# 3rd International Conference on Buildings, Construction and Environmental Engineering (BCEE3-2017)

AA004540

MATEC Web of Conferences Volume 162 (2018)

Sharm el-Shiekh, Egypt
23-25 October 2017

Part 1 of 2

**Editors:**

**T. S. Al-Attar**
**W. S. AbdulSahib**
**M. A. Al-Neami**

ISBN: 978-1-5108-6260-9

**Printed from e-media with permission by:**

Curran Associates, Inc.
57 Morehouse Lane
Red Hook, NY 12571

**Some format issues inherent in the e-media version may also appear in this print version.**

This work is licensed under a Creative Commons Attribution license:
http://creativecommons.org/licenses/by/2.0/
**You are free to:**
**Share** – copy and redistribute the material in any medium or format.
**Adapt** – remix, transform, and build upon the material for any purpose, even commercial.
The licensor cannot revoke these freedoms as long as you follow the license terms.
**Under the following terms:**
You must give appropriate credit, provide a link to the license, and indicate if changes were made. You may do so in any reasonable manner, but not in any way that suggests the licensor endorses you or your use. The copyright is retained by the corresponding authors.

Printed by Curran Associates, Inc. (2018)

For additional information, please contact EDP Sciences – Web of Conferences
at the address below.

EDP Sciences – Web of Conferences
17, Avenue du Hoggar
Parc d'Activité de Courtabœuf
BP 112
F-91944 Les Ulis Cedex A
France

Phone: +33 (0) 1 69 18 75 75
Fax: +33 (0) 1 69 28 84 91

contact-edps@webofconferences.org

**Additional copies of this publication are available from:**

Curran Associates, Inc.
57 Morehouse Lane
Red Hook, NY 12571 USA
Phone: 845-758-0400
Fax: 845-758-2633
Email: curran@proceedings.com
Web: www.proceedings.com

# TABLE OF CONTENTS

## PART 1

## PART 2

# BCEE3- 2017

المؤتمر الدولي الثالث لهندسة البناء والانشاءات والبيئة
23 – 25 تشرين اول 2017
شرم الشيخ – مصر

The 3rd International Conference on Buildings, Construction and Environmental Engineering - BCEE3,
23- 25 October 2017,
Sharm El-Sheikh, Egypt
**www.bcee3.org**
Email: info@bcee3.org

## *Sponsors & Supporters*

## The 3rd International Conference of Buildings, Construction and Environmental Engineering, BCEE3
## 23-25 October 2017
## Sharm el-Sheikh, Egypt

### Introduction

The International Conference of Buildings, Construction and Environmental Engineering is a periodical event and being held every two years. The first conference, BCEE1, was organized in Baghdad, Iraq in 2013 and the second, BCEE2, took place in Beirut, Lebanon in 2015. The BCEE3 was held on October 23-25, 2017 in the Congress Center of the Maritim Jolie Ville Resort, Sharm el-Shiekh, Egypt. This event was organized jointly by the Building and Construction Engineering Department, University of Technology, UOT, Iraq, and the Faculty of Engineering, Menoufia University, MU, Egypt.

### Objectives

1. Providing a forum for researchers to show their researches, conducting a discussion of these scientific researches, exchanging views and experiences, and strengthening the link between researchers and specialists
2. Review of the most important experiences and national developments in the field of Building, Construction and Environment Engineering.
3. Finding modern, eco-friendly alternatives used for the development of construction sector work and resolving obstacles.
4. Exchanging of experiences and encouraging communication between the University of Technology, and local and International Universities in specialized scientific aspects to insure the quality assurance.
5. Benefiting from the experiences of companies and relevant institutions at the work fields of Building and Construction Engineering to identify the problems and work obstacles and give solutions for them.
6. Suggesting scientific strategies to invest the researches and systematic studies in management and development works of Building, Construction and Environment Engineering and how to modify the resources management for the implementation of the engineering projects in the Middle East region.

### Conference Topics

1. Structural Engineering.
2. Building Materials Engineering and Nanotechnology.
3. Geotechnical and Geological Engineering.
4. Water Resources Engineering.
5. Sanitary and Environmental Engineering.
6. Construction Management.
7. Highways and Transportation Engineering.

8. Geomatics Engineering (Surveying, Photogrammetry, Remote Sensing and GIS).
9. Architectural Engineering and Sustainable Developments.

## Conference Committees

### Supreme Committee

1. Prof. Amin D. Thamir, UOT, Iraq/ Chair
2. Prof. Meawad M. El-Kholy, MU, Egypt
3. Prof. Ryiad H. Al-Anbari, UOT, Iraq
4. Prof. Azza E. Lashine, MU, Egypt
5. Prof. Sami Abu El-Non, UOT, Iraq
6. Prof. Alaa A. Atyia, UOT, Iraq
7. Representatives of Supporters

### International Advisory Committee

1. Prof. Riadh S. Al-Mahaidi, **Australia**/Chair
2. Prof. Mohamed Harajli, **Lebanon**
3. Prof. Husham Al-Mansour, **Canada**
4. Prof. Tom Schanz, **Germany**
5. Prof. Siamak Yazdani, **USA**
6. Prof. Andrzej M. Brandt**, Poland**
7. Prof. Caijun Shi, **China**
8. Prof. Muhammad N. Hadi, **Australia**
9. Prof. Abolhassan Vafai, **Iran**
10. Prof. Magdy Abou Rayan, **Egypt**
11. Prof. Sarvesh Chandra, **India**
12. Prof. Omar F. Al-Damluji, **Iraq**
13. Prof. Omar A. Al-Khashman, **Jordan**
14. Prof. Hasan Kurum, **Turkey**

### Organizing Committee

1. Prof. Ryiad H. Al-Anbari, UOT, Iraq / Chair
2. Prof. Tareq S. al-Attar, UOT, Iraq/ Co-Chair
3. Prof. Khaled M. Heiza, MU – Egypt
4. Prof. Aqeel S. Al-Aadili, UOT – Iraq
5. Asst. Prof. Kaiss J. Frayyeh, UOT – Iraq
6. Asst. Prof. Hasan H. Joni, UOT – Iraq
7. Asst. Prof. Mohammed A. Mahmoud, UOT – Iraq
8. Asst. Prof. Maan S. Hasan, UOT – Iraq
9. Asst. Prof. Muzahem A. Alwan, UOT – Iraq
10. Asst. Prof. Wael S. AbdulSahib, UOT – Iraq
11. Dr. Mustafa M. AbdulAama, UOT - Iraq

## Scientific Committee

1. Prof. Abdulrazzaq T. Zibon, UOT – Iraq / Chair
2. Prof. Ashraf S. Zein El-Din, MU – Egypt
3. Prof. Shakir A. Salih, UOT – Iraq
4. Prof. Nabeel A. Jadoo'a, UOT – Iraq
5. Prof. Hussain H. Karim, UOT – Iraq
6. Prof. Mohammed Y. Fattah, UOT – Iraq
7. Prof. Abdulhameed M. Jawad, UOT – Iraq
8. Prof. Wasan I. Khaleel, UOT - Iraq
9. Asst. Prof. Abbas K. Zidan, UOT – Iraq
10. Asst. Prof. Eyad K. Sayhoud, UOT – Iraq
11. Asst. Prof. Waleed A. Abbas, UOT – Iraq
12. Asst. Prof. Faris H. Mohamed, UOT – Iraq
13. Asst. Prof. Falah H. Rahil, UOT – Iraq
14. Asst. Prof. Mahmoud S. Mahdi, UOT – Iraq
15. Asst. Prof. Jaafar S. Maatouq, UOT – Iraq
16. Asst. Prof. Nisreen S. Mohammed, UOT – Iraq

## BCEE3 in numbers:

- No. of received abstracts: 252
- No. of accepted abstracts: 240
- No. of received manuscripts: 217
- No. of accepted manuscripts: 173
- No. of scientific secessions: 23
- No. of keynote lectures: 4
- No. of participating universities: 39 (18 Iraqi + 7 Egyptian + 14 others)
- No. of participating countries: 12 (Algeria, Australia, Egypt, Germany, Iraq, KSA, Malaysia, Russia, Sweden, Turkey, UK and USA)

***Participating Countries***

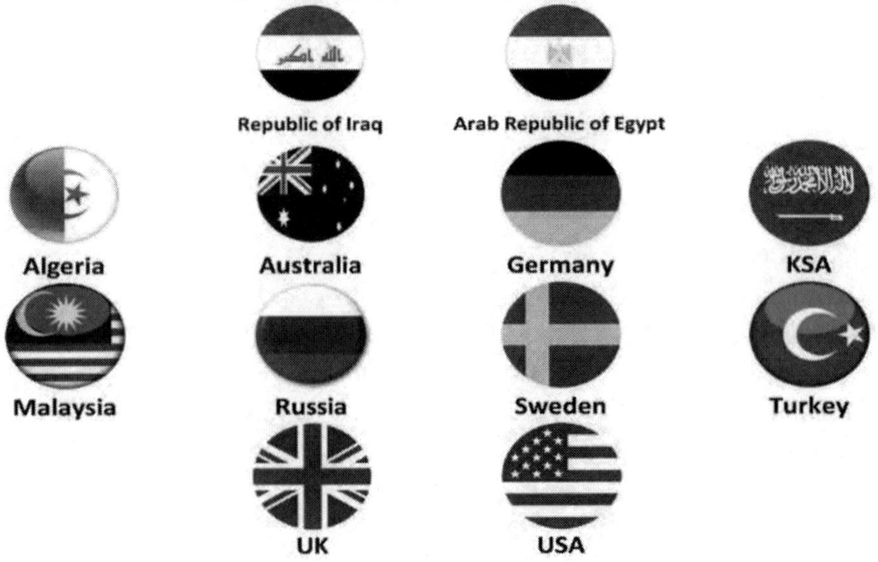

# Participating Universities

| | Republic of Iraq | | Republic of Egypt | | International Universities |
|---|---|---|---|---|---|
| 1 | University of Technology | 1 | Menoufia University | 1 | Swinburne University of Technology –Australia |
| 2 | University of Baghdad | 2 | Ain Shams University | 2 | Ruhr Universitat – Germany |
| 3 | Al-Mustansiriyah University | 3 | Alexandria University | 3 | Umm Al-Qura University – Kingdom of Saudi Arabia |
| 4 | Al-Nahrain University | 4 | Benha University | 4 | King Abdulaziz University - Kingdom of Saudi Arabia |
| 5 | University of Basrah | 5 | Assiut University | 5 | University Putra Malaysia - Malaysia |
| 6 | University of Mousil | 6 | Al-Azhar University | 6 | University Sains Malaysia - Malaysia |
| 7 | University of Kufa | 7 | Millitary Technical College | 7 | University Tenage Nasional – Malaysia |
| 8 | University of Anbar | | | 8 | Moscow State University of Civil Engineering - Russia |
| 9 | University of Tikrit | | | 9 | Lulea University – Sweden |
| 10 | University of Kirkuk | | | 10 | Gazi University – Turkey |
| 11 | University of Al-Qadisiya | | | 11 | Gaziantep University – Turkey |
| 12 | University of Diyala | | | 12 | Heriot Watt University - UK |
| 13 | Al-Qasim Al-Khadra University | | | 13 | Sussex University-UK |
| 14 | University of Karbala | | | 14 | Saint Petersburg State University – USA |
| 15 | Samara University | | | | |
| 16 | Northern Technical University | | | | |
| 17 | Al-Furat Al-Awsat Technical University | | | | |
| 18 | Al-Israa University College | | | | |

edp sciences

MATEC Web of Conferences

# Statement of Peer review

In submitting conference proceedings to MATEC Web of Conferences, I certify to the Publisher that I adhere to the **Policy on Publishing Integrity** of the journal in order to safeguard good scientific practice in publishing.

1. All articles have been subjected to peer review administered by the proceedings editors.
2. Reviews have been conducted by expert referees, who have been requested to provide unbiased and constructive comments aimed, whenever possible, at improving the work.
3. Proceedings editors have taken all reasonable steps to ensure the quality of the materials they publish and their decision to accept or reject a paper for publication has been based only on the merits of the work and the relevance to the journal.

Title, date and place of the conference

The 3rd International Conference on Buildings, Construction and Environmental Engineering, BCEE3
23-25 October 2017
Sharm el-Shiekh, Egypt

Proceedings editor(s):

Prof. Dr. Tareq S. al-Attar
Asst. Prof. Dr. Mohammed A. Al-Neami
Asst. Prof. Dr. Wael S. AbdulSahib

Date and editor's signature

03 / 04 / 2018

# Effect of soil stabilized by cement on dynamic response of machine foundations

*Saad* Al-Wakel[1,*], and *Ahmed* Abdulrasool[1]
[1]Building and Construction Engineering Department, University of Technology, Baghdad, Iraq

**Abstract.** Machine foundations require significant attention from designers. The main goal of the design of machine foundation is to limit the amplitude displacement and not disturb the people who work near the machine. In some cases, if the design of machine foundations does not satisfy the acceptable value of the dynamic response (such as maximum amplitude of displacement), the stabilization of soil under the machine foundation may be used to decrease the amplitude of displacement. This paper outlines effect of stabilized soil under the foundation by cement on the displacement response of machine foundations. Three-dimensional analyses by using finite element method are carried out to investigate the effect of depth of stabilized layer with different percentage of cement content on the dynamic response of the machine foundation. In addition, the effect of area stabilized by cement material on the dynamic response of machine foundation is investigated. The results shown that, the dynamic response of machine foundations generally decreases with increasing the depth of soil layer stabilized with cement. A significant decrease in the displacement of machine foundations is occurred for the stabilized soil layer with a depth of two times of the width of foundation, and the optimum percentage of cement for stabilizing is 6%.

## 1 Introduction

In principle, machine foundations should be designed so that the dynamic forces of machine are transmitted to the soil through the foundation in a way that all kinds of harmful effects are eliminated. The dynamic response of the supporting structure is critical for the in-service stability and safety of machine foundations. In the previous, the simple calculation methods were used, always involving the static loads multiplication by the factor of dynamic. The actual safety factor is not knowledge due to the result being treated as an increased static load. Because of this doubt, the value of dynamic factor was often too high; the exercise showed that the harmful deformations results can be occurred during operation in spite of using such excessive factors. This required a deeper scientific dynamic investigation. Due to the development of machines, a study for more detailed was became urgent. Because increased energy demand around the world in recent years, the control to dynamic response of machine foundations is very important.

Experimental studies on the dynamic behaviour of surface and embedded foundations have been reported by many investigators [1-9] and many others studied the vertical mode of vibration of surface footings, while [10-11] and others studied the vertical mode of vibration of embedded foundations.

Some experimental studies have also been conducted on tensional and sliding modes of vibration [12-14].

All the experimental work cited above has been performed on model or small prototype footings in the field or in the laboratory in soil bins with dimensions in the range of a few feet. The largest size foundations tested in the field were generally foundations for one-story buildings, electric towers, or machine foundations, The largest model foundations used in laboratory tests were comparable with some machine foundations, but in general were about 1/10 to 1/5 of prototype foundations for larger structures such as buildings with several stories.

Some experimental tests of small and rigid model structures of few inches in height, of various shapes and soil densities, have been run on shaking table to study the effect of geometry, soil and embedment on response [15].

The goal of this study is considering the effect of soil stabilized by cement on dynamic response of machine foundations by using three-dimensional finite element modelling.

## 2 Types of machine foundations

The types of Machine foundations based on the shapes of structural as follows (see, Fig. 1) [16]

1. Rigid foundations (Block type)
2. Caisson type or Box
3. Type of wall
4. Type of framed
5. Type flexible or Non-rigid

* Corresponding author: saadfaik231@yahoo.com

© The Authors, published by EDP Sciences. This is an open access article distributed under the terms of the Creative Commons Attribution License 4.0 (http://creativecommons.org/licenses/by/4.0/).

 https://doi.org/10.1051/matecconf/201816201001

In this study, the Machine foundation is considered a rigid foundation.

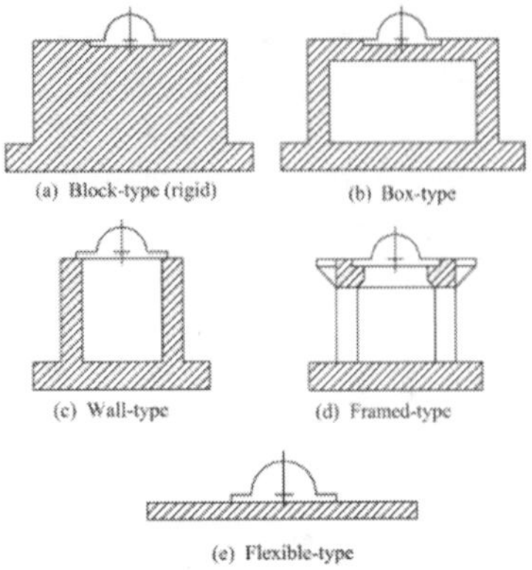


**Fig. 1** Common types of machine foundations.

## 3 Equation of motion of the dynamic system

The general equation of motion for a system is given by Zienkiewicz and Taylor [17]:

$$[M]\{\ddot{u}\} + [C]\{\dot{u}\} + [K]\{u\} = \{F\} \tag{1}$$

where:
$[\boldsymbol{M}]$ = matrix of mass,
$[\boldsymbol{C}]$ = matrix of damping,
$[\boldsymbol{K}]$ = matrix of stiffness,
$\{\ddot{\boldsymbol{u}}\}$ = nodal acceleration vector,
$\{\dot{\boldsymbol{u}}\}$ = nodal velocity vector,
$\{\boldsymbol{u}\}$ = nodal displacement vector, and
$\{\boldsymbol{F}\}$= applied load vector.

## 4 Harmonic excitation of the dynamic force

The dynamic behaviour of foundation plays an important role in providing normal operating conditions for the supported turbo-machine. In this study, the source of the dynamic forces is from a gas turbine machine. The dynamic forces in the vertical direction perpendicular to the longitudinal rotation axis is

$$F(t) = P \sin \omega t \tag{2}$$

where:

$$P = m\, e\, \omega^2 \tag{3}$$

$m$ = proportional part of rotating mass,
$e$ = eccentricity of mass,
$t$ = time,
$\omega$ = circular operating frequency of the machine ($2\pi f$).
$f$= operating frequency.

The maximum amplitude of the dynamic force is 43 kN, the circular operating frequency of the machine 314 red/sec ($f$=50 Hz) at the nominal turbine speed.

## 5 Finite element analysis

A three-dimensional finite element method which is implemented in OpenSees program (Open System for Earthquake Engineering Simulation) [18] is used to analyze the dynamic response of the turbine machine foundation under the effect of harmonic excitation of vertical load. The objective is to investigate the effective depth for the improvement layer under machine foundation, and the effect of the cement percentage to stabilize the soil on the dynamic response of machine foundations.

A concrete foundation is placed at the centre of the model and the dimensions of the foundation are 6 × 6 × 0.5 m. The behaviour of the material of the foundation is assumed linear elastic. Table 1 shows the parameters of the concrete material. The soil boundaries are taken as 212 m wide, 212 m length, and 42 m depth. A linear elastic model is assumed for soil behaviour. The soil properties are assumed uniform throughout the depth of the layer as shown in Table 2.

**Table 1.** Parameters of the concrete material.

| Properties | Value | Units |
|---|---|---|
| Modulus of elasticity (E) | 25742960 | $kN/m^2$ |
| Poisson's ratio (υ) | 0.2 | - |
| Compressive strength of concrete ($f'c$) | 30 | MPa |
| Unit weight of concrete | 24 | $kN/m^3$ |

**Table 2.** Soil properties.

| Properties | Value | Units |
|---|---|---|
| Modulus of elasticity (E) | 10000 | $kN/m^2$ |
| Poisson's ratio (υ) | 0.35 | - |
| Unit weight of dry soil (γ) | 14.5 | $kN/m^3$ |

In this analysis, eight nodes brick element are used. The brick element has eight nodes at the corners with three degrees of freedom for each node. The elementary boundaries with zero-displacement conditions are used at the boundary of the soil model, so that the bottom of the soil is supposed to be fixed. The displacement in X-directions for the nodes at the boundary in Y-Z planes is fixed; as well the displacement in Y-directions for the nodes at the boundary in X-Z planes is constraint. The model of the soil under the machine foundation is assumed homogenous isotropic elastic material due to the small amplitude of vibrations.

The three-dimensional finite element model and the cross-section of the model are shown in Figs 2 and 3, respectively. At the top central point (A) of the foundation (see, Fig. 2), the vertical load is applied and the amplitude of displacements are measured at the same point. In this study, the time period of the analysis is (1000 sec) with time step ($\Delta t$=0.0043 sec). The Newmark method is used for the time integration [19].

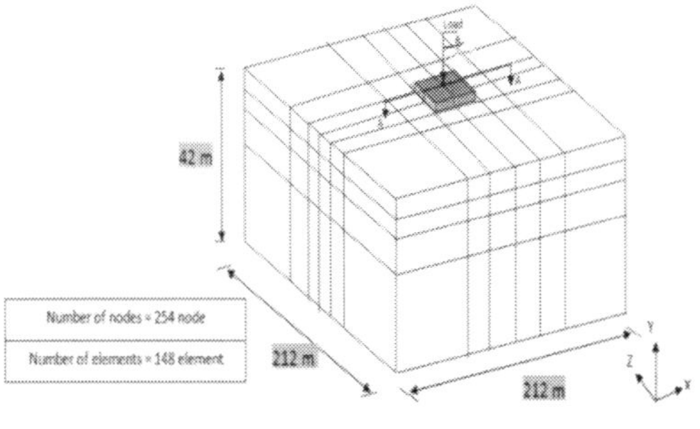


**Fig. 2** The mesh of 3D finite element model.

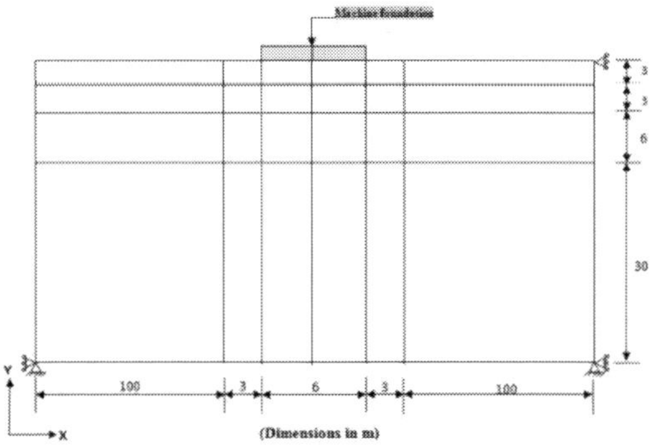


**Fig. 3** Cross-section A-A of machine foundations and soil.

## 6 Effect of cement percentage to stabilize the soil on dynamic response

In this part, the analysis is conducted to determine the vertical displacement of machine foundation that rests on soil stabilized with different percentage of cement (2, 4, 6, 8, and 10).

As a first stage, the thickness of the improved layer for all cases is considered for a depth equal to 2B. The Poisson's ratio, modulus of elasticity and unit weight of dry sand is obtained from previous study [20]. Table 3 shows the soil properties with different percentage of cement.

The relationships between time and amplitude of vertical displacement for the machine foundation constructed on a sand layer stabilized with different percentage of cement is depict in Fig. 4.

**Table 3.** Properties of sand with different percentage of cement.

| **Cement (%)** | **Modulus of elasticity (E) kN/m$^2$** | **Poisson's ratio ($\upsilon$)** | **Unit weight of dry soil ($\gamma$) kN/m$^3$** |
|---|---|---|---|
| 2 | 10906.08 | 0.34 | 19.3 |
| 4 | 11673.09 | 0.33 | 20.48 |
| 6 | 77255.67 | 0.30 | 20.43 |
| 8 | 246080.2 | 0.28 | 21.03 |
| 10 | 849982.9 | 0.24 | 20.50 |

Fig. 5 shows the relationship between the maximum amplitude of displacement of machine foundation and the percent of cement used to stabilize the soil.

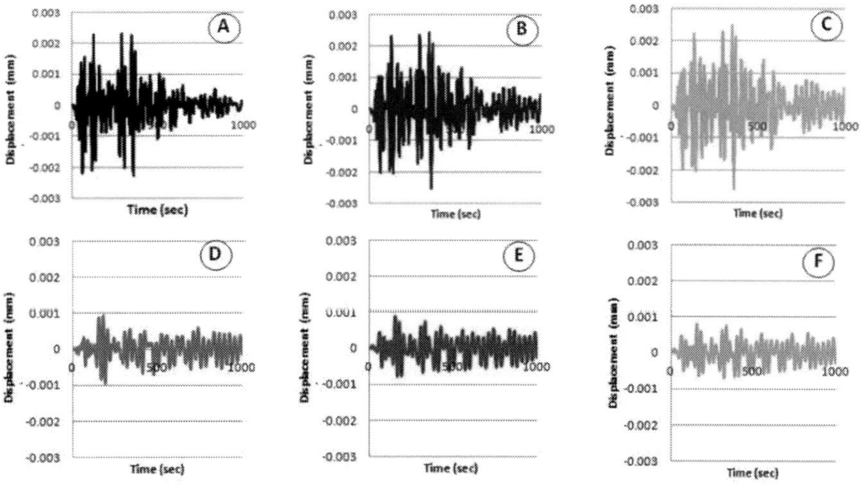


**Fig. 4** Predicted displacement at point A of the foundation rests on stabilized layer at depth 2B with different percentage of cement (A) Soil without treatment; (B) 2% cement; (C) 4% cement; (D) 6% cement; (E) 8% cement; (F) 10% cement.

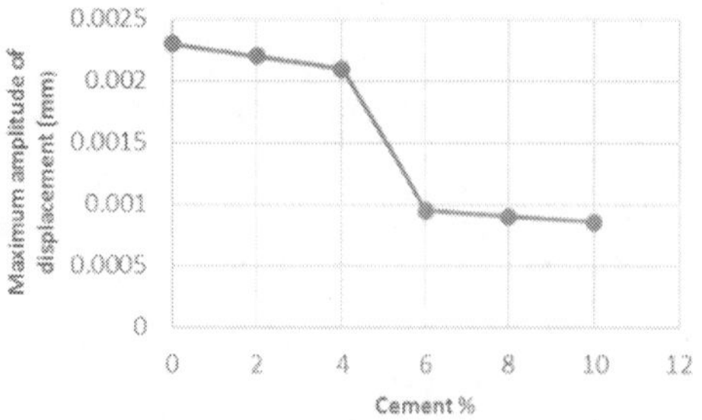


**Fig. 5** Maximum amplitude of displacement vs. cement percentages.

From Fig. 5 it can be stated that, the displacement of the machine foundation decreases with increasing of cement percent used to stabilize the underlying soil and the optimum value is 6%.

## 7 The effect of stabilized layer depth by cement on the dynamic response

The analysis is performed to estimate the optimum thickness of the layer that stabilized with 6% cement for the machine foundation purposes. In this analysis, only the thickness of the stabilized layer is changed with different values of (0.5B, 1B and 2B). The results of the vertical displacement response are presented in Fig. 6.
Fig.7 shows the relationship between the maximum amplitude of displacement of machine foundation and the depth of improvement layer of the soil.
From Fig. 7 it can be seen that the vertical displacement decreases with increase of the thickness of the layer that is stabilized by cement due to the increasing in stiffness of the system (soil-foundation), and the optimum depth for improving the soil is 1B.

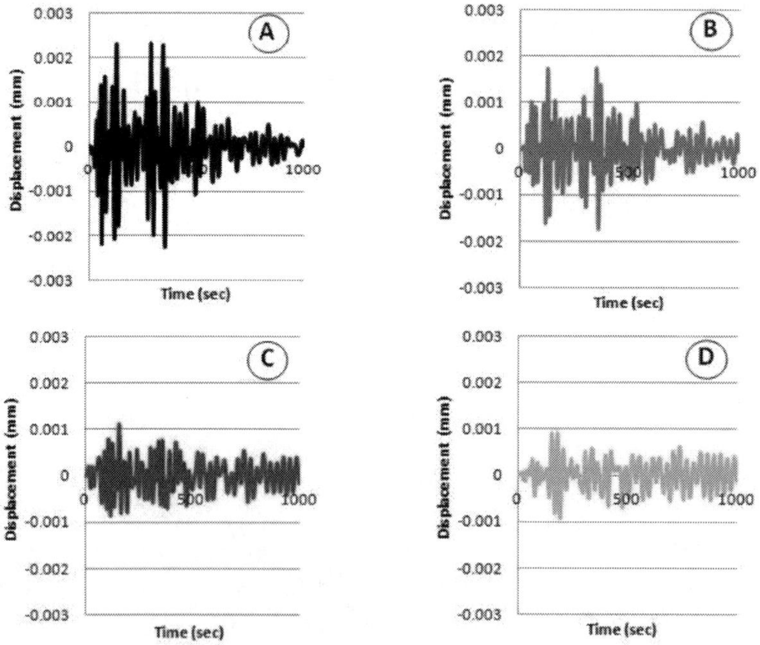


**Fig. 6** Predicted displacement at point A on foundation with different depth of layer stabilized with 6% cement: (A) Soil without treatment (B) 0.5B (C) 1B (D) 2B.

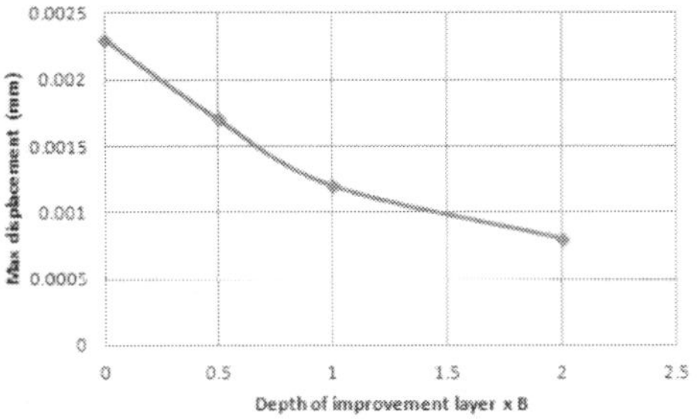


**Fig. 7** Maximum amplitude of displacement vs. depth of improvement layer

## 8 Conclusions

1. Soil stabilization by cement for the soil under machine foundations can be considered as a successful future technique to decrease the displacement of the machine.
2. The maximum amplitude of displacement is affected by the percent cement that used to stabilize the soil due to increasing the stiffness of the foundation, and the optimum value of cement is 6%.
3. The increase in depth layer of the soil improvement with cement leads to a decrease in the displacement response of the machine foundation gradually. However, the depth of improvement upto 0.25B will lead to decrease the dynamic response by 25%.

## 9 References

1. T. K. Al-Azawi, R. K. Al-Azawi, Z. K. Al-Jaberi, Eng. J., University of Baghdad **12**, 429-443 (2006)
2. S. F. A. Al-Wakel, Eng. & Tech. J. **31**, 1447–1458 (2013)
3. S. F. A. Al-Wakel, M. R. Mahmoud, A. S. Abdulrasool, Eng. J., University of Baghdad **20**, 48-61 (2014)
4. K. H. II Stokoe, (PhD Thesis, University of Michigan, 1972)
5. S. M. Erden, (PhD Thesis, University of Massachusetts, 1974)
6. S. N. Varadhi, S. K. Saxena, Geotech. Eng. J. ASCE **106**, 1121-1139 (1980)
7. S. F. A. Al-Wakel, M. R. Al-Qaissy, A. S. Abdulrasool, Eng. & Tech. J. **33**, 1038–1048 (2015)
8. S. F. A. Al Wakel, M. Y. Fattah, H. H. Karim, Eng. & Tech. J. **29**, 2189–2201 (2011)
9. A. S. Abdulrasool, (MSc Dissertation, University of Technology, Iraq, 2012)
10. M. Novak, Soil Mech. & Found. J. ASCE **96**, 337-361 (1970)
11. B. N. Gupta, Geotechnique J. **22**, 129-137 (1972)
12. H. R. Sreekantiah, Geotech. Eng. J. ASCE **108**, 905-917 (1982)
13. A. N. Lin, Earthquake Engineering Research Laboratory (EERL) **82,** 1(1982)
14. R. Henake, F. E. Richart, Jr., R. D. Woods, Geotech. Eng. J. ASCE **109**, 72-88 (1983)
15. A. H. Hadjian, G. E. Haward, C. B. Smith, *Proceedings of the Third International Conference on SMIRT*, K2/5 (1975)
16. N. S. V. Kameswara Rao, "*Foundation Design: Theory and Practice*," John Wiley & Sons (Asia) Pte Ltd, Singapore, 2011.
17. O.C. Zienkiewicz, R. L. Taylor, "*The Finite Element Method*," McGraw-Hill, London, UK, 2005.
18. Open System for Earthquake Engineering Simulation (OpenSees) Manual, 2005.
19. M.G. Katona, O.C. Zienkiewicz, International J. for Numerical Methods in Eng. **21**, 1345-1359 (1985)
20. Z. W. S. Abbawi, K. Y. H. AL-Soudany, Eng. & Tech. J. **33**, 1857–1875 (2015)

# Improving geotechnical properties of clayey soil using polymer material

*Hussein* Karim[1,*] and *Kawther* Al-Soudany[1]

[1] Building and Construction Engineering Department, University of Technology, Baghdad, Iraq

**Abstract.** This study illustrates the application of polymer material for clayey soil stabilization. The article will focus on studying the strength behavior of the clayey soils reinforced with homogenously polymer fiber. In the current research, "polypropylene" was selected as polymer material to reinforce the natural clay soil. This polymer fiber was added to the clayey soil with four different percentages of (0, 1.5, 3, and 5%) by weight of soil. Various tests with different polymer contents were performed to study the effect of using such a polymer as a stabilizing agent on geotechnical properties of clay. As the fiber content increases, the optimum moisture content (OMC) is increased while the specific gravity decreases. For Atterberg's limits, the results indicated increasing liquid limit and plasticity index while decreasing plastic limit with increase in polymer content. The outcomes of the tests also reflected a considerable improvement in the unconfined compressive strength with noticeable improvement in the shear strength parameter (undrained shear strength, $c_u$) of the treated soils. The undrained shear strength obtained from treated soil with 5% polymer addition is more than three times that of the untreated soil. With an increase in polymer content, the consolidation parameters (Compression index $C_c$ and recompression index $C_r$) decreases. Finally, the benefit of the reinforcement is increased with increasing polymer fiber content.

## 1 Introduction

For all engineering structures constructed on weak clayey soils, deformations occur when these soils are exposed to any additional load or due to dissolution of clay when these soils get wet, soaked or leached with water. These deformations may cause a significant failure (cracking, tilting and collapse) of the foundation or earth structures. Many problems regarding foundation such as collapse in road construction projects are associated with the consolidation of the clayey soil layers. Thus, many researchers employ soil stabilization techniques for ground improvement to achieve the required performance [1, 2].

Soil stabilization refers to the addition of a cementing material (s) to the soil to improve one or more of its geotechnical properties. The addition of stabilizing agents to natural soils has a long history either by mechanical mixing to obtain a homogeneous mixture or by adding the stabilizing material to fill the soil voids [3-5].

Several materials (natural, fabricated and byproduct materials) have been used as stabilizers for adaptation of clayey soils. The application of various inorganic or organic additives such as lime, Portland cement, asphalt binder, fly ash, rice husk ash etc., or their combinations often create the modification of soil properties including cementation of soil particles [6, 7].

Clay soils are mostly found in Iraq (more than 20%) particularly in its middle and southern parts and around 35% of the former portions are weak. Thus, the problems of this soil have been taken in consideration by many Iraqi geologists and civil engineers since it is often used as natural foundation bases for structures. There are most complex engineering problems in the clay soil, especially when accompanied by environmental changes in moisture content and undergoes large settlement under long term of loads. Failure of different structures constructed on clay soils in various locations in Iraq were recorded such as Ammara, Nasiriya and Basra that are characterized by their low undrained shear strength (<40 kPa) and compression index as high as 0.3 were reported [2, 8]. Thus, it is necessary to improve large areas covered by weak clay soil that has created problems to roads and airports built on clay soils because of their low bearing capacity.

This research concentrates on investigating experimentally the feasibility of stabilizing and improving the geotechnical properties of soft clay soil using a polymer fiber material in different proportions and to study the polymer effect on the engineering characteristics of the stabilized clay.

## 2 Materials used and methodology

### 2.1 Materials used

#### *2.1.1 Soil used*

The soil used for this study was taken from Alsadda site in Baghdad. At the beginning, the soil natural moisture content was determined immediately. After that, the soil samples were spread out in the laboratory for two weeks

* Corresponding author: husn_irq@yahoo.com

© The Authors, published by EDP Sciences. This is an open access article distributed under the terms of the Creative Commons Attribution License 4.0 (http://creativecommons.org/licenses/by/4.0/).

for air-drying at room temperature to remove the natural water which may influence the analysis. The soil is of low plasticity (CL) classified according to Unified Soil Classification System (USCS). The grain size distribution and engineering properties of the used soil are presented in Figure 1 and Table 1, respectively.

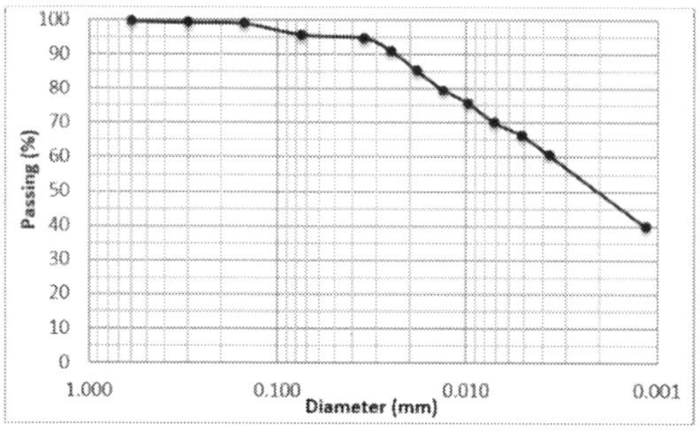


**Fig. 1.** Grain size distribution curve.

**Table 1.** Engineering properties of the used soil.

| Property | Value | Standard |
|---|---|---|
| Liquid limit (L.L. %) | 47 | ASTM D 4318 – 00 [9] |
| Plastic limit (P.L. %) | 23 | ASTM D 4318 – 00 [9] |
| Plasticity index (P.I. %) | 24 | ASTM D 4318 – 00 [9] |
| Specific gravity | 2.69 | ASTM D 854 – 02 [10] |
| Gravel % | 0 | ASTM D 422 – 63 [11] |
| Sand % | 4.2 | ASTM D 422 – 63 [11] |
| Silt % | 29.8 | ASTM D 422 – 63 [11] |
| Clay % | 66 | ASTM D 422 – 63 [11] |
| Activity | 0.41 | ASTM D 422 – 63 [11] |

### *2.1.2 Polymer used*

A commercial product of polypropylene (PP), also known as polypropene, is a thermoplastic polymer utilized in many implementations including textiles and packaging. With the addition of such polymer, the material become tough and abnormally rigid and chemically resistant. Scientifically, polypropylenes are polymers consisting of repeated branched units derived from propane and having the formula $CH_2$ CH ($CH_3$) (Fig. 2). In 2013, the global market for polypropylene was about 55 million metric tons [12].

The properties of polypropylene depend on the molecular weight with its distribution, crystallinity, type and proportion of comonomer [13]. The density of PP is between 0.895 and 0.92 g/cm³, therefore it is plastic with the lowest density. Its Young's modulus is between 3500-3900 MPa. The PP is normally tough and flexible and has good resistance to fatigue and it is reasonably economical and has a melting point that ranges from 160 to 166 °C [16, 17]. At room temperature, PP is resistant to fats and almost all organic solvents. At elevated temperature, PP can be dissolved in non-polarity solvents [18]. The polymer used in this investigation commercially available obtained from Scientific Bureau in Baghdad which is called polypropylene fibers (RHEOFIBRE). These polymer fibers are environmentally accepted and were used to adapt clay soil using different proportions. Figure 3 shows the used polymer with some of its important properties listed in Table 2.

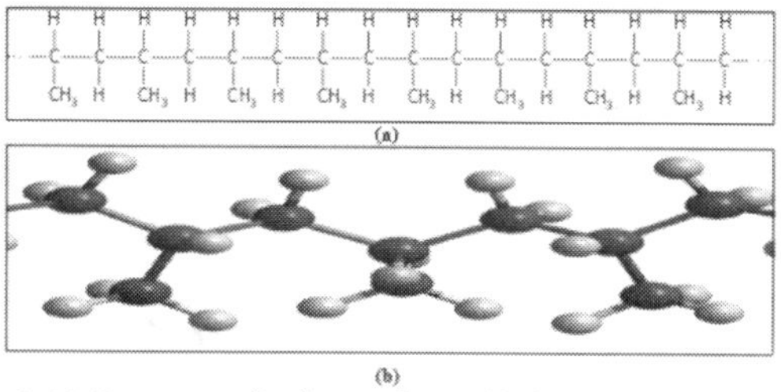


**Fig. 2 (a)** Structures of polypropylene with isotactic tacticity [14]. **(b)** Polypropylene isotactic ball-and-stick model with 3 countable monomers [15].

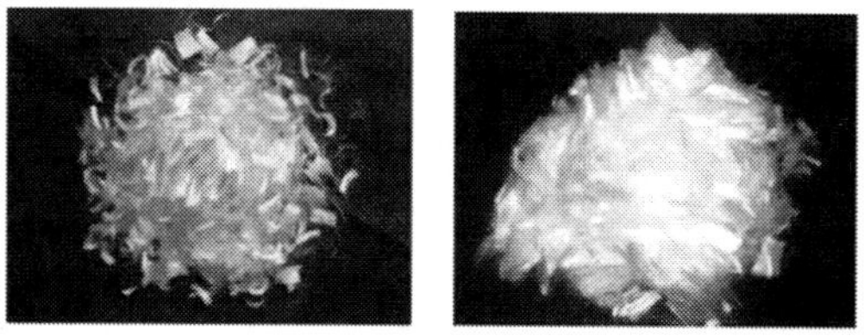

**Fig. 3.** Polymer used (polypropylene fibers- RHEOFIBRE).

**Table 2.** Physical and engineering properties of the polymer used (polypropylene fibers- RHEOFIBRE) [19].

| Property | Value |
|---|---|
| Specific gravity | 0.91 |
| Alkali content | Nil |
| Sulphate content | Nil |
| Air entrainment | Air content will not be significantly increased |
| Chloride content | Nil |
| Constituents | Polymerized Polypropylene |
| Fiber diameter | 18 micron |
| Fiber length | 12 mm |
| Surface area | 230 m²/ kg min. |
| Young's modulus | 3500 – 3900 MPa |
| Tensile strength | Min 350 MPa |
| Melting point | 160 °C |

### *2.1.3 Water*

Tap and distilled water was used in the experiments.

#### 2.1.4 Samples preparation

To investigate the effect of polymer fiber material on the geotechnical properties of the soft soil used, soil-polymer mixtures were prepared using different percentages of polymer material.

At first, the clayey soil was dried and grinded before using it in the mixtures. Then, the required amounts of clayey soil and reinforcement polymer fibers were blended together under dry conditions. Considerable time and attention were spent to get a homogeneous distribution of the fibers in the mixtures. The contents of polymer fibers were chosen as 0, 1.5, 3, and 5 % by total weight of dry soil. It is worth to mention that all the testing
specimens were compacted at lowest density of 1.488 gm/cm$^3$, and the maximum water content of 23.3%, corresponding to the values obtained in the Standard Proctor Compaction Tests [20]. All experimental works were carried out in the Soil Mechanics Laboratory of the Department of Building and Construction Engineering-University of Technology.

## 3 Experimental tests

In this study, laboratory experiments were achieved to assess the effect of polymer addition on the physical properties, compaction, undrained shear strength and consolidation parameters of the untreated (unstabilized) and treated (stabilized) soft clayey soil samples.

Experimental work includes; specific gravity, compaction, optimum moisture content, Atterberg's limits, unconfined compressive, Vane shear, and consolidation tests.

Many geotechnical properties have been studied such as consistency limits (Atterberg's limits: liquid limit L.L, plastic limit P.L., and plasticity index P.I.), maximum dry unit weight ($\gamma_{dry\,max}$), optimum moisture content (OMC), specific gravity ($G_s$), undrained shear strength ($c_u$) and consolidation parameters (compression index $C_c$, and recompression index $C_r$) of the soft clayey soil.

As aforementioned, all the tests have been conducted after choosing the worst case, i.e., by taking the lowest density and maximum water content. Then, static compaction was applied by using the modified proctor mold to study the effect of polymer different percentages on clay by checking the unconfined compressive strength, Vane shear, and consolidation tests.

### 3.1 Effect of polymer on soil properties

For the natural and stabilized samples (after addition 1.5, 3, and 5 % polymer), the above mentioned same tests and procedures were followed.

#### 3.1.1 Specific gravity

The specific gravity of soil was determined according to ASTM D 854-02 [10]- Standard test for specific gravity of soil solids by water pycnometer. Figure 4 shows the specific gravity variation with different polymer contents.

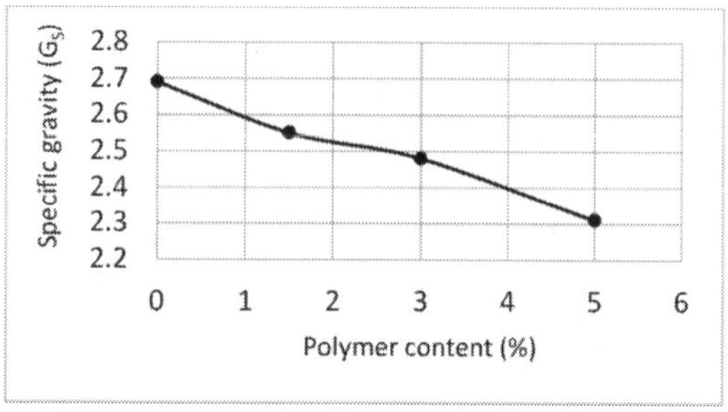


**Fig. 4.** Specific gravity variation with different polymer contents.

#### 3.1.2 Compaction

This test was performed according to ASTM D 1557 [20] to define the relation between the dry unit weight and the moisture content of the soil. Figure 5 shows the optimum moisture content (OMC) obtained from compaction test for untreated (0 % polymer) and treated (1.5, 3 and 5% polymer) soils.

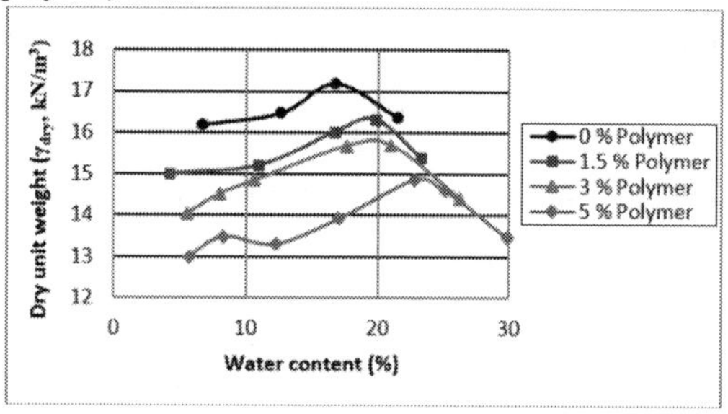


**Fig. 5.** Optimum moisture content (OMC) obtained from the compaction test for untreated and treated soils.

From compaction test, the maximum dry density variation and the optimum moisture content (O.M.C) with different polymer contents are presented in Table 3, and displayed in Figures 6 and 7.

**Table 3.** Maximum dry unit weight and optimum water content.

| Polymer (%) | Max. dry unit weight ($\gamma_{dry}$) (kN/m$^3$) | Optimum moisture content (O.M.C) (%) |
|---|---|---|
| 0 | 17.2 | 16.25 |
| 1.5 | 16.38 | 19.3 |
| 3 | 15.82 | 19.8 |
| 5 | 14.88 | 23.2 |

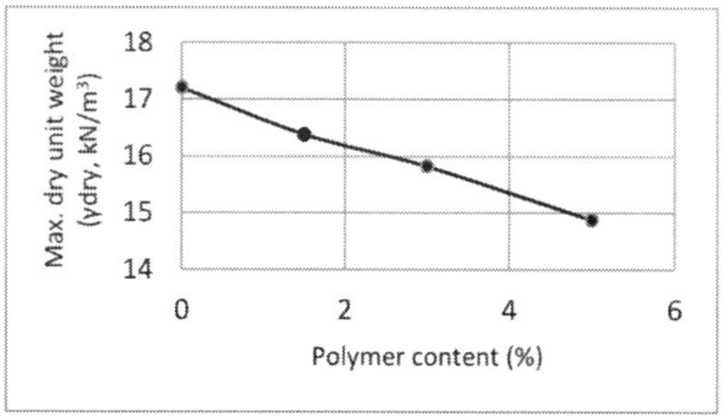


**Fig. 6.** Maximum dry unit weight variation with different polymer contents.

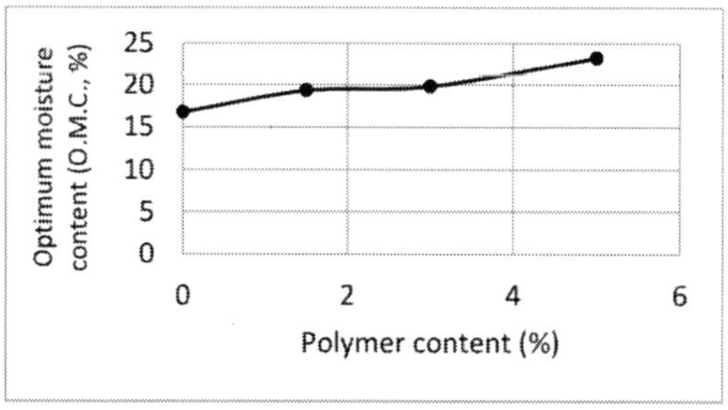


**Fig. 7.** Optimum moisture content variation with different polymer contents.

### *3.1.3 Moisture content*

This test was performed to determine the water (moisture) content of soils. Table 3 and Figure 7 illustrate the optimum moisture content variation with different polymer contents.

### *3.1.4 Atterberg's limits*

A wide variety of soil engineering properties have been correlated with consistency limits (Atterberg's limits). Liquid Limit (L.L) and plastic limit (P.L) were carried out according to the ASTM D 4318-00 [9]. Figure 8 illustrates the variations of these limits with polymer content.

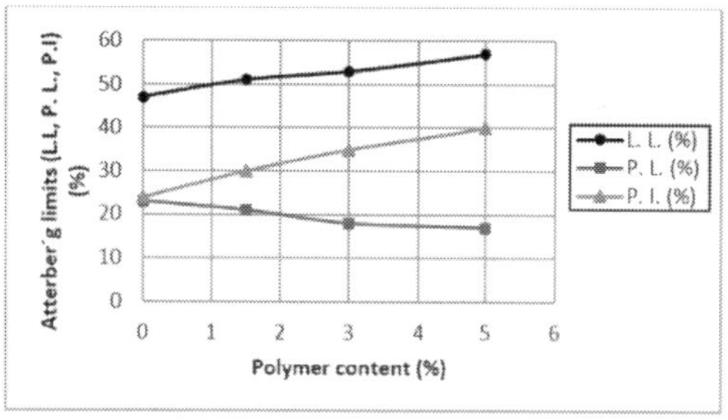


**Fig. 8.** Atterberg's limits variation with polymer content.

### *3.1.5 Unconfined compressive strength*

Unconfined compressive test was applied to determine the unconfined compressive strength (UCS, $q_u$) following the procedure of the ASTM D 2166-00 [21] Standard Test Method for a cohesive soil. Then, the undrained shear strength (or undrained cohesion, $s_u$) of a cohesive soil was calculated which is equal to one-half the unconfined compressive strength ($q_u$). The Vane shear test was also used to estimate the undrained shear strength of fully saturated clays without disturbance according to ASTM D 4767- 04ASTM D2573- 08 [22] Standard Test Method for Field Vane Shear Test in cohesive soil. Figure 9 illustrates the stress-strain relationship from the unconfined compression test for untreated and treated soils. The results of undrained shear strength values for both UCS and Vane shear tests on natural and stabilized soils with different polymer contents are presented in Figure 10.

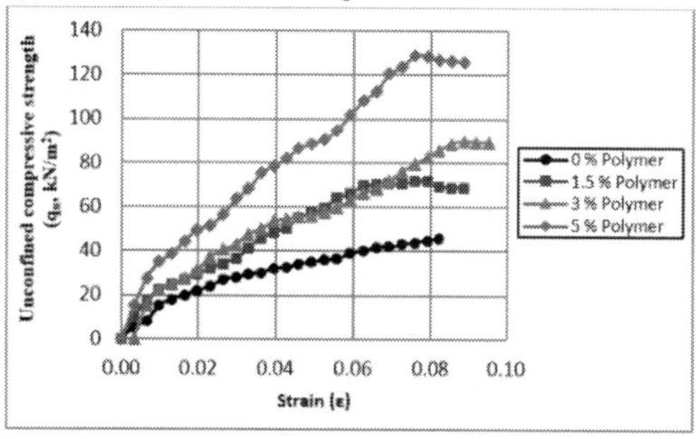


**Fig. 9.** Effect of polymer content on the unconfined compressive strength.

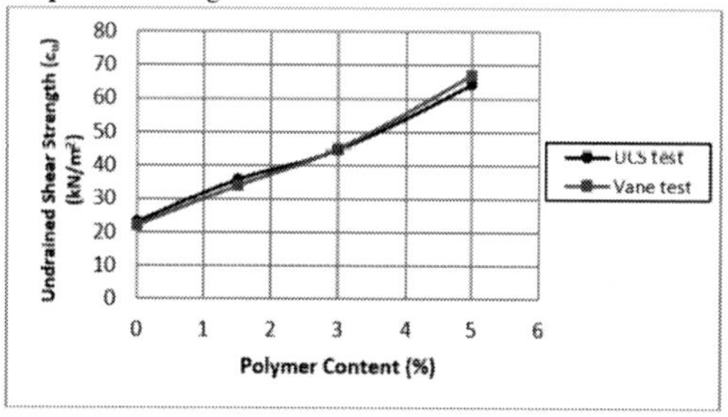


**Fig. 10.** Effect of polymer fibers content on undrained shear strength of soil.

### 3.1.6 Consolidation test

This test is performed according to ASTM D 2435-02 [23] Standard Test Method for One-Dimensional Consolidation Properties of Soils. Loading pressures were 25, 50, 100, 200, 400 and 800 kPa and reloading pressures were 800, 200, 50, and 25 kPa. From e-log p relation, the compression index ($C_c$) and recompression index or swelling index ($C_r$) during unloading were determined at different polymer contents. Figure 11 shows the consolidation test results by relating of e-log p with polymer addition. While, Table 4 shows results of compression index and recompression index values by using different polymer content.

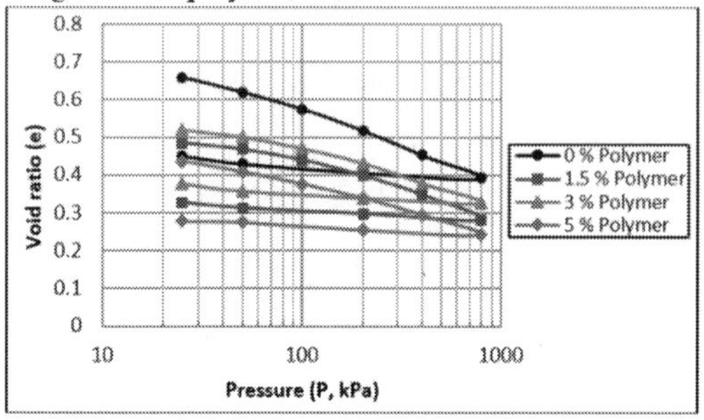


**Fig. 11.** Relation of e-log p for untreated and treated soils with polymer addition.

**Table 4.** Compression index ($C_c$) and recompression index ($C_r$) by using different polymer content.

| Polymer (%) | Compression index ($C_c$) | Recompression index ($C_r$) |
|---|---|---|
| 0 | 0.191 | 0.038 |
| 1.5 | 0.139 | 0.025 |
| 3 | 0.129 | 0.029 |
| 5 | 0.122 | 0.031 |

## 4 Testing results and discussion

As aforementioned, it was desired in this research to evaluate the effectiveness of the clayey soils mixed with polymer fiber material (polypropylene fibers-RHEOFIBRE) as a stabilizer agent. The results of physical and mechanical properties before (0% polymer) and after (1.5, 3, and 5% polymer) shall be discussed in this section.

The addition of polymer fibers affects soil's specific gravity. It is obvious from Figure 4 that with increasing polymer content, the specific gravity of treated soil sharply drops from 2.69 for untreated natural clayey soil to 2.31 after 5 % polymer addition. That is due to the very low specific gravity of the polymer (0.91) compared with that of natural clay soil.

The effect of increasing polymer fibers content addition on soil unit weight is illustrated in Figure 5. It is clearly shown that here is a sharp decrease in dry unit weight and increase in optimum water content with an increase in polymer content. That is due to the light unit weight of the polymer fibers (0.91 gm/cm$^3$). Figure 6 illustrates the influence of polymer addition on the maximum dry unit weight. From this figure, it is clear that for any particular amount of polymer, a sharp decrease in maximum dry unit weight is accompanied due to increasing polymer addition. The explanation is that due to the decline in the average value of the unit weight of the solids in the soil-polymer mixture. Knowing that the maximum dry unit weight of soil ranges from 17.2 kN/m$^3$ for untreated to 14.88 kN/m$^3$ at 5 % polymer addition.

The optimum moisture content (OMC) is increased with increasing the amount of polymer content (Fig. 7). The OMC values gradually increase to 3 % polymer and then considerably increase to 23.2 % at 5% polymer addition compared to 16.75 % for untreated soil.

Concerning the effect of fiber content on Atterberg's limits shown in Figure 8, it was noticed that with increasing polymer content there was an increase in liquid limit and plasticity index from 47 and 24 % for untreated to 57 and 40 % at 5 % polymer addition respectively. While, for plastic limit, the reverse is true, it decreases from 23 to 17 % for untreated and treated (5 % polymer) respectively. As a general trend, this could be due to its higher water absorption affinity (i.e. water needed for hydration).

As regards the results of the unconfined compressive strength and Vane shear tests in case of untreated soil sample, the stress increases gradually with the increase of strain. While for the treated soil, the stress is rapidly increased with the increase of strain. The stress-strain curves of the unconfined compressive test (Fig. 9) showed increase in unconfined compressive strength ($q_u$) from 46.14 kN/m$^2$ for untreated to 128.93 kN/m$^2$ for 5 % polymer addition. By examining stress-strain curves, it can be stated that the addition of polymer (particularly 5%) modify the soil behavior and deformation characteristics in term of brittleness or ductility. Moreover, this addition leads to increase in brittleness behavior of the treated soil and also to increase its stiffness (or the modulus of elasticity, E). Thus, the undrained shear strength obtained from these tests shows linear and sharp increase with polymer content for treated clayey soil particularly for 5 % polymer which is more than three times that of the untreated soil (Fig. 10).

For consolidation test (Fig. 11), the results of compression index ($C_c$) and recompression index ($C_r$) (Table 4) show a decrease in these parameters with increasing polymer content. It has been found that the polymer stabilization markedly reduced the compression index values. The original specimens vary from a void ratio of approximately of 0.659 - 0.391 over a stress range of 25 - 800 kPa. The values of $C_c$ and $C_r$ for the sample without stabilization were found to be 0.191 and 0.038 respectively and reduced gradually to 0.122 for the first parameter and partially reduced to 0.031 for the second one at polymer content of 5 % respectively. The treated specimens have very similar compression behavior, with a void ratio varying according to the polymer contents over the same stress range. The values of compression index reduced gradually and were found to be 0.139, 0.129 and 0.122 at a polymer content of 1.5, 3 and 5% respectively. While, the recompression index partially reduced, i.e., it reduces then increases again at 3 and 5% polymer addition but still with lower values compared to the untreated soil.

The explanation is with increasing polymer content, the consolidation parameters were generally reduced as the polymer generates high tension effect between soil particles, so it is hard to compress the particles and also hard to gain the first situation. Where the $C_c$ was totally reduced while the $C_r$ was partially reduced but still with lower values compared to the untreated soil. Finally, similar results were obtained by Azzam (2014) [1] concerning the increase in shear strength and cohesion as well as the decrease in consolidation parameters.

## 5 Conclusions

This study was conducted to investigate the influence of polymer (polypropylene- RHEOFIBRE) addition on geotechnical properties of clayey soil. From the extensive experimental test results of this study, the following conclusions can be drawn:

1. As the polymer fiber content increases the specific gravity decreases due to low specific gravity of the polymeric material (0.91) compared with that of clay soil (2.69).
2. Compaction results indicated that there were decreasing compaction curves reflecting approximately linear decrease in maximum dry density and approximately linear increase in optimum water content as the polymer content increases.
3. For Atterberg's limits, an increase in both liquid limit and plasticity index, and a decrease in plastic limit were observed with increasing polymer content.
4. Undrained shear strength obtained from both unconfined compression and Vane shear tests show linear increase with a polymer concentration up to 5% for all the clayey soils used as the polymer plays a major role in increasing the tension between soil particles in the soil-fiber mixture.
5. The increase in polymer addition leads to an increase in the brittleness behavior of the treated soil and also to an increase in its stiffness.
6. With an increase of polymer content, the consolidation parameters (compression index $C_c$ and recompression index $C_r$) are generally reduced.
7. The soil stabilization with polymer fibers of polypropylene is considered as a good option as it enhances the soil strength behavior, is environmentally accepted, and is commercially available.

## References

1. W.R. Azzam, Alexandria Eng J **53**, 143-150 (2014)
2. H.H. Karim, Z.W. Samueel, S.F. Ahmed, *Proceedings of the 2nd International Conference of Buildings, Construction and Environmental Engineering, American University of Beirut (BCEE2), 17-18 Oct., 2015, Geotechnical Engineering Part,* 117-121 (2015).
3. W.H. Perloff, *Soil Mechanics: Principal and applications*, (John Wiley & Sons, 1976)
4. B.K.G. Theng, Clay and Clay Minerals **30**, 1–10, (1982)
5. A. Ates, Hindawi Publishing Corporation, Int J Polymer Sci, Vol. 2013, Article ID 356214, 8 P. (2013)
6. K. E. Clare, A. E. Cruchly, Geotechnique **7**, 97-110 (1957)
7. A. Amadi, Leonardo El J Pract Technol **17**, 69-78 (2010)
8. S.F Ahmed, M.Sc. Thesis, University of Technology (2015)
9. ASTM D 4318-00, *Standard test methods for liquid limit, plastic limit, and plasticity index of soils*, (ASTM International, West Conshohocken, PA, United States, 2000)
10. ASTM D 854-02, *Standard test methods for specific gravity of soil solids by water Pycnometer*, (ASTM International, West Conshohocken, PA, United States, 2002)
11. ASTM D 422-63 (Reapproved 2002), *Standard test method for particle size analysis of soils*, (ASTM International, West Conshohocken, PA, United States, 2002)
12. Ceresana: *Market Study: Polypropylene*, (3rd edition), 580 P., available on http://www.ceresana.com
13. D. Tripathi, *Practical guide to polypropylene, Shrewsbury: RAPRA Technology*, (2001)
14. Mediawiki/RicHard-59, File: Polypropylene tacticity.svg, (2012)
15. PakpongICCH444, File:Polypropylene isotactic mini trp.png, (2015)
16. C. Maier, T. Calafut, *Polypropylene: the definitive user's guide and databook*, (William Andrew Inc, USA), 430 P. (1998)
17. W. Kaiser, *Kunststoffchemie für Ingenieure von der Synthese bis zur Anwendung*, (3rd ed., München: Hanser, 2011)
18. S. Koltzenburg, M. Maskos, O. Nuyken, *Polymere: Synthese, Eigenschaften und Anwendungen*, (Springer, 2014)
19. BASF, the Chemical Company, "RHEOFIBRE", Construction Chemicals UAE LLC, (2004), available on http://www.mediawiki.org/wiki/User:RicHard-59
20. ASTM D 1557, *Standard test methods for laboratory compaction characteristics of soil using modified effort (56,000 ft-lbf/ft³ (2,700 kN-m/m³))*, (ASTM International, West Conshohocken, PA, United States, 2000)
21. ASTM D 2166-00, *Standard test method for unconfined compressive strength of cohesive soil*, (ASTM International, West Conshohocken, PA, United States, 2000)
22. ASTM D 4767 - 04ASTM D2573 – 08, *Standard test method for field Vane shear test in cohesive soil*, (ASTM International, West Conshohocken, PA, United States, 2008)
23. ASTM D 2435-02 (ASTM International, West Conshohocken, PA, United States, 2002)

# Investigation of soil-steel interface behavior of Iraqi soil by direct shear apparatus

*Omar* Al-Emami *

*School of Energy, Geoscience, Infrastructure and Society, Heriot-Watt University, United Kingdom

**Abstract.** Soil – structure interaction is an important theme observed in many civil engineering structures like fondations. The interface shear strength plays a significant role in the analysis and design of many structures constructed above or under the ground. In this study, a total of 28 specimens were tested at vertical stresses of 100, 200, and 400 kPa using direct shear apparatus under consolidated drained condition. A silty sand soil, as per USCS classification system was prepared in the laboratory at different water contents (4.5%, 8.8%, and 12.5%) and voids ratios (0.4, 0.6, 0.8, and 1). The frictional resistance of this soil was measured. The soil samples were also sheared against three steel surfaces of different textures (smooth, moderate-rough, and rough). The experimental results showed that the steel surface texture is an effective factor in soil-steel interface shear strength. The interface shear strength of the rough steel surface was found higher than smooth and semi-rough steel surfaces. In addition to the surface roughness, the water content and void ratio also play important roles in interface shear strength.

## 1 Introduction

The interface is usually defined as a contact zone between two bodies. The serviceability of the geotechnical structures is highly dependent on the behaviour of the layer of soil forming close to the structure surface (i.e. interface layer). Numerous previous studies reported the crucial role of the interface behaviour in analysis of the some geotechnical problems like a retaining wall and the skin friction along a pile shaft. Briaud et al. [6] indicated that 43% of the resistance along a rigid concrete pile shaft embedded in stiff clay is provided by skin friction. Whereas, Smith and Ray [7] reported that the frictional resistance generated along a concrete pile shaft contributed about 84% of the total soil resistance subjected to a lateral loading. Extensive researches have been performed to investigate the interface shear strength behaviour between soil and different types of materilas under different test conditions (dry and saturated) [2, 9, 1, 4, 3, 5, 10, 11] using a variety of equipment, e.g., direct shear [2, 8], annular shear device [9], Ring torsion apparatus [4], and simple shear [3]. Their outcomes revealed that the interface shear strength depends on different factors like structure interface condition (i.e. rough, intermediate and smooth), type of construction materials (i.e. concrete, steel, wood and geotextile), moisture content, grain size distribution, different types of soil (i.e. clay, sand and silt), void ratio and normal stress. Potyondy [2] has performed earlier soil structure interface research.

This manuscript was submitted on May 15, 2017. This work is supported by Iraqi Ministry of Higher Education and Scientific Research under grant of scholarship.
O. H. AL-Emami is a PhD student at School of Energy, Geoscience, Infrastructure and Society, Heriot-Watt University, Edinburgh, United Kingdom (e-mail: oha31@hw.ac.uk). Former faculty member, Building and Construction Engineering Department, University of Technology, Iraq

He performed numerous laboratory interface direct shear tests to find out the value of frictional resistance between soil and several types of materials (steel, concrete, and wood). The experimental results showed that the angle of internal friction and cohesion should be considered in calculating of skin friction of cohesive soils. He also indicated that the most important factors influencing on the interface shear strength are surface roughness, applied vertical stress, water content, soil composition, and solid material roughness. Kulhawy and Peterson [12] conducted a series of direct shear tests between sand and various surface roughness of concrete to evaluate the influence of surface roughness on the interface shear strength. They pointed out that the surface roughness has a significant impact on the interface shear strength and that the interface friction angle increased with increasing of surface roughness. Their results showed that the angle of friction of soil is greater than the interface friction angle of both smooth and rough surfaces. In another work, Desai et al., [8] studied the properties of sand against concrete interface with different densities, postulating that the interface behaviour greatly influenced by relative density of sand, magnitude of vertical stress, and rate of shear displacement.

* Corresponding author: oha31@hw.ac.uk

© The Authors, published by EDP Sciences. This is an open access article distributed under the terms of the Creative Commons Attribution License 4.0 (http://creativecommons.org/licenses/by/4.0/).

Tsubakihara and Kishida [14] conducted laboratory tests on friction between cohesive soils and moderate steel using two types of apparatuses : a direct shear apparatus and a simple shear apparatus. Different types of soils were examined in this study, clay and sand-clay mixtures. They stated that there was little difference in the maximum friction coefficient between the tests with different mixture ratios of sand. Paikowsky et al., [15] indicated a series of laboratory tests using a modified direct shear apparatus and a modified simple shear apparatus. The test results showed that at a given applied vertical stress level, the major parameters which controlling the interface shear strength behaviour are particles shape and surface roughness. Recently, the influence of surface roughness on the shear strength behaviour of an interface between silty sand and a cement grout surface by means of direct shear apparatus is investigated by Chu and Yin [16]. Their results indicated that the soil-interface shearing resistance depends primarily on the surface roughness, water content of soil, and magnitude of the applied stress. The influence of soil water content, surface roughness and normal stress on the relationship between shear stress versus horizontal displacement was investigated by Shakir and Zhu [17] using simple shear apparatus. In their study, different concrete surface roughness's (smooth and rough) were used. Clay specimens were compacted at the same dry density with different water content ratios. They stated that the interface shear behaviour for both rough and smooth concrete surfaces governed by the interface sliding occurred within the interface layer between soil and concrete for all specimens except when the water content of specimen is greater than 16% for the case of rough concrete surface where the shear failure occurs within the soil body. Recently, Fleigh et al. [18] classified the interface friction behaviour into three failure modes, namely full sliding at the interface, shear failure within the soil and mixed behaviour where interface sliding and shear deformation of soil specimen proceed simultaneously. In their work, they performed a series of interface direct shear tests to study the effect of surface roughness and texture models on the friction angle between cohesive soils and steel.

## 2 Research Objectives

To the author's knowledge, there is inadequate literature available about the behaviour of interface shear strength of Iraqi soils. Hence, the aim of the present paper is to examine the influence of three major parameters namely surface roughness, soil water content and void ratio on the interface shear strength behaviour between silty sand soil and steel surface using direct shear apparatus.

## 3 Laboratory testing

### 3.1 Material

To achieve the goal of this research, synthetic soil specimens were prepared from sandy soil and kaolin since it could not be possible to find natural soil specimens from Iraq. The soil is prepared in the laboratory according to sieve analysis test results of soil obtained from AL-Rashedia site located in Nineveh governorate (495 km north of Baghdad) and published in a journal paper achieved by Khattab et al., [22]. Standard tests were carried out to determine the engineering properties of the prepared soil. Details are given in Table 1. According to the Unified Soil Classification System (USCS), the soil used in the investigation was classified as silty sand or SM. Fig. 1 shows particle-size distribution curves of the original and prepared soils, which indicate that the soil is composed of 73% sand, 19% silt, and 8% clay. Moreover, a standard compaction test was performed according to (ASTM D1577-02) by using a 2.5 kg rammer and a mould of 1000 $cm^3$. Fig. 2 shows the relationship between the dry unit weight and water content of the soil used. The obtained maximum dry density is 20.20 $kN/m^3$ with an optimum moisture content of 8.8%.

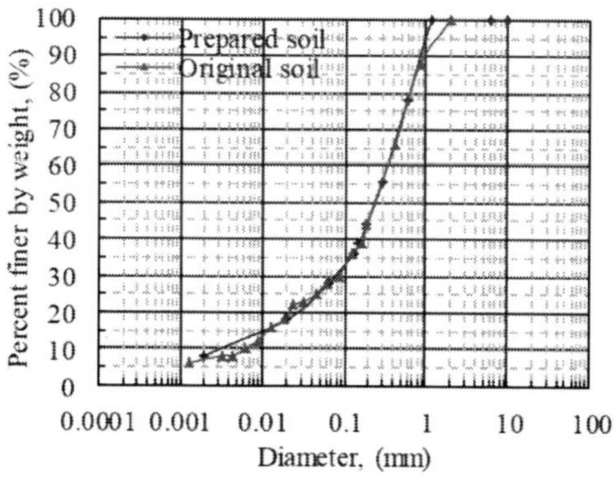


**Fig. 1**. Particle-size distribution curve of the soil

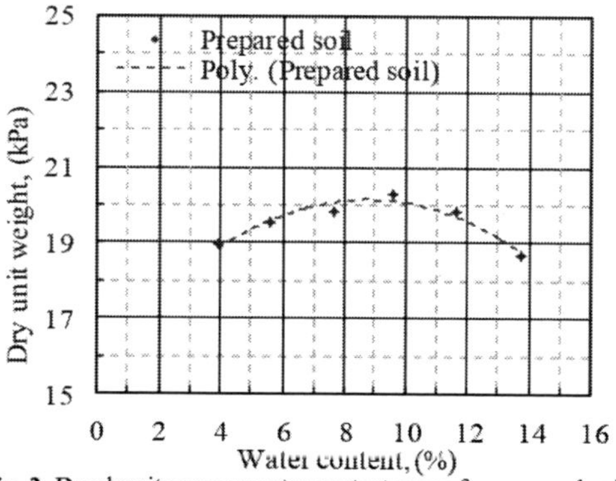


**Fig. 2**. Dry density versus water content curve for prepared soil

### 3.2 Sample preperation

In this research, the soil specimen used in the direct shear test was prepared in the laboratory and placed in the oven at a temperature of 105 °C for 24 hours. Then, the desired quantity of soil was taken and appropriate amount of water was added to the target water content.

The wet soil was wrapped in anti-evaporation plastic bags and stored in containers for 72 hours for water content equalization. For soil to soil specimen, the soil was compacted inside a square mould having inner dimensions of 60 mm by 60 mm and a height of 20 mm. The soil was compacted in four layers of 5 mm thickness of each inside steel mould to the required density. The compaction was done by using a tamping rod. Finally, the mould with the specimen was placed over the upper half of shear box and the specimen pushed slowly inside the shear box until the specimen attached with the lower porous stone. The above procedure was adopted in preparing soil against steel direct shear specimen except that the wet soil was compacted in two layers of 5 mm each. Total number and types of tests performed for this study are presented in Table 2.

**Table 1.** Index properties of prepared soil

| Soil property | Unit | Value |
|---|---|---|
| Specific gravity (Gs) | -------- | 2.74 |
| Liquid limit (L.L.) | % | 21 |
| Plastic limit (P.L.) | % | 20 |
| Plasticity index (P.I) | % | 1 |
| Gravel | % | 0 |
| Sand | % | 73 |
| Silt | % | 19 |
| Clay | % | 8 |
| Standard maximum dry density | kN/m$^3$ | 20.20 |
| Optimum water content | % | 8.8 |
| $D_{60}$ | mm | 0.35 |
| $D_{50}$ | mm | 0.25 |
| $D_{30}$ | mm | 0.075 |
| $D_{10}$ | mm | 0.004 |
| Soil classification (USCS) | -------- | SM |

**Table 2.** Total number and types of tests conducted for this study

| Parameter | Interface | Applied vertical stress (kPa) | | | Test condition |
|---|---|---|---|---|---|
| | | 100 | 200 | 400 | |
| e = 0.4 | Soil | x | x | x | Consolidated drained |
| | Smooth | x | x | x | |
| | Moderat | x | x | x | |
| | Rough | x | x | x | |
| e = 0.6 | Soil | x | x | x | Consolidated drained |
| | Smooth | x | x | x | |
| | Moderat | x | x | x | |
| | Rough | x | x | x | |
| e = 0.8 | Soil | x | x | x | Consolidated drained |
| | Smooth | x | x | x | |
| | Moderat | x | x | x | |
| | Rough | x | x | x | |
| e = 1 | Soil | x | x | x | Consolidated drained |
| | Smooth | x | x | x | |
| | Moderat | x | x | x | |
| | Rough | x | x | x | |
| ω = 4.5% | Soil | x | x | x | Consolidated drained |
| | Smooth | x | x | x | |
| | Moderat | x | x | x | |
| | Rough | x | x | x | |
| ω = 8.8% | Soil | x | x | x | Consolidated drained |
| | Smooth | x | x | x | |
| | Moderat | x | x | x | |
| | Rough | x | x | x | |
| ω = 12.5% | Soil | x | x | x | Consolidated drained |
| | Smooth | x | x | x | |
| | Moderat | x | x | x | |
| | Rough | x | x | x | |

### 3.3 Counterface

In order to investigate the influence of surface roughness on the interface shear stress between soil and structures, three steel plates prepared from a steel sheet were used in this study as a counterface. All steel plates were manufactured having the same dimensions of 100*100*10 mm, which is exactly the same as the size of lower half of shear box. Three different surfaces roughness (smooth, semi-rough, and rough) were performed as shown in Fig. 3. The completely smooth surface polished by fine sandpaper [2]. A value of 0.0025mm was used as a peak to valley height for smooth steel plate [25]. Uesugi and Kishida [3] defined the normalized surface roughness, $R_n$ as a maximum peak to valley height, $R_{max}$ divided by grain size dimeter corresponding to 50% finer, $D_{50}$. Based on the particle-size analysis curve, the normalized surface roughness of smooth, moderate and rough steel plates were equal to 0.01, 4, and 8 respectively.

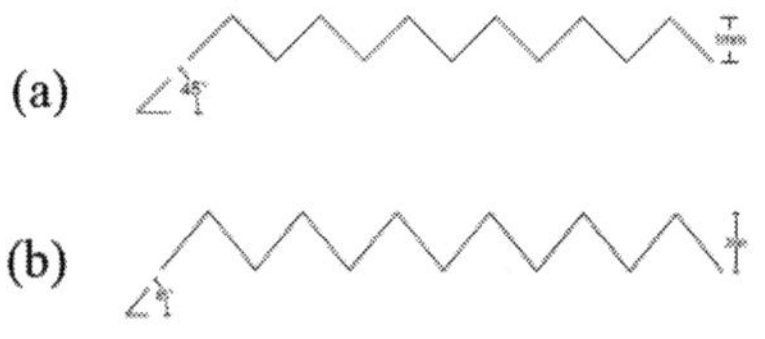


**Fig. 3.** Surface geometry of (a) semi-rough: and (b) rough

## 4 Soil testing method

A fully automated commercially direct shear apparatus was used in this study. All tests were performed according to (ASTM D3080-04). The test procedure consists of three stages :- saturation, consolidation and shearing stages. After placing the soil specimen inside the shear box, the specimen was soaked in water for 24 hours to ensure that the voids are completely filled with water. After saturation, the specimen was consolidated under three applied vertical stresses of 100, 200 and 400 kPa. The consolidation process continues until the vertical deformation versus horizontal displacement curve is levelled off. After consolidation attained, a shearing stage was performed under drained condition with a constant horizontal shearing rate of 0.06 mm/min for all specimens (i.e. soil versus soil and soil versus interface).

## 5 Test results and analysis

### 5.1 Test results of soil to soil specimens

Single stage direct shear tests were performed on initially saturated soil to soil specimens under different applied vertical stresses of 100, 200 and 400 kPa. The effect of different water content and void ratio on the shear strength behaviour of compacted silty sand soil are presented and discussed in the following sections.

#### *5.1.1 Influence of water content on shear strength behaviour*

Typical direct shear test results are presented in terms of shear stress versus horizontal shear stress displacement curves for soil specimens tested at different water contents (4.5%, 8.8% and 12.5%) under 100, 200 and 400 kPa applied vertical stress respectively are shown in Fig. 4a-c. Soil specimens of dry and wet side of optimum were compacted at same dry density (19.2 kN/m$^3$) with different water content (4.5% and 12.5%). As expected, all specimens exhibited a hardening behaviour at beginning of shearing stage until peak shear stress followed by different patterns of shear behaviours (softening and/or hardening). These figures also illustrated that the shear strength increased with the increasing of applied vertical stress. Regardless of the value of vertical stress, specimens compacted at dry side of optimum exhibited higher shear stress compared with those tested at wet side of optimum. These findings coincide with the findings reported by Tiwari and Al-Adhadh, [11]. Fig. 4 also indicates that clear peak shear stress and softening behaviour are obvious for specimens carried out under 400 kPa vertical stress. Fig. 5 depicts Mohr-Coulomb shear envelopes obtained from the three different types of tests. As expected, specimens compacted at optimum water content had higher shear strength compared to those tested at dry and wet side of optimum for all vertical stresses. However, dry side of optimum specimens showed higher shear strength than those compacted at wet side of optimum. Vanapalli et al., [19] and Cokca et al., [20] found the same results. The shear strength parameters along with the water content are summarized in Table 3. As can be seen in Table 3, a significant reduction was observed in the effective cohesion with an increase in water content. In addition, effective friction angle decreased as water content decreased. For example, friction angle decreased from 30° to 29° when water content increased from 4.5% at dry side to 12.5% at wet side of optimum.

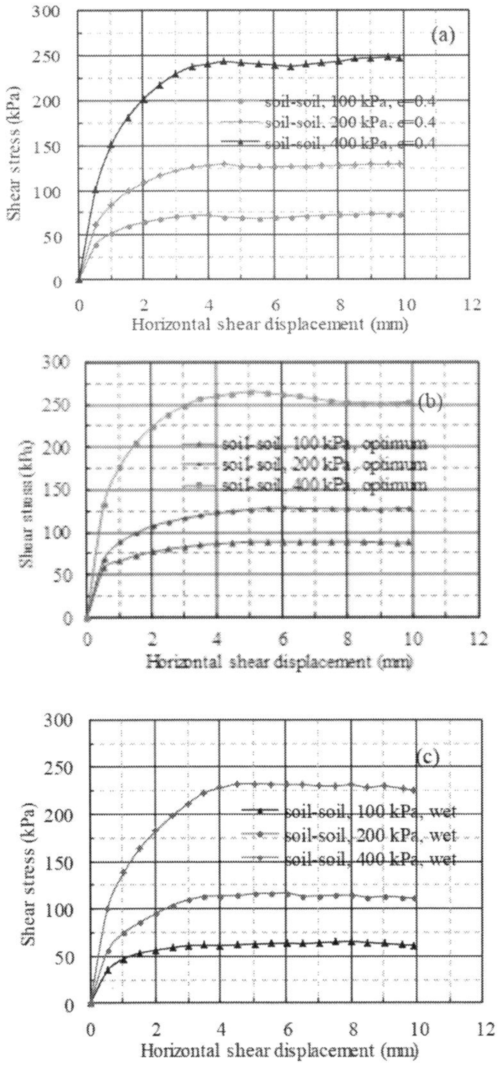


**Fig. 4**. Direct shear test results for soil to soil specimens compacted at different water content (a) dry; (b) optimum; and (c) wet

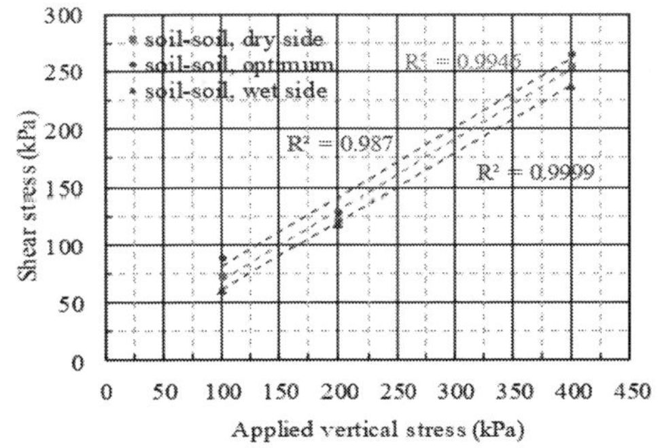


**Fig. 5**. Failure envelopes corresponding to different water content

**Table 3.** Shear strength parameters from different water content tests

| Water content (%) | ϕ′ (deg) | c′ (kPa) |
|---|---|---|
| Dry side = 4.5 | 30 | 10 |
| Optimum = 8.8 | 31.5 | 21 |
| Wet side = 12.5 | 29 | 4 |

### *5.1.2 Influence of void ratio on shear strength behaviour*

Twelve consolidated drained direct shear tests were carried out on the saturated silty sand under three vertical stresses and four different void ratios. Fig. 6a-d shows shear stress versus horizontal displacement curves for soil specimens performed at different void ratios (0.4, 0.6, 0.8, and 1) and water content equal to 7%. It can be seen from Figure 6 that the shear stress decreased with increasing of void ratio for all applied vertical stresses. As can be observed, the maximum shear stress increases as the vertical stress increases for all void ratios. It can be also noticed that the dense specimens (low void ratio) exhibited peak shear stress more clearly than the loose specimens (high void ratio). The values of maximum shear stress from all tests are plotted against applied vertical stress in Fig. 7. The shear strength envelopes for various void ratios are approximately linear. Moreover, the shear strength parameters ϕ′ and c′ decrease with an increasing of void ratio as shown in Table 4. The experimental shear strength result agrees with the findings reported by Tiwari and Al-Adhadh [11].

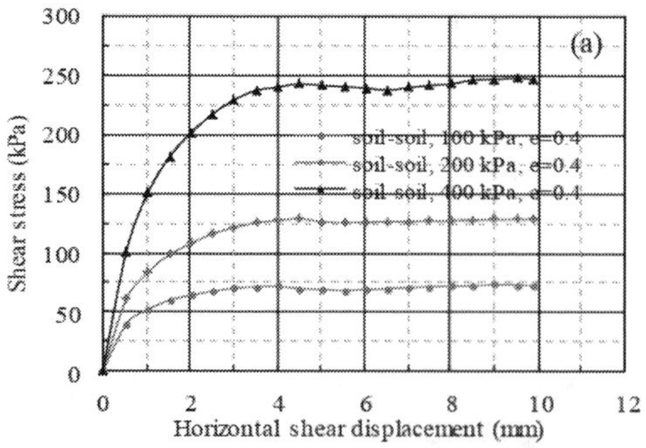


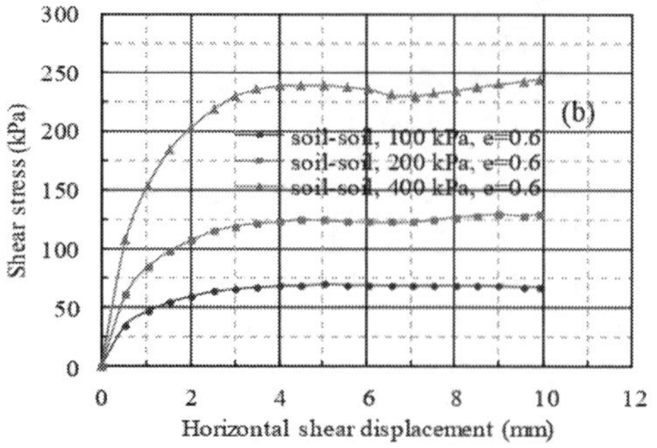


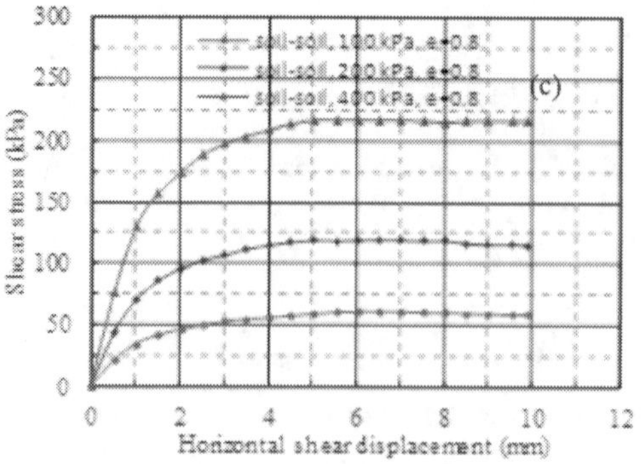


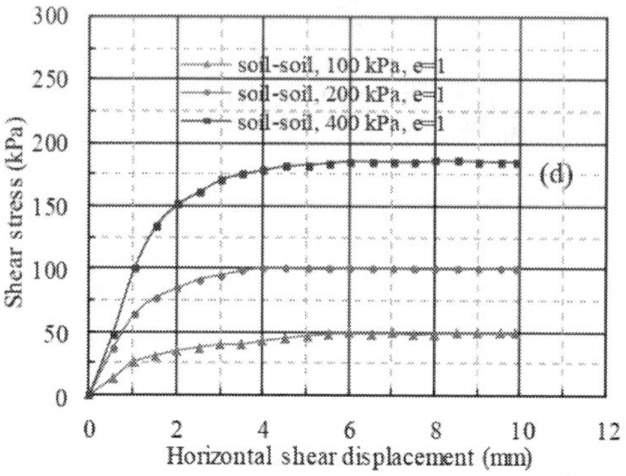


**Fig. 6.** Direct shear test results for soil to soil specimens compacted at different void ratios (a) 0.4; (b) 0.6; (c) 0.8; and (d) 1

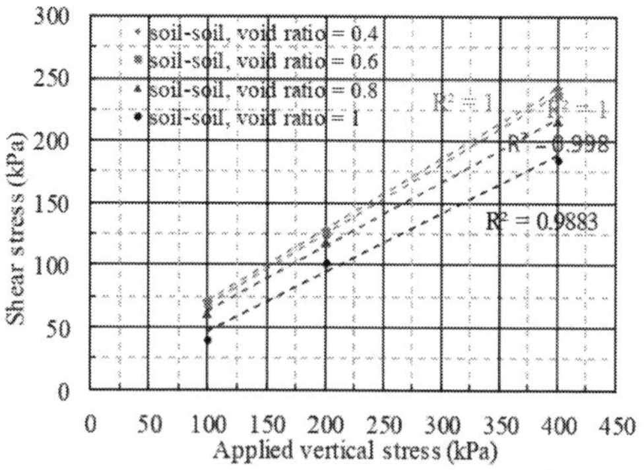


**Fig. 7.** Failure envelopes corresponding to different void ratios

**Table 4.** Shear strength parameters from different void ratio tests

| Void ratio | ϕ′ (deg) | c′ (kPa) |
|---|---|---|
| 0.4 | 29 | 20 |
| 0.6 | 28.5 | 19 |
| 0.8 | 26.5 | 15 |
| 1 | 24 | 0 |

## 5.2 Interface shear tests

This study is set out with the aim of assessing the importance of surface roughness on the shear strength of soil. To do this, a series of interface direct shear tests were carried out to examine the effect of three types of steel surface roughness on the interface shear strength at different void ratios and water content. Curves of shear stress versus horizontal shear displacement for different water content (4.5% and 12.5%) sheared against smooth, semi-rough, and rough steel surfaces under applied vertical stress of 400 kPa are plotted in Fig. 8a-b. Similar behaviour was observed while the specimens were examined under other vertical stresses. It is interesting to note that the shear stress increased with an increase in the surface roughness increases for both dry and wet side of optimum conditions. As can observed from Figure 8a-b that the interface shear strength of specimen with rough steel significantly increased approximately by 1.5 folds and 2 folds for specimens with 4.5% and 12.5% water content, respectively It can be also observed that the curves for rough surface exhibited peak shear stress followed by softening behaviour. As opposed to the rough interface, curves of smooth interface showing no peak shear stress followed by steady state condition. The failure envelopes for soil-steel interfaces performed under different water content are shown in Fig. 9a-b. The failure envelopes are found to be linear. It can be also noticed that there was a remarkable reduction in the interface shear strength compared to the shear strength of soil. The highest reduction was observed for soil sheared against smooth interface compared to the specimens sheared against semi-rough and rough interfaces. This trend was observed for all water content values.

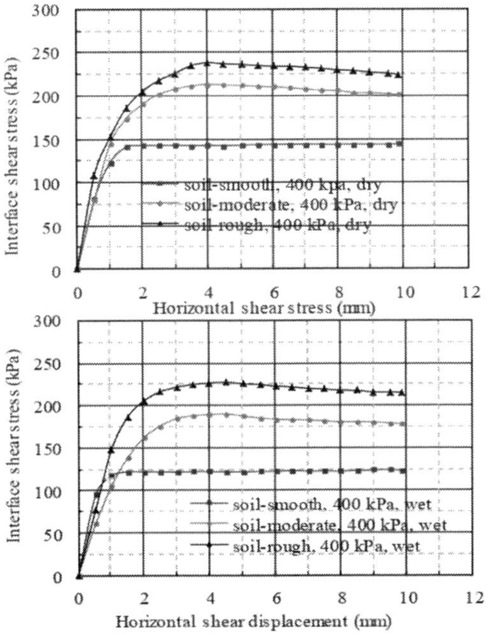


**Fig. 8** Direct shear test results for soil to interface specimens compacted at different water content (a) dry; (b) wet

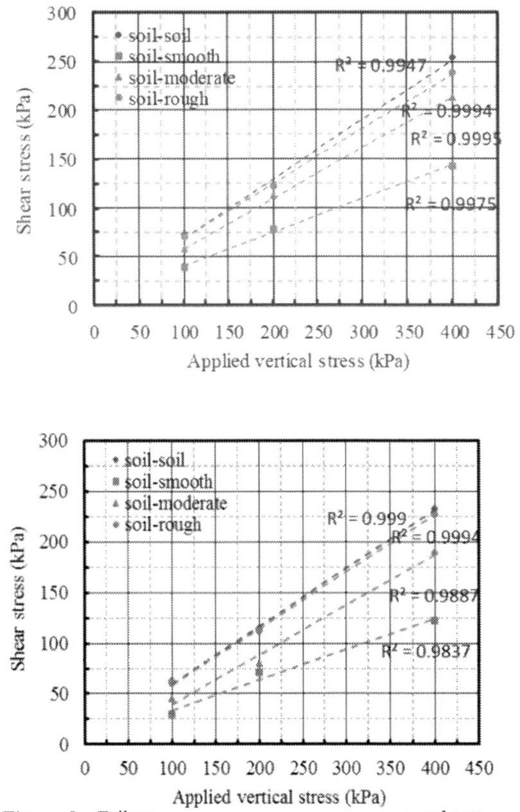


**Fig. 9.** Failure envelopes for the soil to soil and soil to interface for specimens compacted at different water content (a) dry; and (b) wet

These results are consistent with those reported by [21]. The shear strength parameters along with the water content are summarized in Table 5.

**Table 5.** Shear strength parameters from different water content tests

| Interface | Water content (%) | ϕ′ (deg) | c′ (kPa) |
|---|---|---|---|
| Soil | 4.5 | 30.5 | 10 |
| Smooth | | 18.5 | 6 |
| Semi-smooth | | 26.5 | 7.5 |
| Rough | | 28.5 | 12 |
| Soil | 12.5 | 29 | 6 |
| Smooth | | 15.5 | 4 |
| Semi-smooth | | 25 | 0 |
| Rough | | 27 | 5 |

Likewise, the results of interface direct shear tests for the soil specimens with 0.4 and 1 void ratio are presented in Figure 10a-b tested under applied vertical stress of 400 kPa. Similar behaviour was observed while the specimens were carried out under other applied vertical stresses. As evident, there is a consistent increase in the interface shear stress with an increase in the surface roughness irrespective of whether the void ratio was 0.4 or 1. As could observed from Fig. 10a-b that the interface shear strength of specimen with rough steel significantly increased approximately by 3 folds and 1.6 folds for specimens with 0.4 and 1 void ratios respectively. It can be also noticed that the curves for rough surface exhibited peak shear stress followed by softening behaviour. As opposed to the rough interface, curves of smooth interface showing no peak shear stress followed by steady state condition. In general, it is observable that the void ratio has a remarkable influence on the interface shear stress. For example, the maximum shear stress for rough interface decreased by 28% when the void ratio increased from 0.4 to 1. The presented findings seem to be consistent with the results observed by previous researchers such as [2, 12, 10, 21]. The values of peak shear stress form all tests are plotted against the vertical stress in Fig. 11a-b, for different void ratios. The interface failure envelopes for various interfaces roughness are approximately linear. Similar to interface direct shear results under different water content, a significant reduction in the interface shearing resistance compared to the shearing resistance of soil. The highest reduction was observed for soil sheared against smooth interface compared to the specimens sheared against semi-rough and rough interfaces. The shear strength parameters along with the void ratios are tabulated in Table 6. In addition, Figure 12a-b illustrates the influence of surface roughness on the interface angle of friction.

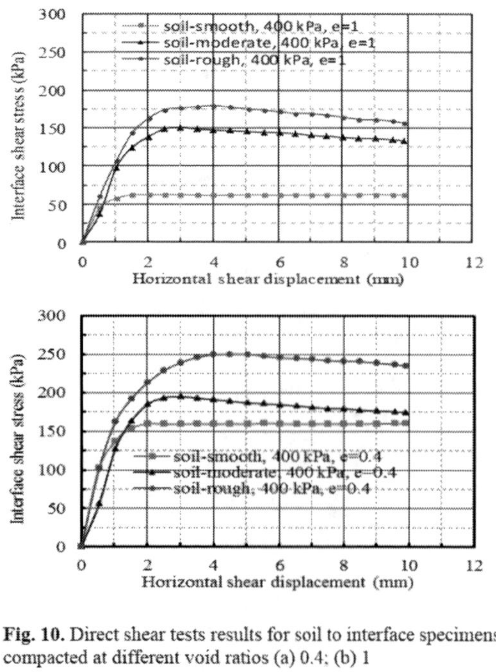


**Fig. 10.** Direct shear tests results for soil to interface specimens compacted at different void ratios (a) 0.4; (b) 1

**Table 6.** Shear strength parameters of soil and interfaces

| Interface | Void ratio | ϕ′ (deg) | c′ (kPa) |
|---|---|---|---|
| Soil | 0.4 | 30.5 | 17 |
| Smooth | | 20 | 8 |
| Semi-rough | | 24 | 13 |
| Rough | | 30 | 16 |
| Soil | 1 | 24 | 7 |
| Smooth | | 5.5 | 23 |
| Semi-rough | | 19 | 0 |
| Rough | | 22.5 | 5 |

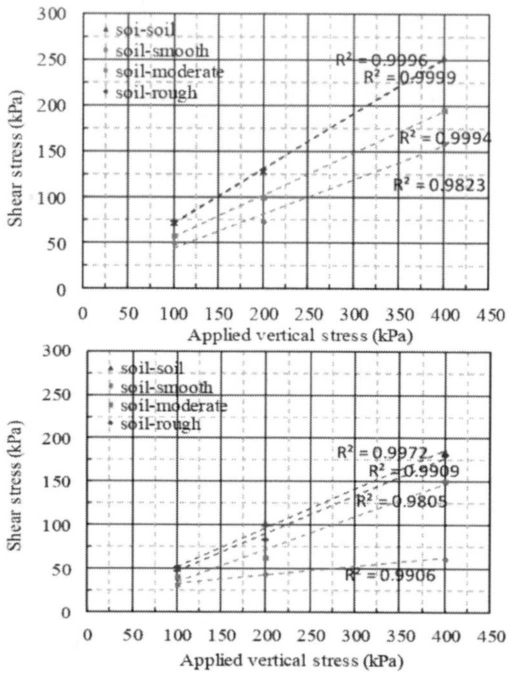


**Fig. 11.** Failure envelopes for the soil to soil and soil to interface for specimens compacted at different void ratios (a) 0.4; and (b) 1

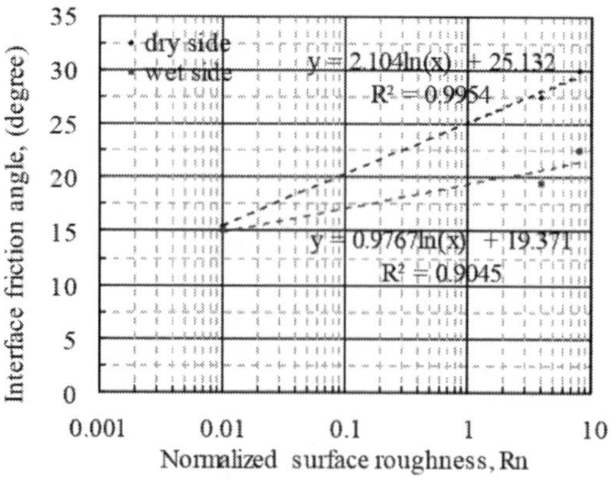

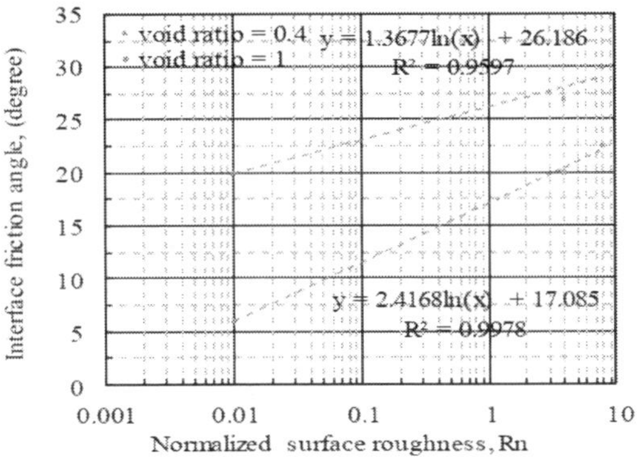


**Fig. 12.** Influence of surface roughness on the interface friction angle

## 6 Conclusions

Consolidated drained direct shear tests were carried out to investigate the influence of water content, void ratio, and surface roughness on the interface shear strength between silty sand and steel plate. Three types of surface roughness (smooth, semi-rough, and rough) were used. The specimens were compacted at different water contents (4.5%, 8.8%, and 12.5%) and void ratio (0.4, 0.6, 0.8, and 1). Based on the results obtained the following conclusions are summarized.

1. The shear strength of soil increases with decreasing both water content and void ratio.
2. Soil shear strength increased with increasing of applied vertical stress. Based on the presented results, it can be seen that soil shear strength was higher than the interface shear strength of smooth, semi-rough and rough interfaces.
3. The interface shear stress increases with increasing the surface roughness. Therefore, from the geotechnical point of view, rough steel surface can be used to increase the interface shear strength of some geotechnical structures such as steel pile foundation and offshore structures.
4. The interface friction angle of different surface roughness increases as the surface roughness increases. Interface friction angle increased by 50% when the surface roughness changed from the smooth to the rough surface for specimens compacted at dry side of optimum.
5. Based on the results, void ratio and water content had an obvious influence on the interface friction angle between soil and steel plates. Referring to the results, high water content and low void ratio produced small friction angle.

## References

1. Y. B. Acar, H.T. Durgunoglu, M. T Tumay, J. of Geo. Eng. ASCE, **108**, GT4, (1982).
2. J. G. Potyondy, Geotechnique, **11**, (1961).
3. M. Uesugi, and H. Kishida, Soils and Found.,**26**, 4, (1986).
4. Y. Yoshimi, and T. Kishida, 10th Int. Conf. on Soil Mech. and Found. Eng., **1**, (1981),
5. K. Fakharian, and E. Evgin, Geo. Tes. J., **19**, 1, (1996).
6. J. L. Briaud, T. D. Smith, and B. Meyer, Symp. on the Press. and Its Marine App., Paris, (1982).
7. T. R. Smith, and B. Ray, Geo. Asp. of Stiff and Hard Clays, Ed. Khera and Lovell, ASCE, New York, (1986).
8. C. S. Desai, E. C. Drumm, and M. M. Zaman, ASCE, J. of Geo. Eng., **111**, 6, (1985).
9. W. F. Brumund, and G. A. Leonards, J. of Test. and Eva., JTEVA, **1**, 2, (1973).
10. A. I. Al-Mhaidib, Eng. J. of the Uni. of Qatar, **19**, 2, (2006).
11. B. Tiwari, and A. Al-Adhadh, Geo. Geol Eng, **32**, (2014).
12. F. H. Kulhawy and M. S. Peterson, 6th Pan-American Conf. on Soil Mech. and Found. Eng., **7**, (1979).
13. Y. Tsubakihara, and H. Kishida, Soils and Found. **33**, 2, (1993).
14. S. G. Paikowsky, C. M. Player, and P. J. Connors, Geo. Test. J., **18**, 2, (1995).
15. L. M. Chu, and J. H. Yin, Geomech. and Geoeng., **1**, 4, (2006).
16. R. R. Shakir, and J. Zhu, J. Fron. of Arch. and Civ. Eng. in China, **3**, (2009).
17. M. Feligha, F. Hammoud, M. Belachia, and M. S. Nouaoria, Geotech Geol Eng, **34**, (2016).
18. E. Cokca, O. Erol, and F. Armangil, Geot. and Geo. Eng., **22**, (2004).
19. S.K. Vanapalli, D.G. Fredlund, D.E. Pufahl, and A.W. Clifton, Can. Geot. J., **33**, (1996).
20. S.E. Mousavi, Transactions B: Applications, **29**, 2, (2016).
21. S.A. Khattab, E. M. Mohammad, and M. G. Jassam, Tik. J. of Eng. Sci., **19**, 3, (2012).
22. T. Hamid, Ph.D. dissert. School of Civ. Eng. and Envi. Sci., Univ. of Oklah., Norman, Okla., (2005).

# Simulation of bearing capacity of bored piles

*Ahmed* Majeed[1,*], *Olla* Haider[1]

[1] Building and Construction Engineering Department, University of Technology, Baghdad-Iraq

**Abstract.** This study focuses on how one can possibly predict the ultimate load for the piles that did not reach failure. This challenge was acquired through Chin- Konder method by which, the estimated settlement that correspond to failure load is well defined. Hence, this research aims to make a comparative study between the results of pile load tests carried out in Al-Basrah sewage treatment plant project, and those results induced from the numerical analysis in term of ultimate pile capacity. Consequently, it may give a clear idea on the ability of numerical simulation in getting close to the actual behavior of piles. In the current study, a numerical study using Plaxis 3D Foundation program has been performed on bored piles by the assistance of site investigations of soil. Mohr- Coulomb and linear elastic models were adopted in the simulation for soil and pile respectively. Ten bored piles were used in this analysis under different values of loading. The diameter and length of pile are 0.6m and 24m respectively. The test results indicate that, an excellent agreement has been found as a response of pile capacity between the field and numerical studies. Also, ideal load- settlement curves were created using Chin- Konder method to predict the failure load of bored piles. Also, the results have demonstrated that, the pile capacity obtained from the simulation process is larger about 51% than that design load estimated before the design of piles. This may present a priority to use the finite element method to be accounted as an effective approach in the primary analysis.

## 1 Introduction

The noticeable disparity in pile design techniques and the uncertainties associated to the traditional theories due to the huge numbers of parameters involved in analysis, necessitates that to alternative analytical ways is essential to account for the whole behavior of piles. In this study, a numerical modeling using Plaxis 3D Foundation software was performed on bored piles placed into layered soil to validate the field test results which has been implemented under same conditions.

Thereupon, the traditional methods for calculating the problems of pile bearing capacity are still insufficient to be accounted in design. As presented, the simplified analysis for the conventional theories ignore some of the facets involved making the solutions inaccurate and deficient to be relied unless verified by certain reliable methods.

Thus, the main role for using the finite element method, is to thoroughly understand and realize some of the practical aspects of bored piles which are hard to obtain using the conventional design methods. These traditional methods have emerged over years to afford conservative solutions seeking the cost and safety, therefore it is important to revise and adjust the bearing capacity with those results obtained from the field load tests especially for strategic projects.

Herein, the pile load test is normally carried out to verify and to check the response of foundations in term of bearing capacity and settlement. The major challenge in this concern is not how to get the design load of pile, but estimation of the behavior of pile that did not reach to failure is still dialectical. Thus, restoring to alternative approaches to expect the actual work load would be greatly required. There are several studies related with this scope, [7] demonstrated that, for 1.2m diameter bored cast in-situ piles executed in residual soils, a reasonable agreement has been gained between the ultimate pile capacity obtained from the finite element analyses in 2D Plaxis and full scale tests. The values of pile settlement in both analysis were 2.6 and 4.9mm under the maximum load equal to 825 tons.

[8] analyzed the results of four static pile load tests and a these results have been validated using Plaxis 3D Foundation program. The model used in the analysis is Mohr coulomb. The pile length and diameter are 9m and 0.62m respectively. Plaxis and in-situ measured load-displacement curves showed a good correlation.

[4] carried out a comparative study in order to assess the behavior of driven piles which have a diameter of 0.35m and embedded length of 10- 12m supported 8 stories-building. The case study compared the field results with

* Corresponding author: 40349@uotechnology.edu.iq

© The Authors, published by EDP Sciences. This is an open access article distributed under the terms of the Creative Commons Attribution License 4.0 (http://creativecommons.org/licenses/by/4.0/).

numerical simulation using Plaxis 2D. The result of finite element analysis shows that the ultimate axial capacity is in good convergence with the axial capacities obtained using empirical method and field results.

[6] Found out that, the simulation of cast in situ piles with a diameter of 620mm and penetration length equal to 15m is an effective method in design. The piles were installed in silt sand, and the numerical analysis has been done using Plaxis 3D software. The test results displayed a pronounced agreement between the two studies in term of load- settlement curves.

[10] Performed a numerical analysis on a bored pile with dimensions of 0.6m diameter and 26m in length, and subjected to axial load using 2D Plaxis software. According to the borehole details, the pile is penetrated in a clayey silt and decomposed of organic matter. The simulation results in 2D Plaxis showed a good agreement between the field and the finite element results.

In summary, this research displays a numerical investigation for the bored piles executed in Al-Basrah sewage treatment project and a comparison of pile capacity is performed with the field results at failure to understand how far the numerical approaches get close to field ones.

## 2 Site investigation and soil details

The project is located at Hamdan region which is a part of Al-Basrah governorate. The field explorations were carried out on the soil layers to obtain their characterization. Ten boreholes were made at different zones, each one describes stratification of soil under a certain structure. According to [2], drilling was performed using flight auger. The undisturbed samples marked (U) were obtained using Shelby tubes. Split spoon samples (SS) were obtained from standard split spoon used in a Standard Penetration Test (S.P.T), which was performed for each test boring at different intervals depending on the stratification of the soil.

The N-Values for S.P.T. and the variation of water content with depth are shown in Figures (1) and (2) respectively. Then, the N values are corrected by taking a percent of 60% ($N_{60}$) and the water content values are somehow useful to calculate the unit weight of the soil in different cases (wet or saturated), which are basically required in the numerical analysis.

The soil investigation along the depth of boreholes exhibits that, the type of brown/gray silty clay or clayey silt with some organic matters or salts at certain depths was shown to be prevailed. Further drilling up to 21m presented sillty sand to dense sand and beyond which the soil is very dense sand with N-values up to 50 according to soil report.

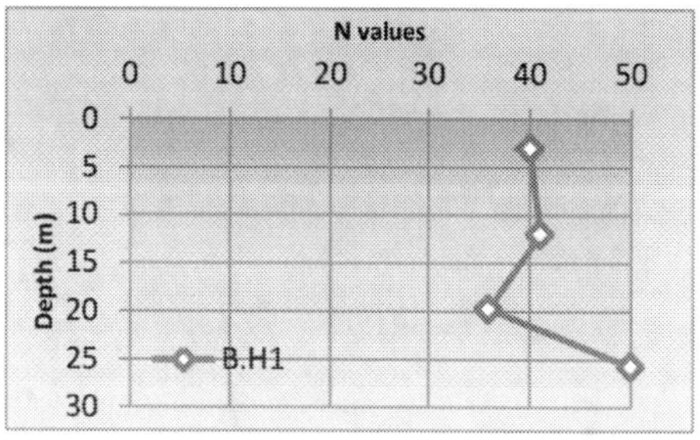


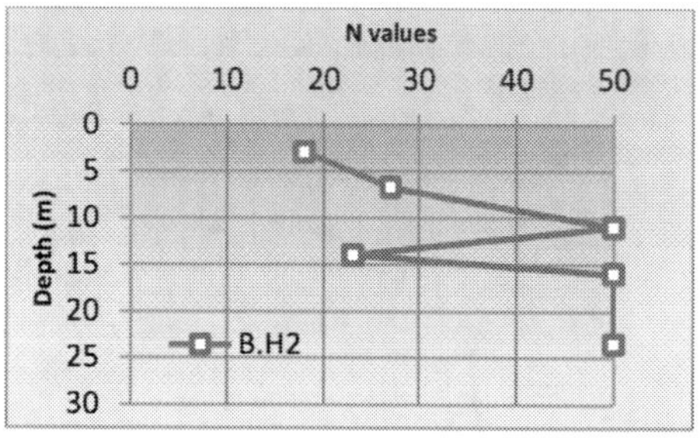


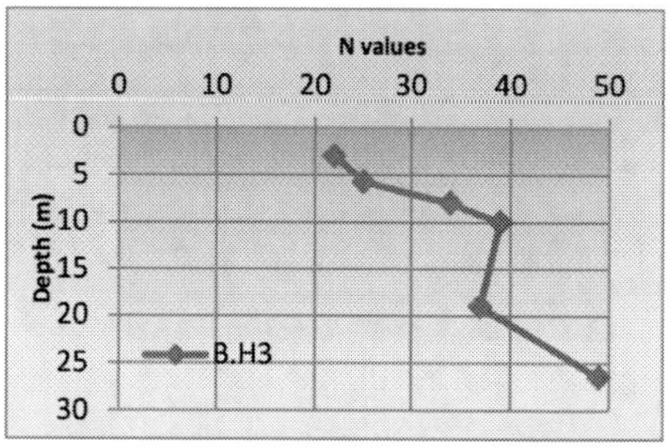


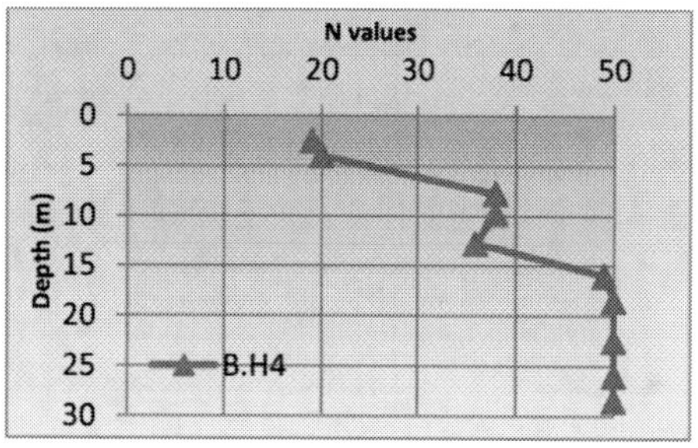

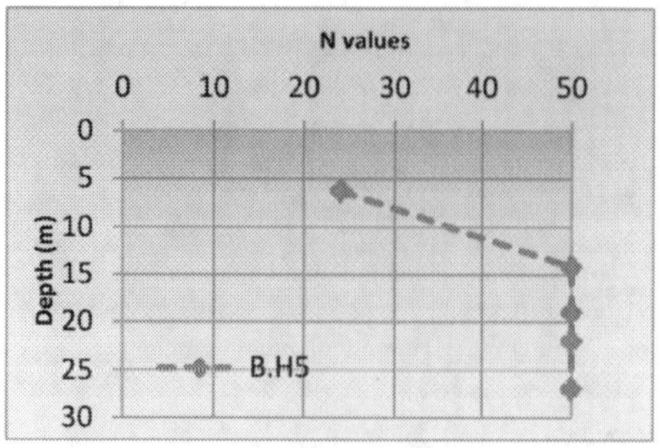


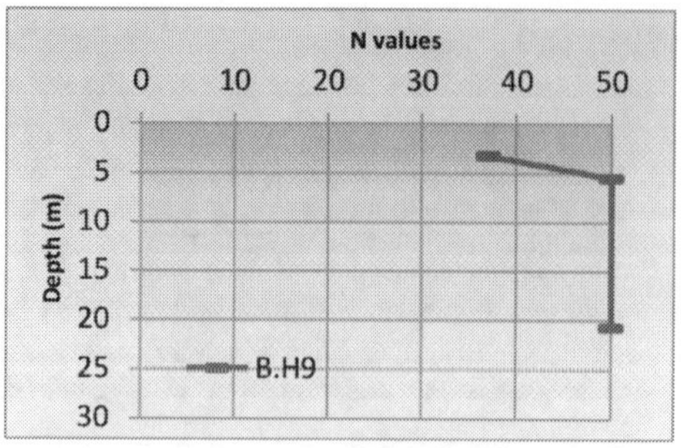


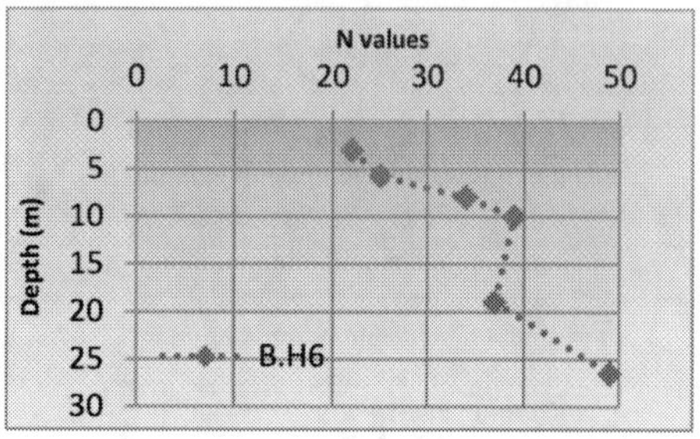


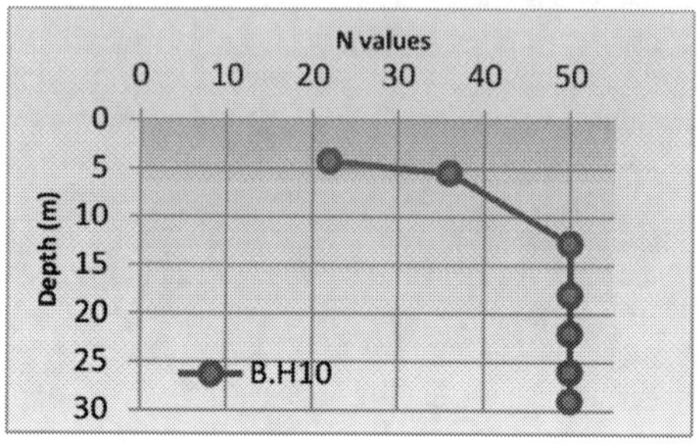


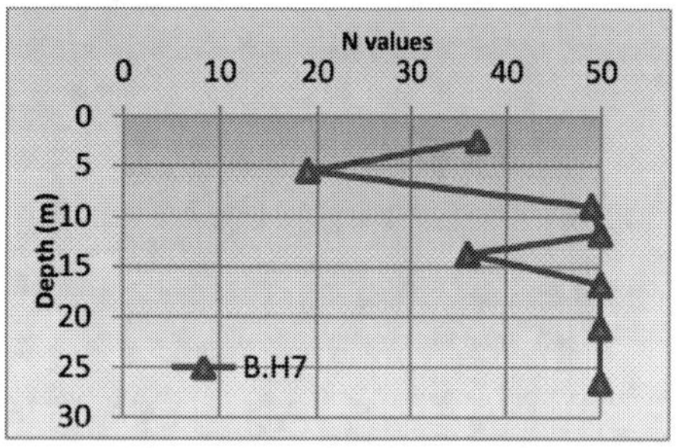


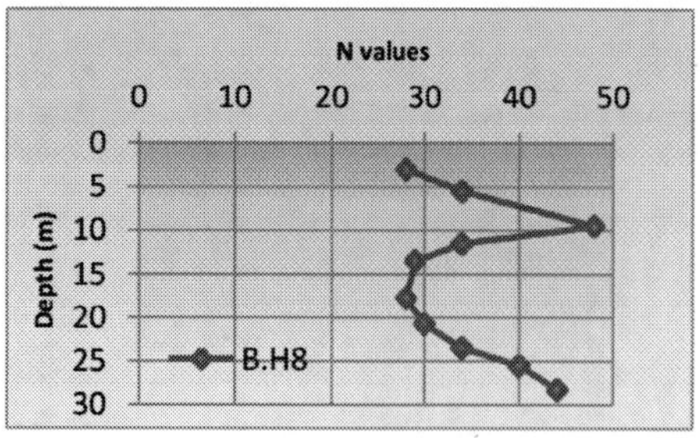


**Fig.1**. Variation of standard penetration values (N) with depth for different boreholes

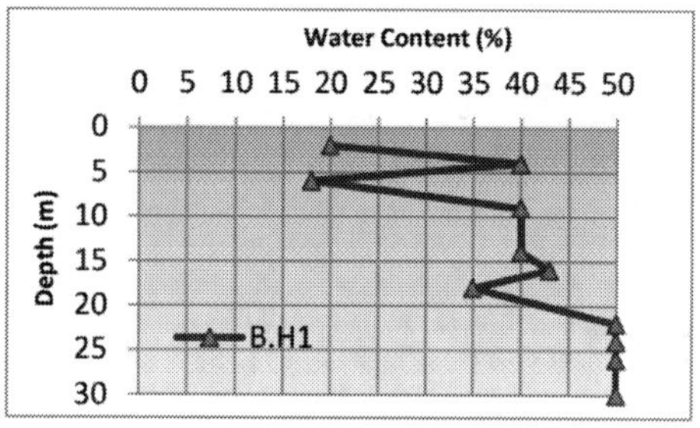


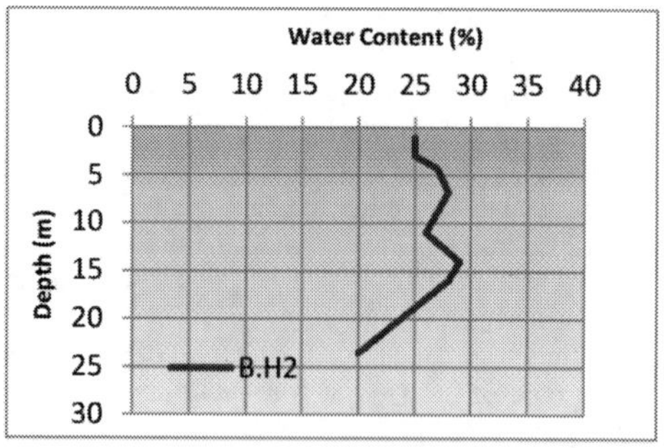

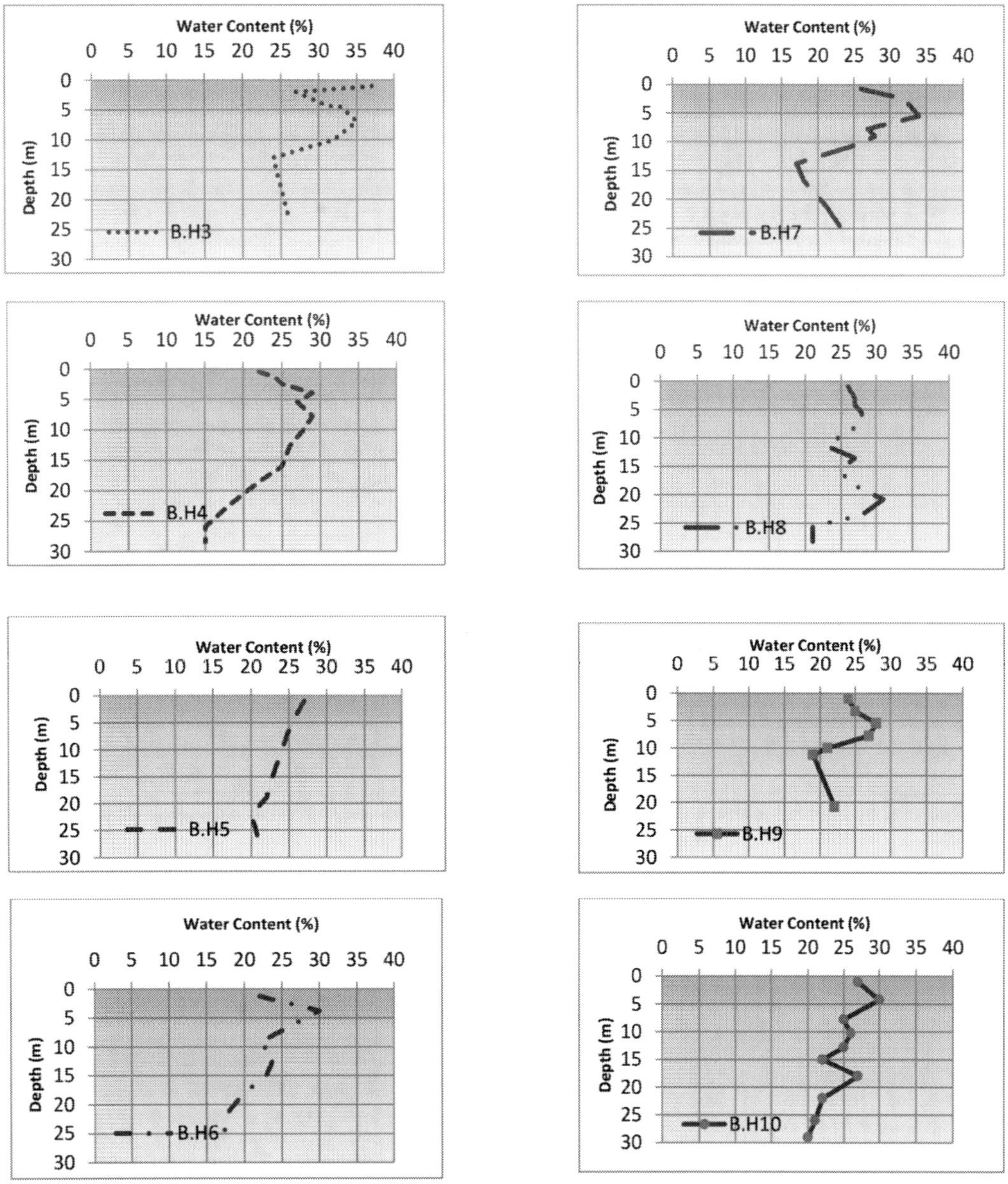


**Fig.2.** Variation of water content values with depth for different boreholes.

An exhaustive analysis was made to obtain the soil properties for each type. The values of specific gravity, void ratio and the unit weights at dry and saturated states for each layer which empanel within the physical soil characterization were calculated based on the visual soil classification with the aid of water content logs. The shear strength parameters which are basically relied on specimens extracted from undisturbed samples were obtained (undrained shear strength $c_u$) especially for the

fine grained soil, nonetheless for some reason there were depths at which no samples had been taken, the shear strength details were induced based on standard values found by [9] which stipulate that the N-values and soil consistency must be known.
The stiffness properties were also computed using triaxial test for the samples taken from undisturbed type, but for the layers at which it is too difficult to extract an intact samples to be tested, feasible magnitudes of modulus of elasticity and Poisson's ratio for each type of soil have been chosen based on soil type and according to [3].
For the coarse grained soil or sand mixed with few quantities of silt or clay, the angle of internal friction φ was determined based on the corrected values of SPT test, for all the cases the $N_{60}$ is predominant. The correlation between the angle of internal friction and $N_{60}$ that is proposed by [5] is expressed as:

$$\phi = \sqrt{20\,Ncor} + 20 \tag{1}$$

The full details of soil properties only for borehole No.1 will be listed in Table 1, just to demonstrate how these parameters have been acquired.

**Table 1.** Borehole log and soil properties No.1

| Property index | Depth m (from- to) | | | | |
|---|---|---|---|---|---|
| | 0-2 | 2-4 | 4-6 | 6-18 | 18-30 |
| Soil Description | Brown silty clay | Silty sand | Stiff silty clay | Clayey sily with organic | Grey silty sand |
| Modulus of elasticity, $E_s$ (MPa) | 6.0 | 14.0 | 4.5 | 45 | 50 |
| Poisson's ratio, ν | 0.3 | 0.3 | 0.4 | 0.3 | 0.3 |
| Angle of internal friction, φ | - | 43° | - | - | 45° |
| Cohesion, c ($kN/m^2$) | 36.0 | - | 47 | 220 | - |
| Angle of Dilation, ψ | - | 13 | - | - | 15 |
| Dry unit weight, γd ($kN/m^3$) | 16.0 | 16.8 | 17.0 | 15.6 | 16.5 |
| Total unit weight, γt ($kN/m^3$) | 20.0 | 19.0 | 20.0 | 19.3 | 19.5 |

## 3 Pile description

Full scale of bored piles in circular cross sections have been evaluated under static compressive load tests. A total number of ten piles were executed at different zones of the project. The structures that were supported by these piles and their function are presented in Table 2. The diameter and length of piles used are 0.8 and 24m respectively. The pile load tests were carried out on working and trial piles by means of Kent ledge dead weights according to [1]

**Table 2.** Summary of piles-supported structures and their function used in this study

| No. | Type of structure | Pile type | Design load (Ton) | Testing load (Ton) |
|---|---|---|---|---|
| 1 | Clarifier Tank TP1 | working | 50 | 100 |
| 2 | Clarifier Tank TP2 | working | 50 | 100 |
| 3 | Aeration Tank | working | 90 | 180 |
| 4 | Clarifier Tank TP3 | Trial | 50 | 150 |
| 5 | Clarifier Tank TP3 | Trial | 50 | 150 |
| 6 | Distribution Chamber | Trial | 50 | 150 |
| 7 | Effluent pumping station | Trial | 50 | 150 |
| 8 | Chlorination Tank | Trial | 50 | 150 |
| 9 | Portable water tank 1 | working | 50 | 100 |
| 10 | Portable water tank 2 | working | 50 | 100 |

## 4 Finite element analysis and mesh generation

The numerical investigation of the bored piles under axial loads were precisely analyzed. Plaxis 3D Foundation software version 1.6 was adopted. The mesh created in simulation process was medium and due to the variation of shear strains near the interface between the pile and soil, the mesh got finer.
The pile and surrounding soil were modeled using linear elastic and Mohr-coulomb models respectively. The first model depended on stiffness parameters of the pile as shown in Table 3. The basic requirements of the latter model are the shear strength parameters which are involved the cohesion c and the angle of internal friction φ as well as the dilation angle ψ beside the modulus of elasticity $E_s$ and Poisson's ratio ν. The pile is assumed as a rigid member, and the interaction at the pile-soil interface must be well defined. Thus, $R_{int}$ value was taken equal to 0.7 due the nature of shear stresses generated along the pile shaft after the fully casting period is reached.

The 15- node wedge element was employed in this study. The 15-node wedge element is composed of 6-node triangles in horizontal direction and 8-node quadrilaterals in vertical direction. The wedge element presents conceivable results and capable to well represent the cases where the strain intensity is localized.

**Table 3.** Pile details used in the current study

| Identification | Value |
|---|---|
| Material Model | Linear elastic |
| Modulus of elasticity, $E_p$ (kN/m$^2$) | 25700000 |
| Poisson's ratio, ν | 0.2 |
| Type of Material | Non porous- concrete |
| Material density (kN/m$^3$) | 24 |

In order to define the cases to be analyzed, a numerical analysis has been performed to simulate the entire behavior of full scale bored piles under compression loads that were carried out on piles of 24m in total length and 0.6m in diameter. The boundaries of mesh used in this analysis were chosen to be (30×30m). Numbers of elements and nodes in generated mesh are 2880 and 8409 respectively. These dimensions are considered satisfactory for eliminating any effects of the boundaries on the test results. The whole system of test corresponding to the mesh layout and soil layers is shown in Figure 3.

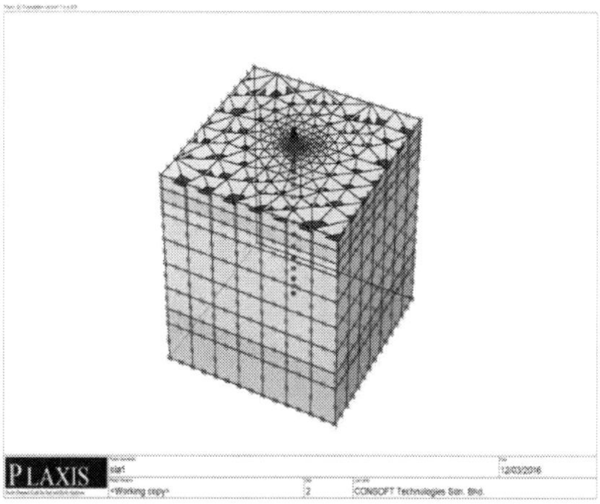


**Fig 3.** Three dimensional mesh for single pile embedded in layered soils

## 5 Presentation of test results

The results obtained from the numerical analysis done using Plaxis 3D foundation program and those gained from the field load tests have been presented. Both studies were performed under same soil and loading conditions, by the assistance of site investigations that were carried out for the required project. Also, the load- settlement curves for the piles examined in the present study will be introduced.

The failure load of piles for the numerical and field results will also be compared, to understand how the finite element technique could be used as a powerful tool in piles design. Figures 4 to 13 present the load- settlement curves of bored piles for the field and numerical studies. The input data adopted in the numerical simulation have exactly relied on the same site investigation of soil, pile properties and the nature of loading that applied on the pile during the load test in the field. This process is to satisfy all the requirements of pile simulation, and to reflect closely as possible the actual behavior of piles

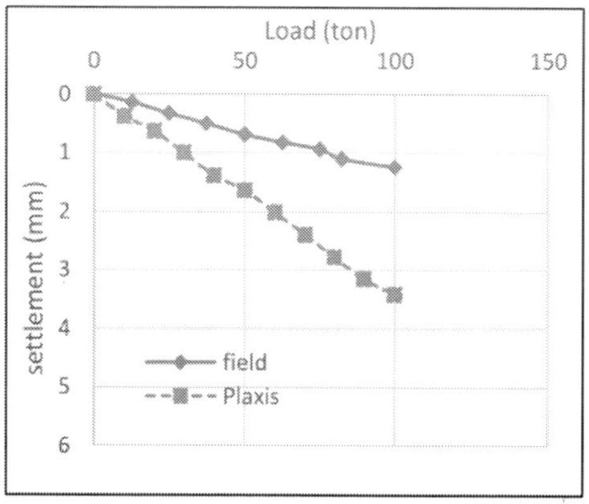


**Fig. 4.** Load-settlement curve for bored pile under Clarifier Tank TP1

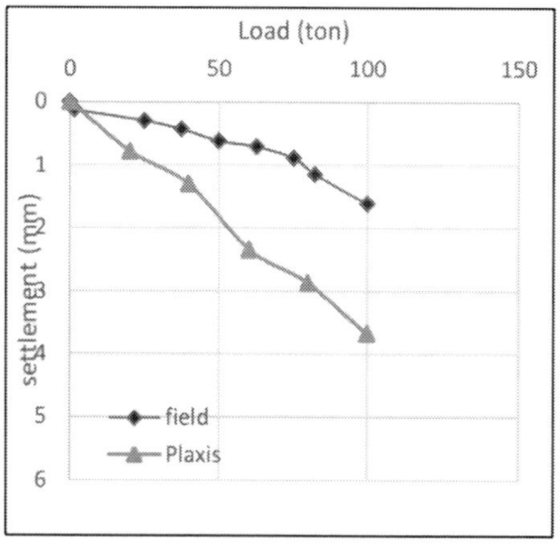


**Fig. 5.** Load-settlement curve for bored pile under Clarifier Tank TP2

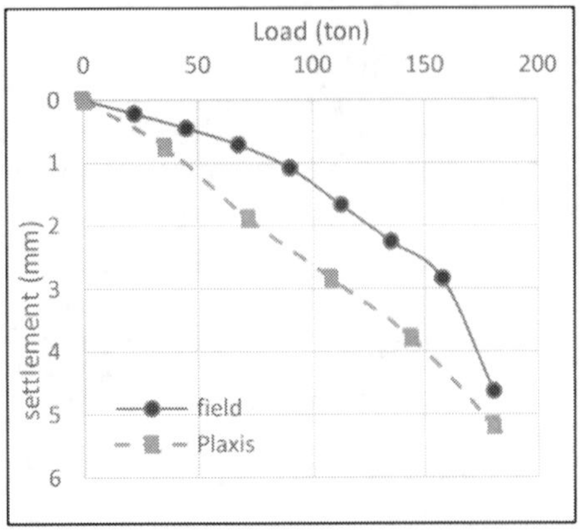


**Fig. 6.** Load-settlement curve for bored pile under Aeration Tank

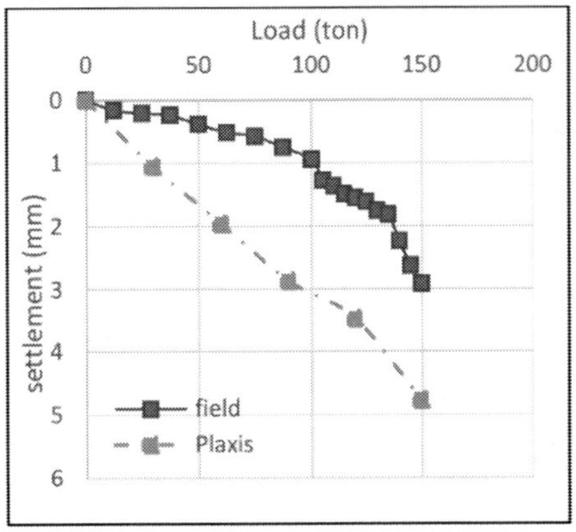


**Fig. 7.** Load-settlement curve for bored pile under Clarifier Tank TP3

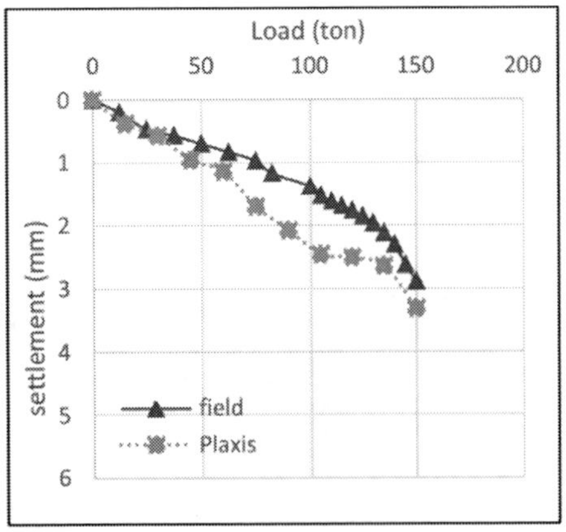


**Fig. 8.** Load-settlement curve for bored pile under Clarifier Tank TP4

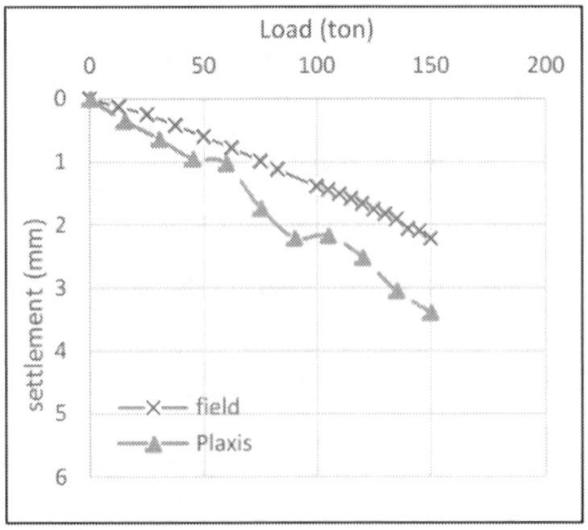


**Fig. 9.** Load-settlement curve for bored pile under Distribution Chamber

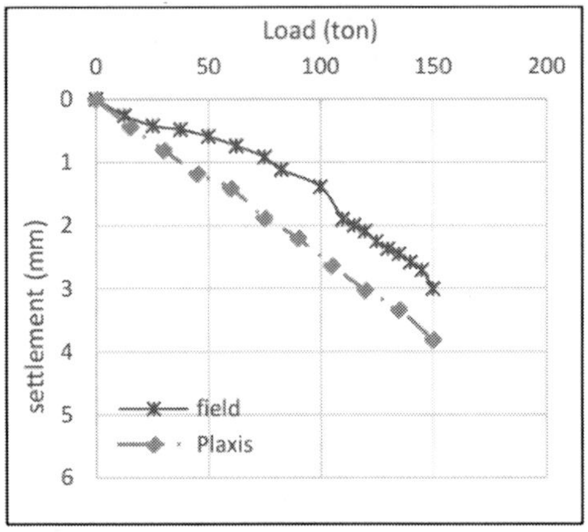


**Fig. 10.** Load-settlement curve for bored pile under Effluent pumping station

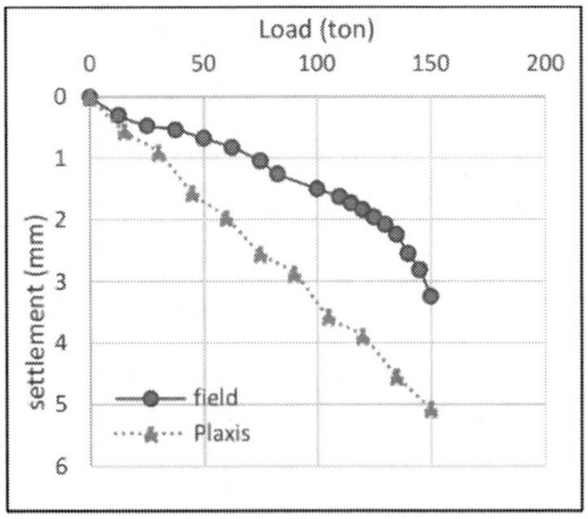


**Fig. 11.** Load-settlement curve for bored pile under Chlorination Tank

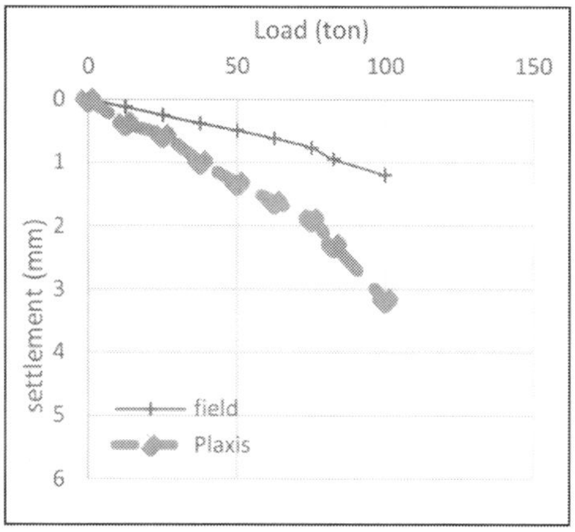


**Fig. 12.** Load-settlement curve for bored pile under Portable water Tank 1

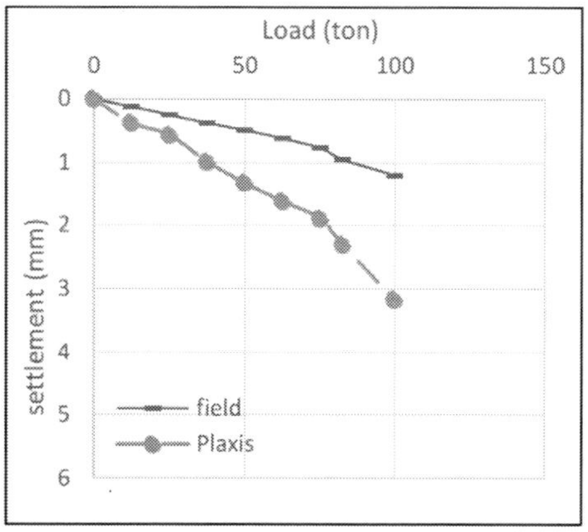


**Fig. 13.** Load-settlement curve for bored pile under Portable water Tank 2

It is noticed that from the figures above, the trend of curves for both studies is approximately similar. This may lead to the fact that, the behavior of piles under an applied load approaches to be same. Also, for all the piles, the values of settlement recorded from the analysis give a good impression on the accuracy and convergence between the numerical and field load test results.

However, the noticeable agreement educed from this analysis may be attributed to the correct employment of modeling details such as inserting actual representative values of soil properties and pile characteristics, in addition to the suitable representation of pile installation in the simulation process, at which the entire behavior of bored pile is significantly affected. A comparison in terms of settlement for the numerical and field tests under testing load is listed in Table 4.

**Table 4.** Values of settlement recorded from the analyses of pile tests

| No. | Type of structure | Settlement values (mm) (Field) | Settlement values (mm) (Numerical) |
|---|---|---|---|
| 1 | Clarifier Tank TP1 | 1.24 | 3.24 |
| 2 | Clarifier Tank TP2 | 1.62 | 3.67 |
| 3 | Aeration Tank | 4.6 | 5.2 |
| 4 | Clarifier Tank TP3 | 2.92 | 4.8 |
| 5 | Clarifier Tank TP4 | 2.89 | 3.31 |
| 6 | Distribution chamber | 2.22 | 4.0 |
| 7 | Effluent pumping station | 3.01 | 3.82 |
| 8 | Chlorination Tank | 3.25 | 5.07 |
| 9 | Portable water tank1 | 1.2 | 3.17 |
| 10 | Portable water tank2 | 1.79 | 3.1 |

### 5.1 Prediction of pile capacity from field and numerical test results

Usually, the aim of pile load test is to obtain the safe capacity of tested pile, and to verify the design calculations computed by traditional methods. So, it is necessary to estimate the pile capacity for the numerical and field studies to withstand on the safe value of pile capacity without failure.

The results of current study demonstrate that, no pile has been reached to the failure. All piles have recorded settlement values less than the allowable limitations for shear failure. This may be referred to numerous facets, for example although the working piles are not loaded to the failure, on the other side, even the trial piles did not fail under the testing load. Therefore, the conventional methods used in piles design give underestimated results.

However, to envision the failure load (ultimate capacity) of piles and since all the tested piles did not reveal a tendency to fail, it is too hard to predict the ultimate capacity in this case, unless an alternative way is used to solve this problem.

A Chin Konder method was adopted in the current study, to expect and track the behavior of pile throughout the stages of loading, until the required level of settlement is satisfied. In this technique, ultimate pile capacity was estimated by using the tangential method. So, the steps used in this method in which the pile capacity has been https://doi.org/10.1051/mateconf/201816201004

achieved, will be displayed as the following: each settlement point is divided with its corresponding load and plot the resulting value against the settlement. The slope of this line is $C_1$ and interception with y-axis is $C_2$. If required the pile load at any level of settlement, an ideal curve can be normally generated using the following equation:

$$Q = \delta / (C1\ \delta + C2) \qquad (2)$$

Where:

Q: the applied load.

δ: settlement and,

C1 and C2: slope of the straight line and y-interception of the straight line respectively.

One example (pile under clarifier Tank TP4) will be reviewed in details, and how the steps of Chin Konder method are applied to obtain the ultimate pile capacity and get an ideal curve. The symbols $C_1$ and $C_2$ are obtained from Figure 14.

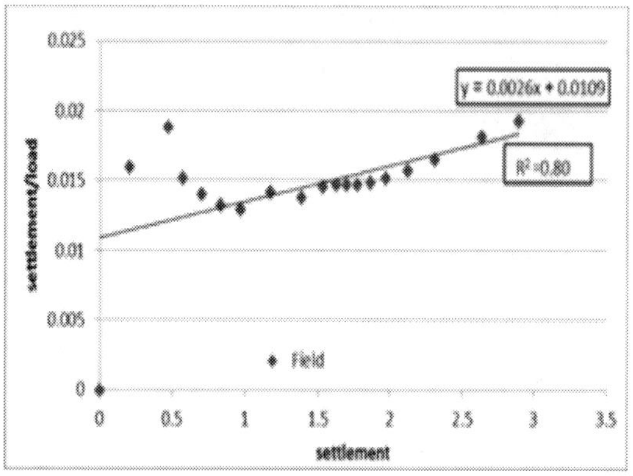


**Fig.14.** Relationship of Settlement /load Vs. Settlement for field test results

So, by applying the equation (2) at any settlement value, and the slope and interception of the straight line are known for each case, the pile capacity is easy to calculate. Thereafter, the ideal curves used to estimate the failure loads of bored piles can be generated.

The challenge of this step for the extrapolated results is to what level of settlement this technique is applicable and reliable. To afford an eligibility for this technique, the field extrapolated test results should be again validated with numerical results at predetermined level of settlement. Hence, the piles are loaded using the finite element analysis to the same pile displacement achieved in Chin-Konder method. For all pile tests, the failure criterion adopted is a tangents method by which the failure load is defined as the load resulted from the intersection of two lines that are parallel to elastic and plastic parts in the load settlement curve. This method is almost applicable for all types of piles and gives rational results. The ideal load-settlement curves generated for the both cases are illustrated in Figures 15 to 24.

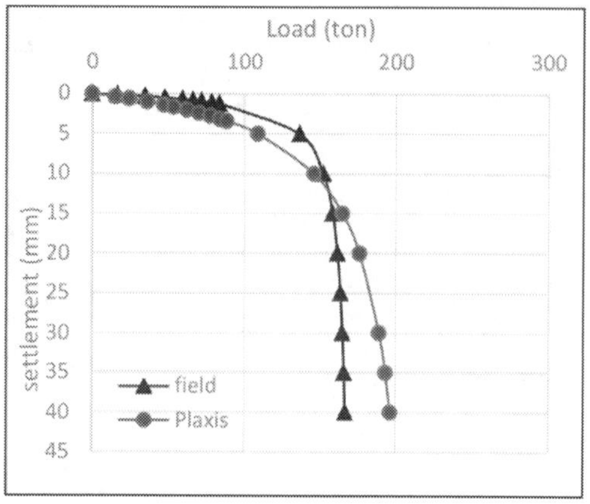


**Fig. 15.** Ideal load-settlement curve for bored pile under Clarifier Tank TP1

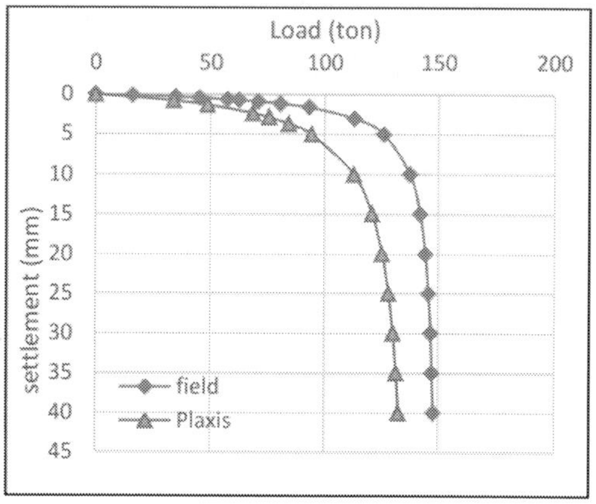


**Fig. 16.** Ideal load-settlement curve for bored pile under Clarifier Tank TP2

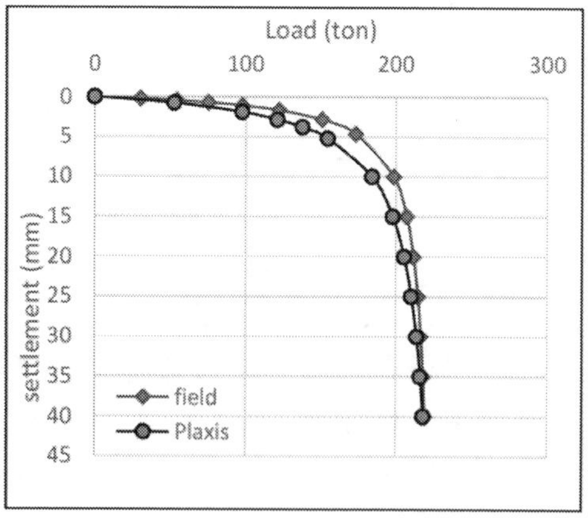


**Fig. 17.** Ideal load-settlement curve for bored pile under Aeration Tank

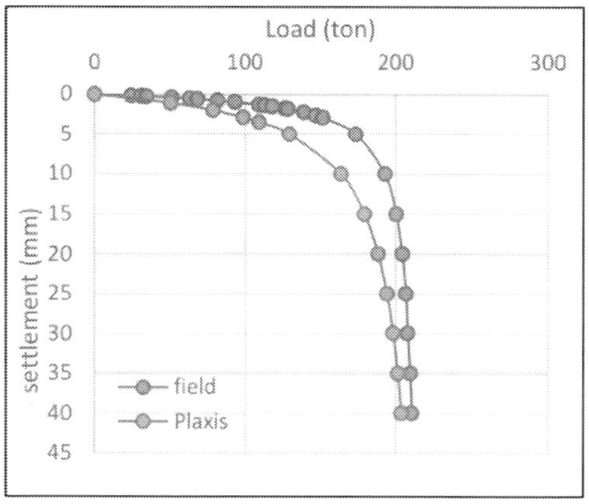


**Fig. 18.** Ideal load-settlement curve for bored pile under clarifier Tank TP3

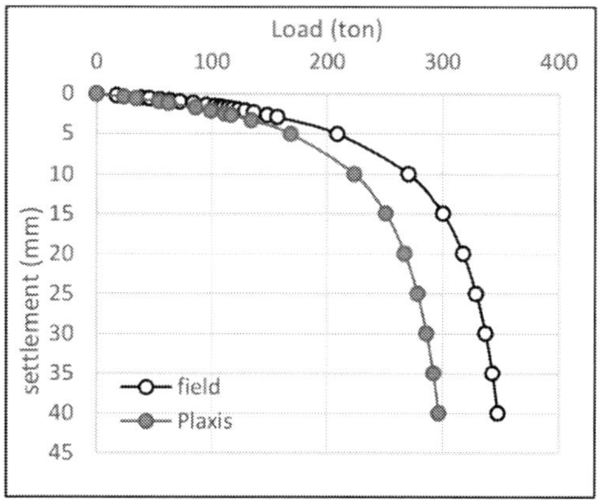


**Fig. 19.** Ideal load-settlement curve for bored pile under clarifier Tank TP4

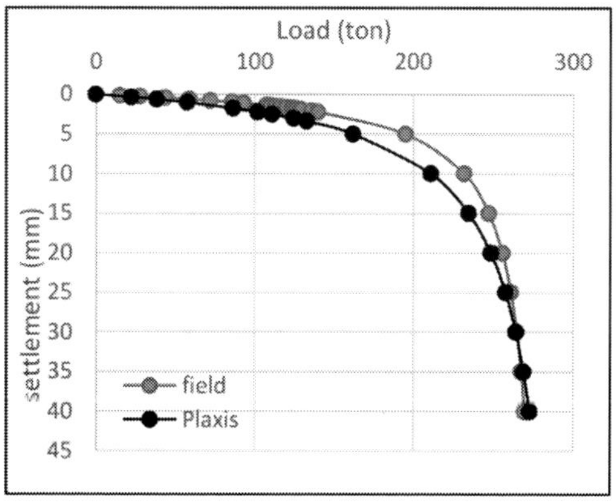


**Fig. 20.** Ideal load-settlement curve for bored pile under distribution chamber

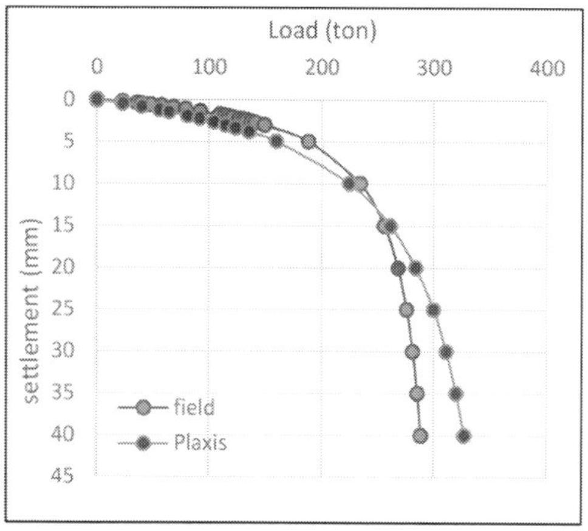


**Fig. 21.** Ideal load-settlement curve for bored pile under Effluent pumping station

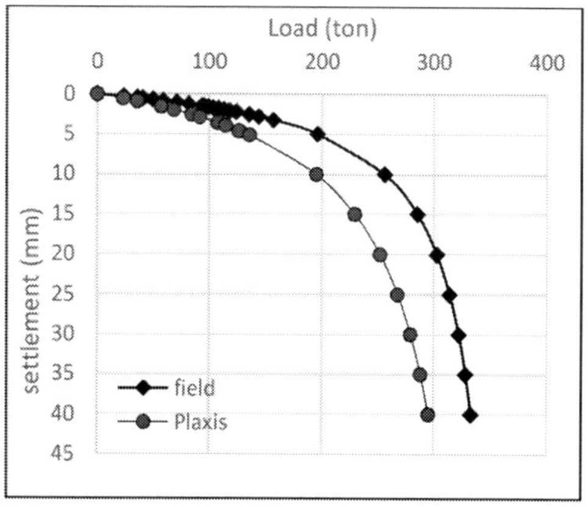


**Fig. 22.** Ideal load-settlement curve for bored pile under chlorination Tank

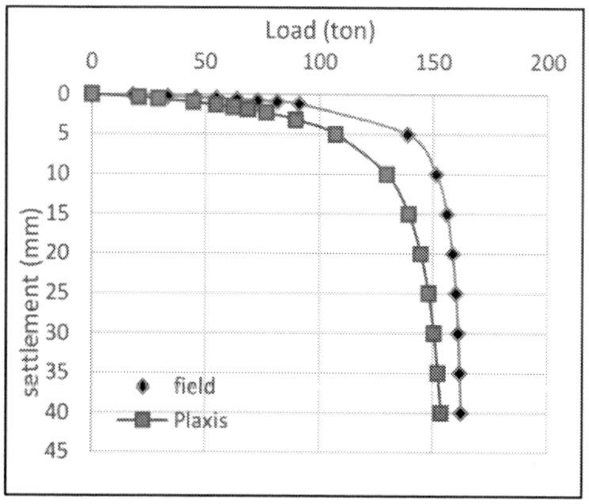


**Fig. 23.** Ideal load-settlement curve for bored pile under Portable water tank 1

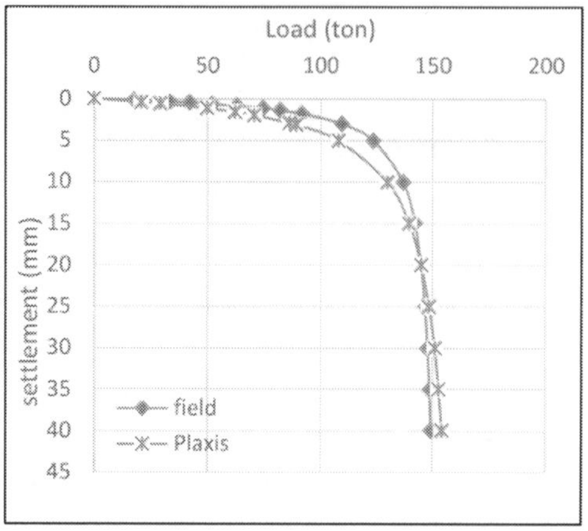


**Fig. 24.** Ideal load-settlement curve for bored pile under Portable water tank 2

It can be noted that, there is a good agreement between the numerical and field test results in term of pile capacity as they appear in Table 5. This absolutely indicates the reliability of the finite element analysis which may be accounted as an alternative approach in design of piles instead of traditional methods that are used in the analysis, in which it presents uneconomical option due to its underestimated response as compared with numerical analysis. The predicted ultimate capacities for all piles used in this study are listed in Table 5.

**Table 5.** Values of pile capacities obtained from the extrapolated field and numerical studies

| No. | Type of structure | Testing Load (Ton) | Pile capacity (Ton) (Field) | Pile capacity (Ton) (Numerical) |
|---|---|---|---|---|
| 1 | Clarifier Tank TP1 | 100 | 155 | 168 |
| 2 | Clarifier Tank TP2 | 100 | 140 | 120 |
| 3 | Aeration Tank | 180 | 210 | 195 |
| 4 | Clarifier Tank TP3 | 150 | 200 | 180 |
| 5 | Clarifier Tank TP4 | 150 | 305 | 255 |
| 6 | Distribution chamber | 150 | 245 | 240 |
| 7 | Effluent pumping station | 150 | 262 | 275 |
| 8 | Chlorination Tank | 150 | 275 | 250 |
| 9 | Portable water tank1 | 100 | 158 | 142 |
| 10 | Portable water tank2 | 100 | 145 | 140 |

Based on the results of pile capacity acquired through extrapolation process, the required new design load for each case can be obtained by dividing the failure load obtained from the ideal load –settlement curve of the field and numerical studies by a suitable factor of safety assumed before design, taking into the account the pile function (trial or working).The findings are outlined in Figure 25.

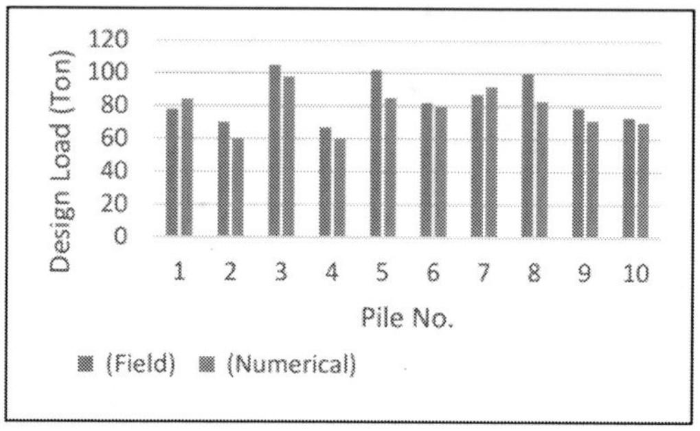


**Fig. 25.** Values of design loads induced from the extrapolated field and numerical results

Now, it is clear to say that; the feasible convergence between the numerical and field test results encourages the numerical methods to be drastically accounted in design instead or beside the conventional approaches especially for the cases when these ways are still incompetent to move on in solving these problems.
The average design load estimated was approximately 50 ton, and at this level of loading the failure mechanism have not plainly revealed. From the numerical analysis, the results exhibited an average design load about 76 ton at a set of ten piles. These new magnitudes of working loads can be utilized in design as a practical step by reducing the number of piles used under the structure, consequently minimizing the quantity of concrete required to execute the piles. Also, construction efforts will be decreased such as the transportation process in the site and time required to accomplish the pile installation also decrease.
Finally, the Plaxis program used in this analysis can be primly relied as a powerful tool in pile design, and by which the response of pile is moving close to the actual behavior of piles.

## 6 Conclusions

Based on the finite element analysis performed using Plaxis 3D Foundation on full scale of bored piles during this study, the following major conclusions can be outlined as:

1. An excellent agreement was gained between the pile capacity obtained from the field and numerical test results.
2. The average design loads estimated from the extrapolated numerical analysis afford economical option, where the pile capacity obtained from this study is higher by about 51% than of that loads predicted using traditional methods.
3. The degree of convergence between pile capacities for both studies is about 90%, this may give a suitable impression on the reliability of test results.
4. A Chin- Konder criterion has successfully proved its reliability in estimation of pile capacity at failure.
5. Finite element analysis using Plaxis program is considered as an effective alternative method in piles design.

## References

1. ASTM D1143, Standard Test Method for Piles under Static Axial Compressive Load, New York, pp. 87-97, (1994).
2. ASTM D5434, Standard Guide for Field Logging of Subsurface Explorations of Soil and Rockl, New York, pp.1-3, (1997).
3. Das. B.M., *Principles of foundation engineering", 6th Edition*, International Student Edition, United States, (2007).
4. Ehsan. M, M., H, N., G and R., N, Comparative Study on Prediction of Axial Bearing Capacity of Driven Piles in Granular Materials", Jr of Tek, (Sciences & Engineering) **61**, 15–20, (2013).
5. Hatanaka. M., and Uchida A. , Empirical Correlation between Penetration Resistance and Internal Friction Angle of Sandy Soils, SAF, **36**, 4, 1-10, (1996).
6. Michal. H., Jakub S., Comparison OF Numerical Analyses with a Static Load Test of a Continuous Flight Auger Pile", SJOCE, **22**, 4, 1 – 10, (2014).
7. Naveen. B.P., T. G. sitharam, and S Vishruth, Numerical Simulation of Vertically Loaded Piles, *Proceedings of Indian Geotechnical Conference, kochi*, (2011).
8. Serhii. Lozovyi, Evhen Zahoruiko "Plaxis Simulation of Static Pile Tests and Determination of Reaction Piles Influence", USATJ, **2**, 68-73, (2012).
9. Terzaghi, K., Peck R.B and Mesri, G., *Soil Mechanics in Engineering Practice*", Third edition, John Wiley and Sons Inc., New York, (1996).
10. Thounaojam, S., and ParbinSultana, Prediction of Bearing Capacity of Bored Cast- in Situ Pile, JOMACE, 1-6, (2015).

# Geotechnical characteristics of some Iraqi gypseous soils

*Tom* Schanz[1] and *Hussein* H. Karim[2,*]

[1] Ruhr Universität Bochum, Bochum- Germany
[2] Building and Construction Engineering Department, University of Technology, Baghdad, Iraq

**Abstract.** In Iraq, especially in the last three decades, extensive developments have been evidenced in the regions of gypseous soils due to the need of construction of many numbers of strategic projects. Failure of different structures constructed on gypseous soil in various regions in Iraq have been noticed. For this purpose, three areas in northern Iraq were selected (Samarra, Tikrit and Baiji) to study their geotechnical characteristics due to their high gypsum contents as well as many engineering problems are faced due to dissolution of gypsum. The experimental work involves testing of many properties such as: scanning electron microscopy (SEM), XRD, chemical, physical, compressibility, collapsibility, shear strength and suction. At low stress level, the test results revealed that, higher collapse potential (CP) is recorded for Tikrit soil. While at low stress level, higher CP is obtained for Baiji soil indicating the increase in CP with decreasing gypsum content. Furthermore, the CP significantly increases with increasing stress level and soaking period at a particular stress level. According to severity classification of the collapse potential, Baiji soil is considered as moderate trouble to slight, while Tikrit soil is considered as trouble to moderate. After soaking, both soils become trouble. As well as, the results showed a reduction in Tikrit soil shear parameters ($\varphi$ and $c$) after soaking period of 6 and 24 hrs as 12.2 to 9.2% in the internal friction angle and 91.5 to 94.2% in cohesion, respectively with respect to dry condition. Maximum total suction is measured for low consistency soils (liquid limit < 30%) represented by Tikrit soil.

## 1 Introduction

Soils which collapse (undergo large consolidation) when unusual amounts of water are added to them exist in many places around the world. There are many types of problematic soils, one of these is the collapsible gypseous soil. Gypseous soil is found in arid and semi-arid regions of the world where gypsum sediments and rocks are widespread such as in Iraq. There are different origins and different definitions for gypseous soil, among which is the definition proposed by Barazanji (1973) [1], who divided this soil into two main groups with respect to gypsum content namely; soil with more than 50% known as gypsiferous soil, and soil containing less than 50% which in turn subdivided into five sub-groups.

Gypseous soils are considered as metastable or collapsible soils defined as any unsaturated soil that goes through a radical rearrangement of particles and susceptible to large abrupt volume changes and sudden collapses take place when such soils are wetted, soaked or leached with water. This can also happen when these soils are inundated under loading [2, 3]. Gypsum mineral ($CaSO_4.2H_2O$) is hydrated calcium sulfate with intermediate solubility ($\approx$2.6 g/l at 25°C) [3-5]. Gypsum presents in soil structure acts as a cementing material when the soil is dry.

Dudley (1970) [6] pronounced the collapse phenomenon in terms of the bonding materials existing at the contact points between soil particles. Collapse can result in a substantial rearrangement of the soil particles, resulting in a significant reduction in total volume of the soil mass. Noticeable collapse of soil demands three main conditions; an open (large void ratio), potentially unstable, partly saturated structure; a high enough value of an applied or existing stress; and a strong soil bonding or cementing agent to stabilize intergranular contacts when dry, while upon wetting will produce collapse.

In general, the settlement of gypseous soils is mainly due to the dissolution of the cementing gypsum which causes high softening of the soil and great loss in strength. So sudden increase in compressibility occurs when these soils are fully or partially saturated accompanied by structure collapse of the soil particularly in sandy gypseous soils. This can cause serious damage and even collapse of the structures founded on or in such soils [7, 8].

Large areas of Iraq are covered with gypseous soils, especially in the west, northwest and southwest regions and covers about 20-30% of Iraq's total area which is equivalent to about 7.3% of the total area of gypseous soils in the world [9-12].

Most hazards which are related to construction on these problematic soils occurred when such soils are subjected to water, causing softening of soils and reducing the shear strength [13-16]. Gypseous soils are highly soluble materials in their nature, and the types of problems associated with it. The failures include collapse and settlement, which can affect all construction including buildings [17-19], roads [20-22] and other

* Corresponding author: husn_irq@yahoo.com

© The Authors, published by EDP Sciences. This is an open access article distributed under the terms of the Creative Commons Attribution License 4.0 (http://creativecommons.org/licenses/by/4.0/).

engineering structures. It is worth to mention that in order to study the
behavior of such soils, it is necessary to perform scanning electron microscopy (SEM) analysis to correlate the morphological and microstructural changes with gypsum crystals dissolution or softening using different magnifications [23].

It is necessary to study the properties of such soils due to the damage caused by these soils to building on or in it. Seepage through sandy gypseous soil causes serious damage to foundation built on such type of soil. Gypsification offers a variety of serious geotechnical hazards. Such a problem causes serious danger in Al-Mosul dam foundation (largest dam located northern Iraq) which is built on highly gypseous soil. The seepage through the soil under foundation causes leaching and dissolution to the gypsum which brought the dam to the danger imminent of collapse, which could kill hundreds thousands of people.

The present study aims to investigate and assess the compressibility, collapse potential, shear strength and suction of gypseous soils. Three sites, located northern Baghdad-Iraq, have been chosen for their high gypsum contents.

## 2 Methodology

### 2.1 Soil's location, sampling and properties

Gypseous soil samples were taken from three sites, northern Baghdad-Iraq; namely Baiji, Tikrit and Samarra areas. These sites were chosen for two reasons; first they are characterized by their high gypsum content, while the second is the foundation problems that have occurred in area close to Baiji where Al-Mosul Dam site is located (Figure 1).

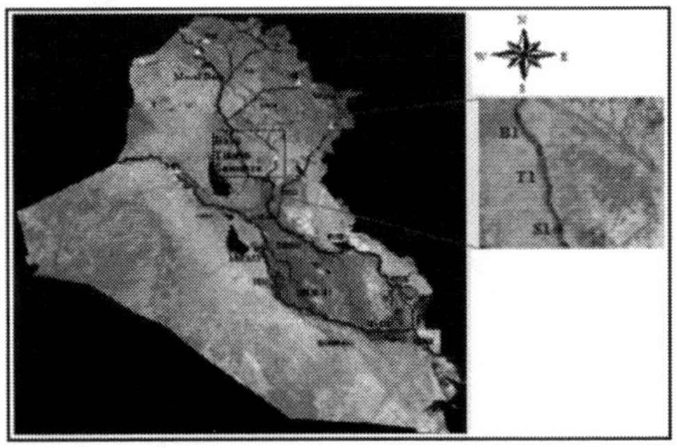

**Fig. 1.** Study Area.

Various tests have been carried out for such soil samples to determine their physical-chemical properties and particle size distribution. X-Ray diffraction test analyses were performed for the three soil sample kinds. The results of the mineralogical composition for the tested soil samples are shown in Table 1. Higher gypsum content is assigned to Tikrit soil sample (73.88%), an intermediate for Samarra soil sample (65.73%) and the lowest is assigned to Baiji soil sample (51.24%).

| Mineral Type (%) | Site | | |
|---|---|---|---|
| | Baiji | Tikrit | Samarra |
| Gypsum | 51.24 | 73.88 | 65.73 |
| Calcite | 10.81 | 9.43 | 12.21 |
| Chlorite | 7.61 | 1.77 | 3.18 |
| Muscovite | 10.8 | 4.23 | 4.88 |
| Plagioclase (Albite) | 9.08 | 3.18 | 5.36 |
| Quartz | 10.47 | 7.51 | 8.64 |

**Table 1.** Mineralogical composition for the tested soil samples.

Physical tests carried out in this study including Atterberg's limits, specific gravity ($G_s$), water content, and grain size distributions are presented in Table 2 with their standards. The Unified Soil Classification System (USCS) was used to classify the tested soils. Other standard mechanical tests performed in this study involve one-dimensional compression, collapse (single collapse test, SCT, and double oedometer test, DOT), and direct shear. A summary of soil's results for some physical properties, grain size distribution and soil classification (USCS) are summarized in Table 3.

**Table 2.** Tests with their standards.

| Item No. | Property | Standard |
|---|---|---|
| 1 | Grain size distributions | ASTM D422 [24] |
| | Specific gravity | BS 1377:6B [25] |
| 2 | Atterberg's limits (L.L and P.L) | BS 1377:2A [26]; BS 1377:3 [27] |
| 3 | Water content | ASTM D2216 [28] |
| 4 | Direct shear tests | ASTM D3080-98 (1998) [29] |
| 5 | One-dimensional consolidation test (Compressibility) | ASTM D 2435-04 (2004) [30] |
| 6 | Collapsibility test (Consolidation tests) | ASTM D 2435, 2004 [30] ; ASTM D5333-03, 2003 [31] |

### 2.2 Compressibility, collapsibility and shear strength of gypseous soils

Collapsible gypseous soils show severe geotechnical and structural engineering challenges around the world. Bonds between soil particles can be formed by capillary forces (i.e. suctions) and/or by cementing of fine materials (i.e. clays or salts). When the applied vertical stresses by means of loading or wetting processes exceed yield strength of these bonding materials, various amounts of collapse potential may abruptly occur [32-34]. Therefore, it is necessary to understand the process of collapse and whether the problems associated with these soils are to be avoided or mitigated. So, the geotechnical properties of gypseous soils cover, compressibility, collapsibility, and shear strength parameters (c and $\varphi$) were studied.

| Soil Type | Water Content (W.C %) | Gypsum Content (G.C %) | Specific Gravity ($G_s$) | Atterberg's Limits | | | Grain Size Distribution % | | | | | | Classification (USCS) |
|---|---|---|---|---|---|---|---|---|---|---|---|---|---|
| | | | | L.L (%) | P.L (%) | P.I (%) | $C_u$ | $C_c$ | Gravel (%) | Sand (%) | Fines (%) | | |
| | | | | | | | | | | | Silt | Clay | |
| Baiji | 1.25 | 51.2 | 2.53 | 37 | 19 | 18 | 3.4 | 1.1 | 6 | 78 | 16 | - | SC |
| Tikrit | 0 | 73.9 | 2.38 | 26 | 21 | 5 | 3.0 | 0.1 | 8 | 74 | 18 | - | SM-SC |
| Samarra | 6.5 | 65.7 | 2.48 | 32 | 25 | 7 | 1.6 | 0.3 | 22 | 61 | 17 | - | SM |

**Table 3.** Summary of soil's results for some physical properties, grain size distribution and classification.

Several researchers have investigated the influence of gypsum on the copressibility characterstics. Such studies reported a decrease in the copression index with increasing gypsum content. Similar contradicting results were reported for the recompression index. Most of the researchers demonstrated an increase in the secondary compression index with increasing gypsum content. This phenomenon is attributed to the continuous dissolution process of gypsum with time as reprted by Seleam (1988) [35] and Nashat (1990) [10]. The same researchers and many others showed that the coefficient of consolidation remains unchanged with increasing gypsum content.

Al-Nouri and Al-Qaissy (1990) [36] stated that the significant reduction in shear strength is mainly due to dissolution of gypsum after wetting. Nashat (1990) [10] and Seleam (1988) [35] found that the collapse potential of gypseous soils (with gypsum content between 20-80%) under a constant stress of 200kN/m$^2$ increases with increasing gypsum content. Petrukhin and Arkelyan (1985) [37] studied the shear strength of gypseous clayey soils with different gypsum content up to 50%. This study showed that $c$ and $\varphi$ parameters are increased with increasing gypsum content upto around 15%, then a decrease was noticed with extra increase in gypsum content.

From the above discussion, the following standard tests were performed in this study including one-dimensional compression; collapse (single collapse tests); and direct shear tests supported by scanning electron microscopy technique.

### 2.3 Collapse potential calculation

The soaking collapse is caused by soaking the dry soil or partly saturated soil with water under specific pressure without flow of water. The factors that affect collapse poential $(CP)$ of gypseous soils are summarized by; soaking pressure ($\sigma_w$), initial water content ($W_o$), initial void ratio ($e_0$), gypsum content (G.C), and duration of sustained load [38].

Gypseous soils are characterized by their collapsible behavior during wetting. The term collapse potential is used to address the hazardous case of collapsibility. Jennings and Knight (1957) [39] and Knight (1963) [40] suggested a double oedometer collapse test to predict the collapsibility of the foundation soil. Two identical samples are tested, one at natural water content and the other after submerging in water for one day. The collapse potential $(CP)$ at a specific stress is defined as [39]:

$$CP = \frac{\Delta e}{1+e_0} \quad (1)$$

where:
$\Delta e$ = the difference in void ratio of the two samples at a specific stress
$e_0$= the natural (initial) void ratio

## 3 Results and discussion

### 3.1 One dimensional compression (Oedometer) and collapse tests results

For consolidation test, the standard oedometer device was used with the stress path of 25, 50, 100, 200, 400 and 800 kPa for loading stage and 400, 100 and 25 kPa for unloading stage following the standard procedure of ASTM D 2435-04 (2004) [30]. The results of this test shown in Figure 2 are presented as e-logσ relation expressing void ratio versus logarithm of vertical stress relationship for the dry soil samples from Baiji and Tikrit (with initial void ratio, $e_0$=0.28 and 0.52, respectively). The figure obviously shows almost a linear trend for the compressibility curves (compression and recompression curves) for both soils which can be attributed to permeability nature of gypseous sandy soil (compared to clayey soil) leading to quick settlement particularly under incremental loading.

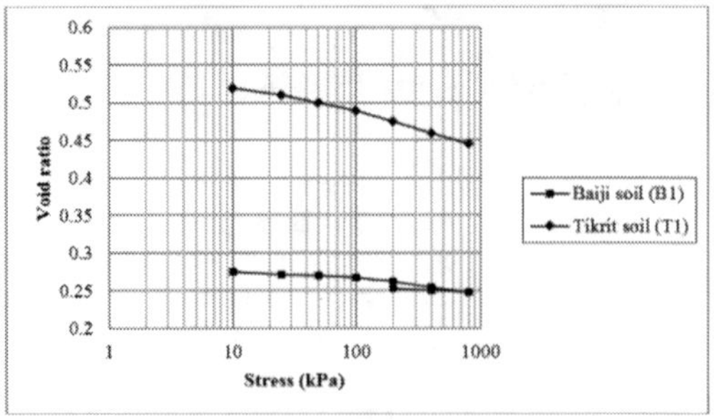


**Fig. 2.** Oedometer test results for Baiji and Tikrit soil samples.

In order to investigate the range of collapse behavior of the studied gypseous soil (Baiji and Tikrit), collapse test was achieved at different conditions (drying and wetting) of soil samples following the standards for consolidation tests [30] and continued as in conventional consolidation tests [31]. The standard oedometer device was used for collapse test following the single (SOT) and double oedometer-collapse (DOT) test procedures applying the same stress sequence for loading and unloading stages for both tests.

Results of the relationship between void ratios versus vertical stress for both soils of single and double collapse-oedometer tests are presented in Figure 3 and Table 4. From the results of these tests as well as the standard oedometer test, the collapse potentials (C.P) are calculated using equation 1.

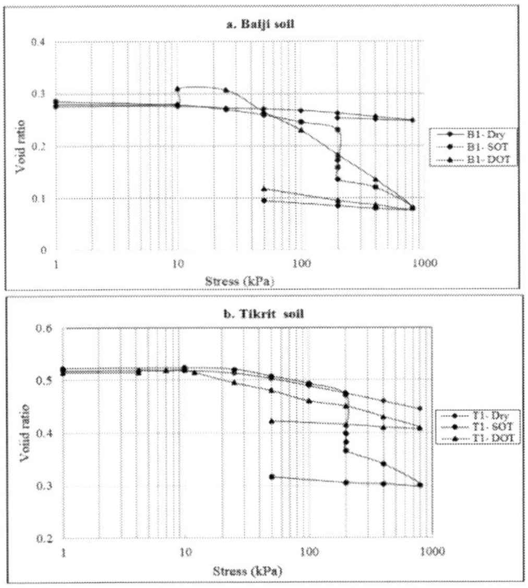


**Fig. 3.** Single and double oedometer tests (SOT and DOT) for: a) Baiji soil, b) Tikrit soil.

**Table 4.** Compression test results of for Tikrit and Baiji soils (Soaked oedometer, DOT).

| Site | Gs | Gypsum Content (%) | $e_0$ | $C_c$ | $C_r$ |
|---|---|---|---|---|---|
| **Baiji soil** | 2.53 | 51.24 | 0.28 | 0.1716 | 0.033 |
| **Tikrit soil** | 2.38 | 73.88 | 0.5163 | 0.055 | 0.011 |

For SOT test results (Figures 3a and b), it can be stated that the response of both soils to wetting is a "collapse" volume change. Furthermore, both natural gypseous soils show a large amount of collapse deformation and sudden volume change at independant entrance of water, where the void ratio is significantly decreased. Otherwise, at high stress level, the gap between the two values of the void ratio at unsaturated and saturated state becomes higher when the inundation took place at high stress level. This behaviour may be due to softening and breaking of gypsum bonds and the reduction of the interparticle contacts. Thus, new particles microstructure is generated with smaller volume due to densification and rearrangement. This finding was also observed by Nashat (1990) [10], Seleam (1988) [35], and Al-Obaidi (2003 and 2014) [12, 34]. The collapse behaviour for both soils is observed when the soil samples are subjected to an overconsolidation state where the soils inundation at vertical stress higher than the maximum preconsolidation pressure (where $\sigma_v$>50 kPa for Baiji soil, and $\sigma_v$>10 kPa for Tikrit soil) as shown in Figures 3a and b.

Regarding DOT test results for Baiji soil (Figure 3a), the matching in the compressibility curves of both SOT and DOT tests is clear except that this soil undegoes a significant swelling at the beginning of the DOT test under loading levels 10, 25 and 50 kPa then the soil starts to collapse after loading level 50 kPa. This swelling volume change occurs as a response for inundation processes at stress level lower than the preconsolidation pressure (i.e. $\sigma_v$<50 kPa), while the collapse behaviour is observed when soil inundation at vertical stress higher than the maximum preconsolidation pressure. This behaviour may be because the soil sample was subjected to an overconsolidation state when inundation was induced at vertical stresses of 25 and 50 kPa. In other words, the volume of soil mass is being expanded when excess water is introduced filling free voids in the soil microstructure. In spite of vertical stress being applied, this did not impede the upward soil movement. Thus, a noticeable collapse settlement is observed as a result of softening and rearrangement of soil particles upon inundation at a stress level higher than the preconsolidation pressure as confirmed by findings of Al-Obaidi (2014) [34] and Lawton et al. (1989) [41]. While, for the DOT test for Tikrit soil (Figure 3b), no matching is observed with SOT test at 200 kPa, because the soil tested in DOT test was disturbed but the soil tested in SOT test was undisturbed and this variation might be due to remolding phenomenon.

By examining the collapse tests results of both soils for double oedometer test (DOT), one may conclude that at low stress level (<50 kPa) the collapse potential of Tikrit soil is higher than Baiji soil due to the higher void ratio (lower density) of the former soil which is due to density effect. While at high stress level (>50 kPa), the results are reversed that is higher collapse potential is obtained for Baiji soil than Tikrit soil which can be attributed to high compression index of soaked Baiji soil.

The comparison between collapse potential obtained by SOT and DOT collapse tests, it is well noticed that both soil samples show a collapse deformation as a result of full saturation at low vertical stress. However, relatively higher collapse potentials was obtained from SOT test than that obtained by DOT test for the same vertical stress level. The explanation for this behavior is in the SOT, all the cementing bonds of gypsum particles exist in the soil structure until sudden inundation is induced at a specified pressure. In this case, large and quick deformations occur as a result of volumetric strain development and cementing bonds softening by gypsum dissolution. While in DOT test, the sample is already stressed and the action of softening and dissolution of gypsum is started before reaching to a specified collapse pressure. Many studies indicate that the maximum collapse occurs when the inundation stress approaches the preconsolidation stress of the natural state compression curve, after that an increase in the degree of saturation directs the compression curve to the saturated soil curve [42, 43].

It is also shown that the collapse potential for both soils increases with increasing stress level as shown in Figure 4. Several collapse tests using DOT conducted by Iraqi studies (such as Nashat, 1990 [10]; Al-Obaidi, 2014 [34]; Seleam, 1988 [35]; Al-Badran, 2001 [38]; and Fattah et al., 2008 [44]) confirm that the collapse potential significantly increases with increasing inundation stress level particularly after the preconsolidation stress of the natural soil. In general, this behavior might be due the_higher stresses (greater than the maximum preconsolidation pressure) applied on soil specimens with looser structure which will destroy and close all cavities in the specimens with an open metastable structure, then denser soil state is obtained. Also from single collapse test, collapse potentials were calculated at 200 kPa stress level for different soaking periods during 4 days (1, 2 and 4 days). It is clear that the collapse potential increases with increasing soaked periods as shown in Figure 5 for Baiji and Tikrit gypseous soil. Depending on the severity classification of the collapse potential when inundation at $\sigma_v$ =200 kPa, Baiji soil is of moderate trouble to slight according to Jennings and Knight (1975) [45] and ASTM D5333 (2003) [31] standard respectively. While, Tikrit soil is considered as trouble to moderate at the same stress level and for the same classification respectively. After soaking, both soils become trouble at $\sigma_v$ =200 kPa.

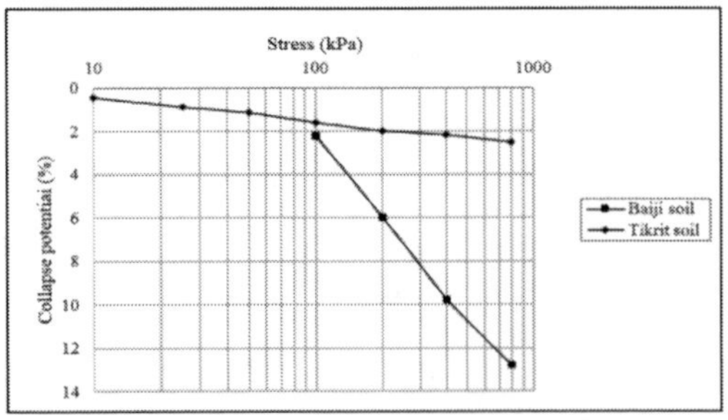


**Fig.4.** Collapse potential versus stress for Baiji and Tikrit soils.

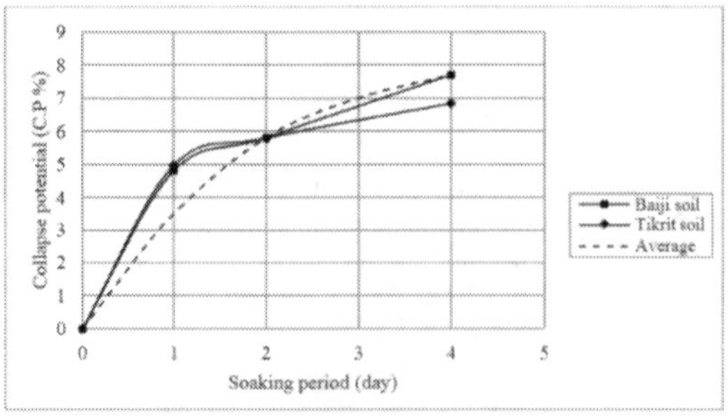


**Fig. 5.** Collapse potential versus time at 200 kPa stress level for Baiji and Tikrit gypseous soils.

## 3.2 Shear strength of gypseous soil

To determine the effective shear strength parameters (cohesion,$c$, and angle of internal friction, $\varphi$), direct shear tests using pneumatic shear device were carried out on dry and soaked soil samples according to ASTM D 3080, 1998. The time of sample soaking in water was 6 and 24 hours. In this series, at low stress level (<50 kPa) the collapse potential of Tikrit soil is higher than Baiji soil, three tests were conducted on 12 soil samples for Tikrit site, 4 of them were tested in dry condition, while the other 8 samples were tested in soaking condition. In order to study the effect of short-term soaking period on shear strength parameters of gypseous soil, the first four samples were soaked for 24 hrs, while the rest samples were soaked for 6 hrs. The normal stresses were 50, 100, 200 and 400 kPa.

Figure 6 and Table 5 present the results of direct shear test of Tikrit soil for the conditions of dry, after 6 hrs soaking and 24 hrs soaking. It can be seen that the percentage of reduction in the angle of internal friction ($\varphi$) and cohesion ($c$) due to soaking were 12.24 %; 91.5 % and 9.2 %; 94.24 % for soaking 6 and 24 hrs respectively. It is well noticed the loss in strength by soaking, basically in cohesion, which is mainly due to softening.

 https://doi.org/10.1051/matecconf/201816201005

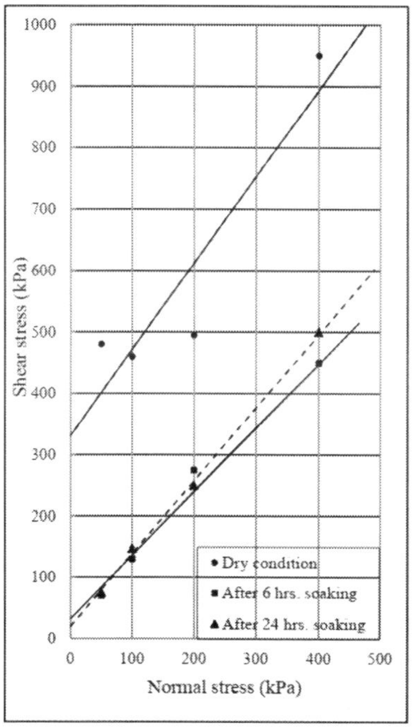


**Fig. 6.** Direct shear test results for Tikrit soil sample for the conditions of dry, after 6 and 24 hrs soaking.

**Table 5.** Shear strength parameters for Tikrit gypseous soil.

| **Shear Strength Parameters** | | | | | |
|---|---|---|---|---|---|
| $c$ **(kPa)** | | | $\varphi$ (°) | | |
| Dry | After soaking (6 hrs.) | After soaking (24 hrs.) | Dry | After soaking (6 hrs.) | After soaking (24 hrs.) |
| 330 | 28 | 19 | 49 | 43 | 44.5 |

To illustrate the behavior of such soils under the effect of both soaking and/or loading on microstructural changes, scanning electron microscopy (SEM) was carried out under different conditions (dry, after 6 and 24 hrs soaking). SEMs of the natural soil samples (Figures 7a, 8a, and 9a) show the presence of gypsum in different shapes (flatty and platy sheets, scattered crystals and as crust encasing soil particles). These gypsum crystals act as a bonding material between soil particles. After soaking (Figures 7b, 8b, and 9b), SEM results point out appreciable microstructural changes which are well identified as voids and solid regions in the micrographs. Besides, these figures show the soil softening by partial dissolution of gypsum particles due to soaking and forming voids between soil particles which may lead to collapsibility. While, when the soil sample is subjected to flooding and loading (after single collapse test) (Figure 7c), the soil grains are weakened by breaking gypsum bond particles due to partial dissolution after soaking and resulting in large voids. With additional load, soil grains may slide one over another resulting in collapse as confirmed by Jefferson and Rogers (2012) [33] and Karim et al. (2015) [23]. Consequently, the tested gypseous soils exhibit a loss of strength basically in cohesion over internal friction angle which is well observed for soaked specimens due to softening bonds destruction and reduction in the inter-particle contacts in the soil after soaking.

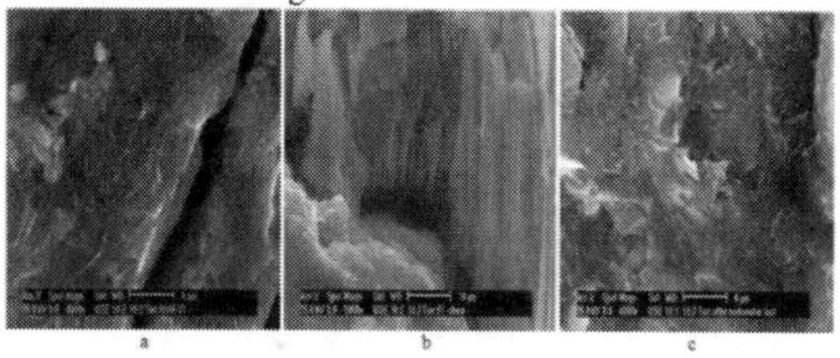

a b c

**Fig. 7.** a) SEM for Tikrit natural soil. b) SEM after short-term soaking. c) SEM after subjecting to single collapse test.

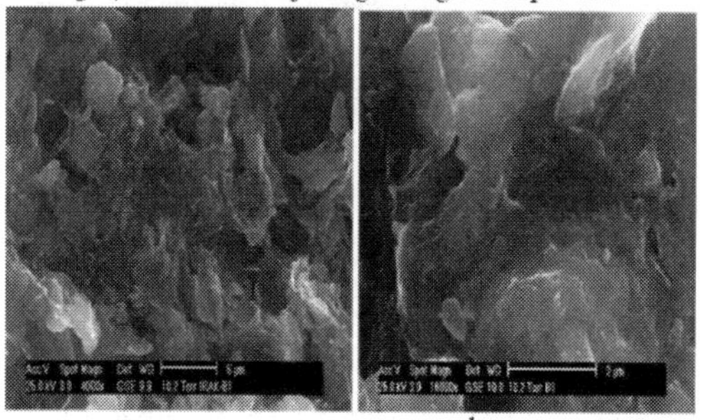

a b

**Fig. 8.** a) SEM for Baiji natural soil. b) SEM after short-term soaking.

a b

**Fig. 9.** a) SEM for Samarra natural soil. b) SEM after short-term soaking.

## 3.3 Suction measurements and related parameters

This test was carried out to investigate the behavior of soils when subjected to suction. The relationships of suction with water content, volumetric water content, degree of saturation and void ratio for Baiji, Tikrit, and Samarra gypseous soils are respectively shown in Figure 10a to d. It is worth mentioning that depsite the suction measurements were incomplete, they may diplay part of these relationships. The test results for the three selected soils clearly show large reductions in their water content, volumetric water content, degree of saturation and void ratio as a result of successive increases in suction values.

https://doi.org/10.1051/matecconf/201816201005

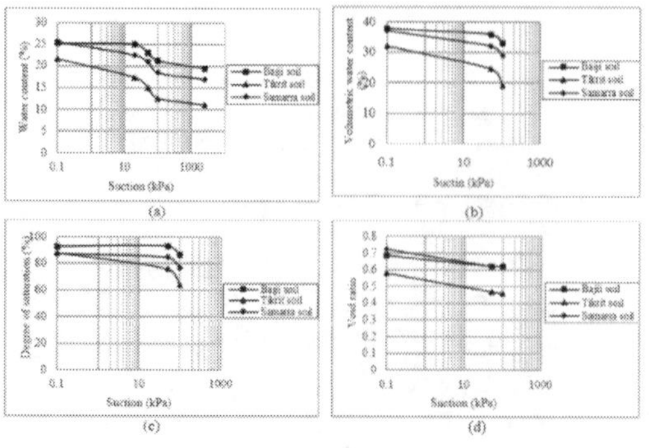


**Fig. 10.** Relationships of suction with water content, volumetric water content, degree of saturation and void ratio for Baiji, Tikrit, and Samarra gypseous soils.

The chilled-mirror hygrometer technique was used to determine and monitor the total suction of soil specimens. It is an indirect method to measure total suction. It uses a dew-point measurement method and the device is a water activity meter by which the relative humidity of the specimens was measured and the total suction was calculated using the thermodynamic relationship between soil suction, or the free energy of the soil-water and the partial pressure of the pore water vapor (Kelvin's equation) as follows:

$$\psi = \frac{RT}{v_{w_o} w_v} ln \left(\frac{\bar{u}_v}{\bar{u}_{v_o}}\right) \quad (2)$$

where: $\psi$=soil suction or total suction (kPa), R=universal (molar) gas constant (i.e., 8.31432 J/(mol K)), $T_K$= absolute temperature (i.e., (273.15+T (K)), T=temperature, °K, $v_{w_o}$=specific volume of water or the inverse of the density of water (i.e., $1/\rho_w, m^3$/kg), $\rho_w$=density of water (i.e., 998 kg/$m^3$ at temperature =200°C), $w_v$ = molecular mass of water vapor (i.e., 18.016 kg/kmol), $\bar{u}_v$=partial pressure of pore water (kPa), and $\bar{u}_{v_o}$ = saturation pressure of water vapor over a flat surface of pure water at the same temperature (kPa). The term $\left(\frac{\bar{u}_v}{\bar{u}_{v_o}}\right)$ is called the relative humidity (RH) and is entered as a function (i.e., water activity) in the above equation.

The total suction of soil samples were measured using the chilled-mirror hygrometer according to ASTM D6836-2008 [46] standard. The total suction-water content relationship calculated from chilled-mirror hygrometer for the three soils is illustrated in Figure 11. The figure indicates that the suction increases with decrease of the water content. The suction curves especially for Baiji and Samarra soils are almost similar to one another, particularly in the higher water content range, while the curve for Tikrit soil is somewhat different. It is shown that the total initial suction at hygroscopic water content for Baiji, Samarra, and Tikrit gypseous soils are 300 000, 180 000 and 1 600 000 kPa, respectively. It is worth mentioning that low liquid limit is obtained for the latter soil. Thus, the maximum suction measured for Tikrit natural soil due to its low consistency (L.L < 30%) as well as its high amount of gypsum (73.88 %) could be the reason for this difference.

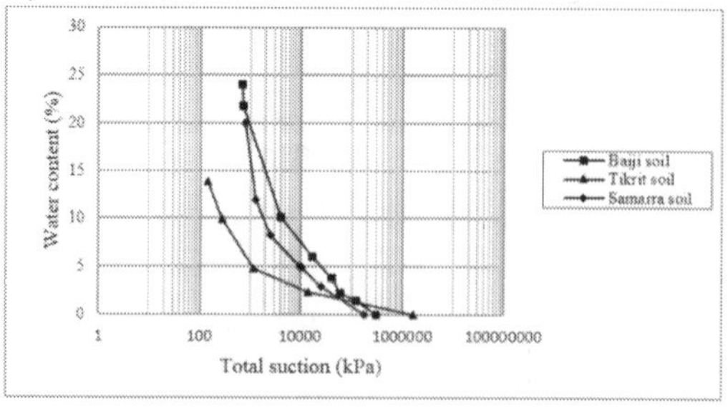


**Fig. 11.** Relation of total suction with water content.

## 4 Conclusions

In this work, a comprehensive laboratory investigation was undertaken to study the collapsibility and shear strength for three Iraqi site locations with different gypsum contents. In the light of experimental test results, the following conclusions can be drawn:

1. The tested gypseous soils exhibit very high shear strength when dry, and by comparing failure envelopes of dry specimens with those of soaked specimens, a loss of strength basically in cohesion is observed in the soaked ones.
2. For Baiji gypseous soil sample, the DOT test showed soil swelling at the beginning of the test under the stress levels 10, 25 and 50 kPa which then started to collapse after loading level 50 kPa.
3. The results of collapse potential-stress relationship show that the CP increases with increasing stress level for both Baiji and Tikrit soil samples. Moreover, Baiji gypseous soil (with lowest gypsum content) exhibits higher CP values at high stress level compared to Tikrit gypseous soil (with highest gypsum content) indicating the increase in CP with decreasing gypsum content.
4. At low stress level (<50 kPa), the collapse potential of Tikrit soil is higher than Baiji soil due to the lower density of the former soil. While at high stress level (>50 kPa), the results are reversed for the soaked Baiji soil which can be attributed to its high compression index.
5. From single oedometer tests (SOT), collapse potentials were calculated at 200 kPa stress level for different soaking periods within 4 days. The results showed that the collapse potential increases with increasing soaking periods for both types of gypseous soils tested. According to severity classification of the collapse potential, Baiji soil is considered as moderate trouble to slight. While Tikrit soil is considered as trouble to moderate. After soaking, both soils become trouble.
6. For Tikrit site, the percent of reduction in the angle of internal friction were 12.2% and 9.2%, while in cohesion were 91.5% and 94.2% due to soaking for 6 and 24 hrs respectively.

7. The maximum suction measured was for Tikrit natural soil due to its low consistency (L.L < 30%) as well as its high amount of gypsum.

This work was supported by Iraqi Ministry of High Education and Scientific Research as part of Ph.D Thesis entitled Geotechnical Engineering Study for Salah Al-Din Governorate Using Geophysical and Remote Sensing Techniques, University of Technology, Iraq. Sincere thanks is expressed to the staff of the Soil Mechanics Laboratory at Bauhaus-Universität Weimar in Germany for their assistance in carrying out the laboratory tests. Deep thanks is also expressed to F.A. Finger-Institute and to the Chemical Laboratories at Bauhaus-Universität Weimar for ESEM, X-Ray and chemical tests.

## References

1. A.F. Barazanji, Soil of Iraq, Ph.D. Thesis, State University of Ghent, Belgium, (1973)
2. S.P. Clemence, A.O. Finbarr, J Geotech Eng Div, Proc. ASCE **107**, 305-317 (1981)
3. Z.M. Mansour, Z. Chik, M.R. Taha, J Appl Sci **8**, 4434-4439 (2008)
4. E. Bock, Can J Chem **39**, 1746–1751 (1961)
5. J. Herrero, J. Porta, Geoderma **96**, 47-61 (2000)
6. J.H. Dudley, ASCE J Soil Mech Found Div **96**, 925-947 (1970)
7. R.N. Arutyunyan, Soil Mech Found Eng **15**, 183-185, (1978)
8. S.S. Razouki, R.R. Al-Omari, I.H. Nashat, H.F. Razouki, S. Khalid, *Proceedings of the symposium on gypsiferous soils and their effect on structures*, NCCL, 7-33 (Baghdad, 1994)
9. FAO, Management of gypsiferous soil, Bulletin No. 62, (Rome, Italy, 1990)
10. I.H. Nashat, Ph.D Thesis, University of Baghdad, Baghdad, Iraq, (1990)
11. A. Al-Mufty, Ph.D Thesis, University of Baghdad, Baghdad, Iraq, (1997)
12. Q.A. Al-Obaidi, M.Sc. Thesis, Al- Mustansiriyah University, Baghdad, Iraq, (2003)
13. S.S. Razouki, O.A. El-Janabi, Q J Eng Geo Hydroge **31**, 87-89 (1999)
14. B.S. Albusoda, M.Sc. Thesis, University of Baghdad, Baghdad, Iraq, (1999)
15. A. Ismail, W.J. Hilo, J App Sci **8**, 4610-4616 (2008)
16. M.A. Al-Dabbas, T. Schanz, M. J. Yassen, Arab J Geosci **22**, 111-119 (2010)
17. R.N. Arutyunyan, A.V. Manukyan, Soil Mech Found Eng **19**, 151-154 (1982)
18. A.H. Cooper, Eng Geol, Special Publication **15**, 265-275 (1998)
19. A.H. Cooper, Q J Eng Geo Hydroge **41**, 409-424 (2008)
20. A.A. Ahmad, M.Sc. Thesis, University of Mosul, Mosul, Iraq, (1985)
21. S.A. Aiban, H.I. Al-Abdul Wahhab, O.S.B. Al-Amoudi, H. R. Ahmed, Constr Build Mater **12**, 329-340 (1998)
22. M. Hua, B. Wang, L. Chen, Y. Wang, V.M. Quynh, B. He, X. Li, Fuel **89**, 1812-1817 (2010)
23. H.H. Karim, T. Schanz, A.N. Ibrahim, *Proceedings of the 2nd international conference of buildings, construction and environmental engineering, American University of Beirut (BCEE2)*, 17-18 Oct., 2015, Geotechnical Engineering part, 95-100 (2015)
24. ASTM D422, Particle–size analysis of soils, (ASTM International, West Conshohocken, PA, United States, 2007)
25. BS 1377:6B, *Method of test for soils for civil engineering purposes, Chemical and electro-chemical tests*, (British Standard Institution, London, UK, 1990)
26. BS 1377:2A, *Method of test for soils for civil engineering purposes*, (British Standard Institution, London, UK, 1990)
27. BS 1377:3, *Method of test for soils for civil engineering purposes*, Chemical and electro-chemical tests, (British Standard Institution, London, UK, 1990)
28. ASTM D2216, *Laboratory determination of water (moisture) content of soil and rock by mass*, (ASTM International, West Conshohocken, PA, United States, 2010)
29. ASTM D3080, *Standard test method for direct shear test of soils under consolidated drained conditions*, (ASTM International, West Conshohocken, PA, United States, 1998)
30. ASTM D2435-04, *Standard test methods for one-dimensional consolidation properties of soils using incremental loading*, (ASTM International, West Conshohocken, PA, United States, 2004)
31. ASTM D5333-03, *Standard test method for measurement of collapse potential of soils*, (ASTM International, West Conshohocken, PA, United States, 2003)
32. S. Houston, W. Houston, C. Zapata, C. Lawrence, Geotech Geol Eng **19**, 333-355 (2001)
33. I. Jefferson, C. Rogers, *ICE manual of geotechnical engineering*, Vol. **1** of ICE manuals (ICE Publishing, London, UK, 2012)
34. Q.A.J. Al-Obaidi, Ph.D Dissertation, Faculty of Civil and Environmental Engineering, Ruhr- Universität Bochum, Bochum, (2014)
35. S.N.M. Seleam, M.Sc. Thesis, University of Technology, Baghdad, Iraq (1988)
36. J.A. Al-Nouri, F.F. Al-Qaissy, *Symposium on recent advances in geotechnical engineering*, 8-14, Singapore, (1990)
37. V.P Petrukhin, E.A. Arkelyan, Soil Mech Found Eng **21**, 23-25 (1985)
38. Y.M. Al-Badran, M.Sc. Thesis, University of Baghdad, Baghdad, Iraq, (2001)
39. J.E. Jennings, K. Knight, *Proceeding 4th international conference on soil mechanics and foundation engineering*, **1**, 316-319 (1957)
40. K. Knight, *Proc. of 3rd regional conference of Africa on soil mechanics and foundation engineering*, **1**, 127-130 (1963)
41. E. Lawton, R. Fragaszy, J. Hardcastle, J Geotech Eng **115**, 1252-1267 (ASCE, 1989)
42. Z. Lin, S. Wang, Eng Geo **25**, 271-282, (1988)

43. A. Al-Mufty, *International conference on geotechnical engineering-Beirut*, (2004)
44. M. Fattah, Y. Al-Shakarchi, H. Al-Numani, Eng Tech J **26**, 1461-1485 (2008)
45. J. Jennings, K. Knight, *Proc. 6th African conf. soil mechanics and foundation engineering, Durban*, 99-105 (1975)
46. ASTM D6836, *Standard test methods for determination of the soil water characteristic curve for desorption using a hanging column, pressure extractor, chilled mirror hygrometer, and/or centrifuge*, (ASTM International, West Conshohocken, PA, United States, 2008)

# Stabilization of soft clayey soils with sawdust ashes

*Hussein* Karim[1,*], *Makki* Al-Recaby[1], and *Maha* Nsaif[1]

[1] Building and Construction Engineering Department, University of Technology, Baghdad, Iraq

**Abstract.** The problems of soft clayey soils are taken in considerations by many Iraqi geologists and civil engineers, because about 35% of the Iraqi clay soils (especially southern Iraq) are weak. Thus, it is necessary to improve the properties of such soils for road construction by means of using of various stabilizers such as sawdust ash. The main goal of the present study is to stabilize soft clay models with sawdust ash (SDA) additive using different percentages (0, 2, 4, 6, 8 and 10% by dry weight of soil). The results revealed that the additive has adverse effects on the property of soil indices by increasing its liquid limit and plasticity index due to clay content. The mixture of sawdust ashes with soft clay soils improves most other physical and mechanical properties of the soil, as expressed by a general reduction in specific gravity and maximum dry density (MDD), as well as a reduction in the compression coefficients ($C_c$ and $C_r$) with an increase in SDA content. While increasing the optimum moisture content (OMC) and the undrained shear strength ($c_u$) with the increase in SDA content. The stabilized soils (with 4 and 10% ash content) resulted in low CBR values (1.6-1.2%) which can be used as sub-base. The SDA can be considered as a cheap and acceptable stabilizing agent in road construction for improving most of the geotechnical properties of the soft clayey soil.

## 1 Introduction

Soft clay soils are recent alluvial deposits presumably formed through the most recent 10,000 years described by their featureless and flat ground surface. These soils are wide spread in Mesopotamian Plain in Iraq particularly in its southern part which are characterized by their low bearing capacity and strength besides their low California bearing ratio (CBR) [1].

Soft clays are characterized by their low undrained shear strength and high compressibility. Such clays are defined as a disturbed cohesive soil whose water content is higher than its liquid limit, such materials display extremely low yield stresses and represent difficult construction conditions. Thus, such types of clays always bring the problems of soil instability and structure settlement. Therefore, the construction of buildings, road, bridges canals and railway in such problematic clayey soils has always been associated with such problems [1, 2].

The designs of earth structures require the knowledge on the behavior and stress-related deformability of soils that will support the structure and the geological conditions of the soil under consideration. Thus, the design should be safe enough against shear failure with reasonable amount of settlement. The geotechnical properties of a soil assessed by proper laboratory or in situ tests are used in predicting the behavior of soil when subjected to loads. However, under certain circumstances, all of the needed parameters cannot be determined or are not determined because of economic or other reasons [3, 4].

Stabilization of soft and compressible soils clarifies the different ways of improvement for the natural soil properties either by using additives through mixing the natural soil with a stabilizing material (lime, cement or organic materials such as ashes) to obtain a homogeneous mixture) [1] or by column type techniques (sand or stone columns) or by both i.e. columns mixed with additives [5]. Nowadays, huge byproduct solid materials (wastes) are available like fly ash, rice husk ash, reeds and papyrus ashes which can be used as soil stabilizer instead of costly lime or cement [6, 7]. There are ongoing research studies concerning the possibility of using other naturally occurring materials such as ashes of agricultural waste products for soil stabilization because of their pozzolanic nature [8]. Saw dust is one of the byproducts from timber industry and wood cutting factories.

The present study investigates whether the sawdust ashes (SDA) can be used as an additive to stabilize soft clayey soils by improving their geotechnical properties which will encourage the use of SDA as a stabilizer for clayey soil to solve road failure problems.

## 2 Materials and methodology

### 2.1 Soil used

Disturbed soil samples used for this study are brought from borrow pits at a depth of 1-1.5 m collected from Al-Maymunah site (Latitude 31° 54′ N and Longitude 47° 2′ E) which is located within Maysan Governorate about 393 km south of Baghdad. Geologically, the

* Corresponding author: husn_irq@yahoo.com

© The Authors, published by EDP Sciences. This is an open access article distributed under the terms of the Creative Commons Attribution License 4.0 (http://creativecommons.org/licenses/by/4.0/).

location is situated within the Mesopotamian basin which is characterized by its recent sediments. Samples were packed in nylon bags. The natural moisture content was determined immediately in the Soil Mechanics Laboratory, University of Technology. After that, the soil samples were spread out in the laboratory for two weeks for air-drying at room temperature to eliminate the natural water which may affect analysis. Then, soils were sieved with sieve No. 4 (4.76 mm opening) to obtain the final soil samples for the tests. After the drying period, lumps in the samples were slightly pulverized with minimal pressure. Thereafter, soil tests have been conducted for this work applying standard procedures. Grain size distribution of the soils used is shown in Figure 1 while the plasticity chart for the natural soil is illustrated in Figure 2. According to the Unified Soil Classification System (USCS) [9], the soil is classified as CL. Table 1 presents the properties of the natural soft soil used in this investigation.

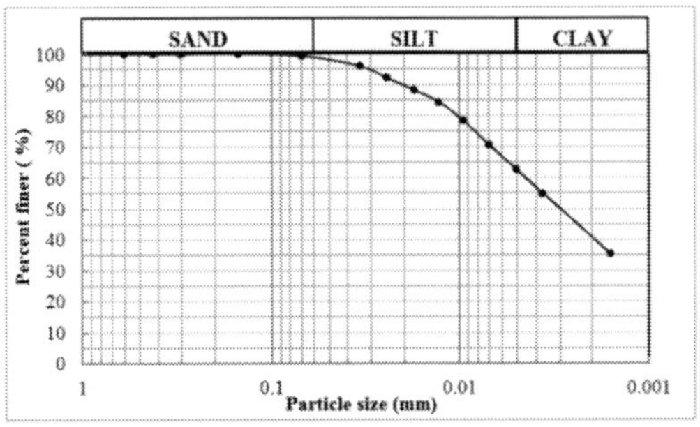


**Fig. 1.** Grain size distribution curve.

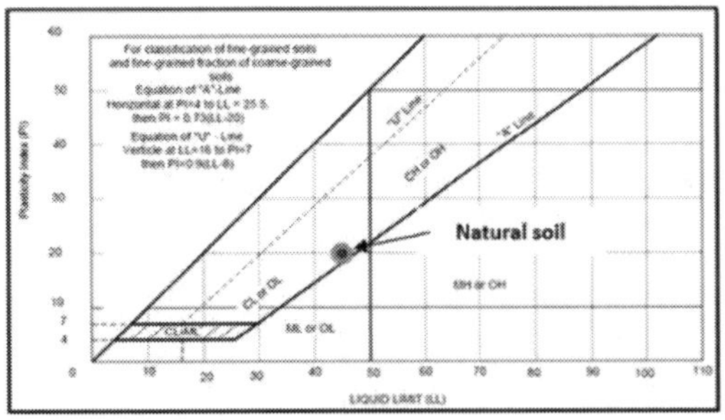


**Fig. 2.** Plasticity chart for the natural soil.

**Table 1.** Physical and chemical properties of natural soil used.

| Property | Value |
|---|---|
| Initial water content (W.C) % | 42 |
| Depth (m) | 1.5 |
| Liquid limit (L.L) (%) | 45 |
| Plastic limit (P.L) (%) | 25 |
| Linear shrinkage (L.S) (%) | 11 |
| Plasticity index (P.I) (%) | 20 |
| Activity ($A_t$) | 0.50 |
| Specific gravity ($G_s$) | 2.68 |
| Gravel (larger than 2 mm) (%) | 0 |
| Sand (0.06 to 2 mm) (%) | 0.5 |
| Silt (0.005 to 0.06) (%) | 37.5 |
| Clay (less than 0.005 mm) (%) | 62 |
| Classification (USCS) | CL |
| Organic material (O.M) (%) | 0.925 |
| Calcium oxide (CaO) (%) | 14.616 |
| $SO_3$ content (%) | 1.108 |
| Total dissolved salt (TDS) (%) | 4.91 |
| pH value | 8.16 |
| Cl | 1.598 |
| OMC (%) | 16.87 |
| MDD (kN/m$^3$) | 18.5 |

## 2.2 Sawdust ashes

Sawdust ashes are loose particles or wood chippings obtained by sawing hard wood into standard useable sizes and then burned. Clean sawdust without a large amount of bark has proved to be satisfactory because of low organic content. Then, the sawdust was spread on the ground and air dried to facilitate its burning. Sawdust was burnt to ashes and was then sieved through a sieve of 75μm to get very fine ash. After that, the sieved ashes were put in oven at 500°C for two hours to ensure removal of any organic matter. Then, it was stored in an air tight container to prevent moisture loss and any form of contamination. Figure 3 shows the burning of sawdust. The chemical composition of sawdust ashes (SDA) are different with various types of wood, but they predominantly comprise silica, alumina and lime [10]. The chemical composition of sawdust used is mainly silica, alumina, and lime besides other oxides with less proportions. The physical properties are tabulated in Table 2.

**Fig. 3.** Burning of sawdust.

**Table 2.** Physical properties and chemical composition of the used sawdust ash.

| Index Property | Index Value |
|---|---|
| Specific gravity | 2.02 |
| Fineness | 75 $\mu m$ |
| Fineness Silica ($SiO_2$) | 76.3 % |
| Alumina ($Al_2O_3$) | 5.8 % |
| Lime (Cao) | 4.7% |
| Iron oxide ($Fe_2O_3$) | 2.9 % |
| $SO_3$ | 1.6 % |
| MgO | 1.2 % |
| Other oxides | 2.5 % |
| Loss in ignition | 3.9 % |

## 2.3 Water

Ordinary water was used in the experimental work.

## 2.4 Preparation of soil mixtures

The selected percentages of ash were at rates 0, 2, 4, 6, 8, and 10% of the dry weight of the natural untreated soil. The soil-additive mixtures were prepared by thoroughly mixing of dry quantities of soil and predetermined amount of stabilizer (ash) in a mixing tray to obtain a uniform color for a minimum of 5 minutes, and the required amount of distilled water was added to the dry mixtures. The prepared soil-ash mixture samples are then mixed carefully until a homogeneous color was obtained.

## 2.5 Experimental tests

Several experimental tests were made to investigate the soil behavior after adding the sawdust ash. These tests are presented in Table 3. All samples for unconfined compressive strength (UCS), California bearing ratio (CBR) tests, and compressibility tests were performed at the optimum moisture content (OMC) and maximum dry unit weight values of the natural soil.

**Table 3.** Conducted tests with their standards.

| Item No. | Property | Standard |
|---|---|---|
| 1 | Specific gravity | ASTM D 854 [11] |
| 2 | Atterberg's limits | ASTM D 4318 [12] |
| 3 | Modified compaction | ASTM D 1557 [13] |
| 4 | UCS ($q_u$) | ASTM D 2166 [14] |
| 5 | CBR | ASTM D 1883 [15] |
| 6 | Compressibility | ASTM D 2435 [16] |

# 3 Test results and discussion

Test results with and without addition of SDA are illustrated in Table 4. These results involved the main preliminary tests (such as Atterberg's limits, specific gravity, particle size analysis, maximum dry density, optimum moisture content, CBR, undrained shear strength and consolidation) before and after the addition of SDA.

**Table 4.** Results of geotechnical properties before and after the addition of sawdust ash.

| Added sawdust ash (%) | L.L (%) | P.L (%) | P.I (%) | $G_s$ | MDD ($kN/m^3$) | OMC (%) | CBR (%) | Unconfined compressive Strength ($q_u$) ($kN/m^2$) | Undrained shear strength ($c_u$) ($kN/m^2$) | Consolidation test | |
|---|---|---|---|---|---|---|---|---|---|---|---|
| | | | | | | | | | | $C_c$ | $C_r$ |
| 0 | 45 | 25 | 20 | 2.68 | 16.87 | 18.5 | 4.9 | 39.2 | 19.6 | 0.24 | 0.02 |
| 2 | 50 | 23 | 27 | 2.67 | 16 | 14.2 | - | 38.5 | 19.3 | 0.12 | 0.02 |
| 4 | 52 | 25 | 27 | 2.65 | 15.7 | 20.76 | 1.6 | 169.23 | 84.6 | 0.14 | 0.03 |
| 6 | 40 | 24 | 16 | 2.8 | 15.4 | 17.5 | - | 98.30 | 49.15 | 0.14 | 0.04 |
| 8 | 50 | 27 | 23 | 2.43 | 15.56 | 19.67 | - | 52.43 | 26.2 | 0.11 | 0.03 |
| 10 | 54 | 27 | 27 | 2.85 | 13.9 | 23.61 | 1.2 | 74.92 | 37.5 | 0.10 | 0.03 |

Based on the specification of maximum liquid limit of 40% by the Federal Ministry of Works and Housing for soils to be used as sub-grade materials, since the average liquid limits of natural clayey soil samples are higher than 40%, therefore soil under consideration are not suitable for road construction unless they are stabilized.

The consistency limits are indicators of clay behavior. The variation in these limits with SDA content is shown in Fig. 4. As a general trend, the whole consistency limits show increase with increasing SDA content. This behavior is confirmed by Adetoro and Faluyi (2015) [17] and Butt et al. (2016) [7]. This could be due to its higher water absorption affinity (i.e. water needed for hydration). According to Whitlow (1995) [18], liquid limits between 45 and 54% indicate that the soil has an intermediate plasticity. It can be also noticed from Table 1 that liquid limit and plasticity index values for both untreated and treated soils are above 35% and 12% respectively, pointing to the high content of clay. Thus, the addition of SDA makes the soil more plastic. It seems that the additive percentages are still not effective due to the high clay content in the soil. Although the SDA addition of 6 % reduces consistency limits, but the general trend of increase in these limits with SDA addition indicating the stabilization of soil to some extent as SDA additive increases. The plasticity indices of samples decreased from 20 to around 16 at 6 % SDA addition may be considered as indication of soil improvement as this causes a removal of some water that can be absorbed by clay minerals [19].

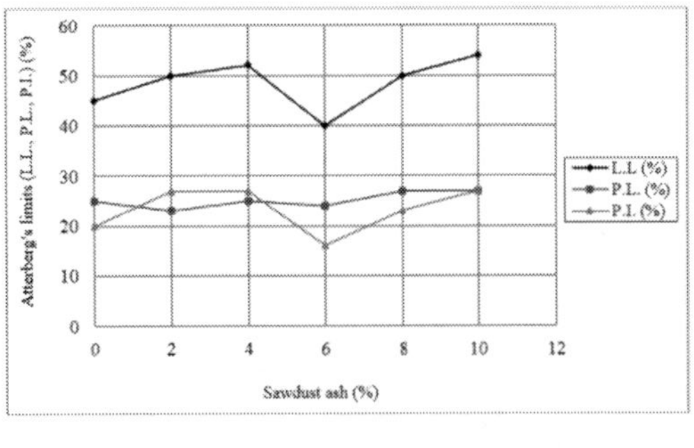


**Fig. 4.** Liquid limit, plastic limit and plasticity index variation with sawdust ash content.

The specific gravities of natural soil and SDA were 2.68 and 2.02 respectively, while the addition of sawdust ash resulted in the reduction of its specific gravity and its unit weight of soft clay-ash mixtures (around 2.43 at 8%; and 2.65 at 2 and 4%) compared to the untreated soft clay soil. This could be considered as an enhancement for the soil property for different geotechnical practices (Figure 5).

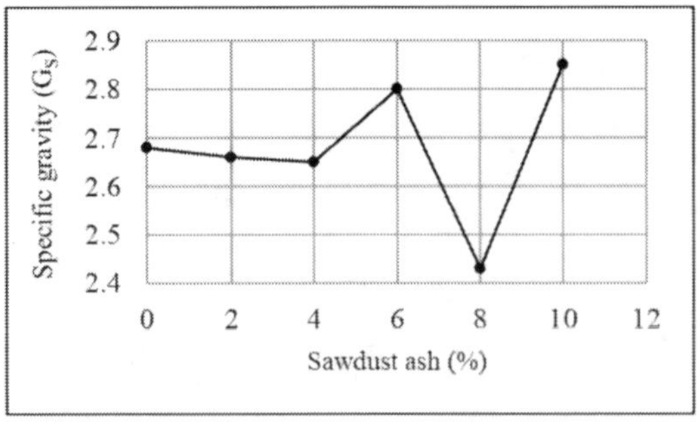


**Fig. 5.** Relation of specific gravity with percent added of sawdust ash.

The optimum moisture content (OMC) for untreated samples is 18.5% with maximum dry density (MDD) of 16.87 kN/m$^3$. With an increase in the SDA percent, an increase in OMC values and a reduction in MDD values were noticed as displayed in Figure 6. The variation in maximum dry density (MDD) and optimum moisture content (OMC) with SDA content (%) is shown in Figures 7 and 8 respectively. The maximum dry density (MDD) in almost all samples slightly decreases with increasing SDA additive which is related to the addition of fine additive with lower specific gravity with respect to the natural soil. But as a general trend, the optimum moisture content (OMC) exhibits an increase with SDA addition which might be a result of water needed for hydration and compaction of soil–ash mixtures. These findings are in conformance to the results obtained by Henry and Ogunribido (2012) [20] and Butt et al. (2016) [7].

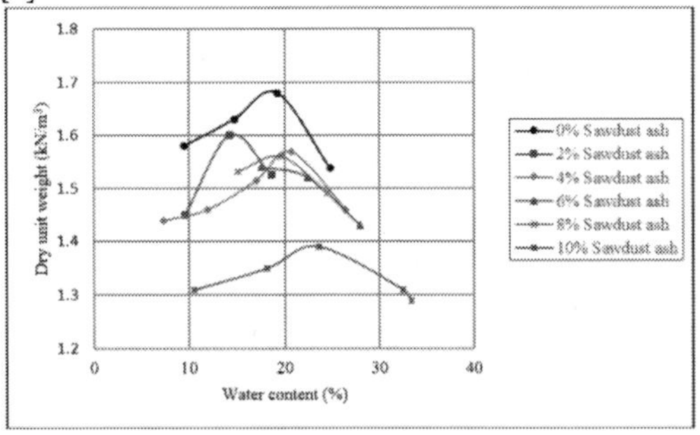


**Fig. 6.** Optimum moisture content (OMC) obtained from the compaction test for untreated and treated soils.

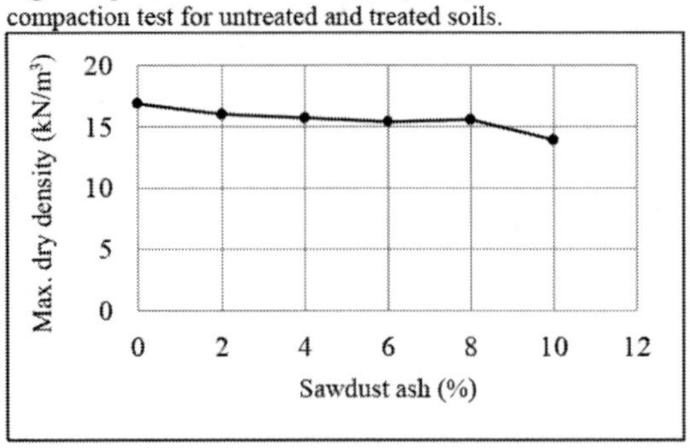

**Fig. 7.** Relation of maximum dry density (MDD) with added sawdust ash.

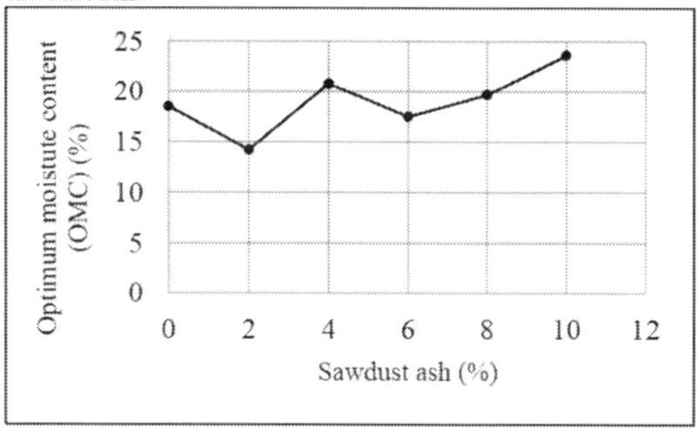


**Fig. 8.** Relation of optimum moisture content (OMC) with added sawdust ash.

The stress-strain curve of unconfined compressive strength (UCS) (Fig. 9) showed sharp increase in unconfined compressive strength ($q_u$) and in turn its undrained shear strength ($c_u$) with increasing SDA content to 4% by more than 4 times (169.2 kN/m$^2$) of the natural soil (39.2 kN/m$^2$). By examining stress-strain curves, it is well demonstrated that the addition of SDA (particularly 4%) modify the soil deformation behavior characteristics in terms of brittleness or ductility by increasing the brittleness behavior of the treated soil and its stiffness which is related to strain softening. While Figure 10 displays the relation of UCS with percent of added sawdust ash. The increase in USC values is due to pozzolanic reactions of SDA. Thus according to Das (2000) [21], the consistency of the natural untreated soil which was very soft clay (<25 kN/m$^2$), and after SDA addition ranges between medium soft (50-100 kN/m$^2$) at 4% SDA to soft clay (<50 kN/m$^2$) for the rest percentages, which is an improvement compared to the untreated soil. It is worth mentioning that almost the same results are obtained for untreated soil and that treated with 2% SDA.

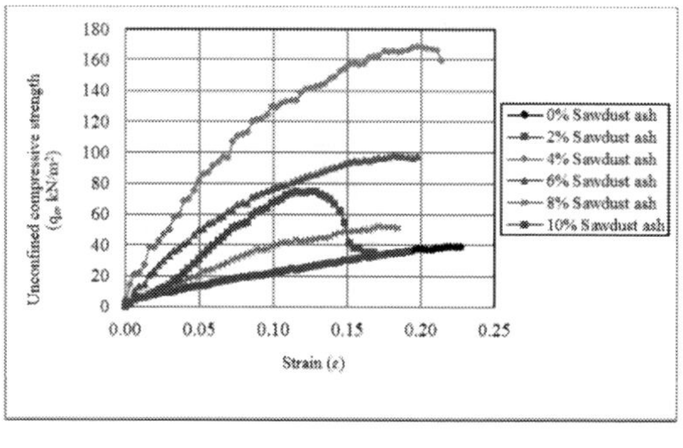


**Fig. 9.** Stress-strain relation of unconfined compression test for treated and untreated soil.

The effect of the SDA addition on soil compressibility is presented in consolidation curves shown in Figure 11. The relation between compressibility coefficients ($C_c$ and $C_r$) with percent added of sawdust ash illustrates that the addition of ashes improved the soft clayey soil by reducing such coefficients particularly compression index, $C_c$, with no noticeable effect in swelling index, $C_r$. As an overall, it can be seen from Figure 12 that the minimum compressibility coefficients were obtained with the addition of 2% ash where $C_c = 0.12$ and $C_r = 0.02$.

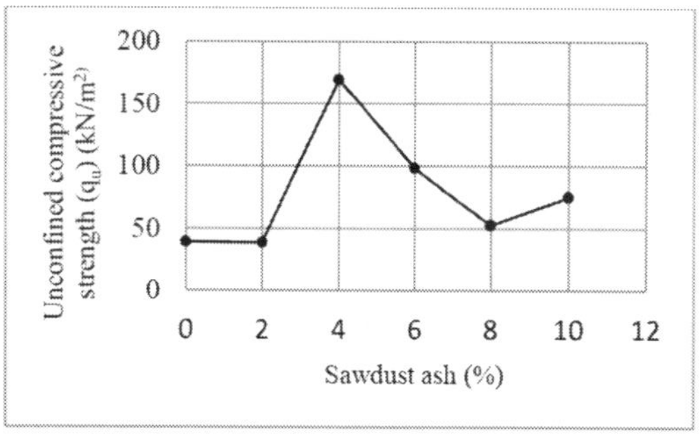


**Fig. 10.** Relation of undrained shear strength (cu) with added sawdust ash.

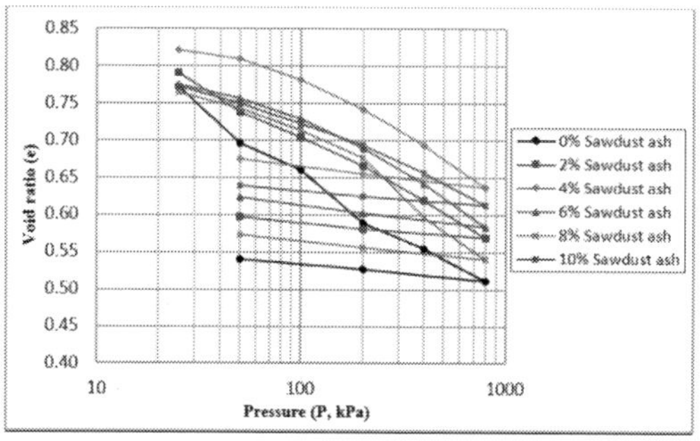


**Fig. 11.** e-log p relation for untreated and treated soils.

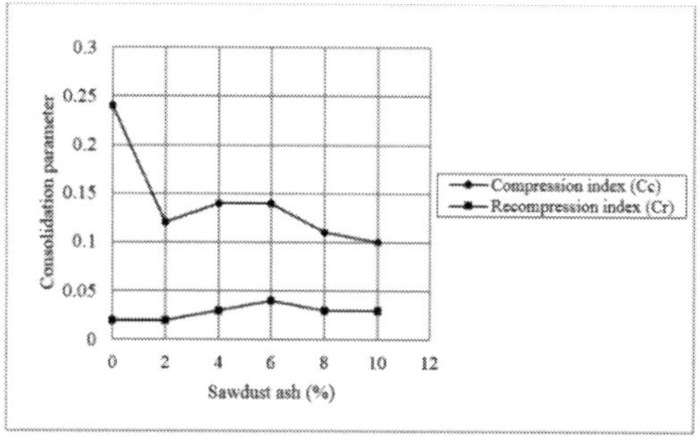


**Fig. 12.** Relation of consolidation parameters (Cc and Cr) with added sawdust ash.

Concerning the California bearing ratio (CBR) tests, although the effect of SDA content on CBR of soil was not tested for all samples (just 0, 4 and 10% SDA addition), but it is important to have an estimate about it. The decrease in CBR in the tested proportions may be due to that such ashes could not be mobilized for the reaction which consequently occupies spaces within the sample. This will reduce the bonds in the soil-ash mixture, so low strength displayed by the SDA addition.

However, according Asphalt Institute (1962) [22], the natural soil sample has CBR=16% which is considered to be a good sub-grade material, and the stabilized soils (with 4 and 10% ash content) resulted in low CBR values (1.6-1.2%) which can be used as sub-base. This decrease in CBR values suggests that the soil samples become weaker indicating that SDA made treated soil high-plastic subbase materials. In general, the higher the plasticity Index, the lower in CBR value. Hence, general rating as sub-grade according to AASHTO (1986) [23] is fair to poor materials.

To summarize, it can be stated that the SDA additives can increase, to some extent, the strength and stability of the soft clay soil in the study area. Thus, both the treated and untreated soil samples are generally regarded as fair to poor for subgrade materials. The additives have adverse effects on the property of soil indices by increase in LL and PI which revealed the increase in fine particles (i.e. clay content) that make the soil less suitable (i.e. no enhancement) for construction purpose.

## 4 Conclusions

Based on the extensive tests conducted and results discussion, the following conclusions can be drawn:

1. Since, the average liquid limits of natural clayey soil samples are higher than 40%, therefore soil under consideration is not suitable for road construction unless it is stabilized.
2. The SDA additive has an adverse effect on the property of soil indices by increasing LL and PI which implied by the increasing clay content.
3. With regard to the influence of SDA on density and compaction, test results show that SDA reduces the compacted dry density and increases the optimum moisture content of soft clay soils.
4. The mixture of SDA with soft clay soils improves the mechanical properties of the soil, as expressed by a reduction in the compression coefficients ($C_c$ and $C_r$) and increasing the undrained shear strength $c_u$.
5. The stabilized soils (with 4 and 10%) resulted in low CBR values (1-2) % which can be used as sub-base material and fair to poor for subgrade materials.
6. In general, the optimum percentage of the SDA for improvement in the strength characteristic of the soil samples is 4-6%.
7. The SDA can be considered as a cheap and acceptable stabilizing agent in road construction for improving most of the geotechnical properties of the soft clayey soil.

## References

1. H.H. Karim, Z.W. Samueel, S.F. Ahmed, *Proceedings of the 2nd international conference of buildings, construction and environmental engineering, American University of Beirut (BCEE2), 17-18 Oct., 2015, Geotechnical Engineering Part,* 117-121 (2015)
2. K.N.M. Yusof, L.C. Lun, P.P. Guan, S.A.A. Tajuddin, *Technology and innovation for sustainable development conference (TISD2006), Faculty of Engineering, Khon Kaen University, Thailand, January* (2006)
3. O.O. Amu, A.R. Origbemide, E.O. Saseun, Res J Appl Sci Eng Tech **2**, 543-546 (2010)
4. O.O. Amu, O.S. Owokade, O.I. Shitan, Int J Eng Tech **3**, 87-94 (2011)
5. H.H. Karim, Z.W. Samueel, M.S. Mohammed, Eng Tech J **34**, Part (A), 1047-1057 (2016)
6. P.P Dahale, P.B. Nagarnaik, A.R. Gajbhiye, El J Geotech Eng **17**, 2443–2461 (2012)
7. W.A. Butt, K. Gupta, J.N. Jha, Geo-Eng **7**, 2-9 (2016)
8. A. Woode, D.K. Amoah, D. Nartey, D. Misefah, D. F.A. Nai, R.A. Nsoh, N.I. Addae, Civ Environ Res **6**, 60-65 (2014)
9. ASTM D 2487-2000, *Standard practice for classification of soils for engineering purposes (Unified Soil Classification System)*, (ASTM International, West Conshohocken, PA, United States, 2000)
10. S. Chowdhury, M. Mishra, O. Suganya, Ain Shams Eng J **6**, 429–437 (2015)
11. ASTM D 854-2002, *Standard test methods for specific gravity of soil solids by water Pycnometer*, (ASTM International, West Conshohocken, PA, United States, 2002)
12. ASTM D 4318-2000, *Standard test methods for liquid limit, plastic limit, and plasticity index of soils*, (ASTM International, West Conshohocken, PA, United States, 2000)
13. ASTM D 1557-2000, *Standard test methods for laboratory compaction characteristics of soil using modified effort (56,000 ft-lbf/ft$^3$ (2,700 kN-m/m$^3$))*, (ASTM International, West Conshohocken, PA, United States, 2000)
14. ASTM D 2166-2000, *Standard test method for unconfined compressive strength of cohesive soil*, (ASTM International, West Conshohocken, PA, United States, 2000)
15. ASTM D 1883-2002, *Standard test method for CBR (California Bearing Ratio) of laboratory-compacted soils*, (ASTM International, West Conshohocken, PA, United States, 2002)
16. ASTM D 2435-1996, *Standard test method for one-dimensional consolidation properties of soils*, (ASTM International, West Conshohocken, PA, United States, 1996)
17. A.E. Adetoro, O.S. Faluyi, Int J Sci Res in Knowledge **3**, 288-296 (2015)
18. American Association of State Highway and Transportation Officials (AASHTO), *Standard specification for transportation materials and methods of sampling and testing* (14th ed.), (USA: Washington DC, AASHTO, 1986)
19. G. Sarkar, M.R. Islam, M. Alamgir, M. Rokonuzzaman, Int J Adv Struct Geotech Eng **01**, 42-49 (2012)

20. T. Henry, T. Ogunribido, Environ Res Eng Manag **2**, 29-33 (2012)
21. B.M. Das, *Fundamental of geotechnical engineering*, (Thomson Learning, USA, 2000)
22. Asphalt Institute, *The asphalt handbook*, Mary Land USA, 176P (1962)
23. R. Whitlow, *Basic soil mechanics,* (Addison Wesley Longman Limited, 1995)

# Performance of soft clay stabilized with sand columns treated by silica fume

*Zeena* Samueel[1], *Hussein* Karim[1,*], and *Mohammed* Mohammed[1]

[1]Department of Building and Construction Engineering, University of Technology, Baghdad, Iraq

**Abstract.** In many road construction projects, if weak soil exists, then uncontrollable settlement and critical load carrying capacity are major difficult problems to the safety and serviceability of roads in these areas. Thus ground improvement is essential to achieve the required level of performance. The paper presents results of the tests of four categories. First category was performed on saturated soft bed of clay without any treatment, the second category shed light on the improvement achieved in loading carrying capacity and settlement as a result of reinforcing with conventional sand columns at area replacement ratio = 0.196. The third set investigates the bed reinforced by sand columns stabilized with dry silica fume at different percentages (3, 5 and 7%) and the fourth set investigates the behavior of sand columns treated with slurry silica fume at two percentages (10 and 12%). All sand columns models were constructed at (R.D= 60%). Model tests were performed on bed of saturated soil prepared at undrained shear strength between 16-20 kPa for all models. For all cases, the model test was loaded gradually by stress increments up to failure. Stress deformation measurements are recorded and analyzed in terms of bearing improvement ratio and settlement reduction ratio. Optimum results were indicated from soil treated with sand columns stabilized with 7% dry silica fume at medium state reflecting the highest bearing improvement ratio (3.04) and the settlement reduction ratio (0.09) after 7 days curing. While soil treated with sand columns stabilized with 10% slurry silica fume provided higher bearing improvement ratio 3.13 with lower settlement reduction ratio of 0.57 after 7-days curing.

## 1 Introduction

Stabilization of soil has become one of the useful solutions to treat the soft soils to achieve the required engineering properties and specification so that the structures can be placed safely without undergo large settlements. The problematic soil is removed and replaced by a good quality material or treated by using mechanical and/or chemical stabilization. Geotechnical properties of such soils may be improved by various techniques. These techniques include compaction, electro-osmosis, ground freezing, chemical stabilization, grouting use of reinforcment such as stone and geotextile waste materials or nontraditional stabilizers, such as silica fume and fly ash, are also applied sometimes for stabilization. Good improvement in engineering properties of soft clay has been noticed by Abd El- Aziz et al. (2014) [1] and Al-Jobouri (2013) [2] when adding mixture of lime and silica fume for stabilizing soft clay. Kalkan (2009) [3] examined the suitability of silica fume to reduce the development of desiccation cracks in compacted clayey liner and over systems. Kalkan (2011) [4] found that silica fume decreases the deformation of expansive clayey soils subjected to cyclic drying and wetting. Karim et al. (2016) [5] studied the behavoir of sand columns (R.D=15%) stabilized with dry silica fume as an additive with different percentages (3, 5 and 7%) and driven in soft clay with $c_u$ between 16-20 kPa. The results showed that the best possible addition of silica-fume in sand-silica fume columne is 7% giving bearing improvement ratio and settlement reduction ratio of 1.56 and 0.5 respectively.

Sand and stone columns, known as column-type technique, are most widely adopted approaches for improving soft and compressible soils due to their high compressive strength and stiffness to the soft soil. The degree of improvement in carrying capacity and reduction in settlement of a soft soil by sand columns is due to two factors: the first one is inclusion of a stiffer sand column in the soft soil. The second factor is the compaction of well graded sand at high relative density. In recent years, many studies and attempts have been carried out to increase the stiffness of stone columns as well as sand columns by mixing additives with backfill materials or by using different patterns of encasement (Abbawi, 2013) [6]. Good improvements in load carrying capacity of sand columns and significant reduction in settlement have been noticed by Al- Saoudi et al. (2014) [7] who used lime as an additive to increase the stiffness of sand columns.

This research focuses on studying the feasibility of improving geotechnical properties of soft clay soil with implemented sand columns (mixed with silica fume using different percentages) and evaluates the degree of improvement in load carrying capacity and settlement reduction ratio.

* Corresponding author: husn_irq@yahoo.com

© The Authors, published by EDP Sciences. This is an open access article distributed under the terms of the Creative Commons Attribution License 4.0 (http://creativecommons.org/licenses/by/4.0/).

## 2 Materials used

### 2.1 Soil

Soil samples used in this study were obtained from Al-Nahrawan area. The soil consists of 16 % sand, 34 % silt and 50 % clay as shown in the grain size distribution (Fig. 1). Atterberg's limits revealed L.L = 37 and P.L = 16. According to the Unified Soil Classification System (USCS), the soil classified as CL (clay of low plasticity). Other physical and chemical properties of the soil are summarized in Table 1. From vane shear test, the cohesion of the soft soil (c) is 35 kN/m$^2$.

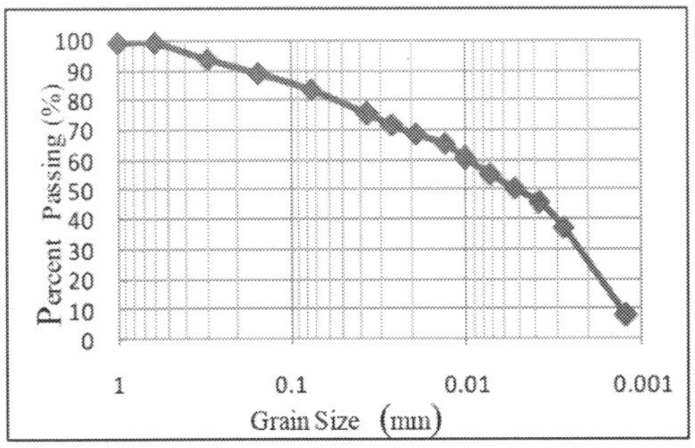


**Fig. 1.** Grain size distribution curve.

**Table 1.** Physical and chemical properties of natural soil used.

| Index Property | Index Value |
|---|---|
| Liquid Limit (L.L) (%) | 37 |
| Plastic Limit (P.L) (%) | 16 |
| Linear Shrinkage (SL) (%) | 12 |
| Plasticity Index (P.I) (%) | 21 |
| Activity (At) | 0.81 |
| Specific Gravity (Gs) | 2.69 |
| Gravel (larger than 2mm) (G) % | 0 |
| Sand (0.06 to 2mm) % (S) | 16 |
| Silt (0.005 to 0.06) (M)% | 34 |
| Clay (less than 0.005mm) (C)% | 50 |
| Classification (USCS) | CL |
| Organic Material (O.M) (%) | 0.39 |
| Calcium Oxide (CaO) (%) | 0.36 |
| $SO_3$ Content (%) | 0.52 |
| Total Dissolved Salt (TDS) (%) | 1.02 |
| pH Value (%) | 9.17 |

### 2.2 Sand

The sand used as a backfill material was brought from Al-Ekhaidhir area. The grain size distribution showed 4% gravel, 94% sand and 2% fines classified as well graded sand as shown in Figure 2. The physical and chemical properties of sand are shown in Table (2). The internal friction ($\phi$) is 44° obtained from direct shear test at dry unit weight 16 kN /m$^3$.

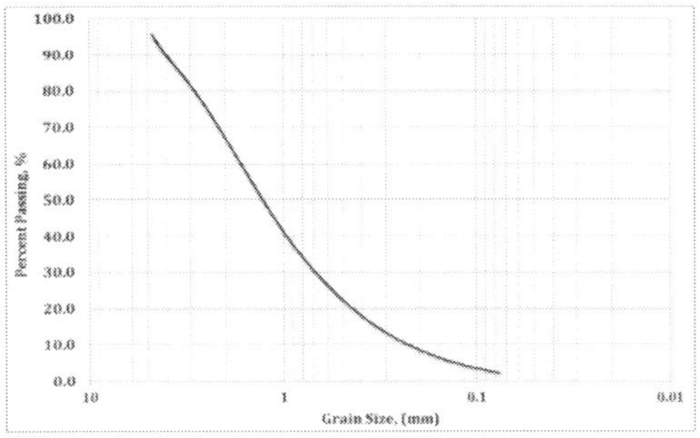


**Fig. 2.** Grain size distribution for the sand used.

**Table 2.** Physical and chemical properties of sand used.

| Index Property | Index Value |
|---|---|
| Max. Dry Unit Weight (kN/m$^3$) | 20.8 |
| Min. Dry Unit Weight (kN/m$^3$) | 16.25 |
| $D_{10}$ (mm) | 0.24 |
| $D_{30}$ (mm) | 0.7 |
| $D_{60}$ (mm) | 1.7 |
| Coefficient of Uniformity ($C_u$) | 7.13 |
| Coefficient of Curvature ($C_c$) | 1.19 |
| Gravel (G %) | 4 |
| Sand (S %) | 94 |
| Fines (%) | 2 |
| Classification | SW |
| Specific Gravity (Gs) | 2.65 |
| Organic Material (O.M) (%) | 0.09 |
| Total Dissolved Salt (TDS) (%) | 0.3 |

### 2.3 Silica fume

The American concrete institute (ACI, 1995) [8] defines silica fume (S.F.) as a "very fine no crystalline silica produced in electric arc furnaces as a byproduct of production of elemental silicon or alloys containing silicon". It is usually a grey colored powder, somewhat similar to Portland cement or some fly ashes. It can exhibit both pozzolanic and cementations properties. Silica fume has been recognized as a pozzolanic admixture that is effective in enhancing the mechanical properties to a great extent. By using silica fume alon with super plasticizers, it is relatively easier to obtain compressive strengths of order of 100-150 MPa in laboratory. The physical and chemical properties of silica fume are shown in Table 3.

**Table 3**. Physical and chemical properties of silica fume used.

| Chemical and physical properties | Composition |
|---|---|
| Chemical composition (%) | |
| $SiO_2$ | > 85 |
| $Fe_2O_3$ | < 2.5 |
| $Al_2O_3$ | < 1 |
| CaO | < 1 |
| $K_2O + Na_2O$ | < 3 |
| C (free) | < 4 |
| $CI^{-1}$ | < 0.2 |
| L.O.I | < 6 |
| Physical Properties | |
| Specific Gravity (Gs) | 2.25 |
| Density ($gm/cm^3$) | 0.75 |
| Moisture content (%) | < 2% |
| Specific Surface ($m^2/gr$) | ~ 20 |

## 3. Experimental work

### 3.1 Steel Container

The model test were carried out by using a steel container with plate thickness of 4mm and internal dimensions 600 mm*600 mm*500 mm.

### 3.2 Model footing

For all model tests, a circular steel model footing with diameter of 64.6mm and thickness 10mm was used.

### 3.3 Loading Assembly

The loading assembly consists of two main parts; a loading frame and a steel container. The loading frame consists of a steel rod with several attachments that host the loading weights. The whole assembly is capable to apply static vertical loads on the footing. Details of the main parts of the loading assembly are shown in Figure 3.

**Fig. 3.** Steel container and loading assembly.

## 4. Sampling and soil mixture preparation

Before stages of preparation of the soil bed, control tests were carried out to obtain a relationship between the shear strength and water content. The undrained shear strength was determined by the Vane shear device. The shear strength of soil decreases with increasing value of water content as shown in Figure 4.

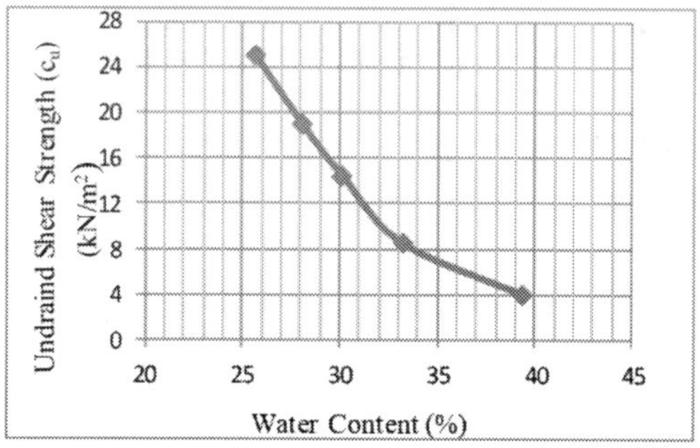


**Fig. 4.** Variation of undrained shear strength with water content.

Following this stage, the natural soil was left for air drying and crushed by a hammer to small sizes and then crushed by using a crushing machine. The air-dried soil was divided into groups, each weight 20 kg. Then, each group was mixed in mixer with sufficient amount of water corresponding approximately to the water range of (27-29%) to get undrained shear strength ($c_u$) between (16-20 $kN/m^2$). After through mixing with water, the soft soil lumps were spread into the container in 5 layers, and tamped with a specially tamping tool to remove any entrapped air. The procedure continued until the required thickness of bed was achieved, the top surface was scraped, level and covered with a polythene sheet and a wooden board of the same size was placed with 5kPa seating pressure. Then, the soil was left for a period of two days to regain its strength by self-weight consolidation. The top surface of the soft soil bed was marked into 4 equal zones and a column was constructed in the center of each quarter by inserting a vertical hollow plastic PVC pipe with external diameter of 50 mm to a depth of 300 mm. The soil inside the column was carefully removed by a hand auger. Later, the sand and sand-silica fume mixture was then poured into the hole in layers. After pouring all the specific amount of the mixture, the full depth of the hole was filled with sand with slight tamping to achieve unit weight of 18.7 $kN/m^3$ at medium state (Dr= 60%). After the completion of the preparation of the bed of soil, a seating load 5 $kN/m^2$ was placed for 24 hours, it was covered with a nylon sheet to prevent any loss of moisture and left for curing period at seven days. Following the curing days for the columns, the static loading system was placed and fixed in position and the footing was incrementally loaded with continuous measurements of the footing up to failure. The same procedure for preparation soft clay was presented in Al-Gharbawi (2012) [9] and Rajab (2013) [10].

https://doi.org/10.1051/matecconf/201816201007

## 5. Experimental results

Prior to the discussion of the model test results, it is important to clarify that the failure is considered as the load corresponding to settlement 10% of the footing diameter. Results and discussion are subdivided into two sections; the first is devoted for discussing the results concerning the improvements in bearing capacity due to the presence of sand columns and sand-silica fume columns. The increase in bearing capacity is determined using the term "bearing improvement ratio" which represents the bearing ratio ($q/c_u$) of the treated soil to that of untreated soil, simply given the notation $(q/c_u)_t/(q/c_u)_u$. The second section of the discussion is devoted to the reduction in settlement gained by each improvement pattern. The term "settlement reduction ration" is explained as the ratio of the settlement of the treated soil to the settlement of the untreated soil and given the notation $S_t$ /$S_{unt}$.

### 5.1 Untreated soil

The first set of model tests was performed on untreated soil to determine the relationship between the applied stresses versus settlement. This relationship is considered as benchmark for comparison purposes of different patterns of improvement. Typical results are shown in Figure 5 which relates the settlement ratio ($S/D_{footing}$) versus bearing ratio ($q/c_u$). The bearing ratio at failure is 4.18 corresponding to 10% settlement ratio.

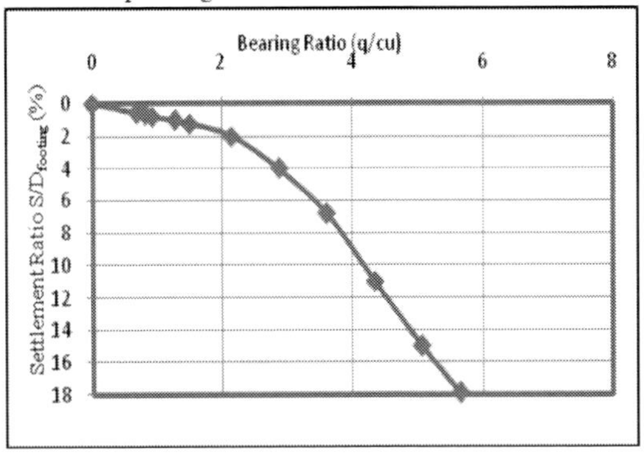


**Fig. 5**. Bearing ratio versus settlement ratio of untreated soil.

### 5.2 Soil treated with sand columns

Figure 6 demonstrates the relationship between $q/c_u$ and $S/D_{footing}$ for soil reinforced with sand column at relative density 60% corresponding to medium state. At initial stress increments up to $q/c_u \approx 2$, the untreated and treated models exhibited approximately the same settlement ratio indicating insignificant influence for the presence of sand columns. At this stage, the applied stress was evenly distributed over the contact area of the composite soil and no sign of stress concentration was noticed. The influence of the sand column becomes noticeable when the bearing ratio exceeding 2 and reaching maximum value at stress levels close to failure with bearing ratio ($q/c_u$) at failure equal to 7.75.

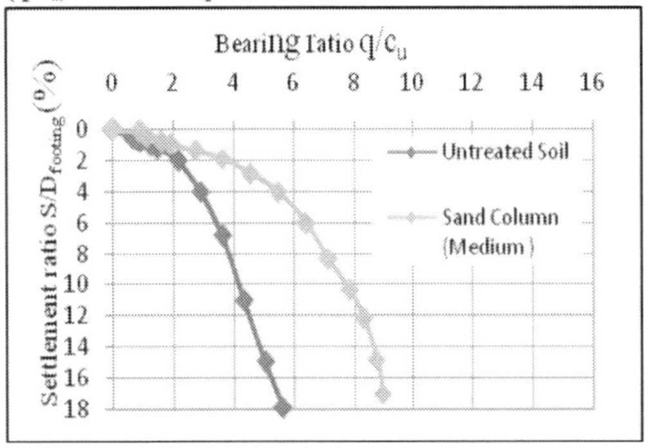


**Fig. 6.** Bearing ratio versus settlement ratio of soil reinforced with sand columns.

The bearing improvement ratio $(q/c_u)_t/(q/c_u)_u$ versus settlement ratio $S/D_{footing}$ at failure is 1.83 as shown in Figure 7.

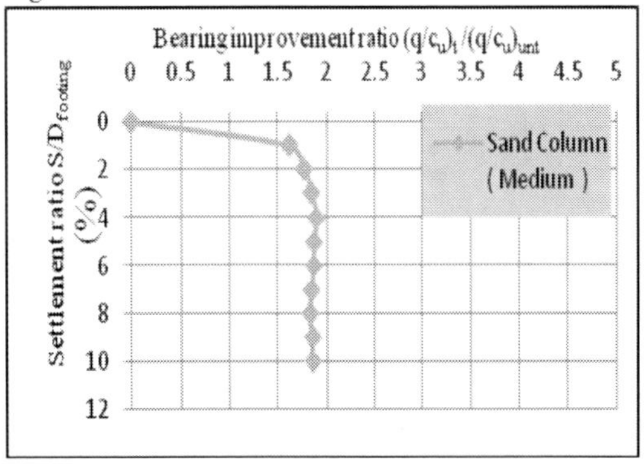


**Fig. 7**. Bearing improvement ratio versus settlement ratio of soil reinforced with sand columns.

Variation of bearing ratio $q/c_u$ versus settlement reduction ratio $S_t$ /$S_u$ is shown in Figure 8. Settlement ratio exhibits a decreasing trend then increasing bearing ratio revealing a settlement reduction ratio of ($S_t$ /$S_u$) equal to 0.29 at failure.

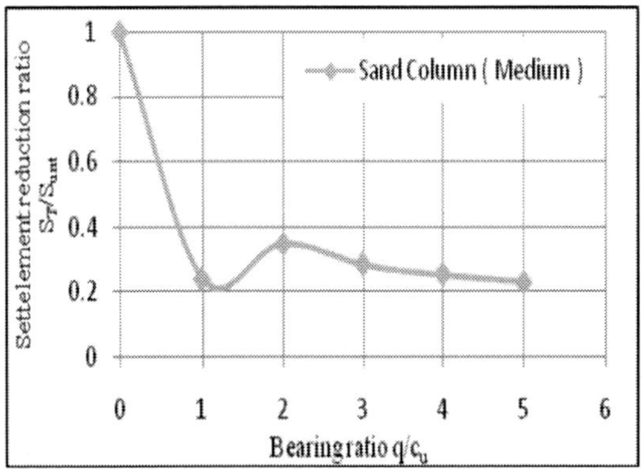


**Fig. 8.** Settlement reduction ratio versus bearing ratio of soil reinforced with sand columns.

### 5.3 Sand column treated with dry silica fume

Six model tests were performed on sand column stabilized with silica fume at medium state (R.D 60%) and tested after curing for seven days. Figure 9 presents the $q/c_u$ versus $S/D_{footing}$ for all percentages (3, 5 and 7%) of silica fume used. At initial stress increments, the results of sand column stabilized with the silica fume showed no deformation compared with untreated sand models. The columns provided sufficient stiffness that led to improve the surrounding soil. The bearing ratio $q/c_u$ at failure are 10.35, 11.15 and 12.7 for 3,5 and 7% silica fume respectively.

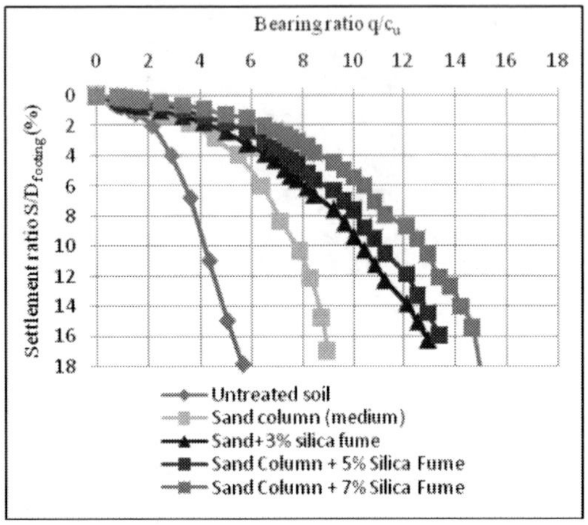


**Fig. 9.** Bearing ratio versus settlement ratio of soil reinforced with sand columns stabilized with silica fume.

The bearing improvement ratio $(q/c_u)_t/(q/c_u)_{unt}$ versus settlement ratio $S/D_{footing}$ is presented in Figure 10. It is shown that $q_t/q_{unt}$ increases with increasing silica fume content. The bearing improvement ratio $(q/c_u)_t/(q/c_u)_{unt}$ at failure = 2.48, 2.66 and 3.04 for 3, 5 and 7% silica fume respectively.

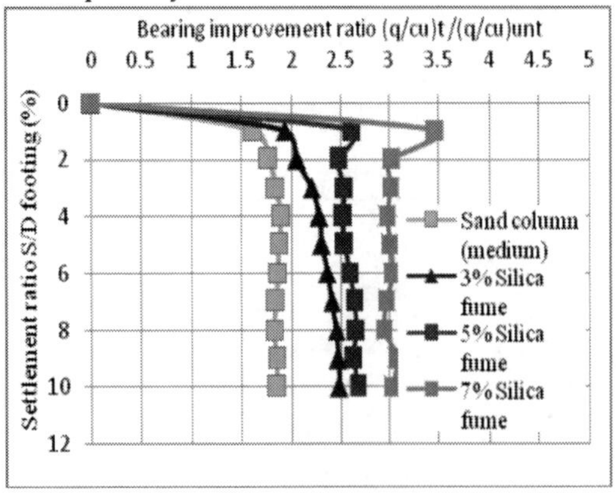


**Fig. 10.** Bearing improvement ratio versus settlement ratio of soil reinforced with sand columns stabilized with silica fume.

Relation between settlement reduction ratio $S_t/S_{unt}$ and bearing ratio $q/c_u$ is shown in Figure 11. It can be shown that 7% silica fume is more efficient in providing the lowest settlement ratio of 0.09. The settlement reduction ratio $S_t/S_{unt}$ are 0.168, 0.121 and 0.09 for 3, 5 and 7% silica fume respectively at $q/c_u$ =5.

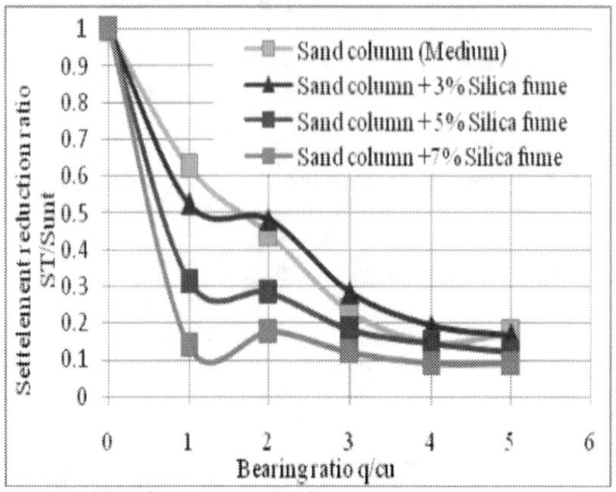


**Fig. 11.** Settlement ratio versus bearing ratio of soil reinforced with sand columns stabilized with silica fume.

### 5.4 Sand column treated with slurry silica fume results

Four model tests were performed on sand column stabilized with silica fume at medium state (R.D 60%) and tested after 7-days curing. Figure 12 presents the $q/c_u$ versus $S/D_{footing}$ for all percentages 10 and 12 % by weight slurry silica fume used. At initial stress increments, the results of sand column stabilized with silica fume showed marginal difference deformation as compared with untreated sand models. These columns provided significant stiffness that led to improve the composite soil. The bearing ratio $q/c_u$ at failure ($S/D_{footing}$ =10%) are 13.12 and 12.02 for 10% and 12% by weight

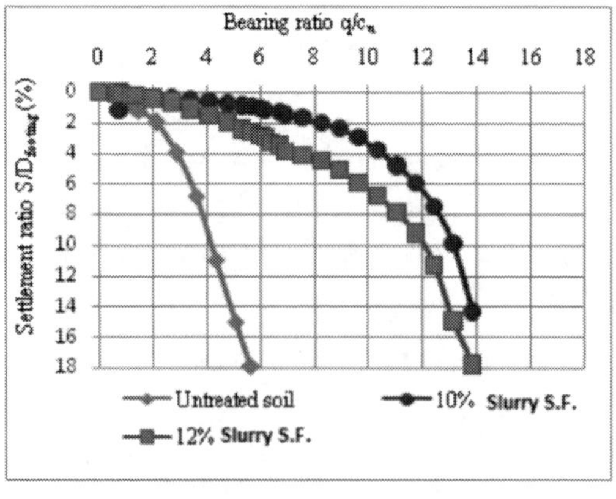


silica fume respectively.

**Fig. 12.** Bearing ratio versus settlement ratio of soil reinforced with sand columns treated with slurry silica fume.

The relationship between bearing improvement ratio $(q/c_u)_t/(q/c_u)_{unt}$ and settlement ratio $S/D_{footing}$ (%) is presented in Figure 13. Results show that the initial peak is about $S/D_{footing}$ =1.5% followed by decreasing then increasing to second peak close to failure. The bearing improvement $(q/c_u)_t/(q/c_u)_{unt}$ at failure =3.13 and 2.88

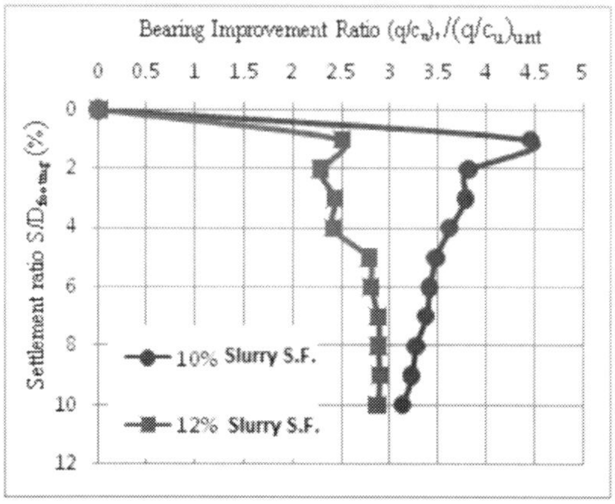


for (10% and 12%) by weight slurry silica fume respectively.

**Fig. 13.** Bearing improvement ratio versus settlement ratio of soil reinforced with sand columns stabilized with slurry silica fume.

Settlement reduction ratio $S_t/S_{unt}$ versus bearing ratio $q/c_u$ are shown in Figure 14. It is shown that at $q/c_u$ =1 settlement reduction ratio is equal to zero indicating that there is no settlement in this stage, but then it starts to increases gradually. The settlement reduction ratio $S_t/S_{unt}$ =0.057 and 0.144 for (10% and 12%) by weight slurry silica fume at $q/c_u$ =5 respectively. Up to 10% of the liquid, the voids are still not filled with liquid. When adding the liquid, the density increases, the percent of voids decreases and the soil strengthens. By increasing the liquid percent more than 10%, the liquid starts to displace (or remove) the soil particles which in turn leads to increasing the percentage of voids and weakness of the soil.

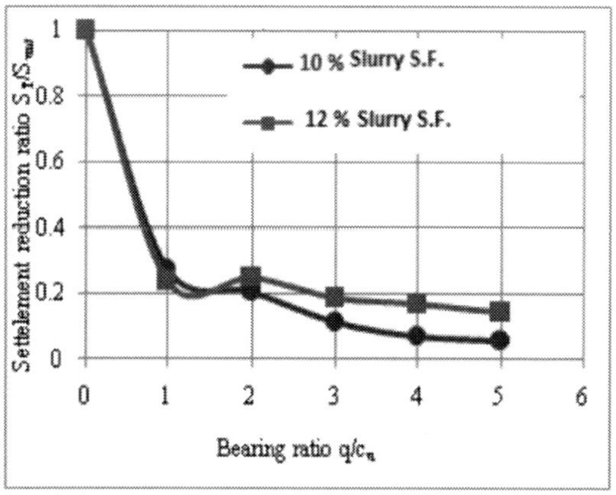


**Fig. 14.** Settlement reduction ratio versus bearing ratio of soil reinforced with sand columns stabilized with slurry silica fume.

## 5.5 Comparison between sand columns reinforced with dry and slurry silica fume

Figure 15 presents the $q/c_u$ versus $S/D_{footing}$ of sand column for dry and slurry. No significant difference was noticed between dry and slurry sand columns at medium state. However, the slurry sand columns provided significant increase in bearing ratio as compared to dry type. Table 4 shows a summary of results for sand columns treated with dry and slurry silica fume types.

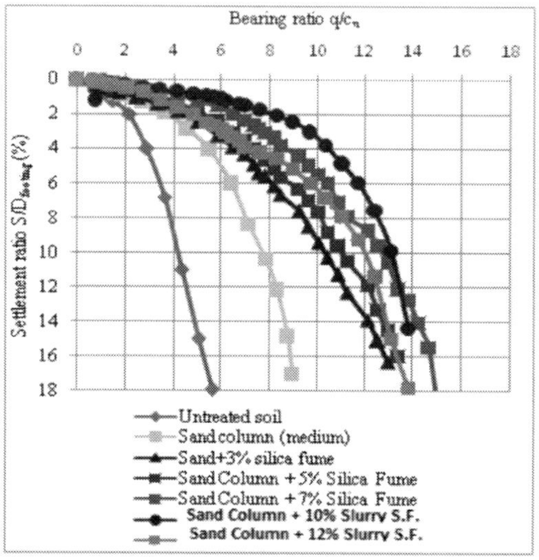


**Fig. 15.** Bearing ratio versus settlement ratio of sand columns treated with dry and slurry silica fume.

**Table 4.** Summary results of sand columns stabilized with dry and slurry type.

| Type of soil treated | $q/c_u$ | $(q/c_u)_t/(q/c_u)_{unt}$ | $S_t/S_{unt}$ |
|---|---|---|---|
| Sand Column +3% Silica Fume | 10.35 | 2.48 | 0.168 |
| Sand Column +5% Silica Fume | 11.15 | 2.66 | 0.121 |
| Sand Column +7% Silica Fume | 12.7 | 3.04 | 0.09 |
| Sand Column+10% (slurry silica fume) | 13.12 | 3.13 | 0.057 |
| Sand Column+12% (slurry silica fume) | 12.02 | 2.88 | 0.144 |

## 6 Conclusions

In the light of experimental tests, the following conclusions can be drawn:

1. The bearing improvement ratio and settlement reduction ratio exhibited by the sand columns only are 1.83 and 0.229 respectively.
2. The results analysis of the model tests indicated an encouraging improvement in load capacity of the sand columns stabilized with silica fume and considerable reduction in the settlement compared to the conventional stone columns.
3. It is worth mentioning that the values obtained for bearing improvement ratio and settlement ratio are limited to type of silica fume used in the test. The best possible addition of silica fume content in sand-silica fume columns is 7% at dry state, giving bearing improvement ratio and settlement ratio of 3.04 and 0.09 respectively. While addition 10% slurry silica fume to sand columns shows bearing improvement ratio and settlement ratio of 3.13 and 0.057 respectively.
4. In general, the use of silica fume as a stabilizing agent with sand provided improvements in terms of bearing improvement ratio and settlement reduction ratio due to increase in stiffness of the columns.

## References

1. M.T. Abd El-Aziz, M.A. Abo-Hashema and M. El-Shoubagy, *The effect of lime-silica fume stabilizer on engineering properties of clayey subgrade, Engineering conference, Faculty of Engineering, Mansoura University,* Sharm *El-Sheikh, Egypt,* April (2004)
2. M.M. Al-Jobouri, M.Sc. Thesis, Civil Engineering Department, University of Baghdad, Iraq, (2013)
3. E. Kallkan, J Appl Clay Sci, **43** (2009)
4. E. Kallkan, J Appl Clay Sci, **52** (2011)
5. H.H. Karim, Z.W. Samueel and M.S. Mohammed, Eng Tech J **34**, Part A (2016)
6. Z.W. Abbawi, Ph.D. Thesis, University of Technology, Baghdad, Iraq, (2010)
7. K.S. Al-Saoudi, M.M. Al-Ksissi and N.A.A. Rajah, Eng Tech J **32**, Part A (2014)
8. ACI Committee 234, ACI Mater J, **92** (1995)
9. A.S.A. Al-Gharbawi, M.Sc. Thesis, University of Technology, Baghdad, Iraq, (2013)
10. N.A.A. Rajab, M.Sc. Thesis, University of Technology, Baghdad, Iraq, (2013)

# Effect of hydraulic conductivity of unsaturated soil on the earth dam performance

*Mohammed* Fattah[1,*], *Mahmood* Ahmed[2] and *Nawar* Ali[2]

[1]Building and Construction Engineering Dept., University of Technology, Baghdad, Iraq
[2]College of Engineering, University of Baghdad, Baghdad, Iraq

**Abstract.** In this paper, the finite element method is uzed to solve the governing equations of flow through earth dams. The computer program Geo-Slope is used in the analysis through its sub-program named SEEP/W. A case study is considered to be Al-Adhaim dam which consists of zoned embankment with a total length of 3.1 km. The dam in its actual design is analyzed. Then, an attempt is made to study the seepage in unsaturated zone of the dam through studying the effect of several parameters including the effect of changing the unsaturated hydraulic conductivity with the degree of saturation of the core soil and changing of water level in the reservoir. A procedure is proposed to define the hydraulic conductivity function from the soil water characteristic curve which is measured by the filter paper method. Fitting methods are applied through the program SoilVision. A parametric study was carried out and different parameters were changed to study their effects on the behavior of partially saturated soil. The study reveals that the rate of flow is decreased by about 20 – 27% when the degree of saturation of the core material is decreased from 100% to 50% at water level 115.75 m, while the exit gradient of flow is decreased by about 13 -15%. This decrease in flow rate becomes 13-15% and 8-9.5% when the reservoir water level is 131.5 m and 143.5 m, respectively, while the exit gradient of flow is increased by about 23-29.5% and 29-29.5% when the reservoir water level is 131.5 m and 143.5 m, respectively. When the state of soil changes from fully saturated S= 100% to partially saturated S= 90%, a rapid increase in head gradient and pore water takes place at the embankment base for different water levels in the reservoir. This decrease plateaus out on further decrease in the degree of saturation.

## 1. Introduction

Seepage problems in geotechnical engineering have been generally solved by sketching flow nets. This method is based on the assumption that water flows only in the saturated zone. This method of solution is practical for simple steady state problems where the boundary of the flow region is clearly defined and the soil conditions are not too complex. However, many seepage problems of practical interest are complex, and a flow-net solution is not feasible. With the development of high-speed digital computers, numerical methods are increasingly used in solving seepage problems. Some of the first attempts at modeling seepage using the finite element method also considered water flow only in the saturated zone. In this "saturated-only" approach, the phreatic line was assumed to be the upper boundary of the flow region for unconfined flow problems. In order to obtain the location of the phreatic line, a trial and error procedure was used to search for a surface that has both zero water pressure along it and zero flow across it. For each trial, the location of the phreatic line was readjusted, and a new mesh is constructed. This method is tedious, and it is also based on the erroneous assumption that the phreatic line was the uppermost flow boundary.

Darcy's law was originally derived for saturated soil, but can equally be applied to the flow of water through an unsaturated soil. The only difference is that for flow through an unsaturated soil, the coefficient of permeability is no longer a constant but is a function of the matric suction of the soil.

Lam et al. (1987) presented the development of a transient finite element seepage model using an approach consistent with that used by geotechnical engineers. The concepts pertinent to the understanding of unsaturated flow were reviewed. The general governing differential equation for transient seepage was derived and the finite element solution to the equation was presented. Three iterative techniques, including the secant method, have also been incorporated to ensure convergence for highly nonlinear flow systems. Changing boundary conditions during a transient process was accommodated by a prescribed step function.

Thieu et al. (2001) presented the application of general-purpose partial differential equation solvers associated soil property functions to the analysis of

* Corresponding author: myf_1968@yahoo.com

© The Authors, published by EDP Sciences. This is an open access article distributed under the terms of the Creative Commons Attribution License 4.0 (http://creativecommons.org/licenses/by/4.0/).

saturated-unsaturated seepage problem. The soil material was assumed to be silt and isotropic with respect to the coefficient of permeability. Experimental data showing the coefficient of permeability versus matric suction, and the volumetric water content versus matric suction were obtained from Ho (1979). The saturated coefficient of permeability was 2.5 x $10^{-7}$ m/sec and the saturated volumetric water content was 0.381. The finite element computer program PDEase2D was used for the analysis of the steady-state seepage problem. A maximum error of 0.1% was specified. The results of the total head distribution were presented in the form of pore-water pressure distribution and flow vectors under steady state seepage. The position of the total head contours and phreatic lines are the same for all permeability functions.

Modelling one-dimensional flow process in unsaturated soils is usually based on the numerical solutions of the Richards's equation which may be expressed in water content form, pressure or mixed form. Bouchemella et al. (2014) tried to test the accuracy of the results and the ability of the different forms of Richards's equation to describe the water flow in unsaturated soils. Implicit finite difference scheme was used in the numerical solution of the water content form and the pressure form of Richards's equation. The solution of the mixed form was obtained using the 1-D Hydrus software. The effect of texture and the initial moisture of the soil on the equation solutions were explored. The results showed a significant difference under some conditions. The choice of the empirical model to describe the hydraulic properties like retention curve and hydraulic conductivity may play an important role on the performance of each form, and can lead to remarkably different simulation results. The accuracy of the results was established by comparing numerical results to the semi-analytical solution of Philip and confronting them to the measured results.

In this study, the aim of experimental work is to define the soil water characteristic curve SWCC by measurement of the soil suction. Soil samples were collected from a site east of Baghdad. The physical properties of this soil were studied by conducting a series of tests in the laboratory, these include: specific gravity, Atterberg limits, grain size distribution by sieve analysis and hydrometer, compaction test and permeability test (falling head). The total and matric suction were measured by the filter paper method at different degrees of saturation. The unsaturated coefficient of permeability was then calculated. These parameters were implemented in the finite element analysis of a zoned earth dam.

## 2. Experimental program

A brown clayey soil was brought from a site east of Baghdad. Standard tests were performed to determine the physical properties of the soil. Details are given in Table 1. Grain size distribution of the soil used revealed 3.3 % sand, 31.7% silt and 65% clay. The samples were subjected to testing program which included the following tests :

The specific gravity for the soil studied was determined according to ASTM D854-00.

The liquid and plastic limit tests were carried out on the soil passing sieve No. 40 according to ASTM D-4318-00. The results are shown in Table 1. Wet sieving (by water), and hydrometer tests were carried out in accordance with ASTM-D422-02. The soil is classified as silty clay according to "ASTM" classification, designated as CL according to the Unified Soil Classification System, and the percentage of clay is summarized in Table 1. The compaction test was carried out on soil passing sieve No. 4 according to ASTM-D698-00 in order to find the dry unit weight and optimum moisture content. The results are summarized in Table 1. Falling head permeability test was carried out on soil samples following the procedure described by Head and Epps (2011).

### 2.1 Total and matric suction of soil measurement by filter paper method

The filter paper method was used for soil suction measurements because of its advantages over other suction measurement devices. The method basically depends on the principle that at equilibrium, the suction value of the filter paper and the soil will be equal. After equilibrium is established between the filter paper and the soil, the water content of the filter paper disc is measured. Then, by using filter paper water content versus suction calibration curve, the corresponding suction value is found from the curve. This is the basic approach suggested by ASTM Standard Test Method for Measurement of Soil Potential (Suction) Using Filter Paper (ASTM D 5298). In other words, ASTM D 5298 employs a single calibration curve that has been used to infer both total and matric suction measurements.

### 2.2 Description of the problem

Al-Adhaim dam is an earth dam located at Al-Adhaim river north of Baghdad in Iraq. It consists of embankment of length 3.1 km from the helm of a major cross-river valley where it is 73.5 m high and go through or underneath spending and channel spillway. Al-Adhaim dam consists of fill soil with sloping core. The initial architecture of the dam has been developed by Bennie and Partners Company. The soil of the foundation consists of sloping layers of sandstone and marl of uneven thickness (Al-Adhaim Dam Report, 1994).

Figure 1 shows section of the dam, which has a base level of about 70 m above the sea level summit dam, and the level at the top of dam was 146.5 m (Al-Adhaim Dam Report, 1994). Al-Adhaim dam is a filling dam consisting of three main zones: shell, core, and filters as shown in Figure 1. The thickness of this filling changes with the height of the dam where it is at greatest value at the base and gradually decreases with increase of the height. The height of the dam depends on the height of land, which varies from a few meters rival parties and up to about 70 m at the confluence with the metaphor of the river, and below is a brief description of the dimensions and slope of the dam:

The shell is composed of sand and gravel in general, the width at top is 12 m, while the slope at the upstream is 1:2.5 and the downstream is 1:2.

The core consists of silty clay sliding towards the upstream of the dam with the slope 1:1.1 and 1:2. The form of core sections is chimney and the kind of slope and the thickness is 8 m at the elevation 143.5 m and this thickness gradually increases until it reaches 33 m at elevation 70 m and the thickness at any point of the core is 50% of the height of water at that point. This thickness is good and impedes water leakage. At elevation 70 m, the core soil merges with the marl soil at the foundation that formed an impediment which prevents leakage of water under the dam (Al-Adhaim Dam Report, 1994). The drainage consists of vertical and horizontal filters. Bennie and Partners implemented that the vertical filter consists of two layers to protect the core, the thickness of the target layer with core is 2 m called filter F and the thickness of other layer is 2.5 m called filter T. The horizontal drainage consists of three layers filter F 0.3 m, filter T 2.5 m, and filter F 0.5 m target with foundation (Al-Adhaim Dam Report, 1994). The specified materials of Al-Adhaim dam are shown in Table 2.

In this paper, the soil from core material has been selected for study and has been subjected to a set of laboratory tests. The soil of core has the properties shown in Table 3, which were calculated from laboratory tests carried out on disturbed samples.

### 2.3 Required relationships for unsaturated soils

Before starting the analysis, there are some relationships required in dealing with partially saturated soil characteristics, which are:

**1. Suction versus degree of saturation**

Total and matric suction of the soil samples are measured by remolding the samples at different degrees of saturation 40%, 50%, 60%, 70%, 80%, and 90% using the filter paper method.

**2. Volumetric water content $\theta_w$**

One of required input data in SEEP/W program is relationship between volumetric water content and pore water pressure. SEEP/W can estimate this relationship from input data such as, volumetric water content at saturated condition, $\theta_s$, and coefficient of volume change, $m_v$, and from closed form solution of van Genuchten (1980), or closed form solution of Fredlund and Xing (1994).

The program Soil Vision is used here to find the soil water characteristic curve and estimated hydraulic conductivity. Mathematical representation of soil behavior is provided by Soil Vision with the fitting of mathematical equations to soil data. Standard equations used to represent soil behavior may be applied to laboratory data. Soil Vision helps application of mathematical equations to laboratory data for provided soil properties (User's Guide Manual of Soil Vision, 2001), including grain-size distribution, soil-water characteristic curve, permeability, compression (consolidation) and compaction, (User's Guide Manual of Soil Vision, 2001):

**3. Unsaturated hydraulic conductivity k**

After the volumetric water content function has been curve-fitted using the method proposed by Fredlund and Xing, (1994), then the hydraulic conductivity function can be predicted over the entire suction range i.e., from 0 to $10^6$ kPa, which removes the need to determine the residual water content, which is usually required for other predictive methods. The difficult task of measuring the unsaturated hydraulic conductivity function directly is often overcome by predicting the unsaturated hydraulic conductivity from either a measured or predicted volumetric water content function.

Additional input data in SEEP/W program is the relationship between hydraulic conductivity and pore water pressure. SEEP/W can estimate this relationship from input data such as saturated hydraulic conductivity and the volumetric water content relationship form solution of Fredlund and Xing (1994). For this reason, the Soil Vision program has been used

In this research, Soil Vision program has been used in order to find the properties of soil such as volumetric water content and unsaturated hydraulic conductivity. After determination of the relation between the volumetric water content and matric suction using the Fredlund and Xing model at every degree of saturation, a relationship between the hydraulic conductivity and matric suction can be estimated from Soil Vision program.

## 3. Results of experiments:

Figures 2 and 3 show the relationship between the total and matric suction and the degree of saturation, respectively. From these figures, it can be shown that the suction of the soil decreases with increase in the degree of saturation and the rate of decreasing in matric suction is not equal to the rate of increase of the degree of saturation.

From the program Soil Vision, and after inputting all the required properties of the soils used in this analysis, i.e., total unit weight, dry unit weight, liquid limit, plasticity index, void ratio, porosity, matric suction value, degree of saturation, and grain size distribution, the soil water characteristic curve is predicted (relation between the gravitation water content and the matric suction through applying fitting methods, such as the method proposed by Fredlund and Xing (1994) and van Genuchten (1980) for fitting the soil water characteristic curve as shown in Figure 4.

The most variable parameter in the Soil Vision program is saturated hydraulic conductivity and Fredlund and Xing fit of soil-water characteristic carve. Finally, the unsaturated hydraulic conductivity curve is predicted in Figure 5 for different degrees of saturation.

From Figure 5, the hydraulic conductivity decreases when the degree of saturation decreases.

A steep permeability function indicates a rapid reduction in the water coefficient of permeability for a small increase in matric suction. In this case, the quantity of water flow into the unsaturated zone is considerably reduced since the coefficient of permeability decreases from $1x10^{-7}$ to $1x10^{-9}$ m/s. Additional input data in SEEP/W program is the relationship between hydraulic conductivity and suction. SEEP/W can estimate this relationship from input data such as saturated hydraulic conductivity and the volumetric water content relationship from the solution of Fredlund and Xing (1994). This relation has been determined as shown in Figure 5.

## 3.1 Seepage analysis of Al-Adhaim dam

The steady state seepage through and under the earth dam is analyzed using the program SEEP/W. The finite element mesh used for the analysis is shown in Figure 6. The mesh includes higher-order eight-noded isoparametric elements for the entire body of the dam and its foundation. The number of elements is 1217 and the number of nodes is 3724. The upstream boundary nodes are designated as head boundaries with total head equals to the water level in the reservoir 131.5 m, and the downstream boundary nodes are designated with total head equal to 93 m. The bottom nodes along the foundation are designated as a zero discharge (no flow). Default boundary conditions are assumed to be ordinary boundary (no flow) in Figure 6. The dam is assumed to be without filters in one case and with filters in the other.

## 3.2 Results of finite element analysis and discussion

At first, the soil that is selected in this research is used in the analysis of the dam as fully saturated soil by the program SEEP/W without and with the presence of filters.

Figure 7 shows the contours of total head and flow path the same water heads. The results of this analysis are summarized in Table 4.

The height of water 143.5 m represents the maximum design water level while the height 131.5 m represents the normal water level and the height 115.75 is the case where the reservoir is 50% filled with water. It can be noticed from Table 4 that the exit gradient is always less than 1.0 which means that the dam is safe in these conditions. Table 5 shows the results of analysis for the earth dam when using the soil of this research. Then, the dam is analyzed considering the core as partially saturated soil with different reservoir water levels 115.75 m, 131.5 m, and 143.5 m with the same degree of saturation without and with the presence of filters. Later, each problem is re-analyzed considering different degrees of saturation. This means that for each case, the degree of saturation is changed from 40%, 50%, 60%, 70%, 80%, to 90% keeping the other parameters constant, and for each degree of saturation, the problem was re-analyzed with the same reservoir water level.

Figure 8 shows the seepage lines and velocity vectors through the dam considering that the core soil is the selected soil in this study for water level of 115.75 m in case of the dam with filters.

The phreatic line resulting from the saturated-unsaturated flow model is in close agreement with the empirical free surface from a conventional flow net construction. This observation supports the assumption that the empirical free surface is approximately equal to a phreatic line. However, water can flow across the phreatic line, as indicated by the nodal flow rate vectors. Water flow across the phreatic line into the unsaturated zone indicates that the phreatic line is not the uppermost flow line, as assumed in the flow net technique.

The water in the saturated and unsaturated zones then flows essentially parallel to the phreatic line in the central section of the dam. The water in the saturated zone then flows across the phreatic line into the unsaturated zone at the toe of the dam.

## 3.3 Effect of height of water in the reservoir

Two cases are considered to study the effect of height of water in the reservoir. The first case includes analysis of the dam without filters. Figures 9 and 10 show the relationship between the quantity of seepage at different heads of water when the dam is constructed without filters for degrees of saturation of the core soil of 40% and 50%, respectively.

Figures 11 and 12 show the relationships between the height of water in the reservoir and exit gradient in the whole dam for conditions of degrees of saturation of 40% and 50%, respectively. It can be noticed that the rate of flow is increased by about 45 - 84% when the height of water is increased from 151.75 m to 134.5 m, while the exit gradient of flow is decreased by about 62.3 - 63.5 %.

Figures 13 and 14 show the relationships between the height of water in the reservoir and head gradient at nodes A and B shown in Figure 6 for conditions of degrees of saturation of 40% 50%, respectively, while Figures 15 and 16 show the relationships between the height of water in the reservoir and the pore water pressure at nodes A and B for different degrees of saturation.

It can be concluded that the pore water pressure at node A which is located at elevation 93.0 m (embankment base) is increased by about 17-27% when the degree of saturation is decreased from 100% to 50% and 40%, respectively. This increase becomes 28 - 39 % at node B located at the center of core base. This means that the effect of unsaturated soil on the values of pore water pressure is clearer at the base of the dam core.

Table 6 summarizes the results of rate of flow, exit gradient of the earth dam at different degrees of saturation while Tables 7 and 8 present water pressure and head gradient at nodes A and B predicted at different degrees of saturation. It can be concluded that when the state of soil changes from fully saturated S= 100% to partially saturated S= 90%, a rapid decrease in head gradient and pore water takes place at node A and node B. This decrease plateau out on further decrease in the degree of saturation S.

At the no-flow, or hydrostatic condition, suction head is distributed linearly because total head is a constant everywhere. The corresponding water content distribution is the soil-water characteristic curve. A minimum water content value occurs at the phreatic line, and the 100% saturation condition occurs at the water table. The air-entry head is the elevation above the water table at which desaturation commences.

Steady state flow causes a downward water flow. The negative pore water pressure increases from the static equilibrium condition. The hydraulic head profile starts with a positive value at ground surface and decreases to zero at the phreatic line. Therefore, water flows downward with a constant, negative flux for steady-state infiltration. Water can flow across the phreatic line, as indicated by the nodal flow rate vectors. Water flow across the phreatic line into the unsaturated zone indicates that the phreatic line is not the uppermost flow line, as assumed in the flownet technique (Fredlund and Rahardjo 1993).

Fattah et al. (2017) found that the water flux decreases linearly with time and with the water level in the reservoir which indicates that the rate of flow in the whole body of the dam shows almost uniform change.

## 4. Conclusions

Based on the experimental results obtained from this research work and analysis of the behavior of a zoned earth dam with unsaturated zone, the seepage and flow analysis carried out on Al-Adhaim dam using the finite element method, the following conclusions can be drawn:

1) The procedure of analysis of the seepage of earth dam including zone of unsaturated soil required a proposed procedure to define the (hydraulic conductivity-suction function. It depends on predicting the SWCC by applying fitting methods with the aid of the program SoilVision, after identifying the basic properties of the soil such as Attereberg limits, particle size distribution, void ratio, porosity, and wet and dry unit weights. The slope of the latter relation can be used to define the $k_{unsat}$ - function. The procedure is found to be successful.
2) The three conditions of water level in the reservoir 143.5 m, 131.5 m and 115.75 m showed that the exit gradient is always less than 1.0 for each degree of saturation. The exit gradient is increased by about 89% and 218% when the water level is raised to 131.5 m and 143.5 m, respectively. This is attributed to increased water head. At the same time, the rate of flow is increased by about 89% and 218%. This means that the increase in rate of flow is approximately equal to the increase of exit gradient due to rising of water level in the reservoir.
3) The rate of flow is decreased by about 20 – 27% when the degree of saturation of the core material is decreased from 100% to 50% at water level 115.75 m, while the exit gradient of flow is decreased by about 13 -15%. This decrease in flow rate becomes 13-15% and 8-9.5% when the reservoir water level is 131.5m and 143.5m, respectively, while the exit gradient of flow is increased by about 23-29.5% and 29-29.5% when the reservoir water level is 131.5 m and 143.5 m, respectively.
4) When the state of soil changes from fully saturated S= 100% to partially saturated, a rapid increase in head gradient and pore water takes place at the embankment base for different water levels in the reservoir. This decrease plateaus out on further degrease in S.

**Table 1**. Index properties of the soil.

| Index property | Value |
|---|---|
| Liquid limit %, LL | 50 |
| Plasticity index %, PI | 25 |
| Specific gravity, Gs | 2.79 |
| Sand % | 3.3 |
| Silt % | 31.7 |
| Clay% | 65 |
| Classification USCS | CL |
| Optimum moisture content | 13.75% |
| Maximum dry unit weight kN/m$^3$ | 17.13 |

**Table 2**: The permeability of Al-Adhaim's dam materials (Al-Adhaim Dam Report, 1994).

| Type of material | | Coefficient of permeability (m/sec) | |
|---|---|---|---|
| | | Horizontal $k_h$ | Vertical $k_v$ |
| The shell | | $1.25 \times 10^{-5}$ | $1.25 \times 10^{-5}$ |
| The core | | $2.25 \times 10^{-7}$ | $1.0 \times 10^{-7}$ |
| The filters | Filter F | $1.2 \times 10^{-3}$ | $1.2 \times 10^{-3}$ |
| | Filter T | $1 \times 10^{-2}$ | $1 \times 10^{-2}$ |
| The found-ation | Marl | $1 \times 10^{-10}$ | $1 \times 10^{-10}$ |
| | Sandstone | $5.5 \times 10^{-6}$ | $5.5 \times 10^{-6}$ |

**Table 3**: Material properties for the assumed core soil.

| Parameter | Value | Unit |
|---|---|---|
| Total unit weight, $\gamma_t$ | 21.017 | kN/m$^3$ |
| Dry unit weight, $\gamma_d$ | 17.13 | kN/m$^3$ |
| Porosity n,% | 38.65 | Degree |
| Void ratio e | 0.63 | — |
| Saturated hydraulic conductivity, $k$ | $1.097 \times 10^{-7}$ | m/sec |

**Table 4**: Results of the basic analysis.

| The original soil of dam core | | | |
|---|---|---|---|
| | Height of water (m) | | |
| | 115.75 | 131.5 | 143.5 |
| Seepage x10$^{-5}$m$^3$/sec/m | 2.3893 | 4.5125 | 7.5975 |
| Exit gradient | 0.11054 | 0.20877 | 0.3515 |
| Max. velocity x10$^{-5}$ m/sec | 0.77028 | 1.4627 | 2.4835 |

**Table 5**: Results of the basic analysis.

| Selected soil | | | |
|---|---|---|---|
| | Height of water (m) | | |
| | 115.75 | 131.5 | 143.5 |
| Seepage x10$^{-5}$m$^3$/day/m | 4.509897 | 7.454249 | 10.35120 |
| Exit gradient | 0.15169 | 0.29172 | 0.48148 |
| Max. velocity x10$^{-5}$ m/sec | 0.69136 | 1.3396 | 2.12017 |

**Table 6**: Summary of pore water pressure and head gradient at nodes A and B predicted at different degrees of saturation.

| a. Height of water (115.75m) | | |
|---|---|---|
| Degree of saturation % | Pore water pressure (kPa) | |
| | Node A | Node B |
| 40 | 192.8 | 396.9 |
| 50 | 191.8 | 397.23 |
| 60 | 193.68 | 398.91 |
| 70 | 194.35 | 398.95 |
| 80 | 194.03 | 406.83 |
| 90 | 194.24 | 398.94 |
| 100 | 178.85 | 394.57 |
| b. Height of water (131.5m) | | |
| Degree of saturation % | Pore water pressure (kPa) | |
| | Node A | Node B |
| 40 | 321.12 | 528.05 |
| 50 | 314.92 | 5.2771 |
| 60 | 318.31 | 528.09 |
| 70 | 319.01 | 528.19 |
| 80 | 317.31 | 527.5 |
| 90 | 319.46 | 528.14 |
| 100 | 320.54 | 527.52 |
| c. Height of water (143.5m) | | |
| Degree of saturation % | Pore water pressure (kPa) | |
| | Node A | Node B |
| 40 | 421.15 | 633.14 |
| 50 | 418.40 | 629.79 |
| 60 | 421.22 | 630.20 |
| 70 | 422 | 630.29 |
| 80 | 422.25 | 630.35 |
| 90 | 422.51 | 630.28 |
| 100 | 313.29 | 632.40 |

**Table 7**: Summary of head gradient at nodes A and B predicted at different degrees of saturation.

| Height of water (115.75m) | | |
|---|---|---|
| Degree of saturation | Head gradient | |
| | Node A | Node B |
| 40 | 0.44189 | 0.2677 |
| 50 | 0.46981 | 0.27364 |
| 60 | 0.46291 | 0.27281 |
| 70 | 0.45692 | 0.27231 |
| 80 | 0.45660 | 0.27162 |
| 90 | 0.45585 | 0.27207 |
| 100 | 0.66071 | 0.92253 |
| Height of water (131.5m) | | |
| Degree of saturation % | Head gradient | |
| | Node A | Node B |
| 40 | 0.92681 | 0.48022 |
| 50 | 0.94768 | 0.48224 |
| 60 | 0.92487 | 0.47991 |
| 70 | 0.92060 | 0.47933 |
| 80 | 0.91781 | 0.47836 |
| 90 | 0.91842 | 0.47970 |
| 100 | 1.1062 | 1.59556 |
| Height of water (143.5m) | | |
| Degree of saturation % | Head gradient | |
| | Node A | Node B |
| 40 | 1.1412 | 0.58895 |
| 50 | 1.1598 | 0.59099 |
| 60 | 1.1404 | 0.5885 |
| 70 | 1.1356 | 0.58812 |
| 80 | 1.1340 | 0.58780 |
| 90 | 1.1331 | 0.58830 |
| 100 | 1.5506 | 1.9252 |

**Table 8**: Results of rate of flow, exit gradient and velocity in the earth dam at different degrees of saturation.

| a. Degree of saturation = 40% | | | |
|---|---|---|---|
| | Height of water (m) | | |
| | 115.75 | 131.5 | 143.5 |
| Seepage (x10$^{-5}$ m$^3$/sec/m) | 3.584909 | 6.565795 | 9.537696 |
| Exit gradient | 0.13123 | 0.23181 | 0.37242 |
| Max. Velocity (x10$^{-5}$ m/sec) | 0.59728 | 1.10814 | 2.0228 |
| b. Degree of saturation = 50% | | | |
| | Height of water (m) | | |
| | 115.75 | 115.75 | 115.75 |
| Seepage (x10$^{-5}$ m$^3$/sec/m) | 3.563482 | 6.5232 | 9.492768 |
| Exit gradient | 0.13094 | 0.23109 | 0.37165 |
| Max. Velocity (x10$^{-5}$ m/sec) | 0.56332 | 1.1253 | 2.0353 |
| c. Degree of saturation = 60% | | | |
| | Height of water (m) | | |
| | 115.75 | 115.75 | 115.75 |
| Seepage (x10$^{-5}$ m$^3$/sec/m) | 3.626726 | 6.571411 | 9.545472 |
| Exit gradient | 0.13194 | 0.23191 | 0.37255 |
| Max. Velocity (x10$^{-5}$ m/sec) | 0.55373 | 1.1063 | 2.0199 |
| d. Degree of saturation = 70% | | | |
| | Height of water (m) | | |
| | 115.75 | 115.75 | 115.75 |
| Seepage (x10$^{-5}$ m$^3$/sec/m) | 3.632947 | 6.582989 | 9.558432 |
| Exit gradient | 0.13205 | 0.23211 | 0.37274 |
| Max. Velocity (x10$^{-5}$ m/sec) | 0.55115 | 1.1026 | 2.0162 |
| e. Degree of saturation = 80% | | | |
| | Height of water (m) | | |
| | 115.75 | 115.75 | 115.75 |
| Seepage (x10$^{-5}$ m$^3$/sec/m) | 3.665261 | 6.514646 | 9.559296 |
| Exit gradient | 0.13242 | 0.23219 | 0.37285 |
| Max. Velocity (x10$^{-5}$ m/sec) | 0.55102 | 1.1022 | 2.0158 |
| f. Degree of saturation = 90% | | | |
| | Height of water (m) | | |
| | 115.75 | 131.5 | 143.5 |
| Seepage (x10$^{-5}$ m$^3$/sec/m) | 3.659299 | 6.476371 | 9.56448 |
| Exit gradient | 0.13319 | 0.232 | 0.37277 |
| Max. Velocity (x10$^{-5}$ m/sec) | 0.55102 | 1.1022 | 2.0158 |

https://doi.org/10.1051/matecconf/201816201008

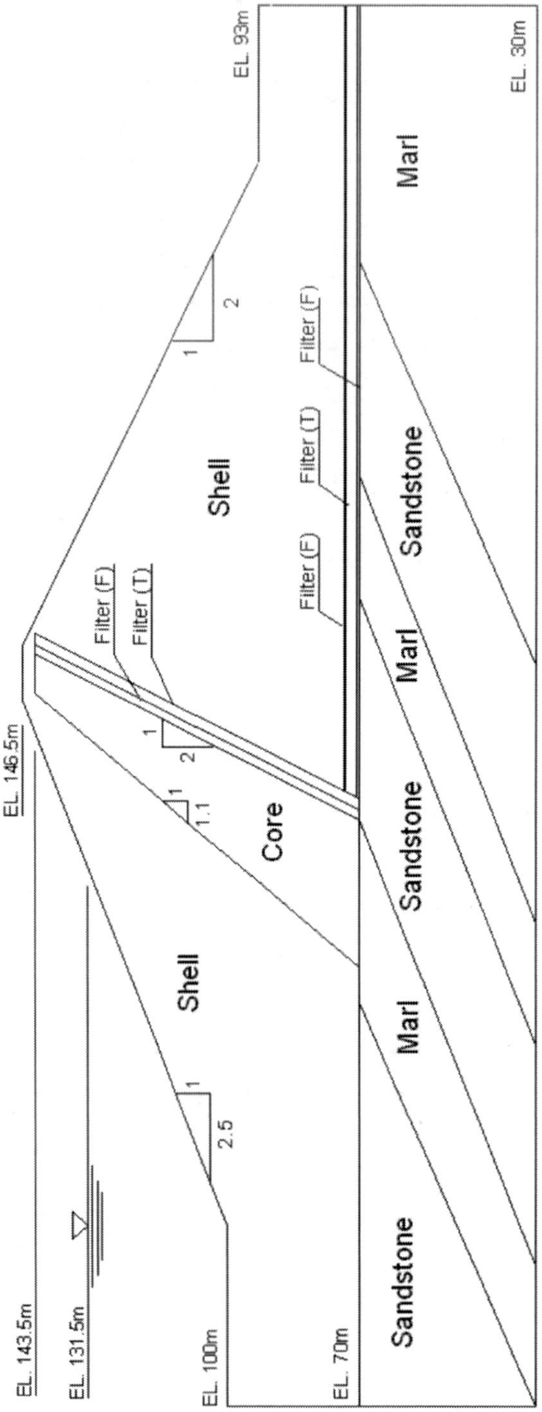


Figure 1: Al-Adhaim Dam – typical cross-section (Al-Adhaim Dam Report, 1994).

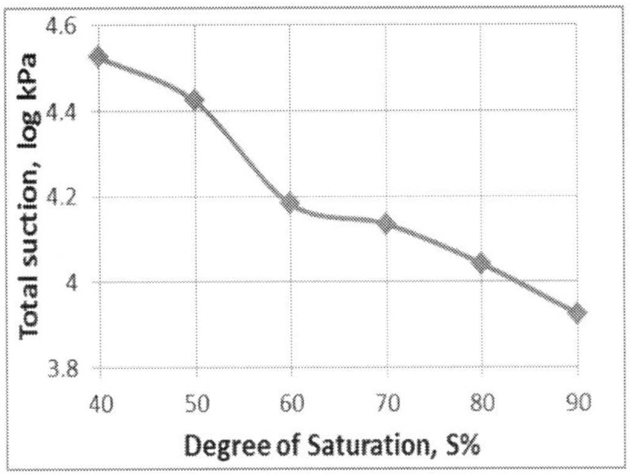


Figure 2: Relationship between the total suction and degree of saturation.

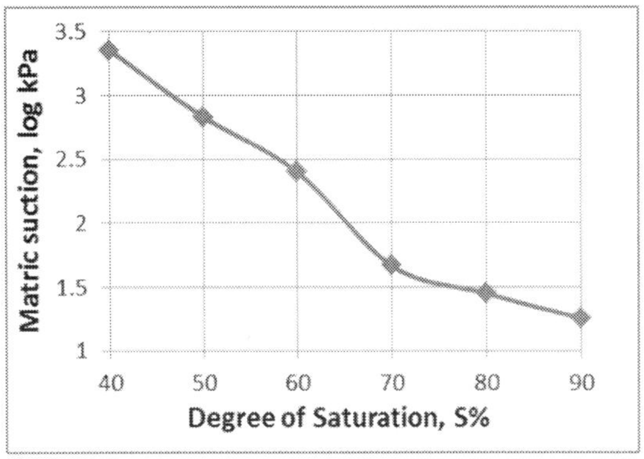


Figure 3: Relationship between the matric suction and degree of saturation.

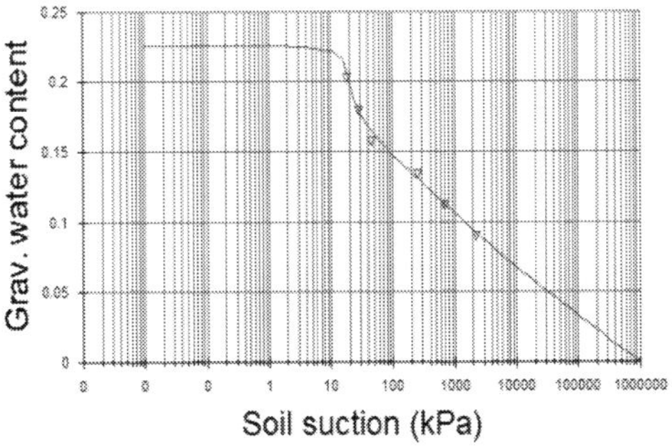


(a) Using Fredlund and Xing (1994) fitting.

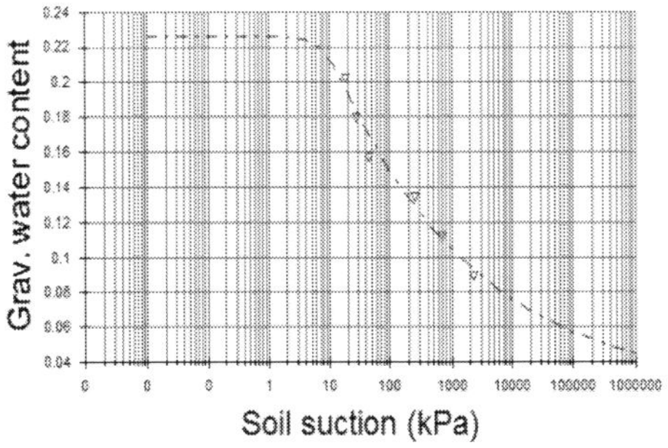


(b) Using Van Genuchten (1980) fitting.

Figure 4: Relationships between the gravitational water content and the matric suction obtained by the program Soil Vision.

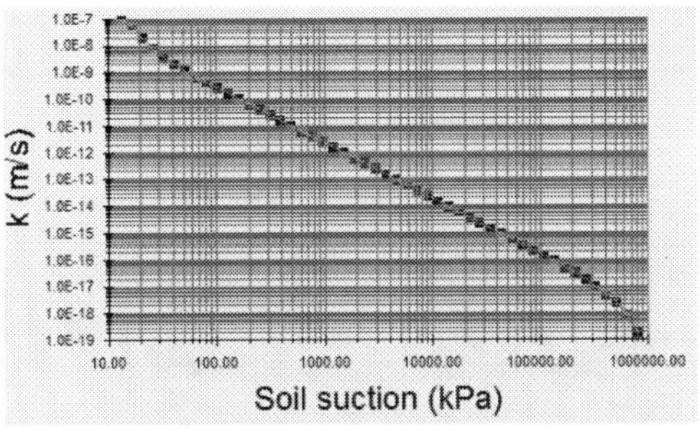


a. hydraulic conductivity at S = 90% .

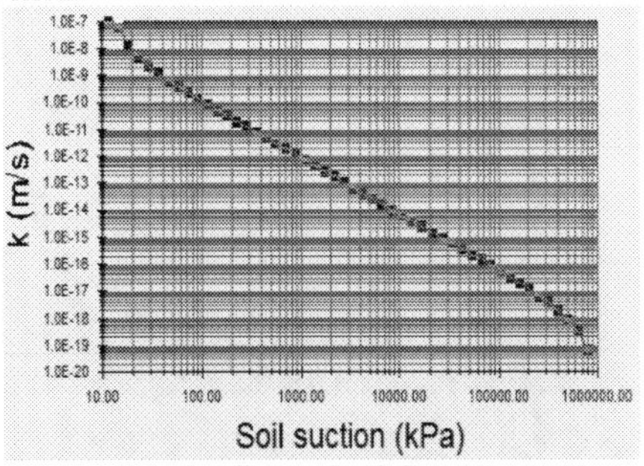


b.hydraulic conductivity at S = 80% .

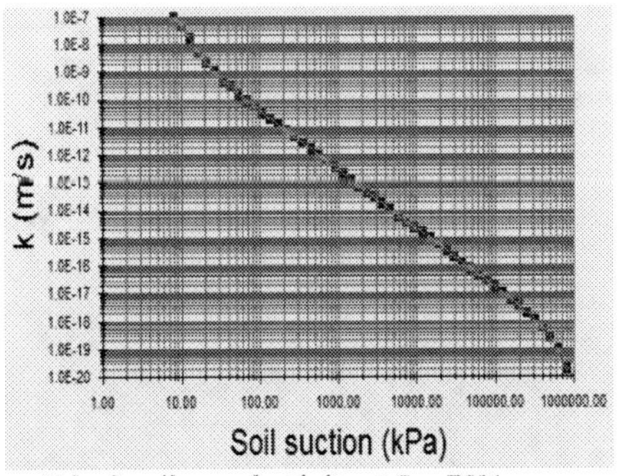


C-hydraulic conductivity at S = 70% .

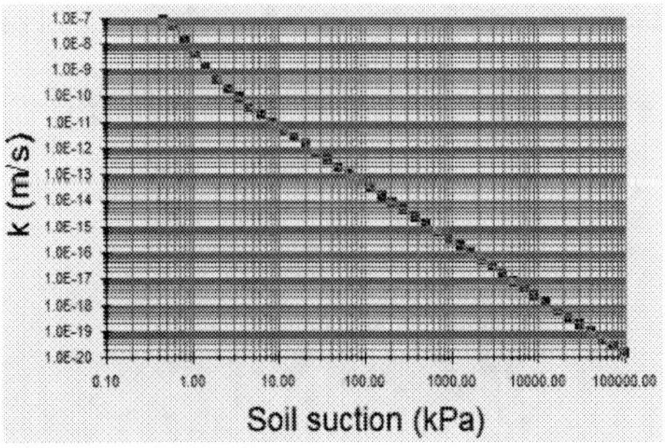


D-hydraulic conductivity at S = 40% .

Figure 5: Relationships between the hydraulic conductivity and the matric suction for different degrees of saturation obtained by the program Soil Vision.

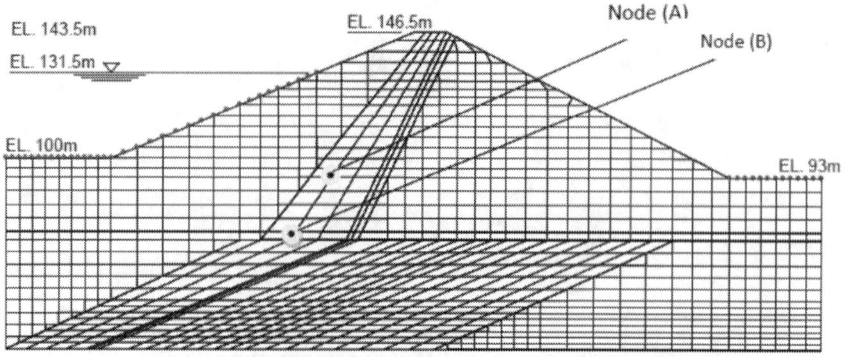


Figure 6: Typical finite element mesh for Al-Adhaim Dam.

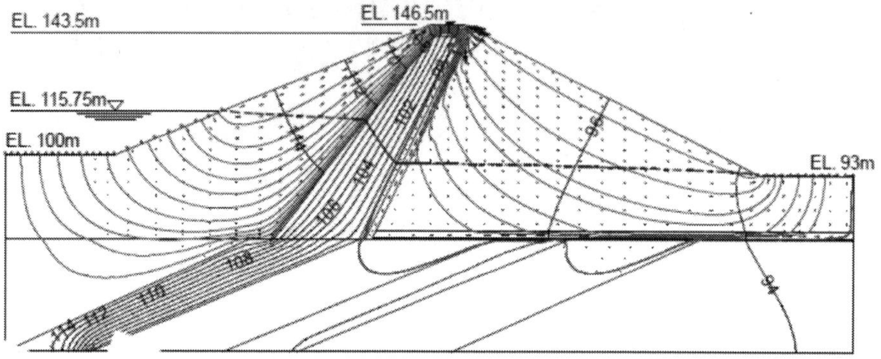


Figure 7: Contours of the total head (in meters) and flow path when the earth dam is without filters and the head of water is 115.75 m.

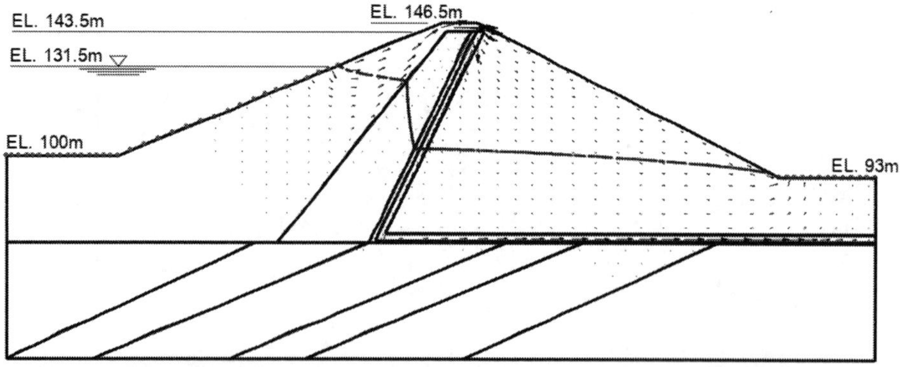


Figure 8: The seepage line and the velocity vectors for Al-Adhaim dam when the head of water is 115.75 m, S =100%, dam with filters.

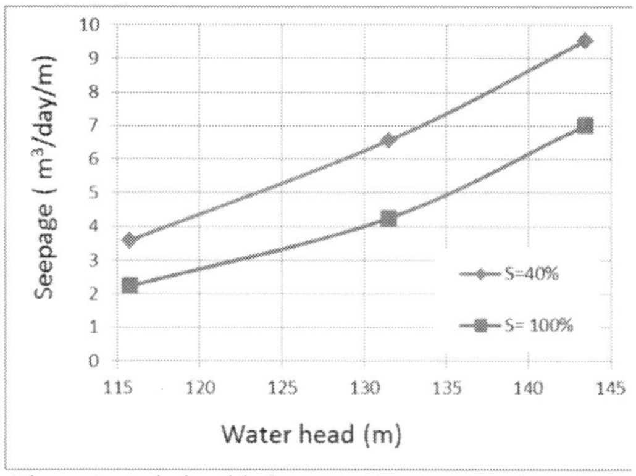


Figure 9: Relationship between the quantity of seepage and head of water of the earth dam without filters, for degree of saturation S=40% and 100%.

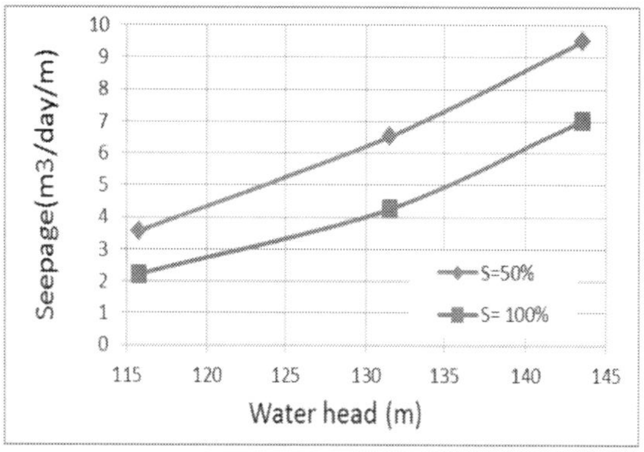


Figure 10: Relationship between the quantity of seepage and head of water of the earth dam without filters, for degree of saturation S=50% and 100%.

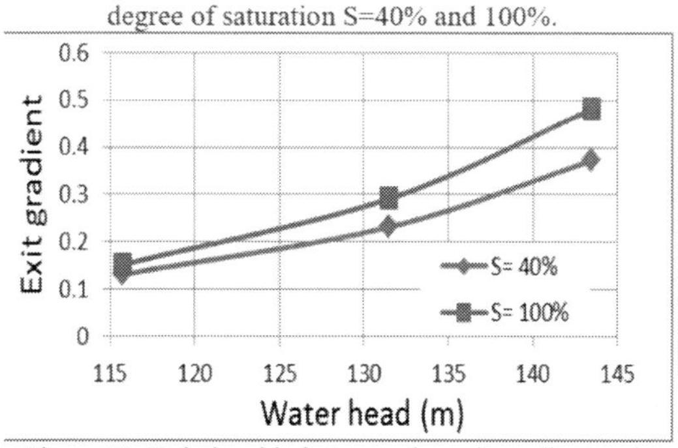


Figure 11: Relationship between the exit gradient and head of water of the earth dam without filters for degree of saturation S=40% and 100%.

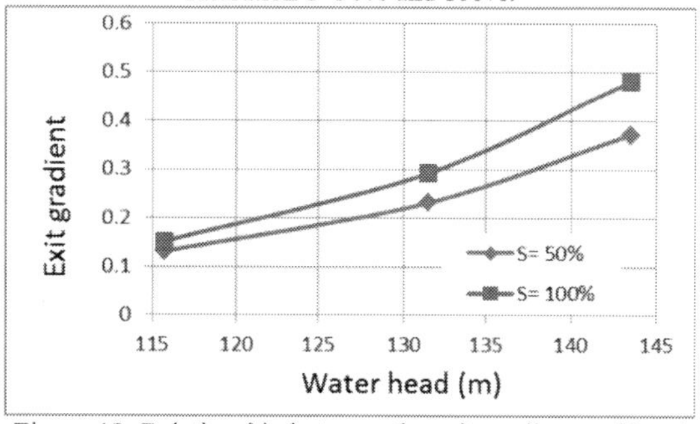


Figure 12: Relationship between the exit gradient and head of water of the earth dam without filters for degree of saturation S=50% and 100%.

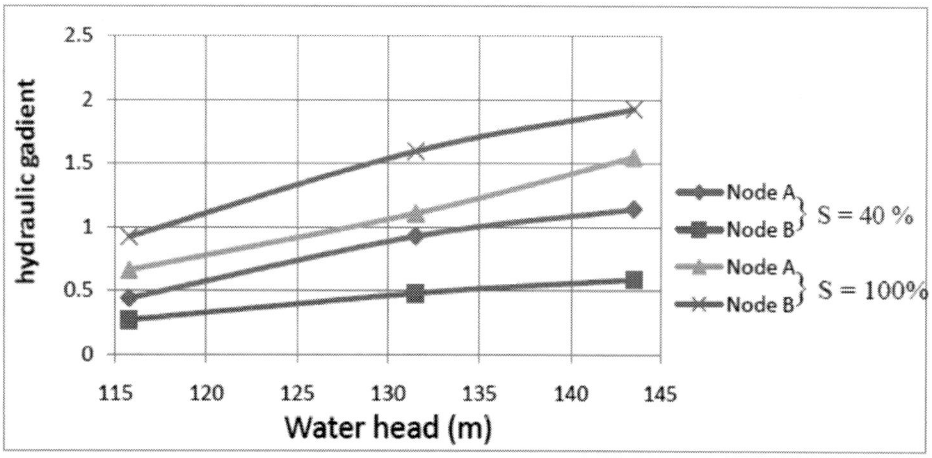


Figure 13: Relationship between the head gradient and head of water of the earth dam at nodes A and B without filters, for degree of saturation S= 40% and 100%

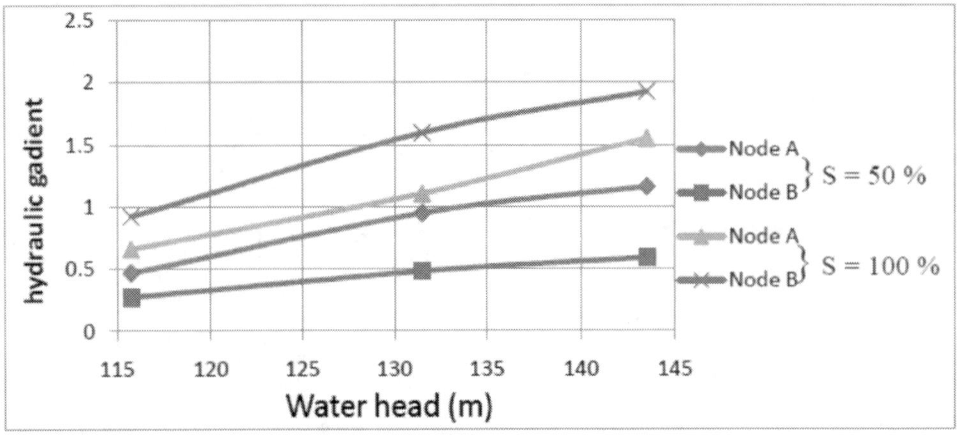


Figure 14: Relationship between the head gradient and the head of water of earth dam at nodes A and B without filters, for degree of saturation S=50% and 100%.

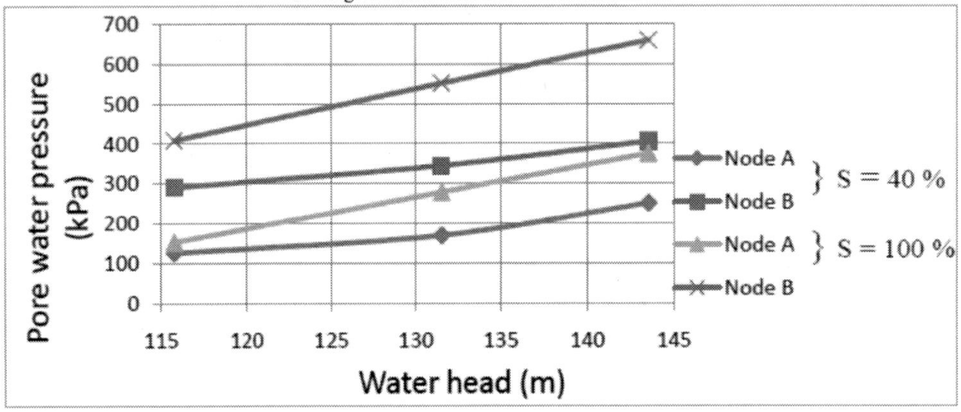


Figure 15: Relationship between the pore water pressure and the head of water of the earth dam at node A and B without filters, for degree of saturation S=40% and 100%.

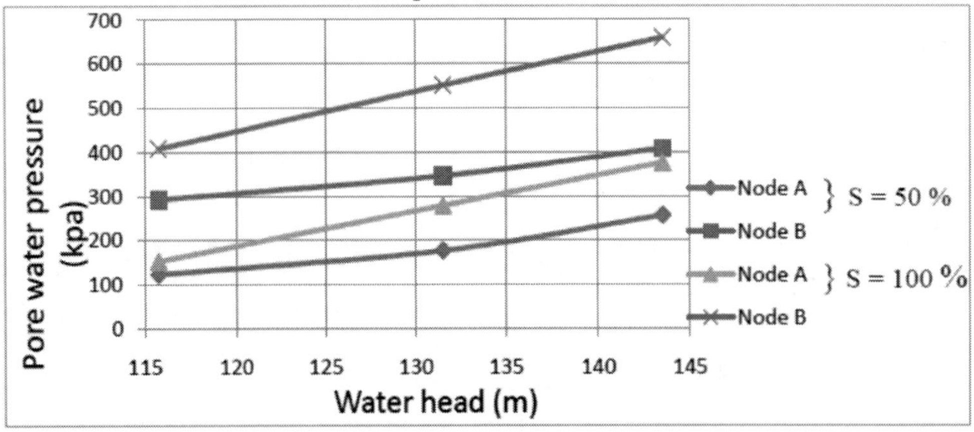


Figure 16: Relationship between the pore water pressure and head of water of the earth dam at node A and B without filters, for degree of saturation S=50% and 100%.

## References

1. Al-Adhaim Dam Report, Final Report of Al-Adhaim Earth Dam, Unpublished Report, Engineering Consultancy Bureau, University of Baghdad, Iraq, (1994).
2. ASTM-D422-00, Annual Book of ASTM Standards, **04.08**, Soil and Rock, pp. 1 – 8.
3. ASTM-D698-00, Annual Book of ASTM Standards, **04.08**, Soil and Rock, pp. 1 – 11.
4. ASTM-D854-00, Annual Book of ASTM Standards, **04.08**, Soil and Rock, pp. 1 – 7.
5. ASTM-D4318-00, Annual Book of ASTM Standards, **04.08**, Soil and Rock, pp. 1 – 16.
6. ASTM-D-5298-03, Annual Book of ASTM Standards, **04.08**, Soil and Rock, pp. 1 – 6.
7. Bouchemella, S., Seridi, A., Alimi-Ichola, I., European Journal of Environmental and Civil Engineering, Vol. 18, June, DOI: 10.1080/19648189.2014.926294, Taylor and Francis, (2014).
8. Fattah, M. Y., Omran, H. A., Hassan, M. A., Acta Montanistica Slovaca , **22**, 1, pp. 43-57, (2017).
9. Fredlund, D.G., Rahardjo, H., Soil Mechanics for Unsaturated Soils, John Wiley & Sone Inc. New York, United States of America, (1993).
10. Fredlund, D.G., Xing, A., Canadian Geotechnical Journal Vol. 31, No. 4, pp. 521 – 532, DOI, 10.1139/t94-060, (1994).
11. Head, K. H., Epps, R. J., 3rd edition, Whittles Publishing, UK, (2011).
12. Ho, P.G., B.Sc. Thesis, University of Saskatchewan, Saskatoon, Canada, (1979).
13. Lam, L., Fredlund, D.G., and Barbour, S.L., Canadian Geotechnical Journal, **24**, 4, pp. 565–580, DOI 10.1139/t87-071, (1987).
14. Thieu, N. T. M., Fredlund, M. D., Fredlund, D. G., and Vu, H. Q., Proc., Int. Conf. on Management of the Land and Water Resources, Hanoi, Vietnam, pp. 49–56, (200).
15. Usre's Guide Manual of Soil Vision, Soil Vision Systems Ltd. Saskatoon, Saskatchewan, Canada, (2001).
16. Van Genuchten, M. T., Soil Science Society America Journal, 44, pp. 892 – 898, (1980).

# Improvement of clayey soil characteristics by using activated carbon

*Kawther* Al-Soudany[1], *Ahmed* Al-Gharbawi[1*], *Marwa* Al-Noori[1]

[1]Building and Construction Engineering Department., University of Technology, Baghdad, Iraq

**Abstract.** The clay soil is weak and unable to carry the applied loads as a result of the weight of buildings or vehicles on the load performing on the soil. In this research, clay soil was grained and mixed with different percentages of activated carbon additives to investigate its performance. One type of clay soil from Al-Taji city was used. The percentages of activated carbon 3, 5, 7 and 9% were added to the soil and the influence of the admixture was observed by comparing the results with the untreated soil. The selected properties for this comparison were specific gravity, consistency limits, compaction, static compaction, CBR, consolidation, swelling and unconfined compressive strength. The results showed that the plasticity index, maximum dry weight and specific gravity decreased as the percentage of additives increased. The unconfined compressive strength increased as the percentage of additives and curing periods (1, 7, 14 and 28) days increased. The amount of increase in soil strength was even more than 100% for the 9% activated carbon. The results showed that the addition of activated carbon has a positive effect to the geotechnical properties.

## 1 Introduction

For all engineering structures constructed on clayey soils, deformations occurr when these soils are exposed to any additional load or due to dissolution of clay when these soils are wet, soaked or leached with water. These deformations may cause a significant failure (cracking, tilting and collapse) of the foundation or earth structures.

Engineering properties such as strength, texture, workability, and plasticity are improved when adding some commercially manufactured additives to the soil in suitable quantities. The operation of blending and mixing some materials with soil, or blending soils together to improve certain properties, is called Stabilization. The benefit of mixing and addition is to achieve the required gradation or change gradient, texture, plasticity, and can act as a binder for cementation the soil. The methods of stabilization are divided into two methods as follows [1]:

a- Mixing the soils with several gradients to obtain material conforming to the required specifications can achieve the mechanical stabilization. Soil mixing can be carried out at the construction site, the central station, or the borrowing area. After that by using traditional methods, the mixture is spread and impacted to required density.

b- The addition of additives to the soil with different percentages of additives such as cement, lime, fly ash, bitumen, or a combination of these materials, will achieve additional mechanical stabilization. The soil classification and degree of improvement in required soil quality determines the percentage and type of additives to be used. Generally, to improve the characteristics of the soil such as gradation, workability, and plasticity requires a smaller amount of additives. Larger quantities of additive are used when needed to increase the strength and durability of the soil. And after the additives are mixed with the soil, spreading and compaction are achieved by conventional means.

This research concentrates on investigating experimentally the feasibility of stabilizing and improving the geotechnical properties of soft clay soil using activated carbon material in different proportions and studying The effect of activated carbon on the engineering characteristics of the stabilized clay.

* Corresponding author: ahmedsgharbawi@yahoo.com

## 2 Experimental work

### 2.1 Materials used

#### *2.1.1 Soil used*

© The Authors, published by EDP Sciences. This is an open access article distributed under the terms of the Creative Commons Attribution License 4.0 (http://creativecommons.org/licenses/by/4.0/).

A standard test was carried out on a brown soil fromAl-Taji site (east Baghdad) to determine its physical and chemical properties. The soil's grain size distribution was 2.8 % sand, 40% silt and 57.2% clay as shown in Figure (1). According to USCS the soil is classified as CL.

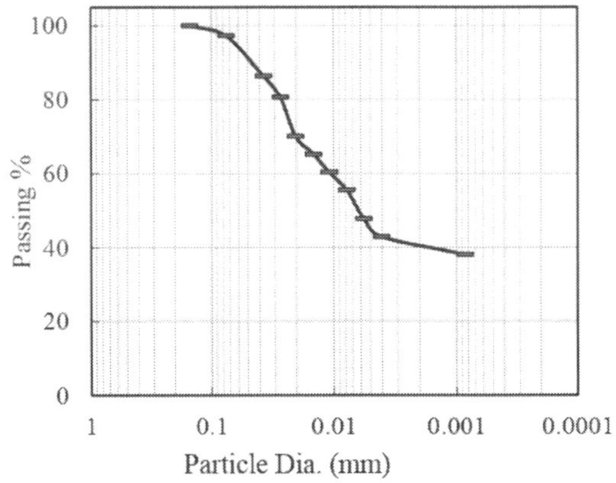


**Fig. 1.** Grain size distribution of Al-Taji soil

### 2.1.2 Activated carbon

The structure of many materials have a high amount of carbon content that can potentially be activated. Coal is one of the raw materials which can stabilize soils and contains (bituminous, anthracite and lignite), wood (both soft and hard), coconut shells, peat and petroleum based residues. Other raw materials have been evaluated such as peach pits, walnut shells, nutshell and but invariably their commercial limitation lies in raw material supply. The ranges for internal surface area and degree of porosity for the most of carbonaceous materials are between 10-15 $m^2$/gm over activation, by controlled oxidation of carbon atoms - usually achieved by the use of steam at high temperatures, the internal surface becomes highly porous and extended. After activation, depending on the plant operating conditions the internal surface area for carbon will be between 700 and 1,200 $m^2$ [2].

**Fig. 2.** Activated Carbon

#### 2.1.2.1 The Manufacturing of Activated Carbon

There are two method of manufacturing activated carbon the first being treating raw material by chemical and the second by steam [3].

*1- Chemical method:*
The chemical activation is dependent on the raw materials used such as sawdust. The most commonly used activating agent used is phosphoric acid, whereas that there are good results for zinc chloride and sulfuric acid. The materials Which were used in past are calcium hydroxide, calcium chloride, manganese chloride and sodium hydroxide, and they are considered as dehydrating agents.

The raw material and reagent are mixed into a paste, dried and carbonized in a rotary furnace at 600 °C. When phosphoric acid is the activating agent the carbonized product is further heated at 800-1000 °C during which stage the carbon is oxidized by the acid. The acid is largely recovered after the activation stage and converted back to the correct strength for reuse. The activated product is washed with water and dried.

Activity can be defined that the change limits between the raw material to activating agent, about 1:05 to 1:4. By increasing the concentration of the activating agent the same objective of the Activation can be achieved by increasing the control of furnace temperature and residence time.

*2- Steam method :*
The steam can be used for activation that applied to virtually all raw materials. The development of the all methods have the same principles of initial carbonization at 500-600 °C and followed by steam to react the activation at 800-1,100 °C.

#### 2.1.2.2 Properties of activated carbon

The properties of activated carbon that we are concerned in this project [3]:

*1- Surface area::*
The surface area of activated carbon reaches from 500 $m^2$ to more than 2000 $m^2$ and the surface area will increase by the increase of activated carbon finesse and it will decrease by the increase of its granular size.

*2- Apparent density:*
The activated carbons have density for the solid or skeletal between 2.0 and 2.1 $gm/cm^3$ (125–130 lbs /cubic foot). Nonetheless , if the air volume of activated carbon takes a large space, then the actual or apparent density will be lower, typically 0.4 to 0.5 $gm/cm^3$ (25–31 lbs/cubic foot). The good quality of activated carbon is provided at the high density which that provide a great volume.

*3- Hardness/abrasion:*
The measurement of activated carbon friction resistance is called hardness and abrasion and it is

considered a very significant indicator for activated carbon to take part of its physical integrity and resist against frictional forces imposed by backwashing, etc. depending on the raw material and activity level.

4- *Effect of oven temperature and adding water on activated carbon:*

Activated carbon will not react with added water, it will adsorb the water because activated carbon has high porosity, and oven temperature of 110 °C in our experiment will not affect activated carbon.

5- *Particle size distribution:*

Activated carbon has fine particles size, which makes better access to the surface area and faster rate of adsorption kinetics. In vapor phase systems against pressure drop needs to be considered, which will affect energy cost. To provide significant operating benefits particle size distribution must be considered.

6- *Structure*

The ability of activated carbon can be dissected by appreciation of its structure. Much of the literature quotes a modified graphite like structure; the modification results from the existence of micro crystallites. It is formed during the carbonization process, which during activation have their proportionate bonding, disrupted causing free valences which are very reactive. In addition, the presence of impurities and process conditions effect the formation of interior vacancies, in the microcrystalline structures. Existence of pores generally explains faults in crystalline structures.

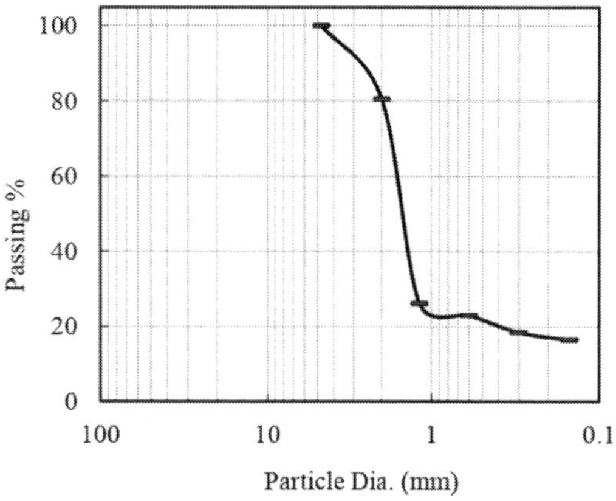


**Fig. 3.** Grain size distribution of AC used

7- Carbon tetrachloride activity

The adsorption of saturated carbon tetrachloride vapour is the measurement of the porosity of an activated carbon.

The adsorption is the process in which atoms, ions or molecules from a substance (it could be gas, liquid or dissolved solid) adhere to a surface of the adsorbent. Adsorption is a process which is based on surface and that is different from absorption which is the process in which a fluid is dissolved by a liquid. In adsorption a film of adsorbate is created on the surface while absorption involves the entire volume of the absorbing substance.

### *2.1.2 Laboratory tests*

The testing program of the samples of the clayey soil were tested to calculate their physical and chemical properties in the natural state. And the testing program of the samples of the clayey soil, were stabilized with different percentages (0, 3, 5, 7, and 9%) of activated carbon materials (AC), included Atterberg's limits, compressibility, Specific Gravity, Unconfined compressive strength natural and improved soil when activated carbon content increase with curing (0, 1, 7, 14, 28) days, consolidation and CBR test [4].

**Atterbeg limits:**

- Liquid limit test: The test of liquid limit was carried out on samples passing sieve (No. 40) 0.425mm, by using Casagrande's apparatus liquid limit apparatus as per the procedures laid down in ASTM D 4318 with clayey soils with (0, 3, 5, 7 & 9%) of activated carbon additive.
- Plastic limit test: The test of plastic limit was carried out on samples passing 0.425mm (No. 40) sieve, by using per the procedures laid down in ASTM D 4318 with clayey soils with (0, 3, 5, 7 & 9%) of activated carbon additive.

**Specific gravity:** The specific gravity (Gs) test was carried out on the soil according to ASTM 854.

**Compaction:** This test was carried out according to the specifications ASTM D 1557. The specimen's diameter was of 102mm and 116mm height. The degree of compaction of soil was effected by the engineering soil characteristics such as compressibility, stiffness, swell potential, permeability and shrink. So that, it is important to achieve the desired degree of relative compaction and it's necessary to meet the required soil characteristics.

**Static compaction:** To prepare samples for unconfined compressive strength UCS and consolidation tests, air dry soil was mixed with the required amount of distilled water to give initial moisture content. The soil was mixed thoroughly by hand and stored in a desiccators for18 hours to obtain uniform moisture distribution.

The static compaction method was followed the preparation of soil samples using a special mold and a compression testing machine. Load was applied statically to the soil sample at a compression rate of 0.04mm/sec. until the depth of penetration approach the required height. The load was maintained on the specimen for about five minutes after reaching the required height to prevent the occurrence of rebound after static load removal.

**UCS:** the unconfined compressive test was carried out according to the specifications ASTM D 2166. The sample of UCS test had 38mm in diameter and 76mm in length. The dry unit weight of 17.6 kN/m$^3$ value of the natural soil at the moisture content 17 %, and improved soil when activated carbon content increase with curing (0, 1, 7, 14, 28)days.

**CBR:** This test was carried out according to the specifications (ASTM D 1883). The sample of CBR had 152mm in diameter and 126mm in length. The testes were performed at the optimum moisture content (OMC) and maximum dry unit weight (MDD) values of the natural soil.

**Swelling:** Consolidation test was carried out according to the specifications (ASTM D 2435). The setup was used for determining the cyclic. The sample which used had 75mm in diameter and 19mm in height. The samples were prepared at Proctor's dry densities.

## 3 Presentation and discussion of test results

### 3.1 Effect on Atterberg's limit

#### *3.1.1 Liquid limit (L.L.)*

Figure (4) is shown the variation of liquid limit for soil stabilized with different percentages of activated carbon. There were decreases in liquid limit at all soil-activated carbon combinations. The cause of the ash is that the activated carbon coats and binds all clay particles which possesses little cementitious value and large particles.

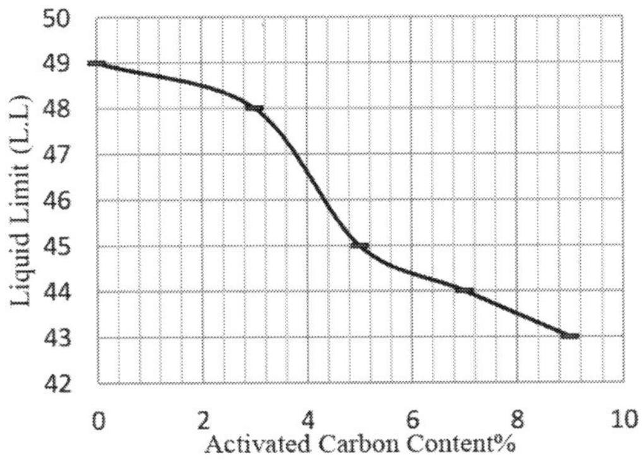


**Fig. 4.** Liquid Limit of Activated carbon mixed soil

#### *3.1.2 Plastic limit (P.L.)*

Plastic limits shown in Figure (5), it was noticed that with increasing AC content there was increase in plastic limit. It is very clear the increase in the plastic limit when the activated carbon content is increased in the combinations. The exchange in cationic reaction adduced for the liquid limit is also applicable here.

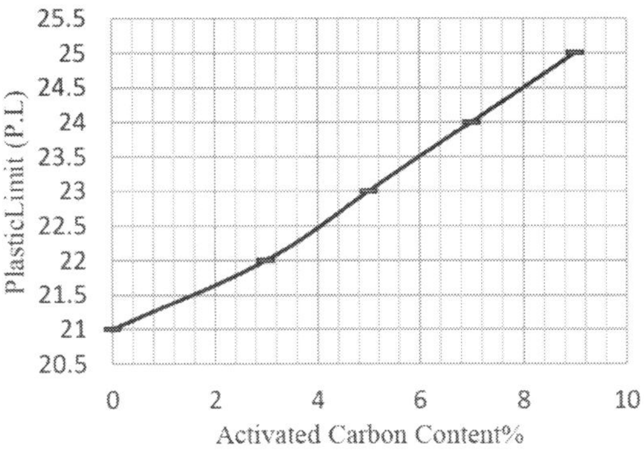


**Fig. 5.** Plastic Limit of Activated carbon mixed soil

#### *3.1.3 Effect on plasticity index (P.I.)*

The Figure (6) clarified the changes of P.I for the different percentages of soil – activated carbon combination. It was observed that increasing activated carbon content at all percentages for both stabilized samples of soil-activated carbon leads to decreases in plasticity index. The results obtained by other investigators (Sridharan et al. 1986, Athanasopoulou 1995) [5-6] agree with this. They observed a significant reduction in the liquid limit and plasticity index when treating a fine-grained soil with additives.

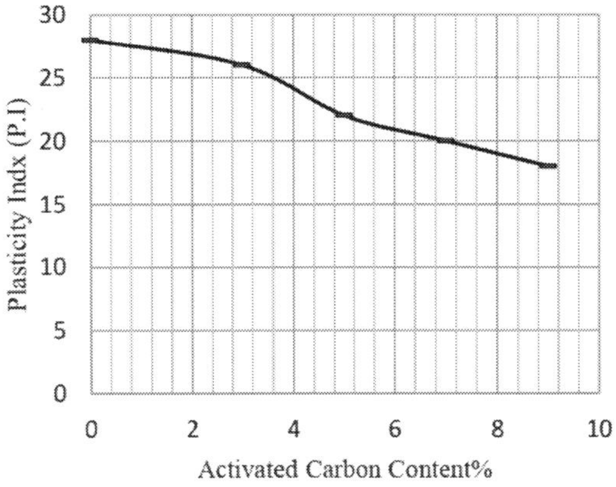


**Fig. 6.** Plasticity Index of Activated Carbon mixed soil

### 3.2 Effect on specific gravity

Figure (7) is shown the variations of specific gravity of the soil with different percentages from activated carbon combinations. It is observed that the increase of AC to the soil made a decreasing in results of the specific gravity value. This decrease in specific gravity can be attributed to the lower specific gravity value of AC (i.e. 2.05).

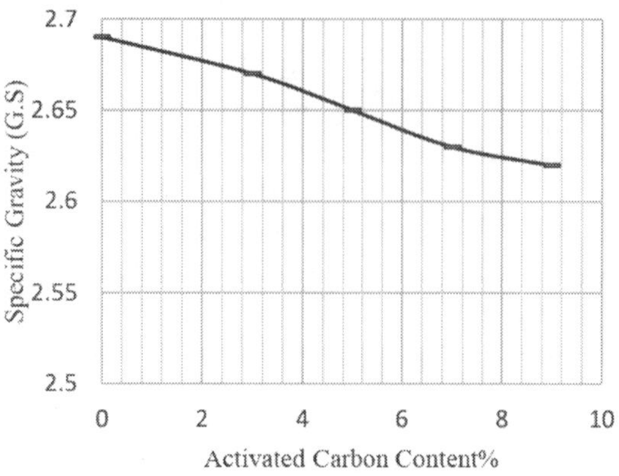


**Fig. 7.** Specific gravity of Activated Carbon mixed soil

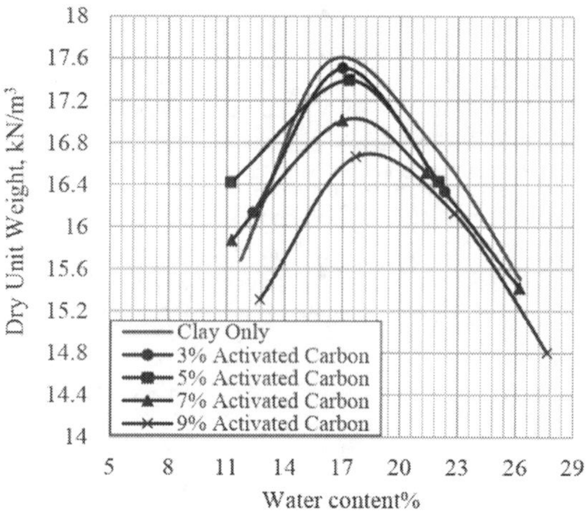


**Fig. 8.** Variation of MDD and OMC with AC content for clay soil.

## 3.3 Effect on compaction

Figure (8) is shown the variation of compaction characteristics for the soil with different percentages of AC between optimum moisture content (OMC) and maximum dry density (MDD. Figure (9) illustrates the relation between MDD and AC content and the results shows that the MDD decreases with the increase in the AC content. Figure (10) illustrates the relation between OMC and AC content and the results shows that when increase in AC content, the OMC is increased too.

The decrease in the MDD can be attributed to the replacement of soil and by the AC in the mixture which has relatively lower specific gravity (2.04) compared to that of the soil which is 2.69 (Ola 1975; Osinubi and Katte 1997) . It may also be attributed to coating of the soil by the AC which result to large particles with larger voids and hence less density (Osula 1991). The decrease in the MDD may also be explained by considering the AC as filler (with lower specific gravity) in the soil voids.

The reason behind increasing in optimum moisture content (OMC) with increase in AC contents is that increase in AC content causes adsorption of a high amount of water which expel the clay particles and the water takes its place.

When the soil is mixed with additives that lead to improvement in variations of the engineering characteristics and improve the physical and chemical properties of the soil such as carbonation, flocculation of clay and cation exchange (Jawad et al., 2014; Zumrawi and Hamza, 2014) [7-8].

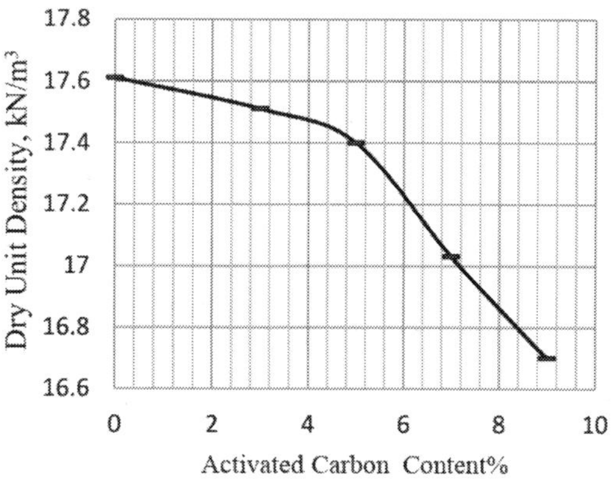


**Fig. 9.** MDD of AC Mixed Soil

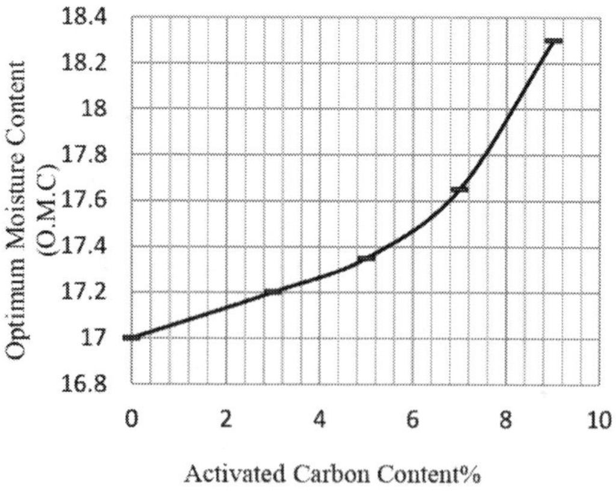


**Fig. 10.** OMC of AC Mixed Soil

https://doi.org/10.1051/matecconf/201816201009

### 3.4 Effect on unconfined compressive strength

Compressive strength (UCS) of a soil is an important factor to assess the stress of the soil and to assess the materials which can use in pavement and constructions. It is one of the important tests which used to determinate of amount of additives required to improve the soil (Singh and Singh 1991). UCS tests were performed on natural soil with (0%, 3%, 5%, 7% and 9% AC) as shown in Figures (11, 12, 13, 14 and 15) respectively. There is no clear variation in UCS for the clay only without AC. With increase of the curing time periods from 0 to 28 days and changes in UCS with the increase in AC from 3% to 9% while increasing curing period from 0 to 28 day are shown in Figure (16) which illustrates two important points. The first is the increase in stiffness of the soil with increasing the percentage of AC as an additive, this is clearly shown in the gradual change of the shapes of curves with increasing AC content for four curing periods. The second point is the increase in stiffness of the model test cured for 28 days as compared to the corresponding model test cured for 7 days; this increase is due to hardening process that developed during 28 days curing period. The compaction method – (i.e. modified test instead of the used standard Proctor's test) would have a significant impact on the strength values (Ghosh 2013) [9].

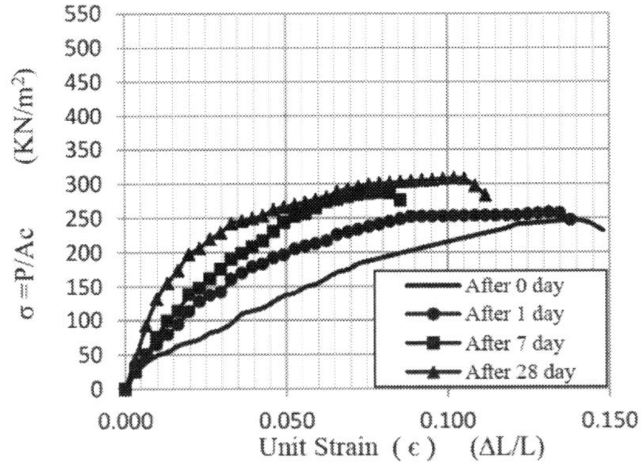


**Fig. 11.** Stress-strain relationship from unconfined compression test for samples treated by 0% AC.

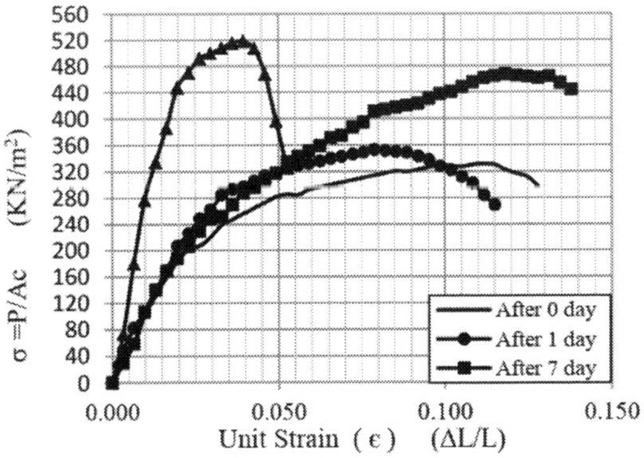


**Fig. 12.** Stress-strain relationship from unconfined compression test for samples treated by 3% AC.

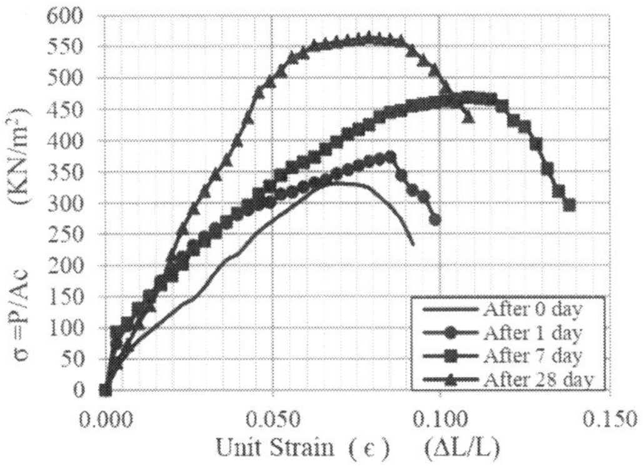


**Fig. 13.** Stress-strain relationship from unconfined compression test for samples treated by 5% AC.

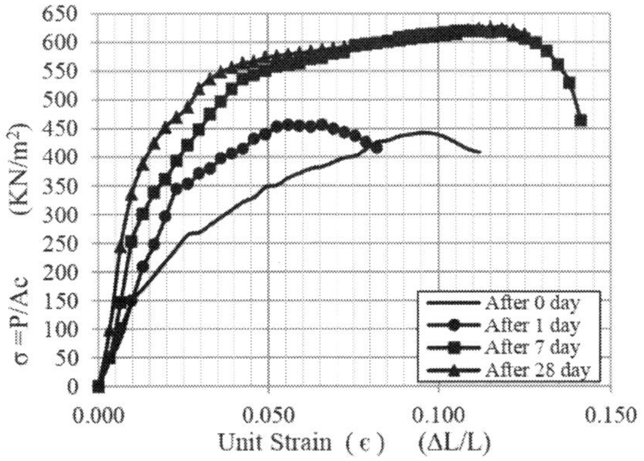


**Fig. 14.** Stress-strain relationship from unconfined compression test for samples treated by 7% AC.

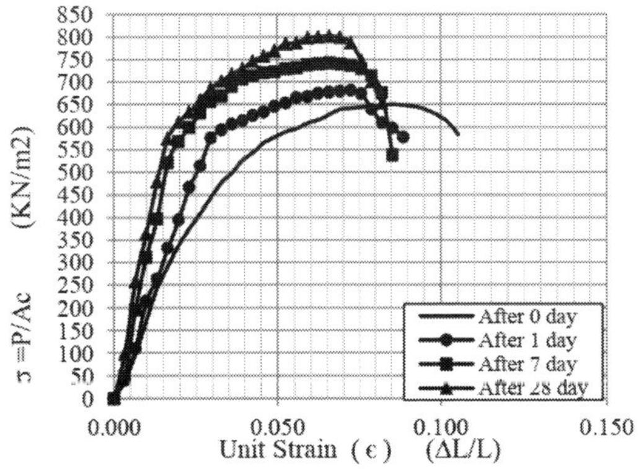


**Fig. 15.** Stress-strain relationship from unconfined compression test for samples treated by 9% AC.

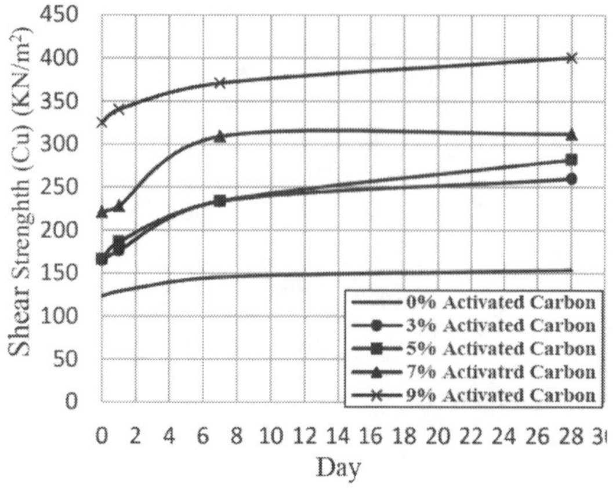


**Fig. 16.** Variation of UCS with AC

**Fig. 17.** CBR of AC Mixed Soil

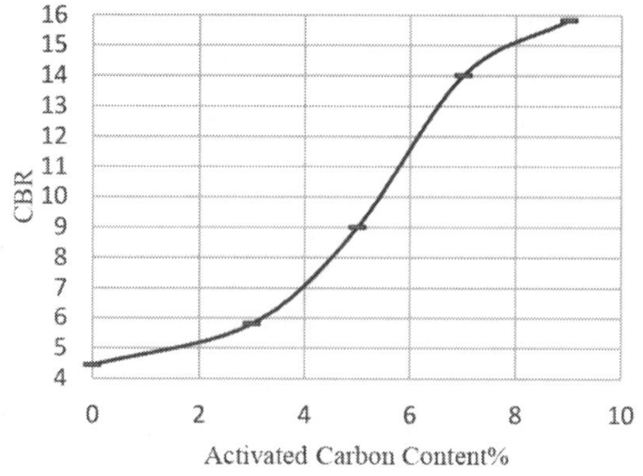


**Fig. 18.** CBR with AC Percentages

### 3.5 Effect on California bearing ratio (CBR)

The California bearing ratio (CBR), is an indicator of bearing capacity of the soil and the strength of the soil. This type of test is used to facilitate the design of the lower layers of the pavement such as base layer and sub-base layer.

The test is performed by measuring the pressure required to penetrate a soil sample with a plunger of standard area. The pressure of soil which is determined from the test is divided by pressure of crushed rock for the same penetration. The CBR test is described in ASTM Standards D1883-05 (for laboratory-prepared samples) and D4429 (for soils in place in field) which this research depends on is AASHTO T193. This test, which is simple, fast and reliable have been used to verify the stabilization of weak soils by adding physical and chemical additives (Raymond et al 2007 and Joel Beeghly et al 2010) [10-11]. Figure (17) is shown the test results of CBR with each percentage of activated. Figure (18) is shown the increase in CBR value of the soil with increase of percentages of AC content. It can be seen that addition of AC increased the CBR of the soil.

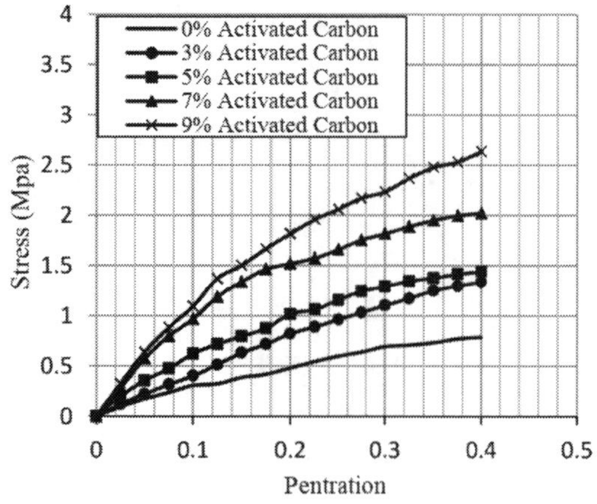


### 3.6 Effect on compressibility characteristics

A consolidation ring of 7.5 cm diameter and 1.9 cm height was used in testing. All samples of soil were compacted by static compaction at $\gamma_{dry}$ of 17.6 kN/m$^3$ and water content of 17%. Every sample was soaked in water after the test. Figure (19) is shown the variation of in void ratio with effective stress of consolidation in semi log scale for the soil with each percentage of AC content.

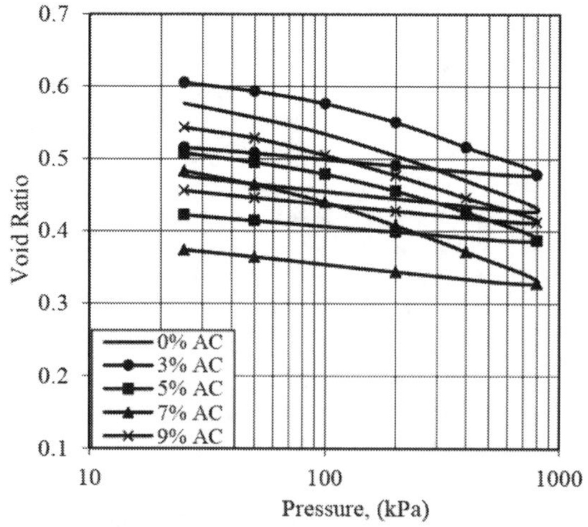


**Fig. 19.** e-Log Relationship for different percent of AC

### 3.7 Effect on swell and swell pressure for prepared soil

The soil samples used in this test were prepared in the laboratory by mixing natural soil with 30% percentage of bentonite and improved by adding 0, 3, 5, 7 and 9% AC powder (passing from 0.6 mm). Static compaction method was followed for preparing the soil samples using a special mold and a compression testing machine. Load was applied statically to the soil sample at a compression rate of 0.04mm/sec. until the depth of penetration approach the required Height of 1.5 cm, the samples were prepared at a dry unit weight of 17.61 $kN/m^3$ and moisture content of 17 %. The load was maintained on the specimen for about five minutes after reaching the required height to prevent the occurrence of rebound after static load was removed. Figures (20), (21) and (22) show the effect of the soil stabilized with different percentage of AC content on the free swell and swell pressure. From these figures we can see that there is rapid decrease in free swell and swell pressure with the increasing percentages of AC content. The reduction in the swelling and swelling pressure values of the improved soils may be due to the addition of non expansive silt size particles to the expansive soil and the interaction between the soil and activated carbon particles. Consequently, the specific surface area and water affinity of the inspected soil samples decrease, which refers to the reduction in the swell-shrinkage values such as free swell and swelling pressure.

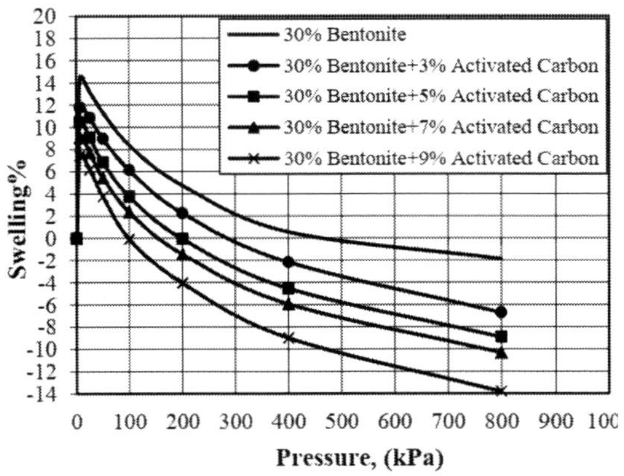


**Fig. 20.** Relationship between swelling and pressure for prepared soil with 3, 5, 7 and 9% Ac.

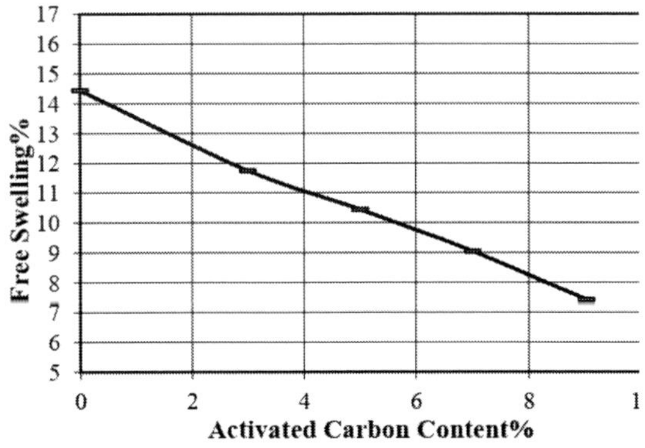


**Fig. 21.** Relationship between free swelling and AC content.

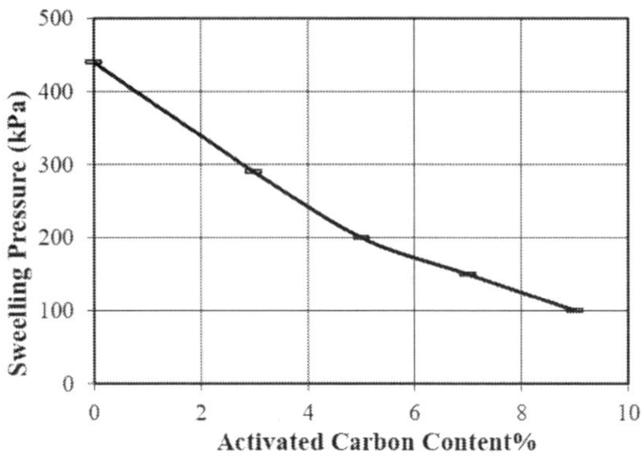


**Fig. 22.** Relationship between swelling pressure and AC content.

## 4 Conclusions

Based on the results obtained it can be concluded that:

1. The specific gravity decreases with increase of AC percentage.
2. The liquid limit slightly decreases with increase of AC percentage.
3. The plastic limit slightly increases with increase of AC percentage.
4. The plasticity index decreases with increase of AC percentage.
5. There is a clear decrease in maximum dry unit weight with the increase of percentage of the AC content.
6. Optimum moisture content increases with increase of AC percentage.
7. The unconfined compression strength was increased by increase both of activated carbon percent and the period of curing.
8. Free swell and swell pressure highly increases with increase of bentonite percentage. While free swell and swell pressure rapidly decrease with the increase of the percentage of the AC content.

## References

1. K. Terzaghi R.B. Peck, *Soil Mechanics in Engineering Particle*, 2nd edition, john Wiley and Sons, New York (1976).
2. P. Buringh, Netherlands by H. Veenman and Zonen N.V ,Wageningen (1960).
3. B.M. Das, *Principles of Foundation Engineering*, 5th Edition Thomoson learning Academic Resource, United State of Amarican (2004).
4. J.E. Bowels, *Foundation Analysis and Design*, 5th ed. Mc Graw-Hill book company, New York (1996).
5. A. Athanasopoulou, Application to Soils Encountered in the Area of Thrace, Ph.D. Dissertation, Department of Civil Engineering, Democritus University of Thrace (1995).
6. A. Sridharan, S.M. Rao, N.S. Murthy, ASTM

https://doi.org/10.1051/matecconf/201816201009

Geotechnical Testing Journal (1986).
7. I.T. Jawad, M.R. Taha, Z.H. Majeed, T.A. Khan, Research Journal of Applied Sciences, Engineering and Technology (2014).
8. M.M.E. Zumrawi, O.S.M. Hamza, International Journal of Science and Research (IJSR) (2014).
9. R. Ghosh, Journal of Civil Engineering and Construction Technology (2013).
10. N. Raymond, Yong, Ouhadi. R. Vahid, Applied Clay Science (2007).
11. H. Joel, Beeghly, Michael. Schrock, International Journal of Soil, Sediment and Water (2010).

# Decrease in failure load of inclined driven piles by using Allpile program

*Ahmed* Al-Gharbawi[1*]

[1]Building and Construction Engineering Department, University of Technology, Baghdad, Iraq

**Abstract.** One of the problems in construction piles in field is driven inclined which maybe become by error during putting pile or driven. This type of problem in construction is difficult to determine bearing capacity. In this research, studying the determination of bearing capacity of inclined pile in deferent angles and different types of soil as well as deferent types of piles shapes by using AllPile program. The results demonstrate that the reduction in bearing capacity of 30° inclined pile less than 20% from the non inclined pile.

## 1 Introduction

The load at the pile head gradually increases from a very small value to the limit load ($Q_L$) required for pile to penetrate into the ground at a constant rate (Salgado, 2006)[1]. Sometimes piles are driven or constructed inclined in civil structures by mistake or for some reason. Many codes and standards do not prefer to use inclined pile in structures especially in seismic regions.

(Li and Luo, 2011)[2] studied the development of pile foundation by using group of piles constructed inclined with a center pile constructed in the different angles of inclination. The result showed that the optimum inclination angle for the center pile is located between 8-12°.

A significant amount of work (Mylonakis et al. 1997)[3] has been carried out in reliable prediction of the response of structures with vertical piles exposed to seismic excitations as vertical piles are undoubtedly a popular choice for a wide range of structures. Inclined piles (sometimes named 'batter piles' or 'raked piles'), on the other hand, are particularly desired for structures demanding substantial lateral stiffness and these piles were used in hydraulic structures, towers, and bridges. The major problem of inclined piles are the mechanism of transfer the vertical load from the structure through the piles cap to piles then the bed of the soil, so the bearing capacity is reduced. In recent years, evidence has been accumulating that inclined piles may, in certain case, be beneficial rather than detrimental both for the structure they support and the piles themselves, (Gazetas and Mylonakis, 1998)[4]. Studies by (Sadek and Shahrour, 2004)[5] showed that supporting by using inclination piles improved some sided of the seismic reaction of the structures.

## 2 AllPile program

The program AllPile for Windows analyzes pile load capacity efficiently and accurately. AllPile can handle all types of piles: drilled shaft, driven pile, auger-cast pile, steel pipe pile, H-pile, timber pile, tapered pile, bell pile, shallow foundation, etc. You can define new pile types and input customized parameters based on local practices and experience. The program is capable of performing the following calculations:

- Lateral capacity and deflection.
- Vertical capacity and settlement.
- Group vertical and lateral analysis.
- FHWA SHAFT program.
- Static and cyclic conditions.
- Negative and zero friction.
- Shallow footing.
- Tower foundation.

The main AllPile window is shown in Figure (1). And the AllPile operations can be divided into three main steps as shown in Figure (2) (Allpile manual, 2014)[6].

* Corresponding author: ahmedsgharbawi@yahoo.com

© The Authors, published by EDP Sciences. This is an open access article distributed under the terms of the Creative Commons Attribution License 4.0 (http://creativecommons.org/licenses/by/4.0/).

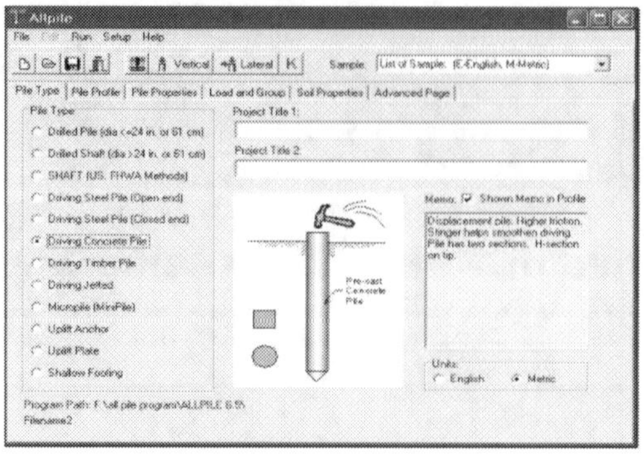

**Fig. 1.** AllPile window.

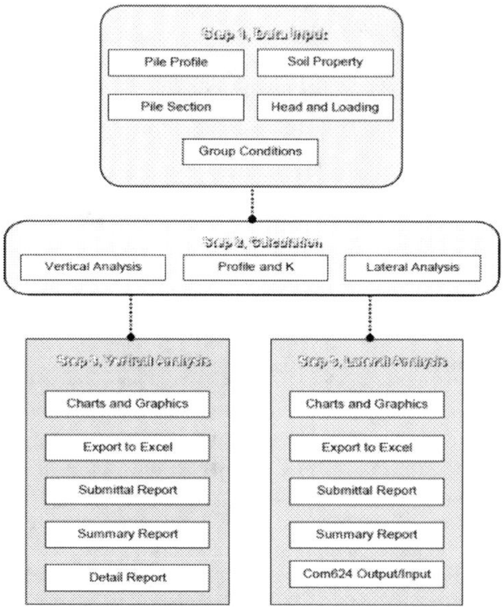


**Fig, 2.** Program Flowchart.

## 3 Properties of soil used

### 3.1 Soil

In this research two types of soil were used (Sand and Clay). In sand soil two states dry and fully saturated and fully saturated clayey soil were used. The properties of soil used is shown in Table (1).

**Table 1.** Soil properties.

| Soil Type | Properties |
|---|---|
| Soft Clay | Undraind Shear Strength (cu) = 25 kN/m$^2$<br>Angle of Internal Friction (Ø) = 0°<br>Submerged Unit Weight (γ) = 8 kN/m$^3$ |
| Sand | Cohesion (c) = 0 kN/m$^2$<br>Angle of Internal friction (Ø) = 33°<br>Relative Density (Dr) = 33%<br>Dry Unit Weight (γ) = 18.1 kN/m$^3$<br>Submerged Unit Weight (γ) = 8.5 kN/m$^3$ |

### 3.2 Pile

In this research four types of piles were used (square driven concrete, circle driven concrete, steel driven, and hollow steel driven pile) each of them had length (h) 12 m and diameter (or width) (d) 30 cm. The properties of piles is shown in Table (2). The sketch of pile in the soil with different angles of inclination are shown in Figure (3).

**Table 2.** Pile properties

| Pile Type | Properties |
|---|---|
| Concrete | Moment of Inertia (I) = 67500cm$^4$ for square Piles<br>Moment of Inertia (I) = 39760.8 cm$^4$ for circular Piles<br>Modulus of Elasticity (E) = 20683 MPa |
| Steel | Moment of Inertia (I) = 39760.8 cm$^4$<br>Modulus of Elasticity (E) = 199938 MPa<br>Yield Modulus (Fy) = 310 MPa |

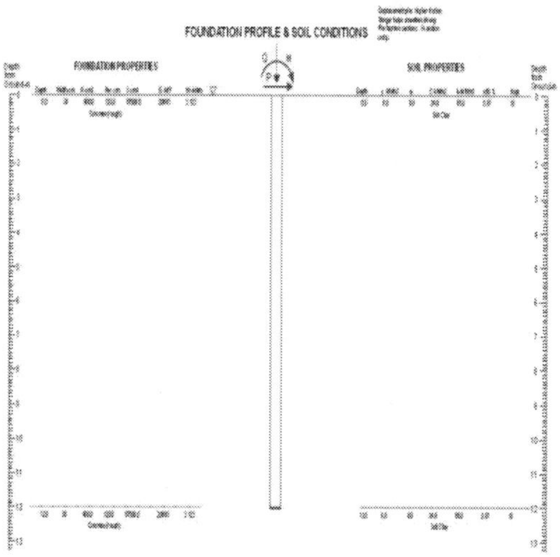


(a)

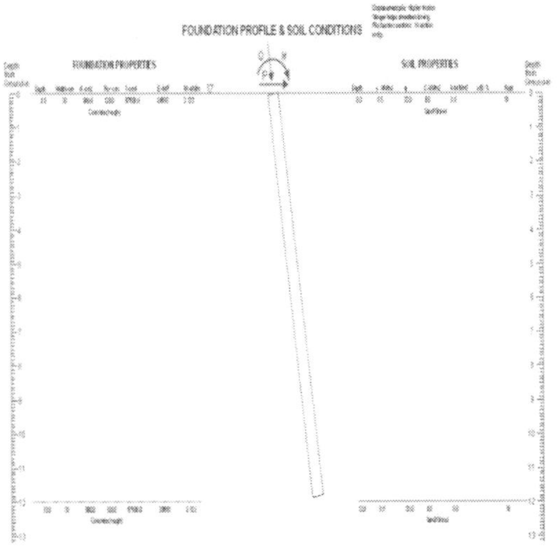


(b)

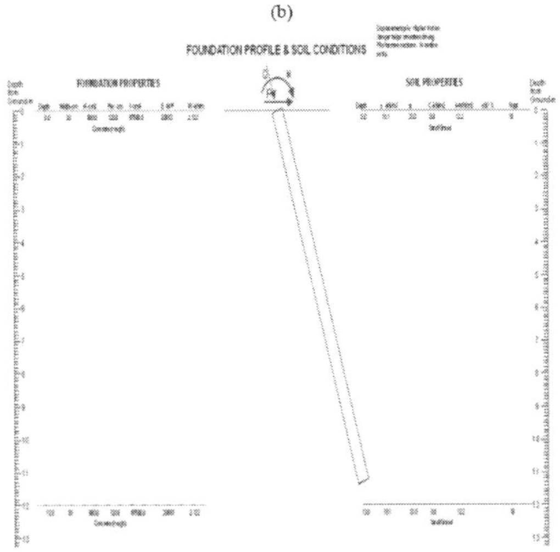


(c)

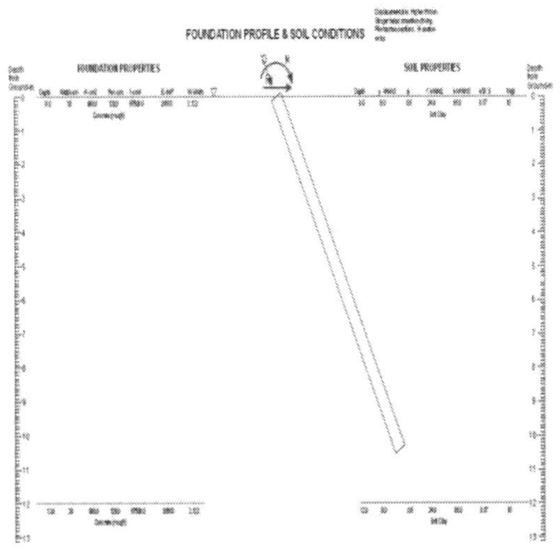


(d)

**Fig. 3.** The sketch of pile in the soil with different angles of inclination (a) 0° (b) 10° (c) 20° (d) 30°

## 4 Presentation of test results

Failure of model tests is defined as the stress required to cause settlement corresponding to 10% of the footing diameter (width) depending on the proposal given by (Terzaghi, 1947).

The analysis of results of all model tests regarding the compression load and the corresponding settlement is illustrated in terms of corresponding load (Q) versus settlement (X).

### 4.1 Soft clay soil

#### *4.1.1 Driven concrete square pile*

4 virtual model tests were performed in this type of piles with inclination 0, 10, 20 and 30°, the result demonstrated the decrease in failure load with increase the inclination. Figures (4) and (5) are shown the relation between load and settlement for bearing effect only and friction effect only for pile, respectively. Figure (6) is shown the relation between load and settlement. The loads corresponding 10% settlement ratio are shown in Table (3).

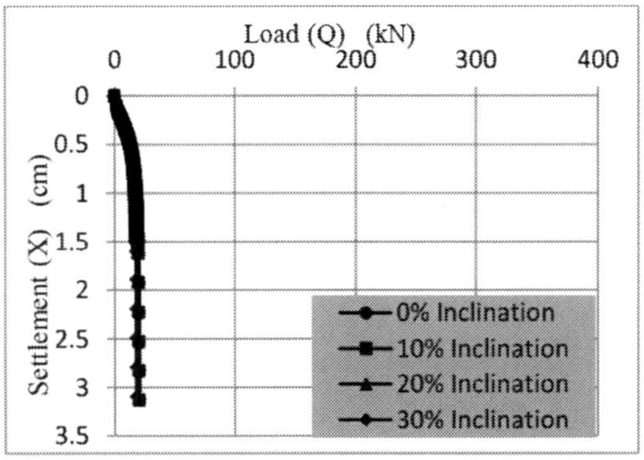


**Fig. .4.** Load-Settlement curve for a square concrete pile driven in soft clay soil for bearing effect only

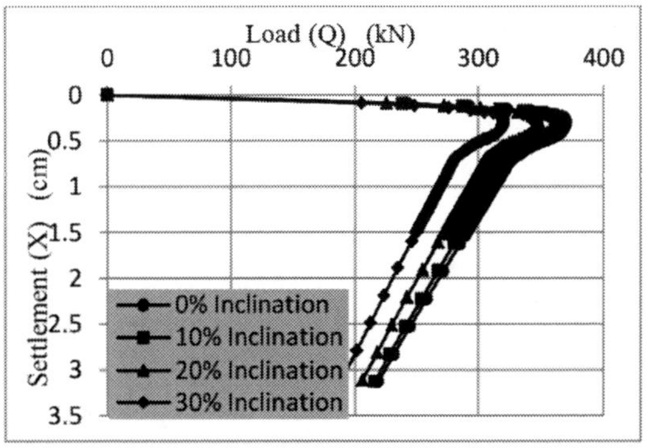

**Fig. 5.** Load-Settlement curve for a square concrete pile driven in soft clay soil for shaft friction effect only

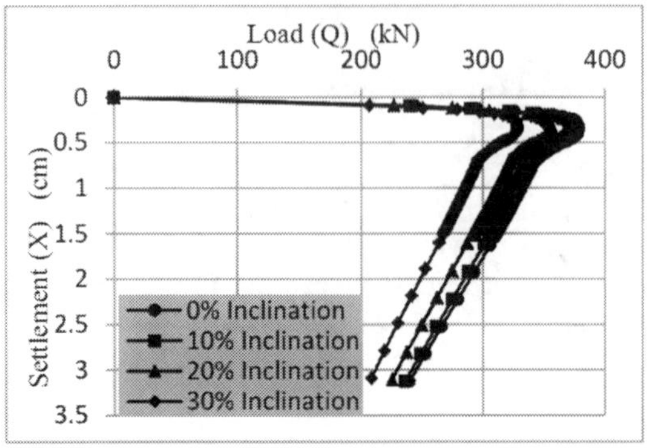


**Fig. 6.** Load-Settlement curve for a square concrete pile driven in soft clay soil

### *4.1.2 Driven concrete circular pile*

4 virtual model tests were performed in this type of piles with inclination 0, 10, 20 and 30°. Figures (7) and (8) are shown the relation between load and settlement for bearing effect and friction effect only for pile, respectively. Figure (9) shows the relation between load and settlement. The loads corresponding 10% settlement ratio are shown in Table (3).

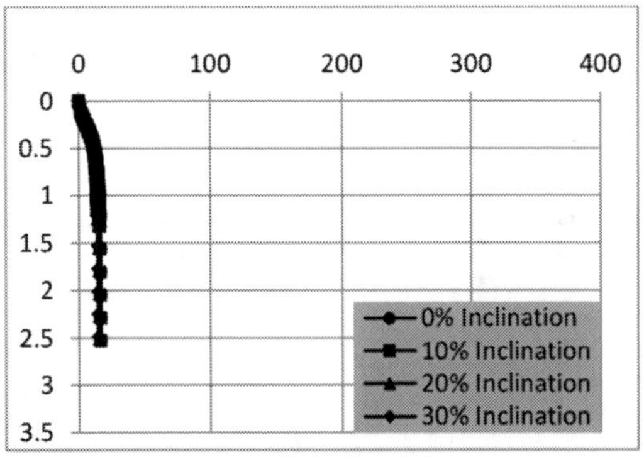


**Fig. 7.** Load-Settlement curve for a circular concrete pile driven in soft clay soil for bearing effect only

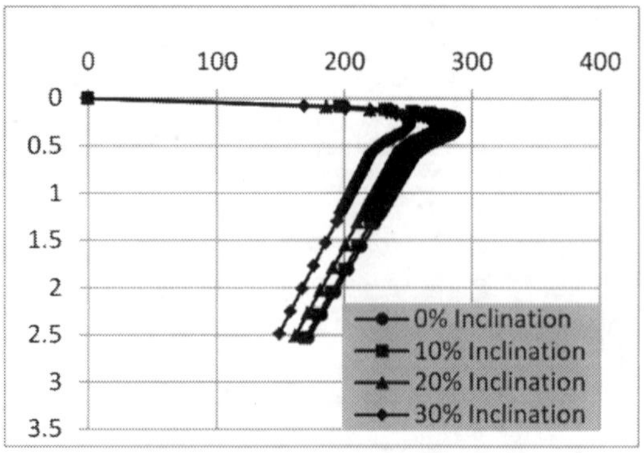


**Fig. 8.** Load-Settlement curve for a circular concrete pile driven in soft clay soil for shaft friction effect only

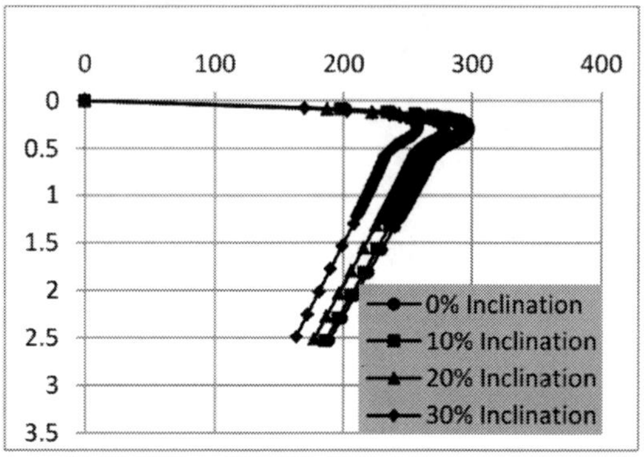


**Fig.9.** Load-Settlement curve for a circular concrete pile driven in soft clay soil with different angles

### *4.1.3 Driven steel circular pile*

4 virtual model tests were performed in this type of piles with inclination 0, 10, 20 and 30°. Figures (10) and (11) show the relation between load and settlement for bearing effect and friction effect only for pile, respectively. Figure (12) shows the relation between load and settlement. The loads corresponding 10% settlement ratio are shown in Table (3).

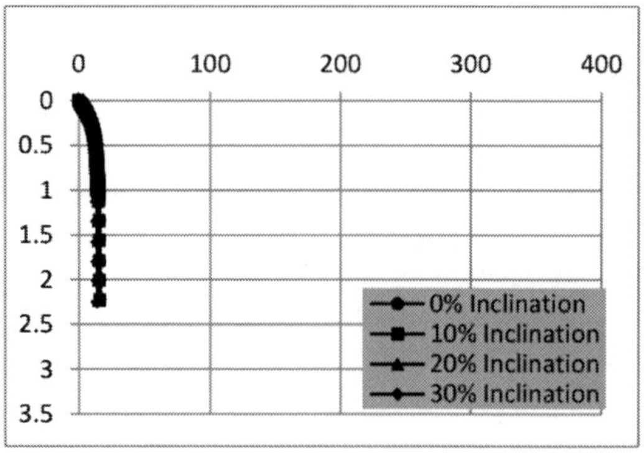

**Fig. 10.** Load-Settlement curve for a circular steel pile driven in soft clay soil for bearing effect only

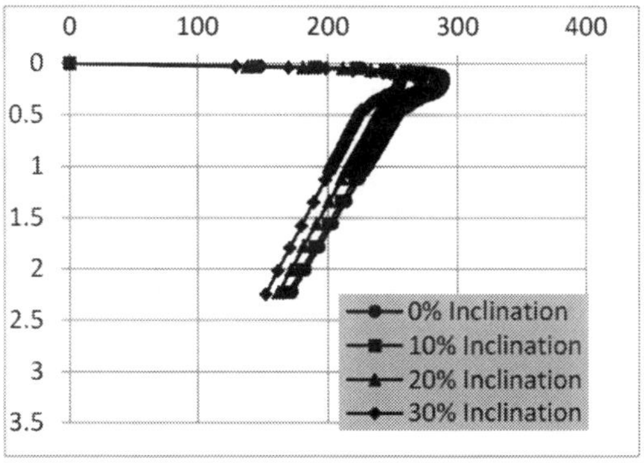


**Fig. 11.** Load-Settlement curve for a circular steel pile driven in soft clay soil for shaft friction effect only

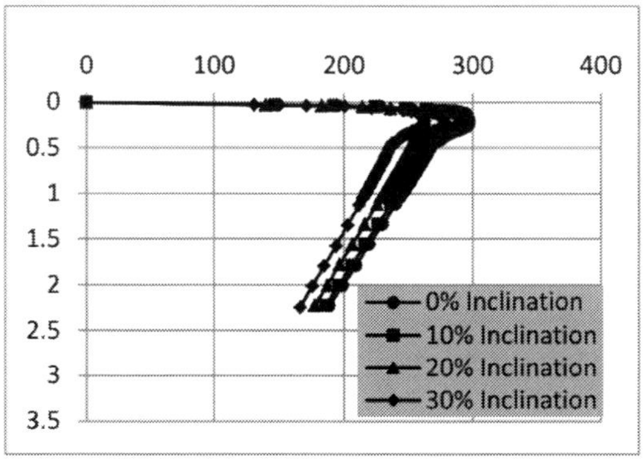


**Fig. 12.** Load-Settlement curve for a circle steel pile driven in soft clay soil with different angles

### *4.1.4 Driven hollow steel circular pile*

4 virtual model tests were performed in this type of piles with inclination 0, 10, 20 and 30°. Figures (13) and (14) show the relation between load and settlement for bearing effect and friction effect only for pile, respectively. Figure (15) shows the relation between load and settlement. The loads corresponding 10% settlement ratio are shown in Table (3).

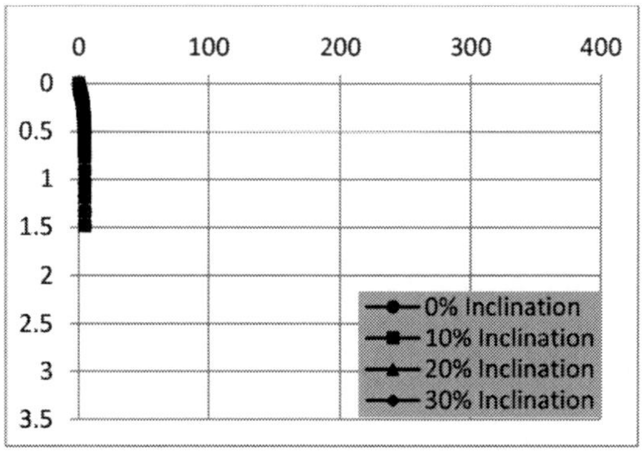


**Fig. 13.** Load-Settlement curve for a hollow circular steel pile driven in soft clay soil for bearing effect only

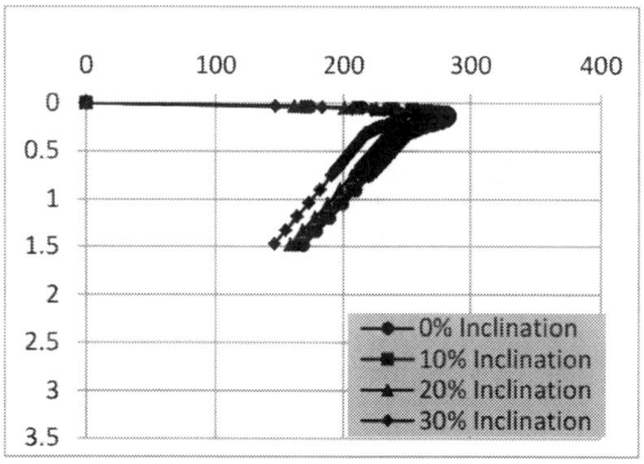


**Fig. 14.** Load-Settlement curve for a hollow circular steel pile driven in soft clay soil for shaft friction effect only

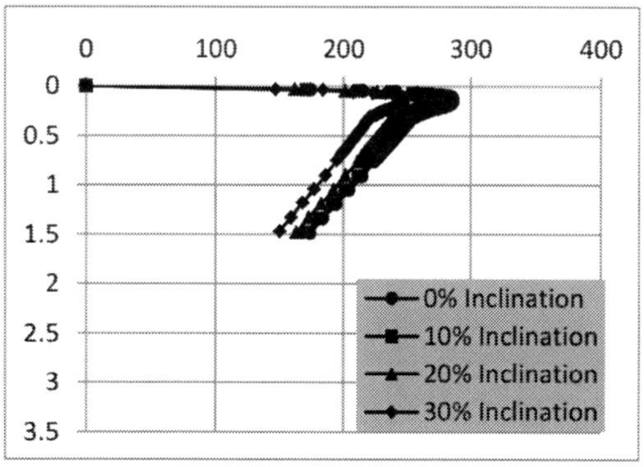


**Fig. 15.** Load-Settlement curve for a hollow circular steel pile driven in soft clay soil

## 4.2 Fully saturated sandy soil

### *4.2.1 Driven concrete square pile*

4 virtual model tests were performed in this type of piles with inclination 0, 10, 20 and 30°. Figures (16) and (17) show the relation between load and settlement for bearing effect and friction effect only for pile, respectively. Figure (18) shows the relation between load and settlement. The loads corresponding 10% settlement ratio are shown in Table (3).

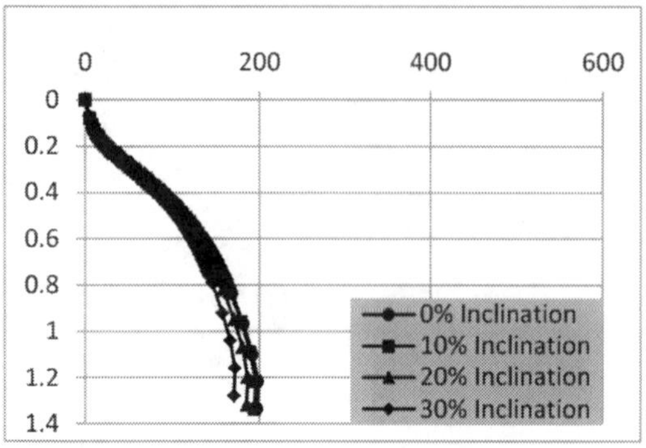


**Fig. 16.** Load-Settlement curve for a square concrete pile driven in fully saturated sandy soil for bearing effect only

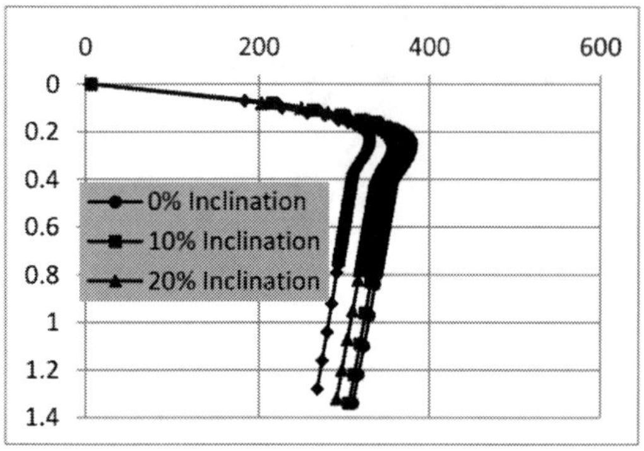


**Fig. 17.** Load-Settlement curve for a square concrete pile driven in fully saturated sandy soil for shaft friction effect only

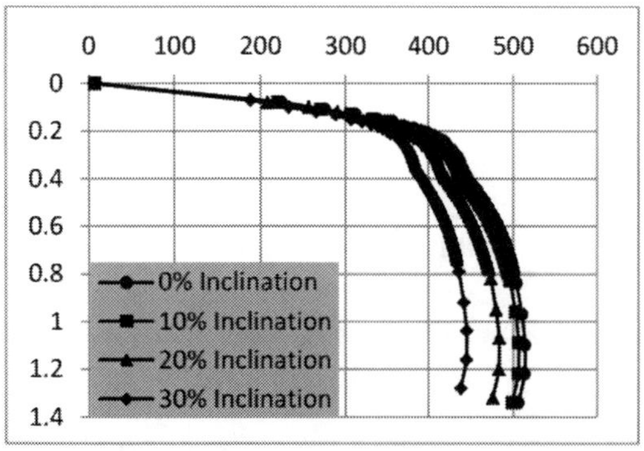


**Fig. 18.** Load-Settlement curve for a square concrete pile driven in fully saturated sandy soil

*4.2.2 Driven concrete circular pile*

4 virtual model tests were performed in this type of piles with inclination 0, 10, 20 and 30°. Figures (19) and (20) show the relation between load and settlement for bearing effect and friction effect only for pile, respectively. Figure (21) shows the relation between load and settlement. The loads corresponding 10% settlement ratio are shown in Table (3).

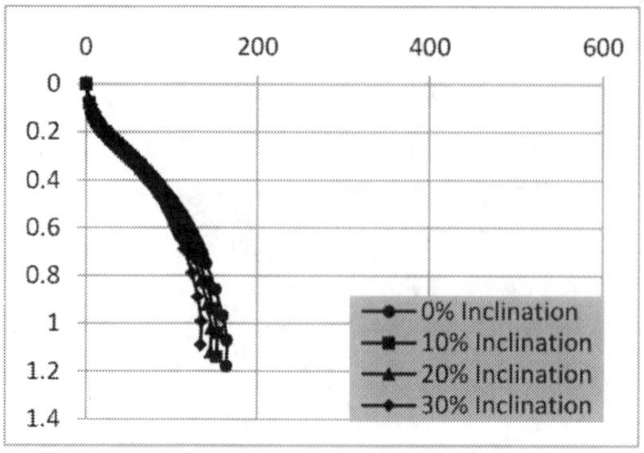


**Fig. 19.** Load-Settlement curve for a circular concrete pile driven in fully saturated sandy soil for bearing effect only

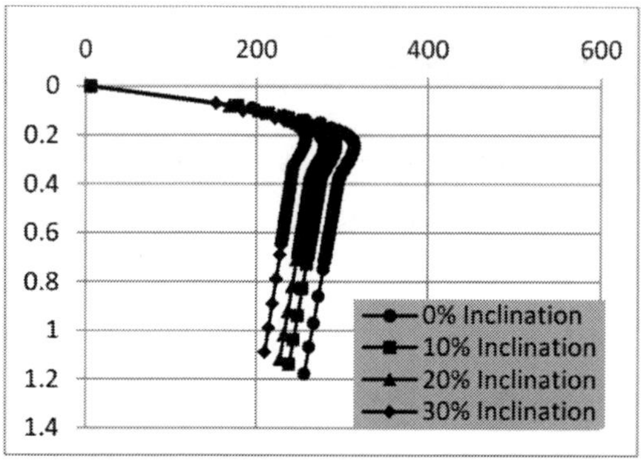


**Fig. 20.** Load-Settlement curve for a circular concrete pile driven in fully saturated sandy soil for shaft friction effect only

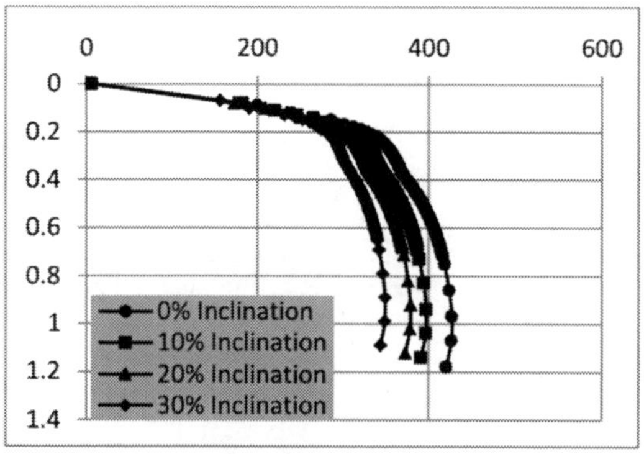


**Fig. 21.** Load-Settlement curve for a circular concrete pile driven in fully saturated sandy soil

*4.2.3 Driven steel circular pile*

4 virtual model tests were performed in this type of piles with inclination 0, 10, 20 and 30°. Figures (22) and (23) show the relation between load and settlement for bearing effect and friction effect only for pile, respectively. Figure (24) shows the relation between load

and settlement. The loads corresponding 10% settlement ratio are shown in Table (3).

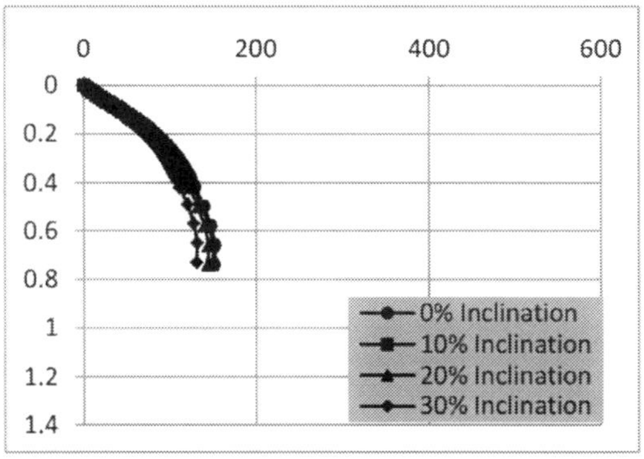


**Fig. 22.** Load-Settlement curve for a circular steel pile driven in fully saturated sandy soil for bearing effect only

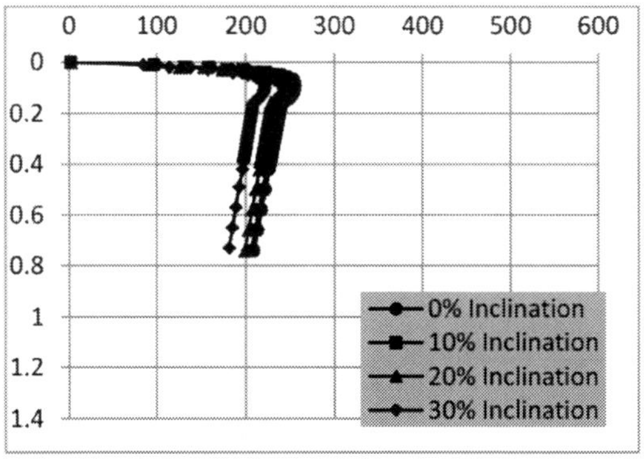


**Fig. 23.** Load-Settlement curve for a circular steel pile driven in fully saturated sandy soil for shaft friction effect only

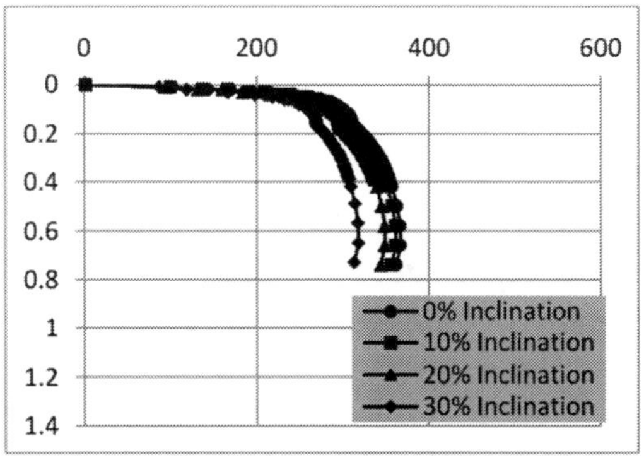


**Fig. 24.** Load-Settlement curve for a circular steel pile driven in fully saturated sandy soil

#### *4.2.4 Driven hollow steel circular pile*

4 virtual model tests were performed in this type of piles with inclination 0, 10, 20 and 30°. Figures (25) and (26) show the relation between load and settlement for bearing effect and friction effect only for pile, respectively. Figure (27) shows the relation between load and settlement. The loads corresponding 10% settlement ratio are shown in Table (3).

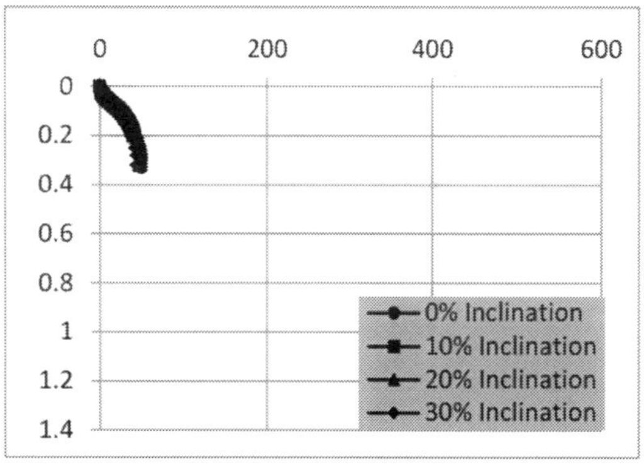


**Fig. 25.** Load-Settlement curve for a circular hollow steel pile driven in fully saturated sandy soil for bearing effect only

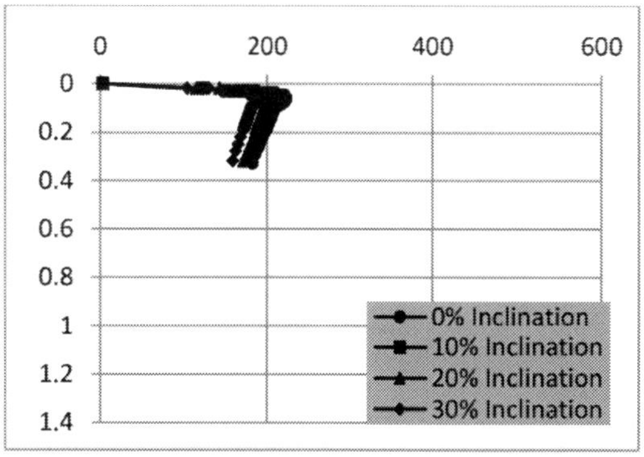


**Fig. 26.** Load-Settlement curve for a circular hollow steel pile driven in fully saturated sandy soil for shaft friction effect only

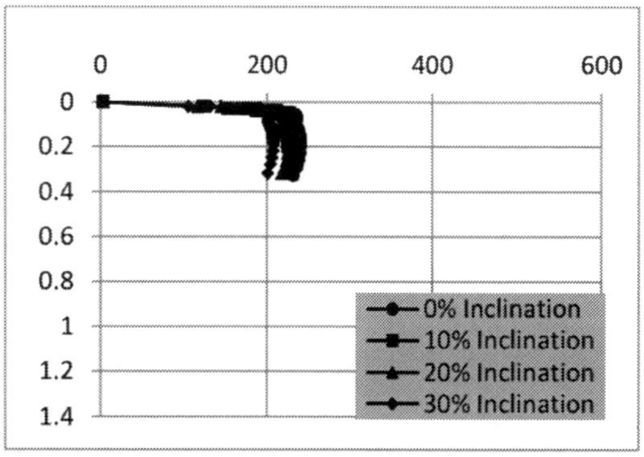


**Fig. 27.** Load-Settlement curve for a circular hollow steel pile driven in fully saturated sandy soil

### 4.3 Dry sandy soil

#### *4.3.1 Driven concrete square pile*

4 virtual model tests were performed in this type of piles with inclination 0, 10, 20 and 30°. Figures (28) and (29) show the relation between load and settlement for bearing effect and friction effect only for pile, respectively. Figure (30) shows the relation between load and settlement. The loads corresponding 10% settlement ratio are shown in Table (3).

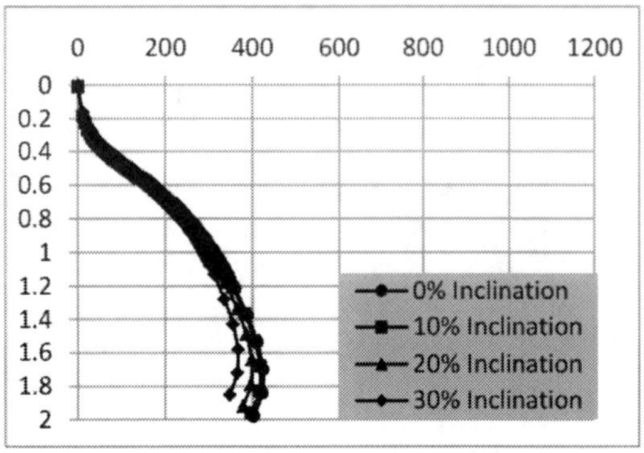


**Fig. 28.** Load-Settlement curve for a square concrete pile driven in dry sandy soil for bearing effect only

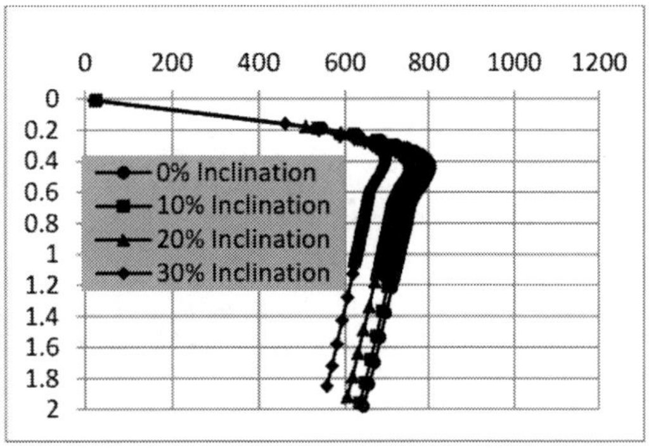


**Fig. 29.** Load-Settlement curve for a square concrete pile driven in dry sandy soil for shaft friction effect only

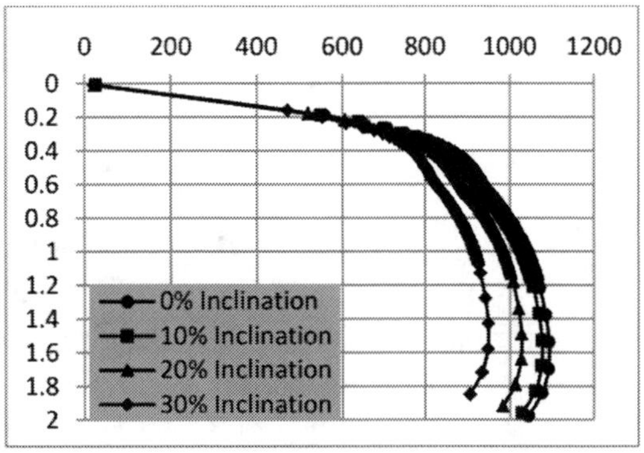


**Fig. 30.** Load-Settlement curve for a square concrete pile driven in dry sandy soil

#### *4.3.2 Driven concrete circular pile*

4 virtual model tests were performed in this type of piles with inclination 0, 10, 20 and 30°. Figures (31) and (32) show the relation between load and settlement for bearing effect and friction effect only for pile, respectively. Figure (33) shows the relation between load and settlement. The loads corresponding 10% settlement ratio are shown in Table (3).

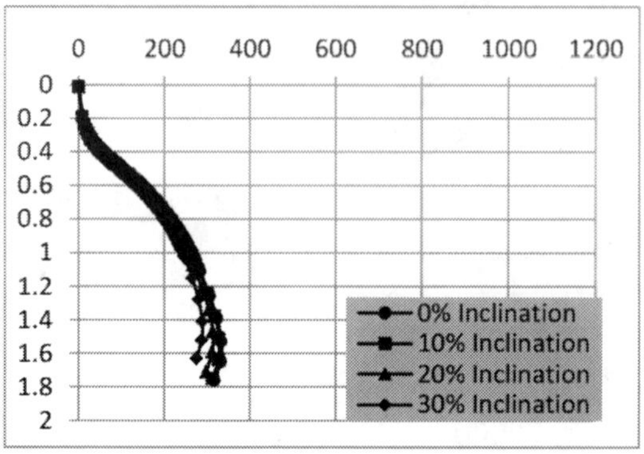


**Fig. 31.** Load-Settlement curve for a circular concrete pile driven in dry sandy soil for bearing effect only

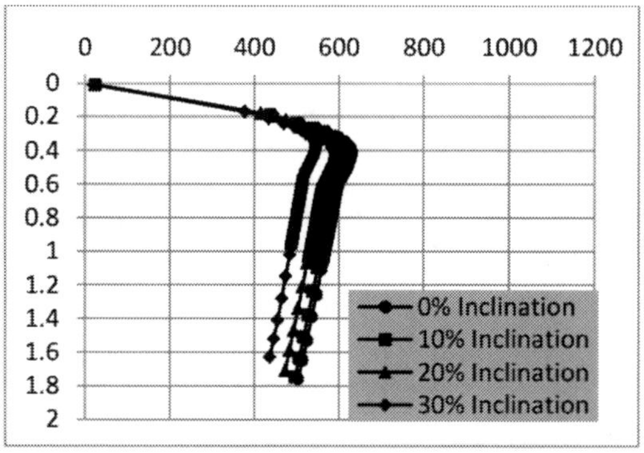


**Fig. 32.** Load-Settlement curve for a circular concrete pile driven in dry sandy soil for shaft friction effect only

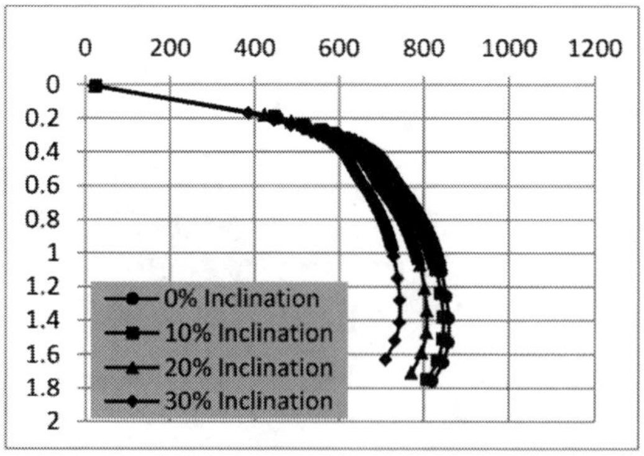


**Fig. 33.** Load-Settlement curve for a circular concrete pile driven in dry sandy soil

#### *4.3.3 Driven steel circular pile*

4 virtual model tests were performed in this type of piles with inclination 0, 10, 20 and 30°. Figures (34) and (35) show the relation between load and settlement for bearing effect and friction effect only for pile, respectively. Figure (36) shows the relation between load and settlement. The loads corresponding 10% settlement ratio are shown in Table (3).

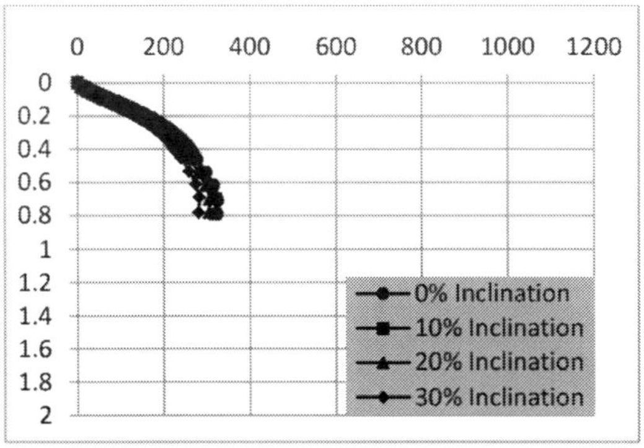


**Fig. 34.** Load-Settlement curve for a circular steel pile driven in dry sandy soil for bearing effect only

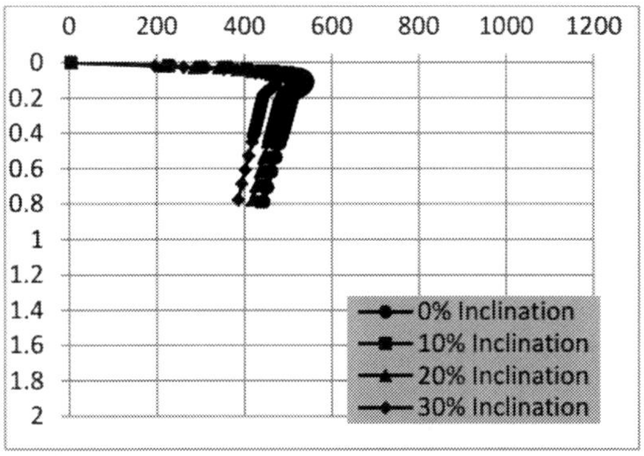


**Fig. 35.** Load-Settlement curve for a circular steel pile driven in dry sandy soil for shaft friction effect only

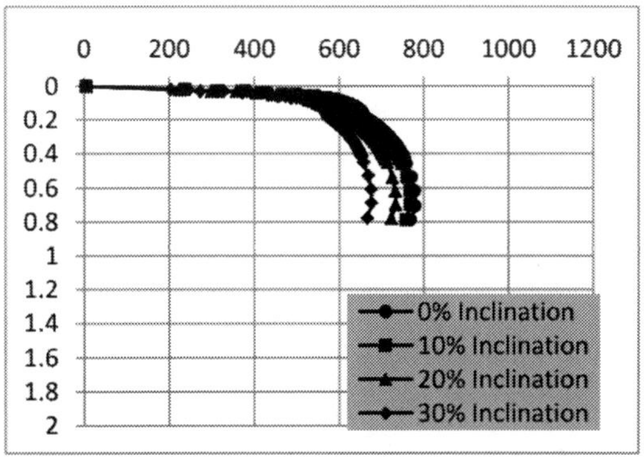


**Fig. 36.** Load-Settlement curve for a circular steel pile driven in dry sandy soil with different angles

#### *4.3.4 Driven hollow steel circular pile*

4 virtual model tests were performed in this type of piles with inclination 0, 10, 20 and 30°. Figures (37) and (38) show the relation between load and settlement for bearing effect and friction effect only for pile, respectively. Figure (39) shows the relation between load and settlement. The loads corresponding 10% settlement ratio are shown in Table (3).

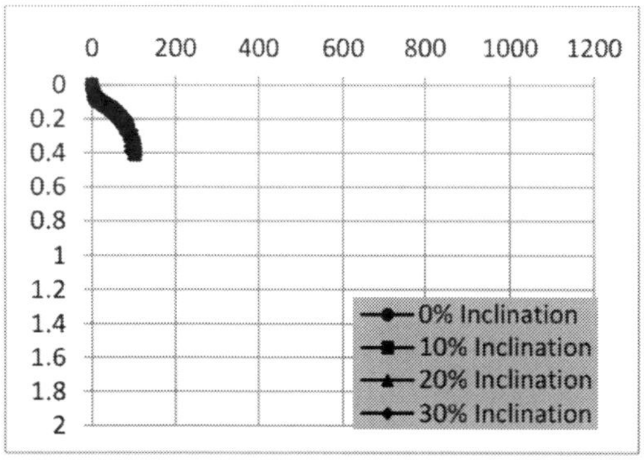


**Fig. 37.** Load-Settlement curve for a circular hollow steel pile driven in dry sandy soil for bearing effect only

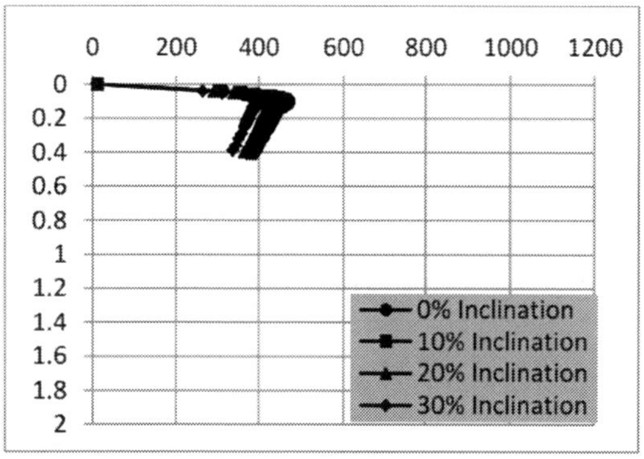


**Fig. 38.** Load-Settlement curve for a circular hollow steel pile driven in dry sandy soil for shaft friction effect only

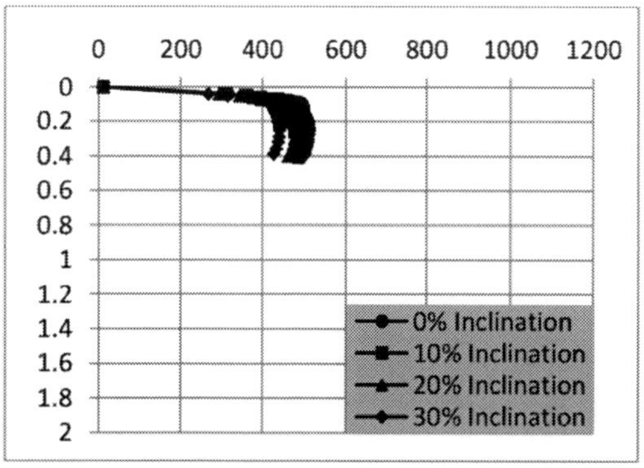


**Fig. 39.** Load-Settlement curve for a circular hollow steel pile driven in dry sandy soil

**Table 3.** Summary of test results.

| Pile Type \ Soil Type | Angles | Soft Clay | Fully Saturated Sand | Dry Sand |
|---|---|---|---|---|
| Square Concrete | 0° | 250kN | 495 | 1100 |
| | 10 | 240 | 488 | 1080 |
| | 20 | 226 | 462 | 1040 |
| | 30 | 212 | 435 | 940 |
| Circular Concrete | 0 | 192 | 423 | 846 |
| | 10 | 186 | 392 | 825 |
| | 20 | 180 | 372 | 800 |
| | 30 | 165 | 344 | 744 |
| Circular Steel | 0 | 195 | 352 | 758 |
| | 10 | 189 | 345 | 736 |
| | 20 | 177 | 336 | 704 |
| | 30 | 162 | 304 | 645 |
| Circular Hollow Steel | 0 | 177 | 230 | 486 |
| | 10 | 170 | 225 | 474 |
| | 20 | 162 | 215 | 460 |
| | 30 | 150 | 197 | 420 |

## 5 Conclusions

- The reduction ratio in failure load for the piles driven in soft clay soil are between 15-16% when the inclination increases from 0° to 30°.
- The reduction ratio in failure load for the piles driven in fully saturated sandy soil are between 13-19% when the inclination increases from 0° to 30°.
- The reduction ratio in failure load for the piles driven in dry sandy soil are between 13-15% when the inclination increases from 0° to 30°.
- Type of soil is the main effect on bearing capacity of piles and the degree of saturation of sand is plays an important role on bearing capacity but did not effect on clayey soil according to Tomlison's equation for piles in clayey soils.
- Shaft resistance is more effective than bearing resistance for clayey and sandy soils and all types of pile because of dimensions and depth of piles.

## References

1. R. Salgado, 1st ed. Mc Graw-Hill book company, New York (2006).
2. L. Li, S. Luo, ICTE 2011, pp. 1643-1648 (2011).
3. G. Mylonakis, A. Nikolaou, G. Gazetas Earthquake Engineering and Structural Dynamics (1997).
4. G. Gazetas, G. Mylonakis, Geotechnical Earthquake Engineering and Soil Dynamics III, ASCE, Geotechnical Special Publication (1998).
5. M. Sadek, I. Shahrour, Soil Dynamics and Earthquake Engineering (2004).
6. *AllPile manual*, version 7, **1** and **2**, USA, (2014).

# Studying of the combined salts effect on the engineering properties of clayey soil

*Anwar* Al-Obaidi[1,*], *Alaa* Ihssan[1], and *Hamed* Allawi[1]

[1]Building research directorate, Ministry of housing and construction, Baghdad, Iraq

**Abstract.** In recent years, a number of studies had been performed to investigate the effect of pore water chemistry on the strength and compressibility characteristics of soil. Although the effect of chloride and sulfates salts separately in pore fluids on the geotechnical properties of soil seems to be well understood, but the influence of combined effect of sulfates and chlorides in pore water on the behavior of soil is still unclear mostly due to the limited numbers of studies as well as the complexity of processes that may occur in soil (with the presence of salts) in pore water-soil interaction. Southern regions of Iraq, especially Basra suffers from low water levels in the summer season in addition to the lack of rain water, which causes a significant increase of salt in the Shatt al Arab. Water salinity continues to increase with time. To investigate the combined impacts of water salinity on the behavior of clayey soils, the basic characteristics of the soil brought from Al-Nahrawan site was studied. Chemical methods were done with three types of water (distilled, water of highly saline as Shatt Al-Arab water and water of Tarmiya as moderate saline water). The effect of water salinity on the geotechnical properties of fine grain soil was investigated. Different laboratory tests such as Atterberg limits, standard compaction, consolidation and shear strength of soil .Results showed that the presence of perceptible amounts of dissolved salts in water can lead to changes in the engineering properties of the soil.

## 1 Introduction

Soil is a mixture of several compounds such as soil solid, air voids and pore water. Water is well known to be strongly attracted to the soil surface. Mitchell (1993) [6] in their soil investigations stated that fabric and bonding both characterise the structure of soils. Micro structural consideration considers that the fabric is mainly referred to soils solid particle arrangement and its flocculation process, yet bonding is dominated by the inter particle forces. Pore properties of the fluid have a direct impact on both the fabric and the bonding mechanism. This is related to the water-soils interaction of clays (Mitchell and Soga, 2005[7]). The term 'salinity' represents the soluble salts total concentration in the soil. Those salinity salts come in different forms including calcium , magnesium , chlorides, sulphates or even carbonates of sodium, Pore water salinity influence on the behavior of solid soil particles as a result to the chemical interaction that may occur between the soil particles and the ions in the pore water. This is clearly observed in fine-grained soils. Studies have proved that many material macroscopic characteristics (e.g., shear stiffness, compressibility, etc.) are affected by pore water salinity concentrations. Yet salinity influence on the geotechnical properties of soil is still yet to be further investigated. The current study looks into the relationships bonding between salt content that fills the gap and the geotechnical properties of fine-grained soils.

### 1.1 Clay water interaction

Clays are known to be hydrophobic colloid. Their behaviors is generally dominated by the surface forces. Surface forces play an important role in the formation of soil microstructure and water-soil interactions (van Olphen, 1991 [9]; Mitchell and Soga, 2005 [7] ).

Negative charged surface and adjacent charges are termed to the diffuse double layer. Ions distribution among the diffuse double layer may cause unbalanced force field near the clay surface. This can generate inter particle forces when two soil particles get closed. The double layer forces may change the movement of soil particles and impact soil behaviors.

* Corresponding author: alobeydi_63@yahoo.com

© The Authors, published by EDP Sciences. This is an open access article distributed under the terms of the Creative Commons Attribution License 4.0 (http://creativecommons.org/licenses/by/4.0/).

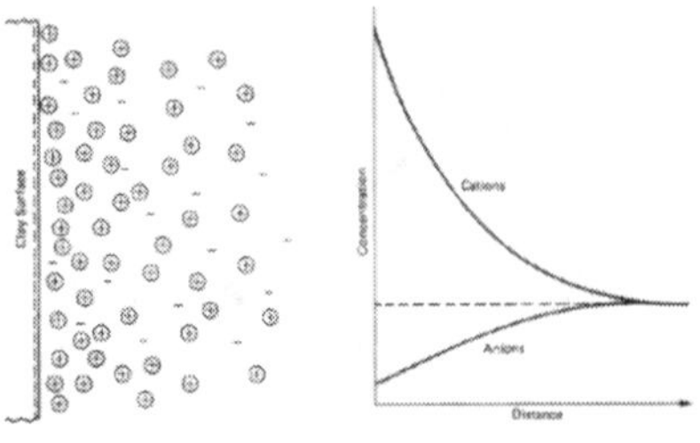

**Fig. 1.** Ions concentrations in the diffuse double layer phase (after Mitchell and Soga, 2005) [7]

### 1.2 Effect of water salinity on soils geotechnical properties

In recent years, a number of studies have been performed to look into the effect of the chemistry of pore water and its impacts on the geotechnical characteristics of soil. Anandhanarayanan and Murugaiyan,(2014) [3] investigated the effect of sea water and different inorganic salt solutions (as permeant liquids) on the compaction characteristics, liquid limit, free swelling, consolidation characteristics and hydraulic conductivity of different types of clay-bentonite mixtures. Sea water and salt solution increased the maximum dry density and decreased the optimum water content of mixtures .Higher cation valance leads to higher increase in the maximum dry density and higher decrease in optimum water content, by increasing the sea water and salt concentration the swelling volume and liquid limit decreases. Higher cation valance and salt concentration cause higher decrease in liquid limit by increasing the sea water and salt water concentration, the compression index decreases the hydraulic conductivity of the samples increases by adding the salt water and sea water.

The salinity of the water has been shown to effects on the geotechnical properties of fine soils used in Korchay dam core had been studied Ajalloeian, et al.,(2013) [1]. Several soil measurments as ( Atterberg limits, compaction, consolidation, direct shear and dispersion (pinhole and chemical) ) have been carried out by the use of waters of different salt concentrations ( distilled, half saline and saline waters ) . The charges in the geotechnical properties of the soil were neglected as there were a some percent of active clay minerals found in the soil. Results revealed the values of the Atterberg limits, compression index and swelling index decrease and consolidation coefficient and shear strength parameters increase as the amount of pore water salinity increases. When comparing those results with other tests carried out using water that has been desalinized, the effect of water salinity on consolidation under high pressures is not consideration. Dispersion tests also revealed that there was no dispersion in the soil when contacted with waters of high salinity.

Aksoy, et al.,(2008) [2] examined the sea water effects on consistency limits and compressibility properties of clayey soils of low and high plasticity.The results revealed that sea water impacts were almost negligible on the tested consistency limitsand compressibility characteristics of soils when they have liquid limits up to 110%. The sea water effect is mostly noticed on the consistency limits and compressibility of Na-bentonites.

## 2 Experimental works

### 2.1 Soil samples

The following tests were done to determine the physical and engineering properties, salinity and pH of soil sample collected from Al-Nahrawan site.

1- Specific Gravity (ASTM D854)
2- Atterberg Limit (ASTM D 4318)
3- Grain Size Analysis (ASTM D 422)
4- Moisture Content (ASTM D 2216-90)
5- Unit Weight (ASTM D 1556-00)
6- Shear Strength (ASTM D 2166)
7- Consolidation (ASTM D 4186)
8-pH (ASTM D4972-01)

The soil samples have been collected from Al-Nahrawan brick factory. Figure (2) shows the distribution of the grain sizes of the soil materials of the collected samples. In this graph it is seen that the soil distribution contained of 44% silt and 56% clay. According to (liquid limit and plasticity index) and in reference to grain size distribution, the soil can be classified as lean clay (CL) using unified soil classification system. Chemical methods were done with three types of water (distilled, Shatt Al-Arab ( highly saline water) and water of Tarmia ( moderate saline water).Concentrations of cations and anions of the waters utilized in this investigation had the specifications listed in Table (1) . The cation charge showed to have played an important part in increasing double layer thickness and in decreasing interparticle attractive force. The effect of three types of water (i.e. saline, half saline and distilled water) on geotechnical properties of fine grained soil had been investigated in this research. Tests including hydrometer Atterberg limits, standard compaction, consolidation, direct shear test and unconfined compression were also performed.

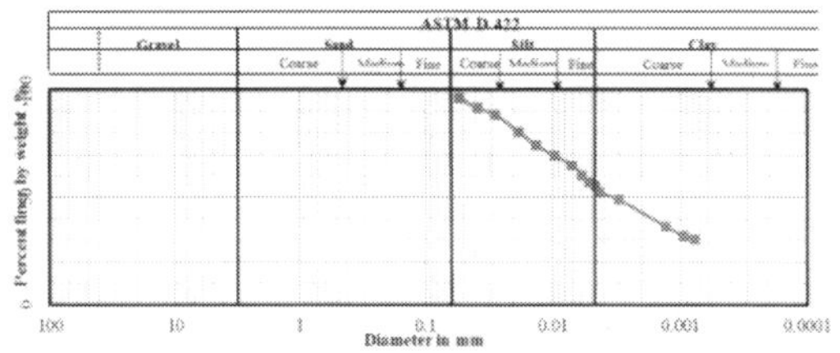


**Fig. 2.** Grain size distribution of Al-Nahrawan site

**Table 1.** Chemical analysis of Waters studied

| Type of water | pH | $Cl^{-1}$ | $SO_4^{-2}$ | $Na^{+1}$ | $K^{+1}$ | $Mg^{+2}$ | $Ca^{+2}$ |
|---|---|---|---|---|---|---|---|
| Distilled | 7.2 | 0.73 | 12.1 | -- | --- | --- | --- |
| Al-Tarmia | 7.6 | 6.39 | 26.5 | -- | --- | --- | --- |
| Shatt-Al-Arab | 8.0 | 65.3 | 45.9 | 30.9 | 5.0 | 14.7 | 28.2 |

## 3 Results and discussions

### 3.1 Effect on consistency limits

Fig. 3. presents the results of the liquid limit (LL), plastic limit(PL) and plasticity index(PI)with different types of water.

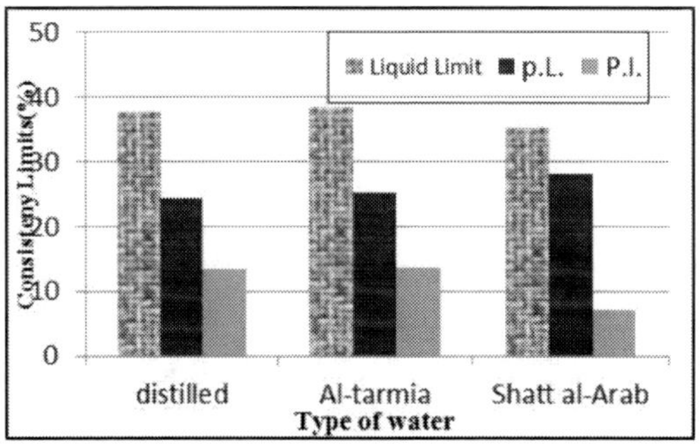


**Fig. 3.**Effect of Salinity of water on consistency limits

The results shows that the liquid limit of soil when mixed with distilled water is 37.6% and it does not seem to increase significantly when mixed with moderate saline water then decreased when mixed with highly saline water (shatt al-Arab).Plastic limit is increased in all types of saline water.Increasing the salt concentration, the cationvalence decreases the inter-particle distances, resulting in the decrease of the liquid limit. Ajalloeian et al., (2013) [1] noticed that time of exposure has no effect on changing Atterberg limits.

### 3.2 Variation of compaction characteristics with Salinity

Fig.4. shows the results of compaction tests performed with waters of different salinity. Relaying on Ajalloeian et al., (2013) [1] results on compaction, Atterberg and exposure time showed that these are not effective factors in the results. In Fig. (4) there were no distinguished differences between these curves. Yet the results have shown that the optimum moisture content has decreased and maximum dry unit weight of the soil has increased when mixing with saline water and the results is more pronounced when the salinity is moderate.

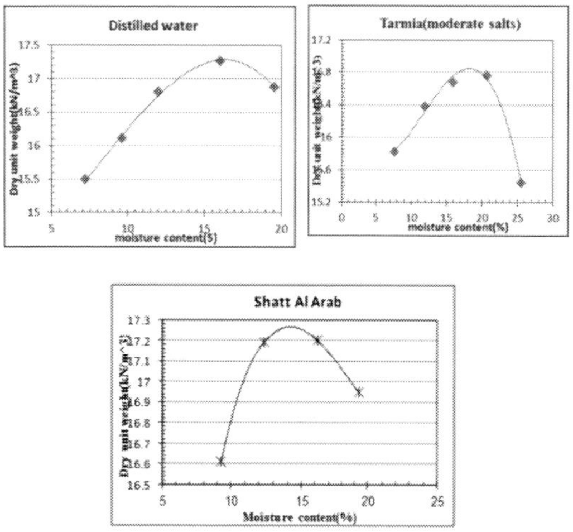


**Fig. 4.** Effect of salinity of water on compaction characteristics

### 3.3 Effect of salinity on consolidation characteristics of soil

Fig.5. shows results of the one dimension consolidation test of three times performance with all types of water as well as compacted soil. Logarithmic vertical stress relationship against void ratios was used to determine the Compression index (Cc) and swelling index (Cs). Average values of compression index (Cc) of distilled water, half saline water and saline water were 0.159, 0.1067 and 0.095 respectively as seen in Fig.6. The results come in agreement with other studies. Compression index value reduction can be related to the reduction of double layer thickness and increasing of attractive force between particles as pore fluid concentration increases.

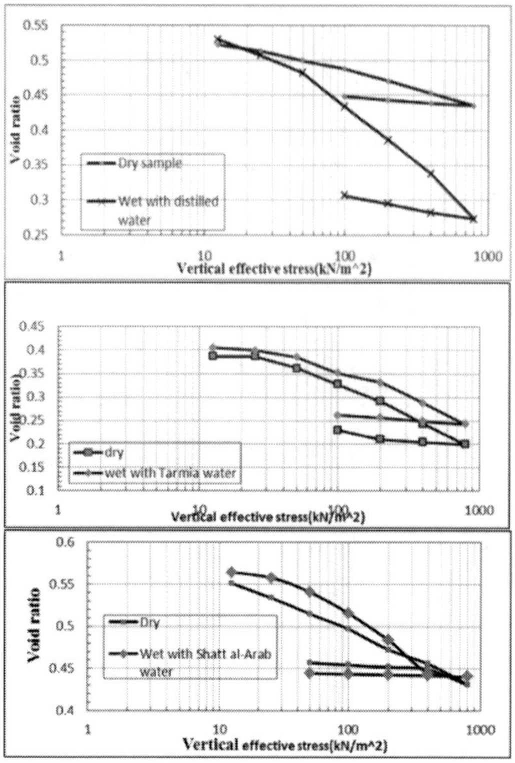


**Fig.5.** Effect of salinity of water on consolidation

It can be concluded that high concentration of the salt concentrations in the water has increased the attraction forces present between the particles. Moreover, this established a better bonding among the particles. Consequently flocculation structure has formed in the soil. Also salt crystals may sediment in the pores of soil and play role of cement. In this condition, subsidence of soil and compression index will be lower.

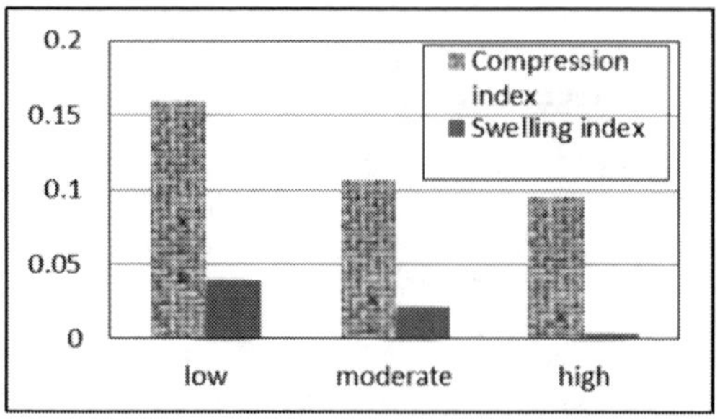


**Fig.6.** Compression index and swelling index versus salinity of water

It was observed that the swell index (Cs) has also decreased in respect to water salinity increment . The average values of the swell index (Cs) with distilled water, half saline water and saline water have been found to be 0.039, 0.0215 and 0.0033 respectively. The changes can be related to the differences of the osmotic pressure. High concentration of the half saline and saline water, osmotic pressure difference between water within layers of clay and water in the pores of soil has been observed to be lower in comparison with the distilled water. So after unloading, less water enters to the space between layers of clay to balance osmotic pressure. Hence Cs has decreased as salinity of water increases.

In Fig.7. illustrates the effect of salinity of water on modulus of volume compressibility (mv) in the stress $200 kN/m^2$.It is seen that the modulus of volume compressibility decreases with increasing salinity of water. This change is due to decrease in double layer thickness as concentration of water increases.

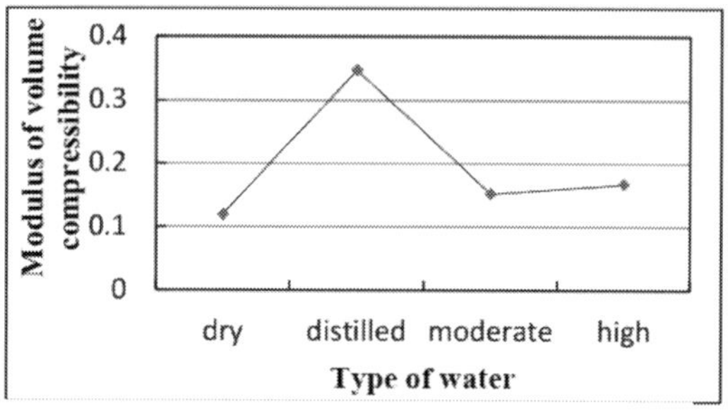


**Fig.7.** Modulus of volume compressibility Vs . Water salinity

## 3.4 Effect of salinity on undrained shear strength of the soil

### *3.4.1 Unconfined compressive tests*

The effect of water salinity on the unconfined compressive strength for the Al-Nahrawan soil is presented in Fig.8. The ultimate strength of samples decreased significantly with increasing water salinity as shown. Results also showed that mixing saline water with soils decreased its undrained compressive strength characteristics as shown in Fig.9. In general, the soils displayed low unconfined compressive strength with increasing salinity than with distilled water. This may be attributed such a strength decrease to the fact that the salt forms brine with water; this brine forms a film around each particle, which decreases the strength of the particles since brine film is weak in surface tension than ordinary moisture, thus it decreases the cohesion of the particles.

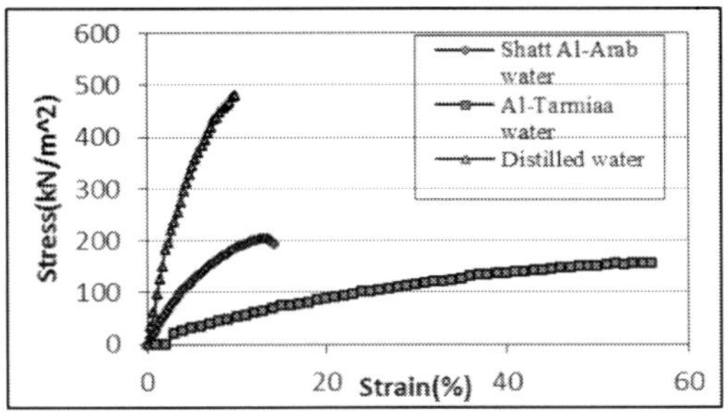


**Fig.8.** Effect of salinity of water on the unconfined compressive strength of Al-Nahrawan soil

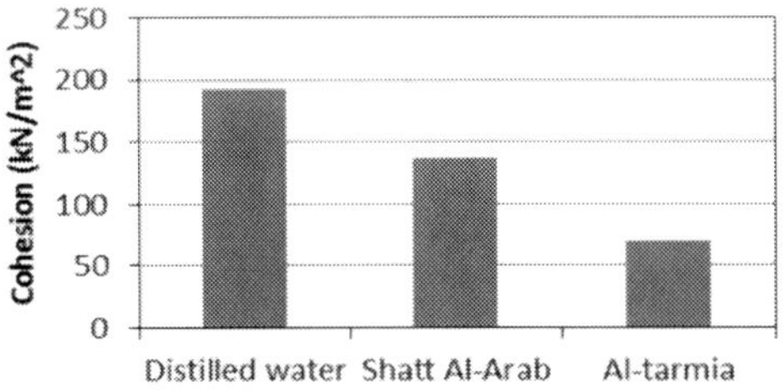


**Fig. 9.** Effect of salinity of water on undrained shear strength(Cu)

### *3.4.2 Direct shear test*

The plot of stress-strain curve obtained after conducting the shear tests is shown in Fig. 9. for as compacted specimens with different water salinity and Fig. 10. for specimens repaired as compacted and then immersed with different water salinity. The fit line was drawn starting from the undrained shear resistance versus percent salt content for direct shear tests to study the effect of different salinity solution on undrained shear strength of the soil.

In general (both as compacted and saturated) the cohesion is decreased with increasing salinity. These results are coinciding with the results obtained from unconfined compression tests.The surfaces of soil particles hold negative or positive charges due to isomorphous substitution process. These surface charges lead to electronic interparticle forces and affect the behavior of soils. The charged surfaces of soil particle have interaction with the dissolved ions in the pore water and influence soil behaviour. Generally, most of the soil particle surfaces hold negative charges caused by isomorphous substitution. Cations are attracted by the negative surface charges and held on the surface of soil particle. Due to the absorbed cations, the electronic forces between soil particles are influenced. In this way, the ion concentration (the salt content) of pore water affects the behavior of soil particles and the properties of soils.

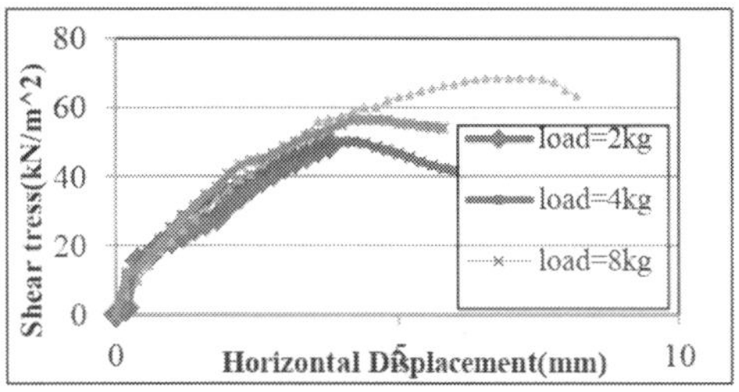


**(a)**Distilled water

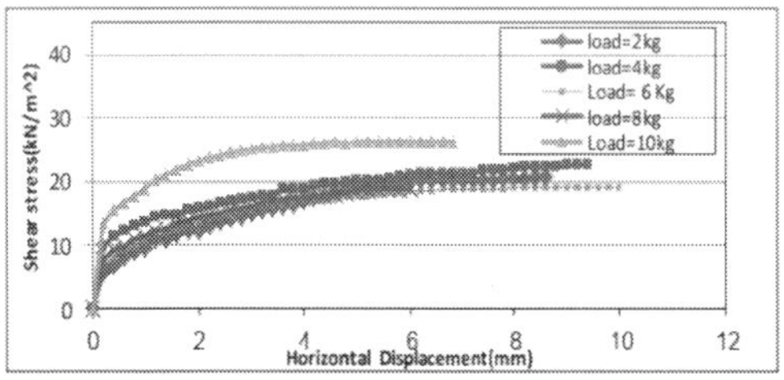


**(b)**Al-Tarmia water

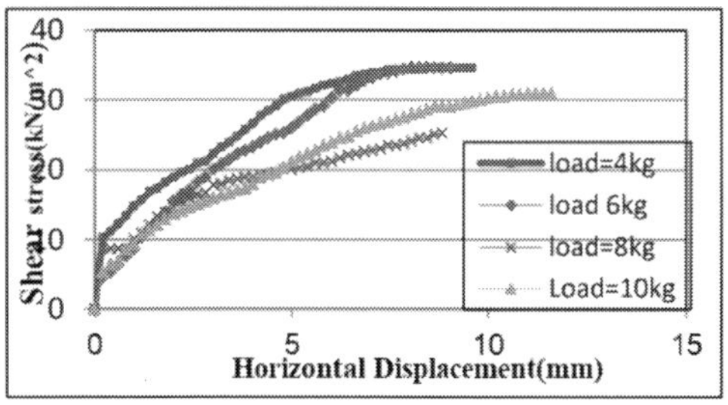


**(c)** Shatt Al-Arab water

**Fig. 10**. Effect of salinity of water on stress strain curve for as compacted specimens

https://doi.org/10.1051/matecconf/201816201011

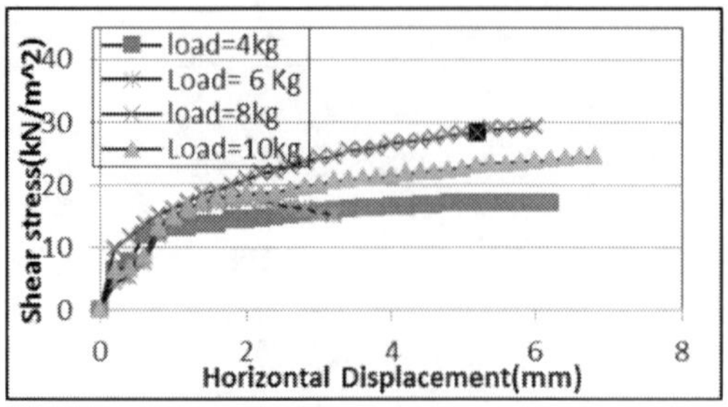


(a)Immersed with distilled water

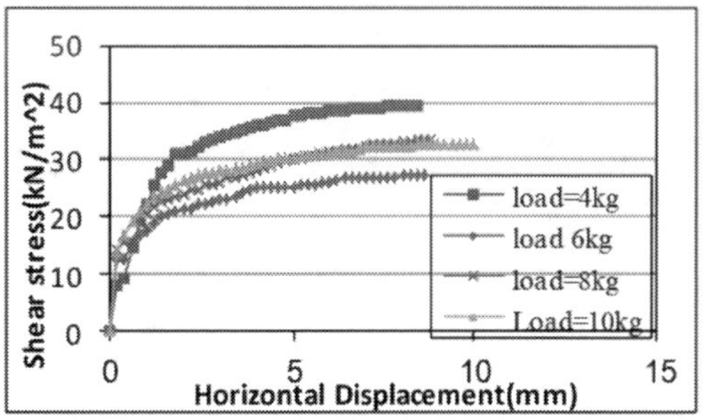


b)Immersed with Al-Tarmia water

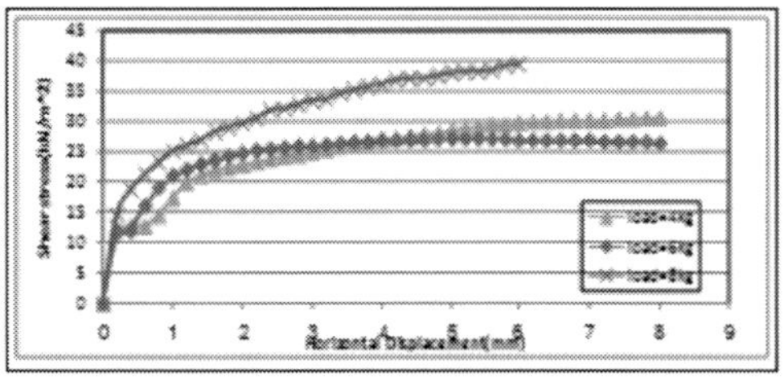


( c) Immersed with Shatt-Al-Arab water

**Fig. 11.** Effect of salinity of water (using in immersed specimens) on stress-strain curve for as compacted specimens

## 4 Conclusions

Based on the experimental results, the following conclusions can be drawn:

1. Salt solution increment can decreases the undrained shear resistance of compacted soil due to the decreased in attractive forces as compared with distilled water.

2. A comparison between direct shear and unconfined compression tests, results showed that as the salinity of water increases, the cohesion decreases and these parameters obtained by unconfined tests may give higher value than direct shear.

3. The test results demonstrated that the salinity has a pronounced effect on compressibility. The compression index and sweelling index decreased with increasing salt concentration

4. The modulus of volume compressibility decreases with increasing salinity of water.
5. Optimum moisture content showed to decrease and maximum dry unit weight of soil showed to increase when mixing with saline water .
6. Liquid limit as well as plastic limit decreases with increasing salinity.
7. Finally, from above results care must be taken when using saline water instead of distilled water or tap water or when the soil is exposed to saline water since the salinity weaken the soil.

## References

1. R. Ajalloeian, H. Mansouri, and A.H.deghpour, EJGE, **18** (2013).
2. Y.Y.Aksoy, A.Kay, and A.H. Oren, Eng. Geology **1 02**, P.P. 54–61(2008).
3. G. Anandhanarayanan, V. Murugaiyan, IJERT, **3** Issue 3, March (2014).
4. ASTM, *Annual Book of the ASTM Standards*, **04-08**(2006).
5. D.Kolymbas, & E.Bauep, GTJ, **6(2),** P.P. 263-270(1993).
6. J.Mitchell, *Fundamentals of soil behavior,* NY: Wiley (1993).
7. J.Mitchell, & K.Soga, *Fundamentals of soil behavior.*3rd edition, New York (2005).
8. R. Ajalloeian, H. Mansouri, and A. Sadeghpour EJGE, **18**, P.P.1419-1434 (2013).
9. V. Olphen, TGSC, *An introduction to clay colloid chemistry*, Reprint version, New York, John Wiley and Sons, (1991).

# Influence of cyclic loading on performance of steel piles in sandy soil

*Mohammed* Al-Neami [1,*], *Mariam* Wasmi [1]

[1]Building and Construction Engineering Department, University of Technology, Baghdad, Iraq

**Abstract.** This paper introduces an experimental study to clarify the response of steel pile models exposed to the cyclic loading. Thirty six models of two types of steel piles are tested (open ended pile and H-pile) with lengths equal to (30, 40, and 50) cm. Three diameters (2.5, 3.5, and 4.1) cm for open ended pipe pile and three flange widths (2.6, 3.6, and 4.4) cm for H-pile are investigated. Jacking technique is employed to installed piles models in dry sandy samples with two different relative densities (60% for medium sand 80% for dense sand). It is found that the pile geometry (diameter and length) with sand density have a high impact on the number of cycles. Analysis of results showed that increasing of pile diameter and relative density cause a reduction in the number of cycles when the length of steel pile models are fixed while variety of diameters of open ended pipe pile has a small effect on the number of cycles. It was found that pipe piles with open ended have more resistance to the cyclic loading compared with H piles under the same geometric conditions (pile diameter, embedded length and sand density) especially in medium sand. Finally, if the testing conditions are the same, number of cycles is decreased with increasing in amplitude loading.

## 1 Introduction

Many permanent deformations are developed in soil when exposures to cyclic loading which causes serious problems as well as excessive settlement or liquefaction due to a build-up of excess pore water pressure (di Prisco and Wood 2012)[1]. Pile foundations are the most populist constitute of deep foundation utilized alike for offshore and onshore structures. Piles foundation may subject to the cyclic loading as a result of machine foundation vibration, traffic loads, vibration induced due to pile driving, wave action in offshore structures or the naturally imposed vibrations in soil strata due to earth tremors. The effect of cyclic loading may be transmitted to the soil surrounding piles resulting significant rotation or deformation during the piles life which not taking in consideration in design.

Many researchers have been studied the behavior of deep foundation under cyclic loading, some of them are representing in following.

Puech et.al, (1982) [2] conducted a cyclic tensile load on steel piles embedded in silts and loose sands. These tests were performed on 27.3cm diameter and 130 cm long piles. The outcomes of these tests showed that the displacement of piles under cyclic loading divided into a transitory stage, which refers to an immediate response of the soil to the cyclic stresses and a cyclic creep stage. For reduced load levels (less than 50%$Q_{ult}$), the displacement rate decreases, indicating stability and for higher load levels (greater than 50%$Q_{ult}$), the rate of displacement increases loading to failure, similar to those observed by Chan and Hanna (1980).

Poulos (1984, 1989a and 1989b) [3, 4, and 5] made experimental and analytical studies on the small piles models in sand. The results of these studies showed that the cyclic deterioration of skin friction is controlled by the amplitude of cyclic displacement, relative to a critical displacement. This critical displacement is equal to the static displacement required for full slip.

Craig and Sabagh (1994)[6] concluded from testing the aluminum pipe installed in dry sand by jacking method that the shaft capacity degraded significantly in the smaller-diameter piles when the cyclic displacement exceeded a threshold which was in the order of displacement required to completely mobilize the shaft resistance.

Jardine and Standing (2000, 2012) [7, 8] demonstrated that axial cyclic loading affects the behavior of full scale piles driven in the sands. Low–level tension axial cycling (with load cyclic amplitude $Q_{cyclic} < \sim 20\%$ the operational static shaft capacity QT) led to minimal cyclic displacements that accumulated at stabilizing rates with almost no impact on pile stiffness and potentially enhancement of the pile's static tension capacity.

Yu et al. (2006) [9] studied the behavior of jacking H-pile embedded in dense sand under cyclic loading. They showed that the effect of previous loading cycles on pile-soil stiffness and load distribution was insignificant.

Yu (2008) [10] conducted a cyclic field test on H pile installed in dense sand. He stated that the base resistance carried about 60% of the overall capacity and the shaft

* Corresponding author: Al_Neami74@yahoo.com

© The Authors, published by EDP Sciences. This is an open access article distributed under the terms of the Creative Commons Attribution License 4.0 (http://creativecommons.org/licenses/by/4.0/).

resistance was contributed mainly by the friction between the pile shaft and the completely or highly decomposed granite.

Jardine and Standing (2012) [8] reported cyclic tests on a 45.7 cm diameter of six pipe piles with open ends driven in medium to dense marine sands. One cyclic test pile had been driven to (L/D = 21.9) and others five to about (L/D = 41.6). Subsequently, most of the cyclic loading suites delimit the influences of the prior cyclic loading. Figure1 summarized the cyclic parameters investigated in a normalized (non-dimensional) axial cyclic interaction diagram whereas the three zones of response, Stable (St), Metastable (MS) and Unstable (US), were characterized that defined the piles capacity performance.

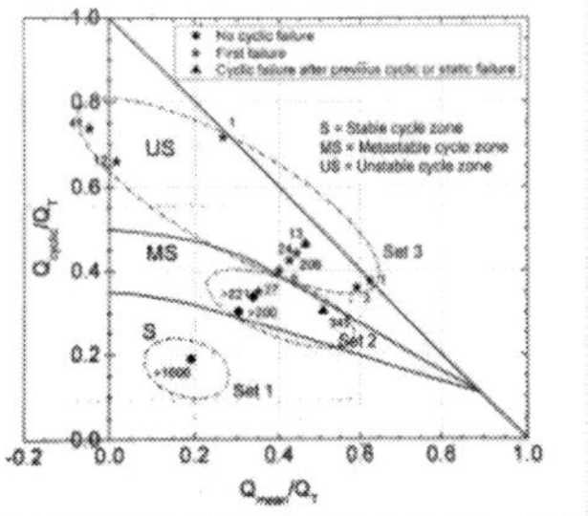


**Fig.1.** Interaction diagram on axial cyclic response of displacement piles (Jardine & Standing 2012).

Li et al. (2013) [11] studied the impact of installation methods on the jacked piles performance subjected to the cyclic vertical loads. They showed that the jacking pile is stiffer to cyclic response than bored pile. The stiffness of pile head reduces with increases the number of cycles and a bored pile showed a more settlement than jacked pile.

The main objective of this work is to study the behavior of steel piles (open ended pipe piles and H piles) installed by the jacking method in dry medium and dense sand under cyclic load and making a clear picture of the effect of amplitude load on the number of cycles.

## 2 Methodology

### 2.1 Materials used

Air-dried sand brought up from east of Baghdad in Iraq was used in this study. The soil is classified as poorly graded sand (SP) according to USCS as shown in Figure 2. Series of laboratory tests were conducted to distinguish sand accordance with ASTM Specification. Two relative densities were used (medium sand and dense sand) and the direct shear test gave that the friction angles are 31.5° and 38.7°, for medium sand (DR =60%) dense sand (DR = 80%) respectively. The outcomes of experimental testing are summarized in Table1.

Piles models used in this study are divided into groups (Figure 3); the first one is open ended steel solid pipe with diameters (2.5, 3.5, and 4.1) cm. secondly is H piles with three flange widths (2.6, 3.6, and 4.4) cm. Eighteen models are used for each model with three lengths (30, 40, and 50) cm.

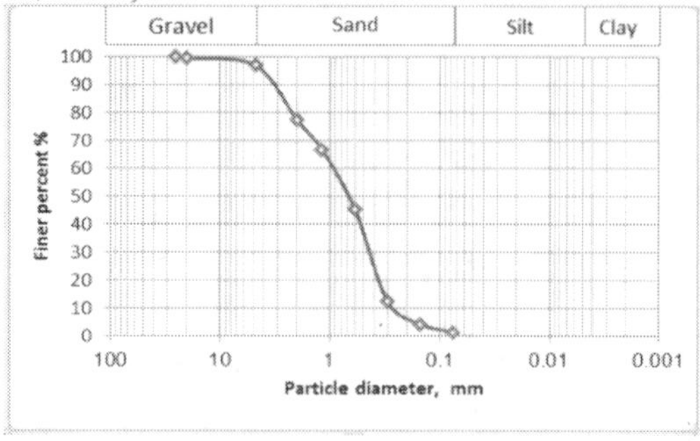


**Fig. 2.**Grain size distribution of sand used.

**Table 1.** Physical properties of the tested sand.

| Property | Value | | Specification |
|---|---|---|---|
| Effective sizes, D10, D30, D50, D60, mm | 0.29, 0.45, 0.7, 0.9 | | ASTM D2487 -11 |
| Coefficient of uniformity, Cu | 3.10 | | |
| Coefficient of curvature, Cc | 0.78 | | |
| Classification (USCS) | SP | | ASTM D2487-11 |
| Specific gravity, Gs | 2.66 | | ASTM D854 - 14 |
| Maximum unit weight, $\gamma_d$ (max.) kN/m³ | 19.4 | | ASTM D4253 - 14 |
| Minimum unit weight, $\gamma_d$ (min.) kN/m³ | 15.8 | | ASTM D4254 -14 |
| Maximum void ratio, $e_{max}$ | 0.68 | | |
| Minimum void ratio, $e_{min}$ | 0.371 | | |
| Relative density, % | Medium = 60 | Dense = 80 | |
| Dry unit weight ($\gamma_d$) kN/m³ | 17.75 | 18.5 | |
| Void ratio | 0.55 | 0.44 | |
| Angle of internal friction (φ), deg. | 31.5 | 38.7 | ASTM D3080 -11 |

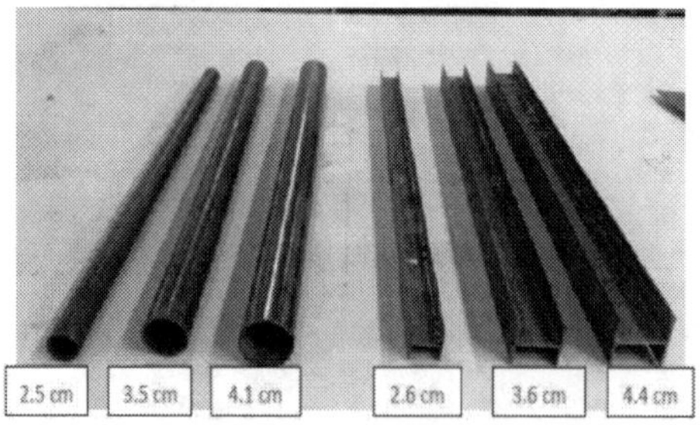


**Fig. 3.** Model piles used.

https://doi.org/10.1051/matecconf/201816201012

## 2.2 Calibration container and sample preparation

All testing of piles model is carried out on dry sand samples compacted inside a steel container has dimensions (100 cm length × 75cm width ×70 cm depth) as depicted in (Figure 4). The container has four wheels move freely on a rail fixed in the ground in form of a reversed T-section. Sand deposit was prepared as layers using a compacted solid plate hammer manufactured for this purpose with dimensions (15cm × 15cm) and thickness of 5mm connected with catcher of 30 cm in length.

According to the previous studies, the stress and displacement modes in the sand modeling are affected by sides of the soil container, additionally, due to the friction between the container walls and soil grains, the vertical stress in the sand can reduce with depth (Kraft, 1991) [12] . So, to eliminate the friction between soil grains and sides of the container; the height of the container (Hc) to its diameter (Dc) should be not more one (in this study Hc/Dc=70/75=0.93) (Tarnet, 1999) [13].

During loading process or pile installation, soil surrounding the pile will be disturbed. This disturbance is extending to the concern zone depending to method of installation and relative density of soil. Previous studies detected that this zone is ranging in between (3D to 8D) where D is the pile diameter (Robinsky and Morrison, 1964) [14]. Therefore, ($8D_{max.}$) are used ($D_{max.}$ is the maximum pile diameter used in this study).

Two values of relative densities are chosen (60% & 80%) to prepare medium and dense sand bed; this means that the weight of sand required in soil modeling is predetermined. Figure 4 explains the stages of sand preparation. Equal weights and lifts of seven layers are used to complete the soil bed; then the top layer is leveled using a ruler with sharp edges (The soil weight of each layer was compacted to desired depth which previously identified by lines drawn on the sides of container).

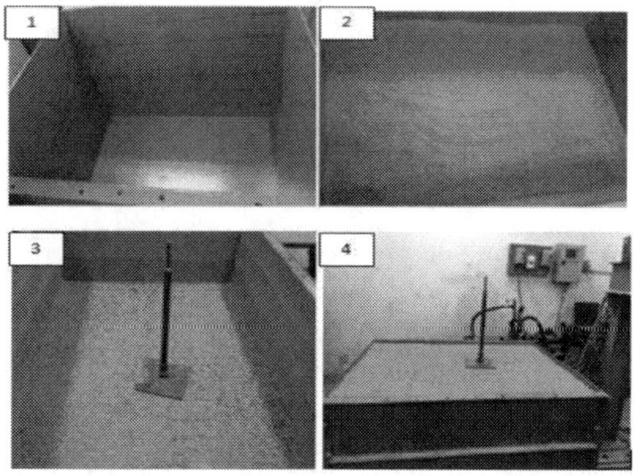


**Fig. 4.** Represents the steps of the sand deposit setup.

## 2.3 Details of pile models

After the completing the sand bed preparation inside the container, pile model is vertically inserted using a steel loading frame had been manufactured by (Ali, 2012) [15] as shown in Figure 5. So, to insure the vertically installation, pile driving system is modified by adding a base plate of (86cm × 20cm) with thickness of 2 cm. A pile holder is added above this system with different diameters (or widths of flange) depending on the pile type. For open ended piles, the pile holder is steel with dimensions (10 cm× 10 cm × 2 cm), while for H-piles, the holder is made from wood with dimensions (15 cm × 15cm × 4 cm) as shown in Figure 6.

The axial loading on model is applied by hydraulic jack having a maximum loading capacity of (1 ton) with a manually adjusted to control on intensity of hydraulic. Load cell with maximum capacity of (2 ton) is used and connected with a digital weighing indicator with an input sensitivity of (0.2 μN/Digit) to record the applied load.

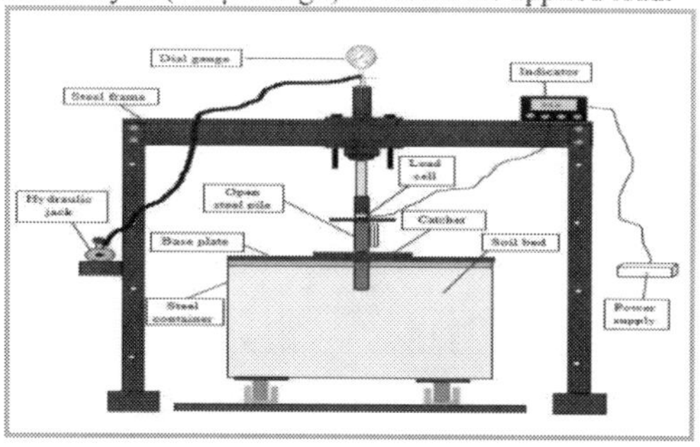


**Fig.5.** Steel loading frame.

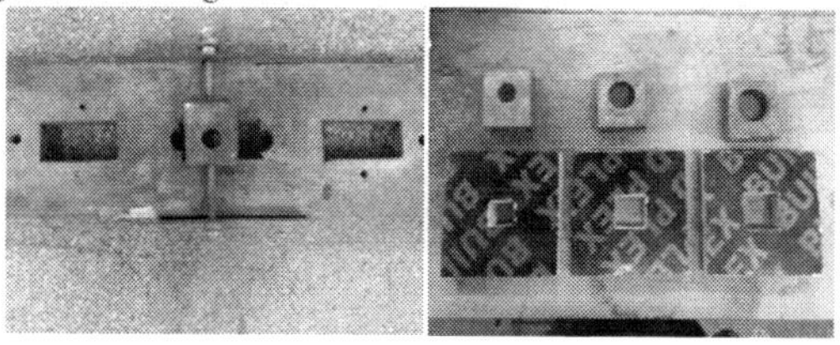

**Fig.6.** Base plates and holders.

## 2.4 Testing program

A total of 36 steel piles models are tested to attain the aim of this study. Eighteen pile models are used for each pile models (pipe piles with ends open and H-piles). To achieve the effect of amplitude load, two relative densities of sand are used (medium sand and dense sand). For each relative density, the load amplitude is chosen corresponding to (0.4) of static ultimate load capacity ($Q_{ult}$), with (2) Hz frequency for each test.

An aluminum plate using as a center guide on the top of the pile was fixed, then the container moved along the rails and fixed in position in such a manner that the center of the aluminum base plate of the piston coincided with the center of the pile model. The static loading was applied gradually through the hydraulic jack that actuating at a controlled displacement of 0.03mm/sec. The load was measured until the developed settlement exceeded 15% of the pile diameter according to ASTM D1143M-07 (2013).

A special testing apparatus manufactured by Rahil (2007) [16] was used (Figure 7). The apparatus has the capability of applied monotonic and cyclic loads. The cyclic loading was applied in a sequence determined beforehand (40% of static load). Tests were leading to at (2) Hz frequency according to the loading value. The process of the loading was continued till the failure occurred or until the recorded settlement exceeded 15% of the pile diameter according to ASTM D1143M-07 (2013), the testing model device.

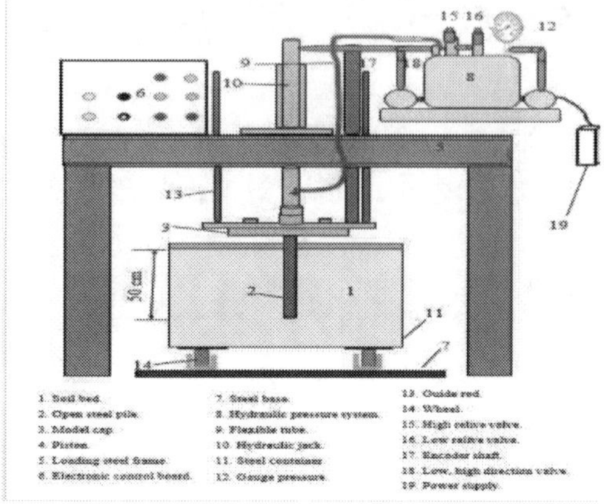

**Fig.7.** Testing model.

## 3 Model Pile Test Results

### 3.1 Load -Settlement Relation

All cyclic tests are depending on the ultimate load ($Q_{ult}$) values obtained from the static tests, therefore, a static test is carried out to concern the magnitude of cyclic loading needing to employ in cyclic testing of the piles models. The load-settlement curves for all piles are illustrated in Figures (8) to (13) and Table 2 represent a summary of the results. It can be see that H-pile carrying a load capacity greater than that of the open ended pipe pile about (13- 136) % and also the settlement of an open ended pipe pile is greater than that of H-pile under the same load and soil conditions. This means that the H-piles have a more resistance than open ends pipe piles.

### 3.2 Embedded Length of the Pile and Number of Cycles Relation

The form of the cyclic loading applied on piles model is of the forming close to the sinusoidal compressive type. All cyclic tests are depending on the ultimate load ($Q_{ult}$) values obtained from the static tests. In cyclic loading the number of cycles of failure is taken as cycles required to causing 15% settlement of the pile diameter at any applied load.

Figures (14) and (15) depict the change of number of cycles with the embedded length of the piles. It can be observed that the number of cycles is linearly changing with embedded length of pile, whereby number of cycles decrease with increasing the length of pile and relative density for the same pile diameter to both piles type (open ended pipe and H-piles) that resulting due to increase in applied load.

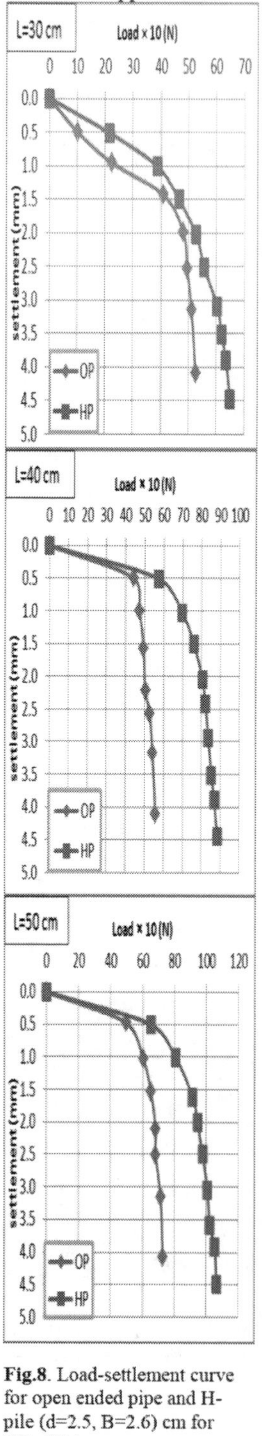


**Fig.8**. Load-settlement curve for open ended pipe and H-pile (d=2.5, B=2.6) cm for DR=60%.

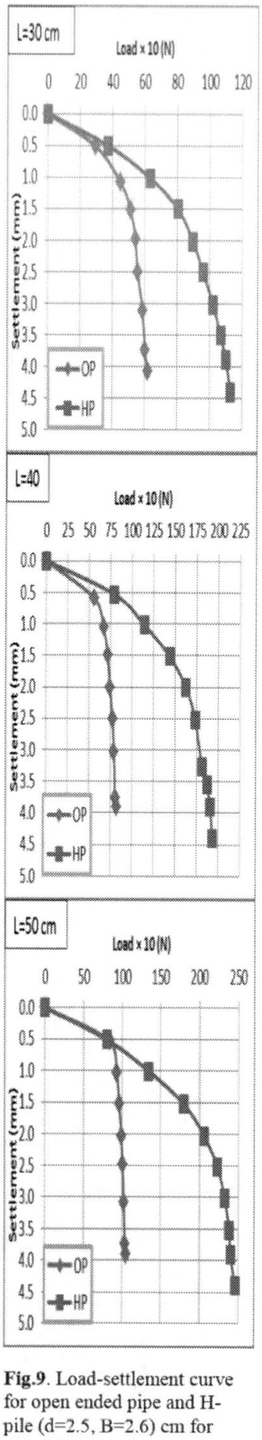


**Fig.9**. Load-settlement curve for open ended pipe and H-pile (d=2.5, B=2.6) cm for DR=80%.

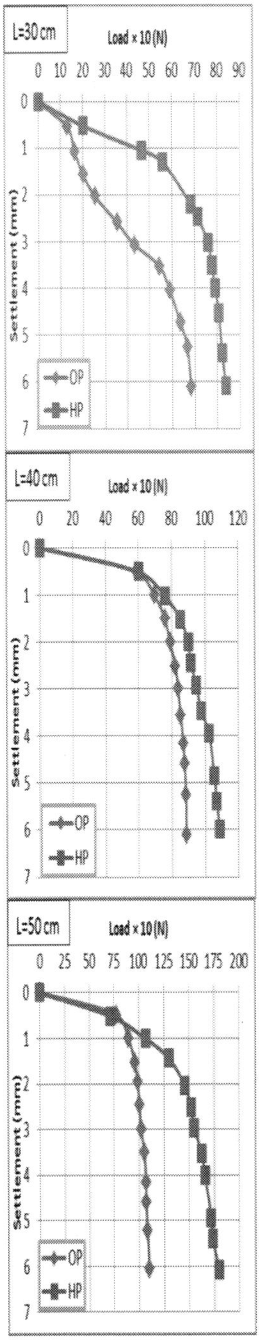


**Fig. 10**. Load-settlement curve for open ended pipe and H-pile (d=3.5, B=3.6) cm for DR=60%.

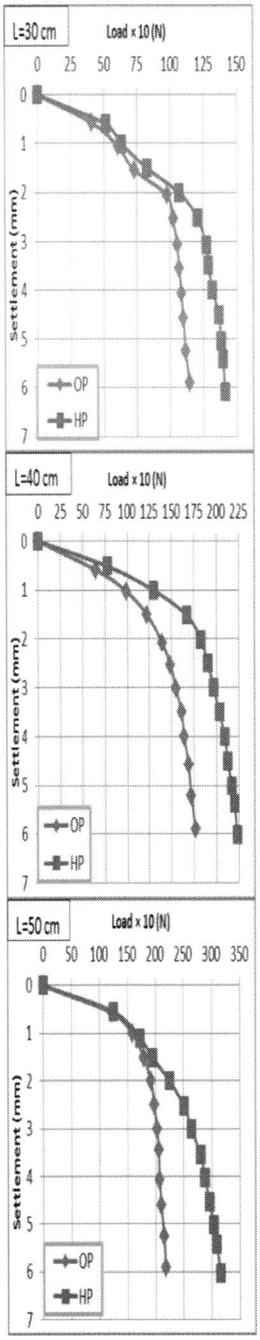


**Fig. 11**. Load-settlement curve for open ended pipe and H-pile (d=3.5, B=3.6) cm for DR=80%.

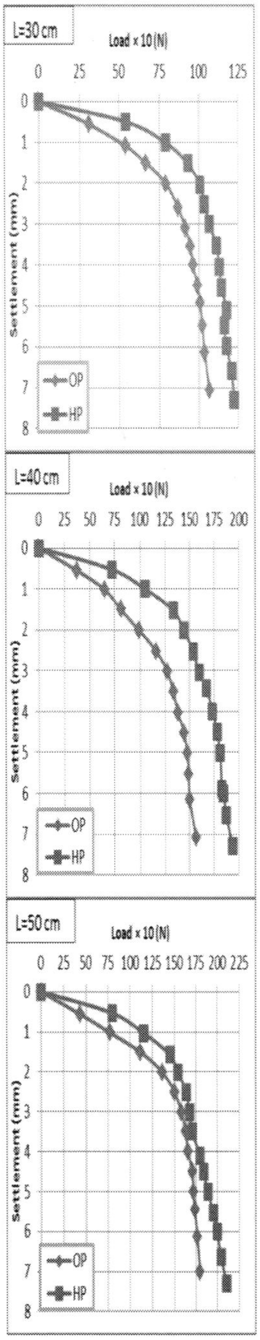


**Fig. 12**. Load-settlement curve for open ended pipe and H-pile (d=4.1, B=4.4) cm for DR=60%.

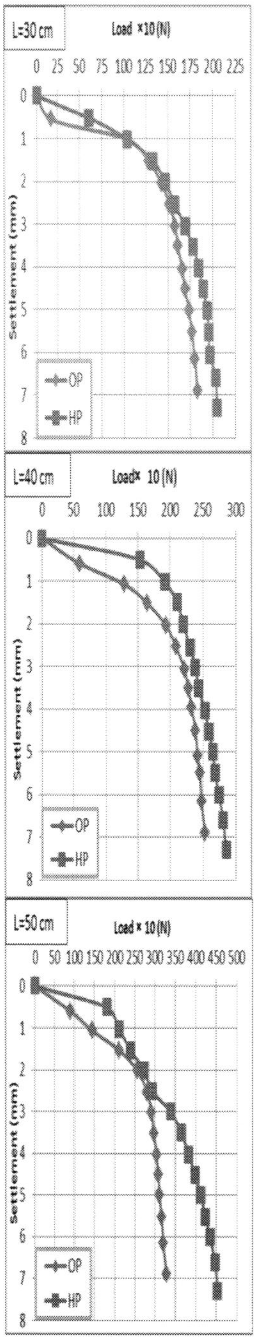


**Fig. 13**. Load-settlement curve for open ended pipe and H-pile (d=4.1, B=4.4) cm for DR=80%.

**Table 2**. Ultimate load capacity of open ended pipe and H pile.

| Pile diameter (d) Flange width (B) (cm) | Length of pile, L (cm) | Qult (N) Open pile | Qult (N) H pile |
|---|---|---|---|
| DR=60% | | | |
| 2.5<br>2.6 | 30 | 511 | 635 |
| | 40 | 543 | 868 |
| | 50 | 713 | 1050 |
| 3.5<br>3.6 | 30 | 666 | 821 |
| | 40 | 883 | 1069 |
| | 50 | 1080 | 1736 |
| 4.1<br>4.4 | 30 | 1038 | 1209 |
| | 40 | 1500 | 1875 |
| | 50 | 1767 | 2040 |
| DR=80% | | | |
| 2.5<br>2.6 | 30 | 605 | 1100 |
| | 40 | 805 | 1906 |
| | 50 | 1040 | 2402 |
| 3.5<br>3.6 | 30 | 1116 | 1395 |
| | 40 | 1720 | 2200 |
| | 50 | 2160 | 3084 |
| 4.1<br>4.4 | 30 | 1800 | 2030 |
| | 40 | 2480 | 2805 |
| | | | |
| | 50 | 3208 | 4464 |

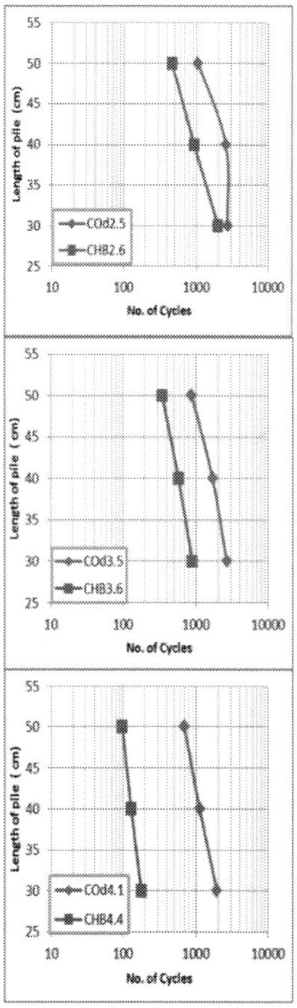


**Fig. 14.** Relationship between length of pile and No. of cycles for DR=60% under (40%$Q_{ult}$).

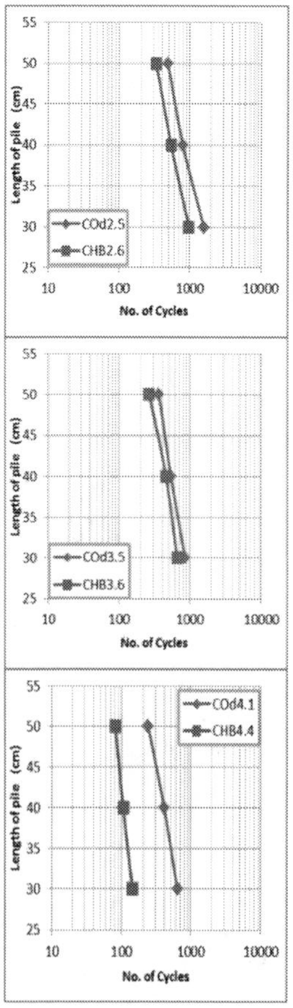


**Fig. 15.** Relationship between length of pile and No. of cycles for DR=80% under (40%$Q_{ult}$).

### 3.3 Pile Diameter and Number of Cycle Relation

Increasing of pile diameter and relative density lead to the decreasing in number of cycles when the lengths of open ended pipe and H-piles are fixed as shown in Figures 16 and 17. The variation of open pipe pile diameter has a small effect on number of cycles because the most shaft resistance is developed during installation process; also the displacement rapidly accumulates over just a few cycles. In H-Pile, a sharply changing is occurred when the width of flange decreased (from 4.4 m to 3.6 m) because the displacements accumulate slowly, then this displacement accumulate rapidly. For the same pile length of both types of piles, the number of cycles decreased with increasing of pile diameter and relative density due to increase in applied load with increase in pile diameter and relative density at static load.

### 3.4 Cycles Number – Settlement Curve

The outcomes of number of cycles and settlement are illustrated in Figures (18) to (23). It can be observed that number of cycles decreased with increasing in the load amplitude, relative density and embedded length of piles for the same pile diameter.

## 4 Conclusions

The conclusions obtained from testing 36 models of piles under cyclic testing can list below:

1. H-Pile afforded a load capacity of about (13 – 136) % greater than that of the open ended pipe piles under the same geometric conditions.
2. Pile geometry (diameter and length) and sand density have highly effect on the number of cycles. The variation of opened end pipe pile diameter has a small effect on number of cycles while decreasing the flange width of H-Pile has more impact on increasing number of cycles.
3. Increasing of pile diameter (flange width) and relative density lead to the decreasing in number of cycles when the lengths of open ended pipe and H-piles are fixed.
4. The number of cycles of the open ends piles is greater than that of the H- piles under the same geometric conditions (pile diameter, embedded length and sand density) especially in medium sand.

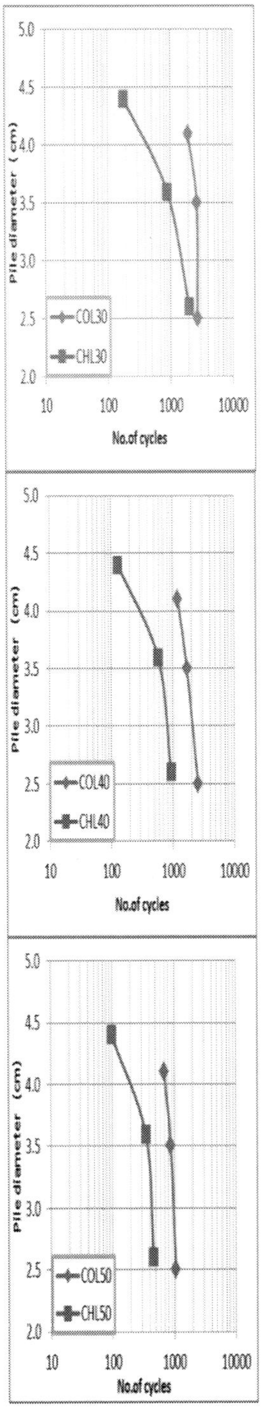


**Fig.16.** Relationship between length of pile and No. of cycles for DR=60% under (40%$Q_{ult}$).

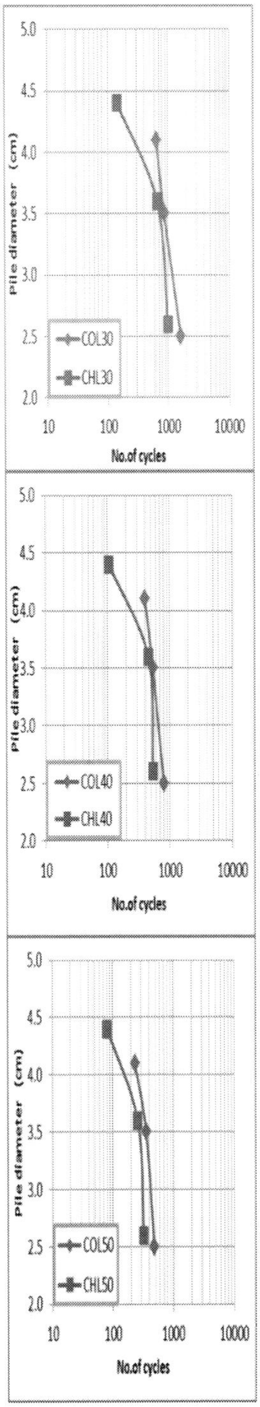


**Fig.17.** Relationship between length of pile and No. of cycles for DR=60% under (40%$Q_{ult}$).

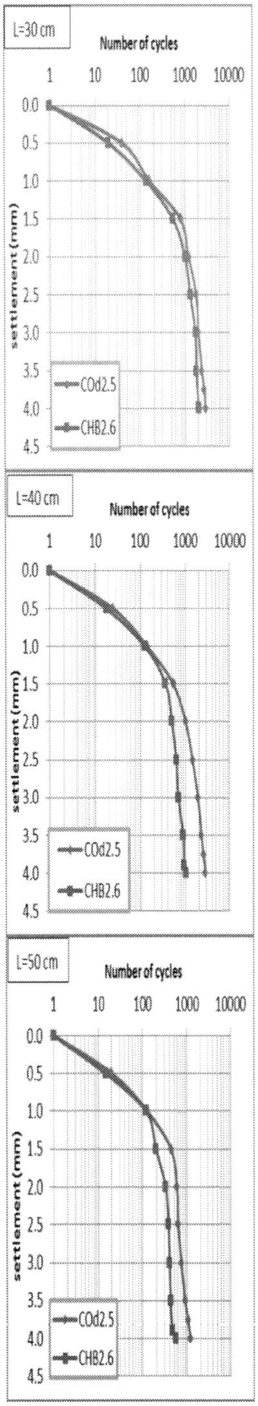


**Fig.18.** Number of cycles-settlement relation for DR=60% (d=2.5, B=2.6) cm.

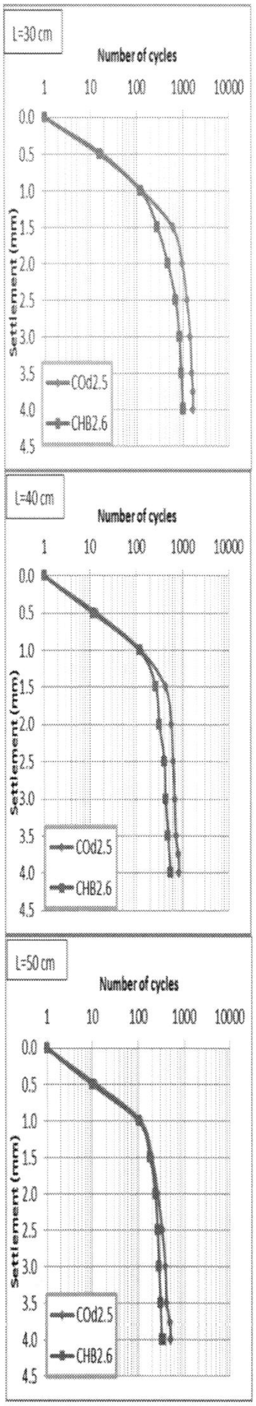


**Fig.19.** Number of cycles-settlement relation for DR=80% (d=2.5, B=2.6) cm.

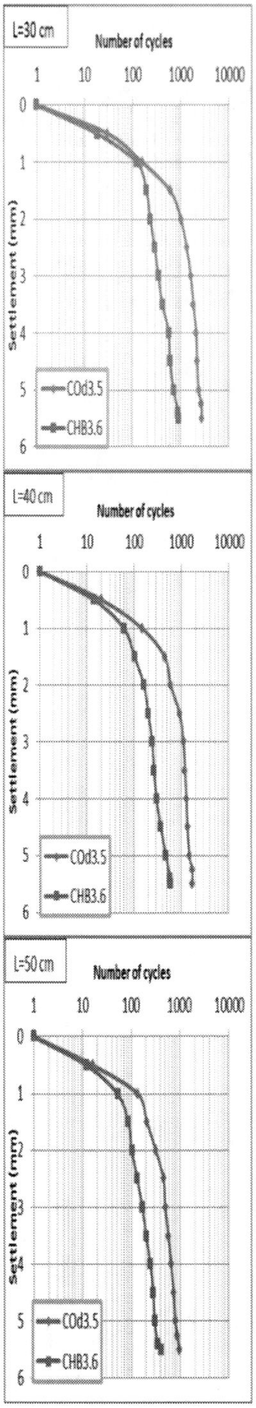


**Fig. 20.** Number of cycles-settlement relation for DR=60% (d=3.5, B=3.6) cm.

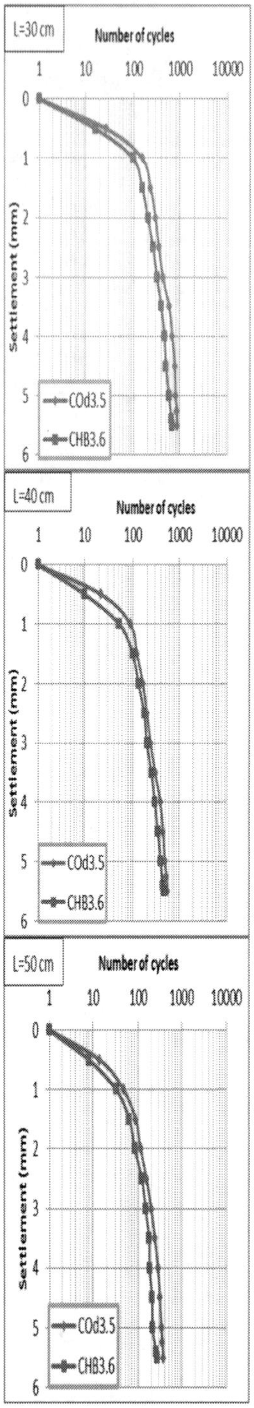


**Fig. 21.** Number of cycles-settlement relation for DR=80% (d=3.5, B=3.6) cm.

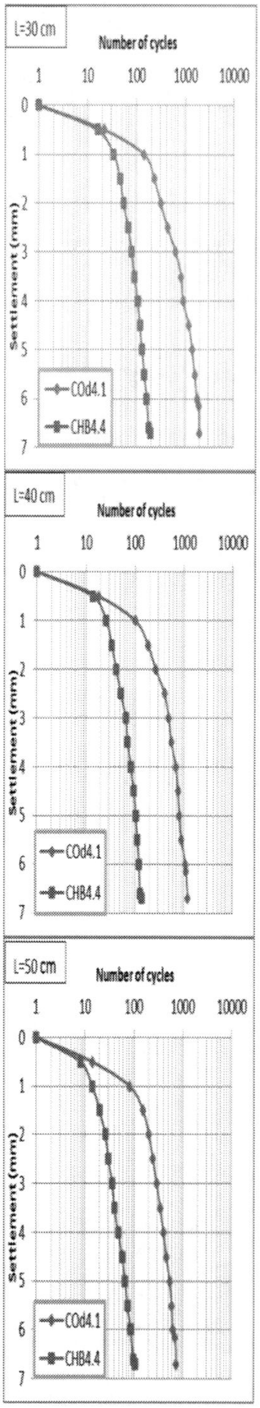


**Fig. 22.** Number of cycles-settlement relation for DR=60% (d=4.1, B=4.4) cm.

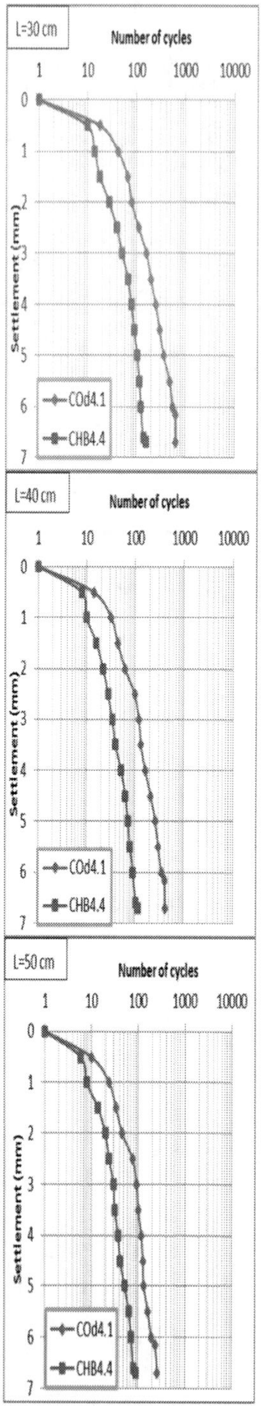


**Fig. 23.** Number of cycles-settlement relation for DR=80% (d=4.1, B=4.4) cm..

## References

1. C. G. di Prisco and D. M. Wood, Springer Vienna, (2012).
2. A. A. Puech, M. Boul on and Y. Meimon, Proc. 2nd Int. conf. on numerical methods in offshore piling 293-312, (1982).
3. H. G. Poulos, *Analysis and design of pile Foundations*, ASCE, 99- 118, (1984).
4. H. G. Poulos, Geot. 39, No. 3, 365 – 41, (1989a).
5. H. G. Poulos, J. Geo. E., ASCE, **115, No.GT6**, 836 – 852, (1989b).
6. W. H. Craig and S. K. Sabagh, Can. Geot. J. **31**, 28-41 (1994).
7. R. J. Jardine and J. R. Standing, **1 & 2.** Offshore Tech. Report, Health and Safety Executive, London: 60 and 200, (2000).
8. R. J. Jardine and J. R. Standing, Soils and Foundations. **52, No. 4**, 723 – 736, (2012).
9. F. Yu, J. Yang, LG Tham, P.K.K. Lee, ST, Geot., **56 No.. 4**, 245-259,(2006).
10. F. Yu, **14**, Bund. A. EJGE. (2008).
11. Z. Li, M. D. Bolton, and S. K. Haigh, Can. Geot. J. **49**: 1074–1087 (2012).
12. L.M. Kraft, J. Geot. E., ASCE, **117 (2)**, 272-296, (1991).
13. O. Ternet, Ph.D. thesis, University of Caen, 184. (1999).
14. E. I. Robinsky, W. L. Sagar, and C. F. Morrison, Can. Geot. J., **1(4)**, 189-204, (1964).
15. A. M. Ali, (2012), M.Sc. Thesis, University of Technology, Iraq, (2012).
16. F. H. Rahil, PhD. Thesis, University of Technology, Iraq, (2007).

MATEC Web of Conferences **162**, 01013 (2018) https://doi.org/10.1051/matecconf/201816201013
*BCEE3-2017*

# Stability analysis of roadway embankments supported by stone columns with the presence of water table under short-term and long-term conditions

*Shaymaa* Tareq Kadhim[1,*], *Ziad* Bashar Fouad[1]

[1]Building and Construction Engineering Department, University of Technology, Baghdad, Iraq

**Abstract.** Use of stone column technique to improve soft foundation soils under roadway embankments has proven to increase the bearing capacity and reduce the potential settlement. The potential contribution of stone columns to the stability of roadway embankments against general (i.e. deep-seated) failure needs to be thoroughly investigated. Therefore, a two-dimensional finite difference model implemented by FLAC/SLOPE 7.0 software, was employed in this study to assess the stability of a roadway embankment fill built on a soft soil deposit improved by stone column technique. The stability factor of safety was obtained numerically under both short-term and long-term conditions with the presence of water table. Two methods were adopted to convert the three-dimensional model into plane strain condition: column wall and equivalent improved ground methods. The effect of various parameters was studied to evaluate their influence on the factor of safety against embankment instability. For instance, the column diameter, columns' spacing, soft soil properties for short-term and long-term conditions, and the height and friction angle of the embankment fill. The results of this study are developed in several design charts.

## 1 Introduction

The most common type of slope instability that is more likely to happen when constructing roadway embankments on soft soil deposits is the global or deep-seated failure. Various ground improvement techniques were proposed to avoid this catastrophic failure; for instance: deeply mixed columns, stone columns, and sand compaction columns [1], [14]. Stone column technique has proven its effectiveness to overcome the embankment instability problem when constructed over a soft soil foundation. Stone columns possess the ability to improve the bearing capacity, minimize settlement, and enhance stability and drainage of improved ground [2], [5], [8], [11], [12], [14].

Embankments built on soft clay soils can fail due to slope instability in local, surficial, toe slope, and deep-seated failure [6], as shown in Figure 1. However, the modes of failure of stone column-supported embankments are more complex since columns can fail in different modes of failure like: rotation, bending, shearing, and sliding [8]. Slope stability analysis can be performed using Limit Equilibrium Method (LEM), Finite Difference Method (FDM), and Finite Element Method (FEM) [4], [7]. Finite Difference Method offers several advantages over Limit Equilibrium Method in computing the factors of safety of slopes like: no assumption should be made for inter-slice forces, the possibility to have different failure surfaces with different shapes and the inclusion of structural elements to simulate different materials rather than assuming equivalent forces [9]. In addition, it was hypothesized that the factor of safety of slope stability computed from Finite Difference Method was 20% lower than that resulted from Limit Equilibrium Method [3]. Reference [6] concluded that the Limit Equilibrium Method overestimates the stability factor of safety of deep-mixed column-supported embankments when compared to Finite Difference Method due to the failure of columns in rotation, tension, or bending.

Reference [1] investigated the stability of embankments supported by granular columns in the undrained (i.e. short-term) condition by means of two methods of equivalency; column wall and equivalent improved ground methods. They recommended a reduction factor of 0.9 for the factor of safety obtained from the equivalent improved ground method if a water table is not considered and 0.92 if the water table is taken into consideration.

In this paper, finite difference numerical analysis was adopted to assess the stability of stone column-supported embankment using FLAC/SLOPE 7.0 software, released

* Corresponding author: shaimakadhim@gmail.com

© The Authors, published by EDP Sciences. This is an open access article distributed under the terms of the Creative Commons Attribution License 4.0 (http://creativecommons.org/licenses/by/4.0/).

by ITASCA Consulting Group. This paper aims to evaluate the stability of stone column-supported embankments in both short-term and long-term conditions with taking into consideration the influence of water table.

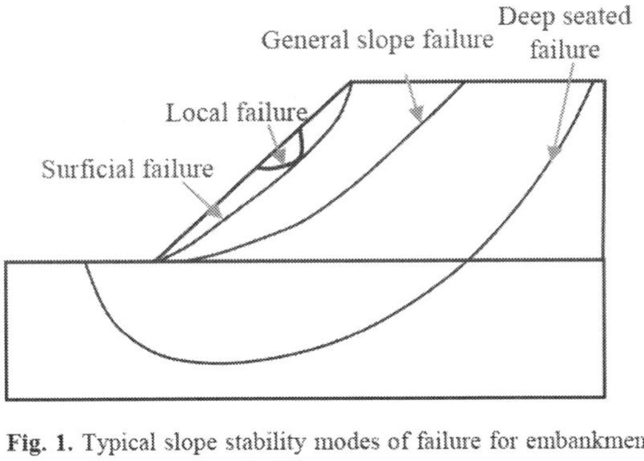


**Fig. 1.** Typical slope stability modes of failure for embankments constructed over soft soils [6].

## 2 Two-dimensional numerical methods

Due to the complexity of three-dimensional (3D) model of embankments, two methods were used to convert that model into plain strain condition (2D), which are: column wall method and equivalent improved ground method.

### 2.1 Column wall method

Reference [13] proposed this method by making an assumption that the width of the column wall is the same as the column diameter as illustrated in Figure 2.

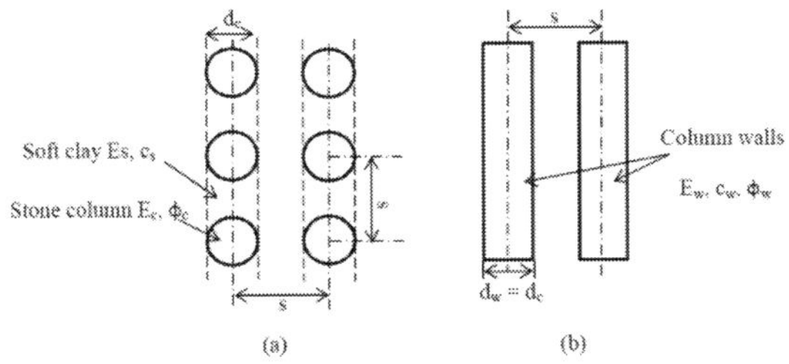


**Fig. 2.** The configuration of stone columns installed in soft clay soil: (a) Individual columns; and (b) Column walls [10].

The shear strength properties of the column wall can be determined as follows [14]:

$$c_w = c_s\,(1 - a_r) \quad (1)$$

$$\phi_w = \arctan(a_r \tan\phi_c + (1 - a_r)\tan\phi_s) \quad (2)$$

$$a_r = \frac{A_c}{d*s} \quad (3)$$

in which $c_w$ and $c_s$ = the cohesion of column wall and surrounding soil deposit, respectively; $a_r$ = the area replacement ratio with respect to the column wall; $\phi_w$, $\phi_s$ and $\phi_c$ = the angle of internal friction of the column wall, soft soil and individual stone column, respectively; Ac= cross-sectional area of individual stone column; d= stone column diameter; s= center to center spacing between columns.

### 2.2 Equivalent Improved ground method

The assumption of this method is based on that both the granular columns and the surrounding soft soil form a composite foundation. The equivalent shear strength parameters of the improved composite foundation are calculated by the following equations [1]:

$$c_{eq} = c_s\,(1 - a_s) \quad (4)$$

$$\phi_{eq} = \arctan(a_s \tan\phi_c + (1 - a_s)\tan\phi_s) \quad (5)$$

$$a_s = \frac{\pi}{4}\left(\frac{d}{s}\right)^2 \quad (6)$$

in which $c_{eq}$ and $c_s$ = the cohesion of equivalent improved ground and soft soil deposit, respectively; $\phi_{eq}$, $\phi_s$ and $\phi_c$ = the angle of internal friction of equivalent ground, soft soil and individual stone column, respectively; $a_s$ = the area replacement ratio of composite foundation. The granular columns were assumed to be installed in a square pattern in this research.

## 3 Two-dimensional numerical modeling

The parameters and dimensions of the stone column-supported embankment for this study were selected from [1] excluding column diameter and center to center spacing. Figure 3 demonstrates the model geometry and mesh for column wall and equivalent improved ground methods. Half of the embankment geometry was adopted in this study utilizing FLAC/Slope 7.0 because of the model symmetry. The composite ground comprised of a soft clay layer of a 10 m thickness placed on top of a 2 m sand layer. The column diameter and length were chosen to be 0.5 m and 10 m, respectively, according to the typical stone column size and length used in practice. The diameter to spacing ratio was assumed to be 20% as spacing measured from center to center of two adjacent columns was equal to 2.5 m. The embankment dimensions were selected with a 5 m height, a crest width of 10 m and a slope equal to 2H:1V. To prevent surficial failure, an exclude area function incorporated in FLAC/Slope 7.0 was activated to ensure that the failure surface did not penetrate a zone thickness of about 0.5 m located at the slope surface. The bottom boundary was restricted in both horizontal and vertical directions since the sand layer was placed above a firm layer of a bedrock. Furthermore, the boundaries on the two sides of the model were fixed in the horizontal direction and permitted to move freely in the vertical direction.

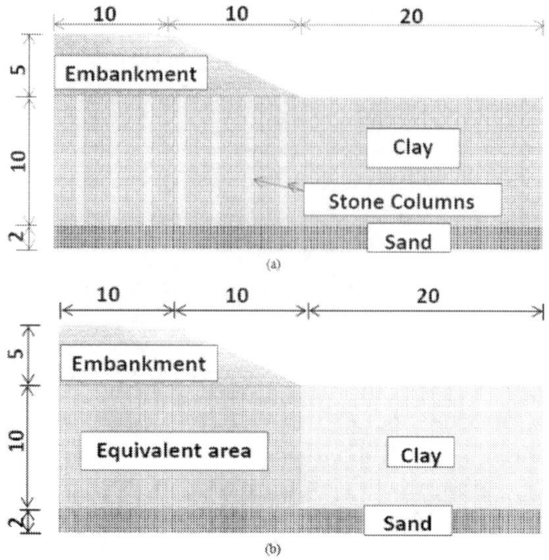


**Fig. 3.** FLAC models (unit: m): (a) Column wall; and (b) Equivalent improved ground.

The soft foundation soil, stone columns, and embankment fill were modeled as linearly elastic perfectly plastic materials with the Mohr-Coulomb failure criteria. Table 1 demonstrates the material properties of embankment fill, foundation soil, and stone columns in short-term condition (i.e. undrained condition), while Table 2 depicts the material properties in long-term condition (i.e. drained condition).

## 4 Results and discussion

The stability of embankments constructed on stone column-improved ground was explored in short-term and long-term conditions in the presence of water table via two methods of equivalency: column wall and equivalent improved ground. Various parameters were investigated in this research as follows:

### 4.1 Diameter of stone column

Figure 4 shows the effect of column diameter on embankment stability in short-term and long-term conditions. Three diameters were chosen for this study: 0.25 m, 0.5 m, and 0.75 m, which simulated diameter to spacing ratios equal to 10%, 20%, and 30%, respectively using a 2.5 m center to center spacing. It is evident that the factors of safety for the two aforementioned methods and for the short-term condition were increased when using larger sizes of stone columns.

**Table 1.** Properties of embankment, soft soil and stone column for a 0.5 m column diameter and a 2.5 m center to center spacing for short-term condition.

| Item | Height (m) | Cohesion (kPa) | Friction angle (Deg.) | Unit weight (kN/m$^3$) |
|---|---|---|---|---|
| Embankment | 5 | 0 | 32 | 18 |
| Clay | 10 | 20 | 0 | 16 |
| Sand | 2 | 0 | 30 | 18 |
| Column | 10 | 0 | 38 | 17 |

**Table 2.** Properties of embankment, soft soil and stone column for a 0.5 m column diameter and a 2.5 m center to center spacing for long-term condition.

| Item | Height (m) | Cohesion (kPa) | Friction angle (Deg.) | Unit weight (kN/m$^3$) |
|---|---|---|---|---|
| Embankment | 5 | 0 | 32 | 18 |
| Clay | 10 | 5 | 25 | 16 |
| Sand | 2 | 0 | 30 | 18 |
| Column | 10 | 0 | 38 | 17 |

In short-term condition, the factor of safety values calculated from the equivalent improved ground method were greater than those computed from the column wall method. This may be attributed to the fact that the equivalent improved ground method provides continuous failure surface in such a way that every point located on that slip surface reached its shear strength at the same time unlike the column wall method where no continuous slip surface was observed and not all the points mobilized their shear strengths [14]. While, the computed factors of safety for both methods were identical for long-term condition. This is consistent with reference [14] findings in long-term condition since all failure points located on the slip surface have mobilized their shear strengths at the same time. For long-term condition, the effect of increasing the column diameter had no influence on the factor of safety values since the effective cohesion and effective friction angle of soft clay were introduced in long-term case so that the slip surface would not enter the strengthened ground zone. In addition, short-term condition yielded lower factors of safety than long-term condition.

### 4.2 Spacing between stone columns

The impact of the center to center spacing on the factor of safety in both short-term and long term is depicted in Figure 5. The chosen values of center to center spacing were varied from 1.67 m to 5 m to match diameter to spacing ratios from 30% to 10% [1]. A 0.5 m column diameter was used for all models. The factors of safety were found to decrease with the increase in stone column spacing for the two methods in short-term condition. As discussed earlier, higher values of factor of safety were

 https://doi.org/10.1051/matecconf/201816201013

obtained from the equivalent improved ground method compared with the column wall method in short-term condition. Whereas for long-term condition, the effect of column spacing was insignificant because the stability failure is dependent on the embankment material properties rather than the improved ground properties.

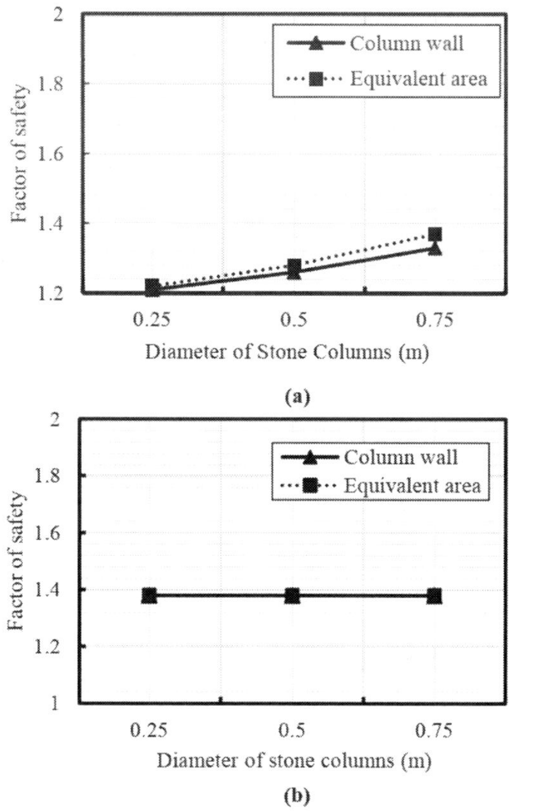


**Fig. 4.** Influence of the stone column diameter: (a) short-term condition; and (b) long-term condition.

### 4.3 Cohesion of soft soil

Figure 6 depicts the effect of soft soil cohesion on the factor of safety for both column wall and equivalent improved ground methods in short-term condition only (i.e. undrained cohesion). It is evident that the factor of safety is significantly increased once the soil cohesion changes in a range from 10 to 25 kPa. However, the effect of increasing the cohesion greater than 25 kPa has a minimal influence on the computed factor of safety by the two methods because the strengthened soft clay soil was able to prevent the slip surface from developing within the foundation area and to generate within the embankment zone instead. Also, the factors of safety obtained from equivalent improved ground method were greater than those resulted from column wall method because of the continuous slip surface developed in equivalent area method.

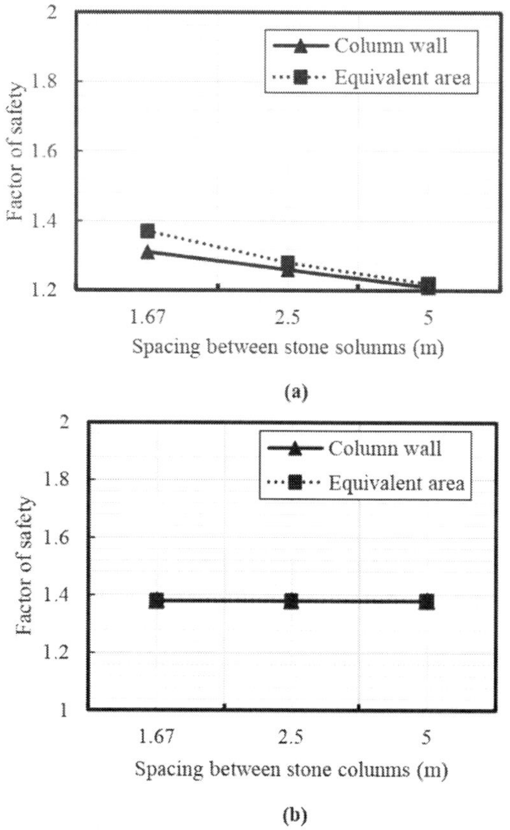


**Fig. 5.** Influence of center to center spacing between granular columns: (a) short-term condition; and (b) long-term condition.

### 4.4 Friction angle of soft soil

Figure 7 illustrates the factor of safety against the angle of internal friction of the soft clay soil in long-term condition while keeping the effective cohesion equal to 5 kPa for all cases. There is an increase in the factor of safety with the increase of the friction angle of soft soil for both column wall and equivalent improved ground methods. While, increasing the friction angle beyond a 20° had an insignificant effect since the strengthened foundation ground had prevented the slip surface from penetrating throughout the composite ground. In other words, the failure surface was influenced by the embankment properties only. The computed factors of safety using column wall method and the equivalent improved ground method were identical in long-term condition as mentioned earlier.

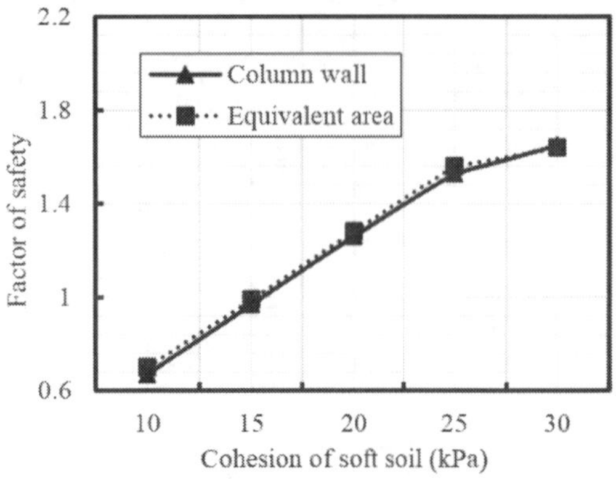


**Fig. 6.** Effect of soft soil cohesion for the short-term condition.

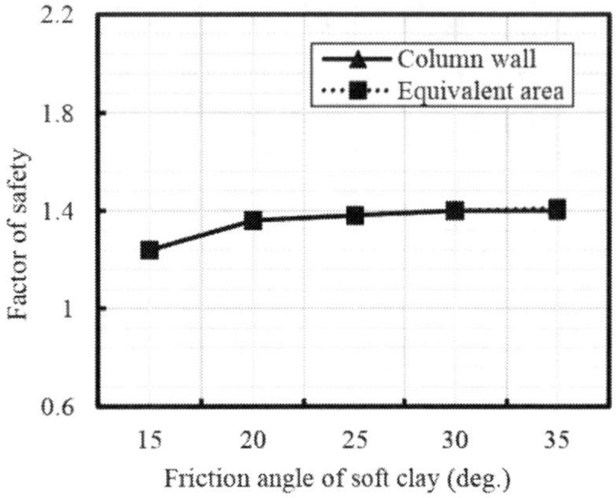


**Fig. 7.** Effect of friction angle of soft soil for the long-term condition.

### 4.5 Height of fill embankment

Figure 8 shows the effect of embankment height on the embankment stability for both short-term and long-term conditions represented by column wall and equivalent improved ground methods. It is obvious that increasing the height of embankment from 2 m to 8 m resulted in a remarkable decrease in the values of factor of safety for both short-term and long-term conditions. Equivalent improved ground method generated slightly higher values of factor of safety than the column wall method in short-term condition. Similarly, the factors of safety computed from aforementioned methods were identical in long-term condition.

### 4.6 Friction angle of fill embankment

In this study, the material of embankment was modeled as a cohesionless material. Figure 9 depicts the effect of the friction angle of the embankment on the stability factor of safety using the column wall method and the equivalent improved ground method in short-term and long-term conditions. In the short-term, the influence of improving the quality of embankment fill (i.e. increasing the friction angle) had an insignificant effect on the embankment stability if the friction angle increased above 30°. This finding is supported by reference [1] who concluded that increasing the friction angle of the embankment material beyond 30° had no influence on the stability factor of safety. This may be associated to that the potential slip surface penetrated deeper throughout the composite foundation and highly influenced by the foundation contribution rather than the embankment material properties. Whereas in long-term condition, there was a continuous increase in the factor of safety values as the angle of the internal friction of the embankment increased.

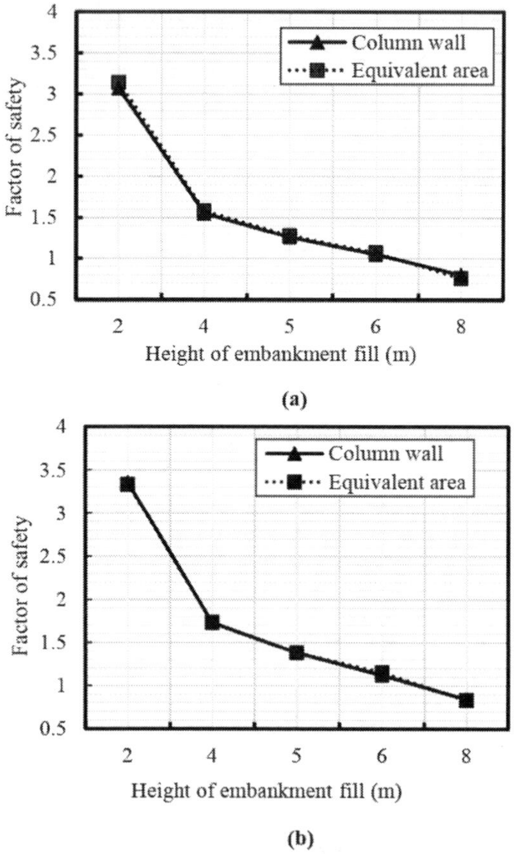


**Fig. 8.** Effect of embankment height: (a) short-term condition; and (b) long-term condition.

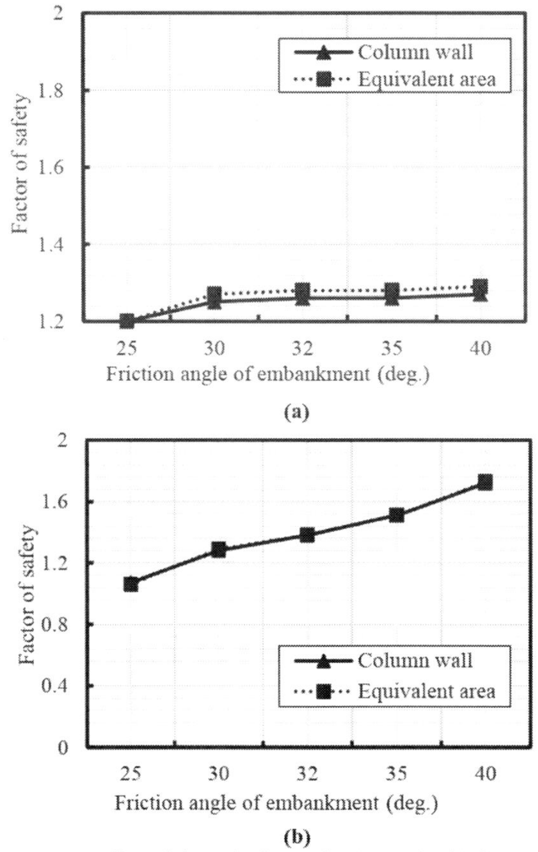


**Fig. 9.** Effect of the embankment friction angle: (a) short-term condition; and (b) long-term condition.

## 5 Conclusions

In order to assess the factor of safety against general deep-seated failure for embankments constructed on granular column-improved ground with the presence of water table, a two-dimensional finite difference numerical analysis was performed using FLAC/SLOPE 7.0 software. A parametric study was conducted in short-term and long-term conditions. Some conclusions might be drawn from this research:

1. The stability of stone column-supported embankments was investigated in short-term and long-term conditions. The results of the numerical analysis showed that the stability factors of safety of the embankment in short-term condition were less than those computed in long-term condition since the short-term condition is considered the worst case scenario in the field.
2. The factors of safety obtained from equivalent improved ground method were found to be higher than those from the column wall method in short-term condition because of the continuous failure surface that was developed in the equivalent improved ground method as compared with the column wall method.
3. The stability factors of safety of embankments constructed on the stone column-improved ground were identical when using column wall and equivalent improved ground methods in long-term condition. This is mainly attributed to that all the points located on the potential slip surface have reached their shear strengths at the same time.
4. A considerable decrease in the stability factors of safety occurred after increasing the spacing measured from center to center between ordinary stone columns. While the embankment stability was improved when using stone columns of larger diameters for short-term condition.
5. The factor of safety considerably increased when the cohesion of the soft soil was less than 25 kPa for short-term condition and when the angle of the internal friction of the soft soil was less than 20° for long-term condition because the slip surface could not penetrate the strong composite ground.
6. Increasing the height of the embankment caused a significant decrease in the stability factor of safety for both short-term and long-term conditions.
7. No remarkable effect on the factor of safety was noticed when increasing the friction angle of the embankment beyond 30° in short-term condition. For long-term condition, improving the quality of embankment fill (i.e. using a material with the higher angle of the internal friction) had a substantial role in the stability of stone column-supported embankment.

## References

1. S. Abusharar, J. Han, Eng. Geo., **120,** 103–110, (2011)
2. A. P. Ambily, S. R. Gandhi, ASCE J. Geotech. Geoenviron. Eng., **133**, 405–415, (2007)
3. M. Cala, J. Flisiak, *FLAC and Numerical Modeling in Geomechanics*, Ecully-Lyon, France, (2001)
4. Y.M. Cheng, C.K. Lau, *Slope Stability Analysis and Stabilization: New Methods and Insight*, Routledge, UK, (2008)
5. J. Han, S. L. Ye, *1st Young Asian Geotechnical Engineers Conference*, Thailand, 197-206, (1991)
6. J. Han, R. L. Parsons, A.R. Sheth, J. Huang, *Deep Mixing Conference*, Sweden, **1.2**, 231–236, (2005)
7. J. Han, D. Leshchinsky, *8th Int. Geosynthetics Conference*, 1347–1350, Yokohama, Japan, (2006)
8. J. Han, *Ground Improvement and Ground Control*, **1**, 99–113, (2012)
9. ITASCA, FLAC/SLOPE User's Guide Version 6.0, Itasca Consulting Group, Inc., License, (2008)
10. S. Kadhim, R. L. Parsons, J. Han, *IFCEE*, 2318-2327, (2015)
11. S. R. Lo, R. Zhang, J. Mak, Geotext. Geomembr., **28**, 292-302, (2010)
12. D. Mc Kelvey, V. Sivakumar, A. Bell, J. Graham, Geotech. Eng., **157,** 137–149, (2004)

13. S.A. Tan, S. Tjahyono, K.K. Oo, Geotech. Geoenviron. Eng., **134**, 185–194, (2008)
14. Z. Zhang, J. Han, G. Ye. Eng. Geology, **168**, 104-113, (2014)

# Prediction of soil water characteristic curve using artificial neural network: a new approach

*Abdul-Kareem* Esmat Zainal[1,*] and *Shaimaa* Hasan Fadhil[2]

[1]College of Engineering, Baghdad University, Baghdad, Iraq
[2]College of Engineering, Al-Mustansiriyah University, Baghdad, Iraq

**Abstract.** Soil–Water Characteristic Curve (SWCC) is an important relationship between matric suction and volumetric water content of soils especially when dealing with unsaturated soil problems, these problems may include seepage, bearing capacity, volume change, etc. where the matric or total suction may have a considerable effect on unsaturated soil properties. Obtaining an accurate SWCC for a soil could be cumbersome and sometimes it is time consuming and needs effort for some soils, either through laboratory tests or through field tests. Accurate prediction of this curve can give more precise expectations in design or analysis that include some unsaturated soil properties, which can save more effort and time. This work will concentrate on proposing a new approach for determining the SWCC using Artificial Neural Network (ANN) depending on some soil properties (air-entry point and residual degree of saturation) through computer software MatLab as a tool for ANN. The new approach is to plot the SWCC curve points instead of obtaining the parameters used in Brooks and Corey (BC) Model (1964), van Genuchten (VG) Model (1980), or Fredlund and Xing (FX) Model (1994). Results showed close agreement in determination of the SWCC by verification of the ANN results with an additional curve sample.

## 1 Introduction

The SWCC provides a conceptual relation between the mass (and / or volume) of water in a soil and the energy state described as (usually) matric suction of the water phase. The SWCC has proven to be an interpretive model that utilizes the capillary model to provide an understanding of the distribution of water in the voids. The effects of soil texture, void ratio, and gradation became part of the interpretation of measured laboratory SWCC data (i.e. these soil properties are used as parameters in prediction of SWCC).

The SWCCs have an important role in the determination of unsaturated soil property functions (e.g. shear strength, volume change, etc.). The procedures that have been proposed for unsaturated soil properties are approximate but satisfactory for analyzing unsaturated soil mechanics problems [1]. Neural Networks are widely used in experimental researches that need some soil parameters determination either in field or laboratory, and to describe some complex soil behavior.

## 2 Previous work

Many researchers had been working on using Artificial Neural Network (ANN), Genetic Programming (GP), and Genetic Based Neural Network (GBNN) in predicting the SWCC.

Lee, et al., 2005 Considered the change of void ratio during the SWCC tests. Additionally, a method reasonably predicting the SWCC for Korean weathered granite soils was suggested based on the test results obtained from the experiments conducted in their study. A method to estimate the parameters used in Fredlund and Xing's equation were proposed using an ANN (artificial neural network). The particle size distribution, compacted water content and void ratio were used as input data in the ANN model for predicting the parameters.

Johari, et al., 2006 Used Genetic-Based Neural Network (GBNN) is employed to predict the soil-water characteristic curve of unsaturated soils. A three-layer network has been trained by genetic algorithm and its topology is determined by trial and error. The network has five input neurons, namely, initial void ratio, initial gravimetric water content, logarithm of suction normalized with respect to air pressure, clay fraction and silt content.

Johari, et al., 2011 Investigated Genetic-Based Neural Network (GBNN) and Genetic Programming (GP) in determining the SWCC. These two models have identical set of input parameters. The authors utilized some soil properties as parameters to predict the SWCC, these parameters include void ratio, initial water content, clay fraction, silt content and logarithm of suction normalized with respect to air pressure. Assessment of the results indicates that predictions from GBNN model

* Corresponding author: kareem_esmat@yahoo.com

© The Authors, published by EDP Sciences. This is an open access article distributed under the terms of the Creative Commons Attribution License 4.0 (http://creativecommons.org/licenses/by/4.0/).

have relatively higher accuracy as compared to GP model.

Johari, and Javadi, 2011, Used a neural network which predicts the soil water characteristic curve of unsaturated soils. Their network has five input neurons, namely, initial void ratio, initial gravimetric water content, logarithm of suction normalized with respect to the atmospheric pressure, clay fraction, and silt content.

Johari, and Nejad, 2015, used Gene Expression Programming (GEP) in their research which was employed as an artificial intelligence method for modeling of the SWCC curve. The principal advantage of the GEP approach is its ability to generate powerful predictive equations without any prior assumption on the possible form of the functional relationship. GEP can operate on large quantities of data in order to capture nonlinear and complex relationships between variables of the system. The selected inputs for modeling are the initial void ratio, initial gravimetric water content, logarithm of suction normalized with respect to atmospheric air pressure, clay content, and silt content. The model output is the gravimetric water content corresponding to the assigned input suction.

Nikhil et al., 2016 Used Fredlund and Xing equation that consists of four fitting parameters. The SWCC (fitting parameters) being a function of pore size distribution and stress state vary across different sites, and thus necessitating the need for setting up a site-specific database for reasonable landslide hazard predictions under an extreme rainfall condition. Therefore, a SWCC fitting parameter estimation model has been developed via an artificial neural network (ANN) using soil samples collected from eight regions in Korea.

## 3 Present work

In this study, examples of ten SWCC curves were taken from experimental tests already published in literature, eight of these curves were used in the learning process of the ANN to obtain the curve points, and another two SWCC curves were used for verifying the reliability of the ANN. These curves are used just to demonstrate the proposed method; larger sets of curves may be used to obtain more accurate results that cover wider ranges of SWCC.

Depending on the typical SWCC shape as shown in Fig. 1 and 2, some ideas are discussed to express the author's point of view in determining the SWCC parameters required for the ANN.

### 3.1 Basic assumptions

In this work, the main difference from other solutions proposed in literature studying how SWCC were obtained is that the main factors (or parameters) that are considered herein are reduced to only two parameters, viz.:

1. The air-entry point value, where the degree of saturation is 100% and the value of Matric Suction is considered as the first input data in kPa.
2. The Sr (residual degree of saturation) point, that is considered with its value of Sr in %, and $(u_a-u_w)_r$ (Matric Suction value at Sr in kPa).

The four input data values are assumed to be the main parameters that can give a reliable SWCC points. The part of SWCC before air-entry point is considered flat, and the part of SWCC after Sr is also considered flat.

Although the SWCC vary between coarse grained soils Fig. 1 and fine grained soils Fig. 2. Since the coarse grained soils have a well-defined air entry point most of the times and a well-defined residual saturation Sr point. In fine grained soils, these points are curvy and not well-defined, the curve takes a smoother shape which leads to using extrapolation in the curve during its transient stage to obtain a defined points for air–entry and the residual stage.

In addition, the domain of the Soil Suction values are smaller for most coarse grained soils (about few hundreds kPa) and much larger for fine grained soils (up to $1\times10^6$ kPa).

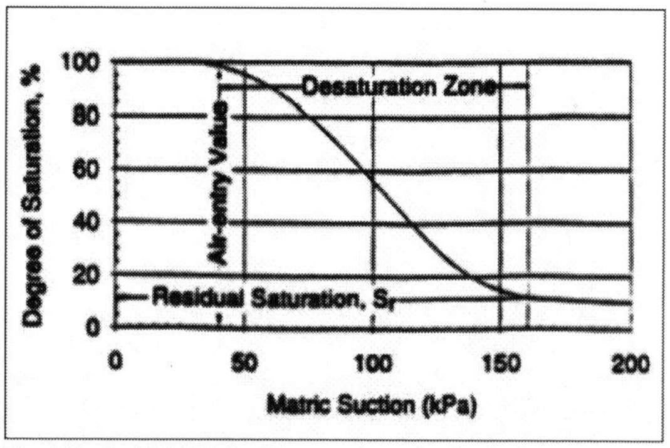


Fig. 1 Typical SWCC for most coarse grained soils [2]

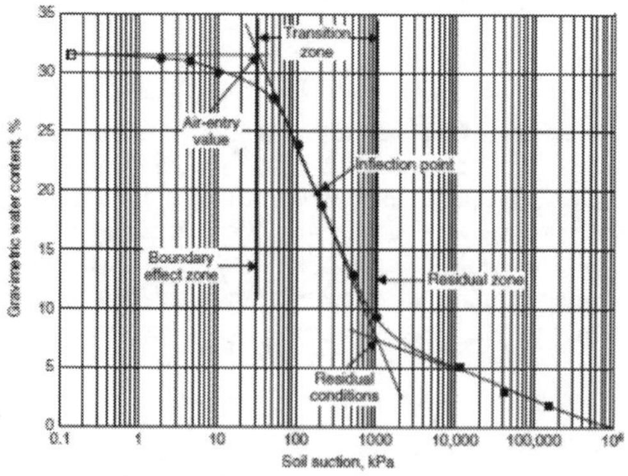


Fig. 2 Typical SWCC for most fine grained soils [1]

In this work, the air-entry point is considered as it is for coarse grained soils and for fine grained soils, it is taken as the mostly representative point near to S=100%. For coarse grained soils, the residual degree of saturation point Sr determined as the point with almost flat part at the end of the curve is considered. Where for fine grained soils, the most end flat point at the curve is considered, the purpose of this consideration is to utilize the ANN to provide the shape of the curve as it is, and the user then can define the needed points by using extrapolation.

### 3.2 Methodology

Ten samples of experimentally obtained SWCC were considered in this study, eight of them were used for the learning process, and the other two SWCC were used to verify the reliability of the ANN.

For each curve, the main points of the air-entry point and the Sr point is defined by their degree of saturation (S) in percent, and the Matric Suction ($u_a$–$u_w$) in kPa.

For each SWCC, the curved part between these two points (including these main points) is digitized into 21 equally spaced points (as possible) in the horizontal domain, and the values of the matric suction and the degree of saturation is obtained via a digitizing computer program (Obtaining more points give more accurate results but take more time in digitization). The two main points mentioned (with their four readings) are used as the input data to the Neural Network.

Fig. 3 shows a typical ANN where the input layer, the hidden layer(s) and the output layer are shown in sequence (A, B, C, and D symbols represent Artificial Neurons). Two networks were used to determine the SWCC. The first network has an input layer with four input neurons each represents one input value as follows:

i- Air-entry degree of saturation (in %),

ii- Air-entry Matric Suction (in kPa),

iii- Residual degree of saturation (in %), and

iv- Residual Matric Suction (in kPa).

The hidden layer will contain 20 neurons using logsig function, and the output layer will contain 21 neurons that represent the degree of saturation using purelin function.

The second network has the same input layer, this time it feeds another hidden layer with 20 neurons with logsig function, and connected to another output layer that represents the Matric Suction corresponding to the degrees of saturation obtained from the first network with 21 neurons using purelin function.

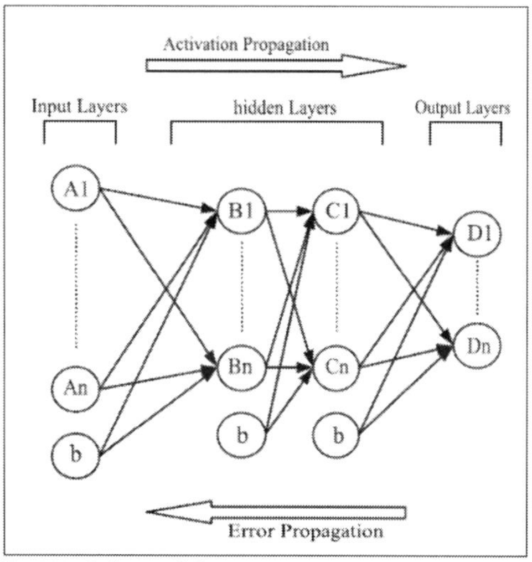


Fig. 3 Typical ANN [4]

### 3.3 Working procedure

The ten SWCC are shown in Fig. 4 to Fig. 9, with the reference of each one, curve 8 was taken in the training process while curves 9 and 10 were taken for the verification process.

All curves were digitized and the data obtained from digitizing these curves are illustrated in Table 1 and Table 2. Table 1 shows the values of the matric suction ($u_a$–$u_w$) in kPa for each curve, where Table 2 shows the corresponding values of the degree of saturation against each value of matric suction. The data are shown in this form as a requirement for the computer software to represent the target values. Table 3 shows the input main points for the curves (air-entry point, and the Sr point).

After the training process was complete, the two ANN (one to represent the matric suction, and the other to represent the degree of saturation) can simulate the verification data.

Results of regression for the matric suction ANN and the degree of saturation ANN are shown in Fig. 10 and Fig. 11 respectively.

The main parameters (four data points) for each curve (namely curve 9 and curve 10) was supplied to each ANN, the input data are shown in Table 4, and the results obtained are shown in Table 5.

Three curves were drawn to represent the results, the first two are the results of verifying data against the digitized data, and the third one is to represent one of the curves already used in the training process. These graphs are shown in Fig. 12, Fig. 13, and Fig. 14.

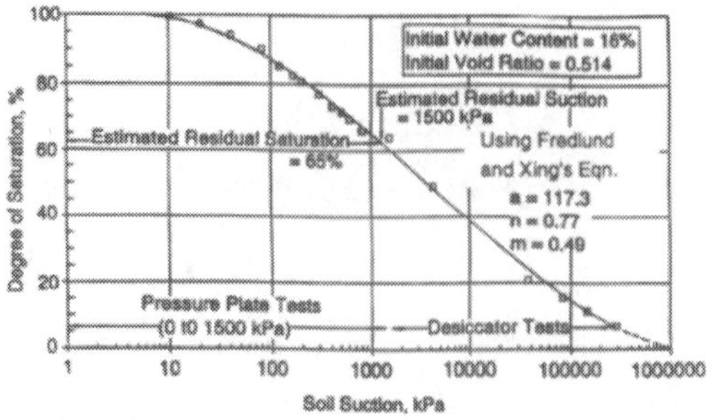


Fig. 4 SWCC for curve 1 after [2]

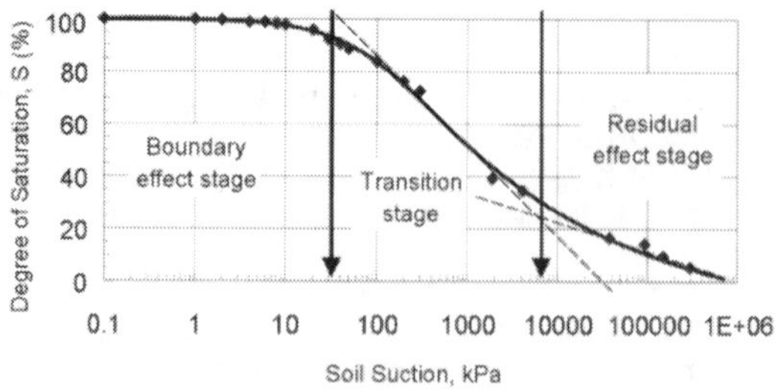


Fig. 8 SWCC for curve 7after [13]

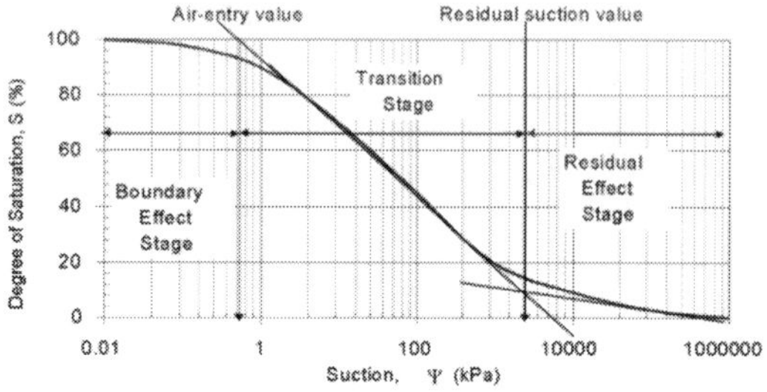


Fig. 5 SWCC for curve 2 after [13]

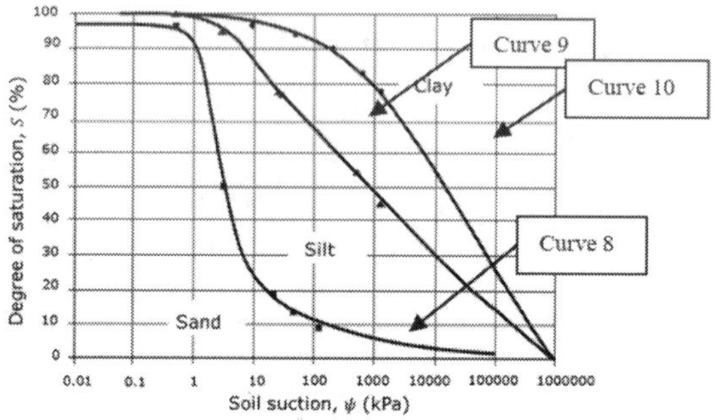


Fig. 9 SWCC for curve 8after [10]

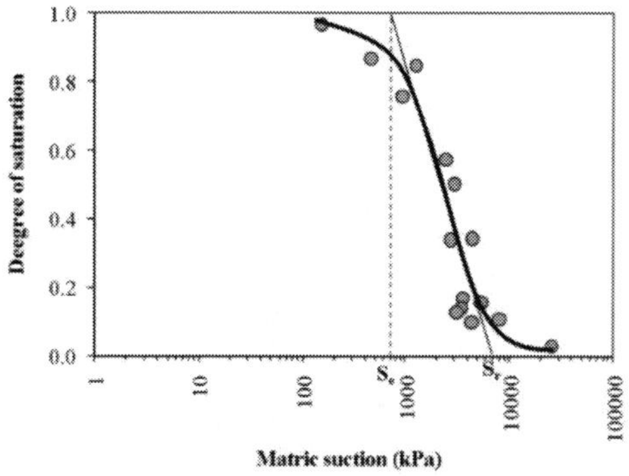


Fig. 6 SWCC for curve 3 after [11]

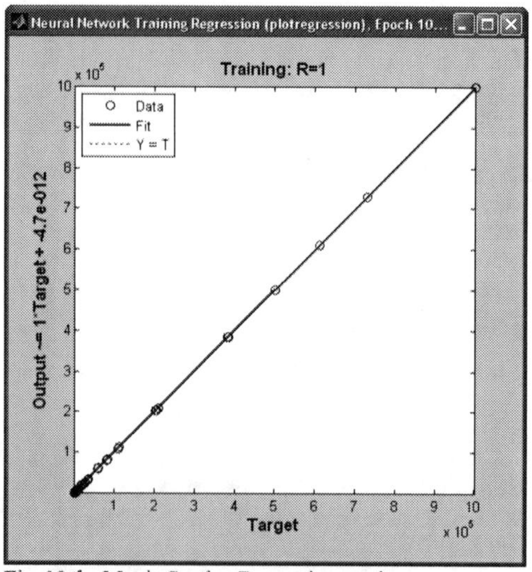


Fig. 10 the Matric Suction Regression result

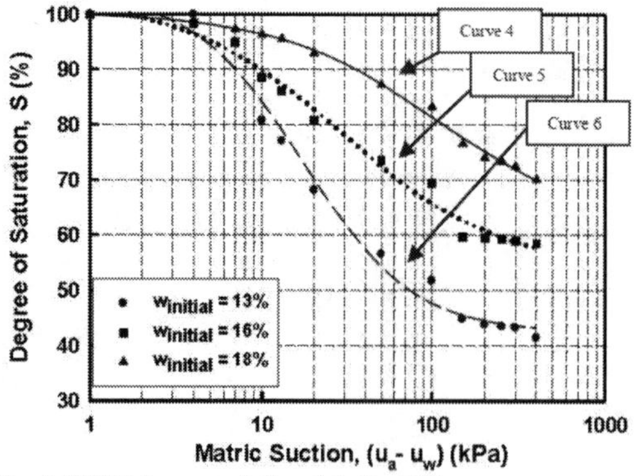


Fig. 7 SWCC for curve 4, 5, and 6 after [12]

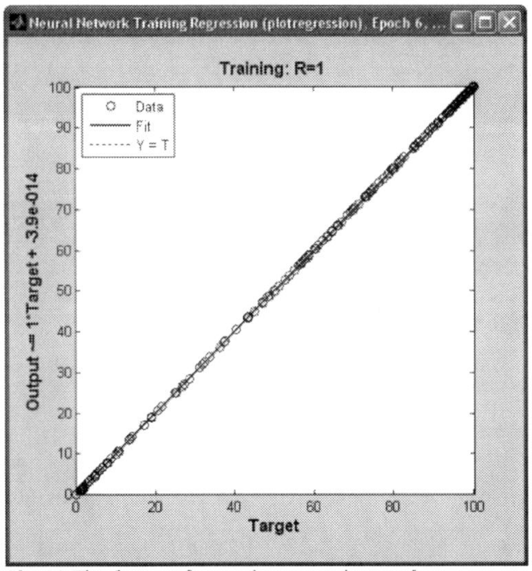


Fig. 11 The degree of saturation regression result

## 4 Discussion

The original curves were obtained from experimental data, the curved lines were digitized to represent the best fit for these experimental points. Results obtained and shown in Fig. 12 to Fig. 14 deviate slightly from the original curved data (as in experimental data points). The best fit lines shown are more representative in showing the curve variation.

Curve 8 best fit line can be represented by (1):

$$S = 74.14\Psi^{-(0.079+0.03536\Psi)} \quad (1)$$

with $R^2$=0.837

Also curve 9 best fit line can be represented by (2):

$$S = 95.12\Psi^{(0.07386-0.02383Ln(\Psi))} \quad (2)$$

with $R^2$=0.9613

Finally, curve 10 best fit line can be represented by (3):

$$S = 104.5\Psi^{(-0.01816\Psi^{0.2011})} \quad (3)$$

with $R^2$=0.9736

where:

Ψ= matric suction (kPa), and

S= Degree of Saturation in percent.

The trend of these result points is the same as in the original data. It is recommended to use more curves and more data points digitized which can give more accurate curves.

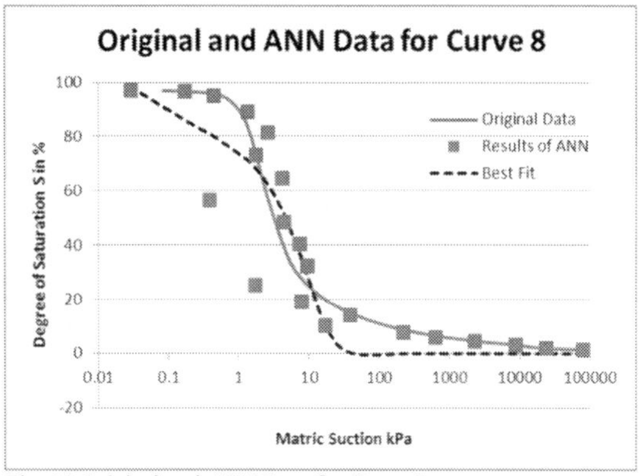


Fig. 12 Original and ANN Data for Curve 8

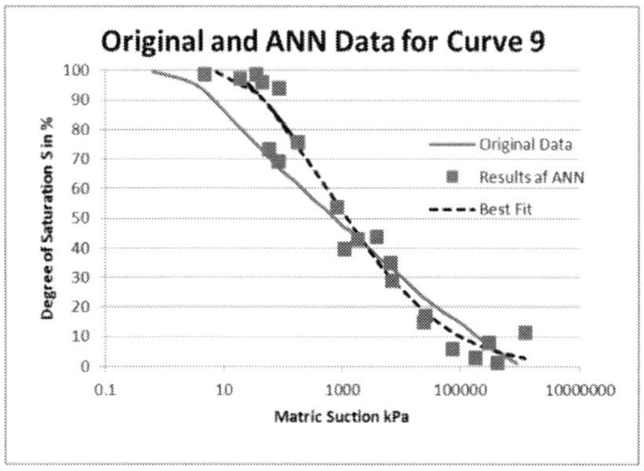


Fig. 13 Original and ANN Data for Curve 9

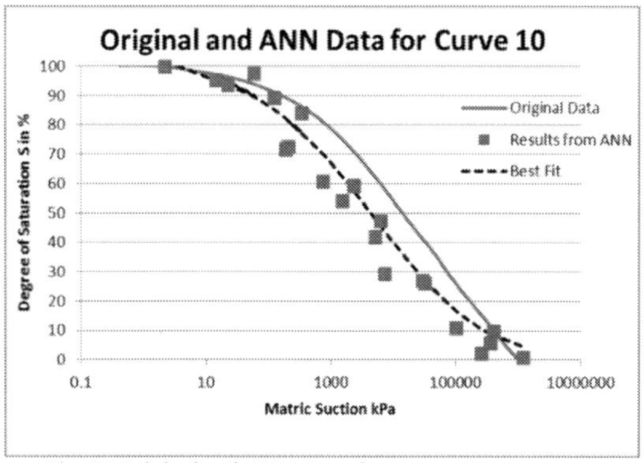


Fig. 14 Original and ANN Data for Curve 10

## 5 Conclusions

The new proposed method of representing the SWCC is found to be satisfactory in representing the SWCC curved segment. Further work in needed to represent the SWCC more accurately and more sample curves are needed.

Other soil properties may be included along with more representatives (i.e. void ratio, soil classification, effective diameter D10, and others).

## References

[1] D. G. Fredlund, H. Rahardjo, and M. D. Fredlund, "*Unsaturated Soil Mechanics in Engineering Practice*", John Wiley & Sons, Inc. (2012).

[2] D. G. Fredlund, S. K.Vanapalli, D. E. Pufahl, "Predicting the shear strength function for unsaturated soils using the soil-water characteristic curve", Unsaturated soils : proceedings of the first International Conference on Unsaturated Soils, UNSAT '95, Paris, France, 6-8 September. Vol. **3**. (1995).

[3] A. Johari, A. Hooshmand Nejad, , "Prediction Of Soil-Water Characteristic Curve Using Gene Expression Programming", IJST, Transactions of Civil Engineering, Vol. **39**, No. C1, pp. 143-165. (2015)

[4] A. Johari, A. A. Javadi, "Prediction of soil-water characteristic curve using neural network", Unsaturated Soils – Alonso & Gens (eds) Taylor & Francis Group, London, pp. 461-466. (2011).

[5] A. Johari, G. Habibagahi, A. Ghahramani, "Prediction of a Soil-Water Characteristic Curve Using a Genetic-Based Neural Network", Scientia Iranica, Vol. **13**, No. 3, pp 284-294. (2006).

[6] A. Johari, G. Habibagahi, A. Ghahramani, "Prediction of SWCC using artificial intelligent systems: A comparative study", Scientia Iranica 8 (**5**).pp 1002–1008. (2011).

[7] S. R. Lee, Y.K. Kim, S. J. Lee, "A Method to Estimate Soil-Water Characteristic Curve For Weathered Granite Soil", Proceedings of the 16th International Conference on Soil Mechanics and Geotechnical Engineering. (2005).

[8] N.V. Nikhil, Yoon Seok, Seung-Rae Lee, Deuk-Hwan Lee, "ANN Based Estimation of SWCC Fitting Parameters for Korean Weathered Soil Considering In-Situ Characteristics", The 2016 world congress on advances on Civil, Environmental, and Materials Research. (ACEM16). (2016).

[9] Won Taek Oh, Sai K. Vanapalli, "Modelling the Stress versus Settlement Behavior of Model Footings in Saturated and Unsaturated Sandy Soils", The 12th International Conference of International Association for Computer Methods and Advances in Geomechanics (IACMAG) 1-6 October, Goa, India.( 2008).

[10] N. Pérez-García, "Determination of Soil-Water Characteristic Curve with pressure plate test". in Groundwater - Contaminant and Resource Management book, Safandila, Querétaro, México: Secretaria de Comunicaciones y Transportes, Mexican Transportation Institute;. 54 p. Publicación Técnica No. 313 (in Spanish). (2008)

[11] Anuchit Uchaipichat, , "Variation of Pile Capacity in Unsaturated Clay Layer with Suction", Ejge, Vol. **17**, pp. 2425-2433. (2012).

[12] S. K .Vanapalli, Taylan Z.N, "Design of Single Piles Using the Mechanics of Unsaturated Soils", Int. Jour. of GEOMATE, March, Vol. **(2)**, No. 1 (Sl. No. 3), pp. 197-204.(2012).

[13] S. K., Vanapalli, A. Wright, D.G. Fredlund, , "Shear Strength Behavior of a Silty Soil Over the Suction Range From 0 to 1,000,000 kPa", 53th Canadian Geotechnical Conference, at Moterial, Quebec,Canada, October. (2000).

**TABLE 1**. MATRIC SUCTION FOR THE 8 CURVES

| 1 | 2 | 3 | 4 | 5 | 6 | 7 | 8 |
|---|---|---|---|---|---|---|---|
| 8.57553 | 0.055602 | 48.1437 | 1.672545 | 1.694723 | 2.120678 | 1.434355 | 0.084768 |
| 21.06522 | 0.134151 | 137.1151 | 2.653551 | 2.235451 | 2.688841 | 3.182805 | 0.197618 |
| 32.65756 | 0.324312 | 223.0924 | 3.320424 | 2.987829 | 3.593593 | 7.78326 | 0.572914 |
| 62.53066 | 0.756323 | 361.177 | 4.617148 | 4.154408 | 5.198023 | 14.65592 | 1.07374 |
| 117.8381 | 1.671164 | 561.8744 | 5.6271 | 5.700876 | 6.541884 | 27.77551 | 1.485428 |
| 215.9374 | 3.51968 | 809.7844 | 7.134633 | 7.227838 | 8.897422 | 163.9119 | 1.838764 |
| 388.6442 | 7.222818 | 1068.71 | 9.410904 | 9.533255 | 10.73145 | 263.0179 | 2.228015 |
| 689.7889 | 14.55798 | 1319.747 | 12.25082 | 12.40947 | 12.36917 | 428.4782 | 2.746242 |
| 1212.124 | 28.93492 | 1573.873 | 18.62726 | 16.15307 | 15.63406 | 702.5664 | 3.428708 |
| 2117.359 | 56.71134 | 1846.009 | 23.68593 | 22.45854 | 20.86181 | 1151.982 | 4.373265 |
| 3683.958 | 110.2674 | 2143.724 | 28.52934 | 27.73189 | 26.29097 | 1980.802 | 5.922081 |
| 6422.307 | 218.2891 | 2472.965 | 36.11855 | 34.24346 | 32.4627 | 3262.017 | 9.679343 |
| 11218.4 | 421.8968 | 2838.589 | 42.37131 | 42.84465 | 40.79334 | 5645.563 | 19.59148 |
| 19596.84 | 1129.149 | 3269.114 | 51.56069 | 55.03947 | 46.95129 | 9898.265 | 46.6512 |
| 34436.92 | 2545.303 | 3783.747 | 62.65208 | 66.1953 | 56.54949 | 17543.03 | 123.0946 |
| 61363.22 | 5959.629 | 4445.367 | 91.29107 | 97.02138 | 80.60604 | 31771.46 | 343.3754 |
| 112443.7 | 14207.29 | 5489.569 | 132.4373 | 138.5041 | 107.5743 | 58543.65 | 978.5758 |
| 209809.3 | 34140.99 | 7415.32 | 194.6825 | 197.7232 | 164.0409 | 109047.6 | 2812.824 |
| 383060 | 83032.35 | 11011.51 | 254.1643 | 293.6675 | 253.8486 | 204000.5 | 8154.716 |
| 611672.6 | 203560.6 | 17446.28 | 313.848 | 387.3668 | 317.6452 | 384117 | 23742.62 |
| 1000000 | 501059.7 | 26079.9 | 403.2 | 478.3629 | 397.4737 | 729544.6 | 80976.37 |

**TABLE 2.** DEGREE OF SATURATION IN % FOR EACH CURVE

| 1 | 2 | 3 | 4 | 5 | 6 | 7 | 8 |
|---|---|---|---|---|---|---|---|
| 100.1479 | 99.16128 | 100 | 100 | 99.66008 | 100.0255 | 99.60786 | 96.87191 |
| 96.74303 | 97.34698 | 97.73097 | 99.09568 | 98.91423 | 99.65047 | 99.54323 | 96.50971 |
| 94.94827 | 95.10542 | 95.78695 | 98.90598 | 97.98307 | 97.23914 | 97.82207 | 95.09384 |
| 90.41395 | 91.45666 | 93.39106 | 98.34405 | 95.94096 | 93.71603 | 96.99501 | 89.29865 |
| 85.64171 | 86.30029 | 90.02373 | 98.15491 | 94.26917 | 90.93591 | 94.3787 | 81.36319 |
| 80.2874 | 79.98811 | 85.39129 | 97.59489 | 92.59903 | 86.36563 | 77.48374 | 73.0326 |
| 74.53619 | 73.14829 | 79.71971 | 96.47899 | 90.00295 | 82.78463 | 70.42706 | 64.73494 |
| 68.54696 | 66.03213 | 73.41563 | 95.46616 | 87.7772 | 80.08857 | 63.60235 | 56.56898 |
| 62.37258 | 58.71499 | 66.90828 | 93.95669 | 84.99637 | 74.96475 | 56.87706 | 48.46888 |
| 56.14542 | 51.22199 | 60.31059 | 92.38928 | 81.47409 | 69.96314 | 50.11863 | 40.40171 |
| 49.86542 | 43.60339 | 53.69032 | 91.19304 | 79.06441 | 64.6338 | 43.2609 | 32.4004 |
| 43.58533 | 36.16061 | 47.07007 | 89.9752 | 76.65473 | 61.11399 | 37.62905 | 25.12348 |
| 37.33166 | 28.46663 | 40.42723 | 88.53282 | 73.87473 | 57.03875 | 31.99731 | 18.96608 |
| 31.13102 | 19.06192 | 33.78439 | 86.82241 | 71.2792 | 55.04169 | 26.59754 | 14.12578 |
| 24.98315 | 14.18188 | 27.14154 | 85.46153 | 69.42514 | 52.73532 | 21.56227 | 10.50379 |
| 19.02026 | 10.58337 | 20.52125 | 82.0205 | 66.08678 | 48.86303 | 17.15659 | 7.869608 |
| 13.61293 | 7.638194 | 14.17199 | 78.49733 | 63.30403 | 47.19182 | 13.34738 | 5.893974 |
| 8.761576 | 5.069951 | 8.748892 | 75.34393 | 60.52129 | 44.98333 | 9.869533 | 4.445176 |
| 3.805091 | 2.727831 | 4.681239 | 73.28257 | 58.84785 | 43.59684 | 6.689906 | 3.259796 |
| 1.982576 | 1.039097 | 1.923897 | 71.24294 | 58.10199 | 43.48938 | 3.808504 | 1.909779 |
| 0 | 1.009049 | 1.674576 | 69.94256 | 57.72755 | 43.29969 | 1.324733 | 1.448798 |

**TABLE 3.** INPUT DATA

| | 1 | 2 | 3 | 4 | 5 | 6 | 7 | 8 |
|---|---|---|---|---|---|---|---|---|
| ir-entry point matric suction | 8.57553 | 0.055602 | 48.1437 | 1.672545 | 1.694723 | 2.120678 | 1.434355 | 0.084768 |
| air-entry point S | 100.1479 | 99.16128 | 100 | 100 | 99.66008 | 100.0255 | 99.60786 | 96.87191 |
| Sr point matric suction | 1000000 | 501059.7 | 26079.9 | 403.2 | 478.3629 | 397.4737 | 729544.6 | 80976.37 |
| Sr S | 0 | 1.009049 | 1.674576 | 69.94256 | 57.72755 | 43.29969 | 1.324733 | 1.448798 |

**TABLE 4**. VERIFYING DATA

| | 9 | 10 |
|---|---|---|
| air-entry point matric suction | 0.651432 | 0.410554 |
| air-entry point S | 99.53902 | 100.1317 |
| Sr point matric suction | 868462.8 | 950007.8 |
| Sr S | 0.987817 | 0.329272 |

**TABLE 5.** VERIFYING DATA RESULTS

| Matric Suction kPa | | Degree of Saturation S% | |
|---|---|---|---|
| 9 | 10 | 9 | 10 |
| 4.728393 | 2.223661 | 98.71132 | 99.64459 |
| 36.89925 | 60.00188 | 98.6911 | 97.28264 |
| 19.64693 | 14.42446 | 97.28464 | 95.17319 |
| 46.64776 | 22.40805 | 95.88645 | 93.42735 |
| 88.63865 | -129.106 | 93.829 | 88.91717 |
| 179.7074 | 345.2875 | 75.47142 | 83.76707 |
| 86.4868 | 197.1986 | 69.00925 | 71.41278 |
| 60.14028 | 210.1119 | 73.1 | 72.32742 |
| 811.1791 | 752.5994 | 53.5159 | 60.65005 |
| 1075.403 | 1510.868 | 39.50234 | 53.91856 |
| 1853.102 | 2333.282 | 42.78961 | 59.19435 |
| 3840.476 | 5252.347 | 43.77121 | 41.71416 |
| 6600.492 | 6502.615 | 35.08857 | 47.15224 |
| 7228.543 | 7531.47 | 28.70058 | 28.96275 |
| 24685.07 | 33397.71 | 14.77886 | 25.99016 |
| 26587.29 | 30670.66 | 16.9301 | 26.65161 |
| 74943.67 | 102013.3 | 5.839024 | 10.75616 |
| 178921.8 | 260958.9 | 2.814036 | 2.149345 |
| 298830.6 | 407437.3 | 7.694084 | 9.607634 |
| 409398.9 | 366794.4 | 0.961794 | 5.576248 |
| 1190123 | 1190970 | 11.33338 | 0.769058 |

# Compressibility of soft Iraqi soil stabilized with traditional Iraqi stabilizers (cement and lime)

*Husam* Baqir [1,*], *Aqeel* Al-Adili[1] *and Ali* Sharef[2]

[1] University of Technology, Baghdad, Iraq
[2] Baghdad Governorate, Directorate of Education, Baghdad, Iraq

**Abstract.** This study shows an improvement of two types of clay soil brought from different parts of Iraq. The first soil (A) from Al - Zaafaraniya site in Baghdad governorate. The second soil (B) from Garma Ali site in the Al Basra governorate, Iraq. Soft clayey soils were treated by a combination of sulphate resistance Portland cement (PC) and Quicklime (LQ) to modify and stability. PC was added in percentages of 2,4,6,8 and 10%, as well as, LQ was added to 2 and 4%, of dry weight. Laboratory tests to determine specific gravity, Atterbergs limits and standard proctor test were conducted. Also, the main objective of this research is the concentrating on compression ratio (CR), the Rebound (Swelling) ratio (RR) and the stiffness during the modulus of elasticity (Es) for treated and natural soils procreation from consolidation test. The results from laboratory tests shows high ability on the enhancing in terms of reduction in plasticity index (greatly increased workability), reduction in compression ratio (CR), reduction in the Rebound (Swelling) ratio (RR), increase in the modulus of elasticity (Es). The change in moisture-density relationships resulting in lower maximum dry densities, higher optimum water content, and less variation of dry density from the maximum over a much wider range of water contents.

## 1 Introduction

The Soft clay soils are late alluvial deposits presumably formed through the 10,000 years covered by their featureless and flat solid ground surface. Brand and Bernard, (1981)[1] identified, such clay by their low undrained shear strength (Cu < 40 Kpa) (B.S.C.P.8004:1986)[2] and high compressibility (CC between 0.19 to 0.44). They are found in high natural moisture content typically ranging from (40 – 60) % with the plasticity index, ranging from (45 – 65) % (Broms, 1987)[3]. Soils with such characteristics create serious problems to geotechnical engineering associated with stability and settlement problems. Clay soils cause cracking and fracture of pavement, railways , highway embankment , roadways, foundations and channel or reservoir linings (Cokça, 1999)[4]. Solutions to the problems presented by clay subgrades include excavating and replacing a thickness of clay with a select fill material and increasing the base thickness layer to decrease subgrade stresses and minimize moisture changes. Both solutions are inherently expansive and wasteful because they require disposal of significant quantities of inferior materials or the use and transportation of additional virgin materials. For low – volume roads, this expense is especially high because, frequently, the distance of the road project to sources of suitable in situ materials is quite far (Prusinski and Bhattacharja, 1999). To proceed with this alternative , soil stabilization using lime and / or hydraulic binders, etc. is a widespread technique employed to improve the workability and hydromechanical properties of soils (Obuzor et al., 2011)[5]. Stability with lime and cement are mainly and commonly used in the above-mentioned works and they have been under continuous development since its launch in the middle of the last century. Lime is used primarily to dewater the soil in order to improve the workability and its bearing capacity. This Lime modification is widely used for building embankment and subgrade of clayey soil since the effect is rapid and modifies the geotechnical characteristics of the soil like the plastic limit, the shear strength and the soil compaction characteristics. Later the pozzolanic reaction between the soil minerals and lime in the presence of water leads to the formation of secondary cementitious products (C-S-H, C-A-S-H,...) increasing the soil cohesion and its resistance (Pomakhina et al., 2012)[6]. The effects of the pozzolanic reaction are mainly effective in long term. If the mechanical resistance of the material is essential in short term, the soil stabilization goes through the use of cement. Cement stabilization is quick, does not need mellowing time and provides a non-leaching platform (Sariosseiri and

* Corresponding author: husamhb65@yahoo.com

© The Authors, published by EDP Sciences. This is an open access article distributed under the terms of the Creative Commons Attribution License 4.0 (http://creativecommons.org/licenses/by/4.0/).

Muhunthan, 2009)[7]. Cement can be used for stabilization of a wide range of soils and the best performances of soils treated with cement have been observed on silt as well as on coarse-grained materials (Currin et al., 1976)[8]. Furthermore, in practice, in the world, a mixed treatment with lime and cement is used since the mixture allows to facilitate the workability without disturbing the effect of the cement in the gain of resistance in the long term.

The addition of lime and cement exerts impacts on the material microstructure. The flocculation/aggregation of clay particles after adding lime or cement modifies, for one thing, the material particle size distribution (Osula, 1996)[9] and, consequently, influences pore distribution over a very short term (Bin et al., 2007)[10], although this distribution is evolving during the curing period as well as in jointly with both the pozzolanic and hydraulic reactions (Metelkova et al., 2012)[11]. Locat et al. (1990)[12] followed up by representing the lime distribution in a soil sample.

## 2 Experiment of investigation

### 2.1 Material used

#### *2.1.1.Soils*

In this study two soil samples (A and B) were used obtained from two different locations. The first soil sample (A) from Al - Zaafaraniya site in Baghdad city. The second soil sample (B) from the Garma Ali site about 538 km south of Baghdad in Al Basra city. The engineering properties of clayey soils are presented in Table 1 and Figure (1) shows the grain size distribution of soils used.

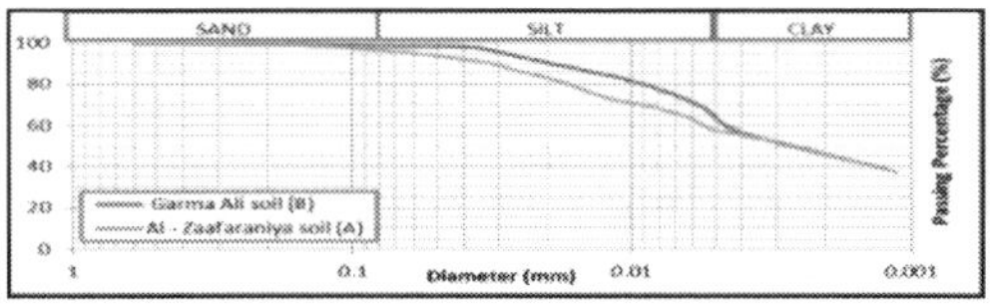


**Fig. 1.** Particle Size Distribution of Soft Soil used

**Table 1.** Physical and Chemical Properties of Natural Soils Used.

| Index Property | Test Standard | Index Valve | |
|---|---|---|---|
| | | Soil A | Soil B |
| Natural water content (%) | ASTM D 2216 | 26 | 31 |
| Depth (m) | --- | 0.5 -1.5 | |
| Liquid Limit (L.L) (%) | ASTM D 4318 | 48 | 58 |
| Plastic Limit (P.L) (%) | ASTM D 4318 | 21 | 27 |
| Plasticity Index (P.I) (%) | ASTM D 4318 | 27 | 31 |
| Activity of Clay (At) | Skempton formula At= (PI/ {percent of clay < 0.002mm}) | 0.6 | 0.56 |
| Specific Gravity (G.s) | ASTM D 854 | 2.72 | 2.75 |
| Gravel (larger than 4.75mm) (G) % | ASTM D 422 | 0 | 0 |
| Sand (0.075 to 4.75 mm) (S) % | ASTM D 422 | 4 | 2 |
| Silt (0.005 to 0.075 mm) (M) % | ASTM D 422 | 38 | 35 |
| Clay (less than 0.005mm) (C) % | ASTM D 422 | 58 | 63 |
| Classification (USCS) | ASTM D 2487 | CL | CH |
| Organic Material (O.M) (%) | ASTM D 2974 | < 0.01 | < 0.01 |
| Calcium Oxide (CaO) (%) | Chemical Analysis | 18.38 | 21.12 |
| SO3 Content (%) | Chemical Analysis | 0.82 | 0.38 |
| Total Dissolved Salt (TDS %) | Chemical Analysis | 2.21 | 1.73 |
| Total Solved Salt (TSS %) | Chemical Analysis | 8.3 | 6.89 |
| PH Value (%) | ASTM D 4972 | 7.95 | 8.69 |
| MDD (KN/m3) | ASTM D 698 | 17.55 | 16.5 |
| OMC (%) | ASTM D 698 | 16 | 20.5 |

#### *2.1.2 Types and properties of additives*

The cement used to be sulfate resistance Portland cement (type V) manufactured by (Al Jessir) factory made in Iraq. The type of lime was used quicklime (unhydrated) manufactured from limestone by the Alnoora factory in Kerbella governorate. Also, quicklime made in Iraq. Table (2) and (3) shows the Physical and chemical properties of the cement and quicklime, respectively. The tests have been conducted in Iraq Geological Survey/Central Laboratories Department.

**Table .2.** The Physical and Chemical Properties of the Cement.

| Index Property | Index Value |
|---|---|
| Physical Properties | |
| Specific gravity (G.S) | 3.15 |
| Compressive strength after 3 days (MPa) | 17 |
| Compressive strength after 7 days (MPa) | 26 |
| Time of initial setting (minute) | 93 |
| Time of final setting (hour) | 4.28 |

| Chemical Properties | |
|---|---|
| $SiO_2$ % | 19.79 |
| CaO % | 63.8 |
| MgO % | 3.19 |
| $SO_3$ % | 2.15 |
| $C_3A$ % | 3.27 |
| L.O.I % | 0.89 |

**Table 3.** The Physical and Chemical Properties of the Quicklime.

| Index Properties | Index Value |
|---|---|
| Physical Properties | |
| Specific gravity (G.S) | 2.99 |
| Retained on Sieve # 30 (%by weight) | 0 |
| Retained on Sieve # 200 (%by weight) | 11 |
| Chemical Properties | |
| CaO % | 92.3 |
| Free Water (%) | 0.08 |
| IR (%) | 2 |
| $SO_3$ (%) | 0.07 |
| L.O.I (%) | 25.3 |

## 2.2 Laboratory tests

A series of laboratory tests consisting of Atterberg limits, compaction and consoildation were conducted on the two selected clayey soils. The combinations of Portland cement and quicklime (PC and LQ) were used for stabilization of the two soils. The percentages of PC were 0,2,4,6,8 and 10%. While the percentages of LQ were 0,2 and 4%. A total of 11 combinations based on soil A and soil B with mixed modes of stabilizers were studied (Table 4).

**Table 4.** Stabilizer combination scheme for stabilizing soils

| Designation | Sample mixture (%) | | |
|---|---|---|---|
| | Soil | PC | LQ |
| PC 0% LQ 0% | 100 | 0 | 0 |
| PC 2% LQ 2% | 96 | 2 | 2 |
| PC 4% LQ 2% | 94 | 4 | 2 |
| PC 6% LQ 2% | 92 | 6 | 2 |
| PC 8% LQ 2% | 90 | 8 | 2 |
| PC 10% LQ 2% | 88 | 10 | 2 |
| PC 2% LQ 4% | 94 | 2 | 4 |
| PC 4% LQ 4% | 92 | 4 | 4 |
| PC 6% LQ 4% | 90 | 6 | 4 |
| PC 8% LQ 4% | 88 | 8 | 4 |
| PC 10% LQ 4% | 86 | 10 | 4 |

### *2.2.1 Atterberg limits tests*

According to ASTM D4318 (2000) can be obtain on plastic limit (PL), liquid limit(LL) and plasticity index (PI).Where studied in this research the plasticity index(PI) whether be treated or untreated soils by used a composite from PC and LQ. initially, the dry soils by airs (passing # 40 sieves) and mix with the predetermined quantity of a composite (PC & LQ) in a dry state together. add distilled water to the soil admixture. Permit water penetration into the soils and the mixing, the paste was allowed stand in an air tight container for about 24 hours to examining. After this calm, the paste was remixed with each stabilizer thoroughly for at least 15 min before the performance the first test. The plastic limit tests conducted on material prepared for the liquid limit test the plastic limit was determined as the average of the three water contents and rounded to the nearest whole number.

### *2.2.2 Compaction tests*

The method given in the ASTM D558 (2000) was applied to determine the maximum dry density and water content which represents (OMC) of the soils. The soil mixtures, with and without additives, were thoroughly mixed for 0.5 hour before process the compaction the first of compaction tests were aimed at determining the compaction properties of the soils sample only. Secondly, tests were performed to determine the proctor compaction properties of the treated soil with varying amount of PC & LQ.

### *2.2.3 Consolidation test*

Odometer tests were performed according to ASTM D 2435. The simple size with a diameter of 7.5 cm and height 1.9 cm were put in the consolidation cells. The soil was mixed with different a composite content mentioned in the table (4). The soil samples were enclosed in a metal ring and placed into the consolidometer. The sample was sandwiched between two filter plates allowing a two-way drainage of the sample. Water was added into the cell around the sample, so the sample remained saturated during the test. Seven vertical static load increments were applied. The changes in the displacement of the sample against time were recorded during each loading increment.

## 3 Results and discussion

### 3.1 Atterberg limits

Figure (2) and (3) illustrate the influence of both compounds on the plasticity index of soil A and B respectively. The addition of 2%LQ and PC at a maximum value of 10%, resulting in a decrease in liquid limit from 48 % to 41.3 % for soil A and from 58% to 49 % for soil B, plastic limit increased from 21 % to 32 % and from 27 % to 38.6 % in soil A and soil B respectively. Hence, plasticity index decreased from 27 % to 9.2 % and from 31 % to 10.3 %. On the other hand, the addition of 4%LQ and PC at a maximum value of 10% result in also a decrease in liquid limit from 48 % to 40 % for soil A and from 58% to 47.1 % for soil B, plastic limit increased likewise from 21 % to 33.7 % and from 27 % to 40.4 % for soil A and soil B respectively. Furthermore, plasticity index decreased as well from 27 % to 6.3 % and from 31 % to 6.7 % for soil A and soil B respectively. The reduction in the plasticity index is attributed to the transformation in soil nature (granular nature after flocculation and agglomeration) and the resulted soil is as crumbly as sandy clay soil. Many researchers illustrated that the decrease in the liquid limit of additive – treated soil depends on the soils type (Jawad et al, 2014)[15]. Table (5) Summaries of Atterberg limits values of soil A and B mixed with different percentages of composite.

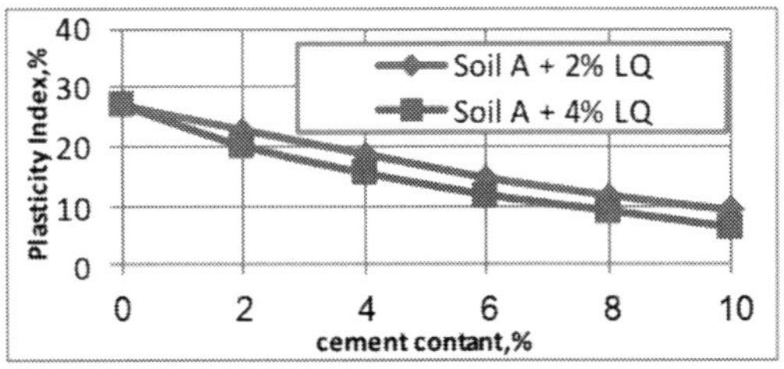


**Fig. 2.** Plasticity index for natural and treated soil A with different (PC, 2% LQ and 4% LQ)

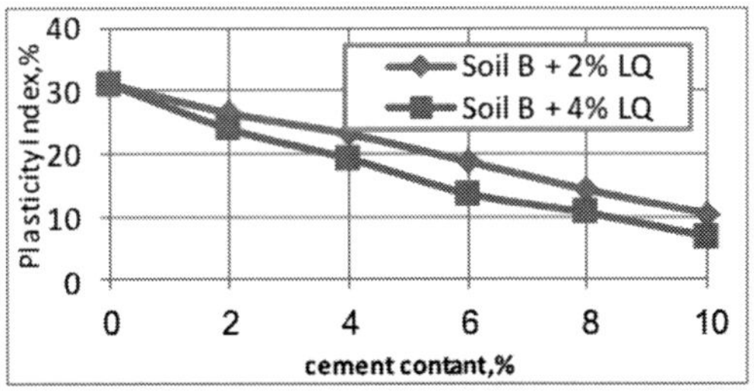


**Fig. 3.** Plasticity index for natural and treated soil B with different (PC, 2% LQ and 4% LQ)

**Table 5.** Atterberg limits values of soil A and B mixed with different percentages of composite

| The Additive | | Soil A | | | Soil B | | |
|---|---|---|---|---|---|---|---|
| **% LQ** | **% PC** | % P.L | % L.L | % P.I | % P.L | % L.L | % P.I |
| **0** | **0** | 21 | 48 | 27 | 27 | 58 | 31 |
| **2** | **2** | 24 | 46.8 | 22.8 | 30.9 | 57.3 | 26.4 |
| **2** | **4** | 26.3 | 45 | 18.7 | 32.7 | 55.8 | 23.1 |
| **2** | **6** | 29 | 43.7 | 14.7 | 35.1 | 53.8 | 18.7 |
| **2** | **8** | 30.9 | 42.6 | 11.7 | 37.4 | 51.6 | 14.2 |
| **2** | **10** | 32 | 41.3 | 9.2 | 38.6 | 49 | 10.4 |
| **4** | **2** | 26 | 46.1 | 20.1 | 32.2 | 56.1 | 23.9 |
| **4** | **4** | 28.9 | 44.4 | 15.5 | 34.8 | 54 | 19.2 |
| **4** | **6** | 30.9 | 42.7 | 11.8 | 37.6 | 51.2 | 13.6 |
| **4** | **8** | 32.7 | 41.8 | 9.1 | 39.3 | 49.8 | 10.5 |
| **4** | **10** | 33.7 | 40 | 6.3 | 40.4 | 47.1 | 6.7 |

### 3.2 Compaction Characteristics

Twenty two from both natural, untreated soils and treated soils with different percentages of PC and LQ were prepared to recognize the effect of compound admixture on the compaction properties. Figures (4) and (5) show the relationship between the maximum dry unit weight (MDD), moisture content (OMC) and different PC content for soils A and B with a 2% LQ, while Figures (6) and (7) present the effect commingle of 4% LQ and 2 to 10% PC on the optimum water content and the maximum dry unit weight (MDD) of the two treated soils. When the adding the composite (PC & LQ) can be seeing that increase the optimum moisture content and decreased the dry unit weight with increasing composite (PC & LQ) content. Similar behavior was also observed in the literature in the case of composite (PC & LQ) stabilized clayey soils. The showing of this behavior is maybe as a result of the following reasons: (1) the composite (PC and LQ) causes aggregation of the particles to occupy large spaces and hence change the effective category of the soils. (2) The pozzolanic reaction between the clay present in the soils and the composite is responsible for the increase in OMC.

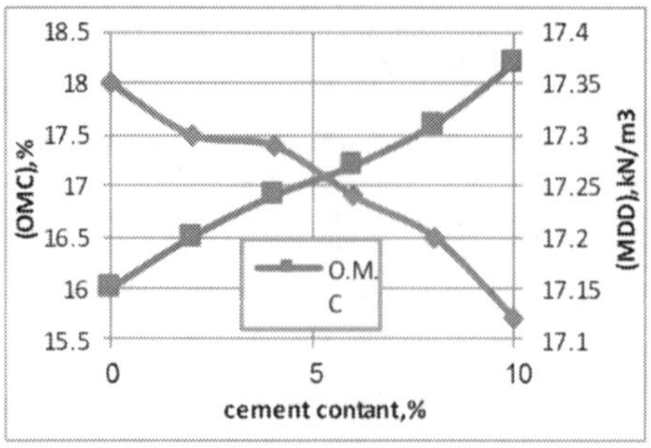


**Fig .4.** Relationship between MDD and OMC soil A with different PC and 2% LQ

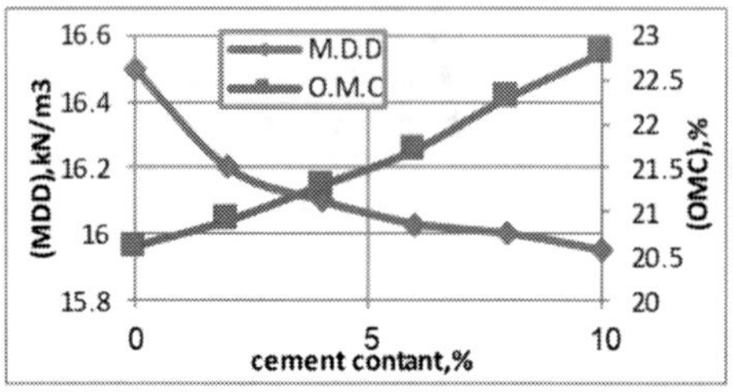


**Fig. 5.** Relationship between MDD and OMC soil B with different PC and 2% LQ

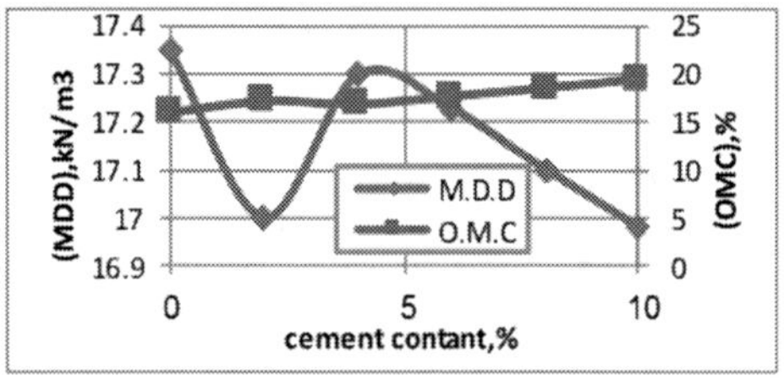


**Fig. 6.** Relationship between MDD and OMC soil A with different PC and 4% LQ

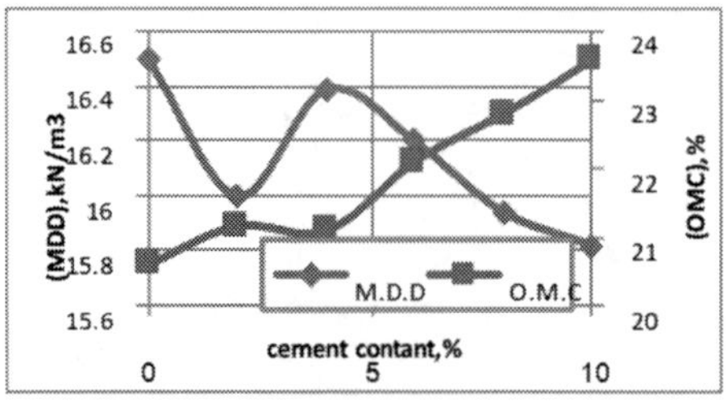


**Fig. 7.** Relationship between MDD and OMC soil B with different PC and 4% LQ

### 3.3 Consolidation test

The (The relation between compression and compression index is not a direct relation, and the compression index (Cc) does not represent the compressibility of the soil. Wesly, (1988)[13] mentioned that the compression ratio (CR) is more realistic to find out the compressibility value, which it is;

$$CR = Cc / (1 + e_o)$$

In contrast, the Rebound (Swelling) ratio (RR) is more accurate than the swelling index (Cr), Where;

$$RR = Cr / (1 + e_o)$$

The confined compressibility

$$(mv) = (((1+v)(1-2v)) / (1-v) Es)$$

The confined (one dimensional) compressibility is also referred to as the coefficient of volume compressibility or the coefficient of volume decrease and the symbol (mv) is widely used to indicate the value of this compressibility.

Results of consolidation tests for the twenty two samples. Figures (8) and (9) show the relationship between compression ratio (CR) and rebound (swelling) ratio (RR) with different PC and 2% LQ for soils A and B respectively. It can be noticed that the additive of 2% LQ content can improve the compressibility of soft clay soils by reducing the compression ratio CR from 0.1264 to 0.0167 and from 0.1106 to 0.034 for soil A and B respectively. And the rebound (swelling) ratio RR decreased from 0.03 to 0.0027 and from 0.02518 to 0.0024 for soils A and B respectively.

Figures (10) and (11) show the relationship between compression ratio (CR) and rebound (swelling) ratio (RR) with different PC and 4% LQ for soils A and B respectively. Results of consolidation test under the compound stabilizer indicate that composite at 4% LQ can decrease the compression ratio CR of soft clay from 0.1264 to 0.0083 and from 0.1106 to 0.0058 for soil A and B respectively. While the rebound (swelling) ratio RR decreased from 0.03 to 0.00207 and from 0.00343 to 0.00094 for soil A and B respectively. This is due to the flocculation and cementitious phenomenal effects by the pozzolanic reaction that forms the secondary minerals (CSH). And it may also due to the formation of inter particle cementation bonds that improved the strength and reduced the compressibility. Furthermore, according to Baquir (1990)[14] this behavior could be attributed to the presence of additive (cement + lime) within the soil, whose crystallization structure has extremely strong structural bonds. This indicates the increased tendency of clay – additive blends to resist the compression and expansion. It is also attributed to the aggregate formation of soil treated with an additive which resulted in stronger particle aggregates and gave higher resistance to compression. Also, because used cement type V (sulfate resistance) causes for the decrease in the compressibility index Cc and swelling index Cr values may be attributed due to sulfate content ($SO_3$). Where Sulfate could possibly interact with the free lime and hence expansive minerals (ettringite crystals) will be formed. It could destroy the structure of soft soils and hence the soil structure becomes weaker, more compressible and a potential to expand (Nicholson, 2015)[16]. Through the figures and the valuses of CR and RR above and also observed bahaviour of untreated and treated soils assume poisson ratio for untreated soil is 0.4 and 0.2 for treated soil through these imposed values for poisson ratio have been calculate the modulus of elasticity (Es).

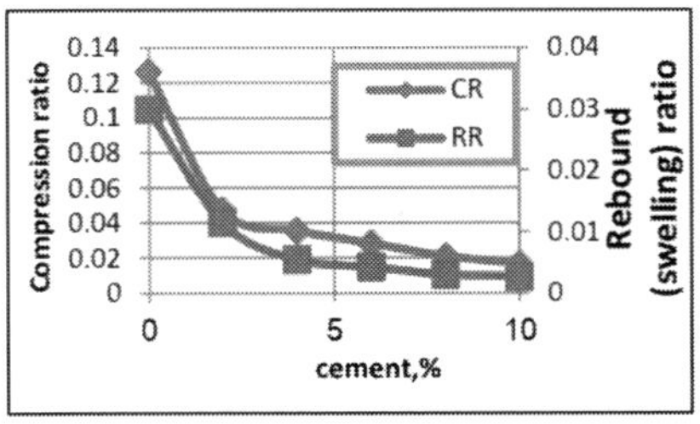


**Fig. 8.** The relationship between compression ratio (CR) and rebound (swelling) ratio (RR) with different PC and 2% LQ for soils A

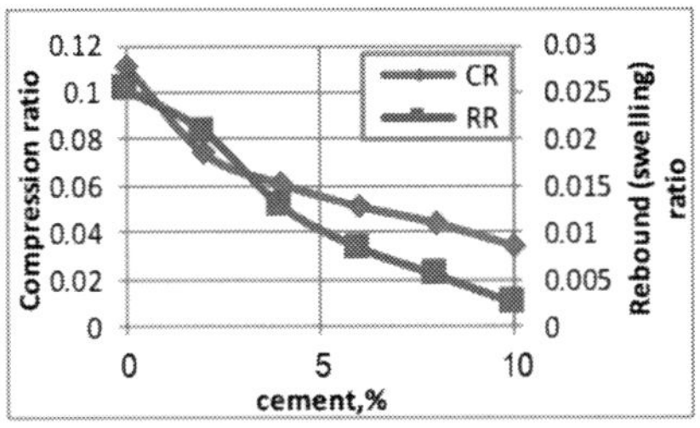


**Fig. 9.** The relationship between compression ratio (CR) and rebound (swelling) ratio (RR) with different PC and 2% LQ for soils B

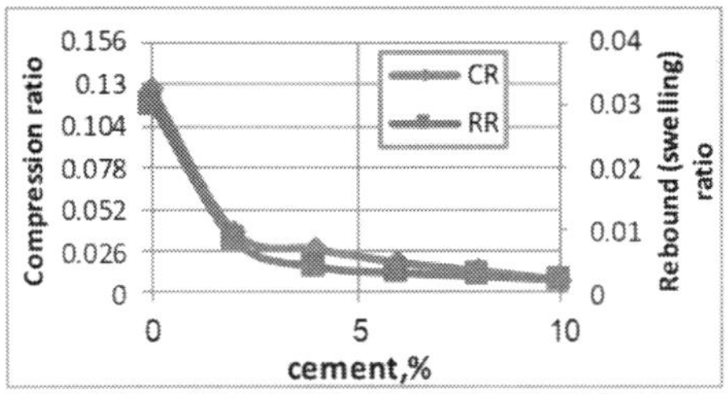


**Fig. 10.** The relationship between compression ratio (CR) and rebound (swelling) ratio (RR) with different PC and 2% LQ for soils A

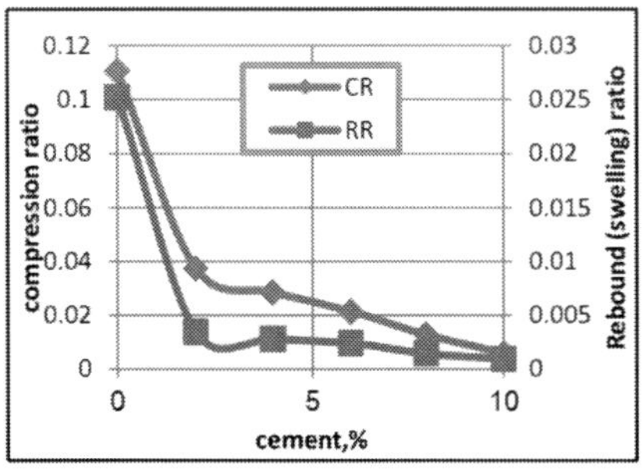


**Fig. 11.**The relationship between compression ratio (CR) and rebound (swelling) ratio (PR) with different PC and 2% LQ for soils B

This decrement shows that the soil samples were effectively stabilized, which indicates an improvement in the engineering properties of the soils by making them good non – plastic subgrade materials. This will considerably reduce the subbase and surfacing materials needed for road construction work. Table (6) summarized the results of consolidationtest.

**Table 6.** Results of the compression index (Cc), Compression ratio(CR) swelling index (Cr) ,Rebound (Swelling) ratio (RR), The confined compressibility (mv) and the modulus of elasticity (Es) for soils mixed with different percentages of Additive (PC & LQ) (A (soil A) and B (soil B))

| The Additives | | Soil A | | | | | |
|---|---|---|---|---|---|---|---|
| PC % | LQ% | Cc | Cr | CR | RR | mv ($m^2$/kN) *$10^{-5}$ | Es (kN/$m^2$) |
| 0 | 0 | 0.209 | 0.052 | 0.1264 | 0.03 | 9.55 | 4815 |
| 2 | 2 | 0.078 | 0.019 | 0.0471 | 0.0114 | 3.58 | 25140 |
| 4 | 2 | 0.059 | 0.009 | 0.0356 | 0.00543 | 2.7 | 33337 |
| 6 | 2 | 0.047 | 0.0076 | 0.0283 | 0.00421 | 2.14 | 42053 |
| 8 | 2 | 0.035 | 0.0052 | 0.0209 | 0.003 | 1.58 | 56911 |
| 10 | 2 | 0.028 | 0.0043 | 0.0167 | 0.0027 | 1.2 | 71170 |
| 2 | 4 | 0.068 | 0.014 | 0.037 | 0.00839 | 2.86 | 31448 |
| 4 | 4 | 0.047 | 0.007 | 0.027 | 0.0041 | 2.06 | 43696 |
| 6 | 4 | 0.031 | 0.005 | 0.0189 | 0.00312 | 1.422 | 63262 |
| 8 | 4 | 0.022 | 0.0041 | 0.0136 | 0.00262 | 1.024 | 87859 |
| 10 | 4 | 0.011 | 0.003 | 0.0083 | 0.00207 | 0.62 | 143161 |

(B)

| The Additives | | Soil B | | | | | |
|---|---|---|---|---|---|---|---|
| PC % | LQ% | Cc | Cr | CR | RR | mv ($m^2$/kN) *$10^{-5}$ | Es (kN/$m^2$) |
| 0 | 0 | 0.196 | 0.044 | 0.1106 | 0.02518 | 8.326 | 5543 |
| 2 | 2 | 0.131 | 0.037 | 0.0737 | 0.0208 | 5.5 | 19629 |
| 4 | 2 | 0.108 | 0.022 | 0.0607 | 0.01275 | 4.58 | 23429 |
| 6 | 2 | 0.091 | 0.014 | 0.05103 | 0.0083 | 3.84 | 27116 |
| 8 | 2 | 0.078 | 0.0096 | 0.04372 | 0.00538 | 3.32 | 34979 |
| 10 | 2 | 0.061 | 0.003 | 0.0340 | 0.002402 | 2.57 | 38865 |
| 2 | 4 | 0.066 | 0.006 | 0.0373 | 0.00343 | 2.81 | 32011 |
| 4 | 4 | 0.051 | 0.0052 | 0.028 | 0.0028 | 2.14 | 42011 |
| 6 | 4 | 0.038 | 0.0039 | 0.02128 | 0.002408 | 1.6 | 55509 |
| 8 | 4 | 0.023 | 0.0026 | 0.01268 | 0.001452 | 0.958 | 44160 |
| 10 | 4 | 0.014 | 0.0017 | 0.0058 | 0.00094 | 0.44 | 204748 |

## 4 Conclusions

This study presents the effect of PC and LQ on Atterberg limits, compaction and consolidation of soft clay (cohesive) soils. On the basis of the test results, the following conclusions can be drawn:

1. Observes to reduce the rate of dehydration of the soil when adding small amounts of the compound (PC & LQ). Also, notes the effect of the dehydration rate increase soil when adding high levels of the compound in the initial period during the first half hour at add the mixture to the soil.
2. The addition of a few percentage PC and LQ cause low decreased in plasticity index, while a higher percentage of the composite (PC & LQ) led to a reduction in plasticity index. Thus, PC & LQ added soils have better workability.
3. The addition PC & LQ increased optimum water content and decreased maximum dry unit weight of the soils.
4. The addition of the composite (PC and LQ) to the sample soils can improve the compressibility of soft clay soils by reducing the compression ratio (CR) from 0.1185 to 0.00705 (as averages for both soils), also reduction in rebound (swelling) ratio from 0.0276 to 0.0015 (as averages for both soils) was observed with increasing composite (10% PC and 4% LQ).
5. Through observe increment in the modulus of elasticity (Es) from 5179 kN/m$^2$ to 173954.5 kN/m$^2$ . PC & LQ treated soils exhibited much more brittle behavior compared with non – treated soils.

## References

1. E.W. Brand, and R.P., Brenner, Elsevier Scientific Publish Company, Amsterdam. (1981)
2. British Standard B.S: C.P. 8004 "Code of Practice for Foundation ", British Standard Institution, London. (1986)
3. B.B., Broms,. Proceeding 5th International Geotechnical Seminar 2 – 4 Dec. (1987)
4. E., Cokça, Electronic Journal of Geotechnical Engineering, Paper 9904. Oklahoma State University, USA. (1999)
5. G.N., Obuzor, J.M., Kinuthia and R.B. Robinson, Eng. Geol. 122 (3–4), 334–338. (2011)
6. E., Pomakhina, D., Deneele, A.-C., Gaillot, M., Paris, G., Ouvrard, Res. 42 (4), 626–632. (2012)
7. F. Sariosseiri, and B. Muhunthan, Journal of Engineering Geology Vol.104, pp.119-125. ,(2009)
8. D.D., Currin, J.J., Allen, D.N., Little, Report No. FJSRL-TR-0006. Frank J. Seisler Research Laboratory, United States Air Force Academy, Colorado. (1976)
9. D.O. Osula.., Eng. Geol. 42 (1), 71–81. (1996).
10. S., Bin, L., Zhibin, Yi, C., Xiaoping, Z.,.. Civ. Eng. 19 (1), 99–104. (2007)
11. Z., Metelkova, J. Bohac, , R., Prikryl, I., Sedlarova, Appl. Clay Sci. 61, 37–43. (2012)
12. J. Locat, , M.-A., Bérubé, M., Choquette Can. Geotech. J. 27, 294–304. ,(1990)
13. Wesley, D Laurence.". Journal of Geotechnical Engineering 114.6 718-723. (1988)
14. H.H., Baquir, "Stabilization of Fao Clay", M.Sc. thesis, University of Technology. (1990)
15. I.T., Jawad, , Research Journal of Applied Sciences, Engineering and Technology, Maxwell Scientific Organization, Vol.8, No. 4, pp:510-520. (2014)
16. P.G., Nichloson, Elsevier Scientific Publish Company, Amsterdam. (2015),

# Effect of foundation embedment on the response of a multi-story building to earthquake excitation

*Mohammed* Hamood[1,*], and *Sajjad* Abid[1]

[1]Building and Construction Engineering Department, University of Technology, Baghdad, Iraq

**Abstract.** During an earthquake, the behaviour of any structure is affected not only by the superstructure response, but also by the response of the soil beneath. Recent structural failure patterns have indicated the significance of soil-structure interaction (SSI) effects. The present study focuses on SSI analysis considering the embedment depth of the foundation of a symmetric six stories reinforced concrete (RC)space bare frame building resting on stiff soil and subjected to seismic loading. The finite element analysis software ANSYS v17.2 is used. Time history (TH) analysis has been adopted. The Response in terms of lateral displacements, base shear forces, base moments and variation in natural time periods are calculated from the analysis of the soil foundation structure interaction (SFSI) model. Results are compared with that obtained from conventional method assuming rigid support at the base (fixed base) of the structure. The results show that the SFSI considering different embedment depths are significant in altering the seismic response of the multi-story building (MSB).

## 1 Introduction

Over the last decades, the dynamic Soil Foundation Structure -Interaction (SFSI) has attracted an intensive interest among researchers and engineers in the fields of soil dynamics, structural dynamics and wave mechanics over the world. The investigation methods include experimental and analytical studies, since, the experimental investigations are very expensive and time consuming [1], calibration of present computer models with a similar previous experimental research can be a certainty, with the methods of analysis and the availability of computer programs, it is in principle possible to evaluate the dynamic response of any structure foundation system to any excitation of the base. Such evaluations, however, can be no better than the assumptions underlying the analyses. Many researchers like Milind [2], Faramarz et al [3] and Ola [4] recommended the inclusion of SSI effects on multi-story buildings. In this work the general Finite Element Analysis (FEA) program ANSYS has been used to evaluate the structural response.

## 2 Soil foundation structure Interaction

SFSI is defined as when an earthquake excitation occurs, the structure and the ground vibrate with influencing each other .This phenomenon is called "Dynamic Soil Structure Interaction (DSSI)" and is recognized as being very important for the seismic design of structures [5] . Soil structure interaction (SSI) has an impact on the structure response mainly in four basic aspects. These effects can be epitomized as first increased lateral displacements of the structure, second lengthened natural time period of the system, also, higher damping in the system and finally changing the base shear force depending on the dynamic characteristics of the soil also the structural system and frequency content of the excitation motion [6], degree of influence of SSI on response of building depends on foundation system type, stiffness of ground and dynamic characteristics of building itself, that is, damping factor and natural period [5].

## 3 Modelling of soil

In this research, Mohr-Coulomb criterion is used to model the soil; Mohr-Coulomb model is applicable to three-dimensional stress space models while being a simple model (Refer Figure (1)) with only two strength parameters to describe the plastic behavior. By means of true triaxial tests, researchers have indicated that stress combinations causing failures in real soil sample quite well agreed with the hexagonal shape of the failure pattern [7].

Mohr-Coulomb criterion is a first-order model shown in Fig.s (1 and 2), it is an elastic-perfectly plastic model which is often used to model the behavior of soil. In general stress state, the model's stress-strain behaves linearly in the elastic range, with two defining parameters from Hooke's law (Young's modulus, E and Poisson's ratio, $\nu$). Failure criterion is defined by two parameters (the angle of friction $\phi$, and cohesion, c) and also a parameter to describe the flow rule (dilatancy angle, $\psi$ which comes from the use of non-associated flow rule

* Corresponding author: mjh_smm@yahoo.com

© The Authors, published by EDP Sciences. This is an open access article distributed under the terms of the Creative Commons Attribution License 4.0 (http://creativecommons.org/licenses/by/4.0/).

which is used to simulate the realistic irreversible change in volume caused by shearing).

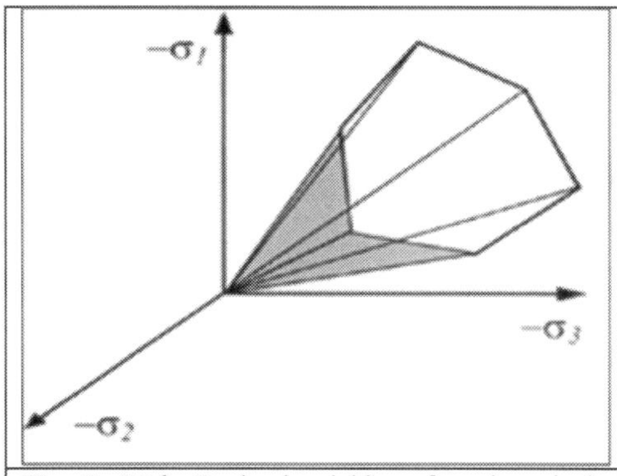


**Fig. 1.** Mohr-Coulomb yield surface in principal stresses [9].

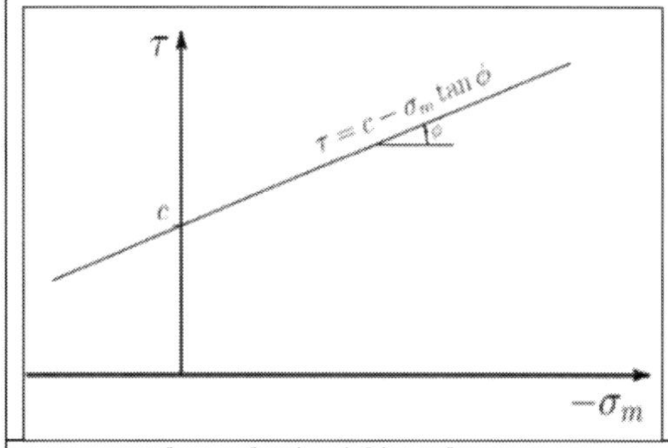


**Fig. 2.** Mohr-Coulomb yield surface as mean stress versus shear [10].

## 4 General description of the structure

The studied building is designed using structural design and analysis software ETABS 2016, according to building code requirements for structural concrete (ACI 318M-14), seismic design category D is used, the building consists of two bays in both directions with a spacing of 6 m in each bay, the building comprising of a six stories with a height of 3 m as shown in Fig.s 3and 4, the girders dimensions are b=0.4 m, h=0.5 m while the columns have a square section of 0.5 m, the longitudinal rebar have 16 mm diameter with four bars at the corners for both beams and columns, stirrups with a 12 mm diameter used in both the columns and beams with a spacing for both columns and beams is 150 mm.

## 5 Loading on the structural models

The building models in the present study were loaded vertically (superimposed dead load, live loads and self-weight of the members) and laterally (seismic loads). It was assumed a live load of (2 kN/m2), superimposed dead load (1.5 kN/m2) in addition to self-weight and seismic load, these loads are used to analyze all FEM - models.

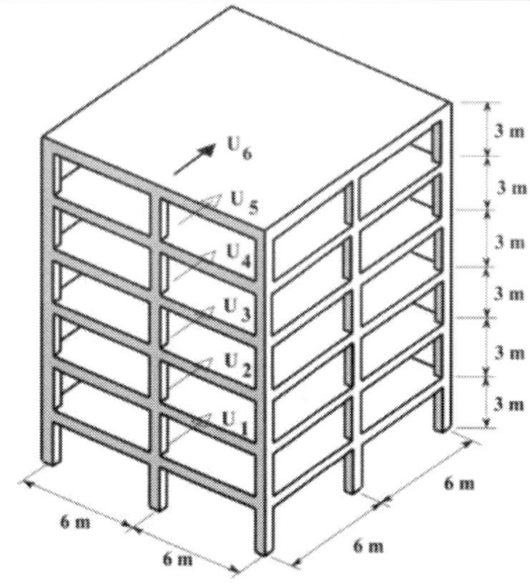


**Fig.3.** Building sketch.

**Fig.4.** FEM model for the Building

## 6 Types of modelling elements with their constitutive models

A finite element modeling is done for the superstructure along with the supporting soil using available finite element software ANSYS. The elements described below have been used in modeling of the structure and the soil:

### 6.1 SOLID65

Is an element used for modeling of 3-D solids without or with reinforcing bars (rebar). The element has an ability to crushing in compression and cracking in tension. The concrete material has the properties of crushing, plastic deformation, cracking (in three orthogonal directions), and creep [8]. Drucker-Prager yield criterion has been used as a constitutive model for concrete.

### 6.2 BEAM189

This element is 3-D quadratic solid with three-nodes, it has six degrees of freedom at each node; these include rotations about the x, y, and z directions and translations in the x, y, and z directions, creep, elasticity, plasticity and other nonlinear material models are supported. A cross-section referencing more than one material associated with this element type can be a built-up. BEAM189 is based on beam theory of Timoshenko, which is a first-order shear-deformation theory: assuming transverse-shear strain is constant through the cross-section (that is, cross-sections remain undistorted and plane after deformation) [8] bilinear isotropic hardening model has been used.

### 6.3 SOLID45

This is an element used for modeling of 3-D solid materials. The element is defined by eight nodes with three degrees of freedom at each node: translations in the nodal x,y, and z directions. The element has creep, plasticity, stress stiffening, swelling, large strain, and large deflection capability [8], Mohr-coulomb model is used for soil formulation.

## 7 Properties of materials:

Properties of Materials are summarized in the table 1.

## 8 Input seismic excitation:

As it is impossible to predict ground motion characteristics that may happen in the future at a construction site as the property of the ground motion is described by many factors such as seismic wave propagation from the source to site, fault mechanism and the amplification characteristics of ground. The significant factors of ground motions affecting the structure's response results are frequency contents, peak foundation acceleration, shapes of waveform and duration of ground motion in this study The 1940 El Centro earthquake has been selected see figures (5) and (6) for more details.

**Table 1**. Properties of materials used.

| Material | Element type | Properties | |
|---|---|---|---|
| Concrete | SOLID65 | Density( $kg/m^3$ ) | 2408.2 |
| | | Poisson Ratio | 0.2 |
| | | Modulus of Elasticity (MPa) | 29386 |
| | | Drucker-Prager Base | |
| | | Uniaxial compressive strength (MPa) | 38.1 |
| | | Uniaxial tensile strength (MPa) | 3.5 |
| | | Biaxial compressive strength (MPa) | 44.2 |
| Rebar | BEAM189 | Density( $kg/m^3$ ) | 7849 |
| | | Poisson Ratio | 0.3 |
| | | Modulus of Elasticity (MPa). | 199100 |
| | | Bilinear Isotropic Hardening | |
| | | Yield Stress (MPa) | 578 |
| | | Tangent Modulus (mPa) | 1991 |
| Soil | SOLID45 | Density( $kg/m^3$ ) | 1939 |
| | | Poisson Ratio | 0.28 |
| | | Shear Modulus (MPa) | 623.41 |
| | | Mohr-Coulomb | |
| | | Initial Inner Friction Angle (degree) | 40 |
| | | Initial Cohesion (Pa) | 5000 |
| | | Dilatancy Angle | 10 |
| | | Residual Friction Angle (degree) | 38 |
| | | Residual Cohesion (Pa) | 0 |

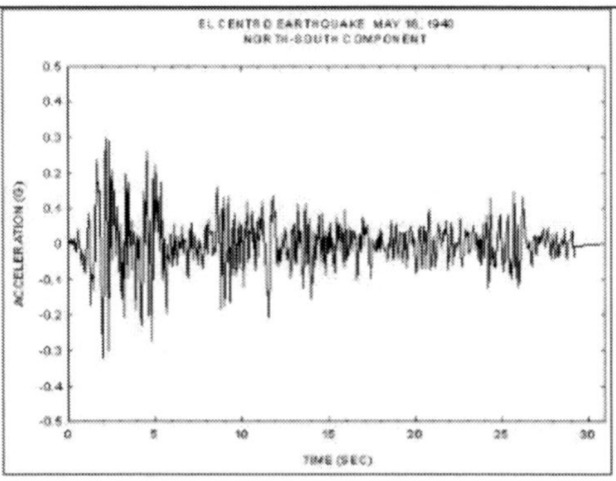


**Fig.5.** Earthquake acceleration time history [10].

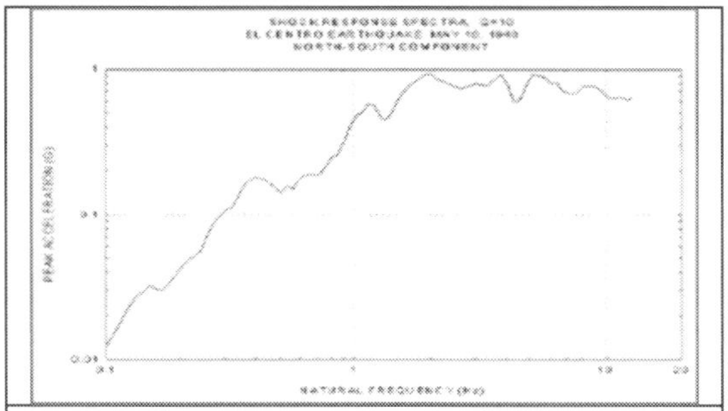

**Fig. 6.** Earthquake response spectrum [10].

## 9 Objectives

1- The present work focuses on the analysis of SFSI of six stories reinforced concrete bare frame for the following cases of analysis:

- Without considering the beneath soil i.e. without SSI effect (fixed case)
- With subsoil condition i.e. soil structure interaction analysis for different embedment condition with raft foundation.

2- Dynamic linear time history analysis method is carried out for all the models.

## 10 Results and discussion

### 10.1 Displacement response of the building:

SFSI, especially for multi-story buildings resting on relatively soft soils may significantly amplify the inter-story drifts and lateral displacements. This amplification of lateral deformations may result in changing the performance level of the structure. Thus, an extensive dynamic analysis to estimate the realistic performance level of a structure should consider the effects of SFSI on the model, the effect of SSI on the story lateral displacement of six stories building has been studied using three different embedment depths of the building, the lateral displacements with time history are presented in Fig.(7). It is observed that the displacement increase occurs in SSI models, the displacement increases more from embedment depth of one story to embedment depth of three stories even though the embedment depth have increased as could be seen clearly from the Table (2).Many researchers indicated this like Janardhan [11], Nithya [12], also he indicated that the low to mid rise buildings is mostly affected by SSI effects, the increased lateral displacement is due to soil flexibility.

### 10.2 Base shear

One of the primary inputs considered for seismic designs is seismic base shear. Base shear is defined as; the maximum expected lateral force that is estimated to occur at the base of a structure as result of seismic ground excitation. The effect of SSI on base shear of building frames with different embedment depths are presented in Table (3) and Figure (8). It is observed that base shear of the building is increased by 52% to 256% due to SSI effect.

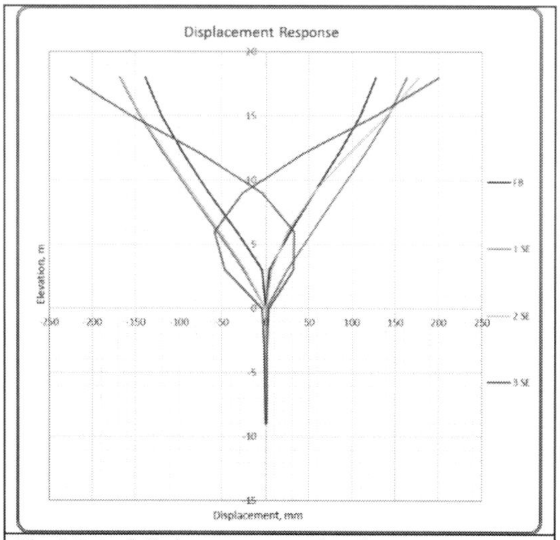


**Fig. 7.** Lateral displacement versus elevation.

**Table 2.** Maximum and minimum top displacement response summary.

| **Displace-ment response** | **Max. Displace-ment (mm)** | **Min. displac-ement (mm)** | **% of Max. FB respon se** | **% of Min. FB respons e** |
|---|---|---|---|---|
| FB (fixed base) | 128 | 140 | 100 | 100 |
| Embedm-ent depth of 1 story | 164 | 171 | 128 | 122 |
| Embedm-ent depth of 2 story | 177 | 168 | 138 | 120 |
| Embedm-ent depth of 3 story | 204 | 228 | 159 | 162 |

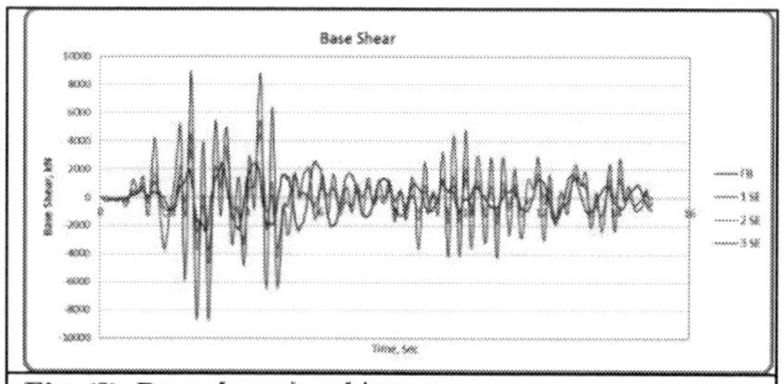


**Fig. (8).** Base shear time history.

 https://doi.org/10.1051/matecconf/201816201016

**Table (3).** Maximum and minimum base shear summary

| Base Shear (kN) | Max. Shear | Min. Shear | % of Max. FB Shear | % of Min. FB Shear |
|---|---|---|---|---|
| FB | 2560 | 2460 | 100 | 100 |
| 1 SE | 4478 | 3751 | 174 | 152 |
| 2 SE | 5450 | 4712 | 212 | 191 |
| 3 SE | 34581 | 34034 | 95 | 104 |

## 10.3 Base moment

The base moment of a multi-story building was computed for different embedment depths based on time history analysis. The base moment obtained from the multi-story building with flexible base is compared with that obtained from the multi-story building with FB. The variations of base moment evaluated from different embedment depths are shown in Table (4) and Fig.(9). It is found that the moment of base of the multi-story building increases with increased embedment depth of foundation. It is also seen that the base moment is maximum in multi-story building with FB as compared to SSI case. the base moment of multi-story building obtained from the 1 SE building is less than that obtained from the 1 and 3 SE building, the overall base moment is less than that with FB as can be seen in the Table 4.

**Table 4.** Maximum and minimum base moment summary.

| Base Moment (kN.m) | Max. Moment | Min. Moment | % of Max. FB Moment | % of Min. FB Moment |
|---|---|---|---|---|
| FB | 36046 | 32487 | 100 | 100 |
| 1 SE | 29578 | 29783 | 82 | 91 |
| 2 SE | 27357 | 30616 | 75 | 94 |
| 3 SE | 34581 | 34034 | 95 | 104 |

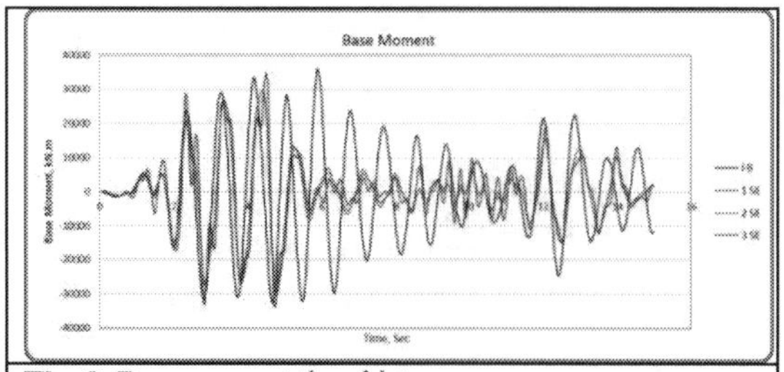


**Fig. 9.** Base moment time history.

## 10.4 Multi-story displacement response

Multi-Story displacement response of each floor level is an important factor that has been evaluated. The Story displacement response is investigated for studied multi-story building using different embedment depths for the foundation. It can be seen from Fig. (7), that the lateral displacement of the building is increasing even though the embedment depth of the foundation has been increased to reach a maximum increase in response of 102% compared to FB case, while the cases of 1 and 2 stories embedded have not shown that much difference in response compared to each other.

## 10.5 Max inter-story drifts

Story drift ratio is the maximum relative displacement of each floor divided by the height of the same floor is an important parameter that has been evaluated. The story drift ratio response is investigated for studied multi-story building using different embedment depths for the foundation. It can be concluded as shown in the figure(10) that the maximum inter-story case is at the base of the building when the SSI is considered regardless of the embedment depth, while in the case of fixed base (FB) the second story is the governing inter-story drift.

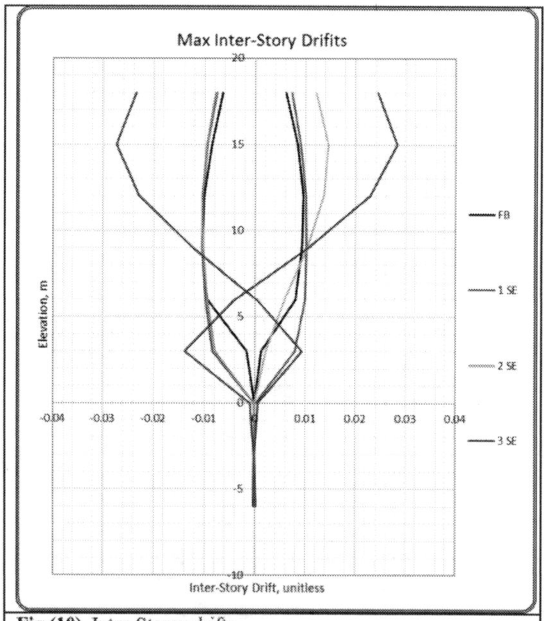


**Fig.(10).** Inter-Storey drifts

## 10.6 Fundamental period

The period of vibration is one of the fundamental parameters in the force-based design of structures as this parameter determines the spectral acceleration and turn the base shear force to which the building should be designed [14]. The fundamental period of vibration, T, is a function of the stiffness of the lateral load resisting system and the building mass [15]. The fundamental periods of the building are summarized in the table below.

**Table (5).** Fundamental period.

| Fundamental Period for Case: | Frequency (Hz) | Period (sec) | % of FB |
|---|---|---|---|
| FB | 1.1815 | 0.8464 | 100 |
| 1 SE | 1.0164 | 0.9839 | 116 |
| 2 SE | 1.064 | 0.9398 | 111 |
| 3 SE | 1.0233 | 0.9772 | 115 |

## 11 Conclusions

This study aimed to investigate the effects of Soil-Structure Interaction (SSI) with the foundation embedment depth of the building. Based on the results from this study, the following conclusions can be drawn:

• Due to soil flexibility, lateral displacement is increased as compared to fixed base, the lateral displacement is increased up to twice of fixed base lateral displacement response, the increase embedment showed no beneficial effect on displacement response of the multi-story building.

• While many codes around the world suggest methods to reduce the base shear when considering SSI, this study confirmed recent studies that are not the case always and the accurate response of structures is influenced by the media on which the structures is founded.

• The SSI tends to reduce the overall base bending of the multi-story building to some extent.

• The maximum story drift in the case of fixed base occurs at first story while on the case of SSI with varying embedment depths is at the base, i.e. it behaves closer to pinned base.

• The natural period of the structure is usually longer when it is analyzed with flexible base systems.

## References

1. B. Furkan and F. Birdal, Nat. Hazards Earth Syst. Sci., (2012).
2. V. M. Milind, International Journal of Scientific & Engineering Research, vol. 6, no. 10, (August-2015).
3. F. Khoshnoudian, E. Ahmadi, F.A. Nik, pp. 453-464, (2013).
4. O. A. Hussien, University of Technology, Bagdad/Iraq, (2015).
5. K. Miura, Graduate School of Engineering Hiroshima University, (2009).
6. P. W. John and J. D. Andrew , Foundation Vibration Analysis, Elsevier, (2004).
7. G. M, "True triaxial tests on dense sands," Balkema, Rotterdam, (1984).
8. ANSYS, Inc., ANSYS Mechanical APDL Theory Reference, Pennsylvania, United States: ANSYS, Inc., (1970).
9. D. F. Craig , A. S. Ahmed and D. L. Michael, The American Society of Mechanical Engineers, (May 20, 2013).
10. T. Irvine, "El Centro Earthquake," May 1940. [Online] Available, http://www.vibrationdata.com/elcentro.htm. [Accessed 15 2 2017].
11. J. Shanmugam, International Journal of Computer Applications, (2015).
12. N. Chandran, A. Rajan, International Journal of Emerging Technology and Advanced Engineering, vol. 4, no. Issue 11, 11, (November 2014).
13. R. Ditommaso, M. Vona, M. R. Gallipoli, M. Mucciarelli, Nat. Hazards Earth Syst. Sci., (4 July 2012).
14. A. R. Shehata , International Journal of Advanced Structural Engineering, (03 September 2013).
15. J. Raju, SSRG International Journal of Civil Engineering, pp. 135-141, (April 2015).

# Remediation of Clayey Soil Using Silica Fume

*Kawther* Al-Soudany[1,*]

[1]University of Technology, Baghdad, Iraq

**Abstract.** This paper evaluates the use of silica fumes as modification of fine-grained soil in order to alter undesirable properties of the native soil and create new useful soils. Silica fume as well as clay material, are used in changing the engineering properties to be compatible and satisfying this is due to their pozzolanic reactivity. The study aims to investigate the uses of these materials in geotechnical engineering and to improve the properties of soils. Four percentages of silica fumes were used in the present study, which is 0, 3, 5 and 7%. Classification, specific gravity, compaction characteristics, swell and swell pressure, CBR and compressive strength tests had been conducted on the prepared and modified soils. Results clarified that the silica fume increasing leads to decrease the plasticity index and liquid limit. Increasing in silica fume causes an increasing in plastic limit and optimum water contents while the maximum dry unit weight values decrease. The compressive shear strength, California Bearing Ratio (CBR), swell and swell pressure is improved by using silica fume so that silica fume can be considered as a successful material in improving the soil properties.

## 1 Introduction

A great challenge faces civil engineering especially when the subgrade soil is found to be clayey soils, these soils usually have a tendency to change in volume if there is any change in moisture content [1]. The changes in moisture content may be due to floods, rains, leaking sewer lines or water evaporation this due to the fact that soils is covered by a pavement or a building. Remarkably, these cases causes the cracking and breaking up of pavements, highway, railways and embankments, foundations and reservoir linings or channe [2, 3].

Construction of this type of soils needs to be improved by using additives. Improvement of subgrade soils has traditionally depended on treating the soils with cement, lime, or waste material such as slag, Silica fume [4, 5, 6]. The silica fume found to be abundant material, and it is about 40% cheaper than other material such as Portland cement [4] but there is insufficient data on the effect of this material on the compressive strength of clayey soil.

There is little data available to define the relationship of shear strength of clay with different percent of silica fume. [7] used silica fume-clay for liners and the main conclusion that using silica fume-clay causes a reduction" in permeability and swelling pressure with high shear strength for landfills, "found that a higher attaching strength between clay and silica fume particles resulted in higher internal shear strength in the clay silica fume mixture. There are very complex engineering problems in the clay soil, especially when accompanied by environmental changes in moisture content and undergoes large settlement under long term of loads. Failure of different structures constructed on clay soils in various locations in Iraq were recorded such as Ammara, Nasiriya and Basra that are characterized by their low undrained shear strength (<40 kPa) and compression index as high as 0.3 were reported [8, 9]. Thus, it is necessary to improve large areas covered by weak clay soil that has created problems to roads and airports built on clay soils because of their low bearing capacity.

This research concentrates on investigating experimentally the feasibility of stabilizing and improving the geotechnical properties of soft clay soil using a silica fume material in different proportions and to study the silica fume effect on the engineering characteristics of the stabilized clay.

## 2 Materials used

### 2.1 Soil

The soil used was taken from a site east of Baghdad. The soil has low plasticity classified (CL) according to Unified Soil Classification System (USCS). The physical and chemical properties are listed in Table 1, the grain size distribution is shown in Figure 1.

* Corresponding author: k.y.alsoudany@gmail.com

© The Authors, published by EDP Sciences. This is an open access article distributed under the terms of the Creative Commons Attribution License 4.0 (http://creativecommons.org/licenses/by/4.0/).

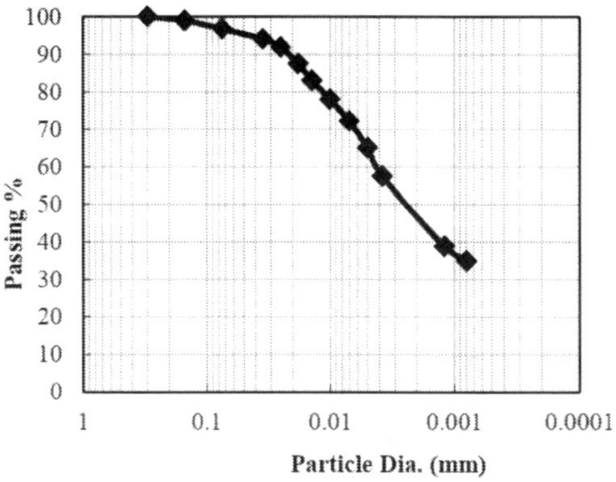


**Fig. 1.** Grain size distribution of soil used.

**Table 1.** Engineering properties of the used soil.

| Index property | Value | Specification | Bentonite |
|---|---|---|---|
| Liquid limit %(LL) | 45 | ASTM D 4318 | 512 |
| Plastic limit %(PL) | 18 | ASTM D 4318 | 38 |
| Shrinkage limit %(SL) | 27 | ASTM D 427 | - |
| Plasticity index %(PI) | 16.5 | ASTM D 4318 | 474 |
| Activity ($A_c$) | 0.45 | Skempton formula* | - |
| Specific gravity (Gs) | 2.69 | ASTM D 854 – 02 | 2.26 |
| Gravel (larger than 2mm)% | 0 | ASTM D 422 – 63 | - |
| Sand (0.06 to 2mm)% | 3.3 | ASTM D 422 – 63 | - |
| silt (0.005 to 0.06mm)% | 31.7 | ASTM D 422 – 63 | - |
| Clay (less than 0.005mm)% | 65 | ASTM D 422 – 63 | - |
| Gypsum content % | 2.92 | Chemical Analysis | - |
| $SO_3$ content % | 1.36 | Chemical Analysis | - |
| Organic matter O.M % | 0.43 | ASTM D 2974 | - |
| PH value | 8.7 | ASTM D 4972 | - |
| Soil symbols (USCS) | CL | ASTM D 2487 | - |

### 2.2 Silica fume material

During the manufacture of silicon and ferrosilicon alloys, Silica fume (SF) which is also known as micro-silica, is produced as a secondary product from reduction of high-purity quartz with coal in electric oven. It could be made from other silicon alloys such as ferrochromium, ferromanganese, ferromagnesium, and calcium silicon [10]. Silica fume is composed of very fine hyaline particles, about 100 times smaller than the average of cement particles with approximated surface area of 20,000 m2/kg (215,280 ft2/lb) depending on techniques nitrogen absorption measurement. Because of a high silica content and its extreme fineness it is considered a very effective pozzolanic material [10, 11]. One of its main uses is to improve concrete properties by increasing concrete bond strength, compressive strength, abrasion resistancein addition to reducing its permeability; therefore,

It helps in protection of the reinforcing steel from abrasion. In the present work, the chemical composition of the used silica is given Table 2, which is applicable according to the [10] requirements.

**Table 2**. Chemical composition of silica fume used in the tests

| Composition | Silica fume |
|---|---|
| $SiO_2$ | 98.87 % |
| $Al_2O_3$ | 0.01% |
| $Fe_2O_3$ | 0.01 % |
| CaO | 0.23 % |
| MgO | 0.01 % |
| K2O | 0.08 % |
| Na2O | 0.00 % |

## 3 Results and discussion

### 3.1 Atterberg limits

Concerning the effect of silica fume content on Atterbergs limits shown in Figure (2), (3) and (4). It was noticed that with increasing SF content there was a decrease in liquid limit and plasticity index. This could be because SF coats and binds all clay particles, which possesses large particles with little cementitious value. This effect is called a pozzolanic reaction between SF and aluminous material as stated by [6].

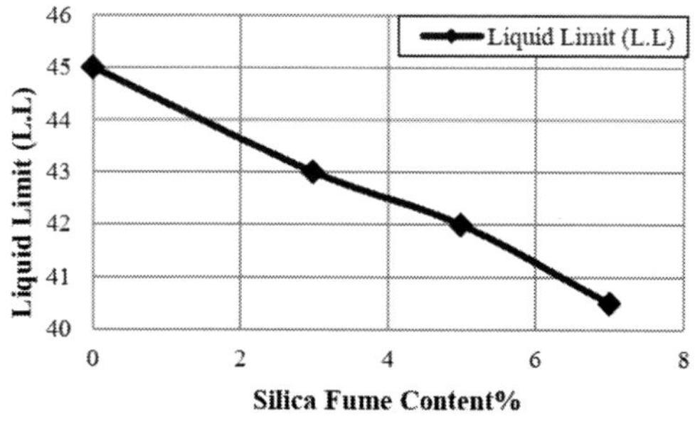


**Fig. 2.** Liquid limits behavior of clayey soil with increasing silica fume.

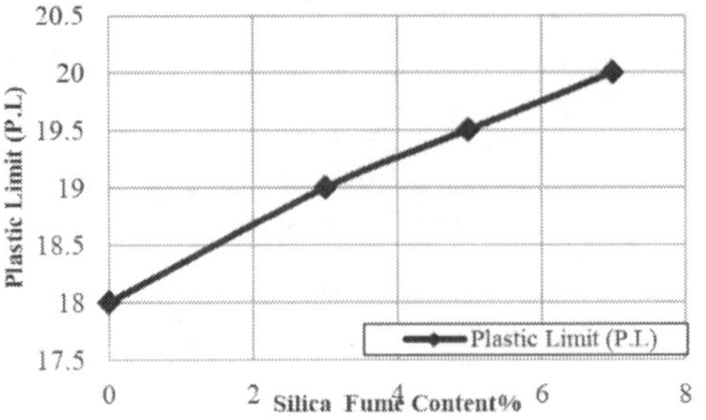


**Fig. 3.** Plastic limits behavior of clayey soil with increasing silica fume.

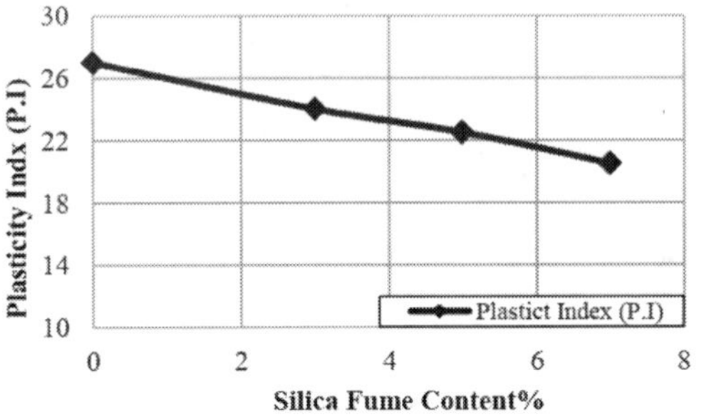


**Fig. 4.** Plasticity behavior of clayey soil with increasing silica fume.

## 3.2 Specific gravity

The addition of SF content affects soil's specific gravity. It is obvious from Figure 5 that with increasing SF content, the specific gravity of treated soil sharply drops from 2.69 for untreated natural clayey soil to 2.62 after 7 % SF addition. That is due to the very low specific gravity of the SF (2.2) compared with that of natural clay soil.

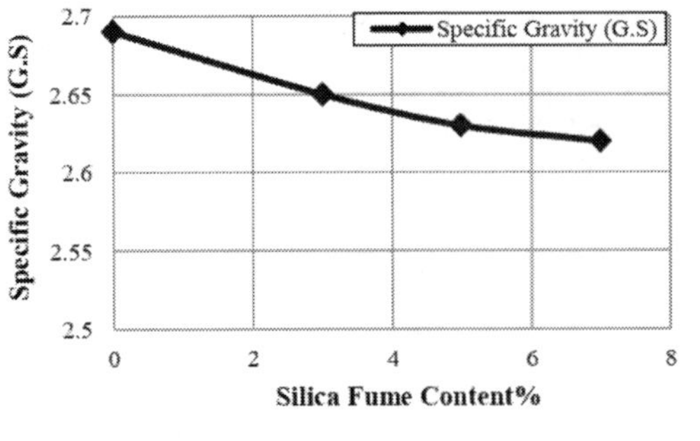


**Fig. 5.** Specific gravity of clayey soil with increasing silica fume.

## 3.3 Compaction test

The dry unit weight relationship with water content for different SF values, is shown in Figure (6). It is clearly shown that due to reduction in the parallel orientation to the clay particles at 7% SF there is a decrease in compaction effect. The optimum moisture content (OMC) is increased with increasing the amount of SF content as shown in Figure (7). While Figure (8) illustrates the effect of SF on the maximum dry unit weight.

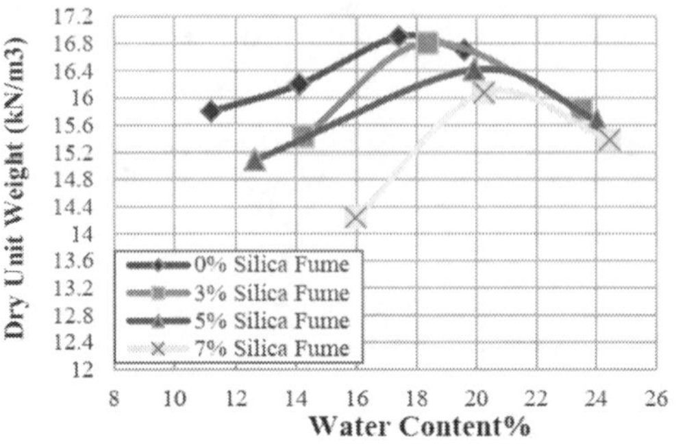


**Fig. 6** Variation of the dry unit weight with moisture content for different SF percents.

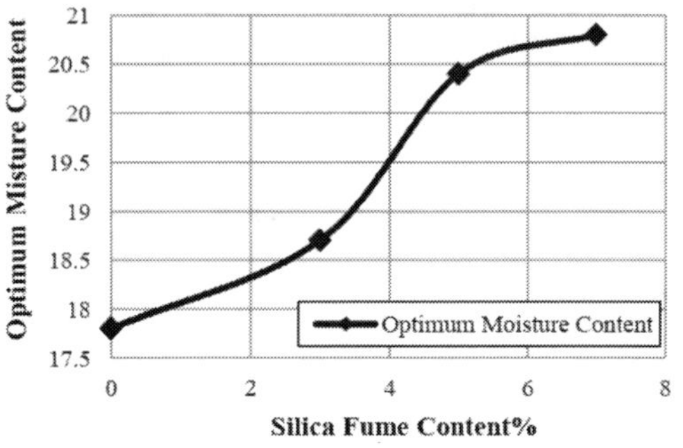


**Fig. 7.** Optimum moisture content variation with different SF contents.

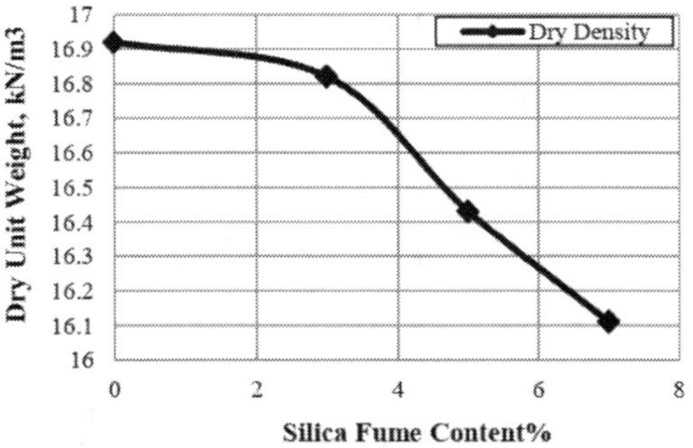


**Fig. 8.** Variation of the maximum dry unit weight with SF percent.

Generally, the optimum moisture content is increased with the increasing of SF percent from 17.8% to 20.8% at SF 7% as could be seen in Figure (7). This may be the effect of SF, which decreases the free silt clay fraction amount, and coarser material with forming of a large surface area, the water is needed here to make these operations take place. This also hinted that the compacting of soil-SF mixtures needs more water as mentioned in [12]. In the same way, can be as result of replacement of soil by silica fume in the blend, which has a lower specific gravity (2.2) in comparison to soil (2.69) the decrease in the maximum dry unit from 16.92 to 16.11 kN/m3, as shown in Figure (8).

### 3.4 Shear strength tests

This test is performed according to ASTM D 2435-02 [23] The most common and adaptable method to evaluate the strength of stabilized soil is the Unconfined Compressive Strength (UCS) test for the compacted soil. It is recommended to perform this test to define the required additive amount to use in operation of soil stabilization [13]. UCS tests were performed on natural soil with (3%, 5% and 7% SF). All the unconfined compressive tests were carried out for specimens with 24% moisture content and 15.5 kN/m3 dry unit weights by static compaction.

UCS testing for soil-SF mixtures was carried out after curing time at 20° C for 2 days and 100% relative humidity. The specimens remain in molds as suggested by ref. [14].

Figure (9) and (10) shows the relationship between the unconfined compressive strength and SF percents while Table (3) shows the relationship between the undrained angle of internal friction and SF percents.

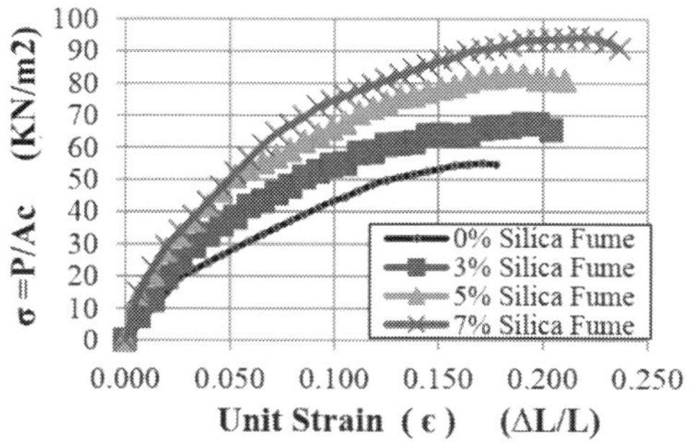


**Fig. 9.** Stress-Strain relationship from unconfined compression test for samples treated by SF.

**Table 3.** Shear strength parameters for soil stabilized with SF.

| SF % | 0% | 3% | 5% | 7% |
|---|---|---|---|---|
| cu, kPa | 27.5 | 33.7 | 41.3 | 47 |
| $\phi_u$, | 4 | 17 | 20 | 22 |

$\phi$u, * These values are calculate from angle of failure plane, $\theta = 45 + \phi u/2$

It can be noticed that UCS and the undrained angle of internal friction increases with increase of SF percents. The reason for these results of development strength is due to the pozzalanic reaction acceleration. The stages of preparation samples are illustrated in Figures (11), (12) and (13).

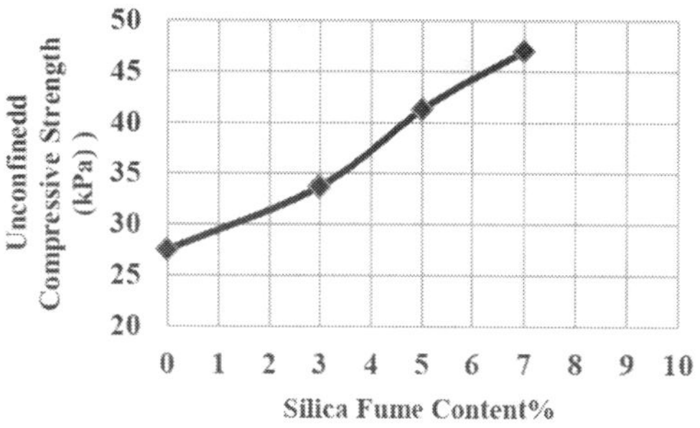


**Fig. 10.** Effect of silica fume on the unconfined compressive strength of soils.

**Fig. 11**. Preparation of Samples.

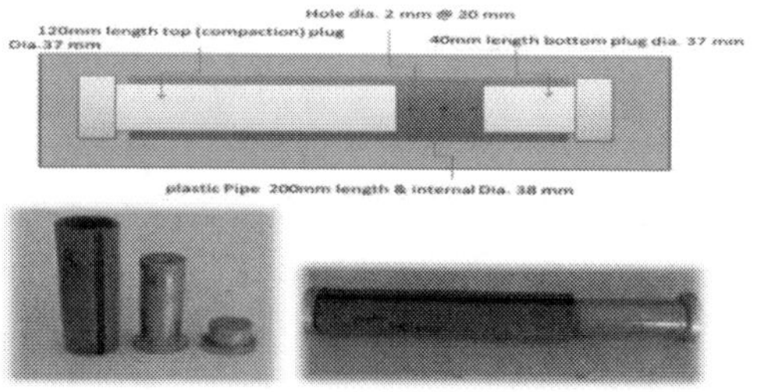


**Fig. 12.** The details of the plastic pipe and the metal plugs.

**Fig. 13.** Sample under shearing in unconfined compression apparatus.

## 3.5 California bearing ratio (CBR)

The adding effect of silica fume on California Bearing Ratio (CBR) values of clayey soil, shown in figures (14), (15). This simple, fast and reliable test had been used to verify the stabilization of weak soils by adding physical and chemical additives [15, 16]. The CBR test is described in ASTM Standards D1883-05 (for laboratory prepared samples) and D4429 (for soils in place in field) which this research depends on it, AASHTO T193. In general it could be seen that the addition of SF increase the CBR value of the soil. The optimum percent of SF mixtures was carried out after curing time at 20° C for 2 days, which gives a higher improvement of CBR value was found when adding 5% and 7%. Thus, there is an enhancement in CBR values with addition SF and finally a good relationship between the values of UCS and CBR shown in Figure (16).

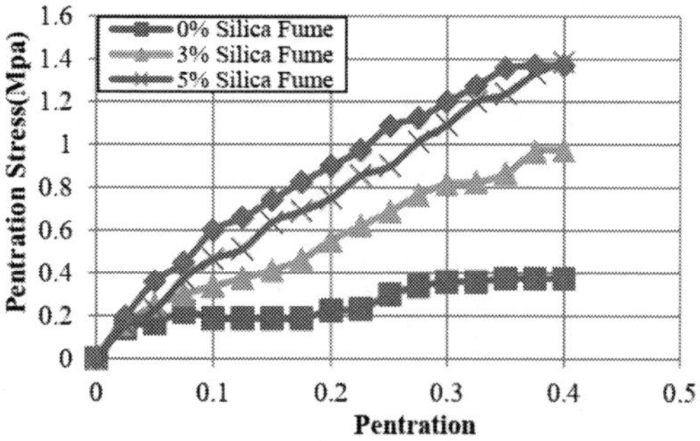


**Fig. 14.** Effect of addition of different percentages of SF on CBR.

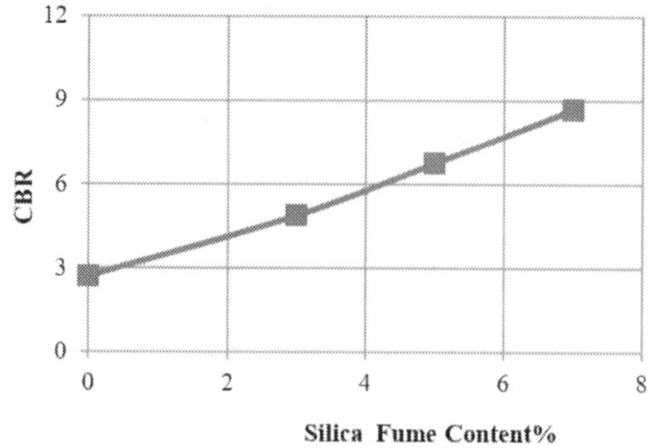


**Fig. 15.** Effect of addition of different percentages of SF on CBR.

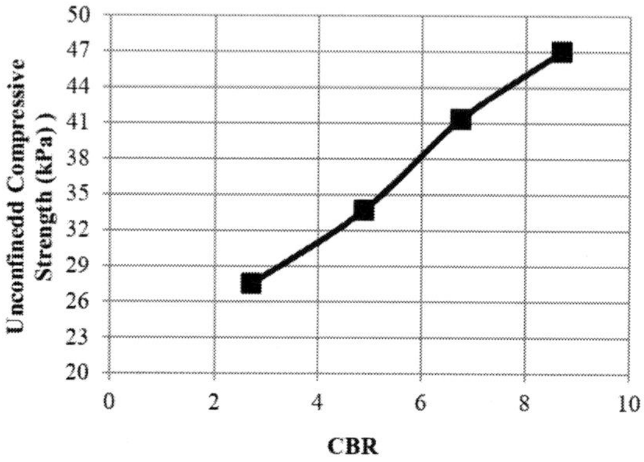


**Fig. 16.** Relationship between CBR and UCS.

## 3.6 Silica fume content effect on swelling

The soil samples used in this test were prepared in the laboratory by mixing natural soil with 50% of bentonite. Static Compaction Method was followed for the preparation of soil samples using a special mold and a compression testing machine. Load was applied statically to the soil sample at a compression rate of 0.04mm/sec. until the depth of penetration approached the required height (1.5cm), the samples were prepared at a dry unit weight of (16.92 $kN/m^3$) and moisture content of (17.8%). The load was maintained on the specimen for about five minutes after reaching the required height to prevent the occurrence of rebound after static load removale. Figure (17), (18) and (19) Show the silica fume effect in different percentages on the free swell and swell pressure of the prepared soil .From these figure it can be observed that the free swell and swell pressure decrease fast with increasing percentage of silica fume. The reduction in the swelling and swelling pressure values of the improved soils may be due to the addition of non-expansive silt-size particles to the expansive soil and the interaction of soil with silica fume particles. In this experiential study the silica fume consists mainly of spherical non-crystal silicate, aluminum, and iron oxides compounded with some microcrystal material and unburned carbon. Consequently, this decreased the specific surface area. This affinity of the inspected soil samples. This refers to the reduction in the swell-shrinkage values such as swelling pressure and free swell. And obtained that 5% silica fume was the perfect percentage to decrease the swell and swell pressure.

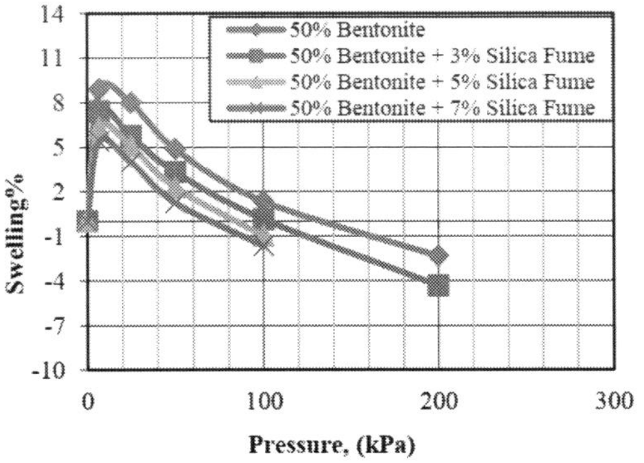


**Fig. 17.** Relationship between swelling and pressure for prepared soil with 3, 5 and 7% silica fume.

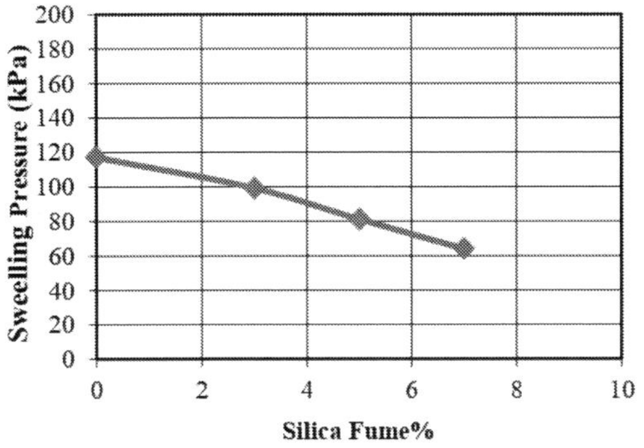


**Fig. 18.** Relationship between swelling pressure and silica fume content for soil with 3, 5 and 7 % silica fume.

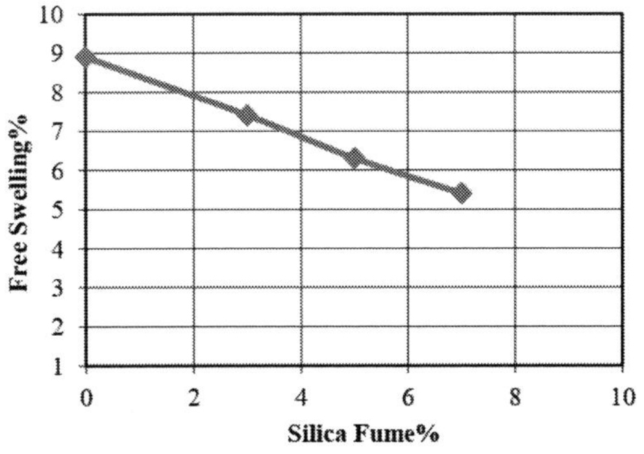


**Fig. 19.** Relationship between free swelling and silica fume content for soil with 3, 5 and 7 % silica fume.

## 4 Conclusions

Based on the results obtained, it can be concluded as follows:

1. For Atterbergs limits, decreasing in both liquid limit and plasticity index, increasing in plastic limit were observed with increasing SF content.
2. The specific gravity decreases with increasing of SF percentage.
3. The silica fume changes compaction parameters of clayey soil. The maximum dry unit weight decreases with the increase of the SF content, while optimum moisture content increases with increasing SF content.
4. Increasing SF improves the Unconfined Compression Strength (UCS).Silica Fume percentages at 5% & 7% gave UCS 41.3% and 47% respectively, after 2 days of curing compared to UCS of untreated soil 27.5%.
5. CBR increases with increasing SF content.
6. Free swell and Swell pressure decrease with increasing of SF percentage. And 5% was the perfect percentage to decrease free swell and swell pressure.

## References

1. F.H. Chen, *Foundation on Expansive Soils*, Development in Geotechnical Engineering, Elsevier Scientific Publishing Company, New York, USA, 463 (1988).
2. E. Cokca, EJGE. **4**, 1-4 (1999).
3. C.R. Wilson, J.G. Davis and N.Mejia, *Landscaping on Expansive Soils*, www. ext. colostate. Edu, (2001)
4. E.J. Yoder, and M.W. Witczak, *Principles of Pavement Design*, 2nd Edition, John Wiley & Sons, Inc., USA, 699 (1975).
5. A.S. Muntohar, Regional Seminar at Islamic University of Indonesia Sept. 4th, Yogyakarta, Indonesia, HYPERLINK (1999).
6. Abd El-Aziz., M. Abo-Hashema , and M. El-Shourbagy, *Fourth Mansoura International Engineering Conference (4th IEC)*, Faculty of Engineering, Mansoura University, Egypt, April (2004).
7. E. Kalkan, and S. Akbulut, , Eng. Geol. Journal, **73**, 145-156 (2004).
8. S.F Ahmed, M.Sc. Thesis, University of Technology, (2015).
9. H.H. Karim, Z.W. Samueel, S.F. Ahmed, *Proceedings of the 2nd International Conference of Buildings, Construction and Environmental Engineering, American University of Beirut (BCEE2), 17-18 Oct., 2015, Geotechnical Engineering Part,* 117-121 (2015)

10. ACI Committee 226, *Silica Fume in Concrete: Preliminary Report*. ACI Materials Journal March-April, PP.66-158 (1987)
11. M.D. Luther, *High-Performance Silica Fume (Microsilica)-Modified Cementitious Repair Materials*, 69th Annual Meeting of the Transportation Research Board, paper no. 890448 (January), 88-94 (1990)
12. K. Harichane, M. Ghrici. And S. Kenai, Int J Civ Eng, Vol. **9**, 2 (2011)
13. R.K. Goswami, and B. Singh, Ground Improvement, **9**, 175-182 (2005)
14. D. N. Little, , *Handbook for Stabilization of Pavements, Subgrades, and Base Course with Lime*, Kendall/Hunt Publishing Company, Dubuqe, Iowa (1995).
15. N. Yong Raymond, V. R. Ouhadi., *Experimental study on instability of bases on natural and lime/cementstabilized clayey soils*, Applied Clay Science (2007).
16. H. Beeghly Joel, S. Michael, Int J Soil Sediment Water, **3** (2010)

# Favorable of grouted micropiles for the load transfer in weak sandy soils

*Ahmed* Al-Obaidi[1,*], and *Ansam* Al-Karawi[2]

[1]Tikrit University, Tikrit, Iraq
[2]Cihan University, Erbeil, Iraq

**Abstract.** Micropiles are defined as small diameter piles, implemented as a cast-in-place replacement or injected grout. Generally assumed with a diameter less than 300 mm. The axial capacity of micropiles develops primarily through the bond between gravity grouted (Type A) or post grouted (Type B, C or D) and soil in bonded zone of the micropile. Because of this, micropiles are useful in a variety of applications. Micropiles (Type D) indicate a two-step process of grouting, neat cement grout is placed under gravity head, then, additional grout is injected via a sleeved grout pipe at a specified pressure. In this research, a numerical model was developed to simulate the properties of the micropile (Type D), in addition to micropile (Type A). A general finite element program ABAQUS was selected for the numerical analysis method and for generating the solution. The parameters that affect the load transfer and ultimate capacity of the micropile foundation, which includes micropile length, diameter, grouted length and weak soil type, were investigated. The main results show that the pressure grouting micropile provides better load transfer mechanism as it makes surrounding soil strength to be fully mobilized upon axial loading. Micropiles (Type D) show more favorable in a capacity where the increment range is up to 50% as compared with ground grouted micropiles (Type A). In addition, the results indicate that the ultimate capacity of micropile increases as the grouted length increases.

## 1 Introduction

Micropiles first appeared as a cost effective way to rehabilitation of existing historical structures. Recently, micropiles have gained popularity all over the world and are being used for buildings, bridges, antenna towers, slope stability and residential construction. Micropiles is superior in case of difficult drilling conditions where other deep foundation methods are not reasonable and consist of any combination of rebar, grout, hollow bar, steel pin, and steel casing.

Typical micropile design procedures are covered by Federal Highway Administration [1], where the bond stress method was developed solely. The use of finite element method for the micropile analysis was adopted by many authors, [2], [3], and [4]. Simulation of micropiles Type D differs from micropiles Type A in which that micropile Type D indicates a two-steps process of grouting. The primary grout is placed under gravity, and after hardening of initially placed grout, additional grout is injected via tube a Manchette at a pressure of 2 to 8 MPa. A packer may be used so that specific levels can be treated several times, if required. In cohesionless soil, the pressure grout, a certain amount of permeation and replacement to loosened soil takes place and causes the formation of cake-like cement paste along the grout soil interface that improves the bond. Pressure grouting also appears to cause a re-compaction or re-densification of the soil around the borehole, [5]. The effective micropile diameter (after post grouting) depends on magnitude of grouting pressure and type of soils; the radial expansion occurs so drilled hole diameter extend from (D) to ($\alpha c$ D)

The effective diameter is estimated using a correction factor ($\alpha c$) [6] that allows for radial expansion due to pressure grouting typical values are shown in Table (1). Many researchers suggested the form of micropiles after the execution in the site with various shapes, as illustrated in the Fig. (1).

In the present study, a general finite element program ABAQUS/CAE is selected to appraise the capacity of micropile in the weak soil, involving the soil-structure interaction which is considered as important topic in developing the performance of the different type of micropiles.

## 2 Parametric study

In the analysis of micropile, the axisymmetric model is built to simulate the soil-micropiles interaction and load

* Corresponding author: dr.obaidi.a.h@tu.edi.iq

© The Authors, published by EDP Sciences. This is an open access article distributed under the terms of the Creative Commons Attribution License 4.0 (http://creativecommons.org/licenses/by/4.0/).

transfer of two micropiles type A and D in weak soil. Also, the micropiles case one under axial loading were used with different lengths and diameters ranging from 5m to 20m and 0.1m to 0.3m, respectively and embedded in weak soil and resting on strong soil. In addition, the effect of post grouted on ultimate micropiles capacity are studied with different grouted length ($L_G$) for micropiles type D. The grouted length was taken as (2, 4…….etc.) m until all micropile length became post grouted, i.e. ($L_G/L$) = 1.0

The convincible shape used to simulate these micropiles in ABAQUS/CAE program are shown in Fig. (2). The magnitude of (αc) was chosen from 1.4 to 1.5 depending on the type of soil.

**Table 1.** Correction coefficient $\alpha_c$ for micropiles type D [6]

| Soil Type | $\alpha_c$ |
|---|---|
| Gravel | 1.8 |
| Sandy gravel | 1.6-1.8 |
| Gravely sand | 1.5-1.6 |
| Coarse sand | 1.4-1.5 |
| Medium sand | 1.4-1.5 |
| Fine sand | 1.4-1.5 |
| Silty sand | 1.4-1.5 |
| Silt | 1.4-1.6 |
| Clay | 1.8-2.0 |
| Marl | 1.8 |

## 3 Numerical models using ABAQUS / CAE program

In ABAQUS/CAE, the block of model consists of parts, in this work two parts used to simulate the soil-micropiles system and the model is created with axisymmetric, deformable body with a shell, base feature.

The boundaries condition of this model extends in the lateral direction by a range equal (10.5 D), D, of micropile diameter from the centre of micropiles, and (L+10D), L, the length of micropiles below the tip in the vertical direction, [7]. Also, the boundary conditions assumed to be hinged at the end of the lower soil to prevent the horizontal and vertical movements, a roller inside of soils to allow a vertical movement and prevent horizontal movement, and symmetry/ antisymmetry / ancastre for the micropile's side.

Elasto – perfectly plastic model with Mohr – Coulomb failure criterion, usually named as Mohr-Coulomb model, is widely used in finite element analysis of geotechnical engineering [8], [9], and [10]. In the present work, the Mohr-Coulomb criterion is used to simulate the soil material due to its simplicity and sufficient accuracy. Table (2) shows the properties of weak soils used in this study.

The micropile is modeled as linear elastic according to [11], [8], and [9]. The value of modulus of elasticity for the micropile is (14.2 GPa) according to [12], and the Poison's Ratio is assumed to be 0.2.

Within a model, the sequence of one or more analysis steps can be defined. In this research, the steps must be used to analyze the behavior of a single pile embedded in weak soil including initial, geostatic and loading steps. The initial step, it allows defining boundary conditions, predefined field, and interactions that are applicable at the very beginning of the analysis. In the geostatic (First) step the gravity load is included, and the acceleration is specified to be 9.81 m/sec2 in the vertical direction and zero in the horizontal direction, but the pressure loading applied in the pile top in the loading (second) step in high magnitude to reach to failure.

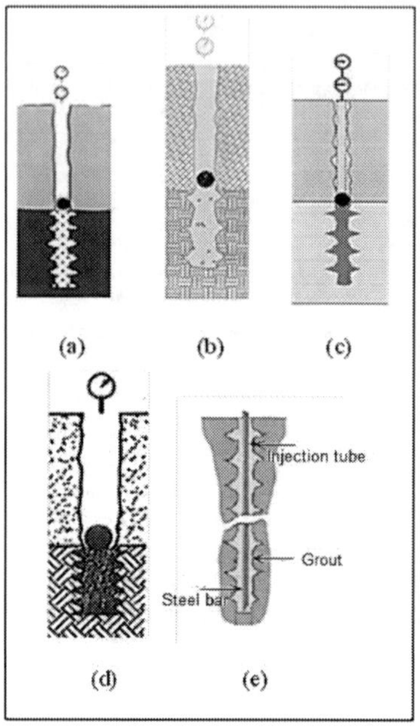


**Fig.**1. Suggested shapes of micropile type (D) by different Authors

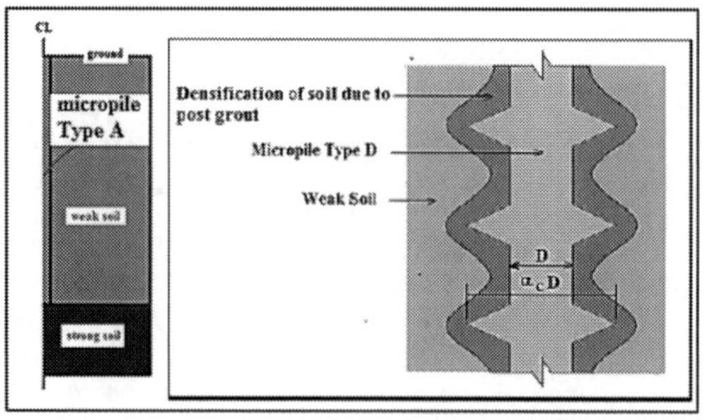


**Fig. 2.** Developed micropile type (D) and (A) model

## 4 Interaction between micropiles and soils

In this study, interface modeling has three steps:

- Defining the contact surfaces that could potentially be in contact.

**Table 2.** Soil Properties used in this Study

| Angle of Internal Friction, Ø° | Density | Unite weight, γ kN/m³ | Modulus Elasticity of Soil, Es kN/m² | Poisson's ratio, ν |
|---|---|---|---|---|
| 25 | Loose | 15 | 25000 | 0.3 |
| 30 | Medium | 16 | 37500 | 0.3 |
| 35 | Dense | 17 | 50000 | 0.32 |
| 40 | V. Dense | 18 | 80000 | 0.35 |

- Starting which surfaces interact with one another (master and slave).
- Defining the mechanical and thermal properties of the surface (tangent and normal).

All the parts of the model are modelled using the Second-order continuum axisymmetric element (CAX4) with reduced order numerical integration available in ABAQUS (6.11).

In the surface based contact approach, two surfaces have to be defined based on their rigidity and the assignment of "master" and "slave" roles to the contact surfaces is used to model the interaction between micropile and soil, (Fig.3).

The contact behaviors at the pile-soil surface include load transfer mechanisms both in the normal and tangent direction. In this research a surface-to-surface contact with finite-sliding formulation was used to define the interaction between the micropile sides and the surrounding soil, the soil is assumed to be slave surface and the pile is master surface [13].

The friction along the pile shaft was found using Coulomb's friction law, [9], and [14]. The tangential behavior was defined using "penalty" friction formulation, with friction coefficient, μ, and the shaft resistance stress is

$$f_s = \mu\, \sigma'_x = \sigma'_x \tan \varphi'_i \qquad (1)$$

Where

$\mu$= the coefficient of friction

$\sigma'_x$= the lateral effective stress

$\varphi'_i$= the soil-pile interfacial effective friction angle.

"Hard" contact model was used to define contact pressure in normal behavior, [7].

## 5 Ultimate capacity of micro piles type (D)

Micropiles Type (D) constructed with post grouted pressure resulting with non-uniform diameters. Therefore, the skin friction increased leading to increasing the ultimate capacity. In this part, the ultimate capacity of micropiles type (D) is investigated. Also, the effect of grouted length ($L_G$) on the increment of ultimate capacity is examined.

The typical load-settlement curves were found as shown in Fig. (4). The ultimate capacity for the micropile was established using the method proposed by [15], where the ultimate load can be determined as the abscissa part of the load-settlement curve changes to a falling straight line.

Figures (5) and (6) show the relationship between the ultimate capacity of micropiles with grouted length ($L_G$), the grouted length is increased from zero (micropile Type (A)) with 2m steps until the full length are grouted.

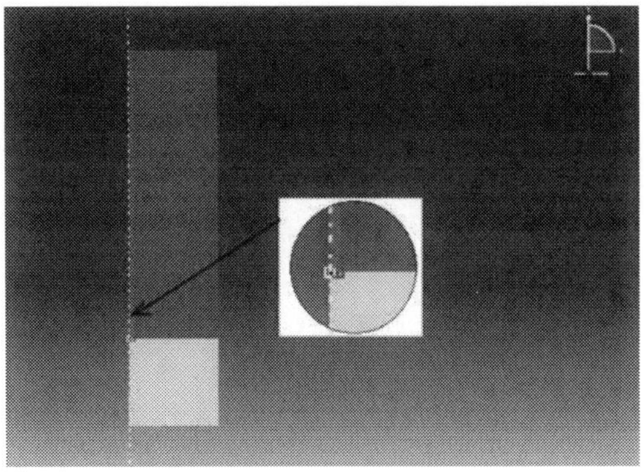

**Fig. 3.** Interaction between soil and micropiles

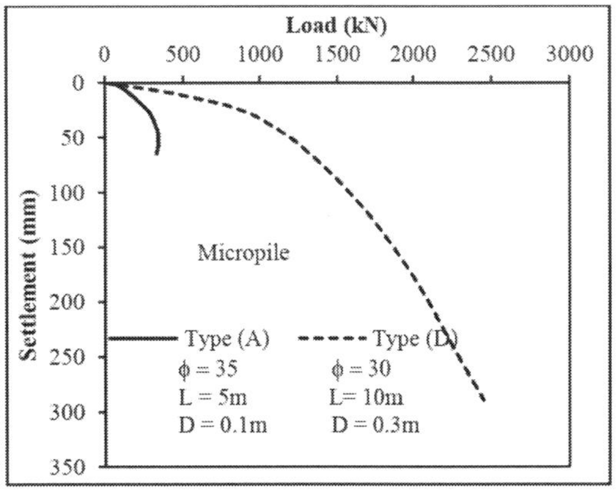


**Fig. 4.** Typical load – settlement curve obtained from the results for micropiles Type (A) and (D)

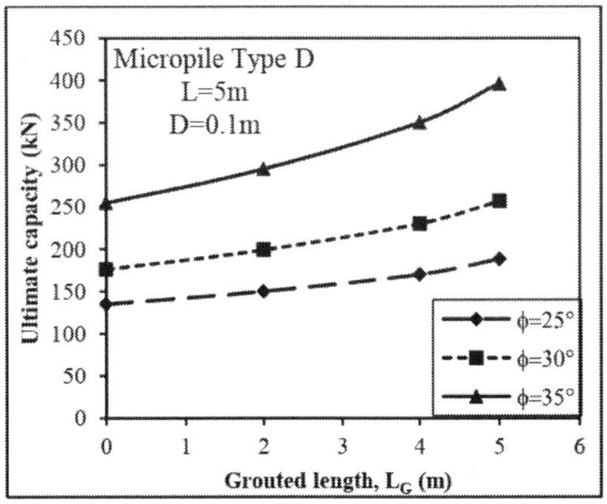


**Fig. 5.** Ultimate micropile capacity versus grouted length for different soil friction and constant diameter, D=0.1m and length = 5m

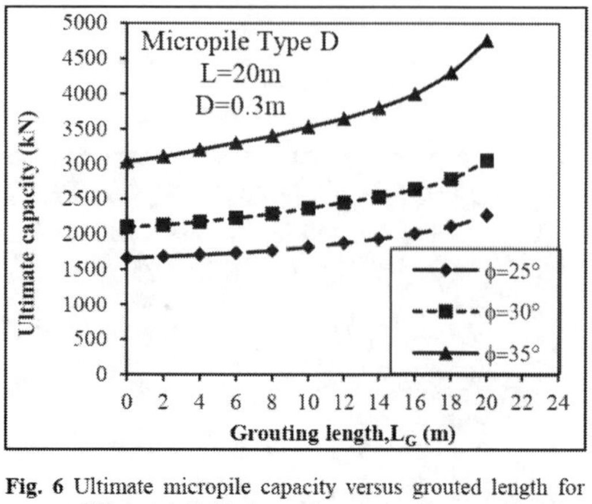


**Fig. 6** Ultimate micropile capacity versus grouted length for different soil friction and constant diameter, D=0.1m and length = 20m

The ultimate capacity of micropile increases as the grouted length increased, the trend of the curve notice to be linear when the grout length less than two third of the micropile length measured from the base. Then the trend became cone curve up; this behavior reflects the effect of the skin friction at the top layers of the soil, also the bonding zone length with toothed edges increases. Thus, it is recommended to post grouted full length of micropile to get better ultimate increment capacity. When the soil friction increased the effect of post grouted on the ultimate capacity are became pronounced more than less soil friction angle, herein the effect of interaction between the grouted cement and the soil increased led to increasing the shaft resistance and so on increased the ultimate capacity of micropile

## 6 Ultimate capacity ratio of micro piles type (D)

The ultimate capacity of micropile Type (D) was studied by using the ultimate capacity ratio for micropile type (D). This ratio represented the ultimate capacity of micropile grouted at different length to the ultimate capacity micropile type (A) at same length, diameter, and angle of internal friction of soil, $(U.C.R)_{LG}$

Figures (7) to (18) show the relationship between the ultimate capacity ratio for micropile Type (D), $(U.C.R)_{LG}$ and the grout length percentage ($L_G$/L). In these figures, the zero value of grout length percentage ($L_G$/L) at x-axis represents micropile type (A). In these figures, the ultimate capacity ratio $(U.C.R)_{LG}$ increases as the grout length percentage increases; the increment is found to be running high as the ($L_G$/L) % increases, where the bonding zone is increased. The value of the ultimate capacity ratio appears to be close to the grout length percentage not exceeding 20% for a different angle of internal friction of soil and micropile diameter.

The ultimate capacity is ratio increased by increasing the angle of soil friction increases. Also, the grouted pressure densify the soil around the micropile and lead to increasing the coefficient of earth pressure. Higher coefficients of earth pressure mean higher shaft capacity because of higher pressure exerted by the soil. The effect will be stronger for longer micropiles due to the increase in the effective stress of the soil.

In micropiles embedded in the weak soil, an increase in the length (L) resulted in a noticeable increase in the ultimate unit skin friction, as revealed in Table (3). For a length increase from L=5m to L=20m with a constant diameter (0.1m), the ultimate capacity increased from 188 kN to 684kN, thus, an elevation of 73%, and 264% for the same length increase but for micropile diameter 0.3m. Increasing might be due to an increase in the bonding pressure. For a certain micropile diameter, the ultimate capacity ratios give the impression to be much closed for micropiles with full grouted length, but for $L_G$/L less than 100% of these values are departed.

## 7 Load transfer through soil for micro piles type (D)

The load transfer behavior of micropile Type (D) embedded in weak soil and resting on the dense soil during loading is shown in Figure (19).

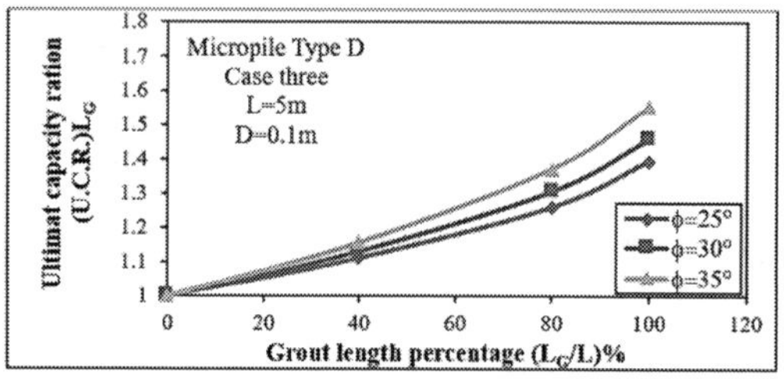


**Fig. 7.** Ultimate capacity ratio versus grout length percentage for micropile with 5m length and 0.1m diameter

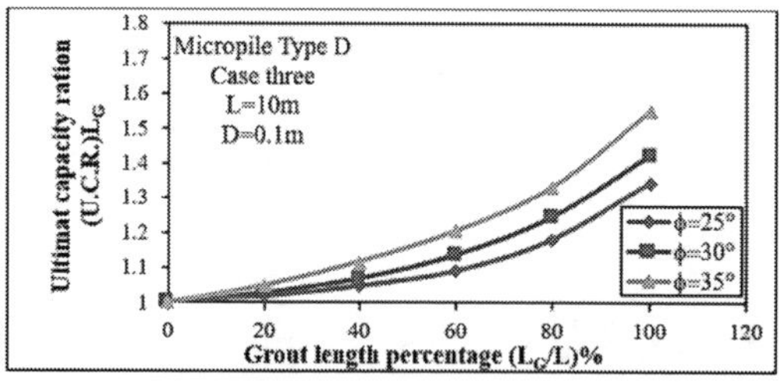


**Fig. 8.** Ultimate capacity ratio versus grout length percentage for micropile with 10m length and 0.1m diameter

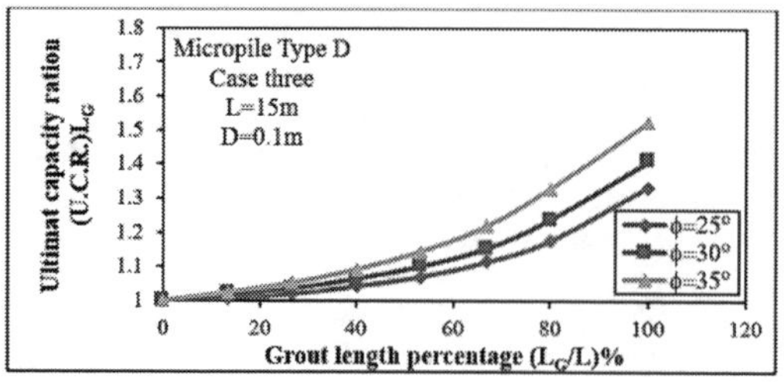

 https://doi.org/10.1051/matecconf/201816201018

**Fig. 9.** Ultimate capacity ratio versus grout length percentage for micropile with 15m length and 0.1m diameter

It is clearly from this figure that skin resistance of the micropile shaft start to transfer the load to the surrounding soil and after the amount of loading increment the end bearing reaction start carrying the load. During loading the load transfer to the skin near the micropile then to the surrounding, soil far away. Before the failure, the shaft resistance is fully mobilized before the base resistance, but the load carrying by the latter is greater than that for micropile type (A).

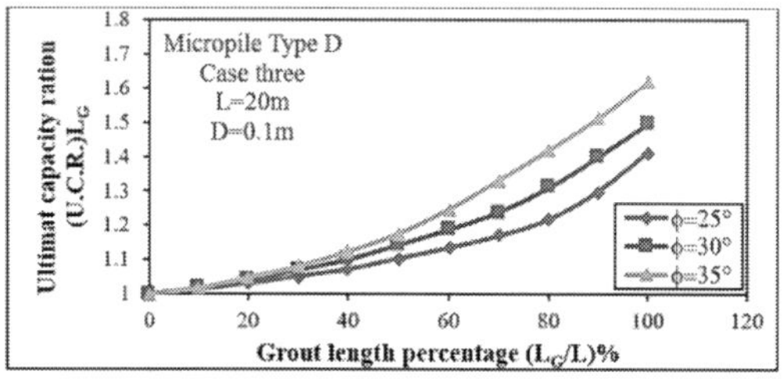


**Fig. 10.** Ultimate capacity ratio versus grout length percentage for micropile with 20m length and 0.1m diameter

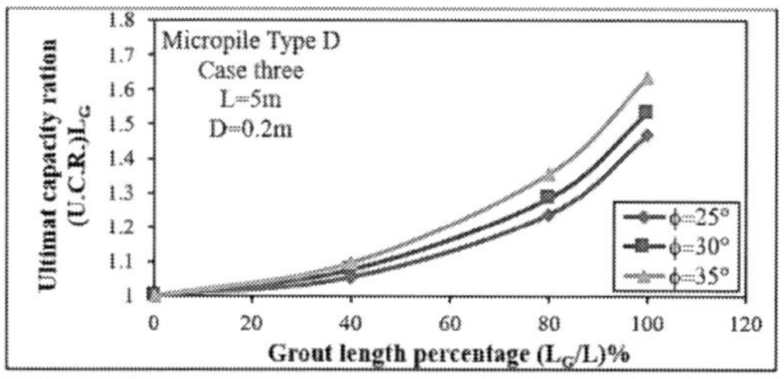


**Fig. 11.** Ultimate capacity ratio versus grout length percentage for micropile with 5m length and 0.2m diameter

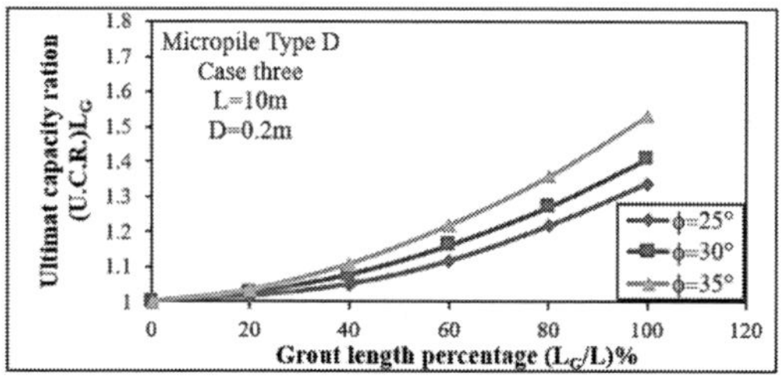


**Fig. 12.** Ultimate capacity ratio versus grout length percentage for micropile with 10m length and 0.2m diameter

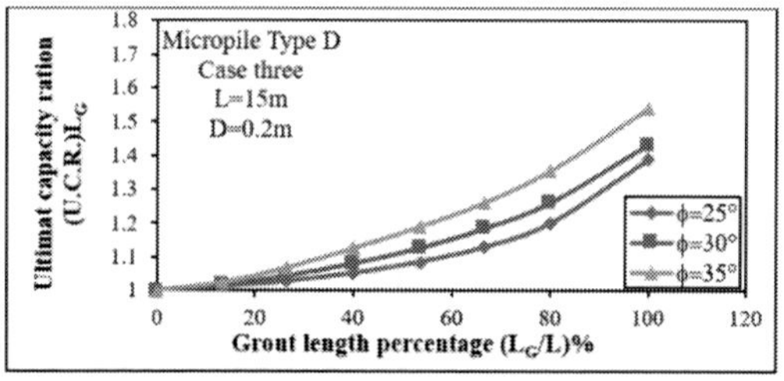


**Fig. 13.** Ultimate capacity ratio versus grout length percentage for micropile with 15m length and 0.2m diameter

## 8 Conclusions

Different behaviors are shown for Type (A) and Type (D), in terms of the mechanism of load transfer and the amount of load carrying capacity. In contrast, the effect of changing micropile length, diameter, and soil friction angle had given a similar behavior, where increasing any of these parameters increases the ultimate capacity of the micropile, with a different percentage.

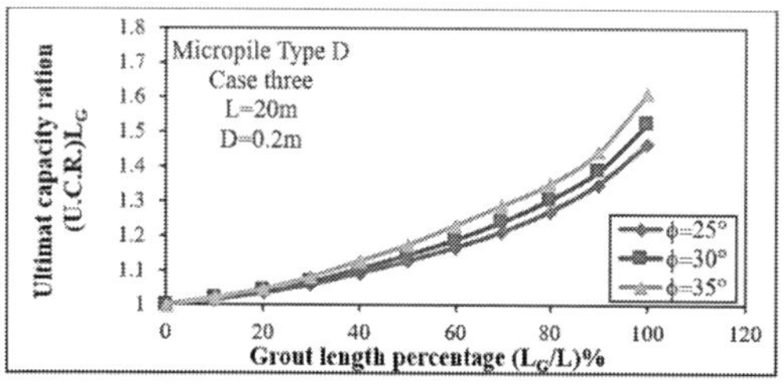


**Fig. 14.** Ultimate capacity ratio versus grout length percentage for micropile with 20m length and 0.2m diameter

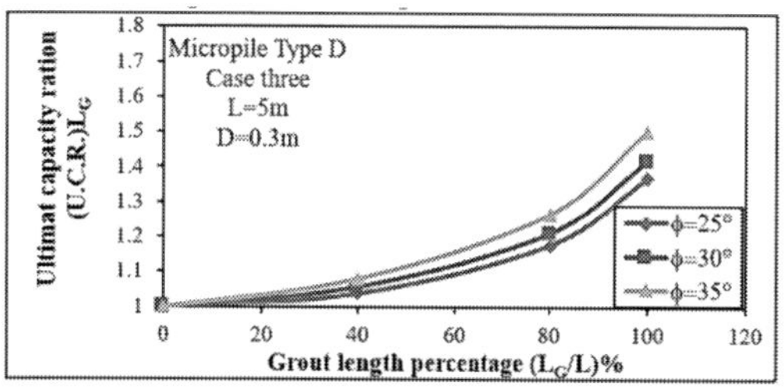


**Fig. 15.** Ultimate capacity ratio versus grout length percentage for micropile with 5m length and 0.3m diameter

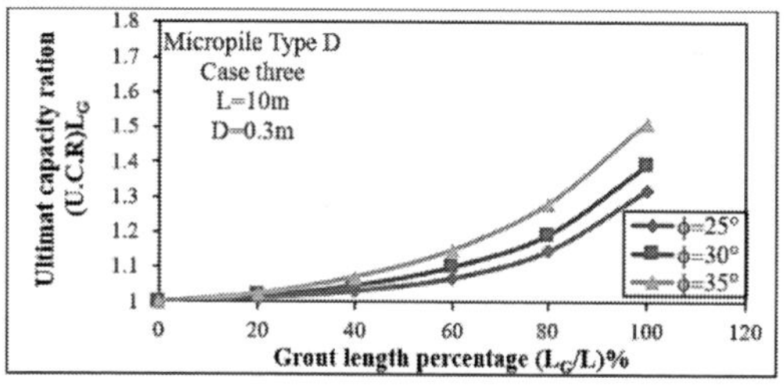


**Fig. 16.** Ultimate capacity ratio versus grout length percentage for micropile with 10m length and 0.3m diameter

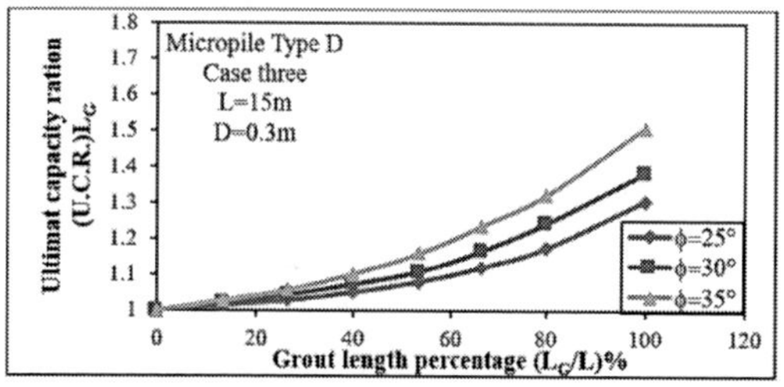

**Fig. 17**. Ultimate capacity ratio versus grout length percentage for micropile with 15m length and 0.3m diameter

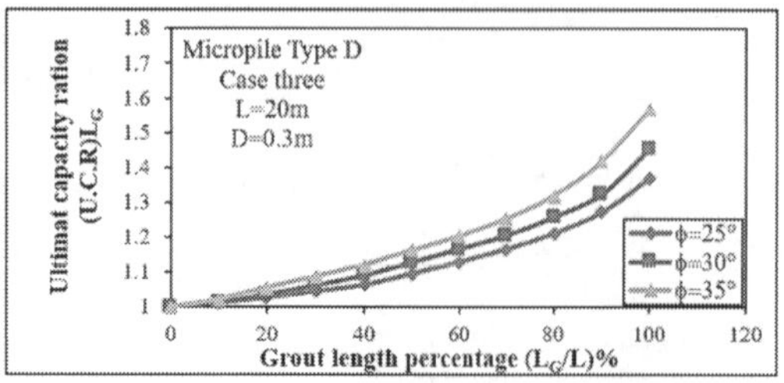


**Fig. 18**. Ultimate capacity ratio versus grout length percentage for micropile with 20m length and 0.3m diameter

It can be shown from the analysis results that the micropiles Type (D) gives the higher values of ultimate capacity as compared to micropiles Type (A). The ultimate capacities ratio for micropiles Type (D) ranges from (1.03) to (1.63) as compares with Type (A). Thus, it is preferable to the designer to think of employing micropile Type (D).

**Table 3.** Ultimate capacity of micropiles Type (D) (full grouted length)

| Micropile | | Angle of Soil friction, | Ultimate capacity, |
|---|---|---|---|
| Length, L (m) | Diameter, D (m) | ϕ (°) | (kN) |
| 5 | 0.1 | 25 | 188 |
| 5 | 0.1 | 30 | 257 |
| 5 | 0.1 | 35 | 396 |
| 10 | 0.1 | 25 | 370 |
| 10 | 0.1 | 30 | 520 |
| 10 | 0.1 | 35 | 790 |
| 15 | 0.1 | 25 | 533 |
| 15 | 0.1 | 30 | 740 |
| 15 | 0.1 | 35 | 1158 |
| 20 | 0.1 | 25 | 684 |
| 20 | 0.1 | 30 | 973 |
| 20 | 0.1 | 35 | 1513 |
| 5 | 0.2 | 25 | 431 |
| 5 | 0.2 | 30 | 585 |
| 5 | 0.2 | 35 | 940 |
| 10 | 0.2 | 25 | 803 |
| 10 | 0.2 | 30 | 1091 |
| 10 | 0.2 | 35 | 1660 |
| 15 | 0.2 | 25 | 1173 |
| 15 | 0.2 | 30 | 1590 |
| 15 | 0.2 | 35 | 2441 |
| 20 | 0.2 | 25 | 1487 |
| 20 | 0.2 | 30 | 2020 |
| 20 | 0.2 | 35 | 3126 |
| 5 | 0.3 | 25 | 674 |
| 5 | 0.3 | 30 | 890 |
| 5 | 0.3 | 35 | 1308 |
| 10 | 0.3 | 25 | 1267 |
| 10 | 0.3 | 30 | 1670 |
| 10 | 0.3 | 35 | 2541 |
| 15 | 0.3 | 25 | 1800 |
| 15 | 0.3 | 30 | 2370 |
| 15 | 0.3 | 35 | 3712 |
| 20 | 0.3 | 25 | 2271 |
| 20 | 0.3 | 30 | 3050 |
| 20 | 0.3 | 35 | 4758 |

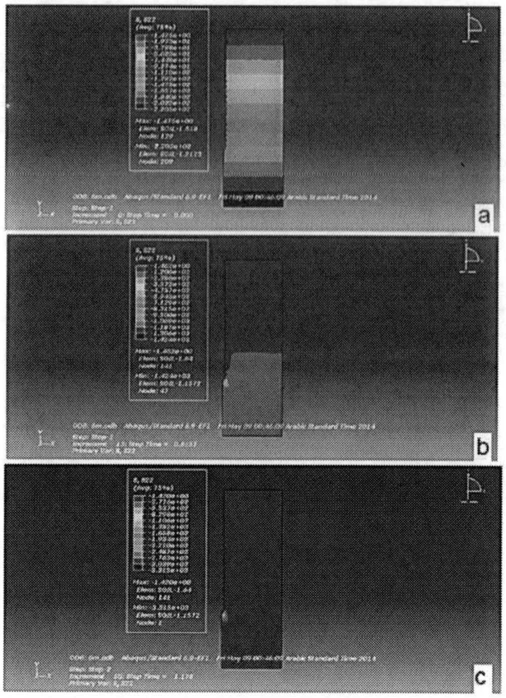


**Fig. 19.** Load transfer for micropile Type (D), (a) geostatic condition, (b) part of load applied, (c) before failure

## References

1. FHWA, Micropiles Design and Construction Reference Manual, Sabatini, P.J., Tanya, B., Armour, T., Groneck, P. and Keely, J., Report No. FHWA-NHI-05-039, (2005).
2. J. Frassetto, C., Performance of Micropiles, M.Sc. Thesis, Concordia University, Canada, (2004).
3. A. Elsalfiti, K., Skin Friction of Micropiles Embedded in Gravelly Soils, M.Sc. Thesis, Concordia University, Canada, (2011).
4. D. Barron, Finite Element Modeling of Micropiles and the Influence of Steel Casing on Load Transfer Mechanisms, M.Sc. Thesis State University of New York at Buffalo, USA, (2016)
5. L. Shong, S. and F. Chung, C., Design and Construction of Micropiles, Geotechnical Course for Pile Foundation Design and Construction, pp: 1-49, (2003).
6. M. Bustamante, and B. Doix, A Method for Calculating Injected Anchors and Micropiles, Bull Liaison Lab. Highways, Paris, (1985)
7. Y. Zhan, H. Wang, and F. Liu, Modeling Vertical Bearing Capacity of Pile Foundation by using ABAQUS, EJGE, **17**, pp. 1855-1865, (2012).

8. G. Wang, and N. Sitar, Numerical Analysis of Piles in Elasto- Plastic Soils under Axial Loading, 17th ASCE Engineering Mechanics Conference, University of Delaware, Newark, pp: 8-14, (2004)
9. S. Salgado, M. Prezzi, and H. Seo, Advanced Modeling Tools for the Analysis of Axially Loaded Piles, Advances in Deep Foundation, International Workshop on Recent Advanced of Deep Foundations (IWDPF07), Port and Airport Research Institute, Yokosuka, Japan, (2007).
10. H. Seo, S. Jeong, and Y. Kim, Load Transfer Analysis of Rock-Socketed Drilled Shaft by Coupled Soil Resistance, Computer and Geotechnics, 36, Issue **3**, April, pp: 446–453, (2009).
11. J. Lee, H., and R. Salgado, Determination of Pile Base Resistance in Sands, Journal of Geotechnical and Geoenvironmental Engineering, Vol. **125**, No.8, pp: 673-683, (1999).
12. H. Elarabi, and M. Alhadi Alshareef, Estimation of Micropile Capacity Grouted Under Pressure, American Journal of Engineering, Technology and Society, Vol. **1**, No. **2**, pp. 11-14, (2014).
13. FHWA, Use of Micropiles for Foundations of Transportation Structures, Seo, H. and Prezzi, M., Final Report, FHWA/IN/JTRP-2008/18, (2008)
14. A. Stuedlein, W., M. Gibson, D. and G. Horvitz, E., Tension and Compression Micropile Load Tests in Gravelly Sand, 6th International Conference on Case Histories in Geotechnical Eng., Arlington, pp: 11-16, (2008)

# Effect of stabilizers on the shear strength of residual soil

*Lamyaa* Najah Snodi [1,*], *Yahya* Atemimi [2] and *Fauziah* binti Ahmad [3]

[1] Civil Engineering Department, College of Engineering, Tikrit University, Tikrit, Iraq
[2] Civil Engineering Department, College of Engineering, Babylon University, Hilla, Iraq
[3] School of Civil Engineering, Universiti Sains Malaysia, Pulau Penang, Malaysia

**Abstract.** In this paper, chemical stabilizer was used for soil characteristics improvement. Styrene Butadiene Rubber (SBR) polymer and lime is effectively used as an inexpensive and easily applied material for soil stabilization and its response performance was evaluated. This research studied improving the engineering properties of Residual soil by mixing it with varying percentages of chemical stabilizer. An evaluation of the effectiveness and performance of SBR and Lime as a soil stabilizer was performed over a series of laboratory tests for geotechnical soil properties. Residual soil used in this study was lateritic soil. The soil was mixed with various amounts of stabilizer for polymer (SBR) {2.5, 5, 10 and 12.5 %} and for lime {1.25,2.5,5 and 10%} by weight and then compacted at the optimum moisture content (OMC) and maximum dry unit weight (MMD). The original soil and the stabilized samples were subjected to unconfined compression test to determine their strength at different curing times (7,14 and 28 day). The results showed a reduction in the plasticity index. The results showed an increase in strength.

## 1 Introduction

In order to solve soil problems, the researchers have paid serious attention in use of a variety of chemical additives to improve the geotechnical properties. Chemical stabilizers can be classified into traditional stabilizers and non-traditional soil additives. Traditional stabilizers could be cement , cement dust [1] and lime [2]. Nontraditional stabilizers can be classified into: sulfates, salts, petroleum resins, tree resins, and polymers [3,4].

There is no study that has thoroughly investigates selecting the type of stabilizer and no clear procedure to follow these materials has been provided [5,6].

Styrene Butadiene Rubber (SBR) is one example of liquid additive nontraditional stabilizer, derived from styrene and butadiene monomers [7]. Lime is hydrated high calcium as one of traditional stabilizers.

Residual soils are separated over a wide range of tropical and semi-tropical region in the world. Therefore, it is commonly used in various engineering practices [8]. A lateritic residual soil has a reddish color; this type of soil is unsuitable for use in construction practices because of its low strength or excessive volume change due to varying moisture content. In these years, the types of deserted soils have started to fall; therefore, the construction companies have turned to un-desert soil, which needs to be improved. Therefore, soil stabilization is one of the most important solutions to improve soil characteristics [9].

This study investigates the strength of residual soils to evaluate the effect of stabilizer on it.

## 2 Materials and methods

Residual soil used in this study has been collected from (USM), Penang, Malaysia.

Distributed soil samples were collected from a depth of 1.0 m from the ground surface and then air- dried. The soil were classified into SM, according to the Unified Soil Classification System (USCS), Table1

**Table 1.** Soils properties

| Properties of Soil | Values |
|---|---|
| $G_s$ | 2.66 |
| Color | Reddish yellow |
| LL% | 44 |
| PL% | 32 |
| PI% | 11.8 |
| % finer<0.075 mm | 32 |
| pH | 4.9 |
| Organic content (OC)% | 6 |
| MMD g/cm$^3$ | 1.85 |
| OMC % | 13.7 |

Styrene Butadiene Rubber (SBR) was obtained from the Company in Malaysia under its commercial name NBT-I. Also, Lime (hydrated high calcium [Ca $(OH)_2$]) was used. Tests were conducted on the stabilizer (polymer

* Corresponding author: dr.lamyaanajah@tu.edu.iq

© The Authors, published by EDP Sciences. This is an open access article distributed under the terms of the Creative Commons Attribution License 4.0 (http://creativecommons.org/licenses/by/4.0/).

and Lime) to identify the properties such as density, shear strength, and pH [10] and these are reported in Table 2.

**Table 2.** Chemical composition of the stabilizer

| Test | Property | | specification |
|---|---|---|---|
| | SBR | Lime | |
| PH | 5.5 | 11.8 | BS.1377: Part3:1990 |
| Density | 1.05 g/cm$^3$ | - | |
| Strength of Shear | increase with time | | ASTM - D196-99 |

The quantities of additives and curing time used in this study are listed in Table 3. All samples were dried in oven at (105 -110°C) over a 24 hour period.were added as a percentage of weight , for SBR blending with distilled water at its optimum moisture content (OMC) then added to the soil. Lime was added to the soil then mixed by hand-, until it reached a uniform color. The samples were then stored in a plastic bag to maintain its moisture content. A basic property test for soil was conducted on control soil. The engineering test properties were conducted in a mixture of soils with an allowance for curing time.

**Table 3.** Program for Mix design

| Soil : SBR ratio | Soil : Lime ratio | Curing time (Days) |
|---|---|---|
| 1:0.025 | 1:0.0125 | 7,14,28 |
| 1:0.05 | 1:0.025 | 7,14,28 |
| 1:0.1 | 1:0.05 | 7,14,28 |
| 1:0.125 | 1:0.1 | 7,14,28 |

A gradation test was conducted on the soil used in this experiment. Hydrometer and Wet sieving test were carried out using distilled water according to BS1377: part 2: 1990. Soil was passed through sieve (4.75mm) before the test.

British standard compaction was used. Dried soil samples by oven were passed through sieve with a 4.75 mm, mixed with 0, 2.5, 5, 10 and 12.5% SBR and with 0, 1.25, 2.5,5 and 10% Lime of dry soil were used. This test was carried out in accordance with BS 1377: part 4: 1990 [11].

The cylindrical soil specimens were having dimensions of 152 mm diameter and 127 mm height. The samples were prepared in accordance with BS 1377: part 4: 1990 clause 7.2.4. By using the machine for compaction, the load was applied with strain rate of (1mm/min) and a cell load of 5kN having steel road (50 mm) diameter fixed at the top of the frame of device was used to record the load applied to the sample. Moreover, a surcharge load of 50N was used as recommended by BS 1377: part 4: 1990: clause 7.4.2.3, and 7.4.3.4.

UCS test was performed on cylindrical samples with a height of 76 mm and a diameter of 38mm, according to B.S 1377: part 7: 1990: clause 7. The required weight of the soil used was calculated using equation:

$$W_d = \gamma_{d\,max.} \times V \quad (1)$$

$W_d$: dry soil weight in (gram),
$\gamma_{d\,max.}$ : maximum dry density( g/cm$^3$),
$V$: volume of mold in cm$^3$.

The chemical stabilizers (SBR, lime) mixed with a quantity of water equal to the optimum moisture content (OMC) of the specimen which was obtained from the compaction test. The mixture was then compacted to the maximum dry density based on the Modified Proctor compaction test.

# 3 Results and discussion

## 3.1 Gradation

Figure 1 shows the gradation result for the residual soil used. The results showed the soil SM has 32% fine grain soil with 100% passing through a No.4 sieve as a silty sand soil.

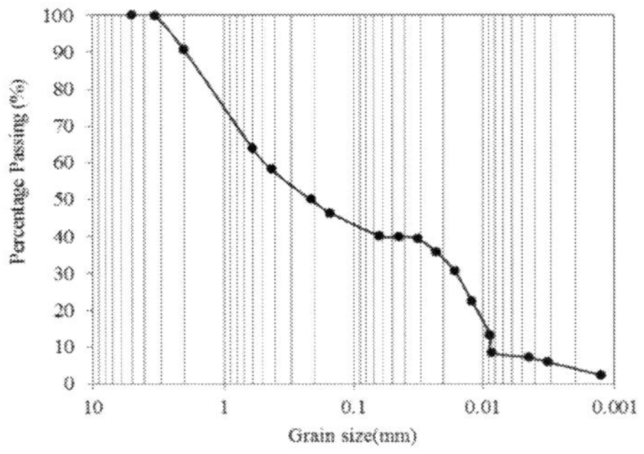


**Fig. 1.** Distribution curve for soil used

## 3.2 Index properties

From the plasticity chart for soil as shown in Figure 2 and Figure 3, the soil treated with SBR and Lime at different percentages exhibited a slight change in the position of all SM – (SBR , Lime) percentage mixtures and were classified as remaining at low plasticity under the A- line. Identical trends were reported by [12, 13].

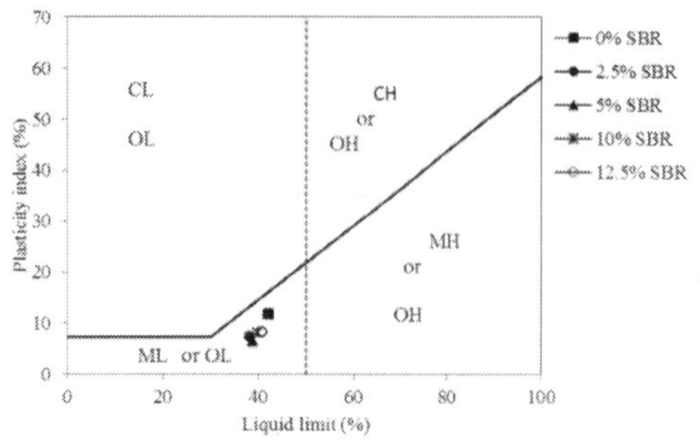


**Fig. 2.** Effect of SBR on consistency limit.

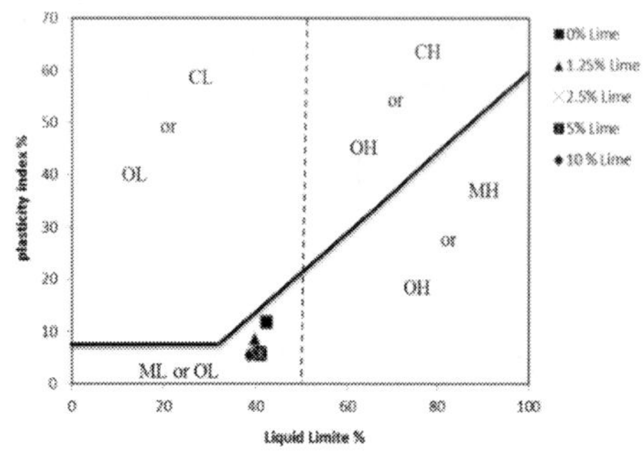


**Fig. 3.** Effect of lime on consistency limit.

Figure 4 and Figure 5 shows a reduction in the plasticity index a maximum reduction of 5% .The soil exhibited a different behavior for SBR. Increase in the plasticity index can be attributed to the liquid stabilizer (SBR) which filled the voids of the soil. The Atterberg limits of this soil when untreated can be attributed to its mineralogical composition with respect to residual soils.

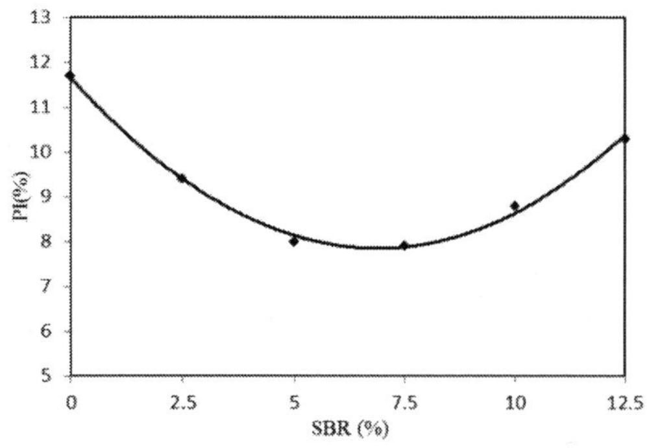


**Fig. 4.** Effect of SBR on PI of soil.

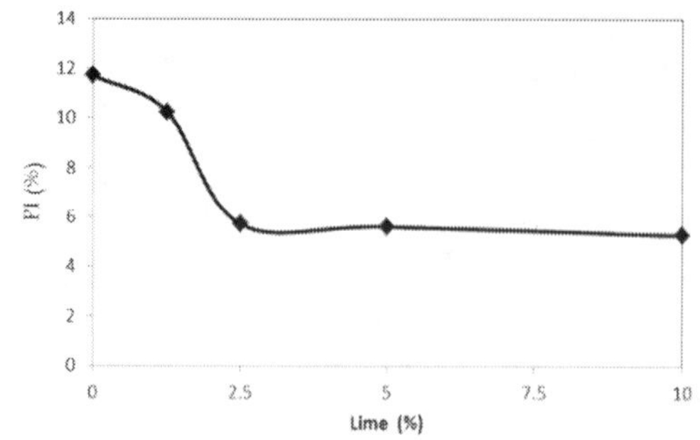


**Fig. 5.** Effect of lime on PI of soil.

### 3.3 Compaction

Results of the compaction test are shown in Figures 6 and Figure 7. It is shown that the soil exhibits a reduction in maximum unit weight compared to the control soil density and this decreased can be attributed to the flocculation and the agglomeration polymer effect, caused by the increase in the soil porosity. In addition, the OMC increased at 5% then began to reduce.. Decreasing the dry density shows it needs a low combustive energy (CE) to achieve a maximum dry density. As a result, compaction costs are reduced [14]. Similar findings have been reported by several researchers [15-17].

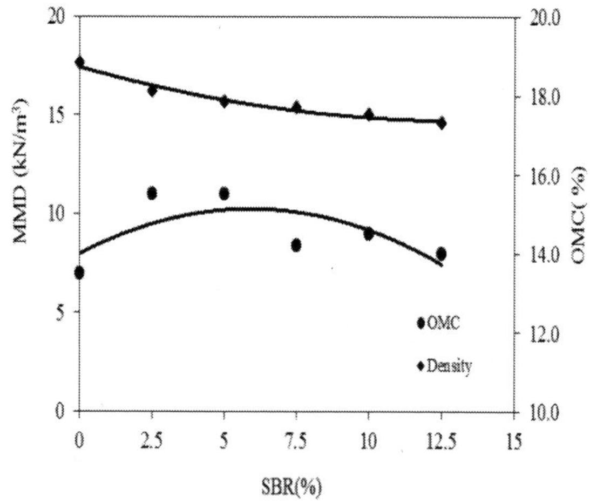


**Fig. 6.** Maximum unit weight and OMC at different SBR%.

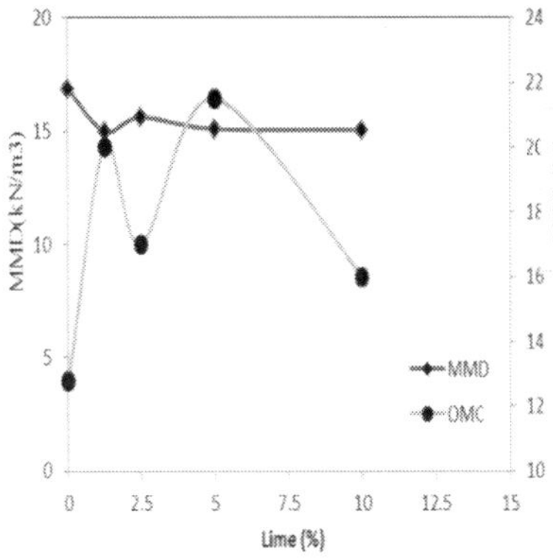


**Fig. 7.** Maximum unit weight and OMC at different Lime%.

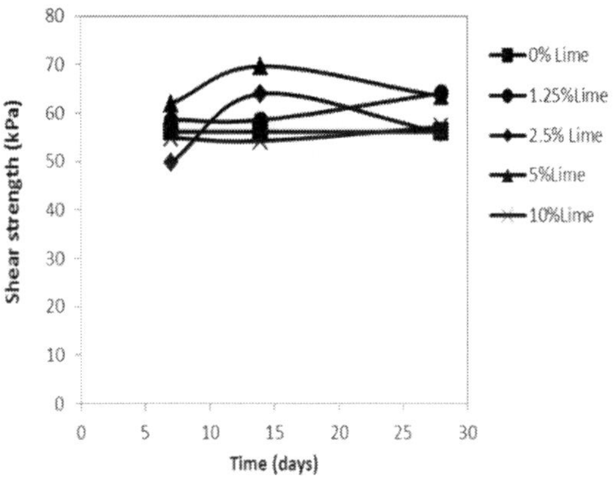


**Fig. 9.** Effect of curing time on strength of soil at different Lime%

### 3.4 Unconfined compression strength test results

Figure 8 shows effect of curing time on strength at different percentages of SBR. The mixture of soil had a strength below that of the original soil during the first day after mixing and then began to increase as the curing time increased. The 5% SBR exhibited a maximum strength when compared to all other mixtures. The strength increased in the soil at 5% SBR after 28 days. A similar behavior trend was found trend [18].
Figure 9 shows effect of curing time on strength of soil at different percentages of lime with soil. The results showed a strength increase of 5% Lime.

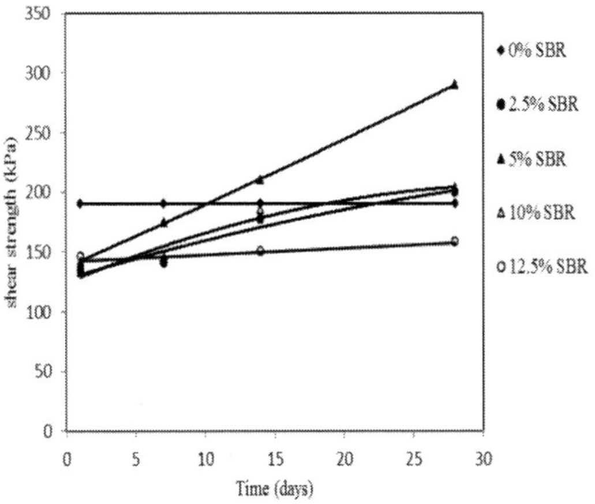


**Fig. 8.** Effect of curing time on strength of soil at different SBR%

## 4 Conclusions

The following conclusions are observed based on the results :

1. The chemical stabilizers affect the plasticity index.
2. Due to the difference in densities between additives and soil we cannot depend on the compaction results.
3. The soil strength is increased.
4. Curing time had an effect on all geotechnical soil properties.
5. The soil texture affects the additives percentage.

## 5 References

1. G. Miller, S. Azad. "Influence of Soil type on Stabilization with Cement Kiln Dust". Construction and Building Materials Journal,**14**, 89-97(2000).
2. N. Attoh. "Lime Treatment of Laterite Soils and Gravels- Revisited" Construction and Building Materials Journal ,**9**, 283-287(1995).
3. S. Lahalih, N. Ahmed. "Effect of New Soil Stabilizers on the Compressive Strength of Dune Sand". Construction and Building Materials Journal, **12**, 321-328(1998).
4. J.Tingle, K. Newman, S. Larson, C. Weiss, J. Rushing. "Stabilization Mechanisms of Nontraditional Additives". Transportation Reaearch Recordion : Journal of the Transportat Research Board **2**, 59-67(2007).
5. S. Naeini, B. Naderinia, E. Izadi. "Unconfined Compressive Strength of Clay Soils Stabilized with Waterborn Polymer". KSCE Journal of Civil Engineering **16**, 943-949(2012).
6. N. Naderi, S. Naeini. "The Influence of Polymer Inclusion and Plasticity Index on the

Unconfined Compression Strength of Clays". 2nd International Conference on New Developments in Soil Mechanics and Geotechnical Engineering, Near East University, Nicosia, North Cyprus. (2009).

7. D. Matzen, E. Straube. "Mechanical Properties of SBR-Networks: I. Determination of Crosslink Densities by Stress-Strain-Measurements". Colloid and Polymer Science **270**, 1-8(1992).
8. B. Badmus. "Plasticity and Compressibility Characteristics of Lateritic Soil from Southwestern Nigeria". Research journal of soil and water management **1**,10-14(2010).
9. Y. Atemimi , F. Ahmad. "Influence of New Liquid Stabilizer on strength of Residual soils". The Iraqi Journal for Mechanical and Material Engineering, Special Volume, Babylon First International Engineering Conference, Issue (C), 436-449 (2015).
10. ASTM. Annual Book of ASTM standards. ASTM International, West Conshohocken, PA, www.astm.org. ,04-08(2005).
11. BS. Methods of Test for Soils for Civil Engineering Purposes 1377 (1990). British Standards Institution 2 park street London W1A 2BS (2000).
12. M. Muazu. "Evaluation of Plasticity and Particle Size Distribution Characteristics of Bagasse Ash on Cement Treated Lateritic Soil". Leonardo Journal of Sciences **10**,137-152(2007).
13. A. Olarewaju, M. Balogun, S. Akinlou. "Suitability of Eggshell Stabilized Lateritic Soil as Subgrade Material for Road Construction". EJGE **16**,899-908(2011).
14. Y. Bagheri, Y. Atemimi, F. Ahmad, M. Ismail. "Soil Strength improvement by using a new compound stabilizer". Caspian Journal of Applied Sciences Research, **2** (AICCE'12 & GIZ' 12), 84-90, (2013)
15. A. Rahman. "The Potential of Some Stabilizers for the Use of Lateritic Soil in Construction". Building and environmental Journal, **21**,57-61(1986).
16. S. Nayak, B. Sunil, S. Shrihari. "Hydraulic and Compaction Characteristics of leachate-Contaminated Lateritic Soil". Engineering Geology **94**,137-144(2007).
17. M. AL-Tai, A. Mohamed Ali, B. Albdean. Improvement of Selected Parts of Basrah Governorate Soil Using Resorcinol Resin. Basra Journal of Science, **27**,1- 16(2009).
18. Y. Atemimi, Y. Bagheri, F. Ahmad. "The Influence of Soil Stabilization Techniques by Styrene Butadiene Rubber on the Soil Strength". Awam International Conference on Civil Engineering (AICCE'12) Geohazard Information Zonation (GIZ'12) Park Royal Penang Resort, 28$^{th}$ – 30$^{th}$(2012).

# Geotechnical properties of reinforced clayey soil using nylons carry's bags by products

*Nahla* Salim[1,*], *Kawther* Al-Soudany[1], and *Nora* Jajjawi[1]

[1]University of Technology, Baghdad, Iraq

**Abstract.** All structures built on soft soil may experience uncontrollable settlement and critical bearing capacity. This may not meet the design requirements for the geotechnical engineer. Soil stabilization is the change of these undesirable properties in order to meet the requirements. Traditional methods of stabilizing or through in-situ ground improvement such as compaction or replacement technique is usually costly. Now a safe and economic disposal of industrial wastes and development of economically feasible ground improvement techniques are the important challenges being faced by the engineering community. This work focuses on improving the soft soil brought from Baghdad by utilizing the local waste material for stabilization of soil, such as by using "Nylon carry bag's by product " with the different percentage and corresponding to 1 %, 3% and 5% ( the portion of stabilizer matters to soil net weight) of dried soil. The results indicated that as Nylon's fiber content increases, the liquid limit decreases while the plastic limit increases, so the plasticity index decreases. Furthermore, the maximum dry density decreases while, the optimum moisture content increases as the Nylon's fiber percentage increases. The compression index ( decreases as the Nylon's fiber increases and provides a maximum of 43% reduction by adding 5% nylon waste material. In addition, the results indicated that, the undrained shear strength increases as the nylon fiber increases.

## 1 Introduction

Stabilization is the method of blending different soils or adding additives to soil to improve the characteristic of the soil such as strength, durability, gradation and plasticity which making the soils more stable. The main objective of stabilization can be listed:

- Decrease cohesiveness (Plasticity)
- Decrease volume expansion or compressibility
- Increase strength.

There are different methods of soil stabilization, which can be categorized into:

a)- Mechanical stabilization
b)-Chemical stabilization
c)- Combined between mechanical and chemical stabilization.

Mechanical stabilization can be achieved either by compaction which causes an interlocking of the soils particle , or by dewatering of wet soil (if water drains out of soil, the soil becomes strong)or by changing in gradation size of soil by adding different types of material such as fine or aggregate and compacting the mixture .

Chemical stabilization, which depends mainly on the chemical reaction between the additive material (cementation) and the soil minerals (pozzolanic material) to a chieve desired effect. Cement, lime, fly ash, bitumen and combination of these can be used for this type of stabilization. Usually stabilized soils have higher strength and lower compressibility than the native soils. Many researchers used these types of stabilization and .

Soil reinforcement is one method of soil improvement in geotechnical engineering application. The main issues of reinforcement first developed by using tensile resisting materials into the soil to improve the strength of soil and increase its stability. In general, soils can withstand high compressive strength but cannot withstand the tensile strength. So the main objective of soil reinforcement is to make soils to absorb tensile loads and shear stresses by using reinforcement elements. So the stability of geotechnical structure can be achieved by randomly using reinforcement and distributed throughout the soil element. This method was used early in ancient times when reeds, sops, straws, and timber were used in reinforcing the soils element by mixing this material with soils to enhance the soil properties. Soil reinforcement can be categorized into Macro-reinforcement and Micro-reinforcement.

The Macro-reinforcement included woven and non-woven reinforcement and widely used in construction industries Micro-reinforcement were used in which conduct by using randomly small reinforcing elements into soil mass with the uniform distribution .

## 2 Soil fiber reinforcement

In general soils have low tensile strength and highly affected by environmental conditions. Soil reinforcement can be defined as a method to improve the physical and

* Corresponding author: nahla_salim2007@yahoo.com

© The Authors, published by EDP Sciences. This is an open access article distributed under the terms of the Creative Commons Attribution License 4.0 (http://creativecommons.org/licenses/by/4.0/).

mechanical properties of the soil. The primary purpose of reinforcing soil mass is

1- to improve its stability,
2- to increase its bearing capacity, and
3- to reduce settlements and lateral deformations [12].

There are different soil reinforcement methods, such as:

stone columns [13] , root piles, soil nailing and reinforced earth. Reinforced earth is a composite material consisting of compacted backfill and man-made reinforcing materials such as fibrous materials (geosynthetic and randomly distributed fibers). The idea of reinforcing soils using fibers is one that has been in use for a long time[12]. Initial developments in soil reinforcement led to the use of plant roots and straws in walls made from soil bricks to improve their mechanical properties. Fiber reinforcement in soils acts as homogeneous materials in composites which are fibers, and have high tensile strength included in a soil matrix. Tensile resistance in the fibers affects the shear stress in the soil and contributes greater strength to the composites soil.

Nylon fibers were used to reinforce silty clay soil with different degree of compaction by [14] they studied the relative strength gained in terms of unconfined compression and concluded that the effect of adding nylon fibers is significantly higher on the residual strength of soil rather than on the peak strength. The results obtained showed that the strain at the peak increases with the increasing the amount of fibers. The result indicated that the peak and the residual strength of the soils compacted up to 93% are significantly more than the samples compacted at the higher densities. Reinforcement by fiber was used in waste containment soil liners by [15 ]. They illustrated the role of fiber reinforcement in reducing crack development in compacted clay samples and their effect on the hydraulic conductivity. The maximum crack reduction observed was approximately 90% for a fiber content of 0.8%. It was found that the hydraulic conductivity of clay liners increased with increasing fiber content.

Reference [16] performed a laboratory test to study the properties of using waste of nylon carpet fiber in reinforcing sandy silt soil. The results revealed that compressive strength increased by 204% using 3% carpet fibers and the ductility of the soil was mentioned by the authors. In addition, field tests have shown that shredded carpet waste fibers of 70 mm long can be used into the soil with special equipment. The availability and the low cost fibers from carpet waste could lead to wider use of fiber reinforced soil and more cost-effective in construction.

Reference[17] conducted the experimental works mainly on the compressive strength and swelling properties of clay reinforced with metal fiber. Different percentages of fibers mixed with soil were used to prepare specimens to be tested. It was found that both peak strength and strain at failure increased with increasing fiber content while swelling pressure and swelling potential decreased with the increase of fiber content. The suitable percentage of fiber content was found to be 8%.

Reference [18] improved the property of expansive soil by the combined application of lime and geosynthetic reinforcement using experimental approach. A series of unconfined compression tests have been conducted on soil mixed with different percentages of lime (2, 4, 8, 12 and 16%) and for different curing periods (3, 7, 14 and 28 days). Significant improvements in strength with increase in lime-content have been observed. At 2, 4, 8, 12 and 16% of lime content the strength improvements were found to be 2.5, 3.5, 8.4, 8.7 and 8.5 times respectively than that of untreated soil. Reference [19] pointed out that, based on the total and effective stresses, the increase in the content of nylon fiber causes an increase in the shear strength parameters of clay.

Significant improvements in strength with increase in lime-content have been observed. At 2, 4, 8, 12 and 16% of lime content the strength improvements were found to be 2.5, 3.5, 8.4, 8.7 and 8.5 times respectively than that of untreated soil. Reference [19] pointed out that, based on the total and effective stresses, the increase in the content of nylon fiber causes an increase in the shear strength parameters of clay.

# 3 Laboratory studies

In this part, experimental work was conducted using standard procedures, the untreated samples were tested for their classification and index properties, as well as their consistency properties. Furthermore, the testing program conducted on the clayey soil samples stabilized with different percentages of recycled nylon strips, included Atterberg's limits, Specific Gravity, Static compaction, vane shear test and consolidation test.

The Atterberg Limits: The Atterberg limits are determined as follows:

Liquid limit (L.L): Casagrande Method according to ASTM [20]. The liquid limit test was conducted on samples passing 0.425mm (No.40) sieve, Clayey soils with (0, 1, 3, 5)% waste material taken from carry bag manufactured by-product, using Casagrande's liquid limit apparatus.

Plastic limit (P.L): ASTM [20]

The plastic limit test was performed on samples passing 0.425mm (No.40) sieve, Clayey soils with (0, 1, 3, 5)% nylon carry bags by product.

Specific Gravity: (ASTM [21] ) The specific gravity (GS) test was carried out on the soil in accordance with ASTM [21]

Compaction Test: (ASTM [22] ) This test is used to determine the relationship between the moisture content and the dry density of a soil using standard proctor method.

Vane shear test: ASTM [23] "Standard Test Method for Field Vane Shear Test in Cohesive Soil.

Consolidation: (ASTM [24]) This test is carried out according to ASTM D 2435 – Standard test.

# 4 Materials used

### 4.1 Soil

The soil sample used for this study was brought from Baghdad city and transported to the soil mechanics laboratory, University of Technology where soil test has been conducted. The soil grain size distribution is shown Figure (1). The chemical and physical properties of natural soil are listed in Table (1).

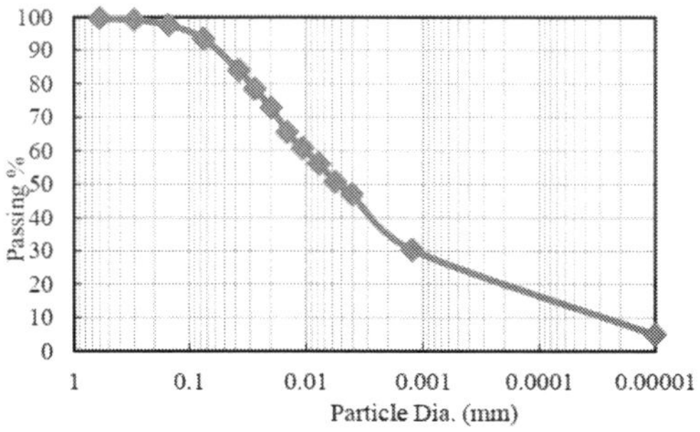


**Fig. 1.** Grain size distribution of clayey soil used

**Table 1.** Engineering properties of the used soil.

| Index property | Value |
|---|---|
| Liquid limit %(LL) | 40 |
| Plastic limit %(PL) | 19 |
| Plasticity index %(PI) | 21 |
| Specific gravity (Gs) | 2.69 |
| Gravel (larger than 2mm)% | 0 |
| Sand (0.06 to 2mm)% | 6.5 |
| silt (0.005 to 0.06mm)% | 43.5 |
| Clay (less than 0.005mm)% | 3.3 |
| Gypsum content % | 50 |
| SO3 content % | 2.92 |
| Soil Symbols (USCS) | CL |

### 4.2 Nylon fiber materials

This material commonly used for the shopping bag, storage and marketing for various purposes due to its most advantage character of less volume and weight. Most of these nylon carry bag by-products are especially made for spot use, having short life span and are being discarded immediately after use. With the few reasons cited above, it is very important that we find ways to re-utilize these nylon carry bags by- product wastes. Therefore, the investigation and attempt have been carried out to demonstrate the potential of reusing these nylon wastes as soil reinforcement for improving the subgrade soils. The nylon waste was collected from nylon carry bag factory as the by-product of carry bags and made into strips using a secateurs and mill machine of different aspect ratios as shown in Figure (2). The used Nylon's fiber properties shown in Table (2).

**Fig. 2.** Nylon's fiber used in experiments.

**Table 2.** Physical and engineering properties of the nylon's fiber used.

| Index Properties | Index Values |
|---|---|
| Specific gravity | 0.98 g/cm^3 |
| Alkali conten | |
| Sulphate content | Nil |
| Air entrainmen | 0.23 % Air content will not |
| Chloride content | Nil |
| Constituents | Polymer Polypropylene |
| Fibre diamete | 17 micron |
| Fibre length | 6 mm |
| Surface area | 225 m^2/kg |
| Young modulus | 3400 – 3800 MPa |
| Tensile strength | Min 350 MPa |
| Melting point | 160 C |

## 5 Presentation and discussions of results

### 5.1 Effect on compaction characteristics

The effect of adding randomly different percentage of nylon fiber on compaction behavior of improved soils was clearly shown in Figures (3), (4), (5) and (6). Figure (3) shows the change of dry unit weight with water content of the intact soil. The compaction curve indicated that the maximum dry unit weight of natural soil is 18.04 kN/ m3 corresponding to 14.62% optimum water content. Figure(4) shows the effect of adding a different percentage of Nylon fiber on the maximum dry unit weight and the optimum moisture content. The outcome reveals that as the Nylon fiber increases the maximum dry unit weight decreases and the optimum moisture content increases. The reduction in maximum dry unit weight can be attributed to the replacement of soil by the nylon's fiber in the prepared soils, which has relatively lower specific gravity compared to that of the soils.

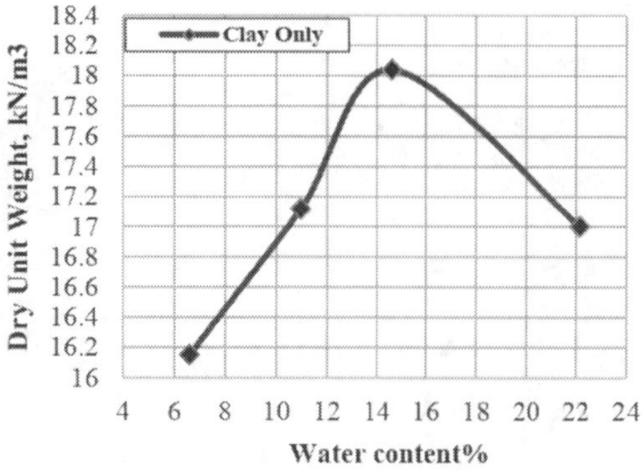


**Fig. 3.** Dry unit weight versus water content.

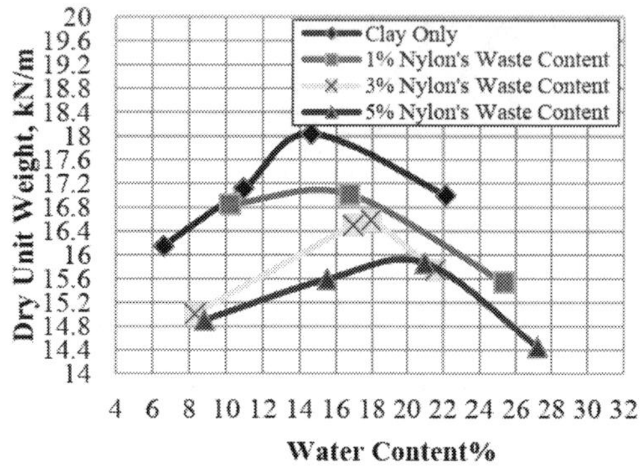


**Fig. 4.** Dry unit weight- water content at different percentage of nylon's fiber.

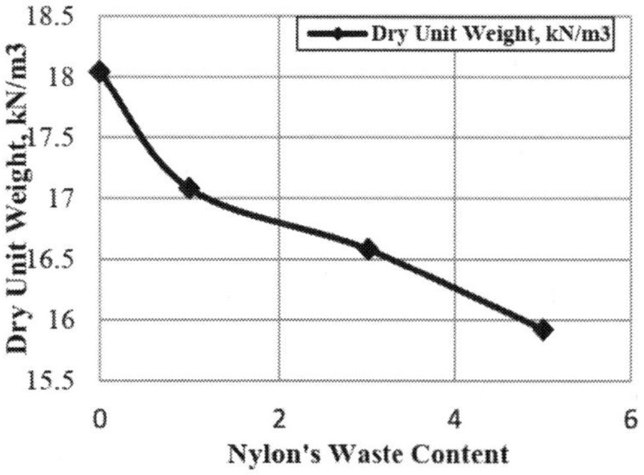


**Fig. 5.** Variation of the maximum dry unit weight with nylon's fiber percent.

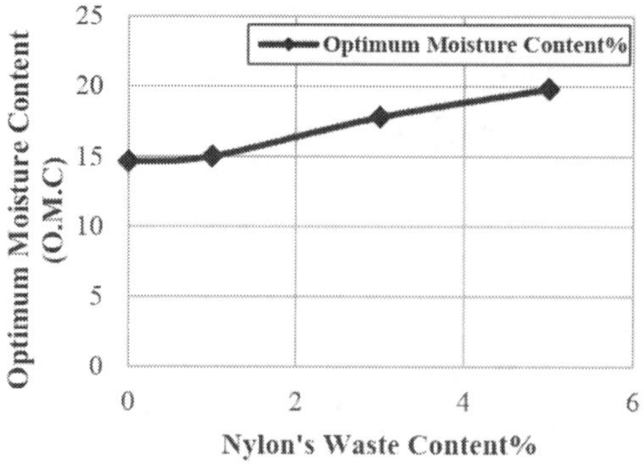


**Fig. 6.** Variation of the optimum moisture content with nylon's fiber percent.

## 5.2 Effect on Atterberg's limits

Figures (7), (8) and (9) show the effect of adding a different percentage of nylon fiber on consistency limits (LL, PL and PI), respectively. It is clearly shown, that the liquid limits (LL) decrease with increase in percentage of nylon fiber while the plastic limits (PL) increase with increase in nylon's plastic fiber. So as a result to change in liquid limit and plastic limit, noticeable decrease in the plasticity index as the nylon plastic fiber increases. This behavior ensured that this additive is considered as a granular binder.

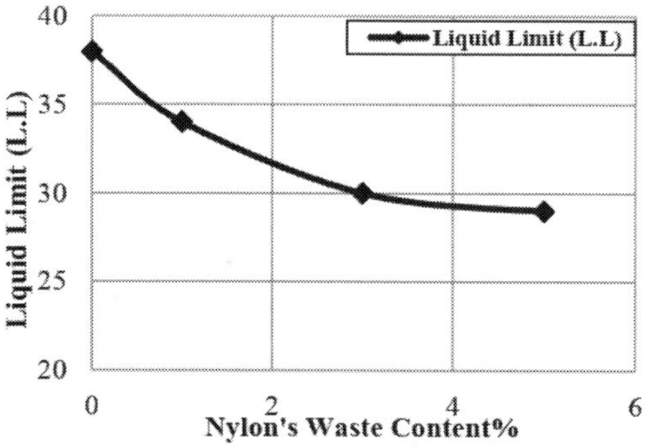


**Fig. 7.** Effect of nylon's fiber additive on liquid Limit.

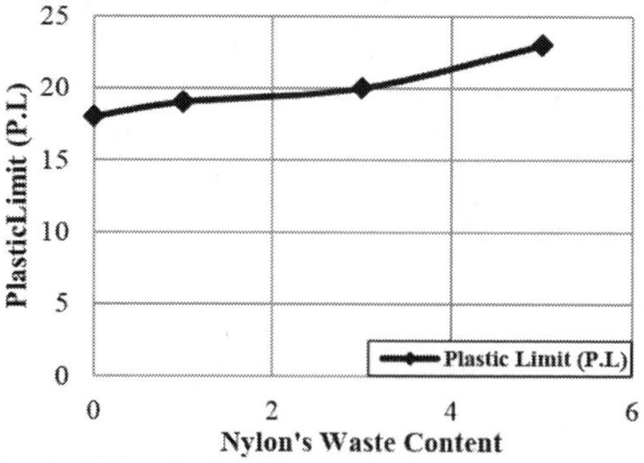


**Fig. 8.** Effect of nylon fiber on plastic limits

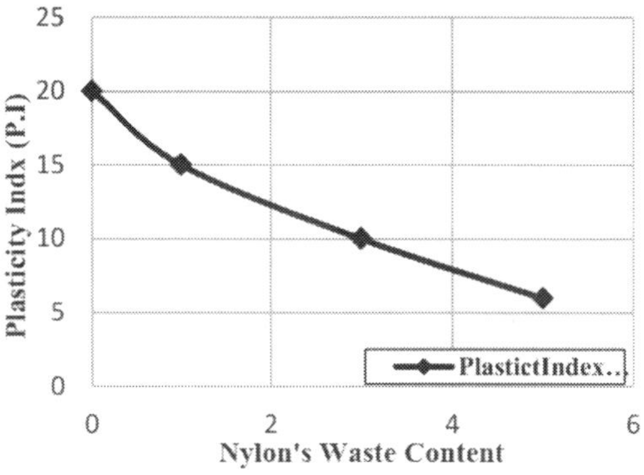


**Fig. 9.** Effect of nylon's fiber on plasticity index.

### 5.3 Effect on specific gravity

Figure (10) demonstrates the specific gravity values of the soil-mixed with the different percentage of nylon's fiber. It is clearly shown a reduction in specific gravity with the an increase in nylon fiber due to the low values of the specific gravity of nylon fiber.

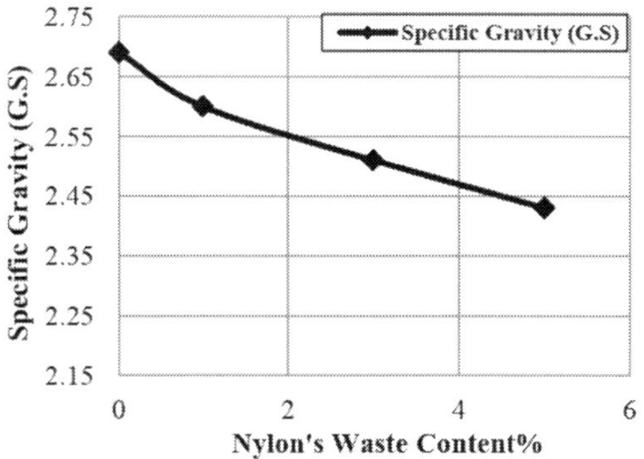


**Fig. 10.** Specific gravity with different % of nylon's carry bags by product.

### 5.4 Effect on unconfined compressive strength

The results of vane shear on stabilized soils with (1, 3, 5) % nylon fiber are presented in (Figure 11). Undrained shear strength obtained from these tests show its linear increase with nylon fiber concentration up to 5% for all of the clayey soil used. The Vane shear test was also used to estimate the undrained shear strength of intact fully saturated clays. The test is:relatively simple, quick and provides a cost-effective way of estimating the soil shear strength; therefore, it is widely used in geotechnical investigations. The results of the test are not reliable if clay contains silt or sand. To prepare samples for vane shear test the specimens were prepared at a dry unit weight of (15 kN/m3) and moisture content of (25%), air dry soil was mixed with the required amount-of-distilled water-to give initial water content. The soil was mixed thoroughly by hand and stored in a desiccator for 24 hours to get uniform -moisture-distribution. Static compaction method was followed for preparation of soil samples using a special mold and a compression testing machine. Load was applied statically to the soil sample at a compression rate of 0.04mm/sec. until the depth of penetration approached the required height. The load was maintained on the specimen for about five minutes after reaching the required height to prevent the occurrence of rebound after static load removal.

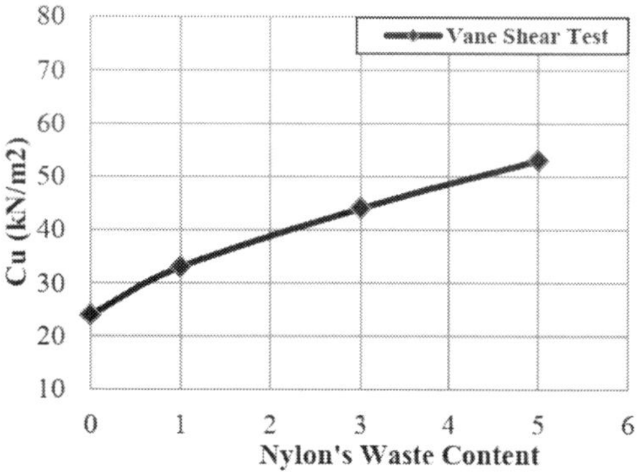


**Fig. 11.** The effect of nylon's fibers content addition on undrained shear strength of soil.

### 5.5. Effect on compressibility characteristics

In this study, the compressibility tests of clayey soils were carried out using one-dimensional consolidation test (Oedometer test). The results of standard consolidation test of soils are presented as a relation between the void ratios versus the logarithm of effective stress curve as shown in Figures (12). The loading compression curves and the recompression curves are an approximate linear. The values of compression index (Cc) and recompression index (Cr ) are given in Table (3). For consolidation test the specimens were prepared at a dry unit weight of (15 kN/m3) and moisture content of (25%) using static compaction, air dry soil was mixed with the -distilled water-to give initial moisture content.

**Table 3.** Results of consolidation test for treated soil.

| Nylon's Fiber % | $C_c$ | $C_r$ | $C_c / C_r$ |
|---|---|---|---|
| 0% | 0.157995 | 0.03336 | 4.736111 |
| 1% | 0.142193 | 0.044138 | 3.221569 |
| 3% | 0.116362 | 0.039413 | 2.952381 |
| 5% | 0.101969 | 0.035086 | 2.90625 |

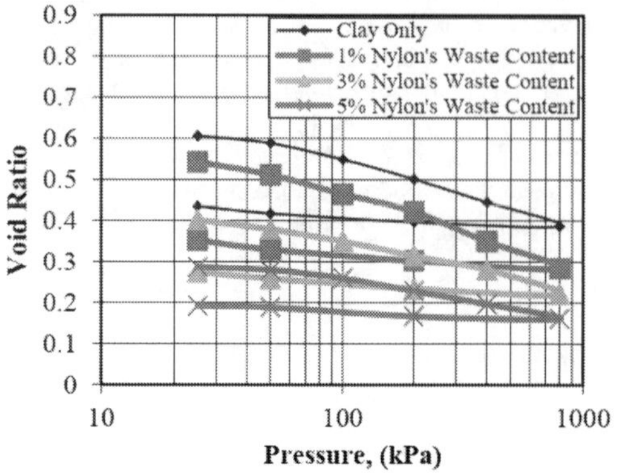


**Fig. 12.** Relation of e-log p for untreated and treated soils with nylon addition.

## 6 Conclusions

The application of recycled fiber has become the new issue due to environmental benefits and cost effect of geofiber. This study focuses on the effect of inclusion randomly nylon carry bag by-products as a reinforcing material on physical and mechanical properties of clayey soils. And by analyzing the experimental results, the following conclusions were made:

1- The addition of randomly distributed nylon waste resulted in reducing liquid limits and slightly increasing plastic limits. These behaviors cause a reduction in plasticity index, this means increase in the workability which enhances the physical properties of soils.
2- The inclusion of nylons waste to the clay soil causes a reduction in the maximum dry unit weight and an increase in the optimum moisture content, the compaction curves go down and shift to the right side.
3- The nylon waste material mixed with clayey soils causes a reduction in specific gravity as the percentage of waste material increase.

4- The compression index ($C_c$) decreases as nylon fiber increases and the maximum decrease is approximately 35% by adding 5% of nylon's waste.

## References

1. B.M. Daz (2002) "*Principles of Geotechnical Engineering*", 5th Edition, Wadsworth Group, California, USA
2. D.N. Little (1995)"*Handbook for Stabilization of Pavement Subgrades and Base Courses with Lime*", Kendal/Hunt, Lowa.
3. N. L. Dallas, Summary of Finding Web Address; (WWW.lime.org/SOIL.PDF), **1** (1999)
4. C.M. Geiman, M.Sc. thesis, Virginia Polytechnic Institute andvState University, at Blacksburg, VA (2005)
5. N. K..Al-Saoudi , F. H. Rahil & K. Y. AL-Soudany, Uotechnology Journal, **29**,15 (2011)
6. K. Harichan, M. Ghrici, S. Kenai, and K. Grine, Geotech Geol Eng, **29** (5), 759 (2011b)
7. N. Kumar, S. Swain, U. Sahoo Geotech Geol Eng, **24**, 2 DOI 10.1007/s10706-012-9532-3 (2012)
8. H. Vidal, "*The Principle of Reinforced Earth*", Highway Research Record, **282**, 1-16 (1969)
9. H. Gregory , S.Chill, In: 6th Int conf geosynt, Atlanta, USA (1998)
10. R.M.Koerner, GCLs" Fourth Edition, Prentice Hall, New, Jersey (1999)
11. I.M.C.F.G.Falorca, M.I.M.,Pinto, "Effect of short, randomly distributed polypropylene microfibres on shear strength behaviour of soils" Geosynthetics International, **18**, 2-11(2011)
12. D.R. Freitag, J. Geotech . Geoenviron Eng., ASCE, **112** (1986)
13. M.Y. Fattah,, B.S.Zabar, and H.A. Hussan,IJG ASCE, **16**, 4 (2016)
14. S.Kumar, and E.Tabor, EJGE, **8** (2003)
15. J. Miller , S.Rifai, J Environ Eng ASCE, 130:981-5 (2004)
16. J.Murray ,D. Frost, Y.Wang, Trans Res Rec, 1714:9 17 (2004)
17. S.I. Al-Azzo,S.M., and T.A. Salim, "Compressive strength and swelling properties of randomly distributed fiber reinforced clayey soil"(2007)
18. D. Baglari, and S.K. Dash, "*Improvement of Expansive soil by lime and reinforcement*" Proceedings of Indian Geotechnical Conference December 22-24, Roorkee (2013)
19. A.R.Estabragh, A.T.Bordbar, and A.A. Javadi, Geotech Geol Eng 31:501-510 (2013)
20. ASTM D 4318-00, "*Standard test methods for liquid limit, plastic limit, and plasticity index of soils*", American Society for Testing and Materials.
21. ASTM D 854-00: "*Standard test method for specific gravity of soil solids by pycnometer*", American Society for Testing and Materials.
22. ASTM D698-00: "*Standard test methods for laboratory compaction characteristics of using standard effort (600 kN-m/m3)*", American Society for Testing and Materials.
23. ASTM D2573- 01: "*Standard Test Method for Field Vane Shear Test in Cohesive Soil*"
24. ASTM D2435: "*Standard Test Method for One-Dimensional Consolidation Properties of Soils*", American Society for Testing and Materials.

# Dynamic response of machine foundation resting on sand-granulated tyre rubber mixtures

*Falah* Hassan[1,*] and *Salwan* Ahmed[1]

[1]Building and Construction Engineering Department, University of Technology, Baghdad, Iraq

**Abstract.** This paper examines the dynamic response of machine foundation resting on sand mixed with granular rubber of sizes (0.07-3) mm to improve the vibration attenuation and to reduce the settlement of the foundation. Granular rubber used is the product of hashing scrap tyres to small pieces in a tyre factory of Babylon in Iraq. Forty Model tests were carried out in a steel box of size (1200x1200x900) mm using a steel square footing of size (200x200x20) mm and a vertically acting rotating-mass type mechanical oscillator to apply a harmonic vertical mode of vibration. Karbala sand was used in this study with two relative densities (loose and medium) mixed with granular rubber ratios of (4%, 8% and 12%) of dry sand for different depths of mixing of (0.25 B, 0.5 B and 1B). The effect of relative density, frequency, depth of mixing and mixing ratio were studied. The results showed that the displacement amplitude and the settlement of machine foundation resting on dry sand granular rubber mixtures decreased with increasing mixing ratio and mixing depth. The optimum mixing ratio and mixing depth for loose state are (8% and 0.5B) at 69 Hz frequency while they are (12% and 1B) at 80 Hz frequency. For medium state they are (8% and 1B) for both frequencies. The settlement became constant after few seconds of vibration application when the foundation resting on sand mixed with optimum granular rubber content.

## 1 Introduction

The volume of used tyre rubbers in the world is significantly increasing every year due to the developing industry and a growing population and their disposals, therefore, they become a major environmental problem worldwide. The large number of scrap tyres is either dumped in landfills or stockpiled across the landscape in huge volume [1, 2, 3]. Tyres in landfills are problematic because they take up large amount of space, which are not compatible, and can 'float' to the top, potentially damaging or breaking the landfill cap [1, 4].

The use of tyre shreds or granular rubber recycling from waste tyres in construction projects, such as embankment roads and improvement the soil under foundation may be feasible as an alternate way to consume the huge stockpile of scrap tyre, with a better understanding of the behavior of rubber-soil mixture. Use of these waste materials in different applications can provide an alternative way to consume it from all over the world and consequently can be cost-effective and reduce some environmental problems from waste tyres and give these wastes an ecological value [3].

The rubber-soil mixture was reported by various researchers [4, 5, 6, 7]. They showed that recycled rubber mixed with soil is a potential composite material, which can be advantageously employed in improving the dynamic behavior of new composite materials, particularly in the case of roads highway and embankments.

A laboratory model test of tyre chips embankment was conducted to generate deformation response data of various constructions under repetitive loads. A better performance in decreasing the surface plastic displacement was noticed, when the tyre-chips covered by a relatively thick soil cap compared to using the tyre-chips in the whole of the fill [5].

The dynamic properties (shear modulus and damping ratio) of specimens constructed by different percentages of granulated tyre rubber and Ottawa sand using torsional resonant column test, were strongly influenced by the percentages of rubber. The shear modulus decreases and the damping ratio increase with increasing the rubber content [7].

The behavior of recycled materials (tyre buffing's) for highways under static and dynamic loads was investigated. The tire buffing may be used to modify the strength, deformation and dynamic behavior of sand and their use is economically feasible [6].

The maximum load carrying capacity associated with less value of rebound deflection of flexible pavement on expansive soil subgrade was obtained for rubber mixed with gravel subbase compared to rubber mixed with fly ash subbase [8].

This paper aims to study the dynamic response of machine foundation resting on sand-granulated tire

* Corresponding author: ffhhhzmr@yahoo.com

© The Authors, published by EDP Sciences. This is an open access article distributed under the terms of the Creative Commons Attribution License 4.0 (http://creativecommons.org/licenses/by/4.0/).

rubber mixtures through study the vibration amplitude and settlement of the foundation and compare the results with that of the soil alone. Also the use of the granular tyre rubber in this field has another benefit contribution solving the problem of the pollution.

## 2 Materials used

### 2.1 Sand

Karbala sand is used in the present study. Standard tests are performed to determine the physical properties. The details of these properties are listed in Table (1). The grain size distribution is shown in Figure (1). The soil is classified as poorly graded sand (SP) according to the unified soil classification system.

**Table 1**. Physical properties of sand used

| Property | Value | Specification |
|---|---|---|
| Specific gravity, Gs | 2.64 | ASTM D854 [9] |
| Coefficient of curvature, Cc | 0.87 | ASTM D422-63 and ASTM D2487-06 [10,11] |
| Coefficient of uniformity, Cu | 2.37 | |
| D10, D30, D60 (mm) | 0.16, 0.23, 0.38 | |
| USCS-soil type | SP | |
| Maximum dry unit weight, γdmax, kN/m3 | 18.82 | ASTM D4253-0 [12] |
| Minimum dry unit weight, γdmin, kN/m3 | 15.32 | ASTM D4254-00 [13] |

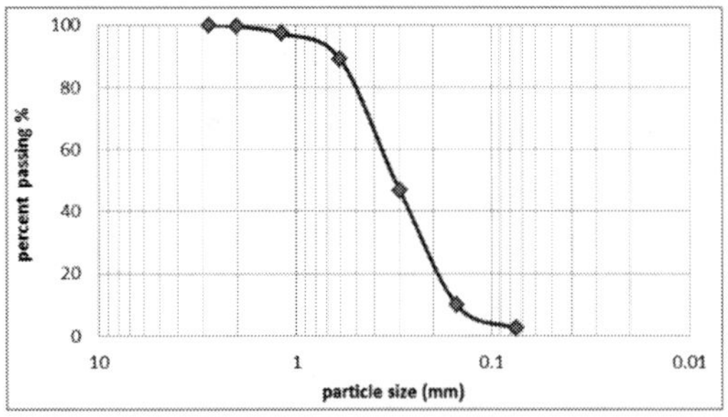


**Fig.1.** Grain size distribution of Karbala Sand

### 2.2 Granular rubber

Granular rubber used is the product of hashing scrap tyres to small pieces in a tyre factory of Babylon in Iraq. The engineering properties of tire rubber are listed in Table (2), these properties are taken from the tyre factory. The grain size distribution of granulated rubber is shown in Figure (2). The specific gravity of this material is determined according by ASTM D 854, and its value 0.88. According to ASTM D6270-98 and since the grain size of the waste tyre rubber is (0.07-3) mm, the rubber is specified as the granular rubber.

**Table 2.** Engineering properties of tire rubber

| Property | Value | Specification |
|---|---|---|
| Void ratio, e | 1.5-2.5 Uncompacted | ASTM D6270-98 [14] |
| | 0.9-1.2 Compacted | |
| Elastic Modulus, E | 1241-5171 kPa | |
| Poisson's Ratio, μ | 0.5 | |
| Water Absorption Capacity | 2%-4% | |

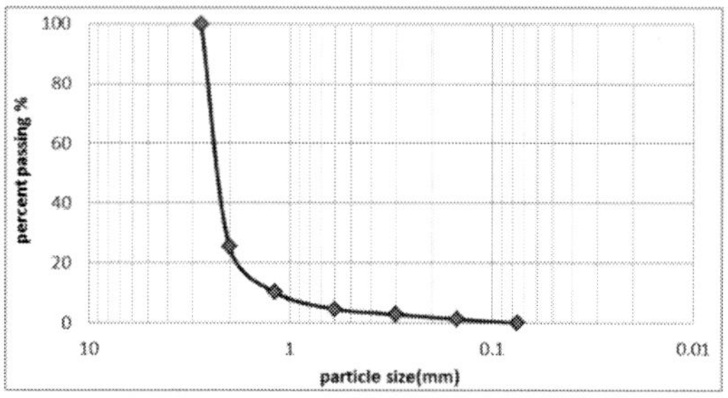


**Fig.2.** Grain size distribution of granular rubber

## 3 Model preparation and setup used

A steel container of size (1200×1200×900) mm is used for preparation of sand. A predetermined weight of sand is poured in a testing tank in layers of 100 mm thick. Each layer is compacted to maintain a uniform density throughout the model and to achieve the desired relative densities (35% and 60%). To study the effect of granular rubber in depth, three depths of sand-granulated rubber of (0.25B, 0.5B and 1B) are used, where B is the width of the footing. The percentages of granulated rubber by weight of sand are (4%, 8% and 12%). The final layer is leveled and then steel square footing of size (200×200×20) mm fixed with the mechanical oscillator on the top of the footing is placed carefully in the center of the tank as shown in Plate (1). A piezoelectric type vibration pickup is placed on the top of the footing to measure the displacement amplitude and connected to a data acquisition system. LVDT is placed on the edge of the footing to measure the settlement of the foundation. A speed control unit is connected to the oscillator to achieve the desired frequencies 69 Hz and 80 Hz.

## 4 Testing program and testing procedure

Forty model tests are performed. They are divided into 20 tests for each relative density (35% and 60%). Two of which are carried out on the sand without granular rubber for comparison. Three percentages of granular rubber by weight of sand are used (4%, 8%, and 12%) for the soil under the footing at a depth of (0.25B, 0.5B and 1B) from the footing. For each relative density two frequencies are used 69 Hz and 80 Hz. After the model preparation a harmonic load was applied at the foundation using a mechanical oscillator having a maximum speed of 6500 rpm. The operation is continued to achieve a stable reading. The results of displacement amplitude and settlement of the foundation are recorded with time on a PC connected to a data acquisition system.

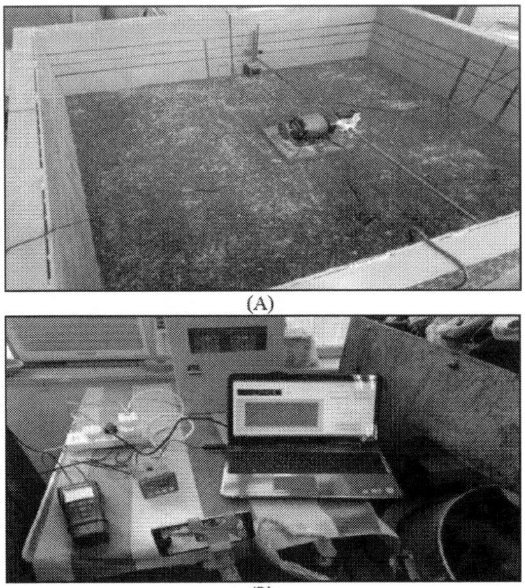


**Plate.1.** (A) Model setup and test apparatus (B) Data acquisition system

## 5 Results and discussion

The results of experimental study are presented in this section. The presentation focusses on the effect of two main parameters, mixing ratio of granular rubber and the depth of an improved layer on displacement amplitude and settlement of machine foundation for the two relative densities under the effect of the two frequencies 69 Hz and 80 Hz .

### 5.1 Effect of mixing ratio (R.D=35%)

#### *5.1.1 Displacement amplitude*

The variation of displacement with time was measured for each test by using a vibration meter; where-as the displacement amplitude (Az) was recorded at the steady state, which occurred after (500) sec from the operation of the oscillator.

The variation of displacement with time for 69 Hz frequency and (0.25B) depth of mixing is shown in Figure (3). It can be noticed from this figure that the displacement of sand is more than the displacement of sand granular rubber mixture for different ratios of mixing. The optimum mixing ratio at this depth of mixing that give the minimum value of (Az) is found at the mixing ratio (8%), beyond which the value of (Az) decreases.

Figure (4) demonstrates the value of ( Az) with mixing ratio and depth of mixing for 69 Hz frequency. It is clear from the figure that the best reduction occurred at (0.5B) mixing depth for (8%) mixing ratio.

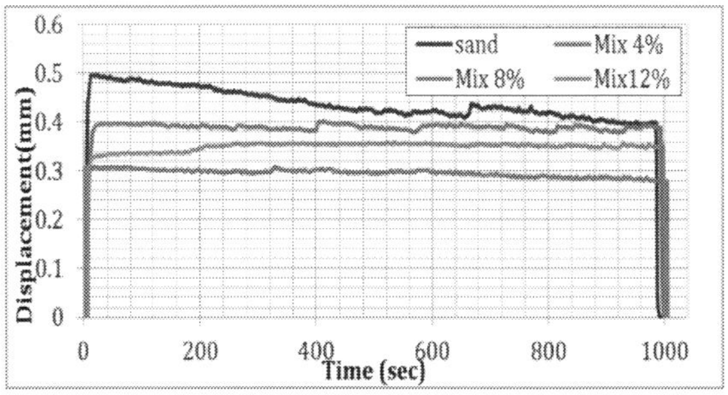


**Fig.3.** The relationship between displacement and time for fr 69 Hz and depth of mixing 0.25B

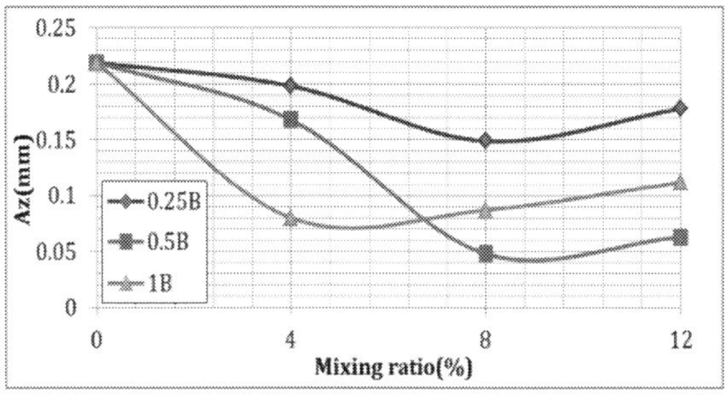


**Fig.4.** Variation of displacement amplitude (Az) with mixing ratios for different depths of mixing and fr 69 Hz

The results of displacement for (0.25B) depth of mixing and 80 Hz frequency are shown in Figure (5). The addition of granular rubber decreases the displacement and the best ratio is (12%). Figure (6) shows the variation of (Az) with a mixing ratio for different depth of mixing. It is clear from the figure that the best mixing ratio is (12%) at a depth of mixing (1B).

The above results showed that the optimum mixing ratio increases with an increase in frequency.

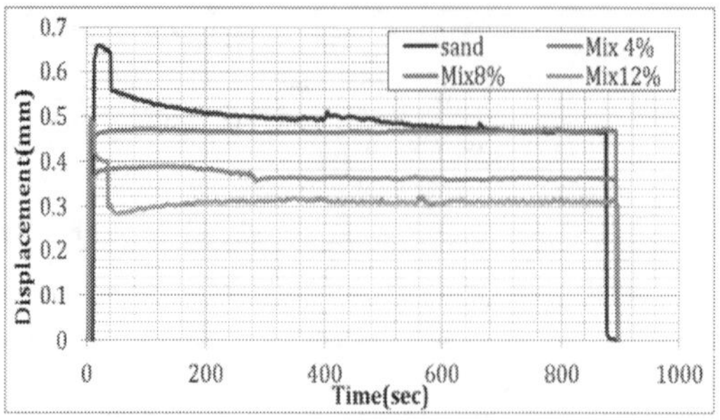


**Fig.5.** The relationship between displacement and time for fr 80 Hz and depth of mixing 0.25B.

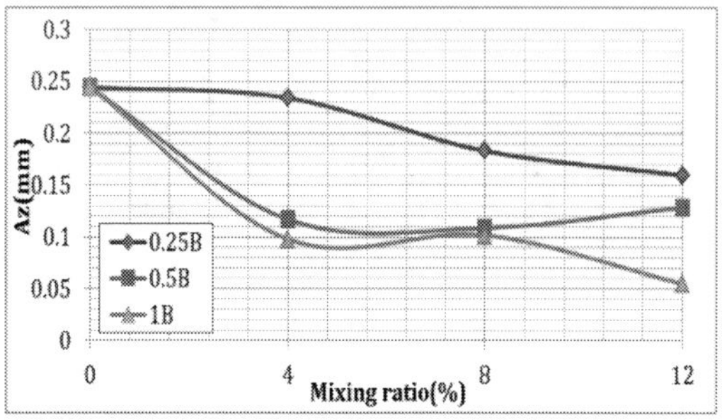


**Fig.6.** Variation of displacement amplitude (Az) with mixing ratios for different depths of mixing and fr 80 Hz.

### *5.1.2 Settlement of foundation*

The variation of settlement with time for the different mixing ratios and 69 Hz frequency is shown in Figure (7). It is clear from the figure that the settlement decreases with increase of mixing ratio and after a short time of vibration application the value of settlement becomes constant, when the mixing ratio is equal to or greater than (8%). This means that the use of optimum percent of granular rubber stopped the plastic deformation of the sand under the foundation.

Figure (8) illustrates the variation of a final settlement (Stf) with a mixing ratio for 69 Hz frequency and different depths of mixing. It can be observed that the settlement decreases with increasing mixing ratio and the best values of mixing ratio is (12%).

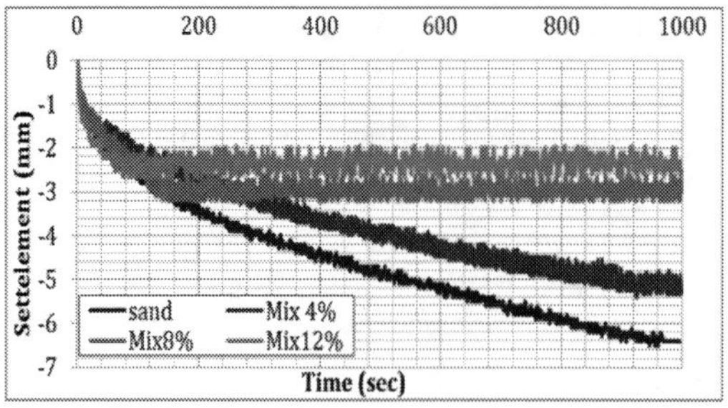


**Fig.7.** The relationship between settlement and time for fr 69 Hz and depth of mixing 0.25B.

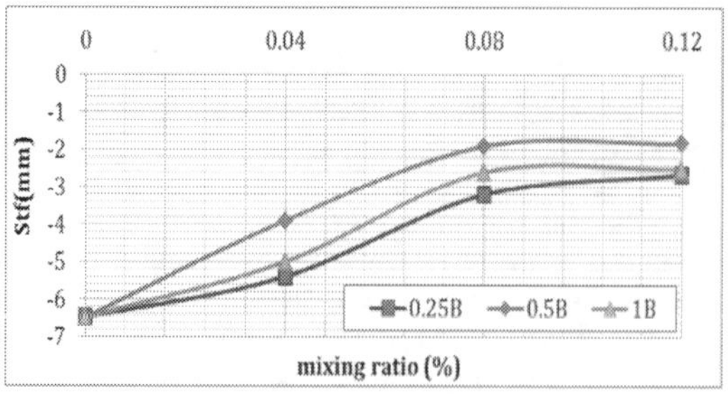


**Fig.8.** Variation of settlement with time for fr 69 Hz and depth of mixing 0.25B.

The results of settlement for the 80 Hz frequency, (0.25B) mixing depth and different mixing ratios are shown in Figure (9). It is clear that the plastic deformation decreases remarkably when using the rubber and the settlement stops after a very short time from the vibration application.

Figure (10) demonstrates the variation of settlement with a mixing ratio for different depths of mixing. The settlement decrease with an increase in mixing ratio.

From the above results it can be concluded that the settlement increases with increase in frequency for the same mixing ratio.

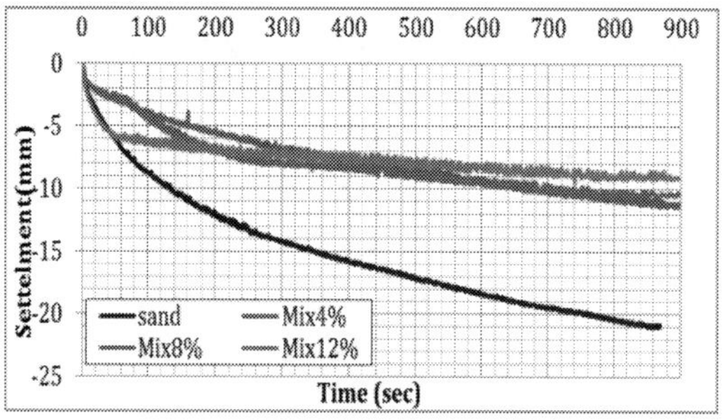


**Fig.9.** The relationship between settlement and time for fr 80 Hz and depth of mixing 0.25B.

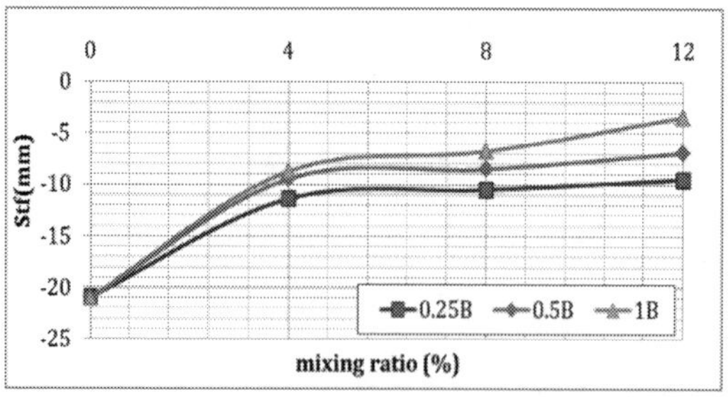


**Fig.10.** Variation of settlement with time for fr 80 Hz and depth of mixing 0.25B.

## 5.2 Effect of mixing depth (R.D=35%)

### *5.2.1 Displacement amplitude*

The relationship between displacement amplitude (Az) and the depth of mixing for different ratios are shown in Figures (11) and (12) for frequencies 69 Hz and 80 Hz, respectively. It is clear that the best depth is (0.5B) for both frequencies.

### *5.2.2 Settlement of foundation*

The variation of settlement with depth of mixing for 69 Hz and 80 Hz frequencies are shown in Figure (13) and Figure (14), respectively. The best depth for controlling the settlement is (0.5B) for the 69 Hz frequency, while the settlement decreases with increasing depth of mixing for 80 Hz frequency.

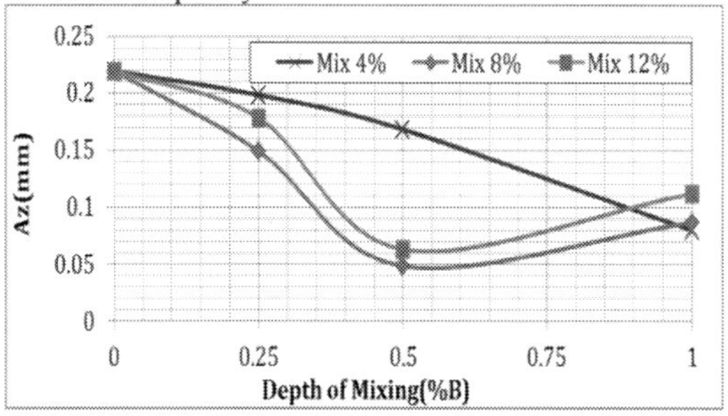


**Fig.11.** Variation of Az with depths of mixing for different mixing ratios and fr 69 Hz.

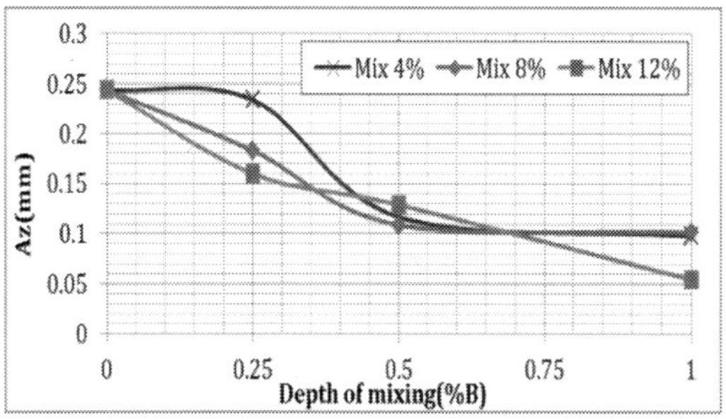


**Fig.12.** Variation of Az with depths of mixing for different mixing ratios and fr 80 Hz.

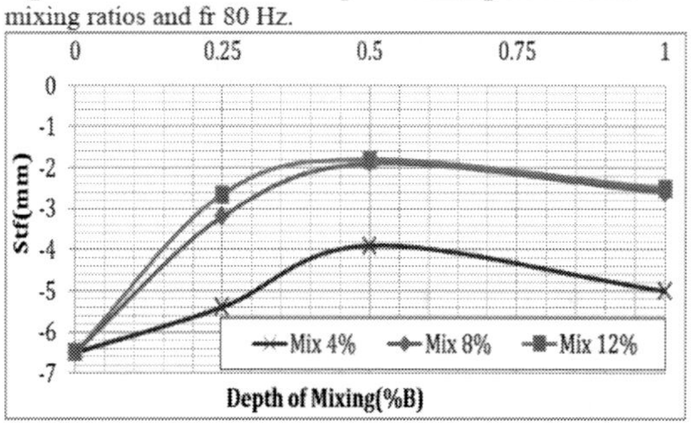


**Fig.13.** Variation of Stf with depths of mixing for different mixing ratios and fr 69 Hz.

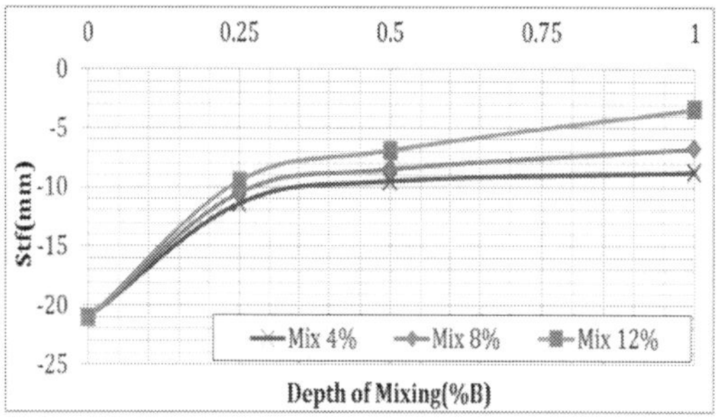


**Fig.14.** Variation of Stf with depths of mixing for different mixing ratios and fr 80 Hz.

## 5.3 Effect of mixing ratio (R.D=60%)

### *5.3.1 Displacement amplitude*

Figure (15) represents the relationship between displacement and time for (0.25B) depth of mixing and for different mixing ratios. It is clear from the figure that the best mixing ratio is (4%) and the displacement amplitude stays nearly constant for the ratios above (4%), while the best value of the ratio is (12%) when the frequency increases to 80 Hz for the same depth as shown in Figures (17) and (18).

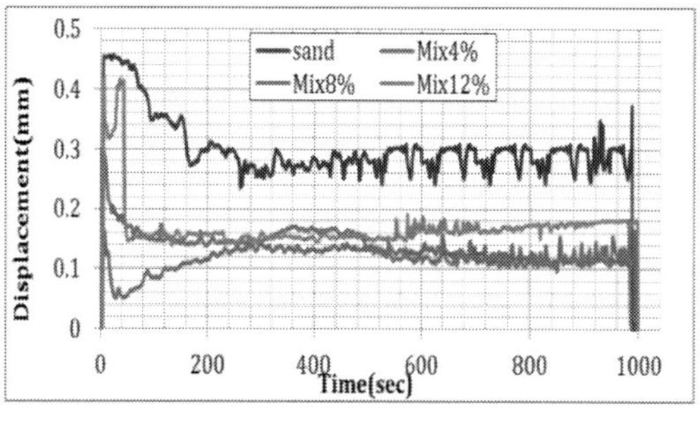


**Fig.15.** The relationship between displacement and time for fr 69 Hz and depth of mixing 0.25B.

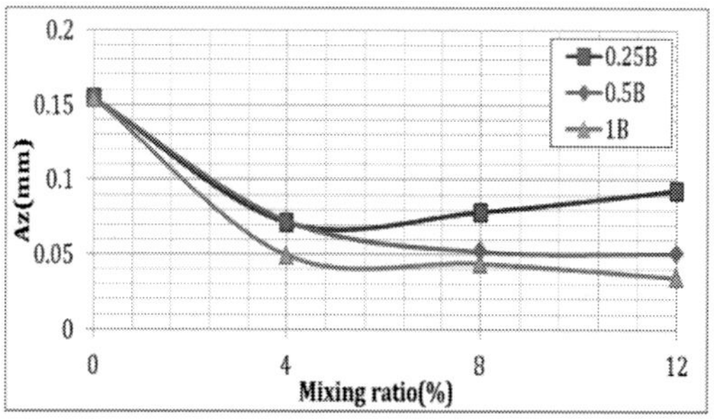


**Fig.16.** Variation of displacement amplitude with mixing ratios for each depth of mixing and fr 69 Hz

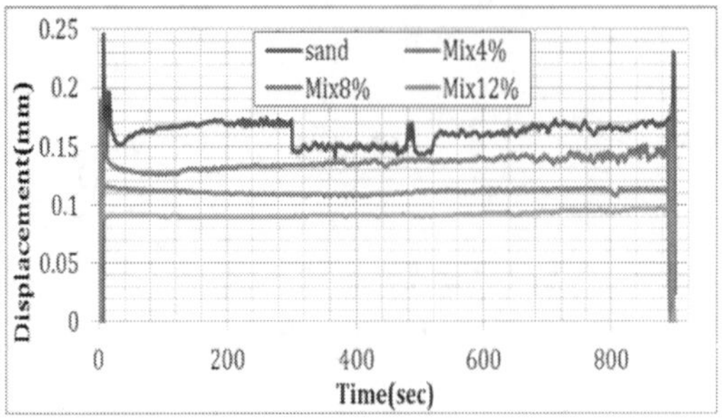


**Fig.17.** The relationship between displacement and time for fr 80 Hz and depth of mixing 0.25B

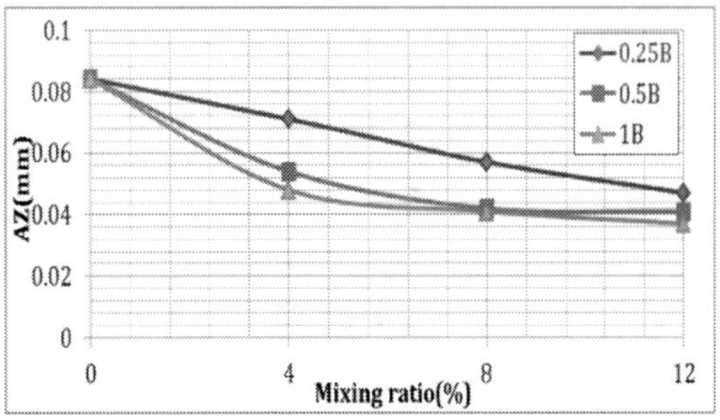


**Fig.18.** Variation of displacement amplitude with mixing ratios for each depth of mixing and fr 69 Hz.

The above results showed that the displacement amplitude for medium sand is less than that of loose state under the same condition, due to increase in the stiffness of sand.

*5.3.2 Settlement of foundation*

The relationship of settlement with time for (0.25B) depth of mixing and 69 Hz frequency and different mixing ratios is shown in Figure (19). The settlement is more uniform compared to the same models of relative density (35%). The results of settlement increase with an increasing mixing ratio until (8%) then become constant as shown in Figure (20).

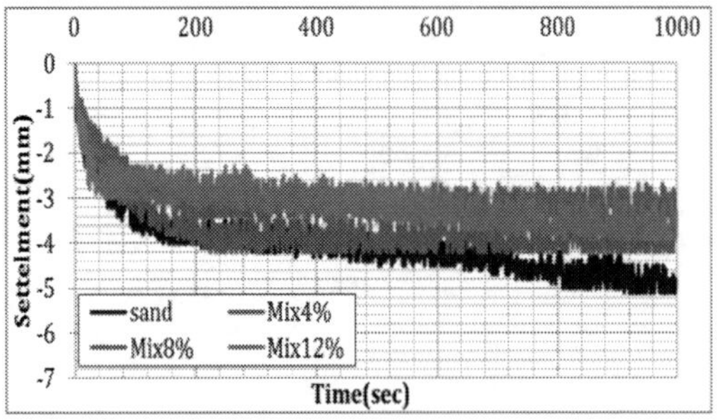


**Fig.19.** The relationship between settlement and time for fr 69 Hz and mixing depths 0.25B

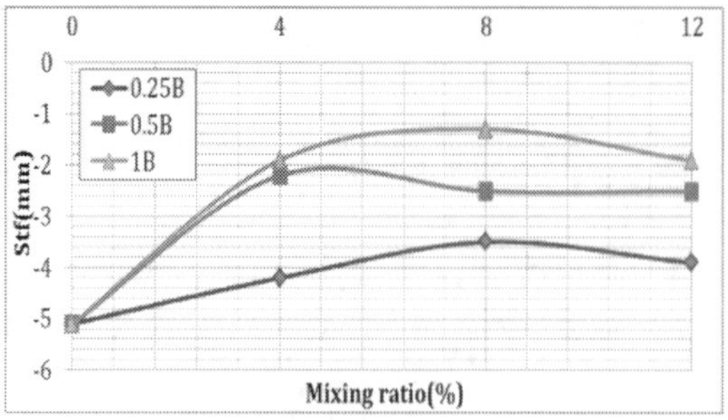


**Fig.20.** Variation of final settlement with mixing ratios for each depth of mixing and fr 69 Hz

In the same way, the results of the settlement with time for frequency of (80 Hz) is presented in Figure (21) for depth of mixing (0.25B). It is clear that the values of settlement stays constant after a very short time from the operation of the machine. Also the settlement decreases with increase in mixing ratio until (8%), then becomes constant as shown in Figure (22). This means that the effect of frequency decreases with increase in the relative density.

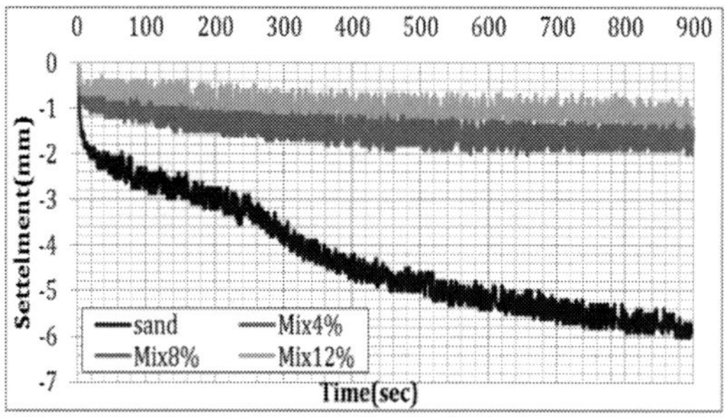


**Fig.21.** The relationship between settlement and time for fr 80 Hz and mixing depths 0.25B

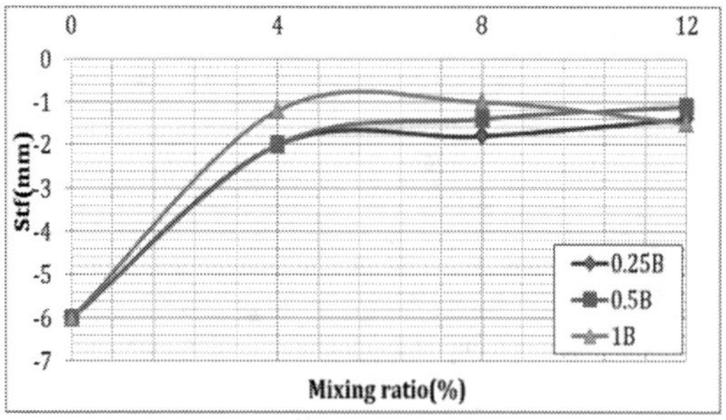


**Fig.22.** Variation of final settlement with mixing ratios for each depth of mixing and fr 80 Hz

## 5.4 Effect of mixing depth (R.D=60%) :

*5.4.1 Displacement Amplitude*

The variation of displacement amplitude (Az) with a depth of mixing under the effect of 69 Hz and 80 Hz frequencies for different mixing ratios are shown in Figure (23) and (24). It is clear from the figures that the best depth of improvement is (0.5B) for the both frequencies.

*5.4.2 Settlement of foundation*

the results of the settlement with mixing depth for the two frequencies are shown in Figures (25) and (26). In general the settlement decrease with increase in the depth of mixing and frequency, but for sand without rubber the settlement increases with increase in the frequency. The preferred depth of mixing layer is (1B) for both frequencies, since at this depth the final settlement value is the minimum.

When mixing the granular rubber with sand, the settlement becomes approximately constant with time and the mixture behaves as an elastic material.

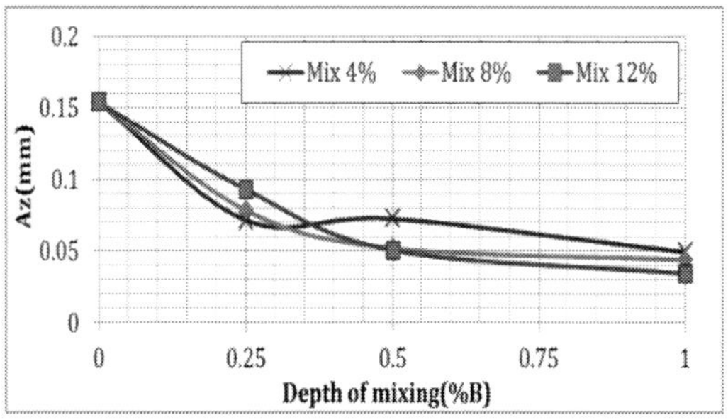


**Fig.23.** Variation of Az with depths of mixing for different mixing ratios and fr 69Hz

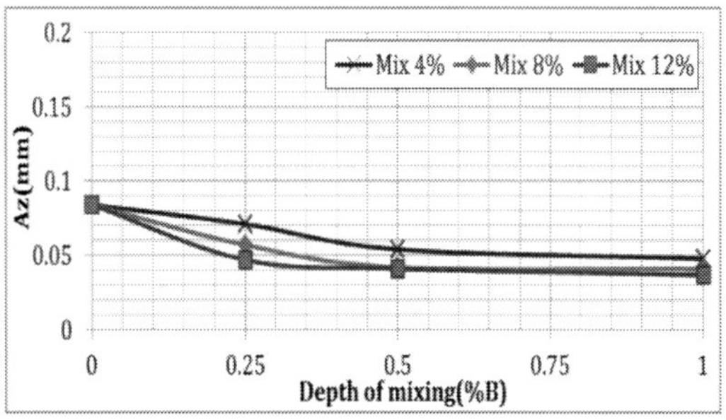


**Fig.24.** Variation of Az with depths of mixing for different mixing ratios and fr 80Hz

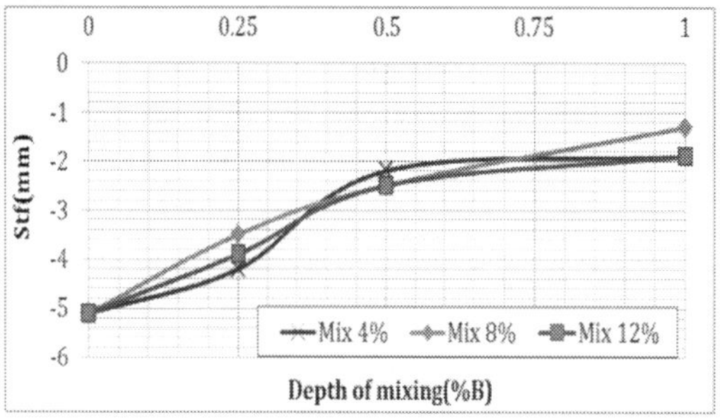


**Fig.25.**Variation of (Stf) with depths of mixing for different mixing ratios and fr 69 Hz.

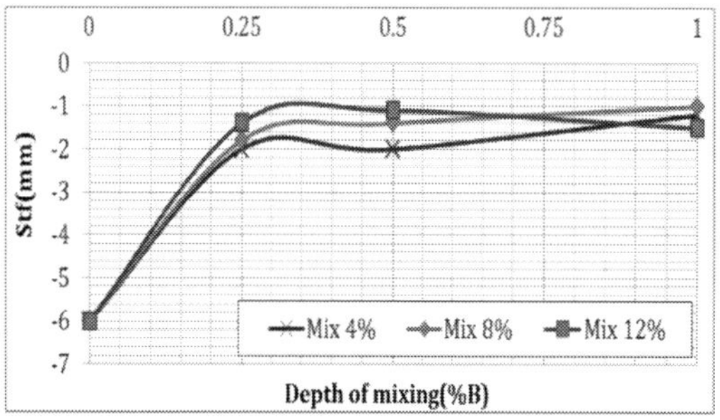


**Fig.26.**Variation of (Stf) with depths of mixing for different mixing ratios and fr 80 Hz.

## 6 Conclusions

- For dry loose sand-granular rubber mixtures, the displacement amplitude increases with increase in the frequency of the machine, and decreases with increase in the mixing ratio and depth of mixing. The optimum mixing ratio is (8%) and the best depth of mixing is (0.5B) for the 69 Hz frequency, while the optimum mixing ratio is (12%) and the best depth of mixing is (1B) for the 80 Hz frequency.
- For dry medium sand-granulated rubber mixtures, the displacement amplitude decrease with increasing frequency, increasing mixing ratio and depth of mixing. The optimum mixing ratio is (8%) and the preferred mixing depth is (1B).
- For dry loose sand-granular rubber mixtures, the final settlement increases with increase in frequency and decrease with increase in mixing ratio and depth of mixing. The optimum mixing ratio and mixing depth are (8% and 0.5B) for frequency 69 Hz, while they are (12% and 1B) for the 80 Hz frequency.
- For dry medium sand-granulated rubber mixtures, the settlement decreases with increase in mixing ratio and mixing depth. The optimum mixing ratio and depth of mixing are (8% and 1B) for the two frequencies.

- For both relative densities used, the most effective depth of mixing is (1B) and the preferred mixing ratio is (12%).
- The settlement of the foundation resting on sand mixed with optimum granular tyre rubber ratio becomes approximately constant after a short time of machine vibration and the mixture behaves as an elastic material .

## References

1. Cetin, H., Fener, M. And Gunaydin, O. Eng. Geol. **88**(1-2), 110-120, (2006).
2. Chiu, C.T. J.Res. Conserv. Recy. **52**(3), 522-532, (2008).
3. Wu, J.Y. And Tsai, M.,Waste. Manage. **29**(2) ,636-642, (2009).
4. Lee, J., Dodds, J. And Santamarina, J.C., J. Mater. Civil.Eng. - ASCE, **19**(2), 179-184, (2007).
5. Bosscher, P.J., Edil, T.B. and Kuraoka, S., J.Geotech. Geoenviron. Eng. - ASCE, **123**(4), 295-304, (1997).
6. Edincliler, A., Baykal, G. and Dengili, K., Resour. Constr. Recy., **42**(3), 223-237, (2004).
7. Feng, Z.Y. and Sutter, K.G., Geotech. Test.J., **23**(3), 338-344, (2000).
8. Prasad, D.S.V. and Prasada Raju, G.V.R., ARPN. J. Eng. Appl. Sci., **4**(6), 89-92, (2009).
9. American Society of Testing and Materials (ASTM).ASTMD854,WestConshohocken, Pennsylvania, USA, (2006).
10. American Society of Testing and Materials (ASTM). ASTM D422-63 (2002), West Conshohocken, Pennsylvania, USA, (2006).
11. American Society of Testing and Materials (ASTM).ASTM D2487-06, West Conshohocken, Pennsylvania, USA, (2006).
12. American Society of Testing and Materials (ASTM). ASTM D4253-00 (2006), West Conshohocken, Pennsylvania, USA, (2006).
13. American Society of Testing and Materials (ASTM). ASTM D4254-00 (2006), West Conshohocken, Pennsylvania, USA, (2006).
14. American Society of Testing and Materials (ASTM).ASTM D6270-98, West Conshohocken, Pennsylvania, USA ,P.40-63, (2004).

# Behavior of a strip footing on reinforced soil subjected to inclined load

*Jawdat* Abbas [1,*]
[1]Civil Engineering Department, University of Tikrit, Tikrit, Iraq

**Abstract.** This study investigates the behavior of a strip footing under inclined load on reinforced sandy soil by using experimental model. The effect of the load inclination angle (α), number of geogrid layers (N) and the relative density ($R_D$) on the bearing capacity, settlement and horizontal displacement were studied. The results showed that by increasing the number of reinforcement layers (N) , the bearing capacity increased, but there is an optimum value (N=4-5) depending on relative density of supporting soil. Also the settlement and horizontal displacement of footing decreasing with increase number of reinforcement layers.

## 1 Introduction

The load from structure is transformed to the supporting soil by foundations. Foundations usually are constructed with a larger area to minimize the pressure. Many types of footings are used for different applications. A strip footing which is largely used to support bearing walls is used in this research. The strip footing is rectangular in form but its length is much greater than its width. Strip footing can be analyses in two dimensions.(plane strain conditions). Footings of building are often under inclined loading such as footings under a vertical and horizontal loads, structure's nature or earthquake[1].

The load inclination leads to reduce the bearing capacity of the supporting soil by sliding the footing and heaving the supporting soil. This could be avoided either by building the footing with larger dimensions to minimize the contact pressure which causes uneconomical design or by raising the "bearing capacity" of the soil under footings.

Many researchers have written on the effective utilization of soil reinforcement as a economical way to enhance the bearing capacity of soil beneath foundations. This could be done by replacing the poor field soil up to a certain depth by the soil or fills of the same soil back with place as horizontal sheets of geosynthetics at different depths beneath the footings. So that, with the potential advantage of utilizing soil reinforcement both the type and the size of footing may be changed leading to an economic design[2,3,4,5,6,7,8,9].

There are many types of geosynthetics according to function and application, (geotextiles, geogrids, geonets, geomembranes, geopipes, geofoam, geocomposite, etc.). Geogrids are plastic made into very open, grid like configuration with large apertures between individual ribs in the machine and cross machine directions. The opening are usually (12-100 mm) in length and/or depth; geogrids are transported to the site in (1- 4 m) width of rolls. Geogrids are formed in various ways[10].

The utilization of geogrid sheets could be particularly convenient when the mechanical properties of the soil under a foundation would suggest the designer in utilizing other solutions. Recently, the utilization of geogrids as a soil reinforcement has become prevalent, because geogrids are dimensionally stable and combine features such as high tensile modulus (low strain at high load), open grid structure which provides improved soil reinforcement interaction, positive shear connection properties, light weight, and more service life[11].

In this paper the bearing capacity , settlement and horizontal displacement of a strip footing subjected to inclined loading resetting on reinforced soil using experimental modeling was investigated.

## 2 Laboratory model tests

The testing devices consist of four parts, the test tank, model footing, the loading system, and vibratory system [12].

### 2.1 Test box

The test box used in this study made of steel plate with inside dimensions 900mm× 900mm and 550mm in height. The sides and the bottom were made of 6 mm thickness plate, the plate was supported by four steel channels, with 150 mm high from the base of the steel box. The internal surfaces of the box were oiled (in order to reduce the friction which may develop during the process). A mark lines were drawn to give the desired thickness of the layers and the position of reinforcement layer, plate (1) shows the test box[12].

* Corresponding author: dr.Jawdatkhadim@tu.edu.iq

© The Authors, published by EDP Sciences. This is an open access article distributed under the terms of the Creative Commons Attribution License 4.0 (http://creativecommons.org/licenses/by/4.0/).

**Plate (1):** Test box

## 2.2 Model of footing

The test footing was a strip steel channel 80mm × 800mm. The load applied to the footing was counted using proving ring of 5 kN capacity, while the vertical deflection and horizontal displacement of the footing was counted by three dial gauges (0.01 mm/ division) as shown in Plate(2). The dimensions of footing model was selected depending on the size of the test box and the influence zone[12].

**Plate (2:)** Strip footings model

## 2.3 The loading frame

The test box was placed over 1100×1100 mm strong steel base of 80mm thick. The base was connected to a stiff loading frame, which was locally manufactured. As shown in Plate (3). The frame consists of two columns of steel channels 1520 mm height, which intern bolted to a loading platform. The platform was designed to slide along the columns and can be fixed at any desired height by means of slotted spindles and holes provided at different intervals along the two columns. The two steel columns were fixed by four short steel angle pieces connected to the lower plate in the frame[12].

**Plate (3)**: Loading system

## 2.4 The loading system

The load was applied by means of mechanical arrangement technique that was employed for the test. The proving ring was attached to a cylindrical steel toothed shaft device of 550mm long and 40mm diameter, which transfers the load to the footing and help to adjust the height of ring to any position required before or after test. A steel plate was made for each one of the footings, as shown in plate (4) which was attached at the end of the proving ring and work to transfer loading as equally distributed line load. Three dial gauges were attached to the footing and fixed to measure the footing vertical and horizontal displacement.

**Plate (4):** A steel plate model which transfer distributed line load on the strip footing.

## 2.5 Vibration system

To achieve the required relative densities, it is easier to use an electrical vibrator. This method based on placing the soil in the box in layers each layer of thickness (50mm) then placing a plate (700×700mm) then moving the vibrator over the whole area of the plate in a specified time .The time needed to reach the required relative density was achieved by performing a series of trials with different measured time. In each trail, the densities were calculated by collecting samples in small aluminum cans of known volume put at different locations in the test tank[12].

### 2.6 Experimental setup

The footing was installed in position and the load was applied to it through the proving ring. The load was applied until failure happened. Vertical settlement and horizontal displacement of the footing for each load increment applied were recorded with the usage of dial gauges.

## 3 Materials properties

### 3.1 Sand properties

The soil used in this study, is sandy soil with the properties shown in Table(1). All the laboratory tests were conducted based on ASTM.[13]. The grain size distribution is shown in Figure (1).

**Table 1.** Soil properties used in this study.

| Properties | | Value |
|---|---|---|
| Moisture content , (ω)% | | 1 |
| Specific gravity, ($G_s$) | | 2.579 |
| Coefficient of uniformity Cu | | 3 |
| Coefficient of curvature Cc | | 1 |
| Unified Soil Classification System | | SP |
| M.I.T classification | Sand (%) | 100 |
| | Silt & Clay (%) | 0 |
| Minimum dry unit weight,( $kN/m^3$) | | 14.873 |
| Maximum dry unit weight,( $kN/m^3$) | | 17.975 |
| For relative density 60%,( $kN/m^3$) | | 16.594 |
| For relative density 80%,( $kN/m^3$) | | 17.255 |

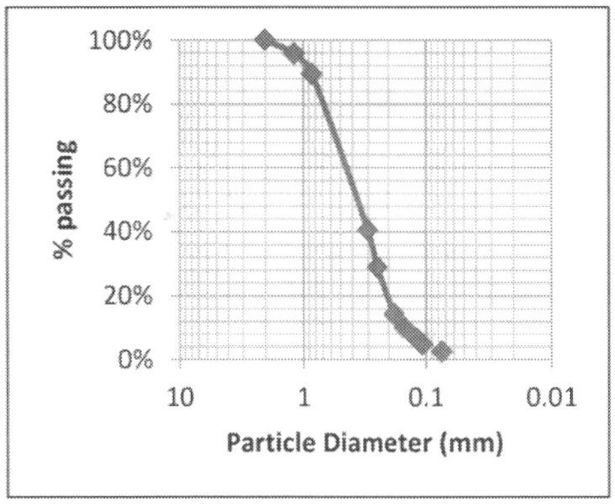


**Fig.1.** Particle Size Distribution of the sand

### 3.2 Reinforcement layers

The reinforcement material used in this study is Geogrid type "TriAx® TX140" Geogrid product from a punched polypropylene layer. The characteristics of a Geogrid sheet are shown in Table (2).

**Table 2.** Characteristics of Geogrid Reinforcement sheets used in this study

| Index Properties | Longitudinal | Diagonal | Transverse | General |
|---|---|---|---|---|
| • Rib pitch[2], mm (in) | 40 (1.60) | 40 (1.60) | - | |
| • Mid-rib depth[2], mm (in) | - | 1.2 (0.05) | 1.2 (0.05) | |
| • Mid-rib width[2], mm (in) | - | 1.1 (0.04) | 1.1 (0.04) | |
| • Rib shape | | | | rectangular |
| • Aperture shape | | | | triangular |
| **Structural Integrity** | | | | |
| • Junction efficiency[3], % | | | | 93 |
| • Aperture stability[4], kg-cm/deg @ 5.0kg-cm [2] | | | | 3.0 |
| • Radial stiffness at low strain[5], kN/m @ 0.5% strain | | | | 225 |
| (lb/ft @ 0.5% strain) | | | | (15,430) |
| **Durability** | | | | |
| • Resistance to chemical degradation[6] | | | | 100% |
| • Resistance to ultra-violet light and weathering[7] | | | | 100% |

## 4 Test program

There are (48) tests conducted to reach the aim of the study. Eight tests were conducted on unreinforced soil. These tests are a reference to comparison and to know how much the improvement in the bearing capacity as a result of using the reinforcement by Geogrid. Also, they are used to determine the effect of variation in the angle of load inclination( α ) on the bearing capacity of unreinforced soil for relative densities (60% & 80%). The remaining (40) tests which were representing the fundamental part of this research, show the effect of the multi-reinforcement layers on the bearing capacity , settlement and horizontal displacement.

## 5 Results and discussion

There are several parameters, which affect the behavior of a strip footing subjected to inclined loading resting on reinforced sandy soil. In this study, most of these parameters were varied within their reasonable ranges in order to explore their effects on the bearing capacity and on each other. The effects of the following parameters : load inclination angle (α) (0°, 5°, 10°, and 15°), number of geogrid layers (N) from (1 to 5 layer), and for two relative densities (60% and 80%) to represent medium dense and dense sand are studied. The depth of first layer (U) is taken equal to 0.35B while the vertical distance between reinforcement layers (h) is taken equal to 0.25B.[12]

The expression bearing capacity ratio (BCR) is presented to calculate the effect of Geogrid reinforcement on the bearing capacity and it can be written as:

$$BCR = \frac{q_{ur}}{q_u} \quad (1)$$

Figure (2) show the major reinforcement parameters of strip footing subjected to inclined load on reinforced sandy soil.

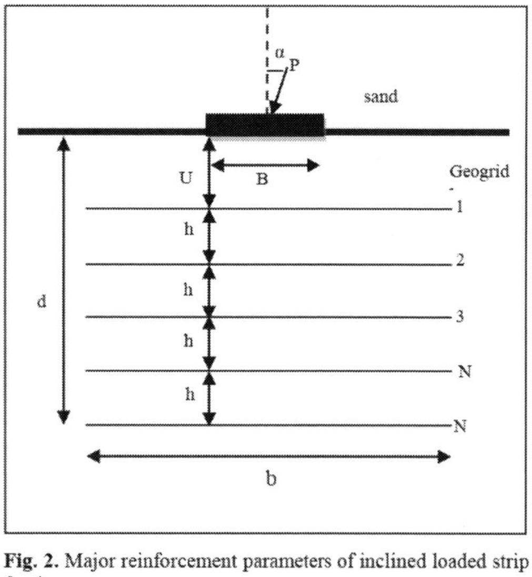


**Fig. 2.** Major reinforcement parameters of inclined loaded strip footing.

failure pressure decreases. Besides the decrease in settlement, the values of the ultimate bearing capacity decreases with an increase in the load inclination too. This happened because of increasing the horizontal load component. The horizontal component of the load increases the lateral stress on the surrounding zone, leaving the less resistance to support the vertical stress generated by the load vertical component [Sowers and Sowers, 1951].

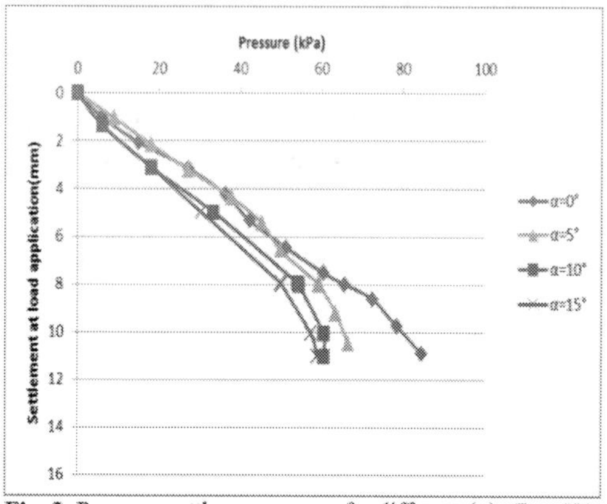


**Fig. 3.** Pressure-settlement curves for different (α) , $R_D$=60%, N=0

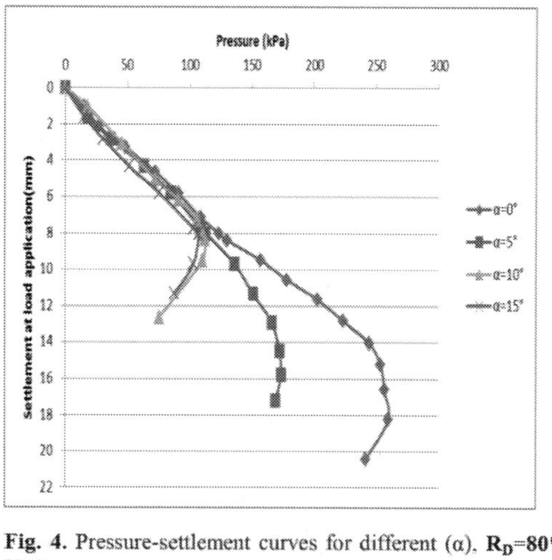


**Fig. 4.** Pressure-settlement curves for different (α), **$R_D$=80%, N=0.**

### 5.1 Effect of load inclination (α)

In order to investigate the effect of the variation of load inclination angle (α) on the ultimate bearing capacity of strip footing, (8) tests have been conducted for several cases.

Figures (3) and (4) show the bearing pressure-settlement curves for different load inclination (α) for two relative densities on unreinforced sand. It is evident from the pressure-settlement curves, that as the inclination of the load is increased, the settlement due to failure pressure decreases.

Figures (5) and (6) show the bearing pressure-horizontal displacement curves for different load inclination (α) for two relative densities on unreinforced sand. By observing the curves of pressure-horizontal displacement of all cases, it can be concluded that the horizontal displacement is increased with an increasing of the load inclination (α).

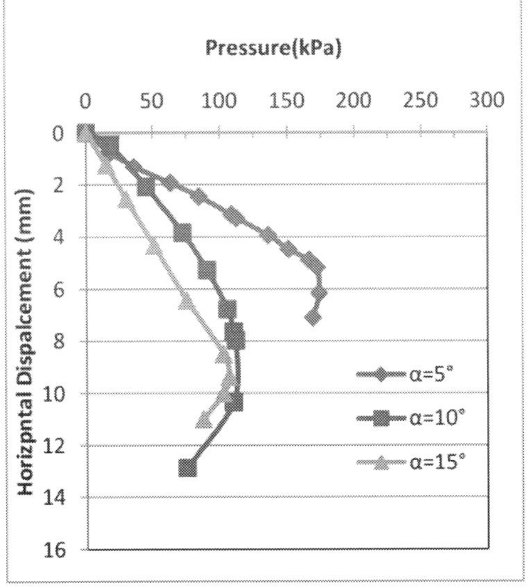


**Fig. 5.** Pressure-horizontal displacement curves for different (α) $R_D$=60%, N=0

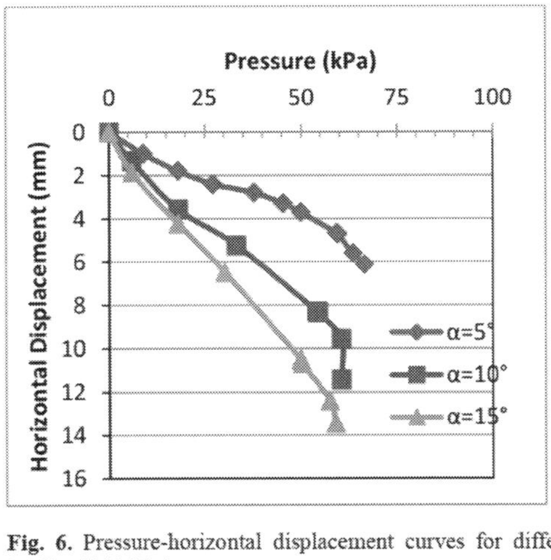


**Fig. 6.** Pressure-horizontal displacement curves for different (α) $R_D$=80%, N=0

## 5.2 Effect of Geogrid Layers

In order to investigate the effect of the change the number of geogrid layers (N) on the ultimate bearing capacity, (36) tests have been conducted for several cases. Figures (7) and (8) show the relationship between the number of geogrid layers (N) and the bearing capacity ratio (BCR) for different values of load inclinations (α) for relative densities 60% and 80%, respectively. It is noticed that the (BCR) significantly increased with the increase of the number of geogrid layers. In addition, it is noticed that there is an optimum value of (N) after which little increase in the value of (BCR) is observed. The reason of this behavior can be clarified as the number of geogrid increase the length of the shear failure surface increased. The second reason is that by increasing the numbers of geogrid layers, the tensile stress which can be carried by the geogrid will increased too. As the results of both reasons, the resistance of the soil against shear failure will increase too.

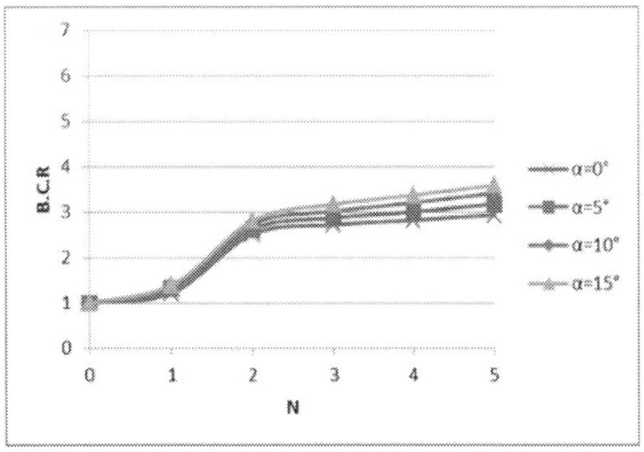


**Fig. 7.** Bearing capacity ratio (BCR) versus Number of reinforcement layer (N) for (α = 0°, 5°, 10° and 15°) and (RD=60%).

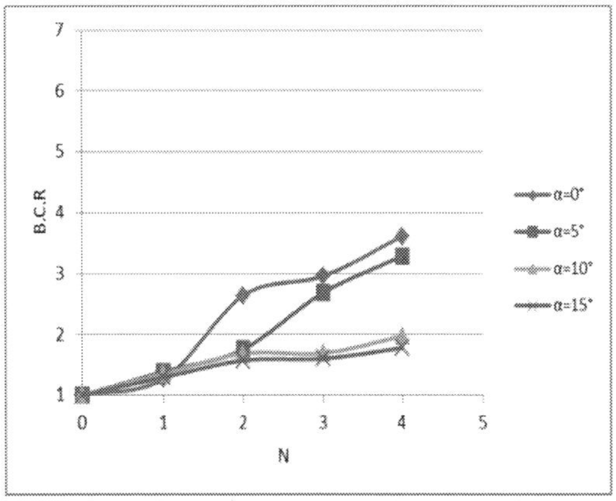


**Fig. 8.** Bearing capacity ratio (BCR) versus Number of reinforcement layer (N) for (α = 0°, 5°, 10° and 15°) and (RD= 80%).

Figure (9) and (10) show the effect of number of reinforcement sheets (N) on the horizontal displacement of the footing which is due to load inclinations (α). It can be seen that the increasing of the reinforcement (N) decreases the horizontal displacements for different values of load inclination (α) for relative densities 60% and 80%, respectively.

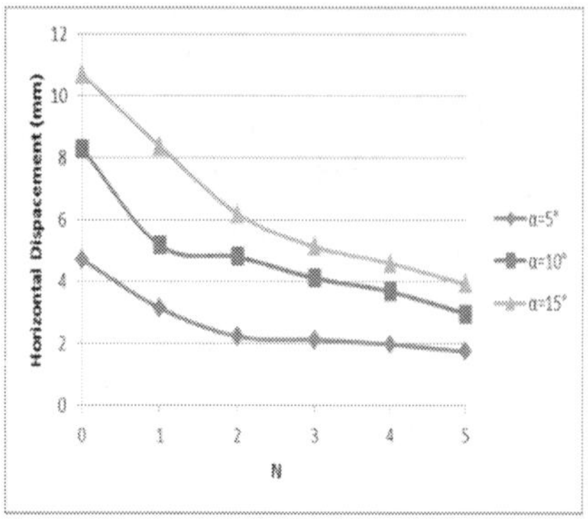


**Fig. 9.** Horizontal displacement (mm) versus Number of reinforcement layer (N) for different values of (α) (RD= 60%).

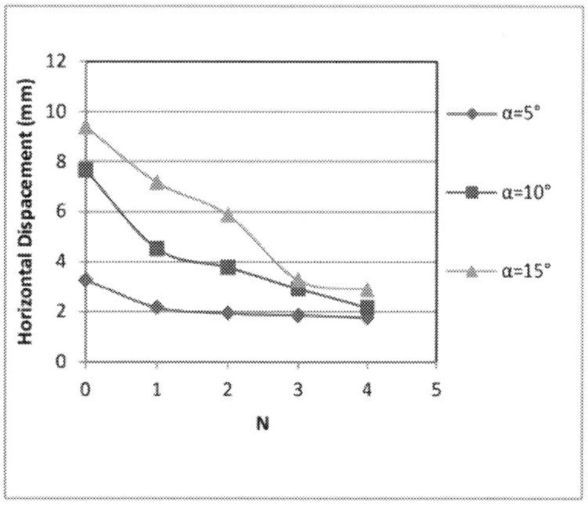


**Fig. 10.** Horizontal displacement (mm) versus Number of reinforcement layer (N) for different values of (α) and (RD= 80%).

Figures (7) to (10) show the influence of the relative density ($R_D$) on the bearing capacity ratio (BCR) and on horizontal displacements for different values of load inclination (α) for relative densities 60% and 80%, respectively. It can be seen that the increase of the relative density ($R_D$) decreases the (BCR) and decreasing the optimum number of reinforcement (N) and reduce the horizontal displacement of the footing which is due to load inclination (α).This could be explained because the soil has been improved and it was observed for the two chosen values of ($R_D$). In addition, it is noticed that the (BCR) for the medium dense sand is larger than that for dense sand. This means that the reinforcement is more sufficient for medium dense sand than for dense sand considering the unreinforced loaded case for each of them as a reference, according to the definition of (BCR).

It should be mentioned that Figures (7) to (10) could be used by practicing engineers as design charts to obtain the number of reinforcement layers required to cancel or to reduce the effect of load inclination or even to increase the factor of safety. Table (3) show the magnitude of enhancement in the bearing capacity of strip footing when use geogrid reinforcement layer for different values of load inclination angle and for two relative densities.(60% and 80%).

**Table 3.** Effect of Geogrid Multi-reinforcement layers on Bearing Capacity Ratio

| $R_D$ % | α | $q_u$ | N | | | | | |
|---|---|---|---|---|---|---|---|---|
| | | | 0 | 1 | 2 | 3 | 4 | 5 |
| 60 | 0° | 65.7 | 1.00 | 1.497 | 2.512 | 2.715 | 2.817 | 2.934 |
| | 5° | 59.5 | 1.00 | 1.575 | 2.621 | 2.969 | 2.997 | 3.17 |
| | 10° | 54.4 | 1.00 | 1.680 | 2.733 | 3.011 | 3.201 | 3.413 |
| | 15° | 50.2 | 1.00 | 1.758 | 2.810 | 3.167 | 3.367 | 3.585 |
| 80 | 0° | 123.3 | 1.00 | 1.273 | 2.642 | 2.959 | 3.607 | |
| | 5° | 112.5 | 1.00 | 1.394 | 1.759 | 2.690 | 3.281 | |
| | 10° | 112.5 | 1.00 | 1.363 | 1.687 | 1.697 | 1.971 | |
| | 15° | 106.9 | 1.00 | 1.298 | 1.577 | 1.605 | 1.780 | |

## 6 Conclusions

From the experimental results and their discussion stated in the above sections, the major conclusions are outlined below:

1. The results showed that using geogrid as a reinforcement material has a significant increased on the ultimate bearing capacity .
2. Bearing capacity increase with increasing in the number of reinforcement layers (N), but there is an optimum value after which little effect is observed. This optimum value is varied (N=4-5) for relative densities (80% and 60%), respectively.
3. Increasing the number of reinforcement layers (N) lead to decreasing the horizontal displacement of the footing.

## References

1. A. Amen, M.Sc. Thesis, Tikrit University, Iraq. (2008)
2. M. El Sawwaf, J. Geotechnical. and Geological Eng., Vol. **135**, No. 10, pp. 1509-1518, (2009).
3. M. Omar, J. Pure and Applied Sciences, University of Sharjah, Vol. **3**, No.2, pp. 35-51,( 2006).
4. C. Patra, R. Behera, N. Sivakugan, and B. Das, Int. f- Geotechnical. Eng., 6(2): 342-352, (2012).
5. R. Purkayastha, and R. Char, j. Geotechnical. Eng. Div., 103(6), 647-651.( 1977).
6. S. Saran, S. Kumar, and A. Kumar, J. Geotechnical and Geological Eng, 25, 123-137,(2007).

7. E. Shin, and B. Das, Geosynthetics Int., Vol. **7**, No. 1, pp. 59-71.( 2000).
8. E. Shin, B. Das, E. Lee, and C. Atalar, J. geotechnical and geological engineering, 20(2), pp.169-180,( 2002).
9. Al-Taay, M.Sc. Thesis, Tikrit University, Iraq, (2010).
10. R. Koerner, *Designing with Geosynthetics*, 5th Edition, Pearson Prentice Hall, USA, (2005).
11. C. Patra, B. Das , M. Bhoi, and E. Shin, J. Geotextiles and Geomembranes, 24(4), 254-259, ( 2006).
12. M. Al-Zandi, M.Sc. Thesis, Tikrit University, Iraq , (2013).
13. ASTM, Vol. 04.13 Geosynthetics, West Conshohocken, United State ,(2004).

# Three-dimensional analysis for piled raft machine foundation embedded in sand

*Mahmood* Mahmood[1,*], *Saad* Al-Wakel[1], *Ihab* Abdulwahhab[1]

[1]Building and Construction Engineering Department, University of Technology, Baghdad, Iraq

**Abstract:** Three-dimensional analysis for the dynamic response of a piled raft foundation subjected to vertical vibration is presented in this study. The analysis considers several factors affecting the amplitude of displacement for deep foundation such as pile cap embedment, pile cap thickness, relative density of the sand and the boundary effect. A validation for an experimental piled raft model depending on a scale factor of (20) using at (Plaxis 3D) computer program was performed. The sand is simulated using Mohr-Coloumb model while the concrete is simulated as linear elastic material. It has been found that embedding the pile cap in the soil and increasing its thickness lead to decrease the maximum amplitude of displacement. Furthermore, the predictions showed that increasing the distance between the foundation and the boundaries and increasing the relative density of the sand can significantly minimize the dynamic response of the foundation.

## 1 Introduction

The use of the finite element method in dynamic foundation problems varies with other applications such that soil stratum of infinite medium that should be performed by a model of a specified size. This type of model creates a 'box' effect which is restricting the energy of the system and deforming its dynamic behavior. To eliminate this problem, lateral boundaries that absorb waves are introduced to contain the energy radiation into the outer region not existed in the model [1].

There are many issues that need careful examination before using finite element computer software such as modeling capabilities, analysis capabilities and processing but the most important one is the validation of results. Validation of the results is necessary before accepting the results [2].

Novak, 1974 [3] computed a vertical response of a machine and its foundations. The foundation consisted of a rectangular block of concrete (4.8 x 3 x 2.4 m). Two cases were considered; embedded (0.60 m) into the soil and having no embedment. It was supported on (10.5 m) long, fixed-top timber piles in medium stiff clay. The machine weight was (10 tons).

Manna and Baidya, 2010 [4] studied the influence of nonlinearity on the dynamic response of cast-in-situ reinforced concrete piles subjected to strong vertical excitation. Forced vibration experiment of single piles (L/d = 10, 15, 20) and (2×2) pile groups with (s/d = 2, 3, 4 for each L/d) were conducted in the field for two embedment conditions of pile cap. From the recorded nonlinear response curves, the effective pile–soil system mass, damping and stiffness were obtained and the nonlinear response curves were back-calculated depending on the theory of nonlinear vibration. The results of the test were compared with Novak continuum approach with dynamic interaction factor approach by both linear and linear-equivalent numerical methods. A good match between the predicted and recorded response was found for linear-equivalent methods by introducing a weak boundary zone around the pile to approximately account for the nonlinear behavior of pile-soil system. The test data were used to create the empirical relationship to estimate the extent of soil separation around the pile with soil subjected to vertical vibration.

Padron et al., 2012 [5] studied the accuracy and effectiveness of the superposition method to assess the coefficients of the dynamic stiffness and damping of embedded footings supported by vertical piles set in uniform visco-elastic soil. A Comparison between these coefficients of pile embedded footings and those obtained by superposing the separate coefficients of the corresponding pile groups and embedded footings reveals that the average of the relative differences is about (10–30%). The results were presented in a set of normalized charts and simple expressions, which can be used to estimate the dynamic stiffness and damping of piled embedded footings, on a condition that the coefficients of the two separate components were known. Since such impedance functions for both embedded footings and pile groups were available for a wide range of cases, the superposition approach studied here was attractive.

Abdulrasool, 2012 [6] studied the dynamic analysis of deep foundations on uniform dry sand experimentally

* Corresponding author: mahmoudal_qaissy@yahoo.com

© The Authors, published by EDP Sciences. This is an open access article distributed under the terms of the Creative Commons Attribution License 4.0 (http://creativecommons.org/licenses/by/4.0/).

and numerically by the finite element method. The numerical analysis involves the displacement response under the effect of dynamic loads of harmonic vertical mode of vibration. The aim of the study is to analyze the dynamic response of machine foundations by simulating the machine on deep foundation numerically using the finite element method. It was concluded that when the piles length and the number of piles increase this will lead to a decrease in the displacement response of the pile foundation because the increase in the mass of foundation. The dynamic response of the deep foundation is influenced by the spacing between piles.

## 2 Numerical analysis

In this study, (Plaxis 3D) has been utilized to simulate the piled raft foundation with four piles in order to study the effect of various factors on the dynamic response of the foundation. Furthermore, it is transformed to a prototype model depending on a scale factor of (20). It is recommended to use a scale factor greater than (10) to eliminate the overestimated results [3]."Table 1" illustrates the scales of centrifugal modeling which is adopted in this analysis where (n) is the scale factor.

Table 1.Scales for centrifugal modeling [9]

| Quantity | Full Scale | Centrifugal Model |
|---|---|---|
| Linear dimension | 1 | 1/n |
| Area | 1 | $1/n^2$ |
| Volume | 1 | $1/n^3$ |
| Time | 1 | 1/n |
| Mass | 1 | $1/n^3$ |
| Force | 1 | $1/n^2$ |
| Density | 1 | 1 |
| Frequency | 1 | n |

### 2.1 Soil geometry and properties

The scaled dimensions of the soil are (15 × 15) m and the depth of the soil is (11) m are specified by single borehole at the first corner of the model. The boreholes locations and profile in the model as same as the information on the soil profile. If multiple boreholes are specified, the program automatically interpolates between the boreholes [8]. "Fig. 1" shows the soil model.

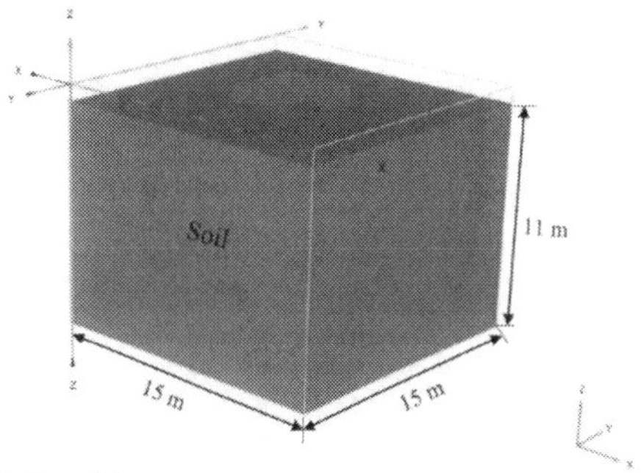


Fig.1. Soil model.

In order to simulate the soil behavior, an appropriate material model and parameters must be assigned to the geometry. The (Mohr-Coloumb) model is chosen for the soil. The material properties of the soil which are taken from the experimental model are listed in "Table 2".

Table 2. Soil material properties.

| Parameter | Value | Unit |
|---|---|---|
| Material model | Mohr-Coulomb | - |
| Drainage type | Drained | - |
| Dry unit weight | 16.87 | $kN/m^3$ |
| Voids ratio | 0.535 | - |
| Young's modulus | $30 \times 10^3$ | $kN/m^2$ |
| Poisson's ratio | 0.28 | - |
| Cohesion | 0 | $kN/m^2$ |
| Friction angle | 37.5 | degree |
| Lateral earth pressure coefficient $K_0$ | 0.4 | - |

The lateral earth pressure coefficient $K_0$ is obtained automatically using the following equation:

$$\mathbf{K_o = 1 - sin\emptyset} \quad (1)$$

The deep foundation consists of two parts; the first is the pile cap which is simulated as a volume with the dimensions of (4 m × 4 m) and a thickness of (0.8 m). The volume is created by extruding a created surface which has the pile cap same dimensions using the extrude tool. The soil material data sets can be assigned to volumes. The data set of the concrete listed in "Table 3" is assigned to the pile cap volume.

Table 3. Concrete material properties

| Parameter | Value | Unit |
|---|---|---|
| Material model | Linear Elastic | - |
| Drainage type | Non-Porous | - |
| Dry unit weight | 24 | $kN/m^3$ |
| Young's modulus | $27.6 \times 10^6$ | $kN/m^2$ |
| Poisson's ratio | 0.1 | - |

The second part of the deep foundation is the four piles which have the same dimensions of (0.42) m diameter and (8) m length. Each pile is simulated as embedded pile. The skin resistance is simulated as linear. The input is defined by the skin resistance at the pile top $T_{top\ max}$ and the skin resistance at the pile bottom $T_{bot\ max}$. This way of defining the pile skin resistance is mainly applicable to piles in a homogeneous soil [9]. The tip bearing is defined by $F_{max}$. The total pile bearing capacity $N_{pile}$ is given by:

$$N_{pile} = F_{max} + 0.5\,L\,(T_{top\ max} + T_{bot\ max}) \quad (2)$$

Where: $L$ is the pile length.
The pile properties are shown in "Table 4".

Table 4. The Piles properties

| Parameter | Value | Unit |
|---|---|---|
| Young's modulus | $27.6 \times 10^6$ | $kN/m^2$ |
| Unite weight | 24 | $kN/m^3$ |
| Predefined pile type | Massive Circular | - |
| Diameter | 0.42 | M |
| Skin resistance type | Linear | |
| $T_{top\ max}$ | 0 | kN/m |
| $T_{bot\ max}$ | 94.6 | kN/m |
| $F_{max}$ | 1589 | kN |

The type of connection of the top of the pile is selected as rigid which makes the rotation and displacement both coupled with the rotation and displacement of the structural or soil element at which the top of the pile is located. On the other hand, the type of the connection at the bottom of the pile is selected as free which allows the connection point to rotate and move relatively to the soil. The pile interaction with the surrounding soil is due to the special interface elements."Fig. 2" shows the simulated piled raft foundation with four piles.

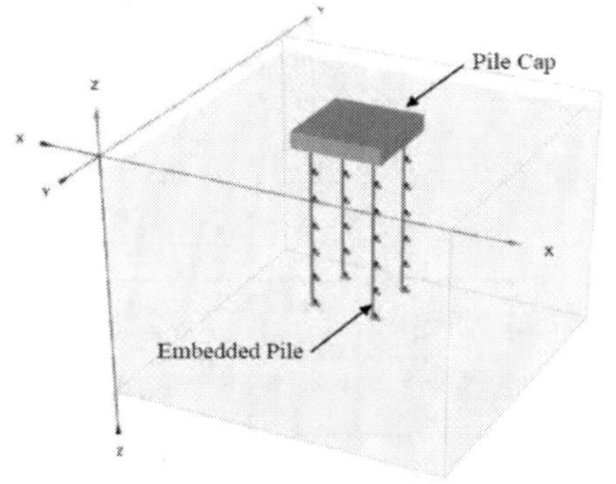


Fig.2 The simulated piled raft foundation with for piles

A surface representing the base of the machine is created on the top of the pile cap volume with dimensions of (2 × 2) m subsequently the static and the dynamic loads are created on it as shown in "Fig. 3".

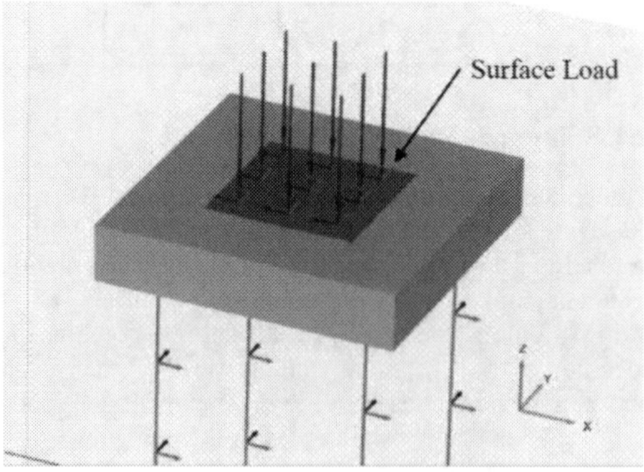


Fig. 3.The surface of machine base

The static load is representing the scaled weight of the machine. It is defined as surface load on the surface which represents the base of the machine whereas the dynamic load which is calculated due to the scaled parameters is defined using a harmonic load multiplayer as shown in "Fig. 4".

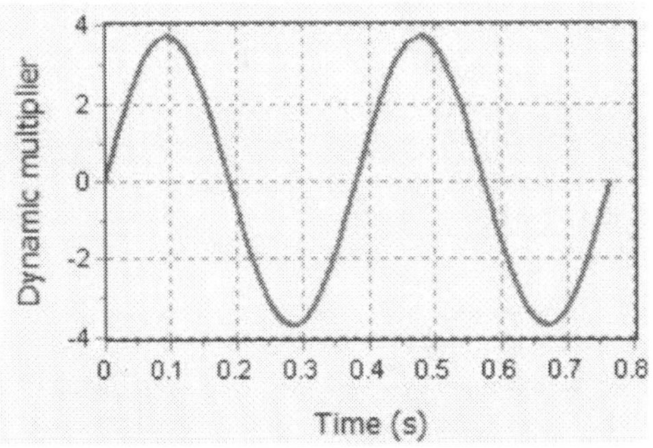


Fig. 4.The harmonic load.

The soil volume is modeled using (10) node tetrahedral elements which are created in an automatic mesh generation procedure. The geometry is divided into volume elements and structure elements if excited. The mesh generation takes in account the position of the volumes in the geometry model. Furthermore, the exact position of loads and structures is taken in consideration in the finite element meshing. A local refinement is considered in the pile cap volume. The created mesh consists of (2871) node. The finite element discretization of the piled raft model is shown in "Fig. 5".

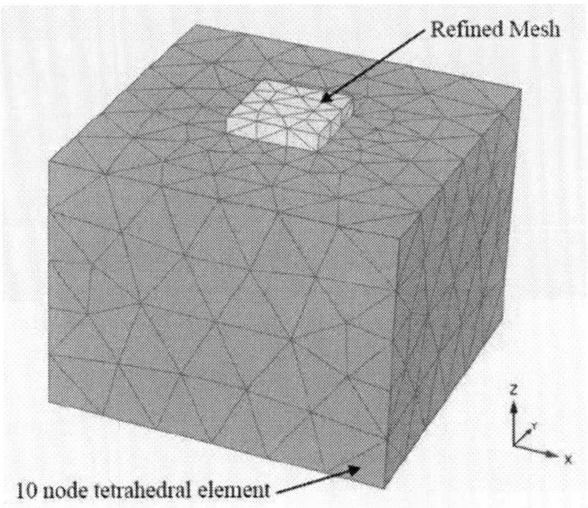


Fig. 5.The finite element discretization of the piled raft model.

The calculation consists of multiple phases defined due to sequence of structures construction and loading in the staged construction mode. This case includes initial phase, static loading phase and dynamic loading phase.

The Initial phase represents the initial conditions of the soil. The initial conditions include the initial geometry configuration and the initial stress state. The soil and the pile cap volumes in addition to the embedded piles are activated in this phase to simulate the initial condition of the model. $K_o$ procedure is selected as a calculation type for this phase. The static and dynamic loads are deactivated during this calculation phase. The static load which represents the weight of the machine is activated in the static loading phase and $K_o$ procedure is also selected as calculation type.

The dynamic load is activated in the next phase and the calculation type is defined as dynamic with a time interval of 5 seconds consisting of 256 step of calculations. Special boundary conditions called viscous boundaries have been defined to account the reflected waves from the model boundaries.

After the creation of the phases, the node at the center edge of the pile cap is selected to be considered in the curves then the calculation process is executed.

## 3. Results and discussion

### 3.1. Validation of the experimental model

The piled raft model and embedded piled raft model with four piles configuration at the frequency of (52.3) Hz. have been simulated in three dimensional finite element modeling using (Plaxis 3D) software where the maximum amplitude of dynamic force $F_{dy.max}$ equals (14.8) kN. In order to validate the experimental model using a scale factor of 20, Novak, 1987 [9] recommended using a scale factor greater than 10 to eliminate the overestimated results.

The Predicted displacement with time of the piled raft model and embedded piled raft model are shown in "Fig. 6" and "Fig. 7", respectively. A good agreement has been achieved between the predicted and the measured results from Abdul wahhab, 2017 [10] as shown in "Table 5". This means that the numerical modeling which is adopted in this study to simulate the machine foundations on dry soil is satisfactory.

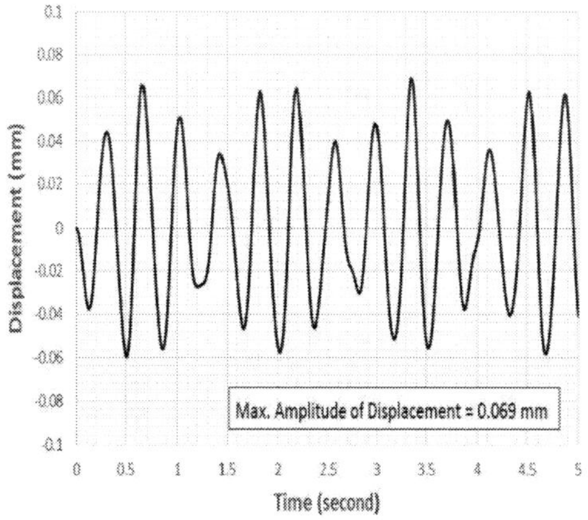


Fig. 6. Predicted vertical displacement with time for the piled raft model with four piles configuration.

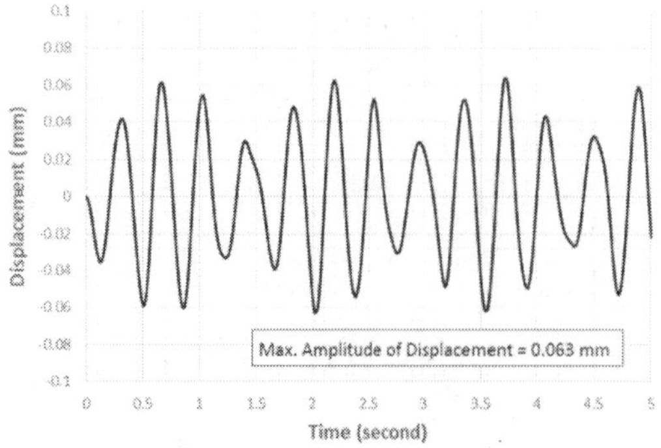


Fig. 7. Predicted vertical displacement with time for the Embedded piled raft model with four piles configuration.

Table 5. The measured and predicted amplitude of displacement of the piled raft foundation.

| Typeof Foundation | Measured Amplitude (mm) | Predicted Amplitude (mm) |
|---|---|---|
| Piled Raft | 0.07 | 0.069 |
| Embedded Piled Raft | 0.061 | 0.063 |

### 3.2. Effect of the pile cap embedment on the dynamic response

To investigate the effect of the pile cap embedment on the maximum amplitude of vertical displacement of the deep foundation, three different elevations of the pile cap (On the soil surface, 0.4 m under the soil surface and 0.8 m under the soil surface) were considered. The dynamic response of the foundation for different elevations of the pile cap is shown in "Fig. 8", while "Table 6" shows the maximum amplitude of the vertical displacement and the percentage of reduction. The reduction in the dynamic response due to embedding the pile cap is approximately linear as shown in "Fig. 9".

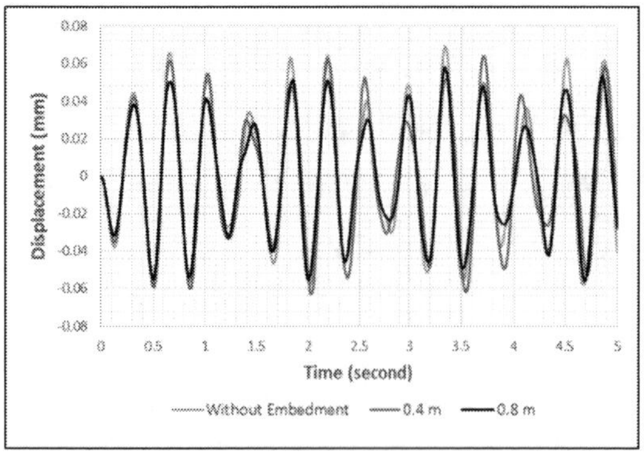


Fig. 8. Effect of the pile cap embedment on the dynamic response

Table 6. The maximum amplitude of displacement and the percentage of reduction for different elevations of the pile cap.

| Depth of Pile Cap Embedment | Max. Amplitude of Displacement | Percentage of Reduction |
|---|---|---|
| No Embedment | 0.069 | - |
| 0.4 | 0.063 | 9 |
| 0.8 | 0.058 | 16 |

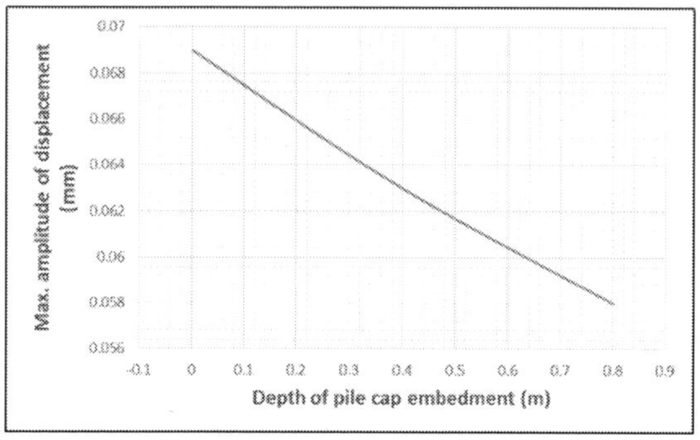


Fig. 9. Reduction in the maximum amplitude of displacement due to increasing the depth of pile cap embedment.

The results showed that embedding the pile cap decreases the maximum amplitude of displacement due to the side friction of the pile cap and the increase of the damping due to increasing the area which absorbs the waves transmitted from the foundation. This behavior agrees with Novak, 1974 [3] who state that embedding the pile cap in a soil layer increases the damping.

### 3.3. Effect of the boundaries of the model on the dynamic response

In this part, three different model dimensions considered (15×15) m, (17.5×17.5) m and (20×20) m to study the effect of the boundary condition on the dynamic response of the foundation, while the depth of the model is remained constant as (11) m. All the dimension of the piles and the pile caps and other parameters remained constant in this study.

"Fig.10"shows the dynamic response for the three models. The percentage of reduction of the maximum amplitude of the displacement due to change in boundary condition is illustrated in "Table 7".

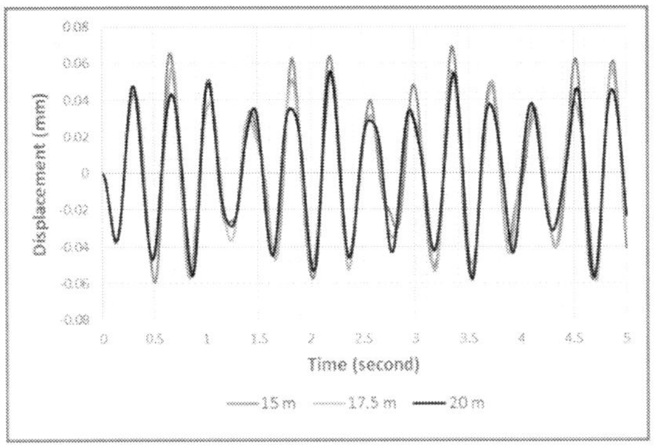


Fig. 10. Effect of the boundary condition on the dynamic response.

Table 7. The maximum amplitude of the displacement and percentage of reduction due to change in the boundary condition.

| Width of the Model | Max. Amplitude of Displacement | Percentage of Reduction (%) |
|---|---|---|
| 15 | 0.069 | - |
| 17.5 | 0.065 | 15 |
| 20 | 0.056 | 19 |

The results showed that the maximum amplitude of vertical displacement decreases by increasing the model dimensions but the reduction is less using larger models due to the less reflected waves from the boundaries as shown in "Fig. 11".

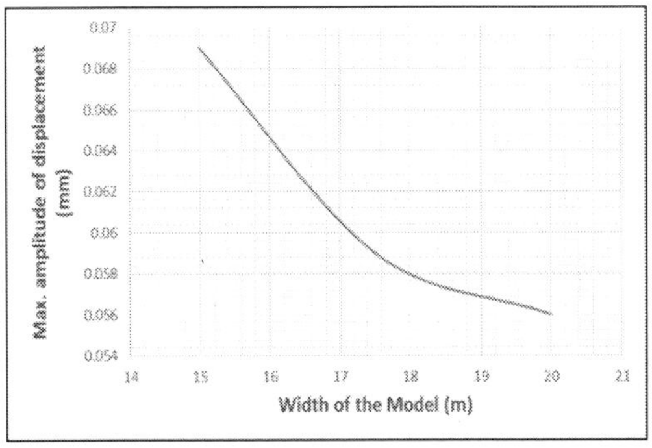


Fig. 11. Reduction in the maximum amplitude of displacement due to increasing the dimensions of the model.

## 3.4. Effect of the relative density on the dynamic response

Three soil profiles have been selected to investigate the dynamic response of the machine foundation with different soil properties. The value of Poisson's ratio (υ) and values for the Young modulus (Es) are adopted from Das and Ramana, 2011 [11].The properties of soil are shown in "Table 8".

Table 8. The maximum amplitude of displacement and the percentage of reduction due to increasing the number of piles.

| Material Properties | Loose | Medium | Dense |
|---|---|---|---|
| Relative Density, Dr (%) | 29 | 50 | 77 |
| Modulus of Elasticity, Es | 25000 | 30000 | 35000 |
| Poisson's Ratio, υ | 0.32 | 0.28 | 0.24 |
| Dry Density ($kN/m^3$) | 16.2 | 16.87 | 17.8 |

The dynamic response for the soil types are shown in "Fig. 12". The maximum amplitude of displacement and the percentage of reduction due to using different soil types are illustrated in "Table 9".

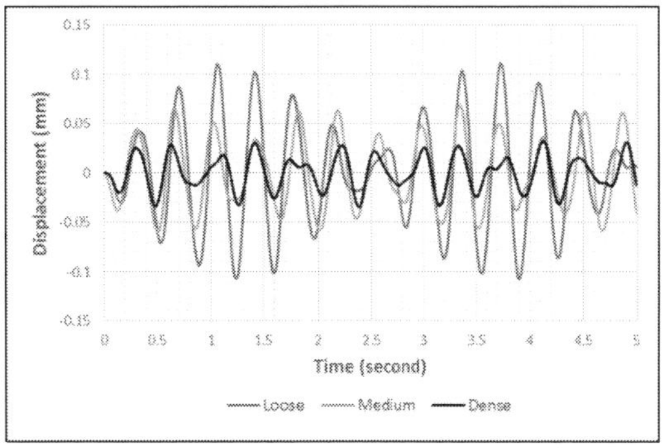


Fig. 12. Effect of the relative density on the dynamic response.

Table 9. The maximum amplitude of displacement and the percentage of reduction due to increasing the relative density.

| Relative Density (%) | Max. Amplitude of Displacement (mm) | Percentage of Reduction (%) |
|---|---|---|
| 29 | 0.112 | --- |
| 50 | 0.069 | 38 |
| 77 | 0.035 | 63 |

From the results of the numerical analysis for the machine foundation embedded in the soil with different soil properties, it can be seen that the maximum amplitude of displacement decreases in the medium sand model. Furthermore it can be noticed that the minimum dynamic response occurred in the dense sand model due to the increase in soil damping using dense soils as shown in "Fig. 13". This behavior agrees with Abdulrasool, 2012 [6] who stated that the maximum amplitude of vertical displacement of pile machine foundation under dynamic loads occurs in the loose sand.

Table 10. The maximum amplitude of the displacement and the percentage of reduction due to increasing the pile cap thickness.

| Thickness of the Pile Cap (m) | Max. Amplitude of Displacement (mm) | Percentage of Reduction (%) |
|---|---|---|
| 0.4 | 0.07 | - |
| 0.8 | 0.069 | 1.4 |
| 1.2 | 0.068 | 2.9 |

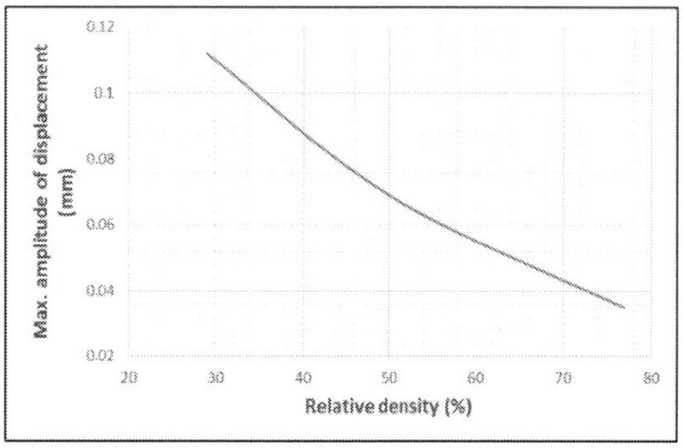


Fig. 13. Reduction in the maximum amplitude of displacement due to increasing the relative density of the soil.

### 3.5. Effect of the pile cap thickness on the dynamic response

Another approach in this part is investigated on the effect of increasing the thickness of the pile cap. The bearing capacity of the pile cap remains the same in a result for keeping the other dimensions constant. Three models have been numerically analyzed with three thicknesses of the pile cap (0.4 m, 0.8 m and 1.2 m).

The dynamic responses of the machine foundation with different pile cap thickness are illustrated in "Fig. 14". The maximum amplitude of the displacement and the percentage of reduction resulted from increasing the pile cap thickness as shown in "Table 10".

The results of the numerical analysis showed that the maximum amplitude of vertical displacement decreases by increasing the thickness of the pile cap due to the weight excess in the pile cap. This means that the stiffness is increased. The reduction in the maximum amplitude of displacement is linear as shown in "Fig. 15"

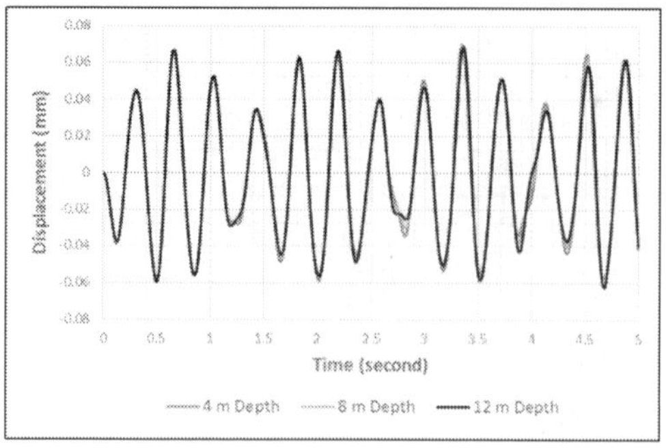


Fig. 14. Effect of the pile cap thickness on the dynamic response.

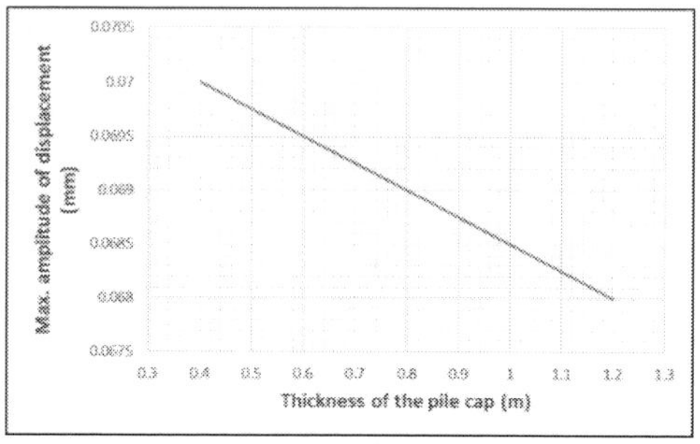


Fig. 15. Reduction in the maximum amplitude of displacement due to increasing the thickness of the pile cap.

## 4 Conclusions

From the study, several conclusions have been made:

1. The dynamic response results of the numerical model and the measured results of the experimental model showed a good agreement. This means that the numerical modeling which is adopted in this study to simulate the machine foundations on dry soil is satisfactory.
2. Embedding the pile cap in the soil decreases the maximum amplitude of displacement due to the side friction of the pile cap and the increase of the damping due to increasing the area which absorbs the waves transmitted from the foundation.
3. The maximum amplitude of vertical displacement decreases by increasing the model dimensions but the reduction is less using larger models due to the less reflected waves from the boundaries.
4. The maximum amplitude of displacement decreases in the medium sand model. Furthermore, it can be noticed that the minimum dynamic response occurred in the dense sand model due to the increase in soil damping using dense sand.
5. The maximum amplitude of vertical displacement decreases by increasing the thickness of the pile cap

due to the weight excess in the pile cap. This means that the stiffness is increased.

## Abbreviations

A list of symbols should be inserted before the references if such a list is needed

| | |
|---|---|
| | Pile length |
| $D$ | Pile Diameter |
| $S$ | Spacing between piles |
| $\varphi$ | friction angle |
| $T_{top\ max}$ | Skin resistance at the pile top |
| $T_{bottom\ max}$ | Skin resistance at the pile bottom |
| $F_{max}$ | Tip bearing |
| $K_0$ | Lateral earth pressure coefficient |
| $F_{dy.max.}$ | The maximum amplitude of dynamic force |

## References


1. G. Gazetas, Soil Dynamics and Earthquake Engineering, , **2**, No. 1, pp. 2-42 (1983).
2. K.G. Bhatia, "Foundations for Industrial Machines" D-CAD Publishers, 158, Vardhman Grand Plaza, Mangalam Place, Sector **3**, Rohini, New Delhi I 10085(2008).
3. M. Novak, Canadian Geotechnical Journal, National Research Council of Canada, Vol. **II**, No. 4, Toronto, Ontario, Canada, pp. 574-598(1974).
4. B. Manna, and D.K. Baidya , Elsevier Ltd, Soil Dynamics and Earthquake Engineering, **30**, Issue 6, June 2010, pp 456-469 (2010).
5. L. A. Padron, G. Mylonakis and D.E. Beskos, "Simple Superposition Approach for Dynamic Analysis of Piled Embedded Footings", International Journal for Numerical and Analytical Methods in Geomechanics, Vol. **36**, pp. 1523–1534(2012).
6. A.S. Abdulrasool, "Vibration analysis of deep foundations in dry sand" M.Sc. thesis, University of technology, Baghdad (2012).
7. Plaxis 3D, (2013). *"Scientific Manual"* pp 40.
8. Plaxis 3D, (2013). *"Tutorial Manual"* pp 10.
9. M. Novak, "Experimental Studies of the Dynamic Behavior of Piles and Pile Groups" Dynamic Behavior of Foundations and Buried Structures, Elsevier Applied Science Publishers, London, pp. 156 (1987).
10. I. G. Abdulwahhab, "Experimental and Numerical Approach for the Behavior of Machinery Piled Raft Foundation Embedded within Cohesionless Soils" M.Sc. Thesis, University of Technology, Baghdad (2017).
11. B. M. Das, and G. Ramana, "Principles of Soil Dynamics", Cl-Engineering.

# Assessment of natural and technogenic conditions of the area of the water-supply system upgrade in Dudinka (Krasnoyarsk Region, Russia)

*Fedor* Bryukhan[1,*]

[1]Moscow State University of Civil Engineering, Yaroslavskoe shosse, 26, Moscow, 129337, Russia

**Abstract.** Territories of the subarctic forest-tundra of Taymyr are characterized by an insufficient state of exploration of their natural conditions. Another feature is vulnerability of natural landscapes to man-made impacts. Along with this, the increased public attention to preservation of the ecological status of these territories necessitates detailed studies of environment components during construction activities. This work is devoted to the geoecological substantiation of the water-supply system upgrade in Dudinka (Taimyr Dolgan-Nenets Autonomous District of the Krasnoyarsk Region, Russia), considering specific natural conditions of the sub-arctic forest-tundra and technogenic impacts of the upgrade project. This work describes primary results of engineering and environmental surveys, which provided necessary and sufficient basic data. It was established that due to a localized nature of technogenic impacts during environmental activities, the combined impact of technogenic factors is insignificant and will not cause environment degradation. A set of recommendations and proposals for environment protection and ecological monitoring organization was developed.

## 1 Introduction

Territories of the subarctic forest-tundra of Taymyr are characterized by an insufficient state of exploration of their natural conditions and vulnerability of natural landscapes to man-made impacts. Particularly, due to technogenic impacts on landscape components geochemical changes occur in soils, bottom sediments, surface waters and vegetation cover in these territories [1]. It should be noted that at large industrial facilities in most cases environmental impacts are observed at all stages of technogenic activity, especially during their construction and operation. Therefore, construction activities in these territories should comply with the "green" construction concept [2, 3] and consider quality improvements in geoecological substantiation of such projects [4].

Taking into account the above considerations, acquisition of adequate baseline data for ecological design during construction development of the subarctic forest-tundra territories shall include appropriate composition and scope of engineering and environmental surveys.

The purpose of the study is a comprehensive geoecological assessment of the territory of the reconstructed water supply system in Dudinka (Taimyr Dolgan-Nenets Autonomous District of the Krasnoyarsk Region, Russia), which is necessary for the development of measures to protect the environment.

## 2 General characteristics of the project and surveyed territory

Figure 1 shows an outline map of the surveyed territory. The main water-supply source of Dudinka is the Lake Samsonkino. The first stage pumping station (point A in Figure 1) is located on its south bank. The project includes:

- upgradation of the first stage pumping station;
- construction of treatment facilities with the second stage pumping station (point B in Figure 1);
- upgradation of the existing main water pipelines and construction of new ones;
- construction of a 6 kV power transmission line from the main step-down substation (point C in Figure 1) for power supply to the designed facilities.

The surveyed territory is situated outside the residential zone of the city and includes (refer to Figure 1):

• an area within the sanitary protection zone (SPZ) of the Lake Samsonkino (shaded);

• sites of the first stage pumping station, treatment facilities and the main step-down substation;

• a 20 m wide strip on the both sides of infrastructure facilities.

The city lies on the right bank of the Yenisey, near the confluence point of the Dudinka river, its right-bank tributary. The distance between the city and the Yenisey estuary is about 430 km.

* Corresponding author: pniiis-gip@mail.ru

© The Authors, published by EDP Sciences. This is an open access article distributed under the terms of the Creative Commons Attribution License 4.0 (http://creativecommons.org/licenses/by/4.0/).

https://doi.org/10.1051/matecconf/201816201024

The water-supply source – Lake Samsonkino – is located 3 km north-east from the city. The catchment area is 8.4 $km^2$, the lake length – 1,960 m, average width – 450 m, average depth – 5.2 m, volume – $4.6\times10^6$ $m^3$, water-surface area – 0.89 $km^2$. The lake is supplied by snow melting and rainfalls. According to the data of long-term hydrological observations at the Lake Samsonkino, the highest flood occurs in mid-May – early June. The minimum water level is observed in March-April.

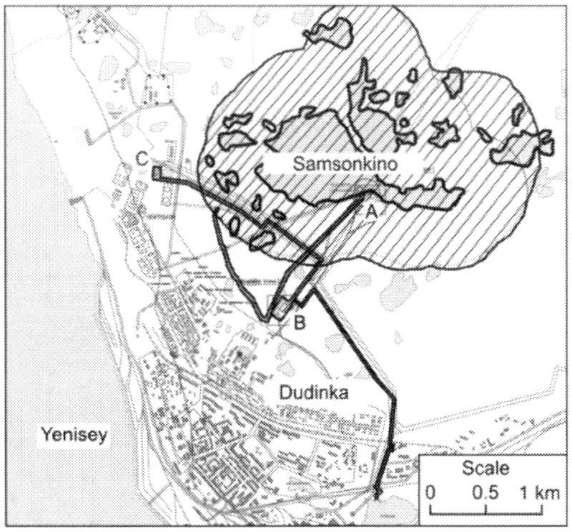


**Fig. 1.** Outline map of the surveyed territory.

The territory is a vast low gently rugged plain with diverse and composite topography and pervasive cryogenic structures [5]. Specific landscape forms are shown in Figure 2. The surveyed territory is characterized by numerous small lakes and boggy areas. The territory is dissected by shallow valleys of small creeks running dry in summer.

The territory belongs to the transitional zone between Arctic tundra and taiga. The dominant form of the vegetation cover is shrubby (including alder, rowan, briar) [6, 7]. Suppressed trees with crooked trunks and irregular crowns occur. Many berry plants grow near the Lake Samsonkino – blueberry, cloudberry and, cowberry. There are high blooming sally and hogweed thickets. Large areas of the territory are covered with lichen. Water birds, mainly ducks, nest in thick grasses (sedge and cotton grass) near the lake.

A reconnaissance survey of the territory showed that it is characterized by a complex combination of peaty-gley tundra soils and gley-podzolic soils, specific for the northern taiga. Based on the results of the reconnaissance it was established that the humus thickness is insignificant, only several centimeters.

**Fig. 2.** Specific landscape forms. Near the Lake Samsonkino (a), one of numerous small lakes (b).

## 3 Basic geoecological works

Survey works were carried out by specialists of Scientific & Industrial Association Gidrotekhproekt and the National Research University – Moscow State Construction University in September of 2016 according to requirements and recommendations [8, 9].

Preparatory works included collection, processing and analysis of data on environment conditions within the target area, including materials of similar surveys previously performed in this territory.

The composition and scope of field works and laboratory studies are shown in Tables 1 and 2. During radio ecological studies – measurement of the gamma Equivalent Dose Rate (EDR), radionuclide analysis of soil samples, measurement of radon flux density – special attention was paid to areas of prolonged exposure for the personnel.

Cameral works included geoecological generalization of field and laboratory research results, predictive assessment of the environment condition, development of proposals for protection measures and ecological monitoring organization.

**Table 1.** Composition and scope of field works.

| Types of works | Work scope |
|---|---|
| Field reconnaissance | Route length – 10 km |
| Engineering and environmental survey of the territory (detection of visual signs of pollution and conditions of landscape components of the environment) | Route length – 10 km, description of 48 typical local points |
| Gamma survey | Survey area – 460 ha |
| Radon hazard survey | 35 measurement points |
| Noise level measurement | 10 measurement points |
| Electromagnetic radiation measurement | 10 measurement points |
| Soil sampling for chemical analysis | 6 samples |
| Soil sampling for radionuclide analysis | 6 samples |
| Soil sampling for microbiological and parasitological analyses | 6 samples |
| Water sampling for chemical analysis | 4 samples |
| Water sampling for microbiological analysis | 4 samples |

**Table 2.** Composition and scope of laboratory studies.

| Test subject | Number of samples | Types of laboratory analyses |
|---|---|---|
| Soil | 6 | Multi-element chemical analysis |
| Soil | 6 | Gamma-spectrometric analysis of radionuclide activity |
| Soil | 6 | Microbiological and parasitological analyses |
| Water | 4 | Quality chemical analysis |
| Water | 4 | Microbiological analysis |

## 4 Results

During the field reconnaissance and engineering and environmental survey, the territory was walked over with execution of ecological descriptions of typical points. A part of the territory, surveyed during reconnaissance and located mostly beyond the SPZ limits, has insignificant signs of technogenic disturbance. The reconnaissance survey did not reveal any signs of pollution within the Lake Samsonkino's SPZ. Generally, the landscape condition can be assessed as satisfactory, with separate areas of technogenic disturbance.

Information on disturbance and pollution levels of environmental components in the surveyed territory, obtained during engineering and environmental surveys, is summarized in Table 3.

**Table 3.** Information on disturbance and pollution of environmental components in the surveyed territory.

| Environmental components | Disturbance and pollution levels |
|---|---|
| Landscape conditions | Generally satisfactory, with separate small disturbed areas |
| Radiation conditions | Radiation characteristics:<br>- gamma EDR at the territory is not more than 0.22 μSv/h;<br>- soil radioactivity level does not exceed 22.5 Bq/kg;<br>- radon flux density is less than 47 mBq/($m^2$×c).<br>The characteristics meet sanitary standards |
| Noise | Noise level is less than 50 dBA and does not exceed the maximum permissible value |
| Electromagnetic fields | Electric and magnetic field strength are respectively 0.19 kV/m and 3.8 A/m and does not exceed the maximum permissible value |
| Atmospheric air conditions | Concentrations of main pollutants do not exceed relevant maximum permissible values |
| Surface water quality | Water of the Lake Samsonkino meets sanitary standards |
| Soil conditions | Chemical contamination category – permissible; no bacteriological or parasitological contamination |

Laboratory analyses of water samples from Lake Samsonkino and other water bodies (Lake Mishkino, the Three Lakes System and the Dudinka river) show that for organoleptic, generalized indicators, the content of harmful ingredients, microbiological and radiation criteria, the water is suitable for drinking supply. After the cleanup envisaged by the project of reconstruction of the water-supply system the water entering into the plumbing from the treatment facilities will improve its quality.

The results of analyses of soil samples and the criteria of the existing Russian sanitary standards allowed to establish that:

• the category of chemical contamination in soils is permissible;

• the category of epidemic danger is pure;

• the level of radioactivity meets the established standards of radiation safety for building materials.

The performed calculations of concentrations fields for the most significant polluting agents showed that the impact of the water-supply system facilities will not lead to noticeable effects on the quality of atmospheric air at the border of the living area of Dudinka.

A comprehensive analysis of landscape components shows that the cumulative state of the natural environment in the areas of reconstruction of the facilities and the surrounding territories as a whole is assessed as safe.

Based on the field and laboratory studies, forecasting was made regarding changes in the ecological conditions of the territory during construction works for the facility upgrade and operation, as well as emergency situations. Potential changes in the ecological conditions in case of technogenic impacts on the atmospheric air, surface and ground waters, environmental impacts of waste, impacts on the soil cover, the geological environment, flora and fauna, influence of the facility on conditions of especially protected natural territories, natural monuments, historical and cultural monuments. It was established that due to a localized nature of technogenic impacts during construction activities, their limited period (about one year) and during environmental activities, the combined impact of technogenic factors is insignificant and will not cause environment degradation. Analysis of technogenic impacts established that during operation of the water-supply system their combined influence will be insignificant too. Protection measures for emergency situations (fires, emergency power cut-offs, water pipeline ruptures, pumping equipment failures, etc.) will also ensure ecologically safe operation of the facility.

Based on predictive assessments of the changes in landscape components a set of recommendations and proposals for prevention and mitigation of the upgrade project impacts was developed. For identification of tendencies in quantitative and qualitative changes of the environment condition proposals for an ecological monitoring organization were developed.

## 5 Conclusions

- The comprehensive environmental assessment of the territory adjacent to Dudinka provided the design process for the water-supply system upgrade with necessary and sufficient baseline data on the environmental conditions.
- Analysis of landscape component conditions established that the general environment conditions at facility upgrade sites and adjacent territories are considered satisfactory.
- Radiation characteristics (gamma EDR, soil radioactivity level, radon flux density), the noise level and electromagnetic pollution of the territory, the state of the atmosphere, surface waters and soils meet the current sanitary standards.
- It was established that due to the localized nature of technogenic impacts during environmental activities, the combined impact of technogenic factors is insignificant and will not cause environment degradation.
- A set of recommendations and proposals for environment protection and ecological monitoring organization was developed.

The author thanks the Scientific & Industrial Association Gidrotekhproekt for financing the research works, as well as the EcoStandard Group for assistance in field measurements and laboratory tests.

## References

1. F.F. Bryukhan, V.V. Lebedev, Kriosfera Zemli, **16** (2012) – in Russian.
2. V.I. Telichenko, Promyshlennoe i grazhdanskoe stroitel'stvo, No 2, 47-51 (2011) – in Russian.
3. V. Telichenko, A. Benuzh, G. Eames et. al., Procedia Engineer., **153** (2016)
4. I.V. Lancova, G.V. Tuljakova, Promyshlennoe i grazhdanskoj stroitel'stvo, No 11, 3-5 (2012) – in Russian.
5. *Atlas Arktiki* (Central Administration of Geodesy and Cartography under the USSR Council of Ministers, Moscow, 1985) – in Russian.
6. Yu.P. Parmuzin, *Middle Siberia: the essay on the nature* (Mysl' Publishing House, Moscow, 1964) – in Russian.
7. Yu.P. Kozhevnikov, Botanicheskiy zhurnal, **82** (1997) – in Russian.
8. *SP 11–102–97. Engineering and Environmental surveys for construction* (Gosstroy of Russia, Moscow, 1997) – in Russian.
9. *SP 47.13330.2012. Engineering surveys for construction, basic principles: Updated edition of SNiP 11–02–96* (Ministry of Regional Development of Russia, Moscow, 2012) – in Russian.

# Ultimate capacity of piles penetrating in weak soil layers

*Ahmed* Al-Obaidi [1,*], and *Pinar* Mahmood[2]

[1]Civil Engineering Department, Tikrit University, Tikrit, Iraq
[2]Ministry of Construction and Housing, Baghdad, Iraq

**Abstract.** A pile foundation is one of the most popular forms of deep foundations. They are routinely employed to transfer axial structure loads through the soft soil to stronger bearing strata. Piles generally used to increase the load carrying capacity of the foundation and reduce the settlement of the foundation. On the other hand, many cases in practice where piles pass through different layers of soil that contain weak layers located at different depths and extension, also some time cavities with a different shape, size, and depth are found. In this study, a total of 96 cases is considered and simulated in PLAXIS 2D program aiming to understand the influence of weak soil on the ultimate pile capacity. The piles embedded in the dense sand with a layer of weak soil at different extension and location. The cross section of the geometry used in this study was designed as an axisymmetric model with the 15-node element; the boundary condition recommended at least 5D in the horizontal direction, and (L+5D) in the vertical direction where D and L are the diameter and length of pile, respectively. The soil is modeled as Mohr-Coulomb, with five input parameters and the behavior of pile material represented by the linear elastic model. The results of the above cases are compared with the results found in a pile embedded in dense soil without weak layers or cavities. The results indicated that the existence of weak soil layer within the surrounding soil around the pile decreases the ultimate capacity. Furthermore, it has been found that increase in the weak soil width (extension) leads to reduction in the ultimate capacity of the pile. This phenomenon is applicable to all depth of weak soil. The influence of weak layer extension on the ultimate capacity is less when it is presentin the upper soil layers.

## 1 Introduction

Pile foundation is one of the most popular forms of deep foundations. They are routinely employed to transfer axial structure loads through the soft soil to stronger bearing strata. Piles generally used to increase the load carrying capacity of the foundation and reduce the settlement of the foundation. These purposes are accomplished by transferring loads through a soft stratum to a stiffer stratum at a greater depth, or by distributing loads through the stratum by friction along the pile shaft, or by some combination of these two[1].

Soil can be grouped into two categories depending on the method of deposition, residual soils are formed from the weathering of rocks and transported soils are those materials that have been moved from their place of origin. The method of deposition and transportation has a significant effect on the properties of resulting soil mass [2]. So many large land areas have been formed with transported soils which deposited primarily by one of the transportation methods and may be considered as weak soil. Also, the weak soil may be a man-made soil like bad filling or accumulated debris.

Weak soil may initiate many geotechnical difficulties including inadequate bearing capacity, the potential for unacceptable settlement and slope instability [3]. So the existence of weak soil is a complex problem and difficult due to the lack of understanding of phenomena of soil-pile interaction and the limited quantity and inexact quantity of soil information that can be provided for analysis [4].

There are many cases in practice where piles pass through different layers of soil that contain weak layers of soils located at different depths and locations, and some time cavities with a different shape, size, and depth.

A laboratory testing is carried out by [5] to established the performance of laterally loaded piles embedded in the soil, which contains cavities. In this study, five groups were carried out; one group included embedded pile in soil without cavities. The second and third groups are implemented on pile embedded in soil with a single cavity located in front of pile surface, in the second group, the cavity exists in the back pile surface in the third one. Group four is performed on the pile with the existence of two cavities located in front of pile surface. While the fifth group is carried out on the pile with the existence of three cavities located in front and at a variable distance from pile face. In all tests, the pile tip was free and subjected to a horizontal load. The results indicate that the number of cavities and their location

*Corresponding author: dr.obaidi.a.h@tu.edu.iq

© The Authors, published by EDP Sciences. This is an open access article distributed under the terms of the Creative Commons Attribution License 4.0 (http://creativecommons.org/licenses/by/4.0/).

have a mutual effect on the behavior of laterally loaded pile, the effect of cavities found in front of the pile is insignificant at X/D > 8 where X is the spacing between cavity and pile and D is the diameter of the pile.

[6], studied the interaction between cavities and adjacent axially loaded pile in clayey soil using the 3D-FE program. A comparison of the behavior of load settlement curve, shear stress on the pile and lateral stress on the pile, is made for cases of the variation in the cavity positions in the vertical direction. It has been concluded that the effect of the cavity on the shear stress along pile length and on the lateral soil stress on the pile is ignored for the depth greater than 20% of the pile length from the ground level.

[7], performed a numerical study using (Plaxis 2D) to analyze the performance of the piles subjected to axial loads and to find the effect of existing weak soils and cavities on ultimate pile capacity with different locations, thickness, and soil properties. The results show that the variation between the ultimate capacity ratios with weak soil layer thickness seems to be linear for all cases for weak soil layer properties. Moreover, presence of cavity reduced the ultimate capacity ratio for the pile, the most dangerous situation when the cavity is located near the pile base. In this study, the research focused on presenting a numerical model using (PLAXIS 2D) as a finite element method to simulate and analyzed the behavior of the pile under axial loads and to find the ultimate pile capacity regards to presence of weak soil layer and cavity at different locations and different extension.

[8] Found that the agreement between numerical results of end bearing resistance of drilled shafts and available experimental and empirical relations is satisfactory. It can be concluded that the elastoplastic Mohr–Coulomb constitutive law with stress-dependent elastic parameters is suitable for numerical modeling of end bearing capacity of drilled shafts in sand and also in sandy soils, with an increase in the pile embedment depth, the end bearing capacity increases at a decreasing rate. The Numerical analyses show that at the pile tip displacement equal to 5% of pile diameter, the end bearing capacity of drilled shafts with different diameters are approximately equal. This means that the end bearing capacity is less dependent on the changes in drilled shaft diameter provided the associated tip displacement be equal to 5% pile diameter.

## 2 Behavior of a single pile in layered soil

According to [9] only punching shear failure occurs in deep foundations to attach importance to the density of the soil so long as the depth-width ratio Lid is greater than four where L = length of pile and d = diameter (or width of pile), the failure surfaces do not revert to the shaft

$$Q_u = Q_b + Q_f = q_0' N_q A_b + P \int_0^L \bar{q_0'} \bar{K_S} \tan\delta \, \Delta L \quad (1)$$

$Q_u$ = load at failure applied to the pile
$Q_b$ = base resistance
$Q_f$ = shaft resistance
$q_0'$ = effective overburden pressure at the base level of the pile
$N_q$ = bearing capacity factors which take into account the shape factor
$A_b$ = bearing area of the base of the pile
P =Circumference
$\bar{q_0'}$ = average effective overburden pressure over the embedded depth of the pile
$\bar{K_S}$ = average lateral earth pressure coefficient
$\delta$ = angle of wall friction

## 3 Parametric study

In the present work, two cases are simulated in PLAXIS 2D program and investigated aiming to understand the influence of weak soil on the ultimate pile capacity.

In this study, a total of 97 cases are studied that were divided into two groups; the first one included the piles embedded in dense soil only, the second group included piles embedded in the dense sand with a layer of weak soil at different extension and location (96 cases). Details of these cases are shown in Figure (1) and (2).

## 4 Numerical model, step-by-step procedure in Plaxis program

### 4.1 Boundary condition

The cross section of the geometry used in this study was designed as an axisymmetric model, and according to [10], the boundary condition recommended at least 5D in the lateral direction, and (L+5D) in the vertical direction where D and L are the diameter and length of piles, respectively. These boundaries condition are necessary to minimize the boundary effect on zone around the shaft, not affecting the deformations and providing sufficient accuracy for the analysis.

### 4.2 Material and properties

In this study, the soil is modeled as Mohr-Coulomb, with five input parameters; Young's moduli, Poisson's ratio, cohesion, the angle of shearing strength, and dilatancy angle. The parameters of soil used in this study are shown in Table (1).

The material of the pile foundation used in the present work that is represented by the linear elastic model. The value of the modulus of elasticity for the pile is (30 GPa), unit weight, $\gamma_c$=25 kN/m$^3$, and a Poisson's ratio, v=0.2.

 https://doi.org/10.1051/matecconf/201816201025

### 4.3 Mesh

In this study, 15-node element was chosen because it is in an accurate element that can produce high-quality stress results for different problems.

### 4.4 Interface

When using 15-node soil element, the corresponding interface element is defined by five pairs of nodes, whereas for 6-node soil elements the corresponding interface elements are defined by three pairs of nodes.

### 4.5 Calculations

After the definition of the model comes the generation of initial effective stress where the pile is deactivated, and there is only soil material. The procedure considers only soil weight and calculates effective stresses in soil elements and interfaces. The initial horizontal effective stress $\sigma'_{ho}$is related to the initial vertical effective stress by the coefficient of earth pressure at rest $k_o$ ($\sigma'_{ho} = k_o . \sigma'_{vo}$). The default $k_o$ value is based on Jaky's formula (1-sin Ø) for normally consolidation soils. When the K0 procedure is adopted, vertical stresses are generated in equilibrium with the self-weight of the soil.

The next step is the calculation process, which is divided into calculation phases, much like project phases. Each calculation phase is divided into some calculation steps because the non-linear behavior of the soil requires loading to apply in small proportions, called load steps.

There are many options available for calculations phase in the program; the most appropriate for the case studied is the staged construction where the user can specify a new state that is to reach at the end of calculation phase. It is possible to change the geometry and load configurations (activate pile and activate loads).

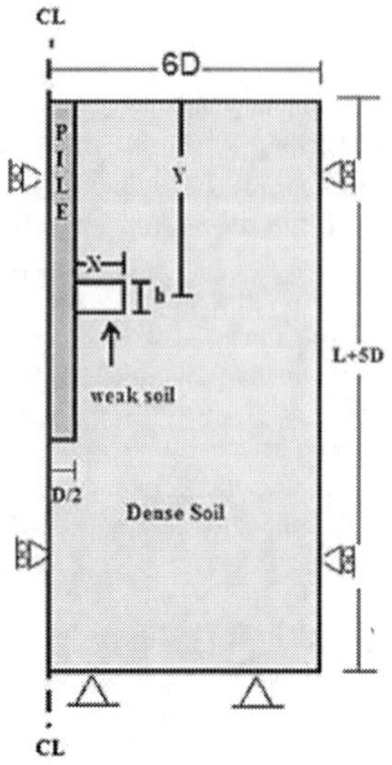


**Fig. 1.** Boundary condition model with existing weak soils

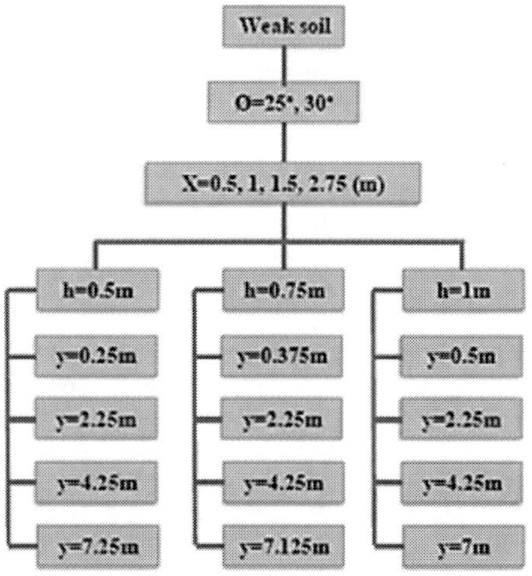


**Fig. 2.** Parametric study of existing weak soils

**Table 1.** Soil properties used in this study

| Soil properties | Dense Sand | Weak soils | |
|---|---|---|---|
| | | 1 | 2 |
| Modulus of elasticity , E kN/m² | 80000 | 37500 | 23500 |
| Unit weight , γ kN/m³ | 18 | 16 | 15 |
| Poisson's ratio , μ | 0.35 | 0.3 | 0.3 |
| Internal friction , φ Degree | 40 | 30 | 25 |
| Cohesion , c kN/m² | 1 | 1 | 1 |
| Angle of dilatancy , ψ Degree | 10 | 0 | 0 |
| Interface reduction factor , $R_{inter}$ | 1.0 | 1.0 | 1.0 |

## 5 Results and discussion

In the present study, a failure load defined as that which causes a settlement equal to 10 percent of the pile diameter to study the effect of weak soil on the ultimate pile capacity.

- In many soil profiles, it occasionally exists weak soil layers; these layers may be extended with limit width. In this research, the weak layer width is considered to change as (0.5m), (0.75m), (1.5m), and fully extended, measured from pile side surface .
- Figures (3 to 5) show the effect of weak soil layer extension on the ultimate capacity ratio, (UCB) and the angle of internal friction of this layer is 25°. From these figures, it can be observed that the increase in the weak soil width (extension) leads to reduction of the ultimate capacity of the pile. This phenomenon is applicable to all depths of the weak soil. The influence of weak layer extension on the ultimate capacity is less when it's present in the upper soil layers. This is due to the fact that amount of the load transfer from pile to soil increased, where more load transfer at lower depths .
- The behavior of the curves in these figures indicates that ultimate pile capacity decreases with increase in weak soil layer extension (width). This is true for all weak layer thickness and depth, but the reduction in the ultimate pile capacity becomes very little and may have fixed values for certain weak soil layer width. For existing of weak soil layer at upper mid-half of the pile length, the fixed values of ultimate pile capacity may be found at the width of (1.5) m for weak soil layer, measured from the pile side. Existing of weak soil layer at lower depths may have constant values at more than (2.5) m width.
- Figures (6 to 8) show the effect of weak soil layer extension on the ultimate capacity ratio, the angle of internal friction of this layer is 30°.
- Same behaviour and effect of existing of weak soil layer as in the previous article (ϕ = 25°), except that, the ultimate pile capacity is greater for weak layer soil with (ϕ = 30°).
- Figures (9 to 11) show the effect of depth weak soil layer centre (ϕ = 25°) on the ultimate pile capacity, figures (12 to 14) same as above but for weak soil layer with the internal friction of (ϕ = 30°).
- The trend of behaviour of curves in all the figures seems to be similar, where the increasing depth of weak soil layer led to decrease the ultimate capacity ratio, regardless of the thickness of this weak soil layer. This reduction increases as the width of weak soil layer increases.
- The minimum ultimate pile capacity was found to be close to pile base, as shown in Tables (2) and (3), the reduction ranges between (0.96 – 0.92) for weak soil layer with internal friction of (ϕ = 25°) and between (0.97 – 0.94) for weak soil layer with internal friction of (ϕ = 30°).

## 6 Conclusions

Based on the results of the numerical study on the pile penetrate weak soil layers, the following conclusions can be drawn. Increasing of the weak soil width (extension) leads to reduction in the ultimate capacity of pile. These phenomena are applicable to all depths of the weak soil. The influence of weak layer extension on the ultimate capacity be less when it is present in the upper soil layers. For existing of weak soil layer at upper mid-half of the pile length, the fixed values of ultimate pile capacity may be found at width of (1.5) m for weak soil layer, measured from the pile side. Existing of weak soil layer at lower depths may have constant values at more than (2.5) m width.

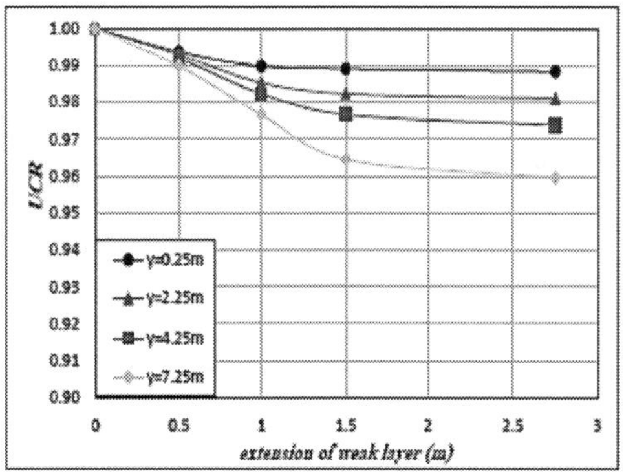


**Fig. 3.** Relationship between UCR and weak soil layer with thickness of (0. 5m) and friction angle (ϕ = 25°) at different depths

 https://doi.org/10.1051/matecconf/201816201025

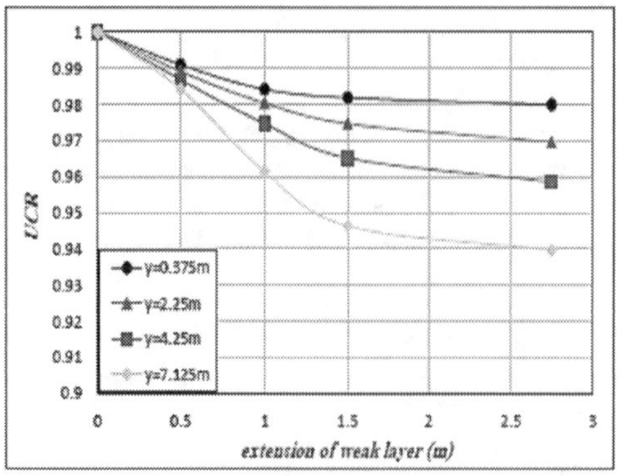


**Fig.4.** Relationship between UCR and weak soil layer with thickness of (0.75m) and friction angle ($\phi$ = 25°) at different depths

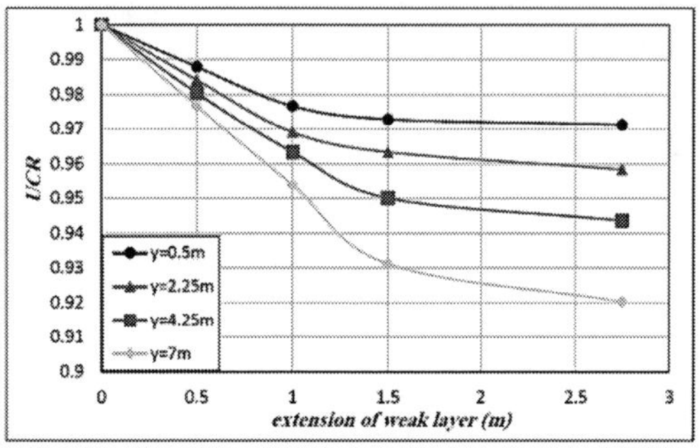


**Fig.5.** Relationship between UCR and weak soil layer with thickness of (1.0 m) and friction angle ($\phi$ = 25°) at different depths

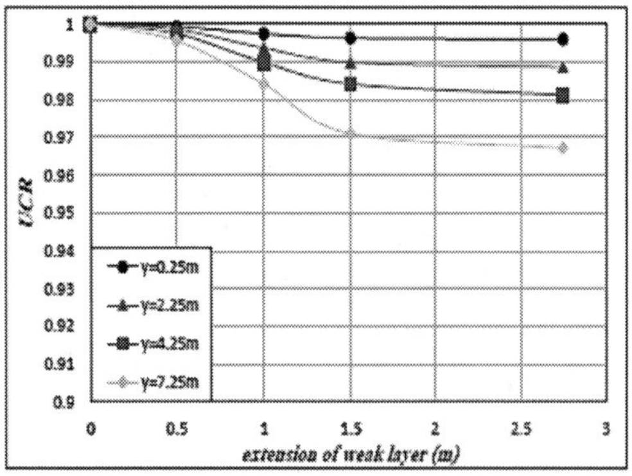


**Fig.6.** Relationship between UCR and weak soil layer with thickness of (0.5 m) and friction angle ($\phi$ = 30°) at different depths

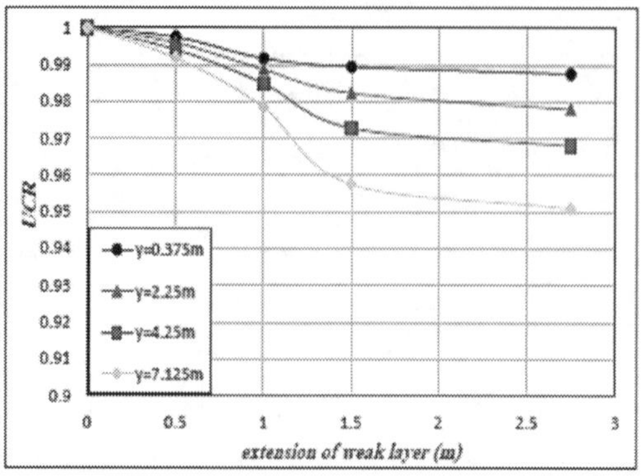


**Fig.7.** Relationship between UCR and weak soil layer with thickness of (0.75 m) and friction angle ($\phi$ = 30°) at different depths

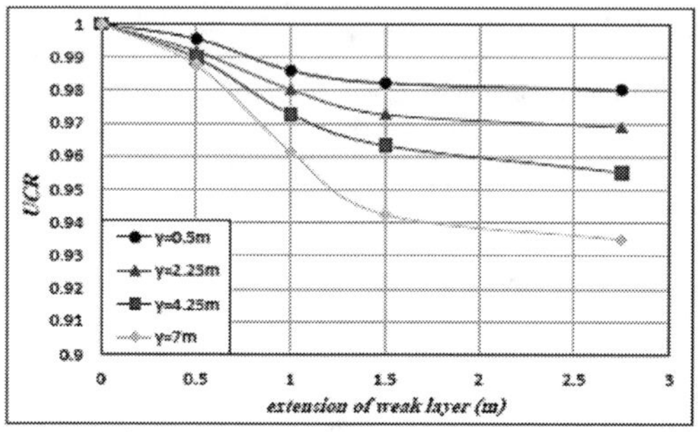


**Fig.8.** Relationship between UCR and weak soil layer with thickness of (1.0 m) and friction angle ($\phi$ = 30°) at different depths

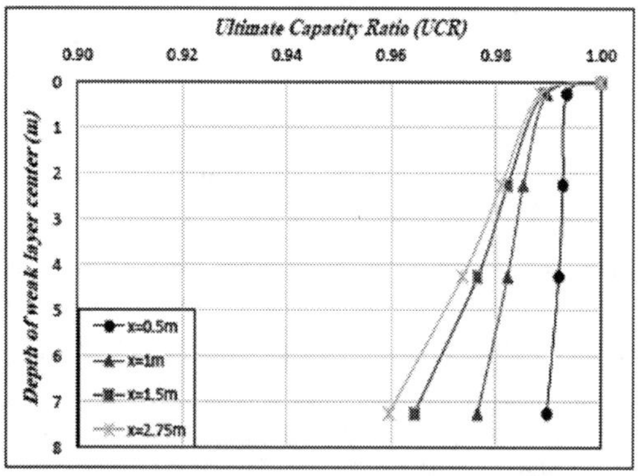


**Fig.9.** Relationship between UCR and Depth of weak soil layer with (0.5m) thickness and friction angle of ($\phi$ = 25°)

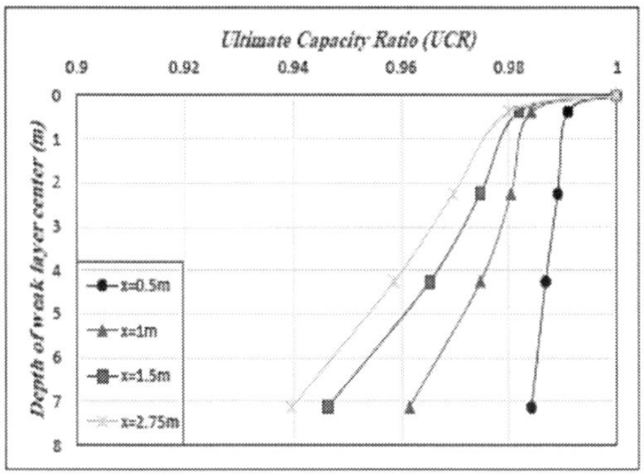


**Fig.10.** Relationship between UCR and Depth of weak soil layer with (0.75m) thickness and friction angle of (ϕ = 25°)

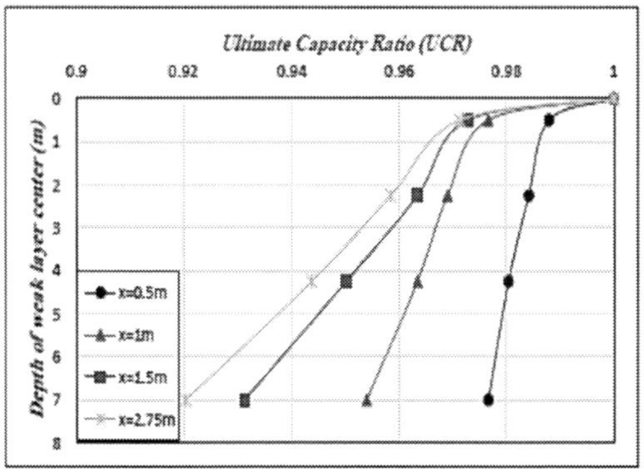


**Fig.11** Relationship between UCR and Depth of weak soil layer with (1.0m) thickness and friction angle of (ϕ = 25°

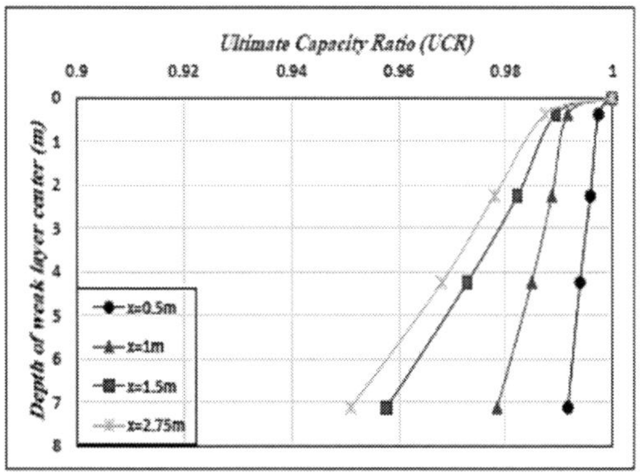


**Fig.12.** Relationship between UCR and Depth of weak soil layer with (0.5m) thickness and friction angle of (ϕ = 30°)

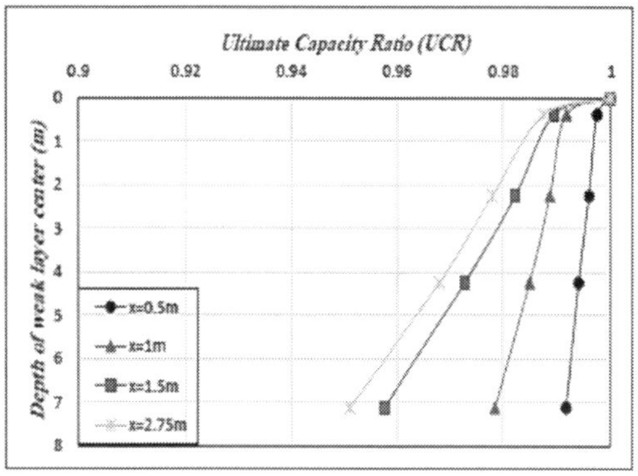


**Fig.13.** Relationship between UCR and Depth of weak soil layer with (0.75m) thickness and friction angle of (ϕ = 30°)

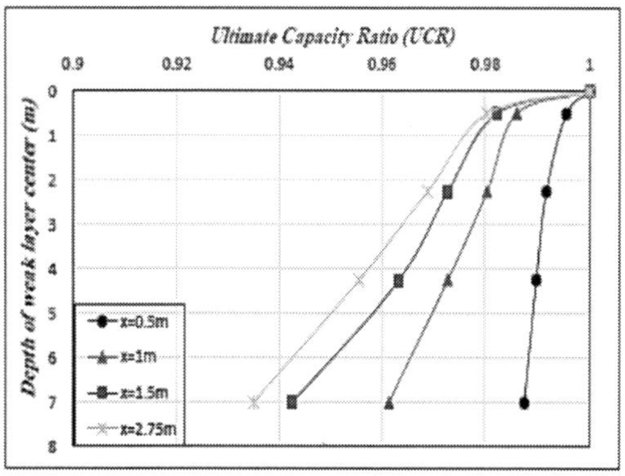


**Fig.14.** Relationship between UCR and Depth of weak soil layer with (1.0m) thickness and friction angle of (ϕ = 30°)

**Table 2.** Effect of weak soil layer (ϕ = 25°) depth, thickness, and extension on the ultimate capacity of pile (kN).

| Depth of weak layer (m) | Thickness of weak soil layer (m) | Extension of weak layer (m) | | | |
|---|---|---|---|---|---|
| | | 0.5 | 1.0 | 1.5 | 2.75 |
| 0.25 | 0.5 | 2625 | 2615 | 2613 | 2611 |
| 2.25 | | 2623 | 2603 | 2595 | 2591 |
| 4.25 | | 2621 | 2595 | 2580 | 2572 |
| 7.25 | | 2615 | 2580 | 2548 | 2535 |
| 0.375 | 0.75 | 2618 | 2600 | 2594 | 2589 |
| 2.25 | | 2613 | 2590 | 2575 | 2561 |
| 4.25 | | 2607 | 2575 | 2550 | 2532 |
| 7.125 | | 2600 | 2540 | 2500 | 2482 |
| 0.5 | 1.0 | 2610 | 2580 | 2570 | 2565 |
| 2.25 | | 2600 | 2560 | 2545 | 2531 |
| 4.25 | | 2590 | 2545 | 2510 | 2492 |
| 7.0 | | 2580 | 2520 | 2460 | 2431 |

**Table 3.** Effect of weak soil layer ($\phi = 30°$) depth, thickness, and extension on the ultimate capacity of pile (kN).

| Depth of weak layer (m) | Thickness of weak soil layer (m) | *Ultimate capacity of pile (kN)* | | | |
|---|---|---|---|---|---|
| | | Extension of weak layer (m) | | | |
| | | 0.5 | 1.0 | 1.5 | 2.75 |
| 0.25 | 0.5 | 2640 | 2635 | 2632 | 2631 |
| 2.25 | | 2638 | 2625 | 2615 | 2611 |
| 4.25 | | 2635 | 2615 | 2600 | 2592 |
| 7.25 | | 2630 | 2600 | 2565 | 2555 |
| 0.375 | 0.75 | 2635 | 2620 | 2614 | 2609 |
| 2.25 | | 2630 | 2612 | 2595 | 2583 |
| 4.25 | | 2626 | 2602 | 2570 | 2557 |
| 7.125 | | 2620 | 2585 | 2530 | 2512 |
| 0.5 | 1.0 | 2630 | 2605 | 2595 | 2589 |
| 2.25 | | 2620 | 2590 | 2570 | 2559 |
| 4.25 | | 2615 | 2570 | 2545 | 2523 |
| 7.0 | | 2609 | 2540 | 2490 | 2470 |

## References


1. L. Reese, C., W. Isenhower, M. and S. Wang, Analysis and Design of Shallow and Deep Foundations, John Wiley and Sons, New York, (2006).
2. D. McCarthy, F., Essential of Soil Mechanics and Foundations, Prenticle Hall, Upper Saddle River, (2006).
3. B. Slocombe, C., Deep Compaction of Problematic Soils", Problematic Soils, Thomas Telford (London), pp: 163-181, (2001).
4. S. Akoobi, I. Effect of Stress Level of Surrounding Soil on Bored Pile Capacity in Sand ", MSc. Thesis University of Technology, The Building, and Construction Engineering Department, Geotechnical Engineering, (2012).
5. M. Al-Mosawe, J, Y. Al-Shakarchi, J. and M. Al-Taie, Pile Embedded in Sandy Soil with Cavities, Journal of Engineering, No.**1**, Vol. **13**, Baghdad university, Iraq, (2007).
6. S. Kadhim, T., Studying The Behavior of Axially Loaded Single Pile in Clayey Soil with Cavities", Eng. & Tech, Journal, No. **8**, Vol. **29**, (2011).
7. A. Al-Obaidi, A, P. Mahmood, N., and A. Al-Karawi, A., Assessment of Vertical Pile Capacity in Soil with Weak Layers and Cavities", 2nd International Conference on Developments in Civil and Computer Engineering Applications, Erbil, Iraq, (2016).
8. G. Ahmadi I and S. Khabbazian M End Bearing Capacity of Drilled Shafts in Sand: A Numerical Approach" Geotechnical and Geological Engineering, Vol. **4**, No. **27,** pp: 195-206, (2008).
9. A. Vesic, S. A Study of Bearing Capacity of Deep Foundations, Final Report, School of Civil Eng., Georgia Inst. Tech., Atlanta, U.S.A, (1967).
10. B. Naveen, P., T. Sitharam, G. and S. Vishruth, Numerical Simulation of Vertically Loaded Piles, Proceedings of Indian Geotechnical Conference, Kochi, No. **N-118**, (2011).

# Improvement ratio and behavior of bearing capacity factors for strip skirt footing model resting on sand of different grain size distribution

*Mahmood* Rashid Mahmood*

Building and Construction Engineering Department, University of Technology, Baghdad, Iraq

**Abstract.** Plain strain model tests were performed on beds of sands with different particle size distribution (Coarse, Medium and Fine) prepared at loose state (Relative density Dr. of 30%). A strip footing model with skirt was placed on the bed of sand and loaded vertically up to failure at different ratios of skirt depth to width D/B of (0.5, 1.0, 1.5, 2, and 3). The applied stress increments and the corresponding settlements were measured. The improvement ratio due to different skirt depth and the behavior of bearing capacity parameters Nγ and Nq at each depth were evaluated and compared with some theoretical approaches. The test results revealed that the improvement ratio increased linearly up to D/B of 1.5 then reduced. Two factors were introduce into the general bearing capacity equation where used to evaluate bearing capacity of skirt footing, there values are about 1.6 for skirt ratio ranged between 0.5 to 1.5, and 1.25 for skirt ratio more than 1.5. Also, it is found that the Nγ parameter for D/B=0 were very close to Vesic proposal for fine and medium grain size distribution, while it's close to Biarez proposal for coarse sand. The behavior of Nq parameter with different skirt ratio shows slight increase up to D/B of 1.5 then decrease with increasing D/B ratio for different grain size distribution. While the behavior of theoretical Nq parameter (depending on angle of internal friction values) shows a linear increase with skirt ratio for different grain size distribution.

## 1 Introduction

Ultimate bearing capacity theories for shallow footings are available since the proposals of Terzaghi in (1948), and its modifications by Meyerhof (1951, 1963), Hansen (1970), Vesic (1973) etc. These theories depend on the shear resistance of the soil beneath the footing. It is known that from the mechanism of shear failure for a footing resting on a homogenous layer the bearing capacity increase with the increasing total length of the failure surface to give higher shear resistance. This can be produced by increasing the width or the depth of foundation. Skirt foundation can be used to increase the depth of footing then increase the length of failure surface that would develop under central vertical loading condition and improve the bearing capacity of the foundation. These skirts are expected to constrain the soil between them and thereby improve foundation performance equivalent to that a foundation embedded to the depth of the skirt tips. Skirts forms closed space at which soil is restrained laterally; it works with the skit as one unit to transfer the loads to the soil at the level of the skirt tip.

So, the skirted foundation transfers the load to a high depth, in comparison to a surface foundation, which mobilizing higher load bearing capacity. Skirted foundations were first used in early 1970's as a supporting unit for floating structures in offshore hydrocarbon projects. Nowadays, skirted foundation is adopted for bridges, marine structures and where water scouring is a major issue.

The ultimate bearing capacity of a centrally loaded strip footing of width (B) is normally calculated from the well known Terzagh's formula[1] as shown in figure (1), which in the case of cohesionless soils (c=0) gives:

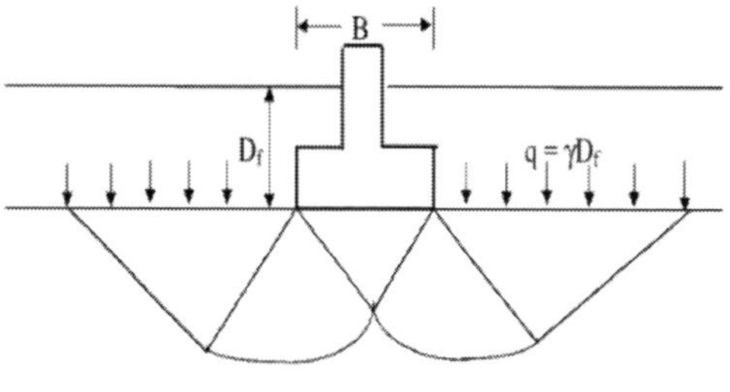


Fig. 1. Mechanism of bearing capacity failure in soil under a rough rigid strip footing subjected to vertical load proposed by Terzaghi (1943)

$$Q_{ult} = \gamma D_f Nq + 1/2\ B\ \gamma\ N\gamma \quad \text{-------------------- 1}$$

Where: $Q_{ult}$ and γ are the ultimate bearing capacity and unit weight of the soil respectively; Nq and Nγ are bearing capacity factors which depend on the angle of

*Corresponding author: mahmoudal_qaissy@yahoo.com

© The Authors, published by EDP Sciences. This is an open access article distributed under the terms of the Creative Commons Attribution License 4.0 (http://creativecommons.org/licenses/by/4.0/).

internal friction. $D_f$ represents the depth of the foundation base below ground level.

Most of geotechnical engineers accept the method of superposition which adopted for estimation the ultimate bearing capacity. For rough continuous foundation Meyerhof[2] solution for estimation Nq parameter is:

$$Nq = e^{(\pi \tan\varphi)} \tan^2\left(45 + \frac{\varphi}{2}\right) \quad \text{---------------- } 2$$

This is found an acceptance from most of Geotechnics. While there is a wide controversy considered over the theoretical values of Nγ Meyerhof suggested that Nγ could be approximate as:

$$N\gamma = (Nq-1)\tan(1.4\varphi) \quad \text{-------------------- } 3$$

Hansen[3] proposed an approximate relationship for Nγ in the form:

$$N\gamma = 1.5(Nq-1)\cot\varphi \tan^2\varphi \quad \text{----------------- } 4$$

Vesic[4] suggested that:

$$N\gamma = 2(Nq+1)\tan\varphi \quad \text{-------------------- } 5$$

Biarez et al.[5] also recommended the following relationship for Nγ:

$$N\gamma = 1.8(Nq-1)\tan\varphi \quad \text{-------------------- } 6$$

He shows that the reason for several theories used to estimate Nγ factor and the few of correlation with the experimental values is due to the difficulty in choosing a representative value of the soil angle of internal friction (φ) for estimating the bearing capacity. So, it is suggested plain strain soil angle of internal friction ($\varphi_p$) instead of ($\varphi_t$) (triaxial) to estimate the bearing capacity.

In 2004, Aghbari and Mohamedzein[6] performed laboratory model test on skirted strip foundation on dense sand and found that the use of structural skirts can improve the bearing capacity by a percent of 1.5 to 3.9 depending on the geometrical and structural properties of the skirts and foundations, soil characteristics and interface condition of soil skirt foundation system as in figure(2). They suggested adding a skirt factor, Fγ, to Terzaghi bearing capacity equations (see equation 7). They mentioned that skirt factor is a function of foundation base friction factor, skirt depth factor, skirt side roughness factor, skirt stiffness factor, and soil compressibility factor. Nevertheless, they reported that the skirt factor ranges from 1.3 to 1.5. They mentioned also, no factor is included the first part (Nq term) of the general equation because the effect of the skirt can be accounted for by the skirt depth.

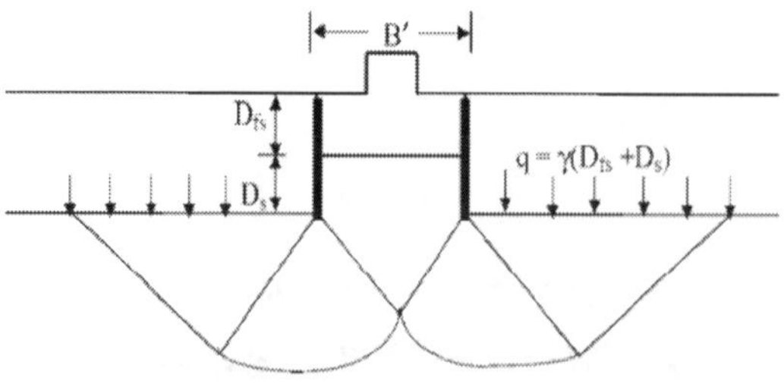


Fig. 2. Mechanism bearing capacity failure in soil under strip footing with skirt subjected to vertical central load (Based on Terzaghi (1943) assumptions)

$$Q_{ult} = \gamma(D_{fs} + D_s)\, Nq + 1/2\, B'\, \gamma\, N\gamma\, F\gamma \quad \text{------- } (7)$$

Where: $D_{fs}$ represents the depth of foundation base below ground level, $D_s$ denotes the depth to the lower edge of the skirt at footing tip, $B'$ shows the total width of foundation including skirts and the skirt thickness.

Sawwaf and Nazer (2005)[7] studied the effect of skirt confinement on the bearing capacity of sand and have found an improvement in bearing capacity as 17 times as that without skirt confinement.

In 2007, Aghbari and Mohamedzein[8] performed tests on circular skirted footing models resting on sand. The results showed that the use of skirts improved the bearing capacity by a factor up to 3.

Eid et al. 2009[9] performed laboratory study on footing load test in sand and concluded that the existence of walls, which surrounded a surface foundation, significantly reduces its settlement and increases its bearing capacity.

El-Wakil 2013[10] reported that 1.3 times increase in bearing capacity of thin wall foundations in dense sands when the wall to width ratio is increased from 0.5 to 1.

Amr Z. EL Wakil (2013)[11] performed eighteen laboratory tests on circular footings of different diameters and different skirt lengths. They found enhancement in bearing capacity of footing with the skirt was increased by about 6.25 times than the footing without the skirt. They studied the effect of relative density of sand and skirt length on bearing capacity of shallow footings

Mahmood and Huda (2013)[12] present study to investigate the behavior of skirted strip model footing founded on beds of soft clays and to evaluate the improvement values in bearing capacity. A series of laboratory model tests were performed to determine the optimal vertical capacities of skirted foundation within different skirted embedded depth (D/B) ratio ranged between (0.0-2.5). It is found that the appropriate skirted embedded depth as the skirt depth to foundation width of 0.5 (D/B=0.5).

P.K. Jain et. al., (2014)[13] presents the result of load tests conducted on a circular footing resting on confined sand to study the influence of cylindrical skirts with different heights and diameters in medium dense sand. The comparison was made between the results of unconfined case and confined case. It is concluded that the pressure corresponding to 5 mm settlement (i.e. 10% S/D ratio), increases appreciably by confining the sand with skirts and it is due to that skirt restricts the lateral displacement of the sand underneath the footing, thus resulting in appreciable improvement.

Pavan and Oruna (2016)[14] presents experimental results by utilizing 50 mm square footing on the sand which is maintained at a relative density of 47% and confining cells of varying length (50 mm, 75 mm, 100 mm, 125 mm & 150 mm) are used. The obtained results indicate that load carrying capacity of footing increases as the length of confinement increases.

In order to verify the effect of grain size distribution incorporated with the behavior of bearing capacity factors through loading skirt footing with the different skirted ratio, model tests were performed simulating the plain strain condition to avoid any effect of the shape of the footing.

## 2 The soil used

Poorly graded sand of rounded particles was used as a natural soil in the present study. The sand was sieved to obtain different grain size distribution (fine, medium and coarse) according to (ASTM D 422-02). Table (1) shows physical soil properties for the sand. Figure (3) shows the grain size distribution curve, according to the Unified Soil Classification System (USCS), the sand is classified as poorly graded sand with symbol (SP).

**Table (1):** Physical properties of the soils used.

| The Properties | Fine sand | Medium Sand | Coarse Sand | Specification |
|---|---|---|---|---|
| Effective size $D_{10}$, (mm) | 0.18 | 0.56 | 1.90 | ASTM D 422[4] |
| **D30, (mm)** | 0.23 | 0.76 | 2.15 | |
| **D60, (mm)** | 0.27 | 1.02 | 2.25 | |
| Coefficient of curvature, $C_c$ | 1.08 | 1.01 | 1.08 | |
| Coefficient of uniformity, $C_u$ | 1.50 | 1.82 | 1.18 | |
| Soil classification (USCS) | SP | SP | SP | |
| Specific gravity, $G_s$, @20°C | 2.68 | 2.65 | 2.61 | ASTM D 854[3] |
| Maximum dry unit weight $\gamma_{max}$, (kN/m$^3$) | 16.6 | 17.45 | 16.41 | ASTM D 4253[5 |
| Minimum dry unit weight $\gamma_{min}$, (kN/m$^3$) | 13.6 | 14.8 | 13.8 | ASTM D 4254[6 |
| Dry unit weight ($\gamma_d$) (kN/m$^3$) at D.R. 30% | 14.5 | 15.5 | 14.4 | |
| Max. void ratio, $e_{max}$ | 0.94 | 0.75 | 0.85 | - |
| Min. void ratio, $e_{min}$ | 0.61 | 0.49 | 0.59 | - |
| Angle of internal friction( Ø°) at D.R. 30% | 32.5 | 33.3 | 31.7 | |

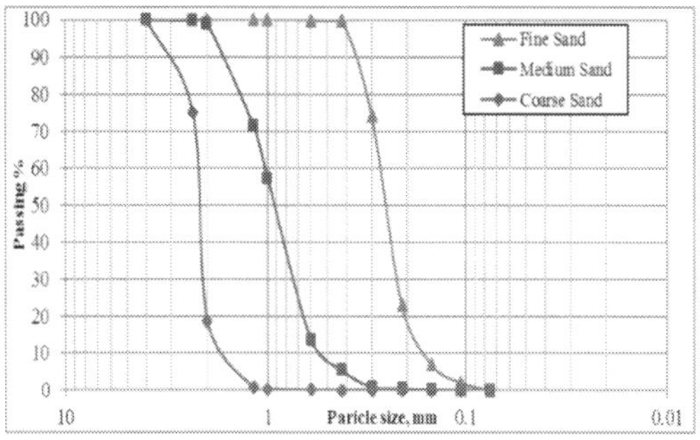


Fig. 3. The different grain size distribution of the soils used

## 3 The Model setup

The apparatus of the testing model and other accessories were performed to achieve the requirements of this research. The apparatus consists of, mainly; steel loading frame with axial loading system, steel container of (600 × 600 ×750) mm, steel tamping hammer, skirt strip footing, load cell, digital weighing indicator dial gauge and magnetic base holder, these components are shown in Plate (1).

Plate 1. The component of the testing model set up

## 4 Testing program

### 4.1 Preparation of soil bed:

The sand was poured into the test container in layers of 100mm thickness to maintain a uniform condition. Corresponding to the loose state of packing each layer was compacted by a steel tamping hammer with uniformly distributed blows to obtain desired relative density of 30%. The soil surface is scraped and leveled by a sharp edge tool to get flat surface.

### 4.2 Foundation model:

Foundation model has divide into two parts; strip model footing and the skirted structure. A mild steel plate of 4 cm width, 40 cm length which gives L/B 1:10 (strip footing), and 1cm in thickness was employed to simulate the proposed shallow strip model footing. The plane strain condition was assumed in all model tests carried out during this study. The structural skirted introduced here to embracing the shallow foundation model was acted via utilizing a jacket made from rigid steel plate with a length of 25 cm, 4 cm width and 0.3 cm in thickness. It is fixed to the edges of steel strip footing through a threaded hole in the center of the footing to allowed rigid connection with the auxiliary of four screw bolts as shown in plate(2).

The steel jacket was holed vertically at an equal spacing to get different (D/B) ratios of (0,0.5,1.0,1.5,2, and 3) with various skirted depth were take into consideration via conversion the location of the screw bolts.

Plate 2. The steel jacket skirt footing model

## 5 Presentation and discussion of test results

### 5.1 Bearing Capacity Ratio

Three loading tests were performed for each type of different graine size distribution with different D/B ratio. Average load-settlement values been revealed the behavior of skirted foundation rested on loose sand. The normal foundation case without skirted that employed as a reference case by which a comparative study can be done. Ultimate bearing capacities at failure detected to show the variation of bearing capacities due to existence skirted structure. The load–settlement measurement continuous beyond failure point and followed by unloading stages. Failure commonly pointed as the $q_{ult}$ recorded by settlement of 10% regarding foundation width.

Figures (4, 5, and 6) deduced that the $q_{ult}$ at failure for different grain size distribution (Coarse, Medium, and Fine sand) at different skirt ratio of (0.0, 0.5, 1.0, 1.5, 2.0, and 3.0) respectively. It is represented a comparative figures summarized the vertical bearing capacities for various cases considered.

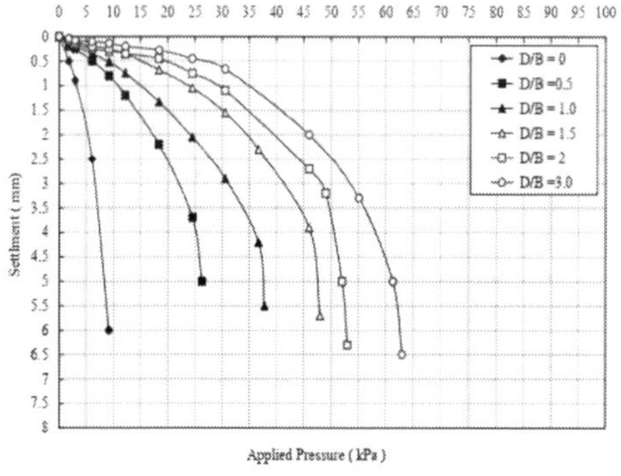


Fig.4. Applied pressure vs settlement for **loose Coarse** sand at different skirted ratio

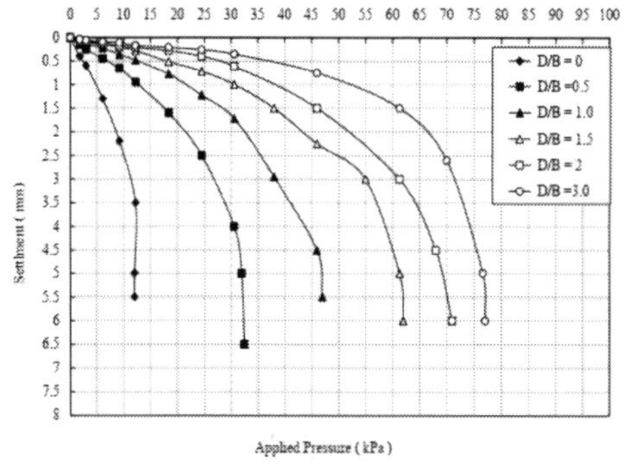


Fig. 5. Applied pressure vs settlement for **loose Medium** sand at different skirted ratio

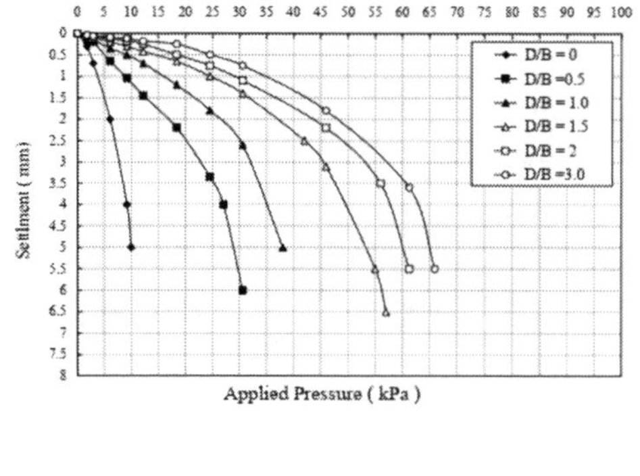

Fig.6. Applied pressure vs settlement for **loose Fine** sand at different skirted ratio

The theoretical improvement ratio of ultimate bearing capacity due to increasing in depth (same as the depth of skirt ratio) shows in figure (7). The figure shows a liner increase in bearing capacity ratio due increase in depth for all the particle size distribution. Coarse graded sand demonstrate higher improvement ratio, while medium graded sand demonstrate lower improvement ratio.

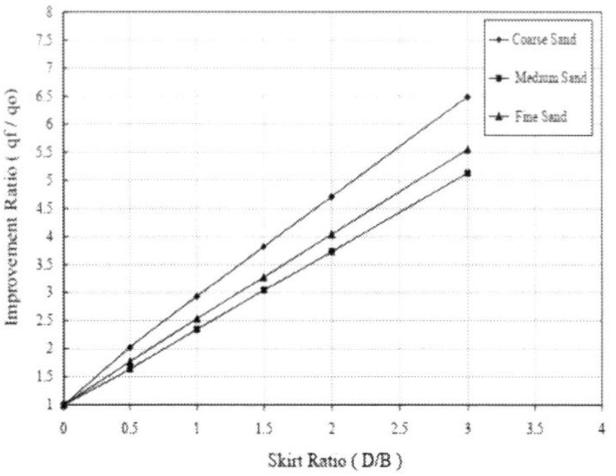


Fig. 7. Theoretical Improvement bearing capacity ratio vs. different depth ratio

The measured improvement ratio as a consequence to the presence of a skirt evaluated for different skirted depth with different grain size distribution at the same relative density shows in figure(8). The figure shows that the improvement ratio (ultimate bearing capacity of skirt foundation to the ultimate bearing capacity of footing at the soil surface) increasing linearly, up to D/B about 2.0 then the improvement ratio was reduced. The coarse sand of the lowest angle of internal friction revealed higher improvement ratio than medium and fine grain size. Same as the fine sand of angle of internal friction lower than medium grain size revealed a higher improvement ratio. That means the soil of lower angle of internal friction when packed through the skirt revealed a higher improvement.

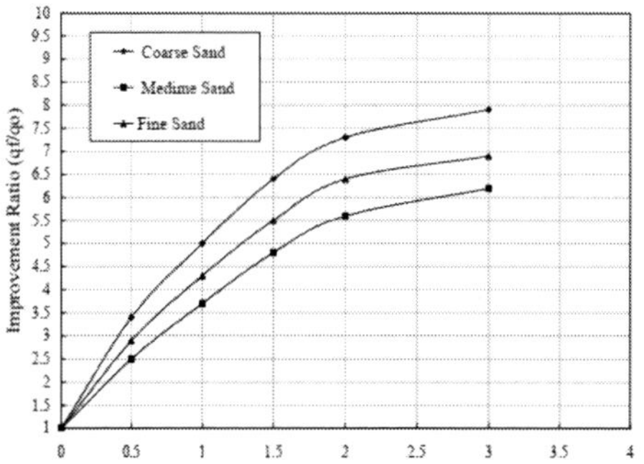


Fig. 8. Improvement bearing capacity ratio vs. skirted depth ratio

Figure (9) shows the net improvement ratio of using skirt than that without skirt at the same depth ratios of the skirt. The figure shows a linear increase of ultimate bearing capacity when using skirt up to skirt ratio about 1.8, at which is the optimum increment ratio occurs, after that the improvement ratio will be decreased. Coarse graded sand gives a higher net improvement ratio, about 2.7 times than that footing without skirt at (D/B) ratio of 1.5. While medium graded sand gives a lower improvement ratio about 1.9 times than that footing without skirt at the same (D/B) ratio.

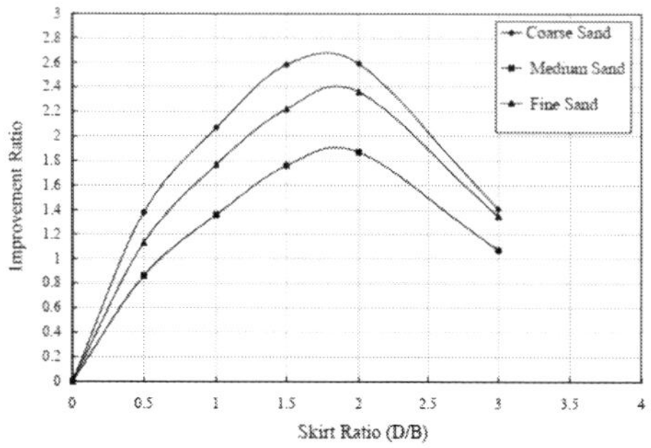


Fig. 9. Net improvement ratio of bearing capacity for skirt footing than that without skirt at the same depth ratio

From the above analysis of improvement ratio due to using skirt foundation it's concluded to introduce skirt improvement factor in order to determine the ultimate bearing capacity of strip footing with skirts. A factor of 1.6 will introduce to the general bearing capacity equation for skirt ratio (D/B) up to 2.0 and 1.25 for skirt ratio more than 2.0.

Figure (10) shows the improvement ratio for the strip footing model without skirt (theoretical improvement bearing capacity ratio as shown in fig.7) multiplied by skirt factor of 1.6 for D/B less than 2.0 and 1.25 For D/B more than 2.0. The figure shows that the values of improvement ratio as approximately the same as that of the improvement ratio for skirt strip footing from experimental results as shown in (fig. 8).

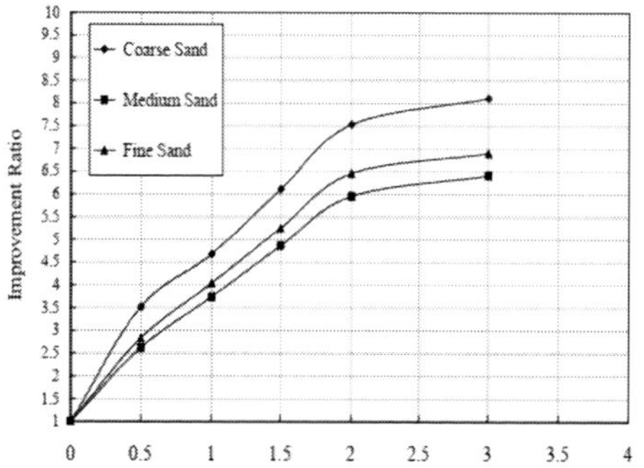

Fig. 10. Theoretical Improvement bearing capacity ratio vs. different depth ratio multiplied by skirted factors

## 5.2 Behavior of bearing capacity factors

### *5.2.1 Variation of $N_\gamma$ :*

Figure (11) shows the variation of theoretical Nγ with the angle of internal friction φ of different particle size distribution for different authors according to equations no. (3, 4, 5, and 6) comparing with the measured values from the experimental model for D/B=0. The results show that the measured Nγ parameters were very close to Vesic proposal for fine and medium grain size distribution, while it's close to Biarez proposal for coarse sand.

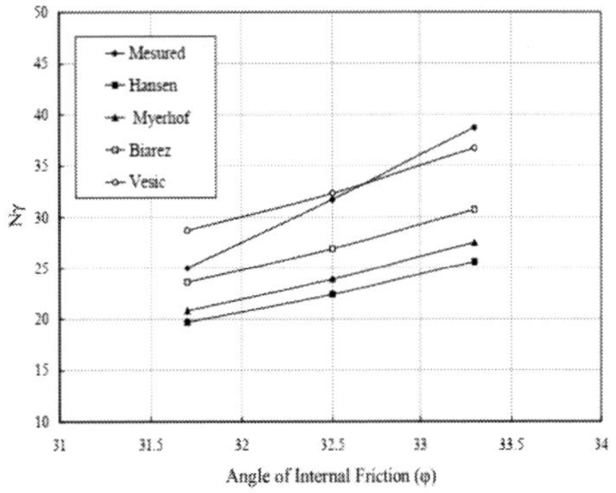


Fig. 11. Behavior of measured and theoretical Nγ parameter of different authors with different angle of internal friction

In general the theoretical Nγ values are all below the experimental values. The general remark that may be drawn is that most of the theories provide conservative results. At the present stage and because measured values are close to Vesic proposed, it will be adopted for calculating Nq parameter according to eq.(5).

Figure (12) shows the behavior of Nγ parameter with skirt ratio, the figure demonstrate that there is a little effect on Nγ values at skirt ratio up to 1.5 but when skirt ration increase Nγ reduced.

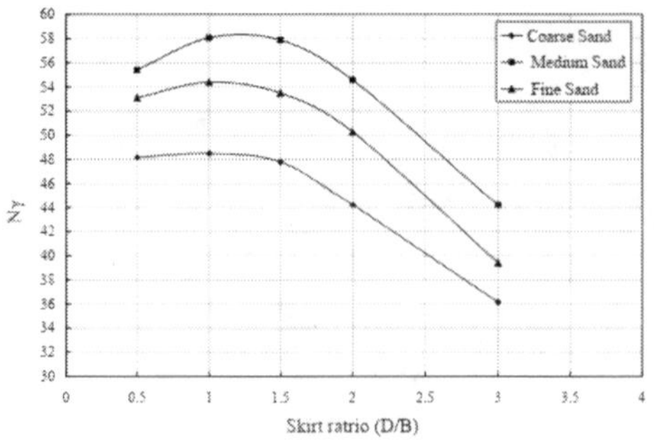


Fig. 12. Behavior of measured Nγ parameter with different skirt ratio for different grain size distribution

### *5.2.2 Variation of Nq :*

Figure (13) shows the variation of Nq versus angle of internal friction (φ) for coarse, medium, and fine graded sand determined from the ultimate bearing capacity of experimental model tests for different skirt ratio (D/B) of (0.5, 1.0, 1.5, 2.0 and 3) by adopting equation no. (1) and Vesic proposal for Nq. The figure shows a conservative value of Nq when compared with those obtained from the model tests. Also it shows that Nq values for skirt ratio of 0.5 to 1.5 are close to each other and reduced with increasing this ratio.

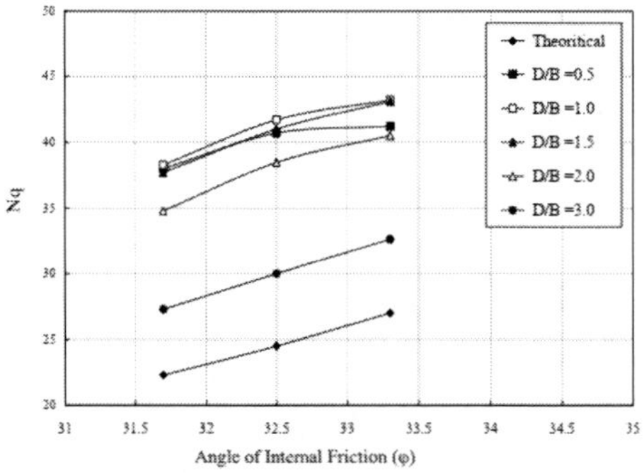


Fig.13. Behavior of $N_q$ parameter for different skirt ratio with angle of internal friction

Figure (14) shows the behavior of Nq parameter with different skirt ratio for different grain size distribution. The figure shows that a little change of Nq parameter for skirt ratio range between 0.5 to 1.5 for all the different grain size distribution, while it decrease with increasing this ratio. This is may attributed to the increase the stress outside the footing area with increasing the skirt depth, where Nq parameter is concerned with the soil above the skirt tip.

At the present state, the figure indicates that no effect of Nq values obtained from the model tests on the skirt ratio. It shows that a slight increase up to

D/B=1.5 then Nq value will tend to decrease close to the theoretical values.

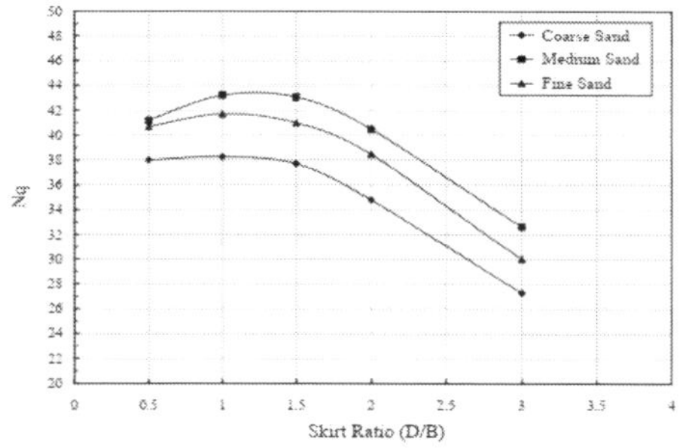


Fig. 14. Behavior of measured Nq parameter with different skirt ratio for different grain size distribution

## 6 Conclusions:

The following conclusions were drawn from the model tests of strip footing with skirt resting on loose sand of different particle size distribution:

1. The theoretical improvement ratio of ultimate bearing capacity due to increase in depth shows a liner increase in bearing capacity ratio for all particle size distribution of the sand used.
2. The improvement ratio of skirt footing increase linearly, up to D/B of 1.5 then the improvement ratio decreased.
3. The net improvement ratio of using skirt foundation than that without skirt at the same depth ratio of the skirt shows a linear increase of ultimate bearing capacity up to skirt ratio of 1.7, at which is the optimum increment ratio occurs, after that the improvement ratio will be decreased.
4. Coarse graded sand gives a higher improvement ratio than that of fine and medium sand, about 2.6 times than that footing without skirt at (D/B) ratio of 1.5. While medium graded sand gives a lower improvement ratio about 1.8 times than that footing without skirt at the same (D/B) ratio.
5. To determine the ultimate bearing capacity of strip footing with skirts rested on loose sand, multiply the general bearing capacity equation by skirt factor (fs) of 1.6 for skirt ratio (D/B) ranged between 0.5 to 1.5 and by 1.25 for skirt ratio more than 1.5.
6. The measured Nγ parameters were very close to Vesic proposed for fine and medium grain size distribution, while it's close to Biarez proposed for coarse sand.
7. Nq values obtained from the model tests for skirt ratio of 0.5 to 1.5 are close to each other and reduced with increasing this ratio.
8. Behavior of Nq parameter has a little effect for skirt ratio of 0.5 to 1.5 for all different grain size distribution, while it decreases with increasing this ratio.

## References

1. K. Terzaghi, "Theoretical soil mechanics". New York: John Wiley (1943).
2. G.G. Meyerhof, Canadian Geotech. J. **15(4)**: 593.( 1978).
3. J.B. Hansen, Bulletin No. 28, Danish Geotechnical Institute, Copenhagen(1970).
4. A.S. Vesic, J. Soil Mech. Found. Div., ASCE, **99(1)**: 45. (1973).
5. J. Biarez, M. Burel, and B. Wack. "Contribution à l'étude de la force portante des fondations", in Proc., V Intl. Conf. Soil Mech. Found. Eng., Paris, France, 1: 603. (1961).
6. M.Y. Al-Aghbari, and Y.E-A. Mohamedzein, Geotechnical and Geological Engineering,**22(1)**:43-57(2004)..
7. M.EI. Sawwaf, and A. Nazer, Journal of Geotechnical and Geoenvironmental Engineering, ASCE, **131(3)**, 359-366(2005)..
8. M.Y. Al-Aghbari ,the Journal of Engineering Research , **4**, No.1 11-16 (2007).
9. T. Hisham, O. Alansari, A. Omar, M. Abdelfattah, N. Mohannad and A. Husam , Canadian Geotechnical Journal,vol. **46**, 438-453(2009).
10. A. Z. El.Wakil. "Horizontal capacity of skirted shallow footings on sand." Alexenderia Eng. Journal (2010).
11. A. Z. EL Wakil , Alexandria Engineering journal vol.**52**, pp.359-364(2013).
13. M. R. Mahmood and H. Muwafak "Experimental Study the behavior of Skirted Foundation Rested on Soft Clayey Soils" 1st International Conference for Geotechnical Engineering and Transportation ICGTE in15-24/4/2013 Eng. & Tech. Journal , Vol. **31** , Part (A),No. 20,( 2013).
13. P.K. Jain et.al" Effect of skirt on pressure settlement behavior of model circular footing in medium dense sand " Chandrawanshi, et al., International Journal of Advanced Engineering Technology E-ISSN 0976-3945.(2014)
14. H. M. Pavan, T. Aruna. International Research Journal of Engineering and Technology (IRJET) e-ISSN: 2395 -0056 Volume: 03 Issue: 07 | July- (2.16).

# Improvement of engineering soil properties using non - traditional additives

*Mohanned* Waheed [1,*] and *Noor* Asmael [2]

[1] Building and Construction Eng. Department, University of Technology, Baghdad, Iraq
[2] Highways and Transportation Department, Mustansiriyah University, Baghdad, Iraq

**Abstract.** Laboratory experiments are conducted to evaluate the effect of some non-traditional additives on the engineering properties of clayey soil, which show problematic phenomenon when used as a construction material. The conducted tests covered the influence of these additives on various parameters like consistency limits, compaction characteristics and CBR value. Two nontraditional stabilizers are selected in this study, polymers and phosphoric acid at three different percent which are (1%, 3% and 5%) of the dry soil weight. It is concluded that addition of the polymer to the clayey soil results in a slight increase in plastic limit while the liquid limit is not affected accompanied by a marginal decrease in the dry unit weight while the optimum moisture content remains unaffected. The addition of phosphoric acid to the clayey soil has no effect on its Atterberg limits. In general, it is observed that polymer is found to be ineffective as a stabilizer to improve clayey soils, especially in small amounts of about (3%). The phosphoric acid treated soil gained better improvement for all amounts of additive used. For (3%) acid treated soil the CBR is about (360%) compared to that of untreated soil, for that, it can be concluded that the improvement using phosphoric acid in the clay soils is a promising option and can be applied to solve the geotechnical stabilization problems.

## 1 Introduction

Improvement of soil properties or soil stabilization is a technique introduced for the purpose of improving poor soil properties to make it acceptable for construction in the engineering projects. The effect of additives on the engineering properties of soil has been studied by many researchers. The wide use of traditional additives such as cement, lime, fly ash, and asphalt emulsion etc. for improvement of soil led to an attempt to select other non-traditional additives, especially, a few studies have been conducted in the field of soil improvement using non-traditional additives. The variety of non-traditional stabilization additives dictated some attempts to classify these different additives according to their active composition.

A synthesis of potential stabilizers that is identified by the US Army Corps of Engineers and contract researchers for the period between (1946 -1977) was presented by Oldham et al. [1]. The results of their investigation identified cement, lime, resins, asphalt, silicates, salts, and other products as potential stabilizers demonstrating varying degrees of engineering success. The report classifies the product performance according to type of soil and demonstrated that the action was different in varying soil types.

Ajayi-Majebi et al. [2] performed an experimental work to study the effect of a mixture of an epoxy resin (bisphenol A/epi-chlorohydrin) and a polyamide hardener for stabilization of clayey silt soils. The mixing ratio of epoxy resin to polyamide hardener is (1:1 ratio). The liquid limit of the soil ranging from (37 to 45) so that the plasticity index is between (13 to 18). It was observed that mixing of (4 %) stabilizer produced large increment in unsoaked CBR values.

Laboratory tests were carried out by Santoni et al. [3] for stabilization a silty-sand soil using some stabilizers. The stabilizers include enzymes, acids, polymers, tree resins, lignin-sulfonates, and petroleum emulsions. The results obtained showed excellent potential of polymers for the improvement, while enzyme and acid are failing to produce significant action for stabilization.

Zandieh and Yasrobi [4] studied stabilization of sandy soil using two different polymers. The results achieved using these polymers showed that there was improved in the compressive strength from (0.03 MPa) for control sample to (5.2 MPa) for improved sample.
The intended task of this paper is studying the effect of two non-traditional additives on the engineering properties of clayey soil, especially, those soils that show a trend to be more problematic during geotechnical construction. The additives selected in this study, include polymers, and phosphoric acid. The investigation covered the influence of these additives on various soil

* Corresponding author: muhannad1978@yahoo.com

© The Authors, published by EDP Sciences. This is an open access article distributed under the terms of the Creative Commons Attribution License 4.0 (http://creativecommons.org/licenses/by/4.0/).

properties as consistency limits and compaction characteristics.

As the California Bearing Ratio test (CBR) is widely used by the engineers to determine the suitability of a soil as a subgrade or subbase for highway and runway design and construction, therefore, it was considered in this study to evaluate the strength improvement of the soil.

## 2 Materials and experimental program

### 2.1 The soil

The soil used is extracted from Baghdad city in Iraq, standard tests were performed to determine the physical properties of the soil, and the results are presented in Table (1), while the grain size distribution is shown in Figure (1), The soil is brown silty clay and it is classified as, low plasticity clay (CL) according to the unified soil classification system.

**Table 1.** The properties of the soil used.

| Property | Index Value | Specification |
|---|---|---|
| Liquid limit (L.L) | 48 | ASTM-D4318-2010 [5] |
| Plastic limit (P.L) | 21 | ASTM-D4318-2010[5] |
| Plasticity index (P.I) | 27 | ASTM-D4318-2010[5] |
| Specific gravity (Gs) | 2.69 | ASTM-D854-2010[6] |
| Sand (4.75mm to 0.075mm) | 4 % | ASTM-D422-2010[7] |
| Silt (0.075 mm to 0.005mm) | 45 % | ASTM-D422-2010[7] |
| Clay (less than 0.005 mm) | 51 % | ASTM-D422-2010[7] |
| Soil Classification | CL | USCS |

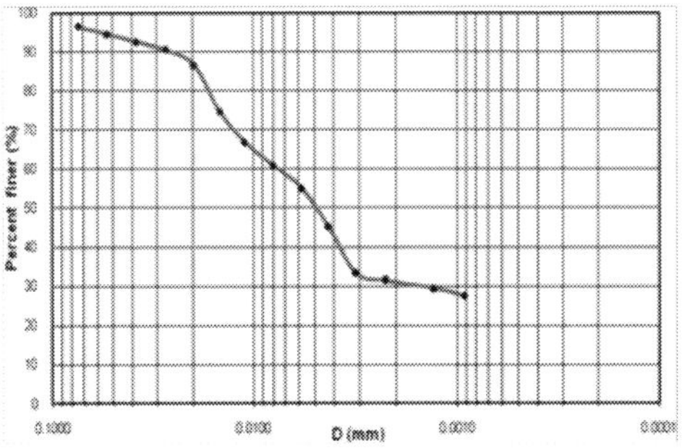


**Fig. 1.** Soil grain size distribution.

### 2.2 Polymers

The polymer used in this investigation is polyvinyl alcohol (PVA), which is commercially available in a granular powder state and it was odorless and tasteless, translucent white or cream colored. It is produced commercially from polyvinyl acetate, usually by a continuous process. The chemical formula is ($CH_2CH(OH)$). It has solubility in water and slightly soluble in ethanol, but insoluble in other organic solvents. Typically (5%) solution of PVA exhibits a pH value of about (5 to 6.5) and has a melting point equal to about (180°C), while a molecular weight of between 26,300 and 30,000, so that the degree of hydrolysis ranged from ( 86.5 - 89%).

### 2.3 Phosphoric acid

Phosphoric acid has varied uses (HSDB, 1995) [8]. For the industry, it is a chemical intermediator reagent in the production of numerous phosphate fertilizers, soaps, waxes, polishes, agricultural feeds, and detergents. Phosphoric acid is added to foods as a preservative, acidifying agent, flavor enhancer, and clarifying agent. It is also used in processes such as the coagulation of rubber latex, electro-polishing and as a catalyst in the production of propylene and butane polymers, ethylbenzene, and cumene .The summary of the chemical properties of Phosphoric acid according to (HSDB, 1995; and HSDB 1999[9]) is shown in Table (2).

**Table 2.** Chemical properties of Phosphoric acid.

| Property | Specification |
|---|---|
| Description | Clear syrupy liquid or unstable crystals; odorless |
| Molecular formula | $H_3PO_4$ |
| Molecular weight | 98 |
| Boiling point | 213°C |
| Melting point | 42.35°C |
| Vapor pressure | 0.03 torr @ 20°C |
| Solubility | Very soluble in hot water; 548 g/100 ml cold water; soluble in alcohol |
| Conversion factor | 4.0 $\mu g/m^3$ per ppb at 25°C |

### 2.4 Laboratory tests

The laboratory tests carried out included the determination of the Atterberg limits (liquid limit, plastic limit, and plasticity index), compaction test and CBR test. The details of these tests are illustrated in the following sections.

#### *2.4.1 Atterberg test*

Atterberg limits tests were carried out based on ASTM-D4318-2010. These tests were conducted to investigate the effect of the additives in varied percentages on the consistency limits of the soil.

#### *2.4.2 Compaction test*

Compaction tests are conducted to determine the effect of stabilizers on maximum dry density (MDD) and optimum moisture content (OMC). Compaction test was performed according to ASTM (D1557–2012) [10] , it is performed in a mold with an internal diameter of 152.4 mm and a height of 116 mm. The specimens are compacted in (5) layers and number of blows equal to (56) for each layer according to modified proctor test.

The first series of compaction tests are conducted in order to find the compaction properties of the untreated soil. Secondly, tests are carried out to determine the effect of additives with varying amounts on the compaction properties. The additive is dissolved in water and then thoroughly mixed with soil. A curing period of (24) hour is allowed for the soil-additive mixture prior to compaction.

#### *2.4.3 CBR test*

The CBR test was considered in this research in order to study the effect of the additives on strength improvement of the soil. CBR test was conducted according to BS (1377-1990)[11]. The samples prepared for CBR tests are compacted to a maximum dry density at the optimum moisture content determined according to modified proctor compaction tests and tested directly after sample preparation .

The preparations of the specimens were achieved according to the procedures in ASTM (D1557–2012). The CBR test is performed in a standard mold of CBR with a (4.5 kg) rammer dropped from a height of (457 mm). The soil was compacted in (5) layers with (56) hammer blows applied per lift. The test is conducted under a seating load of (4.5 kg) and a penetration rate equal (1.27 mm/sec) .

The CBR value is calculated as:

$$CBR = P/Ps * 100\ (\%) \qquad (1)$$

where: P : plunger-load in kN for tested soil.

Ps : plunger-load in kN for a standard soil (which is equal to 13.2 kN for 2.5mm and 20 kN for 5 mm penetration), where the highest of the two values is then considered as CBR value.

## 3 Results and discussions

The effectiveness of the stabilizer for soil strength improvement was studied by mixing varying amounts of additive with the soil samples, and performing the tests in the laboratory. Three different amounts of additive for each of phosphoric acid and polymer are selected for this study, ( i.e., 1, 3 and 5%) of the dry soil weight. The effect of each additive on soil properties is described as follows.

### 3.1 Effect of polymer

The effect of polymer addition on Atterberg limits is shown in Figure (2), it can be seen that there is a small decrease in plasticity index due to a slight increase in plastic limit as the polymer content increased, while the liquid limit showed a negligible increase.

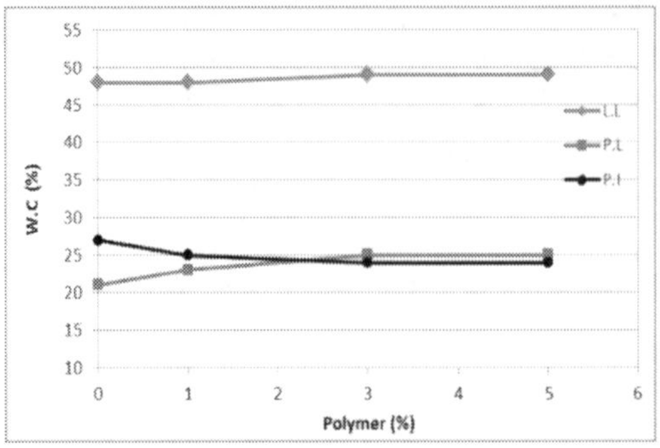


**Fig. 2.** The effect of polymer on Atterberg limits.

The compaction curves for untreated and treated soil with different percentages of polymer are shown in Figure (3). It is noticed that increasing the polymer contents marginally decreases the dry density except for the addition of (5%), where a marked decrease in the dry density value was observed, while the optimum moisture content is not affected in all cases. The decrease in the dry density may be attributed to the clay polymer interaction, as it is believed that the polymer flocculates the particles and spreads them apart resulting in a volume increase of the sample and so reducing its dry weight.

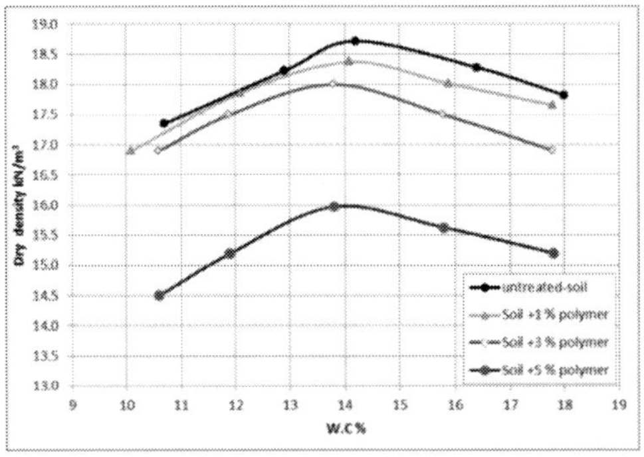


**Fig. 3.** The relation between dry density and water content for different percentages of polymer used.

The load-penetration curves of CBR test for untreated and treated soil with polymer are shown in Figure (4), and the CBR values of untreated and treated specimens using the two additives are given in Table (3), it can be observed, in general, that the addition of polymer led to a significant increase in the CBR values only in case of addition (5%).

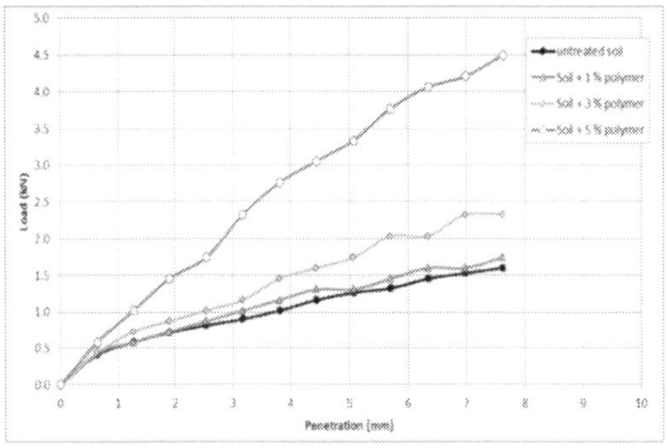


**Fig. 4.** The load-penetration curves of CBR test for untreated and treated soil with polymer.

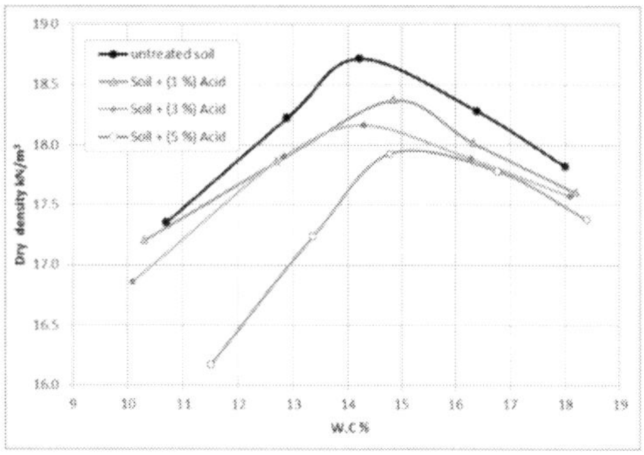


**Fig. 6.** The relation between dry density and water content for different percentages of Phosphoric acid used.

## 3.2 Effect of phosphoric acid

The effect of adding phosphoric acid on Atterberg limits is shown in Figure (5). it is observed that there is no noticeable effect on the liquid limit, however, there is a slight decrease in the plasticity index due to a slight increase in plastic limit. In general, it is found out that the effect of acid is almost absent on Atterberg limits.

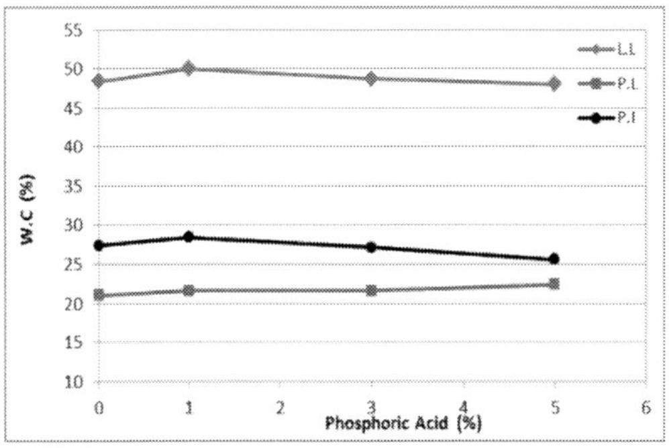


**Fig. 5.** The effect of phosphoric acid on Atterberg limits.

The effect of adding phosphoric acid on the compaction characteristics is shown in Figure (6). It is observed that increasing the phosphoric acid content decreases significantly the dry density and increases slightly the optimum moisture content , however , OMC arise from (14.3 %) for untreated to about (15.4 %) of the soil treated with ( 5 %) acid, while the density declines from (18.7 kN/m$^3$) to (17.9 kN/m$^3$) respectively.

The effect of Phosphoric acid on the load-penetration curves of CBR test for untreated and treated soil are shown in Figure (7), it can be seen, in general, that the addition of acid causes a significant increase in the CBR values for the three different percentages of addition, moreover, the next paragraph shows the percentage of strength improvement in CBR for the three amounts for the two different additives used.

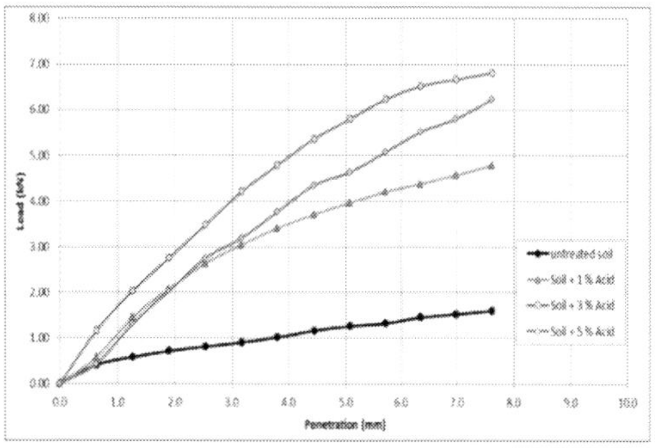


**Fig. 7.** The load-penetration curves of CBR test for untreated and treated soil with Phosphoric Acid.

## 3.3 Effect of stabilizer type on strength improvement

For comparison purposes, the strength improvement in the CBR in percent to that of the soil treated with polymer are drawn together with those treated with phosphoric acid in Figure (8), at the three different percent. It can be observed from this figure, that the addition of (1-3 %) polymer to the clayey soils causes an increase of (8-40 %) in the CBR value, while it increases to about (160 %) for addition of (5 %) polymer.

In general, it can be said the stabilization with polymer in clayey soil gives poor results with respect to strength improvement, especially in less amount.

On the other hand the CBR has improved between (200 to 360 %) due to the addition of phosphoric acid and the maximum value corresponds to a (3%) of phosphoric acid addition. It can be concluded that the improvement using phosphoric acid in the clay soils is a promising option and excellent results can be expected.

Similar results were reported by (Lyons and McEwan [12] and Medina and Guida [13], however, it was pointed out that using phosphoric acid for treatment of kaolinite clays has been brought several benefits in soil properties such as improved strength and workability .

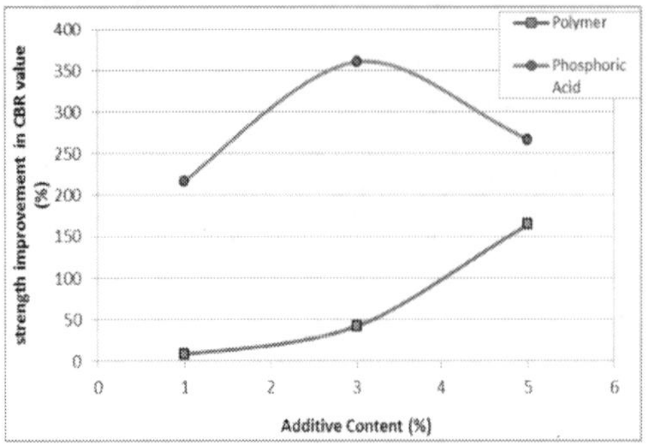


**Fig. 8.** strength improvement in the CBR value of the soil treated soil with polymer and Phosphoric Acid.

**Table 3.** Values of CBR of Untreated and Treated Specimens.

| Case | CBR (At 25 mm penetration) | CBR (At 50 mm penetration) | CBR |
|---|---|---|---|
| Untreated Soil | 6.15 | 6.31 | 6.31 |
| Soil + 1% polymer | 6.59 | 6.53 | 6.59 |
| Soil + 3% polymer | 7.69 | 8.7 | 8.7 |
| Soil + 5% polymer | 13.18 | 16.68 | 16.68 |
| Soil + 1% Acid | 19.99 | 19.87 | 19.99 |
| Soil + 3% Acid | 26.36 | 29 | 29 |
| Soil + 5% Acid | 20.87 | 23.2 | 23.2 |

## 4 Conclusions

Based on the results of the tests conducted throughout this task, it can be concluded that:

1. Adding polymer to clayey soil causes a slight increase in plastic limit while the liquid limit is not influence, and marginally decreases the dry density while the optimum moisture content was not affected in all cases.
2. In general, it can be said that the effect of phosphoric acid is almost négligeable on Atterberg limits, however, there was a slight decrease in the plasticity index due to a slight increase in plastic limit .
3. Increasing the phosphoric acid content decreases significantly the dry soil density and slightly increases the optimum moisture content.
4. It is found out that the polymers are inefficient in soil stabilization, especially when added at low percent.
5. The phosphoric acid-treated soil gained more strength improvement for all amounts of additive used, which can be up to 360% relative to untreated soil for the addition of 3% acid.
6. The stabilization using phosphoric acid in the clay soils is a promising option and excellent results can be expected to solve the geotechnical stabilization problems.

## References

1. J.C. Oldham, R.C. Eaves, D.W. White. Miscellaneous Paper S-77-15, U.S. Army Engineer Waterways Experiment Station, Vicksburg, MS, September (1977).
2. A. Ajayi-Majebi , W.A. Grissom, L.S. Smith, E.E. Jones. National Research Council, Washington, D.C. (1991).
3. R.L. Santoni, J.S. Tingle, S.L Webster. TRB paper 02-3756, TRB January (2002).
4. A. Zandieh, S. S. Yasrobi , Geotech. Geol. Eng. **28**:139–145 (2010).
5. ASTM-D4318, Standard Test Method for Liquid Limit, Plastic Limit, and Plasticity Index of Soil, (2010).
6. ASTM-D854, Standard Test Method for Specific Gravity, (2010).
7. ASTM-D422, Standard Test Method for Particle Size Analysis of Soils, (2010).
8. Hazardous Substances Data Bank. National Library of Medicine, Bethesda, Maryland. Denver, CO: Micromedex, Inc. (1995).
9. HSDB. Hazardous Substances Data Bank. Available online at http://sis.nlm.nih.gov, (1999).
10. ASTM-D1557, Standard Test Methods for Laboratory Compaction Characteristics of Soil Using Modified Effort (2012).
11. B.S. Methods of test for soils for civil engineering purposes, 1377-4, (1990).
12. J.W Lyons, G.J. McEwan, Highway research board bulletin, Washington DC, **318**, 4–14, (1962).
13. J. Medina, H.N. Guida, Journal of Geotechnical & Geological Engineering 13:199–216 (1995).

MATEC Web of Conferences 162, 01028 (2018) https://doi.org/10.1051/matecconf/201816201028
BCEE3-2017

# Volume change behavior of bentonite-sand mixture under constant suction condition

*Yasir* Al-Badran[1,*] and *Tom* Schanz[1]

[1]Faculty of Civil and Environment Engineering, Ruhr-Universität Bochum, Germany

**Abstract.** The unsaturated state of soil can raise many geotechnical problems upon wetting-drying cycles such as volume change resulting in collapse or swell, change in the shear strength, and change in the hydraulic conductivity. The classical principles of soil mechanics for saturated soil are often not suitable for explaining these phenomena. An experimental work was carried out to study the volume change behavior of Bentonite-Sand mixture under constant suction condition. A mixture of 30% Calcigel bentonite and 70 % Haider sand (30B) were tested in this study. The applied suction was kept constant through the tests. Thirteen 30B specimens were tested under constant suction condition. Two different initial states (slurry and loose) were chosen to examine the effect of initial state on the volume change at unsaturated state. The results of first group (initially slurry state) show that once the suction was applied the compressibility of the specimens in this group followed the isotropic NCL up to 255 kPa suction. Subsequently, the slope of compression curve decreased. The results of increasing suction for initially slurry state show that the slope of compression lines became near to zero for suction higher than 600 kPa. Moreover, when applied suction is lower than the air-entry value, $s_{aev}$, the soil behaviors as the saturated soil. Therefore, the effective stress equation can be applied. The results of second group (initially loose state) show that the unsaturated preconsolidation pressure increases as the constant suction increases. The slopes of the unsaturated over consolidated paths in all tests were closed to the slope of saturated over consolidated path ($C_r$), while the slopes of the unsaturated normal consolidated (yield) state are higher than the slope of saturated NCL. Finally the results show that the slope of the unsaturated NCL is higher (steeper) than the slope of saturated NCL.

## 1 Introduction

The soil forming in shallow layers in the arid and semi-arid regions and the compacted soil used in earth dams, highways, embankments, airport runways, and the sealing and buffer materials engineering barriers e.g., used for the nuclear repository are in unsaturated conditions. Whenever the water interacts with a soil that is in a state of unsaturated condition volume change may occur. In nuclear (or radioactive) and high-level toxic waste disposal repositories, temperature of the waste elevates due to radioactivity processes which causes drying of the soil and alters the degree of saturation of the soil. The safe disposal of high-level radioactive nuclear waste generated from nuclear power plants is of great concern in many countries due to nuclear waste long-term (at least 10,000 years) detrimental impact on humans if released into the environment. In many countries, the chosen emplacement depth for the waste repositories has been found to vary between 500 to 1000 m (e. g., [1] and [2]). A geologic disposal facility generally consists of metallic canisters containing the radioactive nuclear wastes, buffer materials between the waste canister and bedrock, and backfill materials that would fill the spaces in the excavated access tunnels and pits. The basic functions of the barrier materials used in the waste repositories require a very low hydraulic conductivity in order to provide long-term stability to the barrier system under the high overburden pressure, and the materials should possess a high radionuclide adsorption capacity and a high heat transfer capacity [3]. Therefore, it is important to investigate the materials that have been proposed (such as bentonite-sand mixtures) as these materials are to be used as sealing and buffer elements for the repository beside the studies performed on the host rock site characterization.

The unsaturated condition of the repository induces hydro-mechanical processes of its buffer materials engineering barriers. To precisely understand this behavior the volume change behavior of bentonite-sand mixture under constant suction condition should be investigated. The volume change behavior in this study

* Corresponding author: husn_irq@yahoo.com

© The Authors, published by EDP Sciences. This is an open access article distributed under the terms of the Creative Commons Attribution License 4.0 (http://creativecommons.org/licenses/by/4.0/).

represents the compression curves of bentonite-sand mixtures under constant suction condition. Two initial states are chosen: slurry and loose density state for each soil mixture.

## 2 Materials used and samples preparation

The materials used in this study are pure bentonite 100B and a mixture of bentonite and quartz sand, 30% bentonite and 70% sand mixture 30B. The bentonite and sand mixture` with compositions of 30B was prepared. The bentonite was mined from Bavaria, Southern part of Germany. Table (1) summarizes the properties of bentonite and sand used in this study based on ASTM standards, ASTM (1997) and DIN standards, DIN (1987). The Liquid limit for (30B) mixture is 30%. The slurry soil was prepared by mixing the soil with water equal to 1.2 of liquid limit (L.L.).

To cover all the possible states of unsaturated volume change, the initial conditions of the specimens used are divided into two states: ((i) Slurry (Sl) when drying path was applied and (ii) Loose (Lo) when wetting path was applied. The initial slurry state was used to study the unsaturated volume change of drying path, whereas the initially loose state was used to study the unsaturated volume change of wetting path. The slurry specimens were prepared by mixing the soil with water content equal to 1.2 times its liquid limit (i.e., 36 % for 30B). The soil was mixed daily for at least one week to produce homogenous mixture. The loose specimens were prepared to produce maximum void ratio of unsaturated state. Water content of range (65 %-10%) to achieve suction ranges (12 MPa-3.5 MPa) brings into being maximum initial void ratio for 30B soil.

**Table1**. Summary of soil properties for materials used, 30B soil.

| Properties of soils | 30B |
|---|---|
| Specific gravity | 2.695 |
| Liquid limit (%) | 30 |
| Plastic limit (%) | - |
| Plasticity index (%) | - |
| Shrinkage limit (%) | - |
| Clay content (%) | 8.1 |
| Fine content (%) | 28.5 |
| Total specific surface area ($m^2/g$)* | 120 |

*determined using the Ethylene glycol monethyl ether (EGME) method [23]

## 3 Experimental program

The study amides to investigate the effect of suction on volumetric yielding behavior under different initial conditions for fine-grained soils. Therefore, the net stress is increased through the test while the suction is kept constant. Thirteen 30B specimens were tested under constant suction condition. The initial states for constant net stress test and the stress paths for each test in (net stress-suction) space are shown in Table (2) and Figure (1), respectively. The following nomenclatures will be used as abbreviation in this work for the mixture percent, type of test, initial condition (e.g. 30B-S39-Sl = 30B mixture with 39000 kPa constant suction test for slurry initial condition): **30B** = 30% bentonite (calcigel) and 70% sand; SX = constant suction test of **X** MPa; **Lo** = loose initial state; **Sl** = slurry initial state.

The tests 30B-S0.1-SL, 30B-S0.45-Sl, 30B-S0.05-Lo, and 30B-S0.1-Lo1 were loaded under constant suction condition then followed by wetting process (reduction in suction under constant net stress) at 0.91 MPa, 1.6 MPa, 0.92 MPa, and 0.925 MPa, respectively. The maximum constant applied suction in the cerise was 39000 kPa (30B-S39-Sl test) because of the possibility to increase the preconsolidation pressure (yield stress) more than 24000 kPa (the maximum limit of the used equipment).

The state of the applied net stress in study is mainly one-dimensional using; the modified controlled-suction oedometer cells, UPC-Barcelona cell, UPC-Isochoric cell, and high stress oedometer cell, in the soil laboratory of Ruhr-Universität, Bochum, Germany. All these cells have a specimen with the following dimensions: 50 mm diameter and 20 mm height. Only one test for each soil was carried out by applying isotropic stress using triaxial device to locate the position of isotropic normal consolidation line (iso-NCL). The specimen has dimensions 50 mm diameter and 40 mm height. The procedure followed for each test depends on the type of the test. Based on the pressure-deformation characteristic of all devices, the measured volume (i.e., dry density and void ratio) at each loading and unloading steps during the compression tests were corrected. The axis translation technique (ATT) and vapor equilibrium technique (VET) were used to apply (control) suction in the used one-dimensional compression cells. For suction range less than 1500 kPa, the axis-translation technique (ATT) was used, while for suction higher than 2000 kPa, the vapor equilibrium technique (VET) was adopted.

## 4 Experimental results and discussion

The results of the volume change test under constant suction condition are presented in this section. Thirteen 30B specimens (8 specimens for initially slurry state and 5 specimens for initially loose state) were tested under constant suction condition. Two different initial states were chosen to examine the effect of initial state (slurry and loose) on the volumetric yielding at unsaturated state. The initial states for constant suction test and the stress paths for each test in (net stress-suction) space are shown in Table (2) and Figure (1), respectively. The initial point is denoted as point A, whereas point B represents the beginning of applying suction. Point C represents the beginning of loading under constant suction condition, and point D represents the end of constant suction condition. In the Figures (2 and 14-16.), the applied constant suction was equal to the initial suction, therefore A, B, and C correspond to the same point. Figures (5.23 and 5.39) were tested under zero suction condition (consolidation test).

Figures (2-11) show the results of first group (initially slurry state) of constant suction test for 30B soil. Within these tests, all the specimens start at slurry condition (point A) then the suction increases under seated load (4-7 kPa), (point B), till reaching a specific suction value (point C). Beyond that the net stress increases under constant suction condition.

The preconsolidation pressure are 7, 200, 300, 400, 600, 800, 1000, and 8000 kPa under 0, 50, 100, 255, 300, 400, 450, and 39000 kPa constant suction conditions respectively as shown in Figures (2a-10a). The increase in the preconsolidation pressure is due to (i) increase in the density before loading (as result of increasing the applied suction) and (ii) hardening of soil material as a result of increase in suction. Most of curves in this group (except 30B-S39-Sl test) joined the 1D-NCL as net stress increases. This behavior is attributed to increase in the degree of saturation during the loading that made the specimens reaching the saturation zone. The preconsolidation pressures, for tests with constant suction range from 0 to 450 kPa, locate on the saturated 1D-NCL and the values of these pressures increase as a result of increasing the density due to applying the suction. Note the values of the void ratio before start loading, Figures (2b-10b). The compression curve of 30B-S39-Sl test (Figure 8a.30a) had undergone a high reduction in volume during the initial part of loading (between point C and 100 kPa net stress). This behavior can be attributed to separation of the sample from the confining ring due to high applied suction (39000 kPa) and the incremental response of the sample changes from oedometric to uniaxial. Therefore, initially the sample had high reduction in volume until it retain back, by filling the space between the sample and the ring with soil material due to lateral deformation, to the oedometric loading condition. As the loading increased the compression curve passed the 1D-NCL till reaching 8000 kPa net stress. Beyond this point the state of the specimen changes from over consolidated to normal consolidated (yield) state with slope higher than the slope of saturated NCL. Once the suction was applied (after point B), the compressibility of the specimens in this group followed the isotropic NCL up to 255 kPa suction. Subsequently, the slope of compression curve decreased. The results of increasing suction for initially slurry show that the slope of compression lines became near to zero (no compression with increasing the suction) for suction higher than 600 kPa. The same behavior was observed in case of 30B-S39-Sl test.

The gravimetric water content, of constant suction tests for initially slurry sate, decreased as suction increased along with drying path of SWCC, Figures (3c-10c). The gravimetric water content decreased slightly as the net stress increased for over consolidation (before the preconsolidation pressure). The reduction in the gravimetric water content during the loading at the over consolidated state is believed to be due to change of the state from drying path (due to increasing the suction before loading) to wetting path (due to increase in the degree of saturation as a results of compression the specimen during loading). The results of controlled-suction collapse tests carried out by Sun et al. (2007) showed similar behavior during the over consolidated state. Beyond the preconsolidation pressure it seems to be that the gravimetric water content-suction relationship depends on the degree of saturation. If the degree of saturation is equal or higher than the degree of saturation at air-entry value (about 0.85 for 30B soil), as the case of 0, 25, 100, and 255 kPa constant suction conditions, the gravimetric water content decreased as net stress increased to keep the specimen in saturated zone. Whereas in case that the degree of saturation was moderately lower than the degree of saturation at air-entry value, as the case of 300, 400, and 450 kPa constant suction conditions, the gravimetric water content increased slightly as net stress increased till the degree of saturation reaches value around 0.85 (the degree of saturation at air-entry value), then the relationship followed similar behavior of first case. The third case is when the degree of saturation was significantly lower than the degree of saturation at air-entry value, as the case of 30B-S39-Sl test under constant suction of 39000 kPa. In this case the results show that the gravimetric water content remained unchanged regardless of increasing the density due to loading. Generally, the degree of saturation decreases with increasing the suction, then increases as the net stress increases, Figures (3d-10d).

For the initially slurry specimens under values for constant suction up to 255 kPa, the specimens remain in saturated state (except one point before loading in the 30B-S0.255-Sl test). The soil before the air-entry value, $s_{aev}$, is in the saturated zone and its behavior as the saturated soil. Therefore, the effective stress (Eq. (1) with $\chi = 1$) can be applied for void ratio and gravimetric water content, Figure (11). In other tests, the effective stress can be applied for range of suction before air-entry value (80 kPa) only.

$$\sigma' = (\sigma - u_a) + \chi(u_a - u_w) \qquad (1)$$

where $\sigma'$ = effective stress; $\sigma$ = total stress; $u_a$ = pore air pressure; $u_w$ = pore water pressure; $\chi$ = effective stress parameter, which has a value of 1 for saturated soils and 0 for dry soils.

Figures (12-17) show the results of second group of tests under constant suction condition for 30B soil: namely the initially loose state. The preconsolidation pressures are 5, 4, 100, 150, and 800 kPa under 50, 100, 4300, 10000, and 39000 kPa constant suction conditions respectively, Figures (12a-17a). The values of the void ratios in this group before loading (points C's) was 1.0, 1.1, 1.44, 1.05, and 1.05 under 50, 100, 4300, 10000, and 39000 kPa constant suction conditions respectively. These values can give some explanations about the relationship between the initial void ratio (initial density) and the unsaturated preconsolidation pressure. Again the increase in the preconsolidation pressure is due to the higher density before loading and hardening of soil material as suction increases. However, the yield states of 30B-S0.05-Lo and 30B-S0.1-Lo tests were achieved immediately when the suction was reduced from 5000 and 3500 kPa to 50 and 100 kPa, respectively. The results of 30B-S4.3-Lo, 30B-S10-Lo, and 30B-S39-Lo tests show that the unsaturated preconsolidation pressure increased from 100 kPa to 800 kPa as the constant suction increased from 4300 kPa to 39000 kPa. Moreover, Figure (17a) shows that the normal consolidated (yield) paths for both 30B-S10-Lo and 30B-S39-Lo tests are the same and the difference in the preconsolidation pressure is due to difference in the initial density. In other words, the results show that, for range of void ratio from 1.44 to 1.5, the unsaturated preconsolidation pressure remains almost constant when the suction was changed from 4300 kPa to 10000 kPa. The slopes of the unsaturated over consolidated path in all tests were closed to the slope of saturated over consolidated path ($C_r$), while the slopes of the unsaturated normal consolidated (yield) state are higher than the slope of saturated NCL. The specimen of 30B-S0.05-Lo test, Figure (12), had initially 5000 kPa suction (point A), and then the suction was reduced to 50 kPa under seating load (point B) which caused an increase of the gravimetric water content from 0.85 to 0.205 (point C), Figure (12b), and reduction in void ratio from 1.56 to 1.04, Figure (12c). While the initial suction for 30B-S0.1-Lo test, Figure (13), was 3500 kPa (point A), then it was reduced to 100 kPa under seating load (point B) which caused reduction in the gravimetric water content from 0.1005 to 0.190 (point C), Figure (13b), and reduction in void ration from 1.6 to 1.06, Figure (13c). The applied constant suction values of the other tests of this group (30B-S10-Lo, 30B-S4.3-Lo, and 30B-S39-Lo) were equaled to the initial suction. The gravimetric water content values remain almost unchanged during the yield state for high suction value (s > 4300 kPa), whereas the gravimetric water content decreased slightly in case of 50 kPa and 100 kPa constant suction conditions, Figure (17c). The results show that all the tests, during yield state had gravimetric water content-suction relationship follows the wetting path of SWCC.

The observed behavior, the slope of the unsaturated NCL is higher than the slope of saturated NCL, is similar to the results of ([4]; [5]; [6]; [7]; [8]; [9]; [10]; [11]; [12]; [13]; [14]; [15]; [16] and [17]). This behavior confirms the assumption of some unsaturated soil models (e. g. [5]; [7]; [13 and [18]), while it disagrees with other unsaturated soil models, where they are assumed that the slope of the unsaturated NCL is lower than the slope of saturated NCL (e. g. [19] and [20]).

*Reference* [21] pointed out that the apparent pre-consolidation pressure of the 50 % bentonite-sand mixture specimens generally increases with increasing suction indicating that the material hardened as suction increases. This behavior is generally similar to the behavior of tested bentonite-sand mixtures (30B) in this study. *Reference* [22] presented results showing that during the compression process at constant soil suction for an unsaturated soil, the degree of the saturation always increases with an increment of net vertical stress. *Reference* [14]cited that the contours of constant suction NCLs for suctions between 10 and 100 kPa appear to be slightly curved and converge towards the fully saturated (zero suction) NCL at high vertical stresses. But the NCLs for higher suction showed straight line behavior. It is expected that the NCLs for higher suction would also converge toward the fully saturated NCL if the samples are loaded to high enough net stresses. Moreover, his results show that all the contours of constant suction NCLs have a slope larger than the saturated NCL. The behavior of soil in the work of *Reference* [14] is very close to the observed behavior of tested bentonite-sand mixtures in this study.

## 5 Conclusions

Regarding to the results of experimental program for a mixture of 30% Calcigel bentonite and 70 % Haider sand (30B) with two different initial states (slurry and loose) and wide range of net stresses under different values of constant suction used in study to examine the volume change behavior of Bentonite-Sand mixture under constant suction condition, the following points can be achieved:

1. The results of first group (initially slurry state) show that once the suction was applied the compressibility of the specimens in this group followed the isotropic NCL up to 255 kPa suction. Subsequently, the slope of compression curve decreased. The results of increasing suction for initially slurry state show that the slope of compression lines became near to zero for suction higher than 600 kPa. Moreover, when applied suction is lower than the air-entry value, $s_{aev}$, the soil behaviors as the saturated soil. Therefore, the effective stress equation can be applied.
2. The results of second group (initially loose state) show that the unsaturated preconsolidation pressure increases as the constant suction increases.
3. The position and the slope of NCL's depend on both net stress and suction.
4. For low suction value the NCL has void ratio lower than the saturated NCL (associated with flatter slope than the saturated NCL).
5. For higher suction value the NCL has void ratio higher than the saturated NCL (associated with steeper slope than the saturated NCL).

## Acknowledgment

The first author would like to acknowledge the financial support for this research provided by Deutsch Akademischer Austausch Dienst, DAAD (German Academic Exchange Service)..

## References

1. Swedish Nuclear Fuel and Waste Management Company, "Final storage of spent nuclear fuel-KBS-3, III barriers". Swedish Nuclear Fuel Supply Company, Division KBS Technical Report, 1992: 9:1–16:12.
2. JNC, H12: Project to establish the scientific and technical basis for HWL disposal in Japan. Supporting Report 2. "Repository design and engineering technology{", JNC TN1410 2000-003. Japan Nuclear Cycle Development Institute, Tokai, Japan. 2000.
3. Brookins D., "Geochemical aspects of radioactive waste disposal". Springer-Verlag, 1984: 267-278
4. Maswoswe, J., "Stress path for a compacted soil during collapse due to wetting". PhD thesis, Imperial College, London. 1985.
5. Josa, A., Balmaceda, A., Gens, A., and Alonso, E.E., "An elasto-plastic model for partially saturated soils exhibiting a maximum of collapse". Proc. 3rd Int. Conference Computational Plasticity, Barcelona, 1: 1992: 815-826.
6. Sivakumar, V., "A critical state framework for unsaturated soil". PhD thesis, University of Sheffield, UK. 1993.
7. Wheeler, S.J. and Sivakumar, V., "An elasto-plastic critical state framework for unsaturated soils". Géotechnique, 45 (1), 1995: 35–53.
8. Wheeler, S. J., and Karube, D., " Constitutive Modelling". Proc. 1st Int. Conf. Unsaturated Soils, Paris, France, 1996: 1323-1356.
9. Sharma, R. S., "Mechanical behaviour of unsaturated highly expansive clays". PhD thesis, University of Oxford, UK. 1998.
10. Matsuoka, H., Sun, D., Kogane, A., Fukuzawa, N., and Ichihara, W., "Stress-strain behaviour of unsaturated soil in true triaxial tests". Canadian Geotechnical Journal, 39, 2002: 603-619.
11. Wheeler, S. J., Gallipoli, D., and Karstunen, M.," Comments on use of the Barcelona Basic Model for unsaturated soils". International Journal Numer. Anal. Meth. Geomech., 26, 2002:1561–1571.
12. Gallipoli, D., Gens, A., Sharma, R. and Vaunat, J., " An elasto-plastic model for unsaturated soil incorporating the effects of suction and degree of saturation on mechanical behaviour". Géotechnique, 53, No. 1, 2003: 123–135
13. Georgiadis, K., " Development, implementation and application of partially saturated soil models in FE analysis". Ph.D. thesis, Imperial College of Science, Tech. & Medicine, Uni. of London. 2003.
14. Jotisankasa, A., "Collapse behaviour of compacted silty clay". Ph.D. thesis, Imperial College of Science, Technology and Medicine, University of London. 2005.
15. Casini, F., Vassallo, R. , Mancuso, C. and Desideri, A., "Interpretation of the behavior of compacted soils using Cam-clay extended to unsaturated conditions". 2nd Int. Conference Mech. of Unsaturated Soil. Weimar, Germany, 7-9 March, 2007.
16. Benatti, J.C.B., Miguel, M.G., Rodrigues, R.A., and Vilar, O.M., " Collapsibility study for tropical soil profile using oedometric tests with controlled suction". 5th Int. Conf. of Unsaturated Soils-Alonso & Gens (eds), Barcelona, Spain,6-8 Sep 2010: 193-198.
17. Al-Badran, Yasir M. H., "Volumetric yielding behaviour of unsaturated fine-grained soils". PhD Dissertation. Faculty of Civil and Env. Eng., Ruhr-Universität Bochum, Germany. 2011.
18. Gallipoli, D., Wheeler, S. J. and Karstunen, M., "Modelling the variation of degree of saturation in a deformable unsaturated soil". Géotechnique, 53, No. 2. 2003: 105-112
19. Alonso, E. E., Gens, A. and Josa, A., "A constitutive model for partially saturated soils". Géotechnique, 40 (3), 1990: 405-430.
20. Kohgo, Y., Nakano, M., and Miyazaki, T., "Theoretical aspects of constitutive modeling for unsaturated soils". Soil Mech. Found. Eng. (Engl. Transl.), 33(4): 1993: 49–63.
21. Agus, S.S., "An experimental study on hydro-mechanical characteristics of compacted bentonite-sand mixtures". PhD Diss. Faculty of Civil Eng., Bauhaus-Universität Weimar, Germany. 2005.
22. Pham Q.H. and Fredlund, D., "A volume-mass constitutive model for unsaturated soils". Proc. Of the 58th Canadian Geotechnical Conf., Saskatchewan, Canada. V. 2. 2005.
23. Cerato A.B. and Lutenegger A.J., " Determination of surface area of fine-grained soils by the ethylene glycol monoethyl ether (EGME) method". Geotechnical Testing Journal, ASTM, 25(3), 2002: 315-321.

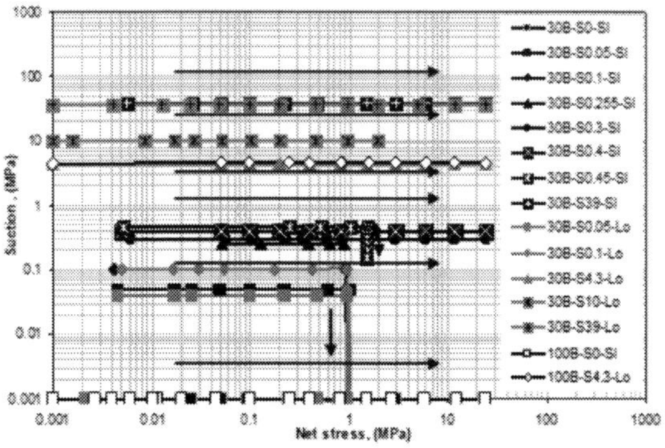


**Fig. 1.** Stress paths of constant suction tests for 30B soil.

 https://doi.org/10.1051/matecconf/201816201028

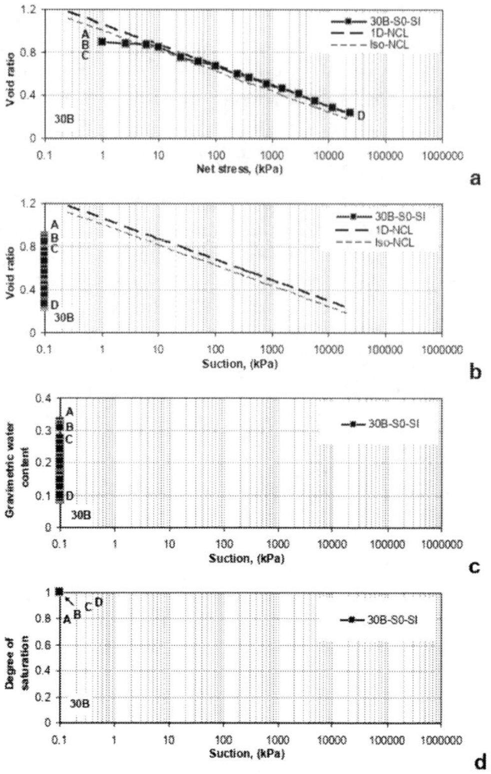


Fig. 2. Void ratio, gravimetric water content, and degree of saturation results of constant suction = 0 kPa condition, saturated, (for log-scale of suction it is given as 0.1 kPa) for initially slurry 30B soil.

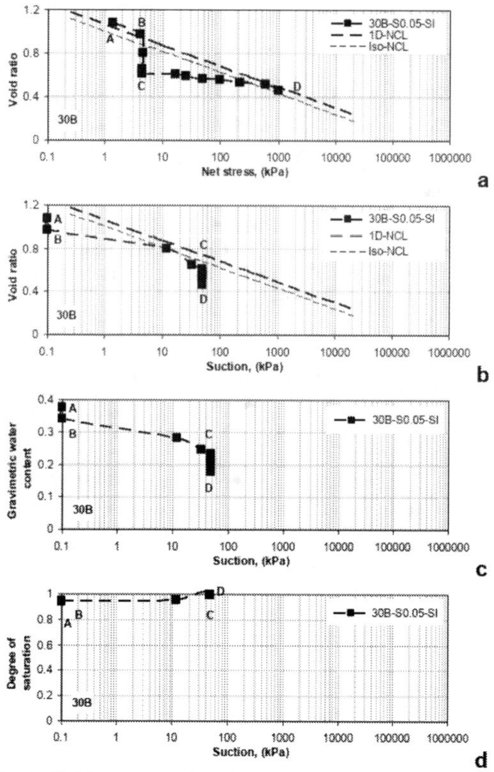


Fig. 3. Void ratio, gravimetric water content, and degree of saturation results of constant suction = 50 kPa condition for initially slurry 30B soil.

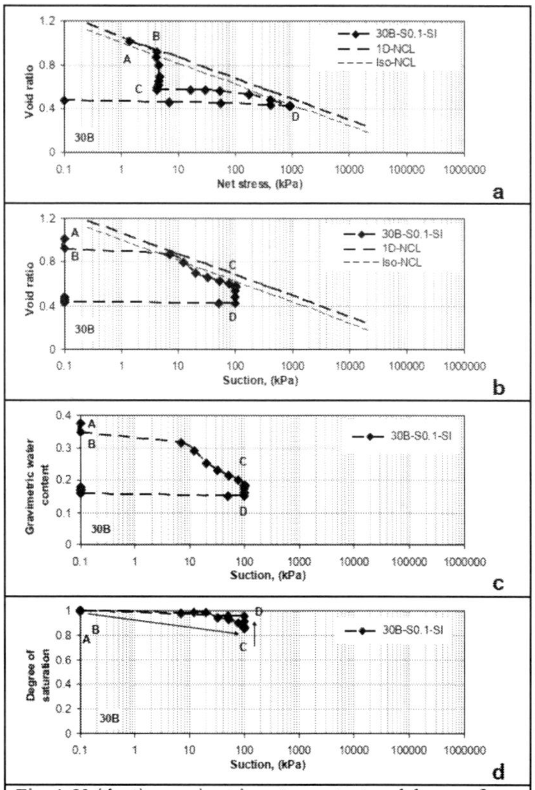


Fig. 4. Void ratio, gravimetric water content, and degree of saturation results of constant suction = 100 kPa condition for initially slurry 30B soil.

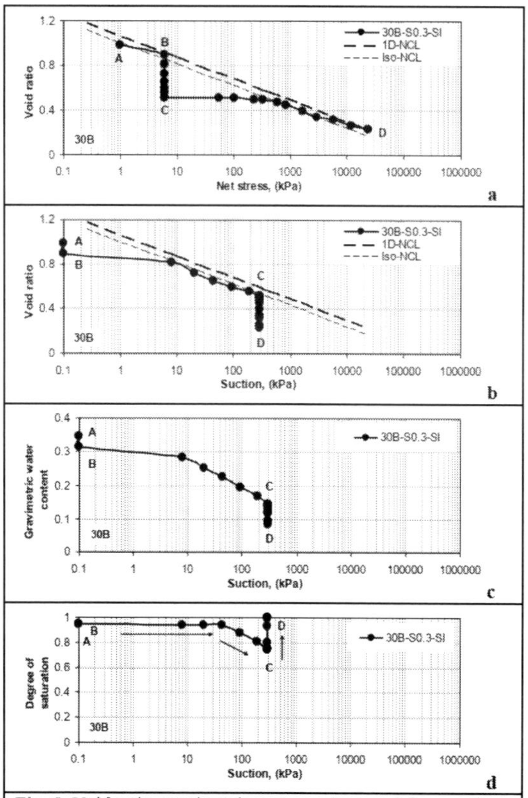


Fig. 5. Void ratio, gravimetric water content, and degree of saturation results of constant suction = 255 kPa condition for initially slurry 30B soil.

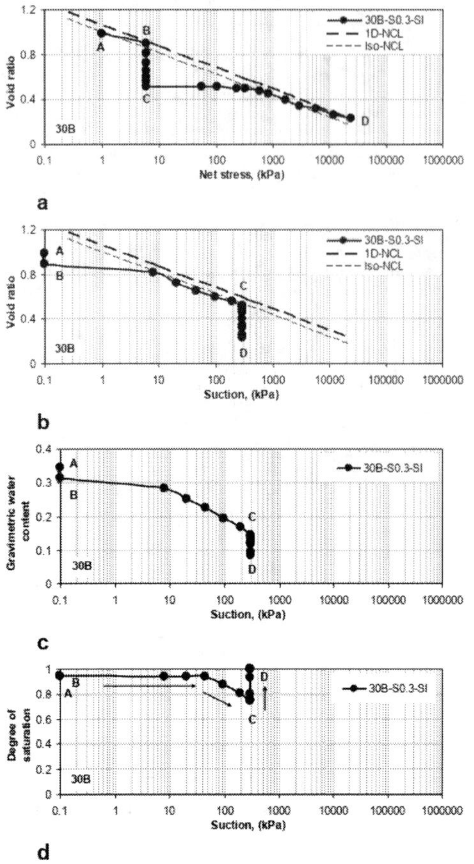


Fig. 6. Void ratio, gravimetric water content, and degree of saturation results of constant suction = 300 kPa condition for initially slurry 30B soil.

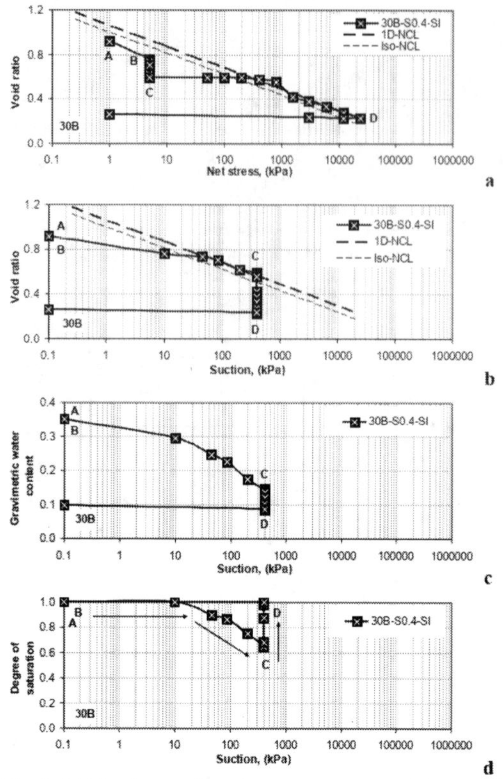


Fig. 7. Void ratio, gravimetric water content, and degree of saturation results of constant suction = 400 kPa condition for initially slurry 30B soil.

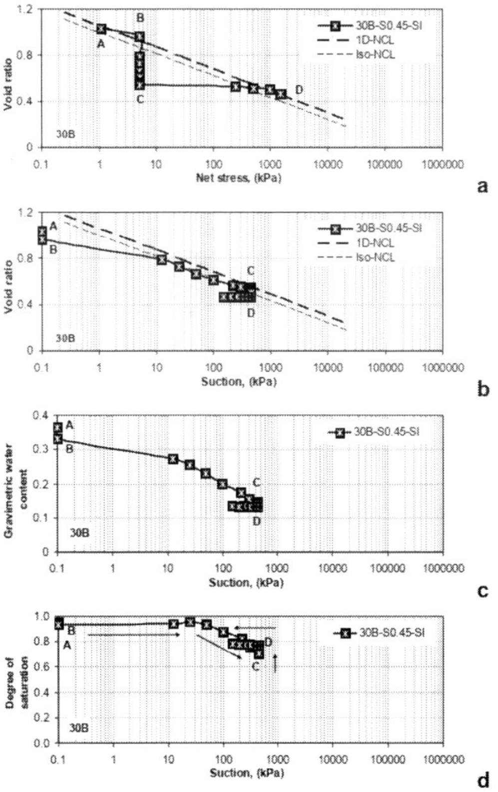


Fig. 8. Void ratio, gravimetric water content, and degree of saturation results of constant suction = 450 kPa condition for initially slurry 30B soil.

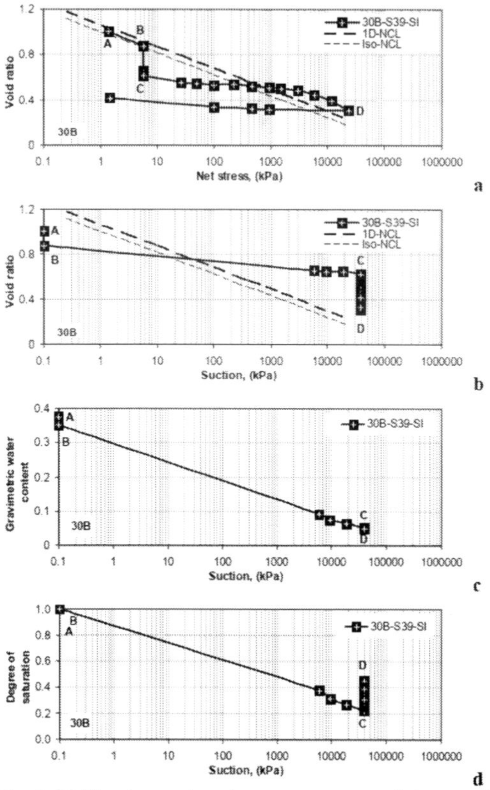


Fig. 9. Void ratio, gravimetric water content, and degree of saturation results of constant suction = 3900 kPa condition for initially slurry 30B soil.

https://doi.org/10.1051/matecconf/201816201028

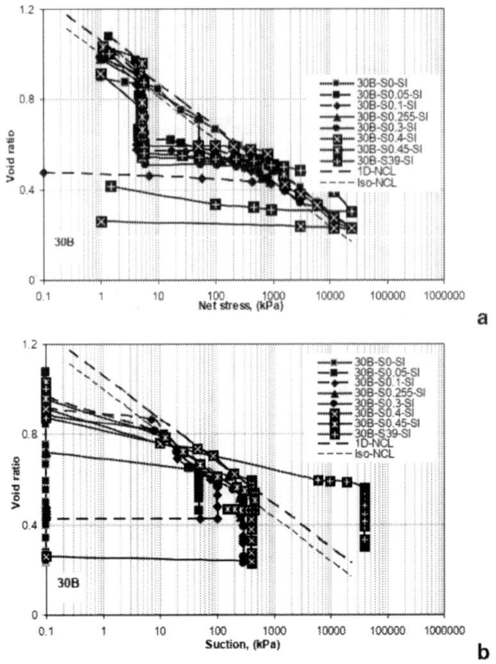


Fig. 10-1. Void ratio, gravimetric water content, and degree of saturation results of all constant suction condition for initially slurry 30B soil.

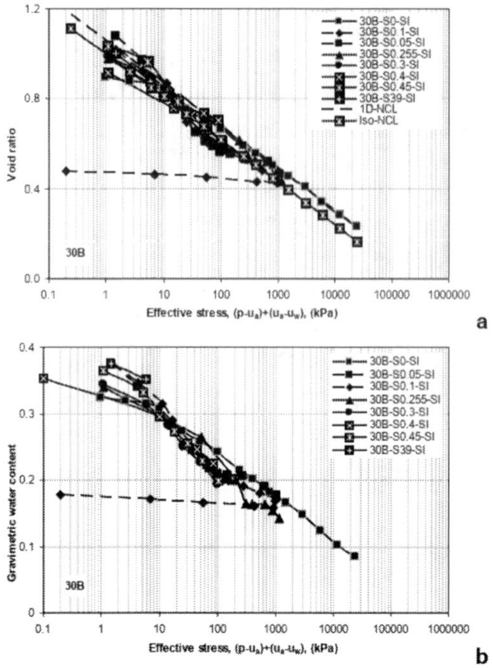


Fig. 11. Void ratio and gravimetric water content verses effective stress relationships for saturation zone of all constant suction condition for initially slurry 30B soil.

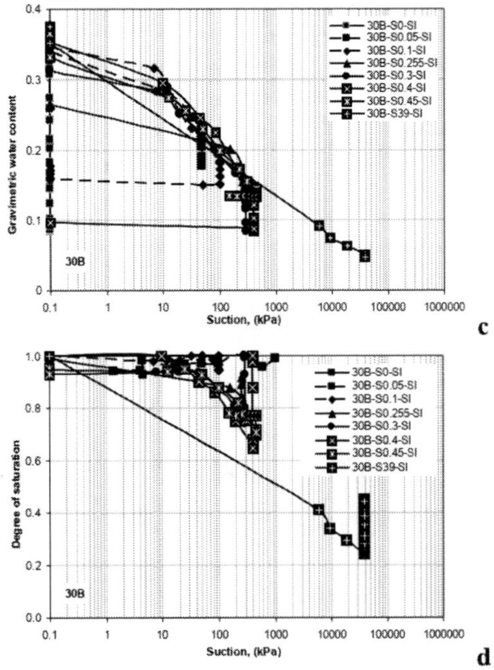


Fig. 10-2. Void ratio, gravimetric water content, and degree of saturation results of all constant suction condition for initially slurry 30B soil.

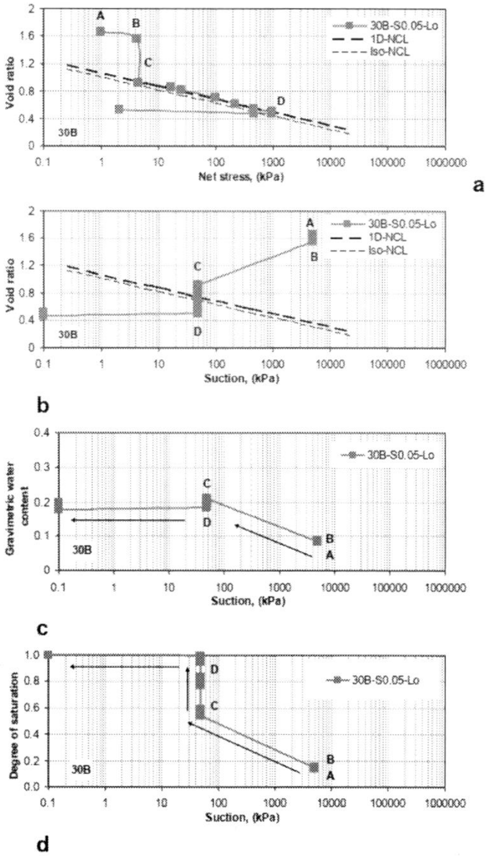


Fig. 12. Void ratio, gravimetric water content, and degree of saturation results of constant suction = 50 kPa condition for initially loose 30B soil.

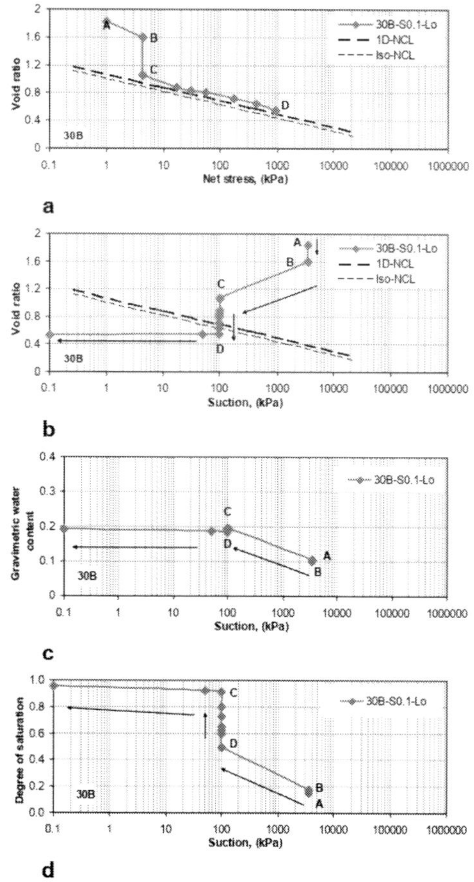


Fig. 13. Void ratio, gravimetric water content, and degree of saturation results of constant suction = 100 kPa condition for initially loose 30B soil.

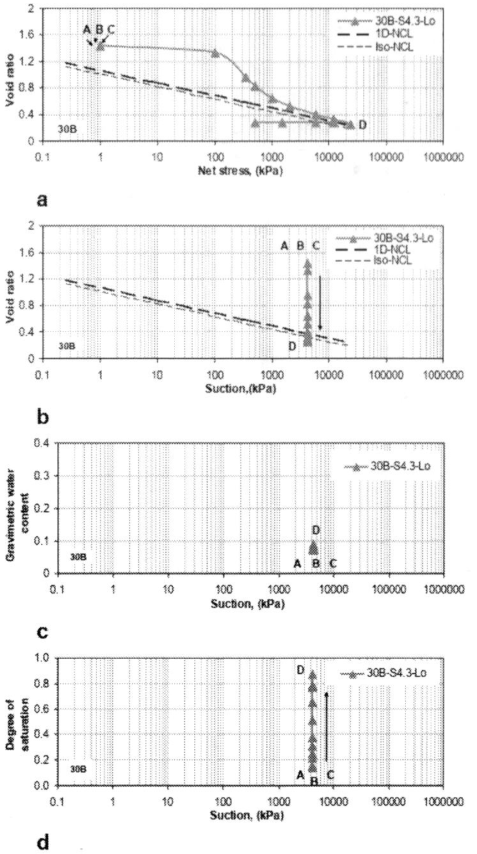


Fig. 14. Void ratio, gravimetric water content, and degree of saturation results of constant suction = 4300 kPa condition for initially loose 30B soil.

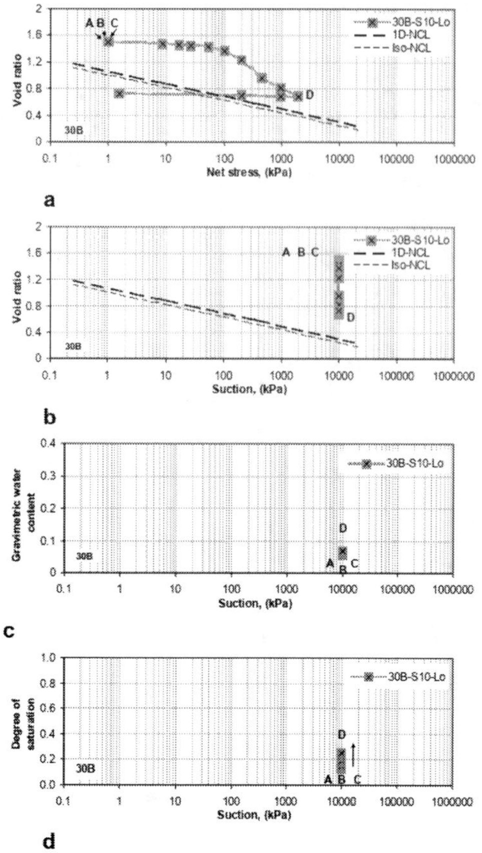


Fig. 15. Void ratio, gravimetric water content, and degree of saturation results of constant suction = 10000 kPa condition for initially loose 30B soil.

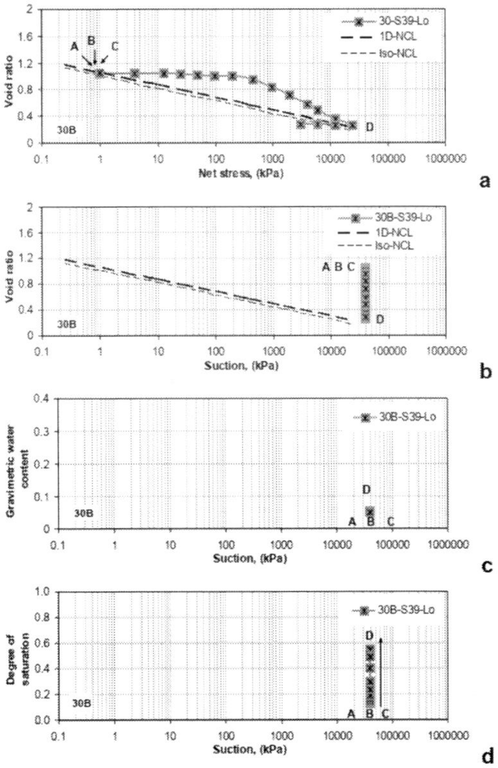


Fig. 16. Void ratio, gravimetric water content, and degree of saturation results of constant suction = 39000 kPa condition for initially loose 30B soil.

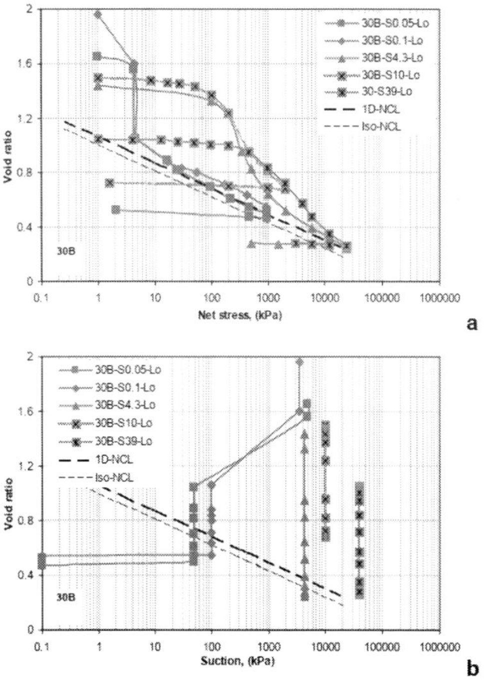


Fig. 17-1. Void ratio, gravimetric water content, and degree of saturation results of all constant suction condition for initially loose 30B soil.

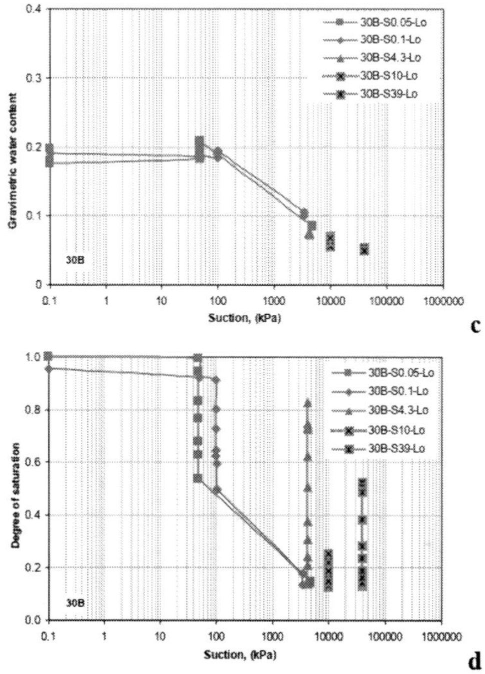


Fig. 17-2. Void ratio, gravimetric water content, and degree of saturation results of all constant suction condition for initially loose 30B soil.

# Field observations and finite element 3-D analysis of soil displacements close to unsupported excavation

*Ala* Nasir Aljorany[1,*] and *Ghusoon* Sadiq Al-Qaisee[2]

[1]College of Engineering, Baghdad University, Baghdad, Iraq
[2]Institute of Technology, Middle Technical University, Baghdad, Iraq

**Abstract.** Soil displacements due to unsupported deep excavation may cause severe damages to the nearby structures and foundations systems. Such excavations affect the state of stresses and displacements field of the surrounding soil. In this study, the soil displacements at five observation points were continuously monitored for23 days, which was the time period of excavation of about 7 m deep open tunnel. The reference points were installed on the ground surface at horizontal distance ranging from 1.25 to 3.25 m from tunnel excavation edge. The construction work was related to the project of developing the Army Channel/ Zeyouna section in the Eastern part of Baghdad City. The field observations indicate upward vertical displacement and outward lateral displacements during the first 10 days when the excavation depth was within 4 m. After that, the displacement trend was generally reversed. The displacement values were within 12 mm and vary from one observation point to another depending on the lateral distance of each point from excavation edge. The finite element package PLAXIS 3D was used to simulate the problem after obtaining the required soil parameters by an extensive site investigation. The analysis results in general, compare well with the field observations in terms of soil displacements at the reference points, especially during the first 10 days. This finding may reveal the reliability of the analysis results at other locations in the surrounding soil.

## 1 Introduction

Open excavations for constructing high rise buildings in crowded inhabited areas have become increasingly a common practice in civil engineering. Such excavations are often deeper than the foundations level of the adjacent buildings. Unless there is a well-designed shoring and sound support, these excavations will necessarily cause lateral and vertical movements of the soil mass surrounding the excavated pits, [1-8]. Such movements may cause severe damages to the adjacent structures. The design of these excavations should therefore include an estimation of the ground movement as well as a stability check of the adjacent buildings, [2-6]. For example, a deep foundation pit nearby a subway line in Taipei was excavated and caused the line tunnel damages and great economic losses, [2]. Fig. 1 displays a case study of a collapsed 13-floor building in Shanghai, China due to an adjacent deep excavation and a heavy rainfall, [2& 9].

The patterns of movement of a soil mass surrounding an open excavation can be explained as follows:

Removing the excavated soil mass is eventually unloading the soil from its geostatic stress state. The resulting displacements due to this unloading are related to the soil elastic rebound. The surrounding soil will, generally, be displaced upward while the pit's sides tilt outside the excavation center in the horizontal direction.

The other pattern of soil mass movement is related to the loss of lateral support upon excavation. The pit sides will consequently be displaced laterally towards the pit center as the lateral soil pressure will change from its at rest ($k_o$) condition to zero, for the case of unsupported excavation.

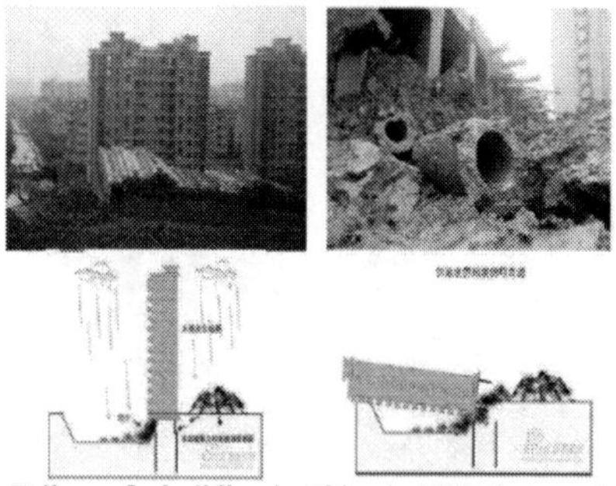

**Fig. 1.** Failure of a building in China in 2009 that was initiated by a nearby deep excavation and a heavy rainfall, [2 and 9].

These two patterns of soil movements are interlocked and usually occur simultaneously as the excavation process proceeds.

In this paper, the soil displacements at five observation points adjacent to an open, unsupported excavation were continuously monitored throughout 23 days period of

* Corresponding author: alaljorany@gmail.com

© The Authors, published by EDP Sciences. This is an open access article distributed under the terms of the Creative Commons Attribution License 4.0 (http://creativecommons.org/licenses/by/4.0/).

excavation of about 7 m deep open tunnel. The finite element package PLAXIS 3D is then used to analyze the problem and compare the measured results.

## 2 Field work

The project of developing the Army Canal in the Eastern part of Baghdad city comprises of excavation of a number of open tunnels. Zayona tunnel, which is taken as the case study in this paper, is a part of this project. The tunnel occupies an approximate area of about (45x28) $m^2$. The proposed unsupported excavation depth is about 7 m. The soil profile at the tunnel location consists of about 20 m deep clayey silt to silty clay layer of low plasticity (CL). The water content is ranging from 15 to 18%, the liquid limit is about 33 and plasticity index of 18. Five observation points were installed, before starting the excavation process, close to two perpendicular sides of the tunnel boundaries as shown in Fig. 2. The red line in this Fig represents the excavation boundaries. Plate 1 displays the earth works after completing the excavation. The point coordinates and elevations are measured using total station before and throughout the excavation period. Table 1 displays the initial positions of the observation points, the horizontal distance of each point from the excavation edge and the coordinates before starting the excavation process. The excavation works started in 26 June 2013 and continued for 23 days to reach a final excavation depth of 7.0 m. Points coordinates and elevations were continuously monitored throughout the excavation period, [10 and 11].

## 3 Results of field work

The field observations of soil displacements are measured by means of surface trigonometry using the surveying instruments such as Theodolite and Levels. Fig. 3. displays the variation of vertical displacement with time at each of the five observation points. It can be noticed that the vertical displacement is gradually increased in the upward direction (positive) until it reached its maximum value after about 10 days when the excavation depth became 4m. After that, the vertical displacement tends to decrease and after about 15 days, downward movement (negative) is detected. In general, the vertical displacements of the points (1 and 3) whether it is positive or negative are greater than that of the other points because these points are closer to the excavation edge.

Fig. 4. and Fig. 5. show the variation of the lateral displacements in X and Y directions at each of the observation points. It is important to note that the tunnel local coordinates axes make an angle of about 45° with the global coordinate axes. In order to figure out the lateral displacements whether they are outward or towards the excavation center, the following should be considered; for points 1 and 2, the resultant of positive X and positive Y displacements indicates outward lateral movement. For points 3, 4, and 5, the resultant of positive X and negative Y displacements indicates the outward movement. In general, the trend of variation of lateral displacement is almost similar to that of vertical displacement especially for observation point 1.

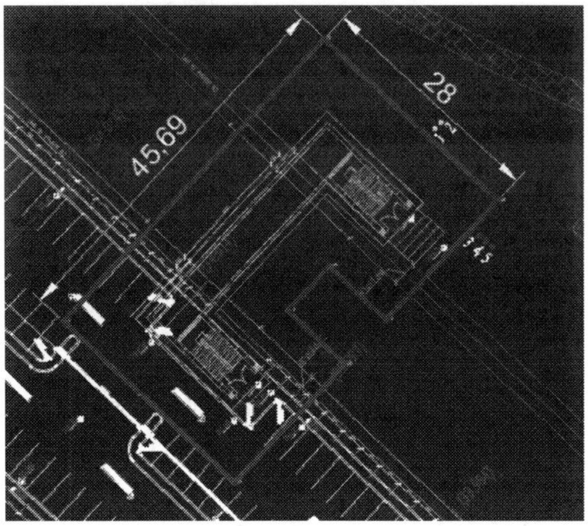


**Fig. 2.** The plan of Zayona Tunnel at Army Canal and Observation Point positions.

**Plate 1.** A Photo of the excavation work of Zayona Tunnel at Army Canal Project.

Table 1. Location and global coordinates before excavation of observation points

| point | Perpendicular Distance from excavation face, m | x | y |
|---|---|---|---|
| 1 | 1.25 | 450493.517 | 3688077.891 |
| 2 | 2.25 | 450494.225 | 3688078.589 |
| 3 | 1.25 | 450498.727 | 3688065.220 |
| 4 | 2.25 | 450499.461 | 3688064.558 |
| 5 | 3.25 | 450500.164 | 3688063.851 |

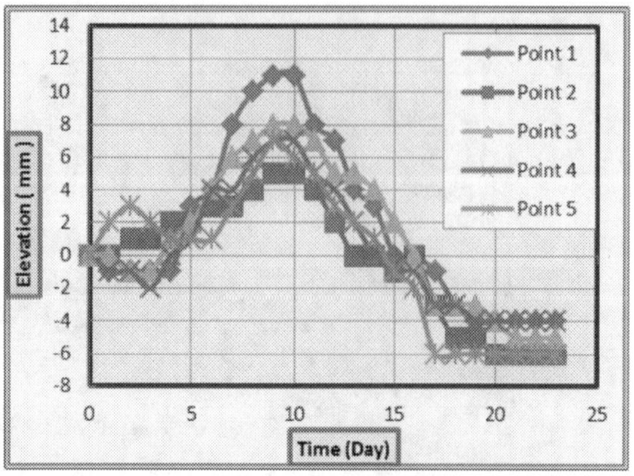


**Fig. 3.** Vertical displacements of the observation points

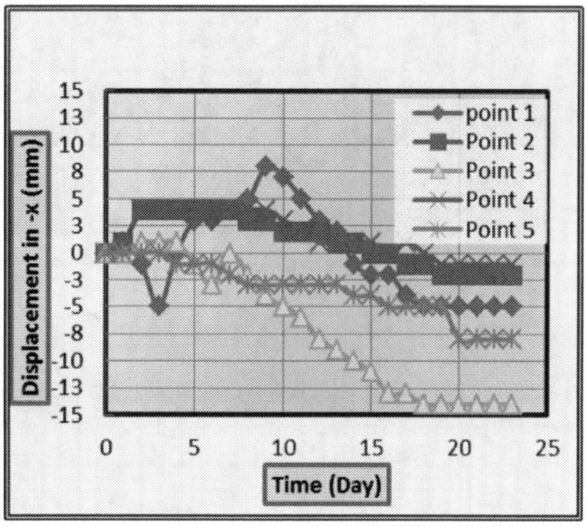


**Fig. 4.** X-coordinates variation of the observation points.

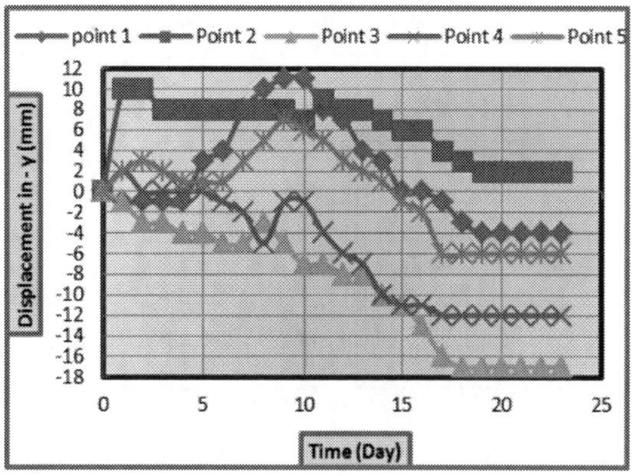


**Fig. 5.** Y- coordinates variation of the observation points.

## 4 Finite element modelling

A series of 3D finite element analyses are implemented, by using PLAXIS 3D Foundation program [12], to analyze the ground movement at the location of observation points. A 10 nodded tetrahedron element is used in the analyses. The Silty Clay layer is modeled as a Hardening soil model and the relevant soil properties are listed in Table 2.The excavation process is modeled as five stages. The relevant excavation depths of these stages are: 1, 2, 3, 5 and 7 m, respectively. Fig. 6 displays the model of the tunnel excavation in PLAXIS 3D Foundation program. Fig. 7a, b and c expose the distribution of vertical ground movement for excavation depth 1m, 5m and 7m respectively. In the early excavation stages, upward vertical ground surface movement (positive) is observed then changes to downward ground surface movement (negative) for all the observation points. The positive (upward) vertical movement at the central area of the tunnel bed increases with increasing the excavation depth. The positive vertical movement at the observation points increases with increasing depth of excavation down to 3.0 m deep and then decreases and becomes downward (negative). The upward (positive) vertical movement is in the range of 5 to 40 mm in central area of tunnel pit while it is range from 3mm (positive) to 25 mm (negative) at location of observation points. Fig. 8 illustrates the distribution of lateral ground movement at each of the excavation depths. For excavation depths less than 3 m, the lateral grounds surface movements is in the outward direction and increase with increasing excavation depth. After that depth of excavation, the lateral grounds surface movements indicate a reverse direction and the negative lateral grounds movements (towards the excavation center) are developed below the level of ground surface. The measured displacements at the observation points are compared with those obtained by the finite element analysis. Fig. 9 displays the measured and calculated vertical displacements at four of the observation points. In general, the trend of variation is almost similar. Good agreements are noticed particularly during the first 10 days when the excavation depth less than 5 m. The comparison between measured and calculated lateral displacements at two of the observation points is displayed in the Fig 10. It can be noticed that all the measured and calculated values of lateral displacements are within the range of +8 mm to -17 mm with almost similar trend of variation.

Table 2. Materials properties of in situ soil

| Parameter | Name | Value | UNIT |
|---|---|---|---|
| Material model | Model | Hardening soil model | - |
| Type of material behavior | Type | Drained | - |
| Unit Weight of soil | $\gamma_{unsat}$ | 18.0 | kN/m$^3$ |
| Young's Modules | $E^{ref}_{50}$ | 6500 | kN/m$^2$ |
| | $E^{ref}_{oed}$ | 6500 | kN/m$^2$ |
| | $E^{ref}_{ur}$ | 15000 | kN/m$^2$ |
| Poisson's ratio | ν | 0.30 | - |
| Cohesion | $C_{ref}$ | 75.0 | kN/m$^2$ |
| Friction Angle | ϕ | 15.0 | ° |

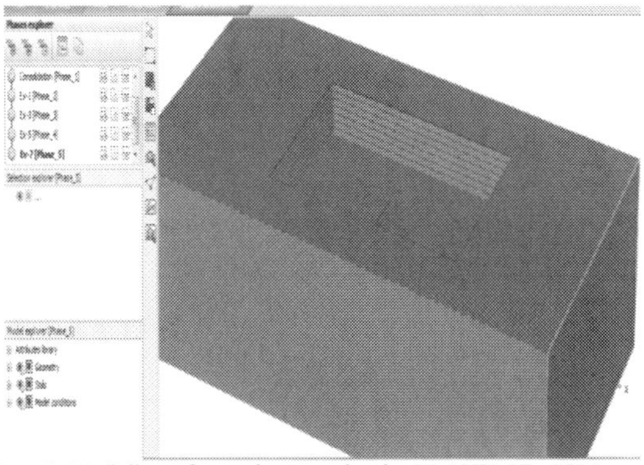

**Fig. 6.** Modeling of tunnel excavation in PLAXIS 3D, 2013

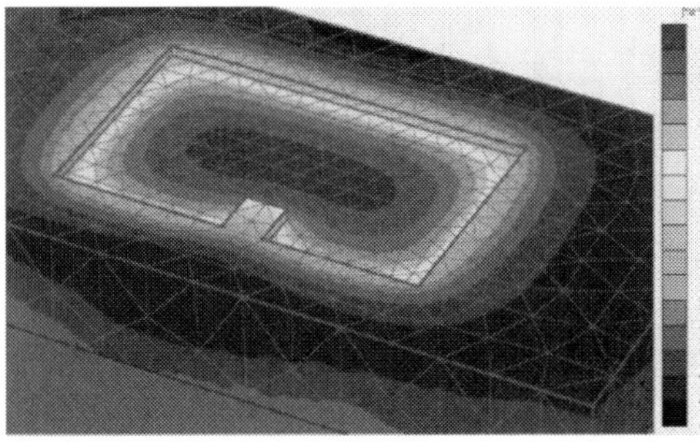

**Fig. 7a** Vertical ground movement distribution at excavation depth 1.0 m

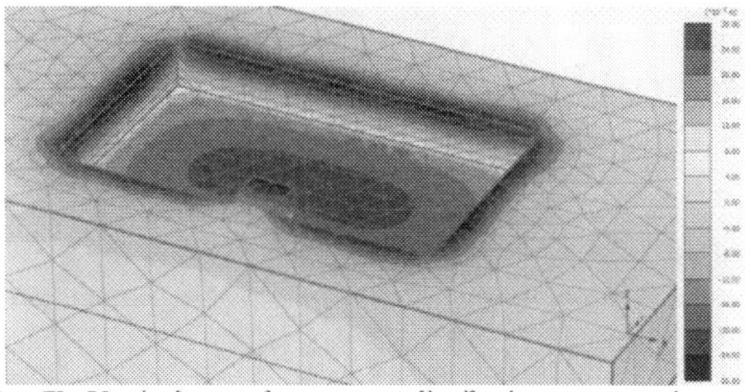

**Fig. 7b** Vertical ground movement distribution at excavation depth 5.0 m

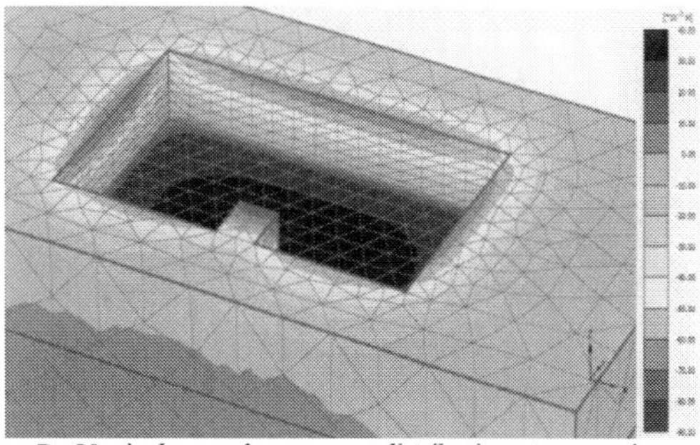

**Fig. 7c** Vertical ground movement distribution at excavation depth 7.0 m

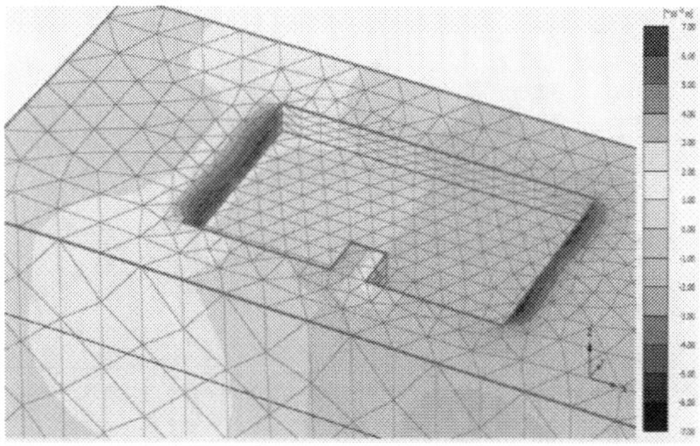

**Fig. 8a** Lateral ground movement distribution at excavation depth 3.0 m

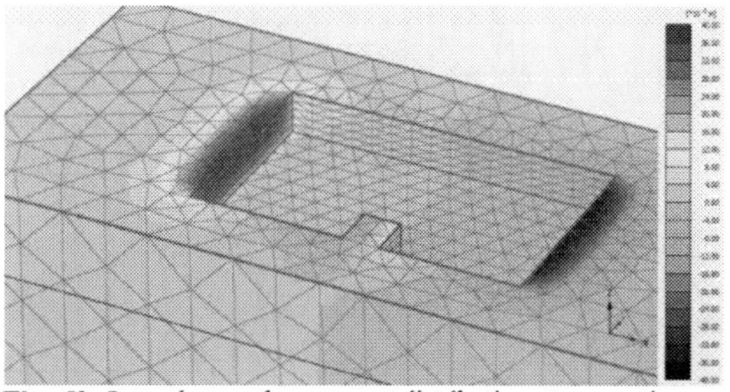

**Fig. 8b** Lateral ground movement distribution at excavation depth 5.0 m

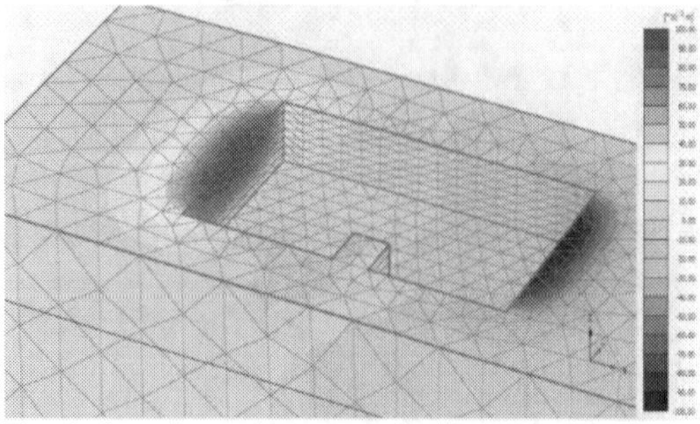

**Fig. 8c** Lateral ground movement distribution at excavation depth 7.0 m

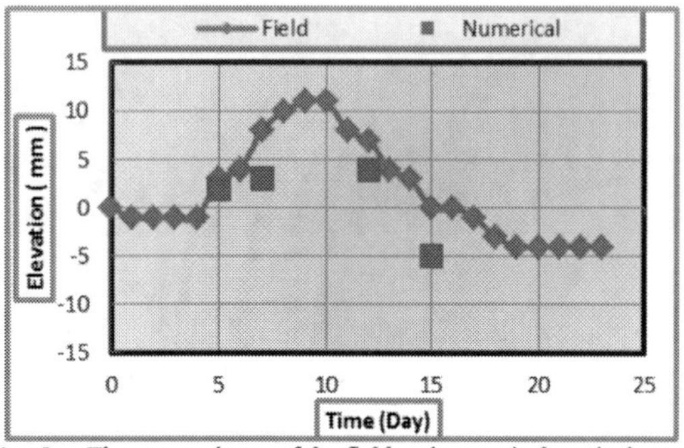


**Fig. 9a.** The comparisons of the field and numerical vertical ground movement at observation point no.1

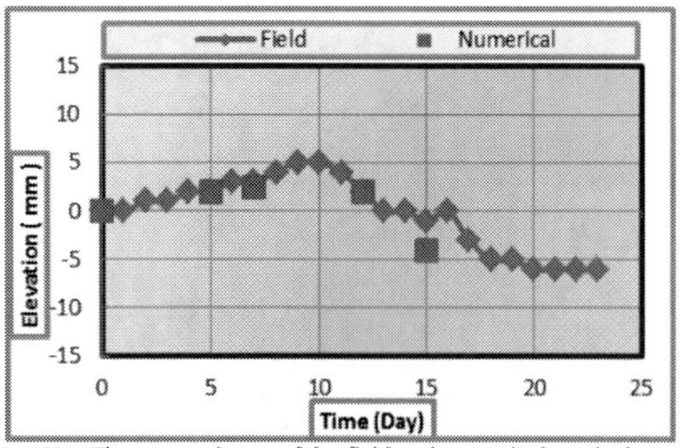


**Fig. 9b.** The comparisons of the field and numerical vertical ground movement at observation point No.2.

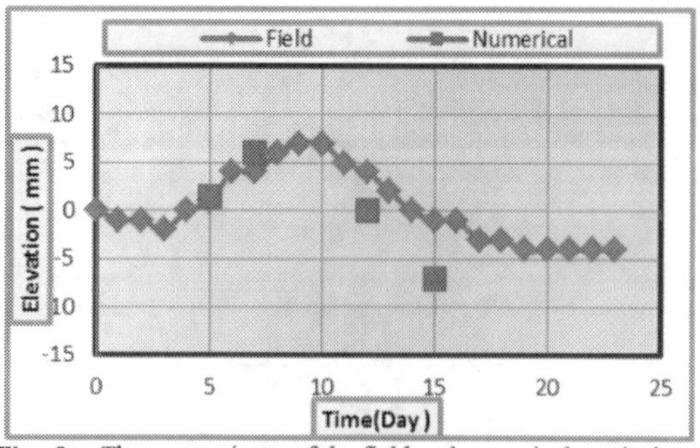


**Fig. 9c.** The comparisons of the field and numerical vertical ground movement at observation point No.4.

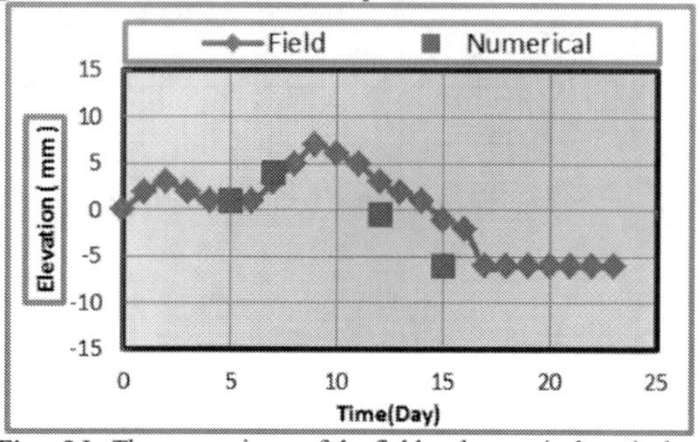


**Fig. 9d.** The comparisons of the field and numerical vertical ground movement at observation point No.5

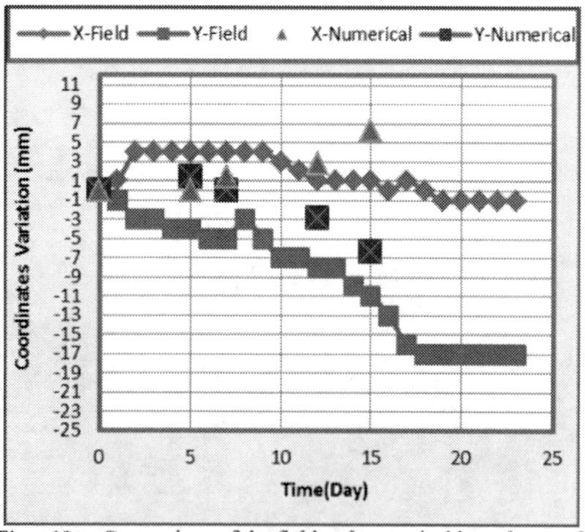


**Fig. 10a.** Comparison of the field and numerical lateral ground movement at observation point No.3.

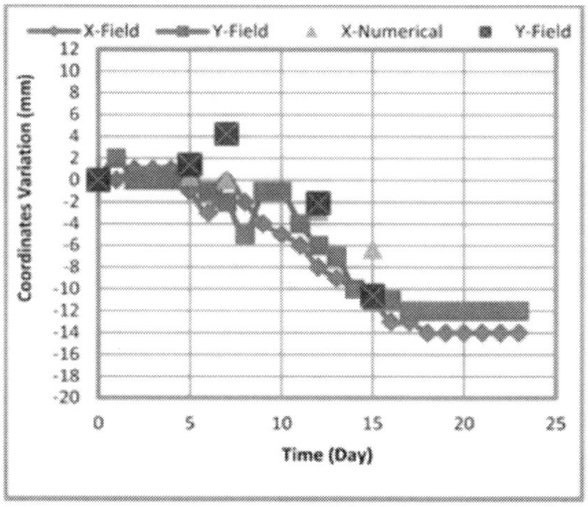


**Fig. 10b.** Comparison between field and numerical values of lateral ground movement.

## 5 Conclusions

All the observation points exhibit upward (positive) vertical displacement with increasing depth of excavation down to a depth of 4.0 m (at the tenth day of the excavation process). After that, downward vertical (negative) displacements are monitored.

In general, the values of vertical displacement at the observation points whether they are positive or negative are greater for the points that are closer to excavation boundaries.

The values of displacements obtained by the finite element analysis reveal good agreements in terms of variation trend and magnitudes with the measured ones especially for excavation depth about 5 m (less than the 10 days). The differences between measured and calculated values become more pronounced for excavation depths greater than 5 m.

## References

1. M. Korff, Deformations and damage to buildings adjacent to deep excavations in soft soils, (2009)
2. S. A. Ahmed, Geotechnical Engineering Group, DOI: 10.13140/RG.2.1.3966.9284, (2014).
3. X. Liu and S. Liu, Geotechnical and Geological Engineering, **30** (2), (2012).
4. J. H. Wang; Z. H. Xu and W. D. Wang, Journal of Geotechnical and Geoenvironmental Engineering, **136** (7), (2010).
5. C.Y. Ou, J.T. Liao, and W. L. Cheng, Géotechnique, **50** (3), (2000).
6. W. Allen Marr and M. Hawkes, Displacement-Based Design for Deep Excavations, Earth Retention Conference (ER), (2010).
7. M. Korff, and R.J. Mair, Proceedings of the 18th International Conference on Soil Mechanics and Geotechnical Engineering, (2013)
8. M. Korff, and R.J. Mair and, F. A. F. Van Tol, Journal of Geotechnicaland Geoenvironmental Engineering, **142** (8), (2016)
9. Chai, J.; Shen, S.; Ding, W.; Zhu, H. & Carter, J. China, Computers and Geotechnics, **55**, (2014).
10. A. N. Aljorany and Gh. S. Al-Qaisee, Journal of Engineering, **22** (5), (2016).
11. Gh. S. Al-Qaisee, Effects of Unsupported Excavation on a Nearby Pile-Soil System, PhD Thesis, Department of Civil Engineering, University of Baghdad, Iraq, (2015).
12. Plaxis 3D foundation; Material models manual V1.6 (2013).

# Some traffic characteristics of rural roads in Iraq

*Hamid* Athab Al-Jameel [1,*], and *Ali* Jihad Kadhim [2]

[1] University of Kufa, Najaf, Iraq
[2] Al-Mustansiriyah University, Baghdad, Iraq

**Abstract.** Now a days, modeling traffic characteristics are very important in evaluating traffic facilitates such normal section, weaving and merging sections. Lane changing and lane utilization are important traffic characteristics that are widely studied. Therefore this study is focused on both of these. The lane changing behavior affects the capacity and safety of roads. The distribution of vehicles among lanes of roadway is important in terms of different aspects such as safety, traffic management and even the structural design of roads. Field data have been collected from different rural roads. Two types of roads were studied; two-lane section and three-lane section. The results of analysis for collected data indicate that the driver behaviour is different in terms of lane utilization in Iraq from other countries and the same behaviour was found for lane changes also. Compared to other countries, Iraq has the lowest lane utilization and highest frequency of lane changes. These data are urgently needed nowadays for calibrating and validating the sophisticated programs such as simulated microscopic programs. These data are so important to mimic the reality for driver behaviors in rural roadways.

## 1 Introduction

Studying driver behavior is very important factor in modeling traffic stream [1]. Lot of studies have proved that field data could be used to model different driver behaviors such as car-following, lane changing, lane utilization and gap acceptance [2-4]. In Iraq, however, approximately no such field data have been collected for rural roads in terms of lane change and lane utilization. [5] has focused on studying the driver behavior for three lane normal section for both urban and suburban roads. Therefore, this study has been devoted to collect and analyze field data for rural roads.

## 2 Lane changing behavior

The lane changing (LC) behavior could be expressed as the desire of a driver to change the lane in order to avoid loss of his/her desired speed or to reach his/her destination. According to previous literatures, the LC could be classified as mandatory and discretionary. The first type is about when a driver tries to reach his/her destination; whereas the second type is how a driver avoids loss his desired speed because of slowing vehicles ahead [3, 5].

## 3 Lane utilization behaviour

Lane utilization (LU) represents how the vehicles are distributed among different lanes. [6] showed that the LU for the three lane normal section in the UK could be summarized as indicated in Figure 1. This figure demonstrates clearly that when the flow is low, most of the vehicles occupy the first lane and to some extent the second lane; whereas the third lane is almost free from vehicles excepting those vehicles who need to overtake.

Generally, as the flow increases from low to moderate, the usage of the third lane increases from approximately zero to approximately 50%. On the other hand, the LU for both the first and second lane decreases until reach the minimum value of the capacity as indicated in Figure 1. This behavior could be interpreted by using the first lane widely at low flow and as flow increases the drivers try to avoid the reduction in their speed by shifting to the third lane until it reaches the capacity.

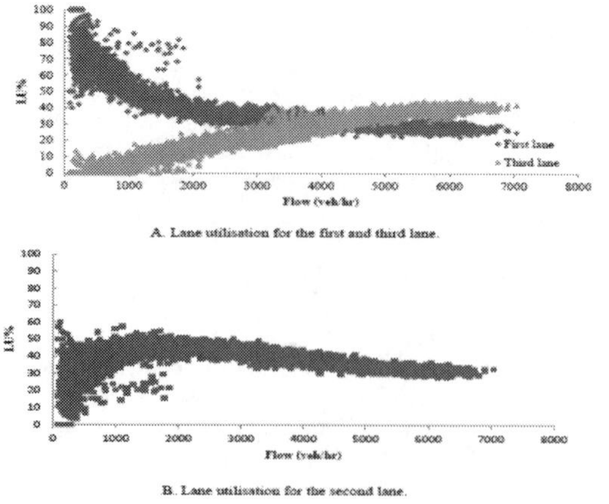


**Fig. 1.** LU for three lane normal section.

*Corresponding author: hamid.aljameel@uokufa.edu.iq

© The Authors, published by EDP Sciences. This is an open access article distributed under the terms of the Creative Commons Attribution License 4.0 (http://creativecommons.org/licenses/by/4.0/).

## 4 Data collection and analysis

Different sites have been selected during this study. These sites are on different rural roads with two and three lanes in each direction. Table 1 indicates these sites with the duration of data collection. Two Sony video cameras have been used to capture the field data. The selection of a vantage point was one of the most challenging things in line with getting security permission that was faced during the data collection stage. Footbridges have been selected as the best vantage point to monitor the traffic stream.

**Table 1.** Different sites of data collection in Iraq.

| Site No. | Site name | Duration | Date |
|---|---|---|---|
| 1 | The Expressway No.1/R4 | 5 hours | 22-9-2016 |
| 2 | R8 (Baghdad - Al-Mahmoudiyah) | 2 hours and 15 minutes | 10-10-2016 |
| | | 3 hours and 15 minutes | 17-10-2016 |
| | | 3 hours | 28-12-2016 |
| 3 | R6 (Baghdad - Al Kut). | 1 hours and 30 minutes | 27-10-2016 |
| 4 | R6 (Baghdad - Al Kut). | 2 hours and 20 minutes | 27-10-2016 |
| 5 | R2(Baghdad - Diyala) | 3 hours and 10 minutes | 23-01-2017 |
| | | 4 hours and 15 minutes | 26-01-2017 |

### 4.1 Site no.1 (expressway no.1- section R4)

The Expressway No.1 is the major freeway in Iraq that extends from the Syrian and Jordanian borders to the Kuwaiti border, and connects various parts of Iraq. It is divided into two parts. The first part, which is about 583 Km long, leads from Baghdad to Safwan; the second part is about 615.9 Km long and leads from Baghdad to the Syrian and Jordanian borders [7].
The study area is a part of section R/4B which starts at station 49+000 Km located at the beginning of section R/4B at AL-Latifiya city. The pavement section has six lanes; three in each direction, with a standing lane, each direction is separated by an island as indicated by Figure 2 and Figure 3.

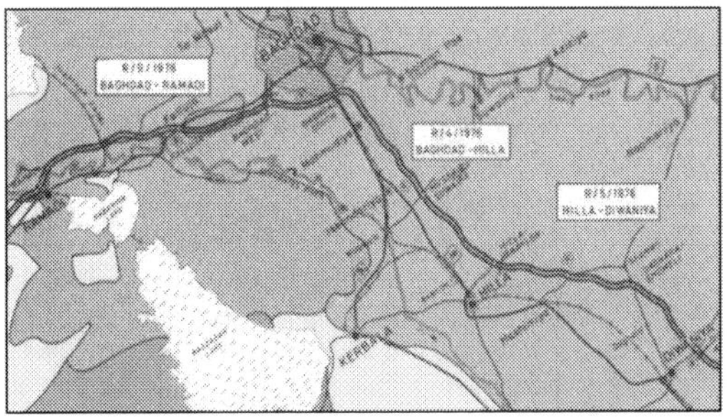


**Fig. 2.** Location of Section R4/Expressway No.1 in Iraq.

**Fig. 3.** Layout of Site No.1.

The most important thing observed on this road is the absence of pavement marking along the road. This mainly makes the task more difficult to track the changing vehicle from one lane to another. Moreover, the absence of these markings makes the drivers to be unstable or restricted by the boundary of his/her lane. This finally leads to high number of LC than the case of roads with such markings.

### 4.2 Site no.2 (Baghdad to Mahmoudiyah R8)

This site represents the link between Baghdad and the southern provinces sites. It consists of the six-lanes; three lanes in each direction. The road surface contains markers and this helps in the analysis of the results as shown in Figure 4.

**Fig. 4.** Details of Site No.2.

### 4.3 Site no.3 (Baghdad to Kut R6)

This site belongs to a road linking the province of Baghdad and Kut, with three lanes in each direction and with a standard width of 12ft as demonstrated in Figure 5. This road starts from the borders of the new city of Baghdad until it reaches a checkpoint in the region of Al-Luj, where this region is within the boundaries of Wasit province. This road is important so that it is considered the main way to travel and exchange goods between Wasit province and Baghdad.

**Fig. 5.** Geometric design of Site No. 3.

## 4.4 Site no.4 (Baghdad to Kut R6)

This site belongs to a road linking the province of Baghdad and Kut, a two-lane in each direction and with the standard width of lane 12ft as shown in Figure 6. This road begins after the checkpoint of Wasit province towards Baghdad governorate until the checkpoint reaches the region of the Al-Luj, and then turns into a section with three lanes. It is used by many of the neighboring population in their movement and is also a link between the southern provinces (ie Basra, Amarah and Dhi Qar) With Baghdad, where many large truck drivers use it to transport goods coming from Basra or from border crossings with Iran .

**Fig. 6.** Details of geometric design for Site No.4.

## 4.5 Site no.5 (Baghdad to Diyala R2)

This is the road linking the province of Baghdad and Diyala province, a two lanes in each direction with a width measuring 12ft as demonstrated by Figure 7. It is one of the important roads linking the provinces of the center and south, northern governorates.

**Fig. 7.** Layout of Site No.5 (Source: from video record).

# 5 Data analysis

Having reported the collection data from the different sites mentioned above, the analysis of these data has taken a lot of efforts to classify the vehicles. Moreover, checking the lane changing for each vehicle is so difficult because the vehicle should be tracked from the time of entering the boundary of determined section by two lines drawn on the screen of the laptop to the time of exiting. Tracking each vehicle is consuming time.

Figure 8 shows the number of LC with the variation of flow from about 300 veh/hr to about 2000 veh/hr. The maximum number of LC is about 1150. It was observed from field data that the overtaking from right which is considered as violation is predominated. However, the number of LC for such section is lower than similar section.

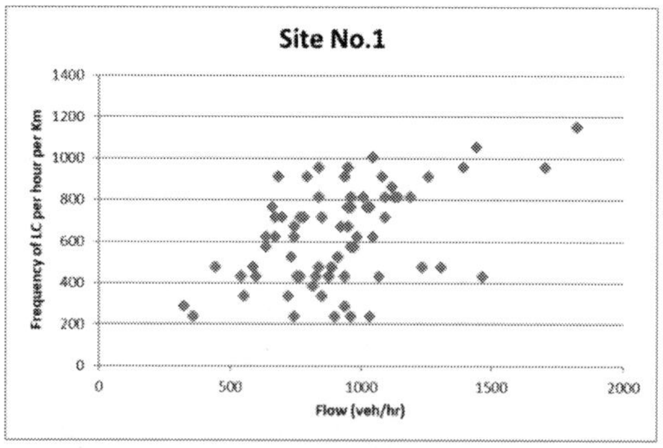


**Fig. 8.** Frequency of LC with flow Site No.1.

The lane utilization (LU) is another factor which represents driver behavior. Figure 9 indicates that drivers prefer to be in the second lane under low to moderate flow. Whereas, the third lane represents the second option for drivers. The first lane has still the lowest flow of vehicles. The maximum value for the LU for the first lane does not exceed 20 at any flow. This could be attributed to the high distresses spreading in this lane such as rutting and other types of damages resulting from heavy vehicles and absence of maintenance.

The percentage of HGV, as indicated in Figure 10, is higher comparing with other normal lanes. This percentage is about 40% under moderate flow. This high value could be the cause that interpreting the high value of LC comparing with other countries such as the UK.

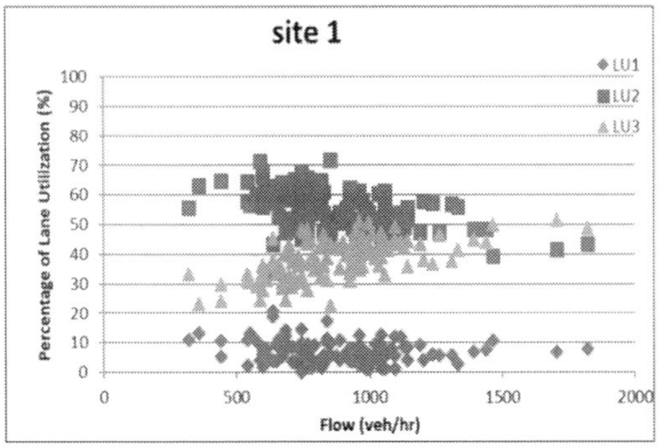


**Fig. 9.** Percentage of LU with flow.

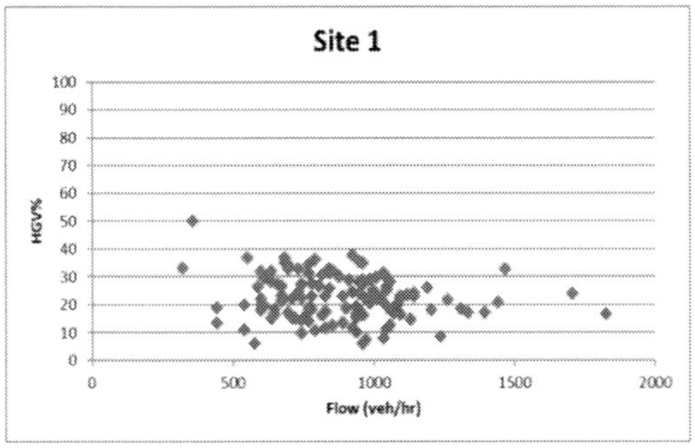


**Fig. 10.** Percentage of HGV with flow.

Figure 11 shows how the drivers are distributed among the three lane normal section. Again as observed in the Site No.1, the drivers focus at the second lane up to moderate flow then the high percentage of vehicles concentrates in the third lane. This behavior is approximately similar to the behavior observed at the Site No.3 (see Figure 14) and the UK sites as indicated in Figure 1.

At the Site No.2 it was observed that the frequency of the LC, as indicated in Figure 12, is similar in its behavior to that reported by Al-Jameel [6] in the UK (there is an increase up to around 3000 veh/hr then there is a reduction in the frequency of LC towards the capacity. However, the maximum value is beyond that observed in the UK. Another factor has an effect on this behavior is the percentage of HGV as indicated in Figure 13.

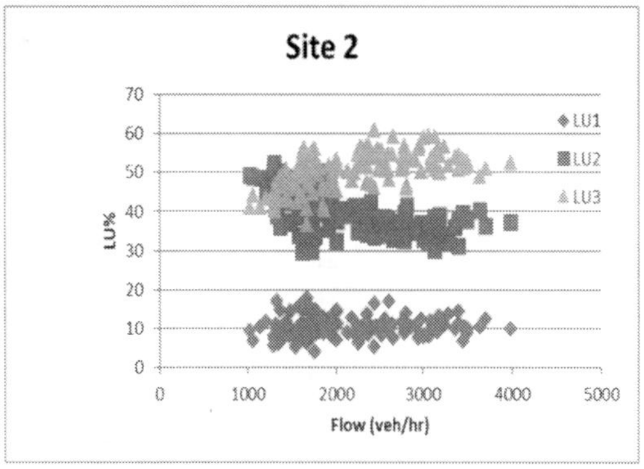


**Fig. 11.** Percentage of LU with flow.

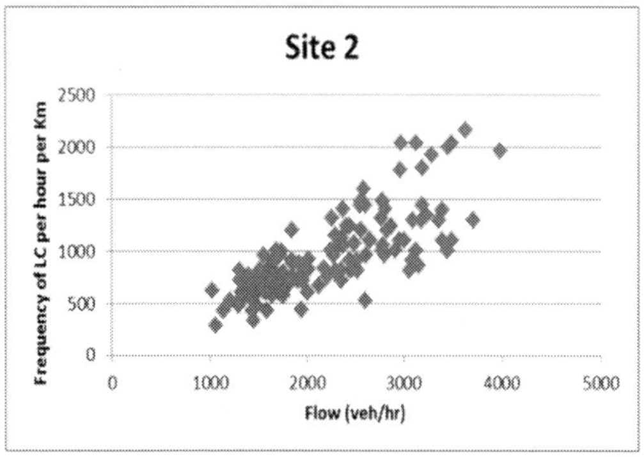


**Fig. 12.** Frequency of LC with flow Site 2.

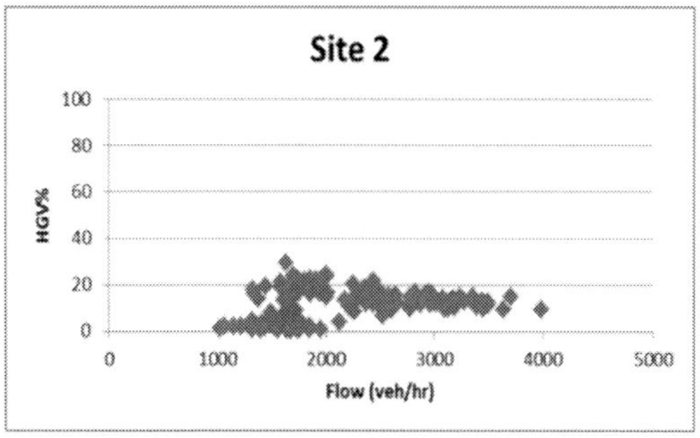


**Fig. 13.** Percentage of HGV with flow.

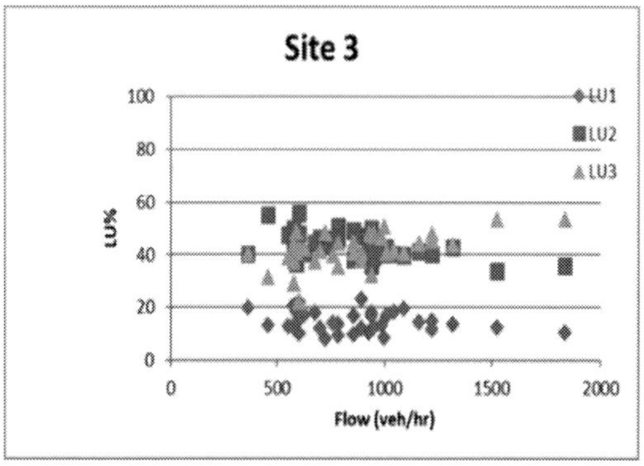


**Fig. 14.** Percentage of LU with flow Site 3.

The fourth and fifth sites are approximately different from the previous sites in terms of number of lane. They are two lane sections. The main factor has been dissected is the LU. It was observed that the drivers prefer to be in the second lane at different flow rate as demonstrated in Figures 15 and 16.

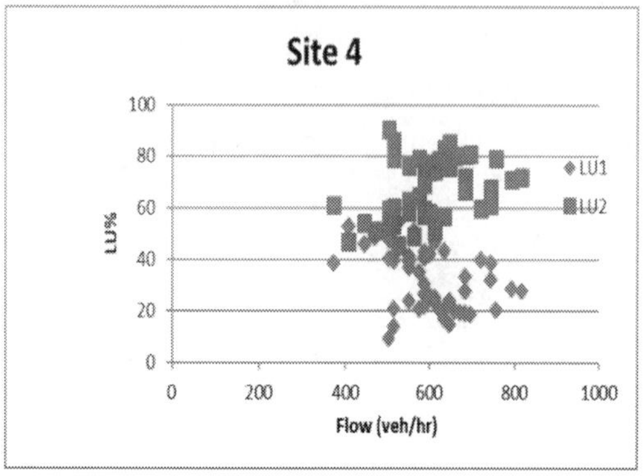


**Fig. 15.** Percentage of LU with flow Site 4.

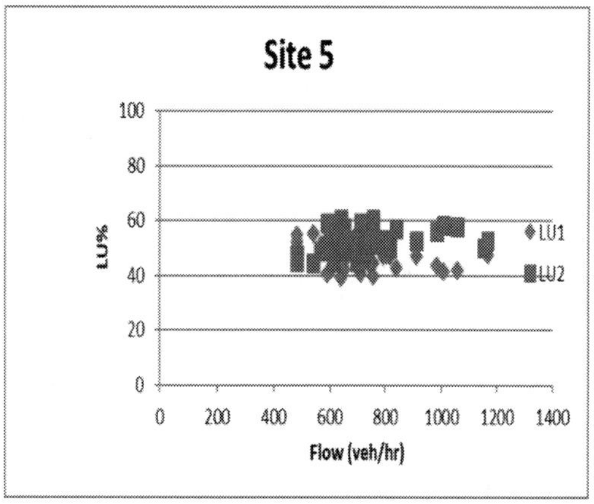


**Fig. 16.** Percentage of LU with flow Site 5.

From the above data for all sites, it was found there is an urgent need for installing loop detector in rural roads. Firstly, to obtain accurate data for some traffic characteristics such as flow, speed, headway and occupancy. Secondly, to get enough amount of data required for LU and other characteristics.

In the light of above, these data could be used for calibration and validation of simulation models in terms of the LC behavior for both two and three lane sections. Besides, the LU is another calibrated and validated factor which has an important factor to correct the behavior of simulated model.

## 6 Conclusions and recommendations

The main conclusions extracted from this study could be summarized as:

- The frequency of lane changes in Iraq is higher than that for other countries such the UK. This could be attributed to illegal LC (from left lane).
- The LU for the three lane section indicates that the passing lane does not use for overtaking the faster vehicles but used by all vehicles. On the other hand, most vehicles avoid to use the first lane because of high distresses.
- The absence of pavement markings does not only affect the FLC but also affects the traffic operations and safety.
- There is a need for correcting the driver behavior in terms of overtaking in illegal way and putting loop detector to get more accurate picture of driver behavior.

## References

1. H. Al-Jameel, Developing a simulation model to evaluate the capacity of weaving sections,University of Salford (2012).
2. J.Wang, A merging model for motorway traffic The University of Leeds (2006).
3. S. Lee, Development of a Three Dimensional Lane Change Model, University of Illinois at Urbana-Champaign (2008).
4. H. Al-Jameel, Kufa journal of Engineering,**5**(2) (2014).
5. H. Al-Jameel, Journal of Babylon,. vol.**25**(1), Iraq (2017).
6. H. Al-Jameel, The 2nd International Conference on Buildings, Construction and Environmental Engineering BCEE2, 17-18 October, Beirut, Lebanon., (2015).
7. M. Hamid, Backcalculation of Flexible Pavement Moduli Based on Genetic Algorithm Technique. Highway & Transportation Engineering Department, Al-Mustansiriya University (2012).

https://doi.org/10.1051/matecconf/201816201031

# Toward converting some individual parks into smart parks in Kerbela city

*Hamid* Athab Al-Jameel[1,*]

[1]University of Kufa, Najaf, Iraq

**Abstract.** The city of Kerbela is from the famous holy cities in Iraq and in the world where the holy shrine of Imam Hussein there. This city suffers from lack of suitable parking area for the vehicles coming to this city. This study selects a surrounding section which is approximately the same as other sections surrounding the city center. Knowing the problems in this section give a comprehensive picture for parking problems in the whole city. The study area has serious parking problems; therefore five off-street parks have been surveyed for two days during ordinary day and the other one in religious day (Thursday at day and night). Other types of data have been collected such as vehicular flow and pedestrian flow in order to select the best management for this area. The location of these five parks is on the north side of the city facing the main arterial street surrounding the holy city and in the opposing direction to the main garage called Al-Garage Al-Muhad from which different trips end and start. The analysis results of field data show that parking behavior differs in the ordinary day from the day of religious issues. It is a specific location has been assigned for one park with multi-stories in the same area as alternative of these separated parks. In addition, it was found that there is a need for managing pedestrian movement related to the parking vehicles in this area. This study also suggests using smart parking system as first step towards building intelligent transportation system to manage the parking facilities.

## 1 Introduction

The holy city of Kerbela is one of the famous holy cities (Al-Najaf, Baghdad and Samara) in Iraq in terms of number of attracting tourists from Iraq and other parts in the world. The number of visitors mainly exceeds 10 million persons for more than three times in year [1]. Therefore, a lot of vehicles used to transfer those visitors. Consequently, there is an urgent need for parking area for those vehicles.

Parking facilities are a necessary component of any transportation system to store vehicles when they are not in use. However, as parking facilities grow large, advanced methods of parking management are needed to reduce inefficiencies in finding available parking spaces [2].

Parking lots usually accompany buildings that house members of institutions (e.g. schools, churches) or organizations (e.g. business establishments) that operate motor vehicles. These plots of land can have capacities ranging from a few vehicles to hundreds of vehicles. Lots are normally paved and have paint marks that outline the space designated for each vehicle. Parking spots are normally arranged in rows, and aisles provide access to the spaces in each row [2]. Smart parking systems (SPS) have arisen in high density population areas to help patrons easily find parking in a crowded facility [2].

As mentioned above, there is a need to manage parking area. SPS is one of these methods as indicated in the next section.

The aim of this study is to evaluate some specific isolated parks distributed arbitrary along section of road up to 500m in length. Then, study the characteristics of these parks, pedestrians' movements and flow rate for the arterial road in the area under study.

## 2 Smart parking system (SPS)

As parking facilities grow larger and demand increases for parking, systems to help drivers find available parking become necessary to better manage the time and resource costs associated with the process of parking. These SPSs are able to monitor parking space usage and direct drivers to available spaces. The primary method of collecting usage information has come in the form of sensors at entrances and exits or sensors at individual parking spaces. Several methods of spreading availability information have been developed such as variable message signs and mobile/internet interfaces [3,4].

Chinrungrueng and Sunantachaikul [5] described a proposed SPS that utilizes optical wireless sensors as its base technology for detecting the movement of vehicles and space occupancy. The authors solved the problem of false positives caused by pedestrians and other non-

* Corresponding author: hamid.aljameel@uokufa.edu.iq

© The Authors, published by EDP Sciences. This is an open access article distributed under the terms of the Creative Commons Attribution License 4.0 (http://creativecommons.org/licenses/by/4.0/).

https://doi.org/10.1051/matecconf/201816201031

vehicular bodies by installing two optical heads, spaced apart, connected to one sensor. In this manner, the optical heads work together to judge the size of the passing object. Only those disturbances fitting the size of a motor vehicle trigger the sensor to register the presence of the vehicle.

Yan-Zhong et al. [6] developed an SPS that utilizes individual sensors for every parking space. These sensors are connected to a wireless network to inform drivers of a lot's real-time capacity and the location of empty spots. Sensors are deployed at every parking space and monitor the space's occupancy. This information is dispersed to the sink node that collects the information and relays it to a management station as indicated in Figure 1. The management system sends out the information to a sign at the entrance informing drivers of the parking utilization. It also updates the appropriate guiding nodes located at every major intersection. These guide nodes direct drivers toward available parking.

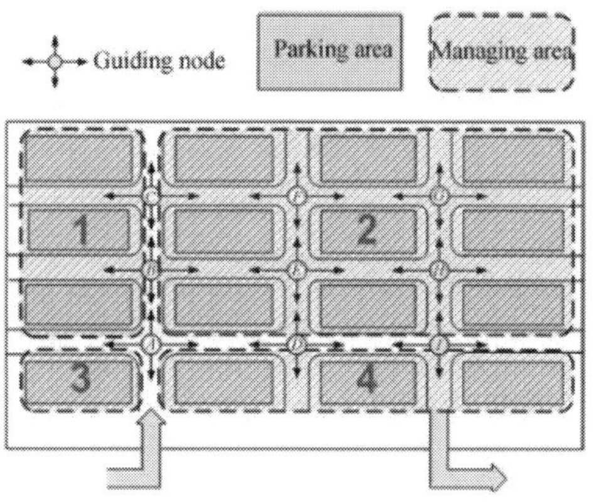


**Fig. 1.** Parking management system using guiding nodes and grouped parking spaces [6].

The guiding nodes do not direct drivers to individual spots, but rather, they guide drivers to parking areas with unused capacity. This is an efficient scheme and is depicted in Figure 1. However, the technology (e.g. sensors, routing protocol) is not perfect based on preliminary experiments conducted by Yan-Zhong et al. [6]. Improved methods for improving data transfer reliability, among other issues, are recommended bythe authors for further testing. Lu et al. [4] published a paper on a smart parking scheme for large parking lots called SPARK (i.e. Smart Parking).

A new type of SPS has been proposed by Geng and Cassandras [7]. This system exploits technologies for parking space availability detection and for driver localization and which assigns optimal parking spots to drivers instead of only supplying guidance to them as in Figure 2. They also described the system infrastructure and basic "smart parking" procedure.

Adhatarao, et al. [8] conducted their study to test the application of SPS using wireless sensor networks (WSNs). They successfully built and test SPS using light sensors to automatically determine the use of parking spaces in a parking area and providing occupied and free parking spaces to drivers. This will help drivers in finding suitable parks with low cost and time.

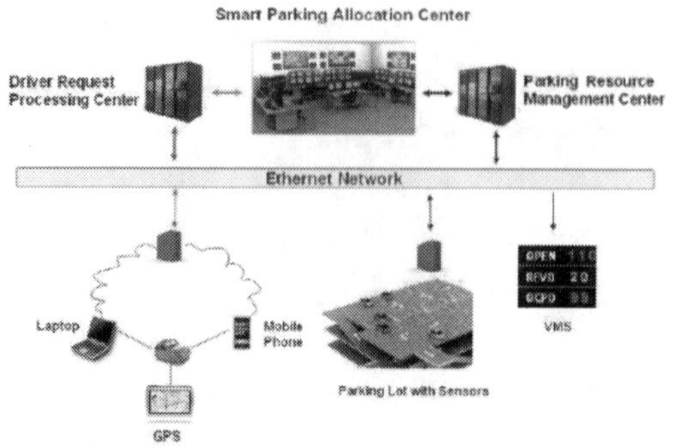


**Fig. 2.** "Smart Parking" Infrastructure (Geng and Cassandras , 2012).

Srotyr, et al. [9] proposed telematics system to provide information for truck drivers about the predicted occupancy for the nearest parking lot. They tested their model and there was encouraging results.

Majidi et al. [10] propose an exponential smoothing and multi-objective decision-making by using cloud-based methods to find the best parking place, taking into consideration of the park-cost for the driver as shown in Figure 3.

**Fig. 3.** An example of inter-vehicle communication [10].

The proposed system is capable of calculating i) a parking place for a specific location (the popularity of a parking place), ii) the possibility of vacancy of a parking place at the time, iii) anticipating the situation of a parking at a particular time, and iv) performing the multi-criteria decision-making among the criteria that the driver or the system supplies. With the assistance of the objective function according to the chosen criteria and the current parking situation, the system points the best places in order of priorities of the driver.

## 3 The parks understudy

Generally, in Kerbela city one could notice two demands for parking: one in ordinary day and the other in day which there is a religious activity (every Thursday and Friday in each week during a year). Figure 4 shows the location of area understudy and Layout A (see Appendix A) indicates the land use for area understudy.

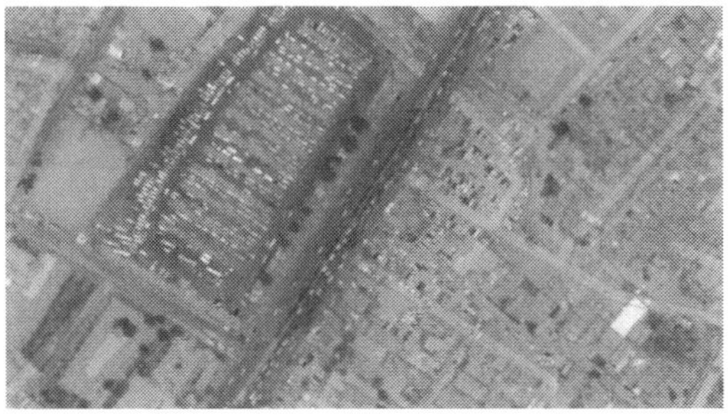

**Fig. 4**.Study area with five parking lots.

Moreover, Figure 5 demonstrates the location of proposed muti-story park in the study area as suggested by the authorities in the city. The importance of study area currently is due to two reasons: firstly, it is the closest to city center where holy shrines are located which are the major trip generators and distributors. Secondly, it is similar to other surrounding parts of city center.

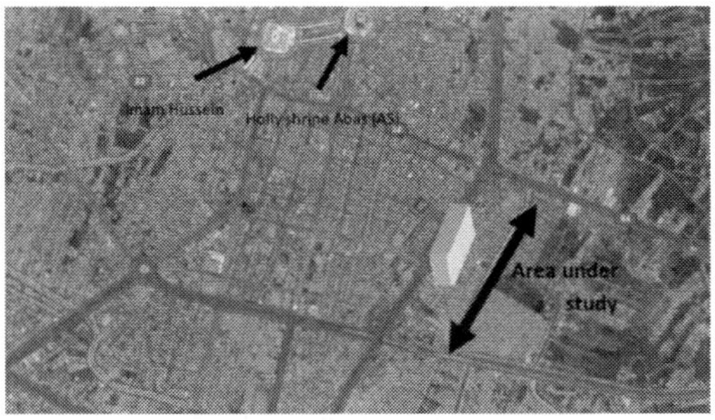


**Fig. 5.** Location of proposed Multi-story Park.

## 4 Data collioection

To understand the whole vision in the study area and know more about the parking behaviour, five parks have been selected in addition to determine the pedestrian and vehicle flow for two days (Wednesday and Thursday).
Three types of parking surveys have been adopted: the number of vehicles entering and leaving, the origin and destination survey (interviews) and turnover calculations. These surveys have been conducted by a group of engineers. They have been distributed to cover different activities such as parking survey, counting vehicles using the main road in the study area, and pedestrian crossing the main road within parking area only.

## 5 Data analysis

As mentioned above, several surveys have been conducted in this study such as flow rate, pedestrian flow and parking surveys.

## 5.1 Flow rate

Flow data have been collected every 15min by the observers. Enumerators have been distributed at the entrance of the main road with the boundary of study area as indicated in Figure 6. According to flow data, it was observed that the peak hour (3700 veh/hr) was from 8:30 to 9:30 as indicating in Figure 7 on Wednesday. Whereas, Figure 8 shows flow on Thursday. This period represents the time from 2:30 P.M until 10:00 P.M. It was observed that the peak hour (3500 veh/hr) for this period from 5:30-6:30 P.M. The flow has been determined just for the direction coming to the city (entering the city) and is in the side of parking area. Then, the maximum value of the peak flow is on Wednesday.

**Fig. 6.** Enumerators collecting field data.

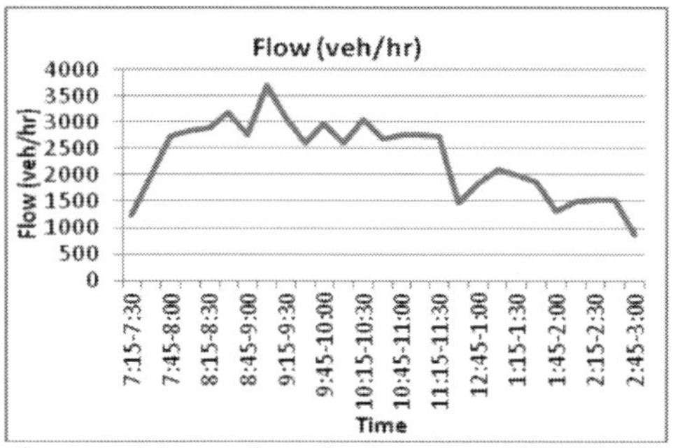


Fig. 7. Peak hour volume for the main road/ Wednesday.

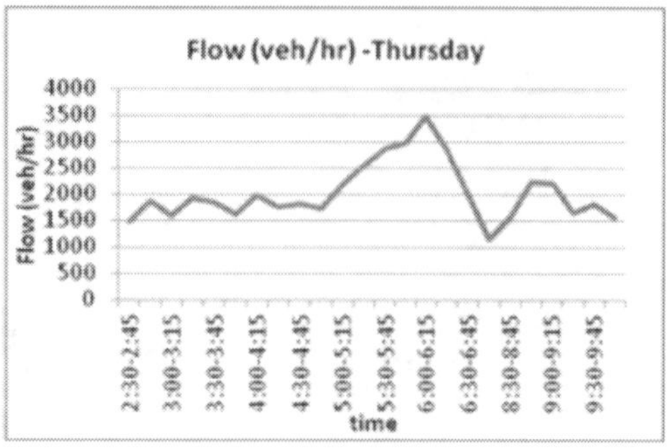


**Fig. 8.** Peak hour volume for the main road–Thursday.

## 5.2 Pedestrian flow

According to the survey for two days, it was noticed that the highest flow for pedestrian crossing the main road along the study area was on Thursday. Figure 9 demonstrates the number of pedestrian crossing in both directions. Thereupon, this high flow rate of pedestrians could be attributed to the high number of visitors crossing the street either coming from taxi/public transportation or parking vehicles in the isolated parks (parks under study). This leads to high interference with the vehicular flow.

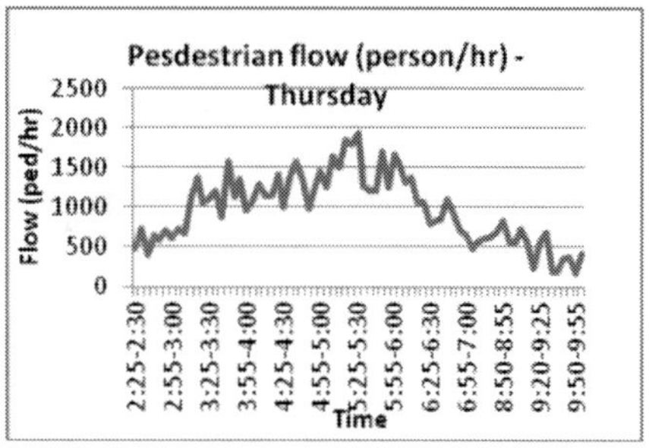


**Fig. 9.** Pedestrian flow crossing the main road.

It has been found that the number of pedestrian crossing the main road at the peak hour is 1900 ped per hour. This leads to high interaction between pedestrian and traffic stream because there is no specific area for pedestrian to cross. This results in high congestion and creates a lot of bottlenecks along this section as indicated in Figure 10.

**Fig. 10.** Pedestrian interactions with traffic stream.

## 5.3 Parking survey

Parking surveys have been conducted for entering and exiting vehicles to obtain accumulated vehicles. Figures 11 and 12 state the accumulative curve for Imam Ridah Park. The number of spaces in Imam Ridah Park is 90. Table 1 shows that the number of vehicles in and out the park. The survey started at 7:30 A.M and the number of parked vehicles was 30. Then, the occupancy for each park has been determined according to in-out accumulated vehicles.

Moreover, the average turnover has been evaluated by determining the number of spaces and the number of vehicles using these spaces. This has been conducted by checking the plate number of each vehicle for each park every 15 minutes by the observers. If the average turnover is 1 this means that the spaces has only occupied by one vehicle and if it 2 means 2 vehicles have used it. The calculated turnover for this park is 1.25 for Wednesday.

Figure 11 shows that the percentage of occupancy for Wednesday which reaches up to 65% from the capacity of park. This value has been found that the same results for the several days (except Thursday and Friday) in previous weeks according to data recording for these parks for each individual park form security purposes. These trips were found as a result of coming to governmental offices there. Whereas, on Thursday, the purpose of these trips was to visit the holy shrines in the city that means they were for religious issues. Figures from 13 to Figure 16 demonstrate the percentage of accumulated parking vehicles for Al-Gadeer, Al-Anwar, and Bab Tureag.

In fact, all other types of parks behave the same during the Thursday night (i.e. at 6:00 P.M reach their capacity). However, Al-Gadeer Park is closed during Friday and Saturday. Moreover, this park is closed up to 2:00 P.M. the number of spaces is 41. The turnover is 1.

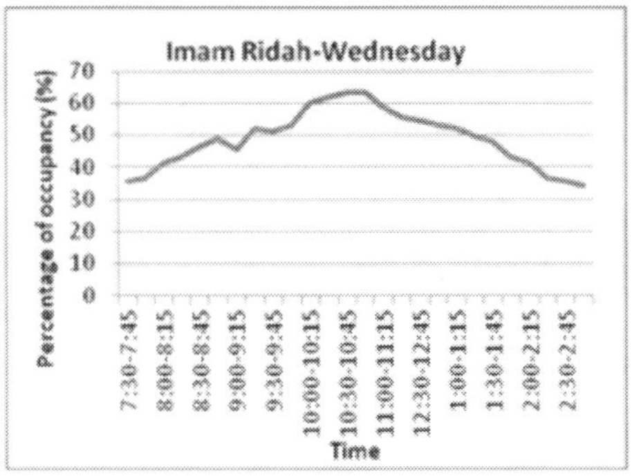


**Fig. 11**. Accumulated parking vehicles for Imam Ridah-Wednesday.

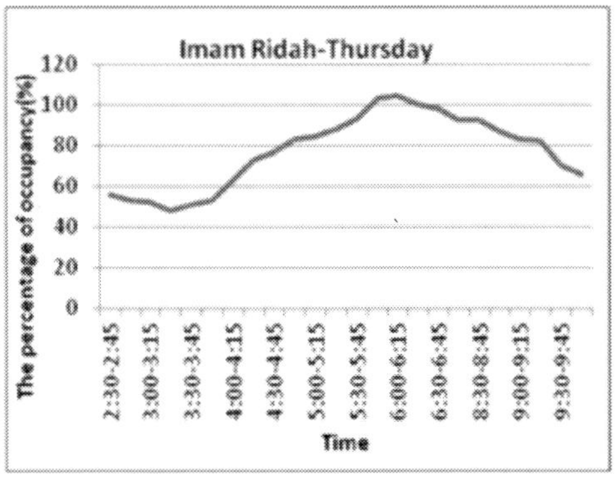


**Fig. 12**. Accumulated parking vehicles for Imam Ridah-Thursday.

Al-Anwar and Baghdad parks were considered as one park because they share in one entrance. The total number of spaces is 85. It was noticed that the accumulated flow during Wednesday operated under capacity by 40%. Whereas on Thursday, the capacity was reached after 6:30 P.M. Then, parking volume starts decreasing due to returning of the visitors back to their origin. The average turnover is 1.35 and 1.4 for Al-Anwar and Baghdad parks, respectively.

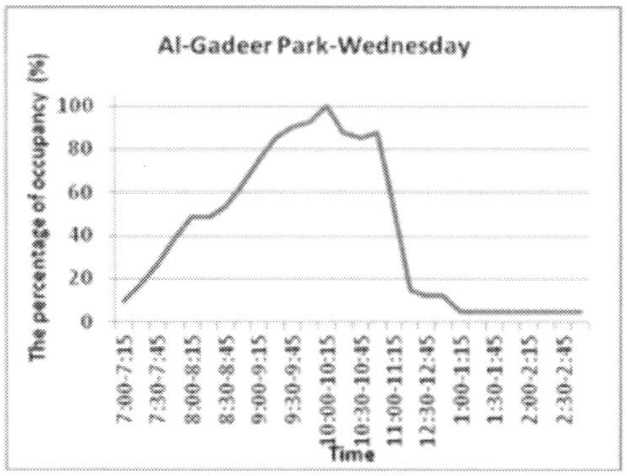


**Fig. 13**. Accumulated parking vehicles for Al-Gadeer Park-Wednesday.

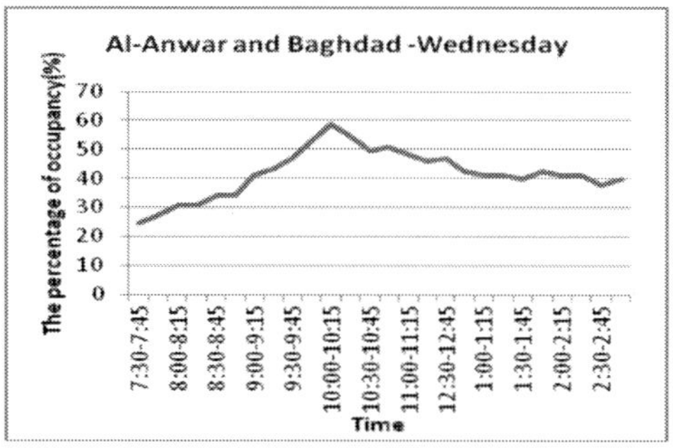


**Fig. 14**. Accumulated parking vehicles for Al-Anwar and Baghdad Parks-Wednesday.

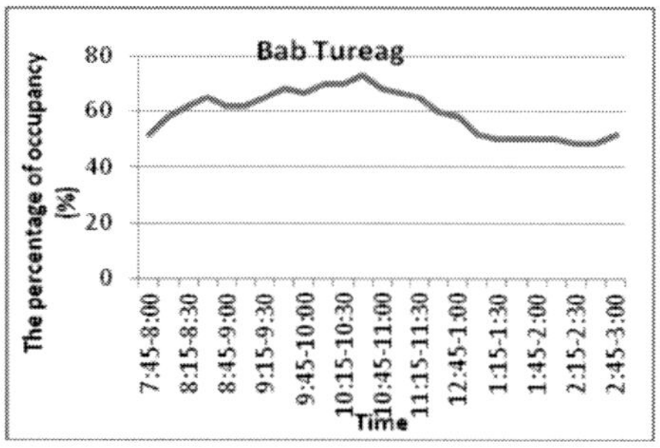


**Fig. 15**. Accumulated parking vehicles for Bab Tureag Parks-Wednesday.

**Table 1.** Accumulated parked vehicles in Imam Ridah Park.

| TIME | IN | OUT | ACCUMULATED VEHICLES | TIME | IN | OUT | ACCUMULATED VEHICLES |
|---|---|---|---|---|---|---|---|
| 7:30-7:45 | 3 | 1 | 32 | 11:15-11:30 | 1 | 1 | 53 |
| 7:45-8:00 | 4 | 3 | 33 | 11:30-11:45 | 2 | 2 | 53 |
| 8:00-8:15 | 5 | 1 | 37 | 11:45-12:00 | 0 | 0 | 53 |
| 8:15-8:30 | 3 | 1 | 39 | 12:00-12:15 | 0 | 0 | 53 |
| 8:30-8:45 | 4 | 1 | 42 | 12:15-12:30 | 1 | 1 | 53 |
| 8:45-9:00 | 5 | 3 | 44 | 12:30-12:45 | 2 | 3 | 52 |
| 9:00-9:15 | 3 | 6 | 41 | 12:45-1:00 | 1 | 3 | 50 |
| 9:15-9:30 | 8 | 2 | 47 | 1:00-1:15 | 1 | 2 | 49 |
| 9:30-9:45 | 3 | 4 | 46 | 1:15-1:30 | 0 | 2 | 47 |
| 9:45-10:00 | 4 | 2 | 48 | 1:30-1:45 | 0 | 2 | 45 |
| 10:00-10:15 | 8 | 2 | 54 | 1:45-2:00 | 0 | 4 | 41 |
| 10:15-10:30 | 6 | 4 | 56 | 2:00-2:15 | 2 | 4 | 39 |
| 10:30-10:45 | 3 | 2 | 57 | 2:15-2:30 | 2 | 6 | 35 |
| 10:45-11:00 | 3 | 3 | 57 | 2:30-2:45 | 2 | 3 | 34 |
| 11:00-11:15 | 2 | 6 | 53 | 2:45-3:00 | 0 | 1 | 33 |
| | | | | | | | |

## 5.4 Origin and destination survey

Origin and destination study has been conducted by distributing the engineers at the entrance of each park. They ask the person involved in the trip about their origin, destination and the purpose of their trips. One of important thing has been noticed during the survey that all drivers used the parks were forced to answer these questions by the park office due to instructions from the authorities in the city for security issues.

Figure 17 indicates the percentages trips coming from different origins. Most of trips were interviewed from these isolated parks for the area under study demonstrates that Al-Najaf city at the first of other cities in terms of generating trips then Al-Basrah and Baghdad. These trips were classified as trips assigning for just visiting the holy shrines there (for religious purpose only during Thursday. On the other hand, Saef Saad quarter represents the highest percentage of trips coming to the city center as indicating in Figure 18. These trips represent the trips coming for either to the governmental offices or marketing.

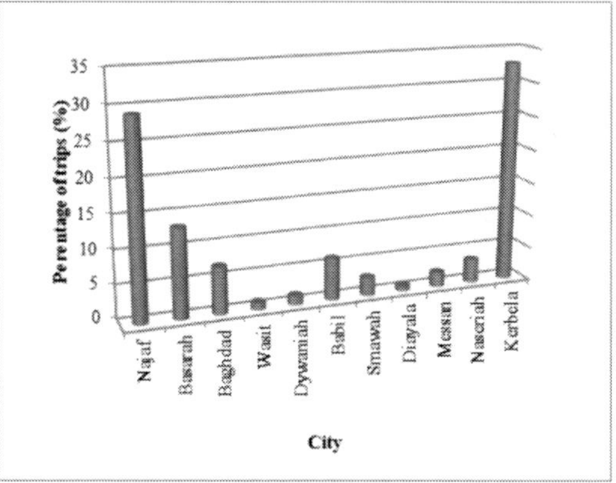


**Fig. 16.** Origin for trips on Thursday.

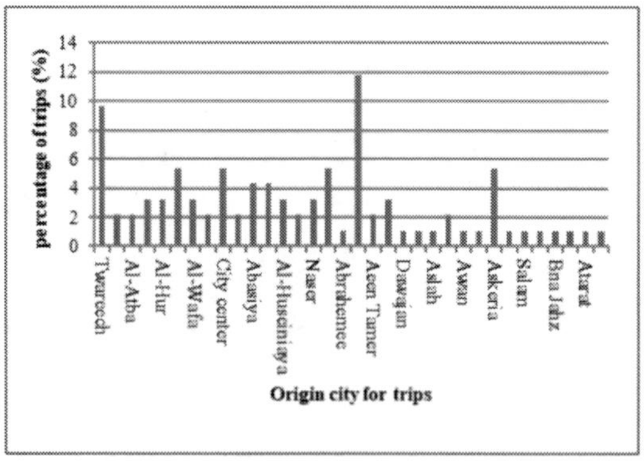


**Fig. 17.** Origin for internal trips on Wednesday.

## 6 The results analysis

According to the analysis of the results for five parks, it was found that most parks operated under their capacities on Wednesday except Al-Gadeer parks as indicated in Figures 11 to 16. Whereas on Thursday, all parks are oversaturated except Al-Gadeer, which was closed after 2:00 P.M.

At this site, the peak period for looking at park starts after 5:30 P.M and about 6:30 all parks are overcapacity (100%). A lot of vehicles can't get a space for parking for more than 1 hour. Other vehicles park even on arbitrary location as on-street parking in the road networks existing in the area under study. In addition, this period is the period of evening pray in the holy shrine for the visitors coming to performance their pray at Thursday evening. Moreover, the peak hour for pedestrian crossing the main road is about 5:30 P.M as shown in Figure 8.

Regarding to the peak flow rate, it was observed that the peak hour occurred on Wednesday morning. However, at this time the parks still worked under their capacities. This could be attributed that for morning flow rate, there are several destinations for these purpose such as just passing for another city because of existing Al-Garage Al-Mud but for Thursday all trips have one destination is for religious issues coming to the holy shrines there. So, all spaces in the area under study were occupied by even illegal parks with high number of pedestrian crossing the main roads towards the city center. The accessibility to these parks is so limited and created congested problems on the main road.

## 7 The suggested smart park

This study has just determined the location of proposed one multi-story park from Kerbela Municipality. The location of this multi-story park has been suggested in the center of these five parks multi-stories. In fact, the suggested location is so suitable for some reasons. Firstly, its location is in the center of studied parks. Secondly, it is located in the closest site to the holy shrines (the destination for the most visitors coming through this direction). The number of spaces is briefly introduced in this study.

The total number of spaces suggested for the multi-story park ranges from 700 to 850 for different layouts. For example, for option no.1 (850), the number of spaces is 100 for base, 140 for ground floor, 232 for each first and second floor, and 148 for roof. So the total number of spaces for multi-story park (850) is about double for the current total spaces for all parks (440).

Referring to the mentioned literatures about smart parks in the above sections, this study suggests using such smart park with using sensors for each parking space and connecting that with variable message in order to indicate the availability of spaces in the multi-story proposed park as indicated in Figures 1, 2 and 3. Specific sensors could be implemented for all spaces and connected with the variable messages which are placed on the main roads to indicate the availability of spaces in the park. This will help more in saving time and reducing congestion in the main road.

On the other hand, the management of routes entering the proposed multi-story park has different options. These options start from current case of entering and leaving from the main road in front of this park. The other option is from behind the park location (i.e. the parked vehicles entering or leaving the park without entering the main road as indicated in Figure 19. The VISSIM simulation model has been used to test the route of traffic in case of coming and leaving the multi-story park from the main road. This option shows that the main road suffers from congestion which is so close to what happened in reality. Another option is summarized using routes as indicated in Figure 19. This option allows no congestion on the main road because the coming and leaving demand for the park comes from determined routes in the figure which do not connect with the main road.

**Fig. 18.** Managing entering and leaving the proposed park using VISSIM model.

There is a need for effective steps which should be implemented such as:

- Followed effective parking management strategies as used in the developed countries (as mention in Section 2).
- Regulating the parking area in order to encourage the transit use and create much vibrant street life.
- Avoiding accumulative of parking spaces over a small area resulting in chaos.

## 8 Conclusions and recommendations

The main conclusions are:

1. The evaluated isolated off-street parks operate under their capacity on Wednesday but these parks suffer from overcapacity on Thursday due to religious issue. This also accompanies with high pedestrian volume crossing the roads.
2. The demand for parking every Thursday night is high around the city center. However, the high demand during special event has been excluded in this study. Therefore, there is a need for constructing multi-story parks in the area understudy and applying smart park principle for it and using variable message on the main road showing the occupancy for parks.
3. Managing pedestrian movement crossing the main road using footbridge or tunnel taking into consideration pedestrian movement resulting from parking area.

Using SPS for the proposed multi-story park by connecting sensors in the park with variable messages on the main arterial road.

## References

1. Middleeasteye.net,http://www.middleeasteye.net/news/arba-karbala-pilgrims-iraq-581628397(2016).
2. G. Surpris, *Evaluating the Effect of Smart Parking Technology on Campus Parking System Efficiency using Discrete Event Simulation* (PhD Thesis, Embry-Riddle Aeronautical University, Florida, 2012).
3. R. Maccubbin, *Evaluating ITS parking management strategies: A systems approach* (Unpublished master's thesis). University of Virginia, 2000).
4. R. Lu, X. Lin, H. Zhu, and X. Shen, *IEEE INFOCOM*, 1413-1421, (2009).
5. J. Chinrungrueng, and U. Sunantachaikul, *International Symposium on Applications and the Internet Workshops* (2007).
6. B. Yan-Zhong, S. Li-Mun, Z. Hong-Song, Y. Ting-Xin, and L. Zheng-Jun, Acta Automatica Sinica, **32(6)**, 968-977(2006).
7. Y. Geng, and C. Cassandras, Social and Behavioral Sciences **54** 1278 – 1287(2012).
8. S. Adhatarao, O. Alfandi, A. Bochem, and D. Hogrefe, IEEE (VNC, 2014).
9. M. Srotyr, M. Jerabek, Z. Lokaj, and T. Zelinka, (Smart Cities Symposium Prague, IEEE, 2015).
10. A. Majidi, H. Polat, and A. Çetin, IEEE, 4th ICSG,(2016).

MATEC Web of Conferences **162**, 01032 (2018) https://doi.org/10.1051/matecconf/201816201032
*BCEE3-2017*

# Improvement of traffic performance at intersections in Karbala city

*Gandhi* Sofia[1], *Abdulhaq* Al-Haddad[1] and *Israa* Saeed Al-Haydari [1, *]

[1]Mustansiriyah University, Baghdad, Iraq

**Abstract**. This study deals with the evaluation of traffic performance at the road network around the old city of Karbala from the southern side, at Al-Tarbia Street and Fatima Al-Zahraa Street, through the evaluation of intersections' performance. This is followed by suggestions of some improvement proposals, which vary from changing timing plan, geometric improvement, to change intersection type completely.
The study area composed of seven intersections, three intersections are four legs signalized, and the other four are roundabouts. The video recording technique is used to collect the traffic data for 27 approaches. These data are abstracted from video films using EVENT program, and processed by prepared EXCEL sheets. While, the spot speed data for each entire link in the network are collected using pavement marking method.
SYNCHRO software was used for evaluation and analysis of signalized intersections, suggestion of best timing plan, and coordination. SIDRA INTERSECTION software was used for evaluation and analysis of both signalized intersections and roundabouts. The best proposal was also evaluated for the target year.

## 1 Holy Karbala city

Karbala is one of the special cities in the Islamic world. It is famous due to its religious aspect. The old city, or city center, is holding the holy shrines of Imam Hussain (the grandson of the Holy Prophet of Islam) and his stepbrother Abbas. It lies between (41°, 10') to (44°, 20') longitude and (32°) to (31°) latitude, departed at about 110 km south-west away from Baghdad. It is surrounded by Al-Anbar city from the north and west, Al-Najaf city from the south, and Al-Hilla (Babylon) city from the east [1].

Karbala province's population increased to (852963) in 2007. Around 53% of them are living in Karbala city. The car ownership also increased by about 133% from 2002 to 2007 [2].

The Holy shrine, as located in the Central Business District area (CBD) represents the core area of the city. It is surrounded from the north and the western north by farms and groves of palm trees, and western desert from the south side. Therefore, the growth of the city extended towards the south side due to the availability of vacant land [3].

### 1.1 The city problem

The rapid growth of the city with the increased number of pilgrims as well as population and auto vehicles together with road network expansion, led to many traffic problems and intolerable congestion. Although the local governorate authorities implemented some roadway resurfacing and traffic improvements. it does not solve most of the traffic congestion problems, especially at Central Business District (CBD) area.

The new growth in Karbala traffic congestion has been recognized as a serious problem in the city, which affects the economy, travel time, driver behavior and causes discomfort to drivers and visitors. The number of vehicles in Karbala increases rapidly without a considerable increase in the capacity of the road network, which leads to increase in delay times and lowers the level of service (LOS).

The study area as shown in Figure (1) represents the nearest road network (from the south side) to the old city, so it is strongly affected by pilgrims that the city attracts. Karbala city attracts around 15 million pilgrims per year during four major occasions and on Thursday night's (weekend). Traffic congestion in this area was increased with increasing traffic demand due to the presence of many activity centers such as schools, governmental buildings, shopping centers, and religious places.

These traffic conditions need to be studied and evaluated using modern software. This is to set the required traffic and geometric proposals to attain efficient traffic flow.

## 2 Study objectives

The main objectives of this study are to evaluate the traffic performance in the studied area, and suggest the required traffic and/or geometric solutions to alleviate the congestion problem at the selected network.

© The Authors, published by EDP Sciences. This is an open access article distributed under the terms of the Creative Commons Attribution License 4.0 (http://creativecommons.org/licenses/by/4.0/).

## 3 Data collection and abstraction

To achieve the study objectives, it is necessary to follow a designed procedure as in Figure (2) to gather the required data.

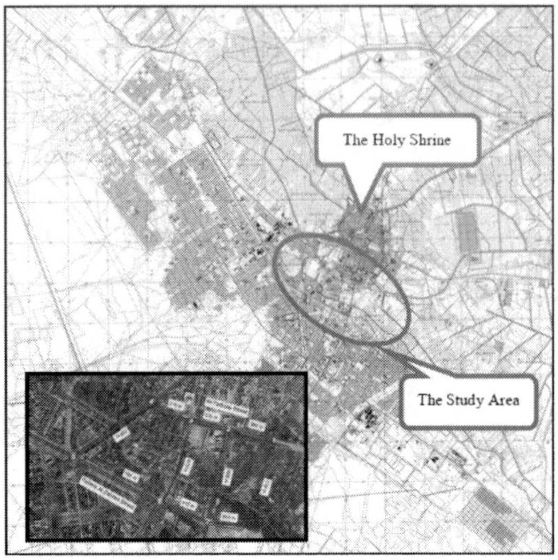


**Fig. 1.** The Road Network at Karbala City

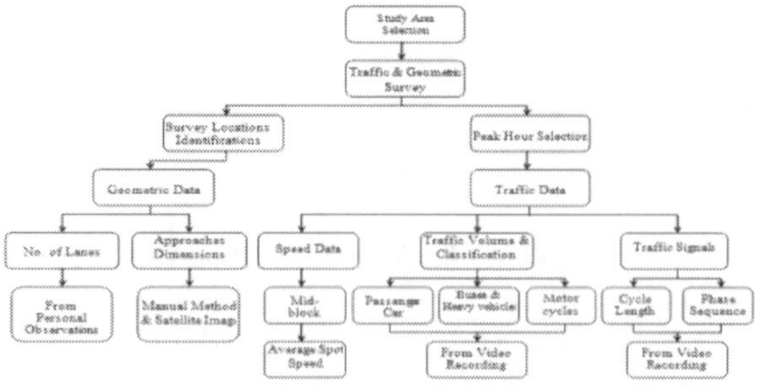


**Fig.2.** Traffic and Geometric Survey Program

### 3.1 Geometric data

Geometric data have been collected by using GIS tools in map measurements depending on the available satellite photographs with an accuracy of 0.6m, updated to 2007 and available at Karbala municipality, like intersection spacing, right of way. Field measurements are conducted for additional data like geometric characteristics of approaches, number of lanes per approach at each intersection, lane width, and splitter island width for each intersection. In addition, some of the geometric features for roundabouts like entry and exit radius, island diameter, circulating width, and so on are drawn to scale and measured, and other features that cannot be measured in the field are also obtained, such as entry angle and effective flare length. Field survey is used to obtain the geometric features that cannot be drawn from satellite image due to unavailability of the updated one. The main geometric features of the selected intersections are shown in Table (1).

**Table 1.** Geometric Features for The Selected Intersections

| Nodes | Approach | Entry Width (We) | No. of Entry lanes ($n_e$) | Exit Width | Entry Radius (Re) | Entry Angle | Effective Flare Length (l') | Circulating Width (Wc) | No. of Circulating Lanes ($n_c$) | Splitter Island Width | Inscribed Circle Diameter (ICD) | Central Island Diameter (ID) |
|---|---|---|---|---|---|---|---|---|---|---|---|---|
| | | (m) | No. | (m) | (m) | (Ø) | (m) | (m) | No. | (m) | (m) | (m) |
| 1* | E/W | 14 | 4/3 | 10.5 | 17 | --- | --- | --- | --- | 3 | --- | --- |
| | N/S | 12.3 | 3/3 | 14 | 10 | --- | --- | --- | --- | 3.5 | --- | --- |
| 2** | E/W | 18.6 | 4/4 | 22.6 | 28.8 | 36 | 18 | 20 | 4 | 8.3 | 67 | 27 |
| | N/S | 12.9 | 2/0 | 11.7 | 20.3 | 36 | 9 | 20 | 4 | 5.3 | 67 | 27 |
| 3** | E/W | 18.6 | 4/4 | 23 | 28.8 | 36 | 18 | 20 | 4 | 8.3 | 67 | 27 |
| | N/S | 12.9 | 2/3 | 11.7 | 20.3 | 36 | 9 | 20 | 4 | 5.3 | 67 | 27 |
| 4** | E/W | 16.1 | 4/4 | 17.9 | 50 | 56 | 17.4 | 17 | 4 | 5.8 | 102 | 68 |
| | N/S | 16.1 | 3/3 | 18.4 | 55 | 56 | 18.2 | 17 | 4 | 5.3 | 102 | 68 |
| 5* | E/W | 9 | 3/3 | 12.3 | 14.5 | --- | --- | --- | --- | 12.9 | --- | --- |
| | N/S | 12.5 | 3/3 | 15.5 | 58.5 | --- | --- | --- | --- | 17 | --- | --- |
| 6** | E/W | 11 | 2/2 | 10 | 30 | 30 | 0 | 16 | 2 | 11.7 | 53 | 21 |
| | N/S | 8.8 | 2/2 | 14.5 | 36 | 30 | 0 | 16 | 2 | 11 | 53 | 21 |
| 7* | E/W | 12.5 | 4/3 | 12.8 | 24.5 | --- | --- | --- | --- | 7.9 | --- | --- |
| | N/S | 11 | 3/3 | 12.8 | 16 | --- | --- | --- | --- | 4.8 | --- | --- |

* Signalized Intersection
** Roundabouts

https://doi.org/10.1051/matecconf/201816201032

## 3.2 Traffic data

The traffic flow data are recorded by using a video camera. Data are recorded on 8mm video camera tapes and later copied onto solid disk. The study area is characterized by a morning peak period of (7:00-9:00 A.M.) and evening peak period of (5:00-7:00 P.M.) for typical weekdays (Monday, Tuesday, and Wednesday) and Thursday.

The abstraction process of the required data is carried out with the aid of EVENT software, which turns the computer into a data capturing device and provides a digital representation of the observed data. Data abstraction was based on sessions of 15 minute periods of recorded data.

After data abstraction stage, data processing is made with the aid of a spreadsheet program. Excel program is used in abstracting data files, obtained from EVENT program. These programs are used to measure the main traffic parameters that are illustrated in Table (2).

**Table 2.** A Summary of Abstracted Data

| Major Category | Data Type |
|---|---|
| Traffic volume data | • Through and turning traffic volume counts<br>• Vehicle classification |
| Driver performance characteristics data | • Headways and saturation flow rate |
| Signal timing control | • Cycle length<br>• Phase sequence |

The output files from EVENT program are opened with the aid of EXCEL program. This program has an option which converts text data to columns. This method separates the recording time into a column and the associated character in the next column. This arrangement makes the dealing with the data easier by sorting them into digits and characters.

For traffic volume data calculation, Pivot table was used to calculate the number of each character presented in the selected column. Other time-related parameters (saturation headway, phase and cycle length) were calculated by arranging EVENT output file in a set of columns according to their characters. Then, subtracting of one digit from another to find the value (time) for the subjected parameters. The traffic data that have been collected are as follows:

### *3.2.1 Traffic Volume Data*

This data includes counting the traffic volumes abstracted from video recording for each approach at the intersections in the selected network, also, traffic composition, and the volume of turning movements. The network average traffic volume (veh/hr) is shown in Figure (3) for typical weekdays.

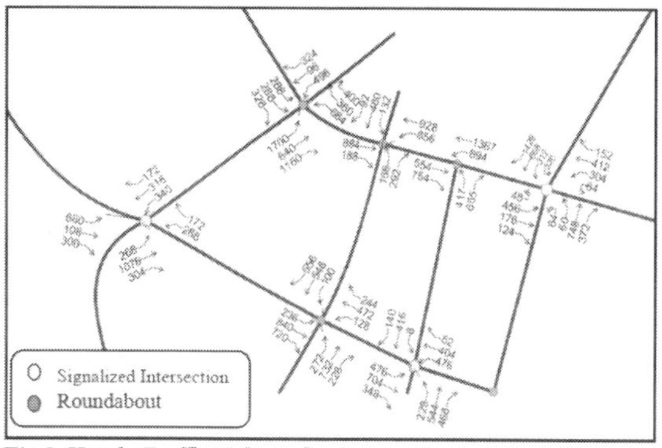


**Fig.3.** Hourly Traffic Volume for Average Typical Weekdays

Roundabouts turning movement estimates were derived from collected traffic volume data as recommended by the Federal Highway Administration [4]. The FHWA procedure requires the entering, exiting, circulating, and right turns for each leg of the traffic circle, as in Figure (4).Traffic passing 'through' the roundabout and 'left turning' vehicles were determined from Equations (1), and (2), respectively, and the definition of each term is shown graphically in Figure (4) [4,5].

$$V_{EB}, through = V_{EB}, entry + V_{WB}, Exit - V_{EB}, right - V_{NB}, right - V_{NB}, circ. \quad (1)$$

$$V_{EB}, left = V_{EB}, entry - V_{EB}, throught - V_{EB}, right \quad (2)$$

where:

*$V_{EB}$, through*: traffic volume at the east-bound (West) approach, moving to through direction.

*$V_{EB}$, entry*: traffic volume at the east-bound (West) approach, entering the roundabout.

*$V_{WB}$, exit*: traffic volume at the west-bound (East) approach, exiting the roundabout.

*$V_{EB}$, right*: traffic volume at the east-bound (West) approach, moving to the right direction.

*$V_{NB}$, right*: traffic volume at the north-bound (South) approach, moving to the right direction.

*$V_{NB}$, circ.*: traffic volume upstream to the north-bound (South) approach, moving in the circulatory roadway.

*$V_{EB}$, left*: traffic volume at the east-bound (West) approach, moving to left direction.

* Corresponding author: israa_hwy@yahoo.com,
israasaeed@uomustansiriyah.edu.iq

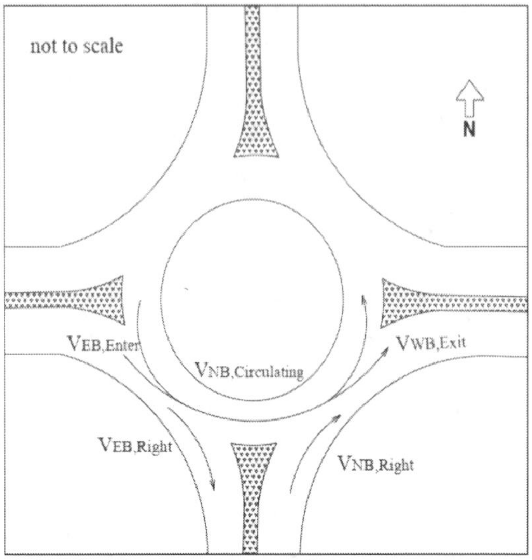


**Fig.4.** Turning Movements per Approach

This numerical method (Algebraic Solution Procedure) was also checked with the method mentioned in [5]. This method named live recording of turning movement patterns using field observer. This is only feasible under low traffic volume conditions where the entire roundabout is visible from one location.

### *3.2.2 Forecasting Traffic Volume Data*

The annual rate of traffic increase in Karbala City is estimated to be (5%) of growth as stated by the [2].

This value is used as an input value to SYNCHRO program in volume tab from network setting. The printout of this stage is illustrated in Figure (5) for forecasted traffic volume.

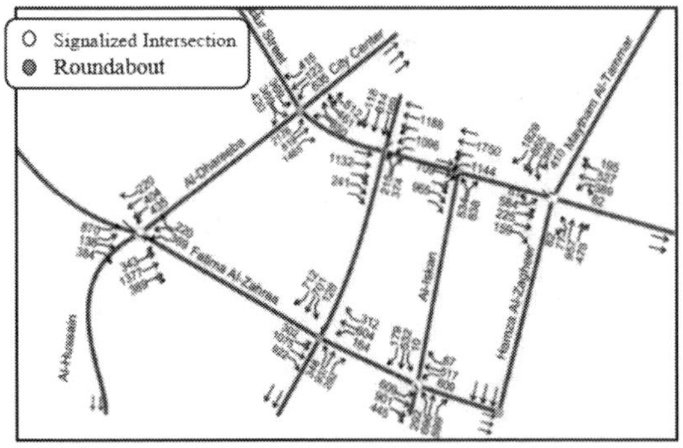


**Fig. 5.** Forecasted Traffic Volume to the Target Year Using SYNCHRO

### *3.2.3 Speed Data*

These data include the measuring of travel times for each internal link within the selected network. The travel time can be converted to the speed as illustrated in Figure (6).

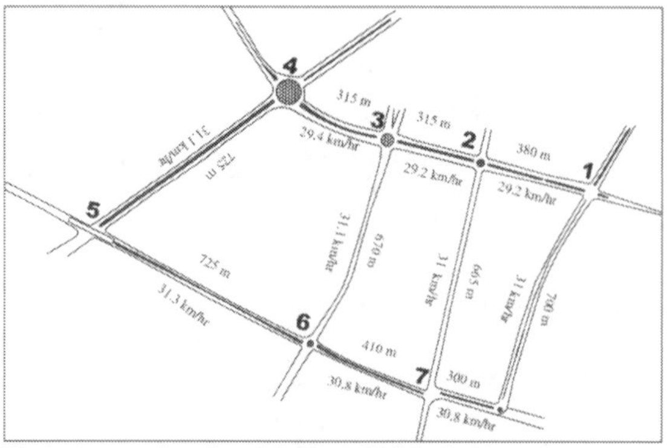


**Fig.6.** Average Spot Speed for Each Link in the Network

### *3.2.4 Traffic Signal Data*

The traffic signals data include the measuring of existing cycle time length, phase length, start up lost time, extension of effective green, and phase sequence.

Cycle length and length of phases have been extracted from video films for the subjected intersections and shown in Table (3). While, the values of both lost time types and extension of effective green are taken as the default values recommended by SYNCHRO and SIDRA software programs, as given in Table (4).

**Table 3.** Phase Order and Cycle Length for Signalized Intersections

| Intersection | Phase Length (sec) | | | | Cycle Length (sec) |
|---|---|---|---|---|---|
| | ↰↑ | ⇄ | ↲↓ | ⇆ | |
| 1 | 25 | 24 | 41 | 23 | 121 |
| 5 | 27 | 21 | 20 | 15 | 91 |
| 7 | 24 | 38 | 14 | 31 | 115 |

**Table 4.** Lost Time and Extension of Effective Green Time for the Selected Software

| *Software* | *Lost time* | *Extension of Effective Green Time* |
|---|---|---|
| **SYNCHRO** | 2.5 sec | 2.5 sec |
| **SIDRA** | 2.0 sec | 2.0 sec |

## 4 Evaluation of existing traffic flow

To investigate the existing conditions of traffic operation schemes and geometry on traffic flow performance, simulation of the actual movement is required. These have been done by applying two software programs as follows:

### 4.1 Application of SIDRA software program

The abstracted and collected data required for this software were fed into the program for each intersection alone. Many runs were implemented to exclude the bias data.

SIDRA INTERSECTION software was used to analyze the existing traffic flow patterns for intersections as isolated at the study area.

Table (5) indicates the output results of simulation runs which include the degree of saturation, total delay, and level of service for each intersection in the study area. All the output results are categorized by medium to high total delays. Thus, the level of service for most of the intersections is (F) according to the LOS categories of the software that based on HCM method.

It is worthwhile to mention that some geometric improvements have been recently made to Al-Tarbia Street. The capacity of this street increased and the level of service (LOS) of the intersections (2&3) are (A and B). But, intersections (1&4) at the beginning and end of this street still suffer from high congestion and low traffic performance characterized by high delay values.

**Table 5.** Total Delay and Level of Service for the Existing Traffic Flow for Isolated Intersections Produced by SIDRA

| *Intersection* | *Max. Degree of Saturation (v/c)* | *Average Delay (sec / veh)* | *LOS* |
|---|---|---|---|
| 1* | 1.89 | 253.6 | F |
| 2** | 0.75 | 9.4 | A |
| 3** | 0.56 | 10.1 | B |
| 4** | 2.80 | 284.8 | F |
| 5* | 1.31 | 97 | F |
| 6** | 1. 57 | 140.5 | F |
| 7* | 1.13 | 76.9 | E |

* Signalized Intersections
** Roundabout

### 4.2 Application of SYNCHRO/Sim traffic software

SYNCHRO/SimTraffic was also applied to simulate the existing traffic flow for all the intersections at the selected network.

Although, SYNCHRO program v.6 is the first version which can analyze multi-lane roundabout, but, it is based on HCM method for roundabout analysis, so only the degree of saturation (v/c) can be obtained from this program, while, SimTraffic have the ability to simulate and analyze both the signalized and unsignalized (roundabout) intersections.

SYNCHRO and SimTraffic have similar delay results for (v/c) less than 0.85, as v/c increases, different results are obtained for the delay. This software has been applied for the selected intersections and the and the analysis results are presented in Table (6).

**Table 6.** Total Delay and LOS for Isolated Intersections Produced by SYNCHRO/Simtraffic for Existing Condition

| | *SYNCHRO* | | | *SimTraffic* |
|---|---|---|---|---|
| **Intersecti on** | **Max. Degree of Saturation (v/c)** | **Average Delay(sec/ veh)** | **LOS** | **Average Delay(sec/ veh)** |
| 1* | 2.08 | 294.3 | F | 160.7 |
| 2** | 1.00 | N/A | N/A | 57.1 |
| 3** | 2.75 | N/A | N/A | 224.8 |
| 4** | 8.80 | N/A | N/A | 408.9 |
| 5* | 1.29 | 109.8 | F | 67.5 |
| 6** | 2.90 | N/A | N/A | 1450.2 |
| 7* | 1.35 | 105.5 | F | 73.3 |

* Signalized Intersections
** Roundabout

The results in Table (6) show that most intersections suffer from an oversaturation condition with high total delay values. SYNCHRO gives unreasonable results for high traffic volume, and large dimension roundabout (intersection 4), While, simTraffic gives unreasonable results at roundabout (intersection 6).

## 5 Improvements

SYNCHRO has the ability to optimize and coordinate the network with signalized intersections. While, SIDRA enables the user to analyze, and evaluates roundabouts with different geometries, so, it will be used for roundabouts improvement.

At the improvement process, the degree of saturation (v/c) was used as an indicator for diagnosing the oversaturated cases. Intersections having v/c ratios equal or greater than 0.9 need to be improved at the first stage.

This section divides the improvement strategies into two parts according to method and software applied. The first part is for signalized intersections (1, 5, and 7). The improvement type is suggested and evaluated by SYNCHRO program. The second part is for all the oversaturated intersections in the studied area. The signalized intersections that cannot be solved by the first part and the roundabouts (4, 6) are involved in the second part. The improvement type is suggested by the researcher and evaluated by SYNCHRO program for signalized intersections, and SIDRA for roundabouts.

### 5.1 Signal timing optimization and coordination

Traffic signal timing optimization is one of the most cost-effective methods for reducing vehicle operating costs and improving traffic flow performance along urban arterials. Arterial signal optimization models are developed to assist traffic engineers in coordinating traffic signal settings along urban arterials and across networks [6].

SYNCHRO contains a number of optimization types. It optimizes cycle length, split times, offsets and phase sequence to minimize driver stops and delay times. These types are applied to the network in the same order shown in Figure (7) [6].

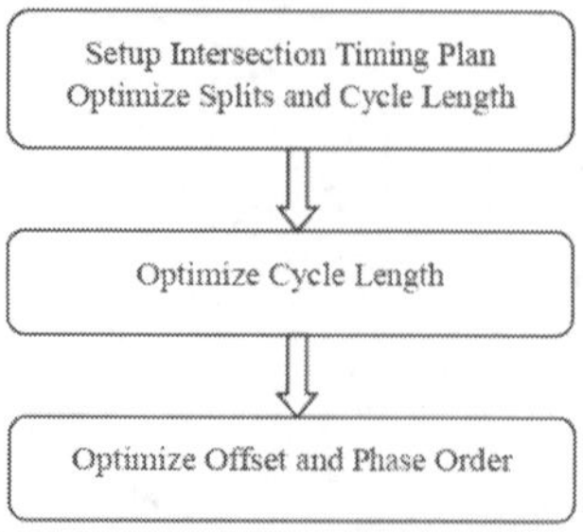


**Fig. 7.** Optimization Steps in SYNCHRO

The range of cycle length recommended by HCM is 60-120 seconds. Long cycle length exceeding 100 seconds will be recommended to accommodate an extra 10% capacity. Long cycle lengths more than 120 seconds may have negative operation aspects including long queues, inefficient use of turning lanes, and blocking. Low cycle lengths may be better to reduce queues which in general increase capacity and provide smoother traffic operation [7]. The optimization results with the HCM recommended cycle range do not result in optimum cycle length, so the default range of SYNCHRO was used.

To calculate the best cycle time and the optimal timing of the existing phase sequences of the observed intersections, SYNCHRO was used. The range of cycle times used in the optimization process was 50-200 seconds, as default values by SYNCHRO. This range was used for each observed isolated intersection. Table (7) provides a summary of the optimization results. This is together with the best cycle time for each intersection in the observed network.

This Table shows that the use of optimization functions for the isolated signalized intersections such as split and offset has a minimal effect on the MOE for intersections (1 and 7) and has no effect at the intersection (5).

The optimization results showed that although the cycle time increased to non-reasonable values, the isolated intersections remain performing under oversaturated conditions, so, it is necessary to improve the performance by coordination.

One of the most common methods for increasing the efficiency of traffic operation is coordination of traffic signals. This is because; the higher increase in the traffic demand will result in a higher increase in the need for coordination. Traffic signal coordination can reduce the traffic congestion in many areas. Substantial improvements in traffic flow and reduction in delay, fuel consumption and stops could be achieved by coordination.

After the coordination process for signalized intersections in the network, the delay times were reduced by 2.8 % and 33.6% for intersections (1) and (5) respectively, while, a reduction of 52.3% is attained at the intersection (7).

Although the optimization and coordination runs improve the network MOE. But, the degree of saturation does not been reduced to a value less than 0.9.

**Table 7.** Measure of Effectiveness for the Improved Intersections with the Optimum Cycle Length Produced by SYNCHRO

| *Intersection* | *Improvement Type* | *Cycle Length* | *Max v/c* | *Total Delay* | | *Total Uniform Stop* | | *Total Fuel Consumption* | | *Un-served vehicles* |
|---|---|---|---|---|---|---|---|---|---|---|
| | | sec | | veh hr/hr | sec/veh | veh-hr/hr | % | Li | km/L | No. |
| 1 | Base Condition | 121 | 1.82 | 213 | 337 | 3937 | 0.69 | 1417 | 2.9 | 1368 |
| | Split | 121 | 1.78 | 215 | 341 | 3935 | 0.69 | 1427 | 2.9 | 1368 |
| | Cycle length | 150 | 1.7 | 207 | 328 | 3986 | 0.70 | 1390 | 3 | 1281 |
| | Network cycle length* | 150 | 1.7 | 207 | 328 | 3986 | 0.70 | 1390 | 3 | 1281 |
| | Phasing Sequence | 150 | 1.93 | 202 | 320 | 4131 | 0.72 | 1374 | 3 | 1171 |
| | Percent Saving | | | 2.8 | 2.7 | -1.2 | -1.4 | 1.9 | -3.4 | 6.4 |
| 5 | Base Condition | 91 | 1.28 | 110 | 123 | 2860 | 0.71 | 724 | 4.7 | 520 |
| | Split | 91 | 1.28 | 110 | 123 | 2860 | 0.71 | 724 | 4.7 | 520 |
| | Cycle length | 138 | 1.11 | 74 | 82 | 3069 | 0.76 | 615 | 5.5 | 105 |
| | Network cycle length | 112 | 0.97 | 73 | 82 | 3056 | 0.76 | 615 | 5.5 | 98 |
| | Phasing Sequence* | 73 | 0.75 | 16 | 14.5 | 2352 | 0.58 | 418 | 8.1 | 0 |
| | Percent Saving | | | 85.5 | 88.2 | 18.3 | 15.2 | 42.3 | -72 | 100 |
| 7 | Base Condition | 115 | 1.43 | 153 | 181 | 3043 | 0.71 | 763 | 2.7 | 677 |
| | Split | 115 | 1.37 | 166 | 197 | 2988 | 0.70 | 8.5 | 2.5 | 520 |
| | Cycle length | 150 | 1.31 | 157 | 186 | 3134 | 0.73 | 778 | 2.6 | 657 |
| | Network cycle length* | 150 | 1.31 | 73 | 82 | 3065 | 0.76 | 615 | 5.5 | 98 |
| | Phasing Sequence | 150 | 1.16 | 208 | 246 | 3320 | 0.78 | 949 | 2.1 | 209 |
| | Percent Saving | | | 52.3 | 54.7 | -0.7 | -7.0 | 19.4 | -103.7 | 85.5 |

* Best improvement type
The percent saving is measured for the best improvement type.

Another trial was made by changing the number of phases and the phasing sequence. The phasing order in Iraqi`s four-leg intersection is four split phases for through and left traffic with permitted right turn and yields only for pedestrians. The leading and lagging and any other protected phase are not used in this system, so the driver is unfamiliar with such type of movement. It may cause problems with head-on collisions between oncoming left turns. Due to these reasons, it will not be applied in this study.

At the intersection (5), due to the presence of overpass for through traffic at the east-west approach, changing the phase sequence with the lead/lag option is applied. The result shows 85% delay saving and 100% vehicles in the intersections are served.

## 5.2 Geometric Improvement

Since the coordination process does not satisfy the aforementioned objectives, so it is necessary to improve the geometries of the oversaturated approaches at intersections (1, 4, 6, and 7).

The improvement strategies will be as follows:

1. Re-marking of pavement.
2. Pavement widening.
3. Change the intersection type.

The MOE produced for each strategy is illustrated in Table (8) for signalized intersections analyzed by SYNCHRO, and Table (9) for roundabouts analyzed by SIDRA.

The first step of the geometric improvement is to apply pavement remarking. From the measurement, it can be noticed that the number of lanes in many approaches of these intersections can be increased by decreasing the lane width (with keeping the width of lanes greater than a minimum value of 2.4 m) according to HCM which is actually observed to be used in the field.

Increasing the number of lanes will increase the saturation capacity of the affected approaches, which leads to improving traffic performance MOE for some intersections at the selected network. Therefore, in this step consideration was given to remark the approaching lanes so that rearranging of queuing vehicles at stop line is obtained.

Also, separating the movement with high traffic volume by exclusive lane is also applied. Optimization run after each improvement type is applied for the signalized intersections; results are shown in Table (8).

**Table 8.** Measure of Effectiveness for the Improved Intersections with the Optimum Cycle Length Produced by SYNCHRO

| *Intersection* | *Improvement Typ* | *Cycle Leng* | *Max v/c* | *Total Dela* | | *Total Uniform St* | | *Total Fuel Consumptio* | | *Un-served vehicles* |
|---|---|---|---|---|---|---|---|---|---|---|
| | | sec | | veh hr/hr | sec/veh | Veh-hr/hr | % | Li | Km/L | No. |
| 1 | Base Condition | 121 | 1.82 | 213 | 337 | 3937 | 0.69 | 1417 | 2.9 | 1368 |
| | Pavement marking | 124 | 1.09 | 78 | 124 | 4084 | 0.72 | 830 | 5 | 276 |
| | Prohibition of U-turn | 100 | 1.04 | 51 | 74 | 3528 | 0.68 | 644 | 5.9 | 44 |
| | Pavement Widening | 104 | 0.98 | 41 | 59 | 3592 | 0.69 | 603 | 6.3 | 0 |
| | Overpass | 102 | 0.83 | 26 | 27 | 1948 | 0.51 | 354 | 7 | 0 |
| | Roundabout* | --- | 0.66 | 12 | 11 | 3779 | --- | 265 | --- | --- |
| | Percent Saving | | | 94.4 | 96.7 | 4.01 | 26.1 | 81.3 | -141 | 100 |
| 7 | Base Condition | 115 | 1.43 | 153 | 181 | 3043 | 0.71 | 763 | 2.7 | 677 |
| | Pavement Remarking | 104 | 0.87 | 40 | 47 | 3251 | 0.76 | 379 | 5.1 | 0 |
| | Roundabout* | --- | 0.47 | 9 | 7 | 3677 | --- | 267 | --- | --- |
| | Percent Saving | | | 73.9 | 74.0 | -6.8 | -7.0 | 50.3 | -88.9 | 100 |

* Analyzed by SIDRA program

Even the pavement remarking is applied to each approach of Al-Hakeem Intersection (intersection 1). This method improves the MOE but not to the desired level.

A suggestion of prohibition of U-turning vehicles at the intersection area by providing U-turn opening upstream for all approaches is applied. This reduces the best cycle time to 100 seconds, and reduces all the measure of effectiveness but not to the desired level.

Another trial was made by widening the east approach due to the presence of wide sidewalk at this section taking into consideration pedestrian volume. Also, shifting of the central island to meet the design criteria is applied at this approach as shown in Figure (8). The obtained results do not reduce the v/c to value less than 0.9. This supports the suggestion of Updated Karbala Master Plan [1] to adopt an overpass at north-south direction. This will reduce v/c to 0.83.

The delay times at the west approach does not reach the desired level, especially for left turning movement at the mentioned intersection, so another suggestion is made by changing the intersection type to a roundabout. Since roundabouts serve left turning movement. Results obtained by SIDRA and illustrated in Table (8) show an increase in all MOE and reduce v/c to a maximum value of 0.66 for the north approach.

The traffic problem at Al-Sayed Jawda Intersection (intersection 7) is solved by dividing the approach width into four lanes; each lane has its exclusive movement.

The traffic lanes that have low traffic volume are assigned to be shared. After the optimization process is applied, the optimum cycle length is set to be 104 sec and v/c is 0.87. So the intersection is in under-saturated condition.

Changing the four-leg signalized intersection to roundabout intersection with circulatory width equals to 14m and island diameter of 30m reduce v/c drastically to 0.47, and the average delay to 7 secs, as seen in Table (8). Figure (9) shows the intersection before and after geometric improvement.

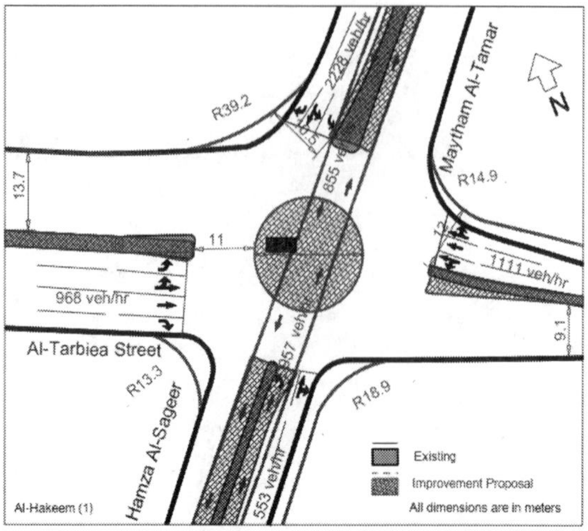


**Fig. 8.** Improvement Proposal for Al-Hakeem Intersection

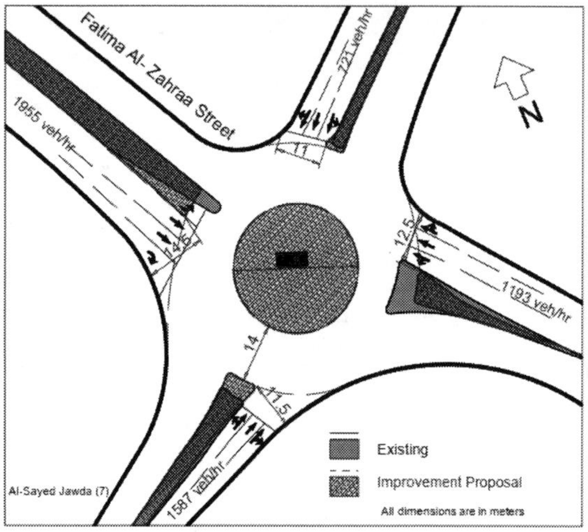


**Fig.9.** Improvement Proposal for Al-Sayed Jawda Intersection

Al-Zahraa square (intersection4) suffers from a high degree of congestion, especially from the East approach. Remarking of the pavement alleviate the P.I by about 75%, but it is still under the oversaturated condition, and needs further modification to achieve the objected delay and performance index.

Since the width of circulating roadway has a great effect on capacity, thereby delay [8]. The circulatory width is suggested to be increased by trimming the central island diameter by 8m. In addition to the adjusting splitter islands shape to provide an extra pavement area for entering and exit approaches, as shown in Figure (10). The results in Table (9) show the degree of saturation is still over the desired limit.

Installing metering signal at West approach controlled by the demand for East approach is proposed. This is to create gaps in the circulating flow that assists the controlling approach traffic (East approach) to enter the roundabout.

The use of metering signals is a cost-effective measure to avoid the need for a fully-signalized intersection treatment [9,10,11,5]. Roundabout metering signals are installed on selected roundabout approaches and used on a part time basis since they are required only when heavy demand conditions occur, during peak periods [10,11].

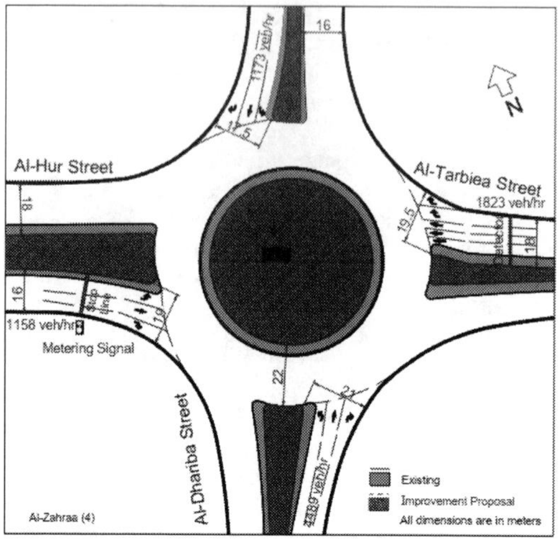


**Fig.10.** Improvement Proposal for Al-Zahraa Square

The use of metering signal can provide beneficial operational benefits during these peak periods. In some cases, metering may be a more cost-effective solution as compared to the geometric improvements, particularly if the traffic condition requiring metering is of a short duration. A basic metering system consists of two components [5]:

A queue detector, located on the downstream entry, is experiencing excessive delays and queues. The queue detector should be placed relatively far back on the downstream entry to detect when there is a long queue. When a long queue is detected, the signal controller activates the metering signal.

A metering signal on the dominant approach, preferably set far back enough from the entry to minimize misunderstanding with the yield sign. If the metering signal cannot be set back adequately, some countries (e.g., Australia) use a special changeable message sign that exhibits a yield sign but can be changed to read "Stop on Red Signal".

**Table 9.** MOE for the Improved Intersections Produced by SIDRA

| *Intersection* | *Improvement Type* | *Max v/c* | *Total Delay* | | *Total Uniform Stop* | *Total Fuel Consumption* | *Performance Index* |
|---|---|---|---|---|---|---|---|
| | | | *veh – hr/hr* | *sec/veh* | *veh /hr* | *Li/hr* | *P.I* |
| 4 | Base Condition | 2.8 | 801 | 485 | 34774 | 1970 | 1482 |
| | Pavement Re-marking | 1.57 | 159 | 77.3 | 10974 | 727 | 456.4 |
| | Widening Wc | 1.39 | 144 | 69.6 | 16370 | 701 | 475.7 |
| | Widening We | 1.13 | 98 | 59.8 | 14356 | 687 | 456.7 |
| | Metering Signal | 0.99 | 50 | 24.3 | 11602 | 556.1 | 325.9 |
| | Percent Saving | | 93.7 | 94.9 | 66.6 | 71.7 | 78.0 |
| 6 | Base Condition | 1.57 | 228 | 140.5 | 20621 | 722 | 591.2 |
| | Pavement Re-marking | 1.78 | 155.5 | 95.8 | 13214 | 575.9 | 404.6 |
| | Widening We | 1.03 | 41.7 | 25.7 | 8640 | 406.3 | 216.5 |
| | Widening Wc | 1.03 | 40.9 | 25.2 | 8550 | 404.9 | 214.3 |
| | Curb-shifting | 0.89 | 24.0 | 14.8 | 7161 | 374.9 | 174.2 |
| | Percent Saving | | 89.5 | 89.5 | 65.3 | 48.1 | 70.5 |

Pavement remarking at Al-Safeena intersection (intersection6) results in lower delay times, stops, and P.I. this with an increased value of v/c. So the widening of the South approach is necessary to add an extra lane. For the North approach, adding 2.5 m to the circulating width to be 12m was applied, as shown in Figure (11).

The entry radius for the East, South, and West approaches are increased. The increase in entry radius is recommended to increase flare length, and entry width and consequently increase the approach capacity.

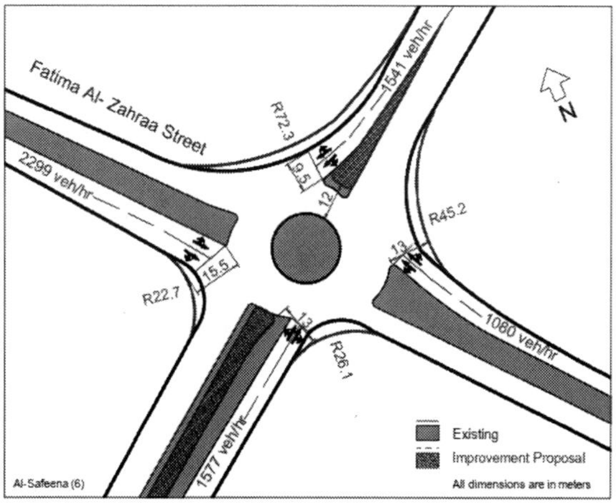


**Fig.11.** Improvement Proposal for Al-Safeena Intersection

## 6 Forecasted network

The growth factor (i.e. traffic forecast factor TFF) was introduced for the application of forecasting traffic volume on the improvement proposals, for the signalized intersection. The growth rate together with the analysis period is applied with SIDRA for the application of forecasting traffic volume on the improvement proposals, for roundabouts (2, 3, 6, and 7), and roundabouts with metering signals (4).

Table (10) shows the MOE obtained from SYNCHRO and SIDRA optimization runs after the best improvement strategy for the target year.

**Table 10.** MOE Produced by SYNCHRO & SIDRA after Best Improvement Strategy for the Base Year, and Target Year

| *Intersection* | *Base Year* | | *Target Year* | |
|---|---|---|---|---|
| | **Average Delay (sec/veh)** | **LOS** | **Average Delay (sec/veh)** | **LOS** |
| **1**** | 11.0 | B | 18.6 | B |
| **2**** | 9.4 | A | 13 | B |
| **3**** | 10.1 | B | 18.4 | B |
| **4***** | 24.3 | C | 80 | F |
| **5*** | 14.5 | B | 25.9 | C |
| **6**** | 14.8 | B | 36.4 | D |
| **7**** | 7.3 | A | 14.2 | B |

* signalized intersection
** Roundabout
*** Roundabout with metering signals at peak periods

This Table indicates that five of the intersections will perform well till five years later, except Al-Zahraa square (intersection 4). Opening the North approach for traffic at the intersection (3) may reduce the right turning vehicles of the East approach by 50%. This may reduce the future delay. Also installing the metering signal at the West approach, and leading the controlling approach in the South approach instead of the East. This will reduce the average future delay time to 29.6, and LOS C.

## 7 Conclusions

The main conclusion that can be drawn from this research is summarized as follows:

1) The optimization and coordination process slightly improve the selected network. This may be attributed to the presence of roundabouts between the signalized intersections that may disturb the platoon movements.
2) Pavement remarking and lane designation improves the intersections performance by an average of 50%.
3) Pavement widening, in addition to pavement marking, improve MOE of the intersections by an average of 66%.
4) Converting of closely spaced intersections to roundabouts, as in intersections (1, and 6), provides lower speed without stopping which is necessary to improve safety, especially in urban area. In addition to

the aesthetic view.

The authors would like to appreciate Dr. Raid R. Al-Muhanna at Karbala University due to his great support during the data collection stage, and providing necessary information for the research.

## References

[1] المركز العالمي للأبحاث الفنية ICTR " *تحديث التصميم الأساسي لمدينة كربلاء المقدسة والحر، تقرير مرحلة إعداد البدائل وتقويمها* "، وزارة البلديات و الأشغال العامة، المديرية العامة للتخطيط العمراني،تشرين الثاني (2007 ) .

[2] مكتب الاستشارات الهندسية, كلية الهندسة/ جامعة بغداد ، " *مشروع الطريق الحولي لمدينة كربلاء المقدسة وربط المدينة بطريق المرور السريع رقم (1)* "، المرحلة الأولى الدراسة الأولية، وزارة الإعمار والإسكان، الهيئة العامة للطرق والجسور، أيار، (2007).

[3] المكتب الاستشاري الهندسي, كلية الهندسة/ جامعة الكوفة ، " *الدراسة الشاملة لتخطيط النقل الحضري لمدينة كربلاء المقدسة* "، التقرير الفني رقم (3)، الاستطلاع الميداني،( 2002).

[4] FHWA Federal Highway Administration, "*Roundabouts: An Informational Guide*", US Department of Transportation, FHWA RD-00-067, June (2000).

[5] NCHRP National Cooperative Highway Research Program, "*Roundabouts: an Informational Guide*", NCHRP Report 672, Transportation Research Board of the National Academies, Second Edition, Washington D.C. (2010).

[6] D. Husch, J. Albeck, "*Traffic Signal-User Guide, Synchro Studio7*", (2006).

[7] TRB, Transportation Research Board," *Highway Capacity Manual*", Washington D.C. Updated (2005).

[8] G.G. Sofia, A.H. Alhaddad, and I.S. Al-haydari, JEAD, **17**, 1, March (2012).

[9] R. Akçelik, the 2nd Urban Street Symposium, USA, July (2003).

[10] R. Akçelik, the ITE 2009 Annual Meeting, USA, August (2009).

[11] R. Akçelik, Akçelik & Associates Pty Ltd, December (2010).

# Assessment of Al-Amarah street within the Al-kut city using pavement condition index (PCI) and GIS technique

*Mohammed* Al-Neami[1,*], *Rasha* Al-Rubaee[1], *and Zainab* Kareem[1]

[1]Building and Construction Engineering Department, University of Technology, Baghdad, Iraq

**Abstract.** The capabilities of geographical system and their spatial analysis is considered the most appropriate tools to enhance pavement management operations, with features such as graphical display of pavement condition. In Iraq, most of transportation agencies do not have a tool that is used as a database for road deteriorations, so there is a need for road surveying and storing the collected information in GIS to know the condition of every road with details. Furthermore, these data can be used for maintenance process and estimation of prior cost. This research has been carried out to estimate of flexible pavement condition through visual surveys using the Pavement Condition Index (PCI) method; so it can provide an easy way to calculate the PCI based on GIS data with Micro PAVER software 5.2. Al-Amarah Street, which is internal road in Al-Kut city in the eastern part of Iraq, is used as a case study. The average pavement condition index of the selected case study is found to be "64" using Micro PAVER 5.2 software which mean "Fair" pavement condition. Arc Map 9.3 has been applied in this study to make an integrated maintenance system for each road in the region demonstrating the annual road deteriorations and the resulting change in the PCI values which occurs every year. The study provides an easy and simplified way of presentation the details of deteriorations on the satellite or the geographical map of the road in which each type of distress has been symbolized with specific sign and each PCI value has been represented with specific color.

## 1 Objective of the research

The aim of this research is to produce a thematic map for distress types in the city network with their completely information about severity, quantity, (x,y) coordinates for each distress type. Taking advantage of the ability of GIS tools to store the data and display it in any time needed; it is useful to use GIS as a database to make a decision for maintenance and then to achieve best results at lowest possible costs.

In this research, visual surveys are used to determine the Pavement Condition Index (PCI) of the selected roads according to ASTM D6433-07, 2016 [1].

Many road institutions in Iraq suffer from lack in road documentation and do not have a database program for storing; this reflects a poor system of maintenance. Therefore, the use of GIS software can effectively solve that.

## 2 Background

The functions of highway facilities are to provide travel quickly, safely, and comfortably. The lack of any of these characteristics indicates functional deficiencies therefore; the causes of the functional deficiency are measured to assess the pavement performance properly (Yoder, and Witczak, 1975) [2].

Evaluation of surface distress is the one that has historically been characterized by a lack of uniformity in data collection practices, since there are currently no standards accepted by the entire transportation community (Haas et al., 1994; Flintsch and McGhee, 2009) [3 and 4].

As distress increases in amount and severity, roughness, and safety hazards increase (Metal, 1980) [5]. Deterioration is a result of complex distress as pavement cracking through fatigue under repeated loadings and environmental cycles (Gary et al., 2009) [6].

To simplify the collection of distresses data, Geographic Information System (GIS) and remote sensing through Global Positioning System (GPS) are proposed by Salih (2006) [7], to be used together with a large amount of quantitative and qualitative information regarding pavement management. A PCI model through using a stepwise regression technique was developed to predict pavement condition for local flexible pavement.

In the recent past (GIS) has acquired tremendous importance in various applications; for transportation planning (GIS) is implemented to solve transportation problems (Didigwu, 2010) [8].

Ibraheem and Falih (2012) [9] pointed out to use GIS as tool for the pavement management decision making. A

* Corresponding author: Al_Neami74@yahoo.com

© The Authors, published by EDP Sciences. This is an open access article distributed under the terms of the Creative Commons Attribution License 4.0 (http://creativecommons.org/licenses/by/4.0/).

GIS based system was built, which provides information that can be used as a platform for all aspects of the Pavement Maintenance and Management System (PMMS) process. 23 sections were selected that distressed with PSI element to draw a map that connected between roadway and their values of PSI. The range of PSI for these sections was low and they suggested a treatment for each section based on the types of distresses and PSI value.
One of most modern programs in geographic information systems is ArcGIS 9.3 which is one of the families of geographic information systems that have contributed to the completion of the GIS (Liu et al., 2010) [10].
To assess pavement condition, PAVER system used by Mohamed, (2015) [11] to perform pavement condition analysis for Section R4 that is a segment from Expressway No. 1 which connects between Hilla and Baghdad city in the middle of Iraq. The PCI factor was used with other measure to estimate the main performance indicators affecting asphalt pavement maintenance alternatives.
ASTM D6433-07 (2016) [1] provides a manner to establish Pavement Condition Index (PCI) in order to state the service level for flexible and rigid roads.

### 2.1 Pavement condition survey

One of the most important functions of pavement engineering is that dealing with evaluation of in-service pavement. It is necessary to know the condition of pavement from standpoint of setting up design criteria and for establishing maintenance and priority (Yoder and Witczak, 1975) [2]. It will serve several purposes including, establishing need for including design requirements of procedures, and pointing out special conditions influencing overlay design (Finn and Monismith, 1984) [12]. The PCI is a quick method of comparing the overall condition of pavement and magnitude of rehabilitation needs. Figure (1) shows how pavement condition typically deteriorates over time Johnson, 2000) [13].

### 2.2 Micro PAVER program

Micro PAVER is developed to provide engineers with a systematic approach for determining maintenance and rehabilitation needs and priorities for pavement management (Shahin and Walther, 1990) [14]. Micro PAVER is used to manage roads, streets, parking lots, and airfield pavement. The PAVER system is based on the Pavement Condition Index (PCI) survey and rating procedure (Norlela, et al., 2009) [15].

PAVER for Windows is an automated pavement management system (PMS). It is a decision making tool for the development of cost effective maintenance and repair alternatives for roads and streets, parking lots, and airfields. PAVER provides many important capabilities (U.S Army Corps of Engineers, 2011) [16].

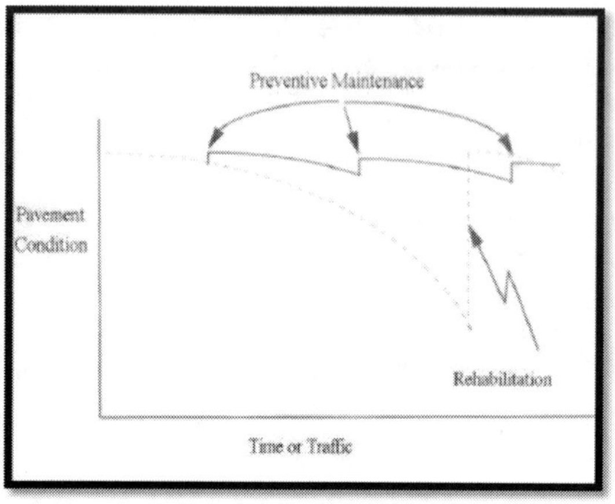


**Fig. 1.** Performance of preventive Maintenance treatments.

## 3 Methodology

### 3.1 Case study

Al-Amarah Street is a collector road in Al-Kut city that is located in the eastern part of Iraq as shown in Figure (2).This road collects from most minor streets in the city. It extends from the entire part to the end of Al-Kut city toward Al- Amarah city. It is a divided multilane with two lanes in each direction. The lane width is 3.5 m with a median of 1 m width. Many heavy vehicles use it toward Al-Amarah city, which in turn leads to deterioration of this road most of the time.

### 3.2 Map digitizing

The ArcView9.3 is full-featured Geographic Information System (GIS) software for visualizing, managing, creating, and analyzing geographic data (Haldar, 2012) [17]. Through the inspection survey, the GPS (Global Positioning System) type of (GEO-XT) was used to identify the location of each distress point in the road as shown in Figure (3).

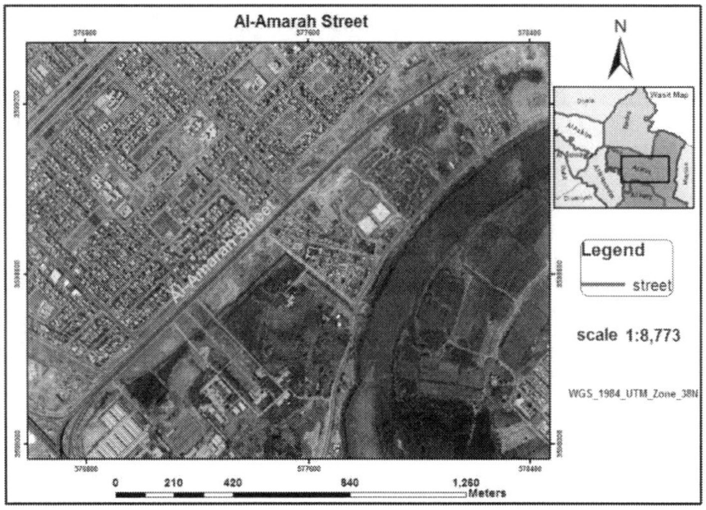


**Fig.2.** Location of Al-Amarah Street.

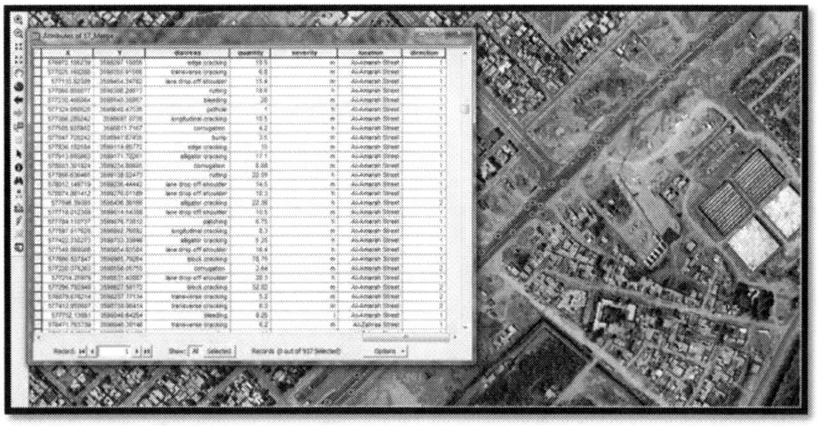

**Fig. 3.** Database of distresses shape file.

Depending on coordinates from GPS, each distress point must be interred in the GIS Software with its type, severity, quantity and location; in which these details about deformation are required to be used as a database. For more readily, the type of distress can be classified as shown in Figure (4) based on the symbol types in the legend of the map. For example, a sign of red triangle refers to alligator distress and a circle with red color refers to corrugation.

Through activated Hyperlink bottom in GIS Arc Map a feature of each distress point can be also displayed as presented in Figure (5). The features with the input data can integrate information about each deformation in the road.

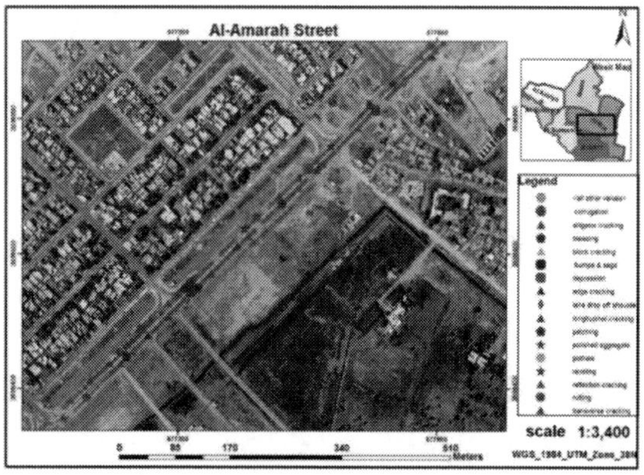


**Fig. 4.** Distress types in Al-Amarah Street.

**Fig. 5.** Connect digital image to failure exist in the roads using hyperlink.

### 3.3 PCI calculation

The following steps are needed for PCI calculation: - (Shahin, 2005) [18]

1- Determine the total number of the sample units in the pavement section (N) by dividing the section area by the sample unit area which is ($225\pm90$ $m^2$),

2- Determine the minimum number of sample units (n) that must be examined in order to estimate the PCI value for the selected road section. Equation (1) was used to obtain (n)

$$n=N.s^2/[(e^2/4)(N-1)+s^2] \quad (1)$$

Determine the sampling interval (i) by dividing the total number of the sample unit (N) by the minimum *number* of the sample units (n).
Where:

*N*= total number of sample units in the pavement section

*e* =allowable error in the estimate of the section PCI (*e* was equal to 5)

*s*= standard deviation of the PCI between sample units in the section (s=10 was used; initially inspected).

### 3.4 Calculations for the case study

The length of Al-Amarah Street is1700m and 7m width, the area of the section is 11900 $m^2$;
N=11900/231=51.5(total sample unit).

Using Eq. (1); n= 12.4 (minimum number of sample units to be surveyed)
i=N/n= 4.15≅4 (as mentioned in Shahin, 2005 [18]).

Length of sample unit to be surveyed= $\frac{231}{7}$= 33 m

That means that the sample units to be surveyed and entered in PAVER are 1, 5, 9, 13... 49.
The dividing process of the section into sample units is conducted using GIS Arc Map as shown in Figures (6) and (7). The selected sample units are with blue color as shown in Figure (7) with their lengths, widths and area of each inspected units. The points of distresses in each selected sample unit are entered in Micro PAVER software with their information as saved in GIS program and presented in Figure (8).
Each side of direction in the road is calculated and entered in Micro PAVER independently.

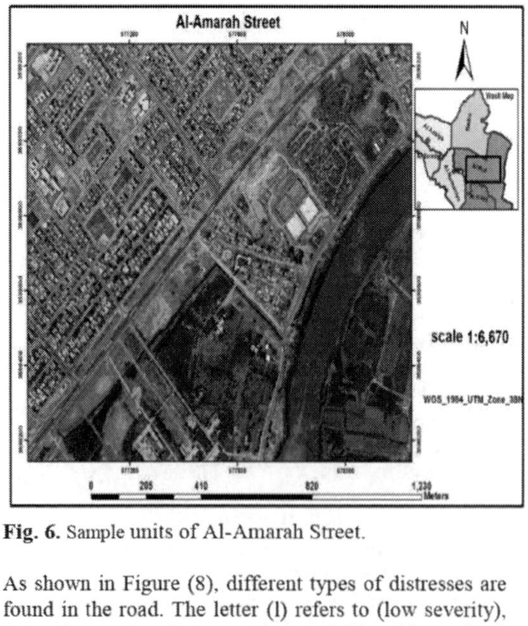


**Fig. 6.** Sample units of Al-Amarah Street.

As shown in Figure (8), different types of distresses are found in the road. The letter (l) refers to (low severity), (m) refers to (medium severity) and (h) refers to (high severity). The detection level of severity of each distress that found in the case study relied on ASTM D6433-07, 2016 [1].

 https://doi.org/10.1051/matecconf/201816201033

## 4 Analysis of results

### 4.1 PCI determination

After entering each distress point in each sample unit, Micro PAVER program gives the PCI value directly as shown in Figure (9) and (10). These figures illustrated that the pavement condition index is 65 and 63 for side 1 and side 2 respectively. This means that the condition of the road is fair as mentioned in ASTM D6433-07 (2016) [1] and shown in Figure (11), also the program indicates the fairness condition.

The case study is within the critical PCI (Mohamed, 2015) [11] that is defined as the PCI value at which the rate of PCI loss increases with time or the cost of applying localized preventive maintenance increases significantly. The typical critical PCI is ranging in between 55 to 70.

**Fig. 7.** Information of sample units of Al-Amarah Street.

**Fig. 8.** The points of distresses in each selected sample unit as entered in Micro PAVER software

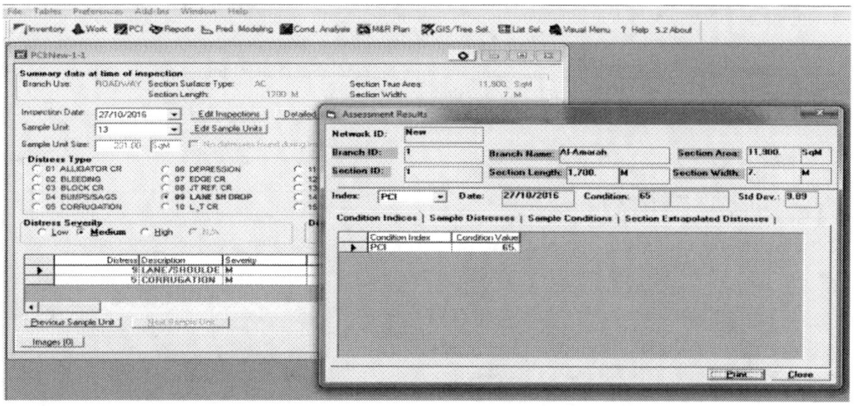

**Fig. 9.** Automated PCI calculation from the Micro PAVER System for Al-Amarah Street, side1.

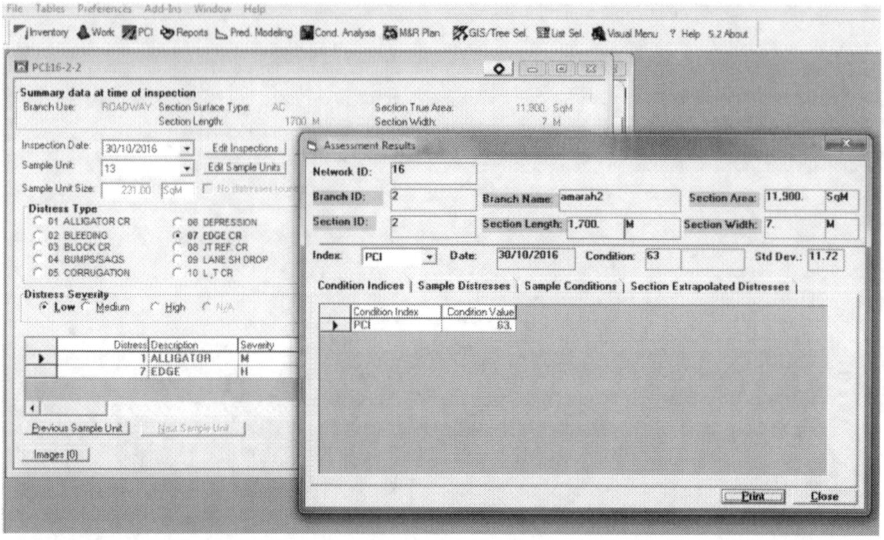

**Fig. 10**. Automated PCI calculation from the Micro PAVER system for Al-Amarah Street, side2.

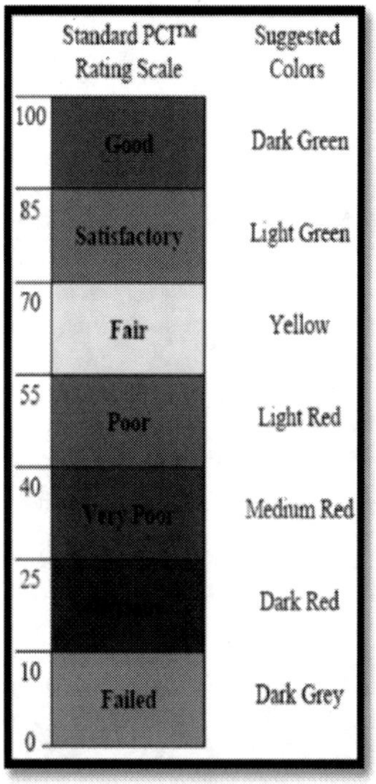


**Fig. 11.** Pavement Condition Index (PCI), rating scale, and suggested colors (ASTM D6433-07, 2016).

Micro PAVER also displays the condition of each sample unit as presented in Table 1.

Figure (12) and (13) presents PCI value for each sample unit, in which it can be noticed that the highest PCI value is 81 and 85 for Side 1 and 2 respectively, while the lowest value for PCI is 53 in Side1and 49 in Side2. Figure (14) and (15) shows the sample units that entered in GIS Arc Map program with their values of PCI for both sides for Al-Amarah Street.

**Table 1.** Condition indices for Al-Amarah Street, side1 &2.

| Sample No. | Sample Type | Sample Size | Units | PCI-Side1 | PCI-Side 2 |
|---|---|---|---|---|---|
| 1 | Random | 231 | $m^2$ | 75 | 54 |
| 2 | Random | 231 | $m^2$ | 57 | 85 |
| 3 | Random | 231 | $m^2$ | 77 | 65 |
| 4 | Random | 231 | $m^2$ | 69 | 49 |
| 5 | Random | 231 | $m^2$ | 81 | 71 |
| 6 | Random | 231 | $m^2$ | 60 | 75 |
| 7 | Random | 231 | $m^2$ | 56 | 55 |
| 8 | Random | 231 | $m^2$ | 57 | 62 |
| 9 | Random | 231 | $m^2$ | 74 | 55 |
| 10 | Random | 231 | $m^2$ | 69 | 55 |
| 11 | Random | 231 | $m^2$ | 53 | 53 |
| 12 | Random | 231 | $m^2$ | 53 | 80 |
| 13 | Random | 231 | $m^2$ | 70 | 53 |

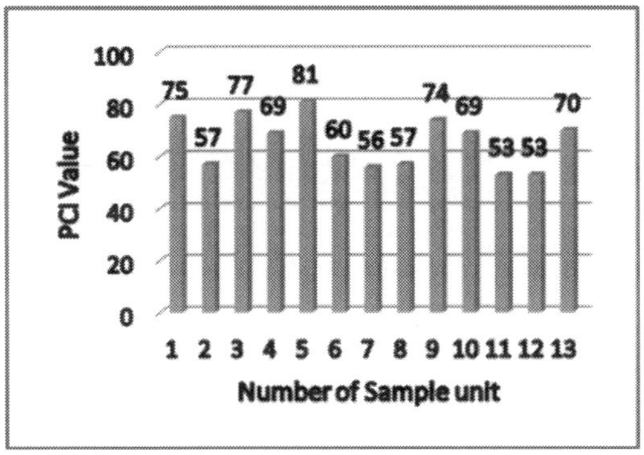


**Fig. 12.** PCI value for each sample units in Al-Amarah St. Side1.

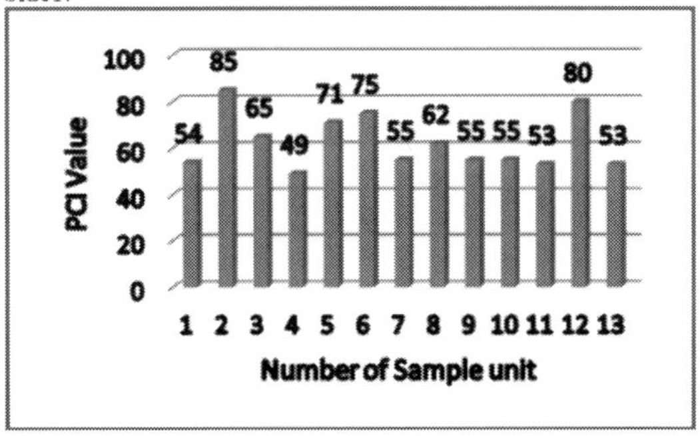


**Fig. 13.** PCI Value for each sample Unit in Al-Amarah st., side2.

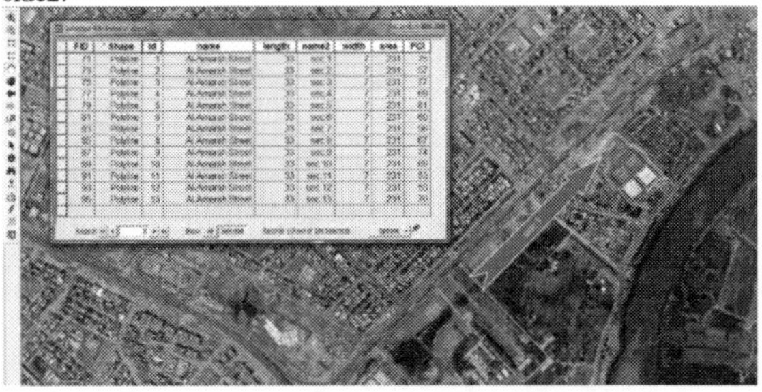

**Fig. 14.** Sample units for side 1 with their PCI values.

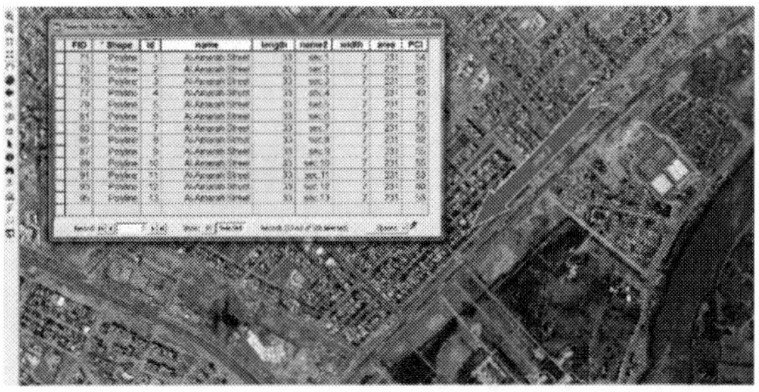

**Fig. 15.** Sample units for Side 2 with their PCI values.

Depending on PCI value, classifying each sample unit with specific color is based on ASTM D6433-07, 2016 [1] as presented in Figure (11).

Figures (16) and (17) show that each sample unit has its own color, based on the color in the legend of the map, in which the difference in color represents the range of PCI value. For example the PCI value between (0-10) is symbolized with gray color which indicates that there are many distresses or high distress intensity within the assessed section, while (86-100) is symbolized with dark green color which in turn indicates that the selected section is with low distresses or low distress intensity.

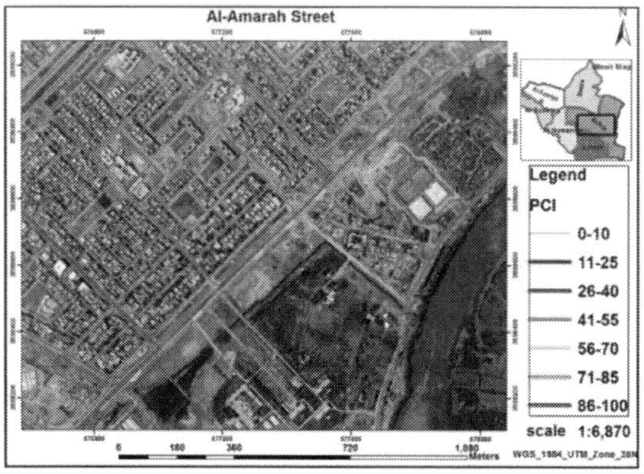


**Fig. 16.** PCI value scale of sample units for side 1 in Al-Amarah St.

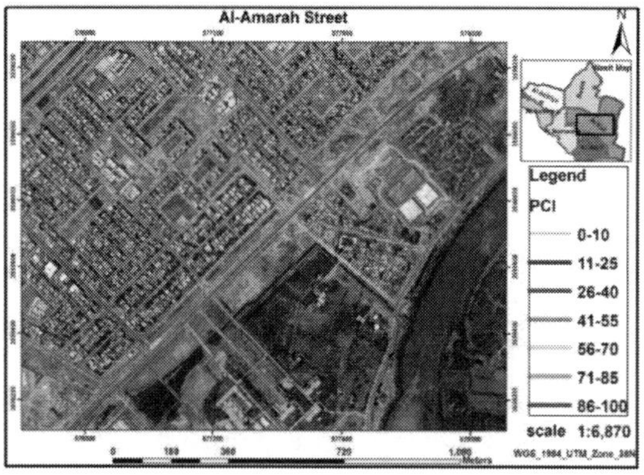


**Fig. 17.** PCI value scale of sample units for side2 in Al-Amarah St.

The total value of PCI for both sides of the road is presented in Figures (9) and (10). The values are 65 and 63 which is located within the range of yellow color region as shown in Figure (18).

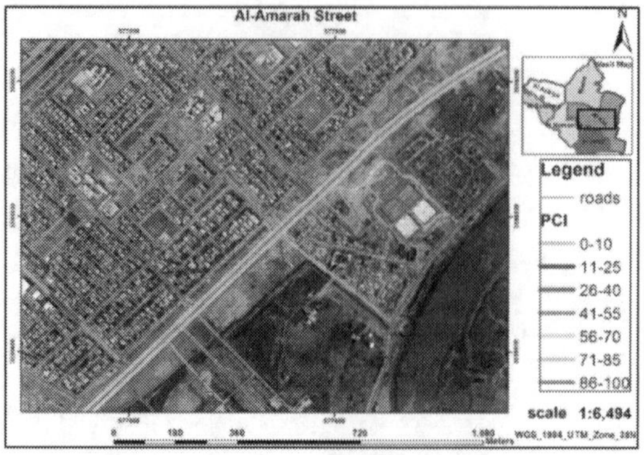


**Fig. 18.** Total PCI value for side1and 2.

## 5 Conclusions

Based on the assessment of results, the following conclusions are drawn

1- To avoid expected errors related to conventional method (hand calculation), pavement condition index (PCI) has been estimated using Micro PAVER 5.2 software. The average pavement condition index of the selected case study is found to be "64" which mean "Fair" pavement condition and critical PCI.

2- Arc Map 9.3 has been applied in this study to a road in a region that has not been surveyed before. Therefore, it is essential to make an integrated maintenance system for each road in the region demonstrating the annual road deteriorations and the resulting change in the PCI values which occurs every year.

3- The study provides an easy and simplified way of presentation the details of deteriorations on the satellite or the geographical map of the road in which each type of distresses has been symbolized with specific sign and each PCI value has been represented with specific color.

4- It is a good way to use the PCI value of road sections for road maintenance prioritization because the adopted methodology depends on the visual condition survey and the calculation of this index depends on direct measurements of each distress of pavement not on just visual rating.

## References

1. ASTM D6433-07, American Society for Testing and Materials (2016).
2. E. J. Yoder and M. W. Witczak, *Principles of Pavement Design,* 2nd Edition, John Wiley & Sons, USA (1975).
3. R. Haas, W. R. Hudson, J. Zaniewski, *Modern Pavement Management,* Malabar, Florida: Krieger Publishing (1994).
4. G. W. Flintsch and K. K. McGhee, NCHRP Synthesis of Highway Practice 401, **8** 152 (2009).
5. L. J. Matel, , T.E.J **106**, 1 (1980).
6. M. Gary, Z. Hao, C. Qinghin, JCEM, **135**, 353 (2009).
7. S. E. Salih, M.Sc. Thesis, Highway & Transportation Department, University of Mustansiriyah, Iraq (2006).
8. A. U. S. Didigwu, Inter. Journal of Arch. and built Envi., **2** (2010).
9. A. T. Ibraheem and D. A. Falih, Eng. Jour., EJ, **4**, pp.44-54 (2012).
10. X. Liu, W. Lu, J. Mao, Y. Liu, 1683-1687 (2010).
11. M. J. Mohamed, M.Sc. Thesis, Highway & Transportation Department, University of Mustansiriyah, Iraq (2015).
12. F. N. Finn and C. L. Monismith, NCHRP Synthesis of Highway Practice, 116 (1984).
13. A. M. Johnson, *Best Practices Handbook on Asphalt Pavement Maintenance*, (No. MN/RC-2000-04). Professional Eng. Services, Ltd. 2 I3 'K'ownes Lane Wayzata, Minnesota 55391, (2000).
14. M. Y. Shahin and J. A. Walther, U.S.A. CERL Tech. Rep. M-90/05, Champaign, IL., (1990).
15. I. Norlela, I. Amiruddin, A. Riza, O. K. Rahmat, *Proc. of the Regional Eng. Postgrad. Con., Malaysia*, (2009).
16. U.S. Army Corps of Engi.-USACE, *Micro Paver 6.5 User Manual*, USA, (2011).
17. S. K. Haldar, *Mineral Exploration: Principles and Applications*, Newnes, (2012).
18. M. Y. Shahin, 2nd edition, Springer Science Business Media, Inc., New York. NY. U.S.A. Smith, R.E, (2005).

# Effect of curing age on the self-sensing behavior of carbon-based engineered cementitious composites (ECC) under monotonic flexural loading scenario

*Ali Majeed* Al-Dahawi[1,*]

[1]University of Technology, Baghdad, Iraq

**Abstract.** The potential effects of curing age on the self-sensing (piezoresistivity) capability of carbon-based Engineered Cementitious Composites (ECC) specimens are under focus in the present study. This non-structural feature can be regarded as one of the best solutions for continuously monitoring of infrastructures in terms of damage and deformations. Carbon fibers which are a micro-scale electrically conductive material were added to the ECC matrix and well dispersed to create the electrically conductive network. This network is responsible for sensing the applied loads on the prismatic specimens. The self-sensing behavior of electrically conductive prismatic specimens under four-point monotonic flexural loading was investigated and compared with dielectric ECC specimens at four ages of curing (7, 28, 90 and 180 days). The results showed that the developed multifunctional cementitious composites can sense the changes in the applied flexural stresses and the resultant mid-span deflection along the adopted curing ages with an improvement in the later ages.

## 1 Introduction

The concrete, which is the main material in the construction of several structures and infrastructures around the world, suffers from of a lot of risks resulting from the applied loads, as well as, from the surrounding weather conditions, which caused, as a whole, a decline in the serviceability index of these structures. Cementitious composites have demonstrated outstanding performance in many civilian facilities such as roads, bridges, dams, sanitary facilities and others. In order to continue working with these materials within the economic boundaries, it was incumbent on the civil engineer to accept the challenge of the conditions referred above in order to ensure the continuity of the work of concrete structures and infrastructures along the economic construction age at the lowest cost. This cannot be achieved easily without incessant monitoring of the structural performance of these cementitious composite constructions in terms of stress, strain, and damage indications to remove the potential defects quickly and economically by the application of the rehabilitation systems opportunely.

To achieve the above goal, and as an initial step, many types of sensors have been used either to be embedded inside the structural members, or attached with them [1]. The only advantage that has been obtained through the use of these sensors is to achieve control in the area of concrete structures. On the other hand, the high cost, the non-homogeneity of the cementitious composites, the limited monitoring of concrete members hampered the use of these sensors. Moreover, the limited durability and the weak resistance of these devices to weather conditions in addition to their contribution in reducing the mechanical properties of concrete infrastructures/structures are the main reasons that led to search for effective, precise, durable and cheap structural health monitoring tools.

The self-sensing technology that appeared in the nineties of the last century is an alternative technology to continuously monitor the behavior of carbon-based cementitious composites effectively without any negative impact on their mechanical properties. This technique is based on the employment of non-structural properties of the cementitious composites to self-sense the distresses suffered by the structure by embedding electrically conductive carbon-based materials within these composites to create semi-homogeneous conductive networks. This gives the ability to sense the development of distresses in terms of strain, stress and damage by means of changes in the electrical resistivity. So, in this case, the cementitious composite material will sense itself without embedded or attached sensors.

This behavior has been found to be more effective in cement-based paste and mortar than in concrete [2]. Many researchers in this field studied the effects of static compression loadings on the self-sensing cementitious composites, besides few studies highlighted on the behavior of these composites under static flexure. It is

* Corresponding author: alimk_ceg@yahoo.com

© The Authors, published by EDP Sciences. This is an open access article distributed under the terms of the Creative Commons Attribution License 4.0 (http://creativecommons.org/licenses/by/4.0/).

important to mention that most of the previous studies concentrated the efforts on clarifying the self-sensing behavior under loadings for only one curing age, and this factor is not included in those studies except the study that was carried out by Fu and Chung [3]. The self-sensing behavior of carbon fiber reinforced mortar was studied under static compression up to failure for the curing ages from 7 to 28 days. It was found in that study that, after 7 days of curing, the electrical resistance increased monotonically with the increase in compressive strain versus decrease in the electrical resistance after 14 and 28 days of curing ages. It was concluded that the reason of this behavior is due to the weakening of fiber-cement interconnection with the progress of curing age. It was presented figuratively that the fractional change in electrical resistance reached up to around 225% at compressive strain of 1.9×10-3 and compressive stress of 39 MPa at 7 days of curing age. These results can be seen, also, within the study of Chung [4].

On the other hand, many studies have applied their tests at only one age of curing. Chen and Chung [2] examined the capability of mortars reinforced by 0.2-4.2 vol.% and concretes contain 0.2-1.1 vol.% of short carbon fiber (CF) to act as a strain sensor via electrical probing under static and cyclic compression, tension and flexure loadings at 7 days age. It was reported that the fractional change in electrical resistivity (FCER) increases with the applied stresses by up to 2100% under static compression up to failure, 5.3% under static tensile loading up to failure and 18.4% under static flexure loading up to failure. It was concluded that the small increase in FCER in the cases of loading under tension and flexure is due to much lower ductility under tension and flexure than compression.

Han and Ou [5] used a 6mm CF with carbon black (CB) as conductive materials to prepare a cement-based mortar and explore its piezoresistivity under single compressive loading after 28 days of curing. The manufactured specimens have dimensions of 50×40×30mm3. Embedded four gauzes for each specimen were used as electrodes to measure the 4-probe electrical resistivity in the perpendicular direction with respect to load direction. It was noted that the FCER decreased with the increase in the applied stress/strain to reach up to 27% at an ultimate stress and strain of 40MPa and 0.0022 respectively.

The damage in carbon fiber reinforced concrete was monitored under uniaxial compression and three-point bending by electrical resistance measurement within a study carried out by Chen and Liu [6]. Pitch-based carbon fibers with 5mm length were used to fabricate the conductive cubic specimens ($70.7\times70.7\times70.7mm^3$) and prismatic specimens ($40\times40\times160\ mm^3$). DC with 4-probe method was used to measure the electrical resistivity. The electrical contacts were copper wires attached around the whole perimeter of the cubes or around the four planes of the prisms with the aid of silver paint. The four wired planes of the four electrical contacts were perpendicular to the stress axis in the case of cubes and parallel to the stress axis in the case of beams. It was observed that the fractional change in resistance with respect to cubes under uniaxial compression decreases when the stress is increased at low level of stress which was equal to the 57% of the fracture stress. The fractional change in resistance at this point reached up to around 38%. After this point, when the damage in the specimen begins due to high level of stress, the fractional change in resistance increased rapidly and reached up to 50% with stress increasing. The researchers divided the resulted load-displacement-FCER curve into three zones. The first one refers to the elastic stage with a deflection range of greater than zero and less than 0.01mm. Within this zone, it was noted that with the increase of deflection, the FCER increased up to around 1.2%. In the second zone, when the deflection is bounded between 0.010mm and 0.015mm, visible cracks occurred and the FCER increased from 1.2% to 1.5%. The end of this zone represented the point of maximum applied load. When moving to the third zone with a deflection of more than 0.015mm, it was seen that the FCER remained approximately constant due to breaking of the conductive network.

Yeh et al. [7] used a 15mm chopped CF to fabricate a cement-based self-sensing material for measuring the strain and detecting the damage of civil engineering structures. The researchers tested the specimens under compression and bending at 28 days age. To measure the electrical resistivity, 4-probe method was used by attaching copper wires around the perimeters of the specimens with the aid of silver paint. Cylindrical samples with dimensions of (100mm diam.×200mm h.) and prismatic samples with dimensions of ($550\times150\times150mm^3$) were prepared for uniaxial compression and 3-point bending tests, respectively. The experimental results of uniaxial compression test showed that the fractional change in electrical resistance reached up to 14% corresponding with $4.2\times10^{-3}$ compressive strain. The experimental results of cyclic loadings of all fabricated types of three-point bending specimens showed that the relationships between fractional change in electrical resistance and strain measured at top and bottom of specimens were linear. At a cyclic tension strain level of $0.05\times10^{-3}$, the fractional change in electrical resistance reached up to around 3.5%, whereas, the value of this parameter reached up to around 5% at a cyclic compression strain level of $0.08\times10^{-3}$.

In the present study, the effects of curing age on the self-sensing behavior of cementitious composites were experimentally investigated under four-point monotonic flexural test at 7, 28, 90 and 180 days of curing. Moreover, electrically conductive chopped carbon fibers (CF) in micro-scale were used to fabricate the multifunctional cementitious composites (MCC).

## 2 Experimental program

### 2.1 Ingredients of carbon-based cementitious composites

The fabricated mixtures, in the present work, comprise

 https://doi.org/10.1051/matecconf/201816201034

mainly the common ingredients which are used in the composition of the cement-based mortar. These are: CEM I 42.5R ordinary Portland cement (PC) (similar to ASTM Type I); Class-F fly ash (FA); fine silica sand with maximum aggregate size of 0.4 mm, specific gravity of 2.60 and water absorption capacity of 0.3%; potable mixing water and polycarboxylate ether based high range water reducing admixture (HRWRA). The chemical and physical properties of Portland cement, FA and silica sand are illustrated in Table 1, and the particle size distributions of these ingredients are represented in Figure 1.

**Table 1.** Chemical and physical properties of PC, FA and silica sand.

| Chemical composition, % | PC | FA | Silica sand |
|---|---|---|---|
| CaO | 61.43 | 7.98 | 0.02 |
| $SiO_2$ | 20.77 | 52.22 | 99.79 |
| $Al_2O_3$ | 5.55 | 16.58 | 0.06 |
| $Fe_2O_3$ | 3.35 | 6.60 | 0.02 |
| MgO | 2.49 | 2.10 | 0.01 |
| $SO_3$ | 2.49 | 0.02 | - |
| $K_2O$ | 0.77 | 1.53 | 0.01 |
| $Na_2O$ | 0.19 | 0.86 | 0.02 |
| Loss on Ignition (LOI) | 2.20 | 10.36 | 0.07 |
| $SiO_2 + Al_2O_3 + Fe_2O_3$ | 29.37 | 75.40 | 99.87 |
| Physical properties | | | |
| Specific gravity | 3.06 | 2.10 | 2.60 |
| Blaine fineness ($cm^2/g$) | 3250 | 2690 | - |

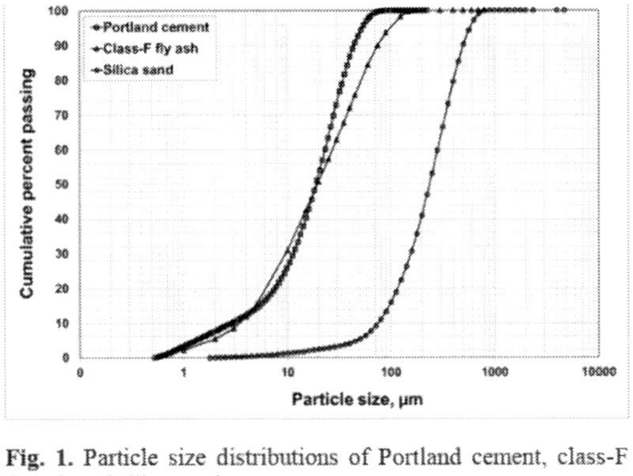


**Fig. 1.** Particle size distributions of Portland cement, class-F fly ash and silica sand.

The carbon-based electrically conductive materials, which were employed to create an electrically conductive network within the cementitious composites, were chopped carbon fibers with 12mm length and aspect ratio of 1600 (CF12). CF were with the tensile strength of 4200 MPa, elastic modulus of 240 GPa, elongation of 1.8%, density of nearly 1.7-2.0 g/cm3 and diameter of 7.5 μm.

Polyvinyl Alcohol fibers (PVA) of 8mm length, 39 μm diameter and specific gravity of 1.3 were also used to enrich the mechanical properties of the developed materials against the applied loads, in spite of its negative impact on reducing the electrical conductivity of the developed MCC because it is a dielectric material. The nominal tensile strength and elastic modulus of these fibers were 1620 MPa and 42.8 GPa, respectively.

In order to remove any undesirable bubbles that may be created due to the use of HRWRA, a defoamer powder was used as a percentage by weight of cementitious materials.

## 2.2 Electrical measuring setup

To measure the electrical resistivity of the specimens under 4-point flexural tests, four brass plates were embedded within 360×75×50mm3 prismatic specimens. The positions of electrodes and the test configuration setup can be seen in the last study carried out by the author [8]. Alternating current (AC) with 2-probe method and direct current (DC) with 4-probe method were used to measure the electrical impedance between the electrodes in the case of AC and to measure the voltage values between the internal electrodes due to a constant applied current on the external electrodes in the case of DC. Engineered Cementitious Composites (ECC) mixtures were used in the present study. ECC was firstly developed by Li [9] to be the unique construction material that has a flexible application together with low cost and weight, isotropic properties and high performance in terms of strength, ductility, fracture toughness and pseudo strain-hardening. ECC is formed from cement-based mortar with a high modulus fiber such as PVA. Since its composition is similar to that of conventional types of concrete and fiber reinforced concrete (FRC), the ECC can be regarded within the non-electrically conductive materials. To add the electrical conductivity property, as an additional function to the ECC without losing its properties, carbon-based materials, which are essentially electrically conductive materials, can be utilized.

## 2.3 Mixture proportions and mixing methods

Water to cementitious materials (PC + FA) ratio (W/CM) and fly ash to Portland cement ratio (FA/PC) were kept constants as 0.27 and 1.2, respectively during the preparation of mixtures. The dosage of CF as it was reported by Al-Dahawi et al. [10]. This dosage was 1.00% by volume of mixture. Moreover, for the same reasons that were previously mentioned, the PVA fiber was used with a 2.00% dosage by the volume of the mixture [9]. In addition to that, the high range water reducing admixture (HRWRA) was used to get the same liquidity of the MCC. It is well known that the differences in the microstructural properties of the carbon-based ECM that were utilized, in the present study, led to difficulties in obtaining the same workability by the use of the same amount HRWRA. To keep the specimens in the same acceptable workability, different dosages of HRWRA were, therefore, used for the different carbon-based MCC. A defoamer dosage of 0.3% by weight of cementitious materials was utilized to remove the undesirable air bubbles.

During the present study, the best methods that were suggested by Al-Dahawi et al. [11] to mix the micro-scale materials (CF) were utilized here for the achievement of the best distribution of the electrically conductive carbon-based materials in order to obtain the best self-sensing capability. Mixing of CF within the cementitious matrix was made in accordance with the findings of that study to get the highest electrical conductivity. Accordingly, CF was first mixed with the dry raw materials (PC, FA and silica sand) in a 20-liter-capacity mortar mixer at 100 rpm for 10 minutes. After slowly adding the mixing water at 100 rpm over 10 seconds, speed was increased to 300 rpm, all of the HRWRA was added over 30 seconds, and mixing of all materials was continued for additional 10 minutes at 300 rpm in the mortar mixer.

## 3 Preparation and testing of specimens

The fabricated specimens were poured, according the mixing procedures that have been recommended by Al-Dahawi et al. [11], in the oiled molds and have been placed in their molds at 50 ± 5% RH, 23 ± 2 °C and covered with plastic sheets for 24 hours. After demolding process, all the specimens were, then, moved into isolated plastic bags to be cured at 95 ± 5% RH, 23 ± 2 °C until 24 hours before the pre-specified test ages, regarding to the specimens that were tested within 28 days. To remove the potential effects of the internal moisture on the electrical measurements due to polarization, the specimens were dried in an oven at 60 °C for 24 hours before testing [12, 13]. The prepared specimens to test after 90 and 180 days have been removed from the plastic bags after 28 days of curing and then kept at 50 ± 5% RH, 23 ± 2 °C till their pre-specified testing dates. All specimens were tested at room temperature.

## 4 Results and discussion

### 4.1 Mechanical properties of the developed MCCs under flexure

The most important mechanical features of the prismatic specimens under the flexural loading are the ultimate flexural strength (also known as a modulus of rupture) and the deflection hardening that can be obtained from the multiple cracking behavior (vs single cracking and localization) as mentioned by Naaman [14] to define the high performance of the fiber reinforced cement composites. The typical stress/deflection curve of the prismatic specimens that behave with the deflection hardening contains many important points. These are the first cracking position; the points of load-carrying capacity greater than the load at the first cracking point and the ultimate flexural strength point which is represent the modulus of rupture value as mentioned above and as defined by the ASTM C 78 – 02. All these points were described in detail in the previous ASTM standard C 1018 – 97. However, the researchers faced a difficulty in correctly distinguishing the first cracking position and the corresponding strength for the materials exhibited a deflection hardening behavior according to the new ASTM standard C 1609 [15]. Therefore, in the present study, the method that was described in ASTM standard C 1018 – 97 and suggested by kim et al. [15] was utilized to identify this point at the stress/deflection curve. This method takes into account the fact that the point which is located at the end of the straight line and the beginning of the non-linear line of the stress/deflection curve is the point of the occurring of the first crack. The simplest way that was explained by Naaman [14] to represent the deflection hardening behavior was used in the present study. He showed that the deflection hardening of the prismatic specimens can be measured by utilizing the ductility index (DI) which is the ratio of the mid-span deflection at the ultimate strength point ($\delta$ult.) and the mid-span deflection at the first cracking point ($\delta$fc) based on a condition that the ultimate stress ($\sigma$ult.) must be greater than the first cracking stress ($\sigma$fc).

The stress/deflection curves of the developed MCC, in the present study, for different curing ages together with the self-sensing behavior can be seen in Figures (2 and 3). In this section, the discussion is limited to demonstrate the mechanical behavior of the developed prismatic mixtures, while the next section will cover the self-sensing behavior exhibited by them. From the stress/deflection curves for the different curing ages above, the previously mentioned desirable characteristics were specified and plotted. Within Figure 4, the mid-span deflection values at the first cracking positions of the different manufactured mixtures at different curing ages can be seen. On the other hand, the mid-span deflection values of the corresponding mixtures at the ultimate flexural strength were demonstrated within Figure 5. The corresponding stresses at the first and post cracking points were presented in Figures 6 and 7, while the ductility index and the ratio of the ultimate strength to the first cracking strength were showed in Figures 8 and 9, respectively. From Figure 4 it can be noticed that the general behavior trend of the first cracking deflection values is the starting with small values of around 0.10 to 0.15mm at 7 days old with an increase up to around 0.20 and 0.30mm at the 28 and 90 days, respectively followed by some decrease at 180 days of curing ages. This trend encompasses the manufactured mixtures except the control mixture at 7 days old. The same trend of deflection values can also be seen in Figure 5 but with the deflection values at the ultimate strength positions except the behavior of the control specimens, which exhibited the inverse trend. This case reflects the enhancement characteristics of adding the micro-scale carbon-based material within the cement-based matrices with the progression of their ages to extend their ductility with multiple cracking behaviors under flexural loadings. Like what has been observed in Figure 4, the trend of the corresponding stress values at the first cracking positions were similar as can be noticed in Figure 6. The stress values at these positions for the earlier ages (7 and 28 days) were around 3 MPa while those at late ages were around 5 MPa. Unfortunately, the high values of the first crack deflections and strengths are not desirable for achieving the deflection hardening behavior and this is what will be seen in the next figures. With approximately constant values of the control specimens against increasing values with the progression of their curing ages with respect to the carbon-based MCC, Figure 7 clears this behavior of the ultimate strengths of the manufactured mixtures. As previously mentioned, the control specimens get only few flexural strength development across their curing ages to reach up to 18% at the end of 180 days old. Whereas the CF - based MCC get gains reached up to 36% of their flexural strength at 180 days old from their original 7-days of curing age values. In brief, it can be concluded that the CF-based MCC registered an enhancement in the flexural strength of about 20% compared with the control specimens at the end of 180 days.

From the calculated values of ductility index (DI) (Figure 8) it can be seen easily that all of the fabricated mixtures exhibited a deflection hardening behavior at all the adopted curing ages. In addition to the DI criterion, it can be seen in Figure 9 another representation of the developed mixtures in terms of deflection hardening which is the ratio of the ultimate flexural strength to the flexural strength at the first cracking position. This figure reflects the stress development with the curing ages together with Figure 7. All mixtures crossed the limit of 1, which is the condition of deflection hardening behavior of the prismatic specimens under flexural loading. As an average, all mixtures reached up to around 1.5 at all the adopted curing ages.

### 4.2 Self-sensing behavior under flexural test

Generally, the FCER/mid-span deflection curves can be divided into three parts according to their slopes that have been obtained from the results shown in Figures (2 and 3) which are due to the progressive load levels. It can be seen that the first slow slope is bounded between the starting point of loading and the first cracking position which is located at the end of the linear section of the stress/mid-span deflection curve. The second part of the FCER/mid-span deflection curve can be specified from the end of the first part up to the end of the load carrying ability of the stress/mid-span deflection curve before occurrence of failure which can be distinguished by fast loss in the ability of the specimen to carry the applied flexural stress. The slope of this part of the FCER/mid-span deflection curve is higher than the first part of the same mixture. The third part of this curve begins at the end of the second part up to the end of the loading process which has generally the highest rate of change in FCER.

Figure 10 abstracts the average values of FCER of the different mixture specimens at the considered curing ages at the first crack position of each specimen, which can be regarded as the most important part within the loading process. From this figure, it can be noticed clearly the ability of CF-based MCC on diagnosing and sensing the first crack position in terms of FCER at most curing ages. At 7 days old, the developed CF-based MCC exhibited an improvement of 171% in FCER as compared with the control specimens. After 28 days of curing, it can be seen that the control specimens achieved the highest value of FCER. An improvement in FCER of 103% was registered at 90 days of curing for

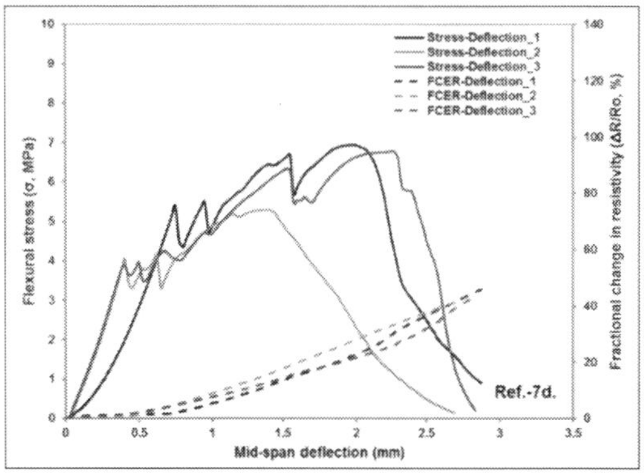


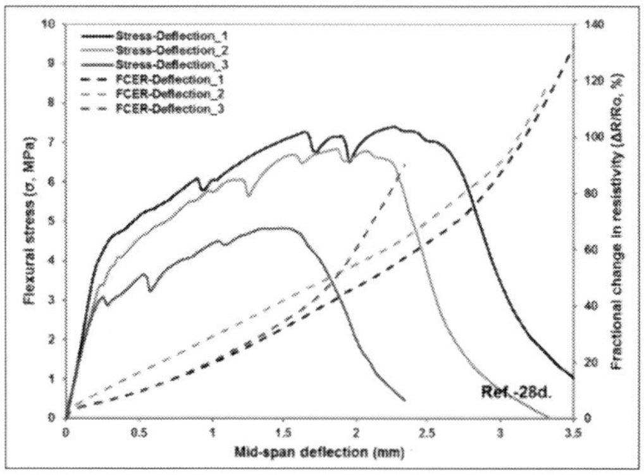


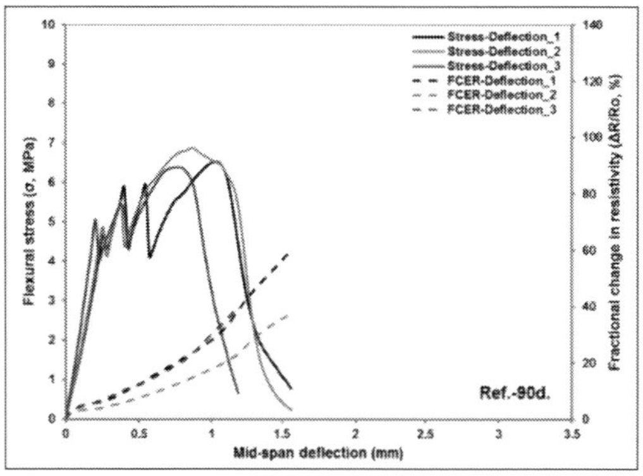

https://doi.org/10.1051/matecconf/201816201034

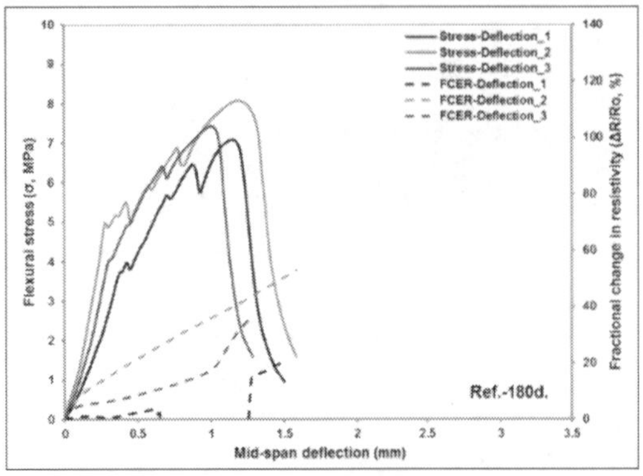


**Fig. 2.** Flexural stress and FCER versus mid-span deflection of the plain mixtures at different curing ages.

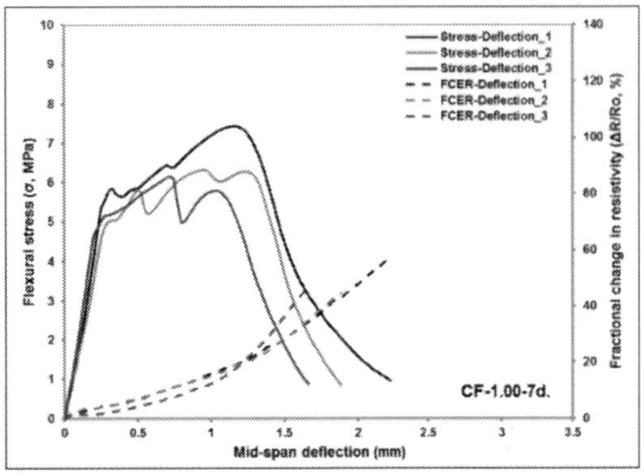


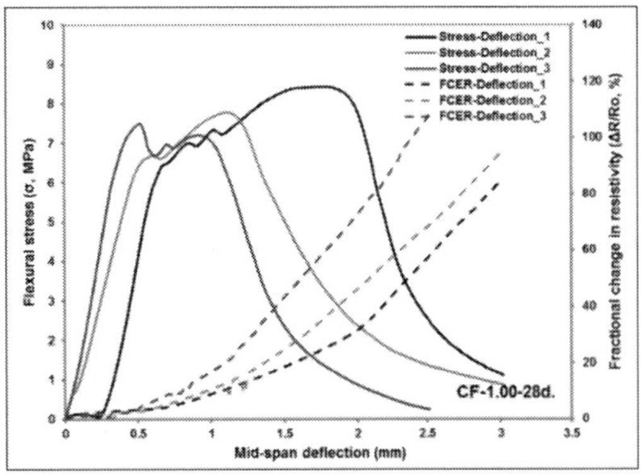


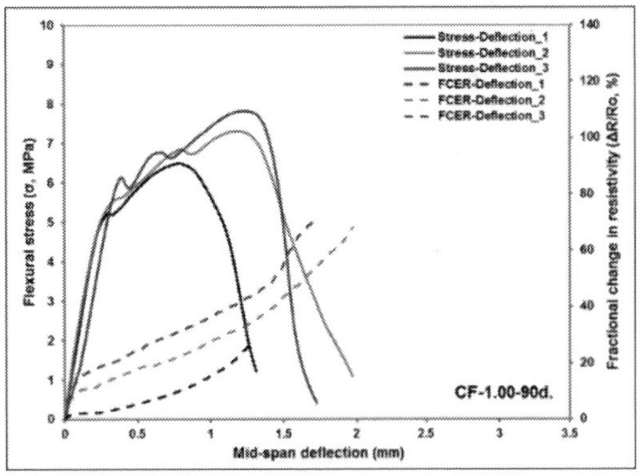


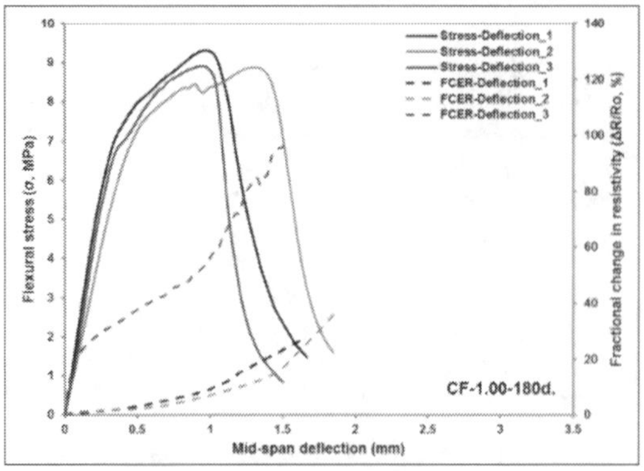


**Fig. 3.** Flexural stress and FCER versus mid-span deflection of the CF-based mixtures at different curing ages.

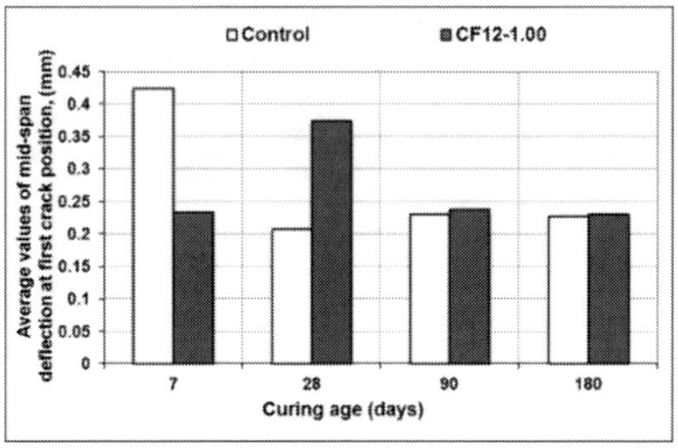


**Fig. 4.** Mid-span deflection (average values) at first crack position for the adopted curing ages.

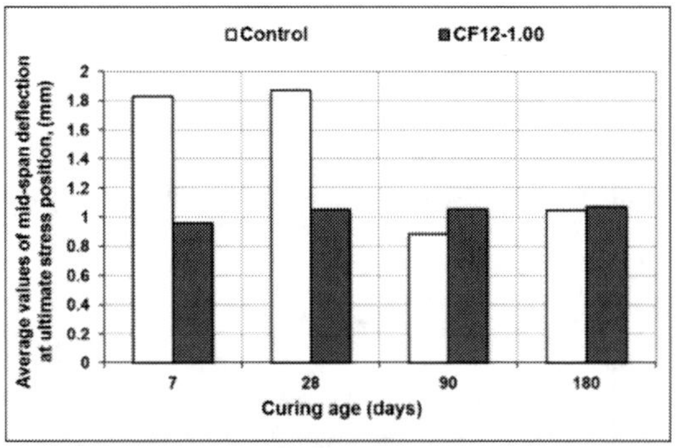


**Fig. 5.** Mid-span deflection (average values) at the ultimate stress position for the adopted curing ages.

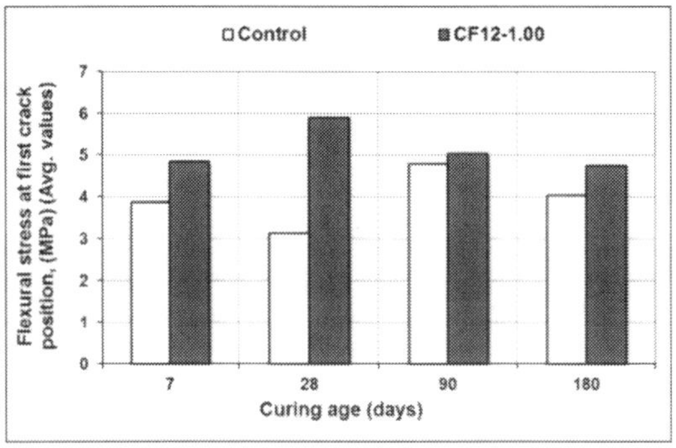


**Fig. 6.** Flexural stress (average values) at the first crack position for the adopted curing ages.

the CF-based MCC. Moreover, at the end of 180 days, the improvement in the FCER was 73% for the carbon-based cementitious composite. As a general behavior, it can also be concluded from this figure that the self-sensing ability in the beginning of the loading process of the developed MCC was magnified with the progressive curing ages.

The changes in the resultant values of FCER due to the same applied flexural stress but with different ages of curing are due to the development of hydration process with the age progression and the effects of the hydration products on the electrically conductive networks within the cement-based matrices which lead to these

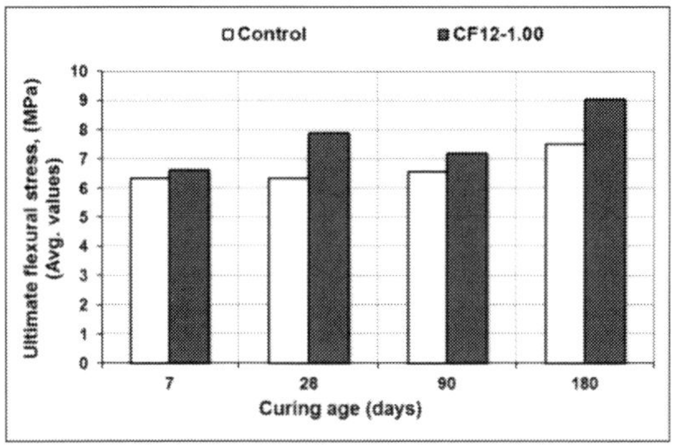


**Fig. 7.** Ultimate flexural stress (average values) for the adopted curing ages.

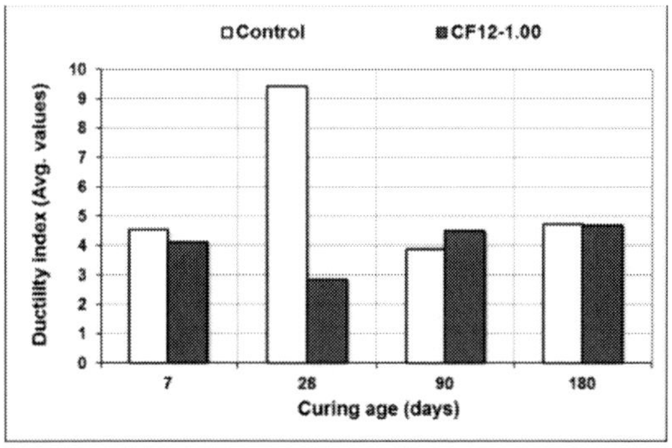


**Fig. 8.** Ductility index (DI) (average values) for the adopted curing ages.

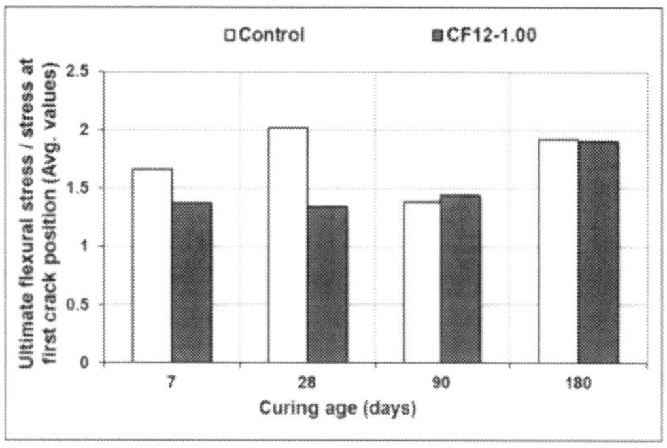


**Fig. 9.** Ratio of the ultimate flexural stress/stress at first crack position (average values) for the adopted curing ages.

deviations. This fact clears the big changes that have been occurred in the mixtures during hydration process between 7 and 28 days. After 28 days, no big differences can be seen in the FCER values which indicate the bounded activity of the hydration process at the advanced ages of curing after this age. The effect of adding the conductive materials within the cementitious composites on the self-sensing ability can also be noticed clearly. It can be pointed out that the self-sensing capability in terms of FCER improves with age, which is important in order to ensure the continuity of monitoring the structural behavior of the monitored structures/infrastructures. This gives an indication of the durability of the developed composites to use as a stress and damage sensor.

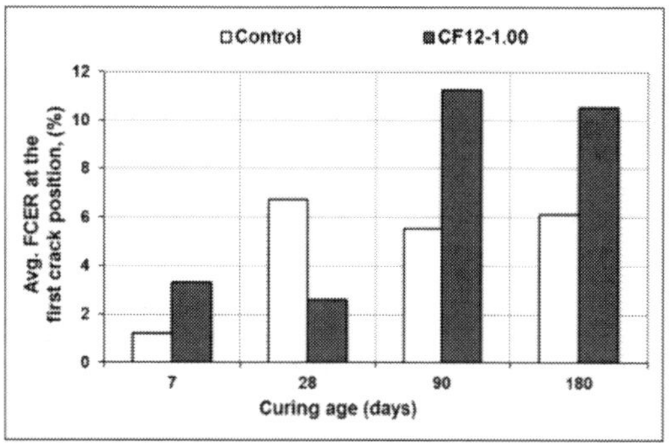


**Fig. 10.** FCER (average values) at different curing ages.

## 5 Conclusions

Mechanical and self-sensing behaviors of cementitious composites incorporating micro-scale CF were investigated under monotonic four-point bending. The following conclusions were drawn:

- The manufactured prismatic plain cementitious composite specimens and those reinforced with micro-scale CF exhibited a deflection hardening behavior under monotonic four-point flexural loading.
- The CF-based cementitious composite specimens enhanced the flexural strength characteristics at all the

adopted curing ages with 20% of enhancement compared with the plain specimens at the end of 180 days of curing.

- The incorporation of CF within the ECC matrix has an important role in improving the self-sensing ability of the monitored structures/infrastructures in terms of stress and damage diagnosis especially in later ages.
- The self-sensing ability of the developed MCC under monotonic four-point bending was magnified with the progress in curing ages.

The author gratefully acknowledge the financial assistance of the Scientific and Technical Research Council (TUBITAK) of Turkey provided under Project: 114R043.

## References

1. V.M. Karbhari, F. Ansari, Structural health monitoring of civil infrastructure systems (Woodhead Publishing in materials, 2009)
2. P.-W. Chen, D.D.L. Chung, Compos. Part B: Eng. **27**, 1, 11-23 (1996)
3. X. Fu, D.D.L. Chung, Cem. and conc. Res. **27**, 9, 1313-1318 (1997)
4. D.D.L. Chung, Jour. of Int. Mats. Sys. & Struc. **13**, 9, 599-609 (2002)
5. B. Han, J. Ou, Sens. and Act. A: Phys. **138**, 2, 294-298 (2007)
6. B. Chen, J. Liu, Const. & Build. Mat. **22**, 11, 2196-2201 (2008)
7. F. Yeh, C. Hsu, M. Lai, K. Chang, W. Liao, in *Life-Cycle of Structural Systems: Design, Assessment, Maintenance and Management* (CRC Press/ Taylor & Francis Group: London, UK. 341-347, 2015)
8. A. Al-Dahawi, G. Yıldırım, O. Öztürk, M. Şahmaran, Const. & Build. Mat. **145**, 1-10 (2017)
9. V.C. Li, in *Fiber reinforced concrete: Present and the future* (Canadian Society for Civil Engineering, Montreal: Canada, 64-97, 1998)
10. A. Al-Dahawi, M.H. Sarwary, O. Öztürk, G. Yıldırım, A. Akın, M. Şahmaran, M. Lachemi, Smart Mat. & Struct. **25**, 10, 1-15 (2016)
11. A. Al-Dahawi, O. Öztürk, F. Emami, G. Yıldırım, M. Şahmaran, Const. & Build. Mat. **104**, 160-168 (2016)
12. H. Li, J. Ou, H. Xiao, X. Guan, B. Han, in *Nanotechnology in Civil Infrastructure* (Springer, 131-173, 2011)
13. M.-q. Sun, R.J. Liew, M.-H. Zhang, W. Li, Const. & Build. Mat. **65**, 630-637 (2014)
14. A.E. Naaman, in *CBM-CI international workshop, Karachi, Pakistan* (Citeseer, 2007)
15. D.j. Kim, A.E. Naaman, S. El-Tawil, Cem. & conc. comp. **30**, 10, 917-928 (2008)

# Simulation of traffic flow in unsignalization intersection using computer software SIDRA in Baghdad city

*Ali* Mohammed [1,2,*], *Hassan* Jony[1], *Alaa* Shakir[2], and *Kamarudin* Bin Ambak [2]

[1]University of Technology, Baghdad, Iraq
[2]Faculty of Civil and Environmental Engineering Universiti Tun Hussein Onn Malaysia(UTHM), Parit Raja, Johor, Malaysia

**Abstract.** Computer simulation is a vital means for the examination of expressways and urban lanes and streets. Transportation experts concentrate on the arrangement and dissemination of traffic jams on roadways. The objective of this study is to calculate and analyses the travel time, delay time, degree of saturation, and level of service and travel speed in Jordan intersection in Baghdad, Iraq using Sidra software. The number of vehicles passing Jordan intersection was recorded by the author of this work from 7am to 3pm for four days. A simulation model has been used to assess the performance of the current intersection. Results demonstrated that the level of service (LOS) for Jordan intersection is D with (35 sec/veh) average delay and degree of saturation (0.996 v/c). It was concluded that Jordan intersection needs further developments like using intelligent transportation system application (ITS) to regulate traffic signals, whereby reducing traffic jam, using closed circuit TV (CCTV) might assist traffic office in identifying jam point to reduce traffic jam.

## 1 Introduction

Traffic jam happens when a volume of Traffic produces demand for more space as compared to the accessible street limit; this point is generally expressed as the degree of saturation. There are various particular circumstances that cause traffic jams; the vast majority of them decrease the limit of a street at a given point or over a specific length. Around half of U.S. traffic jam is repeating [2], and is ascribed to sheer weight of traffic; a major of the other half is attributed to the movement to traffic Ali used SIDRA software to assess the traffic flow in Roundabout and intersection in Bandar Baru Bangi [2], Al-Mimany, used Sidra to analyze and assess the traffic network in Kirkuk city [3]. Ibrahim investigated the assessment of Models for roundabout capacity and performance estimation for using SIDRA, ARCADY, RODEL, SYNCHRO models and Highway Capacity Manual model (HCM 2010) to carry out the analyses [4]. Examined the evaluation and enhancement of traffic operation for Al-Zeoat Intersection in Al-Ramadi city, using SIDRA traffic program [5]. investigated the urban street network in kut city, using geometrics techniques like (GPS, GIS) for detecting the traffic flow patterns of the selected network [6]. HCM (Highway Capacity Manual), HCS and SYNCHRO 6 computer program for operational analysis of Wadi-Saqra signalized intersection in Amman [7]. used aaSIDRA computer Software to Optimize the Traffic Flow in Bandar Baru Bangi Roundabout and intersection in Selangor, Malaysia [9].

The Jordan intersection in Baghdad city, which connects Al-Mansour quarter and Hay Al -Jameeah quarter could be described as the busiest intersection with the maximum traffic jams intersections in Baghdad connects the most vital areas in Baghdad nearby 14th Ramdhan intersection which is a commercial district having numerous shopping centers, governmental facilities like passports office, Iraqi citizenship office, and traffic office. Therefore, it encounters massive jams in the day. Besides, Jordan intersection is linked with Al-Nessor roundabout which is a congested area.

## 2 Data collection

Data has been collected from the field using a video camera with high storage capacity (80 GB) and a long-life battery (battery charging lasts for 10 hours). This camera has been installed on the alrashed bank building. a good vantage points for collecting data. The time of recording started from off-peak (7:00 AM) till another off-peak (3:00 PM) and was evaluated every 15 minutes during two periods the day which is (7am-12pm) and (12pm-3pm). Table1: shows the traffic volume data expressed by a number of cars for all direction, Table1 shows the input worksheet required for analysis of the existing geometric, traffic, and signalized conditions so this data was conducted over three days. Fig 1and 2 shows the existing peak traffic flow determined by SIDRA computer program.

SIDRA computer software has been created by the

© The Authors, published by EDP Sciences. This is an open access article distributed under the terms of the Creative Commons Attribution License 4.0 (http://creativecommons.org/licenses/by/4.0/).

Australian Road Research Board (ARRB), Transport Research Ltd., as a guide for outline and assessment of crossing points, for example, signalized convergences, roundabouts, two-way stop control, and caution sign control crossing points [10,11,12]. In assessing and comparing behavior of roundabouts, there are a few advantages that the SIDRA model has over some other programming model; it gives solid appraisals of geometric delays and related breakdown impacts for the different crossing point sorts. This property of the software is essential to the computation of new options for the intersection treatments consistently.

**Table 1.** Traffic Volume at Jordan Intersection for all Approaches each Quarter Hour.

| Time | From Street Jordan Road | | | From Yarmouk Road | | | From Baghdad Tower Road | | | Total |
|---|---|---|---|---|---|---|---|---|---|---|
| | L | ST | R | L | ST | R | L | ST | UTURN | |
| 7:00-7:15 am | 30 | 33 | 35 | 41 | 33 | 38 | 29 | 50 | 65 | 354 |
| 7:15 - 7:30 | 115 | 55 | 60 | 67 | 62 | 71 | 77 | 70 | 78 | 655 |
| 7:30-7:45 | 179 | 196 | 211 | 122 | 178 | 183 | 197 | 211 | 198 | 1675 |
| 7:45-8:00 | 280 | 379 | 97 | 250 | 250 | 259 | 240 | 269 | 324 | 2348 |
| 8:00-8:15 | 178 | 154 | 165 | 170 | 154 | 169 | 187 | 187 | 203 | 1567 |
| 8:15-8:30 | 103 | 120 | 172 | 133 | 155 | 134 | 130 | 150 | 102 | 1199 |
| 8:30-8:45 | 64 | 77 | 73 | 65 | 68 | 69 | 72 | 77 | 71 | 636 |
| 8:45-9:00 | 69 | 66 | 65 | 55 | 64 | 63 | 71 | 72 | 77 | 602 |
| 9:00-9:15 | 62 | 60 | 69 | 72 | 72 | 71 | 68 | 55 | 59 | 588 |
| 9:15-9:30 | 58 | 49 | 66 | 52 | 45 | 55 | 63 | 68 | 70 | 526 |
| 9:30-9:45 | 55 | 34 | 51 | 58 | 42 | 52 | 58 | 55 | 79 | 484 |
| 9:45-10:00 | 44 | 49 | 54 | 53 | 48 | 47 | 58 | 59 | 66 | 478 |
| 10:00-10:15 | 49 | 66 | 77 | 69 | 73 | 66 | 68 | 65 | 62 | 595 |
| 10:15-1:30 | 42 | 55 | 63 | 61 | 65 | 63 | 69 | 70 | 72 | 560 |
| 10:30-10:45 | 45 | 44 | 59 | 67 | 72 | 67 | 65 | 59 | 60 | 538 |
| 10:45-11:00 | 48 | 48 | 53 | 55 | 69 | 56 | 58 | 64 | 60 | 71 |
| 11:00-11:15 | 44 | 44 | 32 | 44 | 59 | 66 | 70 | 79 | 75 | 74 |
| 11:15-11:30 | 40 | 40 | 37 | 42 | 61 | 68 | 76 | 80 | 79 | 76 |
| 11:30-11:45 | 47 | 47 | 41 | 52 | 58 | 61 | 74 | 79 | 81 | 79 |
| 11:45-12:00 pm | 19 | 19 | 46 | 55 | 59 | 65 | 77 | 83 | 85 | 87 |
| 12:00-12:15 | 39 | 39 | 50 | 54 | 61 | 68 | 79 | 86 | 86 | 75 |
| 12:15-12:30 | 41 | 53 | 57 | 67 | 60 | 84 | 83 | 89 | 88 | 622 |
| 12:30-12:45 | 55 | 57 | 60 | 70 | 68 | 80 | 83 | 86 | 83 | 642 |
| 12:45-01:00 | 46 | 55 | 62 | 73 | 71 | 82 | 86 | 83 | 79 | 637 |
| 01:00-01:15 | 49 | 60 | 69 | 70 | 74 | 78 | 88 | 90 | 93 | 671 |
| 01:15-01:45 | 47 | 66 | 55 | 73 | 70 | 77 | 84 | 85 | 89 | 646 |
| 01:45-02:00 | 49 | 66 | 55 | 73 | 70 | 79 | 83 | 85 | 85 | 645 |
| 02:00-02:15 | 47 | 68 | 53 | 69 | 73 | 70 | 84 | 88 | 79 | 631 |
| 02:15-02:30 | 45 | 70 | 55 | 74 | 70 | 73 | 83 | 85 | 89 | 644 |
| 02:30-02:45 | 44 | 72 | 55 | 75 | 70 | 69 | 81 | 85 | 89 | 640 |
| 02:45-3:00 | 89 | 87 | 98 | 77 | 88 | 71 | 79 | 86 | 94 | 769 |

**Fig. 1.** Jordan intersection study area.

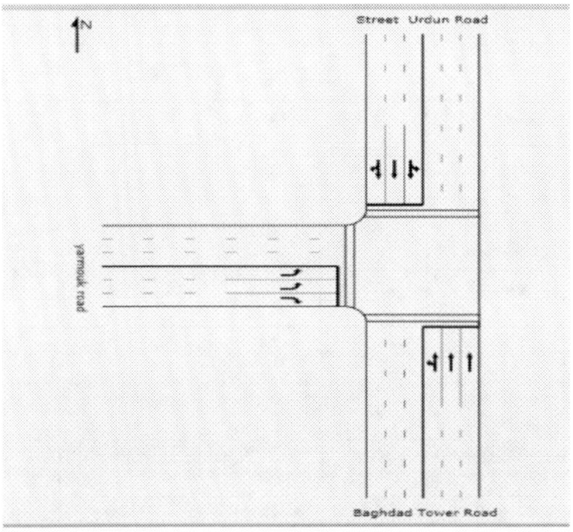


**Fig.2.** Existing Peak Traffic Flow as Shown by SIDRA

**Table 2.** Level of service definitions based on delay hcm (for vehicles) [11]

| LOS | Control Delay (signalized) sec/veh | Delay sec/veh (unsignalized) |
|---|---|---|
| A | ≤ 10 sec | ≤ 10 sec |
| B | 10-20 | 10-15 |
| C | 20-35 | 15-25 |
| D | 35-55 | 25-35 |
| E | 55-80 | 35-50 |
| F | >80 | >50 |

## 3 Results and discussion

### 3.1 Peak hour volumes

Traffic flow shown in Table 1was analyzed using excel program to identify the peak hour. The peak hour at Jordan intersection observed at 8:00 a.m. The total traffic volumes within this hour were recorded as (4023 pc/h) whereas the lease number of vehicles passing Jordan intersection were observed at10.00 am where the total traffic volumes at that hour were (962 pc/h) as shown Fig3. Total traffic flow percentage is illuminated in Fig.4 traffic flow percentage is evaluated by dividing the number of cars in each direction by the total number of cars. Total traffic flow from Yarmouk Street approach is 33%, from Baghdad Tower is 37% and from Jordan intersection is 30%.

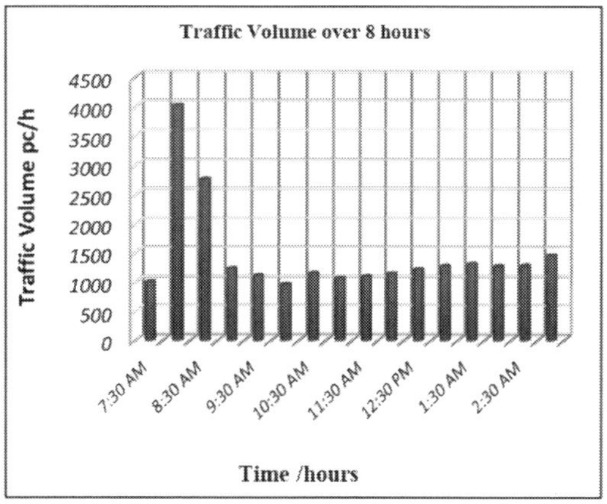


**Fig.3**. Total traffic volumes at Jordan intersection each one hour

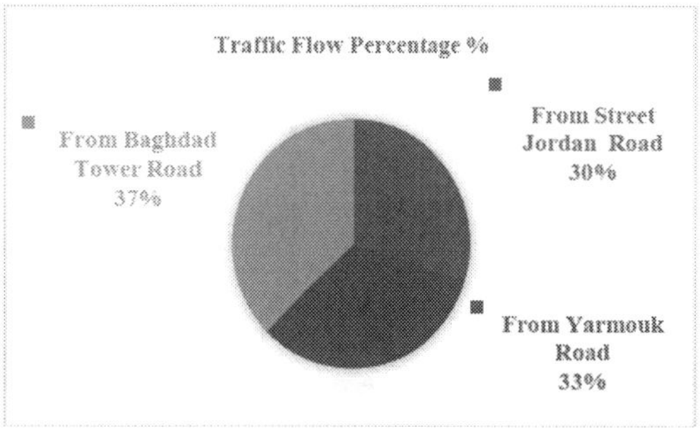


**Fig.4.** Total traffic flow Percentage at Jordan intersection

**Table 3.** Traffic Volume At Jordan Intersection For All Approaches Each Half Hour

| Time | From Street Jordan | From Yarmouk Road | From Baghdad Tower Road | Total |
|---|---|---|---|---|
| 07:30 am | 328 | 312 | 369 | 1009 |
| 08:00 | 1342 | 1242 | 1439 | 4023 |
| 08:30 | 892 | 915 | 959 | 2766 |
| 09:00 | 414 | 384 | 440 | 1238 |
| 09:30 | 364 | 367 | 383 | 1114 |
| 10:00 | 287 | 300 | 375 | 962 |
| 10:30 | 352 | 397 | 406 | 1155 |
| 11:00 | 304 | 389 | 379 | 1072 |
| 11:30 | 239 | 400 | 463 | 1102 |
| 12:00 pm | 260 | 394 | 494 | 1148 |
| 12:30 | 294 | 419 | 507 | 1220 |
| 01:00 | 335 | 444 | 500 | 1279 |
| 01:30 | 346 | 442 | 529 | 1317 |
| 02:00 | 338 | 434 | 504 | 1276 |
| 02:30 | 341 | 431 | 512 | 1284 |
| 03:00 | 274 | 236 | 259 | 769 |

The data was collected for each line for all the directions (left, right and through) and summed up during half an hour to recognize the number of cars passing the intersection during half an hour, the total number of cars passing Jordan intersection was 1,342 cars at 8:00 am whilst the total number of cars passing the intersection at 1:30 pm was recorded as 346 cars, while the number of cars passing al Yarmouk street was around 1,242 cars at 8:00 am , while the number of cars passing the same street was 444 cars at 1:30 pm. The number of cars passing Baghdad Tower at 8:00 am was 1,439 cars, whilst in 1:30 pm were recorded as 529 cars as shown in Table 3.

The percentage of cars passing the intersection each half an hour was calculated from Table 3 by dividing the total number of cars passing Yarmouk Street, Baghdad Tower and Jordan intersection during half an hour by the total number of cars shown in Table 3 to arrive at the percentage of cars passing

**Table 4.** Traffic percentage volume at Jordan intersection for all approaches for each in half an hour

| Time | From Street Jordan | From Yarmouk Road | From Baghdad Tower Road | Total |
|---|---|---|---|---|
| 07:30 am | 33% | 31% | 37% | 100% |
| 08:00 | 33% | 31% | 36% | 100% |
| 08:30 | 32% | 33% | 35% | 100% |
| 09:00 | 33% | 31% | 36% | 100% |
| 09:30 | 33% | 33% | 34% | 100% |
| 10:00 | 30% | 31% | 39% | 100% |
| 10:30 | 30% | 34% | 35% | 100% |
| 11:00 | 28% | 36% | 35% | 100% |
| 11:30 | 22% | 36% | 42% | 100% |
| 12:00 pm | 23% | 34% | 43% | 100% |
| 12:30 | 24% | 34% | 42% | 100% |
| 01:00 | 26% | 35% | 39% | 100% |
| 01:30 | 26% | 34% | 40% | 100% |
| 02:00 | 26% | 34% | 39% | 100% |
| 02:30 | 27% | 34% | 40% | 100% |
| 03:00 | 36% | 33% | 39% | 100% |
| Max | 34% | 36% | 43% | |
| Mini | 22% | 31% | 34% | |
| Average | 29% | 33% | 38% | |

the intersection. It was demonstrated that the highest percentage was recorded from 7:30 am to 9:30 am with 33%, while the percentage of cars passing Jordan Street were as high as 36% at 3:00 pm. However, the percentage at Yarmouk Street was significantly different as the highest percentage was recorded at 11:00 am with 36% and 35% at 1:00 pm. The percentage of cars passing Baghdad Tower was 42% at 11:30 am and 40%

https://doi.org/10.1051/matecconf/201816201035

at 1:30 pm and 2:30 pm as shown in Table 4. The percentage of vehicle passing the intersection each half hour for 8 hours is indicated in Fig.5

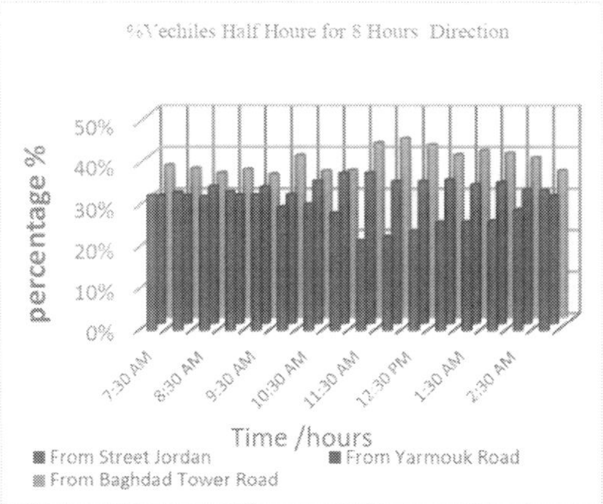


**Fig.5.** %Vehicles per Half hours for 10 hours' direction

**Table 5.** PHF values for Jordan intersection approaches

| Approach | Movement | PHF |
|---|---|---|
| From Street Jordan Road | L | 0.2 |
| | St | 0.032 |
| | R | 0.3 |
| From Yarmouk Road | L | 0.3 |
| | St | 0.3 |
| | R | 0.3 |
| From Baghdad Tower Road | L | 0.4 |
| | St | 0.3 |
| | U turn | 0.3 |

Peak Hour Factor (PHF) is the ratio between total volume and the maximum volume per 30 min rate of flow within the hour [Garber 2010]. Table 5 show PHF values for all approaches at both squares

**Table 6.** Results of the analysis by SIDRA

| Approach | Average Delay sec/veh | LOS |
|---|---|---|
| From Street Jordan Road | 36.0 | D |
| From Yarmouk Road | 4.3 | A |
| From Baghdad Tower Road | 61.6 | E |

## 3.2 Existing LOS at Jordan intersection

Level of Service (LOS) Method is described based on the US Highway Capacity Manual (HCM) a subjective measure used to relate the efficiency of traffic flow. LOS is utilized to dissect highway by sorting traffic flow and doling out quality levels of movement in view of execution measure like speed, density [Highway Capacity Manual HCM, 2010]. LOS definitions based on delay HCM (for vehicles) is indicated in Table 6. Existing LOS was investigated in this study using Computer Simulation Sidra Software. Results illustrated that Jordan Street LOS is (A) while the Yarmouk Road LOS is (E). However, Baghdad Tower LOS of the intersection is (D) as shown in Fig.6, hence it is might be described as unstable flow. Speed is adversely correlated with traffic flow. Average delay is computed using SIDRA software and is related with LOS as given in Table 6

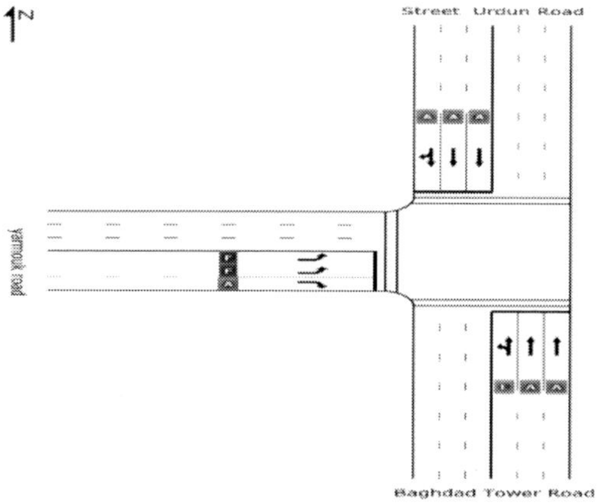

https://doi.org/10.1051/matecconf/201816201035

Fig.6. LOS for Jordan intersection approaches

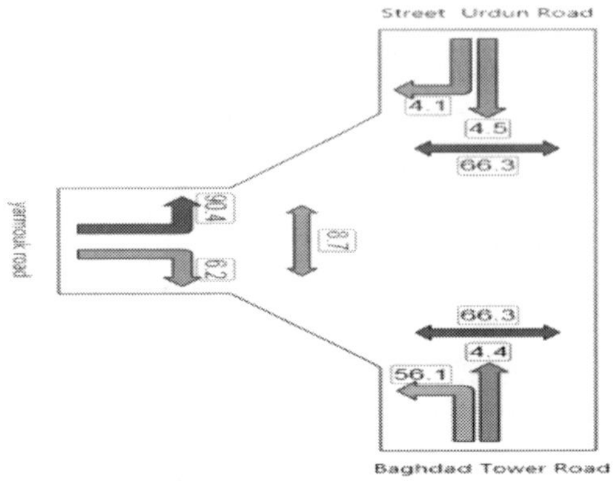


**Fig.7.** Details of LOS for Jordan intersection approaches

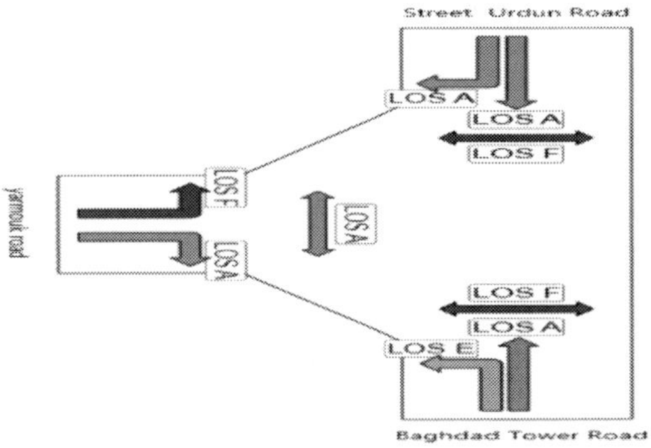


**Fig.8.** Results for average delay for Jordan intersection approaches

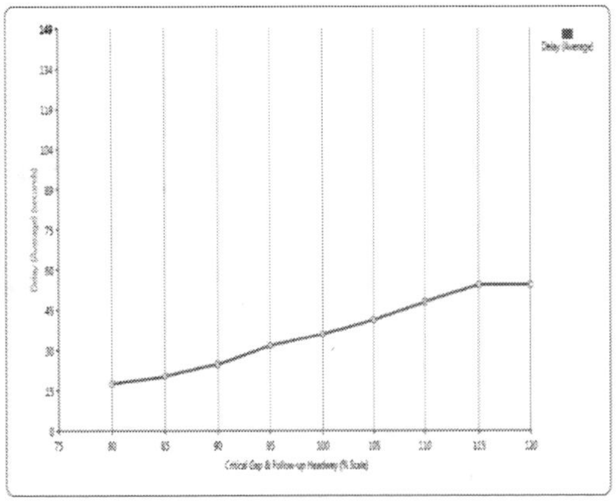

**Fig.9.** Relation between average delay and follow up headway

### 3.3 Average delay

The average delay is a crucial factor in forecasting travel time. Results for the delay is indicated in Fig.8 which shows that both Jordan Street and Yarmouk Street have two colours red and green, red refers to LOS F and green refers to LOS A which has the minimum delay. Delay is adversely related to LOS when the delay is minimized LOS is enhanced [Turley, 2007]. Whilst, Baghdad Tower Road has two colours orange and green, green refers to LOS A and orange refers to LOS E. Relation between average delay and follow up headway is illustrated in Fig.9, It is obvious to note that the follow up heady increase with the increase of average delay. Total delay for Jordan intersection was recorded as 35.4 sec

### 3.4 Degree of saturation

The level of saturation (%) is a proportion of demand to limit on each lane to deal with the intersection, with an estimation of 100% implying that demand and limit are equivalent and no further movement can advance through the intersection. Values more than 85% are normally viewed as torment from movement clog, with lines of vehicles starting to shape. according to colour code based on a degree of saturation. Each degree has a specific colour. Results for the degree of saturation is clarified in Fig.10. Green colour refers to less than 0.6 degrees of saturation, it was observed that Al Yarmouk Street has pink colour means that degree of saturation approach (0.8-0.9), whilst Baghdad Tower road has an orange colour, which refers to (0.9-1) degree of saturation. A relation between Degrees of Saturation and follow up headway is illustrated in Fig.11.The degree of saturation for Jordan intersection is 0.996 v/c

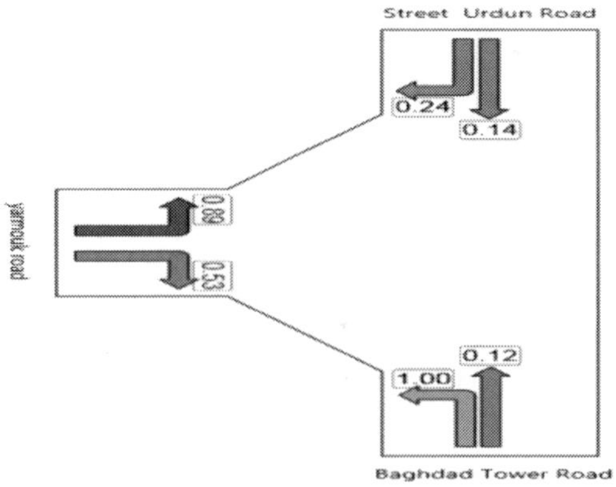

**Fig.10.** Results for degree of Saturation for Jordan intersection approaches

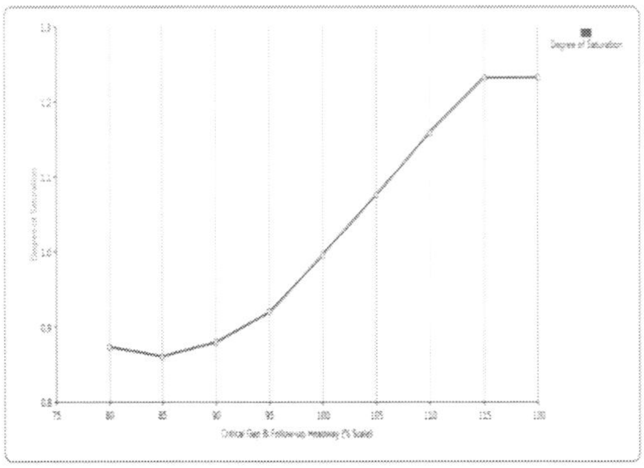

**Fig.11.** Relation between Degree of Saturation and follow up headway

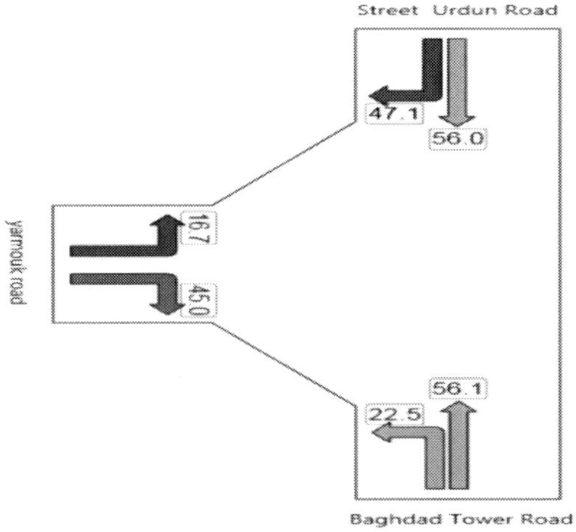


**Fig.12.** Results for Travel Speed for Jordan intersection approaches

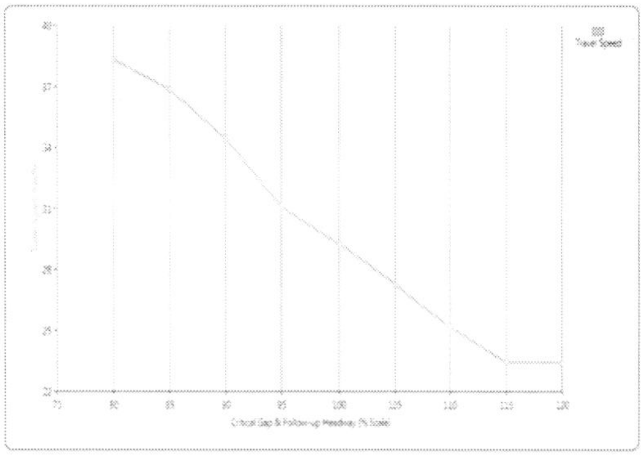

**Fig.13.** Relation between travel speed and follow up headway

### 3.5 Travel speed

Results for travel speed is indicated in Fig.12. According to colour code based on Speed Efficiency light blue colour and dark blue colour that the Speed Efficiency Ratio between (0.8-0.9) and (0.7-0.8).It was noted that Yarmouk Road has red colour and a pink colour which means the Speed Efficiency Ratio is between from (0-0.3) and (0.5-0.7) respectively. Whereas, Baghdad Tower road has a yellow colour and Light blue colour which means that the Speed Efficiency Ratio is between (0.3-0.5) and (0.8-0.9) as shown in Fig.12. A relation between travel speed and follow up headway is illustrated in Fig.13. It is significant to observe that the travel speed is adversely correlated with follow up headway when the travel speed increases the follow-up headway decreases. Travel speed for Jordan intersection is 29.3 km/h sec.

## 4 Conclusions

Conclusions based on the investigation done in this study are below:

- The peak hour at Jordan intersection was recorded at 8:00 am with (4023 pc/h).
- The average delay is adversely correlated with LOS. Total delay for Jordan intersection was recorded as 36 sec; hence LOS of Jordan intersection is D.
- The degree of Saturation increases with follow up headway. The degree of saturation for Jordan intersection is 0.996.
- Travel speed is adversely correlated with follow up headway. Travel speed for Jordan intersection is 29.3 km/h sec

## References

1. A. A. Mohammed, A.A. Shakir, Journal of Engineering Science and Technology. **8**, 351-363 (2013)
2. A. A. Mohammed, Eng. & Tech. **31**, 774-792 (2013)
3. D. A. Al-Mimany, Eng. & Tech. **5**, 542-549 (2015)
4. N. J. Garber, L.A. Hoel, *Traffic and Highway Engineering*, Thomas Learning, Fourth Edition, (2010)
5. H. A. Awad, H.A. Mohammed, Anbar Journal for Engineering, **2**, 3, 46-57 (2010)
6. M. R. Mahmoud, M.Sc. thesis, University of Technology, (2015)
7. M. A. bojaradwh, M. Msallam, B. Jew, Jour. of Environment & Earth Science. **4**, 12, 156-169 (2013)
8. D. Y. Karim, M.R. Mohamad, S. Abdullah. the Eastern Asia Society for Transportation Studies. **5**, 5, 945-958 (2005)
9. A. A. Mohammed, *Tikrit Jou of Eng Sciences*, **20**, 10-25 (2013)
10. R. Akcelik, SIDRA User Guide.Akcelik and Associates Pty Ltd, Melbourne, Australia. 341-347, 2010)
11. HCM Highway Capacity Manual, Transportation Research Board's (TRB) fifth edition of the Highway Capacity Manual HCM, Washington D.C. USA. 2-10 (2010)
12. B. H. Al-Omari ,M.M.Taamneh. Jordan Jour of Civil Engineering, **4**, 375-392 (2007)

# Drainage investigation of surface runoff for highway pavement

*Aqeel* Al-adili.[1,*], *Rasha* Abdulamir [1],and *Hawra* Abbas[1]

[1] Building and Construction Engineering Departent, University of Technology, Baghdad, Iraq

**Abstract.** The aim of this study is to establish the effect of heavy rainfall and the chosen pavement layers on the drainage design, material selection and rutting resistance of the flexible pavement. The test in present study was started with wheel track passing without load and without rain falling on the pavement for a period of time, and it was noticed that no distress appeared on the surface of the pavement. Then, the load is gradually added by using wheel track load of 106 psi for five tests without rain falling and five other tests with gradually increasing rain fall duration and intensity. Deterioration and distresses appeared on the pavement when increasing the wheel track load to (150 psi) under high intensity rain and long term duration of rain fall. By increasing the number of days, which is 103 days of study, when the pavement is saturated, the extra amount of the water will runoff. The clogging material which caused a decrease in the water seepage, increases the time of runoff ending. The clogging materials of fine particles that get deposited on the surface of the pavement resulted by passing the wheel track loading and wear & tear of the pavement surface, and other clogging materials such as salt in the water will penetrate to the pavement and seal the voids and decrease its water seepage. The water seepage decreases by increasing number of days, so the amount of the absorbed water decreases by 89% after 71 days of testing for high rain intensity (116 ml/min.).

## 1 Introduction

This study includes the results which are conducted from the laboratory model test and discussion of the results. The model was subjected to a duration of rainfall and the results of the amount of water absorbed through the pavement layer, under the effect of the simulate track loading weight, has been recorded. Analysis and discussions of the changes and distresses that obtained in the surface pavement are presented in this paper.

## 2 Selection of pavement layer thickness

In this study, a silty clay soil (with 400 mm thickness) has been used with a subbase type B (with 300mm thickness), concrete base (with 150 mm thickness) and an asphalt superpave pavement (with 50 mm thickness). The thickness of the pavement layer was chosen according to (***Iraqi standard specification for Roads & Bridges, 2003***) and ***(Huang, 2004)***. As shown in Figure 1.

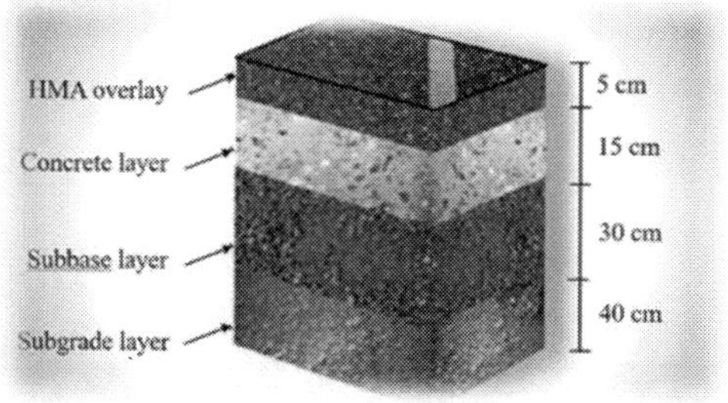


**Fig.1.** Pavement Layer Thickness

## 3 Model preparation

The second phase in this study is the prototype laboratory simulation model. The details for the preparation and manufacturing of the model are described below.

### 3.1 Steel box

Steel container with external dimensions of (150 cm) length, (80 cm) width and (90 cm) depth is constructed by the researcher to simulate the pavement layers (subgrade, subbase, and concrete pavement and asphalt surface) as shown in Fig.2. The steel plate thickness is 4 mm which was used to manufacture each part of the box.

* Corresponding author:aqeeladili@hotmail.com

© The Authors, published by EDP Sciences. This is an open access article distributed under the terms of the Creative Commons Attribution License 4.0 (http://creativecommons.org/licenses/by/4.0/).

**Fig.2.** The Steel Box

Holes were distributed randomly around the steel box to allow the absorbed water by layer to drain out as shown in Fig.3. and four iron pipes as in Fig.4 welded above the edge side of the steel container to install distributed water pipe over it.

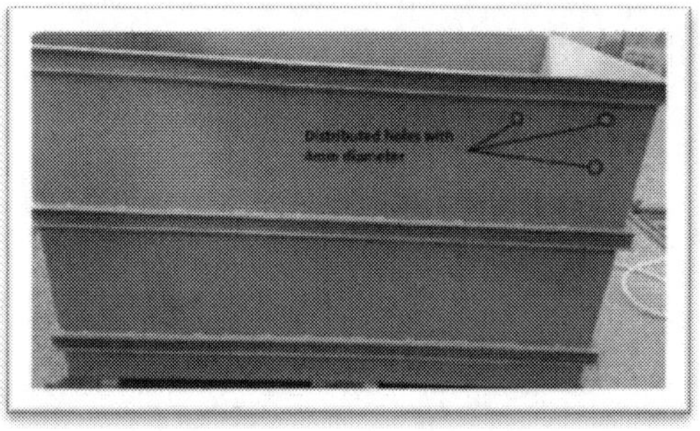
**Fig.3.** Holes Distributed Around the Steel Box

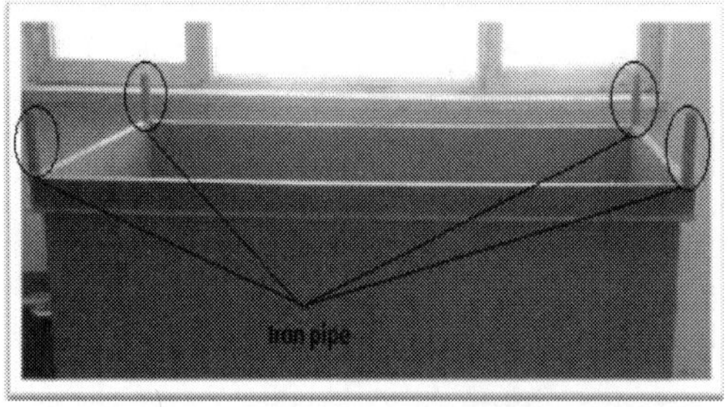

**Fig.4.** Iron Pipe Used to Install Distributed Water Pipe

### 3.2 Distribution water pipe

The distribution water pipe which was manufactured from 12 iron pipes with (1/2") inch (diameter?) and 70 cm length, are connected to frame as a water network with 150 cm length and 80 cm width and 150 cm height from the steel box top and installed over the steel box by screws.

There are holes of 2 mm in diameter distributed along the pipe and the distance between the holes is 3 cm to simulate the rainfall over the highway in the model as shown in **Fig.5**.

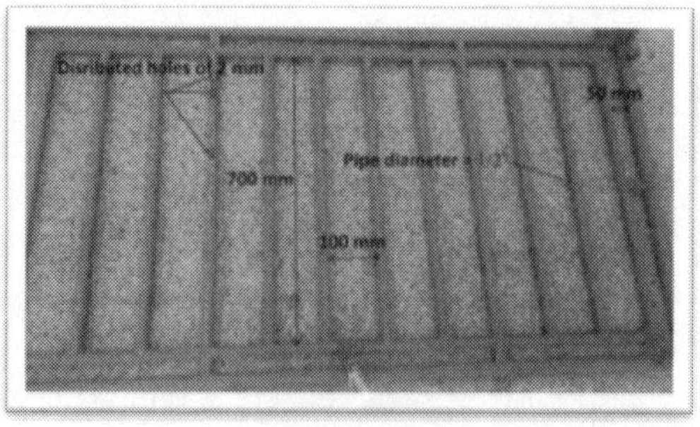

**Fig.5.** Network for Rainfall Simulation

There is a slot in the long side of the water network connected with plastic pipe to the water pump which let the water to flow from the water tank to the water network.

### 3.3 Water pump

An electrical water pump is used which is joined to the water tank to draw water from the water tank to the distribution water pipe by plastic water pipe, the intensity of the water pump is 30 l/min. and with a power of 550W.

### 3.4 Water tank

For this study, a cylindrical water tank with volume of 500 litres is used. The water tank is connected to the electrical water pump from the bottom and it contained a slot of 3/4" at the top of the tank.

### 3.5 Wheel truck

A framework of iron rectangular with 200 cm long and a 150 cm high has been manufactured. This frame contains ball bearing to facilitate the wheel movement. The wheel is manufactured by rubber tyre with 250 mm diameter and 80 mm width and the contact area between the surface of the pavement and the tyre is 17.7 $cm^2$ (2.454 $in^2$). The rubber tyre is moved by three phase's motor with 0.2 m/s speed which is connected to the rubber by toothed wheels and fixed under iron plate which is used to carry the loads.

For the loads, two metal cylinders filled with concrete mix in order to obtain to the required weight have been used. The weight of each cylinder is 80 Kg, and the weight of the motors is 90 Kg, it will add to the load of the cylinders which generate (149 psi) load on the pavement in the contact area between the tyre and the pavement. The loads are placed above the iron plate of the wheel as shown in Fig.6. And Fig.7 show the loaded laboratory model test.

**Fig.6.** Simulate Track Loading

**Fig.7.** Loaded Laboratory Model Test

## 4 Selection of rain intensity for long and short terms

For this study, two types of rain intensity were selected: high intensity rain with 116 ml/min., and low intensity rain with 58 ml/min.

Two aspect of drainage control have been considered in this research, namely a short-term and a long-term intensity rain. The short-term intensity rain takes at periods of time such as 30 minutes, one hour and two hours in the day.

The long-term intensity rain started at the same time of the short term rain but continued for four months. During this periods, the effect of the rain on the pavement and it's deflection, if happened, and its effect on the drainage has been studied.

## 5 Starting the test

The test began with turning on the water pump and opening the faucet to the required water intensity which was previously selected, then the water pump withdraw the water from the water tanker in which its amount of water was calculated before starting the test, then the water starts to flow through the plastic pipe to arrive to the water distribution network in order to simulate the rain fall in the pavement, which was executed and explained previously. In the meantime the wheel track which is loaded with two cylindrical loads is run along the pavement through the test time, and then the changes on the pavement layer during the rain and under the influence of wheel track loading are recorded.

During the test, the water drains and moves toward a channel which was set on the side slope of the pavement, and returns back to the water tanker through the larger plastic pipe and pours on the top of the tanker.

## 6 Amount of the absorbed water through the pavement layer

After ending the rain trial for short and long term and for the required period of time, the amount of water that goesout from the water tanker, the amount of water that is drained and returned back to the tanker and the amount of water that is absorbed through the pavement layer should be determined. In order to calculate the amount of the water that is drained and that is absorbed through the pavement layer, the amount of the water in the water tanker before beginning the test needs to be calculated.For this purpose, and before filling the tanker with water, the amount of added water to the tanker is determined.

After ending the test by one or two hours, the height of water has been recorded and the amount that was indicated on the tanker has been checked and then the result has been obtained. The amount of absorbed water can be then calculated by taking the differences between the amount of water before starting rainfall and the amount of water in the water tanker after rain falling.

## 7 Test result for high intensity rain

This step represents the study result of the amount of absorbed water, ending time of seepage and vertical drainage for high intensity rain (116 ml/min.) during the study period.

### 7.1 Amount of absorbed water

The rain intensity test was conducted for two different times. Before beginning the test, the amount of the water in water tanker has been calculated, then after two or one hour rain falling and after ending the test, the amount of water has been calculated again. The differences between the amount of water before starting the test and amount of water after ending the test is referred to the amount of absorbed water by the pavement layer.

Amount of absorbed water = Amount of water before rain falling (wb) – Amount of water after ending the rain fall (wa) ............... (1)

Table 1 shows how to number the days of testing according the date of the test.

Table 2 shows the results which has been recorded after two hours (120 min) rain falling for (116 ml/min.) intensity rain.

**Table 1.** Days numbering according to the date.

| Intensity ml/min | Time of Test (minute) | Date of test | Days Numbering |
|---|---|---|---|
| 116 | 60 | 17/04/2016 | 1 |
| 116 | 60 | 18/04/2016 | 2 |
| 58 | 120 | 26/04/2016 | 10 |
| 58 | 120 | 27/04/2016 | 11 |
| 58 | 120 | 28/04/2016 | 12 |
| 58 | 60 | 28/04/2016 | 12 |
| 58 | 60 | 04/05/2016 | 18 |
| 58 | 60 | 05/05/2016 | 19 |
| 116 | 120 | 08/05/2016 | 22 |
| 116 | 120 | 09/05/2016 | 23 |
| 116 | 120 | 10/05/2016 | 24 |
| 116 | 60 | 12/05/2016 | 26 |
| 116 | 120 | 17/05/2016 | 31 |
| 116 | 120 | 24/05/2016 | 38 |
| 116 | 120 | 26/05/2016 | 40 |
| 116 | 60 | 31/05/2016 | 45 |
| 116 | 120 | 15/06/2016 | 60 |
| 116 | 120 | 19/06/2016 | 64 |
| 116 | 60 | 26/06/2016 | 71 |
| 58 | 120 | 03/07/2016 | 78 |
| 58 | 120 | 17/07/2016 | 92 |
| 58 | 120 | 18/07/2016 | 93 |
| 58 | 120 | 24/07/2016 | 99 |
| 58 | 120 | 25/07/2016 | 100 |
| 58 | 60 | 26/07/2016 | 101 |
| 58 | 60 | 27/07/2016 | 102 |
| 58 | 120 | 28/07/2016 | 103 |

Fig.8 shows the relationship between the amount of absorbed water and the time for two hours raining under track loading for high rain intensity which is 116 ml/min. This figure shows that the amount of the water that is absorbed through the pavement layer slightly degreases with the increase in time . At the first day of the test the pavement layer needed more amount of water to be saturated and after the saturation, the amount of the water degreases. The amount of absorbed water reached its highest point (45.5 letters) at the first day of test. At the final day of test for this rain intensity, the amount of water records its lowest value which is 10 letters with percentage of 78% after 64 days of testing as shown in Table 2. The amount of water that seeped through the pavement layer decreases rapidly by increasing clogging materials of fine particles deposited on the surface of the pavement which resulted in passing the wheel track loading and the wear of the pavement surface, so the amount of absorbed water decreases which is the same as mentioned by ***(Fang and ONG, 2005)***.

**Table 2.** Test result for amount of absorbed water for (116ml/min.) rain intensity for two hours.

| Amount of Absorbed Water (letters) | Days Numbering | Date of Test | Time of Vertical Drainage Ending (Sec) |
|---|---|---|---|
| 45.5 | 22 | 08/05/2016 | 188 |
| 35.5 | 23 | 09/05/2016 | 188 |
| 25.5 | 24 | 10/05/2016 | 188 |
| 20.5 | 31 | 17/05/2016 | 240 |
| 20.5 | 38 | 24/05/2016 | 241 |
| 15.5 | 40 | 26/05/2016 | 241 |
| 15.5 | 60 | 15/06/2016 | 243 |
| 10 | 64 | 19/06/2016 | 245 |

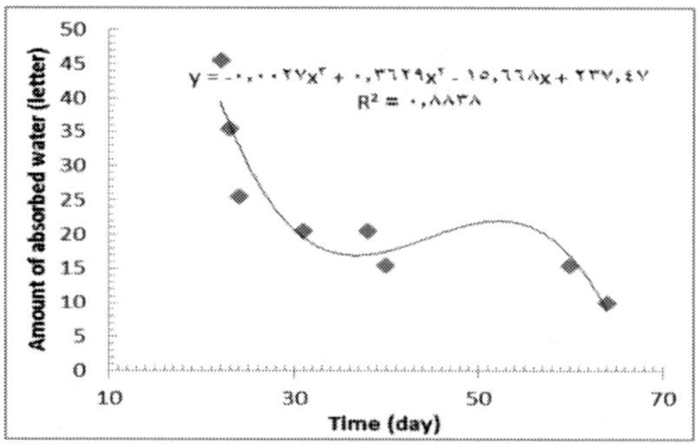


**Fig. 8.** The Relationship Between The Amount of Absorbed Water and The Time (day) for Rain Intensity (116ml/min.) for Two Hours.

Fig.9 shows the relationship between the amount of absorbed water and the time of vertical drainage ending for (116ml/min.) rain intensity for two hours. By passing the days and increasing rain falling test under wheel track repetitions on the pavement, the clogging materials and other distress that appeared on the pavement causes the accumulation of water on the surface which needs more time to evaporate and seep.

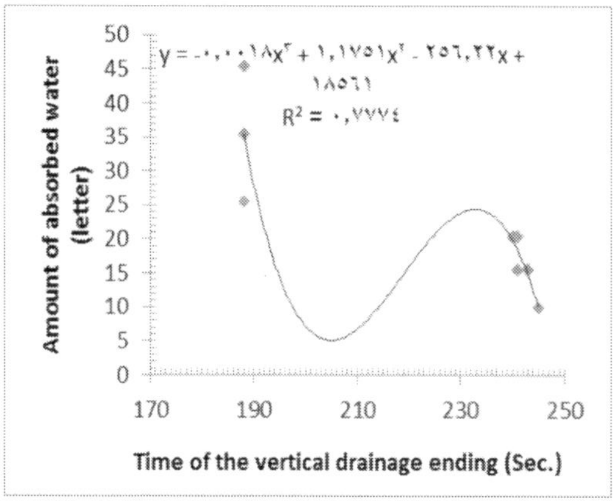


**Fig.9.** The Relationship Between Amount of Absorbed Aater and Time of Vertical Drainage Ending for (116ml/min.) Rain Intensity for Two Hours.

Table 3 and Fig.10 show the results of amount of absorbed water that is obtained after one hour (60 min) of rain fall.

**Table 3.** Test result for amount of absorbed water for (116ml/min.) rain intensity for one hour.

| Date of Test | Days Numbering | Amount of Absorbed Water (letter) | Time of Vertical Drainage Ending (Sec) |
|---|---|---|---|
| 17/04/2016 | 1 | 50.5 | 120 |
| 18/04/2016 | 2 | 47.5 | 180 |
| 12/05/2016 | 26 | 10.5 | 240 |
| 31/05/2016 | 45 | 15.5 | 241 |
| 26/06/2016 | 71 | 5.5 | 247 |

Fig.10 shows that the amount of absorbed water decreases as time increases and its highest value which is 50.5 letters at the first day of testing (17 April) and arrived to its lowest value, 5.5 letters, at the end of the third month of the test at (26 June) with percentage of 90% after 71 days, as shown in Table 3. By increasing the number of days and increasing clogging materials of fine particles deposited on the surface of the pavement which result in passing the wheel track loading and wearing the pavement surface, the water seepage decreases.

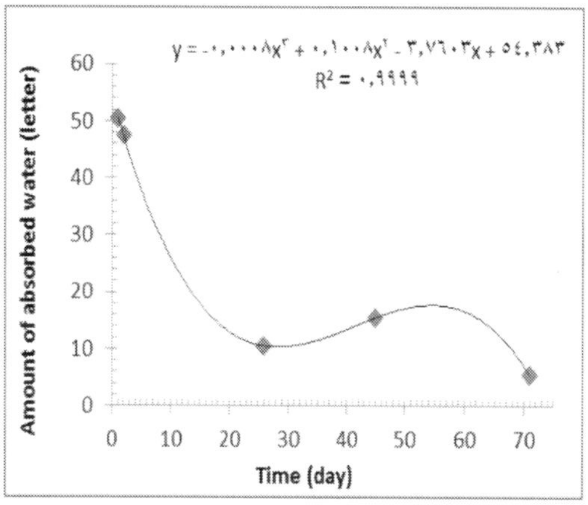


**Fig. 10**. The Relationship Between The Amount of Absorbed Water and The Time for Rain Intensity (116ml/min.) for One Hour.

Fig.11 shows the relationship between amount of absorbed water and time of vertical drainage ending for (116ml/min.) rain intensity for one hour.

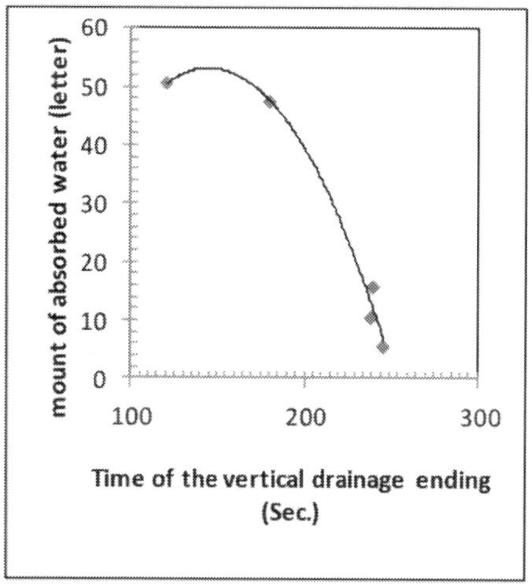


**Fig.11.** The Relationship Between Amount of Absorbed Water and Time of Vertical Drainage Ending for (116ml/min.) Rain Intensity for Two Hours.

The amount of absorbed water in the long duration of rain fall is less than that for the short duration of rain fall because more of the water is runoff, which is the same as indicated by (***Day et al., 1981; Hunt et al., 2002; Valavala et al., 2006; Collins et al., 2008***) that was taken for permeable pavement. Moreover, by increasing the clogging materials that penetrated to the pavement by increasing number of days, the water seepage decreased.

## 8 Test result for low intensity rain

This step represents the result study of amount of absorbed water, ending time of seepage and vertical drainage for low intensity of rain (58 ml/Min.) as selected previously.

### 8.1 Amount of absorbed water

Figures compiled of more than one sub-figure presented side-by-side, or stacked. If a multipart figure is made up of multiple figure types (one part is lineart, and another is grayscale or color) the figure should meet the stricter guidelines.
Table 4 and Fig.12 show the results for the amount of absorbed water through the pavement layer during long term and short term period for the chosen low rain intensity.
The amount of the in letter was calculated before beginning the test, and then after two or one hour rain fall and at the end of the test. Then the difference between the amount of water before starting the test and amount of water after ending the test is referred to the amount of absorbed water by pavement layer.

**Table 4.** Test result for amount of absorbed water for (58ml/min.) rain intensity for two hours

| Date of Test | Days Numbering | Amount of Absorbed Water (letters) | Time of Vertical Drainage Ending(Sec) |
|---|---|---|---|
| 26/04/2016 | 10 | 55.5 | 180 |
| 27/04/2016 | 11 | 50.5 | 180 |
| 28/04/2016 | 12 | 45.5 | 180 |
| 03/07/2016 | 78 | 5.5 | 247 |
| 18/07/2016 | 93 | 5.5 | 248 |
| 24/07/2016 | 99 | 10.5 | 250 |
| 25/07/2016 | 100 | 5.5 | 250 |
| 28/07/2016 | 103 | 5.5 | 253 |

Fig.12 shows the relationship between the amount of absorbed water and the time for continuously two hours rain under track loading for low intensity of rain (58 ml/min.). This Figure shows that the amount of water that is absorbed through the pavement layer decreases with increase in the time. For the first test day of this rain intensity the amount of water was 55.5 letters and then decreases to reach 5.5 letters after 78 days of test. The amount of absorbed water reached its highest point at the first day of test.
When leaving the model without testing, the amount of water increases, because the pavement layers need to absorb more quantity of the water to be saturated, and after saturation it absorbs less water quantity.
The water seepage decreases rapidly by increasing clogging materials of fine particles deposited on the surface of the pavement which result in passing the wheel track loading and the wear of the pavement surface and clogging materials of salts and fine sands and other adulterant materials of the water, so the amount of absorbed water decreases.

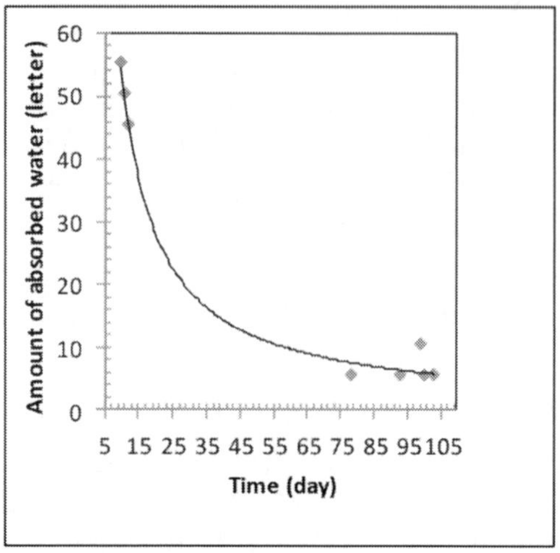


**Fig.12** The Relationship Between The Amount of Absorbed Water and The Time for Rain Intensity (58ml/min.) for Two Hours.

Fig.13 shows the relationship between amount of absorbed water and time of vertical drainage ending for (58 ml/min.) rain intensity for two hours. By passing the days, the amount of water absorbed will decrease. The clogging materials and other distress that appear on the pavement causes the accumulation of water on the surface which needs more time to evaporate and seep so the time of vertical drainage ending increases.

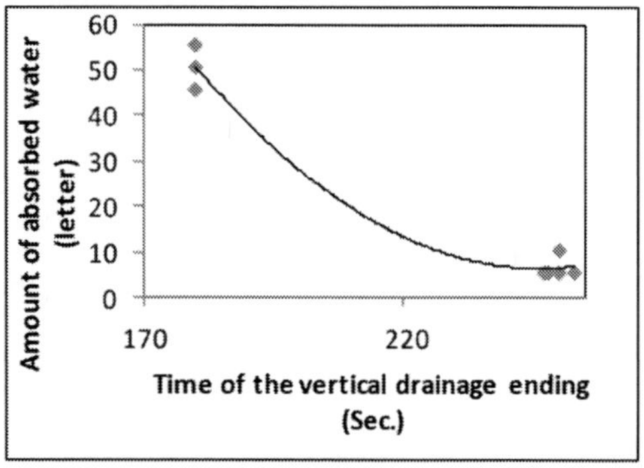


**Fig.13.** The Relationship Between Amount of Absorbed Water and Time of Vertical Drainage Ending for (58 ml/min.) Rain Intensity for Two Hours.

Table 5 and Fig.14 show the result of amount of absorbed water for one hour rain intensity for low intensity rain.

https://doi.org/10.1051/matecconf/201816201036

**Table 5**. Test result for amount of absorbed water for (58ml/min.) rain intensity for one hour

| Date of Test | Days Numbering | Amount of Absorbed Water (letters) | Time of Vertical Drainage Ending (Sec) |
|---|---|---|---|
| 28/04/2016 | 12 | 25.5 | 180 |
| 04/05/2016 | 18 | 35.5 | 188 |
| 05/05/2016 | 19 | 30.5 | 189 |
| 26/07/2016 | 101 | 2.5 | 253 |
| 27/07/2016 | 102 | 2.5 | 253 |

Fig.14 shows that the amount of absorbed water decreases with increase in test days and reaches to its lowest value 2.5 letters at the ending day of the test as shown in Table 5. At the first test day of this rain intensity the amount of absorbed water was 25.5 letters but after leaving it for six days it increases 10 letters to reach 35.5 letters. So after leaving the pavement without rain fall the amount of absorbed water increases.

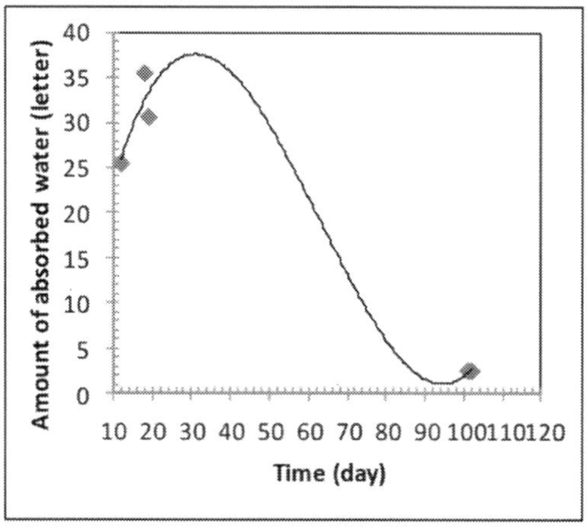


**Fig.14** The Relationship Between The Amount of Absorbed Water and The Time for Rain Intensity (58ml/min.) for One Hour

Fig.15 shows the relationship between amount of absorbed water and time of vertical drainage ending for (58 ml/min.) rain intensity for one hour.

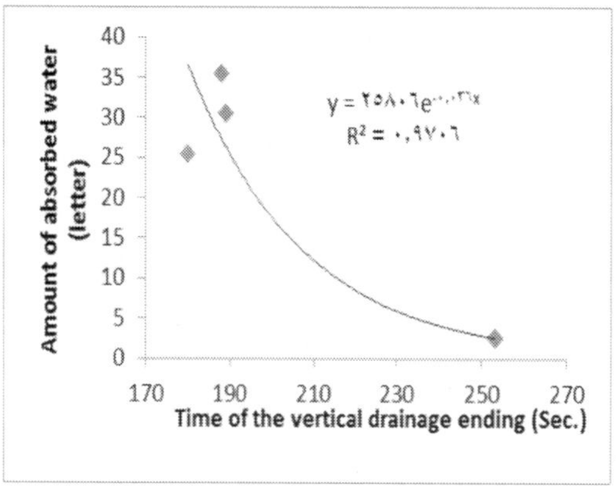


**Fig.15.** The Relationship Between Amount of Absorbed Water and Time of Vertical Drainage Ending for (58 ml/min.) Rain Intensity for One Hour.

Pavement saturation and the clogging material affects the amount of absorbed water. By increasing time, it causes a decrease in absorbing water, the amount of absorbed water in the high rain intensity and for long duration of rain fall is less than that for low intensity rain for short term rain fall. In addition, if there was continuous rain fall for two different intensities of rain with its two different durations of time, saturation pavement and clogging materials have greater effect on absorbing water.

## 9 Rutting result

The wheel passes 240 repetition of passage in one hour and 480 repetition of passage in two hours with 0.2 $m/Sec$ under rain intensity for the required period of time. In the analysis of potential rutting, the rut depth appeares after 960 wheel repetition, as in Table 6 and increases slightly 0.5 after 2640 of wheel passing and increases randomly, then produced 22.5 mm rut depth after 11040 wheel repetition of passages. The test results plotted in Fig.16 show the relationship between the number of load passages and rut depth for the pavement of two different rain intensities. Initially the rut is slowly developed in more than 10 mm until it reaches around 5760 wheel repetition pass.

**Table 6.** Rutting depth result

| Intensity ml/min | Time of Test (minute) | No. of wheel repetition | Rut depth (mm) | Date of test | No. of days |
|---|---|---|---|---|---|
| 116 | 60 | 240 | 0 | 17/04/2016 | 1 |
| 116 | 60 | 480 | 0 | 18/04/2016 | 2 |
| 58 | 120 | 960 | 0.5 | 26/04/2016 | 10 |
| 58 | 120 | 1440 | 1 | 27/04/2016 | 11 |
| 58 | 120 | 1920 | 1.5 | 28/04/2016 | 12 |
| 58 | 60 | 2160 | 2 | 28/04/2016 | 12 |
| 58 | 60 | 2400 | 2.5 | 04/05/2016 | 18 |
| 58 | 60 | 2640 | 3 | 05/05/2016 | 19 |
| 116 | 120 | 3600 | 5.5 | 09/05/2016 | 23 |
| 116 | 120 | 4080 | 6.5 | 10/05/2016 | 24 |
| 116 | 60 | 4320 | 7 | 12/05/2016 | 26 |
| 116 | 120 | 5280 | 9 | 17/05/2016 | 31 |
| 116 | 120 | 5760 | 10 | 24/05/2016 | 38 |
| 116 | 120 | 6240 | 12 | 26/05/2016 | 40 |
| 116 | 60 | 6480 | 14 | 31/05/2016 | 45 |
| 116 | 120 | 6960 | 14.5 | 15/06/2016 | 60 |
| 116 | 120 | 7440 | 15.5 | 19/06/2016 | 64 |
| 116 | 60 | 7680 | 16 | 26/06/2016 | 71 |
| 58 | 120 | 8160 | 17 | 03/07/2016 | 78 |
| 58 | 120 | 9120 | 19 | 18/07/2016 | 93 |
| 58 | 120 | 9600 | 20 | 24/07/2016 | 99 |
| 58 | 120 | 10080 | 20.5 | 25/07/2016 | 100 |
| 58 | 60 | 10320 | 21 | 26/07/2016 | 101 |
| 58 | 60 | 10560 | 21.5 | 27/07/2016 | 102 |
| 58 | 120 | 11040 | 22.5 | 28/07/2016 | 103 |

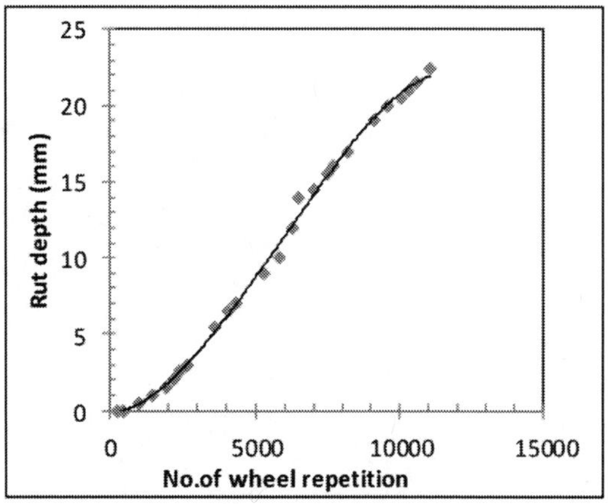


**Fig.16.** Rut Depth Results of Wheel Repetitions.

## 10 Conclusions

The following conclusions can be made from this study:

1) The amount of water of the absorbed water decreases from 50.5 letters on the first day of rain fall to 5.5 letters on the final day to collect 89% decrease after 71 days of testing for high rain intensity. And decreases from 55.5 letters to 2.5 letters after 92 days rain fall to record 95 % decreasing for 58 ml/m rain intensity.
2) Although by increasing the clogging materials of fine particles that are deposited on the surface of the pavement that result of passing the wheel track loading and wear of the pavement surface, and other clogging materials such as salt in the water which penetrate to the pavement and seal the voids and affect its water seepage and causes decrease in the amount of absorbed water.
3) The rut depth appeared after 960 wheel repetitions which was 0.5 mm and increased to 19 mm after 9120 repeated wheel passing to record 97 % increase after 93 days of testing and rain fall under different rain intensity with two different time durations.

## References

[1] FWA .Tien Fang and ONG. Ghim Ping. Analysis and Design of Vertical Drainage Geosynthwtic-Reinforced Porous Pavement for Roads and Car parks. Journal of the Eastern Asia Society for Transportation Studies, Vol. 6, pp. 1286 – 1301 (2005).

[2] G. E. Day, D. R. Smith, and J. Bowers,. Runoff and pollution abatement characteristics of concrete grid pavements. Bulletin 135, Virginia Polytechnic Institute and State University (1981).

[3] B. Hunt, S. Stevens, and D. Mayes,. Permeable pavement use and research at two sites in Eastern North Carolina. In *Global Solutions for Urban Drainage*. Proc., 9*th* Int. Conf. on Urban Drainage. Portland, Oreg.:ASCE (2002).

[4] S. Valavala, , F. Montes, and L. M. Haselbach, Area-rated rational coefficients for Portland cement pervious concrete pavement. *Journal of Hydrologic Engineering*11(3): 257-260 (2006).

[5] K. A. Collins, W. F. Hunt, and J. M. Hathaway. Hydrologic comparison of four types of permeable pavement and standard asphalt in eastern North Carolina. *Journal of Hydrologic Engineering* (in press) (2008a).

[6] E. Z. Bean, W. F. Hunt, and D. A. Bidelspach. Field survey of permeable pavement surface infiltration rates. *Journal of Irrigation* (2007a).

# Evaluation performance of base course mixture containing reclaimed asphalt pavement (RAP)

*Karim* Al helo [1,*], *Zaynab* Qasim [1], and *Ahmed* Majeed [1]

[1] Building and Construction Engineering Department, University of Technology, Baghdad, Iraq

**Abstract.** This paper presented the effect of addition of Reclaimed Asphalt Pavement (RAP) on performance of mixture with (20%, 30%, 40% and 50%) RAP as proportion of asphalt mix. To compare with virgin mixture marshall flow and stability test was used to evaluated durability performance with (0,1,3, and 7) days immersion and Wheel Truck test was used to evaluated the resist to rutting. The best gradation and optimum asphalt content was selected according to Superpave system. Superpave Gyratory Compactor (SGC) was used to compact mixture with 100-mm diameter. The test results indicated that addition of RAP to mixes showed significant increase on resistance of Durability and Rutting.

## 1 Introduction

Recycling Hot Mix Asphalt (HMA) material results in a reusable mixture of aggregate and asphalt binder known as Reclaimed Asphalt Pavement (RAP) , [1]. .Asphalt paving technologists reacted to this situation by developing recycling methods to reduce the demand on asphalt binder and thereby, reducing the costs of asphalt paving mixtures. Many practices that were initially developed during that period are still in use today and have become part of routine operations for pavement construction and rehabilitation, [2]. The asphalt pavement recycling methods include hot Mix recycling, cold and hot in place recycling, and full depth reclamation, [3]. There are three major causes for preferring the RAP on the virgin materials: the increase in asphalt binder and crude oil cost, reduction of quality aggregates, and the pressing demand to save the environment, [4]. For performance tests Li et al. in 2005 studied the properties of low and high temperature of RAP binder by using two fresh asphalt grades; (PG 58-28) and (PG 58-34) and mixes with three contents of RAP (0, 20%, and 40%), [5].

It found that G*/sinδ value is increasing with increase in RAP content which means its more stiffer than mix without RAP. At low temperature, e.g. -24 °C, both the stiffness from failure temperature and the BBR test increase and the m-value is decreases when percentage of RAP increases. Also, they indicated that the fresh asphalt has less stiffness than the RAP asphalt. Only when the percentage of RAP is higher than 20% by weight a considerable variation is shown. The preceding researches on mechanical properties of RAP mixtures indicated that using RAP give a good performance to resist rutting but lower resistance to low temperature cracking due to affected by stiff asphalt of the RAP. The change in properties of asphalt can be significant only when using high content of recycled asphalt pavement (more than 20-30%). Pradyumna, et al. in 2013 indicated that hot mix recycling with 20% RAP can have better performance to those of new conventional hot mix asphalt mixtures, [6]. Various performance tests such as Retained Stability, Indirect Tensile Strength (ITS) and wheel tracking test were adopted. For Retained Marshall Stability for virgin mix was (85.6%) and for RAP mix was (88.1%). While rut depth for virgin mix was obtained 8.20 mm, whereas for 20 % RAP mix, it was 7.6 mm only after 20,000 passes. Van et al. in 2015 designed surface mixtures (1/2 in) content high RAP and binder graded PG 70-22, [7]. Three Hot Mix Asphalt (HMA) test sections with RAP amounts of 30.0, 35.5, and 39.2% were constructed. RAP materials were fractionated to meet Superpave mix design requirements. After 20,000 passes were applied using a Hamburg wheel tracking device, rut depths were less than 3 mm for all field mixtures. Izaks et al., in 2015 studied two RAP sources, first source of RAP is (A7); the second source of RAP is (A6) to be compared with control mixture (AC16), [8].

They noted that the largest average rut depth of 3.4 mm for the AC-16base/bin. (A6) with 30% RAP mixture with lowest asphalt content shows high resistance to permanent deformations, having an average rut depth value of 1.8 mm. Celauro et al., in 2010 used high content of reclaimed asphalt (0%, 40%, and 50%), which were investigated in a laboratory to design surface course (types U3 and U5) with thickness (3 and 5 cm, respectively), binder courses (types BiK2 and BiK3 and base course (type BiK1) with different binder content (K0, K1, K2, and K3). Rutting test was adopted for surface course only (U5), [9]. Marshall Stability for all mixtures was studied, it can be noted that each mixture which was taken into account in that study no only met the minimum levels for Marshall Stability according to

* Corresponding author: karim.alhelo@yahoo.com

© The Authors, published by EDP Sciences. This is an open access article distributed under the terms of the Creative Commons Attribution License 4.0 (http://creativecommons.org/licenses/by/4.0/).

Italian Specifications but also with higher performances pursued by this study.

Also showed that the rut depth is decreasing with increase in RAP content which is affected by hardening the total binder in the mixture as well as it can be observed the mixture with recycled materials have good rutting resistance even though with high asphalt content lees than rut depth for control mix.

## 2 Materials

### 2.1. Asphalt cement

The grade of binder that was used (AC 40-50) or PG (64-16) was obtained from the Al-Dora Refinery, south-west of Baghdad. The testing results conformed to Iraqi specification, [10] and ASTM Requirement. The physical properties of binder are show in Table 1.

**Table 1.** Properties of Asphalt Cement

| Property | ASTM designation | Test results | SCRB specification |
|---|---|---|---|
| Penetration, (25°C, 100 g, 5s), 0.1 mm | D-5 | 47.3 | 40-50 |
| Softening Point. (◦C) | D-36 | 51.5 | - |
| Ductility at 25 C, 5cm/min, (cm) | D-113 | >100 | >100 |
| Specific Gravity | D-70 | 1.03 | - |
| Rotational Viscometer, at 135°C (Pa.sec) | D-4402 | 0.537 | - |
| Rotational Viscometer, at 60°C (Pa.sec) | D-4402 | 0.15 | - |

### 2.2. Aggregate

The crushed quartz aggregate used in this work was obtained from Al-Sedour quarry; this aggregate widely used in local asphalt paving in Baghdad. The physical properties of used aggregate were showed in Table 2.

### 2.3. Mineral Filler

The filler used in this work was Portland cement brought from Al-Mas Company which provided from local market. The physical properties of Portland cement are presented in Table 3.

### 2.4 Reclaimed Asphalt pavement

The reclaimed asphalt pavement materials (RAP) was brought from stoke of Reclaimed Asphalt for Mayoralty of Baghdad-project office at Altalbia-region in Baghdad city. Extraction test was conducted on the reclaimed asphalt pavement to extraction the asphalt from aggregate and filler. The testing has been procedure according to (ASTM-D2172). The percent of asphalt cement was (4%) and gradation of RAP aggregate is determined and added to new materials according to (Iraqi specifications SCRB-R9), as shown in Table 4, [11].

**Table 2.** Physical properties of aggregate

| Laboratory Test | ASTM designation | Test results | SCRB Specification |
|---|---|---|---|
| Coarse aggregate | | | |
| Apparent specific gravity | C-127 | 2.66 | - |
| Bulk specific gravity | C-127 | 2.61 | - |
| Water absorption, % | C-127 | 0.66 | - |
| Percent wear by (Los Angeles abrasion), % | C-131 | 14.6 | 35-45 Max |
| Soundness loss by sodium sulfate solution, % | C-88 | 3.12 | 10-20 Max |
| Fractured pieces, % | D 5821 | 96% | 95 Min |
| Fine aggregate | | | |
| Apparent specific gravity | C-128 | 2.66 | - |
| Bulk specific gravity | C-128 | 2.62 | - |
| Water absorption, % | C-128 | 0.68 | - |

**Table 3.** physical properties of Portland cement

| Property | Test results |
|---|---|
| Specific Gravity | 3.2 |
| %Passing Sieve No.200 or (0.075 mm) | 100 |

## 3 Mix design

Superpave design system was adopted to select design aggregate structure and design binder content for getting optimum asphalt pavement. Aggregate gradation was selected according to ASTM and Iraqi specifications (SCRB, R9) for base course. Figure 1 shows the selection gradation of blend. Four content of binder were selected as (3.45%, 3.95%, 4.45%, and 4.95%). The design asphalt binder content is established at 4.0% air voids the design asphalt binder content was 4.42%. All other mixture properties were checked at the design asphalt binder content to verify that they meet specifications for base course.

**Table 4.** Selected gradation of RAP aggregate

| Standard Sieves (mm) | % Passing by weight, (SCRB-R9) | %Passing by weight (RAP) |
|---|---|---|
| 19 | 100 | 100 |
| 12.5 | 90-100 | 99 |
| 9.5 | 76-90 | 97 |
| 4.75 | 44-74 | 73 |
| 2.36 | 28-58 | 54 |
| 0.3 | 21-5 | 30 |

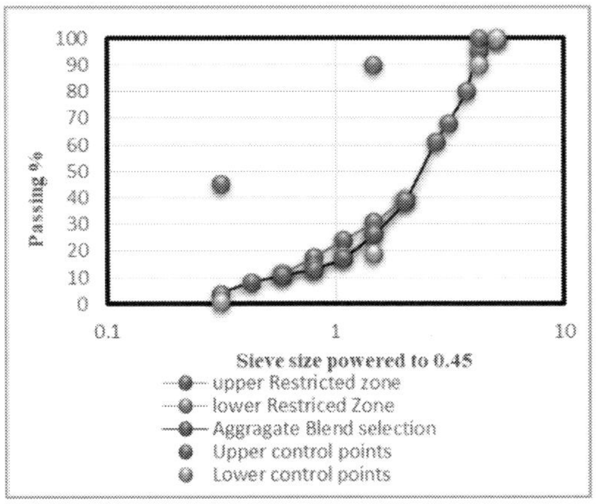


**Fig. 1.** Selected blend of aggregate gradation

### 3.1 Sample preparation

Specimens are required to be mixed and compacted under equiviscous temperature conditions corresponding to viscosities of 0.17±0.02 and 0.28±0.03 Pascal-seconds (Pa·s), respectively. Depending on traffic level in Baghdad ESALS ranged (3 million to < 10 million), [12]; each specimen compacted using Superpave Gyratory Compactor (SGC) to design number of gyrations (100 cycle) and diameter of samples was 100 mm.

The Sample was prepared as following:

- Heating the Reclaimed asphalt pavement at 110°C for (1-2) hour.
- Heating the Aggregate to (170°C).
- Heating the asphalt at (160°C).

After heating the aggregate and asphalt mixed together and then heating reclaimed asphalt pavement is added to mixture. The percent of reclaimed asphalt pavement (RAP) was added are (20%, 30%, 40%, and 50%) of total mix weight.

### 3.2 Durability of hot asphalt mixtures

Durability of an asphalt mixture is defined as its resistance to weathering and the abrasive action of traffic. In terms of its application to asphalt paving materials, durability can be defined as the ability of the materials in the asphalt pavement structure to resist the effects of environmental conditions, such as water, ageing and temperature variations without any significant deterioration for an extended period for a given amount of traffic loading, [13].

To evaluation durability, a mixture is subjected to environmental conditioning, and a mixture property associated with load-related or environmental distress is measured before and after the conditioning process.

Abrasion characteristics of the aggregate in the mixture must also be considered in the evaluation of durability. The greater the protection by asphalt concrete, more durable the mix will be. The fewer air voids in the Total mix, the slower will be the deterioration of the asphalt concrete itself. Theoretical Approach: It's including three steps as following:

a. Marshall Stability

Marshall Stability is calculated from the following equation:

$$So = o * R * T \qquad (1)$$

Where, So = stability numeral, kN
o = stability timepiece reading on Marshall test, kN
R = Proving ring calibration
T = the matter test correction factor

b. Retained Marshall Stability (RMS)

The Retained Marshall Stability is calculated from the following equation:

$$RMS=(Si/So) * 100\% \qquad (2)$$

Where: RMS = Retained Marshall Stability (%)
Si = maximum stability in conditioned set based on times series
So = maximum stability in unconditioned set (0 days)

c. Durability Index (DI)

The formula which used to calculate durability index is shown in following equation:

$$DI = \left(\frac{1}{2tn}\right)\sum_{t=1}^{n-1}(S_i - S_{i+1}) * \left(2tn - (t_{i+1} - t)\right) \qquad (3)$$

Where: Si+1 = percent retained strength at time ti+1
Si = percent retained strength at time ti
ti, ti+1 = immersion time (from beginning of test).

Durability Index was defined as the average strength loss area enclosed between the durability curves. Figure 2 shows a schematic description of durability curve and Figure 3 shows samples for testing procedure.

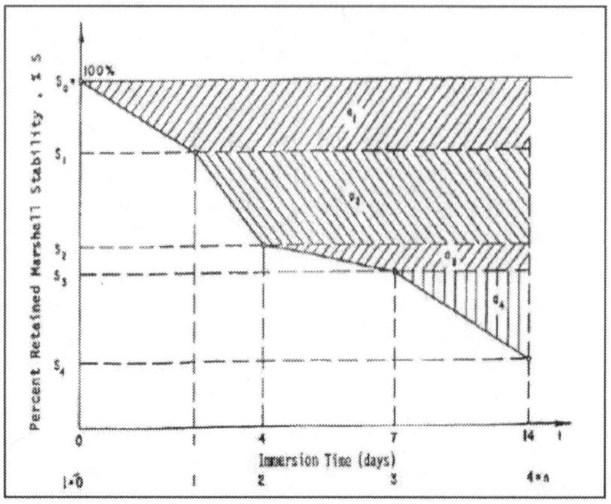


**Fig 2.** Schematic Description of Durability Curve

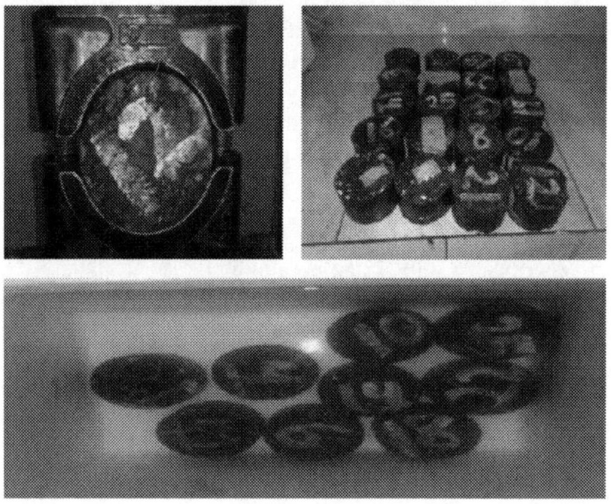

**Fig. 3.** Tested Procedure Samples

# 4 Results

## 4.1 Marshall stability and flow results

Figure 4 shows the relation between stability and RAP percent heated to 110°C. It can be concluded that stability is increased with increase RAP content. When addition of 30% RAP content, the stability is increased with 21.6% and for addition of 40%; stability is increased with 41.9 %.

Figure 5 shows the relation between flow and RAP percent heated to 110°C. It can be illustrated that the increase in inclusion of the RAP causes a decrease in flow for example the addition of 30% RAP, the flow is decreased with 2.7%, and an addition of 50% RAP, flow is decreased with 8.1%.

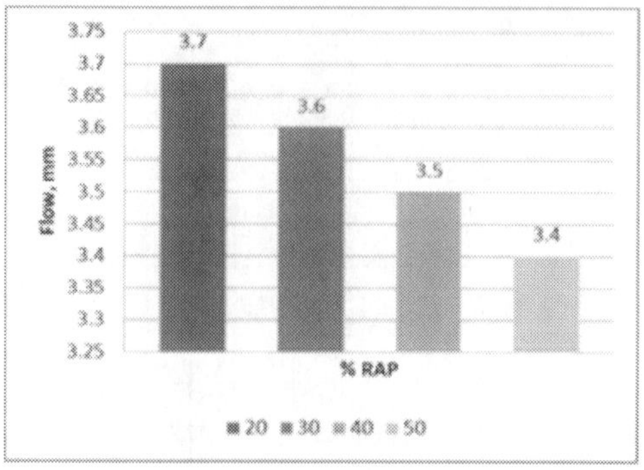


**Fig.4.** Stability with % RAP for 1 day

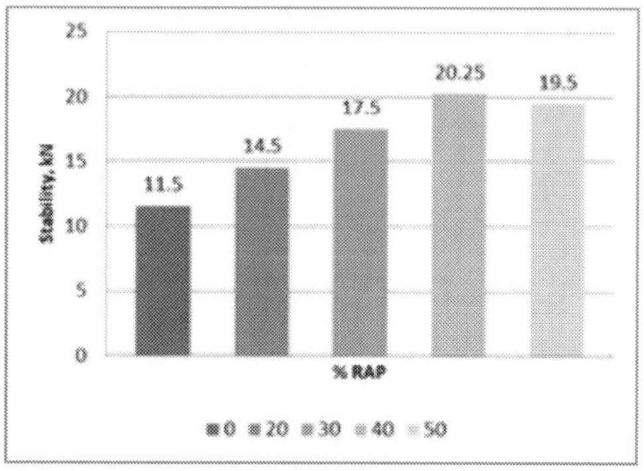


**Fig.5.** Stability for RAP heated to 110°C

## 4.2 Durability test results

The Retained stability (RS) is decreased with increase immersion day for all percent addition. For example, for 20% RAP content the RS value is 98% for one day emersion and is 82.4% for seven days immersion as shown in Table 5. It can be observed that RS is increased with increase RAP content; e.g., the RS for 7 days is 82.4% for 20% addition and is 87.8% for 50% RAP content.

**Table 5.** Retained stability (%) with days

| % RAP | RS for 1 day | RS for 3 day | RS for 7 day |
|---|---|---|---|
| 20 | 98.0 | 92.6 | 82.4 |
| 30 | 97.2 | 93.3 | 85.6 |
| 40 | 96.4 | 92.9 | 88.1 |
| 50 | 95.1 | 92.7 | 87.8 |

Figures 6 and 7 illustrated the relation between stability and immersion day for all percent of RAP additions. It can be noticed that stability for all RAP percent is higher than the control mix. Stability is increased with increase in RAP content until 40% then it decreases for 50% but still above the stability values for 30% RAP content for all immersion days, e.g. the stability value with one day immersion day is increased with (22.3%, 48.7%, 73.5% and 41%) for (20%, 30%, 40% and 50%) RAP content respectively. And the stability value with three immersion days is increases with (26%, 52.1%, 76% and 69.5%) for (20%, 30%, 40% and 50%) RAP content, respectively. While stability value with seven days immersion days is increased with (24.4%, 57%, 88.7% and 83.6%) for (20%, 30%, 40% and 50%) RAP content, respectively.

The addition of RAP that forms from asphalt and aggregates in the "RAP Mix" have developed the Marshall stability and the Retained of Marshall Stability above the "Control Mix". It is confirmed that this can be attributed to the fact that RAP has hardened asphalt, which will cause increased in the stability due to higher asphalt viscosity. The virgin asphalt is having less viscosity than RAP Asphalt "Hardened asphalt" if they have the similarity in grade. For that, when the mixes are subjected to immersion, the stability of the "RAP Mix" will be less affected by hot water than that of the "Control Mix".

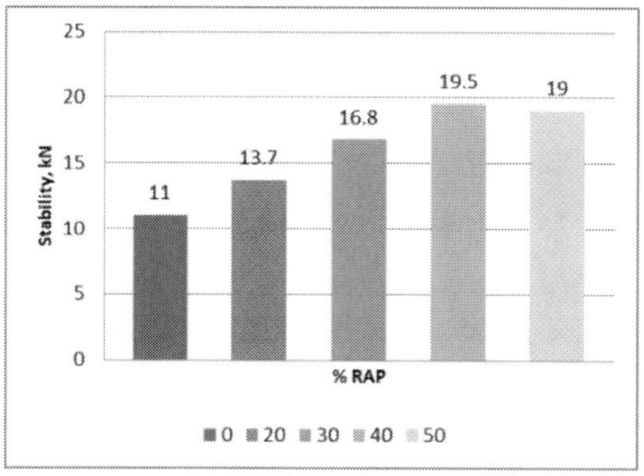


**Fig. 6.** Stability with % RAP for 3 day

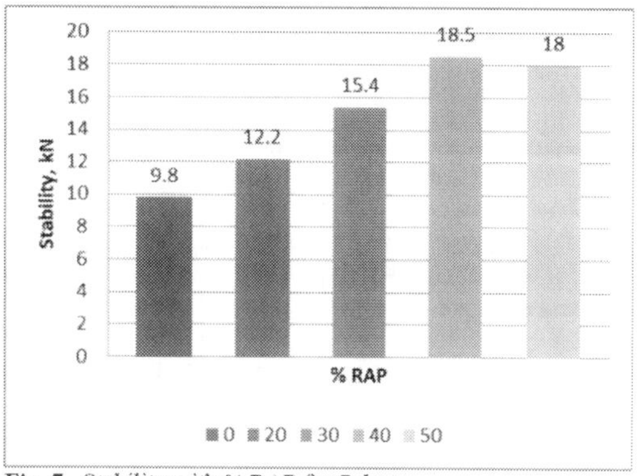


**Fig. 7.** Stability with % RAP for 7 days

### 4.3 Durability index

Figure 8 illustrates the durability index for RAP heated to 110°C with RAP content, it can be noted that durability index DI decreases with increases RAP content, for example DI decrease with (17.3%) and (45.1%) for 20% and 50%RAP content, respectively as shown in Table 6.

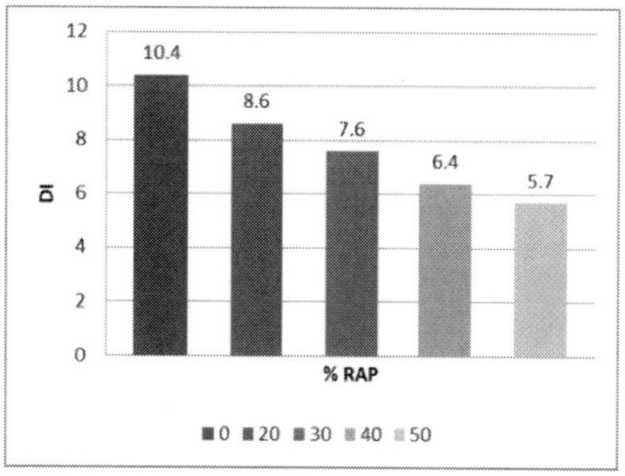


**Fig.8.** DI for RAP Heated to 110°C

**Table 6.** Relation Between Durability Index and RAP content

| RAP % | DI |
|---|---|
| 0 | 10.4 |
| 20 | 8.6 |
| 30 | 7.6 |
| 40 | 6.4 |
| 50 | 5.7 |

### 4.4 Rutting test results

The resistance to permanent deformation increased with increases of RAP content, e.g. Rut depth is deceased with increase RAP content. Figure 9 illustrated that the Rut depth is decreased by (15.4%) when the

RAP is increases from (20% to 30%) percent, while it is decreased by (31.3%) and (35.3%) when the RAP is increased from (20% to 40%) and up to 50%, respectively. Table 7 illustrates the values of rut depth at end of the test.it can be observed that the rut depth is lower than control mix (0% RAP) for all RAP content, e.g., the rut depth with 20%, 30%, 40% and 50% RAP content is decreased by (8.8%, 22.9%, 37.4% and 41.4%), respectively. Table 8 illustrated the ratio of decreases in rut depth with increases RAP content compared with the original mix. This refer to the addition of Reclaimed Asphalt Pavement is improving the resistance of rutting for the mix. The RAP containing mixes become stiffer compared to the mix without RAP and thus has better resistance to permanent deformation.

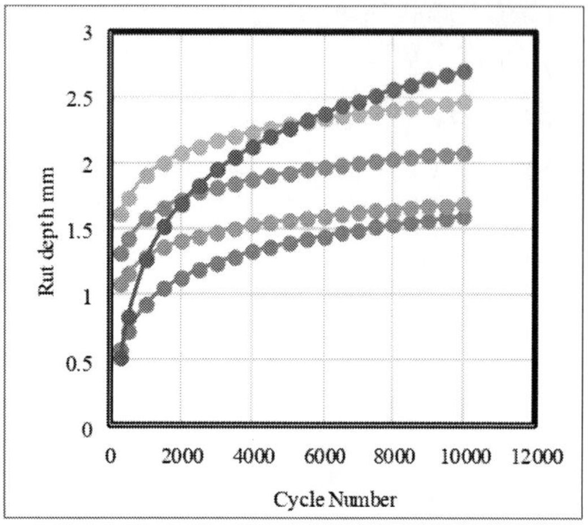


**Fig. 9**. Relation between Cycles number with Rut depth for RAP heated to 110°C

**Table 7.** Relation between rut depths with % RAP

| RAP % | Max rut depth RAP heated to 110°C, mm |
|---|---|
| 20 | 2.46 |
| 30 | 2.08 |
| 40 | 1.69 |
| 50 | 1.59 |
| 20 | 2.46 |

**Table 8.** Decrease Ratio of Rut Depth with RAP

| RAP % | Decrease Ratio for RAP heated to 110 °C |
|---|---|
| 20 | 8.8 |
| 30 | 22.9 |
| 40 | 37.4 |
| 50 | 41.4 |

## 5 Conclusions

- Marshal Stability results show that the increasing RAP will increase the stability for RAP heated to 110° C or 140° C. The stability increased by (22.3%) (48.7%), (73.5%) and (69.4%) for (20%), (30%), (40%) and (50%) RAP content respectively for RAP heated to 110°C.
- The Durability results for RAP heated to 110°C show that increase in RAP causes an increasing in resistance to moisture damage for all percent of RAP. The stability increases by (26%), (52%), (76%) and (69.5%) for one day emersion and increases by (24.5%), (52.7%), (77.2%) and (72.7%) for three days and increases by (24.4%) (57.1%) (88.7%) and (83.6%) for seven days; for (20%), (30%), (40%) and (50%) of RAP content, respectively. The DI values are (8.6) (7.6) (6.4) (5.7) for (20%), (30%), (40%) and (50%) RAP content respectively
- Durability Index results decrease with increase in RAP content
- Wheel-Tracking permanent deformation results for asphaltic base course Mixes indicate that increase in RAP will decrease the permanent deformation the rut depth decreased by (8.8%), (22.9%), (37.4%) and (41.1%) for (20%), (30%), (40%) and (50%) RAP content, respectively.

## References

1. I. L. Al-Qadi, Q. Aurangzeb, S. H. Carpenter, W. J. Pine, J. Trepanier, ICT R27-37, Champaign (2012)
2. West, R.., National center for asphalt technology Report, Auburn University, Alabama (2010)
3. L. Wang, E. Masad, *Proceedings of the Symposium on the Mechanics of Flexible Pavements*, ASCE, **176** (2007)
4. N. Sabahfar, Department of Civil Engineering, Kansas State University (2012).
5. X. Li, M.O. Marasteanu, R.C. Williams, and T.R Clyne, TRB **2051**, 90–97 (2008)
6. T. A. Pradyumna, A. Mittal, P. Jain, PSBS, **104**, 1149-1157 (2013)
7. V. Winkle, H. Lee, R. C. Williams, S. Schram, TRB **94** (2015)
8. R. Izaks, V. Haritonovs, I. Klasa, M. Zaumanis, PE. **114**, 676-684 (2015)
9. C. Celauro, C. Bernardo, B. Gabriele, J. RCR, 54, 337-347 (2010).
10. SCRB, *General Specification for Roads and Bridges, Section R/9*, Revised Edition, State

Corporation of Roads and Bridges, Ministry of Housing and Construction, Iraq (2003).

11. K. H Al Helo, Z. I. Qasim, A. D. Majeed, GJESRM, ISSN 2349-4506 (2016).
12. Z. I. Qasim, A.H. Abed, *the 2nd International Conference on Buildings, Construction and Environmental Engineering (BCEE2)*, Lebanon, (2015).
13. M. Y. Fattah, Z. I. Qasim, Y. A. Zuhier, GJESRM, ISSN 2349-4506 (2017).

# Demand estimation of bus as a public transport based on gravity model

*Noor* Asmael[1*] and *Mohanned* Waheed[2]

[1]Faculty of Engineering, Mustansiriyah University, Baghdad, Iraq
[2] Building and Construction Engineering Department,University of Technology, Baghdad, Iraq

**Abstract.** Bus as a public transport is a suitable service to meet the travel demand between any two zones. Baghdad faced with severe traffic problems along with the development in city size and economy. Passengers have to wait lots of time during commutation to work because of the serious traffic jams. In the last years, rate of car ownership has increased as income levels have gone up and cars have become a preferable mode of transport. Bus, as the only public mode of transport available, is suffering from inconvenience, slowness, and inflexibility. A big emphasis must be given to the public transport system because it introduces an active utilization of limited resources, energy and land. This study determines the demand of public routes for buses using boarding / alighting values to generate a model and assign these demand values to the bus network. Five public routes were selected to collect the required data. Ride check and Point check survey was conducted for each selected route. The results of this study were public demand assigned to the selected bus routes, dwell time, load factor and headway. It is observed that R1 and R3 have the heaviest travel demand; they need special study to improve bus performance and make better transit. The model developed with only limited data available to predict travel demand will assist transportation planners and related agencies in decision making.

## 1 Introduction

Baghdad, one of the most populated cities in Iraq with a population of about 7.5 million in 2016 according to AL-Khatib and Alami in 2012 [1]. So it is important to provide enough public transport as sustainable modes to overcome traffic problems. Public transport is very important in Iraq; not only to fulfills the mobility demand in a growing city, but a bigger share of bus mode have positive effects including decrease in pollution, private mode share and energy. When more people travel by public bus transport, it leads to less people travelling by other modes. Thus, road congestion eventually decreases, road safety increases and fuel cost decreases.

Transportation demand has increased quickly in urban areas due to rapid population growth and alteration in travel patterns. It is encouraged to plan bus systems to meet the needs of users and service providers within the limited resource constraints [2]. For the proper functioning of the transportation network, it is important that demand and supply for mobility are aligned properly [3]. Estimates of bus passenger demand have always been important to bus operators and to transport planners [4]. Prediction of public demand using conventional transportation models has disadvantages, as they need huge data and are unable to deal with changes in generation rates. This has led to the need for a quick, simple model with low cost to inspect different public transport alternatives [4].

The bus planning process is a very complex process that contains five activities: Network Design, Setting Frequencies, Timetable Development, Bus Scheduling and Driver Scheduling [5]. The bus demand is affected mainly by spatial factors, such as the existence of the central business district (CBD), shops, educational institutes, and offices. These important spatial facilities attract many passengers [6]. Very few works concentrated on predicting demand on bus routes, some of these works can be abstracted as follows:

Gahlot et al [7] presented a GIS based methodology for selection of BRT alignment. The spatial distribution of transit trips is considered the base for selection of a BRT corridor in the city for the target year. The developed model utilizes transit trip, land use and the demographic characteristics to select the high BRT ridership corridor. Jäger et al [8] introduced a method to analyze taxi systems and forecast a model to predict the taxi demand. Siridhara [9] investigated public transportation in Greater Nakhon Ratchasima using service characteristics, quality, mode share and ridership. His work identified the current mode share and the effect of public transport role to other modes. Kalaanidhia and Gunasekaran [10] evaluated the urban transportation accessibility and its effect on

* Corresponding author: noor_moutaz@uomustansiriyah.edu.iq

© The Authors, published by EDP Sciences. This is an open access article distributed under the terms of the Creative Commons Attribution License 4.0 (http://creativecommons.org/licenses/by/4.0/).

the ridership of the bus system. They focussed to demonstrate accessibility Indices, which can assist in evaluating the transit facilities. Carpio-Pinedo [11] used multiple linear regressions to forecast bus demand model at a stop level. Bus ridership is predicted using spatial variables, urban environment and network accessibility indicators . Simon and Furth [12] generated route origin/destination matrix from boarding/alighting data and was compared to actual O-D data. This method is recommended as simple, and inexpensive method of developing route O-D matrices. AlSahili• and• Sadeq [13] try to study intercity• bus• demand, evaluate• current• services,• and forecast• future bus• ridership in• the• Palestiniancity. A• relatioship was established between• public demand• and• socioeconomic,• operating• variable s. They found variables obtained from onboard• survey• of• bus• riders which influence• ridership• demand. Li et al [14] try to forecast ridership at station level using back propagation neural network and use a new predictor like population per distance band. The result of their study leads to the adequacy of the proposed models. Zhou et al [15] present research to forecast passenger bus demand by utilizing three models: weighted time varying poisson model, autoregressive integrated moving average and time varying poisson model. Smit [6] develop bus demand models using traditional four step model which included trip generation, distribution and assignment models. Herijanto and Thorpe [16] concluded that when there is attention to the quality and quantity of trip attraction data, then the combined application between singly constrained gravity model and fluid analogy has considerable power in developing countries. Asmael [17] develops an O-D trip matrix between bus terminals based on GIS in Baghdad city and try to suggest future scenarios alternative to meet the requirements of future demand.

## 2 Field survey types

### 2.1 Ride check survey

It is one of techniques used to provide information on passenger load along transit routes. Boarding and alighting is the main data source of this survey. A surveyor travels during the journey of the bus and write the number of boarding and alighting values, dwell time, and the number of passengers on board for every stop. This survey provides an actual service route; journey time and average speed.

Ride check survey includes boarding and alighting counting to observe the change in load profile and estimate load factor. Load profile enables the planner to monitor variation in the load over the transit stops [18]. Load factor equals to the values of "passenger-km" divided by the "capacity-km". Load factor gives information about the effectiveness and service frequency of the bus route. The survey gives information based on demand characteristics to determine which stop can be eliminated or relocated along transit routes. From the ride check survey, dwell time can be estimated at each stop. Dwell time is interval time measured at a bus stop when the bus stops for boarding and alighting passengers [19].

### 2.2 Point check survey

It is one of data collection methods used to provide information on load at specific points along the transit route. From ride check survey results, it can determine the position of maximum load point along each route, which represents the input to point check survey. The data recorded in point check survey include route name, vehicle number, passenger load and vehicle capacity. Then average load per route is determined by analyzing the load of various buses and aggregating by time periods.

## 3 Data collection

Baghdad is the study area of this work. It has the largest concentration of urban population in Iraq. It is the center of Iraq's finance, commerce, and manufacturing. Public transport is the only means of transport for a large segment of population in Baghdad. Most public bus routes are centralized in the main CBD areas, buses travel along the fixed routes and stop at every bus stop according to demand. Five public routes were selected that travel inside the city center. Table (1) and Figure (1) show public routes selected for data collection:

**Table 1.** Five bus public routes selected

| Route Name | Route symbol |
|---|---|
| Bab- Moatham - Shaab | R1 |
| Bab- Moatham – 7Ebkar | R2 |
| Mawwal- Aden Square | R3 |
| Hay-AlJameah- Bab- Moatham | R4 |
| Bayyaa- Bab- Moatham | R5 |

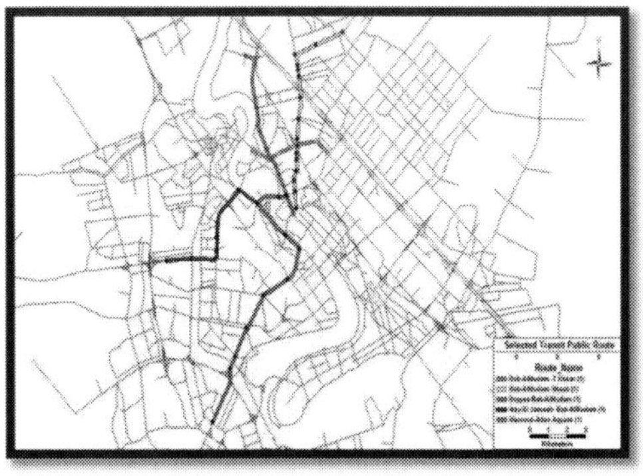

**Fig. 1.** Selected five bus public routes

For each route, ride and point check was conducted. For ride check survey, two observers were riding two vehicles in the same period. The observer was seated at the end of the vehicle for clear vision. When passengers get on/off the bus, the observer records number of boarding and alighting for each stop. For point check survey one observer is sufficient to count passenger at each stop. Only the number of passengers and number of vehicles at specific stops is counted. The travel time, dwell time, boarding and alighting count, number of vehicles were recorded in two peak hours, the first peak hour was from 8-9 AM and the second peak hour was 2-3 PM for three typical days.

## 4 Generation origin-destination based on gravity model

Conventional transportation models require big data collection for predicting public transport demand, which include expensive surveys, and gathering data specific to public transport and road networks. The generated models give inaccuracy in predicting transit demand. Therefore, there is a need for simple, inexpensive method to predict transit demand, and using the OD data from on/ off data, it is able to determine passenger loads and assignment flows on the transport network. If number of transit trips in a certain stop are known, then each origin to each destination trip can be determined. This can be done using a gravity model, which uses the distance between zones and the number of trips from and to the stops to estimate the number of trips between the stops. The gravity distribution model was calibrated to find number of passengers for each bus stop pair with a direct connection. To generate (origin-destination) transit matrix, then at each bus stop, the number of boarding and alighting was entered. Load between all stops are computed in this step and are used to generate an Origin-Destination matrix which gives an overview of how many trips from each stop to all other stops are undertaken.

## 5 Results and analysis

The following represent data results of this work:

### 5.1 Determination of bus route attributes

Table (2) shows summary of data analysis for five selected bus public routes. Headway is the difference in time between two vehicles arriving at a specific stop. It represents an important factor to identify transit service reliability [20].

**Table 2.** Summary of route data analysis

| Route Symbol | Travel Time Min. | Distance Km | Dwell Time Sec. | Headway Min. | Load Factor |
|---|---|---|---|---|---|
| R1 | 38 | 11 | 17 | 0.4 | 0.7 |
| R2 | 24 | 8.5 | 11 | 5 | 0.57 |
| R3 | 14.5 | 4.6 | 16 | 0.9 | 0.65 |
| R4 | 35 | 10 | 9 | 3 | 0.5 |
| R5 | 56.5 | 15 | 17 | 1.4 | 0.55 |

### 5.2 Route load profile analysis

For trip assignment, the shortest path method was used and the following Figures from (2) to (6) show a trip assignment for each bus public route selected during peak hour.

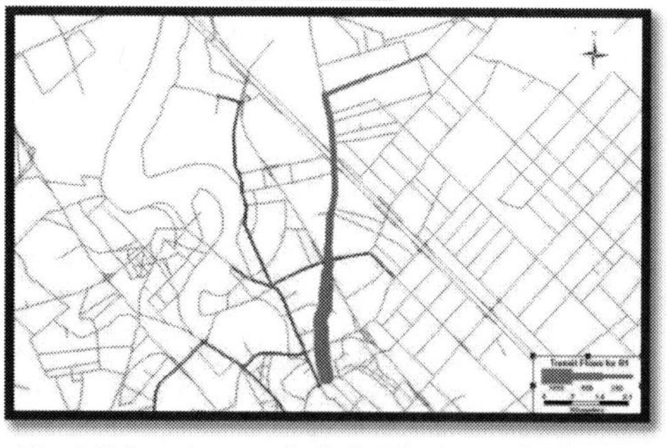

**Fig. 2.** Trip assignment for Bab- Moatham - Shaab route.

**Fig. 3.** Trip assignment for Bab- Moatham – 7Ebkar route.

**Fig. 4.** Trip assignment for Mawwal- Aden Square route.

**Fig. 5.** Trip assignment for Hay-AlJameah- Bab- Moatham route.

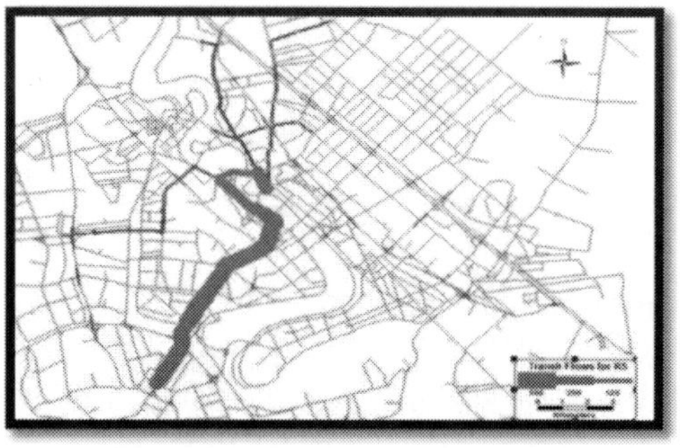

**Fig. 6.** Trip assignment for Bayyaa- Bab- Moatham route.

### 5.3 Boarding and alighting profile analysis

The following Figure(7) shows the results of boarding / alighting count for each bus stop for five selected bus public routes.

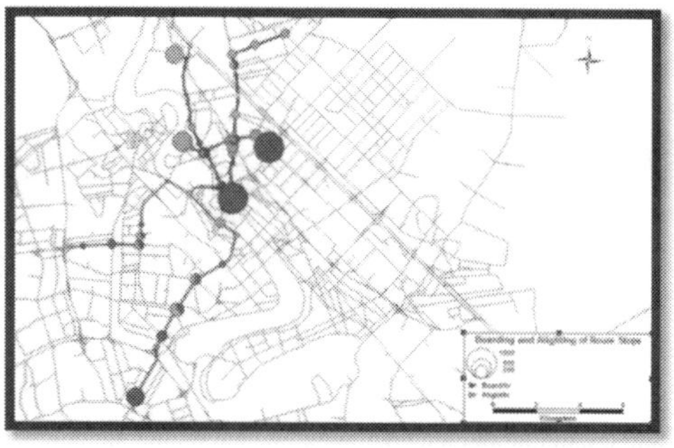

**Fig.7.** Boarding / alighting count of route stops.

### 5.4 Determination of bus demand routes

From trip assignment results the following table (3) represents demand for each bus public route.

**Table 3.** Demand for bus public routes

| Route Symbol | Demand (Passenger per hour) |
|---|---|
| R1 | 2552 |
| R2 | 1778 |
| R3 | 2364 |
| R4 | 407 |
| R5 | 1392 |

From the table results, it is shown that R1 and R3 have the heaviest demand along route, these results are consistent with assignment results of origin-destination matrix developed by [21].

## 6 Conclusions

Ridership analysis represents an important role in providing best transit service and evaluates financial profit for transit agencies. The count of boarding and alighting passengers can provide useful information for planners to determine the bus demand for every public route. Accordingly, improve transit operational strategies. Some of these strategies include increasing or decreasing daily vehicles and or increase/reduce headway to satisfy transit demand. Stop level ridership provides information for planners to identifying the most bus stops with a large number of boarding, and then adjust transit service to meet transit demand to achieve better service quality. This work uses limited data to estimate public demand. The developed model with only limited data provides a good way of estimating public demand based on a stop level and aids in identifying potential future adjustments and the required fleet size. This is a flexible method and easily available data provide a way to improve transit operation to meet transit demand and to achieve better service quality and then to encourage people to use public transport.

## References


1. AL-Khatib and Alami, *Baghdad Comprehensive city development plan*, Report, Amanat Baghdad, Iraq (2012)
2. M. Advani and G. Tiwari, Indian J. Transp. Manag. **30**, 4, pp. 363–391, (2006)
3. E. Klok, *Determinants of the demand for bus transport*, Master Thesis, University of Twente, (2010)
4. P. Mullen, Transportation (Amst)., **4**, 3, pp. 231–252, (1975).
5. A. Ceder and N. Wilson, Transp. Res. Part B: Methodological **208**,4, (1986)
6. Marijn Smit, *Exploring the demand for bus transport*, Master Thesis, University of Twente, (2014)
7. V. Gahlot, B. L. Swami, M. Parida, and P. Kalla, Int. J. Sustain. Built Environ., **1**, 1, pp. 102–109, (2012)
8. B. Jäger,, M. Wittmann, and M. Lienkamp, J. Traffic Logist. Eng. **4,** 2, (2016)
9. S. Siridhara, J. Traffic Logist. Eng., **2**, 2, pp. 120–125, (2014).
10. S. Kalaanidhi and K. Gunasekaran, Procedia - Soc. Behav. Sci., **104**, pp. 885–893, (2013)
11. J. Carpio-Pinedo, Procedia - Soc. Behav. Sci., **160**, pp. 205–214, (2014).
12. B. J. Simon and P. G. Furth, J. transp. Eng., **111**, 6, pp. 583–593, (1985)
13. K. Al-Sahili and A. Sadeq, J. Public Transp., **6**, 2, pp. 19–35, (2003)
14. J. Li, M. Yao, and Q. Fu, Discret. Dyn. Nat. Soc., **2016**,(2016)
15. C. Zhou, P. Dai, and R. Li, Proc. - *IEEE 13th International Conference on Data Mining Workshops*, pp. 1069–1076, (2013)
16. W. Herijanto and N. Thorpe, J. East. Asia Soc. Transp. **6**, pp. 1708–1723, (2005)
17. N. Asmael , Int. Res. J. Eng. Technol. **3**,10, pp. 1-5, (2016)
18. A. Ceder, *Public Transit Planning and Operation: Theory. Modeling and practice*. 0xford: Elsevier, (2007)
19. V. Vuchic,*Urban transit: operations, planning, and economics*, (2005)
20. C. Liao, and H. Liu, Transp. Res. Rec. 2143 ,34–43,(2010)
21. N. Asmael,*A GIS-Assisted Optimal Route Selection Based on Transportation Network Design (Baghdad Metro Case Study)*, Ph.D Thesis, University of Baghdad (2015)

# Assessment at Al-Ameer signalized intersection in Samawa city

*Hasan* Joni[1], *Mafaz* Hikmatt[1]

[1] Building and Construction Engineering Department, University of Technology, Baghdad, Iraq

**Abstract.** The aim of this paper is to evaluate traffic performance of Al-Ameer signalized intersection in Samawa city, which fits with the prevailing conditions and geometric properties of the intersection. The technique of video recording has used for collection the traffic volume data for all approaches. These data are abstracted from video films. **SYNCHRO 8** software was used to evaluate and analyze the intersection and choose the best suggestion. The evaluation process result showed that the intersection is operated with level of service (LOS F). By suggestion of several strategies which vary from signal optimization to geometric improvements. The best solution has been found by suggestion an overpass at the east-west direction, and the level of service has improved from (LOS F) to (LOS C), this result is considered an acceptable and economical solution for the existing problems at intersection.

## 1 Introduction

Most of governmental agencies require studies for traffic impact, also known as analyses of traffic impact and assessments of traffic impact, for the developments that produce a significant amount of new traffic or changing in existing traffic patterns.

The network of road transportation of any country plays an essential role in the economical development, then, ensure that there are road networks connecting all parts of country and that the transportation infrastructures is in good, the perfect physical conditions is very critical. Lacking in suitable transportation network reaching to viable parts of the country becomes a huge economical challenge [1].

The systems of transportation are the motor of the economical activities in whole urban societies the entire world, consequently sustaining a livelihood of a people living in. The standard facilities of urban transportation which contain roads, railways, airways and waterways. Through these, the major proportion is road network. Logically, most of researches and planning studies have concentrated on the road systems. In substance, system of road transportation is major factor in economic activities of the most urban areas. In current times, a large increase in traffic and transport demand have seen in many cities, which has thus, lead to decay in the capacity and inactive performance of traffic system [2].

## 2 Intersection

An intersection is the area where two or more roadways connect. It includes the roadway and roadside facilities available for traffic movement.

For many years, measurement of the performance level of signalized intersections has primarily focused on vehicle delay. Besides vehicle delay, there are another performance measurements like the spatial volume of queues, and the number of vehicle stops at intersection approaches have been found to play an important part in the performance of evaluation of signalized intersections in developing countries, where most of the times the intersections remain near or at oversaturated status. Hence, situation of go and stop is very frequent at the intersection. These measurements not only related to the level of service (LOS) that provided to the drivers, but, also to the air pollution and fuel consumption level which generated by the vehicles that crossing the signalized intersection. particularly, estimations of vehicle stop play an essential factor in determining consumption of vehicle fuel and emission at intersection approaches, estimation of queue length is significant not only for designing of lanes, but also to guarantee that the operations of traffic signal do not influence in queues which spill backing into the upstream intersection [3].

Signalized intersections could be defined are the basic elements in the system of urban transportation and carry the heavy traffic for motorized and non-motorized vehicles and pedestrian, which, create many conflicts between turning and crossing, merging of maneuvers. For a diversity of reason like economic, auto ownership growth and population, increasing of traffic demand could exceeding the capacity of the intersection through the peak hour periods. Consequently, deterioration of traffic conditions deteriorates and safety worsens. Congestion and dangerous of traffic condition lead to increase emission, accidents, consumption of fuel, and noise, thus, a quality of life, resources of world energy and global atmospheric conditions lead to deteriorate service [4].

An intersection is any site or place where many users mix and compete for time within the same area. Intersection takes many shapes and forms, range from complex junctions to driveways to the meeting of two ways. They are oftentimes defined by their operations and layout: traffic signals, T-junctions and roundabouts.

© The Authors, published by EDP Sciences. This is an open access article distributed under the terms of the Creative Commons Attribution License 4.0 (http://creativecommons.org/licenses/by/4.0/).

Compactness, Simplicity, low speed and eye contact are favored in intersection design [5].

Performance of Intersection as calculated by the delay is an action of different factors include, turning movement demands, signal timing plan, traffic stream compositions, intersection geometry, pedestrians volume, temporary variation in traffic demand, driver characteristics, headway distributions at each traffic flow, conditions of road surface, visibility and weather. Some of these factors are constant such as (signal timing planning and geometry) while others differ like (traffic demand, weather, etc.) [6].

## 3 Delay time types

There are different types of delay time that could be calculated at intersections, and every one fills an alternate need to the transportation engineer. Intersection delay may include two components: stop (queue) delay and control delay. Stop delay, or queue delay, is difficult to quantify due to its stochastic nature that is affected by random arrival. The sophisticated techniques could work better in estimation of queue delay, but are often impractical for planning models due to this requirement of intense data. It is often hard to find a well-balanced queue delay model for integration into planning models. Estimation procedure of the signalized intersections capacity and LOS are worked around the idea of moderate control delay per vehicle. The control delay could be defined as a part of the total delay assigned to the operations of traffic signal for signalized intersections [7, 8]. The control delay (overall delay) can be grouped into decelerations delay, accelerations delay and stopping delay:

a- Stopping delay is simpler to be measured, typically, ~~a~~ transportation experts define the stopping delay as the delay that happens when the vehicles are completely stopped or immobilized,

b- The delay that happens by an acceleration or deceleration of vehicle is categorized as acceleration and deceleration delay, respectively.

In HCM2000 (TRB 2000), the control delay is included in initial deceleration delay, queue move up time, stopping delay, and final acceleration delay, however, in prior versions it included just the stopping delay. Besides the control delay, there are other kinds of delay time which vehicles experience at the signalized intersections. This kind of delay is known as "geometric delay" which is the time lost because of the geometrical layouts of intersection. Geometric delays might be huge for the turning movement. So, the total delay of a vehicle is the whole sum of control delay plus to geometric delay [9].

## 4 Al-Ameer signalized intersection as a case study

This intersection in Samawa City is a four leg signalized intersection which is operates as a fixed time signal control and has four signal phases, and represents one of the most important intersections in the city center, where many lively activities happen at this intersection. The south approach suffers from high traffic volume due to many shopping centers, offices, cafes and others activities exist in this street and the west approach leads to neighborhoods and residential areas, and as shown in the Figure 1.

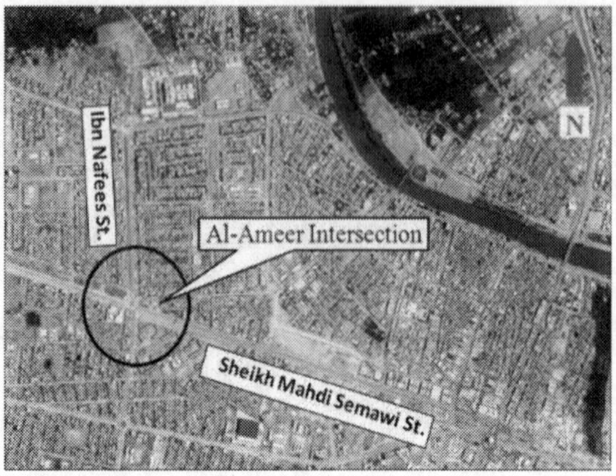


**Fig. 1.** The Study Area and the Selected Intersection

The layout diagram of this intersection is shown in Figure (2) as in the following.

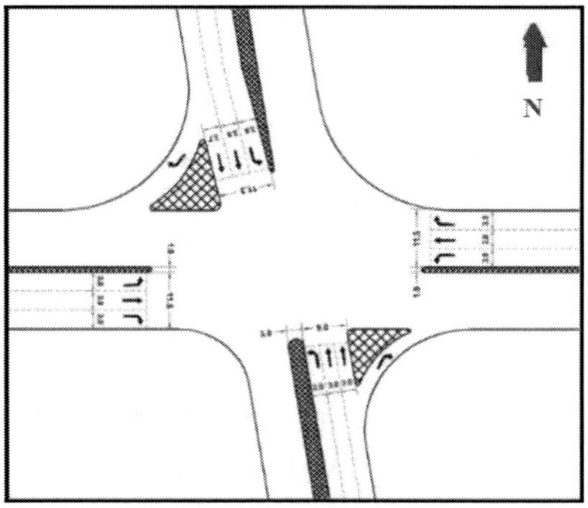


**Fig. 2.** The layout diagram of Al-Ameer Intersection

## 5 Signal phasing and timing

In Samawa City, to control the traffic, fixed time traffic signals are being used for a significant number of years. Unfortunately, rather than using any traffic engineering knowledge, these signals have been timed by traffic police from arbitrary judgment only. As a result, they became ineffective to serve the purpose properly and efficiently. Therefore, this research assumed that the selected intersection work actuated system by traffic police and four phase signals (each approach with one phase) and use the maximum actual green time.

## 6 Major parameters for evaluating intersection operations

Capacity, delay and level of service are the main parameters for evaluating the operations of intersections.

## 6.1 Capacity

The Highway Capacity Manual (2000) demonstrates the capacity of facility as " the maximum hourly rate that a person or a vehicle can reasonably be expected to traverse a point or uniform section of a lane or roadway through a given time period under the prevailing roadway, traffic and control conditions " [10].

## 6.2 Delay

Delay is one of the most important factors that are used at the optimization procedure of traffic signal timing. Moreover, delay is the main parameter in calculating the level of service (LOS) providing to the drivers at signalized intersection. Delay, is a factor that is difficult to be estimated because it contains the delay related to the deceleration to the stopping, the stopping delay and the delay related to the acceleration from the stopping [11].

## 6.3 Level of service

Level of service (LOS) is defined in terms of the average total vehicles delay of the whole movements through an intersection. Vehicle delay is a way of evaluation many intangible parameters which include driver discomfort, loss of travel time and frustration. In particular, levels of service (LOS) criteria are expressed in terms of average delay per vehicle through a specific time period (such as, the AM peak hours). Vehicle delay is a complicated measurement based on several variables, include signal phasings (i.e., progress of movements during the intersection), length of signal cycle and volumes of traffic with respect to capacity of intersection [7].

## 7 Methodology of data collection

To fulfill the objectives of the study, it was important to go ahead with the designed procedures to gather the required data. These procedures are classified into four steps as explained below:

1. Describing the study area limitations.

2. Defining the peak hour periods.

3. Defining the specified input data for the selected software.

## 8 Geometric data

The main objective behind collecting the geometrical data for the signalized intersections is to evaluate and enhance the traffic through increasing the width of a lane, adding a new lane, creating an overpass or underpass.

Some of geometric data have been collected via using GIS techniques by using map measurements and depending on the available satellite images for Samawa city with precision of 60cm, and updated to 2013. The slopes of all intersections are done by using the level instrument. Field measurements are obtained by using measuring tape to get such as, the median width and lane width, some of geometric features are obtained by observation such as; existence of exclusive LT or RT lanes, number of lanes per approach and Parking.

## 9 Traffic volume data

The traffic volume data are collected for two week by video recording technique for three different periods: morning peak period of (7:00- 9:00 A.M.), noon peak period of (1:00-3:00 P.M.) and evening peak hour of (5:00 -7:00 P.M.), after many personal observations, and pilot surveys have been done in the study area. Also, many personal interviews are done with the interested people such as traffic policemen at the study area and several road users in order that determining the peak hour periods for collecting the traffic data.

The video recording is made in normal conditions, good weather (dry), and good visibility, and by using a video camera SONY HANDYCAM lens 8mm, and later copied onto solid disk. From all of these considerations, it was concluded that Al-Ameer intersection has morning peak hour of (7:30 – 8:30) A.M. as shown in Figure 3.

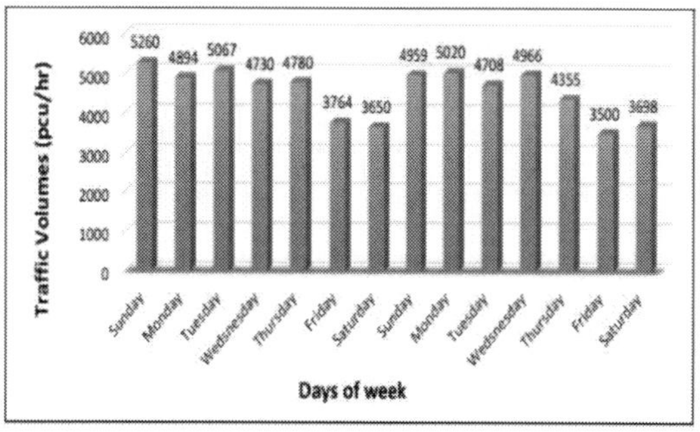


**Fig. 3.** Total Traffic Volume at Al-Ameer Intersection

## 10 Traffic volume data for intersection

This data include calculation of the traffic volumes abstracted from the video recording for each approach at intersection, also, composition of traffic (passenger car ,private car ,heavy vehicle, bus and motorcycle), and volumes of turning movement (though, left ,right, and U-turn). The traffic volume is transformed to passenger car units (PCU) by multiplying every vehicle kind by its factor according to HCM 2000.

## 11 Traffic signal data

For the signalized intersections the cycle length, phase length, green time and all red time are measured from the video film.

- Cycle length: is the whole (total) time for the signal to finish one cycle, and represented by second.
- Phase length: is a portion of the cycle length specified to any group of the traffic movements that receive the right of way together through one or more interval.
- Yellow time : is the value of time of the yellow interval , generally, the value of time must to be between 3 to 5 second, based on width of street ,speed of approach and the local standards .The minimum value of time permitted by the SYNCHRO 8 is 2 second and the maximum is 10 second [12].

• All red time: is the value of time for all red intervals that follow the yellow intervals. All red time must be of a suitable interval to allow time at the intersections to clear before the traffic is released [12].

## 12 Data abstraction

Data abstraction was based on sessions of 15 minute periods of recorded data. The choice of a 15 minute period was depended on these considerations:

1. To guarantee that the sample is adequate to give meaningful results.
2. The traffic flow should be effectively constant during the period without incidents.
3. Decreasing of observer fatigue and subsequent mistakes. To get the required data from a video film, it was important to play back the film many times. To guarantee that the video film begins at the same point in time.

## 13 Application of SYNCHRO 8 software

SYNCHRO 8 was applied to simulate the existing traffic flow for the selected intersection. It is an entire programming package for modeling, optimization, and simulation traffic system. The results in Table 1 show that most the intersection suffers from an oversaturation condition with high total delay values and unacceptable level of service (LOS F).

**Table 1.** Performance Analysis for Al-Ameer Intersection.

| Intersection Name | Approach | Traffic volume | LOS | Average Delay |
|---|---|---|---|---|
| Al-Ameer | North | 890 | F | 84.0 |
| | South | 1682 | F | 286.0 |
| | East | 1302 | F | 154.1 |
| | West | 1381 | F | 184.5 |
| Summary | | | | |
| **Average Delay = 191.5 sec LOS = F V/C = 1.97** | | | | |

## 14 Improvement of intersection

This section presents the improvement strategies applied to the traffic flow at the selected intersection, through the application of the calibrated software SYNCHRO 8.

This section divides the improvement strategies into two parts according to method and software applied. The first part could be named as, "the cycle length optimization", and include cycle length optimization, splits optimization, and phasing sequences. The second section could be name marking, pavement widening and construction an overpass.

### 14.1 Cycle length optimization method

The optimization of Traffic signal timing considered one of the most cost-effective methods to reduce the costs of vehicle operating and to improve traffic flow performance along urban streets. SYNCHRO 8 includes a number of optimization kinds. It optimizes the cycle lengths, phase sequences and split times in order to decrease the delay time and stop. These types are applied to the selected intersection in the same order and as shown in Figure 4. [13].

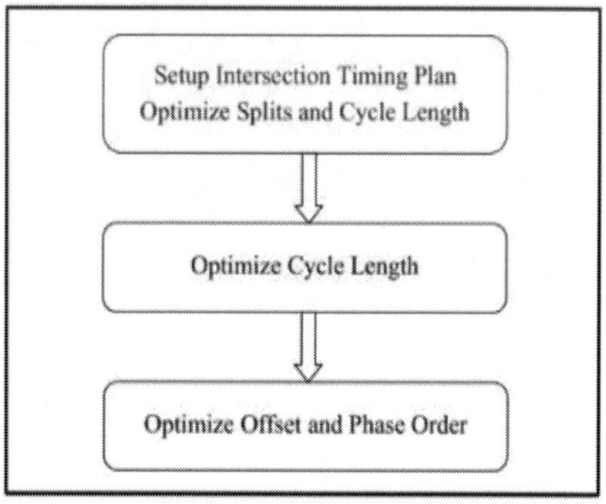


**Fig. 4.** Optimization Steps in SYNCHRO 8

SYNCHRO 8 begins with the shortest cycle length, and then optimizes the splits for that cycle length. If the split for every phase is not capable to clear out the percentile critical traffic, SYNCHRO 8 will have to try a higher cycle length till the percentile critical traffic is cleared.

The optimize splits order will automatically set the splits for all the phases. Time is splited based on every lane group, and the volume of traffic is divided by its adjusted saturation flow rate. Splits optimization will estimate minimum splits sittings for every phase whenever potential.

The range of cycle length recommended by HCM is 60-120 seconds. Long cycle length exceeding 100 seconds will be recommended to accommodate an extra 10% capacity. Long cycle lengths more than when the cycle length exceeded 100 second, it will be recommended to accommodate a further more 10% capacity. The long cycle lengths more than 120 second might have negative operation sides; include inefficient use of turning lanes, long queues and blocking. Low cycle length might be better to reduce the queue which in general raise the capacity and provide a smoother traffic operation. The range of cycle times used in the optimization process was 50-200 seconds, as a default values by SYNCHRO.

**Table 2.** Effectiveness Measure for Al-Ameer Intersection created by SYNCHRO 8

| Improvement type | Cycle Length | V/C | Average delay | LOS | Fuel consumed | Unserved vehicles |
|---|---|---|---|---|---|---|
| Base condition | 162 | 1.97 | 191.5 | F | 942 | 1108 |
| Cycle Length | 191 | 1.71 | 192.2 | F | 932 | 1091 |
| Splits | 162 | 1.91 | 208.8 | F | 981 | 1162 |
| Phase Sequences | 185 | 1.68 | 167.5 | F | 816 | 947 |
| Percent saving | | | 12.5% | | 13.4% | 14.5% |

Table 2. shows that, the use of optimization functions for Al-Ameer signalized intersection has no effect, the level of service remained the same level (LOS F) each time that used the optimization functions, according to the degree of saturation (V/C), there are a few differences with this functions.

Another trial was done by changing the number of phases and the phasing sequence. The phasing order in Iraq is four-leg intersection and divided into through and left traffic with permitted right turn and yields only for pedestrians. The leading and lagging and any other protected phases are not used in this order, so the drivers are unfamiliar with this type of movements. It might cause problems with head on collisions between oncoming left turns, and also by applying this method there is no acceptable change happened.

The optimization results showed that although the cycle time increased to non-reasonable values, the intersection remains performing under oversaturated conditions, so, it is necessary to improve the performance by trying the second part of interaction strategies.

## 14.2 Geometric improvements

Since the cycle length optimization process did not improve the intersection, so it is necessary to improve the geometries of the oversaturated approaches at Al-Ameer intersections by the strategies which mentioned before, and as shown in table 3.

**Table 3.** Effectiveness Measure of Al-Ameer Intersection that Produced SYNCHRO 8

| Improvement type | Cycle Length | V/C | Average delay | LOS | Fuel consumed | Unserved vehicles |
|---|---|---|---|---|---|---|
| Base condition | 162 | 1.97 | 191.5 | F | 942 | 1108 |
| Pavement Remarking | 191 | 1.67 | 191.8 | F | 930 | 1115 |
| Pavement Widening | 171 | 1.32 | 94.5 | F | 544 | 376 |
| Overpass | 117 | 0.93 | 29.2 | C | 214 | 0 |
| Percent Saving | | | 84.8 % | | 77.3 % | 100 % |

The first step of the geometric improvement is to apply pavement remarking. In this step consideration was given to remark the approaching lanes so that rearranging of queuing vehicles at stop line is obtained. Also, separating the movement with high traffic volume by exclusive lane is also applied. Optimization run after each improvement type is applied for the signalized intersections.

The second trail was with the pavement widening, from the measurement, it can be noticed that the number of lanes in many approaches of these intersections can be increased by decreasing the lane width (with keeping the width of lanes greater than a minimum value of 2.4 m according to HCM) which is actually observed to be used in field. Increasing in number of lanes will increase the saturation capacity of affected approaches, which leads to improve traffic performance measure of effectiveness (MOE) for some intersections.

The pavement remarking is applied to the intersection, but this method did not improve the measure of effectiveness (MOE), the level of service (LOS) is still with unacceptable level (LOS F), and the delay results that computed from this method for intersection is considered unsatisfied results.

By applying the second strategy the pavement widening, this method has improved the delay value at Al-Ameer intersection with 51.4 % saving percent, and did not change the level of service (LOS F) were is the results shown in Table (3), using this method required enough areas and widths inside the location that intersection located in, and there is a flexibility to widen the lanes width. Al-Ameer intersection, located at restricted area.

Suggestion an overpass was the third trial, this method has applied on Al-Ameer intersection, because this intersection suffers from high traffic volume especially the through movement, and also, there are enough space for construction. Adopting an overpass at the east-west direction, reduced the delay to 29.2 sec with accepted level of service (LOS C), also rearrange the cycle length to be 117 sec as shown in Table 3 with some geometric changes and as illustrated in Figure 4 when compared with Figure 5.

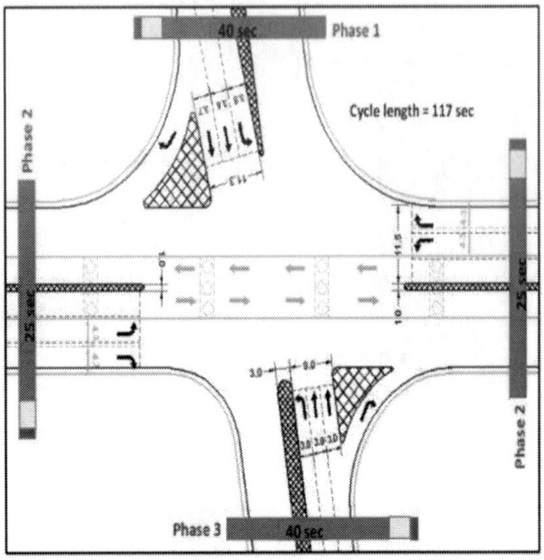


**Fig. 5.** Improvement proposal for Al-Ameer Interection

## 15 Conclusions

1. The local case study of Al-Ameer intersection suffer from highly congestion on four approach which lead to level of service to be at minimum requirements (level F) according to Highway Capacity Manual.
2. Using of signalization timing optimization by using SYNCHRO 8 software including splits for each phase and phase sequences shows not effective solution to improve the traffic operation performance due to high traffic volume.
3. The pavement widening strategy of geometric improvement has improved the delay value at intersection with 50.7 % saving percent, and did not change the level of service (LOS F).
4. The adopting an overpass as a best strategy at the east-west direction, reduced the delay to 29.2 sec with accepted level of service (LOS C), also rearrange the cycle length to be 117 sec.

## References

1- NCC, *National Council for Construction,* (2003)
2- M. L. Chiguma, in Analysis of side friction impacts on Urban Roads: Case Study Dar-es-Salaam. Diss .KTH (2007).
3- H. Rakha, Y. Kang and F. Dion, in Estimating Vehicle Stops at under saturated and Oversaturated Fixed-Time Signalized Intersections , (2001).
4- X. Yu and G. Sulijoadikusumo, in Assessment of Signalized Intersection Capacity in Response to Downstream Queue Spillback. (2012).
5- NACTO, in *Urban Street Design Guide Overview*. (2012).
6- H. Bruce and Z. Abdy, in Signalized Intersection Analysis and Design: Implications of Day-to-Day Variability in Peak Hour Volumes on Delay, Journal of Transportation Engineering 134.7 (2008): 307-318.
7- HCM, *Highway Capacity Manual*, (2000).
8- A. K. Jameel, in Estimating Delay Time at Palestine Street Intersection in Baghdad City Using HCM and SIDRA Models, Al-Qadisiya Journal for Engineering Sciences,**4** No.1 (2011).
9- Y. Darma, M. R. Karim, J. Mohamad, and S. Abdullah, in Control Delay Variability at Signalized Intersection Based on HCM Method, **5**, pp. 945 – 958 (2005).
10- NRC, "*Estimation of Highway Capacity* ", Rev.1, (2011)
11- Y. S. King, in Delay, Stop and Queue Estimation for Uniform and Random Traffic Arrivals at Fixed-Time Signalized Intersections *(2000).*
12- Trafficware, Ltd. in *Traffic Signal-User Guide, Synchro Studio 8*, (2011).
13- D. Husch, J. Albeck, in *Traffic Signal-User Guide, Synchro Studio7*, (2000).

https://doi.org/10.1051/matecconf/201816201040

# Application of mechanistic empirical approach to predict rutting of superpave mixtures in Iraq

*Zaynab* Qasim [1,*], *Hamed* Hamdou [2], and *Namir* Alkawaaz [3]

[1] Building and Construction Engineering Department, University of Technology, Baghdad, Iraq
[2] University of Baghdad, Baghdad, Iraq
[3] University of Al-Mustansiriyah, Baghdad, Iraq

**Abstract.** In Iraq rutting is considered as a real distress in flexible pavements as a result of high summer temperature, and increased axle loads. This distress majorly affects asphalt pavement performance, lessens the pavement useful service life and makes serious hazards for highway users. Performance of HMA mixtures against rutting using Mechanistic- Empirical approach is predicted by considering Wheel-Tracking test and employing the Superpave mix design requirements. Roller Wheel Compactor has been locally manufactured to prepare slab specimens. In view of study laboratory outcomes that are judged to be simulative of field loading conditions, models are developed for predicting permanent strain of compacted samples of local asphalt concrete mixtures after considering the stress level, properties of local material and environmental impacts variables. All in all, laboratory results were produced utilizing statistical analysis with the aid of SPSS software. Permanent strain models for asphalt concrete mixtures were developed as a function of: number of passes, temperature, asphalt content, viscosity, air voids and additive content. Mechanistic Empirical design approach through the MnPAVE software was applied to characterize rutting in HMA and to predict allowable number of loading repetitions of mixtures as a function of expected traffic loads, material properties, and environmental temperature.

## 1 Introduction

Various inconvenient types of distresses regularly impact on the asphaltic paving mixture during its service life. These distresses are caused by construction practices, load, climate, and /or lacking materials. One of the genuine distresses is rutting; or permanent deformation which at last may lead to complete failure of pavement. Such distress will lessen the performance of asphalt pavements, which makes second rate ride quality drivers as well as yields higher life-cycle cost, [1].

FHWA in 2003 described rutting as longitudinal surface depression in the wheel path, which may have associated transverse displacement, [2]. Furthermore, Huang in 1993 stated that the permanent deformation is an imperative factor in design of flexible pavement, [3].

With expansion in activity load and tyre weight, most of the rutting occurs in the upper layers as opposed to that in subgrade. Permanent deformation may likewise start different types of asphalt pavement distress, for example, cracking, [4]. Such distress is related with the asphalt- cement binder and can be controlled essentially by modifying the binder material with chemical additives, [5]. In Iraq under the impact of heavy traffic loading, moisture damages and high temperature, specific requirements are expected to control the material quality of highway pavement in order to increase the durability, [6].

Accordingly and to accomplish a reliable assessment of the asphalt pavement rutting, one must consider material, loading and environmental factors at which permanent distortion accumulate for finding out pavement layer. Furthermore, conditions experimental tests which satisfactorily simulate the field conditions could also be considered. In particular, the issue of development of performance tests and measure of rutting performance with regard to Mechanistic Empirical approach has become the focus of current research.

To achieve the main goals of this study, available local materials are used consisting of asphalt binder, aggregates, mineral filler and additive contents. Asphalt binders of (40-50 or PG 64-16) and (85-100 or PG 58-22) from Al-Daura refinery in Baghdad were obtained, and the aggregate from Al-Nibaie quarry whereas the mineral filler brought from lime factory in Karbala, the additive used in this study is SBR polymer (Styrene Butadiene Rubber) is brought from Najaf Tyres Factory in Iraq. The Superpave mix design is adopted with varying volumetric compositions. The Superpave Gyratory Compactor SGC was utilized at the NCCL laboratory for preparing 24 cylindrical specimens of asphalt concrete of (15 cm diameter and 14.5 cm height) for carrying out volumetric design criteria. For proof testing the ability of HMA to resist permanent deformation, Wheel Tracking devices have gained a great consideration as potential candidates during the

* Corresponding author: Zaynab.2010@yahoo.com

© The Authors, published by EDP Sciences. This is an open access article distributed under the terms of the Creative Commons Attribution License 4.0 (http://creativecommons.org/licenses/by/4.0/).

implementation phase of Superpave. It is considered the most widely used simulative test, the rut depth criteria for the WTT test were developed by AASHTO TP-63 for its employment, [7]. The basic principle of WTT is to simulate the stress conditions during which 108 slab samples were tested under repeated, moving-wheel loads as they actually occur on in-service pavements.

For preparing the asphalt concrete slabs, Roller Wheel Compactor Machine has been locally manufactured to prepare slab specimens of (32 cm length×28 cm width×4 cm height). The application of Mechanistic-Empirical approach is presented using MnPAVE, 2011 program and its output results with that of the developed models are compared. The Mechanistic-Empirical Pavement Design Guide MEPDG is developed under the NCHRP a project represents a major advancement to pavement design and analysis. It uses site specific traffic, climatic conditions and materials properties to predict rutting and cracking performance of flexible pavement structures, [8].

## 2 Wheel tracking tests

In this study, compacted asphaltic slabs are prepared at air voids equal to (4%) using Roller Compactor Device according to (EN12697-Part 33:2003) and Superpave system (AASHTO Designation: T 312-2010). The dimensions of the compacted slabs used in this work are 320 mm (12.8 inch) in length and 280 mm (11.2 inch) in width and 40 mm (1.6 inch) in height as proposed by (EN 12697-Part 22:2003), [9]. Proportion of aggregate and asphalt binder are used for mixing, curing, and compacting, [10]. The fractions of aggregate are separated into nine sizes, as retained on each of the conventional sieves.

The desired mineral filler or Limestone is added according to the gradation requirements. The aggregate was combined into batch of (9700 g for slab specimen) in the mixing bowl and heated to the mixing temperatures prior to mixing with heated asphalt corresponding to each binder. Mixing and compaction temperatures for both types of asphalt binder were determined using Superpave Rotational viscometer, it were 160 ±3°C , 146 ±4 °C for asphalt binder PG (64-16) grade and 152±2°C , 140±5°C for asphalt binder PG (58-22).

## 3 Results and analysis

In this study outcomes of the wheel tracking test were plotted on a diagram showing rut depth (normally in millimetres) versus the number of passes for each test. A curve is plotted for the loaded wheel for the test running, an examination of the graph can reveal the number of passes to failure, the maximum rut depth occurring. The rut depth development monitored with an automatic displacement measuring device by setting the machine motion and taking readings of the vertical displacement initially, r0, and then after every (25±1) load cycles. The tracking is continuing for 10000 load cycles or until a rut depth of 25 mm is reached, whichever is shorter. The maximum rut depth for loaded wheel is displayed prominently along with a box declaring whether the test passed or failed. The experiment design for the rutting test is a full factorial with; three asphalt contents, two asphalt types, two air voids contents, three temperatures, and three percent of SBR additives, resulting in a nominal total of 108 tests. Based on results and analysis of permanent deformation tests, Table 1 summarizes and provides a qualitative description of the Influence of Temperature, Performance Grade and Mix Components on Plastic Parameters. As shown in this Table the permanent deformation is highly dependent on temperature, Performance grade and air voids, while moderate and low for polymer content and asphalt content respectively.

**Table 1.** Influence of Different Parameters on Plastic Parameters

| Variable | | Rut Depth, mm | TR, μm/cycle | WTR, μm/cycle | WTS mm/ $10^3$ cycles |
|---|---|---|---|---|---|
| Temp., °C | 40 to55 | 19.44 | 9.4 | 9.7 | 14.65 |
| | 55 to 70 | 30.46 | 26.2 | 26 | 34.58 |
| Asphalt Content,% | 4.1 to4.7 | -8.25 | -32 | -33 | -24 |
| | 4.7 to5.3 | 19 | 67 | 68 | 31 |
| Performance Grade | PG58-22to PG 64-16 | -30 | -25 | -26 | -29 |
| Air Voids Content,% | 4.0 to 7.0 | 56 | 73 | 74 | 96 |
| Polymer SBR Content,% | 0 to 1.5 | -23 | -17 | -17 | -28 |
| | 1.5 to 3 | -9 | -8 | -8 | -12 |

## 4 Statistical models

The statistical techniques are used for the model development required for evaluating permanent deformation in asphalt concreting mixtures. In statistical modelling, the overall objective is to develop a predictive equation correlate the permanent deformation in the form of permanent strain with the independent variables. These variables include number of passes, asphalt grade, air voids, asphalt content, testing temperature, and additive content.

Numerous studies have been made to characterize the permanent deformation response of asphalt concrete materials due to repeated load applications, [11]. The repeated compressive load test is used to investigate the variables that influence the relationship between the number of load applications, (N), and the permanent strain, (εp). For this research, relationship between the permanent deformation in terms of plastic strain (εp) of asphalt concrete mixtures is presented in equation (1):

$$\varepsilon p = RD / h \qquad (1)$$

Where: εp = Vertical Permanent Strain at Mid Thickness (h) of the Layer, (%),

 https://doi.org/10.1051/matecconf/201816201040

RD = Rut Depth at (N) Number of Load Cycles (mm), and, h = Thickness of Asphalt Concrete Sublayer.

Analysis of results, calculation of standard error, coefficient of variation for the permanent strain ($\varepsilon p$), permanent deformation (RD) and allowable number of repetitions (Nf) models are presented in Table 2.

**Table 2.** Summary Results of Final Models

| Model | $R^2$ | SER |
|---|---|---|
| $RD = 2.335\ LnN + 5.752\ Ln\ P_{Seff} + 9.73\ LnT - 0.309\ LnP_M - 28.02\ \eta + 1.789\ A_v - 49.68$ | 0.88 | 0.231 |
| $\varepsilon_p = 0.058\ LnN + 0.144\ Ln\ P_{Seff} + 0.243\ LnT - 0.008\ Ln\ P_M - 0.701\eta + 0.045\ A_v - 1.242$ | 0.88 | 0.231 |
| $N_f = 7.23*10^6 * RD^{2.476} * P_{Seff}^{-1.031} * T^{-1.947} * P_M^{0.06} * \eta^{2.411} * A_v^{-1.7}$ | 0.89 | 0.207 |

## 5 Mechanistic-empirical analysis approach

The NCHRP 1-37A project conveyed the latest M-E based strategy that fuses broadly adjusted models to anticipate distinct distresses instigated by traffic load and weather conditions. The NCHRP 1-37A methodology likewise joins vehicle class and load distributions in the design plan, a stage forward from the Equivalent Single Axle Load (ESAL) approach utilized as a part of the AASTHO design equation and different methods, [8].

M-E design methods depend on the mechanics of materials that relate input information, for example, wheel load and material properties to an output of pavement reaction, for example, stress or strain. In the M-E Design Guide method, the asphalt pavement is viewed as a multi-layered elastic system. The materials in each of these layers were described by modulus of Elasticity (E*) and Poisson's ratio (v).

This technique requires the determination of critical stress, strain, or deflection in the asphalt pavement by some mechanistic method and the estimation of resulting damages by some empirical failure criteria before the thickness design and remaining life of the current pavement which will be assessed. In the M-E design process, the multi-layer structure is analysed mechanistically to evaluate the critical strains created within the structure. These values of strain are utilized to appraise the structural capacity in terms of repeated traffic loading by utilizing the empirically derived transfer functions.

The outcomes are compared with the results obtained from laboratory based statistical models and field section cores to validate the mechanistic component of asphalt concrete mixtures. Powell and Leech in 1983 showed that the dynamic stiffness of the mixture increases by 30 % if the void content of the material is reduced by 3%. Linear elastic analysis of the construction as a whole shows that, by reducing void content, the thickness of the construction can be reduced by 8%, [12].

The other advantage of adequate compaction is the increase in the resistance to the permanent deformation. An increase of 3% in compaction leads to a reduction of the permanent deformation of about 50% after 1000 passes, measured with a pneumatic tyre wheel tracking machine, [13].

On account of the perpetual flexible pavement enough stiffness in the upper pavement layers is expected to keep away from rutting and enough total pavement thickness and flexibility in the lowest layer to oppose fatigue cracking from the pavement structure bottom. Since the HMA pavement is tailored to oppose specific distresses in each layer, the materials selection, mix design, and performance testing should be specialized for every material layer.

## 6 Application of MnPAVE program

MnPAVE is a computer program developed through a joint effort by the Minnesota Department of Transportation and the University of Minnesota version 6.1 on October, 2011. MnPAVE combines known empirical relationships with a portrayal of the mechanics and physics behind flexible pavement behaviour. The mechanistic portions of the program depend on finding the tensile strain at the asphalt layer bottom, the compressive strain at the top of the subgrade, and the maximum principal stress in the middle of the aggregate base layer.

MnPAVE comprises of three information modules: Traffic, Climate, and Structure, and three design levels: Basic, Intermediate, and Advanced. The level is chosen based on the amount and quality of information known about the material properties and traffic data.

Material information consists of layer thickness, Poisson's ratio, dynamic modulus, and an index indicating the slip degree among layers. MnPAVE assumes zero slip at all layer interfaces. Other inputs include load and evaluation locations. Loads are characterized by pressure and radius. MnPAVE output includes expected life of the pavement and the damage factors. Reliability also has been incorporated into the latest version.

The M-E PDG requires an expansive arrangement of material properties. Three parts of the design procedure require material properties: the climate model, the pavement response models, and the distress models. Climate related properties are utilized to investigate temperature and moisture variations inside the pavement structure. To process the state of stress/strain at critical locations in the structure due to traffic loading and temperature changes; rectified as suitable for temperature and moisture impacts; the models of pavement response uses material properties. These structural responses were utilized by the distress models along with correlative material properties to predict pavement performance.

The asphalt concrete dynamic modulus defined as one of the material properties required in the M-E PDG which is considered innovative for pavement design methods. Temperature dependency and Loading time of

https://doi.org/10.1051/matecconf/201816201040

asphalt mixtures are characterized by the dynamic modulus, |E*|. The dynamic modulus master curve models the variation of asphalt concrete stiffness because of the rate of loading and temperature variation (hardening with low temperature/high frequency and softening with high temperature/low frequency). The non-linear elastic behaviour of unbound granular materials is modelled by a stress-dependent resilient modulus included as factors input. The complex dynamic modulus |E*| is the principal material property input for asphalt concrete.

It is a function of mixture characteristics: (binder, gradation of aggregate and volumetric properties), temperature, age, and rate of loading. For information, the dynamic modulus master curve is constructed based on time-temperature superposition principles by shifting laboratory frequency sweep test data, [14] and [15].

Binder viscosity measured using the Superpave Rotational Viscometer (RV) is also a required input. For inputs, the dynamic modulus master curve is developed in view of an empirical predictive equation. The |E*| predictive equation is an empirical relationship between |E*| and mixture properties as shown below:

$$\log E^* = 3.750063 + 0.02932\,.\rho_{200} - 0.001767.(\rho_{200})^2$$
$$-0.002841\,.\rho_4 - 0.058097.V_a - 0.802208.\left(\frac{V_{beff.}}{V_{beff.} + V_a}\right) +$$
$$\frac{3.871977 + 0.0021\,.\rho_4 + 0.003958\,.\rho_{38} - 0.000071.(\rho_{38})^2 + 0.005470.\rho_{34}}{1 + e^{(-0.603313 - 0.313351.\log(f) - 0.393532.\log(\eta))}} \quad (2)$$

Where: E* = dynamic modulus, 105 psi;
η = binder viscosity, 106 Poise;
f = loading frequency, Hz;
Va = air void content, %;
$V_{beff}$ = effective binder content, % by volume;
ρ4 = cumulative % retained on the 4.75-mm sieve;
ρ34 = cumulative % retained on the 19-mmm sieve;
ρ200 = % passing the 0.075-mmm sieve;
ρ38 = cumulative % retained on the 9.5-mm sieve.

### 6.1 Binder viscosity values

The binder's viscosity values used in this research are estimated utilizing the ASTM viscosity-temperature equation (NCHRP Report 1-37A, 2004), given by the relationship as equation (3):

$$\log \log \eta = A + VTS^*\log TR \quad (3)$$

In which:
η = bitumen viscosity, cP;
A = regression intercept;
TR = temperature, Rankine (TR = T Fahrenheit + 460);
VTS= regression slope of viscosity temperature susceptibility.

The values of viscosity for binders and test temperatures relevant to this study are presented in Table 3.

**Table 3.** Binder Viscosity According to ASTM Viscosity-Temperature Relationship

| Binder Grade | Test Temperature | Viscosity, Pa.sec |
|---|---|---|
| PG 64-16 | 40°C | 2.87 |
| PG 64-16 | 55°C | 2.25 |
| PG 64-16 | 70°C | 1.79 |

## 7 MnPAVE results and data analysis

The recent software provides capabilities for the analysis and performance prediction of various input parameters for local conditions. In MnPAVE, the Research Mode provides the components important to finish an asphalt pavement design, more features in entering information. The allowed repetitions for rutting were determined by simulating the strains because of traffic loads, according to equation (4):

$$NR = K_{R1}\,\varepsilon v^{KR2} \quad (4)$$

Where: NR = number of repetitions to rutting failure (Allowed Repetitions), cycle
KR1, KR2 = MnPAVE coefficients
εv = vertical strain at the top of the subgrade (in/in)
In order to evaluate the effect of input variability on the calculated reliability; the influence of binder type, air voids, binder content and temperature on dynamic modulus is evaluated, as described in the following articles :

### 7.1 Effect of air voids

One of the essential parameters which have a pivot role on the performance of asphalt pavement is air voids. At two levels of air voids; the dynamic modulus of asphaltic mixture is evaluated: (a) 7% air voids level and (b) 4% air voids level with binder of PG (64-16) and temperatures of 40°C, 55°C and 70°C. The results are shown in Figure 1. It is found that increasing of air voids from 4% to 7% will decrease E* and reliability by 11.2% and 12% respectively for asphalt of PG (64-16). Figure 2 show that reliability value of 93 % is reached for rutting and the allowable repetitions of loading greater than 8300 cycle through running of verification; so this design is conservative. Same behaviour can be achieved for the other type of asphalt binder.

### 7.2 Effect of asphalt binder type

One of the basic components in asphaltic mixtures is asphalt binder. The performance of an asphalt mixture is straightforwardly identified with binder mechanical characteristics. Subsequently, there is a need to assess the relationship between the binder's properties and asphaltic mixtures such that an appropriate comprehension and selection of an asphalt binder can be made to enhance the asphaltic mixture performance. In this study, the impacts of asphalt binder properties on asphalt mixtures at various temperatures were assessed. Figure 3 shows the comparison of two types of binders; PG (64-16) and PG (58-22), High temperature and low frequency related with the rutting resistance of the asphalt mixture. The outcomes demonstrate that the PG grade, particularly the high PG grade impacts the dynamic modulus at high temperature; higher PG grade leads to higher E*. It is found that a change of binder of PG (64-16) to PG (58-22) causes 5.5 % decrease in E* and 20 % decrease in reliability at 40°C testing temperature. Figure 4 shows that reliability value of 93% is reached for rutting and the allowable repetitions of loading greater than 8300 cycles for PG (64-16).

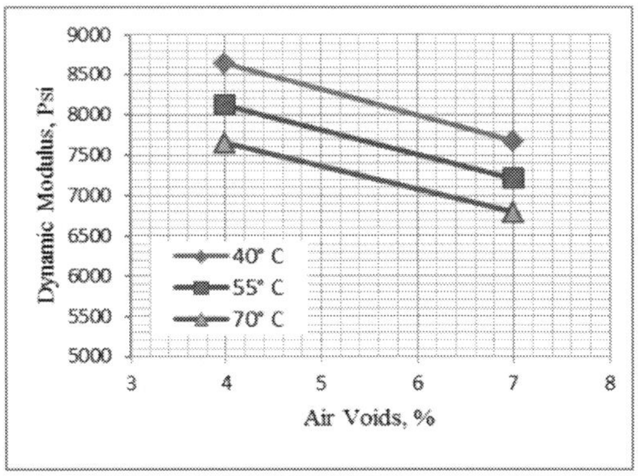


**Fig. 1.** Effect of Air Voids on Dynamic Modulus for PG (64-16)

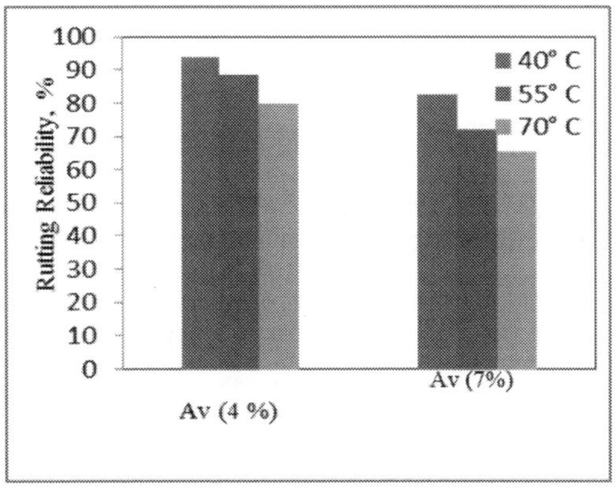


**Fig. 2.** Effect of Air Voids on Reliability

### 7.3 Effect of asphalt content

The effect of asphalt content on Dynamic modulus was evaluated; Figures 5 and 6 show the comparison of dynamic modulus and reliability for three various asphalt contents. It is noted that the increase of asphalt content from 4.1% to 4.7 % causes 4.9 % decrease in E* at 40° C

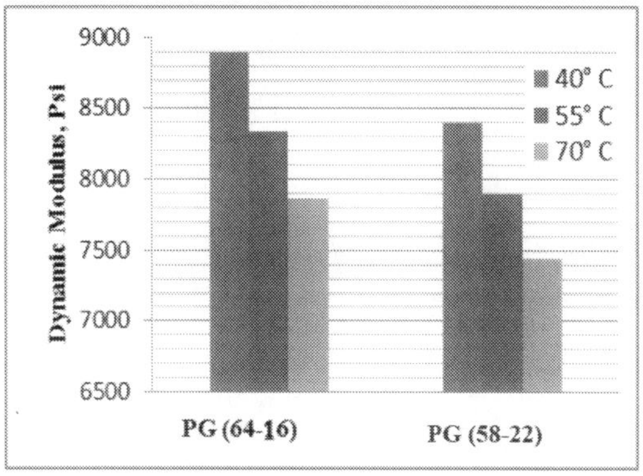


**Fig.3.** Effect of Asphalt Binder Type on Dynamic Modulus

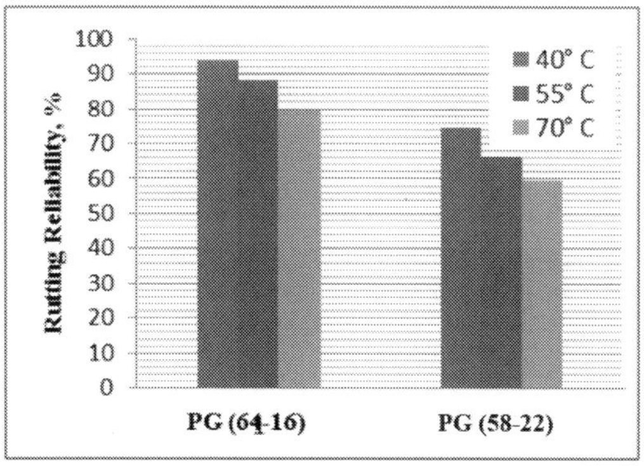


**Fig.4.** Effect of Asphalt Binder Type on Reliability

test temperature, while an increase of asphalt content from 4.7% to 5.3 % causes 5.5 % decrease in E* at 70° C test temperature. It is viewed as that the asphalt binder is softer and less influential regarding the load bearing capacity at high temperatures; rather, aggregate skeleton assumes a more important part. Asphalt content has some effect on E* and allowable number of repetitions.

## 8 Comparisons between the laboratory model and MnPAVE results

A comparison between the structural performances of MnPAVE software results with Superpave laboratory model is carried out. The pavement volumetric and mechanical properties were used as an input data for MnPAVE Software.

The WTT allowable number of repetitions results for rutting performance of mixtures utilized in this research and developed for various degree of test temperatures are compared with MnPAVE results. Figures 7 to 9 show the relationships between WTT's allowable number of repetitions model which was developed previously and stated in Table 3, Good relationship can be observed unmistakably between these two types of outcomes.

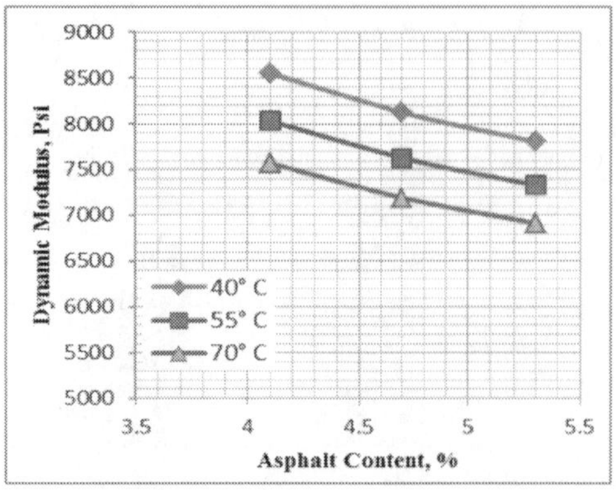


**Fig.5.** Asphalt Content with Dynamic Modulus

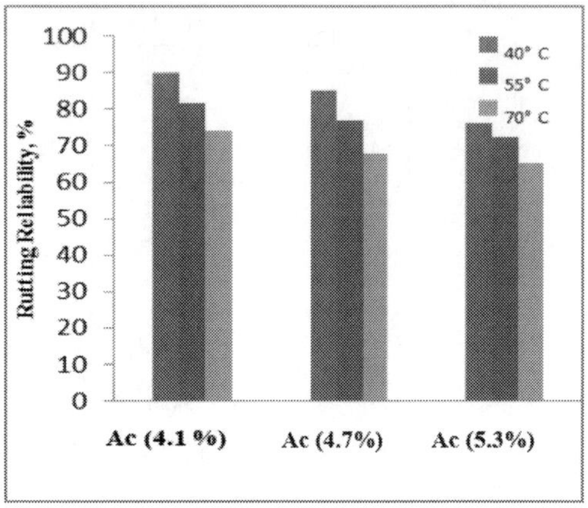


**Fig. 6.** Asphalt Content with Reliability

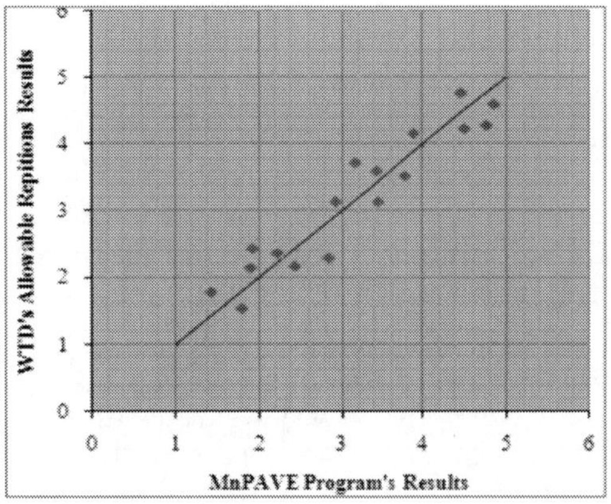


**Fig. 7.** WTT's Allowable Repetitions vs. MnPAVE Results at 40°C

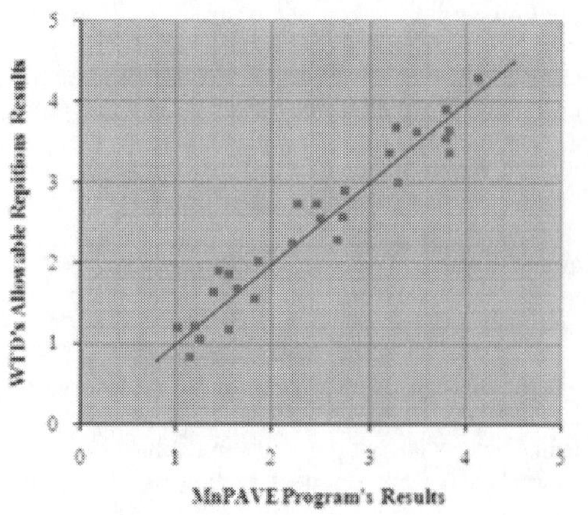


**Fig. 8.** WTT's Allowable Repetitions vs. MnPAVE Results at 55° C

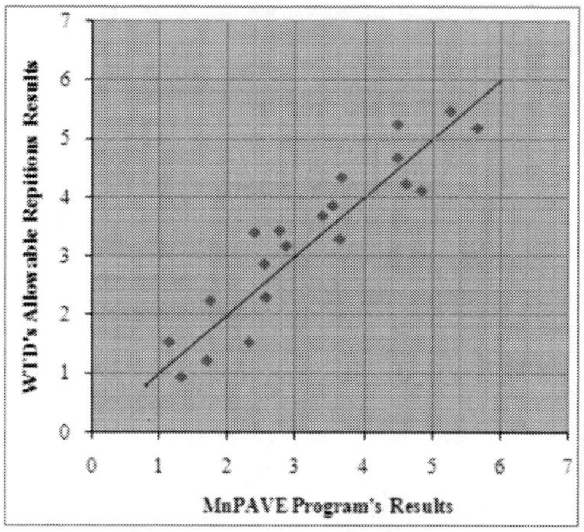


**Fig. 9.** WTT's Allowable Repetitions vs. MnPAVE Results at 70° C

## 9 Conclusions

In light of wheel-track test comes about, a model was developed to predict permanent strain of local asphalt concrete mixtures of wearing course after short- term aging mixtures for different mix properties and test conditions utilizing statistical technique, as equation 5:

$$\varepsilon p = 0.058\ LnN + 0.14\ Ln\ PS + 0.24\ LnT - 0.008 LnPM - 0.701\eta s + 0.045\ Av - 1.242 \quad (5)$$

Where: εp = Vertical Permanent Strain at Mid Thickness (H) of the Layer; N = Number of Loading Cycles; PS = Effective Asphalt Content, (%) by Weight of Total Mix; T = Test Temperature in Degree Centigrade (°C); ηs = Viscosity at Testing Temperature (Pa.sec); Av = Air Void of Asphalt Mixtures (Av from 4.0-7.0 %), and; PM = Additive (SBR- Polymer) Content, (%) by Weight of asphalt binder.

Mechanistic Empirical design approach through MnPAVE 2011 software, used to characterize rutting in HMA and to predict allowable number of loading repetitions as a function of expected traffic loads, material properties, and environmental conditions, has been employed to verify the developed model with satisfactory result. The influence of variables on dynamic modulus E* and rutting reliability is evaluated. It is viewed that increasing of air voids from 4% to 7% causes a decrease E* and reliability by 11.2% and 12% respectively for asphalt of PG (64-16).While a change of binder of PG (64-16) to PG (58-22) causes 5.5 % decrease in E* and 20 % decreases in reliability at 40°C testing temperature. Furthermore, an increase of asphalt content from 4.1 to 4.7 % causes 4.9 % decrease in E* at 40° C test temperature, while a change of asphalt content from 4.7 to 5.3 % causes 5.5 % decrease in E* for PG (64-16).

The authors might want to recognize the National Centre for Construction Laboratories (NCCL) and the College of Engineering of Baghdad University in Iraq for their support during the preparation of this research.

## References

1. K. P. Khosla, V. Kadayam, K.I. Harikrishnan, FHWA/NC/2002-030, Final Report Submitted to NCDOT, (2007).
2. J.S. Miller, W.Y. Bellinger, Final Report Submitted to U.S. Transportation Dep., FHWA-RD-03-031, (2003).
3. Y. Huang, *Pavement Analysis and Design*, 2nd Edition Prentice Hall College Div, New Jersey, USA. (1993).
4. R. Garba, Norwegian university of Science and Technology, *NVF conference*, 28 June-2 July, Reykjavik, Iceland, (2000).
5. H. Hamdou, N. Ahmed, Z. I. Qasim, *IMS/7 International Conference on Advances in Engineering Materials*, UAE, (2014).
6. Z. Qasim, A. Abbas, Z. Qasim, NJES, **20** pp.183 - 193, (2017).
7. *AASHTO Guide for Design of Pavement Structures*, Washington, D. C., USA. (2010).
8. C. Olidis, D. Hein, NCHRP Final Report ERES Consultants Division, Quebec, (2004).
9. European Standards EN 12697: CEN/TC 227 "Road materials", August. (2005).
10. Asphalt Institute, *Asphalt Hand Book*, Six Edition, Kentucky, USA. (2007),
11. M. Fattah, K. Al Helo, Z. Qasim, RMPD Journal, DOI: 10.1080/14680629.2015.1119711, Taylor and Francis Group, (2015).
12. W. Powell, D. Leech, Transport Research Laboratory, Crowthorne, LR 1102, (1983).
13. H. Henny, A. André, Civil Engineering & Management Department, University of Twente, Netherlands, 7500 AE. (2000).
14. Y. Huang, *Pavement Analysis and Design*, 2nd Edition, Prentice Hall, Englewood Cliffs, New Jersey, USA, (2004).
15. T. Pellinen, J. Song, S. Xiao, *8th Conference of Asphalt Pavements in South Africa*, CAPSA-04, September 12-16. (2004).

# Impact of the Superpave hot mix asphalt properties on its permanent deformation behavior

*Zahra* Qasim [1], *Alaa* Abed [2], and *Zaynab* Qasim [3,*]

[1]University of Baghdad, Baghdad, Iraq
[2]University of Al-Nahrain, Bagdad, Iraq
[3]University of Technology, Baghdad, Iraq

**Abstract.** In Iraq, the severity of rutting has increased in asphalt pavements possibly due to the increase in truck axle loads, tyre pressure, and high pavement temperature in summer. As of late, Superpave has been accounted as an enhanced system for performance based design, analysis of asphalt pavement performance prediction for asphalt concrete mixes. In this research the development of permanent deformation in asphalt concrete under repeated loadings was investigated, Wheel-Tracking apparatus has been used in a factorial testing program during which 44 slab samples were tested to simulate actual pavement. The objectives of the present research include; investigating the main factors affecting rutting in asphalt concrete mixture, quantifying the effect of SBS polymer and steel reinforcement on asphalt concrete mixtures in addition to studying the effect of variables on the asphalt concrete mixes against moisture sensitivity. It has been determined that that increasing of compaction temperature from 110 to 150 °C will decrease the permanent deformation by 20.5 and 15.6 percent for coarse and fine gradation control asphalt mixtures, respectively. While the permanent deformation decreases by 21.3 percent when the compaction temperature is increased from 110 to 150 °C for coarse gradation SBS modified asphalt mixtures.

## 1 Introduction

Rutting is the major distress in flexible pavements in Iraq as a result of increasing axle loads, and high summer temperature, [1]. Rutting reduces the useful service life of the pavement. By affecting vehicle handling characteristics, it creates serious hazards for highway users; the gradual effects of weathering and the action of vehicle traffic are the two major elements contribute to asphalt pavement deterioration as sited by [2].

Rutting has been a major concern for flexible asphalt pavement. It is also one of the most common pavement permanent deformations due to repetitive traffic loads which accumulate small deformations of pavement materials appearing as longitudinal depressions in the wheel paths of the roadways.

Early detection and repair of pavement defects is the most important preventive maintenance procedure. There are five areas of distress for which guidance is needed: fatigue cracking, (wheel path) rutting, thermal cracking, friction, and moisture susceptibility, all of these distresses can result in loss of performance, but rutting or permanent deformation is the one distress that is most likely to be a sudden failure as a result of unsatisfactory hot mix asphalt or asphalt mixture, other distresses are typically long term failures that show up after a few years of traffic [3].

Up-to-date, Superpave has been widely known as an improved system for perform based design, Investigation about asphalt concrete mixes and asphalt pavement performance prediction. It will be an organized methodology comprising of choice for materials, determination for design aggregate structure, asphalt binder content, and evaluation of moisture sensitivity.

In Iraq, the severity of rutting has been increased in asphalt pavements possibility due to the increase in track axle loads, tyre pressure; and high pavement temperature during summer, [4].

## 2 Research objectives

The main objectives of this study is to investigate the main factors affecting rutting in asphalt concrete mixture in Iraq such as; mix properties, types of filler, loading and temperature conditions by using superpave mix design system, investigate the effect of additive (SBS) on the asphalt concrete mixes improvement and quantify the effect of steel reinforcement in asphalt concrete slabs using Wheel-Tracker testing.

## 3 Materials properties

The materials used in this research are widely available and currently used in road paving in Iraq except steel reinforcement which is brought from local market.

* Corresponding author: zaynab.2010@yahoo.com

© The Authors, published by EDP Sciences. This is an open access article distributed under the terms of the Creative Commons Attribution License 4.0 (http://creativecommons.org/licenses/by/4.0/).

### 3.1 Binder

The binder used in this research is classified as (40-50) penetration grade or PG (64-16) performance grade; it is brought from Daurah Refinery in Baghdad. Table 1 shows the physical properties of the used Asphalt Cement.

### 3.2 Aggregate

The (crushed) aggregate used in this study is brought from Al-Nibaie quarry. The physical properties and chemical composition of the aggregate (coarse and fine) are shown in Table 2 and 3. Two types of mineral filler are used: limestone dust brought from the lime factory in Karbala governorate and Portland cement is from factory of Kubesa in Iraq. It is thoroughly dry and free from aggregations of fine particles or lumps.

**Table 1.** Properties of asphalt cement, according to ASTM requirement and Iraqi specifications.

| Test | ASTM Designation | Test Results | SCRB/ R9 2003 |
|---|---|---|---|
| Penetration, (100gm, 25°C,5 sec., (0.1mm) | D-5 | 45 | 40-50 |
| Ductility (25°C, 5cm/min), cm | D-113 | >100 | >100 |
| Flash Point, °C | D-92 | 289 | Min. 232 |
| Specific Gravity (25°C) | D-70 | 1.043 | …… |
| Residue from thin film oven test, D-1754 | | | |
| %Retained Penetration,1/10 mm | D-5 | 67.4 | >55% |
| Mass Loss, % | D-1754 | 0.38 | < 0.75 |
| Ductility of Residue, cm | D-113 | >100 | > 25 |

**Table 2.** The Physical properties of al-Nibaie aggregate

| Test | ASTM desig. | Test result | Superpave Specificatio n |
|---|---|---|---|
| Superpave coarse aggregate angularity, % | ASTM D-5821 | 97 | 90/95 Min |
| Fine aggregate angularity, % | T-304 | 61 | 45 Min. |
| Percent flat and elongated particles, % | ASTM D-4791 | 1.6 | 10% Max. |
| Clay content by Sand equivalent,% | T-176 | 51 | 45 Min. |
| Toughness, by (Los Angeles Abrasion) | T-96 | 15.2 | 35-45 Max. |
| Soundness loss by sodium sulfate solution,% | T-104 | 2.46 | 10-20 Max. |
| Deleterious materials, % | T-112 | 0.5 | 0.2-10 Max. |

**Table 3.** Chemical composition of al-Nibaie aggregates

| Chemical Compound | Content,% |
|---|---|
| Silica, Sio2 | 84.73 |
| Lime, CaO | 3.37 |
| Magnwasia, MgO | 0.53 |
| Sulphuric Anhydride, SO | 2.9 |
| Alumina, AlO | 0.62 |
| Ferric Oxide, FeO | 0.58 |
| Loss on Ignition | 6.25 |
| Total 98.98 | 98.98 |
| **Mineral composition** | |
| Quartz | 81.2 |
| Calcite | 10.02 |

Two types of aggregate gradation have been used in this research as shown in Table 4, The aggregate is sieved and recombined in the satisfactory proportions to declaration the gradation of wearing course as headed by SCRB specifications [5]. The aggregate maximum size gradation of 19 mm is used in this study. The aggregate fractions are separated into nine sizes using dry sieve analysis. Mineral filler (Limestone and Portland cement) are added according to requirements of desired gradation. Figure 1 shows the gradation curve for the aggregate. Four lines are presented: the lower, the upper curves of the Iraqi specifications of SCRB in addition to the control points according to Superpave system.

**Table 4.** The selected aggregate gradations as a percent passing by weight

| Sieve Size | Gradation % Passing | | Iraqi Specification SCRB,R9, 2003 | |
|---|---|---|---|---|
| (mm) | Coarse Blend | Fine Blend | Max. | Min. |
| 19 | 100 | 100 | 100 | 100 |
| 12.5 | 98 | 92 | 100 | 90 |
| 9.5 | 88 | 78 | 90 | 76 |
| 4.75 | 67 | 48 | 74 | 44 |
| 2.36 | 47 | 32 | 58 | 28 |
| 1.18 | 23 | 16 | --- | --- |
| 0.6 | 15 | 12 | --- | --- |
| 0.3 | 17 | 8 | 21 | 5 |
| 0.075 | 5 | 5 | 10 | 4 |

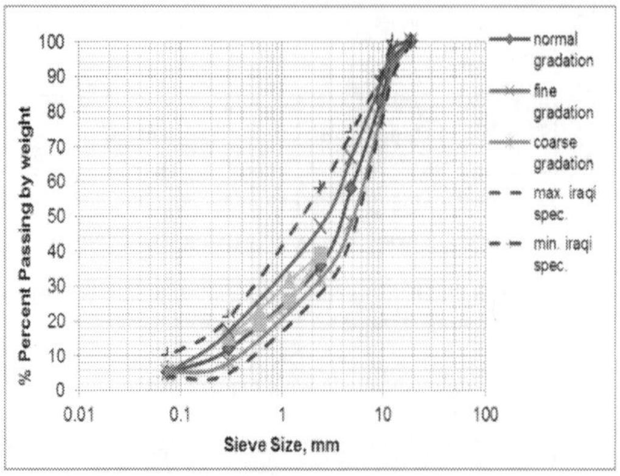


**Fig. 1.** Selected gradation for used aggregated

### 3.3 Additives

One type of polymer has been used in this research as an additive to the asphalt binder. The additive is known as SBS polymer (Styrene Butadiene Styrene), it's widely accepted asphalt binder modification agent used by the whole bitumen industries. 3% SBS is added by weight of asphalt binder.

Improvement in fatigue damage, thermal cracking, rutting resistance, temperature susceptibility, and stripping have led SBS polymer modified binders to be a substitute for non-modified, neat bitumen in many paving and maintenance applications. Table 5 shows the physical properties of modified binder.

**Table 5.** Properties of modified binder

| Test | Original 40-50 | Modified +3% SBS | SCRB/R9 2003 |
|---|---|---|---|
| Penetration, (100gm, 25°C,5 sec., (0.1mm)) | 45 | 68 | 40-50 |
| Ductility (25°C, 5cm/min), cm | >100 | >200 | >100 |
| Flash Point, °C | 289 | 300 | Min.232 |
| Residue from thin film oven test, D-1754 | | | |
| % Retained Penetration,1/10 mm | 67.4 | 50 | >55% |
| Mass Loss, % | 0.38 | 0.36 | < 0.75 |
| Ductility of Residue, cm | >100 | >200 | > 25 |

### 3.4 Mesh steel reinforcement

Configuration of the Mesh Steel Reinforcement used in this research consists of diamond shaped hole of size 40 by 30 by 1 mm, as shown in Figure 2. This mesh is protected against corrosion by a zinc coating. These meshes are available in local market and proposed to be used in this research as a reinforcement interlayer.
As a part of the experimental work, forty-four asphalt slab samples, with and without mesh steel reinforcement were prepared and tested in the laboratory. The reinforcement were cut and placed in the middle of asphaltic mixtures in the rectangular mold of the Roller Compacter machine with dimensions of 300 mm by 400 mm.

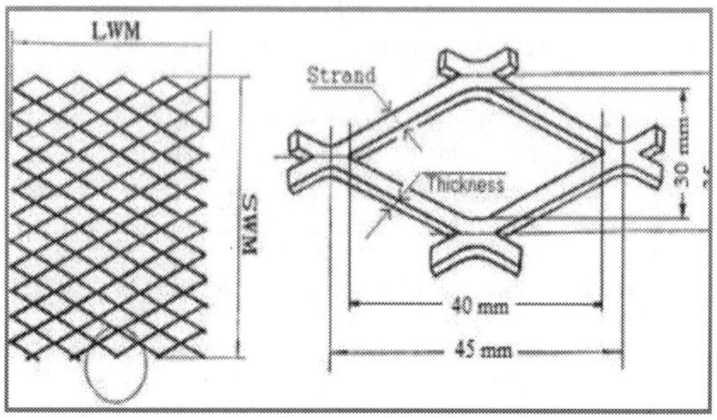


**Fig.2.** Configuration of mesh steel reinforcement

## 4 Mix design

In this research the volumetric properties of the HMA were found in accordance with Superpave Volumetric Mix Design (AASHTO M-323), [6]. Obviously, the design air voids (4%) for all Superpave mixes meet the specifications for percent air voids, [7]. Optimum binder content for different mixtures of the surface layer is determined using Superpave method. Four mixtures were tested to determine the optimum asphalt content: two gradations (coarse and fine gradation) and with two types of asphalt (origin and modified asphalt with SBS polymer). The Superpave Gyratory Compactor was used to prepare 24 cylindrical specimens of asphalt concrete for carrying out volumetric design according to Superpave system (AASHTO Designation: T 312-2010). Specimens were prepared by mixing at the appropriate mixing temperature, which is 160° C for the selected asphalt binder PG (64-16) and 180° C for modified asphalt binder. The specimens were then short-term aged by placing the loose mix in a flat pan in a forced draft oven at the compaction temperature 148° C and 165° C, for 2 hours for origin and modified asphalt respectively . Finally, the specimens were then removed and either compacted or allowed to cool loose (for Gmm determination).

## 5 Evaluation of moisture susceptibility

Moisture damage in reality can be defined as the loss of strength and durability in asphaltic mixtures caused by the presence of water, [8]. Moisture damage is induced by the loss of bond between the asphalt cement or the mastic (asphalt cement, the mineral filler and small aggregates) and the coarse and fine aggregate. Moisture damage accelerates as moisture permeates and weakens the mastic, making it more susceptible to moisture during cyclic loading. For evaluating the moisture damage in asphalt concrete, specimens with dimensions of (150 mm) (6") in diameter and (95 mm) (3.8") height for Superpave asphalt design mixes are used. Specimens are prepared according to AASHTO T 283-07, left to cool at room temperature for 24 hours. The number of desired gyrations is determined to obtain test specimens with approximately 7 percent ±1 of air voids using different number of gyrations that is (75) ± 3 for asphalt coarse gradation mixtures, as shown in Figure 3.

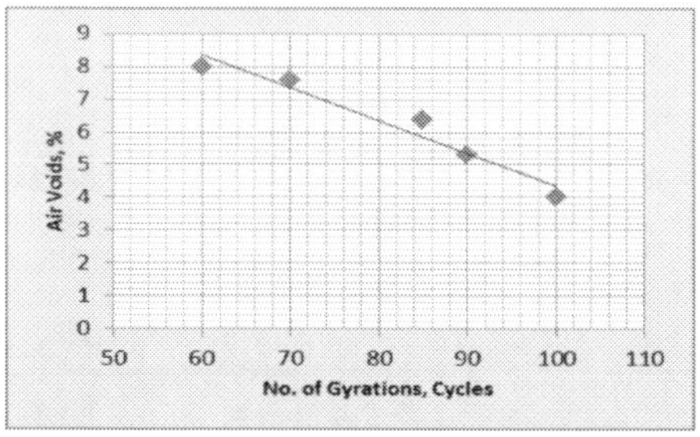


**Fig.3.** Relation between no. of gyrations and air void percentage

Eight subset of test specimen are produced in this research. Specimens are compacted in Superpave Gyratory Compactor to a specimen height of 95 mm with air void content approximately seven percent as presented in Figure 4. Each subset containing six specimens prepared for each mix with optimum asphalt content.

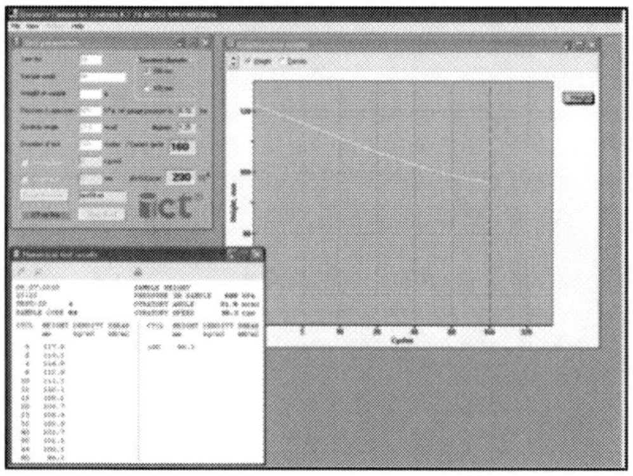

**Fig. 4.** Moisture susceptibility test by superpave gyratory compactor

Six specimens (three conditioned and three unconditioned specimens) are tested by Versa Tester Machine, as shown in Figure (5) at a rate of (50.8 mm/min.) (2"/min.) until the maximum load is reached and the specimen completely fractures then the load is recorded. Indirect Tensile Stress is calculated as follows:

$$St = 2P/\pi tD \quad (1)$$

where: St = tensile strength, kPa (psi); P = maximum load, N (lbf);
t = specimen height immediately before tensile test, mm (in); D = specimen diameter, mm (in.).
Then the Tensile Strength Ratio is calculated as follows:

$$TSR = (S_{tm}/S_{td})*100 \quad (2)$$

Where: TSR= tensile strength ratio in percent; $S_{tm}$ = average tensile strength of the moisture conditioned subset, kPa; and $S_{td}$ =average tensile strength of the dry subset, kPa.

**Fig. 5.** Indirect tensile test device

## 6 Sample preparation

The roller compacter apparatus can compact asphalt slabs to a target density using loads per unit roll width, which are consistent to those of pavement rollers used in the highway construction. The roller compactor provides a pneumatically powered means of compacting slabs of asphaltic material in the laboratory under conditions, which simulate in-situ compaction, as shown in Figure 6.

In this research, compacted asphaltic slabs for rutting test are prepared at air voids equal to (4%) using Roller Compactor Device according to (EN12697-Part 33:2003) and Superpave system (AASHTO Designation: T312-2010). The dimensions of the compacted slabs used in this work are of (400 mm by 300 mm by 50±6 mm) as proposed by (EN 12697-Part 22:2003)[9].

Proportion of aggregate and asphalt binder are used for mixing, curing, and compacting. The fractions of aggregate are separated into 9 sizes, as retained on each of the following sieves, 3/4", 1/2", 3/8", No.4, No.8, No.16, No.30, No.50, and No.200) using dry sieve analysis. The aggregate retained on the 3/4" sieve is discarded. Mineral filler (Limestone) and cement have been added according to the desired gradations requirements. The aggregate is combined into batch of (13400 gm for slab specimen) on the mixing bowl and heated to the mixing temperatures prior to mixing with asphalt binder which heated to the mixing temperatures corresponding to each binder.

**Fig. 6.** Compacted slab by roller wheel compactor machine

**Fig.7.** Wheel-tracking device and tested Sample

## 7 Wheel-tracking testing

The Wheel Tracker of pavement is a device used for testing the asphalt mixes wearability by simulating roadway conditions, this test was performed according to EN 12697-22:2003 [9] and AASHTO Designation: T 340-10. The test provides information about the rate of permanent deformation from a moving the sample under concentrated load. It uses a Linear Value Displacement Transducer (LVDT's) to measure the deformation or rutting of the specimen. The loaded wheel applies about 700 N (158 pounds) of load at contact points and passes repetitively over the asphaltic sample for up to 10,000 cycles.

Wheel-tracking machine is constructed so as to enable the test specimen in its cradle to be moved backwards and forwards under the loaded wheel in a fixed horizontal plane. The centre-line of the tyre track is (5 mm) from the theoretical centre of the specimen. The centre of the contact area of the tyre describes a simple harmonic motion with respect to the centre of the top surface of the test specimen with a total distance of travel of (230±10) mm and a constant loading frequency of (26.5±1.0) load cycles per 60 seconds for the test device in approximately 10,000 load cycles or 20 mm maximum allowed deformation is reached.

The experiment design for the permanent deformation testing is a full factorial with; two asphalt contents, three compaction temperatures, two asphalt types: original and modified with SBS, with and without mesh steel reinforcement, and two types of gradations, resulting in a nominal total of 44 slab tests.

The compacted specimens, which are 30 cm in width, 40 cm in length and 5 cm in height, are cooled to room temperature for a period of 24 hours in accordance with (EN-12697-22). The specimens are placed in mold and then placed on the carriage table of WTD for testing. The specimens in the mold are labelled with information mix type.

The holder of the displacement transducer is disengaged and reference plate is adjusted such that the transducer probe is compressed approximately 70% of its total travel. This allowed having sufficient travel available to measure the track formation on the sample.
If the maximum allowed deformation is reached before 10,000 passes, the wheel is lifted off the failed sample. Test results are compiled in a Microsoft Access database application which provides several means of reporting results, as shown in Figure 7.

## 8 Results and analysis

Permanent deformation and vertical permanent strain ($\varepsilon p$) were measured at testing temperature of 40 °C, frequency level 53 passes per minute, two compaction temperatures of 110 °C and 150 °C, two asphalt contents, two types of gradation, two types of filler, with and without mesh steel reinforcement and SBS polymer was used as modifier to asphalt binder.

### 8.1 Effect of compaction temperature

This study investigates the influence of compaction temperature on coarse and fine asphalt mixes for modified and unmodified binders.

For this, Superpave mix designs for two compaction temperatures of 110 °C and 150 °C and two asphalt binders (control, 3% styrene butadiene styrene (SBS) modified) were carried out. A total of 44 specimens were manufactured with a short-term age for two hours at the mixture compaction temperatures prior to test.

Figure (8) shows effect of compaction temperature on Permanent deformation (i.e. rut depth, RD). It can be seen that permanent deformation decreases when compaction temperature changes from 110 to 150 °C as shown in Table (6). It can be observed that the rut depth (RD) decreases about 20.5 and 15.6 percent when the compaction temperature increases from 110 to 150 °C for coarse and fine gradation control asphalt mixtures respectively. While the rut depth (RD) decreases about 21.3 percent when the compaction temperature increases from 110 to 150 °C for coarse gradation SBS modified asphalt mixtures as shown in Figure 9.

The results from this study showed that the compaction temperatures significantly affected the volumetric properties of the SBS modified mixes. The mixtures containing SBS-modified binders, the rut depth decreased by 21.3 percent and the air–void contents significantly decreased with an increase in compaction temperature because the binder in modified asphalt mixtures is stiffer than in conventional mixtures; therefore, there is a need for a higher compaction temperature.

**Table 6.** Percentage of change in permanent deformation with compaction temperature

| **Rut Depth , mm** | **Effect of Temperature, 110° to 150 °C (%)** | |
|---|---|---|
| | **Coarse gradation** | **Fine gradation** |
| Control asphalt Mixtures | -20.5 | -15.6 |
| Modified asphalt Mixtures | -21.3 | - |

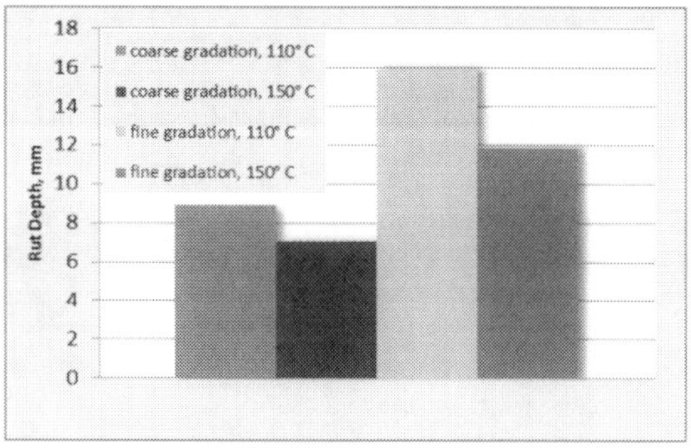


**Fig. 8.** Effect of compaction temperature of control mixtures on permanent deformation

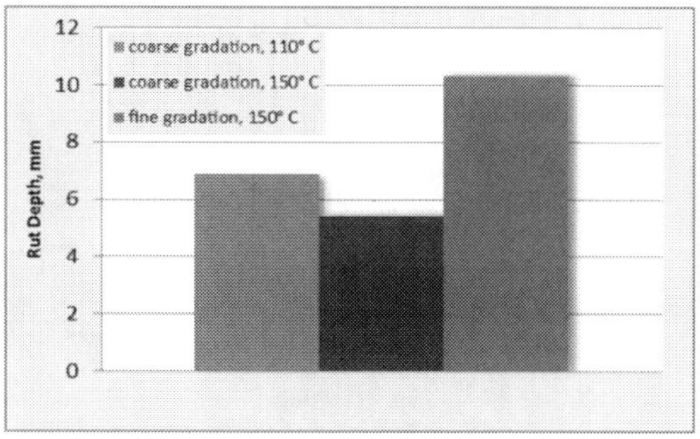


**Fig. 9.** Effect of compaction temperature on permanent deformation of modified mixtures

### 8.2 Effect of asphalt / binder content

Based on the data shown in Figures (10) to (12), it appears that the examined asphalt content has influence on the plastic response of the material. The plastic strain is increased with the increases in asphalt content from optimum AC to opt. +0.5 percent asphalt content for conventional and modified asphaltic mixtures. For (RD) value, it can be found that it increases about 10.6 percent when asphalt content increases from 4.6 to 5.1 percent for conventional coarse asphalt mixture, while it increases about 8 percent when asphalt content from 4.9 to 5.4 percent for SBS modified asphalt mixture, as shown in Table (7).

It can be concluded for Superpave mixtures, the total asphalt content plays an important significant role in controlling overall rutting resistance of the conventional and modified asphaltic mixtures, i.e rutting performance is highly influenced by total asphalt content, voids in mineral aggregate (VMA), voids filled with asphalt , (VFA), and dust-to-binder ratio.

**Table 7.** Percentage of change in permanent deformation with original asphalt content

| Variable | Original Asphalt | Modified Asphalt |
|---|---|---|
| | 4.6 % to 5.1 % | 4.9 % to 5.4 % |
| Rut Depth, mm | 10.6 | 8.0 |

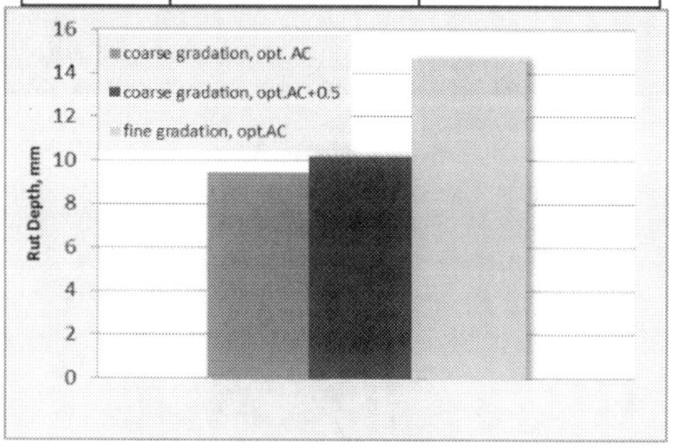


**Fig.10.** Effect of asphalt content of conventional mixture on Permanent deformation

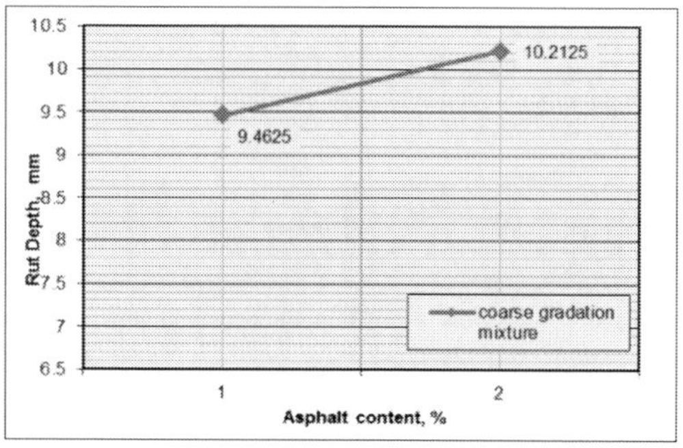


**Fig. 11.** Change in permanent deformation with asphalt content

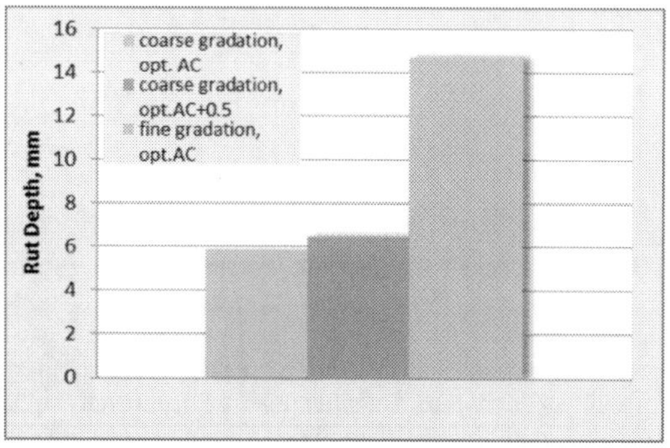


**Fig.12.** Effect of asphalt content of modified mixture on permanent deformation

### 8.3 Effect of aggregate gradation

Aggregate is present as a major portion of asphalt concrete mixture. It is found that researchers have come to different conclusions with regard to the effect of aggregate gradation on resistance to permanent deformation of asphalt mixtures.

Two gradations of aggregate that are typically used to produce hot mix asphalt in Iraq were used in this research. They are: coarse and fine gradation, both of them passing below the restricted zone (on the Superpave gradation chart). Figure (13) shows the results showed effect of aggregate gradation of mixture on permanent deformation; it concluded that rutting resistance of asphalt paving mixes is affected by the mix gradation of aggregate. Coarser gradation had higher resistance to rutting from fine gradation of aggregate by 65.6 percent as shown in Figure (14).

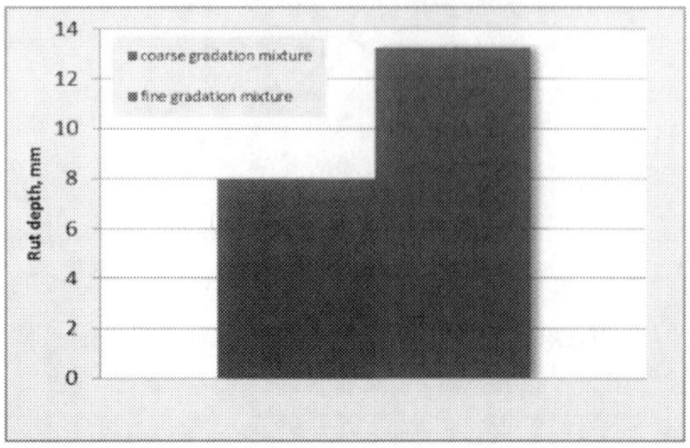


**Fig.13.** Effect of aggregate gradation of mixture on permanent deformation

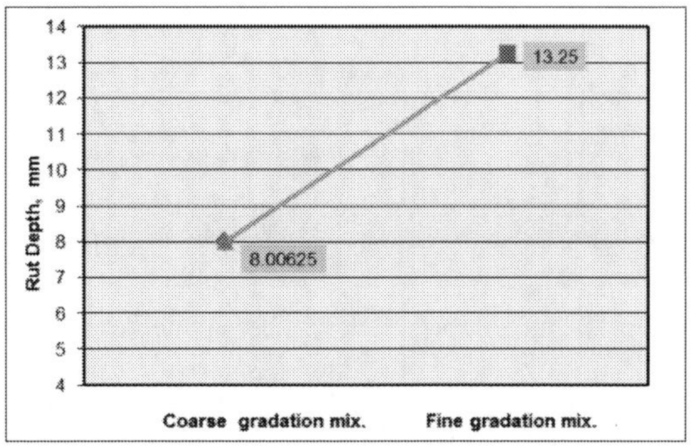


**Fig. 14.** Change in permanent deformation with aggregate gradations

## 8.4 Effect of type of filler

The resistance of asphalt mixture to permanent deformation is related to the stiffness of asphalt binder, mixture volumetrics, and the bonding interaction between asphalt binders and aggregate. Mineral filler is usually added into asphaltic mixture to stiffen asphalt binder and improve asphaltic mixture density and strength. These fillers are typically fine powders with particle sizes in the range of 0–100 μm, two types of filler were used in this research, there are: limestone and cement. However, the properties of mineral fillers of the compacted mixtures can influence on the permanent strain for conventional and modified mixtures as depicted in Figures (15) and (16). The percentage of change in Permanent Deformation is presented in Table (8). It can be observed that the rut depth (RD) decreases about 23.5 and 30.6 percent when the cement filler is used in asphaltic mixtures instead of limestone filler for coarse and fine gradation mixtures respectively. The effect of type of filler on the mixture rutting potential was more significant for the fine mixture than the coarse mixture.

**Table 8.** Permanent deformation change with type of filler

| Rut Depth , mm | Effect of Type of Filler From (lime) to (cement), % |
|---|---|
| Coarse gradation | -23.5 |
| Fine gradation | -30.6 |

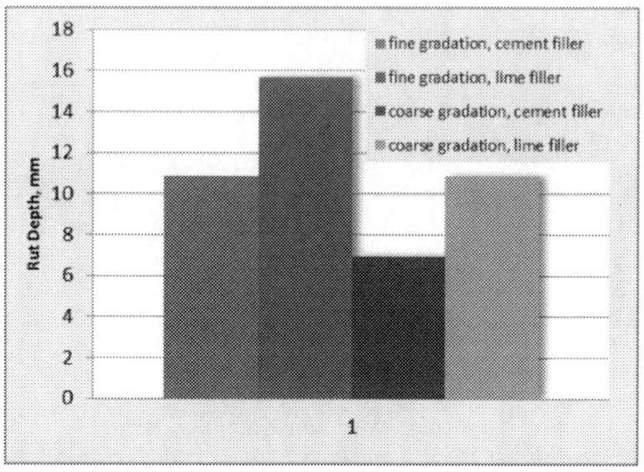


**Fig. 15.** Effect of type of filler in mixture on permanent deformation

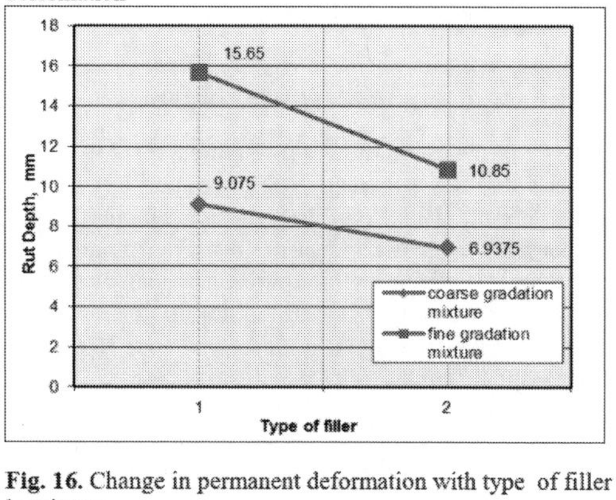


**Fig. 16.** Change in permanent deformation with type of filler in mixture

## 8.5 Effect of steel reinforcement

Reinforcing is a structural measure of increasing strength against the variety of stresses and improving its strength characteristics. Several studies have been conducted in order to investigate the effect of reinforcement materials on asphalt pavements. In this research one type of reinforcement material is used to improve the strength properties of asphaltic concrete mixtures and reduce the shear strain.

In this research an improvement was observed in rutting behavior of metal reinforced samples compared with those without reinforcement. Moreover, samples containing expanded mesh reinforcement performed better against rutting. It was observed that the permanent deformation of the asphalt concrete pavement is decreased in the presence of metal reinforcement; Figure (17).

The reinforcement stiffness and its interlocking with the asphalt concrete contribute to the restraining effect. The developed strains in the reinforcement around the vicinity of the loading area manifested the restraining effect of reinforcement. Moreover, there was a reduction in settlement over the loading area of reinforced samples compared with that of those without reinforcement.

As a result, when reinforced samples are achieved compared to those of control ones, it appears that permanent deformation of asphaltic pavements decreases by 41.1 and 19.7 percent for coarse and fine aggregate gradation respectively as shown in Figure (18).

**Table 9.** The Percentage of change in permanent deformation with steel reinforcement

| Variable | Effect of Steel Reinforcement, (%) | |
|---|---|---|
| | Coarse gradation | Fine gradation |
| Rut Depth , mm | -41.1 | -19.7 |

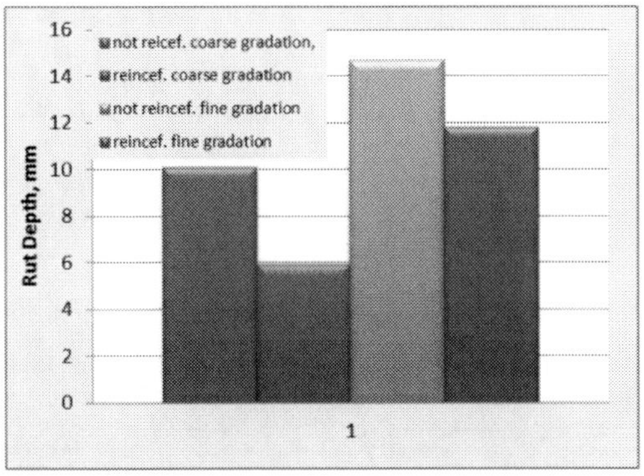


**Fig. 17.** Effect of steel reinforcement on permanent deformation of mixtures

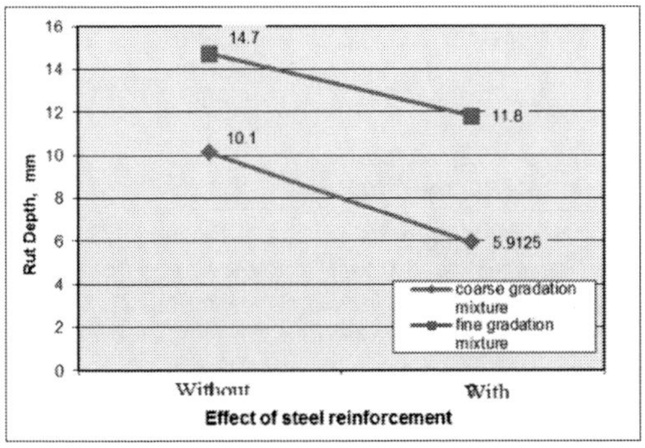


**Fig. 18.** Change in permanent deformation with steel reinforcement

## 8.6 Effect of SBS polymer additive

It was noted that the mixes with the polymer-modified binder were more difficult to mix and compact during the preparation of the samples. Any cooling of the mix greatly increases the viscosity of the asphalt and the stiffness of the mix.

In this research, the addition of additive (SBS) polymer tends to improve the mix properties. In Superpave mix design, the results showed that rutting performance of the asphaltic samples has improved for the polymer modified cases.

The percentage of change in permanent deformation with Polymers Content was shown in Table (10). It can be observed that the permanent displacement (RD) decreases about 37.22 and 29.5 percent when SBS polymer is used for coarse and fine gradation mixtures, respectively, as shown in Figures (19) and (20).

Based on the above evaluation, the use of asphalt modified by SBS polymer gives the wearing course better rutting resistance than the fine gradation of the same asphalt performance grade.

At the testing temperature of $40^{\circ}$ C±5, the asphaltic mixtures appeared to maintain stiffness levels. However, it should be noted that when SBS polymer modifier is used, the HMA become stiffer at the high testing temperature. It can be concluded that SBS polymer-modified binder made the polymer modified mixtures stiffer than control mixes with unmodified asphalt binder at high temperature.

**Table 10.** Percentage of change in permanent deformation with polymers content

| Variable | Effect of Polymer (SBS) Content (%) | |
|---|---|---|
| | Coarse gradation | Fine gradation |
| Rut Depth , mm | -37.22 | -29.5 |

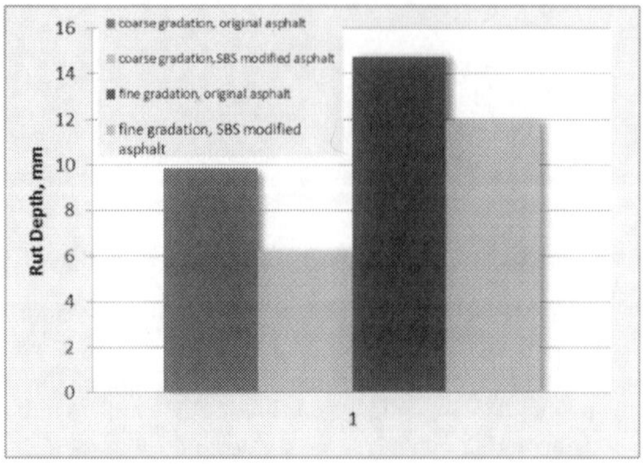


**Fig. 19.** Effect of SBS polymer on permanent deformation of mixtures

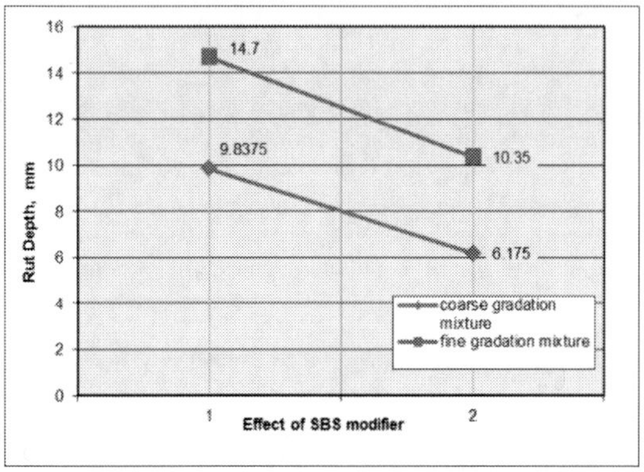


**Fig. 20.** Change in permanent deformation with SBS polymer modifier

## 9 Conclusions

Considering all results of laboratory tests and analysis, the following conclusions are presented:

- Wheel-track permanent deformation results for asphaltic wearing course mixes indicate that increasing of compaction temperature from 110 to 150 °C will decrease the permanent deformation by 20.5 and 15.6 percent for coarse and fine gradation control asphalt mixtures, respectively. While the rut depth (RD) decrease about 21.3 percent when the compaction temperature increases from 110 to 150 °C for coarse gradation SBS modified asphalt mixtures.
- The permanent deformation is increased with the increases in asphalt content from optimum AC to opt. +0.5 percent asphalt content for conventional and modified asphaltic mixtures. For (RD) value, it can be found that it increases about 10.6 percent when asphalt content increases from 4.6 to 5.1 percent for conventional coarse asphalt mixture, while it increase about 8 percent when asphalt content from 4.9 to 5.4 percent for SBS modified asphalt mixture.
- Wheel-track permanent deformation results showed effect of aggregate gradation of mixture on permanent deformation, it concluded that rutting resistance of asphalt paving mixes is affected by the mix gradation of aggregate. Coarser gradation had higher resistance to rutting from fine gradation of aggregate by 65.6 percent for conventional asphalt mixtures.
- The permanent deformation decreases about 23.5 and 30.6 percent when the cement filler is used in asphaltic mixtures instead of limestone filler for coarse and fine gradation mixtures, respectively. The effect of type of filler on the mixture rutting potential was more significant for the fine mixture than the coarse mixture.
- The permanent displacement (RD) decreases about 37.22 and 29.5 percent when SBS polymer is used for coarse and fine gradation mixtures respectively. Based on the results, the use of an asphalt modified by SBS polymers gives the coarse wearing course better rutting resistance than the fine gradation of the same asphalt performance grade.
- The permanent deformation decreases when reinforced samples are achieved compared to those of control ones, it appears that permanent deformation of asphaltic pavements decreases by 41.1 and 19.7 percent for coarse and fine aggregate gradations respectively.
- The impact of factors on tensile strength ratio TSR and moisture susceptibility is assessed. It is discovered that utilizing cement as filler, TSR is increased by (5 and 3.5) percent for coarse and fine gradations at optimum design of asphalt binder content, while TSR increases by 2.07 and 2.5 percent for coarse and fine gradations at modified optimum design of asphalt binder content.
- Also when utilizing the SBS polymer modifier for coarse and fine gradations, it can be presumed that TSR is increased by 6.1 and 4.5 percent at optimum design asphalt of binder content for lime filler, respectively, while TSR increased by 3.2 and 3.4 percent for coarse and fine gradations at modified optimum design of asphalt binder content for cement filler, respectively.

## References

1. H. M. Hamdou, N. G. Ahmed, Z. I. Qasim, IMS/7 *International Conference on Advances In Engineering Materials*, UAE, 2014.
2. Y.Huang, *Pavement Analysis and Design*, Prentice Hall, Englewood Cliffs, New Jersey, USA. (1993).
3. I. Ishai, and J. Craus, "Transportation Research Record 1530, National Research Council, Washington, D.C., pp. 75-86 (1996).
4. Z. I. Qasim, A.S. Abbas, Z. I., Qasim, NJESl, 2017.
5. *General Specification for Roads and Bridges*, SCRB/ R9, Iraq (2003).
6. AASHTO Guide for Design of Pavement Structures, *The American Association of State Highway and Transportation Officials*, Washington, D. C., USA. (2010).
7. M. Y. Fattah, K. H. Al Helo, Z. I. Qasim, JRMPD, (2015).

8. Z. I. Qasim, A.H. Abed, *The 2nd International Conference on Buildings, Construction and Environmental Engineering* (BCEE2), Lebanon, (2015).
9. European Standards EN 12697: *Bituminous mixtures Test methods for hot mix asphalt*, CEN/TC 227, (2005).

# Chemical recycling of polyethylene terephthalate (waste water bottles) for improving the properties of asphalt mixture

*Mohammed* Alzuhairi [1, *], *Ahmed* Al-Ghaban [1], *Shams* Almutalabi[1]

Department of Materials Engineering, University of Technology, Baghdad, Iraq

**Abstract.** One of the most prevalent of waste materials is Polyethylene Terephthalate (PET) which is used mainly to produce the drinking water bottles. In this research, the waste plastic bottles, which are normally made from PET was used to investigate the possibility of using this material as an additive in asphalt concrete mixtures. Six different proportions (w/w %) of Degradated Polyethylene Terephthalate (DPET) (2, 4, 6, 8, 10, and 12%) have been added to bitumen to prepare the specimens. The tests include Marshall Method of mix design and coating with asphalt. The results indicated that the mixture property modification increased as the content of (DPET) increases. This additive gives maximum flexibility and rigidity of the asphalt, according to ductility and penetration tests. Marshall Method gives better resistance against permanent deformations and better engineering properties in terms of stability, flow value, air voids and water absorption comparing with non-modified mixtures.

## 1 Introduction

Plastic is a commonly available material which is used widely all over the world and it has many advantages such as: Lightweight, flexibility, robustness and rigidity. Although it is used in different applications like bags, food packages, containers and bottles, the problem raised here is how to get rid of it after use, because it's made of polymer chemicals which are not biodegradable. Polymers can be divided into two major classifications including thermoplastic materials (foams, adhesives, embedding resins and elastomers) and thermosetting materials (or Thermosets) [2]. Plastic is also an inexpensive, lightweight and durable material, which can easily be molded into a variety of products that can be found use in a wide range of applications. As a result, the production of plastics has increased significantly over the last 60 years [3]. Several possible ways to get rid of plastics have been investigated. Recycling is one of these methods. It involves processing of plastic materials (waste) to turn them into new materials as an essential way to keep environment green. The new product of recycled plastics can be used as an additive. The plastic addition to bitumen in the road construction has increased its smoothness and lifespan and also makes it environment friendly and economical. Plastic waste is used as a modifier of asphalt to enhance many properties of asphalt roads. Plastic water bottles, that are mainly made from Polyethylene Terephthalate (PET), though do not have any direct danger on the environment, but because of the large quantities of plastic waste dumped in the environment and the sea, it would be a serious problem to the environment[4].

From the environmental and sustainability point of view, chemical recycling is the most suitable process because, in chemical recycling operations, new raw materials are produced that in turn can be reused again in the production of chemical and in petrochemical or refineries. It is also a partial solution to the solid PET waste issue [5].

There are different ways to improve asphalt mixture properties, first is constructing road pavement of higher thickness, the second is using different types of additives as a modifier (e.g. different types of fibers and polymers) in the asphalt mixture. Constructing high-thickness pavement will cause considerably higher construction cost. Thus, using additives might be a better solution to overcome the pavement deterioration problem [6].

In recent years, a serious environmental problem due to the amount of waste plastic bottles is increasingbeing increased steadily and has become a global problem. So, the world focuses on solving this problem by many methods such as recycling. It is one of the most important efforts currently available to reduce this problem by providing opportunities to reduce the use of oil, carbon dioxide emissions and the quantity of waste that requires disposal. This work aims at:

- Analyzing the effect of additive degraded PET on the properties of mixed asphalt for road paving and coating applications.
- Studying the effect of temperature on the mechanical and physical properties of modified asphalt and finding the optimum percentage of plastic waste (PET) through making comparison between conventional and modified asphalt.

* Corresponding author: dr.alzuhairi@gmail.com

© The Authors, published by EDP Sciences. This is an open access article distributed under the terms of the Creative Commons Attribution License 4.0 (http://creativecommons.org/licenses/by/4.0/).

This research is very important in many ways - one of them is the recycle of tens of thousands of tons used for asphalt improving and second – it is a creative method for chemical degradation of PET in a simple and low cost manner.

## 2 Materials

### 2.1. Bitumenous

The asphalt used in this research is petroleum asphalt brought from Al-Obaidy asphalt plant, refined at 40/50 penetration-grade asphalt cement was used for this investigation. Hot mix asphalt (HMA) used in this work is available in large quantities locally in Iraq. The properties of asphalt binder based on the values of Iraqi specification for Roads and Bridges (2016) [7].

### 2.2 Aggregates

Aggregates are the main material in the asphalt mixture. The crushed coarse and fine aggregates were brought from Al -Nibaee quarry in the north of Baghdad; the coarse aggregate consists of strong, hard, durable pieces. According to specification (2016)[7] the gradation of coarse type aggregate ranges between ¾ in. (19.0 mm) and No.4 sieve (4.75 mm). While the gradation of fine type aggregate ranges between 4.75 mm (No.4) sieve and 0.075 mm (No.200) sieve.

**Tabel 1.** Properties of asphalt (bitumen) binder of al-obeidi asphalt plant*.

| Test | ASTM | Specification | |
|---|---|---|---|
| | | Min | Max |
| Penetration (0.1 mm) | ASTM D5 | 40 mm | 50 mm |
| Ductility (cm) | ASTM D113 | >100 | <100 |
| Flash & Fire Point (°c) | ASTM D 92 | 175 | 250 |
| Ring & Ball Softening Point °C | ASTM D36 | 50 | 60 |
| Specific Gravity | ASTM D70 | 1.01 | 1.06 |
| Viscosity at 135 °C | ASTM D2170 | >210 | ....... |

*These tests had been done in asphalt technology laboratory, University of Technology.

The fine type consists of tough grains free of clay, loam or other deleterious materials. The physical properties of Al-Nibaee aggregate listed in the table (2).

### 2.3 Filler

The filler used consists of Portland cement and limestone powder, these fillers added to hot mix asphalt, to increase the density and strength of the mixture. These fillers should pass through 75 μm SI sieve (International System of Units), as shown in Table 3.

**Tabel 2.** Physical properties of al-nibaee aggregates*.

| Property | ASTM | Coarse Aggregate | Fine Aggregate |
|---|---|---|---|
| Bulk Specific Gravity | ASTM C127 | 2.646 | 2.63 |
| Apparent Specific Gravity | ASTM C127 | 2.656 | 2.667 |
| Percent Water Absorption | ASTM C127 | 0.47 | 0.523 |
| Percent Wear (Los-Angeles Abrasion) | ASTM C131 | 22.69 | ....... |

*These tests were made by the national center for construction labs and researches in Baghdad /Iraq, as well as table (3).

**Tabel 3.** Physical properties of cement and limestone*

| Filler | Specific Gravity |
|---|---|
| Cement | 3.12 |
| Limestone | 2.73 |

### 2.4 Water bottles (PET)

In this study the waste plastic water bottles (PET) were added as a modifier of asphalt to achieve a service life of road pavement. It is cut into many small sized pieces. This material is present everywhere in huge quantities due to its wide use and low cost. This research includes adding Ethylene Glycol (EG), which was used as a solvent for glycolysis and zinc acetate particles were used as a catalyst acting as an analytical reagent grade to PET. Table (4) shows the properties of waste plastic, and Table (5) shows the main properties of ethylene glycol and zinc acetate [8].

**Tabel 4.** Properties of waste (PET).

| Density | Melting point | Molar weight |
|---|---|---|
| 1.38 g/cm$^3$ ,20 °C | 260 °C | 192.2 g/mol |

**Tabel 5.** Ethylene glycol & zinc acetate properties (purity 99.9%).

| Material | Producer | *M. Wt.* | *P, g/cm$^3$* |
|---|---|---|---|
| Ethylene Glycol | Barcelon (Spain) | 62.07 | 1.1132 |
| Zinc Acetate | Sigma Aldrich (Germany) | 219.5 | 1. 735 |

## 3 Experimental work

### 3.1 Close system

In the first step, mixture of 100 gm of PET, 116 ml of ethylene glycol (EG.) (4:1, EG: PET, molar ratio) with 0.5% zinc acetate based on the weight of PET as catalyst were heat treated at 190°C for 5 hours until the mixture came in the resin state. Heat treatment involves complete condensation (reflux) of rising vapors in close system-no material is lost, through use of glass condenser which uses water as a cooler, then unreacted (EG) solution was separated from the previous mixture leaving behind the final form of the Degradated Polyethylene Terephthalate (DPET) as shown in Figure (1).

**Fig. 1** Degradation of Polyethylene Terephthalate (DPET) after a liquid separation.

Close system of chemical recycling for PET is shown in Figure (2), with U-tube manometer above apparatus to ensure no loss of any vapors and materials from the system [9].

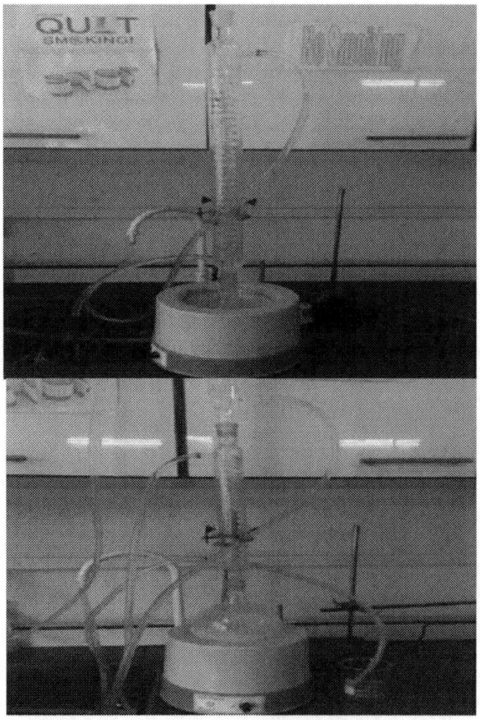

**Fig. 2** Close system of chemical degradation.

In the second step, the product DPET, was mixed with bitumen by (2, 4, 6, 8, 10, and 12) (w/w %) at 200 °C for 2 hours, with mechanical mixing in 250 Revolution per Minute (RPM), as shown in the Figure (3). Finally product poured into a mold to cool at room temperature.

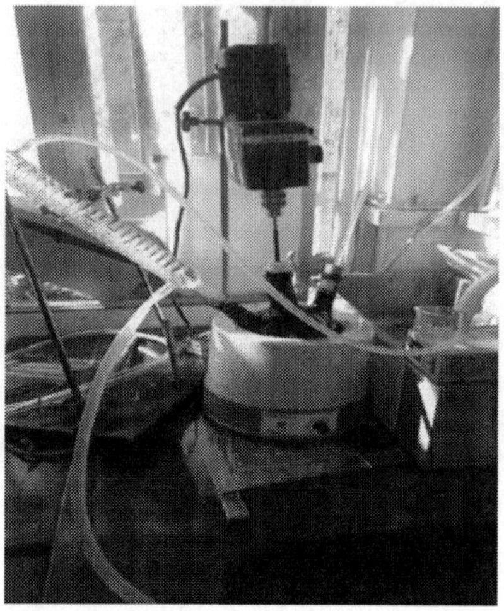

**Fig. 3** Mechanical mixing of bitumen and DPET.

### 3.2 Marshall mix design

The grades of aggregates and their quantities used for preparing Marshall Samples were used according to Iraqi specification for Roads & Bridges (2016). Suitably designed mix of bitumen will determine an economical blend through several trial mixes and resist the strong loads of traffic under negative environmental conditions such as (heavy rain, humidity and high temperature) also fit the characteristics of pavement surface and structure. This method determines the resistance to plastic flow within the cylindrical shape samples of bitumen paving mixture loaded on the lateral surface of Marshall apparatus according to ASTM (D1559) [ASTM, 2003]. This method includes preparing of cylindrical sample which is 2.5 inches (63.5mm) in height, and (4 inches) (101.6 mm) in diameter. The optimum bitumen content was (5%) by asphalt mix weight [10]. The samples of asphalt concrete mixtures of the additive type DPET were prepared to measure the adding effect of DPET to the mixture. The samples were prepared by the addition of six different percentages (2, 4, 6, 8, 10 and 12%) added to the content of optimum asphalt. Three samples were measured for each proportion to determine the unit weight, flow and stability and air voids. The apparatus of this test includes: Marshall mold, compaction hammer, and spatula are heated on a hot plate to a temperature between (100-110) °C, the procedure is done by applying the addition of DPET that mixed previously to the bitumen by mechanical mixture and placed into the oven and heated until it became liquid at (100-110) °C, (45) minutes and heating (1200gm) of aggregates and filler of each sample until it reaches a (160-180) °C, then mixed together. The asphalt mixture was placed in the preheated mold, and it was then speculated mechanically for 15 min at (150 -170) °C. Two filter papers were placed on the bottom of the mold filled with mixture in preparation to place the asphalt mix in the suitable mold. 72 blows were applied with the compaction hammer for each surface. After compaction was finished, the mold containing sample had to be cooled in air till enough cohesion had improved to result in the cylindrical proper shape, then the sample had to be extruded by a hydraulic jacking device to provide a constant and gentle pressure. Finally, samples immersed in a water bath at 60 °C ready for the test[10].

## 4 Results and discussion

### 4.1 The stability

In this method, the stability values in asphalt mixtures are higher than the stability values in conventional mixture: the value in conventional mixture is (12.1 KN) with filler cement, and (9.6 KN) with filler dust limestone, while the maximum stability value at the added percentage of DPET (8 %) is found nearly (13.8 KN) with filler cement, and (11.8 KN) with filler dust limestone at 150°C, Also the stability values in asphalt mixtures are higher than the stability value in conventional mixture: (11.5 KN) with filler cement, and (9.8 KN) with filler dust limestone, while the maximum stability value at the percentage (8 %) of DPET is found nearly (14.7 KN) with filler cement, and (11.9 KN) with filler limestone at 170°C.This increase demonstrates that an improvement has been achieved with regard to the compressive strength of asphalt mixture as the percentages of DPET additions indicate, as shown in Figures (4) and (5). This shows that the results (0-12%) of the additive are within the range of Iraqi specification for Roads and Bridges (2016) which determine the minimum value of stability test (8 KN). DPET works to increase the viscosity, stiffness and hardness of the bitumen, which leads to reduction in penetration and softening point as mentioned above, and the Marshall value for the mixture with cement filler is higher than the value mixture with dust limestone because mixture gives high density and lower air voids.

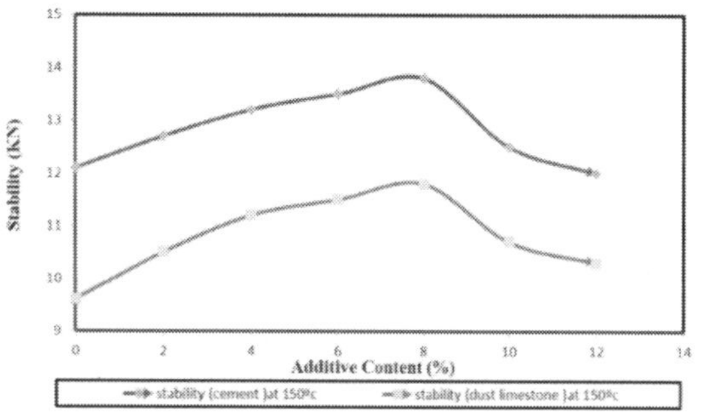


**Fig. 4** DPET effects on Asphalt mixture stability at (150 °C).

### 4.2 The flow

In the modified asphalt, the flow values in asphalt mixture are higher than those in a conventional asphalt mixture. The flow values decrease continuously as the DPET percentage increases, as shown in Figures (6) and (7). When adding % DPET, the flow values extend from (2.7mm) till they reach (4.1 mm) with the filler cement, while the flow values extend from (3.2 mm) till they reach (4.8 mm) with the filler dust limestone at 150°C. But when the temperature is 170°C, the flow values extend from (2.9 mm) till they reach (4.2 mm) the filler cement, while the flow values extend from (3.7 mm) till they reach (4.8 mm) with the filler limestone. All the results, were in the range of Iraqi specification for Roads and Bridges (2016) [the specification values extend from (2 to 4 mm)]. Flow increases that means the elastic behavior of the flow is related to the increase in ductility.

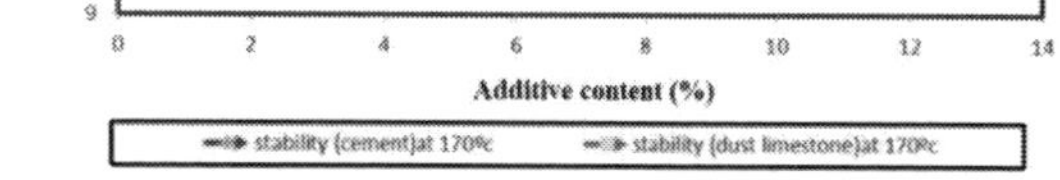


Fig. 5 DPET effects on Asphalt mixture stability at (170 °C).

 https://doi.org/10.1051/matecconf/201816201042

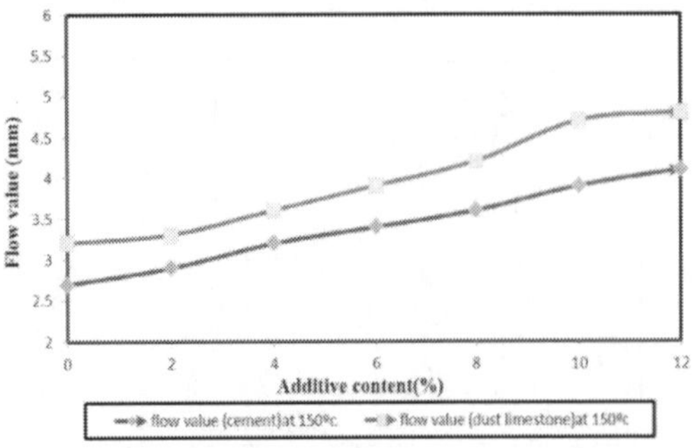


**Fig. 6** DPET effects on Asphalt mixture flow at (150 °C).

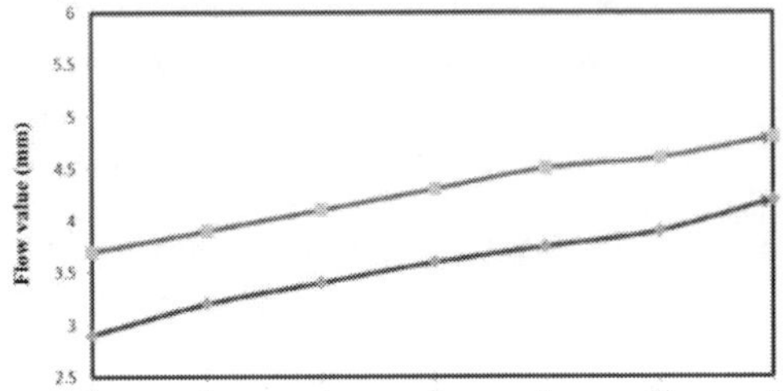


**Fig.** 7 DPET effects on Asphalt mixture flow at (170 °

### 4.3 The bulk density

The bulk density decreases as DPET content increases, when adding % DPET, the bulk density values extend from (2.32 mm) till they reach (2.278 mm) with the filler cement, while the flow values extend from (2.281 mm) till they reach (2.243 mm) with the filler dust limestone at 150°C. When the temperature is 170°C, the flow values extend from (2.38 mm) till they reach (2.32 mm) with the filler cement, while the flow values extend from (2.27 mm) till they reach (2.28 mm) with the filler dust limestone. This decrease in bulk density is caused due to low density of PET added instead of asphalt, the effect of DPET in the asphalt mixture on the bulk density is shown in Figures (8) and (9). The results agree with the values of Iraqi specifications for Roads and Bridges (2016), which extend from (2.1 - 2.45 gm/cm$^3$). Bulk

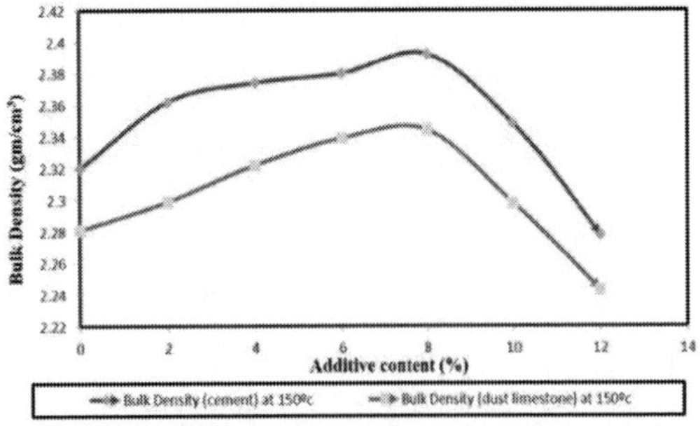


**Fig. 8** DPET effects on bulk density of Asphalt mixture at (150

density of mixture with cement filler is higher than mixtures with dust limestone filler due to specific gravity of cement filler which is (3.12) and specific gravity of limestone filler which is (2.73), this difference in specific gravity values lead to this result.

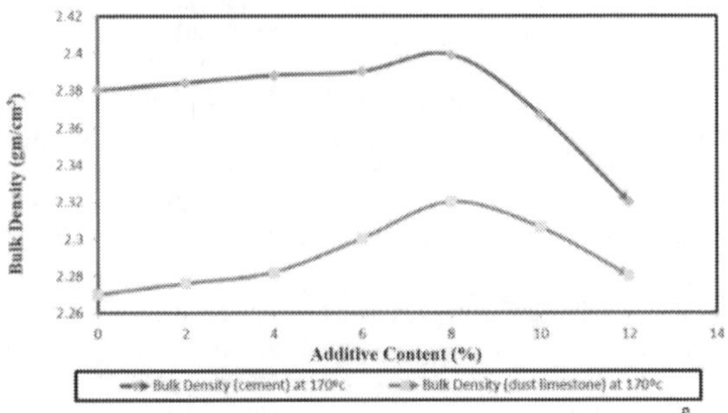


**Fig. 9** DPET effects on bulk density of Asphalt mixture at (170 °C).

### 4.4 Air voids

The percentages of air voids (during the addition method) decrease as DPET percentage increase, when adding % DPET, the AV. values extend from (6.2 mm) till they reach (3.4 mm) with the filler cement, while the flow values extend from (6.4mm) till they reach (4.4 mm) with the filler dust limestone at 150°C. When the temperature is 170°C, the flow values extend from (5.8 mm) till they reach (4.2 mm) with the filler cement, while the flow values extend from (6.1mm) till they reach (4.5 mm) with the filler limestone, this decrease occurs due to increase in asphalt percentages which has a lower density than DPET, values of air voids with different DPET content percentages are shown in Figures (10) and (11) that deviate from the Iraqi specification for Roads and Bridges (2016) (values that extend from 3% to 5%). Air voids in cement are less than those in dust limestone due to the high smoothness in cement as compared to the limestones, which consequently have effect the asphalt mixtures. Mixture with cement filler gives a high bulk density that leads to lower porosity due to high softness of cements.

### 4.5 Coating tests (dilution)

The diluents and mixture of bitumen & DPET that meet concrete specifications consist of 30 (v/v) of toluene as the solvent for bitumen and DPET mixture. Toluene is an organic solvent, colorless with a special smell related to paint, thinners, one of the aromatic hydrocarbons which has density 0.87 g/cm$^3$, pb is 95°C, and molar weight equal to 92.14 g/mol. Toluene widely used as substitute for gasoline is chemically toxic. It can be added to the fuel and used as an argument for a large number of compounds, including drugs, dyes, detergents, perfumes, explosives and synthetic resins [10, 11].

### 4.6 Preparation of the samples

https://doi.org/10.1051/matecconf/201816201042

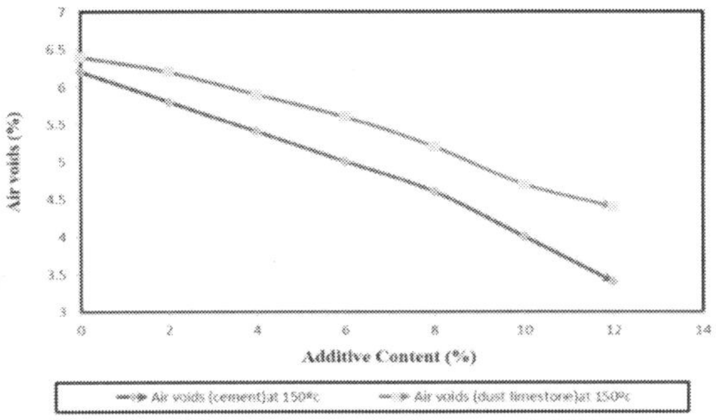


**Fig. 10** DPET effects on air voids of Asphalt mixture at (150 °C).

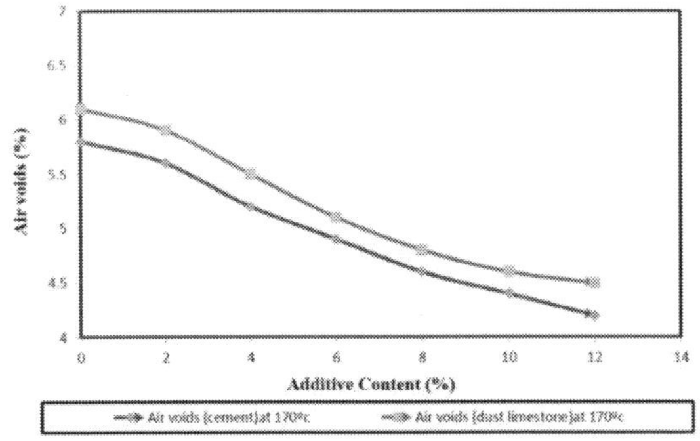


**Fig. 11** DPET effects on air voids of Asphalt mixture at (170 °C).

The different samples of concrete are prepared at (1:2.75)' cement: sand with water cement ratio (0.495) and composting size (5cm×5cm×5cm) of the mold. Then, the concrete is repainted with the diluted bitumen and left in the air for three days to dry completely. Finally, the repainted concrete is immersed in water for seven days at different temperatures (0, 20, 40, and 60 °C) to investigate the water absorption as shown in Figure (12).

**Fig. 12** Specimens of coated

### 4.7 Waterproofing

When the percentages of DPET content increase, the percentages of water absorption decrease. The waterproofing in the concrete coated with modified asphalt is more than the uncoated one, but due the temperature increases the porosity of modified asphalt will go bigger as a result of stretching. This leads to increasing in the penetration and absorption, consequently leading to increase the need for more % DPET in the coating. Figure (13) shows the values of water absorption of different percentages of DPET content for various ranges of temperatures agreed with Iraq weather.

## 5 Conclusions

In this work the results are going to be useful and helpful in the road paving and coating applications by waste plastics recycling, So the following conclusion could be listed:

1. Waste plastic can improve the waterproofing and performance of roads.
2. The cost of materials used in this research is reduced because the cheapness and affordability of the materials.
3. The DPET additives show a great improvement in asphalt specification in terms of the flow, stability, bulk density, and air voids.
4. The coating with DPET modified asphalt has shown that the percentage of water absorption has decreased at various temperature ranges in Iraq.

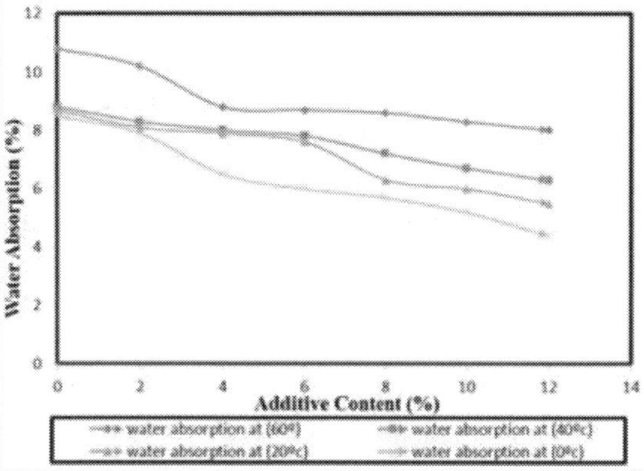


**Fig. 13** DPET effects on water absorption of Asphalt mixture different temperature.

## References

1. M. Alzuhairi, A.Al-Ghaban, S. N. Almutalabi. "Chemical Recycling of Polyethylene Terephthalate (PET) as Additive for Asphalt." ZANCO J. of Pure and Applied Sciences. 28, 2; 675-679 (2016).
2. M. Sulyman,, Haponiuk, K. Formela "Utilization of Recycled Polyethylene Terephthalate (PET) in Engineering Materials: A Review." Int. J. of Envir. Sc. and Develop., 7, 2, 100. (2016).
3. W. Aishah, S. Mohd Rawi, and N. Desa, "Utilization of Plastic Bottle Waste in Sand Bricks." J. of Basic and Applied Scientific Research, 2090-4304 (2015).

 https://doi.org/10.1051/matecconf/201816201042

4. T.. Mathew,"Pavement materials: Bitumen" Lecture notes in Transportation Systems Engineering, (2009).
5. S. AbdulWahab, Iraqi specification for Roads and Bridges, Planning Ministry Centeral of Orgnization for Standerization and Quility Control (2016).
6. Mosadeghzad, et al."Preparation and properties of acacia sawdust/UPR composite based on recycled PET"; Malaysian Polym. J., 4, 1, 30-34 (2009).
7. M. Alzuhairi,, K. Shabbeb, S. Alsa,edy, "Heat and Chemical Treatments for Sawdust/UPE Composites", Eng. & Tech. J., 34 Part (A),.8 (2016).
8. F. Awaja and D. Pavel, "A Review of Recycling of PET" Eur. Polym. J., 41, 1453- 1477 (2005).
9. D. Torre, "Feasibility overview of a solar water heater made of polyethylene terephthalate bottles for rural areas in Guatemala.", Institute of Science and Technology for Development, Rafael Landivar University, (2011).
10. A. Raheem, and L. Uyigue, "The conversion of post-consumer PolyEthylene Terephthalate (PET) into a thermosetting polyester resin." Arch. Appl. SCI. Res 2 240-254. (2010).
11. A. Mohammad,, and L. Shbeeb, "The use of polyethylene in hot asphalt mixtures." American Journal of Applied Sciences 4, 6 390-396. (2007).

# Performance of railway track system under harmonic loading by finite element method

*Ammar* Shuber[1] , *Mohammed* Hamood [1,*] and *Walaa* Jawad[1]
[1]Building and Construction Engineering Department, University of Technology, Baghdad, Iraq

**Abstract.** The performance of railway track under dynamic loading depends on many parameters such as type of sleepers, ballast, soil – structure interaction and the relation of contact forces between the track and railway. The rail track classified as low and high speeds, where speed is a very important factor in design. Many theoretical models were derived and some developed for dynamic analysis of railway track system with assumptions that simplified or matched actually the real behavior especially the support of sleepers as linear or nonlinear and the contact between railway and track. In the present paper, a Finite Elements Approach is one of the numerical analysis by ANSYS software that adopted here to study the performance of harmonic analysis of railway track system. Harmonic ranges were applied with constant applied loading based on the European code for the steel rail.

## 1 Introduction

Rail is the most important part of the whole system of train railway, made from steel. The functions of rail are to transmute the weight of the train to the lower components, guide the wheels and give smooth surface for running. Sleepers are another part typically made from timber, concrete, steel and other alternate materials, the functions of sleepers are to cushion and transfer external loads to the ballast. Ballast used to fix and anchors the sleepers, resists the lateral and vertical forces which come from train, that means provides stability, also distributes the loads to the subgrade layers. Subgrade layers are designed to resist all loads from the components of railway system and rest on hard soil to prevent any unexpected failure. Rail joints that are made from angle bars are used to hold the ends of the rail in the right place and worked as beam and to prevent vertical and horizontal movements of the rail.

In 2009, Jabbar Ali Zakeri and He Xia [1], adopted (FEA) to analyze two dimensional beam element (Timoshenko beam) that represented railway and concerned the train – track interaction under the effect of dynamic loading. In 2009, Valeri Markine et al [2], looked closely at the relationships between railway pads, sleeper's pad and ballast foundation with applied dynamical loading from tracks, and investigated the relationship between wheel, rail and dynamic loading by using FEM. In 2010, A. Gomes Correia et al [3], studied many softwares that deal with Finite Elements approach to simulate and evaluate the performance of railway in case of high speed of track train and used the data in all softwares from the available experiments. In 2011, Khanh Nguyen et al [4], reported and studied the loading on ballast under the effects of dynamic loading in case of high speed. In 2011, Huan Feng [5], studied the effects of many parameters on the design of railway system under dynamic loading. Finite Elements Approach was considered in the study and many models were simulated that described the relation between beams, soil, ballast, sleepers and railway track system. Euler-Bernoulli principle was adopted to simulate rail and sleepers; spring model was used to represent the relationship between sleepers and ballast. In 2012, Puguh B. Prakoso [6], used conventional method and compared the results by using another approached as (FEM) to model the whole track system. Beam rest on the elastic foundation that adopted the concept of Winkler's theory is the conventional approach that was classified as one dimensional as it still has lack of applications for railway analysis. FEM also adopted by ANSYS software to model the railway track system and compared the results with Zimmermann Method. In 2014, D. Kishore Kumar and K. Sambasivarao [7], studied the performance of pre-stressed concrete sleepers under static and impact loading by modeling the problem using Finite Elements Method (FEM). Natural frequency was calculated under various supports condition such as free or fixed. In 2014, A. Ortega García [8], investigated the dynamic behavior of railway track. The frequency due to the vibrations caused by track was analyzed using numerical approach and divided in two parts, first, simulated the railway track using FEM, and the second, checked out the safety of the track according to vertical displacements due to velocity.

* Corresponding author: mjh_smm@yahoo.com

© The Authors, published by EDP Sciences. This is an open access article distributed under the terms of the Creative Commons Attribution License 4.0 (http://creativecommons.org/licenses/by/4.0/).

In 2016, Zhenxing He and Xinwen Yang [9], studied the performance of sleepers by coupled vehicle-track system under dynamic loading. Three dimensional models were used for two-block sleeper. All external loading that developed from car, frames and wheel was assumed rigid.

Too many approaches were considered and adopted to analyze the railway system. In present work Finite Elements Analysis (FEA) has been adopted. Analysis has been done using two dimensions that requires to make some assumptions to reduce the complexity of the problem. Three dimensions simulation using finite elements analysis represent the actual behaviors of the problem by taking into account all parameters that affects the performance of the railway system. By selecting suitable elements to represent the real characteristics of each parameters, elements size, real input data, suitable solver and good accuracy lead to accurate results, so selection of three dimensions was done to simulate the railway system to ensure that the problem gets solved correctly.

## 2 Modeling

Railway track, sleepers, ballast and soil layers were taken into account to simulate the whole railway track system. The model consists of a discrete system of springs, dampers and masses that take into account the parameters effects on the performance of railway track system.

Numerical approach is adopted for railway track system performance. Finite Elements Approach depends on the assumed strain field and the degree of freedoms for each nodes when analysis is three dimensional and then the equations are formed in matrix form to find the unknowns.

In two dimensional approach of (FEA), the rail track represented as beam element has two degrees of freedom at each node at the ends, vertical displacement and rotation (u, ϑ) as follow:

$$\{y_i\} = \{u_i\ \vartheta_i\} \qquad ...(1)$$

The relationship between the forces and the displacements is expressed by the equation below:

$$\{F\}^e = [k]^e \{y\}^e \qquad ...(2)$$

Where, (k) is the stiffness matrix for each element and then after assembly in one stiffness matrix as global matrix. The mass matrix formed by the sum for all masses defined as:

$$[m] = \begin{bmatrix} m_1 & 0 & \dots & 0 \\ 0 & m_2 & \dots & 0 \\ \dots & \dots & \dots & \dots \\ 0 & 0 & \dots & m_n \end{bmatrix} \qquad ...(3)$$

In each part of the railway system there are damping (c) which defined as damping matrix as follow:

$$[c]^e = \lambda [m]^e + \gamma [k]^e \qquad ...(4)$$

Where, (λ) and (γ) are damping confections. The general equation that collects all parameters based on the Newton's second law, so:

$$[M]\{y_r''\} + [c]\{y_r'\} + [k]\{y_r\} = \{q\} \qquad ...(5)$$

Where, [M], is the sum of all masses, [c], is the sum for all damping coefficients, [k], is the sum for all stiffnesses and {q} is the sum for all foresees, {yr}, {yr'} and {yr''} are the displacement, velocity and acceleration of the rail, respectively. The relations between all elements as finite elements after construction of the assembly matrix that contains all parameters. The three dimensional finite elements differ from the two dimensions because the degrees of freedom become more and the problem becomes more complex.

### 2.1 Assumptions

The assumptions adopted in present study are listed below:

- Concrete as sleepers, steel as railway and all others materials are modeled as isotropic and homogeneous.
- Steel was assumed to be an elastic-perfectly plastic material and identical in tension and compression (no hardening).
- Initially plane sections remain plane after loading (no nonlinear geometry).
- The maximum compressive strain in the concrete is assumed to be 0.003 mm/mm.
- Perfect bond exits between different materials.
- Self-weight of all materials is ignored.

The ANSYS [10] program is one of the most powerful and versatile that is adopted in the present study. It is effective and efficient, especially for harmonic load analysis. The reinforcement in the discrete model uses one dimensional bar that is connected to same concrete mesh nodes so that makes integrity and there is no friction between them. Full displacement compatibility between the reinforcement and concrete is a significant advantage of the discrete representation, see Figure (1).

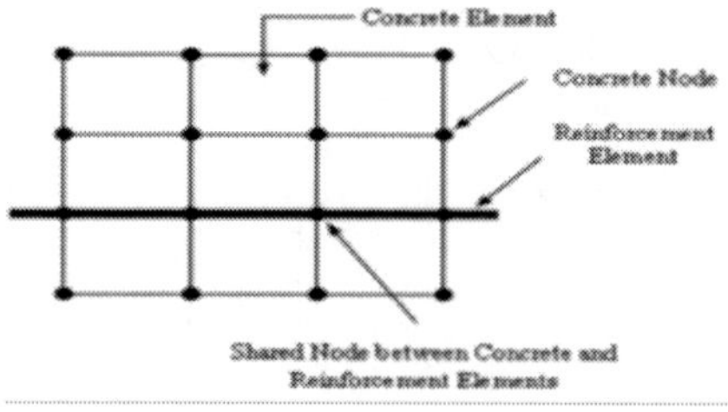


**Fig.(1).** Reinforcements representation

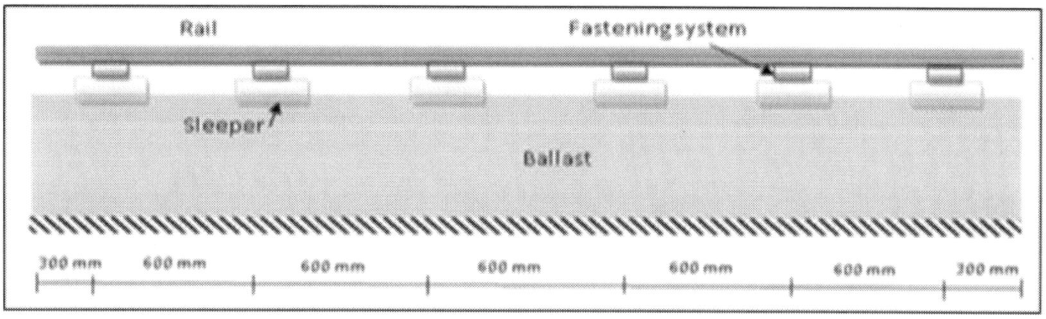


**Fig. (2).** Section in railway track system

## 2.2 Simulations and loading

For railway steel I section, SOLID185 is selected. The element is defined with eight nodes having three degrees of freedom at each node – translations in the nodal x, y, and z directions. The SOLID65 element is used to model concrete sleepers. This element has eight nodes with three degrees of freedom at each node – translations in the nodal x, y, and z directions. Link180 element is used to model steel reinforcement. This element is a 3D spar element and it has two nodes with three degrees of freedom at each node – translations in the nodal x, y and z directions. The use of a square (aspect ratio equal to unity) or rectangular mesh is recommended for good accuracy and saving time, so that the mesh is set up such that square or rectangular elements are created.

The magnitude of applied loading based on the load adopted by European code at the nodes along the railway that represents the train loading. Fixed supports are assumed at the bottom of sleepers to simulate the ballast layer and rock bed.

Harmonic response analysis is a linear analysis and the full method is adopted here which considers all matrix elements and calculates all displacements and stresses in a single pass.

# 3 Definition of the model

A section of a railway track system as shown in Figure (2) is modeled using (FEA) by ANSYS. The reduced number of elements in the model allows simplifying the numerical works performed by the (FEA), so that only one single railway will consider.

Table (1) lists the symbols mark, loading type, frequency range, and sleeper's status.

## 3.1 Geometry

The model is made up of (12000 mm) total classic railway track represented with varied structural element in three dimensional models. Each element has material and geometrical properties according to their function in the superstructure. The cross sectional for railway is UIC – 60 as shown in the Figure (3), the type of connections between railway and the sleepers are rigid connection, the sleeper cross section shown in the Figure (4).

**Table (1).** Model classifications

| Model mark | Loading | Frequency range (Hz) | Loading (kN) | Sleepers status |
|---|---|---|---|---|
| RW1 | Static | --- | Euro loading | Normal |
| RW2 | Static | --- | Euro loading | Two sleepers fail |
| RW3 | Static | --- | Euro loading | Three sleepers fail |
| RW4 | Static | --- | Euro loading | Four sleepers fail |
| RW5 | Low | 0-40 | Euro loading | Normal |
| RW6 | Low | 0-40 | Euro loading | Two sleepers fail |
| RW7 | Low | 0-40 | Euro loading | Three sleepers fail |
| RW8 | Low | 0-40 | Euro loading | Four sleepers fail |
| RW9 | Medium | 40-400 | Euro loading | Normal |
| RW10 | Medium | 40-400 | Euro loading | Two sleepers fail |
| RW11 | Medium | 40-400 | Euro loading | Three sleepers fail |
| RW12 | Medium | 40-400 | Euro loading | Four sleepers fail |

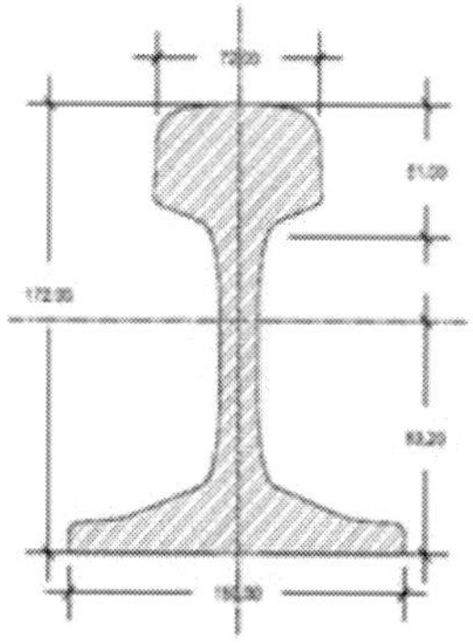

**Fig. (3).** UIC-60 cross section

* Corresponding author: mjh_smm@yahoo.com

https://doi.org/10.1051/matecconf/201816201043

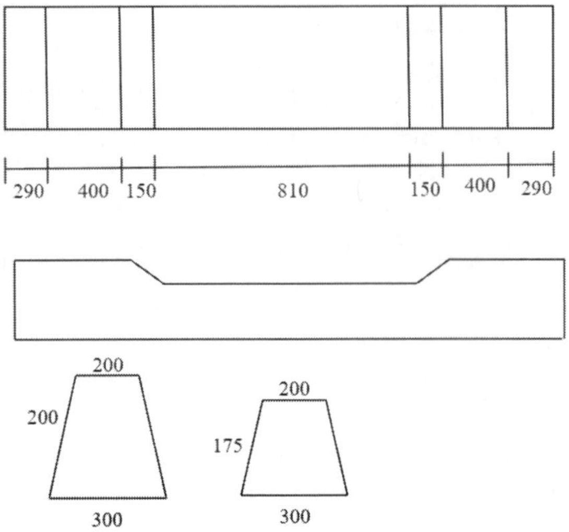


**Fig. (4).** Sleeper cross section. (All dimensions in mm)

### 3.2 Elements type

For railway steel I section, SOLID185 is select. The element is defined with eight nodes having three degrees of freedom at each node – translations in the nodal x, y, and z directions. The geometry and node locations for this element type are shown in Figure (5).

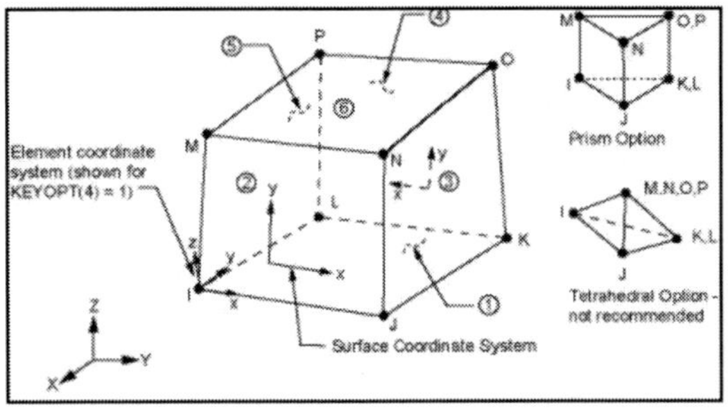


**Fig. (5).** SOLID185 element

The Solid65 element is used to model concrete sleepers. This element has eight nodes with three degrees of freedom at each node – translations in the nodal *x*, *y*, and *z* directions. This element is capable for cracking and crushing in three normal directions. A schematic of the element is shown in Figure (6).

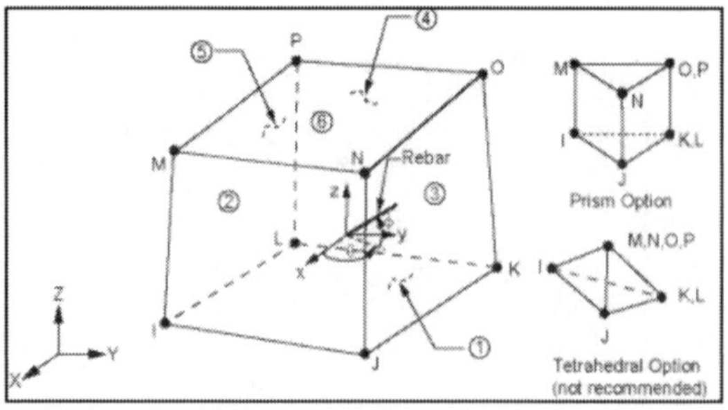


**Fig. (6).** SOLID65 element

Link180 element is used to model steel reinforcement. This element is a 3D spar element and it has two nodes with three degrees of freedom at each node – translations in the nodal x, y and z directions see Figure (7).

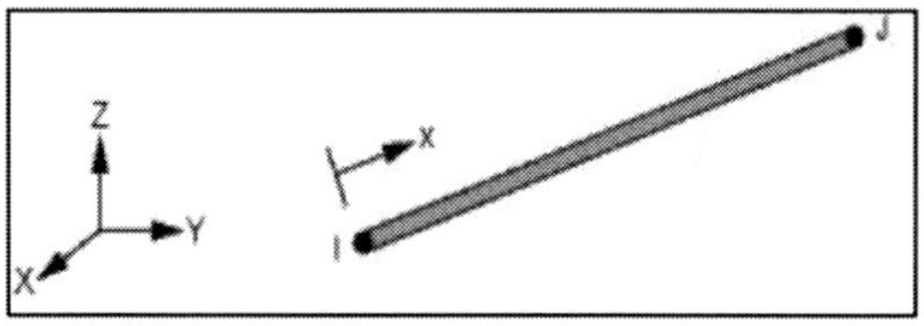


**Fig. (7).** Link180 element

The real constant and material properties like modulus of elasticity, Poisson's ratio for each element are used as input data to complete the model.

### 3.3 Meshing

The use of a square (aspect ratio equal to unity) or rectangular mesh is recommended for good accuracy and time saving, so that the mesh is set up such that square or rectangular elements were created. A convergence study was carried out to determine an appropriate mesh density. No need to mesh the reinforcement because of the link element created at the same nodes of concrete.

### 3.4 Loading

The loading of the track comes from the moving wheel, which is loaded by a constant force dead load of the car. The vertical actions used for Railway Bridge according to EURO code are defined by means of load models as shown in Figure (8).

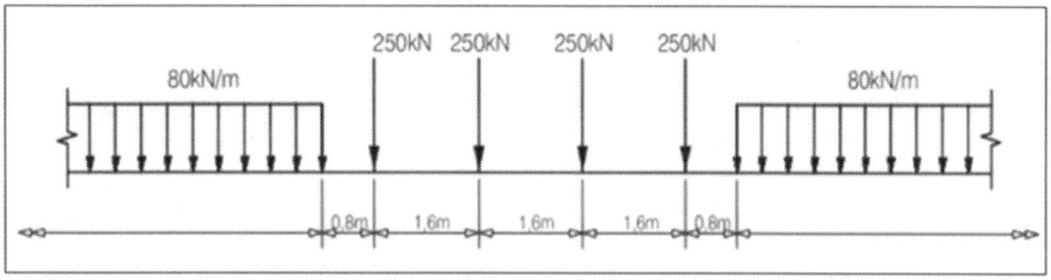


Fig. (8). Load Model and characteristic values for vertical loads

## 4. Sleeper's failure

The ehaviour of the soil is not exactly the same under all sleepers; also the loading of the track is irregular, so that the amount of settlement will differ from one sleeper to another. Because of the reasons mentioned above the sleepers are not always fully supported, and some sleepers may even become completely unsupported (voided) and a gap appears between the sleeper and the ballast layer. At the zone of failure, the track geometry starts to deteriorate, the variations of the train/track interaction force increase, and this speeds up the track deterioration rate.

## 5. Analysis

The finite elements model for normal case (no sleepers fail) is shown in Figure (9), and the finite model is shown in Figure (10). Figure (11) shows the three dimensional railway track system in case of two, three and four sleepers failure locations.

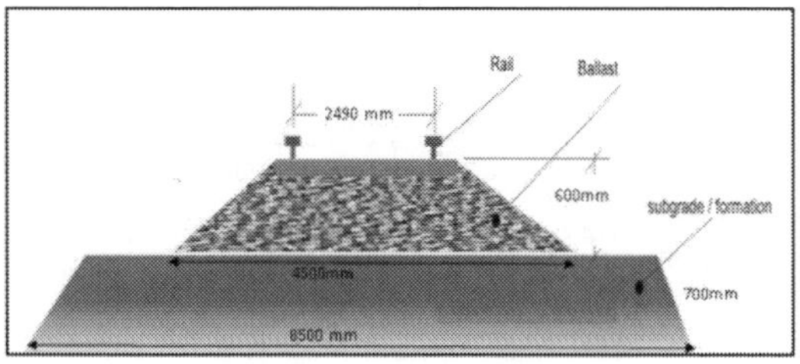


**Fig. (9).** The actual geometry of the railway system

**Fig. (10).** Three dimensional view of normal railway system

( a )

( b )

* Corresponding author: mjh_smm@yahoo.com

( c )

**Fig. (11).** Three dimensional and of two, three and four sleepers failure locations for railway system

The analysis results from ANSYS modeling shown in Figures (12) to (14) while the modeling represent the full performance of whole railway track system for three and four that fail under static loading.

In case of dynamic – harmonic loading analysis, the deflection – frequency for railway system in case of no sleep failure is shown in Figure (15). Figures (16) to (19) represented the performance of railway system under low harmonic loading in the range 0f (0-40 Hz). In case of medium frequency range (40-400 Hz), Figures (20) to (25) showed the full performance of the railway system for all cases.

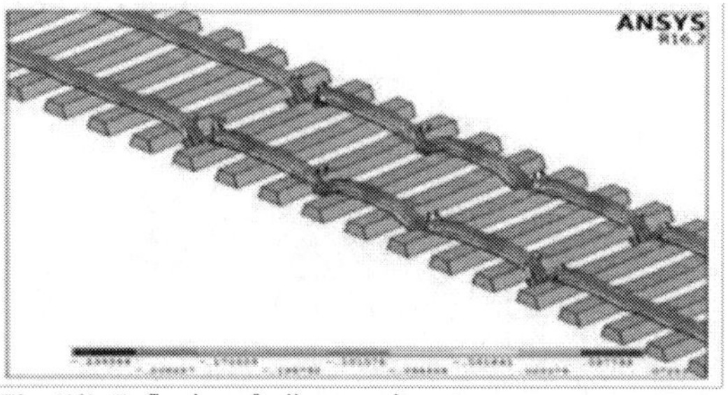


**Fig. (12).** Deflection of railway track system

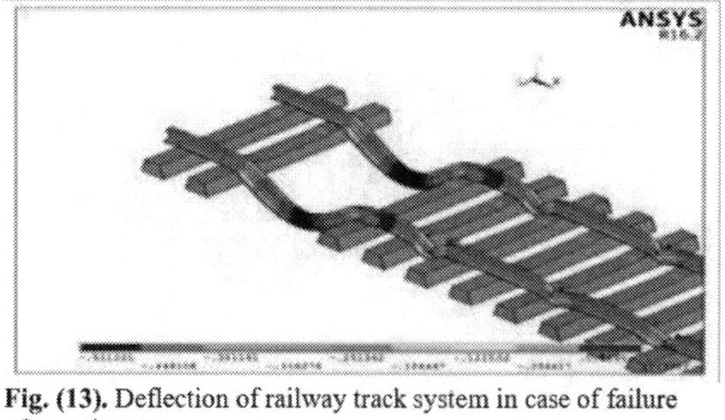


**Fig. (13).** Deflection of railway track system in case of failure of two sleepers

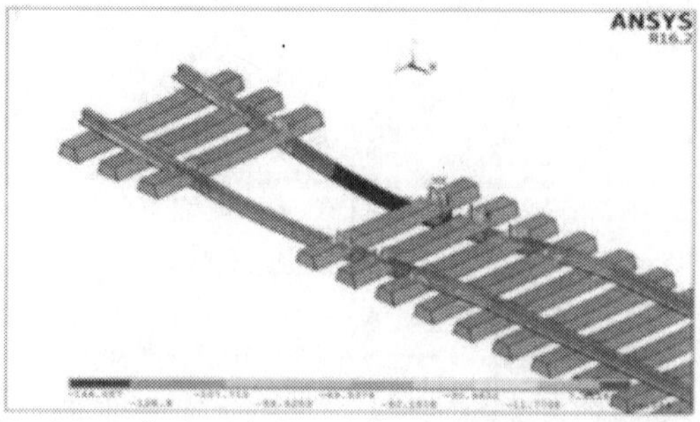


**Fig. (14).** Deflection of railway track system in case of failure of three sleepers

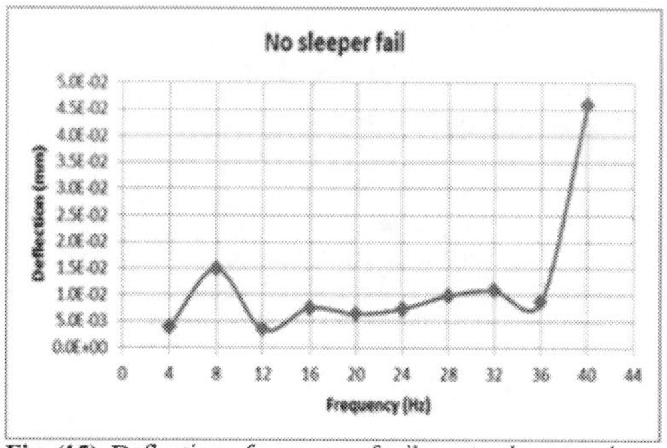


**Fig. (15).** Deflection –frequency of railway track system in case of failure of no sleeper failure

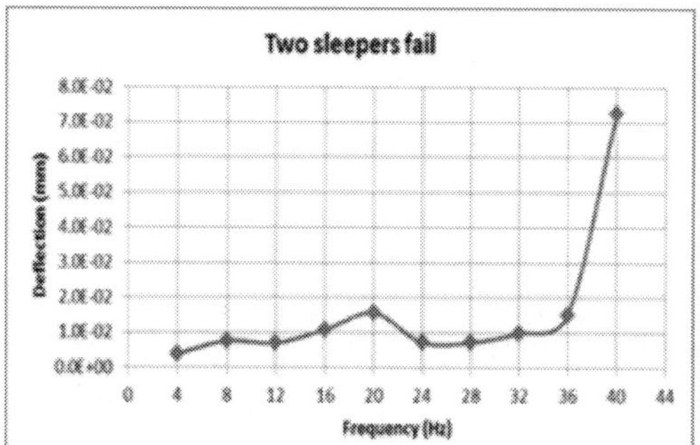


**Fig. (16).** Deflection –frequency of railway track system in case of failure of two sleeper failure

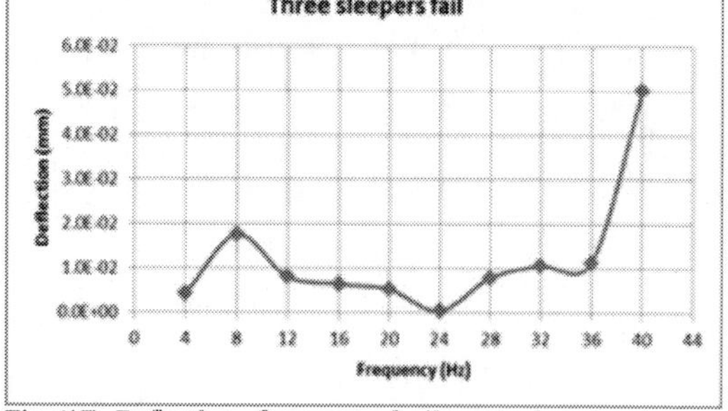


**Fig. (17).** Deflection –frequency of railway track system in case of failure of three sleeper failure.

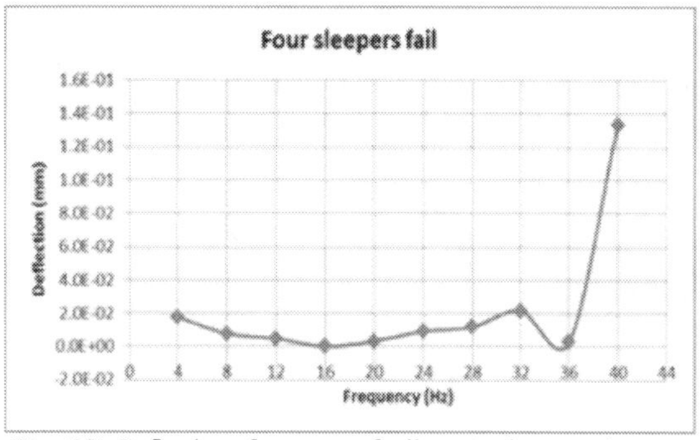


**Fig. (18).** Deflection –frequency of railway track system in case of failure of four sleeper failure.

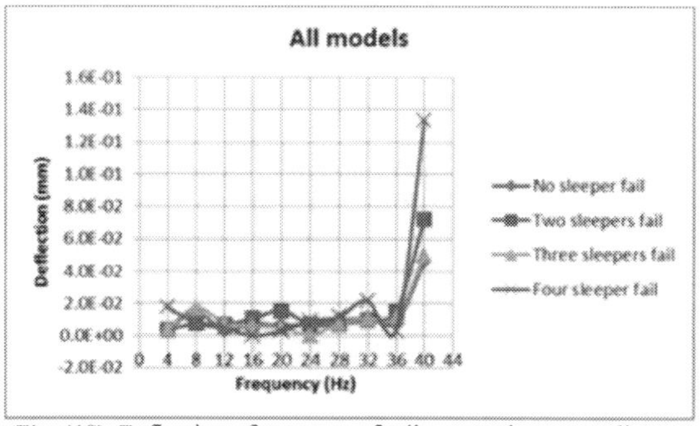


**Fig. (19).** Deflection –frequency of railway track system all models

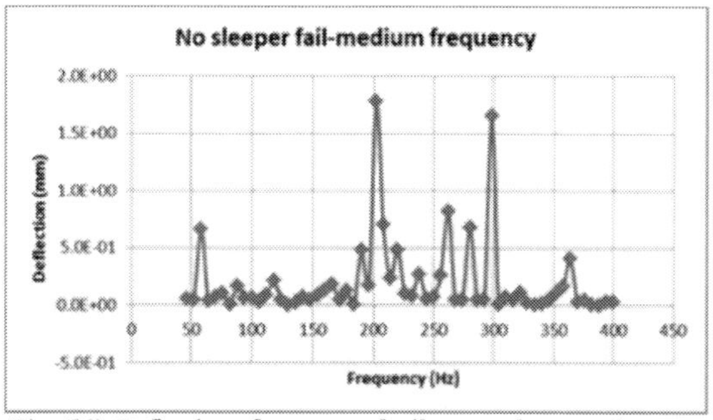


**Fig. (20).** Deflection –frequency of railway track system in case of failure of no sleeper failure.

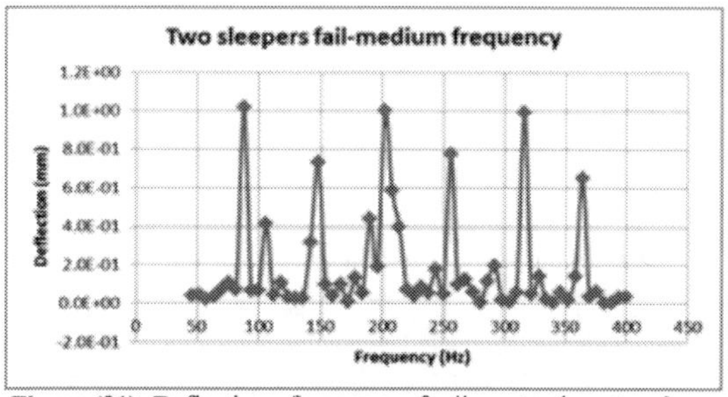


**Figure (21):** Deflection –frequency of railway track system in case of failure of two sleeper failure

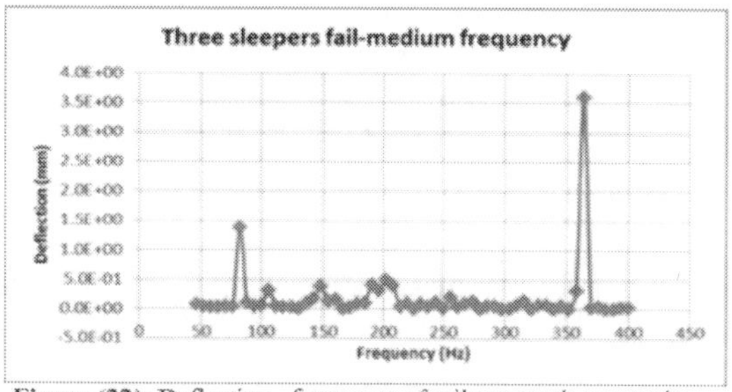


**Figure (22):** Deflection –frequency of railway track system in case of failure of three sleeper failure

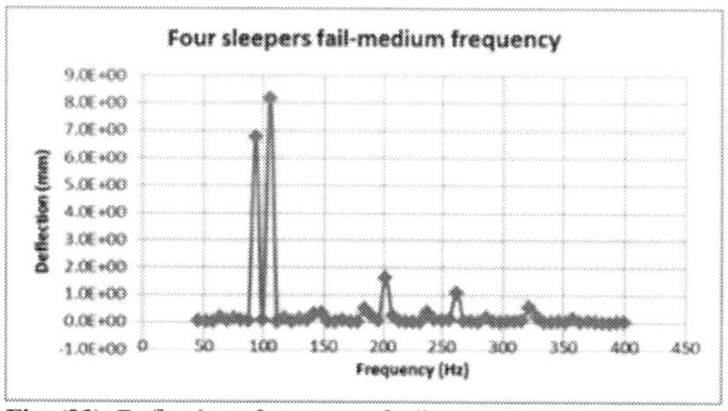


**Fig. (23).** Deflection –frequency of railway track system in case of failure of four sleeper failure

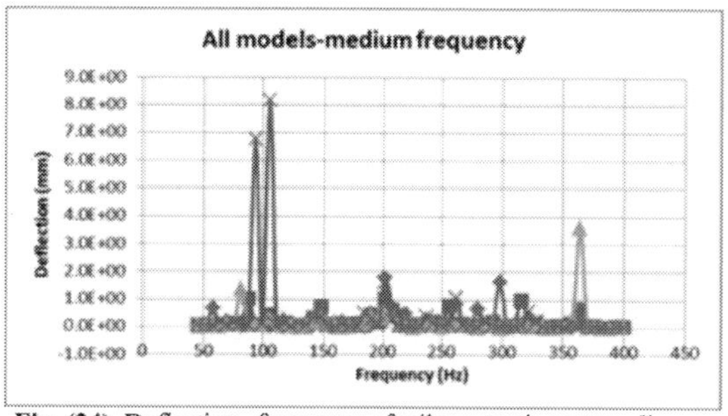


**Fig. (24).** Deflection –frequency of railway track system all models

## 6. Conclusions

Based on the finite elements analysis by ANSYS, the following points may be drawn:

- The static analysis of the whole railway track system successful for normal and two failures of sleeper cases because of the deflection within limit (less than 1.67 mm if assumed behaved as simply supported beam).
- In case of three and four failures of sleeper cases, the railway fails because of increase in deformation values, so that it concluded that the fail will occur in case of span failure equal and more than 1.8 m.
- In case of harmonic loading, the low range at frequency equal to (40 Hz) the deflection becomes higher but less than the limit in case of static loading.
- In case of harmonic loading, the low range at frequency equal to (40 Hz) the deflection becomes higher

but different in values based on the number of failures of sleepers.

- In case of harmonic loading, the medium range at frequency equal to (200, 200 with other ranges, 370, and 110 Hz) in case of normal, two, three and four failure of sleepers, the deflection becomes higher but less than the limit in case of static loading.

## 7. References

1. Jabbar Ali Zakeri and He Xia, "Application of 2D-infinite beam elements in dynamic analysis of train-track interaction", Journal of Mechanical Science and Technology, 23, (2009), 1415~1421.
2. Valeri Markine et al, "A dynamic model for analysis of damage of railways switches", International Symposium on dynamics of vechiles on roads and tracks , 2009, Sweden, 1-13.
3. A. Gomes Correia et al, " Dynamic analysis of rail track for high speed trains. 2D approach", Foundation for Science and Technology (FCT), (project POCI/ECM/61114/2004 - Interaction soil-railway track for high speed trains), 2010.
4. Khanh Nguyen et al, "Dynamic Analysis of High Speed Railway Traffic Loads on Ballasted Track", Advances in Environmental Vibration, Fifth International Symposium on Environmental Vibration, Chengdu, China, October 20-22, 2011.
5. Huan Feng, "3D models of railway track for dynamic analysis", Master Degree Project, Division of Highway and Railway Engineering, Department of Transport Science, School of Architecture and the Built Environment, Royal Institute of Technology, SE-100 44 Stockholm, Stockholm 2011.
6. Puguh B. Prakoso, "The Basic Concepts of Modeling Railway Track Systems Using Conventional and Finite Element Methods", INFO
7. D.Kishore Kumar and K.Sambasivarao, "Static and Dynamic Analysis of Railway Track Sleeper", International Journal of Engineering Research and General Science Volume 2, Issue 6, October-November, 2014
8. A. Ortega García, " Numerical and experimental analysis of the vertical dynamic behavior of a railway track", Master of Science Thesis, University of Technology, 24 April 2014.
9. Zhenxing He and Xinwen Yang, "Dynamic Response Analysis of an Asymmetric Coupled Vehicle-Track System Generated by Voided Elastic Two-Block Sleeper", Hindawi Publishing Corporation, Shock and Vibration, Volume 2016, Article ID 4985265, 11 pages
10. ANSYS, Inc. Help Manual: is a UL registered ISO 9001:2000 Company. Version (16.2), 2016, USA.

# Developing optimized prioritizing road maintenance

*Hussein* Ali Ewadh[1, *], *Raid* Almuhanna[2], and *Saja* Alasadi[2]

1College of Engineering, University of Babylon, Hilla, Iraq
[2] College of Engineering, University of Karbala, Karbala, Iraq

**Abstract.** Increased demand for efficient maintenance of the existing roadway system needs optimal usage of the allocated funds. The paper demonstrates optimized methods for prioritizing maintenance implementation projects. A selected zone of roadway system in Kerbala city represents the study area to demonstrate the application of the developed prioritization process. Paver system PAVER integrated with GIS is used to estimate and display the pavement condition index PCI, thereby to establish a priority of maintenance. In addition to simple ranking method by PCI produced by the output of PAVER, the paper introduces PCI measure for each section of roadway. The paper introduces ranking by multiple measures investigated through expert knowledge about measures that affect prioritization and their irrespective weights due to a predesigned questionnaire. The maintenance priority index (MPI) is related to cost of suitable proposed maintenance, easiness of proposed maintenance, average daily traffic and functional classification of the roadway in addition to PCI. Further, incremental benefit-cost analysis ranking provide an optimized process due to benefit and cost of maintenance. The paper introduces efficient display of layout and ranking for the selected zone of roadway system based on MPI index and incremental BCR method. Although the two developed methods introduce different layout display for priority, statistical test shows that no significant difference between ranking of all methods of prioritization.

## 1 Introduction

There is increase demand for new construction as well as efficient rehabilitation and maintenance of the existing roadway system. To preserve the investment spent on this huge network of pavement, extensive maintenance and repair activities are necessary, with the intention of using funds optimally [1]. Accordingly, for some nation's interstate highway system in developed countries as well as for the case of restricted budgets in developing countries, the focus is shifting from new construction to maintaining, preserving, and rehabilitating highway assets. However, the resources made available for road maintenance are limited in most countries [2].

According to the World Bank Report, "The developing countries have lost precious infrastructure worth billions of dollars through the deterioration of roads. The cost of restoring these roads is going to be three to five times greater than the bill would have been for timely and effectively maintenance" ,hence there is a need to manage the network more efficiently in a scientific manner [3].

The Pavement Maintenance & Management Systems (PMMS) have a two main components; first one, all importance's database are identify, such underway and historical information on pavement condition index, pavement structure layer, and traffic value. The second component is a collection of tools that allows us to identify current and future pavement conditions, forecasting financial needs, and identify and prioritize pavement project [3].

The paper demonstrates methods for prioritizing maintenance implementation projects. Due to budget and other constraints, it is believed that site of lowest pavement condition index (PCI) does not always justify maintenance project first.

The term "prioritization" is refer to a review of potential maintenance projects for building and preparation an ordered list of selected sites by maximizing benefits according to budget and other qualification depended on the score of ranking and optimization processes. "Ranking" indicate to an ordered list of projects or project alternatives based on given factors or project benefits and costs.

Prioritization methods presented in this paper are firstly serviceable to developing optimal maintenance projects program across multiple sections of a roadway (in a selected zone) or for an entire roadway system.

### 1.1 PAVER and MicroPAVER

PAVER and Micro PAVER [4] are developed to extend engineers with a systematic process in order to determine maintenance and rehabilitation (M&R) needs and priorities for pavement management. While Micro PAVER executes on a microcomputer; PAVER is the mainframe version. The PAVER is developed to optimize the use of funds allocated for pavement M&R.

* Corresponding author: ewadh57@yahoo.com

© The Authors, published by EDP Sciences. This is an open access article distributed under the terms of the Creative Commons Attribution License 4.0 (http://creativecommons.org/licenses/by/4.0/).

Micro PAVER is utilized to manage all types of driveway (roads, streets), parking lots, and airfield pavement. The PAVER system is depended on the PCI survey and evaluation procedure. also, needs to create of a database in the network inventory to performed network and project analysis, the last one expand the users with detailed of current PCI survey information, possible alternatives for M&R. It is used for current year or near term needs. Network analysis which is used for forecasting long-term M&R needs supply the users with the future PCI, budget planning and project priorities. The PAVER system is written in FORTRAN and C languages and design reside in IBM or compatible with personal computer [5]. The PCI is an evaluation process that is determined according to the procedures included in ASTM D 5340, Standard Test Method for PCI Survey. This procedure is used worldwide to provide a measurement of the condition of pavements taking into account the functional performance with effects of structural performance. cyclic PCI definitions on the same pavement will appear the deferent in performance level with time. The PCI ranges of 0 "Failed" to 100 "Excellent", with an "Excellent" condition equivalent to a pavement at the starting of its life cycle, and a "Failed" condition equivalent to a severely deteriorated pavement without virtually remaining life.

## 2 Maintenance priority

After application of PAVER, a list of pavement condition for the network sections is prepared. In most cases, limited financial resources make it impossible to perform all sections of low pavement condition indices. in the list. funding needs more than available funding. When this happens, one of the methods for prioritizing and optimizing will be needed in order to prepare a maintenance and rehabilitation program. The following is a list of methods for establishing priorities; however, alternate methods can be developed based on an agency's policies and administrative decisions [6].

• The matrix method can be depended on many factors such as traffic and condition; i.e., the pavements in the worst condition and has a heaviest traffic is given the highest priority.

• The condition index method can be depended on relative scores which ranked from 0 (for worst) to 100 (for best). In order to develop a final list of projects, Priorities can combine condition score with many factors such as traffic or functional class.

• In the benefit-cost ratio process, the sections would have the highest priority when it has the highest benefit-to-cost ratio. Whereas the previous methods are likely to favor a "worst-first" policy, the benefit-cost rationale can provide high priorities for pavements in fair-to-poor condition rather than always starting with the pavements in the worst condition.

• The cost-effectiveness procedure is the same to benefit-cost except that the function is to maximize the performance of the sections while considering cost. Performance, in this case, is a measure of the effectiveness of a particular strategy on a segment over time. Each segment in the agency's network can then be ranked against each other to arrive at a list of maintenance and rehabilitation options. This method does not require a "worst first" approach.

• The maximum benefits procedure is inherent in most optimization methods. This methods can also be developed with life cycle costs and prioritization. For example, that set of projects from all candidate projects, which maximizes the combined benefit-cost ratio or cost effectiveness for a specific budget, would be selected for M&R treatments. In this research a set of maintenance prioritization methods are proposed to achieve efficient alternatives to maximize cost effectiveness for a limited financial resource.

## 3 The proposed maintenance prioritization methods

Three prioritization methods are proposed in this research as follows:

1) Simple ranking by PCI measure.
2) Ranking by multiple measures.
3) Incremental benefit-cost analysis ranking

Simple ranking by PCI measure provides a prioritized list of projects depended on a pavement condition index criterion. Other measures in addition to PCI are included in the process of ranking by multiple measures. Investigation of expert knowledge about measures that affect prioritization of sites for maintenance is achieved through a predesigned questionnaire. The irrespective scores of each measure and its importance are also investigated. Ranking by multiple measures is improved alternate for condition index method and the matrix method developed by Washington State Department of Transportation.

Further, incremental benefit-cost analysis ranking provide an optimized process of ranking by taking PCI measure as benefit (when maintenance of the roadway section is achieved) and cost of maintenance (according of the suitable maintenance strategy for a specific distress in that section). Hence, this process considers the effect of budget constraints and allocated resources in developing an optimized maintenance project set. Also, this process is related to the benefit-cost ratio process and the cost-effectiveness procedure developed by Washington State Department of Transportation.

## 4 Methodology of research

This study in the done an application of paver system integrated with GIS to establish a priority of maintenance for a selected zone in Karbala city. Due to a restricted budget for a new construction as well as highway demand to maintenance they roadway system,

https://doi.org/10.1051/matecconf/201816201044

the study make use of pavement management existing system to demonstrate the efficient tool of maintenance by PAVER.

A selected zone of roadway system in Kerbala city, represents the study area to demonstrate the application of prioritization process. Karbala is located in the Central Region of Iraq on the edge of the Eastern Plateau Bank, west of the Euphrates River, and specifically between longitudes 43.33 north. A selected zone of urban streets for seven neighborhoods within the CBD area in Karbala city is shown in Figure (1-a). These streets are classified into: major arterial, minor arterial, collector, and local as shown in Figure (1-b).The study area include many activity centers such as; schools, governmental buildings, and religious please.

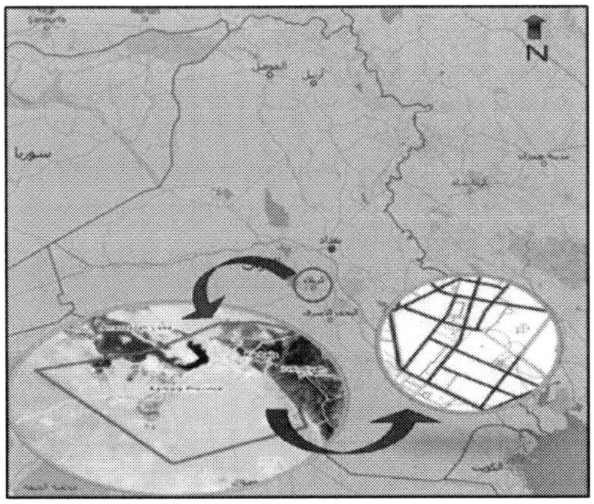


(a)

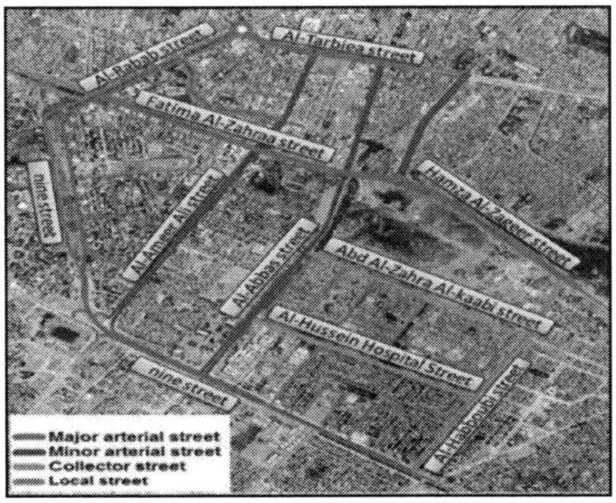


(b)

Fig. 1. The selected zone of road network in Karbala

## 5 Application of the proposed prioritization process

Site inspection is achieved for all sections in the selected zone in the study area. According to the output of PAVER and other factors affecting pavement condition, the following articles demonstrate the application of the prioritization process.

### 5.1 Simple Ranking by PCI Measure

Simple ranking by PCI depends mainly on the output of PAVER. PCI measure for each section of roadway represents the degree of need to maintenance. Consequently, all sections of the roadway inspected are ranked from lowest PCI to the highest one.

PAVER 6.5.7 allows the pavement manager to maintenance and rehabilitation (M&R) sections roads and to determine work priorities from the set of critical PCI value. The priority developed afford to the user to limit priorities according to critical PCI value, facility use, or pavement type that used by the M&R module, to improve a prioritized list of bespoke pavement perfection projects.

The priority typical scale of PCI that can be used to manage M&R operations is shows in Plate (2). However, users can see the critical PCI to show the behavior of different sections of their pavement depended on their own experience. For example, with a pavement suffering from a high distress rate, crack sealing (preventive maintenance) may not be cost-efficient and therefore the critical PCI value may be adjusted [7].

On the other hand, and depending on PCI values from PAVER output and using GIS tools to layout the maintenance priority for the selected sections as shown in Plate (3).

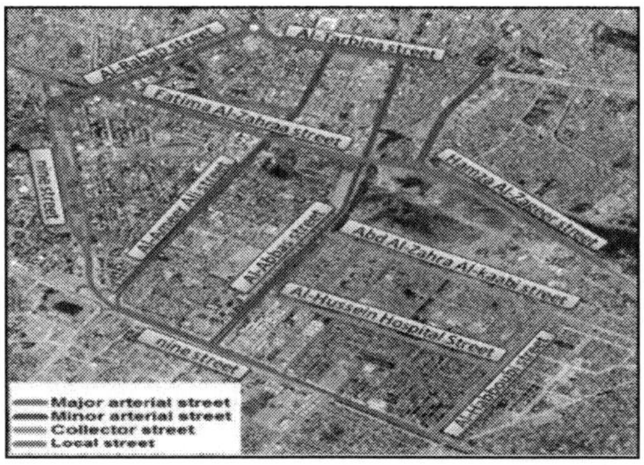


Plate 1.

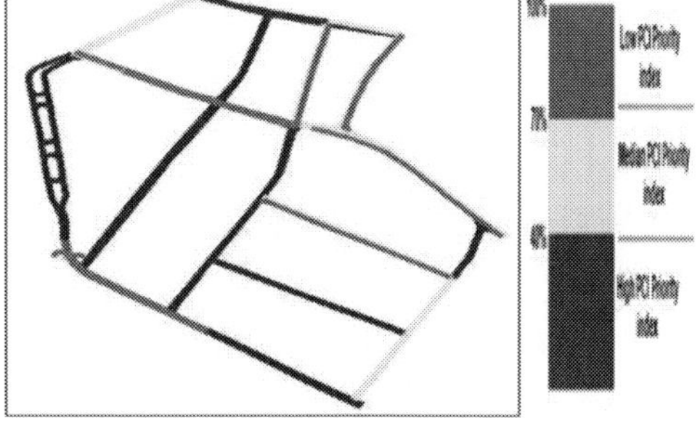

Plate 2. Priority of Maintenance and Rehabilitation for Selected Roadway Zone in Karbala City.

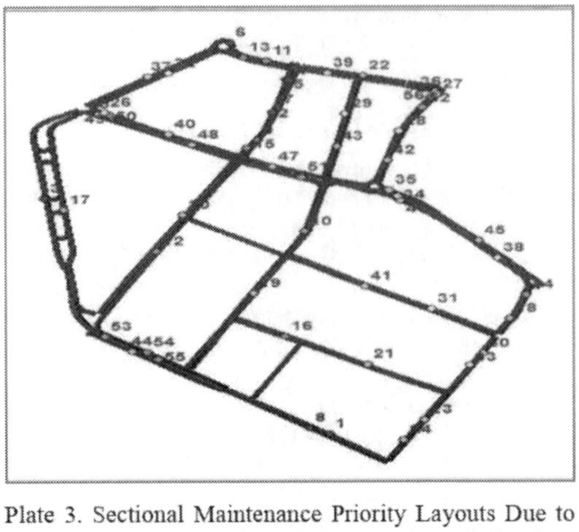

Plate 3. Sectional Maintenance Priority Layouts Due to PCI Values for Selected Roadway Zone in Karbala City

M&R operations is shows in Plate (2). However, users can see the critical PCI to show the behavior of different sections of their pavement depended on their own experience. For example, with a pavement suffering from a high distress rate, crack sealing (preventive maintenance) may not be cost-efficient and therefore the critical PCI value may be adjusted [7].

On the other hand, and depending on PCI values from PAVER output and using GIS tools to layout the maintenance priority for the selected sections as shown in Plate (3).

## 5.2 Ranking by Multiple Measures

Investigation of expert knowledge about measures that affect prioritization of sites for maintenance is achieved through a predesigned questionnaire. The irrespective scores of each measure and its importance are also investigated. Table (I) sample of the questionnaire are used.

The maintenance priority index (MPI) (as a dependent variable) is related to other independent variables in addition to PCI determined from PAVER system. Other independent variables include; Cost of suitable proposed maintenance, Easiness of proposed maintenance, Average Daily Traffic (ADT) and functional classification of the roadway. An equation is developed to estimate maintenance priority index as follows:

MPI=

$$(a * PCI) - (b * C) + (c * E) + (d * ADT)(e * F.C.) \quad (1)$$

Where:

MPI: Maintenance priority index of section.

PCI: Pavement condition index of section.

C: cost of maintenance for each section per square meter.

E: easiness of work through maintenance sections.

ADT: Average Daily Traffic for each section.

F.C.: Functional Classifications of each section.

a,b,c,d,e: Coefficients represent the irrespective weights value for the corresponding independent variables due to experts response.

Table 1. Sample of the questionnaire for the irrespective variables affecting weights of variables for priority of pavement maintenance

| No. | DEFINITION | Irrespective Weights* | Scale Reference |
|---|---|---|---|
| 1 | How costly would it be to implement maintenance to address a good pavement condition? | – | 1. Not Costly<br>5. Very Costly |
| 2 | What impact has PCI measure on pavement condition? | + | 1.Not Impact<br>5. Highly Impact |
| 3 | How easy would it be to implement the maintenance? | + | 1. Not Easy<br>5. Very Easy |
| 4 | What impact has the functional classification of the section on priority of maintenance? | + | 1. Not Impact<br>5. Highly Impact |
| 5 | What impact has the average daily traffic (ADT) on priori of maintenance? | + | 1. Not Impact<br>5. Highly Impact |

*Weights (from 0 to 10), Experts are asked to give the irrespective weight for each item

Statistical analysis is made for 35 questionnaire samples (sample of the questionnaire is shown in Table (I) to check the normality for data of Irrespective weights. Consequently, the average values of the irrespective weight for each variable are estimated as: a = 7.6, b = -6.4, c = 5.0, d = 8.3, e = 6.9.

The weights of independent variable in equation (1) are found during the inspection round of the roadway section. During the site inspection of pavement condition, data of ADT and the function classification of section roads ranking from 1 up to 5 (1:local, 2:local, distributor, 3:collector, 4:minor arterial, 5:majer arterial) are collected. Further, maintenance cost and work easiness for each distress type in flexible pavement are estimated through investigation of experience of experts for each type of maintenance process as shown in Table (II). This cost contain aftertime overlays and/or upgrading made necessary when the riding quality of pavement lowering to a definite low level of acceptability [8].

The maintenance priority of sections in study area (using data for 46 major and minor arterial sections, and 10 collector sections) is estimated by using equation (1). Plate (4) shows a layout of maintenance priority for the selected sections using GIS tools. Multiple priority

maintenance MPI can be classified into three ranges corresponding to the scale.
On the other hand depending on MPI values and using GIS tools to layout the maintenance priority for the selected sections as shown in Plate (5).

Table 2. The maintenance cost for flexible pavement distress and work easiness

| Distress Type | Severity Levels | Description | Financial unit cost (ID) | E |
|---|---|---|---|---|
| Alligator, Block Cracking, Patching, Potholes, Depression | Low | Surface Patching | 15,000 | 4 |
| | Medina | Deep Patching | 30,000 | 3 |
| | High | Deep Patching | 40,000 | 2 |
| Long ,Transverse , Edge Cracking | Low | Do Nothing | 0 | 5 |
| | Medina | Crack Sealing | 3,000 | 4 |
| | High | Slurry seal, Thin Overlay | 15,000 | 3 |
| Rutting ,Shoving, Swell | Low | Do Nothing | 0 | 5 |
| | Medina | Milling & Repave | 25,000 | 3 |
| | High | Deep Patching | 40,000 | 2 |
| Asphalt Bleeding | Low | Do Nothing | 0 | 5 |
| | Medina | Do Nothing | 0 | 5 |
| | High | Milling & Repave | 25,000 | 3 |
| Weathering / Raveling | Low | Do Nothing | 0 | 5 |
| | Medina | Slurry Seal | 15,000 | 4 |
| | High | Thin Overlay | 20,000 | 3 |
| Bumps& Sags ,Corrugations, Slippage Cracks | Low | Do Nothing | 0 | 5 |
| | Medina | Surface Patching | 15,000 | 4 |
| | High | deep Patching | 40,000 | 2 |
| Reflection Cracks | Low | Do Nothing | 0 | 5 |
| | Medina | Refill | 8,000 | 4 |
| | High | Refill | 10,000 | 4 |
| Lane/Shoulder Drop | Low | Do Nothing | 0 | 5 |
| | Medina | Surface Patching | 3,000 | 4 |
| | High | deep Patching | 4,000 | 4 |

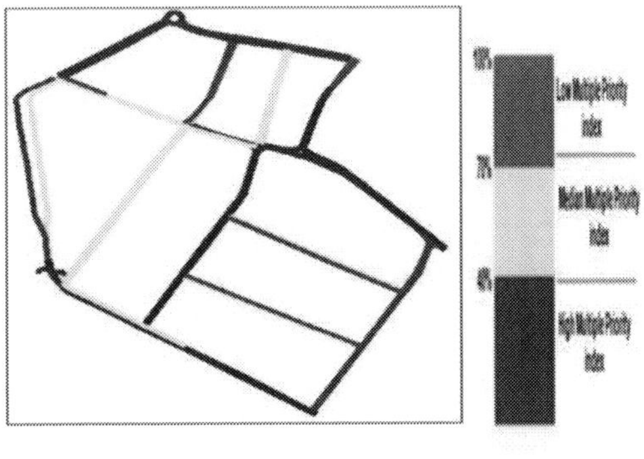

Plate 4. Priority of Maintenance and Rehabilitation Due to Multiple Priority Index for Selected Roadway Zone in Karbala City

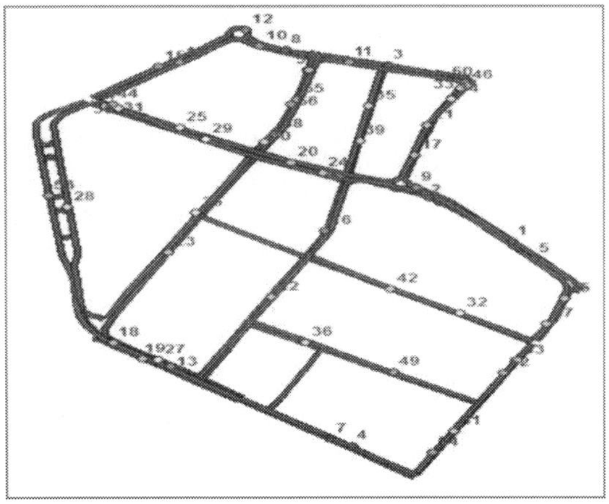

Plate 5. Sectional Maintenance Priority Layout Due to Multiple Measures for Selected Roadway Zone in Karbala City

### 5.3 Incremental Benefit-Cost Method

This method represents an extension of the benefit/cost ratio. The benefit gained from pavement maintenance is assumed to equal decrease of PCI out of hundred (100 – PCI). On the other hand, the ratio of benefit to cost of pavement maintenance represents BCR depends on two variable benefit value (PCI-100). PCI determined from PAVER system, and maintenance cost of each section per square meter is estimated as explained in the multiple measure priority method. In this paper, input data (benefit and cost) are used for arterial and collector section roads (in study area). The sections are arranged depending on cost-effectiveness with (from low to high cost), a BCR greater than 1.0 in increasing order depended on their estimated cost. The section which involves with a lowest cost is listed first one. starting at the up of the list, determine the variance between the benefits of the first and second project's. Similarly determine the variance between the first and second costs of the projects.

The variance between the two values for each of benefits or cost projects are used to compute the BCR for the incremental investment by the following equation (AASHTO, 2010):

Incremental BCR = (PVbenefits 2 – PVbenefits 1) / (PVcosts 2 – PVcosts 1) (2)

Where:
PVbenefits 1 = Present value of benefits for lower cost of pavement maintenance.
PVbenefits 2 = Present value of benefits for higher cost pavement maintenance.

PVcosts 1 = Present value of cost for lower cost pavement maintenance.
PVcosts 2 = Present value of cost for higher cost pavement maintenance.

If the incremental investment BCR is greater than 1.0, the higher cost between the two project is compared with the following project in the list. If the the incremental investment BCR is less than 1.0, the lower cost between the two project is compared with the following project in the list. Return this process. The project selected in the last pairing is considered as the best economic investment. To find the ranking of all projects, the complete evaluation is iterative without the projects already determined to be the best economic investment until the ranking of every project is determined [9].

In order to perform a most efficient and effective application of this method, MATLAB software program is developed. Furthermore, GIS Geostatic Analyses Tools is used to layout the priority result of this method as shown in Plate (6).

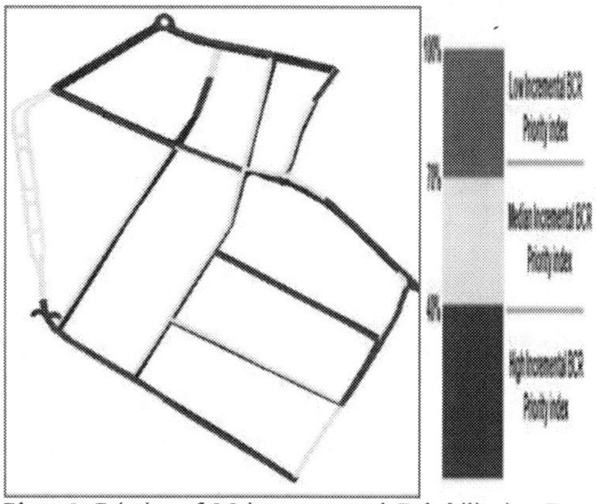

Plate 6. Priority of Maintenance and Rehabilitation Due to Incremental BCR Index for Sections of Karbala City

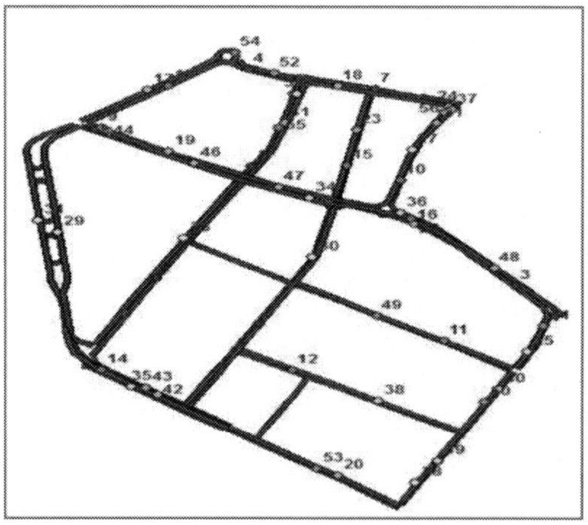

Plate 7. Sectional Maintenance Priority Layout Due to Incremental BCR Index for Sections of Karbala City

On the other hand, and depending on ranking values through incremental BCR priority index and using GIS tools to layout the maintenance priority for the selected sections as shown in Plate (7).

## 6 Maintenance priority result analysis

Maintenance priority rank result from each: PAVER PCI, matrix method, and incremental BCR method are analysis by using Wilcoxon signed-rank test (Matched Pairs Test) in SPSS software. The testing equality of means of two continuous allocation that are clearly non normal, and independent samples (i.e., there is no pairing of observations), the Wilcoxon rank-sum test or Wilcoxon two-sample test is an appropriate alternative to the two-sample [10]. The test is achieved assuming that (null hypothesis) there is no significant difference between ranking of each method of priority of maintenance. As shown in Table (III), the test shows that to a 0.05 degree of significance, there is no significant difference between rankings of each method.

Table 3. Results of Wilcoxon matched Paris test

A-Descriptive Statistics

| | N | Mean | Std. Deviation | Minimum | Maximum |
|---|---|---|---|---|---|
| PCI priority | 56 | 28.5000 | 16.30951 | 1.00 | 56.00 |
| EQ priority | 56 | 28.5000 | 16.30951 | 1.00 | 56.00 |
| BCR priority | 56 | 28.5000 | 16.30951 | 1.00 | 56.00 |

B- Wilcoxon Signed Ranks Test

| | | N | Mean Rank | Sum of Ranks |
|---|---|---|---|---|
| EQpriority – PCIpriority | Negative Ranks | 27[a] | 30.19 | 815.00 |
| | Positive Ranks | 29[b] | 26.93 | 781.00 |
| | Ties | 0[c] | | |
| | Total | 56 | | |
| BCR_priority – PCIpriority | Negative Ranks | 31[d] | 25.73 | 797.50 |
| | Positive Ranks | 23[e] | 29.89 | 687.50 |
| | Ties | 2[f] | | |
| | Total | 56 | | |
| BCRpriority – EQpriority | Negative Ranks | 30[g] | 26.63 | 799.00 |
| | Positive Ranks | 25[h] | 29.64 | 741.00 |
| | Ties | 1[i] | | |
| | Total | 56 | | |

a. EQpriority < PCIpriority b. EQ_priority > PCI_priority c. EQ_priority = PCI_priority
d. BCRpriority < PCIpriority e. BCRpriority > PCIpriority f. BCRpriority = PCIpriority
g. BCRpriority < EQpriority h. BCRpriority > EQpriority i. BCRpriority = EQpriority

C- Test Statistics[a]

| | EQpriority - PCIpriority | BCRpriority - PCIpriority | BCRpriority - EQpriority |
|---|---|---|---|
| Z | -.139-[b] | -.474-[b] | -.243-[b] |

| | | | |
|---|---|---|---|
| Asymp. Sig. (2-tailed) | .890 | .636 | .808 |
| a. Wilcoxon Signed Ranks Test | b. Based on positive ranks. | | |

## 7 Conclusions

1. Depending on PAVER output integrated with GIS, the paper demonstrates two developed optimized methods of prioritization in addition to simple ranking of PAVER.
2. Investigation of expert knowledge about measures that affect prioritization of sites for maintenance results that in addition to PCI determined from PAVER system, other independent variables include; cost of suitable proposed maintenance, easiness of proposed maintenance, average daily traffic (ADT) and functional classification of the roadway.
3. The paper introduces efficient display of layout and ranking for the selected zone of roadway system based on multiple priority maintenance (MPI) index based on multiple factors affecting prioritization of sites for maintenance results.
4. The paper introduces efficient display of layout and ranking for the selected zone of roadway system based on incremental BCR priority index.

Although the two developed methods introduce different layout display for priority, statistical test shows that no significant difference between ranking of all methods of prioritization.

## References

1. Vepa T.S., George K. P., and Shekharan A. R., "Prediction of Pavement Remaining Life" Transportation Research Record, No. 1524, pp 137-144, 1996.
2. Bent Thagesen, Technical University of Denmark, "Highway and Traffic Engineering in Developing Countries". contact the Promotions Department, E & FN Spon, 2–6 Boundary Row, London SE1 8HN, UK. Tel: International +44 71–865 0066, 2005.
3. NIJU.A, "GIS based Pavement Maintenance & Management System (GPMMS)". Department of Civil Engineering, National Institute Of Technology Calicut, Calicut, Kerala 673 601, 2006.
4. Shahin, M.Y. and Walther, J.A., "Pavement Maintenance Management for Roads and Streets using the PAVER
5. Norlela I., Amiruddin I., and Riza A. , "An Overview of Expert Systems in Pavement Management", European Journal of Scientific Research ,Vol.30 No.1, pp.99-111, 2009a.
6. Washington State Department of Transportation (WSDOT), "A Guide for Local Agency Pavement Managers", Washington State Department of Transportation – Northwest Technology Transfer Center, Olympia, WA, 1994.
7. "A Best Practice For The Implementation of A Pavement Management System for Small and Medium Airport", Architectural & Engineering Services, Civil Engineering Directorate, Airport Engineering, ATR-045, 2002.
8. American Association of State Highway and Transportation Officials (AASHTO), Guide for Design of Pavement Structures, AASHTO, Washington, D.C., 1993.
9. AASHTO, "Guide for the local calibration of the mechanistic-empirical pavement design guide". Washington, DC: American Association of State Highway and Transportation Officials, 2010.
10. Ronald E. Walpole, Raymond H. Myers, Sharon L. Myers, Keying Ye., "Probability & Statistics for Engineers & Scientists". Ninth Edition,University of Texas at San Antonio, 2012.

# Evaluation of using crushed brick as coarse aggregate in concrete layer within rigid highway pavement

*Ali* Alwash[1], *Fatimah* Al-Khafaji[1]

[1]University of Babylon, Babylon, Iraq
[2]University of Babylon, Babylon, Iraq

**Abstract.** Most of the present studies related to the field of highway pavement construction technique tend to make use of the local available materials as substitutes for the imported and necessary materials for some of the practical application. For this reason this research aims at looking for the prospect of used locally available aggregate such as crushed clay bricks for the aim of producing proper concrete with suitable thermal and mechanical properties. Experimental investigations have been carried out to asses the effect of partial replacement of coarse aggregate by free manually crushed Brick with percentages (10, 20, 30 and 40)% of virgin coarse aggregate in concrete mix for highway rigid pavement. While the percentage (0)% replacement represent reference mix. Mix proportion based on the target of compressive strength for all replacement percentage of (33) MPa at (28) days to achieve AASHTO requirement for highway concrete rigid pavement .The results of flexural strength, modulus of elasticity, density and thermal conductivity refers to better performance (less thickness of concrete layer with large spacing between contraction or expansion joints and less stresses due to warping induced concrete layer) for concrete mix with 20% crushed brick as replacement of coarse aggregate.

## 1 Introduction

Concrete is the widely used paving material due to suitable features such as durability, strength, cast effectives and availability.

The clay brick from demolished construction wall. [1]

From the last years, it has been established as the growing problem of the waste from the demolished construction walls which exist in every location of the world. The sustainable use of this waste can solve large environmental problems.

Accumulation of these waste is considered as real problem. Therefore these demolished construction waste should be recycled and reused as raw materials for new application like rigid pavements recycled crushed brick can be used for manufacturing concrete mix with suitable features for ensure sustainable management for this operation. [2]

Recycling materials is defined as reusing of old ones in new application. By this operation, the consumption of raw materials is reduced and the waste is eliminated.

Recycled crushed brick as fine aggregate can be used for producing mortar as well as concrete mix.

The use of this waste materials have good effect on economy. Some application needs utilization of low cost materials without any bad impact on the environment.

Almost, at the preparation of concrete mix, the aggregate is used more than cement because it's the cheapest and the strength of aggregate is very significant for the final concrete strength. Therefore the percentage of aggregate equal to 70-80% of the total volume of concrete mix. [2]

The use of crushed brick aggregate must be under investigation for important application. A review on the subject indicated that there was another information of the utilization of such waste however the crushed brick aggregate is different from that of crushed concrete aggregate due to its porosity. [3]

Some researchers used good burnt crushed brick as coarse aggregate in concrete for achieving high strength concrete. This was done by finding mechanical properties of brick aggregate concrete without needing that of crushed brick aggregate itself.

Other researchers used clinker brick after crushing as coarse aggregate. It was found only 7% loss in compressive strength of concrete in comparison to that with natural aggregate. But there is a decrease in density of crushed brick concrete by about 9.5%. [3]

Recycled crushed brick aggregate has an absorption values 22-25% by weight in comparison with dry material. For the study of this property, it obtained that the saturation time in water is 30 min and the absorption of waste aggregate concrete is higher by about 20% than that of control cement with variation according the type and composition of waste aggregate. But the use of

© The Authors, published by EDP Sciences. This is an open access article distributed under the terms of the Creative Commons Attribution License 4.0 (http://creativecommons.org/licenses/by/4.0/).

crushed brick aggregate with fine gradation leading to higher cement content in concrete mix. [1,2]
Therefore the major problem of using crushed brick after recycling as aggregate its high water absorption. [4]

# 2 Experimental works

## 2.1. Materials

### 2.1.1. Cement:

Sulfate resistance cement was used. The specific gravity of the cement used was 3.15.
*2.1.2. Fine aggregate:*
Normal weight natural sand from Kerbala region was used as fine aggregate. The grading of the requirement of Iraqi specification No.45/1984, grading area (3).

### 2.1.2. Coarse aggregate

#### Natural gravel

The grading was within the limits of gravel by British Standard (B.S..882:1973)

#### Crushed brick:

Free manually crushed brick (as demolished building material). The grading is made in a manner similar to grading of natural gravel aggregate.

### 2.1.3. Concrete Mix:

The concrete mix and water-cement ratio based on target C 30 an portable water was used in all mixes.

### 2.1.4. Curing

After casting, the specimens were separated and kept in the water bath for a curing period of 28 days for proper hydration process under laboratory temperature.

## 2.2. Tests:

- The compressive strength is fabricated according to B.S: (1881, part 116, 1989).
- The modulus of rupture and static modulus of elasticity are performed according to (ASTM C78-2002) and (ASTM C469-2002) respectively.
- The linear coefficient of thermal expansion is calculated according to AASHTO TP60-2004 while thermal conductivity as a measure of the rate at which heat passes through unit thickness for a temperature difference of one degree, is measured according to (ACI 122R-02)

# 3 Results and Discussion

## 3.1. Dry density

Table (1) shows the densities for all specimens. It is noticed that the density decreases by replacing natural gravel by crushed brick. Also this light weight aggregate (LWA) presents a reduction in density from normal weight concrete ranging between 9% to 20%.

## 3.2. Compressive Strength:

The results reported are the average of 3 specimens as shown in Table (2) at the age of 28 days. There is a reduction in strength that may belong to two causes:**[5]**

- Recycled brick have no proper bond with cement in concrete mix.
- High porosity of crushed brick surface leading to higher water requirement for workability.

Increasing the substation content with waste aggregate leading to decrease in 28 days compressive strength with range of 5-11% when compared with control concrete. This variation in decrease depends on the strength of original brick. Therefore we can estimate the strength of recycled aggregate concrete from the strength of original brick. [2, 6].
However, the strength of wasted concrete agree with requirement of pavement applications in highway (above 30 MPa).

## 3.3. Flexural Strength:

Flexural strength is defined as the stress breaking or deforming permanently and the material and the result as shown in Table (2). The angularity of the crushed brick aggregate and its surface roughness leading to proper bond between aggregate and the cement paste that could increase the flexural strength as shown for 20% replacement that aggregate with [2, 7]. Although of that mentioned above, flexural strength of crushed brick concrete is about 15% lower than that of control concrete.

## 3.4. Modulus of Elasticity:

Elastic modulus of concrete is related to the elastic modulus of the aggregate and the cement paste and their relative parentage. The modulus of elasticity of concrete with crushed brick as an aggregate is about 12-23 % lower than the one of a normal concrete as shown in Table (2). With the increase percentage of substitution of natural aggregate with crushed brick aggregate, the modulus of elasticity decreases which agreed with [2].

## 3.5. Thermal Conductivity and Coefficient of Thermal expansion:

Thermal conductivity is the character of material that shows its possibility for heat conductivity. The thermal

behavior of concrete is indicated by means of the thermal conductivity. The results of test are shown in Table (3).
Thermal conductivity is also defined as the quality of heat gradient per unit time between entering face and leaving at the opposite face at a temperature difference of about one degree.
Usually the mineralogical property of aggregate, water with air voids content and temperature with moisture state of concrete, affect significantly on the thermal conductivity of control concrete. Therefore thermal conductivity decreases significantly by variance of moisture content in the concrete state. While the source of aggregate significantly influences the thermal conductivity of concrete.
Almost, small length change due to expansion and contraction of materials related to the rising or falling temperature. Therefore the coefficient of thermal expression is the best measure of them. This coefficient is the main parameter to be adopted in the rigid pavement design. The average value of the coefficient is already used for pavement design. The variance in type of aggregate used in the concrete mix causes numerous values of coefficients. But this operation leads to a wrong assumption of thermal movement and any possible distress.
The coefficient of thermal expansion of recycled crushed brick concrete, is lower than that of control concrete that agreed with **[2,8]**. This decrease is due to low coefficient of thermal expansion of waste materials. The result is shown in Table (3).

**Table 1.** Results of density for different concrete mix (different percentage of coarse aggregate replacement by crushed brisk) at age 28 days.

| Mix type | (0% rep.) | (10% rep.) | (20% rep.) | (30% rep.) | (40% rep.) |
|---|---|---|---|---|---|
| Density kg/m3 | 2324 | 2115 | 2040 | 1899 | 1870 |

**Table 2.** Results of compressive strength, modulus of rupture and modulus of elasticity for different concrete mixes (different percentages of coarse aggregate replacement by crushed brick) at age 28 days.

| Mix type | Compressive strength (MPa) | Modulus of rupture (MPa) | Modulus of elasticity (GPa) |
|---|---|---|---|
| Reference mix (0% replacement) | 36.80 | 5.60 | 29.6 |
| (10% replacement) | 34.71 | 4.75 | 26.0 |
| (20% replacement) | 33.90 | 5.35 | 25.5 |
| (30% replacement) | 33.20 | 4.91 | 24.1 |
| (40% replacement) | 33.04 | 4.70 | 22.8 |

**Table 3.** Results of thermal conductivity, coefficient of thermal expansion for different concrete mixes (different percentage of coarse aggregate replacement by crushed brick) at age 28 days.

| Mix type | Thermal conductivity W/m Kelvin | Coefficient × 10-6 expansion mm/mm/Co |
|---|---|---|
| Reference mix (0% replacement) | 2.19 | 13.87 |
| (10% replacement) | 1.51 | 13.19 |
| (20% replacement) | 1.35 | 13.03 |
| (30% replacement) | 1.28 | 12.69 |
| (40% replacement) | 1.27 | 12.95 |

## 4 Design thickness of concrete layers:

Rigid highway pavement is concrete layer with thickness (h) on base (or subbase) layer. The strength is needed for the beam action of the layer due to the high modulus of elasticity of concrete. Thickness nomograph has a wide range in values of flexural strength and modulus of elasticity **[9]**. Table (4) shows a comparison which is based on layer thickness and different concrete mixtures.

**Table 4.** Calculated thickness values of concrete slab for various concrete mixtures types.

| Mix type | $W_{18}$* $10^3$ | K (pci) | MR (psi) | E*10^ 6 (psi) | H (in) |
|---|---|---|---|---|---|
| Reference mix (0 replacement) | 4000 | 200 | 812 | 4.29 | 7.9 |
| (10 replacement) | 4000 | 200 | 688 | 3.77 | 8.4 |
| (20 replacement) | 4000 | 200 | 776 | 3.71 | 8.0 |
| (30 replacement) | 4000 | 200 | 712 | 3.49 | 8.3 |
| (40 replacement) | 4000 | 200 | 682 | 3.30 | 8.7 |

Where:
$W_{18}$: Predicted number of 18 Kip equivalent single axle load application.
MR: modulus of rupture (psi).
K: modulus of subgrade reaction (pci)
E: modulus of elasticity for Portland cement concrete (psi)

The thickness design results indicate that the concrete mixtures containing 20% coarse aggregate replacement by crushed brick render the slab concrete thickness of rigid more slightly using reference concrete mixtures .

### Stresses due to warping

Usually, if the concrete layer is subjected to temperature grading with depth, layer surface is warping (which means concave the layer edges down during day and concave up during the night). But the effect of temperature grading is resisted by the weight of layer which will hold the layer in its first state.

Table (5) illustrates the edge and interior stresses due to warping of concrete slab for various concrete mixtures type adopted in this research in case of using these mixes in construction of rigid pavement highway.[10]

a: $\sigma_{ix} = (E * \alpha * \Delta T * 0.5) * \{(C_x + \mu C_y/(1 - \mu^2)\}$

as $\sigma_{ix}$: internal stress in x direction due to warping

b: $\sigma_{ex} = (C_x * E * \alpha * \Delta T)/2$

as $\sigma_{ex}$ : edge stress in x direction due to warping.

as I = $\{(E * h^3)/(12k * (1 - \mu^2)\}^{\wedge 1/4}$

where:

Radius of relative stiffness (inch)
Pavement layer thickness (from Table 4)
Poisson's ratio of the pavement taken 0.15 for concrete
Linear coefficient of thermal expansion (mm/mm/°F) (From Table 3)
From Bradbury curve, (as $L_x$/I & $L_y$/I)

**Table 5.** Calculated stresses due to warping of concrete slab for various concrete mixture types

| concrete type | Stress due to warping psi | |
|---|---|---|
| | Interior | Edge |
| 0% | 440.7 | 391.0 |
| 10% | 393.1 | 360.7 |
| 20% | 382.8 | 351.8 |
| 30% | 370.7 | 337.3 |
| 40% | 365.2 | 335.0 |

For interior warping stresses, the results indicate that coarse aggregate with 20% replacement by crushed brick gives a decrease in warping stresses by (13%) compared to reference concrete mixtures for the two joint distance and for edge warping stresses to the results indicate that coarse aggregate with 20% replacement by crushed brick give a decrease in warping stresses by (10%) compared to reference concrete mixtures for the two joint distances. This behavior is due to the low coefficient of thermal expansion and thermal conductivity of coarse aggregate with 20% replacement by crushed brick reference mix.

## 5 Conclusions

After reviewing of main properties of concrete made with crushed clay brick as an aggregate and according to results of experimental investigation that are shown in the research, following facts are concluded:

- Compressive strength decreases slightly when using clay brick as coarse aggregate with up to 40% replacement and achieving the pavement requirements.
- Modulus of rapture decrease when using clay brick as coarse aggregate but this decrease became very less at 20% replacement.
- The values of modulus of elasticity with modulus of rapture at 20% coarse aggregate replacement by clay brick, reflected on concrete layer thickness same to that of reference mix (lower very slightly)
- Density decreases significantly when using clay brick as coarse aggregate for all replacement percentage and this decrease imposed a clear increase in joint spacing.
- The use of clay brick as coarse aggregate improve the thermal insulation properties of concrete layer, for all replacement percentage.
- It seems that, the use of clay brick waste which has a very high absorption of water is a very effective technique for sustaining cement hydration and increases the volume of hydration products that fill pores within the concrete and increase its durability.

## References

1- J. Yang, D. Qiang, B. Yiwang, Construction and Building Materials **25**: 1935–1945 (2011).
2- K. Ivana, N. Ivanka, & B. Dubravka, Technical Gazette, **15** (3), 35-40 (2008).
3- M. G. Osama. **17** (4): 1022-1038 (2011).
4- S. R. Laith & M. R. M. Laith. **2** (5): 45-52( 2013)
5- M. S. Haifa, Tikrit Journal of Engineering Sciences. **19** (3): 33-40 (2012)
6- M. A. Haider, Journal of Babylon University, Pure and Applied Sciences **19** (1): 226-239 (2011)
7- K. M. Hamoud "Effect of addition of steal fibers on compressive and tensile strength of structural light weight concrete made of broken bricks". IJCE-7th issue. 99-118. (2007)
8- A. Shakir and E. Maha, Eng and Tech. **26** (12):1508-1513 (2008)
9- AASHTO *Guide for design of pavement structures*. (1993)
10- V. Tom Mathew and K. V. Krishna Rao. " *Introduction to Transportation Engineering*". NPTEL. **16**: 1-6. (2007).

# Effects of waste PVC addition on the properties of (40-50) grade asphalt

*Nawal* Salman[1,*] and *Zeena* Jaleel[1]

[1]Building and Construction Engineering Department, University of Technology, Baghdad, Iraq

**Abstract.** Disposal of plastic waste poses a serious threat nowadays. Effective solutions are required to reduce or eliminate this problem. One of these solutions is to use these materials in asphalt mixtures. PVC (Poly Vinyl Chloride) during melting has adhesive properties which can be used with asphalt to reduce the bitumen mixture costs. Investigation of physical characteristics of asphalt cement (40-50) mixed with PVC is presented in this paper. The main objective is to study the change in bitumen properties after mixing with PVC of percent (2.5, 5, 7.5, 10, 12 and 15) % by weight of bitumen. Penetration, ductility, loss on heating, softening point, flash and fire tests were conducted for each percent. It is concluded that penetration decreases by 62.8% on addition of 15% PVC. Ductility also drops when PVC dosage increases. Softening point increases when PVC dosage increases by 6 % for 15% PVC.

## 1 Introduction

There is an urgent requirement to improve the road performance in view of the increase in traffic volume on roads and overloading of transport vehicles along with an absence of effective legislation to protect the highways, and also due to the significant difference in temperatures during day and night and across seasons. Hence the need to improve the properties of asphalt material. Population increase and industrial growth throughout the world result in creation and production of new types of non-biodegradable plastic waste. Recent studies have shown that plastic waste can stay on earth for 4500 years unchanged and without degradation. [1]

In order to reduce pollution due to plastic waste an effective solution for recycling must be found. One such solution is the use of plastic waste materials in flexible pavement mixture. This process can reduce the cost of construction. The use of plastic wastes with asphalt mixture in pavement construction is not a new work. The idea began to spread across the world in the last two decades, where researchers have put together various ideas and researches to find alternatives to the use of plastic waste with asphalt.

Rajasekaran et al. [2] investigated possibility of Polyethylene polypropylene polystyrene in coating aggregate, which is showing better aggregate properties that can be used in flexible pavement construction. Marshall Mix sample prepared using the coated aggregate. Marshall Stability improve in range of (18-20) kN. This process is environment friendly and helps reduce costs.

Rahman et al. [3] used two types of plastic waste, polyethylene and PVC to improve asphalt mixture properties using Marshall Mix design. The tests includes measurement of unit weight, stability, flow, and V.T.M, the study concluded that polyethylene modifier up to 10 % and PVC up to 7.5 %, can be used for flexible pavement construction".

Chavan [4] utilized waste plastic bags in coating the aggregate then used in the asphalt mixture to enhance properties of asphalt mix besides solving the plastic waste problems"

Gawandea etal. [5] Reviewed developed techniques to use plastic wastes for construction of flexible pavement. Bitumen material can be modified by using plastic waste, so the modified mixture can be used on the wearing surface of the pavement. The modified mixtures show good stability, flow and moisture resistance.

Jassim et al. [6], investigated stability of Marshall and retained strength index to evaluate plastic waste particles properties such as thickness, size and content percent. Basis results of the tests, it was found that the plastic waste reduced to a thin size of fine particles, thin thickness resulted in developing stability of Marshall and water resistance to damage, and could also solve environmental problems.

## 2. Materials

### 2.1 Asphalt material

The asphalt material used for this study is (40-50) penetration graded bitumen, which is supplied by Al-Dura refinery in Baghdad city. Al-Dura refinery produces asphalt cement graded from refined crude oil petroleum, which is used for pavement construction. Usually the penetration graded asphalt classified to (40-

*Corresponding author: 40096@uotechnology.edu.iq

© The Authors, published by EDP Sciences. This is an open access article distributed under the terms of the Creative Commons Attribution License 4.0 (http://creativecommons.org/licenses/by/4.0/).

50), (50-60), (60-70), (85-100), (120-150), (200-300). Standard ASTM tests were performed on the original bitumen getting the following properties shown in Table 1.

**Table 1.** Test Results for the Origin Bitumen Material

| Property | Units | ASTM Designation | Test Results | SCRB Specification |
|---|---|---|---|---|
| Penetration at 25 $^0$C ,100 gm ,5 sec. | 0.1 mm | D-5-86 | 43 | 40-50 |
| Flash point | $^0$C | D-92 | 342 | Min.232 |
| Fire point | $^0$C | D-92 | 345 | - |
| Softening point | $^0$C | D-36-86 | 52.5 | (51-62) |
| Ductility at 25 $^0$C , 5 cm /min. | cm | D-113-79 | 105 | >100 |
| Loss on heating (% max.) | % | D-1754-83 | 0.3 | 0.75 % |

## 2.2 Poly Vinyl Chloride (PVC)

Plastic waste may be grouped depending on their physical properties to thermoplastic and thermosetting materials. Thermoplastic materials can be formed into any desired shape under the effect of pressure and heat. Thermoplastic materials become solid when cooled. Some examples of thermoplastic and thermosetting materials are tabulated below in Table 2. [1]. Plastics can also be classified according to their chemical sources. According to sources of plastic, there are six types: Cellulose Plastics, Synthetic Resin Plastics, Protein Plastics, Natural Resins, Elastomers and Fibres. Table 3 gives the source of waste plastic generation. [1]. The essential raw materials for PVC are derived from salt and oil. PVC is manufactured from two starting materials;

- 57% of the molecular weight derived from common salt
- 43% derived from hyrdocarbon feedstocks (increasingly ethylene from sugar crops is also being used for PVC production as an alternative to ethylene from oil or natural gas). [7]

**Table 2.** Typical Thermoplastic and Thermosetting Resins [1]

| Thermoplastic | Thermosetting |
|---|---|
| Polyethylene Teryphthalate(PET) | Bakelite |
| Polypropylene (PP) | Epoxy |
| Poly Vinyl Acetate (PVA) | Melamine |
| Poly Vinyl Chloride (PVC) | Polyester |
| Polystyrene (PS) | Urea - Formaldehyde |
| Low Density Polyethylene (LDPE) | Alkyd |
| High Density Polyethylene (HDPE) | |

**Table 3.** Waste Plastic & its Source [1]

| Waste Plastic | Origin |
|---|---|
| Low Density Polyethylene (LDPE) | Carry bags, sacks, milk pouches, bin lining, cosmetic and detergent bottles |
| high density Polyethylene (HDPE) | Carry bags, bottle caps, house hold articles etc |
| Polyethylene Terephthalate (PET) | Drinking water bottles etc |
| Polypropylene (PP) | Bottle caps and closures, wrappers of detergent, biscuit, wafer packets, microwave trays for readymade meal etc.. |
| Polystyrene (PS) | Yoghurt pots, clear egg packs, bottle caps. Foamed Polystyrene: food trays, egg boxes, disposable cups, protective packaging etc. |
| Polyvinyl Chloride (PVC) | Mineral water bottles, credit cards, toys, pipes and gutters, electrical fittings, furniture, folders and pens, medical disposables, etc. |

PVC is a polymer having only aliphatic (linear) carbon atoms in their backbone chains. [8]. Table (4) shows some of physical properties for PVC. [8]

**Table 4.** Physical Properties for PVC

| Density gm/cm$^3$ | Tensile strength MPa | Elongation at break % | Flexural Modulus GPa | Thermal coefficient expansion |
|---|---|---|---|---|
| 1.3-1.6 | 40-50 | 2-80 | 2.1-3.4 | $80*10^{-6}$ |

PVC is chemically resistant to acids, salts, bases, fats, and alcohols. It is also resistant to some solvents for example, PVC is resistant to fuel and some paint thinners. Some solvents may only swell it or deform it but not dissolve it, but some of them like tetrahydrofuran or acetone, may damage it. [8]

# 3 Plastic PVC preparation

PVC collected from factories which produce PVC doors and windows to house construction. The plastic waste is cleaned well to ensure no unwanted material.

## 3.1 PVC Shredding

Shredding is cutting the PVC into small particle sizes between 2.5 to 3.0 mm by shredding machine from the seller where the waste PVC supplied.

## 3.2 Preparation of blend

First step to mix PVC with asphalt was heating the PVC particles until they soften then added to hot asphalt. It was found that PVC particles soften at (180-200) $^o$C. This process would fail if the PVC particles are gathered and clustered. Particles were added gradually to hot asphalt of (180-200) C$^o$. PVC mixed with percent of (2.5, 5, 7.5, 10, 12, and 15) % by weight of bitumen until plastic waste mixture with bitumen became homogeneous. Figure (1) shows the shredded PVC. Figure (2) shows mixing process with bitumen.

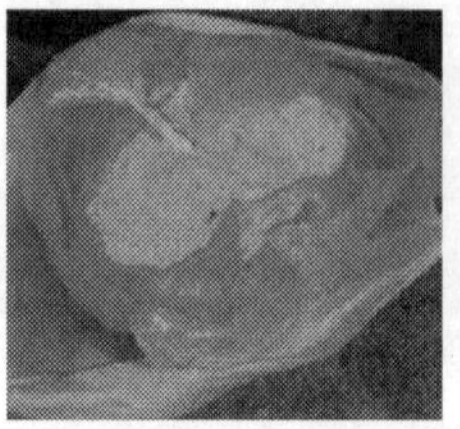

**Fig. 1.** Shredded PVC

**Fig. 2.** Mix PVC with Bitumen

## 4 Experimental works

Standard ASTM tests were conducted for the PVC and asphalt cement mixture. These tests included penetration, ductility loss on heating, softening point, flash and fire test. PVC added to (40-50) graded asphalt with percent of (2.5, 5, 7.5, 10, 12 and 15) by weight of bitumen. Test results are shown in Table (5). Penetration Index (P.I) also calculated for each percent to detect asphalt temperature susceptibility. P.I calculated from equation 1:

$$\frac{20 - P.I}{10 + P.I} = 50 * \frac{\log(800) - \log(penteration.)}{T_{rb} - T_{penteration}} \quad (1)$$

Where:
P.I = penetration index
Trb= softening point temperature
Tpenteration= penetration temperature = 25° C
P.I range from (-2, 2) which indicate normal temperature susceptibility. P.I greater than 2 indicates low temperature susceptibility, while P.I less than -2 indicates high temperature susceptibility. [9]

## 5 Results and Discussion

The Penetration values decrease significantly when (40-50) bitumen is mixed with the modifier (PVC). The original grade decreases by 62.8 % on adding 15 % of PVC by weight of bitumen. Figure (3) shows the relationship between penetration and PVC content. It may be seen that the ductility values for bitumen modified with 12 and 15 percent are very low compared to original binders. It drops from 105 cm to 16 cm at 15 % PVC. The ductility values decrease with the increase in percentage of modifier (PVC) because asphalt becomes harder.

Figure (4) shows the relation between Ductility and PVC percent. Penetration and ductility values decrease with the increase in dosage of PVC. Plastic materials are organic polymers which are usually solid so penetration

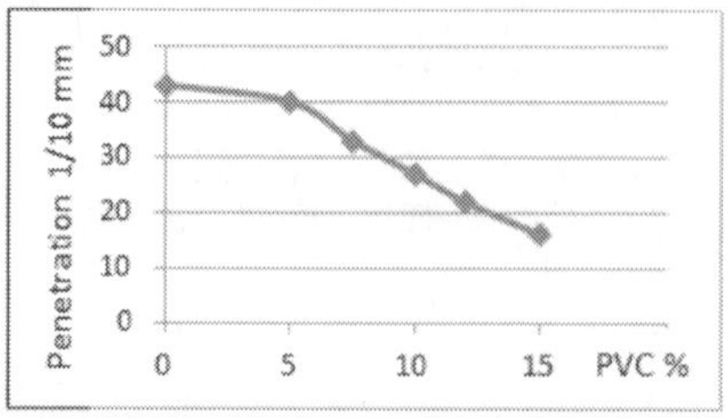


**Fig. 3.** Relation between Penetration and PVC

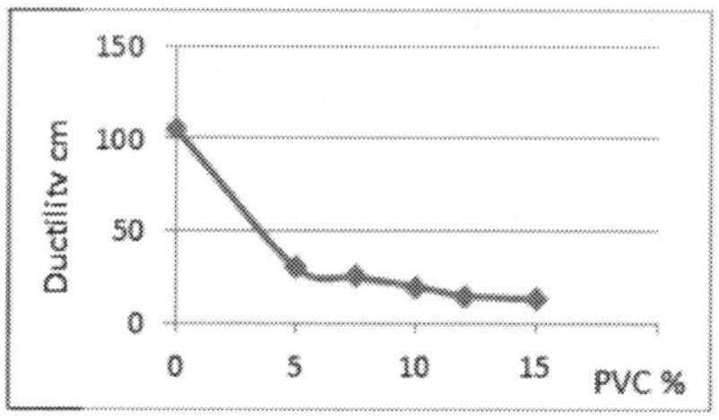


**Fig. 4.** Relation between Ductility and PVC

The softening point increases with the increase in percentage of PVC dosage because the bitumen becomes increasingly more brittle. Softening point of bitumen increase to 55.8 °C by the addition of 15% PVC. It's increased by 6 %. Figure (5) shows softening point with PVC%. Penetration Index P.I was calculated for each percent to detect temperature susceptibility for asphalt material after adding PVC. The calculation results are illustrated in Table (5) which shows that penetration index decreases with increase in PVC %. The penetration index from 2.5% to 12% indicate normal susceptibility to temperature, while at 15% PVC asphalt material become high susceptibility. Figure (6) shows the relation between P.I and PVC%.
Flash and fire points decrease with PVC increase, this can be shown in Figure (7). Flash point decreases from 345 ° C to 286 ° C when adding 15 % PVC, which meets the Iraqi standards that specifies minimum 232 ° C as a flash point. Flash point decreases by 23 %., while fire point decrease by 22% on addition of 15 % PVC.

**Table 5** Properties of Asphalt with PVC

| PVC % | Penetration (1/10) mm | Ductility cm | Softening Point C° | Flash point C° | Fire point C° | Loss on Heating | Penetration Index | Penetration Index Reflection |
|---|---|---|---|---|---|---|---|---|
| 0 | 43 | 105 | 52.5 | 342 | 345 | 0.3 | -0.934 | Normal |
| 2.5 | 42 | 50 | 52.7 | 340 | 343 | 0.35 | -0.937 | Normal |
| 5 | 40 | 30 | 53 | 337 | 340 | 0.4 | -0.972 | Normal |
| 7.5 | 33 | 25 | 54 | 320 | 325 | 0.4 | -1.14 | Normal |
| 10 | 27 | 20 | 54.5 | 300 | 308 | 0.4 | -1.416 | Normal |
| 12 | 22 | 15 | 55.6 | 282 | 285 | 0.6 | -1.552 | Normal |
| 15 | 16 | 13 | 55.8 | 263 | 268 | 0.8 | -2.017 | High |

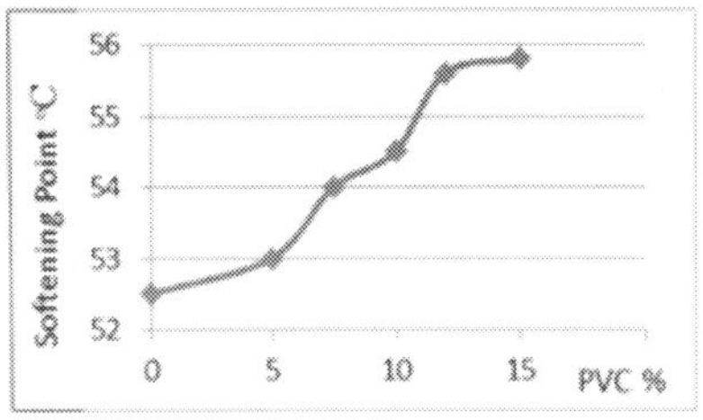


**Fig. 5.** Relation between Softening Point & PVC

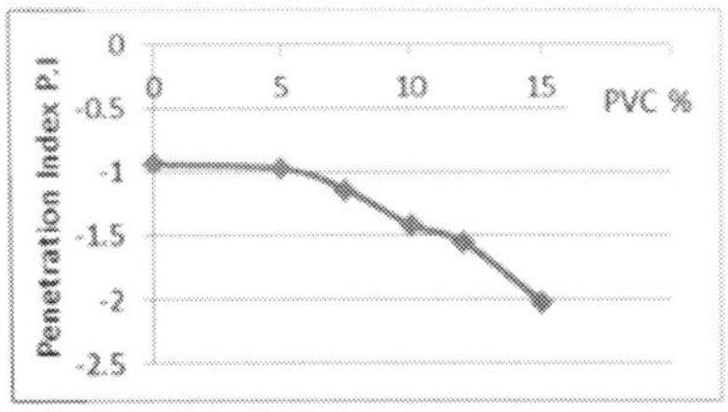


**Fig. 6.** Relation between P.I & and PVC

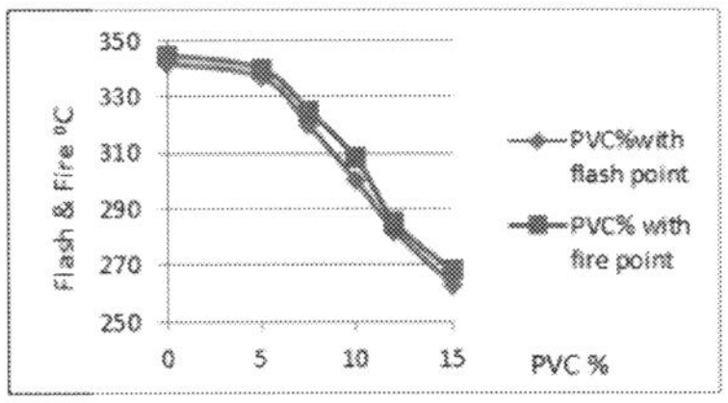


**Fig.7.** Relation between Flash & Fire Point & PVC

Loss on heating test shows convergent results, at 2.5 % of PVC the loss was 0.35 %, at (5, 7.5, 10) the loss value was 0.4%, at 12 % the loss was 0.6 %, all these results meet the Iraqi standards which permits maximum loss value as 0.75 %. The last percent of PVC % (15) recorded 0.8% loss value, which was out of the range of specification. Figure (8) represents the loss on heating value.

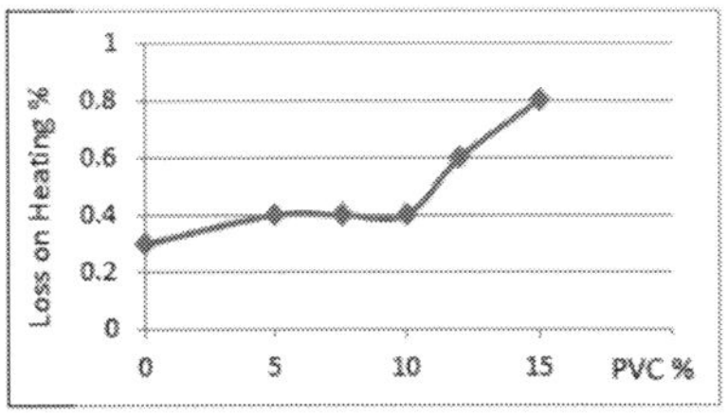


**Fig. 8.** PVC% with Loss on Heating Value

## 6 Conclusions

This study investigates the use of PVC with bitumen as additives to study the change in physical properties of (40-50) graded asphalt. From the experimental work, the following can be concluded:

1. Penetration decreased by 62.8 % on addition of 15 % PVC. Generally asphalt became harder with the increase of PVC dosage.
2. Ductility value decreased with increase in PVC dosage. Ductility value dropped from 105 cm with 0 % PVC to 16 cm on adding 15 % PVC.
3. Softening point increased with the increase in PVC dosage. It increased by 6 % on adding 15 % PVC.
4. Penetration Index (P.I) decreased with an increase in PVC dosage. With PVC percent of (5, 7.5, 10, 12) % asphalt cement has normal susceptibility to temperature, while with PVC dosage of 15 % the asphalt cement becomes highly susceptible to temperature.
5. Flash point decreased by 23 % when PVC % dosage increased.
6. Fire point decreased by 22 % when PVC % dosage increased.
7. Loss on heating value increased from (0.3 to 0.6) % when PVC dosage increased. 15 % PVC recorded 0.8% loss on heating value which is outside the range of the specifications.

## 7 Recommendations

Further research may be carried out on softer grade of asphalt like (85-100), (120-150) and also on other types of PVC like plastic bottles, carrying bags, pipes...... etc.

## References

1. *Indian Roads Congress*, (2013).
2. S.Rajasekaranl, R. Vasudevan, S. Paulraj, AJER e-ISSN: 2320-0847 p - ISSN: 2320-0936, **2**, 11, 01-13, (2013).
3. Md. N. Rahman, M. Ahmeduzzaman, M. A. Sobhan, T. U. Ahmed, AJCEA, **1**, 5, 97-102, (2013).
4. A. J. Chavanl, (IJAIEM), **2**, 4, April (2013).
5. A. Gawandea, G. Zamarea, V.C. Rengea, S. Taydea, and G. Bharsakaleb, JERS, **3**, 2, (2012).
6. H. M. Jassim , O. T. Mahmood , S. A. Ahmed, (IJETT), **7** , 1 , (2014).
7. http://www.bpf.co.uk/plastipedia/polymers/PVC
8. https://www.britannica.com/science/plastic
9. J. Read, D. Whiteoak, *The Sell Bitumen Handbook*, Sixth Edition, Shell Bitumen, UK, (2015).

# Behavior of high performance artificial lightweight aggregate concrete reinforced with hybrid fibers

*Wasan* Khalil[1*], *Hisham* Ahmed [1], *and Zainab* Hussein [2]

[1] professor in the Building and Construction Engineering Department, University of Technology, Baghdad, Iraq
[2] lecturer in College of Engineering, Al Mustansiriyah University, Baghdad, Iraq

**Abstract.** In this investigation, sustainable High Performance Lightweight Aggregate Concrete (HPLWAC) containing artificial aggregate as coarse lightweight aggregate (LWA) and reinforced with mono fiber, double and triple hybrid fibers in different types and aspect ratios were produced. High performance artificial lightweight aggregate concrete mix with compressive strength of 47 MPa, oven dry density of 1۸۲۸ kg/m3 at 28 days was prepared. The Fibers used included, macro hooked steel fiber with aspect ratio of 60 (type S1), macro crimped plastic fiber (P) with aspect ratio of 63, micro steel fiber with aspect ratio of 65 (type S), and micro polypropylene fiber (PP) with aspect ratio of 667. Four HPLWAC mixes were prepared including, one plain concrete mix (without fiber), one mono fiber reinforced concrete mixes (reinforced with plastic fiber with 0.75% volume fraction), one double hybrid fiber reinforced concrete mixes (0.5% plastic fiber + 0.25% steel fiber type S), and a mix with triple hybrid fiber (0.25% steel fiber type S1+ 0.25% polypropylene fiber + 0.25% steel fiber type S). Fresh (workability and fresh density) and hardened concrete properties (oven dry density, compressive strength, ultrasonic pulse velocity, splitting tensile strength, flexural strength, static modules of elasticity, thermal conductively, and water absorption) were studied. Generally, mono and hybrid (double and triple) fiber reinforced HPLWAC specimens give a significant increase in splitting tensile strength and flexural strength compared with plain HPLWAC specimens. The percentage increases in splitting tensile strength for specimens with mono plastic fiber are, 20.8%, 31.9%, 36.4% and 41%, while the percentage increases in flexure strength are 19.5%, 37%, 33.9% and 34.2% at 7, 28, 60, 90 days age respectively relative to the plain concrete. The maximum splitting tensile and flexure strengths were recorded for triple hybrid fiber reinforced HPLWAC specimens. The percentage increases in splitting tensile strength for triple hybrid fiber reinforced specimens are 19.5%, 37%, 33.9% and 34.2%, while the percentage increases in flexure strength are 50.5%, 62.4. %, 66.8% and 62.2% at 7, 28, 60 and 90 days age respectively relative to the plain concrete specimens.

Keywords: Artificial Lightweight Aggregate, Plastic Fiber, Hybrid Fiber, High Performance LWA.

## 1 Introduction

High performance structural lightweight concrete (HPSLWC) is an advanced in concrete technology which has the properties of High Performance Concrete (HPC) and Lightweight Concrete (LWC) such as, high strength, low permeability, low density, and high thermal insulation properties which are appropriate for many applications.

High performance lightweight concrete typically has a compressive strength from 34 - 69 MPa and an air dry density not exceeding 2000 kg/m$^3$ [1, 2].

Lightweight Aggregate (LWA) used in concrete reduces the weight of different structural members which leads to a decrease in the load transmitted to the foundations, and less work and efforts are required to transport these members. The occurrence of voids and pores in LWA provides good thermal and acoustic insulating properties and a good fire resistance to concrete [3].

In Iraq, local natural LWA is very limited, such as porcelinite aggregate. The large consumption of natural lightweight aggregate leads to production and use of artificial lightweight aggregate, also the use of waste material as lightweight aggregate is the main aim of many researchers. Artificial lightweight aggregate can be produced from a wide variety of raw material and production procedures [4].

Several investigations have been carried out to study the influence of mono plastic fiber, polypropylene, and

* Corresponding author: wasan1959@yahoo.com

© The Authors, published by EDP Sciences. This is an open access article distributed under the terms of the Creative Commons Attribution License 4.0 (http://creativecommons.org/licenses/by/4.0/).

steel fiber on properties of artificial lightweight aggregate concrete [5, 6]. Very little works have been done to investigate the properties of hybrid fiber reinforced lightweight concrete containing natural or expanded clay lightweight aggregate.

The behavior of hybrid fiber high strength lightweight aggregate was investigated by Daneti and Wee [7]. Hooked end macro steel fiber (30mm length with aspect ratio of 55), straight micro steel fiber (12mm length and aspect ratio of 120mm), and micro polypropylene fiber (12mm length with aspect ratio of 109) were used. The concrete mixes containing mono macro steel fiber, double hybrid fiber (0.3% macro steel fiber + 0.3% polypropylene fiber) and triple hybrid fiber (0.2% macro steel fiber + 0.2% micro steel fiber + 0.2% polypropylene fiber) with constant total fiber volume fraction of 0.6%. Coarse lightweight expanded clay aggregate was used. The compressive strength of single and hybrid fiber showed almost equal or slightly lower strength compared with plain lightweight aggregate concrete. However, splitting and flexural tensile strengths of single and hybrid fiber were increased in the range of 16% - 52% and 1% - 28% respectively compared to plain LWAC. Lightweight aggregate concrete containing mono and triple fiber showed higher splitting and flexural strengths when compared with double hybrid fibers.

Widodo et al. [8] examined the effect of hybrid polypropylene-steel fiber on properties of lightweight concrete made of pumice breccia as coarse aggregate. Monofilament Polypropylene with 12 mm length and 18 $\mu$m diameter and macro hooked end steel fiber with 60 mm length and 0.75mm diameter were used. Fiber addition was done using fixed polypropylene fiber volume fraction of 0.1% and different steel fiber additions of 0.0%, 0.5%, 1.0%, 1.5%, and 2.0% by volume into the mixture. Different properties including density, compressive strength, modulus of elasticity, splitting tensile strength, modulus of rupture and the flexural load-displacement behavior of the lightweight aggregate concrete with hybrid fiber addition were studied. The results demonstrate that the addition of steel fiber to LWAC causes a decrease in the workability but the density increased. The compressive strength and modulus of elasticity of lightweight concrete increased. The splitting tensile strength and modulus of rupture for hybrid LWAC with 0.1% polypropylene + 1.5% steel fiber improved by about 222.28% and 187.47% respectively. Flexural performance is much better in hybrid fiber reinforced LWAC of 0.1% polypropylene +1.5% steel compared to the plain concrete mixes.

## 2 Research significance

Research significance can be summarized as follows:

- Producing high performance LWAC containing artificial sustainable LWA produced from locally available materials and studying its properties.
- Investigate the effect of inclusion mono and hybrid fibers on the properties of the produced high performance LWAC.

## 3 Experimental programs

### 3.1 Materials

- Ordinary Portland cement (Type I) from Bazian Company in Iraq was used. Test results show that the adopted cement satisfies the Iraqi Specifications No. 5/1984 [9].
- Normal weight natural sand with maximum size of 4.75 mm was used in this investigation. The results demonstrate that the grading of the sand, physical properties and sulfate content are within the requirements of the Iraqi Specifications No. 45/ 1984 [10].
- An artificial lightweight aggregate was produced from bentonite clay and sodium silicate liquid waste from glass plant through this study, as mentioned by the research (11). Bentonite clay was screened through 75μm sieve. Many trials were carried out to select the proper amount of sodium silicate that produces a paste of adequate plasticity. Sodium silicate liquid waste with 50% mixing ratio by weight (clay: sodium silicate, 1:1) was selected. The clayey paste was rounded by handed to form balls, then they were left to dry at laboratory temperature for 24 hours to avoid the formation of shrinkage cracks. The clayey balls were then dried in an oven for 24 hours at 100 °C to ensure complete drying, and then they were burned at different high temperatures. It was found that burning these balls at temperature between 750 - 800 °C for two hours produces lightweight aggregate with properties conforming to ASTM C 330 Specifications [12]. Finally, the furnace was opened and the specimens were left inside the furnace until they were cooled to laboratory temperature. The clayey balls were crushed manually to smaller size by means of a hammer, and then screened on standard sieve series in order to prepare lightweight aggregate with grading which conforms to ASTM C 330 Specifications [12]. Table 1 shows the properties of the produced lightweight aggregate.

**Table 1**. Properties of the produce lightweight aggregate [13]

| Properties | Specification | Test Results |
|---|---|---|
| Specific gravity | ASTM C127 [13] | 1.53 |
| Absorption % | ASTM C127 [13] | 12.9 |
| Dry loose unit weight, $kg/m^3$ | ASTM 29/C29M[14] | 538.22** |
| Dry rodded unit weight, $kg/m^3$ | ASTM 29/C29M [14] | 543.21 |
| Aggregate crushing value, (%) | BS 812-part 110-1990 [15] | 51.6 |
| Sulfate content (as $SO_3$), % | BS 3797-part 2-1981[16] | 0.97 |

* Physical analysis was conducted by National Center for Construction Laboratories and Researches (NCCLR)
* * Within the limit of ASTM C 330, ≤ 880 $kg/m^3$

- Two types of high performance LWAC admixtures were used in this work including, chemical admixture based on modified Polycarboxylic ether (Sika – Viscocrete - 5930). It is a third generation superplasticizer meeting the requirements of ASTM

C494M / 04[17] types F and silica fume produced by Sika Company. The results indicate that the silica fume used satisfy the physical and chemical requirements of the ASTM C1240 [18].

- Four types of fiber were used in this investigation including:

a- Macro hooked steel fibers (type $S_1$) with 30 mm length and 0.5 mm diameter (aspect ratio l / d = 60), the ultimate tensile strength for individual fibers is 1180 MPa and density is 7800kg/m$^3$.

b- Straight steel fibers (type S) with 13 mm in length and 0.2mm in diameter (aspect ratio l / d = 65), the ultimate tensile strength for individual fibers is 1180 MPa and the density is 7800 kg/m$^3$.

c- Crimped plastic fiber (P) with 50 mm length, 0.8 mm diameter (aspect ratio l /d = 63), and minimum tensile strength between 250- 350 MPa.

d- Micro polypropylene fiber (PP) with 12mm length, 18 micron diameter (aspect ratio l/d = 677), and minimum tensile strength of 350 MPa.

### 3.2 Concrete mixes

Control LWAC mix (plain) and many of LWAC mixes reinforced with mono and hybrid fiber was prepared in this work. Table 2 explains the details of concrete mixes studied in this work.

**Table 2.** Details of lightweight aggregate concrete mixes

| Mix Symbol | Fiber Volume Fraction (%) | | | | Mix proportion |
|---|---|---|---|---|---|
| | $S_1$ | P | S | PP | |
| MAR | · | · | · | · | 1:1.18:0.73 by weight (Cement:Sand:LWA), Cement Content 550 kg/m$^3$, w/c =0.25, HRWRA= 3L/ 100kg of cement, silica fume 10% as a replacement, by weight of cement |
| MAP | · | ·.75 | 0 | 0 | |
| $MAH_4$ | 0 | 0.50 | 0.25 | 0 | |
| $MAH_5$ | ·.25 | · | 0.25 | 0.25 | |

### 3.3 Preparation, casting and curing of specimens.

After conducting workability test (slump and inverted slump cone tests) the concrete specimens were prepared by casting the concrete in different standard molds. The molds were well cleaned and oiled before casting of concrete to avoid the adhesion of hardened concrete to the internal surfaces of the molds. The fresh concrete was placed in the molds with layers according to the standard specifications for each test and compaction by means of vibration table. The top layers of concrete specimens had been smoothed by steel trowel, and then the specimens were covered with nylon sheets for 24 hours to prevent the evaporation of water. After that the concrete specimens were demoulded and fully immersed in tap water until the time of testing.

### 3.4 Experimental tests

The following experimental tests were carried out to investigate the effect of artificial LWA on some properties of HPLWAC reinforced with fiber:

- Slump test according of ASTM C-143 [19]
- Fresh density test according ASTM C 567-05a [20].
- Inverted slump cone test according to ASTM C995[21].
- Oven dry density test according to ASTM C 567-05a [20] (using150×300 mm cylindrical specimens).
- Compressive strength test according to B.S. 1881[22] (using 100 mm cube specimens).
- Splitting tensile strength test according to ASTM C496 [23](using 100×200 mm cylindrical specimens).
- Flexural tensile strength test according to ASTM C78 [24] (using 100×100×400 mm prism specimens).
- Static Modulus of Elasticity test according to ASTM C469 [25] (using 150×300 mm cylindrical specimens).
- Ultrasonic pulse velocity test according to ASTM C597 [26]( using 100 mm cube specimens).
- Thermal conductivity test was carried out on specimens with dimensions 200×100×100 mm (length × width × thickness).
- Water absorption test according to ASTM C642 [27] (using 100 mm cube specimens).

## 4 Results and Discussion

### 4.1 Selection of mix proportions for high performance LWAC

Reference lightweight aggregate concrete mix was designed in accordance with ACI 211.2 [28] without any admixtures with compressive strength of 20 MPa at age 28 days. The mix proportion of this mix is 1:1.18:0.73 (cement: sand: LWA) by weight, w/c ratio of 0.42 to have slump value of 100 ± 5mm, and cement content of 550 kg/m$^3$. Several trial mixes were achieved in order to select the optimum silica fume content and the optimum dosage of superplasticizer, while w/c ratio was modified to have the same slump value of the plain mix (without fiber). The results demonstrate that the optimum dosage of superplasticizer was 3.0 Liter per 100 kg of cement. The oven dry density and the compressive strength at 28 days age for the selected reference mix are 1815 kg/m$^3$ and 41.6 MPa respectively. The details of the trial concrete mixes with different dosages of superplasticizer are shown in Table [3]. Different dosages of silica fume as a partial replacement by weight of cement (5%, 8%, and 10 %) were used with the selected mix containing the optimum dosage of superplasticizer. Table [4] shows the effect of silica fume content on workability, compressive strength and density of LWAC mixes. It can be noticed that the workability of the concrete mix is slightly decreased with the increase of silica fume content. This is because of the high surface area of very fine particles of silica fume which leads to absorbed water. The compressive strength and oven dry density of the selected mix containing 10 % silica fume as a replacement by weight of cement and 3 liter of superplasticizer /100 kg of cement are 47 N/mm$^2$ and 1828 kg/m$^3$ respectively at 28 days age.

### 4.2 Workability

The workability test results including slump test and inverted slump cone test are shown in Table [5]. The slump test did not give accurate results for workability of fiber reinforced concrete mixes [29]; hence inverted slump cone test is recommended for this type of mixes. Generally, the results indicate that the inclusion of fibers reduces the workability. This is due to the large surface area of fibers that increases the viscosity of the mixtures [30]. All HPLWAC mixes reinforced with fibers were produced without any fibers balling. Despite the reduction in workability, visible inspections show that the addition of fibers to HPLWAC enhances the uniformity and stability for fresh mixes. This is because fibers can make network structure in fresh concrete that can effectively restrain the segregation of LWA; this was also indicated by **Libre et al.** [31].

**Table 3**. Details of the trial mix with different dosages of HRWRA

| w/c Ratio | HRWRA (L/100 kg of cement) | Slump (mm) | Oven dry density kg/m³ | Compressive strength (MPa) | |
|---|---|---|---|---|---|
| | | | | 7 days | 28 days |
| 0.42 | 0 | 105 | 1748 | 29.7 | 34.5 |
| 0.40 | 1.0 | 105 | 1765 | 30.4 | 35.7 |
| ٠,٣٥ | 1.5 | 100 | 1775 | 30.7 | 36.9 |
| 0.30 | ٢.0 | 98 | 1787 | 32.2 | 37.0 |
| ٠,٢٧ | 2.5 | 100 | 1800 | 33.4 | 39.0 |
| ٠,٢٥ | 3.0 | 100 | 1815 | 34.5 | 41.6 |
| 0.23 | 3.0 | 105 | 1808 | 33.1 | 40.7 |

**Table 4**. Details of trial mix with various dosages of silica fume

| Silica fume (% ) replacement by Weight of Cement) | w/c Ratio | Slump (mm) | Oven Dry Density kg/m³ | Compressive Strength (MPa) | |
|---|---|---|---|---|---|
| | | | | 7 day s | 28 days |
| 0 | 0.25 | 100 | 1815 | 34.5 | 41.6 |
| 5 | 0.25 | 100 | 1818 | 35.1 | 45.8 |
| 8 | 0.25 | 98 | 1823 | 38.9 | 46.5 |
| 10 | 0.25 | 95 | 1828 | 40.0 | 47.0 |

## 4.3 Fresh density

The fresh densities of HPLWAC are listed in Table 5 Results indicate that concrete reinforced with plastic fiber has lower fresh density compared with the plain concrete. The inclusion of steel fibers (0.25% volume fraction) in double hybrid fiber concrete mix ($MAH_4$) increases the fresh density relative to concrete mix reinforced with plastic fiber (MAP) but it is still lower than that for plain concrete mix. The high volume fraction of steel fiber (0.5%) in triple hybrid fiber reinforced concrete mix ($MAH_5$) explains highest fresh density. This is because of the low specific gravity of plastic fiber and the high specific gravity of steel fiber.

## 4.4 Ovendry density

The oven dry density of HPLWAC mixes prepared in this work is shown in Table 5. The inclusion of plastic fiber also causes a reduction in oven dry density of HPLWAC containing artificial LWA (mix MAP). Concrete specimens reinforced with double hybrid fibers show increase in densitites compared with those containing mono plastic fiber with the same total volume fraction but it is lower than plain specimens. The high volume fraction of steel fiber (0.5%) in triple hybrid fiber reinforced concrete mix ($MAH_5$) gives concrete with highest dry density. This is due to the high density of steel fibers used as compared to plastic fiber.

**Table 5**. Workability and density for HPLWAC

| Mix | Slump cone (mm) | Inverted slump cone | | Fresh density (kg/m³) | Oven Dry density (kg/m³) |
|---|---|---|---|---|---|
| | | Sec. | cm* | | |
| MAR | 95 | 18 | -- | 1935 | 1828 |
| MAP | 85 | 105 | --- | 1923 | 1821 |
| $MAH_4$ | 80 | 120 | 2 | 1930 | 1823 |
| $MAH_5$ | 65 | 120 | 6 | 1950 | 1835 |

*The distance from top of concrete level to the top of cone is recorded when the elapsed time is more than 120sec.

**Equilibrium density is equal to oven dry density plus 50 kg/m³ according to ASTM C567-05a[20].

## 4.5 Compressive Strength

The compressive strength development with age for all HPLWAC mixes are listed in Table [6]. Generally, the compressive strength increases with age for all concrete specimens. The compressive strength of concrete specimens containing artificial LWA and reinforce with mono plastic fiber at 7, 28, 60 and 90 days decreases by about 6.2%, 8.8%, 12.6% and 13.4% respectively relative to plain concrete (without fiber). This may be attributed to the high content of air and the presence of voids which reduce the compressive strength of concrete. Concrete specimens reinforced with double hybrid fibers ($MAH_4$) show higher compressive strength than those reinforced with mono plastic fiber (MAP), but it is still less than those for plain concrete specimens. This is attributed to the inclusion of steel fibers that slightly increases the compressive strength of concrete. The inclusion of triple hybrid fiber in HPLWAC ($MAH_5$) slightly improves the compressive strength relative to the plain concrete specimen (MAR).

**Table 6**. Compressive strength for HPLWAC

| Mix symbol | Compressive strength (MPa) | | | |
|---|---|---|---|---|
| | 7 days | 28 days | 60 days | 90 days |

https://doi.org/10.1051/matecconf/201816202001

| | | | | |
|---|---|---|---|---|
| MAR | 40.0 | 47.0 | 51.8 | 53.5 |
| MAP | 37.7 | 43.2 | 46.0 | 47.2 |
| $MAH_4$ | 39.0 | 45.6 | 48.4 | 49.2 |
| $MAH_5$ | 41.8 | 48.5 | 53.9 | 54.8 |

### 4.6 Ultrasonic pulse velocity

The UPV test is used to estimate the uniformity and quality of HPLWAC. The results of UPV for HPLWAC in Table 7 are in the range of 4.22- 4.65 km/sec. These values represent the good quality of concrete specimens prepared in this investigation [32].

**Table 7**. Ultrasonic pulse velocity for HPLWAC

| Mix symbol | UVP (km/sec) | | | |
|---|---|---|---|---|
| | 7 days | 28 days | 60 days | 90 days |
| MAR | 4.40 | 4.51 | 4.57 | 4.61 |
| MAP | 4.22 | 4.34 | 4.45 | 4.53 |
| $MAH_4$ | 4.29 | 4.41 | 4.48 | 4.60 |
| $MAH_5$ | 4.35 | 4.53 | 4.60 | 4.65 |

### 4.7 Splitting tensile strength

The development of splitting tensile strength with age for HPLWAC containing artificial lightweight aggregate is shown in Table 8 and Fig. 1. The results demonstrate that the inclusion of mono plastic fibers causes an increase in splitting tensile strength relative with plain concrete. The percentages of increase in splitting tensile strength are 20.8%, 31.9%, 36.4% and 41% at ages 7, 28, 60 and 90 days age respectively as compared to the plain concrete specimen. This is because the action of fiber in arresting the cracks [٧]. All hybrids fiber reinforced HPLWAC ($MAH_4$ and $MAH_5$) demonstrate higher splitting tensile strength compared with concrete mix reinforced with mono plastic fiber (MAP). The maximum splitting tensile strength of HPLWAC with artificial aggregate is recorded for concrete specimen reinforced with triple hybrid fiber ($MAH_5$). The percentages of increase in splitting tensile strength for specimen reinforced with triple hybrid fiber are 34.6%, 56%, 63.6 % and 68.1% for 7, 28, 60 and 90 days age respectively compared to the plain concrete specimen (MAR). This is because of the synergy between macro and micro fiber in arresting crack.

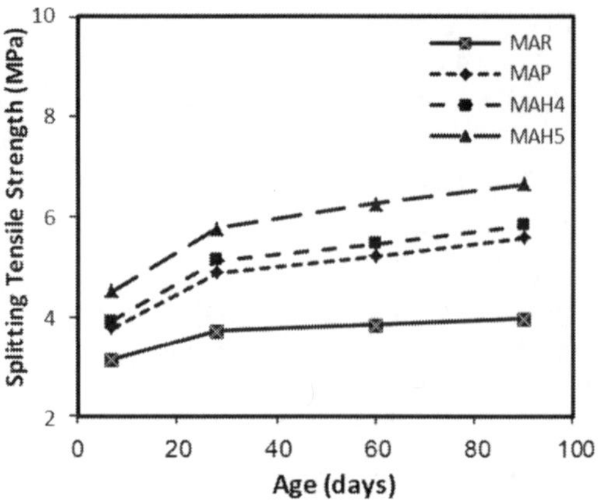


**Fig.1.** Splitting tensile strength development with various ages for lightweight concrete specimens.

**Table 8**. Splitting tensile strength for HPLWAC

| Mix Symbol | Splitting Tensile Strength (MPa) | | | |
|---|---|---|---|---|
| | 7 Days | 28 Days | 60 Days | 90 Days |
| MAR | 3.12 | 3.70 | 3.82 | 3.95 |
| MAP | 3.77 | 4.88 | 5.21 | 5.57 |
| $MAH_4$ | 3.92 | 5.13 | 5.46 | 5.83 |
| $MAH_5$ | 4.50 | 5.77 | 6.25 | 6.64 |

### 4.8 Flexural tensile strength (modulus of rupture)

The influence of fiber on flexural strength is listed in Table 9 and Fig. 2. The results demonstrate that the addition of fibers to HPLWAC increases the flexural strength for both mono and hybrid (double and triple) fiber reinforced concrete specimens compared to plain concrete specimens. The results of HPLWAC containing artificial LWA indicate that the inclusion of mono plastic fibers shows increase in flexural strength compared with reference concrete. The percentages of increase in flexural strength for concrete mix reinforced with mono plasticfiber (MAP) are 19.5%, 37%, 33.9% and 34.2% at 7, 28, 60 and 90 days age, respectively compared to plain concrete (MAR). This is due to the action of fiber that arrested the cracks. Hybrid fiber reinforced HPLWAC specimens ($MAH_4$ and $MAH_5$) demonstrate higher flexural strength relative to concrete specimen with mono plastic (MAP). The highest flexural strength is recorded for concrete specimens reinforced with triple hybrid fiber ($MAH_5$). The percentages of increase in flexural strength for these specimens are 50.5%, 62.4%, 66.8% and 62.2% at 7, 28,

60 and 90 days age respectively compared to plain concrete. This is due to the high tensile strength of steel fiber in triple hybrid fiber concrete mix and the synergy between micro and macro fibers in arresting micro and macro cracks respectively [33].

**Table 9** Flexural strength of HPLWAC

| Mix symbol | Flexural Strength (MPa) | | | |
|---|---|---|---|---|
| | 7 days | 28 days | 60 days | 90 days |
| MAR | 4.10 | 4.87 | 5.18 | 5.47 |
| MAP | 4.90 | 6.67 | 6.92 | 7.34 |
| $MAH_4$ | 5.25 | 6.83 | 7.36 | 7.85 |
| $MAH_5$ | 6.17 | 7.91 | 8.64 | 8.87 |

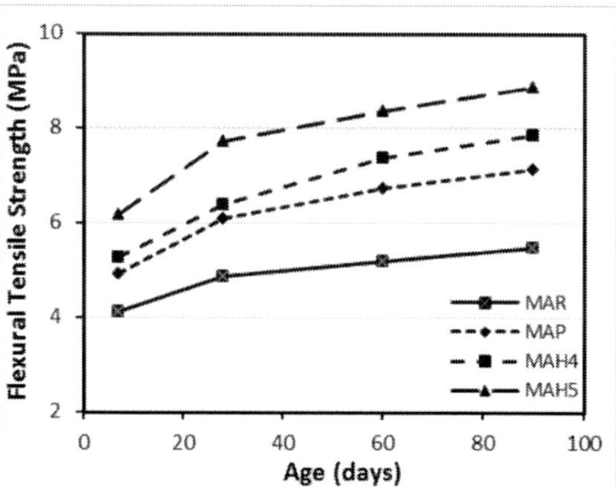


**Fig. 2.** Flexural tensile strength development with varies ages for lightweight concrete specimens.

## 4.9 Static modulus of elasticity

Static modulus of elasticity for all HPLWAC specimens is illustrated in Table 10. The results indicate that the inclusion of mono plastic fiber in concrete causes a decrease in modulus of elasticity by about 11.7 % and 5.1% at 28 and 90 days age respectively compared with plain concrete. All hybrid fibers reinforced HPLWAC ($MAH_4$ and $MAH_5$) show increase in modulus of elasticity in comparison with concrete specimen reinforced with mono plastic fiber (MAP). The highest modulus of elasticity is recorded for concrete specimen reinforced with triple fiber ($MAH_5$). This is due to the use of macro and micro steel fibers combined with other fibers (polypropylene fiber) in this specimen.

**Table 10.** Static modulus of elasticity of HPLWAC

| Mix symbol | Static modulus of elasticity (GPa) | |
|---|---|---|
| | 28 Days | 90 Days |
| MAR | 19.073 | 20.844 |
| MAP | 17.081 | 19.520 |
| $MAH_4$ | 17.528 | 20.143 |
| $MAH_5$ | 20.436 | 21.024 |

## 4.10 Thermal Conductivity

The results of the thermal conductivity of all HPLWAC specimens are shown in Table 11. Thermal conductivity of plain HPLWAC containing artificial lightweight aggregate is 0.9278 W/ (m. K). Thermal conductivity of normal weight concrete is in the range of 1.98 - 2.94 W/ (m. K) depending on the type of aggregate used [34]. The low thermal conductivity of lightweight aggregate concrete is attributed to the presence of pores in lightweight aggregate that resist the heat flow through the concrete and affects the conductivity of the material. The low unit weight and thermal conductivity combined with the ability to cast in any desired shape enable this structural lightweight concrete to be suitable material for casting different thermal insulated structural members. The inclusion of mono plastic fiber in concrete specimens (MAP) causes a slight reduction in thermal conductivity. This reduction may be attributed to the low thermal conductivity of lightweight plastic fiber which increases the insulation property of the concrete specimens. Concrete specimens reinforced with hybrid (double and triple) fibers ($MAH_4$ and $MAH_5$) show increase in thermal conductivity relative to plain concrete (MAR). This is because the high thermal conductivity of steel fiber.

## 4.11 Water absorption

The results of water absorption for HPLWAC specimens are listed in Table 11. The water absorption of all HPLWAC specimens are less than 10 percent by weight. This shows the good quality of HPLWAC specimens prepared in this work [32]. Also, HPLWAC specimens with artificial lightweight aggregate reinforced with mono or hybrid fibers indicate a decrease in water absorption relative with plain concrete specimen (MAR). This is because of the role of fibers which effectively increases the crack resistance that reduces the permeability of concrete.

**Table 11.** Thermal conductviy and water absorbtion of HPLWAC

| Mix symbol | Water absorption at 28 days (%) | Thermal conductivity at 28 days [W/(m. k)] |
|---|---|---|
| MAR | 5.85 | 0.9278 |
| MAP | 5.74 | 0.9132 |
| $MAH_4$ | 5.57 | 0.9379 |
| $MAH_5$ | 5.34 | 0.9987 |

# 5 Conclusions

From the experimental results presented in this work, the following conclusions can be drawn:

[1] The produced sustainable lightweight aggregate concrete containing artificial LWA has a

compressive strength of 47 MPa and oven dry density of 1828 kg/m$^3$ at 28 days age.

[2] The addition of plastic fiber causes a reduction in oven dry density, while steel fiber causes increses in oven dry density. The addition of hybrid fiber (double and triple) shows increase in densitites compared with specimens containing mono plastic fiber with the same total volume fraction.

[3] The compressive strength of concrete specimens at different ages decreases with the addition of mono plastic fiber in the range from (6% -13%), while the inclusion of double and triple hybrid fiber ( mixes $MAH_4$ and $MAH_5$) slightly improves the compressive strength relative to the plain concrete specimens.

[4] The inclusion of mono plastic fibers causes an increase in splitting tensile strength compared to plain concrete. Double and triple hybrid fiber reinforced specimens ($MAH_4$ and $MAH_5$) show higher splitting tensile strength compared with specimens reinforced with mono plastic fiber (MAP).

[5] Mono plastic fibers show increase in flexural strength compared with plain concrete. Hybrid fiber reinforced specimens ($MAH_4$ and $MAH_5$) show higher flexural strength relative to specimens reinforced with mono plastic fiber (MAP).

[6] The inclusion of mono plastic fiber in concrete specimens causes a decrease in modulus of elasticity. Double and triple hybrid fiber reinforced concrete specimens ($MAH_4$ and $MAH_5$) show slight increase in modulus of elasticity in comparison with specimens reinforced with mono plastic fiber (MAP). The highest modulus of elasticity is recorded for specimens reinforced with triple hybrid fiber reinforced specimens ($MAH_5$).

[7] Thermal conductivity of plain specimens is 0.9278 W/ (m. K), while for specimens reniforecd with mono plastic fibers it is slightly lower. Specimens reinforced with double and triple hybrid fiber ($MAH_4$ and $MAH_5$) show higher thermaal conductivity relative to both plain and plastic fiber reinforced specimens.

[9] The water absorption of all HPLWAC specimens reinforced with mono and hybrid fibers decreases compared with plain specimens.

## References


1. Eurocode No.2, part I, General Rules and Rules for Buildings, Final Text,( October 1991)
2. Hoff, G.C., US Army Corps of Engineers, Engineer Research and Development Center, (August, 2002).
3. De Gennaro, R., Cappelletti, P., Gerri, G., De Gennaro, M., Doudi, M., and Langella, A., Applied Clay Science, Vol. **28,** No. **1**, pp.309-319, (2005)
4. Chi, J.M., Huang, R., Yang , C.C., and Chang , J.J., Cement and Concrete Composites,Vol.**25**, Issue **2**, pp.197-205,(2003).
5. Cabral, A., Schalch, V., Molin, D., and Ribeiro, J., Construction and Building Materials, Vol. **24**, pp. 421–430, (2010)
6. Sabin, G. M., Taylor and Francis Group, (201).
7. Daneti, S.B., and Wee, T. H., Magazine of Concrete Research, Vol.**63**, Issue **11**, pp.871-881, (2011).
8. Widodo, S., Satyarno, I. and Tudjono, S., Civil Engineering, Vol. **23**, No. **8**, pp. 1211–1219, (2011).
9. Iraqi Standard, NO.**5**, "Portland Cement", The Central Organization for Standardization and Quality Control, (1984), (in Arabic).
10. Iraqi Specification, No.**45**, "*Aggregate from Natural Sources for Concrete and Constructions*", (1980), (in Arabic).
11. Khalil, W. I., Ahmed, H. K., and Hussein, Z. M, 17[th] International Conference on Building Science and Engineering, Berlin Germany Conference, Sep 14-15, Vol. **19**, No.**(9)** Part IV, (2015).
12. ASTM C330-03, "*Standard Specification for Lightweight Aggregates for Structural Concrete*" Annual Book of ASTM Standards, Vol.**04-02,** Concrete and Aggregates, United States, ( 2003).
13. ASTM C127, "*Standard Test Method for Density Relative Density (Specific Gravity) and Adsorption of Coarse Aggregate*", Annual Book of ASTM Standards, Vol. **04.02**, pp. 68–73, (2004).
14. ASTM C29/C29M, "*Standard Test Method for Bulk Density (Unit Weight) and Voids in Aggregate*", Annual Book of ASTM Standards, Vol. **04.02**, pp. 1- 4, (2004).
15. BS 812, Part 110, "*Method for Determination of Aggregate Crushing Value (ACV)*", British Standards Institution, pp. 8, (1990).
16. BS 3797, "*Specification for Lightweight Aggregates for Masonry Units and Structural Concrete*", British Standards Institution, pp. 2, (1990).
17. ASTM C494/C494M, "*Standard and Specification for Chemical Admixtures for Concrete*", Annual Book of ASTM Standards, Vol. **04.02**, pp. 271–279, (2004).
18. ASTM C1240 "*Standard Specifications for Silica Fume Used in Cementitious Mixtures*", Annual Book of ASTM Standards, Vol.**04**, pp. 200-208, (2007).
19. ASTM C143/C143M–03, "*Standard Test Methods for Slump of Hydraulic–Cement Concrete*", Annual Book of ASTM Standards, Vol. **04.02**, pp. 95–98, (2004).
20. ASTM C567, "*Standard Test Methods for Determination Density of Structural Lightweight Concrete*", Annual Book of ASTM Standards, Vol. **04.02**, pp. 302–304, (2004).
21. ASTM C955/C995, "*Standard Test Methods for Time of Flow of Fiber-Reinforced Concrete through Inverted Slump Cone* ", Annual Book of ASTM Standards, Section 6, Vol. **04.02**, (2001).
22. BS. 1881, Part 116, "*Method for Determination of Compressive Strength of Concrete Cubes*", British Standards Institution, 1881; pp. 3, (1989).
23. ASTM C496/C496M, "*Standard Test Methods for Splitting Tensile Strength of Cylindrical*", Annual Book of ASTM Standards, Vol. **04.02**, pp. 283–287, (2004).
24. ASTM C78-02, "*Standard Test Method for Flexural Strength of Concrete (Using Simple Beam with*

*Third-Point Loading)*", Annual Book of ASTM Standards, American Society for Testing and Materials, Vol. **04-02**, pp.1-3, (2007).

25. ASTM C469, "*Standard Test Methods for Static Modulus of Elasticity and Poisson's Ratio of Concrete in Compression*", Annual Book of ASTM Standards, Vol. **04.02**, pp. 256–259, (2004).
26. ASTM C597, "*Standard Test Methods for Pulse Velocity through Concrete*", Annual Book of ASTM Standards, **Vol. 04.02**, pp. 310-313, (2004).
27. ASTM C642," *Standard Test Method for Density, Absorption, and Voids in Hardened Concrete*", Annual Book of ASTM Standards Volume: **04.02**, (2013).
28. ACI Committee 211, "*Standard Practice for Selecting Proportion for Structural Lightweight Concrete*", ACI Manual of Concrete Practice, Part 1, pp. 211–218, (2004).
29. Topcu, I.B., and Canbaz, M., Construction and Building Materials, Vol.**21**, Issue 7, pp.1486–1491, (2007).
30. Chen, B., and Liu, J., Cement and Concrete Research, Vol.**35**, No.6, pp.2458-2464, (2005).
31. Libre, N.A., Shekarchi, M., Mahoutian, M., and Soroushian, P., C construction and Building Materials, Vol. **25**, No.12, pp. 2458-2464, (2011).
32. Neville, A. M., "*Properties of Concrete*", Fourth and final edition, Longman Group, Ltd., (2005).
33. Kim, D. J., Park, S.H., Ryu, G.S. and Koh, K.T., Journal of Construction and Building Materials, Vol.**25**, No.5, pp.4144- 4155, (2011).
34. Milovanovic, B., Pecur, I.B., and Gabrijiel, I., Novel Development and Innovation in Cementitious Materials, 12-13, Imperial College, London, United Kingdom, pp.1-5. ( 2011).

# Acidic solution effects on no-fines concrete produced by using recycled concrete as coarse aggregate

*Ikbal* Gorgis[1*], *Whab* Faleh Abd[1] and *Shaker* Al-Mishhadani[1]

[1] Building and Construction Engineering Department, University of Technology, Baghdad, Iraq

**Abstract.** This paper investigates durability of no fine concrete containing demolished concrete as coarse aggregate after crushing to different sizes. Different no fine concrete mixes were considered using Portland cement type I with two types of coarse aggregates, crushed demolished concrete and crushed natural gravel were used with two ratios by weight (1:5 and 1:7) C/Agg. Graded aggregate and single size were used with a maximum size of 20 mm. W /C ratio was kept as 0.4 for all mixes and super-plasticizer was required to keep the same flow and compaction factor value for all mixes. Cube specimens with 150mm were cured and divided to two parts, the first part was exposed to 60 cycles of freezing- thawing; the second part of the sample was immersed in Nitric Acids solution with pH of 3.5 for (7, 28, 90 and 180 days) and then tested for compressive strength. The results indicated that it is possible to produce homogenous and workable mixes by using demolished crushed concrete as coarse aggregate. The compressive strength after cycles of freezing- thawing and immersing in Nitric acid (HNO3) at (7, 28, 90 and 180) days was decreased for samples made with crushed demolished concrete. Also it is found that the performance of concrete mixes containing graded coarse aggregate and 1:5 cement/aggregate ratios was better than other mixes.

## 1 Introduction

No fine concrete is a method of producing lightweight concrete by removing the sand from conventional concrete. This concrete is consisting of only cement, coarse aggregate and water [1]. Very often only grading and single sized coarse aggregate of size passing through sieve (20 mm) retained on sieve (10 mm) is used. Because of some of the advantages it possesses over the conventional concrete, no fines concrete is becoming popular.

Safeguarding of natural resources, recycling of waste materials and the protecting the environment have become of paramount importance. To reduce the problem of cumulative waste, dumping should be used in an environmentally safe method either as economical and beneficial purposes or as raw materials for other products. Lightweight concrete has become commonly used recently because of huge development in the construction techniques and methods of mixing. Moreover, the use of demolished concrete for the purpose of manufacture of lightweight concrete can represent the double benefits of using this waste, thus cleaning the environment, and also by the production of good concrete which can be used [2]. Durability of concrete may be defined as the ability of concrete to resist weathering action, chemical attack and abrasion while maintaining its desired engineering properties.

Wen et al [3] showed the effect of using waste concrete as coarse aggregate in new concrete pavements. Variables included the source of the recycled coarse aggregate (RCA) percent replacement of coarse natural aggregate with waste concrete as aggregate (0% to 45%). Replacement of rubble concrete as aggregate for natural coarse aggregate by up to (45%) by volume had no significant effect on any of the concrete properties. These results indicate that high quality waste concrete as aggregate can be used as a replacement for a portion of the coarse natural aggregates in new Portland cement concrete pavements.

Witchayangkoon et al [4] showed the effect of using waste concrete as a replacement of natural coarse aggregate to produce concrete paving blocks. The experiment was conducted to monitor concrete strength of the specimen samples. Waste concrete as aggregate has been used in different proportions (50%, 70%, and 100%) as replacement of natural coarse aggregate. The mix concrete uses W/C ratio of (0.47) with design slump (100 ± 25) mm. Average concrete strengths at 28 days were (66.5, 45.6 and 35.0) MPa. Thus, this work showed possibility that dumped concrete could be recycled with full (100%) replacement of coarse aggregate in making concrete paving blocks.

Qasrawi et al [5] studied the effect of using rubble concrete as aggregates on the basic properties of normal concrete. Results showed that the use of demolished concrete as recycled aggregates has an adverse effect on the workability of concrete. Such an effect can be easily retained by using plasticizers. The ratio of aggregate replacement used was (25, 50, 75 and 100%) by weight

* Corresponding author: iqbalkorkess@yahoo.com

© The Authors, published by EDP Sciences. This is an open access article distributed under the terms of the Creative Commons Attribution License 4.0 (http://creativecommons.org/licenses/by/4.0/).

of coarse aggregate and water/cement ratio (0.45, 0.55 and 0.65). Concrete strength has been reduced by (5% to 25%) depending on the percent of the normal aggregate replaced by waste concrete as recycled aggregate and the W/C ratio. With respect to the split tensile strength, recycled concrete as aggregate was slightly lower. For example, the use of (100%) replacement reduced the compressive strength of the mixes by (25%) for W/C ratio (0.45), while in the case of tensile strength this value was (14%).

The aims of this research was producing no fine concrete by using demolished concrete as recycled coarse aggregate and comparing the durability of this concrete with no fine concrete produced by using natural aggregate. It was aimed to use local material to maintain the economic side thus, providing an incentive and encouraging using no fine concrete in practical applications as road pavement.

## 2 Experimental works

Materials used to complete the experimental work were:

### 2.1 Cement

Ordinary Portland cement type (I) manufactured in Iraq with a commercial name of (Karasta) was used for no fine concrete mixes throughout the present work. The physical properties and chemical analysis for cement are present in Tables 1 & 2. Results showed that the cement conformed to the Iraqi Standard Specification IQS No.5/1984[6].

### 2.2 Coarse Aggregate

Two types of coarse aggregates were used in this study to produce no fine concrete, they were:-

- Natural Crushed Coarse Aggregate: Natural crushed coarse aggregate of (20mm) single and grading size, brought from (Al-Soddor source) region. The grading and properties of natural crushed coarse aggregate stratify the Iraqi specification IQS No.45/1980[7], as shown in Tables 3&4.

- Demolished Concrete Aggregate: A rubble concrete was used as coarse aggregate. It was brought from waste found in the university laboratories (chosen carefully to contain no fiber or lightweight aggregate) and used as 100% replacement of coarse aggregate. The waste concrete was crushed into smaller sizes manually by a hammer to obtain the required sizes. The grading and properties of demolished concrete as coarse aggregate conform to the Iraqi specification IQS No.45/1980[7], as shown in Table 3 and Table 4. The aggregate grading and single size were similar to that of natural aggregate used in this study.

### 2.3 Water

Ordinary tap water was used for mixing and curing.

### 2.4 Admixture

The superplasticizer used in this work to keep constant W/C ratio for all mixes; commercially known as (Glenium 54), Table 5 shows the typical properties of it. This admixture was identical to the requirement of the ASTM C494-03(8) type F.

**Table 1**. Chemical composition and main compounds of the cement

| Oxide | Content % | Limits of Iraqi Specification No.5/1986[6]. |
|---|---|---|
| CaO | 66.11 | - |
| $SiO_2$ | 21.93 | - |
| $Al_2O_3$ | 4.98 | - |
| $Fe_2O_3$ | 3.10 | - |
| MgO | 2.0 | <5 |
| $SO_3$ | 2.25 | < 2.8 |
| Insoluble Residue I.R | 1.29 | <1.5 |
| Loss on ignition L.O.I | 2.39 | <4.0 |
| Lime Saturation Factor, | 0.93 | 0.66-1.02 |
| Main Compounds(Bogue's equations) | | |
| C3S | 64.57 | - |
| C2S | 14.15 | - |
| C3A | 7.95 | >3.5 |
| C4AF | 9.43 | - |

**Table 2**. Physical properties of cement.

| Physical properties | Test results | Limit of Iraq specification ( No. 5-1984)[6] |
|---|---|---|
| Specific surface area (Blaine method), $m^2/kg$ | 376 | ≥230 |
| Setting time (Vicate apparatus),<br>Initial setting time, hrs: min<br>Final setting time, hrs: min | 1:05<br>4:00 | ≥45 min.<br>≤ 10 hrs |
| Soundness (Autoclave ) method, % | 0.12 | ≤0.8 |

**Table 3**. Grading and physical properties of single size coarse aggregate

| No | Sieve size (mm) | Percentage passing | |
|---|---|---|---|
| | | % Passing | (Limits of Iraqi specification No.45/1980)[7] |
| 1 | 20 | 100 | 85-100 |
| 2 | 10 | 14 | 0-25 |
| 3 | 5 | 1 | 0-5 |
| 4 | 2.36 | - | - |
| Sulfate content ( % ) | | 0.08 | ≤ 0.1% |
| Specific gravity | | 2.69 | |
| Absorption (%) | | 0.55 | |

**Table 4.** Grading and physical properties of coarse aggregate.

| No. | Sieve size (mm) | Percent passing % | | |
|---|---|---|---|---|
| | | Natural aggregate | Demolished aggregate | Limits of Iraqi Specification No. 45/1980(7) |
| 1 | 20 | 100 | 100 | 85-100 |
| 2 | 10 | 30 | 30 | 30-60 |
| 3 | 5 | 0 | 0 | 0-10 |
| 4 | 2.36 | - | - | - |
| Sulfat content (%) | | 0.08 | 0.08 | ≤0.1% |
| Specific gravity | | 2.69 | 2.48 | |
| Absorption (%) | | 0.55 | 4.5 | |

 https://doi.org/10.1051/matecconf/201816202002

**Table 5.** Technical description of high range water reducing admixture

| Technical properties T 25 °C | |
|---|---|
| color | Dark Brown |
| Specific gravity | 1.21@ 25±2°C |
| Chloride content | Nil |
| Flash point | N/A |
| PH | 6.6 |
| viscosity | 128+/-30cps@ 25±2°C |

## 3 Mixing of no fine concrete

Trail mixes were adopted to check the required properties of no fine concrete with suitable amount of w/c ratio and converge on the influence of changing aggregate type on durability of no fine concrete. The details for all mixes used in this search are presented in Table 6. Mixing was done by using (0.1) $m^3$ pan tilting mixer, and then cast in oiled steel molds with 100mm side dimension and compacted by simple rod (25 blows) to prevent any segregation of the mix. All samples were demoulded and immersed in water tank up to age of test.

## 4 Fresh concrete tests

### 4.1 Flow test

This test was conducted for giving the index of the cohesion of the no-fine concrete produced and to show the extent of their tendency to segregation according ASTM C124 -76(9). The increase in the flowing percentage of concrete gives an indication to its higher consistency.

### 4.2 Compacting Factor Test

Compacting factor test is used according to BS 1881: Part 103: 1993(10). The specimen falls from the initial cone and is captured in a second cone. It is then allowed to fall into a test cylinder.

### 4.3 Fresh density

The measurement of concrete density was carried out according to ASTM C138-01(11) in which the average of three cylinders with (100*200) mm masses was considered to determine the fresh concrete density.

## 5 Hardened concrete tests

Durability of no fine concrete with recycled aggregate was required, for which two properties were investigated freezing - thawing as well as to immersing in Nitric acid ($HNO_3$) as a replacement of Acidic rain effect.

### 5.1 Compressive strength

Compressive strength test was executed by British Standard BS 1881 part 116:1983(12) using cubical samples size (100x 100x100) mm and taking the average result of three samples at (7, 28, 90 and 180) days age for each mix.

**Table 6**, Constituents of no-fine concrete mixes.

| No. | Mix description | Mix Details | Mix proportion by volume cement: aggregate | w/c | S.P |
|---|---|---|---|---|---|
| 1 | NSA5 | Natural single size aggregate | 1:5 | 0.4 | - |
| | RCSA5 | Recycled concrete single size aggregate | 1:5 | 0.4 | - |
| 2 | NSA7 | Natural single size aggregate | 1:7 | 0.4 | 0.36 |
| | RCSA7 | Recycled concrete single size aggregate | 1:7 | 0.4 | 0.36 |
| 3 | NGA5 | Natural grading size aggregate | 1:5 | 0.4 | - |
| | RCGA5 | Recycled concrete grading size aggregate | 1:5 | 0.4 | - |
| 4 | NGA7 | Natural grading size aggregate | 1:7 | 0.4 | 0.36 |
| | RCGA7 | Recycled concrete grading size aggregate | 1:7 | 0.4 | 0.36 |

### 5.2 Freezing - Thawing

This test method covers the determination of the resistance of concrete specimens to rapidly repeated cycles of freezing- thawing in the laboratory. Freezing-and-thawing cycle's according to ASTM C666- 03(13) (Procedure A) was used. The freezing-thawing apparatus consists of a suitable chamber in which the cube specimens may be subjected to the specified freezing-and-thawing cycle together with the necessary refrigerating and heating equipment and controls to produce continuously and automatically reproducible cycles within the specified temperature requirements. The specimens were completely surrounded by air during the freezing phase of the cycle and by water during the thawing phase. This test method shall consist of alternately lowering the temperature of the specimens from 4.4°C to -17.8°C and raising it from -17.8°C to 4.4°C within 2 h. Each specimen was exposed to 60 cycles and then checked for compressive strength.

### 5.3 Immersed in nitric acid ($HNO_3$)

Acidic rain is one of the factors which cause deterioration of concrete. Acid rain, or acid deposition, is a broad term that includes any form of precipitation with acidic components, such as sulfuric or nitric acid that fall to the ground from the atmosphere in wet or dry forms. This can include rain, snow, fog, hail or even dust that is acidic. In this study Nitric acid ($HNO_3$) with pH (3.5) was used. All no fine concrete specimens (cube specimens 100mm) were kept in plastic bags sealed until the age of 28 days then put in plastic tank to be immersed in Nitric acid (HNO3) for PH (3.5- 4). The concentration of acid was checked every week till the test day.

# 6 Test results

## 6.1 Fresh properties (Flow and Compacting Factor)

The results of the fresh properties of all no fine concrete mixes (flow and compacting factor) are indicated in Table 7. The results showed that it is possible to have a homogeneous mix using demolished concrete as replacement of coarse aggregate with cement/aggregate (C/A) ratios of 1:5 and 1:7 and w/c ratio of 0.4. Superplasticizer was used for mixes with C/A ratio of 1:7 to keep the flow and impact factor nearly the same for all no fine concrete mixes.

The results indicated that using demolished concrete as coarse aggregate causes a slight reduction in compaction factor and flow test compared to natural crushed aggregate. This is due to harsh surface of recycle aggregate. The aggregate / cement ratio had a significant effect on the flow test and compaction factor, in which increasing the aggregate/ cement ratio increased both test results. The type of aggregate size (single or graded) of the used aggregate had significant effect on compaction factor and flow test of mixture productive. Using single size aggregate led to a reduction in flow and compaction factor more than grading size.

## 6.2 Density Test Results

The results of the fresh and oven dry density at (1 and 28) days for all no fine concrete mixes were listed in Table 8. Where fresh density gives an early expectation of all the hardened properties of concrete, which gives an indication whether the mix will satisfy the requirements of no fines concrete to be used as structural or insulation members. Besides, it gives an early solution to overcome any unsatisfactory property. The fresh density ranges between (1680-1950) kg/m3 for all mixes.

Using demolished concrete as coarse aggregate exhibits a reduction in unit weight (fresh and oven dry densities) compared with natural crushed aggregate no-fine concrete. This reduction of fresh density attributed to the demolished concrete aggregate which require large amounts of water to get suitable workability which was avoided by using S.P. Hence, the porous nature of recycle concrete aggregate produces mixes with lower density compared with natural crushed coarse aggregate. In addition the harsh nature of the demolished concrete particles required higher water to overcome the friction problem between the particles leading to more voids and lower density. The mixes with cement/aggregate C/A ratio of 1:7 present lower fresh density as compared with 1:5.

Table 8 also indicates the result of oven dry density at (1 and 28) days for all no-fine concrete specimens. In no-fine concrete the oven dry density would be more important because the concrete sample contain large amounts of voids and pores that retain water (mixing water + curing water) inside them.

## 6.3 Compressive strength test results

The average compressive strength results at (7, 28, 90 and 180) days age for all no fines concrete mixes are listed in Table 9. The results showed that the compressive strength increased with the age, which is due to the increasing in the hydration process with time. The compressive strength for no fine concrete containing demolished concrete as coarse aggregate was lower than concrete made of natural coarse aggregate. This is due to interfacial transition zone (ITZ) representing the bond between aggregate and paste and is often weaker than either the aggregate or hydrated cement paste. In no fine concrete with natural aggregate, the (ITZ) occurs between aggregate and cement paste only, while in no fine concrete cast with demolished concrete aggregate, there are two different (ITZs) instead of one, an old (ITZs) occurring between the original aggregate and the old mortar and another new (ITZ ) happening between the demolished concrete aggregate and new cement paste. The weakness in rubble concrete aggregate cause a reduction in compressive strength of no fine concrete.

Fig.1 shows the relationship between C/A ratio and compressive strength, which indicates that the compressive strength increases with decreases percentage of coarse A/C ratio. The cement content also has a significant role in compressive strength ,where concrete mix with high cement content exhibit higher compressive strength than concrete mix with low cement content due to increase cement paste around aggregate particles leads to improve cohesive strength and bond between aggregate particles and cement paste. Gradation size of aggregate effect on compressive strength led to increase in compressive strength compared with single aggregate type.

## 6.4 Freezing - Thawing results

The cube specimens were cured in water for 14 days and then they were subjected to 60 cycles of freezing-thawing. Table 10 represented the compressive strength for all no fine concrete specimens before and after being subjected to 60 cycles of freezing- thawing, respectively. The compressive strength decreased with increasing the cycles of freezing- thawing. Due to reduction in rate of hydration during the period of freezing that led to reduction in the compressive strength for no fine concrete specimens, on the other hand freezing process increased micro-cracks in no fine concrete. Specimens cast with demolished concrete as recycled coarse aggregate has more effect on strength as compared with normal coarse aggregate. Due to nature cellular and high absorption for waste concrete as aggregate, therefore. Freezing water in this type of aggregate is more harmful as increases pores volume. Fig. 2-a indicates the relationship between cement/aggregate ratio and freezing-thawing. The results indicate that the compressive strength for no fine concrete specimens that subject to freezing- thawing decrease with increase percentage of coarse aggregate/cement ratio.

**Table 7**. Fresh properties of all no fine concrete mixes

| Set No. | Mix | C/A | S.P.by cement wt. (%) | Flow (%) | Compacting Factor |
|---|---|---|---|---|---|
| 1 | NSA5 | 1:5 | 0 | 61 | 0.88 |
| | RCSA5 | 1:5 | 0 | 60 | 0.85 |
| 2 | NSA7 | 1:7 | 0.35 | 64 | 0.87 |
| | RCSA7 | 1:7 | 0.35 | 62 | 0.86 |
| 3 | NGA5 | 1:5 | 0 | 63 | 0.89 |
| | RCGA5 | 1:5 | 0 | 61 | 0.87 |
| 4 | NGA7 | 1:7 | 0.35 | 65 | 0.90 |
| | RCGA7 | 1:7 | 0.35 | 64 | 0.88 |

**Table 8**. Fresh density and oven dry density of all no fine concrete mixes.

| Set No. | Mix | Fresh density kg/m$^3$ | Oven dry density at 1days kg/m$^3$ | Oven dry Density at 28days kg/m$^3$ |
|---|---|---|---|---|
| 1 | NSA5 | 1850 | 1740 | 1790 |
| | RCSA5 | 1735 | 1650 | 1705 |
| 2 | NSA7 | 1780 | 1685 | 1730 |
| | RCSA7 | 1680 | 1610 | 1660 |
| 3 | NGA5 | 1950 | 1800 | 1860 |
| | RCGA5 | 1800 | 1710 | 1768 |
| 4 | NGA7 | 1835 | 1720 | 1775 |
| | RCGA7 | 1735 | 1680 | 1710 |

**Table 9**. Average compressive strength result for no fine concrete mixes.

| Set No. | Mix | Compressive strength MPa | | | |
|---|---|---|---|---|---|
| | | 7 days | 28 days | 90 days | 180 days |
| 1 | NSA5 | 6.80 | 8.10 | 9.82 | 13.40 |
| | RCSA5 | 4.80 | 5.70 | 7.10 | 9.70 |
| 2 | NSA7 | 6.65 | 7.85 | 9.20 | 13.00 |
| | RCSA7 | 4.55 | 5.50 | 6.60 | 9.40 |
| 3 | NGA5 | 7.30 | 8.70 | 11.10 | 14.00 |
| | RCGA5 | 5.10 | 6.30 | 8.10 | 10.40 |
| | RCGA7 | 4.80 | 6.10 | 7.8 | 9.80 |

**Table 10**. The compressive strength for all no fine concrete specimens subjected to 60 cycles of freezing - thawing.

| Set No. | Mix | Compressive strength MPa | | Percentage of reduction |
|---|---|---|---|---|
| | | No cycle (28 days) | 60 cycles | |
| 1 | NSA5 | 8.10 | 7.6 | 6.2 |
| | RCSA5 | 5.70 | 4.5 | 21.1 |
| 2 | NSA7 | 7.85 | 6.8 | 13.4 |
| | RCSA7 | 5.50 | 4 | 27.3 |
| 3 | NGA5 | 8.70 | 7.92 | 8.9 |
| | RCGA5 | 6.30 | 4.8 | 23.8 |
| 4 | NGA7 | 8.60 | 7.5 | 12.8 |
| | RCGA7 | 6.10 | 4.5 | 26.2 |

Fig. 2-b indicates the effect of gradation type of aggregate (single size and graded) on the freezing - thawing for no fine concrete. Using graded size led to increase in the compressive strength for specimens that are subjected to freezing-thawing as compared with using single size of coarse aggregate particles, due to reduction in the volume and number of voids and pores.

### 6.5 Immersed in nitric acid [compressive strength of no fine concrete]

Results for all no fine concrete specimens immersed in Nitric acid ($HNO_3$) are presented in Table 11. The results of compressive strength for mixes show slight increase with age due to hydration of cement. Although the results showed a reduction in strength for all specimens at age of (7, 28, 90 and 180) days as compared with specimens cured in water, see Fig. 3. This reduction was due to transformation of $Ca(OH)_2$ to a highly soluble nitrate salt resulting in greater deterioration as shown by equation ($2HNO_3 + Ca(OH)_2 \rightarrow Ca(NO_3)_2 + 2H_2O$) (Calcium Nitrate). With the help of water, these soluble salts may easily be transported to the outer parts of cement. In this situation, continuous reactions increase the porosity of cement paste and increased pore volume speed up the rate of reaction.

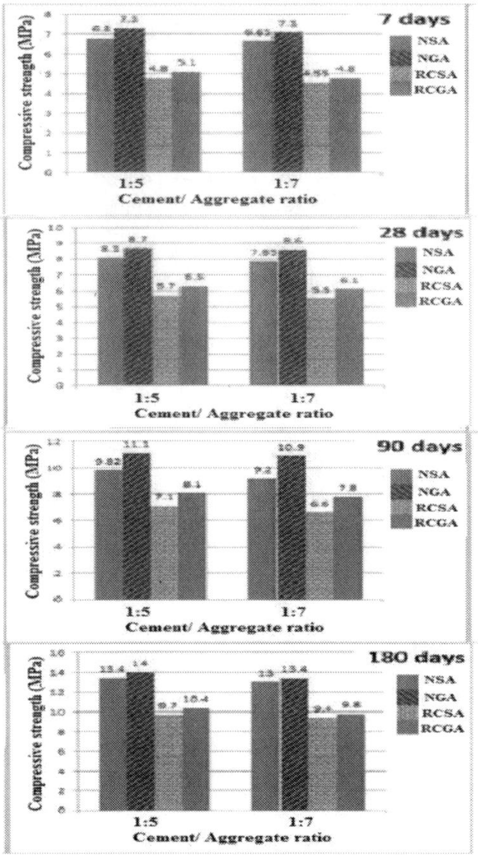


**Fig. 1** :Compressive strength and C/A ratio relationships at (7, 28, 90 and 180) days age.

The Ordinary Portland cement is a highly alkaline material (pH above 12.5) and thus, as pH decreases the equilibrium of the cement matrix is disturbed and hydration compounds are easily changed by hydrolytic decomposition. From results of different test ages indicating slightly increased in compressive strength. Results of no fine concrete with demolished concrete as recycled coarse aggregate were lower than natural coarse aggregate in all ages due to the weakness in properties for this type of aggregate. On other hand, rubble concrete aggregate be surrounded by old cement leading to more interaction causing reduction in the hydration rate.

Reduction in compression strength results increased as the percentage of cement ratio /coarse aggregate increased. This is mainly due to the low mortar content and the high porosity. Using graded size aggregate led to increase in compressive strength as compared with using single aggregate size. Due to increase in the bond strength between cement paste and aggregate particles.

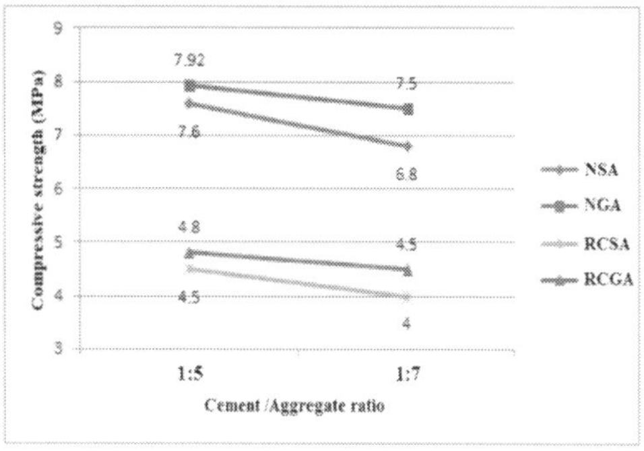


a- Compressive strength and C/Aggregate ratio relation.

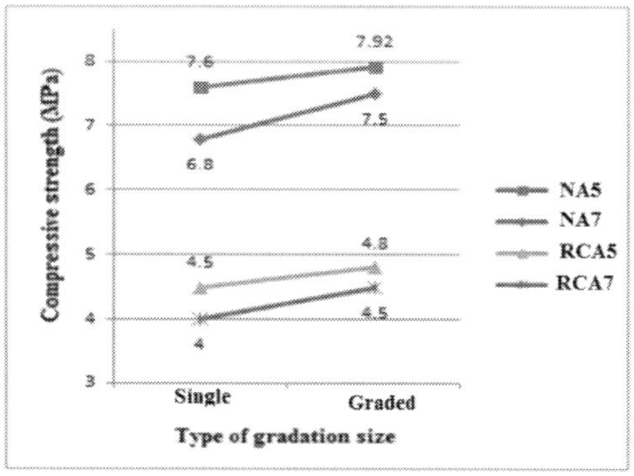


b- Compressive strength and type of gradation.

**Fig. 2** Relationship between compressive strength and C/Agg and type of gradation after 60 cycles of freezing - thawing.

**Table 11**. Compressive strength of all no fine concrete mixes specimens immersed in nitric acid ($HNO_3$).

| Set No. | Mix Designation | Compressive Strength (MPa) | | | |
|---|---|---|---|---|---|
| | | 7 days | 28 days | 90 days | 180 days |
| 1 | NSA5 | 4.1 | 4.6 | 6.3 | ٧.2 |
| | RCSA5 | 2.2 | 2.8 | 4.15 | 5.4 |
| 2 | NSA7 | 3.9 | 4.15 | 5.1 | 6.6 |
| | RCSA7 | 2 | 2.6 | 3.3 | 3.9 |
| 3 | NGA5 | 6.1 | 6.8 | 7.5 | ٨.1 |
| | RCGA5 | 3.3 | 4 | 5.1 | 5.8 |
| 4 | NGA7 | 4.4 | 6.2 | 6.9 | 7.3 |
| | RCGA7 | 3 | 3.85 | 4.4 | 5.2 |

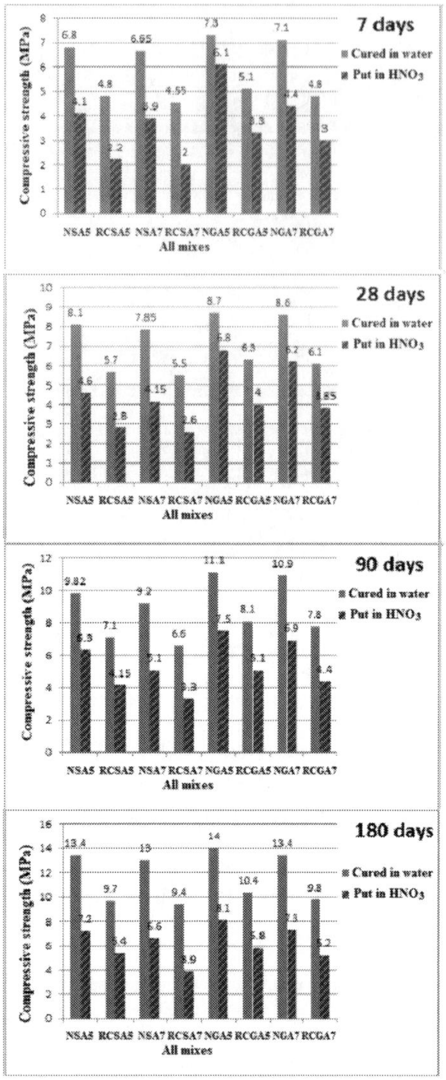


**Fig. 3** The change in the compressive strength for specimens cured in water and these kept in acid solution (HNO3) for all ages.

## 7 Conclusions

Based on the results of this research, the following conclusions may be drawn:

1- The workability of no fine concrete produced by demolished concrete as coarse aggregate is less than the workability of no fine produced by natural crushed aggregate. The flow test and compaction factor of no fine concrete produced with C/A ratio of 1:5 was less than no fine concrete produced with 1:7. The results of flow test and compaction factor of no-fine concrete produced with single aggregate type is less than that of no-fine concrete produced by graded aggregate by about 2%.

2- The hardened properties compressive strength and oven dry density for no-fine concrete produced by demolished concrete as coarse aggregate were less than that of no-fine produced by natural crushed aggregate. The hardened properties for no-fine concrete produced with C/A 1:5 were higher than no fine concrete produced with C/A 1:7. The harden properties for no fine concrete produced by single aggregate size was less than no fine produced by graded aggregate.

3- Using graded aggregate produce higher compressive results for all tests than specimens with single graded aggregate this was due to less voids in the mix.

4- Results for both freezing-thawing cycles and immersing in nitric acids indicated that the compressive strength increases with decreasing the percentage of cement ratio /coarse aggregate. This is mainly due to the high mortar content and the low porosity.

5- The compressive strength for specimens immersed in acidulous solution ($HNO_3$) shows a slight increase with age up to 180 days. No fine concrete produced with 1:5 C/A ratio exhibits higher compressive strength, than that of no fine concrete with 1:7 C/A ratio by about (5-38.7) %.

## References

1. A. M.Neville A. M., "Properties of Concrete", fifth Edition, First published 2010.
2. Abadjieva & Sephiri ,Dept. of Civil Eng., University of Botswana, Private Bag 0061, Gaborone, Botswana,pp.1-6.
3. H. Wen, I. D. McLean and K. Willoughby, 2015 TRB Annual Meeting and Publication in the Transportation Research Record: J. T. R. B.
4. B. Witchayangkoon , M. Pattanasuwan, P. Nakarin and P. Jampatong. P, ATEAS, http://TuEngr.com/ATEAS.
5. H. Qasrawi, I. Marie and H. Tantawi. Conference: Proceedings of the 5th Jordanian International Civil Engineering Conference, At Amman, Jordan, January 2012, pp.1-7.
6. IQS. No.5-1984, "Portland cement", Iraqi Standard Organization, COSQC.
7. IQS. No. 45-1980, Iraqi Standard Organization, COSQC.
8. ASTM C494-04, Annual Book of ASTM Standards, ASTM, Vol. **04**-02, pp. 1-8, 2004.
9. ASTM C124–76, Annual Book of ASTM Standards, ASTM, Vol. 04-02, pp. 1-8, 2004.
10. BS 1881: Part 103: 1993, British Standards Institution, 1881; pp. 3.

11. ASTM C138–01, Annual Book of ASTM Standards, pp. 4, 2001.

12. BS 1881: Part 116: 1989, British Standards Institution, 1881; pp. 3.

13. ASTM C666-03, Annual Book of ASTM Standards, ASTM, Vol. 04-02, pp. 1-8, 2004.

14. K. Josiphiah, D. Yusuf. And T. Kassar.T, Civil and Environmental Research, ISSN 2224-5790 (Paper) ISSN 2225-0514 (Online), Vol.7, No.12, 2015.

# Fresh and some mechanical properties of sifcon containing silica fume

*Shakir* Salih[1], *Qais* Frayyeh[1,*], and *Manolia* Ali[2]

[1] Building and Construction Engineering. Department, University of Technology, Baghdad, Iraq
[2] Material Engineering Department, Collage of Engineering, Al-Mustansiriya University, Baghdad, Iraq

**Abstract.** Slurry infiltrated fiber concrete (SIFCON) is one of the recently developed construction material. SIFCON could be considered as a special type of fiber concrete with high fiber content. The matrix of SIFCON consists of flowing cement mortar or cement slurry. SIFCON has a very good potential for application in area where resistance to impact and high ductility are needed especially in designing the seismic retrofit, in the structures under impact and explosive effects and repair of structural reinforced concrete element. The main objective of this paper is to determine the effect of steel fiber content and silica fume (SF) cement replacement on the mechanical properties of SIFCON concrete. The percentage of SF replacement was 10% by weight of cement in SIFCON slurry, and three different volume fractions of hooked ended steel fiber (6, 8.5, and 11) % were used. The tested properties of SIFCON were compressive strength and splitting tensile strength which were carried out on standard size of cubes and cylinders respectively at the age of 7and 28 days. It was observed that the mechanical properties of SIFCON were affected in a positive manner by using silica fume as a partial replacement of cement and by adding steel fiber reinforcement in different percentages. The compressive and splitting tensile strength up to 83.7 MPa and 17.3MPa, respectively were obtained at the age of 28 days.

## 1 Introduction

Slurry Infiltrated Fibrous Concrete (SIFCON) is relatively a new high performance and advanced material [1] that differs from ordinary Fiber Reinforced Concrete (FRC) in terms of fabrication and composition [2].Fiber content in FRC generally varies from (1 to 3) % by volume, while the fiber content of SIFCON can be anywhere from (4 to 20)% even though the current practice ranges from (4 to 12%, and when using hooked ended steel fiber with length more than 30 mm the fiber content range between (4 to10)%. SIFCON matrix has no coarse aggregates but a high cementitious content. It may contain fine or coarse sand and mineral admixture such as silica fume, fly ash and latex emulsions [3]. So, the matrix of SIFCON is either cement paste or flowing cement mortar as opposed to regular concrete used in FRC. Also, the production of SIFCON far different from FRC which is produced by adding the fiber to the wet or dry concrete mix. While SIFCON is produced by first sprinkling fibers into a mold until it is completely filled, then the fiber network of SIFCON is infiltrated by cement slurry [2, 4, 5].

SIFCON is a unique construction material exhibiting new behavioral phenomenon, that of Fiber lock, the fibers are subjected to mechanical and frictional interlock, in addition to the bond with the matrix, which is believed to be responsible for its outstanding stress-strain properties. SIFCON also possesses high strength as well as large ductility and far excellent potential for structural applications when accidental loads are encountered during services [1].

Many researchers study the mechanical behavior of SIFCON produced with fly ash or ground granulated blast furnace slag replacement. Also, there is little information about SIFCON characteristic produced using silica fume as additional ingredient to cement paste or slurry

Sashidhar [6] examined the compressive strength characteristics of SIFCON produced with low tensile strength of steel fiber, ultimate tensile strength of 390 MPa, using mineral admixtures like metakaoline and silica fume in the cement mortar. Fibers with 8 to 12% volume fraction and with aspect ratio of 40 to 60 were used. The study reported an 88% increase in compressive strength for SIFCON when compared to conventional concrete. It is further reported that the use of metakaoline and silica fume in the matrix cause increase of (11 to 14) % in the compressive strength of SIFCON.

Kim and Choi[7] studied the mechanical properties of SIFCON using hooked –end steel fiber with volume friction varying from 4% to 10%. The addition of silica

*e-mail: jwd_ks@yahoo.com

© The Authors, published by EDP Sciences. This is an open access article distributed under the terms of the Creative Commons Attribution License 4.0 (http://creativecommons.org/licenses/by/4.0/).

fume was 10% by weight of cement, and superplastisizer of 0.5% to improve the workability of cement slurry .The water /cement ratio was kept constant as 0.4.The test results show that the compressive strength of SIFCON was increased to about (1.59 to 2.68) times when compared with the cement paste, while the increase in tensile strength was about (2.51 to 8.77) times than that of the cement paste. They also observed that the ductility and toughness of SIFCON were increased with increasing the volume fraction of steel fiber.

The flowable slurry with rich cement is the binder in SIFCON products. Using high amount of cement not only affects the production cost, but also has negative effects on heat of hydration and may cause shrinkage problems, therefore, the incorporation of mineral admixtures as a partial replacement of cement seems to be a suitable solution to overcome these problems, furthermore, the use of mineral admixtures may have a positive effect on mechanical properties and durability of SIFCON products.

There is very limited information about the behavior of SIFCON containing silica fume as a replacement of cement .Therefore, the main aim of this research is to evaluate the effect of silica fume as a partial replacement (10% by weight of cement)and the effect of volume fraction of steel fiber ranging from (6 to 11%) on fresh and mechanical properties of SIFCON.

# 2 Experimental Work

## 2.1 Materials

### *2.1.1 Cement*

Ordinary Portland cement (type I) which is commercially known as (Krasta) was used in this research, its physical and chemical properties conformed to the Iraqi specification NO.5/1984[8].

### *2.1.2 Fine aggregate*

River sand passing through 1.18 mm sieves, to ensure complete infiltration through steel fiber, was used as a fine aggregate. It conformed to the Iraqi specification NO.45/1984[9], zone (2). Its specific gravity, sulphate content and absorption are 2.6, 0.34%, and 2%, respectively.

### *2.1.3 Water*

Ordinary tap water was used in this work for all mixing and curing of SIFCON specimens

### *2.1.4 Silica fume (SF)*

Densified silica fume, which is commercially known as MEYCO/MS610 from BASF Company with fineness of (21000 m$^2$/kg), was used in this work as a partial replacement of cement. It conformed to the ASTM C1240-05[10]

### *2.1.5 High range water reducer (HRWR)*

For the production of SIFCON high range water reducing admixture was needed to improve the workability (flowability) of the slurry, which should be liquefied enough to flow through the dense fiber bed without leaving honeycombs. In this work (Glenium 54) from BASF Company was used. It conformed to the ASTM C494 type F [11].

### *2.1.6 Steel fiber*

Hooked end steel fiber with diameter of 0.7mm and length of 35 mm, giving aspect ratio of 50 and tensile strength of 1100 MPa was used in this investigation .Fibers were randomly orientated in the matrix.

## 2.2 Mix proportion

The first step in this work was preparing many trial slurry mixes to find a suitable mix that has the desirable fresh properties satisfying the requirements of SIFCON slurry with respect to viscosity, fluidity and filling ability without segregation or pore pockets in the fiber network. While the second step was selected the minimum and maximum volume fraction of steel fiber to be used after the trial mixes of SIFCON. Two types of SIFCON matrix ($M_1$ and $M_2$) were prepared with the same mix proportion (1:1) (cement: fine aggregate) by weight depending on literature review findings which show that the proportion of sand to cement is equal to 1 in most cases, and the addition of sand beyond cement to sand ratio of (1:1) decreases the required compressive strength. The cement content was 885 kg/m$^3$ and (water /binder) ratio was kept constant at 0.3 for the two mixes ($M_1$ and $M_2$). The first SIFCON matrix ($M_1$), control mix, was obtained by using a superplastisizer which is (Glenium 54). While the second type of SIFCON matrix ($M_2$) was prepared from the control mix by a partial replacement by weight of cement with 10% of silica fume (SF). The details of the two SIFCON mixes are shown in Table (1)

**Table 1.** Mix proportion used for SIFCON slurries

| Mix Symbol | w/b* ratio | Cement kg/m$^3$ | Sand kg/m$^3$ | SF (10% rep.) kg/m$^3$ | Water L/m$^3$ | SP** % by wt. of cement |
|---|---|---|---|---|---|---|
| $M_1$ | 0.3 | 885 | 885 | 0.0 | 265.5 | 1.2 |
| $M_2$ | 0.3 | 796.5 | 885 | 88.5 | 265.5 | 2.4 |

*w/b = water /(cement + silica fume)
**SP dosage was regulate to obtain similar flow diameter and viscosity (without bleeding and segregation)

Each mix was prepared with three volume fractions (6%, 8.5% and 11%) of steel fiber. The volume fraction of 6% was a minimum practical limit that could fill the mold without using vibration while 11% volume fraction was the maximum practical limit that fills the mold with intense vibration to obtain complete penetration of the slurry into the fiber network. Volume fraction of 8.5% was selected as an intermediate value which required only light vibration

### 2.3 Test of fresh SIFCON

Testing of SIFCON in its fresh state is of serious importance for the production of SIFCON. Its matrix must be liquid enough and have sufficient fineness to flow through the dense fiber bed. The mini slump flow and V-funnel test, according to EFNARC [12], were used to determine the flowability, filling ability and viscosity of the slurry. The mini flow test represents the flowability, segregation resistance and uniformity of the slurry. Base diameter, top diameter, and height of mini slump flow test apparatus used are 100, 70 and 60 mm, respectively. A value between (240-260) mm spread diameters is required for SIFCON mortar. The other test to assess the viscosity of the slurry was V-funnel test, a flow time between (7-11) seconds is considered appropriate [12]. The details of these two tests could be found in many references [12, 13, 14]. Results of the fresh properties of SIFCON mixes are given in Table (2) and shown in Fig. (1)

**Table 2.** Fresh properties of SIFCON mix

| Mix Symbol | Mini slump flow (mm) | V-funnel (s) | SP. % by wt. of cement |
|---|---|---|---|
| $M_1$ | 258 | 8 | 1.2 |
| $M_2$ | 257 | 9.5 | 2.4 |

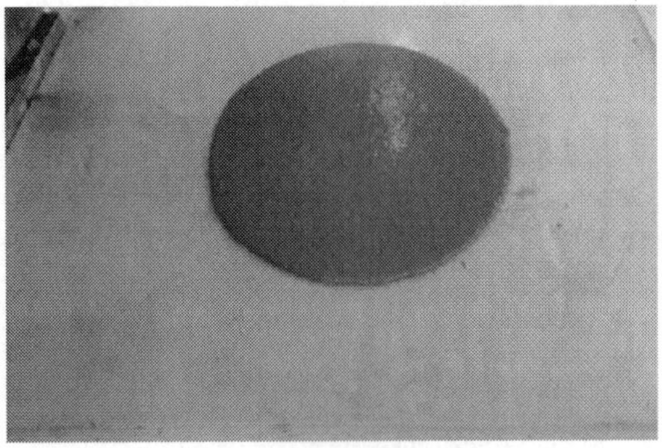

**Fig.1**. mini slump flow test for SIFCON slurry

### 2.4 Preparing, casting, and curing of the test specimens

The first step in preparing test specimens is placing steel fibers into the molds, and then they were infiltrated with the cement slurry which must be flowable and liquid enough to infiltrate thoroughly the dense matrix in the fiber -filled form.

In this study, casting of the two- layer technique was used for incorporating fiber in the mortar slurry ,which proved effective during the casting of SIFCON specimens and found to be easier in actual practice than the single layer technique especially in case of high content of steel fiber

This technique involved initial placing of the fibers in the mold only up to the half depth, followed by infiltration of the slurry up to this level. Then the contents in the mold were vibrated. The process was repeated until the entire mold was filled with the required fiber content. No vibration was required with volume faction(Vf)of 6%,while alight vibration for(6-10)sec by table vibrator was applied in case of Vf of 8.5% and intense vibration for(15-20)second in case of the maximum Vf (11%) to ensure that SIFCON slurry will penetrate completely into the fiber pack , as shown in Fig.(2).

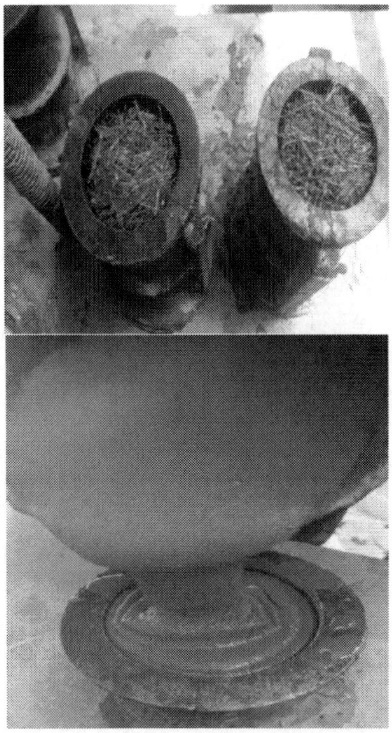

**Fig.2.** preparing and casting of SIFCON specimens

The weight of steel fiber to be used in each mold depends on the dimensions of the mold, density and the volume fraction of the steel fiber used. After a bout 24 hours from casting time, the specimens were demolded, marked and immersed in tap water tank to be cured for 7 and 28 days.

### 2.5 Testing of hardened SIFCON specimens

#### *2.5.1 Compressive strength*

This test was conducted in accordance to B.S.1881: Part 116:1989[15] by using 100 mm cube. The compressive strength value of each mix represents the average of three specimens. They were tested at 7 and 28 days of water curing.

#### *2.5.2 Splitting tensile strength*

This test was carried out according to ASTMC496 04[16] by using (100*200 mm) concrete cylinder. The specimens were tested at 7 and 28 days of water curing. The average of three specimens was used for each variable in this test.

## 3 Results and Discussion

### 3.1 Fresh properties of SIFCON

Table (2) shows the results of mini slump flow and V-funnel test that satisfied the criteria of appropriate flowability , adequate viscosity and filling ability for the two slurry mixes ($M_1$ and $M_2$). It can be seen from the results of the reference mixture ($M_1$) that the addition of HRWR is very necessary to obtain a mixture with low w/c ratio and in the same time satisfying the homogeneity and the other mentioned requirement.

The method followed to find the superplastisizer(SP) amount needed for $M_1$ is by checking first the slump flow with different amount of SP until reaching the aimed value of (240-260)mm ,then checking the two other requirement by V-funnel test. This is because the slump flow test is the simplest and the fastest one.

The great or positive effect of SP on fluidity and cohesiveness of cement slurry was because the polycarboxylate superplasticizer absorbs onto and disperses the cement particles; thus improving workability and reducing friction, and if the particles tend to flocculate then the multi-ion polymer interferes. These particles remain frictionless, so maintaining the dispersing effect for longer time after mixing [13].

The resulted feature of using silica fume (SF), as cement partial replacement, to reference mix ($M_1$) is shown in Table (2) as mix ($M_2$). It was noticed that this replacement caused a decrease of workability making the mix cohesive and sticky and to obtain similar viscosity and fluidity, the SP dosage of mixture has to be increased. Insufficient fluidity or viscosity was observed for lower SP dosage than the values given in Table (2). On the other hand, greater dosage caused bleeding and segregation behavior of mixtures. The lower workability and stickiness of the mixtures by SF replacement is a result of its very large surface area. The superplasticizer when mixed in concrete are absorbed by cement and silica fume particles, make them mutually repulsive as a result of anionic nature of superplasticizer, which causes the cement and silica fume particles to become negatively charged [13,17]

It can be seen from the results that SF incorporation reduces the spread diameter of mini slump and increases the V-funnel time. In other words SF replacement causes relatively high viscosity. This can be attributed to increased cohesiveness with SF. Sonebi *et al.* [18] mentioned that the increased dosage of SF reduced the spread diameter of the mini slump and volume of penetrated slurry into the fiber mass.

### 3.2 Hardened properties of SIFCON

#### *3.2.1 Compressive strength*

The compressive strength test results for SIFCON cubes (100) mm cured at 7 and 28 days for two different types of mortar ($M_1$ and $M_2$) with different volume fraction of steel fibers (6%,8.5% and 11%) are presented in Table(3) and plotted in Fig. (3), Fig. (4) , Fig. (5) and Fig. (6).The results show that the compressive strength increases with the increase of steel fiber volume fraction for both the two groups (Group 1 and Group 2).

**Table 3.** Compressive strength of SIFCON mixes

| Group No. | Mix symbol | Compressive strength (MPa) | |
|---|---|---|---|
| | | 7 days | 28 days |
| *1* | $M_1F$-6 | 63 | 70.7 |
| | $M_1F$-8.5 | 66.3 | 74.8 |
| | $M_1F$-11 | 70.6 | 76.5 |
| *2* | $M_2F$-6 | 69.5 | 78.5 |
| | $M_2F$-8.5 | 71.4 | 81 |
| | $M_2F$-11 | 74.7 | 83.7 |

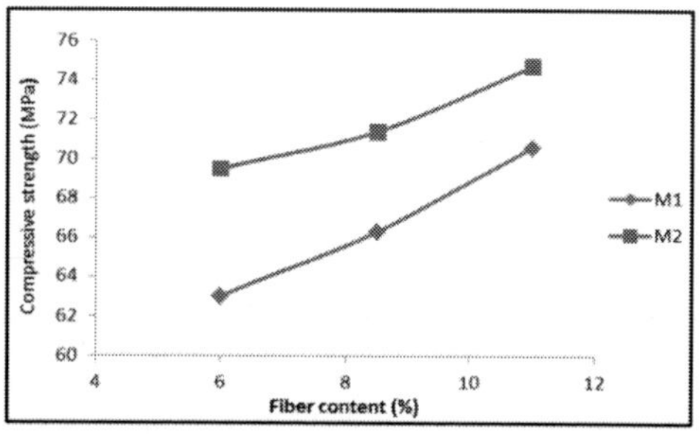


**Fig.3**.The effect of steel fiber content on the compressive strength for all SIFCON mixes at age of 7 days

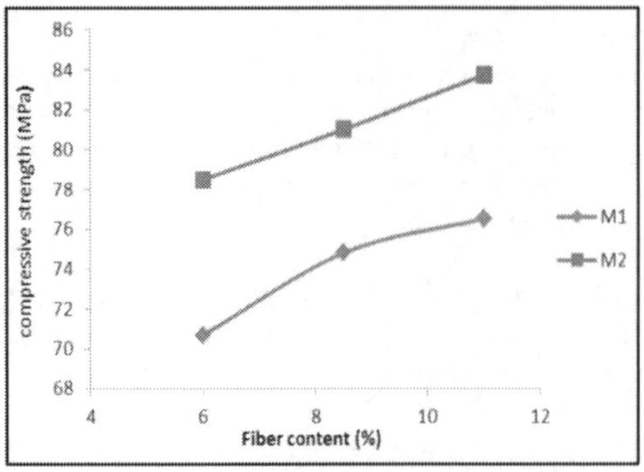


**Fig.4.** The effect of steel fiber content on the compressive strength for all SIFCON mixes at age of 28 days

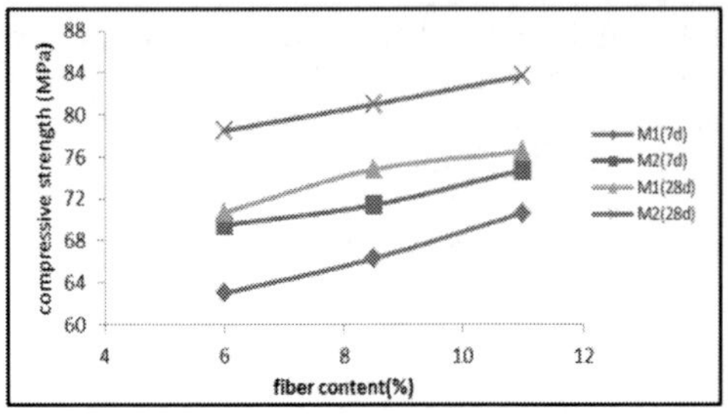


**Fig.5.** The effect of steel fiber content on the compressive strength for all SIFCON mixes at age of (7 and 28) days

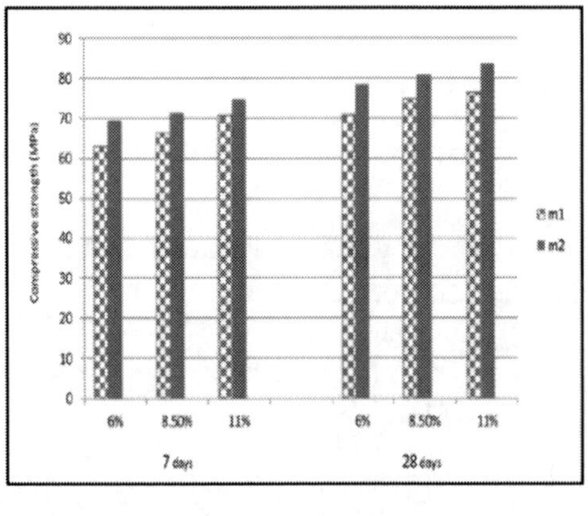


**Fig.6.** Compressive strength development with time for SIFCON mixes with different fiber content

For example, in Group (1),the increase in fiber content from 6% to (8.5 % and 11%) leads to increase in compressive strength of SIFCON mixes ($M_1$F-8.5 and $M_1$F-11) to (5.2 and 12%) at age of 7 days , and (5.8 and 8.2%) at age of 28 days, respectively compared with $M_1$F-6.This increase in compressive strength is due to the higher gained bond between fibers/ matrix interfaces by increasing fiber volume fraction up to 11%.These findings are agreed with other researchers [19] and [20]. The results also show that all SIFCON mixes in Group 2 have a slight increase in the compressive strength due to the incorporation of silica fume (as a partial replacement by weight of cement) .which causes an increase by about (10.3%, 7.7%, 5.8%) for 7 days as well as (11%, 8.2%, 9.4%) at the age of 28 days for $M_2$F-6, $M_2$F-8.5 and $M_2$F-11mixes in Group2 compared with their counterpart mixes in Group 1,as shown in Fig.(5) and Fig.(6).

The effect of SF on improving the compressive strength of SIFCON mixes is due to the high surface area of SF which fills in the spaces between cement grains. This phenomenon is referred to as (particle packing). Also SF accelerates the hydration of cement by providing nucleation sites where the products of cement hydration can more readily precipitate from solution result in more densified microstructure and also improve the cohesive force between fiber and matrix. SF is a very reactive pozzolana material which reacts with calcium hydroxide to form additional binder material called calcium silicate hydrate which is similar to the calcium silicate hydrate from Portland cement [13]

### *3.2.2 Splitting tensile strength*

The tensile strength investigations of SIFCON mixes with two different types of mortar ($M_1$ and $M_2$) and different volume fractions of steel fiber (6,8.5 and 11%) were carried out at the ages of 7, and 28 days. The results are presented in Table (4) and plotted in Fig. (7), Fig. (8) and Fig. (9).

**Table 4.** Splitting tensile strength results for SIFCON mixes

| Group No. | Mix symbol | Splitting tensile strength (MPa) | |
|---|---|---|---|
| | | 7 days | 28 days |
| 1 | $M_1$F-6 | 9.15 | 11.2 |
| | $M_1$F-8.5 | 11.3 | 12.8 |
| | $M_1$F-11 | 13.2 | 14.6 |
| 2 | $M_2$F-6 | 11.1 | 12.9 |
| | $M_2$F-8.5 | 14.4 | 16.2 |
| | $M_2$F-11 | 16.1 | 17.3 |

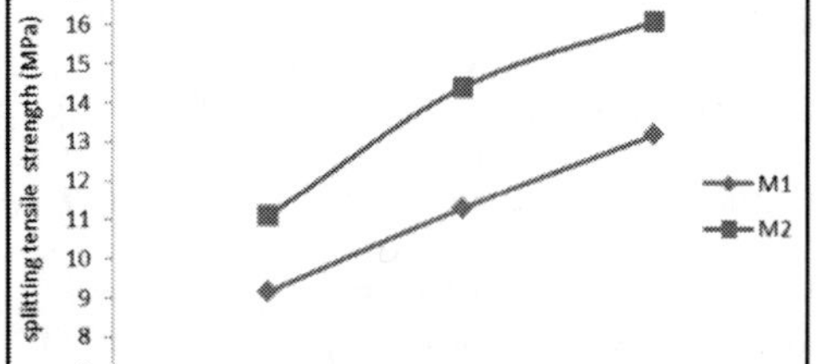


**Fig.7.** The effect of steel fiber content on the splitting tensile strength for all SIFCON mixes at age of 7 days

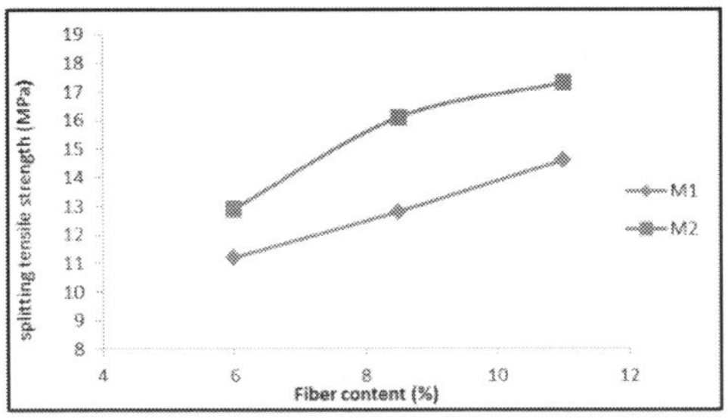


**Fig.8**.The effect of steel fiber content on the splitting tensile strength for all SIFCON mixes at age of 28 days

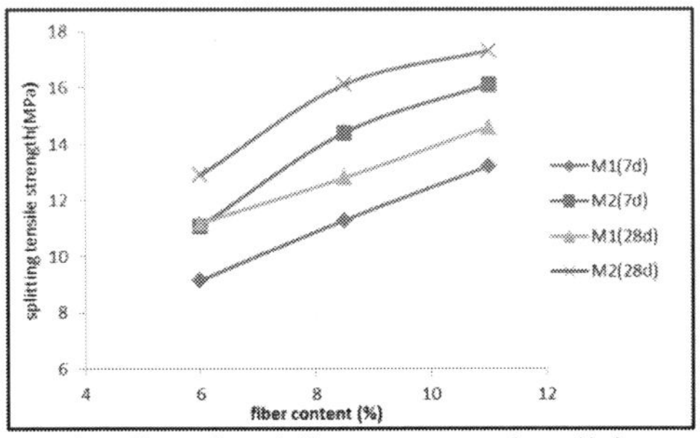


**Fig.9**.The effect of steel fiber content on the splitting tensile strength for all SIFCON mixes at age of (7 and 28 days)

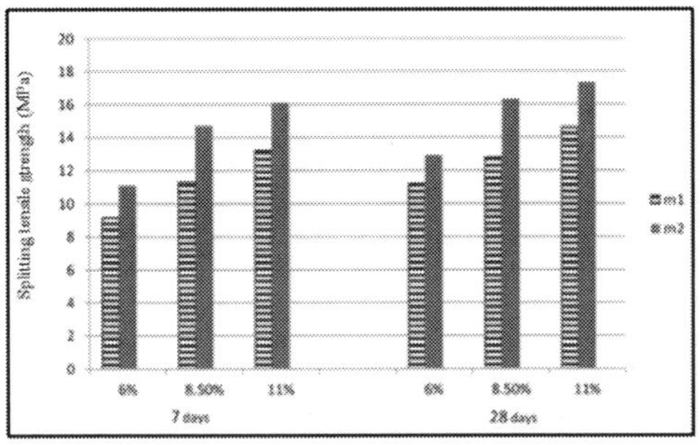


**Fig.10.** Splitting tensile strength development with time for SIFCON mixes

The results show that splitting tensile strength increases with the increase in volume friction of steel fiber for both the Groups (Group 1 and Group 2).Also the effect of steel fiber content on increasing the tensile strength was more than its effect on compressive strength

As shown in Group (1),the increase in fiber content from 6% to (8.5 % and 11%) increases the splitting tensile strength of SIFCON mixes ($M_1$F-8.5 and $M_1$F-11) to (23.5 and 44.3%) for the age of 7 days , and (14.3 and 30.4%) for the age of 28 days ,respectively compared with $M_1$F-6.And in Group (2) the increase in splitting tensile strength of SIFCON mixes ($M_2$F-8.5 and $M_2$F-11) to (29.7 and 45%) at age of 7 days, and (25.5 % and 34.1%) at age of 28 days, respectively compared with $M_2$F-6.The reason behind this behavior is due to arresting the macrocracks as well as microcracks by effect of steel fiber, also using hooked end SF leads to an improved bond which in turn leads to an improved mechanical properties.

The results show that Group 2 SIFCON mixes have an improved tensile strength due to the incorporation of silica fume (SF) that results in an increase by about (21.3%, 27.4%, and 22%) for 7 days as well as (15.2%, 26.5 %, and 18.5%) at the age of 28 days for ($M_2$F-6, $M_2$F-8.5, and $M_2$F-11) mixes in Group2 compared with SIFCON mixes in Group1,as shown in Fig(10). It was mentioned that the filler effect, pozzolanic effect and the enormous surface area enhance the bond between fiber and matrix interface in addition to the growth restricting for microcracks, this explains the improved mechanical properties in Group 2.

## 4 Conclusions

1-The test results show that SF replacement with cement caused increase in the viscosity of slurry (decreased the spread diameter of mini slump and increased the V-funnel time),which can be controlled by using proper dosage of SP. Therefore, SIFCON slurry having proper filling ability and flow ability properties can be produced with SF replacement

2-The mechanical properties of SIFCON (compressive strength and splitting tensile strength) were increased with increase in volume fraction of steel fibers. The increase in volume fraction of steel fibers from (6% to 11%) will increase the compressive and splitting tensile strength of SIFCON mixes ($M_1$F-11) to 12% and 44.3%, respectively at the age of 7 days.

3- The use of silica fume (as a partial replacement by weight of cement) in the matrix resulted in an increase in the mechanical properties of SIFCON. The increases in compressive and splitting tensile strength were about 9.4% and 18.5%, respectively at age of 28 days for $M_2$F-11 mix.

## References

[1] D. Shanthini , E.K. Mohanraj, International Journal of Science and Engineering Research (IJOSER), **3**, Issue 10, (.2015).

[2] D.R Lankard, Concrete International,.**6**, No.12, pp.44-47,(1984).

[3] R. Giridhar ,P. Rama , M. Rao, International Journal for Technological Research in Engineering , **2**, Issue 7, (2015).

[4] D. R. Lankard. , J.K.Newell, SP-81, Fiber Reinforce Concrete-International Symposium, American Concrete Institute, Detroit, Mich, pp. 287-306,(1984).

[5] M. Sonebi, L. Svermova , J.M. Bartos., ACI material journal ,pp.136-145,( 2004).
[6] C. Sashidhar, Ph.D Thesis, Jawaharlal Nehru Technology University, Hyderabad, (2005).

[7] S. kim , J.choi, journal of the Korea concrete institute, **18**, No.5, pp703-708, ( 2006).

[8] Iraq Standard Specification (IQS) No.5,"Portland cement", Baghdad, 8p ,( 1984).

[9] Iraq Standard Specification (IQS) No.45,"Natural sources of aggregate used in building and concrete", Baghdad, 13p,(1984) .
[10] ASTM C 1240 – 05, "Standard Specification for Silica Fume Used in Cementitious Mixtures", American Society for Testing and Material International, **04.02**, (2005).

[11] ASTM C 494-05, "Standard Specification for Chemical Admixtures for Concrete", American Society for Testing and Material International, (2005).

[12] EFNARC, "Specifications and Guidelines for Self-Compacting Concrete", EFNARC, UK (www.efnarc.org) , pp. 1-32, (2005).

[13] B.S. Mahdi, Ph.D. thesis, Building and Construction Engineering, University of Technology, Iraq, (2009).

[14] H. Yazici , S. Aydın, H. Yiğiter, M. Y. Yardımc , G. Alptuna, journal of material in civil engineering ,**22**, No.11 , pp.1093-1101,(2010).

[15] B.S 1881: Part 116, "Method for Determination of Compressive Strength of Concrete Cubes". , British Standards Institution, (1989).

[16] ASTM C496-04."Standard Test Method for Splitting Tensile Strength of Cylindrical Concrete Specimens", American Society for Testing and Materials, (2004).

[17] V. Ramakrishnan , V. Srinivasan, ACI Publications SP-79, pp. 797-812, (1984).

[18] M. Sonebi, L. Svermova. , P, J. M. Bartos, Materials' and Structures, pp.79-86, (2005).

[19] V.S. Parameswaran, T.S. Krishnamoorthy, K. Balasubramanian, , S. Gangadar, Structural Engineering Research Center, Taramani, Madras, India

[20] M.G. Krishman, , D. Elevaras, International Journal of Engineering research and technology, **3**, Issue5, pp.534-537, (2014).

# Shrinkage of high performance lightweight concrete exposed to hot-dry weather conditions

*Hisham* Ahmed [1], *Wasan* Khalil [2] and *Nada H. Jumaa** [3]

[1] Civil Engineering Department, Al-Esra'a University College Baghdad, Iraq
[2] Building and Construction Engineering Department, University of Technology, Baghdad, Iraq
[3] Ministry of Higher Education and Scientific Research, Baghdad, Iraq

**Abstract.** The main aim of this investigation is to study the combined effect of hot-dry weather conditions on the plastic and drying shrinkage of high performance lightweight aggregate concrete (HPLWAC) specimens, along with the other properties including, workability, setting time compressive and splitting strength. The experimental program including the use of fixed mix proportions and was carried out in a typical Iraqi summer day (under actual conditions) of different times during the day. The results indicate that the use of lightweight aggregate play a main role in decreasing the effect of hot dry weather on plastic shrinkage, as well as the plastic shrinkage strain of HPLWC specimens cast at 12:00 pm. less than that of normal weight aggregate concrete specimens by about 36.7%. While drying shrinkage of HPLWC specimens at initial ages up to 7 days was low and it increases with a higher rate at later ages.

## 1 Introduction

High temperature, low RH and high wind speed alone or a combination of these factors encourage rapid evaporation of bleeding water that results in shrinkage, thereby inducing tensile stresses in concrete. If stresses induced due to shrinkage, are greater than the tensile strength of concrete, cracks are formed [1] [2]. The use of high performance concrete in hot dry weather conditions normally requires special precautions against excessive water evaporation, because it has high-cement content and low-water to cementitious materials ratio, the low bleeding rates may result in early age cracking, which is associated with self-desiccation and autogenous shrinkage. In order to avoid such a risk of cracking, it is necessary to prevent the internal relative humidity from decreasing as the hydration of cement takes place [3] [4].

Due to the rapid development of very tall buildings, long span and large– size concrete structures, the requirements for better concrete performance are, higher strength, lightweight, higher toughness and others, therefore, high performance lightweight concrete (HPLWC) has been used for structural purposes.

## 2 Experimental program

### 2.1 Materials

#### *2.1.1 Cement*

Ordinary Portland cement (type 1) manufactured by Mass Bazian Factory was used in this study. Both the chemical composition and physical properties of the cement conformed to I.Q.S No. 5/1984[5].

#### *2.1.2 Silica fume (SF)*

Silica fume used was commercially known as (Sika fume-HR), it was brought from Sika Company. The physical properties and chemical composition of the silica fume show that the silica fume used satisfies the requirements of ASTM C 1240-06 [6].

#### *2.1.3 Fine aggregate*

Natural sand brought from Al-Ukhaider region was used as fine aggregate passing through sieve 4.75 mm. The results demonstrate that the fine aggregate grading and the sulfate content conform to the I.Q.S No.45/1984 [7] Zone 2. Also, lightweight fine aggregate (Pumice) was used as a partial replacement by volume of the natural sand, this lightweight aggregate was crushed by crusher machine. The gradation of pumice fine aggregate was prepared as that of the natural sand. Specific gravity, absorption and dry loose - unit weight of pumice fine aggregate was 1.32, 17.3%, and 968 $kg/m^3$, respectively.

#### *2.1.4 Coarse aggregate*

Natural crushed gravel of nominal maximum size of 12.5mm was used as coarse aggregate. It was brought

* Corresponding author: nadaalrubai@gmail.com

© The Authors, published by EDP Sciences. This is an open access article distributed under the terms of the Creative Commons Attribution License 4.0 (http://creativecommons.org/licenses/by/4.0/).

from Al- Nebai quarry region. The grading of coarse aggregates met ASTM C33/03 requirement [8]. Lightweight coarse aggregate (pumice) was also used as partial replacement of normal coarse aggregate. Grading of pumice coarse aggregate falls in the size designation of 12.5 to 2.36 mm in accordance to ASTM C 330-04 [9]. The specific gravity, absorption and dry loose - unit weight of pumice coarse aggregate are 0.93, 28.6%, and 672 kg/m$^3$, respectively.

#### *2.1.5 High range water reducing admixture (Superplasticizer)*

A high range water reducing agent was used (commercially known as Sika Visco Crete -5930), and meets the requirements for superplasticizer (Sp) according to ASTM C -494 type F [10].

### 2.2 Concrete Mixes

Mix design is carried out in accordance with volumetric method of ACI 211.2 [11]. The reference concrete mixture is designed to have a characteristic compressive strength of 40 MPa at 28-days. Only one mix with 1: 1.33: 0.84 (cement: fine aggregate: coarse aggregate) by weight was used throughout this investigation. The design was carried out to conform to the requirements of structural lightweight aggregate concrete, according to ACI Committee 213 classification [12].

Concrete mixtures contained 572 kg/m$^3$ cement, 10% by weight of cement silica fume, superplasticizer (1.5% by weight of cement), and 50% lightweight fine aggregate was used as partial replacement to normal weight fine aggregate (sand), and 80% lightweight coarse aggregate used as partial replacement to crushed gravel.

Mixing process was carried out in the field under actual hot-dry weather conditions in a typical Iraqi summer days at different hours of the day at (7:30am, 9:30am, 12:00 pm, 3:00pm), concrete mixes were denoted by (M 7:30, M 9:30, M 12:00, M3:00), respectively. Before mixing, all concrete ingredients and mixer as well as molds were kept under sun radiation. Other specimens were prepared in shadow site at (12:00) pm. viz. (M 12:00 Sh) and others in laboratory denoted by (M Ref).

Another concrete specimen denoted by (M-N.W.A) was prepared with the same mix proportions mentioned above but it contains only normal weight fine and coarse aggregate which was cast at 12:00pm. under sun radiation.

### 2.3 Mixing Procedure

Before mixing, the pumice aggregates were pre-soaked for 30 minutes in water and used after removal of excess surface water by spin–drying for 10-15 minutes in a centrifuge to produce the aggregate particles in state saturated surface dry (SSD) condition. A pan mixer of 0.1 m$^3$ capacity was used. Coarse aggregate was added first to the drum followed by fine aggregate. All these materials were first mixed for 1 minute to get homogenous materials. The required cement and SF were mixed manually until a homogenous mixture was obtained, then they were added to the rotary drum. Two thirds of the water was added to the dry mixture and mixed for 1 minute. Finally the superplasticizer and one third of the remained water were stirred and added to the mixture. After that the mixture was mixed for 6 minutes.

### 2.4 Curing

All the specimens were demolded after 24 hours from casting and water cured for 13 days, and then they were left in the same conditions of casting until the time of test.

### 2.5 Experimental Tests

The following tests and measurements were conducted on prepared mixtures:

#### *2.5.1 Workability*

The workability of all concrete mixes was measured at different times by using slump test immediately after mixing in accordance with test method ASTM C-143 [13].

#### *2.5.2 Setting time*

According to the method described in ASTM C 403 [14], the initial and final setting time of concrete was determined, by means of penetration resistance measurements on mortar sieved from the concrete mixture.

#### *2.5.3 Compressive strength*

The compressive strength test was carried out on cubes of 100 mm according to BS.1881: part 116[15].

#### *2.5.4 Splitting tensile strength*

The splitting tensile strength test was carried out on cylinders of 100 ×200 mm according to ASTM C 496 [16].

#### *2.5.5 Length change*

The test method involves the measurement of length change of HPLWAC specimens, which permits assessment of the potential for volumetric changes of concrete specimens. The wooden molds 100 ×100 ×400 mm were used for casting the specimens. After 24 hours from casting, the specimens were demolded and left in air without any curing for a period of 6 months. A stainless steel demec points for 150 mm were fixed at selected positions on the sides of the specimen to measure the length change by using mechanical extensometer. The accuracy of the dial gauge of the measuring device is 0.002 mm.

### *2.5.6 Plastic shrinkage strain and plastic shrinkage cracking*

The concrete specimens utilized for evaluating the effect of hot weather exposure conditions on the plastic shrinkage strain and plastic shrinkage cracking were 1000 ×1000×50 mm. Wooden forms were used to cast the concrete specimens. The concrete mixing process was conducted in the site in actual hot weather conditions at different times of the day (9:30 am., 12:00 pm., and 3:00 pm.) and others were cast in shadow site at 12:00 pm., after casting the specimens were left outside and exposed to atmospheric conditions. To measure plastic shrinkage strain, two aluminum strips measuring 25×70×2 mm were placed on the two corresponding sides of the specimen as shown in Figure1. The movement of each strip, due to shrinkage, was recorded through two linear variable differential transducers (LVDTs) that were connected to a data acquisition system. Shrinkage displacements were recorded every 30 minutes during 24 hours [17]. Also the concrete specimens were visually monitored for 6 hours to detect the onset of plastic shrinkage cracks. The lengths of cracks were recorded and average width was measured by using optical microscope.

**Fig. 1.** Plastic shrinkage test

## 3 Results and discussion

### 3.1 Workability

The relation between the workability of HPLWAC represented by the slump and the actual hot-dry weather conditions is showed in the Table 1, where the workability is decreased by high temperature and decreased relative humidity. It has been noted that the initial concrete temperature effects on the workability of concrete and the amount of water required for a given slump due to evaporation.

**Table 1.** Slump of HPLWAC specimens cast at different hours of the day in hot dry weather conditions

| Mix designation | w/cm (%) | Slump (mm) | Air temp. (°c) | RH (%) | Concrete temp. (°c) |
|---|---|---|---|---|---|
| M Ref | 0.30 | 115 | 22 | 58 | 21.7 |
| M 7:30 | 0.34 | 90 | 39 | 17 | 33.3 |
| M 9:30 | 0.34 | 75 | 48 | 13 | 37.6 |
| M 12:00 | 0.34 | 55 | 54 | 9 | 48.5 |
| M 3:00 | 0.34 | 45 | 56 | 7 | 49.7 |
| M 12:00(Sh) | 0.34 | 105 | 45 | 13 | 42.4 |

### 3.2 Initial and Final Setting Time of Concrete

The penetration strength as a function of time were plotted to determine the initial and final setting time of HPLWAC specimens. They were determined using penetration resistance values of 3.5 MPa and 27.6 MPa, respectively as illustrated in Figure 2 and Table2.

**Table 2.** Initial and final setting time of concrete specimens cast at different hours

| Mix designation | Initial setting (I.S) min (h: min) | Final setting (F.S) min (h: min) |
|---|---|---|
| M Ref | 295 (4:55) | 380 (6:20) |
| M 7:30 | 217 (3:37) | 285 (4:45) |
| M 9:30 | 200 (3:20) | 268 (4:28) |
| M 12:00 | 180 (3:00) | 235 (3:55) |
| M 3:00 | 193 (3:13) | 257 (4:17) |
| M 12:00(Sh) | 210 (3:30) | 275 (4:35) |

It can be seen that both setting times of (M12:00) are less than other specimens. This is attributed to the high initial temperature of these specimens, it is about 48.5°C, however, (M 3:00) had higher initial temperature of 49.7°C, but the setting time is more than that of (M12:00). This may be it exposed to sun radiation for a less period. On the other hand, both the initial and final setting times are reduced with the rise in temperature and the reduction in RH which accelerates the rate of concrete stiffening (i.e. slump loss) and increase the rate of temperature rise inside the concrete. Also it can be seen from pervious study [18] that the initial and final setting time of normal weight concrete cast at 12:00 pm. under actual hot-dry weather conditions was less than that of current study. It is about (2:15) hrs. and (3:39) hrs., this may be attributed to the internal reservoir of water, which is stored in porous lightweight aggregate (pumice), that causes a delay the setting time of concrete.

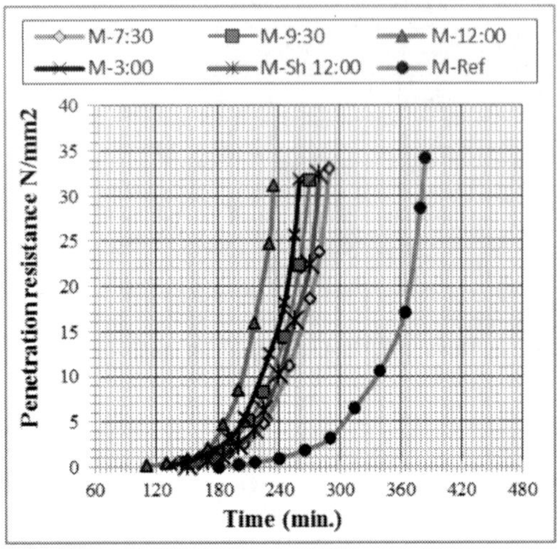


**Fig. 2.** The relationship between penetration resistance and time of HPLWAC specimens

### 3.3 Compressive Strength

The compressive strength results at age (3, 7, 28, 90 and 180) days of HPLWAC specimens that were cast at different hours of the day under actual hot-dry weather conditions are shown in Table 3.

The results show that the percentage increase in compressive strength at age 7 days of specimens cast at (7:30am., 9:30 am, 12:00pm. and 3:00pm) was (5.2, 6.5, 6.8, 6.2)%, respectively, compared with the 3-day strength, while for (M Ref) specimens the percentage increase is 4.6%. Whereas the percentage increase in 28-day strength of specimens M (7:30, 9:30, 12:00, and 3:00) compared with the 7-day strength, is about (21.3, 23.3, 23.8, and 22.9)%, respectively, while for (M Ref) is 37.3%, also for the same specimens, the 90-day strength increased by about (8.7, 8.2 ,7.9, and 8.6)% compared to 28-day strength, whereas (M Ref) the percentage increase is higher, it is about 15.3%. The results also have shown that the percentage of increase in strength during 2 months is less than that during 21 days. These values are typically less than that for normal weight concrete. The strength development in lightweight concrete is limited by the inherent strength of aggregate. These results are in agreement with those found by Al-Khaiat and Haque [19]. Also the percentage increase in 180-day strength compared to the 90-day is (5.1, 4.5, 4.4, and 4.7)% and for (M Ref) is 5.3%, this means that the gain in strength at 28 day and later ages for (M Ref) is higher than that for specimens cast and cured under hot weather conditions. This is in agreement with other researchers[1,20,21] which reported that "Concretes mixed, placed, and cured at elevated temperatures normally develop higher early strengths than concrete produced and cured at lower temperatures, but strengths are generally lower at 28 days and later ages"

**Table 3.** Compressive strength of HPLWAC specimens cast in hot dry weather conditions

| Mix designation | Compressive strength ( MPa) | | | | |
|---|---|---|---|---|---|
| | 3 days | 7 days | 28 days | 90 days | 180 days |
| M Ref | 34.6 | 36.2 | 49.7 | 57.7 | 60.8 |
| M 7:30 | 36.6 | 38.5 | 46.7 | 50.8 | 53.4 |
| M 9:30 | 38.7 | 41.2 | 50.8 | 55 | 57.5 |
| M 12:00 | 39.8 | 42.5 | 52.6 | 56.8 | 59.3 |
| M 3:00 | 37.3 | 39.6 | 48.7 | 52.9 | 55.4 |
| M 12:00( Sh) | 37.6 | 39.9 | 48.2 | 52.2 | 54.7 |

### 3.4 Splitting Tensile Strength

The relationship between splitting tensile strength and age of HPLWAC specimens that were cast at different hours under actual hot weather conditions is shown in Figure 3. Generally the results indicate that the pattern of the behavior of all specimens in splitting tensile strength is similar to that of the compressive strength.

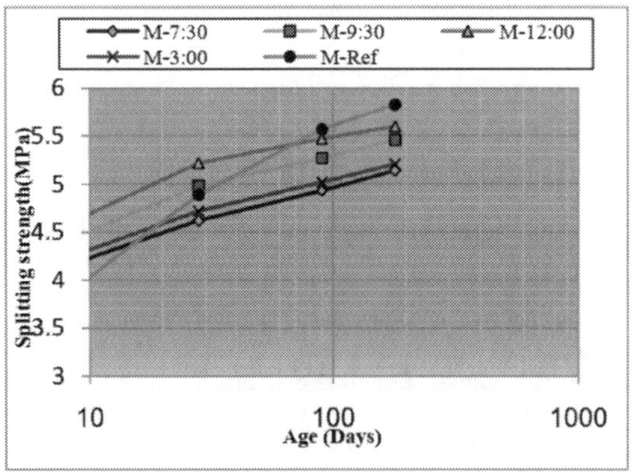


**Fig. 3.** Relationship between splitting tensile strength and age of HPLWAC specimens casted at different hours under actual hot-dry weather conditions

The figure display also that the splitting tensile strength at early ages (3, and 7) days of specimens M (7:30, 9:30,

12:00, and 3:00) was higher than that of reference specimens, the percentage increase at 3 days was (7.9, 13.2, 17.3, and 9.8) % respectively, and at 7 days the percentage increase was (9.6, 16.3, 20.6, and 11.7) % respectively, compared to reference specimens. On the other hand, the splitting tensile strength of specimens (MRef) at 180 day age was higher than that of specimens (M 7:30, M 9:30, M 12:00, and M 3:00) by about (11.8, 6.3, 3.9, and 10.6)%, respectively.

## 3.5. Length Change of HPLWAC Specimens

The data in Table 4 and Figure 4 indicate that drying shrinkage up to 7 days for specimens casted at (7:30 am., 9:30am., 12:00 pm., and 3:00 pm.) is low, and not exceeding (108, 92, 59, and 162) micro strain respectively. This behavior may be attributed to the process of internal curing in pumice concrete due to the supply of absorbed water from (pre-soaked pumice) aggregate pores into the finer capillary pores of cement paste that can reduce the early age drying shrinkage of concrete. Also, the evaporation process is reduced due to the formation of dense hardened paste and the segmented capillary pores. Because of this, the core of the prisms resists the surface contraction on drying, sometimes resulting in surface shell drying of the specimens [22].

From Fig. 4 it can be seen also that short-time exposure of fresh concrete is expected to reduce shrinkage, such reduction was observed in concrete specimens cast at 3:00 pm. which was exposed to intensive drying for a short time, maximum (peak) shrinkage was 616 micro strain at age 56 days, while peak shrinkage for specimens cast at (12:00 pm.) was higher than that of specimens (M-3:00) it is about 902 micro strain. Whereas in specimens M (7:30, and 9:30) the shrinkage at 56 days is less than that of specimens M (12:00, and 3:00), it is (102, and 300) micro strain. The effect of exposure time on shrinkage is due to the effect of drying on the structure of the concrete which, in turn, affects differently on the strength and shrinkage. While at a later age after 84 days, it can be seen that the shrinkage reduces for specimens M (12:00, and 3:00), it is of magnitude (524, and 410) micro strain. This may be due to the fact that there are no further volume changes in concrete, and the internal cracking occurs, the presence of cracks, including internal crack, reduces shrinkage because some of the induced strains are taken up by the cracks and are not reflected, hence, there is a reduction in measured shrinkage [23]. While the peak shrinkage of specimens cast at (9:30 am., and 7:30 am.) was (735, and 767) micro strain, respectively at age 168 days, and for specimens cast in shadow site at 12:00 pm. was (670) micro strain at age 126 day.

On the other hand, it can be noted also from the figure that the maximum swelling occurred to the specimens during 180 days of exposure period of (200, 340, and 359) micro strain for specimens cast at (7:30 am., 9:30am., and 12:00pm) at age (70) days, respectively, and about 464 micro strain for specimens (M-3:00) which happened at age 98 days. This is because of the sudden changes in relative humidity due to the rain storms which results in an increase in the volume of the paste due to a rewetting of the specimens (i.e. swelling).

**Table 4.** Length change of HPLWAC specimens cast under hot dry weather conditions

| Age (days) | Length change $\times 10^{-6}$ of HPLWAC specimen | | | | |
|---|---|---|---|---|---|
| | M-7:30 | M-9:30 | M-12:00 | M-3:00 | M-12:00(Sh) |
| 4 | 32 | 27 | 16 | 0 | 27 |
| 5 | 54 | 48.6 | 38 | 14 | 156 |
| 6 | 92 | 75.6 | 48 | 22 | 403 |
| 7 | 108 | 91.8 | 59 | 162 | 485 |
| 14 | 437 | 315.3 | 459 | 119 | 331 |
| 21 | 296 | 111.5 | -43 | 569 | 189 |
| 28 | - 87 | -190.4 | -257 | -162 | 125 |
| 42 | -119 | -173.4 | -281 | -342 | 45 |
| 56 | 102 | 299.6 | 902 | 616 | 259 |
| 70 | -200 | -340.4 | -359 | -194 | 43 |
| 84 | 702 | 572.6 | 524 | 410 | 11 |
| 98 | -195 | -340.4 | -114 | - 464 | 451 |
| 112 | 11 | -43.4 | 183 | -194 | 281 |
| 126 | 232 | 204.6 | -54 | 184 | 670 |
| 140 | 16 | -102.4 | 361 | -86 | 238 |
| 154 | 497 | 237.6 | -200 | 281 | 200 |
| 166 | 767 | 734.6 | - 6 | -32 | 465 |
| 182 | -38 | -70.4 | -52 | 0 | 105 |

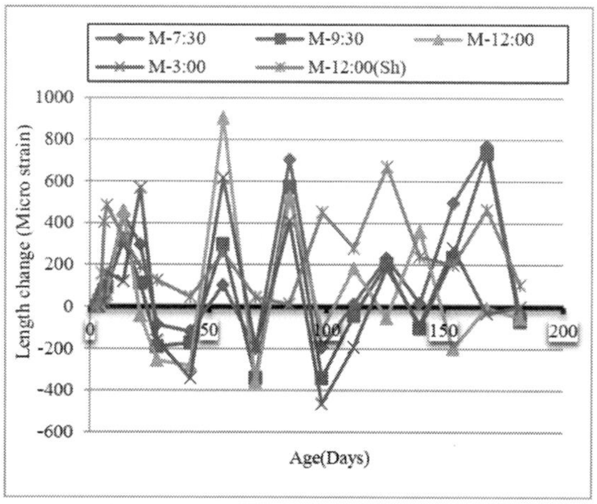


**Fig. 4**. Length change of HPLWAC specimens cast at different hours in hot dry weather conditions

## 3.6 Plastic Shrinkage Strain and Plastic Shrinkage Cracking

The effect of actual hot dry weather conditions on plastic shrinkage strain of HPLWAC specimens is depicted in Figure5, which illustrates the variation of plastic shrinkage strain over a period of 24 hrs. for different cast times of specimens. It can be observed from the figure that the plastic shrinkage strain increases with the period of exposure to ambient conditions, and it increased almost approximately linearly with time up to about 4 hrs., after that the rate is decreased. It can be noticed also from Figure5 that the maximum plastic shrinkage strain is for specimens (M3:00), and around 809 micro strain after 24 hrs. from casting. This is because of the excessive evaporation of mixing water due to the high ambient air temperature, and low RH at the time of cast; they are about 58°C, and 10% respectively, leads to an initial concrete temperature of 46°C. This refers to the point that increased rate of evaporation and mix concrete temperature lead to excessive tendency for plastic shrinkage and cracking.

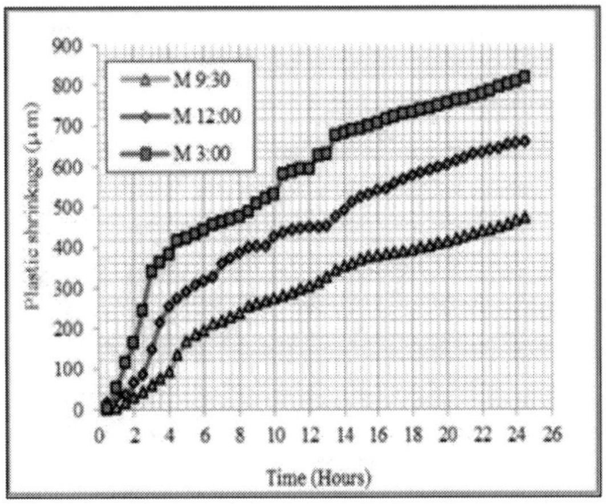


**Fig. 5.** The effect of actual hot dry weather conditions on plastic shrinkage strain of HPLWAC cast specimens at different hours

From Table 5, it can be seen that the number of hair cracks of specimens (M-3:00) are more than those of other specimens. 8 cracks with width ranged from 0.1 to 0.65 mm can be observed. While maximum plastic shrinkage strain for specimen cast at 12:00 pm. is 656 μm over a period of 24 hrs. from cast, which is about is 19% less than that of specimen cast at 3:00 pm. This may be due to the fact that initial concrete temperature of specimen cast at12:00 pm. was less about 44°C, since air temperature was 55°C and RH was 13%, and the specimens (M-12:00) have less number of hair cracks than those of the specimens (M-3:00), which are 4 cracks, with width ranging from 0.06 to 0.7 mm. Whereas specimen cast at 9:30 am. has a maximum plastic shrinkage strain of 463 micro strain, that is (43 and 29.5) % less than those of specimens cast at (3:00 and 12:00) pm., respectively. This may be because the air temperature at time of cast specimens (M-9:30) is low at about 42°C and the RH is 17% and the initial concrete temperature was 38°C, this means that the plastic shrinkage strain decreases with reducing temperature and increasing RH, also plastic shrinkage cracks of specimens (M-9:30) are short ranged from 5 to 10 cm, and with a small number of hair cracks nearly 4 cracks. The results of the maximum plastic shrinkage strain of current work are in agreement with the findings of Al-Amoudi et al. [17].

**Table 5.** Plastic shrinkage cracks of HPLWAC specimens cast at different hours in actual hot weather conditions

| **Mix designation** | **M-9:30** | **M-12:00** | **M-3:00** | **M- Sh** | **M-N.W.A** |
|---|---|---|---|---|---|
| **Exposure condition** | under sun radiation | | | shadow site | under sun radiation |
| **Time of cast** | 9:30 am. | 12:00 pm. | 3:00 pm. | 12:00 pm. | 12:00 pm. |
| **Air temp. (°C)** | 41.9 | 55 | 58 | 48 | 55 |
| **Initial concrete temp.(°C)** | 38.2 | 44.3 | 46.4 | 41.6 | 47.3 |
| **Time of first crack(after cast)(hr:min)** | 4:40 | 3:10 | 2:30 | 4:00 | 2:00 |
| **Width of first crack (mm)** | 0.04-0.7 | 0.08-1.2 | 0.1-0.8 | 0.02-0.6 | 0.6-1.8 |
| **Length of first crack(mm)** | 140 | 300 | 220 | 280 | 340 |
| **Cracks Number (crack width)** | 4(most of them short and hair cracks) | 4 ranged ( 0.06-0.7)mm | 8 ranged (0.1-0.65) | 7 ranged (0.1-0.5) most of them hair crack | many cracks distributed throughout the specimen |

Figure 6 shows that specimen (M12:00Sh) has plastic shrinkage strain at 24 hrs. less than that of the specimens cast under sun radiation by 18.5 %. The increase plastic shrinkage strain for concrete specimen (M-12:00) may be attributed to the low bleeding [17]. The absence of bleed water in the fresh concrete enhances the evaporation of pore water from surface concrete layers under hot dry weather. The evaporation of mixing water from fresh concrete induces shrinkage stress in it [17].

It can be seen also from Figure 6 that the plastic shrinkage strain of specimens (M- N.W.A) is 1037 μm at 24 hrs., which is higher than that of specimen (M-12:00) by 58%. This may be attributed to the fact that the lightweight concrete specimen had low modulus of elasticity about 24.4 GPa [24]. The lower the modulus of elasticity, the lower will be the amount of the induced elastic tensile stress for a given magnitude of shrinkage [20].

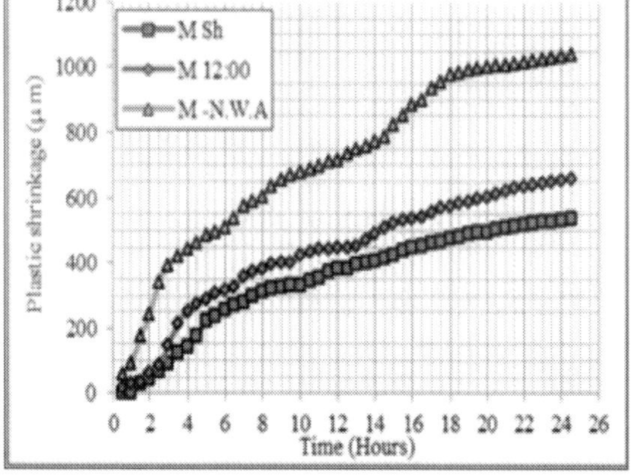


**Fig.6.** The variation of plastic shrinkage strain of concrete specimens cast at 12:00 pm.

## 4 Conclusions

The following conclusions can be drawn from the results and discussions presented from this investigation:

1.Using pre-soaked lightweight aggregates (pumice) as internal water reservoirs for producing HPLWAC under actual hot-dry weather condition played positive role in improvement of concrete properties by compensating the evaporation of water due to the rising temperature and decreasing relative humidity, and provide additional moisture in concrete for a more effective hydration of the cement.

2.The compressive strength and splitting tensile strength performance results of specimens (M 12:00) cast and cured under actual hot-dry weather and exposed to sun radiation was more than that for specimens of shadow site at all ages. While strength performance of reference specimens was slightly more than that of specimens (M 12:00) at later ages.

3.The plastic shrinkage strain of HPLWAC specimens increases almost linearly with time up to about 4 hours, after that the rate of increase in plastic shrinkage decreases until 24 hours. The highest plastic shrinkage strain was 809 micro m in the concrete slab cast at 3:00 pm. under hot dry weather conditions. While the lowest plastic shrinkage strain was 463 μm in concrete slab cast at 9:30 am. Also plastic shrinkage strain of HPLWAC specimens cast at 12:00 pm. is less than that of normal weight aggregate concrete specimens by about 36.7%.

4.The drying shrinkage of HPLWAC specimens at initial ages up to 7 days is low. It does not exceed (108, 92, 59, and 162) micro strain for specimens cast at (7:30 am., 9:30am., 12:00 pm. and 3:00pm.), respectively. Drying shrinkage increases with a higher rate at later ages. Maximum drying shrinkage for specimens M1 (7:30, and 9:30) at age 168 days was (767, and 735) micro strain respectively, while for specimens M1 (12:00, and 3:00) was (902, and 616) micro strain respectively at age of 56 days. Whereas, specimens cast in shadow site at 12:00 pm. was 670 micro strain at age 126 days.

## References

1. A. M. Neville, 4th ed., Longman Group Ltd., (2005).
2. A.A. Almusallam, J., **23**, pp.353-361, (2001).
3. S. Zhutovsky, K. Kovler, A.Bentur, Cem. Concr. Res. J, **54**, pp. 36–42, (2013).
4. P. F. de J. Cano-Barrita, B. J. Balcom, T. W. Bremner, M. B. MacMillan, W. S. Langley, Mater. Struct. J., **37**, pp. 522-531, (2004).
5. Iraqi Specification, No. 5, *Portland cement*, (1984).
6. ASTM C1240, *Standard specification for use of silica fume as a mineral admixtures in hydraulic-cement concrete, mortar and grout*, pp. 1-8, (2006).
7. Iraqi Specification, No.45, *Natural sources of aggregate*, (1984).
8. ASTM C33, *Standard specification for concrete aggregates*, **04.02**, pp. 1-11, (2003).
9. ASTM C 330, *Standard specification for lightweight aggregates for structural concrete*, **04.02**, pp.1-4, (2004).
10. ASTM C494, *Standard specification for chemical admixtures for concrete*, **04.02**, pp. 271–279, (2004).
11. ACI Committee 211, *Standard practice for selecting proportion for structural lightweight concrete*, pp. 1–20, (2004).
12. ACI Committee 213, *Guide for structural lightweight aggregate concrete*, (2003).
13. ASTM C143, *Standard test method for slump of hydraulic-cement concrete*, Annual Book of ASTM Standards, **04.02**, pp. 1–3, (2004).
14. ASTM C 403, *Time of setting of concrete mixtures by penetration resistance*, **04.02**, pp. 1–6, (2004).
15. British Standard 1881, Part 116, *Method for determination of compressive strength of concrete cubes*, pp. 3, (1989).
16. ASTM C496, *Standard test methods for splitting tensile strength of cylindrical*, **04.02**, pp. 283–287, (2004).
17. O.S.B. Al-Amoudi, M. Maslehuddin, M. Shameem, M. Ibrahim, Cem. Concr. Comp. J, **29**, pp. 690–699, (2007).
18. H.K. Ahmed., Build. Res. Mag., **1**, pp.5-19, (1982).
19. H. Al-Khaiat, M.N. Haque, Cem. Concr. Res. J, **28**, pp. 859–866, (1998).
20. P.K. Mehta, J. M. P. Monteiro, pp. 1-684, (2006).
21. K.A. Schindler, B.F. McCullough, A paper offered for publication at the Annual Meeting of the

Transportation Research Board. Washington, D. C., Jan. (2002).

22. L. Gunduz, Construct. Build. Mater. J., **22**, pp.721–728, (2008).
23. I. Soroka, London, U.K.: E&FN Spon (1993).
24. N. Jumaa, Ph.D. dissertation, Build. Construct. Engine. Depart. Tech. Univ., Iraq, (2017).

# The effect of using synthetic fibers on some properties of modified juss

*Nada* Mahdi Fawzi [1,*], *Luma* Abdul Ghani Zghair [2], *Hind* Hussein Hamad [3]

[1] College of Engineering, Baghdad University , Baghdad, Iraq
[2] Civil Engineering Department, Al- Mustansiriyah University, Baghdad, Iraq
[3] College of physical education and sports science, Baghdad University, Baghdad, Iraq

**Abstract.** This paper examines the mechanical properties of a composite material made of modified Iraqi gypsum (juss) reinforced with polypropylene fibers. The modified juss was prepared by adding two percentages of cement (5, 10) %. Two percentages of polypropylene fibers were used, to reinforce the modified juss (1, 2) %. The water/dry compound ratio used was equal to 0.53%. The composite was evaluated based on compressive strength, flexural strengths, absorption percentage, density, acoustic impedance, ultra – pulse velocity, longitudinal shrinkage and setting time tests. The results indicated that the inclusion of cement on to juss increases the compressive strength, absorption percentage, density, acoustic impedance, ultra – pulse velocity, longitudinal shrinkage and a reduction in flexural strength and setting time were observed by adding the cement. In addition, the inclusion of polypropylene fiber was significant in improving mechanical performance of the composite material, it shows a great improvement in longitudinal shrinkage, modulus of rupture and absorption percentages.

Index Terms— modified juss, compressive strength, flexural strengths, absorption percentage, density, acoustic impedance, longitudinal shrinkage

## 1 Introduction

Plaster is a very cheap building material, it's used since ancient time. Plaster has a lot of applications such as fire proofing plasterboard and decoration. The only most important factor likely to affect the strength of the gypsum binder is the water. The existence of atmospheric moisture even at low relative humidity is enough for bring about a significant reduction in strength of gypsum binder. For this reason, gypsum products are used in the building industry as a surface finish on interior walls.

Many studies have been carried out to enhance the mechanical strength of the plaster Dalmay, et al., [1] presented a study about the use of different natural fibers namely (flax and hemp) to enhance the mechanical properties of plaster. Results showed that the behavior of composite matrix was modified from brittle to a non-linear one. Moreover, even a small addition of fibers enhance the flexural strength. The result also showed that when fibers were added to the plaster, hemp fibers delayed the setting time as compared to flax fibers. Research by Carvalho, et al.,[2] showed that calluses pulp fibers improved the modulus of rupture compared to gypsum paste without calluses pulp fibers. Coalk, [3] used a mixture of Portland cement –natural pozzolana to study its impact on density, strength development and durability of gypsum. The result showed that an increase in the density and strength development of gypsum – Portland cement and gypsum – Portland cement – pozzolana compared to gypsum only. Moreover, the density and strength development, decreases with increase in the percentage of natural pozzolana they also concluded that natural pozzolana can be effective in improving the resistance of gypsum–Portland cement blends to water. Some other researchers show different results from Coalk, [3] Bentur, et al., [4] reported that Mixing of Portland cement and gypsum is usually infeasible because it can result in the formation of ettringite, which leads to deterioration. A combination of Portland cement and silica fume, may provide conditions to eliminate the formation of ettringite, and perhaps make it feasible to produce a gypsum-Portland cement-pozzolana blend which owns the advantages of both gypsum and Portland cement, but the researchers also reported that increasing the percentage of pozzolana material can be effective in improving the strength of gypsum. Gencel, et al., [5] used two additives [expanded vermiculite and polypropylene fibers] to improve the physical and mechanical properties of the gypsum. The gypsum composite was analyzed based on unit weights and compressive strength. They reported that the pp

* Corresponding author: Naljalawi@yahoo.com

© The Authors, published by EDP Sciences. This is an open access article distributed under the terms of the Creative Commons Attribution License 4.0 (http://creativecommons.org/licenses/by/4.0/).

fibers enhanced the mechanical properties, while, both expanded vermiculite and the fibers reduced the unit weight of the gypsum composites. Abdulhussein, [6] used resole resin to improve the properties of local Iraqi gypsum. The prepared composites of elasticity. On the other hand a reduction in bending strength and swelling % were observed when the proportion of resole resin in gypsum composite was increased. Al –Ubaidi,[7] and Frieh, et al.,[8] prepared different gypsum composite samples to improve the Iraqi gypsum. Hydrated lime, limestone, metakaoline and kaoline were used in this work. The result shows that the setting time and the compressive strength increased by using hydrated lime and limestone while the compressive strength decreases for some weight percent of adding by 19% and 13.6% for metakaoline and kaoline respectively, in addition, the modulus of rapture decreased by adding these admixtures. Sern, et al., [9] used different forms and percentages of waste rubber to improve the elastic behavior of gypsum. The result indicated that the addition of waste rubber particles lead to decreases in bending and compressive strength and increases in elastic modulus, they also reported that a reduction in toughness was observed and this may cause a worse behavior in crack resistance. In the present work gypsum composites containing cement and cement - polypropylene fibers were prepared to examine their physical and mechanical properties.

## 2 Research significant

Synthetic fibers are already used in different types of materials like concrete, plastics and lime-based products. They show different features like the combination of good mechanical, acoustic and thermal properties that allow these types of materials to be used for different applications. The main defect associated with juss is its brittleness, especially under tensile stress. Therefore, it is interesting to investigate different methods that could enhance the mechanical properties of juss. Adding fibres and cement to obtain a composite material is one way to improve the behaviour of the product, especially after the failure of the matrix. The aim of this work was to the study the effects of adding synthetic fibers, namely polypropylene, on the setting time and the mechanical properties of the composite matrix.

## 3 Materials

### 3.1 Local juss

A local Iraqi juss was used in this study. The consistency and the setting time were measured following the IQS NO. 27, [10] and IQS NO.28, [11] standard, respectively. Table 1 illustrates the property of this juss.

**Table 1.**
The chemical and physical properties for the various binders

| Oxide | Cement | Juss |
|---|---|---|
| CaO | 62.96 | |
| $SiO_2$ | 20.51 | 27.48 |
| $Fe_2O_3$ | 2.96 | 0.59 |
| $Al_2O_3$ | 5.57 | 6.65 |
| MgO | 2.35 | 0.8 |
| $SO_3$ | 1.9 | 34.75 |
| Loss on agent L.O.I% | 0.93 | 7.25 |
| I.R% | 0.83 | |
| L.S.F | 0.93 | |
| Fineness ($m^2$/g) | 3.817 | |
| Vicat set (hr:min) Initial Final | 3:00 < 10 | |
| Mortar Compressive Strength ( N/mm$^2$) at 3 days 7 days | 30> 15 32 > 23 | |

### 3.2 Cement

The used cement was ordinary Portland cement Type (1), which conformed to IQS No. 5[12]. Table 1 shows the properties of this cement.

### 3.3 Polypropylene fibers pp

Crimped Polypropylene fibers (pp) with (30 mm) length, (0.47 mm) diameters and Specific gravity of 1.1 were used.

## 4 Experimental work

### 4.1 Modified juss

The modified juss was prepared by adding two percentages of cement (5, 10)% to the local Iraqi juss. For preparing modified juss reinforced with (1, 2)% of pp fiber, the dry compounds (fibers, cement and juss) were first mixed by hand until the fibers were coated with cement and juss. Then, the dry mix was slowly added to water and mixed for a proper time by hand. The water/dry compound ratio used was equal to 0.53 to ensure a sufficient fluidity even for the highest percentages of fibers. The mix is then poured into moulds and compacted by thin rod. Table 2 shows the details of juss mixes, local Iraqi juss was used in this study. The consistency and the setting time were measured following the IQS.

**Table 2.** Set and description of mixes

| Mix notation | Descriptions of mixes |
|---|---|
| G | reference juss |
| GC5 | juss blended with 5% cement |
| | |
| GC10 | Juss blended with 10% cement |
| GF2 | juss blended with 2% pp fibers |
| GC5F1 | juss blended with 5% cement and 1% pp fibers |
| GC5F2 | juss blended with 5% cement and 2% pp fibers |
| GC10F1 | juss blended with 10% cement and 1% pp fibers |
| | |

## 5 Result and Discussion

### 5.1 Setting Times

The setting time of the gypsum and the gypsum composite samples were measured following the Iraq standard NO.28,[11]. Table 3 shows the result of the setting time

**Table 3** Setting time of juss specimens

| Mix notation | Setting Time min |
|---|---|
| G | 12:30 |
| GC5 | 10:35 |
| GC10 | 8:55 |
| GF2 | 10:20 |
| GC5F1 | 10:25 |
| GC10F1 | 8:40 |
| GC5F2 | 10 |
| GC10F2 | 8 |
| | |

It was shown that with constant water/dry compound, the setting times of both composites made with cement and pp fibers were reduced by about (16, 30, 17, 16.6, 31, 19, and 35) % for GC5, GC10, GF2, GC5F1, GC10F1, GC5F2 and GC10F2, respectively comparable with (G), as shown in Fig 1. This may be due to the reduction in water content because it is already consumed by the cement paste during continuous hydration, as a result, setting times is reduced

In all cases, the setting times of both composites made with cement and pp fiber mixtures conformed to the Iraqi specifications.

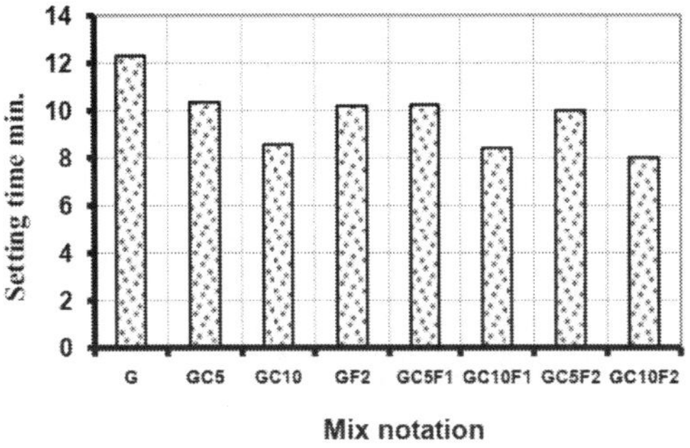


**Fig. 1.** Setting time of juss samples

### 5.2 Compressive Strength

The compressive strengths of the gypsum and the gypsum composite samples were measured following the ASTM C472-99 [13] standard. Table 4 shows the result of the compressive strength. The results show that the composite made with a matrix of juss and cement, had a compressive strength higher than for composite made with a matrix of juss, cement and pp fibers. The increase of the compressive strength indicate good adhesion matrix of juss and cement. The increase for mixes was (15.5, 37.3, 6, 25, 1 and 20) % for GC5, GC10, GC5F1, GC10F1, GC5F2 and GC10F2, respectively, comparable with reference G.

The results also show that the composite made with a matrix of juss and 2% pp fibers (GF2%) has a lower compressive strength as shown in Fig 2 comparable with (G). This may be attributed to the presence of voids, where the addition of fibers may increase the voids in juss sample.

**Table 4.** Compressive strength and modules of rupture of juss specimens.

| Mix notation | Compressive strength ($N/mm^2$) 14 DAYS | Modules of rupture ($N/mm^2$) 14 days |
|---|---|---|
| G | 4.95 | 4.3 |
| GC5 | 5.72 | 3.6 |
| GC10 | 6.8 | 3.3 |
| GF2 | 4.23 | 10.9 |
| GC5F1 | 5.25 | 5.1 |
| GC10F1 | 6.2 | 4.9 |
| GC5F2 | 5 | 7.2 |
| GC10F2 | 5.94 | 6.9 |
| | | |

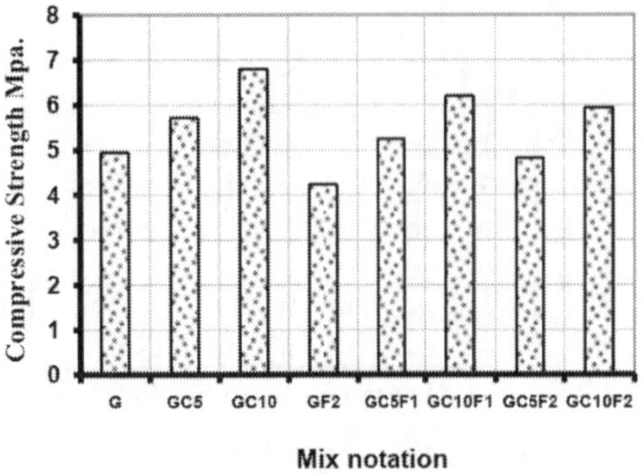


**Fig. 2**. Compressive strengths of juss samples

### 5.3 Modulus of Rupture test

The samples were analyzed based on modulus of rupture tests with three -point loading, following the recommendations of the ASTM C293-02 [14]. Table 4 and Fig. 3 show the results of modulus of rupture for juss samples. The addition of pp fibers causes an increase in modulus of rupture by (153, 18.6, 13.9, 67.4, 60.4) % for GF2, GC5F1, GC10F1, GC5F2 and GC10F2, respectively compared to (G). The increase in modulus of rupture for GF2 samples is very significant as shown in Fig.3. This increase may be due to the high tensile strength of the pp fibers which tries to arrest the micro cracks that occur under increased loads. On the other hand, the addition of cement causes a decrease in modulus of rupture by (16 and 23)% for GC5 and GC10, respectively. This may be attributed to the following:

1. In the case of the compressive strength tests, the formation of ettringite which can form stress as a result of the interaction between the cement and juss, as reported by Bentur, et al., [4] was less pronounced and this may due to the low percentage of cement.

In the case of modulus of rupture tests the forms of stresses may increase the deflection and as a result, reduce the modulus of ruptures value.

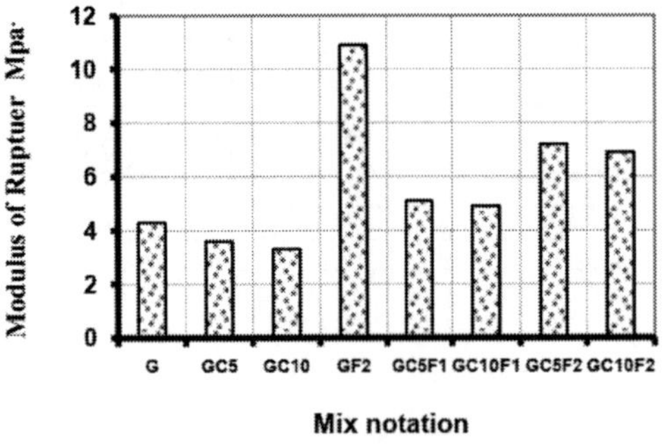


**Fig. 3**. Modulus of rupture of juss samples.

### 5.4 Absorption test

The Table 5 and Fig. 4 show the results of absorption tests for juss samples. The addition of cement causes an increase in absorption by (4 and 9)% for GC5 and GC10, respectively compared to (G). This increase may be due to the absorption of cement paste during the hydration process thus increasing the absorption %. While the addition of pp fibers reduce the absorption percentage for gypsum samples. This may be due to the water proof property of pp.

**Table 5.** Absorption and longitudinal shrinkage percentages of juss specimens

| Mix notation | Absorption % | Longitudinal Shrinkage % |
|---|---|---|
| G | 0.23 | 2 |
| GC5 | 0.24 | 2 |
| GC10 | 0.25 | 3 |
| GF2 | 0.2 | 0.8 |
| GC5F1 | 0.2 | 1.4 |
| GC10F1 | 0.22 | 1.9 |
| GC5F2 | 0.22 | 1 |
| GC10F2 | 0.23 | 1.6 |
| | | |

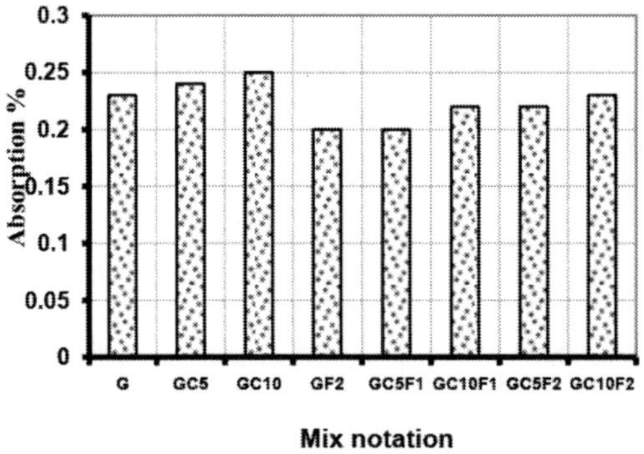


**Fig. 4**. Absorption percentage of juss samples

### 5.5 Longitudinal shrinkage

Table 5 and Fig. 5 show the results of longitudinal shrinkage for juss samples. The addition of pp fibers causes a decreases in longitudinal shrinkage by (60, 30, 50, 5, 50 and 20) for GF2, GC5F1, GC10F1, GC5F2 and GC10F2, respectively compared to (G). Using pp fibers can be effective in retaining the cementitous matrix integrity and reducing the Longitudinal Shrinkage which can occur after the hydration of cement paste. On the other hand, the addition of cement (10)% causes an increase in longitudinal shrinkage by (50 % ) for GC10, this may be attributed to reduce in water content because it is already consumed by the cement paste during continuous hydration, consequently, longitudinal shrinkage is increased.

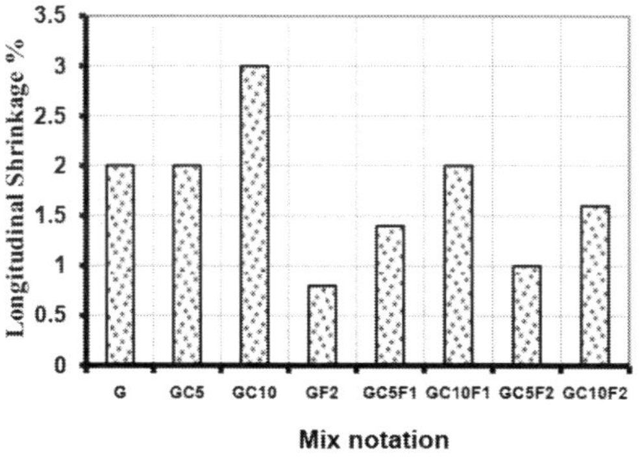


**Fig. 5**. Longitudinal shrinkage percentage of juss samples

### 5.6 Density test

Table 6 and Fig. 6 shows the results of density tests for juss samples. A general increase in the density of juss–Portland cement blends is observed compared to reference Juss. The increase_for mixes was (21, 24, 20, 29, 24 and 31) % for GC5, GC10, GC5F1, GC10F1, GC5F2 and GC10F2, respectively, compared to reference G. This result indicates that a uniform density can be obtained for juss -Portland cement blends. This finding is supported by Coalk, [4]. The results show that the composite made with a matrix of Juss, cement and a reinforcing pp fibers, had a density higher than for composite made with a matrix of Juss and cement only. The increase of the density indicates good adhesion matrix of Juss, cement and pp fibers. On the other hand GF2% (with juss and 2% pp fibers) has a slight loss of density. This may be due to the low density of PP fibers.

**Table 6.** Density, ultra – pulse velocity
And acoustic impedance of juss specimens

| Mix notation | Density kg/m$^3$ | Ultra – pulse velocity Km/sec | Acoustic Impedance kg/s.m$^2$ |
|---|---|---|---|
| G | 1.11 | 1.5 | 1.66 |
| GC5 | 1.34 | 1.54 | 2.1 |
| GC10 | 1.38 | 1.64 | 2.26 |
| GF2 | 1.08 | 1.26 | 1.6 |
| GC5F1 | 1.33 | 1.66 | 2.21 |
| GC10F1 | 1.43 | 1.8 | 2.57 |
| GC5F2 | 1.38 | 1.76 | 2.42 |
| GC10F2 | 1.45 | 1.82 | 2.63 |

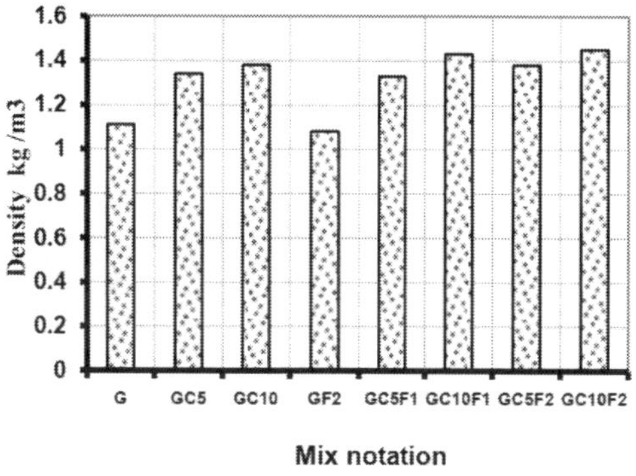


**Fig.6.** Density of juss samples

### 5.7 Ultra – pulse velocity test

The samples followed the recommendations of the ASTM C597-02[15]. Table 6 and Fig.7 show the results of ultra – pulse velocity tests for juss samples. It appears that the Ultra – pulse velocity value increases with the addition of cement and PP fibers compared to reference juss. The increase for mixes was (3, 9,11, 20, 17 and 21) for GC5, GC10, GC5F1, GC10F1, GC5F2 and GC10F2 respectively, comparable with reference G. This may be due to the increases in specimen density and reduction voids and discontinuity point. On the other hand GF2% (with juss and 2% pp fibers) has a slight reduction in ultrasound velocity wave, the reason of this anomaly may be due to the fact that pp fibers are good acoustic insulator, so they act as obstacles to disperse waves.

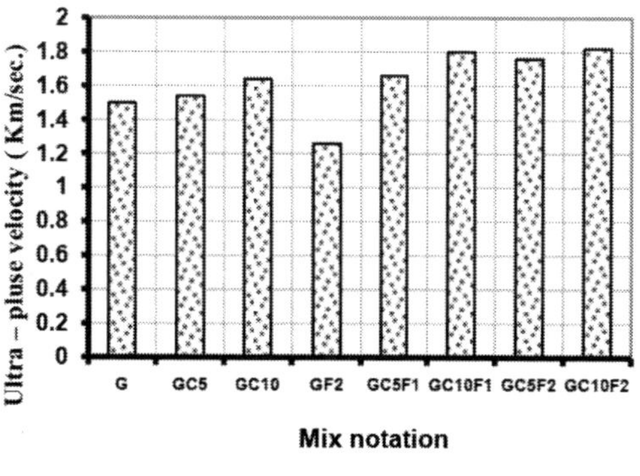


**Fig 7**. Ultra – pulse velocity of juss samples

### 5.8 Acoustic impedance test

The acoustic impedance test of the gypsum and the gypsum composite samples were measured as per the following equation [16]

$$\text{Acoustic Impedance test} = \text{velocity} \times \text{density} \quad (1)$$

Table 6 and Fig. 8 show the results of acoustic impedance tests for juss samples. It appears that the acoustic impedance value increases with the addition of cement compared to reference juss. The increase for mixes was (27, 36, 33, 55, 46 and 58)% for GC5, GC10, GC5F1, GC10F1, GC10F2 and GC5F2, respectively, compared to reference G. This may be due to the increases in specimen density and reduction voids and discontinuity point.

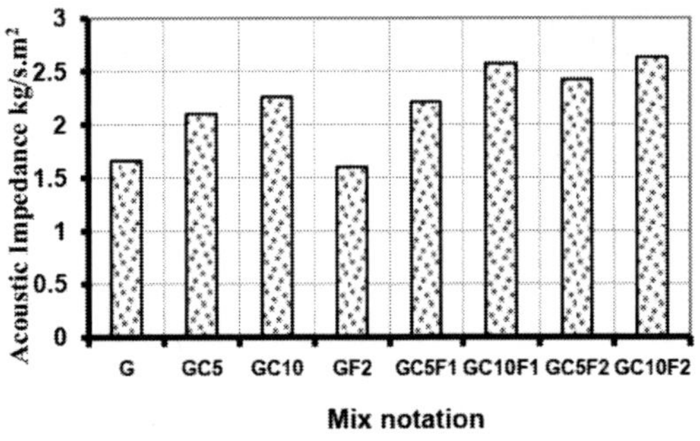


**Fig 8**. Acoustic impedance of juss samples

On the other hand GF2% (with juss and 2% pp fibers) has a slight reduction in acoustic impedance, the reason of this anomaly may be due to decreases in specimen density

## 6. Conclusion

The findings from this study are summarized as follows:

1. The density of juss–Portland cement-pp fiber blends ranges from about 1.35 to 1.45 kg/m$^3$. Density variations are usually caused by the changes in mix proportion. Lower densities can be obtained by increasing the juss or reducing the cement content in the mix. In addition, the use of pp fibers only in juss mixtures, led to a decreases in density by (3)% lower than reference mixes.
2. The Acoustic Impedance value and the Ultra – pluse velocity increase with the addition of both cement and pp fibers. On the other hand, a decrease in Acoustic Impedance value and Ultra – pluse velocity in mixes using pp fibers only, by (3.6 and 20)%, respectively lower than reference mixes.
3. pp fibers (2)% are effective in reducing the longitudinal shrinkage and absorption percentages. While adding 5% to 10% Portland cement by weight of binder dramatically increases the longitudinal shrinkage and absorption percentages when compared to paste without Portland cement.
4. pp fibers and Portland cement were found to accelerate the setting time of juss when the water/dry compound was constant, comparable with reference juss and in all cases, the setting time of the composite material conformed to the Iraqi specification criteria.
5. The addition of pp fibers causes an increase in modulus of rupture while the addition of cement causes a decrease in modulus of rupture comparable with reference juss.
6. A blend of juss and a mixture of Portland cement and pp fibers can provide a system with strength levels higher than those of juss only.

## References

1. P. Dalmay, A. Smith, T. Chotard, P. S. Turner, V. Gloaguen, and P. Krausz,, Journal of Materials Science Vol. **45**, PP.793–803,( 2010)
2. M. A. Carvalhoa, C. C. Júniora, H. S. Juniorb, R.Tubinoc,, and M. T. Carvalhod, Materials Research, Vol. **11**, No. 4, PP. 391-397 , (2008)
3. A. Colak, Cement and Concrete Research , Vol. **32** , PP. 109– 115, (2002)
4. A. Bentur, K. Kovler, and A. Goldman, Advances in Cement Research, Vol. **6**, No. 23, PP. 109-116,( 1994)
5. O. Gencel, J. J. Diaz, M. Sutcu, F. Koksal, F.P.A Rabanal, G. M. Barrerae, and W. Brostow, Energy and Buildings , Vol.**70** , PP. 135–144,( 2014)
6. A. A. Abdulhussein, Journal of Kerbala University , Vol. **10**, No.2, PP. 211-220. (2012)
7. L. S. Al – Ubaidi, MSc, University of Technology, PP.1-98 .( 2004)
8. k. J. Frieh, W. A. AL –Qasi, and L. S. Al – Ubaidi, Eng .&Tech. Journal,Vol.**24**, No .9, PP. 380 -393,( 2005)
9. A. Sern, M. Rio, J. G. Palomo,., and M. Gonzalez, Construction and Building Materials , Vol. **35**, PP. 633–641, (2012)
10. Iraqi Specification, , Physical examinations of juss for construction , Central Organization for Standardization and Quality Control No. 27 ,( 1988)

11. Iraqi Specification, No. 28, juss for construction, Central Organization for Standardization and Quality Control (2010)
12. Iraqi Specification, No. 5, "Portland cement"(1984).
13. ASTM C472-99 Standard Test Methods for Physical Testing of Gypsum, Gypsum Plasters and Gypsum Concrete (1999)
14. ASTM C293-02 , Standard test method for flexural strength of concrete (using simple beam with center-point loading) , Annual Book of ASTM Standard, Vol. **04**-02,. .(2002)
15. ASTM C597-02, Standard test method for pulse velocity through concrete , Annual Book of ASTM Standard, Vol. **04-02** .(2002)
16. E. K. Lawrence, $4^{th}$ edition ,John Wiley & Sons Inc,( 2000)

# Utilization of brick waste as pozzolanic material in concrete mix

*Raghda* Resin[1], *Abeer* Alwared[1], *Suhair* Al-Hubboubi[2]

[1]Environmental Department, College of Engineering, Baghdad University, Baghdad, Iraq
[2]Building Research Directorate, Ministry of Construction and Housing, Baghdad, Iraq

**Abstract.** Recycling of construction and demolition waste and recycling of industrial by-products into building materials has an enormous potential. Brick waste can be treated by reusing for construction. Recently, the utilization of pozzolanic materials in cement and concrete have increased considerably due to their diverse benefits such as lesser use of cement , saving in production costs, improvement in the durability of concrete and so on. Ground bricks was added to concrete mix as partial replacement of cement at (5, 10, 15) % by weight. It was evaluated as a pozzolanic material in conjunction with high range water reducing admixture superplasticizer (HRWR)-The study confirmed that the use of 10 % of ground brick improved compressive, flexure strength of concrete by about (23.3%) and (5.92%) respectively at 240 days as compared to reference concrete and decreased the depth of water penetration under pressure by about 43.75% at 150 days, also the water absorption decreased by about 2.52%. The uses of 10% of grinding brick in conjunction with high range water reducing admixture reduced corrosion of steel reinforcement by about 18%.

## 1 Introduction

Preservation of the environment and conservation of rapidly diminishing natural resources should be the essence of sustainable development. Continuous industrial development poses serious problems in construction and demolition waste disposal [1].Construction and demolition (C&D) waste could be broken concrete, bricks from buildings, or broken pavement. Brick waste is not the major amount of waste in construction, but it contributes largely in demolition activities. These waste materials are directly dumped into the environment and it causes environmental pollution as well as scarcity of landfill .Thus, waste utilization has become an alternative solution to overcome the problem for disposal of waste instead of minimizing the brick waste produced, it can increase the strength and performance of concrete production. Further ,it also helps in minimizing the land fill scarcity problem [2] .Concrete is one of the most widely used construction materials in the world. However, the production of Portland cement, an essential constituent of concrete, leads to the release of significant amount of $CO_2$, a greenhouse gas (GHG). The production of one ton of Portland cement produces about one ton of $CO_2$ and other GHGs. The environmental issues associated with GHGs, in addition to natural resources issues, will play a leading role in the sustainable development of the cement and concrete industry during this century [3]. Research studies revealed that pozzolanic materials significantly improve the properties of concrete [4][5][6].Material transport properties, especially permeability, affect the durability and integrity of a structure. High permeability, due to porosity or cracking, provides ingress for water and other corrosive agents. If such agents reach reinforcing bars within the structure, the bars corrode, thus compromising the ability of the structure to withstand loads, which eventually leads to structural failure. Therefore, there has been an interest in determining the permeability of pozzolanic material – concrete compared with conventional concrete [7] The aim of this study focuses on the feasibility of using bricks of locally available construction waste for establishing economically-friendly and affordable green environment through evaluating its pozzolanic activity and checking the suitability as partial replacement of cement to produce high performance concrete.

© The Authors, published by EDP Sciences. This is an open access article distributed under the terms of the Creative Commons Attribution License 4.0 (http://creativecommons.org/licenses/by/4.0/).

## 2 Experimental work

### 2.1 Materials

*Cement*: Ordinary Portland cement conforming to IQS No. 5/1984 was used in all types of concrete mixes in this work. Tables (1 and 2) represent the chemical and physical properties of the used cement respectively.

*Fine aggregate*: Natural sand of zone (1) conforming to IQS No.45/1984 was used as fine aggregate. The gradation of the used fine aggregate and its physical and chemical tests were shown in Tables (3 and 4).

*Coarse aggregate*: Crushed river gravel with 20 mm maximum size was used as coarse aggregate. Table (5) shows the properties of coarse aggregate which conforms to Iraqi standards No.45/1984.

*High Range Water Reducing Admixture:*

High range water reducing admixture (HRWR) super-plasticizer, known as (PC2000) complies with ASTM C 494 type N was used. It decreases water/cement ratio to improve the mechanical property of concrete and durability. It was used in dosages between (0.5 to1.5) percent of the cement weight. The properties of this admixture are listed in Table (6).

*Waste Brick Powder:*

The waste bricks obtained from the garbage of a demolished building were grinded in Jet mills machine and then sieved to get the desired particle size (100μm, sieve No.150), Fig. (1). its chemical and physical properties are summarized in Tables (7) and (8) All the materials used in this work were supplied by the Iraqi local market. The properties of cement, fine an respectively. X-ray diffraction, Fig. (2), shows that brick composed of quartz (being the main component), illite, albite, and hematite. However, the intensities of XRD patterns of bricks are different and suggest that the bricks were manufactured differently by using materials of high clay content. [8].

*Steel Reinforcement:*

For half-cell potential measurement, deformed steel bars of 12.5 mm diameter and 400 mm length were used to reinforce test specimens; the steel bars were cleaned thoroughly using steel brush for removing any rust deposits from their surfaces. Table (9) represents the physical properties of the steel bars.

**Fig. 1.** Brick powder passing sieve (No. 150)

**Table1.** Chemical properties of cement

| Compound composition | Abbreviation | Percent by weight | Limit of IQS No.5/1984 |
|---|---|---|---|
| Lime | CaO | 62.73 | - |
| Silica | $SiO_2$ | 20.41 | - |
| Alumina | $Al_2O_3$ | 4.91 | - |
| Iron oxide | $Fe_2O_3$ | 3.21 | - |
| Magnesia | MgO | 2.6 | ≤ 2.85% |
| Sulfate | $SO_3$ | 2.8 | ≤ 5% |
| Loss on ignition | L.O.I | 3.17 | ≤4% |
| Insoluble residue | I.R | 1.05 | ≤ 1.5 |
| Lime saturation factor | L.S.F | 0.93 | 0.66-1.02 |
| Main compounds % by weight | | | |
| Name of compounds | | Abbreviation | Percent by weight |
| Tri calcium silicate | | $C_3S$ | 49.57 |
| Di calcium silicate | | $C_2S$ | 21.36 |
| Tri calcium aluminate | | $C_3A$ | 7.58 |

**Table2.** Physical properties of cement

| Physical properties | Limits of cement | Limits of IQS No. 5/1984 |
|---|---|---|
| Fineness Blaine $m^2/kg$ | 359.7 | ≥ 230 |
| Initial setting time (h:min) | 1:35 | ≥ 00:45 |
| Final setting time (h:min) | 4:40 | ≤ 10:00 |
| Compressive strength (MPa)<br>3 days<br>7 days | 28<br>34 | ≥ 15<br>≥ 23 |

**Table3.** Gradation of fine aggregation

| Sieve size (mm) | Percentage passing % | Limit of IQS No.45/1984 , zone (1) |
|---|---|---|
| 10 | 100 | 100 |
| 4.75 | 97 | 90-100 |
| 2.36 | 77.9 | 60-95 |
| 1.18 | 51.6 | 30-70 |
| 0.6 | 30.3 | 15-34 |
| 0.3 | 10.9 | 5-20 |
| 0.15 | 2 | 0-10 |

**Table 4.** Physical and chemical properties of fine

| Properties | Values | Limits of IQS No.45-1984 |
|---|---|---|
| Sulfite % ($SO_3$) | 0.18 | ≤ 0.5% |
| Fineness modulus | 3.3 | --- |
| Absorption % | 1.5 | ---- |

**Table5.** Gradation for coarse aggregate

| Sieve size (mm) | Percentage passing % | Limit of Iraqi standards No. (45/1984) |
|---|---|---|
| 20 | 100 | 100 |
| 10 | 23.5 | 0-25 |
| 5 | 1.15 | 0-10 |

**Table 6.** Properties of HRWR

| | |
|---|---|
| Chemical content | Poly-carboxylate based |
| Color, appearance | Brown liquid |
| PH | 0.5±6 |
| Density | 1.03±0.02 g/cm3 |
| Chloride content (%) | Without chloride ion |

**Table 7.** Chemical Properties of waste brick powder

| Oxide composition | Oxide content % | ASTM C618 Class C limitation |
|---|---|---|
| $SiO_2$ | 39.35 | Min 50% |
| $Al_2O_3$ | 17.77 | |
| $Fe_2O_3$ | 2.40 | |
| $SO_3$ | 1.60 | Max. 5% |
| CaO | 23.2 | |
| MgO | 7.5 | |
| L.O.I | 2.59 | |
| Others | 5.5 | |
| Total | 99.91 | |

**Table 8.** Physical properties of brick waste

| Physical properties | Brick | ASTM C618 Class C Limitation |
|---|---|---|
| Fineness: Amount retained when wet-sieved on 45 µm (No. 325) sieve, % | 20 | Max 34 |
| Specific gravity | 2.98 | |

**Table 9.** Physical properties of the steel deformed bars

| Item | Properties | limits of the American standard (ASTM A615) of steel type (420) |
|---|---|---|
| Tensile strength ($N/mm^2$) | 619.2278 | 620 Min. |
| Yield strength ($N/mm^2$) | 542.064314 | 420 Min. |
| Elongation (%) | 16.182573 | 9% for dia.12 Min. |

## 2.2 Concrete mix proportions

Concrete mixture consists of 400 $kg/m^3$ cement, 750$kg/m^3$ sand, 1035 $kg/m^3$ gravel, and a w/c was ranged within (0.41-0.51), 1.2% HRWR by weight for reference concrete and 1.4% for brick –modified concrete and it was designed to achieve compressive strengths of 50 MPa at 28 days. The slump was kept within the range (80-90) mm as shown in Table (10).

According to a typical mixing procedure, a mixer of a 20 kg capacity was used to prepare the concrete mix: First, gravel and sand wee mixed together and followed by cement, initially mixed dry material for 1.5 min; finally, the addition of HRWR and 30% of water together. After 1.5 min of mixing, the remaining water was added to the mixer. Then all batches were mixed for about 2 min. The water was adjusted to maintain a slump within the range .To prevent fresh concrete from segregation. The mixing duration was kept as low as possible. For the testing, specimens of the hardened properties were prepared by direct pouring of concrete into molds with compaction by vibrating table. High performance concrete was mixed by a similar procedure to reference concrete except that the required quantity of brick powder was mixed with dry cement very well .Then mixed with other ingredients in the same manner as for reference concrete.

**Table 10.** Concrete mix

| Concrete mix | Cement kg/m³ | Waste added kg/m3 | ercentage of waste % | Fine aggregate kg/m³ | Coarse aggregate kg/m³ | w/c ratio | HRWR adm.,% |
|---|---|---|---|---|---|---|---|
| R | 400 | 0 | 0 | 750 | 1035 | 0.41 | 1.2 |
| B5 | 380 | 20 | 5 | 750 | 1035 | 0.51 | 1.4 |
| B10 | 360 | 40 | 10 | 750 | 1035 | 0.47 | 1.4 |
| B15 | 340 | 60 | 15 | 750 | 1035 | 0.48 | 1.4 |

R: reference concrete Mix; B5, B10, B15: 5, 10, 15%brick,

## 2.3 Curing procedures

Concrete casting was performed in two layers. Each layer was compacted by using a vibrating table until no air bubbles emerged from the surface of concrete and the concrete was adjusted to the top of the mold smoothly (ASTM C192-88). The specimens then were kept in the laboratory for about (24) hrs. The specimens were covered with nylon sheets to prevent evaporate on of water from the concrete. After that the specimens were remolded carefully, immersed in tap water in the curing basin at room temperature until the time of testing.

## 2.4 Experimental program

Tests for hardened concrete were; strength activity index (ASTM C311) [9], compressive strength B.S 1881: Part 116 [10], flexural strength (ASTM C293-79) [11], depth of water penetration under pressure (BS EN 12390-8:2009), absorption after immersion (ASTM C642-97) [12] and The electrochemical potential of the reinforcing steel was measured relative to a copper-copper sulfate electrode (CSE). The relationship between half-cell potential and corrosion of steel embedded in concrete are shown in Table (11) .The half –cell potential measurements are interpreted according to ASTM C 876-99 [13].The compressive strength test was done with 60 cubes at the ages of 7, 28, 90, 150 and 240 days. The flexural strength and absorption after immersion tests were done with 12 specimens for each test at the age of 150 and 240 days. The depth of water penetration under pressure was done with 6 cubes at the age of 150 day. The half-cell potential measurements was done with 6 specimens and monitoring the current at different period of time.

**Table 11.** Relationship between half – cell potential and corrosion of steel embedded in concrete (ASTM c 876-99)

| Half-cell potential (mV) | Interpretation |
|---|---|
| > -200 | Low probability of corrosion |
| -200 to -350 | Uncertain corrosion activity |
| < -350 | high probability of corrosion |

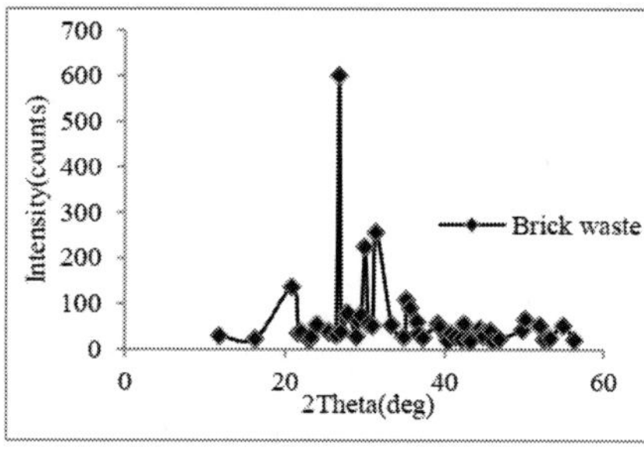


**Fig. 2 .** X-Ray analysis for brick waste

# 3 Results and discussion

## 3.1 Strength activity index (SAI)

SAI is used to determine if pozzolana results are within the acceptable level of strength development when used with hydraulic cement in concrete. The results may not provide a direct correlation of how pozzolana will contribute to strength in concrete, due to conduct the test with mortar. The strength activity index (SAI) of waste brick powder was determined with and without HRWR and the obtained values are presented in Table (12).The SAI of brick powder was within the limits of ASTM C618 class C at 7 days and 28days with or without HRWR however using the superplasticizer provide good spread of brick powder through the mixture and reduce the water requirement which led to a significant increase in strength. The total percentage of $SiO_2$ + $Al_2O_3$ + $Fe_2O_3$ of brick powder was 59.52 % which was within the limits of ASTM C618 class C,

(Min50%), due to high fineness of the brick powder. It filled the voids between cement particles and the sand and contributing to greater than 112 % strength activity index, which could have contributed by the pozzolanic reaction of brick powder with $Ca(OH)_2$, producing C-S-H and increasing the strength.

**Table 12.** Strength activity index (SAI)

| | Water requirement | | Strength Activity Index at 7 days,% | | Strength Activity Index at 28 days,% | |
|---|---|---|---|---|---|---|
| | Without HRWR | With 1% HRWR of cementitious materials | Without HRWR | With 1% HRWR of cementitious materials | Without HRWR | With 1% HRWR of cementitious materials |
| Brick | 103 | 84 | 75 | 97 | 82 | 112 |
| ASTM C618 Class C Limitation | Max 105% of control | | Min 75 | | Min 75 | |

### 3.2 Compressive strength

The effect of different content (0, 5, 10 and 15%) by weight of brick waste as a partial replacement of cement on the compressive strength of concrete at various ages (7, 28, 90 ,150 and 240) days is shown in Fig.(3). It can be seen from this figure that the compressive strength of reference concrete was improved by the replacement with brick waste to a percentage less than 10 % by weight at all curing ages but it start to decline as cement percentage reached 15% by weight at all curing ages but it was still higher than that of the reference concrete. This may be attributed to pozzolanic reaction [14]. $SiO_2$ reacts chemically with alkalis in cement and form cementitious product that contributes to the strength development, due to very fine powder of brick waste, also brick waste may be effectively filling the voids and giving rise to dense concrete microstructure. The maximum increase in compressive waste was 23.3% for 10% replacement of cement at 240 days curing age.

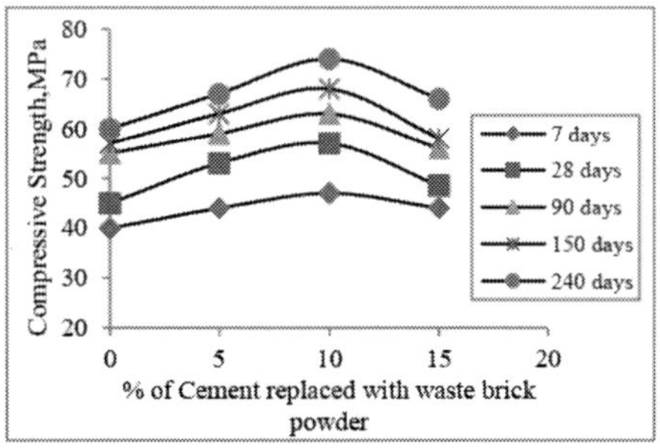


**Fig. 3.** Effect of various content of waste brick powder on compressive strength as compared with reference concrete

### 3.3 Flexural strength

The flexural strength test results for 10% by weight of brick waste are presented in Table (13). The flexural strength for reference concrete was improved by the replacement of 10% by weight brick waste as well as it tend to increase with increasing curing age from 150 to 240 days. This shows that significant pozzolanic reaction had taken place during this period of time. Similar observations were reported by [15][16].

**Table 13.** Flexural strength

| Curing ages(days) | Reference concrete | 10% brick - concrete |
|---|---|---|
| 150 | 3.88 | 4.38 |
| 240 | 5.24 | 5.55 |

### 3.4 Premeability

The results for water absorption test or the porosity test for 10% replacement brick waste–concrete and reference concrete at 150 and 240 day curing ages were plotted in Fig.(4), which indicated that the percentage water absorption for 10% brick waste– cement was less than that of reference concrete at all curing age, also it decreased with increasing curing ages from 150 to 240 days. The maximum reduction of water absorption was 13.87% as compared with reference concrete. This reduction in water absorption and water accessible porosity is attributed to the pozzolanic reaction of brick waste, which can refine the pore structures and decrease in the connectivity. The transport of water is governed by factors like pore connectivity and pore size distribution, other than water accessible porosity, indicating that pozzolanic reaction can refine the capillary pores for water transport due to more fineness property of particles than cement and the capacity of the pozzolanic towards fixing $Ca(OH)_2$ generated during the reactions of hydration of cement.[17][18]These results are in agreement with the finding of [19]. The water penetration depth for 10% replacement brick waste–concrete and reference concrete are calculated for 150 days curing age and results present in Fig. (5), this figure shows that the depth of water penetration for brick - concrete was less than that for reference concrete. The incorporation of brick waste as replacement of cement content may improve the pore structure in the transition zone, and thereby reduce water permeability. The results are in agreement with those obtained by [20] who reported a reduction in the water depth penetration when using pozzlanic materials as cement replacement.

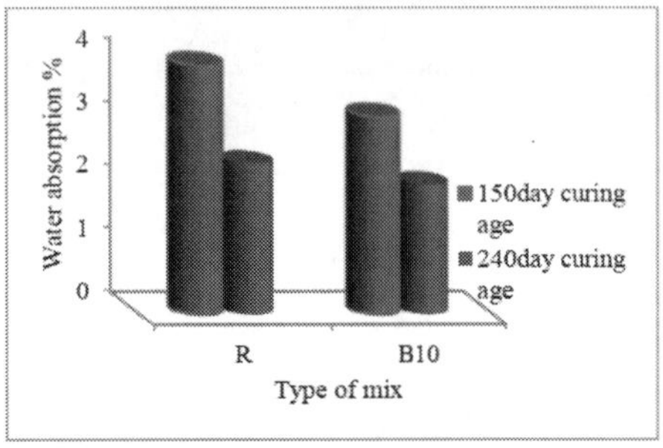


**Fig. 4.** Water absorption test

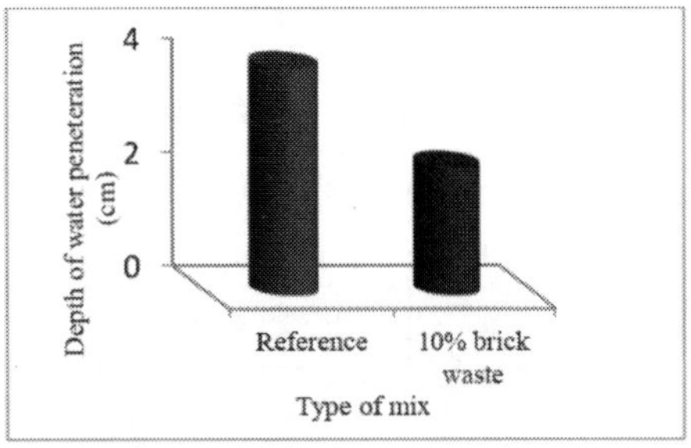


**Fig. 5.** Water penetration for the concrete mixes

### 3.5 Half–cell potential

The results of half-cell potentials of steel as referenced to a copper- copper sulphate electrode (CSE) for 10% replacement brick - concrete and reference concrete embedded in 3.5% NaCl solution were shown in Fig. (6).It is obvious that the potentials of steel in all types of concretes were shifted towards negative potential at the beginning of exposure period. The highly negative potential is not an indicative of high corrosion activity. This may be due to the higher moisture content of concrete after 28 days curing in tap water. The half-cell potential of steel increases with increasing moisture content of concrete [21]. At 150 days specimens of 10% by weight brick- concrete showed that the steel potential was (-265 mV) which was lower than that of reference concrete (-338 mV). This behaviour may be proposed on the basis of the better adsorptive capacity resulting in a barrier film difficult to break or penetrate by the corrosive chloride ions [22] after that it started to fluctuate and appeared to shift to more negative potential. At the end of 300 days of exposure the average potential in reference concrete reached -356 mV, while the average potential of brick concrete reached (-337mV). It should be noted that all B10 concrete of steel potential was significantly below the threshold value stated by ASTM C876,Table 9. This pozzolanic reaction reduces the porosity and permeability of cement paste making it stronger and significantly more durable [23]. Use of superplasticizer in concrete lead to reduced water content and further to low permeability, thereby, reducing the penetrability of the aggressive ions.

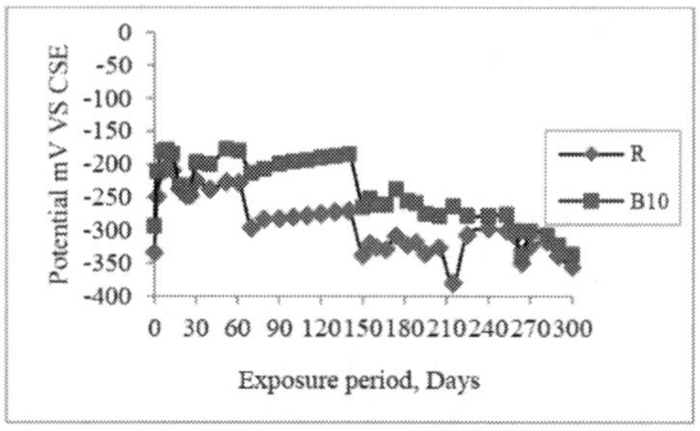


Fig. 6. Potential versus time behavior for 10% brick material – concrete with reference concrete

## 4 Conclusions

This study concluded the performance of brick waste powder as supplementary cementitious materials to produce high performance concrete by using these materials in conjunction with high range water reducing admixture and minimizing the threat of these types of solid wastes for environmental protection. This study confirmed that the use of 10 % of grinding brick improved strength of concrete and decreases the permeability of water under presser and reduced corrosion of steel reinforcement of about 18%.

## References

[1]T.B. Ilker, G.F.Nedim, 25,7(1995)
[2]B.V. Bahoria, D.K. Parbat and P.B., Naganaik J. of Envi. Res. And Deve. 7,4 (2013)
[3]T.R Naik ,. (2005)
[4] J. Bijen, Constr. &Build. Mat., **10** (1996)

[5], L. O. Ettu, N. N. Osadebe, and, M. S. W. Mbajiorgu, IJMER **3**, 2 (2013).
[6]A A, Aliabdo, A M Abd-Elmoaty and H. H, Hassan, Alex. Eng J. **53** (2014)
[7]J.Rapoport, C.M., Aldea and S. P. Shah. Permeability of Cracked Steel Fiber-Reinforced (2002).
[8] T.Kavas, and A.Olgun, Ceramics **52**,1 (2008).
[9] ASTM C311-02, pp. 9 (2002)
[10]BS 1881: Part 116- 1989, British Standards Institution, (2003).
[11] ASTM C293-79 , **04**, pp.162-163 (1979).
[12] ASTM C642-97, pp.1-2.1997.
[13]ASTM C876-99, p. 6. (2013).
[14] A. Heidari, and B.Hasanpour, Asian J. of Civil Engineering **14**, 5 (2013)
[15] S Shruthi., S, Chandrakala and G. Narayana, IJRET: Inter. J. of Res. in Eng. and Tech. **4**, (2015)
[16] T.Subramani, and S.B., Ram, IOSRJEN **05**,05 ( 2015).
[17] V.M Malhotra and P.K. Mehta, London.
[18] S. Ghosal and S., Moulik, Int. J. of Eng.Res.**4**,9, (2015 )
[19] D. Hongjian, H.J., Gao, and W. Li, ACMSM23 **I**, pp. 267-272. (2014).
[20]B., Menadi, S.Kenai, J. Khatiband, A. Al-Mokhtar, Con. & Build. Mat. **23**, pp 625-633.(2009)
[21] B.B. Hope and A.K., IP, **15**, 3, pp.525-534. (1985).
[22]A.Phanasgaonkar, B.Cherry, and M., Forsyth, pp.6. (1997)
[23] R.Lewis, L., Sear, P. Wainwright, and R., Ryle, chapter3, pp. 66, (2003)

# XRD analysis for hydration products of different lime-pozzolan systems

*Nahida* Kadum[1], *Zeyad* Al-Azzawi[2], and *Tareq* al-Attar[2,*]

[1]Ministry of Youth and Sport, Baghdad, Iraq
[2]University of Technology, Baghdad, Iraq

**Abstract.** This study represents a part of an extensive experimental program devoted to study the properties of different Lime-Pozzolan systems as a sustainable binder to replace Portland cement in concrete. These systems were Lime-Silica fume, Lime-Silica fume-Fly ash and Lime-Metakaolin. Four powder samples for X-ray diffraction, XRD, analysis was prepared from the studied systems. These samples were chosen to monitor the effect of curing age on the hydration process and to study the difference in hydration products when using different binder systems. According to the defined minerals by XRD analysis, the hydration products of all investigated Lime-Pozzolan mixtures are closely similar to those of Portland cements. The current XRD analyses have been employed to build recommendations for the forecast performance for each system.

## 1 Introduction

Utilizing of natural Pozzolan, such as volcanic ash, in combination with Lime to produce hydraulic concretes was first done by the Romans. Two millenniums old structures, including the Pantheon, are a good proof to the strength and durability of this ancient material [1]. Lime-natural Pozzolan cements were widely used for all kinds of construction as the earliest important building materials. The invention of Lime and Lime-natural Pozzolan cements backs to the Neolithic period, 7000 BC. They were widely used in the masonry construction of aqueducts, arch bridges, retaining walls and buildings during Roman times [2]. In the middle and late 19th century, such binders were commonly used for plain concrete structures, often in combination with crushed bricks as coarse aggregate for Dutch military works [3]. The use of Lime-natural Pozzolan cements, however, had not been continued due to some disadvantages such as: long setting time and low early-age strength. On the other hand, Lime-natural Pozzolan cements have a number of advantages including low cost and more importantly long-term engineering properties such as very low heat evolution, decreased permeability and increased chemical resistance. The invention of Portland cement in the 19th century resulted in a drastic reduction in use of Lime-natural Pozzolan cement because of the faster setting time and higher early-age strength of Portland cements. In recent decades, the environmental side-effects of Portland cement production process have caused a growing interest in the demand of Lime-Pozzolan binders [2]. The manufacture of Portland cement is an energy and carbon intensive process and it is responsible for 5 to 8 percent of the total global carbon emissions [4].

X-Ray Diffraction, XRD, is a technique used to determine the mineral properties of a crystalline solid. When a crystalline solid is exposed to X-Rays of a particular wavelength, layers of atoms diffract and produce a pattern of peaks, which characterize the minerals present in the material [5]. Powder XRD has numerous advantages like non-destructive nature, high sensitivity, reliability, easy sample preparation, system is user friendly, operational procedure is convenient, easy data interpretation that could be used for both qualitative and quantitative analysis, in wide range of applications. Since every material has its unique diffraction patterns so materials and compounds can be identified by using a database of diffraction patterns [6].
Lime-Pozzolan systems could be considered as promising new binders; therefore, the aim of this study is to investigate the XRD patterns for these mixtures. This investigation will enrich the available database and could give an indication about the effect of the chemical composition on the behavior of these mixtures in their service life.

## 2 Experimental work

### 2.1 Lime-Pozzolan mixture

In an extensive experimental program, three different Lime-Pozzolan systems were investigated. The program was mainly devoted to study these systems as a sustainable binder to replace Portland cement in

* Corresponding author: 40076@uotechnology.edu.iq

© The Authors, published by EDP Sciences. This is an open access article distributed under the terms of the Creative Commons Attribution License 4.0 (http://creativecommons.org/licenses/by/4.0/).

concrete, for more details see Kadum [7]. The systems were Lime-Silica fume, LS, Lime-Silica fume-Fly ash, LSF, and Lime-Metakaolin, LMK. Tables 1and 2 show the chemical composition and physical properties of hydrated Lime, Silica fume, Fly ash and Metakaolin, respectively. Table 3 lists the mix proportions for the produced concrete throughout this study.

**Table 1.** Chemical and physical properties of hydrated lime

| Test | | Results |
|---|---|---|
| Chemical Composition, % | CaO | 86.444 |
| | MgO | 2.14 |
| | $CO_2$ | 2.47 |
| Chemical Factor | | 84.2 |
| Blaine Fineness, $m^2/kg$ | | 1200 |
| Pozzolanic Receptivity Index, % | | 101.04 |

**Table 2.** Chemical composition and physical properties of used Silicafume, Fly ash and Metakaolin.

| Pozzolana Type | | SF | FA | MK |
|---|---|---|---|---|
| Chemical Composition, % | $SiO_2$ | 90.64 | 65.65 | 51.35 |
| | $Al_2O_3$ | 0.02 | 17.69 | 22.4 |
| | $Fe_2O_3$ | 0.01 | 5.98 | 2.3 |
| | CaO | 1.21 | 0.98 | 3 |
| | MgO | 0.01 | 0.72 | 0.17 |
| | $SO_3$ | 0.23 | 0.19 | 0.15 |
| | LOI | 2.86 | 3.1 | 7.8 |
| Blaine Fineness, $m^2/kg$ | | 20000 | 773 | 1100 |
| Specific Gravity | | 2.35 | 2.26 | 2.48 |

**Table 3.** Mix proportions for different Lime-Pozzolan systems.

| Mix | LS | LSF | LMK |
|---|---|---|---|
| Lime, $kg/m^3$ | 166 | 200 | 200 |
| SF, $kg/m^3$ | 166 | 80 | --- |
| FA, $kg/m^3$ | --- | 120 | --- |
| MK, $kg/m^3$ | --- | --- | 200 |
| Fine aggregate, $kg/m^3$ | 680 | 704 | 704 |
| Coarse aggregate, $kg/m^3$ | 1020 | 1056 | 1056 |
| Water, $kg/m^3$ | 166 | 160 | 198 |
| HRWR, % of L+P | 2.5 | 1 | 2.5 |

### 2.2 XRD test

The test was conducted in the Nanotechnology and Advanced Materials Research Center, University of Technology. Figure 1 shows the X-ray Diffraction apparatus. Four powder samples were prepared for the test. They were chosen as follows:

a. Two samples from the Lime-Silica fume mixture at ages of 7 and 90 days to monitor the effect of curing age on the hydration process.

b. Two samples from mixes Lime-Silica fume-Fly ash and Lime-Metakaolin mixtures at the age of 90 days to study the difference in hydration products when using different binders.

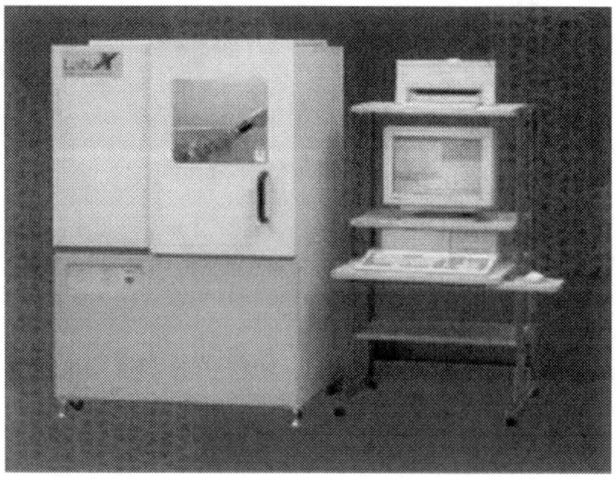

**Fig. 1.** X-ray diffraction apparatus

## 3 Analysis and discussion of XRD results

Table 4 and Figures 2 - 5 display the result of the XRD of powder samples taken from mixes LS, LSF and LMK. Data of XRD powder, listed in Appendix 3 in reference No. 8, were adopted to define the minerals represented by peaks in Figures 2 - 5. According to the defined minerals, it could be concluded that the hydration products of Lime-Pozzolan mixtures are closely similar to those of Portland cements.

**Table 4.** Summary of results of the XRD of powder samples taken from mixes LS, LSF and LMK.

| Mix | Age, days | High Peaks | Peak No. | 2 theta, deg. | d, A° | Rel. Int., % |
|---|---|---|---|---|---|---|
| LS | 7 | 1 | 4 | 26.7 | 3.34 | 100 |
| | | 2 | 1 | 11.7 | 7.53 | 73 |
| | | 3 | 5 | 29.39 | 3.04 | 58 |
| | 90 | 1 | 7 | 29.5 | 3.02 | 100 |
| | | 2 | 1 | 11.8 | 7.49 | 66 |
| | | 3 | 3 | 21.0 | 4.21 | 52 |
| LSF | 90 | 1 | 6 | 29.6 | 3.02 | 100 |
| | | 2 | 1 | 11.9 | 7.45 | 55 |
| | | 3 | 5 | 26.8 | 3.32 | 52 |
| LMK | 90 | 1 | 14 | 29.7 | 3.01 | 100 |
| | | 2 | 9 | 23.2 | 3.83 | 81 |
| | | 3 | 12 | 26.9 | 3.31 | 69 |

### 3.1 Effect of binder type on hydration products

Figures 2 and 3 show the XRD patterns for LS mix, which incorporate Silica fume as a Pozzolanic material. The only detected phases in the examined powders are the calcium silicate, CS, calcium hydroxide, CH, and quartz, Q. The occurrence of other phases such as: calcium aluminate, CA, calcium ferrite, CF, and calcium sulphoaluminate, CAŠ have not been detected. This observation is expected since Silica fume has the highest $SiO_2$ content in all Pozzolan. These patterns were in agreement with the work of Chow and Barbhuiya [5] Bottrill [9] and Bernd [10].

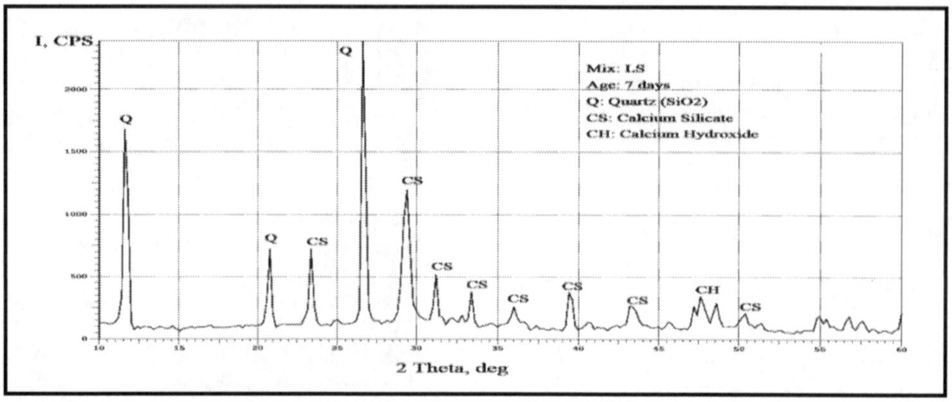


**Fig. 2.** XRD pattern for mix LS at age of 7 days.

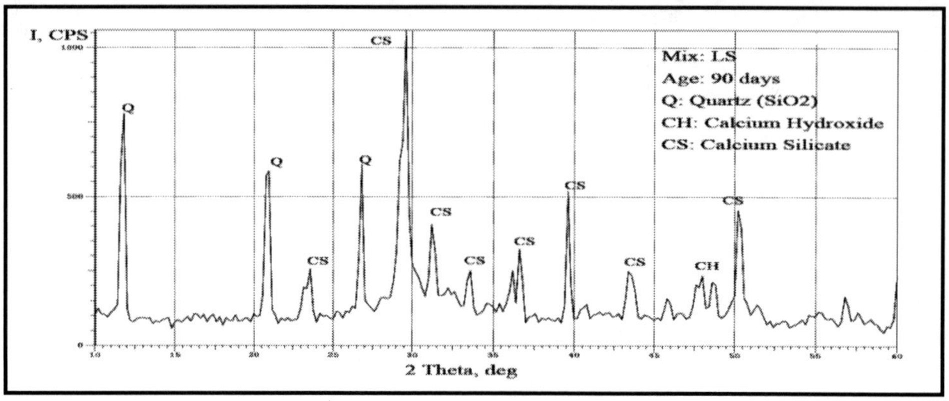


**Fig. 3.** XRD pattern for mix LS at age of 90 days.

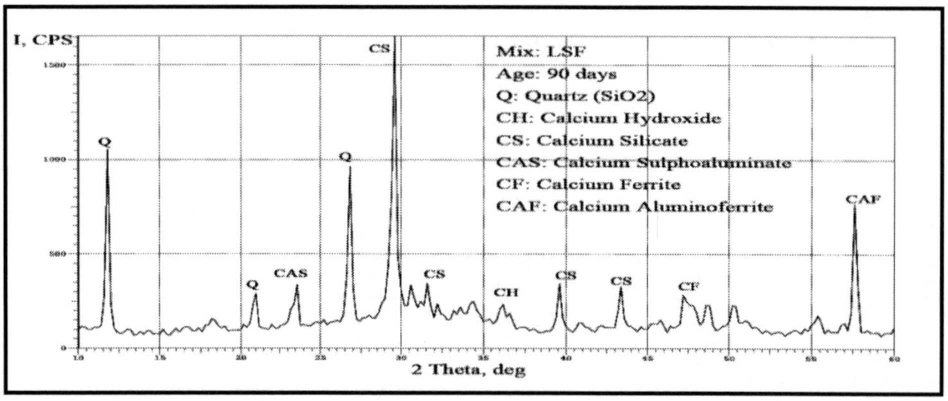


**Fig. 4.** XRD pattern for mix LSF at age of 90 days.

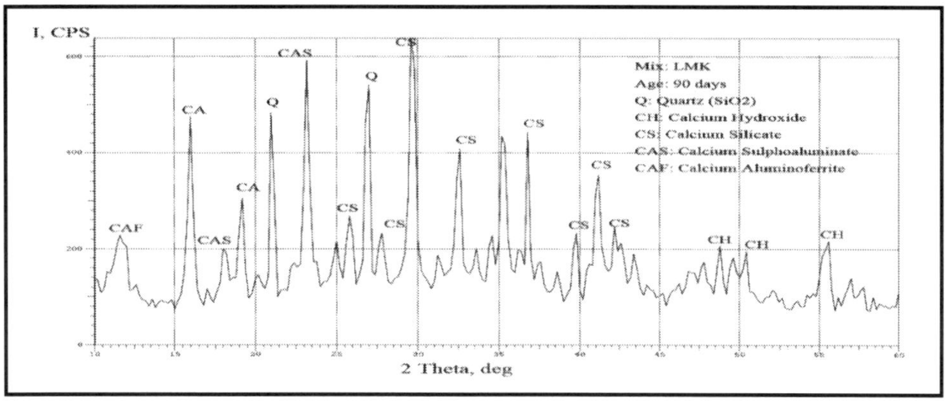


**Fig. 5.** XRD pattern for mix LMK at age of 90 days.

Figure 4 shows the XRD pattern for LSF mix at the age of 90 days. Due to the existence of $Al_2O_3$, 18 percent, and $Fe_2O_3$, 6 percent, in Fly ash, calcium aluminoferret, CAF, calcium ferrite, CF, and calcium sulphoaluminate, CAŠ have been detected in the powder in addition to the calcium silicate, CS, calcium hydroxide, CH, and quartz, Q phases. Similar patterns had been reported by Ward and French [11]. No traces of calcium aluminate, CA, phase was found. Although this observation gives the impression that this mixture is immune to sulfate attack, but the durability of LSF mixture to such attack should be investigated. Moreover, calcium aluminate has a passivity action on steel reinforcement corrosion process in reinforced concrete; therefore, the resistance of this mixture to steel corrosion should also be investigated.

When examining Figure 5 that displays the XRD pattern for the mix LMK at 90 days, the main observation is the existence of calcium aluminate, CA, phase in addition to all mentioned above phases.

That could be resulted due to the effect of both the higher content of $Al_2O_3$, 33 percent, and the lower content of $Fe_2O_3$, 2 percent, in Metakaolin. The same precautions that are recommended for Portland cements against sulfate attack should be taken. The examined XRD pattern for mix LMK was in agreement with the work of Gameiro *et al.* [12], Mlinárik and Kopecskó [13] and Santos Silva *et al.* [14].

Based on the above mentioned data, it could be concluded that the hydration products of the Lime-Pozzolana systems depending on the silica to alumina ratio in the Pozzolanic material. Also it could be outlined here that the obtained products are closely similar to those of Portland cements, i.e. calcium silicate, calcium aluminate, calcium hydroxide and calcium sulfoaluminate.

### 3.2 Effect of curing age on hydration products

A comparison between the XRD patterns for mix LS in Figure 2, for the age of 7 days, and in Figure 3, for age 90 days, could be made to detect the effect of curing age on hydration products. The rise of calcium silicate, CS, peaks is the main difference between the two figures. In contrast to that, the quartz, Q, peaks have been dropped down due the involvement of the $SiO_2$ in the reaction with CaO to produce calcium silicate, CS.

## 4 Conclusions

1. According to the defined minerals by XRD analysis, the hydration products of the Lime-Pozzolana systems depending on the silica to alumina ratio in the Pozzolanic material. Also it could be outlined here that the obtained products are closely similar to those of Portland cements, i.e. calcium silicate, calcium aluminate, calcium hydroxide and calcium sulfoaluminate.
2. The only detected phases in the examined Lime-Silica fume powder are the calcium silicate, CS, calcium hydroxide, CH, and quartz, Q. The occurrence of other phases such as: calcium aluminate, CA, calcium ferrite, CF, and calcium sulphoaluminate, CAŠ have not been detected.
3. In the powder of Lime-Silica fume-Fly ash system, calcium aluminoferret, CAF, calcium ferrite, CF, and calcium sulphoaluminate, CAŠ have been detected in addition to the calcium silicate, CS, calcium hydroxide, CH, and quartz, Q phases. No traces of calcium aluminate, CA, phase was found.
4. For the mix LMK, the XRD pattern showed the existence of calcium aluminate, CA, phase in addition to all mentioned above phases.

## References

1. R. W. Steiger, Aberdeen Group. Pub. No. **C950949** (1995)
2. A. Allahverdi, and J. Ghorbani, J. Ceramics-Silikáty **50**, 4 (2006)
3. H. A. Heinemann, O. Çopuroğlu, T. G. Nijland, Delft University of Technology (2010)

4. E. R. Grist, K. A. Paine, A. Heath, J. Norman, H. Pinder, J. case studies in construction materials, Elsevier **1** (2014)
5. P. L. Chow, S. Barbhuiya, ISEC Press (2014)
6. A. Chauhan, P. Chauhan, J. Analytical and Bioanalytical Techniques, **5**, 5 (2014)
7. N. M. Kadum, MSc Thesis, University of Technology, Iraq (2015)
8. F. M. Lea, Edward Arnold, UK (1970)
9. R. S. Bortrill, Tasmania Department of Resources and Energy - Report **1991/12**, (1991)
10. F. Bernd, Proc. 10th Int. Inorganic-Bonded Fiber Composites Conference, IIBCC São Paulo, Brazil (2006)
11. C. R. Ward, D. French, Proc. World of Coal Ash, WOCA, Kentucky, USA (2005)
12. A. Gameiro, A. Santos Silva, R. Veiga, A. Velosa, MATEC9, **46**, 2, (2012)
13. L. Mlinárik, K. Kopecskó, Civil Engineering & Architecture, **56**, 2 (2013)
14. A. Santos Silva, A. Gameiro, J. Grilo, R. Veiga, A. Velosa, Applied Clay Science, **88**, (2014)

# The effects of adding waste plastic fibers on some properties of roller compacted concrete

*Adil* Abed[1,*], *Abdulkader* Al-Hadithi[2], and *Ahmed* Salih Mohammed[1]

[1]University of Anbar, College of Engineering, Civil Engineering Department, Ramadi, Anbar, Iraq
[2]University of Anbar, College of Engineering, Dams and Water Resources Engineering Department, Ramadi, Anbar, Iraq

**Abstract.** An attempt to produce of roller compacted concrete (RCC) improved by adding waste plastic fibers (WPFs) resulting from cutting the PET beverage bottles was recorded in this study. The method which is used for production of RCC is an approved design method for ACI committee (5R-207,1980)[1]. WPF was added by volumetric percentages ranging between (0.5 to 2 %) and reference concrete mix was produced for comparison reason. Many tests were conducted on the models produced by rolling compacted concrete like compressive strength, flexural strength, modulus of elasticity, dry density, water absorption and ultrasonic pulse velocity. The analysis of the results showed that the use of plastic waste fibers (1%) had led to improvement in the properties of each of the compressive strength and flexural strength compared with reference concrete. Results also showed that the addition of these, fibers increase water absorption and reduce the speed of Ultrasonic pulse velocity.

## 1 Introduction

Roller compacted concrete (RCC) differs from conventional concrete principally in its requirement for a zero – slump consistency at the time of mixing and placement. This allows the material to be placed in successive layers using earth moving construction methods, When vibratory compaction adds to these individual layers, the resulting material exhibits comparatively high compressive strengths relative to the quantities of Portland cement used in the mixture , In addition, RCC mixtures are often able to utilize marginal aggregates which could not be tolerated in conventional concrete mix designs [2]. Batayneh et al. [3] Studied the effects of ground plastic on the properties of concrete. Concrete mixes containing plastic particles which were partially proportioned replaced with fine aggregate up to 20%. They concluded that the addition of plastic particles led to a reduction in the strength properties. Many researchers [4], [5] studied the effects of adding waste plastics as aggregates or as fibers on the properties of RRC. All these researches proved that, the addition of WPFs led to an increment in compressive strength because of the fibers increase of interlocking among the matrix and reduce cracking development during loading. The aim of this research is studying the effect of adding waste plastic fibers, which are produced from the cutting by hand of plastic beverage bottles on the properties of RCC.

* Corresponding author: adil_eng2005@yahoo.com

## 2 Experimental program

### 2.1 Materials

Ordinary Portland cement has been used in casting all specimens throughout the experimental work. Physical and chemical analysis and tests proved that, this kind of cement conforms to the Iraqi Specifications I.Q.S. 5/1984[6]. For preparing mixes, natural sand is used with a maximum aggregate size of 4.75mm. The sieve analysis and physical properties of this aggregate conforms to the Iraqi Specifications I.Q.S. No.45/1984 zone 2[7]. Crushed gravel is used for preparing mixes with a maximum size of aggregate equal to 14mm . The sieve analysis and physical properties of this aggregate proved that, this aggregate conforms to the Iraqis specifications I.Q.S. No.45/1984.

The water, which is used for both mixing and curing is potable water from the water-supply network system (tap water). The geometrical characteristics of WPFs throughout the experimental work are illustrated in Table(1).

© The Authors, published by EDP Sciences. This is an open access article distributed under the terms of the Creative Commons Attribution License 4.0 (http://creativecommons.org/licenses/by/4.0/).

 https://doi.org/10.1051/matecconf/201816202008

**Figure 1.** Waste domestic plastics are made into fiber

**Table 1.** Characteristics of plastic fibers

| Type of Fibers | Average Length (mm) | Average Width (mm) | Average Thickness (mm) | Specific gravity |
|---|---|---|---|---|
| Plastic fibers | 30 | 4 | 0.30 | 1.12 |

## 2.2 Testing moulds

Standard Cylindrical mould of 150mm diameter and 300mm height is used to prepare the test specimens .Cubic moulds of 100mm and 150 mm have been used for some of other testing specimens .Prisms moulds of (100×100×500mm) are used for the flexural strength testing. The results are modified to obtain equivalent values as for cylinder specimens.

## 2.3 Equipments which used for production of RCC samples

Modified hammer conforming to ASTM D-1557 (modified proctor, weighted 4.5 kg) was used for compaction of RCC materials, into testing moulds. The vibrating compaction hammer with a minimum mass (without tamping plate) of 10 ± 0.2kg a shown in Fig (2).It also has minimum power input of 900 Watt and capable of providing at least 2000 impacts /min. Tamping plate: A circular steel plate attached to a metal shaft, which insert into the vibrating hammer chuck. The diameter is 140±3mm and a mass of 3±0.1 kg.

**Figure 2.** Vibrating compaction Hammer with tamping plate

## 2.4 Mixes

In order to select the mixture proportion for RCC, the design method recommended by ACI committee 207-5R, 1980[8], is adopted .The weights of the materials per cubic meter of RCC are:

| Material | Weight (kg/m$^3$) |
|---|---|
| Ordinary cement | 202-355 |
| Water | 114-160 |
| Aggregate | 2020-2090 |

Therefore it is decided to use the mix which contains 329 kg/m$^3$ cement and water/cement ratio 0.4 as fixed mix proportions, at the same time the aggregate content is 2060 kg/m$^3$ where the gravel is 927kg/m$^3$ and the sand is 1133kg/m$^3$ [8].

## 2.5 Preparation of specimens

The specimens of RCC are prepared by using Cylindrical steel moulds of size (150×300mm), cubes of size (100 × 100 × 100 mm) and prisms of size (100 ×100×500 mm) for most testing which is carried out throughout this work. Before casting, the moulds are cleaned, rigidly tightened, and lightly oiled to prevent concrete sticking to them. After mixing, the materials are placed by filling these cylinders, cubes and prisms in three layers and compacted by placing the hammer with tamping plate on concrete. After that, vibrating by the hammer will be started allowing the concrete to compacted under the tamping plate. The observation should be taken of the concrete in the annular space, in the region between tamping plate edge and in the insider wall of the mould. When the concrete consolidates, mortar should fill in the annular space between the outer edge of the tamping plate and the inside mould wall.
Until the mortar forms a ring round the total primeter of the tamping plate, this mortar should be observed. The vibrating hammer should be stopped, when the mortar ring forms complete the tamping plate according to ASTM C-1435[10].

# 3 Results and discussions

Figure 3. illustrate the dry density and moisture content relationships for the same samples. It can be seen that the dry density increases with increasing moisture content until it reaches an optimum, then the dry density decreases with further increase in moisture content. 15% cement content (as percentage of total dry aggregate weight) gives the maximum dry density of 2.324 g/cm$^3$ at 5.5% moisture content.

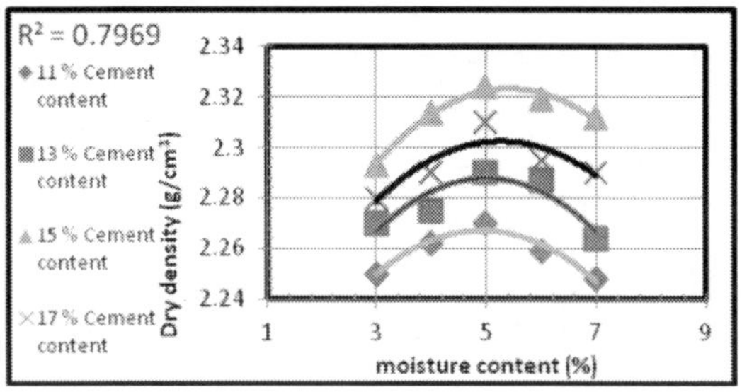


**Figure 3.** Relationships between dry density and moisture content.

## 3.1 Effect of volume fraction of plastic fibers on the mechanical properties of RCC.

The compressive strength is determined by using cylinder specimens according to ASTM C-39 2003[11]. Figure 4. shows the relationship between compressive strength with WPFs. The compressive strength increases from plastic fibers ratio of (0%) up to ultimate compressive strength at plastic fibers of (1%) then the strength decreases as plastic fibers ratio increases at any curing age. The reduction in compressive strength beyond the optimum (plastic fibers) of (1%) may be due to decrease in workability factor, this leads to resistance in that the compaction process and not to obtain fully compacted mixture. The ultimate compressive strength at (7,14 and 28) days is higher by (41.2, 48.5 and 52.153%) respectively, than the compressive strength at plastic fibers ratio of (0%).

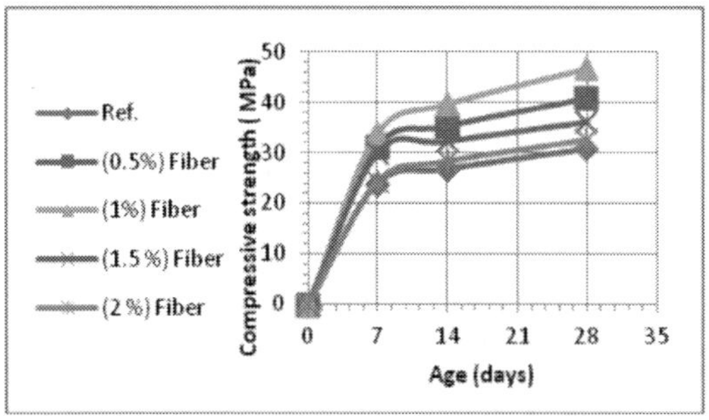


**Figure 4.** The relationship between compressive strength of RCC mixes with age.

Determination of the modulus of elasticity for each mix was done according to ASTM-C-469 2002[12]. Figure 5. shows the effect of volume fraction of plastic fibers on the static modulus of elasticity of RCC at 28 days. It can be seen that there is a significant increase in the static modulus of elasticity with the increase in the volume fraction of plastic fibers up to (1 %) by volume. The average increase is higher by (44.73%) than the static modulus of elasticity at plastic fibers ratio of (0 %).

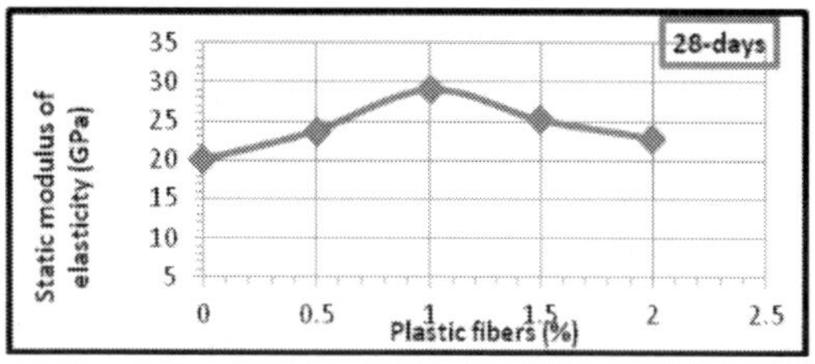


**Figure 5.** Variation in static modulus of elasticity with WPFs in (28 days)

The flexural strength of the (100×100×500 mm) specimens are determined by applying center point loading of sample beam according to ASTM C-293 2003[13].The effect of volume fraction of plastic fibers on the Flexural Strength (ultimate flexural Strength) of RCC specimens at (7, 14 and 28) days is shown in Figure 6. It can be noticed that the flexural strength increases with the increase of plastic fibers up to (1%) .The ultimate flexural strength at 28 days is higher by (17.86%) than the flexural strength at (plastic fiber) ratio of (0%).The flexural strength increases because the plastic fiber works bridging among materials during applied load.

## 3.2 Absorption of water

According to ASTM C-642 1997[14], The water absorption test is carried out using cubic moulds (150mm).The relationship between the plastic fibers and absorption of RCC mixes is presented in Figure 7. It can be noticed that the absorption increases with the increase of fiber content to (1.5%). The reason for this is the irregular distribution of fibers when increasing the volumetric ratios of fiber and this leads to roll and increase air voids under the lamellar body of the waste plastic fibers[15]. As shown in the same figure, it can be seen that, after approximately (1.5%) plastic fibers, the mix begins to have absorption less than of the mix that has (1.5%) plastic fibers where the rate of increase until(1.5%) lies between (16.95% to 35.37%) from the reference mix.

## 3.3 Ultrasonic pulse velocity (UPV)

This test was performed according to ASTM C597 - 09[16]. Figure 8. shows that the pulse of RCC decreases with the increase of the plastic fiber ratio. The decrease in pulse velocity is slight, where this decrease is because finding plastic fibers in the mix where this fiber works on the extinction the wave and this gives rise to deficiency in time of wave transfer between specimens terminal. As shown in the same Figure it can be seen at approximately (2%) that plastic fiber is decreased (1.45%) than the mix that has (0%) plastic fiber.

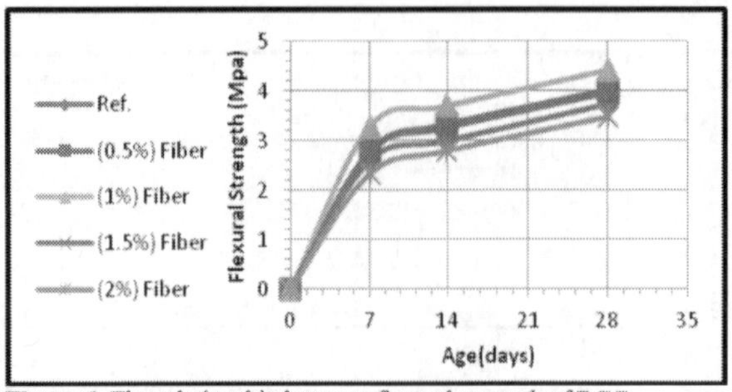


**Figure 6.** The relationship between flexural strength of RCC mixes with age.

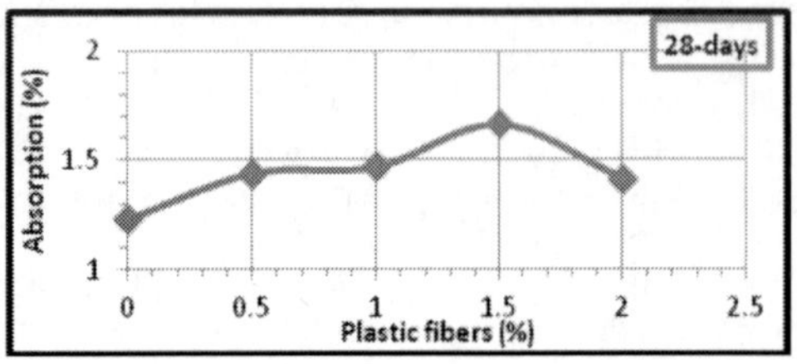


**Figure 7.** Variation in absorption with a WPFs ratio of RCC mixes in (28 days).

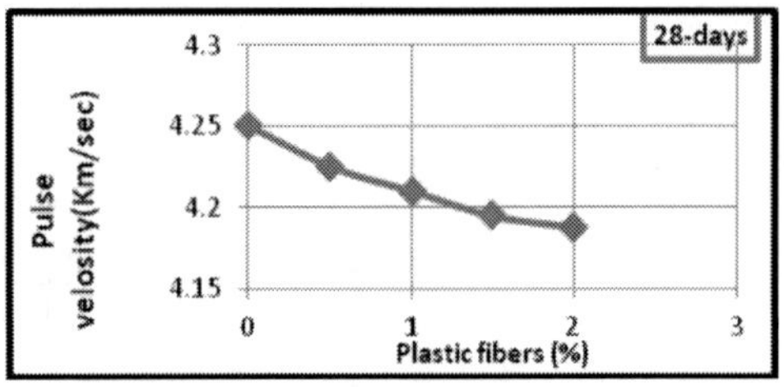


**Figure 8.** Variation in pulse velocity with WPFs ratio of RCC mixes in (28 days).

## 4 Conclusions

The compressive strength of RCC depends mainly on the water/cement ratio, cement content and degree of compaction.

- It is shown that the optimum water/cement ratio is (0.4) for the mix adopted throughout this investigation.
- Compressive strength for RCC mixes increases with the addition of plastic fiber to the reference mix as a percentage of the mix. Compressive strength increases until (plastic fiber) ratio reaches an optimum value (1 %) and it is higher by (52.15%) than the compressive strength of (plastic fiber) of 0 % at 28 days.
- The flexural strength increases with the an increment of WPFs ratio, and the ultimate flexural strength at 28 days is higher by (17.86%) than the flexural strength at (plastic fiber) ratio (0%).
- The porosity and absorption increases with the increase of the percentage of plastic fiber where the rate of increase until (1.5%) plastic fiber between (10.12 to 52.4 %) and (16.95 to 35.37 %) for porosity and absorption respectively.
- The pulse velocity of RCC decreases with the increase of the plastic fiber ratio; at approximately (2%) plastic fiber was decreased to (1.45 %) than the mix that has 0 % plastic fiber.

## References

1. ACI Committee 207, (ACI-207 5R-80),ACI, 1980.
2. K. D. Hansen, Concr Int J, Vol. 9, No. 2, pp.15-17, Feb. ,1987.
3. M. Batayneh, I. Marie and I. Asi, Waste Manag . Vol. 12, No. 27, pp. 1870–1876, 2007.
4. Piggott, R.W, PCA R & D serial No.2261, Portland cement Association, Skokie, Illinois , USA, PP. 62, 1999.
5. E. Reid and J. Marchand, Nova Scotia, 1998.
6. *Iraqi specifications* - No(45), Ministry of Planning, Baghdad, Iraq,1999.
7. *Iraqi standard specification*- No(5), Ministry of Planning, Baghdad, Iraq, 1999.
8. ACI Committee 207-5R-96, *ACI Manual of Concrete Practice*, part (1), 1-49, 1996.
9. ASTM D1557, *Annual Book of ASTM Standards*, Vol. (04.08), West Conshohocken, PA., United States, 2002.
10. ASTM D1435, *Annual Book of ASTM Standards*, Vol. (04.08), West Conshohocken, PA., United States, 2009.
11. ASTM C39/C39M-01, Vol. 04.02), West Conshohocken, PA., Issued October 2003.
12. ASTM C469-02, *Annual Book of ASTM Standards*, Vol. (04.02), West Conshohocken, PA., Issued October 2003.
13. ASTM C293, *Annual Book of ASTM Standards*, Vol. (04.02), West Conshohocken, PA., Issued October, 2002.
14. ASTM C642 (1997), "Standard Test Method for Density, *Annual Book of ASTM Standards*, West Conshohocken, PA., Issued October.
15. A. I. Al-Hadithi and W. K. Frhaan, Iraqi J. of Civil Eng., Vol. 11, No. 1, pp.1-20,2017.
16. ASTM C597-09, *Annual Book of ASTM Standards*, West Conshohocken, PA., 2009.

# Performance of self-compacting concrete containing hybrid fibers in aggressive environment

*Mohammed* Salman [1], *Qais* Frayyeh [2*], *Luma* Zghair [1]

[1] Civil Engineering Department, Al- Mustansiriyah University, Baghdad, Iraq
[2] University of Technology, Baghdad, Iraq

**Abstract.** Self-compacting concrete can be used in many concrete applications some of it susceptible to acid attacks. The aims of this study are to investigate the resistance of different SCC mixture designs to sulphuric acid attacks. The test variables included the type of hybrid fibers (steel, plastic and polypropylene fibers). The powder content of the mixes was kept constant. (500)Kg/m3. The slump flow, L-box, and V-funnel were performed for mixes in their fresh state. In the present work, the specimens were immersion in sulphuric acid solution at concentration of (0.5%) up to 289 days after normal curing for 28 days. After concrete has hardened, two types of test are performed before and after immersion in sulphuric acid solution. Firstly destructive tests are conducted including (compressive strength, splitting tensile strength, and modulus of rupture). Secondly non-destructive tests are performed including (mass loss). The result obtained from this work, shows that the use of hybrid fiber was significant in improving the resistance to sulphuric acid solution in the case of splitting tensile strength, mass loss and modulus of rupture on the other hand, the use of hybrid fibers was insignificant in reducing the concrete compressive strength loss.

Index Terms— self-compacting concrete, hybrid fibers, sulfuric acid

## 1 INTRODUCTION

Deterioration of concrete members prone to aggressive acid environments are durability issue that affects the life cycle performance and maintenance of infrastructure.[1] Concrete can be attacked by liquids with pH value below 6.5 but the attacks are severe only at a pH below 5.5 [2] Sulfuric acid fluids are classified as the most aggressive of natural threats to concrete structures. Generally, they arise from industrial operations, but they can be caused by urban area activities. Large amount of acids exist in sewage systems. Acid attack is influenced by the processes of disintegration and leaching of cement paste constituent. A very important quantity of admixtures in SCC paste can negatively or positively affect its resistance to acid aggression. [3] .The deterioration of concrete in acidic environments is influenced by several factors such as the type of cement used, permeability of concrete, and the surrounding environmental conditions. Furthermore, the solubility of calcium salt produced from the acid – base reaction of the cementitious paste and attacking acid is an important factor. [4] Many studies about the performance of concrete exposed to sulphuric acid solution have been completed.

An experimental program was carried out by **Senhadjiet al.** [5] to study the influence of different supplementary cementitious materials (silica fume, natural pozzolan) and limestone fine at various replacement levels. The results showed that the addition of limestone powder or natural pozzolan can improve the acid resistance of mortar, but at different rates depending on the proportion of supplementary cementitious materials. On the other hand, mortars with silica fume are seriously damaged in the sulphuric acid environment. They reported that despite the improving SF the compressive strength and reduced the porosity of concrete, this improvement was not effective for enhancing the resistance of concrete to sulphuric acid attacks. In other words, converting C–H into calcium silicate hydrate and refining the porosity of concrete did not improve its resistance to severe sulphuric acid attacks. They also reported that C-S-H which formed from pozolana reaction may be chemically unstable in acidic environments.

**Fan et al** .[6] estimated the tensile properties of normal concrete (283 kg / $m^3$ cement) exposed to acid rain, in this study the pH level of acid rain was below 5.6. Based on the test result the following was observed

At the initial damage stage, both the elastic modulus and the tensile strength of the concrete samples exhibit a slight increase. When the damage continues, the tensile strength and the elastic modulus of the specimen gradually decrease after 70 days immersion in acid solution.

**Bassuoni and Nehdi**[1] examined the resistance to sulfuric acid solutions (pH of 2.5 and 1.0) of a different

*Corresponding author: jwd_ks@yahoo.com

© The Authors, published by EDP Sciences. This is an open access article distributed under the terms of the Creative Commons Attribution License 4.0 (http://creativecommons.org/licenses/by/4.0/).

range of SCC mixtures containing cement only or cement with different pozolana material (silica (fume, slag and fly ash) with or without hybrid fiber reinforcement macro steel + micro polypropylene). The results indicated that:

1. The resistance of SCC to aggressive sulfuric acid solution was improved using binder with pozolana.
2. Using hybrid micro – and macro – fibers can be effective in retaining the cementitous matrix integrity and controlling disruptive pressures resulting from voluminous reaction product. **Saleh et al.** [7] used different types of fibers (carbon and polypropylene) to minimize the corrosive action of sulfuric acid solution on concrete industrial water tanks. Compressive strength test, ultrasonic pulse velocity, bulk density, and water absorption tests were done to the samples after curing in normal water for (7, 28, 60, 90, 180) days and then they were exposed to dilute sulfuric acid solution (10% concentration) for (7, 30, 60, 90, 180) days. The results indicate that using carbon fibers was more effective for enhancing the resistance of concrete to sulphuric acid attacks.

## 2 Research significant

Several concrete elements are susceptible to chemical attacks of sulfuric acid, including, industrial floors of chemical plants, superstructures, sewage pipe systems, etc., even though there has been an increased use of SCC in many concrete applications, a comprehensive review of literature indicates that there is lack of information on the role of hybrid fibers, which can be inserted in SCC in such aggressive environment.

## 3 Materials

The binders used included ordinary Portland cement (OPC), (ASTM C150 – Type 1) [8]. The cement was tested and checked according to IQS 5:1984[9], silica fume (SF) of class Npozzolans (ASTM 618-03)[10], limestone powder (chalk powder ) was used throughout this investigation. Table (1) shows the chemical and physical properties for the various binders.

Three types of fibers were used for the purpose of this study. These types are steel, plastic and polypropylene fibers. Table (2) illustrates the properties of these fibers. The binder content was kept constant at 500 kg/m$^3$ compatible to the guidelines for SCC mixture design [e.g. EFNARC] [11]. The fine aggregate has a fineness modulus of 2.6, and specific gravity of 2.65. The maximum nominal size for crushed coarse aggregate was 10 mm, with a specific gravity of 2.62. The constituents of the selected SCC mixes are given in Table 3.

**Table 1** . The chemical and physical properties for the various binders

| Oxides | cement | Silica fume | Chalk powder |
|---|---|---|---|
| $SiO_2$ | 20 | 96.68 | 2.24 |
| $Fe_2O_3$ | 3.9 | 0.069 | 0.12 |
| $Al_2O_3$ | 4.5 | 0.20 | 0.42 |
| CaO | 62 | 0.54 | 68.73 |
| MgO | 2.43 | 0.12 | 0.70 |
| $SO_3$ | 2.03 | 0.61 | <0.07 |
| specific gravity | 3.2 | 2.13 | 2.42 |
| specific surface area m$^2$/g | 4.37 | 0.157 | 3.17 |

**Table 2** . Properties of fibers

| Types of fiber | Plastic | Poly propylene | Micro steel |
|---|---|---|---|
| Description | crimped | monofilament | straight |
| Length of fiber (mm) | 30 mm | 12 | 15 |
| Tensile strength (MPa | 250-300 | 350 | 2600 |
| Aspect Ratio | 63 | 66.7 | 75 |

**Table 3** . The constituent of the selected SCC mixes

| Mix notation | SP% | Fibers by volume % | | |
|---|---|---|---|---|
| | | Steel | plastic | Polypropylene |
| SCC LP SF | 8.7 | - | | |
| SCC SF LP SFHR1 | 11 | 0.3 | - | 0.2 |
| SCC SF LP SFHR2 | 13 | 0.3 | 0.2 | - |

Note: for all mixes: water = 170l/m$^3$, sand =778 kg/m$^3$, gravel = 890 kg/m$^3$, limestone powder =150 kg/m$^3$, cement =315 kg/m$^3$and silica fume = 35 kg/m$^3$

To improve flowability of the SCC mixtures, copolymer-based Superplasticizer, (SP) designed for the production of SCC (Glenium 51) with relative density of 1.1 at 20°C was incorporated in all mixtures. The dosages of SP was modified to maintain a slump flow of 600 - 750 mm, T50 (4 to 10) sec., L box index (>0.75) (3 Ø10 mm with 50 mm gaps) and V-funnel flow (3 to 25) sec. The fresh properties were evaluated and compared with EFNARC] [11] and the European Guidelines for SCC [12]. The result of fresh properties of SCC mixtures are summarized within Table 4.

**Table 4.** Fresh properties of concrete mixtures

| Mix notation | Slump depression mm | T50 Sec. | L-box index | V-funnel Sec. | Sp % wt. of cement per $m^3$ |
|---|---|---|---|---|---|
| SCC LP SF | 680 | 7.34 | 1 | 12 | 8.7 |
| SCC LP SF HR1 | 600 | 10 | 1 | 20 | 11.6 |
| SCC LP SF HR2 | 700 | 4 | 1 | 16 | 12 |

## 4 Preparation of acid solution

In the present work, chemical immersion tests were adopted in order to assess the resistance of SCC. All concrete specimens were cured in water for 28 days, after which they were immersed in acid solution. The initial pH (2.3) of the solution increased quickly. A digital portable pH meter was used for monitoring the pH levels of the sulfuric acid solutions. Specimens were fully immersed for 41 weeks, before chemical immersion, the concrete specimens were left to dry under room temperature. Each group of mixtures had its own acid bath. This is to provide similar acidic environments for the different binder mixtures in each group of mixtures.

## 5 Harden concrete properties

### 5.1. Compressive strength of concrete

Table 5. presents the result of compressive strength at 28 days for each concrete as well as the compressive strength of concrete cubes after 90, 182 and 289 days immersion in sulfuric acid solution. The 28 days compressive strength of each concrete is used as a criterion for comparison of the compressive strength of the cubes after immersion in the sulfuric acid solution. Fig. (1) shows the relative change of compressive strength against the immersion time upto 289 days for all concretes mixtures.

**Table 5.** Compressive strengths of concrete mixes

| Mix notation | Mix No. | Compressive strength 28 days before immersion ( Mpa) | Compressive strength after immersion ( MPa) | | |
|---|---|---|---|---|---|
| | | | 90 days | 182 days | 289 days |
| SCC LP SF | M1 | 69.0 | 41.5 | 53.0 | 44.1 |
| SCC LP SF HR1 | M2 | 68.5 | 30.0 | 28.0 | 25.6 |
| SCC LP SF HR2 | M3 | 67.5 | 45.0 | 45.5 | 35.6 |

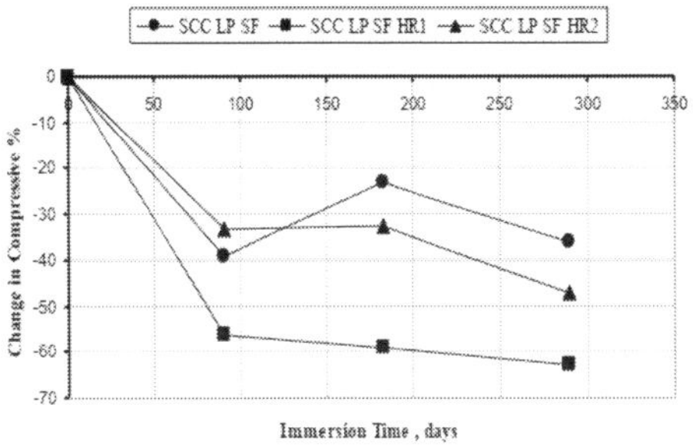


**Fig1.** The relative change in compressive strength after immersion in $H_2SO_4$ solution

From Fig (1), the relative change of compressive strength after 289 was (36, 62.6 and 47) % compressive strength loss for M1, M2 and M3, respectively. It was observed that at the age of 180 days some specimens exhibit loss of compressive strength as shown in SCC LP SF and SCCLPSF HR2 while in mixes SCCLPSFHR1 there was an increase in concrete compressive strength. The compressive strength loss at this stage probably due to the micro cracks occurs between the matrix content after the diffusion of sulfuric acid. The mass gain can be attributed to saturation of the specimens it was observed that the use of hybrid fibers was insignificant in reducing the concrete compressive strength loss, it may increase the voids in concrete matrix and this may cause further diffusion of sulfuric acid and result further corrosion of concrete that is exposed to sulfuric acid solution.

### 5.2. Mass loss of concrete specimens

Table 6 presents the result of density at 28 days for each concrete as well as the density of concrete cubes after 90, 182 and 289 days immersion in sulfuric acid solution. After immersion in $H_2SO_4$ solution, the relative change of mass against the immersion time up to 289 days was (11.76, 7.74 and 7.8) % mass loss for M1, M2 and M3 respectively, as shown in Fig. 2.

**Table 6.** Mass loss of concrete samples

| Mix notation | Mix No. | Density of concrete specimens 28 days before immersion ( kg / $m^3$) | Density of concrete specimens after immersion ( kg / $m^3$) | | |
|---|---|---|---|---|---|
| | | | 90 days | 182 days | 289 days |
| SCC LP SF | M1 | 2516 | 2535 | 2505 | 2220 |
| SCC LP SF HR1 | M2 | 2577 | 2520 | 2545 | 2378 |
| SCC LP SF HR2 | M3 | 2500 | 2385 | 2570 | 2305 |

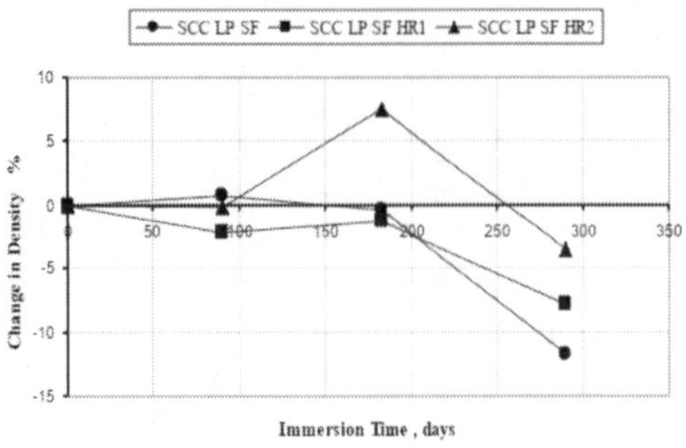


**Fig 2.**The relative change of density after immersion in $H_2SO_4$ solution

It was observed that at the age of 90 days some specimens exhibit loss of mass as shown in SCC LP SF HR1 and SCCLPSF HR2 while in mixes SCCLP SF there was an increase in concrete mass. The mass loss at this stage probably due to the micro cracks occurs between the matrix content after the diffusion of sulfuric acid. The mass gain can be attributed to saturation of the specimens or that is, in the initial conditioning age, the porosity in the concrete decreases which results in the mass increase. With continued exposure, most of the specimens exhibit mass loss. The reaction between sulfuric acid and the cement constituent of concrete results in the conversion of $Ca(OH)_2$ to calcium sulfate (gypsum) which, in turn, may be converted to calcium sulfoaluminate (ettringite). Each of these reactions involves an increase in volume of the reacting solids by a factor of about two. The formation of calcium sulfate leads to softening (decrease in density) of the concrete. Since mass depends on both volume and density, the initial mass gain of the test specimens is probably due to the relative increase in volume being greater than the relative decrease in density [13]

## 5.3 Splitting tensile strength of concrete specimens

Table 7 shows the results of 28 day splitting tensile strength for all mixes as well as the splitting tensile strength of concrete specimen after 90, 182 and 289 days immersion in sulfuric acid solution. From Table 7, the relative change of splitting tensile strength against the immersion time up to 289 days for M1, M2 and M3 was (48, 35 and 35)% respectively, as shown in Fig. 3. With continued exposure, all of the specimens exhibit tensile strength loss. This was also observed by Fan et al [6]. At the end of the test (282 days), it was observed that the use of hybrid fibers in mixes (SCC LP SF HR1) and (SCC LP SF HR2) enhanced the resistance of SCC to sulfuric acid solution comparable with (SCC LP SF).

**Table 7.** Splitting tensile strengths of concrete samples

| Mix notation | Mix no. | Splitting tensile strength 28 days before immersion (MPa) | Splitting tensile strength after immersion (MPa) | | |
|---|---|---|---|---|---|
| | | | 90 days | 182 days | 289 days |
| SCC LP SF | M1 | 3.8 | 3.6 | 3.5 | 2 |
| SCC LP SF HR1 | M2 | 4.3 | 3.9 | 3.3 | 2.8 |
| SCC LP SF HR2 | M3 | 4.9 | 3.6 | 3.8 | 3.2 |

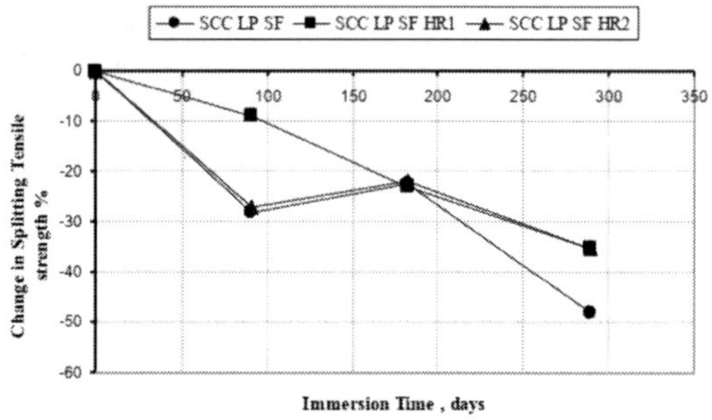


**Fig 3.**The relative change of splitting tensile strength after immersion in $H_2SO_4$ solution

## 5.4 modulus of rupture of concrete specimens

Table 8 shows the results of 28 day modulus of rupture for all mixes as well as the modulus of rupture of concrete specimen after 90, 182 and 289 days immersion in sulfuric acid solution. The relative change of modulus of rupture at 289 days comparable with result of 28 day modulus of rupture, was (41, 24 and 25)% loss for M1, M2,and M3 respectively, as shown in Fig. 4.

**Table 8** Modulus of rupture of concrete samples

| Mix notation | Mix no. | Modulus of Rupture 28 days before immersion (MPa) | Modulus of Rupture after immersion (MPa) | | |
|---|---|---|---|---|---|
| | | | 90 days | 182 days | 289 days |
| SCC LP SF | M1 | 6.1 | 8.0 | 6.0 | 3.6 |
| SCC LP SF HR1 | M2 | 6.6 | 6.5 | 7.5 | 5.0 |
| SCC LP SF HR2 | M3 | 6.7 | 6.3 | 7.5 | 5.0 |

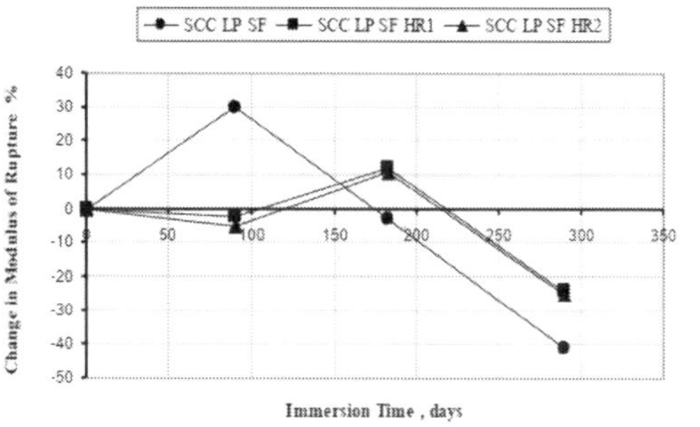


**Fig4.** The relative change of modulus of rupture after immersion in $H_2SO_4$ solution

After 90 and 182 days immersion in sulfuric acid solution some specimens exhibit decrease or increase in modulus of rupture. The decreases in modulus of rupture probably due to the micro cracks occur between the matrix content after the diffusion of sulfuric acid. The increases in modulus of rupture can be attributed to saturation of the specimens or reduction in porosity due to the chemical reaction between the sulfuric acid and the paste which results in increase in the modulus of rupture.

Finely, most of the specimens exhibit loss in modulus of rupture, at the end of the test (282 days); it was observed that the use of hybrid fibers affects the (fr) in the same manner as (ft).

## 6 Conclusion

After immersion in sulfuric acid solution. As a result, the following conclusions were obtained:

1. SCC concrete mixes were susceptible to sulfuric acid attacks but differed in the level of deterioration depending on the type of concrete mixes.
2. In the case of concrete specimens subjected to compressive strength load, the use of hybrid fibers in mixes M2 and M3 was insignificant in reducing the concrete compressive strength loss comparable with M1 the reduction was (62.6 and 47)% respectively while for M1, the reduction was (36)%.
3. Concretes made with silica fume only have a higher mass loss comparable with M2 and M3 (specimens with hybrid fibers) the reduction was (11.76)% and for mixes with hybrid fibers the reduction was (7.74 and 7.8)% respectively.
4. In the case of concrete specimens subjected to tensile strength load the use of hybrid fibers in mixes M2 and M3, enhanced the resistance of SCC to sulfuric acid solution comparable with M1 the reduction in tensile strength was (35 and 35)% respectively while for M1, the reduction in tensile strength was (48)%.
5. It was observed that the use of hybrid fibers affects the modulus of rupture in the same manner as splitting tensile strength.

## References

1. M.T. Bassuoni , and M. L. Nehdiet, Cement and Concrete Research **37** pp 1070-1084 , (2007)
2. P. S. Sesha , S. T. Sehadri , R. Srinivasa, Sravana , P. Sarika , Volume: **02,** Sep-(2013)
3. H.Siad , H. A. Mesbah , H. Khelafi , B. S. Kamali , and M. Mouli , Can. Journal in civil engineering , **37** pp 441- 449, (2010)
4. M.T. Bassuoni M. L. Nehdiet, and M. Amin , : Proceedings of the ICE - Construction Materials, Volume **160**, August, PP 113 –123, (2007 )
5. Y.Senhadji , G. Escadeillas , M. Mouli , H. Khelafi , and Benosman , Powder Technology **254** 314–323, (2014)
6. Y.F. Fan., Z.Q. Hu, and H.Y. Luan, Interaction and Multiscale Mechanics, Vol. **5**, No. 1 , pp.41-56, (2012)
7. S. A. Saleh , Z. A. Al –Din , and R. Z. Jasim , Eng .&Tech. Journal,Vol.**30** No . 11,pp.259 -272,( 2012)
8. ASTM C150, "Specification for Portland Cement" Annual Book of ASTM Standard, Vol. **04-02**, (2002)
9. Iraqi Specification, No.5., "Portland cement".(1984)
10. ASTM C618-03, "Standard Specification for Coal Fly Ash and Raw or Calcined Natural Pozzolan for Use in Concrete", Annual Book of ASTM Standard, Vol. **04-02**, February, p. 3. ( 2003)
11. EFNARC , "Specification and Guidelines for Self-Compacting Concrete" , , February, pp.32 (2002)
12. Self-compacting concrete European project group, "The European Guidelines for Self-Compacting Concrete" May, pp 43-56,( 2005)
13. E. K. Attiogbe , and S. H. Rizkalla , ACI Materials Journal *I* November-December PP: 481-488,( 1988)

# Effect of high temperature and type of cooling on some mechanical properties of cement mortar

*Faisal* Abdulhussei[1], *Waleed* Abbas[2] and *Qais* Frayyeh[2,*]

[1]Madenat al-Elem University College, Baghdad, Iraq
[2] Building and Construction Engineering Department, University of Technology, Baghdad, Iraq

**Abstract.** Mortar of cement as construction materials subjected sometimes to high temperature. Some of properties of this mortar being studied after this effect. The effect of high temperature 100, 200, 400 and 700 °C (exposed for two hrs.) on some mechanical properties (compressive and flexural strength) of two groups of cement mortar samples (with and without the addition of crushed bricks and superplasticizer as modifying materials) has been studied. Two methods of cooling samples by air and by water for 1/2 hr. was used, then tested after 3, 7 and 28 days. The results showed that the compressive and flexural strength for reference mix exposed to 700°C and water cooling decreased by 65.3 % and 64.7%, respectively, compared with their reference mix tested at 20°C in 28 days. While mixes containing 100% of crushed brick as an additive and air cooling decreases by 12.3% and 9% of their compressive and flexural strength, respectively compared with the mixes tested at 20°C in 28 days. Also showed that the decreases in flexural strength for no sand mixes containing 100% of crushed brick and 4% of superplasticizer exposed to 700°C and then water cooling was 28.2% compared to those for reference mixes tested at 20°C.

## 1 Introduction

Cement mortar as a material in service is exposed to high temperatures during fire. The main properties of cement mortar after such an exposure to fire are of great importance in terms of the serviceability of the materials.

Fire represents one of the most sever risks to buildings and structures. To find the resistance of construction materials for fire, test is performed torch for each country according to fire-fighting systems in place there. It has been found through experiments that cement mortar had bad properties in fire resistance; so many researches had been done to limit and improve the ability to withstand high temperatures.

Aydin [1] studied the effect of high temperature up to 900°C on the compressive strength of cement pumice mortars containing different amounts of ground granulated blast furnace slag (GGBFS). Aydin concluded that at 900°C, the mortar containing 80% GGBFS exhibited only 23% 23% and 28% compressive strength loss when cooled in air and water, respectively, whereas mortars without GGBFS lost almost 70% of their strength.

Aydin and Baradan [2] investigated the effects of different high temperatures on the mechanical properties of cement pumice mortars containing fly ash. They concluded that the cement pumice mortar containing 60% of fly ash revealed the best performance particularly at 900°C. This mixture did not show any loss in compressive strength at all test temperatures when cooled in air.

Hossain, Islam and Karim [3] studied the effect of high temperature during fire on the bond and compressive strength of cement mortar containing high volume fly ash without using any water reducing admixtures. They indicated that the mortar containing 50% fly ash as a replacement of cement exhibits greater resistance to high temperature. Also, compressive and bond strengths of mortar containing different percentages of fly ash initially increase with the increase in temperature but after 200°C they decrease with the further increase in temperature.

## 2 The objective

Studying the effect of high temperature on compressive and flexural strength of cement mortar, and how to decrease these effects by replacing with crushed bricks as waste materials.

*Corresponding author: jwd_ks@yahoo.com

© The Authors, published by EDP Sciences. This is an open access article distributed under the terms of the Creative Commons Attribution License 4.0 (http://creativecommons.org/licenses/by/4.0/).

# 3 Experimental works

## 3.1 Materials

### *3.1.1 Cement*

Ordinary Portland cement type (I) manufactured in Iraq with trade mark of (AL. MASS) has been used. The test results show that the cement conforms to provisions of (I.Q.S. No. 5 / 1984) [4]. Tables 1 and 2 show the chemical analysis and physical properties of cement.

### *3.1.2 Fine Aggregate (sand)*

Sand used after it has been sieved by sieve size (4.75 mm) with specific gravity of 2.64, sulfate content is 0.35%, and percentage of fine materials less than 0.075 mm was 3.4%. The grading of fine aggregate was shown in Table 3. It conformed to (I.Q.S.NO. 45 – 1984) [5].

### *3.1.3 Crushed bricks*

The bricks that have been used in this research were waste material from Iraqi factory in Al-Nahrwan. The waste bricks were crushed to get same grading as fine aggregate used as partial or complete replacement of sand. The grading of crushed brick was shown in Table 4. It was shown that the grading of crushed brick within the same zone of sand. Also, the chemical test results indicated that specific gravity was 1.84, sulfate content in crushed bricks was 0.34% and fine materials less than 0.075 was 3.8%.

### *3.1.4 Superplasticizer*

Superplasticizer SP 703S type A according to (ASTM C494 – 2004) [6] was used as a high range water reducing admixture by weight of cement of (1%, 2%, and 4%). Table 5 shows the chemical properties of the super plasticizer used according to manufacturer.

### *3.1.5 Water*

Drinking water was used for all the mixes and for the curing process of samples.

**Table 1.** Chemical analysis of cement*.

| Components | | Test Result | Limits of I.Q.S. No. 5 / 1984 |
|---|---|---|---|
| Lime (%) | CaO | 61.89 | - |
| Silica (%) | $SiO_2$ | 21.37 | - |
| Alumina (%) | $Al_2O_3$ | 4.6 | - |
| Iron Oxide (%) | $Fe_2O_3$ | 3.35 | - |
| Sulfate (%) | $SO_3$ | 2.42 | $\leq$ 2.8% |
| Magnesia (%) | MgO | 3.05 | $\leq$ 5% |
| Potash (%) | $K_2O$ | 0.36 | |
| Soda (%) | $Na_2O$ | 0.27 | |
| Loss on Ignition (%) | L.O.I | 2.16 | $\leq$ 4% |
| Lime Saturation Factor | L.S.F | 0.91 | 0.66 – 1.02 |
| Insoluble Residue (%) | I.R | 0.6 | $\leq$1.5 |
| Main compound (Bogue's equation) | | | |
| Tricalcium Silicate | $C_3S$ | 46.95 | - |
| Dicalcium Silicate | $C_2S$ | 25.85 | - |
| Tricalcium Aluminate | $C_3A$ | 6.52 | - |
| Tetracalcium Aluminoferrite | $C_4AF$ | 10.19 | - |

**Table 2.** Physical properties of cement*.

| Physical Property | Test Result | Limits of I.Q.S. No. 5 / 1984 |
|---|---|---|
| Specific Surface Area ($m^2$/kg) | 369 | 230 (min) |
| Setting time | | |
| Initial ( hrs. : min.) | 2:00 | 0:45 (min.) |
| Final ( hrs. : min ) | 3:35 | 10:0 (max.) |
| Compressive strength ( N/$mm^2$) | | |
| 3 days | 24.9 | 15 (min.) |
| 7 days | 29.7 | 23 (min.) |

*Test done at NCCL.

**Table 3.** Grading of fine aggregate (sand).

| Sieve Size (mm) | % passing | Limit of I.Q.S. No. 45 – 1984 - Zone 2 |
|---|---|---|
| 4.75 | 98 | 90 – 100 |
| 2.36 | 83 | 75 – 100 |
| 1.18 | 65 | 55 – 90 |
| 0.6 | 47 | 35 – 59 |
| 0.3 | 18 | 8 – 30 |
| 0.15 | 6.0 | 0 – 10 |

**Table 4.** Grading of crushed bricks

| Sieve Size (mm) | % passing | Limit of I.Q.S. No. 45 – 1984 |
|---|---|---|
| 4.75 | 100 | 90 – 100 |
| 2.36 | 88 | 75 – 100 |
| 1.18 | 70 | 55 – 90 |
| 0.6 | 61 | 35 – 59 |
| 0.3 | 19 | 8 – 30 |
| 0.15 | 7 | 0 – 10 |

**Table 5.** Properties of superplasticizer SP 703S

| | |
|---|---|
| Appearance | Dark brown / black liquid |
| Specific gravity | 1.235 @ 25 ± 2 °C |
| Chloride content | Nil |
| Flash point | N / A |

## 3.2 Mix Proportioning

The flow of the mortar was considered to mix design so that the w/c ratio was determined according to ASTM C1437-04 [7]. The reference mortar mix was designed to have flow 80 ± 5 mm, the cement: sand ratio was 1:3 by weight and w/c ratio was 0.6. Seven mortar mixes were prepared. Its details are shown in Table 6.

## 3.3 Preparation, Casting and Curing of Specimens

The mix of the raw materials was done according to ASTM C 305 – 99[8]. After conducting flow test, the specimens were prepared by casting in different standard molds. The standard molds were cleaned and oiled before casting to avoid the adhesion to sides of the molds. The mortar was placed in the molds with layers according to the standard specifications for each test and compaction by means of vibration table. The top layers of mortar specimens had been smoothed by steel trowel, and then the specimens were covered with nylon sheets for 24 hours to prevent the evaporation of water. After that the specimens were demolded and fully immersed in tap water for 28 days. At age 28 days the specimens subjected to different constant temperatures (100, 200, 400 and 700 °C (exposed for two hrs.) in two ways. First one when the samples were oven dry and second one where the samples were fully saturated.

**Table 6.** Mixes proportion details (by weight)

| Mix Symbol | Cement | Sand | Crushed bricks | SP % from cement | w/c to give flow of 80 ± 5 mm | Oven Dry density kg / cm$^3$ at 28 days age |
|---|---|---|---|---|---|---|
| $B_0$ | 1 | 3 | - | - | 0.6 | 1980 |
| $B_1$ | 1 | 2.25 | 0.75 | - | 0.62 | 1840 |
| $B_2$ | 1 | 1.5 | 1.5 | - | 0.66 | 1760 |
| $B_3$ | 1 | - | 3 | - | 0.72 | 1680 |
| $B_4$ | 1 | 2.25 | 0.75 | 1 | 0.58 | 2080 |
| $B_5$ | 1 | 1.5 | 1.5 | 2 | 0.55 | 2120 |
| $B_6$ | 1 | - | 3 | 4 | 0.52 | 2180 |

# 4 Results and Discussion

## 4.1 Oven dry density

This test was done according to ASTM C 567-05a [9] using 50×100 mm cylindrical specimens. The results of dry density for different mortars demonstrated in Table 6 and Fig. 1. From the results, it is clear that the replacement of sand by crushed brick with (0.25%, 50% and 100%) reduce the dry density of mortars with (7.07%, 11.11% and 15.15), respectively. The reasons of these reductions may be due to the use of supplementary material (crushed brick) which has a specific gravity much lesser than that of sand and its coarse grains. Also, the results indicated that the use of superplasticizer increase the dry density. These may be due to reduction in the w/c which needs to give flow of 80 ± 5 mm.

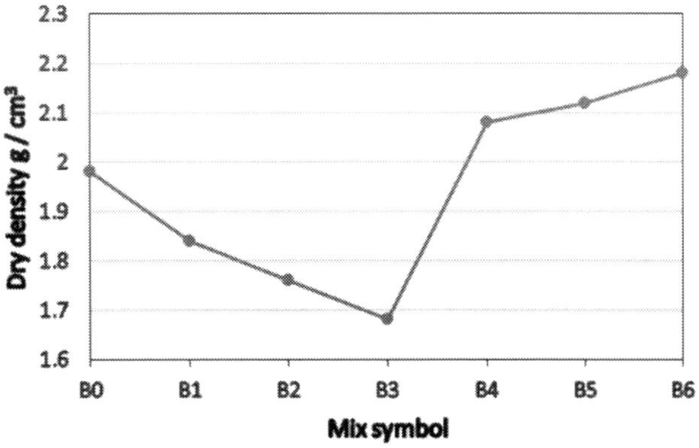


**Fig. 1.** Oven dry Density for different mixes tested at (28) day and curing in water at 20 °C

## 4.2 Compressive Strength

This test was done according to ASTM C 109-02 [10] using 50 mm cube specimens. From the results, it's clear that the use of water curing is better than air curing methods and this is in conformity with the findings of the earlier researchers. The use of super plasticizer for specimens ($B_4$, $B_5$ and $B_6$) with percent of (1%, 2% and 4%), respectively reduces the ratio of (w/c) which increases the compressive strength to ensure good workability (Shirkavand and Baggott, R,1995[11]) (Fig. 2. & 3.). Also, the results indicated that the compressive strength of specimens without replacement of sand by crushed bricks when exposed to 700°C dropped by (37.1%) (Mix $B_0$), while this drop reduced to (19.5%) when the total sand was replaced by crushed bricks (Mix $B_6$). This reduction of drops referred may be related to the use of refractory materials (crushed bricks) which had stability when exposed to high temperatures compared with sand.

From Fig. 3, the values of compressive strength were achieved when tested at age of (28) days for specimens ($B_6$) and exposed to heating of (700) °C which represents an increase of (163.15%) compared to reference mortar specimens ($B_0$) when exposed to heating of (700) °C and tested at age (28) days. These results are the same researchers results of (Ahmed, A. H and Abdul Kaream, O.M,2010[12]) and these may be due to reduction of required water which increases strength and reduces voids.

The results showed there is a big loss in compressive strength for the reference mortar specimen ($B_0$) when exposed to high temperature tested at (28) days as shown in Fig. 2 & 3 and this suits the result archived by the researcher (Aydin & Baradan , ,2007[2]) which included "The loss of the mortar exposed to high heating temperature treated with water is greater than the loss in compressive strength for the mortar at the same condition treated with air".

Fig. 4 represents the relationship between compressive strength with dry density for cement mortar for different mixes tested at (28) days and cured in water at 20°C. It's clear to observe that the relations between compressive strength and dry density are positive.

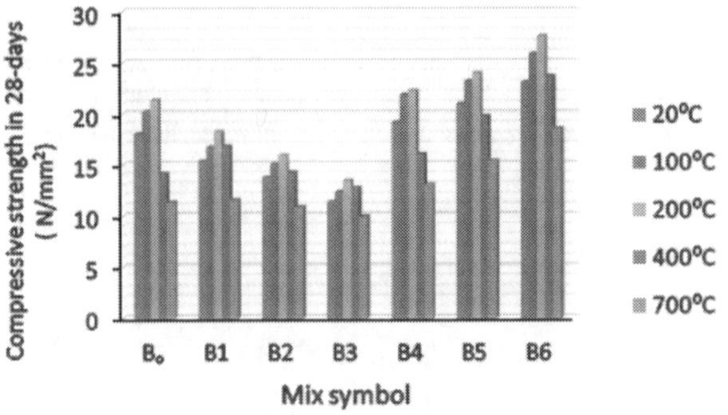


**Fig. 2.** Compressive strength in 28-days for different mixes of mortars cooled with air and exposed to different temperature

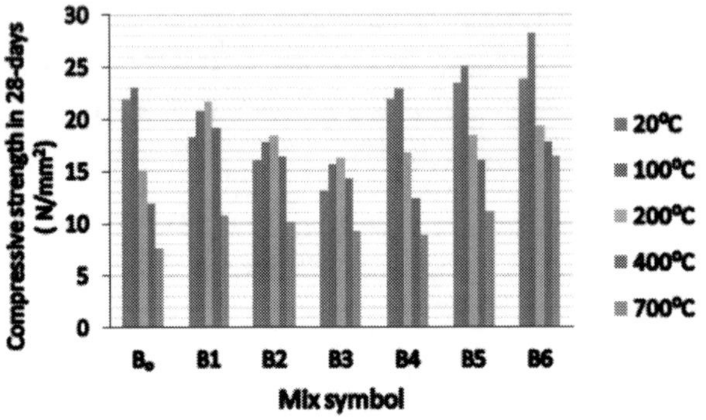


**Fig. 3.** Compressive strength in 28-days for different mixes of mortars cooled with water and exposed to different temperature

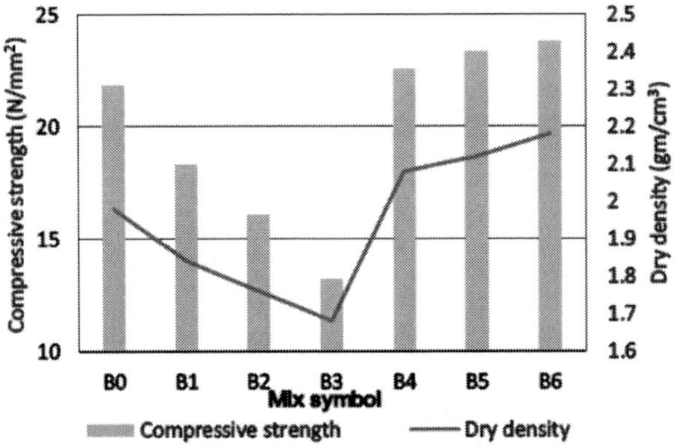


**Fig. 4.** The relationship between compressive strength and the dry density for different mixes tested at (28) day and cured in water at 20°C

### 4.3 Flexural Strength

This test was done according to ASTM C348-02 [13] using 40×40×160 mm prism specimens. The result values for flexural strength for all mixes exposed to different heating temperatures for half hour after treatment with air or water and then tested are shown in Fig. 5. & 6.

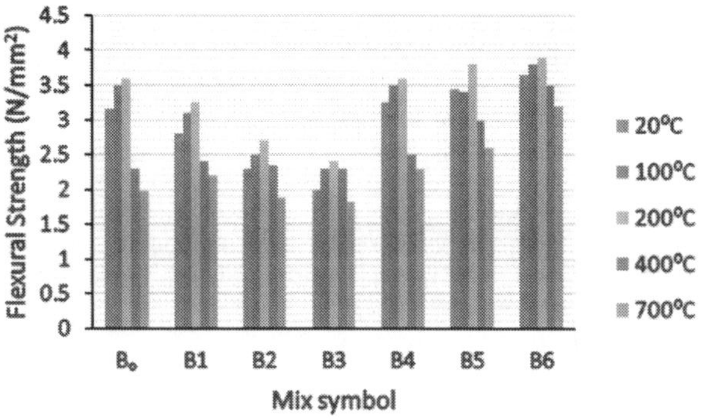


**Fig. 5.** Flexural Strength in 28-days for different mixes cooled by air and exposed to different temperature

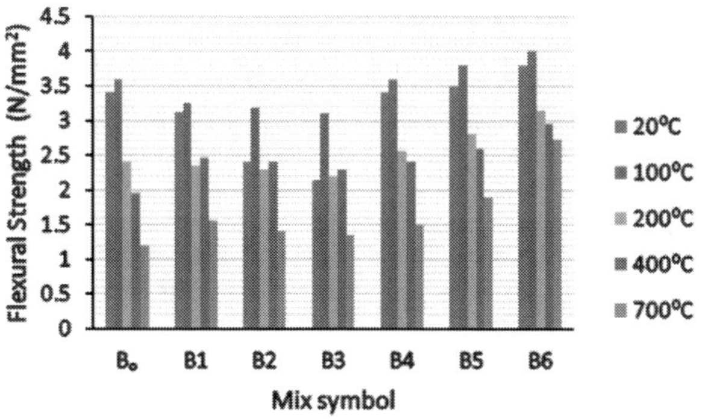


**Fig. 6.** Flexural Strength in 28-days for different mixes cooled by water and exposed to different temperature

From the results, it's clear that the use of water curing gives good results compared with air curing methods and this is also in conformity with the findings of the earlier researchers as for compressive strength.

For the specimens exposed to different heating temperatures for half hour after being treated with water and tested, the results showed the loss in flexural strength was huge (between 28.2% - 64.7%), especially for those specimens exposed to (700)°C and tested at (28) days for specimens $B_0$ to $B_6$ respectively, which was about double the loss for the same specimens treated with air after being exposed to heating degree as shown in Fig. 5 & 6 this result doesn't counter the research (Aydin, 2007[2]) .Many authors (Georgali B. ,and Taskiridis P. E,2005[14], Fu Y.F. ,et all ,2004[15], Serdar A. ,et all ,2008 [16]and Fares , Noumoure ,and Remond,2004[17]) agree on the fact that strength decreases with increasing temperature of exposure.

From Fig. 6, the values of compressive strength were achieved when tested at age of (28) days for specimens ($B_6$) and exposed to heating of (700) °C which represents increase by percent of (127. 5%) compared to reference mortar specimens ($B_0$) when exposed to heating of (700) °C and tested at age (28) days. These may be due to reduction of required water which increases strength and reduces voids.

Fig. 5 shows the loss of strength at highest heating degree of temperature (700) °C tested at age (28) days with percent of (37.3%, 9% and 12.3%) for specimens ($B_0$ , $B_3$ and $B_6$), respectively compared to the same specimens when tested at the same age with heating degree of (20)°C, while Fig. 8 showed the relation of the percentage of loss of strength at highest temperature (700)°C between the specimens treated with air and others treated with water.

## 5 Conclusions

From the experimental results presented in this research, the following conclusions can be drawn:

1- All specimens exposed to a temperature of 700°C showed a decrease in their compressive and flexural strengths. The difference in the strength decrease was relatively different between specimens treated with water and those left in air to dry compared to the reference specimens tested at 20°C after (3, 7 and 28 days).

2- The decrease in both compressive and flexural strength was more obvious in specimens subjected to

700°C than that treated in water and for all references tested at 20°C.

3- The minimum compressive strength loss was achieved when specimens ($B_3$) were exposed to 700°C and air treated then tested after 28 days, the decrease in strength was (9.1%) compared to that of reference specimens tested at 20°C and the percentage loss in flexural strength was (9%) for the same above case.

4- The minimum loss in compressive and flexural strength for specimens ($B_3$) was (29.6%, 37.3%), respectively for the specimens at 700°C by water treated and tested after 28 days when comparing those results relative to the reference samples at 20°C.

## References

1- Aydin, S. (2008). Fire Safety Journal, **Volume 43, Issue 8**, November 2008, Pages 610 – 617

2- Aydin, S. Bardan, B. (2007). Cement and Concrete Research, **Volume 37, Issue 6**, 2007, Pages 988 – 995

3- Hossain, M.A., Islam, M.N., and Karim, M.R. (2006) *31st conference on our world in concrete & structures*, 2006, Singapore

4- Iraqi standard Specification, **ISQ No. 5:1984**, *Properties of Ordinary Portland Cement*, 1984, Iraq

5- Iraqi standard Specification, **ISQ No. 45:1984**, *Properties of Natural Aggregate Resources*, 1984, Iraq

6- American Society for Testing and Materials, ASTM C494-99a, (2004), *Standard Specification for Chemical Admixture for Concrete,* Annual Book of ASTM Standards, **Vol. 04-01**, USA

7- American Society for Testing and Materials, ASTM C1437-, (2004), *Standard Specification for Test Method for Flow of Hydraulic Cement Mortar,* Annual Book of ASTM Standards, **Vol. 04-02**, USA

8- American Society for Testing and Materials, ASTM C305-99, (2004), *Standard Practice for Mechanical Mixing of Hydraulic Cement Pastes and Mortars of Plastic Consistency*, Annual Book of ASTM Standards, **Vol. 04-01**, USA

9- American Society for Testing and Materials, ASTM C0642-97, (2004), *Standard Specification for Test Method for Density, Absorption, and Voids in Hardened Concrete*, Annual Book of ASTM Standards, **Vol. 04-02**, USA

10- American Society for Testing and Materials, ASTM C0109-02, (2004), *Standard Specification for Test Method for Compressive Strength of Hydraulic Cement Mortars,* Annual Book of ASTM Standards, **Vol. 04-01**, USA

11- Shirkavand, M. and Baggott, R. (1995, Cement and Concrete Research, **Volume 25, Issue 7**, 1995, Pages 1512 – 1522

12- Ahmed, A. H. and Abdul Kaream, O.M. (2010). Al Rafidain Engineering J., **Vol. 18, No. 4**, Pages 43-54

13- American Society for Testing and Materials, ASTM C348-02, (2004), *Standard Test Method for Flexural Strength of Hydraulic-Cement Mortars*, Annual Book of ASTM Standards, **Vol. 04-01**, USA.

14- Georgali B.and Taskiridis P. E. (2005). Cement and Concrete Composites, **Vol. 27**, Pages 255 – 259

15- Fu, Y.F., Wong, Y.L. and Poon C.S,et al., (2004). Cement and Concrete Research, **Volume 33**, 2004, Pages 789 – 797

16- Serdar A, Halit Y. and Bulent B. (2008). Construction and Building Materials, **Vol. 22**, 2008. Pages 504- 512.

17- Fares H., Noumoure A., and Remond S. (2009). Cement and Concrete Research, **Vol. 39**, 2009. Pages 1230 – 1238.

# Freezing and thawing resistance of internally cured high performance concrete

*Muthana* Saadi[1,*], *Tareq* al-Attar[2], and *Shatha* Hasan[2]

[1]Ministry of Planning, Baghdad, Iraq
[2]University of Technology, Baghdad, Iraq

**Abstract.** The behavior of internally cured high performance concrete, HPC, exposed to freezing and thawing cycles, was investigated. Two saturated curing agents, Limestone dust and powder of Porcelanite rock, were used to facilitate internal curing for concrete. These agents were used as partial replacements of fine aggregate in two volumetric percentages, 20 and 30 percent. The cast concrete specimens were separated in two groups according to curing method: water-cured and sealed (only internally-cured) specimens. The concrete specimens were subjected to three exposure systems, $F_0$: without freezing and thawing, and $F_1$ and $F_2$: with 50 and 100 cycles of freezing and thawing, respectively. The freezing and thawing test was done as stipulated by the ASTM C666. The conducted tests for each exposure were: compressive and flexural strengths. The results revealed that internal curing does not enhance the concrete resistance to freezing and thawing cycles. Using saturated agents has increased the moisture content of concrete and makes it more vulnerable to frost action deterioration. Sealed specimens for all investigated mixes showed lower reductions in strength than water-cured ones. The lesser water content of these mixes may be the reason for that behavior.

## 1 Introduction

Cyclic freezing and thawing deterioration is known to be the major source of deterioration of concrete in most of the places experiencing severe cold weather. The extent of the problem can be appreciated from the fact that almost one-third of all concrete poured in Nordic European countries is susceptible to freezing and thawing deterioration [1]. Infrastructures in northern Japan have also suffered extensively due to this phenomenon [2]. Damage due to freezing and thawing deterioration can take various forms, the most common being scaling and micro-cracking [3]. Both these forms of damage are known to potentially progress individually within the same structural member [4].
Other forms of damage may include large-scale spalling and crumbling of concrete and crossing fatigue-like cracks visible on the surface. All these symptoms are generally accompanied by loss in strength and serviceability of concrete. Scaling is a surface phenomenon and is usually responsible for the wearing off of the outer few millimeters of concrete. Light scaling can cause concrete surfaces to become rough, and more severe scaling can expose coarse aggregate, sometimes even scaling off more than 10 mm of cover concrete. It has been found that processes like improper troweling, insufficient curing and bleeding lead to the formation of a highly porous zone of concrete near the surface that is much more susceptible to scaling [5].

Though its mechanism isn't very clear yet, the use of deicing salt, generally used to remove snow from pavements, bridge decks, runways, sidewalks is generally held responsible for intensified and faster scaling of concrete surface, especially in concretes with low air contents. Tensile pressures generated within concrete, during cyclic freezing and thawing, are known to generate cracks within concrete, reducing its strength and integrity. Microscopic studies verify the growth of around 1 - 10 μm wide micro-cracks upon subjecting concrete to freezing and thawing cycles. These cracks have been found to pass through cement paste and follow the boundaries of most large aggregate particles [6].
Many factors are found to affect the propagation of micro-cracks within concrete, with air-entrainment, water/cement ratio and strength. Curing and saturation are believed to have major influences on durability of high performance concrete when exposed to freezing and thawing cycles.

## 2 Experimental work

### 2.1. Materials and concrete mix

The cement used throughout this investigation is sulfate resisting Portland cement. Test results showed that the cement conforms to the provisions of Iraqi Standard IQS 5/1984- Type V [7]. Natural sand with fineness modulus

* Corresponding author: muthana_abdalkader@yahoo.com

© The Authors, published by EDP Sciences. This is an open access article distributed under the terms of the Creative Commons Attribution License 4.0 (http://creativecommons.org/licenses/by/4.0/).

of 3.19 was used as fine aggregate. The coarse aggregate was crushed gravel with maximum size of 19 mm. The grading and sulfate content of both used aggregates are conforming to the provisions of the Iraqi Standard IQS 45/1984[8]. Silica fume conforming to the ASTM C1240 [9] was the used supplementary cementitious material throughout this investigation. It has a specific surface of $20m^2/g$.

Porcelanite and limestone dust were adopted as internal curing agents. They were graded as the fine aggregate and used as partial replacement for fine aggregate in 20 and 30% by volume.

Many trial mixes were done in this work to obtain an average required compressive strength, $f_{cu}$, at 28 days equals to 60 MPa. The weight proportions of the chosen mix were: 1: 1.4: 1.9: 0.25 (binder: fine aggregate: coarse aggregate: W/CM ratio). The total binder content was 500 $kg/m^3$.

Five mixes were produced to investigate the effect internal curing agents, limestone dust and Porcelanite, on resistance to freezing and thawing. These mixes were:

$M_0$: reference mix made without fine aggregate replacement.

$M_{20L}$: a mix made with 20% fine aggregate replacement by limestone dust.

$M_{30L}$: a mix made with 30% fine aggregate replacement by limestone dust.

$M_{20P}$: a mix made with 20% fine aggregate replacement by Porcelanite.

$M_{30P}$: a mix made with 30% fine aggregate replacement by Porcelanite.

### 2.2 Curing procedures

After casting and demolding of concrete specimens, two curing procedures were adopted. In the first; the specimens were completely immersed in tap water until the time of testing (water-cured specimens denoted with **w** letter). The curing water was replaced at least once a month. Second method of curing is to put the specimens in plastic bags and prevent any water to penetrate the bags (sealed specimens denoted with **s** letter). The second method was done to illustrate the impact of internal curing on the behavior of tested specimens.

### 2.3 Freezing and thawing exposure

The freezing and thawing test was carried out according to ASTM C666-08[10].The exposure conditions were as fallow:

1. F0: mixes were not exposed to freezing and thawing cycles.
2. F1: mixes were exposed to 50 cycles of freezing and thawing.
3. F2: mixes were exposed to 100 cycles of freezing and thawing.

According to the above mentioned standard method of test the specimens were cubes (100×100×100) mm and prisms (100×100×400) mm. They were put in the test chamber at the age of 14 days. Specimens in the pre-test period were kept in water for water-cured specimens and in plastic bags for sealed specimens.

The compressive strength test was considered as an indication for concrete resistance to the above described exposure conditions. This test was carried out according to the BS EN 12390 [11] for cubes with dimensions of (100×100×100) mm at ages of 14, 28 and 60 days.white, and this should be taken into account when preparing them.

## 3 Results and discussions

Table 1 displays the effect of different freezing and thawing exposures on compressive strength of the studied concrete mixes. Figure 1 shows the compressive strength variations of mix $M_0$ for both water-cured and sealed specimens when being exposed to different freezing and thawing exposures. It could be observed that the rate of gain (or reduction) in strength is almost the same (the lines are almost parallel). This behavior may be a result of that the two groups of specimens have suffered the same exposure conditions. The only difference between them is that the water-cured group has been in water till the ages of 14 days; meanwhile the second group was kept in bags from the date of casting till the age of test. This difference seems to have no effect on later behavior, especially if one keeps in mind the low permeability characteristic of studied concrete.

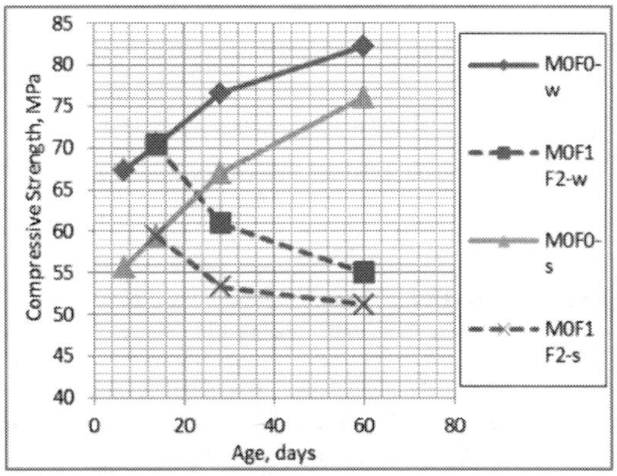


**Fig. 1.** Compressive strength variations of mix $M_0$ for both water-cured and sealed specimens when being exposed to different freezing and thawing exposures

Figures 2 and 3 illustrate the effect of replacement of fine aggregate by Limestone dust on concrete resistance to freezing and thawing cycles for water-cured and sealed specimens, respectively. The use of saturated Limestone dust has adversely affected the compressive strength and the reduction in strength increased with the increased replacement percentage and increased number of freezing and thawing cycles. It is worthy to note that when the divergence between the solid and dotted lines, of the same mix and at the same age, increases it means the reduction in compressive strength will increase.

 https://doi.org/10.1051/matecconf/201816202011

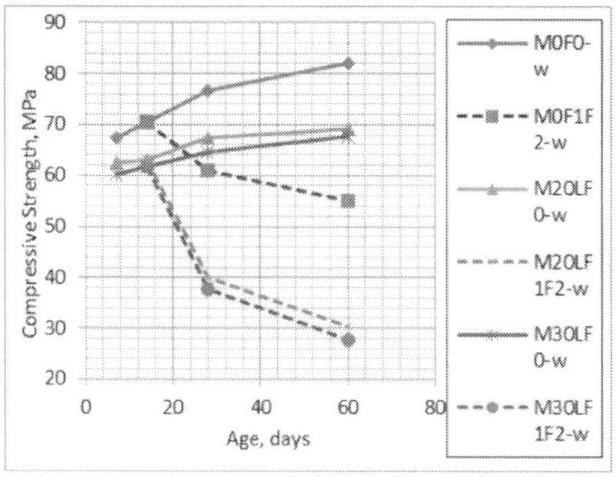


**Fig. 2.** Compressive strength variations of mixes $M_0$, $M_{20L}$ and $M_{30L}$ for water-cured specimens when being exposed to different freezing and thawing exposures

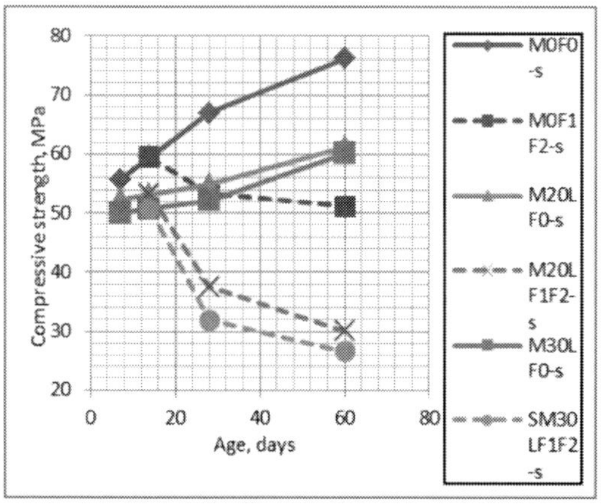


**Fig. 3.** Compressive strength variations of mixes $M_0$, $M_{20L}$ and $M_{30L}$ for sealed specimens when being exposed to different freezing and thawing exposures

For water-cured mix $M_{20L}$, when making comparisons between F0, F1 and F2, the reductions in strength will be 41 and 56 percent for 50 and 100 cycles, respectively. And for 30 percent replacement, mix $M_{30L}$; these reductions will be 42 and 59 percent for 50 and 100 cycles, respectively. Moreover, when making comparisons between mixes $M_0$ and $M_{20L}$, the reductions in compressive strength will be 34 and 45 percent for 50 and 100 cycles, respectively. For mixes $M_0$ and$M_{30L}$; these reductions will be 38 and 49 percent for 50 and 100 cycles, respectively.

These results could reveal the following conclusions:

a. The two adopted replacement percentages have the same effect in magnitude on concrete durability and cause almost the same degradation in compressive strength.

b. The extra carbonation bond between limestone and cement paste could create a saturated stiff matrix with low permeability. This situation prevents the water to escape and to exert higher pressure on pore walls leading to more microcracking and more strength reduction [12].

Sealed specimens for the same mentioned mixes, $M_{20L}$ and $M_{30L}$, showed lower reductions in strength than water-cured ones. The lesser water contents of these mixes may be the reason for that behavior [13].

Figures 4 and 5 display the behavior of mixes $M_{20P}$ and $M_{30P}$ when being exposed to freezing and thawing cycles for water-cured and sealed specimens, respectively. For water-cured specimens, using saturated Porcelanite has negatively affected the compressive strength and the reduction in strength was negatively proportioned to the replacement percentage and positively proportioned to the number of freezing and thawing cycles. On contrast to that, for sealed specimens the reduction in compressive strength increased with the increased replacement percentage and increased number of freezing and thawing cycles. Sealed specimens showed lower reductions in strength than water-cured ones. The lesser water contents of these mixes may be the reason for that behavior.

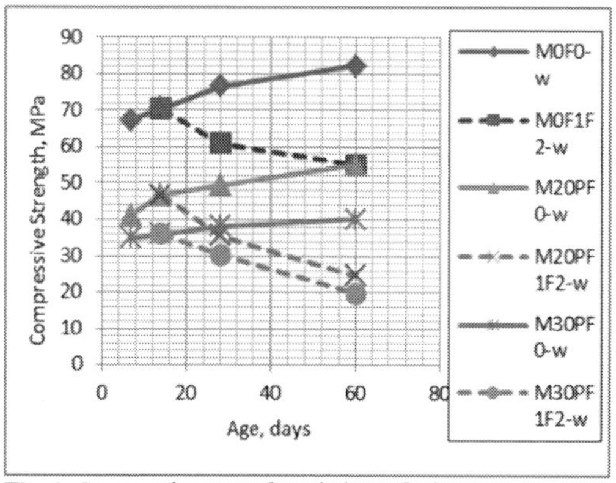


**Fig. 4.** Compressive strength variations of mixes $M_0$, $M_{20P}$ and $M_{30P}$ for water-cured specimens when being exposed to different freezing and thawing exposures

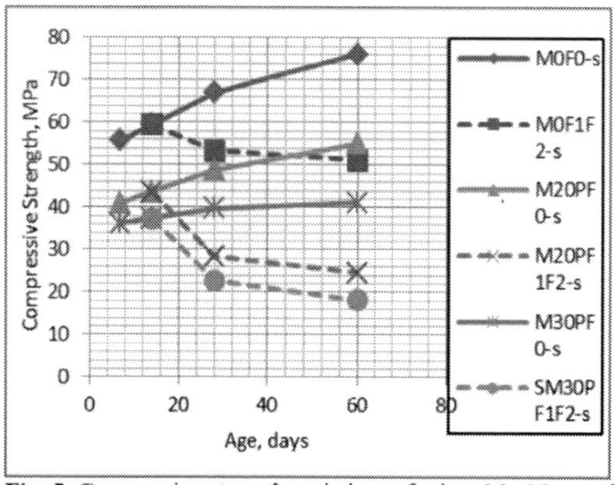


**Fig. 5.** Compressive strength variations of mixes $M_0$, $M_{20P}$ and $M_{30P}$ for sealed specimens when being exposed to different freezing and thawing exposures

When making comparisons between F0, F1 and F2, for water-cured mix $M_{20P}$, the reductions in strength will be

28 and 55 percent for 50 and 100 cycles, respectively. For 30 percent replacement, mix $M_{30P}$; these reductions will be 20 and 51 percent for 50 and 100 cycles respectively. Moreover, when making comparisons between mixes $M_0$ and $M_{20P}$, the reductions in compressive strength will be 42 and 55 percent for 50 and 100 cycles, respectively. For mixes $M_0$ and $M_{30P}$; these reductions will be 50 and 64 percent for 50 and 100 cycles, respectively. The last comparison leads to the conclusion that the high water absorption of Porcelanite is the reason for the higher deterioration caused by freezing and thawing cycles.

## 4 Conclusions

1. In the present work, the F1 and F2 (50 and 100 cycles of freezing and thawing respectively) exposures seem effective enough to investigate the behavior of internally cured concrete.
2. The internal curing does not enhance the concrete resistance to freezing and thawing exposures F1 and F2 and that may be due to:
a. The standard method of test (ASTM C666) stipulated that specimens shall be cured for 14 days in water prior to testing and such period is not enough for concrete with low permeability (high performance concrete) to be saturated. Therefore, the results of mix $M_0$ (without fine aggregate replacement) showed almost the same rate of reduction for both water-cured and sealed specimens.
b. Using saturated agents has increased the moisture content of concrete and makes it more vulnerable to deterioration resulted from frost action.
3. Sealed specimens for all investigated mixes showed lower reductions in strength than water-cured ones. The lesser water contents of these mixes may be the reason for that behavior.
4. Higher water absorption of Porcelanite than limestone dust caused higher deterioration by freezing and thawing cycles.

## References

1. V. Penttala, F. Al-Neshawy, Cement and Concrete research, **32**, No. 9 (2002)
2. T. Ohta, Proc. Second Canada/Japan Workshop on Low Temperature Effects on Concrete (1991)
3. P. Mehta, P. Monteiro, *Concrete microstructure, properties, and materials* (Indian Concrete Institute, 1997)
4. J. Marchand, R, Pleau, R. Gange, Mat. Sci. Conc. (1995)
5. M. Pigeon, C., Talbot, J., Marchand, H. Hornain, Cem. Conc. Res. **26**, 10 (1996)
6. S. Jacobsen, J., Marchand, H. Hornain, Cem. Conc. Res., **25**, 8 (1995)
7. IQS No. 5, COSQC, Iraq (1984) (in Arabic)
8. IQS No. 45, COSQC, Iraq (1984) (in Arabic)
9. ASTM C1240-15, Annual Book of ASTM Standards, USA (2015)
10. BS EN 12390- P: 3, British Standard Institution, UK (2002)
11. ASTM C666-15, Annual Book of ASTM Standards, USA (2015)
12. T. Rønning, Ph.D. Thesis, Nor. Inst. Tech. (2001)
13. A. M. Neville, *Properties of concrete*, (Prentice Hall, UK, 2011)

**Table 1.** Compressive strength results for all mixes at different freezing and thawing exposures for water-cured and sealed specimens

| Mix | Compressive Strength of Concrete Mixes, MPa, for Different Exposures at Ages: | | | | | | | | | |
|---|---|---|---|---|---|---|---|---|---|---|
| | for water-cured specimens | | | | | for sealed specimens | | | | |
| | F0 | | | F1-50 cycles | F2-100 cycles | F0 | | | F1- 50 cycles | F2 -100 cycles |
| | 7 d | 28 d | 60 d | | | 7 d | 28 d | 60 d | | |
| $M_0$ | 67.3 | 76.6 | 82.2 | 61.0 | 55.0 | 55.7 | 67.0 | 76.1 | 53.3 | 51.2 |
| $M_{20L}$ | 62.5 | 67.4 | 69.1 | 40.1 | 30.3 | 52.3 | 54.1 | 61.2 | 37.5 | 30.1 |
| $M_{30L}$ | 60.1 | 64.7 | 67.7 | 37.7 | 27.8 | 50.0 | 52.1 | 60.1 | 31.9 | 26.6 |
| $M_{20P}$ | 41.2 | 49.3 | 55.0 | 35.5 | 24.7 | 41.0 | 48.8 | 55.1 | 28.3 | 24.6 |
| $M_{30P}$ | 34.9 | 38.1 | 40.1 | 30.3 | 19.8 | 36.2 | 39.7 | 41.2 | 22.7 | 18.1 |

# Properties of foamed concrete reinforced with hybrid fibres

*Waleed* Abbas [1,*], *Eethar* Dawood[2], and *Yahya* Mohammad[2]

[1]Building and Construction Engineering Department, University of Technology, Iraq
[2]Building and Construction Engineering Department, Technical College of Mosul, Iraq

**Abstract.** The properties of foamed concrete reinforced with carbon fibres and hybrid fibres of carbon with polypropylene fibres has been studied. Various volumetric fractions of carbon fibres (0.5, 1 and 1.5%), hybrid fibres of carbon fibres (CF) with polypropylene fibres (PPF) as (1% CF + 0.5% PPF) & (0.5% CF + 1% PPF), also the mono polypropylene fibres as 1.5% PPF were used to reinforce foamed concrete mix. Fresh and hardened properties of all mixes included flowability, density, absorption, compressive strength, splitting tensile strength, and flexural strengths has been tested. Results showed that inclusion of carbon fibres up to 1% volumetric fraction may increase the compressive strength by about 36% higher than that of control mix. Whereas, the use of 1.5% carbon fibres exhibit significant increase in splitting and flexural tensile strengths by about 47 and 114%, respectively, compared to the reference mix. On the other hand, the hybridization of 1% CF + 0.5% PPF increased the splitting tensile strength and flexural strengths by 53% and 114%, respectively, compared with plain foamed concrete mix.

Keywords: Foamed concrete, Carbon fibres, Hybrid fibres, Economical Efficiency

## 1 Introduction

Lightweight Foamed Concrete (LFC) can be represented as one of the recent advancement of concrete technology in civil engineering that can be used in a wide range of construction projects [1]. Significant improvements over the past 20 years in production equipment and better-quality surfactants (foaming agents) has enabled the use of foamed concrete on a larger scale [2].

The use of structural lightweight foamed concrete (SLWFC) which possesses good mechanical properties and durability ensures many advantages compared to normal weight concrete. Thus, SLWFC is a relatively new construction material which provides more efficient strength-to-weight ratio in structural elements [3]. Concrete has an inherently brittle nature and some disadvantages such as poor deformability and weak crack resistance in the practical usage. The fibers are used to improve the brittle nature of lightweight foamed concrete [4].

Fibers used in concrete can be classified into two types, Low modulus with high elongation fibers such as, polypropylene and polyethylene. this type has capable of large energy absorption properties and do not improve the compressive strength. On the other hand, high modulus with high strength fibers such as steel, glass and carbon produce strong composites [5].

The combination of different fibers may offer potential advantages in improving concrete properties. Hybrid fibers can provide reinforcement at all ranges of strains. Combination of low and high modulus fibers can arrest cracks at micro level as well as macro level [6].

Therefore, this study was prepared to induce carbon fibers and also combination of carbon and polypropylene fibers with foamed concrete. These different mixes are tested and evaluated from the view of economic efficiency.

## 2 Materials and mix proportion

### 2.1 Materials

The materials used in the present work are: cement, sand, water, silica fume foam agent, polypropylene fibers and carbon fibers.

Ordinary Portland cement (OPC) type (I) commercially known as AL-Mass cement factory (Sulaimaniyah governorate of Iraq) was used in this study. The physical characteristics of ordinary Portland cement is showed in Table 1 and conformed to IQS: 5/1984 [7]. Whereas the chemical compositions for the cement are shown in Table 2 and they conformed to ASTM C150 [8].

*Corresponding author: waleedalqaisi@yahoo.com

© The Authors, published by EDP Sciences. This is an open access article distributed under the terms of the Creative Commons Attribution License 4.0 (http://creativecommons.org/licenses/by/4.0/).

**Table 1.** Physical properties of cements.

| **Physical properties** | **Results** | **Limits of IQS: 5/1984**[7] |
|---|---|---|
| Initial setting time (minute) | 100 | $\geq 45$ |
| Final setting time (minute) | 320 | $\leq 600$ |
| Fineness (Blaine $m^2$/ kg) | 300 | $\geq 230$ |
| Soundness by Autoclave Method (%) | 0.02 | Not more than 0.8 |
| Compressive strength (MPa) | | |
| 3 days | 21 | $\geq 15$ |
| 7 days | 27 | $\geq 23$ |

**Table 2.** Chemical composition of cement.

| **Composition** | **Abbreviation** | **Percentage by weight** | **Limits of IQS: 5/1984**[7] |
|---|---|---|---|
| Lime | $CaO$ | 61 | - |
| Silica | $SiO_2$ | 19.84 | - |
| Alumina | $Al_2O_3$ | 5.08 | - |
| Iron Oxide | $Fe_2O_3$ | 4.8 | - |
| Sulphate | $SO_3$ | 2.49 | $\leq 2.8$ |
| Potash | $K_2O$ | 0.1 | |
| Soda | $Na_2O$ | 0.3 | |
| Equivalent $Na_2O$ | $Na_2O+0.658K_2O$ | 0.36 | $\leq 0.6\%$ |
| Magnesia | $MgO$ | 2.48 | $\leq 5.0\ \%$ |
| Loss on ignition | L.O.I. | 3.8 | $\leq 4.0\ \%$ |
| Insoluble residue | I.R. | 0.40 | $\leq 1.5\ \%$ |
| Main Compounds (Bogue's equations) | | | |
| Tri calcium Silicate | $C_3S$ | 49.45 | - |
| Di calcium Silicate | $C_2S$ | 19.57 | - |
| Tri calcium Aluminate | $C_3A$ | 5.34 | - |
| Tetra calcium Aluminate –Ferrite | $C_4AF$ | 14.61 | - |

Silica fume (Sika Fume HR) was used as a partial replacement of cement, the material properties are shown in Table 3.

**Table 3**. Material properties of Silica fume.

| **Form** | **Agglomerated** |
|---|---|
| Particles Color/ Appearance | Grey |
| Specific Gravity | 2.20 |
| Size of particles | 0.1 µ |
| Dosage | 2 - 10 % by weight of cement |
| Chloride content | Nil |

The natural sand used as fine aggregate was supplied from AL- Ukhaider region. The specific gravity and fineness modulus of sand are 2.65 and 3, respectively. The grading limits are according to ASTM C 33-02 [9] and given in Table 4.

**Table 4.** Grading of fine aggregate.

| **Sieve No. (mm)** | **Passing (%)** | **Limits of ASTM C 33-02[9]** |
|---|---|---|
| 4.75 | 95 | 95-100 |
| 2.36 | 80 | 80-100 |
| 1.18 | 59 | 50-85 |
| 0.6 | 44 | 25-60 |
| 0.3 | 18 | 5-30 |
| 0.15 | 4 | 0-10 |

Sika Lightcrete 02 was used as a foaming agent to obtain lightweight foamed concrete by entraining a controlled amount of air bubbles to concrete mix. The foaming agent was diluted in 30 parts of water before using it. Carbon fibers with 8mm length were used with the lightweight foamed concrete, the properties of the carbon fibers are listed in Table 5.

**Table 5.** Properties of carbon fibers.

| **Fiber properties** | **Quantity** |
|---|---|
| Fiber length | 8 mm |
| Diameter | 7 ± 2 micron |
| Aspect ratio | 1140 |
| Tensile strength | 3.5 GPa |
| Young's Modulus | 230 GPa |
| density | 1.7 $g/cm^3$ |
| Chemical Resistance | High |
| Absorption | Nil |
| Melt Point | 3500°C |
| Shape | chopped strand |

Whereas, Monofilament polypropylene fibers (Sika fiber) were used in the lightweight foamed concrete mixes. Chopped strand shape fibers are 12 mm long with 18 µm of diameter and density of 0.9 $g/cm^3$. The properties of the polypropylene fibers are listed in Table 6.

**Table 6.** Properties of polypropylene fibers.

| Fiber properties | Quantity |
|---|---|
| Fiber length | 12 mm |
| Diameter | 18 microns |
| Aspect ratio | 670 |
| Tensile strength | 500 MPa |
| Young's Modulus | 3.5 GPa |
| density | 0.9 g/cm$^3$ |
| Chemical Resistance | High |
| Absorption | Nil |
| Melt Point | 160 °C |
| Shape | chopped strand |

### 2.2 Mix proportions

The proportions of the foamed concrete mixes (C0-C6) as shown in Table 7 were prepared using different volumetric fractions of fibers. However, the control or reference mix (C0) has been prepared to be in the flowability range of 110±5%.

Besides, the foamed concrete reinforced with carbon fiber mixes were designated by the mixes as C1-C3 due to the use of 0.5, 1 and 1.5% of carbon fibers, respectively. Thus, the flow value was recorded for each mix proportion.

Furthermore, the hybridizations of carbon fibers (CF) and polypropylene fibers (PPF) were presented in the mixes C4-C5. Such hybrid fibers were as 1% CF + 0.5% PPF and 0.5% CF + 1 % PPF, respectively. And again, the flow test was done for these mixes.

Lastly, the use of 1.5% PPF was used to prepare the mix of C6. The flowability was also measured for this mix.

## 3 Experimental work

The molds of 100 mm cubes were used for testing the compressive strength of the foamed concrete mixes according to BS 1881 Part:116; 1983 [10]. The density and absorption for all mixes was also examined using100 mm cubes according to ASTM C 642 [11]. The flexural strength of foamed concrete mixes was done using prismatic molds of 100 × 100 × 400 mm according to ASTM C78 [12]. Whereas, the cylindrical molds 100×200 mm were used for splitting tensile strength according to ASTM C567 [13].

## 4 Results and discussion

### 4.1 Effect of fibers on Flowability

The effect of fibers on flow of the concrete mixes is depicted in Table 7 and Figure 1. It can be noticed that the inclusion of carbon fibers with foamed concrete mix reduces the flowability and thus, the flowability decreased from 110 % to 60% due to 1.5% carbon fibers. On the other hand, each replacement of carbon fibers by polypropylene fibers leads to better flowability. In other words, the effect of polypropylene fibers on flowability or workability is much less than that of carbon fibers [14].

### 4.2 Compressive strength

Table 8 shows the results of compressive strength for specimens at ages of 7 and 28 days after water curing. The results of the compressive strength of foamed concrete mixes indicate that the increase in carbon fibers increases the compressive strength due to the reduction of porosity and an improvement in mechanical bond strength [15]. The percentage of compressive strength increase from using 1.0 % of carbon fibers in the foamed concrete was about 35% for 28 days. However, the use of 1.5% reduces the percentage of the increase.

Alternatively, the results of the hybrid fibers-specimens also show that the use of 1% CF + 0.5% PPF fibers gave higher value of compressive strength than that of 0.5%CF + 1 %PPF, this is probably due to the ability of carbon fibers to enhance the ultimate strength capacity due to the higher stiffness of carbon fibers than that of polypropylene [16].

Figure 2 shows the percentages of increase of the compressive strength of foamed concrete due to different percentages of fibers.

**Table 7.** Mix proportions of foamed concrete mixes reinforced with fibers.

| Mixes | Mix proportion | | | | | fibers % | | Flow % | Fresh density kg/m$^3$ |
|---|---|---|---|---|---|---|---|---|---|
| | Cement | Sand | w/c % | Silica fume | Foam kg/m$^3$ | CF | PPF | | |
| **C0** | 0.9 | 1.9 | 0.34 | 0.1 | 1 | --- | --- | 110 | 1820 |
| **C1** | 0.9 | 1.9 | 0.34 | 0.1 | 1 | 0.5 | --- | 80 | 1800 |
| **C2** | 0.9 | 1.9 | 0.34 | 0.1 | 1 | 1 | --- | 75 | 1810 |
| **C3** | 0.9 | 1.9 | 0.34 | 0.1 | 1 | 1.5 | --- | 60 | 1820 |
| **C4** | 0.9 | 1.9 | 0.34 | 0.1 | 1 | 1 | 0.5 | 80 | 1800 |
| **C5** | 0.9 | 1.9 | 0.34 | 0.1 | 1 | 0.5 | 1 | 80 | 1800 |
| **C6** | 0.9 | 1.9 | 0.34 | 0.1 | 1 | --- | 1.5 | 90 | 1800 |

**Table 8.** Properties of foamed concrete mixes reinforced with fibers.

| Mixes | Compressive strength (MPa) | | Splitting tensile strength (MPa) | | Flexural strength (MPa) | | Absorption (%) | Oven dry density kg/m$^3$ |
|---|---|---|---|---|---|---|---|---|
| | 7 Days | 28 Days | 7 Days | 28 Days | 7 Days | 28 Days | 28 Days | 28 Days |
| C0 | 11.2 | 17.1 | 1.72 | 1.9 | 1.7 | 2.1 | 12.2 | 1670 |
| C1 | 11.5 | 17.3 | 1.8 | 2.0 | 2.6 | 3.8 | 9.9 | 1745 |
| C2 | 13.8 | 23.1 | 2.0 | 2.6 | 3.1 | 4.3 | 7.4 | 1805 |
| C3 | 12.0 | 21.4 | 2.1 | 2.8 | 3.4 | 4.5 | 7.5 | 1820 |
| C4 | 10.0 | 21 | 2.2 | 2.9 | 2.9 | 4.5 | 9.2 | 1775 |
| C5 | 12.3 | 19.1 | 2.0 | 2.1 | 3.3 | 4.0 | 8.8 | 1770 |
| C6 | 7.0 | 13.8 | 1.9 | 2.0 | 2.8 | 3.8 | 8.9 | 1715 |

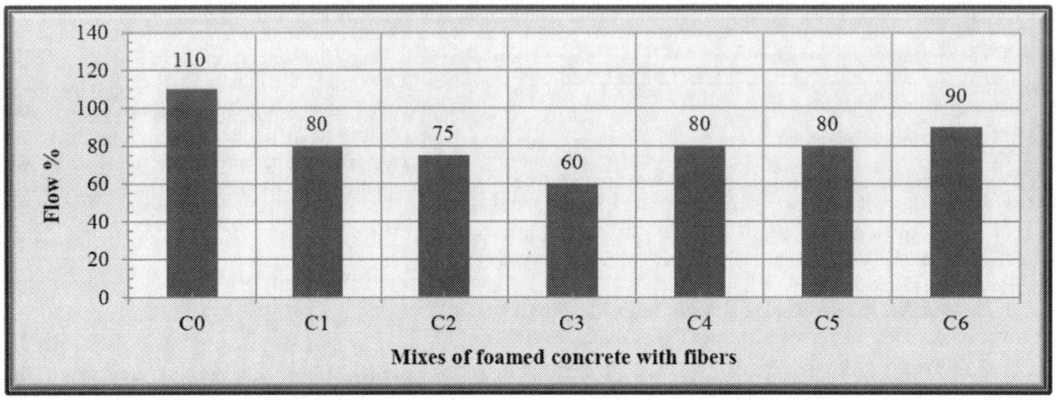


**Fig. 1.** Effect of fibers on the flow of the concrete mixes.

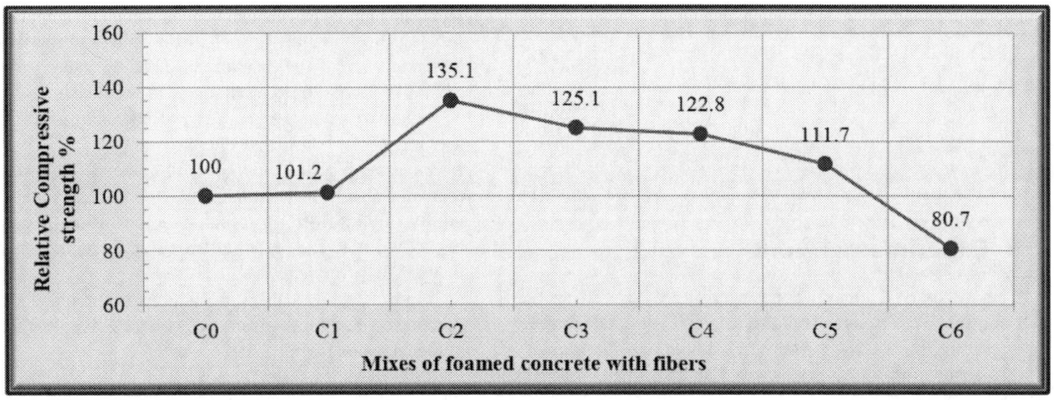


**Fig. 2.** Effect of fibers on the compressive strength for 28 days.

## 4.3 Splitting Tensile Strength

The results of splitting tensile strength for specimens at ages of 7 and 28 days after water curing are shown in Table 8.

The results show that the inclusion of carbon fibers in the foamed concrete mix increases the splitting tensile strength of foamed concrete. As the volumetric fraction of carbon fiber increases, the splitting tensile of foamed concrete increases, too.

Thus, the highest value recorded due to the use of 1.5% CF in the foamed concrete mix and the percentage of increase is about 47% higher than that of reference foamed concrete for 28 days. This is related to the high tensile strength of carbon fibers that may promote the tensile strength of the foamed concrete and the ability of the fibers to arrest the cracks enhancing the homogeneity of the foamed concrete [17].

On the other hand, the hybridization of 1% CF with 0.5% PPF increases the splitting tensile strength significantly. The percentage increases of the latter case

(C4) was found to be higher than that of the control mix (C0) by about 53%. This can be attributed to the ability of foamed concrete with two fibers to bridge the cracks effectively, thus, the micro-mechanical feature of cracks bridging is operative from the stage of damage evolution to beyond ultimate loading.
Figure 3 indicates the percentages of increase of the splitting tensile strength of foamed concrete due to different percentages of fibers.

### 4.4 Flexural strength

The results of flexural strength for specimens at ages of 7 and 28 days after water curing are shown in Table 8.

The results of flexural strength of foamed concrete reinforced with carbon fibers show that the inclusion of fibers increases the flexural strength of foamed concrete. The highest value has been obtained due to the use of 1.5% carbon fiber in the foamed concrete mix and the percentage of increase is about 114% higher than that of plain foamed concrete.

Also, the hybridization of 1% CF with 0.5% PPF increases the flexural tensile strength significantly. The percentage increases by such use (C4) was also found to be higher than that of the control mix (C0) by about 114%.

This effective increase in flexural strength may have resulted from better compaction and homogenous distribution of fibers in mortar mixes and the ability of different types of fibers to restrain and bridge the cracks [18].

Figure 4 shows the percentages of increase of the flexural strength of foamed concrete due to different percentages of fibers.

### 4.5 Absorption

Table 8 shows the results of absorption for specimens at ages of 28 days after water curing.

It can be noted that the use of carbon fibers may significantly reduce the absorption of foamed concrete. The comparison between C0 with C3 shows that the use of 1.5% vol. of carbon fibers lowered the absorption from 12.2% to 7.5%. This could be due to the fact that carbon fibers reduce the porosity of the foamed concrete by using these fibers [19].

On the other hand, the absorption reduction by using polypropylene fibers illustrates that these fibers have less ability to decrease the absorption than that of carbon fibers.

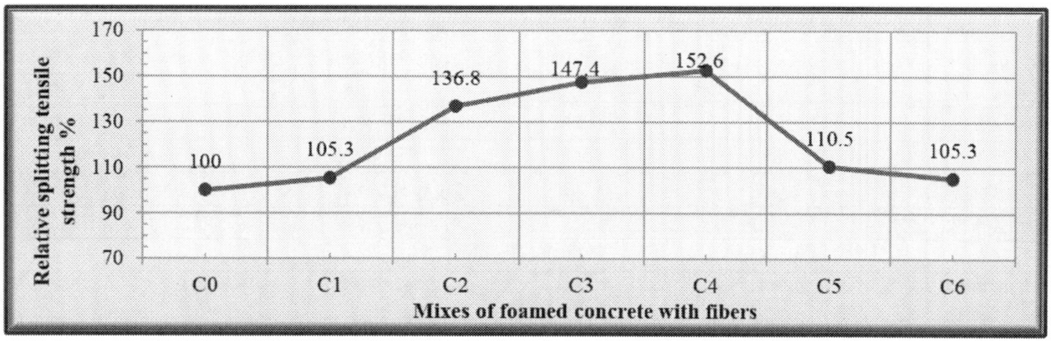


**Fig. 3.** Effect of fibers on the splitting tensile strength for 28 days.

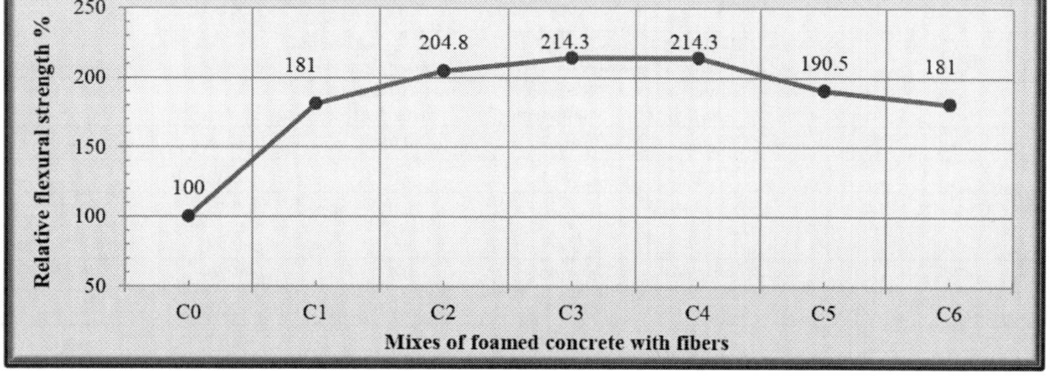


**Fig.4.** Effect of fibers on the flexural strength for 28 days.

### 4.6 Oven dry density

Table 8 shows the results of absorption for specimens at ages of 28 days after water curing.

The test results of foamed concrete show that the increase in Carbon fibers leads to a significant increase in the oven dry density. The highest amount of carbon fibers used in the study (1.5%) produces an increase in the density from 1670 kg/m$^3$ to 1820 kg/m$^3$. These results are also supported by other researchers [20]. Whereas, the unit weight of concrete increases with the increasing fiber content.

On the other hand, the effect of polypropylene fibers on the density of foamed concrete was found to be much less than that of carbon fibers. Therefore, the use of 1.5% of polypropylene fibers increases the density from 1670 kg/m$^3$ to 1715 kg/m$^3$.

## 5 Economic Efficiency criteria

Some logical steps were adopted to evaluate the performance of foamed concrete mixes reinforced with fibers and to study the economic efficiency of these fibers. For each mix, the cost of the mix per cubic meter of concrete was calculated. These steps are listed as follows:

**Step1:** Find the maximum value in each column as shown in Table 9.

**Step2:** Classify the columns into two types; maximum columns and minimum columns, according to the requirement of each property.

**Step3:** Find the coefficient for each value of the property, as shown in Table 10.

**Step4:** Find the average coefficient for each mix, the largest average coefficient presents the best mix.

Therefore, the results of the highest coefficient indicate that foamed concrete reinforced with hybrid fibers as 1 % CF + 0.5 % PPF exhibit the best performance compared with other mixes. In other words, in-spite of the increase of the cost of foamed concrete by the inclusion of fibers, the enhancements obtained make this mix as the best effective one.

Also, the comparison between C2 and C6 illustrates that the use of 1.5% CF is more effective than that of 1.5% PPF. So, the selection of costly fibers may be the suitable choice due to the requirements of the best performance.

**Table 9.** Maximum value for each property.

| Mixes | Compressive strength (MPa) 28 Days | Splitting tensile strength (MPa) 28 Days | Flexural strength (MPa) 28 Days | Oven dry density kg/m$^3$ | Cost \$/m$^3$ |
|---|---|---|---|---|---|
| **C0** | 17.1 | 1.9 | 2.1 | 1670 | 100 |
| **C1** | 17.3 | 2.0 | 3.8 | 1745 | 121 |
| **C2** | 23.1 | 2.6 | 4.3 | 1805 | 143 |
| **C3** | 21.4 | 2.8 | 4.5 | 1820 | 164 |
| **C4** | 21.0 | 2.9 | 4.5 | 1775 | 147 |
| **C5** | 19.1 | 2.1 | 4.0 | 1770 | 130 |
| **C6** | 13.8 | 2.0 | 3.8 | 1715 | 114 |
| **Maximum value** | **23.1** | **2.9** | **4.5** | **1820** | **164** |

**Table 10.** Coefficient calculation.

| Column type | Maximum (value / max. value from table 9) | | | Minimum 1 - (value / max. value from table 9) | | Average coefficient |
|---|---|---|---|---|---|---|
| **Mix type** | **Compressive coefficient** | **Splitting coefficient** | **Flexural coefficient** | **Oven dry density coefficient** | **Cost coefficient** | |
| C0 | 0.74 | 0.66 | 0.47 | 0.08 | 0.39 | 0.47 |
| C1 | 0.75 | 0.69 | 0.84 | 0.04 | 0.26 | 0.52 |
| C2 | 1.00 | 0.90 | 0.96 | 0.01 | 0.13 | 0.60 |
| C3 | 0.93 | 0.97 | 1.00 | 0.00 | 0.00 | 0.58 |
| C4 | 0.91 | 1.00 | 1.00 | 0.02 | 0.10 | 0.61 |
| C5 | 0.83 | 0.72 | 0.89 | 0.03 | 0.21 | 0.53 |
| C6 | 0.60 | 0.69 | 0.84 | 0.06 | 0.31 | 0.50 |

## 6 Conclusions

This paper describes the properties of foamed concrete reinforced with carbon and a combination of carbon fibers (CF) and polypropylene fibers (PPF). Some conclusions are observed as follows:

1. The use of carbon fibers may significantly decrease the flowability or workability of foamed concrete and have more effect than that of polypropylene fibers.
2. The use of 1% carbon fibers increases the compressive strength by about 36% compared with reference mix. Whereas, the use of 1.5% of carbon fibers increases the splitting and flexural strengths by about 44 and 116%, respectively, compared with normal foamed concrete.
3. The hybridization of 1% CF with 0.5% PPF fibers improves the compressive strength by about 23% compared with reference mix. Whereas, the percentages of increase for splitting and flexural strengths are about 48 and 116%, respectively, compared with normal foamed concrete.
4. The economical factor efficiency can be calculated to evaluate the mixes depending on best performance with least cost. Thus, the use of 1% CF + 0.5% PPF fibers, gives the highest coefficient of such evaluation.

## 7 Acknowledgements

The work described in this paper is a part of Master research program of the 3rd author. The authors express there thanks to the Building and Construction Department staff in the University of Technology-Baghdad for supporting their research. Special thanks to the technical staff in Concrete and Materials Lab in the department.

## References

1. S.K. Lim, C.S. Tan, X. Zhao, T.C. Ling, K.S. of Civil Engineers, **19**, 7, 2191-2197 (2015).
2. S.W. Dhengare, A.L. Dandge, H.R. Nikhade, J.I.A.A.T.S, **16** (2015).
3. O. A. EL-Nawawy, A.H. Zaher, A. A. Talaat, A. S. Mostafa, I.J.E.I.T, **4**, 10 (2015).
4. N. K. Praful, I.J.T.E.E.E. **3**, 4, (2015).
5. Yoo-Jae Kim, Jiong Hu, Soon-Jae Lee, Byung-Hee You, Hindawi Publishing Corporation Advances in Civil Engineering, Article ID 549642, 8 pages (2010)
6. J. S. Darole, V.P. Kulkarni, A.P. Shaikh, B.E. Gite I.J.E.R.A. **3**, 4,1408-1411 (2013).
7. Iraqi Standard Specification (IQS), **5** (1984).
8. ASTM C 150, Annual book of ASTM standards, **04.01** (2007).
9. ASTM C 33, Annual book of ASTM standards, **04.02** (2002).
10. British Standard Institution, B.S. 1881, Part 116, (1983).
11. ASTM C 642, Annual book of ASTM standards, **04.02** (2006).
12. ASTM C78, Annual book of ASTM standards, **04.02** (2002).
13. ASTM C567, Annual book of ASTM standards, **04.02** (2000).
14. E.T. Dawood, M. Ramli, Construction and Building Materials, **25**, 5, 2240–2247 (2011).
15. P.N. Balaguru, S.P. Shah, *Fiber-Reinforced Cement Composites*, Engineering series, McGraw-Hill International Editions (1992).
16. R.V. Balendran, F.P. Zhou, A. Nadeem, A.Y. Leung, Building and Environment **37**, 12, 1361–1367 (2002).
17. E.T. Dawood, M. Ramli, Construction and Building Materials **28**, 1, 193–200 (2012).
18. M. Sahmaran, I.O. Yaman, Construction and building materials **21**, 1, 150-156 (2007).
19. G. E. Nawy, *Fundamentals of high strength high performance concrete*, Longman House, Burnt Hill, Harlow, England, (1996).
20. F. Koksal, F. Altun, I. Yigit, Y. Sahin, Construction and Building Materials **22**, 8,1874–1880 (2008).

# Relationship between splitting tensile and compressive strengths for self-compacting concrete containing nano- and micro silica

*Ali* Jaber[1,*], *Iqbal* Gorgis[1] *and Maan* Hassan[1]

[1]Building and Construction Engineering Department, University of Technology, Baghdad, Iraq

**Abstract.** This paper describes the relationship between splitting tensile strength and compressive strength of self-consolidating concrete using data collected from laboratory specimens tested at standard conditions. The results were then compared with some expressions published in international literature. The investigated variables included: type of cement, percentage of nanosilica and percentage of microsilica as a cement replacement by weight. In spite of concrete not being designed to resist direct tension the knowledge of tensile strength is needed to estimate the cracking load. In the absence of test results an estimate of the tensile strength may be obtained by using the relationship proposed. The verification of the proposed formula based on experimental data was estimated by means of the integral absolute error (IAE). The output of this study has provided a better understanding of the correlation between splitting and compressive strengths of SCCs and the effect of some related variables on the resultant behavior, which has therefore, helped to generate new expression with better accuracy.

## 1 Introduction

Compressive strength is the main measure of structural quality of concrete. Structural design depends mainly on the compressive strength for some purposes the tensile strength is of interest. In the design of highway and airfield slabs, shear strength, and resistance to cracking requires information about the tensile strength. The compressive and splitting tensile strength are closely related depending on the nature strength of concrete. Concrete is not normally designed to resist direct tension; the knowledge of tensile strength is used to estimate the load under which cracking will develop. This is due to its influence on the formation of cracks and their propagation to the tension side of reinforced concrete flexural member. There is no direct relation between the compressive and the splitting tensile strengths. It was noticed that with the increased compressive strength, the tensile splitting strength is also increased but at a decreasing rate [1][2].

The test process of splitting tensile strength involves loading the cylinder specimens to induce transverse tension. Two strong parallel plates distribute and apply a stress in two diametrically opposite points on the cylinder diameter leading to the propagation of cracks through the paste, due to high tensile stresses that cause splitting of the specimen along the vertical plane [2 & 3].

The equation that was used to calculate the splitting tensile stress (fspt) which causes the failure of the specimen according to ASTM C496 / C496M-11 [3] is:

$$fspt = \frac{2F}{\pi DL} \quad (1)$$

where: (F) the applied force; (D) the diameter of the sample; and (L) the sample length. The empirical formula relating $(fspt)$ and $(fc')$ had been suggested by researchers (see Table I) having the following type:

$$fspt = k\,(fc')^{n} \quad (2)$$

where: (k) and (n) are constant coefficients; and (fc′) is the compressive strength of concrete cylinder. The values of (n) between (0.5) and (0.85) have been suggested. The former value is used by the ACI 318-2014 [4], different values of the experimental coefficients (k) and (n) were proposed by other researchers which are indicated in Table 1. Table 1 shows the experimental parameters (k) and (n) proposed by various concrete design codes.

* Corresponding author: aliattiea@gmail.com

© The Authors, published by EDP Sciences. This is an open access article distributed under the terms of the Creative Commons Attribution License 4.0 (http://creativecommons.org/licenses/by/4.0/).

**Table 1.** Values of the constant coefficients (k) and (n)

| Sources | k | n |
|---|---|---|
| ACI 318 [4] | 0.56 | 0.5 |
| ACI 363R [5] | 0.59 | 0.5 |
| Gardner [6] | 0.47 | 0.59 |
| Nihal [7] | 0.387 | 0.63 |
| JCI [8] | 0.13 | 0.85 |
| JSCE [9] | 0.23 | 0.67 |
| CEB-FIB [10] | 0.3 | 0.67 |
| Raphael [11] | 0.313 | 0.667 |
| Ahmad and Shah [12] | 0.462 | 0.55 |
| Oluokun et al. [13] | 0.294 | 0.69 |

Some studies suggest that ACI 318-2014 [4] coefficients underestimate the splitting tensile strength for high strength concrete and overestimate it for low strength concrete [14]. Other research findings have consistently indicated that the 0.5 power relation adopted in ACI 318-2014 [4] does not agree particularly well with the test results. Consequently, researchers have proposed several alternative relations.

## 2 Research significance

This investigation was intended to present a relationship between compressive strength and splitting tensile strength for SCC containing nanosilica and microsilica as a cement replacement by weight and cast using two types of cement (type I and V). Two ages (28 and 90 days) were studied in this analysis to predict the constant coefficients (k) and (n) using regression analysis. For this purpose, the existing relationship between the compressive strength and splitting tensile strength of concrete was examined.

The evaluation of the constant coefficients (k & n) and the derived constant coefficients from regression analysis models were estimated through the use of Integral Absolute Error (IAE) described by:

$$IAE = \frac{\sum[(Q_i - P_i)^2]^{0.5}}{\sum Q_i} \quad (3)$$

where: (Oi) the experimental splitting tensile strength; and (Pi) the predicted splitting tensile strength from regression analysis.

## 3 Materials and experimental program

### 3.1 Materials

Two types of Commercial Iraqi cements were used. The first was Ordinary Portland Cement (OPC Type I) and the second was Sulfate Resisting Portland cement (SRPC Type V). Both of them conform to ASTM C150/C150M-15 [15] and Iraqi Specification No. 5/1984 [16]. Two types of silica were considered:

- Densified microsilica (mS) MEYCO MS 610 conforming to ASTM C 1240-15 [17], supplied by BASF, with density of 2.2 g/cm$^3$ and $SiO_2$ content more than 90%.
- Colloidal nanosilica (nS) produced by Jinan Yinfeng Silicon Products Company, with slight blue and transparence, pH value of 9.55 and density 1.204 g/cm$^3$.

Local natural fine (FA) and crushed coarse aggregate (CA) were used, they were within the requirements of ASTM C33/C33M-13 [18] and also conform to the Iraqi Specification No. 45/1984 [19]. The specific gravity and bulk density were 2.65 and 1670 kg/m$^3$ for fine aggregate, 2.63 and 1685 kg/m$^3$ for coarse aggregate, respectively. A constant amount of 100 kg/m$^3$ ground limestone powder (LP) was applied as filler for all SCCs mixes.

Superplasticizer (SP) was used to increase the workability of SCC for this work, unique polycarboxylic ether with long lateral chains based SP with the specific gravity of 1.07 was used as a high range water reducer (GLENIUM 54 produced by BASF company and conforming to ASTM C494/C494-15 Type F [20] depending on the dosages used) to improve the performance of SCC. Table 2 describes the various physical and chemical characteristics of the cement types and each mineral admixture used in this study.

**Table 2.** Chemical composition and physical properties of cement and admixtures

| Oxides | Percent chemical composition (%) | | | | |
|---|---|---|---|---|---|
| | Cement type I | Cement type V | Micro-silica | Nano-silica | Lime stone filler |
| CaO | 66.11 | 63.32 | 0.92 | --- | 65.32 |
| $SiO_2$ | 21.93 | 21.95 | 95.14 | 30.1 | 1.02 |
| $Al_2O_3$ | 4.98 | 3.76 | 0.71 | --- | 0.61 |
| $Fe_2O_3$ | 3.10 | 4.66 | 0.46 | --- | 0.32 |
| $SO_3$ | 2.25 | 1.10 | 0.95 | --- | 0.12 |
| MgO | 2.00 | 2.15 | --- | --- | 0.31 |
| Insoluble material | 1.29 | 0.49 | --- | --- | |
| Loss on ignition | 2.39 | 1.93 | 1.41 | --- | 31.5 |
| Lime saturation factor | 0.93 | 0.85 | --- | --- | |
| Physical properties | | | | | |
| Specific gravity | 3.12 | 3.08 | 2.2 | 1.204 | 2.7 |
| Surface area- cm$^2$/g | 3760 | 3400 | 20750 | 50000 | 3900 |

### 3.2 Experimental program

#### *3.2.1 Properties of fresh SCC*

Fresh state properties such as filling ability, viscosity and passing ability of concrete, Slump Flow and T50 cm, V-Funnel and V-Funnel at T5 minutes, J-Ring and L-Box Tests, were conducted according to European

Federation of National Associations (EFNARC) for SCC specifications [21].

Using the provided materials, fourteen SCC mixes were casted according to the mix proportions listed in Table 3. A total time of mixing was 5 minute in a compulsory mixer; fresh properties according to EFNARC were done.

### *3.2.2 Properties of hardened SCC*

Three specimens (cubes, 100 mm side length) were cast for each mix of SCC to determine the compressive strength.

**Table 3.** Mixture proportions of the investigated concretes.

| Mix* | Cement $kg/m^3$ | mS $kg/m^3$ | nS $kg/m^3$ | FA $kg/m^3$ | CA $kg/m^3$ | LP $kg/m^3$ | Water $L/m^3$ | SP $L/m^3$ | W/Cem | W/P |
|---|---|---|---|---|---|---|---|---|---|---|
| SCC Ref. I | 400 | 0 | 0 | 850 | 850 | 100 | 152 | 5 | 0.38 | 0.3 |
| SCC 3% mSI | 388 | 12 | 0 | 850 | 850 | 100 | 152 | 5.4 | 0.38 | 0.3 |
| SCC 4.5 % mSI | 382 | 18 | 0 | 850 | 850 | 100 | 152 | 5.8 | 0.38 | 0.3 |
| SCC 6% mSI | 376 | 24 | 0 | 850 | 850 | 100 | 152 | 6.4 | 0.38 | 0.3 |
| SCC 3% nSI | 388 | 0 | 12 | 850 | 850 | 100 | 152 | 11.82 | 0.38 | 0.3 |
| SCC 4.5 % nSI | 382 | 0 | 18 | 850 | 850 | 100 | 152 | 15.21 | 0.38 | 0.3 |
| SCC 6% nSI | 376 | 0 | 24 | 850 | 850 | 100 | 152 | 18.74 | 0.38 | 0.3 |
| SCC Ref. V | 400 | 0 | 0 | 850 | 850 | 100 | 152 | 5 | 0.38 | 0.3 |
| SCC 3% mSV | 388 | 12 | 0 | 850 | 850 | 100 | 152 | 5.4 | 0.38 | 0.3 |
| SCC 4.5% mSV | 382 | 18 | 0 | 850 | 850 | 100 | 152 | 5.8 | 0.38 | 0.3 |
| SCC 6% mSV | 376 | 24 | 0 | 850 | 850 | 100 | 152 | 6.4 | 0.38 | 0.3 |
| SCC 3% nSV | 388 | 0 | 12 | 850 | 850 | 100 | 152 | 11.82 | 0.38 | 0.3 |
| SCC 4.5% nSV | 382 | 0 | 18 | 850 | 850 | 100 | 152 | 15.21 | 0.38 | 0.3 |
| SCC 6% nSV | 376 | 0 | 24 | 850 | 850 | 100 | 152 | 18.74 | 0.38 | 0.3 |

*SCC 3% nSV means self-consolidating concrete with 3% nanosilica as a replacement of cement type V; W/Cem, ratio of water to cementitious materials; W/P, ratio of water to powder materials.

The cubes were cured sealed for the first day, demoulded after 24 hours and cured in water till the age test (28 and 90 days). The average of three cubes was used to calculate the compressive strength. The compressive strength test was performed according to B.S 1881: Part 116: 1989 [22] ( BS EN 12390-3:2009). Cylinders conforming to the ASTM C 496/C 496M – 11 [3] were cast to calculate the splitting tensile strength after 28 and 90 days.

## 4 Test results and discussion

### 4.1 Results of fresh SCC

Preliminary results given in Table 4, concern fresh concrete properties. The experimental results show that, type of cement has no significant effect on fresh properties. The SCCs mix presents a good filling ability (slump-flow ranged from 695 to 740 mm and T50 from 2.8 to 3.6 second and V-funnel flow times were between 7.4 to 9.1 second) and an acceptable passing ability (L-box results ranged between 0.86 and 0.95 and J-ring 695 to 720 mm). Therefore, fresh SCC properties complied well with EFNARC guidelines, which is an indication of a good workability. The mix with colloidal nanosilica gave closer flow result to the lower limit. The plasticizer content was changed (as indicated in Table 3) to stay within the limit of fresh tests specified standards. The supper plasticizer dosage depends on the surface area and the mineralogy and the fine-ness of particles. The supper plasticizer requirement is rather related to changes in the surface area of the mix than to the concentration (by weight) of the nanosilica addition, as presented by Sobolev et al. [23][24].

Test results of fresh SCC mixes with different percentages of different $SiO_2$ indicated that, there is a remarkable reduction in slump flow for both types of cement when using micro or nanosilica. Consequently, it could be caused by higher surface area for both macro and nanosilica that require higher w/c. Also, the reduction by microsilica was less than nanosilica for the same added percent.

Test results for V-funnel time are presented in Table 4 and indicate that using different percentages of $SiO_2$ has less/no effect on the results. Besides, test results of L-box for all SCC mixes, clarify that using nanosilica cause an increase in the flowability and viscosity of the mixes as compared to micro silica.

### 4.2 Results of hardened SCC

#### *4.2.1 Compressive strength*

The cube specimens were cured and tested at 28 and 90 days' age, they gave an indication to an enhancement that occurred as a result of $SiO_2$ nano particles incorporation on all SCC mixes and the influence of the cement type, which are presented in Table 5. The results of cube compressive strength (fcu) listed in Table IV were equated by cylinder (fc') using factor equal (0.8) multiplying by (fcu) [22].

The cylinder compressive strength (fc') presented in Fig. 1 displayed an improvement in SCCs mix with OPC and SRPC using nS and mS. It was indicated that using nS developed higher strength than mS for specimens with two types of cement and both ages 28 and 90 days, but the improvement for specimens with OPC were higher than specimens with SRPC. The behavior of collided $SiO_2$ nano particles was responsible for the enhancement in the strength due to acting as nucleus sites to tightly bond with C-S-H gel particles, resulting in more intensified microstructure. During the development of hydration process with time, the mix contained colloidal nanosilica indicated higher increase in the compressive strength. Several investigations in concrete containing nanosilica concluded that the pore structure was denser and lower coarse capillary pores than Portland cement concretes [25][26].

**Table 4.** Concrete properties in fresh stat

| Mix description | Slump flow (mm) | Flow time (sec.) | J-ring (mm) | L-box ($H_2/H_1$) | V-funnel (sec.) |
|---|---|---|---|---|---|
| EFNARC requirements[21] | 650 - 800 | 2- 5 | 600 –750 | 0.8 -1.0 | 6 -12 |
| SCC Ref. I | 740 | 3.5 | 695 | 0.95 | 8.0 |
| SCC 3% mSI | 717 | 3.6 | 703 | 0.94 | 8.5 |
| SCC 4.5 % mSI | 710 | 3.2 | 696 | 0.93 | 9.0 |
| SCC 6% mSI | 705 | 2.9 | 702 | 0.91 | 9.4 |
| SCC 3% nSI | 705 | 3.2 | 710 | 0.93 | 8.7 |
| SCC 4.5 % nSI | 700 | 2.9 | 698 | 0.92 | 9.3 |
| SCC 6% nSI | 695 | 3.5 | 716 | 0.90 | 9.8 |
| SCC Ref. V | 735 | 3.4 | 704 | 0.94 | 8.2 |
| SCC 3% mSV | 720 | 3.2 | 720 | 0.93 | 8.8 |
| SCC 4.5% mSV | 710 | 2.9 | 713 | 0.92 | 9.0 |
| SCC 6% mSV | 706 | 3.0 | 699 | 0.90 | 9.2 |
| SCC 3% nSV | 710 | 3.3 | 717 | 0.92 | 9.0 |
| SCC 4.5% nSV | 705 | 3.5 | 722 | 0.91 | 9.3 |
| SCC 6% nSV | 700 | 2.8 | 690 | 0.89 | 9.7 |

**Table 5.** Concrete properties in hardened state

| Mix description | Compressive Strength (MPa) | | | | Tensile Splitting Strength (MPa) | |
|---|---|---|---|---|---|---|
| | 28 days | | 90 days | | 28 days | 90 days |
| | fcu | fc' | fcu | fc' | fc' | f c' |
| SCC Ref. I | 52.3 | 41.84 | 55.4 | 44.32 | 4.12 | 4.92 |
| SCC 3% mSI | 55.9 | 44.72 | 58.8 | 47.04 | 4.63 | 5.24 |
| SCC 4.5 % mSI | 57.8 | 46.24 | 60.3 | 48.24 | 4.36 | 5.11 |
| SCC 6% mSI | 58.3 | 46.64 | 61.5 | 49.2 | 4.14 | 4.96 |
| SCC 3% nSI | 63.9 | 51.12 | 66.1 | 52.88 | 4.96 | 5.53 |
| SCC 4.5 % nSI | 67.8 | 54.24 | 69.7 | 55.76 | 4.59 | 5.48 |
| SCC 6% nSI | 68.1 | 54.48 | 71.2 | 56.96 | 4.21 | 5.28 |
| SCC Ref. V | 45.6 | 36.48 | 47.8 | 38.24 | 3.91 | 4.50 |
| SCC3% mSV | 50.2 | 40.16 | 53.1 | 42.48 | 4.45 | 4.93 |
| SCC4.5%mSV | 55.3 | 44.24 | 58.5 | 46.8 | 4.25 | 4.88 |
| SCC 6% mSV | 55.8 | 44.64 | 58.9 | 47.12 | 4.01 | 4.64 |
| SCC 3% nSV | 57.6 | 46.08 | 60.5 | 48.4 | 4.81 | 5.41 |
| SCC 4.5% nSV | 62.2 | 49.76 | 66.3 | 53.04 | 4.65 | 5.22 |
| SCC 6% nSV | 63.1 | 50.48 | 67.5 | 54 | 4.23 | 5.12 |

### *4.2.2 Tensile splitting strength*

The average 28 and 90 days' age tensile splitting strength results are presented in Table V and plotted in Fig. 2. It was indicated that the reference specimens with OPC gave higher tensile splitting strength than SRPC, however, the improvement in using nS and mS for SRPC was higher than OPC. The results of specimens with nS gave higher splitting strength than specimen with mS for both types of cement. The higher splitting tensile strength was for SCC specimens with 3% mS and nS for both ages as indicated in Fig.2. This was due to the stronger bond between the hardened paste and the aggregate. Mixes containing nanosilica produce an enhancement in the quality of the interfacial zone (ITZ), this is due to precipitation of smaller and stronger (higher stiffness) C–S–H gel and an accelerated rate of hydration of the paste, as reported by K. Audenaert and G. De Schutter [25].

This study was presented to investigate the applicability of existing relationships between the compressive strength and splitting tensile strength for SCC mixes containing mS and nS. The applicability of existing relationships shown in Table 6 between the two strengths were examined to estimate the Integral Absolute Error (IAE) and listed in Table 6. The results indicated that the relation presented by Gardner [6] and Nihal [7] gave approximately the smaller values of (IAE%) for both ages.

The Nonlinear relationships between the experimental results and predicted splitting tensile strength from all relations indicated in Table 1 for all SCCs specimens are shown in Figs. 3 to 6. In view of the numerous factors influencing the relationships of the strengths of concrete, it is not surprising that no simple exact relation is applicable. However, these correlations are felt to be representative in lieu of specific testing for concrete design and evaluation.

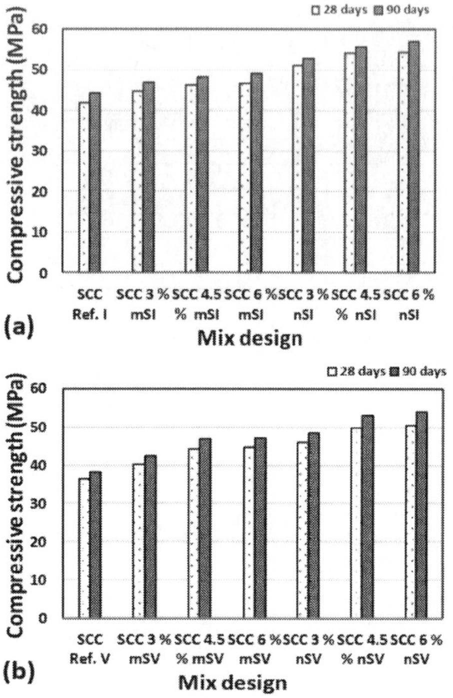


**Fig. 1** Average compressive strength for SCCs at 28 and 90 days, (a) Ordinary Portland cement and (b) Sulfate Resisting Portland cement.

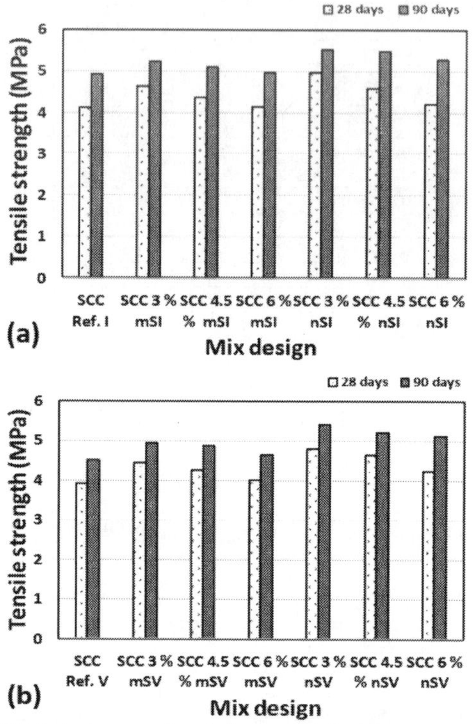


**Fig. 2.** Average tensile splitting strength for SCCs at 28 and 90 days, (a) Ordinary Portland cement and (b) Sulfate Resisting Portland cement.

All theoretical study presented above gave higher values for splitting strength (when applying their relations), the reason for this is that they ignore flaws, and all materials contain flaws.

Based on this research, the relationship between the compressive and splitting tensile strengths of SCCs mixes containing nano and micro silica for 28 and 90 days was studied. A new equation was predicted by regression analysis, in which for simplicity a square root was used as given in most research and ACI-318-2014 [4]:

$$fspt = 0.65 * \sqrt{(f'c)} \tag{4}$$

Using equation (4) to calculate the values of splitting tensile strength and IAE% as presented in Table VI, it complements well with experiment results and gave lesser values for IAE%.

## 5 Conclusions

Experimental results of SCC mixes made with both types of cement, OPC and SRPC and replacements of cement with (3-6) % of micro silica or nano silica increased concretes compressive strength compared to the reference SCC mixes. The enhancement in strength was more visible in concretes made with OPC.

As previously stated, there is little information in the literature concerning the accuracy and validity of the equations used for the purpose of estimating splitting tensile strength from compressive strength. This is especially true for normal strength concrete without any fiber or additives. To assess the accuracy of other power function relationships, the IAE concept was used for the experimental data reported by the various researchers.

The relationship between the compressive and splitting tensile strengths of SCCs mixes used in this research containing nano and micro silica for 28 and 90 days was found to be $fspt = 0.65 * \sqrt{f'c}$ . The IAE results for 28 and 90 days results were between 5 % and 10.8% respectively, which were almost the lesser value.

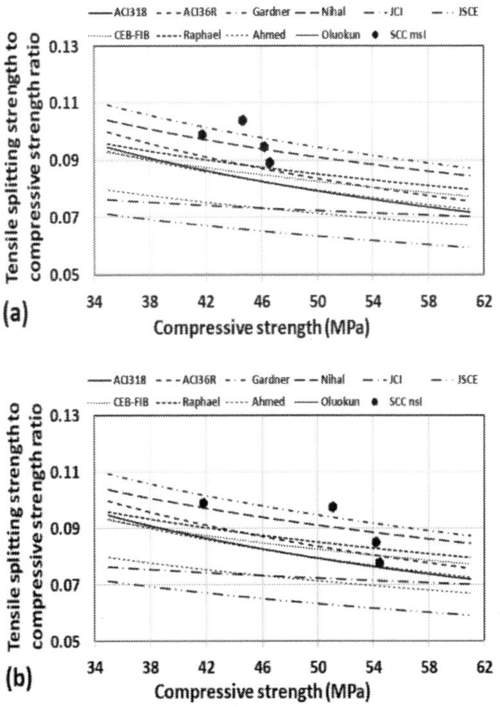


**Fig. 3.** Comparison of the relationship for splitting tensile strength and compressive strength for SCCs mixes with OPC at 28 days, (a) with mS and (b) with nS.

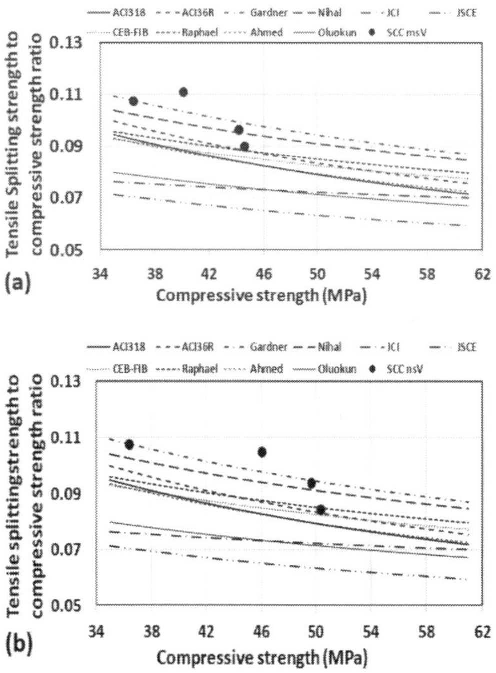


**Fig. 4.** Comparison of the relationship for splitting tensile strength and compressive strength for SCC mixes with SRPC at 28 days, (a) with mS and (b) with nS.

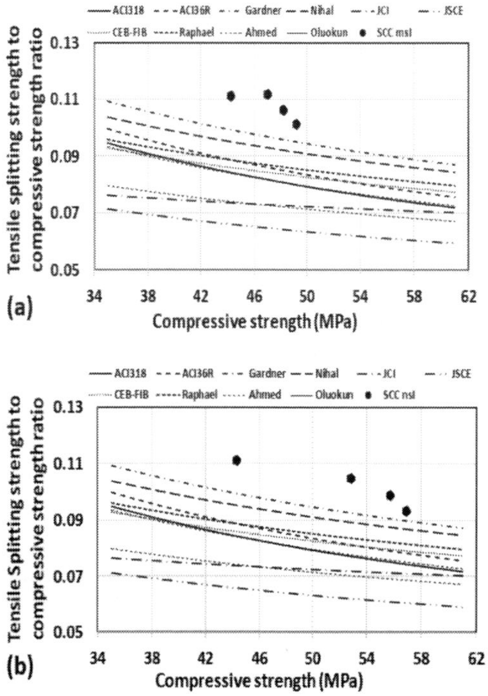

**Fig. 5.** Comparison of the relationship for splitting tensile strength and compressive strength for SCC mixes with OPC at 90 days, (a) with mS and (b) with nS.

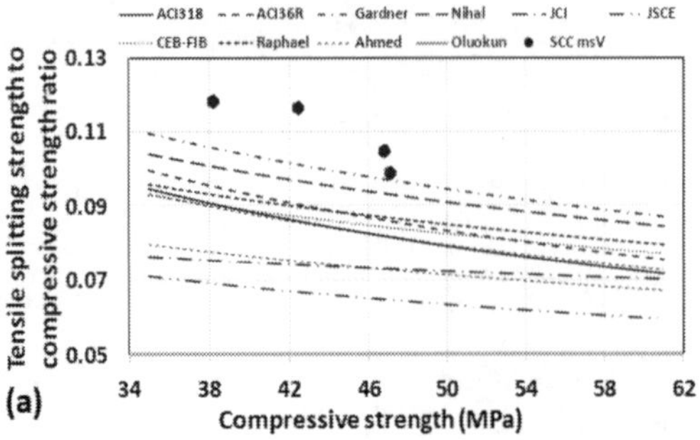


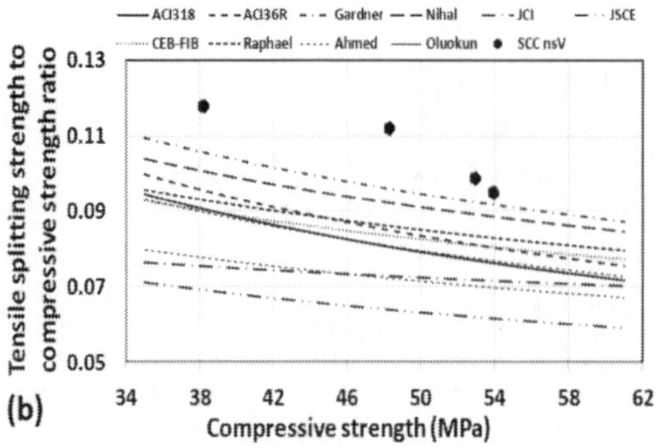


**Fig. 6.** Comparison of the relationship for splitting tensile strength and compressive strength for SCC mixes with SRC at 90 days, (a) with mS and (b) with nS.

**Table 6.** Values of the integral absolute error (IAE)

| Sources | IAE % (28 days) | | | | IAE % (90 days) | | | |
|---|---|---|---|---|---|---|---|---|
| | OPC | | SRPC | | OPC | | SRPC | |
| | msI | nsI | msV | nsV | msI | nsI | msV | nsV |
| ACI 318 [4] | 13.04 | 11.47 | 13.00 | 14.36 | 27.06 | 27.70 | 23.59 | 26.34 |
| ACI 363R [5] | 8.39 | 8.26 | 8.49 | 9.69 | 22.45 | 22.89 | 19.09 | 21.67 |
| Gardner [6] | 5.15 | 8.30 | 5.03 | 5.25 | 10.92 | 9.79 | 8.54 | 9.69 |
| Nihal [7] | 3.97 | 6.90 | 5.43 | 6.52 | 14.94 | 13.53 | 12.71 | 13.63 |
| JCI [8] | 23.57 | 19.09 | 25.26 | 24.32 | 36.07 | 33.87 | 34.35 | 34.53 |
| JSCE [9] | 31.82 | 29.73 | 32.09 | 32.95 | 45.02 | 45.11 | 41.96 | 44.16 |
| CEB-FIB [10] | 11.07 | 9.33 | 12.27 | 12.06 | 24.31 | 23.04 | 22.03 | 23.05 |
| Raphael [11] | 8.27 | 7.93 | 9.57 | 9.83 | 21.53 | 20.11 | 19.33 | 20.23 |
| Ahmad and Shah [12] | 13.24 | 11.13 | 13.53 | 14.46 | 27.02 | 27.16 | 23.87 | 26.16 |
| Oluokun et al. [13] | 6.27 | 7.12 | 7.91 | 8.85 | 19.11 | 17.28 | 17.16 | 17.68 |
| Proposed Equation | 4.21 | 6.61 | 4.53 | 4.92 | 11.71 | 11.29 | 9.42 | 10.85 |

## References


[1] A. M. Neville, 4th Edition, Longman Group Ltd., Essex, 1995, 844 pp."

[2] M. S. Hassan, Z. M. Al-azawi, and M. J. Taher, *Arab. J. Sci. Eng.*, vol. 41, no. 10, 2016.

[3] ASTM C496/C496M-11, "Standard Test Method for Splitting Tensile Strength for Cylindrical Concrete Specimens," *ASTM Int. West Conshohocken, PA*, vol. 04.02, p. 5, 2011.

[4] A. C. I. 318, "Building Code Requirements for Structural Concrete (ACI 318-14): An ACI Standard: Commentary on Building Code Requirements for Structural Concrete (ACI 318R-14), an ACI Report," 2015.

[5] ACI Committee 363 R, "State-of-the-Art Report on High-Strength Concrete (ACI 363R-92)," *Am. Concr. Institute, Farmingt. Hills, Mich.*, p. 55, 1992.

[6] N. J. Gardner, *ACI Mater. J.*, vol. 87, No.1, pp. 68–78.

[7] N. Arioglu, Z. Canan Girgin, and E. Arioglu, *ACI Mater. J.*, vol. 103, no. 1, pp. 18–24, 2006.

[8] JCI 2008, "Guidelines for Control of Cracking of Mass Concrete," *Japan Concr. Inst.*, 2008.

[9] JSCE 2007, "Standard specifications for concrete structures-2007," *Mater. Constr. Japan Soc. Civ. Eng.*, 2008.

[10] CEB-FIB 1991, "Model Code for Concrete Structures. Evaluation of the Time Dependent Behavior of Concrete," *1991*.

[11] J. M. Raphael, in *Journal Proceedings*, 1984, vol. 81, no. 2, pp. 158–165.

[12] S. H. Ahmad and S. P. Shah, *PCI J.*, vol. 30, no. 6, pp. 92–119, 1985.

[13] F. A. Oluokun, E. G. Burdette, and J. H. Deatherage, *Mater. J.*, vol. 88, no. 2, pp. 115–121, 1991.

[14] H. S. S. Ros, *Concr. Inst. Proc.*, pp. 427–432, 2013.

[15] ASTM C150/C150M-15, "Standard Specification for Portland Cement," *ASTM Int. West Conshohocken, PA*, vol. 04.01, p. 7, 2015.

[16] COSQC Iraq, "Standard Specification for Portland Cement," *Iraqi Specif. No. 5*, 1984.

[17] ASTM C1240-15, "Standard Specification for Silica Fume Used in Cementitious Mixtures," *ASTM Int. West Conshohocken, PA*, vol. 04.02, p. 9, 2015.

[18] ASTM C33/C33M-13, "Standard Specification for Concrete Aggregates," *ASTM Int. West Conshohocken, PA*, vol. 04.02, p. 11, 2013.

[19] COSQC Iraq, "Standard Specification for

Concrete Aggregates," *Iraqi Specif. No. 45*, 1984.
[20] ASTM C494/C494M-15, "Standard Specification for Chemical Admixtures for Concrete," *ASTM Int. West Conshohocken, PA*, vol. 04.02, p. 10, 2015.
[21] EFNARC, "The European Guidelines for Self Compacting Concrete, Specification, Production and Use," *Eur. Proj. group, http//www.efnarc.org, May*, 2005.
[22] B.S 1881 Part 116, "Method for Determination of Compressive Strength of Concrete Cubes," *Br. Stand. Inst.*, 1989.
[23] K. Sobolev and S. P. Shah, ACI, 2008.
[24] M. S. Hassan, I. N. Gorgis, and A. H. Ali,
[25] K. Audenaert and G. De Schutter, 2009.
[26] R. T. Abdulkareem, M. S. Hassan, and I. N. Gorgis, *Eng. &Tech.Journal*, vol. 34, no. 3, pp. 527–538, 2016.

# Effect of different curing conditions on the mechanical properties of reactive powder concrete

*Mazin* Abdulrahman[1,*], *Alyaa* Al-Attar[2] *and Marwa* Ahmad[2]

[1]Civil Engineering Department, Tikrit University, Tikrit, Iraq
[2]Technical College, Kirkuk, Iraq

**Abstract.** Reactive Powder Concrete (RPC) is an ultra-high performance concrete which has superior mechanical and physical properties, and composed of cement and very fine powders such as quartz sand and silica fume with very low water/ binder ratio and Superplasticizer. Heat treatment is a well-known method that can further improve the performance of (RPC). The current research including an experimental study of the effect of different curing conditions on mechanical properties of reactive powder concrete (compressive strength, modulus of rupture and splitting tensile strength), the curing conditions includes three type of curing; immersion in water at temperature of 35 °C (which is considered as the reference-curing situation), immersion in water at temperature of 90 °C for 5 hours daily and curing with hot steam for 5 hours daily) until 28 days according to ASTM C684-99 [8]. This research includes also the study of effect of adding silica fume as percentage of cement weight on mechanical properties of reactive powder concrete for different percentage ratios (5%,10% and 15%). Super plasticizer is also used with ratio of (1.8%) by weight of cementitious material; constant water cement ratio (0.24) was used for all mixes. For each reactive concrete mix, it has been cast into a cubes of (150*150*150) (to conduct the compression test), a cylinders of 150mm diameter with 300mm height (to conduct split test) and prisms of (500*100*100)mm to conduct the modulus of rupture test. The results showed that the best method of curing (according to its enhancing the RPC mechanical properties) is the method of immersion in hot water at temperature 90 °C for the all silica fume percentages, and the best used silica fume percentage was (10%) for the all used curing methods.

## 1 Introduction

The conventional concrete has a compressive strength of usually less than 40 MPa. It is relatively brittle and its resistant to plastic deformation is relatively low and have low tensile strength and the cracks appears fast. Also, conventional concrete is heavy somewhat and have low resistance to vibration loads. Thus, the designers and construction professionals sought to introduce new ideas and approaches that would make the concrete more efficient. Recently, a new generation of concrete with excellent characteristics and properties called High Performance Concrete (HPC)have been developed [1].

The Reactive Powder Concrete RPC, which was manufactured in France in the 1990s is a new type of high performance concrete which has good mechanical properties and durability [2] [3]. The major difference between reactive powder concrete and conventional concrete is that no coarse aggregate is involved in reactive powder concrete, but crushed quartz and quartz sand are used instead, with a high dosage of silica fume. So that all materials are granular size less than 600 micrometers.

The advanced mechanical and physical properties of RPC are obtained by optimizing packing density of concrete mixture with fine gradation of all mix particles, and by using highly refined silica fume to improve the microstructure of hydrated cement paste through the pozzolanic reaction. The granular size less than 600 micrometers and can be widely used in structural engineering applications, especially in locations where large loads generate tension stresses and high cutting forces.

Furthermore, researchers reported that application of different heat cure processes improve mechanical properties of RPC [4] The aim of the heat treatment of concrete is to increase the hydration speed and thus decrease the setting and hardening period. Besides conventional water curing (standard) of setting concrete, resistance can be gained in a shorter time with different curing methods. These kinds of cures that facilitate the resistance gaining are generally categorized as heat treatment [5]. Thus, heat treatment is a well-known method that can further improve the performance of (RPC).

* Corresponding author: dr.mazinburhan @tu.edu.iq

© The Authors, published by EDP Sciences. This is an open access article distributed under the terms of the Creative Commons Attribution License 4.0 (http://creativecommons.org/licenses/by/4.0/).

## 2 Objective of research

1- Studying the possibility of producing reactive powder concrete.
2- Studying the effect of different curing methods on the properties of hardened concrete (compressive strength, split strength and the modulus of rupture). The treatments used include ordinary treatment by immersion in water at 35°C and immersion in hot water treatment at 90 °C for 5 hours per day and steam treatment for 5 hours per day according to ASTM C684-99 [8] until testing of samples.
3- Study the effect of the addition of silica fume at different ratios (5%, 10%, 15%) by cement weight on the mechanical properties of (RPC).

## 3 Experimental program

### 3.1 Material used

#### *3.1.1Cement*

Local Iraqi cement was used (Type mass), which is an ordinary Portland cement. The results of the test were in accordance with the Iraqi standard specifications (IQS No.5/1984) [9]. The physical and chemical properties of this cement are shown in Table (1).

**Table 1.** Physical and chemical properties of cement

| Properties | Value | Limits of IQ Specification 5/1984[9] |
|---|---|---|
| Specific Surface Area | 259 | 230 lower limit |
| Initial Setting Time (min.) | 127 | Not less than 45 min. |
| Final Setting Time (hr.) | 5 | Not more than 10 hrs. |
| Compressive Strength (MPa) at age of 3 days. | 23.6 | (15 Mpa) |
| CaO | 61.44% | - |
| $Al_2O_3$ | 4.1% | 8% upper limit |
| $SiO_2$ | 3.93% | 5% upper limit |
| $SO_3$ | 1.21% | 4% upper limit |
| $Fe_2O_3$ | 3.93% | 5% upper limit |
| MgO | 2.46% | 2.5% upper limit |
| L.O.I | 1.26% | 4% upper limit |
| I.R | 0.78% | 1.5 % upper limit |
| $C_3S$ | 58.17% | - |
| $C_2S$ | 16.34% | - |
| $C_3A$ | 6.7% | - |
| $C_4AF$ | 13.28% | - |

#### *3.1.2 Fine aggregate*

The sand used in the concrete mix is known as quartz sand of granular size (150-600) micrometer. The sand sieve analysis and its properties are shown in Tables 2 and 3 and it conforms to the British Specifications (B.S.882) [10].

**Table 2** Grading of fine aggregate

| Sieve Size (mm) | Cumulative Passing % | Limits of B.S.882 (zone III) |
|---|---|---|
| 4.75 | 100 | 90-100 |
| 2.36 | 92 | 85-100 |
| 1.18 | 83 | 75-100 |
| 0.6 | 62 | 60-79 |
| 0.3 | 23 | 12-40 |
| 0.15 | 5.25 | 0-10 |

**Table 3.** physical and chemical properties of fine aggregates

| properties | Test result | Limit of Iraqi specification No.5/1984 |
|---|---|---|
| Specific gravity | 2.7 | ------- |
| Absorption | 0.7 | ------- |
| Sulphate content | 0.13 | 0.5 (max) |
| Materials finer than 0.075 mm | 1.24 | 5 (max) |

#### *3.1.3 Water*

Ordinary tap water was used for mixing and curing all concrete mixes of this investigation.

#### *3.1.4 Silica fume*

Silica fume used in this research is type (silica fume HR). Silica fume is produced as a by-product of the silicon industry, which is 100 times finer than the Portland cement. When used in concrete, it is used as a filler and as cementitous material, where the Silica Fume's small particles fill the spaces between cement particles and between cement and aggregates. Silica fume also reacts with Calcium Hydroxide resulting from the process of cement hydration to form extra calcium silicate during the reaction, thus producing a denser, stronger and less permeable concrete. It was used in the form of addition to the cement as percentages of cement weight by 5%, 10%, 15%. The used silica fume is of high quality, and is dependable as concrete additive according to the US specification (ASTM C1240-03) [11], Tables (4) and (5) illustrate the physical and chemical properties of this type of silica fume respectively.

**Table 4**. Physical properties of silica fume

| property | Value |
|---|---|
| Density | 300 kg/m$^3$ |
| form | powder |
| Appearance/ colour | Grey powder |

**Table 5.** Chemical properties of silica fume

| property | Test Value | Limits According to ASTM |
|---|---|---|
| SiO2 | 90 % | ➤ 85 % |
| L.O.I | 2.86 % | < 6 % |

*3.1.5 Superplasticizer*

In this study, the super-plasticizer PC260 is used to enhance the early and final resistance of the concrete. The additive is highly effective for the production of reactive powder concrete. It gives good workability and flow-ability. This additive does not contain chlorides or any compounds that help to iron corrosion. Table (6) shows the specifications for this additive [12]. The additive is commercially known as PC260, which is different from the standard super-plasticizers, which consists of long chain polycarboxylic ether polymers, which makes the concrete smooth and gives high compressive strength and the plasticizer is suitable for use with ordinary Portland Cement.

**Table 6**. Specification of superplasticizer

| Main action | Super-plasticizer properties |
|---|---|
| Appearance | Light yellow liquid |
| Specific Gravity | 1.1±0.02 |
| Viscosity | 22cp,20C |
| Chloride Content | 0.1% |
| Chemical Base | Polycarboxylic either polymers |

## 3.2 Design of concrete mix

The key to the production of Reactive Powder Concrete (RPC) is the high cement content, glass sand (150-600) micrometer, and the ratio of water to cement is very low with high content of super plasticizer, where the mortar consisting of quartz sand, cement and silica fume have a great effect on providing high resistance. The concrete mix of reactive powder concrete is designed in order to obtain the highest compressive strength according to the ACI code ACI-211-91 [13]. The following quantities of contents are used: 990 kg/m$^3$ of ordinary Portland cement, 1090 kg/m$^3$ of the quartz sand content, 237.6 kg/m$^3$ the water content and 1.8% per 100 kg of cement of the plasticized ratio. Three ratios of silica fume as additives to cement has been used (5%, 10%, 15%) in addition to the silica-free reference concrete. This plasticized ratio of 1.8% was used because it is the optimum ratio that gives the best operational reliability through the precipitation test. The mixing of concrete components is important for the required workability and the homogeneity of mix, and concrete mixing process affects the quality of the concrete at the hardened state and the mixture materials should be distribute regularly during concrete for reactive concrete powders. The steps of mixing are as follows: A mixer with a capacity of 0.1 m$^3$ used, where cement and quartz sand are mixing together in there dry state for 3 minutes and then add silica fume and mixing there for another 5 minutes. The water and plasticizer are mixed separately and gradually added to the pre-mixed dry ingredients and mix all for 3 minutes.

**Table 7.** Ingredients of concrete mix

| Ingredients kg/m3. | | | | |
|---|---|---|---|---|
| Water(kg/m$^3$) | Fine aggregate (kg/m$^3$) | Cement (kg/m$^3$) | Silica fume (kg/m$^3$) | Super plasticizer % |
| 237.6 | 1090 | 990 | 0 | 1.8 |
| 237.6 | 1090 | 990 | 49.5 | 1.8 |
| 237.6 | 1090 | 990 | 99 | 1.8 |
| 237.6 | 1090 | 990 | 148.5 | 1.8 |

## 3.3 Cast of samples

After completion of the mixing process, the contents of the mix shall be emptied and poured into the molds using the iron molds set for this purpose. All molds shall be cleaned before casting. The inner faces of the molds shall be greased to prevent adhesion of the hardened concrete with the inner surface of the molds. The cubes and cylinders are casted in three layers whereas the prisms are casted in two layers and each layer is vibrated by the electric vibrator. After casting, the samples are placed in a dry place and covered with plastic sheeting to prevent plastic cracking due to shrinkage for 24-hour and then cured. The following samples where casted: three cubes of dimensions (150*150*150) mm , three cylinders of dimensions (500*300) and three prisms of dimensions (100*100*500)mm. Thus, the number of models are (72) cubic, (72) cylinder and (72) prism, Fig. (1).

**Fig. 1.** Casting of concrete specimens

## 3.4 Curing of samples

Curing is the process of controlling the rate of moisture loss by providing continuous moisturizing of the concrete in order to obtain good concrete quality. This is created by providing a suitable environment for the curing to obtain the appropriate strength because the quantity and rate of humidity depends on the temperature and humidity of the concrete sample. Three kinds of curing were used: the curing with immersed in normal water at temperature 35°C , curing by immersing in hot water at temperature 90°C for a period of five hours per day at the age of 28 days according to specifications ASTM C684 [8], where a metal basin is equipped with two electric heaters and the basin is equipped with a metal cover to maintain the temperature of water, and finally the curing by hot steam for 5 hours continuously daily and for 28 days according to specifications ASTM C684 [8], where a thick cloth room is equipped with dimensions of cubic meter connected to a steam generator that works on electric power and generates relative humidity of 95% and temperature degree 95°C.

### 3.5 Studied variables

As mentioned previously, the variables studied in the current research are the method of concrete curing as well as the percentage of silica fumet, where three ratios of silica fume were used for each curing method in addition to the reference reactive powder concrete free of silica fume, see Table (8).

**Table 8**. Studied variables of RPC specimens

| Group No. | Specimen No | curing | Silica fume add % |
|---|---|---|---|
| G1 | S1 | 35 $^0$C | 0 |
| | S2 | 90 $^0$C | 0 |
| | S3 | steam | 0 |
| G2 | S1 | 35 $^0$C | 5 |
| | S2 | 90 $^0$C | 5 |
| | S3 | steam | 5 |
| G3 | S1 | 35 $^0$C | 10 |
| | S2 | 90 $^0$C | 10 |
| | S3 | steam | 10 |
| G4 | S1 | 35 $^0$C | 15 |
| | S2 | 90 $^0$C | 15 |
| | S3 | steam | 15 |
| G5 | S1 | 35 $^0$C | 0 |
| | S2 | 90 $^0$C | 0 |
| | S3 | steam | 0 |
| G6 | S1 | 35 $^0$C | 5 |
| | S2 | 90 $^0$C | 5 |
| | S3 | steam | 5 |
| G7 | S1 | 35 $^0$C | 10 |
| | S2 | 90 $^0$C | 10 |
| | S3 | steam | 10 |
| G8 | S1 | 35 $^0$C | 15 |
| | S2 | 90 $^0$C | 15 |
| | S3 | steam | 15 |

## 4 Hardened concrete tests

### 4.1 Compressive strength test

The compressive strength was conducted according to BS 1581: Part 116: 1983] [14] for cubes (150 * 150 * 150) mm where it was loaded axially using a (U-Test) machine of capacity (2025 KN) with loading rate of (7kN / sec), so that the load is on the opposite sides of the cube and vertical to the direction of casting, the rate of three cubes strength was adopted at the age of (7, 28) days considered from the time of adding water to the mixture, Fig. (2).

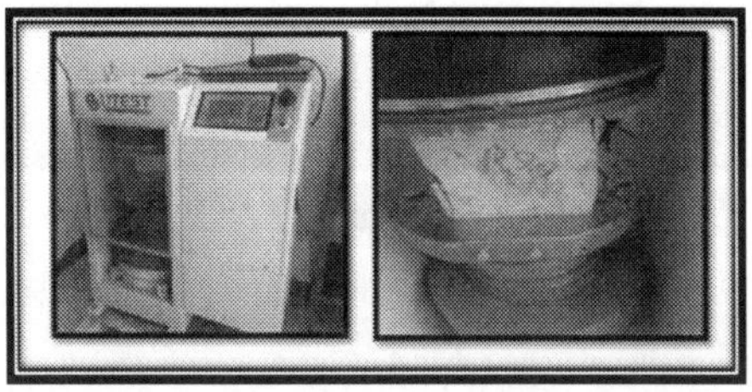

**Fig. 2.** Compressive Test of Concrete Cubes

### 4.2 Splitting strength test

ASTM C 496-96 [15] was adopted to conduct the splitting strength test on 28-day cylindrical specimens with standard dimensions (300*150) using an electrical test machine (U-TEST). The average of three specimens for each type of RPC mix was adopted, Fig. (3).

**Fig. 3.** Split Test of concrete cylinder

### 4.3 Modulus of rupture test

Concrete prisms of dimensions (500 * 100 * 100) mm were used to measure the modulus of rupture ($f_r$) of RPC by using the third point loading method according to (ASTM C78-02) [16]. The average of results of three prisms for each RPC mix at the age of (28) day was adopted. The test is conducted using a (U-TEST) machine with a capacity of 300 kN and a load ratio of 1 MPa/min, Fig. (4).

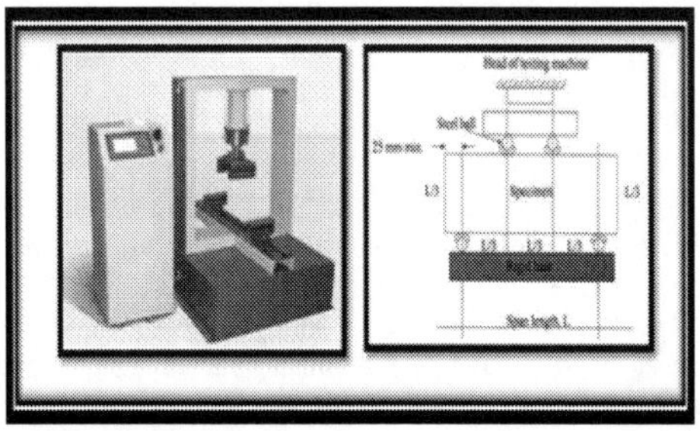

**Fig. 4.** Prism Test

## 5 Results and discussion

### 5.1 Compressive strength

The average of three cubes strength $f'_c$ was adopted for the compressive strength tests. Figures (5) and (6) shows the results for the compressive strength at 7 and 28 days age for different silica fume ratios under different curing conditions. Results showed that increasing of silica fume ratio significantly increases the compressive resistance of concrete cubes to all concrete mixes compared with the reference concrete mix which is free of silica dust, where the curing by immersing in water with 35°C with 5% SF increases the compressive strength by (4.1%, 7.07%) for concrete cubes at age of 7 and 28 days respectively, and for 10% S.F, the increase was 9.7% and 10.08% at 7 days and 28 days respectively. When adding 15% of the S.F, this leads to decrease in compressive strength by (2% ,2.5%) for 7 and 28 days age respectively.

When using the curing by hot water immersing at 90 °C method with adding 5% of S.F, this leads to increases in compressive strength by (34.5% , 32.7%) for ages of (7, 28) days respectively. Adding 10% of silica dust increases $f'_{c\ by}$ (41.9%, 37.1%) for ages of (7 , 28) days respectively. Adding 15% of silica fume results in a decrease in compressive strength of 26% and 19.3% for concrete cubes with 7 days and 28 days respectively.

For the steam method of curing with adding 5% S.F, this will lead to an increas in compressive strength by (2% , 8.7%) for ages of (7 , 28) days respectively. When adding 10% of S.F, this leads to an increas in compressive strength by (10.6%,14.2%) for ages of (7 , 28) days respectively, and when adding 15% of S.F, the increasing in compressive strength will be (25%,35%) for ages of (7, 28) days respectively. The increasing of compressive strength values with increasing of S.F percentages can be attributed to the reason that the S.F is a very fine material that penetrates between the granules of cement and the quartz sand, reduces voids and increase the density. Also, the pozzolanic activity of this material causes more hydration process which leads to more strength and producing the gel, which increases the density of concrete, and thus, increasing the compressive strength. For the effects of different treatments on the compressive strength for constant S.F ratios, and considering the curing by water at 35°C is the reference curing method, the effects is as follow: When the S.F ratio is 0%, the accelerated curing contributes to the increased compressive strength of the concrete by ratios (24%, 14.6%) for curing by hot water of temperature degree 90°C and for steam curing respectively for 7 days age, and the percentage increase is (13.6%, 17.6%) for curing by hot water of temperature degree 90°C and for steam curing respectively for 28 days. For 5% S.F ratio, accelerated treatment contributes to an increased compressive strength as the percentage increase is (17.4%, 42) for curing by hot water of temperature degree at 90°C and for steam curing respectively for 7 days age, and the percentage increase is (22.8%, 44.22%) for hot water curing at 90 ° C and steam treatment respectively at 28 days. When the percentage of S.F is 10%, accelerated treatment contributes to an increased compressive strength as the percentage increase is (41%, 63.4%) for hot water curing 90 ° C and steam curing respectively at 7 days, and with increasing percentage of (23.8%, 42.75%) for hot water curing 90 ° C and steam curing respectively at 28 days, At the rate of addition of S.F 15%, the accelerated curing contributes to the increase of compressive resistance but with very slight percentage increase of (0.05%, 1%) for hot water curing 90 ° C and steam curing respectively at 7 days, either in case of the accelerated curing at the ages of 28 days, the accelerated curing has a negative effect on compressive strength as the percentage of reduction is (25%, 20% for hot water curing 90 ° C and steam curing respectively. From the above values, it can be concluded that accelerated curing using immersing in 90°C hot water method improves the compressive strength compared with steam-accelerated curing and reference curing method.

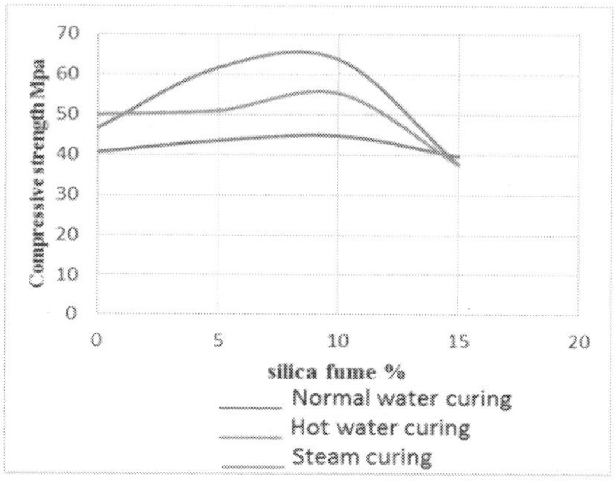


**Fig. 5.** Relationship between silica percentage and compressive strength at 7 days

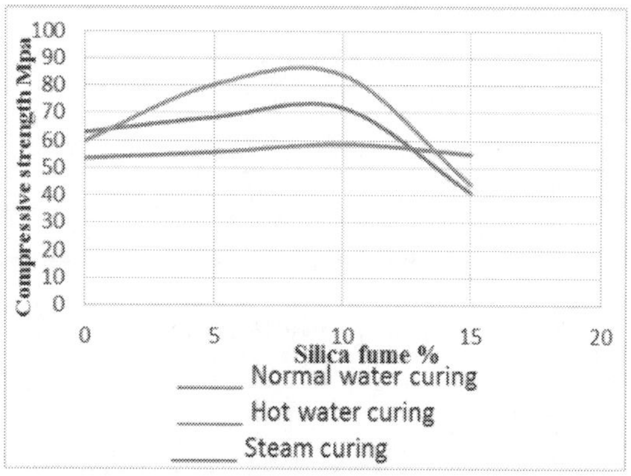


**Fig. 6.** Relationship between silica percentage and compressive strength at 28 days

## 5.2 Split tensile strength

The results showed, Figure (7), that the addition of S.F results in an increase in the tensile strength of the RPC cylinders. Considering that the curing with water at 35°C temperature is the reference curing, the effect of silica fume on splitting strength is as follows: adding S.F increases the tensile strength by (14%, 137.5%, 80.3%) for S.F percentages (15%, 10%, · 5%), respectively comparing with RP contains no S.F. When applying a hot water at 90°C curing, the addition of S.F increases the tensile strength by (25.5%, 38.4%) for the S.F addition (5%, 10%). But it decreases by about (40%) when adding 15% of S.F.

Either by steam curing, this will leads to increases of tensile strength by (214.1%, 243.9%, 33.3%) for the addition of S.F (5%, 10% and 51%), respectively. As for the effect of different curing methods on the tensile strength values, with constant S.F ratio, when the concrete is free of S.F, the percentage of increase in the splitting tensile strength was (1%, 135.9%) for hot water curing at 90°C and steam curing respectively. In the case of 5% S.F ratio, the increase in tensile strength was (63.77%, 38.8%) for hot water curing of 90°C and steam curing respectively.

For 10% of S.F ratio, the percentage of increase in splitting tensile strength is (35.7%, 15.27%) for hot water curing of 90°C and steam curing respectively and for 15% of S.F, then the increase splitting tensile strength is 34.4% for hot water curing of 90°C, but in steam curing method, there was a decrease in splitting tensile strength by about 7%. The results indicated that increase addition of S.F from 5% to 10% had a positive effect on the splitting tensile strength $f_t$.

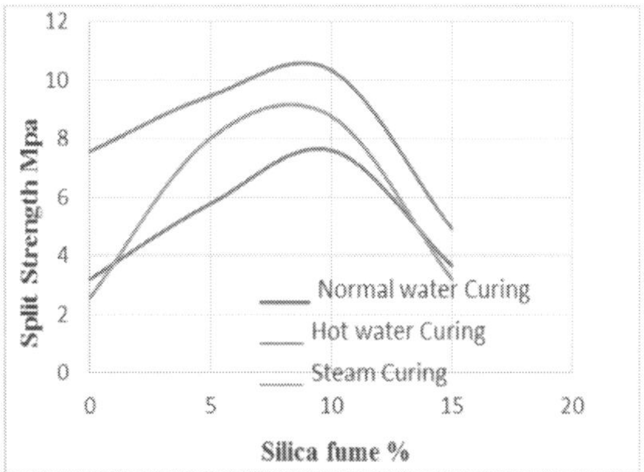


**Fig. 7.** Split tensile strength results

### 5.3 Modulus of rupture

The experimental program results, Figure (8) showed that the effect of the change in the addition of S.F and the effect of the change in the curing methods is as follows: when the immersion in water at 90°C curing method is used, then the increase S.F ratio contributes to the increase in modulus of rupture by (36.9% and 26.7%) for the addition ratio (5% and 10%) respectively. However, a decrease of (8.9%) is occurred for S.F ratio of 15% , and when the hot water at 90°C curing method is used, then the increase S.F ratio contributes to the increase of the modulus of rupture by (36.9%, 31%, 26.7%) for the S.F addition ratios (15%, 10%, 5%) respectively, and for the steam curing method, the increase in the silica dust additive content contributes to the increase in the modulus of rupture by (51.4%, 21.3%, 12.6%) for the S.F addition ratios (15%, 10%, 5%) respectively.

Either the effect of different treatment methods on modulus of rupture values is as follows: when the concrete is free of S.F and cured with hot water at 90°C and steam curing, the modulus of rupture values increase by (1%, 48.7%), where the increase in $f_r$ due to steam curing method was very slight compared to increase due to hot water curing method. At the percentage of 5% of S.F and when the RPC is cured by hot water at 90°C and cured by steam method, the values of modulus of rupture increased by (44.4%, 0%), and as noted, the steam curing method have no effect in $f_r$ increase. At the percentage of 10% of S.F and when the RPC is cured by hot water at 90°C and cured by steam method, the values of modulus of rupture increased by (12.1% ,96.5%) respectively, and when 15% ratio of S.F is used, and the RPC is cured with hot water at 90°C and with steam method, the values of modulus of rupture is increased by (22.3% ,30.9%) respectively.

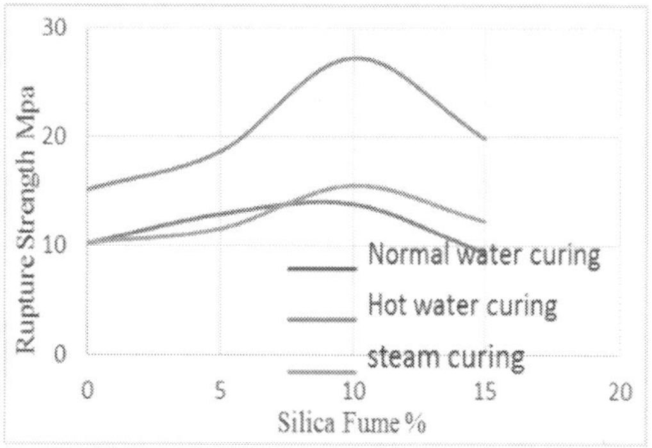


**Fig. 8.** Modulus of rupture results

**Table. 9.** Compressive strength, split tensile strength and modulus of rupture at 28 days age

| Curing Method | Silica Fume % | Compressive Strength for cubes $f_{cu}$ (MPa) | Split Tensile Strength for Cylinders $f_t$ (MPa) | Modulus of Rupture $f_r$ (MPa) |
|---|---|---|---|---|
| Immersion in water 35°C | 0 | 53.55 | 3.2 | 10.22 |
| | 5 | 55.78 | 5.77 | 12.95 |
| | 10 | 58.77 | 7.6 | 13.9 |
| | 15 | 55.1 | 3.66 | 9.4 |
| Immersion in hot water 90°C | 0 | 59.79 | 7.55 | 15.2 |
| | 5 | 80.45 | 9.45 | 18.7 |
| | 10 | 83.9 | 10.32 | 27.3 |
| | 15 | 44 | 4.92 | 19.9 |
| steam | 0 | 63 | 2.55 | 10.3 |
| | 5 | 68.5 | 8.01 | 11.6 |
| | 10 | 71.9 | 8.76 | 15.6 |
| | 15 | 41 | 3.2 | 12.3 |

## 6 Conclusions

1. (Reactive Powder Concrete can be product by using silica fume , quartz sand with mixing ratio (1:1.1) (cement:quartz sand) with cement content of 990 kg /m$^3$ , w/c equal to 0.24 ratio, and an addition ratios of silica fume (0%, 5%, 10%, 15%) from the weight of cement and super-plasticized content 1.8% from the weight of cementations materials.

2. Accelerated methods of curing using hot water at 90°C and steam for 5 hours per day allow a better contribution in improvement of the mechanical properties of hardened reactive powder concrete (compressive strength, indirect tensile strength $f_t$, modulus of rupture $f_r$) compared with water-immersed curing at a temperature of 35°C. The immersion in hot water at 90°C is better than steam curing method to improve the mentioned properties.

3. For the various curing methods mentioned , the best addition ratio of silica fume in improving the mechanical properties of concrete (compressive strength, indirect tensile strength $f_t$, modulus of rupture $f_r$) is the ratio of (10%).

## References

1. P. Richard and M. H. Cheyrezy, *Proceedings of the V. Mohan Malhotra Symposium, ACI*, SP–144, S. Francisco. No.4, pp.507–518, (1994)

2. D. Jeong and H. Lee, , *international Conference on sustainable Building* ,Asia, Seoul.

3. W. Shuxin, and C. Victor Li., ACI Materials Journal, pp. 97-105, (2006)

4. M. Dashti Rahmatabadi, International Journal of Structural and Civil Engineering Research **Vol. 4**, No. 4, pp.354-358,( 2015)

5. F. Hattatoglou and A. Bakis, World Academy of Science, Engineering and Technology, International Journal of Civil, Environmental, Structural, Construction and Architectural, **Vol.10**, No.4, pp.503-507, (2016)

6. N. Al-.Jubory, civil Eng. department, University of Mosul (2015)

7. N.P. Lee and D.H. Chisholm, Study Report No. 146, Building Research Levy, New Zealand, pp. 1-29, (2005)

8. ASTM C684-(2003), Standard Test Method for Making, Accelerated Curing, and Testing Concrete Compression Test Specimens, http://www.astm.org/Standards/C684.htm

9. Iraqi specification, No.5/1984., "Portland cement"

10. B.S.882, "Specification for Aggregates from Natural Sources for Concrete", British Standards Institution, (1992)

11. ASTM C1240 –03, "Standard Specification for Use of Silica Fume as a Mineral Admixture in Hydraulic-Cement Concrete, Mortar, and Grout" (2003)

12. Superplasticizer ,pc260, for DCP company, product data sheet, http://www.dcp-int.com/psmenaenglish.

13. ACI Committee 211-91 ""Standard Practice for Selecting Proportions for Normal, Heavyweight, and Mass Concrete" (ACI 211.1-91) (Reapproved 2009), American Concrete Institute, (1991)

14. B.S. 1881 Part 116 "Method for Determination of Compressive Strength of Concrete Cubes" British Standards Institution, (1989)

15. ASTM C 496 – 96 "Standard Test Method for Splitting Tensile Strength of Cylindrical Concrete Specimens manual book of ASTM standards، **Vol. 04**, No.2, pp-36-38, (2004)

16. ASTM designation C78-02،"Standard test method for flexural strength of concrete (using samples beam with third point loading) (2002)

# Study some mechanical properties of self-compacting concrete with nano silica under severe saline environment conditions

*Ghalib* Habeeb[1] and *Zahraa* Hashim[1]

[1] University of Babylon, Hilla, Iraq

**Abstract.** The main aim of this research is to evaluate the performance of Nano silica self-compacting concrete which is subjected to severe saline conditions that contain sulfates and chlorides at concentrations similar to those existing in the soils and ground water of the middle and southern parts of Iraq. For this purpose, ordinary and sulfate resistant Portland cement without and with 3% Nano silica addition by weight of cementitious materials were used. Splitting tensile strength, flexural strength, static modulus of elasticity and ultrasonic pulse velocity were investigated for all exposure conditions and all types of mixes of self-compacting concrete at ages of 28, 60, 90, 120 and 180 days. Test results revealed that the inclusion of Nano Silica in concrete mixes improved clearly the mechanical properties of self-compacting concrete compared with reference concrete.

## 1 Introduction

Self-compacting concrete (SCC) has been defined as the concrete that does not require vibration for placing and compaction. It is capable to flow under its own weight, completely filling formwork and achieving full compaction even in the presence of congested reinforcement. The hardened concrete is dense, homogeneous and has the same engineering properties and durability as traditional vibrated concrete (**European Project Group, 2005**). SCC has been developed in Japan to get durable concrete structures due to increase of the homogeneity, which is attributed to improve the characteristics of SCC in fresh state; filling ability, passing ability and segregation resistance, in addition to enhancing the hardened properties of SCC. On the other hand, the durability of concrete structures is closely associated with the permeability of the surface layer. Therefore, the absence of compaction of the surface layer due to the difficulties of vibration in narrow spaces between the formwork and the re-bars, which can be characterized as a key factor of poor durability performance of reinforced concrete structures subjected to aggressive environments. Overcoming this was one of the main reasons for the original development of SCC in Japan (**Singh and Thombre, 2013**).

## 2 Research significance

Sulfate ions have been found in the groundwater and soil in the middle and southern parts of Iraq. In the regions of these parts, the concrete structures can be exposed to sulfate attack and likely suffer from two types of deterioration; loss of strength of the matrix resulting from degradation of calcium silicate hydrate (C-S-H), and volumetric change leading to expansion due to formation of gypsum or ettringite that leads to cracking. Protection against sulfate attack requires reducing the permeability of concrete; hence, the low permeability can prolong the service life of the structures that are subjected to severe exposure condition (**AL-Rawi, 1985**).

The key for enhancing the durability of SCC subjected to sever solution environments is a good compaction of microstructure with reducing the porosity of the mortar matrix and interfacial transition zone (ITZ) (**Safiuddin et al., 2008**). Thus, the addition of Nano Silica to concrete leads to improve the hardened properties of concrete because of its high pozzolanic activity and additional (C-S-H) are produced for generation more strength and reducing the free calcium hydroxide. In addition, the microstructure of concrete is improved and the permeability is reduced with utilizing Nano Silica (Nano filler and anti-leaching effects), the cohesiveness between the particles of concrete will be improved and segregation and bleeding will be decreased, thus, making concrete durable **(Li, 2004)**, **(Quercia et al., 2012)**.

## 3 Materials

### 3.1 Cement

Tasluja-Bazian Ordinary Portland cement (OPC) (ASTMC150 Type I) and Tasluja-Bazian sulfate resistance Portland cement (SRPC) (ASTMC150Type V)

© The Authors, published by EDP Sciences. This is an open access article distributed under the terms of the Creative Commons Attribution License 4.0 (http://creativecommons.org/licenses/by/4.0/).

https://doi.org/10.1051/matecconf/201816202015

manufactured in Iraq were used for concrete mixes throughout the present work. This cement complied with the Iraqi specification (IQS, No.5:1984). The physical properties and chemical analysis of the cement used are given in Tables (1) and (2).

### 3.2 Fine Aggregate (Sand)

The fine aggregate from Al-Eakhadir region was used throughout this work. The test results indicate that the adopted fine aggregate conforms to the Iraqi specifications (IQS No.45/ 1984) for grading. It's some physical and chemical properties are shown in Table (3).

### 3.3 Coarse Aggregate (Gravel)

Rounded gravel of 14 mm maximum size from Al-Nebai quarry was used as a coarse aggregate in all mixes. Some physical properties and sulfate content of coarse aggregate are shown in Table (4).

### 3.4 Water

Potable water from the water-supply network system was used for both mixing and curing process of concrete.

### 3.5 Superplasticizer

To achieve high workability needed to produce SCC, super plasticizer known as Glenium 54 was used. Glenium 54 is free from chlorides and complies with (ASTM C494/ C494M, 2001) Type F. Table (5) shows the typical properties of Glenium 54.

### 3.6 Micro Silica

Micro silica, also known as Silica fume or condensed silica fume, is a byproduct of producing silicon metal or ferrosilicon alloys. It consists of silicon dioxide ($SiO_2$) usually more than 85% in non-crystalline (amorphous) form. Densified micro silica fume under commercial name (Sika® Fume S 92 D) had been used as replacement of cement with 5% by weight. Physical properties of silica fume are shown in Table (6).

### 3.7 Nano Silica

Nano Silica is not an industrial byproduct. It is manufactured under special conditions so as to have a high degree of purity and the well-defined range of diameter for the particles usually between (10-50) nm.

Nano Silica used in this work was produced in China by precipitation method in powder form making Nano Silica to have a high degree of purity about 99.96% and diameter for the particles (35nm). Table (7) presents its properties. X-ray diffraction (XRD) of Nano Silica was investigated as illustrated in Figure (1). As it can be seen, the XRD patterns show an approximately broad peak centered about 2Θ = 20° which demonstrates the amorphous structure of utilized Nano Silica. The utilized Nano Silica particles are expected to have high pozzolanic reactivity due to their amorphous structure.

**Table 1.** Physical Properties of Cement.

| *Physical properties* | *OPC* | *SRPC* | *Limits of (IQS No. 5/1984)* | |
|---|---|---|---|---|
| | | | *OPC* | *SRPC* |
| Fineness, Blaine, $m^2/kg$ | 334 | 360 | ≥230 | ≥250 |
| Setting Time (Vicat's method):<br>Initial (minute)<br>Final (hours: minute) | 190<br>3:30 | 150<br>4:30 | ≥45 minute<br>≤ 10 hours | |
| Compressive strength (MPa)<br>3days<br>7days | 22.05<br>28 | 24.90<br>34.80 | ≥15<br>≥23 | |

**Table 2.** Chemical Composition and Compounds of Cement.

| *Chemical Composition* | *Percent by Weight of cement* | | *Limits of (IOS NO.5/1984) for* | |
|---|---|---|---|---|
| | *OPC* | *SRPC* | *OPC* | *SRPC* |
| CaO | 61.37 | 62.26 | / | / |
| $SiO_2$ | 21.24 | 21.40 | / | / |
| $Al_2O_3$ | 5.96 | 4.02 | / | / |
| $Fe_2O_3$ | 3.32 | 4.64 | / | / |
| $SO_3$ | 2.30 | 2.25 | ≤ 2.8% If $C_3A$>5 % | ≤ 2.5% If $C_3A$<5% |
| MgO | 2.71 | 2.41 | ≤ 5% | ≤ 5% |
| Free CaO | 0.84 | 1.06 | / | / |
| L.O.I. | 1.69 | 1.5 | ≤ 4% | ≤ 4% |
| I.R. | 0.62 | 0.79 | ≤ 1.5% | ≤ 1.5% |
| L.S.F. | 0.87 | 0.9 | 0.66 – 1.02 | 0.66 – 1.02 |
| ***Main compounds (Bogue's equation) percentage by weight of cement*** | | | | |
| *Main compounds (Bogue's equation)* | *percentage by weight* | | *Limits of (IQS NO.5/1984) for* | |
| | *OPC* | *SRPC* | *OPC* | *SRPC* |
| ($C_3S$) | 37.00 | 50.70 | / | / |
| ($C_2S$) | 33.06 | 23.19 | / | / |
| ($C_3A$) | 10.18 | 2.81 | / | ≤3.5 |
| ($C_4AF$) | 10.09 | 14.11 | / | / |

**Table 3.** Physical properties and sulfate content of fine aggregate

| *Properties* | *Test result* | *Limits of Iraqi Specification No.4/1984* |
|---|---|---|
| Specific gravity | 2.6 | ----- |
| Fineness modulus | 2.51 | ----- |
| Absorption, % | 1.7 | ----- |
| Passing sieve size 75µm, % | 2.5 | ≤ 5% for natural fine aggregate |
| Sulfate content $(SO_3)$, % | 0.25 | ≤ 0.5% |

**Table 4.** Physical Properties and Sulfate Content of Coarse Aggregate.

| *Properties* | *Test result* | *Limits of Iraqi Specification No.45/1984* |
|---|---|---|
| Specific gravity | 2.65 | ----- |
| Dry rodded density $(kg/m^3)$ | 1640 | ----- |
| Absorption, % | 0.8 | ----- |
| Sulfate content $(SO_3)$, % | 0.02 | ≤ 0.1% |
| Passing sieve size 75µm, % | 0.8 | ≤ 3% |

**Table 5.** Typical Properties of Glenium 54.

| | |
|---|---|
| Form | Whitish to straw colored liquid |
| Relative density | 1.07 at 25℃ |
| Chloride content | Free |
| pH | 6.5 |
| Storage | Should be stored in original containers and at above 5℃ . if frozen, gradually thaw and agitate until completely reconstituted. |

**Table 6.** Physical Properties of SF.

| *Physical properties* | *SF* | *Limit of specification requirements ASTM C-1240* |
|---|---|---|
| Specific surface, min, $(m^2/g)$ | 24 | ≥ 15 |
| Accelerated pozzolanic activity index (%) at 7 days | 131 | ≥ 105 |
| Bulk density $(kg/m^3)$ | 400 | ----- |
| Percent retained on 45µm (No.325) sieve, max, % | 5 | ≤ 10 |

**Table 7.** Properties of NS.

| *Properties* | *Description* |
|---|---|
| $SiO_2$ % | 99.96 |
| Physical form | Powder |
| Color | White |
| Particle size, mm | 35 |
| Surface area $(m^2/g)$ | 270 |

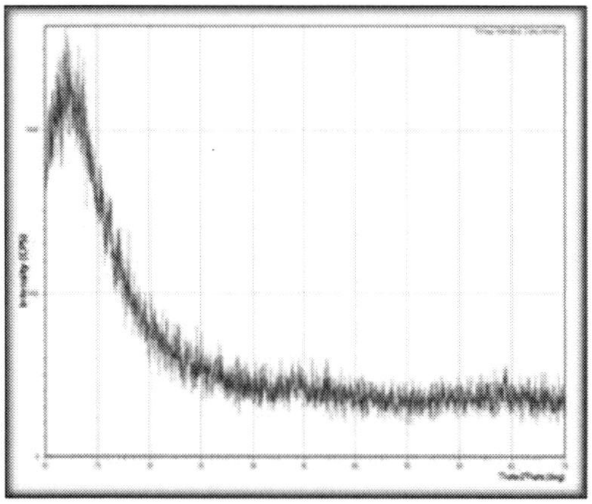

**Fig. 1.** X-ray diffraction patterns of NS.

## 4 Concrete mixes

Four mixes were made, all based on the same control mixture. These mixes were divided into two series each one with different type of cement (I or V). The concrete mixes were designed according to (**European Project Group, 2005**). The mix proportions were shown in Table (8).

## 5 Mixing placing and curing procedures

Mixing process was conducted in pan mixer with a capacity of $0.05m^3$. The 3/4 dosage of superplasticizer was added to mixing water and mixed about 30 second in a small pan mixing separately from other material then the Nano Silica was stirred with this solution and mixed about 3 minutes in order to disperse uniformly and avoid agglomerations of Nano Silica since it has a high surface energy before 2 minutes from the end of previous mixing, the cement and silica fume had been added in pan mixer and mixed for 1minute in dry state. The fine aggregate was added to the drying materials and mixed for 1minute. The prepared solution of (superplasticizer, water and Nano Silica) was added and mixed for 3 minutes. Thereafter, gravel was added and mixed for 1 minute, then the remaining dosage 1/4 of super plasticizer is added and mixed for 2 minutes. The mixture is then discharged, tested and placed. The total time of mixing process was about 8 minutes.

All specimens were demolded after 24 hours and placed in tap water for 28 days as initial curing. After 28 days of curing in tap water, all the concrete specimens were divided into three groups. The first group was partially submerged in ($SO_4^{--}+Cl^-$) solution, the second group was partially submerged in ($SO_4^{--}$) solution, while the third group was submerged in tap water for comparison. These aggressive solutions at concentration identical to those existing in soil and ground water of the middle and southern parts of Iraq.

## 6 Preparation of aggressive solutions

The salts of sulfates and chlorides used in preparing the solutions and providing the suitable concentrations of the ions were pure $MgSO_4.7H_2O$, $Na_2SO_4$, $CaCl_2.2H_2O$ and NaCl. Tables (9) and (11) show the types and concentrations of salts that used. Tables (10) and (12) illustrate the actual inions and cations provided by such salts.

**Table 8.** The main details of SCC Mixes used in the present study.

| Mix Notation | Cement ($kg/m^3$) | Silica fume ($kg/m^3$) | Cementations Material ($kg/m^3$) | Nano Silica ($kg/m^3$) | Sand ($kg/m^3$) | Gravel ($kg/m^3$) | SP % | Water ($kg/m^3$) | w/b ratio |
|---|---|---|---|---|---|---|---|---|---|
| **OPC-0%NS (OR)** | 475 | 25 | 500 | --- | 780 | 820 | 0.7 | 180 | 0.36 |
| **OPC-3%NS (ON3)** | 475 | 25 | 500 | 15 | 780 | 820 | 2.0 | 180 | 0.35 |
| **SRPC - 0%NS (SR)** | 475 | 25 | 500 | --- | 780 | 820 | 0.7 | 180 | 0.36 |
| **SRPC - 3%NS (SN3)** | 475 | 25 | 500 | 15 | 780 | 820 | 2.0 | 180 | 0.35 |

**Table 9.** Salts types & concentration used in sulfate solution

| *Types of salts* | *Chemical form* | *Concentration* | | *Salt content % by wt. of cement* |
|---|---|---|---|---|
| | | *ppm* | *gm/l* | |
| Sulfate salts | $MgSO_4.7H_2O$ | 15000 | 15 | 1.5 |
| | $Na_2SO_4$ | 3250 | 3.25 | 0.325 |

**Table 10.** Ions concentration of ions used in sulfate solution.

| *Anions* | | *Cations* | |
|---|---|---|---|
| *Type* | *Concentration, ppm* | *Type* | *Concentration, ppm* |
| $SO_4^{--}$ | 8041 | $Mg^{++}$ | 1480.5 |
| | | $Na^+$ | 1053 |

**Table 11.** Salts types & concentration used in sulfate-chloride solution.

| *Anions* | | *Cations* | |
|---|---|---|---|
| *Type* | *Concentration, ppm* | *Type* | *Concentration, ppm* |
| $SO_4^{--}$ | 8041 | $Mg^{++}$ | 1480.5 |
| $Cl^-$ | 33718.9 | $Na^+$ | 20713 |
| | | $Ca^{++}$ | 1908.2 |

**Table 12.** Ions concentration used in sulfate-chloride solution.

| *Types of salts* | *Chemical form* | *Concentration* | | *Salt content % by wt. of cement* |
|---|---|---|---|---|
| | | *ppm* | *gm/l* | |
| Sulfate salts | $MgSO_4.7H_2O$ | 15000 | 15 | 1.5 |
| | $Na_2SO_4$ | 3250 | 3.25 | 0.325 |
| Chloride salts | $CaCl_2.2H_2O$ | 7000 | 7 | 0.7 |
| | NaCl | 50000 | 50 | 5 |

## 7 Tests of hardened concrete properties

The hardened properties of concrete which were studied: splitting tensile strength, flexural strength, modulus of elasticity and ultrasonic pulse velocity test (U.P.V.). After the concrete had been cast and cured, splitting tensile strength of (104) cylinders with dimensions (100×200)mm, flexural strength of (104) prisms with dimensions (100x100x400) mm, static modulus of elasticity of (104) cylinders with dimensions (150x300)mm and U.P.V. of (156) cubes with an edge length (100)mm were examined. The average of two specimens was taken for each test except U.P.V., the average of three cubes was taken.

## 8 Results and discussion

### 8.1 Splitting Tensile strength

The results of splitting tensile strength of SCC specimens cured in tap water and other Aggressive solution exposer are shown in Table (13) and Figure (2-a,b,c). All specimens cured in tap water exhibited a continuous

 https://doi.org/10.1051/matecconf/201816202015

development in splitting tensile strength with age progress. While SCC specimens containing Nano Silica (ON3 and SN3) exposed to ( $SO_4^{--} + Cl^-$) solution revealed the increment in splitting tensile strength up to 180 days of exposure, due to the presence of Nano Silica, which has increased the strength of concrete. while the reference SCC specimens (without Nano Silica) reveal the increment in splitting tensile strength up to 90 days and 120 days of exposure for (OR and SR) concretes respectively; after that a decrease in strength occurred up to 180 days. This reduction is due to the attack of sulfate ions, which leads to form the gypsum and ettringite that leading to expansion, spalling, cracking and loss of strength of concrete; thereby the deterioration in concrete occurs.

As illustrated in Table (13) and Figure (2-c), SCC specimens with Nano Silica (SN3) exposed to ( $SO_4^{--}$) solution offered the increment in splitting tensile strength up to 180 days of exposure, due to the presence of Nano Silica. While (ON3) specimens exposed to ( $SO_4^{--}$) solution show that the increases in splitting tensile strength is stopped in 120 days of exposure; then a stability in strength is observed at 180 days, because of the presence of pure sulfate ions and the use of OPC in Nano Silica concrete. In contrast for (OR and SR) concretes, the increment in splitting tensile strength is observed up to 90 days and 120 days of exposure respectively; then a decrease in strength is occurs up to 180 days.

**Table 13.** Results of Splitting Tensile Strength of SCC Mixes.

| *Mix Designation* | *Before exposure at 28 days* | *60 days* | *90 days* | *120 days* | *180 days* |
|---|---|---|---|---|---|
| *Splitting tensile strength (MPa) (Tap water)* | | | | | |
| OPC- 0%NS (OR) | 4.24 | 5.15 | 5.57 | 5.78 | 6.08 |
| OPC- 3%NS (ON3) | 6.06 | 7.17 | 7.60 | 7.86 | 8.25 |
| SRPC- 0%NS (SR) | 5.00 | 6.29 | 7.13 | 7.58 | 7.94 |
| SRPC- 3%NS (SN3) | 6.73 | 8.23 | 8.96 | 9.22 | 9.90 |
| *Splitting tensile strength (MPa)( $SO_4^{--} + Cl^-$)* | | | | | |
| OPC- 0%NS (OR) | 4.24 | 4.74 | 4.97 | 4.91 | 4.55 |
| OPC- 3%NS (ON3) | 6.06 | 7.00 | 7.35 | 7.56 | 7.69 |
| SRPC- 0%NS (SR) | 5.00 | 6.04 | 6.41 | 6.68 | 6.63 |
| SRPC- 3%NS (SN3) | 6.73 | 8.16 | 8.69 | 9.07 | 9.61 |
| *Splitting tensile strength (MPa) ( $SO_4^{--}$)* | | | | | |
| OPC- 0%NS (OR) | 4.24 | 4.60 | 4.63 | 4.24 | 3.97 |
| OPC- 3%NS (ON3) | 6.06 | 6.90 | 7.27 | 7.48 | 7.48 |
| SRPC- 0%NS (SR) | 5.00 | 5.76 | 6.14 | 6.22 | 5.79 |
| SRPC- 3%NS (SN3) | 6.73 | 8.03 | 8.59 | 8.91 | 9.42 |

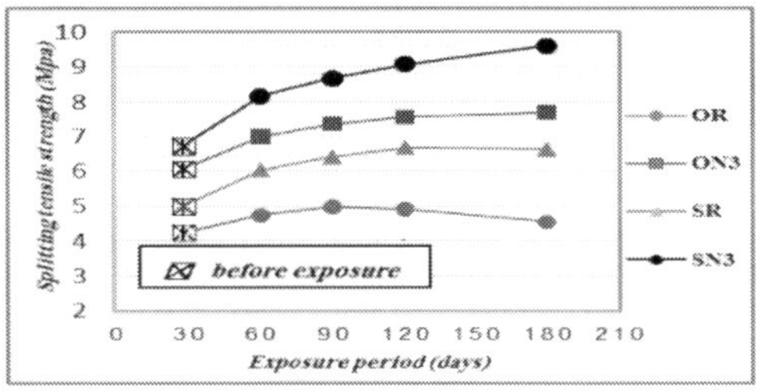


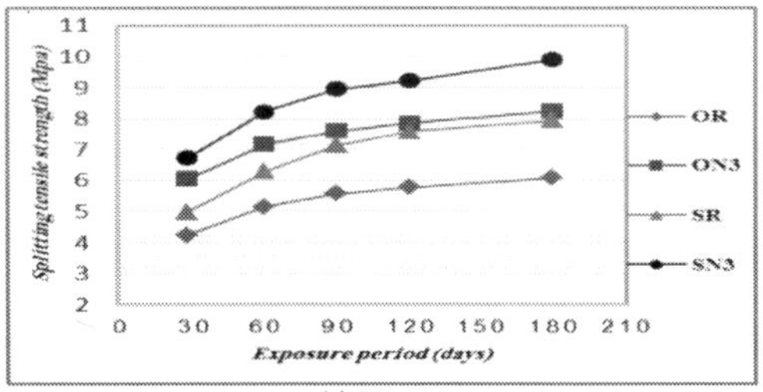


(a) Tap water

(b) ( $SO_4^{--} + Cl^-$) solution.

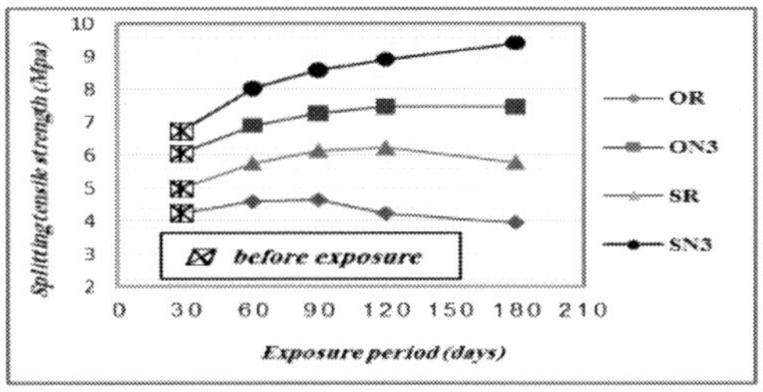


**Fig. 2.** Splitting Tensile Strength Development of SCC Mixes Exposed to various exposure conditions.

This reduction is caused by the existence of pure sulfate ions, which will consume more amount of portlandite (CH) from hydration of cement and $C_3A$ compared with combined sulfate ions; thereby, the gypsum and ettringite are formed. Also, the presence of $Mg^{+2}$ cations associated with sulfate salts, that convert a further portlandite in the case of pure sulfate attack (when compared with available portlandite in combined sulfate with chloride solution) to gypsum by the simultaneous formation of Brucite (magnesium hydroxide), which is insoluble and reduces the pH of the system. In the absence of the portlandite, (C-S-H) gel is no longer stable and is also attacked by magnesium sulfate, (C-S-H) gel will be converted to a cohesion less, porous, reticulated magnesium silicate hydrate (M-S-H) gel, resulting in loss of cohesiveness and reduction in strength, as reported **(Asrar et al., 1997)**, **(Prasad et al., 2006)**.

The reason of Nano silica reduces the loss in splitting tensile strength of SCC specimens exposed to aggressive solution solutions and improves the splitting tensile strength of SCC specimens cured in tap water is attributed to Nano Silica enhances the tensile strength of hardened cement mortar by two mechanisms. The first of which is the packing effect of Nano particles as filler hardened microstructure of cement mortar and thus increasing its density as well as the strength. The second is the pozzolanic effect, Nano Silica reacts with $Ca(OH)_2$ and produces C-S-H. Therefore, it reduces the formation of gypsum and ettringite which leads to reduce the internal tensile stress forming from expansion of ettringite so increasing tensile strength. The same conclusion is reported as **(Sobolev et al., 2006)**, **(Qing et al, 2007)**.

From the results, it is also clear that SRPC mixes have higher splitting tensile strength than OPC mixes at all ages and for all exposure conditions and therefore, enhancing sulfate resistance is achieved. This behavior is due to a low content of $C_3A$ compound and high ratio of $C_3S/\ C_2S$ in SRPC than OPC. Also, it is clear that combined sulfate attack reduces the deterioration in splitting tensile strength than pure sulfate. These trends were also reported by **(Al-Amoudi et al., 1995)**.

## 8.2 Flexural Strength

The results of flexural strength of SCC specimens are illustrated in Table (14) and Figures (3- a, b, c). The data in Table (14) and Figure (3-a) indicate that all specimens cured in tap water demonstrate a continuous increasing in flexural strength with age. While the specimens exposed to ( $SO_4^{--} + Cl^-$ ) solution which contained Nano Silica (ON3 and SN3) exhibit an improvement in flexural strength up to 180 days of exposure, relative to corresponding reference specimens without Nano Silica (OR and SR) which show a continuous increase in flexural strength up to 90 and 120-days of exposure respectively. Those results are also summarized in Figure (3-b).

On the other hand, all SCC specimens which were subjected to ( $SO_4^{--}$ ) solution have the same behavior that in ( $SO_4^{--} + Cl^-$ ) solution, except for (OR) specimens which exhibit a noticeable strength increase at 60 days of exposure, beyond this period a reduction in flexural strength is observed. This increase in flexural strength at 60 days may be attributed to the filling up of pores by reactions between products and crystallization of salts. This phenomenon has also been stated by **(Ganjian and Pouya, 2005)**.

The enhancement in flexural strength for Nano Silica concretes can be mainly due to addition of Nano Silica particles producing a homogeneous microstructure, characterized by compact and small sized CSH gel. So, a denser ITZ was produced that means the bonding strength at the interface zone is increased and the formation of micro cracking is decreased. Hence, the micro cracking initiation occurs at a higher stress level. The addition of Nano Silica caused a refinement of the microstructures (less interconnected pore structure) and induced precipitation of small sizes C-S-H gel having high stiffness. The improvement of the microstructure was reflected by the mechanical properties due to the fact that the pozzolanic gel structure presents better mechanical properties than the C-S-H gel precipitated in standard OPC concrete **(Sobolev et al., 2006)** and **(Khanzadi et al.,2010)**. As a results of Sulfate attack, the results also material into interstitial spaces inside the skeleton of show a decrease in flexural strength for SCC mixes immersed in saline solutions. It is clear that the deterioration in flexural strength for SCC specimens exposed to pure sulfate solution is more than that exposed to combined sulfate solution. Also, the SRPC mixes showed higher flexural strength than OPC mixes.

**Table 14.** Results of Flexural Strength of SCC Mixes.

| *Flexural strength (MPa) (Tap water)* | | | | | |
|---|---|---|---|---|---|
| *Mix* | *Before exposure at 28 days* | *60 days* | *90 days* | *120 days* | *180 days* |
| **OPC- 0%NS (OR)** | 8.41 | 9.81 | 10.38 | 10.84 | 11.25 |
| **OPC- 3%NS (ON3)** | 11.80 | 13.07 | 13.81 | 14.33 | 14.54 |
| **SRPC- 0%NS (SR)** | 9.46 | 11.73 | 12.49 | 12.94 | 13.54 |
| **SRPC- 3%NS (SN3)** | 13.08 | 14.42 | 15.63 | 16.00 | 16.40 |
| *Flexural strength (MPa)( $SO_4^{--} + Cl^-$ )* | | | | | |
| *Mix* | *Before exposure at 28 days* | *60 days* | *90 days* | *120 days* | *180 days* |
| **OPC- 0%NS (OR)** | 8.41 | 9.01 | 9.44 | 9.38 | 8.85 |
| **OPC- 3%NS (ON3)** | 11.80 | 12.91 | 13.34 | 13.64 | 13.77 |
| **SRPC- 0%NS (SR)** | 9.46 | 11.24 | 12.04 | 12.23 | 12.12 |
| **SRPC- 3%NS (SN3)** | 13.08 | 14.34 | 15.45 | 15.76 | 16.20 |
| *Flexural strength (MPa) ( $SO_4^{--}$ )* | | | | | |
| *Mix Designation* | *Before exposure at 28 days* | *60 days* | *90 days* | *120 days* | *180 days* |
| **OPC- 0%NS (OR)** | 8.41 | 9.34 | 9.33 | 9.14 | 8.27 |
| **OPC- 3%NS (ON3)** | 11.80 | 12.80 | 13.00 | 13.44 | 13.48 |
| **SRPC- 0%NS (SR)** | 9.46 | 11.00 | 11.89 | 12.11 | 12.00 |
| **SRPC- 3%NS (SN3)** | 13.08 | 14.34 | 15.33 | 15.50 | 16.06 |

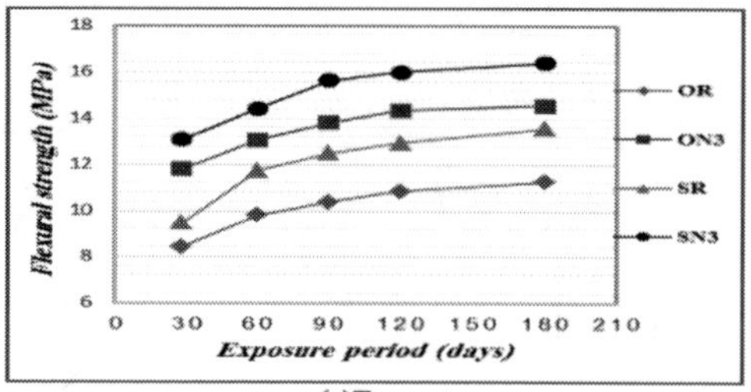

(a)Tap water

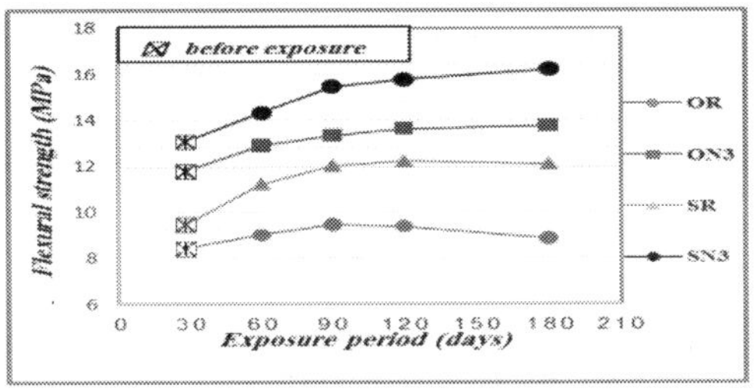


(b) ($SO_4^{--}$ + $Cl^-$) solution

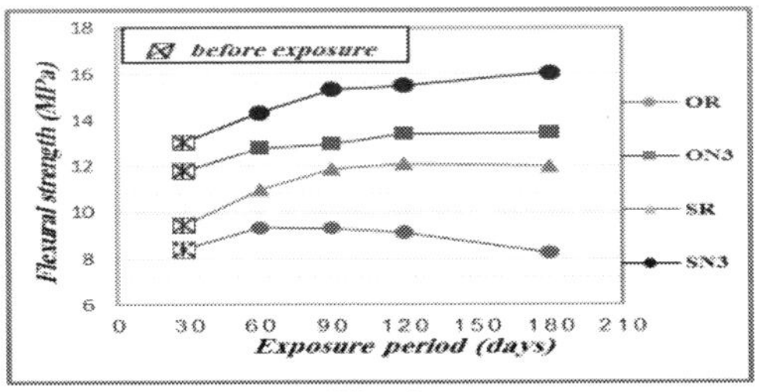


(c) ($SO_4^{--}$) solution.

**Fig. 3.** Flexural Strength Development of SCC Mixes Exposed to various exposure conditions.

## 8.3 Static Modulus of Elasticity

The results of modulus of elasticity of SCC specimens cured in tap water are illustrated in Table (15) and plotted in Figure (4-a). While, the results for both sulfate exposure conditions for all SCC specimens are also shown in Table (15) and Figures (4- b and c). It is observed from these results that SCC specimens tends to have lower modulus of elasticity, the static modulus of elasticity of SCC specimens ranged between (30.23-48.28) GPa. This behavior is due to the high content of paste and low content of aggregate at SCC. From these results, it is observed that all SCC specimens exposed to either ($SO_4^{--}$ + $Cl^-$) solution or ($SO_4^{--}$) solution exhibited the percentages of decrease in elastic modulus at all ages of the test compared with those cured in tap water. The reason of this decrease is attributed to the decrease in modulus of elasticity of the matrix caused by the degeneration of the interfacial bond strength between bulk cement paste and aggregate.

Generally, it is clear that the decrease in elastic modulus in SRPC mixes was less than that in OPC mixes. This is due to low content of $C_3A$ compound and high ratio of $C_3S/C_2S$ in SRPC compared with those in OPC. Also, it is observed that the decrease in the modulus of elasticity for SCC mixes exposed to ($SO_4^{--}$) solution is higher than those subjected to ($SO_4^{--}$ + $Cl^-$) solution, which may be due to: *(i)* the increased solubility of ettringite and gypsum in chloride-bearing solution Which can reduce the amount of sulfate products formed in cement paste, *(ii)* the ettringite formed in the sulfate chloride solution being less expansive, and *(iii)* the higher diffusivity of chloride ions compared with sulfate ions permittings the chloride ions to react with $C_3A$ to form Freidel's salt, which will reduce the conventional sulfate attack. This agrees with **(Al-Amoudi et al., 1995).** As well as, it is also clear that Nano silica addition reduces the reduction in modulus of elasticity at all SCC specimens. On the other hand, the modulus of elasticity increases with adding Nano Silica at all mixes and at all ages.

**Table 15.** Results of static modulus of elasticity of scc mixes.

| *Modulus of Elasticity (GPa) (Tap water)* | | | | | |
|---|---|---|---|---|---|
| *Mix* | *Before at 28 days* | *60 days* | *90 days* | *120 days* | *180 days* |
| OPC-0%NS (OR) | 31.30 | 33.21 | 35.21 | 36.00 | 36.65 |
| OPC-3%NS (ON3) | 39.00 | 40.40 | 42.75 | 43.65 | 44.28 |
| SRPC-0%NS (SR) | 33.78 | 35.71 | 38.36 | 39.56 | 40.74 |
| SRPC-3%NS (SN3) | 40.83 | 43.12 | 45.78 | 46.90 | 48.28 |
| *Modulus of Elasticity (GPa) ($SO_4^{--}$ + $Cl^-$)* | | | | | |
| *Mix* | *Before at 28 days* | *60 days* | *90 days* | *120 days* | *180 days* |
| OPC-0%NS (OR) | 31.30 | 32.19 | 33.06 | 32.66 | 30.67 |
| OPC- 3%NS (ON3) | 39.00 | 40.18 | 42.33 | 43.00 | 43.17 |
| SRPC- 0%NS (SR) | 33.78 | 35.40 | 37.47 | 37.64 | 36.90 |
| SRPC- 3%NS (SN3) | 40.83 | 42.80 | 45.53 | 46.35 | 47.60 |
| *Modulus of Elasticity (GPa) ($SO_4^{--}$)* | | | | | |
| *Mix Designation* | *Before exposure at 28* | *60 days* | *90 days* | *120 days* | *180 days* |
| OPC- 0%NS (OR) | 31.30 | 31.75 | 32.23 | 32.20 | 30.23 |
| OPC- 3%NS (ON3) | 39.00 | 40.00 | 42.19 | 42.80 | 42.01 |
| SRPC- 0%NS (SR) | 33.78 | 35.24 | 36.53 | 36.57 | 36.30 |
| SRPC- 3%NS (SN3) | 40.83 | 42.68 | 45.36 | 45.76 | 46.63 |

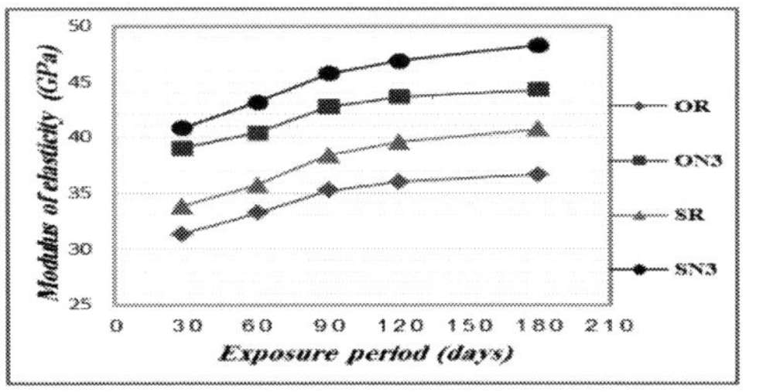


(a) Tap Water.

https://doi.org/10.1051/matecconf/201816202015

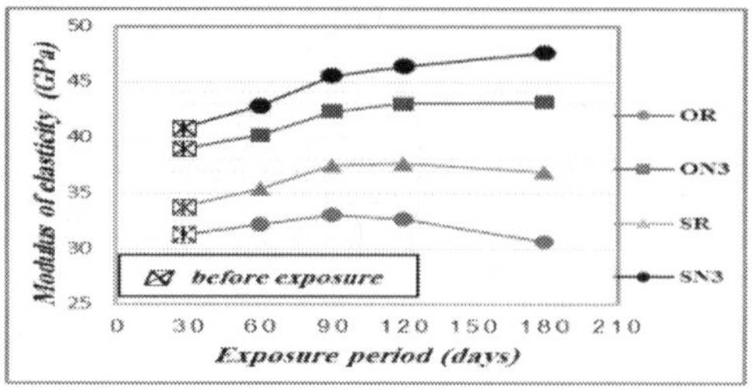


(b) ( $SO_4^{--}$ + $Cl^{-}$) solution.

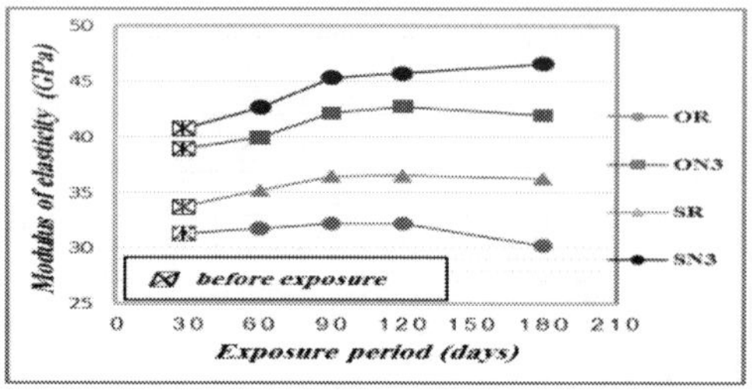


(c) ( $SO_4^{--}$ ) solution.

**Fig. 4.** Modulus of Elasticity Development of SCC Mixes Exposed to various exposure conditions.

### 8.4 Ultrasonic Pulse Velocity (U.P.V.)

The results of U.P.V. test was measured on 150 mm cubes at different curing ages (28, 60, 90, 120 and 180 days). These results are illustrated in Table (16) and Figures (5- a, b and c).

The results show that the U.P.V. for SCC cured in tap water exhibit a continuous increasing in U.P.V. with age progress. The recorded results of U.P.V. for SN3 ranged between (5.321-5.837) km/sec, ON3 ranged between (5.264-5.738) km/sec, SR ranged from (4.912-5.539) km/sec, and OR ranged from (4.857-5.398) km/sec. The behavior is mainly attributed to the increase in density of concrete with age progress. The same conclusion is reported by **(Hadi, 2009)**. The test results also indicate that all SCC specimens subjected to pure sulfate or combined sulfate solution demonstrate a reduction in U.P.V. values at all exposure periods relative to corresponding SCC cured in tap water. Moreover, the SCC specimens exposed to pure sulfate solution elucidate more reduction in U.P.V. compared with combined sulfate solution. This reduction can be related to the formation of micro-cracks resulted by sulfate attack which formed with a large quantity in pure sulfate solution compared with combined sulfate solution, and these micro-cracks connecting to each other to form larger cracks over time due to exposure. In such a case, the pulse will be diffracted around the larger cracks, thereby increasing the travel path and travel time, leading to the decrease in pulse velocity. This is also reported by **(Mahmoud, 2012)** and **(Kazem, 2013)**.

The U.P.V. values increase with adding of Nano Silica for all SCC mixes. This behavior may be attributed to the effect of Nano Silica Which increase the density and filled the pores of the cement matrix with new CSH gel, this makes the pulse velocity increases for SCC.

**Table 16.** Results of U.P.V. of SCC Mixes.

| *U.P.V. (km/sec) (Tap water)* | | | | | |
|---|---|---|---|---|---|
| *Mix* | *Before at 28 days* | *60 days* | *90 days* | *120 days* | *180 days* |
| **OPC- 0%NS (OR)** | 4.857 | 5.067 | 5.245 | 5.325 | 5.398 |
| **OPC- 3%NS (ON3)** | 5.264 | 5.39 | 5.575 | 5.635 | 5.738 |
| **SRPC- 0%NS (SR)** | 4.912 | 5.229 | 5.386 | 5.463 | 5539 |
| **SRPC- 3%NS (SN3)** | 5.321 | 5.543 | 5.698 | 5.759 | 5.837 |
| *U.P.V. (km/sec) ($SO_4^{--}$ + $Cl^{-}$)* | | | | | |
| *Mix* | *Before at 28 days* | *60 days* | *90 days* | *120 days* | *180 days* |
| **OPC- 0%NS (OR)** | 4.857 | 5.050 | 5.165 | 5.174 | 5.154 |
| **OPC- 3%NS (ON3)** | 5.264 | 5.381 | 5.550 | 5.584 | 5.607 |
| **SRPC- 0%NS (SR)** | 4.912 | 5.217 | 5.296 | 5.383 | 5.394 |
| **SRPC- 3%NS (SN3)** | 5.321 | 5.537 | 5.667 | 5.727 | 5.745 |
| *U.P.V. (km/sec) ($SO_4^{--}$)* | | | | | |
| *Mix Designation* | *Before exposure at 28 days* | *60 days* | *90 days* | *120 days* | *180 days* |
| **OPC- 0%NS (OR)** | 4.857 | 5.000 | 5.074 | 5.097 | 5.087 |
| **OPC- 3%NS (ON3)** | 5.264 | 5.375 | 5.467 | 5.532 | 5.542 |
| **SRPC- 0%NS (SR)** | 4.912 | 5.210 | 5.276 | 5.300 | 5.324 |
| **SRPC- 3%NS (SN3)** | 5.321 | 5.516 | 5.624 | 5.688 | 5.740 |

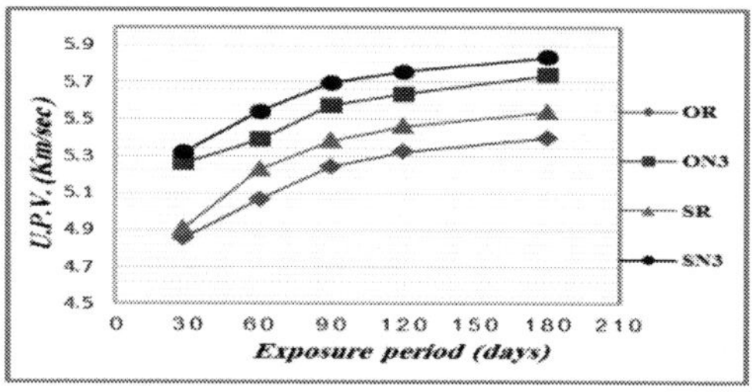


(a) Tap Water

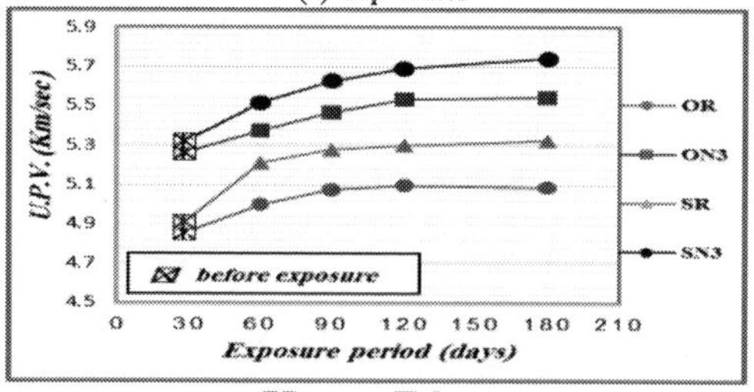


(b) ( $SO_4^{--}$ + $Cl^{-}$) solution.

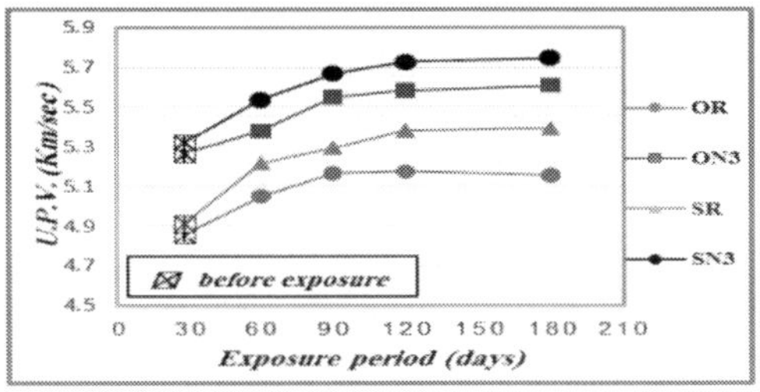


(c) ( $SO_4^{--}$ ) solution.

**Fig. 5.** U.P.V. Development of SCC Mixes Exposed to various exposure conditions.

## 9 Conclusions

- SCC specimens exposed to ($SO_4^{--}$+ $Cl^-$ ) or ($SO_4^{--}$) solution demonstrated a significant reduction in splitting tensile strength, flexural strength, modulus of elasticity and U.P.V. when compared with the specimens cured in tap water.

- The utilizing of 3% Nano Silica as an addition by weight of cementitious materials was found to be effective in improving the resistance of SCC to external sulfate attack. Also, the using of Nano Silica was found to be more benefit in OPC than SRPC.

- The deterioration in the mechanical properties, non-destructive tests was higher in ($SO_4^{--}$) solution compared with ($SO_4^{--}$+ $Cl^-$ ) solution.

## References

[1] Al-Amoudi, O. S., Maslehuddin, M., and Abdul-al, Y. A., (1995), Construction and Building Materials, Vol.9, Issue 1, February 1995, pp.(25-33).
[2] Al-Rawi R. S., (1985), ACI-RILEM, Joint Symposium, Monterey, Mexico, pp. 543 .
[3] Asrar, N., Malik, A.U., Ahmad, S., and Andijani, I.N., (1997), Research and development center, presented in second acquired experience symposium on desalination plants, Sept. 29, 1997.
[4] ASTM C494/C494M, (2001), *"Standard Specification for Chemical Admixtures for Concrete"*, American Society for Testing and Materials.
[5] European Project Group, (2005), *"The European Guidelines for Self-Compacting Concrete: Specification, Production and Use"*, may, 2005, PP.63.
[6] Ganjian, E., and Pouya, H.S., (2005), *Cement and Concrete Research, Vol. 35, pp. (1322-1343).*
[7] Hadi, A.M., (2009), M.Sc. Thesis, University of Babylon.
[8] Iraqi Organization of Standards (IQS), No.45/1984, *"Aggregate from Natural Sources for Concrete and Construction"*.
[9] Iraqi Organization of Standards (IQS), No.5/1984, *"Portland Cement"*.
[10] Kazem, Z.H., (2013), M.Sc. Thesis, University of Babylon.
[11] Khanzadi, M., Tadayon, M., Sepehri , H., and Sepehri, M.,(2010), HBRC Journal, vol.20, No. 20, April 2010, p.p. (1-13).
[12] Li, G. (2004). Cement and Concrete Research, vol. 34, No.6, Jun. 2004, pp.(1043-1049).
[13] Mahmoud, R.H.A., (2012), M.Sc. Thesis, University of Babylon.
[14] Prasad, J., Jain, D.K., and Ahuja, A.K., (2006), Asian Journal of Civil Engineering, Vol. 7, No. 3 (2006), pp. (259-268).
[15] Qing, Y., Zenan, Z., Deyu, K. and Rongshen, C., (2007), Construction and Building Materials, Vol.21, No.3,2007, pp. (539-545).
[16] Quercia, G., Spiesz, P., Husken, G., & Brouwers, J. (2012), In Proceedings of the International Congress on Durability of Concrete (ICDC 2012), (pp. 18-21).
[17] Safiuddin, M. D., West, J.S., and Soudki, K.A., (2008), Journal of Applied Sciences Research, Vol.4, No. 12, pp. (1834-1840).
[18] Singh, S., and Thombre ,S., (2013) International Journal of Latest Research in Science and Technology , ISSN 2278-5299 , Vol.2 , Issue.2, April (2013) ,pp.(87-90).
[19] Sobolev, K. , Flores, I. and Hermosillo, R.,(2006), "Proceedings of ACI Session on, November 7, Denver, USA 2006, pp. (91-118).

# Mechanical properties of papercrete

*Harith* Zaki[1,*], *Iqbal* Gorgis[1] and *Shakir* Salih [1]

[1] Building and Construction Engineering Department, University of Technology, Baghdad, Iraq

**Abstract.** This paper studies the uses, of waste paper as an additional material in concrete mixes. Papercrete is a term as the name seems, to imply a mixture of paper and concrete. It is a new, composite material using waste paper, as a partial addition of Portland cement, and is a sustainable, building material due to, reduced amount of waste paper being put to use. It gains, latent strength due to presence of hydrogen bonds in microstructure of paper. Papercrete has been, reported to be a low cost alternative, building construction, material and has, good sound absorption, and thermal insulation; to be a lightweight and fire-resistant material. The percent of waste paper used (after treating) namely (5%, 10%, 15% and 20%) by weight of cement to explore the mechanical properties of the mixes (compressive strength, splitting tensile strength, flexural strength, density), as compared with references mixes, it was found that fresh properties affected significantly by increasing the waste paper content. The compressive strength, splitting tensile strength, flexural strength and density got decreased with increase in the percentage of paper.

## 1 Introduction

The world has been faced, with environmental problems such as global warming, forest destruction and lack of resources. Recycling of resources help solve, these environmental problems. Principally, paper; is one of the most effective recycling, resources to prevent global warming. According to Resource; Information System Incorporated (RISI), paper recycling is being done to the extent of about 60% ~ 80% in many countries, See Table 1[1]. Waste management refers to awareness regarding the collection, transport, treatment, recycling and storage, of waste. The term usually refers to materials resulting from human activities and to the reduction; of their effect on the people's health, the environment, or the aspect of a habitat. Waste management may also contribute to saving natural; resources through the reuse of recoverable parts [2].

Waste paper has been used as building materials for decades, especially in cementitious; matrices, and since then a lot of research has been conducted to develop; the mechanical properties of the composite like; compressive, tensile, flexural strength, and etc. [3].

Use of wastepaper in structural' concrete could; become economical and environment friendly, which helps to minimize harmful effects of construction process [4 &5]. Paper is the most frequent; type of waste found in all activity areas and exemplifies an important source of cellulous fibers. In different forms (quality paper, mixed paper, newspapers and journals) paper exemplifies about 41% of all household waste produced today. Paper is a recyclable material; that after reaching the garbage dumps; has a biodegradation period of 3-12 months. That is why the benefits for nature through paper waste recycling are very important. Paper can be recycled in the technological flow; that has produced it for the manufacture of paper or the produce of cardboard and pasteboard. Paper can only be recycled 6-10 times, because with each recycling, the length of cellulose fiber is minimized, resulting in a decrease of the mechanical strength and quality (appearance, color) of the manufactured paper and an increase of technological losses.

**Table 1.** Recycling ratio of paper in some countries [1]

| | *Korea* | *Japan* | *China* | *Taiwan* | *HongKong* |
|---|---|---|---|---|---|
| *Recycling Ratio (%)* | *75.4* | *72* | *34* | *68* | *80* |
| | *U.S.A* | *Canada* | *Germany* | *U.K* | *France* |
| *Recycling Ratio (%)* | *52* | *68* | *75* | *65* | *64* |

Cellulose fiber; insulation uses recycled newspaper as a raw material, from which raw fibers are extracted and subsequently treated with special additives for fire proofing, as well as for preservation against pests. The manufacturing technology includes the application by injection of cellulose fibers. It can be applied for thermal insulation; of walls, floors, as well ceilings and roofs and also good sound insulation [2]. More than 450 million; tons of paper is generated worldwide per annum and it is predict that the demand for paper will extent to 500 million tons per annum by the end of 2020 [6].

Papercrete was produced by blending sufficient amount of waste paper with Portland cement type I,

* Corresponding author: almasrafharith@gmail.com

© The Authors, published by EDP Sciences. This is an open access article distributed under the terms of the Creative Commons Attribution License 4.0 (http://creativecommons.org/licenses/by/4.0/).

water and aggregate, and then slump and strength values were compared with those of conventional concrete. The slump minimized when a higher amount of paper; pulp content is added to concrete. The slump increased up-to 5% replacement by weight of cement, above 5% the slump minimized as the paper pulp content in the concrete mixtures increased. Many factors are accountable for adverse effects on the workability of paper pulp; concrete such as amount of paper pulp replacement, paper pulp physical properties and the carbon content of the paper pulp would be the central reasons for the lowering of concrete workability. The spotted compressive, splitting tensile and flexural strength raised up to 10% addition of waste paper pulp and further increase in waste paper pulp minimizes the strengths progressively. So the most appropriate mix proportion was using 5 to 10% replacement of waste paper pulp by weight of cement [7].

Paper is fundamentally wood cellulose, which is considered a fibrous' material. Cellulose is the second most generous material on earth after rock. Cellulose is a natural polymer, a long chain of associated sugar molecules made by the associating of smaller molecules. The links in the cellulose chain are a type of sugar: ß-D-glucose. The cellulose chain bristles with polar-OH groups. These groups compose many hydrogen bonds with OH groups on approaching chains, bundling the chains together Fig. 1. The chains also pack orderly in places to form hard, stable crystalline regions that allow the bundled chains even more balance and strength. This hydrogen bonding produces the basis of papercrete strength [8]. The raw dry paper sludge includes silica and calcium oxide, followed by alumina and magnesium oxide. Cement blended with 10% and 20% calcined paper display a smaller reduction in compressive strength than the control cement [9].

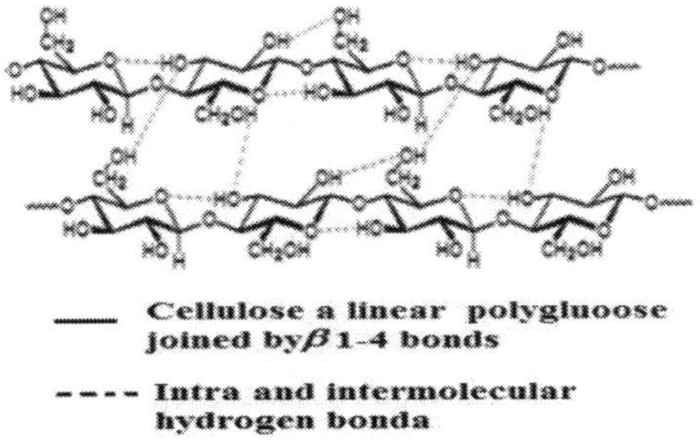


**Fig. 1.** Chemical structure of paper (Cellulose hydrogen bond)[10].

Papercrete is a complex material made of Portland cement, waste paper, water and/or sand. Papercrete has been reported to be a cheap alternative building material; to have good sound absorption and thermal insulation; to be a light weighted and fire- resistant material. It should be noted that papercrete is a relatively new concept with limited scope. Papercrete is famous by alternative names such as fibrous concrete, Padobe and Fidobe [11].

This paper summarizes the behavior of concrete with the waste paper pulp by adding cement in the range of 5%, 10%, 15% and 20% which may help to minimize the disposal problem of sludge and reinforce the properties of concrete.

The different objectives of the study of Papercrete are: -

- Study the properties of Papercrete to define its potential as a sustainable building material.
- Limit the Ideal Papercrete Mix.
- Formulate the samples of the locally available materials.
- Determine the strengths of various Papercrete mixes.
- Do prorated analysis of Concrete and Papercrete.

## 2 Materials and properties

### 2.1 Cement

Ordinary Portland cement (Type I) manufactured by (Lafarge company/ Bazian) commercially known as (Krasta) was used throughout the work. The physical properties are, given in Table 2, whereas the chemical composition and main compounds of cement are explained in Table 3. The test results of cement present that the cement conforms to Iraqi Specification No.5/1984 [12]. Tests were carried out by National Center for Construction Laboratories and Researches (NCCLR).

### 2.2 Fine aggregate

Al-Ekhaider natural sand was used as fine aggregate throughout the experimental work. Table 4 and Fig. 2 present physical and chemical properties of it. The test results were presented the gradation of the sand lies in zone [2] and the sulfate content (0.37%) which are within the requirements of the Iraqi specification No. 45/ (1984) [13]. The tests were conducted at the Construction Materials Laboratory at Civil Engineering Department at University of Technology.

### 2.3 Coarse aggregate

Al-Soddor crushed coarse aggregate was used as coarse aggregate through the experimental work. It was tested to determine the sieve analysis as shown in Table 5 and Fig. 3 Also physical and chemical properties of gravel. The test results presented the gradation of the gravel and the sulfate content which are within the requirements of the Iraqi specification No. 45/ (1984) [13].

### 2.4 Waste papers

Waste paper used in this study was collected from schools, administration offices and libraries. The physical properties of the waste paper are shown in Table 6. The papers were cut into small pieces by using paper cutter as shown in Fig. 4. These small dimensions prevent the paper from clumping when waste paper is assorted with water so the paper expands evenly in the papercrete mix. The paper was then soaked in water for three days Fig. 5 and then the paper sludge was placed on non- absorbent plate for squeezing out extra water content, after that the paper is subjected to grinding by using kitchen blender to obtain the required size of paper sludge. The paper sludge was collected from the grinder and stored for casting of papercrete specimens. Fig. 6 shows the paper mass after grinding in the wet grinder.

### 2.5 Water

Water is an important ingredient of papercrete as it is contributory in the chemical reaction with cement. Potable water should be used for both soaking and mixing of papercrete. It should be free from organic matter and the pH value should be between 6 and 7.

**Table 2.** Physical properties of cement

| Physical Properties | Test results | Limits of Iraqi specification No.5/1984 (12) |
|---|---|---|
| **Fineness (Blaine method), ($cm^2/gm$)** | 3760 | ≥2300 $cm^2/gm$ |
| **Setting time (Vicate apparatus),**<br>**a. Initial setting, hrs:min**<br>**b. Final setting, hrs:min** | 2:05<br>4:00 | ≥45min<br>≤10hrs |
| **Compressive strength, MPa**<br>**For 3 days**<br>**For 7 days** | 20.0<br>25.0 | ≥15 MPa<br>≥23 MPa |
| **Soundness Autoclave method%** | 0.12 | <0.8 |

**Table 3.** Chemical composition and main compounds of cement

| *Oxides Composition* | *Composition Content %* | *Limits of Iraqi specification No.5/1984* |
|---|---|---|
| *Lime, CaO* | 66.11 | - |
| *Silica, $SiO_2$* | 21.93 | - |
| *Alumina, $Al_2O_3$* | 4.98 | - |
| *Iron oxide , $Fe_2O_3$* | 3.10 | - |
| *Magnesia ,MgO* | 2.00 | 5.0 % Max |
| *Sulfate, $SO_3$* | 2.25 | 2.8 % Max |
| *Loss on Ignition, (L.O.I.)* | 2.39 | 4.0 % Max |
| *Insoluble residue, (I.R)* | 1.29 | 1.5 % Max |
| *Lime Saturation Factor, (L.S.F.)* | 0.93 | 0.66-1.02% |
| ***Main compounds (Bogues equations)*** | | |
| *$C_3S$* | 58.16 | - |
| *$C_2S$* | 19.00 | - |
| *$C_3A$* | 7.95 | > 5% |
| *$C_4AF$* | 9.43 | - |

**Table 4.** Properties of fine aggregate

| Sieve size (mm) | Passing by weight % | Limits of IQS (No.45/1984) Zone (2) [13] |
|---|---|---|
| 9.51 | 100 | 100 |
| 4.75 | 98 | 90-100 |
| 2.36 | 89 | 75-100 |
| 1.18 | 77 | 55-90 |
| 0.6 | 55 | 35-59 |
| 0.3 | 21 | 8-30 |
| 0.15 | 5 | 0-10 |
| Fineness Modulus =2.8 | | |
| Property | Test Results | Limits of IQS No.45/1984) (13) |
| Specific gravity | 2.6 | - |
| Bulk Density (kg/m3) | 1712 | - |
| Sulfate content% | 0.37 | ≤0.5 |
| Absorption% | 3.2 | - |

**Table 5.** Properties of coarse aggregate.

| Sieve size (mm) | Passing by weight % | Limits of IQS (No.45/1984)(13) |
|---|---|---|
| 20 | 100 | 95-100 |
| 14 | - | - |
| 10 | 42 | 30-60 |
| 5 | 3 | 0-10 |
| Property | Test Results | Limits of (IQSNo.45/1984)(13) |
| Specific gravity | 2.68 | - |
| Bulk Density (kg/m3) | 1712 | - |
| Sulfate content% | 0.098 | ≤0.1 |
| Gypsum % | 0.210 | ≤ 0.215 |
| Fine material % | 0.56 | ≤ 3 |
| Absorption % | 0.43 | - |

**Table 6.** Physical properties of waste paper.

| | |
|---|---|
| Moisture Content % | 2.67 |
| Specific Gravity ( SSD) | 0.98 |
| Density ($kg/m^3$) | 800 |
| Absorption % | 89 |
| Organic Materials % | 70 |
| In-organic materials % | 30 |

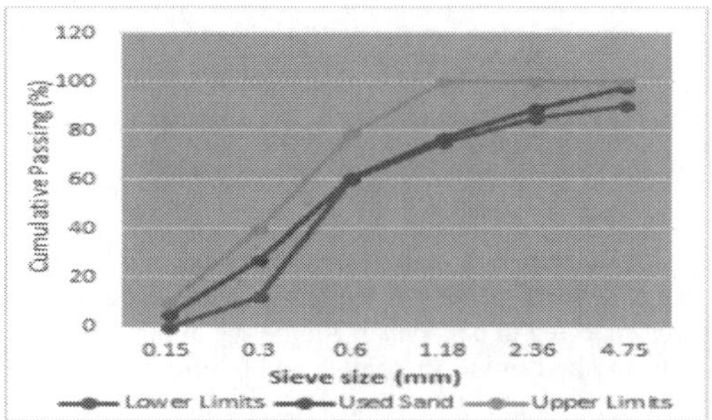


**Fig. 2** . Grading curve for used fine aggregate.

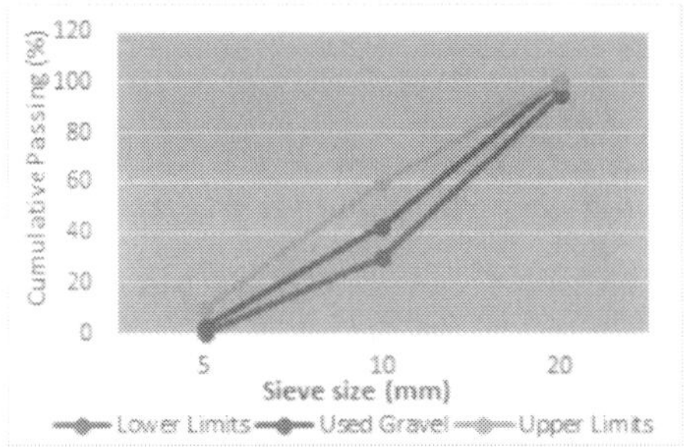


**Fig. 3**. Grading curve for used coarse aggregate.

**Fig. 4**. Sample of paper shredder

**Fig. 5**. Paper pieces soaked in water

**Fig. 6**. Paper mass after soaked and grinding

# 3 Experimental program and test procedures

## 3.1 Proportion of the mix.

The mix proportion was designed to get approximately about 25MPa compressive strength of normal concrete. Trial mixes were prepared for workable mix while adding waste paper. At last the mix proportion by weight was [1:1.9:2.5] cement: fine aggregate: coarse aggregate, water cement ratio was kept constant as w/c = 0.35 to keep the slump value the same by adding super plasticizer called Sika. The weight of cement, sand and gravel that were used in all mixes were 400 (kg/m3), 760 (kg/m3) and 1000 (kg/m3) respectively. Four Papercrete mixes were then prepared by adding paper sludge by weight of cement as 5%, 10%, 15% and 20% as shown in Table 7.

## 3.2 Fresh concrete test

### *3.2.1. Slump Test*

Workability is one of the important features defining the fresh properties of concrete and can be defined as a measure of the ability of concrete to be mixed, handled, transported and placed without loss in homogeneity and with less air voids. A slump test is a suitable test to determine the workability for all types of concrete mixes; the test was performed according to ASTM C143 [14].

### *3.2.2. Fresh Density*

This test is conducted according to ASTM C567-05a [15]. The average of three cubes of (100) mm is considered to determine the fresh concrete density.

## 3.3 Hardened Concrete Test

- Density of hardened concrete is determined using air dried cubes of sizes (100x100x100) mm according to B.S 1881: part 114: 1989 [16]. The specimens were tested after picking the cubes out of the curing water after drying their surfaces by a piece of cloth. Density (ρ) is the mass of a unit volume of hardened concrete expressed in kilograms per cubic meter.

- Compressive Strength was determined according to B.S 1881: part 116, 1989 [17]. The average compressive strength of three cube specimens was recorded. This test was conducted at 7, 28 and 60 days of age.

- Splitting Tensile Strength was done according to ASTM C496 –07 [18] and the average splitting tensile strength of three specimens were recorded. This test was conducted at ages 7, 28, and 60 days.

- The modulus of rupture was determined by using (100×100×400) mm prism specimens; in conformity with ASTM C78 -04 [19] the prisms were subjected to two third point loading system. The loading was applied at a rate of 0.02 MPa/sec, the specimens were tested at age of (7, 28, and 60) days. The average of two specimens in each age was adopted.

- Absorption test is determined using cubes of sizes (100x100x100) mm to define the absorption capacity in accordance with the ASTM C642-03 [20].

- Fire resistance test, papercrete cubes cured for 28 days and air-dried for 1 hour were burnt to a temperature of 800°C for 10 minutes in a furnace. The cubes were allowed to cool for 1 hour in the open air before the determination of their residual compressive strength.

 https://doi.org/10.1051/matecconf/201816202016

## 4 Results and discussion

### 4.1 Fresh Concrete

The slump test and fresh density results are explained in Table 7. Together slump and density decreased when a higher amount of paper pulp content was included. The as-received pulp presented a high water-absorption capability. Thus, when a higher amount of paper pulp was included in the mixture, it required more water to achieve a given slump. The workability of concrete containing paper-mill residual was improved by the addition of super-plasticizer to achieve good workability with reinforced strength of papercrete. While density decreased by the addition of paper. Several factors could lead to apposite effects on the workability and density of papercrete. The amount of paper pulp addition, paper pulp physical properties, and the carbon content of the paper pulp would be the main reasons for the reduction of concrete workability and density. The increase in water demand becomes larger with an increase in the paper pulp content to about 20%.

**Table 7**. Results of slump and fresh density

| *Mix ID* | *% of pulp paper mass by weight of cement* | *Admixture (Sika) (%) by weight of cement* | *Slump (mm)* | *Fresh density (kg/m³)* |
|---|---|---|---|---|
| *R* | 0 | 0.75 | 45 | 2445 |
| *M5* | 5 | 1 | 48 | 2469 |
| *M10* | 10 | 1.5 | 50 | 2203 |
| *M15* | 15 | 2.5 | 46 | 2183 |
| *M20* | 20 | 3 | 42 | 2135 |

### 4.2 Hardened concrete

The dry density and water absorption test of the papercrete was done by using pulp paper and the results are shown in Table 8 and Fig. 7. It was noted that the bulk density and water absorption of the papercrete change inversely with the waste paper content of the mixtures. The water absorption of the cubes expanded with the expanding waste paper content. High measure of water ingestion is because of the nearness of cellulose materials that effectively assimilates water and holds it for long time. The average test results of splitting tensile strength, compressive strength and flexural strength done at 7, 28 and 60 days age are recorded in Tables 8 and 9. The test results demonstrated an expanding with curing age because of proceeding of hydration process. The compressive, splitting tensile and flexural strength of concrete blends with paper pulp were less than reference mixes for all test ages aside from the blend with (5%) expansion of paper pulp by weight of cement. The results showed reduction when higher paper pulp contents were involved in the concrete mixes. Fig. 8 presents the compressive strengths of all 5 mixes at 7, 28 and 60 days. The paper pulp content in the concrete mixtures performed a great role in the mechanical properties. Yet, as the paper mass content rises, the strength falls down.

The same trend was seen in Fig. 9 for splitting tensile strength. This is because of loss in cohesion and also the poor binding of calcium- hydrate – silicate (C-S-H) gel on cellulosic material. Besides adding waste paper will take part of concrete mix from the volume and any contraction will decrease the concrete mix which will debilitate the concrete. On the other hand the flexural strength results shown in Fig. 10, it can be seen that flexural strength of concrete mixes with paper pulp were as much as reference mixtures for all test ages. The fire resistance was depicted as a measure of the residual compressive strength of papercrete cured for 28 days, dried and subsequently subjected to heat at 1000°C. The 28-day compressive strength result for papercrete cubes, before and after burning, is introduced in Fig. 11. There is a general lessening in their compressive strength after burning. The crevice between the compressive strength lines of the papercrete progressed toward becoming smaller with increase in waste paper content. This demonstrates that the fire resistance of papercrete increased with increase in waste paper content.

**Table 8**. Results of hardened concrete mixes properties

| Mix ID | Dry density (kg/m³) | Water absorption (%) | Compressive Strength (MPa) | | |
|---|---|---|---|---|---|
| | | | *7 days* | *28 days* | *56 days* |
| R | 2140 | 6.3 | 32.6 | 39.7 | 48.7 |
| M 5 | 2159 | 5.9 | 33.2 | 42.1 | 49.9 |
| M 10 | 2015 | 8.4 | 21.7 | 31.4 | 37.3 |
| M 15 | 1989 | 12.4 | 17.4 | 22.2 | 26.9 |
| M 20 | 1894 | 14.6 | 11.3 | 16.8 | 21.2 |

**TABLE 9**. Average tensile strength of mixes.

| Mix ID | Splitting Tensile Strength (MPa) | | | Flexural Strength (MPa) | | |
|---|---|---|---|---|---|---|
| | *7 days* | *28 days* | *56 days* | *7 days* | *28 days* | *56 days* |
| R | 2.5 | 3.3 | 3.5 | 4.4 | 5.5 | 5.8 |
| M 5 | 2.7 | 3.5 | 3.6 | 4.3 | 5.4 | 5.7 |
| M 10 | 2.2 | 2.8 | 3.3 | 3.4 | 4.4 | 4.8 |
| M 15 | 1.8 | 2.4 | 2.7 | 3.2 | 4.1 | 4.3 |
| M 20 | 1.5 | 1.9 | 2.4 | 2.4 | 3.4 | 3.6 |

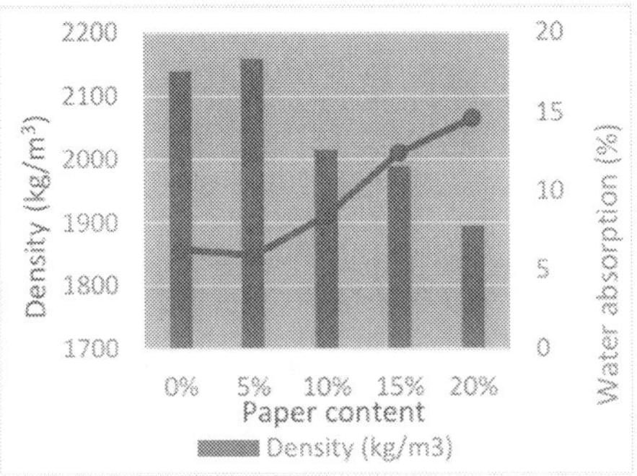


**Fig. 7**. Variation in density with water absorption

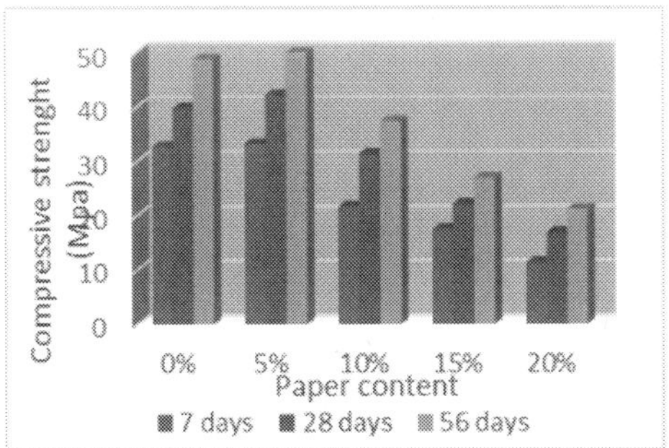


**Fig. 8**. Results of compressive strength of concrete

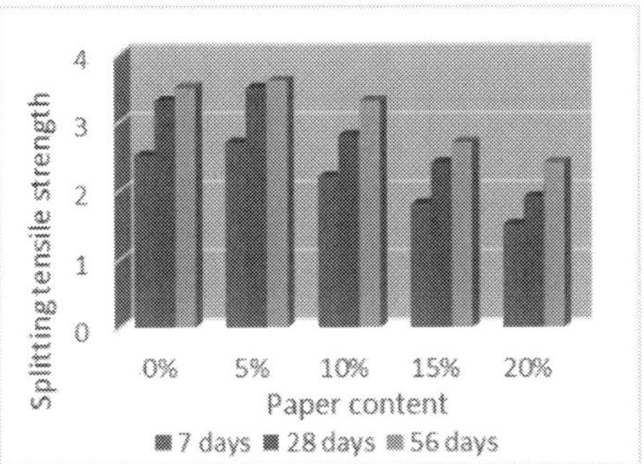


**Fig. 9**. Results of Splitting Tensile Strength of Concrete

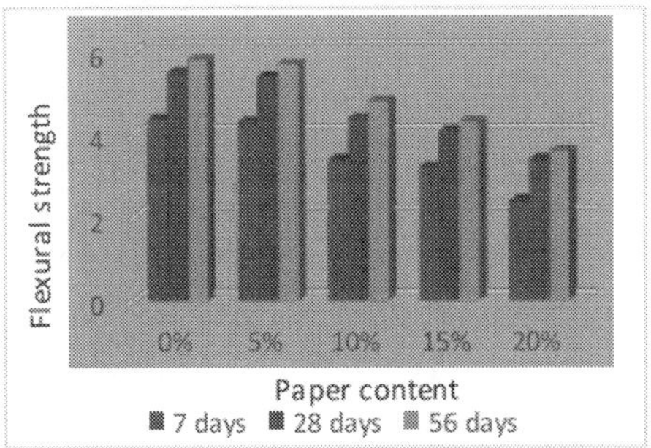


**Fig. 10**. Results of flexural strength of concrete.

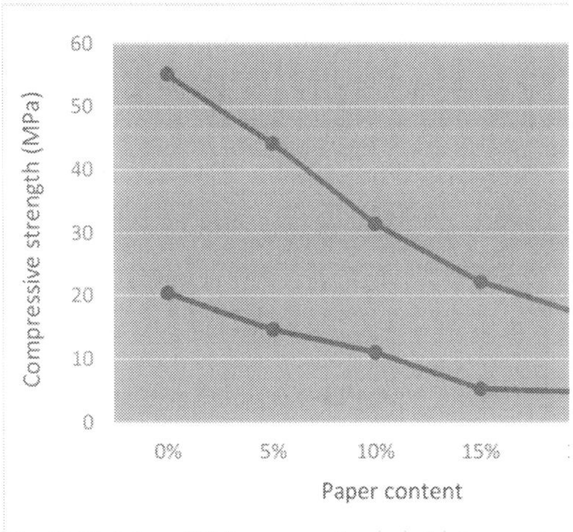


**Fig. 11**. Variation of 28-day compressive strength of papercrete (before and after burning)

## 5 Conclusions

Based on the experimental work results in this investigation, the following conclusions can be drawn:

1. The low bulk density of papercrete demonstrates that waste paper could be an important alternate sustainable material for production of light weight concrete.
2. The reuse of wastes is essential from various perspectives: It helps to save and sustain the natural resources which are not replenished; it decreases the pollution of the environment and it also helps to save and recycle energy in the production process. Additionally, suitable landfill sites are becoming more difficult to find as urban areas expand.
3. It has been confirmed that adding waste paper has a distinct antagonistic impact on the slump and fresh density. For which request higher water or higher chemical admixture dosages to keep the slump values as far as possible.
4. Adding waste paper to concrete mix prompted an increase in water absorption and a decrease in dry density for all mixes used aside from the mixture with 5% of paper pulp. The results of dry density demonstrate that lightweight concrete could be created by adding waste paper.
5. Results of compressive strength, and splitting tensile strength for (10%, 15% and 20%) mixes decreased with the increase of the amount of wastepaper. While in the mixture with (5%), the strength was higher than the reference since waste paper contains considerable amount of alumino-siliceous material that is combined with calcium, leading to the improvement in its strength. The advancement of this strength is predominantly inferable to the hydraulic and pozzolanic activity of waste paper that is activated by the alkalis and to some extent, $Ca(OH)_2$

which is released from the hydration process. While results of flexural strength were less than reference mixes for all test ages.

6. Papercrete should not be utilized for external walls and near-ground walls due to its high water absorption capacity. If it has to be used for external walls, the surface of the walls must be waterproof.

7. The reductions in compressive strength of the papercrete cubes after burning were less than 1 N/mm$^2$. This might be ascribed to the thermal insulation of the interior of the burnt papercrete cubes.

8. The optimum percent of paper is 20% and can be utilized to produce bricks with light weight and good fire resistance.

## References

1. J.H. Chung, B.H. Kim, H.k. Choi & C.S. Choi, Int. Con. on S. B. **14**, 317-318 (2015).
2. C. Aciu, D.A. Ilutiu, N. Cobirzan, & A. Balog, Int. Con. Int. in Eng., 295-300 (2014).
3. R.S. Gallardo & M.A. Adajar. Sym. on Inf. D. & E. 2006, 7-8 (2006).
4. V.K. Vishwakarma, & P.K. Patil, Int. J. for Sc. R. & D. **4,** 561 (2016).
5. J. Singh & P.K. Patil, Int. J. for Sc. R. & D. **4,** 784 (2016).
6. A. Hashmi, H.N. &, N. Baig, Int. J. of Eng. **11**, 337-340 (2013).
7. S. F. Seyyedalipour, D. Y. Kebria, N. R. Malidarreh, & G. Norouznejad, Int. J. of Eng. R. & A. **4**, 115-122 (2014).
8. S. Khandelwal, K. L. Prajapat, M. Kumar, L. Bhantia, A. Sharma, V. Sharma, Int. J. of C. R. & D. **4,** 668 (2015).
9. I. Vegas, J. Urreta, M. Frías, & R. García, *C. & B. M., 23(8)* (2009).
10. G.V.S. Prasad, P.P. Reddy, M.Swathi, P.D.V.Kumar, T.Praveenraja, M.Naveen, In. J.of Eng. R., 3 (2015).
11. B J Fuller, AFafitis and J L Santamaria. ASCE Civil Engineering 75, 72 -77 (2006).
12. IQS. No.5-1984, "Portland cement", Iraqi Standard Organization, Central Organization for Standardization and Quality Control.
13. IQS. No. 45-1984," Aggregate from Natural Sources for Concrete", Iraqi Standard Organization, Central Organization for Standardization and Quality Control.
14. ASTM C143, "Standard Test Method for Slump of Hydraulic Cement Concrete", 2000.
15. ASTM C567–05a, "Standard Test Methods for Determination Density of Structural Lightweight Concrete" ASTM International, Annual Book of ASTM Standards, American Society for Testing and Materials, **04.02**, United States, 2006.
16. B.S.1881, Part 114, "Methods for Determination of Density of Hardened Concrete", British Standards Institution, 1983.
17. B.S 1881: Part 116, *''Method for Determination of Compressive Strength of Concrete Cubes''.* , British Standards Institution. , **3**, 1989.
18. ASTM C496/C496M, "Standard Test Method for Splitting Tensile of Cylindrical Concrete Specimens", Annual Book of ASTM standard, **04.02**, 283-287 (2004).
19. ASTM C78-02.*''Standard Test Method for Flexural Strength of Concrete"*, Annual Book of ASTM standard, **04.02** (2002).
20. ASTM C642–03a, "Standard Test Method for Density, Absorption, and Voids in Hardened Concrete", Annual Book of ASTM Standards, American Society for Testing and Materials, **04-02** (2003).

 https://doi.org/10.1051/matecconf/201816202017

# Effect of internal curing on performance of self-compacting concrete by using sustainable materials

*Nada* Aljalawi[1], and *Amar* Yahia AL-Awadi[1]

[1]Engineering College, University of Baghdad, Baghdad, Iraq

**Abstract.** This paper is devoted to investigate the effect of internal curing technique on the properties of self-compacting concrete. In this study, self-compacting concrete is produced by using limestone powder as partial replacement by weight of cement with percentage of (5%), sand is partially replaced by volume with saturated fine lightweight aggregate which is thermostone aggregate as internal curing material in three percentages of (5%, 10%, 15%) for self-compacting concrete, and the use of two external curing conditions which are water and air. The experimental work was divided into three parts: in the first part, the workability tests of fresh self-compacting concrete were conducted. The second part included conducting compressive strength test and modulus of rupture test at ages of (7, 28 and 90) days. The third part included doing the shrinkage test at age of (7, 14, 21, 28) days. The results show that internally cured self-compacting concrete has the best workability and the best properties of hardened concrete which include (compressive strength, modulus of rupture) of externally cured self-compacting concrete with both water and air as compared with reference concretes. Also, the hardened properties of internally cured self-compacting concrete with percentage of (5%) with thermostone aggregate is the best as compared with that of percentages (10% and 15%) in both external curing conditions. In general, the results of shrinkage test have shown reduction in shrinkage of internally cured self-compacting concrete as compared with reference concretes and this reduction increases with increase in the thermostone aggregate content-within-self-compacting-concrete.

## 1 Introduction

Self-compacting concrete (SCC) is a type of concrete that gets compacted under its self-weight. It is commonly abbreviated as SCC and defined as the concrete that is capable of self-compacting and occupying all the spaces in the formwork without any vibration; purely by means of its self-weight, thereby eliminating the need for either external energy input from vibrators or any type of compacting effort. The use of SCC is spreading worldwide because of its very attractive properties in the fresh state as well as after hardening. The use of SCC will lead to reduce the technical costs of in situ concrete constructions and eliminate some of the potential human error. It will replace manual compaction of fresh concrete with a modern semi-automatic placing technology and in that way, improve health and safety in and around the construction site [22].

Term internal curing means "supplying water throughout a freshly placed cementitious mixture using reservoirs, via pre-wetted lightweight aggregates that readily release water as needed for hydration or to replace moisture lost through evaporation or self-desiccation" according to the definition provided in American Concrete Institute. Also, the hydration of cement continues because of the availability of internal water that is not part of the mixing water, refers to internal curing technique [1].

## 2 Literature review

The concept of SCC was first proposed by Professor Hajime Okamura, in 1986 in Japan as a

© The Authors, published by EDP Sciences. This is an open access article distributed under the terms of the Creative Commons Attribution License 4.0 (http://creativecommons.org/licenses/by/4.0/).

solution to concrete's concerns. There are many advantages of using SCC, these include [7] [15]:

- Elimination of the need for vibration.
- Reducing the noise pollution.
- Improving the filling capacity of highly congested structural member.
- Reducing the construction time and labor cost.

Self-compacting concrete (SCC) is characterized by high resistance to segregation that can be cast without compaction or vibration. It flows like honey, de-aerates, self-compacts and has nearly a horizontal concrete level after placing [14]. Internal curing has been discussed as an added advantage in concrete research. It has wider prospect in many countries and it is possible to get benefit from the internal curing instead of traditional external curing. Internal curing has a significant contribution in shrinkage reduction. In the advancement of concrete technology internal curing is found to be beneficial in terms of enhancing concrete performance as well as environment friendly [20].

The main properties of lightweight aggregate (LWA) are its high porosity, high absorption, low specific gravity and cellular structure. These properties make LWA a suitable material for internal curing technique [19].

By replacing a portion of the normal weight aggregates with pre-wetted lightweight aggregates (LWA), additional internal curing water is provided to the concrete mixture. During the hydration of the cement paste within the concrete mixture, this internal curing water will be drawn from the LWA into the hydrating paste, maintaining a high degree of saturation (water-filled pores) in the cement paste and avoiding or at least reducing shrinkage stresses and their accompanying autogenous deformations [9] [6].

## 3 Expermental work

### 3.1 Materials

Limestone powder ($CaCO_3$) was used as a filler material and as percentage replacement of cement for producing SCC. The fineness (Blaine specific surface) of the used limestone powder (LSP) is (200 $m^2/kg$) and the chemical composition of LSP is given in Table (4).

Ordinary Portland Cement (O.P.C) (ASTM Type I). This cement complied with the Iraqi specification [16]. The coarse aggregate was brought from Al-Nibaii region with a nominal size of (14) mm. Al-Ekhaider natural sand is used as fine aggregate in concrete mixes. Table (1) and Table (2) show the sieve analysis and properties for the sand used throughout this work. The grading and properties of the used sand and coarse aggregate satisfies the requirements of the [17].

Glenium 51 (G51) is used in this research as chemical admixture and complies with [5] type F.

Thermostone aggregate is considered as one type of lightweight aggregate and used in this study as internal curing material. Thermostone was gained from Karbala thermostone factory as waste and broken into smaller size particles like sand. Then, the crushed thermostone is washed and cleaned with water afterward dried by spread in air. The crushed particles are sieved and partially replaced by volume with the same size of sand with a certain percentage to have the same grading as the used sand which satisfies the grading requirements of the [17].

Later, the thermostone aggregate is soaked in water for (24) hours to bring the aggregate particles to saturated condition. Table (3) shows the physical properties of the used thermostone aggregate.

### 3.2 Mix Design and Proportions

The mix design method of the used SCC in this study is according to the European guidelines for SCC [12], and then the proportions of materials were modified after obtaining a satisfactory self-compactability by evaluating through fresh concrete tests. The mix proportions of the SCC which was used throughout this research are shown in Table (5).

### 3.3 Tests of Fresh Concrete

Slump flow test, T50cm test, V- funnel test and L-box test are used as test methods for workability properties of SCC. These methods are given in the European Federation dedicated to specialist construction chemicals and concrete systems SCC guidelines [12].

### 3.4 Tests of Hardened Concrete

Tests of hardened concrete in this research are shown below:

- Compressive strength test was conducted according to the British Standard [8] using (Tinius Olsen testing machine) compression device of (1000 kN) maximum capacity. Three cubes (100*100*100) mm were tested for each mix at each age of (7, 28 and 90) days for the determination of compressive strength by using two sets of mixes for SCC, one of these sets is cured in water and the other is cured in air along the test period.
- Modulus of rupture test was performed on two (100x100x400) mm prisms according

to [4] with span of (300) mm at age of (7, 28 and 90) days and the average of two specimens of each mix was adopted.

- Shrinkage test was conducted according to [3]. A micro – meter dial gauge with (0.001) mm reading accuracy was used in this test. Pins were fixed on prisms (100x100x400) mm after casting the specimens. The shrinkage test was conducted by using two sets of prisms specimens for SCC, the first set is cured for 7 days in water, after that it is taken out from water tank and left in air (laboratory conditions) for 21 days, the second set is cured in air along the test period for 28 days. The length change was calculated at age of (7, 14, 21 and 28) days by using the following eq. (1):

$$L = [(L_x - L_i)/G] \;\ldots\ldots\; (1)$$

where:
L: Shrinkage strain at (x) age.
$L_x$: Comparator reading of specimen at (x) age minus comparator reading of reference bar at (x) age (mm).
$L_i$: Initial comparator reading of specimen minus reading of reference bar at that same time (mm).
G: Nominal gauge length (mm).

# 4 Results and discussion

## 4.1 Test Results of Fresh Concrete

The results of the slump flow test, V- funnel test and L-box test are shown in Table (6). These results indicate that in general, the workability of fresh SCC improves with increasing thermostone aggregate as partial sand replacement percentage as compared with reference concrete. This is due to that the pre-wetted fine lightweight aggregate (LWA), which provides a set of water-filled reservoirs within the concrete as additional moisture and in turn improves the workability of fresh SCC [13] [22]. These results are within the acceptable criteria for SCC [11] and also indicate excellent deformability and filling ability without any segregation, bleeding and blocking.

## 4.2 Test Results of Hardened Concrete

The results of the hardened properties which include (compressive strength, modulus of rupture and shrinkage) of SCC in this research are shown in Tables (7) to (12), and represented in Figures (1) to (6). From these results, it can be seen that internally cured SCC with thermostone aggregate has the best hardened properties which include (compressive strength, modulus of rupture) of externally cured SCC with both water and air as compared with reference concretes which didn't contain thermostone aggregate, as a result of internal curing technique. The highest increase in compressive strength reaches 16.72% and 12.67% for SCC internally cured with thermostone aggregate as partial sand replacement with percentage of (5%) and cured in water and air, respectively.

In general, the results of shrinkage test have shown reduction in shrinkage of internally cured SCC with thermostone aggregate as compared with reference concretes which didn't contain LWA.

The increased strength may be due to increase in the degree of cement hydration as a result of internal curing water, which leads to increase the hydration products, improve the interfacial transition zone by filling internal voids of concrete, reduction of shrinkage induced micro cracking and decrease the porosity of SCC. This complies with studies carried out by [18] [10] [2].

# 5 Conclusions

Depending on the results of this investigation, the following conclusions can be drawn:

1. It is possible to produce self-compacting concrete (SCC) by using thermostone aggregate as partial sand replacement with three different percentages (5%, 10% and 15%) by volume as internal curing material to internal curing technique.
2. It is possible to get benefit from the internal curing instead of traditional external curing because internal curing is easy to use and has a significant contribution to shrinkage reduction, enhancing durability, sustainability and hence, improving overall SCC performance.
3. Improving the workability and the hardened properties which include (compressive strength and modulus of rupture) of internally cured SCC with increasing thermostone aggregate as partial sand replacement as compared with reference concrete.
4. The SCC mixes are internally cured with thermostone aggregate and externally cured in both water and air exhibit very low shrinkage as compared with reference concrete.

# References

[1] ACI Committee 213R-03, 2003, Guide for Structural Lightweight Aggregate Concrete, Reported by ACI committee 213, ACI Manual of Concrete Practice, pp. 213R-1-38.
[2] AL-Awadi, A.Y., 2013, M.Sc. Thesis, University of Baghdad.
[3] ASTM C490-00, 2000, Use of Apparatus for the Determination of Length Change of

Hardened Cement Paste, Mortar, and Concrete, Annual Book of ASTM Standards C490 – 00.

[4] ASTM C293-02, 2002, Flexural Strength of Concrete (Using Simple Beam with Center – Point Load), Annual Book of ASTM Standards C293 – 02, pp. 1-3.

[5] ASTM C494-05, 2005, Standard Specification for Chemical Admixtures for Concrete, Annual Book of ASTM Standards.

[6] Bentz, D.P., 2007, Internal Curing of High-Performance Blended Cement Mortars, ACI Materials Journal, Vol. (104), No. (4), pp. 408-414.

[7] Bouzoubaa, N., and Lachemi, M., 2001, Self-Compacting Concrete Incorporating High-Volumes of Class F Fly Ash: Preliminary Results, Cement and Concrete Research, Vol. 31, No.3, pp413-420.

[8] British Standards Institution. B.S 1881, Part 116, 1985, Method for Determination of Compressive Strength of Concrete Cubes.

[9] Cusson, D., and Hoogeveen, T., 2005, Nantes, France, pp. 579-584.

[10] Cusson, Z., and Lounis, L., 2010, .

[11] EFNARC, 2002, Specification and Guidelines for Self-Compacting Concrete, pp. 32, www.efranice.org.

[12] ERMCO, 2005, The European Guidelines for SCC, pp. 63, www.efca.info.

[13] Friggle, T., and Reeves, D., 2008, American Concrete Institute, Farmington Hills, MI, pp. 71-80.

[14] Holm, T.A., and Bremner, T.W., 2000, Engineer Research and Development Center , ERDC/SL TR-00-3, C.116, Vicksburg.

[15] Horta, A., 2005, M.Sc. Thesis, Georgia Institute of Technology, pp.228.

[16] Iraqi Specifications, IQS No.5, 1984, The Portland Cement, Central Apparatus for Standardization and Quality Control. (Translated from Arabic).

[17] Iraqi Specifications, IQS No.45, 1984, The Used Aggregate from Natural Sources in Concrete and Building, Central Apparatus for Standardization and Quality Control. (Translated from Arabic).

[18] Lura, P., 2003, Ph.D. thesis, Technical University of Delft.

[19] Lura, P., Bentz, D.P., Lange, D.A., Kovler, K., and Bentur, A., 2004, Proceeding International RILEM Symposium on Concrete Science and Engineering, Northwestern University, Evanston, Illinois. RILEM Publications S.A.R.L., pp. 137–151.

[20] Munaz, A.N., Bushra, I., and Rahman, S., 2011, BUET-Japan Institute of Disaster Prevention and Urban Safety, Bangladesh University of Engineering and Technology, Dhaka, Bangladesh, pp. 22-28.

[21] Selvamony, M.S., Ravikumar, S.U., and Basil, G., 2010, .

[22] Villareal, V.H., 2008, Eds. D. Bentz and B. Mohr, American Concrete Institute, Farmington Hills, MI, pp. 45-56.

**Table 1**. Sieve analysis of sand and thermostone aggregate

| Sieve size (mm) | Percent passing (%) | I.Q.S.45: 1984 Limits Zone (2) |
|---|---|---|
| 10 | 100 | 100 |
| 4.75 | 99.8 | 90 - 100 |
| 2.36 | 84.4 | 75 - 100 |
| 1.18 | 65.6 | 55 - 90 |
| 0.60 | 41.8 | 35 - 59 |
| 0.30 | 11 | 8 - 30 |
| 0.15 | 2.2 | 0 - 10 |
| Fineness modulus = 2.95 | | |

**Table (2): Physical and chemical properties of sand**

| Property | Test result | I.Q.S.45: 1984 Limits |
|---|---|---|
| Apparent specific gravity | 2.55 | ------ |
| Absorption, % | 2.95 | ------ |
| Bulk density (kg/m3) | 1710 | ------ |
| Sulphate content (SO3) | 0.19% | 0.50% (max) |

**Table (3): Physical properties of thermostone aggregate**

| Property | Test result |
|---|---|
| Shape | Crushed |
| Apparent specific gravity | 1.14 |
| Bulk density ($kg/m^3$) | 675 |
| Absorption, % | 48 |

**Table (4): Chemical composition of limestone powder**

| No. | Constituents | Quantity (%) |
|---|---|---|
| 1 | CaO | 56.21 |
| 2 | $SiO_2$ | ---- |
| 3 | $Al_2O_3$ | ---- |
| 4 | $Fe_2O_3$ | ---- |
| 5 | L.O.I | 43.78 |
| 6 | $SO_3$ | ---- |
| 7 | MgO | ---- |

**Table (5): Mix proportions of SCC mixes***

| Thermostone Aggregate (fine) $kg/m^3$ | | | Coarse Aggregate $kg/m^3$ | Sand $kg/m^3$ | Limestone Powder (5%) $kg/m^3$ | Cement $kg/m^3$ | Index of Mixes |
|---|---|---|---|---|---|---|---|
| 15% | 10% | 5% | | | | | |
| --- | --- | --- | 850 | 825 | 25 | 475 | Mix-R |
| --- | --- | 16.3 | 850 | 783.75 | 25 | 475 | Mix-T5% |
| --- | 32.6 | --- | 850 | 742.5 | 25 | 475 | Mix-T10% |
| 48.9 | --- | --- | 850 | 701.25 | 25 | 475 | Mix-T15% |

*Water (kg/m$^3$) = 185
Glenium 51 (liter per 100kg of cementitious materials) = 1.1

**Table (6): The results of the slump flow test, V- funnel test and L-box test**

| Mixes | Compressive strength (MPa) | | |
|---|---|---|---|
| | 7 days | 28 days | 90 days |
| Mix-R | 40.76 | 42.88 | 45.71 |
| Mix-T5% | 44.04 | 47.66 | 53.35 |
| Mix-T10% | 41.98 | 45.98 | 50.82 |
| Mix-T15% | 37.84 | 40.73 | 42.04 |

**Table (7): Test results of compressive strength of SCC internally cured with thermostone aggregate and externally cured with water**

| Index of Mixes | Slump flow | | V-Funnel | | L-Box | |
|---|---|---|---|---|---|---|
| | D (mm) | T50cm (sec) | TV (sec) | TV5min (sec) | Blocking Ratio (H2/H1) | T40cm (sec) |
| Mix-R | 736 | 3.5 | 10 | 12 | 0.9 | 3.5 |
| Mix-T5% | 747 | 3.3 | 9.5 | 11.6 | 0.91 | 3.4 |
| Mix-T10% | 760 | 3.1 | 8.5 | 10.5 | 0.92 | 3.2 |
| Mix-T15% | 772 | 2.8 | 7.2 | 9.3 | 0.93 | 3 |
| | | | | | | |

**Table (8): Test results of compressive strength of SCC internally cured with thermostone aggregate and externally cured with air**

| Mixes | Compressive strength (MPa) | | |
|---|---|---|---|
| | 7 days | 28 days | 90 days |
| Mix-R | 38.27 | 40.22 | 42.63 |
| Mix-T5% | 40.21 | 43.48 | 48.03 |
| Mix-T10% | 38.96 | 41.92 | 45.95 |
| Mix-T15% | 35.13 | 37.83 | 40.89 |

**Table (9): Test results of modulus of rupture of SCC internally cured with thermostone aggregate and externally cured with water**

| Mixes | Modulus of rupture (MPa) | | |
|---|---|---|---|
| | 7 days | 28 days | 90 days |
| Mix-R | 5.66 | 5.87 | 6.04 |
| Mix-T5% | 5.98 | 6.31 | 6.65 |
| Mix-T10% | 5.80 | 6.04 | 6.28 |
| Mix-T15% | 5.63 | 5.75 | 5.88 |

**Table (10): Test results of modulus of rupture of SCC internally cured with thermostone aggregate and externally cured with air**

| Mixes | Modulus of rupture (MPa) | | |
|---|---|---|---|
| | 7 days | 28 days | 90 days |
| Mix-R | 5.32 | 5.51 | 5.65 |
| Mix-T5% | 5.57 | 5.85 | 6.07 |
| Mix-T10% | 5.45 | 5.66 | 5.86 |
| Mix-T15% | 5.31 | 5.43 | 5.52 |

**Table (11): Test results of volume change of SCC internally cured with thermostone aggregate and externally cured with water for 7 days and with air until 21 days**

| Mixes | Shrinkage × $10^{-6}$ | | | | | |
|---|---|---|---|---|---|---|
| | Water curing | | | Air curing | | |
| | 1 day | 4 days | 7 days | 14 days | 21 days | 28 days |
| Mix-R | 0 | +20 | +25 | -40 | -135 | -180 |
| Mix-T5% | 0 | +30 | +35 | -30 | -40 | -50 |
| Mix-T10% | 0 | +35 | +40 | -20 | -30 | -35 |
| Mix-T15% | 0 | +40 | +50 | -5 | -10 | -15 |

**Table (12): Test results of volume change of SCC internally cured with thermostone aggregate and externally cured with air until 28 days**

| Mixes | Shrinkage × $10^{-6}$ | | | | | |
|---|---|---|---|---|---|---|
| | Air curing | | | | | |
| | 1 day | 4 days | 7 days | 14 days | 21 days | 28 days |
| Mix-R | 0 | -105 | -190 | -220 | -240 | -255 |
| Mix-T5% | 0 | -25 | -50 | -65 | -75 | -85 |
| Mix-T10% | 0 | -15 | -30 | -45 | -50 | -65 |
| Mix-T15% | 0 | -10 | -20 | -35 | -40 | -50 |

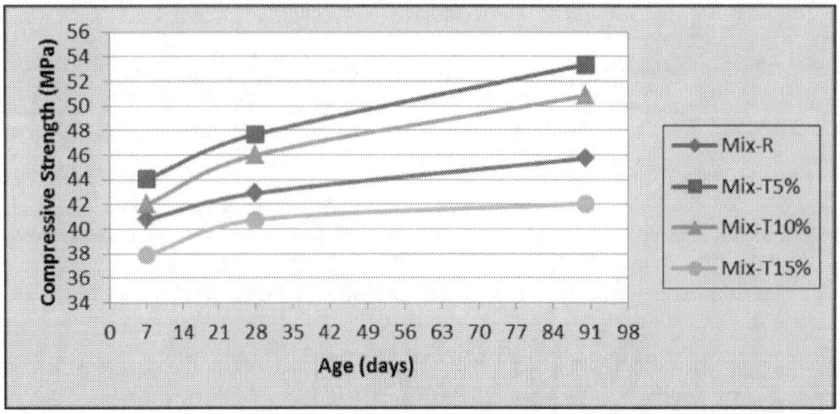


**Figure (1): Compressive strength development of SCC internally cured with thermostone aggregate and externally cured with water**

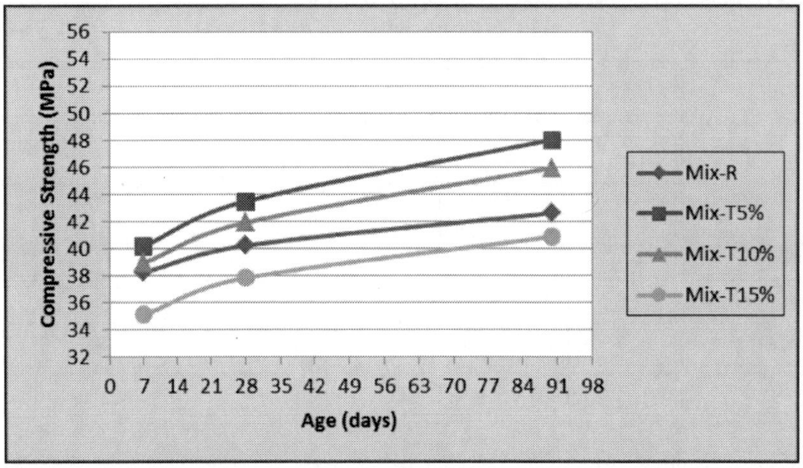


**Figure (2): Compressive strength development of SCC internally cured with thermostone aggregate and externally cured with air**

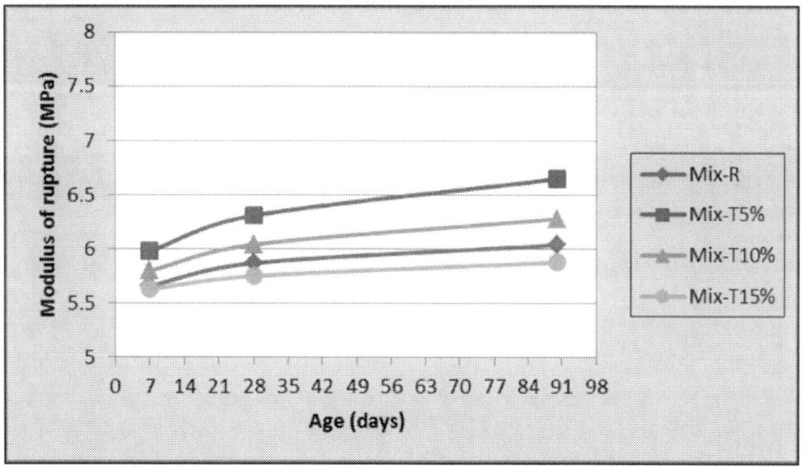


**Figure (3): Modulus of rupture development of SCC internally cured with thermostone aggregate and externally cured with water**

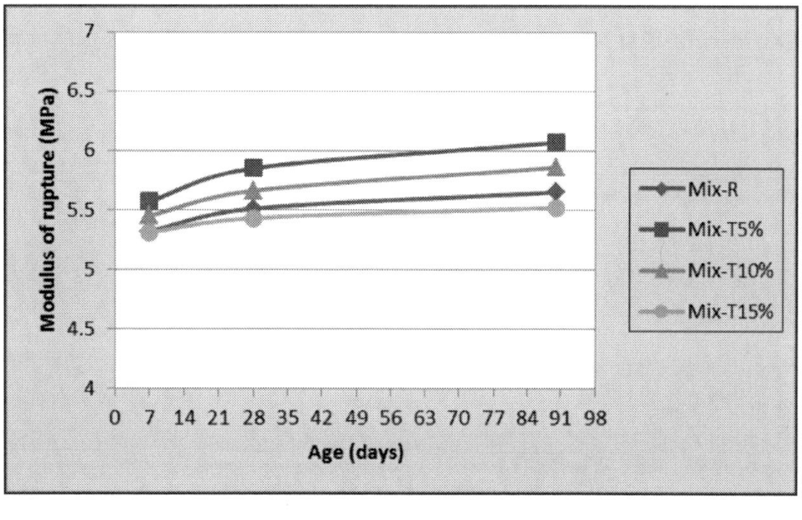


**Figure (4): Modulus of rupture development of SCC internally cured with thermostone aggregate and externally cured with air**

https://doi.org/10.1051/matecconf/201816202017

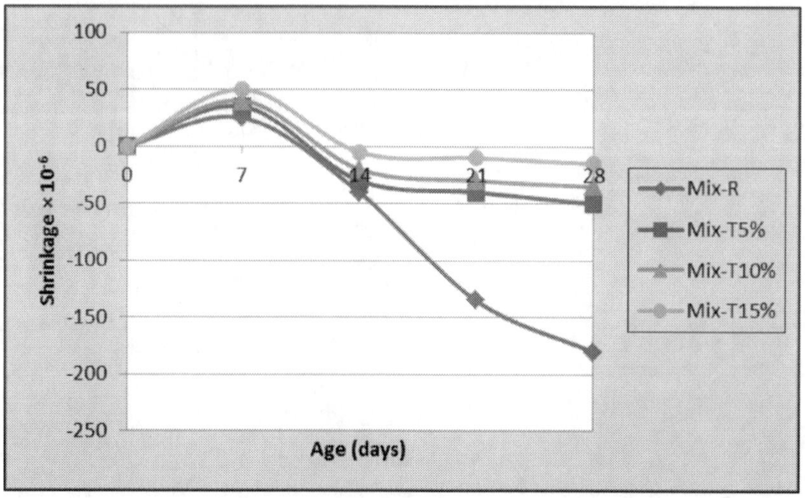


**Figure (5): Volume change development of SCC internally cured with thermostone aggregate and externally cured with water for 7 days and with air until 21 days**

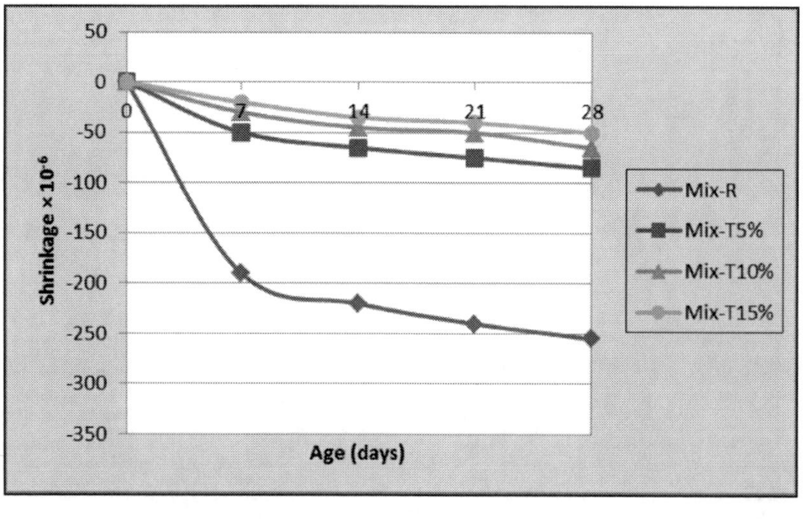


**Figure (6): Volume change development of SCC internally cured with thermostone aggregate and externally cured with air until 28 days**

 https://doi.org/10.1051/matecconf/201816202018

# Compressive strength improvement for recycled concrete aggregate

*Dhiyaa* Mohammed[1], *Sameh* Tobeia[1], *Faris* Mohammed[1] and *Sarah* Hasan[1]

[1] Building and Construction Engineering Department, University of Technology, Baghdad, Iraq

**Abstract.** Increasing amount of construction waste and, concrete remnants, in particular pose a serious problem. Concrete waste exist in large amounts, do not decay and need long time for disintegration. Therefore, in this work old demolished concrete is crashed and recycled to produce recycled concrete aggregate which can be reused in new concrete production. The effect of using recycled aggregate on concrete compressive strength has been experimentally investigated; silica fume admixture also is used to improve recycled concrete aggregate compressive strength. The main parameters in this study are recycled aggregate and silica fume admixture. The percent of recycled aggregate ranged from (0-100) %. While the silica fume ranged from (0-10) %. The experimental results show that the average concrete compressive strength decreases from 30.85 MPa to 17.58 MPa when the recycled aggregate percentage increased from 0% to 100%. While, when silica fume is used the concrete compressive strength increase again to 29.2 MPa for samples with 100% of recycled aggregate.

## 1 Introduction

Concrete is the premier construction material across the world and the most widely used in all types of civil engineering works, including infra-structure, low and high-rise buildings, defense installations, environment protection and local/domestic developments. Concrete is a manufactured product, essentially consisting of cement, aggregates, water and admixture(s). Among these aggregates and, inert granular materials such as sand, crushed stone or gravel form the major part. Traditionally aggregates have been readily available at economic prices and of qualities to suit all purposes [1]. However, in recent years the wisdom of the continued wholesale extraction and use of aggregates from natural resources has been questioned at an international level. This is mainly because of the depletion of quality primary aggregates and greater awareness of environmental protection. In light of this, the availability of natural resources to future generations has also been realized [2].

In fact, many countries throughout the world have now introduced various measures aimed at reducing the use of primary aggregates and increasing the use of recycled aggregates as a partial or full alternative (if any possible) of natural aggregate to produce recycled aggregate concrete, where it is technically, economically, and environmentally acceptable [3].

## 2 Aim of work

This work aimed to experimentally investigate the recycling old demolished concrete into recycled coarse aggregate and the effects of this aggregate on concrete properties, especially compressive strength and the possibility to improve it by using certain admixtures like silica fumes.

## 3 Recycled concrete aggregate productions

Waste concrete sources represented by, demolition of old and deteriorated buildings and traffic infrastructure, and their substitution with new ones, is a frequent phenomenon today in a large part of the world. The main reasons for this situation are changes of purpose, structural deterioration, rearrangement of a city, expansion of traffic directions and increasing traffic load and, natural disasters (earthquake, fire and flood) [3,4].

Recycled aggregate is generally produced by two-stages, the first one represented by crushing of demolished concrete, while the second is the screening and removal of contaminants such as reinforcing steel and other embedded items, and care must be taken to prevent contamination by other materials that can be troublesome, such as asphalt, soil and clay balls, chlorides, glass, gypsum board, sealants, paper, plaster, wood, and roofing materials. The quality of concrete with recycled aggregate is very dependent on the quality of the recycled material used [5-8].

## 4 Experimental work

In this work, several concrete cubes have been made

© The Authors, published by EDP Sciences. This is an open access article distributed under the terms of the Creative Commons Attribution License 4.0 (http://creativecommons.org/licenses/by/4.0/).

by using natural coarse aggregate as well as recycled coarse aggregate, cured and tested in order to investigate the effects of recycled aggregate on concrete compressive strength. Admixtures used also to improve recycled concrete aggregate performance.

## 4.1 Materials

### *4.1.1 Cement*

Ordinary Portland cement was used throughout this work. It was conforming to the Iraqi Standard No. 5/1984[9]. The cement is made in Iraq and is in the form of 50 kg bags.

**Table 1.** Ordinary Portland cement test results

| Test description | Test result |
|---|---|
| Initial setting time | 114 minutes |
| Final setting time | 292 minutes |

### *4.1.2 Fine Aggregate (Sand)*

Natural sand from Al-Ukhaider region was used in this work. The test results were conforming to Iraqi specification (IQS 45/1984) [10].

### *4.1.3 Recycled Coarse Aggregate*

Concrete mass from demolition buildings sites collected in different sizes, crushed to produce several smaller sizes and grades, sieve analysis has been made as shown in Fig. (1), and the maximum size of recycled coarse aggregate is (25) mm.

### *4.1.4 Natural Coarse Aggregate (Gravel)*

Natural coarse aggregate with a maximum size of (25) mm from Al-Nibaae region was used as coarse aggregate. The grading test results is shown in Fig. (1).

### *4.1.5 Mineral Admixtures (Silicafume)*

Silicafume, with 10 percent replacement by weight of cement, has been used in order to improve the performance of concrete with recycled aggregate.

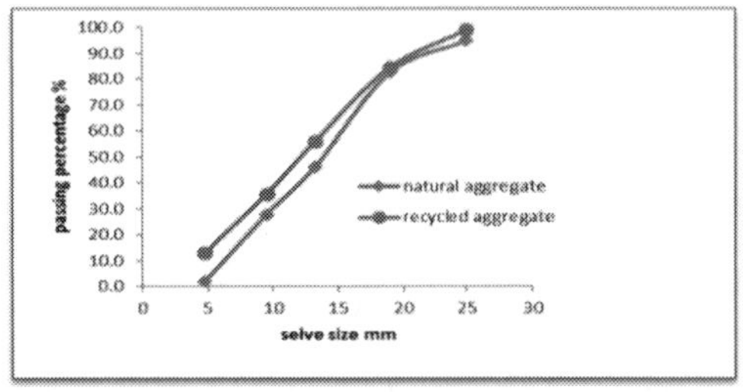


**Fig. 1.** Gradation curves for natural and recycled aggregate

## 4.2 Specimen preparation

In this work, several concrete cubes with (150) mm have been made with concrete mix ratio 1:2:4. These cubes have been divided into five groups according to recycled aggregate and admixtures percentage, the replacement percentage of recycled aggregate studied in this work are 0%, 50% and 100%, while admixtures percentage added are 0% and 10%. The five groups can be described as follows:

i) Group A: Concrete cubes made with 0% replacement percentage of recycled aggregate (100% natural aggregate), 0% added admixtures and slump value about 10mm.

ii) Group B: Concrete cubes made with 50% replacement percentage of recycled aggregate and 50% of natural aggregate, 0% added admixtures and slump value about 7mm.

iii) Group C: Concrete cubes made with 100% replacement percentage of recycled aggregate (0% natural aggregate), 0% added admixtures and slump value about 5mm.

iv) Group D: Concrete cubes made with 50% replacement percentage of recycled aggregate and 50% of natural aggregate, 10% added admixtures and slump value about 12mm.

v) Group E: Concrete cubes made with 100% replacement percentage of recycled aggregate (0% natural aggregate), 10% added admixtures and slump value about (12) mm.

## 4.3 Tests of Concrete Specimens

All concrete cubes specimens have been tested for compressive strength at 28 days. Hydraulic test machine (ELE International ADR 3000) as shown in Figure (2) is used in these tests. Figures (3) and (4) show concrete cubes during test procedure.

Table (2) summarizes the concrete compressive strength test results with corresponding replacement percentages of recycled aggregate and Silica Fume admixture.

The effects of replacement percentage of recycled aggregate and the use of silica fume admixture on concrete compressive strength have been illustrated in Figure (5) and Figure (6).

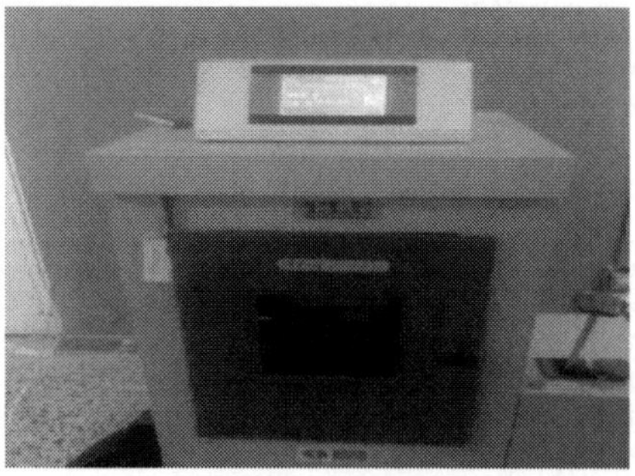

**Fig. 2.** Compressive strength test machine

**Fig. 3.** Group A concrete cube during the test

**Fig. 4.** Group B concrete cube during the test

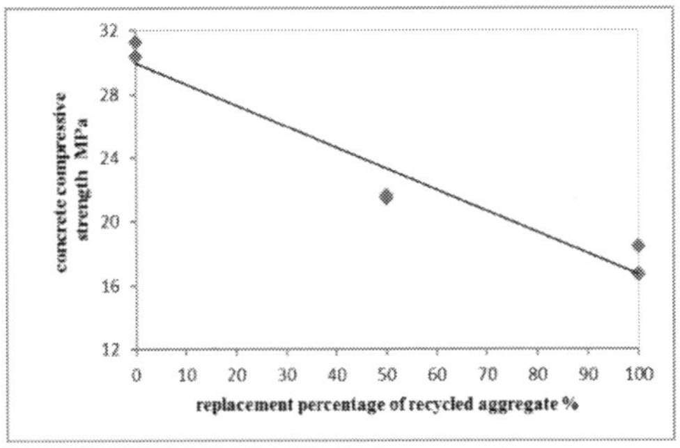


**Fig. 5.** Concrete compressive strength VS. Replacement percentage of Recycled aggregate (without silicafume)

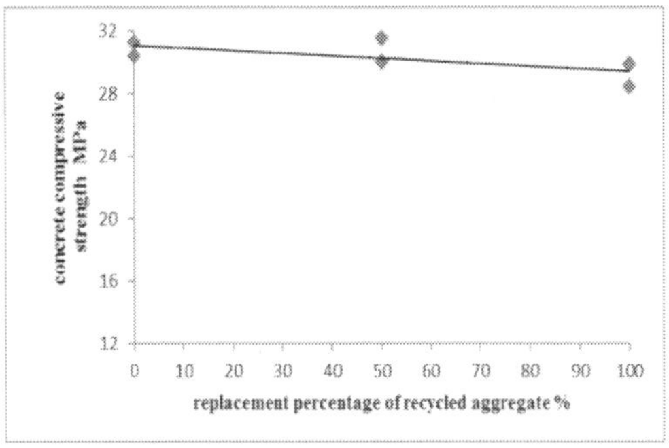


**Fig. 6.** Concrete compressive strength VS. Replacement percentage of Recycled aggregate (with silica fume)

## 5 Results and discussion

The main objective of this work is to determine the compressive strength of concrete. Concrete mixes with compressive strength of 30MPa are made using recycled aggregate with percentage (0, 50, and 100) % of the total coarse aggregate with and without adding silica fume. The use recycled aggregate as an aggregate cause a reduction in the compressive strength of concrete in comparison with concrete made with natural aggregate only, these percentages of reduction are (30, and 43) % for (50, and 100)% replacement of natural aggregate. This might be due to weakness of the recycled aggregate that the failure being in it.

Addition of silica fume to the recycled concrete aggregate cause an increase in the concrete compressive strength. This increase in the concrete compressive strength is due to its pozzolanic action which coming from its reaction with calcium hydroxide in the matured cement leading to production of cementitious material.

## 6 Conclusions

The main conclusions obtained from this work are summarized as follow:

1) The use of recycled aggregate in construction reduces the impact of waste material on environment, also reduces the excavation and transportation of natural aggregate from natural resources.

2) The compressive strength of natural aggregate concrete was higher than the compressive strength of recycled aggregate concrete.

3) The decrease in compressive strength with 50% replacement of aggregate concrete is 30%, while this percentage of decrease is 43% for 100% replacement of aggregate concrete.

4) The increase in compressive strength is observed to be about 42.5% with 50% replacement of aggregate concrete due to addition of 10% silica fume, while this increase in compressive strength is 65% for 100% replacement of aggregate concrete.

## References

1) A. M. Neville and J. J. Brooks, second edition, (2010).
2) Mirjana Malesev , Vlastimir Radonjanin and Snezana Marinkovic, Sustainability, **2**, (2010).
3) Nik. D. Oikonomou, Cement & Concrete Composites **27**, (2005).
4) How-Ji Chen, Tsong Yen and Kuan-Hung Chen, Cement and Concrete Research, **33**, (2003).
5) Qiong Liu, Jianzhuang Xiao, Zhihui Sun, Cement and Concrete Research, **41**, (2011).
6) M. Etxeberria, E. Vazquez, A. Mari, M. Barra, Cement and Concrete Research, **37**, (2007).
7) V. Letelier, E. Tarla, G. Moriconi, Sustainable Civil Engineering Structures and Construction Materials, SCESCM, (2016).
8) K. Zielinski, Modern Building Materials Structures and Techniques, MBMST, (2016).
9) Iraqi standard no. 5/1984, the Central Organization for Standardization and Quality Control "COSQC', Baghdad, (1984).
10) Iraqi standard no. 4, the Central Organization for Standardization and Quality Control "COSQC', Baghdad, (1984).

**Table 2.** Concrete compressive strength test results

| No. | Group | Replacement percentage of recycled aggregate % | Silicafume admixture % | Average concrete compressive strength (MPa) |
|---|---|---|---|---|
| 1 | A | 0 | 0 | 30.9 |
| 2 | B | 50 | 0 | 21.6 |
| 3 | C | 100 | 0 | 17.6 |
| 4 | D | 50 | 10 | 30.8 |
| 5 | E | 100 | 10 | 29.2 |

# Relationship between amorphous silica in source materials and compressive strength of geopolymer concrete

*Basil* Al-Shathr[1], *Mohamed* Shamsa[1,*], and *Tareq* al-Attar[1]

University of Technology, 10066, Baghdad, Iraq

**Abstract.** Geopolymer is a new sustainable binding material. It was developed to reduce $CO_2$ footprint of existing Portland cement concrete. One ton of Geopolymeric cement generates 0.18 ton of $CO_2$ from combustion carbon-fuel. This figure is 6 times less than the emission of Portland cement manufacture. The relationship between the compressive strength of Geopolymer concrete and the percentage of amorphous silica in the source material has been studied in the present work. Six mixes with different source materials were investigated to verify this relationship. The used Pozzolanic materials were three types of Fly ash, two types of Metakaolin and one type of ground granulated blast furnace slag. Geopolymer concrete samples were cured by heating for 72 hours. The testing ages for compressive strength were 7, 14, 28, and 60 days. The results showed that a noticeable relationship between compressive strength and amorphous silica was observed. The microstructure of the six mixes was studied in detail through the SEM and XRD analysis.

## 1 Introduction

Davidovits [1] stated that the geopolymerization process involves a reaction between alkaline liquids and compounds containing alumina and silica to produce a binder. The binder is termed "Geopolymer". Unlike ordinary Portland /Pozzolanic cements, Geopolymer do not form calcium-silicate-hydrates (CSHs) for matrix formation and strength, but the aluminosilicate gel forms by Geogeopolymerization binds the aggregates and provides the strength to Geopolymer concrete. The strength of Geopolymer depends on the composition of its constituent materials [2]. Natural or by-product materials containing silicon (Si) and aluminum (Al) in amorphous form could be used as source materials for Geopolymer.

Palomo *et al.* [3] indicated that this new cementitious material is an amorphous alkali silicoaluminate (containing small proportions of quartz, mullite, and maghemite from the original Fly ash), similar to that obtained in the alkali activation of Metakaolin.

During the first step of the reaction, the amorphous portion of the Fly ash begins to dissolve; in accordance with the FTIR results, the Al in tetrahedral coordination reacts but the Al in octahedral coordination does not.

Duxson *et al.* [4] showed that the fundamental chemical and structural characteristics of Geopolymers derived from Metakaolin, Fly ash and slag are explored in terms of the effects of raw material selection on the properties of Geopolymer composites. It was shown that the raw materials and processing conditions are critical in determining the setting behavior, workability and chemical and physical properties of Geopolymeric products. The XRD diffractogram labeled b in Fig.1 is of a Geopolymeric gel with nominal composition $NaAlSiO_4$, synthesized from Metakaolin (diffractogram a) in a 35 wt% water activating solution at 40 _C, 1 day after synthesis. Although this material appears amorphous to XRD initially, evidence of the same reaction mechanism being responsible for the formation of Geopolymer and crystalline zeolite can be observed, with the formation of zeolite observed within 7 days of synthesis (diffractogram c). Similar results obtained when Fly ash (diffractogram d) is used as activated material, as seen in Fig.1 (diffractogram e and f).

The compressive strength of Geopolymer concrete is not dependent only on the presence of silica in raw materials, whereas the alumina has main function and active in strength acquisition. De Silva *et al.* [5] showed that setting time of the Geopolymer systems was mainly controlled by the alumina content. Essentially, the setting time increased with increasing $SiO_2/Al_2O_3$ ratio of the initial mixture. Up to a certain limit, the $SiO_2/Al_2O_3$ ratio also found to be responsible for observed high-strength gains at later stages. An increase in the $Al_2O_3$ content, i.e. for low $SiO_2/Al_2O_3$ ratio, led to products of low strength.

Saidi *et al.* [6] studied the effect of Si/Al and Na/Al ratios on mechanical properties of Metakaolin based Geopolymer. There results indicated that an increase of Si/Al ratio improves the Geopolymer mechanical properties due probably to the formation of Si-O-Si bond in the Geopolymer at the expense of Si-O-Al one. It was also shown that an introduction of NaOH in the Geopolymer composition enhances the mechanical properties until a molar ratio of Na/Al=1. Beyond this value, the excess of Na+ cation causes fragility of the

* Corresponding author: mhshamsa990@gmail.com

© The Authors, published by EDP Sciences. This is an open access article distributed under the terms of the Creative Commons Attribution License 4.0 (http://creativecommons.org/licenses/by/4.0/).

Geopolymer structure confirmed by a decrease of the mechanical strength.
The objective of this paper was to study the effect of amorphous silica in different source materials on the strength development of Geopolymer concrete.

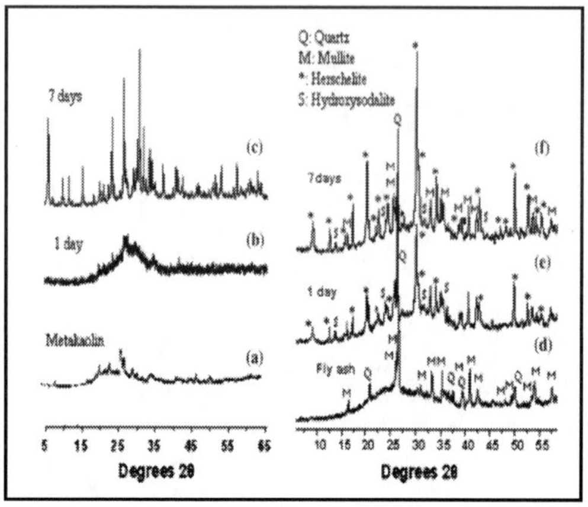


**Fig. 1.** XRD diffractogram of (a) typical Metakaolin (with slight muscovite impurity), (b) Geopolymer with nominal composition $NaAlSiO_4$ obtained after 1 day and (c) 7 days after synthesis, showing peaks corresponding to zeolite X, (d) typical Fly ash, (e) inorganic polymer from Fly ash at 20 h, 85°C and (f) 7 days at 85°C [4]

## 2 Experimental work

### 2.1. Materials

The following materials have been used in the production of Geopolymer concrete:

### 2.1.1 Source materials (Pozzolanic materials)

- Three types of Fly ash used (FA1) from China and (FA2, FA3) from electrical power plant in Turkey.
- Two types of Metakaolin (MK1, MK2); the M.K. derived from kaolin by calcination at 700°C for 2 hours. The raw kaolin brought from Al-Ramadi region in Iraq.
- Ground granulated blast furnace slag (GGS).

The chemical composition of the Pozzolanic materials determined by XRF analysis is given in Table (1). Fig. 2. shows the XRD patterns of these source materials.

### 2.1.2 Alkaline solution

Alkaline solution was obtained by blending sodium silicate and sodium hydroxide solutions.
- Sodium silicate with chemical composition of $Na_2O$ = 13.5%, $SiO_2$ = 32.5%, $H_2O$ = 54%.
- Sodium hydroxide NaOH flakes with 97-98% purity used. The sodium hydroxide solution was prepared by dissolving the NaOH flakes in tap water to produce an 8M solution.

### 2.1.3 Aggregate

Crushed gravel with 12.5 mm maximum size used, as coarse aggregate, natural graded sand was the fine aggregate according to ASTM C33 [7], as shown in Tables 2 and 3. $SO_3$ contents were 0.39 and 0.07 % for fine and coarse aggregate respectively.

### 2.1.4 High-range water reducer

A high range water reducer superplasticizer (KUT PLAST SP400) based on modified sulphonated naphthalene formaldehyde condensate was used to enhance workability of Geopolymer concrete.

**Table 1.** Chemical composition of the used Pozzolanic materials.

| Oxides % | FA1 | FA2 | FA3 | MK1 | MK2 | GGS |
|---|---|---|---|---|---|---|
| $SiO_2$ | 50.5 | 60.6 | 63.0 | 53.30 | 56.77 | 30.7 |
| $Al_2O_3$ | 22.7 | 31.7 | 27.1 | 34.63 | 30.85 | 13.3 |
| $Fe_2O_3$ | 6.33 | 3.98 | 4.12 | 2.21 | 2.48 | 0.35 |
| CaO | 13.8 | 0.98 | 1.20 | 0.94 | 0.58 | 42.4 |
| MgO | 1.22 | 0.72 | 0.74 | 0.32 | 0.59 | 6.89 |
| Others | 5.42 | 1.98 | 3.71 | 8.6 | 8.73 | 6.32 |

**Table 2.** Sieve analysis of coarse aggregate

| Sieve size ( mm) | Passing (%) | ASTM C-33[6] Requirements |
|---|---|---|
| 19 | 100 | 100 |
| 12.5 | 98.5 | 90-100 |
| 9.5 | 63.9 | 40-70 |
| 4.75 | 2.7 | 0 – 15 |
| 2.36 | 0.2 | 0 – 5 |
| Percentage materials less than 75 micron | 0.66 | 1 % upper limit |

**Table 3.** Sieve analysis of fine aggregate.

| Sieve size ( mm) | Passing (%) | ASTM C-33[6] Requirements |
|---|---|---|
| 9.5 | 100 | 100 |
| 4.75 | 98.2 | 95 – 100 |
| 2.36 | 90.8 | 80 – 100 |
| 1.18 | 73.3 | 50 – 85 |
| 0.6 | 52.3 | 25 – 60 |
| 0.3 | 15.1 | 5 – 30 |
| 0.15 | 4.2 | 0 – 10 |
| Percentage materials less than 75 micron | 1.98 | 3% upper limit |

### 2.2 Mixing and specimen preparation

Six series of Geopolymer concrete mixes were produced by varying the Pozzolanic material type, as shown in Table 4.
First the alkaline activators with 8 M (sodium silicate, sodium hydroxide, water and superplasticizer) were

prepared. The proportion of NaOH to $Na_2SiO_3$ was 1:1.5 for FA and GGS mixes, while it was 1:2 for MK mix to obtain workable mixes. The dry materials (Pozzolanic material, fine and coarse aggregates) were mixed in a mixer for 4 mins to have the required homogeneity. The alkaline activators were then poured into the mixer and mixed continuously for approximately 4 mins until the concrete was well-mixed. The fresh concrete then was cast in two layers in steel cylinder molds of 100 mm diameter and 200 mm height and vibrated for two mins for each layer on vibrating table to remove entrapped air.

**Table 4.** Mix proportions of Geopolymer concrete.

| Materials (kg/m$^3$) | FA1, FA2, FA3 | MK1, MK2 | GGS |
|---|---|---|---|
| Pozzolanic material | 400 | 400 | 400 |
| NaOH | 19 | 26 | 36 |
| Sodium Silicate | 111 | 200 | 150 |
| Water | 55 | 74 | 65 |
| Fine aggregate | 650 | 650 | 650 |
| Coarse aggregate | 1200 | 1200 | 1200 |
| Superplasticizer | 12 | 12 | 12 |

### 2.3 Curing and testing procedures

The molded specimens were placed in the oven for 24 hours, after that they were taken out and removed from their molds. After that, they additionally cured in the oven for another 48 hours. Fly ash based Geopolymer, GGBS based Geopolymer cured at 70±5° C and the Metakaolin based Geopolymer cured at 45±5° C, according to the recommendations of Alhifadhi [8] and Al-Shathr *et al.* [9]. The specimens removed out and allowed to cure at room temperature (20±5°C) until testing their compressive strength using a digital compression testing machine at 3, 7, 28, and 60 days age. The reported compressive strength is average strength of three specimens.

## 3 Results and discussion

### 3.1 Compressive strength

Table 5 and Fig. 2 indicate the compressive strength of Geopolymer concrete mixes. In all cases, the strength of the concrete increased with age. The rate of strength gain is high at early ages and gradually decreases at later ages.

Five types of Geopolymer concrete studied to know the effect of amorphous silica content in the raw materials on compressive strength development. The three types of Fly ash based Geopolymer, FA1*, FA2* and FA3*, have differences in their compressive strength at all ages. Fly ash used in mixes FA2* and FA3* have approximately the same chemical compositions and surface area, as shown in Tables 1 and 6. Therefore, it was expected that both mixes should have a similar strength.

**Table 5.** Compressive strength of Geopolymer concrete mixes.

| Age days | Compressive strength of concrete mixes containing pozzolanic material type, MPa | | | | | |
|---|---|---|---|---|---|---|
| | FA1 | FA2 | FA3 | MK1 | MK2 | GGS |
| 7 | 44.7 | 34.49 | 40.56 | 19.38 | 22.70 | 40.00 |
| 14 | 53.92 | 46.21 | 52.12 | 22.12 | 27.84 | 48.40 |
| 28 | 65.40 | 52.28 | 61.02 | 26.57 | 35.44 | 52.66 |
| 60 | 65.27 | 52.35 | 61.75 | 27.36 | 36.10 | 51.96 |

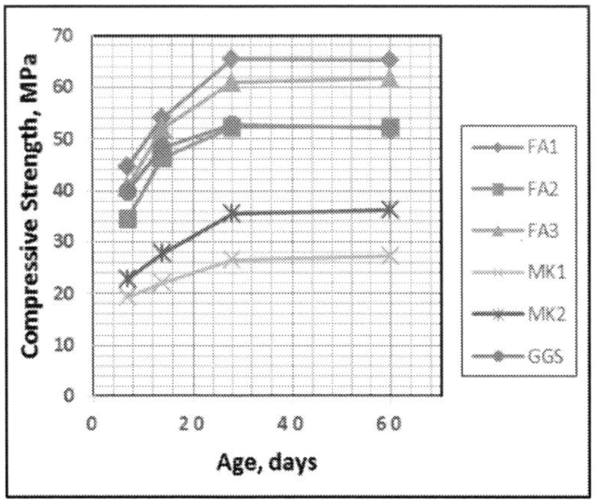


**Fig. 2.** Effect of age on compressive strength of Geopolymer concrete mixes

From the XRD results for raw materials, shown in Fig. 3, there is a difference in the amount of amorphous silica which is indicated by the hump in the chart. It is proposed that these differences could be the result of varying furnace temperatures or changes in the feed coal composition. The Si/Al ratio in FA2 mix is lower than FA3, so according to De Silva *et al.* [5] the $SiO_2/Al_2O_3$ ratio could be responsible for the observed high-strength gain. On the other hand, the higher content of calcium compounds in FA1 compared to others is thought to be the second factor that was attributing to the faster strength development of mix containing FA1. The formation of calcium silicate, calcium aluminate hydrates [10], and calcium–silico–aluminates [11] is possible, and in many cases, the high early strengths achieved with blended Geopolymeric cements ascribed to the formation of these compounds. Further evidence that supports this statement is the slag based Geopolymers, GGS*, which contains both amorphous and semi crystalline newly formed phases contained calcium.

The GGS* mix showed higher strength than FA2* mix especially at early age, 14 days, and that could be due to high content of calcium compounds in GGS.

**Table 6.** Specific surface area of the used Pozzolanic materials.

| Material | FA1 | FA2 | FA3 | MK1 | MK2 | GGS |
|---|---|---|---|---|---|---|

https://doi.org/10.1051/matecconf/201816202019

| | | | | | | |
|---|---|---|---|---|---|---|
| Fineness ($m^2/kg$) | 765 | 773 | 778 | 16800 | 17250 | 681 |

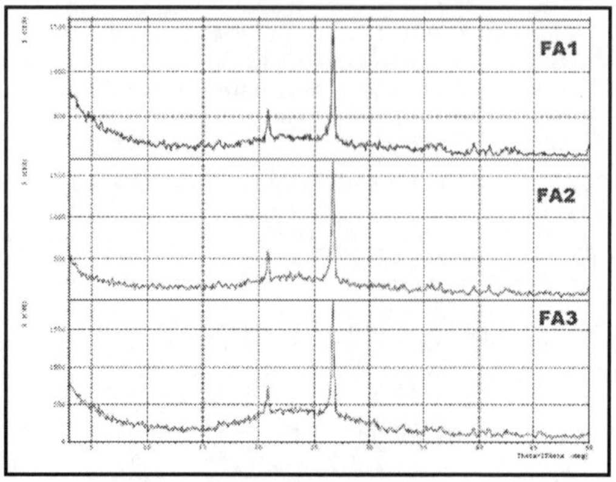


**Fig. 3.** XRD pattern of the used Fly ashes.

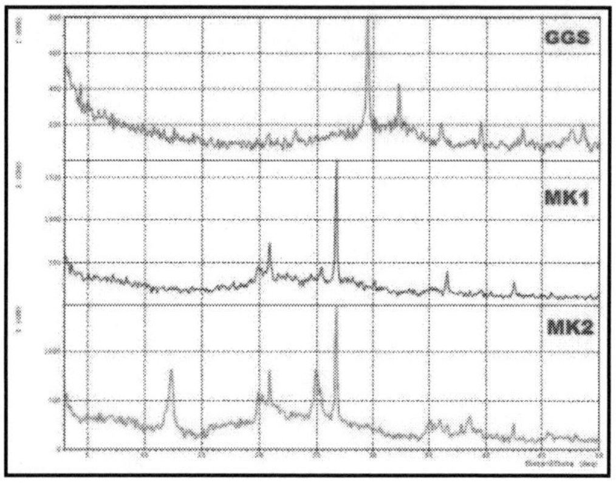


**Fig. 4.** XRD pattern of the used Metakaolin and GGS.

In MK1 and MK2 mixes [6] the results of XRD analysis of Metakaolin as shown in Fig. 4 displayed a revealing hump pattern in the range of 15° to 30° (2θ). This hump indicates that these materials are amorphous.

Some peaks related to crystallized phases were also detected, which is attributed to quartz and elite initially present in the kaolin and which did not contributed to the geopolymerization. The diffractogram of the MK2 shows some peaks which are not present in the MK1 spectra which are indicating more orderly phases in this material.

### 3.2 Geopolymerization products and microstructure

The products of geopolymerization of concrete mixes can be generally traced by using XRD and EDX analysis combined with scanning electron microscope, SEM, images. Figs. 5 and 6 illustrate the XRD analysis of Fly ash, Metakaolin, and GGS based Geopolymers in 60 days age.

Main crystalline phases observed in the Fly ash Geopolymer samples are quartz, mullite and are largely unaffected by the geopolymerization process. By linking the results between XDR analysis and ESM images and EDX analysis, it could be noticed that the FA1* sample showed crystalline needles Fig. (7-a, b), especially in early ages, similar to the needles formed during the reaction of the Portland cement. This may be due to the high CaO ratio, which was 14% in the XRF analysis, Table 1. Meanwhile, its ratio was in the EDX analysis for Geopolymer concrete is 3.25, which is a high ratio compared to the samples FA2* and FA3*, which were 1.83 and 1.62 respectively.

From the XDR analysis at 60 days for mix FA1* Geopolymer concrete sample Fig. 5, it is worthy to point out to the emergence of peaks representing the formation of $C_2S$ gel in 2θ equals to 28° and 46.5°.

From the XDR analysis of the raw material, Fig. 3, the sample FA2, had higher percentage of the crystallized silica than FA3. After two months, the XRD analysis shows the high rise in the intensity of the quartz for FA2* Geopolymer concrete. The SEM images for FA2* in Fig.7-c, d showed the appearance of the crystallized material clearly and the emergence of many voids in early ages that may reduce the compressive strength.

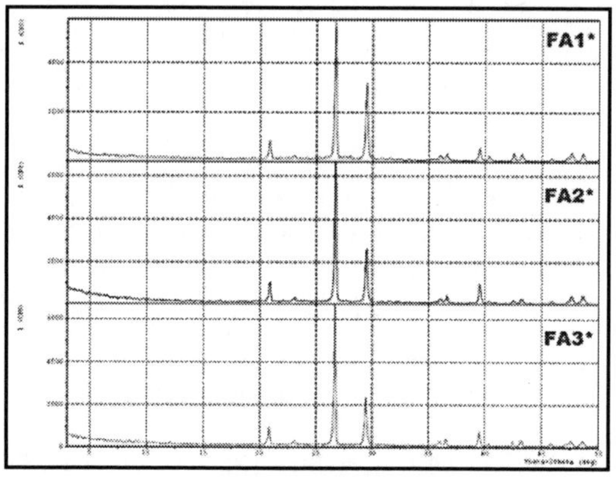


**Fig. 5.** XRD of Fly ash Geopolymer concrete after 28 days.

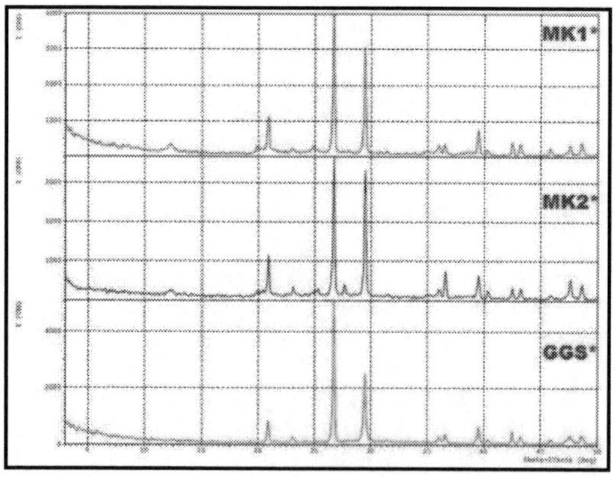


**Fig. 6.** XRD pattern of Metakaolin and GGS Geopolymer concrete at age of 28 days

The increase of the elements (Si, Al, Na, and Ca) in the EDX analysis indicates an increase in compressive strength, where, at low Si, low compressive strength achieved only if the silica already present in the crystalline material is not involved in the geopolymerization process. When comparing the FA2* sample FA3* samples, it could be noted that the ratios of the aforementioned elements in the EDX analysis of FA3* sample are higher, resulting in increased compressive strength during the age of the examination.
The XRD analysis of the raw Metakaolin samples in Fig. 4 showed that MK2 has a higher ratio of the amorphous silica, which is clear from the hump in the pattern. When analyzing two month old Geopolymeric concrete samples, the crystallization intensity of MK1* is much higher than that of MK2*. This indicates that silica, which crystallized in MK1, did not contribute to geopolymerization, resulting in reduced compressive strength as observed in Table 5 in all ages.
SEM images in Fig. 8 a-d showed a clear manifestation of the amount of silica crystallized in the MK1* sample. From the EDX analysis, it could be noted that the ratio of elements Si, Al, Na, and Ca in the MK2* sample is higher than MK1* at the age of two weeks. In a system with a high Si concentration, condensation starts with the formation of oligomeric silicates, leading to 3-D rigid polymeric structures. Low Si concentration results in poly-sialate polymer structures, and confirms that crystallized silica originally present in MK1 did not contribute to geopolymerization so the Si percent was high in two months old. In addition, the Si / Al ratio may be considered as the second indicator on the increase in strength, it is clear from the EDX analysis that the ratio in the sample MK2* at the age of 2 months was about 3.43 and about 1.92 in the MK1* sample.
For the GGS* sample, calcium oxide plays a major role in the process of strength development. Moreover, the absence of quartz in the XRD analysis of GGS material in Fig. 4 is evidence on the process of consumption of amorphous silica. According to SEM images and EDX analysis of the GGS* Geopolymer concrete samples, Figs. 10-e, f, there is no quartz and low Si ratio during the age, where 19.01 at two weeks to 9.92 at the age of two months. The $C_3S$ and $C_2S$ gel as evident in the XRD analysis in Fig. 7 as well as the calcium aluminate hydration gel appearance at ($2\theta \approx 42.5°$).

## 4 Conclusions

The following conclusions can be drawn from this study:
1. The ratio of amorphous silica in the raw Pozzolanic material has a significant effect on the compressive strength of the copolymer concrete.
2. The ratio of quartz (crystalline silica) found in the raw Pozzolanic material does not contribute to the process of geopolymerization and therefore will be high also after the formation of Geopolymer concrete.
3. The ratio of Ca in the raw material has an effect on increasing compression resistance in early ages, as it contributes to the formation of gel-like gel resulting from the hydration of cement.
4. Increasing the Si, Al, Na, Ca elements in the Geopolymer concrete has an indication of increased compressive strength.
5. The high Si ratio of EDX analysis of the Geopolymer concrete is not necessarily indicative of increased compressive resistance unless the silica present in the raw material is amorphous silica.
6. The properties of a Geopolymer can be significantly by relatively minor changes in Si and Al concentrations during synthesis, where increasing the Si / Al ratio to a certain extent leads to increased compressive resistance.

## References

1. J. Davidovits, *Geopolymer chemistry and applications*, (Inst. Geop., France 2011)
2. S. Wallah, B. Rangan, Curtin Uni. Tech. (2006)
3. A. Palomo, M. Grutzeck, M. Blanco, Cem. Conc. Res, **29**, 8 (1999)
4. P. Duxson, A. Fernández-Jiménez, J. Provis, G. Lukey, A. Palomo, J. van Deventer, J. Mat. Sci., **42**, 9 (2007)
5. P. De Silva, K. Sagoe-Crenstil, V. Sirivivatnanon, Cem. Conc. Res. **37**, 4 (2007)
6. N. Saidi, B. Samet, S. Baklouti, Int. J. Mat. Sci., **3**, 4 (2013)
7. ASTM C33-15, Annual Book of ASTM Standards (2015)
8. M. Alhifadhi, MSc thesis, Uni. Tech., Iraq (2015)
9. B. Al-Shathr, T. al-Attar, Z. Hasan, J. Babylon Uni. **24** (2016)
10. M. Luxán, M. Sánchez de Rojas, M. Frías, Cem. Conc. Res., **19**, 1 (1989)
11. J. Davidovits, J. Mat. Edu.**16** (1994)

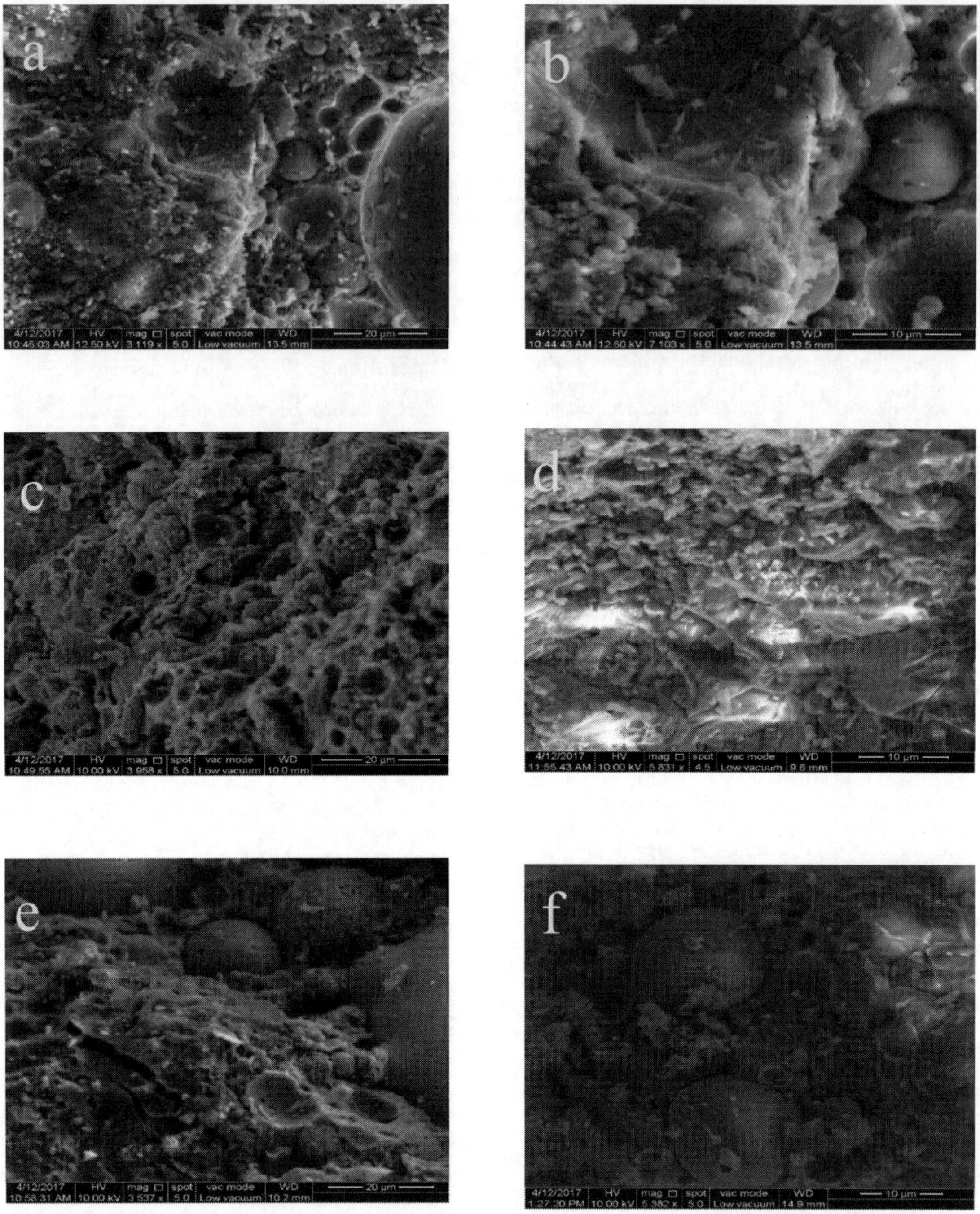


**Fig.7.** SEM pictures for fly ash based Geopolymer concrete; (a) FA1* at 14 days, (b) FA1* at 60 days, (c) FA2* at 14 days, (d) FA2* at 60 days, (e) FA3* at 14 days, and (f) FA3* at 60 days.

https://doi.org/10.1051/matecconf/201816202019

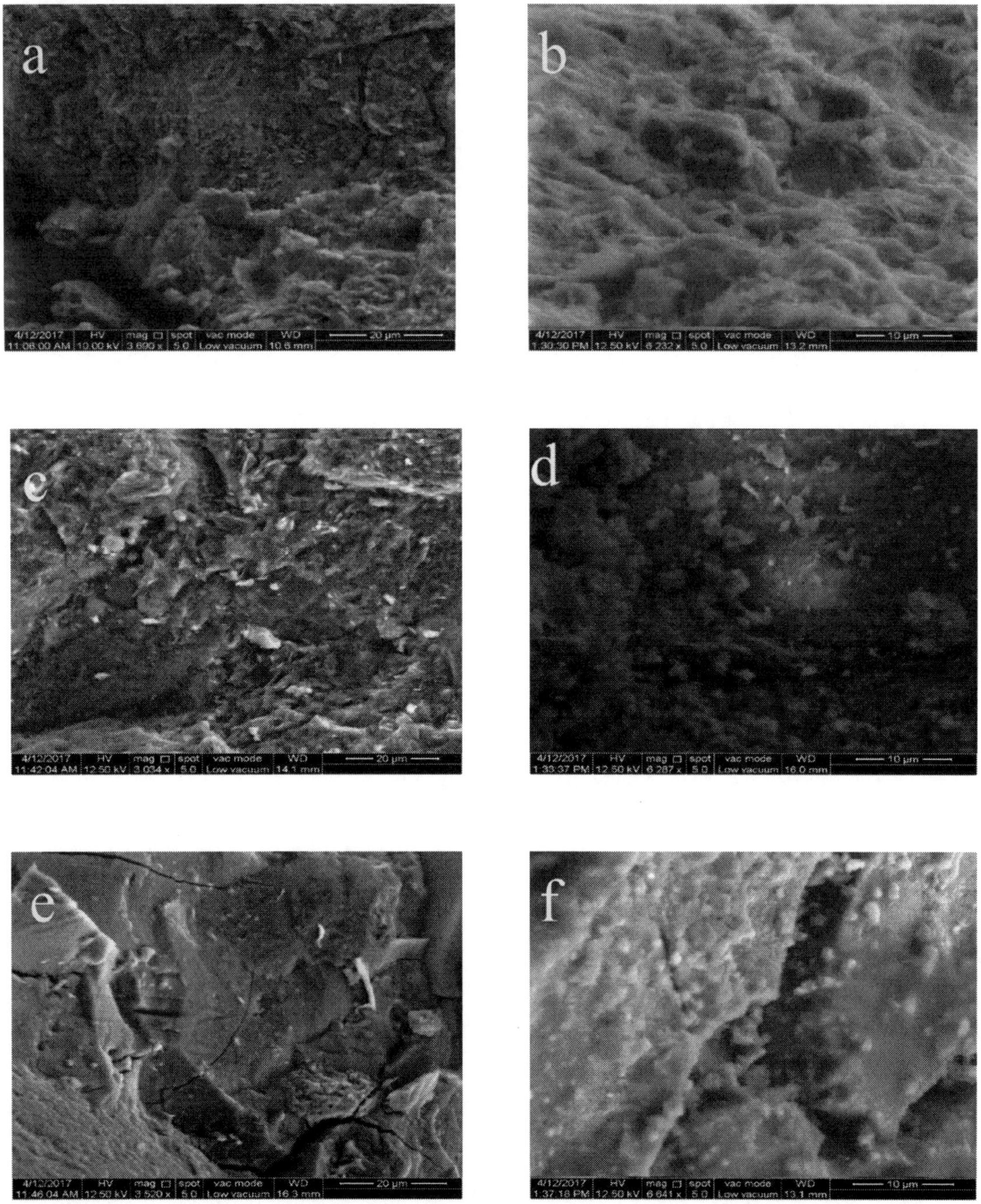


**Fig.8.** SEM pictures for metakaolin and GGS based Geopolymer concrete; (a) MK1* at 14 days, (b) MK1* at 60 days, (c) MK2* at 14 days, (d) MK2* at 60 days, (e) GGS* at 14 days, and (f) GGS* at 60 days.

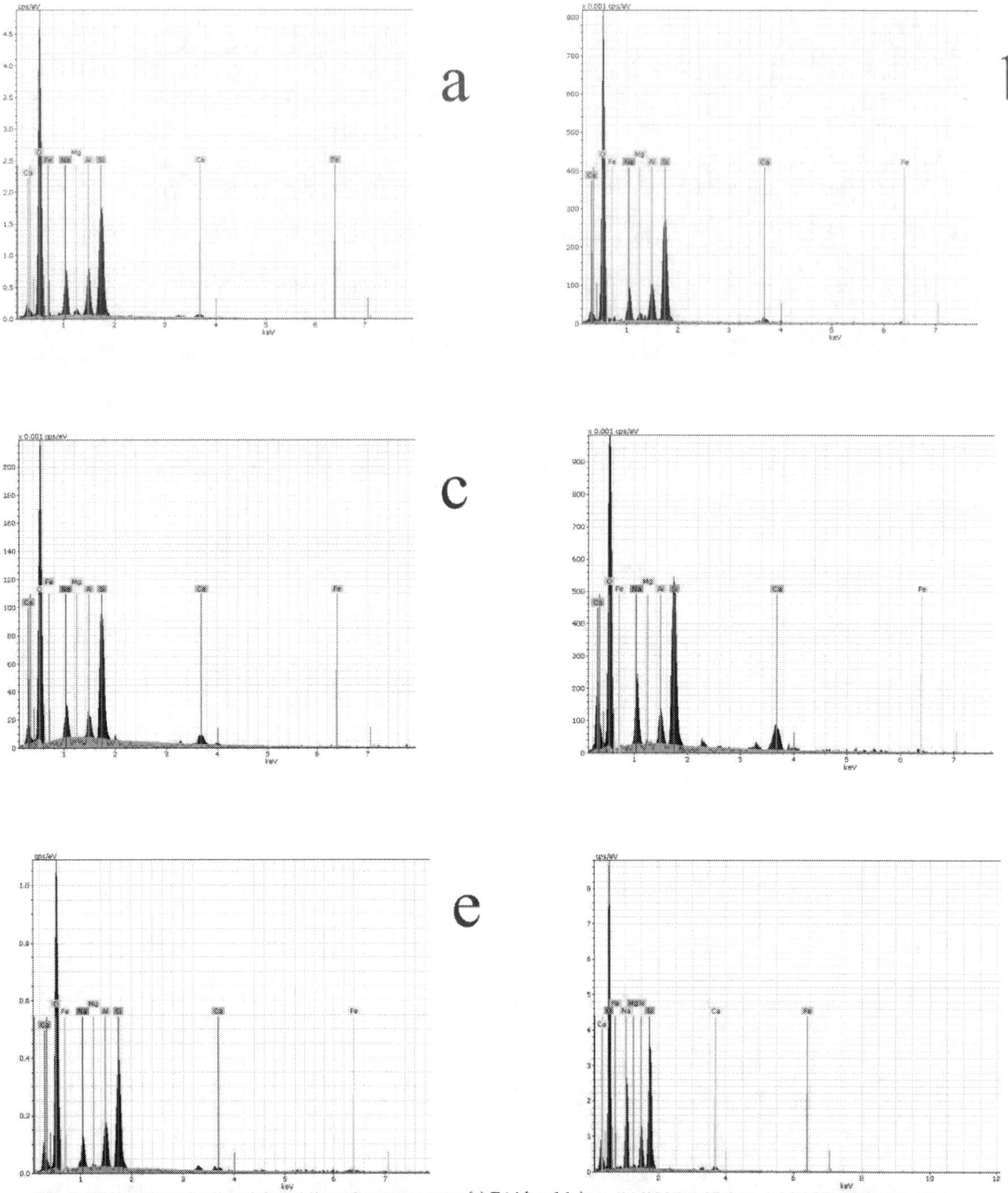


**Fig. 9.** EDX analysis for fly ash based Geopolymer concrete; (a) FA1* at 14 days, (b) FA1* at 60 days, (c) FA2* at 14 days, (d) FA2* at 60 days, (e) FA3* at 14 days, and (f) FA3* at 60 days.

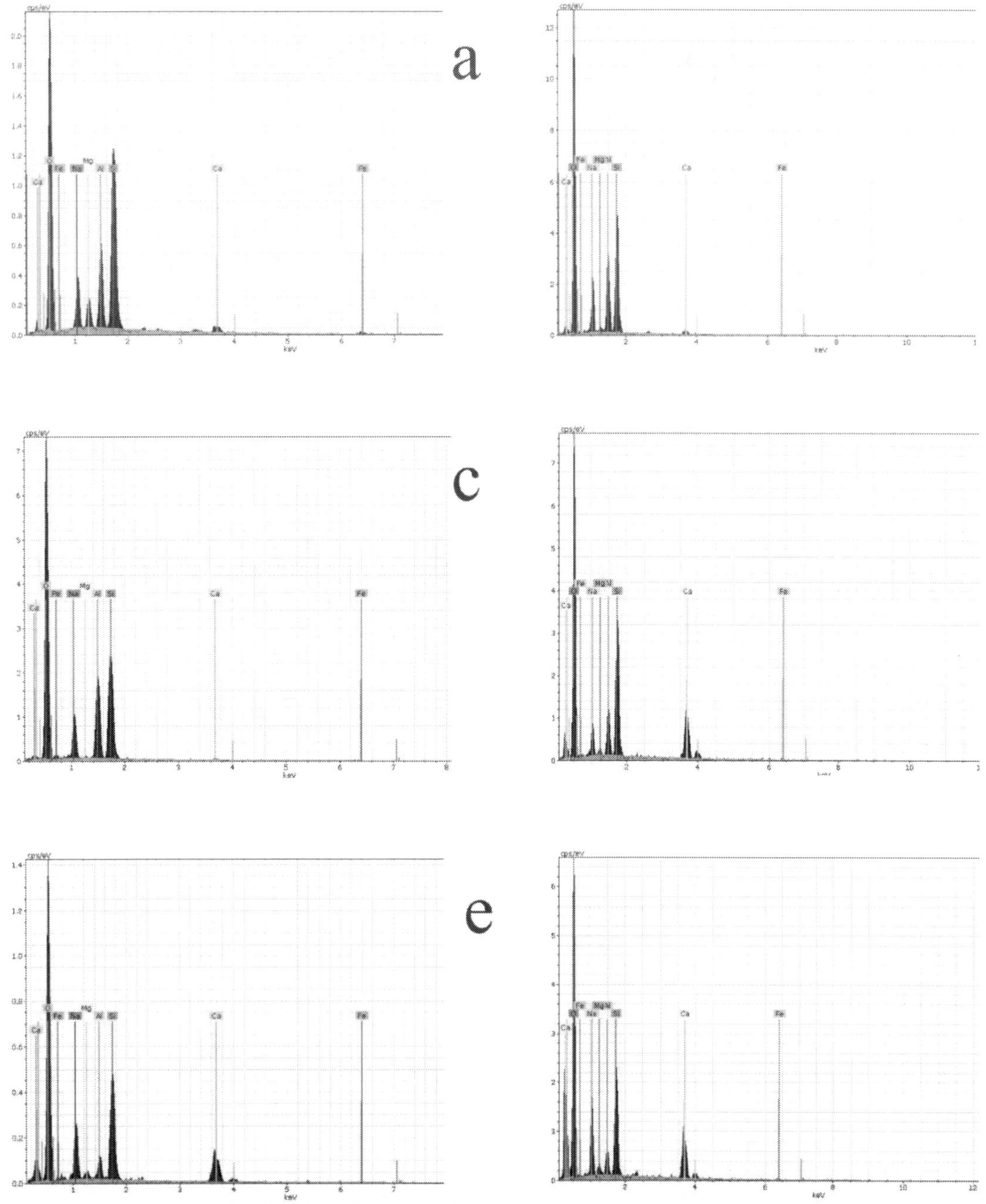


**Fig. 10.** EDX analysis for metakaolin and GGS based Geopolymer concrete; (a) MK1* at 14 days, (b) MK1* at 60 days, (c) MK2* at 14 days, (d) MK2* at 60 days, (e) GGS* at 14 days, and (f) GGS* at 60 days.

# The influence of recycled concrete aggregate on the properties of concrete

*Nisreen* Mohammed[1], *Kaiss* Sarsam[1] and *Mazin* Hussien[1]
[1] Building and Construction Engineering Department, University of Technology, Baghdad, Iraq

**Abstract.** Use of Recycled Coarse Aggregate (RCA) in concrete can be described in terms of environmental protection and economy. This paper deals with the mechanical properties of concrete compressive strength, splitting tensile strength, modulus of elasticity, and modulus of rupture. Three kinds of concrete mixtures were tested, concrete made with Natural Coarse Aggregate (NCA) as a control concrete and two types of concrete made with recycled coarse aggregate (50% and 100% replacement level of coarse recycled aggregate). These kinds of concrete were made with different targets of compressive strength of concrete f 'c (35MPa) and (70 MPa). Fifty specimens were tested of the fresh and hardened properties of concrete. The waste concrete from laboratory test cubes was crushed to produce the Recycled Coarse Aggregate used in recycled concrete. A comparative between the experimental results of the properties for fresh and hardened concrete is presented in the paper. Recycled aggregate concrete (RCA) had a satisfactory performance despite the replacement ratios. It was found using the size of Recycled Coarse Aggregate (RCA) of (5-14) mm has quite similar in performance with the same size of Natural Coarse Aggregate (NCA), it is necessary to use high quality of recycled concrete (with low levels of impurities). Recycled aggregate as an alternative to natural aggregates -seems quite successful.

## 1 Introduction

USE of recycled aggregate (RA) in concrete is to be considered as a trend for environment protection and to support the economy. The application of recycled aggregate as being used in construction since the end of World War II, by demolished concrete pavement as recycled aggregate in stabilizing the base course for road construction (Olorusongo, F.T., 1999) [1]. These are many advantages of using recycle aggregate in construction industry such as environmental friendliness and economic sense. The wastes from demolition and construction works represent a big problem as it gets
accumulated gradually and increases with time, so the recycled aggregate the perfect solution for this problem. The sustainable concrete construction is to be considered as a main strategy for the construction industry. It is important to reduce energy consumption that is associated with carbon dioxide and gas emissions, which cause ‘‘ greenhouse effect”, Portland cement production releases huge amount of carbon dioxide and other greenhouse gases (GHG) in the atmosphere (Malhotra and Bilodeau, 1999) [2]. The construction waste produced from building demolition alone is estimated to be 123 million tons per year in the USA [3]. In Iraq the amount of the construction waste amounts to 1,111,788 tons/year (the Iraqi environmental ministry). Since the materials waste is increases with the growing population and increasing urban cities, an investigation and research on the usage of waste construction materials is significant. Therefore, many research and analysis have been made on recycled aggregate. Also the recycled aggregate is less expensive and easy to obtain as compared with normal aggregate. In order to reduce the usage of natural aggregate, recycled aggregate can be used as the replacement materials (N.Sivakumar et al. 2014) [4].

## 2 Literature Review

Many researches are involved interests in previously used processing of demolished concrete, the mixture design including the physical and the mechanical properties in addition to the durability process. Most of these researches are widely being studied and tested such as Hansen [5,6] and ACI 555 reported [7], which showed that some Characteristics of Recycled Aggregate Concrete (RAC) may be normally lesser than the ordinary concrete, but in spite of that, it gives satisfactory result for some practical usages related to Civil Engineering. The key mechanical properties for (RAC) are represented by compressive strength, flexural strengths, modulus of elasticity, and the tensile strength of such concrete. In present years, many investigations have also been performed for the Mechanical properties of (RAC). Bairagi and Kishore [8], found a comparable ratio of strain–stress relationship for recycled aggregate concrete to that of normal concrete. Poon and Lam [9], showed that a

© The Authors, published by EDP Sciences. This is an open access article distributed under the terms of the Creative Commons Attribution License 4.0 (http://creativecommons.org/licenses/by/4.0/).

particular amount of the mortar and cement paste from original concrete when smashed concrete is crushed and remains stuck to the stone particles of the recycled aggregate, thus the attached mortar is the main cause for lower quality of (RCA) compared to (NCA).. The density of (RCA) is decreased to 10% compared with the density of (NCA). (Rahal, 2007) [10] , Refer to the range of water absorption of coarse (RCA) which reaches from 3.5% to 9.2%. (Lopez-Gayarre et al., 2009) [11], Showed that the resistance of the abrasion (Los Angeles test) for (RCA) is lowered up to 70% as compared that for (NCA). (Sanchez de et a. 2004) [12].
Noticed that the recycled coarse aggregate has lesser specific gravity and bulk density than the natural coarse aggregate. (Yang, K.H. et al.,2008) [13]. The main conclusions that can be drawn are the concrete with recycled coarse aggregate decrease up to (25%, 10%, 45%) for compressive strength, flexural tensile strength and modulus of elasticity, respectively as compared to those with concrete containing natural coarse aggregate. (BS 8500-1, 2006) [14], explain Recycled Aggregate (RA) is defined as the generic term for aggregate result from the reprocessing of inorganic material previously used in construction. American Concrete Institute (ACI 318) [15], specifies that aggregates must meet the requirements of ASTM C33 [16]. Neither (ACI 301) [17] nor (ACI 318) included specific provisions on using RCA in concrete. Guidance on use of recycled materials in concrete is mainly provided by ACI Technical Committee 555[18], in the report "Demolition and Reuse of Hardened Concrete".

## 3 Experimental work

The principal objective of this research work, the recycled aggregate collected from the waste concrete of laboratory test cubes (the debris have been selected, cleaned and sieved in laboratory). This paper deals with the study of compressive and splitting tensile strengths as well as modulus of elasticity, and modulus of rupture. Five types of concrete mixes were casted and made with f 'c (35 and 70) MPa, three for natural coarse aggregate (NCA) and two types of concrete made with recycled coarse aggregate (RCA) (50% and 100% replacement level of coarse recycled aggregate). Fifty specimens were tested of the fresh and hardened properties of concrete.

### 3.1 Material testing and Mix design

#### *3.1.1 Cement*

The cement used in casting all the specimens in this study is Ordinary Portland Cement (OPC) type (I), well-known (Mass), Al-Sulaimaniya, Iraq. It was tested according to (Iraqi Standard Specification (I.Q.S No.5:1984) [19] and complied with (ASTM C150-02), [20] physical properties of cement in Table (1).

**Table 1.** Physical Properties of Cement

| Physical Properties | Test result | Limits of Iraqi specification No.5:1984 |
|---|---|---|
| Fineness ($cm^2/g$) by Blaine method | 2632 | >2300 |
| Setting time (Vicat's method) Initial setting (min) Final setting(min) | 168<br>245 | >45 min<br><10 hrs |
| Compressive strength for cement mortar cube (70.7)mm at, MPa<br>3 days<br>7 days | 16.9<br>24.5 | >15<br>>23 |
| Soundness using Auto clave% | 0.2 | <0.8 |

#### *3.1.2 Fine aggregate*

Natural sand from AL-Ukhaider region, Iraq, was used in this study. The fine aggregate grading was checked according to the I.Q.S (No. 45/1984) [21] Zone 2, and the British Standard Specification (B.S. 882:1992) [22]. Table (2) shows the physical properties of fine aggregate.

**Table 2.** Physical properties of fine aggregate

| Physical properties | Test result | Limit of Iraqi specification No.45:1984 |
|---|---|---|
| Specific gravity | 2.63 | - |
| Sulfate content | 0.39% | 0.5% (max) |
| Absorption | 1.75 % | - |
| Fineness modulus | 2.9 | - |

#### *3.1.3 Coarse aggregate*

The maximum size of (14 mm) crushed natural gravel was used in this work as saturated surface dry (SSD) condition, the limits confirm to the I.Q.S (NO.45/1984) and specifications (ASTM C33-02) [23]. Table (3) shows the physical properties of the natural coarse aggregate according to Iraqi specification (No.45:1984) [22].

**Table 3.** Physical properties of Natural Coarse Aggregate

| Physical properties | Test result | Limit of Iraqi specification No.45:1984 |
|---|---|---|
| Specific gravity | 2.65 | - |
| Absorption | 0.7% | - |
| Crushed value | 15% | - |
| Sulfate content | 0.06% | 0.1% (max) |

#### *3.1.4 Recycled coarse aggregate*

Recycled coarse aggregate used in this study was obtained by crushing waste concrete test cubes used in previous works to get small pieces with maximum size (14 mm) of recycled aggregate (the debris have been selected, cleaned and sieved in laboratory). The sieve analysis of recycled coarse aggregate (RCA) according to the I.Q.S (No.45: 1984), and (ASTM C33-02) specifications. Table (4) shows the physical properties of Recycled Coarse Aggregate according to the I.Q.S (No.45:1984). The recycled coarse aggregate is shown in Figure 1.

**Table 4.** Physical properties of recycled coarse aggregate

| Physical properties | Test result | Limit of Iraqi specification No.45:1984 |
|---|---|---|
| Specific gravity | 2.4 | - |
| Absorption | 3.6% | - |
| Crushed value | 21% | - |
| Sulfate content | 0.08% | 0.1% (max) |

**Fig. 1.** Recycled Coarse Aggregate

**Table 5.** Details of mixes

| Specimen | Compressive strength | Code | Cement $kg/m^3$ | Sand $kg/m^3$ | Gravel ($kg/m^3$) | | Water $l/m^3$ | Glenium51 $l/m^3$ | Silica fume $kg/m^3$ |
|---|---|---|---|---|---|---|---|---|---|
| | | | | | Normal | Recycled | | | |
| N35 | 35 MPa | 35N.C | 400 | 745 | 990 | - | 185 | 1.6 | - |
| | | 50%R35 | 400 | 745 | 495 | 495 | 185 | 2.1 | - |
| | | 100%R35 | 400 | 745 | - | 990 | 185 | 2.5 | - |
| H70 | 70 MPa | 70N.C | 450 | 750 | 1000 | - | 126 | 12.5 | 50 |
| | | 50%R70 | 450 | 750 | 500 | 500 | 126 | 16 | 50 |
| | | 100%R70 | 450 | 750 | - | 1000 | 126 | 19 | 50 |

**Table 6.** Casting Samples of Concrete Properties

| Concrete Properties | Sample | No. of samples | Total No. of Samples |
|---|---|---|---|
| Compressive strength | Cubs (150×150×150)mm | 6 | 18 |
| Modulus of rupture | Prism (100 x 100 x 500)mm | 3 | 9 |
| Modulus of elasticity | Cylinder 150 mm diameter x 300 mm height | 3 | 9 |
| Splitting tensile strength | Cylinder 150 mm diameter x 300 mm height | 3 | 9 |

Note: The average values were obtained from the three measured values.

#### *3.1.5 Water*
Normal potable water has been used in the process of mixing concrete for the production and curing of concrete.

#### *3.1.6 Admixtures (Superplasticizer)*
A superplasticizer used throughout this work was "Glenium 51" with nominal dosage of (0.5 liter per 100 kg of cement) as recommended by the manufacturer. This material was classified as type (A) and (F) in ASTM C494-05 [24]. The main virtue of (Glenium 51) is its ability of water reduction and keeping the workability of fresh concrete. Superplasticizers increase the early and ultimate compressive strength. The relative density of Glenium 51 (1.1 g/cm$^3$ @ 20° C) and pH=(6.6).

#### *3.1.7 Mix Design*
Designing the Mix as in many experimental mixes have beenmade in accordance with the standards of the ACI 211.1-97[25]. Concrete mixture has been designed in order to realize cylinder strength of (35 MPa, and 70 MPa) at (28) days (obtain the normal, and high strength concrete respectively) Mixture details are given in Table (5). It was found that the used mixture produces good workability as well as uniform mixing of concrete without segregation, the slump test was carried out according to (ASTM C143-00) [26]. The slump was approximately (±100 mm).

### 3.2 Casting, curing and testing work
Fifty specimens were tested the fresh and hardened properties through the pouring of casting, curing and testing work for each mix and for each type of coarse aggregate. six cubes of (150 × 150 × 150) mm in size were used according to (BS 1881:part116) [27], for testing the compressive strength of concrete at 7 days and 28 days for three types of concrete mixes for NCR and two replacement level of coarse recycled aggregate (50% and 100%). Each of the three cylinders of (150 mm diameter and 300 mm height) of each type of mix was used to find the split tensile strength of concrete at 28 days according to (BS 1881: part 117) [28] and the modulus of elasticity of concrete at 28 days as recommended by (ASTM C469-02a)[29]. Moreover, three prisms size (100mmx100mmx500mm) were cast using steel molds to determine the modulus of rupture at 28 days according to (ASTM C78-02) [30] for each type. The samples were kept at room temperature for 24 hours. After that, the samples were demolded and put in water for treatment. After (7) and (28) days, they were removed from the water to air dry environment for (24) hours and then tested in accordance with the standard specifications to determine the concrete properties of hardened concrete, the following standard samples were cast as illustrated in Table (6).

## 4 Results and discussion
### 4.1 Compressive strength
The Compressive strength is most important property of the hardened concrete. According to (BS 1881:part116), standard cubes of (150 x150x150 mm) were used, they were poured and tested to compute ($f'_c$) at 7 days and 28 days of age. A hydraulic universal machine with 3000 kN capacity was used in test. The results are presented in Table (7).

**Table 7.** Concrete Compressive Strength of Specimens

| Specimen | Code | $f'_c$ (MPa – 7 days) | $f'c$ (MPa – 28 days) |
|---|---|---|---|
| N35 | 35N.C | 29 | 36 |
| | 50%R35 | 28 | 35 |
| | 100%R35 | 26 | 33 |
| H70 | 70N.C | 60 | 70 |
| | 50%R70 | 56 | 68 |
| | 100%R70 | 50 | 60 |

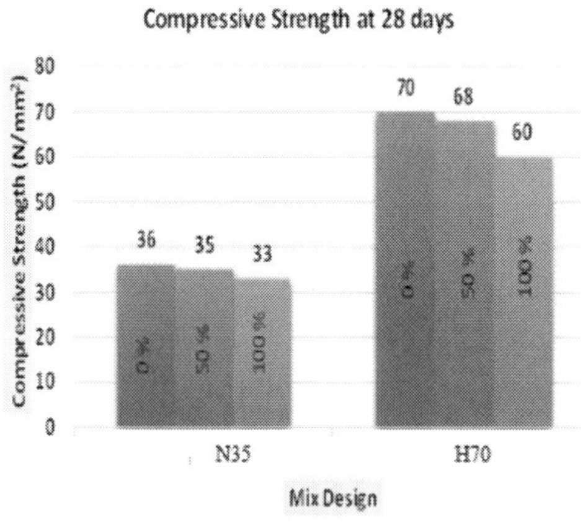


Figure (2): Compressive strength at 28 days

The results show that the lowest compressive strength was achieved for concrete specimens containings high percentage of recycled aggregate. In general, the $f'_c$ of concrete at 7 days was 80% of these at 28 days. Fig.(2) shows that the $f'_c$ for 50% contain of RCA at 28 days has dropped around 2.8% for both types. Even upto 100% replacement of recycled aggregate, the compressive strength was reduced only to a maximum of 8.34% for type(35 MPa) and 14.29% to type (70 MPa), with respect to that of control concrete. The compressive strength of the concrete specimens for 50% recycled aggregate is (35)N/mm$^2$, which meets the target strength of (35)N/mm$^2$ to the first kind from the results, it is clear that there is a possibility to use 50% recycled coarse aggregate in applications like normal concrete.

### 4.2 Splitting tensile strength

According to (BS 1881: part 117), a (150*300) mm has been used for the splitting tensile test. The cylinders are moist cured in water and tested at age of 28 days by a compressive load. The split tensile test indicates a decreasing trend of split tensile strength at 28 days of curing, when the percentage of recycled aggregate is increased. Table (8) represents the tensile strength values for mixes at 28 days.

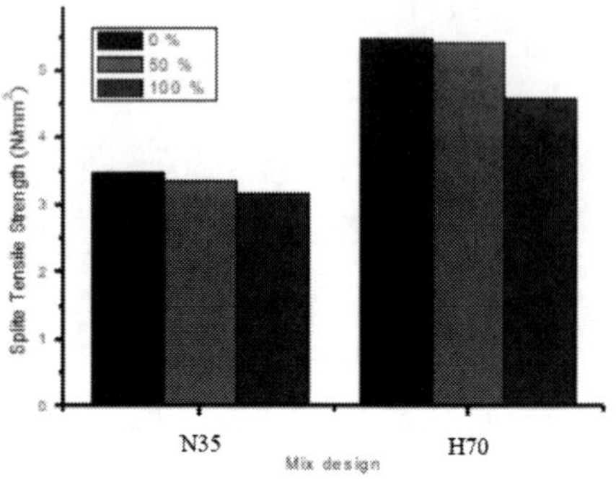


Figure (3): Splitting tensile strength at 28 days

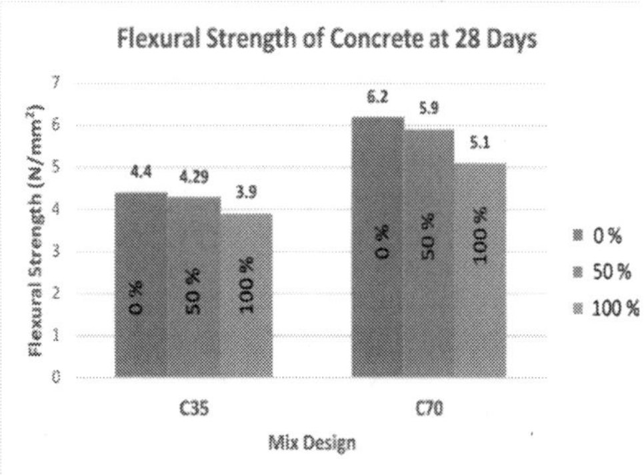


Figure (4): Flexural strength of concrete at 28 days

Figure (3) shows the splitting tensile strength for two kinds of mixes (N35 and H70). The lowest tensile strength was achieved by concrete specimen with 100% recycled coarse aggregate, for both kinds but acceptable through the difference does not exceed (8.86%) for first type and (16.2%) to second kind, between 0% replacement and 100% of recycled coarse aggregate, However, the concrete specimen with the 50% recycled aggregate has the maximum tensile strength. It just drops (3.7%), which still remainsed (96.3%) when compared to 0% recycled aggregate concrete specimen for first type, and (1.3%) only drops with the second type of mix design.

**Table 8.** Splitting tensile strength at 28 days

| Specimen | Code | Split tensile strength -MPa – at 28 days |
|---|---|---|
| N35 | 35N.C | 3.5 |
| | 50%R35 | 3.37 |
| | 100%R35 | 3.19 |
| H70 | 70N.C | 5.5 |
| | 50%R70 | 5.43 |
| | 100%R70 | 4.61 |

### 4.3 Flexural strength

Using (100*100*400) mm prism to express the modulus of rupture by two point load of hydraulic machine with loading capacity of (50 KN) mentioned in (ASTM C78-02). The flexural strength capability for the RAC and NAC mixes during 28 days of curing has been shown in Table (9). The concrete samples with 100% of replacement recycled aggregate have the lowest amount of flexural strength compared to specimens with 50% of the recycled aggregate concrete. Fig.(4) shows that there is a reduction in flexural strength about 2.3% and 11.4% to the concrete samples with 50% and 100% coarse aggregates, respectively for first mix and the drop is 4.8% and 17.75% to the concrete specimens with 50% and 100% replacement, respectively with the second kind of mix design.

**Table 9.** Flexural strength of concrete at 28 days

| Specimen | Code | Flexural strength -MPa – at 28 days |
|---|---|---|
| N35 | 35N.C | 4.4 |
| | 50%R35 | 4.29 |
| | 100%R35 | 3.9 |
| H70 | 70N.C | 6.2 |
| | 50%R70 | 5.9 |
| | 100%R70 | 5.1 |

### 4.4 Modulus of elasticity

According to (ASTM C469-02a) the chord-modulus method has been used to find the modulus of elasticity. The modulus of elasticity can be calculated from the stress-strain diagram. The chord-modulus is the slope for the linear part of curve between the point in which the strain will be 0.00005, and the point where stress equals 40% of the ultimate stress. All cylinders sample are tested at age of (28 days). The results of mixes are given in Table (10). The specimen of NCR has the highest value of modulus of elasticity compared with the specimens of 100% recycled aggregate. From the experimental results, the modulus of elasticity of full NCA was 31.378 GPa to specimen (N35) while the full RCA (100% replacement level) was 27.208 GPa as indicated in Fig. (5). It is a loss of 4.17 GPa, so the percentage of variation is 13.3% between 0% and 100% recycled coarse aggregate. For the second specimen (H70) the modulus of elasticity of NCR specimens was 42.03 GPa, while the modulus of elasticity of 100% recycled coarse aggregate specimens was 36.566 Gpa, so, the drop percentage was 13%.

**Table 10.** Modulus of Elasticity of Concrete at 28 days strength

| Specimen | Code | Modulus of Elasticity -GPa – at 28 days |
|---|---|---|
| N35 | 35N.C | 31.378 |
| | 50%R35 | 29.496 |
| | 100%R35 | 27.208 |
| H70 | 70N.C | 42.03 |
| | 50%R70 | 39.852 |
| | 100%R70 | 36.566 |

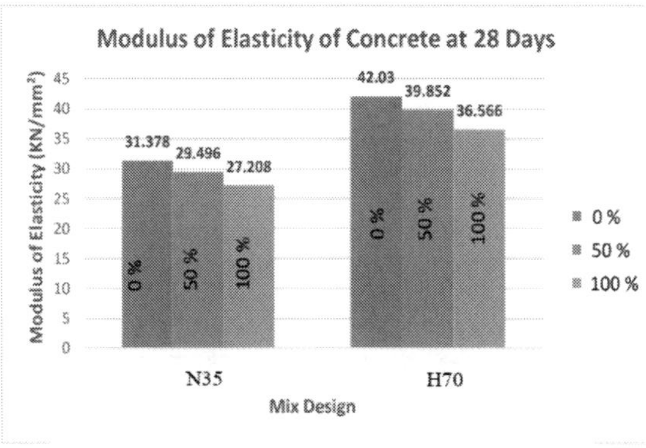


Figure (5): Modulus of Elasticity of Concrete at 28 days

## 5 Conclusions

This paper deals with the experimental tests of fresh and hardened Properties of recycled aggregate concrete. According to the results mentioned in this paper, there are some of the assumptions listed below.

1- The percentage of RCA Water absorption is more than the NCA level, because the older mortar stays adhered to the aggregate surface which leads to lessening the strengths.

2- The values of compressive strength to RCA and NCA for both mixing (N35 and H70) and two kinds of replacement aggregate (50%, and 100%) is clearly being noticed the RCA is lower than NCA, so, the strength of concrete decreases whenever the percentage of RCA increases.

3- Tensile stress of RC specimens showed about 8.86 % reduction compared to that of NC specimens for first type (N35) and 16.2% for the second type (H70).

4- Test results showed that the flexural strength of the recycled aggregate concrete is found to be lower than the natural aggregate byabout 11.4% in the first type (N35) and 17.75% in the second kind (H70).

5- For the modulus of elasticity results the values decrease by about 13 % by replacing 100% of recycled aggregate.

6- Based on the test results, it is recommended that the RCA can be replaced with natural aggregate 50% is sufficient than the replacement 100%. The replacement RCA for 100% which need more investigate with high strength concrete so the designing engineering should take into account the decrease of strength when using recycled aggregate concrete.

7- The main trends detected specify that RCA can be used in high strength concrete mixes, if the engineering characteristics are consistent with acceptable compressive strength, flexural strength and modulus of elasticity.

8- There is a necessity to specify and identify new standards for recycled aggregates that can be used effectively under different conditions.

9- With the adoption of the developed recycling systems, it is possible to recycle concrete waste to lead to high efficiency in reducing both the cost and the environmental impact.

## References

1- Olorusongo, F.T. (1999). Proceeding of the International Seminar on Exploiting Wastes in Concrete held at the University of Dundee, Scotland, UK on 7 September 1999, pp. 163-170.

2- Malhotra VM, Bilodeau In: Mehta PK, editor. Proc int symp on 'Concrete technology for sustainable development in the twenty-first century'. New Delhi (India): Radha Press; 1999. pp. 43–64.

3-Transportation Applications of Recycled Concrete Aggregate—FHWA State of the Practice National Review 2004; U.S. Department of Transportation Federal Highway Administration: Washington, DC, USA, 2004; pp. 1-47.

4-N.Sivakumar, S.Muthukumar, V.Sivakumar, D.Gowtham, V.Muthuraj. , International Journal of Engineering And Science Vol.4, Issue 01 (January 2014), PP. 27-36.

5-T.C. Hansen, second state-of-the-art report, developments from 1945–1985, Mater. Struct. 19 (111) (1986) 201– 246.

6- Hansen, T.C., Ed.; Taylor and Francis, Oxfordshire, UK, 1992; pp. 316.

7- ACI Committee 555, Removal and reuse of hardened concrete, ACI Mater. J. 99 (3) (2002) 300– 325

8- N.K. Bairagi, R. KishoreResour. Conserv. Recycl. 9 (3) (1993) 109– 126

9- Poon, C.S., Lam, C.S., 2008. Cement and Concrete Composites 30 (4), 283–289.

10- Rahal, K., 2007, Building and Environment 42 (1), 407–415.

11-Lopez-Gayarre, F., Serna, P., Domingo-Cabo, A., Serrano-Lôpez, M.A., Lopez-Colina,C., 2009. Waste Management 29 (12), 3022–3028.

12-Sanchez de Juan, M.; Gutierrez, P.A. In Proceeding of the International RILEM Conference: The Use of Recycled Materials in Building and Structures, Barcelona, Spain, 8–11 November 2004; pp. 545-553.

13- Yang, K.H.; Chung, H.S.; Ashour, A. , ACI Mater. J. 2008, 3, 289-296.

14- BS 8500-1 (2006), Volume 2, Section 3, Part 9 BA 92/07. (BRE Digest 433), Volume 2, Section 3, Part 9 BA 92/07.

15- ACI Committee 318, "Building Code Requirements for Structural Concrete", ACI 318-11, Committee 318 Report, American Concrete Institute, Farmington Hills, Michigan, 2011, 443 pages.

16- ASTM C 33 – 11a, "Standard Specification for Concrete Aggregates", *Annual Book of ASTM Standards*, American Society for Testing and Materials, West Conshohocken, Pennsylvania, USA, 2011, 11 pages

17- ACI Committee 301, "Specifications for Structural Concrete", ACI 301-10, Committee 301 Report, American Concrete Institute, Farmington Hills, Michigan, 2010, 81 pages.

18- ACI Committee 555, "Removal and Reuse of Hardened Concrete," ACI 555R-01, Committee 555 Report, American Concrete Institute, Farmington Hills, MI, 2001, 26 pages.

19- Central Organization for Standardization and Quality Control; "Iraqi Standard Specification for the Portland Cement" ;IQS (5) ,1984, Baghdad, Iraq.

20- ASTM Designation C150-02 "Standard specification for Portland cement", 2002 Annual Book of ASTM Standards, American Society for Testing and Material, Philadelphia, Pennsylvania, Section 4, Vol. (4.02), pp 89-93, 2002.

21- Iraqi Specification No. 45, "Natural Sources for Gravel that is Used in Concrete and Construction", Baghdad, 1984.

22-British Standard Specification (B.S.882:1992) Aggregates from Natural Sources for concrete", 1992.

23- American Society of Testing and Material (ASTM), "Standard Specification for Concrete Aggregates", ASTM C33-02, West Conshohocken, PA, 2002.

24- Standard Specification for Chemical Admixtures for Concrete. ASTM-C494-05, American Society for Testing and Material, 2005.

25- ACI 211.1-97, (Re-approved 1997), "Standard practice for selecting proportions for normal, heavy weight and mass concrete", ACI Manual of Concrete Practice, American Concrete Institute, Detroit, 1997.

26- ASTM Designation C143-00 "Standard specification for test method for slump of Portland cement concrete", 2000 Annual Book of ASTM Standards, American Society for Testing and Material, Philadelphia, Pennsylvania, Section 4, Vol. (4.02), pp 1-3, 2000.

27- BS 1881: Part 116:1983, "Method for Determination of Compressive Strength of Concrete Cubes".

28- BS 1881: Part 117:1983, "Method for Determination of Tensile Splitting Strength".

29- ASTM Designation C469-02a, "Static Modulus of Elasticity and Poisson's Ratio of Concrete in Compression" 2002 Annual Book of ASTM Standards, American Society for Testing and Materials, Section 4, Vol.04.02, pp.236-239.

30- ASTM Designation C78-02, "Flexural Strength of Concrete".

# The role of ambient temperature variation on drying shrinkage development of self-compacting Portland-limestone cement concrete

*Basil* Al-Shathr[1], *Ammar* Abdulameer[1,*] and *Tareq* al-Attar[1]

[1]Building and Construction Engineering Department, University of Technology, Baghdad, Iraq

**Abstract.** The ambient temperature records in Iraq show a large variation between day and night reaching 20 C, depending on the season, whether it is summer or winter. For this reason, the aim of this research is to study the effect of these conditions on the drying shrinkage of self-compacting concrete produced by using Portland-Limestone cement (ASTM C595 – Type IL). SCC mixes were designed to attain compressive strengths of 40 and 60MPa at 28days with and without silica fume respectively. Same mixes were reproduced with ordinary Portland cement (ASTM C150 – Type I) for comparisons. Two maximum sizes of aggregate 10 and 20 mm were incorporated in this work. The drying shrinkage was measured for 180 days after 7 days of water curing. The range of ambient (outdoor) temperature variation was from - 4 to + 39 °C and the relative humidity ranged from 15 to 60 %. The results of this exposure were compared to that of specimens kept in the shrinkage chamber, with a temperature of 21 °C and relative humidity 35%. The current results showed that due to the irreversible nature of shrinkage strain, the drop of ambient temperature and the rise of atmosphere moisture or relative humidity would not reverse the shrinkage strain. It is important to figure the final total accumulated strain when dealing with ambient temperature variation. The drying shrinkage characteristics for concrete made with Type IL cement, are found similar to that for concrete produced with Type I cement.

## 1 Introduction

Drying shrinkage is a major cause of the concrete deterioration especially in Middle East region, thus the acknowledgement of the shrinkage deformation needs to be evaluated. Ingram and Daugherty [1] indicated that in many countries some limestone are allowed to be added to Portland cement products, and by controlling the properties of the cement they will be able to produce products that are both ethically and economically viable. Sivilis et al. [2] concluded that limestone cements, have lower paste water demand than the relative pure cements. Holt and Leivo [3] stated that drying shrinkage is greatly affected by environmental conditions such as wind and humidity in the first days of curing. Sagaraa and Pane [4] revealed that the shrinkage initially tends to be higher for high strength concrete; however the ultimate shrinkage strain is larger for normal strain concrete. Khan and Montgomery [5] concluded that the use of milled limestone in self- compacting concrete with suitable content may reduce drying shrinkage of concrete. Rozie're et al. [6] reported that shrinkage cracking susceptibility increases by increasing the volume of paste and that reducing the volume of paste may affect flowability and surface quality, but it is a way to reduce shrinkage and risk of cracking without affecting strength. Kalkan and Lee [7] had pointed out that the use of SCC makes a beam more vulnerable to restrained shrinkage cracking. Heirman et al. [8] introduced test results which revealed that the SCC mixtures exhibit higher shrinkage and creep deformations compared with the limestone cement concrete mixtures. Leemann et al. [9] attributed the higher shrinkage of SCC mixtures is due to their higher paste volume. For high performance lightweight concrete, Al-Attar and Al-Sakini [10] showed experimentally that using higher lightweight aggregate to binder ratio has negative effect on shrinkage. The results of Gupta et al. [11] gave a conclusion that shrinkage strain of high strength concrete with 10% fly ash and silica fume replacement were higher by (6-10) % than concrete without them.

The aim of this research is to investigate the effect of temperature variation on the rate of drying shrinkage development and on ultimate recorded values.

## 2 Experimental work

### 2.1 Materials

Two types of Portland were used throughout this study. They were ordinary Portland cement, confirms to the ASTM C150-Type I [12], and Portland-limestone cement IL, confirms to the ASTM C595-Type IL [13]. Silica fume with an activity index of 120% was used as supplementary cementitious material in producing the

* Corresponding author: aalrahamy@yahoo.com

© The Authors, published by EDP Sciences. This is an open access article distributed under the terms of the Creative Commons Attribution License 4.0 (http://creativecommons.org/licenses/by/4.0/).

self-compacting concrete mixes. The silica fume was conforming to the ASTM C-1240 [14] and was added by 10% as partial replacement of cement for high strength concrete mixes. Table (1) shows the chemical composition and some physical properties for both types of cement and for silica fume.

**Table (1).** Chemical and physical properties of used ordinary and limestone Portland cement and silica fume

| No. | property | | OPC | IL | SF |
|---|---|---|---|---|---|
| 1 | Oxide content,% | CaO | 60.8 | 61.3 | < 1 |
| | | $SiO_2$ | 19.9 | 17.7 | >85 |
| | | $Al_2O_3$ | 4.7 | 4.1 | - |
| | | $Fe_2O_3$ | 3 | 4.7 | - |
| | | MgO | 1.5 | 2.8 | - |
| | | $K_2O$ | 2.3 | 0.5 | - |
| | | $Na_2O$ | 0.46 | 0.1 | - |
| | | $SO_3$ | 0.11 | 2.3 | < 2 |
| 2 | Loss on Ignition (L.O.I) | | 2.7 | 5.8 | |
| 3 | Fineness (Blain $m^2$/kg) | | 309 | 368 | > 15000 |
| 4 | Specific gravity | | 3.15 | 3.07 | 2.3 |
| 5 | Compressive strength, MPa at 28 days | | 42 | 48.4 | - |

Limestone dust passing sieve (75 μm) was added to produce SCC, Table (2) shows the properties of the used limestone dust.

**Table (2).** Chemical composition of limestone dust

| Oxide | $SiO_2$ | $Fe_2O_3$ | $Al_2O_3$ | CaO | MgO | $SO_3$ | L.O.I | $CO_3$ | $CaCO_3$ |
|---|---|---|---|---|---|---|---|---|---|
| Content, % | 0.21 | 3.36 | 0.03 | 50.84 | 2.99 | 0.08 | 42.2 | 0.19 | 98 |

Natural sand with fineness modulus of 2.97 and SO3 of 0.3% was used. Natural crushed coarse aggregate with maximum sizes of 10mm and 20mm were used. High performance superplasticizer (Sika ViscoCrete -5930) confirming to ASTM C494 Types F [15] was used for producing high strength self- compacted concrete.

### 2.2 Concrete mixes and testing program

Four SCC mixes, AI40 and BI60 for Type I cement and AIL40 and BIL60 for Type IL cement, as shown in Table 3, were prepared according to the recommendations of ACI Committee 237 [16].

**Table (3).** Mix details

| Mix | | Binder, kg/$m^3$ | | Limestone Dust, kg/$m^3$ | Aggregate, kg/$m^3$ | | Max Size of Agg, mm | Water, kg/$m^3$ | ViscoCrete L/$m^3$ |
|---|---|---|---|---|---|---|---|---|---|
| | | Cement-Type | Silica Fume | | F A | C A | | | |
| AI40 | | 400 – I | 0 | 100 | 764 | 800 | 20 | 152 | 8 |
| AIL40 | | 400 – IL | | | | | | | |
| BI60 | | 450 – I | 50 | 50 | | | 10 | 155 | 13 |
| BIL60 | | 450 - IL | | | | | | | |

Fresh concrete mixes were tested for self-compactability. 40 MPa was the average required compressive strength at 28 days of cylinders, d=100mm and h=200mm, for mixes AI40 and AIL40. With w/c ratio of 0.38 and an aggregate maximum size of 20 mm, the attained strength values were 44 and 41 MPa respectively. Mixes BI60 and BIL60 were high strength concrete. They were designed to yield 60 MPa as average required compressive strength at 28 days and they attained 67 and 62 MPa respectively. These mixes were cast with a water/binder ratio of 0.31 and a maximum size of aggregate of 10mm. For drying shrinkage, prisms of 100x100x400 mm were exposed to indoor controlled conditions, 21 o C and 35 % relative humidity, and outdoor conditions, - 4 to +39 o C and 15 to 75% relative humidity. The drying shrinkage strain was measured according to the ASTM C 157 [17] for a period of 180 days after 7 days of moist-curing for the studied mixes.

## 3 Results and discussion

### 3.1 Fresh SCC results

Table (4) lists the results of the self-compactability carried out test for the studied SCC mixes. The results revealed that the produced mixes were all conforming to the requirements of the international EFNARC [18]. The replacement of Portland cements with silica fume has enhanced the passing and filling abilities of the tested mixes. This behavior could be attributed to the high surface area of silica fume.

**Table (4).** Fresh SCC test results

| Type of mixes | Slump flow | | V-funnel | L-Box |
|---|---|---|---|---|
| | D | $T_{500}$ | | |
| | mm | Sec. | Sec. | (H2/H1) |
| AI40 | 750 | 3 | 8 | 0.93 |
| AIL40 | 750 | 3 | 8 | 0.93 |
| BI60 | 790 | 2 | 6 | 0.98 |
| BIL60 | 790 | 2 | 6 | 0.98 |
| EFNARC Requirements 2002 [18] | 600-800 | 2-5 | 6-12 | 0.8-1 |

## 3.2 Drying shrinkage behavior for indoor controlled exposure

Table (5) and Figure (1) show the values and drying shrinkage strain development for the studied mixes when exposed to the indoor controlled exposure, temperature of 21 °C and 35% relative humidity.

**Table (5).** Drying shrinkage strain results in the indoor controlled exposure for the studied SCC mixes

| Indoor shrinkage strain, × $10^{-6}$ (Temp = 21 °C, R. H. = 35%) | | | | |
|---|---|---|---|---|
| Age, days | Mix | | | |
| | AI40 | AIL40 | BI60 | BIL60 |
| 3 | 49 | 49 | 38 | 33 |
| 7 | 100 | 85 | 85 | 75 |
| 14 | 150 | 120 | 103 | 96 |
| 21 | 181 | 170 | 155 | 137 |
| 28 | 225 | 205 | 200 | 187 |
| 60 | 293 | 281 | 268 | 237 |
| 90 | 350 | 336 | 316 | 300 |
| 120 | 372 | 359 | 350 | 338 |
| 150 | 427 | 406 | 391 | 363 |
| 180 | 475 | 466 | 438 | 425 |

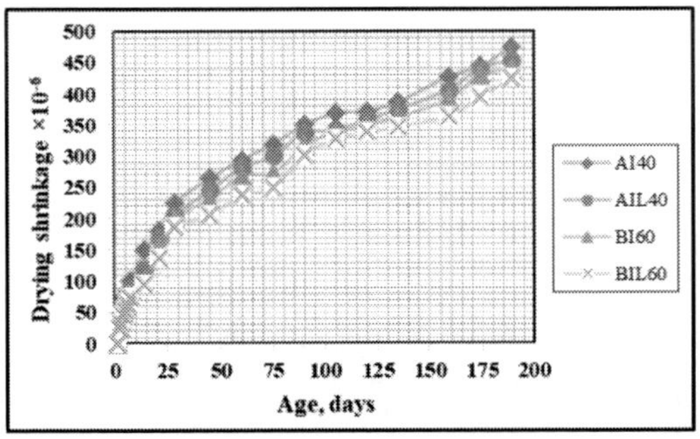


**Figure (1).** Drying shrinkage behavior of indoor controlled exposure for the studied SCC mixes

Generally, high strength concrete recorded lower drying shrinkage values than normal strength concrete. The ratios of highest shrinkage, at 180 days, for B series mixes to A series mixes were: 92 and 91 % for Types I and IL respectively. This behavior could be explained by that high strength mixes, BI60 and BIL60, have higher bond between aggregate and cement paste because of using aggregate with smaller maximum size, i.e. with higher surface area. This good bond makes the aggregate more effective in restraining the shrinkage strain [19]. The denser interface zone due to the inclusion of silica fume and to using lower w/c ratio could be another factor [20].

Small differences could be observed between shrinkage values for the studied mixes when regarding the type of used cement. Mixes made with Type IL cement showed always lower shrinkage values than those made with Type I cement. The ratios of highest shrinkage strain of Type IL cement to Type I cement mixes were: 98 and 97 for series A and B respectively. The difference may be due to the smaller clinker to cement ratio of Type IL cement [21].

With respect to rate of shrinkage development, all the studied mixes have almost the same rate and the parallelism of curves shown in Figure (1) could be considered as an indication. According to Figure (1), it could be concluded that the recorded shrinkage values at the end of 180 days, could not be considered as the ultimate shrinkage. The curves at this age do not level at this age and it is expected for these curves to attain higher values at later ages. This observation could be attributed mainly to the controlled temperature and relative humidity which did not allow faster evaporation rate [22].

### 3.3 Drying shrinkage behavior for outdoor uncontrolled exposure

Figure (2) display the temperature variation for the outdoor uncontrolled exposure. This variation was recorded for six months in Baghdad city through the period from 19/10/2016 to 19/04/2017. The relative humidity in this period had ranged from 15 to 60%. Three intervals could be distinguished in this period: from the starting date to the age of 60 days the average temperature had fallen from 25.5 to 4.5 O C, from 60 to 120 days ages the average temperature stayed with small variation, 4.5 to 11 O C and finally from 120 days to the end of testing period the average temperature raised from 4.5 to 30 O C.

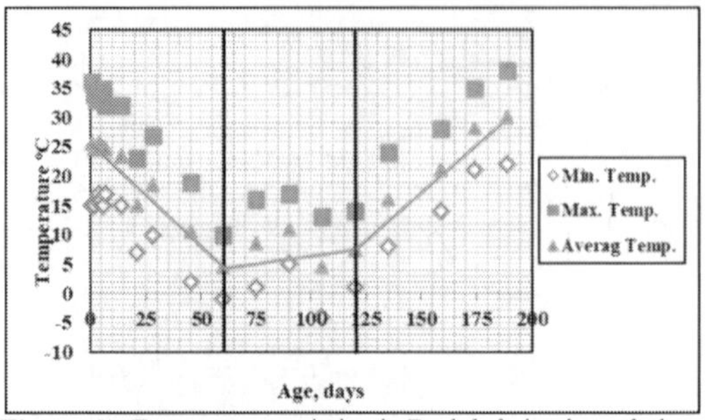


**Figure (2).** Temperature variation in Baghdad city through the period from 19/10/2016 to 19/04/2017 (outdoor uncontrolled exposure).

Table (6) and Figure (3) show the results of drying shrinkage development in the outdoor uncontrolled exposure for mixes AI40, AIL40, BI60 and BIL60.

**Table (6).** Drying shrinkage strain results in the outdoor uncontrolled exposure for the studied SCC mixes

| Outdoor shrinkage strain, × $10^{-6}$ | | | | | | | | |
|---|---|---|---|---|---|---|---|---|
| Age, days | Mix | | | | Ambient Temperature, °C | | | R.H., % |
| | AI40 | AIL40 | BI60 | BIL60 | Min. | Max. | Av. | |
| 3 | 112 | 101 | 50 | 43 | 15 | 36 | 25.5 | 43 |
| 7 | 196 | 182 | 112 | 100 | 17 | 32 | 24.5 | 40 |
| 14 | 219 | 198 | 137 | 127 | 15 | 32 | 23.5 | 37 |
| 21 | 250 | 235 | 180 | 150 | 7 | 23 | 15 | 40 |
| 28 | 300 | 280 | 210 | 200 | 10 | 27 | 18.5 | 42 |
| 60 | 425 | 390 | 281 | 270 | -1 | 10 | 4.5 | 45 |
| 90 | 550 | 483 | 344 | 315 | 5 | 17 | 11 | 60 |
| 120 | 679 | 660 | 485 | 415 | 1 | 14 | 7.5 | 44 |
| 150 | 723 | 715 | 573 | 513 | 14 | 28 | 21 | 42 |
| 180 | 775 | 759 | 631 | 600 | 22 | 38 | 30 | 15 |

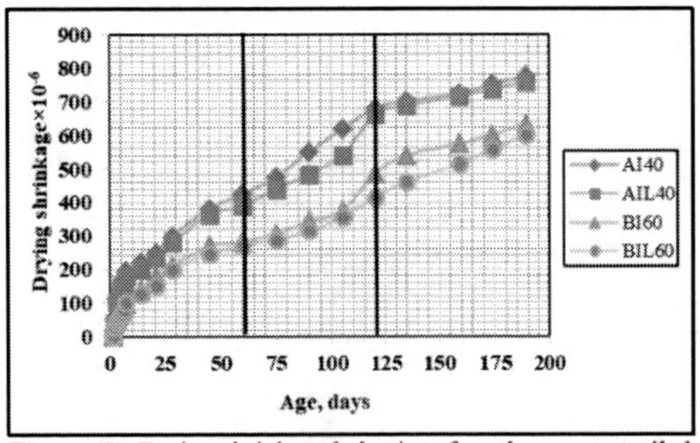


**Figure (3).** Drying shrinkage behavior of outdoor uncontrolled exposure for the studied SCC mixes

According to the tabulated results it could be assigned that in this exposure the values of shrinkage strain at the age of 180 days were higher than those recorded in the indoor controlled exposure for all mixes. At this age, 180 days, the ratios of ultimate strain of outdoor to indoor exposure were: 163, 163, 144 and 141% for mixes AI40, AIL40, BI60 and BIL60 respectively. This may be due to the faster rate of drying in the outdoor exposure. This faster rate caused the concrete mixes to attain their ultimate strain values earlier. Although of the ambient temperature fluctuation in this exposure, the total drying shrinkage values were higher. This could be considered

as an indication to the irreversible nature of most of shrinkage strain. The drop of ambient temperature and the rise of atmosphere moisture or relative humidity would not reverse the shrinkage strain. It is important to figure the final total accumulated strain when dealing with ambient temperature variation.
The ambient temperature fluctuation has affected the rate of shrinkage development. The three above mentioned intervals for temperature variation can be easily assigned on the shrinkage development rate as shown in Figure (3).
The Type IL cement mixes exhibited lower shrinkage strain than Type I cement mixes but the differences were small and the rate was similar. The ratios of ultimate shrinkage strain for Type IL cement mixes to Type I cement mixes were: 98 and 95 for series A and B respectively. Therefore, it could be concluded that the drying shrinkage characteristics for concrete made with Type IL cement, ASTM C595, are mostly similar to that produced with Type I cement, ASTM C150.

## 4 Conclusions

1. Portland-limestone cement, ASTM C595 - Type IL, could be used to produce self-compacting concrete with the required filling and passing abilities.
2. For indoor controlled exposure, 21 $^0$C and 35% RH, the recorded shrinkage values at the end of 180 days, could not be considered as the ultimate shrinkage. This observation could be attributed mainly to the controlled temperature and relative humidity which did not allow faster evaporation rate. Meanwhile, at this age in the outdoor uncontrolled exposure concrete mixes attained their ultimate strain values.
3. Although of the ambient temperature fluctuation in the outdoor exposure, the total drying shrinkage values were higher than those of the steady controlled temperature and relative humidity exposure.
4. Due to the irreversible nature of shrinkage strain, the drop of ambient temperature and the rise of atmosphere moisture or relative humidity would not reverse the shrinkage strain. It is important to figure the final total accumulated strain when dealing with ambient temperature variation.
5. The drying shrinkage characteristics for concrete made with Type IL cement, ASTM C595, are mostly similar to that produced with Type I cement, ASTM C150.

## References

[1] K. D. Ingram and K. E. Daugherty, Cement and Concrete Composites, **13**, 165-170, (1991).
[2] S.T. Sivilis, E. Chaniotakis and E. Badogiannis, Cement and Concrete Composites, **21**,107-116, (1999).
[3] E. Holt and M. Leivo, Cement and Concrete Composites, **26**, 521-530, (2004).
[4] A. Sagaraa and I. Pane, Procedia Engineering, **125**, 1087 – 1093,(2015)
[5] V. Khan and D. Montgomery, Proceedings of the 1st International RILEM Symposium on Self-Compacting Concrete, Stockholm, Sweden, September (1999)
[6] E. Rosziere, S. Granger, Ph. Turcry and A. Loukili, Cement and Concrete Composites,**29**, 626–636, (2007).
[7] I. Kalkan and J. H. Lee, KSCE Journal of Civil Engineering, (2013).
[8] G. Heirmana, L. Vandewalle, D. Van Gemert , V. Boel, K. Audenaert, G. De Schutter, B. Desmet and J. Vantomme, Engineering Structures, **30**, 2945–2956, (2008).
[9] A. Leeman, P. Lura and R. Loser, Construction and Building Materials, **25**, 2283-2289, (2011)
[10] T. S. Al-Attar and J. S. Al-Sakini, International Journal of Civil & Environmental Engineering, **11**, No.6, (2011)
[11] S.M. Gupta, V.K. Sehgal and S.K. Kaushik, World Academy, Engineering and Technology, **50**, (2009).
[12] ASTM C150 – 15,Specifications for Portland Cement, Annual Book of ASTM Standards, (2015).
[13] ASTM C595 – 15, Specifications for Blended Hydraulic Cements, , Annual Book of ASTM Standards, (2015).
[14] ASTM C1240 – 15, Specifications for Silica Fume Used in Cementitious Mixtures, Annual Book of ASTM Standards, (2015).
[15] ASTM C494 – 13, Specification for Chemical Admixtures for Concrete, Annual Book of ASTM Standards, (2015).
[16] ACI Committee 237, Self-Consolidating Concrete, ACI237R-07, ACI Manual of Concrete Practice, USA, (2013)
[17] ASTM C157-08, Standard Test Method for Length Change of Hardened Hydraulic-Cement Mortar and Concrete , Annual Book of ASTM Standards, (2015).
[18] EFNARC, Specifications and Guidelines for Self-Compacting Concrete, UK, (2002)
[19] T. S. Al-Attar, Engineering & Technology Journal, **26**, 146-153,(2008).
[20] P. K. Mehta and P. J. M. Monteiro, *Concrete Microstructure, Properties and Materials*, 3rd ed. Mc-Grew Hill, USA, (2006).
[21] T. J. Barrett, H. Sun and W. J. Weiss, Indiana Department of Transportation and Purdue University, USA, (2013).
[22] A. M. Neville, *Properties of Concrete*, 5th ed. Prentice Hall, UK, (2011).

# Effect of silica fume/binder ratio on compressive strength development of reactive powder concrete under two curing systems

*Mohammed* Abed[1], *Mohammed* Nasr[1,*] and *Zaid* Hasan[1]

[1]Babylon Technical Institute, Al-Furat Al-Awsat Technical University, 51015, Babylon, Iraq

**Abstract.** This paper aims to investigate the influence of Silica fume proportion ratio in respect to the total amount of binder on compressive strength of reactive powder concrete cured in two curing systems. Four ratios of Silica fume (0%, 15%, 25% and 35%) as replacement of cement weight were considered. After de-molding, two curing systems were used: the first included immersing the cubic specimens in water at 24 ± 2 °C until the test. In the second, the specimens were immersed in hot water at 105 ± 5 °C (accelerated curing) for 48 hours, then in water at 24 ± 2 °C until the test. The results show that mix which contains 25% Silica fume imparts more enhancement on compressive strength as compared to the control mix. Also, it was found that the second system of curing has more influence on compressive strength development than the first one, especially at earlier ages.

## 1 Introduction

Reactive powder concrete (RPC) that was developed in Paris in 1990 is a special type of concrete that has two main properties: high strength and high performance. The gradation in its materials gives a low percent of voids with high density when compared with traditional concrete [1]. Coarse aggregate is absent in reactive powder concrete, therefore, the microstructure is far from non-homogenous between the cementitious materials and fine aggregate [2].

Different curing systems were applied in the literature in the production of RPC. For activity accelerating, heat treatment may be adopted to improve the microstructure and enhancing the earlier strength of RPC when silica fume is used in the mix [3]. Talebinejad et al. [4] indicated that the heat curing at (90-200)°C modified the pozzolanic activity, while, Cheyrezy et al. [5] showed that heat curing was essential at (200-250) °C for getting excellent compressive strength, especially at earlier ages. Treatment using static pressure was also used and led to enhance the compressive strength by about 33% in comparison to others without the static pressure [6]. A study by Yazıc et al. [7], found that some reduction in bonding (paste and fiber) happened after using of autoclave and steam curing when compared with normal curing.

The goal of this study is to investigate the influence of silica fume/ binder ratio on compressive strength of RPC using two curing systems: accelerated and normal curing. Additionally, due to that previous work [8] showed that the smaller the maximum particle size and the better the packing degree of fine aggregate, the higher the compressive strength of RPC, so, sand with a maximum grain size of 0.3 mm is used to achieve the aim of this study.

## 2 Experimental Program

### 2.1 Materials

Ordinary Portland cement, commercially called "Kar", was used in this study. The chemical composition of the cement, which conforms to Iraqi specifications No.5/1984 [9], is illustrated in Table 1. Silica fume purchased from "BASF" company was used as a pozzolanic material, see Table 2. Natural sand with single size (0.15 - 0.3 mm) was used as fine aggregate. In order to enhance workability, third generation superplasticizer (Gelnium 54), based on polycarboxylic ether was added.

### 2.2 Mix proportions, mixing and casting

A total of four mixes were made to achieve the goals of this investigation. Three of them include replacing of cement (by weight) with silica fume in proportions of 15%, 25%, and 35%. The other mix, without adding, was considered as control. The sand/ binder and water/ cementitious ratios for all mixes were 1:1 and 0.18, respectively. The details of mix proportions per one cubic meter are shown in Table 3.

Due to the low water content and high cementitious materials that are usually used in reactive powder concrete, it required relatively a long mixing time. The depended mixing procedure in this research was as follows:

* Corresponding author: msn_alamar@yahoo.com

© The Authors, published by EDP Sciences. This is an open access article distributed under the terms of the Creative Commons Attribution License 4.0 (http://creativecommons.org/licenses/by/4.0/).

- All dry materials were fed into the mixer and mixed at a low-speed rate (140 r/ min) for one minute;

- The mixing water and superplasticizer, which were previously mixed, were then added, and the all materials were mixed for one minute at low-speed rate;

- After that, the mixer was stopped for one minute and the speed was changed to a moderate rate (285 r/ min);

- Then, the mixer was operated at this rate for eight minutes.

After the mixing was complete, the fresh RPC was placed in standard cube molds. The following casting and tamping procedures were as described by ASTM C109 standard [10].

**Table 1**. Chemical composition of cement.

| Oxides Composition | Content, % | Iraqi specifications (No.5/1984), % |
|---|---|---|
| CaO | 61.8 | - |
| $SiO_2$ | 20.3 | - |
| $Al_2O_3$ | 4.16 | - |
| $Fe_2O_3$ | 3.9 | - |
| MgO | 3.4 | <5 |
| $SO_3$ | 2.4 | <2.5 |
| Free lime | 3.01 | - |
| L.O.I. | 0.03 | <4 |
| L.S.F. | 0.96 | 0.66- 1.02 |
| Insoluble residue | 1 | <1.5 |

**Table 2.** Properties of silica fume.

| Property | Value or description |
|---|---|
| Unit weight | 2300 kg/m$^3$ |
| Fineness (blain) | > 15 m$^2$/g |
| $SiO_2$ | > 85 % |
| CaO | < 1 % |
| Chloride content | < 0.1 % |
| Activity index | 112 % |

**Table 3.** Details of mix proportions.

| Mix symbol | Cement, kg/m$^3$ | Silica fume, kg/m$^3$ (%) | Sand, kg/m$^3$ | Water, kg/m$^3$ | Superplasticizer, % of binder |
|---|---|---|---|---|---|
| M0 | 1150 | 0 | 1150 | 207 | 2.58 |
| M15 | 977.5 | 172.5 (15) | 1150 | 207 | 2.58 |
| M25 | 862.5 | 287.5 (25) | 1150 | 207 | 2.58 |
| M35 | 747.5 | 402.5 (35) | 1150 | 207 | 2.58 |

### 2.3 Curing

After 20-24 hours of casting, the specimens were de-molded. Thereafter, two systems of curing were applied: the first (SC1) included immersing the specimens in water at °24 ±2 until the test. In the second (SC2), the specimens were cured in hot water (accelerated) at °105 ±5 for 48 hours, then immersed in water at °24 ±2 until the test. The curing systems were not applied until the end of 24 hours from casting.

## 3 Tests

### 3.1 Compressive Strength Test

The compressive strength test was performed using standard 50×50×50 mm cubic specimens at three ages: 7, 14 and 28 days. Three specimens were taken into account for each mix at the specified age, so, a total of 72 specimens was made in this study.

## 4 Results and discussion

### 4.1 Compressive Strength Results

Compressive strength results of the RPC mixes at 7, 14 and 28 days ages are clarified in Figures 1 to 5. In general, results showed that mixes which contained silica fume awarded higher strength at the specified age than without adding mix for SC1 (except M35 at 28 days age) and SC2 curing systems. The pozzolanic reaction, well packing in between cement particles and the densifying of the microstructure that provided by silica fume, could interpret this behavior.

At 7 days ages, for SC1 curing, maximum compressive strength (81.28 MPa) was reached at 25% replacement of silica fume compared with control mix. For SC2 curing, results showed that the higher the silica fume/ binder ratio the lower the compressive strength of RPC. Where, higher strength (95.73 MPa) was gained by replacing 15% of cement with silica fume, while minimum strength (87.37 MPa) was obtained at 35% replacement. At 14 days age, mix that contained 25% of silica fume gave maximum compressive strength in comparison with other mixes for both SC1 and SC2 curing systems (86.87 MPa and 96.6 MPa, respectively). Also, this behavior could be seen in the M25 mix at 28 days age for SC2 curing system, where it folded maximum arising in compressive strength (21%) with respect to without adding mix. While, for the SC1 system, the increment in compressive strength decreased with increasing of silica fume/ binder ratio. The compressive strength of M35 mix was less than the control mix by about 2%.

For comparison between the two curing systems, results indicated that SC2 system, which included hot water curing, had more effect on strength development at 7 days age than later ages. This might be as a result of temperature increase which speeded up the reactions, and consequently the hydration products formation, and

https://doi.org/10.1051/matecconf/201816202022

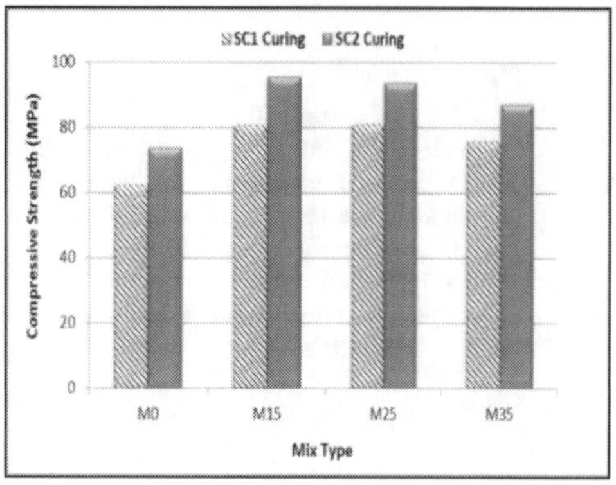


**Fig. 1.** Compressive strength results at 7 days.

led to increasing the early strength [11]. Similar behavior was also found in the literature [12]. Furthermore,

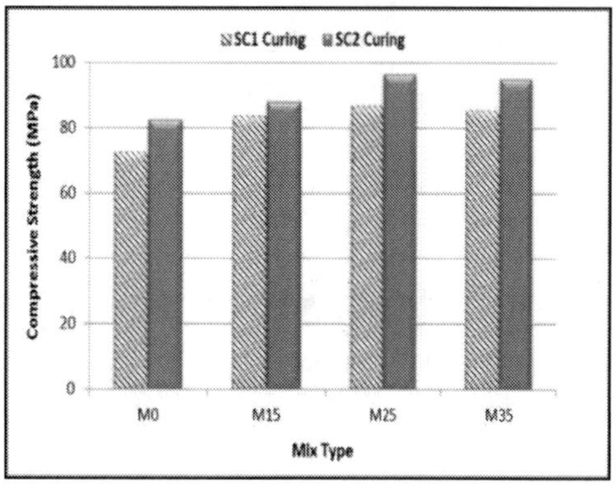


**Fig. 2.** Compressive strength results at 14 days.

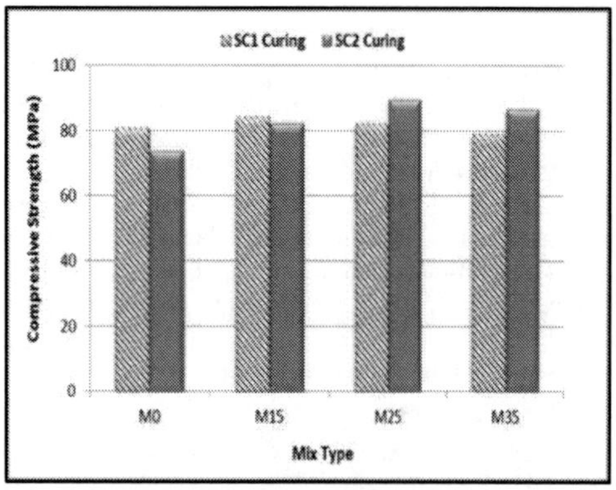


**Fig. 3.** Compressive strength results at 28 days

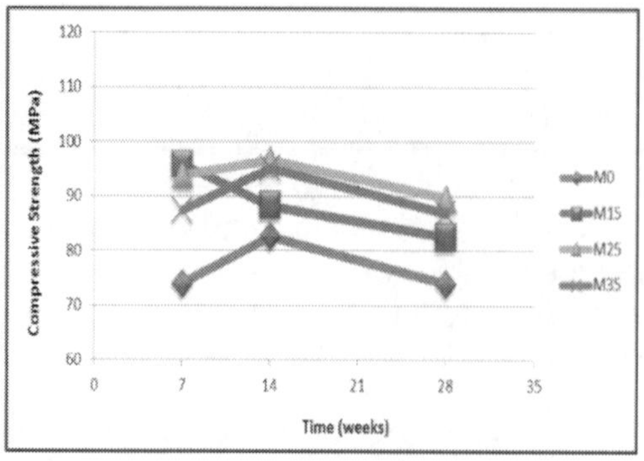


**Fig. 4.** Compressive strength development with time (accelerated curing system).

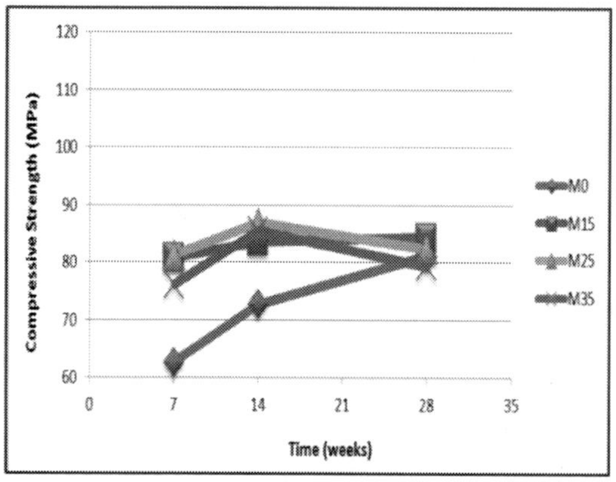


**Fig. 5.** Compressive strength development with time (normal curing system).

though that compressive strength showed higher initial values at early ages (7 and 14 days), it tended to decrease at later ages (28 days), but it remained higher than control mix. At 28 days, for M0 and M15 mixes, compressive strength values of the normal curing system (SC1) were higher than that in the SC2 system. Also, the influence of accelerating curing (SC2) was more pronounced for higher replacements of silica fume (25% and 35%) than that for lower replacement (15%).

Figures 4 and 5 illustrate the compressive strength development with time for all mixes. It can be seen that, for accelerated curing systems (Figure 4), all mixes except M15 showed strength increasing at 14 days age in comparison with beginning age (7 days) then tended to decrease at the later age (28 days). This action could result from the non-homogeneous distribution of hydration products within the matrix due to the accelerated effect of hot water curing that caused, finally, the strength to be decreased at later ages. On the other hand, the M15 mix showed permanent decrease with time for all considered ages. For normal curing system (Figure 5), the results revealed that control and M15 mixes had a comparable exhibition, continual increase in strength with time. However, the other mixes (M25 and M35) showed similar behavior as for the accelerated curing system.

## 5 Conclusions

Based on the results of this investigation, the following conclusions can be obtained:

1. Replacing of cement with silica fume up to 35% enhances the compressive strength of RPC, but in different rates with time.
2. The best compressive strength result, for SC1 and SC2 curing, is obtained when replacing of cement by 25% of silica fume.
3. The influence of accelerated curing is more obvious on strength development of RPC at early ages than later ages, especially for M25 and M35 mixes.
4. The influence of SC2 curing becomes more clarity at 25% and 35% replacements of silica fume than that for 15%.

## References

1. Rahmatabadi M.A.D., Int. J. Struct. Civ. Eng. Res., Vol. **4**, No. 4, pp. 354-358, (2015)
2. Sujatha T. and Basanthi D., IJEAR, Vol. **4**, Issue: spl -2, pp. 85-87, (2014)
3. Sumith K.K., Gururaj A., and Siddesh R.K.M., JETIR, Vol. **2**, Issue: 6 pp. 1749-1757, (2015)
4. Talebinejad I., Bassam S.A., Iranmanesh A. and Shekarchizadeh M., International Symposium on Ultrahigh Performance Concrete, pp.133–41, (2004)
5. Cheyrezy M., Maret V. and Frouin L., Cem. Concr. Res., Vol. **25**, No. 7, pp. 1491–1500, (1995)
6. Helmi M., Hall M.R., Stevens L.A. and Rigby S.P., Cons. Build. Mat., Vol. **105** pp. 554–562, (2016)
7. Yazıc H., Yardımc M.Y., Aydın S., and Karabulut A.S., Cons. Build. Mat., Vol. **23**, pp. 1223–1231, (2009)
8. Zhu Z., Li B., Liu J. and Lv X., Trans Tech Publications, Vols. **548-549**, pp. 247-253, (2014)
9. Iraqi Standard Specifications No. 5, Portland Cement, Central Organization for Standardization and Quality Control, Iraq, (1984)
10. ASTM C 109/C 109M – 13, Standard Test Method for Compressive Strength of Hydraulic Cement Mortars (using 2-in. or [50-mm] cube specimens), ASTM, West Conshohocken, PA, USA, (2013)
11. Yunsheng Z., Wei S., Sifeng L, Chujie C., Jianzhong L., Cem. Concr. Comp., Vol. **30**, pp. 831–838, (2008)
12. Arel H.S., Vol. **6**, pp. 664–674, (2016)

# Performance of super-absorbent polymer as an internal curing agent for self-compacting concrete

*Suhair* Al-Hubboubi [1,*], *Tareq* al-Attar [2], *Haider* Al-Badry[1], *Samir* Abood[3], *Rawaa* Mohammed[1] and *Buthaina* Haddhood[1]

[1] Building Research Directorate, Iraq
[2] University of Technology, Iraq
[3] Al-Mutasim State Company for Construction Contracts, Iraq

**Abstract.** Internal curing agent by using super-absorbent polymer was present in this study, its effect on the properties of self-compacting concrete was evaluated .The SAP content in the concrete mix was 0.5 % by weight of cement. Three procedures for curing were adopted; curing in water, curing in water and air and curing in polyethylene sealed bags. Fresh concrete tests conducted to assess the self-compactability of the produced concrete. Moreover, compressive and splitting strength tests were carried out. The testing program had been extended to the age of 90 days.The use of super-absorbent polymer did not affect the fresh state characteristics of the studied SCC and achieved an increase in both compressive and tensile strengths as compared to the reference concrete mix.

## 1 Introduction

Okamura [1] defined Self-Compacting Concrete (SCC) as concrete that is able to flux in the interior of the formwork, flow in a natural manner and then pass through the reinforcing bars and other obstacles under the action of its own weight .In order to achieve the required flowability, high strength concrete and low water to cementitious materials ratio (w/cm), high-range water reducer (HRWR) admixtures are essential [2]. Self-desiccation will arise with mixtures having low water-cementitious materials ratios (w/cm) [3].The aim of internal curing is to prevent self-desiccation and the accompanying autogenous stresses and strains that may lead to early age cracking by maintaining saturated conditions within a hydrated cement paste

The efficiency of internal curing, preserve saturated conditions within a hydrating cement paste in order to avoid self-desiccation and the accompanying autogenous stresses and strains that may lead to early age cracking.

depending on the measurement of a wide sets of performance properties like restrained shrinkage internal relative humidity, restrained shrinkage and creep, degree of hydration, and compressive strength development ,the efficiency has been estimated for the internal curing for example by using saturated lightweight aggregates (LWA) or super-absorbent polymers (SAP) [4-7].

Super-absorbent polymers (SAP) are generally white sugar-like hygroscopic materials also SAP are hydrophilic networks that can absorb and retain large amounts of water or aqueous solutions. [8]. Super-absorbent polymers made from partially neutralized, lightly cross-linked poly (acrylic acid) are proved to give the best performance versus cost ratio. Dried polymers are milled in to granular white solids while in water they swell to a rubbery gel that in some cases can be up to 99% water. These polymers are manufactured at low solids levels for both quality and economic reasons [9]. Very efficient internal water curing can be ensured by using SAP, which is defined as "incorporation of a curing agent serving as an internal reservoir of water, gradually releasing it as the concrete dries out [10]. From a strength point of view, there are two opposite effects for the addition of SAPs to concrete, first it generates voids in the concrete and thus reduces strength;second enhances the degree of hydration due to the internal water curing provided by the SAP and that way increases the strength. Water-cement ratio (w/c), the maturity of the concrete, and the amount of SAP addition determines which of these two effects is dominant depends on the water-cement ratio (w/c), the maturity of the concrete, and the amount of SAP addition. The total effect seems to be described well with existing models, such as the gel-space ratio concept. In particular, at a high w/c (more than 0.45), also there is very little effect on hydration through the addition of SAP which generally reduces compressive strength. At a low w/c (less than 0.45), SAP addition may increase the compressive strength [11].

## 2 Material and Experimental work

### 2.1 Materials

Cement: In all types of concrete mixes Portland cement conforming to IQS No. 5/1984- type I [12], was used . Tables 1and 2 show the chemical and physical properties of the cement

* Corresponding author: author@e-mail.org

© The Authors, published by EDP Sciences. This is an open access article distributed under the terms of the Creative Commons Attribution License 4.0 (http://creativecommons.org/licenses/by/4.0/).

**Table 1.** Chemical properties of cement

| Compound composition | Abbre-viation | Percent by weight | Limit of IQS No.5/1984 |
|---|---|---|---|
| Lime | CaO | 62.02 | - |
| Silica | $SiO_2$ | 21.13 | - |
| Alumina | $Al_2O_3$ | 5.33 | - |
| Iron oxide | $Fe_2O_3$ | 3.07 | - |
| Magnesia | MgO | 1.71 | ≤ 2.85% |
| Sulfate | $SO_3$ | 2.65 | ≤ 5% |
| Loss on ignition | L.O.I | 4 | ≤ 4% |
| Insoluble residue | I.R | 1.04 | ≤ 1.5 |
| Lime saturation factor | L.S.F | 0.88 | 0.66-1.02 |
| **Main compounds % by weight** | | | |
| Compounds | Abbreviation | Percent by weight | |
| Tri calcium silicate | $C_3S$ | 44 | |
| Di calcium silicate | $C_2S$ | 27.6 | |
| Tri calcium laminate | $C_3A$ | 8.9 | |
| Tetra calcium alumino ferrite | $C_4AF$ | 9.3 | |

**Table 2.** Physical properties of cement

| Physical properties | Limits of cement | Limits of IQS No. 5/1984 |
|---|---|---|
| Fineness m2/kg | 340.2 | ≥ 230 |
| Initial setting time (h:min) | 1:40 | ≥ 00:45 |
| Final setting time (h:min) | 4:15 | ≤ 10:00 |
| Compressive strength( Mpa)<br>3 days<br>7 days | <br>33<br>39 | <br>≥ 15<br>≥ 23 |

**Table 3.** Gradation of fine aggregate

| Sieve size (mm) | Percentage passing % | Limit of IQS No.45/1980, zone( 2) |
|---|---|---|
| 10 | 100 | 100 |
| 4.75 | 90 | 90-100 |
| 2.36 | 74.5 | 75-90 |
| 1.18 | 63 | 55-90 |
| 0.6 | 44.4 | 35-55 |
| 0.3 | 14.8 | 8-30 |
| 0.15 | 3 | 0-10 |
| Sulfite % (SO3) | 0.047 | ≤ 0.5% |

**Table 4.** Gradation of coarse aggregate

| Sieve size(mm) | Percentage passing % | Limit of Iraqi specification NO. (45/1984) |
|---|---|---|
| 20 | 100 | 100 |
| 10 | 37 | 30-60 |
| 5 | 2.8 | 0-10 |
| Sulfite content | 0.07 | ≥0.1% |

Fine aggregate: Natural sand conforming to IQS No.45/1984- Zone 2 was used [13]. Table (3) shows the gradation of the used sand and its main properties.

Coarse aggregate: 20 mm maximum size of crushed river gravel was used, conforming to Iraqi specification to IQS No.45/1984 [13], and was shown in Table (4).

Silica fume: Silica fume (exported from Elkem material company) is a highly reactive material used in this research as Pozzolanic materials to produce SCC. Silica fume conforms to the ASTM 1240 [14]. Physical and chemical properties for this by-product are listed in Table 5and 6.

Limestone powder (LSP): Limestone powder (locally known as Al-Gubra). It has been used as filler for concrete production for many years. It was added as 9 % replacement by weight of cementitious materials to enhance the workability of SCC mixes. Chemical properties for this material are listed in Table 7.

High Range Water Reducing Admixture:

Superplasticizer known commercially as (BETONAC®-1030) was used. It conforms to the ASTM C 494- type G. SCC (self-compacting/ consolidating) concrete was produced by designing high efficiency acrylic copolymer-based superplasticizer, with extremely high levels of workability without segregation. The SP content in the concrete mixes was 8.750 Kg/m3. Its properties (as claimed by the manufacturer ) are listed in Table 8.

Crushed Brick: Pieces of clay bricks, conforming to IQS No.25/1984-Class B [15], was used. The water absorption for this brick was about 21.16%.The brick was crushed and graded as the grading curve of sand (Zone 2). It was used by two percentages (15 and 20) % as a partial volumetric replacement of sand. Before using the crushed brick in the mix, it was soaked in water for (24) hours to bring the particles to saturated condition.

Super-absorbent Polymer, SAP: Commercial SAP, white sugar-like was used. It is described as an ionic polyacrylamide and used efficiently in agriculture to reduce watering intervals. SAPs has the ability to gain and preserve aqueous solutions up to several hundred times of its own weight as can be seen in Figure 1. It was added (0.5%) by weight of cement. The dry particle of SAP carefully mixed with sand before adding the mixing water.

**Table 5.** Physical and Chemical Properties of Silica Fume

| Compound composition | Percent by weight | Limit of ASTM C1240 |
|---|---|---|
| Loss on ignition | 2.14 | ≤6 |
| $SiO_2$ | 93.47 | ≥85 |
| $Al_2O_3$ | 2.15 | |
| $Fe_2O_3$ | 0.65 | |
| $SO_3$ | Nil | |
| Total | 99.79 | |

**Table 6.** Physical properties of Silica Fume

| Physical properties | results | Requirement of ASTM C1240 |
|---|---|---|
| Retaining on sieve 45μm,% | 6 | ≤10 |
| strength activity index | 125 | ≥105 |

**Table 7.** Chemical Properties of LSP

| Compound composition | Percent by weight |
|---|---|
| Loss on ignition | 43 |
| $SiO_2$ | 1.83 |
| $Al_2O_3$ and $Fe_2O_3$ | 0.63 |
| $SO_3$ | 0.2 |
| MgO | 0.58 |
| CaO | 53.71 |

**Table 8.** Properties of HRWR

| Chemical content | Polycarboxylate based |
|---|---|
| Color, appearance | transparence |
| pH | 7.5 at 20°c |
| Density | 1.108 g/ml |

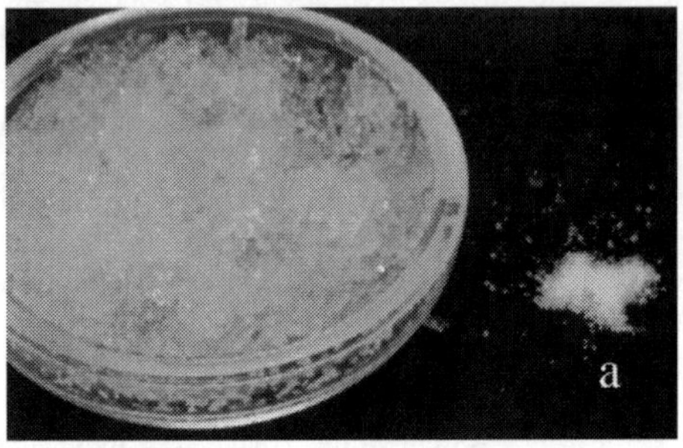

**Fig. 1.** (**a**) Dry SAP powder and (**b**) swollen SAP.

## 3 Concrete Mix Proportions

For the production of SCC, the mix proportioning should be performed so that the predefined properties of the fresh concrete are reached. The total powder content was 500kg/m$^3$, consisting of cement 405kg/m$^3$, silica fume 50kg/m$^3$and limestone powder 45kg/m$^3$. The fine and coarse aggregate contents were (880 and 900) kg/m$^3$ respectively. Water content was 172 kg/m$^3$ for reference mix (R) and for mixes containing crushed brick (B 15 and B 20%). Meanwhile the mix containing 0.5% SAP had 173 kg/m$^3$ of water. According to these proportions, the reference mix yielded 60 MPa compressive strength at 28 days. The concrete was mixed using a drum mixer of 50 L capacity. Mixing procedure followed the laboratory procedure outlined by Emborg [16] as follows: Fine aggregate was added to the mixer with 1/3 of water and mixed for 1-minute following that, the powder mixture of cement, limestone and Silica fume was added. After that, the coarse aggregate was added with the last 1/3mixing water and 1/3 of superplasticizer, and mixing for 1.5 minutes; then, the remaining 2/3 of the superplasticizer was added and mixed for 1.5 minutes; the mixture was then discharged and cast in moulds.

Curing Procedures

After the demoulding of specimens, three methods of curing were used. The first one was by immersing specimens in tap water until the time of testing (water-cured). The second method was by curing the specimens in water for 7 days and then the specimens were put in a wetted burlap bags and all covered with plastic sheet (moist-cured). The last method was by curing the specimens in water for 7 days and then left in laboratory conditions (air-cured).

## 4 Experimental Program

The fresh SCC tests include; the slump flow, L-box. The tests were conducted according to the standard methods cited by international literature [17] and the results were recorded.

Tests for hardened concrete were; compressive strength B.S 1881: Part 116 [18], splitting tensile strength (ASTM C496) [19] and ultrasonic pulse velocity (ASTM C597) [20]. The compressive strength test was done cubes at 7, 28 and 90 days curing age. The splitting strength and UPV tests were done on cylinders at the age of 90 days.

## 5 Results and Discussion

### 5.1 Fresh SCC

Desired fresh properties of SCC include suitable flowability, good passing and filling abilities, isolation resistance and stability, which are achieved by properly proportioning the constituent materials and related admixtures. Table 9 shows the results of fresh SCC tests. The tested concrete has proven itself to have self-compactability.

**Table 9.** Fresh Tests results for all fresh mixes.

| Mix | Details of replacement | Slump flow | | L – Box H1/ H2 |
|---|---|---|---|---|
| | | D(mm) | $T_{500}$(sec) | |
| Ref | No replacement | 650 | 5 | 1 |
| B15% | 15 % brick by wt of sand | 705 | 4.5 | 1 |
| B20% | 20% brick by wt. of sand | 700 | 3.5 | 1 |
| SAP0.5% | 0.5 % SAP by wt. of cement | 700 | 4.5 | 1 |

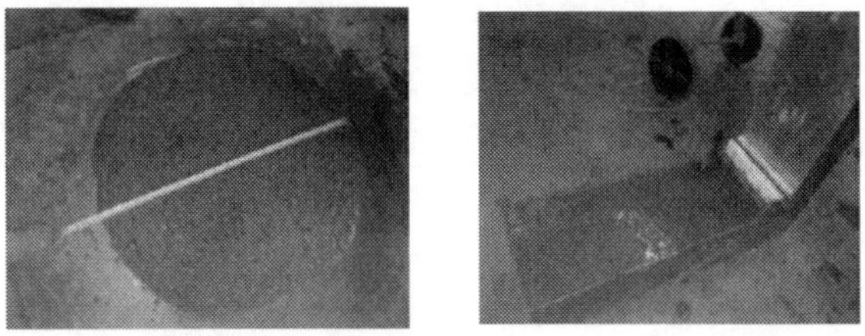

**Fig. 2.** (a) **Slump flow** (b) **L – Box**

**Table 10**. Compressive strength results

| Mix Description | Curing method | Compressive strength, N/mm$^2$ | | |
|---|---|---|---|---|
| | | 7 days | 28 days | 90 days |
| Ref | Water cured | 34 | 60 | 66 |
| | Moist cured | - | 58 | 67 |
| | Air cured | - | 62 | 64 |
| B15% | Water cured | 33.4 | 55 | 63 |
| | Moist cured | - | 56 | 68 |
| | Air cured | - | 59 | 64 |
| B20% | Water cured | 31 | 58.2 | 65 |
| | Moist cured | - | 60.7 | 69.2 |
| | Air cured | - | 62 | 69 |
| SAP0.5% | Water cured | 35.4 | 59 | 69 |
| | Moist cured | - | 61 | 71 |
| | Air cured | - | 66 | 73 |

## 5.2 Hardened SCC

There is continuous increase in strength for all concrete specimens with the increasing of curing age as can be seen in Table 10, which is attributed to the increase of the bond strength between the concrete ingredients and the continuous hydration process which increased the dense hydrated calcium silicate in concrete with increased curing age.

Results also showed that the compressive strength of SCC increased by curing with water or moisture, however using internal curing enhancing this increment especially at later ages.

For water-cured specimens, the Ref mix always showed higher strength except the case of mix SAP.5% at 90 days curing age.The change in compressive strength at 28 days is -8, -3, and -2 for mixes B15%, B20% and SAP.5% respectively. This change for 90 days is -5, -2 and +4 % for mixes B15%, B20% and SAP.5%.

For moist and air curing procedures, the mixes B20% and SAP.5% showed their best enhancement for compressive strength. There were increments in strength for 28 and 90 days. At 28 days the change in strength is +5 and +5 % for moist-curing and is 0 and +7% for air-curing for mixes B20% and SAP.5% respectively. At 90 days these figures change to +3 and +6% for moist-curing and +8 and +14% for air-curing for mixes B20% and SAP.5% respectively. It could be concluded that when there is a lack in curing water the internal curing by SAP is the best solution for this problem.

The improvement of the interfacial transition zone, raise hydration because of internal curing, and absence of shrinkage-induced microcracking will enhanced in strength. During mixing a 0.5% SAP addition permits an active control of geometric and thermodynamic properties of the water phase. The water is free in the formed SAP inclusions , and the initially added SAP particles governed the size and shape of the inclusions [21].The moisture can be seen clearly in SAP-concrete at early ages of curing (7days-broken cube), Figure 4.

Results of splitting tests of air-cured SCC showed the importance of using internal curing. The splitting strength increased by about 14%, 47% and 40% in B15%, B20% and SAP0.5%, respectively as compared to the reference mix. An enhancement was also observed in the Ultrasonic Pulse Velocity test and modulus of elasticity due to the high density of internally cured SCC concrete mixes.

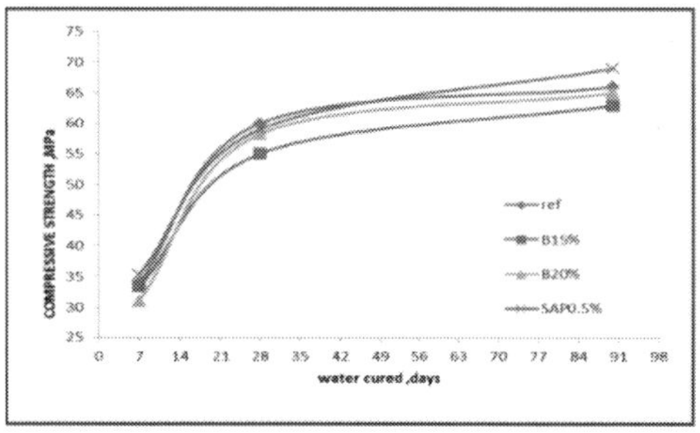


Fig.3. Compressive strength for all concrete mixes cured by water

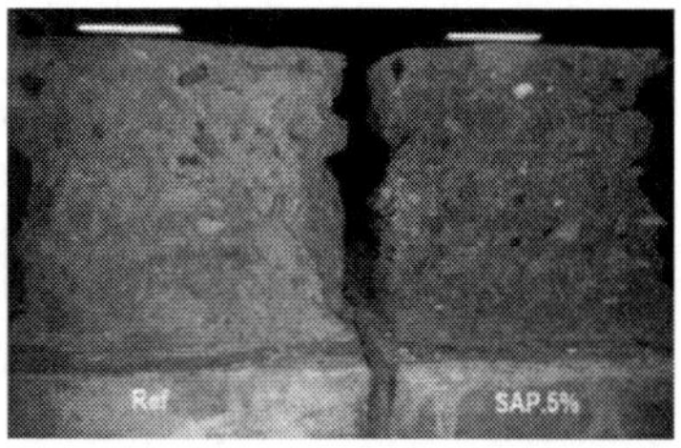


Fig. 4. Moisture in concrete due to SAP addition

**Table 11.** Splitting Strength, UPV, Density and Modulus of Elastisity

| Concrete Mixes | Splitting strength (MPa) 90 days | Ultrasonic Pulse Velocity 90 days | Density kg/m3 | Modulus of elasticity GPa |
|---|---|---|---|---|
| Ref | 2.5 | 4.47 | 2358.7 | 47.13 |
| B15% | 2.86 | 4.37 | 2368.4 | 45.229 |
| B20% | 3.69 | 4.545 | 2334.5 | 48.223 |
| SAP0.5% | 3.5 | 4.545 | 2361.4 | 48.779 |

## 6 Conclusions

Based on the obtained results, the followings conclusions can be drawn:
1. Saturated Crushed brick has no adverse effect on self-compactability tests of fresh SCC. SAP concrete needed a small addition of water due to high water absorption capacity of the SAP.
2. The replacement of fine aggregate by 20% crushed brick as internal curing material caused better enhancement in mechanical properties (strengths) of SCC than the case of 15% replacement.
3. The compressive strength of HPC containing 0.5% SAP yielded higher strength than HPC mixtures with or without saturated crushed brick.
4. Increments in splitting strength by about (14 and 47) % were achieved due to adopting internal curing by 15% and 20% of crushed brick and 40% in SAP concrete.
5. Application of internal curing, results in higher ultrasonic pulse velocity values and density when compared with the normally cured concrete.
6. The most important benefit of using SAP as an internal curing is when the concrete is produced in hot weather; an increase in compressive strength by about 14% was achieved due to the application of internal curing by SAP for concrete left in air.

## References

1. Okamura, H., "Self-Compacting High Performance Concrete," ACI Concrete International, Vol. 19, No. 7, pp.50-54 July (1997).
2. Erkmen, B., Shield, C. K., and French, C. E." Self-Compacting Concrete (SCC) for Prestressed Bridge Girders," Department of Civil Engineering University of Minnesota, October (2008).
3. ACI Committee 308,"Guide to curing concrete, 308R-01", ACI manual of concrete practice, (2001).
4. Bentz, D.P., Halleck, P.M., Grader, A.S., and Roberts, J.W., "Four-Dimensional X-ray Microtomography Study of Water Movement during Internal Curing" Proceedings of the International RILEM Conference – Volume Changes of Hardening Concrete: Testing and Mitigation, Eds. O.M. Jensen, P. Lura, and K. Kovler, RILEM Publications S.A.R.L., Bagneux, France, 11-20,(2006).
5. Youssef, D., "The use of lightweight sand for internal curing and its effect on performance of HPC used for concrete infrastructures," Sherbrooke, Québec, Canada, June, (2013).
6. Cusson, D., "Effect of blended cements on effectiveness of internal curing in HPC," ACI SP-256-9, (2009).
7. Mohammed, L. M., "improvement SCME properties of concrete with internal curing by using lightweight aggregate" MSc thesis, University of Baghdad, 2011.
8. Zohuriaan-Mehr, M. J. and Kabiri, K., " Superabsorbent Polymer Materials: A Review," Iranian Polymer Journal, 17 (6), 2008, 451-477.
9. Elliott, M., "Superabsorbent Polymers," Product Development Scientist for SAP, BASF, http://chimianet.zefat.ac.il/download/Super absorbant_polymers.pdf.
10. Hasholt, M., Seneka-Jespersen, M., and Jensen, O." Mechanical Properties of Concrete with SAP Part I: Development of Compressive Strength," In O. Jensen, M. Hasholt, & S. Laustsen (Ed.), International RILEM Conference on Use of Superabsorbent Polymers and Other New Additives in Concrete. Bagneaux: RILEM Publications S.A.R.L. 2010.
11. Mechtcherine, V. and Reinhardt, H. W., "Application of superabsorbent polymers (SAP) in concrete construction,"state of the art report prepared by technical committee 225-sap, Springer, 2011.
12. IQS No.5,"Portland cement", Central Organization for Standardization and Quality Control,Iraq,1984.
13. IQS No.45," Aggregate from natural sources for concrete and building Construction, "Central Organization for Standardization and Quality Control,Iraq,1984.
14. ASTM C1240," Standard Specification for Silica Fume Used in Cementitious Mixtures", Annual Book of ASTM standard, Vol.04.02, 2005.
15. IQS No.25,"Clay building brick", Central Organization for Standardization and Quality Control, Iraq, 1988.
16. Emborg M., "Mixing and Transport ", Brite EuRam, Task 8.1, June 2000, pp.11-13.
17. EFNARC,"The European Guidelines for Self-Compacting Concrete Specification, Production and Use", 2005.
18. BS 1881: Part 116, (1989), "Method for Determination of Compressive Strength of Concrete Cubes", British Standards Institution, (2003).
19. ASTM C496 "Standard Test Method for Splitting Tensile Strength for Cylindrical Concrete Specimens", American Society for Testing and Materials, (2004).
20. ASTM C597 "Standard Test Method for Pulse Velocity Through Concrete", American Society for Testing and Materials, (2002).
21. Jensen, O. M., "Use of Superabsorbent Polymers in Concrete", Concrete international, January, 2013.

# Production of lightweight Geopolymer concrete using artificial local lightweight aggregate

*Waleed* Abbas[1], *Wasan* Khalil[1], *and Ibtesam* Nasser*[2]

[1]Building and Construction Engineering Department, University of Technology, Baghdad, Iraq
[2]Middle Technical University, Baghdad, Iraq

**Abstract.** Due to the rapid depletion of natural resources, the use of waste materials and by-products from different industries of building construction has been gaining increased attention. Geopolymer concrete based on Pozzolana is a new material that does not need the presence of Portland cement as a binder. The main focus of this research is to produce lightweight geopolymer concrete (LWGPC) using artificial coarse lightweight aggregate which produced from locally available bentonite clays. In this investigation, the binder is low calcium fly ash (FA) and the alkali activator is sodium hydroxide and sodium silicate in different molarities. The experimental tests including workability, fresh density, also, the compressive strength, splitting tensile strength, flexural strength, water absorption and ultrasonic pulse velocity at the age of 7, 28 and 56 days were studied. The oven dry density and thermal conductivity at 28 days age are investigated. The results show that it is possible to produce high strength lightweight geopolymer concrete successfully used as insulated structural lightweight concrete. The 28-day compressive strength, tensile strength, flexural strength, dry density, and thermal conductivity of the produced LWGPC are 35.8 MPa, 2.6MPa, 5.5 MPa, 1835kg/m3, and 0.9567 W/ (m. K), respectively.

## 1 Introduction

Environmental pollution is the most severe problem that the world is facing today. In the construction industry, the production of ordinary Portland cement (OPC) causes emission of pollutants which result in environmental pollution[1], are being made in the construction industry to address this by developing alternative binders in concrete.

The use of geopolymer technology is one of these alternatives. Geopolymer binders produced from a chemical reaction between silica and alumina molecules that contained in an active pozzolanic material (such as fly ash) react in highly alkaline conditions [3]. The resulting gel binder reacts to produce geopolymer concrete (GPC). The gradual growth of a global effort to minimize greenhouse gas emissions has grown because it has been found that one ton of OPC production needs about 2.8 ton of raw materials, such as limestone and coal, besides releases about 0.7 ton of carbon dioxide ($CO_2$) into the earth's atmosphere due to the decarbonation of lime in the kiln [4]. Approximately 12% of the total $CO_2$ emissions into the earth's atmosphere emitted from the consumption of fossil fuels for commercial and residential heating accounts. In addition, huge amounts of $CO_2$ are emitted from facilities cooling. As a result, the improvement of energy saving systems and new renewable energy sources has become one of the major issues in building structures. The use of lightweight concrete as a building material is very effective in saving energy because of the increase in thermal insulation capacity through the flagging thermal conductivity of lightweight aggregates [5]. Furthermore, the application of structural lightweight concrete has some useful advantages such as, the reduction of the dead load, and a lower density of concrete allows for using smaller and lighter weight structural member that increases the available space and enhances the seismic resistance capacity of the upper structures. In addition, the smaller and lighter elements of precast concrete members are used to make the lifting and transporting system less expensive [6]. Promote effects are expected when fly ash binder and lightweight aggregates are combined to produce environmentally eco-friendly concrete because of the great advantages of both materials. First significant effect is the excellent reduction in $CO_2$ emission from concrete building structures by the use of fly ash (FA) binder with an energy-saving effect due to the use of lightweight aggregates. Thus, precast concrete can be produced with supreme quality and economic efficiency from an early higher strength development capacity of FA paste and a lower density of aggregates.

Since the construction industry is considered to be one of the rapid growing industries, minimizing the heat loss in buildings by increasing its thermal insulation

* Corresponding author: ibtesammhabbaba@yahoo.com

© The Authors, published by EDP Sciences. This is an open access article distributed under the terms of the Creative Commons Attribution License 4.0 (http://creativecommons.org/licenses/by/4.0/).

properties is necessary, as it would enable energy efficient buildings and improved environmental sustainability. Besides, the utilization of industrial waste would be an added advantage [7]. Today, geopolymer is greatly diffuse in industry sector and considered acceptable alternative and has been used in many field applications like precast beams, boat ramp, pavement, bricks, retaining wall, water tanks, precast bridge, decks,…etc.,[8 ,4]. Sachin[8] reports that using LECA as an artificial aggregate gives only 8 $N/mm^2$ without ambient curing temperature. Most researchers agreed that the curing temperature for curing Fly ash geopolymer concrete must be more than 60°C to increase ability to dissolve Si and Al in fly ash.

The aim of this research is to produce lightweight geopolymer concrete (LWGPC) with high strength, using artificial coarse lightweight aggregate which is produced from byproducts.

## 2 Research significance

A lot of researches examined the mechanical properties of GPC and indicated this material as a green or eco friendly material. Very little work was carried out on some properties of lightweight aggregate GPC. In this investigation high strength lightweight aggregate GPC was produced for the first time in Iraq from fly ash and local artificial lightweight aggregate with high compressive strength and low dry density. The beneficial consequence of producing high strength lightweight geopolymer concrete is to reduce the dead load of buildings, and enhance their thermal insulation, which makes this method a future construction alternative.

## 3 Experimental program

### 3.1 Materials

#### *3.1.1 Fly ash*

The fly ash used is recovered as a result of coal combustion during production of electricity from ISKENment-Turkey power station. It is a fine and glassy powder. The results indicate that the fly ash adopted satisfies the requirements of ASTM C 618[9]. The chemical composition of the used fly ash is presented in Table (I). The fly ash has a specific gravity of 2.33 and surface area of 773 $m^2/kg$.

**Table 1**. Chemical analysis of Fly ash*

| Oxide | Content, percent |
|---|---|
| $SiO_2$ | **59.56** |
| $Al_2O_3$ | **29.33** |
| $Fe_2O_3$ | **3.36** |
| CaO | **2.20** |
| MgO | **0.66** |
| $SO_3$ | **0.67** |
| $Na_2O$ | **0.21** |
| $K_2O$ | **2.02** |
| L.O.I | **1.99** |

* Chemical composition analysis was carried outby IraqGeological Survey Central Laboratories Department.

**Table 2.** Chemical requirement for (ASTM C 618, 2005).

| Oxide composition | Pozzolan class F | Fly ash |
|---|---|---|
| $SiO_2$+$Al_2O_3$+$Fe_2O_3$,min. percent | 70 | 89.73 |
| $SO_3$, max. percent | 5 | 0.67 |
| Loss on ignition, max. percent | 6 | 2.77 |

### 3.1.2 The alkali-activated solutions

The alkali-activated solution was used as a combination of two materials: sodium silicate ($Na_2SiO_3$) solution and sodium hydroxide (NaOH) solution to form the binder between the aggregates and other unreacted materials in the geopolymer concrete. Sodium hydroxide in the form of flakes (NaOH with 98 % purity), and sodium silicate solution ($Na_2O$ = 13.0 %, $SiO_2$ = 32.0 % and density = 1.5 g/ml at 20°C, silicate viscosity (CPS) 20°C was 600) were used in this study. For preparing the solution of 1M, 40 g of NaOH flakes (molecular weight of NaOH = 40) was dissolved in one liter of distilled water. In this study different molar concentrations were produced depending on the ratio of caustic soda flakes to water. The solution was prepared a day before it was mixed with sodium silicate in order to be cool, then it was mixed with liquid sodium silicate in a weight proportion of (1: 2.5) (sodium hydroxide: sodium silicate) just before the mixing process.

#### *3.1.3 Fine aggregate*

Al-Ekhaider natural sand was used as fine aggregate of maximum size 4.75 mm. Its gradation zone was (2). The results display that each of physical properties, sand grading and sulfate content were within the requirements of the Iraqi Specification No.45/1984. Its absorption, specific gravity, density and sulfate content are 1.8 %, 2.6, 1670 $kg/m^3$, and 0.12%, respectively.

#### *3.1.4 Admixtures*

A high-range water reducing admixture DARACEM 19CFMQ was used. This type of super-plasticizer is liquid based sulphonated naphthalene and complies with the ASTM C494-2005 Types F[10].

#### *3.1.5 Coarse aggregate*

The coarse lightweight aggregate (LWA), used in this research was locally manufactured artificial lightweight aggregate produced from bentonite clay and water glass (sodium silicate) [10]. The manufacturing method of LWA has four main steps, the raw materials preparation, mixing of raw materials in proper ratios, firing the mixture to an adequate temperature and eventually the cooling process. Bentonite clay was mixed with sodium silicate liquid (clay: sodium silicate, 1:1 by weight). The clayey paste was twirled to form balls, and then they were dried at laboratory temperature for 24 hours, after that they were dried in an oven for 24 hours at 100 °C to ensure the complete drying. The rounded clay balls were burnt in an electrically heated furnace up to 800 °C firing temperature for soaking period of 2 hours. The fired balls were broken manually to smaller size by means of a hammer, and then screened on standard sieve series to fit the grading of coarse lightweight aggregate that conforms to ASTM C 330 Specifications[11]. The manufacturing steps that were carried out to produce the artificial LWA are shown in Fig. (I).

**Figure 1.** Preparation of aggregates

### *3.1.6 Water*

The water used in mixing the geopolymer concrete is potable water from the water supply network system. It was free from organic materials and suspended solids.

## 3.2 Mixing, Casting, and Curing of Lightweight Geopolymer Concrete specimens

Lightweight aggregates and natural sand were moistened for 24 h and then air-dried for another 24 h to imitate the saturated surface dried-state. The coarse and fine aggregates were mixed in the rotary drum mixer for about 3 min, and then the fly ash was mixed with the aggregate for another 5 minutes. The alkaline liquid (Sodium silicate and the sodium hydroxide solutions) were maintained at room temperature for at least 24 h and then it was added to the mixture and mixed for not less than two minutes.

The super-plasticizer and the remaining water were finally mixed for two minutes and added to the mixture. Mixing was continued for another 4 minutes [11]. Then the initial slump was tested and the mixture was poured into varied steel molds to measure the mechanical properties.

**Table 3.** The properties of the produced coarse lightweight aggregate*

| Properties | Test Results of Coarse aggregate | Specification |
|---|---|---|
| Absorption % | 11.5 | ASTM C127 |
| Specific gravity | 1.63 | ASTM C127 |
| Aggregate crushing value % | 773.31 | BS 812-part 110-1990 |
| Dry loose unit weight, (kg/m$^3$) | 768.52** | ASTM 29/C29M |
| Dry rodded unit weight, (kg/m$^3$) | 51.6 | ASTM 29/C29M |
| Sulfate content ( $SO_3$), % | 0.97*** | BS 3797-part 2-1981 |
| Stain intensity**** | No stain | |
| Stain index | 0 | ASTM 641 |

*Physical analysis was conducted by National Center for Construction Laboratories and Researches .

**Within the limit of BS 3797 part 2 ≤ 1.0%.

*** Within the limit of ASTM C330 ≤ 880 kg/m$^3$.

****Staining material test was done in National Center for Construction Laboratories and Researches

The mixture was compacted to remove the entrapped air using vibrating table. Immediately after casting, the surface of the mixture was leveled and the specimens were covered using plastic film to reduce water loss. The specimens were kept in ambient condition for 24h. Then the specimens were cured in an oven for 48h at 60°C. Afterwards, the specimens were taken out of the oven and kept at ambient condition with an average temperature of 28°C. The procedure of mixing has a major effect on workability and strength of geopolymer concrete.

Many researchers stated that geopolymer concrete can be manufactured by adopting the conventional techniques used in the manufacturing of Portland cement concrete [12][13].

## 3.3 Experimental Tests

Several tests were adopted in this investigation including:

- Slump test according to ASTM C143[14].

- Fresh density test according to ASTM C127[15].
- The compressive strength test according to (BS. 1881: Part 116:)( Cube specimens of 100 mm) [16].
- The splitting tensile strength test according to (ASTM 496)(Cylinder specimens of 100 ×200 mm) [17].
- Flexural strength test according to ASTM C78 (Prism specimens of 100 ×100×400 mm) [18].
- Water absorption test conducted according to ASTM C642 (Cube specimens of 100 mm) [19].
- Dry density test according to ASTM C642(Cube specimens of 100 mm) [19].
- Ultrasonic Pulse Velocity (UPV) According to the ASTM C597 (Cube specimens of 100 mm) [20].

# 4 Results and discussion

## 4.1 Selection of Mix Proportions for LWGPC

Davidovits, [14] mention that geopolymer concrete attain most of its strength at early age, usually 7 days. Thus, in this investigation several trial mixes were adopted to select the LWGPC mix with proper density and compressive strength at 7 days age. The curing of fly ash lightweight geopolymer concrete is mostly carried out in previous researches at elevated temperatures; this is because the reaction of fly ash based geopolymer martial is so slow and usually show both slow setting and (slow?) strength development. All previous researchers conducted that curing time and curing temperatures significantly affect the compressive strength of geopolymer concrete [8, 9]. The rise in curing temperatures and longer curing time increase the compressive strength. Table (4) shows the ingredients of all trial mixes prepared in this investigation, their fresh density, and compressive strength at 7 days. Different mix proportions of fly ash: sand: lightweight aggregate (Fa: S: LWA) were used. The fly ash content is varied between 342 and 540 kg /m$^3$; the super-plasticizer was used as 1.9% from the weight of fly ash. The total water in the mix includes the water used in the activator solution and the water added to concrete. The water to fly ash ratio (w/Fa) used is between 0.25 and 0.61, while the ratio of the activator solution to fly ash (A/Fa) is between 0.35 and 0.7 by mass. The molarity of NaOH solution used in the trial mixes is in the range of 10-16, while sodium silicate to sodium hydroxide solutions ratio remains at 2.5 for all trial mixes. The results demonstrate that geopolymer concrete mix (M8) has the highest compressive strength of 29.7 at 7 days, but its fresh density is high (1980 kg / m$^3$), so mix M10 is selected as lightweight geopolymer concrete with compressive strength of 27.3 MPa at 7 days
and fresh density of 1951 kg/m$^3$ with fly ash content of 500 kg /m$^3$. The compressive strength for all trial mixes are shown in Fig. (2). In this research, the effect of alkali activator molarity and curing temperature on the compressive strength of the selected LWGPC mix was investigated. Three groups of LWGPC mixes were prepared using different activator molarities of 12, 14, 16 , each group is subjected to different curing temperatures of 60, 80 and 90 °C , as shown in Table (IV) and Fig.(3). The results show that as the alkali activator molarity increases, the density and compressive strength of LWGPC are increased too. Also the excess of curing temperature increases the compressive strength. It can be concluded that the selected mix has an alkali activator molarity of 16 and cured in temperature of 90 °C for 48 hours with compressive strength and fresh density of 32.8 MPa and 1951 kg/m$^3$ at 7 days age[21]. The same conclusion for curing temperatures was found by other researcher[22]. Other researcher [23] reported that the increase in curing temperature improves the activation of fly ash, which raise the compressive strength and the increase in compressive strength is minor with the increase in duration of curing time from 24hrs to 48hrs. Also, he specifies that the compressive strength of concrete will increase, whenever the molar concentration of sodium hydroxide liquid is increased.

## 4.2 Properties of fresh lightweight Geopolymer concrete

### *4.2.1 Workability and fresh density*

Fly ash has hollow and spherical shape particles with smooth surface texture that lead to increase in the workability[24]. Fly ash lightweight geopolymer concrete prepared in this investigation has a good workability with slump value of 245 mm and fresh density of 1951 kg/m$^3$.

| Mix Symbol | Fa. (kg/m$^3$) | Mix proportion Fa: S: LWA | A/Fa. | W/Fa. | NaOH molarity | Fresh density (kg/m$^3$) | Compressive strength at 7 days ( N/mm$^2$) |
|---|---|---|---|---|---|---|---|
| M1 | 342 | 1 : 2.40 : 1.41 | 0.59 | 0.61 | 12 | 1800 | 10.2 |
| M2 | 350 | 1 : 2.55 :1.62 | 0.72 | 0.53 | 10 | 1810 | 8.3 |
| M3 | 375 | 1 : 2.26 :1.33 | 0.4 | 0.34 | 14 | 1800 | 15.5 |
| M4 | 460 | 1 : 2.50 : 0.60 | 0.5 | 0.25 | 14 | 1850 | 18.2 |
| M5 | 480 | 1 : 0.73 : 0.66 | 0.35 | 0.34 | 14 | 1800 | 24.8 |
| M6 | 460 | 1 : 2.0 : 0.60 | 0.4 | 0.34 | 12 | 1830 | 14.5 |
| M7 | 490 | 1 : 1.32 :1.38 | 0.60 | 0.35 | 10 | 1900 | 9.2 |
| M8 | 490 | 1 : 1.53 :1.22 | 0.4 | 0.25 | 14 | 1980 | 29.7 |
| M9 | 490 | 1 : 1.73 :1.32 | 0.4 | 0.25 | 12 | 2020 | 18.4 |
| M10 | 500 | 1 : 1.50 :1.10 | 0.4 | 0.35 | 16 | 1951 | 27.3 |
| M11 | 500 | 1 : 1.60 : 1.20 | 0.6 | 0.35 | 16 | 2000 | 23.3 |
| M12 | 500 | 1 : 1.60 : 1.10 | 0.5 | 0.34 | 16 | 1950 | 21.2 |
| M13 | 540 | 1: 0.78 : 1.42 | 0.4 | 0.31 | 14 | 1980 | 18.7 |
| M14 | 540 | 1 : 1.40 : 0.92 | 0.4 | 0.31 | 14 | 2020 | 20.2 |
| M15 | 540 | 1 : 1.10 : 0.80 | 0.4 | 0.31 | 14 | 1950 | 16.5 |

**Table 4.** Trial mixes for lightweight GPC

Note1: The mix proportion is by weight; Fa= fly ash, S= sand, LWA=lightweight aggregate, A=activator, W= water.
Note 2: Curing temperature =60 °C for two days, superplasticizer dosage =1.9% by weight of fly ash, sodium silicate: sodium hydroxide solutions =2.5 for all trial mixes.

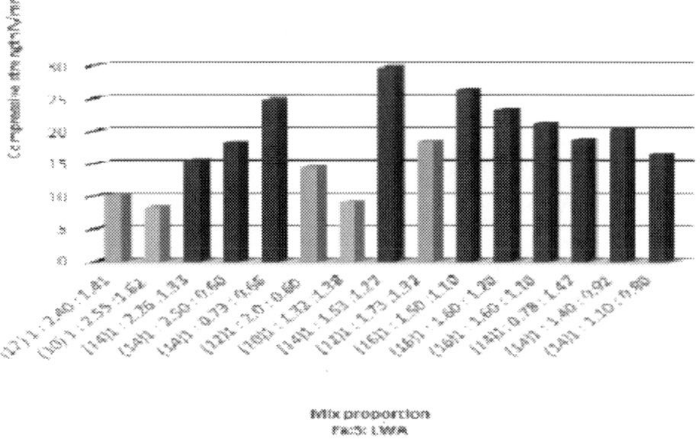


**Fig 2.** The compressive strength for all trial mixes

**Table 5.** Effect of activator molarity and rising curing temperatures on the compressive strength of LWGPC

| Mix symbol | Fa. (kg/m$^3$) | Mix proportion Fa: S: LWA | Alkali activator molarity | Fresh density (kg/m$^3$) | Compressive strength at 7 days after curing in 60°C for 48 hours (N/mm$^2$) | Compressive strength at 7 days after curing in 80°C for 48 hours (N/mm$^2$) | Compressive strength at 7 days after curing in 90°C for 48 hours (N/mm$^2$) |
|---|---|---|---|---|---|---|---|
| GPLW1 | 500 | 1: 1.5:1.1 | 12 | 1943 | 18.2 | 20 | 27.3 |
| GPLW2 | 500 | 1: 1.5:1.1 | 14 | 1945 | 25.6 | 27 | 30.2 |
| GPLW3 | 500 | 1: 1.5:1.1 | 16 | 1951 | 27.3 | 29.2 | 32.8 |

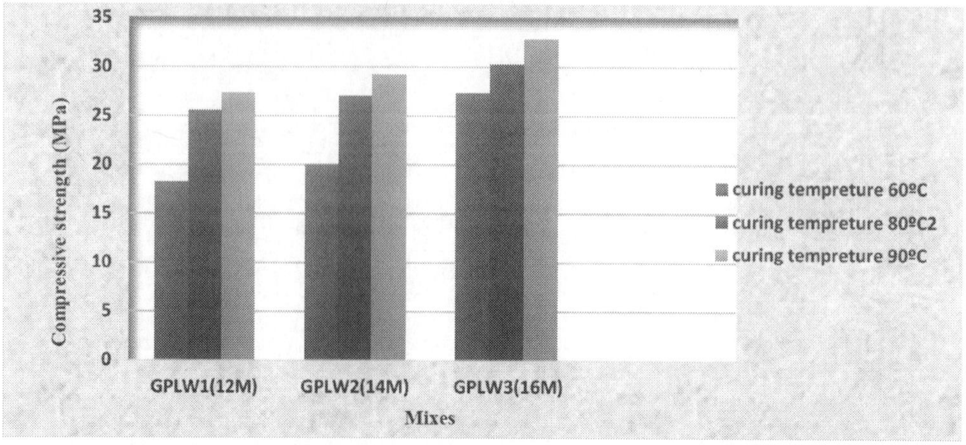


**Fig. 3.** The effect of rising the curing temperature on the compressive strength of LWGPC

**Table 6.** Properties of lightweight geopolymer concrete

| Mix proportion | Age (days) | Compressive strength (MPa) | Splitting tensile strength (MPa) | Flexural strength (MPa) | Oven dry density (kg/m$^3$) | Water absorption (%) | Thermal conductivity (W/m.K) | Pulse velocity (m/sec) |
|---|---|---|---|---|---|---|---|---|
| 1: 1.5: 1.1 Fa: S: LWA (Fa ) contain 500 kg/m$^3$ curing temperature 90° | 7 | 32.8 | 2.22 | 4.6 | - | 8.27 | - | 4.15 |
| | 14 | - | - | - | - | - | - | |
| | 28 | 35.8 | 2.59 | 5.5 | 1835 | 7.5 | 0.9567 | 4.25 |
| | 56 | 36.7 | 2.8 | 5.61 | - | 6.6 | - | 4.35 |

## 4.3 Properties of hardened lightweight Geopolymer concrete

### *4.3.1 Compressive Strength*

Table (5) and Fig (4) present the compressive strength of lightweight geopolymer concrete at 7, 28, and 56 days. Generally the compressive strength increases with age. The percentage increases are 9.5 % and 12 % at 28 and 56-day age, respectively relative to that at 7 days age. This is due to the continuous geopolymerization process.

### *4.3.2 Splitting Tensile Strength*

Table (6) and Fig (5) present the splitting tensile strength of lightweight geopolymer concrete at 7, 28 and 56 days. Generally, the splitting tensile strength increases with age. The percentage increases are 16.7 % and 26 % at 28 and 56 days age, respectively relative to that at 7 days age.

https://doi.org/10.1051/matecconf/201816202024

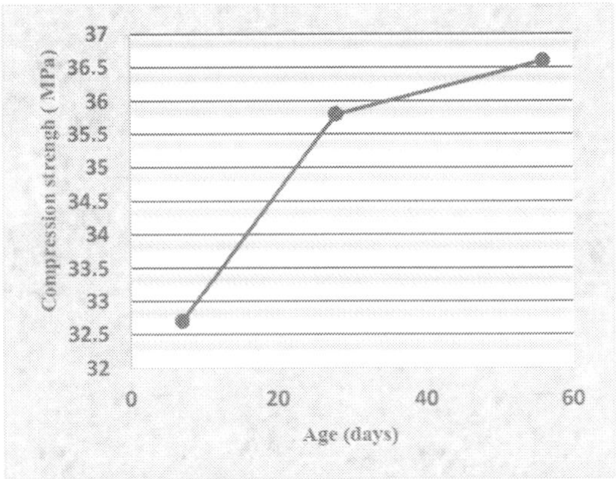


**Fig. 4.** Compressive strength development of LWGPC with age

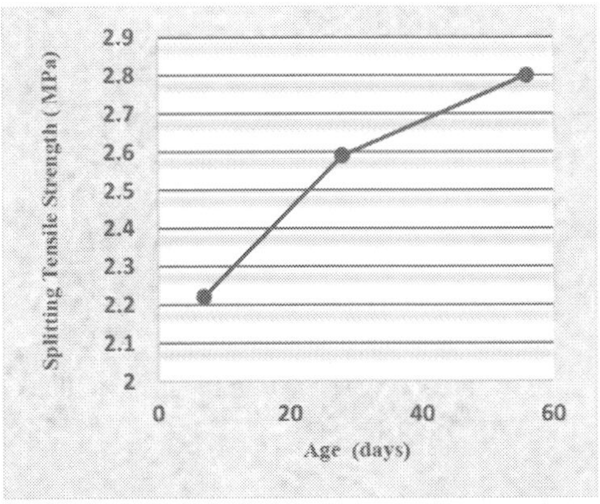


**Fig. 5.** Effect of age on the splitting tensile strength of LWGPC

#### *4.3.3 Flexural Strength*

Table (6) and Fig (6) present the flexural strength of lightweight geopolymer concrete at 7, 28, and 56 days. Generally, the flexural strength increases with age. The percentage increases are 19.5 % and 22 % at 28 and 56 days age, respectively relative to that at 7 days age.

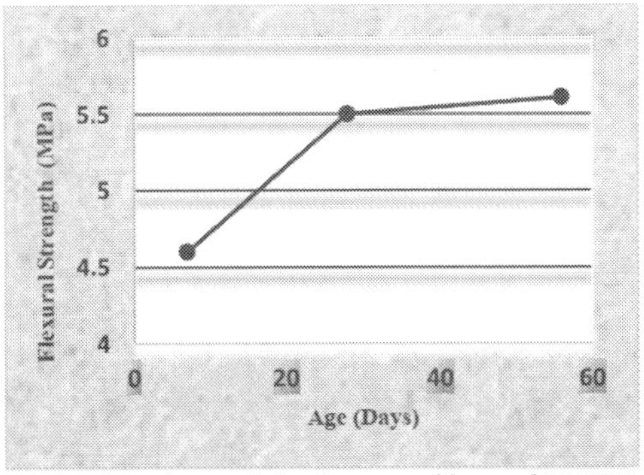


**Fig. 6.** Flexural strength development with age of LWGPC

#### *4.3.4 Water Absorption:*

The water absorption was conducted at different ages (7, 28 and 56) days. Water absorption decreases with the increase in age as shown in Table (6). This is because of hardening of the structure of fly ash geopolymer concrete with time. At all investigated ages, absorption values were less than 10 percent. According to Neville [5], most good concretes have an absorption value below 10 percent by mass. In addition to that, the absorption has a negative relationship with age and that is a good indication to the continuous geopolymerization process.

#### *4.3.5 Oven-dry density*

Oven-dry density for fly ash geopolymer concrete prepared in this investigation is 1835 kg/m³ at 28 days age, thus, it satisfied the requirements of lightweight concrete.

#### *4.3.6 Ultrasonic Pulse Velocity (UPV)*

The UPV test is usually used to estimate the uniformity and quality of GPLWAC, specify the presence of voids, cracks, beside its evaluate the depth of the crack. The UPV values for geopolymer concrete specimens with artificial local lightweight aggregate are in the range of 4.15- 4.35 km/sec as shown in Table (6). According to Neville[5] these values are considered as good quality concrete. The values of UPV obviously show the same trend for the compressive strength of LWGPC.

#### *4.3.7 Thermal conductivity*

The thermal conductivity of LWGPC is 0.9567 W/(m. K) at 28 days age. Normal weight concrete have thermal conductivity in the range of 1.98 - 2.94 W/(m. K) depending on the type of aggregate used [25]. Lower thermal conductivity for lightweight geopolymer concrete is attributed to the presence of voids in lightweight aggregate. The low thermal conductivity of the produced LWGPC made it a good thermal insulating material.

## 5 Conclusions

The main findings of this research are as follows:

1. LWGPC can be produced using fly ash activated by alkali activator (sodium hydroxide:sodium silicate in a weight proportion of 1: 2.5) and locally artificial lightweight coarse aggregate.
2. The density and compressive strength of high strength LWGPC increase as the alkali activator molarity, curing temperature, and the age increase.

3. The selected LWGPC mix, was containing fly ash of 500kg/m$^3$ with alkali activator molarity of 16 and cured at temperature of 90°C.
4. The compressive strength of LWGPC increases with age. Its value is 32.7, 35.8, and 36.8 MPa at 7, 28, and 56 days, respectively. Whereas the percentage increases in splitting tensile strength is 16.7 and 26% at age 28 and 56 days, respectively relative to that at 7 days age. Also the flexure strength increases with age. The percentage increase is 19.5 and 22% at 28 and 56 days, respectively relative to that at 7 days age.
5. The produced high strength LWGPC has oven dry density and thermal conductivity of 1835 kg/m$^3$ and 0.9567 W/ (m. K) at 28 days age, respectively.
6. The produced high strength LWGPC is classified as insulated structural lightweight concrete, in addition it is an environmentally eco-friendly concrete.

## References

[1] A. S.Pratik, B. shinde, and Amit, D. chougule., International Research Journal of Engineering and Technology (IRJET), **3**, (2016).

[2] A. D. Najmabadi, Faculty of Civil Engineering Universiti Teknologi Malaysia JANUARY, Master of Engineering (Civil – Structure), (2012).

[3] E. Ivan Diaz-Loya, E. N. Allouche, and S. Vaidya, ACI materials journal, **108**, (2011).

[4] E. Gartner, Cement and Concrete research, **vol. 34**, pp. 1489-1498, 2004.

[5] A. M. Neville, *Properties of concrete*, Fourth and final edition, Longman Group, Ltd., (2005).

[6] O. Kayali, Construction and Building Materials, **22**, pp. 2393-2399, (2008).

[7] S.-C. Ng and K.-S. Low, Energy and Buildings, **42**, pp. 2452-2456, (2010).

[8] S. P. G. Babu2, Transactions on Engineering and Sciences . **2**, October (2014).

[9] ASTM C618, *Standard Specification for Coal Fly Ash and Raw or Calcined Natural Pozzolan for Use in Concrete,* American Society for Testing and Materials. (2002)

[10] ASTM C494 / C494M, *Standard Specification for Chemical Admixtures for Concrete*, American Society for Testing and Materials.(2005)

[11] ASTM C330-03, *Standard Specification for Lightweight Aggregates for Structural Concrete*, American Society for Testing and Materials. ( 2003).

[12] N. A. Lloyd, and V. Ranjan, Proceedings of the 10th ACI International Conference on Recent Advances in Concrete Technology and Sustainability Issues, Seville, **ACI SP- 261**, pp. 33-54., (2009).

[13] S. H. a. K., and Sanni, R.B., International Journal of Research in Engineering and Technology, pp. 366-371, (2013).

[14] ASTM C143-03, *Standard Test Method for Slump of Hydraulic-Cement Concrete*, ASTM International, American Society for Testing and Materials, **04-02**, United States, (2004).

[15] ASTM C127, *Standard Test Method for Density Relative Density (Specific Gravity) and Adsorption of Coarse Aggregate*, American Society for Testing and Materials, 04-02, United States, (2004).

[16] BS 1881, Part 116, *Method for Determination of Compressive Strength of Concrete Cubes*, British Standards Institution, 1881.( 1989).

[17] ASTM C496, *Standard Test Method for Splitting Tensile Strength for Cylindrical Concrete Specimens*, American Society for Testing and Materials. **04-02**, (2004) .

[18] ASTM C78–02, *Standard Test Methods for Flexural Strength of concrete (Using Sample Beam with Third Point Loading)*, American Society for Testing and Materials. Vol. **04.02**, ( 2004).

[19] ASTM C642-97, *Standard Test Method of Specific Gravity, Absorption and Voids in Hardened Concrete*, American Society for Testing and Materials. Vol. **04-02**, 1997.

[20] ASTM C597, *Standard Test Method for Pulse Velocity through Concrete,* American Society for Testing and Materials, Vol. **04-02**,( 2015).

[21] Mohammed R. Nagral, Tejas Ostwal, and Manojkumar V. Chitawadagi, International Journal of Computational Engineering Research (IJCER),**4**,( 2014).

[22] A. Palomo, M.W. Grutzek , and M.T. Blanco, Cement and Concrete Research,, vol. **29**, pp. pp. 1323- 1329,( 1999).

[23] S. V. Joshi, and M. S. Kadu, International Journal of Environmental Science and Development, **3**, (2012).

[24] J. Davidovits, "*Geopolymer Chemistry and Applications*",3rd edition, Institut Géopolymère, France., (2011) .

[25] B. Milovanovic, I.B. Pecur, and I.Gabrijiel, Novel Development and Innovation in Cementitious Materials, Imperial College, London, United Kingdom, **12**, pp.1-5, (2011).

# Physical and mechanical characteristics of sustainable self-consolidating concrete incorporating high volume fly ash and cement kiln dust

*Mohammed* Aldikheeli[1,*], *Shakir* Salih[2], and *Faiq* Al-Zwainy[3]

[1]College of Engineering, Kufa University, Najaf, Iraq
[2]Department of Building and Construction Engineering, University of Technology, Baghdad, Iraq
[3]College of Engineering, Al-Nahrain University, Baghdad, Iraq

**Abstract.** In recent years, sustainable development has become a global issue and several industries (including concrete industry) have made every effort to reduce the high material cost, high energy depletion and $CO_2$ emission in addition to soundly eliminate industrial waste residue. The purpose of this study is to investigate the possibility of production sustainable self-consolidating concrete (SSC) with green materials: Portland limestone cement (PLC), high volume class F fly ash (HVFA) and locally available cement kiln dust (CKD) depending on the limitations of fresh properties that is specified by ACI 273R-07 and EFNARC and to assess the effects of these green materials on the hardened properties. Moreover, microstructure of this concrete was also investigated by means of scanning electron microscopy (SEM). The results displayed that by replacing high levels of cement by type F fly ash and locally available CKD, it was conceivable to make eco-friendly SCC, satisfying the rheological demands for self-consolidating, with agreeable 28-day compressive strength.

## 1 Introduction

Self-consolidating concrete (SCC) is a very significant advance in the concrete technology in recent times. It is a modern kind of "high performance concrete" with the capability of flowing under its self-weight and without the necessity of vibrations [1]. A better working environment can be provided by using SCC by removing the vibration noise, so it can be contributed for the constructions' sustainability [2]. Civil engineering meeting one fundamental challenge that is to perform projects in harmony with nature using the impression of sustainable development. The concrete industry may be considered to be unsustainable due to the huge production and consumption cycles of concrete have substantial environmental influences [3]. Every year the production of Portland cement is the reason for about 7% of carbon dioxide ($CO_2$) releases in addition to further detrimental emissions such as cement kiln dust. Generality of these emissions are imputed to the clinker production of Portland cement [4, 13]. The industry has transferred to blended cement in order to reduce the dangerous effects of cement production, where in this cement part of the clinker is replaced for other materials such as ground granulated blast furnace slag, fly ash, natural and industrial pozzolana or limestone [5, 14]. Many standard around the world were standardized the blended cements such as (ASTM C 595-12 [6], EN 197-1 [7] and CSA A3001-08 [8]). To get better concrete sustainability and even to attain green or ecological concrete various separate strategies or in combination have been followed [9]. One way is the use of Portland-limestone cement (PLC) where it has a smaller carbon footprint [10].

Both technical and economical benefits can be achieved by using Portland-limestone cements. It is foreseeable that the production of Portland-limestone cement around the world in the future will be constantly increased [11]. Another way to improve concrete sustainability is by using an increased ratio of supplementary cementing materials (SCMs) (such as silica fume, fly ash, slag and natural pozzolan) or by-product materials. Energy saving, environment protection and natural resources conservation can be obtained by using SCMs and other alternative/waste materials in concrete [12]. In this research the by-product materials used were fly ash with high volume fraction and Iraqi cement kiln dust (CKD). The costs due to the production and removal of kiln dust raised up and due to these high costs in addition to future storage space problems, and environmental concerns, the reduction and use of kiln dust has become an important effort.

## 2 Research significance

To produce high volume cement past that required in SCC, high cement content is wanted (450–600 kg/m$^3$) [15, 16] which is not desired neither technically nor

* Corresponding author: mohammedr.aldikheeli@uokufa.edu.iq

© The Authors, published by EDP Sciences. This is an open access article distributed under the terms of the Creative Commons Attribution License 4.0 (http://creativecommons.org/licenses/by/4.0/).

https://doi.org/10.1051/matecconf/201816202025

economically/environmentally. The hope is to make the construction industry proceed toward the sustainability concept and to satisfy this hope the substituting additives such as filling materials and mineral admixtures can be used, especially in SCC [20]. The study of "green", "sustainable" or "eco-efficient" concrete has advanced rising attention among the major contemporary publications about concrete because the affairs concerning the industrial wastes recycling, durability of concrete, environment and the cost will place a pressure on the employment of waste materials [17]. The current study aims at highlighting the possibility to produce sustainable self-consolidating concrete (SSC) with green materials: Portland limestone cement (PLC), high volume class F fly ash (HVFA) and locally available cement kiln dust (CKD) and to evaluate hardened characteristics of SCC produced with these materials.

## 3 Materials and experimental program

### 3.1 Materials

#### *3.1.1 Cement*

In the present study, the cement used was local Portland-lime stone cement (PLC) available in the markets, Karasta CEM II/A-L 42.5 R, produced by Lafarge Company in Al – Sulaymaniyah city in Iraq. It complies with European Standard EN 197-1 [7] and Iraqi industrial license No: 3868. The physical and chemical characteristics of cement used in this study are presented in Table (1).

#### *3.1.2 Fine and coarse aggregate*

As fine aggregate natural sand brought from Al-Najaf quarry was used in this work. The grading, physical and chemical properties of the sand used are shown in Table (2). It has a fineness modulus of 2.5 and within the grading zone 3. A Crushed gravel with a maximum size of 20mm was used as a coarse aggregate, the grading, physical and chemical properties of gavel used are shown in Table (3). Both types of aggregate were conformed to the Iraqi specification No.45 / 1984[23].

#### *3.1.3 Chemical admixture*

A high performance superplasticizer based on modified polycarboxylic ether which is commercially famous (GLENIUM 54) was used, for the liquefaction of the concrete mixtures to achieve the desired workability, throughout this study as a "high range water reducing admixture" (HRWRA). It complies with ASTM C494 [24]. Typical addition proportions were provided by the manufacturer (Basf Chemical Company) as 0.5 - 2.5 liters per 100 kg of cement (cementitious material).

#### *3.1.4 Fly ash*

Fly ash is generally captured from the chimneys of electric power generation facilities. Fly ash used in present study was obtained from Turkey. The physical and chemical properties of fly ash as provided by the manufacturer are tabulated in Table (4). It can be seen from Table (4) that the fly ash used is considered as class F fly ash as per ASTM C618 standard [25]. As shown in Fig. (1), Pozzolanic Activity Index (PAI) test was conducted on the fly ash used and it was 85%.

#### *3.1.5 Cement kiln dust*

Cement kiln dust (CKD) is a by-product of cement production. CKD is a fine-grained, particulate material easily entrained in the combustion gases moving through the kiln [18]. The chemical composition of cement kiln dust differs from one cement plant to another because it depends on many factors, such as the composition of the raw materials, the kiln process incorporated for cement manufacturing, and the fuel used to heat the kiln, while the particle size distribution is determined by the collection method employed. Table (5) indicates the chemical composition and Fig. (2) shows the Scanning Electron Microscopy (SEM) of the cement kiln dust used in this research.

**Fig. 1.** The pozzolanic activity index (PAI) test.

**Fig. 2.** SEM for CKD used

## 3.2 Experimental program

### 3.2.1 Mix proportions

Because of SCC mixtures highly reliant on the properties and the composition of its ingredients, it can be considered a delicate mix. Two disagreeing properties should be found in each SCC mix, and these are the high flow-ability and the high segregation resistance. For powder type SCC the equilibrium between these two properties can be achieved by the dispersal effect of high-range water-reducing admixture (super-plasticizer) and cohesiveness created by the addition of the filler materials due to the high concentration of fine particles [26]. Yet, there is a continuous improvement in the mix design ways and developing procedures for testing SCC. Mix design criteria are generally concentrated on the mixture proportions and type of the ingredients. One of the chief characteristics in selection the mix proportions of SCC mixtures is the amendment of superplasticizer dosage and the water / powder ratio [26]. In the present work, the reference SCC mixture (REF) was designed according to Okamura and Ouchi [1] taking into account the recommendations of the EFNARC [21] and ACI 237 [22]. The used water/powder ratio (w/p) was kept constant as 0.36, while the dose of used super-plasticizer was changed to attain the fresh properties that satisfy the limitations of EFNARC [21] and ACI 237 [22], for this, for each mixture several trial batches were produced. Seven different SCC mixes were investigated in this study. One reference concrete mixtures made with 500 $kg/m^3$ of (PLC) alone (REF), three mixtures had 40%, 50%, and 60% Class F fly ash that replace cement (40, 50, 60 FA) and two mixtures had a cement replacement of 20% and 30% by cement kiln dust (20, 30 CKD). In addition to one binary mixture (50B) had cement replacement of 50% of the two materials (20% CKD plus 30% FA). Table (6) shows the mixture proportions of these mixes.

### 3.2.2 Mixing sequence and samples preparation

The batching sequence and mixing process impact the consistency of production of SCC. In this study drum type mixer of 0.1 $m^3$ capacity was used to mix the concrete ingredients. The dry constituents of concrete mixes were placed in the mixer such that the cement or (cement plus powder materials) is placed between two layers of sand followed by two layers of gravel, this prevents spillage of cement in air, the dry materials were well mixed for about 3 minutes to attain uniform mix. Then, about 80 % of the required quantity of tap water was added and mixed thoroughly for another 3 minutes. Finally, The HRWR diluted with the residual mixing water was then presented through 30 second, and the concrete was mixed for 2.5 minutes [30]. The concrete stayed at rest in the mixer for 1 minute to permit any big air bubble captured throughout mixing to rise to the surface. The concrete was then remixed for 1 minute [27]. After that the concrete was casted in the moulds without any vibration and instantaneously covered with plastic wrap and kept undisturbed for 24 hr. in laboratory circumstances. After 24 hr., specimens were removed from the moulds and placed in curing tank until testing.

### 3.2.3 Test procedure

#### 3.2.3.1 Tests on fresh SCC

To calculate and evaluate the fresh features of SCC there are several test methods that had been developed around the world. Among these test methods there is no single test that can be used alone to assess all of the main parameters, so a combination of tests is necessary to totally describe a SCC mix. In this work, the three main characteristics of SCC listed below are performed according to the methods mentioned in EFNARC [21] and/or ACI 237R-07[22]:

i- Filling ability: ability to fill a mould or form under its own weight. This property can be assessed by slump–flow and $T_{50}$ test. To execute the test, the fresh concrete mix was filled into the slump cone till full and then the slump cone was raised upwards to permit the concrete mix to flow. When the flow ended, two perpendicular diameters of the concrete patty made were measured and averaged as the slump-flow as in Fig. (3-a). While the slump-flow test is carrying out, viscosity is indirectly measured by $T_{50}$ test. Any sign of segregation was observed, as indicated by the formation of a strip of paste/mortar without coarse aggregate at the edge of the concrete patty.

ii- Passing ability: ability to flow through reinforcing bars or other obstacles without segregation and without mechanical vibration. This property can be assessed by J-ring test. According to ACI 237R-07 [22] the difference between the conventional slump flow value and the J-ring slump flow value is calculated and it is an indication of the degree to which the passage of SCC through reinforcing bars is restricted. According to EFNARC [21] the passing ability can be indicated by measuring the differences in height between the concrete inside and that just outside the J-Ring as shown in Fig. (3-b).

iii-Resistance to segregation: ability of fresh concrete to remain homogenous in composition during transport and placing. This property can be assessed by Column segregation test as shown in Fig. (3-c). This test was done according to ACI 237R-07 [22] and the following equation is used to determine the percent segregation:

$$S\% = \left[\frac{CAB - CAT}{\frac{CAB + CAT}{2}}\right] \times 100 \quad if\ CAB > CAT \qquad (1)$$

$$S\% = 0 \qquad if\ CAB < CAT$$

Where: S = percent segregation; CAT = mass of coarse aggregate in the top section of the column; and CAB =

mass of coarse aggregate in the bottom section of the column.

**Fig. 3.** Tests of fresh (SCC)

### *3.2.3.2 Tests on hardened SCC*

i- Compressive strength test:
Concrete compressive strength test was performed according to the BS EN 12390-3 [28] on 100 mm cubes, by using ELE digital compression machine of 2000 KN as shown in Fig. (4-a).

ii-Splitting tensile strength test:
Splitting tensile strength of the concrete was carried out in according to ASTM C496 – 04 [29] on cylinders of (100 mm × 200 mm) by using the same machine used for testing the compressive strength. The specimen was placed horizontally between the plates of testing machine and the load was increased at a rate of (0.94 KN/s.) until failure by splitting along the vertical axis takes place as shown in Fig. (4-b).

iii-Static modulus of elasticity test:
The static modulus of elasticity was determined according to ASTM C469 – 02 [40] using (d=150 mm, h=300 mm) cylindrical specimens and mechanical strain gauges (ELE) of effective length equal to 150 mm as shown in Fig. (4-c). The chord modulus was used in this study and in this modulus, the slope of a line drawn from a point representing 50μ€ to the point corresponding to 40% of the ultimate stress and it is calculated as follow:

$$Ec = S2 - S1/\epsilon 2 - 0.00005 \qquad (2)$$

where:
$E_c$= chord "Young" modulus of elasticity, (MPa)
$S_2$= stress corresponding to 40% of ultimate load, (MPa)
$S_1$= stress corresponding to a longitudinal strain (0.00005), (MPa
$\varepsilon_2$ = longitudinal strain produced by stress $S_2$

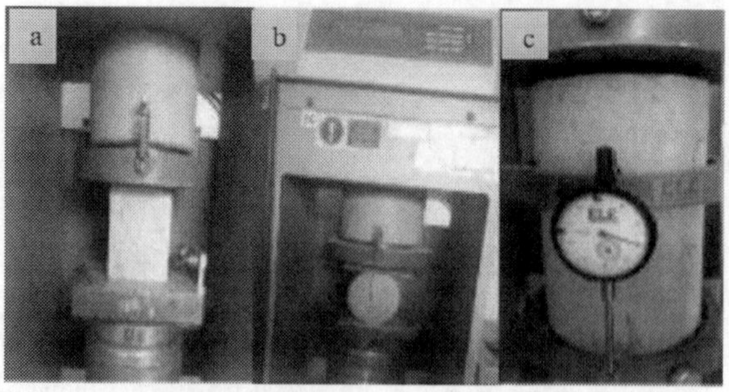


**Fig. 4.** Tests of hardened SCC

## 4 Results and discussion

### 4.1 The effect of replacement level of fly ash and cement kiln dust on the fresh properties of SCC

According to EFNARC [21] a concrete mixture can only be classified as SCC if the requirements for filling, passing, and segregation resistivity characteristics are fulfilled. Fig. (3) shows the fresh SCC after conducting the slump flow, J-ring (flow and height differences) and column segregation tests. Table (7) presents the fresh properties namely (filling ability, passing ability and segregation resistance) of the studied mixes accompanied by the acceptable criteria proposed by EFNARC [21] and ACI 234R-07[22]. In terms of filling ability, it can be seen from table (7) and Figure (5) that the slump flow diameter ($d_s$) for all mixtures ranges between (668-752) mm which is a sign of good deformability conforming to the recommendations of EFNARC [21] (class $SF_2$) and ACI [22]. Similarly, the $T_{50}$ (which may be considered as in direct method to assess plastic viscosity) was always in the range specified (2-5) sec and according to EFNARC [21] all mixes within $VS_2$ class. The slump flow diameter ($d_s$) for SCC with (40, 50, 60) % fly ash replacement was (727, 740, 752) mm with an increase of (4.4, 6.1, 8.2) % respectively in comparison to the reference SCC mix though they have lower super-plasticizer dosage. The slump flow diameter ($d_s$) of CKD mixes were lower than that of FA and REF mixes and ($d_s$) reduced as CKD replacement increase. This reduction in filling ability may be imputed to absorbing mixing water by CKD due to the free lime that is found in CKD which rapidly reacts with water in addition to the higher fineness of CKD [19]. In binary mix (50B) which contains CKD together with FA showed better filling and passing abilities than mixes with CKD alone. This can be due to spherical shape of FA particles which induces a ball bearing effect so the poor fresh properties of CKD can be overcome by the incorporation of FA together with CKD to make binary system.

For fly ash series ($T_{50}$) increase with the increasing Fly ash content and this is probably because of lower specific gravity of fly ash compared with cement and this will lead to increasing paste volume which results in greater plasticity and better cohesiveness [32]. Also for this reason the stability of FA concrete mixtures increases with an increase of the FA ratio used. With respect to the stability the different SCC mixes operate well, as all these mixtures showed a segregation resistance according to ACI 237R [22] less than 10 mm as shown in Fig. (6).

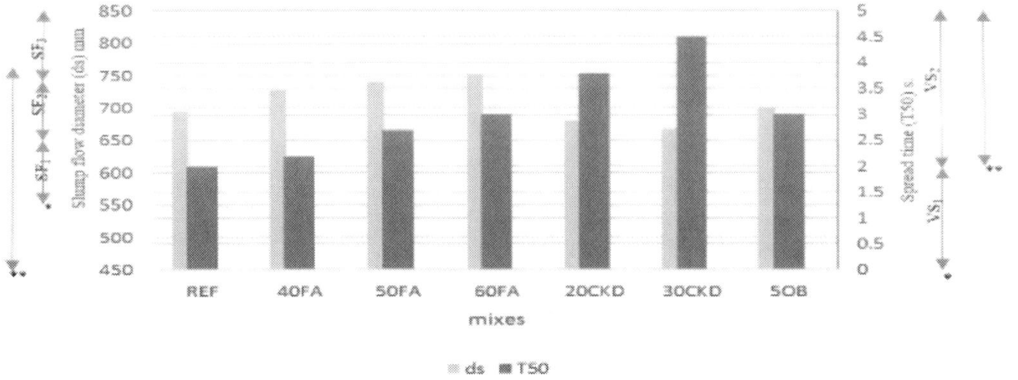


**Fig. 5.** filling ability for SCC mixes with acceptance criteria of EFNARC* [21] and ACI ** [22].

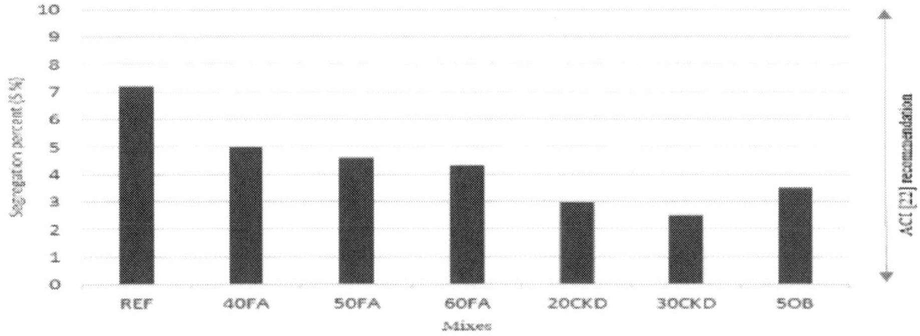


**Fig. 6.** Segregation resistance (static stability) for SCC mixes with acceptance criteria of ACI237R [22].

In the J-ring test as shown in Fig. (7) and Table (7) all SCC mixes exhibited passing ability in terms of the difference in heights of the concrete inside and outside of the ring according to EFNARC [21] in the range of (2-6.3). Also, it was observed that when passing ability expressed by the difference of values between the slump flow without and with the J-ring according to the ACI 237R [22] ($d_s - d_J$) to be with (0-25) mm except for 30CKD mix which was around 38 mm so this mix may exhibit minimal to noticeable blocking and this may be due to the incorporation of high level of CKD, increase in the fines amount of the mix and thus, making the mixture very sticky.

With respect to the dosage of super plasticizer (SP) used, it can be seen from Table (7) that FA mixes consume lower SP dosage compared with CKD mixes and according to Mehta [31] this may be due to three mechanisms. First, the fine fly ash particles prevent cement flocculation thus, the cement particles are dispersed effectively and do not trap large amounts of water. Also, since the fly ash particles have a smooth, spherical surface so that inter-particle friction is reduced. Finally, there is more efficient particle packing within the paste. Moreover, the higher dosage of SP in CKD mixes can be compensated with the incorporation of FA together with CKD as in mix (50B) which had the same SP dosage as 20 CKD mix but with higher slump flow diameter.

From the fresh properties it can be seen that all the SCC mixes had fallen in desired range as per EFNARC[21] and ACI 234R[22] recommendations, so sustainable SCC can be produced by using HVFA and CKD. Hooton et al. [11] stated that the replacement of clinker with limestone filler in the production of blended cement decreases the required amount of water and increases the workability so the significant role of PLC in obtaining these results must be considered.

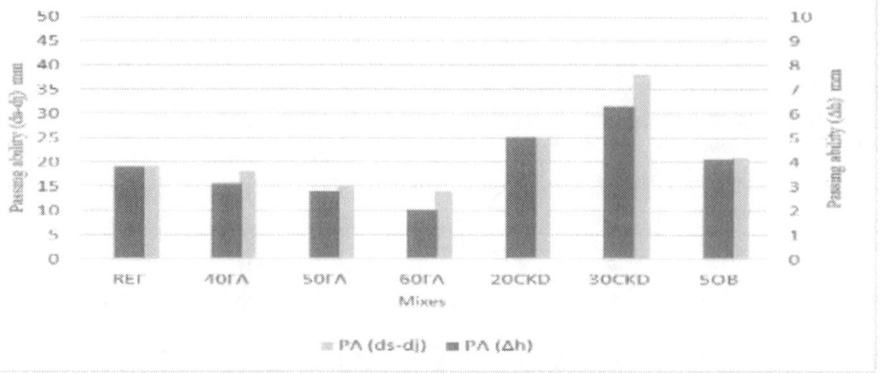


**Fig. 7.** Passing ability (PA) for SCC mixes according to EFNARC [21] (Δh) and according to ACI 237R [22] ($d_s$-$d_j$)

## 4.2 The effect of replacement level of fly ash and cement kiln dust on the hardened properties of SCC

### *4.2.1 Compressive strength*

The average compressive strength of all SCC mixes at 7, 28, 56 and 90 days are shown in Fig. (8) and Table (8). As the curing time increased the compressive strength of all SCC mixes also increased. The REF SCC mix gives the higher compressive strength compared to other SCC concrete mixes at all ages as shown in Fig. (8) and this may be due to high Portland limestone cement (PLC) used where according to Leemann et al. [33] the concrete that is produced with (PLC) will have a slightly lower interfacial transition zone (ITZ) thickness. This can be interpreted as an effect of grain-size dependent particle packing. Besides, REF mix containing higher super-plasticizer (SP) dosage compared to FA mixes and it is generally agreed that when (SP) is used to lower water requirement, an increase in compressive strength will occur.

In terms of FA SCC mixes that contain (40, 50, 60) % replacement levels, the results showed that at all test ages, as the fly ash amount increased, the compressive strength of these mixes decreased and it was lower than the REF SCC control mix, especially at early ages as shown in Figure (8). According to Yazici et al. [34] because of the dilution effect in addition to slow pozzolanic reaction of fly ash this reduction in compressive strength occurs in binary blended Portland cement comprising fly ash, but in this study there was a slight reduction in compressive strength at early ages compared to REF mix and this may be due to the use of (PLC) instead of ordinary Portland cement, where Weerdt et al. [35] indicated that when compared to limestone free cements the existence of limestone powder can be led to the creation of hemi- and monocarbonate and to a stabilization of ettringite, where a part of the ettringite (AFt) transformed to monosulphate (AFm). So, an increase in the volume of the hydrates can happen and an increase in compressive strength may occur. This effect was amplified for the fly ash-limestone blended cements due to the additional alumina provided by the fly ash reaction. Thus, the problems of low early compressive strength in addition to demolding problem (due to long setting time) associated with HVFA concrete were overcome in this study.

However, the compressive strength of the binary SCC mix (50B), which contains (30% FA+20% CKD) is still lower than that of the REF control mix and this could be attributed to the dilution effect, which cannot be compensated through pozzolanic reaction of fly ash and micro filler effect of CKD. Though, when compared the (50B) SCC mix with FA and CKD SCC mixes, the results indicate that this mix has higher compressive strength compared to the others as shown in Figure (8) and this may be due to two reasons; the first reason is the remarkable synergistic effect of the pozzolanic reaction by FA and micro filler by CKD (higher packing density). The second reason is the activation of FA by CKD, this activation was also reported by many studies [36, 37] where the calcium hydroxide created through CKD hydration (due to free lime content) and high sulphate content in CKD (which worked as an activator in the existence of alkalis) stimulates the pozzolanic reaction in fly ash containing system and results in the creation of (C-A-S-H).

For FA SCC mixes the compressive strength is higher than that of CKD SCC mixes as shown in Figure (8) and this may be due to the higher amount of alumina that is found in the mixtures that containing fly ash which will cause an increment in $C_3A$ amount, moreover, there is an increment in "calcium silicate hydrate (C-S-H)" content that is generated due to the pozzolanic reaction of fly ash [38]. Whereas for CKD SCC mixes the lowering in compressive strength may be attributed to the high amounts of alkalis, chlorides and sulphates, and also due to the dilution effects (due to lowering cement content) which influence the amount of (C-S-H) and aluminate hydrates, leading to the creation of significant amounts of hydrated sulpho-aluminates and chloro-aluminates that aid expansion and softening.

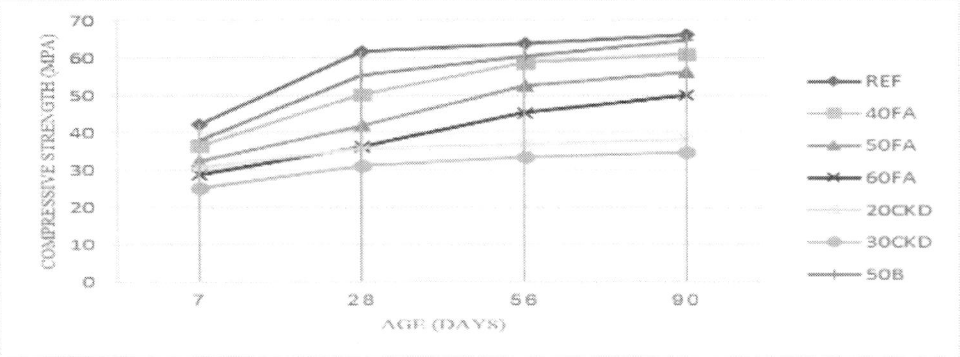


**Fig. 8.** development of compressive strength for SCC mixes

### *4.2.2 Splitting tensile strength*

Broken specimens from split tensile strength test were also used for another assessing the segregation and regularity in the allocation of the aggregates as shown in Fig. (9). The splitting tensile strength governs the cracking behavior and affects other properties such as stiffness and durability of concrete. In this research, the splitting tensile strength was measured at 28 and 90 days and the values for all types of SCC mixes are presented in Table (8) and Fig. (10). The test results showed that in general, all concrete mixes exhibit a continuous increase in the splitting tensile strength from 28 to 90 days. The REF SCC mix showed superior performance in splitting tensile strength compared to other SCC mixes (40, 50, 60) % FA, (20, 30) % CKD and 50B and the percentage increase was (28.6%, 60.2%, 80.6%, 93.3%, 114% and 15.3%) at 28 days and (12.5%, 30%, 38.5%, 80%, 111% and 7.2%) at 90 days, respectively as shown in Fig. (10), this may be due to high amounts of cement and superplasticizer used causing an improvement in the interfacial transition zone (ITZ) and enhancing the bond between the aggregate and cement paste.

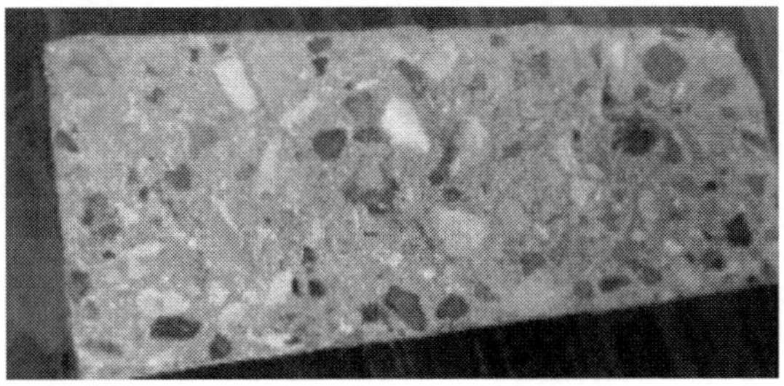

**Fig. 9.** broken specimen from splitting tensile test.

For FA SCC mixes the results showed that the splitting tensile strength decreased as fly ash content increased in both ages but these mixes exhibit better splitting tensile strength compared with CKD SCC mixes as shown in Fig. (10), this behavior may be attributed to the pozzolanic reaction of fly ash with calcium hydroxide $Ca(OH)_2$ being liberated during the hydration of cement, thus, the capillary voids are reduced in size and dense crystals of $Ca(OH)_2$ are replaced with additional (C-S-H) (the main strength-contributing compound) of low permeability as a result of that strengthening the "interfacial transition zone" and lowering the micro cracking which causes a significant increase in tensile strength. Whereas the reduction in tensile strength in CKD mixes is due to poor bonding between aggregate and cement paste.

For binary mix (50B) showed superior performance compared to FA and CKD mixes with a very slight reduction compared to REF mix and this may be due to the activation of fly ash by CKD, where a collection of activators for alumina-silicates system such as CaO, $CaSO_4$, and different forms of alkalis) are found in CKD and this will result in a binding system with enhanced mechanical characteristics when used together with fly ash due to the development of ettringite, monosulfate and syngenite phases in addition to the (C-S-H) formation [37]. So all these products in addition to the process of pore-size and grain size refinement greatly reduced the porosity in the (ITZ) zone thus, reducing the micro-cracking and improving the splitting tensile strength.

### *4.2.3 Modulus of elasticity*

Table (8) shows that the average values of elastic modulus range between (27.4 to 38.24) GPa and (29.6 to 41.25) GPa at 28 and 90 days respectively. It is clear from Fig. (11) that the modulus of elasticity of the mixes produced increased with age and decreased with higher replacement level. The results indicated that the modulus of elasticity for FA mixes at 28 days decreased when FA content in the mixture increased whilst at 90 days there was not a significant difference among these mixes. The increment in the modulus of elasticity at later ages for FA mixes can be attributed to the densification of the paste structure due to the pozzolanic action in addition to improvement in the bond between the aggregate particles and the paste.

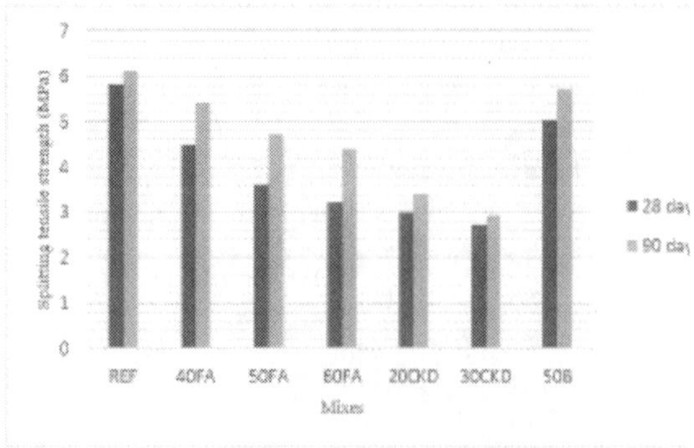


**Fig. 10.** splitting tensile strength for SCC mixes.

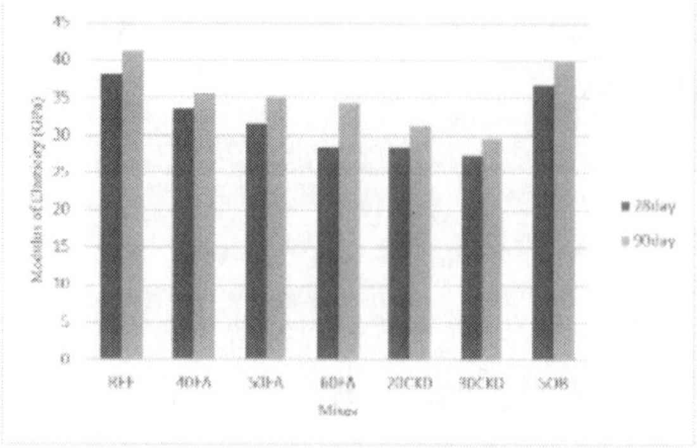


**Fig. 11.** Modulus of elasticity of SCC mixes.

For binary mix (50B) the percentage increase in modulus of elasticity compared to FA mixes were (13.8, 21, 34) % and (16, 17.5, 20.5)% at 28 and 90 days, respectively. This may be explained with the effect of the un-reacted fly ash particles in FA mixes while these particles were activated by CKD when used together in binary mix (50B) so the number of these particles will be reduced and additional hydration products will be formed which will reduce the porosity, especially in the "interfacial transition zone" thus, enhancing the modulus of elasticity.

With respect to the CKD mixes, it showed lower modulus of elasticity compared to other SCC mixes and the modulus of elasticity decreased as CKD increased and this may be due to very high fineness of CKD such that as its content increases, greater shrinkage will occur and thus, a greater number of micro-cracks in the aggregate paste interface will appear that will reduce the modulus of elasticity.

### 4.3 Relationships between hardened properties

The relationship between the values of splitting tensile strength and modulus of elasticity with the compressive strength for SCC mixes was shown in Fig. (12, 13), respectively. Each of these figures, includes equation and correlation coefficient ($R^2$) in addition to equations provided by ACI 318[41] and European standard CEB [42] for estimating the splitting tensile strength and modulus of elasticity based on compressive strength for normal strength concrete. The high value of correlation coefficient indicates that the splitting tensile strength and modulus of elasticity have a strong relationship with compressive strength. This relationship looks to be unaffected by the cement replacement nature and content for SCC mixes.

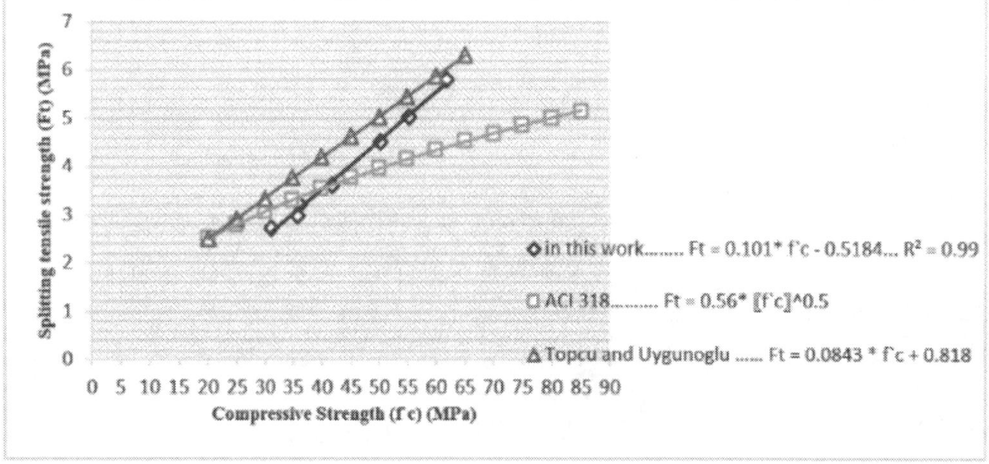


**Fig. 12.** The relation between splitting tensile strength and compressive strength for all mixes; ACI 318 [41] and Topcu and Uygunoglu [43].

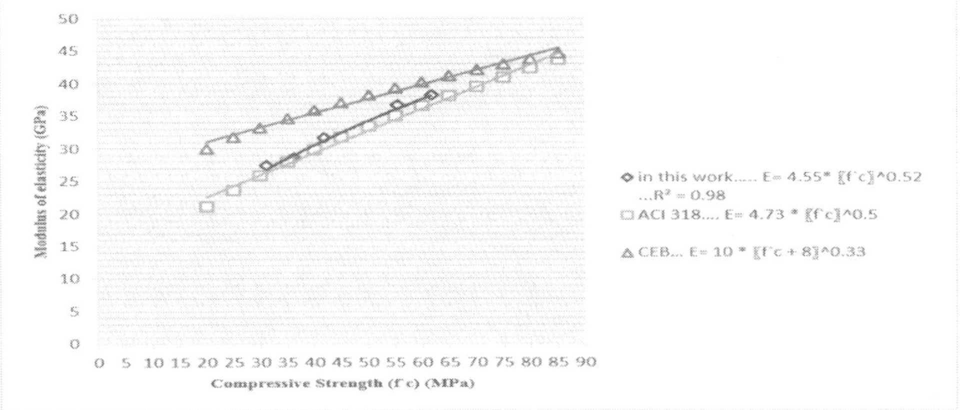


**Fig. 13.** The relation between modulus of elasticity and compressive strength for all mixes; ACI 318 [41] and CEB [42]

### 4.4 Microstructure investigation from Scanning Electronic Microscopy (SEM)

Fig. (14) shows microstructural characteristics from SEM imaging of the studied mixes. For REF SCC mix that contain only (PLC), the SEM images Fig. (14-a) indicated that its microstructure was uniform and compacted with low porosity. Lothenbach et al. [39] reported that high and low forms of carbo-aluminates will be formed as a result of the reaction of ground $CaCO_3$ with various calcium aluminate hydrates during Portland cement hydration. Calcium hemi-carbo-aluminate generates an early hydration product in calcite containing ordinary Portland cement blends. After about 28 days, it transforms approximately entirely to calcium monocarbo-aluminate, a stable AFm phase. Thermodynamic calculations and experimental observations showed that mono-carbo-aluminate creation is preferred to mono-sulfo-aluminate. The existing sulfate reacts with water and calcium hydroxide, crystalizing as ettringite. Due to further ettringite creation, the whole volume of the hydrated phase increases, and the total porosity decreases.

For mixes containing (40, 50, 60) % FA Fig. (14-b, c, d) respectively, it can be seen that there are some un-reacted spherical fly ash particles whereas in binary mix (50B) the amount of these particles was reduced that is an indication to the activation that happened to fly ash by cement kiln dust as discussed previously. Moreover, the formation of the hydration product "ettringite" as a needle-like shape can be recognized.

For binary mix (50B) the synergy between the inclusion materials (PLC, FA, and CKD) can produce a higher packing density that can lead to a denser microstructure as shown in Figure (14-g).

For CKD SCC mixes the SEM images, Figure (14-e, f), showed lower compacted microstructure compared to other SCC mixes and this may be due to the larger amount of chloride present in cement kiln dust causing assortment of crystallization of hydration products which result in an opening of the pore system and this confirms the reduction of strength of these mixes.

## 5 Conclusions

- Self-consolidating concrete mixtures can be produced with Portland limestone cement, high-volume class F fly ash, cement kiln dust and a low dosage of superplasticizers without the use of any viscosity modifying admixtures and with mechanical performance suitable for structural applications.
- One of the principal justifications of incorporating Portland limestone cements has been to enhance concrete sustainability by lowering related carbon emissions and energy usage. Although the sustainability advantages of this approach are impressive, this research investigates the premise that the PLC may also enhance concrete performance through beneficial synergies (interaction of elements that when combined exceed the sum of their individual contributions).
- Because of the unavailability of fly ash in Iraq the substitution of less-expensive CKD as a partial cement replacement improves the sustainability of SCC with reasonable strength. Where it reduces cost and environmental pollution from the disposal of CKD.
- The decreased properties during the short term of high volume fly ash concrete are effectively recovered by the incorporation of CKD and PLC (binary mix).
- When fly ash is used with CKD in binary mix, it intrinsically alleviates the higher water request of CKD (higher requirement for super-plasticizers) and improve the strength.
- The replacement of cement by CKD alone lowered the compressive strength of concrete particularly with a high amount of replacement.
- A good relationship was found between hardened properties and this relationship seems

https://doi.org/10.1051/matecconf/201816202025

to be unaffected by the cement replacement type and content for SCC mixes.

- The addition of fly ash resulted in a denser and uniform microstructure compared to the addition of cement kiln dust.
- Micrograph scanning showed a good agreement with mechanical test results.

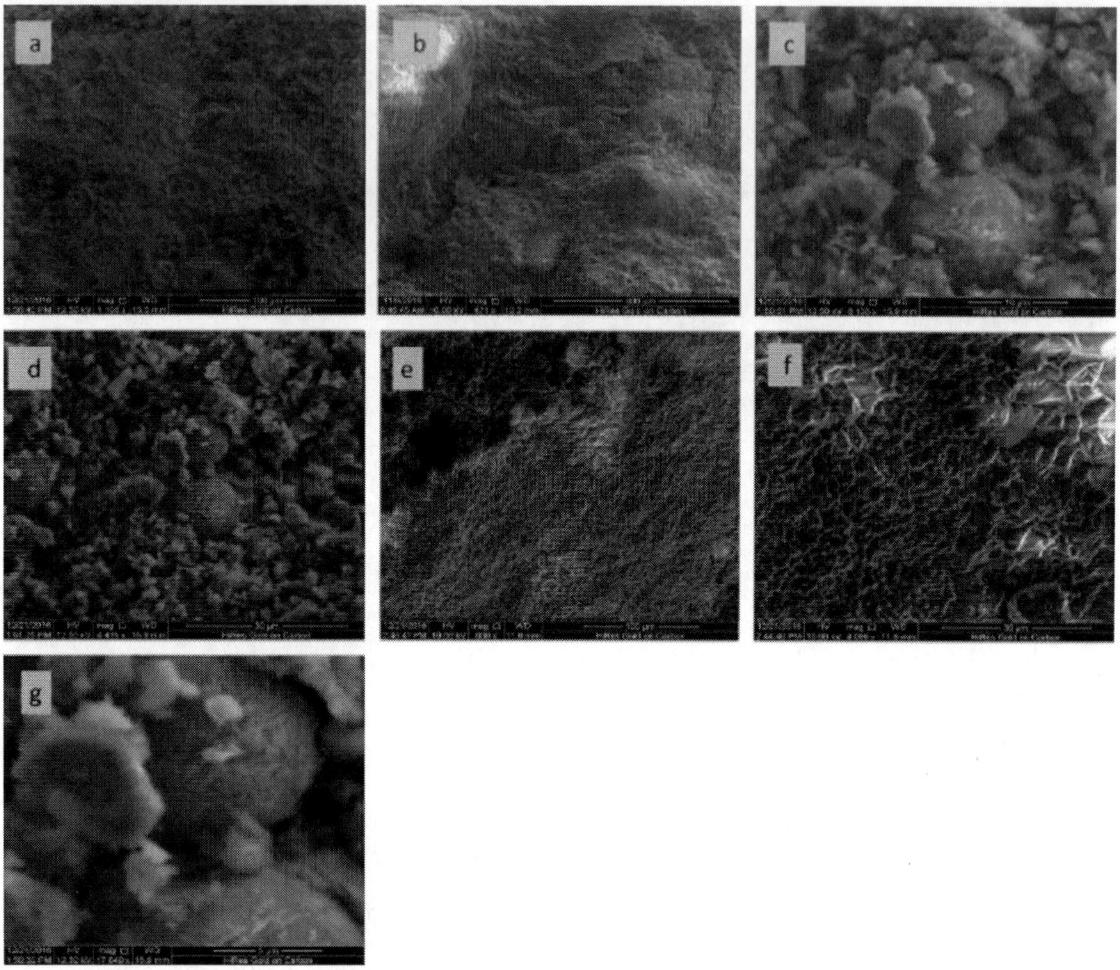


**Fig. 14.** SEM images for SCC mixes. (a, b, c, d, e, f, g) for (REF,40%FA,50%FA,60%FA,20CKD,30CKD and 50B) respectively.

## References

[1] H. Okamura and M. Ouchi , J. of Advan Concr Tech. Vol. **1**, pp. 5-15. (2003).

[2] J. Hu, I. Souza and F. Genarini, J. of Sust Cem-Based Mater. Vol. **6.** (2016). http://dx.doi.org/10.1080/21650373.2016.1230901

[3] International Energy Agency [IEA] and World Business Council for Sustainable Development [WBCSD]. (2009). " Cement Technology Roadmap 2009 - Carbon emissions reductions up to 2050 [Online].Available: http://www.iea.org/papers/2009/Cement_Roadmap.pdf.

[4] A. Gursel, M. Eric, H. Arpad and S. Alex. Cem and Concr Comp. Vol. **51**. pp.38–48. (2014)

[5] P. Pipilikaki, M. Katsioti and JL. Gallias, Const and Build Mater. Vol. **23**, pp. 1042-1049. (2008) http://dx.doi.org/10.1016/j.conbuildmat.2008.05.001

[6] American Society for Testing and Materials, ASTM C595-12, "Standard specification for blended hydraulic cements ". ASTM International. (2012).

[7] European Standard, Cement - Part 1: Composition, Specifications and Conformity Criteria for Common Cements, EN 197-1, (2000).

[8] Canadian Standards Association (2008). "A3001-08: Cementitious materials for use in concrete".

[9] C. Meyer, Cem and Concr Comp. Vol. **31**, pp. 601–605. (2009)

[10] N. Voglis, G. Kakali, E. Chaniotakis and S. Tsivilis. Cem and Concr Comp. Vol. **27**, pp. 191–196. (2005). http://dx.doi.org/10.1016/j.cemconcomp.2004.02.006

[11] D. Hooton, M. Nokken and M. Thomas, Cement Association of Canada. (2007). www.cement.ca

[12] V. M. Malhotra and P. K. Mehta, Ottowa, Canada, *Supplementary Cementing Materials for Sustainable Development*. (2008).

[13] B. Mobasher, In: *CBM-CI International Workshop*, Karachi, Pakistan, pp. 73–90. (2008).

[14] M. Nisbet, M. Marceau and M. VanGeem, PCA R&D Serial No. 2137a. The Portland Cement Association. (2002).

[15] X. Gao, S. Kawashima, X. Liu and P. Surendra , Cem and Concr Comp. Vol. **34**, pp. 478–485. (2012).

[16] H. Dehwah , Const and Build Mater. Vol. **26**, pp. 547–551. (2012).

[17] R. Kraus, T. Naik, B. Ramme and R. Kumar, Const and Build Mater. Vol. **23**, pp. 3439–3442. (2009).

[18] American Society for Testing and Materials, ASTM D 5050-08, "Standard Guide for Commercial Use of Lime Kiln Dusts and Portland Cement Kiln Dusts". Annual Book of ASTM Standards (2008).

[19] P. Kunal, R. Siddique and A. Rajor, Const and Build Mater. Vol. **63**, pp. 49–55. (2014).

[20] P. Domone , Cem and Concr Comp. Vol. **28**, pp.197–208. (2006).

[21] EFNARC. (2005) "Specification and guidelines for self-compacting concrete ". European Federation of Producers and Applicators of Specialist Products for Structures.

[22] ACI Committee 237R-2007. "Self-Consolidating Concrete ". Farmington Hills: American Concrete Institute.

[23] Iraqi Specification, No.45/1984 "Aggregates from Natural Sources for Concrete and Construction". National Centre for Construction Laboratories and Researches, (2001).

[24] American Society for Testing and Materials, ASTM C 494-05, "Standard Specification for Chemical Admixtures for Concrete". Annual Book of ASTM Standards. (2005).

[25] American Society for Testing and Materials, ASTM C 618-05, "Standard Specification for Coal Fly Ash and Raw or Calcined Natural Pozzolan for Use in Concrete". Annual Book of ASTM Standards. (2005).

[26] B. Felekoglu, S. Turkel and B. Baradan , Bulding and Environment. Vol. **42**, pp.1795–1802. (2007).

[27] W. Long, K. Khayat, G. Lemieux, S. Hwang and N. Han , MDPI, Mater J. Vol. **7**, pp. 2474-2489. (2014).

[28] BS EN 12390-3. Testing hardened concrete. Compressive strength of test specimens. (2009).

[29] American Society for Testing and Materials, ASTM C 496-04,"Standard Test Method for Splitting Tensile of Cylindrical Concrete Specimens". Annual Book of ASTM Standards. (2004).

[30] H. Thanh and HM. Ludwig , Mater. and Desi. Vol.**89**, pp.156-166. (2015). http://dx.doi.org/10.1016/j.matdes.2015.09.120

[31] P.K. Mehta, *Proceedings of the International Workshop on Sustainable Development & Concrete Technology*. Beijing, China, pp. 3-14. (2004).

[32] ACI Committee 232.2R-03. "Use of fly ash in concrete ". Farmington Hills: American Concrete Institute. (2003).

[33] A. Leemann, R. Loser and B. Munch , Cem and Concr Comp. Vol. **32**, pp.116–120. (2010). http://doi.org/10.1016/j.cemconcomp.2009.11.007

[34] H. Yazici, M. Yardimci, S. Aydin and A. Karabulut, Const. and Build. Mat. Vol. **23**, pp.1223–1231. (2009). http://doi.org/10.1016/j.conbuildmat.2008.08.003

[35] K. Weerdt, M. Haha and et al. , Cem Concr Res. Vol. **41**, pp. 279–291. (2011). http://doi.org/10.1016/j.cemconres.2010.11.014

[36] R. Siddique and A. Rajor , Reso, Cons and Recy. Vol. **61**, pp. 59–68. (2012). http://doi.org/10.1016/j.resconrec.2012.01.006

[37] P. Chaunsali and S. Peethamparan, ACI Mater J. Vol. **110**, pp. 297–304. (2013).

[38] P.K. Mehta, and P.J. Monteiro. "*Concrete: Microstructure, Properties, and Materials* " Third Edition, McGraw -Hill. (2006).

[39] B. Lothenbach, G. Saout, E. Gallucci and K. Scrivener, Cem Concr Res. Vol. **38**, pp. 848-860. (2008).

[40] American Society for Testing and Materials, ASTM C 469-02, " Standard Test Method for Static Modulus of Elasticity and Poisson's Ratio of Concrete in Compression " Annual Book of ASTM Standards. (2002).

[41] ACI Committee 318 "Building Code Requirements for Structural Concrete". Farmington Hills: American Concrete Institute. (2005).

[42] Euro International Committee of Concrete. CEB-FIP model code 1990. London: Thomas Telford; 1993, as sited in reference [43].

[43] B. Topcu and T. Uygunoglu, Const and Build Mater. Vol. **24**, pp. 1286–1295. (2010)

Table 1. Chemical and physical characteristics of Portland limestone cement (PLC) used in this study*.

| Oxides or Property | PLC test results | Requirement of EN 197-1 [7] | Requirement of Iraqi industrial license No: 3868** |
|---|---|---|---|
| $SiO_2$ | 18.8 | - | - |
| $Al_2O_3$ | 4.8 | - | - |
| $Fe_2O_3$ | 2.7 | - | - |
| CaO | 61.9 | - | - |
| MgO | 2.5 | - | $\leq$ 5.0% |
| $SO_3$ | 2.6 | $\leq$ 4.0% | $\leq$ 2.5% if $C_3A$ less than 5%<br>$\leq$ 2.8% if $C_3A$ more than 5% |
| $Na_2O$ | 0.2 | - | - |
| $K_2O$ | 1.1 | - | - |
| $(Na_2O)_{eq.}$*** | 0.92 | - | - |
| L.O.I | 4.5 | - | - |
| Fineness ($m^2$/Kg) | 390 | - | - |
| Initial setting time (min.) | 128 | $\geq$ 60.0 | $\geq$ 45.0 |
| Final setting time (hr.) | 3.3 | - | - |
| 2 days compressive strength (MPa) | 23 | $\geq$ 20.0 | $\geq$ 20.0 |
| 28days compressive strength (MPa) | 49 | $\geq$ 42.5 | $\geq$ 42.5 |

* Chemical analysis and physical properties were carried out in the laboratory of Al – Kufa cement mill.

** Limit by ICOSQC (Iraqi Central Organization for Standardization & Quality Control).

*** $(Na_2O)_{eq.} = Na_2O + 0.658\ K_2O$

Table 2. Grading and some physical and chemical properties of sand used.

| Sieve size (mm) | Cumulative passing % | Limits of Iraqi specification No.45/1984 [23], zone 3 |
|---|---|---|
| 4.75 | 97 | 90-100 |
| 2.36 | 89 | 85-100 |
| 1.18 | 75 | 75-100 |
| 0.60 | 60 | 60-79 |
| 0.30 | 22 | 12-40 |
| 0.15 | 5 | 0-10 |
| Fineness modulus = 2.5 | | |
| **Property** | **Test results** | **Limits of Iraqi specification No. 45/1984[23]** |
| Specific gravity | 2.58 | - |
| Sulfate content ($SO_3$) % | 0.21 | ≤ 0.5 |
| Absorption % | 1.82 | - |
| Materials finer than sieve No. 200 | 1.4 | ≤ 5.0 |

Table 3. Grading and some physical and chemical properties of gravel used.

| Sieve size (mm) | Cumulative passing % | Limits of Iraqi specification No. 45/1984[23] |
|---|---|---|
| 37.5 | 100 | 100 |
| 20 | 100 | 95-100 |
| 10 | 39 | 30-60 |
| 5 | 3 | 0-10 |
| Property | Test results | Limits of Iraqi specification No. 45/1984[23] |
| Specific gravity | 2.62 | - |
| Sulfate content (SO3) % | 0.03 | ≤ 0.1 |
| Absorption % | 0.7 | - |

Table 4. chemical analysis and physical properties of the fly ash*

| Oxides or Property | Fly ash | ASTM C618-05[25] class F requirement |
|---|---|---|
| $SiO_2$ | 50.5 | SiO2 + Al2O3 + Fe2O3 > 70 |
| $Al_2O_3$ | 22.7 | |
| $Fe_2O_3$ | 9.3 | |
| CaO | 10.8 | |
| MgO | 1.2 | - |
| $Na_2O$ | 1.0 | - |
| $K_2O$ | 0.8 | - |
| $TiO_2$ | 0.7 | - |
| $SO_3$ | 1.5 | 5.0 max. |
| Loss on Ignition | 1.2 | 6.0 max. |
| Specific gravity | 2.12 | - |
| Specific surface area ($m^2/kg$) | 420 | - |

* According to the supplier.

 https://doi.org/10.1051/matecconf/201816202025

Table 5. Chemical composition and physical properties of cement kiln dust used throughout this work*

| **Oxides** | CaO | $SiO_2$ | $Al_2O_3$ | $Fe_2O_3$ | MgO | $SO_3$ | $Na_2O$ | $K_2O$ |
|---|---|---|---|---|---|---|---|---|
| **Content (percent)** | 44.5 | 16.7 | 4.5 | 2.0 | 1.3 | 5.5 | 0.3 | 3.7 |
| **Property** | L. O. I. | | Fineness ($m^2$/kg) | | | | | |
| **Test result** | 20.0 | | 565 | | | | | |

* Chemical analysis were carried out in the laboratory of Al – Kufa cement mill

Table 6. Mix proportions of the concrete mixes.

| Mix ID | Mixture proportions (kg/$m^3$) Cement | Fly ash | Cement kiln dust | Water | Sand | Gravel | W/P* | SP** |
|---|---|---|---|---|---|---|---|---|
| REF | 500 | - | - | 180 | 800 | 800 | 0.36 | 0.8 |
| 40FA | 300 | 200 | - | 180 | 800 | 800 | 0.36 | 0.7 |
| 50FA | 250 | 250 | - | 180 | 800 | 800 | 0.36 | 0.6 |
| 60FA | 200 | 300 | - | 180 | 800 | 800 | 0.36 | 0.55 |
| 20CKD | 400 | - | 100 | 180 | 800 | 800 | 0.36 | 0.9 |
| 30CKD | 350 | - | 150 | 180 | 800 | 800 | 0.36 | 1.1 |
| 50B | 250 | 150 | 100 | 180 | 800 | 800 | 0.36 | 0.9 |

* W/P : water / powder : water / (cement + FA +CKD)

** Sp : superplasticizer : (Lit/100Kg cementitious material)

Table 7. Fresh properties of SCC mixes

| Mix ID | | Filling ability: Slump flow ($d_s$) mm | Filling ability: Spread time ($T_{50}$) S. | Passing ability (J-ring test): Differences in heights (mm) | Passing ability (J-ring test): Flow ($d_j$) | Passing ability (J-ring test): ($d_s$-$d_j$) | Segregation resistance % |
|---|---|---|---|---|---|---|---|
| REF | | 695 | 2 | 3.7 | 676 | 19 | 7.2 |
| 40FA | | 727 | 2.3 | 3.1 | 709 | 18 | 5 |
| 50FA | | 740 | 2.7 | 2.8 | 725 | 15 | 4.8 |
| 60FA | | 752 | 3 | 2 | 738 | 14 | 4.5 |
| 20CKD | | 680 | 3.8 | 5 | 655 | 25 | 3 |
| 30CKD | | 668 | 4.5 | 6.3 | 630 | 38 | 2.5 |
| 50B | | 700 | 3 | 4.1 | 679 | 21 | 3.5 |
| Acceptance criteria of SCC suggested by | ACI [22] | 450 – 760 | 2 – 5 | - | | 0 – 25 | 0 - 10 |
| | EFNARC [21] | 550 – 850 | 2 – 5 | 0 – 10 | | - | - |

Table 8. Hardened properties of SCC mixes

| Mix ID | Compressive strength (MPa) | | | | Splitting tensile strength (Mpa) | | Modulus of elasticity (GPa) | |
|---|---|---|---|---|---|---|---|---|
| | 7 days | 28 days | 56 days | 90 days | 28 days | 90 days | 28 days | 90 days |
| REF | 42.1 | 61.7 | 63.8 | 66.2 | 5.80 | 6.12 | 38.24 | 41.15 |
| 40FA | 36.4 | 50.2 | 58.7 | 61.1 | 4.51 | 5.44 | 33.58 | 35.52 |
| 50FA | 32.3 | 41.8 | 52.7 | 56.3 | 3.62 | 4.71 | 31.60 | 35.11 |
| 60FA | 28.7 | 36.2 | 45.3 | 50.1 | 3.21 | 4.42 | 28.52 | 34.24 |
| 20CKD | 30.6 | 35.7 | 36.8 | 38.1 | 3.00 | 3.4 | 28.30 | 31.22 |
| 30CKD | 25.2 | 31.1 | 33.4 | 34.7 | 2.71 | 2.90 | 27.41 | 29.60 |
| 50B | 38.0 | 55.4 | 60.6 | 64.8 | 5.03 | 5.71 | 36.74 | 39.85 |

 https://doi.org/10.1051/matecconf/201816202026

# Effect of discontinuous curing and ambient temperature on the compressive strength development of fly ash based Geopolymer concrete

*Bayrak* Almuhsin[1,*], *Tareq* al-Attar[1], and *Qais* Hasan[1]
[1]Building and Construction Engineering Department, University of Technology, Baghdad, Iraq

**Abstract.** In the current study, 6 mixtures of Geopolymer concrete have been studied. The effect of discontinuous curing in oven and in atmosphere ambient temperature has been inspected by exposing the Geopolymer concrete specimens to temperature in the oven for few hours then stopping the oven to let it cool down to the ambient temperature. The compressive strength test of 100x200 mm cylindrical specimens for each mixture has been performed at different ages. It was found that the ambient temperature has vast effect on the compressive strength of the Geopolymer concrete in discontinuous curing. Discontinuous curing saves energy and can be a good replacement to the continuous curing when the ambient temperature is 40° C or more. Specimens that were cured continuously in ambient temperature of 43° C has resulted in compressive strength of 23 MPa at age of 40 days; to enhance the compressive strength, it is advised to impose few hours of discontinuous oven curing. It was also found that the rest period (starts when pouring concrete in forms and ends when imposing oven curing to the Geopolymer) has a positive influence on the compressive strength of Geopolymer concrete, but when no rest period is allowed, the later ages compressive strength is remarkably higher.

## 1 Introduction

Geopolymer concrete or so-called mineral-polymer concrete is one of the promising innovations in the concrete industry [1]. It has mainly attracted the attention of the scientists and the concrete industry entities as an alternative to conventional concrete because of being much more environmental friendly than the latter. Large amount of carbon dioxide is being released into the air during the manufacturing of Portland cement [2, 3]. For producing one ton of Portland cement, one ton of Carbon dioxide is emitted to the atmosphere [4]. Knowing that concrete is the second most consumed material, just below water, with an annual consumption of (4180) metric tons/ year (as for 2014) [5]. The result is (4180) metric tons/ year of Carbon dioxide is being added to the atmosphere every year from the production of Portland cement that will be used for the manufacturing of the conventional concrete.

Portland cements paste gains strength from the process of hydration that results in C-S-H gel. The mechanism of developing strength in Geopolymer is explained by the poly-condensation of the precursors of silica and alumina.

Mineral-polymer results from the polymerization of materials that alumina and silica constitutes most of its contents. To enforce the polymerization of the alumina-silica enriched materials, an alkaline activator such as sodium hydroxide and sodium silicate solution is to be added. The addition of the alkaline solution results in the dissolving of the oxide minerals from the silica-alumina oxides and finally poly-condensation to form the final stable silico-aluminate product [6]. The production of the Geopolymer can be as described below:

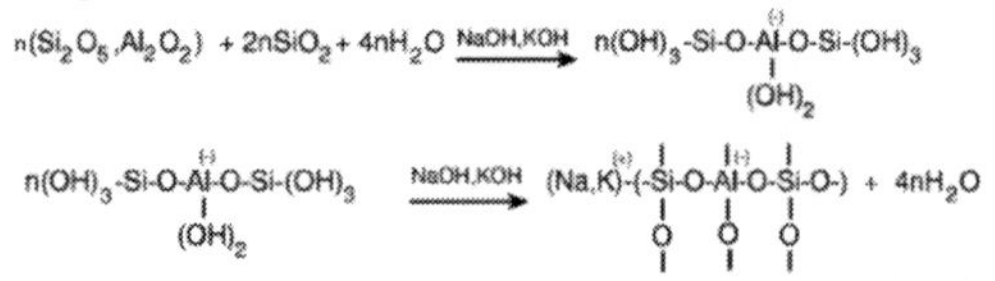

In spite of the fact that since Davidovits has published his innovations in Geopolymer binders and cements and many researchers have studied the physical and chemical properties of the Geopolymer and the production of it for various uses, none of the concrete standardization entities have adopted standard procedures for producing Geopolymer concrete. Thus, more information are required to provide more knowledge for the database in the field to be enriched with more aspects resulting in better understanding of the physical and chemical properties that contribute to the Geopolymer concrete properties.

The advantages of Geopolymer concrete to the environment and the durability of it are considered much higher than those of the Portland cement concrete and make it a good replacement to the conventional Portland cement.

One of the innovative developments in concrete industry is the high range water reducing agents (HRWRA) which is necessary for workable and durable concrete. HRWRA enhance the fresh properties of the concrete, and improve hardened concrete properties,

* Corresponding author: bayrak.almuhsin@yahoo.com

© The Authors, published by EDP Sciences. This is an open access article distributed under the terms of the Creative Commons Attribution License 4.0 (http://creativecommons.org/licenses/by/4.0/).

particularly, the durability of the hardened concrete [7-11].

Due to the advantages of the HRWRA to the industry of concrete, the Geopolymer concrete, as a potential alternative to Portland cement concrete, needs to be investigated for the appropriate type of HRWRA for Geopolymer concrete. The current study used naphthalene-based superplasticizer to get the required workability and to enhance the compressive strength of the Geopolymer concrete.

## 2 Experimental Work

For the production of the Geopolymer for the research program, the following materials were used: fly ash class F, sodium hydroxide flakes, sodium silicate liquid, crushed gravel, sand, superplasticizer, and tap water.

### 2.1 Materials

#### *2.1.1 Fly ash*

Electrical power generation plants that use coal as a fuel in the boilers extract large amounts of fly ash as a byproduct for the combustion process [12]. Fly ash is a fine powdered material consisting of fine particles that are driven out of the boiler by the flue gases through the chimney. ASTM C 618 classifies fly ash based on chemical requirements and physical requirements into Class F and Class C [13]. The requirements for class F is shown in **Table 1**.

**Table 1.** Chemical requirements [13]

| | Class F |
|---|---|
| $SiO_2$+$Al_2O_3$+$Fe_2O_3$, min, % | 70.0 |
| (SO3), max, % | 5.0 |
| Moisture content, max, % | 3.0 |
| Loss on ignition, max, % | 6.0 |

The fly ash used in the current study is manufactured at Iskenderun power station in compliance with DIN EN 450 – fly ash for concrete and is classified as class F in accordance to ASTM C 618. The product specifications as reported by the manufacturer is shown in **Table 2**.

**Table 2.** Fly ash properties

| Parameter | Average |
|---|---|
| Loss On Ignition, % | 2.6 |
| Fineness > 0.045 mm, % | 15.1 |
| Chloride (CI), % | 0.012 |
| Gross Density, kg/m$^3$ | 2330 |

The specific surface area of the fly ash is 0.356 gm/m$^2$, the maximum grain size is 350 μm, the mass percentage retained on 45 μm (no. 325) sieve is 64.25% which exceeds ASTM C 618 limit of 34% for fly ash designated to be used as a Pozzolan material in normal concrete. The percentage of fine materials was disregarded due to the fact that the fly ash is being used as a cementitious material and thus, finer particles results in more chemically active material. The particle size distribution is shown in **Fig. 1**.

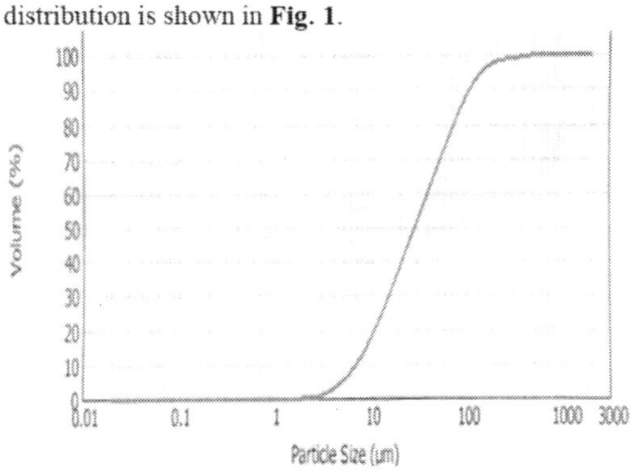


**Fig. 1**. Fly ash particle size distribution

The chemical composition analysis was conducted in the General Laboratories Department of the Iraq Geological Survey in Baghdad. The detailed results are tabulated in **Table 3**.

**Table 3.** Fly ash chemical composition

| Oxide | $SiO_2$ | $Al_2O_3$ | CaO | $SO_3$ | LOI | $Na_2O$ | $K_2O$ |
|---|---|---|---|---|---|---|---|
| Content | 58.13 | 18.89 | 1.60 | 0.18 | 3.28 | 1.27 | 1.46 |

#### *2.1.2 Sodium Hydroxide*

Commercially available sodium hydroxide (NaOH), known as caustic soda, was used for the current study. The physical appearance is white solid smooth shiny flakes. The purity of the sodium hydroxide was measured in the General Laboratories Department of the Iraq Geological Survey in Baghdad and was reported as 98.8%. The product is hydroscopic thus; it was kept in sealed airtight bags in order to protect the flakes from absorbing moisture from the air. Sodium hydroxide reacts with carbon dioxide ($CO_2$) from the air to form sodium bicarbonate (NaHCO) and this is another reason for protecting sodium hydroxide from the exposure to air.

#### *2.1.3 Sodium silicate*

Sodium Silicate ($Na_2SiO_3$) has been supplied from the local market. The consistency is a liquid of 0.55 water content by mass.

#### *2.1.4 Coarse aggregate*

The coarse aggregate that was used for the current study was normal weight crushed aggregate with maximum size of 12.5 mm and conforming to ASTM C 33 [14] as shown in **Fig. 2**. The bulk density of the coarse aggregate was found to be 2665 kg/m$^3$, the absorption is 0.5% and the sulfate content is 0.08%.

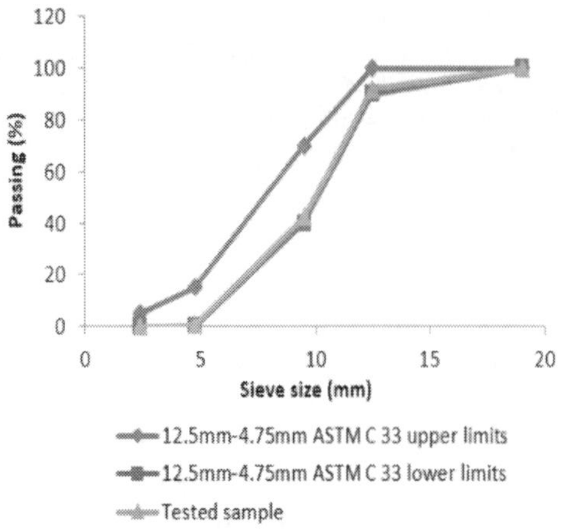


**Fig. 2.** Course aggregate sieve analysis

### *2.1.5 Fine aggregate*

Clean and washed river sand was used in the current study. ASTM C 33-03 specifications [14] were used for the standardization of the sand; it was found that the tested sand conformed to the specifications as shown in **Fig. 3**. The bulk density of the fine aggregate was found to be 2600 kg/m$^3$ and the sulfate content is 0.13%.

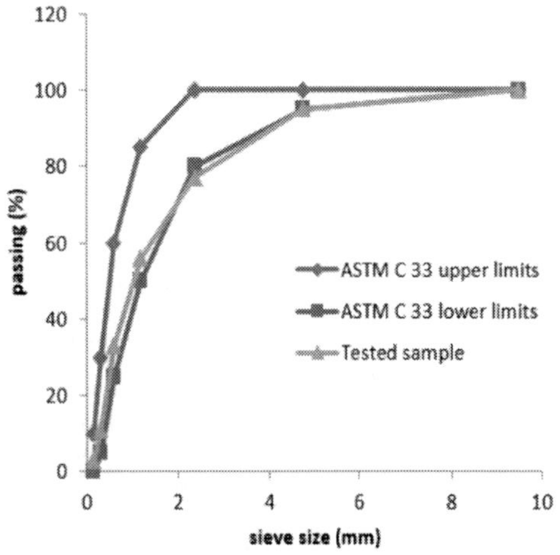


**Fig. 3.** Fine aggregate sieve analysis

### *2.1.6 Superplasticizer*

The superplasticizer used is the KUT PLAST SP 400 high range water reducing admixture of relative density 1.22-1.24 which is a solution based on modified sulfonated naphthalene formaldehyde condensate. The product was purchased from the local market and was stored in a place protected from extreme temperature.

### *2.1.7 Water*

Clean tap water was used for the sodium hydroxide solution preparation and as an extra mixing water for more workable Geopolymer concrete.

## 2.2 Mixing procedure

The first step of mixing was to prepare the alkaline solution by mixing the water with the sodium hydroxide flakes with stirring by a rod for 30 seconds. The solution then was left for at least 1 hour for the complete dissolving of the NaOH flakes. Then the sodium silicate liquid was added to the NaOH solution with stirring for 1 minute and left for at least 30 minutes in the laboratory. The fine and coarse aggregate were mixed together in the electric pan mixer for 2 minutes then the fly ash was added and mixed for 1 minute. The HRWA superplasticizer was added to the alkaline solution not later than 2 minutes of the addition of the alkaline solution to the dry materials. The HRWA superplasticizer was mixed for 20 seconds with the alkaline solution, then directly added to the dry materials in the mixer. The mixing lasted for 4-7 minutes in the mixer. When extra-water was required, it was added to the wet mix at least 1 minute after the mixing of the alkaline solution with the dry materials.

## 2.3 Curing procedure

The curing of the Geopolymer concrete differs from that of the Portland cement concrete. The former is cured by heat, while the latter is cured by water. The specimens were cured by two methods: One of the methods is the oven discontinuous curing, which expose the Geopolymer concrete specimens to temperature in the oven for few hours then stopping the oven to let it cool down to the ambient temperature. This method was applied to mix 3 through mix 6. The curing regime is illustrated in **Fig. 4**, 5, and 6. The oven temperature is 90°C for mix 3, 4, and 5 and is 80°C for mix 6. The second curing method is exposing of the specimens to ambient temperature only.

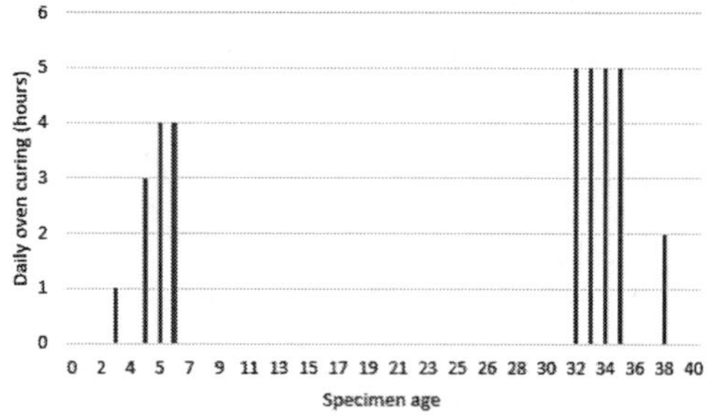


**Fig. 4.** Discontinuous oven curing of mix 3

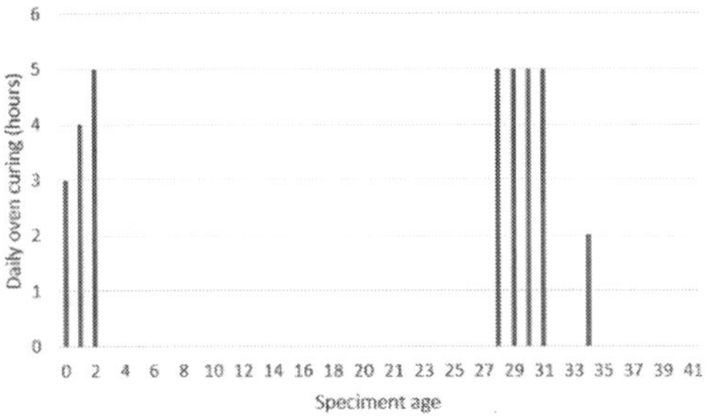


**Fig. 5.** Discontinuous oven curing of mix 4

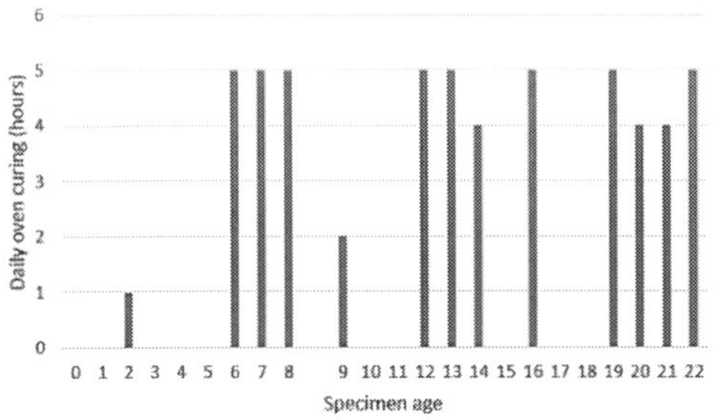


**Fig. 6.** Discontinuous oven curing of mix 5

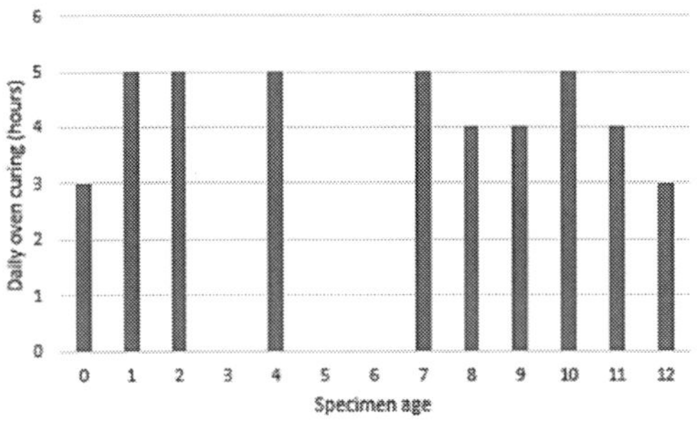


**Fig. 7.** Discontinuous oven curing of mix 6

## 3 Experimental study

### 3.1 mixture characteristics

The selection of the parameters that was based on their influence on the fresh and hardened properties of the Geopolymer concrete. Geopolymerization of the precursors of the mineral-polymer is series of chemical processes that include reactions, forming, and arranging to result in the Geopolymer structure. It has been found by many researchers that both the alkaline solution properties and the oxides in the fly ash have vast impact on the properties of the mineral-polymers [15-20]

In the current study NaOH molarity was kept constant at 12 M. The effect of the molarity of NaOH is controversial. Some studies indicated that the higher the NaOH concentration, the higher the compressive strength [21]. Some studies show that high NaOH concentration results in lower compressive strength [17]. The molarity of the caustic soda solution affects the setting time of the Geopolymer concrete [16].

The percent of $Na_2O$ in the solution can be controlled by the concentration of NaOH and the ratio of sodium silicates to sodium hydroxide (SS/SH), see **Fig. 8**. The fly ash used in the current study has 1.27% by weight of $Na_2O$. Depending on the molarity and SS/SH ratio of the mixtures shown in **Table 4**, $Na_2O$ from fly ash can contribute in a range of (5-15)% of the total $Na_2O$ weight of the Geopolymer.

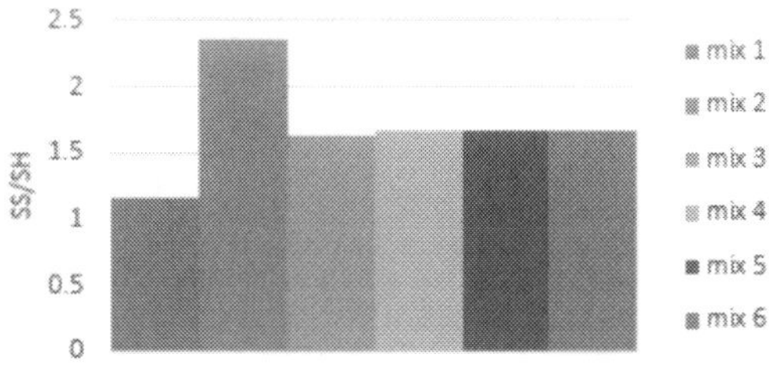


**Fig. 8.** The ratio of sodium silicate to sodium hydroxide in the mixes

The weight of the fly ash in the mix is an important factor for the compressive strength of the Geopolymer concrete. The ratio of Fly ash/($Na_2O$+$SiO_2$) and Fly ash/(Fly ash+$Na_2O$+$SiO_2$) is considered to be utilized for studying the contribution of fly ash quantity relative to $Na_2O$, and $SiO_2$ in the various mixes.

**Table 4.** Parameters of the Geopolymer of current study

| Mix No. | Water/Fly ash weight ratio | Water/Solids weight ratio | HRWRA type | HRWRA/Fly ash (l/100 kg) | Molarity | $Na_2SiO_3$/NaOH weight ratio | Fly ash/($Na_2O$+$SiO_2$) (%) | Fly ash/(Fly ash+$Na_2O$+$SiO_2$) (%) | $SiO_2$:$Na_2O$ weight ratio (%) |
|---|---|---|---|---|---|---|---|---|---|
| 1 | 0.35 | 0.30 | K* | 2.28 | 12 | 1.17 | 614 | 86 | 77 |
| 2 | 0.32 | 0.27 | K* | 2.57 | 12 | 2.35 | 665 | 87 | 113 |
| 3 | 0.32 | 0.26 | K* | 2.94 | 12 | 1.64 | 504 | 83 | 94 |
| 4 | 0.32 | 0.26 | K* | 3.33 | 12 | 1.66 | 498 | 83 | 94 |
| 5 | 0.32 | 0.26 | K* | 3.33 | 12 | 1.66 | 498 | 83 | 94 |
| 6 | 0.32 | 0.26 | K* | 4.12 | 12 | 1.66 | 498 | 83 | 94 |

* KUT PLAST SP 400

### 3.2 Mixture proportions

The mix proportions for the six mixtures of the current study is shown in **Table 5**. The quantities are for one cubic meter of Geopolymer concrete.

**Table 5.** Mixture proportions

| Mix no. | Fly Ash (kg/m$^3$) | Extra water (kg/m$^3$) | Fine aggregate (kg/m$^3$) | Coarse aggregate (kg/m$^3$) | Sodium Silicate (kg/m$^3$) | NaOH solution(kg/m$^3$) | Superplasticizer (kg/m$^3$) |
|---|---|---|---|---|---|---|---|
| 1 | 452 | 46 | 751 | 858 | 104 | 89 | 13 |
| 2 | 410 | 43 | 768 | 987 | 107 | 45 | 13 |
| 3 | 414 | 11 | 723 | 1000 | 130 | 79 | 15 |
| 4 | 413 | 9 | 721 | 998 | 131 | 79 | 17 |
| 5 | 413 | 9 | 721 | 998 | 131 | 79 | 17 |
| 6 | 412 | 9 | 719 | 995 | 131 | 79 | 21 |

The curing type, curing temperature and ambient temperature are detailed in **Table 7**

### 3.3 Testing Methods

#### *3.3.1 Workability*

Slump test as an indicator for the concrete workability was conducted in compliance with ASTM C 43 specifications [22]

#### *3.3.2 Compressive strength*

Cylinders with d=100 cm and h=200 cm were used to shape the Geopolymer concrete specimens. The specimens were tested using 2000 kN compression testing machine. The test specimens and procedures were conducted in compliance with ASTM C 39-05 specifications [23].

## 4 Results and discussion

### 4.1 Compressive strength

The compressive strengths of the tested cylinders are shown in **Table 7**. It can be seen that the compressive strength of mix 1 which was cured in ambient temperature of 43°C has a progressive strength with time and reached 29.21 MPa at 40 days. While for mix 2 which was also cured at ambient temperature of 43°C the compressive strength is 23.45 MPa at 40 days. Referring to **Table 5**, this increase in strength can be explained by higher percentage of the fly ash that leads to higher binder in the matrix. Another factor that led to the higher strength for mix 1 is the $Na_2SiO_3$/NaOH weight ratio which when increased from 1.17 to 2.35 negatively affected the compressive strength. **Table 4** shows that water to fly ash ratio for mix 2 is 0.32 which is lower than that for mix 1 (0.35). This decrease in water was not enough to compensate the negative effect of the high SS/SH ratio. This leads to a conclusion that SS/SH ratio is a vital factor when designing Geopolymer concrete.

Mixes 3 through 6 were cured discontinuously in the oven. For mix 3, the specimens were left atambient temperature of 42°C for the first three days after casting. Then they were exposed to oven curing for 1 hour at 90 °C then left at ambient temperature of 42°C until they were tested for the compressive strength at the fourth day (see **Fig. 4**). The development of strength was noticeable if compared to mix 1 and mix 2. One hour of discontinuous curing was enough to result in a compressive strength after only four days that is close to the compressive strength of mix 1 after 7 days and to mix 2 at 40 days. Discontinuous curing for mix 3 resulted in 40.58 MPa at 40 days age.

Discontinuous curing of 12 hours within 3 days for mix 4 (see **Fig. 5**) resulted in a compressive strength which is almost equal to that of mix 3 at 21 days. In spite of this improvement in the early strength of mix 4, the strength improvement at later ages is more obvious for mix 3 than mix 4. Both mix 3 and mix 4 were cured discontinuously for 40 days with total oven curing of 34 hours, but the compressive strength of mix 3 was 40.58 MPa which is 23% more than that of mix 4 at age 40 days. Everything is the same for mix 3 and mix 4 except the curing regime. To explain the reason behind the divergence in the compressive strength of the two mixes, it is important to mention that mix 3 was left in the ambient temperature for 3 days as a rest period prior to any oven curing. Mix 4 was directly cured in the oven after the casting in the molds. Curing regime has enhanced the geopolymerization of the goepolymer concrete through the activation of the chemical reactions of the fly ash with the alkaline solution in the early age of mix 4, but mix 3 has developed 40 days compressive strength of 23% higher than mix 4.

Mix 5 is identical to mix 4 and was left in the ambient temperature for two days of rest period then cured at 90°C for one hour then left in the ambient temperature for three days and discontinuous oven curing was then resumed. The compressive strength at 21 days of mix 5 is almost equal to that of mix 4 in spite of the fact that mix 5 oven curing time is remarkably more than that of mix 4. The effect of the rest period can be observed from the need of mix 5 for more oven curing to reach the compressive strength of mix 4 at the same age.

The influence of the rest period was also inspected in mix 6. No rest period was allowed for mix 6 (which is identical to mix 4 and mix 5). It is obvious that the progress of the compressive strength for mix 6 is higher than that for mix 5 in spite of the drop in the ambient temperature to 36°C which was 38°C in the case of mix 5. Again, this divergence in compressive strength is demonstrated by the influence of the rest period.

**Table 6** compares the compressive strength of mix 3, mix 4, mix 5, and mix 6 for specific curing times and shows that for mixes without rest period, a higher early compressive strength can be obtained by the same oven curing time and in earlier age than the mixes with the rest periods.

**Table 6.** Early strength comparison of mixes 3 through 6

| Mix no. | Rest period | Age (days) | Oven curing (hours) | Compressive strength (MPa) |
|---|---|---|---|---|
| 3 | Yes | 7 | 12 | 23.60 |
| 4 | No | 3 | 12 | 27.00 |
| 5 | Yes | 8 | 16 | 24.15 |
| 6 | No | 7 | 18 | 31.74 |

## 5 Conclusions

One of the main expenses associated with the production of the Geopolymer concrete is the oven curing costs. The following conclusions summarize the effect of the investigated factors on both compressive strength and the cost of the Geopolymer concrete:

1. The ratio of SS/SH is an important factor that influences the compressive strength of the Geopolymer concrete and thus, trail mixes of various SS/SH for optimization are needed to result in the required compressive strength with the least oven curing. A value of 1.17 was found the best for ambient cured Geopolymer concrete.
2. Discontinuous oven curing, even for short period of 1 hour significantly improves the compressive strength. An increase of 56% in the compressive strength was achieved by one hour of oven curing.
3. In the case of discontinuous curing, the rest period when allowed to the Geopolymer concrete upon casting results in higher 40 days compressive strength than that of Geopolymer concrete which is directly oven cured. The strength improvement is remarkably evident within the period from 28 days to 40 days.
4. The early strength improvement of Geopolymer without rest period excitation is higher than when rest period is allowed. In one example, it has shown a compressive strength of 23% higher at 3 days age than the Geopolymer concrete without rest period excitation.
5. The ambient temperature plays a major role in the Geopolymerization process of the Geopolymer concrete in discontinuous curing. Within the experimental tests temperature range of (43-36)°C, the higher the ambient temperature the higher the compressive strength.

**Table 7.** Compressive strength of the tested cylinders

| Mix no. | Curing type | Curing temperature °C | Ambient temperature °C | Age at test (day) | Hours in the oven | Compressive strength (MPa) |
|---|---|---|---|---|---|---|
| 1 | A* | 43 | 43 | 7 | 0 | 22.90 |
| | | | | 14 | 0 | 24.62 |
| | | | | 40 | 0 | 29.21 |
| 2 | A* | 43 | 43 | 7 | 0 | 15.13 |
| | | | | 14 | 0 | 16.00 |
| | | | | 40 | 0 | 23.45 |
| 3 | O** | 90 | 42 | 4 | 1 | 22.00 |
| | | | | 7 | 12 | 23.60 |
| | | | | 18 | 12 | 25.10 |
| | | | | 21 | 12 | 27.10 |
| | | | | 25 | 12 | 35.77 |
| | | | | 40 | 34 | 40.58 |
| 4 | O** | 90 | 41 | 3 | 12 | 27.00 |
| | | | | 13 | 12 | 30.00 |
| | | | | 21 | 12 | 30.78 |
| | | | | 40 | 34 | 32.97 |
| 5 | O** | 90 | 38 | 8 | 16 | 24.15 |
| | | | | 14 | 18 | 29.27 |
| | | | | 21 | 55 | 30.35 |
| 6 | O** | 80 | 36 | 7 | 18 | 31.74 |
| | | | | 11 | 40 | 32.09 |
| | | | | 12 | 43 | 33.83 |

A* Cured in ambient temperature
O** Cured in an electrical oven

## References

[1] M. Albitar, M. M. Ali, P. Visintin and M. Drechsler, *Construction and Building Materials,* **136**, (2017)

[2] C. Horsley, M. H. Emmert and A. Sakulich, **184**,(2016)

[3] S. Ruan and C. Unluer, *Journal of Cleaner Production,* **137**,(2016)

[4] A. Palomoa, M. Grutzeckb and M. Blancoa, *Cement and Concrete Research,* **29**, (1999)

[5] U.S. Geological Survey, Reston, Virginia, (2015)

[6] S. Saha and C. Rajasekaran, *Construction and Building Materials,* **146**, (2017)

[7] S. Demie and M. F. Nuruddin, *Construction and Building Materials,***41**, (2013)

[8] R. Derabla and M. L. Benmalek, *Construction and Building Materials,* **66**, (2014)

[9] M. R. Hall and K. B. Najim, *Construction and Building Materials,* **73**, (2014)

[10] K. Ma, J. Feng, G. Long, Y. Xie and X. Chen, *Construction and Building Materials,* **143**,(2017)

[11] Okamura, Hajime and M. Ouchi, *Journal of Advanced Concrete Technology*, **1**, 1, (2003)

[12] R. Rani and M. K. Jain, *Powder Technology*, (2017)

[13] ASTM C618, ASTM International, (2015)

[14] ASTM C33, ASTM International, (2015)

[15] C. B. Cheah, M. H. Samsudin and M. Ramli, *Journal of Cleaner Production*, (2016)

[16] Y.-K. Cho, S.-W. Yoo, S.-H. Jung, K.-M. Lee and S.-J. Kwon, *Construction and Building Materials*, **145**, (2017)

[17] M. N. Hadi, N. A. Farhan and M. N. Sheikh, *Construction and Building Materials*, **145**, (2017)

[18] A. Nazari, H. Khanmohammadi, M. Amini, H. Hajiallahyari and A. Rahimi, *Materials and Design*, **41**, (2012)

[19] A. Sharmin, U. J. Alengaram, M. Z. Jumaat, M. O. Yusuf, S. A. Kabir and I. I. Bashar, *Construction and Building Materials*, **144**, (2017)

[20] A. A. Siyala, K. A. Azizlia, Z. Mana and H. Ullaha, *Procedia Engineering*, **148**, (2016)

[21] K.-W. Lo, K.-L. Lin, T.-W. Cheng, Y.-M. Chang and J.-Y. Lan, *Construction and Building Materials*, **143**, (2017)

[22] ASTM C143, ASTM International, (2015)

[23] ASTM C39, ASTM International, (2015)

# Assessment the thermal properties lightweight concrete produced by using local industrial waste materials

*Osama* AbdulAmeer

Building and Construction Engineering Department, University of Technology, Baghdad, Iraq

**Abstract.** In this research study of the thermal properties of light weight concrete produced from using two locally industrial wastes materials, to evaluate the thermal insulation, and some physical properties like strength of concrete (compressive and flexural tensile strength and variation of density of light weight concrete. Two types of wastes admixtures were used in this study, (chopped rubber, and wood sawdust) with (5%, 10%, 15% and 20%) percent of each one. Thermal conductivity, compressive strength, flexural tensile strength and variation of density have been examined for each specimen at all percentages of admixtures, and compare with the reference concrete specimens. Experimental test results indicated that using these types of wastes as replacement of fine aggregate in concrete significantly affects the thermal conductivity, compressive strength, flexural tensile strength (rise or fall out), especially at (5%) of adding materials, as well as utilization of additives in concrete to produce low density mixture with wood sawdust or as high density concrete when using rubber chopped in the concrete mixture.

## 1 Introduction

Thermal insulation and some of physical properties play an important role in constriction and designing of modern building, especially when considering wall properties such as weight and cost. Loss of heating from cement buildings occupies an important factor in air conditioning science. One of the main factors that affect cooling load in air conditioning space, is the thermal characteristics of building material. Increasing thermal insulation is the dominant factor in reducing heat that could be transfer of the building [1, 2]. The best uses of waste materials depends on their use being economically competitive with the alternative natural material, these costs are primarily made up of handling, proceeding and transportations. For these mentioned reasons, many research works were endeavoured to make use of the enormous quantity of waste like, rubber tyres (chopped worn-out tyres), iron splinters (scrap), wood sawdust, ash rice husk and silica gel, to investigate the basic properties of concrete mixes with different admixtures, produce lightweight concrete, improve strength of concrete, and reduce costs of construction, as well as, reduce environmental pollution and preventing the accumulation of the raw materials. Industrial waste admixtures of concrete, generally have two main categories, organic wastes, and in-organic wastes (urban wastes) [3].

The objective of this work is to investigate some properties of concrete mixes with different additives, e.g., thermal conductivity, compressive strength, flexural strength, density (as light-weight concrete) and their influences on concrete behaviour. The additives included in this work are, rubber wastes (chopped worn-out tyres), and sawdust which were added as replacement of fine aggregates to concrete with mixing ratio of concrete is (1:2:4).

## 2 Materials

### 2.1 Cement

The ' Tasloga' ordinary Portland cement was used throughout this work. It is produced by factory in Sulymania. It is stored in air tight plastic containers to avoid exposures to different atmospheric conditions .The chemical & physical properties are shown in Tables(1) & (2) respectively. The tables conform to the limitations of Iraqi specification No. (5/1984) [4].

* Dr.osamaeedan@gmail.com

© The Authors, published by EDP Sciences. This is an open access article distributed under the terms of the Creative Commons Attribution License 4.0 (http://creativecommons.org/licenses/by/4.0/).

**Table 1.** Chemical composition of the cement *

| Compounds | % by weight | Limits of Iraqi spec. No.5/1984 |
|---|---|---|
| Ca O | 61.5 | ----- |
| $SiO_2$ | 22.3 | ----- |
| $Al_2O_3$ | 4.4 | ----- |
| $Fe_2O_3$ | 2.74 | ---- |
| $SO_3$ (%) | 2.65 | Less than 2.8 |
| Mg O (%) | 2.5 | Less than 5.0 |
| L.O.I (%) | 1.9 | Less than 4.0 |
| $C_3A$ | 2.6 | Less than 3.5 |

* These tests were made in (NCCLR).

Table 2. Physical properties of the cement

| physical properties | Test results | Limits of Iraqi spec. No.5/1984 |
|---|---|---|
| Specific surface area blaine Method $m^2/kg$ | 375 | >230 |
| Setting time (Vicats method)<br>Initial setting (min)<br>Final setting(min) | 196<br>283 | < 60<br>< 600 |
| soundness | 0.2% | <0.8% |
| Compressive strength of Mortar (Map)<br>3-days<br>7-days | <br>15.3<br>27.5 | <br>>15<br>>23 |

## 2.2 Fine aggregate

Natural sand from "El-Akhaider" was used for concrete mixes. All Tests of it indicate with conform to the limitations of No. (45/1984) specifications [5]. The (absorption, $SO_3$ content, and the percentage of the materials finer than sieve No. 200sieve of the sand were (1.19, 0.22., and 0.8) % respectively, and the specific gravity of that is 2.62 .The grain size distribution of the fine aggregate as shown in Table (3).

**Table 3.** Seive analysis of sand

| Sieve size (mm) | Accumulated percentage passing | Limit of Iraqi speci. No. 45/1984, Zone (3) |
|---|---|---|
| 4.75 | 100 | 90-100 |
| 2.36 | 95 | 85-100 |
| 1.18 | 90 | 75-100 |
| 0.6 | 75 | 60-79 |
| 0.3 | 38 | 12-40 |
| 0.015 | 7.5 | 0-10 |

Fineness modulus = 2.43

## 2.3 Coarse aggregate

(5-19) mm natural crushed- gravel, was used in this study. Sieve analysis of it indicates that the conformance with the limitations of No. (45/1984) Iraqi specifications [5].

## 2.4 Water

In mixing and curing, potable water was used.

## 2.5 Admixtures

Industrial wastes (saw dust and rubber chopped) were used in this work as admixture. The chemical analyses of these admixtures are shown in Table (4 and 5).

**Table 4.** Chemical properties of Tires (rubber) cutting *

| composition | content% |
|---|---|
| Rubber hydrocarbon | 48 |
| Carbon black | 31 |
| Acetone extract | 15 |
| Ash | 2 |
| Residue chemical balance | 4 |

* Testing was made at Babylon Tyres Factory.

**Table 5.** Chemical analysis of wood saw dust*

| Oxide composition | Content % |
|---|---|
| $SiO_2$ | 86.2 |
| $CaSO_3$ | 2.8 |
| $Al_2O_3$ | 4.5 |
| Mg O | 2.5 |
| $SO_3$ | 2.2 |
| Loss on ignition | 1.8 |

*Analysis was made at X-Ray Diffract -Meter, type XRD-6000

# 3 Mix proportion

The mixing process was performed in pan type mixer of (0.1) $m^3$capacity. The aggregate and saw wood dust were used in a saturated surface dry condition. The dry materials were well mixed for about 1.5 minutes to satisfy uniform mix, then the amount of required tap water was added and the mix elements were mixed for another 1.5 minutes.

All molds used, were cleaned, and the surface of it was oiled and filled with concrete in layers and compaction was made by vibrating table for a period of 20 seconds/layer. Care was taken to avoid concrete segregation. Then the top faces of the molds were troweled. After casting, the specimens were covered with polyethylene sheets to avoid cracks associated with the plastic shrinkage and maintained in moist cabinet at a temperature of 23 ± 1° C and the relative humidity of about 90%. The specimens were then remolded after 24 hours from casting, then the specimens were placed in tap water for 28 days after this period the specimens were removed from water and left inside the Laboratory for 3 days to allow for drying.

# 4 Experimental tests

## 4.1 Thermal conductivity

Thermal conductivity measure of the quantity of the heat that flows through a material per unit time. Thermal conductivity of most materials found to change with the percentage of impurities or admixtures. From the Furriers steady –state heat condition equation (thermal conductivity) is determined as follows:

$$K= (Q)(\Delta X)/(A)(\Delta T) \quad (1)$$

*Where:*
*K: Thermal Conductivity ( w /m.℃ ).*
*Q: The Heat Transfer rate (kj ).*
*A: Area ($m^2$).*
*T: Temperature (℃ ).*

The Thermal Conductivity was run in the laboratories of Material Engineering Department, University of Technology. The samples were prepared to determine the thermal conductivity of concrete and the test apparatus -Less disc apparatus - type ( Griffin and George) as shown in Plates (1and 2) [6].

## 4.2 Physical properties tests

Experimental work have been carried out according to Iraqi specification Code 1989(CBRI, 1989) [7], and B.S. (1881-1983) [8], for compressive strength tests and density, and according to ASTM (C192-07) [9], for flexural tensile strength tests. The admixtures were used with different weight percentages from cement content (5%, 10%, 15%, and 20 %) for rubber particles and wood saw dust, the mixing ratios of reference concrete are 1:2:4, with water cement ratio (w/c) of (0.5). Average of three specimens was taken for each ratio.

# 5 Discussion of results

## 5.1 Effect of adding wastes on concrete strength and density

The tests results provide the behavior for concrete for each type of admixtures. This research revealed that the compressive strength of concrete (in 28 days ) increased with adding (5%) of different types of admixtures as shown in Table (6) , and Fig.(1)

**Table 6.** The compressive strength results of the concrete samples with different additives types

| Admixtures % (by wt. of cement) | Compressive strength of concrete in (28 days) containing Rubber cutting N/mm$^2$ | Compressive strength of concrete in (28 days) containing wood saw dust N/mm$^2$ |
|---|---|---|
| 0 | 25.53 | 25.51 |
| 5 | 28.76 | 28.3 |
| 10 | 23.36 | 21.4 |
| 15 | 16.88 | 17.3 |
| 20 | 15.52 | 24.08 |

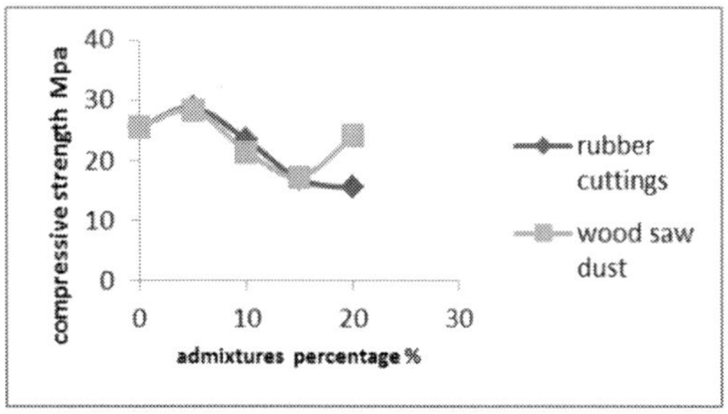


**Fig. 1.** Compressive strength of with different types and percentages of admixtures.

Concrete modified by rubber chopped with 5% have been noticed to be able to increase the compression strength as shown in Fig.(1). This increase is about (8.4) with respect to reference concrete, while decrease for other admixture percentages. This could be attributed to the inclusion of cutting inside existing voids and pores through mixture, which behaves as bond material .While, when increasing the addition ratio (more than 5%), the rubber will act as weakness regions .Wood dust increases the concrete compressive strength of about (13%), by using (5%) addition percentage, compared to reference concrete as shown in Fig.(1) .This can be attributed to the high content of ($SiO_2$) in saw dust ,and the texture of this material, which reflect in concrete mixture .The concrete density with additives show significant decreasing in comparison with normal concrete. Table (7) shows that the rubber cutting gives density less than wood saw dust, also see Fig.(2).

**Table 7.** Dry density of concrete with different percentages of wastes.

| Admixtures % (by wt. of cement) | Dry density of concrete containing of rubber chopped (kg/m$^3$) | Dry density of concrete containing of wood saw dust (kg/m$^3$) |
|---|---|---|
| 0 | 2400 | 2400 |
| 5 | 2403 | 2442 |
| 10 | 2369 | 2436 |
| 15 | 2285 | 2398 |
| 20 | 2177 | 2374 |

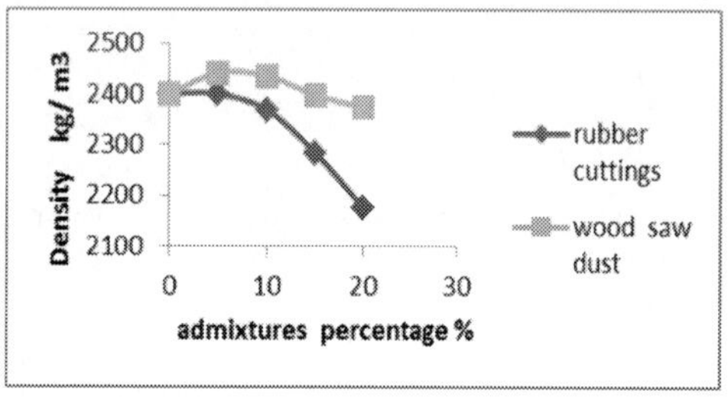


**Fig. 2.** Dry density of with different types and percentages of admixtures.

Table (8) and Figure (3) , shows the behavior flexural strength of concrete (with size of specimens used 100*100*400mm) modified by addition of the different percentage of additives (rubber cutting and saw wood dust). They shows that the flexural strength is decreased with increasing of rubber cuttings while, it is increased with increasing saw wood dust .The maximum flexural strength was provided by adding (10%) of saw dust.

**Table 8. Flexural strength of concrete**

| Admixtures % (by wt. of cement) | Flexural strength of concrete containing rubber cuttings (N/mm$^2$) | Flexural tensile strength of concrete containing wood saw dust (N/mm$^2$) |
|---|---|---|
| 0 | 5.6 | 5.6 |
| 5 | 5.12 | 5.65 |
| 10 | 3.77 | 6.15 |
| 15 | 2.88 | 5.92 |
| 20 | 1.83 | 5.80 |

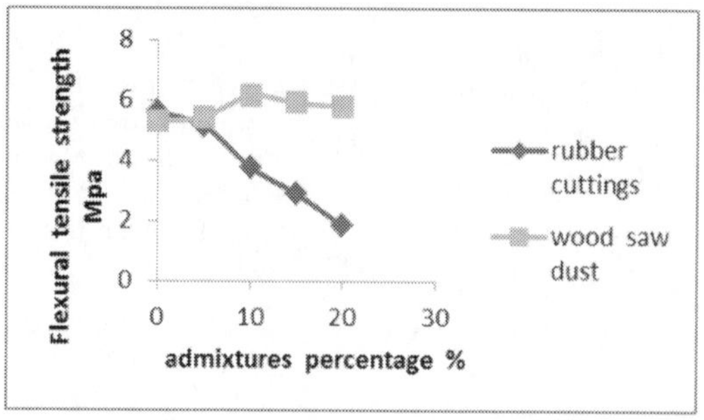


**Fig. 3.** Flexural strength of with different types and percentages of admixtures

## 5.2 Effect of additives on thermal conductivity of concrete

Table (9) shows the thermal conductivity of rubber and wood saw dust and, Fig.(4) shows the relationship between thermal conductivity and admixtures percentages. It is clear that the increase of additives leads to decrease in thermal conductivity value. Fig.(4) shows the effect of wood ratio on the thermal conductivity of concrete .It is clear that the increase in wood ratio leads to a decrease in thermal insulation , this is due to the presence of voids within the concrete matrix which affects thermal conductivity of the produced material.

**Table 9.** Thermal conductivity of concrete

| Admixtures % (by wt. of cement) | Thermal conductivity of concrete containing rubber cuttings (w/m.k) | Thermal conductivity of concrete containing wood saw dust (w/m.k) |
|---|---|---|
| 0 | 0.704 | 0.704 |
| 5 | 0.638 | 0.539 |
| 10 | 0.550 | 0.441 |
| 15 | 0.480 | 0.322 |
| 20 | 0.420 | 0.218 |

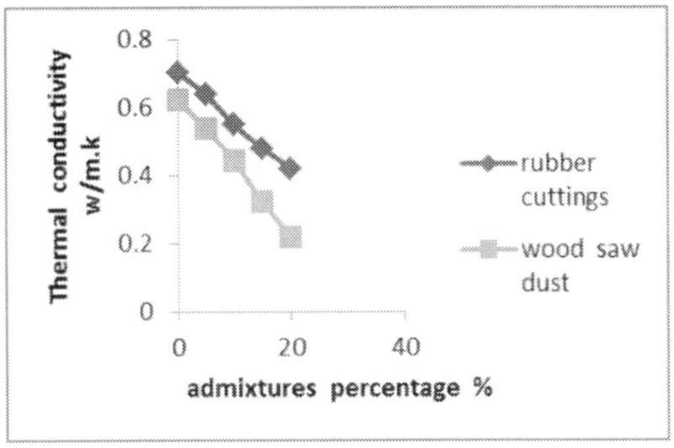


**Fig. 4.** Thermal conductivity of concrete

**Plate 1. The concrete samples**

## 6 Conclusions

This study has been carried out to investigate the behavior of concrete addition of organic and non-organic wastes mixtures .The study concludes the following points .

1. Adding 5% chopped rubber or saw dust wood gives increase in compressive strength of concrete by 11.23%, and 9.78% ,respectively.
2. Flexural strength of mixed concrete with rubber cutting decreases considerably by increasing the rubber additives, while wood saw dust increased flexural strength.
3. The thermal conductivity of concrete mixes decreases, with addition of the rubber cutting or saw dust wood.
4. The use of waste industrial materials as a replacement of aggregate (which represents the main salt sources) reduces the salt concentrations in the concrete mixture; this may be lead to improvement the durability of concrete.

**Plate 2**. Lees Disc apparatus

## Appendix

**Plate 1.** The concrete samples

## References

1. P. Turgut, and B. Yesilata, ",(Energy & Building ,pp ,679-688, 2008) .
2. M.S. Solemez , (Building and env. ,pp1-5, 34 1991).
3. ACI committee 213, guide for structural light weight aggregate concrete, ACI manual of concrete practice ,part-1 ,materials and general properties of concrete (1999).
4. I.Q. Iraqi Specifications No. 5, "Portland Cement", Baghdad, (1984).
5. I.Q. Specifications No. 45, "Natural Sources for Gravel that is Used in Concrete and Construction", Baghdad, 1984.

6. CBRI, Central Building Research Institute coconut pith cement concrete for thermal insulation , Building Materials, No.9, Rookie, India, (1993).

7. I.Q. Specification code, Ministry of planning, Iraq, (1989).

8. B.S. British standard institute, Method of making and curing test specimens, BS 1881, part-3, London, (1983).

9. ASTM, C192-2007 ,Standard practice for making and curing test specimens in the Laboratory ,Annual book of ASTM standard ,Philadelphia, vol. 04 , (2007).

# Mechanical properties of fly ash geopolymer mortar reinforced with carbon fibers

*Qais* Frayyeh [1*], and *Ahmed* Swaif [1]

[1] Building and Construction Engineering. Department. University of Technology. Baghdad, Iraq.

**Abstract.** Portland cement industry releases high amounts of CO2, so efforts have been made for alternative binders. Geopolymers are one of the highlighted elective binders. The aims of this study are to produce fly ash Geopolymer boards reinforced with carbon fibers and investigate the mechanical properties of several blends of fly ash Geopolymer mortar reinforced with carbon fibers (FGC). Carbon fibers (CF) were added at different percentages (WF) of 0.2%, 0.5%, 1%, and 1.5% by weight. The sodium hydroxide condensation of 8 molars, sodium silicate to sodium hydroxide weight ratio of 2.0, the alkaline liquid/binder proportion of 0.47 and the dry heat curing of 70°C for 24 hours were utilized as a piece of blends. Fresh and hardening state properties such as Flow capacity, flexural strength and compressive strength, and physical properties such as bulk density were determined from the 3-day FGC samples after the curing process. The results obtained from this study, show that the uses of CF were significant in improving the compressive and flexural strength due to the bonding characteristics of fibers with Geopolymer pastes. Then again, the used of CF results in slightly decrease of the bulk density. The optimal compressive and flexural strength outcomes were at weight percentage 1% of CF.

## 1 INTRODUCTION

Another type of substances with bond attributes designated geopolymer has been created as of late [1]. Alkaline activation, normally alluded to as geopolymerization, is characterized as a process which changes the shiny constituents in all around compacted and concrete highlighted composites. An intensive base is substantial to expand the surface hydrolysis of the particles of crude substances with a disintegration comprising of a specific sum of silica and alumina found in substances amid polymerization. This medium can be gotten by single or joined base arrangements alluded to as activators [2].

Alkali activator solutions assume a critical part in the disintegration of Si and Al oxides. Henceforth, geopolymers are combined by blending the most well-known alkaline activators that are strong alkaline solutions, for example, sodium hydroxide (NaOH), potassium hydroxide (KOH), potassium silicate or sodium silicate with aluminosilicate receptive materials [3] [4].

The concentration of the activator significantly affects the compressive qualities of the geopolymers. The perfect condensation of the activator expands the quality of the geopolymer. Separated from the perfect condensation, a few misfortunes may happen in the mechanical characteristics of the substance because of the free OH particles in the alkali-activated framework which can alter the geopolymer structure of the substance. Curing and age temperature of the geopolymers are different factors which affect the compressive quality of the samples. Be that as it may, for these factors to be compelling, adequate concentration of activator must be accessible in the medium amid geopolymerization in light of the fact that NaOH concentration affects the quality esteems than on the curing temperature and curing time esteems [2] [5][6].

Geopolymers are by and large formed by fly ash initiated at low temperatures. Fly fiery remains a mechanical waste with pozzolanic characteristics acquired from warm power plants and utilized rather than cement in concrete development. SiO2 and Al2O3 in the fly ash react with calcium hydroxide shaped because of hydration of Portland type and begin a pozzolanic response. Hence, fly fiery remains can be utilized rather than Portland type [1].

Fly ash has large amounts of shapeless silica and alumina furthermore, is frequently blended with base solution for break down them. Silicates start to break up when the fly fiery remains come into contact with alkali Solution. Fly fiery remains can be utilized as a source substance in the era of geopolymers since geopolymerization is basically in view of the alumina-silicate chain. In any case, the sort and condensation of the base solution assume an essential part in the arrangement of the fly fiery remains. The dissolvability

*Corresponding author: jwd_ks@yahoo.com

© The Authors, published by EDP Sciences. This is an open access article distributed under the terms of the Creative Commons Attribution License 4.0 (http://creativecommons.org/licenses/by/4.0/).

of Al3+ and Si4+ ions in sodium hydroxide solvency is higher than in potassium hydroxide arrangement [7] [8]. Along these lines, NaOH importantly affects the framework and the compressive quality of the geopolymers [9]. Si, Al and other minor ions start to break down when the fly fiery remains contact with NaOH. The measure of disintegration relies on upon the NaOH condensation and disintegration time [1]. The NaOH fixation impacts the deterioration method and the official of strong particles in the geopolymeric framework's last structure of the watery stage [9]. The blend of fly powder with 10 molarity (M) of NaOH is appropriate for the geopolymer blend [1].

The readied blend can be exposure to curing like room temperature or like a given temperature. Aluminosilicate responsive substances break up in strong alkaline arrangements and free SiO4 and AlO4 tetrahedral structure forms. In any case, the reaction of the fly ash debris in the creation of geopolymers is low at ambient temperatures [3]. A long these lines, the geopolymerization procedure through which the material can accomplish high compressive strength at given temperatures in the vicinity of 40 and 95 °C can create [6] [9] [10] [11]. The microstructure of the fly ash geopolymers contains aluminosilicate gel, unreacted fly ash and other precious stone phases [12].

Past reviews have presumed that Class F fly ash is a decent hotspot for geopolymers [13] and that NaOH is superior to other activators for the initiation of the fly ash in the creation of geopolymers.

At the point when NaOH is utilized together with water glass, the compressive strength of the geopolymer material is higher than when just NaOH is utilized. The explanation for this is the way that in the geopolymerization handle, water glass expands the response items in which the Si substance is higher, and gives more mechanical quality [14]. Different reviews demonstrate that when utilizing the water glass with base activators, compressive and flexural quality increment [15]; however, when water glass is utilized together with NaOH, rather than utilizing just NaOH, tests increment huge quality even in one day [16] [17]. In any case, water glass in powder shape displays brings down execution than water glass in fluid frame [15].

A few reviews in the literature have tried blends of various curing times and temperatures with NaOH fixations.

Among these reviews, analysts discovered both geopolymers in which fly ash and bottom ash were utilized [18], and additionally geopolymers which were orchestrated with fly ash, kaolin, sodium silicate solution and NaOH [11]. Furthermore, different reviews in the literature are delivered light inorganic polymers by incorporating Class F fly ash, metakaolin and lightweight aggregates with NaOH sodium silicate solution [19] or created polymers by utilizing fly ash and sodium silicate solution [20], readiness and portrayal of new geopolymer-epoxy sap mixture mortars [21] and some other studies combined 7-day fly powder based geopolymers [22] and created 7-day fly ash-based geopolymers mortars [23]. The past concentrates for the most part centered on geopolymer paste. In these reviews, geopolymer materials were subjected to long curing times, and drawn out stretches of time were considered for the discovery of their properties. In any case, in the present review, NaOH and sodium silicate solution (water glass) were utilized to initiate the Class F fly ash to decide the physical and mechanical properties of the orchestrated geopolymer materials in a brief timeframe.

The adjustments in the geopolymerization procedure and the properties of the last item were explored by curing geopolymer mortars, which were readied utilizing alkali-activated fly ash and crushed sand, with varying NaOH concentrations, temperatures, also, periods of time. Along these lines, the adjustments in the properties of the geopolymer mortars in brief timeframes were uncovered by distinguishing the connections between the alkali solution concentrations, curing temperature and curing time.

## 2 Research significance

The target of this examination is to survey the impacts of utilizing CF at various weight portions in a blend of (FGM) and concentrate some mechanical characteristics of this mortar.

## 3 Materials

The raw materials used to synthesize fly ash geopolymers include Turkish hard coal fly ash from power station Iskenderun (ASTM class F fly ash), the composition of the fly ash utilized is shown in Table 1 and Figure 1shows the detailed particle size distributions determined by the standard test method ASTM D422 [24],the alkaline activator was a mixture of chemical grade NaOH solution 8 molarity (99% purity quotient, local market)and liquid sodium silicate solution with the chemical composition of SiO2=28.75%, Na2O=13.75% and H2O=57.5% (local market), natural sand was used, A high range water reducer superplasticizer based on modified sulfonated naphthalene formaldehyde condensate was used to enhance workability of fly ash Geopolymer mortar and fibers used in this study were Carbon fibers with 12 mm in length, their properties shown in Table 2.

**Table 1.** Fly Ash Chemical Composition

| Component | Mass % |
|---|---|
| $Fe_2O_3$ | 5.98 |
| $SiO_2$ | 65.65 |
| $Al_2O_3$ | 17.69 |
| CaO | 0.98 |
| $SO_3$ | 0.19 |
| LOI | 3.1 |
| $Na_2O$ | 1.35 |
| MgO | 0.72 |
| $K_2O$ | 2.99 |

Chemical composition data from the data sheet and materials safety provided by the suppliers.

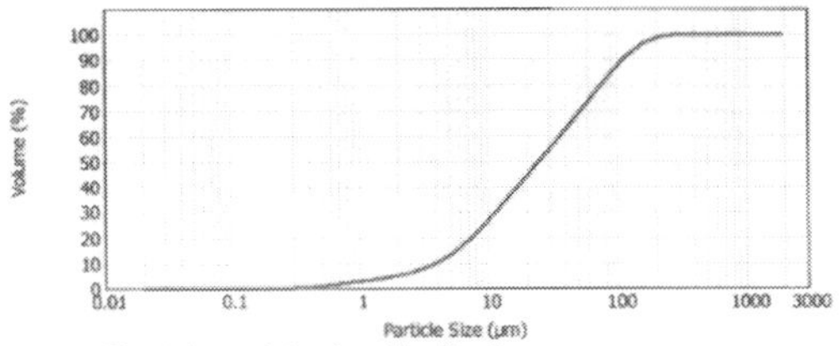

**Fig. 1.** Fly Ash particle size distribution

**Table 2.** Mechanical properties of Carbon fiber

| Property | Carbon fiber |
|---|---|
| Specific gravity | 1.79 |
| Tensile strength, MPa | 3900 |
| Length mm | 12 |
| Diameter μm | 10 – 13 |
| Elongation at break (%) | 15 |

Mechanical properties from the data sheet and materials safety provided by the suppliers.

## 4 Mix proportion, mix detail and curing

The mortar proportions are shown in Table 3. A total of five mortar mixtures were prepared, one of them without fibers and four mixtures with CF. The fibers were uniformly added into the mortar mixture at four different contents (0.2 %, 0.5%, 1% and 1.5% by weight of fly ash). The alkaline liquid was prepared one day before using by blending sodium silicate and sodium hydroxide arrangements, and then the base liquid was mixed with superplasticizer prior to starting of the mixing process. The dry materials (Fly Ash, Sand and Fibers) were firstly mixed for three minutes. From that point forward, the blended alkaline liquid with superplasticizer were added to dry materials and mixed for another three minutes until the mixture was homogenous. The dry heat curing system of 70°C for 24 hours was adopted for all specimens for this study[6] [9] [10] [11].

**Table 3.** The mortar mix proportions

| Material | Quantity (kg/m$^3$) |
|---|---|
| Fly Ash | 900 |
| Sand | 1331 |
| Sodium silicate solution | 247 |
| sodium hydroxide solutions(8 M) | 174 |
| superplasticizer | 19 |

## 5 Mechanical and physical properties of mortar

### 5.1 Flow

The impact of fibers on the stream of the five mortar blends is highlighted in Table 4. It is easy to note that the inclusion of CF in FGC blends reduces the flowing ability. On the other hand; a higher amount of CF makes loss flow ability [25]. In summary, the effect of CF on the flowing capacity is less than the control blend by 20% as shown in figure 2.

### 5.2 Compressive strength

The compressive strength outcomes of mortar blends are illustrated in Table 5. The compressive strengths of geopolymer mortars containing CF were clearly superior to that of control mix. At 28 days, the compressive qualities of mortars containing 1% Carbon fibers increased 30.15% to 34.1MPa compared with 26.2 MPa of the control mortar as illustrated in figure 3. The outcomes took after an indistinguishable inclined from those of the compressive quality, and affirmed with different examines [26] that expansion of fibers enhanced the quality of mortar and concrete. The expansion more than 1% CF, regardless, began to antagonistically influence the quality of geopolymer. The compressive qualities at 28 days of 1.5% of Carbon fibers reduced to 31.3 MPa. The compressive quality of the pastes were fundamentally enhanced because of the change in the mechanical security quality between the filaments and mortar, where the strands give the capacity to defer development of miniaturized scale splits and capture their engendering a while later up to a specific degree of the strands' volume portion.

### 5.3 Flexural strength

The flexural strength of the mortar mixes is shown in Table 6. The increase in the flexural strength of the blends containing a 1% CF increased44.42% to 4.76 compared with 3.29 MPa of the control mortar, which might be on account of the better compressed and homogeneity of the fiber appropriation as appeared in figure 4. It can likewise be noticed that the expansion over 1% CF somewhat diminishes the change in flexural quality, which might be on account of the physical challenges in giving a homogenous appropriation of the filaments inside the mortar blend [27].

### 5.4 Bulk Density

Table 7 shows the results of density for all mixes, and from the outcomes about, it can be seen that the expansion of CF in the mortar blends slightly decreases the bulk density, and this can be ascribed to the particular gravity of fiber which reduce the overall density of mortar [28] as shown in figure 5.

**Table 4.** The flow of mortar blends

| Mix name | Flow % |
|---|---|
| The control | 34 |
| CF (0.2% WF) | 32.7 |
| CF (0.5% WF) | 31.3 |
| CF (1% WF) | 29.1 |
| CF (1.5% WF) | 27.2 |

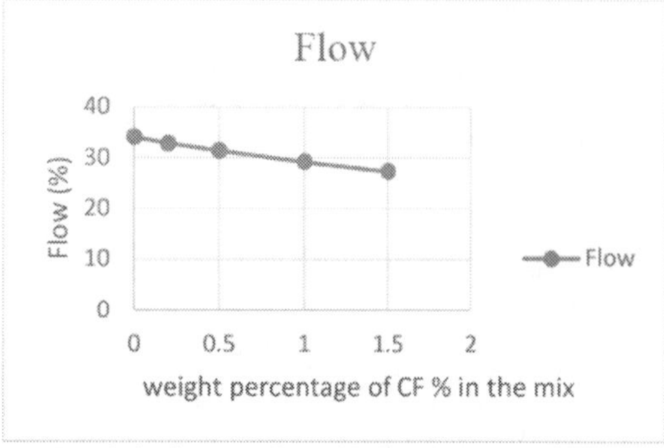


**Fig. 2.** Relation between flow with fiber content

**Table 5.** The compressive strengths of mortar blends

| Mix name | Compression strength (age3days) MPa | Compression strength (age7days) MPa | Compression strength (age28days) MPa |
|---|---|---|---|
| The control | 16.23 | 24.33 | 26.2 |
| CF (0.2% WF) | 17.3 | 25.8 | 28 |
| CF (0.5% WF) | 21.1 | 28 | 30.2 |
| CF (1% WF) | 24.4 | 31.9 | 34.1 |
| CF (1.5% WF) | 21.7 | 28.8 | 31.3 |

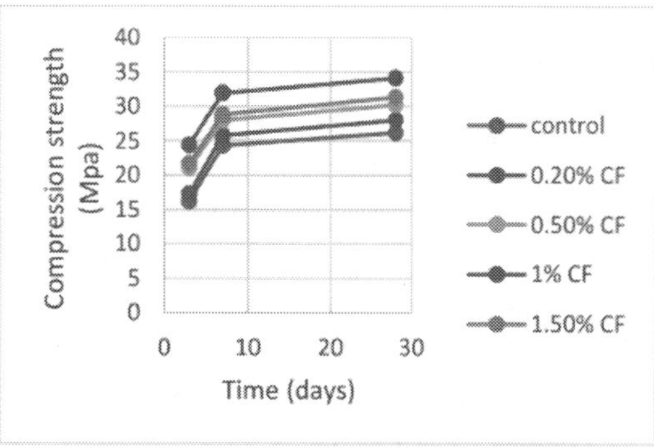


**Fig. 3.** Relation between compression strength with time

**Table 6.** The flexural strengths of mortar blends

| Mix name | Flexural strength (age7days) MPa | Flexural strength (age28days) MPa |
|---|---|---|
| The control | 3.919 | 3.296 |
| CF (0.2% WF) | 3.302 | 3.601 |
| CF (0.5% WF) | 3.5 | 3.94 |
| CF (1% WF) | 4.2 | 4.76 |
| CF (1.5% WF) | 4.06 | 4.701 |

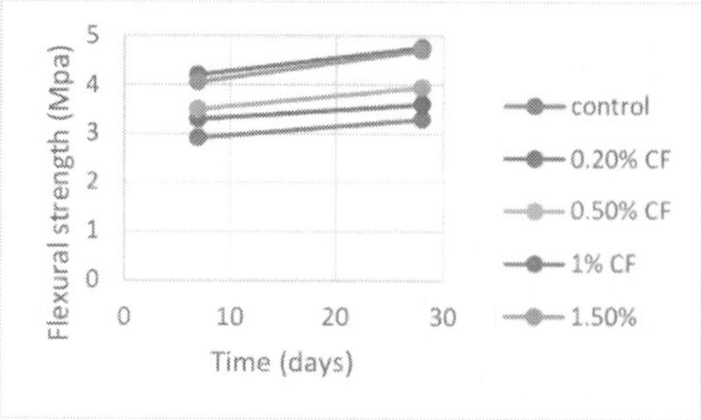


**Fig. 4.** Relation between flexural strength with time

**Table 7.** The density of mortar blends

| Mix name | Density (age 28 days) kg /m$^3$ |
|---|---|
| The control | 2175 |
| CF (0.2% WF) | 2173 |
| CF (0.5% WF) | 2169 |
| CF (1% WF) | 2163 |
| CF (1.5% WF) | 2156 |

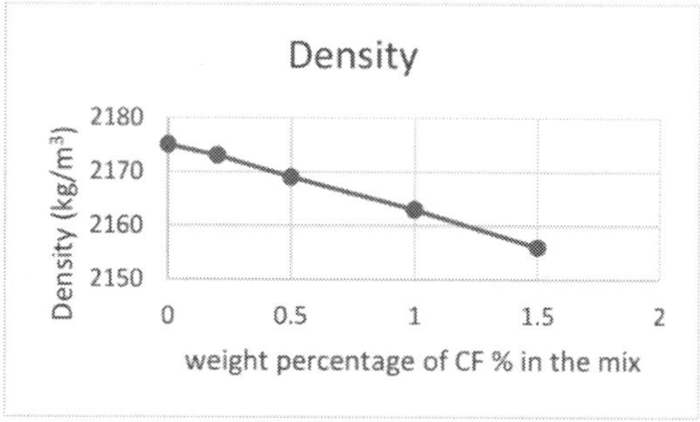


**Fig. 5.** Relation between bulk density with fiber content

## 6 Conclusions

In light of the aftereffects of this review, the take after conclusions could be drawn.

1. The possibility of production of fly ash geopolymer boards reinforced with CR with very acceptable properties which encourage to utilize the boards in several application.
2. The utilization of CF at little weight portions (1% Carbon fibers) has clear effects on the properties mortar.
3. The use of CF affects the workability of fiber reinforced mortars; CF produce harsh mixes significantly with increase the weight percentage of CF.
4. The compressive quality outcomes demonstrate that the utilization of the CF increases the compressive quality and that the best change is gotten when the CF are 1% of the blend. This is

because of the mechanical bond between the paste and the CF.

5. The flexural quality outcomes demonstrate that there is a good increment in flexural quality by expanding the fiber content. The utilization of1% CF as a weight portions gives the most noteworthy increment, however past this rate there is a slight diminishing in this change.
6. The use of CF slightly reduces the density, and this may be not considered as a very significant factor if it is compatible with acceptable ranges of compressive strength.

## References

[1] Rattanasak, Ubolluk, and PrinyaChindaprasirt. Minerals Engineering 22, no. 12 (2009): 1073-078.

[2] Vargas, Alexandre Silva De, Denise C.c. Dal Molin, Antônio C.f. Vilela, Felipe José Da Silva, Bruno Pavão, and Hugo Veit. Cement and Concrete Composites 33, no. 6 (2011): 653-60. doi:10.1016/j.cemconcomp.2011.03.006.

[3] Somna, Kiatsuda, Chai Jaturapitakkul, PuangratKajitvichyanukul, and PrinyaChindaprasirt. " Fuel 90, no. 6 (2011): 2118-124. doi:10.1016/j.fuel.2011.01.018.

[4] Brough, A.r, and A. AtkinsonCement and Concrete Research 32, no. 6 (2002): 865-79. doi:10.1016/s0008-8846(02)00717-2.

[5] Hu, Mingyu, Xiaomin Zhu, and Fumei Long. " Cement and Concrete Composites 31, no. 10 (2009): 762-68. doi:10.1016/j.cemconcomp.2009.07.006.

[6] Bakharev, T. Cement and Concrete Research 35, no. 6 (2005): 1224-232. doi:10.1016/j.cemconres.2004.06.031.

[7] Jaarsveld, J. G. S. Van, and J. S. J. Van Deventer. Industrial & Engineering Chemistry Research 38, no. 10 (1999): 3932-941. doi:10.1021/ie980804b.

[8] Deventer, J.s.j. Van. " Miner Eng 2012;29:89–104." Handbook of Low Carbon Concrete, 2017, 217-62. doi:10.1016/b978-0-12-804524-4.00010-5.

[9] Panias, Dimitrios, Ioanna P. Giannopoulou, and Theodora Perraki. Colloids and Surfaces A: Physicochemical and Engineering Aspects 301, no. 1-3 (2007): 246-54. doi:10.1016/j.colsurfa.2006.12.064.

[10] Chindaprasirt, P., T. Chareerat, and V. Sirivivatnanon. Cement and Concrete Composites29, no. 3 (2007): 224-29. doi:10.1016/j.cemconcomp.2006.11.002.

[11] Swanepoel, J.c., and C.a. Strydom. Applied Geochemistry 17, no. 8 (2002): 1143-148. doi:10.1016/s0883-2927(02)00005-7.

[12] Lecomte, I., C. Henrist, M. Liégeois, F. Maseri, A. Rulmont, and R. Cloots. Journal of the European Ceramic Society 26, no. 16 (2006): 3789-797. doi:10.1016/j.jeurceramsoc.2005.12.021.

[13] Schmücker, Martin, and Kenneth J.d. Mackenzie. "Microstructure of sodium polysialatesiloxogeopolymer." Ceramics International 31, no. 3 (2005): 433-37. doi:10.1016/j.ceramint.2004.06.006.

[14] Criado, M., A. Palomo, and A. Fernandezjimenez. Fuel 84, no. 16 (2005): 2048-054. Doi:10.1016/j.fuel.2005.03.030.

[15] Pacheco-Torgal, Fernando, João Castro-Gomes, and Said Jalali. Construction and Building Materials 22, no. 7 (2008): 1315-322. doi:10.1016/j.conbuildmat.2007.03.019.

[16] Fernández-Jiménez, A., and A. Palomo. Cement and Concrete Research 35, no. 10 (2005): 1984-992. doi:10.1016/j.cemconres.2005.03.003.

[17] Fernández-Jiménez, A., and A. Palomo. Fuel 82, no. 18 (2003): 2259-265. doi:10.1016/s0016-2361(03)00194-7.

[18] Li, Qin, Hui Xu, Feihu Li, Peiming Li, Lifeng Shen, and JianpingZhai. Fuel 97 (2012): 366-72. doi:10.1016/j.fuel.2012.02.059.

[19] Wu, Hwai-Chung, and Peijiang Sun. Construction and Building Materials 21, no. 1 (2007): 211-17. doi:10.1016/j.conbuildmat.2005.06.052.

[20] Provis, John L., Chu Zheng Yong, Peter Duxson, and JannieS.j. Van Deventer. Colloids and Surfaces A: Physicochemical and Engineering Aspects 336, no. 1-3 (2009): 57-63. doi:10.1016/j.colsurfa.2008.11.019.

[21] Colangelo, Francesco, Giuseppina Roviello, Laura Ricciotti, Claudio Ferone, and Raffaele Cioffi. Materials 6, no. 7 (2013): 2989-3006. doi:10.3390/ma6072989.

[22] Jaarsveld, J.g.s. Van, J.s.j. Van Deventer, and G.c. Lukey. Materials Letters 57, no. 7 (2003): 1272-280. doi:10.1016/s0167-577x(02)00971-0.

[23] Temuujin, J., A. Van Riessen, and K.j.d. Mackenzie. Construction and Building Materials 24, no. 10 (2010): 1906-910. doi:10.1016/j.conbuildmat.2010.04.012.

[24] Standard test method for particle-size analysis of soils. West Conshohocken, PA: ASTM International, 2007.

[25]Makhlouf, Said, SouadKhedache, DihiaDjefel, and Gilles Lefebvre. " Key Engineering Materials 550 (2013): 91-98. doi:10.4028/www.scientific.net/kem.550.91.

[26] Ramli, M., and E. Dawood. Slovak Journal of Civil Engineering XIX, no. 3 (2011). doi:10.2478/v10189-011-0013-0.

[27] Dawood, EetharThanon, and MahyuddinRamli. ISRN Civil Engineering 2012 (2012): 1-5. doi:10.5402/2012/718549.

[28] Li, Bao Jia, GuoZhong Li, and Chao Ning. Advanced Materials Research 146-147 (2010): 38-41. doi:10.4028/www.scientific.net/amr.146-147.38.

# Some properties of sustainable concrete containing two environmental wastes

*Wasan* Khalil [1,*], and *Nazar* Al Obeidy [2]

[1] Building and Construction Engineering Department, University of Technology, Baghdad, Iraq
[2]National Center for Construction Laboratories, Baghdad, Iraq

**Abstract.** This investigation includes the use of 15% of glass wastes as a partial substitution to cement in combination with plastic wastes as volumetric replacement to natural coarse aggregate to produce sustainable concrete. Different volumetric replacements of plastic waste to natural coarse aggregate (25%, 50%, 75%, and 100%) were used in concrete containing 15% glass powder as a replacement by weight of cement. Generally, the results show that the inclusion of 15% glass powder improves the compressive strength, splitting tensile strength and flexural strength by about 13.3%, 36.3%, and 34.7%respectively at 60 day age in comparison with reference concrete without wastes, also the results show a decrease in water absorption and an increase in dry density. The inclusion of plastic waste aggregate in the presence of 15% glass powder leads to a decrease in the compressive strength, flexural strength, splitting tensile strength, dry density, ultrasonic pulse velocity, and thermal conductivity. The percentage reductions are 59.8%, 46.3%, 43.6%, 20.5%, 28.6%, and 54.4% respectively for concrete specimens that including 100% plastic waste coarse aggregate in comparison with concrete specimens without plastic waste aggregate.

## 1 Introduction

The concept of sustainability is now very important in the construction industry. It is known as ensuring the complacency of the present requirements without permuting the ability of future generation [1]. Concrete is one of the main materials used in construction activities in the world. Annually the production of concrete all over the world is about twelve billion tons that consume nearly 1.6 billion tons of Portland cement, ten billion tons of rocks and one billion ton of water [2]. The production of one ton of Portland cement creates about 1 tons of $CO_2$ gas that causes environmental pollution. It is clear that the concrete industry significantly affects the environment. It is not reasonable to confine the use of concrete, but there are several manners to reduce its environmental influence [2, 3]. Huge quantities and types of glass waste and plastic are generated all over the world due to the large use of these products. Most of these solid wastes have been buried in landfill sites; this land filling of wastes is unfavorable because these wastes are non-biodegradable, which makes this process environmentally less friendly. So, it is required to start a new solution for sustainable development. One of the best solutions of sustainable solid waste management is the ability to Reduce, Reuse and Recycle of these wastes to the maximum as possible [4]. To produce eco-friendly concrete, some researchers studied the replacement of cement by other materials (supplementary cementitious materials). The chemical composition of these materials mainly contains reactive silica that reacts with Ca $(OH)_2$ (one of the hydration products of cement) to produce additional gel, which improves some properties of concrete [5, 6]. Some types of these materials are by–products from industry. In addition, there are natural materials, namely natural pozzolans such as metkaolin that is produced from kaolin clay after firing this clay to a temperature between 600 °C - 800°C [7, 8].

In Iraq the by–product supplementary cementitious materials such as, silica fume, fly ash, etc. are not available, while metkaolin as a natural Pozzolan can be produced from kaolin clay, which exists in large deposits in Iraq. However, the use of metkaolin as a replacement to cement consumes the kaolin clay from environment; in addition, it consumes a large amount of energy during the burning process that makes this material less sustainable. Therefore, it is very important to find alternative sustainable material that can be used as a replacement to cement in concrete industry. The use of different waste materials (recyclable Portland cement concrete, plastic, brick and others) as aggregate in concrete industry can consume huge quantities of these materials. In this case, the problems of consume aggregate in construction sites, and environmental pollution problems can be solved. The inclusion of plastic waste as aggregate in concrete can significantly improve some properties of concrete, because plastic has low thermal conductivity, low density, good abrasion resistance and high toughness.

The aim of this study is to investigate the effect of using

* Corresponding author: wasan1959@yahoo.com

© The Authors, published by EDP Sciences. This is an open access article distributed under the terms of the Creative Commons Attribution License 4.0 (http://creativecommons.org/licenses/by/4.0/).

glass waste powder and plastic waste, on some properties of concrete.

## 2 Research significance

Research significance can be summarized as follows:

- Recycling of glass and plastic wastes (collection, washing, and preparing) and classifying them based on ASTM.
- Investigating the properties of concrete containing the 15% of glass waste powder as a replacement to cement by weight and different percentages of crushed plastic as replacement by volume of natural coarse aggregate.

## 3 Experimental program

### 3.1 Materials

- Ordinary Portland cement type (1) was used in this investigation. Test results show that the taken cement has satisfied the Iraqi Standard No. 5/1984.
- Normal weight natural sand with maximum size of 4.75mm was used. The results demonstrate that the gradation of the sand, $SO_3$ content and physical properties conform to the Iraqi Standard No. 45/1980.
- Natural crushed gravel with maximum size of14mm was used in this study. The grading and sulphate content of the coarse aggregate conform to the Iraqi Specifications No. 45/1984.
- Potable water from the water supply network was used for mixing and curing of concrete.
- High, range water reducer admixture with a commercial mark of GLENIUM 54.0 was utilized. It is compatible with ASTM C494 type F.
- Glass powder waste and plastic waste aggregate were used after preparation and classification.

### 3.2 Preparation of glass waste

White glass waste was collected, washed and broken into small pieces as shown in Figure (1-a). The glass was then crushed by using a crusher type Retsch shown in Figure (1-b). The particle size of the glass waste after the crushing process varied from 0-0.5 mm as shown in Figure (1-c). After that it was grinded by a grinder type Retsch shown in Figure (1-d). The grinding was carried out at different time periods (1, 5, 10, and 15 minutes). Glass powder sample after grinding process is shown in Figure (1-e). Finally, the glass powder was screened on sieve No.200 (75 microns), as shown in Figure (1-f).

### 3.3 Experimental test on glass powder

The following tests were conducted on the prepared glass powder to make sure that this powder had properties which conform to the properties of natural pozzolans according to ASTM C618; all results are illustrated in Table (1):

•Fineness test was carried out on the glass powder after each grinding period (1, 5, 10, and 15 minutes) in accordance with ASTM C204.

•Strength Activity Index (S.A.I.) test at 7 and 28 day age was conducted according to ASTM C 311.

•The Remaining on the Sieve 45 Microns (No. 325) test was carried out based on ASTM C 430.

•The Chemical Analysis (Quantitative and Qualitative) was conducted using X-ray diffraction by EDX-7000 device [9].

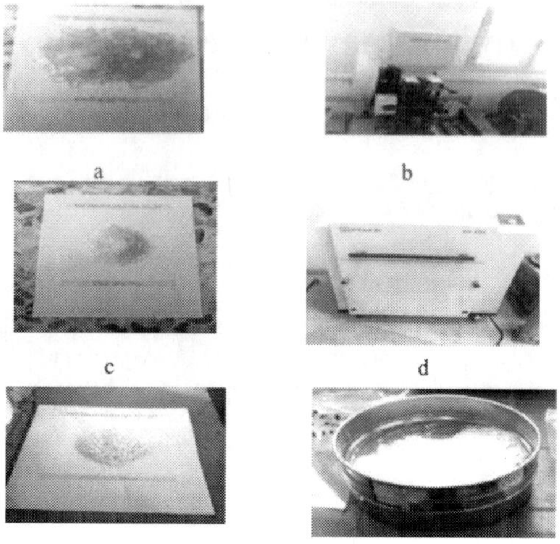


**Fig. 1**. Preparation of Glass Waste Powder

**Table 1**. Properties of Glass Powder

| Some of chemical composition of components | Value (%) | Requirements of ASTM C-618Specifications Class N |
|---|---|---|
| $SiO_2$ | 84.862 | ≥ 70 |
| $Al_2O_3$ | 5.529 | |
| $Fe_2O_3$ | 0.311 | |
| CaO | 7.934 | --- |
| $SO_3$ | 0.036 | ≤ 4 |
| $K_2O$ | 1.251 | --- |
| $TiO_2$ | 0.037 | --- |
| $ZrO_2$ | 0.011 | --- |
| $Cr_2O_3$ | 0.010 | --- |
| SrO | 0.008 | --- |
| Retained on a Sieve 45 μm (%) | 9.88 | ≤ 34 |
| Fineness(Blain Method) After 10 Minutes Grinding ($cm^2/gm$) | 7340 | --- |
| Specific Gravity | 2.265 | --- |
| L.O.I (%) Loss on Ignition | 0.41 | ≤ 10 |
| Strength Activity Index*2(%)<br>7 days<br>28 day | <br>80.2<br>83.0 | <br>≥ 75<br>≥ 75 |

### 3.4 Preparation of plastic waste

Vegetable containers were collected, washed with water, and then crushed to small pieces by a crusher in Bub Al -

Sham region in Baghdad. The crushed plastic waste was sieve on standard sieves then prepared with gradation that agrees with Iraqi Standard No.45/1980 and corresponding to natural aggregate used in this study, as shown Figure (2).

a. Plastic Waste Aggregate. b. Natural Crushed Aggregate

**Fig. 2**. Coarse Aggregate Used in this Investigation.

## 3.5 Experimental test on plastic waste

### *3.5.1 Fourier Transforms Infrared Spectroscopy Analysis*

Fourier transforms infrared spectroscopy (FTIR)[10] analysis is a subset of spectroscopy that deals with the infrared region of the electromagnetic spectrum. It can be used to identify compounds or investigate the composition of a material. Figure (3) shows the analysis of plastic waste used in this study. By observing the results of the analysis and comparing them with the results of the standard plastic analysis by FTIR for each of polyethylene, polypropylene, polystyrene, and polyethylene chloride, the test result shows that the plastic is closer to the polyethylene.

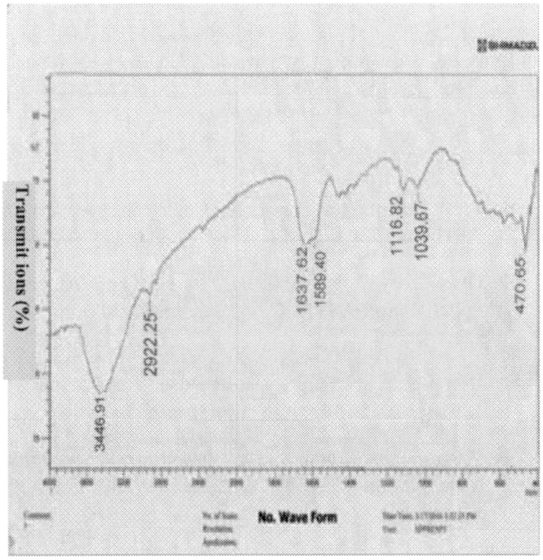


**Fig. 3**. FTIR Analysis of Plastic Waste

### *3.5.2 Sieve analysis*

The coarse plastic aggregate was prepared to satisfy the requirements of IQS 45/1980 for natural coarse aggregate with max size of 14 mm. The grading of plastic waste aggregate used in this investigation is shown in Table (2). It is the same as that for crushed natural coarse aggregate used in this study.

**Table 2.** Grading of Plastic Waste Aggregate

| Sieve size (mm) according to Iraqi standard | Passing (%) | Limits of Iraqi Standard No. 45 /1980 (5-14)mm |
|---|---|---|
| 14 | 100 | 90-100 |
| 10 | 70 | 50-85 |
| 5 | 2 | 0-10 |

- Density test was carried out based on ASTM D792 [11].

- Flexural Strength test was conducted based on ASTM D790 [12].

- Compressive Strength test based on ASTM D695 [13] was carried out.

- Tensile Strength test was carried out according to ASTM D 638[14].

- Thermal conductivity was carried out using Lee's Disc Method [15].

- Impact Test was carried out based on ISO 179[16].

All test results are illustrated in Table (3).

According to the properties of plastic waste in Table (3) and the Figure 3 of FTIR analysis, the plastic waste used in this investigation can be classified as polyethylene plastic. Generally polyethylene plastic is classified in to three types, low density (0.910-0.925 gm/cm3), medium density (0.926-0.94 gm/cm3), and high density (0.941-0.959 gm/cm3) according to ASTM D3350 [17], so plastic waste used in this investigation is high density polyethylene (HDPE).

**Table 3**. Physical Properties of HDPE Plastic Waste Used in this Investigation

| Item | Property | Value | ASTM Specifications |
|---|---|---|---|
| 1 | Density( $gm/cm^3$) | 0.949 | ASTM D792[11] |
| 2 | Flexural Strength ($N/mm^2$) | 878.34 | ASTM D 790[12] |
| 3 | Compressive Strength ($N/mm^2$) | 26.46 | ASTM D638[14] |
| 4 | Tensile Strength ($N/mm^2$) | 7.7 | ASTM D638[14] |
| 5 | Thermal of Conductivity ( W/m.K) | 0.118 | Lee's Disc Method |
| 6 | Impact Charpy ($kJ/m^2$) | 16.27 | ISO-179[16] |
| 7 | Color | Black | --- |

### 3.6 Selection of mix proportion for the reference concrete

Reference concrete mix was designed according to British Method for concrete mix design [18], to get concrete compressive, strength at least of 40 MPa at 28 day age without any additives. The mix of cement: sand: crushed coarse aggregate is in the ratio 1:1.4:1.8 by weight with cement content of 500 kg/m$^3$, slump test value of 100±5.0 mm and w/c ratio of 0.42. Many trial mixtures were conducted to choose the optimum dose of high, range water reducer admixture (HRWRA). The w/c ratio was modified to give the constant workability of the reference mixture (slump test value of 10.0±5.0 mm). The major task of utilizing HRWRA is to decrease the amount of water mix, whereas keeping the constant workability of reference mixture. The details of the designed reference concrete mix containing various dosages of superplasticizer (HRWRA) are given in Table (4). The experimental results in this study mentioned that the optimum dose of high range water reducing admixture is 1.5 L / 10.0 kg of cement that gives reduction in water by about 35.71% and maximum compressive, strength of 59.6 MPa at 28 day age.

**Table 4**. Experimental Concrete Mixtures Containing Different Dosage of Superplasticizer

| Mix proportion by weight | Dosage of HRWR (liter/100 kg of cement ) | w/c ratio | Slump (mm) | Water reduction (%) | Compressive strength (MPa) | |
|---|---|---|---|---|---|---|
| | | | | | 7 days | 28 day |
| 1:1.4:1.8 (cement :sand :gravel)cement content of 500kg/m$^3$, w/c=0.27, HRWRA =1.5 Liter/100 kg | 0 | 0.42 | 95 | --- | 28.9 | 40.46 |
| | 0.5 | 0.36 | 102 | 14.28 | 38.1 | 48.6 |
| | 1.0 | 0.30 | 104 | 28.57 | 41.5 | 53.8 |
| | 1.5 | 0.27 | 102 | 35.71 | 47.7 | 59.6 |
| | 2.0 | 0.28 | 102 | 33.34 | 43.2 | 55.2 |

### 3.7 Concrete mixes

After the selection of optimum dosage of HRWRA, several concrete mixes were prepared. Concrete mixes prepared in this investigation are shown in Table (5).

### 3.8 Mixing, preparation, and curing of concrete specimens

The processes for all concrete mixes were performed in an electrical rotary vessel mixer of 0.1m$^3$ capacity according to ASTM C 192/C192M. After removing the concrete specimens from molds, all the specimens were completely immersed in water storage tank containing water mixed with 3gm/litter of calcium hydroxide according to ASTM C511. Concrete specimens were cured till the time of test.

**Table 5.** Details of Concrete Mixes Used in this Investigation

| Group | Concrete mixes | Mix symbol | Glass powder waste content (%) | Plastic waste content (%) | Age (day) |
|---|---|---|---|---|---|
| 1 | Reference | G0P0 | 0 | 0 | 60 |
| 2 | Concrete Containing 15% Glass waste Powder as a Replacement to Cement with natural coarse aggregate | G15P0 | 15 | 0 | |
| 3 | Concrete containing 15% Glass Powder and different percentage of plastic waste as volumetric replacement to natural coarse aggregate | G15P25 | 15 | 25 | |
| | | G15P50 | 15 | 50 | |
| | | G15P75 | 15 | 75 | |
| | | G15P100 | 15 | 100 | |

### 3.9 Experimental tests

The following experimental tests were carried out to study the influence of two wastes on some properties of concrete:

- Slump test based on American Specification ASTM C143.
- Fresh density test based on American Specification ASTM C 138.
- Oven dry density test based on American Specification ASTM C 642 (using100×100 mm cube specimens).
- Compressive, strength test based on British Standard 1881 (utilizing 10.0 mm cube specimens).
- Water absorption test based on American Specification ASTM C642 (using 100 mm cube specimens).

 https://doi.org/10.1051/matecconf/201816202029

- Splitting tensile strength test, based on American Specification ASTM C496 (using 10.0×200 mm cylindrical specimens).
- Flexural tensile strength test based on American Specification ASTM C78 (utilizing 10.0×10.0×400 mm prism specimens).
- Ultrasonic pulse velocity test according to ASTM C597 (using 100 mm cube specimens).
- Thermal conductivity test (using cube specimens of 100 mm) [19].

## 4 Results and discussion

### 4.1 Properties of concrete containing glass powder waste

#### *4.1.1 Fresh concrete properties*

1) Workability

The experimental results in Table (6) indicate that there is a slight increase in workability with incorporation of 15% glass powder waste for the same slump compared with reference without wastes, because the glass powder has low water absorption compared with cement. Also it can be seen that there is a slight increase in workability with the increment in plastic waste coarse aggregate in concrete. The improvement in workability of concrete mixes with plastic waste aggregate is attributed to the very low absorbent characteristic of plastic waste in comparison with natural aggregate this correspond with that mentioned by Abduel Majid [20].

2) Fresh density

The results of fresh density for concrete mixtures containing different percentage of HDPE plastic and 15% glass powder waste are shown in Table (6) and Figure (4). It can be observed that the use of 15% glass powder only leads to a decrease in the fresh density of concrete compared with reference (without glass powder) this is due to the low specific gravity of glass powder (2.265) compared with specific gravity of cement (3.15). Also, the increase in plastic waste aggregate content decreases the fresh density. This is attributed to the low density of plastic aggregate (0.949kg/m$^3$) compared with that of the natural aggregate (1.715 kg/m$^3$) in concrete this is compatible with explanation by Kou et al. [21].

**Table 6.** Effect of Different Contents of Plastic Waste and 15% Glass Powder Waste on Workability, Fresh, Oven Dry Density, and Water Absorption of Concrete

| Symbol of mixes | Content of plastic waste (%) | Reduction in water relative to the reference (G0P0) without any wastes (%) | Fresh density (kg/m$^3$) | Oven dry density ( kg./m$^3$) | Water absorption (%) |
|---|---|---|---|---|---|
| G0P0 | --- | -- | 2496.4 | 2303.4 | 0.73 |
| G15P0 | 0 | 3.7 | 2484.6 | 2341.9 | 0.66 |
| G15P25 | 25 | 6.6 | 2366.7 | 2285.0 | 0.82 |
| G15P50 | 50 | 9.3 | 2275.3 | 2107.1 | 0.91 |
| G15P75 | 75 | 13.1 | 2143.7 | 2023.0 | 0.98 |
| G15P100 | 100 | 17 | 2055.5 | 1862.4 | 1.07 |

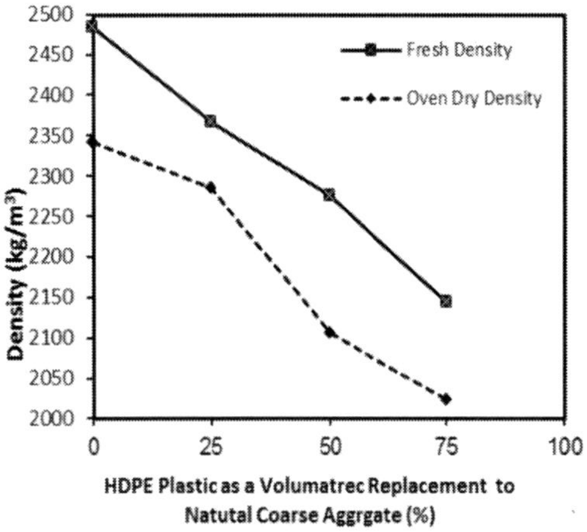


**Fig. 4.** Fresh and oven dry density of all concrete specimens

#### *4.1.2 Hardened concrete properties*

1) Oven dry density

The test results at 60 day age presented in Table (6) and Figure (4) show a slight increase in dry density for specimens containing 15% glass powder by about 1.671% relative to reference (G0P0). This increment is due to the change of calcium hydroxide ions that have specific gravity of 2.24 to extra gel with specific gravity of 2.3-2.6 [22], that fills the space between particles in the microstructure of concrete. The inclusion of plastic waste aggregate in concrete leads to a significant decrease in the oven dry density of concrete because of the lower density of plastics relative to the density of natural coarse aggregate. The results indicate that concrete specimens containing 100% HDPE plastic waste aggregate and with 15%, glass powder waste has a dry density of 1862.4kg/m$^3$

2) Water absorption

Water absorption of all concrete specimens prepared in this investigation is listed in Table (6). It's clear that the water absorption is decreased for concrete containing 15% glass powder by 9.8% in comparison with the reference specimens (without any wastes), this is due to the pozzolanic reaction between the silica in glass powder and calcium hydroxide. This reaction produces extra gel [5], which fills the spaces between particles, as well as segment the continuous capillary pores in the microstructure of concrete, that reduces the permeability consequently and decreases the absorption. The incorporation of plastic waste as coarse aggregate in concrete increases the water absorption with the increase in plastic waste content in concrete specimens. This is because of the shape of plastic aggregate (fluky and elongation), that increases the continuous path between pores and increases porosity. Also the low density of plastic waste leads to unsuitable compaction then more pores are formed.

The water absorption of each concrete mixes is very low and less than 10 %. This displays the good quality of all concrete mixes in this investigation [23].

3) Compressive strength

The effect of 15% glass powder waste and HDPE Plastic waste coarse aggregate on the compressive, strength of concrete is explained in Table (7) and Figure (5).The test results demonstrate that the concrete specimens without any wastes (G0P0) is high strength concrete according to ACI 363 [24] that has a compressive strength of 64.7 N/mm2 at 60 day age. The compressive strength of concrete containing 15% of glass powder at 60 day age is higher than reference because of the pozzolanic reaction that occurs between the reactive silica in glass powder and calcium hydroxide produced from the hydration of cement. This reaction produces extra gel that increases the strength at later ages [5]. The inclusion of plastic waste coarse aggregate significantly decreases the compressive strength of concrete. It's clear that the percentage reductions in compressive strength for concrete specimens with 25%, 50%, 75%, and 100% volumetric replacement of plastic waste coarse aggregate are 25.6%, 37.1%, 53.2%, and 59.8% respectively relative to concrete specimens without coarse plastic waste aggregate (G15P0). The reducing in compressive strength is because of the decrease in the bond strength, between the cement paste and HDPE plastic waste aggregate this corresponds with mentioned by Kinda et al. [25]. In addition, the natural aggregate is stronger than plastic aggregate, and as the most strength of concrete is from the strength of aggregate because approximately three quarters of the volume of concrete an occupied by aggregate [26], therefore the quantity and the strength of aggregate play an important role in the strength of concrete.

**Table 7**. Compressive, Splitting, and Flexural Strengths of Concrete Containing Different Percentage of Plastic Waste and 15% Glass Powder Waste

| Symbol of mixes | Content of plastic waste as a volumetric replacement to natural coarse aggregate (%) | Compressive strength at 6o day (MPa) | Splitting tensile strength at 60 day (MPa) | Flexural strength at 60 day (MPa) |
|---|---|---|---|---|
| G0P0 | --- | 64.7 | 3.97 | 5.68 |
| G15P0 | 0 | 73.3 | 5.41 | 7.65 |
| G15P25 | 25 | 54.5 | 4.38 | 6.83 |
| G15P50 | 50 | 46.1 | 3.63 | 5.84 |
| G15P75 | 75 | 34.3 | 3.19 | 4.92 |
| G15P100 | 100 | 29.5 | 3.05 | 4.11 |

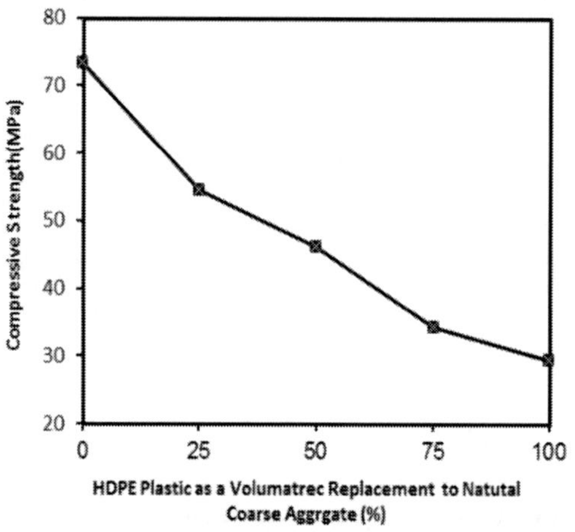


**Fig. 5**. The effect of plastic waste content on the compressive strength of concrete

4) Splitting tensile strength

The splitting tensile strength values for concrete specimens containing 15% glass powder waste and different percentages of coarse plastic waste aggregate at 60 day age are presented in Table (7) and Figure (6). It can be observed that there is an increase in splitting tensile strength for concrete containing 15%, glass powder (G15P0) of about 36.3%, compared with the specimens without glass powder (G0P0). This is due to the formation of extra gel from the pozzolanic reaction, in addition the glass powder fills the space between particles and increases the bond between the components of concrete [5, 27]. Generally, the splitting tensile strength for all concrete samples containing plastic waste aggregate decreases with the increase in plastic content relative to natural crushed coarse aggregate replacement by volume. The percentage decreases in splitting tensile strength are 19.0%, 32.9%, 41.1%, and 43.6% for specimen containing 25%, 50%, 75%, and 100% volumetric content of plastic waste aggregate relative to the concrete specimens without plastic aggregate (G15P0). This is due to the lower bonding strength between the surfaces of plastic aggregate and cement paste compared with natural aggregate and cement paste, also natural coarse aggregate is stronger than plastic aggregate, that leads to more compatibility between aggregate and cement paste, then improves the interfacial transition zone between aggregate and cement paste.

5) Flexural strength

The test results of flexural strength (modulus of rupture) for all concrete specimens are listed in Table (7), and shown in Figure (6). An increase in flexural strength was recorded for concrete containing 15% glass powder as a partial replacement by weight of cement by about 34.7 % compared with the specimens without glass powder (G0P0). This increase is because glass powder improves the transition zone between aggregate and cement paste consequently due to the pozzolanic reaction, as well as its filling ability to the spaces between particles by extra gel produced from this reactions, therefore the porosity that is inversely proportional to strength is decreased. Generally, the use of plastic waste aggregate as a replacement to natural coarse aggregate in concrete composite leads to a significant decrease in flexural strength for the same reasons mentioned in the compressive and splitting tensile strengths. The percentage reductions are 10.7%, 23.7%, 35.7%, and 46.3% for concrete specimens with 25%, 50%, 75%, and 100% plastic aggregate respectively relative to concrete specimens without plastic aggregate (G15P0).

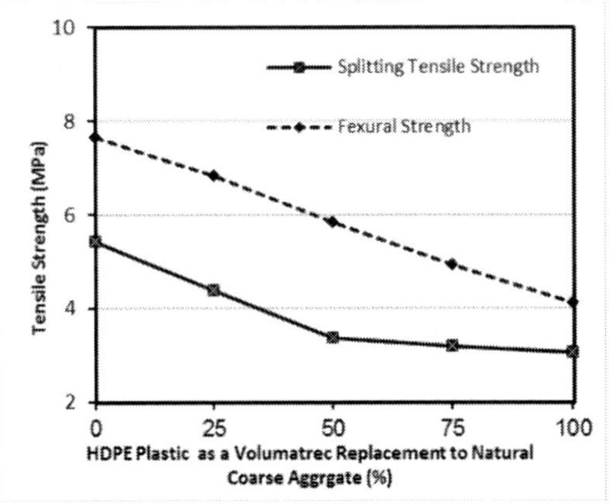


**Fig. 6**. The effect of plastic waste content on splitting and flexural strength of concrete.

6) Thermal conductivity

Thermal conductivity results of all concrete specimens at 60 day age are given in Table (VIII). It can be seen that there is a significant decrease in thermal, conductivity factor with the rise in plastic waste aggregate content. This is because the formulation of a large numbers of pores and holes in the texture of concrete in presence of plastic waste aggregate. Also due to porosity that considered main factor that affect the thermal conductivity factor of concrete and the closed pores with air that has low thermal, conductivity factor [28]. Also, plastic waste aggregate has lower thermal conductivity (0.1175W/m.K) in comparison with natural coarse aggregate (2W/m.K); this is compatible with Semiha et al. [29].

7) Ultrasonic pulse velocity

The results of UPV for concrete specimens with and without HDPE plastic are given in Table (8). This non-destructive test is very useful to investigate the homogeneity of concrete containing polymer and plastic waste, i.e. more homogeneity mean increase transmission of velocity through the concrete. The reason of decrease in ultrasonic pulse velocity with the increment in plastic waste is because more non-homogeneity of concrete with plastic as coarse aggregate as a result of different shapes, low modulus of elasticity, also the plastic is considered very poor for transition wave compared with natural aggregate.

**Table 8**. Thermal Conductivity and Ultrasonic Pulse Velocity of Concrete Containing Different Percentage of Plastic Waste and 15% Glass Powder Waste

| Symbol of mixes | Percentage of plastic waste as a volumetric replacement to natural coarse aggregate (%) | Thermal conductivity (W/m.K) | Ultrasonic pulse velocity ( km/sec) |
|---|---|---|---|
| G0P0 | --- | --- | --- |
| G15P0 | 0 | 2.3402 | 5.35 |
| G15P25 | 25 | 2.1699 | 5.03 |
| G15P50 | 50 | 1.6651 | 4.69 |
| G15P75 | 75 | 1.3029 | 4.35 |
| G15P100 | 100 | 1.0663 | 3.82 |

## 5 Conclusions

From the experiential results and materials used offered in this investigation, the following conclusions can be drawn:

1. The workability of fresh concrete is improved with the use of glass powder or plastic or both; this is important to reduce the water demand of the mixture by increasing workability with the same water cement ratio.

2. There is a decrease in fresh density with the use of 15%glass powder waste or with the increase in plastic waste aggregate content.

3. There is a significant decrease in oven dry density with the increase in plastic waste aggregate, while there

https://doi.org/10.1051/matecconf/201816202029

is a slight increase when 15% glass powder waste is only used.

4. The incorporation of 15%glass powder leads to a decrease in water absorption of concrete. On the other hand the increase in plastic waste content as coarse aggregate in concrete leads to increase in water absorption.

5. The inclusion of 15% glass powder in concrete improves strength of concrete. While, the inclusion of plastic waste coarse aggregate significantly decreases the compressive strength of concrete. The percentages reduction in compressive strength for concrete samples with 100% plastic waste coarse aggregate is 59.8% relative to concrete specimens without any plastic waste coarse aggregate.

6. The inclusion of plastic waste aggregate in side concrete leads to significant decrease in splitting and flexural strengths. The maximum reduction in splitting and flexural strengths is 43.6% and 46.3% for concrete specimens with 100% plastic aggregate respectively.

7. There is a significant reduction in thermal conductivity and ultrasonic pulse velocity with the increase in plastic waste aggregate content.

8. It's possible to produce structural lightweight concrete with good thermal insulation property using 100% HDPE plastic waste replacement to natural coarse aggregate by volume, e.g. the compressive strength, oven dry density, and thermal conductivity are 29.5 MPa, 1862.4 kg/m3, and 1.0663W/m.K respectively.

## References

[1] K., Sobolev, T.R., Naik, Int. Canada, 295-305, (2005).

[2] S.P., Shah, Wang, K., Development and Concrete Technology, China, 15-23, (2004).

[3] D., Jevtic, D., Zakic, A., Savic, Mechanical Testing and Diagnosis J., **1**, 22-39, (2012).

[4] Oyenuga, A.A., Bhamidimarri, R., Int. J. of Social Behavioral Educational, **9**, 1131- 1137, (2014).

[5] Mehta, P.K., and Monterio, P.J.M., McGraw-Hill Companies, Inc., New York, (2006).

[6] M.C.G. Juenger, R., Siddique, Cement and concrete research J., **78**, 71-80, (2015).

[7] A.M., Rashad, J., **41**, 303-318, (2013).

[8] Z., Abdul Rahim, Ph.D. Thesis, Jawaharlal Nehru Technological University, Hyderabad, India, (2010).

[9] Instruction Manual, EDX-7000, Shimadzu Corporation, Japan, (2008).

[10] IRAFFINITY-1 Fourier Transform Infrared Spectrophotometer "*Manual* ", Shimadzu Corporation, Japan, (2008).

[11] ASTM D-792, "*Standard Test Methods for Density and Specific Gravity (Relative Density) of Plastics by Displacement*", Annual Book of Standards, **0801**, (2003).

[12] ASTM D790, "*Test Method for Flexural Properties of Unreinforced and Reinforced Plastics and Electrical Insulating Materials*", Annual Book of ASTM, **0801**, (2003).

[13] ASTM D695, "*Standard Test Method for Compressive Properties of Rigid Plastics*", Annual Book of ASTM Standards, **0801**, (2002).

[14] ASTM D638, "*Standard Test Method for Tensile Properties of Plastics*", Annual Book of ASTM Standards, **0801**, (2010).

[15] Griffin and George:, Griffin and George Ltd., UK. (2002).

[16] European Standard EN ISO 179-1 Standard, "*Plastics Determination of Charpy Impact Properties*", (2001).

[17] ASTM D3350,"*Standard Specification for Polyethylene Plastics Pipe and Fittings Materials*", Annual Book of ASTM standards, **0802**, (2007).

[18] Building Research Establishment, "*Design of Normal Concrete Mixes*", London, (1997).

[19] Quick Thermal Conductivity Meter, "*Operation Manual, QTM-500*", KYOTO Electronics Manufacturing Co., LTD, Japan, (2014).

[20] K.N., Abduel Majid, A.B., Emhemmad, M.B.K., Milad, Int. J. of Chemical and Environmental Eng., **4**, 399-402, (2013).

[21] S.C., Kou, G. Poon Lee, W.L., Lai, Waste Management J., **29**, 621–628, (2009).

[22] S. Mindess, J.F., Young, D., Darwin, "*Concrete*", Second Edition, Prentice Hall, U.S.A., **66**, (2003).

[23] Neville, A.M., "*Properties of Concrete*", , Longman Group Ltd., United Kingdom, (2011).

[24] American Concrete Institute Committee ACI 363, "*State-of-the-Art Report on High-Strength Concrete*", ACI 363R-92, Detroit, (1992).

[25] H., Kinda, S. K., Bernard, W., Prince, Waste Management J., **30**, 2312–2320, (2010).

[26] Zongjin, Li, "*Advanced Concrete Technology*", John Wiley and Sons, Inc., New Jersey, 2011.

[27] Thomas, M., Taylor and Francis Group, London, (2013).

[28] List of Thermal Conductivities, 2016. Available at: https://en.wikipedia.org/wiki/List_of_thermal_conductivities.

[29] A., Semiha, A., Kubilay, and C.A., Duran, Composites Part B: Engineering J., **45**, 721–726, (2013).

# Effects of nano and micro size of clay particles on the compressive and tensile strength properties of self-consolidating concrete

*Baydaa* Alrashedi[1,*], and *Maan* Hassan[1]

[1]Building and Construction Engineering, University of Technology, 10066, Baghdad, Iraq

**Abstract.** Many past studies concerned about using of nanoclay (nC) particles as an active pozzolan to concrete and their influence on the physical and mechanical properties. In this study, the effects of various nanoclay particle sizes and dosages on the compressive and flexural tensile strengths of SCCs were investigated. Progressive nC percentages of 2%, 5% and 8% were replaced with cements and the produced SCCs were evaluated and compared with similar replacement levels of the metakaolin MK which have comparable chemical compositions. The produced SCCs were tested for compressive, flexural tensile and splitting tensile strengths in 28 and 90 days. Results indicated that nC replacement harmed all the studied mechanical properties at 28 days age. After 90 days, however, both compressive and tensile strengths of nC concretes show superior strengths than control concrete and also exceeded MK concretes made of equivalent replacement levels. This behavior demonstrates the pozzolanic activity of the nC particles at later ages and proved to be significantly more effective than early ages.

## 1 Introduction

Self-compacting concrete (SCC) is an innovative concrete that does not require vibration for placing and compaction. It is able to flow under its own weight, completely filling formwork and achieving full compaction, even in the presence of congested reinforcement [(1,2)]. It was first launched in Japan in the late 1980's by Okamura to solve problems of pouring and setting concrete in high rebar densities structures [(3)]. The high workability of SCC results in a well compacted microstructure with reduced porosity in mortar matrix and interfacial zone, and thus, improves the electrical resistivity and transport properties of concrete. This is the key to enhanced durability performance of SCC [(4)].

There are two ways of achieving SCCs, first by increasing the powder content, and second by incorporating chemical admixture. However, the chemical admixtures are expensive and their use may increase the cost of materials while the use of fine materials can ensure the required concrete properties without increasing the cost [(5, 6)].

Nanoclay (kaolin) is a fine, white, clay mineral that has been traditionally used in the manufacture of porcelain. Kaolinite is the mineralogical term that is applicable to kaolin clays. Kaolinite is defined as a common mineral, hydrated aluminum disilicate, the most common constituent of kaolin [(7)].

Nowadays, in concrete design, concrete researchers and developers are taking advantage of secondary cementations materials to give concrete greater strength. One of the newest technologies to break into the concrete design is the use of pozzolanic nano-particles in the concrete matrix. By using pozzolanic nano-particles, the development of the strength bearing crystals of cement paste can be increased or controlled [(8)]

Nanotechnology is a very active research field and has applications in a number of areas. Currently this technology is being used for the creation of new materials, devices and systems at molecular, nano- and micro-level [(10)]. Nanoclay particles have shown promise in enhancing the mechanical performance, the resistance to chloride penetration, and the self-compacting properties of concrete and in reducing permeability. Clay and the properties of clay that are important as a mineral admixture to cement exist on the nanoscale, [(11-15)].

In this study, two clay particle sizes: metakaolin (MK) and nanoclay (nC) were mixed in progressive replacement ratios with SCCs. The main aim was to assess the effects of nano size clay particles against alternative micro size metakaolin on the mechanical properties of SCCs. In terms of economical and industrial applications, the results of this study may provide more insights into the utilization of new but costly nanoscale mineral (i.e. nC) instead the more widely used MK.

## 2 Experimental program

### 2.1 Preparation of raw material

* Corresponding author: baydaa.engineer@yahoo.com

© The Authors, published by EDP Sciences. This is an open access article distributed under the terms of the Creative Commons Attribution License 4.0 (http://creativecommons.org/licenses/by/4.0/).

The preparation of raw material in this work is explained briefly in this section. Ordinary Portland cement (C) with trade mark of (Cresta) made in Iraq was used that conforms to Iraqi specifications number 5-2001 (16). Fine aggregate (S) was brought from river sand and sieved using a 4.75mm sieve to remove any possible large and clogging particles bigger than 4.75 mm. Meanwhile, the 12mm coarse aggregate (G) was selected the grading curves conform to ASTM C33(17).

To maintain the freshness of SCC the sika based on modified polycarboxylic (viscocrete-5930) (SP) was used which conform to ASTM C 494 Type F. (18). Table (1) shows the chemical composition of ordinary Portland cement, nanoclay(nC), kaolin (K), metakaolin (MK) and limestone powder (LSP) used in this study.

**Table 1**: Chemical composition of SCC fine materials

| constituent | C% | nC% | K% | MK% | LSP% |
|---|---|---|---|---|---|
| silica | 21.9 | 38.64 | 50 | 49.64 | 1.12 |
| Alumina | 4.97 | 13.15 | 32 | 10.7 | 0.64 |
| Iron oxide | 3.2 | 2.25 | 1.4 | 3.7 | 0.3 |

The nanoclay powder used in this study was a hydrophilic Montmorillonite materials. It was brought from (Nanocor® Inc, USA) and was used as received without any type of further treatment or manipulation. The Montmorillonites materials are high purity aluminosilicate minerals. It is sometimes referred to as phyllosilicates. Montmorillonites have a sheet-type or plate like structure. Even though their dimensions in the length and width directions have a scale of hundreds of nanometers, the plate thickness is only one nanometer. As a result individual sheets have aspect ratios ranging from 200-1000, with a majority of platelets in the 200-400 range after purification. The theoretical formula for montmorillonite is:

$$(Na, Ca)_x (Al, Mg)_2 (SiO_{10}) (OH)_2.nH_2O \quad (1)$$

No chemical treatment is needed for this type of nanoclay to be dispersed in water which makes it easy to use in both concrete and mortar preparation.

### 2.2 Mixing and curing

The ACI 237R-07 procedure was followed to design SCC mix. It provides a guideline for proportioning mixtures and makes use of batches as trial mixes and then adjust the proportions based on the fresh and hardened testing results. The mix proportions of the ingredients used in this study are kept constant and only the nC or MK contents were changed as a cement replacement by weight (see Table-2).

The typical batching sequence that has been used to minimize balling of fine materials as below (19):

1) Adding the coarse aggregates to the mixer with more than 50% of mix water and mixing for 1 minute; 2) after performing the dry mix for the powders (cement, limestone, nanoclay or microclay particles) followed by adding the fine aggregate to the previous ingredients and mixing for 1 minute. As mentioned by other researcher (20) nano materials were vigorously dry mixed with cement for 30 min or more to disperse nano particles with cement; 3) adding remaining water with 1/3 of SP and mix for 1 1/4 minutes after that adding the 2/3 of the leftover of SP dosage with 2 3/4 minute mix time; 4) the mixture is then discharged, tested and casted.

In the present study the above procedure was followed. The total mix time takes 5 minutes. The concrete was prepared using ordinary Portland cement with a partial replaced by nC and MK. The ingredients were homogenized on mixer to assure complete homogeneity and then water was added. The molds used in this study were cubes of (100 *100mm) size and cylinders (100*200mm) size. The samples were kept wet in molds, and then cured in water until testing age.

## 3 Results and discussion

### 3.1 Compressive strength

The compressive strength test was carried out using cubes (100mm) according to BS 1881: part 116-2004(21). The cubes tested by using (ELE) machine. The average of three cubes was recorded. This test was conducted at 28, 90 days of ages and presented in Table (4) and Fig. (1).

As can be seen, the compressive strength of concretes containing nC reduced at 28 days compared to control. Reduction rate was increased as the replacement percentages increased and reached a maximum reduction rate of 23.3% at 8% nC replacement. In contrast, progressive replacement percentages of MK lead to improve the compressive strengths and the improvements were a function of the replacement values of MK. This could be attributed to the higher pozzolanic activity of MK particles due to burning process during the preparation stage. These results suggest that at early ages up to 28 days, the positive effect of clay particles burning overrides the effect of using lower particle size even if it is in nano scale.

**Table 2**: Details of concrete mixes used in the present work (kg/m$^3$)

| Mixes | C | nC | MK | LSP | P | W/C | G | S | SP |
|---|---|---|---|---|---|---|---|---|---|
| SCC | 400 | 0 | 0 | 100 | 500 | 192 | 840 | 840 | 4 |
| SCC-nC2% | 392 | 8 | 0 | 100 | 500 | 192 | 840 | 840 | 5 |
| SCC-nC5% | 380 | 20 | 0 | 100 | 500 | 192 | 840 | 840 | 6 |
| SCC nC 8% | 368 | 32 | 0 | 100 | 500 | 192 | 840 | 840 | 8 |
| SCC- MK 2% | 392 | 0 | 8 | 100 | 500 | 192 | 840 | 840 | 5 |
| SCC- MK 5% | 380 | 0 | 20 | 100 | 500 | 192 | 840 | 840 | 6 |
| SCC- MK 8% | 368 | 0 | 32 | 100 | 500 | 192 | 840 | 840 | 8 |

SCC=self-consolidation concrete, nC= nanoclay, MK=metakoalin,C=cement, p= total powder,

LSP=limestone powder, S= sand, G= crushed gravel, SP=superplasticizer**

**Table 3**: Compressive strength test results for SCCs

| Group No. | Mix Description | Compressive strength (MPa) | |
|---|---|---|---|
| | | 28 days | 90 days |
| Control | SCC | 33.4 | 44.2 |
| 1 | SCC-nC2% | 33.1 | 49.2 |
| | SCC-nC5% | 30.1 | 51.7 |
| | SCC-nC8% | 25.6 | 54.1 |
| 2 | SCC- MK 2% | 34 | 47 |
| | SCC- MK 5% | 43.4 | 49.4 |
| | SCC- MK 8% | 53 | 52.6 |

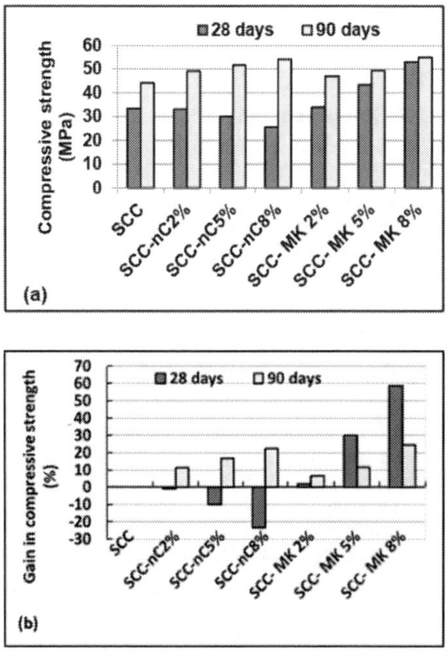


**Fig. 1**. Compressive strengths for SCCs contained MK and nC at 28 and 90 day (a) and gain in strengths (b)

At later ages (90 days), however, compressive strength of concretes containing nC particles observed to have higher level compared to control and also exceeded concretes of similar replacement percentages of MK. The gain in strength reached 22% at 8% replacement value. These results reflect the beneficial effect of nano scale particles at later time. They also illustrate the slow reaction rate of non-burned clay particles to actively participate in improving the cement past microstructure through better pore refinement, more consumption of Ca $(OH)_2$, micro filling action, and higher pozzolanic reactions.

To explain the differences in gain of strengths due to using nC against MK SCCs in this study, the ratios of 90 days to 28 days are presented in Fig. 2 below. In the case of nC concretes, the ratios increased as the percentages of replacement increased indicating higher activity in later ages compared with earlier ages. In the case of MK concretes, however, 90 d/ 28 d ratios of similar replacement levels were less than nC concretes indicating that concretes made with MK reached their maximum rate of strength at ages up to 28 days. After 90 days they almost became of constant level of strength and no more gains are achieved.

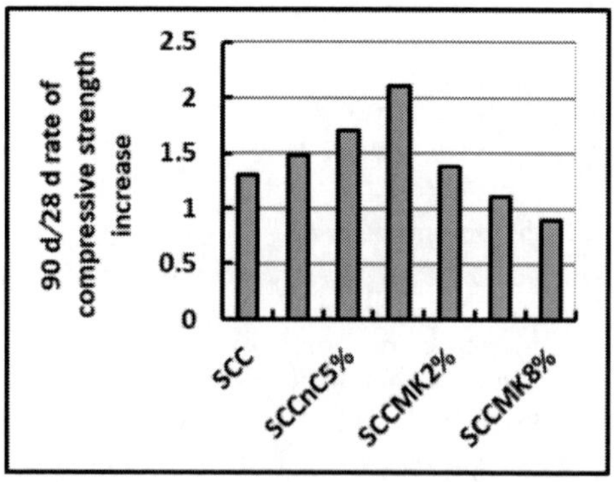


**Fig. 2**. Ratios of increase in compressive strengths (90 d/ 28 d) for SCCs with MK or nC cement replacements.

### 3.2 Tensile strength results

The splitting and flexural tensile strength tests were carried out according to ASTM C496-2004[22] and ASTM C1609 [23] respectively, at ages of 28, and 90 days. The results are shown in Table 4 and presented in Figs. 3 and 4.

**Table 4**: Splitting and flexural strength test results for SCCs.

| Group No. | Mix Description | Splitting strength (MPa) | | Flexural strength (MPa) | |
|---|---|---|---|---|---|
| | | 28 days | 90 days | 28 days | 90 days |
| **Control** | SCC | 2 | 2.4 | 2.9 | 2.9 |
| 1 | SCC-nC2% | 1.8 | 2.7 | 2.7 | 3 |
| | SCC-nC5% | 1.8 | 2.7 | 2.6 | 3.7 |
| | SCC-nC8% | 1.9 | 2.8 | 2.6 | 3.9 |
| 2 | SCC- MK 2% | 2.0 | 2.5 | 3 | 3.1 |
| | SCC- MK5% | 2.3 | 2.7 | 3.1 | 3.4 |
| | SCC- MK 8% | 2.5 | 2.7 | 3.5 | 3.8 |

In general, the incorporation of nanoclay particles into concrete cement past led to reduction in both splitting and flexural tensile strengths for all SCC specimens at 28 days ages, as shown in Figs. 3 and 4, respectively. The addition of 2%, 5%, 8% nanoclay reduced splitting strength by 10%, 10%, 5%; and flexural strength by 7%, 10%, 10%, respectively, compared to control specimens. The reduction in mechanical properties can be attributed to the poor dispersion and agglomeration of the nanoclay in the cement paste at higher replacement percentages. Other researchers reported that nanoclay contents more than 1% could create weak zones in the form of micro voids as stress concentrators [24-28]. This behavior is correlated with the compressive strength results.

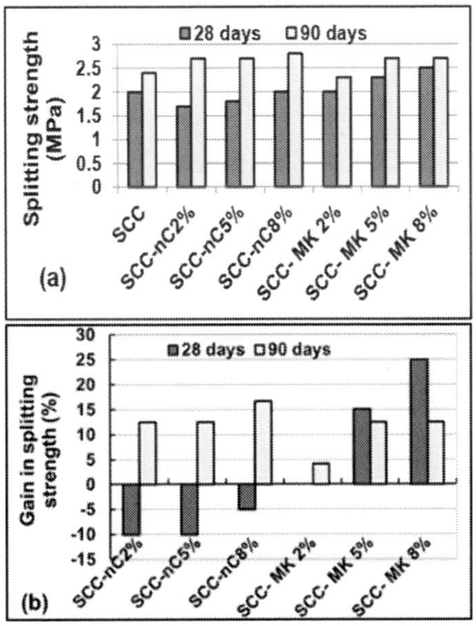


**Fig. 3**. Splitting strengths for SCCs contained MK and nC at 28 and 90 day (a) and gain in strengths (b)

After 90 days, both splitting and flexural tensile strengths were increased. Progressive values of nC replacements resulted in enhancements in tensile strengths of the investigated SCCs.

The use of MK led to noticeable improvement in both 28 and 90 days. Concretes with 2%, 5%, 8% MK replacement have 0%, 15%, 25% higher at 28 days and 4%, 12.5%, 12.5% higher at 90 days of splitting strengths, respectively. The improvement in flexural strengths in similar replacement levels were 3%, 7%, 20% higher at 28 days and 7%, 17%, 31% higher at 90 days, respectively. The results suggest that using nC with cementitious materials may reduce early ages strengths in comparison to MK, but it could have significant improvements during the next few months and even exceed the MK improvement rates.

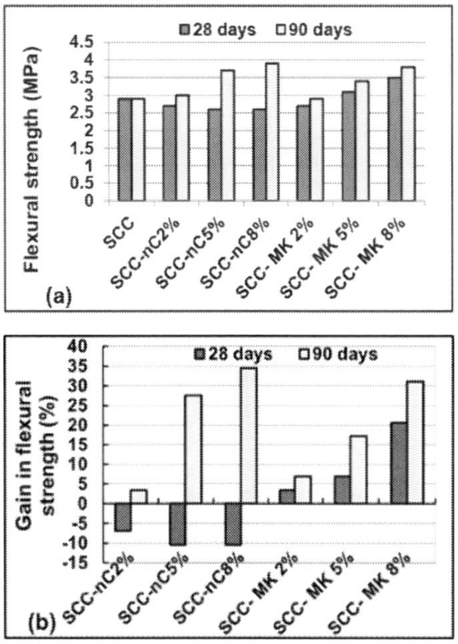


**Fig. 4**. Flexural strengths for SCCs contained MK and nC at 28 and 90 day (a) and gain in strengths (b)

## 4 Conclusions

Based on the experimental work results in this research work, the following conclusions can be drawn:

1. For all SCCs, progressively higher cement percentage replacements from 2% to 8% of nC led to significant reduction in the compressive and tensile properties at 28 days age. The reduction was proportioned to the percentage of the cement replacement. On the contrary, use of metakaolin led to enhancement of the compressive and tensile properties at the same dosages and similar age.

The low pozzolanic activity of nC at early ages up to 28 days is probably behind this reduction.

2. At later age (i.e. 90 days), both concretes made with nC or MK showed remarkable increase in the compressive and tensile properties. The gain in strengths of nC concretes was more visible, and the level of strengths exceeded the MK concretes made with equivalent replacement levels.

3. Among various nC and MK dosages used, 8% provided the better later age strength achieved.

Acknowledgments: The authors would like to thank the Building and Construction Engineering Department of University of Technology, Baghdad, Iraq, for extending the facilities and laboratories for the above research work.

## References

1.Okamura,H., and Ouchi,M., Journal of Advanced Concrete Technology Vol. No.1, PP.5-15, April(2003).

2. EFNARC,The European Federation of Specialist Constructionand Concrete System, pp.63, May, (2005).

3. Borroni, M., Proceedings of the 2nd FIB international congress, Naples, Italy, pp. (1-7), (2006).

4. Safiuddin, Md., West, J.S., and Soudki, K.A, Journal of Applied Sciences Research, Vol.4, No.12, pp.(1834-1840),(2008).

5. Tregger NA,Pakula ME,Shah SP, Cem Concr Res 40 (Compendex):384-91, (2010).

6. Ferron R., PhD Thesis, Northwestern University, Evanston, IL; (2008).

7. Rafat Siddique, Department of Civil Engineering, Thapar University,India, (2008).

8. Jonathan S. Belkowitz and Daniel Armentrout, Concrete Sustainability Conference, PP(1-14), (2010).

9. Syeda A. Yousaf and Salamat Ali, Journal of Faculty of Engineering & Technology, PP (11, 12), (2007-2008).

10. Perumalsamy B. and Ken Ch, Proceedings of ACI Session on "Nanotechnology of Concrete: Recent Developments and Future Perspectives", Denver, USA, PP (15), 7 November, (2006)

11. Florence S.,& Konstantin S, Construction and Building Materials, 24 (11), (2010).

12. Chang T-P, Shih J-Y, Yang K-M, Hsiao T-C, J Mater Sci;42(17), PP.(7478–87), (2007).

13. Kuo W-Y, Huang J-S, Lin C-H., Cem Concr Res;36(5) ,PP.(886–95),(2006).

14. Morsy MS, Aglan HA, Abd El Razek M. M., Construct Build Mater; 23(1), PP. (515–21), (2009).

15. He X, Shi X., Transport Res Board Record: J Transport Res Board (2070), PP. (13–21), (2008).

16. Iraqi Standard No. (5) for the year 2001, "Portland Cement" 16. The Central Agency for Standardization and Quality Control.

17. ASTM C33/C33M-13, American Society for Testing and Material International, (2013).

18. ASTM C494-99a, Annual Book of ASTM Standards, American Society for Testing and Materials, Vol. 04-02,(2004).

19. R.T. Abdulkareem, M.S. Hassan, I.N. Gorgis, Eng. & Tech. Journal. 34: 527-538, (2016).

20. Farzadnia, N., Ali, A. A. A., Demirboga, R., and Anwar, M. P., Cement and concrete research, **48**,(2013).

21. B.S 1881: Part 116, British Standards Institution,(1989).

22. ASTM C496-04, American Society for Testing and Materials, 2004.

23. ASTM C1609M – 12, American Society for Testing and Material International, (2012).

24. Morsy MS, Alsayed SH, Aqel M., Constr Build Mater;25(1):145–9,(2011).

25. Li H, Xiao H, Yuan J, Ou J., Composites Part B; 35(1):185–9,(2004).

26. A. Hakamy a, FUA Shaikh b, I.M. Low. Construction and Building Materials 49 ,298–307,(2013)

27. M.S. Hassan, I.N. Gorgis and Ali A. Jaber ARPN Journal of Engineering and Applied Sciences.12 (17), (2017).

28. M.S. Hassan, Z.M. Al-azawi, M.J. Taher., Arab. J. Sci. Eng. **41**. (2016). doi:10.1007/s13369-016-2056-z.

# Lightweight self-compacting concrete with light expanded clay aggregate (LECA)

*Khaled* Heiza, *Fatma* Eid and *Taha* Masoud

Faculty of Engineering, Menoufia University, Menoufia, Egypt

**Abstract**. Lightweight concretes have been successfully applied in building constructions for many years due to their favorable material properties, particularly their low specific weight in connection with a high strength, a high capability of thermal insulation and a high durability. The development leading to lightweight self-compacting concrete (LWSCC) represents an important advanced step within the recent years. This concrete combines the favorable properties of a lightweight concrete with those of a self-compacting concrete. Research work is aimed on development of (LWSCC) with the use of light aggregates "Light expanded clay aggregate (LECA)". In this research, first by specific gravity factor method, twenty different mix designs of (LWSCC) were cast and tested to find out the values of slump flow, J-ring , V-funnel and 28 day compressive strength. Based on the results obtained, the best mix design was selected for further investigation. This paper also focuses on studying the effect of changing the reinforcement ratio on reinforced two way slabs when the dimensions were kept constant.

## 1 Introduction

Lightweight self-compacting concrete (LWSCC) is predicted to produce high workability while not segregation and high durability with reduced weight. The success to production of high quality (LWSCC) lies within the use of aggregates. Expanded clay (LECA) is a ceramic material created by expanding and vitrifying select clay in a rotary kiln. The utilization of expanded clay aggregate with other quality supplementary cementing materials (such as fly ash and silica fume) can provide highly workable and durable( LWSCCs). Expanded clay (LECA) and other light weight aggregates such as: expanded shale, pumice, slate, perlite, bottom ash etc. have been successfully used in the production of lightweight concretes (LWCs) over the decades [1,2 and 3]. Use of those aggregates has property to the sustainable development by protective energy, increasing structural potency and increasing the service lifetime of structural lightweight concrete (LWC). These advantages add to those of (LWSCC) to additional support property development and contribute to projects obtaining Leadership in Energy and Environmental Design (LEED) certification (ESCSI 2004)[4,5]. (LWSCC) is capable of filling up the formwork and encapsulate reinforcement by its self-weight without the need for additional compaction or external vibration. It has excellent segregation resistance, high flowability and passing ability at fresh state as well as better mechanical and durability properties in the hardened state [6, 7].

## 2 Research program

This research was conducted in two phases: Phase (I) focused on the experimental study of the fresh and hardened properties of LWSCC mixes. Twenty concrete mixtures were designed. Three key mix design parameters namely water (w) to binder (b) ratio (w/b) (0.28–0.35), dosage of high range water reducing admixtures (HRWRA) (0.3–1.2 % by total content of binder) and total binder content (B) (410–510 kg/m3) were selected to study the influence on the properties of( LWSCC). The tested (LWSCC) properties were, slump flow, V-funnel flow time, J-ring flow diameter/height difference, unit weight and compressive strength. Phase( II ) based on the results obtained, the best mix which achieve the properties of light weight concrete and self-compacting concrete (so–called standard mix design) was selected for further investigation. By the standard mix the effect of changing reinforcement ratio of two way slabs reinforced by welded wire mesh with variable diameter of wire will be studied.

## 3 Materials

### 3.1 Cement

Portland cement type I (CEMI42.5N), provided by the Suez Cement Company., meeting the requirement of E.S. 7417/2001 [12].

© The Authors, published by EDP Sciences. This is an open access article distributed under the terms of the Creative Commons Attribution License 4.0 (http://creativecommons.org/licenses/by/4.0/).

### 3.2 LECA

The coarse lightweight aggregate used in this study was locally produced from expanded clay type (LECA) from National Cement Company. The maximum nominal size of (LECA) was 16 mm. The sieve analysis of( LECA) is given in Table (1). In this study, the dry (LECA) was pre-soaked in water for 48 hours before mixing in concrete until steady weight of LECA was achieved, to ensure that all voids inside (LECA) particles are fully filled with water. The physical and mechanical properties of coarse aggregate are given in Table (2).

**Table 1.** Sieve analysis of ( LECA)

| Sieve size (mm) | 19 | 16 | 9.5 | 4.75 | 2.36 |
|---|---|---|---|---|---|
| % | 100 | 98 | 79 | 30 | 7 |

**Table 2.** Physical and mechanical properties of (LECA)

| Description | 24 h water absorption: % | Specific gravity | unit weight kg/m3 |
|---|---|---|---|
| LECA | 18.2 | 1.08 | 667 |

### 3.3 Fine aggregates

Natural siliceous sand from El-khatatba was used as a fine aggregate.

### 3.4 Fly ash

Fly ash complying with chemical and physical requirements of American specification [ASTM C618] [13] was used.

### 3.5. Silica fume

Micro silica (silica fume) is a product of ferro silicon alloys industry. The product is a rich silicon dioxide powder where the average particles size is around 0.1micrometers

### 3.6 Viscosity agent

Sika Viscocrete 3425 was used as viscosity enhancing agent (VEA). It contains a polycarboxylic -based copolymer-based mixture and modified cellulose product to achieve the dual action effect of high-range water reducer and viscosity-modifying admixture, respectively. It meets the requirements for super plasticizers according to Swiss specification [ SIA 162(2989)] , European specification [EN934-2] , and American specification [ ASTM- C-494 type G and F). Mechanical and physical properties are given in Table (3).

**Table 3.** Typical Properties of Viscocrete (3425).

| Properties | Value |
|---|---|
| Appearance | Clear liquid |
| Density | 1.08 kg/Lit (ASTM C494) |
| PH Value | 4.0 |
| Solid content | 40% by weight |
| Chloride content | Zero |

### 3.7 Reinforcing steel

Welded wire mesh is arranged in one layer with different diameters of wire in the slab. This type of mesh is made of steel wires with diameter and size of mesh as shown in Table (4) The proof stress is equal to 400 N/mm2 and its ultimate strength is 600 N/mm2 while its modulus of elasticity equal to 170 kN/mm2 .

**Table 4.** Properties of Wire Mesh .

| Diameter (mm) | spacing (mm) | No. of wires / m2 | wire weight kg / m | Weight of mesh kg/m2 |
|---|---|---|---|---|
| 3 | 30 × 30 | 34 ×2 | 0.0555 | 3.778 |
| 4 | 40 × 40 | 25 ×2 | 0.0986 | 4.938 |
| 5 | 50 × 50 | 20 ×2 | 0.1541 | 6.173 |

## 4 Concrete mix design

Six mixes of (LWSCC) were designed using the specific gravity factor method, as described elsewhere [8, 9]. The best six mixes are shown in Table (5). The mix proportions of each mix were calculated by assuming that all its ingredients have to produce a cubic meter of concrete. Various mixes with different w/c ratio, all aggregate types used in this study were weighted on a saturated-surface-dry (SSD) basis, to achieve good moisture control in the concrete mixes. The constituent materials were mixed in a pan mixer with a capacity of 0.08 m3 and a mixing speed of 50 rev/min at ambient temperatures of about 25C. The (SCA) was separately added into the concrete after about 1 min of mixing. Mixing was continued for another 2 min before the mixture was left to rest for approximately 3 min.

**Table 5.** Mixture Design.

| Mix No. | $C_1$ | $C_2$ | $C_3$ | $C_4$ | $C_5$ | $C_6$ |
|---|---|---|---|---|---|---|
| W/c | 0.35 | 0.32 | 0.32 | 0.3 | 0.3 | 0.28 |
| Binder(kg) | 510 | 410 | 500 | 510 | 500 | 510 |
| Cement (kg) | 382 | 325 | 390 | 382 | 400 | 382 |
| FA (kg) | 76 | 62 | 72 | 76 | 72 | 76 |
| S F ( Kg) | 51 | 23 | 27 | 51 | 27 | 51 |
| Water (Lit) | 180 | 130 | 160 | 153 | 160 | 143 |
| SP (Lit) | 3.8 | 1.2 | 6 | 5.1 | 1.5 | 3.8 |
| Sand (kg) | 790 | 840 | 825 | 900 | 830 | 870 |
| LECA (kg) | 390 | 474 | 412 | 450 | 415 | 410 |

### 4.1 Casting of test specimens

a- Cubes with 10cm side length were used for determination of the compressive strength of light weight self-compacting concrete.

b-By using the best mix which achieve the properties of self-compacting and light weight concrete, three slabs were cast using wooden mould with dimension (1000 mm * 1000mm * 60 mm) as shown in figure(1).The thickness of the slabs were kept constant, the thickness of mesh layers and their combinations were varied in order to get higher strength. To prepare the mould for casting, it was first oiled which makes it easy for demoulding process. Initially, the first portion of mortar was evenly placed and the layer of mesh was then placed in position and mortar spacers were used in the areas between these meshes. Table (6) shows the properties of reinforced slabs.

**Table 6.** Properties of reinforced Slabs.

| Slab No. | Dimension (cm ) | Reinforcement | Spacing of mesh (cm) |
|---|---|---|---|
| $A_1$ | 100*100*6 | Welded mesh thick 3mm | 3 * 3 |
| $A_2$ | 100*100*6 | Welded mesh thick 4mm | 4 * 4 |
| $A_3$ | 100*100*6 | Welded mesh thick 5mm | 5 * 5 |

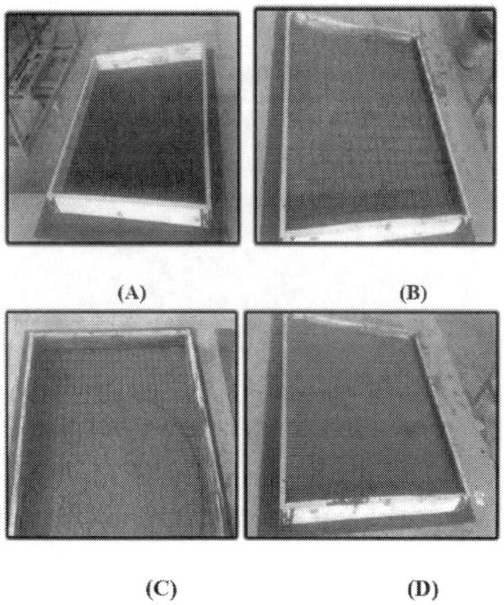

**Fig.1.** Casting of Specimens

## 5 Results and discussion

### 5.1 Fresh characteristics of ( LWSCC)

Immediately after the mixing, the value of slump flow, J-ring and V-funnel test were determined by the following methods Table (7).

#### 5.1.1 Slump flow test

The slump flow test was used to evaluate the free deformability and flowability of (LWSCC) in the absence of obstruction.

The result of slump flow for mixtures showed that an increase in the w/b from (0.28 to 0.35) significantly increased the slump flow However, at fixed( HRWRA% )the slump flow range got limited with the increase of binder content.

Great positive effect of the coupled parameters (w/b and HRWRA) in increasing the slump flow .The slump flow test for (LWSCC ) is shown in Fig. 2.

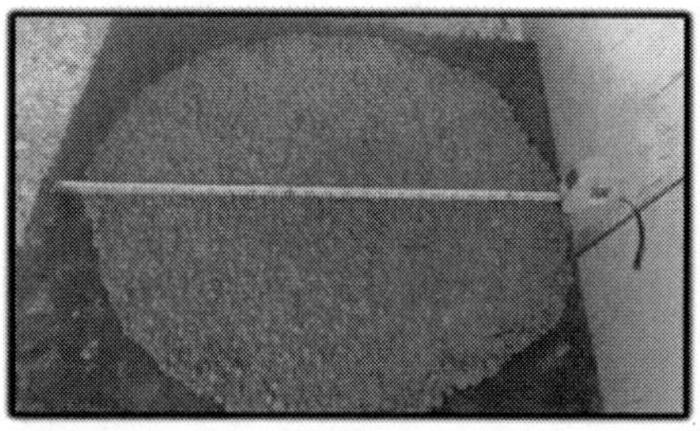

**Fig. 2.** Slump Flow Test.

#### 5.1.2 J-ring test

The J-ring test is used to determine the passing ability of the (LWSCC) (Fig. 3).

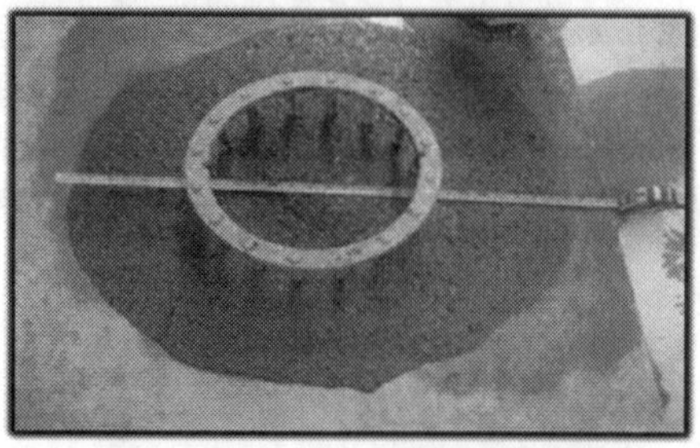

**Fig. 3.** J-ring test

#### 5.1.3 V-funnel flow test

The V-funnel test is used to determine the deformability through restricted area.

An increase of the w/b from (0.28 top 0.35) significantly reduced the V-funnel flow time whereas an increase of (HRWRA) from 0.3 to 1.2 % only slightly reduced the V-funnel flow time. However, combined maximum increase of both (w/b and HRWRA) parameters resulted in a substantial reduction of the V-funnel flow time the test was carried out for (LWSCC) as shown in Fig. 4.

For all mixes, the slump flow and J-ring flow/J-ring height difference were positively influenced by (w/b and HRWRA). An increase of either or both parameters led to an increase in the measured responses/properties. However, an increase in the binder content alone affects the results negatively—showing a decrease in the measured responses. So that mix No C5 and C6 don't achieve values of acceptance of self-compacting concrete [10].

**Fig. 4 .** V-funnel Flow Test.

**Table 7.** Fresh properties of mixtures

| Mix No | Slump flow (mm) | T at 50 cm (sec) | V-funnel (sec) | J-Ring flow (mm) | J – Ring height diff ( mm) |
|---|---|---|---|---|---|
| $C_1$ | 750 | 1 | 7 | 700 | 5 |
| $C_2$ | 650 | 1.5 | 8 | 630 | 8 |
| $C_3$ | 700 | 1 | 7 | 650 | 6 |
| $C_4$ | 690 | 1.2 | 7 | 640 | 6 |
| $C_5$ | 550 | 4 | 13 | 500 | 10 |
| $C_6$ | 450 | - | 15 | 430 | 12 |

### 5.2 Hardened characteristics of (LWSCC)

As noted from the results showed in Table (8) compressive strength increases with decrease of w/c with saturated expanded clay and increases with increase of binder. The best compressive strength mix is C6 but it doesn't meet the requirements of self-compacting concrete .All mixtures unit weight less than 2000 kg /m3 so that it can be light weight concrete.

The best mixture which achieves the requirements of light weight self-compacting concrete (LWSCC) with best strength is mixture no C4.

**Table 8.** Compressive strength and unit weight of different mixes.

| Mix No | Compressive strength (kg/ cm$^2$) | | Unit weight (kg/ m$^3$) | |
|---|---|---|---|---|
| | 7- day | 28-day | fresh | air dry |
| $C_1$ | 162 | 190 | 1820 | 1790 |
| $C_2$ | 130 | 153 | 1850 | 1810 |
| $C_3$ | 155 | 186 | 1880 | 1830 |
| $C_4$ | 167 | 227 | 1910 | 1865 |
| $C_5$ | 152 | 184 | 1860 | 1800 |
| $C_6$ | 172 | 235 | 1900 | 1850 |

### 5.3 Behavior of reinforced light weight self-compacting concrete slabs.

#### 5.3.1 Test Setup of Slabs

A special flexure loading frame was fabricated for making the test of the slabs and the details of the test setup is shown Fig. 5. In order to test the slabs on a four point loading (at 1/3rd span), over an effective span of 850 mm, the middle line of the slab, and also the roller supports were marked and slab was seated on the bottom rollers . Loading was applied manually through a hydraulic

 https://doi.org/10.1051/matecconf/201816202031

jacking arrangement to cause downward deflection (see Fig.5). The load was given through the jack in tiny increments and also the deflection of the slab was recorded up to failure using dial gauges. The proving ring readings and displacement values were observed instantaneously. The proving ring readings are taken at every five division interval and also the corresponding deformation values were determined in the displacement indicator. The loading was continued till the ultimate failure of the slab is reached, and the above measurements were taken at different load levels till final failure. The ductility (deflection) of slabs using different reinforcing mesh thickness (3 mm,4 mm and 5 mm) has been studied initially crack and ultimate failure loads.

**Fig. 5.** Flexure test set-up of slabs.

### 5.3.2 Results

A total of three specimens were tested in this investigation on four point loading tests and the results of the experimental tests carried out are analyzed including the flexural loads and deflection (Fig.6, 8, 9 and 10). The failure of the slab specimen's results from the yielding of wire mesh reinforcement is followed by the crushing of mortar. Initially fine flexural cracks appeared at the bottom of the specimen. With further increase in the load, regularly spaced vertical cracks were observed and they extended from the bottom of the specimen towards top Fig.(7).The load was increased up to ultimate stage.

Slab No. (A3) reinforced with wire mesh 5mm diameter recorded the highest value of ultimate load due to bigger diameter and pitch spacing ( 50mmx 50mm) while slab No (A1) reinforced with wire mesh 3mm diameter obtained the lowest value of ultimate load due to their smaller diameter and pitch spacing (30mm x 30mm)

.

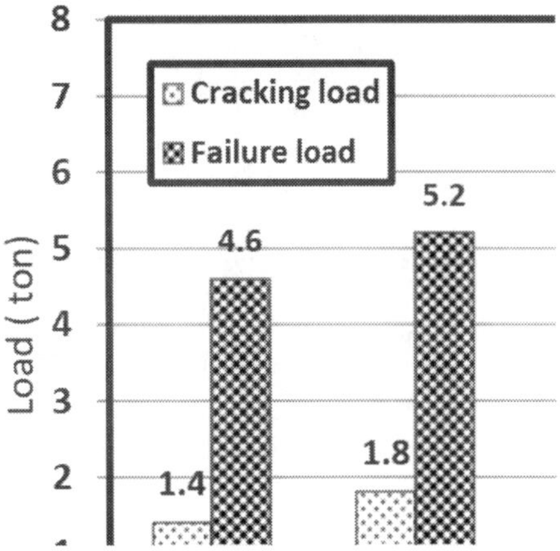


**Fig. 6.** Cracking and failure load for testing slabs

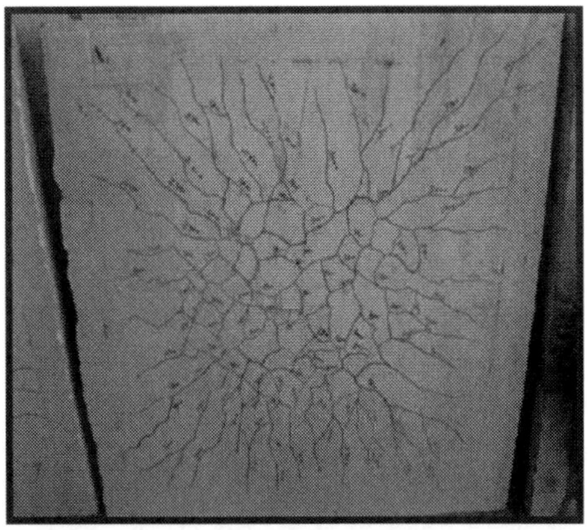

( A1 )

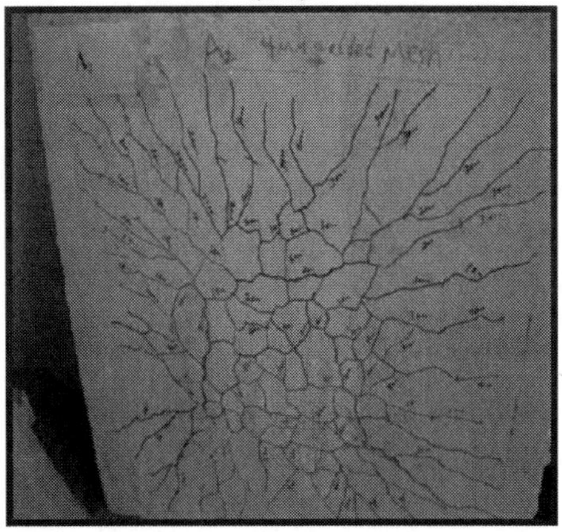

$(A_2)$

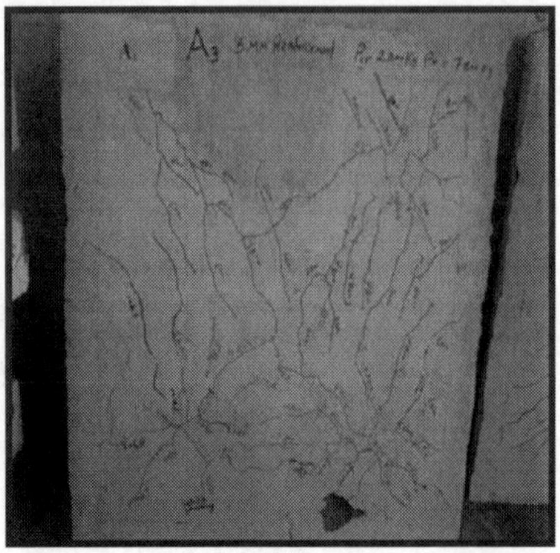
(A3)

**Fig. 7.** Crack pattern for slabs (A1, A2 and A3)

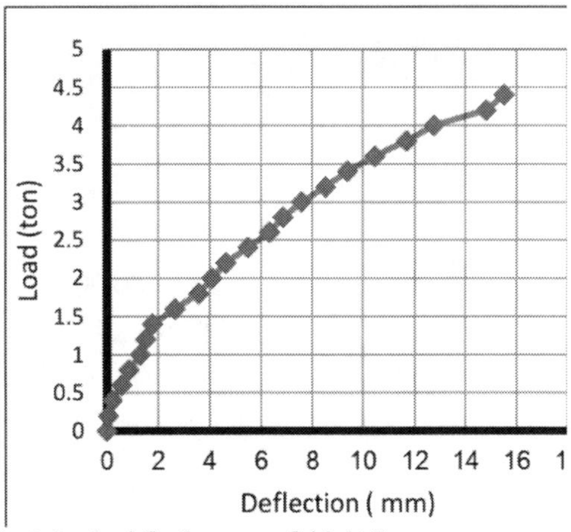


**Fig. 8.** Load – deflection curve of slab (A1)

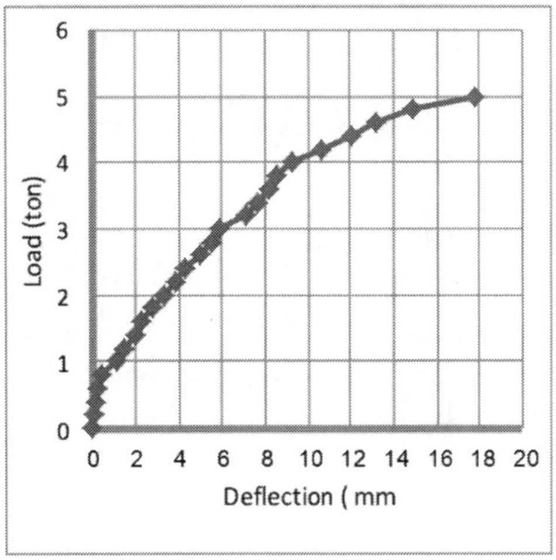


**Fig. 9.** Load – deflection curve of slab (A2)

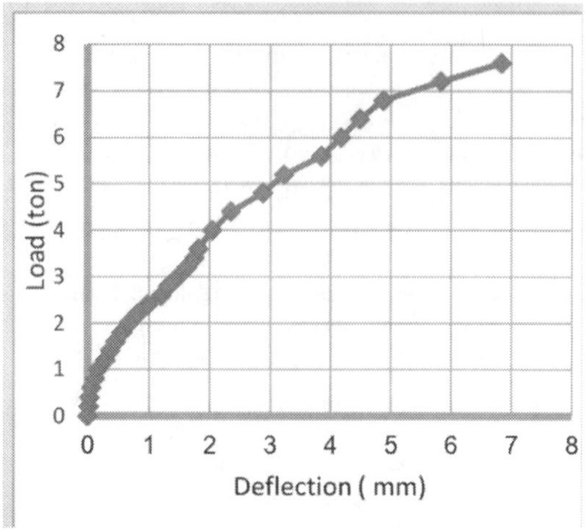


**Fig. 10.** Load – deflection curve of slab (A3)

## 6 Conclusions

Based on the results and observations of the experimental study presented, the following conclusions could be drawn as follows:

1 - It is possible to manufacture a structural light weight concrete with low density and high self-compacting characteristics using expanded clay aggregate ( LECA).

2 - The way of mixing of this concrete is different: light-weight aggregate (LECA) is recommended to be pre-wetted either in the mixing device by batching light weight aggregate (LECA) and water (and adding the rest of materials after a while) or before the process of mixing (in the store or similar). Pre-wetted light weight aggregate (LECA) gives higher stability of rheological behavior of fresh concrete and foaming of cement matrix is better to control.

3- In condition of fresh properties, the passing ability and filling capacity increase with the increases of w/b and there is a great effect when (HRWRA) increased.

4-In condition of fresh properties, the total binder content had influence on workability and static stability (segregation resistance) of (LWSCCs). For a given w/b, the (HRWRA) demand decreased with the increase of total binder content. On the other hand, segregation resistance increased with the increase in total binder content.

5- In condition of hard properties the compressive strength increased with decrease in w/c, and increased with increase in binder amount.

6 - The flexural loads at first crack and ultimate loads depend on thickness of reinforcing mesh layers used in the slab.

7- By increasing the reinforcement ratio from 0 .6 % for slab No (A1) to 0.81 % for slab No (A2), the flexural loads increased by 13 % .

8- By increasing the reinforcement ratio from 0 .81 % for slab No (A2) to 1% for slab No (A3) , the flexural loads increased by 46.15 % .

8 - During exposing slabs to ultimate load, it was observed that the slab did not break into pieces after failing, instead the cracks were found to widen. This is because the welded wire mesh prevents the cement mortar from breaking and falling off. By simply observing the cracks it was inferred that the slab with layer of wire mesh with 4 mm thickness has more crack patterns compared to slab with layer of wire mesh 5 mm thickness. Hence with an increase in thickness of layers of wire mesh we can improve the capacities of slabs.

## References

1. M. G. Stamatakis, M. Bedelean, H., Gorea, D., Alfieris, E., Tziritis, and S., Kavouri, Refractories World forum, **3 , 1**, (2011).
2. Z. Wu, Y., Zhang, J., Zheng, and Y., Ding, Construction and Building Materials, **23, 5**, (2009).
3. C. L., Hwang, and M. F., Hung, Construction and Building Materials, **19, 8**, (2005).
4. K. M., Hossain, Cement and Concrete Research, **34, 2**, (2004).
5. D. Fragoulis, M. G., Stamatakis, E., Chaniotakis and G. Columbus, Materials Characterization, **53, 2**, (2004).
6. U. I. B., Topc , and L. T., Uygunog, Construction and Building Materials, **24, 7,** (2010).
7. ESCSI, Worldwide, Salt Lake City, UT, Publication No. 9349, (2004).
8. Y., Bai, R., Ibrahim, and P. A., Muhammed Basheer, Proceedings of an International Workshop on Sustainable Development and Concrete Technology, Beijing, (2004).
9. K. S., Chia, and M. H. Zhang, Magazine of Concrete Research, **56, 8,** (2004).
10. ACI 211.2, ACI Man. of Conc. Prac., (2013).
11. EFNARC, (2005).
12. E.S.7417/2001, Egyptian Standards, (2001).
13. ASTM C618, American society for testing and materials, ASTM International, (2003).

# Identification and analysis of factors affecting labour productivity in Iraq

*Tareq* Khaleel[1], *and Yasser* Nassar[1,*]

[1] Building and Construction Engineering Department, University of Technology, Baghdad, Iraq

**Abstract.** Productivity is a very important element in the estimation process in construction management. The objective of this research is to identify and analyze the factors which affect labor productivity in construction projects. In this research, 42 effective factors were collected from site survey, interview with engineers and experts, and previous research in the Arab world. These factors are grouped into Nine categories "Workforce, Leadership, Motivation, Supervisors, Safety, Project, Time, Material, and External". A survey questionnaire of 70 respondents was distributed among different experts. A statistical analysis was done using SPSS and EXCEL packages. The Relative Importance Index was used to find out the most significant factors affecting the labor productivity in construction sites. The results accomplished from the survey revealed that the major factors negatively affect the labor productivity (ranked from the worst factor with Relative Importance Index values, respectively) namely, Availability Material (88.571%), Climate status "Weather" (88%), Religious occasions (86.29%), Number of working groups (86%), Ganger experience (85.714%), Workforce surveillance (84.857%), Ganger Age (84%), working at height (82%), Drawings and specifications alteration during execution (81.69%), and Sequence of floor (80.571%). Based upon these findings, this can help the construction professionals to improve the productivity and project performance in Iraq.

## 1 Introduction

The productivity of labor plays a major role in the success of any construction project. Therefore, the construction industry plays an important role in the economy of any developing country. The productivity is associated with the performance of the workers and this is considered most the important factor that has a significant impact on the growth of the construction industry [1]. Thus, improving productivity in construction projects and construction contributes significantly to gross domestic product of any country [1]. However, the productivity may be affected by many factors and variables unexpected difficulties. Such variables include factors related to manpower, materials, tools, equipment, construction methods, political factors, finance and other internal and external factors. In order to improve and increase the productivity of labor, the factors affecting productivity, must be reduced through study, identifying and analyzing the factors that have the highest impact on labor productivity. According to [2], the improved productivity is a major concern for any project aiming at success and profit or optimal use of resources, or as an effective and efficient conversion of resources into marketable products and identify business profitability. Currently, the construction industry in Iraq, suffers significantly from delay in the completion of projects, as well as excessive and increasing costs, which are indicators of the existence of problems in productivity. So this paper aims to identify the most important factors that affect the productivity of labor in construction projects in Iraq.

## 2 Research objectives

This study aims to:

1. Identify factors influencing the productivity of work in construction projects in Iraq

2. Analyze and rank the factors influencing the productivity of labor through the Relative Importance Index in the Iraqi construction projects, these factors can be used by the private and public sectors to improve and increase productivity in projects and control the quality of work performed.

## 3 Productivity-concept and basic definition

There are various definitions of productivity, they all point to comparison between the output and input. In 1950, the Organization for European Economic Cooperation introduced (OEEC) a formal definition of productivity as, " a division product which is obtained by dividing the output on a worker" [3]. As interpreted by the researcher Peles [4], productivity is "the performance

* **Corresponding author: dr.nassar12@yahoo.com**

© The Authors, published by EDP Sciences. This is an open access article distributed under the terms of the Creative Commons Attribution License 4.0 (http://creativecommons.org/licenses/by/4.0/).

achieved by workers". Whereas the researchers Handa and Abdulla [5] defined productivity as "the rate output of goods and / or services on the input of the basic resources, for example, of workers, capital, technology, materials and energy". The researcher Finke's [6] definition of productivity is "the amount of work produced by one worker in an hour and the equipment used during the hour, or the crew during the same hours". The American Society of Engineers defines productivity costs as a "relative measure of the efficiency of the labor, whether it was good or bad this efficiency, compared with the established rules or standards required"[7]. The researchers Horner and Duff [8], define productivity as the "quantity produced per unit with input". ***Relying on the previous definitions, the present paper defines productivity, as "a method to measure the efficiency of work through the exploitation of resources available at the workplace."***. From the above definitions, it is concluded that the productivity is generally defined as the ratio of outputs:

$$Productivity = \frac{Output}{Input} \quad ......(1)\ (\,above\ definitions)$$

## 4 Factors Influencing labor productivity

Over the years, the factors that affect the productivity of labor have been the subject of investigation by many researchers. The importance of these factors varies from one country to another, and from one project to another. Table (1) illustrates the most important results of some researchers conducted in some countries.

**Table 1.** The most influential factors, labor productivity

| Researcher /year | | The most influential factors |
|---|---|---|
| Olomolaiye et al, [9] | | 1. Shortage of materials<br>2. Rework<br>3. Shortage of Equipment's<br>4. Delays in the Notes supervisors<br>5. Absenteeism and inconsistencies in the labor |
| Nigeria | 1987 | |
| Horner et al, [10] | | 1. Experience and skills of labor<br>2. Capacity building<br>3. Quality supervision |
| (England) | 1989 | |
| Lim and Alum, [11]. | | 1. Lack of qualified supervision<br>2. Shortage of skilled labor<br>3. High rate of labor turnover<br>4. Labor absenteeism<br>5. Communication with foreign labors. |
| Singapore | 1995 | |
| Kaming et al, [12]. | | 1. Shortage of materials<br>2. Rework<br>3. Absent laborers<br>4. Lack of appropriate Equipment |
| Indonesia | 1997 | |
| Makulsawatudom *et al*, [13]. | | 1 .Lack of material<br>2. Incomplete drawings<br>3. Incompetent supervisors<br>4. Lack of tools and equipment<br>5. Labor absenteeism<br>6. Poor communication<br>7. Instruction time<br>8. A poor site layout<br>9. Inspection delay<br>10. Rework |
| Thailand | 2004 | |
| Alinaitwe *et al*, [14]. | | 1. Incompetent supervisors<br>2. Lack of skill<br>3. Rework<br>4. Lack of tools/equipment<br>5. Poor construction method |
| Uganda | 2007 | |
| Specimen *et al*, [15]. | | 1. Lag of materials<br>2. Delay in arrival of material,<br>3. Unclear instruction in labor<br>3. Labor strikes Financial difficulties<br>4. Higher absenteeism of Labor<br>5. No supervision Method<br>6. Supervisors absenteeism<br>7. Lag of equipment and design change. |
| Indonesia | 2011 | |
| Gopal Shree Rja et al, [16]. | | 1. Lack of skill<br>2. Experience of the workers<br>3. Late payment<br>4. Poor health of the workers<br>5. Lack of empowerment<br>6. Design changes<br>7. Lack of labor safety<br>8. Poor condition of<br>9. Equipment/tools<br>10. Ignore safety precaution |
| India | 2016 | |

## 5 Factors affecting labor productivity in countries neighbouring Iraq and arab countries

Due to the similarity of some of the factors affecting the productivity of work between Iraq and its neighboring countries. Factors such as, weather, security, economy, availability of materials and other factors affecting the productivity. This study is going to focus on the surrounding Arab countries, which are investigated for factors that affect productivity. In particular, the paper focuses on the top five influential factors influencing productivity, as shown in Table (2).

**Table 2.** Factors influencing labor productivity in countries surrounding Iraq

| Researcher | | The Top Five Most Influential Factors on Productivity |
|---|---|---|
| Jarkas and Biter, [17] | | 1. Clarity of technical specifications;<br>2. The extent of variation/change order execution;<br>3. Coordination level among design disciplines;<br>4. Proportion of work subcontracted; and<br>5. Design complexity level. |
| Kuwait | 2012 | |

| | | | |
|---|---|---|---|
| Zakeri, M., et al, [18] | Iran | 1996 | 1. Material shortage;<br>2. Weather and site conditions;<br>3. Equipment breakdown;<br>4. Drawing deficiencies/change orders; and<br>5. Lack of proper tools and equipment. |
| Serdar Durdyev, et al, [19] | Turkey | 2013 | 1. Poor Communication;<br>2. Poor Site Layout;<br>3. Lack of Tools;<br>4. On-time/Late payments; and<br>5. Reworks; Motivation. |
| Mahamid et al, [20] | Saudi Arabia | 2013 | 1. Lack of labor experience;<br>2. Poor communication and construction coordination;<br>3. Bad relations between labors and management team;<br>4. Payments delay by owner; and<br>5. Misuse of time schedule |
| Ghanim Bekr, [21] | Jordan | 2016 | 1. Poor planning and scheduling;<br>2. Material shortage of project site;<br>3. Equipment shortage;<br>4. Lack of skilled labor; and<br>5. A Poor site management. |
| Nabil Ailabouni1 et al, [22] | UAE | 2007 | 1. Proper Work Timings give a balance between works;<br>2. Leadership Skills of supervisors;<br>3. Salaries on time;<br>4. Technically qualified / educated to the trade; and<br>5. Reasonably well paying job. |
| Remon, Fayek Aziz et al, [1] | Egypt | 2014 | 1. Laborer experience and skill;<br>2. Incentive programs;<br>3. Availability of the materials and ease of handling;<br>4. Leadership and competent; and<br>5. Competency of labor supervision. |
| Enshassi et al, [23] | Gaza Strip | 2007 | 1. Material shortages;<br>2. Lack of labor experiences;<br>3. Lack of labor surveillance;<br>4. Misunderstanding between labor and superintendents; and<br>5. Drawings and specifications alteration during execution; |
| Wa'el Alaghbari et al, [24] | Yemen | 2015 | 1. Laborer experience and skill<br>2. Availability of the materials<br>3. Leadership and competency<br>4. Material availability in the market<br>5. Political and security situation |

## 6 Classification of factors influencing the productive labor

Many researchers in engineering project management in most developed countries have tried to classify the factors affecting labor productivity in construction projects in major global groups. Through the adoption of many techniques and methods to classify the factors affecting, however, the researcher noted that there is no convergence of views and classifications in different research, as shown in Table (3).

**Table 3.** The factors influencing productivity of labor according to groups

| Authors | Group's Influence |
|---|---|
| Shroeder, R.G, [25] | 1. Work Force Group<br>2.Special Work and storage production<br>3. Product-specific Group<br>4. Work and storage Group<br>5. External Group |
| prkopenko, [26] | 1. Job related Group<br>2. Resources Related Group<br>3. Environmental Related Group |
| Herbsman & Ellis, [27] | 1. Technology Group<br>2. Administrative Group |
| Talhouni, [28] | 1. Management Group<br>2. Site Group<br>3. Design Group<br>4. Weather Group |
| Enshassi et al, [23] | 1. Manpower Group<br>2. Leadership Group<br>3. Motivated Group<br>4. Time Group<br>5. Material/Tools Group<br>6. Supervision Group<br>7. Project Group<br>8. Safety Group<br>9. Quality Group<br>10. External Group |
| Brent and Ellis, [29] | 1. Management Group<br>2. Human / Labor Group<br>3. Technology Group<br>4. External Group |
| Robles et al, [30] | 1. Project Group<br>2. Humanist Group<br>3. Management Group<br>4. Materials and Tools Group<br>5. Environmental Group |
| Shree Raja Gopal and Murali, [16] | 1. Workforce Group<br>2. Management team Group<br>3. Psychological Group<br>4. Material / Equipment Group<br>5. Supervision Group<br>6. Schedule compression Group<br>7. Safety Group<br>8. Miscellaneous Group<br>9. External Group |

## 7 Research methodology

This research is based on a site survey aimed at collecting all necessary information in an effective manner. The survey offers (42) influential factors in productivity, which have been collected from previous relevant research, review and revision by the participants in the initial pilot questionnaire. These factors were classified into 42 factors surveyed in nine main groups, namely:

1. Workforce group, 2. Leadership group, 3. Motivation group, 4. Supervision group, 5. Safety group, 6. Project

group, 7. Time group, 8. Material / equipment group, 9. External group.

### 7.1 Design of the questionnaire

The questionnaire included two parts: the first section includes general information about the participant. Also, it includes specialization, current work, years of experience, academic qualifications and the Guild degree; and answers are made by ticking the appropriate choice. The second section includes nine tables, which represent the nine groups of factors influencing the productivity of labor and the scale dish Likert quintet is used in the design of the questionnaire [31]. So the scale extends from (2 very low) to (5 very high). To confirm the consolidated, the results of the questionnaire, a second assessment has been conducted after reviewing and auditing to ensure its effectiveness and suitability for the construction projects in Iraq. The original questionnaire consists of 42 factors affecting labor productivity in construction projects in Iraq. Before distributing the questionnaire, a pilot test (pilot study) has been conducted to ensure that the questionnaire was designed gradually and appropriately simpler. The questions in the questionnaire are forwarded to six of arbitrators (degree expert engineer) as referees for the purpose of benefiting from their expertise in making the necessary adjustments to delete and add to reach the final version of the questionnaire.

### 7.2 Sample size

The sample of target in the search, the respondents are engineers who work as: operational director, project manager, project coordinator, construction manager, site manager, site engineer, superintendent, estimator, supervisor, etc. They work at contractor companies in Iraq both private and government. To get a statistically representative sample of the target, using equations (1),(2) which have been used by many researchers, including Hogg & Tannis, [32].

$$n = \frac{m}{1 + (\frac{m-1}{N})} \dots\dots\dots\dots\dots (1)$$

The m=Sample size of the population is not limited,

N= Sample size from a limited population

To find the value of (m), equation (2) is used as below:

$$m = \frac{Z^2 * p * (1-p)}{E^2} \dots\dots\dots\dots\dots (2)$$

Whereas:

(Z) = value shows the level of confidence (for example, 2.92, 1.575 and 2.245 represent the values of the confidence levels at 99%, 95% and 90% respectively),

(P) = degree of contrast between the target sample elements (0.5)

(E) = choice for point of maximum error.

Using a confidence level of 95% and the level of significance at 5% when the sample size is not specified, the estimated value (m) = clear application of the equation No. (1) As Follows:

$$m = \frac{(1.96)^2 * 0.50 * (1-0.50)}{(0.05)^2} \approx 385$$

### 7.3 Data collection

The target sample of this research is engineers working in the public and private sectors and in various administrative fields. Such as, project manager, supervision engineer, design engineer, road engineer and others. Among the 100 engineers (N=100), the required sample size to succeed in work of the total target sample can be calculated by applying the following equation (2) as follows:

$$n = \frac{385}{1 + (\frac{385-1}{100})} \approx 79$$

To ensure obtaining the required sample size (79 samples), (100 questionnaire) forms were handed out. (79 Questionnaire) forms were filed in time and the number represents the ratio (79%). (9 questionnaire) forms were excluded due to incompleteness, bringing the total questionnaires after the revision, which were used for the analysis to (70 questionnaire), which constitute the ratio (70%). To analyze the data, using the (Relative Importance Index, RII) according to the following equation (3), [20, 17, 23]:

$$RII\,\% = \frac{5*(n5) + 4*(n4) + 3*(n3) + 2*(n2) + 1*(n1)}{5*(n1+n2+n3+n4+n5)} \dots (3)$$

Where: n1, n2, n3, n4 and n5 = the number of respondents who selected:

n1= number of respondents who selected little effect.

n2= number of respondents who selected some effect.

n3= number of respondents who selected average effect.

n4= number of respondents who selected high effect.

n5=number of respondents who selected very high effect.

In contrast, these five expressions are defined by the equal intervals, as the following,

$10.0 \leq$ little effect (**LE**) $\leq 20.0$

$20.0 \leq$ some effect (**SE**) $\leq 40.0$

$40.0 \leq$ average effect (**AE**) $\leq 60.0$

60.0 ≤ high effect (**HE**) ≤ 80.0

80.0 ≤ very high effect (**VHE**) ≤ 100

### 7.4 Measuring the consistency of the questionnaire

Stability is defined as the stability of the scale and lack of contradiction with itself, meaning that the measure would be re-applied to the same sample over a period of time and it will give the same results, the value of consistency between the two values will be zero or one, where the closer the value of one indicates the high stability of the questionnaire and vice versa. In order to measure the stability of the questionnaire, this study uses the stability coefficient (Cronbach's alpha) for equation (4) to ensure the stability of the search tool, and use equation (3) to come up with a valid value (Cronbach's Alpha), where the normal range of Cronbach's coefficient (Alpha) value is between (0.0) and (1.0). The closer the Alpha is to (1) the greater the internal consistency of data [33].

$$\alpha = \frac{K}{K-1}\left[1 - \frac{\sum_{i=1}^{K} S_i^2}{s_t^2}\right] \quad \text{.........(4)}$$

Whereas:
K: number of items in a group.
s2i: the variance associated with the item (I).
s2 t: the variance associated with the sum of all (k) item scores……. (3)

Table (4) demonstrates the values of trust and credibility for each factor in the questionnaire according to Cronbach's Alpha. Chronbach's shows that the Alpha values are in the range of (0.916 to 0.981). This is a high-scale (Excellent), and, therefore, it ensures the reliability and validity of each group in the questionnaire.

**Table 4.** Reliability and Validity of factors affect groups

| Factors Affect Group | No. of Factors | Reliability * | Validity |
|---|---|---|---|
| Workforce Factors | 7 | **0.98** | **0.989** |
| Leadership Factors | 3 | **0.931** | **0.965** |
| Motivation Factors | 4 | **0.965** | **0.982** |
| Supervisors Factors | 4 | **0.968** | **0.984** |
| Safety Factors | 4 | **0.916** | **0.957** |
| Project Factors | 8 | **0.981** | **0.990** |
| Time Factors | 4 | **0.981** | **0.990** |
| Material Factors | 4 | **0.954** | **0.977** |
| External Factors | 4 | **0.95** | **0.975** |
| Total Factors Affect Group | 42 | **0.96** | **0.98** |

*** (Cronbach's Alpha)**

## 8 Results and discussion

In this study, 42 factors that negatively affect the labor productivity in the construction of buildings in Iraq have been identified and ranked according to the Relative Importance Index. These factors have been categorized into nine groups.

To find the results of this research, the following software has been used:

1. (SPSS) version 22 is employed to find the Relative Importance Index and Chronbach's Alpha of the impact strength (1 to 5) for each factor individually according to participants, replies to the questionnaire, and to find frequencies that are used in the calculation of importance coefficient.

2. Excel program (MS Excel 2015) for the application of the formula for the calculation importance coefficient

### 8.1 Workforce group

The Relative Importance Index and ranks of the seven factors are classified under the "Workforce Group" as shown in Table (5).

**Table 5.** Ranking Workforce factors

| First Group Workforce Factors | RII% | Rank | Degree of Effect |
|---|---|---|---|
| Number of working groups | **86** | **1** | **VHE** |
| Ganger experiences | **85.7** | **2** | **VHE** |
| Ganger age | **84** | **3** | **VHE** |
| Number of assistant workers | **75.7** | **4** | **HE** |
| Disloyalty | **72** | **5** | **HE** |
| Lack of competition | **62** | **6** | **HE** |
| Misunderstanding among Workforce | **60.3** | **7** | **SE** |
| **Result of Workforce Group** | **RII%** | **Rank** | **Degree of Effect** |
| | **75.1** | **4** | **HE** |

Table (5) illustrates that the surveyed participants ranked "Number of working groups" as the most important factor that influence labor productivity in this group, with a RII of 86%. This top ranked factors influence is further ranked as the fourth in its effect among all factors explored. In contrast, the factor "Misunderstanding between Workforce" is the least important factor that influences labor productivity in this group, with a RII of 60.3%. This group scored the fourth rank.

### 8.2 Leadership group

The Relative Importance Index and ranks of the three factors are classified under the "Leadership Group" as shown in Table (6).

**Table 6.** Ranking Leadership Factors

| Second Group leadership Factors | RII% | Rank | Degree of Effect |
|---|---|---|---|
| Lack of Workforce surveillance | 84.86 | 1 | VHE |
| Misunderstanding between Workforce and superintendents | 65.14 | 2 | HE |
| Lack of periodic meeting with Workforce | 55.14 | 3 | AE |
| Result of Leadership Group | RII% | Rank | Degree of Effect |
| | 68.38 | 7 | HE |

Table (6) shows that the surveyed participants ranked "Lack of Workforce surveillance" as the most important factor that influences labor productivity in this group, with a RII of 84.86%. This top ranked factors influence is further ranked as the sixth in its effect among all factors explored, contrariwise, the factor "Lack of periodic meeting with Workforce" is the least important factor that influence labor productivity in this group, with a RII of 55.14%. This group scored the seventh rank.

### 8.3 Motivation group

The Relative Importance Index and ranks of the fourth factors are classified under the "Motivation Group" as shown in Table (7).

**Table 7.** Ranking motivation factors

| Third Group Motivation factors | RII% | Rank | Degree of Effect |
|---|---|---|---|
| Payment delay | 78 | 1 | HE |
| Financial motivation system | 72 | 2 | HE |
| Moral Motivation system | 63.43 | 3 | HE |
| Lack of training sessions | 54 | 4 | AE |
| **Result of Motivation Group** | RII% | Rank | Degree of Effect |
| | 66.86 | 8 | HE |

Table (7) illustrates that the surveyed participants ranked "Payment delay" as the most important factor influencing labor productivity in this group, with a RII of 78%. This top ranked factors influence is further ranked as (Twelft) in its effect among all factors explored, contrastively, the factor "Lack of training sessions" as the least important factor that influence labor productivity in this group, with a RII of 54%. This group scored the eighth rank.

### 8.4 Supervisors group

The Relative Importance Index and ranks of the fourth factors are classified under the "Supervisors Group" as shown in Table (8).

**Table 8.** Ranking Supervisors Factors

| Fourth Group Supervisors Factors | RII% | Rank | Degree of Effect |
|---|---|---|---|
| Drawings and specifications alteration during execution | 81.69 | 1 | VHE |
| Rework | 79.71 | 2 | HE |
| Supervisors absenteeism | 77.71 | 3 | HE |
| Inspection delay | 72 | 4 | HE |
| **Result of Supervisors Group** | RII % | Rank | Degree of Effect |
| | 77.78 | 2 | HE |

Table (8) illustrates that the surveyed participants ranked "drawings and specifications alteration during execution" as the most important factor that influence labor productivity in this group, with a RII of 80.29%. This top ranked factors influence is further ranked as the (Nine) in its effect among all factors explored. Contrariwise, the factor "Inspection delay" is the least important factor that influences labor productivity in this group, with a RII of 72%. This group scored the second rank.

### 8.5 Safety group

The Relative Importance Index and ranks of the fourth factors are classified under the "Safety Group" as shown in Table (9).

**Table 9.** Ranking Safety Factors

| Fifth Group Safety Factors | RII% | Rank | Degree of Effect |
|---|---|---|---|
| Working at high place | 82.00 | 1 | HE |
| Accidents | 61.14 | 2 | HE |
| Unemployment of safety officer at construction site | 59.43 | 3 | AE |
| Violation of safety precautions | 57.71 | 4 | AE |
| **Result of Safety Group** | RII% | Rank | Degree of Effect |
| | 65.07 | 9 | HE |

Table (9) illustrates that the surveyed participants ranked "Working at high place" as the most important factor that influence labor productivity in this group, with a RII of 82%. This top ranked factors influence is further ranked as (The eighth) in its effect among all factors explored, In contrast, the factor "violation of safety precautions" is the least important factor influence labor productivity in this group, with a RII of 57.71%. This group scored the ninth ranks.

### 8.6 Project group

The Relative Importance Index and ranks of the eight factors are classified under the "Project Group" as shown in Table (10).

https://doi.org/10.1051/matecconf/201816202032

**Table 10.** Ranking Project Factors

| Sixth Group Project Factors | RII% | Rank | Degree of Effect |
|---|---|---|---|
| Sequence of floor | 80.571 | 1 | VHE |
| Site complication | 76.00 | 2 | HE |
| Design complexity | 74.00 | 3 | HE |
| Level of Communication | 68.86 | 4 | HE |
| Interference | 67.14 | 5 | HE |
| Working in confined spaces | 64.86 | 6 | HE |
| Location of Project | 63.41 | 7 | HE |
| Type of activities in the project | 54 | 8 | AE |
| **Result of Project Group** | RII% | Rank | Degree of Effect |
| | 68.57 | 6 | HE |

Table (10) illustrates that the surveyed participants ranked "sequence of floor" as the most important factor that influence labor productivity in this group, with a RII of 80%. This top ranked factors influence is further ranked (tenth) in its effect among all factors explored. Contrastively, the factor "Type of activities in the project" is the least important factor that influences labor productivity in this group, with a RII of 54%. This group scored the sixth rank.

### 8.7 Time group

The Relative Importance Index and ranks of the eight factors are classified under the "Time Group" as shown in Table (11).

**Table 11.** Ranking Time Factors

| Seventh Group Time Factors | RII% | Rank | Degree of Effect |
|---|---|---|---|
| Misuse of time schedule | 75.14 | 1 | HE |
| Increasing No. of labor in order to accelerate work | 70.0 | 2 | HE |
| Working for 7 days of the week without the holiday | 68 | 3 | HE |
| Work overtime | 66 | 4 | HE |
| **Result of Time Group** | RII% | Rank | Degree of Effect |
| | 69.79 | 5 | HE |

Table (11) illustrates that the surveyed participants ranked "Misuse of time schedule" as the most important factor that influences labor productivity in this group, with a RII of 75.14%. This top ranked factors influence is further ranked (fourteenth) in its effect among all factors explored, while, the factor "Work overtime" as the least important factor influencing labor productivity in this group, with a RII of 66%. This group scored the fifth rank.

### 8.8 Material / Equipment group

The Relative Importance Index and ranks of the eight factors classified under the "Material and Equipment Group" are shown in Table (12).

**Table 12.** Ranking Material and Equipment

| Eight Group Material/Equipment | RII % | Rank | Degree of Effect |
|---|---|---|---|
| Availability of Material | 88.6 | 1 | VHE |
| Unsuitability of materials storage location | 77.14 | 2 | HE |
| Tool and equipment shortages | 74 | 3 | HE |
| Inefficiency of equipment | 62 | 4 | HE |
| **Result of Material Group** | RII % | Rank | Degree of Effect |
| | 75.43 | 3 | HE |

Table (12) illustrates that the surveyed participants ranked "availability of material" as the most important factor that influence labor productivity in this group, with a RII of 88.6%. This top ranked factors influence is further ranked as (the first) in its effect among all factors explored. In contrast, the factor "Inefficiency of equipment" is the least important factor influencing labor productivity in this group, with a RII of 62%. This group scored the fifth rank.

### 8.9 External group

The Relative Importance Index and ranks of the eight factors are classified under the "External Group" as shown in Table (13).

**Table 13.** Ranking External Factors

| Ninth Group External Factors | RII% | Rank | Degree of Effect |
|---|---|---|---|
| Weather changes | 88 | 1 | VHE |
| Religious occasions | 86.29 | 2 | VHE |
| Security | 78 | 3 | HE |
| Impact of neighboring buildings | 73.71 | 4 | HE |
| **Result of External Group** | RII% | Rank | Degree of Effect |
| | 81.5 | 1 | VHE |

Table (13) illustrates that the surveyed participants ranked "Weather changes" as the most important factor influencing labor productivity in this group, with a RII of 88%. This top ranked factor's influence is further ranked as (second) in its effect among all factors explored, while, the factor "Impact of neighboring buildings" as the least important factor influence labor productivity in this group, with a RII of 73.71%. This group scored the first rank.

## 9 The ten Most influential factors on labor productivity (using the fishbone technique)

A Fishbone diagram, also called a cause and effect diagram or Ishikawa diagram, is a visualization tool for categorizing the potential causes of a problem in order to identify its root causes.

Table (14) and Figure (1) show the results of the most influential factor in productivity through the results of the survey. It is found that the highest factor affecting is the "availability of materials in the project" and the value of the Relative Importance of this Index is (88.571%), while the factor that occupies the tenth position in the ranking is the floor of the sequence and the value of the Relative Importance Index (80.571%).

**Table 14.** The ten most influence factors on labor productivity

| Rank | Factors Affect | Degree of Effect | RII % | Related Group |
|---|---|---|---|---|
| 1 | Availability of Material | VHE | 88.571 | Material/Equipment |
| 2 | Weather changes | VHE | 88.00 | External |
| 3 | Religious occasions | VHE | 86.286 | External |
| 4 | Number of working groups | VHE | 86.00 | Workforce |
| 5 | Ganger experiences | VHE | 85.714 | Workforce |
| 6 | Lack of Workforce surveillance | VHE | 84.857 | Leadership |
| 7 | Ganger age | VHE | 84.00 | Workforce |
| 8 | Working at high place | VHE | 82.00 | Safety |
| 9 | Drawings and specifications alteration during execution | VHE | 81.690 | Supervisors |
| 10 | Sequence of the floor | VHE | 80.571 | Project |

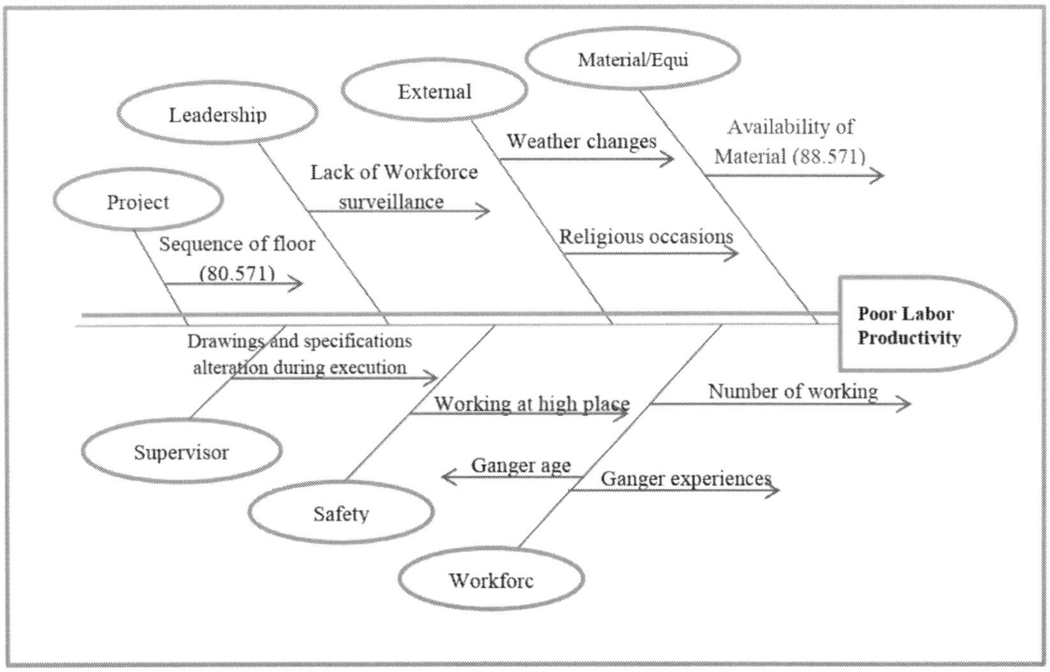


**Fig 1.** shows the ten most influence factors on productivity **(Using a Fishbone Technique)**

## 10 Groups ranking

The 42 identified factors are grouped into nine groups. The ranking of these groups is associated with importance as assessed by engineers who work in the construction sector in Iraq.

Figure (2) indicates that the top important group affecting labor productivity in public construction is External, Supervisors, Material/Equipment, Workforce, Time, Project, Leadership, Motivation and Safety respectively.

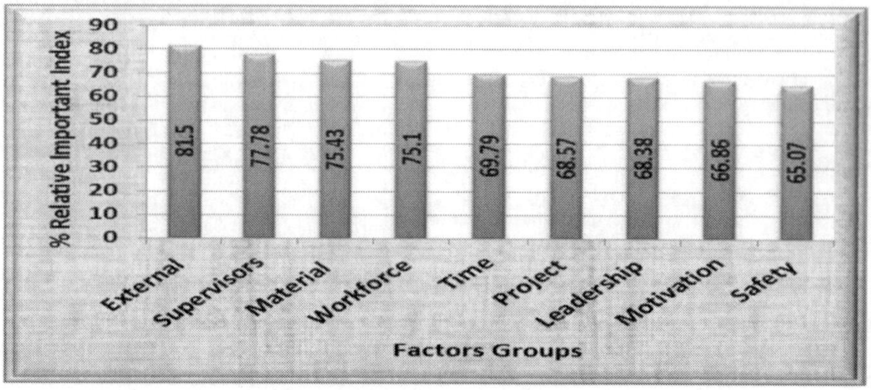


**Fig 2.** Main groups ranking

## 11 Overall ranking factors

Relative Importance Index and ranking of all investigated 42 factors that affect labor productivity in the construction sector in Iraq are listed in Table (15). The analysis clarifies the followings: (1) there are 10 factors with importance index higher than 80%, (2) the minimum importance index is 54%. These indicate that the identified factors are highly relevant to the problem of labor productivity in public and private construction sectors in Iraq. Table (15) shows that the top ten important factors affecting labor productivity in public and private construction projects in Iraq among engineers who work in the construction sector. They are ordered as follows: **(1) Availability of Material**, (2) **Climate status "Weather changes"**, (3) **Religious occasions**, **(4) Number of working groups**, **(5) Ganger experiences**, (6) **Lack of Workforce surveillance**, **(7) Ganger age**, **(8) Working at a high place, (9) Drawings and specifications alteration during execution, and (10) Sequence of floor.**

## 12 Conclusions and recommendations

The objective of this study was to identify the factors that influence labor productivity in construction projects in Iraq. In this study, it was found that forty two factors negatively affect labor productivity.

The most important, namely:

1. Availability of Material;
2. Climate status "Weather changes";
3. Religious occasions;
4. Number of working groups;
5. Ganger experiences;
6. Lack of Workforce surveillance;
7. Ganger age;
8. Working at high place;
9. Drawings and specifications alteration during execution;
10. Sequence of floor;

**Moreover**, 42 factors considered in the study were divided into nine groups, which were ranked according to their Relative Importance Index:

1. External factors group.
2. Supervision factor group.
3. Material / Equipment factor group.
4. Workforce factors group.
5. Time factor group
6. Project factor group.
7. Leadership factors group.
8. Motivation factor group.
9. Safety factor group

Based on the results of this study, the next actions **recommended** are:

1. Provide appropriate scheduling and programming taking into consideration period in demand for the supply of materials.

2. Choosing a suitable storage location for purchasing materials in any project, which should be easy, accessible and close to / near the work sites.

3. In order to have improved site management, the construction teams should have more communication and coordination during all project stages.

4. It is needful to employ project scheduling techniques (such as computer-aided construction project management) in each project to optimize the times of regarding activities, and to ensure that works allow continuous job performance.

5. Site inspections must be conducted permanently during the project design phase to avoid design changes during the implementation phase.

## References

1. S. Hafez, R. Aziz, E. Morgan, M. Abdullah, And E. Ahmed, , *American Journal of Civil Engineering*, **2** (2): 35-40 (2014).
2. S. Wilcox, B. String fellow, R. Harris, and B. Martin, *Transportation research board, committee on management and productivity. Washington*, USA, (2000).
3. D. Sumanth, *McGraw-Hill, New York,* (1984).
4. C. Peles, *Transaction of the American Association of Cost Engineers*, 31st Annual Meeting, Atlanta, USA, (1987).
5. V. Handa, and O. Abdalla, *Construction Management and Economics*, **7** (1): 19–28 (1989).
6. M. R. Finke,. *Journal of Construction Engineering Management*, **124** (6): 490–497 (1998).
7. E. Allmon, C. Hass, J. Borcherding, and P. Goodrum, *Journal of Construction Engineering and Management*, **126** (2): 97–104 (2000).
8. R. M. W. Horner, and R. Duff, *Construction Industry Research and Information Association* (CIRIA), London, UK, (2001).
9. P. Olomolaiye, A. Jayawardane, And F. Harris, *Chartered Institute of Building*, UK, (1998).
10. R. M. W. Horner, B. T. Talhouni, and H.R. Thomas, *Proc. Of the 3rd Yugoslavian Symp. On Construction Management*, Cavtat, **18**–28 (1989).
11. E. C. Lim, and J. Alum, *International Journal of Project Management*, **13** (1): 51-58 (1995).
12. P. Kaming, P. Olomolaiye, G.D. Holt., and F. Harris, *International Journal of Project Management*, **15**(1): 21-30, (1997).
13. A. Makulsawatudom, M. Emsley, and K. Sinthawanarong, *Journal of King Mongkut's Institute of Technology North Bangkok*, **14** (3), 1–6, (2004).
14. H. M. Alinaitwe, J. A. Mwakali, and B. Hansson, *Journal of Civil Engineering and Management*, **XIII** (3): 169-176, (2007).
15. A. Soekiman, K. Pribadi, B. Soemardi, R. Wirahadikusumah, *The Twelfth East Asia-Pacific Conference on Structural Engineering and Construction, proceed Engineering* **14**, 865-873, (2011).
16. T. G. Shree Raja Gopal, K. Murali *International Journal of Recent Scientific Research*. Vol **7**, Issue, 6, pp.11744-11747, (2016).
17. A. Jarkas, and G. Bitar, *Journal of Construction Engineering and Management*, ASCE, **138** (7): 811-820, (2012).
18. M. Zakeri, P. Olomolaiye, G. Holt and F. Harris, *Construction Management and Economics*, **14** (5), 417–26, (1996).
19. S. Durdyev, S. Ismail, and N. Abu Bakar,. *Razak School of Engineering and Advanced Technology, University Technology Malaysia, International Campus, Jalan Semarak*, 54100, Kuala Lumper, Malaysia. DOI: 10.7763/IPEDR. V**55**. 1, (2012).
20. I. Mahamid, *Contractors' Perspective. International Journal of Architecture, Engineering and Construction* (IJAEC), **2** (3): 194 – 202, (2013).
21. G. Bekr, DOI: **10**.5176/2301-394X_ACE16, (2016).
22. N. Ailabouni, K. Gidado, and N. Painting,. *Proceeding Conference for Postgraduate Researchers of the Built and Natural Environment* (PRoBE), Glasgow, United Kingdom, **33**–46, (2007).
23. A. Enshassi, S. Mohamed, Z. Abu Mustafa, And P.E. Mayer, *Journal of Civil Engineering and Management*, **13** (4): 245-254, (2007).
24. W. Alaghbari, A. Salim, K. Dola, and A. Ali, *International Journal of Housing Markets and Analysis*, **5** (1): 41-52, (2012).
25. G. Schroder, Roger, Mc Graaw – Hill Company (1981).
26. J. Prkopenko, , *ILO, Geneva*, (1993).
27. Z. Herbsman, and R. Ellis, *Construction. Management. Econ.*, **8** (1), 49-61, (1990).
28. B.T Talhouni, Ph.D. thesis, Dept. of Civil Engineering, Univ. Of Dundee, Dundee, UK, (1990).
29. G. Brent Hickson and A. Leighton Ellis, *The Journal of the Association of Professional Engineers of Trinidad and Tobago*, **2** (1), 4-11, (2014).
30. G. Robels, *International Journal of Civil, Environmental, Structural, Construction and Architectural Engineering*, **8** (10), 1021-1030, (2014).
31. S. Durdyev, And S. Ismail, *Scientific Research and Essays*, **7** (4): 824-833, (2012).
32. R. Hogg, And E. Tannis,. 8th Ed., *Prentice Hall, Upper Saddle River*, NJ, (2009).
33. S. Abu Shaban,. *MSc Thesis, Islamic University of Gaza, Construction Management Department, Palestine*, (2008).

**Table 15.** The Overall ranking of factors that negatively affecting labor productivity

| Item | Factors Affecting Labor Productivity | Number of Respondents scoring Impact | | | | | RII % | Rank |
|---|---|---|---|---|---|---|---|---|
| | | LE | SE | AE | HE | VHE | | |
| 1 | **Availability of Material** | 1 | 2 | 6 | 18 | 43 | **88.571** | **1** |
| 2 | **Climate status "Weather changes"** | 0 | 1 | 13 | 13 | 43 | **88.00** | **2** |
| 3 | **Religious occasions** | 1 | 1 | 15 | 11 | 42 | **86.286** | **3** |
| 4 | **Number of working groups** | 1 | 1 | 12 | 18 | 38 | **86.00** | **4** |
| 5 | **Ganger experiences** | 1 | 2 | 9 | 22 | 36 | **85.714** | **5** |
| 6 | **Lack of Workforce surveillance** | 1 | 2 | 10 | 23 | 34 | **84.857** | **6** |
| 7 | **Ganger age** | 1 | 4 | 10 | 20 | 35 | **84.00** | **7** |
| 8 | **Working at high place** | 0 | 0 | 20 | 23 | 27 | **82.00** | **8** |
| 9 | **Drawings and specifications alteration during execution** | 0 | 1 | 17 | 28 | 25 | **81.690** | **9** |
| 10 | **Sequence of the floor** | 0 | 0 | 23 | 22 | 25 | **80.571** | **10** |
| 11 | **Rework** | 1 | 2 | 21 | 19 | 27 | **79.71** | **11** |
| 12 | **Payment delay** | 1 | 4 | 23 | 15 | 27 | **78.00** | **12** |
| 13 | **Security** | 1 | 3 | 21 | 22 | 23 | **78.00** | **12** |
| 14 | **Supervisors absenteeism** | 1 | 3 | 24 | 17 | 25 | **77.71** | **13** |
| 15 | **Unsuitability of materials storage location** | 2 | 2 | 25 | 16 | 25 | **77.14** | **14** |
| 16 | **Site complication** | 3 | 4 | 15 | 30 | 18 | **76.00** | **15** |
| 17 | **Number of assistant workers** | 3 | 3 | 22 | 20 | 22 | **75.71** | **16** |
| 18 | **Misuse of time schedule** | 2 | 4 | 24 | 19 | 21 | **75.14** | **17** |
| 19 | **Design complexity** | 1 | 3 | 30 | 18 | 18 | **74.00** | **18** |
| 20 | **Tool and equipment shortages** | 2 | 3 | 25 | 24 | 16 | **74.00** | **18** |
| 21 | **Impact of neighboring buildings** | 2 | 3 | 25 | 25 | 15 | **73.71** | **19** |
| 22 | **Disloyalty** | 3 | 6 | 24 | 20 | 17 | **72.00** | **20** |
| 23 | **Inspection delay** | 2 | 4 | 29 | 20 | 15 | **72.00** | **20** |
| 24 | **Increasing No. of labor in order to accelerate work** | 1 | 4 | 32 | 13 | 17 | **70.00** | **21** |
| 25 | **Financial motivation system** | 3 | 4 | 29 | 16 | 18 | **70.00** | **21** |
| 26 | **Level of Communication** | 4 | 7 | 31 | 10 | 18 | **68.86** | **22** |
| 27 | **Working for 7 days of the week without the holiday** | 4 | 7 | 30 | 15 | 14 | **68.00** | **23** |
| 28 | **Interference** | 5 | 10 | 25 | 15 | 15 | **67.14** | **24** |
| 29 | **Work overtime** | 5 | 7 | 30 | 18 | 10 | **66.00** | **25** |
| 30 | **Misunderstanding between Workforce and superintendents** | 4 | 6 | 37 | 14 | 9 | **65.14** | **26** |
| 31 | **Working in confined spaces** | 2 | 8 | 39 | 13 | 8 | **64.86** | **27** |
| 32 | **Location of Project** | 4 | 8 | 38 | 11 | 9 | **63.71** | **28** |
| 33 | **Moral Motivation system** | 5 | 12 | 31 | 10 | 12 | **63.43** | **29** |
| 34 | **Lack of competition** | 7 | 11 | 28 | 11 | 13 | **63.43** | **29** |
| 35 | **Inefficiency of equipment** | 6 | 10 | 34 | 11 | 9 | **62.00** | **30** |
| 36 | **Accidents** | 4 | 12 | 38 | 8 | 8 | **61.14** | **31** |
| 37 | **Misunderstanding among Workforce** | 8 | 10 | 31 | 15 | 6 | **60.29** | **32** |
| 38 | **Unemployment of safety officer at construction site** | 10 | 7 | 36 | 9 | 8 | **59.49** | **33** |
| 39 | **Violation of safety precautions** | 7 | 14 | 35 | 8 | 6 | **57.71** | **34** |
| 40 | **Lack of periodic meeting with Workforce** | 11 | 12 | 35 | 7 | 5 | **55.14** | **35** |
| 41 | **Lack of training sessions** | 11 | 16 | 30 | 9 | 4 | **54.00** | **36** |
| 42 | **Type of activities in the project** | 10 | 16 | 33 | 7 | 4 | **54.00** | **36** |

# Cost optimization for public school building projects during design stage using value engineering

*Afrah* Hassan[1,*] and *Majid* Yahya[1]

[1]Building and Construction Engineering Department, University of Technology, Baghdad, Iraq.

**Abstract.** During the past ten years, various systems of building components have been applied in public school projects in Iraq, with no systematic method used for selection and evaluation, but only based on the designer's experience. This paper displays evaluation and selection techniques based on value engineering methodology to find the optimal cost for school building projects in Iraq during design stage. The most important criteria for performance, constructability and sustainability criteria, which based on the Leadership in Energy and Environment Design used in this assessment were obtained from a survey of 49 professional designers and consultants, adoption of the Super Decisions Software Program, which uses analytical hierarchy process (AHP) for determining the relative importance of the main criteria and sub-criteria, that allows the decision-makers to evaluate the suitable alternatives of design for the external wall system in Iraq's school buildings was built.

## 1 Optimizing building cost

When making a decision to optimize the design of the building, one must specify the various criteria to judge the optimized solution. The likelihood of any one solution being the best of each criterion is remote. In practice, therefore, one criterion is usually adapted and this on the basis of the use of money, becomes the objective of value maximizing with minimum cost.

There are three elements common in the optimization:

1. The set of alternatives.
2. The value system, which assigns a numerical value for "goodness" so that means "the best" can be known.
3. The means calculating the numerical value of each alternative and comparing this value with each other.

Every owner would like to ensure that many available alternatives have been thoroughly investigated and that alternative selected is the most efficient and economical both on short-term and long-term basis. To accomplish this, the design team performs life-cycle cost analysis, and value engineering as well as model analysis.

### 1.1 Value engineering applied to design

The VE program should be established at a management level where it can effectively challenge design criteria, including established standards, and where it can have access to operational and maintenance cost. This VE program should supplement present method and provide better information on which to make design decisions.

Value engineering studies applied at the design phase of project will realize a great cost saving in practice; it is this stage that receives the most attention. VE plays an important role in both preliminary and detailed design. Preliminary design improvements will probably contribute greatest to initial cost of savings. Whereas VE application to detailed design can eliminate design failure and address itself to maintenance aspects of an engineering system.

Figure 1. shows the decisions that have the significant influence on the expense of funds during the life cycle of the facility. Owners and consultants are key decision makers. To ensure the optimum results, it is necessary to involve the owner and the consultant in the professional assessment process.

With regard to the total costs of the facility, consultant fees represent the little expenditure for all initial costs. Consultants' decisions affect about 50% of the total costs of the facility. Therefore, optimal results can be expected when resources are allocated to environmental equipment at an early stage of the design process, focusing on the impact of the owner and consultant.

* Corresponding author: bce.41593@uotechnology.edu.iq

© The Authors, published by EDP Sciences. This is an open access article distributed under the terms of the Creative Commons Attribution License 4.0 (http://creativecommons.org/licenses/by/4.0/).

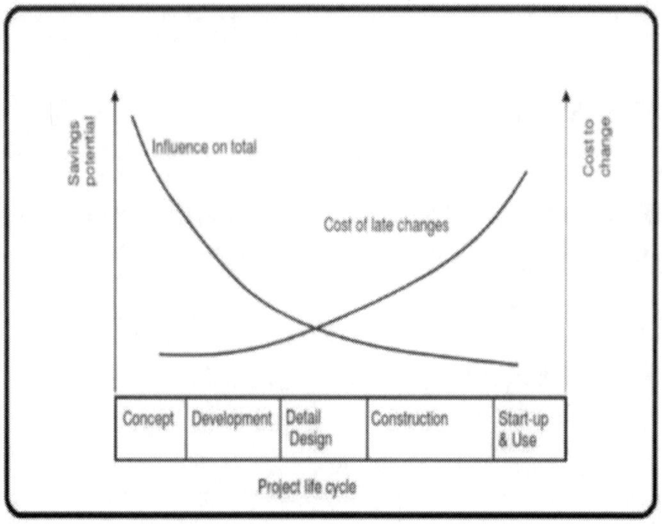


**Fig. 1.** Cost savings potential over project duration

### 1.2 Value index

Dell 'Isola defines the Value Index as the result of the relationship between relative importance and relative cost. This output plays an important role in the application of the value engineering methodology as it will be a guide to the elements for which value engineering is applied and calculated by the following equations:

$$\textit{Value Index} = \frac{\textit{Relative importance}}{\textit{Relative cost}} \text{.....(1)}$$

$$\textit{Value Index} = \frac{(\textit{Function} + \textit{Quality})}{\textit{Cost}} \text{..........(2)}$$

The critical value of the Value Index that can be used to determine the good economic value of the product or service is (1). If the Value Index is greater than or equal to (1) it represents a good economic value, and if the value index is less than (1) it represents a weak economic value.

### 1.3 Value engineering methodology (job plan)

The Job Plan consists of several stages or phases. There is no general agreement on the number or titles of these phases despite the fact that regardless of the terms used or number of stages (which varies from 5 to 8 depending on the source of Job Plan), they all share in the goal of VE studies. In general, this goal is simply to put forward a standard methodology of analysis that is broken down into a series of steps to assist the user in its application and to improve the decision making process for systematically displaying all facts and ideas necessary for an effective analysis.

General overview of Job Plans indicates that at a minimum, any Job Plan must be comprised of the following phases as perceived by the researcher:

1- Information phase.

2- Function Analysis phase.

3- Creative phase.

4- Analytical / Evaluation phase.

5- Development phase.

#### *1.3.1 Information phase*

This is the first phase of the value engineering methodology, in which the system under study is thoroughly understood in all its formal, functional and practical aspects. It collects as much information as possible about the product or system under study and its components (for the current design). The designer, the client and the project manager are required to provide the necessary information, including the following:

1. Design criteria (constructability, conditioning and cooling, load planning)
2. Design calculations
3. Site conditions (topography, soil conditions, surrounding areas,)
4. Detailed design drawings
5. Technical Specifications
6. Operating schedule and its estimated costs
7. The maintenance schedule and the estimated costs
8. Table of power consumption rate
9. Construction Cost Estimates (Bill of Quantities)
10. Services available.

### 1.3.2 Function Analysis phase

In the functional analysis, it is necessary to reach inside the item details by analyzing the components of the system under study and the knowledge of basic and secondary functions.

The functional analysis will be completed as follows:

1. Identify functions.
2. Functions classification.
3. Link the functions to diagram (FAST Diagram).
4. Select which functions can be improved.

#### 1.3.2.1 Identify functions

The application of a function in a value engineering study is a function analysis, the project or product is evaluated by identifying what the item does. The verb is an action verb and the noun, a measurable noun, for example: a light bulb function is a lighting area.

#### 1.3.2.2 Functions classification

The classifications of function includes the following:

1. Basic Function: Is that which is essential to the performance of a user function, or the function describing the primary utilitarian characteristic of a product or design to fulfil a user requirement.
2. Required Secondary Function: A required secondary function is any function that must be achieved to meet codes, standards, or mandatory owner requirements.
3. Secondary Function: If secondary functions are removed from the design, both the basic and required secondary functions can be realized. As such, their worth is zero.

#### 1.3.2.3 Functional Analysis Systems Technique (FAST) Diagram

Using FAST involves a function block diagram based on answers to what? why? What?.

FAST is an effective way to get good answers for three logical reasoning questions:

What is the problem?

Why is the solution necessary?

How can the solution be achieved?

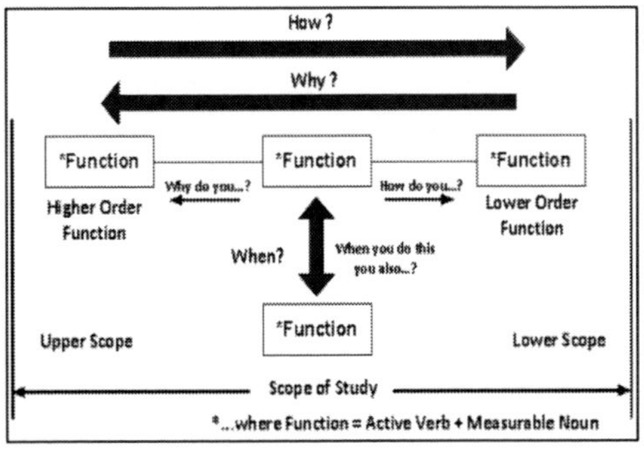


**Fig. 2.** FAST Diagram.

### 1.3.3 Creative phase

The goal of this phase is to generate ideas for alternative solutions to the basic function of the item.

The traditional alternatives are eliminated through the use of all the tools of creative knowledge to solve the problem, including brainstorming sessions and conferences to solve problems used to create an open space of free flow of information and it generates a list of idea.

in the creative phase are not seen as appropriate.

### 1.3.4 Evaluation phase

In this phase, the team examine creative phase and evaluate the feasibility of each idea by identifying its advantages and disadvantages.

In the order of ideas, the following should be considered [22]:

- Did you meet the requirements of aesthetics, performance, quality and reliability?
- Does the proposed idea meet the required functional requirements?
- Will a redesign or excessive delay be created for the project?
- Is there an improvement in operation and maintenance?
- Will life cycle cost savings be achieved?
- Does the idea have a reasonable chance of acceptance and implementation?
- Was the proposed design used in the past?
- Does the idea fundamentally affect the aesthetics of the building or project?

Top ideas rated are selected by the team.

### 1.3.5 Development phase

At the development phase, the best ideas from the evaluation phase are developed into proposals that can be applied. The team develops initial designs, life cycle cost, comparisons of original designs and suggested alternative ideas.

During this phase, the team's technical expertise becomes very important.

Often, it is necessary to consult external experts, vendors and reference sources for additional assessment information before developing available alternatives.

#### 1.3.5.1 Life-Cycle Cost (LCC)

LCC is a method used to compare and evaluate the total cost of competition solutions to meet identical functions based on the expected life of the facility or product to be obtained. In conducting a value study, the LCC analysis is what was carried out in the development phase of the VE Plan to select the least expensive alternative.

The LCC relates to total construction costs over the economic life of the project.

Figure 2 shows how total building costs were incurred.

There are many ways to bring life cycle costs to a comparable time base; present value, interest rate and annual equivalent.

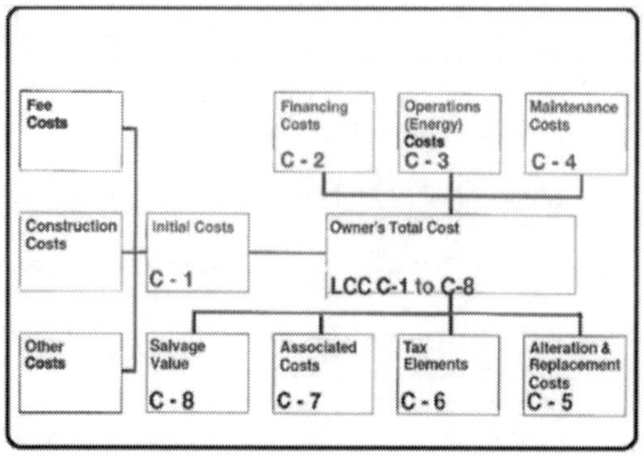


**Fig.3**. Life cycle cost elements

### *1.3.5.2 Weighted Evaluation*

As a last step, economic cost data should be mitigated with human factors such as comfort, appearance, performance, safety and costs (initial operation, maintenance, replacement and rescue).

A weighted evaluation is used to formalize the process. A weighted assessment ensures optimal decisions. Good decisions are made by properly focusing on all criteria. It is important during the evaluation process to discuss and weigh the following areas:

A) Needs versus desires
B) Important vs. unimportant
C) Designing the exchange for the required jobs

The weighted evaluation procedure has been divided into two parts: the criteria -weighted process and analysis matrix. The weighted process of criteria is designed to isolate important criteria and establish them weight and / or relative importance. Figure 4 shows a typical weighted valuation for example of a car purchase.

## 2 Case study for value engineering application in schools design

The researcher selected steel-structure school building to consider as the case study for this research, which was implemented by the Iraqi Ministry of Education / Directorate General for school buildings in the previous years and some have been under implementation, and will be the evaluation of a suitable alternatives for selection model to achieve optimal cost for constructional parts of the building .

From the researcher point of view an important part of the building was studied in terms of cost, different construction methods, materials and other basic functions.

Super-structure (primary elements) was of a greater significance than other parts of buildings that consist the frame, exterior and interior walls, exterior & interior finishes, roofs for the application of the Value Engineering methodology.

Any job plan must consist of the following phases as envisioned by the researcher:

1. Information phase.
2. Creative phase
3. Analytical phase.
4. Evaluation and development.
5. Recommendation phase.

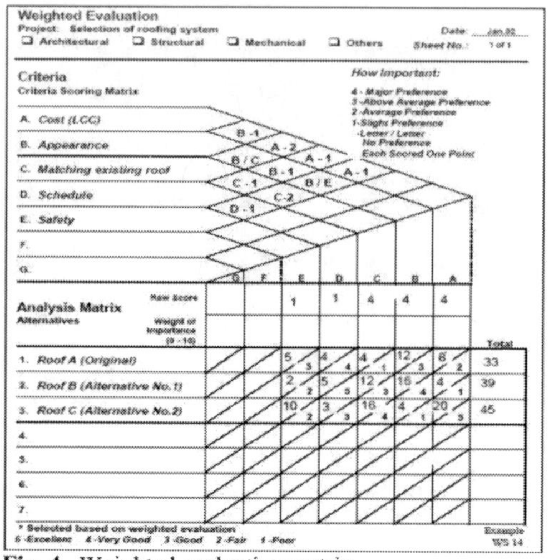


**Fig. 4.** Weighted evaluation matrix

### 2.1 General information about the building

Brief of description as follows:

- A12 class's school with the area of 1400 square meters.
- The yard area is equal to 3000 square meters in average (50x60 or 40x75) m$^2$.
- 20 cm prefabricated light weight concrete fence.
- The school has been designed for three stories (ground +1st. +2nd floor), as shown in the Fig.5.

The Client, Ministry of Education intends to further develop the 12 classes' schools in Baghdad city, seeking for the PC contractor who has the ability for the rapid installation.

Contract is on a Lump-sum basis due to the Bill of quantities, and the itemized break down of the lump sum price- unpriced cleared by weight factor chart.

According to a contract signed between the Engineering Consultancy Bureau of AL-Nahrain University and The General Directorate of School Projects for checking of the 12 Classroom Primary School Steel Structure.

**Fig.5** The shape of school building.

## 2.2 Proposal alternatives for external walls non-load-bearing of School building

All required data and information in this proposal were derived or taken from the bill of quantities for steel structure schools (records, calculations, documentation, and opinions of specialist engineers).

The current system used to provide external walls for the building is a metal-framed sandwich panel with insulation system.

### *2.2.1 Information phase*

External walls for the steel-structure school building constructs from (Composite material, Sandwich panel, studs and high density, moisture, scratch and heat resist Gypsum board) as indicated in the drawings and finishing material table as shown in Fig. 6.

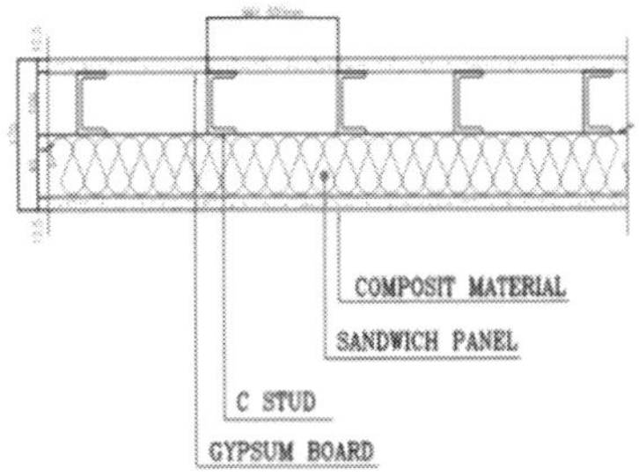


**Fig. 6.** External wall section.

### *2.2.2 Functional Analysis Phase*

The researcher evaluated building components to search for any potential changes for the items of the project. Next, a functional analysis was performed for the exterior wall system for the current system (composite materials), as shown in Table 1

**Table** 1 Functional analysis for current system

| Project: Steel-structure school building Exterior wall system | | | | Item: |
|---|---|---|---|---|
| NO. | Component | Function Verb Noun | Function Level | Original Cost per $m^2$($) |
| 1 | C studs (100mm thickness) | Framing | Basic | 15 |
| 2 | Sandwich panel(45mm thickness) | Thermal comfort | Basic | 10 |
| 3 | Gypsum board high density , moisture 2 Layer 12.5 mm | Sheathing | Secondary | 17 |

### *2.2.3 Speculative Phase*

Brainstorming sessions were led by VE team. Several ideas were recorded for various aspects of the external walls for the building. Table shows a list of some of the ideas for the exterior wall system of
the building; these proposals are:

1. Solid brick wall Brick cavity wall
2. Solid block wall
3. Precast concrete wall panel
4. Light Weight Block Wall

**Table** 2. Brainstorming list of Exterior wall system proposals

| Study Title: Exterior wall system | | Team: VE team |
|---|---|---|
| Generate as many ideas as possible to fulfil the basic function of the item understudy. Do not evaluate the ideas here at all. List everything, judge later. | | |
| NO | System | System assemblies |
| 1 | Brick cavity wall | Consist of brick masonry units. All joints are filled with mortar. |
| 2 | Light Weight Block Wall (THERMOSTON)E | Block masonry units. All joints are filled with mortar. |
| 3 | Precast concrete wall panel | Precast concrete wall panels, cured off-site. |

### *2.2.4 Analytical Phase*

The first part of this phase was used to refine the list of ideas. Possible ideas were identified and retained, while other ideas were ignored. Ideas with potential were studied more closely, so that they could be incorporated with their advantages and disadvantages, as shown in Table 3.

**Table 3** Advantages and disadvantages of proposed systems for Exterior Wall.

| No | Idea | Advantages | Disadvantages |
|---|---|---|---|
| 1 | Metal-framed sandwich panel | Better aesthetics<br>Thermal comfort | Require skilled labor<br>High initial cost of material<br>Long construction time<br>Frequent maintenance |
| 2 | Precast concrete wall | Best aesthetics<br>Permanent materials<br>Fast construction time<br>Best quality control | Transportation arrangement<br>Thermal heat gain |
| 3 | Brick cavity wall system | Less maintenance<br>Low initial cost<br>Materials are available | Large material wastage<br>Large construction time |
| 4 | Light Weight Block wall | Better aesthetics<br>Durable material<br>Thermal comfort<br>Acoustic insulation | Large material wastage<br>Transportation arrangement |

### *2.2.5 Development Phase*

The best ideas and proposal alternatives are used to reach the optimal decision in the following steps:

#### *2.2.5.1 Identifying the main criteria and sub-criteria for external wall systems of school building design*

Three basic concepts of building design were adopted as follows:

a). There are several concepts concerning the constructability which can be applied in the design and configuration, these concepts are aimed at:

1. Choose the design to enable an efficient creation process.
2. Use of parts, materials and work tools available locally as much as possible.
3. Flexibility: Going to determine the design desired results and is not required for ways to reach those results so as to give flexibility to the port to get to the result in the best way.
4. Skills of available labor: The absence of one of two factors, the availability of labor or degree of skilled labor will be very expensive, so the results should be the introduction of this factor in the early design stages.

There are requirements of the performance in the design phase.

b). Performance means "interest of origin after the completion of its creation." So there are several requirements to improve the quality of performance to be considered during the design phase and these requirements:

1-Durability

It is the ability of an item, product or building to maintain its intended function of life expectancy with the intended maintenance levels in the intended use conditions.

2- Safety

Safety is, in any case, of great importance in all stages of the life cycle of the building and starting from the design stage. The terms of danger and risk are frequently used when analyzing the subject of safety. The Hazard: is the recipe for the product that can lead to harmful results the Risk: It is the possibility of injury because of the hazard when the product is turned on by the user.

3- Maintainability

It can be defined as "the function of the design and installation characteristics that affect the programme or under environmental operating conditions Maintenance".

c). At present, strategies have emerged to achieve high performance in the major areas of environmental health and to incorporate the principles of sustainability into the development of construction elements. In 1993 a non-profit organization was established that adopted a voluntary program called "Leadership in the environment and energy design". From it Indoor environmental quality, including control of thermal and acoustic systems and finally the criteria of use of materials and resources, including emphasis on the use of local materials and construction waste management.

#### *2.2.5.2 Assessment of criteria importance (Weights)*

For the purpose of determining the importance of criteria (weights), and also to choose the best alternative from a number of alternative proposals, this research has relied on the field questionnaire, and the improved AHP program (Analytic Hierarchy Process) method is used to determine the weight for different indicators (criteria). The program will identify these important criteria through the pairwise comparison between criteria; as will be explained later, for access to indicators (indexes) for the comparison.

Where selected sample respondents are asked to assess the importance of criteria for buildings design according to a scale ranging from 9 to 1 (9 degree = high importance of criterion) 1 degree = low importance of the criterion.

#### *2.2.5.3 Build an AHP decision model*

Analytic Hierarchy Process (AHP) is one of the most versatile decision-making methods and is one of the most used algorithms for selecting the optimal alternative. This method was designed by Professor Thomas L. Saaty at the University of Pittsburgh in the mid-1970s and can be defined as a method of arranging decision alternatives and selecting the best alternative when a decision maker has multiple objectives or criteria on which the decision is based. While Wang (Et. 2004) defines it as the decision-making tool that analyzes or disassembles the complex problem into a multi-level hierarchical structure of goal, criteria and alternatives. The basic idea of this approach is to transform objective estimates of relative importance into a set of degrees or total weights. By having this method of fundamental property, which is based on the Pairwise Comparison, it complements the various quantitative and qualitative measures to combine them into one comprehensive degree that expresses the order of the alternative between a set of decision alternatives.

#### *2.2.5.4 Use Super Decisions Software to build AHP decision model*

A Super Decisions model consists of clusters of elements (or nodes) arranged in levels. The simplest hierarchical model has nodes connected by lines in them, the clusters contain goal, criteria elements and the alternatives of the decision as shown in Figure 7.

#### *2.2.5.5 Weighted evaluation matrix Using Super Decisions Software (AHP)*

After the results of the questionnaire are analyzed, the weights of criteria are specified for external wall systems of school building design; that will be found by using a program (AHP), which will make pair comparisons between criteria depending on the values of arithmetic mean for criteria.

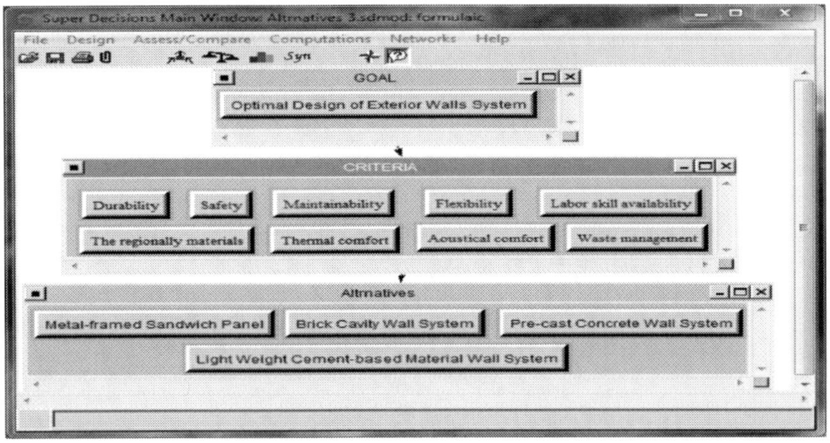


**Fig. 7. Hierarchy model for criteria and alternatives**

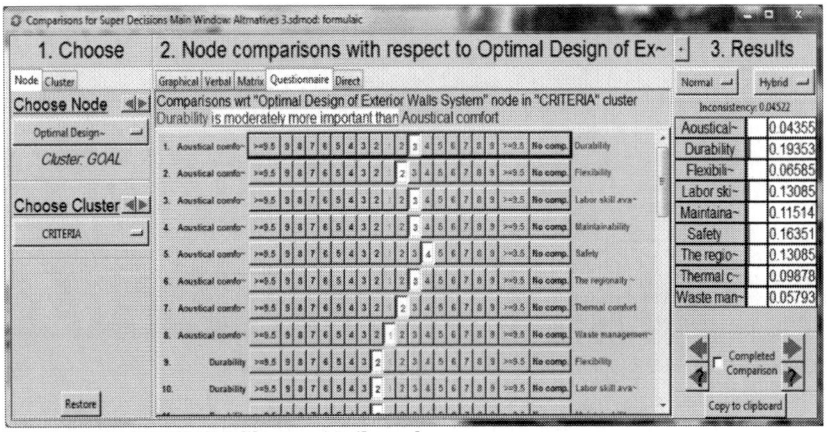


**Fig. 8. Pairwise comparisons for criteria with respect to the goal**

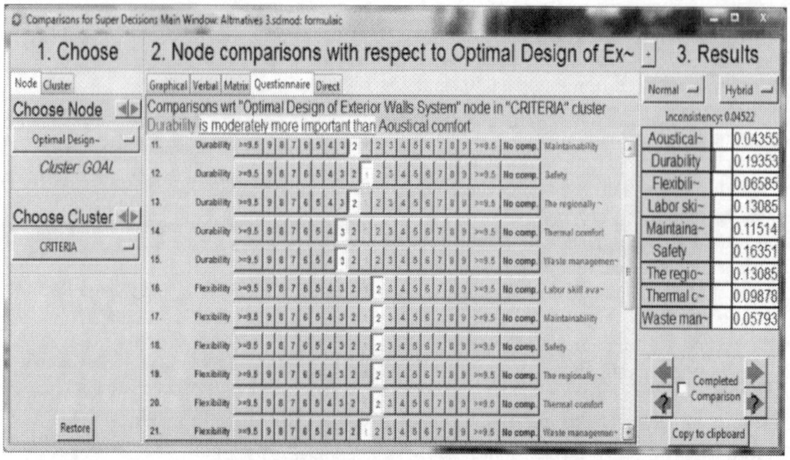


**Fig. 8. (Continued) Pairwise comparisons for criteria with respect to the goal**

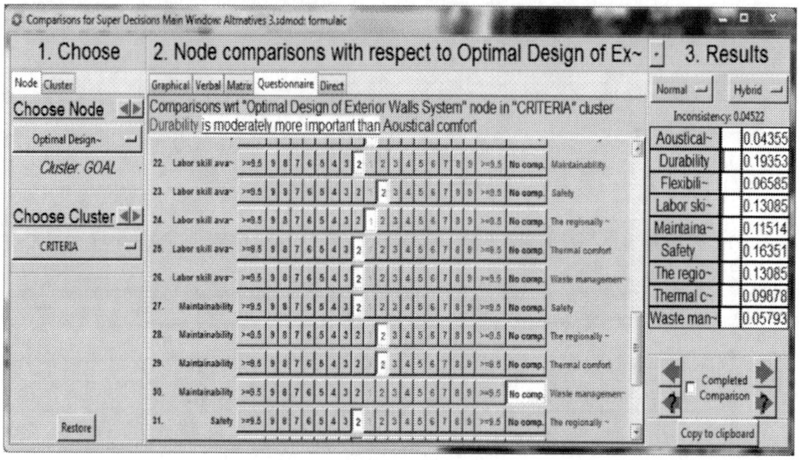


**Fig. 8. (Continued) Pairwise comparisons for criteria with respect to the goal**

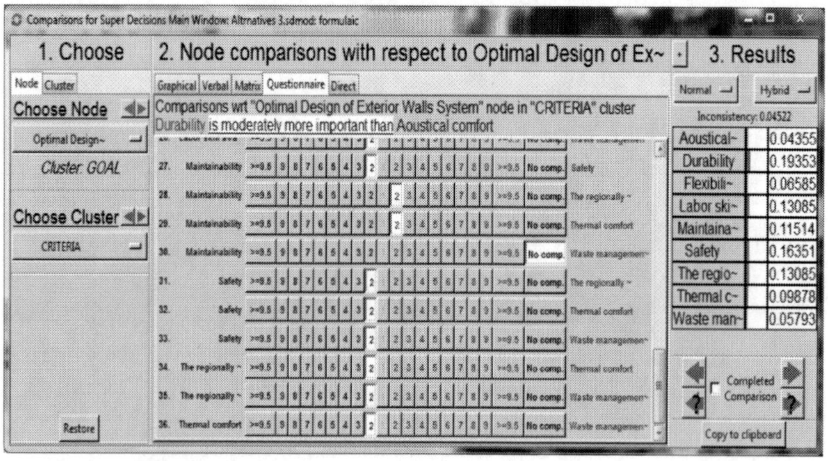


**Fig. 8. (Continued) Pairwise comparisons for criteria with respect to the goal**

Completing the required data in the software program has been guided by civil engineers who were specialists in external wall system for buildings.

After the comparison was completed, the whole inconsistency index is less than 0.10, so the results are fine.

The function index is determined by comparing the alternatives with the main criteria of wall design using the AHP software program. As shown in Fig 9. Then the evaluation matrix is worked out.

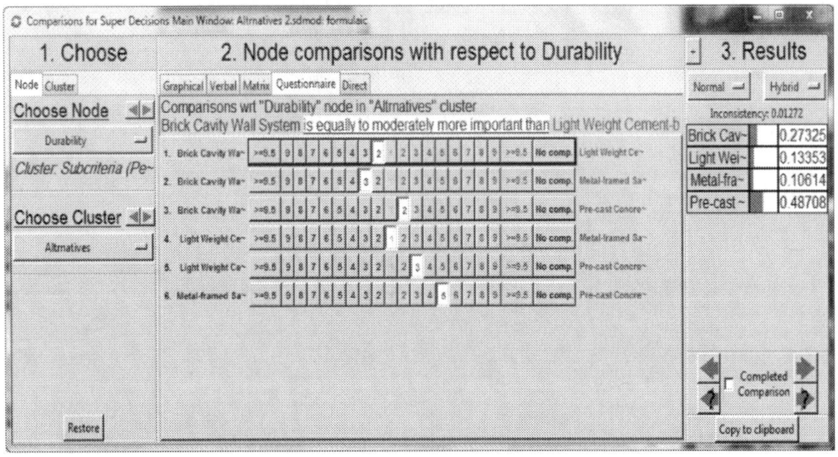


**Fig .9. Pairwise comparisons for alternatives with respect to the Durability criteria.**

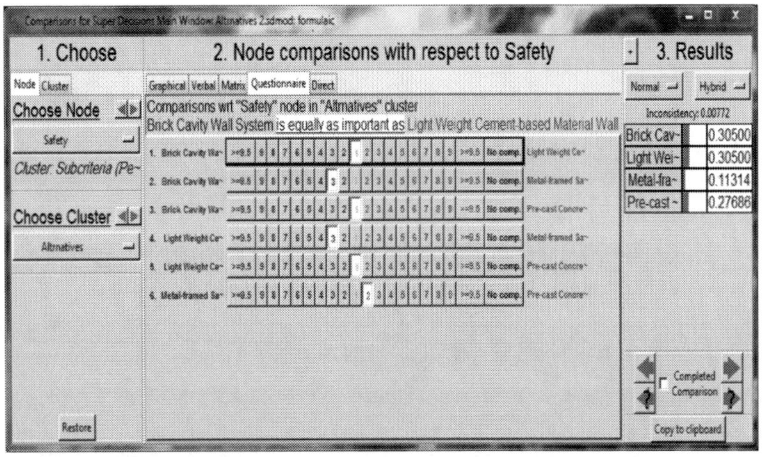


**Fig .9. (Continued) Pairwise comparisons for alternatives with respect to the Safety criteria.**

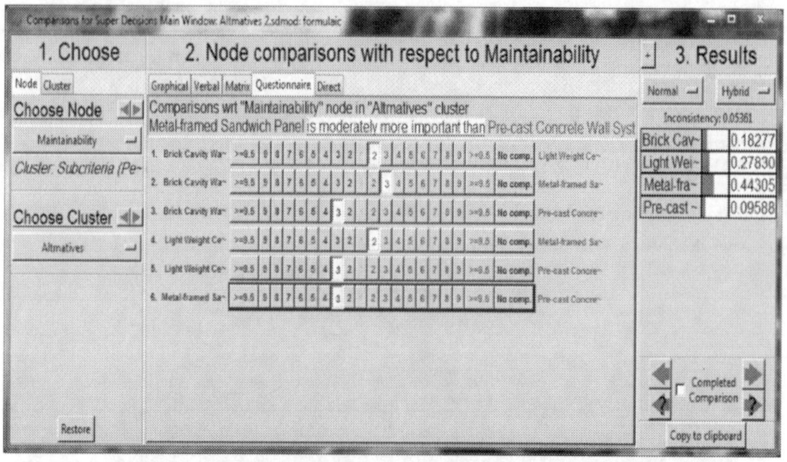


**Fig .9. (Continued) Pairwise comparisons for alternatives with respect to the Maintainability criteria.**

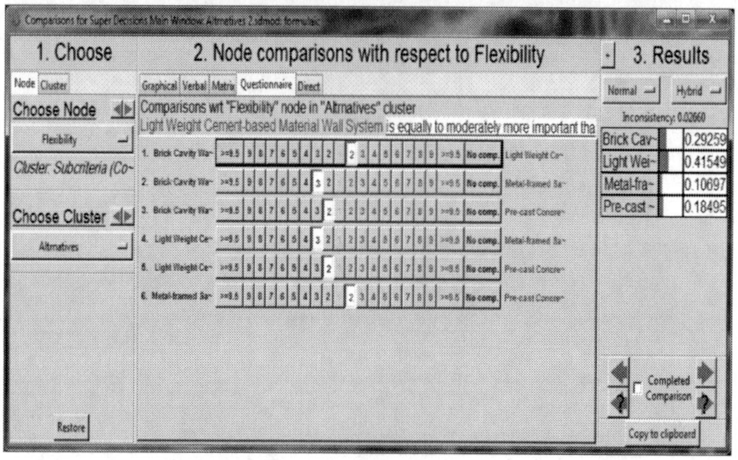


**Fig. 9. (Continued) Pairwise comparisons for alternatives with respect to the Flexibility criteria.**

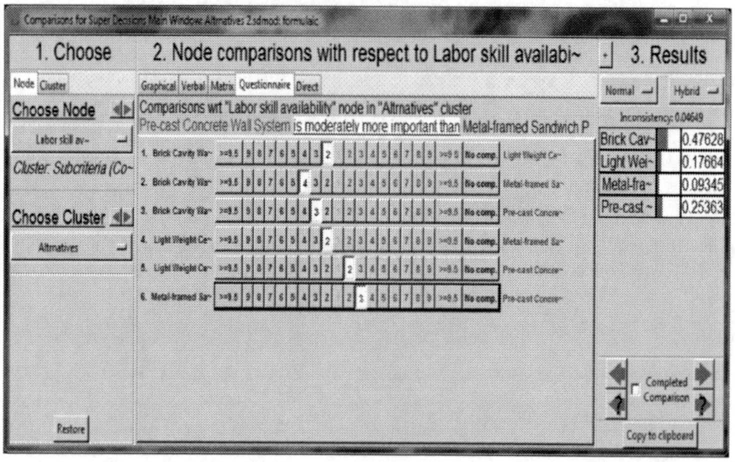


**Fig. 9. (Continued) Pairwise comparisons for alternatives with respect to Labor skill availability criteria.**

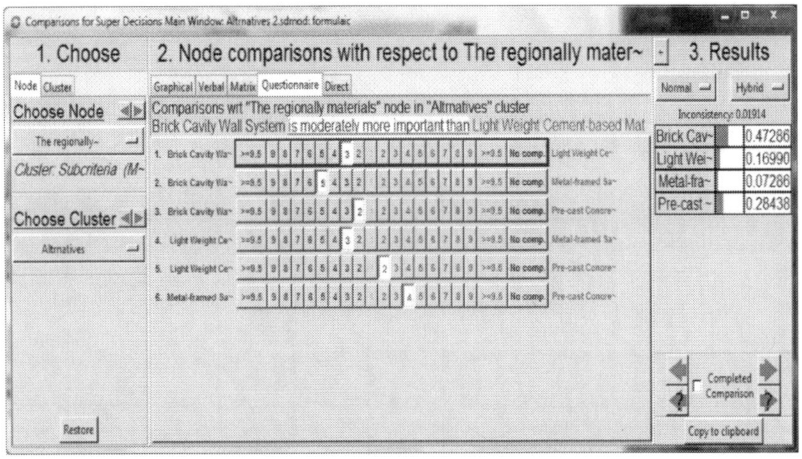


**Fig. 9. (Continued) Pairwise comparisons for alternatives with respect to the regionally materials criteria.**

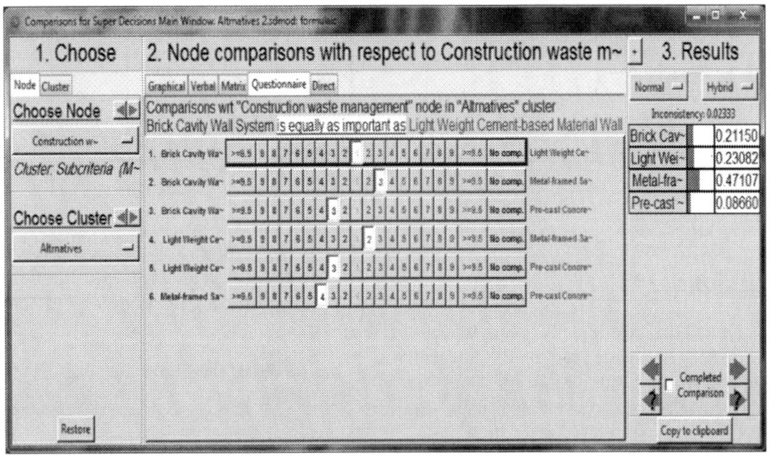


**Fig. 9. (Continued) Pairwise comparisons for alternatives with respect to The Waste Management criteria.**

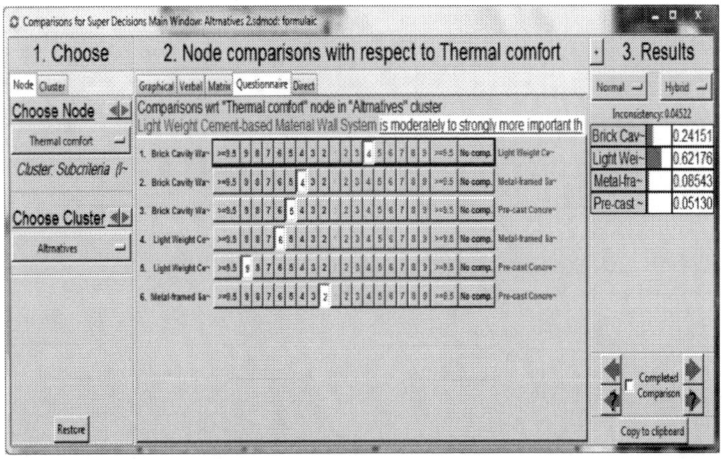


**Fig. 9. (Continued) Pairwise comparisons for alternatives with respect to The Thermal properties criteria.**

**Table 4** Analysis Matrix

| Evaluation criteria | Assigned weight | ALTERNATIVES | | | | | | | |
|---|---|---|---|---|---|---|---|---|---|
| | | Brick Cavity Wall System | | Light Weight Block Wall System | | Precast Concrete Wall Panel | | Metal-framed sandwich panel Wall System | |
| | | Ranking | Score | Ranking | Score | Ranking | Score | Ranking | Score |
| Durability | 0.194 | 0.270 | 0.052 | 0.130 | 0.025 | 0.490 | 0.095 | 0.110 | 0.021 |
| Safety | 0.164 | 0.310 | 0.051 | 0.310 | 0.051 | 0.280 | 0.046 | 0.110 | 0.018 |
| Maintainability | 0.115 | 0.180 | 0.021 | 0.280 | 0.032 | 0.100 | 0.012 | 0.440 | 0.051 |
| Flexibility | 0.066 | 0.290 | 0.019 | 0.420 | 0.028 | 0.180 | 0.012 | 0.110 | 0.007 |
| Labor skill availability | 0.131 | 0.480 | 0.063 | 0.180 | 0.024 | 0.250 | 0.033 | 0.090 | 0.012 |
| The regionally materials | 0.131 | 0.470 | 0.061 | 0.170 | 0.022 | 0.280 | 0.037 | 0.070 | 0.009 |
| Waste Management | 0.058 | 0.210 | 0.012 | 0.230 | 0.013 | 0.090 | 0.005 | 0.470 | 0.027 |
| Thermal properties | 0.099 | 0.240 | 0.024 | 0.620 | 0.061 | 0.050 | 0.005 | 0.090 | 0.009 |
| Acoustical properties | 0.044 | 0.330 | 0.014 | 0.040 | 0.002 | 0.480 | 0.021 | 0.060 | 0.003 |
| Total weight | Score=Weight* Ranking | 0.317 | | 0.258 | | 0.264 | | 0.157 | |

## 3 Conclusions

The evaluation matrix resulted in prioritization of four alternatives among a wide range of alternatives to exterior wall systems in school buildings. They are, in the order of preference: Brick Cavity Wall System, Precast Concrete Wall, Light Weight Block Wall System and Metal-framed sandwich panel Wall System.

This proposed method should help designers and decision maker to select the best external wall system for any particular situation.

In addition, this method can be used for evaluation and choose any construction system by following the value engineering methodology as shown in this paper

## References

1. Dell ' Isola , R.S. Means , USA ,1997.
2. Saaty , T. L. , Int. J Service Science, Vol.1 ,No.1 , 2008.
3. Abudlmohsen Al-Hammad, M.A., Journal of Architectural Engineering , volume: 2, No.3 , September, 1996.
4. Department of Public Works LEED Certification fact sheet_alt layout.doc , USA , 2007.

# Major factors contributing to the construction waste generation in building projects of Iraq

*Tareq* Khaleel [1], and *Ahmed* Al-Zubaidy [1,*]

[1] Building and Construction Engineering Department, University of Technology, Baghdad, Iraq

**Abstract.** Due to the economic growth and improvement of the construction industry witnessed by most countries, there has become a crucial need for employing modern possibilities in the construction sector to build taller, longer and deeper structures. However, one aspect that heads forward with the same intensity is the generation of 100 million tons of construction waste every year. This generation has occurred due to several factors with different levels of importance. Hence, this study reveals 15 factors influencing construction waste generation and categorizes them into 3 groups, (materials management on site), (materials handling, transportation and storage) and (site management and practices). A questionnaire survey of 100 respondents was distributed among different engineers to assess the construction waste factors. Results showed that damage of materials on site, double handling of materials and incompetent contractor's technical staff were the most significant factors of each category with Relative Importance Indexes (RII) of 0.866, 0.844 and 0.83, respectively. These findings will help the practitioners to reduce construction waste quantities in sites and improve waste management performance factors to control the construction waste problems.

## 1 Introduction

The construction industry has a dynamic function in most developing countries. Currently, it is facing many barriers and issues associated with unbelievable quantities generated from construction waste. Undoubtedly, the improvement in the standard of livings, changes in consumption behaviors as well as the natural growth in population lead to increase the amount of construction waste, which is demonstrated significantly in the past two decades [1]. Quantification of construction waste generated from different construction activities was being the first priority of most environmental agencies. To show that, United States Environmental Protection Agency (EPA) has revealed that 170 million tons of construction waste were generated in 2002 [2].

As a developing country, Iraq is confronted with the construction waste problem which has resulted in illegal dumping that adds to environmental pollution and, furthermore, increases the project cost. This problem can worsen if not addressed properly. Design, materials handling, transportation and storage are major contributory activities in generation of construction waste [3]. Such activities would deplete time and energy without adding value to the client, thus, resulting in wastage of construction materials, delay in meeting the specified time and execution of useless activities. Finding out the root cause of construction waste is an effective way to avoid it.

In this context, this research is meant to reveal most frequent construction waste factors of common categories, collected from several methodological approaches that influence the construction waste generation. Furthermore, evaluating those factors using statistical measurements to identify the priority and impact of each factor in order to reduce the project cost deviation and thus, increase the project profits.

## 2 Problem statement

It is observed that the quantities of construction waste are growing rapidly due to the fact that there is an intensive rise in the number of construction projects throughout the world, especially nationwide. The rates of construction waste materials in most projects in Iraq have already exceeded the permissible limits set by the Iraqi Ministry of Construction & Housing (MOCH). Green Middle East in 2007 stated that "Gulf Cooperation Council (GCC) countries produce nearly 80 million tons of waste annually" [4]. This indicates that the generation of construction waste, at this region, is rising expeditiously. Therefore, would motivate the experts for crucial necessity to develop an effective waste management system in sites as soon as possible. Furthermore, would motivate to identify the influencing factors which contribute to increase the construction waste quantities.

* Corresponding author: ahmed_mustafa1993@yahoo.com

© The Authors, published by EDP Sciences. This is an open access article distributed under the terms of the Creative Commons Attribution License 4.0 (http://creativecommons.org/licenses/by/4.0/).

## 3 Literature review

### 3.1 Concept of construction waste

Experts in the construction sector have believed, for long decades, that there is a direct relationship with only the debris disposed from project site and the construction waste produced from different construction activities [5]. Hence, some researchers have defined the waste produced from construction activities, in terms of mathematical analysis, as "the difference between the value of materials delivered and accepted on site, and materials used properly as specified and accurately measured in the work" [6].

Nowadays, waste is defined by means of lean production philosophies as "any inefficiency that results in the use of equipment, materials, labor or capital in larger quantities than those considered as necessary in the production of a building" [7]. Another research work defined waste as "any loss produced by activities that generate direct or indirect costs but do not add any value to the product from the point of view of the client" [8]. In virtual works, some studies reported, through seven sites, that labors consume nearly 46% of working time for the value-adding activities, whereas the remaining majority consumed for non-value adding activities [9]. Other proved even worse conclusions that only from 3 to 20% of steps add-value, and their share of the whole cycle time ignored from 0.5 to 5% [10]. Hence, minimizations of non-value adding activities provide an extensive potential improvement in most operational processes. Besides a clear understanding of the construction waste concept, it is necessary to identify the classification of waste in different perspectives.

### 3.2 Classification of construction waste

Construction waste can be classified into two different categories, namely physical waste and non-physical waste.

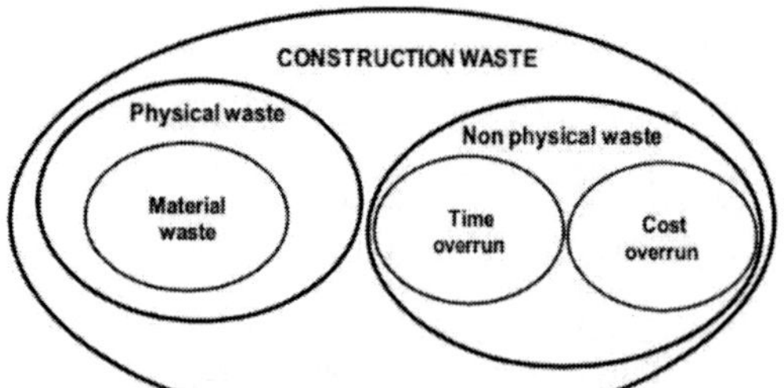


**Fig.1.** Construction waste classification [11]

#### *3.2.1 Physical Waste*

Physical construction waste is denoted as "the waste which originates from construction, renovation and demolition actions including land excavation, civil and building construction, site clearance, roadwork, and building demolition and renovation" [12-15]. In other words, physical waste is recognized as the debris which can be observed at any construction project. It involves exhaustion of construction materials, since they are irreparably damaged or simply lost. The wastage is usually removed from the construction site to landfills. However, physical waste is classified into three sub-divisions: additional material loss, man-hour overproduction and equipment operational time.

#### *3.2.2 Non-physical Waste*

On the contrary, both time and cost overrun are the major forms of non-physical waste which is commonly developed during construction process. Waiting durations, work interruptions, delays in execution and obtaining work permits, reworks and performance of unnecessary works are forms of time wastages [7,8], whereas, cost overrun, which is also referred as "cost increase" or "budget overrun", involves unanticipated costs incurred in excess of the budgeted amounts [16]. Several researchers have reported that both time and cost wastes may cause of material waste generation. For instance, Memon et.al stated that "non-physical waste includes undesired activities, which can cause the physical waste, such as rework, and unnecessary material movements" [17].

### 3.3 Construction waste generation

The construction waste generation has created serious problems, both locally and globally. Construction waste materials are generated when new buildings are built. Buildings include all residential and non-residential structures, as well as public works projects, such as roads, bridges and dams [18]. In 2002, the United Kingdom (UK) generated 200 million tons of total waste. 50% of this quantity was generated by construction, demolition, renovation and refurbishment works. 24% generated in commercial and industrial activities, with households responsible for a further 14% [19]. Whereas EPA, which is located in the United States (US), stated that "approximately 136 million tons of construction waste was generated in 1998 [20], and it raised to 170 million tons in 2003[21]".

The Arab gulf region is facing a serious challenge associated with inadequate management of construction waste, where most countries have the highest per capita waste generation in the world. Industrial growth, changing lifestyles, construction strategies, unsustainable consumption patterns and rapid urbanization have all contributed to the growing of waste problem. Green Middle East in 2007 stated that (GCC) countries produce nearly 80 million tons of waste annually; half of this consists of construction material wastage, and approximately 33% is municipal solid waste while the rest is industrial waste [4]. Others estimate the solid waste quantities in (GCC) at around 120 million tons per year [22]. Hence, the scale of the challenge faced by waste management sectors in the Arab gulf region is even larger. It is obvious that these rates are growing rapidly and create a serious problem which threaten the socio-economy of countries. Therefore; an immediate development of waste management approaches is being a crucial necessity in this region.

### 3.4 Causes of construction waste generation

Among various researches published around the world about construction waste generation and its management, only a few studies can be identified as those that are focused on revealing the factors contributing to construction waste generation. Researchers that covered construction waste generation detected a number of construction waste sources that may rise during the life cycle of a project.

The construction management performance factors are described as the actions which decrease construction waste quantities and contribute to increase recycling at the construction sites. Hence, with the concept of the possibility of decreasing construction waste, several factors were collected through interviews and extensive literature reviews. Subsequently, it was discovered that several factors would have influence on construction waste generation. Thus, 15 factors were categorized by common characteristics: "Materials Management On-Site", "Materials Handling, Transportation and Storage" and "Site Management and Practices". Table (1) lists these 15 factors by categories.

**Table 1.** Construction waste sources

| Category | | Factor |
|---|---|---|
| Materials Management on-Site | 1.1 | Damage of materials on site |
| | 1.2 | Manufacturing defects |
| | 1.3 | Packaging defects |
| | 1.4 | Theft and vandalism |
| | 1.5 | Inadequate control on construction materials in site |
| | 1.6 | Oversized of building elements during execution due to non-straightness of construction works |
| Materials Handling, Transportation and Storage | 2.1 | Wrong handling of materials |
| | 2.2 | Materials storage away from the work site and wastage during transportation |
| | 2.3 | Poor and wrong storage of materials |
| | 2.4 | Double handling of materials |
| | 2.5 | Accidents during handling and transportation |
| Site Management and Practices | 3.1 | Lack of responsibility of waste management by the resident engineer |
| | 3.2 | Absence or weakness of waste management plan |
| | 3.3 | Incompetent contractor's technical staff |
| | 3.4 | Lack of positive incentive that aims to waste reduction |

As mentioned above, these 15 factors indicate the elements which may contribute to decreasing waste quantities at the construction site. As such, if a construction site is assessed based on these 15 factors in terms of execution level of the site, it is possible to assess the performance level of the project. However, since all the factors do not have the same level of importance, it is necessary to determine the magnitude of importance of all the identified factors.

## 4 Methodology

### 4.1 Questionnaire design

A structured questionnaire consists of two sections: the first section describes the demographic data sample of the participated respondents, while the other section identifies the 15 factors that contribute to construction waste generation which have been categorized into 3 groups as shown in Table (1). Then, the respondents were requested to evaluate the potential causes of construction waste by assigning score to each factor wherever observed. The questionnaire measured on a 5-point Likert scale where 1 represents "Ineffective", 2 is "Somewhat Effective", 3 is "Moderately Effective", 4 is "Effective" and 5 is "Very Effective".

### 4.2 Data collection

The Center for Strategy Research recommends that the minimum sample size of a questionnaire survey to accurately estimate the reality is 30 respondents or more [23]. Thus, a total of 100 respondents were elected in this study from different engineers who have different years of experience; 42 respondents included were collected via web and the rest were collected via form. There is a crucial necessity for testing the reliability of results when dealing with questionnaire survey. Hence, Cronbach's Alpha is the most commonly statistical measurement used to estimate the reliability of questionnaire provided with Likert scale. Table (2) illustrates some of descriptive statistics with Cronbach's Alpha of the data sample.

**Table 2.** Statistical analysis of data sample

| Statistic Function | Value |
|---|---|
| N | 100 |
| Minimum | 3.11 |
| Maximum | 4.33 |
| Median | 4.02 |
| Mean | 3.92 |
| Standard Deviation | 0.33 |
| Cronbach's Alpha | 0.848 |

It is observed that Cronbach's Alpha coefficient is 0.848, which rule of thumb considers it as a good indication when it normally lies between 0.8-0.9.

### 4.3 Data analysis

The analysis was aided using Statistical Package for Social Sciences (SPSS) software V24. The rates, linked up with each factor, were exposed for further statistical analysis for the purpose of identifying the priority and the importance of the contributory factors by use of Relative Importance Index (RII). The value of RII mainly lies between (0 to 1), the higher the better. Equation (1) describes the computational method to calculate the RII for each factor separately, whereas Tables (4) to (6) show the analysis results.

$$RII = \frac{\sum W}{A * N} \qquad (1)$$

Where:
$W$: is the attached score of each factor and mainly ranges between (1 to 5), (where "1" is "ineffective" and "5" is "very effective");

$A$: is the highest score (i.e. 5 in this study) and; $N$: is the sample size of the questionnaire (i.e. 100 in this study).

## 5 Results and discussion

### 5.1 Respondent's profile

The sample frame for this study consisted of architects, civil, electrical and mechanical engineers who deal with the construction waste directly at the construction site. The purpose of the respondent's profile diverse is to obtain more opinions about construction waste factors from different perspectives. Table (3) represents the demographic data of this study.

**Table 3.** Demographic Data of the Study

| **Respondent's Characteristics** | **Percentage (%)** |
|---|---|
| ***Engineering Specialization*** | |
| Civil | 81 |
| Architect | 9 |
| Electrical | 1 |
| Mechanical | 9 |
| ***Academic Qualification*** | |
| B.Sc. | 62 |
| Higher Diploma | 1 |
| M.Sc. | 17 |
| Ph.D. | 20 |
| ***Work Field*** | |
| Academic | 16 |
| Site | 66 |
| Both | 18 |
| ***Years of Experience*** | |
| 1-7 | 15 |
| 8-14 | 23 |
| 15-21 | 22 |
| 22-28 | 11 |
| 9-35 | 15 |
| More than 35 | 14 |

### 5.2 Construction waste factors

After analyzing the respondent's profile, it is necessary to analyze the score of the factors concerned at which ranked by respondents. The concerned factors were categorized under 3 different groups as follows:

#### *5.2.1 Materials management on-site factors*

Damage of materials on site is realized to be the highest influential waste factor of materials management category with RII of 0.866. This is obvious, especially in Iraq due to non-compliance with standard storage conditions which have a direct relation to material inventory. Inadequate control on construction materials at site is rated to be the second highest influential waste factor in this group with RII of 0.846 and may occur as a result of engineer's inexperience. Some engineers may ignore or may not take the topicality of construction works or activities seriously such as road leveling before casting, concrete forms and vertical columns straightness, etc. Perhaps those reasons may be one causes of waste generation which is related to the oversize of building elements during execution due to non- straightness of construction works with RII of 0.804.

Theft and vandalism is ranked fourth with RII of 0.798. The influence of this factor could be by employing a guard at the site and building a steel fence (BRC steel) surrounding the construction site. Both packaging and manufacturing defects have occupied the last ranking with RII of 0.738 and 0.712, respectively. Table (4) summarizes the RII and rank of material management on site group.

**Table 4.** Materials management related factors

| No | Factor | RII | Rank |
|---|---|---|---|
| 1 | Damage of materials on site | 0.866 | 1 |
| 2 | Manufacturing defects | 0.712 | 6 |
| 3 | Packaging defects | 0.738 | 5 |
| 4 | Theft and vandalism | 0.798 | 4 |
| 5 | Inadequate control on construction materials in site | 0.846 | 2 |
| 6 | Oversized of building elements during execution due to non- straightness of construction works | 0.804 | 3 |

#### *5.2.2 Materials handling, transportation and storage factors*

Respondents who assessed the factors, have considered that double handling of materials factor to be the most important factor in this group with highest RII of 0.844. When the storage is far away from the construction site, the probability of generating waste increases. Therefore, wastage during transportation has been ranked as the second position with RII of 0.832. Inadequate stacking or insufficient / wrong storage is ranked as the third position with RII of 0.822. Thus, the storekeeper must comply the standards of the factory, which is related to material storage instructions, in order to minimize wastage resulted from this factor Wrong handling of material is ranked as the fourth factor with RII of 0.718. Due to weak probability of accidents during handling and transportation, it makes sense to be as the last factor would affect on wastage with RII of 0.622. Table (5) represents the RII and rank of material handling, transportation and storage category.

**Table 5.** Materials handling, transportation and storage related factors

| No | Factor | RII | Rank |
|---|---|---|---|
| 1 | Wrong handling of materials | 0.718 | 4 |
| 2 | Materials storage away from the work site and wastage during transportation | 0.832 | 2 |
| 3 | Poor and wrong storage of materials | 0.822 | 3 |
| 4 | Double handling of materials | 0.844 | 1 |
| 5 | Accidents during handling and transportation | 0.622 | 5 |

### 5.2.3 Site management and practices factors

Contractor's incompetent technical staff is intended to be the most dominant factor in site management category with RII of 0.83 as shown in Table (6). To reduce waste, waste management plan should be developed. Hence, the absence or weakness of waste management plan is ranked as the second position with RII of 0.818, which may occur due to the lack of responsibility on waste management by resident engineer with RII 0.758. Because of engineers' negligence about construction waste management, the incentive schemes will be discarded by them, which is a convincible matter by ranking it as the last factor with RII of 0.756.

**Table 6.** Site management and practices related factors

| No | Factor | RII | Rank |
|---|---|---|---|
| 1 | Lack of responsibility on the waste management by the resident engineer | 0.758 | 3 |
| 2 | Absence or weakness of waste management plan | 0.818 | 2 |
| 3 | Contractor's incompetent technical staff | 0.83 | 1 |
| 4 | Lack of positive incentive that aims to waste reduction | 0.756 | 4 |

To sum up, Figure (2) illustrates the identified factors with their importance level and priority as a total percentage in a pie chart where the graph legend has been listed from Table (1).

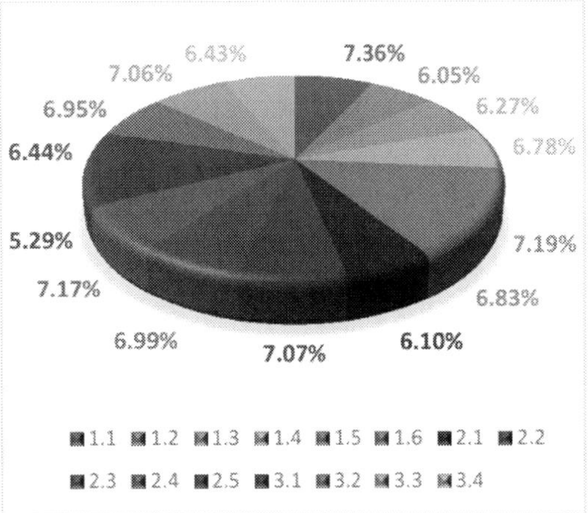


**Fig. 2.** Construction waste priority factors

## 6 Conclusions

The construction industry is a major element in worldwide economy. As the construction industry increases, the construction waste increases too with the same growing intensity. Least priority in waste minimization and management systems in Iraq leads to generation of huge amounts of construction material waste annually. Illegal landfill disposal, project cost deviation are major problems affecting Iraqi construction sector, which are closely related to construction waste. Fifteen factors have been identified and were considered as mainly responsible for construction waste generation through different project stages with different levels of importance.

## References


1. A. Katz and H. Baum, *Waste Management*, **31**, 353-358, (2011).
2. H. Yuan, *Waste Management*, **32**, 1218-1228, (2012).
3. L. Ekanayake and G. Ofori, *in 2nd Southern African Conference of Strategies for a Sustainable Built Environment*, (2000).
4. Green Middle-east, Solid Waste – Solid Growth Prospects", 2017. [Online]. Available: http://www.greenmiddleeast.com/sw/default.html. [Accessed: 07- Apr- 2017].
5. C. Formoso, L. Soibelman, C. De Cesare and E. Isatto, *Construction Engineering and Management*, **128**, 316-325, (2002).
6. L. Pheng and S. Tan, *Construction Management and Economics*, **16**, 621-635, (1998).
7. L. Koskela, Stanford, California, California, USA, (1992).
8. C. Formoso, E. Issato and E. Hirota, *in Method for waste control in the building industry*, (1999).
9. J. Christian and D. Hachey, *Construction Engineering and Management*, **121**, 20-26, (1995).
10. Ciampa, D., Timebased Competition, Blackburn (Ed.), Business One Irwin, Homewood, IL, 273-293, (1991).
11. S. Nagapan et.al., *Faculty of Civil and Environmental Engineering (FKAAS) Digest*, **1**, 15-20, (2012).
12. O. Kofoworola and S. Gheewala, *Waste Management*, **29**, 731-738, (2009).
13. L. Shen, V. Tam, C. Tam and D. Drew, *Construction Engineering and Management*, **130**, 472-481, (2004).
14. C. Poon, *Waste Management*, **27**, 1715-1716, (2007).
15. V. Tam and C. Tam, *Building Research & Information*, **36**, 37-43, (2008).
16. Shanmugapriya, S. & Subramanian, K., *International Journal of Emerging Technology and Advanced Engineering*, **3**, 734-740, (2013).
17. Memon, A.H. Abdul-Rahman, I. and Memon, I., *Life Science Journal*, **11**, 417-424, (2014).
18. M. Abdelhamid, *HBRC Journal*, **10**, 317-326, (2014).
19. Department for environmental Food & Rural Affairs, (2012).
20. US Environmental Protection Agency (EPA). Prepared by Franklin Associates. (1998), Available from: http://www.epa.gov/epaoswer/hazwaste/sqg/c&drpt.pdf.
21. US Environmental Protection Agency (EPA), 2003. [Online]. Available: http://www.epa.gov/epawaste/conserve/rrr/imr/cdm/pubs/cd-meas.pdf. [Accessed 15-Apr-2017].
22. V. Gautam, *South Asia & Middle East, Environmental & Building Technologies Practice*, (2009), Source: www.frost.com
23. Center for Strategy Research, Inc. Sample Size…How Much is Enough?, (2006), available athttp://www.csr-bos.com/Center_for_Strategy_Research_New/Twist/2006_07_Twist.html.

# An investigation into factors causing delays in highway construction projects in iraq

*Bevian* I. Al Hadithi

Building and Construction Engineering Department/ University of Technology, Baghdad, Iraq

**Abstract.** The highways sector is a prominent sector in any country's economy because of its impact on the well-being and safety of its citizens. The transport sector has an impact on social improvement and investment in the nation on the illustration that allows access to markets, production, jobs, health and other social services.This study investigates the causes of delay of highway construction projects in Iraq, which is frequent occurrence. Data was collected using questionnaires which were distributed to the key project participants; contractors, owners and consultants. The data were analyzed using the Frequency index and Spearman's rank correlation. The top seven causes of project delays were observed to be political decisions and political realities, the economic crisis of the country, delays in materials test of and obtaining the results, delay in monthly payments of contractor, failure treatment of the delays when implementing the project, the effects of weather, rain and high temperatures, delay in activities during implementation. It is recommended to establish an appropriate number of laboratories and adopt the field laboratory mechanism for the external and remote screens. Owners should give special attention to pay progress payment to contractors on time. The competent contractor who has prior experience in implementing the high projects should be selected. The contractor must take into consideration the weather conditions when preparing the time plan necessary to implement the project. The project management should identify these reasons and deal with them quickly in order to reduce the total delay of the project.

## 1 Introduction

A lot of construction projects in the world and Iraq in particular are suffering from delay in implementation. It could say that a project that does not suffer from delays is rarely found.

One of the recurring problems in the construction industry is
the issue of delay, which negatively affects the success of the project in terms of time, cost and quality [1]. Delay has a negative effect on owners, contractors and consultants in terms of growth in adversarial relationships, mistrust, litigation, arbitration, cash-flow problems and a general feeling of trepidation toward one another [2]. Project value was found to have a negative correlation with time loss due to the demonization indicating that as project size increases time loss decreases [3]. Delay is mostly common in the traditional type of contracts in which the contract is awarded to the lowest bidder.

This procurement method is mostly practiced in developing
Countries. Ensuring that the project is delivered on time is one of the most significant needs of the clients in the construction industry [4]. The number of projects delayed in Iraq for the period 2006-2007, according to the sample of the research, (571) of the total (1993) project. School projects ranked first with a response rate of 27%, followed by public buildings projects with 24% response and road projects with 21% response. Note that this is due to the large number of projects implemented from these types compared to the projects of electricity and transport, for example, which received the lowest rates [5]. In Iraq, according to a survey finding that many projects suffer from a great deal of delay and due to specific factors, many projects were temporarily or permanently abandoned during the political crisis between 2003 and 2014.This factor escalates disputes to be settled by litigation process. The parties involved in the project use litigation as a last resort to settle disputes [6].

## 2 Literature review

A total of 130 general construction projects was studied in Jordan, which were built during the period 1990-1997. The study found that the main reasons were related to thedisinr or user changes, weather, site conditions, delayed births, economic conditions and increases in quantities[7].Al-Najjar (2008) concluded that the most important factors causing time overruns in building construction projects in the Gaza Strip as perceived by contractors were: strikes, Israeli attacks and border closures, lack of materials in the markets, shortage of construction materials at the site, delays of material deliveries to site, cash shortages during construction, poor

© The Authors, published by EDP Sciences. This is an open access article distributed under the terms of the Creative Commons Attribution License 4.0 (http://creativecommons.org/licenses/by/4.0/).

site management, poor economic conditions (currency, inflation rate, etc.), Shortage of equipment and tools on site, and owner delay in freeing the contractor payments for completing work[8].Kasimu & Abubakar investigate the causes of delay in the Nigeria construction industry. The survey method was adopted for this work through questionnaire distributed to contractors, clients and consultants in the construction industry in Nigeria.The results obtained from ranking analysis show that , improper planning, lack of communication, design errors and shortage of supply is the major cause of delay[9]. Also Surveys were conducted in Vietnam to identify the causes of delay in government projects and to find solutions for dealing with the delay. The five most highly ranked delay factors were: information delays and lack of information exchange between the parties; incompetent owner; incompetent supervision consultant; incompetent contractor; and difficulties in financing project by owner[10].

Delays in road construction projects in Kenya were studied, and the questionnaire was distributed to contractors and consultants.Data were collected and analyzed using the relative significance index and Spearman rank. The payment by the client, the slow decision-making process and bureaucracy in the client organization, inadequate planning, scheduling, and arbitrage are the five main causes of project delays.Client improves their financial management systems so that they can pay contractors in a timely manner, reducing bureaucracy and red tape in client organizations to accelerate slow decision making was one of the main recommendations of this study[11]. The causes of delayed road construction in Egypt were analyzed, a survey questionnaire was distributed. The relative importance index (RII) is calculated according to the highest values, the top 20 and the twenty lowest reasons for the construction projects in Egypt are determined.The test results reveal a good correlation of causes and groups between contractors and site/design engineers and between consultants and site design engineers and a somewhat low correlation between contractors and consultants. So there is no root cause that can be taken for granted to be most or least effective delay causes. The Proposed model for predicting actual road construction project duration was developed[12].

Emphasis has been placed on delays in donor-funded road projects that have become imperative, and an endemic problem in Sri Lanka is hindering the effective use of external assistance to these projects. The main factors that led to the delay were identified. An alarming situation had been revealed that, during the execution process, increase in quantities had a significant effect on the project time and accounted for about 56% of the delays. The errors and omissions in detail design, changes in specifications and scope, were the most prevalent sources of quantity increases. Further, during the planning process, the procurement of work activities was delayed by 23%, mainly due to shortcomings in contract documents and approvals from relevant agencies[13]. Several sample projects implemented over the years 2000-2008 have been analyzed. All projects were managed by the same organization and taken from the Ministry of Rural Works and Public Works of Jordan. By reviewing the literature,19 factors have been identified which are causing delays in road construction projects. Land and weather conditions are the most important factors affecting time and cost in road construction projects in Jordan[14].A study was conducted to investigate the time performance of road construction projects in the West Bank in Palestine to identify the causes of delay and their severity according to contractors and consultants through a questionnaire survey. The field survey included 34 contractors and 30 consultants. A total of 52 causes of delay were identified during the research. The survey concluded that the top five severe delay causes are political situation, segmentation of the West Bank and limited movement between areas, award project to lowest bid price, payment delay by owner, and a shortage of equipment[15].

## 3 Justification for research

After interviewing project managers in companies specialized in the implementation of highway projects construction in Iraq it was concluded that there is a big delay happening in these projects.

Studies on the implementation of high construction projects in Iraq focus on the cost side without consideration of the factors Causing Delays in these projects

## 4 Objectives

The main objectives of this study include the following:

1- To identify the causes of delays in highway construction projects in Iraq based on their frequent occurrence

2- To study the differences in views of the three major parties in any construction project, namely, owners, contractors and consultants.

## 5 Research methodology

The objectives of the research were achieved through a literature review. The field survey included 47 respondents made up to 28 contractors, 6 consultants and13 owners were represented by the General Authority for Iraqi Roads and Bridges. The questionnaire was divided into three parts. The first included education level information and years of experience Fig (1), Fig (2). The second part, which is the most important, focused on causes of construction delay.Questionnaires were delivered to participants in person; in addition an electronic questionnaire was used because of the shortage of number of questionnaires retrieved, which is one of the difficulties encountered in this study.

The respondents were requested to indicate their response to the frequency of occurrence of 64 well-recognized construction delay factors. These causes were categorized into the following seven groups of factors:

- Contractor
- Owner
- Consultant
- Materials

- External factors
- Equipment &labor
- Related to the project

The questionnaire fundamentally based on Likert's scale of 5 ordinal measures from 1 to 5 according to the level of their frequent occurrence:
(5) = always
(4) = often
(3) = sometimes
(2) = rarely
(1) = never happened

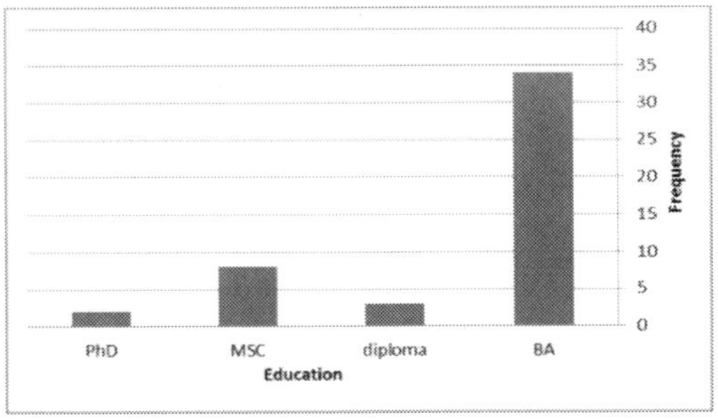


**Fig. 1**. Level of education of questionnaire

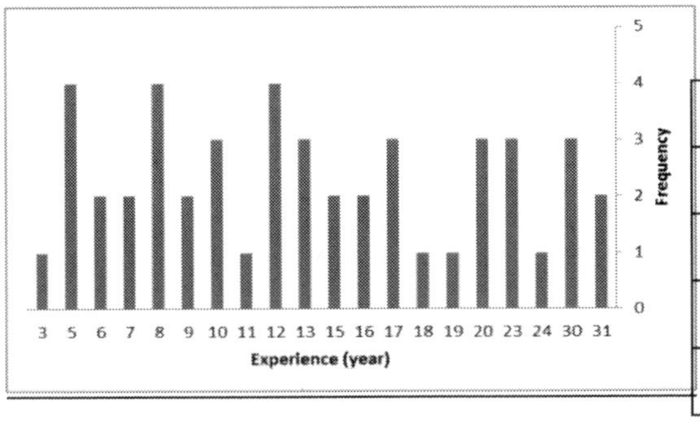


**Fig. 2**. Experience years of questionnaire

## 6 Consistency coefficient for the questionnaire

To calculate the consistency coefficient for the questionnaire based on their frequency of occurrence,using the CronbachAlpha equation, which was implemented using the Statistical Package for Social Sciences (SPSS). The results are shown in the following table:
The results show that Cronbach's coefficient for all items equal (0.923). This range is considered high; the result ensures the reliability of the questionnaire [16].

## 7 Data analysis

Histograms have been used to identify the delay factors that are frequent occurrence in the implementation of highway projects in Iraq from the point of view of contractors, employers and consultants, and the collected data were analyzed through the Frequency Index:

A formula is used to rank the causes of delay, based on frequency of occurrence as identified by the participants.

$$\text{Frequency Index (F. I) (\%)} = \Sigma a (n/N) * 100/5 \quad \text{------} \quad (1)$$

Where (a) is the constant expressing weight given to each response (range from 1 to 5), n is the frequency of the response and N is the total number of responses.
Spearman's rank correlation was used to show the level of agreement between two parties.

$$rs = 1 - 6\Sigma d^2/(n^3 - n) \quad \text{---------} \quad (2)$$

Where 'rs' is the Spearman'srank correlation coefficient;'d' is the difference in ranking between any two parties; and 'n' is the number of factors. The correlation coefficient varies between +1 and –1, where +1 means a perfect positive relationship (agreement), while – 1 means a perfect negative relationship (disagreement). Sample estimates of correlation close to unity in magnitude mean good correlation, while values near zero indicate little or no correlation.

According to the contractors' opinion, the results of the delay factors which has the effect on the implementation of highway projects in Iraq are shown in Table (1), and Fig 3,4,5and 6 respectively.

**Table (1)** results of the delay factors according to the contractors' opinion

| Delay factor | Level of frequent | Percentages % |
|---|---|---|
| the economic crisis of the country | always | 35.7 |
| Delay in monthly payments of the contractor | often happening | 50% |
| The effects of weather, rain and high temperatures | often happening | 50% |
| Delays in materials test of and obtaining the results | often happening | 46.4% |

The overall top six delay factors which are always happening in the implementation of highway projects in Iraq, are shown in Table (2), and Fig 7, 8, 9,10,11and 12 respectively.

**Table (2)** results of the delay factors according to the owners' opinion

| Delay factor | Level of frequent | Percentages % |
|---|---|---|
| the economic crisis of the country | always | 46.2% |
| Ineffective inspection of equipment takes | | 46.2% |
| Not to provide services to workers through the implementation of external roads | always | 38.5% |
| Services for underground construction such as telephone and electricity lines, water pipes, oil pipelines, and rainwater pipes | always | 38.5% |
| Deficiency of skilled and unskilled laborers, take | always | 30.8% |
| Changes in site topography after design, take | always | 30.8% |

The overall top five delay factors which always happening in the implementation of highway projects in Iraq, according to the Consultants' opinion are shown in Table (3), and Fig 13,14,15,16 and 17 respectively.

Table (3) results of the delay factors according to the Consultants' opinion

| Delay factor | Level of frequent | Percentages % |
|---|---|---|
| the economic crisis of the country | always | 100% |
| Political decisions and political realities | always | 66.7%. |
| Cash flow problems faced by the contractor | always | 50%. |
| Changes in the contract by the owner during the construction process | always | 50%. |
| Ineffective inspection of equipment | always | 50%. |

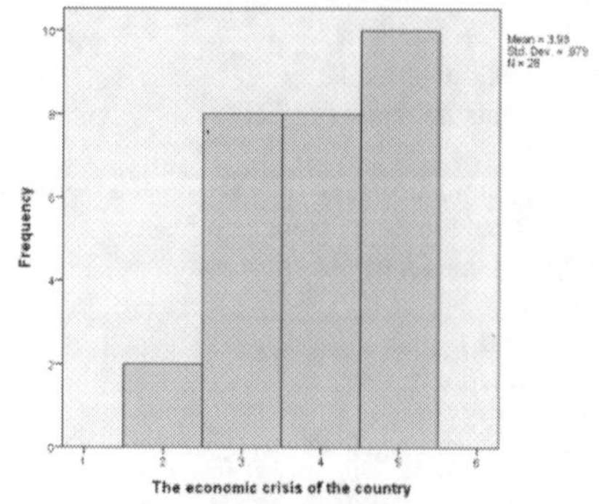


**Fig (3)** frequency occurred of the economic crisis in the country in the projects according to the contractor's opinion

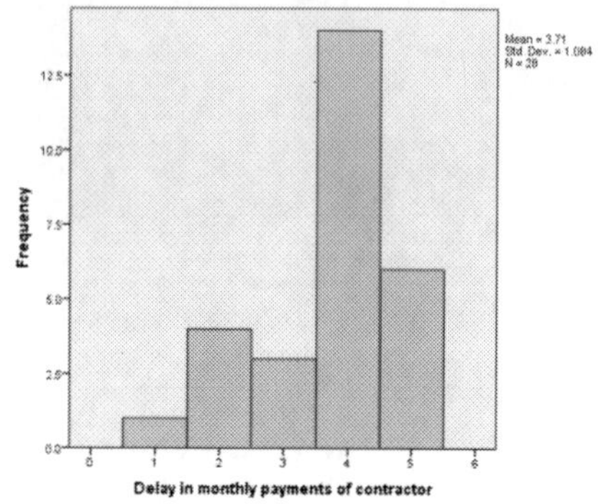


**Fig (4)** the frequency of the delay in monthly payments of contractor

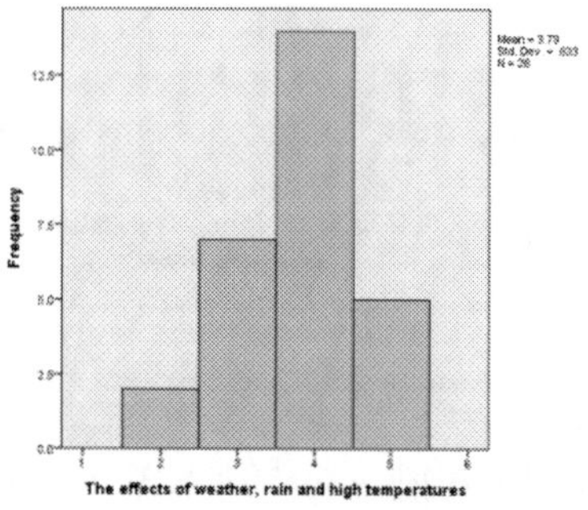


**Fig (5)** the frequency of the effects of weather, rain and high temperatures

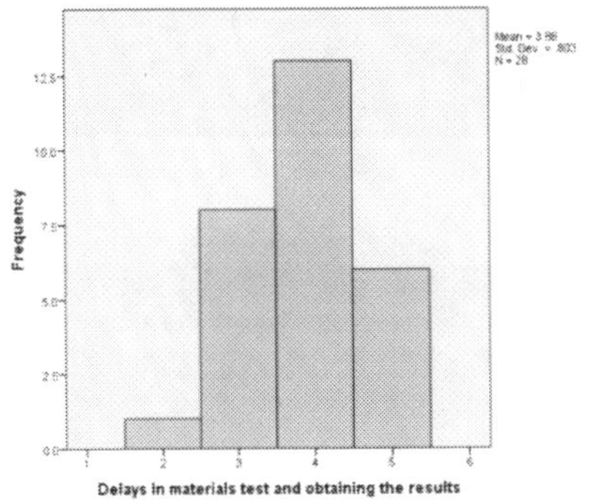


**Fig (6)** the frequency of delays in materials test of and obtaining the results

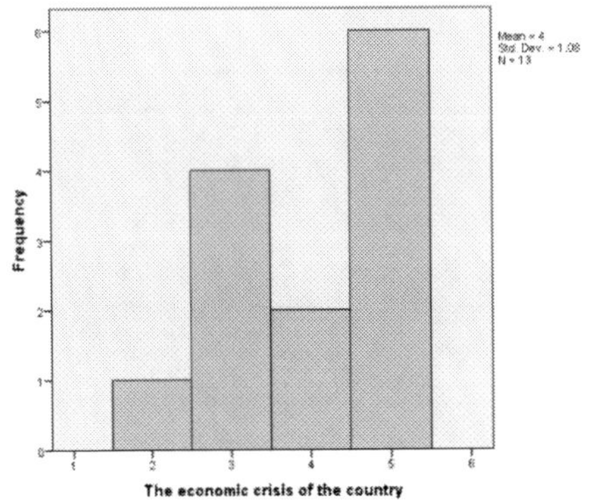


**Fig (7)** frequency occurred of the economic crisis in the country in the projects according to the owners' opinion

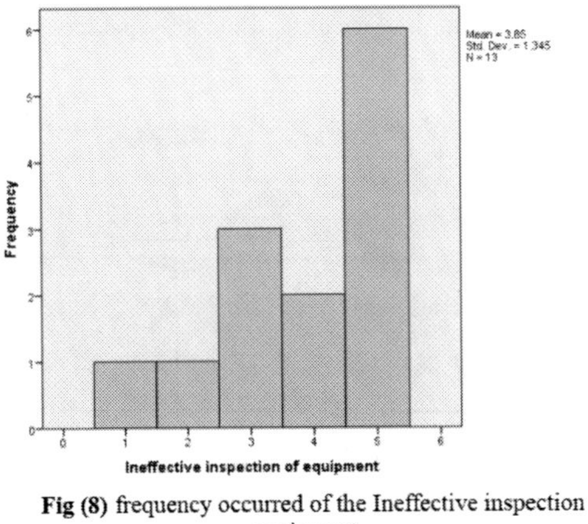


**Fig (8)** frequency occurred of the Ineffective inspection of equipment

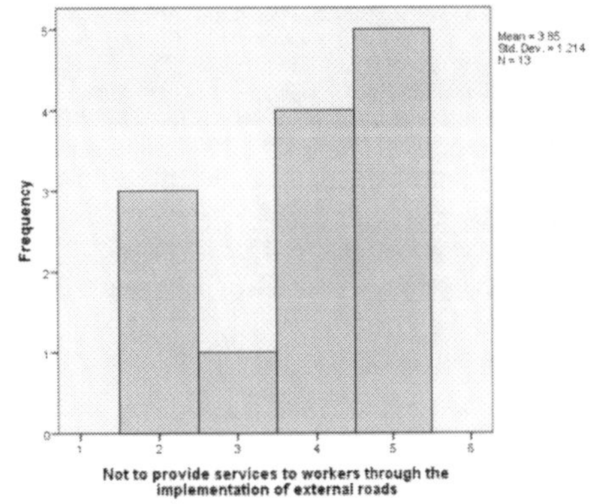


**Fig (9)** frequency occurred not to provide services to workers through the implementation of external roads

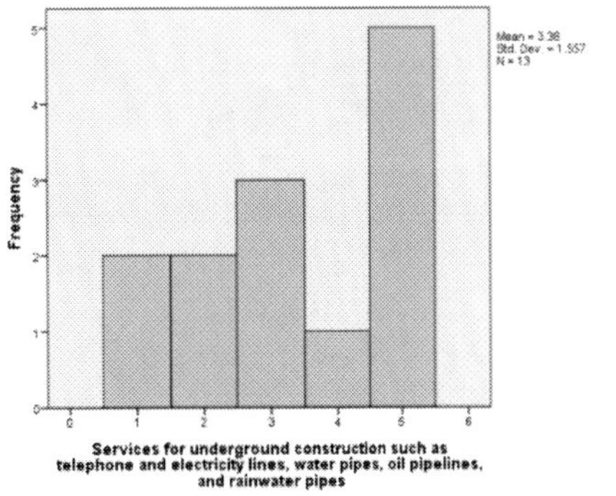


**Fig (10)** frequency occurred Services for underground construction such as telephone electricity lines etc.

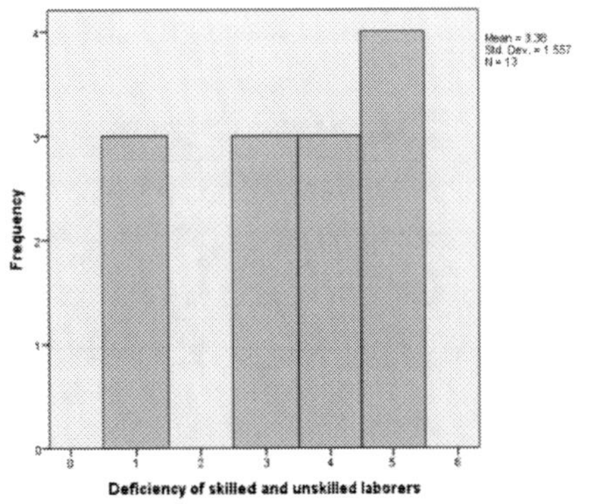


**Fig (11)** frequency occurred Deficiency of skilled and unskilled laborers

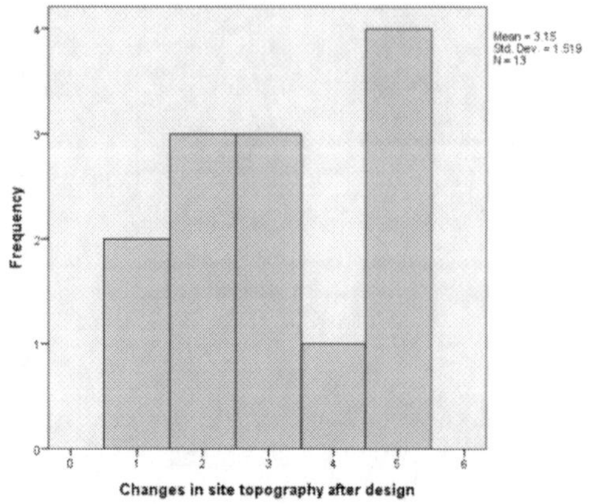


**Fig (12)** frequency occurred Changes in site topography after design

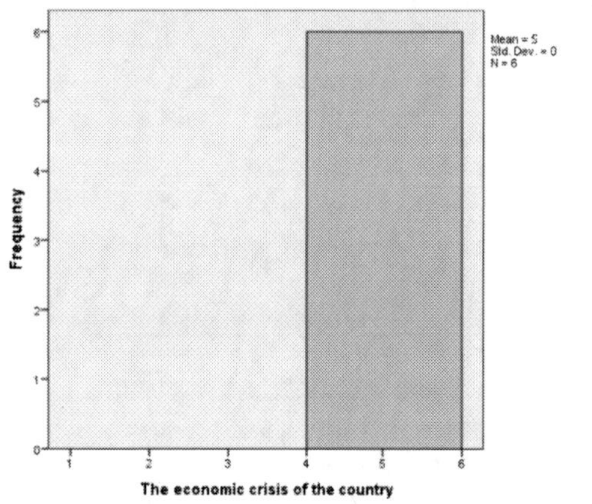


**Fig (13)** frequency occurred of the economic crisis in the country in the projects

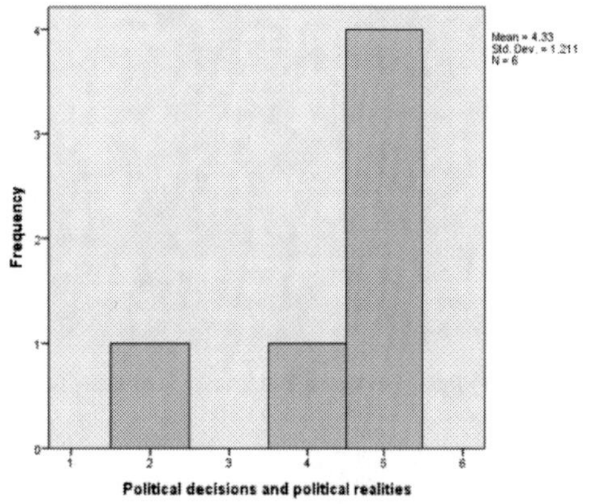


**Fig (14)** the frequency of the political decisions and political realities

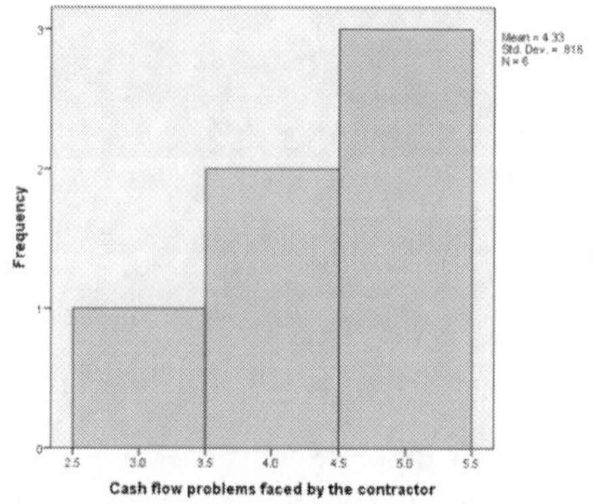


**Fig (15)** the frequency of Cash flow problems faced by the contractor

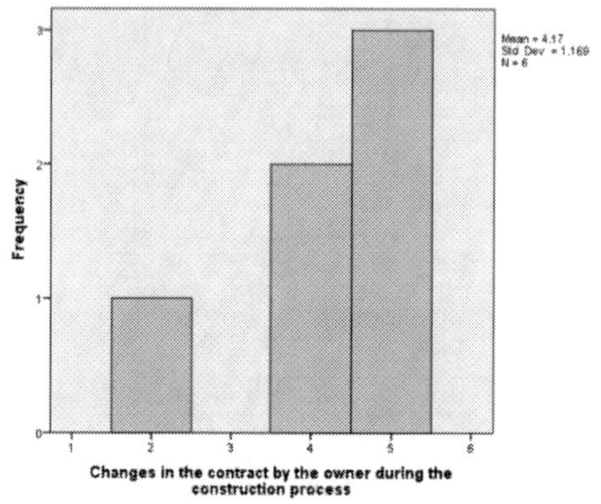


**Fig (16)** the frequency of Changes in the contract by the owner during the construction process

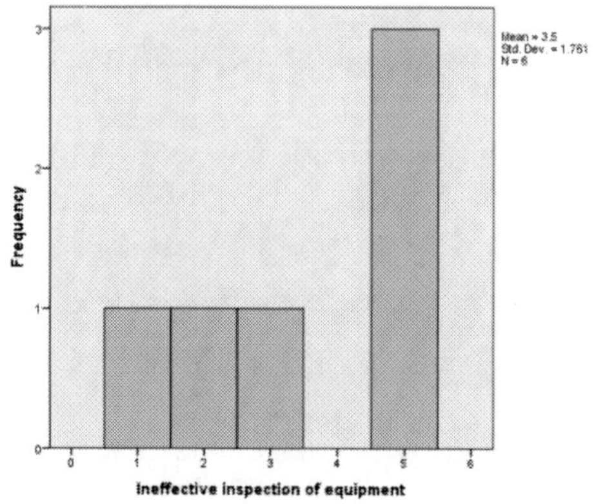


**Fig (17)** the frequency of ineffective inspection of equipment in the projects

### *7.1 The delay causes from the contractor's and owner's point of view*

Table 4shows all causes of delay and frequency of occurrence of each reason and its arrangement from the point of view of the contractor and the owner.
The overall top seven causes of delay identified by both consultants and contractors were:
Political decisions and political realities (F.I=69.79),The economic crisis of the country (F.I=68.94), Delays in materials test of and obtaining the results (F.I=68.94), Delay in monthly payments of contractor (F.I=65.96), Failure treatment the delays when implementing the project (F.I=67.66),The effects of weather, rain and high temperatures (F.I=65.53),Delay in activities during implementation (F.I=61.28).The Spearman's rank correlation coefficient between the contractors and owners was calculated and found to be 0.57 which showed that there was a positive correlation between the views of them.

**Table 4** Delay causes from the contractor and owner point of view

| | (F. I) (%) = Σ a (n/N) * 100/5. | RANK |
|---|---|---|
| **Reasons related to the Contractor** | | |
| Delay in delivering the site | 50.21 | 41 |
| Deficiencies in bidding by contractors/causing calling them to clarify it in the stage of the award of the contract | 47.23 | 50 |
| Failure treatment the delays when implementing the project | 65.53 | 5 |
| Delay in the consent of the owner of the land to expand the road through its | 48.51 | 47 |
| Poor Relationship between the contractor and consultant | 46.38 | 53 |
| Cash flow problems faced by the contractor | 58.72 | 11 |
| Inflation in material prices on the market | 45.96 | 56 |
| Late deliveries of materials from suppliers | 46.38 | 53 |
| Supplier inefficiency | 49.79 | 42 |
| Claims problems | 55.32 | 19 |
| Delay in activities during implementation | 61.28 | 7 |
| Cost estimates are not accurate | 60.43 | 10 |
| Delay penalties | 48.51 | 47 |
| **Reasons related to the owner** | | |
| Changes in the contract by the owner during the construction process | 53.19 | 28 |
| Delay in monthly payments of contractor | 65.96 | 4 |
| Problems in purchasing land belonging to citizens | 53.62 | 27 |
| Misunderstandings with suppliers and subcontractors | 51.49 | 33 |
| Insufficient and incomplete feasibility studies | 57.02 | 15 |
| Design problems | 61.28 | 7 |
| **Reasons related to the consultant** | | |
| Error in site surveys | 56.60 | 16 |
| Error in approving the drawing by the consultant | 56.60 | 16 |
| Inadequate authority for decision-making by consultant | 51.91 | 32 |
| Mistakes in Consultant drawings | 54.04 | 25 |
| Weak consultant experience | 51.06 | 36 |
| Insufficient site information for the consultant | 57.45 | 14 |
| The inaccuracy of dimensions | 55.32 | 19 |
| Reasons related to the materials | | |
| Shortage of materials during implementation | 52.77 | 29 |
| Using materials does not conform to specifications | 54.47 | 23 |
| Unexpected and repeated adjustments in material specifications during construction | 51.06 | 36 |
| Material quality problems | 45.96 | 56 |
| Weakness in the site materials management | 51.06 | 36 |
| Rise in the prices of materials | 54.89 | 22 |
| Insufficient materials at the beginning of the project, causing delays | 51.49 | 33 |
| Do not put the material in its proper place, causing difficulty in transporting it inside the site | 50.64 | 39 |
| Damage of materials | 48.51 | 47 |
| Disposal of road construction waste | 51.49 | 33 |
| Delays in a materials test of and obtaining the results | 67.66 | 3 |
| **Causes of external delay** | | |
| The effects of weather, rain and high temperatures | 65.53 | 5 |
| Low temperatures | 47.23 | 50 |
| Political decisions and political realities | 69.79 | 1 |
| The economic crisis of the country | 68.94 | 2 |
| **Reasons for labor and equipment** | | |
| Deficiency of skilled and unskilled laborers | 55.32 | 19 |
| Poor manpower productivity | 60.85 | 9 |
| Injuries on site | 40.85 | 62 |
| Labor strikes | 35.32 | 64 |
| Absence of equipment | 48.94 | 46 |

| | | |
|---|---|---|
| Delayed delivery of equipment | 49.79 | 42 |
| Shortage in equipment availability | 52.34 | 31 |
| Insufficiency of operators | 54.47 | 23 |
| No adequate spaces for temporary and permanent equipment | 45.11 | 61 |
| Ineffective inspection of equipment | 56.60 | 16 |
| Not to provide services to workers through the implementation of external roads | 52.77 | 29 |
| Asphalt plants are far from work sites | 57.87 | 13 |
| Absence of workers or operators of equipment | 45.53 | 59 |
| Delays related to the project | | |
| Control the traffic on the site | 49.79 | 42 |
| Changes in site conditions | 46.81 | 52 |
| Unexpected Earth Conditions | 50.64 | 39 |
| Insufficient data collection about project site | 54.04 | 25 |
| Changes in site topography after design | 45.96 | 56 |
| Night working conditions and lack of sufficient work lights | 45.53 | 59 |
| Rework due to errors during construction | 49.79 | 42 |
| Theft of machines/ materials | 46.38 | 53 |
| Services for underground construction such as telephone and electricity lines, water pipes, oil pipelines, and rainwater pipes | 58.30 | 12 |
| Work environment / noise | 39.15 | 63 |

### *7.2The causes of delay from the consultant's point of view*

Table 5 shows all causes of delay and frequency of occurrence of each reason and its arrangement from the point of view of the consultant.
The top seven causes of delay identified by the consultants were:
The economic crisis of the country (F.I=12.77), Delays due to Cash flow problems faced by the contractor (F.I=11.06),Changes in the contract by the owner during the construction process (F.I=10.64), Delay in monthly payments of contractor (F.I=10.64),insufficient and incomplete feasibility studies(F.I=10.64),Delays in materials test and obtaining the test results (F.I=9.79),The effects of weather, rain and high temperatures(F.I=9.79).

**Table 5** Delay causes from the consultant' point of view

| | (F.I) (%) = Σ a (n/N) * 100/5. | RANK |
|---|---|---|
| **Reasons related to the Contractor** | | |
| Delay in delivering the site | 5.96 | 55 |
| Deficiencies in bidding by contractors/causing calling them to clarify it in the stage of the award of the contract | 9.36 | 9 |
| Failure treatment the delays when implementing the project | 6.81 | 44 |
| Delay in the consent of the owner of the land to expand the road through its | 6.38 | 51 |
| Poor Relationship between the contractor and consultant | 7.23 | 39 |
| Cash flow problems faced by the contractor | 11.06 | 2 |
| Inflation in material prices on the market | 8.94 | 14 |
| Late deliveries of materials from suppliers | 7.66 | 33 |
| Supplier inefficiency | 6.81 | 44 |
| Claims problems | 7.66 | 33 |
| Delay in activities during implementation | 9.36 | 9 |
| Cost estimates are not accurate | 9.36 | 9 |
| Delay penalties | 8.09 | 21 |
| **Reasons related to the owner** | | |
| Changes in the contract by the owner during the construction process | 10.64 | 4 |
| Delay in monthly payments of contractor | 10.64 | 4 |
| Problems in purchasing land belonging to citizens | 7.23 | 39 |
| Misunderstandings with suppliers and subcontractors | 8.09 | 21 |
| Insufficient and incomplete feasibility studies | 10.64 | 4 |
| Design problems | 8.09 | 21 |
| **Reasons related to the consultant** | | |
| Error in site surveys | 8.09 | 21 |
| Error in approving the drawing by the consultant | 7.66 | 33 |
| Inadequate authority for decision-making by consultant | 5.96 | 55 |
| Mistakes in Consultant drawings | 6.81 | 44 |
| Weak consultant experience | 6.38 | 51 |
| Insufficient site information for the consultant | 7.66 | 33 |
| The inaccuracy of dimensions | 6.81 | 44 |
| **Reasons related to the materials** | | |
| Shortage of materials during implementation | 7.23 | 39 |
| Using materials does not conform to specifications | 8.09 | 21 |
| Unexpected and repeated adjustments in material specifications during construction | 8.09 | 21 |
| Material quality problems | 5.96 | 55 |
| Weakness in the site materials management | 8.09 | 21 |
| Rise in the prices of materials | 9.36 | 9 |
| Insufficient materials at the beginning of the project, causing delays | 7.66 | 33 |
| Do not put the material in its proper place, causing difficulty in transporting it inside the site | 9.36 | 9 |
| Damage of materials | 7.23 | 39 |
| Disposal of road construction waste | 8.51 | 19 |
| Delays in a materials test of and obtaining the results | 9.79 | 7 |
| **Causes of external delay** | | |
| The effects of weather, rain and high temperatures | 9.79 | 7 |
| Low temperatures | 5.96 | 55 |
| Political decisions and political realities | 11.06 | 2 |
| The economic crisis of the country | 12.77 | 1 |
| **Reasons for labor and equipment** | | |
| Deficiency of skilled and unskilled laborers | 8.09 | 21 |
| Poor manpower productivity | 8.94 | 14 |
| Injuries on site | 5.96 | 55 |
| Labor strikes | 5.53 | 61 |
| Absence of equipment | 8.09 | 21 |
| Delayed delivery of equipment | 7.66 | 33 |
| Shortage in equipment availability | 8.09 | 21 |
| Insufficiency of operators | 6.81 | 44 |
| No adequate spaces for temporary and permanent equipment | 5.96 | 55 |
| Ineffective inspection of equipment | 8.94 | 14 |
| Not to provide services to workers through the implementation of external roads | 4.68 | 64 |
| Asphalt plants are far from work sites | 8.94 | 14 |
| Absence of workers or operators of equipment | 6.38 | 51 |
| Delays related to the project | | |

| | | |
|---|---|---|
| Control the traffic on the site | 6.38 | 51 |
| Changes in site conditions | 7.23 | 39 |
| Unexpected Earth Conditions | 8.51 | 19 |
| Insufficient data collection about project site | 6.81 | 44 |
| Changes in site topography after design | 8.09 | 21 |
| Night working conditions and lack of sufficient work lights | 5.53 | 61 |
| Rework due to errors during construction | 6.81 | 44 |
| Theft of machines/ materials | 8.09 | 21 |
| Services for underground construction such as telephone and electricity lines, water pipes, oil pipelines, and rainwater pipes | 8.94 | 14 |
| Work environment / noise | 5.11 | 63 |

## 8 Conclusions and recommendations

The delay causes in highway construction projects in Iraq are discussed in a field survey. It studied the frequency index of the causes of delay from the contractors' and consultants' points of view.64 causes of delay were identified. The identified causes are classified into seven groups. Factors affecting the delay of highway projects were concluded:

### 8.1 Contractor's and owner's view

1- Political decisions and political realities and the economic crisis of the country.These two factors are repeated in the contractorS'and owners'view,delay causes from the consultants'point of view.The reality and the unstable political situation in Iraq caused a volatile security situation, which leads to the closure of some roads and within cities, this situation leads to high cost of materials, lack of resources, delays in the arrival of raw materials to site and delays in the arrival of vehicles that transport asphalt mixtures to the site of the project,thus, will cause delay in the project.

2- Iraq is suffering from a terrible economic and financial crisis because of its dependence on oil as its sole source of foreign currency, in addition to the phenomenon of corruption pervasive in all aspects of the state and the policy of fiscal austerity adopted by the government, which caused the interruption of the implementation and completion of hundreds of infrastructure projects, such as road projects.

3- Delays in materials test and obtaining their results.This is due to the lack of technical devices entrusted with the inspection and the difficulty of obtaining construction materials conforming to the applicable specifications from natural quarries or from the asphalt supply plants. The weakness of quality control over the local product leads to dispersion of the contractor's effort and thus delays the project. On the other hand, it is necessary to establish an appropriate number of laboratories and adopt the field laboratory mechanism for the external and remote screens.

4- Delay in monthly payments of contractor.Difficulties in financing project by owner will result in delays in payment to contractors. This directly affects the completion of the project and cause time overrun as well as give rise to disputes between the various parties so owners should give special attention to pay the contractors on time.

5- Failure treatment the delays when implementing the project. The contractor's lack of experience in identifying the problems that occur during the implementation of the project, which cause delays and slow decisions and guidelines directly affect the delay. Therefore, Therefore, the contractor who has previous experience in the implementation of road projects should be selected.

6- Effects of weather, rain and high temperatures.It is meant to be a total or partial disruption of work due to rain, strong winds, or a significant drop in temperature, which prevents workers from doing their work.Therefore, the contractor must take into consideration the weather conditions when preparing the time plan for the project.

7- Delay in activities during implementation.There are several reasons to delay the project activities depending on the type of effectiveness, so the project manager should identify these reasons and deal with them quickly in order to reduce the total delay of the project.

### 8.2 Causes of delay from the consultant's point of view

In addition to the factors already appearing in clause 6.1,viz.Political decisions and political realities,the economic crisis of the country, delays in materials tests and obtaining their results, delay in monthly payments of contractor and the effects of weather, rain and high temperatures,other factors have appeared, according to the consultants' view which are:

1-Delays due to cash flow problems faced by the contractor.

Difficulties in financing projects develop because of lack of liquidity by the contractor. So the contractor must identify sources of funding and the need to develop a financial plan for each project beforehand. This helps predicting all non-nature conditions that occur during project implementation viz Lack of liquidity by the contractor, high worker wages, material prices or sudden breakdowns in equipment.

2-Changes in the contract by the owner during the construction process.

The existence of changes or amendments to the contract during the implementation (viz the change or modification of the designs or in the way of implementation or adding an item which is necessary or deleting an item which is not important) occurs most often because of the owner, so the impact of these changes on the project time should be assessed.

## References

1. A. A. Salunkhe, R. S. Pail. IJRET, 3 (1), 543-547 (2014).
2. S.Ahmed, S. Azhar, P. Kappagantula,. Proc. 39th Annual Conf. ASC, Clemson Univ.(2003) .

3. Ng S.T., R.M. Skitmore, K.C. Lam, A.W.C. Poon. I. J. P M. 22 (2) 139–146(2004).
4. M. Latham,. "Constructing the Team", HMSO, London, 87-92.(1994).
5. H.A Awad., G.K Hindawi.. Journal of Babylon University,14(4),483-494. 2007)
6.A.G. Bekr. JJCE. 9(2),149-162. (2015).
7.A.H, Al-Momani. International Journal of Project Management, 18 (3) (2000).
8. J. Al-Najjar, Islamic University, Gaza (2008
9. A, M Kasimu, D. I Abubakar, IJCRB. 4(2), 785- 794. (2012)
10. Soo-Yong Kim et al ,6th ICCEPM .(2015)
11. A.S Msafiri. AJCE, 3(3), 51-63. (2015)
12. F. Remon, et al.Alexandria Engineering Journal, 55, (2),1515–1539(2016).
13. Jeyakanthan et al,www.iiirr.ucalgary.ca/files/iiirr/190.
14. A .Nabil, A. S. Zaydoun IJET, 4 (2) 288-293 (2015),.
15. M. Ibrahim, B. Amund, D. Nabil. ASCE,.28:300-310,(2012)
16. U. Sekaran,. Research methods for business: a skill-building approach. New York: John Wiley & Sons (2003).

# An empirical analysis of the relationship between cost of control activities and project management success

*Samiaah* Al-Tmeemy [1,*] and *Bassam* Al Bassam [2]

[1] Surveying Department, Technical Institute, Baquba, Iraq
[2] Civil Engineering Department, Al Israa University College, Baghdad, Iraq

**Abstract.** To achieve the objectives of continuous improvement programs, construction managers must link the achievement of quality with cost. This paper aims to associate project management success (PMS) with cost of control (COC) activities in an empirical manner. Thus, the purpose is to determine the extent to which COC activities impact PMS. Quantitative method was adopted to collect data from Malaysian building companies using postal and email surveys. Hypothesis is tested using correlation and simple linear regression analysis. The findings of this study indicate that COC activities are positively associated with the PMS. The empirical evidences obtained from this research, provides financial justification for all quality improvement efforts. This can assist building contractors to enhance the success of project management by reducing the level of business failures due to poor quality, cost overruns, and delays.

## 1 Introduction

The objective of continuous improvement programs is not only to meet customer's requirements, but to do it at the lowest cost as well. Thus, to achieve the objectives of continuous improvement programs, construction managers must link the achievement of quality with cost [1,2]. This is because the achievement of quality target at a prohibitive cost is of little use [3]. In the same vein, downgrading quality to achieve competitive cost is equally inappropriate [3]. This linkage gives rise to cost of quality to be one of the most effective tools that harmonizes the economics of the company with product quality [4].

Quality costs are associated with two categories of quality-related activities: cost of control (COC) activities and cost of failure control (COFC) activities [5,6]. While COC refers to the cost of doing things right (i.e., prevention and appraisal costs), COFC comprises the cost of getting them wrong (I.e., failure cost). It must be remembered that COC are unavoidable costs that must be borne by the design and construction companies if their products/services are to be delivered right the first time [7]. On the other hand, COFC are almost avoidable in construction, as most originate from ineffective management practices [7]. However, quality improvements are often difficult for managers to justify and accept, mainly because the benefits are not always apparent from a strategic and economic perspective [8].

In general, any serious attempt to improve quality must take into account the costs associated with achieving quality [1,2]. Therefore, managers must employ more quantitative approaches to improve the clarity and precision of the relationship between COC and project management success (PMS).

Several studies showed the positive influence of COC activities on the level of quality and productivity improvement. However, limited research has been conducted to show conclusively the extent to which COC activities are important to project success. In addition, no research has been published to address the impact of quality control activities on project management success. The current research is concerned with the relationship between COC and PMS. Specifically, this research determines the extent to which the COC activities impact PMS in terms of adherence to budget, schedule, and quality target. According to Al-Tmeemy et al. [9], the PMS can be measured by three criteria; adherence to quality target, adherence to Budget and adherence to schedule .

Managing quality costs and understanding the relationships between these costs and project management performance is important to make better decisions related to improving project management performance and enhancing success. This relationship provides justification for increasing COC expenditures to improve the return on quality investments.

* Corresponding author: sam_saf61@yahoo.com

© The Authors, published by EDP Sciences. This is an open access article distributed under the terms of the Creative Commons Attribution License 4.0 (http://creativecommons.org/licenses/by/4.0/).

## 2 Relationship between COC and PMS

A critical factor in successful management of building projects is effective quality management [10]. A poor quality management system connotes that information gathering, reporting and management in a project become uncoordinated, thereby necessitating multiple re-drawing and re-keying of information [7]. Ultimately, wasted time, unnecessary costs, increased errors and misunderstanding rework and non-conformance are accepted as inevitable features of the construction process [7,11-13]. Controlling quality costs leads to reduction in building cost and wasted time in correcting errors and rectifying failures. Thus, the greatest savings can be derived from reducing failure areas and minimizing defects. Therefore, implementing COC activities should improve quality which will ultimately mitigate costs through reduction in internal and external failure costs. Undeniably, reducing failure costs decreases wasted time and materials and such resources are spent prudently according to contract requirements. Eventually, orientations along the COC activities will have consequences for project management performance outcomes. Thus, the following hypothesis is proposed:

H1: There is a significant relationship between COC and PMS.

## 3 Research methodology

The main aim of this research is to establish empirically the extent to which COC activities impact PMS and each of its criteria. Hence, the positivist paradigm –appears to be the most suitable approach for this study, implying that the research approach of deductive reasoning must be used to derive a conclusion from a known premise. As this research adopts the positivist philosophy, the quantitative method was considered necessary for data collection. Meanwhile, a cross-sectional survey has been selected to gather data for the empirical study. A variety of statistical procedures were employed in data analyses, including descriptive statistics, goodness of measurements, correlation and regression.

The population was limited to contractors within Kuala Lumpur. The target respondents of this research were drawn from Construction Industry Development Board Malaysia (CIDB). According to CIDB statistics in 2008, there were 1196 active contractors involved in construction activities in Kuala Lumpur. Of this total, 594 contractors were chosen using stratified sampling method. The sample contained 95 contractors of G6 and 499 contractors of G7.

Data collected from the questionnaire survey was analyzed qualitatively. Various statistical procedures were used; these include descriptive statistics, validity and reliability, correlation and simple linear regression. Correlation refers to the strength of relationship between two or more variables [14]. Pearson's correlation coefficient is useful for determining the strength and direction of the association between two variables [15]. Therefore, Pearson's correlation coefficient was considered appropriate and thus, applied to assess the existence of relationships between COC and PMS.

Even though correlation is a useful research tool for examining the relationships between variables, this test has been noted to provide little information on the predictive power of the individual variables [15]. Nevertheless, regression analysis helps in exploring the effects of independent variable(s) on one dependent variable [16]. Therefore, regression analysis was applied to quantify the relative contribution of COC to enhance project management success.

As this research considers only one independent variable, a simple linear regression was used [17]. By applying simple linear regression with a single independent variable, the coefficient conveys how much the dependent variables are expected to increase (if the coefficient is positive) or decrease (if the coefficient is negative) when that independent variable increases by one.

To apply a valid regression model on the data several regression assumptions must be met [18]. Without verifying if the data have met the regression assumptions, results may become misleading. These assumptions represent the requirements of the underlying statistical theory [18] and a violation of any of these assumptions can affect the performance of the model [19]. The assumptions include: normality, linearity and homoscedasticity [18].

## 4 Results and discussion

### 4.1 Response rate

Of the 594 questionnaires dispatched to the selected sample, 153 were satisfactorily completed, pegging the total response rate at 25.7%. This figure is acceptable according to Akintoye [20] and Dulaimi et al. [21]. Both of whom have stated that normal response rate in the construction industry for postal questionnaires is within the range of 20 -30%.

Frequency distribution is conducted to show four main profiles namely: educational level, age, occupation and period of experience as shown in Table 1. The General Respondents' Demographics (GRD) revealed that the majority of (90.2%) held a Bachelor's degree as shown in Table 1. The highest frequency of respondents (40.5%) was aged between (40 – 49) years. Regarding the occupation of respondents, 'manager' position accounted for the largest percentage (55.6%). Majority of respondents (96.1%) had more than 10 years of experience.

**Table 1.** Descriptive Statistics of GRD.

| GRD | Groups | Frequency | Percent | Cumulative Percent |
|---|---|---|---|---|
| Educational | Diploma | 0 | 0 | 0 |
| | Bachelor | 138 | 90.2 | 90.2 |
| | Master | 13 | 8.5 | 98.7 |
| | PhD | 2 | 1.3 | 100 |
| Age | 20-29 yrs | 12 | 7.8 | 7.8 |
| | 30-39 yrs | 59 | 38.6 | 46.4 |
| | 40-49 yrs | 62 | 40.5 | 86.9 |
| | 50-59 yrs | 10 | 6.5 | 93.5 |
| | 60 + yrs | 10 | 6.5 | 100 |
| Occupation | Quantity Survey | 5 | 3.3 | 3.3 |
| | Project Engineer | 5 | 3.3 | 6.5 |
| | Quality Control | 7 | 4.6 | 11.1 |
| | Site Engineer | 41 | 26.8 | 37.9 |
| | Manager | 85 | 55.6 | 93.5 |
| | Director | 10 | 6.5 | 100 |
| Experience | 5 - 9 yrs | 6 | 3.9 | 3.9 |
| | 10 - 14 yrs | 37 | 24.2 | 28.1 |
| | 15 - 19 yrs | 25 | 16.3 | 44.4 |
| | 20 - 24 yrs | 60 | 39.2 | 83.7 |
| | 25 + yrs | 25 | 16.3 | 100 |

### 4.2 The correlation

The results in Table 2 (r=0.595, P=0.000) offer strong support for accepting H1, which stated that COC is significantly associated with PMS. Therefore, the effective implementation of COC activities will enhance PMS. The project manager is responsible for delivering the project within the agreed time, cost and quality.

**Table 2.** Correlation between COC and PMS.

| | Correlation | Sig. (2-tailed) |
|---|---|---|
| PMS | 0.595* | 0.000 |
| * Correlation is significant at the 0.01 level (2-tailed). | | |

### 4.3 Simple regression analysis

Regression can be used to discover whether the coefficients on the independent variables are truly different from zero. This indicates that independent variables actually have an effect on dependent variables, or alternatively, any apparent differences from zero are merely due to random chance [18]. The relationship between independent and dependent variables is expressed in the form of an equation. This equation yields a predicted or fitted value of dependent variable for each value of independent variable in the data set, holding all other factors fixed. The dependent and independent variables are entered in the proposed model in a linear way. However, the relationship between the variables is not necessarily linear [18].

**Table 3:** Regression Analysis Results for COC and PMS

| Model Summary of PMS | | | | | |
|---|---|---|---|---|---|
| Corr. R | 0.595 | | | | |
| R Square | 0.354 | F | | | 82.880 |
| Adjusted R Square | 0.350 | Sign. F | | | 0.000 |
| St. Error | 1.4092 | | | | |
| Coefficients | | | | | |
| Variable | B | Std. Error | Beta | t | Sig. |
| COC | 0.291 | 0.032 | 0.595 | 9.104 | 0.000 |
| (Constant) | 3.205 | 1.148 | | 2.791 | 0.006 |

The results of this study are in line with those of previous studies [22-24] that supported the philosophy of "zero defect" (i.e., 100% conformance to contracted requirements). These studies confirmed that COC implementation is the most important factor for eliminating failure and adhering to quality target in building projects. The findings likewise strengthen the suggestions offered by Abdul-Rahman et al. [25], who stated that almost all construction-related non-conformances could be avoided either by timely inspections or by the use of more experienced and skilled employees. Moreover, the results are consistent with the finding of Winner et al. [26], who reported that scrap and rework can be reduced by 75% through product and process design optimization. The findings of this study are also in line with Chong and Low [27], who claimed that most of defects in buildings were preventable by implementing COC activities to ensure better design, better workmanship, better materials and better maintenance.

The value of $R^2$ for the generated model is 0.354 (Table 3), implying that COC accounts for 35.4% of the variation in PMS. The value of $R^2$ is close to the value of adjusted $R^2$, indicating that the model can be generalized with shrinkage only by 0.4%. The result for ANOVA shows a significant F statistic (F=82.880, p= 0.00) at 0.01 level. This result indicates that the regression model is statistically significant.

The t test for COC is significant at 0.01 level (t=9.104, p=0.00), as shown in Table 3. Therefore, the null hypothesis is rejected and the alternative hypothesis, which states that coefficient B is significantly different from zero, is accepted. This indicates that COC implementation has a significant impact on PMS. The B coefficient of 0.291 indicates that PMS is predicted to increase by 29% units when COC increases by one unit. This relationship can be expressed by the following equation:

$$PMS = 3.205+0.291\ COC \qquad (1)$$

This model expresses a static relationship and the optimum point will vary according to the nature of the project and environment. Nevertheless, it must recognize the dynamics of changing quality costs over time. It has been noticed that the level of COC expenditures appears to have no immediate impact on internal failure costs [4,28-30]. This is because of the time lag between cause and effect [28], as well as COC's positive correlation with internal failure costs in subsequent months [29]. The increase in COC, over a longer timeframe, continuously decreases failure costs and improves quality levels; subsequently, investment in COC yields a permanent reduction in failure costs in the total amount of COQ. Recognizing these dynamics, the objective is not to determine the optimal level of COC, but to link the identified process improvements through implementation of COC activities and enhancement of PM success through adherence to cost, schedule and quality target.

In construction projects, the timing of implementation of quality management is important [31]. Hendrickson [32] claimed that the decisions regarding quality should be made during the preliminary stages of construction and the design and planning stages. He added that during these preliminary stages, the component configurations, material specifications and functional performance are decided. Abdul-Rahman [31] shared the same point of view when he asserted that quality management should start from the inception up to the commission stages. The earlier implementation of quality management in a project is, the more likely problems can be recognized and savings realized. Abdul-Rahman [31] further explained that at the beginning of a project, the cost of non-conformance is nil and the main contractor/project manager should exploit the situation by implementing sound planning, anticipating problems and difficulties, and building trust between work contractors. Love and Irani [7] likewise asserted that COC must be borne by design and construction organizations if their products/services are to be delivered right the first time.

## 5. Conclusions

This study indicates that COC activities significantly enhance PMS. Consequently, PMS is expected to be affected positively by the implementation of COC activities; specifically, PMS is predicted to increase by 29.1% when COC increases by one unit.

This study provides knowledge and evidence that investment of resources in improving quality could be justified by the dramatic improvement in return on quality investments. This knowledge would enhance the awareness of contractors, managers, engineers and site workers on the importance of COC, providing a unique opportunity to better understand the extent to which the implementation of COC activities can impact PMS. This awareness can enlighten all concerned parties of the significance of COC in eliminating and preventing quality-related problems at an early stage. Subsequently, the company is able to establish a culture of commitment to perform the work correctly at the first time. Without a doubt, this culture will assist in reducing the level of project failure and lead to project management success.

The research has achieved its aim of investigating the relationship between COC activities and PMS and determining the extent to which these activities impact project management success. However, some limitations of this study are worth mentioning, as explained in the following sections. The first limitation is that this research is confined to building construction within Kuala Lumpur, and exclusive focus is placed on the top two grades of construction companies (i.e., G6 and G7). It is possible that different viewpoints may be found among other respondents, such as same-grade companies in other parts of Malaysia. Therefore, the results obtained may not be representative of the population of Malaysian construction and the situation in the rest of the world. However, the results have indicated that the models can be generalized to the population with bias of 0.0%– 0.6%.

The second limitation is that this research uses Juran's model where the appraisal and prevention costs are combined into COC, as it is difficult to decide which activities stand for prevention of quality failures since almost everything is well-managed [33]. However, it is well known that the two costs of appraisal and prevention maintain different behaviors and implications for cost control. Therefore, they are supposed to be considered separately.

The third limitation is that the main concern of this study is on COC and not COFC. Limitation on time and resources, COC was mainly focused in this study. In fact, the COFC helps to identify the improvement opportunities and focus on areas of high expenditure and wastage. Subsequently, this will contribute to PMS. Moreover, the relationship between COC and PMS is dynamic; therefore, it is expected that this relationship will change over time.

The limitations outlined above indicate several aspects where there is potential for improvement in the construction building industry. As an extension to this study, additional studies are needed to assess more accurately the effect of implementation of COC activities on PMS over time. For further studies, it would be interesting to take into account the effect of tracking and recording the failure cost on the performance of the construction projects. An evaluating and reporting quality failure cost is perceived as a useful indicator of project performance and provides opportunities for quality improvements and cost reduction.

## References

1. Abdelsalam, H.M.E.and Gad, M.M. International Journal of Project Management **27**, 501-511 (2009).
2. Schiffauerova, A. and Thomson, V. International Journal of Quality & Reliability Management **23**, 647-669 (2006b).
3. Hall, M. and Tomkins, C. Construction Management and Economics **19**, 727-740 (2001).
4. Chang, C. Journal of the Academy of Business and Economics **II**, 172-187 (2005).
5. Hansen, D.R.and Mowen, M.M. (Thomson, South-Western, United State of America, 2006).
6. Feigenbaum, A.V. Total quality control (McGraw-Hill Inc., New York, 1983).
7. Love, P.E.D.andIrani, Z. Information and Management **40**, 649-461 (2003).
8. Czuchry, A.J., Yasin, M.M.and Ltle, G.S. Industrial Management & Data Systems **99**, 362-366 (1999).
9. Al-Tmeemy, S.M.H.M., Abdul-Rahman, H.andHarun, Z. International Journal of Project Management **29**, 337-348 (2011).
10. Achi, F.O., Onukwube, H.N. and Oluwaseyi M, Atlanta USA, 2007).
11. Abdul-Rahman, H. Construction Management and Economics **13**, 23-32 (1995).
12. Alwi, S., Keith, H.and Sherif, M. Asia Pacific Building and Construction Management Journal **6**, 2-6 (2001).
13. Smallwood, J.and Rossouw, J.-H. in (COBRA ) (Dublin, Ireland 2008).
14. Bewick, V., Cheek, L.andBall, J. Critical Care **7**, 451-459 (2003).
15. Field, A. Discovering (Sage Publications Ltd, London, 2000).
16. Kerlinger, F.N.andLee, H.B. (Harcourt College Publishers, Orlando, 2000).
17. Tabachnick, B.G.and Fidell, L.S. (Pearson Education Inc., United States of America, 2007).
18. Hair, J.F., Black, W.C., Babin, B.J.and Anderson, R.E. Multivariate Data Analysis (Pearson Prentice Hall, New Jersey, 2010).
19. Wang, G.C.S.andJain, C.L. (Graceway Publishing Company, Inc., New York, 2003).
20. Akintoye, A. Construction Management and Economics **18**, 77-89 (2000).
21. Dulaimi, M.F., Ling., F.Y.Y.and Bajracharya, **21**, 307-318 (2003).
22. Crosby, P.B. (McGraw-Hill, New York, 1979).
23. Kazaz, A., Birgonulb, M.T.and Ulubeylia, S. **40**, 1356-1365 (2005).
24. Low, S.P.and Yeo, H.K.C. International Journal of Quality & Reliability Management **15**, 329-349 (1998).
25. Abdul-Rahman, H., Thompson, P.A.and Whyte, I.L. International Journal of Quality & Reliability Management **13**, 48-60 (1996).
26. Winner, R.I., Pennell, J.P., Bertrand, H.E. and Slusarczuk, M.M. (Institute for Defense Analyses, Alexandria, Virginia, 1988).
27. Chong, W.-K.andLow, S.-P. Journal of Performance of Constructed Facilities **20**, 213–221 (2006).
28. Campanella, J. and Corcoran, F.J. (ASQC Quality Press, Milwaukee, WI, 1983).
29. Ittner, C. Operations Research **44**, 114-130 (1996).
30. Love, P.E.D., Mandal, P.andLi, H. Determining the causal structure of rework in construction projects. Construction Management and Economics **17**, 505-517 (1999).
31. Abdul-Rahman, H. International Journal of Quality & Reliability Management **14**, 464-281 (1997).
32. Hendrickson, C. (Prentice Hall, Pittsburgh, 2000).
33. Aoieong, R.T., Tang, S.L.and Ahmed, S.M.A Construction Management and Economics **20**, 179-192 (2002).

# The effects of stakeholder's engagement and communication management on projects success

*ISRAA* Fadhil Alqaisi / *(PMP, CQE)/Master in Surveying Engineer*

*Ministry of Construction and Building / Engineering Construction Office*

**Abstract.** Managing stakeholders' expectations and interests is key to a project's success. So, identifying stakeholders at the beginning of the projects, recognizing and managing their needs and expectations will contribute to the creation of a suitable environment and be catalyst for success. This can be achieved through the establishment of appropriate and timely communication that meets the requirements of stakeholders. This includes providing the decision makers with the required data and receiving feedback to ensure alignment among project objectives and stakeholders expectations. This paper mentions one of the projects which neglected the proactive planning and management of stakeholder's requirements that causes waste in time and resources and many issues that appear as a result of poor planning, and the lessons learned from it.

Key words: Stakeholders, Stakeholder Register, Engagement, Expectation, Strategic Objectives, Communication.

## 1 Stakeholders

### 1.1 Stakeholder Definition

Stakeholders are individuals, groups, or organizations who may get affected by, or perceive themselves to be affected by a decision, activity, or outcome of a project (1). Stakeholders contribute in project process and activity, conceive the idea, develop project plans, execute the plans, make decisions, etc. So, the management of their requirements and expectations is critical to project success.

It is essential to identify key stakeholders at the start of the project and create the stakeholders register that will be updated throughout the life of the project. This register will contain information about stakeholders help in prioritizing them, prioritization is important because it is useless to devote the same time and effort to all stakeholder.

### 1.2 Stakeholder register

Stakeholder register is the central document for tracking project stakeholders' information and should contain as a minimum the following: -

1. Contact information like stakeholder name, location, job title....

2. Category: Stakeholders will be categorized according to their (power, interest, influence and impact)

3. Project Role

4. Status

5. Priority: based on prioritized rating the project manager will determine the amount of efforts and time that should be devoted to each stakeholder

The stakeholder register should be updated as an ongoing activity to add new stakeholder, remove disbanded stakeholder, updates, etc.

The stakeholder register will be the key input to stakeholder engagement plan and communication management plan.

Inputs to communication management plan include (2)

1. Contact information
2. Priority
3. Project role
4. Status

© The Authors, published by EDP Sciences. This is an open access article distributed under the terms of the Creative Commons Attribution License 4.0 (http://creativecommons.org/licenses/by/4.0/).

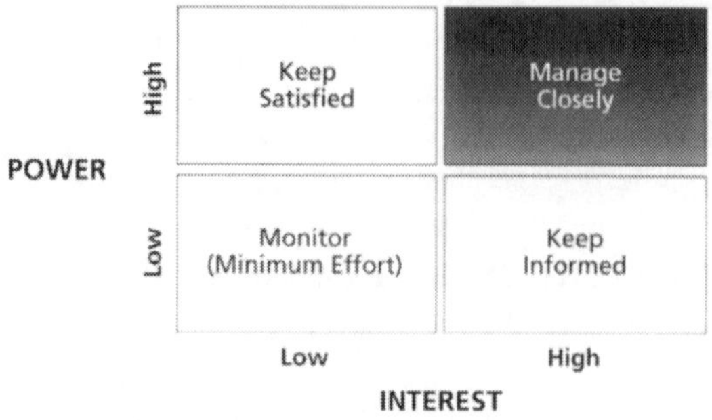


**Fig. 1.** Power and Interest Matrix (1)

**Interest** indicate the level of concern of stakeholder about project outcomes

**Power** is the stakeholder level of authority

### 1.3 Stakeholder Engagement Assessment Matrix

In this matrix the project manager identifies the current status of each stakeholder and the desired status, this information will be included in the stakeholder register. The matrix contains six columns (as shown in Table 1.)

Unaware: The stakeholder has no idea about the project and its outcomes.

Resistant: The stakeholder opposed to the project or its manager.

Neutral: The stakeholder neither resistant nor supportive to the project.

Supportive: -The stakeholder agrees with project goals and its outcomes.

Leading: The stakeholders actively support the project

**Table 1.** Stakeholder Engagement Assessment Matrix (1)

| Stakeholder Name | Unaware | Resistant | Neutral | Supportive | Leading |
|---|---|---|---|---|---|
| Stakeholder | C | | | D | |
| Stakeholder | | | C | D | |
| Stakeholder | | | | D C | |

## 2 Stakeholders Groups

### 2.1 Project team stakeholder

Effective management of a project team starts with a kick off meeting. In this meeting, the project goals and deliverables and their importance to the organization will be clarified (2). In addition, kick off meeting helps in building team relationships and commitment to the project.

Later the project team member will be observed to address any issue that may occur due to poor performance, lack of proper direction or motivation...etc.

The project manager may use personal conversation with individual team member to reach the best solution.

### 2.2 Executive stakeholder

Executive stakeholders approve the projects, fund the projects, terminate the projects and decide if the projects are complete. They have the majority of power on project scope and deliverables. They include:

1. Customer

2. Sponsor

3. Executive council

4. Supplemental executive

Executive stakeholders have a lot to keep track of and some of them may have no experience in project terminology so the best approach is to convert project language into executive language (as shown in Table 2.).

**Table 2.** Convert Project Language into Executive Language (2)

| Attribute | What It Means to the Project Manager | How to Describe It to an Executive |
|---|---|---|
| Work breakdown structure | A decomposition of the project into the smallest assignable tasks. The key document to feed the project management plan, budget, schedule, and so on. | A planning document that helps the project team ensure that all project work is accounted for and organized. |
| Project management plan | The primary document used for project planning and execution. Many sub plans, such as the quality plan and the scope management plan, feed into the project management plan. | The central document to integrate all aspects of project execution. |
| Earned value management | A technique to compare cost of performed work, cost of scheduled work, actual cost of work performed, and other key project metrics. | An early warning system to identify deviations from the planned time line and budget. |
| Project charter | A document that formally permits the project manager to perform certain tasks and make certain decisions. Typically, known as the primary document to launch a new project. | A document to ensure that everyone is in agreement about the project's expected direction and the role of the project manager. |
| Change request | A formal document that can be a corrective action, preventive action, defect repair, or other adjustment to scope, budget, and so on. | A form used to ensure that all the right people approve project changes and are aware of the changes. |

It is critical to understand the important aspect and issue for each executive and this can be achieved through individual conversation to solicit feedback from stakeholders about questions asked by the project manager like:

- What is the most important aspect of this project for you?
- How often would you like to be updated?
- How much detail do you want in the project reports you receive?
- What is the best way to communicate with you (for example, e-mail, telephone, some sort of shared server)?
- What are the watch-outs the project team should be on the lookout for?
- What are your expectations from the project manager and the project team?
- Are there other executives we should talk to about this project?

### 2.3 Other stakeholders

Vendors

Consultants

Governmental agencies

## 3 Communication

Communication can be viewed as a metaphorical 'pipeline' along which information is transferred from one person to another (3). It is the lifeblood of any system of human interaction as without it, no meaningful or coherent activity can take place (4). Communication doesn't mean agreement to what has been said.

Consensus may be facilitated by a variety of communication-enhancing tools & forms of information. Several of the formal documents frequently discussed in project management include the following (5)

- Project charter.
- Communications management plan.
- Project stakeholder register.
- Organizational strategic plan.
- Project management plan.
- Gantt charts and other project time lines.

Each of these documents, and many other project documents like these, serves an important communication function. These documents communicate among the relevant stakeholders' important aspects of the project. Communication may be verbal, non-verbal, written and visual. Each type has its advantage and disadvantage.

### 3.1 verbal communication

While verbal communication (the use of spoken word to send and receive messages) is an important way to discuss difficult to understand issues, verbal communication may lead to misunderstanding of some phrases or words, especially in different culture.

In addition, verbal communication occurs in real time which may require immediate response and answer to questions being asked and the project manager may not be well prepared.

### 3.2 Non-verbal communication

Much of communication happen non-verbally. Non-verbal communication can reinforce the message and it includes tone of voice, facial expression, speed of speech, etc.

Non-verbal communication may be misunderstood due to incorrect interpret because many people do not have the required skill and experience to interpret the body language.

### 3.3 Written communication

Written communication helps to communicate ideas with proper time but if we desire to benefit from this type of communication, it requires from both parties to be engaged at the same time. Also, it is clear that non-verbal communication cannot be presented.

### 3.4 Visual communication

Visual communication includes images, charts, graphs, and a variety of tools used to visually display project information. Specific stakeholders are likely to understand visual communication better than other forms of communication. Visual imagery can communicate some information more concisely than words.

## 4 Managing difficult stakeholders

Projects almost handle difficult cases that should be dealt with but are not so easy. The kind of change can cause discomfort. The skilled project manager has awareness of this and consider project dispute and differences as opportunities to prove his or her excellent skills in project management which include the power of dealing with challenging issues.

So, it is important to know the groups that the stakeholder belongs to (2)

1. Stakeholder whose support is prefer to have.

2. Stakeholder whose support is a crucial for project success.

It is obvious that the support of the stakeholder who belongs to the first groups is not required for the project progress but it is better to gain their commitment and support. Whereas the stakeholder who belongs to the second groups is very important because of their ability to cancel or delay the project, and this make the project manager strive to build good relationship with this stakeholder no matter how complex. The first step is to understand the root cause of their disagreement, sometimes the cause is obvious and other times it requires deep investigation. The Project managers must put themselves in the stakeholder's situation, and he can do this by answering the following few questions

- What are the causes that make the stakeholder's argument valid if it valid?
- What facts and information the stakeholder points to that would support the position of his / her?
- What activities may increase the stakeholder's dispute—and make him or her even more opposed to the project's direction?
- Is there anything that can be done to fairly decrease the dispute?

Project manager can be benefit from engaging the executive sponsor in three ways:

First, the sponsor will be aware of the potential problem and can contribute with the project manager to manage the issue before it become complex.

Second, it is possible that the sponsor may have deferent preview on the project and this can put spot of light on the root causes.

Third, contribution work with the executive sponsor provides important support for the project manager because most of the time project managers do not have formal authority and working with the executive sponsors can make things happen easily if necessary. So, it is helpful to the project manager to involve someone with formal position authority when dealing with difficult stakeholders.

Each stakeholder dispute has unique characteristics. A plan should be developed in order to decrease each dispute. This plan should be developed with the engagement of the executive sponsor and it might include a wide range of choices such as the following:

- Conduct a meeting with each difficult stakeholder in order to reach an agreement.
- Try to find if there are any project details that may not be known by the difficult stakeholder and communicate it to him.
- If there are any project details that the stakeholder cannot understood when he previously informed it must be re-communicate.
- Engage the difficult stakeholder to provide input for resolving the disagreement.
- Engage a third party who have a good relationship to both the project manager and the difficult stakeholder, or one of the Executive Council members.
- Join additional stakeholders to the project who can make a balance against the difficult stakeholder.
- Tactically try to remove those difficult stakeholders from the project.

Finally, the best plan of action for each situation should be developed by the good contribution of the project manager and the executive sponsor using their creativity and awareness.

## 5 Obtaining stakeholders Buy-In

Earning stakeholder buy-in is a continuous process, not a single activity.

The steps as shown in the Circle of Support process is (6)

- Engage stakeholders in the process of the development and implementation of concepts and ideas. People are support what they contribute with.
- Observe stakeholder and recognize if there any different between the verbal and nonverbal signs project and this will help project manager to be aware of the best response.
- Respond to the concerns of the project stakeholder.

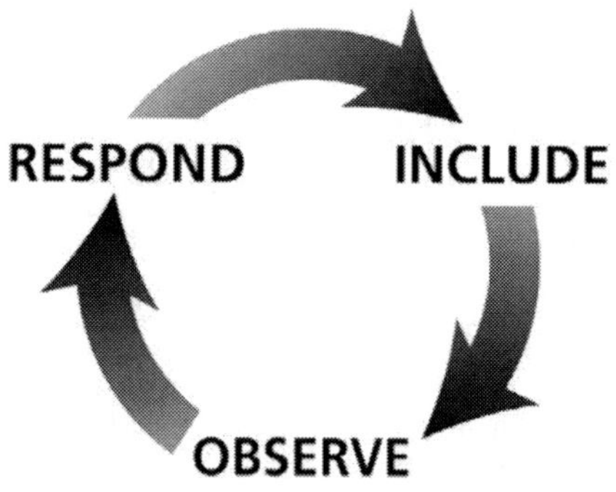


**Fig. 2.** Circle of Support Graphic **(8)**

### 5.1 Involving stakeholders

Two major benefits can be gained from engaging stakeholders in the project:

#### *5.1.1 Personal Ownership*

People in nature prefer their own ideas and concept. So, the stakeholders be able to insert their concept and ideas into the project by possibly modifying the project's planned direction, revising the way of the project execution, or providing input into the project issues and decisions when they involve in the project.
However, it is the project manager responsibility to clarify for each stakeholder the particular areas where their involvement is required and the domains where decisions have been taken. Stakeholders like to know how they can contribute to the project success, and this clarification will assist them focus their efforts on the domains that will be appraised by the project team.

### *5.1.2 Robust Ideas*

Project feedback can be increases by including and engaging additional project stakeholders and this will make the project manager reach to the different viewpoints, cultural diversity, experiences and other assets that each stakeholder has it.

Through these diversity more stable project solutions are developed. On the other hand, there is a risk appear from including additional number of people because of the increasing of the number of communication channels also the complexity of managing the stakeholder's requirements may be increased. And to prevent the large number of stakeholders providing feedback and ideas on each domain of the project, the project manager should behave tactically to deploy each stakeholder.

To benefit from stakeholder's feedback, the project manager should engage individuals from a different background, field of skills and knowledge, and exercised experiences. Project managers should utilize their awareness and understanding of the project environment to make sure that a diverse and comprehensive group of individuals are engaged to be stakeholders of the project and included in the project activity. (7)

### 5.2 Observe Stakeholders

Observe the stakeholders is the second step in the Circle of Support process where the project manager tries to understand the level of support of each stakeholder for project goals and objectives. Observing stakeholders required effort and time which differ from one stakeholder to another depending on the criticality of them to the project support.

Observing should consider both verbal and nonverbal communication.

### 5.3 Respond to Stakeholders

In this final step in the Circle of Support process the project manager will respond depending on the observation result. It will be a big mistake if the project manager responds before including and observing stakeholder's trends and attitudes

Generally, People tend to show one of the three situations shown in Fig. (3) or some combination (e.g., individual may be unsure but tending toward resistance).

The project manager responds differently depending on the stakeholder position on the continuum of support. Also, responses differ within each spot on the scope.

Fig. 3. Resist/Unsure/Support Continuum (8)

### Case study (1)

Poor communication and management of stakeholder expectations were identified in an important project and resulted in negative impacts.

The design and implementation work of an integrated residential complex have been assigned to one of the contracting companies and agreed on a fixed price per square meter based on technical specifications and drawings.

After months of effort, the final designs of the complex were completed (the work was done in the office only) a team of specialists go to the project site for the purpose of marketing products for the project and obtain the required revenue.

The surprise was the final beneficiary (residents of the area and neighboring) unwilling to buy these apartments because of their lack of compatibility with the culture of the region, so the decision is to re-designs to suit the requirements of the final beneficiary and to convert the complex to houses rather than apartments (the end user requirements had been neglected).

After the re-design and for the second time a team of specialists go to the project site for marketing and laying the foundation stone, a number of residents objected that this land belongs to them and the project cannot be started. (This time, the ownership of the planned land has been omitted).

### Case study (2)

One of the projects for training in the field of design

1. The training needs of the design department were determined (in an engineering office (to be able to compete with similar companies and departments and to remain in the market

2. By analyzing these needs, priorities were determined and the initial plan was developed but we were not able to implement this plan because it has been rejected from senior management according to shortage in funding.

3. This plan was replaced with the alternative plan, so the work begins with the analysis of stakeholders who have the power and interest, and the development of the necessary plans and strategies for the purpose of

transferring them to support the implementation of the plan.

4. After implementing the plan and conducting the required contacts, a cooperative relationship was established and the trainers were obtained without charge to train the employees of the design department on the required program and implement the plan to the required extent.

## 6 CONCLUSIONS

1. Lack or weakness of effective planning for all aspects of the project, including managing the expectations of stakeholders and identifying ways of communicating with them, causes major problems in the project, which result into project closure sometimes

2. Due to the absence of an official authority for the project manager and for the purpose of being able to manage the project efficiently and effectively, the project manager should have soft skills in addition to knowledge in the field of project management processes and practices.

3. Working in the project as a team contributes to overcoming many of the obstacles. It is so important to take into account that some members may be from different geographical locations, which requires the development of plans for appropriate communications.

## 7 Recommendations

1. It is highly recommended to involve project managers in training courses in the field of body language as an important part of the communication process as incorrect interpreting may have negative effects on the course of the project.

2. Giving the appropriate importance to stakeholders' engagement as an important part of the project planning process, which starts with the early beginnings of the project and continues throughout the life of the project.

3. Based on the attitudes and status of stakeholders who have the power and influence and try to gain their support for the project and identify their requirements and expectations and work side by side to align them with the goals and objectives of the project.

4. Work on establishing a database of lessons learned for each project that is accessible to avoid falling in the same problems and try to follow successful experiences.

## References

1. PMBOK® Guide, 5th ed. 2013. Newtown Square, PA: Project Management Institute.

2. George Jucan. Managing Project Stakeholders. (2013)

3. Axley, S. 'Managerial and organizational communication in terms of the conduit metaphor'. Academy of Management Review ((1984)

4. Thomason, G. A Textbook of Human Resource Management, Institute of Personnel management, London. (1988)

5. Andrew Dainty, David Moore, Michael Murray. Communication in Construction theory and practice (2006)

6. Brox, Denene. "Power of Persuasion." PM Network, December.(. 2011)

7. Fisher, Roger, and William Ury; 2nd ed. with Bruce Patton, Getting to Yes. 1991.

8. Roeder, Tres.. A Sixth Sense for Project Management. Bloomington, IN: AuthorHouse. 2011

MATEC Web of Conferences **162**, 02038 (2018) https://doi.org/10.1051/matecconf/201816202038
*BCEE3-2017*

# Effective factors to encourage the foreign direct investment in Iraq

*Shahla* Mohammad Ali[1]

[1]Building and Construction Engineering Department, University of Technology, Baghdad, Iraq

**Abstract.** Foreign direct investment in Iraq cannot take its complete role for different reasons, such as: Lack of security, Corruption, Lack of Transparency, Unequipped banking system, undeveloped arbitration law, Intellectual Property Rights (IPR) issue, and internal disputes over oil rights. It was found that Iraq rates as one of the worst places in the world to do business, languishing at 166 out of 183 countries, according to a World Bank report and for starting a business Iraq ranks even lower

## 1 Introduction

Foreign direct investment represents a governing interest in a business enterprise in one country by an entity based on another country.
Foreign direct investment (FDI) is well-known by foreign investment in the portfolio, a negative investment in securities of another country for example equities and public bonds, and the "control" component. Pertaining to the Financial Times, "standard control tariffs adopt a world-wide threshold value of 10 per cent of the voting stock, however this is a gray area where a smaller group of shares will be given control in companies that are exercised on a large scale. Critical inputs can give actual control.
The origin of the investment does not affect the definition as a direct foreign investment, meaning that investment can be "inorganic" by purchasing a company or project in the target country or "organic" throughout the expansion of the operations of an existing company or project in the targeted country.
In general, FDI includes "acquisitions, mergers and acquisitions, the construction of new facilities and the reinvestment of profits from external operations for economic activity and intra-corporate loans."
In a narrow sense, FDI indicates the mere construction of new facilities. It is not possible to compare numerical FDI numbers based on various definitions as a portion of the country's national accounts, and according to GDP equation Y = C + I + G + (sniff) [i.e. consumption + total investment + government spending + exports and imports]. In addition to domestic investment, a foreign investment can also be considered. FDI is also defined as net investment flows (flow minus flow) for an interest of permanent management, which is greater than or equal to 10 percent voting stock, in an institution that operates in an economy excluding an investor's economy in another country.
Foreign direct investment is the sum of capital (i.e. other long-term capital and short-term capital) as shown in the balance of payments. FDI typically includes the contribution in management, management, joint venture, transmission of experience, technology, and skills. The FDI stock can be considered the net cumulative foreign direct investment (FDI) for any given period of time (i.e., INDI minus). Direct investment does not involve investment through the acquisition of shares. Foreign direct investment can illustrates the movements of international factors [1].

## 2 Methodology

### 2.1 Types of FDI

There are many types of foreign direct investment; the most common are listed below [2]:
1. The emergence of horizontal FDI as the company tries to replicate its country activities at the same step of the value chain in the host country through access to foreign direct investment [3].
2. Foreign direct investment in the platform as foreign direct investment from the exporting country to the targeted country for export to a third country.
3. Vertical FDI occurs once a company's money flows through FDI in the upstream or downstream direction in financially, administratively and organizationally various value chains, i.e. companies carry out value-added activities in a fixed time period in a host country.

### 2.2 Importance and difficulties of FDI

* Corresponding author: shahlahikmat@yahoo.com

© The Authors, published by EDP Sciences. This is an open access article distributed under the terms of the Creative Commons Attribution License 4.0 (http://creativecommons.org/licenses/by/4.0/).

There has been a fast growth in the population worldwide since 1950 in most developing countries.

This population rise has been met by increases in GDP more rapidly, and therefore each capita income has increased in most countries and throughout many countries since 1950. Whereas the quality of data collected since 1950 is possibly questioned, the average is estimated. Only significant increases in GDP per capita have been recorded in countries ravaged by wars, crises and countries with further serious external challenges such as Haiti, Somalia and the Niger. The data available to check these situations are freely-accessed.

The rise in foreign direct investment could be linked to improve the growth in economy because of capital inflows, increased tax revenues of the host country and the creation of other revenues. Host countries frequently try to guide foreign direct investment to innovative infrastructure projects and other projects to promote economic growth. The competition between new firms can result in a remarkable improvement in both of productivity and efficiency in the host country and the development of competencies. It has been recommended that the adoption of foreign entity policies in a local subsidiary can develop corporate governance standards.

Moreover, foreign investment can lead to soft skills' transfer through training and providing job opportunities, the availability of more innovative technology for the domestic market, the transfer of expertise and skills, and access to R & D resources. [4]

## 2.3 Identification of foreign direct investment

There are large number of theories that try to clarify the determinants of FDI. These theories can be considered essential steps to develop a systematic agenda for the development of FDI. Nevertheless, their respective potential that can be a self-sufficient general theory that could describe and clarify all types of FDI (FDI, domestic, industry and country-level), has been investigated and questioned in the work of various scientists.

Agrawal (1980) [5] could be given as an example. Daning is one of the most reviewers by authors who work on foreign direct investment. Donning (1993) described three major kinds of FDI based on the motive of investment from the viewpoint of an investment firm. The first form of FDI that seeks to achieve the market and presence in it is called concrete, which aims to assist local, regional and global markets. It is also called horizontal FDI, because it comprises duplication of production services in all sectors of the host country.

One type of FDI is investment in foreign direct investment or substitution of exports represents an alternative to FDI. Horizontal FDI offers better service of the local market for goods and services than domestic production, market size and market development of the host economy play many significant roles as difficulties in accessing domestic markets, for instance taxes, duties and transport costs. The second type form of FDI is called for by the search for new resources: when companies invest abroad to acquire resources that are not available and new in the country of origin, like natural resources, raw materials or low-cost labor. Particularly in the manufacturing and extractive sector, when multinational corporations invest directly for export, considerations of factor costs become essential.

Unlike horizontal FDI, vertical or export-oriented FDI is the transfer of parts of the domestic production chain to the host country. The availability of low-cost labor is a major engine of export-oriented FDI. Of course, foreign direct investment in the extractive resource sector, like oil and natural gas, is concerned to countries with abundant natural endowments. The third form of foreign direct investment, which is called the pursuit of efficiency and skills, it occurs when the company can profit from the joint management of geographically spread activities in the existence of economic scale and scope. In 1998, the World Investment Report [4] analyzed the determinants of FDI and the host country's determinants were characterized in the three groups, which are political and economic factors and facilitated business.

## 2.4 Prospects for foreign direct investment in Iraq

Some positive indicators of business chances are [6]

- Higher increase in th rate of crude oil revenues:

Iraq, according to statistics of the International Monetary Fund and the World Bank, has the growth of gross domestic product by 10% in 2013 linked to the current paid from the production of crude oil to the current (3) billion barrels per day, including about (2.3) million barrels now, Oil to about (90) dollars a barrel, Iraq's production will be about (100) billion dollars a year of revenue. Iraq is the fourth largest crude oil reserves in the world, and seeks to export (6) million barrels per day by 2017, according to the special report issued by the International Energy Agency (2012), Iraq could be the second largest source of crude oil globally over the next two decades to ( 2020 ).

- Increased government spending: The growing need for basic services makes government spending high, which is reflected in turn on economic growth, through funding resulting from crude oil revenues, making the need for investment high

- The laws of foreign investment: The issuance of the Investment Law, No. (13) in the year of 2006 and its amendments, gives the right of foreign ownership of companies in all sectors and by (100%),excluding the extraction of crude oil and minerals. In this context, the law aims to achieve equality between Iraqi and foreign investors in all forms of trade as a result of reforms. Thus, the investment grew from (3087) billion dollars in 2003 to (55.67) billion dollars in 2011, i.e. growth by 40.3% Value Added.

Diversification of the economy: Long sanctions and wars in Iraq have led to a lack of capital, technology and skills for more than two decades, prompting the government to pay attention to investment, especially foreign investment in all sectors of the economy away from oil. The market has been gradually opened to appeal foreign investment and joint ventures in the areas of housing, industry, agriculture, infrastructure, transportation, financial services, tourism and others. The government, through the investment law, exempted foreign companies from taxes for a period of ten years and exempted them for three years from import duties.

- Economic Reforms: Iraq's best reform of institution-building for an open economic environment is the conclusion of the agreement in February 2010 with the International Monetary Fund, which embraces a hydrocarbon law with two main objectives:
A) To draw up a new law allowing Iraq to improve its resources.
(B) To draw up a law and regulations for the fair distribution of revenues from crude oil within the country; Although both sides are still in the realm of controversial political negotiations, foreign oil firms have signed contracts all over Iraq, allowing them ("strategic entry") at the moment to ensure a greater market share in the future.
- The opportunity available in the retail sector, which is Iraq's heavy reliance on imports, may increase the chances for investment in the retail sector.

# 3 Results and discussion

## 3.1 Difficulties that facing foreign investments in Iraq

The difficulties and challenges facing foreign investors in business are very large. Iraq is one of the world's worst countries for commercial investment, with 166 out of 183 countries, consistent with the World Bank report. However, this low classification did not prevent foreign companies from working within the Iraqi market despite facing a range of difficulties and challenges, which are as follows:
Security: Although a major problem but the security environment has improved significantly since 2010, Iraq remains one of the world's most dangerous places to invest with the possibility of renewed violence and political instability. However, the security situation in the Kurdistan region of Iraq has been more stable security.
- Financial and administrative corruption: Iraq is ranked as seventh among the most corrupt countries in the world, according to the International Corruption Index (2012) as this is the case for most economies of developing countries. There is a lack of transparency in government laws and regulations and the participation of government officials in bribery, tenders, contracts and embezzlement of public funds. The lack of the judicial system, the public opinion and the media have caused weakness in the face of corruption and allowed the regulatory sector to play its proper role. Corruption is considered one of the most important factors that hinder the growth of the Iraqi business sector according to the classification of the National Institution for Democracy in its report on the survey of businesses in Iraq (2011)
- Lack of transparency: There is an urgent need to develop transparency in order to improve the business environment and investments in Iraq in the field of procurement and procurement procedures and government contracts, since the illegal activity of these procedures may be at a high level, especially in the award of contracts free and open competition. According to the US Business Trade Directory (2012), the Iraqi government's ability to deliver projects is weak and lacks the means to advance technology in that area.
- An inefficient banking system: Reform of the banking system has become imperative for the purpose of attracting foreign investment, although the banking sector has grown rapidly over the past five years due to high oil revenues. (7) banks are state owned and hold 89% of the country's bank deposits.
The government banks are ineffective and are exposed to bad debts, old losses and the aging of electronic banking systems. In addition to their lack of liquidity, their borrowing rates are low compared to those of the private sector, they suffer from a decrease and a shortage of capital, this is reflected in their activity by borrowing, and these banks require finding equal opportunities for competition as compared to the private sector ones. As government banks rely on Iraqi government spending, restructuring of the organization is more transparent in its policies and financial activities, which may be difficult at all.
Intellectual property rights: Intellectual property rights have specific controls that the Iraqi market lacks control and protection, although the Iraqi government has obliged the adoption of quality control and intellectual property rights on most imported products in July 2011, but counterfeit products in consumer and electronic goods are widely available.
- Arbitration Law: there is lack of arbitration law to settle commercial disputes and lack of commercial arbitration courts in all provinces to settle commercial disputes, which include investors from foreign companies. The Iraqi government is currently drafting a governing law to be entrusted to the United Nations International Trade Law Commission on International Commercial Arbitration.

- Differences of oil rights: One of the issues that affect on the increase in revenues from oil exports is the dispute over the rights of domestic oil and what arose from the dispute over oil rights with the Kurdistan region. The government has finalized the hydrocarbons law and settled disputes with the region, which will provide more clarity to the foreign investor in the oil field; this is the basis for Iraq's economic growth, which requires an investment of $ 25 billion annually to increase its production level.

# 4 Conclusions

Iraq has been assessed as one of the worst places in the world to practice business, languishing at 166 out of 183 countries, consistent with a World Bank report. For starting a business Iraq ranks even lower, at 174. This is due to the following reasons: Lack of security, Corruption, Lack of Transparency, Unequipped banking system, undeveloped arbitration law, Intellectual Property Rights (IPR) issue, and internal dispute over oil rights.

## References

1. Wikipedia, (2014).
2. R. Zelot, "FDI effects on developing country economics", FDI journal, **4**, (2011).
3. R. S. Saunders, "The Determinants of Foreign Direct Investment." *Canadian Journal of Economics,* **28**, pp. 77-84, (2009).
4. UNCTAD, " World Investment Report, (2010).
5. J. Agarwal, "Determinants of foreign direct investment: A survey", Review of World Economics (Weltwirtschaftliches Archiv), **116** (4), pp. 739-773, 1980.
6. S. A. Cheema ,"Investing in Iraq: Prospects and Challenges", New Delhi India (Feb 20, 2013)
7. J. Parletun, "The Determinants of Foreign Direct Investment: A Regional Analysis with Focus on Belarus", (2008) http://biblioteket.ehl.lu.se/olle/papers/0002948.pdf.

# The performance of hydraulic jump over rough bed stilling basin consisting of cubic-shape elements

*Jaafar* Maatooq[1], and *Enass* Taleb [1]

[1]Building and Construction Dept., University of Technology, Baghdad, Iraq

**Abstract.** Three groups of cubic roughness elements non-protruding to the flow were fixed with a specific configuration at the downstream of a stream lined lib sluice gate. The aim was to test the performance of the hydraulic jump under using such elements as appurtenances of stilling basin. The experimental works has been conducting to test; four different intensities, I=8%, 10%, 12% and 16%; three width to height ratios, We/he=0.75, 1.5 and 2; and three length to height ratios, Le/he=0.75, 2 and 4, of cubic elements made from waterproof stiff rubber. The test was performed under the incoming Froude number; F1 ranged between 2.7 to 9. After analysing the experimental data one from the second group configuration gives a higher reduction of sequent depth ratio, y2/y1 as compared with the depth ratio of the un-forced hydraulic jump which calculated by Belanger equation, this decreases arrived 20%. Also greater relative energy has been dissipated, and the gain in dissipation (%GED) ranged between 10.8-22.3%.

## 1 Introduction

Hydraulic jumps are divided into two types according to the bed characteristics. The first is a classical hydraulic jump occur over a smooth bed and has been studied extensively by different researchers such as [4,19,13,14,22]. All these studies were based on comparison with Belanger equation. The second is formed over the so called roughness elements when different appurtenances are used to configure a roughness on stilling basin bed. This type is known as a forced jump and has been studied by many researchers [e.g., 4,23,2,20,8,21,10,5,9]. A hydraulic jump is extremely in turbulent, and the flow along it is rapidly varied. It is characterized by the development of large-scale turbulence, surface waves and spray, and air entrainment. Normally any hydraulic jump should need a stilling basin in order to contain it and to dissipate more energy that mainly is excessive when accompanied with the high velocity through the structure (e.g., under a sluice gate). The stilling basin should be designed to confine the formation of the hydraulic jump, inducing more dissipation of kinetic energy, thereby avoiding or minimizing the damage ability downstream. However, the designer focusing on knowing the height of jump into a stilling basin that has been proposed in order to get a safe height of the side walls with a suitable free board, the less the sequent depth, y2 the shorter height of side wall need. These two advantages (i.e., more dissipation of surplus kinetic energy and reducing the sequent depth ratio) can be achieved by roughing the bed of stilling basin by specific elements. The variation in the expected advantage from the use of these appurtenances depends on many factors such as; type, shape, and configurations along with the hydraulic boundary conditions. Specifically, with the kind of roughness elements have been adopted in the present study, some of the relevant previous studies should review. [20], studied the effect of cubic elements as a roughness installed into bed at intensity I=10% on the length of rectangular hydraulic jump under different flow conditions. The aim was to obtain the optimum length of jump on a rough bed section for both engineering and economics terms. [21], studied the effect of horizontal rectangular roughness elements placed at the downstream of the sluice gate. Different intensities of roughness have been used in experiments ranging between 0 to 100%, beside that different height of roughness he= 0.8, 1.2, 1.6 and 2 cm, and different roughened lengths, Lr ranged between 1.6 to 200.66 cm were adopted. The roughness parameters that produced the minimum length of jump and hence the minimum length of stilling basin from the hydraulic point of view were determined and specified. [10], carried out an experimental study on non-protruding prismatic roughness and he studied the effect of roughness type and arrangement on the length of the jump, energy dissipation, and tail water. The

© The Authors, published by EDP Sciences. This is an open access article distributed under the terms of the Creative Commons Attribution License 4.0 (http://creativecommons.org/licenses/by/4.0/).

investigation was restricted to incoming Froude numbers ranging between 7.3 and 16.6. The results were compared with properties of jump occurring on the smooth bed. The analysis of experimental data showed that the length of the jump was obtained to be 40% less than in traditional (over smooth bed) jump, and the energy losses through the hydraulic jump ranged from 5% to 10% higher than those formed on smooth beds. Moreover, the tail water depth in a roughened bed was 20% lower compared to that formed on a smooth bed. [5], carried out an experimental study on non-protruding prismatic roughness and he has studied the effect of the varying upstream flow parameters, i.e. the initial Froude number, Fr1, and initial depth, y1, with the boundary roughness parameters, as the length ratio, pitch ratio, and roughness arrangement, on the characteristics of the hydraulic jump. [6], measured the characteristics of hydraulic jump (sequent depth and roller length occurring on bed roughness of closely crushed gravel particles cemented to the bottom. A new solution of the momentum equation for sequent depth ratio as a function of the initial Froude number and the proportion between the height of roughness and the initial depth of flow has been introduced. [25], studied, two types of natural bed roughness with crushed sedimentation materials 20 and 27 mm in size to show the effect of these roughness on the characteristics of hydraulic jump, including roller length, jump length, initial and sequent depths, relative energy losses through the jump, boundary stress, and shear stress. The roughness models were tested under 72 experiments with Froude numbers from 5 to 12.7. [11], carried out an experimental study on a rough bed consisting U-shape elements to find the best length and intensity of these elements on hydraulic jump characteristics. These elements fixed at the downstream of vertical gate equal twice the gate opening. The feature of the flow was tested under four different intensities I= 4.5, 8, 12 and 18% and eight different lengths roughness used with a sufficient intensity. The U-shape roughness has been compared with other roughened shapes, and the efficient performance was achieved using the U-shape roughness. The results show that the higher effect roughness intensity was 12.5% at which the relative sequent depth of jump (y2/y1) was reduced by 14-20% and minimized the relative length of the jump by 28-47%. In the present study, four different intensities, I=8%, 10%, 12% and 16%; three width to height ratios, We/he=0.75, 1.5 and 2; and three length to height ratios, Le/he=0.75, 2 and 4, of cubic elements made from waterproof stiff rubber have been adopted to show the effects of these arrangements on the characteristics of the hydraulic jump.

## 2 Experimental works and procedure

The experimental investigations were carried out by using a laboratory flume in the fluid laboratory of Building and Construction Engineering Department, UOT-Baghdad. The cross section of the flume was 0.3m wide, 0.3m deep, with a total length 15m. The steel plate walls added to the inlet part of the flume through 4m length to rising the depth up to 0.6m, in order to ensure access sufficient heads upstream the stream lined lib sluice gate which is used as a control device to get a broad range of Froude number. The flow rate was measured using flow meter (Rotameter type), attached to the supply line of the flume for maximum capacity 102.5 $m^3/hr$. The depth of water was measured using a vernier point gauge with 0.1mm accuracy. The local flow velocity was measured by using a propeller current meter mounted on a wading rod gage. When the propeller rotates the number of revolutions in a given time is counted on a digital counter panel every 10 second, and recorded in Hertz. The average was taken from the recorded data after nearly constant readings. Then the flow velocities obtained after converting the average "Hertz" reading to velocity in cm/s by using an appropriate relationship which was prepared by authors after extracted data from chart issued by "Nixon" company, the manufacturer of the device. Three groups of stiff waterproof rubber cube were used as a roughness element appurtenance in the stilling basin and its dimensions shown in Fig.1.

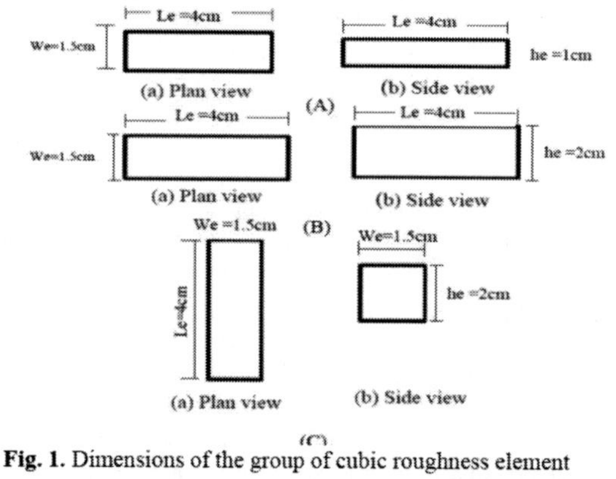


**Fig. 1.** Dimensions of the group of cubic roughness element

Different roughness intensities were considered by using these three groups. The percent roughness intensity (I) is defined as the ratio between the projected area, and the total area of the basin covered by roughness was calculated by;

$$I=100aN/BLr \quad (1)$$

https://doi.org/10.1051/matecconf/201816203001

Where: I= percentage intensity of elements in the basin,
B= width of the basin,
Lr= length of roughened bed,
a= is a plane area of one element, and
N=the number of roughness elements.

The first row of cube elements was at a distance of 18cm downstream from the gate (10cm from the beginning of the basin) followed by the other rows up to the desired length of roughness that gives required intensity. The crest levels of elements at the same level of the upstream bed that gives a non-protruding feature inside the basin to overcome the phenomenon of cavitation. The elements being depress, increase bed shear, thereby, creating more turbulence. The three different configuration groups were installed to show the effect of length, (Le), width, (We) and height, (he) as well as the intensity of these elements on the hydraulic jump characteristics and stilling basin performance. The information about the geometrical characteristics of the three groups of roughness elements and their configurations are listed in Table 1.

After analyzing the results of both groups A and B it was found the best bed roughness intensity was at I=12%, that was recorded by configuration B3. The more dissipation of energy by reducing the sequent depth and lesser length of the hydraulic jump were achieved with this configuration. Accordingly, third group C was conducted with the best intensity (I=12%), and the variation was adopted to show the effect of the element width and length relative to its height on the hydraulic jump characteristics. The arrangement of C1 is shown in Fig.2.

## 3 Results and discussion

The study included fifty one experiments conducted for a basin with the three groups consisting of different configurations and arrangements of roughness elements. As mentioned previously, all of these configurations were used to investigate their impact on hydraulic jump performance specifically on its properties such as; depth ratio (sequent depth to initial depth), energy loss, and gain in energy dissipation. The comparison between groups of roughness elements depends on the benefits which arise from usage. The streamlined lib sluice gate opening refers to the initial depth of the hydraulic jump, y1 ranging between 1cm to 4cm to achieve the incoming Froude number, Fr1 between 9.2 to 2.7 at which the operation discharge for this achievement was ranged between 30.61$m^3$/hr to 72.47$m^3$/hr respectively.

**Table 1.** Geometrical characteristic of the roughness element configurations

| Group | Config. | Element No. | No.of runs | I | Dimensions (cm) | | |
|---|---|---|---|---|---|---|---|
| | | | | | he | Le | we |
| A | A1 | 40 | 6 | 8% | 1 | 4 | 1.5 |
| | A2 | 50 | 6 | 10% | 1 | 4 | 1.5 |
| | A3 | 60 | 6 | 12% | 1 | 4 | 1.5 |
| | A4 | 80 | 6 | 16% | 1 | 4 | 1.5 |
| B | B1 | 40 | 6 | 8% | 2 | 4 | 1.5 |
| | B2 | 50 | 6 | 10% | 2 | 4 | 1.5 |
| | B3 | 60 | 6 | 12% | 2 | 4 | 1.5 |
| | B4 | 80 | 6 | 16% | 2 | 4 | 1.5 |
| C | C1 | 60 | 3 | 12% | 2 | 1.5 | 4 |

**Table 1.** Continued

| We/he | Le/he | No.of rows | No.of col. | X1 (cm) |
|---|---|---|---|---|
| 1.5 | 4 | 5 | 8 | 7.25 |
| 1.5 | 4 | 5 | 10 | 5 |
| 1.5 | 4 | 5 | 12 | 3.5 |
| 1.5 | 4 | 5 | 16 | 1.62 |
| 0.75 | 2 | 5 | 8 | 7.25 |
| 0.75 | 2 | 5 | 10 | 5 |
| 0.75 | 2 | 5 | 12 | 3.5 |
| 0.75 | 2 | 5 | 16 | 1.62 |
| 2 | 0.75 | 5 | 12 | 6 |

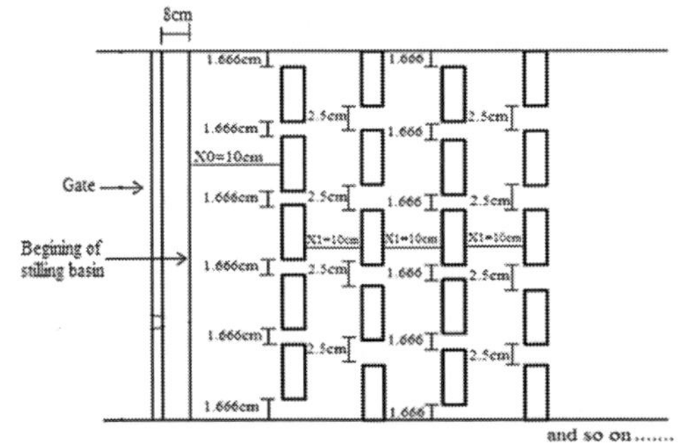


**Fig. 2.** Plan view arrangement of the configuration C1 at best intensity I=12%.

The effect of We/he and Le/he as well as the intensity of roughness elements on the sequent depth ratio was determined. The variation of y2/y1 when using models of group A (We/he=1.5, Le/he=4) and Group B (We/he=0.75, Le/he=2) was sketched with Fr1 as shown in Figs.3 and 4.

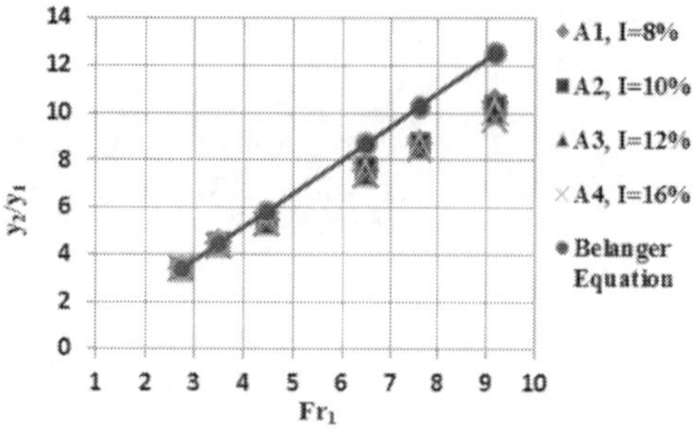


**Fig. 3.** Variation of y2/y1 for group A

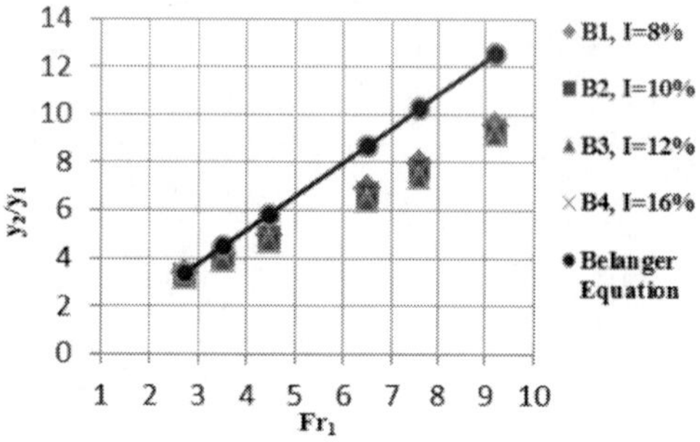


**Fig. 4.** Variation of y2/y1 for group B

These figures illustrate that the sequent depth ratio for both groups roughness A and B are smaller than that for classical jump (Belanger equation). Also as can be observed the variation of intensity does not have an appreciable effect on depth ratio. While for the same intensity the decrease of We/he and Le/he has an effect on decreasing the values of y2/y1 that can be achieved when the height of roughness elements increased. The best roughness intensity was I=12% with model B3, where the minimum sequent depth ratio was achieved as depicted in Fig.4. To study the effect of width/height and length/height ratios of roughness elements a comparison between configuration B3 and configuration C1 (which have the same intensity and roughness height) was conducted. Fig.5 shows the variations of sequent depth ratio as affected by the two geometric ratios of elements. This figure demonstrates clearly that the performance with B3 was better as compared to configuration C1. Moreover, the visual observation shows that the better stability of jump was achieved by using B3 model. The comparison with previous results presented by some researches, at hand, is illustrated in Fig.6. In this figure the results of work done by [5] who used staggered roughness with W/L=4, W/Z=6 where W is the longitudinal spacing between two roughness elements in successive rows, L is the length of the roughness element, and Z is the height of the roughness element, [3] whose used lozenge roughness shape, [25] whose used two types of natural bed roughness with crushed sedimentation materials 20 and 27 cm in size, and [11] whose used U-shaped roughness elements.

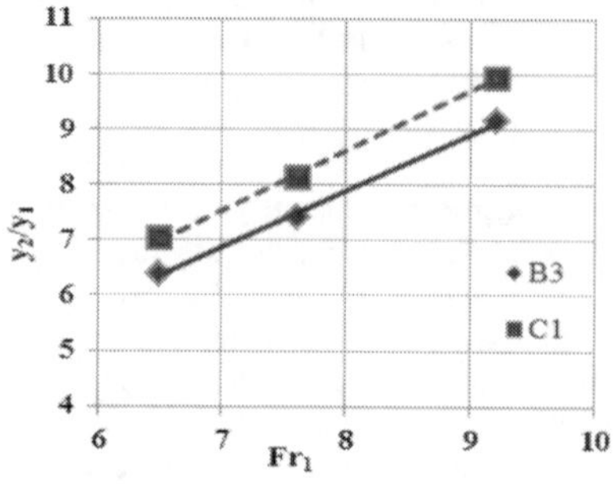


**Fig. 5.** Effect of width and length to height of roughness element

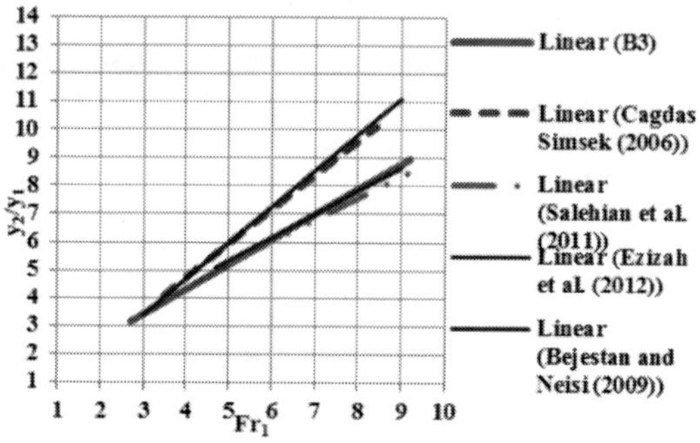


**Fig. 6.** Effect of B3 on y2/y1 as compared with the results of previous studies

This comparison sketched as one set of data together with configuration B3. The trend curves in the figure illustrate that B3 gives a higher reduction in y2 compared to the results of [5] and [11], but gives approximately the same effect on the sequent depth y2 as compared to the results of [3] and [25]. In order to show the amount of difference between the sequent depth of forced jump y2 and sequent depth of classical jump y2*, a dimensionless depth deficit parameter D was used. This parameter indicator was adopted previously by numerous researchers such as [8], [17], [26], [1], and [9] as given below:

$$D = \frac{y_2^* - y_2}{y_2^*} \tag{2}$$

Eq.2 refers that, the higher the D, the lower sequent depth that is associated with a specified configuration. Accordingly, Depth deficit parameter D was computed for all configurations of groups A and B roughness elements and listed in Table 2. The last column in the table refers to the required tail water that needs to achieve a sequent depth, y2 as compared with that calculated by Belanger equation. Fig.7 shows the trend variation of D with Fr1 for all configurations and the higher deficit parameter was recorded with configuration B3. Comparison of the depth deficit parameter that corresponding with configuration B3 of the present study with those proposed by previous studies shows that: the average value with staggered rough elements of W/L=4 and W/Z=6 was D= 0.11 which introduced by [5], whereas the mean value of D = 0.2 was found with configuration B3 and is close to the average value of D= 0.19 which obtained by [25] for crushed gravel materials 27 mm in size.

**Table 2.** Average value of D and Percent reduction of y2

| Configuration | he (cm) | We/he | Le/he | Average value of D | The sequent depth required for forced jump |
|---|---|---|---|---|---|
| A1 | 1 | 1.5 | 4 | 0.09 | $0.91y_2^*$ |
| A2 | 1 | 1.5 | 4 | 0.09 | $0.91y_2^*$ |
| A3 | 1 | 1.5 | 4 | 0.10 | $0.90y_2^*$ |
| A4 | 1 | 1.5 | 4 | 0.12 | $0.88y_2^*$ |
| B1 | 2 | 0.75 | 2 | 0.15 | $0.85y_2^*$ |
| B2 | 2 | 0.75 | 2 | 0.17 | $0.83y_2^*$ |
| B3 | 2 | 0.75 | 2 | 0.20 | $0.80y_2^*$ |
| B4 | 2 | 0.75 | 2 | 0.18 | $0.82y_2^*$ |

In the present study, the measurements of the length of the jump Lj was considered as a longitudinal distance from the beginning of the jump to the section beyond it the water depth did not change. One of the major economic aims of the designer is to present the shortest stilling basin with the best hydraulic performance. The indicator of this goal is a decrease of the hydraulic jump length, Lj. Thus, one of the advantageous of the appurtenances within a stilling basin is to achieve this aim. Based on that, the success model depends on how it performance close of the target.

Fig.8 illustrates the relation between the initial Froude number and the length ratio of the jump Lj/y2* for groups of roughness elements. The We/he, Le/he, as well as the intensity, have been selected as the criteria to show their effects on the length of the jump. It is seen from figure that all configurations of group A have a slightly effect on the length ratio of the jump. While for group B, the increase in intensity caused an appreciable effect in the reduction of the length of hydraulic jump and the maximum reduction was at I= 12%. That achieved with configuration B3. Fig.9 shows the comparison between the results of configuration B3 with results proposed by [5] for staggered roughness and [25] for crushed gravel materials 27 mm in size. It seems from this figure that the performance of B3 gives the best results in reducing the length of the jump compared to those staggered by previous researchers.

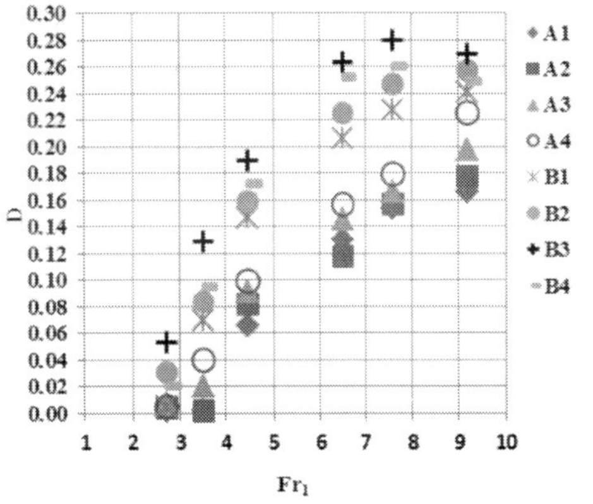


**Fig. 7.** Relationship between D and Fr1

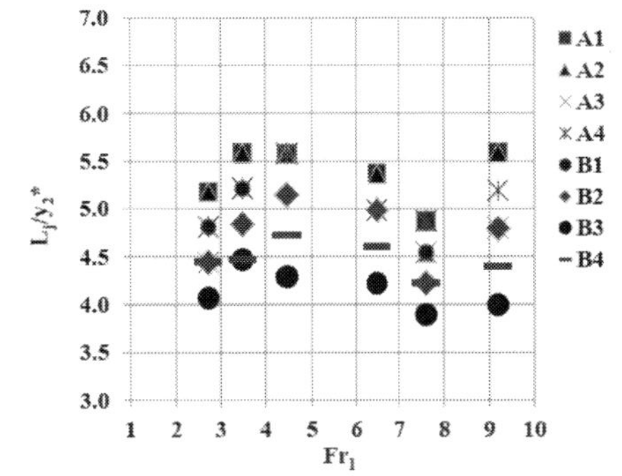


**Fig. 8.** Effect of different configuration of roughness elements on jump length

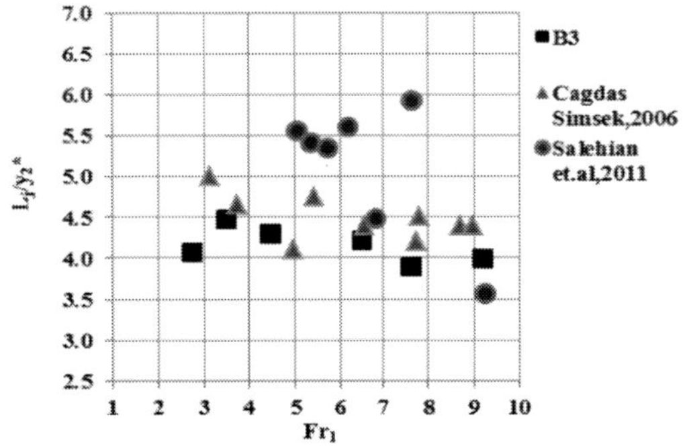


**Fig. 9.** Comparing the results of model B3 with previous studies

Energy dissipation ratio, ΔE/E1, is the difference between the specific energies before and after the hydraulic jump relative to the initial energy. The ΔE hence, reflect the amount of energy lost through the jump when the flow passes from supercritical to subcritical. To determine the effect I, We/he, and Le/he on the energy dissipation ratio, each criterion of these parameters is individually tested as given below;

Fig.10 represents the variation of, ΔE/E1 with the change of intensity of roughness at different Froude numbers for configurations of group A. It was found in the nature of the spread of the data points that, the variation in intensity does not have a significant effect on the rate and amount of energy dissipation along the jump.

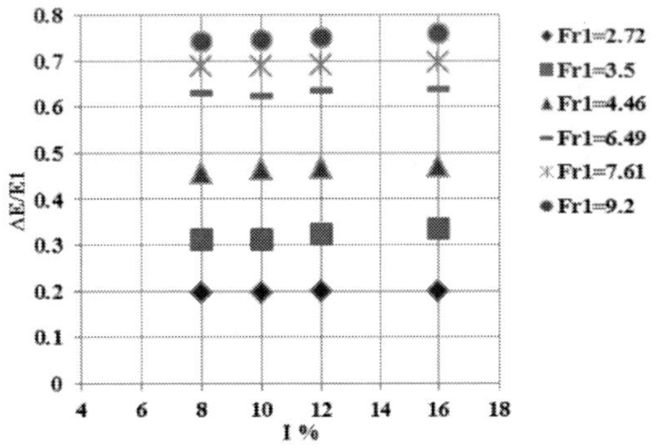


**Fig. 10.** Variation of ΔE/E1 with I for group A configuration

The change of ΔE/E1 as We/he changed for the roughness intensity (I=12%) with different Froude numbers, is shown in Fig.11. It can be concluded from this figure that the We/he has a significant effect on the rate of energy dissipation. The upper and lower dissipation of energy has been recorded with We/he=0.75 and 1.5 respectively.

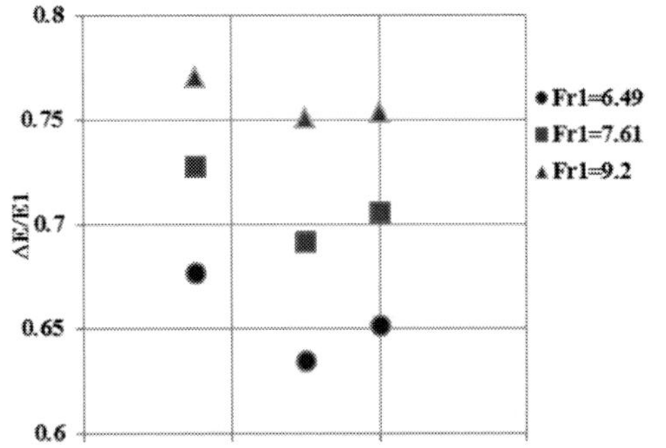


**Fig. 11.** Effect of $W_e/h_e$ on energy dissipation

The variation of ΔE/E1 with Le/he for the roughness intensity (I=12%) and Froude number ranged between 6.49 and 9.2, is shown in Fig.12. This figure indicates that the length of roughness elements relative to its height has a significant effect on the dissipation of energy. The higher dissipation of energy occurred with Le/he=2.

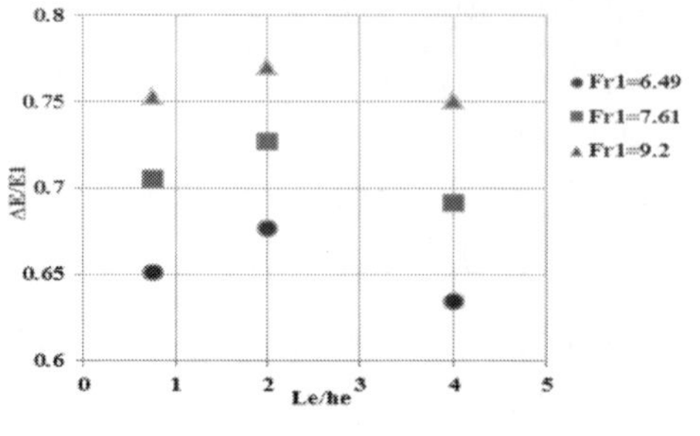


**Fig. 12.**Effect of $L_e/h_e$ on energy dissipation

Fig.13 introduces the relationship between the relative energy losses, ΔE/E1 as the dependent variable with Froude number for different configurations of roughness elements. As mentioned previously, the variation in the intensity has no significant effect on the rate of energy dissipation along the jump, while the change in We/he and Le/he has an appreciable effect. However, the best configuration affected the hydraulic performance was B3, where the energy dissipation was from 22.6% to 77%.

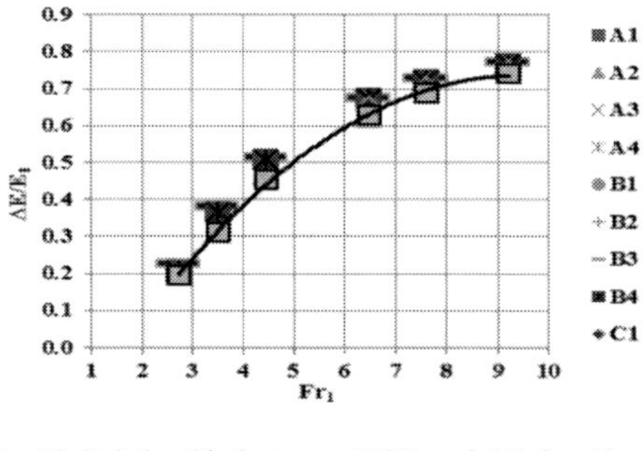


**Fig. 13.** Relationship between ΔE/E1 and Fr1 for all roughness elements

The effect of incoming Froude number is evident on energy dissipation rate. Thus the energy dissipation is basically a function of the initial Froude number Fr1. Based on this concept the relationship between the rate of energy dissipation ΔE/E1, and the Froude number Fr1, at different configurations of roughness elements is defined by the following equation with $R^2$ equals 0.999. It should be referred that Eq.3 can be used with all Froude numbers ranging between 2.72 and 9.2. The general view of jump with using the configuration B3 as roughness elements on bed of stilling basin is shown in Fig.14.

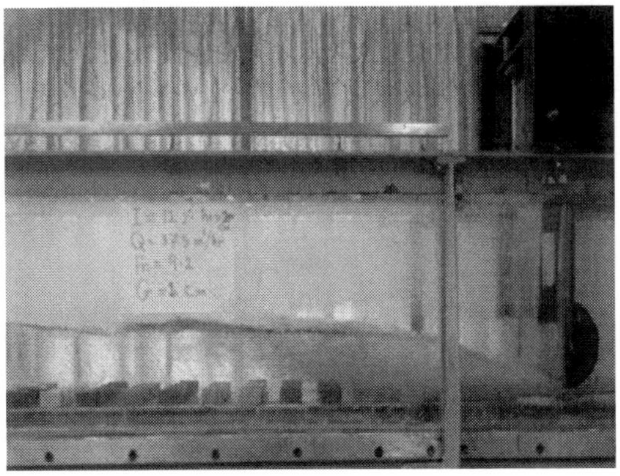

**Fig. 14.** General view of jump with using the configuration B3>

$$\frac{\Delta E}{E1} = -0.0112Fr_1^2 + 0.223Fr_1 - 0.321 \quad (3)$$

The gain in energy dissipation can be defined as the percent of the additional dissipation of energy (when the jump forced by confining it within a stilling basin equipped with baffled appurtenances) to the energy which was dissipated by the hydraulic jump when occurred on a smooth bed (free jump) under identical upstream flow conditions. This parameter was adopted previously by numerous researchers [e.g., 10, 27 and 1] it can be expressed as;

$$G_{ED} = \frac{\Delta E - \Delta E^*}{\Delta E^*} * 100 \quad (4)$$

Where:

$\Delta E$= is the energy loss for a jump formed in a rough bed (forced jump).

$\Delta E^*$= is the energy loss for a jump that occurs on a smooth bed (free jump).

To determine the effects of I, We/he, as well as Le/he on the gain in energy dissipation, the influence of these parameters individually has been investigated as follows;

Fig.15 illustrates the variation of gain in energy dissipation %GED with the intensity of roughness elements I for the Froude number ranging between 2.72 to 9.2 when group A used. It seems from this figure that the variation of roughness intensity has a significant effect on the gain of energy for the same We/he and Le/he (i.e., for the same configuration).

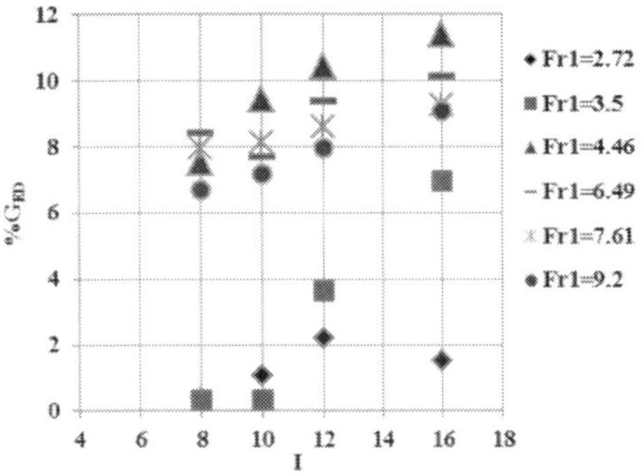


**Fig. 15.** Effect of intensity on the gain of energy dissipation when group A used

The effect of We/he when the roughness intensity (I=12%) on %GED for Froude number ranged between 6.49 and 9.2, is shown in Fig.16. It can be concluded from this figure that the variation in the We/he has a high effect on the gain of energy dissipation. A higher gain was recorded with We/he=0.75.

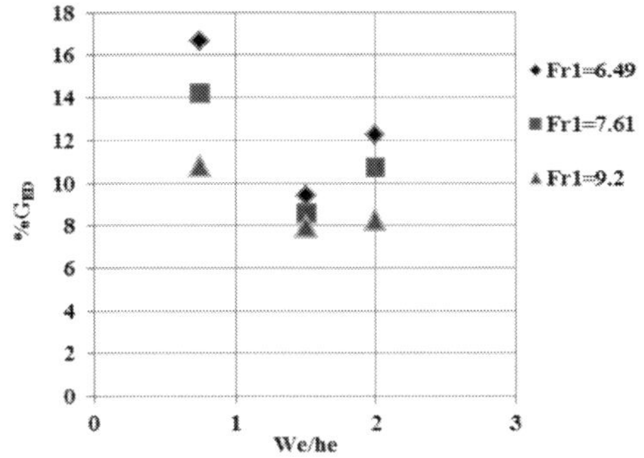


**Fig. 16.** Effect of $W_e/h_e$ on the gain of energy dissipation

The variation of %GED with Le/he for the same roughness intensity (I=12%) and the same range of Froude number (6.49-9.2), is shown in Fig.17. The trend of points in this figure generally shows that the length of the element is directly influenced by the advantageous gain of energy dissipation, and when Le/he=2 a higher dissipation occurred that indicated to a higher percentage of energy gained with this dissipation.

The variation of percent gain in energy dissipation for various configurations of roughness elements is presented in Fig.18. The trends of data points and values confirm that the Configuration B3 is the best in terms higher gain in energy dissipation as compared to other configurations. The gain in energy in case of B3 ranged between 10.8-22.3% larger than that resulted in the free jump.

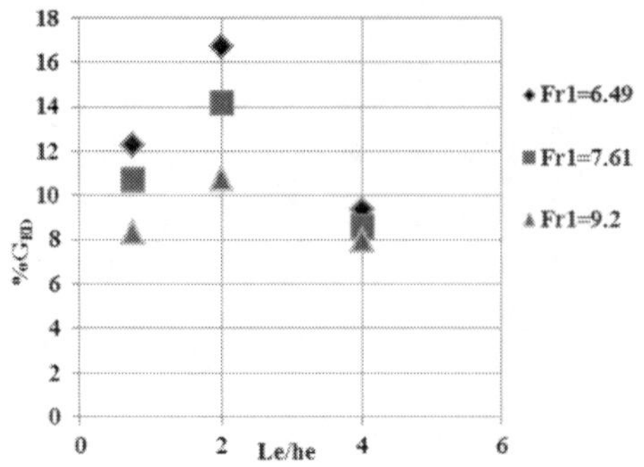


**Fig. 17.** Effect of $L_e/h_e$ on the gain of energy dissipation

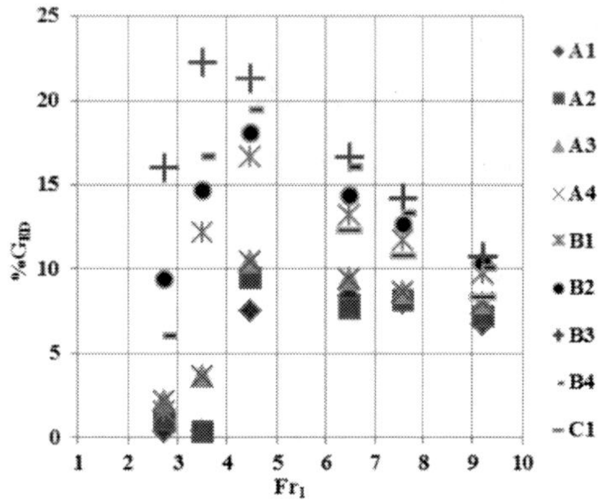


**Fig. 18.** Percent gain of energy with all configurations of roughness elements

## 4 Conclusions

Experimental tests were performed for hydraulic jump stilling basin by adopting three groups of roughness elements installed on a bed of the basin at non-protruding arrangement including numerous configurations at different intensity, width of the element to its height (We/he) and length of the element to its height (Le/he). The types of hydraulic jump were bounded for the incoming Froude number, Fr1 between 2.72 and 9.2. The data thus collected it was processed analytically and the results of the analysis presented in numerous figures. The best configuration of roughness elements that which give minimum sequent depth was B3. The higher reduction in y2/y1 associated with this configuration was 20% compared with a classical jump (Belanger equation). The length ratio of hydraulic jump Lj/y2* for all configurations of roughness elements located below the limitation curve that presented [23]. It was also found that the configuration B3was more effective in reducing the jump length. One of the most parameters used as an indicator of a beneficial performance of hydraulic jump is the dissipation of kinetic energy. Accordingly, the relative energy dissipation ΔE/E1 for all configurations was tested. It was found that the variation in intensity does not have a significant effect on the amount of energy dissipation along the jump, while the change in We/he and Le/he have significant effects on the rate of energy dissipation. However, the higher dissipation of energy has been recorded with We/he=0.75 and Le/he =2. This arrangement is for configuration B3. Also more gain in energy dissipation %GED was associated with configuration B3, which ranged between 10.8-22.3%.

The authors are grateful to the Directorate of Laboratory in Building and Construction Eng. Dep. for approval to use flume and measurement facilities. Appreciation is due to the staff member of Hydraulic Laboratory for assistance in preparing models. Deep acknowledgement is presented to the Building and Construction Engineering Department of the UOT, Iraq for academic assistance.

## References

1. A. Abbaspour, A. Hosseinzadeh, D. Farsadizadeh, A. Sadraddini, J. Hydro-env. Res. **3**,109-117 (2009)
2. J. AL-Baidhani, M.Sc. Thesis, Baghdad Univ, Irrigation and Drainage Engineering (1989)
3. M. Bejestan, K. Nesi, Asian J. of App. Sci., **2**, 436-445 (2009)
4. J. Bradley, A. Peterka, **82**,1401-1406 (1957)
5. S. Çagdas, M.Sc. Thesis, Middle East Tech. Univ., Dept. of Civil Eng., Ankara, Turkey (2006)
6. F. Carolo, V. Feroo, V. Pampalone, J. Hy. Eng., ASCE, **133**,989-999 (2007)
7. S. Chaurasia, J. Irr. and Drai. Eng., ASCE, **129**, (2003)
8. S. Ead, N. Rajaratnan, J. Hy. Eng., ASCE, **128**, 656-663, (2002)
9. A. Ellayn, Zhi-lin SUN, J. of Zhejiang Univ.-Sci., **13**, 519-525, (2012)
10. T. Evcimen, M.Sc. Thesis, Middle East Tech. Univ., Dept. of Civil Eng., Ankara, Turkey, (2005).
11. G. Ezizah, N. Yousif, S. Mostafa, Asian J. Appl. Sci., ISSN 1996-3343, DOI:10.3923, (2012)
12. W. Hager, R. Bremen, N. Kawagowshi, J. Hy. Res., **28**, 591-608, (1990)
13. W. Hager, *Kluwer Acad. Pub.*, Dordrecht, The Netherlands, (1992)
14. W. Hager, R. Bremen, J. Hy. Res., **27**, 565-585, (1989)
15 W. Hughes, J. Flack, J. Hy. Eng., **110**, 1755-1771, (1984)
16. F. Izadjoo, M. Bajestan, 17th Canadian, Hydrot. Conf., Hydrot. Eng., Edmonton, Alberta, Canada, Aug. 17–19, (2005)

17. F. Izadjoo, M. Bajestan, J. App. Sci., **7**, 1164-1169, (2007)
18. A. Kazemiyanzadeh, M. Bajestan, 3rd Conf. mang. of Iran's water res., Tariz, (2008)
19. H. Leutheusser, V. Kartha, J. Hy. Eng., **98**, 1367-1383, (1972)
20. H. Mohamed Ali, J. Hy. Eng., **117**, 83-93, DOI:10.1061/(ASCE)0733-9429(1991)117:1(83), (1991)
21. A. Negm, 5th Int. Conf. on Hydro-Sci. and Eng., Warsaw, (2002)
22. I. Ohtsu, Y. Yasuda, J. Hy. Eng., ASCE, **120**, (1994)
23. A. Peterka, Eng. Monograph No. 25. USBR, Denver, USA, (1958)
24. N. Rajaratnam, K. Hurtig, J. Hy. Eng., **126**, 310-12, (2000)
25. S. Salehian, M. Bajestan, H. Mousavi, H. Kashkhooli, S. Kashefipour, World Appl. Sci. J., **13**, 1005-2011, (2011)
26. A. Samir, Final Res. Report No. 14/427, King Saud Univ., Collegae of Eng. Res. Center (2007)
27. N. Tokyay, In Impacts of Global Climate Change Conf., EWRI, May, Alaska, USA, 408-416. DOI:10.1061/40792(173)408, (2005)

# AUTHOR INDEX

# AUTHOR INDEX

# AUTHOR INDEX

# AUTHOR INDEX

# 3rd International Conference on Buildings, Construction and Environmental Engineering (BCEE3-2017)

MATEC Web of Conferences Volume 162 (2018)

Sharm el-Shiekh, Egypt
23-25 October 2017

Part 2 of 2

**Editors:**

**T. S. Al-Attar**
**W. S. AbdulSahib**
**M. A. Al-Neami**

ISBN: 978-1-5108-6260-9

**Printed from e-media with permission by:**

Curran Associates, Inc.
57 Morehouse Lane
Red Hook, NY 12571

**Some format issues inherent in the e-media version may also appear in this print version.**

This work is licensed under a Creative Commons Attribution license:
http://creativecommons.org/licenses/by/2.0/
**You are free to:**
**Share** – copy and redistribute the material in any medium or format.
**Adapt** – remix, transform, and build upon the material for any purpose, even commercial.
The licensor cannot revoke these freedoms as long as you follow the license terms.
**Under the following terms:**
You must give appropriate credit, provide a link to the license, and indicate if changes were made. You may do so in any reasonable manner, but not in any way that suggests the licensor endorses you or your use. The copyright is retained by the corresponding authors.

Printed by Curran Associates, Inc. (2018)

For additional information, please contact EDP Sciences – Web of Conferences
at the address below.

EDP Sciences – Web of Conferences
17, Avenue du Hoggar
Parc d'Activité de Courtabœuf
BP 112
F-91944 Les Ulis Cedex A
France

Phone: +33 (0) 1 69 18 75 75
Fax: +33 (0) 1 69 28 84 91

contact-edps@webofconferences.org

**Additional copies of this publication are available from:**

Curran Associates, Inc.
57 Morehouse Lane
Red Hook, NY 12571 USA
Phone: 845-758-0400
Fax: 845-758-2633
Email: curran@proceedings.com
Web: www.proceedings.com

# TABLE OF CONTENTS

## PART 1

## PART 2

# Static and dynamic water wells levels of Kirkuk city represented by geo-statistical algorithms in GIS

*Ghadah* Mohammed [1]

[1] Northern Technical University, Kirkuk, Iraq

**Abstract.** Water is great importance for human life as well as for agriculture therefore, different scientific methods are used for curtailed the time and possibilities for well water sources investigation. Fresh water is the most importance for Human life and cannot be healthy without it. In the paper two different interpolation methods were used, which are the ordinary kriging and Inverse Distance Weighting (IDW) to predict static and dynamic water levels of wells in Kirkuk City, and the number of wells was 64. The data of wells were obtained from digging water wells in Directorate of Kirkuk. By using different methods different results were introduced. Furthermore, three wells were used as testing wells for interpolation algorithms.These are office of the Nation Mission, the Emergency department, and Wahid Athar district wells and it was found that the Emergency department well is the most accurate and nearest from the actual levels for both static and dynamic with RMSE of – 2.23 and – 1.35 in kriging method respectively as well as – 2.97 and – 2.65 respectively in IDW method. However, the average of the three testing wells overall IDW interpolation algorithm is the best, accurate and nearest from the true elevations in stable levels of wells.

## 1 Introduction

With the necessary human needs for water, the various scientific methods used to increase saving time and possibilities to search for well water sources. Fresh water considered the most important for human life, as life itself is not possible without it. Also fresh water cannot be substituted by anything else. Fresh water and the various natural surface water bodies had used and consumed by human for a wide range of purposes (1). At the same time, in many Western Asian countries the further extraction of surface and groundwater resources through the subsidization and centralization of large- and small-scale agricultural production result from changing development paradigms and political uncertainties prompted the adoption of national policies to pursue food security through food self-sufficiency (3).

There is an ever-growing demand for information on groundwater hydrology and on the hydraulics of the movement of water in aquifers because groundwater plays an important role in the developing and managing of water resources. (2). Therefore, the search, store and use them, as an alternative is necessary for surface water. The method used wells to extract groundwater from the earth in various depths and the various compounds contained in the water and by the geology of the ground core. Static and dynamic water levels were predicted in this study.

During the few years, the most popular computer program is the Geographic Information System (GIS). By GIS program the data cannot acquire, always updating, as well as we can analyzing and modelling the data by it (5). Many studies by GIS were prepared for
processing region and city center. (4) a Hydrogeological study about Adana city settlement by GIS program. Water quality analyze maps prepared and groundwater level maps. Researches were prepared for aquifers and basins of larger areas with the same way.

In this study, static and dynamic water levels of wells in the Kirkuk City were selected, which are 64 wells. The geographical locations of the wells were converted to the UTM coordinates system by using website to convert them online. Then, I corrected the coordinates of some wells locations by our experience about the study area. For prediction the static and dynamic water level of Kirkuk City, the interpolation analyst in GIS with two difference methods were used, which are Kriging and Inverse Distance Weighting (IDW). Furthermore, I compared between the two interpolation methods after testing them by three wells. The main objectives of this study are generating data for wells for areas that unobserved or do not have available wells as well as comparing different geostatistical interpolation methods and choosing the best one close to the.

## 2 Material and methods

In this paper, geo-statistical algorithms ordinary kriging and Inverse Distance Weighting (**IDW**) used to predict static and dynamic water levels of wells in the selected area (Kirkuk City) which are 64 wells. The data and

© The Authors, published by EDP Sciences. This is an open access article distributed under the terms of the Creative Commons Attribution License 4.0 (http://creativecommons.org/licenses/by/4.0/).

information of wells Kirkuk city obtained from digging water wells Directorate of Kirkuk.

## 3 Study area

Fig. (1) Shows Kirkuk City located in north of Iraq is the capital of the Kirkuk province. A city located at longitude 44° 00′E to 44° 50′ E and latitude 35 ° 13′ N to 36 ° 29′ N with a total area of 96.79 km2 .it is 236 kilometers north of the capital, Baghdad. The average elevation of Kirkuk is 346 meters. Kirkuk is located in a vast area with a very diverse population and dramatic demographic changes in the twentieth century. Kirkuk has multilingual, and the development of distinct ethnic groups was a process that took place throughout the twentieth century of urbanization in Kirkuk.

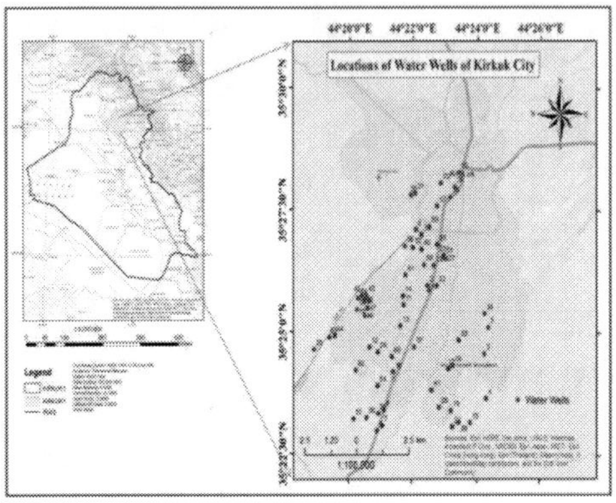


**Fig. 1.** Location map of study area and observation water wells.

## 4 Interpolation methods

### 4.1 Ordinary kriging

Kriging is a geo-statistical interpolation method, which is the statistical models that include autocorrelation based on statistical models—which is, the statistical relationships among the measured points. That is the reason behind the geo-statistical techniques capable to predict the surface of the production in additional to that it provides a certainty measurements or precise predictions.

Kriging's theory presumably that the direction or distance among sampling points reflects a spatial link which, can be used to illustrates the variation in the surface. This tool commensurate with the mathematical function for some points or all points within a specified radius for determining the output value for each location. Kriging tool are a multi-stage operations. It involves modeling of variables, exploratory statistical analysis of data, (optionally) exploring the surface of variance, as well as creating a surface. Kriging is the best tool when there is directional bias in the data or a spatially connected distance. They often used in geology and soil science (6).

$$\hat{Z}(S_0) = \sum_{i=1}^{N} \lambda_i Z(S_i) \qquad \text{..................1}$$

Where:

$Z(S_i)$ = the value measured at the ith location

$\lambda i$ = an unknown weight for the value measured at the ith location

$S_0$ = forecast site

N = the number of values measured

### 4.2 Inverse Distance Weighted (IDW)

Inverse Distance Weighted is the simplest, most readily, and one of the most precise techniques of interpolation. The assumption behind IDW is that the value of non-sampled points can estimated as a weighted average of values at sampled surrounding points As much; the surrounding sampled points are near to the non-sampled point, as the value accuracy of the point increases (7).

$$Z_o = \frac{\sum_{i=1}^{N} z_i d_i^{-n}}{\sum_{i=1}^{N} d_i^{-n}} \qquad \text{..................2}$$

$Z_o$ = the estimation value of variable z in point i.

$Z_i$ = sample value in point i.

$d_i$= distance between sample point and estimated point.

N = A coefficient that determines weigh based on a distance.

## 5 Results and discussion

### 5.1. Interpolation results

#### *5.1.1 IDW interpolation of static elevation*

Fig. (2) Shows the IDW interpolation method for elevation map of Kirkuk City wells. We can classify the stable elevations in to four main classes, which are very high, high, moderate, and low. The first class is starting from 256.2079322 cm to 268.1243654 cm, from 268.1243655 cm to 294.9363403 cm, which represents the highest stable elevation represented by white and light purple colors, and distributes in the north part of study area. The second class is starting from 219.8628107 cm to 240.7165688 cm and from 240.7165689 cm to 256.2079321 cm illustrates the high class, which is the dark purple and dark brown colors and distributes in the center, east, and west parts of study.

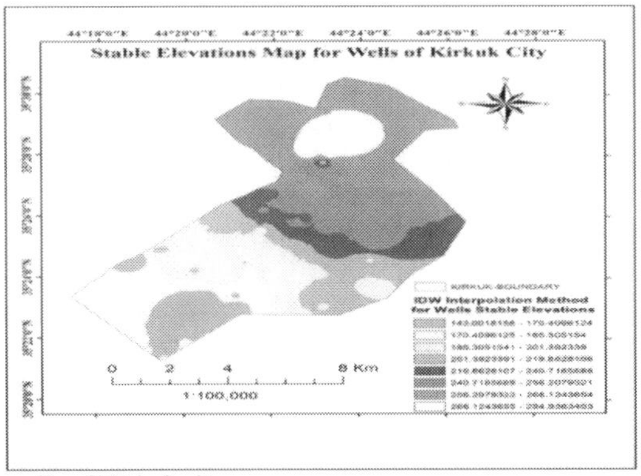


**Fig.2.** IDW interpolation of Stable elevation map for wells of Kirkuk city.

area. The third class is starting from 185.3051541 cm to 201.392339 cm, from 201.3923391 cm to 219.8628106 cm, illustrates the moderate class represented by light brown and light yellow colors, and distributes in the center, and eastern and western south parts of study area.

The fourth class is starting from 143.0018158 cm to 170.4096124 cm, from 170.4096125 cm to 185.3051514 cm and the low class represented by celestial and off-white colors, and distributes in the south and western south parts of the study area map.

### *5.1.2. IDW interpolation of Dynamic elevation*

Fig. (3) shows the IDW interpolation method for dynamic elevation map of Kirkuk City wells. We can classify the stable elevations in to four main classes, which are very high, high, moderate, and low. The first class is starting from 273.3054678 cm to 287.2524524 cm, from 287.2524525 cm to 311.927887 cm illustrates most high stable elevation represented by orange and dark red colors, and distributes in the north part of study area. The second class is starting from 243.8022308 cm to 258.8220604 cm and from 258.8220605 cm to 273.3054677 cm illustrates the high class, which is the yellow and green colors and distributes in the center, east, and west parts of study area. The third class is starting from 209.4711914 cm to 225.5638661cm and from 225.5638662 cm to 243.8022307 cm, clears the moderate class represented by dark blue and celestial colors, and distributes in the center and eastern and western south parts of study area. The fourth class is starting from175.140152 cm to 198.7427415 cm and from 198.7427416 cm to 209.4711913 cm and clears the low class represented by pink and white colors and distributes in the south, eastern and western south parts of the study area map.

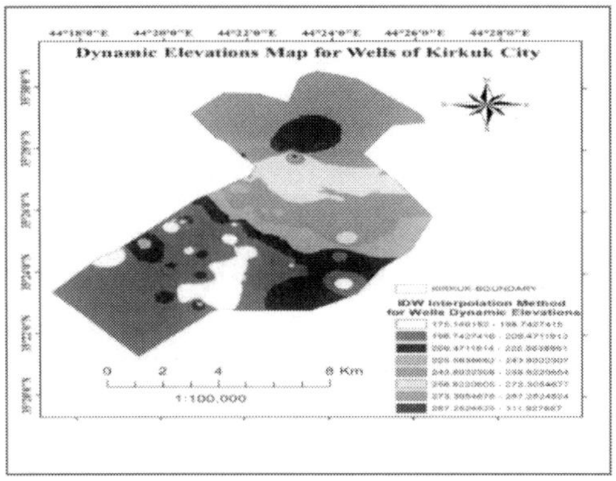


**Fig. 3.** IDW interpolation of dynamic elevation map for wells of Kirkuk city.

### *5.1.3 Kriging interpolation of Stable elevation*

Fig. (4) states the kriging interpolation method for Stable elevation map of Kirkuk City wells. The stable elevations can be classified in to four main classes, which are very high, high, moderate, and low. The first class is starting from 267.7103578 cm to 286.4908393 cm, from 286.4908394 cm to 299.6371765 cm illustrates most high stable elevation represented by red and dark orange colors, and distributes in the north part of study area. The second class is starting from 227.0193141 cm to 247.0518278 cm and from 247.0518279 cm to 267.7103577 cm illustrates the high class, which is the light orange and yellow colors and distributes in the center, east, and west parts of study area. The third class is starting from 184.4502223 cm to 205.1087521 cm and from 205.1087522 cm to 227.019314 cm, clears the moderate class represented by light yellow and light green colors, and distributes in the center and eastern and western south parts of study area. The fourth class is starting from 140.00308823 cm to 160.6616121 cm and from 160.6616122 cm to 184.4502222 cm and shows, the low class represented by light green and dark green colors and distributes in the south part of the study area map.

### *5.1.4. Kriging interpolation of Dynamic elevation*

States the kriging interpolation method for Dynamic elevation map of Kirkuk City wells. We can classify the elevations in to four main classes, which are very high, high, moderate, and low. The first class is starting from 283.7948575 cm to 304.3256078 cm, from 304.3256079 cm to 316.5330811cm illustrates most high stable elevation represented by dark red and dark orange colors, and distributes in the north part of study area. The second class is starting from 246.0626676 cm to 262.1543368 cm and from 262.1543368 cm to 283.7948574 illustrates the high class, which is the light orang and desert colors and distributes in the center, east, and west parts of study area. The third class is starting from 212.2146737 cm to 228.3063428 cm, from 228.3063429 cm to 246.0626675 cm, clears the moderate class represented by grey and light grey colors, and distributes in the center, and eastern and western south parts of study area. Fig. (5) illustrate the fourth class is starting from 175.0373688 cm to 198.3425449 cm and from 198.342545 cm to 212.2146736 cm and clears, the low class represented by light and dark blue colors and distributes in the south, eastern and western south parts of the study area map.

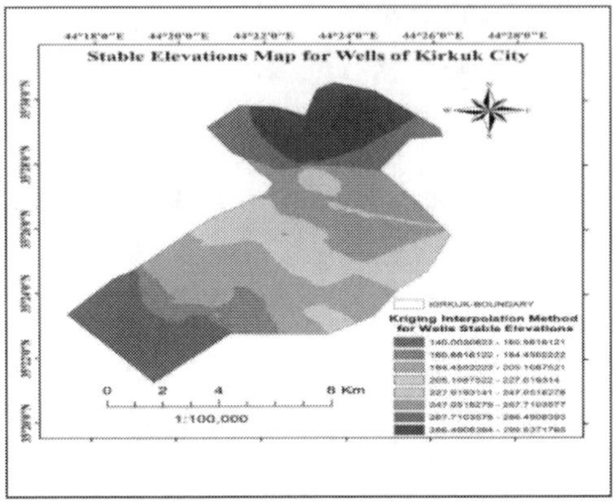


**Fig. 4.** Kriging interpolation of Stable elevation map for wells of Kirkuk city.

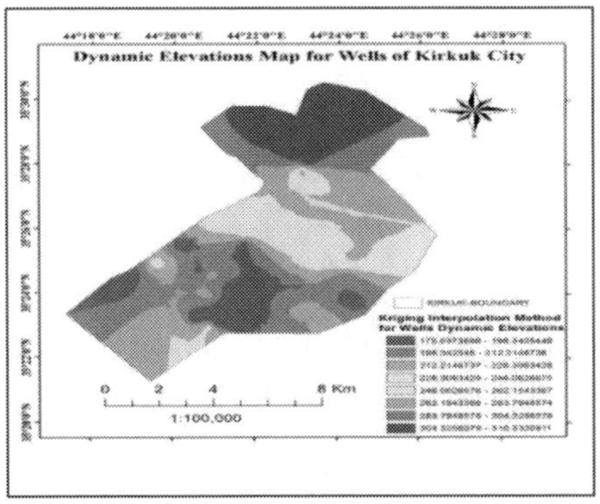


**Fig. 5.** Kriging interpolation of Dynamic elevation map for wells of Kirkuk city.

## 5.2 Validation

Three wells used to test the interpolation methods, which are Office of the United Nations Mission well, the emergency department well, and Wahid Athar district well. The testing wells distributed on study area by homogeneous form. After that, the testing wells checked with the interpolation predict information to examine the validation of each methods i.e. ordinary kriging and IDW methods. The results of testing are illustrated in Table 1 & 2, as follows: in kriging method, the Emergency department well was the most accurate and nearest from the actual levels, where the difference in level between the testing and interpolation result for both static and dynamic with – 2.23 cm and – 1.35 cm, respectively. Likewise the same testing well which is the Emergency department got accurate and closer results from IDW interpolation method, where 2.97 cm and – 2.65 cm for static and dynamic respectively. And so on, the average difference level of the three testing wells overall IDW interpolation algorithm is the best, accurate and closer from the true elevations in stable levels of wells.

**Table 1.** Validation in levels between Testing wells water level and Ordinary Kriging Method.

| ID | NAME | SL | DL | SLK | DLK | DSL | DDL |
|---|---|---|---|---|---|---|---|
| T1 | OOTUNM[1] | 274 | 280 | 264.36 | 284.05 | 9.64 | -4.05 |
| T2 | TED[2] | 192 | 196 | 194.23 | 197.35 | -2.23 | -1.35 |
| T3 | WAD[3] | 192.3 | 216.3 | 181.82 | 183.75 | 10.48 | 32.55 |
| | Average Difference Level | | | | | 17.89 | 27.15 |

**Table 2**. Validation in levels between Testing wells water level and Inverse Distance Weighted IDW Method.

| ID | NAME | SL | DL | SLK | DLK | DSL | DDL |
|---|---|---|---|---|---|---|---|
| T1 | OOTUNM[1] | 274 | 280 | 262.05 | 280.93 | 11.95 | -0.93 |
| T2 | TED2 | 192 | 196 | 194.97 | 198.65 | -2.97 | -2.65 |
| T3 | WAD3 | 192.3 | 216.3 | 187.53 | 198.18 | 4.77 | 18.12 |
| | **Average Difference Level** | | | | | 13.75 | 14.54 |

Where:

**OOTUNM**[1]: Office of the United Nations Mission

**TED**[2]: The emergency department

**WAD**[3]: Wahid Athar district

## 6 Conclusions

Using different interpolation methods for prediction a static and dynamic water levels methods can give more options to indicate the more accurate one. The single interpolator cannot produce chief results for the generation of static and dynamic water levels maps, therefore we always using two-interpolation method. Four maps was produced, two for static water level in two methods ordinary kriging and IDW and they give some difference in spatial distribution of the static level due to the difference in equation and mechanism of each method. Furthermore, another two maps for dynamic water level as well same to previous two maps that produced by two different interpolation methods ordinary kriging and IDW with some difference in results of spatial distribution of dynamic water levels of wells. Generally, by using the validation technique for evaluation, all of the methods gave similar RMSE values. The method used in this study (kriging method) had better performance for pH in surface of soil also the lognormal kriging excellence both IDW and splines to interpolating electrical conductivity in surface soils. To identify most accurate interpolation method, three testing wells had used to validate which method is the best. The Emergency department well was the most accurate and nearest to the truth results of both static and dynamic water levels and even by using both ordinary kriging and IDW methods. However, IDW interpolation algorithm was the best, accurate and nearest than the ordinary kriging in overall average difference between testing and predicting stable and dynamic water levels of wells.

## Acknowledgment

The author would like to express my thanks to the staff of the general commission for groundwater. Especially the Head of Department Dr. Ahmed Nazem, And I have to extend my sincere thanks to Eng- Fouad Mohammed sadiq . For his advice during my research. In addition, I have to extend my sincere thanks to Mr. Ali jassim nayyef and Mr. Sabah Ahmed Khorshid for their advice during work in my research.

## References

1. Shiklomanov, Igor A. "World water resources." A new appraisal and assessment for the 21st century (1998).
2. Bear, Jacob. Hydraulics of groundwater. Courier Corporation, 2012.
3. Brander, Luke M., et al. Ecosystem Services 1.1 (2012): 62-69.
4. Çobanoğlu, İbrahim, Et Al. "Coğrafi Bilgi Sistemi (Cbs) Kullanilarak Adana İli Yerleşim Alaninin Hidrojeolojik Özellikleri Ve Su Kalitesinin Değerlendirilmesi."
5. Çelik, Recep. Arabian Journal of Geosciences 8.6 (2015): 4279-4286.
6. Di, Minin Enrico, et al. "Global priorities for national carnivore conservation under land use change." (2016).
7. Ziary, Y., and H. Safari. Proceedings of FIG Working Week. Hong Kong 13 (2007).

# Evaluation of different types of artificial intelligence methods to model the suspended sediment load in Tigris River

*Mustafa* Al-Mukhtar[1*]

[1]Building and Construction Engineering Department, University of Technology, Baghdad

**Abstract.** Modeling of suspended sediment load in rivers has a major role in a proper management of water resources. Artificial intelligence has been identified as an efficient way to model the complex nonlinear hydrological relationship. In this study, Adaptive Neuro Fuzzy Inference System (ANFIS), in addition to two different kinds of Artificial Neural Network (ANN) i.e. feedforward and radial basis networks were used and compared to model the suspended sediment load (SSL) in Tigris River-Baghdad using the streamflow discharge as input. To this end, an intermittent data of SSL and streamflow were collected over the period 1962-1981 from Sarai station in Baghdad. 70 % of these data was used to calibrate (train) the networks and the remaining 30% for the validation (test). The coefficient of determination ($R^2$), root mean square error (RMSE), and Nash and Sutcliffe model efficiency coefficient (NSE) were used to judge whether the observed and modelled data belong to the same distribution. Results revealed that the ANFIS model outperform the other methods. $R^2$, RMSE, and NSE of ANFIS during the calibration phase were equal to 0.58, 75617, and 0.58, respectively and during the validation were 0.72, 27944, and 0.59, respectively. Therefore, ANFIS approach is recommended to estimate the river suspended sediment load.

## 1 Introduction

Modelling sediment transport accurately is of great importance tool in many fields of water resources management such as water quality, reservoir management, designing of dam and operation. It is well known that higher concentrations of suspended sediment are, the higher risky consequences on river ecology [1]. Therefore, it is of great importance to estimate the inflow amounts of sediment in water systems.

Artificial intelligence approaches such as artificial Neural Network (ANN), support vector regression, fuzzy logic (FL) are used in prediction of suspended sediment [2] as they provide a good alternative to conventional approaches like linear or deterministic models.

Artificial neural network (ANN) is a tremendously parallel- distributed information processing system, which is obtained from research on the nature of human brains [3]. Many advantages of this technique and in particular in the field of hydrology were proved. Foremost, it can manipulate with any arbitrary continuous functions, modeling a nonlinear system considering no prior assumption of processes involved, and ultimately can give a good solution when input data are incomplete or vague [4,5]. Recently, many successful applications of ANN's in the field of sediment transport were reported [6-15].

The Adaptive Neuro Fuzzy Inference System (ANFIS) is a kind of ANN that integrates both neural network and fuzzy logic principles [16]. It is characterized by the capability of capturing the benefits of both items in a single framework. In addition, the set of fuzzy IF–THEN rules that have learning capability to approximate nonlinear functions is adopted in its inference system [17]. Recently, many successful applications of ANFIS in the field of water resources were reported. For example, [18] applied the ANFIS model to modeling groundwater level fluctuation using hydro meteorological data (precipitation and temperature). They compared the results from the ANFIS model with multiple layer perceptron optimized by two different approach i.e. scaled conjugate gradient and Bayesian neural networks. They found that ANFIS outperform the other methods in modeling the noise-free hydro meteorological data. [19] used the ANFIS model to simulate the soil moisture (SM) in an arid area of a basin in China. Their results showed that ANFIS model could give accurate estimation of SM and can be successfully applied for SWC forecasting. [20] used the ANFIS model to estimate the suspended sediment concentration in Mad River Catchment near Arcata, USA. They compared results from ANFIS model with the generalized regression neural networks (GRNN), radial basis neural networks (RBNN) and multi-layer perceptron (MLP) and two different sediment-rating curves (SRC). They found that ANFIS model perform better than the other models in daily suspended sediment concentration estimation.

The objectives of this study is to investigate the capability of two types of artificial neural networks (i.e.

*Corresponding author: mmalmukhtar@gmail.com

© The Authors, published by EDP Sciences. This is an open access article distributed under the terms of the Creative Commons Attribution License 4.0 (http://creativecommons.org/licenses/by/4.0/).

feedforward and radial basis networks) and the ANFIS model to simulate instantaneous suspended sediment load in Sarai station-Baghdad using corresponding streamflow discharge as inputs. Therefore, three codes were prepared with the use of MATLAB software (Release 2008a). To the best of knowledge, no previous studies were conducted to simulate the suspended sediment load in Tigris River-Iraq using these approaches. The positive results from this study could be of use to provide a better insight for the stakeholder about the sediment load in Baghdad.

## 2 Material and methods

### 2.1 Adaptive Neuro Fuzzy Inference System ANFIS

ANFIS is a special type of ANNs which integrates both neural networks and fuzzy logic principles [21]. Thus, it considers a universal approximator as it has the capability of approximating any real continuous function on a compact set to any degree of accuracy using either Takagi–Sugeno or Mamdani fuzzy inference system (if-then rule) [22]. "Takagi and Sugeno's fuzzy if-then rules are characterized as the output of each rule is a linear combination of the input variables plus a constant term, and the final output is the weighted average of each rule's output" [21]. A hybrid-learning rule, which combines the gradient method and the least squares estimate to identify parameters pertaining membership function (MF), was adopted to be used in ANFIS since it is faster and not trapped by local minima as in the basic rule. As a typical example it is assumed a FIS has two parameters i.e. x and y and one output f. The first-order Sugeno's fuzzy model (a typical rule set with two fuzzy If-Then rules) might be written as:

$$\text{Rule 1: If x is } A_1 \text{ and y is } B_1;\ \text{then } f_1 = p_1x + q_1y + r_1 \quad (1)$$

$$\text{Rule 2: If x is } A_2 \text{ and y is } B_2;\ \text{then } f_2 = p_2x + q_2y + r_2 \quad (2)$$

Where $p_1$, $q_1$, $r_1$ and $p_2$, $q_2$, $r_2$ are the variables in the then-part (consequent part) of the first-order Sugeno fuzzy model. The general structure of ANFIS model is shown in Figure 2. A comprehensive details about ANFIS functioning could be found in [21]. In this study, each run with different numbers of member functions (MF) were tried and the one that gave the minimum squared error was selected. Triangular-shaped built-in MFs for the ANFIS models was found to be efficient for modelling SSL.

### 2.2 Artificial neural networks ANN's

#### *2.2.1 Feedforward neural network*

Feedforward neural network *(FF)* is a multi-layer perceptron type in which nodes in one layer are only connected to nodes in the next layer, it has been used in many prediction subjects [23]. "Typically, feedforward networks have one or more hidden layers of sigmoid transfer function followed by an output layer of linear transfer function. Multiple layers of neurons with nonlinear transfer functions allow the network to learn nonlinear and linear relationships between input and output vectors. The linear output layer lets the network produce values outside the range -1 to +1" [24]. A supervised learning technique called backpropagation is used to train the network. The backpropagation computation is derived using the chain rule of calculus [25]. The default algorithm used in optimizing weights between connections is Levenberg-Marquardt backpropagation [25].

#### *2.2.2 Radial basis neural network*

Radial basis network (RB) typically have three layers: an input layer, hidden layer with non-linear radial basis function (RBF), and linear output layer. The RBF estimates the output by using the standard Euclidean distance between the inputs and its corresponding weight [4, 24] . In other words, for each node the Euclidean distance between the center and the input vector is estimated, and then transformed by a non-linear function (Exponential) that determines the output from the hidden layers. The radial basis function has a maximum of 1 when its input is 0. When the distance between the weight (**w**) and input (**p**) decreases, the output increases. Thus, a radial basis neuron acts as a detector that produces 1 whenever the input **p** is identical to its weight vector **w**.

## 3 Model Assessment Criteria

Three most widely used criteria were used in this study in order to test the validity of the evaluated models, which are:

i. Determination coefficient ($R^2$): the determination coefficient (equation 3) "describes the proportion of the variance in observed data explained by the model. Its values range from 0 to 1. Higher values indicate less error variance and typically values greater than 0.5 are considered acceptable" [26, 27].

$$\mathbf{R^2} = \left[\frac{\sum_{i=1}^{n}(O_i-\bar{O})(P_i-\bar{P})}{\sqrt{\sum_{i=1}^{n}(O_i-\bar{O})^2}\ \sqrt{\sum_{i=1}^{n}(P_i-\bar{P})^2}}\right]^2 \quad (3)$$

Where $O_i$ is the observed value, $\bar{O}$ is the average observed value, $P_i$ is the predicted value, and $\bar{P}$ is the average predicted value.

ii. Root mean square error (RMSE): root mean square error (equation 4) is one of the error indices commonly used in model evaluation. The closer value to zero, the better model performance.

$$\mathbf{RMSE} = \sqrt{\frac{\sum_{i=1}^{n}(O_i - P)^2}{n}} \quad (4)$$

iii. Nasch and Sutcliffe Model coefficient efficiency (NSE): The NSE which compares the simulated values with the average observed is calculated from:

$$\text{NSE} = 1 - \frac{\sum_{1}^{n}(O_i - S_i)^2}{\sum_{1}^{n}(O_i - \bar{S})^2} \quad (5)$$

Where: $O_i$ is the observed value; $S_i$ is the simulated value; and $\bar{S}$ is the average value of observations. It ranges from -∞ to 1. When the value approaches to one, the prediction is better. The value of less than or equal to zero means that average values if used as predicted would be better than the model.

## 4 Case study and application

The intermittent instantaneous suspended sediment load and streamflow of Tigris River at Baghdad (Sarai station, latitude 33° 18′ N, longitude 44° 23′ E) operated by the Iraqi ministry of water resources are used in the study. The drainage area at this site is 134000 km$^2$. The 112 daily flow and suspended sediment load data (1962 to 1981 water years) are used to train (calibrate) and testing (validation) the ANFIS and ANN models. The first 80 values were used for calibration and the remaining 32 for testing. Fig. 1 shows the locations of the Sarai sediment station at Baghdad.

Using the ANFIS model, different membership function was tried in order to find the optimum function that represent the observed suspended sediment load based on the assessment criteria ($R^2$, RMSE, and NSE). Thus, it was found that the triangular-shaped built-in membership function was the best representative of model prediction. For modelling the suspended sediment with the ANN's, all the data were standardized to be within [-1, 1].

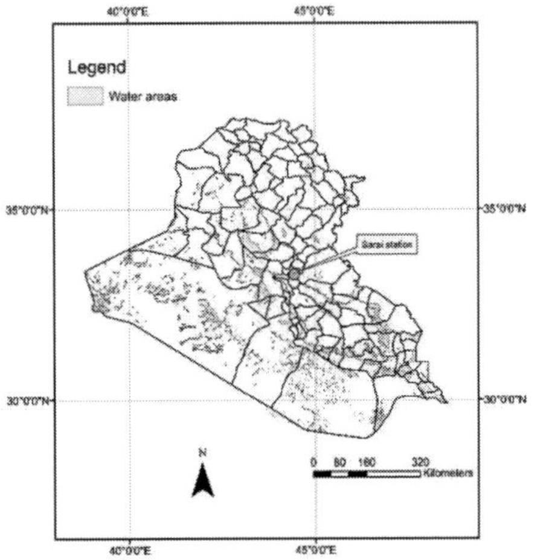


**Fig. 1.** The study area.

## 5 Results and Discussion

In each application of the evaluated methods, two data sets were used. The first was for the calibration and the second for the validation. In order to assess how well the methods performed, three evaluation criteria were used ($R^2$, RMSE, NSE). Table 1 shows the performance criteria of the evaluated methods. As it can be seen that $R^2$, RMSE, and NSE of ANFIS were 0.58, 75617 and 0.58, respectively during the calibration and 0.72, 27944, and 0.59, respectively during the validation. While for the feedforward network were 0.55, 101830, and 0.54, respectively during the calibration and 0.71, 30395, and 0.50 respectively during the validation. Using the radial basis network the values of $R^2$, RMSE, and NSE were 0.64, 89867, and 0.64, respectively during the calibration and 0.61, 35190, and 0.33, respectively during the validation. The descending order of the performance was ANFIS, FFN, and RBN. Thus, it can be concluded that the ANFIS model outperform the other two evaluated methods. this could be attributed to the fact that the ANFIS model combines the linguistic representation of a fuzzy system with the learning ability of the ANN [20]. Thus, can be strongly eliminating pseudo signals (noise).

Figs. 2 and 3 show the observed suspended sediment load versus the modelled with the ANFIS model during the calibration and validation periods, respectively. As it can be seen that the modelled data were in a good agreement with the observed. Moreover, it can be noticed that the peaks values of observations during the both data sets (calibration and validation sets) were in a rather good consistence with those modelled. Figs. 4 and 5 show the observed and modelled suspended sediment load with the feed forward network during the calibration and validation, respectively. It can be seen that the performance during the calibration of the feedforward network was pretty well in matching the observed data set. While, however, during the validation there was a significant difference between the both data sets. As such it could be interpreted the lower values of $R^2$, RMSE and NSE during the validation period with respect to ANFIS (Table 1). The feedforward network was not capable to model the peaks values of suspended load during the validation period. Peak values of sediment load are of significant importance in reservoir and dam management. Figs. 6 and 7 show the observed and modelled sediment load with the radial basis neural network during the calibration and validation, respectively. As with the feedforward, the radial basis network was pretty well capable to represent the observed sediment data during the calibration. While, during the validation, a divergence between the both data sets was observed.

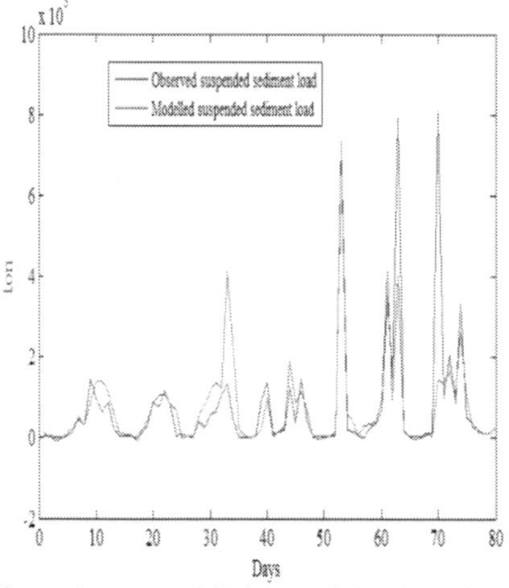


**Fig. 2.** Observed versus modelled suspended sediment load from ANFIS during the calibration.

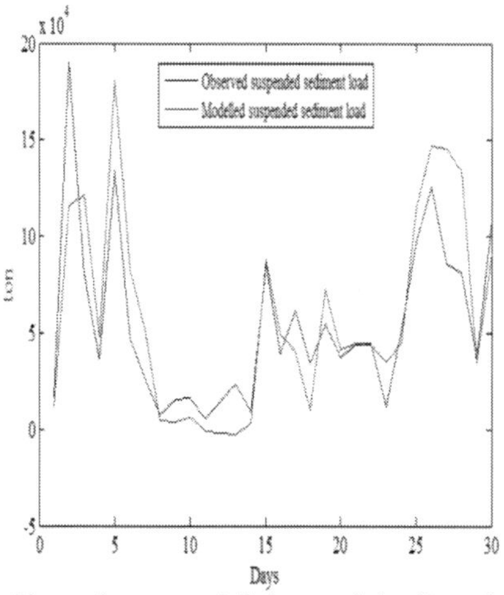


**Fig. 3.** Observed versus modelled suspended sediment load from ANFIS during the validation.

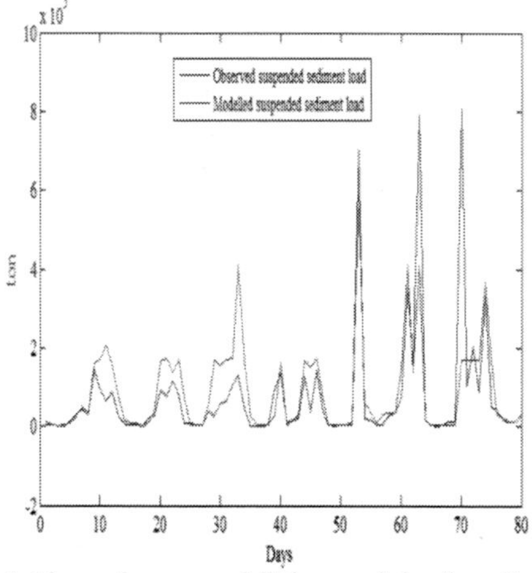


**Fig. 4.** Observed versus modelled suspended sediment load from feedforward network during the calibration.

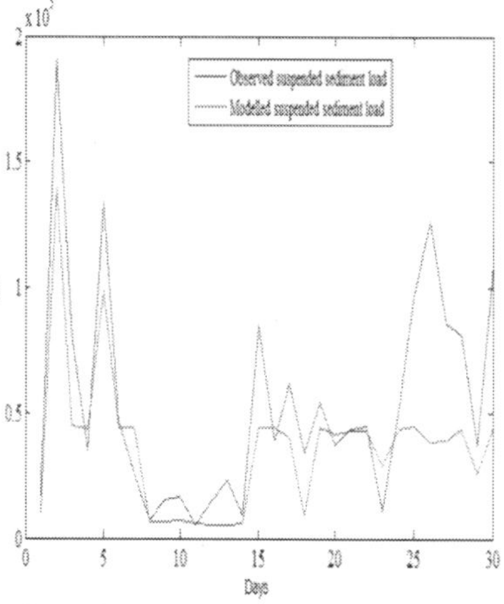


**Fig. 5**. Observed versus modelled suspended sediment load from feedforward network during the validation.

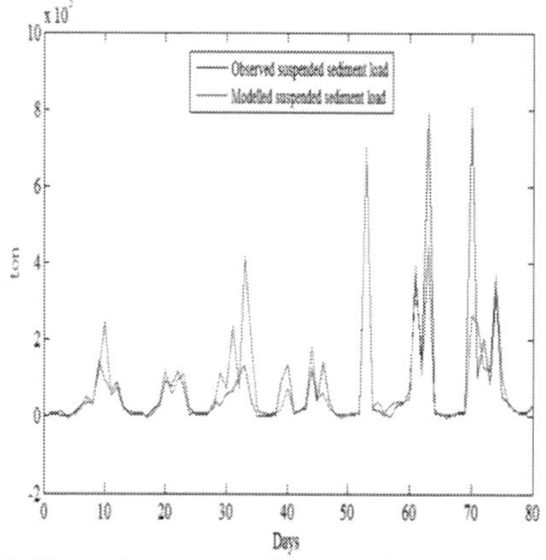


**Fig. 6.** Observed versus modelled suspended sediment load from radial basis network during the calibration.

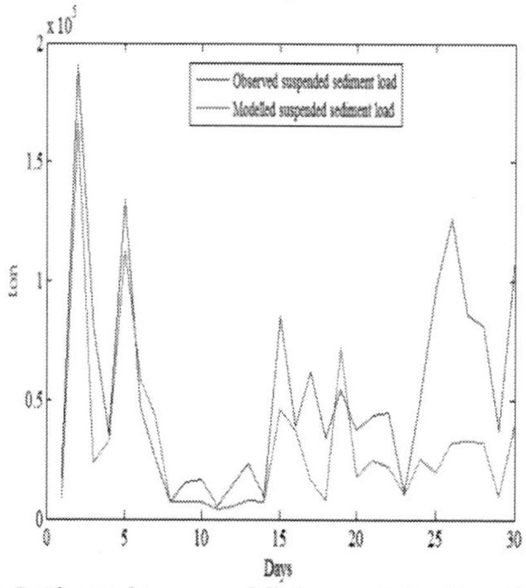


**Fig. 7.** Observed versus modelled suspended sediment load from radial basis network during the validation.

**Table 1.** Performance criteria of the evaluated methods.

| Method | Calibration | | | | Validation | | |
|---|---|---|---|---|---|---|---|
| | R^2 | RMSE | NSE | | R^2 | RMSE | NSE |
| ANFIS | 0.58 | 75617.0 | 0.58 | | 0.72 | 27944.0 | 0.59 |
| FFN | 0.55 | 101830.0 | 0.54 | | 0.71 | 30395.0 | 0.50 |
| RBN | 0.64 | 89867.0 | 0.64 | | 0.61 | 35190.0 | 0.33 |

## 6 Conclusions

This study evaluated three different types of artificial intelligence (ANFIS, FF and RB neural networks) to model the suspended sediment load of Tigris River-Sarai station in Baghdad. Three assessment criteria were used for the evaluation i.e. $R^2$, RMSE and NSE. It was found that the ANFIS outperformed the other evaluated methods. This could be attributed to the ANFIS structure. The ANFIS integrates the advantage of the simplifying function of fuzzy reasoning and the self-learning ability of neural networks and thus gives a strong capability of eliminating noise [14]. This study could be of use to provide a better insight for the stake holders about the amount of suspended sediment load in Tigris River.

## References

1. P. J. Wood and P. D. Armitage, *Environ. Manage.*, **21**, no. 2, pp. 203–217(1997).
2. J. Shiri and Ö. Kisi, *J. Hydrol. Eng.*, **17**, no. 9, pp. 986–1000 (2011).
3. B. Müller, J. Reinhardt, and M. T. Strickland, Springer Science & Business Media (2012).
4. ASCE Task Committee on Application of Artificial Neural Networks, *J. Hydrol. Eng.*, **5**, no. 2, pp. 115–123 (2000a).
5. ASCE Task Committee on Application of Artificial Neural Networks, *J. Hydrol. Eng.*, **5**, no. 2, pp. 124–137 (2000b).
6. M. Alp and H. Cigizoglu, *Environ. Model. Softw.*, **22**, no. 1, pp. 2–13 (2007).
7. F. Anctil, M. Filion, and J. Tournebize, *Ecol. Modell.*, **220**, no. 6, pp. 879–887, ( 2009).
8. H. K. Cigizoglu and M. Alp, *Adv. Eng. Softw.*, **37**, no. 2, pp. 63–68, (2006).
9. M. Çimen, *Hydrol. Sci. J.*, **53**, no. 3, pp. 656–666, (2008).
10. E. Kakaei Lafdani, A. Moghaddam Nia, and A. Ahmadi, *J. Hydrol.*, **478**, pp. 50–62, (2013).
11. H. Kerem Cigizoglu and Ö. Kisi, *J. Hydrol.*, **317**, no. 3–4, pp. 221–238, (2006).
12. Ö. Kisi, *Hydrol. Sci. J.*, **49**, no. 6, pp. 37–41,(2004).
13. Ö. Kisi and M. Zounemat-Kermani, *Water Resour. Manag.*, **30**, no. 11, pp. 3979–3994, (2016).
14. T. Rajaee, S. A. Mirbagheri, M. Zounemat-Kermani, and V. Nourani, *Sci. Total Environ.*, **407**, no. 17, pp. 4916–4927, (2009).
15. Y.-M. Zhu, X. X. Lu, and Y. Zhou, *Geomorphology*, **84**, no. 1–2, pp. 111–125, (2007).
16. J. J.-S. Roger, *AAAI*, , **91**, pp. 762–767 (1991).
17. A. Abraham, *Fuzzy systems engineering*, Springer, pp. 53–83 (2005).
18. S. Maiti and R. K. Tiwari, *Environ. Earth Sci.*,**71**, no. 7, pp. 3147–3160, (2014).
19. J. Si, Q. Feng, X. Wen, H. Xi, T. Yu, W. Li, and C. Zhao, *J. Hydrol.*, **527**, pp. 679–687, (2015).
20. M. Cobaner, B. Unal, and O. Kisi, *J. Hydrol.*, **367**, no. 1–2, pp. 52–61, (2009).
21. J. R. Jang, IEEE, **23,** 3 (1993).
22. J.-S. R. Jang, C.-T. Sun, and E. Mizutani, (1997).
23. H. R. Maier and G. C. Dandy, *Environ. Model. Softw.*, **15**, pp. 101–124, (2000).
24. M. H. Beale, M. T. Hagen, and H. B. Demuth, "Neural Network Toolbox: users guide," (2012).
25. M. T. Hagan, H. B. Demuth, and M. H. Beale, *Neural network design*. (1996).
26. C. Santhi, J. G. Arnold, J. R. Williams, W. A. Dugas, R. Srinivasan, and L. M. Hauck, *J. Am. Water Resour. Assoc.*, **37**, no. 5, pp. 1169–1188, (2002).
27. Ö. Kisi, T. Haktanir, M. Ardiclioglu, O. Ozturk, E. Yalcin, and S. Uludag, *Adv. Eng. Softw.*, **40**, no. 6, pp. 438–444, (2009).

# Possibility of reusing Al-Machraya River for feeding Hawizeh marsh

*Mahmoud* Al-Khafaji [1,*], *Hayder* Al Thamiry [2] and *Ala* Al-Saedi [2]

[1]Building and Construction Engineering Dept., University of Technology, Baghdad, Iraq.
[2] College of Engineering, University of Baghdad, Baghdad, Iraq.

**Abstract.** Al Machraya River was considered as one of the water feeders of Hawizeh Marsh. In 1986, the outlet of this river into the marsh was blocked and the river was used as a main channel for the East Tigris Irrigation Project near Kalat Salih. This causes significant decrease in the available water supply sources, deterioration in the water quality distribution patterns and increasing the stagnation areas within the marsh. This research aims to study the possibility of reusing this river for feeding Hawizeh Marsh. A frequency analysis study was carried out to study the maximum and minimum probable water level (MMPWL) of Tigris River at the upstream of Kalat Salih Barrage. Six statistical models; Normal distribution, Log-Normal type II, Log-Normal type III, Pearson type III, Log- Pearson type III and Gumbel type I distribution were used to estimate the MMPWL. The results show that Pearson type III and Gumbel type I distribution models are the best to fit the maximum and minimum daily water level (WL), respectively, at the upstream of the Barrage. The estimated MMPWL were compared to the required WL in Hawizeh Marsh. The difference between Tigris River and Hawizeh Marsh water levels were found to be not operative to cause a significant flow toward the marsh. Therefore, Al Machraya River cannot be used to feed Hawizeh Marsh.

## 1 Introduction

Al Machraya River was considered as one of the water feeders of Hawizeh Marsh. Its design discharge is 40m3/s. Since 1986, this river is used to irrigate the East Tigris Irrigation Project which is located near Kalat Salih. The intake of this river is located at 4 km upstream of Kalat Salih Barrage Fig.1. This barrage is located at E 716616 N 3490285, about 35 km downstream of Al Amarah City. Kalat Salih Barrage is controlled by Al Amarah Barrage which controlled the water levels in Tigris River downstream of Al Amarah Barrage. Operation of Kalat Salih Barrage can be grouped into three categories according to the period and level of operation; 1979 to 1990, which is the high water level period, 1990 to 2003, which is the low water level period (to reduce the participation of Southern Iraqi marshlands) and 2004 till now, which is the moderate level period (in order to rehabilitate the southern Iraqi marshlands).

The decrease in water resources that feed the marsh causes significant deterioration in the ecological system of the marsh. Sustainability of this system is highly affected by the status of water quality within the marsh in terms of pollution levels and distribution patterns. Location of Al Machraya River outlet into the marsh can contribute effectively to improve the water quality distribution patterns and decrease the stagnation areas within the marsh. In order to reuse this river to feed the marsh, the available water levels in Tigris River at the inlet of this river must be enough to ensure flowing of the required discharge into the marsh. Accordingly, it is of important to study the possibility of satisfying the feeding requirements with the available water level at the inlet and outlet of the river.

This research aims to conduct a comparative study for the maximum and minimum probable water level (MMPWL) in Tigris River at the inlet of Al Machraya River and in the marsh at the outlet of this river. This can be achieved through carrying out a frequency analysis study for the MMPWL at the upstream of Kalat Salih Barrage. For this purpose, statistical distribution models of the maximum and minimum daily water level should be tested to specify and then apply the best fit models. Results of this analysis can be compared with the water levels at the outlet of the river, which are required to achieve the positive operational control of the marsh to ensure the continuity of the ecological system balance of Hawizeh Marsh.

The maximum and minimum daily water levels in Kalat Salih Barrage as measured by Ministry of Water Resources (MoWR) [7] are listed in Table 1.

The water levels in Hawizeh Marsh, which are required to achieve the positive operational control of the marsh to ensure the continuity of the ecological system balance of Hawizeh Marsh are shown in Fig.2, [4].

* Corresponding author: 41100@uotechnology.edu.iq

© The Authors, published by EDP Sciences. This is an open access article distributed under the terms of the Creative Commons Attribution License 4.0 (http://creativecommons.org/licenses/by/4.0/).

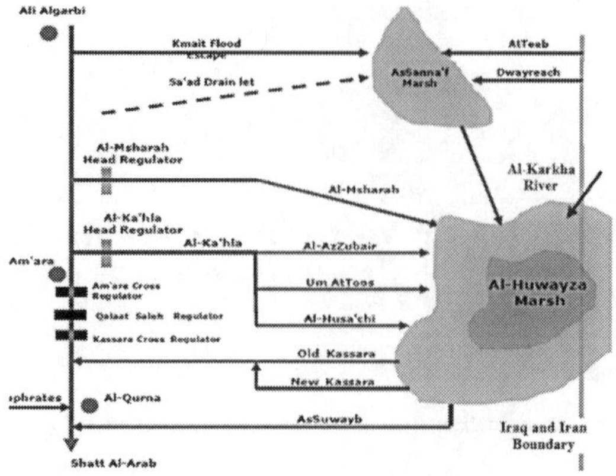


**Fig. 1.** Schematic diagram of the study area [4].

**Table 1**. The Maximum and Minimum daily water levels in Kalat Salih barrage [7].

| Year | Water Surface Elevation (m amsl) | |
|---|---|---|
| | Maximum | Minimum |
| 1996 | 6.40 | 4.00 |
| 1997 | 6.50 | 4.40 |
| 1998 | 6.80 | 4.60 |
| 1999 | 6.80 | 4.40 |
| 2000 | 6.70 | 4.10 |
| 2001 | 6.70 | 4.10 |
| 2002 | 6.30 | 4.40 |
| 2003 | 6.40 | 4.40 |
| 2004 | 7.22 | 5.20 |
| 2005 | 7.06 | 5.61 |
| 2006 | 7.15 | 5.40 |
| 2007 | 5.90 | 4.60 |
| 2008 | 5.12 | 4.12 |
| 2009 | 6.75 | 4.25 |
| 2010 | 5.37 | 4.03 |
| 2011 | 5.30 | 3.99 |
| 2012 | 5.60 | 3.55 |
| 2013 | 4.90 | 3.40 |
| 2014 | 5.42 | 3.77 |
| 2015 | 4.20 | 4.25 |

## 2 Material and methods

### 2.1 Frequency analysis

The MMPWL in Tigris River at the upstream of Kalat Salih Barrage were determined by using six statistical models. These models are; Normal distribution [2], Log-Normal type II [2], Log-Normal type III [2], Pearson type III, Log- Pearson type III [8] and Gumbel type I distribution [6]. These water levels were estimated for return periods of 5, 10, 20, 50 and 100 years. The minimum and maximum daily water levels of 20 years for the period 1996 to 2015 were used to perform the frequency analysis, as shown in Table 1.

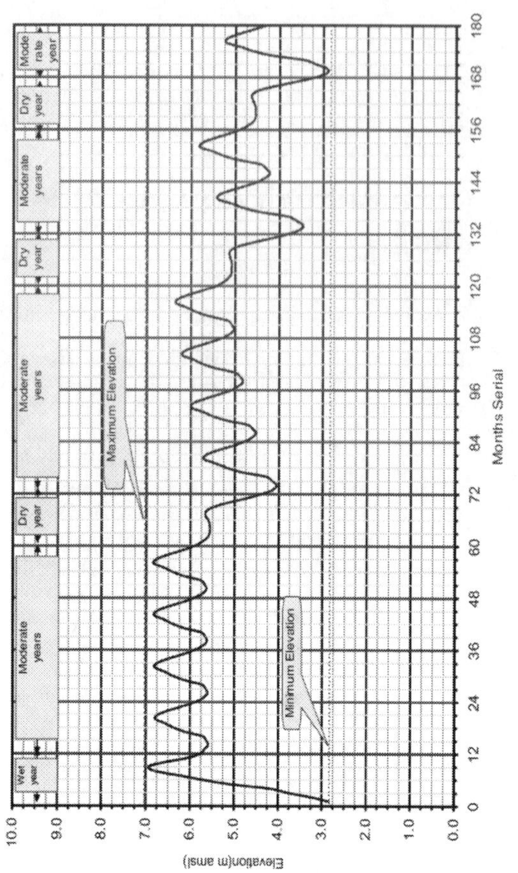


**Fig. 2.** Monthly variation of water level within Al Huwayza Marsh [4].

The parameters of distributions which are; Mean ($\bar{X}$), Standard deviation ($S_d$), Coefficient of variation ($C_v$), Skewness Coefficient ($C_s$) and Kurtosis Coefficient ($C_k$) were estimated by using Method of Moment. These parameters give an indication about which distribution should be accepted [5]. This data has a confidence level between 90-95 % according to the method of reading and the actual situation of existing staff gages in site.

### 2.2 Graphical analysis of fit

There is no exact answer to the question 'Which distribution is best for representing the minimum and maximum water level?'. All the distributions may be fitted and plotted as cumulative density function (CDF) on the normal or log-normal probability papers and may give an answer to such a question [1]. To specify the best statistical distribution models of the maximum and minimum water levels at the upstream of Kalat Salih Barrage, the six statistical distribution models were tested. The cumulative density function (CDF) of these models were fitted and plotted on the normal or log-normal probability papers.

### 2.3 Goodness of fits

A set of data or observations can be described by using some statistical parameters such as the mean and standard deviation. These parameters are estimated from samples. If the samples are unrepresentative then the estimates may be too high or too low. The estimations cannot be used if they differ from expected values by more than certain prescribed limits. Therefore, it is necessary to test the statistics to specify significance of the difference. Such tests are called the test of significance. According to this test, the smallest values of standard error (SE), root mean square error (RMSE) and bias (BIAS) lead to the best fit.

## 3 Results and discussion

The obtained parameters of distribution for the maximum and minimum water levels in Tigris River at the upstream Kalat Salih Barrage are listed in Table 2.

**Table 2.** Obtained parameters of distribution for minimum and maximum water levels at Tigris River upstream Kalat Salih Barrage.

| Parameter | Maximum water level | Minimum water level |
|---|---|---|
| Mean ($\bar{X}$) | 6.13 | 4.33 |
| Standard deviation ( $S_d$ ) | 0.849 | 0.561 |
| Coefficient of variation ( $C_v$ ) | 0.084 | 0.132 |
| Skewness Coefficient ( $C_s$ ) | -0.680 | 0.816 |
| Kurtosis Coefficient ( $C_k$ ) | 2.672 | 3.466 |
| Mean of logarithmic water levels ($\bar{Y}$) | 1.803 | 1.458 |
| Standard deviation for logarithmic water levels ( $S_{dy}$ ) | 0.147 | 0.126 |
| Coefficient of variation for logarithmic water levels ( $C_{vy}$ ) | 0.054 | 0.078 |
| Skewness Coefficient for logarithmic water levels ( $C_{sy}$ ) | -0.796 | 0.386 |
| Kurtosis Coefficient for logarithmic water levels ( $C_{ky}$ ) | 2.771 | 2.897 |

The normal distribution sample estimates of the skewness and kurtosis for the maximum water levels are -0.680 and 2.672, whereas for minimum water levels are 0.816 and 3.466, respectively. These parameters are neither close to zero nor close to 3. Therefore, the Normal distribution may not be a suitable distribution for the maximum and minimum daily water level.

The computed skewness of Log-Normal type II distribution for the maximum and minimum water levels ($Cs = 3Cv + Cv^3$ according to [3]) are 0.40 and 0.25 respectively, which are quite incomparable to Cs = -0.680 and 0.816. Furthermore, $C_{sy}$ for maximum and minimum water levels are -0.796 and 0.386, respectively, which are different from zero. While $C_{ky}$ of logarithmic distribution model for maximum and minimum water levels are 2.771 and 2.897, respectively which are close to 3.000. This distribution is not satisfying all the conditions. Therefore, it is not the suitable distribution.

The accuracy of log-normal type III distribution was checked graphically or from the goodness of fit tests. as this distribution has no limitations.

For Log-Pearson type III distribution, the skewness coefficient of logarithm of data should be greater than zero [3]. Since the sample estimates of the population skewness for maximum and minimum water levels are -0.796 and 0.386, respectively, which are greater than zero for minimum water level, then the Log-Pearson type III distribution can be accepted for the minimum water levels only.

The best values of skewness and kurtosis for Gumbel distribution are 1.14 and 5.4, respectively [3]. The computed sample estimates of the skewness and kurtosis are not close to Gumbels skewness and kurtosis. Also, Gumbel distribution or any of the other distributions cannot refuse until checking the result of those distributions with the available data.

The plotted cumulative density functions (CDF) on the normal or log-normal probability papers for the considered six models are shown in Fig.s 3 to 14. The maximum and minimum water levels at Tigris River upstream Kalat Salih Barrage for different return periods and distributions are given in Table 3. The maximum water level is 7.00, 7.14 and 7.23 m amsl for return period of 20, 50 and 100 years, respectively. These levels may occur in wet years only, during which the water level at Hawizeh Marsh will be at maximum level, 7 m amsl. The total difference between Tigris River and Hawizeh Marsh water levels will not be operative to cause a significant flow towards the Marsh.

Kalat Salih Barrage is controlled by Al Amarah Barrage and there are many irrigation projects along the reach between those two barrages. So, the water levels upstream Kalat Salih Barrage will not reach any extreme values. Additionally, in flood season, Kumait escape (40 km upstream of Al Amara City, with maximum discharge of 400 $m^3$/sec and a weir crest at 8 m amsl) will divert large share of the flood discharge towards Hawizeh Marsh throughout AsSanna'f Marsh.

Consequently Al Machraya River cannot be used as a feeder for Hawizeh Marsh under the present conditions of the Tigris River and Hawizeh Marsh.

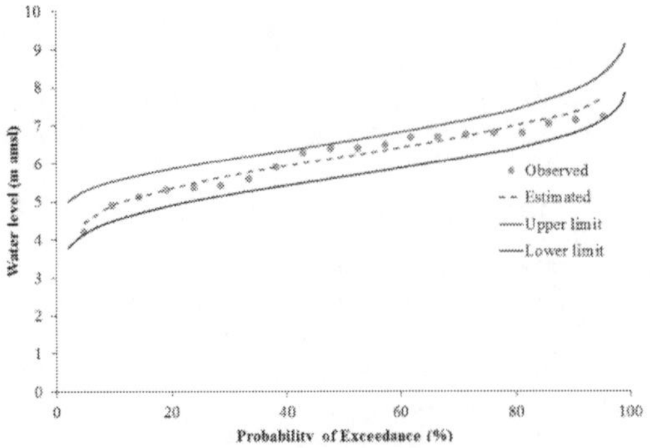


**Fig. 3.** Normal Plot of Maximum Daily Water Level with 95% confidence intervals.

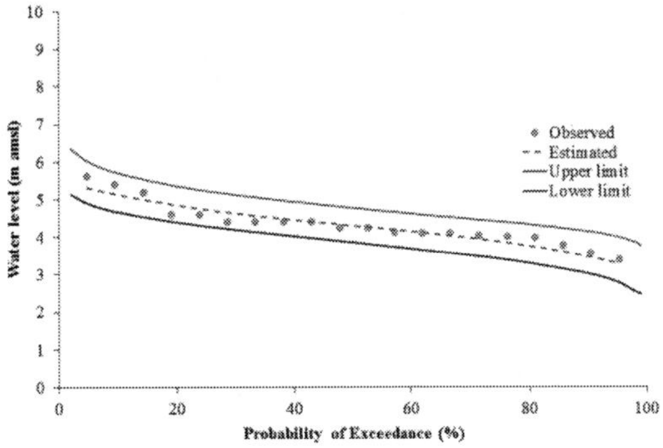


**Fig. 4.** Normal Plot of Minimum Daily Water Level with 95% confidence intervals.

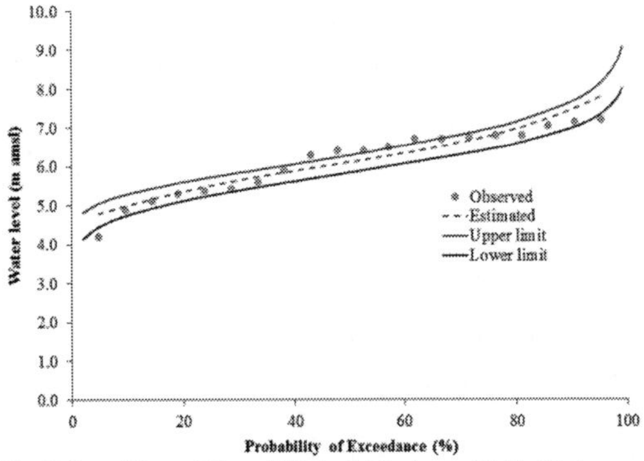


**Fig. 5.** Log- Normal Type 2 Plot of Maximum Daily Water

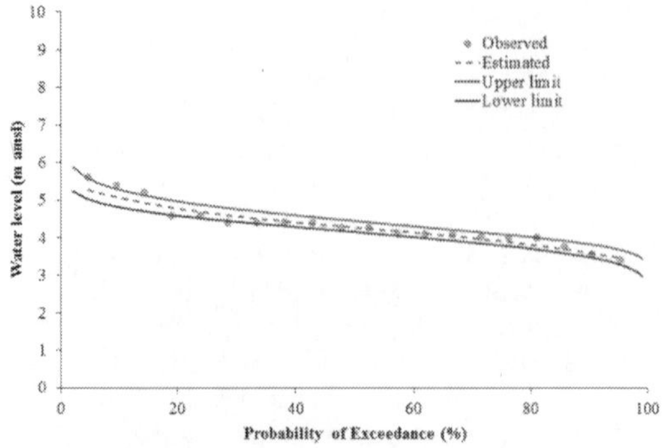


**Fig. 6.** Log- Normal Type 2 Plot of Minimum Daily Water Level with 95% confidence intervals.

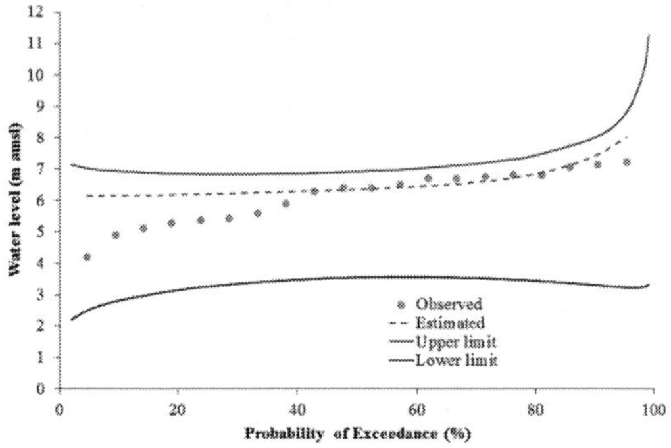


**Fig. 7.** Log- Normal Type 3 Plot of Maximum Daily Water Level with 95% confidence intervals and A=6.1 m amsl.

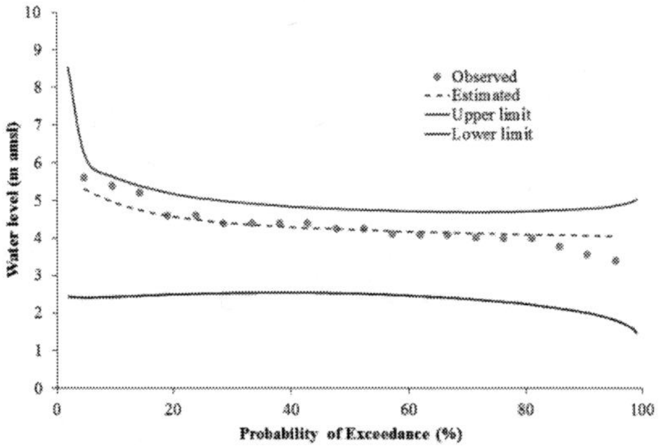


**Fig. 8.**Log- Normal Type 3 Plot of Minimum Daily Water Level with 95% confidence intervals and A=4.02 m amsl.

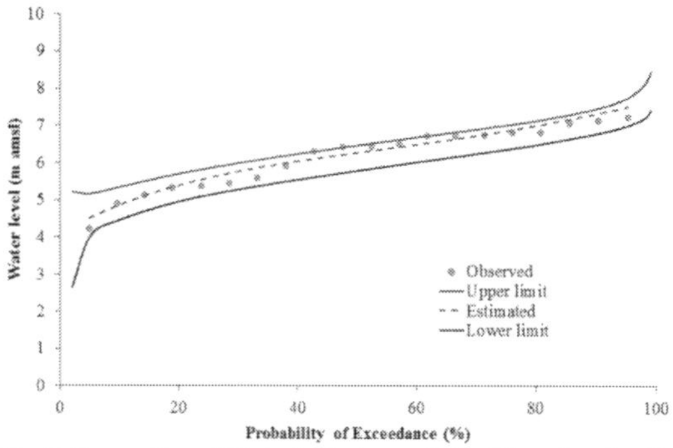


**Fig. 9.** Person Type 3 Plot of Maximum Daily Water Level with 95% confidence intervals.

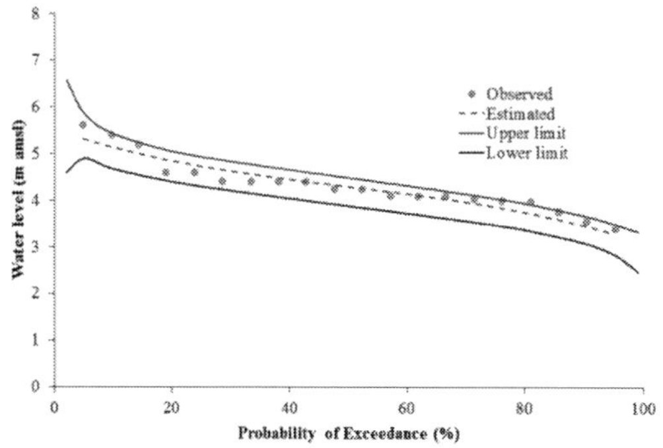


**Fig. 10.** Person Type 3 Plot of Minimum Daily Water Level with 95% confidence intervals.

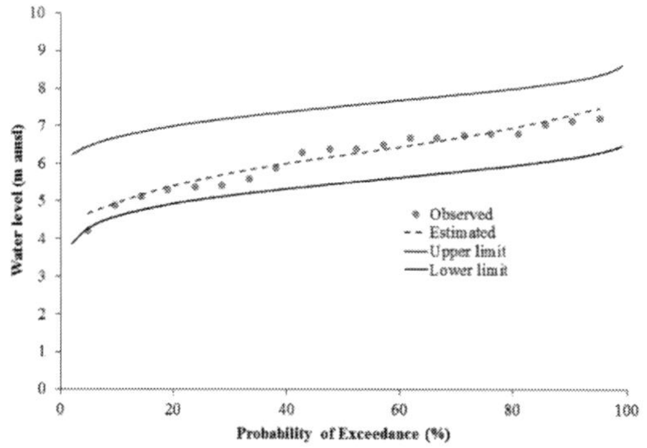


**Fig. 11.** Log-Person Type 3 Plot of Maximum Daily Water Level with 95% confidence intervals.

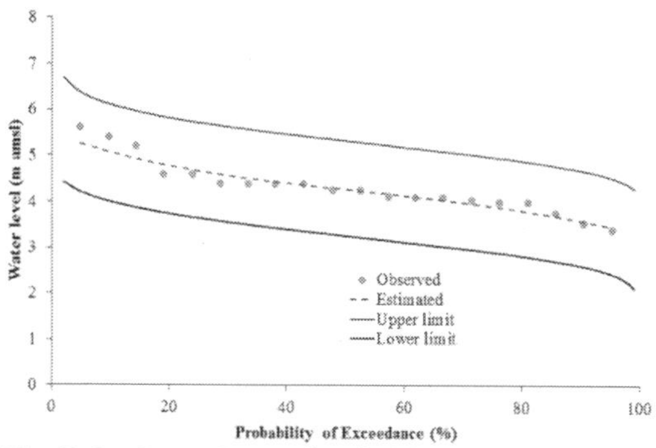


**Fig. 12.** Log-Person Type 3 Plot of Minimum Daily Water Level with 95% confidence intervals

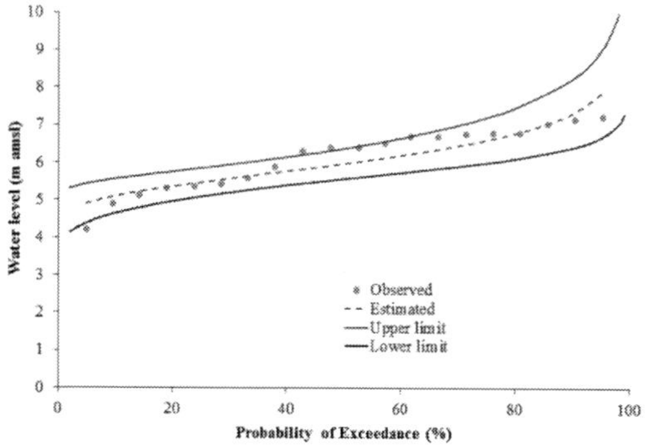


**Fig. 13**. Gumbel type1 Plot of Maximum Daily Water Level with 95% confidence intervals

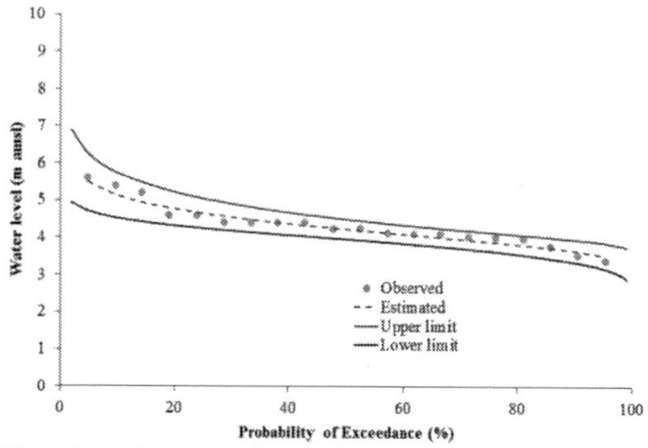


**Fig. 14.** Gumbel type1 Plot of Minimum Daily Water Level with 95% confidence intervals.

**Table 3.** Water Level (mamsl) versus return Periods of six models.

| | Type | Return Period (Year) | | | | |
|---|---|---|---|---|---|---|
| | | 5 | 10 | 20 | 50 | 100 |
| Maximum water level(m amsl) | Normal | 6.66 | 6.92 | 7.14 | 7.38 | 7.55 |
| | Log Normal II | 6.7 | 7 | 7.26 | 7.56 | 7.77 |
| | Log Normal III | 6.27 | 6.57 | 6.98 | 7.71 | 8.45 |
| | Person III | 6.67 | 6.86 | 7 | 7.14 | 7.23 |
| | Log Person III | 6.71 | 6.92 | 7.06 | 7.21 | 7.3 |
| | Gumbel I | 6.59 | 6.94 | 7.27 | 7.71 | 8.03 |
| Minimum water level(m amsl) | Normal | 3.98 | 3.79 | 3.64 | 3.48 | 3.37 |
| | Log Normal II | 3.66 | 3.46 | 3.3 | 3.13 | 3.01 |
| | Log Normal III | 3.83 | 3.8 | 3.79 | 3.78 | 3.77 |
| | Person III | 3.65 | 3.41 | 3.21 | 2.98 | 2.83 |
| | Log Person III | 3.66 | 3.46 | 3.3 | 3.13 | 3.01 |
| | Gumbel I | 3.66 | 3.51 | 3.39 | 3.28 | 3.21 |

## 4 Conclusions

The decision of reusing Al Machraya River for feeding Hawizeh Marsh in terms of hydrology is based on the satisfactory water level to the flow requirements. This can be studied through comparing the water levels in Tigris River at the inlet of Al Machraya River with that in the marsh at the outlet of the river. The probable maximum and minimum water level at the inlet of the river were obtained through carrying out a frequency analysis of the daily maximum and minimum water levels for the period from 1996 to 2015. This analysis involves testing six statistical distribution models to specify the best fit model. It is found that Pearson type III distribution is the best model for estimating the probable maximum daily water levels. While Gumble type I distribution is the best model for estimating the probable minimum daily water levels. The estimated probable maximum water levels at the inlet of the river upstream Kalat Salih Barrage for the return period of 5, 10, 20, 50 and 100 years by using the best fit distributions are 6.67 ± 0.32, 6.86 ± 0.16, 7.00 ±0.12, 7.14 ± 0.13 and 7.23 ± 0.16mamsl. Whereas, the minimum water levels for the same return periods are 3.66 ± 0.31, 3.51± 0.37, 3.39 ± 0.42, 3.28 ± 0.48 and 3.21 ± 0.52m amsl.

The maximum water levels usually occur during the wet years only when the water level at Hawizeh Marsh is at its maximum level, 7 m amsl. The total difference in water levels between the inlet and outlet of the river are not significant to satisfy the requirements of flow from Tigris River to Hawizeh Marsh via Al Machraya River. Accordingly, Al Machraya River cannot be used as a feeder for Hawizeh Marsh under the present conditions of Tigris River and Hawizeh Marsh.

## References

1. R. I. Levin, D. S. Rubin, Wesley Publishing Company(1994).
2. P.B. Bedient, W. C. Huber, Addison–Wesley publication Company (1992).
3. C. Walck, Particle Physics Group Fysikum University of Stockholm (e-mail: walck@physto.se) (2007).
4. CRIMW (Center for the Restoration of the Iraq Marshlands and Wetlands), Ministry of Water Resources, Baghdad, Iraq (2006).
5. D.George, M. Mallery17.0 update (10a ed.) Boston: Pearson, (2010).
6. M. U. Ibrahim, MSc. thesis, College of Engineering. , University of Baghdad (1994).
7. MoWR (Ministry of Water Resources), unpolished documents, Ministry of Water Resources, Baghdad, Iraq (2016).
8. S. K. Gary, New Dehli Khanna (1993).

# Effect of baffle block configurations on characteristics of hydraulic jump in adverse stilling basins

*Ali* Abbas [1], *Haider* Alwash [1] and *Ali* Mahmood [2]

[1] Building and Construction Engineering Department, University of Technology, Baghdad, Iraq
[2] Rafidain General Company for Dams Construction, Ministry of Water Resources, Baghdad, Iraq

**Abstract.** The construction of stilling basin with adverse slope change the characteristics of hydraulic jump such as sequent depth ratio, length of jump ratio, length of roller and energy dissipation ratio, consequently the dimensions of stilling basin are changed, also using baffle blocks with different configurations develop these characteristics. In this study different shapes of baffle block (models (A), (B), (C) and (D)) installed in the stilling basins at adverse slopes (- 0.03, - 0.045, - 0.06) in addition to horizontal bed, all these models are tested in the stilling basin to show their effects on the characteristics of hydraulic jump, the experiments applied for the range of Froude number ($Fr_1$) between 3.99 and 7.48. The baffle block model (D) showed good results when compared with models (B) and (C), therefore it used with arrangement of (single and double row) and compared with baffle block model (A) at slopes (0, - 0.03, - 0.045, - 0.06) to study the effects of baffle blocks on hydraulic jump when bed slopes are changed. In general using baffle block caused a reduction in sequent depth ratio, length of jump ratio and the length of the roller, but the energy dissipation ratio increased.

## 1 Introduction

The hydraulic jump is a phenomenon that occurs when a super critical flow is changed to subcritical flow when the flow obstructed. The rapid change in flow conditions is accompanied by considerable turbulence and dissipation of energy, transferring some of the flow's initial kinetic energy into potential energy [1]. Belanger (1838) predicted the sequent depth ratio ($y_2/y_1$) (for horizontal smooth bed) by using the equation of momentum with the assumption of neglected friction [2], and can be written as:

$$y_2/y_1 = 0.5\left(-1+\sqrt{1+8F_{r1}^{\,2}}\right) \quad \ldots\ldots(1)$$

Where:
$y_2$ = sequent water depth of a hydraulic jump (m).
$y_1$=upstream water depth of hydraulic jump (m).
$Fr_1$= upstream Froude number $=\frac{v_1}{\sqrt{gy_1}}$
$v_1$= upstream velocity (m/s).
g = gravitational acceleration (m/s$^2$).

The characteristics of hydraulic jump affected when the bed conditions of stilling basin are changed, these bed conditions represented by using adverse slope of stilling basin and availability of appurtenances as baffle block.

Hydraulic jump at the adverse bed slope represent an unstable phenomenon which generates some difficulty in the controlling of it, previous studies showed that the hydraulic jump at adverse slopes greater than (- 0.025) is impossible to control, and other studies showed an impossible to keep the jump completely on the adverse slope [4], [5].

The hydraulic jump can be established at $Fr_1 \geq 9$ and for $Fr_1 \geq 4$ the adverse jump still relatively steady without further tailwater adjustments and for $Fr_1 < 4$ (especially as the adverse bed slope increased) continuous adjustment was required even the stabilize position was obtained [6].

The friction in the bed of stilling basin would be responsible for the stability of jump on an adverse slope at a minimum value of upstream Froude number ($Fr_1$), therefore the presence of artificial roughness, baffle blocks or sills are essential to reach to stable condition [7].

The characteristics of the hydraulic jump that represented by dimensionless parameters such as length of jump ratio ($Lj/y_1$), sequent depth ratio ($y_2/y_1$), roller length ratio ($Lr/y_1$) and energy dissipation ratio ($\Delta E/E_1$)

© The Authors, published by EDP Sciences. This is an open access article distributed under the terms of the Creative Commons Attribution License 4.0 (http://creativecommons.org/licenses/by/4.0/).

reduced when the adverse stilling basin is used instead of horizontal bed slope [6], [8], [9].

Stilling basin must be designed perfectly to ensure efficient operating over a wide range of flow. Additional devices may be used to stabilize the jump, reduce the length and height of the jump and increase the energy dissipation. Baffle blocks one of these devices which used to stabilize the jump and dissipate energy as a result of impact action [10].

Baffle block used with different shapes such as cubic and trapezoidal (trapezoidal shape in section), the cubic shape is effective when the best dimensions of height, width, spacing, and the best location in the basin were used. United States Department of the Interior Bureau of Reclamation (USBR) recommended that the corners of baffle block must be not be rounded because the corners are effective in producing of eddies which help in energy dissipation [11].

Previous studies showed that, the model of baffle blocks which have an ability to circulate the jet of water in the vertical transverse direction behave best than others for dissipation of energy, also the rotation in jet of water prevents the jump action to the extent, in this case there is a complete energy dissipation and reduction in the stilling basin length [12].

At adverse bed of stilling basin the specific energy in up stream of hydraulic jump ($E_1$) and in the end of the hydraulic jump ($E_2$) can compute by equations (2) and (3) respectively [3].

$$E_1 = y_1 + \frac{v^2_1}{2g} \quad \text{............ (2)}$$

$$E_2 = y_2 \cos\theta + \frac{v^2_2}{2g} + Lj \sin\theta \quad \text{...... (3)}$$

Where:
($\theta$) = Angle of inclination of bed.
(Lj) = length of the hydraulic jump (m).
The losses in energy due to hydraulic jump compute by the equation:

$$\Delta E = E_1 - E_2 \quad \text{.......... (4)}$$

## 2 Experimental work

The experimental work of this study has been done in the laboratory of fluid mechanics, Building and Construction Department, University of Technology, Baghdad.

The laboratory flume that used were divided into two sections, the first one located in the upstream side of the sluice gate which has a rectangular section with 0.3 m wide, 0.55 m deep and 2.30 m long and the second section located in the downstream side of the sluice gate which has a rectangular section with 0.3 m wide, 0.3 m deep and 10.20 m long.

A plastic platform was installed in the flume to make a different adverse bed slope (-0.03, - 0.045, - 0.06) in addition to horizontal bed. The platform consists from three parts as shown in Figure (1), the first part of platform represents the stilling basin zone and used to adjust the inclination of the bed slope (adverse and horizontal slopes) which has a length of 1 m, the second part represents a horizontal plane has the same level of end of the first part with a length of 1m, the third part represents a transition between the level of end of adverse slope and the level of the bed of the flume.

**Fig. 1** Platform parts.

The models of baffle blocks that used in the experimental work can be described as the following:

1- Baffle block according to USBR recommendations (Model A) with 50 mm height and 37.5mm wide with 50% blockage ratio as shown in Figure (2).
2- Baffle blocks with a trapezoidal shape (in plan) were used with 3 models (B, C and D) as shown in Figure (3) and (4).

Model (B) was made with an apex angle of 53°, model (C) with apex angle of 70° and model (D) with apex angle of 90°, the dimensions of all baffle models summarized in table (1).

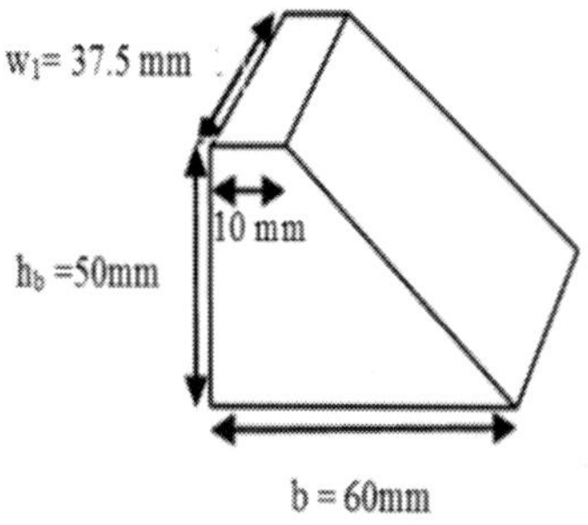


**Fig. 2.** USBR baffle block (Model A).

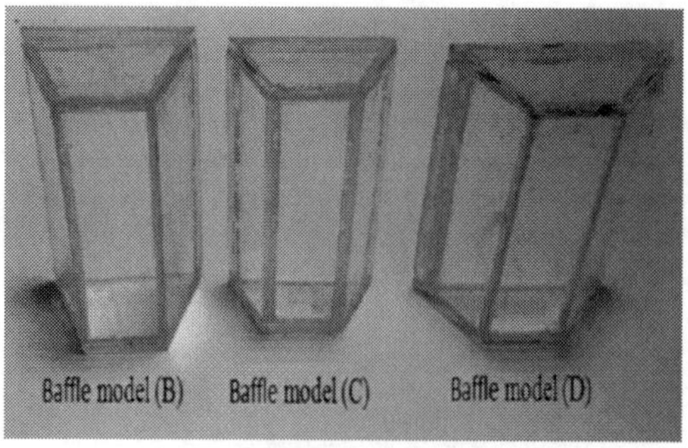


**Fig. 3.** Shape of trapezoidal baffle blocks.

https://doi.org/10.1051/matecconf/201816203005

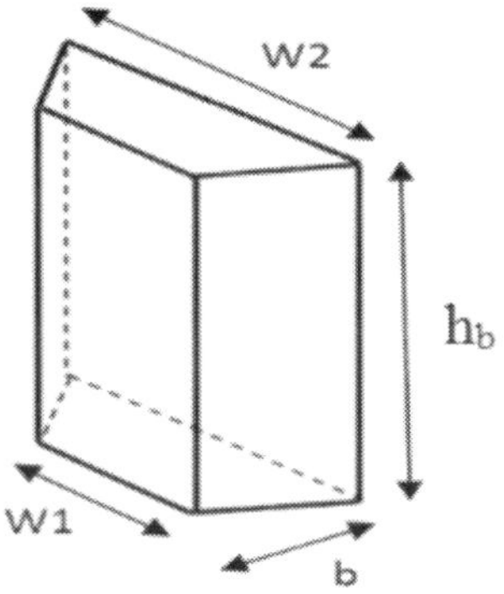


**Fig. 4.** Shape of trapezoidal baffle block.

**Table 1.** Dimensions of baffle block models.

| Baffle models | Height ($h_b$) (cm) | Width (mm) | | Length (b) (mm) | |
|---|---|---|---|---|---|
| | | $w_1$ | $w_2$ | Upper side | Bottom side |
| (A) | 50 | 37.5 | 37.5 | 10 | 60 |
| (B) | 50 | 20 | 37.5 | 17.5 | 17.5 |
| (C) | 50 | 20 | 37.5 | 12.4 | 12.4 |
| (D) | 50 | 20 | 50 | 15 | 15 |

Baffle blocks fixed on the platform with a distance ($X_b$) from the toe of the hydraulic jump, the best stability of hydraulic jump and minimum depth of tailwater occurs when baffle blocks at a location ratio ($X_b/y_2$) =1.87, therefore the baffle blocks fixed at a distance ($X_b$) = 0.32m.

In general, for each experiment the plastic platform installed on the particular slope and the baffle blocks fixed on it (if required). The sluice gate adjusted at a required opening and the flow rate controlled by regulating valve and flow meter.

The hydraulic jump at adverse slope is difficult to be controlled and continually adjusting for the tailgate is essential to reach a stable position. After the free jump is formed and stabilized in its position as shown in Figures (5 to 8), the sequent depth measured by point gage, length of the roller and jump length were measured by measuring tape.

**Fig. 5.** Hydraulic jump at horizontal bed with single baffle model (A), ($Fr_1$= 5.23).

**Fig. 6.** Hydraulic jump at smooth adverse bed slope (-0.03), ($Fr_1$= 3.98).

**Fig. 7.** Hydraulic jump at adverse bed slope (- 0.045) with single baffle model (D), ($Fr_1$= 6.43).

**Fig. 8.** Hydraulic jump at smooth adverse bed slope (- 0.06), ($Fr_1$= 5.83).

## 3 Results

### 3.1 Testing of baffle blocks

#### *3.1.1 Single row of baffle block*

Baffle block models (A, B, C and D) arranged with a single row and tested on slope (- 0.06) to show any which have good result, experimental data of hydraulic jump on slope (- 0.06) with baffle model (D) show a less value for the ratios ($Lj/y_1$), ($y_2/y_1$), ($Lr/y_1$) and high value of ($\Delta E/E_1$) when compared with baffle model (B and C) as shown in Figures (9 to 12), therefore baffle block models (A and D) installed on slopes (0, - 0.03, - 0.045) to study their effect in the characteristics of hydraulic jump when the bed slope is changed.

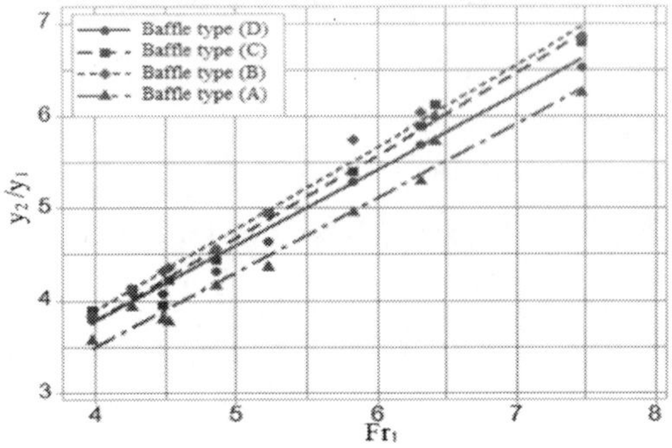


**Fig. 9.** Sequent depth ratio versus $Fr_1$ for baffle models (A, B, C and D) at slope (-0.06).

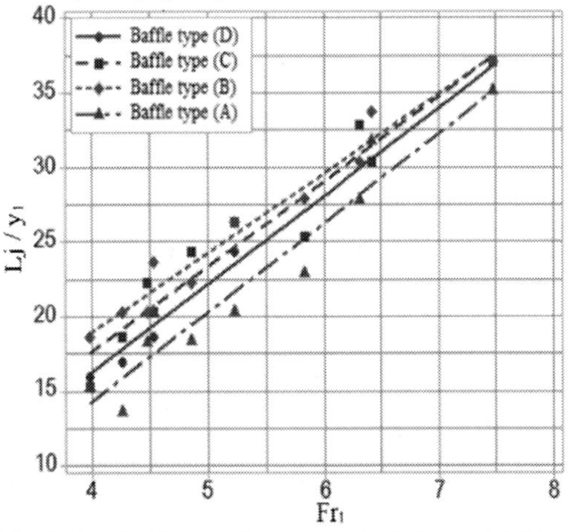


**Fig.10.** Length of jump ratio versus $Fr_1$ for baffle models (A, B, C and D) at slope (-0.06).

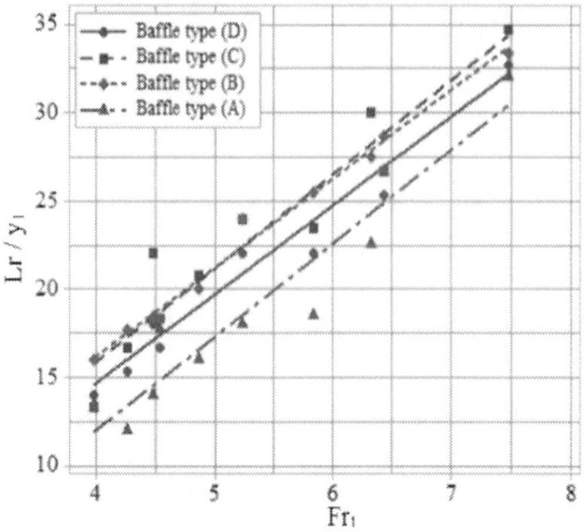


**Fig.11.** Length of roller ratio versus $Fr_1$ for baffle models (A, B, C and D) at slope (-0.06).

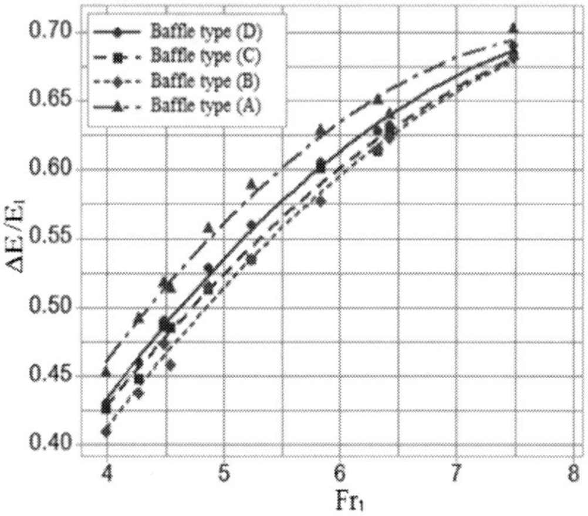


Fig.12. Energy dissipation ratio versus $Fr_1$ for baffle models (A, B, C and D) at slope (-0.06).

**Table (2).** Values of (ẋ/b) ratio for different slopes.

| No. | Bed slopes | (ẋ) mm | (b) mm | (ẋ/b) |
|---|---|---|---|---|
| 1 | 0 | 30 | 15 | 2 |
| 2 | -0.03 | 30 | 15 | 2 |
| 3 | -0.045 | 45 | 15 | 3 |
| 4 | -0.06 | 90 | 15 | 6 |

### *3.1.2 Double row of baffle blocks*

The efficiency of stilling basin increased when the baffle blocks arranged as double row, the distance between first and second row represented by the ratio (ẋ/b), where (ẋ) and (b) represent the longitudinal distance between baffles rows and the length of baffle respectively. The ratio ($\dot{x}$/b) does not the same for all slopes (increase with increasing the adverse slope) due to the effects of the weight of water in adverse slope, table (2) shows the variation of ($\dot{x}$/b) ratio with bed slopes, also Figure (13) show the hydraulic jump at stilling basin with double row of model (D).

**Fig.13.** Hydraulic jump at stilling basin with double row of model (D).

### 3.2 Effects of bed conditions on hydraulic jump characteristics

Characteristics of hydraulic Jump changed when bed conditions varied such as using a smooth bed or using baffle blocks as single or double row with various bed slopes. The results of hydraulic Jump on a smooth bed case compared with that for the cases of single row for baffle models (A and D) and the double row of model (D), all these cases applied at slopes (0, - 0.03, -0.045, -0.06).

#### *3.2.1 Sequent depth ratio*

Experimental results for a certain slope show a reduction in the ratio of ($y_2/y_1$) for the case of double baffle model (D) when compared with results for cases of smooth bed, single row of model (D) and single row of model (A) as shown in Figure (14), the comparison for different cases (at adverse slope (-0.06)) summarized in table (3). Also, using the double baffle model (D) at adverse slope (-0.06) instead of baffle model (A) at the horizontal slope cause a reduction in the ratio of ($y_2/y_1$), the average reduction in this case reaches to about 18.3%. The reduction in the ratio ($y_2/y_1$) when the second row of model (D) used is due to the obstruction of the water particles by the second row and make them to circulate with limited zone, therefore reducing the tailwater is essential to prevent formation of submerged jump.

Using single baffle model (D) at adverse slope (-0.06) gives a good reduction in sequent depth ratio when compared with the case of baffle model (A) at the horizontal slope, the average reduction in this case reaches to about 10.7%. The reduction in the ratio ($y_2/y_1$) lead to reduce in height of the side wall of stilling basin, consequently the cost of construction of stilling basin is reduced.

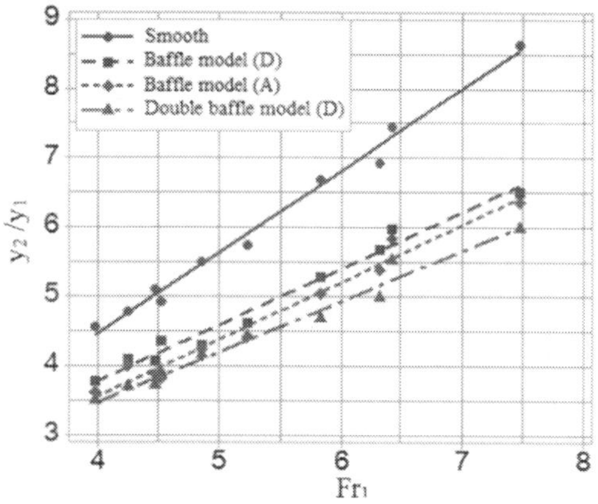


**Fig.14.** Variation in ratio ($y_2/y_1$) due to the change in the bed conditions on slope= (- 0. 06).

**Table (3).** Reduction in ($y_2/y_1$) for different cases on slope (-0.06).

| Bed condition | | Reduction in ($y_2/y_1$) |
|---|---|---|
| Case 1 | Case 2 | |
| Smooth | Double model (D) | 26% |
| Single model (D) | | 8.6% |
| Single model (A) | | 4.4% |

#### *3.2.2 Length of jump ratio*

The length of Jump ratio ($Lj/y_1$) change when the status of basin bed is changed, the results for a certain slope show a reduction in the ratio of ($Lj/y_1$) for the case of double baffle model (D) when compared with results for cases of smooth bed, single row of model (D) and single row of model (A) as shown in figure (15), the comparison for different cases (at adverse slope (-0.06)) summarized in table (4). Also, using the double baffle model (D) at adverse slope (-0.06) instead of baffle model (A) at the horizontal slope cause a reduction in the ratio of ($Lj/y_1$), the average reduction in this case reaches to 38.1%. The reduction in the ratio of ($Lj/y_1$) when the second row of model (D) used is due to the obstruction of the water particles by the second row and restrict its motion, therefore the jump does not extend towards downstream.

Using baffle model (D) at adverse slope (- 0.06) gives a good reduction in length of jump ratio when compared with the case of baffle model (A) at the horizontal slope, this reduction in the ratio ($Lj/y_1$) reaches to about 29.1%, the reduction in length of jump ratio ($Lj/y_1$) means a reduction in the length of stilling basin, consequently the cost of construction of stilling basin is reduced.

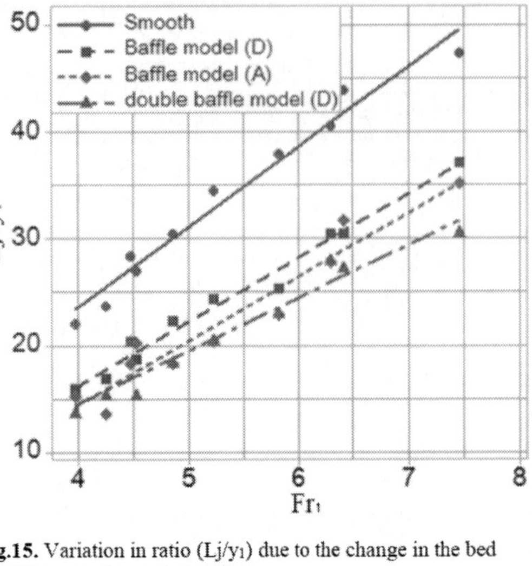


**Fig.15.** Variation in ratio ($Lj/y_1$) due to the change in the bed conditions on slope= (- 0. 06).

**Table (4).** Reduction in ($Lj/y_1$) for different cases at adverse slope (-0.06).

| Bed condition | | Reduction in ($Lj/y_1$) |
|---|---|---|
| Case 1 | Case 2 | |
| Smooth | Double model (D) | 37 % |
| Single model (D) | | 12.7% |
| Single model (A) | | 5.6% |

*3.2.3 Energy dissipation ratio*

The energy dissipation ratio of hydraulic jump change when the bed status is changed, as mentioned previously using smooth stilling basin with adverse slope (-0.06) instead of horizontal smooth bed caused a reduction in the energy dissipation ratio, but using baffle block convert this reduction to gain in the energy dissipation ratio.

The experimental results show that the average gain in the ratio of ($\Delta E/E_1$) reaches to about 10.7% when double baffle of model (D) at slope adverse ( -0.06) is used instead of horizontal smooth bed. Also, experimental results for a certain slope show an increasing in the ratio of ($\Delta E/E_1$) for the case of double baffle model (D) when compared with results for cases of smooth bed, single row of model (D) and single row of model (A) as shown in Figure (16), the comparison for different cases (at adverse slope (-0.06)) summarized in table (5). The additional impact action that produce from the second row of baffle blocks cause the increasing in ratio ($\Delta E/E_1$), also the trapezoidal shape of the baffle model (D) makes the water particles to impact with other and loses its energy.

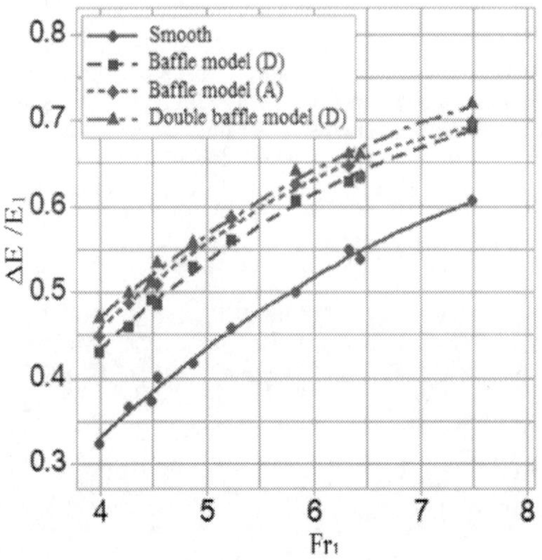


**Fig.16.** Variation in ratio ($\Delta E/E_1$) due to the change in the bed conditions at slope= (- 0. 06).

**Table (5).** Gain in ($\Delta E/E_1$) for different cases at adverse slope (-0.06).

| Bed condition | | Gain in ($\Delta E/E_1$) |
|---|---|---|
| Case 1 | Case 2 | |
| Smooth | Double model (D) | 28.6 % |
| Single model (D) | | 5.7% |
| Single model (A) | | 2.2% |

## 4 Conclusions

• The hydraulic jump in the stilling basin with a single row of model (D) has a less ratio of ($Lj/y_1$), ($y_2/y_1$) and ($Lr/y_1$) and a high ratio of ($\Delta E/E_1$) when compared with models (B) and (C), therefore model (D) represents the best trapezoidal model and used with model (A) on other slopes to show their effects when bed slopes are changed.

• In the arrangement of baffle blocks with double row the distance between the first and second rows (ẋ) does not the same for all slopes, this distance increase with increasing the adverse slope.

• At adverse slope (-0.06) the average reduction in ($y_2/y_1$) ratio and ($Lj/y_1$) ratio reach to about 26% and 37% respectively when the double baffle model (D) is used instead of smooth bed.

•The average reduction in the ratio ($y_2/y_1$) reaches to about 10.7%, while the average reduction in the ratio ($Lj/y_1$) reaches to about 29.1% when the baffle model (D) at adverse slope (- 0.06) used instead of baffle model (A) at horizontal bed.

•The average reduction in ($y_2/y_1$) ratio reaches to 18.3 %, while the average reduction in ($Lj/y_1$) ratio reaches to 38.1% when the double baffle model (D) at adverse slope (-0.06) used instead of the horizontal smooth bed.

• At adverse slope (-0.06) the average gain in ($\Delta E/E_1$) ratio reaches to about 28.6% when the double baffle model (D) is used instead of smooth bed.

•The energy dissipation ratio ($\Delta E / E_1$) reduced when the adverse slope used instead of the horizontal slope, but using the double row of baffle model (D) at adverse slope increasing the efficiency of stilling basin and convert the reduction in energy dissipation to gain, the average gain in ratio ($\Delta E / E_1$) reach to about 10.7% when the double baffle of model (D) at slope (-0.06) is used instead of horizontal smooth bed.

## References

1. S.K. Gupta, R. C. Mehta, and V. K. Dwivedi, Procedia Engineering **51**, 529 – 537(2013).
2. W. H, Hager, Vol.**8**, Zurich, Switzerland, Swiss Federal Institute of Technology (ETH), 10-18 (1992).
3. V.T.Chow, McGraw-Hill Book Company,(1959).
4. N. Rajaratnam, J. Irrig. and Power, India, **23** (2), 137–149(1966).
5. K.M. Abdul, and S. Rajagopal, J. Irrig. Pwr. India, **29**, 77- 82. (1972)
6. J. A. Mccorquodale, and M. S. Mohamed Journal of Hydraulic Research Vol. **32**, 1, 119-130. (1994).
7. H. K. Shayan, and J. Farhoudi, Journal of Hydraulic Structures Vol. **No.II** University of Tehran, Tehran, Iran.( 2013).
8. M. S. Bateni, and F. Yazdandoost, Proceedings of the Institution of Civil Engineers, Water Management 162 Issue WM5 Pages 321–327, research gate. (2009).
9. M. k.Beirami, nd M.R. Chamani, , "Can. j. Civil. Eng. **37**, 535-543 by NRC Research press(2010).
10. M. H.Chaudhry, "Open-Channel Flow" (Second Edition), University of South Carolina, Springer, (2008).
11. A. J.Peterka, (Eighth Edition) United States Department of the Interior, Bureau of Reclamation (1984).
12. A. Bestawy, Asian Transactions on Engineering Volume **03**, Issue 01(2013).

# Characteristics of flow over the free overfall of triangular channel

*Raad* Irzooki[1,*], *Safa* Hasan[2]

[1]Environmental Engineering Department, Tikrit University, Tikrit, Iraq
[2]Civil Engineering Department, Tikrit University, Tikrit, Iraq

**Abstract.** In the present paper, the effects of side slopes, bed slopes and bed roughness on the flow over free overfalls in triangular channels have been studied experimentally. For this purpose, three models of triangular channels with free overfalls have been constructed and fixed in a 6m length laboratory flume. These three models had length of 244cm with different values of side slopes (Z) (0.8(H):1(V), 1:1 and 1.33:1). Each one of these models had four different bed slopes (S) (0, 0.0041, 0.0082 and 0.0123). For each bed slope, the bed was roughened with three particle sizes of sand (ds) (1.18mm, 2.36mm, and 4.75mm). The experimental testing program included sixteen series of experiments for each model. Four of them were for smooth beds and twelve for rough beds. A total of forty eight experiments were tested for different rates of discharge (Q). Experimental results of all models showed that Froude number ($Fr_b$) of flow decreases with the increasing of end depth ratio values ($y_b/y_c$) for different bed roughness, different bed slopes. The relations between the brink depth ($y_b$) and the critical depth ($y_c$) were found to be a simple linear formula for various bottom slope and different bottom roughness. An empirical expression was obtained for the flow over the free overfall in triangular channels for different bed slopes and roughness. The results of the present study have been compared with studies were obtained by other investigators, the comparison shows a very good agreement between them.

## 1 Introduction

The overfall is a weir with broad crest causing sudden change in water depth by free fall of water [1]. Studying the free overfall is important in hydraulic structures subject because of the possibility of using this structure in measuring the discharge in the laboratory in addition to its main function as a device to transport the flow from high level to lower level. The overfall structures can be shown permanently in natural or artificial channels. In natural channels the overfall formed by changing river bed level from a high level to a lower level as a result of erosion of the river bed, but in the artificial channels these structures can be formed by changing the bottom level of these channels for energy dissipation purposes.

Many experimental and theoretical researches were conducted to study the water overfalls. Rouse, 1936, is the first which pointed out the possibility of using the free overfall as a simple structure for measuring the discharge without requiring any calibration [2]. He found that the end depth ratio (EDR) for critical flow in rectangular and horizontal free overfall channel equal to 0.715. The end depth ratio is the ratio of end depth (bring depth) (yb) to the critical depth (yc). Also, Rouse proposed an empirical equation to measure the rate of discharge over free overfall of rectangular channel with Froud Number (Fr) ranging between 1 and 20. The researchers Delleur et al. presented a similar study on a rectangular channel, but with greater values of roughness. The results of this study show that the relative roughness (ks/yc) less than 0.1 (where ks represents height of roughnes) [3]. Also, Delleur's et al. study shows that when the value of (ks/yc) more than 0.1, the EDR value will be smaller than the expected value due to the effect of roughness. (Rajaratnam et al., 1964) [4] present an experimental study on the various shapes of channels sections, they found that the value of the EDR is equal to 0.715 for confined rectangular channels, and for unconfined channels it is equal to 0.705. Also, they obtained that EDR equal to 0.795 for unconfined triangular channels. (Ali and Sykes, 1972)[5] and (Ahmad, 2002)[6] found that (EDR) for the triangular channel equal to 0.795 and 0.762, respectively. (Ramamurthy et al., 2004)[7] studied the pressure head, flow velocity and water surface profile for trapezoidal free overfall channel. (Beirami et al., 2006)[8] presented theoretical study for different free overfall channel sections with horizontal slopes by using momentum equation. He found that the end depth ratio

* Corresponding author: dr.raadhoobi@tu.edu.iq

© The Authors, published by EDP Sciences. This is an open access article distributed under the terms of the Creative Commons Attribution License 4.0 (http://creativecommons.org/licenses/by/4.0/).

(EDR) for rectangular, triangular and parabola channels sections equal to 0.7016, 0.8051 and 0.7642, respectively, and indicated in this study that (Andersn, 1967)[9] and (Murty, 1994)[10] found (EDR) equal to 0.694 and 0.705 for the rectangular channels, 0.762 and 0.795 for the triangular channels and 0.735 and 0.758 for the channels parabola, respectively. (Gue et al., 2008)[11] investigated the channel bed roughness and bed slope effects on the flow over rectangular free overfall. In this study the bed was roughened by discrete square cross section strips with dimensions (6mm×6mm) and various distances between these strips. This study concluded that EDR decreases with increasing the ratio of spacing to height of strips (L/h). (Tigrek et al., 2008)[12] studied, experimentally, the bed slope and Manning's coefficient effects on the properties of flow over rectangular free overfalls. They concluded that EDR increases with increase in the roughness for any bed slope, also, EDR increases with decrease in the bed slope for both sub- critical and super-critical flow. (Mohammed et al., 2011)[13] studied the effect of bed roughness and bed slope in a rectangular overfall channel. Different bed roughness cases were used in this study with two bed slopes. They found six equations to evaluate EDR according to different bed roughness. (Ibrahim, 2016)[14] studied experimentally the bed roughness, side slope and bed slope effects on flow characteristics over free overfall trapezoidal channels. In this study, as done by Gue et al., 2008, the bed was roughened by square strips fixed at different spacing. This study shows that EDR decreases with increasing the bed slope and the average value of EDR is equal to 0.729 for sub-critical and super-critical flow with smooth bed.

The aim of the present study is to investigate the hydraulic characteristics of the free overfalls of the triangular channels with an indication of the influence of the channel bed roughness, longitudinal bed slope and side slope on the depth of water at the edge of overfall as well as the relationship of this depth with critical depth. Also, the present study aims to find empirical equation to calculate the discharge for the cases studied in this research.

## 2 Dimensional analysis

Dimensional analysis proves to be a generally valid method to recognize the information structure in the relationships between physical parameters in a precise and clear way. It starts from the fact that in quantitative natural science, the descriptive quantities have dimensions and can be divided correspondingly into basic quantities and derived quantities.

Dimensional analysis is helpful in laboratory work, because it provides a guide to those variables that significantly influence the phenomena; thus, it indicates the way in which experimental work should progress [15].

In this study a dimensional analysis was applied to predict an empirical equation for measuring the flow discharge over the free overfall of triangular channels as shown in Figure (1).

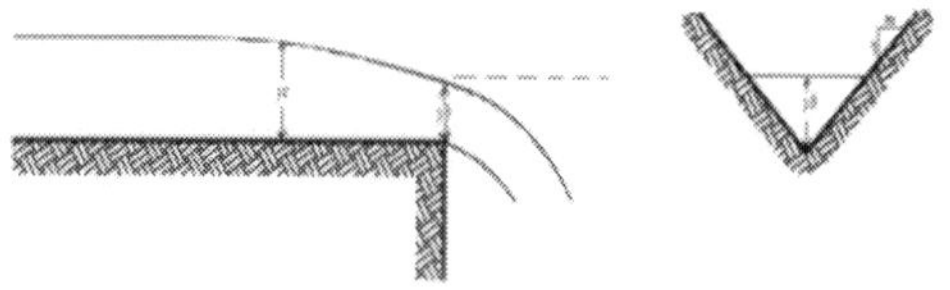

**Fig. 1.** Free overfall for triangular channel.

The possible factors influencing the flow discharge (Q) over free overfall are:

$$Q = f_1(y_b, y_c, S, m, N, g, \rho, \mu) \quad (1)$$

Q= Average flow discharge ($L^3T^{-1}$)
$y_b$ = Bring depth (L)
$y_c$ = Critical depth (L)
S= Longitudinal slope of the channel
m=Side slope of the channel (m (horizontal):1 (vertical))
N= Manning's roughness coefficient
g= Acceleration of gravity ($LT^{-2}$)
ρ= Mass density ($ML^{-3}$)
μ=Water viscosity ($ML^{-1}T^{-1}$)

Using Buckingham's Pi-theorem (π theorem), and choosing ($y_b$, g and ρ) as repeated variables, the following equation of dimensionless parameters can be obtained:

$$\frac{Q}{\sqrt{g y_b^5}} = f_2(\mathrm{Re}, \frac{y_c}{y_b}, S, N, m) \quad (2)$$

As mentioned by Chow, 1959[16], when the flow is turbulent, the viscosity forces become weak relative to inertia forces, therefore, it can neglect Reynolds's number (Re). Rearranging Equation (2), the following equation can be obtained:

$$\frac{Q}{\sqrt{g y_b^5}} = f_3(\frac{y_c}{y_b}, S, N, m) \quad (3)$$

## 3 Experimental work

The experimental work of this study was carried out in Hydraulic laboratory at Environmental Engineering Department in Tikrit University. This work was conducted in 6m long rectangular flume, with cross section of 0.3m width and 0.4m height. The walls of this flume is Perspex glass and the bed is painted iron. The models of triangular free overfall channel of each test were constructed using 6mm thickness Perspex glass, these models were fixed in the working flume by adhesive material to prevent any leakage from sides and bottom. The experimental tests of this study were divided into three groups (A, B and C). Each group represents free overfall triangular channel with side slope, (m), 0.8:1, 1:1 and 1.33:1, respectively. The tests of each group were carried out with four different longitudinal slopes (S) (0 (horizontal slope), 1/244, 2/244 and 3/244). For each of the above tests four different roughness (N) of the channel sides were used, the first one is without using any roughing material on the channel sides, and in the other three types the channel sides were roughened by using three different sizes of gravel (1.18, 2.36 and 4.75mm). For each run, eight different discharges (at least) were passed, therefore 425 runs were carried out in this study.

The discharge which passes through the channel was measured by using 90o V-notch weir with sharp crest. This weir was manufactured from 10mm thickness Perspex glass. The flow rate passing over this weir was calculated using the volumetric method, where for each head above the weir crest, the discharge was measured three times and the rate of these discharges was calculated. Thus, the discharge equation for this weir was developed as follows:

$$Q = 0.012\, H^{2.552} \quad (4)$$

$Q$ = Rate of discharge (ℓ/sec)
$H$ = Depth of water over the weir crest (cm)

Two point gauges with accurately reading of 0.1mm were used for measuring the bring depth (yb) of the triangular overfall channel and the water depth over the weir crest. Figure (2) shows working flume with experimental setup details.

**Fig. 2.** Experimental flume with accessories.

The variables that were measured for each experiment are the water depth at the beginning of the channel ($y_n$), the bring depth at the edge of the free overfall ($y_b$) and the depth of water above the triangle weir ($H$). The critical depth ($y_c$) for each test was calculated by the following equation:

$$\frac{Q^2}{g} = \frac{A_c^3}{T_c} \quad (5)$$

Where :
$A_c$ = The cross section area of flow at critical condition
$T_c$= Top width of water surface at critical condition
=2my$_c$

$$A_c = m y_c^2 \quad (6)$$

$$T_c = 2 m y_c \quad (7)$$

Substituting Equations (6) and (7) in Equation (5), the following equation produced:

$$y_c = \sqrt[5]{\frac{2Q^2}{g m^2}} \quad (8)$$

## 4 Results and discussion

The experimental results of this study were prepared according to parameters that were produced from dimensional analysis as mentioned in previous section.

### 4.1 Relation of bring depth ($y_b$) with critical depth ($y_c$)

Figures (3, 4, 5 and 6) show the relationship between the bring depth ($y_b$) and the critical depth ($y_c$) for (m1) triangular channel free overfall with various bottom slope and different bottom roughness. It can be seen that this relation is defined as simple linear formula as follows:

$$y_b = a_1 (y_c) \quad (9)$$

From these figures it can be seen that ($y_b$) increases with increase in ($y_c$) at a high correlation coefficient for the sub-critical and super-critical flow conditions, as shown in Table (1).

N0 means the bed of the channel without using any roughing material, and N1, N2, N3 represent the bed roughness with using gravel of size 1.18, 2.36 and 4.75mm, respectively. S0 refers to horizontal bed slope and S1, S2, S3 represent bed slope of the triangular channel equal to (1/244= 0.0041), (2/244=0.0082) and (3/244=0.00123), respectively.

Also, the above figures show that the bed slope and bed roughness have a slight effect on the relation between the bring depth and the critical depth. It is worth to mention that same results were produced for the channels with side slopes (m2 and m3).

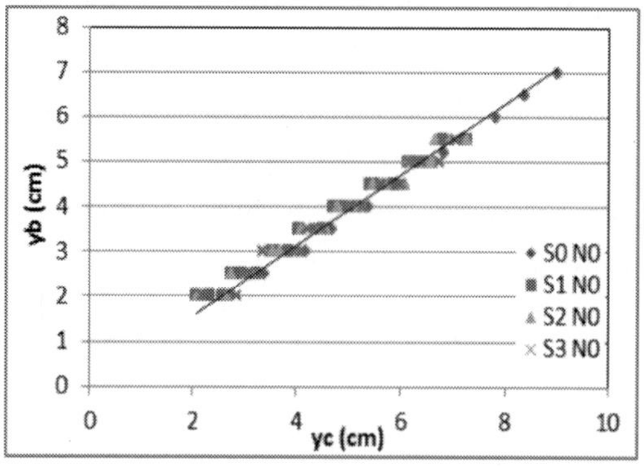


**Fig. 3.** Variation of ($y_b$) with ($y_c$) for triangular channels without roughness material and different bed slopes.

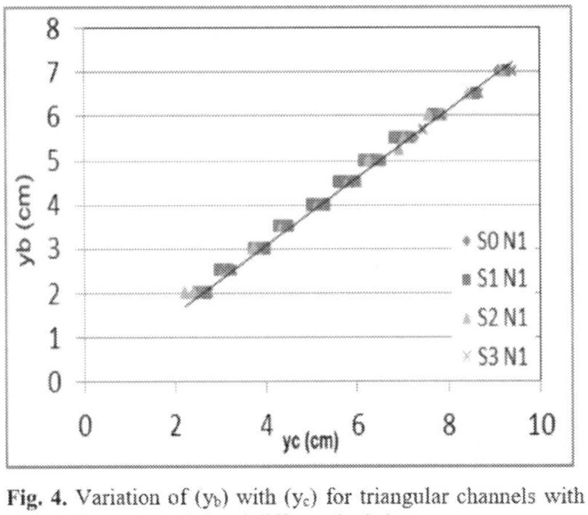


**Fig. 4.** Variation of ($y_b$) with ($y_c$) for triangular channels with N1 roughness material and different bed slopes.

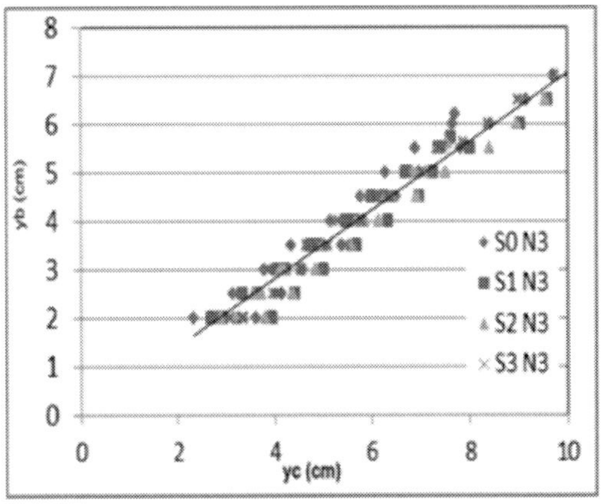


**Fig. 5.** Variation of ($y_b$) with ($y_c$) for triangular channels with N2 roughness material and different bed slopes.

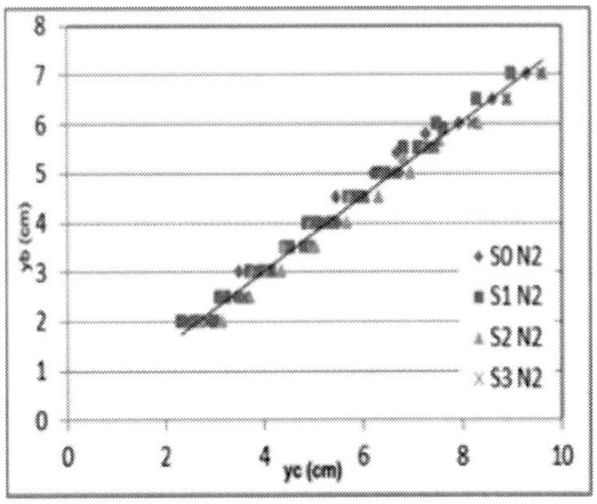


**Fig. 6.** Variation of ($y_b$) with ($y_c$) for triangular channels with N3 roughness material and different bed slopes.

**Table 1.** Magnitudes of (a1) constant and determination coefficients for triangular channels different bed roughness and different bed slopes.

| Roughness | $a_1$=(yb/yc) | $R^2$ | Fr |
|---|---|---|---|
| N0 | 0.788 | 0.978 | 0.45-1.47 |
| N1 | 0.7704 | 0.995 | 0.44-1.13 |
| N2 | 0.7577 | 0.983 | 0.32-1.00 |
| N3 | 0.7048 | 0.955 | 0.31-0.94 |

### 4.2 Relationship of ($Q/\sqrt{gy_b^5}$) with ($y_b/y_c$)

The dimensionless parameter ($Q/\sqrt{gy_b^5}$) represents Froud Number ($Fr_b$) of flow over free overfall triangular channel in terms of bring depth ($y_b$). The variation of this parameter according to the ratio ($y_b/y_c$) was studied for different bed roughness and for all cases of bed slope that were investigated in this study. Figures (7, 8 and 9) show the relation of the parameter ($Q/\sqrt{gy_b^5}$) or ($Fr_b$) with the ratio ($y_b/y_c$) for different bed roughness and S1 bed slope of triangular channel with side slope (m1, m2 and m3), respectively.

Same relations were observed channels with bed slope S2 and S3.

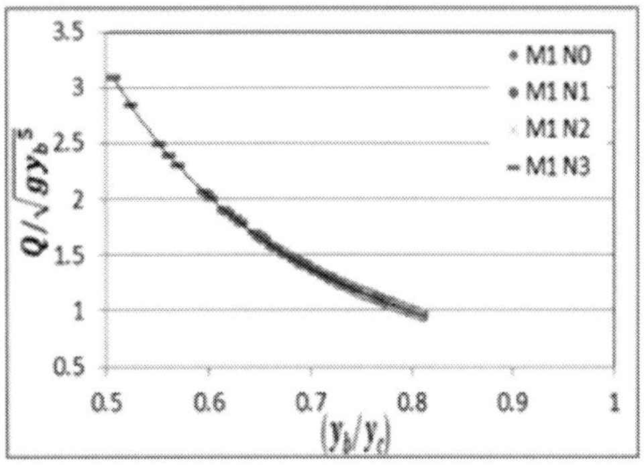


**Fig.7.** Variation of ($Q/\sqrt{gy_b^5}$) with ($y_b/y_c$) for triangular channels with m1 side slope, S1 bed slope and different roughness material.

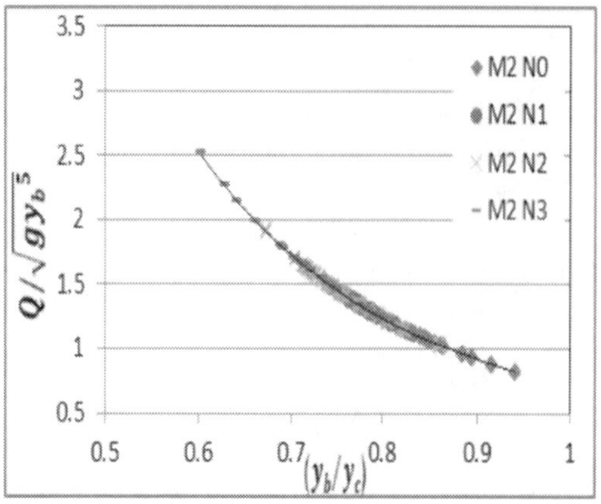


**Fig. 8.** Variation of ($Q/\sqrt{gy_b^5}$) with ($y_b/y_c$) for triangular channels with m2 side slope, S1 bed slope and different roughness material.

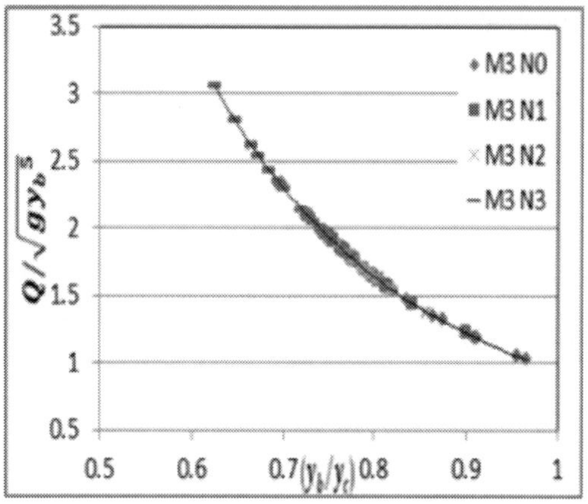


**Fig. 9.** Variation of ($Q/\sqrt{gy_b^5}$) with ($y_b/y_c$) for triangular channels with m3 side slope, S1 bed slope and different roughness material.

From these figures it can be seen that ($Fr_b$) decreases with increase in the end depth ratio ($y_b/y_c$). Also, these figures show that the variation of the side slope effects on the relation between ($Fr_b$) and ($y_b/y_c$), where, for constant ($Fr_b$) the end depth ratio ($y_b/y_c$) increases with increase in (m) (which means decreasing the inclination of the channel sides with the horizontal). In addition to the above, the bed roughness and bed slope of the channel was found to have no or very little effect on this relation for all conditions of flow.

## 5 Discharge equation of triangular channel with free overfall

The experimental data were used to produce an empirical equation for measuring the discharge through triangular channel with free overfall, where approximately 70% of these data can be expressed in to following equation by using a statistical program SPSS 19.

$$Q = 1.252 g^{0.5} m y_b^{2.5} \frac{S^{0.01}}{N^{0.047}} \tag{10}$$

The coefficient of determination of this equation (R2) equal to 0.922.

This equation clearly shows that the influence of bed slope (S) and bed roughness (N) is relatively small on the discharge. This agrees with the analysis of the above results, where these results show that no or little influence of the bed slope and bed roughness on the Froud Number, which is the same on the discharge. The magnitude of the parameter ($\frac{S^{0.1}}{N^{0.047}}$) is approximately equal to unity, therefore, Equation (10) can be reduced to a final form as follows:

$$Q = 1.252 g^{0.5} m y_b^{2.5} \tag{11}$$

The remaining approximately 30% of the experimental data were used to verify the above developed equation. Figure (10) represents the relation between the resulted experimental discharges for different cases of triangular channel with free overfall and the discharges were measured from Equation (11) for the same conditions of these cases. This figure and the calculated standard error ratio for this relation, which is equal to 0.154%, show a good agreement between the experimental and measured results.

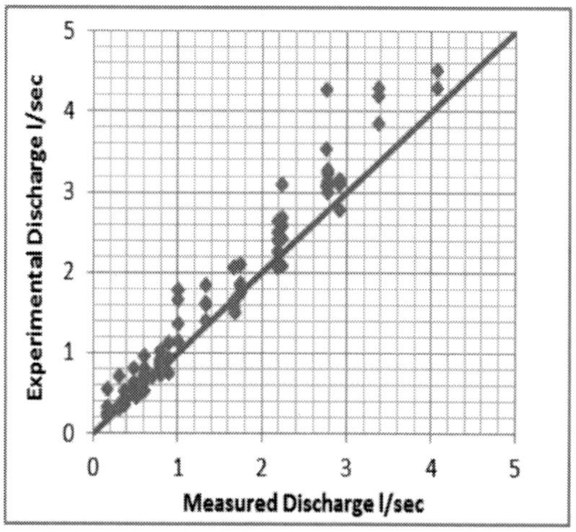


**Fig. 10.** The relation between experimental discharge and calculated discharge by Equation (11) for triangular channels overfalls.

Also, the observed equation of this study coincides with relations that were derived by earlier researchers, for horizontal channels without roughness material, as shown in Table (2). Figure (11) represents the relation between the calculated discharge and bring depth for triangular channel free overfall with side slope (m1) using the different observed equations.

**Table 2.** Observed equations for flow over triangular channel free overfall.

| Researchers | Equation | Difference Ratio % |
|---|---|---|
| Rajaratnam and Muralidhar, 1964 | $Q = 1.2548\, g^{0.5} m y_b^{2.5}$ | 0.0255 |
| Andersen, 1967 | $Q = 1.395\, g^{0.5} m y_b^{2.5}$ | 1.2096 |
| Ali and Sykes, 1972 | $Q = 1.243\, g^{0.5} m y_b^{2.5}$ | 0.0825 |
| Murty, 1994 | $Q = 1.2548\, g^{0.5} m y_b^{2.5}$ | 0.0255 |
| Ahmad, 2002 | $Q = 1.2276\, g^{0.5} m y_b^{2.5}$ | 0.2258 |
| Beirami et al., 2006 | $Q = 1.2158\, g^{0.5} m y_b^{2.5}$ | 0.3372 |
| The present study | $Q = 1.252\, g^{0.5} m y_b^{2.5}$ | - |

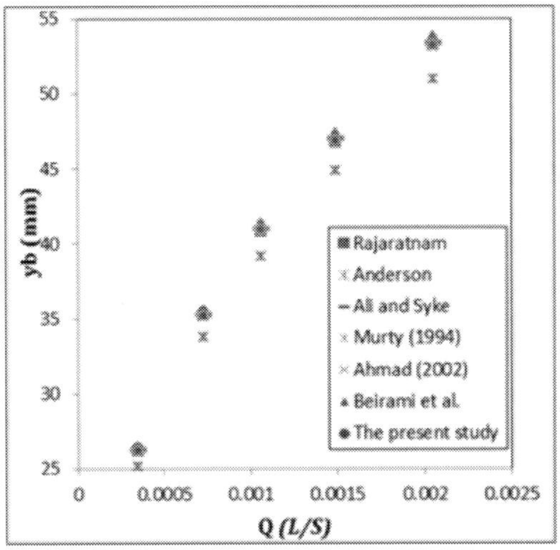


**Fig. 11.** Comparison of present study with earlier researches.

## 6 Conclusions

In the present study, the characteristics of flow in triangular channels with free overfalls were experimentally examined. The experiments were carried out by passing different discharges through channels with different bed slope, side slopes and bed roughness.

Based on the analysis of the resulting data for 425 tests, the following points represent the main conclusions of this study:

1- The bring depth ($y_b$) increases with increase in the critical depth ($y_c$), and the average end depth ratio (EDR) or ($y_b/y_c$) was found equal to 0.755.

2- Froud Number ($Fr_b$) decreases with increase in the end depth ratio ($y_b/y_c$) and this relation is affected by variation in the side slope of the triangular channel, where the same Froud Number, the EDR increases with increase in the magnitude of (m) from 0.8 to 1.33.

3- The bed roughness and bed slope of the channel was found to have no or very little effect on the relation between ($Fr_b$) and EDR.

4- A good equation was developed for measuring the discharge through triangular channel with free overfall.

## References

1. A. Y. Mohammed, Journal of Al-Rafidain Engineering **17**, 1, (2009).
2. H. Rousel, Civil Engineering **6**, 4, 257–260 (1936).
3. J. W. Delleur, J. C. I. Dodge, and K. N. Gent, Journal of Hydraulic Division, ASCE **82**, 4 (1956).
4. N. Rajaratnam, and D. Muralidhar, Journal of Hydraulic Division, ASCE **90**,1, (1964).
5. K. H. M. Ali, and A. Sykes, Journal of Hydraulic Division, ASCE **98**, 5 (1972).
6. Z. Ahmad, Conf. of Hyd., Water Resources and Ocean Eng., 115-119 (2002).
7. A. S. Ramamurthy, C. Zhai, and J. Qu, Journal of Irrigation and Drainage Engineering, ASCE **130**, 5 (2004).
8. M. K. Beirami, S. V. Nabavi, and M. R. Chamani, Iranian Journal of Science and Technology, Transaction B, Engineering **30**, No. B1 (2006).
9. M. V. Andersen, Acta Polytechnica Scandinavica **42**, 1-24 (1967)
10. B. S. Murty, J. Hydr. Res. **32**, 2, 219–232 (1994).
11. Y. Gue, L. Zhang, Y. Shen, and J. Zhang, Journal Hydraulic Engineering, ASCE **134**, 5 (2008).
12. S. Tigrek, C. E. Firat and A. M. Ger, Journal of Irrigation and Drainage Engineering, ASCE **134**, 89 (2008).
13. M. Y. Mohammed, A. Y. Al-Taee, and A. N. Al-Talib, Damascus University Journal **27**, 1 (2011).
14. S. S, Ibrahim and B. M. A. Noori, ZANCO **28**, 2, 127-136 (2016).
15. M. H. Chaudhry, *Open Channel Flow*, 2nd ed. Springer Science and Business Media, LLC, New York (2008).
16. V. T. Chow, *Open-Channel Hydraulics*, International Edition, McGraw-Hill, (1959).

# Impact of using two different sources of water quality for simulation of salt and moisture content in unsaturated zone

*Qassem* Jalut[1,*], and *Anmar* Saleh[1]

[1] Department of Civil Engineering, Diyala University, Baquoba, Iraq

**Abstract.** Monitoring water quality and moisture content throughout the root zone become essential for proper management and for reduction of undesirable and diverse effects caused by using different water quality. In the field level, poor management of using different quality water can cause a huge reduction on the vegetation production. Two field plots were selected in Al-Khalis region in Diyala governorate as a case study. Four scenarios were applied for the data collected for a complete season from a local meteorological station and by using HYDRUS-3D V2.05 software package to simulate water and salt transport through a 1m depth of unsaturated zone. Irrigation scheduling was estimated using FAO method. The net water depth required for a season was 65 mm for 7 days irrigation interval and 21 irrigation events. For depths (20, 40, 60, 80, and100 cm) the salt concentration were simulated for the four scenarios and for the whole season. For the limit of controlling soil salinity at low level in initial and develop stage of plantation, it has been found that the Fourth scenario has to be chosen.

## 1 Introduction

Recently a proper management for irrigation in areas using different water qualities is needed. This task become vital in semi-arid and arid lands ( Iraq as example ) . Availability of good quality waters becomes a real problem facing many of the developing countries , Pereira et al. 2009 ,and Gonçalves et al. 2007 [7], [4], among many others emphasized the importance of monitoring the change in soil moisture distribution and salinity profile in unsaturated zone to the use of different water quality in irrigation.

Short and long term monitoring process of the effect of using different water quality on soil properties, soil environment and hence crop yield draw a very wide attention. Numerical and analytical modelling become an usual practice for many researcher in this area of research (Jarvis , 1994; Ahuja et al., 2000 and Šimu˚nek et al., 2006 ,[6], [1], [10] to mention a few).

Soil moisture and salt distribution in partially saturated zone has been modelled using Richards equation and transport convective – dispersive equation with root water uptake term to take care of the plant stresses on soil water. Because of the non - linearity and complexity of the govern equations, analytical solution seems to be impossible. Feddes and Raats,2004;Šimu˚nekand Hopmans , 2009, [2] , [11] to mention a few used a numerical approach to solve the govern equation of solute transport in unsaturated zone.

A finite element model HYDRUS (2D/3D) was selected in the present study for simulating solute and also water in partially saturated oil with plant water uptake. The model has the ability to display graphical results such as water contents , pressure heads , concentrations , Root water uptake, velocities , and temperatures [12].

## 2 Study area

Two fields of tomatoes were selected in Al-Khalis region of Diyala – Iraq , Al-khuyls field ( 33° 50′ 10″ N and 44° 32′ 50″ E ) and Al-hashmyat field ( 33° 46′ 02″ N and 44° 31′ 35″ E). The climate in study area is mostly dry sub-humid. Summers are so hot and the winters are moderate . With a scarcity of rainfall and a decrease in fresh water sources during the months of years.

* Corresponding author: anmar_shs@yahoo.com

© The Authors, published by EDP Sciences. This is an open access article distributed under the terms of the Creative Commons Attribution License 4.0 (http://creativecommons.org/licenses/by/4.0/).

 https://doi.org/10.1051/matecconf/201816203007

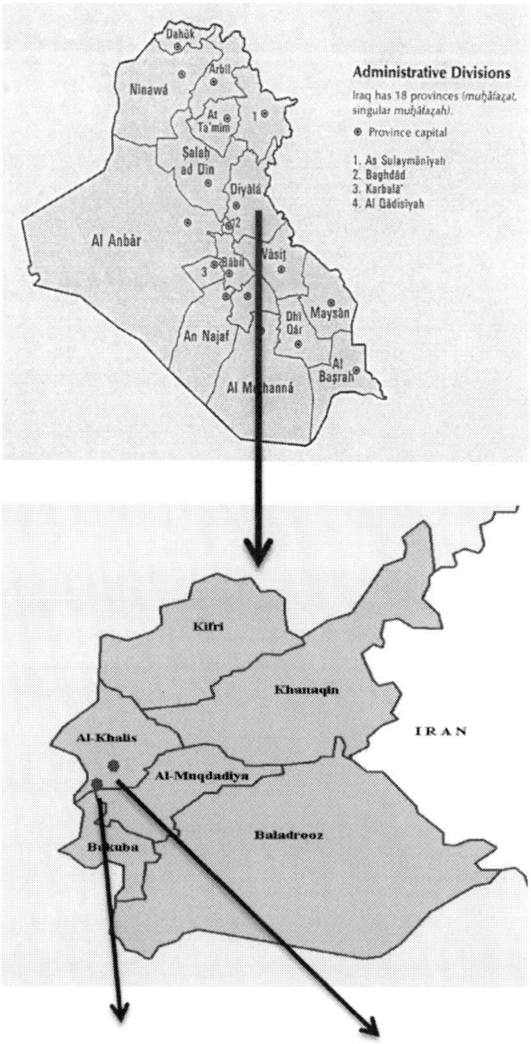


Al-hashmyat Al-khuyls

**Fig. 1.** location of Al-khuyls and Al-hashmyat fields .

## 3 Irrigation water requirement

### 3.1 Crop water requirements

Tomatoes are often planted in Iraq, in the month of April and the season ends in about five months after planting . The actual evapotranspiration can be estimated using equation (1).

$ET_0 \times K_c = ET_c$ (1)

Where $ET_c$ is actual evapotranspiration in mm per day, $K_c$ is factor related crop, and $ET_0$ is potential evapotranspiration in mm per day .

**Table 1** . Growth stages duration.

| Crop | total (days) | Initial stage | devlop. stage | Mid stage | Last stage |
|---|---|---|---|---|---|
| Tomatoes | 150 | 35 | 40 | 50 | 25 |

**Table 2** . The potential evapotranspiration and stages duration from 16 April to 12 September / 2015 .

| Month | Apri. | May | Jun. | Jul. | Aug. | Sep |
|---|---|---|---|---|---|---|
| ET0 (mm//day) | 4.84 | 6.29 | 7.31 | 7.83 | 6.7 | 5.65 |
| Kc(each months) | 0.45 | 0.55 | 0.763 | 1.15 | 1.002 | 0.8 |
| Growth stage) | | S1 | S2 | S3 | S4 | |
| Kc(each stage) | | 0.45 | 0.75 | 1.15 | 0.8 | |
| Duration (month) | | 16 Apr-20May | 21May -29Jun | 30Jun-18Aug | 19Aug-12Sep | |
| Duration(day) | | 35 day | 40 day | 50 day | 25 day | |

The water requirement by crop was 846(mm).

### 3.2 Irrigation scheduling

Irrigation scheduling was estimated using FAO manuals. The net water depth required for a season was 65 mm for 7 days irrigation interval and 21 irrigation events, these values are shown in Table 3 .

**Table 3** . Irrigation scheduling.

| Date | 18 Apr. | 26 Apr. | 3 May | 10 May | 17 May | 24 May | 31 May |
|---|---|---|---|---|---|---|---|
| Amount (mm) | 65 | 65 | 65 | 65 | 65 | 65 | 65 |
| Date | 7 June | 14 June | 21 June | 28 June | 5 July | 12 July | 19 July |
| Amount (mm) | 65 | 65 | 65 | 65 | 65 | 65 | 65 |
| Date | 26 Jul. | 2 Aug. | 9 Aug | 16 Aug | 23 Aug. | 30 Aug. | 6 Sep. |
| Amount ( mm) | 65 | 65 | 65 | 65 | 65 | 65 | 65 |

## 4 HYDRUS (2D/3D) model

HYDRUS is software used for, numerically simulation (2D/3D), solute and water flow in porous media of variably saturated [12]. Below is the description of solution steps of HYDRUS -3D processes used in present research.

### 4.1 Flow of water

The governing water flow equation in unsaturated zone is described using the Richards equation

$$\frac{\partial \theta}{\partial t} = \frac{\partial}{\partial x_i}\left[K\left(K_{ij}^{A}\frac{\partial h}{\partial x_j} + K_{iz}^{A}\right)\right] - S \qquad (2)$$

Where t is time [T], Θ is water content [$L^3 L^{-3}$], $kij^A$ are the anisotropic hydraulic tensor $K^A$, xi is (i=1, 2) are the spatial coordinates [L], h is the pressure head [L , and K is hydraulic conductivity of unsaturated function [$L.T^{-1}$], S is a sink term [$T^{-1}$]

### 4.2 Root uptake

The plant uptake of water (s), is water removed from soil per unit (t) from a unit volume. Feddes et al. in [1978], [3], defined as :

$$S(h) = \alpha(h) S_p \qquad (3)$$

Where α(h) is the water stress of response function ,is a prescribed dimensionless function.

### 4.3 Solute transport

$$-\theta R_k \frac{\partial c_k}{\partial t} - q_i \frac{\partial c_k}{\partial x_i} + \frac{\partial}{\partial x_i}\left(\theta D_{ij,k}\frac{\partial c_k}{\partial x_j}\right) + F_k c_k + G_k = 0 \qquad (4)$$

Where,θ is water content($L^3L^{-3}$),c is the concentration of solute in liquid ($ML^{-3}$) , *Dij* [$L^2T^{-1}$] is an dispersion tensor, q is the volume flux ($L^2 T^{-1}$) ,. $F_k.c_k$ and $G_k$ are coefficients, and $R_k$ constant of universal gas ($ML^2 T^{-2} K^{-1} M^{-1}$).

## 5 Input data

Irrigation fields are divided into basins with dimension of (6*4*1 m) and the figure below shows the typical geometry and finite element mesh used in HYDRUS-3D ,V 2.05.

**Fig. 2.** Geometry of basin and element mesh used in HYDRUS.

### 5.1 Initial conditions and boundary conditions

The initial moisture content of soil for both soil profiles of fields was (0.25) $cm^3.cm^{-3}$ .Initial for the HYDRUS - 3D solute were specified in term the of ($EC_{sw}$ ). All values of chemical and physical characteristic of soil and water use in irrigation are shown in Table 4. Free drainage and atmospheric were put as boundary of bottom and the top surface conditions for each field, respectively and no flow boundary from other sides. Weather conditions were defined using meteorological data from (Al-Khalis meteorological station). As required by model, ETc daily values were split to soil evaporation (E) and crop transpiration (T), and soil evaporation coefficient (Ke) and a basal crop (kcb) are used, where:

$$ETc=(kcb+ke)\times ETo \qquad (5)$$

**Table 4**. Chemical and physical (initial conditions)[a] (Muneam,2015).

| Fields | Al-khuyls | | | Al-hashmyat | | |
|---|---|---|---|---|---|---|
| Depth (cm) | 0–30 | 30–60 | 60-100 | 0–30 | 30–60 | 60–100 |
| Sand(%) | 38.9 | 4.1 | 5.2 | 29.3 | 15.6 | 15.8 |
| Silt(%) | 32.1 | 55.9 | 54.8 | 69.7 | 67.4 | 69.2 |
| Clay(%) | 20 | 40 | 40 | 1 | 17 | 15 |
| Bulk density | 1.43 | 1.55 | 1.6 | 1.61 | 1.68 | 1.5 |
| $EC_{sw}$ (ds $m^{-1}$) | 5.7 | 3.7 | 0.19 | 8.3 | 18.9 | 18.1 |
| $EC_{iw}$ (ground) | (6.51) | | | (2.68) | | |
| $EC_{iw}$ (fresh) | (0.23) | | | (0.23) | | |

a , $EC_{sw}$, electrical conductivity of soil water ,
$EC_{iw}$ electrical conductivity of irrigation water

There are two different criteria by which degree of salinity of soil can be measured, total dissolved solids (TDS) with unit milligrams solute per liter (mg/L), and by electrical conductivity of a saturation paste extract (ECe) with unit deci-simens per meter (ds/m),[formerly micromhos per centimeter,(Mmho/cm) . And can be converting from ($EC_e$) to electrical conductivity of soil water ($EC_{sw}$) by using mathematical relationship Connecting the two values

$k_{EC} \times EC_e = EC_{sw}$ (6)

Where $K_{EC} = EC_{fc} \div EC_e$ the value between (2 to 4) ($EC_{fc}$) in situ water of soil at filed capacity and ($EC_e$) of the water of soil in the saturation extract, [8].In our study we assumed the value, $K_{EC}$ (2).

### 5.2 Hydraulic properties

Hydraulic properties of unsaturated soil, $K(h)$,and $\theta(h)$ in equation (2) are in general highly non-linear functions of the pressure . HYDRUS-3D permits the use of 5 different analytical models to find the properties .In our study we use *van* Genuchten, 1980 ,[13]. The soil water retention, $K(h)$,and $\theta(h)$,functions according to Genuchten :

$$\theta(h) = \begin{cases} \theta_r + \dfrac{\theta_s - \theta_r}{[1+|\alpha h|^n]^m} & h < 0 \\ \theta_s & h \geq 0 \end{cases} \quad (7)$$

Where $\theta s$ – is saturated moisture content [-]; $\theta r$ - is residual moisture content [-];*a, m, n* - empirical parameters [$cm^{-1}$], [-], [-]; *Se* is effective saturation [-];

$$K(h) = K_s S_e^l \left[1 - \left(1 - S_e^{1/m}\right)^m\right]^2$$

Where *l* [-] pore connectivity parameter = 0.5;*Ks* is saturated hydraulic conductivity [$cm.d^{-1}$]; $m=1-(1/n)$, when $n > 1$, all values of parameters are shown in Table 5.

**Table 5**. Soil hydraulic and solute transport parameters of the van Genuchten–Mualem functions.

| Fields | Al-khuyls | | | Al-hashmyat | | |
|---|---|---|---|---|---|---|
| Depth (cm) | 0–30 | 30–60 | 60-100 | 0–30 | 30–60 | 60–100 |
| θr ($cm^3cm^{-3}$) | 0.075 | 0.089 | 0.087 | 0.030 | 0.055 | 0.0599 |
| $\theta_s$ ($cm^3cm^{-3}$) | 0.423 | 0.432 | 0.417 | 0.324 | 0.353 | 0.390 |
| α ($cm^{-1}$) | 0.012 | 0.010 | 0.010 | 0.013 | 0.007 | 0.005 |
| η(–) | 1.437 | 1.403 | 1.393 | 1.491 | 1.528 | 1.643 |
| ℓ (–) | 0.50 | 0.50 | 0.50 | 0.50 | 0.50 | 0.50 |
| $K_s$ (cm $d^{-1}$) | 9.15 | 3.31 | 2.58 | 31.88 | 5.98 | 15.75 |

## 6 Results and discussion

The HYDRUS-3D start simulation on sixteenth of April (2015) in both fields, at the beginning of tomatoes growing season, and the simulation end on twelfth of September (2015).

Four scenarios were used as follows:

The first scenario using groundwater in all irrigation application, the second scenario using one application freshwater and one application groundwater, the third scenario use two application freshwater and two application groundwater, and the fourth scenario use two application freshwater and one application ground water, for all season.

### 6.1 Volumetric water contents

Figure 3 shows a significant change in the water content in the depths near the surface of the soil as a result of the large amount of irrigated water, while this effect reduced in the deeper depths. Also, the large change in moisture content in the nearby depths is due to the evaporation of the water from the soil, which is an effect in the shallow depths.

In addition, in the Al-khuyls field the water content in the depths (20, 40 and 60) cm begins change from the day that the watering begins, but in the depths (80 and 100) cm, the change begins from the 8th day , while in the Al-hashmyat field there was a change in the day of watering in all depths of the soil because of its sandy texture. The movement of water was fast since the soil is sandy with high permeability.

The impact of use of different qualities of water is evident. For instant, the second scenario used groundwater throughout the season is different from other scenarios. Especially the effect is evident in the middle season stage when the plant is more water-intensive than the rest of the stages. The water content is increased because its absorption of salt water is less than fresh water. While in the rest of the scenarios the moisture content is reduced due to the presence of fresh water which cause an increase in the plant's absorption of water.

### 6.2 Overall salinity

Figure 4 shows a significant difference in salinity for both fields, and for all scenarios due to differences in salinity of the irrigation water, as well as the difference in salinity of the initial value of soil water. When applying the second scenario, an increase in the salinity in the Al-khuyls field has been, reaching its peak in the middle stage, whereas in the Al-hashmyat field, a decrease in salinity has been noticed, because the salinity of the soil is higher than the salinity of water used for irrigation water. After this, there is an increase in salinity values

https://doi.org/10.1051/matecconf/201816203007

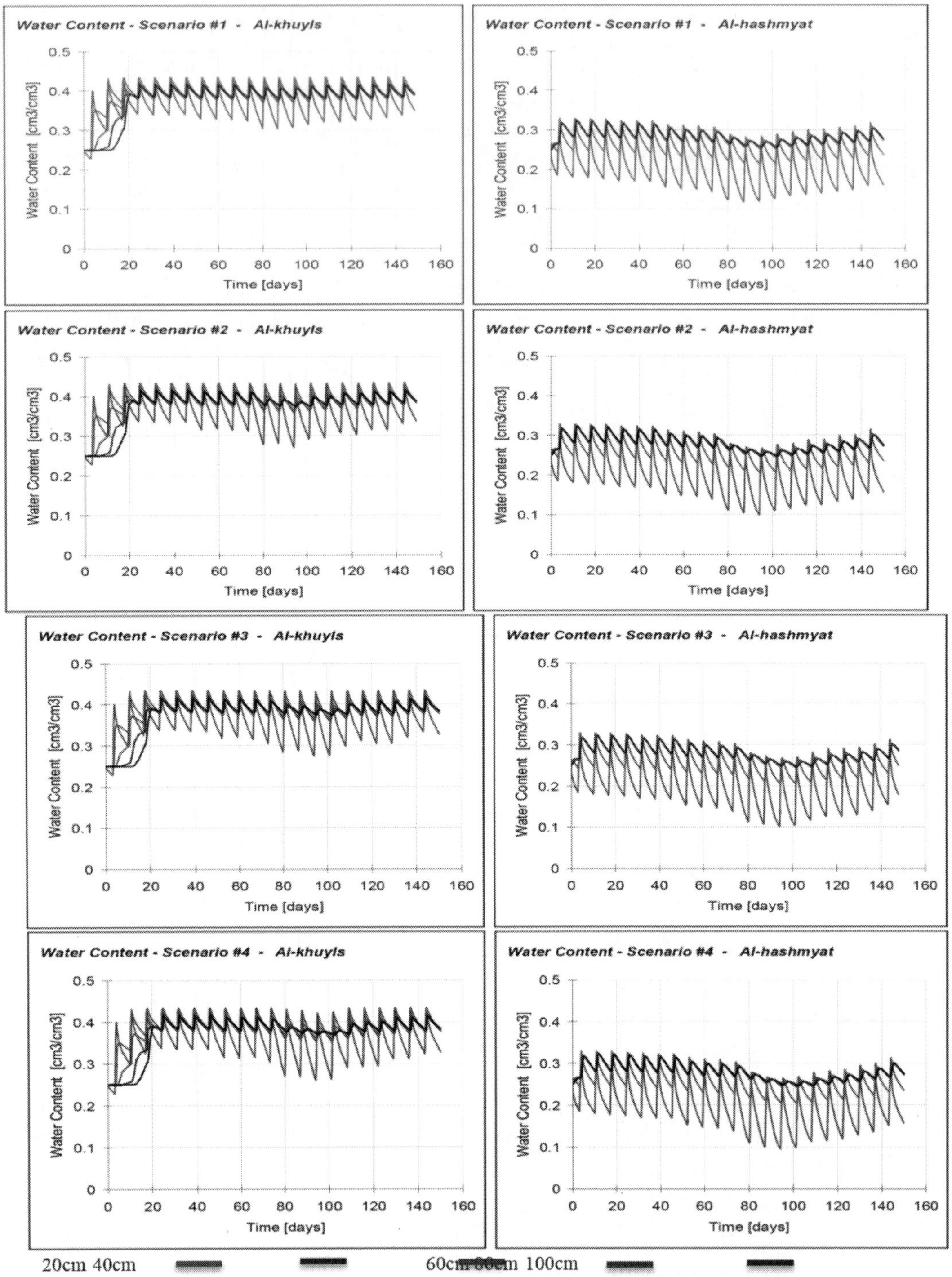


**Fig. 3.** Simulation of water content in both fields at different depth from April[16th] - September[12th] (2015).

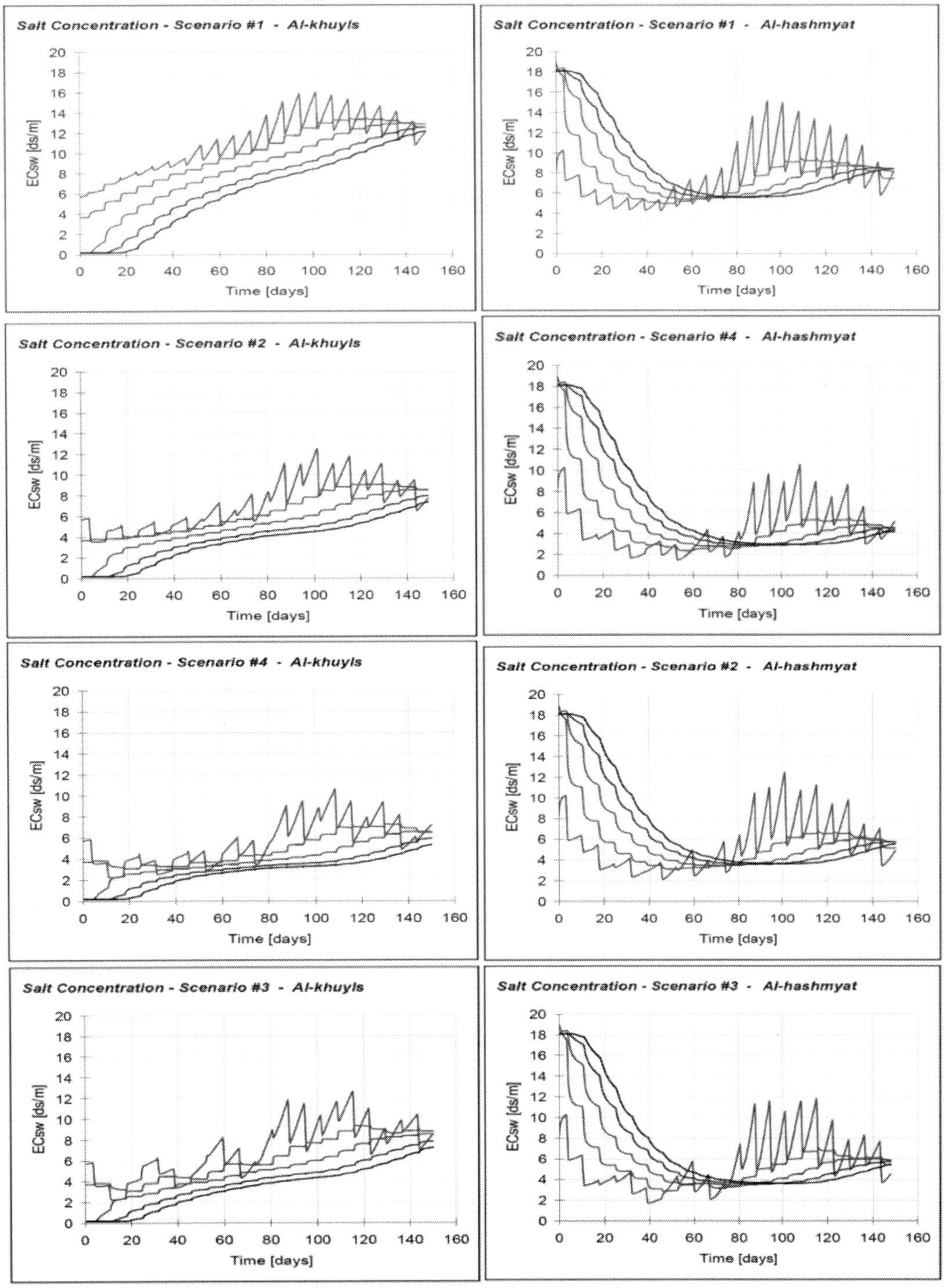


20cm 40cm 60cm 80cm 100cm

**Fig. 4.** Simulation of , $EC_{sw}$ in both fields at different depths from April$^{16th}$ - September$^{12th}$ (2015)

especially in the middle stage due to high evaporation rates. The greatest reduction of salinity was in the fourth scenario. In this scenario the following sequence were used: (freshwater – freshwater –groundwater), in which a 25 % of surface fresh water can be saved.

## 7 Conclusion

1- A significant moisture content fluctuation has been noticed in shallow depths due to irrigation applications. Then chances tend to die out at deeper depths

2-High permeability in Al-hashmyat sandy soil field causes water to move fast by which soil moisture content is changing at all depths in the day of irrigation.

3-The use of high water salinity causes a direct effect on plant water absorption ,Notice the occurrence decrease in the plant
water absorption and hence increase in water content of soil

4-The use of freshwater causes an increase in the plant water absorption and hence decrease in water content of soil

5-The grain size distribution has a direct effect on soil permeability and has a significant effect on soil water content distribution throughout the soil profile.

## References

1. Ahuja , L , Rojas, K.., Hanson, J., Shafeer, M., . (Ed.), 372 (2000).
2. Ayres. R. and D. Westccot ., FAO,29 (1976) .
3. Feddes, R., and P. Raats . 2004. In R. Fedees et al. (eds.),UR Fronts Ser. 6. Kluwer Acad. Press, Wageningen, The Netherlands pp. 95–141.
4. Gonçalves, J. M., Pereira, L. S., Fang, S. X., & Dong, B. Agricultural water management, 94(1-3), 93-108 (2007).
5. Hanson, B., Hopmans, J. W., & Šimůnek, J. *Vadose Zone Journal*, *7*(2), 810-818 (2008).
6. Jarvis, N .,. Department. Soil. Sci., Sweden university. Agri. Sci., Uppsala, Sweden, pp, 51(1994) .
7. Pereira, L. S., Cordery, I., & Iacovides, I. Springer Science & Business Media (2009).
8. Roberts, T ., White , S., Wariick, A.,Thompson, T., 2008Agr. Water Mang. 95 , 669–677.
9. Roberts , T., Lazarovitch, N., Warrick, A., Thompson, T.L.,. Soil Sci. , J. 73 , 233—240(2009).
10. Šimu°nek, J., Van Genuchten, M., Šejna, M., technical manual version - 1.0, pc progress, Prague, republic, pp 21(2006.).
11. Šimu°nek, J., Hopmans, J., . Eco. Modell. 220, 505–521(2009).
12. Šimu°nek, J ., Van-Genuchten, M. Th., Še jna, M., technic manual version-2, pc- progress, Prague, republic. , pp 21(2012).
13. Van Genuchteen, M., . Sci. So. AM. J. 44, pp 892–898(1980).

# Sediment predictions in Wadi Al-Naft using soil water assessment tool

*Imzahim* Abdulkareem Alwan [1*], *Ibtisam* Karim[1], and *Mahmood* Mohamed [2]

[1] Building and Construction Department, University of Technology, Baghdad, Iraq
[2] Ministry of Water Resources, Baghdad, Iraq

**Abstract:** Sediment production is the amount of sediment in the unit area that is transported through the basin by water transfer over a specified period of time. The main aim of present study is to predict sediment yield of Wadi, Al-Naft watershed with 8820 Km$^2$area, that is located in the North-East of Diyala Governorate in Iraq, using Soil-Water Assessment Tool, (SWAT) and to predict the impact of land management and the input data including the land use, soil type, and soil texture maps which are obtained from Landsat-8 satellite image. Digital Elevation Model,(DEM) with resolution (14× 14) meter is used to delineate the watershed with the aid of model. Three Land-sat images were used to cover the study area which were mosaic processed and the study area masked- up from the mosaic, image. The area of study has been registries by Arc-GIS 10.2 and digitized the soil hydrologic group through assistant of Soil Plant Assistant Water Model, (SPAW) which was progressed by USDA, Agricultural, Research Service, using the data of soil textural and organic matter from Food and Agriculture Organization (FAO), the available water content, saturated hydraulic conductivity, and bulk density. The results of average, sediment depth and the maximum upland sediment for simulation period (2010-2020) were predicted to be (1.7 mm), and (12.57 Mg/ha), respectively.

## 1 Introduction

The Soil and Water Assessment Tool (SWAT) was developed by the US Department of Agriculture to predict the impact of land management, practices on water, sediments and agrochemicals yield [1]. It is necessary to minimize requirements data in order to get accurate watershed response. Model accuracy is important for implementing -effective practices of watershed management, [2]. The required watershed data such as: land use / land cover, soil characteristics, topography, or digital elevation model (DEM), require GIS to facilitate modeling of the model parameterization. Hydrological modeling system data should be collected such as soil characteristics, topography, land use / land cover. SWAT is an existing model and requires many data such as: weather variables, soil characteristics, terrain, land cover and land use occurring in the watershed area.

Many researchers have been using SWAT to predict the sediment yield. Radaideh, J, et al (2009) [3] evaluated the quality of rainwater collected and stored in the public water supplies by government agencies in the cisterns and storage tanks. Nayak1,T, (2012) [4] used the SCS-CN method to study the hydrological response in the Development of watersheds to identify potential changes in surface runoff in watersheds as a consequence of planned or unplanned land-use changes Mahasneh, L in (2013) [5]. The effects of water collection interventions on sediments, quality and runoff were studied. The study adopted the SWAT model to predict the effects of water harvesting interventions on biophysical and hydrological parameters.

## 2 SWAT – sediment simulation

SWAT is an extended code, watershed-scale simulation model with daily time- step operation and is planned to evaluate the effect of several administration water practices, agricultural chemical yields as well as sediment. It has physically, established effective computation and is accomplished of simulating a wide level spatial parts. Most parameters of the model for predicting sediment contain: climate, crop growth, hydrology, soil temperature, sediment, pesticides, nutrients and, land management [6].

The model was presented for continuous simulation, as reflected in single event models. Physical procedures related crop growth, flow water, sediment transport, nutrient distribution, etc. are modeled by (SWAT) directly using above mentioned input data. Model

* Dr.Imzahim Abdulkareem Alwan: 40164@uotechnology.edu.iq

© The Authors, published by EDP Sciences. This is an open access article distributed under the terms of the Creative Commons Attribution License 4.0 (http://creativecommons.org/licenses/by/4.0/).

advantages include: simulation of un-gauged watershed, prediction of effect of successive input data such as: changes in management practices, vegetation, water quality and climate. The model also generates weather data as follows: precipitation, relative humidity, solar radiation, wind speed, average temperature of the month and above data variables. It provides useful system to fill in daily missing data in the observed parameters [7]. SWAT watersheds are divided into several sub-watersheds, further subdivided into a single land use / land use known as hydrological response units, HRU. [8].

ArcSWAT divides a catchment into discrete units having the spatial variant of the main physical and hydrological characteristics and processes can be treated as being homogeneous. Landuse/Landcover and the soil maps within sub-basin limits used for generating single group and all groups will be considered as a similar physical property named as Hydrological Response Unit (HRU).

Water balance for each HRUs calculated at a daily time stage. ArcSWAT divided the river basin into sub-divisions having like soil and land properties and that are found in the same sub-basin.

## 3 Runoff modelling

The model selected for the application to the Wadi Al-Naft watershed, is ArcSWAT ver. 2012. The model applies SCS Curve Number (CN) method for estimating direct run-off volume response from rain storms, [9]. This method simulates infiltration loss, although it may account interception surface storage losses through its initial abstraction characteristic [10]. Watersheds have a certain group of soil and fair pasture cover can be classified by various Curve Numbers. These Curves represent rainfall and runoff relationship, such that, [11]:

$$F/S = Q/P \quad (1)$$

$F$: Actual retention (mm).
$S$: Potential retention (mm).
$Q$: Actual runoff (mm)
$P$: Total rainfall (mm).

Initial abstraction,Ia, is all losses before runoff begins. It includes water retention in surface depression and intercepted by vegetation, infiltration, and evaporation as shown in Figure (1).

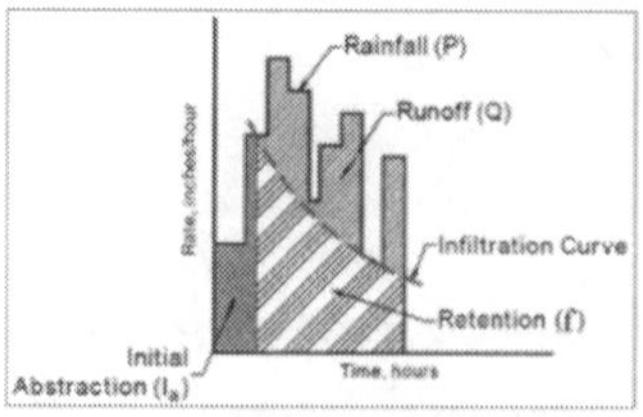


**Fig. 1.** SCS Curve Number (CN) Method, [10].

Ia is subtracted from rainfall P in equation (1) yields

$$(P - Ia - Q) / S = Q / (P - Ia) \quad (2)$$

Solving for $Q$ yields

$$Q = \frac{p-0.25S}{p-Ia+S} \quad (3)$$

## 4 Sediment component

The model calculates sediment yield produced by rainfall and runoff and used the modified universal soil loss equations[12], which is given by equation 4,[13].

$$\text{sed} = 11.8(Q_{surf} * q_{peak} * area_{hru})^{0.56} * K_{USLE} * C_{USLE} * P_{USLE} * LS_{USLE} * CFRG \quad (4)$$

sed: sediment yield (metric tons/day).
$Q_{surf}$: surface runoff rate (mm/ha.).
$q_{peak}$: peak runoff rate ($m^3$/s).
$area_{hru}$: area of each HRU (ha.).
$K_{USLE}$: soil erodibility factor of USLE.
$C_{USLE}$: cover and management factor of USLE.
$P_{USLE}$: support practice factor of USLE.
$LS_{USLE}$: topographic factor of USLE and,
$CFRG$: Factor of coarse fragment.

ArcSWAT uses either the variable storage routing or Muskingum River routing technique to route flow of water. The routing model that simulates the transport of sediments in the channel network includes two components of sedimentation and degradation.

The maximum amount of transferred sediments is a function of the peak velocity of the channel using the following equation, [14].

$$conc_{sed,ch,mx} = c_{sp} * v_{ch,pk}{}^{spexp} \quad (5)$$

In which:
$conc_{sed,ch,mx}$: Maximum transported sediment (kg/L).

$c_{sp}$: Factor defined by the user.
$v_{ch,pk}$: Peak velocity of the channel (m/s), and
$spexp$: Exponent varies between 1 and 2 and was set 1.5.

The maximum concentration of sediment calculated with formula five is compared with the concentration of sediment in the reach at the initial time step, $conc_{sed,ch,i}$ .

- If $conc_{sed,ch,i} < conc_{sed,ch,mx}$ deposition is the prevalent process in reach, and the remaining amount of sedimentation calculated by the formula (6).
- If $conc_{sed,ch,i} > conc_{sed,ch,mx}$ degradation is the prevalent process in reach and the net amount of sediment is computed by the formula (7) [14].

$$sed_{dep} = (conc_{sed,ch,i} - conc_{sed,ch,mx}) * V_{ch} \quad (6)$$

$$sed_{deg} = (conc_{sed,ch,mx} - conc_{sed,ch,i}) * V_{ch} * K_{CH} * C_{CH} \quad (7)$$

$sed_{dep}$:Ssedimentation quantity [metric tons].
$conc_{sed,ch,i}$:Iinitial sediment concentration (kg/L ).
$conc_{sed,ch,mx}$: Ceiling concentration of sediment (kg/L).
$V_{ch}$ :Water volume in the reach part ($m^3$).
$Sed_{deg}$ ; Quantity of sediment entrained in the reach part (metric tons) .
$K_{CH}$ : Parameter of channel erodibility (cm/hr/pa), and
$C_{CH}$ : Parameter of the channel cover.
Ending quantity of the reach sediment is located via the formula (8), [14].

$$\boldsymbol{sed_{ch} = sed_{ch,i} - sed_{dep} + sed_{deg}} \quad (8)$$

$sed_{ch}$: Quantity of reach suspended sediment (metric tons).
$sed_{ch,i}$ : Reach initial quantity of the suspended sediment (metric tons).

The quantity sediment conveyed out of reach is computed as [11]:

$$\boldsymbol{sed_{out} = sed_{ch} \frac{V_{out}}{V_{ch}}} \quad (9)$$

$sed_{out}$: Quantity of sediment, conveyed out of reach (metric tons).
$sed_{ch}$: Reach suspended sediment (metric tons).
$V_{out}$ : Outflow volume during the time step ($m^3$).
$V_{ch}$ : Water volume in the reach segment ($m^3$).

## 5 Wadi Al-Naft Watershed

Wadi Al-Naft watershed is located in the north east of Iraq republic. It lies in the north of Diyala Governorate between (45°00'00" - 46°00'00") E and (33°00'00"-34°00'00") N along LandSat-8 track (36, 37) path and (168) row. The watershed area is about 8820 km². Figure (2) illustrates the study area.

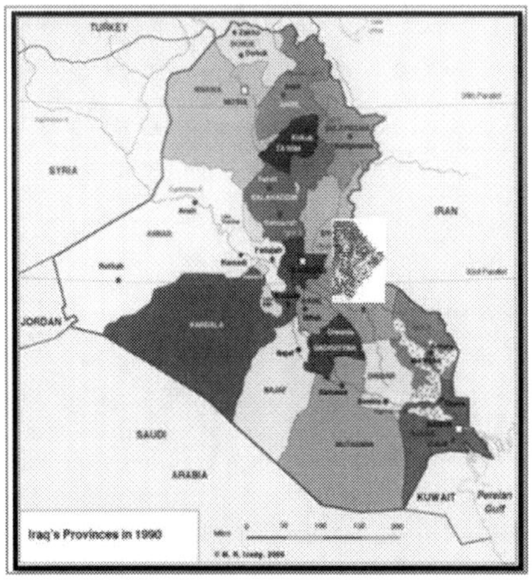


**Fig. 2.** Location of Wadi Al- Naft Region

## 6 Data collection

### 6.1 Climate Data

The Climate data rainfall, minimum and maximum temperature, solar, wind speed and relative humidity were obtained from Iraq Meteorological Organization and Seismology at Kanaqin station, for the period (2000 to 2014). The geographical location of Diyala Governorate is described as dry and hot in the summer months of June, July and August, with a temperature of 31.5 degrees Celsius. The climate of the region is usually windy through winter and the season ranges from December to February. The temperature in winter is about 7.6 degrees Celsius.

Relative humidity in summer to winter in succession was 25.5% to 65.6%. While the evaporation is 53 mm in winter and 329.5 mm in the summer, where the wind speed in winter is 1.2 m / s and in the summer 1.8 m / s. And finally the period of sunshine up to 5.1 and 10.6 hours in winter and summer, respectively.

### 6.2 Digital Elevation Model

Digital Elevation Model for Wadi Al-Naft region is loaded from the USGS website of Landsat8. Figure (3) shows DEM with resolution (14m).

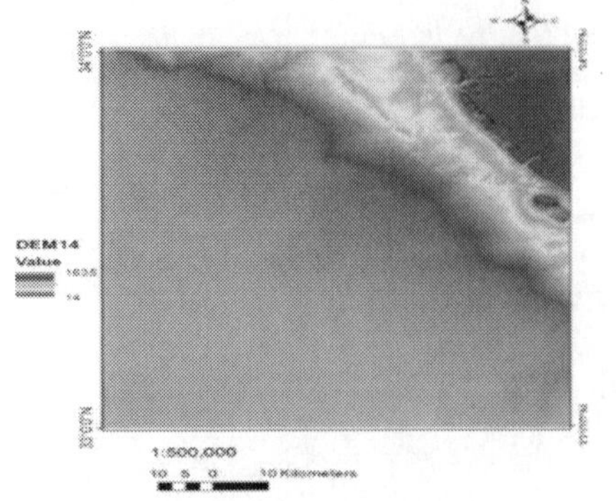

**Fig. 3.** DEM 14.0 mfor Wadi Al-NaftRegion.

### 6.3 Landuse(LU), landcover(LC) and soil digital maps

The satellite imagery, Land-sat 8 can be obtained from the USGS Website, which has three bands with 30.0 m spatial resolution which are used to produce the Landuse (LU) / Landcover (LC) map. Three Landsat images are utilized to cover the area of the valley.

However, the images were mosaic treated and the study area (Wadi Al-Naft) masked up from the mosaic image. Thematic mapping formulated Landuse (LU) / Landcover (LC) categories were classified into three types: agricultural land (34%), forest –mixed (57%) and water (9%). Soil classes were achieved over supervised classification using technique of GIS software. According to exploratory soil map of Iraq 1960 and Land-use map at scale 1,000,000 Table (1).

The area of study has been registered by Arc-GIS 10.2 and has digitized the soil hydrologic group through assistance of Soil Plant Assistant Water Model, (SPAW), using the textural data and organic matter data from Food and Agriculture Organization (FAO), the available water content, saturated hydraulic conductivity and bulk density.

## 7 Delineation digital elevation models

Primarily when using the model of SWAT, it required illustrating the topographic characteristics such as: elevation above mean sea level, incline, flow network of creeks, and distance to nearest creek, and dividing the basin in sub-basin (sub-watershed).

Within SWAT the number of sub-basins to be included in the study, based on the smallest area of all sub-basins. Lesser account provided a large amount detail in the flat areas, greater account yields in sub-basin that were also great in the mountainous areas.

The best sizes of sub-basin were gained by applying a threshold value of 3000 hectares in addition to manually adding more information in the mountainous area, resulting in the delineation in eighty three sub-catchments through used DEM 14.0m., Figure (4).

**Table 1.** Soil Hydrologic Classification

| No | Soil Hydrologic Classification | |
|---|---|---|
| | *Type of soil* | *Percentag e area* |
| 1 | Clay | 1.74 |
| 2 | Clay loam | 9.25 |
| 3 | Silty loam | 0.8 |
| 4 | Loamy | 0.37 |
| 5 | Loamy sand | 49.01 |
| 6 | Sandy | 0.62 |
| 7 | Sandy clay loam | 1.9 |
| 8 | Sandy loam | 7.37 |
| 9 | Silty | 14.73 |
| 10 | Silty clay | 14.21 |

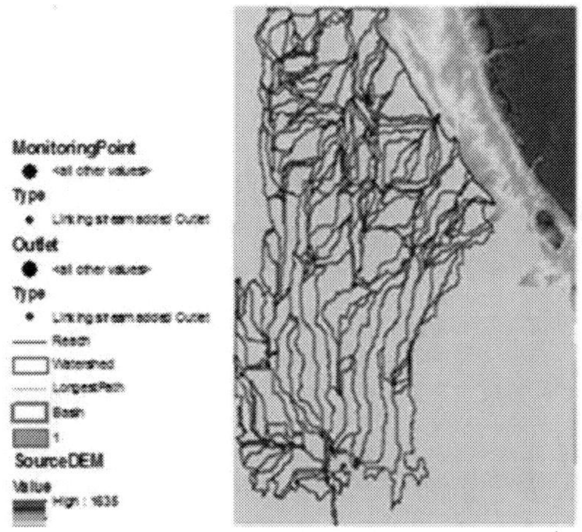


**Fig. 4.** Watershed Delineation and Stream Net.

## 8 Hydrologic response unit analysis

The application of (HRU) by subdividing the catchment area into numerous areas having same land use and soil incorporated assists the model to reflect changes in the hydrologic situations with various land covers, crops and soils to raise the accuracy of load predictions and provide physical description of the water balance.

Water flow is predicted separately for all Hydrology Response Unite and routed to get the whole overland for the watershed. Slope characterization is established upon the DEM defined in the watershed delineation.

Datasets of the Landcover have been connected to the SWAT databases, soil texture and linked the result with SWAT datasets (Soil Hydrologic Groups). Figure (5), (6) and (7) illustrate slope, land cover and soil analysis from DEM (14m) resolution.

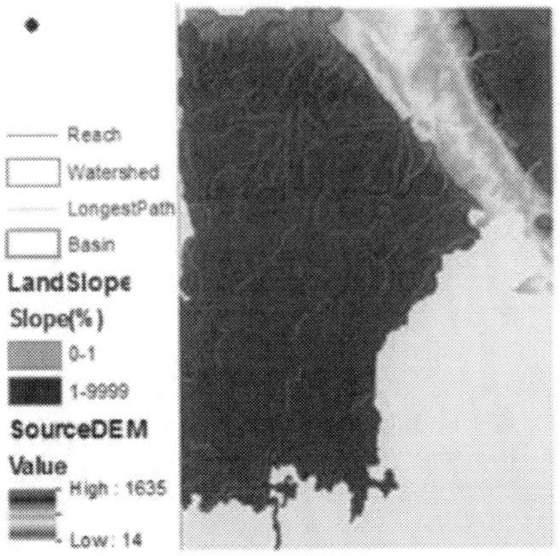


**Fig. 5.** Slopes Reclassified of Study Area.

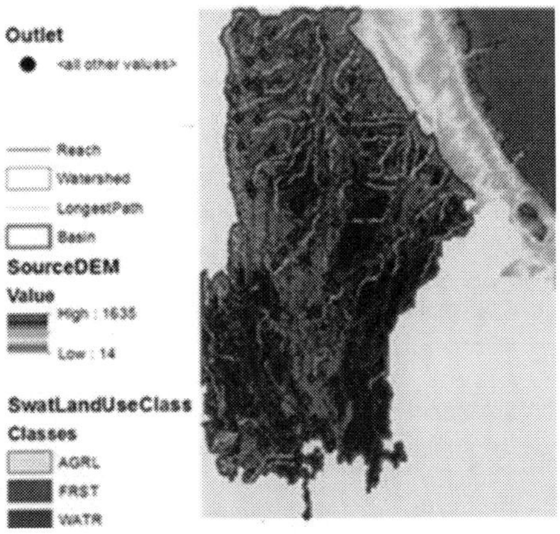


**Fig. 6.** Land cover Reclassified of Study Area.

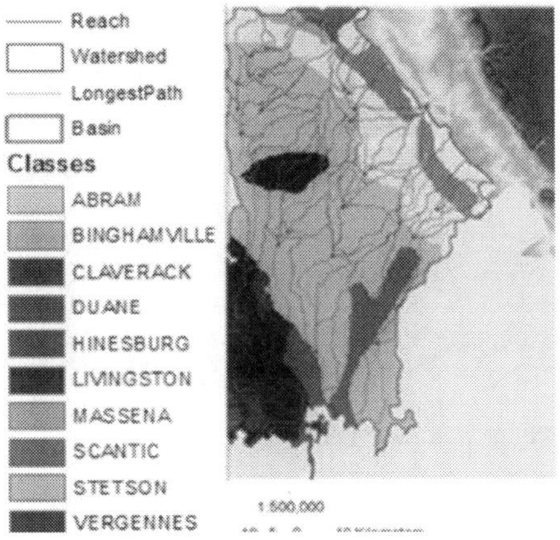


**Fig. 7.** Soil Reclassified of Study Area.

## 9 Results and discussion

To simulate the rate water flow status at the watershed region, net area after delineation step of the watershed DEM 4910.6km$^2$ was considered. Curve Number (CN) values ranged between (60 to 90). Soil of the study area is reclassified by SWAT model as in Fig. 7. The result of average sediment yield for the period of simulation (2010-2020) is found to be (1.7 mm) as illustrated in Table (2). Fig.8 and Fig.9 illustrate the sediment yield and the average maximum upland sediment yield with the value of (12.57 Mg/ha).

**Table 2.** Average Annual Sediment Yield.

| Year | Surface Runoff Depth(mm) | Sediment Yield (mm) |
|---|---|---|
| 2010 | 9.87 | 0.67 |
| 2011 | 30.68 | 2.39 |
| 2012 | 25.48 | 2.3 |
| 2013 | 9.3 | 0.72 |
| 2014 | 28.1 | 2.23 |
| 2015 | 6.5 | 0.44 |
| 2016 | 31.38 | 3.38 |
| 2017 | 30.15 | 2.38 |
| 2018 | 18.5 | 1.4 |
| 2019 | 17.32 | 1.06 |
| 2020 | 23.47 | 1.79 |

https://doi.org/10.1051/matecconf/201816203008

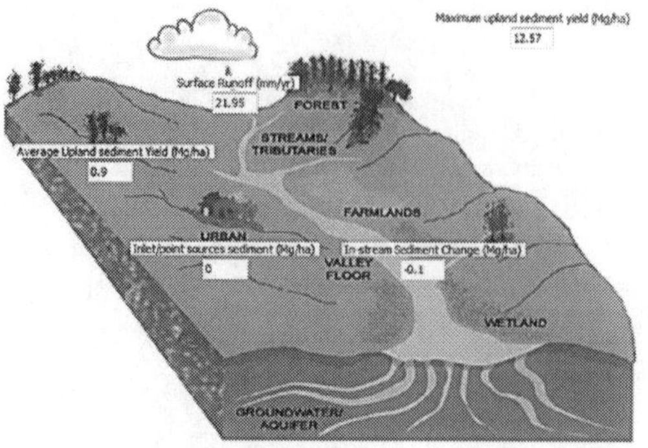


**Fig. 8.** Sediment Load.

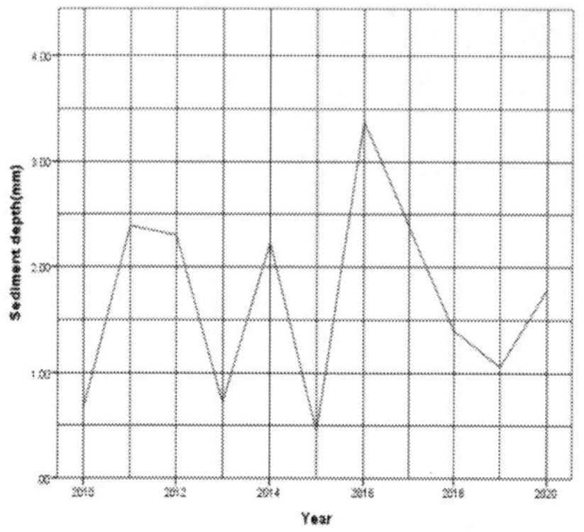


**Fig. 9.** Annual Sediment Depth.

## 10 Conclusions

- Stream flow is taking into account the influent via increase in the number of sub-watersheds because the overland flow is immediately associated with Curve Number, also the stream flow is not influent significantly via the size of the sub watersheds. There is a slight increase in stream flow because of an excess in conveyance gains (inter flow) and a contraction into conveyance losses as sub-watershed size drops.
- The produces of sediment predicted were immediately associated to sub-watershed size which showed a variation due to the sensitivity of overland incline, incline of the channel and drainage density. The variation factors reason for variation in sediment degradation and deposition, and, lastly, with sediment production.
- Primary parameters variance in sub-watershed delineations result in wide variance in the forecast sediment harvests. The results of predictions for sediments become stable for further improvements to the sub basin division, showing that there is a threshold level of division that cannot obtain more accuracy in predictions.

## References

1. E.P. Quernerl and M. van Zanen. Van Hall Larenstein, *University of Applied Sciences* (2013).
2. Chow, V. T., Maidment, D.R., and Mays, L.W., New York, N.Y.: McGraw-Hill.(1988)
3. J.Radaideh ,K. Al-Zboon , A. Al-Harahsheh , and R. Al-Adamat , *Jordan Journal of Earth and Environmental Sciences*, Volume **2**, Number 1 (2009).
4. T. Nayak , M. Verma , and H. Bindu, *International journal of geometrics and geosciences* Volume **3**, Number 1,(2012).
5. L. Al-Mahassneh, *SWAT conference*, France.(2013).
6. S.Neitsch, J. G. Arnold , J.R. Kiniry, and J.R. Williams, Texas, USA: Texas Water Resources Institute.(2011).
7. Hawkins, H. Richard, Timothy, J. Ward, E. Donald Woodward, A. Joseph , and V. Mullen, *American Society of Civil Engineers Publications*.(2008).
8. Ponce, Victor, and H. Richard, Hawkins, *Journal of Hydrologic Engineering*; 11-19.(1996).
9. SCS, Soil Conservation Service, Technical Release 55.(1986).
10. Chow, Ven, Te, New York.(1964).
11. J.R.Williams, and R.W.Hann, Texas: Texas Water Resources.(1978).
12. C.T. Yang . The McGraw- Hill Companies. .New York.(1996).
13. J.R. Williams, Proceedings of the sediment-yield workshop, Oxford, MS, 1972, 244–252, (1975).
14. J.G. Arnold, and J.R. Williams, Agricultural Research Service Black land Research Centre-Texas Argillite Research.(2012).

# Hydraulic performance of abrupt outlet transition structure with strip guide vanes at subcritical flow

*Jaafar* Maatooq[1,*], and *Zainab* Zghaier [1]

[1]Building and Construction Dept., University of Technology, Baghdad, Iraq

**Abstract.** For some stream structures such as barrages, regulators, as well as change of channel cross section, the downstream expansion transitions structures are the common requirement. Eddies, as a result of the flow separation in such structures leads to destruction the bed and sides of the downstream channel. Head loss, on the other hand, produced through the expansion is important because it affects the stage at downstream. This study is restricted to developing the hydraulic performance of abrupt outlet transition at subcritical flow for both decreasing the head losses besides spreading the flow transversely to achieve regular velocity across the width as much as possible. New appurtenances, is adopting to install at specified configuration into a sudden transition. The appurtenances consisted of two successive rows of thin plates used as guide vanes to direct and spread the flow across the width with more uniformity. The guide vanes also keep the head losses at minimum possible. This attempt, through the results, proved that it has considerable hydraulic and economic advantages as compared to using a flared wall. Greater uniformity in velocity distributions of flow across the width, shorter length, and 70% lesser head loss are the results of the hydraulic performance of abrupt expansion transition equipped with two sets of strip vanes as compared to that with the plain one.

## 1 Introduction

When a change of channel shape, a variation of bed elevation, contraction, or expansion of channel width occurs, the transition structure becomes essential to provide a continuous link. Such changes are often required in natural and artificial channels at inlet and outlet structures, syphons, aqueducts, weirs, falls, bridges, barrages, etc. This paper has relevance only to abrupt expansive transitions for subcritical flow. When the flow sections tend to expansive, the flow tends to separate while subjected to the positive pressure gradient associated with flow deceleration, thus resulting in a considerable loss of energy [4]. As a consequence of this separation the eddies are forms adjacent to the boundaries. The present study involves the design of an expansive transition in rectangular open-channel sections at subcritical flow. The design aims are the flow expands with short transition length associated with minimum head loss. The transverse spreading of flow with more uniform distribution to prevent a jet from concentrating at one side wall or at the center is also the major aim of the study. [1,4,9,11] all had agreed when subcritical flow passes through an expansion transition there is an increase in pressure associated with decrease in velocity, which means a conversion from kinetic to potential energy. [11], used a triangle baffles as appurtenances to divide the incoming flow into two parts to reduce the deflection in energy conversion. [1], developed the model by adding two sets of triangular baffles, but in both types of researches, significant eddies occur behind the model. The recommendations proposed by the previous studies in this field are used as a start to develop the model of present study.

## 2 Dimensional analysis

The first step to perform this experimental work is to determine the parameters that have governed influences on hydraulic performance. Dimensional analysis is the most useful tool in experimental fluid mechanics, allowing for the implicit formulation of criteria for dynamic similarity in a simple and direct manner. A prerequisite to run a model with suitable dimensions and boundaries of the experimental work is a proper design of such model with assumptions for magnification length, width, depth, and angle. For the condition of expansion transition structure, the following parameters influencing the hydraulic performance can be grouped and listed herein;

A- Geometrical properties:

y1- water depth at upstream of the transition (L)

*Corresponding Author: 40071@uotechnology.edu.iq; jaafarmaatooq@gmail.com

© The Authors, published by EDP Sciences. This is an open access article distributed under the terms of the Creative Commons Attribution License 4.0 (http://creativecommons.org/licenses/by/4.0/).

https://doi.org/10.1051/matecconf/201816203009

B1-channel width before transition (L)
B2- channel width after transition (L)
B- Flow properties:
V1- flow velocity at upstream section ($LT^{-1}$)
E1- Specific energy of flow at upstream (L)
E2- Specific energy of flow at downstream (L)
g- Acceleration due to gravity ($L\ T^{-2}$)
C- Fluid properties:
ρ - Mass density of the flowing liquid ($ML^{-3}$)
μ - Fluid dynamic viscosity ($ML^{-1}\ T^{-1}$)
The energy of flow at the end of transition will be in dependence with aforementioned parameters; thereby the functional relationship is as follows:-

$$E2=f(\ E1,\rho,\mu,g,V,y1,B1,B2) \qquad (1)$$

The Buckingham $\pi$- theorem is used to create the dimensionless parameters by taking ρ, V1, and y1 as repeating variables. Elimination and delimitation, permit the terms to be restatee as:

$$H_L/y1=f(Fr,B1/y1,B2/B1) \qquad (2)$$

In which the HL= E1- E2. The physical significance of the above parameters visualized that; a head loss is related to inlet water depth (HL/y1), Fr represents a Froude number, and the terms (B1/y1) and (B2/B1) are representing the aspect ratio and the expansion ratio respectively.

## 3 Experimental setup and procedure

The investigation was made on an abrupt expansion in a laboratory flume to develop an outlet transition profile. The approach channel of width, 10 cm, and length, 4 m was constructed of Plexiglas and placed at an upstream portion. Then the headwalls normal to the direction of flow are provided as a sudden expansion to the flume with a width, 30 cm. A tail gate was given at the downstream end of the flume for flow depth regulation. Fig.1, illustrate the model as adopted in the study. Water circulation was continued by a centrifugal pump mounted beneath the flume, lifting water from the floor reservoir tanks running parallel to flume, top of which was used for providing access walk way to the working section. The discharge ranged between 2 L/s – 19 L/s was diverted to the flume via an electromagnetic flowmeter, and its value was recorded directly on a digital control panel.

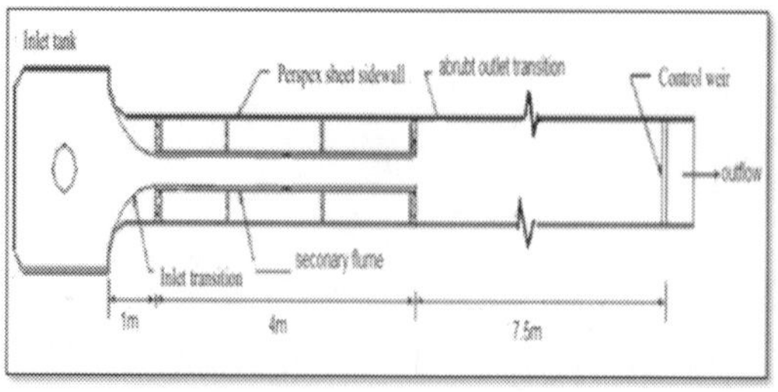


(a) Top view of flume

(b) The flume and abrupt outlet model

**Fig. 1.** Schematic representation and view of model undertaken

To test the transition model, Steps that were followed to find a significant model agreeably accomplished are:

Two sets of vanes have been used, the first set sited at 0.5B1from inlet of transition section and the other set was fixed at 1B1. These locations have been selected by the recommendation provided by [1].

The primary length of each vane was used as recommended by [1] and equal to 0.25B1. The strip vane angle with the flow was 30°as recommended [9]. The trial procedure was used to establish logic and recommendations during searching attempts. The data were collected for a section located at 25 cm upstream of expansion and for six sections within the expansion at which the flow depths and local velocities were measured via point gauge and propeller current meter respectively. The local velocity was measured at 0.6y from the water surface to representing depth average velocity of the section at this point. The velocity for each station X and Z denoted as (Vxz) was measured at 0.1b, 0.25b, 0.5b, 0.6b, 0.75b, 0.8b and 0.9b laterally as shown in Fig.2. Whereas, this velocity was measured vertically from water surface at, 0.2y, 0.4y, 0.6y and 0.8y. All vertical measurements of velocity were at stations along transition center line.

 https://doi.org/10.1051/matecconf/201816203009

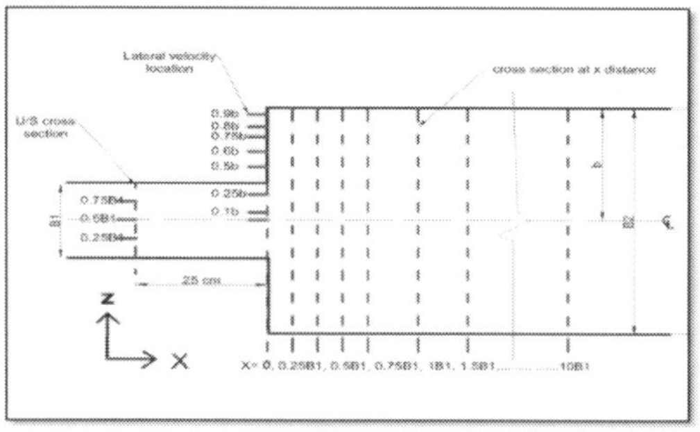


(a) Measurement sections and lateral positions to collect data (X-Z) directions

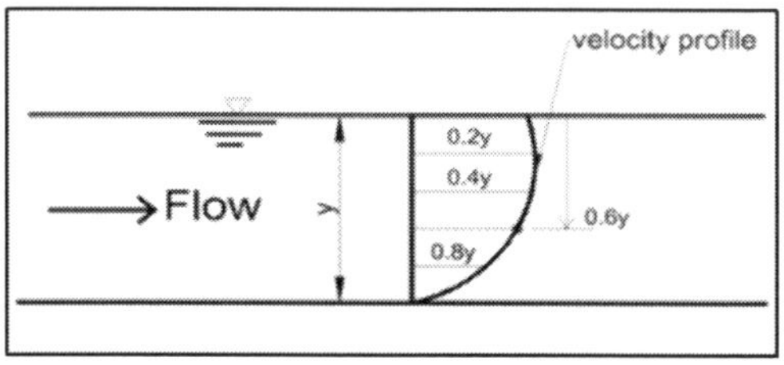


(b) Measuring positions at y-direction (vertically)

**Fig. 2.** Measurement sections and positions for X, Y and Z directions

The transition was tested for two expansion ratios B2/B1 = 2.5 and 3. For each expansion ratio experiments were conducted at three different aspect ratios B1/y1= 0.71, 1 and 1.42. The Froude number F1, at the entrance of transition, were 0.3, 0.5, and 0.7 for each aspect ratio.

## 4 Design and development transition model

Many recommendations of previous researches and data analysis outputs have been reviewed, and some of higher relevance have been used as the initial attempt for present work to design a significant model with practical effect on flow separation and velocity distribution taking energy losses into consideration. Accordingly; the present model was intended as recommended by the previous studies in this field e.g.; [9,1,11]. In the present work, development was made by introducing the stripe vanes as a flow guide by which the inflow divided into three parts, two for sides and one direct along the center. Many trials conducted during the investigation synchronized with changing the limitations of previous researchers. For the first run of the initial design, as shown in Fig.3, a significant separation occurs because of the unsuitability of both the dividing of flow by the first set of the vanes and length of each vane, thereby; the development becomes necessary. If changing the spacing between the strips in the second set that led to no effect on the center flow where, it is responsible for separation phenomenon. Thus, a second trial was conducted by changing the length of the vanes, spacing, and the configuration of installation. The procedure of the second trial was replayed for each strip length checked during the investigation. On the other hand, the changing of strip angle has been adopted according to its effect on flow separation. As a result, the final design was unfolded as displayed in Fig.4.

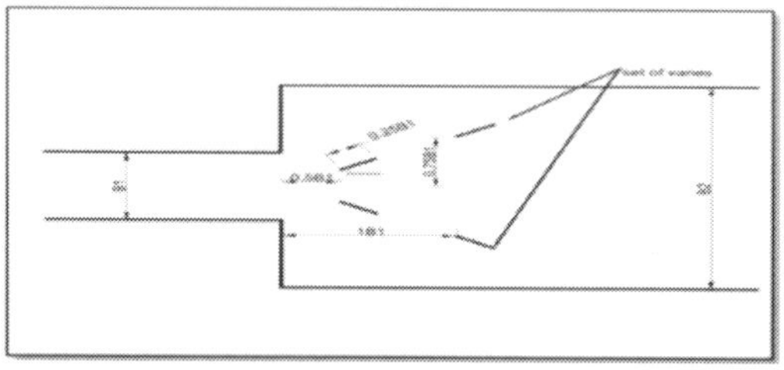

**Fig. 3.** Top view of primary design of model

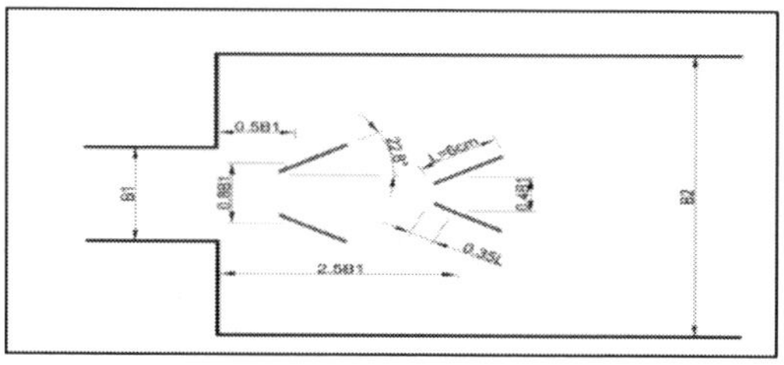

(a) Final design details sketch

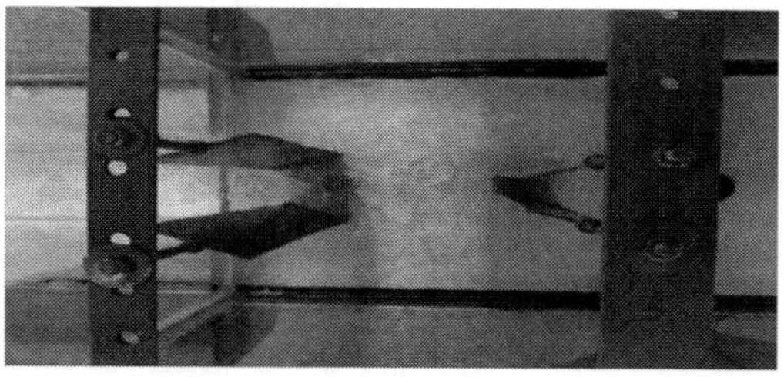

(b) Top view of final design and installation

**Fig. 4.** Final design, configuration, and installation of strip vanes

All sets made individually from galvanized steel stripes gage 18 (0.002m thickness) as adopted by most researchers e.g., [10]. The preparing of these model pieces at the aforementioned configuration and dimensions needs a precise manufacturing equipment, necessary facilities, and technician. Accordingly, the models made by using Computer Numeric Control (CNC) machine.

## 5 Results, analysis and discussion

The collected data was analyzed to show the hydraulic performance pertinent for abrupt outlet transition with strip vanes model compared with conventional shape structures (plain structure). The performance indicators adopted in the present research were the velocity distribution, head loss, and water surface profile. The profile obtained from setting up the strip vanes model on abrupt expansion structure. Using the two flow dividers with a flare angle induced to eliminate the separation and the chances of eddy formation to the less significant amount. Also, transition length ended shorter even before model's length ended and which refers to the dual action of the present model as a distributor and guide reaction with the whole performance.

The current transition (Model C) is tested for performance and compared with the straight wall transition at 22.5° flare angle, which is commonly used in the field, here named as model B. Also the abrupt transition without appurtenances (plain structure) named as model A was tested as a base of performance. Fig. 5 illustrates the variation of transverse velocity distribution at selected sections within a transition for the three models. The good performance related to the uniformity of velocity distribution besides a decrease in its local values was distinguished with using model C. However, long separation reach and concentrated on one side have been corresponded with both models A and B. Moreover, the less transition length at more uniformity in spreading along with less value of local velocity was significantly recorded with model C. The data of Fig. 5 were collected at expansion ratio equal to 3, aspect ratio equal to 1 with Q= 5.50 L/s, so that Froude number at the entrance was 0.55 for all models undertaken.

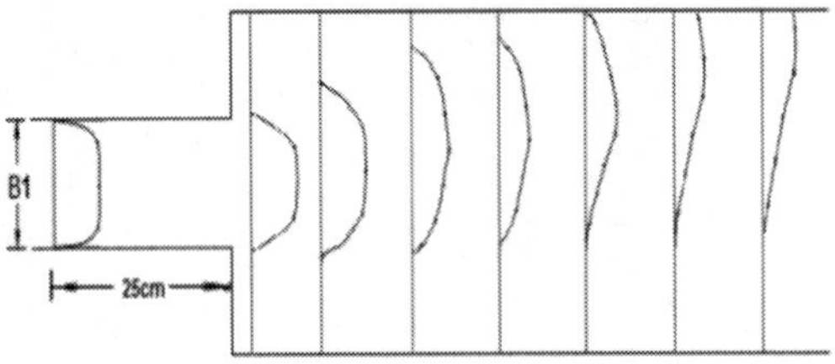


(a)

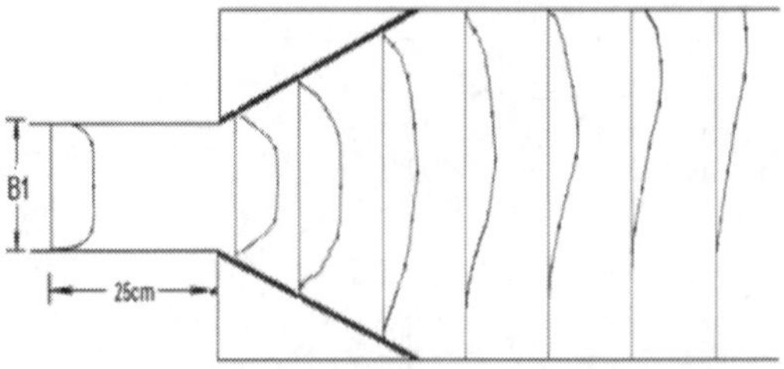


(b)

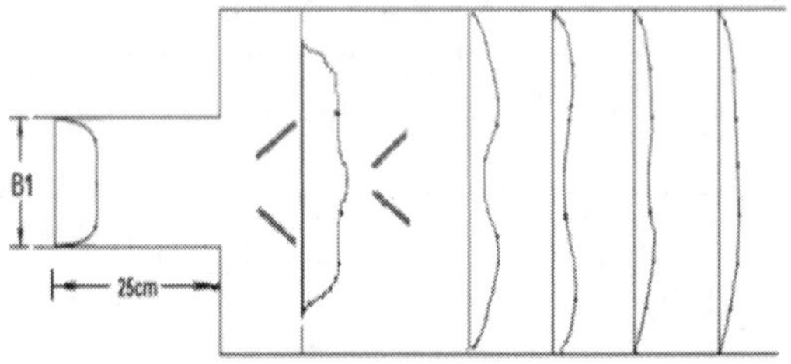


(c)

**Fig. 5.** Velocity distributions across the width and along the length of A, B and C models of transitions

For expansion ratio, B2/B1=3, the greater reduction in head loss was recorded when Fr = 0.55 with model C basis on model A as compared with model B under the same flow conditions. This behavior refers to generate an inverse influential reaction by strips on flow. Thus, this influence reduces the action of eddies which are produced in A and B models. This phenomenon leads to an appreciable increase in losses. On the other hand for expansion ratio at 2.5, the advantage of the reduction in head loss gradually decreases associated with the increasing of Froude numbers which make it the main parameter in this expansion ratio. Generally, head losses recorded for vanes model are less than that for abrupt and competing values for 22.5° models at all Froude number bounded by the present study. The data listed in Table 1 illustrates this behavior, while the percentage of the difference in head losses between model C and model A and that between model C and model B are listed in the last two columns, respectively.

https://doi.org/10.1051/matecconf/201816203009

**Table 1.** Overall head losses data collected for all models of the study

| B2/B1 | B1/y1 | Q(m³/s) | Fr | HL1 (van) (m) | HL2 (abrupt) (m) |
|---|---|---|---|---|---|
| 2.5 | 0.71 | 0.0079 | 0.3 | 0.000601 | 0.001489 |
| 2.5 | 0.71 | 0.01317 | 0.5 | 0.004922 | 0.008694 |
| 2.5 | 0.71 | 0.01844 | 0.7 | 0.014421 | 0.021018 |
| 2.5 | 1 | 0.00469 | 0.3 | 0.000202 | 0.000596 |
| 2.5 | 1 | 0.00781 | 0.5 | 0.001204 | 0.004207 |
| 2.5 | 1 | 0.01094 | 0.7 | 0.005389 | 0.012586 |
| 2.5 | 1.42 | 0.00255 | 0.3 | 6.16E-05 | 0.00016 |
| 2.5 | 1.42 | 0.00425 | 0.5 | 0.000571 | 0.002122 |
| 2.5 | 1.42 | 0.00595 | 0.7 | 0.003024 | 0.006158 |
| 3 | 0.71 | 0.00574 | 0.35 | 0.001791 | 0.001594 |
| 3 | 0.71 | 0.00902 | 0.55 | 0.007163 | 0.009304 |
| 3 | 0.71 | 0.0123 | 0.75 | 0.020299 | 0.022777 |
| 3 | 1 | 0.00347 | 0.35 | 0.000617 | 0.000716 |
| 3 | 1 | 0.0055 | 0.55 | 0.003028 | 0.004587 |
| 3 | 1 | 0.00743 | 0.75 | 0.00991 | 0.01288 |
| 3 | 1.42 | 0.00203 | 0.35 | 0.000354 | 0.000453 |
| 3 | 1.42 | 0.00319 | 0.55 | 0.001833 | 0.00271 |
| 3 | 1.42 | 0.00445 | 0.75 | 0.005201 | 0.00693 |

**Table 1.** (continued)

| HL3 (22.5°) (m) | $\left[\frac{HL1 - HL2}{HL2}\right]\%$ | $\left[\frac{HL1 - HL3}{HL3}\right]\%$ |
|---|---|---|
| 0.000996 | -59.6373 | -39.6586 |
| 0.005019 | -43.3862 | -1.9327 |
| 0.013948 | -31.3874 | 3.3912 |
| 0.000498 | -66.1074 | -59.4378 |
| 0.002174 | -71.3810 | -44.6182 |
| 0.00596 | -57.1826 | -9.5805 |
| 6.16E-05 | -61.5000 | 0 |
| 0.000765 | -73.0914 | -25.3595 |
| 0.002834 | -50.8931 | 6.7043 |
| 0.001495 | 12.3588 | 19.7993 |
| 0.008136 | -23.0116 | -11.9591 |
| 0.018579 | -10.8794 | 9.2578 |
| 0.000617 | -13.8268 | 0 |
| 0.004197 | -33.9873 | -27.8532 |
| 0.011827 | -23.0590 | -16.2086 |
| 0.000354 | -21.8543 | 0 |
| 0.002125 | -32.3616 | -13.7412 |
| 0.00597 | -24.9495 | -12.8811 |

The multi- regression has been following via Minitab-15 Software to construct a deterministic equation for head loss when strip vanes are installed within the abrupt transition to a specified configuration. The data collected have been employed to the functional relationship which resulted from dimensional analysis at Eq. 2, and the following empirical formula restated after regression;

$$\frac{HL}{y_1} = \frac{(B_2/B_1)^{3.55} * Fr^{3.791}}{144.196 * (B_1/y_1)^{1.44}} \tag{3}$$

The statistical indicators for this formula are; ($R^2$=0.976) and standard error of estimate (S = 0.2304). Fig (6) represents the normal probability of collected data. The boundary conditions of application in practice are for Froude number ranging between 0.3 and 0.75 with expansion ratio not exceeding 3.

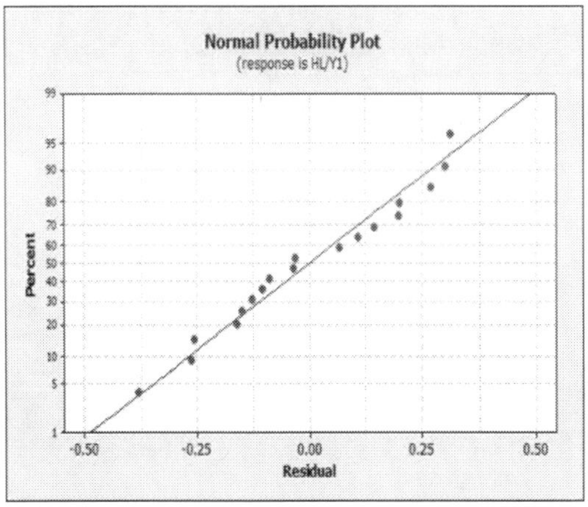


**Fig. 6.** Normal probability plot of data regressed

## 6 Conclusions

Through the experimental program for the present study and after analyzing the data, it can be focusing on the following conclusions:

1. New appurtenances not investigated before were adopted and installed in a specified configuration into abrupt transition.
2. The appurtenances consisted of two successive rows of thin plates used as guide vanes to direct and spread the flow across the width at more uniformity and prevent to prevent loss of more kinetic energy.
3. The hydraulic performance of the appurtenances has been evaluated, and comparisons with conventional (plain structure) proved that it has significant advantages both hydraulically and economically.
4. The experimental results showed that the abrupt transition with strip vanes becomes shorter than and induces less head losses to range between 11% and 71% less than that occurs with the abrupt plain transition.
5. The data collected have been employed to the functional relationship which resulted from dimensional analysis and a deterministic equation for head loss when strip vanes installed within the abrupt transition was restated after regression.

## 7 Notations

The following symbols have been used in this study:
B1= bed width at transition inlet.
B2= bed width at transition end.
Fr= Froude number.
g= gravitational acceleration.
hL= head loss.
L= length of transition.
Q= discharge.
b= bed width at transition end /2.
V= velocity of flow.
x= longitudinal distance from inlet.
y= depth of flow.
y1= depth of flow at inlet.
y2= depth of flow at the outlet.
Δ= difference.

The authors are grateful to the CSED/Water Resources Ministry of Iraq for approving the use of the flume and measurement facilities. Appreciation is due to a staff member of the Hydraulic Laboratory for assistance in preparing models. Deep acknowledgement is presented to Building and Construction Engineering the UOT/Baghdad for academic assistance.

## References

1. Lloyd H. Austin, Gaylord V. Skogerboe and Ray S. Bennett Utah State University, 47-68 (1970)
2. Henderson, F. M., *MacMillan Co., New York, N. Y.*, 235-250 (1966)
3. Hinds, J.,Trans., ASCE, **92**, 1423-1459 (1928)
4. Alauddin, M. and Basak, B.C., J. Civil Eng. **34**(2), 91–101(2006)
5. Chow, V. T., *McGraw-Hill, N.Y.*, 307-318 (1959)
6. Nashta, C.F., and Garde, R.J., J. Hy. Res., **26**(1), 49-65 (1988)
7. Swamee, P. K., and Basak, B. C., J. of Irrg. and Drain. Div., ASCE, **119**(1), 1-17 (1993)
8. Sahar Najmeddin MASc, Civil Eng. College at Concordia Univ. Montreal, Quebec, Canada. 44-71 (2012)
9. Hyatt, M. L., M.Sc. Thesis, Utah State Univ., Logan, Utah, 46-80 (1965)
10. Novak, V. Guinot, A. Jeffrey and D. E. Reeve, *Spon press*, 531-581 (2010)
11. Clifford D. Smith and James N. G. Yu, J. Hy. Division, **92** (2), 1-17 (1966)

# The relationship between the coefficient of discharge and the scour hole depth generated due to the combined flow over weir and below gates

*Saleh* Khassaf [1], *Hassan* Omran[2] and *Fadhel* Abdulabbas[3*]

[1]Department of Civil Engineering, College of Engineering, Al-Basrah University, Basrah, Iraq
[2]Department of Building and Constructions, University of Technology, Baghdad, Iraq
[3]Faculty of Engineering, Kufa University, Najaf, Iraq

**Abstract.** In this research, a study was conducted experimentally to investigate the scour hole dimensions downstream the combined structures which consist of weir and gate due to the effect of the coefficient of discharge ($C_d$) due to flow above the compound weir to the flow under the gate. Fifteen models have been designed, and every model is formed from composite weir consisting of two geometric shapes with rectangular gate of constant dimensions. In this study, the experiment was conducted in a laboratory channel was constructed from blocks and concrete with length of 18 m, 1 m width and depth of 1 m, where the laboratory models were installed after 7 m from the main gate which controlled the passage of water from the main reservoir into the flume. At the beginning, the calibration process was conducted to identify the actual discharge values that pass in the flume, and then seventy-five experiments were conducted in order to derive the formulae for investigating the non-dimensional depth ($SD/d_{50}$) and non-dimensional length ($SL/d_{50}$) of the scour hole due to the effect of ($C_d$). The bed of the flume was spread with sand layer of 30 cm thick for a distance of 4 m downstream combined structure. Two samples of sand were used in the experiments with different median size of particles ($d_{50}$), the first of 0.7 mm and the second of 1 mm. Using the Excel program 2017, Six polynomial relationships were derived to calculate the dimensionless scour depth ($SD/d_{50}$) in terms of the coefficient of discharge ($C_d$), where the resultant coefficients of determinations ($R^2$) from these relationships were high.

## 1 Introduction

Weirs and gates are the common and important structures which are used in controlling and adjusting the flow in irrigation channel. Weirs are widely used for flow measurements. One of the demerits of weirs is that they need to be cleaned of sediment and trash periodically. Sluice gates are used extensively for flow control and water measurement for long time. One disadvantage of the sluice gates is they retain the floating materials.

In order to maximize their advantages, weirs and gates can be combined together in one device, so that water could pass over the weir and below the gate simultaneously.

Fig. 1 shows this structure, this compound device creates a new hydraulic condition in compression with weir or gate, each other alone. The combined weir and gate systems can be used in minimizing sedimentations and depositions [1].

The economic aspect is one of the important factors in the creation of such a project and to reduce the cost of any project, optimal running of the structure without damage in future is taken into consideration.

The non-use of concrete floor at downstream of the structure, which is designed to dissipate energy flow passing over the edge of the weir is one of the most important methods reduce cost. This alternative is less expensive and is available and ensures the lack of access scour, which leads to the occurrence of damage in the structure and thus keeps it fully within the allowable limits.

* Corresponding author: fadhil.alshitali@uokufa.edu.iq

© The Authors, published by EDP Sciences. This is an open access article distributed under the terms of the Creative Commons Attribution License 4.0 (http://creativecommons.org/licenses/by/4.0/).

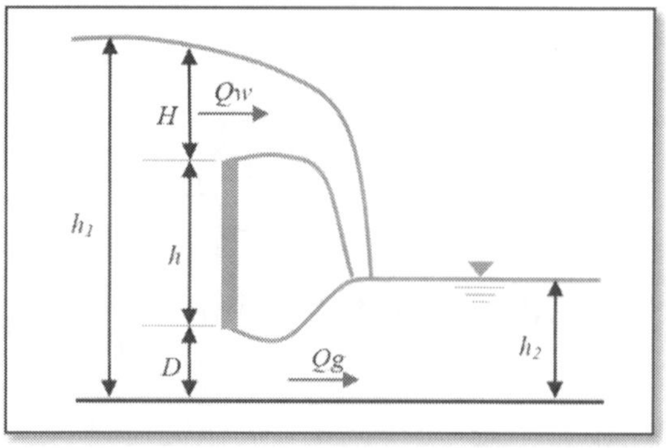


**Fig. 1.** Definition sketch for combined free flow over weirs and under gates.

Ahmed [2], investigated the scour characteristics downstream weirs, gates and combined structures consisting of weir and gate. The study included the measurement of maximum scour depth ($D_s$) and the length ($L_s$) of scour hole downstream these structures. Also, the effects of structure height, under sluice opening height, discharge variation and bed material size ($D_{50}$) on the depth and length of scour hole. Two empirical relationships were obtained to estimate ($D_s/D_{50}$) and ($L_s/D_{50}$) in terms of Froude number ($Fr_0$) , relative water surface fall ($\Delta H_w/P$), relative opening height ($h_0/\Delta H_t$) and relative discharge ($q_u/q_d$) for compound gates with high correlation coefficients.

Dehghani et al. [3] studied the scour characteristics of scour hole downstream of combined free over weir and below gate experimentally. The conceptual model of flow field downstream of combine flow over the weir and under the gate indicates that there are interactions between the flows over the weir and under the gate and the scour hole cuts and fills alternatively. By increasing the Froude number, the maximum depth of scour (hs), length of scour ($l_1$) and sedimentation length ($l_2$) increase.

Sobeih et al. [4] investigated the scour depth downstream weir with holes. An experimental study (with 171 experiments) was carried out in a rectangular flume. The weir used in the research was perforated with different numbers of holes located at different heights (h) with respect to the height of weir (p). The arrangement of opening were included with three cases; first with no hole, second with one hole and third with three holes. The diameter of hole (d) differs from case to another and it was 0.5 inch, 0.75 inch and 1 inch with different height (h) for each case as a percent of weir height (p) and it was 0, 0.25 and 0.5. The research showed that the case when the ratio of height of the hole (h) to the weir height (p) equal to 0.25, the scour depth is minimum and when the same ratio equal to 0.5, the scour depth is the maximum value. Also, at a specific value of (h/p), when the ratio of the diameter of the hole (d) to the height of the weir (p) equal to 0.149, the scour depth is minimum with number of holes equal to one. While, when the same ratio equal to 0.075 with three holes, also the minimum value of scour be realized.

An empirical equation was derived to estimate the scour depth depending on multi variables such as Froude number, height of holes, number and diameter of holes, weir height and the flow condition at downstream side of the weir.

## 2 Laboratory models

Fifteen models were used in the experimental work to investigate the derived formulae describing the relationships between dimensionless scour hole depth and the coefficient of discharge ($C_d$). Laboratory experiments have been conducted by a channel, which was manufactured by the researcher and was built from blocks and concrete, and consisted of the following parts as shown in Fig. 2.

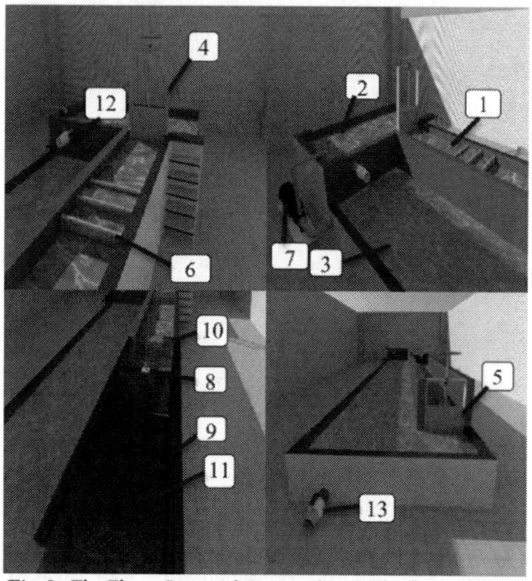


**Fig. 2.** The Flume Parts and Accessories: 1. The flume, 2. Head basin, 3. Lateral basin, 4. Vertical sluice head gate, 5. Vertical. sluice tail gate, 6. Stilling screens, 7. Main pump, 8. Gauge point, 9. Rails, 10. Iron frame, 11. BRC mesh, 12. Overflow valve, 13. Exhausting and cleaning valve.

A calibration process was conducted to the flume by using a standard weir, which is designed according to USBR limitations for standard sharp crested weir with 90° V notch [5]. The laboratory models were divided into four groups with respect to compound weir shape as shown in Fig. 3:

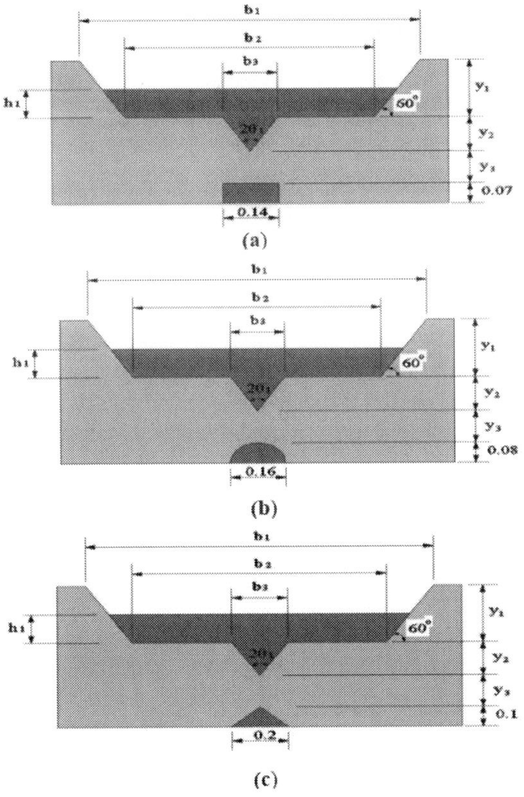


**Fig. 3.** General definition sketch for (a) first five models (b) Second five models (c) Third five models.

The geometrical dimensions for all tested models are listed in Table 1. It should be mentioned that the following geometric parameters are constant for all the tested models:

1) The width (B) = 1 m and the overall height (L) = 0.5 m. for all models.
2) The height of upper part of compound weir ($y_1$) for all models = 0.2 m.
3) The gates dimensions are same for each five models = (0.14 m. width * 0.07 m. height for rectangular gate, 0.16 m. width * 0.08 m. height for semi- circular gate and 0.2 m. width * 0.1 m. height for triangular gate).

**Table 1.** Geometric Properties for Laboratory Models.

| Mod .No. | $y_2$ (m) | $y_3$ (m) | $b_1$ (m) | $b_2$ (m) | $b_3$ (m) | D (m) | $\Theta_1$ | $\Theta_2$ |
|---|---|---|---|---|---|---|---|---|
| 1 | 0.12 | 0.11 | 0.88 | 0.19 | 0.15 | 0.14 | 60 | 30 |
| 2 | 0.09 | 0.14 | 0.76 | 0.36 | 0.18 | 0.14 | 45 | 45 |
| 3 | 0.08 | 0.15 | 0.61 | 0.38 | 0.2 | 0.14 | 30 | 60 |
| 4 | 0.07 | 0.16 | 0.58 | 0.43 | 0.22 | 0.14 | 20 | 70 |
| 5 | 0.06 | 0.17 | 0.53 | 0.46 | 0.26 | 0.14 | 10 | 80 |
| 6 | 0.12 | 0.1 | 0.88 | 0.19 | 0.15 | 0.16 | 60 | 30 |
| 7 | 0.09 | 0.13 | 0.76 | 0.36 | 0.18 | 0.16 | 45 | 45 |
| 8 | 0.08 | 0.14 | 0.61 | 0.38 | 0.2 | 0.16 | 30 | 60 |
| 9 | 0.07 | 0.15 | 0.58 | 0.43 | 0.22 | 0.16 | 20 | 70 |
| 10 | 0.06 | 0.16 | 0.53 | 0.46 | 0.26 | 0.16 | 10 | 80 |
| 11 | 0.12 | 0.08 | 0.88 | 0.19 | 0.15 | 0.2 | 60 | 30 |
| 12 | 0.09 | 0.11 | 0.76 | 0.36 | 0.18 | 0.2 | 45 | 45 |
| 13 | 0.08 | 0.12 | 0.61 | 0.38 | 0.2 | 0.2 | 30 | 60 |
| 14 | 0.07 | 0.13 | 0.58 | 0.43 | 0.22 | 0.2 | 20 | 70 |
| 15 | 0.06 | 0.14 | 0.53 | 0.46 | 0.26 | 0.2 | 10 | 80 |

Each model was examined by five values of actual discharge ($Q_{act}$) which are 0.037, 0.031, 0.026, 0.011 and 0.006 $m^3/s$.

## 3 Sieve analysis for bed material samples

In this study, sand was used as a material for the bed of flume and the samples which were selected included all the grades of sand from coarse to fine and as classified by the USCS classification [6].

To investigate the effect of sediment size as an effective parameter on the scouring process, two samples were selected to conduct the laboratory experiments with two mean diameter, the first of 1 mm and the second of 0.7 mm.Sieve analysis and preparation of the quantities required was conducted by NSGF Company for production of sand and gravel filters. The results of sieve analysis for both samples mentioned above are as shown in Fig. 4 and Fig. 5.

The geometric standard deviation $\sigma_g$ of the sand size equal to 2.27 for the first sample and 2.62 for the second sample, which implies that the sand is of well-graded soil. The $\sigma_g$ is defined as: [7]

$$\sigma_g = \sqrt{\frac{d_{84.1}}{d_{15.9}}} \quad (1)$$

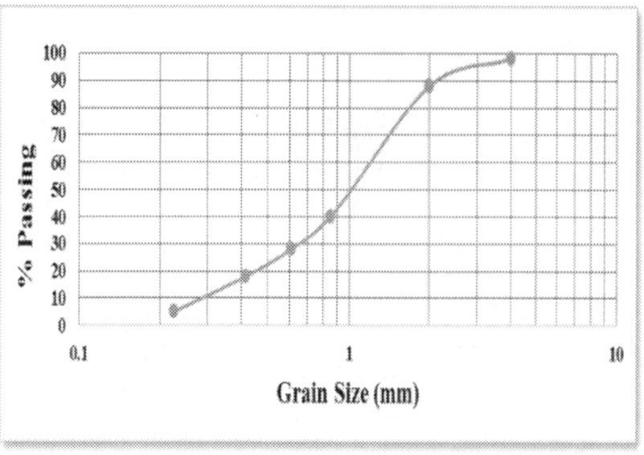


**Fig. 4.** Sieve analysis of bed material for the first sample ($d_{50}$ = 1 mm).

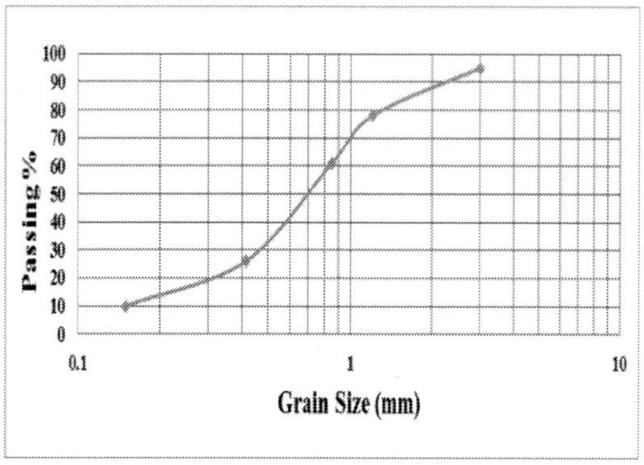


**Fig. 5.** Sieve analysis of bed material for the second sample ($d_{50}$ = 0.7 mm).

## 4 Results analysis and discussions

### 4.1 For the first five models

As known, the increasing in flow section area (increase in $Q_{act}$) causes an increase in the value of the discharge coefficient and vice versa. As a result, the increase in the value of the discharge coefficient causes increase in the value of scour depth.

But whenever the theoretical discharge value increases through the combined section and the actual discharge remains constant, the value of discharge coefficient will decrease.

For the first five models, Figure (6) shows that the maximum value of SD/d50 recorded was 390 when $C_d$ equals to 0.5009 with $d_{50}$ equals to 0.7 mm in model no. 5, while the minimum value of SD/$d_{50}$ recorded was 114 for $C_d$ value of 0.2328 in model no. 1 with $d_{50}$ equals to 1 mm (Figure (7)).

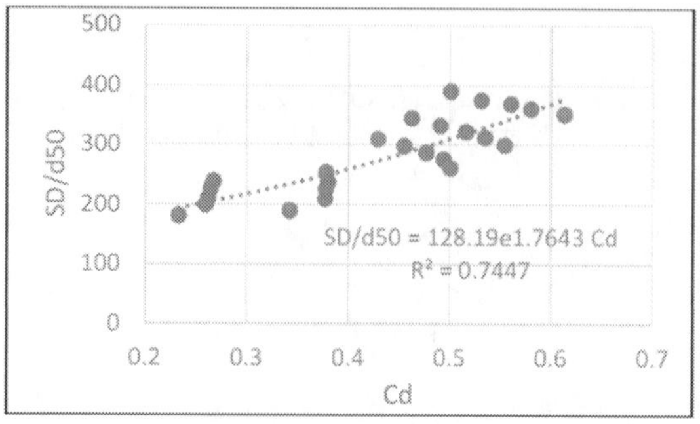


**Fig. 6.** The Relationship between the Non-Dimensional Scour Depth and the Coefficient of Discharge for the First Five Models ($d_{50}$ = 0.7 mm).

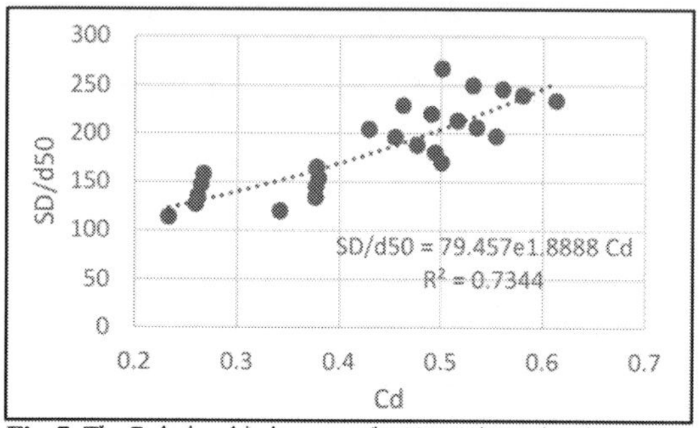


**Fig. 7.** The Relationship between the Non-Dimensional Scour Depth and the Coefficient of Discharge for the First Five Models ($d_{50}$ = 1 mm).

### 4.2 For the second five models

In the second five models, results recorded that the maximum value of SD/d50 of 395.1 for Cd equals to 0.4781 with d50 equals to 0.7 mm in model no. 10 and minimum value of 140 when Cd equals to 0.2215 with d50 equals to 1 mm in model no. 6 as shown in Figures (8) and (9).

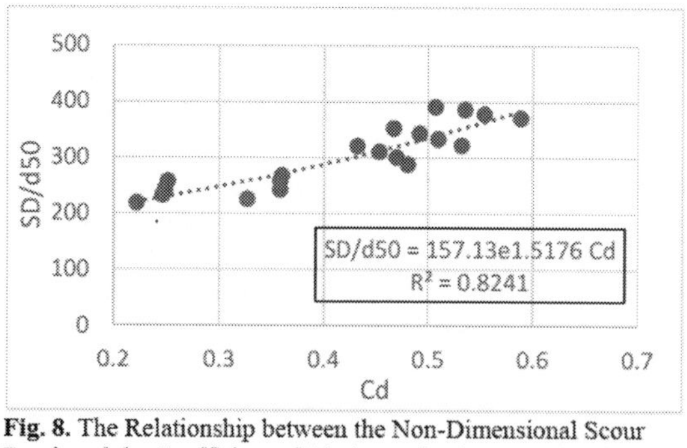


**Fig. 8.** The Relationship between the Non-Dimensional Scour Depth and the Coefficient of Discharge for the Second Five Models ($d_{50}$ = 0.7 mm).

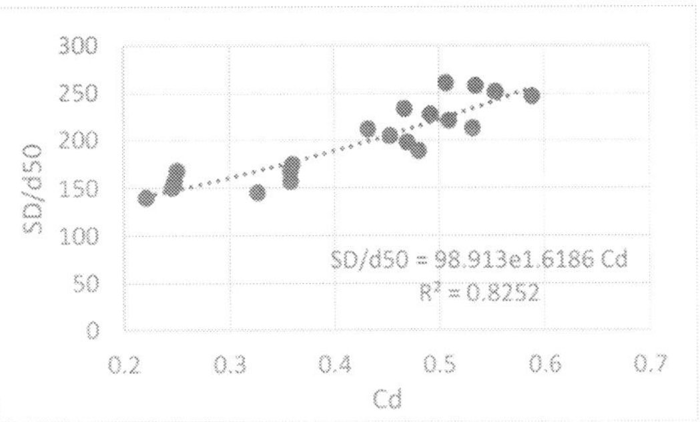


**Fig. 9.** The Relationship between the Non-Dimensional Scour Depth and the Coefficient of Discharge for the Second Five Models ($d_{50}$ = 1 mm).

### 4.3 For the third five models

The third five models, as shown in Figures (10) and (11), results recorded the maximum value of SD/d50 of 379.4 for Cd equals to 0.5217 with d50 equal to 0.7 mm in model no. 15 and minimum value of 107 when Cd equals to 0.2377 with d50 equals to 1 mm in model no. 11.

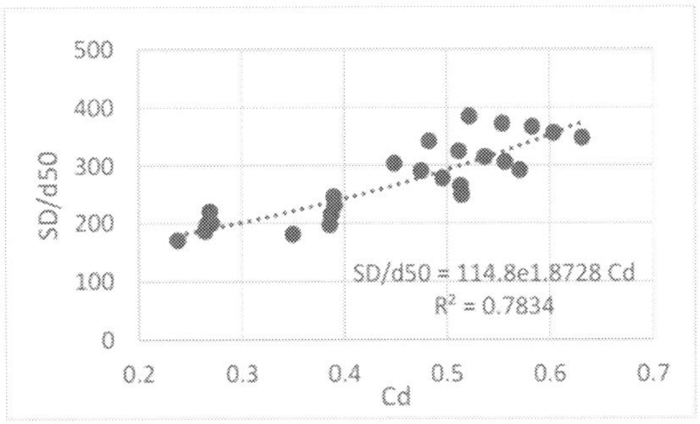


**Fig. 10.** The Relationship between the Non-Dimensional Scour Depth and the Coefficient of Discharge for the Third Five Models ($d_{50}$ = 0.7 mm).

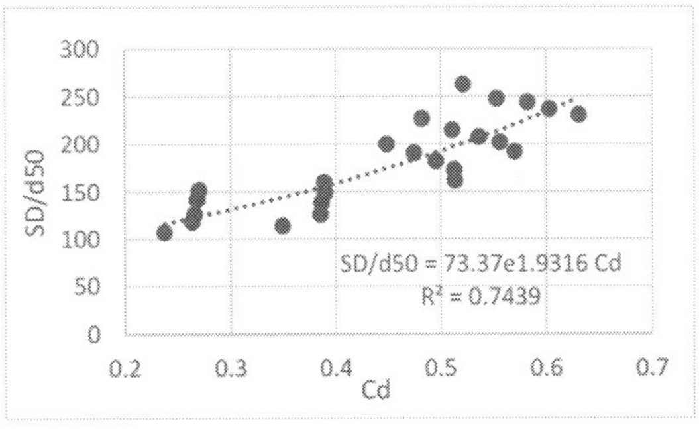


**Fig. 11.** The Relationship between the Non-Dimensional Scour Depth and the Coefficient of Discharge for the Third Five Models ($d_{50}$ = 1 mm).

## 5 Scour hole form

Surfer-ver. 11 program was used to draw a three-dimensional profile to the sandy bed of the flume after scour process is complete. The following figures from (12) to (14) represent the form of scour hole downstream of some combined models, which were used in the laboratory experiments.

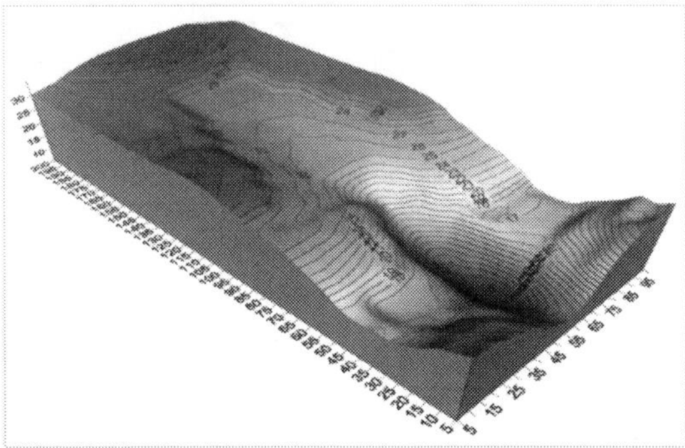

**Fig. 12.** 3D- Scour Profile for Model #1 for Q= 0.037 $m^3/s$ and $d_{50}$ = 0.7 mm.

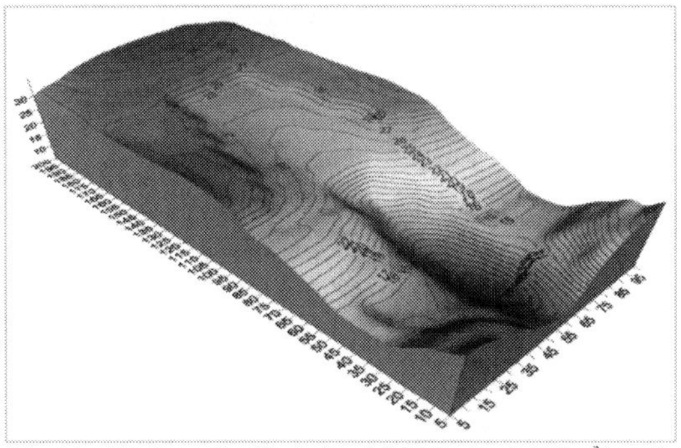

**Fig. 13.** 3D- Scour Profile for Model #6 for Q= 0.037 $m^3/s$ and $d_{50}$ = 0.7 mm.

**Fig. 14.** 3D- Scour Profile for Model #11 for Q= 0.037 $m^3/s$ and

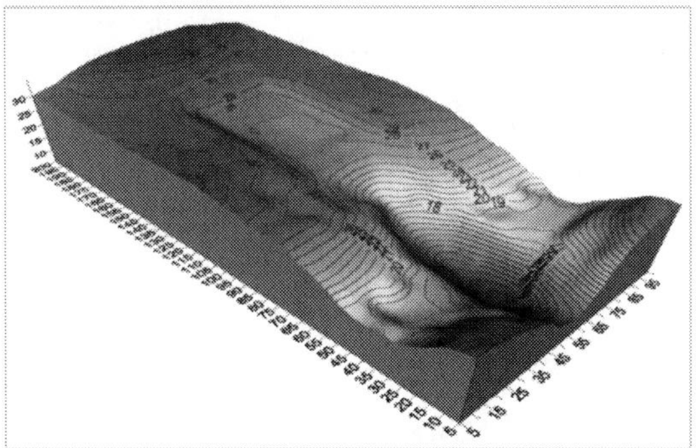

$d_{50}$ = 0.7 mm.

## 6 Conclusions

In this study, the following conclusions are made from $SD/d_{50}$ and $SL/d_{50}$ vs. $Q_r$ for various values of actual discharges $Q_{act}$ and $d_{50}$:

1) The results show that the maximum values of $SD/d_{50}$ for the fifteen models was recorded generally in the second five models and especially in model no. 10 in which the compound weir consisted of trapezoidal upper part and rectangular lower part with a gate semi-circular in shape. The maximum scour depth was recorded as 27.7 cm in sandy bed with $d_{50}$ of 0.7 mm and 30 cm thickness. In addition, the maximum scour depth was recorded against actual discharge of 0.037 $m^3/s$ and relative discharge. While the minimum value of $SD/d_{50}$ was recorded in model no. 11, in which the gate is triangular. The minimum value was 10.7 cm in sandy bed with $d_{50}$ of 1 mm and 30 cm thickness.
2) Both values of SD increase with increase in the values of actual discharge passing through the flume and relative discharge between compound weirs to the flow under the gate.
3) The most controlled hydraulic factor in the value depth of the scour hole is Froude number in terms of the mean size of bed material ($d_{50}$) where always the relationship is positive, that is, when the value of Froude number increases each of SD and SL increased for all models in all groups.
4) The size of the depositions changes with the depth and length of the scour hole where whenever the hole depth increases as a result of the free fall of water from the edge of the compound weir, the deposition of sediments was more, while the flow through the gate is helping to pay these sediments away and make the form seem more flat.
5) Whenever the weir width increases, the scour hole generated becomes more flat even with the increase in the value of discharge passing through it and this is evident by the results of the third and fourth five models.

## References

1. Dehghani, A. A., Bashiri, H., Dehghani, N., River Flow – Dittrich, Koll, Aberle & Geisenhainer, (2010).
2. Ahmed, A. A., University of Mousel, (2007).
3. Dehghani, A.A., Bashiri, H., Shahmirzadi, M.E.M., WSEAS Press, (2010).
4. Sobeih M. F., Helal, E. Y., Nassralla, T. H., Abdelaziz, A. A., International Journal of Civil and Structural Engineering, Volume 3, No. 1 (2012).
5. USBR, U. S. Department of the Interior, Bureau of Reclamation, Revised Reprinted of Third Edition, (2001).
6. Murthy, V. N. S., Marcel Dekker, Inc., First Edition, (2002).
7. Chanson, H, Butter Worth Heineman, Second Edition, (2004).

# Estimate of throughput of bridge transitions and pipe passages built on minor rivers of piedmont areas of Krasnodar Territory-Russia

*Fedor* Bryukhan[1,*], *Aleksey* Vinogradov[2]

[1]Moscow State University of Civil Engineering, Yaroslavskoe shosse, 26, Moscow, 129337, Russia
[2]Saint Petersburg State Forest Technical University, Institutskiy per., 5, Saint-Petersburg, 194021, Russia

**Abstract.** Stability and accident-free operation of engineering road structures including bridge transitions and pipe passages built on mountain rivers mostly depends on a stream regime and lack of obstructions for water flow. Such structures pose a prominent potential hazard being built in piedmont areas of Caucasus Mountains characterized by flash floods and blockage of structures by floating debris, mudflow deposits and wastes of construction. This notwithstanding, the threats caused by these phenomena are poorly studied. The purpose of this study is in estimation of throughput of bridge transitions and pipe passages built on minor rivers of piedmont areas and analysis of hazards caused by floods and obstructions to water flow. The results of calculation of capacities of existing road structures are provided herein. A qualitative assessment of potential emergencies in case of severe flood is also given. A major hazard of possible blockage of waterways that can cause emergency even in regular flood conditions is noted.

## 1 Introduction

In the context of intensive building activities including engineering and construction of roads on the Black Sea Coast of Krasnodar Territory, the man-made hazards of the projects under construction is particularly topical [1]. The floods of mountain rivers and blockage of waterways due to floating debris, mudflow deposits and wastes of construction is one of the sources of both natural and man-made hazards that can cause emergencies [2-4]. An example of the occurrence of the floating debris is shown in Fig. 1.

**Fig. 1.** Floating debris near the bridge transition over watercourse #1 after a flash flood.

This notwithstanding, the respective issues of hydrologic support to building and operating of road structures are poorly studied. Therefore, the purpose of this study is to estimate the throughput of engineering road structures built on minor rivers of piedmont areas and analysis of hazards caused by floods and obstructions to water flow.

## 2 Data and methodology

The low density of gauging stations grid is the feature of the minor rivers of Northern Caucasia. Thus, a maximum runoff of rainfall flood is calculated using the Maximum Rainfall Intensity Method recommended by [5].

The maximum daily precipitation depth of 1 percent exceedance based on long-term weather records for a studied area over a period of 1940-2010 published by the Northern Caucasus Service for Hydrometeorology and Environmental Monitoring was used in calculation.

## 3 Results and discussion

For purposes of the Preliminary Hazard Analysis (PHA), only those hazards to a number of culverts on piedmont rivers that can be caused by dangerous hydrologic phenomena are reviewed. As soon as the accident mechanisms of hydraulic structures and spring-loaded pipe passages outbreaks are identical, the terms given in [6] will be used in accident risk analysis for a certain bridge transition.

* Corresponding author: pniiis-gip@mail.ru

© The Authors, published by EDP Sciences. This is an open access article distributed under the terms of the Creative Commons Attribution License 4.0 (http://creativecommons.org/licenses/by/4.0/).

The waterways of hydraulic structures are reviewed below as the only elements that can be exposed to potential hazard. The results of PHA are given in the Table 1.

The What-If Risk Analysis Technique [6, 7], that became widely used in various theoretical and experimental areas was used at the stage of the hazard identification. The results of the hazards identification are given in the Table 2.

Let us focus on some specific targets. The first conditions to be observed are on road passages through watercourse on the road section Novorossiysk-Sochi between Sequa River and Malaya Plyakho River near Novomikhailovsky Village of Tuapse District.

The watercourse of the area under review belongs to a water-resource region of the Western Transcaucasia [5]. Its water regime depends on average elevation of drainage basin and level difference. The slope exposure has a great impact on distribution of precipitation, and therefore regime of run-off. The major part of reception basin area is forested.

**Table 1.** Results of the PHA in connection with waterways.

| Undesirable phenomenon, process or event that can trigger an accident | Expected impact for people, property and environment | Preliminary estimation of scenario probability |
|---|---|---|
| Flood and blockage of waterway sections by floating debris | Outbreak of hanging dam or its failure with an outburst of breaking wave downstream | Highly probable event due to narrow waterway sections, frequent flood events and presence of floating debris near river channel |
| Mudflow with consequent ablation of supports, reduction of cross-sectional area of waterway sections or blockage of water inlets | Construction failure and outbreak of hanging dam with an outburst of breaking wave downstream | Improbable due to rarity |

**Table 2.** Results of What-If technique use.

| Name of element | Damage source | Type and scale of damage | Effects to people, property and environment |
|---|---|---|---|
| Waterway of bridge transition or pipe passage | Blockage of waterways and sections | Raising of water level, structural instability and failure | Outburst of breaking wave downstream, formation of breaking mudflow, failure of a part of bridgework, failure the flat accommodation and infrastructure facilities, loss of life and substantial damage to environment |

The observed data available at the Northern Caucasus Service for Hydrometeorology and Environmental Monitoring was collected and classified in order to analyze a study degree of the adjacent areas. The best known areas are essentially represented for rivers characterized by 200-500 $km^2$ drainage area and located to the north of the area under study. The rivers of smaller drainage areas are whether not observed at all, or covered by short-term observations. The analysis of the records of peak discharge over many years revealed the peak discharge of the rivers on the area under study is ten times greater than average peak discharge or more. In these conditions, unit discharge increases when a drainage area decreases.

According to the geographical demarcation provided in [8], the rivers in the area under review are classified as the zone of summer floods. However, as soon as each stream located in the area under study is not large and has a drainage basin at the feet of hillside (up to 550 m by Baltic Height System), there is no glacial and sufficient snow feed for it. Thus, maximum runoff of every watercourse is formed only by rainfall floods that are possible in any season. This enables formation of a very high rainfall flood with heavy precipitation covering an entire drainage area and physical and hydrographical properties of a channel networks provides for shortest basin lag. Drainage basins of these rivers can form the streams characterized by up to 20 $m^3/(s \times km^2)$ of unit discharge in case of heavy precipitation.

As previously stated, a formula of Maximum Rainfall Intensity was used in order to calculate a peak discharge of rainfall flood for the drainage basins under study. The comparative results of peak discharge for 1% annual exceedance probability (i.e. maximum calculated discharge with occurrence probability of 1 in 100 years [5]) and capacity of bridge transitions and pipe passages in discharge conditions are given in the Table 3.

Based on calculation and passages observation results it was discovered that most of them are not capable to endure discharging the flood of 1 percent exceedance as confirmed by wash-away of bridge transitions and pipe passages as a result of foods in 2002, 2009 and 2010. If it is remembered that waterways are blocked by stems and branches of trees, as well as construction wastes, then it can be said that none of the structure under study is capable to endure discharging a considerable flooding. Examples of the destructive impact of the floating debris are shown in Fig.s 2 and 3.

The event of a chain of hanging dams resulted from the blockage of culverts and their further sequential outbreaks can cause disastrous consequences. For example, the bridge transition on watercourse # 9 (see Fig. 3) at a five-time margin of its throughput was completely blocked during a common rain shower. It caused a breakthrough of accumulated flood waters onto the road. It is seen that even at such throughput margin accident-free operation of the bridge cannot be ensured when the flow of floating debris is high.

**Table 3.** Comparative results of peak discharge for 1% annual exceedance probability and throughput of bridge transitions and pipe passages in discharge conditions.

| Water-course* | Drain-age area, km² | Peak discharge for 1% annual exceedance probability, m³/s | Water-way section area, m² | Throughput for dis-charge conditions, m³/s |
|---|---|---|---|---|
| # 1 | 1.5 | 30 | 3.5 | 12 |
| # 2** | 15 | 225 | 72 | 235 |
| # 3 | 0.18 | 5.4 | 0.5 | 0.8 |
| # 4 | 0.1 | 3.0 | 0.8 | 1.9 |
| # 5 | 0.13 | 3.9 | 1.8 | 4.9 |
| # 6 | 0.03 | 0.9 | 3.2 | 21 |
| # 7 | 0.46 | 9.2 | 1.8 | 6.4 |
| # 8*** | 8.2 | 164 | 14 | 98 |
| # 9 | 3.4 | 1.51 | 3.2 | 7.7 |
| # 10**** | 762 | 764 | 152 | 780 |

* Watercourses are numbered consecutively – starting from unnamed watercourse # 1 and to Mzymta River (# 10) in the south.
** Malaya Plyakho River.
*** Sequa River.
**** Mzymta River

The bridge transition having a waterway sized at 3.8×3.8 m and built on Sequa River is exposed to the highest risk. If water rises at a waterway to 4 m or more, it flows in a pressurized regime calculated approximately. When water level rises up to 8 m (with water outbreak onto the road), the maximum discharge in the culvert not blocked by floating debris is 98 m³/s, and that is significantly lower than the peak discharge for 1% annual exceedance probability estimated as 164 m³/s. At that, average peak discharge is 34 m³/s. Once in several years the area upstream of the bridge is flooded, and, due to abundant floating debris, the road structure is damaged and outbreaks of flood water occur.

**Fig. 2.** The destructive impact of floating debris at bridge transition on watercourse # 5.

The next target under review is a newly-built bridgework passing Mzymta River, a major watercourse of Sochi Resort. The in-stream support of the bridge (pier) is located in over-water length near Kepsha Settlement close to Krasnaya Skala. The width of a stream valley in its narrowest point where the pier of the cable bridge is erected is up to 60 m. The pier narrows the width of waterways down to 20 m. Although the size of waterway is enough to maintain a regular flood, the probability of its blockage is very high in the event of a great number of floating debris. The approximate volume of water that can be accumulated in the water head of the bridgework under review can reach about 3 MCM. The height of a breaking wave in the first travel kilometer reaches 20 m with a breakthrough rate of discharge at about $10^4$ m³/s [3]. There is a high probability of mudflow in the event of outbreak due to a great volume of debris, detritus and construction waste in a river channel. The volume of such flow can reach 50 MCM with momentary discharge of up to $5\times10^4$ m³/s. Every 5 or 7 years an emergency can be expected for this structure at a calculated discharge of 199 m³/s.

**Fig. 3.** Consequences of a flash flood at bridge transition on watercourse # 9.

According to the results of PHA, the natural and man-caused impact can become the external causes of accidents and emergencies happened to piers of the bridgework under review built on the road near Kepsha Settlement as follows:

- earthquake followed by landslide near the bridge piers;
- flooding followed by debris floating;
- and mudflow.

The high probability of described above scenarios is supported by emergencies repeatedly happened in North Caucasian Federal District (e.g. Krymsk, 2012).

## 4 Conclusions

• The minor rivers of Northern Caucasus Region are characterized by high rainwater floods. The drainage basins of these rivers adjacent to Black Sea Coast are capable to develop the streams characterized by a runoff rate of 20 $m^3/(s\times km^2)$ in the event of heavy precipitation (250 mm per day or more). When this happens, the capacity of larger part of road structures built in recent decades in Northern Caucasia to discharge rare floods raises reasonable doubts.

• Almost every watercourse in this region is highly contracted in places of bridge transitions and pipe passages. In the event of flash flooding bridge waterways and pipe passages would be blocked by floating debris and sediments. The area would be flooded with a high risk of following of outbreak of flood water. The event of a chain of hanging dams resulted from the blockage of culverts and their further sequential outbreaks can cause disastrous consequences.

• The discharge of breaking wave containing a large volume of carrying rock, floating debris and waste to a sea shore can cause a failure of accommodation and infrastructure facilities, blocking of traffic network, loss of life and severe environmental damage.

## References

1. A.A. Goncharov, Sistemy Bezopasnosti, No 1, 365-368 (2014) – in Russian.
2. M. Bollschweiler, M. Stoffel, Prog. Phys. Geog., **34** (2010).
3. A.Yu. Vinogradov, Bull. of the Moscow State Regional University, ser. "Natural Sciences", No 4, 52-56 (2014) – in Russian.
4. M. Stoffel, B. Wyżga, R.A. Marston, Geomorphology, **272** (2016).
5. SP 33-101-2003. (Gosstroy of Russia, Moscow, 2004) – in Russian.
6. Hydraulic structures accidents risk assessment guidelines (All-Russian Scientific Research Institute of Hydraulic Engineering, St. Petersburg, 2005) – in Russian.
7. A.J. Card, J.R. Ward, P.J. Clarkson, J. Healthc. Risk Manag., **31** (2012).
8. Atlas of designed hydrological maps and graphical charts (Gidrometeoizdat Publishing House, Leningrad, 1986) – in Russian.

# Estimation of probable maximum precipitation for tropical catchment

*Saad* Sammen [1], *Thamer* Mohamed [1], *Abd Alhalim* Ghazali [1], *Lariyah* Sideq[2] and *Azlan* Abdul Aziz [1]

[1] Department of Civil Engineering, Faculty of Engineering, University Putra Malaysia. Serdang, Selangor, Malaysia
[2] Department of Civil engineering, College of Engineering, University Tenaga Nasional, Kajang, Selangor, Malaysia

**Abstract.** Probable Maximum Precipitation (PMP) is the maximum precipitation depth for specific region or station within a certain time. The main purpose of PMP estimation is calculate the Probable Maximum Flood (PMF). The PMF is considered necessary for design and manage the hydraulic structures. PMP can be estimate using two methods, either using a physical method or by using statistical method. In this study, statistical approach was used to estimate the PMP for Temengor catchment in Perak state, Malaysia. Extreme value type-1 distribution (EV1) is adopted to estimate the extreme rainfall and Hershfeid method was used to estimate PMP value. Also, intensity duration curve (IDC) was derived for 1, 2 and 3 days storm duration with return period 5, 10, 50, 100, 500 years. The results showed that the values of PMP for 1000 return period are 222.361mm, 311.847mm and 348.307mm for 1, 2 and 3 days respectively.

## 1 Introduction

The PMP has been given the definition as being the greatest depth of precipitation for a particular time period which is meteorologically possible over a specified area or given station (World Meteorological Organization (WMO), 1986). Two techniques are commonly applied for calculating the PMP value, physical method and statistical method. In physical method the PMP for various time periods are calculated by maximize and transposition of actual storm events. In order to produce a PMP value that will satisfy for the lack of a sufficient storm database, storm transposition and envelopment methods are incorporated. For around the last 2 decades, the PMP has been estimated many times for dam in Malaysia using methods of maximization and transposition (NAHRIM, 2008; Desa et al., 2007).

Statistical approach is the second common technique that used for PMP estimation. A statistical method is applied for the estimation of the PMP rainfall at a given area or location which is calculated from the distribution of the frequency fitted to the maximum rainfall records annually. The annual maximum rainfall data are usually used as a key variable in problems related to hydraulic structures design. The philosophy on which this is based is that, if the structure which has been designed is able to attitude the maximum value in one year, it will also be able to attitude other values. The statistical technique is more useful when meteorological data, other than an abundance of rainfall data, are unavailable, such as wind speed and dew point temperatures.

Another kind of statistical technique used to estimate PMP values for areas which are small was developed by Hershfield (1961, 1965). This technique was based on an equation for general frequency presented by Chow (1951). This technique requires a series of maximum annual daily rainfall measurements at a particular observation point as the input data. Hershfield's statistical method has been employed, extensively, in the estimation of PMP for areas which have a long duration of recorded rainfall data. Therefore, Hershfield's method was considered desirable for undertaking in this study to estimating the PMP for 1 day and 3-day durations in the Temengor Catchment stations as their data for daily rainfall is available for long durations.

## 2 Study area and data collection

The study area for this work was at the Temengor catchment station covering a total area size of 3506 $km^2$. The catchment area's elevation ranges from 206m to 2156 m. Lake Temengor covers about a 150 $km^2$ area of the catchment which is otherwise forested. The location map for the catchment area of Temengor is shown in Figure 1.

One of the basic requirements for estimating the PMP is long period rainfall data. Rainfall is a vital element in the hydrological cycle; as such, rainfall needs to be quantified when various water resource projects are being planned and designed. Generally, to quantify rainfall, intensity-duration-frequency (IDF) curves are used (Chow et al., 1988). For this study, seven rainfall stations were used. These stations were located upstream

© The Authors, published by EDP Sciences. This is an open access article distributed under the terms of the Creative Commons Attribution License 4.0 (http://creativecommons.org/licenses/by/4.0/).

of the Temengor dam as shown in Figure 2. The properties of the rainfall stations are presented in Table 1.

## 3 Theory of hershfield method

Chow's (1951) method for analyzing the frequency of rainfall was adapted by Hershfield when developing his technique to estimate the PMP. For a hydrological analysis, Chow (1951) suggested applying the following formula for frequency distributions.

$$x_{PMP} = \bar{X}_n + K_m \cdot \sigma_n \qquad (1)$$

When estimating the PMP, the factor $K_m$ can be calculated by applying the formula:

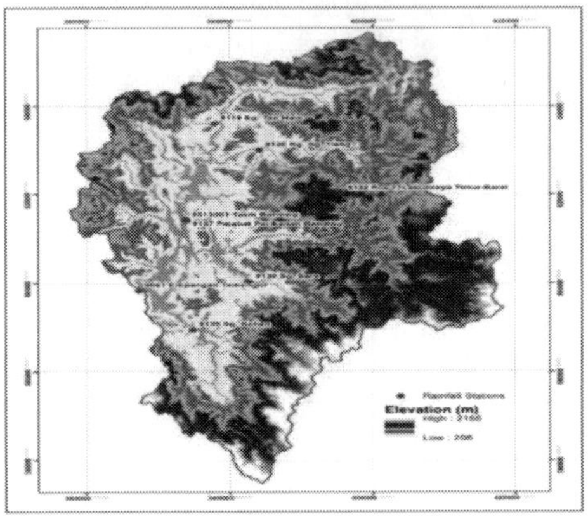

**Fig. 1.** Map of Temengor catchment

**Table 1.** Details of rainfall stations at Temengor catchment

| St. No. | St. Name | Coordinate | | St. Height above S.L MSE (m) |
|---|---|---|---|---|
| | | Lat. | Long. | |
| 9138 | Sg. Sara | 05 25 32N | 101 25 56E | 270 |
| 9132 | Lebuhraya Timur | 05 36 20N | 101 32 54E | 1042 |
| 9120 | Kg. Sungai Tiang | 05 41 38N | 101 26 38E | 260 |
| 9187 | Pejabat Perikanan | 05 33 06N | 101 21 13 E | 285 |
| 9139 | Sg. Kelian | 05 19 37N | 101 21 43 E | 259 |
| 9061 | Empangan | 05 24 23N | 101 17 38 E | 185 |
| 9119 | Sg. Tan Hain | 05 44 51N | 101 23 18 E | 258 |

$$K_m = (\bar{X}_1 - \bar{X}_{n-1})/\sigma_{n-1} \qquad (2)$$

Where

$X_{PMP}$: PMP value for specific station
$\bar{X}_n$: Average of maximum annual rainfall
$\sigma_n$: Standard deviation maximum annual rainfall
$K_m$: Coefficient related to Frequency
$\bar{X}_1$: Maximum value of maximum annual rainfall
$\bar{X}_{n-1}$: Mean of annual maxima, excluding highest
$\sigma_{n-1}$: Standard deviation of annual maxima, excluding highest

A survey conducted on more than 2600 stations worldwide calculated the factor $K_m$, above, as ranging from the highest value of 14.5 down to a value under 3. In Hershfield's work, he adopted the highest value and rounded it up to 15 in order to estimate the PMP.

$$x_{PMP} = \bar{X}_n + 15\sigma_n \qquad (3)$$

Research works undertaken in the USA, Canada, and India have shown that in areas where the annual rainfall is very heavy, the value of $K_m = 15$ tends to result in values of the PMP which are determined to be too large when using other techniques. In addition, in areas where the annual rainfall is low, the values are too small. As a result, rather than using the highest value of $K_m = 15$, used worldwide, a more appropriate $K_m$ value was considered to be one based on stations in Malaysia, which would result in a more reliable estimation of the PMP.

The estimation of the PMP for various locations is not dependent on the single highest value of $K_m$ in that area as was shown by Hershfield (1965). Rather, he found that, the $K_m$ that had been derived, empirically, was different, inversely, from the mean of the series, and directly with the duration of the rainfall. That is to say, when the mean of the series ($X_n$) rose in magnitude, the $K_m$ value tended to become lower. Therefore, each location will have its own $K_m$ value which is dependent on the magnitude of the location's $X_n$. Following that, Hershfield (1965) employed the envelope $K_m$ curve to obtain $K_m$ values for a variety of $X_n$ values. Therefore, it was concluded that, if an appropriate envelope curve could be developed based upon the locations in the study area, it would provide realistic estimates of the PMP for various time periods of rainfall. Consequently, appropriate envelope $K_m$ curves representing the relation between $K_m$ and the mean annual maximum rainfall were found for various time periods of rainfall. However, the natural characteristics of rainfall was realized, leading to $K_m$ versus $X_n$ curves being created for a variety of time periods for the location.

Recently National Hydraulic Research Institute Malaysia (NAHRIM) has conducted a PMP (Technical Research Publication No.1 - Derivation of Probable Maximum Precipitation for Design Flood in Malaysia, 2008) study covering the whole Malaysia where they have established standard method of estimating PMP by both statistical and hydro-meteorological methods. In statistical method, envelop curves for frequency factor ($K_m$) of different duration storm events were developed with respect to their corresponding mean annual maximum precipitation values (Figure 3). Second degree polynomial equations were then derived for each of the envelop curves which are then used to determine the $K_m$

value of any particular location of a catchment or station where the mean of annual maximum rainfall is available. These curves have been found useful for the determination of $K_m$ values for any location in Peninsular Malaysia and hence been used for the Temengor catchment PMP calculation.

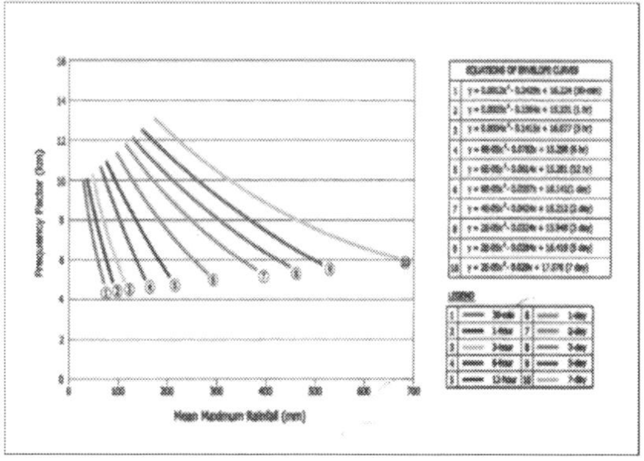

**Fig. 3.** $K_m$ Envelope curves for different durations in west Malaysia (NAHRIM, 2008)

## 4 Methodology

The most widely used statistical method for estimating the PMP is by Hershfield (1961, 1965) which has become one of the standard methods suggested by the World Meteorological Organisation (WMO, 1986) for estimating the PMP at any location. The flow chart for PMP establishment is shown in Figure 4. Therefore, it is considered desirable to carry out a study on the estimation of PMP for 1, 2 and 3 days.

Rainfall analysis was carried out to use Gumble Extreme Value (GEV). The values were determined for 5, 10, 20, 50, 100, 1000 and 10000 year ARI. Frequency analysis was carried out for 1 day and 3 days rainfall storm duration for the Temengor Reservoir. According to Gumble theory of extreme events, the probability of occurrence of an event equal to or larger than a value $X_o$ is

$$P(x \geq x_o) = 1 - e^{-(e^{\wedge}-y)} \quad (4)$$

Where y is dimensionless variable given by

$$y = \alpha(x - a) \quad (5)$$
$$a = x - 0.45005\sigma_x \quad (6)$$
$$\alpha = 1.2825/\sigma_x \quad (7)$$

Where: x is the mean value of the recorded rainfall and $\sigma_X$ is the standard deviation of the recorded rainfall. For the required P, eq. (4) can be rearrange to

$$y_P = - \ln(-\ln(1 - P)) \quad (8)$$

If P is replace by return period ($T_P = 1/P$), $y_T$ is given by

$$y_T = -(\ln.\ln(T_P/T_P - 1) \quad (9)$$

Now rearrange eq. (4), the value of x with the return period ($T_P$) is given by

$$x_{T = x + K.}\sigma_x \quad (10)$$

Where: $x_T$ is the estimated event magnitude and K is given by

$$K = (y_T - 0.577)/1.2825 \quad (11)$$

Note that eqn (10) is of the same form as the general equation of hydrologic frequency analysis.

## 5 Results and discussion

The rainfall intensity is commonly required for planning and designing of various water resource projects. Many sets of relationships have been developed and used in several parts of the world. Rainfall intensity were estimate for storm duration of 1day and eight different return periods of 5, 10, 20, 50, 100, 500 and 1000 years. Frequency analysis techniques are used to estimate the rainfall intensity, for different return periods from rainfall data. Distribution analysis of rainfall frequency is based on Type I extreme value (Gumbel) distributions. Areal precipitation of the study area was obtained by adopting Thiessen polygon method using Geographical Information System (GIS) technique as shown in Figure 4. Also from Thiessen Polygons, The mean area weights or proportion of rainfall that a station contributes to catchment were determined as shown in Table 2. By using this frequency distribution function, the maximum rainfall intensity for durations of 5, 10, 20, 50, 100, 500 and 1000 years return periods, have been determined as shown in Table 3. Intensity duration curve for 1, 2 and 3 days and different return period was shown in Figure 5.

**Table 2.** Weight of rainfall stations to compute main rainfall

| Rainfall Station Name | Area ($Km^2$) | Respective Weight |
|---|---|---|
| 9138 Sg. Sara | 571.417 | 0.166 |
| 9132, Lebuhraya Timur-Barat | 797.926 | 0.232 |
| 9120 Kg. Sg. Tiang | 431.013 | 0.125 |
| 9187 Pejabat Perikanan Banding | 430.361 | 0.125 |
| 9139 Sg. Kelian | 600.925 | 0.175 |
| 9061 Empangan Temengor | 112.374 | 0.033 |
| 9119 Sg. Tan Hain | 490.445 | 0.143 |
| | ∑ Area = 3434.461 | ∑Respective Weight =1 |

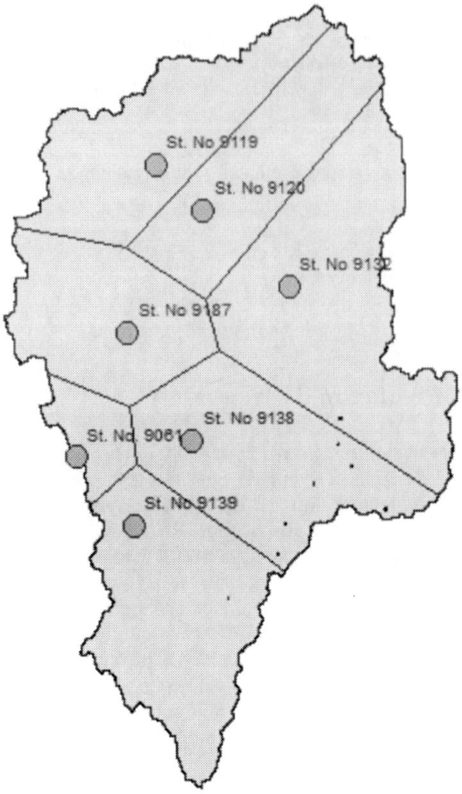


**Fig. 4.** Thiessen polygons derived for the study area

**Table 3.** Rainfall depth (PMP) for different return period

| Return Period (years) | Rainfall Intensity | | |
|---|---|---|---|
| | 1 day | 2 days | 3 days |
| 5 | 99.084 | 137.736 | 158.69 |
| 10 | 116.192 | 161.899 | 185.00 |
| 50 | 153.845 | 215.079 | 242.92 |
| 100 | 169.763 | 237.56 | 267.40 |
| 500 | 206.547 | 289.512 | 323.98 |
| 1000 | 222.361 | 311.847 | 348.30 |

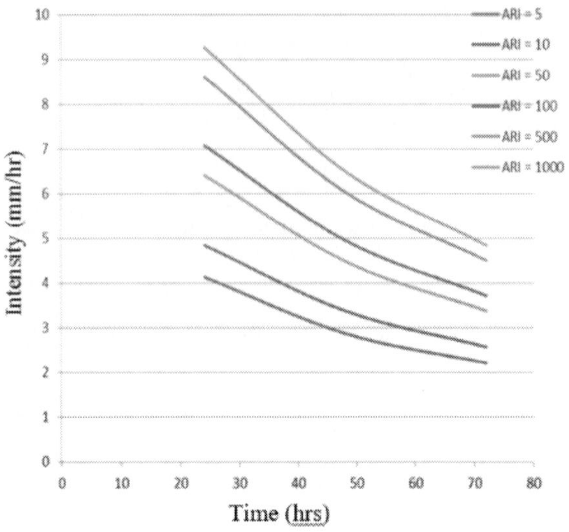


**Fig. 5.** Intensity Duration Curve

## 6 References

1. World Meteorological Organisation, 1986, Operational Hydrology Report No.1, WMO No. 332, 96-108, (Second Edition).
2. National Hydraulic Research Institute of Malaysia (NAHRIM) (2008). Nahrim Technical Research Publication No.1 (TRP 1).
3. Desa, M. N.; Rekhecha, P. R., 2007, Atmos, Res, 84, 84 – 90.
4. Hershfield, D. M., 1961: J. Hyd. Div., Am. Soc. of Civil Eng., 87, Hy5, 99-116.
5. Hershfield, D.M., 1965: J. Am. Water Works Assoc., 57, 965-972.
6. Chow, V. T., 1951: Trans. Am. Geophys. Union, 32, 231-237.
7. Chow V.T.,D.R. Maidment and L.W. Mays (1988). "Applied Hydrology". McGraw-Hill.

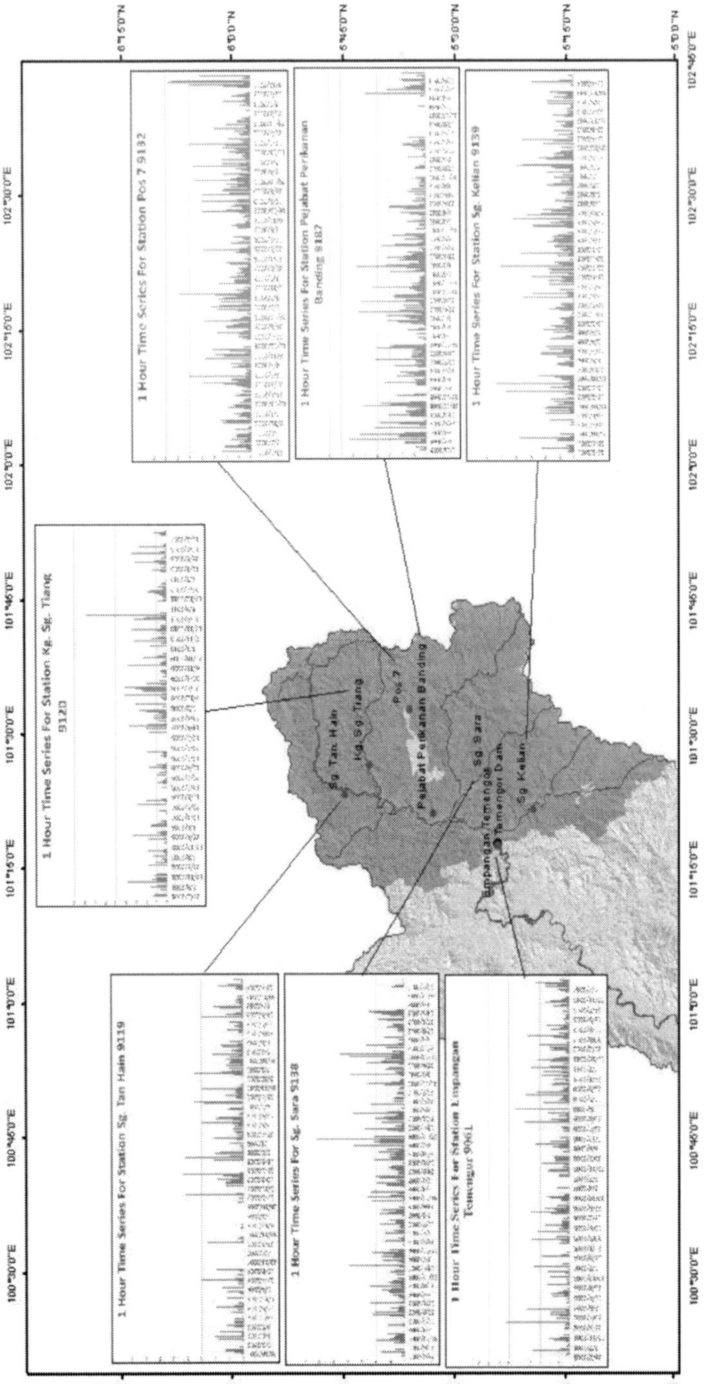


**Fig. 2.** Location of rainfall stations at Temengor catchment ar

MATEC Web of Conferences **162**, 03013 (2018) https://doi.org/10.1051/matecconf/201816203013
*BCEE3-2017*

# Nonlinear analysis of water–soil–barrage floor interaction

*Abdul-Hassan* Al-Shukur [1], *Atheer* Zaki Al-Qaisi [2,*], *Ayaat* Majid Al-Rammahi [1]

[1] College of Engineering, University of Babylon, Babylon, Iraq
[2] Al–Mussaib Technical College, Al–Furat Al–Awsat Technical University, Babylon, Iraq

**Abstract.** Fluid – Structure Interaction (FSI) and Soil–Structure Interaction (SSI) are mostly considered in the non–linear analysis of water–soil–barrage floor interaction. A three dimensional (3D) - section of a barrage is selected and modelled via ANSY 15.0. The ogge barrage floor shape has been taken as a case study of analysis. The non–linear analysis of the structure model is developed through selecting suitable available elements in ANSYS which are related to the case under study and is compared with linear analysis. The results of the analysis are obtained by suggesting different characteristics of concrete, soil, and rock materials as a parametric study. Both closed barrage and in operation are the cases have been considered in analysis. It is concluded from this study that ANSYS/APDL is adequate tool to simulate and analysis the problem that need sufficient experiences to select suitable available elements to get the acceptable results. It is also concluded that the deformation in barrage floor has little difference by (6%) in linear and non–linear analysis in case of no operation. While this deformation in non-linear analysis with operation condition is less by (13%) than of no operation case. The non-linear deformation of barrage floor is more effective by change of density of barrage concrete itself, modulus of elasticity and yield stress of both soil and rock foundation materials, while no sensitive effect of change of Poisson's ratio on the deformation. The maximum seepage flow at exit point downstream barrage floor with upstream and downstream piles is lesser by (67.40%) than without pile. In the view of stability, it is concluded that the barrage floor is already safe in exist gradient and piping problems with or without piles.

## 1 Introduction

Barrage is that type of hydraulic structures which is relatively low–level dams and constructed across a river for the purpose of rising water

As a non–gravity structure, the upstream concrete barrage floor is relatively thin (i.e. it is stable with small load), while the downstream concrete floor is built so as to contain the hydraulic jump [1].

Most hydraulic structures behave as Soil-Structure Interaction (SSI) or Fluid-Structure Interaction (FSI) or both SFSI [2,3,4]. The barrage floor is in contact with water and soil. So it behaves as a fluid–structure and soil–structure interaction problems under effect of different types of loads.

Finite element modelling and analysis are carried out to give the behaviour of barrage floor resting on homogeneous soil condition considering soil-raft-structure system [5]. The non–linear SSI analyses with advances in computer analysis capabilities can ensure cost – effective of an entire soil – foundation – superstructure system [6].

The scope of recent researches is to present a non–linear analysis of water–soil–barrage floor interaction. In present study, the stresses, displacements, uplift pressure distribution, hydraulic gradient, and percentage of seepage under barrage are analysed and the results of these analysis are presented in the forms of tables and figures.

## 2 Numerical modelling of the problem

The ANSYS software is used for finite element modelling. To achieve this purpose, a case study from [7] has been taken to be simulated.

An ogee shape of a typical barrage section (bays 3-4), have been selected for this study as shown in Fig. 1(a and b).

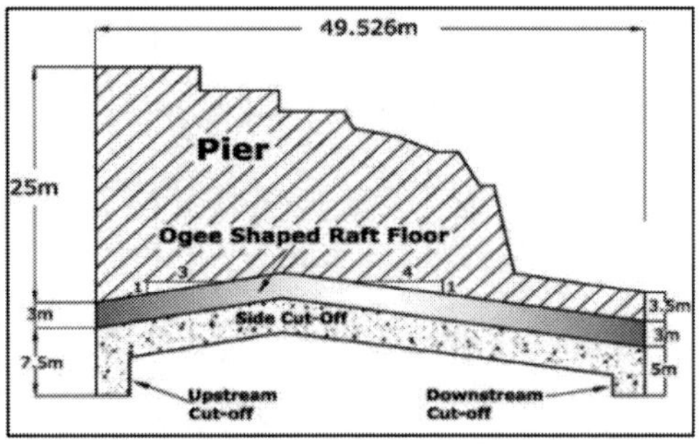


**Fig. 1-a.** Side view of barrage [7].

* Corresponding author: atheermohsin@yahoo.com

© The Authors, published by EDP Sciences. This is an open access article distributed under the terms of the Creative Commons Attribution License 4.0 (http://creativecommons.org/licenses/by/4.0/).

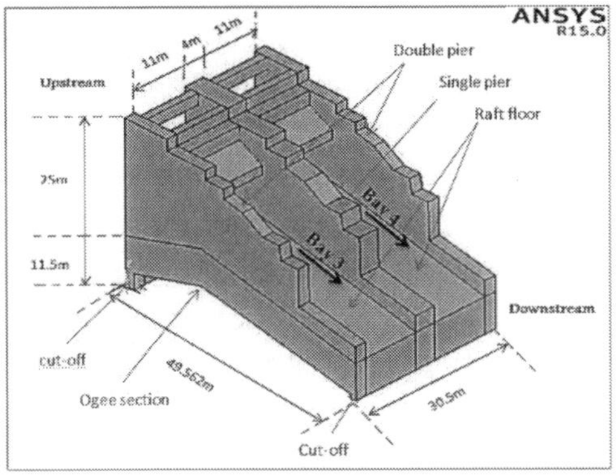


**Fig. 1-b.** 3D section of barrage.

**Fig.1.** Barrage section; a. Side view, b. 3D section [7].

A 3D-shape with 8-Node SOLID 185 element (homogenous structural solid), shown in Fig. (2) which is available in ANSYS 15.0, is a suitable element used for modelling of soil and rock foundation. Such element has three degrees of freedom at each node: translations in the nodal x, y, and z directions. The element has plasticity, hyper elasticity, stress stiffening, creep, large deflection, and large strain capabilities. It also has mixed formulation capability for simulating deformations of nearly incompressible elastoplastic materials, and fully incompressible hyper elastic materials [8].

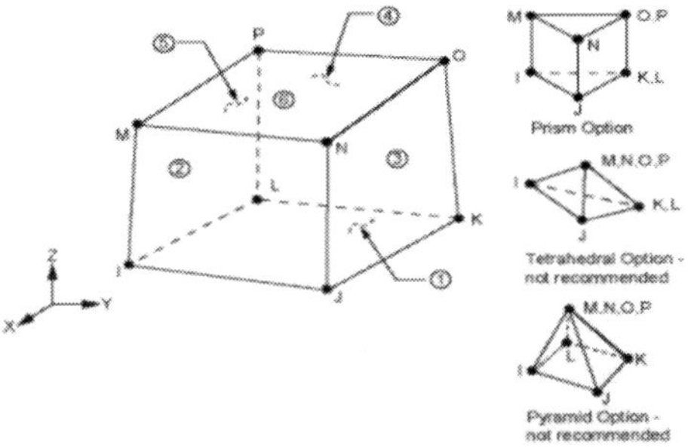


**Fig. 2.** SOLID 185 element geometry [8].

The Extended Drucker -Prager (EDP) material of soil is modeled to represent the non-linear behavior of the soil with structure. There are several topics to define the EDP material model. However, EDP yield criteria and linear plastic flow potential function forms are selected to model the soil. The EDP linear yield criteria form is:

$$f(\sigma, \sigma_y) = \sigma_e + \alpha \frac{1}{3} tr(\sigma) - \sigma_y = 0 \quad (1)$$

Where: $\alpha$ and $\sigma_y$ are pressure sensitivity and uniaxial yield stress, respectively and depend on both shear strength parameters of soil ( $C$ and $\phi$ )which should be defined as constants ($C_1$ *and* $C_2$) in ANSYS/APDL.

In other hand, the linear form of the plastic flow potential is:

$$Q(\sigma, \sigma_y) = \sigma_e + \bar{\alpha} \frac{1}{3} tr(\sigma) \quad (2)$$

Where: $\bar{\alpha}$ is the flow potential pressure sensitivity.

The other soil parameters that also should be defined are modulus of elasticity $\mathbf{E_s}$, Poisson's ratio $\boldsymbol{\vartheta_s}$, and density $\boldsymbol{\rho_s}$. Tables (1 and 2) give values of the soil and rock parameters.

**Table 1.** Soil domain properties [7, 9].

| Material Properties | Symbol in ANSYS | Corresponding Values |
|---|---|---|
| ***Nonlinear properties [9]*** | | |
| Pressure sensitivity, $\alpha$ | $C_1$ | 2.252 |
| Uniaxial yield stress, $\sigma_y$ | $C_2$ | 23.40 KPa |
| Flow potential, Q | LFPOT | 0.59 |
| ***Linear properties [7]*** | | |
| Modulus of elasticity, $E_s$ | EX | $1 \times 10^5$ KPa |
| Poisson's ratio, $\vartheta_s$ | NUXY | 0.3 |
| Density, $\boldsymbol{\rho_s}$ | DENS | **2000 kg/m$^3$** |

**Table 2.** Rock domain properties [7, 9].

| Material Properties | Symbol in ANSYS | Corresponding Values |
|---|---|---|
| ***Nonlinear properties[9]*** | | |
| Pressure sensitivity, $\alpha$ | $C_1$ | 0.875 |
| Uniaxial yield stress, $\sigma_y$ | $C_2$ | 12000 KPa |
| Flow potential, Q | LFPOT | 0.7 |
| ***Linear properties[7]*** | | |
| Modulus of elasticity, $E_r$ | *EX* | $1 * 10^7$ KPa |
| Poisson's ratio, $\vartheta_r$ | *NUXY* | 0.25 |
| Density, $\boldsymbol{\rho_r}$ | *DENS* | **2600 kg/m$^3$** |

About the barrage there are two types of floors, the first being called the gravity type and the second as the raft type [10]. The raft type is recommended and discussed in this study. This type of floor is constructed as reinforced concrete structures. A 3D-8-node SOLID 65 element is a suitable available one via ANSYS to model this type of barrage floor. The description geometry of this element is shown in Fig 3. The solid is capable of cracking in tension and crushing in compression. This element is used for the 3D modelling of solids with or without reinforcing bars (rebar). The element is defined by eight nodes having three degrees of freedom at each node: translations in the nodal x, y, and z directions. This element permits to model the concrete as a linear behavior concrete material as well as a non-linear behavior [8].

In other hand, Solid 187 element is also selected to simulate the barrage itself. It is a higher order 3D, 10-node element having three degrees of freedom at each node: translations in the nodal x, y, and z directions as shown in Fig.4. The element has plasticity, hyperelasticity, creep, stress stiffening, large deflection, and large strain capabilities. It also has mixed formulation capability for simulating deformations of nearly incompressible elastoplastic materials, and fully incompressible hyperelastic materials.

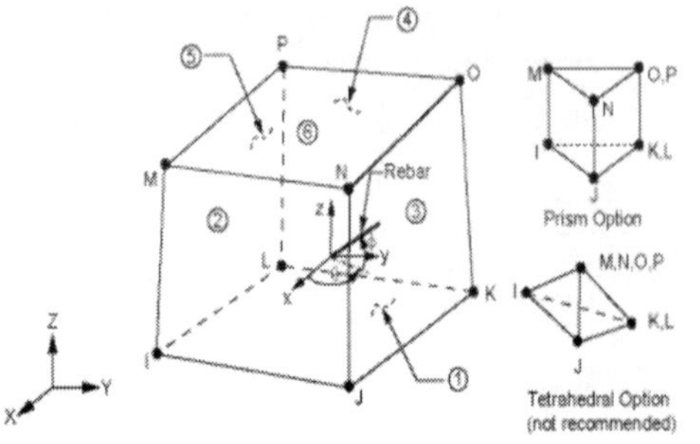


**Fig. 3.** SOLID 65 element geometry [8].

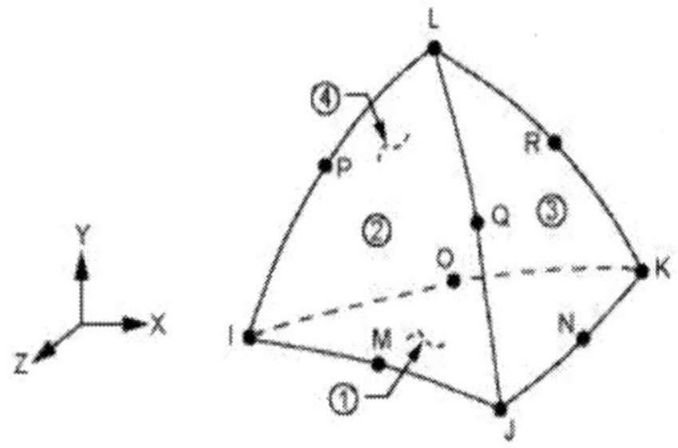

**Fig. 4.** SOLID 187 element geometry [8].

In this study, the behaviour of non-linearity of concrete is considered. In this case, the concrete is capable of plasticity, creep, cracking and crushing [11]. In other hand, the criterion for failure of concrete due to a multiaxial stress state can be expressed in the form [12]:

$$\frac{F}{f_c} - S \geq 0 \qquad (3)$$

Where: **F**= a function of the principlal stress $(\sigma_{xp}, \sigma_{yp}, \sigma_{zp})$,

**S** = failure surfac expressed in terms of principal stresses and five input parameters $f_t, f_{cb}, f_1$, and $f_2$ defind in Table (3),

$f_c$= uniaxial crushing strength, and

$\sigma_{xp}, \sigma_{yp}, \sigma_{zp}$= principal stresses in principlal directions.

If Eq. (2) is satisfied, the material will crack or crush. A total of five input strength parameters are presented in Table (3). So, the material that is used in barrage floor is a mass concrete and the specifications of concrete material that should be defined in ANSYS/APDL are density $\rho_c$, modulus of elasticity $E_c$, and Poisson's ratio $\vartheta_c$. The specifications of C35 grade concrete are as given in Table (3).

**Table 3.** Mass concrete properties [7].

| Material Properties | Symbol in ANSYS | Corresponding Values |
|---|---|---|
| *Pier/Abutment [7]* | | |
| Modulus of elasticity, $E_c$ | EX | 25 Gpa |
| Poisson's ratio, $\vartheta_c$ | NUXY | 0.15 |
| Density, $\rho_c$ | DENS | 2500 kg/m$^3$ |
| *Raft Floor [7]* | | |
| Modulus of elasticity, $E_c$ | EX | 25 Gpa |
| Poisson's ratio, $\vartheta_c$ | NUXY | 0.18 |
| Density, $\rho_c$ | DENS | 2500 kg/m$^3$ |
| *Nonlinear properties for both pier and floor* | | |
| Shear transfer coefficients for an open crack. | $C_1$ | 0.3** |
| Shear transfer coefficients for a closed crack. | $C_2$ | 1** |
| Ultimate uniaxial tensile strength, $f_t$ | $C_3$ | 1.5 Mpa* |
| Ultimate uniaxial compressive strength, $f_c$ | $C_4$ | 30 Mpa* |

* From [13].
** From [14].

The main function of a barrage is to work as a low dam. The retained water is on the upstream and down-stream barrage is considered in this study. FSI involves as the pressure load on the barrage floor. FLUID 30 Element is by ANSYS 15.0 for modelling the fluid medium and the interface in fluid/structure interaction problems. It has 8-corner nodes with four degrees of freedom per node: translations in the nodal x, y and z directions and pressure as shown in Fig. (5). The translations, however, are applicable only at nodes that are on the interface. Typical applications include sound wave propagation and submerged structure dynamics. The governing equation for acoustics, namely the 3-D wave equation, has been discretized taking into account the coupling of acoustic pressure and structural motion at the interface. When there is no structural motion, the element is also applicable to static, modal and reduced harmonic response analyses. This element needs to be defined by some parameters, namely, a reference pressure, and the isotropic material properties. The speed of sound ($\sqrt{K/\rho_w}$) in the fluid is input by SONC where K is the bulk modulus of the fluid (Force/Area) and $\rho_w$ is the mean fluid density (Mass/Volume) (input as DENS). The dissipative effect due to fluid viscosity is also included and input as (MP, VISC). All these materials of water domain could be given in Table (4).

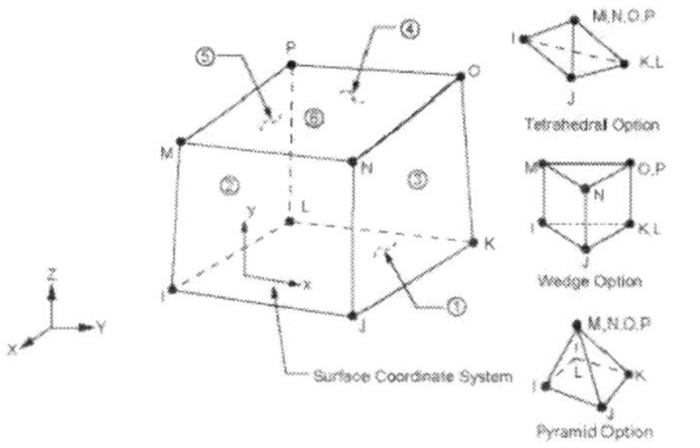


**Fig.5.** FLUID 30 element geometry [8].

**Table 4.** Fluid (water) domain properties [4].

| Material Properties | Symbol in ANSYS | Corresponding Values |
|---|---|---|
| Bulk modulus of the fluid, $K_w$ | *EX* | 2.02 Gpa |
| Reference pressure, $P_w$ | *PREF* | $20 \times 10^{-6}$ Pa |
| Poisson's ratio, $\vartheta_w$ | *NUXY* | 0.2 |
| Sonic velocity, $V_w$ | *SONC* | 1440 m/s |
| Density, $\rho_w$ | *DENS* | 1000 kg/m$^3$ |
| Viscosity, $\mu_w$ | *VISC* | $1.0 \times 10^{-3}$ Pa.S |

To study the contact between two bodies, the surface of one body is conventionally taken as a contact surface and the surface of the other body as a target surface. The "contact-target" pair concept has been widely used in finite element simulations. For rigid-flexible contact, the contact surface is associated with the deformable body; and the target surface must be the rigid surface. For flexible-flexible contact, both contact and target surfaces are associated with deformable bodies. The contact and target surfaces constitute a "Contact Pair".

The interface elements that modeled the interaction among water and soil with barrage floor are CONTA 174 and TARGE170. CONTA174 is used to represent contact and sliding between 3D "target" surfaces (TARGE170) and a deformable surface, defined by this element as shown in Fig. (6). Contact occurs when the element surface penetrates one of the target segment elements (TARGE170) on a specified target surface. Coulomb and shear stress friction is allowed. This element also allows separation of bonded contact to simulate interface delamination.

TARGE170 is used to represent various 3D "target" surfaces for the associated contact element (CONTA 173) as shown in Fig.(7). The contact elements themselves overlay the solid, shell, or line elements describing the boundary of a deformable body and are potentially in contact with the target surface, defined by TARGE170. This target surface is discretized by a set of target segment elements (TARGE170) and is paired with its associated contact surface via a shared real constant set. The translational or rotational displacement, temperature, voltage, and magnetic potential, forces, and moments can be imposed on the target segment element on target elements. For rigid target surfaces, these elements can easily model complex target shapes. For flexible targets, these elements will overlay the solid, shell, or line elements describing the boundary of the deformable target body.

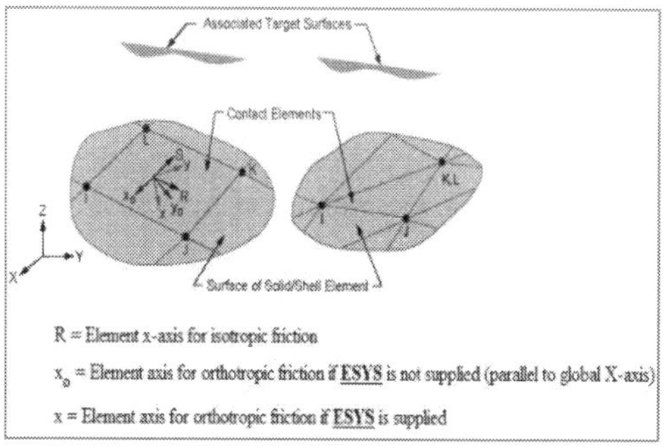


**Fig. 6.** CONTA 173 element geometry [8].

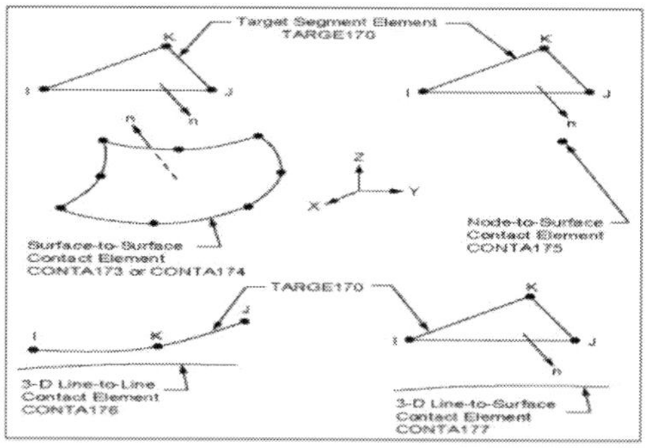


**Fig.7.** TARGE 170 element geometry [8].

The boundary conditions in this model are: the depth of the soil and rock media considered is 80m from the crest level, the extent of the surrounding soil and rock is up to 35m on both sides of the transverse section of the raft and 50m on both in upstream and downstream side equivalent to the length of the raft floor along the flow has been considered [7].

The boundary condition imposed on the finite element models consist of restraining the limiting boundary of the foundation soil and rock in such manner that displacement normal to the boundary surface are restrained i.e. the base of the foundation media at the depth of 80 m is restrained against vertical displacement and at the ends along and across the direction of flow, foundation media is restrained against the horizontal displacement.

The weight of barrage body itself is applied on the foundation as a distributed load (kN/m$^2$) by using (SF) command. The self-weight of the soil and rock media has not been considered as it has been assumed that

 https://doi.org/10.1051/matecconf/201816203013

entire soil and rock media is already settled by its own weight [7]. The water imposed behind barrage is considered as a pressure on the surface of a barrage floor. So it was represented also by (SF) command. In ANSYS 15.0, the fluid-structure interface (FSI) may be automatically flagged if acoustic elements are adjacent to solid structural elements. However, the FSI is flagged as a surface load (SF) on the interface of water-barrage and water – soil by defining fsi =1. The complex impedance value can be used to include damping that may be present at a structural boundary with a sound absorption lining. These impedance boundary conditions may also be applied to a flagged FSI interface. A zero value of IMPD removes the damping calculation. The displacement degrees of freedom (UX, UY and UZ) at the element nodes not on the interface should be set to zero to avoid zero-pivot warning messages if the coupled acoustic elements are used (KEYOPT(2) = 1).

# 3 Results analysis and discussion

## 3.1 Linear analysis

According to linear materials properties given in Tables 1, 2, and 3, the barrage section shown in Fig. (1) is modelled with different structure elements that available in ANSYS. This model is simulated for static linear analysis at empty condition by using 8–nodes Solid 65 and 185 elements additionally of 10-nodes Solid 187 element to make some indication about behaviour the structure by using different mechanism of finite element. The deformation in y-direction (Uy) of barrage floor with distance at ogee section is compared with the results of element given in [7]; as shown in Fig. 8. The results shown in Fig. 8 show the convergence in results between current study and [7] with using Solid 65 and 185 elements, but give some different results when Solid 187 element has been used and this is because of the different geometry of this element from other and this will give some different behaviour of finite element itself.

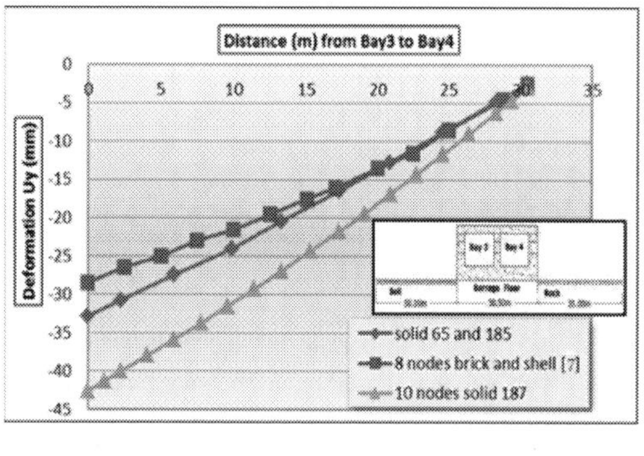


**Fig. 8.** Deformation of barrage by using different elements.

## 3.2 Nonlinear analysis

To make a comparison between linear and non-linear behaviours of barrage floor, the same section shown in Fig. 1 and according to the nonlinear as well as to linear properties given in Tables 1, 2, and 3, the nonlinear analyses in non-operation condition by using solid 65 and solid 185 is performed and the results of deformation are shown in Fig. 9.

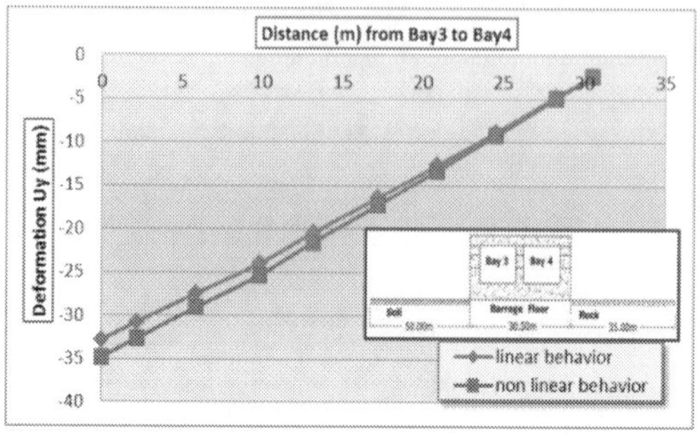


**Fig. 9.** Deformation of barrage by linear and nonlinear analysis at ogee section in non-operation condition.

It is shown from Fig. 9 that the deformation of barrage floor in both linear and nonlinear analyses give close results, where the difference is 6.12% only at the soil side of foundation. This is due to that in fact the design of hydraulic structures is depends on the working stress principle and taken as an uncracked section.

In addition to deformation; the stress at x, y, z directions in model is studded and shown in Fig. 10 ( a, b, and c).

Fig. 10 (a, b, and c) indicates that the stress at y-direction is higher than other directions followed by z-direction. This is due to interaction between soil and barrage floor and the displacement in y-direction. It is noticeable that a structure in contact with rock behaves as a rigid structure than in a flexible state. Therefore, the critical stress is shown at rock side of the modelled section than in the soil side of barrage (modulus elasticity of rock is greater by 100 times than that of soil (Tables 1 and 2)).

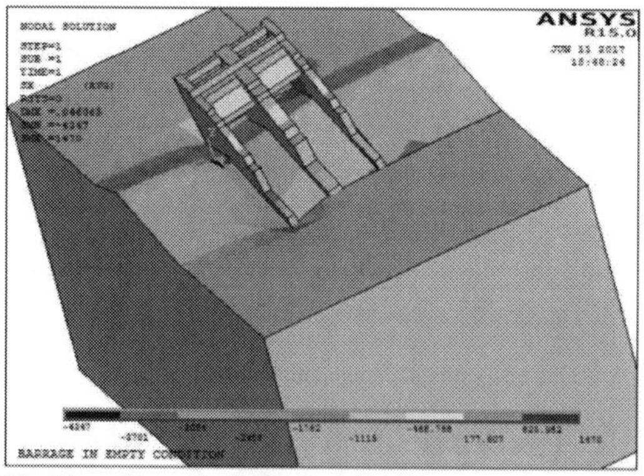


**Fig. 10- a.** Stress distribution in x-direction.

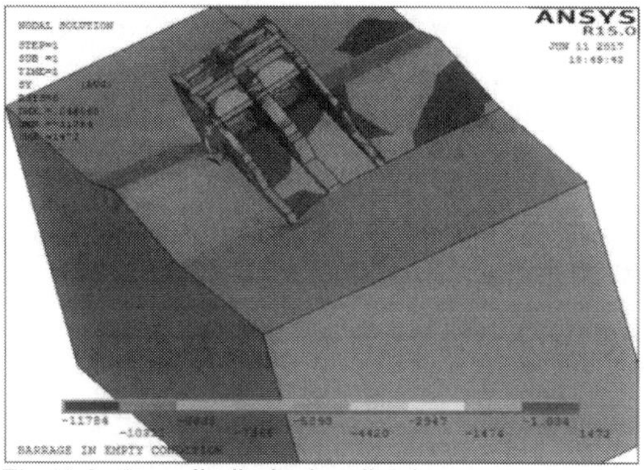


**Fig. 10- b.** Stress distribution in y-direction.

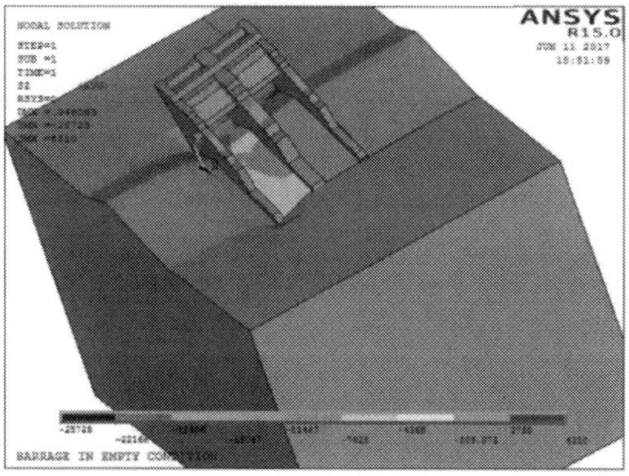


**Fig. 10- c.** Stress distribution in z-direction.

**Fig. 10.** Stress distribution in the considered barrage section; a. in x-direction, b. in y-direction, c. in z-direction.

Also, steady flow condition of barrage in nonlinear analysis is simulated by assuming HFL=8m at upstream and 4m at downstream which is suitable to the height of barrage at upstream and downstream conditions to compare with non-operation condition, as shown in Fig.11.

The operation condition gives result slightly lower than the non-operation condition due to the presence of uplift pressure as shown in Fig. 12. This is happened as a result of balanced forces system equilibrium with the gravity load in the case of flow condition.

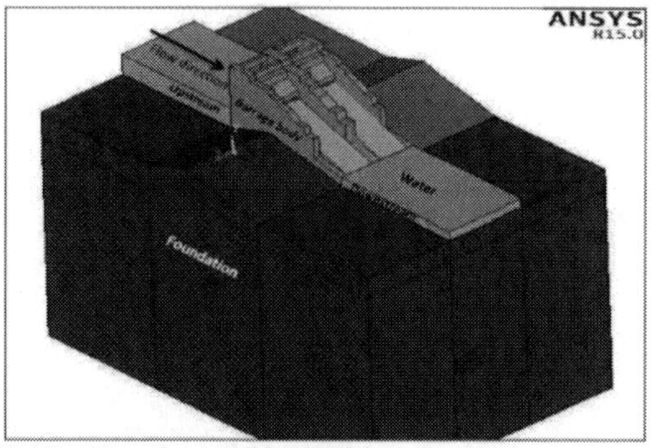


**Fig.11.** Operation condition model.

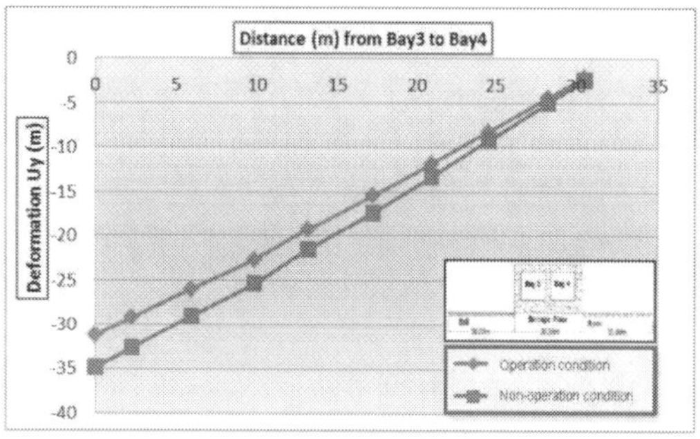


**Fig. 12.** Deformation of barrage with distance in operation and non- operation conditions at ogee section.

### 3.3 Some effective parameters

The properties of materials such as density ρ, modulus of elasticity E, Poisson's ratio $\vartheta$, and yield stress $\sigma_y$ are selected as the effective parameters to give an indication on the behaviours of related structures at empty nonlinear condition for ogee section. The range of these properties are given in [15] as follows:

a- Density for barrage body ($\rho_c$), soil foundation ($\rho_s$), and rock foundation ($\rho_r$); as shown in Fig. (13- a, b, and c). The selected values of these parameters depend on the principle of light to heavy structures.

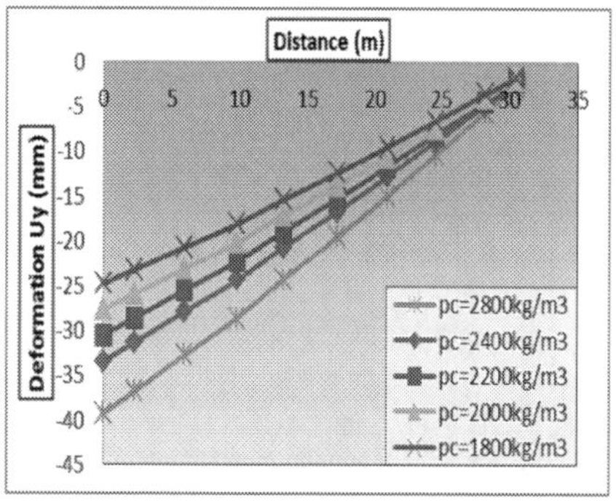


**Fig. 13–a.** Effect of density of barrage body on deformation.

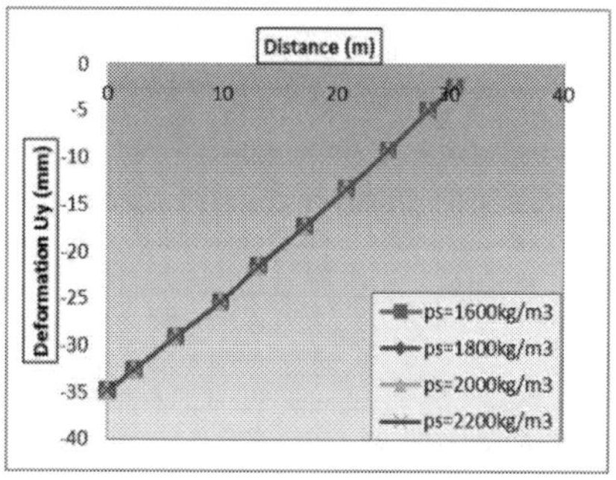


**Fig. 13–b.** Effect of density of soil foundation on deformation.

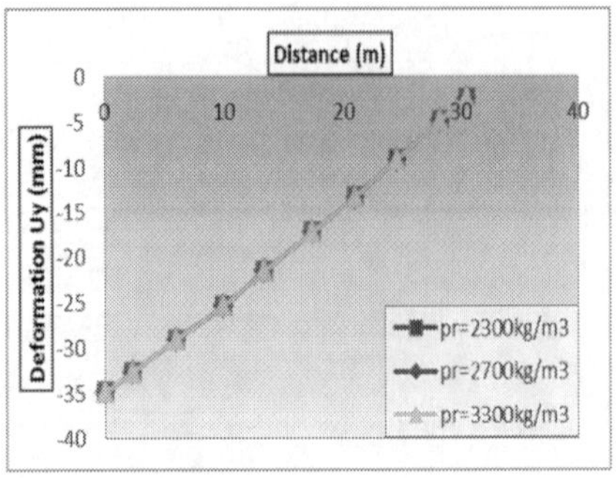


**Fig. 13–c.** Effect of density of rock foundation on deformation.

**Fig. 13.** Effect of density on deformation; a. barrage body, b. soil foundation, c. rock foundation.

b- Modulus of elasticity for barrage body (Ec), soil foundation (Es), and rock foundation (Er); as shown in Figures (14– a, b, and c).

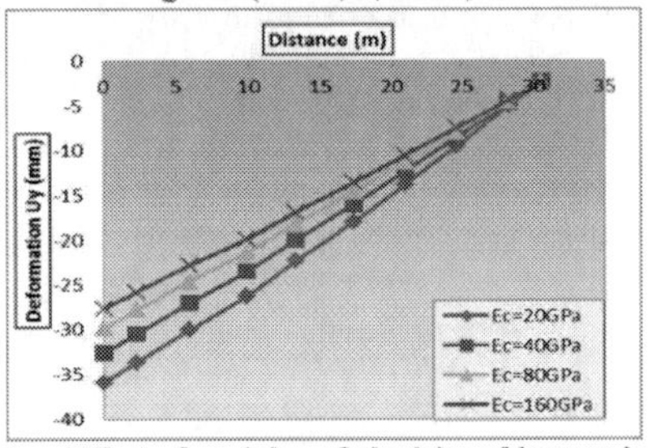


**Fig. 14–a.** Effect of modulus of elasticity of barrage body on deformation.

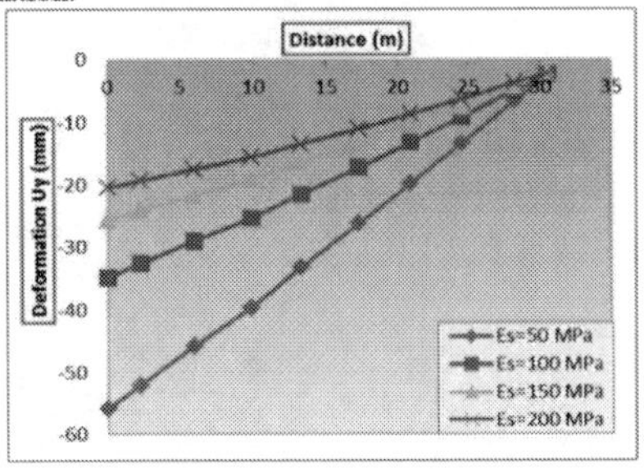


**Fig. 14–b.** Effect of modulus of elasticity of soil foundation on deformation.

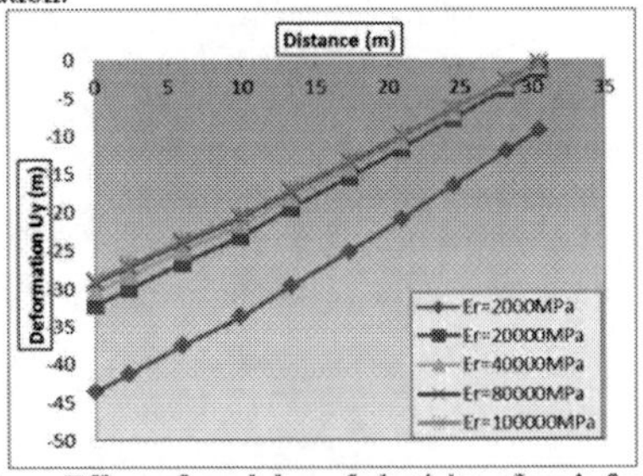


**Fig. 14– c.** Effect of modulus of elasticity of rock foundation on deformation.

**Fig. 14.** Effect of modulus of elasticity on deformation; a. barrage body, b. soil foundation, c. rock foundation.

c- Poisson's ratio for barrage body ($\vartheta_c$), soil foundation ($\vartheta_s$), and rock foundation ($\vartheta_r$); as shown in Figures (15- a, b, and c).

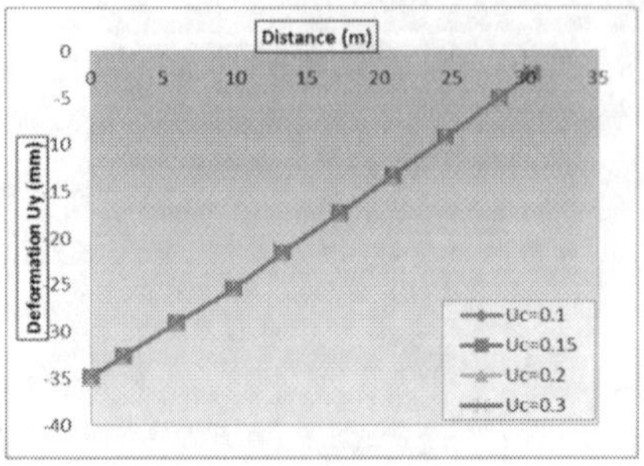


**Fig.15–a.** Effect of Poisson's ratio of barrage body on deformation.

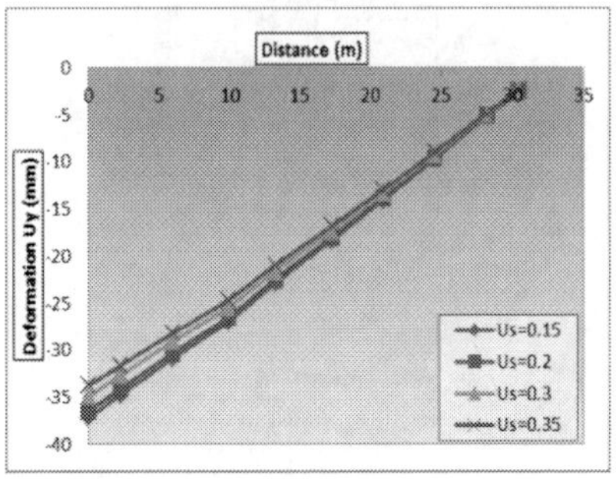


**Fig. 15–b.** Effect of Poisson's ratio of soil foundation on deformation.

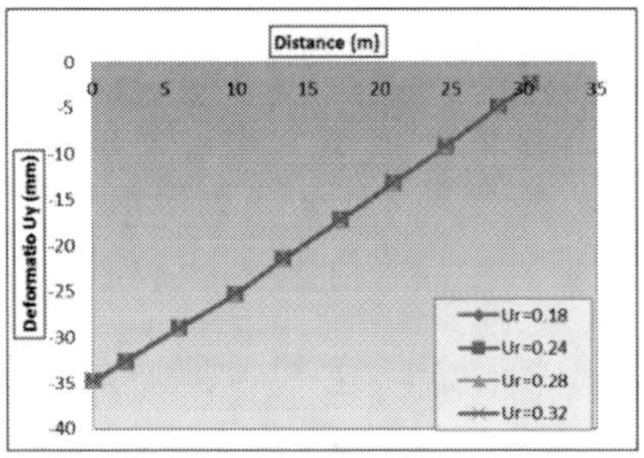


**Fig. 15–c.** Effect of Poisson's ratio of rock foundation on deformation.

**Fig. 15.** Effect of Poisson's ratio on deformation; a. barrage body, b. soil foundation, c. rock foundation.

d- Yield stress, ($\sigma_y$) for soil and rock foundation as showed in figures (16 – a and b).

https://doi.org/10.1051/matecconf/201816203013

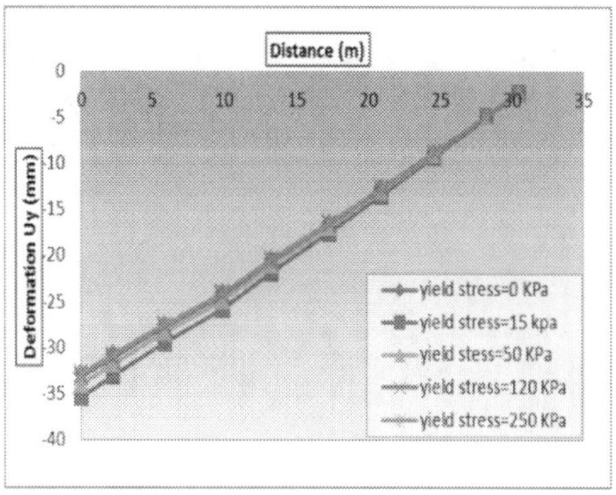


**Fig. 16–a.** Effect of yield stress of soil foundation on deformation.

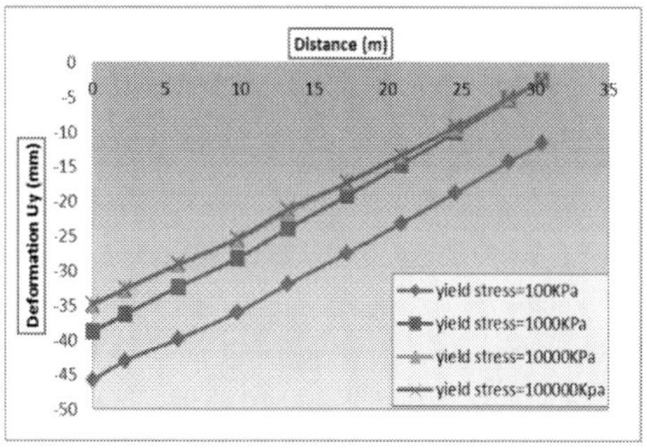


**Fig. 16–b.** Effect of yield stress of rock foundation on deformation.

**Fig.16.** Effect of yield stress on deformation; a. soil foundation, b. rock foundation.

It can be noticed from Figs. (13–a, b, and c through 16– a, b, and c) that the change of parameters may be effect on the deformation of barrage floor rather than others. Figure (13– a) shows that the increasing in density of barrage concrete material ($\rho_c$) will increase the deformation due to increasing of gravity force, where the increasing both density of soil ($\rho_s$), and rock ($\rho_r$) foundation material don't effect on the deformation of barrage floor as shown in Figs. (13–b , and c). The most important parameter affected on the deformation is the modulus of elasticity of soil foundation ($E_s$); Fig. (14–a, b, and c) according to settlement equation given by [16]:

$$S_e = \frac{qBI}{Eu} \quad (4)$$

Where: q is the applied load per unit area (kN/m$^2$), B is the width of loaded area (m), I is the influence factor which is a function of H/B and L/ B, H is the depth of layer, L is the length of structure, and Eu is young modulus of elasticity (kN/m$^2$). In other hand, Poisson's ratio shows little effects on the deformation of barrage floor as shown in Fig. (15–a, b, and c). Moreover, it is shown from Fig. (16–a and b), that the increasing yield stress of soil foundation material from (0–250 kPa), the deformation will decrease so little, while the increasing yield stress of rock foundation material from (100-100000 kPa), the deformation will decrease by (28.57%), where the low value of yield stress of rock make it behave as soft soil.

### 3.4 Seepage analysis

Hassan in his study [17] suggested the different configurations of sheet piles as an effective parameters to find the optimum hydraulic design of barrage. In this study, Four configurations of sheet pile is considered, which is without sheet pile, sheet pile at upstream and downstream, sheet pile at upstream only, and sheet pile at downstream only. The length of sheet pile was 10m for upstream and 6m for downstream according to limit of length in [17]. The permeability (**K**) for the selected type of soil (i.e. Gravelly soils) is equal to (1 cm / s) according to type of soil foundation [15]. The barrage and foundation sections are simulated for FL=8m at upstream only for steady seepage condition using PLANE 77 thermal element that available in ANSYS Software to study effects of presence of sheet piles on the uplift pressure, exit hydraulic gradient, and seepage changes. The Geometry of this element is shown in Fig (17).

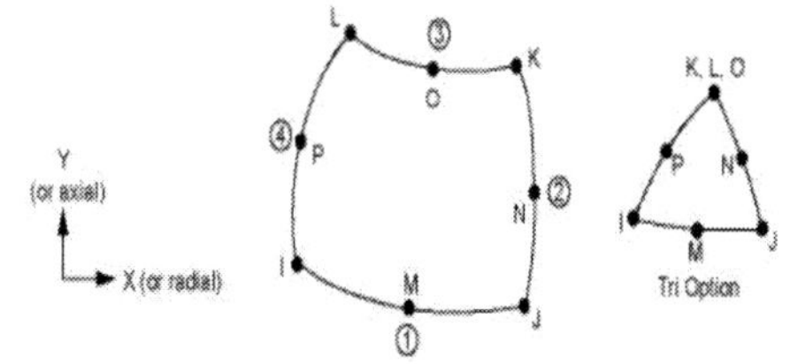


**Fig. 17.** PLANE 77 element geometry [8].

Figs. (18, 19, 20, and 21) illustrates the results of equipotential lines, flow lines and seepage discharge at foundation underneath barrage floor for given four configurations of sheet piles after the numerical simulation has been run. The maximum seepage flow at exit point downstream barrage floor with up and downstream piles is lesser by (67.40%) than without pile. While it is lesser than by (30.71%) and (17.83%) with upstream pile only and with downstream pile only, respectively and these result are close to the results of [18].The results of seepage could be summarized in Table (5).

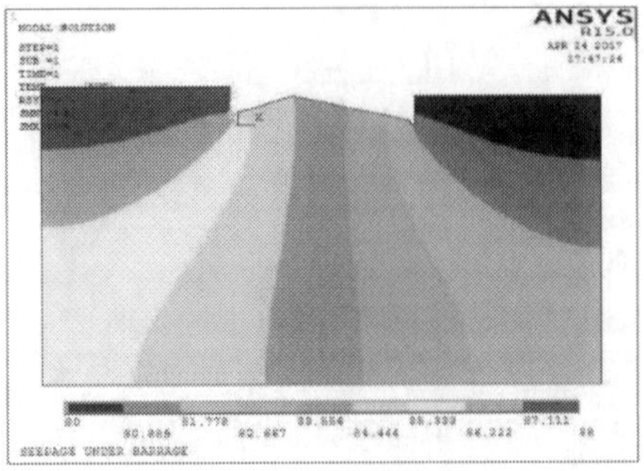


**a.** Equipotential lines.

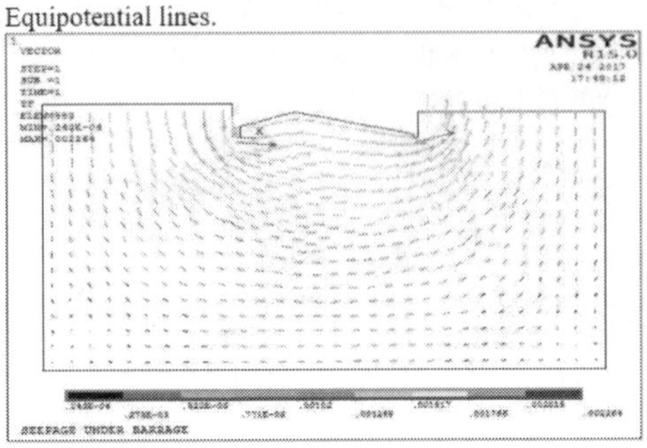


**b.** Flow lines.

Fig. 18. Cross section of barrage foundation under barrage without sheet pile; a. equipotential lines, b. flow lines.

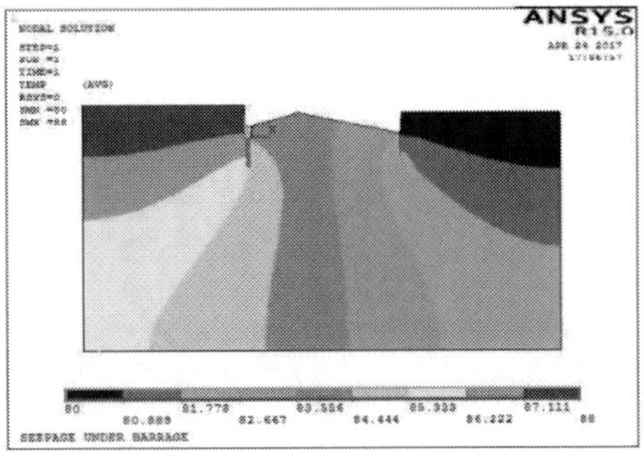


**a.** Equipotential lines.

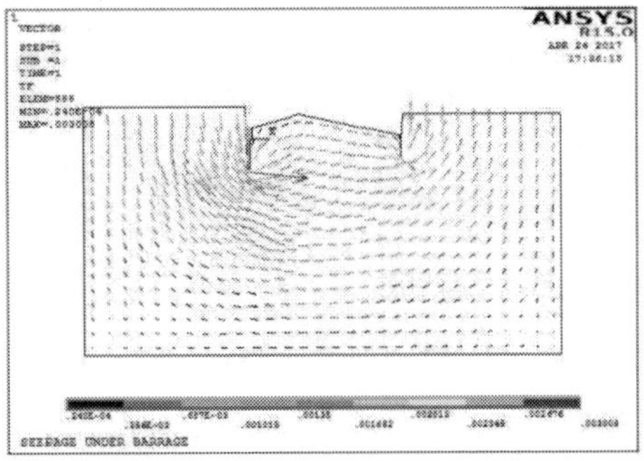


**b.** Flow lines.

**Fig. 19.** Cross section of barrage foundation under barrage with sheet pile at upstream and downstream; a. equipotential lines, b. flow lines.

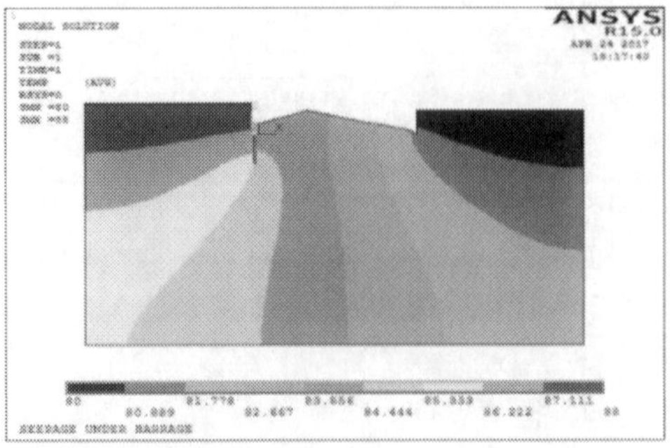


**a.** Equipotential lines

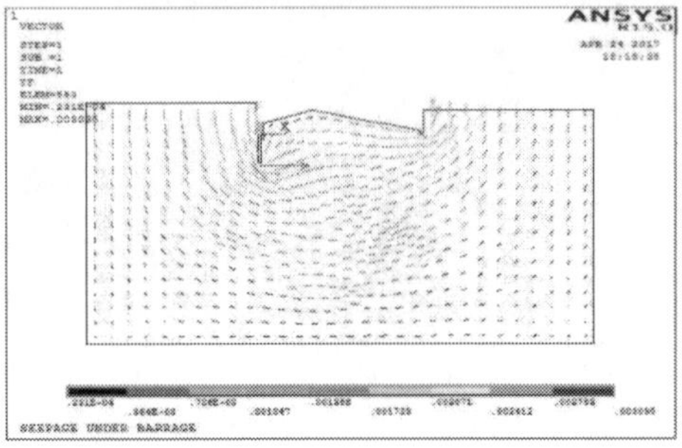


**b.** Flow lines

**Fig. 20.** Cross section of barrage foundation under barrage with sheet pile at upstream only; a. equipotential lines, b. flow lines.

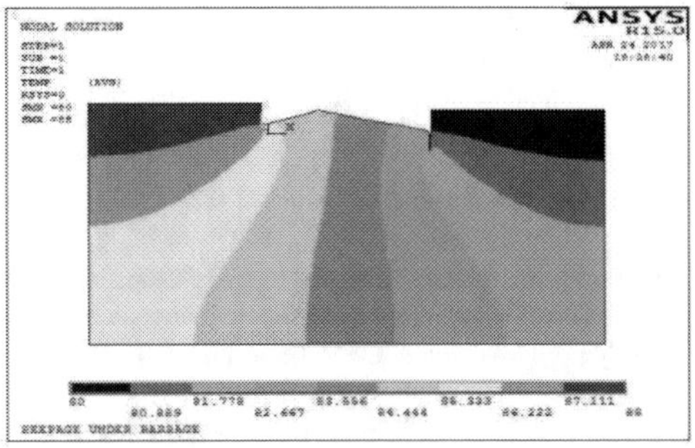


**a.** Equipotential lines.

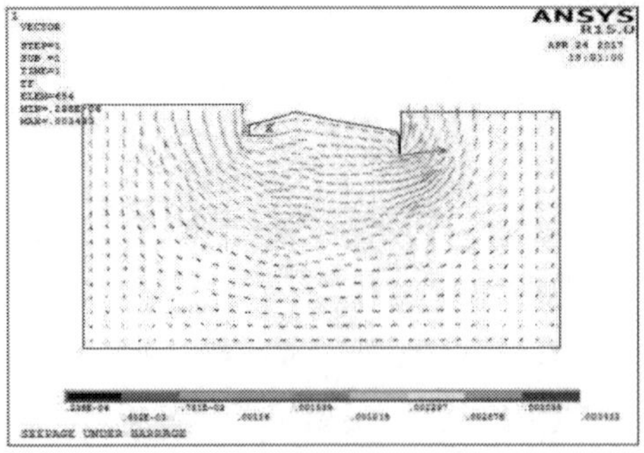


**b.** Flow lines.

**Fig. 21.** Cross section of barrage foundation under barrage with sheet pile at downstream only; a. equipotential lines, b. flow lines.

**Table 5.** Flow seepage quantity underneath barrage floor condition.

| Barrage floor conditions | Flow seepage ($m^2/s/m$) |
|---|---|
| Without piles. | $2.264\times10^{-3}$ |
| With both upstream and downstream piles. | $1.350\times10^{-3}$ |
| With only upstream pile. | $1.729\times10^{-3}$ |
| With only downstream pile. | $1.918\times10^{-3}$ |

The effects of sheet piles on the uplift pressure distribution underneath barrage floor are presented in Fig. 22. It can be noted from this figure that the best condition which give lesser values of uplift pressure only at contact part of barrage floor with foundation is at presence of sheet pile at upstream only.

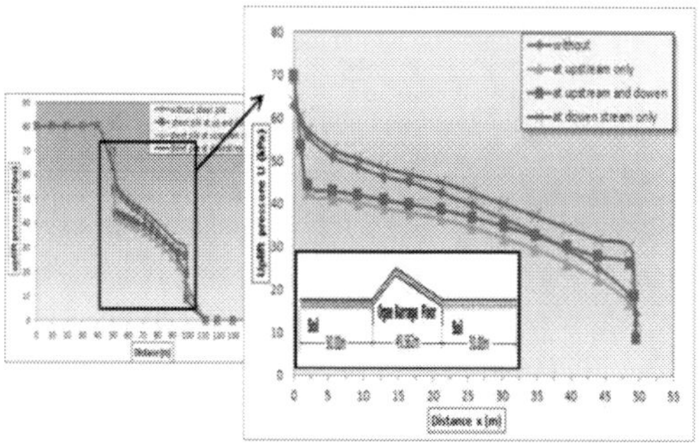


**Fig. 22.** Uplift pressure distribution in the foundation with different conditions of sheet pile.

The effects of sheet pile presence on the exit hydraulic gradient are presented in Fig. 23. It is shown from this figure that the barrage floor without sheet pile is the critical condition which give results closer to the limit of safe exit gradient {i.e., $(^1/_6)$} [17], while the sheet pile at up and downstream are the most safe condition followed by the pile at downstream only and this is close with results of [18].

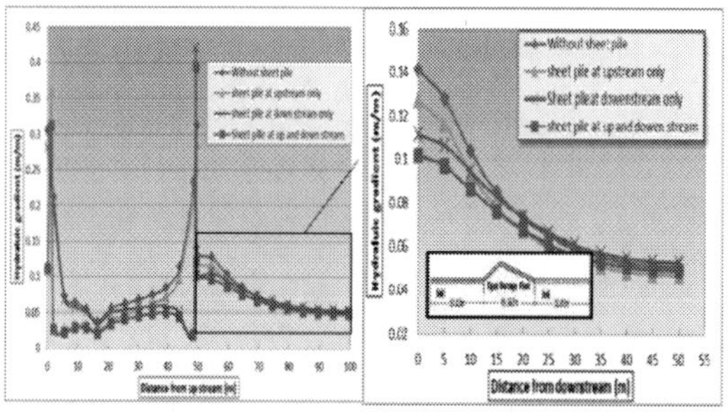


**Fig. 23.** Effects of the sheet pile existing on the changes of the hydraulic gradient

Moreover, the changes of the seepage discharge underneath barrage and foundation is drawn versus the conditions of the sheet pile is presented in Fig. 24.

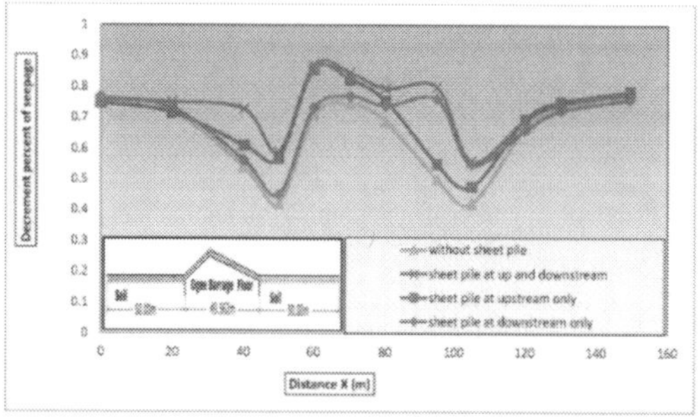


**Fig. 24.** Effects of the sheet pile existence on the changes of the seepage.

According to Fig. 24, the providing sheet piles at both upstream and downstream give maximum decrement to (90%) beneath barrage floor rather than other providing of piles conditions.

## 4 Conclusions

- ANSYS/APDL is adequate tool to simulate and analysis the problem that need sufficient experiences to select suitable available elements to get the acceptable results.
- The deformation in barrage floor has little difference by (6%) in linear and non–linear analysis in non-operation case. While this deformation in non-linear analysis in the flow condition is less by (13%) than of non-operation case.
- The maximum stress in y– direction is more than that in x– and z– directions due to interaction, where the critical section at rock side of foundation.
- The non-linear deformation of barrage floor is more effective by change of density of barrage concrete itself, modulus of elasticity and yield stress of both soil and rock foundation materials, while no sensitive effect of change of Poisson's ratio on the deformation.
- The maximum seepage flow at exit point downstream barrage floor with upstream and downstream piles is less than that without pile by (67.40%). While it is less than that with upstream pile only and with downstream pile only by (30.71%) and (17.83%), respectively.
- The lesser values of uplift pressure only at contact part of barrage floor with foundation is at presence of sheet pile at upstream only.
- The existence of the sheet pile at both upstream and downstream is an appropriate solution for the hydraulic gradient reduction and piping happening under barrage and reducing the quantity of seepage.
- In the view of stability, the barrage floor is already safe in exist gradient and piping problems with or without piles, where the safe exit hydraulic gradient is far away from the critical limited value with pile at upstream and downstream or with downstream only.

• The maximum decrement percentage of seepage under barrage floor reach to (90%) underneath barrage floor with upstream and downstream pile.

The authors are grateful for the support of staff of Al-Hindiya Barrage Project for giving opinions that necessarily to finish this study.

## 5 References

1. P. Novak, A. Moffat , C. Nalluri, R. Narayanan, *Hydraulic Structures*, 4th Edition, Taylor and Francis (2007).
2. A. Mohsin, H. Omran, A.– H. Al–Shukur, IJIRSET, **04**, 8961-8973 (2015).
3. A–H. Al-Shukur, A. Al-Rammahi, IJCIET, **08**, 873–881(2017).
4. A. Al-Qaisi, *Optimal Design of Concrete Gravity Dams on Random Soil*, PhD. Thesis, Building and Constructing Engineering Department, University of Technology, Baghdad, Iraq (2016).
5. K. Venkatesh, A. Pandey, N. Samadhiya, 15th ICRA GEESD, May 24–29, 45, San Diego, California (2010).
6. Z. Lubkowski, M. Willford, X. Duan, A. Thompson, A. Kammerer, 13th WCEE, August 1–6, 1415, Vancouver, B.C., Canada (2004).
7. K. Venkatesh, N. Samadhiya, A. Pandey, 6th ICCH GE, Aug 11th – Aug 16th, Missouri University of Science and Technology (2008).
8. ANSYS. ANSYS User's Manual, *ANSYS Theory Manual*, Version 15.0 ( 2013).
9. American Society for Testing and Materials (ASTM), *Annual Book of ASTM Standards*, Vol. 04.08 Soil and Rock (I): D421 – D5876, Philadelphia, Pa (2004).
10. Indian Institute of Technology (IIT), *Water Resources Engineering*, Kharagpur (2008).
11. R. Taylor, P. Beresford, E. Wilson., IJNME, **10** , 1211-1219 (1976).
12. K. Willam, E. Warnke, ISMES, **19**, (1975).
13. S. Sypedoor, S. Gholizadeh, S. Talebian, CEES, **27**, 295–313(2010).
14. H. Hasan, A. Rahman, H. Ameen, Am. J. Sci. Ind. Res., **03**, 94-98 (2012).
15. D. Prices, *Engineering Geology: Principles and Practice*, deFreitas, Michael (Ed), Chapter 2–Geological Materials, 21-61(2009).
16. D. Bergado, S. Ahmed, C. Sampaco, A. Balasubramaniam, GE J., ASCE, **116**, 136 –155 (1990)
17. Z. Hassan, *Optimal Hydraulic Design of a Barrage Using Genetic Algorithm*, M.Sc. Thesis, College of Engineering, University of Babylon (2012).
18. S. Khassaf, A. Al-Adili, R. Rasheed, 13th IWT Conference, Hurghada, Egypt, 301-310 (2009).

# Derivation of suspended sediment data for Al-Adhiam watershed-Iraq using artificial neural network model

*Mahmoud* Saleh Al-Khafaji[1], *Mustafa* Al-Mukhtar[1] and *Ahmed* Saud Mohena [2]

[1] Building and Construction Engineering Department, University of Technology, Baghdad, Iraq
[2] Ministry of Water Resources, Baghdad, Iraq

**Abstract.** The mean part of river sediments is suspended sediment load, its prediction and simulation has important significance to manage the water resources and environments. In Iraq, most researchers avoid to fighting in sediment researches when related with hydrological models spatially with that need enough observed sediment data for calibration and validation because the sediment data very limitation or scars. The aim of this study is employing the Artificial Neural Network (ANN) model to estimate the suspended sediment load of Al-Adhaim watershed in Iraq from available measured sediment data, identify the suitable pattern of input and target data sampling and obtaining the best nonlinear equation between the river discharge and suspended sediment load. To this end, the ANN model was training and tested with the available sediment data, which was for water year (1983-1984). Two modes were applied for input and target data sampling each mode has two cases, where in the first mode the time series data sampling was used with flow as an input for case one while flow and average precipitation in case two with used suspended sediment as a target variable. For second mode the supervise data sampling was used with the same input and target division in first mode. The performance of the model was evaluated by using Coefficient of determination ($R^2$) and the Nash- Sutcliffe efficiency (NS) and standardization of root mean square error (RSR), the statistical analysis model testing for Al-Adhiam watershed showed satisfactory agreement between observed and estimated daily values for Mode2- Case2. $R^2$, NS and RSR of the testing period were 0.99 and 0.8and 0.2 respectively. The result shows that the conducted ANN model can be used with the best net as a predictor for sediment yield in this watershed. The model was used to predict daily sediment load data for period from 1Oct. 1984 to 31Spt 1985. The predicted daily sediment data was plotted against daily measured flow. The correlation between predicted sediment and measured flow was in good agreement with $R^2$ =0.89 and the best relation was polynomial equation from second degree.

## 1 Introduction

Sediment yield is defined as the total measured sediment load outflow from a watershed at a point of reference in a specified period. Processes of detachment, transportation, and deposition of soil materials induce the sediment outflow from the watershed by rainfall and runoff. Guesses of sediment yield are required in a wide range of problems such as transport of pollutants and sediment in rivers, design of reservoirs and dams, lakes and estuaries, undertaking clean up following floods, design of stable channels, Protection of fish and wildlife habitats, determination of the effects of watershed management [8].

The most reliable method in guessing sediment quantity is the use of its measured records, but sediment sampling is very complex and requires high experience because of its considerable fluctuation within the river section and user-unfriendly measurement tools. Moreover, it is consuming time and costly. These constraints led to low frequency of sediment observation especially in the remote regions.

In Iraq the suspended sediment data is very scars or not available. Many models have been provided to simulate sediment load. Convention sediment rating curve and regression models, in which the system is supposed to be static, are often used to estimate suspended sediment load of a river. [10, 7] employed statistical approach and sediment rating curve to predict sediment load. The results of these studies show that the predicted sediment was underestimated. The hydrological models like deterministic models are useful for sediment yield prediction. The major drawback of these models are being require large amount of input data and computational time [9]. Physics-based increasingly being used to simulate complex water resource systems including simulation suspension sediment load but required enough measured sediment data for conducting the calibration and validation

© The Authors, published by EDP Sciences. This is an open access article distributed under the terms of the Creative Commons Attribution License 4.0 (http://creativecommons.org/licenses/by/4.0/).

processes [16]. Other researchers used Artificial Neural Network (ANN) model to prediction sediment load and compared the results with the sediment-rating curve [1, 14]. The results showed that the ability of ANN model for predicting sediment was better than other regression approaches due to their abilities in capturing nonlinear system among variables and used more than one input variable. [5] developed two different ANN methods to simulate relationship of suspended sediment load with river flow and precipitation by using hydro meteorological data. The obtained results showed that the provided models produced considerably better causes than Multi Linear Regression (MLR). [17] Predicted suspended sediment load by using ANN modeling based on the average of precipitations, precipitation intensity, temperature and flow discharge. This model was applied on the Long Chuanjiang River. It was capable to predict monthly suspended sediment load of flow accurately, [6] compared the estimated suspended sediment using the rating curves and ANN model for El- Kebir catchment. Daily suspended sediment and Daily water discharge data of Ain Assel gauging station, were used as inputs and targets in the models. The feed-forward and cascade-forward back-propagation was the ANN model based techniques by using Levenberg–Marquardt. The models results concluded that these models have the highest performance to re-estimate the global annual sediment yields and daily sediment load. Thus in the view of the superior performance of ANN represent suitable solution to predict sediment load with limitation suspension sediment data.

Al-Adhaim watershed is considered one of the most important catchment that feeding Tigris river in Iraq due to its size. Therefore, it effects remarkably on water quality of Tigris River. The watershed boundaries totally located inside Iraq in the northeast zone. In other words, the watershed is entirely managed by the Iraqi water resources ministry. Because of the lack in suspended sediment data, most of researchers avoid to use the hydrological model for simulating and study the impact of climate on sediment yield spatially in Iraq.

The aim of this study, modeling and driving suspended sediment data for Al-Adhaim watershed, from the available data scars data that was justly for water year 1983-1984, determine the best data input and target sampling and distribution that use in ANN model for this watershed and Deriving the best nonlinear correlation between flow and suspended sediment load.

## 2 Martials and methods

### 2.1 Descriptions of the study site

Al-Adhaim watershed is located on the northeastern border of Iraq, between 35o 42' 24"to 34o 33' 8" N and 43o 41' 9" to 45o 27' 31" E with a total drainage area of 11217 square kilometers Fig. 1. It lies entirely inside Iraqi territories, and it outstretches between the basin of lower Zab northward, and Diyala river southward an, Shwan Secrema-Dagh eastward and Himreen mountains westward [2].

Al-Adhaim Dam was constructed on AL-Adhaim River In 2000 at coordinate 34o 32' 9"N and 44o 30'44" E approximately 134km north Baghdad City and about 65 km southeast of Kirkuk City all the drainage basin of Al-Adhiam is concentrated in Sallahadein, Diyala, Sulamaniya and Kirkuk provinces [4]. It is bounded between Himreen mountains westward and Kirkuk Hill eastward. AL_Adhaim Watershed is an arid area with limited rainfall and no snowfall. Thus, main flow occurs only during the rainy season. The major rainfall storm occurs from October to May. While the weather is relatively dry during the other months of the year. The minimum temperature is -4 °C in winter. It is raised to maximum 49 °C in summer. The annual rainfall of this region is 610 mm.

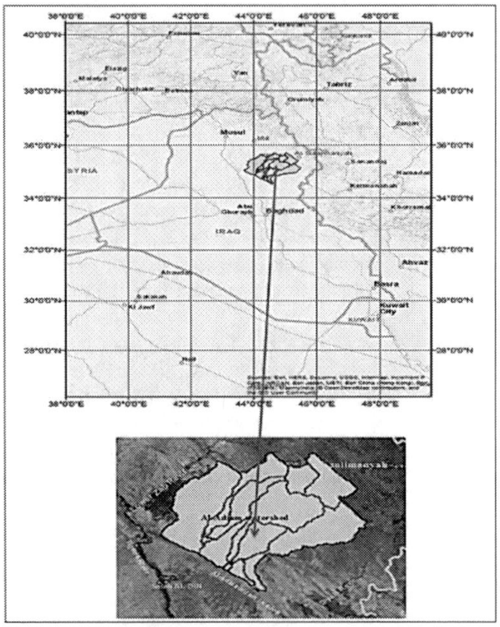


**Fig. 1.** Location map for Al-Adhiam watershed.

### 2.2 Data used

In this research, river discharges and average precipitation were used as input variables, while suspended sediment load was used as target variable .

### 2.3 Hydrological data

Considering the purpose of this study, ANN model needs for daily recorded flow. This data was obtained from the Iraqi Ministry of Water Resources (MOWR)/ National Center for Water Resources Management (NCWRM) for the period from 1983 to 1986.. The data were recorded at Enjana gauge station which located at coordinates 34° 32' 9"N and 44° 30'44" E at the current outlet for the watershed. The daily observed sediment data for the water year 1983-1984 from October 1983 to October 1984 was collected from a study prepared by Al-Ansary at el. (1986) in which they calculated sediment discharge and soluble load for Al-Adhiam River.

### 2.4 Metrological data

Due to the lack in daily precipitation data, the monthly precipitation data was collected for the period from 1985 to 1989 from Iraqi Meteorological Organization and Seismology (MOS). Thus, the Climate Forecast System Reanalysis (CFSR) was used in this study. CFSR dataset consists of hourly weather forecasts produced by the National Centers for Environmental Prediction (NCEP). CFSR Provides weather requirement such as Precipitation, maximum and minimum temperatures [11], the monthly and yearly precipitation data were verified with those counterparts from the observed using statistical merits ($R^2$, RMSE, NS). This data were downloaded on 23March 2017 from (http://globelweather.tamu.edu/) for AL_Adhaim watershed. 12 CFSR weather stations were for Al-Adhiam watershed.

### 2.5 Processing data in ANN

Neural networks are composed of simple components operating in parallel. These components are like that biological nervous system as in nature. The network form is determined largely by the linked between elements. . It can train a neural network to behavior a particular form by modified the values of the weights (connections) between elements. The input values or signal (X1, X2, . . . , Xp) are transmitted to the neuron through the bifurcations that represent different input ways, Fig. 2. Each input way has its own linked weight (W1, W2,. . . ,Wp). These input weights are very critical because they permit to sum and process the signals based on their values and effects on input vectors. An input weight vector is passed through the cell if it gives any real value except zero at the synapse. All the transmitted signals are ordinary integrated by collecting all the inputs. The mathematical model (referred to as activation form) is applied to complete the summation process within the cell body to produce an output signal. Different combinations of data and attached weight are used to obtain the correlation between the input and output within the ANN model using Equation (1), [15].

$$Y = f(I) = f\left[\sum_{i}^{p} W_i X_i + b_k\right] \quad (1)$$

Where;
$X_i$ = input value or signal i
$W_i$ = weight of input value i
P = number of input values
$b_k$ = bias at the cell of the body
Y = output
f=activation function
Many activation functions or equations can be used within the neurons. The Non-linear step function is the most commonly used function.

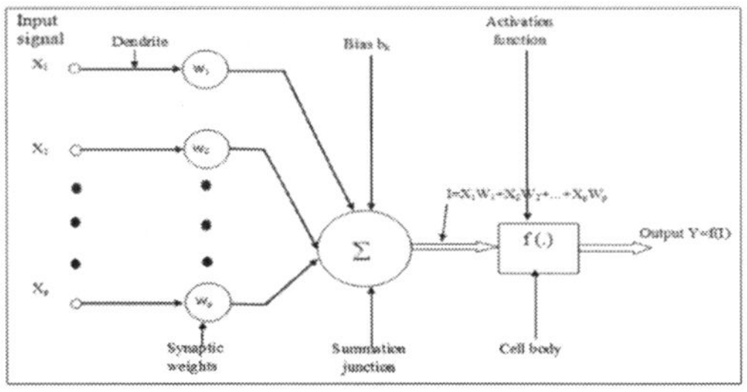


**Fig. 2.** Fondamental parts of a typical neural network [15].

### 2.6 Artificial neural networks structure

The basic neural network used in this study has a five-layer learning network. It consists of the input layer (introducing data to the ANN), the three hidden layers (processing data), and the output layer (producing the ANN results), Fig. 3.

The Levenberg–Marquardt algorithm of training the neural networks was used applied on the feed-forward backpropagation method (FFBP). In this study, the combinations of data partitioning approach, number of hidden layers, transfer function, number of neurons in each hidden layer, error goal, etc., were tried. The presentation of a newff function in relative to inputs, target, and number of neurons has produced a feed-forward network. This function is based on utilizing the units where a biased weighted summation of their inputs is performed. Then, these units transmit this activation level through a transfer function to produce their output, and the units were prepared in a layered feed-forward topology. Once the number of layers and number of elements in every layer, has been chosen, the network's weights and levels have been set so as to minimize the estimate error made by the network. This was the role of the training algorithms [6].

To estimate daily-suspended sediment load applying ANN, at first, measured water discharge (Qw), average precipitation and suspended sediment discharge (Qs) data were collected for the period from 1Oct.1983 to 30 Sept.1984. Before entering the input and target data to the ANN model, the effect of extreme values in the data sets was prevented by normalizing this data between 0 and 1 by using Equation (2), [14].

$$Y_{norm} = \frac{Y_i - Y_{min}}{Y_{max} - Y_{min}} \quad (2)$$

Where, $Y_{norm}$ is the normalized dimensionless variable; $Y_i$ is the measured value of variable; $Y_{min}$ and $Y_{max}$ are the minimum and maximum values of the variables, respectively [14].

The information was modeled utilizing Matlab R2015b software. Different structures of ANN model were built with a variation in hidden layers and joining different equations. Consequently, the optimal structure is found using Trial and Error method. Finally, the network of the obtained optimal structure and its training, was evaluated by using the test package data.

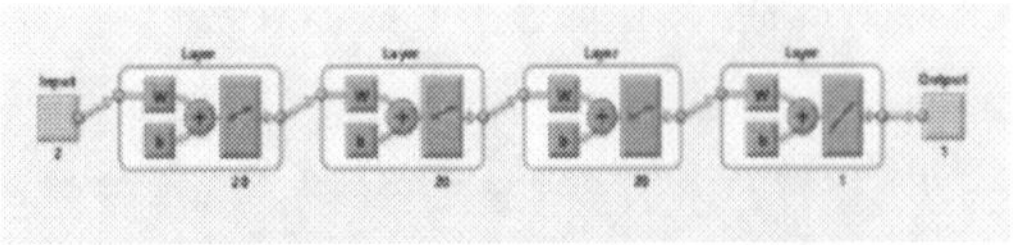

(a)

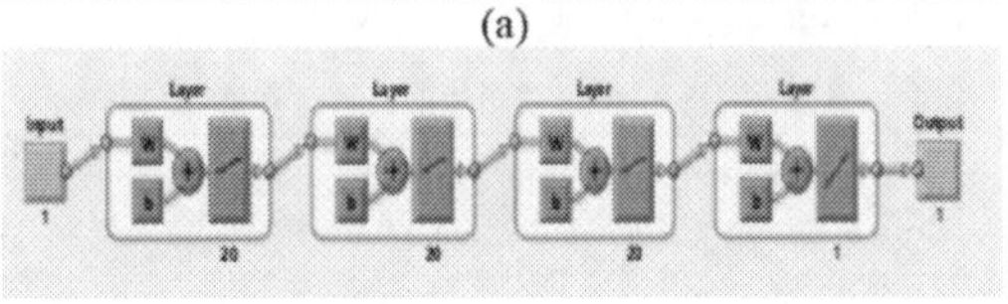

(b)

**Fig. 3**. An exemplary five-layer feed-forward ANN structure; (a): Two input variables feed-forward, (b): One input variables feed-forward.

### 2.7 Input and target data distribution

The obtained data was for one water year from 1 Oct. 1983 to 30 Sep. 1984 with 366 recorded points, Therefore, two modes were used for division and sampling the input variable and target variable as follows:

**i. Mode One:** Using time series sampling for input and target data with two cases as follows:

a. Case one**:** Using flow as an input variable and suspension sediment as a target variable, where the data from 1 Jan. 1984 to 30 Sep. 1984 for training and from 31Oct. 1983 to 31Dec. 1983 for testing.

b. Case two**:** Using flow and average precipitation as an input variable and suspended sediment as a target variable, where the data from 1 Jan. 1984 to 30 Sep. 1984 for training and1 Jan. 1984 to 30 Sep. 1984 for training and from 31Oct. 1983 to 31 Dec. 1983 for testing.

**ii. Mode Two:** using supervised sampling for input and target data along the period from 1 Oct. 1983 to 30 Sep. 1984, with two cases as follows**:**

a. Case One: Using flow as an input variable and suspended sediment as a target variable, with 70% for training and 30% for testing.

b. Case Two: using flow and average precipitation as an input variable and suspended sediment as a target variable with 70% for training and 30% for testing.

### 2.8 Evolution of ANN model

ANN model efficiency was assessed utilizing coefficient of deterministic ($R^2$), standardizes of root mean squared error (RSR), and Nash–Sutcliffe coefficient of efficiency (Ns) [12]. The relationship between the measured and the predicated values within Al-Adhiam watershed was obtained using the correlation coefficient (R). The fundamental function was customized to the respective data input and outputs of the following general form:

$$R^2 = \frac{[\sum_{i=1}^{n}(S_{o,i} - \overline{S_o})(S_{o,i} - \overline{S_s})]^2}{[\sum_{i=1}^{n}(S_{o,i} - \overline{S_o})^2 \sum_{i=1}^{n}(S_{o,i} - \overline{S_s})^2} \quad (3)$$

$$RSR = \sqrt[2]{\frac{\sum_{i=1}^{n}[S_{s,i} - S_{o,i}]^2}{n}} \quad (4)$$

$$NS = 1 - \frac{\sum_{i=1}^{n}(S_o - S_s)_i^2}{\sum_{i=1}^{n}(S_{o,i} - \overline{S_o})_i^2} \quad (5)$$

R: coefficient of determination.
RSR: Standardizes of root mean squared error.
Ns: Nash–Sutcliffe coefficient of efficiency.
$S_o$: the measured value of the suspension sediment variable.
$\overline{S_o}$: mean of the measured values of the suspension sediment variable.
$S_S$: estimated value of suspension sediment variable.
$\overline{S_S}$: mean of the estimated values of suspension sediment variable.
n: number of the considered data points.

The value of R represents the strength of the relation between measured and estimated values. It value is ranged between 0 and 1. The low (high) values indicate a poor (perfect) forecasting capability of the model. The values of RSR is varies from 0, with perfect model simulation where there is no residual variation, to a large positive value. The best model performance corresponding to the lower RSR value. [15].

## 3 Results analysis

### 3.1 Structure characteristics of ANN

The number of neurons in the hidden layer was determined using the trial and error procedure to reach the best characteristics of optimal structure of ANN that were used in this research. Table 1 illustrates the characteristics of optimum structure of ANN.

Table 1. Characteristics of optimum structure of ANN.

| | |
|---|---|
| Number of input layer neuron | 1 |
| Number of first hidden layer neurons | 20 |
| Number of second hidden layer neurons | 20 |
| Number of third hidden layer neurons | 20 |
| Number of output layer neuron | 1 |
| Transfer function | Gaussian-Sigmoid-Sigmoid-Sigmoid |
| Learning rate | 0.001 |
| Momentum coefficient | 0.95 |
| Iteration | 1000 |

### 3.2 ANN model performance

Two modes were applied for data sampling and data distribution for training and validation, to get the optimum distribution with best net. Each Mode has two cases of data sampling, as it discussed earlier. When analyzing the descriptive statistical results that listed in Table 2. $R^2$, NS and RSR were relied upon to evaluating the ANN model performance to determine the best sampling mode for data distributing for training and validation. Also to identifying the optimum ANN model, which can be used for sediment load prediction depend on model performance in training and validation processes. From Table 2, it can be noticed that Mode2, when supervised distribution data used, has the best values of statistical measurements. It has the highest value of $R^2$ and NS with lowest value of RMSE and RSR for training and validation. The model performance in Case2, when used average precipitation (ave pcp) and flow (Q) as an input variable, is better than that of Case1 when only flow was used as an input variable. Fig. 4 shows the comparison between the estimated and measured suspended sediment load for validation period. Fig. 5 shows the correlation between the measured and estimated suspended sediment load of ANN model for the validation period Mode2 Case1 and Case2 respectively.

Table 2 shows the best performance of ANN model was with Mode2 for the two cases when using the supervised sampling for the data used in training and validation with some vantage for Case2. Therefore, the model that obtained from the two cases was applied to predict sediment suspended sediment yield of Al-Adhiam watershed.

Table 2. The best data sampling and their related performance criteria for the two Modes.

| Mode (M) | Input (v) | Targ-et (v) | Training | | | | Validation | | | |
|---|---|---|---|---|---|---|---|---|---|---|
| | | | $R^2$ | NS | RMSE | RSR | $R^2$ | NS | RMSE | RSR |
| M1-cas1 | Q | Sed | 0.68 | 4E-05 | 69264 | 1 | 0.005 | -591718 | 1187155 | 591719 |
| M1-cas2 | ave pcp+Q | Sed | 1 | 0.99 | 723 | 0.01 | 0.038 | -104360 | 5.00E+05 | 104361 |
| M2-cas1 | Q | Sed | 0.99 | 0.99 | 622 | 0.01 | 0.984 | 0.78 | 74580 | 0.22 |
| M 2-cas2 | ave pcp+Q | Sed | 1 | 0.99 | 250.4 | 0.01 | 0.993 | 0.8 | 70480 | 0.2 |

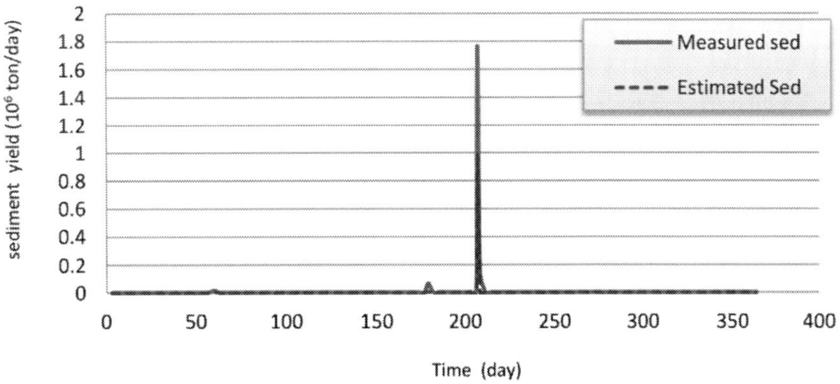


(a)

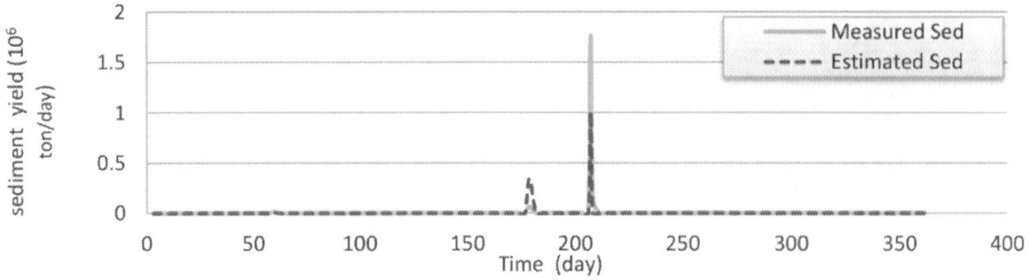


(b):

**Fig. 4.** Measured and estimated suspended sediment of ANN model. for the validation period; (a): Mode2-Case1 (for water year 1983-1984 (122 day)), (b): Mode2-Case2 (for water year 1983-1984 (122 day).

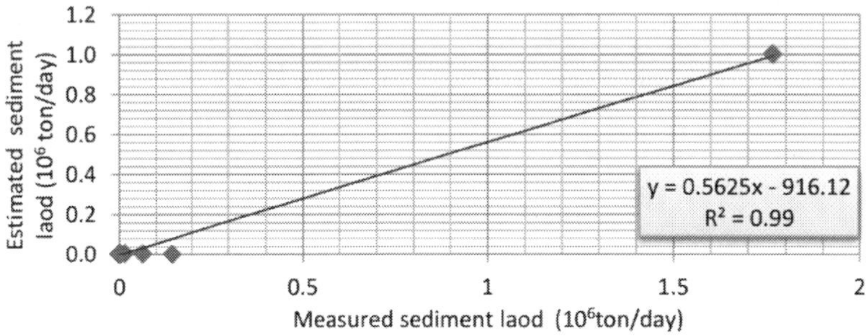


(a)

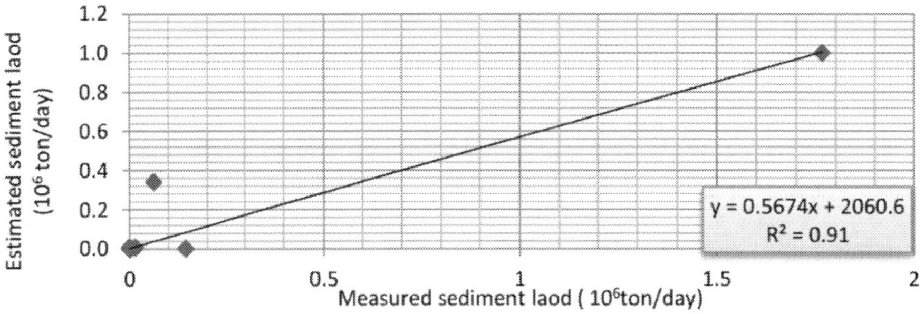


(b)

**Fig. 5.** Correlation between the measured and estimated suspended sediment of ANN model for validation period; (a):Mode2 -Case1, (b):Mod2-Case2.

### 3.3 Prediction of suspended sediment load using ANN model

The two best ANN models that obtained using Mode2 with Case1 and Case2 from training and validating processes were applied to predict suspended sediment yield from Al-Adhiam watershed. These models were applied on the watershed for period from 1 Oct. 1984 to 31 Dec. 1985. Fig.s 6 and 7 show the daily time series of the measured flow and estimated suspended sediment for the period 1 Oct. 1984 to 31 Dec. 1985. As shown in Fig.s 6a and 7b, the relationship between the predicted suspended sediment load and the measured flow for the two Cases can be represented by a nonlinear relationship. The value of $R^2$ for Case2 (using precipitation and flow as input variables) with ANN model was higher than that of Case1 (using only flow as input variable). In addition, results of regression analysis of these cases show the effect of precipitation on the values of predicted suspended sediment load when it used with flow as an input variable to ANN model. In addition, these results show the sensitivity of ANN model to the type and number of input variables. The predicted daily sediment data was plotted against daily measured flow. The correlation between predicted sediment and measured flow was in good agreement with $R^2$ =0.89 and the best relation was polynomial equation from second degree.

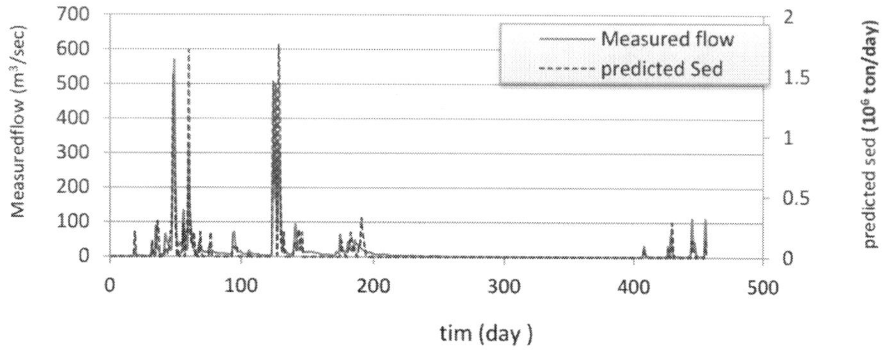


(a)

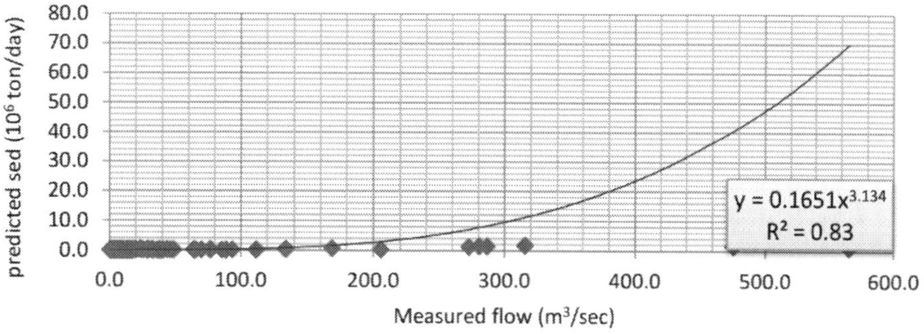


(b)

**Fig. 6.** Correlation between the measured flow and predicted suspended sediment load using ANN model for Mode2-Case1(for the period from 1 Oct. 1984 to 31 Dec. 1985); (a): Daily time series, (b): Regression model.

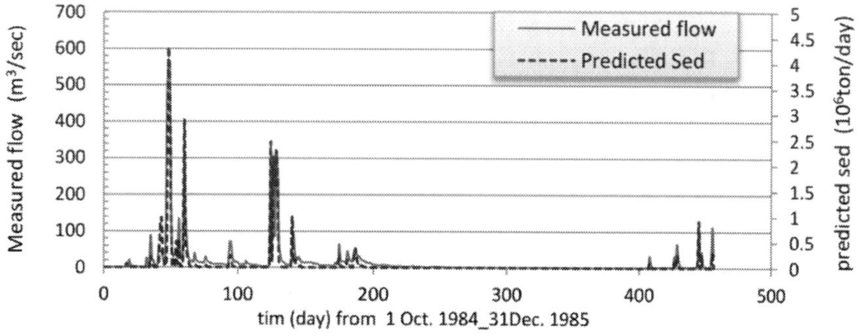


(a)

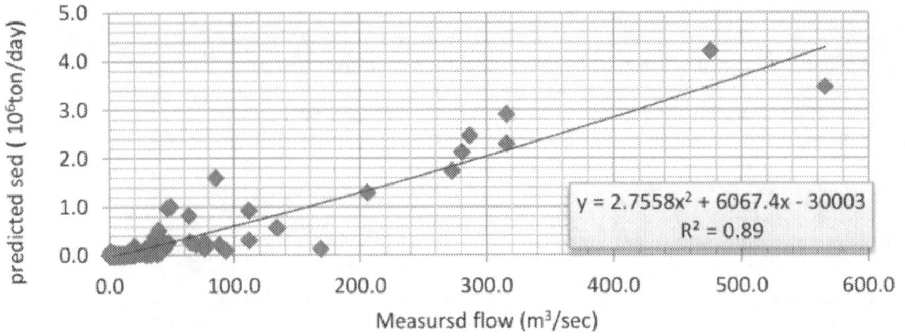


(b)

**Fig. 7**. Correlation between the measured flow and predicted suspended sediment load using ANN model Mode2-Case2 for period from 1 Oct. 1984 to 31 Dec. 1985; (a): Daily time series, (b): Regression model.

## 4 Conclusion

ANN model was training and validated with two modes each mode have two cases. The best performance for the model where the higher value of $R^2$ and NS with lower value of RSR was when applying supervised sampling and using precipitation and flow as input variables and sediment as a target variable. Consequently, ANN model is sensitive to the type and number of input variable. In addition, the predicted daily sediment data was plotted against daily measured flow. The correlation between predicted sediment and measured flow was in good agreement with $R^2$ =0.89 and the best relation was polynomial equation from second degree, it can be concluded the relationship between suspended sediment and flow mostly be nonlinear.

**Acknowledgment.** The authors would like to thank Ministry of Water Resources/ National Center for Water Resources Management in Iraq and to the staff of Building and Construction Engineering Department / Water and Hydraulic Structures Engineering Branch for their efforts to achieve this work.

1. R. J Abrahart and S. M. White, Hydrology, Oceans and Atmosphere **26.1,** 19-24 (2001).
2. N. Al-Ansari A., F. K Al-Shami, H. A Hassan and S. Ali, H, 5th Iraqi Geological Congress (1978).
3. NA Al-Ansari, A Sayfy, GT Al-Sinawi, AA Ovanessian, Water Res, 5.2, 173-187 **(1986)**.
4. A. Al-Kadhimi, M., L. Ahmed, and R. Al-Mphergee, Jordan Journal of Civil Engineering **5.2,** 229-244, (2011).
5. M. Alp, and H. Kerem Cigizoglu, Environmental Modelling & Software **22.1**, 2-13. (2007).
6. Alp, M., and Cigizoglu, H. K, Environmental Modelling & Software, 22(1), 2-13 (2007).
7. Z. Boukhrissa, A., et al., Journal of earth system science **122.5**, 1303-1312 (2013).
8. J. Córdova. R., and M. Gonzalez, Soil technology **11.1**, 57-65 (1997).
9. H. CIĞIZOĞLU KEREM, Turkish Journal of Engineering and Environmental Sciences **26.1**, 16-26 (2001).
10. J. De Vente, , and P. Jean, Earth-Science Reviews **71.1**, 95-125 (2005).
11. R. I. Ferguson, Water Resources Research **22.1**, 74-76 (1986).
12. D. R. Fuka, et al., Hydrological Processes **28.22**, 5613-5623 (2014).
13. J. E. Nash and J. V. Sutcliffe, Journal of hydrology, **10(3)**, 282-290 (1970).
14. National Centers for Environmental Prediction (NCEP), ( http://globelweather.tamu.edu/)
15. M. Shabani and N. Shabani, Australian Journal of Basic and Applied Sciences, **6(12)**, 157-164 (2012).
16. R. M. Wambua, B. M. Mutua, and J. M. Raude, American Journal of Water Resources, **4(2)**, 35-43 (2016).
17. Z. X. Xu, J. P. Pang, C. M. Liu and J. Y. Li., Hydrological Processes, 23(25), 3619-3630 (2009).
18. Y. Zhu, X. X. Lu, and Z. Yue, Geomorphology **84, no. 1**, 111-125 (2007).

# Accuracy evaluation of 2D geometric correction models: Iraq as a case study

*Abdul Razzak* Ziboon [1,*], *Israa* Mohammed [1]

[1] Building and Construction Engineering Department, University of Technology, Baghdad, Iraq

**Abstract.** One of the most important preprocessing methods for remote sensing data and geometrical distrortion is the geometric correction. In this research, several mathematical models were used in two dimensions case and were compared to each other to determine the accuracy of each mathematical model when used in three different regions with different terrains. Three high-resolution satellite images of the QuickBird satellite (a flat area, a hilly area and a mountain area) of Iraq have been used in this work. The flat area is chosen in Baghdad, while the hilly and mountain areas are chosen in Irbil in the north of Iraq. In this research, the mathematical models used are the first and second order polynomials, as well as the projective transform. All of these models were applied to all different topographic areas and their accuracy was assessed based on the Matlab program. The results of the models in the three areas studied indicate that the best precision is achieved with the second order polynomial model, while the worst precision is obtained with the first order polynomial model. On the other hand, the precision of the projective transformation is almost similar to the precision of the first order polynomial.

## 1 Introduction

Remote sensing data is collected using several techniques and extensions. For all these systems there are errors associated with the process of data collection. Therefore, preprocessing of the remote sensing data must be done prior to the beginning of the analysis process to eliminate these errors.
The process of image restoration is usually applied to remove distortion, deformation and reduction of noise that are associated with image processing. The image restoration process produces a corrected image that matches as far as possible with the original scene from geometrical and radiological perspective [1].

To correct remote sensing information, internal and external distortions should be considered. Internal distortions are errors that occur within the sensor. These are usually considered as systematic errors and can usually be removed using a mathematical calibration model.
External distortions occur as a result of deviation of the sensing system, changes in the atmosphere and terrain of the earth. These errors are nonsystematic errors and can be eliminated by using control points that appear in the imaged scenes [2].

## 2 Geometric correction models

Geometric correction models are usually simple and do not need to know the components of the sensor physics. In this sense, these mathematical models must have the ability to link the image coordinates with the ground coordinates [3].
In this paper, several two dimensional models were used for transformation combined with a number of ground control points. Thes\e models are usually available in many image processing programs.
The mathematical models used in this paper are:
- 2D Polynomial Model.
- Projective Transformation Model.

### 2.1 (2D) Polynomial Model

Polynomial models are usually used to transform between two different coordinate systems. The transformation order depends mainly on the distortion of the image, the number of control points and the terrain.
Usually corrections are only first order or second order. Unwanted results from high-order corrections are the result of the use of complex transformation equations. For the purpose of fitting all ground control points there will be significant distortion in the transformed image.
Because the two-dimensional polynomial model does not take into consideration the elevations of ground control points, these mathematical models are more suitable for the flat imaged areas, ie when the image is not affected by terrain. It is usually used in images that have few or small distortions [4]. The following functions represent the general form of the two-dimensional polynomial transformation:
- 1st order polynomial.

* Corresponding author: razzak1956 @yahoo.com

© The Authors, published by EDP Sciences. This is an open access article distributed under the terms of the Creative Commons Attribution License 4.0 (http://creativecommons.org/licenses/by/4.0/).

$$x=a+bX+bY \tag{1}$$

$$y=d+eX+fY \tag{2}$$

- 2[nd] order polynomial.

$$x=a+bX+cY+dXY+eX^2+fY^2 \tag{3}$$

$$y=g+hX+iY+jXY+kX^2+lY^2 \tag{4}$$

Where (x, y) represents the image coordinates of the ground control points of the original input image, while (X,Y) represents the ground coordinates of the ground control points that are observed using GPS and (a, b, c to l) represent the transformation parameters [5].

### 2.2 Projective transformation Model

$$x = (a + bX + cY)/(gX + hY + 1) \tag{5}$$

$$y = (d + eX + fY)/(gX + hY + 1) \tag{6}$$

The above equations are called projective transformation equations. To every point x, y in one plane that represented by image coordinates of the ground control points of the original input image there corresponds a point (X, Y) in the other plane represented by the ground coordinates of the ground control points that are observed using GPS, provided that the common denominator is not zero [6]. If the denominator becomes zero then the corresponding point moves to infinity. It is called vanishing point. The (a, b, c to h) are the eight transformation parameters [7].

## 3 Study areas

### 3.1 Study area (1)

The first study area is chosen in Baghdad city in the middle of Iraq. The area has an elevation range of between (32-47 m), and can be considered as a flat area. Ground control points and check points were distributed as shown in Figures (1) and (2) respectively.

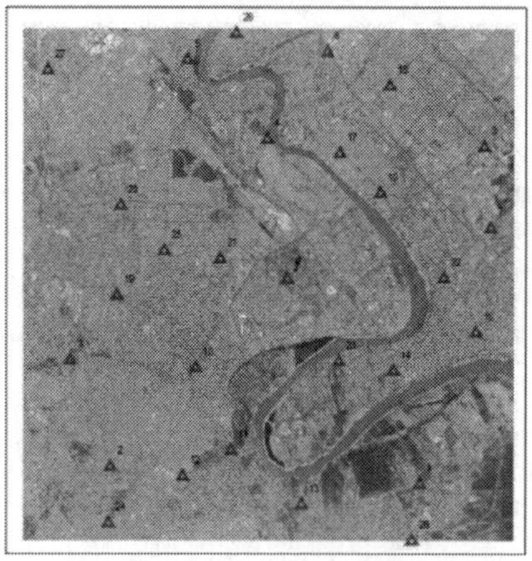

**Fig. 1.** The first study area with GCPs distribution

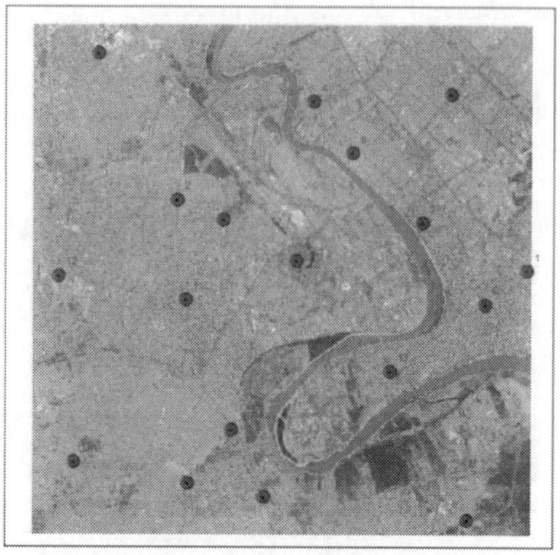

**Fig. 2.** The first study area with CPs distribution

### 3.2 Study area (2)

The second study area is chosen in the center of Irbil city at the north of Iraq. This area has an elevation range of between (377-528 m), and can be considered as hilly area. Ground control points and check points were distributed as shown in Figures (3) and (4) respectively.

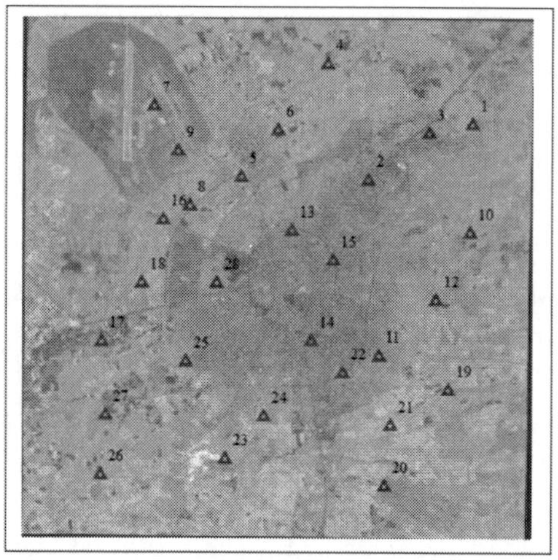

**Fig. 3.** The second study area with GCPs distribution

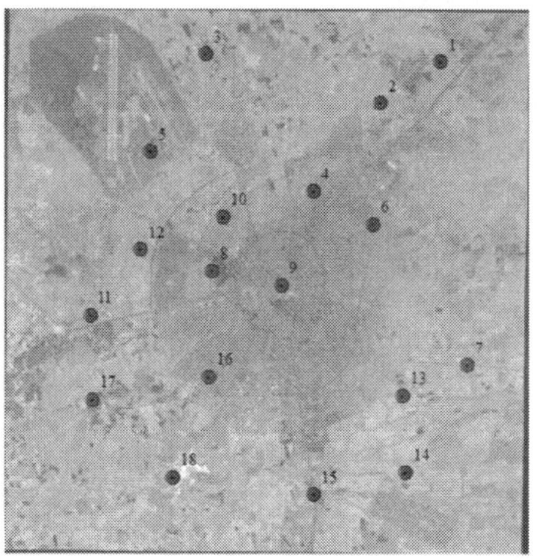

**Fig. 4.** The second study area with CPs distribution

### 3.3 Study area (3)

The third study area is chosen in Soran, Irbil city at the north of Iraq. The area has an elevation range of between (1450-3150 m), and can be considered as mountain area. Ground control points and check points were distributed as shown in Figures (5) and (6) respectively.

**Fig. 5.** The third study area with GCPs distribution

**Fig. 6.** The third study area with CPs distribution

## 4 Results and discussion

In this paper, different geometric correction mathematical models are applied using three high resolution satellite QuickBird images (panchromatic 0.6 m in spatial resolution) for three different terrain areas (flat area, hilly area and mountain area) in Iraq. All the geometric models which are used, utilized the ground control points GCPs in order to determine the transformation parameters between the image and the ground. A total number of (46) GCPs were selected, well distributed over each of the three study areas, (28) points were used as control points and the rest of them were considered as check points. The ground coordinates of all GCPs were collected through the DGPS, type (Leica GPS SR20) with the accuracy of 10mm + 2ppm (Rapid Static). The selection of GCPs targets is as accurate as possible. The (TRMSE) for GCPs and check points have been calculated for all models in order to find the best model. All these models are evaluated through the Matlab environment facilities [8].

The geodetic parameters used in DGPS coordinates characteristics and the images information are given in Table 1, this information was used in all methods.

**Table 1**. The used geodetic parameters and images information

| Parameter | QuickBird (1) | QuickBird (2) | QuickBird (3) |
|---|---|---|---|
| Image type | panchromatic | panchromatic | panchromatic |
| Spatial Resolution | 0.6m | 0.6m | 0.6m |
| Map Projection | UTM | UTM | UTM |
| Datum | WGS 84 | WGS 84 | WGS 84 |
| Zone Number | 38 | 38 | 38 |
| Acquisition Date | 2008 | 2008 | 2008 |
| Measurement Method | DGPS | DGPS | DGPS |

The summary of the results and TRMSE conclusion using the three models for the three study areas can be illustrated in Table 2.
The results of this table show that the 2nd Order 2D Polynomial model is the best model because the best accuracy is achieved with this model. This is due to the nonlinear curve fitting of this model for GCPs distribution. These results ensure that the higher the order of the function, the better is the fit to the GCPs. Also transformation of the 2nd order is nonlinear transformation that can be used to fit the exact location of point [8].

On the other hand, the worst accuracy is obtained with 1st order 2D polynomial model and the accuracy of projective transformation model is nearly similar to the accuracy of 1st order 2D polynomial model.

**Table 2**. The summary of results for the three study areas

| Model | TRMSE (pixel) Flat Area | | TRMSE (pixel) Hilly Area | | TRMSE (pixel) Mountain Area | |
|---|---|---|---|---|---|---|
| | Cont | Chck | Cont | Chck | Cont | Chck |
| 1st Order 2D Polynomial | 0.6042 | 0.6494 | 0.6914 | 0.7528 | 0.8003 | 0.8508 |
| 2nd Order 2D Polynomial | 0.5433 | 0.6153 | 0.6436 | 0.7025 | 0.7113 | 0.7875 |
| Projective Transformation | 0.5878 | 0.6337 | 0.6787 | 0.7206 | 0.7867 | 0.8467 |

## 5 Conclusions

• The results of 2D models in the three studied areas indicate that the best accuracy is obtained with the second order polynomial model, while the worst accuracy is obtained with the 1st order polynomial model. On the other hand, the accuracy of projective transformation is nearly similar to the accuracy of 1st order polynomial. Therefore, in case of using 2D models, the 2nd order polynomial is recommended to achieve high geometric accuracy.
• Due to the best fitting of nonlinear curve to GCPs distribution, the precision of second order polynomial transformation model is better than that of first order polynomial transformation model.
• In the results of 2D polynomial and projection transformation models, a linear increase of TRMSE is found, passing from a flat area to more rugged terrains, reaching a maximum value of about (0.8 pixel) with a polynomial of first order in the mountain area. Therefore, the 2D models are recommended only in case of flat areas, but they failed in the case of more rugged terrains.
• In the process of calculating the coefficients of mathematical models by least square adjustment solution, the results of Gauss elimination method is more accurate than the results of inverse method especially in case of complex models such as 2nd order 2D polynomial. This is due to the more complicated equation and huge matrices of these models that are required in calculating these coefficients.

## References

1. John R. J. *Introductory Digital Image Processing* (Pearson, 2016)
2. T Toutin. IEEE, **40**, 2241, (2002).
3. Ali O. *Accuracy Assessment of the DEM and Orthoimage Generated from ASTER* (METU, 2005)..
4. C. V. Tao,Y. Hu, PE&RS, **67**, 1347, (2001).
5. M. Hosseini J. Amini, ISPRS H. W., (2005).
6. F. Samadzadegan, A. Milanlak, M. Majdabadi *In XXXVI ISPRS* (2005).
7. Edward M. M., James S. B. *Introduction to Modern Photogrammetry* (Wiley, 2001).
8. Ali, B. H. *Enhancing The Accuracy of Geometric Correction of High Resolution Satellite Images* (UoB, 2006).

# Investigation of surface area of lakes and marshes from satellite images by using remote sensing and geographic information system integration in Iraq

*Alaa* Dawood[1,*], *Yousif* Kalaf [1], *Nagham* Abdulateef [1], and *Mohammed* Falih[1]
[1] College of Engineering, Baghdad University, Baghdad, Iraq.

**Abstract.** Water level and distribution is very essential in almost all life aspects. Natural and artificial lakes represent a large percentage of these water bodies in Iraq. In this research the changes in water levels are observed by calculating the areas of five different lakes in five different regions and two different marshes in two different regions of the country, in a period of 12 years (2001 - 2012), archived remotely sensed images were used to determine surface areas around lakes and marshes in Iraq for the chosen years . Level of the lakes corresponding to satellite determined surface areas were retrieved from remotely sensed data .These data were collected to give explanations on lake level and surface area fluctuations. It is important to determine these areas at different water levels to know areas which are being flooded in addition to the total area inundated .The behavior of hydrological regime of these lakes during the period was assessed using an integration of remote sensing and GIS techniques which found that the total surface area of the lakes had diminished and their water volumes reduced. The study further revealed that the levels of the lakes surfaces had lowered through these years.

## 1 Introduction

Lakes and marshes are water-filled areas of variable size, in a basin surrounded by land, apart from any river or other outlet that feeds or drains the lake. Lakes are located on land and not part of the ocean, and are therefore distinct from lakes, which are also larger and deeper than ponds. Docks and swamps can be contrasted with rivers or streams, which usually flow. However, most lakes are fed and drained by rivers. Natural lakes are generally found in mountain areas, fault zones and areas where there is persistent ice. [1][2] The late developments in this field and in satellite image processing have been very helpful in getting a lower coast and a more accurate data. Some of these applications are capable of estimating the percentage of moisture in the soil, the classifications of ground cover, and determining the land use and soil types. It is also possible by the use of satellite images to locate the most threatened areas by floods, this enables us to take the needed measurements to avoid such disasters by giving us the privilege of locating the needed dams to build according to specific standards. Also these remote sensing applications could be very useful in observing rivers flow. [2]

This enables us suggest the appropriate locations and designs for the future water controlling projects to be able to store the excessive flood water and use it back again in dry seasons . For a long time the water issue has been considered as one of the most important components to keep the environmental system in balance to provide all the human needs , especially in dry and semi dry climate countries , such as Iraq , where water is considered one of the most important natural resources , because it plays a great part in controlling population distribution and their economic activities especially in agriculture , so it has an important part in providing the needed food supply for the people . Also many people consider water supplies and their resources in the country as a matter of national security that may maintains the security of the country when water is available.

These reasons and several more show the real importance of water in keeping and maintain the water amount and distribution according to the needs. There for many elaborated and detailed studies should take place for further knowledge to the environmental and natural changes that have occurred and affected water lands during the past decades.[7] Now days in Iraq has developed a good number of water controlling projects, that can be used to regulate the water distribution for several uses so constant and detailed studies on water areas are required . In our research we have studied five different lakes in five different places through Iraq, we have calculated the areas of these lakes through and we have studied two different marshes in two different places through Iraq. In this study the areas of lakes and marshes were calculated by using integration of Geographic Information Systems (GIS). [5]

* Corresponding author: alnurshedi.alaa@gmail.com

© The Authors, published by EDP Sciences. This is an open access article distributed under the terms of the Creative Commons Attribution License 4.0 (http://creativecommons.org/licenses/by/4.0/).

GIS has grown into a powerful tool that can calculate and integrates driven types of spatial data and perform a variety of spatial analysis. This evolution has been driven by significant advances with the developed computer technology. GISs and environmental models function with a board spectrum of geospatial data that are used for diver's applications and spatial analyses at different scales .data are collected through observation, measurement, and inference. The examination and organization of data into a useful form produces information which then enables appropriate analysis and modelling [5]. The main objective of this paper is to estimate the rate and the extent of water loss from Iraqi lakes during the period 2001– 2012. In addition, changes in the aerial extent and the volume of water within these lakes and marshes are further discussed and their potential environmental impacts are assessed.

## 2 Water resources in Iraq

Water bodies are very essential, they are the main working artery in life to keep the environmental system in balance especially in dry and semi dry climates. Especially during the second half of the past century due to four main reasons:

1. The largely decreased amount of falling water in the last decades in our planet affecting our country too, which led to cause scarcity of water in a lot of areas in Iraq. [8]

2. Large areas of Marshes , which represent half of the water surfaces in Iraq and occupy an area of 8900 $km^2$, were dried for war purposes during the 1990 of the past century , the drying process of these marshland considered as an environmental and a social disaster . The southern regions of Iraq are still suffering from the devastating effects of this process. Although a lot of efforts were paid to refresh these marshes by filling them with water. [8]

3. The Iraqi politics and the bad planning regarding plans and future projects to provide the needed amount of water for the country , water projects in Iraq started in 1913 with Al-hindya dam , and then a few more dams were built in 1930 . The main purposes of these projects were, to maintain the water level in Iraqi rivers, prevent seasonal floods, irrigation and generating electricity. The development of life and the growing water needs created new goals for these water projects that were also used for water storage .That required the building of larger water reservoirs that could be used for storing water and then use it back again when needed. [8]

4. Many lakes and rivers were affected by the dams and irrigation projects that were built at the headwaters in neighboring countries such as Turkey, Iran and Syria. [8] Figure (1) shows the expected world map and the water distribution in year 2015 according to a special studies made by the UN. As it shown figure (1), most of the Arabic countries are facing large threats according to their water supplies, most of the Arabic countries, Pakistan and Iran are lying in the very high stress area. The total area of Iraq is 437,072 $km^2$, the land area is 432,162 $km^2$ while the water area is 4,910 $km^2$ .The continuous decreasing of water resources in Iraq has considered one of the major problems that affecting the environmental system in Iraq.

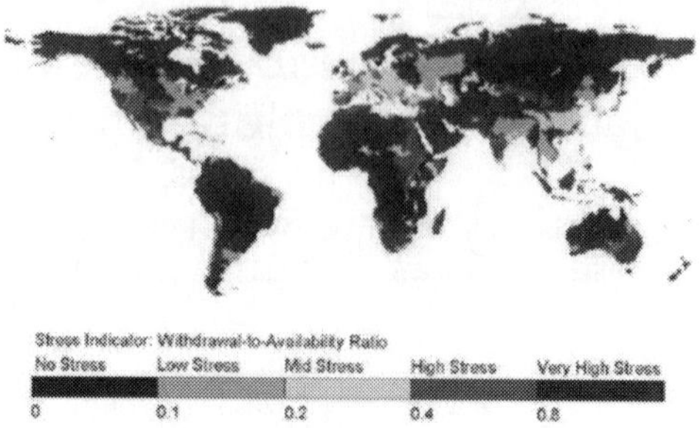


**Fig. 1.** .Expected world map and water distribution in 2015

## 3 Methodology and results

The remote sensing data that were used in this research are Images from Landsat satellite (7, TM, ETM+, and MSS) and (ArcGIS9.3). Digitizing creates a polygon that represents the water body on the satellite image by outlining each water body in a digital image using a mouse cursor. This new polygon can then be used to estimate surface water area and can be compared to all previous and subsequent images The disadvantage to this is the time required to trace all of the pond shorelines in each region for each time frame. The advantage is an accurate representation of the water bodies in an image. [10]

The purpose of the paper is to study the fluctuations of the surface areas of lakes and marshes in Iraq. In order to delineate the boundaries of the lakes and marshes , satellite images was used , Land sat archive images, for the period (2001-2012) for lakes and the period (2005-2012) for marshes , were used as historical records for the situation of each lake and marsh during this periods .So we started by choosing five of the most important lakes of Iraq, (Al-Tharthar , Al-Habanyah ,Derbendekhan, Dukan and Hemrin ), and choosing two of the most important marshes of Iraq, (Al-Hammar, Al-Hawayzah) these lakes and marshes were chosen due to their important role in regulating the water resources in Iraq and for their distributed locations in the center, northern and southern of Iraq to create a sample indicating how the water bodies have changed across different places of Iraq . After locating these seven water bodies we started collecting satellite images for every one of the mover the past 12 years since 2001 to 2012 for lakes and the past 8 years since 2005-2012. We have collected a single satellite image for each year for area measuring.

Those images were all taken at the same time of the year to maintain the accuracy of the comparison and the research. We achieved this research by the use of the GIS technology as a remote sensing instrument, which enabled us to work on large scale areas, and provided the needed accuracy for the work. We used the collected satellite images of the lakes and marshes from the past 12 years and 8 years respectively to draw virtual lines to determine the borders of the lake and marshes in each year. These drawn borders enabled us to calculate the areas of the lakes and marshes for each time period a year and compare these areas with the areas calculated from other satellite images at earlier times. After calculating the areas for each lake we drew a diagram for every lake showing the changes in areas according to their years. In these diagrams the horizontal lines represented the year of the measured area while the vertical lines represented the areas in square kilometers. Figure (2), shows the studied lakes' distribution in Iraq.

## 4 Data integration of GIS

The GIS allows linking or merging of information that is difficult to connect by any other means. Thus, GIS can use sets of variables assigned to construct and analyze new variables. For example, using GIS technology, it is possible to combine agricultural records with hydrographic data to determine which streams will bear certain levels of fertilizer runoff. [3], [4] Agricultural records can indicate the amount of pesticides applied to a plot of land. [11] By identifying these parcels and intersecting them with streams, GIS can be used to predict the amount of runoff nutrients in each stream. Then with convergence streams, total downstream loads can be calculated as the stream enters the lake. [7]

Including artificial and natural lakes and marshes, these artificial lakes and marshes were made to store water of rivers in flood seasons and then to be used later when the water level in rivers is low. So these artificial and natural water reservoirs are very essential for water level to be controlled Figure (2), shows the studied lakes distribution in Iraq.

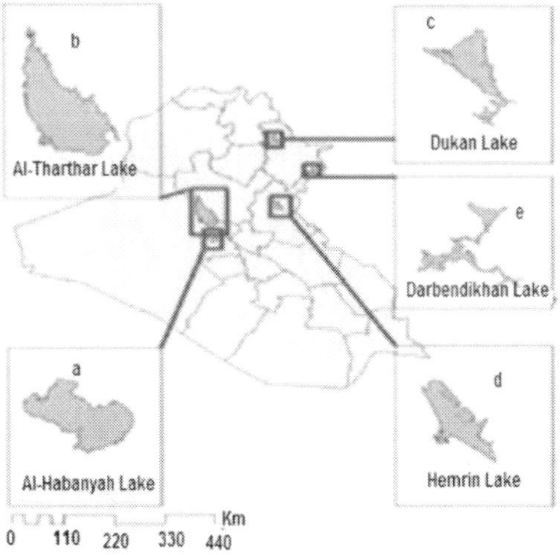


**Fig. 2.** The studied lakes distribution in Iraq.

## 5 The Lakes

### 5.1 Al-Habanyah Lake

It is a shallow natural lake and lies in a depression to the south of the Ramadi city. It faced a lot of changes in those 12 years, in 2001 the total area of the lake was (242.17) $km^2$. When in the upcoming years the lake witnessed a rate of increasing in its area to become in 2012 (292.16) $km^2$. Figure (2a), shows location of Al-Habanyah Lake in Iraq and figures (3), (4) shows satellite images of lake across these 12 years and the area of the lake in each year.

### 5.2 Al-Tharthar Lake

Lake Tharthar, lies in the center of Iraq about 120 km northwest of Baghdad. Between the Tigris and the Euphrates rivers. It is the largest lake in Iraq. The construction of several dams has made it an important source for irrigation. Its size shrinks and expands dramatically with the seasons. The lake is deep and nutrient poor. Al-Tharthar Lake is the largest natural lake in Iraq, and one of the most important ones. The collected satellite images and calculations showed that since 2001 the lake area was increasing at each year until 2003 to achieve its largest average area. In 2001 the lake was 1922.183 $km^2$ in area, and it began to increase at a rate of 21% per year from its original area in 2003, to have the area of 2357.526 $km^2$. After 2003 the lake witnessed an obvious decreases in its area at a rate of 1.8 per year, to have a 1688.47 $km^2$ of area in 2012. Figure (2b), shows location of Al- Tharthar Lake in Iraq and figures (5), (6) shows satellite images of lake across these years and the area of the lake in each year.

### 5.3 Hemrin Lake

Lake Hamrin, also known as the Diyala dam, is a man-made lake about 50 km northeast of Baquba, in the Diyala province of Iraq. It is a source of fish and also provides water to nearby palm groves and other farms, it was reported that due to the Iranian dam of the Lund River, the lake lost about 80 % of its capacity [11]

The satellite images and areal calculations showed a lot of disparity in total area of the lake from one year to the next one, some of the years witnessed an obvious increasing in water level while in other ones the area decreased significantly .The Lake reached its highest level in 2003 when the surface area was 230.387 $km^2$. While reaching its lowest in 2010 making only 80.421 $km^2$ as a surface area. In 2012 the lake area was 130 $km^2$. Figure (2d), shows location of Hamrin Lake in Iraq and figures (7), (8) shows satellite images of lake across these 12 years and the area of the lake in each year.

### 5.4 Dukan Lake

Dukan Lake is the largest lake in Iraqi Kurdištan. It is located near the city of Rania, and the reservoir on the small Zab created through the construction of Dukan dam multi-purpose dam to provide water storage, irrigation and hydropower. [9] The surface area of the lake is 270 $km^2$. As such, the drainage basin of the Dukan Dam is 1700 $km^2$. [6][8]

Dukan Lake also witnessed a less disparity than Hemrin Lake, and decreases in the surface area over the years. The spatial calculations showed that the lake has reached the maximum area of 361.585 $km^2$ in 2002, while reaching its minimum area with only 60.3 $km^2$ in 2008. Figure (2c), shows location of Dukan Lake in Iraq and figures (9), (10) shows satellite images of lake across these 12 years and the area of the lake in each year.

### 5.5 Darbendikhan Lake

The satellite images for this lake showed that the surface area in 2001 was 50.1234 $km^2$, over the following years the lake levels witnessed a lot of ups and downs, and the lake has reached its highest levels in 2004 scoring 95.45 $km^2$. The rate of increasing in 2001 was 82% from its previous area in 2004 to have 95.45 $km^2$ as a surface area, while in 2012 the lake has reached its lowest levels having only 41.546 $km^2$. Figure (2e), shows location of Darbendikhan Lake location in Iraq and figures (11), (12) shows satellite images of lake across these 12 years and the area of the lake in each year.

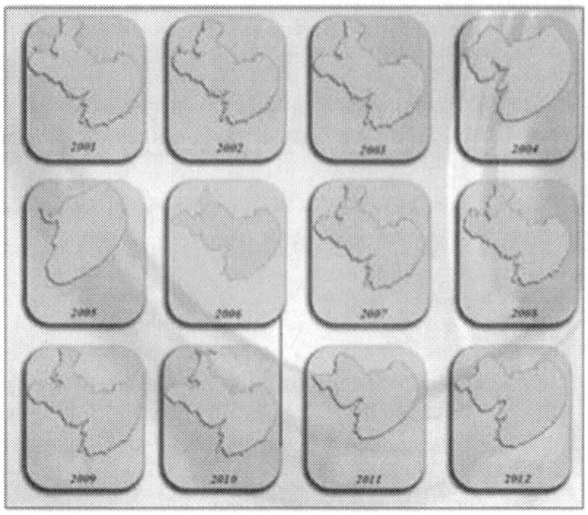

**Fig. 3.** Satellite images of Al- Habanyah Lake.

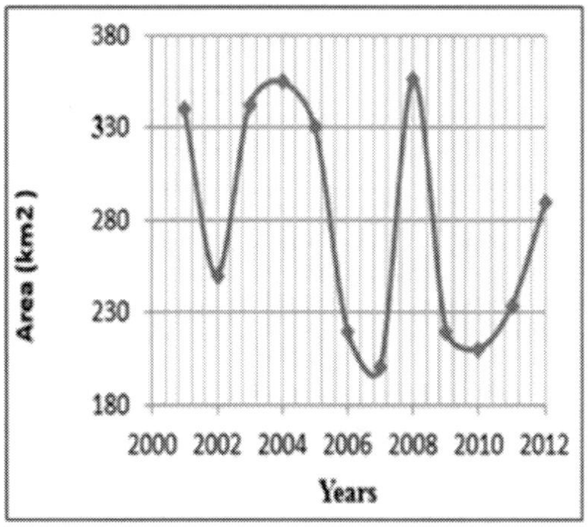


**Fig. 4.** The area of Al- Habanyah Lake.

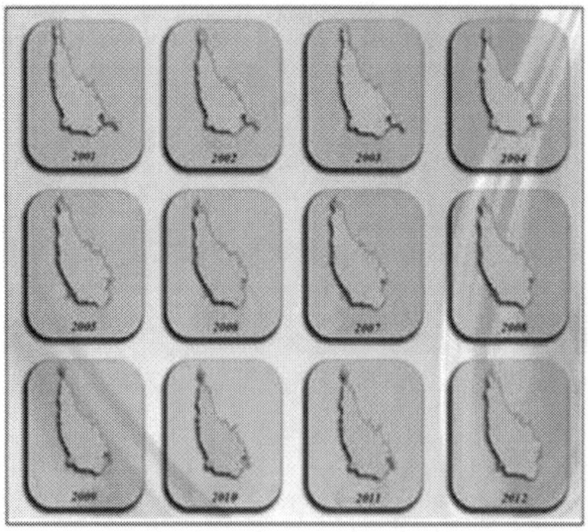

**Fig. 5.** Satellite images of Al- Tharthar Lake.

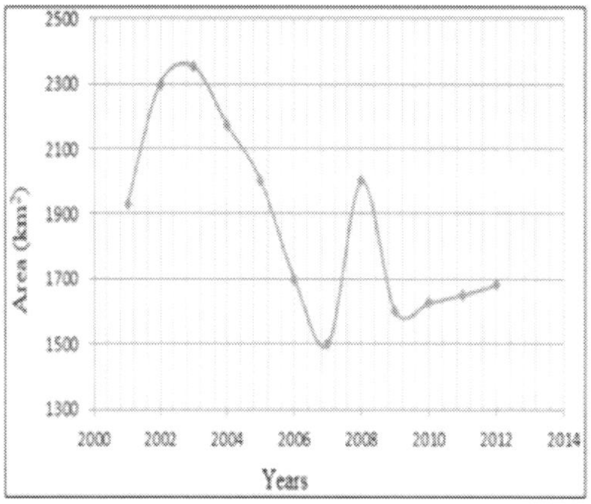


**Fig. 6.** The area of Al- Tharthar Lake

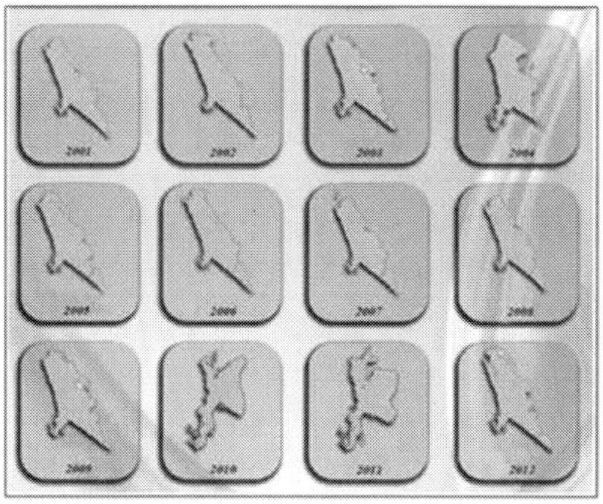

**Fig. 7.** Satellite images of Hemrin Lake.

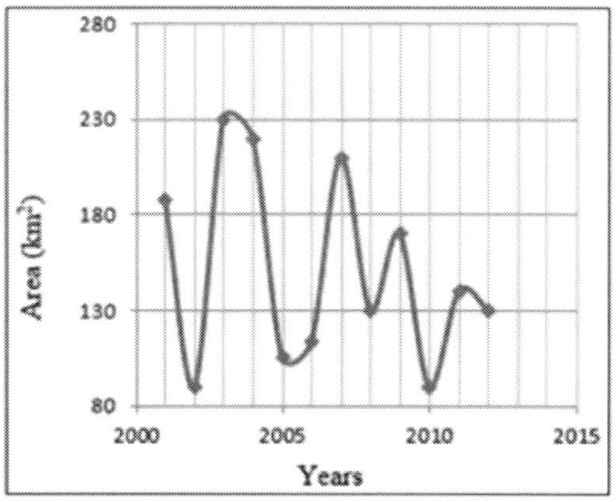


**Fig. 8.** The area of Hemrin Lake.

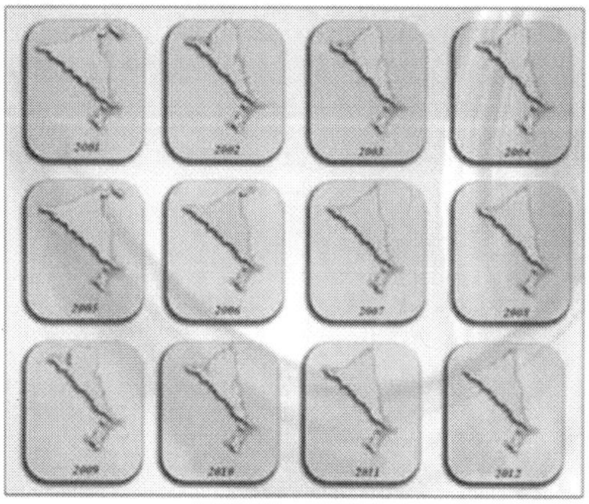

**Fig. 9.** Satellite images of Dukan Lake.

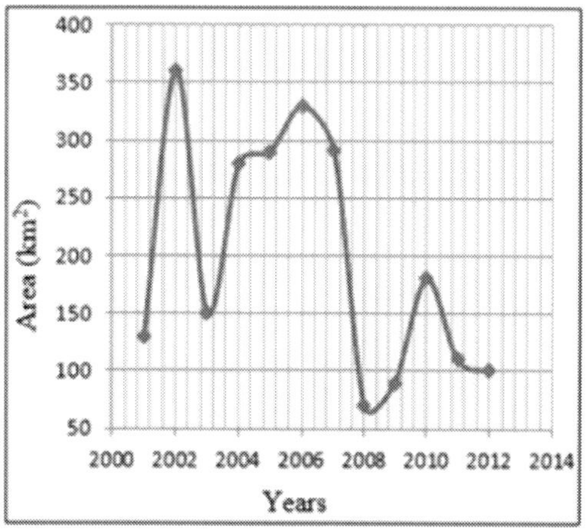


**Fig. 10.** The area of Dukan Lake.

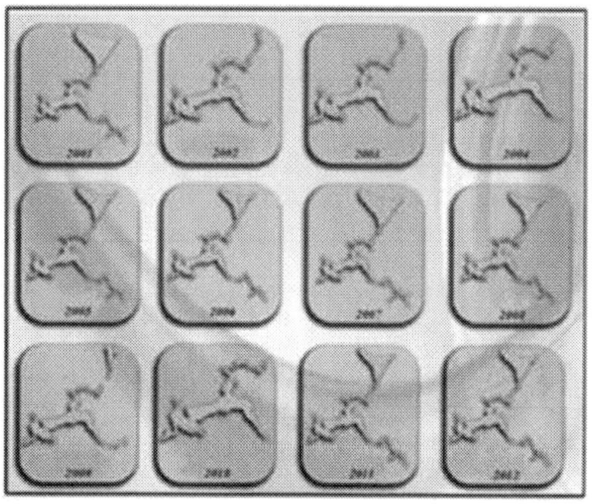

**Fig. 11.** Satellite images of Darbendikhan Lake.

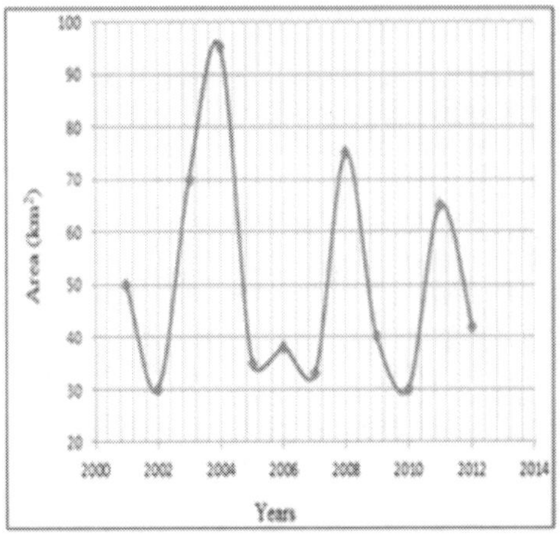


**Fig. 12.** The area of Darbendikhan Lake.

## 6 Marshes

### 6.1 Al-Hammar marsh

This marsh is located south of the Euphrates River in Iraq, and extends from Nasiriya in the west to the outskirts of Basrah at the Shatt al-Arab west. A length of 90 km and width of 25-30 km. and the surface area of the maximum of about 2800 $km^2$ in the flood season goes down to about 600 $km^2$ during the dry season and is a permanent lake this lagoon with a vast area was dried by 94% during the past fifteen years the last was during this period. The spatial calculations showed that the marsh has reached the maximum area of 1092.87 $km^2$ in 2011, while reaching its minimum area with only 611.87 $km^2$ in 2005. Figures (13), (14) shows satellite images of marsh through these 8 years and the area of the marsh in each year.

### 6.2 Al-Hawizeh marsh

The swamps are located in the east of the Tigris and lie in Iran. The Iranian side of the marshes, known as the Great Hor, feeds on the Karkh River, while the Tigris, Mashra and Khela disperse the Iraqi side, with much less water than Karkh. During spring, the Tigris River may flow directly into the swamps.. Al-Hawizeh is one of the marshes of Iraq located in the provinces of Maysan and Basrah. It is bordered by the east of Iran. The Karkh River from Iran is located on the river. 80 km and width 30 km. During the spring flow, the Tigris can flow directly into the marshes. The maximum area of the marshes reaches about 3000 $km^2$. 21% and northern and central parts of the marshes are permanent but the Southern parts become seasonal in normal state. [8]

The satellite images for this marsh showed that the surface area in 2005 was 1600 $km^2$, over the following years the marsh levels witnessed a lot of ups and downs, and the lake has reached its highest levels in 2006 scoring 1671.45 $km^2$. The rate of decreasing to 2001 was 45% from its previous area in 2010 to have 1010 $km^2$ as a surface area, while in 2011and 2012 the lake has increased levels having only 1100.54 and 1190.67 $km^2$ respectively. Figures (15), (16) shows satellite images of marsh through these 8 years and the area of the marsh in each year.

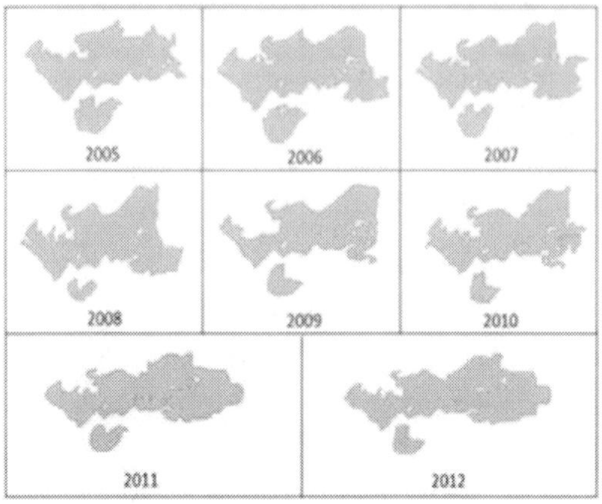


**Fig. 13.** Satellite images of Al-Hammar marsh.

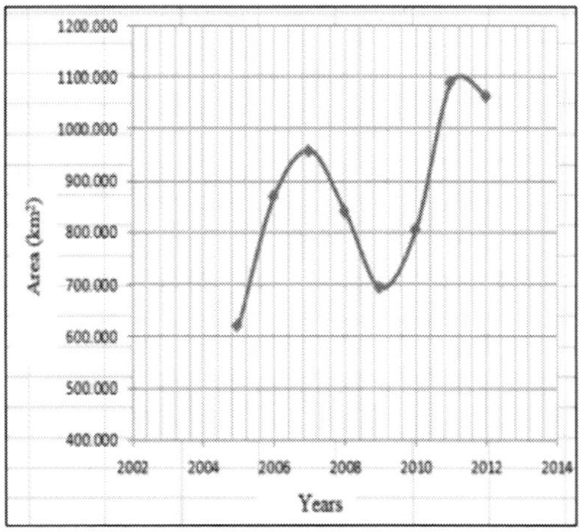


**Fig. 14.** The area of Al-Hammar marsh.

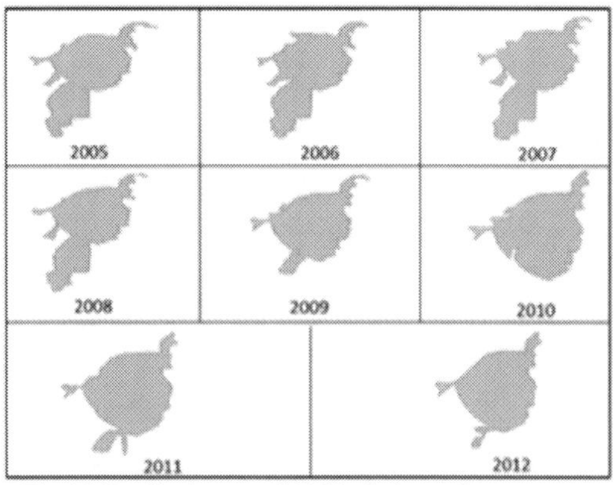


**Fig. 15.** Satellite images of Al-Hawizeh marsh.

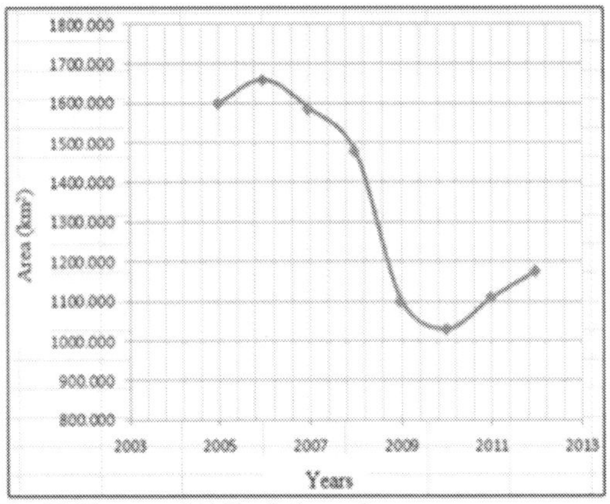


**Fig. 16.** The area of Al-Hawizeh marsh.

## 7 Conclusions

Remote sensing and demographic analysis in order to assess recent surface hydrological and environmental changes in lakes and marshes in Iraq, and concluded that:

1. Due to its location in a highly arid zone, Iraqi lakes are exposed to high evaporation rates and result in an annual loss of about 2.5 m from their water column. Only 10% of this amount is recharged to the aquifers under the lakes.
2. In this research we recommend that we take the measures necessary to make the most of the huge amount of water resources before you lose completely through evaporation.
3. Lakes and marshes in Iraq are the main surface water resources, and given the importance of these lakes and marshes, we conducted a study to determine the changes that have occurred to the areas of these lakes during the past 12 years for lakes and the past 8 years for marshes. These changes are usually influenced by political, economic and climatic conditions.
4. GIS techniques were applied to achieve this study and were calculated for each polygon and compared to the corresponding lakes and marshes of the images at later times. The results indicate that the total area of these five lakes in 2012 decreased by 15.567% of the total area in 2001. And the results indicate that the total area of Al hammar marsh 2012 increased by 85% of the total area in 2005, but the results indicate that the total area of Al Hawizeh marsh 2012 decreased by 34% of the total area in 2005.
5. During these 12 years of our research (2001 to 2012), there was a marked disparity in the lake areas, and in some years the regions were in an increase while in other regions. In general, 2008 recorded the minimum surface area of these lakes in an area of only 1795.1567 $km^2$ while the highest level was in 2003 when the total area was 3455.506 km $^2$.

6. During these 12 years of our research (2001 to 2012), there was a marked disparity in the marshes areas, and in some years the regions were in an increase while in other regions. In general, 2011 recorded the maximum surface area of these marshes in an area of only 2200.1435 km2 while the lowest level was in 2009 when the total area was 1810.436 $km^2$

## References

1. R.G. Bryant, M.P. Rainey , JORSE **12**, 360 (2002)
2. C. Castaneda, J. Herrero, M.A. Castared JAE **63**, 497 (2005)
3. Chang K.T. *Introduction to Geographic Information System* (3d .Ed. McGraw Hill international Edition, 2006)
4. Demers M. *Geographic Information System Modeling in Raster* (New York, Wiley, 2002)
5. ESRI, *Arc GIS 9.3 Help Manual, Environmental System Research Institute Inc.* (Redlands, Californie 2008)
6. A.K. Fink, I.D. Ostrizhnov, PTE **17** ,10 (1984),
7. R.H. French, J.J. Miller, C. Dettling, J.R. Carr, JH **81**, 67 (2006)
8. Albertine J. *Dukan Dam*, (Encyclopédie of the modern Middle East. New York, 1996)
9. Konecny G. *Geo information Remote sensing, Photogrammetric and Geographic Information System*, (Taylor & Francis, London and New York 2003) ngley P.A. Goodchild M.F. Maguire D.J. Rhind D.W., *Geographical Information Systems and Science*, (England, John Wiley & Sons Ltd, USA, 2005)
10. Roaf M. *The Hamrin Sites*, (BSA, London 1992)

MATEC Web of Conferences **162**, 03017 (2018) https://doi.org/10.1051/matecconf/201816203017
*BCEE3-2017*

# Utilization of urban sustainability approaches to deal with ancient cities: a case study of Hit Castle, Iraq

*Thaer* Mahmood[1,*]

[1]College of Engineering, University of Anbar, Ramadi, Iraq

**Abstract.** This research aims to highlight the current situation of some Iraqi ancient cities by taking the Hit Castle as a case study. It is one of the oldest cities in Iraq. Due to the previous circumstances that Iraq has one through, this city was ignored and not looked after. Furthermore, due to the expensive land in this city, many old houses were demolished and rebuilt in a modern style. In order to conserve this ancient city, the urban sustainable approaches are suggested to be used. For this purpose, a study was made and a site observation was carried out. Based on that, it was found that this city could be redeveloped/ rehabilitated by employing the urban sustainability approaches and concepts. Important recommendations have been identified and the governmental bodies will be invited to conserve the ancient cities using these recommendations as guidelines.

## 1 Introduction

The ancient and historical sites are comprise of noteworthy number of heritage sites and thus it necessitates further efforts for conservation and redevelopment for future generations [19]. Globally and most recently, recognizing urban legacy concerning illustrations, and urban destinations as opposed to single monuments has bee
n placed ahead by distinctive authorities, like experts, academics, professionals and global associations [12].Redevelopment and rehabilitation can yield significant improvements to the ancient cities. Indeed, they enhance the caliber of the built environment, encourage the appreciation of the city's history and culture, and support the character and community's personality [14] .Several factors influence the motivation of using the redevelopment and rehabilitation for the ancient cities including population increase, the failure of existing building stock, economic activities and the demand for higher level of living [3, 5, 7, 11].
Sustainability development resides at the nexus of philosophy, practice, politics, and engineering organization [6]. The main concept of sustainability development in cities is harmonized between three elements, namely society, environment and economy [10, 17]. Based on these three elements, a reliable and cohesive foundation, and conditions for sustainable cities can be yielded [13, 18] as reported in their research that cities sustainability development is maintained through the integration between social progress, environmental support and economic development. In another study, it was stated that maintainable advancement to urban communities includes economical progressions of population, maintainable advancement of assets, ecological markets, and economic progress of urban governance [13]. These transformations and the unpredictable
Biological framework of the urban environment, society, and economy end up in adjustments and facilitated a sensible structure, optimized organization, furthermore exceedingly successful operation [8, 15]. In the literature, cities sustainability developments were inspected by several scholars that reflect the sustainability of urban systems [1, 2, 4].
The planning was compatible with the peoples' mentality in the far time. Their view was simple and they were satisfied with their situation in terms of their houses and environment regardless of whether it was right or not. The urban and technical evolution had directed the humans thinking to restudy and re-plan their lives to be fitted with this development in terms of the future expansions and considering the environmental circumstances in the design of their cities [6]. In order to plan the ancient cities to be exist, historical studies should be carried out for the buildings and residents. These cities should be conserved and recovered within the general frame of the new cities and they should be treated as a live body that interacts with the different residents activities. When an ancient city is to be replanted, the following considerations should be taken into account:
Stop the deterioration and disappearance of the ancient city by repairing and maintaining the existing buildings.

* Corresponding author: drthaersh@gmail.com

© The Authors, published by EDP Sciences. This is an open access article distributed under the terms of the Creative Commons Attribution License 4.0 (http://creativecommons.org/licenses/by/4.0/).

Organize the land use by defining each use in a beneficial manner by determining the population density, the building height, and legislative laws that aid in conserving the buildings in the ancient cities.
Reroute the vehicle traffic outside the ancient cities and restrict the car traffic for only necessary uses.
Increase the urban awareness for these city residents about the importance of these unique environments.
The recognition of the ancient buildings as a heritage value that should be conserve them has started in some European countries early the last century particularly after the world wars [4]. This recognition represents by renew and development of the ancient cities as they represent the history and civilization of any country. Later on during the second half of the last century, the developing countries have stated to be interested in renewing the ancient cities that they have. The studies that have been conducted in this regard have established some theoretical fundamentals meanwhile different methods for conserving the ancient cities have been concluded. Iraq is considered to be one of the richest countries all over the world in its heritage and legacy where it can be seen that there are numerous ancient cities that need to be rehabilitated and conserved.
The presented study aims to establish a reliable planning approach including redevelopment/rehabilitation of an ancient city that is older than Babylon city .The ancient Hit city (Hit Castle) is located in Anbar province, Iraq. The significance of this case study "Hit Castle" is owing to its outstanding coordinate which lies on the Euphrates River. In addition, this castle was initiated within the same period of the ancient cities of the Mesopotamia region. Hence, applying the proposed approach is highly recommended for the same heritage monuments that initiated within the same period. The study focuses on the understanding the reality of the situation in Hit Castle and find out the possibility of applying the suitable policy of urban sustainability. The implementation approach for the urban sustainability and the planning indicators are related to the land use and the role of the construction materials used to identify some applicable recommendations to prevent the heritage locations from deterioration and urban and ecological collapse.

## 2 Urban sustainable plan for hit castle

When developing an urban renewal plan for an old traditional Arab city, it is necessary to preserve the implicit standards of the city that guided the planning of this city and gave it this distinctive form. These criteria depend not only on the social, economic and environmental factors, but also on the religious factors in the city since its inception. The mosque in the city (Al Farouk Mosque) is located on the side overlooking the river and not in the center of the city because the city is very old and was Jewish in the past as confirmed by most historical books as shown in the schematic.
The tabulated theoretical concepts abstracted from the state-of-the-art relevant researches [19]. In this research, the spatial analysis approach has been used to study the status of Hit Castle that represents a historical heritage city located in the Mesopotamia region. The spatial analysis approach was used owing to the following reasons:
Nowadays, a noticeable shrinkage of the urban area.
An increasing and expanding in the urban state destruction all over the study area.
Land is buzzing and misapply.
The importance of conserving the historical symbols areas that present a cultural heritage monument.
The necessity of achieving the sustainability of such urban heritage cites.
Facts and data collection were obtained from the original Civil and Department of Municipality of Hit City, and by interviewing the researchers of previous studies [9,14]Additionally, facts and data were collected based on inputs from the former residents of the study area in order to reach the proposed policies for sustainability of urban area of Hit Castle. The data were analyzed using interviews, participating style with allocative groups depending on the superiority of the stakeholders to consider the proposed redevelopment and rehabilitation of an urban sustainable development policy.
A number of scientific conferences have stressed the high desire of the population to renew their ancient cities and to preserve the cultural heritage and the urban cohesive fabric that was ingrained in the traditional Arab cities where civilization and heritage cannot be purchased, but need preservation and sustainability of the heritage remain in the cities. The development of the plan for urban renewal includes several dimensions:
2.1 Desires of the inhabitants of the city

The most important things taken by the plan of urban renewal in the development of its scientific and technical plan is to take into consideration the wishes of the population in the work of urban sustainability and maintain a unified tendency to retain the spiritual and social values derived from Islamic teachings and embody these tendencies in the urban form of the city
We move away from the cloning of modern and western architecture, The latter deviates from the desires and approaches western laws and regulations that are often destructive to the cultural and historical heritage of the city and are far from the urban sustainability plan carried out in preserving the historical and heritage areas which bear the values and history of peoples and tribes with a history of ancient life.

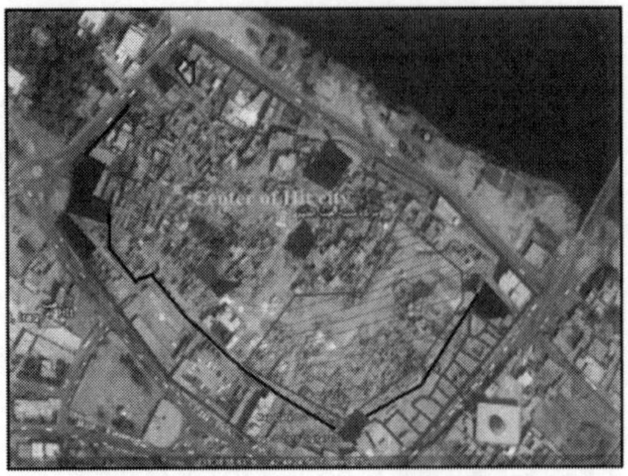

**Fig.1**. Satellite image for ancient Hit city castle

### 2.1 City size

Urban sustainability requires a limited area. Therefore, when developing an urban renewal plan, a governance must be established for the total population to avoid the urban problems resulting from overcrowding, misuse of urban spaces and distortion of the urban structure of the city, etc. The population increase can be planned within the extensions of the municipality of Hit in the east and west of the city, which draws and accommodates population growth and new families.

### 2.2 Urban structure

The clear from the foregoing that the market forces play a prominent role in the formation of a new urban structure for the city of Hit similar to the structure of other cities in the capacity of its streets and the disruption of the urban fabric of the old city and its center in Hit, The urban sustainability plan includes an urban structure derived from traditions and orientation to the social aspects supported by the spiritual values of the city .The technological progress and change in the commercial, economic and security movement in the city will be observed. In the center of the old city of Hit there will be regular elements represented by the integrity of the commercial axis of the city and parallel to the river and around the fence.

## 3 Strategic plan for urban sustainability

The aims of the urban sustainability have been previously highlighted. In addition general and planning concepts and indications about the ancient city of Hit in terms of different land uses and the required areas for these uses are specified. In order to make the studied city attractive for the residents, the planning problems and urban features should be sorted out based on the urban sustainability plan by applying the scientific and planning manners for the urban renewal [20] as demonstrated in Table 1. The approaches of urban sustainability to develop and conserve the Hit Castle can be stated by explaining the urban sustainability polices as follow:

### 3.1 Urban sustainability polices

These deal with the urban structure of the Hit ancient city which in turn represents the urban tissue and planning structure (residential regions and its aspects). In fact, ther are many planning indicatores and concepts for the urban sustainability namely:

1. Sustainable urban planning indicator, which represents the ideal distribution for the land use in the city.
2. Sustainable habitation and housing indicator.
3. Sustainable transport indicators.
4. Sustainable commercial use indicator.
5. Sustainable industrial use indicator.
6. Sustainable under and above services indicator.
7. Green zones indicator.
8. Clean energy indicator.
9. Sustainable and renewable energy indicator.
10. Social sustainability indicator.
11. Privacy in dealing with the historical centers indicator.

**Table 1**. Urban sustainable methods on the Hit castle

| Quarter | Redevelopment | Rehabilitation |
|---|---|---|
| Western door quarter | Develop and improve some of the residential, administrative, and commercial buildings. The ancient city s urban morphology should be maintained in addition to the traditional style of the buildings such as buildings orientation, shadow provision. Modern buildings can be constructed outside the castle and the link between the old and the modern building should be considered and also the orientation of buildings to be compatible with the sun and wind movements. | For a li,ited number of heritage buildings rehabilitate the residual buildings as they have an architectural value. These buildings can be reused for suitable activities to activate the social and economic life in the city center in addition to cultural role and tourism development. In order to achieve, the modifications of traditional urban tissue should be prohibited and also all the deformation should be removed. In general, the entire original designs in terms of architectural, pedestrians, and commercial, etc. should be maintained whenever possible. |

| | | |
|---|---|---|
| Southern door quarter (Al-wesata) | Modern buildings could be constructed to be similar to the old ones that face the river. Same building materials for the original ones should be used in addition to utilizing modern technologies in terms of making the front views for the new buildings to be compatible with the old ones to conserve the historical flavor. The land use should be made to construct multi-story car parks and provide enough spaces to be used for walking and having joy. | Rehabilitate the regions that were destroyed due to the last war which have a historical value in addition to repairing the city infrastructures. When new buildings are constructed they have to have the same height to conserve the street styles and the pedestrians' paths. The new rehabilitated building should maintain the compatibility between the new and old buildings. |
| Eastern door quarter | Construct new buildings while conserving the urban morphology for the traditional city and residential unites such as buildings orientation, shadow provision and constructing new modern buildings outside the ancient city taking into account the buildings orientation in terms of sun and wind direction. | Rehabilitate and repair the heritage buildings and adding new uses for the buildings that do not have heritage vale by using them as car parks and widening the main street in the western parts of the ancient city in addition to rehabilitate the infrastructures. An urban fulfill , by new building, should be made for the areas that surrounding the Sit-Nafesa Mosque and these buildings should be compatible with the existing buildings. |

## 4 Implementation approaches for urban sustainability

Conservation of urban tissue, and planning and urban structure

This can be achieved by conserving the urban tissue for the ancient city in addition to the urban congestion, at the same time conserving the unique planning structure for the ancient city center as shown in Fig 1 . It can be seen that there are impressive solutions for the weather conditions by designing the residential neighborhood, zigzagged alleys planning, internal and external squares, etc. In order to renew the ancient city in an urban manner, minor modification can be conducted to provide some essential services for the house's owner (residents).

### 4.1 Residential use of the Hit castle

The residential units are counted as irregular elements in the urban structure which includes residential buildings. It is planned that 2665 people (approximately 429 families) to be accommodated in 429 residential units; therefore, the number of residential units must be maintained. Table 2 shows the actual situation nowadays of the residential units.

**Table 2**. The urban case analysis for the residential buildings in Hit castle

| The residential unite situation | Number | Percentage |
|---|---|---|
| Engaged | 78 | 18 % |
| Demolished | 296 | 69 % |
| Needed for rehabilitation | 34 | 8 % |
| New | 21 | 5 % |
| Total | 429 | 100% |

From Table 3, it can be noted that 330 residential units need repairing, rehabilitating and modification. There are 99 residential units could be expected from the rehabilitation works as they need just minor maintenance that includes new unites as presented in Table 2; however, these could be removed and redesigned to be compatible to the existing original ones. In general, each unite could be historically and urbanely studied and then a decision can be made whether it can be repaired or not. In fact, there are some unites almost destroyed; therefore, there is no economic feasibility in terms of repairing them. However, these unites could be utilized by redeveloping them urbanely but the compatibility of urban tissue and structure should be carefully considered.

**Table 3. The urban situation for the residential units and the urban sustainable methods for treatment**

| Urban situation | Renewal policy | Future function |
|---|---|---|
| Engaged | Urban conservation | Residential use |
| Demolished | Redevelop and conserve | Residential use and services |
| Needed for rehabilitation | Rehabilitation | Residential use |
| New | Conserve and rehabilitation | Residential use |

### 4. 2 Commercial services

It is one of the regular elements of the urban structure that lay along the ring street that surround the ancient city as illustrated in the schematic diagram (Fig.2). These streets are mostly modern due to being open. These streets tear the urban tissue, especially in the direction of the river. Thanks to the improvement in living standards and time factor, these streets are given strange character in terms of the capacity of streets and building shapes and height, which are three and four floors for some buildings while in the ancient city it did not exceed one floor for commercial use and no more 15 – 32 m2.

The plan for urban renewal contains rebuilding the destroyed urban tissue in the direction of the river and modern markets. This should be combined with providing wide sidewalks for the pedestrians, specific lanes for emergency service vehicles like ambulance, fire, etc. without being used by the residential and visitors' cars. Additionally, the essential services such as water, sewage, lightening, and communications should be served using separated paths in order to provide good services for the residents.

In these areas, excesses have occurred with increasing the height of buildings to three floors, especially, in the street that overlooks the river and the modern market. The other excess is the change in land use from residential to commercial because of the increased price of land in the region. It can be noted that most of the new buildings are reinforced concrete structures and the others were built using concrete blocks whereas the old construction were built using the clay brick and the roofs were built using wood and mud as it will be explained later. The architectural deformation is clearly seen in this area and strict regulation should be legislated to conserve the traditions and values of this city. The main reasons for stopping construction of multistory concrete commercial buildings in this area can be summarized as follows:

1. The street's traffic jams due to multi use buildings.
2. The change in the commercial use from small shops to large store because of the large space
3. The structural construction is considered as strange style and this is easily noted at the first-time it has been seen due to the difference between the old and the new buildings.
4. The multistory buildings caused prejudice in the balance of the old city and damage historical and cultural appearance

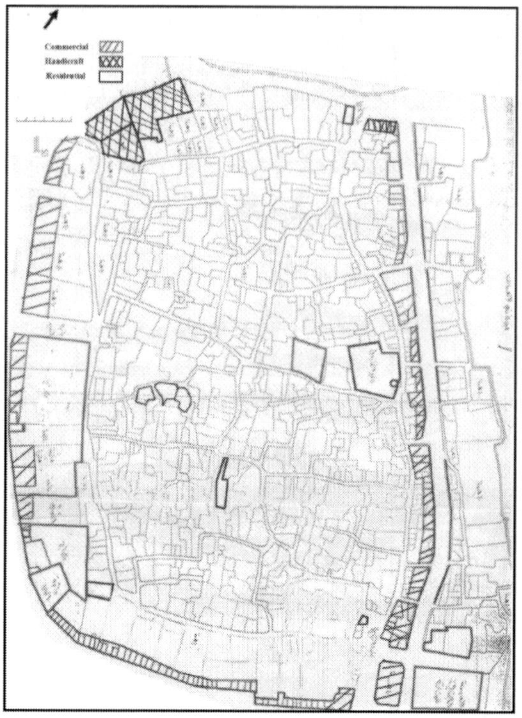


**Fig.2**. Schematic diagram for the urban sustainability for the commercial, handicraft and residential uses in Hit City Castle.

### 4.3 The infrastructure services (above and underground)

From the site visits to the Hit Castle, no services have been seen where there was either sewage or rain discharge network. The electricity services are not sufficient in general. Regarding the drinking water network, a new one has been constructed by the United Nation program, whereas there is no communication services at all. In order to improve the provided services for the residents, the following points should be considered:

Improve the drinking water in the residential, commercial, and social services buildings.

5. Develop a new mechanism for disposal of the solid waste outside the city by providing small cars and small closed vehicles.
6. Construct sewage network for the residential, commercial, and public service buildings.
7. Construct a rain, discharge network or instead provide a surface network if the first is difficult to be constructed.
8. Improve and repair the existing electricity network based on a standard level and utilize the sun power in lightening and generating electricity.
9. Provide a communication network (wire and wireless) based on the available technologies.

 https://doi.org/10.1051/matecconf/201816203017

10. Improve the roads efficiency for both paved and unpaved streets. The unpaved roads, can be paved using the locally-available materials (tar and aggregate). This can be done via the support of the Hit City mayoralty.

### 4.4 The traffic and transportation services

The uses of traffic and transportation in the ancient city contain some limitations that should be included, by the planner, in the renewable plan. The car traffic through the ancient city should be prevented and the passage of cars should be limited to the ring road as shown in Fig 3, where the traffic should be limited to the three entrances as suggested in the urban sustainable plan and the ring and commercial road. There is an increase in the pedestrian paths which are grouped and interrelated to be reached to the city ring road. The vehicles can enter from the Eastern door to reaches to the small mosque (Sit-Nafesa) and from the Western door to Al-Khareef door. A hierarchy in the pedestrians' roads could be conducted till the project road, while the commercial road the parallel to the river can be used to passed the vehicles through the modern market.

The suggested entrance by the researchers is located near to the Hag Omer Hindi garage, where the vehicles can enter till the wall entrance and then only pedestrians can pass due to the narrow paths. Suggested car parks for the car owners who live in the ancient Hit City in addition to the visitors and shoppers can be located near to the Eastern entrance towards Hit bridge and Hag Omer Hindi could be developed to be multi-story car parks. In order to widen the street, the private lands can be bought by the Hit City municipality from the owners.

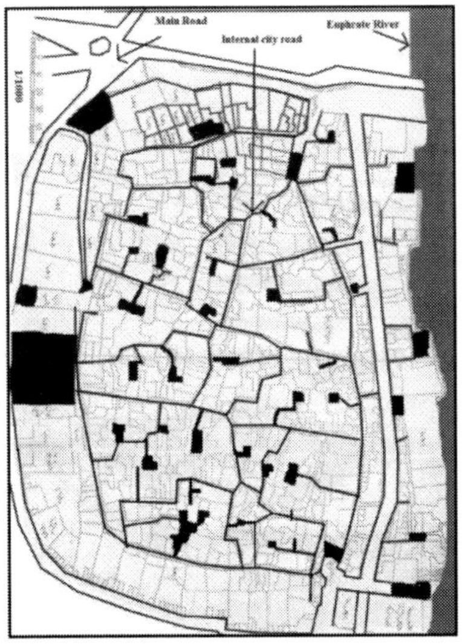


**Fig.3**. Schematic diagram for the urban sustainability for the pedestrians, traffic, and transport

### 4.5 Materials roles in urban sustainability

The used construction materials in Iraqi ancient civilizations such as Mesopotamia, Nineveh, and Sumerian were mainly clay-based materials. This was due to the nature of the geological formation that led to use this material for more than 5000 years ago. The ancient Iraqi cities, especially in the middle and south of Iraq were built on sedimentary lands which made use of clay due to its availability, ease in-manufacturing and building such as Babylon and Orr. The clay (mud) was formed into building units by mixing the clay-soil with water and adding straws from locally-available plants as reinforcement to produce clay brick. These units are known nowadays as brick; however, the main difference is the modern brick is burnt whereas the old units are air/sun dried. As it has been previously mentioned that Hit is one of the oldest Iraqi cities, however, clay brick was not used in building the houses.

A sophisticated technique was developed in ancient Hit City which characterized by using small stones that were bound together using the lime as a binder as shown in Fig. 4. This technique is considered a significant step towards the durable construction as use of the brick neededs to be maintained almost yearly. Whereas, the stone and lime are durable construction materials, therefore, using them to save time and resources. The wall's width was about 50 cm while the foundation is about 1m. The gypsum was used for plastering the inside wall and the lime with sand for outside use.

The roofs were constructed using wood beams in the short direction and some plans residual were used in the other direction to make a covering for the roof with mud and tar as seen in Fig 4. Interestingly, some of the roofs were built using the arches rather than flat roof which means that even at that time there was a fundamental understanding of the 'modern' structural theories. The use of locally-available materials (stone and lime) and the width of the walls in addition to the inside and outside layers of gypsum and lime could provide a unique thermo-physical behavior and thermal insulation for the building (house).

**Fig.4**. Walls and Roofing system in Hit Castle's houses

### 4.6 Rehabilitation of the damaged houses

In order to conserve the historical/heritage value of the Hit Castle and to achieve the sustainability aim, same locally-available materials can be used i.e. reused the same (demolished) materials. Even if there is a shortage in some materials, natural resources (stone, lime, wood, tar) should be used which is widely available in the studied city. The architectural/ structural style should be conserved by using the same material; however, the newly-developed building materials can be used in order to make the repaired houses sustainable and (durable).

## 5 Conclusions

Based on the study conducted and the detailed analysis of the surveyed information, the following conclusions can be highlighted:

- The Hit Castle has a historical value where it was a Jewish city before the Islamic distribution; therefore, it can be counted as a heritage city that should be conserved from the accelerated urban deterioration.
- The increase in demolition works for older houses and rebuilding of these houses using modern materials and style increases the architectural style (urban pollution) and gives a negative impression from the urban and architectural legacy.
- The variety in the land use in Hit Castle provides indicators for using different urban sustainability approaches in addition to applying a group of policies that related to each type based on its structural situation.
- The unique location for a Hit Castle beside the Euphrates River has given it an important architectural indicator that needs to be compatible with the modern construction. Where, it has been noted that modern commercial buildings has taken an important part of the urban structure of the ancient castle.
- The already-existing demolished construction materials can be reused in rehabilitating the residential/ commercial buildings in order to keep the historical value in terms of the materials and styles.
- For foundations and roofing works, locally-available materials such as lime, stone, wood, tar, clay, and residual plants can still be used; therefore, the maintenance and rehabilitation works will be easily implemented. In some situations, the developed construction materials can also be used by keeping the general style compatible with the original buildings.
- In order to bring the attention to the ancient cities that cannot be found everywhere and to highlight the importance of using the urban sustainability approaches, the following recommendations can be emphasized:
- Conduct a comprehensive study (planning and historical) to determine the historical value of the buildings in the Hit Castle.
- Implement the planning instructions and laws that deal with conserving the buildings having heritage value and stopp the demolishing and urban change works.
- Limiting the construction of multi-story commercial buildings on and near to the castle in order to minimize the urban pollution.
- Conserve the variety of land uses in the Hit Castle and attempting to simulate this variety in other locations in the city after employing the most modern technology.

## References

1. J . Breuste, S. Qureshi, Urb. Eco. Sys. J. E 14, 3(2011)
2. S. Cambell, Readings in Planning Theory (John Wiley & Sons, 2016)
3. F. Cheng, S. Geertman,et al., Env .& Urb .Sys. J. E 35,5 (2011)
4. D.L. Childers, M.L.Cadenasso, et al.,Sust. J. E 7,4 (2015)
5. J.P. Ganning, Eco. Dev. Quar. J. E 30, 3 (2016)
6. J.D .Gatrell, R .R. Jensen,et al. Sustainability: Policy and Praxis (McGraw –Hill, 2016)
7. A. Gospodini, Cities, J. E 18, 5 (2001)
8. T. Hak, J. Kovanda,J.Weinzettel,Eco.Ind. J. E 17,1(2012)
9. A.M. Hassan, Damascus Eng. J. E 2,12(2012)
10. Y.R. Jabareen, Planning Edu. &Res. J. E 26,1 (2006)
11. T. Kuyucu,D. Danis, Urb. Aff. Rev. 51,3(2015)
12. P. Pattananurot, Procedia Eng. J. E 32,(2012)
13. R.R. Shaker, I.G. Sirodoev, Habitat Int. 55, 2 (2016)
14. F.Steinberg, Habitat Int. 20,3 (1996)
15. L. Tran, Ecological Indicators, 61,1(2016)
16. H. Wang, Transp. Geog. J. E 40, 2 (2014)
17. N.H. Wong, S.K. Jusuf, C.L.Tan, Land.& Urb. Plan. J. E 100,4(2011)
18. J. Zhang, ACTA Geog. of Chin. J. E 56,1 (2001)
19. B. Yang, L. Shi, Cleaner Production, J. E 141(2017)
20. A. M. Hassan, Dmascus Eng. J. E 2,13(2012)

MATEC Web of Conferences 162, 03018 (2018) https://doi.org/10.1051/matecconf/201816203018
BCEE3-2017

# Selecting optimum railway track using GIS techniques

*Amjed* Al-Hameedawi[1,*], *Mohammed* Salih[1], *Haidar* Mohammed[1], and *Maher* Hassan[2]

[1] Building and Construction Engineering Department, University of Technology, Baghdad, Iraq
[2] College of Engineering, University of Karbala, Karbala, Iraq

**Abstract**. This paper mainly aims at analysing and selecting the best railway route within the Karbala City. The significance of having suitable locations for railway route is recognized as a crucial element in the pursuit to improve the quality of public transport in general. This paper uses technique based on bridging the gap between all the criteria to determine suitable location for the railway. It was also intended to organize non-spatial data for describing the process. The site selection of railway is the process of finding locations that meet the desired conditions set by the selection criteria. For this issue a professional know-how to deal with data depending on multi criteria is necessary to achieve of the task of decision making. Due to the tangle of the issue, a special effort must be done, to have a good usefulness of site selection. The results of spatial model in this paper shows the best railway track is (Karbala–ALhindya).

## 1 Introduction

Rail transportation is a critical method of transport all through the world. Every day, it transports a great many travellers and products from one point to another. The present levels of urbanization inside Karbala have prompted expanded populace inside the city most dwelling in regions around Karbala. This has prompted an extraordinary request of transportation administrations which couldn't be met by the street transport as it were. The consequence of this popularity has constantly growled movement on significant streets and thruways. The administration henceforth must be focused on investigating elective means, including the development and change of the suburbanite rail framework in Karbala and environs. It is basic that workers know all the accessible and advantageous methods of transport to take with a specific end goal to touch base to their goal. With the change of rail framework, workers should have a rail data framework. This realizes the requirement for an open transport data framework. With the accessibility of the data framework, all the data required by the clients will be accessible to them at the snap of a finger. This will likewise put a conclusion to the uncertainty of suburbanites with respect to their vehicle needs. The organizers and supervisors of open transportation will have the capacity to settle on skilful choices about the suburbanite needs. The objective in a site choice venture is to locate the best area with fancied conditions that fulfil foreordained determination criteria. Location selection commonly includes two fundamental stages: (i) site screening (i.e., distinguishing proof of few applicant locales from a wide geographic region and a scope of choice components) and (ii) site assessment (i.e., inside and out examination of every candidate site to locate the most appropriate one).The selection procedure endeavours to upgrade various targets in deciding the appropriateness of a specific site for a characterized travel office. Such enhancement regularly includes a huge number of variables, some of the time negating [1]. A portion of the essential variables that add to the trouble of the best possible decision incorporate the presence of various conceivable choices inside a search for domain, numerous goals, destinations, assorted qualities of gatherings, and absence of quantitative measures of the factors impact, vulnerabilities with respect to effect timing and greatness. In addition to instabilities in regards to government impact on the selected procedure through enactments, vulnerabilities in regards to conceivable postponements of approval and development [2, 1]. Geographic Information Systems (GIS), Multi-Criteria Decision Making techniques (MCDM), and Expert Systems have broadly been utilized as a part of tackling site determination issues throughout the previous two decades [3, 4, and 5]. Be that as it may, each of these systems has its own particular impediments in tending to spatial information, which is crucial when one is managing spatial choice issues, for example, a site choice issue. For instance, the conventional MCDM strategies have been non-spatial. In any case, in a genuine circumstance it can barely be accepted that the whole review zone is spatially homogeneous, in light of the fact that the assessment criteria used to change crosswise over space. A changed approach has continued spreading, in which the three apparatuses are joined in a way so that the weakness of one instrument is supplemented by the quality of another. A GIS system is

* Corresponding author: amjednaser@gmail.com

© The Authors, published by EDP Sciences. This is an open access article distributed under the terms of the Creative Commons Attribution License 4.0 (http://creativecommons.org/licenses/by/4.0/).

used to manipulate the spatial investigation required in the screening phase of candidate locales [1]. There are a few viewpoints to consider when managing the railroad condition. The most self-evident one may be the track, or all the more particularly the track geometry, which must meet the set necessities keeping in mind the end goal to secure a sheltered operation of trains [6].The significance of the region of Karbala from religious and monetary aspects is highly appreciated in Iraq. An extensive number of tourists visit the blessed hallowed places during the specified time as a rule and in particular specific religious events turn into an imperative vacationer nature of the basic financial and come back to the nation obliged to be enhanced and the improvement of foundation in the city. for the region is also known for its strategic importance in Iraq history where it contains holy shrines of Imam Hussain and his brother Imam Abbas Peace Be Upon Them. A conceivable arrangement that this venture tries to address, is the advancement of rail data system which would, in addition to other things, show spatial dissemination of proposed railroad tracks/stations inside Karbala and its environs along the current railroad lines; show reasonable destinations to construct new rail route stations. The objective of this paper is to make use of fundamentals and practical merits of the application of GIS in the railway project management and selection of the most suitable site for the railway. Additionally, this paper can be considered as milestone for such projects. The site must be ideally chosen meaning it must be well disposed with condition, far into rural territories, close to a populated region choosing the best site for a railroad venture is not a simple undertaking. The analysis must consider financial, statistic and physical criteria and the truth of the review zone, and in addition the arrangements of governments and past involvement in this confused undertaking [7, 8]. The principle target is to make a Geodatabase of the railroad lines that will empower capacity of data about rail route lines and to do their spatial investigation to decide their reasonableness according to their area and dissemination.

Specific objectives:

•To map the spatial distribution of the existing and proposed railway lines within the study area.

•To create a comprehensive database containing attributes of the railway lines.

•To produce the suitability of railway maps.

•To propose suitable sites for additional railway lines.

## 2 Study area

The Karbala City lies in the heart of the Iraqi state; it is located in the south-west of Baghdad, and is bordered to the north and from the west Unbar city, while bordering to the south of Najaf province, and is bordered by the city of Babylon from the north and northeast. There are many primary and secondary roads and rural roads, which are used by visitor to reach the city of Karbala, and despite the fact that the majority of the roads are unpaved, are well suited to use. Nevertheless, they are used in many cases especially in times of religious events taking place which draws visitors in large numbers to this province, and calls for the need to embark on many of these roads because of the great momentum in the main roads.

## 3 Methods

The model of a reasonableness site was assembled utilizing a model builder in ArcGIS 10.5. It was constructed in view of the MCE methodology known as the multi-criteria evaluation in which the weight of the relative significance found to every criterion and an aggregate score is then led for every paradigm by multiplying the relative significance percentage by the scale an incentive for that criterion (Figure 1). The appropriate maps were produced by coordinating different models [9, 10, and 11].

The main phases of the methodology used in the study area:

a) Data acquisition through authority printed reports.

b) Converting data into geometric elements through GIS.

c) Data interpretation and analysis.

d) Model development and extraction of measurements from analyses.

While choosing a site for a railroad line, there ought to be sufficient land accessible for the station building, for the proposed line as well as for any future development. The proposed region ought to likewise be with no religious structures. The proposed site ought to ideally be on a fairly level ground with great drainage infrastructures. It ought to be conceivable to give the greatest allowable grade in the yard. The station site ought to ideally have a straight arrangement so that the different signs are plainly unmistakable. The closeness of the station site to a bend creates various operational issues. The station site ought to be effectively available. The site ought to be close to towns and villages. Close-by towns ought to be associated with the station by methods for approach streets for the comfort of travelers. While choosing the site, it ought to be checked that satisfactory water supply is accessible for travelers and operational needs.

The railway determination relies upon number of criteria, including [2]:

•The railroad track would be situated close to residential areas to serve the number of inhabitants in these districts.

•The railroad track would be situated far from agricultural areas.

•Railway track would be situated close to freeways.

•Railway track should be less crossing with an inner street system of the city.

•Railway way should be flat area the tolerated slope is (0-10) %.

•Do not cross a railroad track with crucial establishments or infrastructure.

Data extraction is the way of creating more information from effectively arranged details. In the information input stage, utilizing the GIS need to:

1. Layer represented road network inside the city of Karbala.
2. Expressway network layers.
3. Recent Satellite imagery for Karbala city.
4. Digital Terrain Model.

At that point utilize an ArcMap program and spatial extensions to pick the best area for the track of the railroad by recognizing purposes of arrange framework and include them in the characteristic procedure.

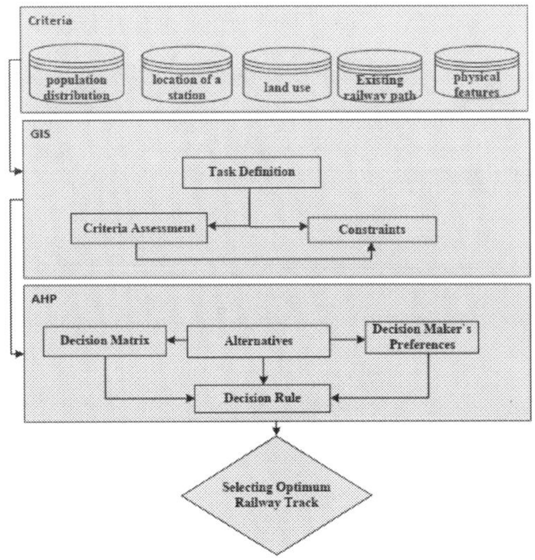


**Fig. 1.** Methodological schema.

## 4 Results and discussion

After extracting required datasets, they were reclassified and last stride the layers were weighted utilizing Analytic Hierarchy Process (AHP). The Analytic Hierarchy Process (AHP) is one of the more well-known basic leadership methods that are broadly used to address Multi-Criteria Decision- Making (MCDM) issues. This technique separates the issue into a chain of command of sub-issues. At that point from the inspired judgments from specialists on the near execution or criticality of the sub-issues, needs are computed. These needs empower the basic leadership identified with sorting, positioning or choosing the most reasonable option in MCDM issues. ArcGIS Spatial Analyst gives a rich arrangement of tools to perform cell-based (raster) investigation. Of the three sorts of GIS data (raster, vector, and TIN), the raster information structure is the most complete displaying condition for spatial investigation. There are two principle approaches to perform separate examination in ArcGIS Spatial Analyst: Euclidean distance and cost distance. The Euclidean distance capacities measure straight-line distance from every cell to the nearest source (the source recognizes the objects of intrigue, for example, wells, streets, or a school). The cost distance capacities (or cost weighted distance) changes Euclidean distance by comparing separation as a cost element, which is the cost to go through any given cell. The Euclidean Distance capacity is utilized habitually as an independent capacity for applications: this can be utilized while making a reasonableness guide, when information relating to the distance from a specific value is required. For this review, the distance of every railroad line was ascertained and the yield was in raster format.

### 4.1 Site selection

This phase in which, the final data representation in the form of thematic maps performed based on modelling process in GIS. The proposed railroad track for indicated criteria have been chosen and illustrated in as a type of maps (Figures 2-5).

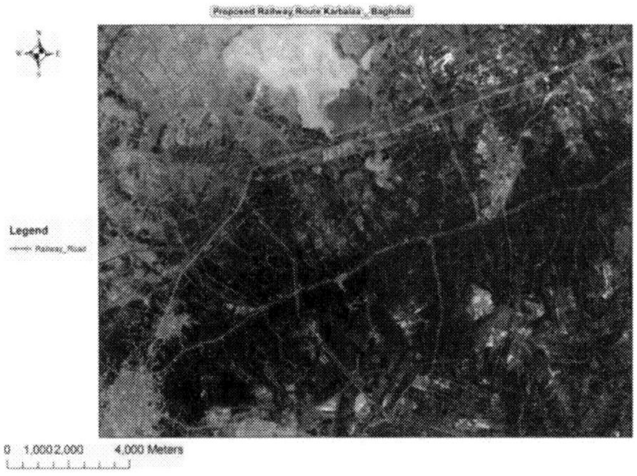


**Fig. 2.** Proposed railway route (Karbala – Baghdad).

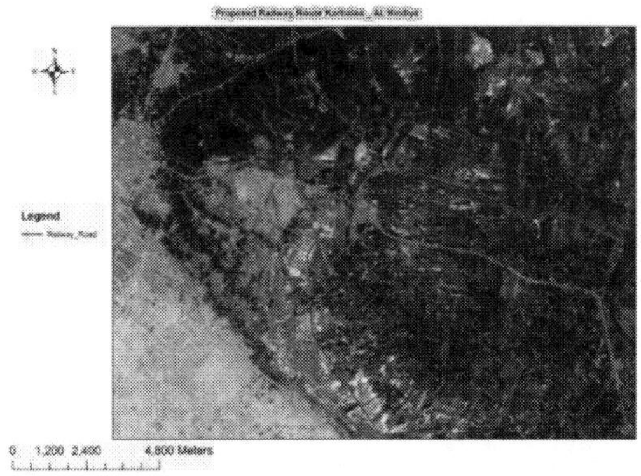


**Fig. 3.** Proposed railway route (Karbala – Hindya).

https://doi.org/10.1051/matecconf/201816203018

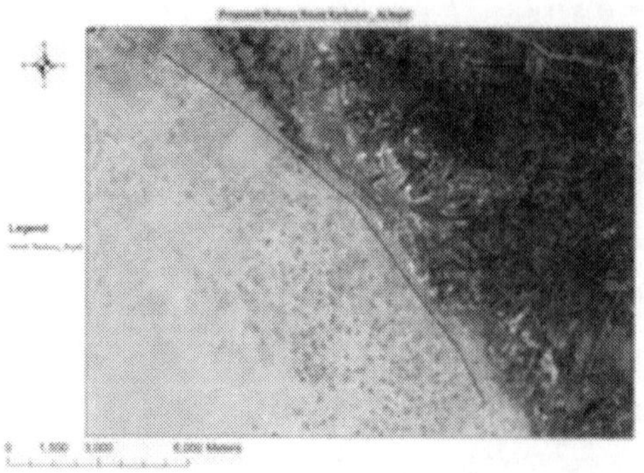

**Fig. 4.** Proposed railway route (Karbala – Najaf).

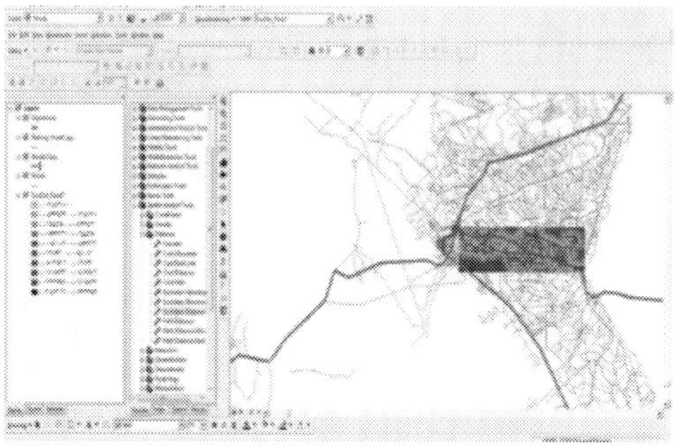

**Fig. 5.** Snapshot of proposed railway route (Karbala – Hindya).

## 5 Conclusions

This paper uses technique based on bridging the gap between all the criteria to determine suitable location for the railway. GIS is effective with tremendous usefulness in its applications when it is connected to decision making process. It is a tool that is extremely helpful in recognizing patterns and issues and finding answers for them. It represents the expressing picture on a guide presented with every one of the parameters related with it. The results of spatial analysis in this project shows the best railway track is (Karbala–ALhindya) because it is going through several residential areas, far from agricultural areas, the lack of a railway track list and is the main corridor for most visitors to the city.

## References

[1] A. Farkas, Inter. Con. **7**, (2009)
[2] W. Roets, (Doctoral dissertation),(2005)
[3] E. Briceño-Elizondo, D. Jäger, M. J. Lexer, J. Garcia-Gonzalo, H. Peltola, S. Kellomäki, Ecol. Indic. **8**, (2008)
[4] S. Kalogirou, Comp, envir. and urban systems **26**, (2002)
[5] I. Tsoy, LWREX. **6**, (2015)
[6] E. Andreas, (2011).
[7] M. Bojesen, L. Boerboom, H. Skov-Petersen, Land use pol., **42**, (2015)
[8] J.Irizarry, E. P. Karan, F.Jalaei, Auto.in Cons. **31**, (2013).
[9] A. N. M. Al-Hameedawi, *Integrative GI Technology Applied to Best-Site Selection for Industrial Areas in Erbil City, Iraq*, TUDpress Ver. der Wiss. Dresden, (2014)
[10] S.Corrente, S. Greco, R. SłOwińSki, Decis. Sup. Sys. **53**, (2012)
[11] D. Ergu, G. Kou, Y. Shi, Com. and Oper. Research, **42**, (2014)

# Sentinel-2 imageries with meteorological data utilize remote sensing and GIS for wheat crop prove production

*Muslim* Sbahi [1,*], *Abdul-Razzak* Ziboon [1] and *Khalid* Hassoon [2]

[1] Building and Construction Engineering Department, University of Technology, Baghdad, Iraq
[2] Ministry of Science and Technology, Baghdad, Iraq

**Abstract.** Agricultural statistics are considered one of the most important topics in the developed countries now. In addition to the irrigation systems used, they relate to the quality and quantity of production and the consequent export and import of agricultural crops in order to achieve food security for the country. Therefore, the present traditional methods used in the collection of statistical data are not sufficient to cover the vast areas of cultivated strategic crops and to determine the effects therein. This is due to many factors, the most important of which is climatic changes. Hence the importance of the application of remote sensing techniques for image and data in the survey, agricultural statistics and monitoring of the quality and quantity of production using the latest satellite images available free from the owner of artificial satellite Sentinel2 which has spatial resolution of 10m instead of 30m available in artificial satellite Landsat8. Research results showed that the temperature can be controlled using modern techniques used in irrigation, which improves the atmosphere and thus contribute to providing a suitable environment for the wheat crop, which helps to increase the amount of production these have been clearly monitored using applicable indicators derived from satellite imageries data.

## 1 Introduction

Remote Sensing (RS) as defined in many scientific sources, is the science and art of obtaining data about an object, area or phenomenon under monitoring. Through specialized software, data obtained without physical contact is analyzed and information is generated about these phenomena [1]. At present, remote sensing techniques in most developed countries have become one of the most important methods used to monitor terrestrial environmental changes, particularly agricultural areas. The importance of monitoring the agricultural areas is due to the vast areas cultivated and the difficulty to control by traditional methods (field visits) and determine the areas and health status of the crops and their need for nutrition. In addition, agricultural production is linked to food security. Using technology, processing and analysis of remote sensing data and modern computers assist surveyors in measuring and monitoring the Earth's natural resources and identify changes periodically [2]. The Sentinel-2 satellite was launched on 23 June 2015 and received the first data within a selected study area on December 19, 2015. It continued to send data until the end of the monitoring process on 8 March 2016 by five satellite imageries. The Sentinel-2 is characterized by spatial resolution of 10 meters, which is suitable in determining the areas and health status of areas estimated at hundreds of donum suitable according to the criteria of remote sensing [3]. The meteorological data obtained from the Iraqi Ministry of Agriculture stations were linked to theoretical results of the imageries analysis in order to enhance the results because of significant impact on the crops. Research works to employ modern techniques of meteorological stations daily and Satellite Sentinel-2 imageries with high spatial resolution improve the the production of the wheat crop.

## 2 Objectives

The research aims to use modern techniques represented by using imageries of satellite Sentinel-2 that has made use of its data for the first time. In addition, integration of the satellite data and daily agricultural meteorologist and atmospheric data network affiliated to the Iraqi Ministry of Agriculture can predict the annual crop of wheat and derive the amount of output for large areas with high accuracy and less cost and effort.

## 3 Study areas

Areas of research and study are located in the Middle Euphrates region of Iraq in the province of Karbala; the first area is east of the province within the borders of Al-Hassainya Nahia and is called Agriculture Division of Imam Aoun. The second study area is west of the province within the borders of Ain al-Tamur Nahia and is called Agriculture Division of Ain al-Tamur (Fig. 1).

* Corresponding author: surveyor.moslem@gmail.com

© The Authors, published by EDP Sciences. This is an open access article distributed under the terms of the Creative Commons Attribution License 4.0 (http://creativecommons.org/licenses/by/4.0/).

Agriculture Division of Imam Aoun depends on surface irrigation using electric pumps and the nature of the distribution of farms is irregular in forms and the soil is not fertile. The Agriculture Division of Ain al-Tamur has fertile soil and is characterized by farms with circle shape depending on the nature of the fixed pivot spraying irrigation, which depends on water artesian wells.

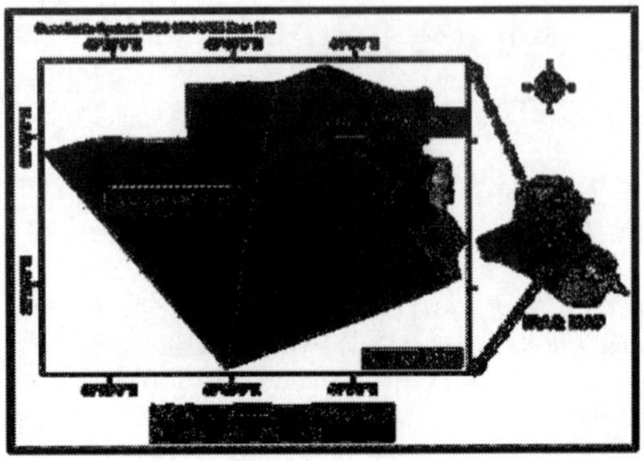

**Fig. 1.** Iraq map with location study areas in the holy province of Karbala

## 4 Theoretical background

### 4.1 Growth stage scales

The chart or table of crop growth stages is very important in order to describe the growth stages of each crop. It is important to understand the growth and development of wheat from sowing to maturity and harvesting for the proper use of fertilizers and pesticides to ensure the safety of the crop [4]. Many of the determinants and signs of increases constrain in use until certain stages of growth have been achieved. A widely used scale in the United States is the scale Feekes [5]. Feekes is distinguished from other measures as it gives a general description of the stages of the wheat crop growth which helps monitor the growth process by using remote sensing techniques to understand the changes taking place.

### 4.2 Growth stage scales

Agricultural meteorological stations of modern techniques are used in the collection of meteorological data. The stations are located in the center of the areas of study (see Fig. 2) for data to be collected in an orderly manner and accurate monitoring. Monitoring is done of several air elements which affect agricultural production. The meteorological stations monitor both degree of minimum and maximum temperature, the intensity of solar radiation as well as relative humidity of minimum and maximum, in addition to the many other elements. Temperature is one of the most important factors affecting the determination of the amount of agricultural production, as well as the production of agricultural crops through their direct and indirect effects. Temperature affects the growth stages of different physiological processes occurring within the crop, as well as determines the places of spread and concentration of the crop in different geographic regions [6]. Thus, three temperature ranges (minimum, optimal, and maximum) for wheat crop were identified as shown in Table 1.

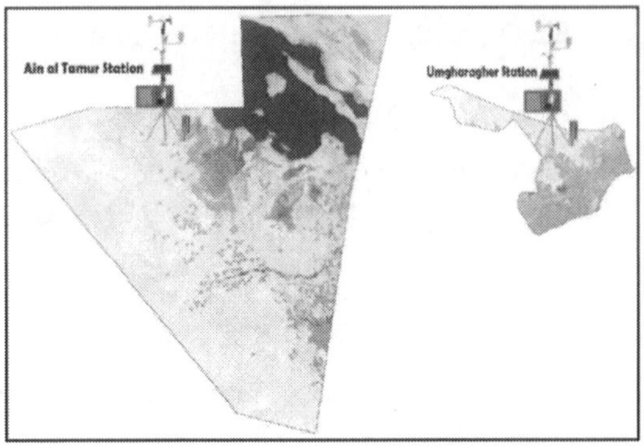


**Fig. 2.** The meteorological stations location for Imam Aoun and Ain al Tamur in the Holy Karbala' province

**Table 1.** Températures limites of the wheat crop

| Wheat Growth Stages | Month | Maximum Temp. C° | Optimum Temp. C° | Minimum Temp. C° |
|---|---|---|---|---|
| Planting - Emergence | November | 30 | 18 | 12 |
| Emergence- Start Tillering | December | 20 | 13 | 3 |
| Vegetative Tillering Growth | January - February | 22 | 17 | 10 |
| Heading | March | 30 | 22 | 15 |
| Ripening | April – May | 40 | 25 | 17 |

### 4.3 Growth stage scales

The Normalized Difference Vegetation Index (NDVI) law see equation (1), is used to calculate the vegetative intensity related to the health status of crops affected by external and internal surroundings of the crop. Moreover, data from remote sensing can be considered the best and fastest method as compared to the conventional way to efficiently manage nitrogen [7]. The equation (2) is used for change detection the contrast between two images for same band. The equation (3) is used for change detection the NDVI between two images at difference date. The logic of image algebra is for change detection using image differencing or band rationing [9].

$$NDVI= (Red - NIR)/ (Red + NIR) \quad (1)$$

Where:
NIR = (Near-Infrared band)
Red = (band visible)

$$\Delta\ BV\ a,\ b,\ d = BV\ a,\ b,\ d\ (1) - BV\ a,\ b,\ d\ (2) + c \quad (2)$$

Where:
Δ BV a, b, d = change in pixel value
BV a, b, d (1) = brightness value on date one
BV a, b, d (2) = brightness value on date two
c = a constant (e.g. 127)
a = line No.
b = column No.
d = a single band (e.g. Sentinel-2 band 13)

$$\Delta\ NDVI\ a,\ b = NDVI\ a,\ b\ (1) - NDVI\ a,\ b\ (2) + c \quad (3)$$

Where:
Δ NDVI a, b = change pixel value at NDVI
NDVI a, b (1) = NDVI value on date one
NDVI a, b (2) = NDVI value on date two

# 5 Methodology

## 5.1 Work steps

Research depended on the following data (Fig. 3):
1. Calculation of crop germination calendar using CROPWAT 8.0 is a tool to support decision-making by the land and water development division of Food and Agriculture Organization (FAO).
2. Sentinel-2 Satellite Data Starting download from 19 December 2015 to 08 March 2016 with five image.
3. Meteorological Data that it has been received from the Meteorological stations affiliated to the Iraqi Ministry of Agriculture. It included data (degree of temperature minimum and maximum, relative humidity minimum and maximum, intensity of solar radiation, wind speed and direction)
4. Field Survey Data include crop growth data, coordinates crop positions, and verification of field sampling data.

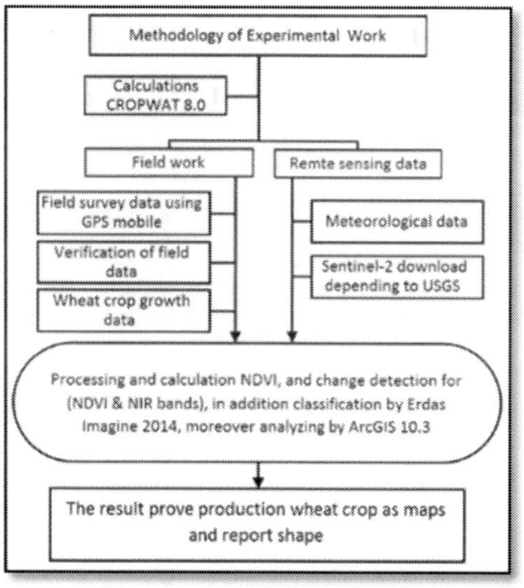


**Fig. 3.** Steps research methodology with study materials

### *5.1.1. Data collection and analysis stage for Agriculture Division of Imam Aoun.*

Farm data (C & E) that depend on surface irrigation, which take irregular forms as in Fig. (4 & 7). Using ERDAS IMAGINE program has been for spectral reflectance calculation of the five samples installed on each farm see Fig. (5 & 8) with the coordinates of taken samples as in the Table (2 & 3). Field visits each month for final samples described as representing the production of the final average farm see Fig. (6 & 9).

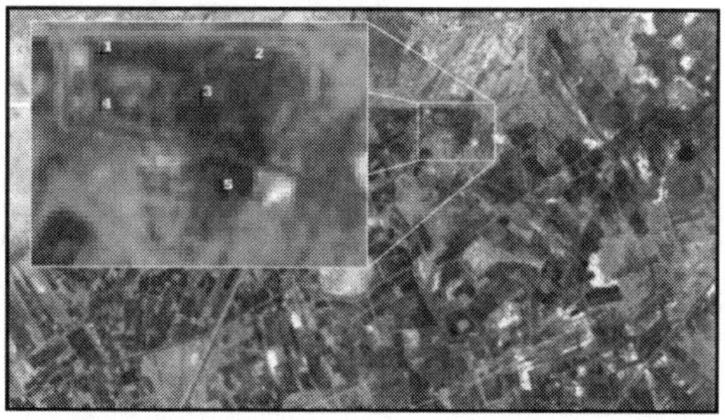

**Fig. 4.** Sentinel-2 image scene on 8-march 2017 shown Farm C with the five samples.

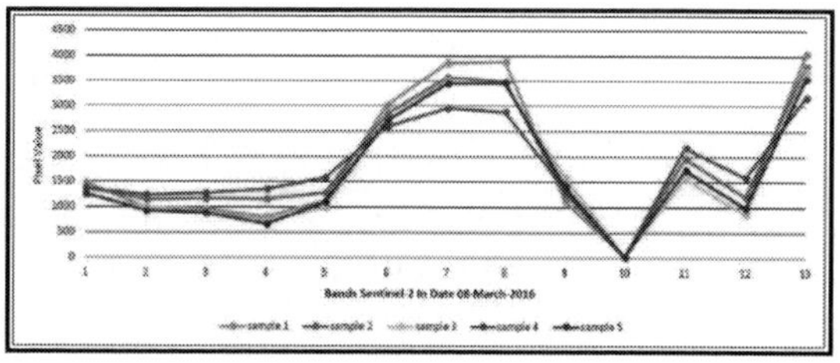

**Fig. 5.** Spectral profile for the five samples at the Farm **C**

**Table 2.** The coordinates of the samples the farm **C**

| No. sample | E | N |
|---|---|---|
| 1 | 423592.998601 | 3625900.39598 |
| 2 | 423905.721972 | 3625887.10349 |
| 3 | 423798.682429 | 3625810.84656 |
| 4 | 423590.899787 | 3625787.7596 |
| 5 | 423842.05793 | 3625625.45127 |
| Coordinates system : WGS 1984 UTM Zone 38N | | |

**Fig. 6.** The final production of the Farm **C**

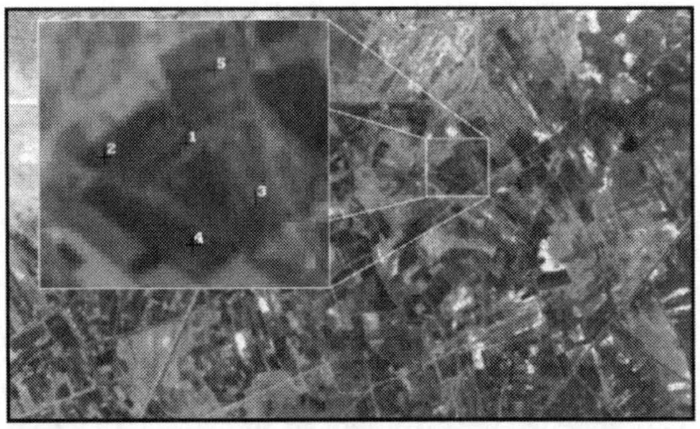


**Fig. 7.** Sentinel-2 image scene on 8-march 2017 shown Farm **E** with the five samples.

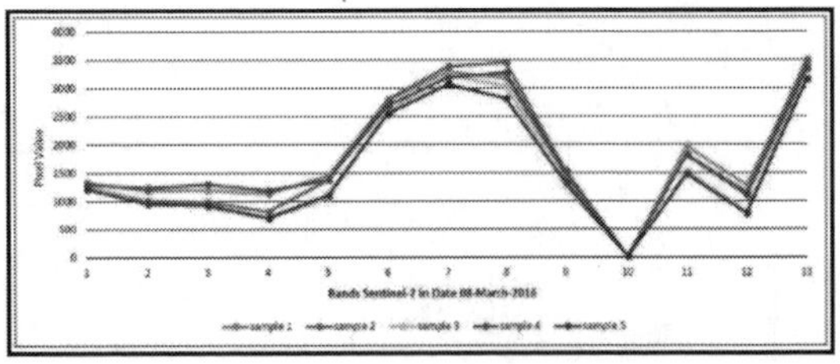

**Fig. 8.** Spectral profile for the five samples at the Farm **E**

**Table 3.** The coordinates of the samples the farm **E**

| No. sample | E | N |
|---|---|---|
| 1 | 424225.612287 | 3625358.47756 |
| 2 | 424061.138242 | 3625334.08787 |
| 3 | 424364.445891 | 3625251.53816 |
| 4 | 424238.119819 | 3625160.85856 |
| 5 | 424282.812659 | 3625501.64803 |
| Coordinates system : WGS 1984 UTM Zone 38N | | |

**Fig. 9.** The final production of the Farm **E**

#### *5.1.2. Data collection and analysis stage for Agriculture Division of Ain al-Tamur.*

Farm data (F & G) that rely on fixed pivot irrigation (sprinkler) and the farms take circular forms of regular sizes (60, 80, 120 donum) depending on the length of the sprayer as in Fig. (10 & 13). Using ERDAS IMAGINE program has been for spectral reflectance calculation of the five samples installed on each farm see Fig. (11 & 14) with the coordinates of taking samples as in the Table (4 & 5). Field visits each month final samples described as representing the production of the final average farm see Fig. (12 & 15).

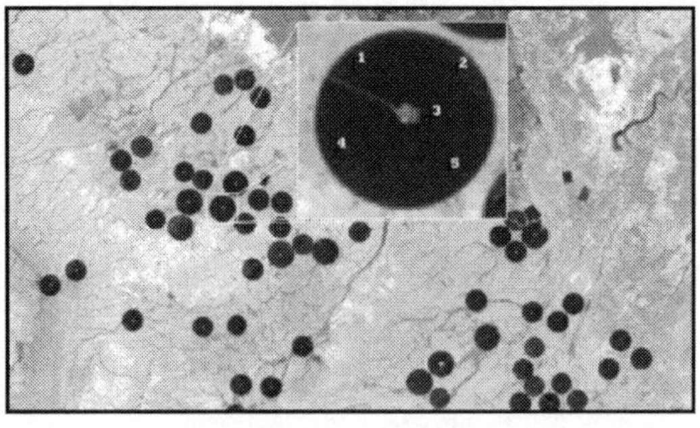


**Fig. 10.** Sentinel-2 image scene on 8-March 2017 showing Farm **F** with the five samples.

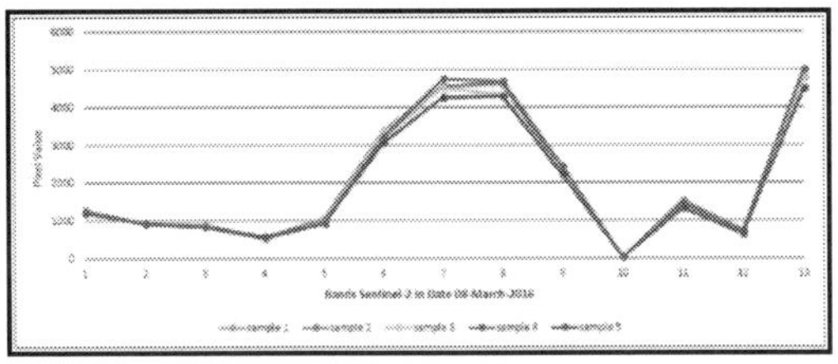

**Fig. 11.** Spectral profile for the five samples at the Farm **F**

**Table 4.** The coordinates of the samples the farm **F**

| No. sample | E | N |
|---|---|---|
| 1 | 354750.993277 | 3598102.4246 |
| 2 | 355107.519445 | 3598083.85552 |
| 3 | 355013.436151 | 3597917.97182 |
| 4 | 354685.382559 | 3597805.31946 |
| 5 | 355079.046869 | 3597734.75698 |
| Coordinates system : WGS 1984 UTM Zone 38N | | |

**Fig. 12.** The final production of the Farm **F**

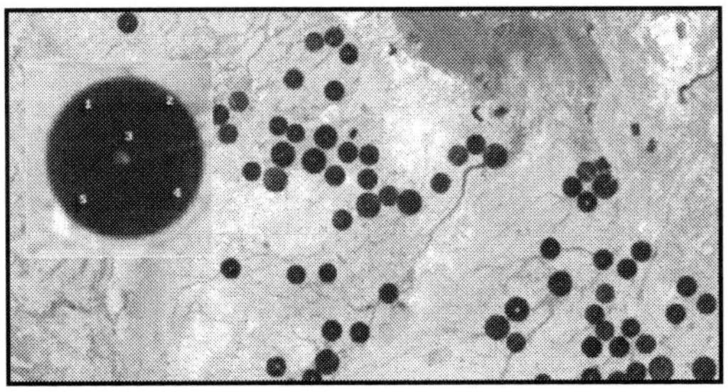

**Fig. 13.** Sentinel-2 image scene on 8-March 2017 showing Farm **G** with the five samples.

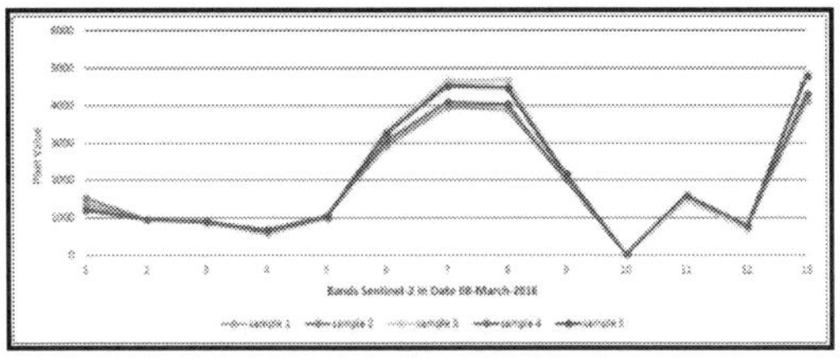

**Fig. 14.** Spectral profile for the five samples at the Farm **G**

**Table 5**. The coordinates of the samples the farm **G**

| No. sample | E | N |
|---|---|---|
| 1 | 353222.651975 | 3597798.6622 |
| 2 | 353471.010036 | 3597806.63091 |
| 3 | 353344.17477 | 3597703.03771 |
| 4 | 353491.595865 | 3597537.68702 |
| 5 | 353205.386441 | 3597517.76525 |
| Coordinates system : WGS 1984 UTM Zone 38N | | |

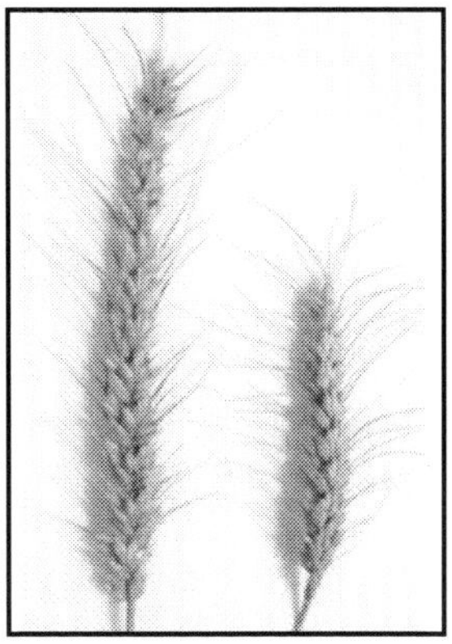

**Fig. 15.** The final production of the Farm **G**

### *5.1.3. Calculation NDVI and change detection (NDVI and NIR band) for Agriculture Divisions.*

The NDVI was calculated with change detection (NDVI & NIR band) for both Agriculture Divisions see Fig. (16 & 17). In addition, the Meteorological data linking theory with the results of indicators pointing to the reasons for the weakness and strength of signal indicators, which affect the quantity and quality of the final crop.

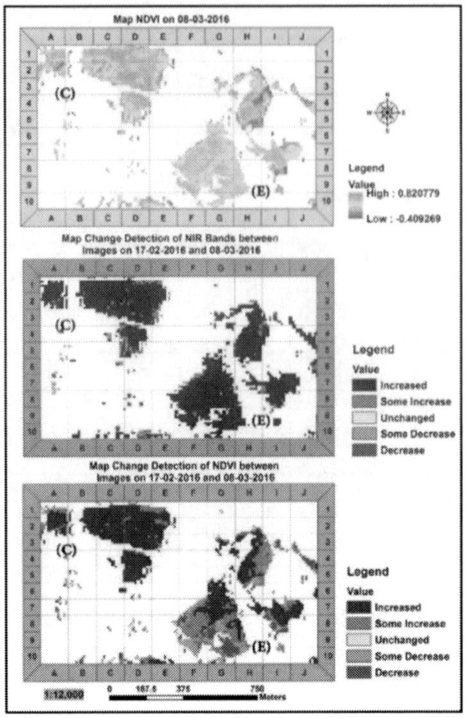


**Fig. 16.** Scenes NDVI on 8-March 2017 and change detection between NDVI & NIR band on (17-02-2017 & 08-03-2017) showing Farms C & E at Agriculture Division of Imam Aoun.

**Table 6.** Supervised classification results of the Agriculture Division of Imam Aoun

| Data | Difference | Maximum likelihood |
|---|---|---|
| Sentinel-2 image on 08-03-2016 | Overall classification accuracy | 94.32% |
| | Overall statistics kappa | 0.9351 |
| | Area of wheat crop | 2814 Donum |

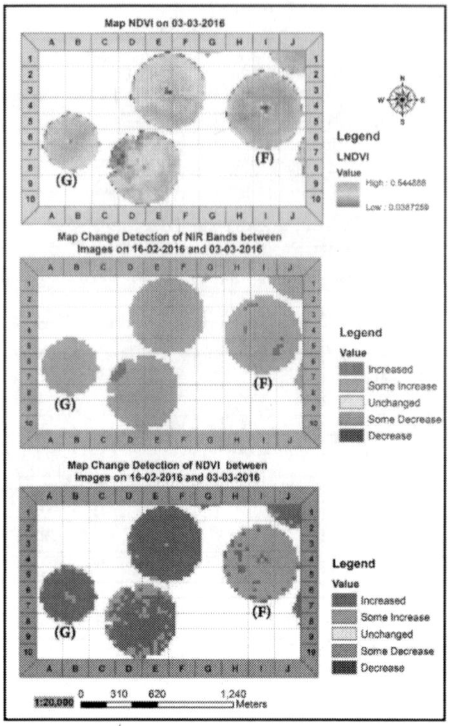


**Fig. 17.** Scenes NDVI on 8-March 2017 and change detection between NIR band & NDVI on (17-02-2017 & 08-03-2017) showing Farms G & F.

### *5.1.4. Classification stage to Agriculture Divisions.*

Classification of both the regions of the study has been using the method of Maximum likelihood. Classification accuracy showed good results as in the Table (6 & 7). The quantity of wheat crop decreased by 686 donum from proven in the agricultural plan at Agriculture Division of Imam Aoun. While the Ain al-Tamur Agriculture Division has seen a marked rise in the quantity of production up to 3200 donum from what was mentioned in Sticky agricultural plan. See Fig. (18 &19). The first study area was classified into eight categories depending on the field visits and the points that were monitored using GPS, which proved the existence of different types of soil, random residential areas, palm trees, wheat and barley crops, as well as some water sources. While the second study, area was classified into five categories because of the lack of ground phenomena.

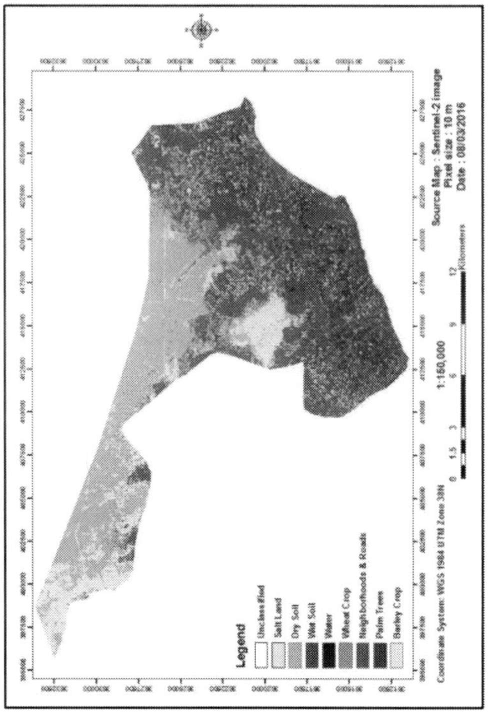


**Fig. 18.** Classification of the Agriculture Division of Imam Aoun by maximum likelihood method for Sentinel-2 image on 08-03-2016.

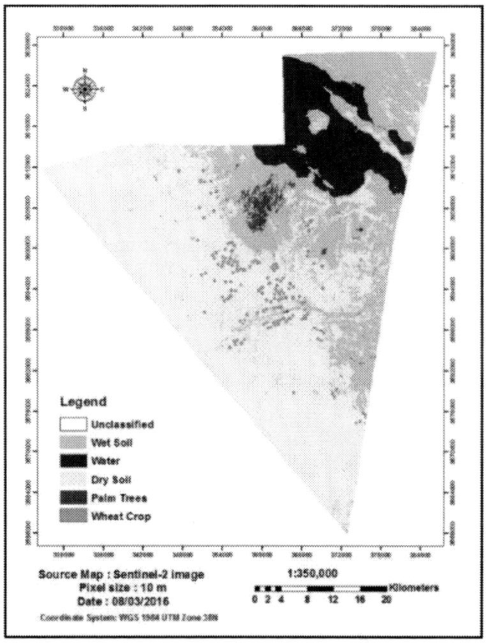


**Fig. 19.** Classification by Maximum likelihood method to Ain al-Tamur Agriculture Division for Sentinel-2 image on 08-03-2016.

**Table 7.** Supervised classification results of Ain al-Tamur Agriculture Division

| Data | Difference | Maximum likelihood |
|---|---|---|
| **Sentinel-2 image on 08-03-2016** | Overall classification accuracy | 97.50% |
| | Overall statistics kappa | 0.9688 |
| | Area of wheat crop | 10739 Donum |

## 6 Conclusions

1. The effectiveness of spatial-resolution 10 meters in agricultural applications.
2. Efficiency indicators used to determine the damage to the phenomenon in the crop in record time too.
3. Production pivotal farms spray twice more than farms that depend surface irrigation.
4. Agricultural Meteorological data contributed to the finding and interpretation of the increase or the decline in farms during monitoring period.
5. Ain al-Tamur farms are considered an important source of wheat due to the clean water quality and the quality of virgin territory and irrigation sprinkler.

## References

1. Isabel Patricia Maria Coppa, "The Use of Remote Sensing data For Broad Acre Grain crop Monitoring in Southeast Australia", PhD Thesis, SET Portfolio, RMIT University, Geospatial Science, School of Mathematical and Geospatial Sciences, Melbourne, Australia (2006).
2. Ghilani C. Wolf P, *"Elementary Surveying an Introduction to Geomatics"*,(13th edition, U.S., The Pennsylvania State University & Environmental Engineering University of Wisconsin–Madison 2012).
3. European Space Agency site 2016. http://www.esa.int/Our_Activities/Observing_the_Earth/Copernicus/Sentinel-2
4. Wise K. Johnson B. Mansfield C. Krupke C, "*Managing Wheat by Growth Stage*" (University PURDUE2013).
5. Kent McVay, Mary Burrows, Fabian Menalled and Kevin Wanner "*Montana Wheat Production Guide*" (Publication, Montana state University 2010).
6. Saeed M. Jassem M. Kadhim F, "*Thermal Limits and Their Effect on Wheat and Barley Crops at Baladruz District*" Diyala Magazine, **66**, 411-426, (in Arabic) (2015).
7. Basso B. Cammarano D. Carfagna E, *"Review of Crop Yield Forecasting Methods and Early Warning Systems" Research Report*, Michigan

# The spatial distribution of remote sensing stations to monitor water quality (Tigris River) within the administrative limits of the capital Baghdad, by using Geomatics technologies

*Suhaib* Mohammed [1,*], *Abdul-Razzak* Ziboon [2]

[1]Ministry of Environment, Baghdad, Iraq
[2] Building and Construction Department, University of Technology, Baghdad, Iraq

**Abstract.** The spatial distribution of remote sensing stations in the rivers is important technique in the detection of pollution in the river water and acceptance of Comparison Environmental determinants, as a result of toxic waste and waste from other service activities. Therefore, the study of the importance of those stations of the spatial distribution using spatial data and availability, which represent how to hold water in complete tests within the ranges of these stations in order to get to the planning ways to address and reduce pollution within the concept of sustainable development. This will address the project to outline the importance of using technology direct remote sensing in the presentation of spatial data of the stations as well as the use of geographic information systems software for the preparation of spatial databases in the presentation of contaminants databases and relying on the global system signature.

## 1 Introduction

The spatial distribution of remote sensing stations in the rivers is one of the important technical issues in the detection of water pollution in the rivers through the dumping of toxic waste through activities (industrial, service, agricultural). Therefore, the importance of these stations from their spatial distribution using spatial data and availability is significant. This study will provide a brief overview of the importance of the use of remote sensing in the presentation of the spatial data of these stations, as well as the use of "GIS" program to prepare spatial databases in the presentation of pollutant databases and relying on the global signature system.

## 2 Study problem

The problem of the study is that there are no spatial databases (spatial distribution) for remote sensing stations to monitor the pollution of the water sources of the Tigris river of Baghdad.
Therefore, modern technologies will be used to detect and identify the pollution areas of the rivers within the administrative boundaries of the city of Baghdad.

## 3 The objective of the study

The study aims to highlight the role that geotechnical techniques can play in the detection and detection of pollution levels through the use of "remote sensing techniques, satellite images, GPS and GIS" through data obtained from remote sensing stations for monitoring water resources in Baghdad, Water treatment projects as well as pumping stations for the irrigation system on the"Tigris" within the administrative border of the city of Baghdad.

## 4 Methodology

The structure and title of the study will address the method of theoretical and library based on the collection of data on the nature of the problem of the study and then describe the modern mechanisms of collection, storage, processing and analysis of these data and show them in the form of information through objective maps of the tests on the water quality of "Tigris" river within the administrative boundaries of the city of Baghdad.

### 4.1 Geomatics engineering

The term Geomatics first appeared in the early 1980's in the Canadian University of "LAVAL", based on the concept that computer technology produced a scientific revolution in surveying or geometric measurements and digitally represented data to fit a large amount of data. The definition of geomantic is thus, "Integrated multidisciplinary science to select the appropriate devices and technologies to collect, store, model,

* Corresponding author: enggissuhaib79@gmail.com

© The Authors, published by EDP Sciences. This is an open access article distributed under the terms of the Creative Commons Attribution License 4.0 (http://creativecommons.org/licenses/by/4.0/).

analyze, retrieve, display and distribute spatial information resulting from several sources of specific accuracy and characteristics in a digital image. [3].

### 4.2 Remote sensing

The latest technology to collect data from a region by advanced devices and these devices send data to researchers automatically "for study without having to visit the area to collect the required data. A famous example of this technique is the satellite imaging where the satellite takes pictures of the surface of the Earth and then sends these images to the earth stations for analysis and interpretation of the surface form, areas of global warming, the temperature of the earth, the heat of the ocean ... etc.
We can divide the remote sensing process into two types according to the method of data collection.

#### 4.2.1. Separate Sensitivity Process.

This process is done by placing the data collection device away from the area to be studied. The device will receive the reflected or emitted electromagnetic waves emitted from that area and convert these waves into numbers or images that the experts can understand and analyze.
Examples of this method are the astronomical observations which collect light rays from the stars and analyze them to identify the components of those stars as well as space photography where the cameras capture the light reflected from the surface of the earth convert to digital images and send to earth stations for use in various studies and research. Figure (1) below shows the Mechanism of remote sensing technique. [1].

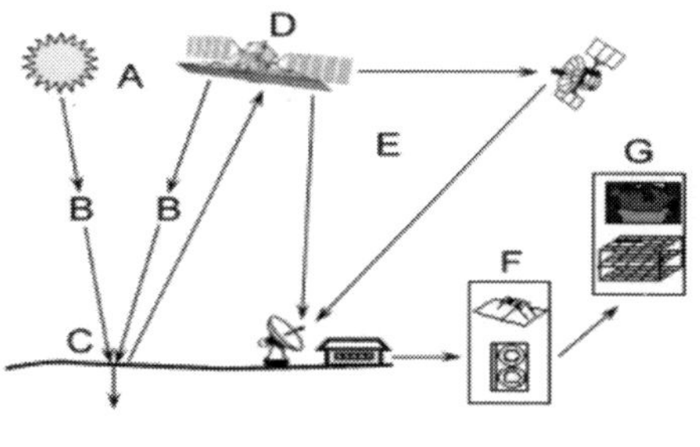


**Fig.1**.Mechanism of remote sensing technique. [1].

#### 4.2.2 Connected Sensitivity Process.

In which the data collection device is located in the area to be studied and left, where the device automatically collects data and sends it to satellites or other data transfer technology (CSD, GPRS) to researchers who are often far from the study area.

#### 4.2.2.1 Connected Sensitivity devices (remote sensing stations)

Figure (2) below shown some of the most important components of these stations:

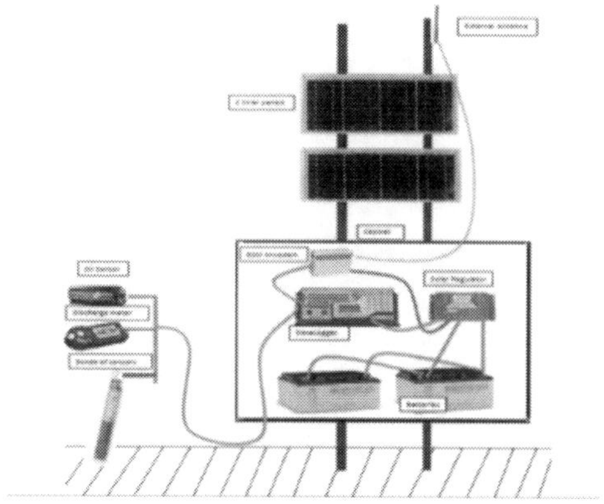

**Fig. 2.** the most important components of these Substation remote sensing. [5].

#### 4.2.2.2 Sensors

They are optical sensors, such as the dissolved oxygen sensor in the water or the ion selective electrode, such as the pH sensor, both of which operate in an electro chemical method. The sensors vary according to the different chemicals that are detected,Figure (3), and can be immersed in the river water completely.

**Fig. 3.** Optical sensors & Ion selective electrode Sensors. [5].

#### 4.2.2.3 Sonde

It's a device connected by more than one sensor.Figure (4). It transmits the entire readings from all sensors to the device that analyzes the data.

 https://doi.org/10.1051/matecconf/201816203020

(sonde)

**Fig. 4**.Sonde transmits the entire readings from all sensors to the device that analyzes the data.[5].

### 4.2.2.4 Data Logger

This device will store the data from the sonde, which by its nature are electrical signals in the unit of measurement (mv) and digitized according to the type of variable read by the sensor and according to the units of measurement adopted globally. The mind of the substation is all the instructions issued from it, for example, the data is sent and received at specific times by instructing the transmitting device to signal the transmission of data and instructing each sensor to take reading at a specific time as well as determine the form of data sent either as tables or in the form of (Text document).

### 4.2.2.5 Signal Transmitter (modem)

It transfers the data analyzed by the data logger from the remote station to the main station where the data is displayed and tabulated.

### 4.2.2.6 Solar panels

To provide continuous power to the substation by charging high-capacity and long-lasting lithium batteries that operate the entire plant components, "which is a mechanism for the use of environmentally friendly renewable energy.

### 4.2.2.7 Main station

This consists of a modern and high-caliber computer calculator. It contains a signal receiving device from the substation and software that displays the readings from the substations and is able to receive data from more than one station. The software then analyzes these readings and displays them appropriately and creates graphs for each variable. The GPRS service has been adopted by mobile phone companies to transfer the signal in the form of Circuit Switch Data (CSD). This is a method of transferring encrypted information. This method has a low cost compared with other signal transmission devices and additionally efficient. Figure (5).represents a remote sensing station in the ( AL Jadiriya) near the Babel Hotel.

**Fig. 5.** Represents a remote sensing station in the ( AL Jadiriya) near the Babel Hotel.. Source (searcher).

## 5 Practical aspect of the study

### 5.1 Phase I (data collection stage).

1- Field survey phase for the purpose of monitoring the locations of remote sensing stations using GPS. Figure (6).data collection.

**Fig. 6.** Part of the on-site monitoring of station coordinates.

2- Desktop survey for the purpose of collecting data for laboratory tests and results of readings of remote sensing devices.

### 5.2 Phase II (design and construction of the GDB)

1-.Field survey coordinates using Arc GIS' 9.3' are summarized on a high-resolution satellite image of '0.6' m.

2-.The geographic database is built using Arc GIS "9.3" and categorized according to the approved standard specifications, according to the 'UTM' Zone(38N) projection system, the WGS 84 Datum and the global Ellipsoid (WGS84).

### 5.3 Phase III (Development of the GDB)

1- The drawing of the symptoms digitizing the Tigris River and points of remote sensing stations and use of the process of topological correction (Topology) of the layers resulting from the drawing process.

2- The process of linking the field survey data and the desk survey with the geographic database for the purpose of accessing the database as final.

### 5.4 Phase IV ( producing a digital map and thematic maps)

1- Production of digital maps in 'mxd' format that can be updated, developed and modified using Arc GIS '9.3'.

2- Production of objective maps on a specific scale for several subjects related to the study provided by the researcher (administrative boundaries of Baghdad governorate, distribution of remote sensing stations in Baghdad, average results of remote sensing readings of the value of a particular element in water, etc.) and high-resolution satellite images.

## 6 Outputs

Outputs are multiple thematic maps in terms of the spatial distribution of stations as well as the descriptive data of the variables of the tests obtained from the stations for monitoring the variables related to the quality of the Tigris. as shown in the Figure (7).

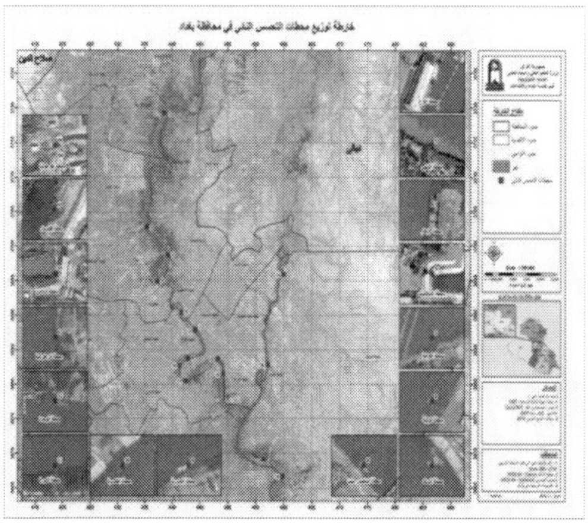

**Fig. 7.**Thematic maps in terms of the spatial distribution of stations. Source (searcher).

## 7 Conclusions

The researcher's approach is to compare between the concentration of Tigris river variables at the time of entering the water of the river, the administrative boundary of the city of Baghdad at the first station, the station( Karkh) as the first monitoring point as well as the station (Madaaen) as the last monitoring station on the Tigris River before leaving the water administrative boundaries of the city of Baghdad. This will focus on the most important etiological factors that affect the change in results in a significant and steady manner.

1-Temperature variable and the measurement unit "c°".

We note that the temperature change between (Karkh) and (Madain) stations is one degree Celsius($\Delta T= +1\hat{c}$ ).

2- PH variable .

We note that the amount of change in the value of hydrogen ion between (Karkh) and (Madain) stations ($\Delta PH = -0.3$ ). It is noted that all the results of the stations fall within the new parameters of the system of river maintenance from pollution ( no. 25) of 1967 and the instructions attached thereto.

3- Total Dissolved Solid (TDS).

We note that the amount of change in the value of the TDS variable between the between (Karkh) and (Madain) stations.($\Delta$TDS) = ( +280) mg / l.

4- Dissolved Oxygen (D.O).

We note that the amount of change in the value of dissolved oxygen concentration (DO) between (Karkh) and (Madain) stations. ($\Delta$DO)=( -1.8) mg/l .

5- Turbidity (Nephlemetric Turbidity Unit "NTU").

We note that the amount of change in the value of turbidity "Turbidity" between (Karkh) and (Madain) stations. ($\Delta$Turb) = (-9.2) NTU.

6- Chlorophyll variable.

We note that the amount of change in the value of the Chlorophyll concentration variable between (Karkh) and (Madain) stations.($\Delta$ Chlorophyll) = ( -1.1)mg/l.

7- Blue green algae "BGA".

We note that the amount of change in the value of the BGA concentration variable between (Karkh) and (Madain) stations. ($\Delta$BGA) = ( -124) Cells/ml.

## 8 Recommendations

1-Finding suitable solutions, including the development and expansion of design energy and address the breakdown of sewage treatment plants located south of the city of Baghdad on the River Diyala (Northern Rustamiyah and Rustamiyah South), as well as the station Karkh streams that flow on the Tigris River south of the city of Baghdad after the convergence of the Tigris River Diyala River where the sewage treatment plant has been out of work since 2003 because of the acts of sabotage that followed the war.

2-Activating all laws and legislations to stop all irregular discharges from stations, for example (Abdel Mohsen El-Kazemi, PN Sidya, TS1), which are mainly designed for "rainwater drainage". The irregular packages have turned them into sewage and industrial sewage stations directly addressed to the river.

3-Follow up all industrial and service activities and impose a remote sensing mechanism on them and develop their environmental awareness and treatment of pollutants through practical and scientific methods and impose appropriate fines on the amount of pollution to the aquatic environment in case they exceed the permissible limits in accordance with the applicable laws and legislation.

The Environmental Protection Act of 2009 has many sections and chapters, including Chapter Three, Article 9, which states: "Those who produce environmental pollution shall comply with the provision of pollution treatment methods and systems using the environmentally clean technologies and their operation and ensure their efficiency and treatment of defects therein. Provide the measurement and control of pollutants and their nature and record the results of measurements and also the adoption of the construction of a database on environmental protection and includes concentrations and levels of pollution. Chapter (V) in this law states that all activities affecting the environment to slavery and in the eighth chapter, he explained the so-called compensation for damages as a result of negligence by complying with environmental laws. In Chapter 9, there were punitive provisions that authorized the Minister to give any warning to any establishment, laboratory or exporting authority for pollution to remove the polluting agent). [4].

4-To work on developing the environmental performance of the control and updating its mechanisms and keeping abreast of the scientific development in this field, by highlighting the role that space images can play in the detection and location of pollution, as well as the daily digital surveillance technology, "digital detectors for field surveying. The enforcement of environmental laws or tax system, which is one of the important tools that can be relied upon in direct activation to address the problem of environmental pollution and will lead to a positive impact to control the values of levels of pollution through the daily use in factories and the Purification plants and power plants located on the Tigris and Diyala rivers in Baghdad.

5-Invest the political and diplomatic momentum in activating the international and bilateral agreements of the shared nature of water resources and maintaining the quantity and quality of the water entering the Iraqi border with each of countries (Turkey, Syria and Iran) because it has a direct and indirect impact on the quality of water and its ability to absorb the pollutants and purify them.

## References

1. A. Abdul-rahman, S. Zlatanova, V. Coors, Innovations in Geographic Information Systems (Springer, 2006)
2. R. Rossi, J. Dungan, L. Beck, Remote Sens. Environ 49, 1 (1994)
3. Gomaa Mohammed, Geomatics Informatics, Saudi Arabia , 2(2014)
4. Env, law of protection and improvement, Ch. III Art, IX(2009)
5. www.ysi.com (accessed 2 April 2015)

# Change detection and building spatial geodatabase for Iraqi marshes

*Oday* Jasim[1,*], *Noor* Hamed[1], and *Tamarra* Abdulgabar[2]

[1]Building and Construction Department, University of Technology, Baghdad , Iraq
[2]Ministry of Labour and Social Affairs, Baghdad , Iraq

**Abstract.**The Iraqi Marshlands has natural and economic potential through its environment rich in various forms of lives. This region has suffered numerous setbacks due to human and natural factors, especially in the last two decades of the last century, which led to significant environmental degradation. The purpose of this paper is to prepare spatial data for the area of the marshes in Iraq as a base (Hour-al Hoveizah and central marshes and Hammar). Also, the other aim is to produce a digital geodatabase for the marshes for the years 1973, 1986, 1999, 2006 and 2016 by using ArcGIS. The process of building geodatabase has been through done in three stages: the first stage is including data collection. The second stage will be by merging the satellite images covering the Iraqi marshes and add to images in order to get the image mosaic process. Also, a georeferencing of satellite images is to be done with all the traditional maps of the same area of the marsh. Finally, complete the full geodatabase for the area of interest by using ArcGIS as the in Cartography Design. The results of this research would be a geodatabase for the Iraqi marshes.

## 1 Introduction

The Iraqi marshes are considered one of the oldest areas with population stability on their lands. The number of population in the marshes area increased during the years of 1977-1997. The population growth rate was 1.9% between the censuses from 1977 to 1987. The population rate decreased to 0.5% between1987 to 1997. While in the period between 1997 -2009 the population growth rate has dropped to -0.6%. The decline in the rate of population growth has occurred due to several reasons. The main reason was the migration of people from the marshes to the province and district centres or other provinces because of the drying and military events and the lack of employment opportunities. The marshes suffered from neglect through poor infrastructure and drying due to no program or development plans. The relationship between man and the environment of social and economic could not be represented, analysed or studied without reference to spatial location. The environment and social and economic activities are closely linked and it is not possible to understand and see the changes taking place or develop without reference to the spatial dimension. GIS provides enormous possibilities to deal with the data within the system of spatial databases. The success of any database depends on the quality and efficiency of the spatial data that form the backbone of any system. Database-building phase needs to be accurate in the collection and incorporation of information in order to get good analysis result. Data is available in multiple formats including spatial information. The spatial information includes maps, aerial photographs or satellite. Also, statistical data has attributes not related to the geographical location. There is sophisticated software that connects between the spatial and statistical data and gives users the possibility of collection, storage, and analysis of data and this so-called term is geoinformation. For the purpose to improve the status of these areas and to prepare or update basic designs to preserve the natural and social environment, the database has to be set up for the Iraqi marshes during different periods depending on the satellite imagery and maps by using geoinformation for its ability to store and update spatial and non-spatial data and find what the relationships between spatial and non-spatial data are.

## 2 Literature review

There are several studies have been published that might be more important due to the aim of this research, such as:

- Aseel Abbas Salman, (2006) has showed selected visible bands in the digital visual interpretation process is considered an optimum technique to observe the soil types using remote sensing techniques. Also, she has been constructed a geographical database and information about layers of soil of the overall studied area (AL-hammar marshes) digitally to represent the geotechnical soil characteristics in associated

* Corresponding author: 40004@uotechnology.edu.iq

© The Authors, published by EDP Sciences. This is an open access article distributed under the terms of the Creative Commons Attribution License 4.0 (http://creativecommons.org/licenses/by/4.0/).

files, and produced digital soil map by using GIS techniques depending on remote sensing data. [2]

- Noor Mahdi alaq, (2011), has assessed the effect of climate factors on marshland of southern Iraq by using remote sensing techniques. She has studied the influence of seasonal climatic factors on floor coverings for the marshes and the possibility of a connection between the marsh area and evaporation using remote sensing techniques and satellite images classified. [3]
- Hussein Kareem Alsaaidi, (2014) has conducted a hydrological study for the marshland of the Shweja, Saadiya, and Dalmaj. He has presented the analysis of physical and chemical properties of marshes to identify causes of decreased amounts of water throughout the drought period and developed proposals for developing the ecological realities of the region.

## 3 Description of study area

The Iraqi marshes located in the central region and the south of the Tigris and Euphrates rivers, along three governorates (Basra, Masan, and Di kar) as shown in figure.1. The most important marshlands are (Al-hammar, the central marshes and Hovaiza). The Iraqi marshes have several advantages including biodiversity of a variety of wildlife, animal, plant, numerous kinds of fish and millions of birds [5].

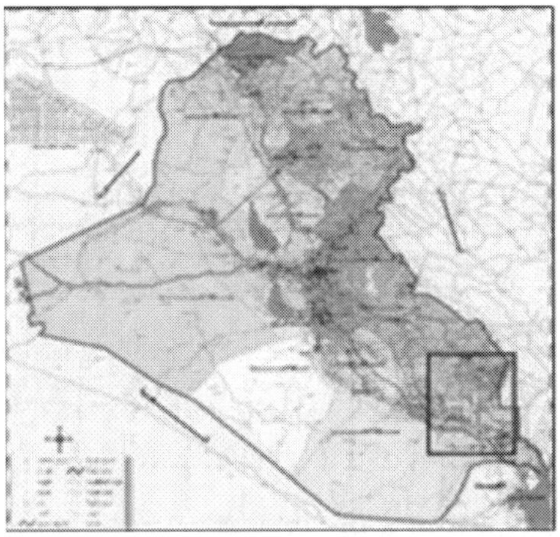

**Fig. 1.** Study area in south Iraq.

## 4 Methodology of work

Requiring of an appropriate spatial data for Iraqi marshlands was represented by collecting four old marsh's maps and downloading five satellite images for study area. The marsh's map sheets were of scale 1/250000 for years 1973, 1986, 1999, and 2006. The five satellites images include two MSS collected in 1973, 1988, two TM collected in 1999, 2006, and one sentinel image collected in 2016. Additional data was collected from government offices. These data include name and area of the marshes, and the factors affecting the low or high water levels. Thus these factors impact the increase or decrease of tracked marshes.

Processing and interpretation of images are basic steps for development of geodatabase and digital maps for Iraqi marshes by using Arc GIS as in the cartography design. The obtained maps are analysed to detect changes in marsh's areas.

### 4.1 Image processing

Mosaicking and matching for all scenes to produce satellite image of Area of Interest (AOI) marsh for each of interest year. All satellite images were geo referenced ,which is the simplest way to link an image to a map projection system by geometric transformation. A transformation is a function that relates the coordinates of two systems. Once such a transformation has been determined, the map coordinates for each image pixel can be calculated. The resulting image is called a georeferenced image. [4]

### 4.2 Building geodatabase

Layers are determined and established to represent features. These features represent various items, such as shallow and deep marshes, road network, administration boundary, villages, and rivers. Also, attribute data has been designed for each feature. These layers will be used in producing the marsh's map with the desired scale and year.

### 4.3 Topology process

If a geospatial vector dataset is derived from a reference dataset either by digitizing or copying geometries, both datasets have a topological relationship. Topology represents a set of rules that prevent interference and overlap of a single or a number of layers. This overlap causes a repeat of spaces. To get high spatial accuracy, it is necessary to hold topology process after screen digitizing process, or topology is the numerical description of the relationships between geographic features, as encoded by adjacency, linkage, inclusion, or proximity [5].

### 4.4 Layout process

Outputs are the final result and can be either in the form of maps, schemas or reports. The scale of the map is an essential factor in this process. Determining of the scale of 1/50000 and adding other maps elements are applied to produce a map of Iraqi marshes for years 1973, 1986, 1999, 2006, and 2016. Brief Technical Methodology for building geodatabase, production map for Iraqi marshes with scale 1/50000, and detecting changes in their area by using GIS and remote sensing techniques are presented in steps. The main steps for that methodology are shown in Table 1.

**Table 1.** Production and updating map with generalization

| Item | Description |
|---|---|
| 1 | Collect, prepare and enter spatial data for the Marshlands area (satellite images and traditional maps) |
| 2 | Combine the satellite images covering the study area with (7) images and making them as unified image in the process of mosaics |
| 3 | Geo reference process for raster images. |
| 4 | Create spatial geo data base through Arc catalog window |
| 5 | Correct interpretation of the imagery or map based on the map legend |
| 6 | Start digitizing of the features from satellite images and traditional maps through the Arc map window with activation of SNAPPIN |
| 7 | Conversion of the digitized features into the geo data base GIS layers |
| 8 | Carry out topology process and fill attribute data Production of digital maps of the Iraqi Marshlands for the years (1973, 1986, 1999, 2006, 2016) with scale1/ 500000 |

And for all these items descriptions the practical works with the geomatics software such as ERDAS, ARC GIS 9.3 occurs as shown in figures 2, 3, 4, 5, 6, and 7.

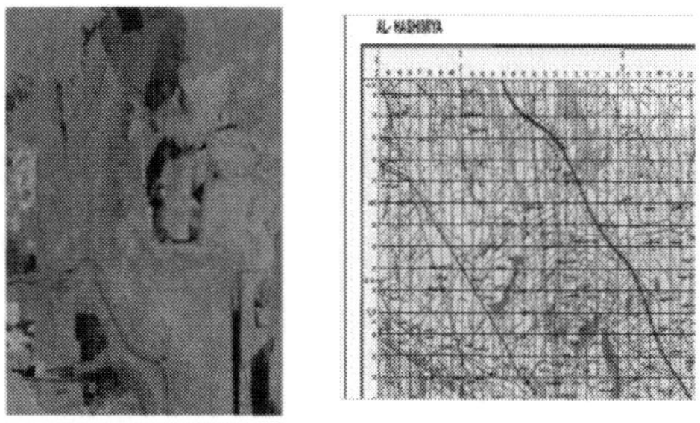

(a) (b)

**Fig.2.** The collected data: (a) satellite image, (b) traditional map.

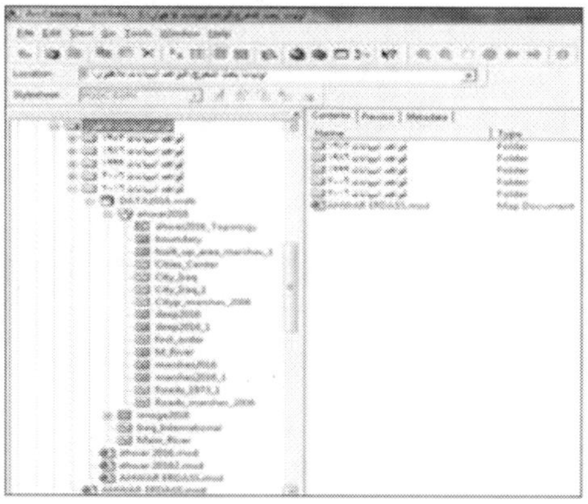

**Fig.3.** Shows creating geodatabase.

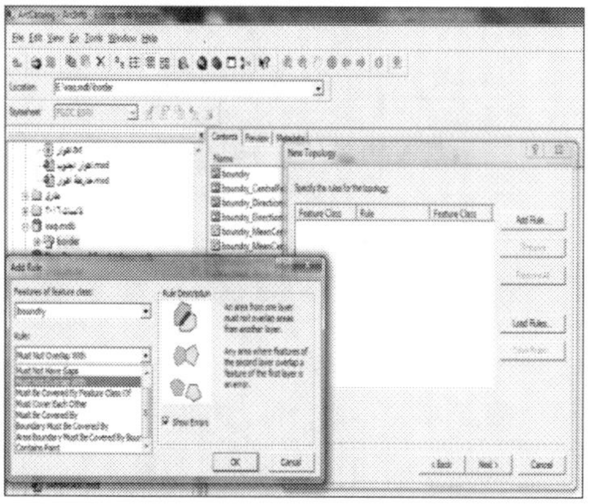

**Fig.4** .Shows digitizing of features.

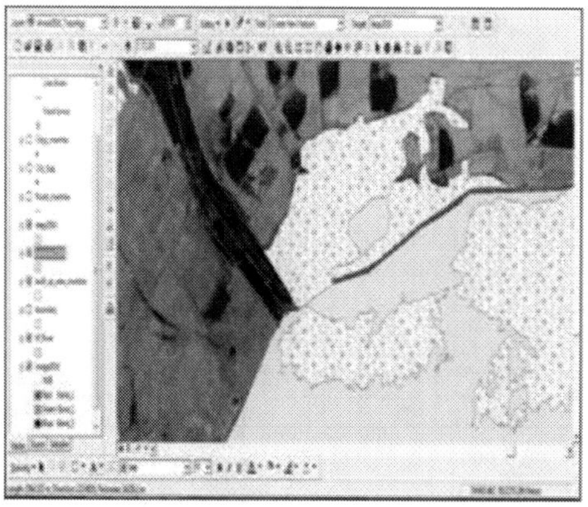

**Fig.5.**Shows the topology.

https://doi.org/10.1051/matecconf/201816203021

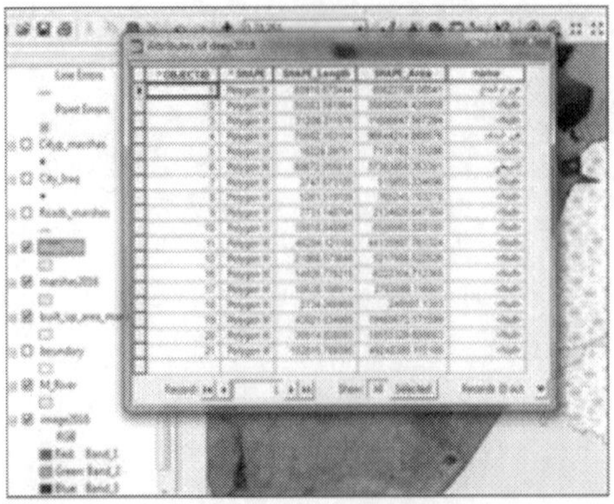

**Fig. 6.** Shows attributes data of Iraqi marshes.

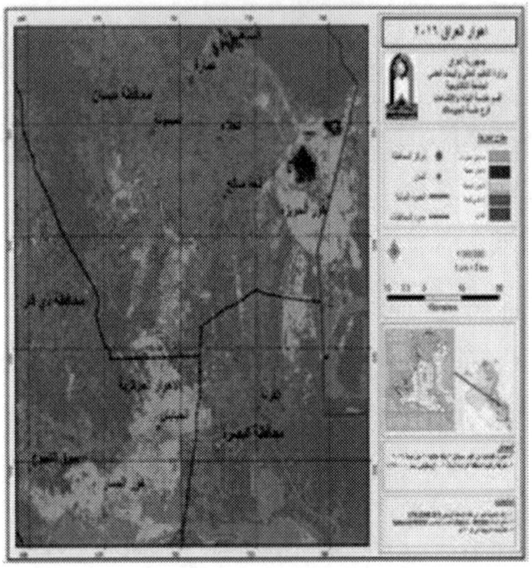

**Fig.7.**Shows Iraqi marshes map for year 2016.

## 5 Summary and Conclusions

The produced maps and calculated area for the Iraqi marshes during different decades by using Arc GIS software have shown the presence of the area of these marshes, which reached at their highest level in 1973 of about 892.383 km$^2$ and the swamp (flood area) of about 9076 km$^2$. Since 1986, and during the Iran-Iraq war, these marshes area was reduced to about 1150 km$^2$and the swamp area was reduced to 6442 km$^2$. In 1999 marshes areas reached to the lowest levels due to the continuous drying processes during the drying period which lasted more than a decade where many of the marshes in those areas disappeared, and the marsh areas reduced to 419 km$^2$and swamp marshes reduced to 769 km$^2$.

In 2006, after attempts to revitalize the marshes, these marshes have begun increasing these areas from what it was at the end of the nineties. The deep marsh areas have amounted to about 782 km$^2$ while shallow marshes reached 3333 km$^2$. From 2016 until now, the study area is about 926 km$^2$while swamp marshes are about 2466 km$^2$.

Remote sensing data is primary source, extensively used for change detection in recent decades. The main information on land use /land cover in the form of maps and statistical data is very vital for spatial planning. In this research, remote sensing and geographic information system (GIS) were used in order to study land use /land cover changes for the period of marshes in the history in Iraq. The use of remotely sensed data (satellite and aerial) to detect changes in land use as well as precise and accurate analysis using GIS is widely preferred over other conventional survey techniques because the method is very efficient for assessing the change or degrading trends of a region.

Finally, planning and decision making processes are dependent on geomatics, especially with engineering works. Geomatics must be recognized for the public to communicate with world technology.

## References

1. A. Salman, (Master thesis), (2006)
2. N..Alaq, (Master thesis), (2011)
3. H. Alsaaidi, (Doctoral dissertation), (2014)
4. Wim H. Bakker, Lucas L. F. Janssen, *Principles of Remote Sensing An introductory textbook* (ITC Educational Textbook Series 2001)
5. Shahab F. *GIS basics* (Published by New Age International Ltd, New Delhi, 2008)

# Assessing the performance of commercial Agisoft PhotoScan software to deliver reliable data for accurate 3D modelling

*Ahmed* Jebur[1,*], *Fanar* Abed[2] and *Mamoun* Mohammed[1]

[1] Engineering Technical College, Central Technical University, Baghdad, Iraq
[2] College of Engineering, University of Baghdad, Baghdad, Iraq

**Abstract.** 3D models delivered from digital photogrammetric techniques have massively increased and developed to meet the requirements of many applications. The reliability of these models is basically dependent on the data processing cycle and the adopted tool solution in addition to data quality. Agisoft PhotoScan is a professional image-based 3D modelling software, which seeks to create orderly, precise n 3D content from fixed images. It works with arbitrary images those qualified in both controlled and uncontrolled conditions. Following the recommendations of many users all around the globe, Agisoft PhotoScan, has become an important source to generate precise 3D data for different applications. How reliable is this data for accurate 3D modelling applications is the current question that needs an answer. Therefore; in this paper, the performance of the Agisoft PhotoScan software was assessed and analyzed to show the potential of the software for accurate 3D modelling applications. To investigate this, a study was carried out in the University of Baghdad / Al-Jaderia campus using data collected from airborne metric camera with 457m flying height. The Agisoft results show potential according to the research objective and the dataset quality following statistical and validation shape analysis.

## 1 Introduction

Photogrammetry is a suitable tool to equip information about man-made and natural objects such as terrain, vegetation and land cover urban features. Photogrammetry and digital image-based processing techniques have an important role to make visualization technology practical and provide low cost workflow when compared to traditional and geodetic survey. The photogrammetric process offers various benefits over the traditional land surveying methods[1]. It represents the elastic job which any project data requires and can be accomplished at the same time it would be long lasting, and corresponding with the permanent cost of image acquisition. It can also be supported with data of higher resolution for digital mapping products such as maps; digital elevation models (DEMs) and orthoimages. Due to this capability, it is convenient and practical for many applications, such as archaeology, monitoring issues and 3D city modelling [2].

A 3D city model is a digital representation of an urban environment with 3D geometry of urban features and structures, however, buildings have the most feature print phenomenon [3]. 3D city models can be technically generated from different digital techniques such as photogrammetry, laser scanning and others. 3D city model has become ready and available for several purposes beyond visualization, and are used in a large number of fields including Engineering [4]. The 3D city model, simulation is done by using true and reliable information via photogrammetry which can be provided with the only economic means to obtain truly 3D city data [5]. With regards to precision and accuracy, 3D city models can be used for simulation and analysis from visualization and animation [6, 7].

Reference [7] discuss about using the Leica Photogrammetric Suite (LPS) software to deliver 3D data in order to create (DTM) from Aerial Photographs. Reference [8] discuss about 3D building modelling from Close range Photogrammetry by using PhotoModeler scanner software and found to be a reliable software for accurate derivable for short-range images. However, [9] also discuss about the use of digital camera for Close Range Photogrammetry (CRP) by using PhotoModeler scanner software. Their result shows that CRP technique is capable to generate the 3D photorealistic model and structural building with photo texturing. On the other hand, [10] introduces a modelling method from aerial images for 3D city modelling. They used a commercial photogrammetric tool and deliver promising outputs showing the validity in terms of time, labor, and reusability. Further, [11] discuss the production of 3D city model for Al-Yarmouk University by using Photogrammetry techniques and geographic information system (GIS). In this study they have discussed the corresponding required production workflow for building 3D GIS Model of a complex site of interest using Digital Photogrammetry.

* Corresponding author: *kareemahmed1980@yahoo.com*

© The Authors, published by EDP Sciences. This is an open access article distributed under the terms of the Creative Commons Attribution License 4.0 (http://creativecommons.org/licenses/by/4.0/).

In a 3D city model development, it is important to choose suitable data with high reliability, appropriate methods and high-efficiency software for production [9]. These data can be derived from aerial images, terrestrial photos and laser scanning data etc. However, the high-efficiency software such as (LPS) Leica photogrammetry suite, Agisoft Photoscan or PhotoModeler scanner or others, has to be investigated and validated for individual data. As the majority of the research depends on various softwares, this research discusses the possibility of using the Agisoft Photoscan software to deliver reliable data for accurate 3D modelling applications. It also features other different methods to evaluate the results obtained through the Agisoft Photoscan software, compared to the Leica photogrammetry suite (LPS) which has been referred to in this work.

## 2 Study Area and Dataset

The study area comprises the boundaries of the college of Engineering / University of Baghdad / IRAQ which equals to (1.33329 sq. km). It is located within (33° 16′ - 33°16′39″ N) latitude and (44° 22′10″- 44°23′05″ E) longitude, respectively as shown in Figure (1).

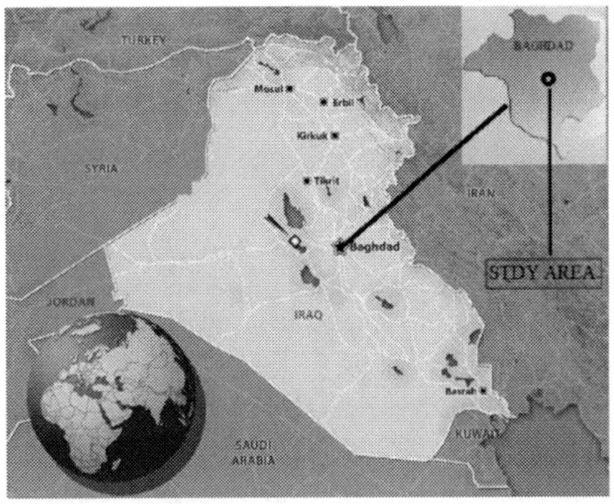


**Fig.1.** Study Area.

The dataset consists of eight aerial photos captured by metric camera format of (230 x 230 mm) in 1983. This dataset is an interesting one to be investigated because it was the first photogrammetry project to cover the majority of Baghdad University. The panchromatic images are consisting of two flight lines with 20% side lap and more than 60% end lap. These images were captured with flying height of 457 m above ground and camera focal length of 152.16 mm as shown in Figure (2).

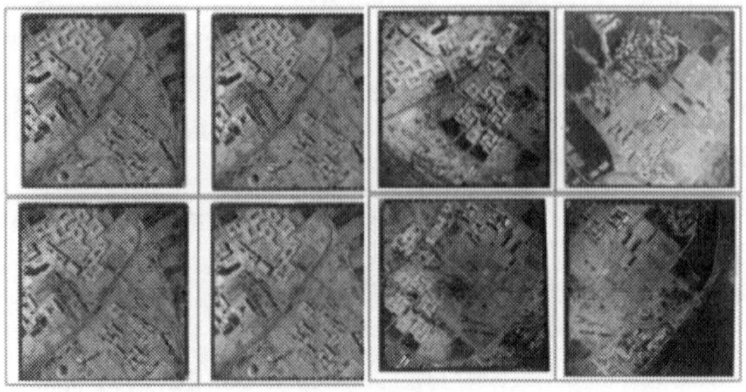

**Fig.2.** Represents The Block of Photos of The Dataset Including Flight lines 1 and 2.

## 3 Methodologies

This research was carried out in four basic stages. The first stage was accomplished by collecting GCPs and adjusting the network delivered. The second stage consisted of photo processing and triangulation using least squares solution. The third stage consisted of implemented procedures to carry out the spatial analysis to create raster surface. And the last stage included the statistical analysis, assessment, and validation of the generated 3D surface.

### 3.1 Ground Control Points (GCPs) Collection

The ground control points were obtained by using differential global positioning system (DGPS) technique, and later adjusted accurately following robust least squares scenario. The type of the GPS device used to collect the data was Topcon GR-5, and the working mode was the real time kinematic (RTK). The coordinates of 29 ground points were collected and located on well-defined features with good and homogeneous distribution over the study area. These were measured with reference to the UTM projection reference. From all collected points, only 20 were used as GCPs and the remaining 9 points were used as Check Points as show in Figure (3).

**Fig.3.** Distribution of GCPs and Check Points in Study Area.

A high number of Check Points were used in this research because the reliability of the accuracy estimations depends among other variables, on the number of Check Points and their distribution [12]. Later a network of 3D coordinates are delivered, however, it cannot be post-processed for accurate later stages without adjustment. Therefore, 3D network adjustment was implemented to adjust all of the 38 GCPs using non-linear solution (variation of coordinate's model). Least squares routine was adopted for accurate final derivable. The adjusted coordinates of the GCPs was used later on for the next stage of processing.

## 3.2 Photo Processing by Simultaneous Solution (Bundle Adjustment)

The photo processing with aerotriangulation workflow is also named as bundle adjustment technique because it is based on the idea that many light rays which come from the light source (camera sensor) are forming a bundle of rays all intersects to generate spatial positions. The bundles from all involved images are later adjusted simultaneously so that corresponding light rays intersect at positions of ground features [13]. Collinearity equations are the formula used as a mathematical model to adjust these bundles using least squares. However, these equations are not linear and thus, need to be linearized for an optimal solution. This can be obtained mathematically by adopting the first order terms of Taylor's theorem using initial values to the unknowns [14]. 3D dense point clouds are generated from this step as input for the next further steps.

### *3.2.1 Bundle Block Adjustment using LPS software*

Leica Photogrammetry Suite (LPS), provided a flexible environment for aeritriangulation and bundle adjustment. LPS is a commercial digital photogrammetric solution which allows fast processing of many types of image-based solutions such as triangulation or orthorectification with accurate and precise outputs. It can manipulate any kind of image collected from different sources of cameras and satellite sensors. It can easily model various camera and sensor types by generating the image geometry at the time of exposure and deliver the orientation parameters for individual images at the time of collection. Later, the software computes the mathematical relationship between the bundle of images to compute the geospatial coordinates of the ground points based on the collinearity condition equations assumption [15]. This can be applied by following space intersection and space resection techniques for the bundle of images simultaneously. However, to do so it is important to engage some additional techniques during the process of image capturing, these are the GPS and the INS techniques. They both normally deliver initial measurements to the exterior orientation; however, these measurements need to be adjusted for further processing in the triangulation modelling due to high level accuracy requirement [15]. A bundled solution is adopted to calculate the exterior orientation parameters of individual images in the block and the final geospatial coordinates of the unknown points. The block of eight images (2 flight lines) was included in this research and was simultaneously treated in one solution. A least squares adjustment is applied to compute the unknowns in the entire block following multi iterations to decrease individual errors. It is obvious that block triangulation is the optimal operation to define the mathematical relationship between the images in a certain block and the ground. Once the relationship is defined correctly, accurate geospatial positions of ground features can be extracted.

### *3.2.2 Bundle Block Adjustment using Agisoft photoscan software*

Agisoft PhotoScan is professional image-based 3D modeller software seeking at creating orderly precision 3D content from fixed images [16]. It works with arbitrary images those qualified in both controlled and uncontrolled conditions. In Agisof PhotoScan software, the images can be captured from any position, in a way that the object can be reconstructed from at least two photos taken from different positions. All the process including image alignment and 3D model reconstruction are fully automated as shown in Figure (4). Following the recommendations of many users all around the globe, Agisoft PhotoScan, has become an important source to generate precise 3D data for different applications.
Agisoft photoscan Project Manager also uses the bundle block adjustment for triangulation method. However, Agisoft PhotoScan has a major issue regarding photo orientation as compared to the LPS. This becomes obvious when the photos are highly oriented from the Nadir (in Kappa direction). This can be interpreted due to the lack of the PhotoScan software to apply photo calibration including distortion parameters during the adjustment process. This could be a major issue if the photos were not originally in digital form and has been transformed into a digital form through scanning process which is the case in this research.

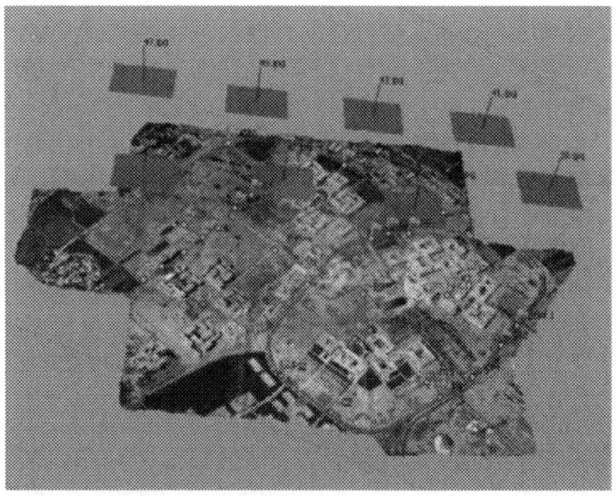

**Fig.4.** Illustrate Image Alignment and 3D Model Reconstruction.

## 3.3 Spatial Analysis (Interpolation)

Interpolation is the procedure used to foretell cell values in locations that Lack sample points. It is based on the principle of spatial self-correlation or spatial dependence between the near and distant organism. Spatial analyst methods which are considered to be one of the important tools in GIS, these tools are used to create maps. Spatial analysis method is a public term for all manipulations of spatial data with the aim of improving one's understanding of the geographical phenomena represented by the data [17]. In this research the spatial analysis was carried out using the ArcGIS® Geostatistical Analyst toolbar to generate 3D model. The spatial analysis was conducted using multiple interpolation methods which are available in the ArcGIS® Geostatistical Analyst toolbar. The spatial interpolators assessed in this work were: kriging method, IDW method, and spline method. These methods were chosen because they are widely used and approved in the literature.

#### *3.3.1 Spatial Analysis Methods*

##### *3.3.1.1 Kriging Method*

Kriging is a geostatistical gridding method that has proved its validity and efficiency in many applications. This method produces visually attractive maps of irregularly spaced data. Kriging is a very flexible gridding method and can produce very accurate surfaces of any kind of data. It also can be custom-fit to a data set, by specifying the appropriate variogram model. Within Arc Map, Kriging can be either a smoothing or an exact interpolator, depending on the specified parameters [18].

##### *3.3.1.2 Inverse Distance Weighting (IDW) Method*

The Inverse Distance is a weighted average interpolator, which can be either smoothing or exact interpolate data. Data are weighted with this method to a certain power during the interpolation process, so that the influence of one point, relative to another, decreasing with distance from the node of grid. The greater the power of weighting when far removed from the grid node, lesser effect on the points, during interpolation [19]. By using a linear -weighted combination IDW determines the values of the cells for a set of sample points. The weight assigned is considered as a function of the distance from a certain input point to the output cell location. The greater the distance, lesser the effect on the output value [17].

##### *3.3.1.3 Nearest Neighbours Method*

The nearest neighbours (NN) method foresees the value of the attribute at an un sampled point based on known values of the nearest sampled points. This can be achieved by drawing perpendicular bisectors between sampled points (n), forming such as Thiessen polygons. This produces one polygon for individual samples and also in the centre of the polygon located in the sample. All points are nearer to its enclosed sample point than to any other sample points in each polygon [20]. Within each polygon all points (or locations) are assigned the same value (weights) [21].

### 3.4 Statistical Analysis and Validation Assessments

In order to determine the difference between both softwares which have been used, comparison between the produced data from both softwares has been. The method that was used for the comparison is statistical, which depends on some of the standards. In order to assess the surface generated by different interpolation methods, some field measurements has adopted for this assessment, which is regarded as a reference, such as length, width or building area as shown in Figure (5). In order to verify the accuracy and reliability of the surface formed by the methods of interpolation, comparison was made with field measurements.

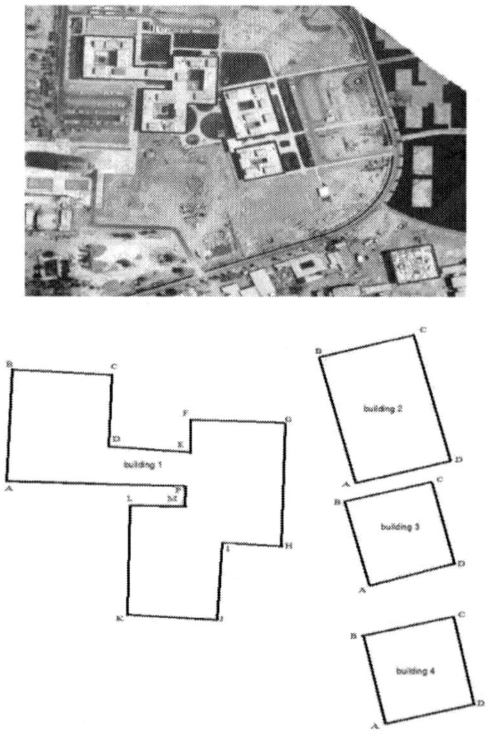

**Fig.5.** Illustrates Field Measurements for Validation Assessment.

## 4 Results

### 4.1 Result of GCPs Adjustment

Table (1) shows the corrected coordinates of ground control points using least squares method after applying 3D network adjustment as stated in section 3.1.

**Table 1.** The Corrected 3D Coordinates of the GCPs.

| | Point ID | $X_m$ | $Y_m$ | $Z_m$ | Type |
|---|---|---|---|---|---|
| 1 | P7 | 441820.02 | 3681854.52 | 36.12 | Control |
| 2 | Point 2 | 442057.74 | 3681729.71 | 35.63 | Control |
| 3 | Point 6 | 442122.97 | 3681835.03 | 36.60 | Control |
| 4 | C2 | 442387.34 | 3681975.24 | 35.05 | Control |
| 5 | P12 | 442426.57 | 3681818.94 | 35.54 | Control |
| 6 | P3 | 442169.90 | 3681468.58 | 36.78 | Control |
| 7 | Point 5 | 442168.13 | 3681271.72 | 35.92 | Control |
| 8 | P5 | 442400.08 | 3681334.27 | 36.00 | Control |
| 9 | 1200 | 442024.10 | 3681031.63 | 35.71 | Check |
| 10 | C7 | 442018.14 | 3681032.84 | 35.72 | Control |
| 11 | C9 | 441833.53 | 3681595.12 | 34.90 | Check |
| 12 | C11 | 441846.08 | 3681868.01 | 36.15 | Check |
| 13 | Point 1 | 441747.75 | 3682054.30 | 33.82 | Control |
| 14 | P9 | 442106.40 | 3682106.20 | 35.01 | Check |
| 15 | C15 | 442103.99 | 3682051.10 | 35.20 | Check |
| 16 | C13 | 441840.13 | 3682077.14 | 35.74 | Control |
| 17 | P13 | 441619.08 | 3681622.60 | 34.79 | Control |
| 18 | Point 3 | 442550.16 | 3681361.10 | 35.94 | Control |
| 19 | 1000 | 441759.13 | 3681296.50 | 34.99 | Check |
| 20 | C3 | 442174.52 | 3681618.74 | 36.51 | Control |
| 21 | C14 | 441986.11 | 3682110.09 | 35.59 | Control |
| 22 | C1 | 442181.34 | 3681837.57 | 36.89 | Check |
| 23 | C12 | 441749.52 | 3681854.89 | 36.42 | Control |
| 24 | Point 4 | 441892.90 | 3681582.47 | 35.63 | Control |
| 25 | P8 | 441774.15 | 3681630.71 | 35.41 | Control |
| 26 | C4 | 442097.56 | 3681256.32 | 36.25 | Control |
| 27 | C5 | 442339.65 | 3681249.01 | 36.39 | Check |
| 28 | C16 | 442324.72 | 3681318.42 | 36.69 | Control |
| 29 | C6 | 442482.30 | 3681346.87 | 36.10 | Check |

## 4.2 Result and Analysis of The bundle Block Adjustment

Triangulation has been applied by using LPS and Agisoft photoscan software. In individual overlapped areas at least three GCPs must be specified for each stereo pair, and on stereo image, several tie points were measured. The triangulation method is performed to estimate the ground coordinates (X, Y, Z) locations of tie points in stereo models. It also computed the exterior orientation parameters (EOP) of individual images. Figure (6) illustrates the distribution of the GCPs, tie points and check points in the adjusted stereo models. However, Figure (7) illustrates the distribution of the GCPs and check points in the adjusted stereo model by Agisoft photoscan software.

In triangulation, a less number of control points were used in the block for both softwares. The results of the triangulation are 3D point's clouds as shown in Figures (8 and 9). The number of the generated point's clouds was equal to 1784 by LPS software points, however, more than 7519294 points was generated by Agisoft photoscan software.

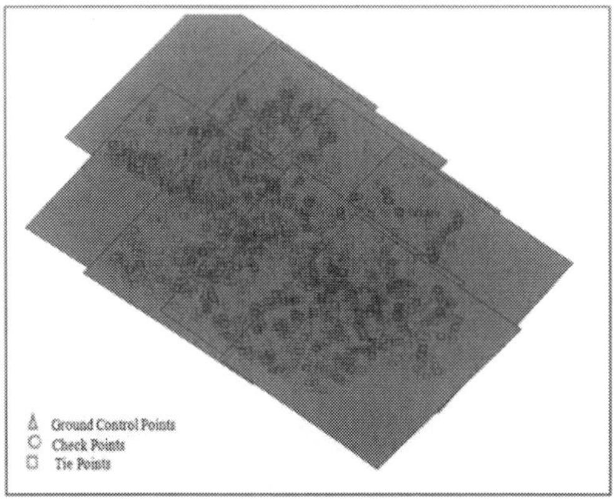


**Fig.6.** The Distribution of the GCPs, Tie Points and Check Points in the Adjusted Stereo Model by LPS software.

**Fig.7.** Distribution of Ground Control Points and Check Points in the Adjusted Stereo Model by Agisoft Photoscan software.

After the process of triangulation has been completed, RMSE was calculated. Nine check points were used to compare the triangulated coordinates from photogrammetry with those measured by DGPS and adjusted later. The results are listed in Tables (2 and 3) for both softwares.

**Fig.8.** Illustrates the Triangulated Points Clouds Generated From LPS Software.

**Fig.9.** Illustrates the Triangulated Points Clouds Generated From Agisoft Photoscan Software.

### 4.3 Result of Interpolation Methods

For interpolation purposes, points were exported from LPS and Agisoft Photoscan softwares to Arc Map for spatial analysis and to generate a surface for the study area. The results are as shown in Figures (10, 11 and 12) and Figures (13, 14 and 15) for both the softwares, respectively.

### 4.4 Result of Statistical Analysis and Validation Assessment

Statistical analysis was performed on the points clouds that were generated by Agisoft photoscan and LPS software. In this research, statistical analysis was conducted by using one-sample T-test to find error values. One-sample T-test stating an assumption about a population with large sample sizes, generally meaning $\geq$ 30, from a continuous population, the distribution of sample is well approximated by a normal distribution. For this research there is one value to accept or reject which is the null hypothesis: p-value. Whereas, P-value referred to significance level, it is used as a standard for accepting or rejecting the null hypothesis. For this study, 95% CI was applied according to NSSDA approach whereas P-value is 0.05. After finding the value of one-sample T-test, the P-value from the statistical table can be easily determined. Further it cannot reject the null hypothesis when the P-value is greater than 0.05, and it should accept alternative hypothesis when P-value is less than 0.05. The following Figures (16 and 17) show the results of the statistical analysis.

**Table 2.** RMSE Delivered from LPS Software.

| | Reference coordinates of check Points (m.) | | | Calculated coordinates of check Points (m.) | | |
|---|---|---|---|---|---|---|
| **Point ID** | **X** | **Y** | **Z** | **X** | **Y** | **Z** |
| 1200 | 442024.99 | 3681031.63 | 35.71 | 442024.82 | 3681031.48 | 35.15 |
| C9 | 441833.53 | 3681595.12 | 34.90 | 441833.79 | 3681594.96 | 35.17 |
| C11 | 441846.08 | 3681868.01 | 36.15 | 441846.12 | 3681867.75 | 36.25 |
| P9 | 442106.40 | 3682106.20 | 35.01 | 442106.36 | 3682106.23 | 34.99 |
| C15 | 442103.99 | 3682051.10 | 35.20 | 442104.59 | 3682051.16 | 34.54 |
| 1000 | 441759.13 | 3681296.50 | 34.99 | 441759.41 | 3681296.52 | 34.78 |
| C1 | 442181.34 | 3681837.57 | 36.89 | 442181.64 | 3681837.36 | 36.40 |
| C5 | 442339.65 | 3681249.01 | 36.39 | 442339.31 | 3681248.94 | 36.39 |
| C6 | 442482.30 | 3681346.87 | 36.10 | 442482.20 | 3681346.65 | 36.100 |
| Mean Square Error in check points (m) | | | | | | |
| RMSE X=0.3001 | | RMSE Y 0.1569 | | | RMSE Z = 0.3527 | |

**Table 3.** RMSE Delivered from Agisoft Photoscan Software.

| | Reference coordinates of check Points (m.) | | | Calculated coordinates of check Points (m.) | | |
|---|---|---|---|---|---|---|
| **Point ID** | **X** | **Y** | **Z** | **X** | **Y** | **Z** |
| 1200 | 442024.99 | 3681031.63 | 35.71 | 442025.12 | 3681031.93 | 35.99 |
| C9 | 441833.53 | 3681595.12 | 34.90 | 441833.64 | 3681595.05 | 35.17 |
| C11 | 441846.08 | 3681868.01 | 36.15 | 441846.58 | 3681868.38 | 36.49 |
| P9 | 442106.40 | 3682106.20 | 35.01 | 442106.19 | 3682106.07 | 34.49 |
| C15 | 442103.99 | 3682051.10 | 35.20 | 442103.76 | 3682051.08 | 35.31 |
| 1000 | 441759.13 | 3681296.50 | 34.99 | 441759.54 | 3681296.26 | 34.68 |
| C1 | 442181.34 | 3681837.57 | 36.89 | 442181.89 | 3681837.14 | 37.09 |
| C5 | 442339.65 | 3681249.01 | 36.39 | 442339.55 | 3681249.30 | 36.19 |
| C6 | 442482.30 | 3681346.87 | 36.10 | 442482.23 | 3681347.19 | 35.88 |
| Mean Square Error in check points (m) | | | | | | |
| RMSE X=0.310 | | RMSE Y =0.276 | | | RMSE Z =0.294 | |

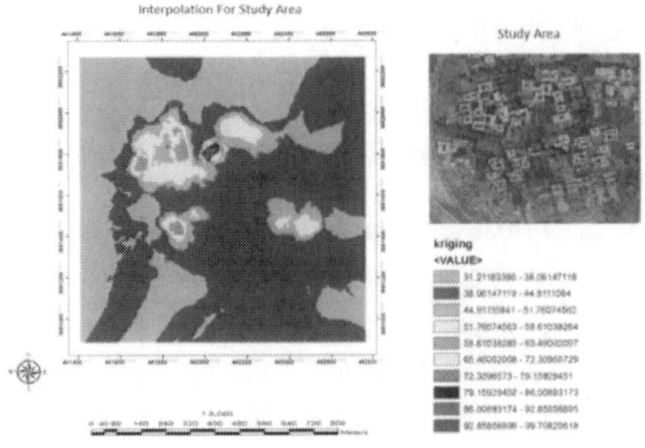


**Fig.10.** Interpolation by kriging Method (Results from LPS Software).

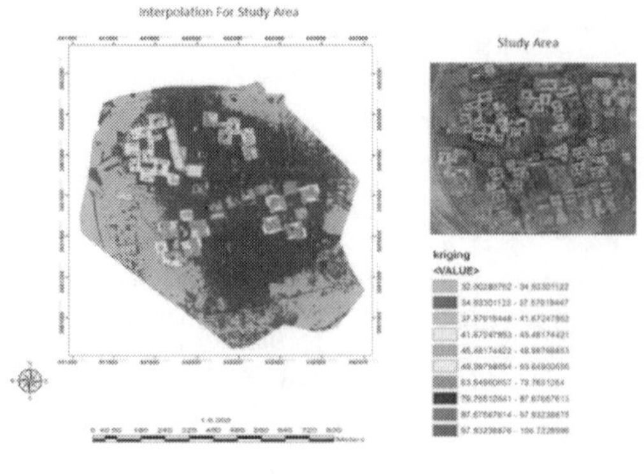


**Fig.13.** Interpolation by Kriging Method (Results from Agisoft PhotoScan Software).

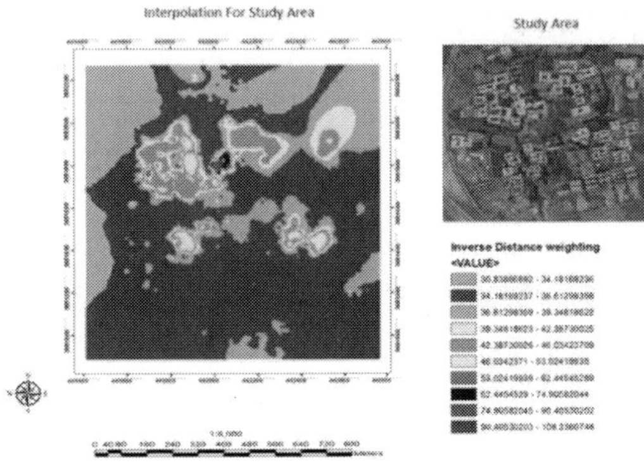


**Fig.11.** Interpolation by IDW Method (Results from LPS Software).

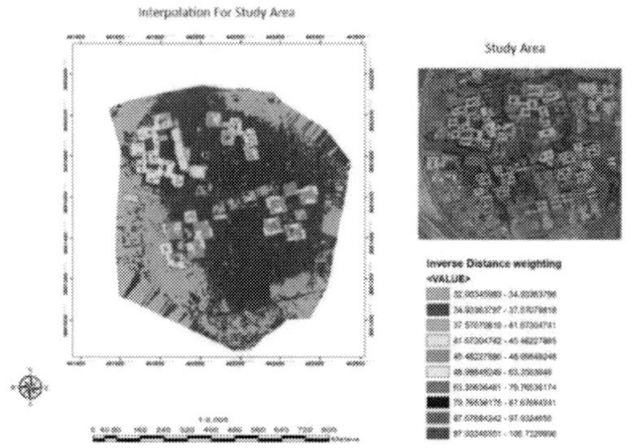


**Fig.14.** Interpolation by IDW Method (Results from Agisoft PhotoScan Software).

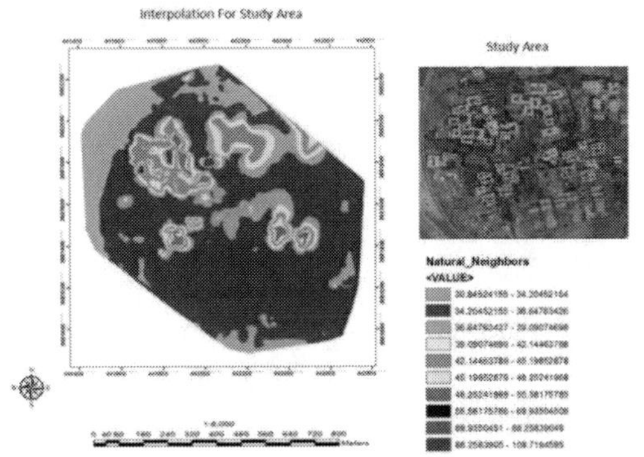


**Fig.12.** Interpolation by Nearest Neighbours Method (Results from LPS Software)

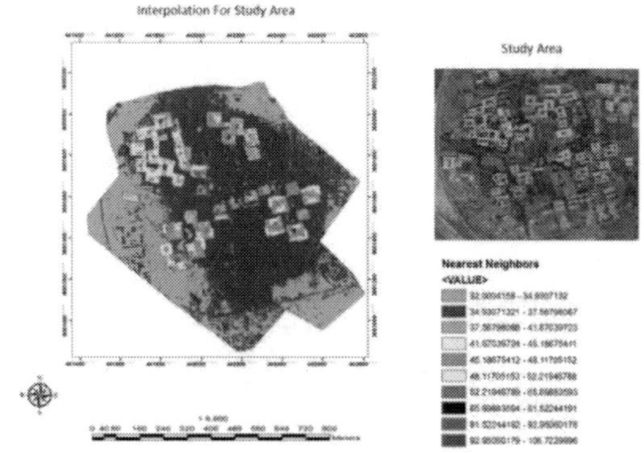


**Fig.15.** Interpolation by Nearest Neighbours Method (Results from Agisoft PhotoScan Software).

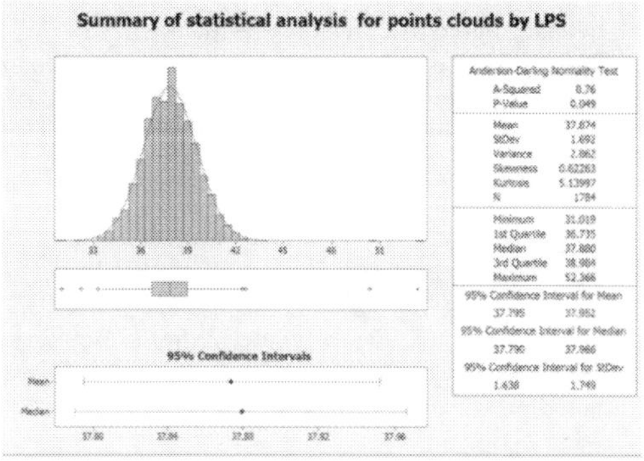


**Fig.16.** Summary Results of Statistical Analysis for LPS data.

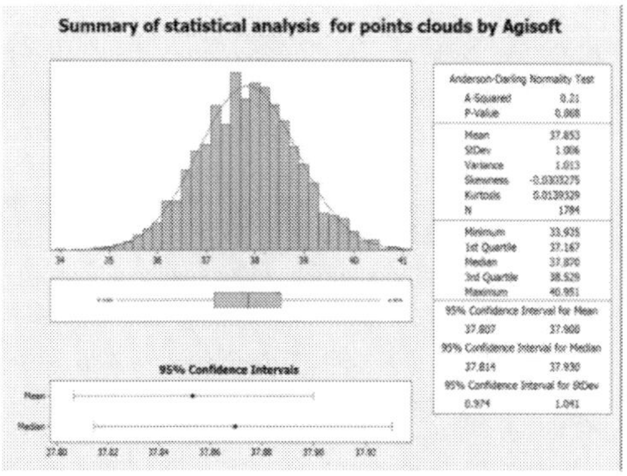


**Fig.17.** Summary Result of Statistical Analysis for Agisoft Photoscan data.

In order to evaluate the generated surfaces from the Agisoft Photoscan software, interpolated surfaces were assessed. Based on the standard deviation delivered, it was found that the kriging method is more reliable than other methods. Therefore, the Kriging method was selected to apply in the validation process with the reference measurements. The validation process was based on the shape geometric measurements for some accurate and generic objects, such as the length, width and the area of selected features. The results are shown in Tables 4 and 5 below.

**Table 4.** Illustrate the Validation Results of Lengths and Width Measurements.

| Interpolation Method | Building | Side | Reference distance (m) | Computed distance (m) | Diff (m) |
|---|---|---|---|---|---|
| | | AB | 70.31 | 72.43 | 2.12 |
| | | BC | 64.22 | 62.12 | -2.1 |
| | | CD | 48.09 | 46.17 | -1.92 |
| | | DE | 49.97 | 51.18 | 1.21 |
| | | EF | 20.97 | 20.7 | -0.27 |
| | | FG | 58.23 | 59.11 | 0.88 |
| | Building 1 | GH | 78.84 | 80.25 | 1.41 |
| | | HI | 35.79 | 36.53 | 0.74 |
| | | IJ | 48.08 | 49.97 | 1.89 |
| | | JK | 58.2 | 56.54 | -1.66 |
| | | KL | 69.75 | 70.94 | 1.19 |
| | | LM | 36.18 | 33.85 | -2.33 |
| | | MP | 13.98 | 12.79 | -1.19 |
| | | PA | 114.61 | 111.41 | -3.2 |
| Kriging | | | | | |
| | | AB | 82.08 | 84.58 | 2.5 |
| | Building 2 | BC | 58.22 | 60.6 | 2.38 |
| | | CD | 82.08 | 84.59 | 2.51 |
| | | DA | 58.22 | 60.69 | 2.47 |
| | | | | | |
| | | AB | 52.13 | 55.5 | 3.37 |
| | Building 3 | BC | 52.3 | 54.14 | 1.84 |
| | | CD | 52.13 | 55.46 | 3.33 |
| | | DA | 52.3 | 54.17 | 1.87 |
| | | | | | |
| | | AB | 54.64 | 57.87 | 3.23 |
| | Building 4 | BC | 54.97 | 56.96 | 1.99 |
| | | CD | 54.64 | 57.89 | 3.25 |
| | | DA | 54.97 | 56.97 | 2 |
| | | | | SD = 2.234m | |

**Table 5.** Illustrate the Validation Results of Area Measurements.

| | Building | Reference area $m^2$ | Computed area $m^2$ | Diff $m^2$ |
|---|---|---|---|---|
| | Building 1 | 13783.88 | 14094.54 | 310.67 |
| Kriging | Building 2 | 4778.70 | 5130.27 | 351.58 |
| | Building 3 | 2726.40 | 3035.08 | 308.68 |
| | Building 4 | 3003.56 | 3320.92 | 317.35 |
| | SD = 372.432 $m^2$ | | | |

## 5 Discussion and Conclusions

In this paper, the ability of Agisoft PhotoScan software to derive reliable data from aerial images was assessed and analyzed accordingly and to subsequently show the potential of this software for accurate 3D modelling. To investigate this, a study was carried out in the University of Baghdad / Al-Jaderia campus using data collected from airborne metric camera with 457m flying height. The Agisoft results show potential according to the research objective and the dataset quality following statistical and validation shape analysis. The reliability of the data delivered depends on different variables: the quality of the used imagery, image orientation, adopted scenario, GCPs availability and distribution, etc. The GCPs was measured and adjusted accurately following robust least squares scenario. The study is based on assessing the Agisoft results with those delivered from Leica Photogrammetric Suite software as a reference. Multiple interpolation methods were adopted, assessed and compared to find the most reliable surface created from the triangulated data delivered from the Agisoft software. Statistical analysis was applied and the results were interpreted accordingly for individual interpolation methods. Shape geometric measurements were conducted in the field following accurate and generic object selection in order to validate the generated 3D models. The following 3D RMSE values of 0.310m, 0.276m and 0.294m were delivered from Agisoft software. However, Leica Photogrammetry Suite has resulted 0.290m, 0.157m and 0.353m 3D RMSE values in the X, Y, and Z respectively.

The results delivered indicate that the accuracy of the Aigesoft Photoscan software in the process of triangulation is high compared to the LPS software through the RMS outputs in terms of accuracy. From the methods of interpolation that depend on the data which was obtained from both softwares, it was found that interpolation results delivered from the Aigesoft Photoscan data is better than those delivered from the LPS data. This is because the number of points that intensified by the Aigesoft Photoscan are much more than the points intensified by LPS because of the ability of Aigesoft Photoscan to intensify the points during the process of triangulation to the number of millions of points contrary to the LPS software. Further, the results from the statistical analysis indicate that the method of kriging is the best interpolation method to intensify 3D points which is apparent from the standard deviation value delivered. Therefore, the following conclusions can be made upon the case study results:

- ➢ The results show the reliability and the validation of Agisoft Photoscan software in the triangulation process and thus, in 3D modelling applications.
- ➢ The great potential of Agisoft Photoscan software to intensify points during triangulation process in comparison with the LPS software for accurate 3D modelling applications.
- ➢ The easy use of the Agisoft photoscan software comparing to the LPS software as the former doesn't need any approximate values to the unknowns including the exterior orientation parameters and the control points.
- ➢ Kriging interpolation method is the best algorithm to generate surface from 3D data for accurate modelling.

## References

1. A. Gruen, X. Wang, *In Photogramm. Week* (1999)
2. K. Hammoudi, *Contributions to the 3D city modeling: 3D polyhedral building model reconstruction from aerial images and 3D facade modeling from terrestrial 3D point cloud and images* (Université Paris-Est, 2011)
3. M. Lancelle, D.W. Fellner, *In Image Vision Comput.* 363, (2004)
4. R. Billen, et al. *3D City Models and urban information: Current issues and perspectives* (EDP sciences, 2014)
5. I. Suveg, G. Vosselman, ISPRS J. Photogramm. **58**, 3 (2004)
6. N. Yastıklı, Z. Alkış, O. Emem, *In CIPA XIX Conf. Proc.* (2003)
7. Y. Al-kinani, Eng. Tech. J. **8,** 33 (2015)
8. S. Singh, K. Jain, V. Mandla, *In Geomatrix-12 Conf. Proc.* (2012)
9. A. Rasam, et al. *In 4th IEEE ICSGRC Conf. Proc.* (2013)
10. A. Ali, C. Berebbia, *Digital Architecture and Construction* (WIT Press, 2006)
11. A. Abdul-rahman, S. Zlatanova, V. Coors, *Innovations in 3D Geo Information Systems* (Springer, 2006)
12. A. Aguilar, et al. *In IMProVe Conf. Proc.* (2011)
13. P. Wolf, B. Dewitt, *Elements of photogrammetry: with applications in GIS* (McGraw-Hill, 2000)
14. S. Ghosh, Photogrammetria. **31,** 3 (1975)
15. I. Erdas, *LPS project manager user guide* (Erdas Inc., 2008)
16. L. Agisoft, *Agisoft PhotoScan user Manual* (Agisoft LLC, 2014)
17. C. Childs, E. Services, Arc user **32**, 7 (2004)
18. C.. Yang, et al. *In XX ISPRS Congress* (2004)

19. Y. Hu, *Mapping monthly precipitation in Sweden by using GIS* (Gothenburg, 2010)
20. R. Rossi, J. Dungan, L. Beck, Remote Sens. Environ **49**, 1 (1994)
21. R. Webster, M. Oliver, *Geostatistics for environmental scientists* (John Wiley & Sons, 2007).

MATEC Web of Conferences 162, 03023 (2018) https://doi.org/10.1051/matecconf/201816203023
BCEE3-2017

# Building archaeology geodatabase in Iraq using GIS

*Abbas* Kalaf [1], *Hisham* Al Sharaa [1,*], *Israa* Hatem[1], *and Hussein* Mohammed[1]

[1] Building and Construction Department, University of Technology, Baghdad, Iraq

**Abstract.** Geomatics has been an important tool in archaeology. The combination of Geomatics and archaeology adopters have been considered a perfect match, since archaeology often involves the study of the spatial dimension of human behavior over time, and all archaeology carries a spatial component. Since Iraqi archaeology becomes one of the main victims of destruction by negligence and terror attacks, makes our great heritages forgotten. Hence, it is necessary to build a secure database for all Iraqi archeological sites with their two main types (investigated and uninvestigated) and rely on digital system by creating digital maps for each Governorate with their archeological database system. Results of archaeological studies are rich in spatial information. GIS is adept at processing these large volumes of data especially those that are geographically referenced. It is effective, accurate and a fast tool. The tools made available through GIS help in data collection, its storage and retrieval, its ability for customization and, finally, the display of the data so that it is visually comprehensible by the user. The most important aspect of GIS in archaeology lies, however, not in its use as a pure map-making tool, but in its capability to merge and analyze

## 1 Introduction

The Iraqi archeology with all its classifications is one of the most important sources of historical information. Therefore, this research focuses on the adopters its effects and saving this heritage. Iraq possesses a large percentage of these monuments, including the civilization of Ur and Babylon, as well as the existence of many sites and archaeological remains; Iraq is a symbol of great cultural prosperity. From this simple description, the need to introduce modern digital techniques from geomatics has become urgent in an attempt to manage the databases of Iraqi archeological sites and use them in the correct geometric form to prevent them from loss. The objective of mapping archaeological sites is to ensure that, they are not affected, and spatial database can be updated and analysed easily.

This research includes an overview of the spatial data problem of any project accomplished using digital techniques in geomatics in terms of combination, geometrical correction and spatial data links and display by information such as the numerical objective station. This will contain the collection of data needed for this subject of the project after the processing of sources. In addition, Maps, satellite images of the study area and raw spatial data collected are then processed and equipped to prepare spatial databases.

## 2 Problem statement

The problem is in the abundance and dispersion of data pertaining to all Iraqi archeology sites, gaps in database documentation for many years, dependence on paper documentation and Excel program, which are prone to loss and damage and are not easy to use / refer to them. The field survey for the archeology sites around the country presents the scarcity of documentation and updating of the geographical information of these archaeological sites for several years, during which the site changes due to changing weather conditions, which can cause sites location, to be lost, especially non-prospecting sites.

## 3 Area of interest

The study area covers the whole of Iraq from east to west and from north to south. Iraq is located in the heart of the ancient and modern world. Iraq is located within three zones (37, 38 and 39). Its center is Baghdad, which is located at 45° longitude E and 33° latitude N. Iraq is bordered by Turkey (Anatolia) in the north, Iran (Persia) in the east, Kuwait and Saudi Arabia (the Arabian Peninsula) in the south and Jordan and Syria (Levant) in the west. The governorates of Dhi Qar and Qadissiya were selected as study area for the research output spatial data and within the archaeological sites surveys for the year 2010-2015.

* Corresponding author: hishamalsharaa@live.com

© The Authors, published by EDP Sciences. This is an open access article distributed under the terms of the Creative Commons Attribution License 4.0 (http://creativecommons.org/licenses/by/4.0/).

## 4 Methodology

The theoretical aspect of collecting descriptive data from multiple sources, official departments and field surveys, as well as spatial data collection to use them in spatial information techniques to arrange and organize these data both spatially and descriptively as a practical aspect by constructing a spatial database using ArcGIS 10.3 for building and analysing the existing archaeological sites.

## 5 Geomatics Techniques

There are many techniques involved in the field of geomatics. However, in this paper a brief explanation of the theoretical part of digitalizing, its contribution, preparation and construction of spatial databases, framed to find a digital system that will be reliable in all spatial analysis, and will represent the best way to achieve any engineering goal.

These techniques can based on how spatial databases are prepared based on:

1. Display data and digital processing software.
2. Geographic information systems.
3. Maps.
4. Satellite system.

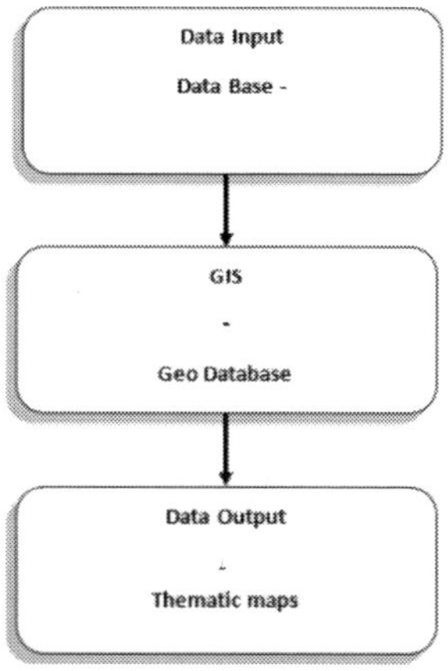


**Fig. 1**. Database building system

## 6 Spatial Geo Database

One important term in GIS system is the installation of geographical locations of the archaeological sites precisely by using Global Positioning System (GPS). Because of the difficulties to reach some archaeological sites, the researcher entered the coordinates sent by the Archaeological Authority in each Governorate to the General Archaeological Authority and converted these coordinates to maps and turned them into databases containing location data. This will help in terms of better and faster decision-making, lesser conflicts related to infrastructure sites and taboos, leading to lesser changes in the selection of project sites and objective production of maps using the information available in the database.

### 6.1 Steps to build a spatial geodatabase:

The spatial database is built using Arc Catalog, which is part of the ArcGIS10.3 software package, where the project database is prepared according to the Geographical Coordinate System and the WGS84 support platform and WGS84.
The database was designed according to standard specifications as shown in the figure (2).

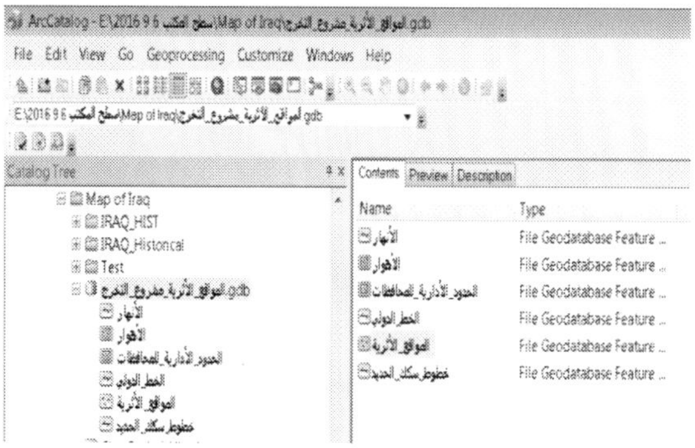


**Fig. 2**. Steps to build a spatial geodatabase

The Draw points represent the effects depending on the boundaries of the provinces using the ArcMAP program and the distribution of points by archaeological sites, taking into account the name of the archaeological site on each point as shown in Figure (2&3).

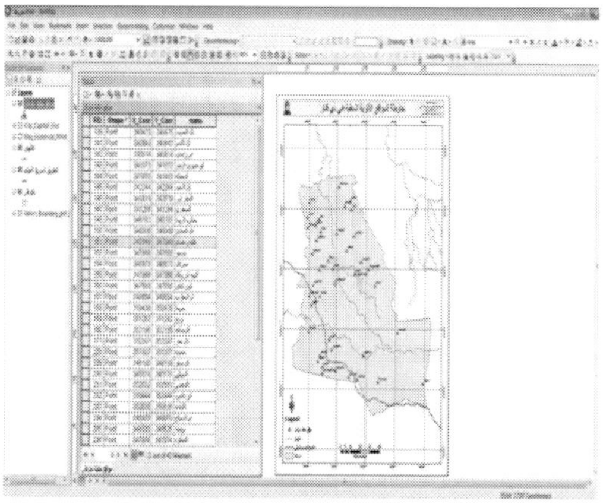

**Fig. 3.** The descriptive table of the archaeological attribute site

## 7 Maps Generations

The map below shows all points of archeology after completion of the process of drawing points representing the effects distributed across the provinces

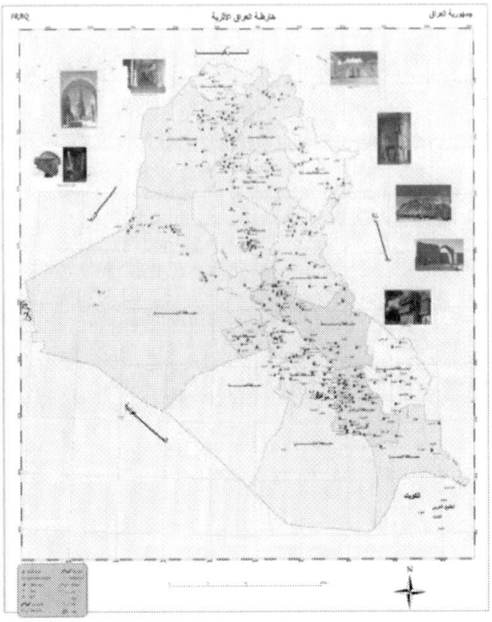

**Fig. 4.** Complete the process of drawing points that represent the archaeology sites

## 8 Topology

### 8.1 Process

This process is the topological correction stage, which is considered one of the important stages in the process of drawing and preparing the geographical data base and is necessary in all the work of geographic information systems. It is the process of organizing spatial relations and dealing with all the problems and overlaps that occur between the different layers in the geographic data base, which are organized using special parameters to be installed during the preparation of the spatial database.. For this purpose, it is necessary not to overlap the symptoms drawn with each other for administrative boundaries. For example, the borders of the Governorate of Nazareth should not interfere with any of the borders of the neighboring governorates. (Must not Have Gaps) This parameter is necessary to avoid gaps during the process of drawing the symptoms of the administrative boundary, for example, they should not make a gap between the boundaries of the province and other Governorate needs to be adjacent without interference. This is meant to be the point of the effects within the boundary of the administrative boundary layer, in other words it must not be located outside the administrative boundaries of the provinces as it does not use the other determinants because the class effects of the points in the geometric shape do not need other determinants.
The process of checking and detecting errors: After completing the process of entering the parameters of the Topology process, we use another program, the Arc MAP program for the purpose of auditing and detection of errors and overlaps are identified in the process of adding parameters using the tool Topology Tools. By using Error Inspector, errors, if any, are detected in the graphics, and are handled and corrected and the changes are stored on the original layer, whether the administrative boundary or the layer effects and then the layers are ready for the process of display in maps or in the digital versions.

### 8.2 Output

After the completion of preparing the geographical data base in its final form, it is possible to display the information that was entered in multiple formats according to the need. The final output is objective specific for each possible effect, can be extracted in the form of maps or any files when clicking on any effect. The entrance to the system enables all information and offers advantage of the search feature and query in ArcGIS program in the search for the impact of a particular Auadh effects as well as possible sort effects by prospecting and by the excavating and by Governorate will show ... etc any in general, the database is characterized by flexibility and oversight E dealt with where the researcher produce thematic maps concerning the effects of which will be mentioned below.

## 9 Results and desiccations

Production of a map for the undeclared effects of Dhi Qar Governorate according to the 2010 surveys and using the quadratic coordinate system according to Figure (5).Production of objective maps derived from the map of the distribution of antiquities Figure (6) below shows the production of a thematic map of the archaeological sites declared in the province of Qadisiyah.
Production of objective maps showing the distribution of the effects in Qadisiyah Governorate Figure (7) below Production of a thematic map concerning archaeological sites in Qadisiyah Governorate and Production of full geodatabase system as illustrated in Figure (8).

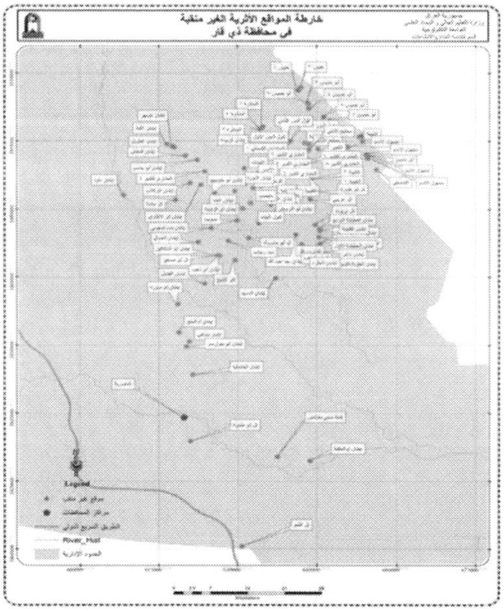

**Fig. 5.** Map of the undiscovered archaeological sites of Dhi Qar governorate according to the 2010 survey

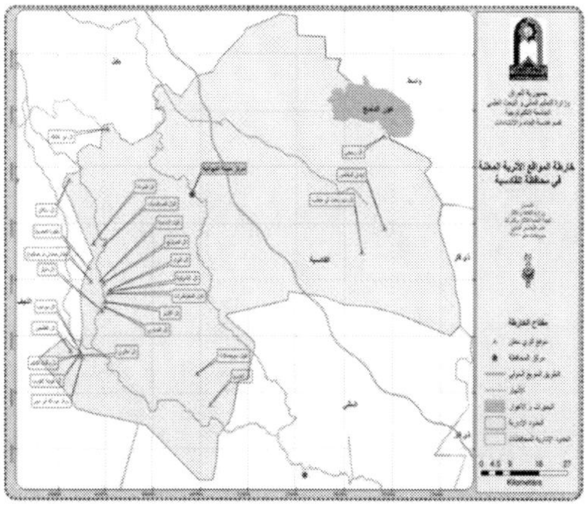

Fig. 6. Production of a thematic map concerning the archaeological sites declared in Qadisiyah governorate

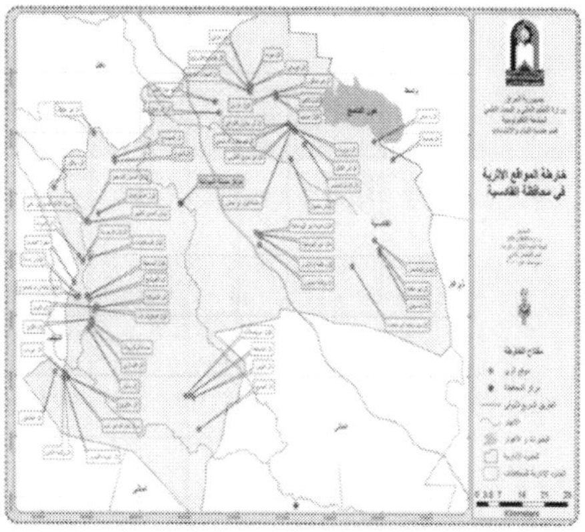

**Fig. 7.** Production of a thematic map concerning archaeological sites in Qadisiyah governorate

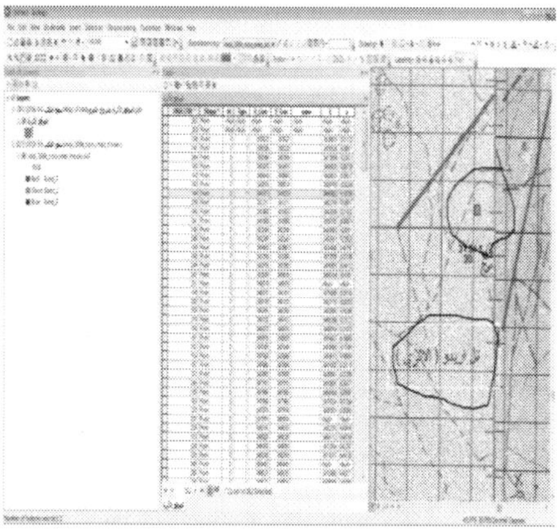

**Fig. 8. Production of full geodatabase system**

# 10 Conclusions

1. The maps produced on Iraqi provinces illustrated the excavated archaeological areas.
2. Many archaeological sites are undeveloped and unprotected.
3. Weak statistics and follow-up of data collection by the relevant authorities, which negatively affect the process of input, analysis and production.
4. The high possibility of GIS programs in the field of measuring distances and spaces and high-resolution, geographical locations and in the coding and coloring. The final output maps are in a manner impossible to achieve by traditional methods.
5. The presence of some areas of relics within the submerged marshes.
6. The possibility of preparing a geographic database by the GIS, regardless of size and storage within the computer and then referring to them at any time and making adjustments to them according to the latest changes result in high accuracy in a very short time compared to the traditional methods and then producing

maps showing any recent change in the study area.

7. The accuracy of the spatial data of archaeological sites sent to the General Authority for Antiquities was observed. There is a creep in the coordinates of the site between geographical coordinates and square.
8. It is possible to benefit greatly from the geographical phenomena plotted by the GIS programs after the establishment of the database to install them together and find spatial relations between them, impact each other through the spatial analysis of the same program and the production of new maps.

## 11 Recommendations

1. Control the indiscriminate construction and prevent excesses on uninhabited areas by the state and to follow modern techniques in prospecting.
2. The use of geographic information systems (GIS) programs in many analytical methods to produce digital analytical maps of many phenomena that are of interest to decision makers.
3. The use of GIS programs to produce all the maps needed by planners, researchers and decision makers according to the built-in database such as using the ARCGIS software to produce thematic maps and all maps according to user needs and coordinates and updating them.
4. Building geographic databases that contain urban architectural data and keep them up-to-date.
5. For the purpose of increasing the accuracy of the produced map, it is preferable to use modern, correct and high-resolution satellite images, by increasing accuracy, determining the projection system, coordinates and conducting the geo-referencing process.
6. Choose the appropriate method to connect the AutoCAD drawings drawn above the corrected spatial images of the required study area and perform the geo-referencing process to minimize or eliminate errors resulting from the linkage such as crawling based on GPS ground control points.
7. Adopting real coordinates at the drawing using Autocad using the corrected images of the study area and converting the Autocad maps to the ARC GIS programs to obtain the required accuracy and facilitate the work to build a geographical database containing all data and information belonging to the study area.
8. The production of three-dimensional maps of the archaeological areas and the construction of geographical databases containing all the contents of the place of built areas and areas and internal methods as well as building materials used and build a database of all the contents of the explanatory and conduct analysis and linking parameters and conduct relationships in the future.
9. Updating land surveys of archaeological areas and adopting this as a national project to preserve the cultural history of Iraq.

## References

1. Hussein H.M. *"Building Archaeology Geodatabase in Iraq Using GIS,"* (UOT, 2016).
2. H. M.J. Al Sharaa, J. Eng. Sustain. Dev.,**20**,1-13(2016)

MATEC Web of Conferences **162**, 03024 (2018) https://doi.org/10.1051/matecconf/201816203024
*BCEE3-2017*

# Proposal of building a standard geodatabase for urban land use

*Haifaa* Husein[1,*], *Oday* Jasim[1], and *Saja* Mahmood[2]

[1] Building and Construction Engineering Department, University of Technology, Baghdad, Iraq
[2] Ministry of Planning, Baghdad, Iraq

**Abstract.** The digital technologies represented by digital processing programs, local data bases and satellite receivers system is a set of technologies made to organize the process of building descriptive field database, and conveying them to computers, in order to store, analyze and manifest data as well as upgrade and extract it in the form of digital maps, which represent an asset in any engineering and planning study. For a variety of types of mixed land uses and others land uses in the urban city, which cause great difficulty in how to build spatial Geodatabase in terms of frequency of the land use and overlap between land uses, the paper will find the proposed standard Geomatics techniques to get rid of these difficulties. The paper will tackle the engineering and survey methods which could reinforce manifestation of uses of logical database in urban areas. Lastly, the research concludes that relying on important field survey and digital references in preparing suggested criteria for field uses, in addition to the recommendations provided in this field for the post graduate and undergraduate students as well governmental specialized departments.

## 1 Introduction

The importance of digital technologies has emerged as an advanced means of handling spatial and the descriptive data in terms of input to the computers, method of storage, processing and analysis of these data, updating and archival on paper and screen in a digital format for the study of various topics, including the preparation of rules standard spatial data for urban land uses. One of the most important of these technologies is GIS technology, Which showed a clear development in the 1980s. One of the most important results is the creation of geographic information systems that can be used through personal computers. The decade of the 1990s has been characterized by the widespread use of geographic information systems (GIS) technologies around the world and the emergence of many manufacturers of cartridges, and a number of complementary technologies for information systems such as global signature system and remote sensing. Using remote sensing techniques, geographical information system algorithms and the global signature system,
Geomatics involves a variety of digital techniques that are dedicated to the surface of the earth and the surrounding environment using computers that play an important role in linking digital technology, space and aerial images in such a way as to facilitate access to information for taking decisions. This science deals with many techniques, including the techniques that specialize in the preparation and construction of databases, like spatial sensing, photogrammetry, remote sensing, field survey, geographic information system (GIS) and global positioning system (GPS). This project includes two main aspects - first involves the preparation of a standard spatial database using GIS, while the second includes urban land use. The importance of geographic information systems has emerged as an advanced means of dealing with data in various fields of scientific information, and geographic information systems have helped solve many of the problems of information regulation using the computer and linking it to a geographic perspective, as geography is the link between spatial data, attribute data and statistics. The main objective of the research is to demonstrate the importance of showing the overall structure and important steps to build standards rules spatial data for urban land uses using digital techniques in geomatics to be supplied to Architectural Planning libraries specialized in such courses.

## 2 Description of study area

The project provides a high-resolution aerial location for an urban residential complex (mahala 614) in Baghdad for the year 2009 within the district of Karkh in the direction of Mansour and bounded between latitudes 33°17′24" and 33°17′21"north and longitudes 44°19′28″ and 44°20′53″ east. The temporal boundary was limited to the current state of land use in the study area during the year 2016 based on the digital

* Corresponding author: author@e-mail.org

© The Authors, published by EDP Sciences. This is an open access article distributed under the terms of the Creative Commons Attribution License 4.0 (http://creativecommons.org/licenses/by/4.0/).

maps of the image and the preparation of the field survey for the study area is shown in figure (1) through access to below the area of the district in Baghdad city.

**Fig.1.** Study area in Baghdad

## 3 Land cover land use classification

A land use and land cover classification system which can effectively employ orbital and high-altitude remote sensor data should meet the following criteria, The minimum level of interpretation accuracy in the identification of land use and land cover categories from remote sensor data. The accuracy of interpretation for the several categories should be about equal, Multiple or mixed uses of land should be recognized when possible. Aggregation of categories must be possible and finally The classification system should be suitable for use with remote sensor data obtained at different times of the year. [1]

The term land use refers to the activities that man performs in a specific spatial and at a specific time. It is known as activities carried out by humans on the ground to which a direct link is related to Classification of land uses. Classification is defined as combining similar observations of individual characteristics into categories and thus, isolating them from other categories . Since, the study of land use and its representation in detail is not an easy process, the classification of these uses is required because it is a way to arrange the uses of the land in certain categories by a specific property for the purpose of achieving the desired goal of the school after obtaining information and data from the field registration of land use survey and updating information for events and services, and placing them locally, these events and information are organized into files within the calculator that are in light of cartographic representation, taking into account the overlap between the uses such as housing, commercial , industry and others services. There are several methods and systems for the classification of land uses such as, Coding System, which is a regulatory tool for recording land uses and is a fast way to access the land uses data and this system will organize these data into spatial groups [2] ,the coding system have a great the importance during Field inventory or field survey as converting data from field records to spatial file records as well as its importance in helping facilitate the task of determining the subtypes of the uses for example, a housing code of 10 is used for residential use of one floor, 11 and 12 for two floors and so on. Colouring System is also one of the most important and most widely used systems in the study of land use patterns, but with particular difficulty in terms of the cost of producing the map of land use and introduces with this system the development of certain symbols of coordination with each colour intended for specific use [3]. .Graphic Symbol System is also help to understand the pattern of land use according to the symbols of knowledge, After the system is selected to classify land uses, the information can be simplified and shown on three main groups:

1. Area represent of the land use, housing, industry, commerical ,adminstrative, education etc..
2. Lines represent road networks and lines of all infrastructures.
3. Symbols represent the actual services like the fire station, the water towers, etc.

The methods and systems used in the classification of land uses in some countries are as follows [2]:

• In USA ,they used in the past a colour scheme in coding of the land uses .

• At present , the municipalities of Dubai , Egypt and Saudi Arabia have used the symbols and code system to mark land uses in order to standardize the planning indicators, to facilitate their classification, storage, , retrieval and utilization in a unified scientific method that avoids criticism, supports integration, saves time and effort. It establishes an integrated database that supports various development decisions and helps to spread knowledge awareness.

• Based on the Colour System in America and the system of code and symbols tracked in the countries mentioned above, and with the use of Anderson classification of land cover and land use, the researcher will prepare a proposal for the construction of land use databases in the urban city of Iraq using the main Geomatic techniques.

### 3.1 Data collection

The data collection process includes both spatial and attributes types and can be summarized as follows: Spatial Data: Spatial data include high-resolution satellite image of the Area of intreset. This is where the coordinates of the image are corrected as making each point on the map similar to the one on the earth through the process of georeferencing of the aerial image Georeference, which is the first major process to deal with GIS and in this process the metadata will be converted to digital formats that the computer can handle. Attribute Data Collection.This stage was divided into the following two sub stages.

• Stage one: Conducting a ground survey of the area of study in order to determine the land uses for which information must be collected and found such as

Residential use, commercial use, health use, educational use, Religious use, Recreational use, mixed use, roads, light industries

• Stage two: Designing a questionnaire for the land use survey.

### 3.2 Design and construction of GDB

The Building of Geodatabase of land cover was completed by using the coding of Enderson classification from the interpretation of the image, as well as the Building of Geodatabase for land use by Arc GIS 9.3 and using the Arc catalog where a folder containing two databases of Personal Geodatabase One was called Land use and the other Land cover as shown in figure (2).

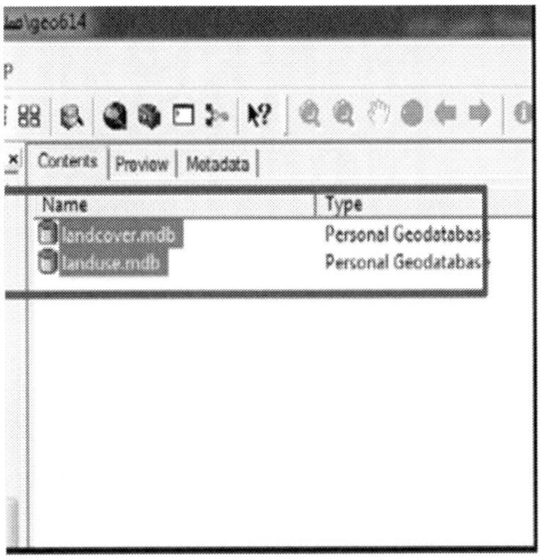


**Fig. 2.**The two databases land cover, land use

### 3.3 Preparing the Spatial Database

At this stage, the researcher start to draw all features in the land cover and land uses using Arc GIS9.3 program and based on the high resolution image. This geodatabase include a store of geographic data as in feature data set with the set of intgrity of rules (topology) and with a tables represent nongeographic objects.and also these features and data represent as in a domain and subtype stages which is very important in the classifications of the land cover and land use . Figure (3) show the layers of geodatabase of the study area. After collecting the attribute data through the survey of land uses, the data will be entered into the land use database. The sub-type and domain, which is one of the most important features of the personal database, will be used. It facilitates the classification of data and save time and effort of re-writing it, as well as preventing errors while entering the data. Figure (4,5) shows the land cover,land use (LCLI) of the study area.

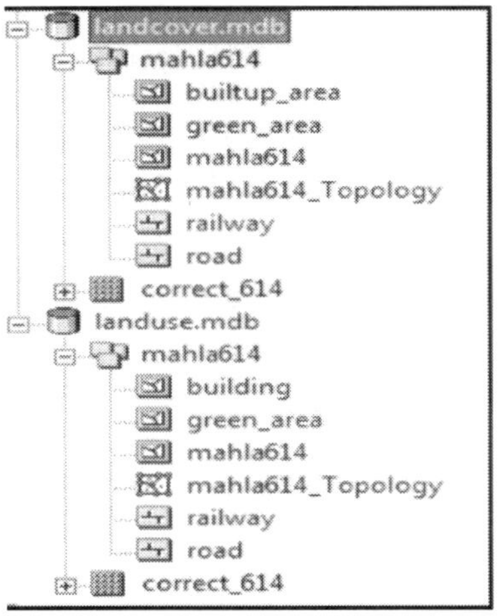


**Fig.3.** Show the layers of the geodatabase.

**Fig.4.** Show the Land cover of the study area.

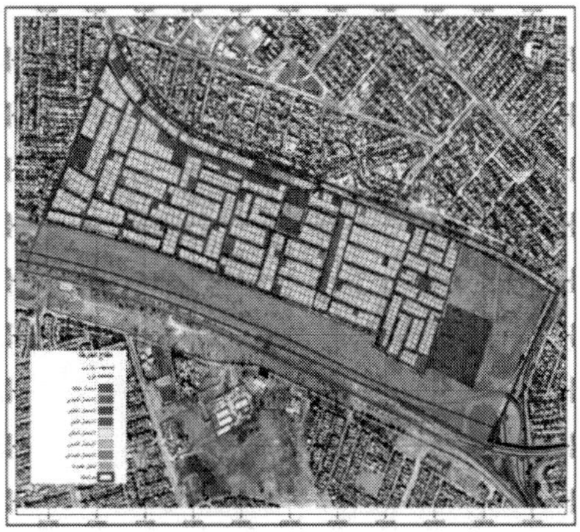

**Fig.5.** Land use of the study area.

The subtype is used by the researcher to choose from, and each subtype will have a domain of selections, and this is done in the following steps:

1. The second level table data will be the field to be served with its subtype while the other fields will be domain of this subtype.
2. Chose the layer of buildings in the land use database and select all of the Selected Properties and then select Field and add the fields where data will be entered with Consider field type.
3. Repeat the previous step but choose Subtype and then from the field Subtype choose Landuse-level two
4. Enter the data in the Subtype fields
5. When open a layer that has its Subtype function using the Arc map9.3 note that Choose any type of work to have Subtype and thus save the time and effort of rewriting again.
6. Using the Arc catalog, add the Domain fields by pressing (RC) on the Land use Geodatabase and select Domain and begin adding the third level data with consideration Select the type of short integer field and the code is coded values.
7. Repeat the previous step to create the Domain for the field use-level four.
8. Link the Subtype field to the Domain fields, by selecting the layer of buildings in the land use data and press RC on the properties and then select Properties then choose Subtype after linking the fields.
9. After completing the linking process open the layer using the program Arcmap9.3 note that the existing properties of the Domain that is associated appears with the same field as the Subtype and when the type of the Subtype is changed The Domain field also changes. This is the idea of the Domain & Subtype for the proposed geodatabase of the Urban land use

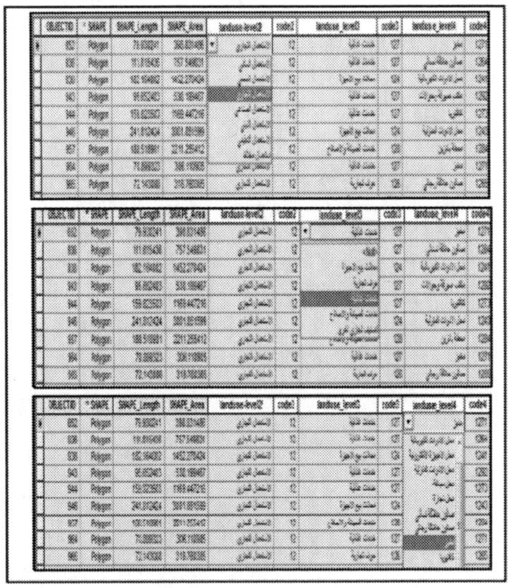

as shown in figure (6).

**Fig.6.** Proposal geodatabase of urban land use

### 3.4 Production of a digital map and thematic maps

The final outputs are accompanied by digital maps to produce a thematic map of classification Land Cover and Land Use by designing the map according to the standard map production standards using Arc map 9.3 and going to the Layout view With the Insert tool,

then add the map title, chart, and map information as shown in figure (7)

**Fig.**7. Final layout of the urban land use map

## 4 Summary and Conclusions

Despite the importance of field survey in the accuracy of access to data, the delay in completing Land surveys and the associated high cost of access to one building in detail, prevent its application in a manner which leaves learners and planners with important data on the reality on the ground and this may lead to the wrong conclusions in these courses and plans. So it is important to introduce digital techniques into Geomatics in the preparation of spatial databases for urban land use. The difficulty of identifying the details of the urban land use has brought this research to necessity to adopt planning criteria to ensure easy access or recognition of the quality of urban land use. The project has reached an understanding of all the standard mechanisms used recently in developed countries and has drawn from them what can be incorporated into digital techniques in Geomatics and digital forms. High-resolution data is one of the best ways to get data and information for any area where colourful aerial images give a discriminating ability to locations and spatial phenomena. The ability of digital techniques in Geomatics to provide the possibility of handling and processing a large volume of data, will help to increase the efficiency of urban land use.

## References

1. Anderson, J., Hardy, E., Roach, R. Witmer, USGS, Prof. Pap 964, 5(1983)
2. Stuart F. Chapin, Edward J. Kaiser, Urban Land use planning (university of Illinois,1972)
3. De C. Joseph, Koppelman L, Planning Design Criteria, (Published by Van Nostrand Reinhold 1969)

# Accuracy assessment of cadastral maps using high resolution aerial photos

*Imzahim* Alwan[1,*], *Noor* Hamed[1] and *Haifaa* Husien[1]

[1] Building and Construction Engineering Department, University of Technology, Baghdad, Iraq

**Abstract.** A cadastral map is a map that shows the boundaries and ownership of land parcels. Some cadastral maps show additional details, such as survey district names, unique identifying numbers for parcels, certificate of title numbers, positions of existing structures, section or lot numbers and their respective areas, adjoining and adjacent street names, selected boundary dimensions and references to prior maps. In Iraq / Baghdad Governorate, the main problem is that the cadastral maps are georeferenced to a local geodetic datum known as Clark 1880 while the widely used reference system for navigation purpose (GPS and GNSS) and uses Word Geodetic System 1984 (WGS84) as a base reference datum. The objective of this paper is to produce a cadastral map with scale 1:500 (metric scale) by using aerial photographs 2009 with high ground spatial resolution 10 cm reference WGS84 system. The accuracy assessment for the cadastral maps updating approach to urban large scale cadastral maps (1:500-1:1000) was ∓0.115 meters; which complies with the American Social for Photogrammetry and Remote Sensing Standards (ASPRS).

## 1 Introduction

The cadastre is one of the most important country's - registers which shows individual cadastral units of parcels and land ownership [1]. The Cadastre is a methodology arranged for public inventory of data concerning properties within a district based on survey of their boundaries; these boundaries are shown on large scale maps [2].

The urban area in all countries has witnessed many changes in terms of urbanization and growth of population. Because of the daily changes in an urban area, updating cadastral maps can be considered as continuous problem with no end. Those cadastral maps were required to be updated but no updating was carried out in last five to six decades. Consequently, the cadastral maps in Iraq are out of today's development and are unable to serve the urban requirements. Maps in Iraq/ Baghdad Governorate are largely out dated and don't reflect the realities of the ground.

Iraq's maps referenced to Clark 1866 which was the first local geodetic system accomplished during 1930s. The Iraqi cadastral maps were referenced to this system. In 1967 a new system called English system or Nahrwan 67 appeared and was referenced to Clark 1880[3].

In Iraq specifically in Baghdad city, the existing cadastral maps produced for the center of Baghdad only by the Union of Surveying and Cartographic Enterprises of Poland represented by (Pole Service) between 1974-1979 with scale 1:500 [4].The coordinates for these cadastral maps have been computed in UTM Universal Transfer Mercator System on Clark 1880 ellipsoid. The point where the ellipsoid coincides with the geoid is Karbala 79 [4]. Clark ellipsoid 1880 was designed to fit the shape of Iraq, this employed as a local coordinate system. There is a shift between Clark 1880/ Karbala 79 datum and WGS84 because of the difference between local and geocentric datum [3].

## 2 Study Area

The study area is a Neighborhood No.512 located inside the boundary of Baghdad Municipality (Sadr City). This neighborhood has a rectangular shape, as shown in Figure1.

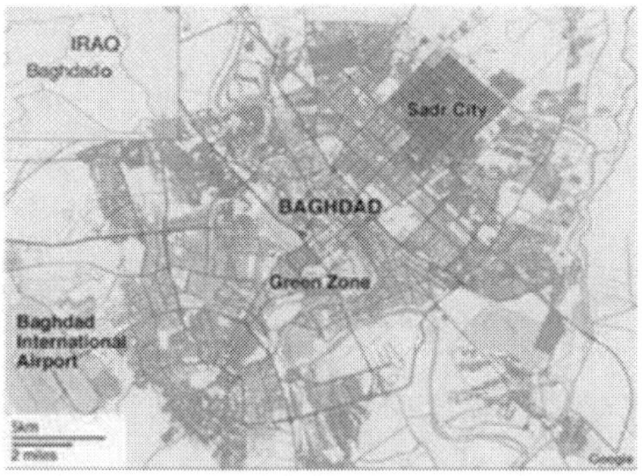


**Fig. 1.** Baghdad municipality boundary

## 3 Methodologies

* Corresponding author: 40164@uotechnology.edu.iq

© The Authors, published by EDP Sciences. This is an open access article distributed under the terms of the Creative Commons Attribution License 4.0 (http://creativecommons.org/licenses/by/4.0/).

## 3.1 Data collection and pre-processing

In this paper, the geospatial data sources are: twenty traditional cadastral maps represent the study area, the map scale is1:500 as shown in Figure 2, with their index as shown in Table 1. These maps were obtained from Baghdad Municipality the main official office deals with urban cadastral maps that were carried by Pole Service Company. The datum of these maps is Clark 1880/ Karbala79.

**Fig. 2.** Traditional cadastral map

**Table 1.** Index of traditional cadastral maps

| No. | Index of map | Map code |
|---|---|---|
| 1 | 3690-444 | 18-3-1 |
| 2 | 3690-444 | 17-2-1 |
| 3 | 3690-444 | 17-2-3 |
| 4 | 3690-444 | 17-2-4 |
| 5 | 3690-444 | 17-1-2 |
| 6 | 3690-444 | 17-1-4 |
| 7 | 3690-444 | 17-4-1 |
| 8 | 3690-444 | 17-4-2 |
| 9 | 3690-444 | 17-4-4 |
| 10 | 3690-444 | 12-3-1 |
| 11 | 3690-444 | 12-3-2 |
| 12 | 3690-444 | 12-3-3 |
| 13 | 3690-444 | 12-3-4 |

The aerial photo captured in 2009 with ground sample distance (GSD) of 10 cm considered as base map as shown in figure 3. This photo was geo-rectified to WGS84 UTM coordinate system. Five ground control points were selected to measure through used DGPS. These five control points were used to generate coefficients for first order polynomial and a nearest neighbor method was applied to resample the photo. The root mean square errors were found to be less than 0.4 pixels. The preprocessing was applied by using ERDAS 9.0 and ArcGIS 9.2 software; Table (2) shows the first order polynomial coefficients; Table (3) shows the coordinates of ground control points.

**Table 2.**1st Order Polynomial Coefficients

| | | | |
|---|---|---|---|
| $a_0$ | -4434970.826 | $b_0$ | -36882532.76 |
| $a_1$ | 9.996679716 | $b_1$ | -0.000855515 |
| $a_2$ | -0.009542856 | $b_2$ | 10.00421759 |

**Fig. 3.** The aerial photo of Baghdad city/neighborhood No.512

## 3.2 Digitizing

An aerial photo represents the basic reference for the data which is required for updating urban cadastral maps, because of the fact that the photo represents the current reality of all existing artificial man- made features. Consequently, the aerial photo represents the best reference for detecting and checking any changing that might happen [6].A photomap was produced from the aerial photo . The photomap would be considered as a base map and consisted of all characteristics of base map, then this base map can be used as a reference map to update the cadastral maps. The digitizing process of the map features was carried using ArcGIS9.2 software. The updated maps for the Neighborhood No.512 have been developed according to projection WGS84 UTM because of the fact that the map was produced using ArcGIS 9.2 software base on the aerial photo as shown in Figure 4.

**Fig. 4.** The updating maps of neighborhood No.512

**TABLE 3.** Coordinates of ground control points

| GCP No. | Longitude | Latitude | Easting (m) | Northing (m) | X-Residual (m) | Y-Residual (m) | RMSE (m) | Ellipsoid Height (m) |
|---|---|---|---|---|---|---|---|---|
| 1 | 44°26'37.725'' | 33°18'41.707'' | 448226.347 | 3685968.217 | -0.203 | -0.790 | 0.816 | 31.114 |
| 2 | 44°26'57.938'' | 33°18'19.016'' | 448745.340 | 3685266.604 | -0.154 | -0.045 | 0.161 | 31.822 |
| 3 | 44°26'18.489'' | 33°18'52.072'' | 447730.683 | 3686290.066 | -0.017 | 0.407 | 0.407 | 31.563 |
| 4 | 44°26'12.149'' | 33°18'41.485'' | 447564.978 | 3685964.887 | -0.155 | 0.078 | 0.173 | 31.687 |
| 5 | 44°27'08.435'' | 33°18'26.750'' | 449018.029 | 3685503.370 | -0.220 | 0.350 | 0.413 | 31.821 |

### 3.3 Geo-referencing

The principles of manual Georeferencing of traditional cadastral maps and aerial photo have been used through the selection of a number of well-defined features in the cadastral maps as well as the aerial photo. The coordinates in a cadastral map are representing the source and the coordinates of an aerial photo are the reference. Minimum twenty (20) points were used in georeference processes. The Georeferencing takes into consideration the resultant updating map would have the optimum precision through the application of weighting system to the maps through which starting with the higher weight map and then gradually done to the lowest weight map. The problem was that there were no tied points connect between the maps as there was between photos, therefore, to reduce the edge demerges between twenty maps covering the neighborhood the weighting system was used. This system depends on the number of surrounding maps. There are four types of different weights (4, 3, 2, and 1) as illustrated in Table 4.

**TABLE 4.**Maps weighting

| Index of map | Map number | weight |
|---|---|---|
| 3690-444 | 12-3-1 | 2 |
| 3690-444 | 12-3-2 | 2 |
| 3690-444 | 12-3-3 | 2 |
| 3690-444 | 12-3-4 | 3 |
| 3690-444 | 17-1-2 | 2 |
| 3690-444 | 18-3-1 | 1 |
| 3690-444 | 17-2-1 | 2 |
| 3690-444 | 17-2-3 | 4 |
| 3690-444 | 17-2-4 | 2 |
| 3690-444 | 17-4-1 | 2 |
| 3690-444 | 17-4-2 | 4 |
| 3690-444 | 17-4-4 | 1 |

The map has higher weight 4 with map index 17-2-3 that is surrounded by other four maps. The georeference of this map through the use of a minimum twenty (20) of well-defined features (points) which were represented by houses sides and streets, taking into consideration that those points must be well distributed on the map which must include a minimum of two points on each side closed to edge boundaries of the map to avoid rotation and translation as shown in Figure 5 and the result of georeferenced map illustrated in Table 6. The map index 17-4-2 was georeferenced by the same approach.

All maps having a weight equal to three then two and finally one have been georeferenced. The Georeferencing between communed edges of maps was done by selecting more than two well-defined features in common edge then choosing a number of well distributed points inside. The well-defined features must be presented first by house- sides and then by street, therefore the priority should be given to the houses and then the streets. The final results of all georeferenced maps are shown in Figure 6 and TRMS obtained is listed in Table 5.

**Table 5.** TRMS of twenty maps

| Map index | TRMS | Map index | TRMS |
|---|---|---|---|
| 17-2-3 | 0.26560 | 17-1-2 | 0.20488 |
| 17-4-2 | 0.10217 | 12-3-3 | 0.25965 |
| 12-3-4 | 0.25861 | 12-3-2 | 0.19990 |
| 17-2-4 | 0.24099 | 12-3-1 | 0.25528 |
| 17-4-1 | 0.24910 | 18-3-1 | 0.22448 |
| 17-2-1 | 0.30271 | 17-4-4 | 0.14207 |

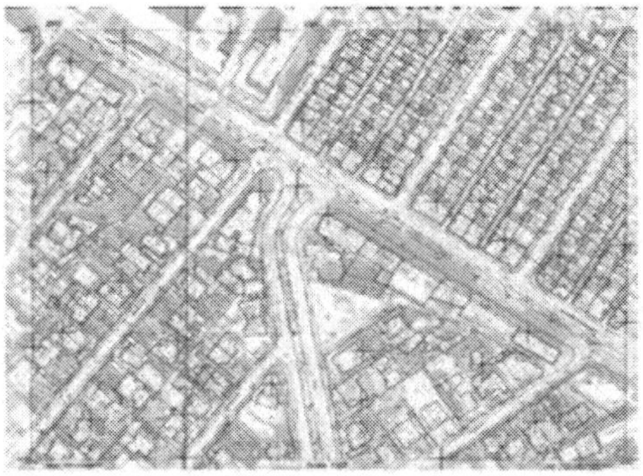

**Fig. 5.** Georeferencing of map 17-2-3

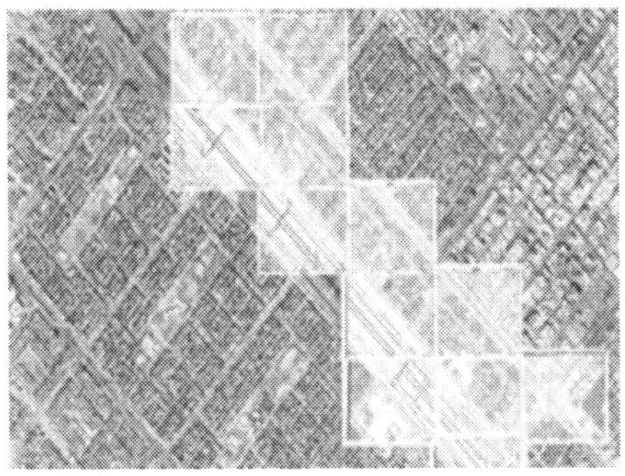

**Fig. 6.** Georeferencing maps of neighborhood No.512

**Table 6.**Georeferencing results of map 17-2-3

| Point ID | X-source | Y-source | X-reference | Y-reference |
|---|---|---|---|---|
| 1 | 448800.631 | 3685662.745 | 448512.315 | 3685943.020 |
| 2 | 449104.752 | 3685508.320 | 448812.780 | 3685786.700 |
| 3 | 449090.173 | 3685684.775 | 448801.830 | 3685963.740 |
| 4 | 448996.087 | 3685698.916 | 448707.837 | 3685978.055 |
| 5 | 448800.041 | 3685784.979 | 448510.620 | 3686065.456 |
| 6 | 449160.709 | 3685719.707 | 448872.467 | 3685198.775 |
| 7 | 449146.356 | 3685638.059 | 448858.806 | 3685917.078 |
| 8 | 449001.690 | 3685517.155 | 448714.320 | 3685795.711 |
| 9 | 449159.978 | 3685538.385 | 448873.025 | 3685816.778 |
| 10 | 449048.668 | 3685696.912 | 448760.349 | 3685926.341 |
| 11 | 448966.495 | 3685500.085 | 448678.754 | 3685778.699 |
| 12 | 449073.118 | 3615523.063 | 448785.877 | 3685801.478 |
| 13 | 449060.950 | 3685505.740 | 448775.891 | 3685784.104 |
| 14 | 449104.752 | 3685508.320 | 448812.780 | 3685786.700 |
| 15 | 449122.141 | 3685213.045 | 448834.987 | 3685491.277 |
| 16 | 449194.942 | 3685226.979 | 448907.497 | 3685505.692 |
| 17 | 449162.067 | 3685256.652 | 448874.591 | 3685535.250 |
| 18 | 449108.489 | 3685200.747 | 447721.009 | 3685479.338 |
| 19 | 449198.800 | 3685362.449 | 448911.995 | 3685641.515 |
| 20 | 449146.749 | 3685499.879 | 448859.364 | 3685776.905 |

## 4 Precision assessments

Unfortunately most if not all the literature uses the term accuracy and true value, since as a matter of fact nobody can determine the true value of any measurement at all. For this reason, we are actually dealing with the precision and not the accuracy, as well as the most probable value and not the true value [5].

The Planimetric accuracy of the land parcels was carried out by comparing length and area of selected plot boundaries from aerial photo with in-situ field observations using total station and field measurement parcel [6]. The accuracy of field boundaries prepared from the image was validated by the following method.

Different existing houses have been chosen randomly in Neighborhood No.512; the length of seven well defined side of these houses in the photomap and field survey used (EDM) Electronical Distance Measurement has been measured as shown in Table 7, RMSE was∓0.115 m.

**Table 7.** Precision assessment

| House No. | Ground Distance (m) | Map Distance (m) | Difference between map and Ground Distance (m) |
|---|---|---|---|
| 1 | 11.333 | 11.211 | -0.093 |
| 2 | 22.944 | 23.054 | 0.110 |
| 3 | 22.645 | 22.525 | -0.120 |
| 4 | 6.452 | 6.552 | 0.100 |
| 5 | 11.537 | 11.640 | 0.100 |
| 6 | 23.184 | 23.053 | -0.131 |
| 7 | 22.935 | 23.136 | 0.201 |

## 5 Results and conclusion

Cadastral map with scale 1:500 (metric scale) was produced using aerial photo 2009 has a resolution 10 cm with a WGS84 system as shown in Figure7. A precision assessment of the updating approach for Urban Large Scale Cadastral Maps (1:500-1:1000) was ∓0.115m which complies with the American Society for Photogrammetry and Remote Sensing Standards (ASPRS) [7].

The originality of this research is through use of the old cadastral maps and the new aerial images with high resolution (10 cm) to update and produce the cadastral maps with sufficient map scale 1/500.

The ArcGIS and ERDAS software were used in the updation of cadastral updated maps methodology.

**Fig. 7.** ArcGIS cadastral maps

## References

1. W. Effenberg, .*Spatial Cadastral Information Systems themaintenance of digital cadastral map*, Doctorate thesis ,(Univercity of Melbourne, Department of Geomatics, 2001).
2. P.M. Brown and D. Divid Moyer, . *Multipurpose Land Information Systems*, (The GuideBook, The Federal Geodetic Control Committee, 1989).
3. J.Gazdzicki, H. Kwiatkowshi, *New Geodeticcontrol Network In Iraq: Desgin, Surveys, And Data Processing*, (Geokart –Poland. 1977)
4. A. Jaronski, R. Pazus.,. The Measurment of Geodetic Control In Republic Of Iraq , *State Enterprise for Survey and Cartography*, (Warszawa, Jasns 2/4 ,Poland. 1976).
5. J.M. Anderson and E.M. Mikhail,.*Surveying Theory and Practice*, (WCB McGraw-Hill, 7th Edition 1998)
6. S. Kay, P. Spruyt, and K. Alexandrou,. Geometric Quality Assessment of Orthorectified VHR Space Image Data. *Photogrammetric Eng. and Remote Sensing*, **69**, 484,(2003).
7. ASPRS .Positional Accuracy Standards for Digital Geospatial Data. *Photogrammetric Engineering & Remote Sensing* Vol. **81**, A26,( 2015)

# Utilization of thermal bands of Landsat 8 data and geographic information system for analysis of urban heat island in Baghdad governorate 2016

*Alaa* Khalaf[1,*]

[1]Ministry of Science and Technology, Baghdad, Iraq

**Abstract.** This study aimed to determine Urban Heat Island in Baghdad Governorate capital of Iraq by thermal bands of Landsat 8 images and Geographic Information System. Two cloud free Landsat 8 scenes covering the study area were selected for analysis. These scenes were located at path 168/37 and path 169/37. Images were acquired during daytime at 10:33 AM and it were dated August 19, 2016 and August 26, 2016, respectively. All images mentioned above were registered in Universal Transverse Mercator projection zone 38. Temperatures have been extracted from thermal bands 10 and 11 for each scene and then these results have been merged by mosaic process to produce one scene for band 10 and band 11. Finally, urban heat island has been extracted for study area by taking the average temperatures of band 10 and band 11. The final results shown that Baghdad Governorate contains four heat islands: water body, built-up area, green area, and bare land area. The Heat Island of bare land was highest, followed by green area, built-up area and water body. The highest summer daytime Heat Island was found on bare land (57°C), followed by green area (45 °C),built-up area (43 °C), and water body (37 °C).

## 1 Introduction

Most urban areas have been created at a quickened rate all over the world, in both developing and developed nations. Urban development and improvement reduces the normal land cover and replaces it with another front of synthetic components (e.g. structures, roads, modern regions, and so forth.). As need be, the nearby atmosphere of these refers to changes by new land utilization and cover modifications which has made the Urban Heat Island (UHI) [1].

It is outstanding and recorded that urbanization can affect nearby climate. An urban warmth island is a metropolitan region which is essentially hotter than its encompassing provincial ranges. Higher urbanization prompts more particular urban warmth island with tremendous temperature contrasts amongst urban and provincial regions [2].The climate of the city impressions of natural warmth created is because of the variety in the dissemination of surface temperatures between the scopes of the city and complexities. The qualities of the stream of warm radiation, which drives low spillage warm radiation of the surface into space, as a result of the gradual addition of structures or developed region to build the heat of the roads. Then again getting to be focused on urban communities and the urban region inside which are hotter than its edges in the appearance is known as Urban Heat Island [3].

The power of the UHI depends on a few parameters. The most imperative variable is the warmth affected from sun, while the built up is extremely hot during the evening contrasted and day time. Other factors include atmospheric conditions; when the environment has cloudless skies and sunlight, winds conditions, etc. The framework or geometry of the city and geology of the land assume a critical part, yet primarily in the spatial appropriation of temperatures in the city. At last the emanation of anthropogenic warmth consequences for the UHI power though its effect is weaker and it depends upon the season [4]. Many studies have demonstrated that air temperatures in downtown amid the night are by and large higher than air temperatures in encompassing country regions. Urban Warmth Island is commonly present at daytime and evening time, yet have a tendency to be higher amid the night. On a midyear day, the sun can warm dry, uncovered urban surfaces, for example, asphalts and rooftops, to a temperature more blazing than the air temperature. The soggy surfaces in country environment remain mostly near air temperatures. The normal contrast in daytime temperature between the urban and provincial zones is 5°C to 10°C, and the distinctions in evening time temperature is 10°C to 15°C considered higher than the daytime distinction [1].This study has been used in the Landsat 8, and GIS for

* Corresponding author: alaaspace1@yahoo.com

© The Authors, published by EDP Sciences. This is an open access article distributed under the terms of the Creative Commons Attribution License 4.0 (http://creativecommons.org/licenses/by/4.0/).

deriving urban heat islands and to determine the thermal island peak more effectively in Baghdad Governorate in 2016.

## 2 Reasons for urban heat island

There are various causes which add to the arrangement of Urban Heat Island. The causes which assume critical part in the formation of UHI are described below [5].

### 2.1 Low albedo materials

Albedo is assessed by the proportion of the reflected sun powered vitality to the occurrence sunlight based vitality. On the off chance that the albedo of the urban surface is low, it will store more sun based vitality and the expanding of urban temperature.

### 2.2 Human gathering

As the human get-together is noticeable at the downtown areas inferable from the accessibility of different offices, discharge of $CO_2$ is more in these zones. $CO_2$ stores warm upgraded climatic temperature. A definitive impact is that it aids the arrangement of Warmth Island as it were.

### 2.3 Destruction of trees

Trees capture the sun oriented warmth and furthermore assimilate $CO_2$ for their own photosynthesis, making the earth cool. With the annihilation of vegetation, the productivity of chilling framework goes drastically off causing formation of the procedure.

### 2.4 Urban canopy

The heat reflected by a building is caught by the close-by taller structures which are known as the urban canopy. UHI is exacerbated with the arrangement of urban canopy.

### 2.5 Wind blocking

Because of the nearness of thickly arranged structures, wind speed is lessened. Accordingly, the cooling impact by convection reduces, consequently the warmth caught, can't be extinguished, resulting in escalation of the impact.

### 2.6 Air pollutant

In the urban territories, particularly in the downtown area, air contamination is prominent. Fumes from vehicles and mechanical contaminations discharged in the earth, trap sun powered radiation along these lines leading to higher temperatures and the microclimate impact ends up noticeably more grounded. The causes are summarized in the Figure (1).

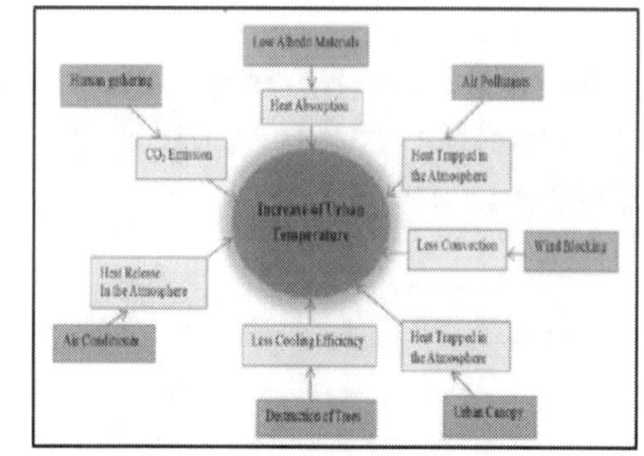


**Fig.1.** Urban Heat Island formation

## 3 Methodology

The urban heat island map of study area has been produced by the following procedures:

### 3.1 Determination satellite image data

The study area is covered by two scenes of the Landsat 8, the first scene is located at path168/row37 and the second scene is located at path169/raw37.These scenes are acquired during daytime at 10:33 AM and it is dated August 19, 2016 and August 26, 2016 respectively.

### 3.2 Registration of satellite images

Satellite images (two scenes) are geometrically corrected in the datum The World Geodetic System 1984 and UTM projection zone 38N using the first order of polynomial function.

### 3.3 Mosaic images and extract study area

The satellite images are merged by using mosaic process and then the study area is extracted from them.

### 3.4 Producing urban heat island

ARCGIS 10.2 spatial model tool is used to produce Urban Heat Island (UHI) for study area depending on thermal bands 10 and 11 for Landsat .

### 3.5 The model is performed by the following steps

#### *3.5.1 Conversion from Digital Number (DN) to radiance*

Landsat thermal bands 10 and 11 are converted from DN to Radiance for all scenes (mentioned above in step 1) which depend on its metadata files.

#### *3.5.2 Conversion from radiance to temperature degrees*

Results in the last step is converted to temperature Kelvin degrees and then subtract 273 from the Kelvin degrees to convert it into Celsius degrees

#### *3.5.3 Band 10 (Celsius degrees) for path 168/37 and path 169/37 is merged by mosaic process*

#### *3.5.4 Band 11 (Celsius degrees) for path 168/37 and path 169/37 is merged by mosaic process*

#### *3.5.5 Final step is performed by taking average of results of last two steps and extract final Urban Heat Island for study area*

## 4 Study area

The study area is Baghdad Governorate in center of Iraq. The study area is bounded by the coordinate (from 43° 50′ to 44° 55′) East and (from 32° 48′ to 33° 45′) North in zone 38N according to Geographic coordinate system. The selected area could be recognized in Figure (2).

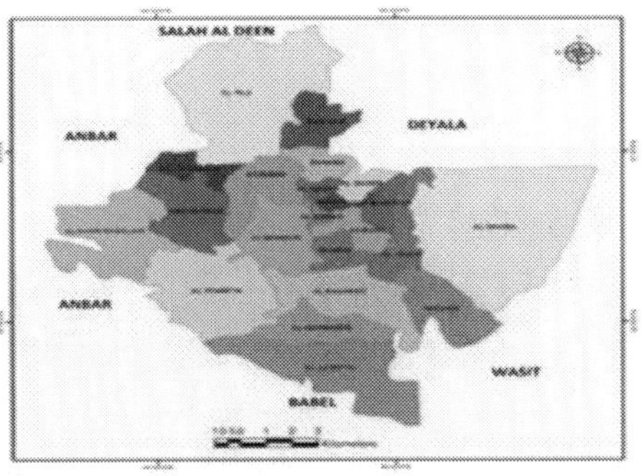


**Fig.2.** Map of study area modify by author (source surveying foundation 2010)

## 5 Materials and methods

### 5.1 Soft-ware packages

ArcGIS 10.2 has been used to process, manage, and analyze the raster and thematic data sets in this research .

### 5.2 Satellite Images data

The study area is covered by two scenes of the Landsat 8 as illustrated in Figures 3 and 4. Multispectral Landsat (path 168 / row37 and path 169 / row37) are dated August 19, 2016 and August 26, 2016 respectively; imagery remotely sensed data-sets were assembled to Urban Heat Island analysis in the study area by using thermal bands 10 and 11 with spatial resolution 100 m.

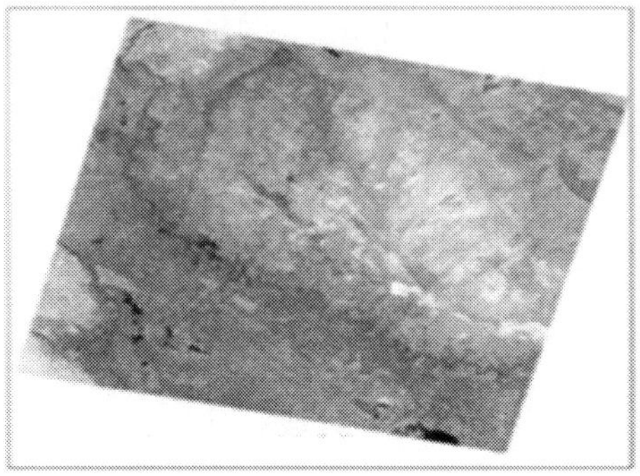

**Fig.3.** first scene of study area 2016.

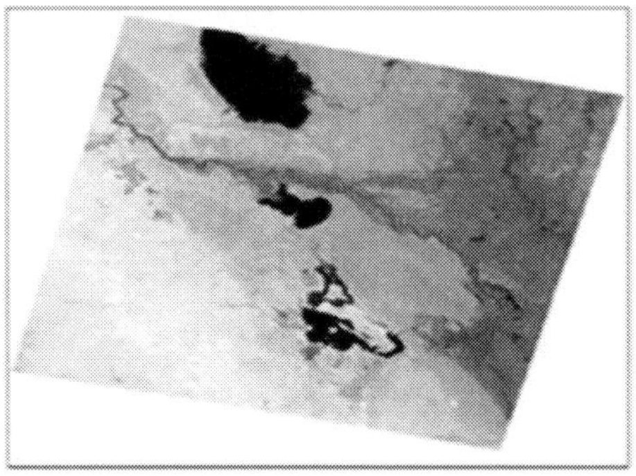

**Fig.4.** second scene of study area 2016

### 5.3 Pre-processing

The data preprocessing is performed utilizing ArcGIS 10.2 program .Each Landsat 8 scene is composed of the independent single layer images. Thus, it is firstly used to combine the single-layer images to a multi-layer image of Landsat 8 using a layer composite tool. Secondly, the geometric correction was made for the Landsat 8 images, by which each point on the image would have latitude and longitude geographical coordinates with datum WGS84. This is the most important step in pre-processing. Finally, the image subset function was utilized to cut our study area as outlined in Figure 5.

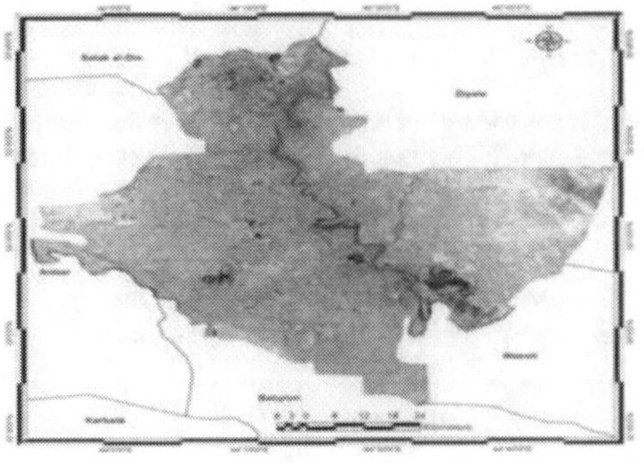

**Fig.5**. image of study area august 2016 (color composite 6, 4, 3)

### 5.4 Image processing and analysis

In satellite images, each pixel value is represented as digital numbers (DN) changing for each sensor type. In determination of Urban Heat Island for study area from Landsat 8, the digital number of image pixels should be changed into spectral radiance utilizing the sensor adjustment information in order to acquire helpful values. As one of the most widely used software packages in the field of digital image processing, the spatial modeler in ARCGIS package gives a powerful tool to fulfill most image processing functions. The spatial model to recover UHI from Landsat 8 information utilizing ARCGIS spatial model apparatus is composed. In this model, the Landsat 8 thermal bands 10 and 11 are used. These two bands are then converted to Urban Heat Island as illustrated in Figures 6 and 7 according to USGS 2013 Equation 1; digital numbers were converted to spectral radiances (Lλ).

$$L\lambda = ML * Q\,cal + AL \qquad (1)$$

Where:
$L\lambda$ = Spectral radiance [$W/(m2 * sr. * \mu m)$]
$ML$ = Radiance multiplicative scaling factor for the band (Radiance Multi Band n from the metadata).
$AL$ = Radiance additive scaling factor for the band (Radiance Add Band n from the metadata).
$Q\,cal$ = L1 pixel value in DN

In the final stage of the processing, urban heat island (UHI) was calculated with the simplified formula (2) (USGS, 2013).

$$T = K2 / \ln(K1/L\lambda + 1) \qquad (2)$$

Where:
$T$ = TOA Brightness Temperature, in Kelvin.
$L\lambda$ = Spectral radiance [$Watts/(m2 * sr. * \mu m)$]
$K1$ = Thermal conversion constant for the band ($K1$ constant band n from metadata)
$K2$ = Thermal conversion constant for the band ($K2$ constant band n from metadata)

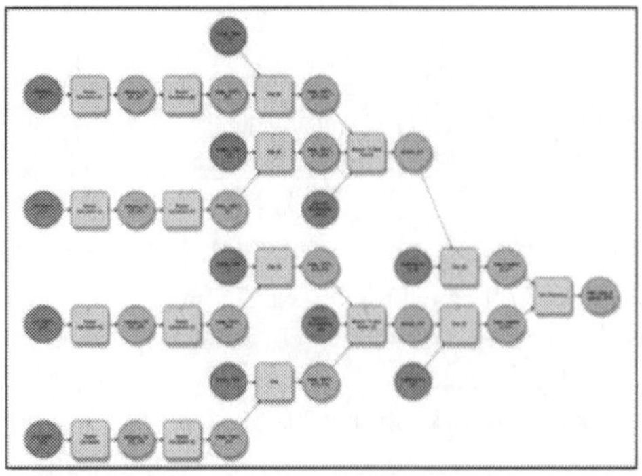

**Fig.6.** spatial model processing to convert DN to UHI

## 6 Results and discussion

The study has utilized thermal band of Landsat 8 for extracting urban heat island for region of study, GIS incorporated with satellite images helped in distinguishing the urban heat Island tops in Baghdad Governorate. For the most part Urban Heat Island force has been discovered based on distinction between the temperature of the urban zone and the surrounding zones. The results have shown that Baghdad Governorate contains four heat islands: water body, built-up area, green cover and non-vegetation or bare land cover as illustrated in Figure 7. The Heat Island of bare land was highest, followed by green area, built-up area and water body was lower. The highest summer daytime Heat Island was found on bare land (57°C), followed by green area (45°C), built-up area (43°C) and water (37 °C).

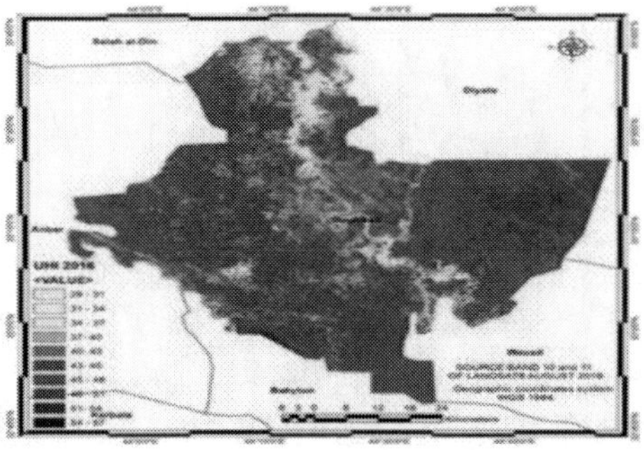

**Fig.7.** Urban Heat Island for Baghdad Governorate August 2016

## 7 Conclusions

1. GIS Spatial modeler gives a capable apparatus to achieve most image processing capacities.
2. Urban Heat Island helps to remove merging problem between built-up and bare land area.
3. The results illustrate that study area contains four heat islands viz.: water body, built-up area, green cover and non-vegetation or bare land cover.

4. The results illustrate that Heat Island of bare land was highest, followed by green area, built-up area and finally, water body. The highest summer daytime Heat Island was found on bare land (57°C), followed by green area (45°C), built-up area (43°C) and water (37 °C).

## References

1.N. Khalid, Urban Heat Island in Erbil City, M.sc Thesis, Department of Physical Geography and Ecosystems Science, Lund University, Sweden (2014)

2 .L. Liu and Y. Zhang, J. Remote Sens. **3**, 1536 (2011)

3. M. Almutairi, J. Geos. & Envi.Protec. **3**, 19 (2015)

4. J. Montavez, A. Rodriguez, J. Jimenez, Int. J. Climatol. **20**,900 (2000)

5. Md. Nuruzzaman, In. J. Environmental Monitoring & Analysis, **3**, 68 (2015)

# 3D Digital modeling for archeology using close range photogrammetry

*Abbas* Khalaf[1,*], *Tariq* Ataiwe[1], *Israa* Mohammed[1], and *Ali* Kareem[1]

[1] Building and Construction Engineering Department, University of Technology, Baghdad, Iraq

**Abstract.** This research is to evaluate the feasibility of applying three-dimensional modelling of the close-range photogrammetry in documenting archaeological monuments by using digital photogrammetry image processing software and digital consumer camera. The digital camera used was Nikon D3100, the processing software was (AgiSoft PhotoScan) and (ArcGIS, ArcScene extension). The study area was selected in the centre of Baghdad province by choosing one of the archeological monuments in it, namely the Abbasid alace. A set of camera locations represent the locations of the images, and as a result of the processing, 81 digital images were arranged in a sequence in which the results of this step were verified. The points cloud after processing were 1,082,617 points. Six control points were selected, used as distances constrained. The validity of the fixed location of the points can be ascertained by checking the data. The program provide the error and accuracy for each image, where a total error in the scale bar was 0.005253 meters, a total error of marks points was 0.010957 meters and the accuracy for all six points was 0.005 meters.

## 1 Introduction

The importance of archaeological sites come from that it represents an important dimension of civilization and legacy, and as a means of identification for originated peoples and originality confirmation on the identity, and attracting source for tourism, and a means to communicate with the different communities by attract the interested researchers and scholars to these vestiges. Hence, the importance of preserving this heritage has emerged through sustainability documented in various ways and styles to bring out all the details and content of the simplified manner and attractive to the attention of the recipient and show the evident as it is [1,2].
he means of documentation of archaeological sites become more diverse, varied studies aimed at documenting archaeological sites, traditional studies which requires a long time and big effort or required unconventional expertise and skills at the same time a high cost, so it necessary to find ways to achieve a low-cost, ease to implement , with simple skills,less time than previous methods , new style to attract and interactive and is able to give a real impression, all of that can be found through the three-dimensional modeling with close range Photogrammetry[3,4].
Through the development of different technologies, operating systems and application systems, the users are provided with tools and means to easily handle the size of the information obtained from the authentication and treatment process. In order to reach the goal of the study, the research dealt with the construction of a three-dimensional model of one known archaeological site and the definition of its history and its structure. The definition the photogrammetry, types and its divisions and the mechanism of work, and the possibilities of the analytical elements of features and acquire the necessary documentation of information through use of multiple images of the site.
Three-dimensional modelling from multiple images without using Ground Control Points , this can be done using a specialized software in the processing of multiple images and generating three-dimensional model (such as a Agisoft PhotoScan program)[5,6].

## 2 Research goals

The aim of the research is to document the heritage buildings as an input to the process of preserving the heritage environment by presenting the different methods of documentation and finding the least complicated and least expensive way. This is done by presenting a three dimensional and interactive model including all the details. The feasibility of forming a database and information for all the effects as a basic reference for documentation and restoration with the possibility of the introduction of these data and models for display on the Internet and specifically to serve researchers and those interested in this area.

## 3 Study area

The study area was selected in the center of Baghdad province by choosing one of the archeological

* Corresponding author: khalaf1955@yahoo.com

© The Authors, published by EDP Sciences. This is an open access article distributed under the terms of the Creative Commons Attribution License 4.0 (http://creativecommons.org/licenses/by/4.0/).

monuments in it, namely the Abbasid palace, as shown in Figure 1, within the coordinates below which dates back to the 5th century AH, which consists of two-storey building with internal dimensions (21.5 x 20 m), each floor has many rooms and various uses with a large central courtyard and main entrance, we choose this archaeological site because it is one of the most important landmarks in Baghdad.

The coordinates of the site:

- WGS_1984_UTM_Zone_38N
- P1 = ( E = 442632.195 , N = 3689511.394 ) m
- P2 = ( E = 442646.826 , N = 3689451.54 ) m

**Fig. 1.** Abbasid palace

## 4 Research materials and methodology

The Materials used in this research include the following:

1-A digital camera type Nikon D3100 For the purpose of taking pictures of the study site with different locations and angles.

2- A software package specializing in three-dimensional modeling and analysis of Close Range Photogrammetry (AgiSoft PhotoScan)

3 - Using ArcScene, one of the extensions of ArcGIS

- The methodology of the work can be summarized in the following igure (2) :

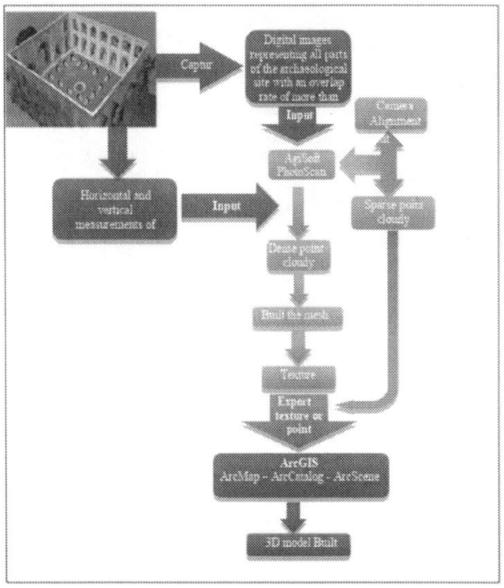


**Fig. 2.** The methodology flowchart

## 5 Create a 3D model using Agisoft PhotoScan

After collecting the images that were needed for the modeling process, we review images and remove distorted and unclear ones, which can be quickly identified before introducing to the program to get the results faster and better and then add the required images to the program and start the steps of treatment as follows:

• The first step is the process of organizing images and rearranging them to be sequential (Camera Alignment). In this step, the program will search for the common points in the images and work on matching them, in addition to finding the location of the camera and for each image as in Fig. 3 and Fig. 4 and finding the calibration parameters of the camera. A set of camera locations represent the locations of the images, and as a result of the processing, 81 digital images were arranged in a sequence in which the results of this step were verified.

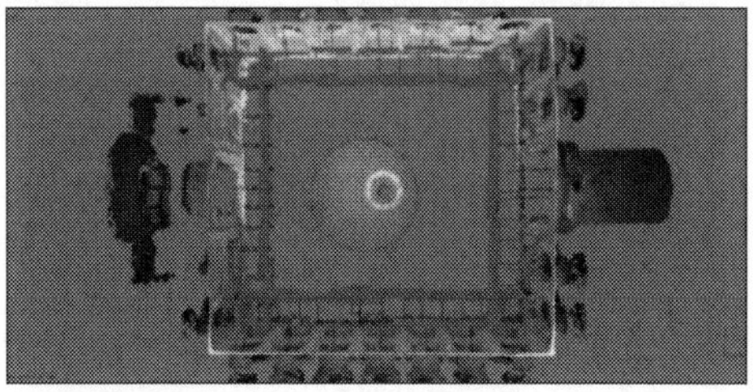

**Fig. 3.** Camera Alignment

**Fig. 4.** Image capture locations on one of the sides for the corresponding side

This sparse point cloud figure (5), consisting of points greater than 1 million points (1,082,617 points), it is the result of image rearranging. It is not directly used in the creation of the 3D model, and can be exported for use in external programs. Example of using sparse point cloud is to be used in edit triangular or reference formats, and the camera location can be used to create 3D models.

**Fig. 5.** Sparse point cloud

• In this step and through the tools available in the program we place the markers to fix the real scale – this is done by inserting distances measured between points as shown in figure (6). As the number of horizontal and vertical points increase, we can obtain better results. In our search, 6 points were selected used as distance constrained. The validity of the fixed location of the points can be ascertained by checking the data. The program provides the error and Accuracy for each image, where a total error in the scale bar was 0.005253 meters, a total error of marks points was 0.010957 meters and the accuracy for all six points was 0.005 meters.

**Fig. 6.** The location of markers

• In this step, the dense point cloud as depicted in Figure (7) is constructed. The number of points formed is more than thirty-four million points (34,813,976 points). It is constructed depending on the locations of the cameras. The program creates a dense cloud that can be edited or classified before being exported and processed to create the network for the three-dimensional model.

**Fig. 7.** Dense point cloud

• The next step is to build the mesh as shown in igure (8), where the program creates a three-dimensional mesh representing the body's surface based on the thick point cloud created in the previous phase. In addition, it is possible to create a geometric model and a polygonal network faster by creating it on the scattered points cloud alone.

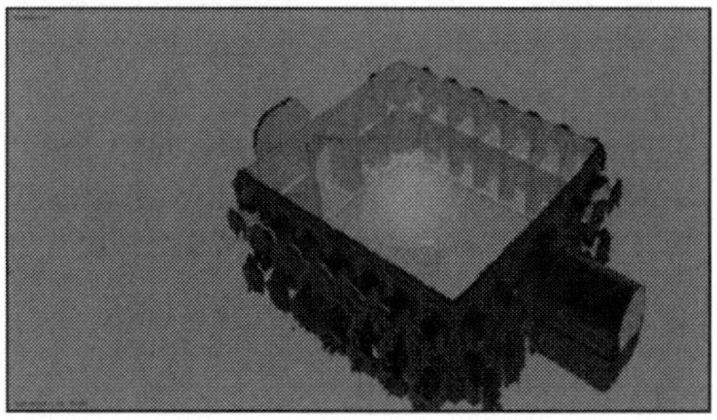

**Fig. 8.** Built the mesh

• After the construction mesh the construction of the outer texture as shown in figure (9) and igure (10) or generation of orthophoto can be completed [7].

**Fig. 9.** Outer texture

**Fig. 10.** Final outer texture

## 6 Create a 3D model using ArcGIS

Using ArcGIS 10.3 with its main extensions: ArcMap, ArcCatalog, ArcScene

After exporting the points from the AgiSoft PhotoScan program in text file format, we use ArcMap, import the points file figure (3-16), and convert the points into a shapefile figure (3-17) which can be used, and through the 3D data for the points (X, Y , Z) to build a 3D model in the ArcScene format (3-18)

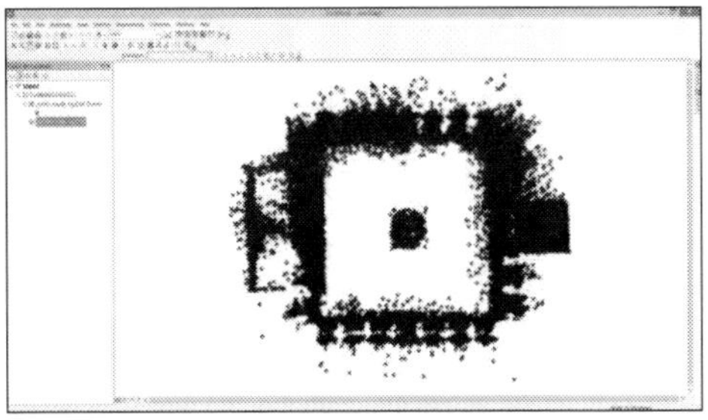

**Fig. 11.** Points in ArcMap

**Fig. 12.** Shapefile

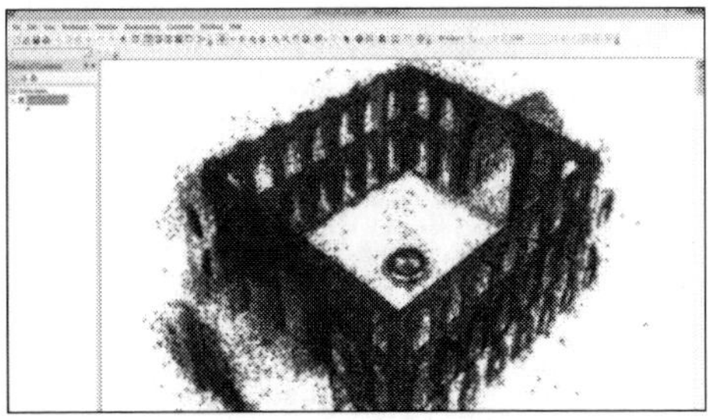

**Fig. 13.** 3D Points in the ArcScene format

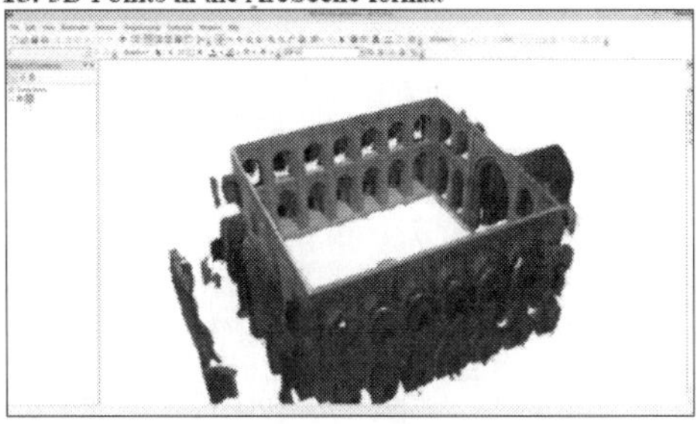

**Fig. 14.** Build a 3D model in the ArcScene

## 7 Conclusions:

Based on the results of the study, we can summarize the conclusions as follows:

- Development of imaging equipment, image capture and processing software has helped the possibility of building full-scale digital engineering models using low-cost digital cameras and personal computer software,
- The methodology used in this research saves money and time and does not require special imaging expertise compared with laser scanning, the price of the camera did not exceed 500 USD, the processing time was approximately 24 hours and the photos gathered by non-professional photographer, but required to follow the guidelines of the work plan in the manner of taking multiple images for the purposes of photogrammetry and is especially effective in areas with limited spaces.

## 8 Recommendations:

- The need to use of electronic means in the process of electronic documentation to obtain accurate results.
- Perform documentation using 3D modeling for all sites and incorporate them into a comprehensive database.
- Increasing awareness of the importance of heritage as a cultural and tourist source.

## References

1. F. Saleh, R. Bahgat, *In 29th CAA conf. Proc.* (2001).
2. Paul R.W., Bon A. D., Benjamin E. W., *Elements of photogrammetry: with applications in GIS* (McGraw-Hill, 2014)
3. Mario A. G., *Basics of geomatics* (Springer, 2010).
4. H. Lin, J. Zhu, B. Xu, W. Lin, Y. Hu, J. Lee, S. Zlatanova *3D Geo-Information Sciences* (Springer, 2009).
5. Stephen R. *Design and Implementation of a Spatially Enabled Panoramic Virtual Reality Prototype* (UNB, 2002).
6. G. Olga, K. Dimitrios, K. Georgios, P. Petros, *In 29th CAA conf. Proc.* (2001).
7. L. Agisoft, *Agisoft PhotoScan user Manual* (Agisoft LLC, 2016)

# Calculating of adjusted geoid undulation based on EGM08 and mean sea level for different regions in Iraq

*Ali* Fanos[1,*], *Rusul* Tahir[2], *Suad* Mohammed[1], and *May* Mahmood[1]

[1] Ministry of Water Resources, State Commission of Survey, Baghdad, Iraq
[2] Civil Engineering Department, Al Esraa University College, Baghdad, Iraq

**Abstract.** In last decades Global Navigation Satellite System (GNSS) or as known Global Positioning System (GPS) technique is considered a revolutionary technique in the field of geodetic survey in comparison with traditional techniques (level, theodolite and total station). The height obtained from GNSS technique is ellipsoid height and to have a physical meaning in a surveying or engineering application it must be transformed to orthometric height. Therefore, a geoid model has to be used to do this transformation process. In Iraq there is no specific geoid that can be used in order to get proper orthometric height. This research aims to calculate adjusted geoid undulation based on Earth Gravitational Model 2008 (EGM08) through observation of Iraqi official vertical network using GNSS technique. Different regions in Iraq have been chosen to perform this research. The result of this research can assist a lot to enhance the accuracy of elevations obtained from GNSS and support the establishment of Iraq geoid.

## 1 Introduction

Most surveying, engineering and scientific work, orthometric heights related to the geoid, a close approximation to mean sea level (M.S.L), are needed [1]. Combining of data acquired by the global navigation satellite system (GNSS), leveling and geoid information has been a key process for different geodetic applications. Despite the fact that these three height types are significantly varies, basically in terms of physical meaning, reference surface realization/definition, accuracy and observational methods, they should fulfill the simple geometrical relationship [2]

$$N = h - H + \varepsilon \quad (1)$$

In which $h$ is ellipsoidal height derived by GNSS technique, $H$ is orthometric height obtained by spirit leveling and gravimetric, $N$ is geoid height and $\varepsilon$ is small quantity because of the curvature of the plumb line and the deflection of the vertical.

Normal height is height above the mean sea level and it is one of many height types that are all computed by slightly different methods. The others are: orthometric height and dynamic height. The normal height of a point is computed from geopotential numbers by dividing the point's geopotential number, for instance, its geopotential variance with which of sea level, by the mean, normal gravity computed along the plumbline of the point. Therefore, normal height is dependent upon the chosen of reference ellipsoid. Normal heights figure prominently in the theory of the Earth's gravity field. The reference surface which normal height is measured from is called the quasi-geoid. The representation of M.S.L similar to the geoid and close to it, but lacking the physical interpretation of an equipotential surface.

The distance H along a plumb line from the point to the geoid is the orthometric height of a point. Orthometric height is for practical purposes height above sea level (Fig.1).

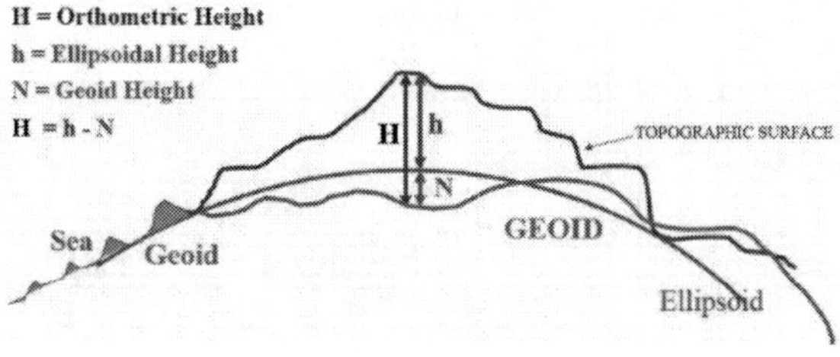


**Fig.1.** The difference between the orthometric and ellipsoidal height.

The geoid is the shape that the sea's surface would take under rotation and the effect of earth's gravity, in the other effects absenteeism such as winds and tides. The surface is expanded through the continents. The points on the geoid have the similar potential energy of gravity (centrifugal potential energy and the sum of gravitational potential energy). Gravity force acts everywhere perpendicular to the geoid, which means water levels parallel to the geoid and plumb lines point perpendicular [3].

Particularly, the geoid is the equipotential surface which could correspond with the mean sea surface if the atmosphere and sea were in equilibrium [4].

* Ali Mutar Fanos: engalim87@gmail.com

© The Authors, published by EDP Sciences. This is an open access article distributed under the terms of the Creative Commons Attribution License 4.0 (http://creativecommons.org/licenses/by/4.0/).

The geoid surface is lower than the reference ellipsoid wherever there is a negative gravity anomaly (mass deficit) and higher than the reference ellipsoid wherever there is a positive gravity anomaly (mass excess) [5].

Ali (2016) [6], calculated the geoidal height for different ground control points (GCPs) distributed in side Baghdad University campus based on the difference between the ellipsoidal height and the orthometric height of the same point. EGM08 has been used to calculate the maximum and the minimum values of the geoid undulation. However, the calculation of geoidal height or geoid undulation needs to be performed for wide area with well points distribution.

In last decade, GNSS technique has been widely used in Iraq. However, regarding the elevation, the local elevations in Iraq are related to the mean sea level (M.S.L) and the zero point is located in Al Faw south of Iraq. Meanwhile, there is no specific geoid that can be used in order to get proper elevation or orthometeric height. Thus, this research is trying to find the adjustment value of geoid undulation based on EGM08 through observe the local vertical network using GNSS.

## 2 Geodetic referencing in Iraq

The most significant datum in Iraq are: Karbala 1979 and The Iraq Geospatial Reference System (IGRS).

### 2.1. Karbala 1979

A completely new geodetic network was established in the 1970s by the Cartography GEOKART and Polish State Enterprise for Geodesy (Polservice) covering all of Iraq. This was a conventional astro-geodetic control network which consist of around 2780 points with an average inter-point distance of 15 km. A new spirit-leveled vertical network, tied to two tide gauges at the port of Al-Faw, supplemented the horizontal network. Observations of gravity were made along all precise leveling lines. The original point of the horizontal network was moved to Karbala instead of Nahrwan. The coordinates change of any point in Iraq is important: in the Baghdad region the shift between Karbala 1979 and Nahrwan coordinates is around 400 meters. Karbala 1979 was utilized in combination with the UTM map projections system and with a dedicated TM projection as well.

### 2.2 The Iraq Geospatial Reference System (IGRS)

USA armies after the first Gulf War establishes the beginnings of a new geodetic network, the Iraqi Geospatial Reference System (IGRS), an effort in which the Iraqi Ministry of Water Resources / State Commission on Survey played a key role. Several points of the Karbala 1979 network that still existed were observed utilizing high accuracy GNSS technique and many new points were established. A series of six Continuously Operating Reference Stations (CORS) is acting as the backbone of the IGRS, that have been utilized to extend the coverage of the IGRS primary control network.

The CORS network (Fig.2) is operated jointly by the US National Geodetic Survey, which is part of the US National Oceanographic and Atmospheric Administration (NOAA) (NGS) and the Iraq Ministry of Water Resources / State Commission on Survey. High Accuracy Reference Network (HARN) points have been established relative to the CORS network forming the IGRS control network consists.

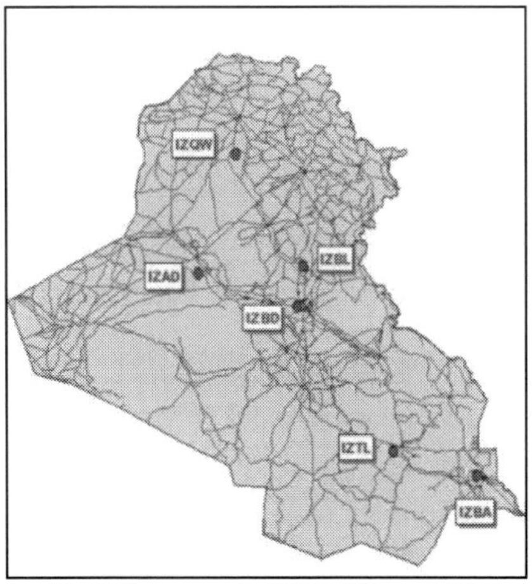


**Fig.2.** The distribution of initial CORS stations in Iraq.

Initially the geodetic surveying work of the IGRS network was managed by British and US Army engineers. In this phase a series of 64 HARN points have been established in southern Iraq (Fig.3). The aim of this work was to make a beginning for a nationwide HARN network with a distance of about 50 km between each two points.

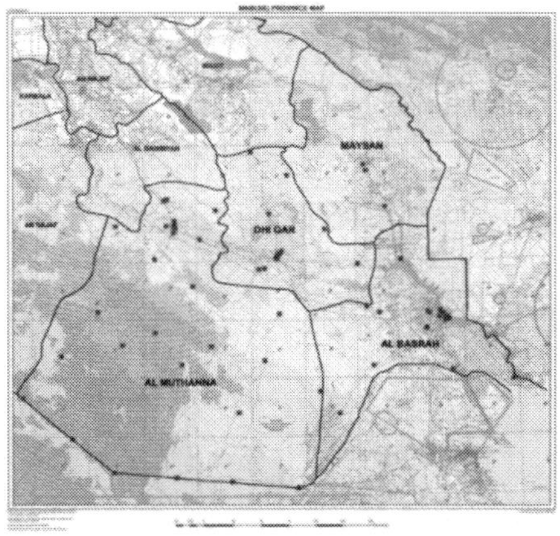


**Fig. 3.** The distribution of 64 HARN points in southern Iraq.

After that Ministry of Water Resource / State Commission on Survey installed 7 CORS stations (Fig.4) and continued establishing the HARN points for whole Iraq provinces.

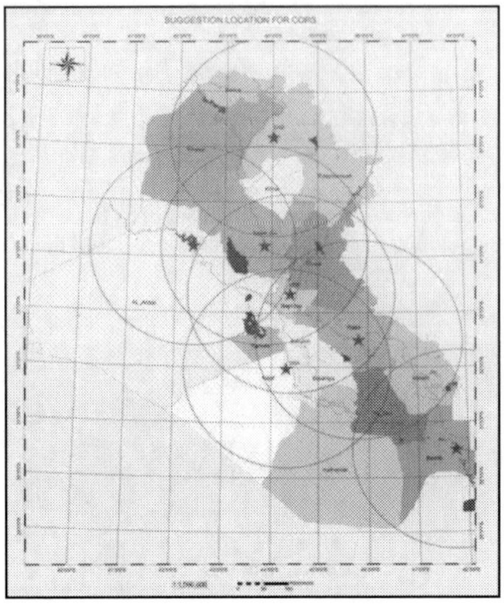

**Fig. 4.** The distribution of CORS stations in Iraq.

At this phase of IGRS the orthometric height of points was obtained by subtracting geoid undulation based on the EGM96 model from the ellipsoidal height that derived using GNSS technique. In that time, EGM96 model was the most accurate global geoid. Geoid undulation accuracy based on EGM96 is estimated to be not better than 0.5 meters. Therefore, orthometric height accuracy derived via EGM96 will be of the same magnitude order as the model itself (0.5 meter). This is much larger than the accuracy of ellipsoidal height that can be obtained through GNSS technique.

## 3 Study area

The methodology of this research has been performed in different regions in Iraq and focus on the most important provinces namely: Babylon, Karbala, Najaf, Kut, and Dioania (Fig.5). GNSS technique has been widely used in these regions due to the increase of projects that have been carrying out in these provinces in terms of oil, construction, industrial and other projects. That is why this research focus on these regions.

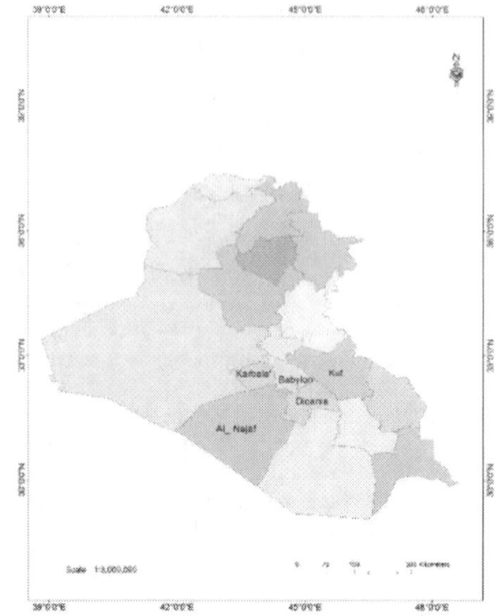


**Fig. 5.** The study area.

## 4 Methodology and material

The methodology of this research can be divided into two aspects the first one is the reconnaissance and observation of the local vertical network by GNSS technique, the second one is measuring of HARN points by GNSS and by precise level based on local vertical points.

### 4.1 Reconnaissance and observation of the local vertical network

The local vertical network is based on Faw mean sea level and consist of many leveling lines covering whole Iraq (Fig.6). Each line has many points and each point has a number printed on it. The number consists of two parts the first one refers to the line number and the other refers to the point number (for instance, 35 - 10 it means line number 35 and the tenth point in the line). The shape of the points are pyramid on a cube (Fig.7).

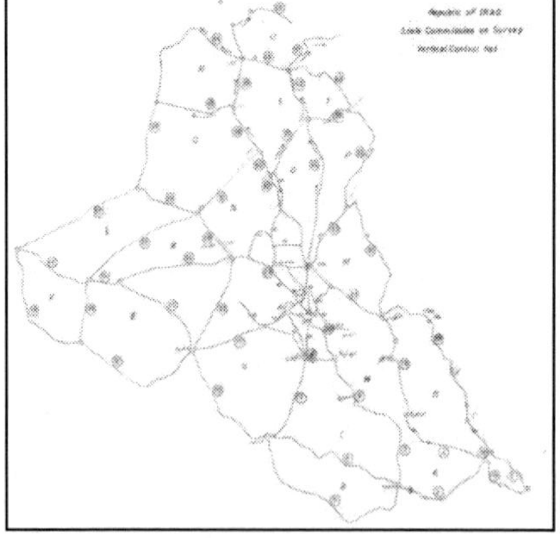

https://doi.org/10.1051/matecconf/201816203028

**Fig. 6.** Precise leveling lines.

**Fig.7.** A precise leveling point.

Reconnaissance has been performed to the lines in concerning province based on points sketch that prepared previously by the company that established the vertical network (Polservice). This process has been carried out in order to know the existing points because some points have been damaged or removed. After that, a plan has been made to specify which points will be observed in a specific line. Then the points have been measured using GNSS technique (Topcon GR3 and Trimble 5700) (Fig.8). Each point has been observed for five hours. Then the raw data has been sent to NGS after one month from the observation date to do post-processing based on Iraq CORS station and get very accurate information in terms of longitude, latitude, and altitude in addition to x,y,z. Then EGM08 has been used in order to get geodetic height or orthometric height which supposed to be same the Faw M.S.L. After that, the difference between the height obtained by GNSS and the actual height for each point has been calculated which represents the adjustment value of geoid undulation.

**Fig. 8.** Observation process for local vertical points.

## 4.2 Measuring of HARN points

State Commission on Survey has established many HARN points distributed in all Iraq (Fig.9). This process has been performed based on standard design that NGS follows (Fig.10), and also based on maximum and minimum distance between each two points.

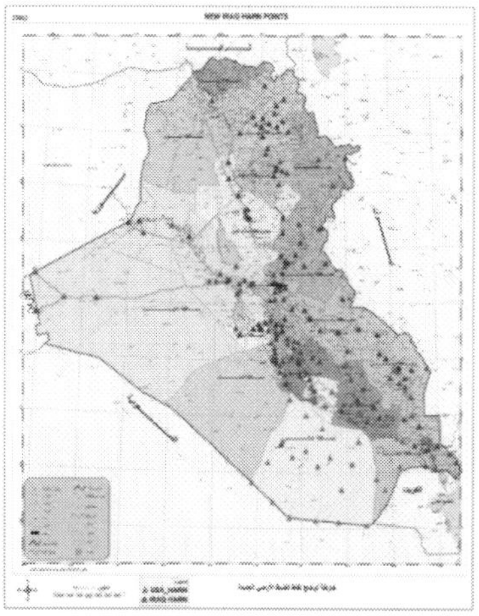

**Fig. 9.** The distribution of HARN points which established by S.C.S.

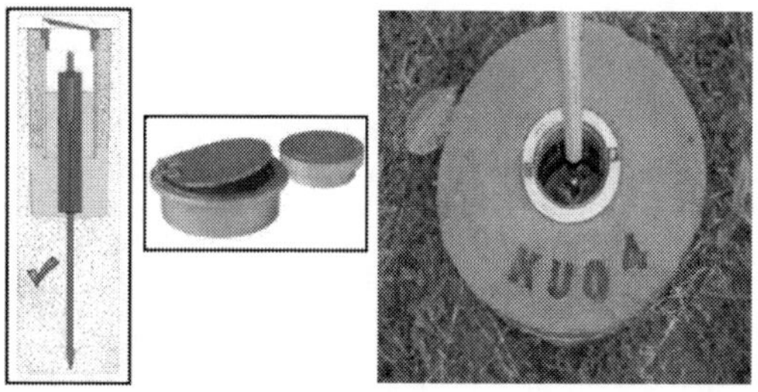

**Fig. 10.** Design of HARN points.

All HARN points have been observed using dual frequency GNSS receiver (Fig.11), in two sessions each

https://doi.org/10.1051/matecconf/201816203028

one for 5.5 hours and the data processing has been carried out by OPUS program through NGS website.

**Fig. 11.** Observing of HARN points.

Almost all the HARN points (Appendix 1) have been measured based on the nearest local vertical point. This process has been carried out using two precise level and two operators and do the double check to get accurate measurements. Then the difference between the orthometric height obtained by GNSS and precise level has been calculated for each HARN point and added to the geoid undulation (N) which obtained by applying EGM08 to get adjusted N for each point. After that, the average of the difference has been calculated to represent the adjustment value of N obtained using GNSS in each province.

Root mean square error (RMSE) is a frequently used measure of the differences between values (measured or calculated value and actual value). RMSE has been applied in order to check the accuracy

$$RMSE = \sqrt{\frac{\sum_{t=1}^{n}(h_{Faw} - h_{EGM08})^2}{n}} \qquad (2)$$

In which $n$ is the number of points, $h_{Faw}$ is the orthometric height based on Faw mean sea level, and $h_{EGM08}$ is the orthometric obtained from GNSS based on EGM08.

## 5 Results and discussion

Figure (12) illustrates the difference between ellipsoidal height, orthometric height based on Faw mean sea level and orthometric height obtained by GNSS using EGM08. This figure shows the difference for the five provinces (Babylon, Karbala, Najaf, Kut and Dioania). In all provinces, the orthometric height based on Faw mean sea level and EGM08 is almost same with a small difference. The difference between orthometric height (both Faw and EGM08) and ellipsoidal height is cleared in this figure. The smallest difference between orthometric height and ellipsoidal height is in Karbala and the highest difference is in Kut.

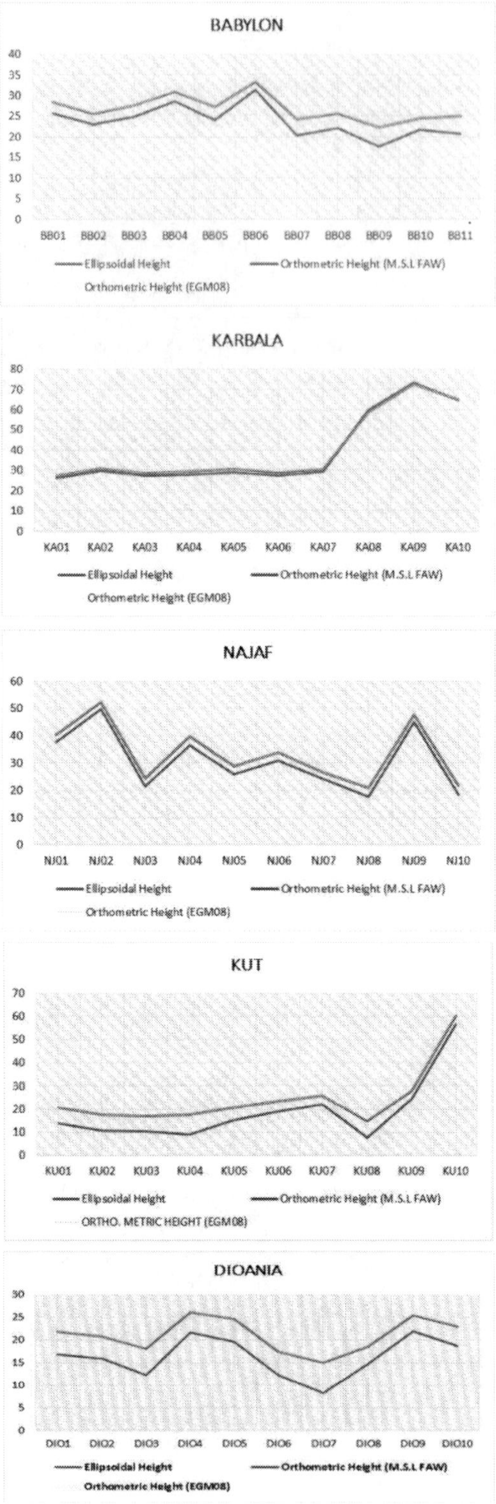


**Fig. 12.** The difference between orthometric height based on Faw M.S.L and EGM08 and ellipsoidal height

The root mean square error (RMSE) has been calculated for each province based on Faw orthometric height and the orthometric height obtained from GNSS-based on EGM08. Table (1) shows the value of RMSE for each province. The highest value is in Kut province (0.1361) and the lowest value is in Najaf province (0.0283).

**Table 1.** The RMSE of Each Province.

| No. | Province | RMSE |
|---|---|---|
| 1 | BABYLON | 0.0438 |
| 2 | KARBALA | 0.0416 |
| 3 | NAJAF | 0.0283 |
| 4 | KUT | 0.1361 |
| 5 | DIOANIA | 0.0600 |

According to RMSE results, one must not rely on just EGM08 without considering the correction value of each province. Thus, in order to get accurate elevation, this correction has to be added.

Figure (13) illustrates the geoid undulation based on EGM08 and the adjusted one based on the observation related to Faw mean sea level in each province. In Babylon province, the adjustment value is ranging around (-0.02m to -0.07m). In Karbala province, it is ranging (-0.05m to 0.05m). In addition, the figure clears that the geoid undulation can be positive or negative which means the geoid can be higher than ellipsoid and vice versa. It is ranging from (-0.05 to 0.01) in Najaf province and from (-0.08 to -0.18) in Kut province and from (-0.01 to -0.08) in Dioania province.

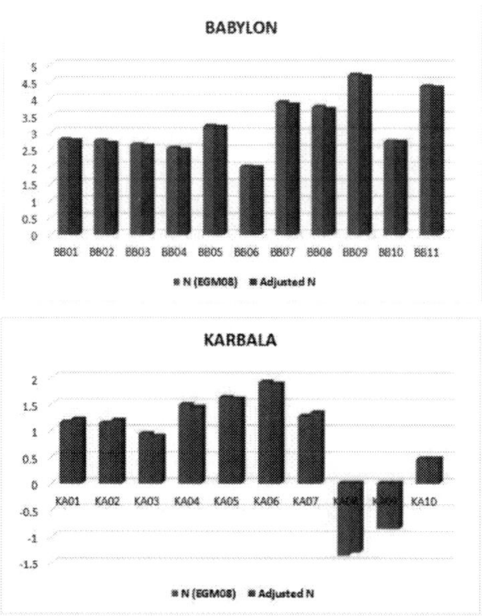


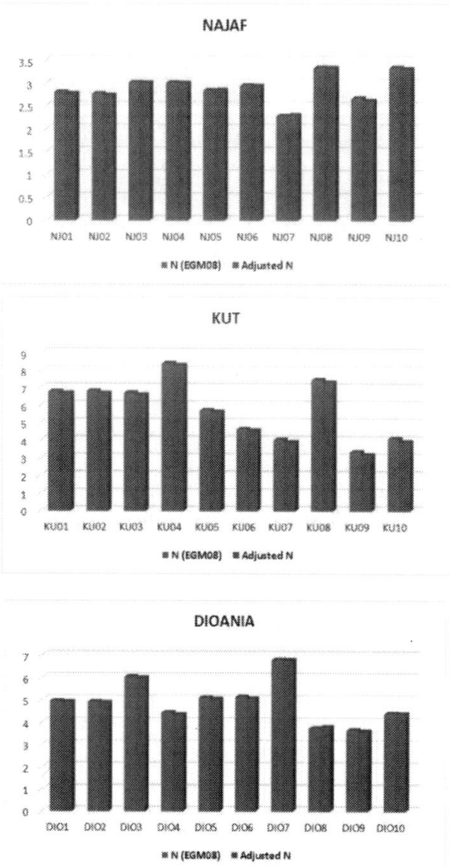


**Fig. 13.** Geoid undulation based on EGM08 and Faw M.S.L.

Table 2 clarifies the average adjustment value for each province that should be added to the geoid undulation (N) obtained by GNSS technique based on EGM08. These values based on the observation of local vertical network and the HARN points that measured by GNSS technique and by precise level related to the nearest local vertical point (Faw M.S.L). As a result, EGM08 can be used as second order in Karbala and Najaf province and as third order in the Iraq middle provinces. In the northern provinces (Erbil, Duhok, and Sulimania), EGM08 cannot be used due to the high differences.

**Table 2.** The Average Adjustment Value for each province.

| Province | Average ajustement value |
|---|---|
| BAGHDAD | 0.058 |
| BABYLON | -0.044 |
| KARBALA | 0.007 |
| NAJAF | -0.019 |
| KUT | -0.134 |
| DIOANIA | -0.045 |
| MUTHANA | 0.110 |
| THI QAR | 0.067 |
| MISSAN | 0.149 |
| BASRAH | 0.148 |
| ERBIL | -0.293 |
| DUHOK | 0.323 |

## 6 Conclusions

Global Navigation Satellite System (GNSS) is considered as an advanced technique in the field of geodetic surveying due to the capability of this technique over the conventional techniques (theodolite, level, and total station) in terms the number of operators, time-saving, intervisibility, and accuracy. However, regarding the height, this technique measures the ellipsoidal height whereas the most significant height in all application is the orthometric height. To do this transformation (ellipsoid to orthometric height) a geoid model has to be used.

In Iraq there is no specific geoid that can be used to get proper orthometric height which supposed to be based on Faw mean sea level. In this research, EGM08 has been used to calculate the adjusted geoid undulation (N) based on the difference in orthometric height related to Faw M.S.L and the one obtained based on EGM08. The result shows that EGM08 can be utilized as second order in Karbala and Najaf province and as third order in the other provinces.

The adjusted geoid undulation has been calculated in this research for different provinces in Iraq. The results of current research can improve the accuracy of the orthometric height or elevation obtained by GNSS technique and can support the establishing of Iraq geoid.

For future work, gravity observation can be combined with the height observation to improve the accuracy and establishing Iraq geoid. In addition, this research can be improved by increase the number of observations for local vertical network.

We acknowledge Mr. Wisam Abdulkadhim Hussein for his support our research.

## References

1. P. Banerjee, G.R. Foulger, C.P. Dabral, JGEOD **73**, 79 (1999)
2. C. Kotsakis, M.G. Sideris , JGEOD **73**, 412 (1999)
3. T. Statella, S.M. Figueiredo, P. Pina, *In EPSC, Proc.* (2015) in Nantes, France, **10**, 722
4. C. Conrad, "Lecture 3: Earth's figure, gravity, and geoid", GG612 Accelerated Introduction to Geology II, SOEST - University of Hawaii (2012)
5. C.A. Stein, Eos **72**, 427 (1991)
6. A.M. Ali, ARJ **2**, 403 (2016)

**Appendix 1.** The coordinates of observed HARN points

| Point | Easting | Northing |
|---|---|---|
| BB01 | 445659.061 | 3592819.7 |
| BB02 | 443736.413 | 3584152.909 |
| BB03 | 442835.171 | 3597549.892 |
| BB04 | 443602.3 | 3613534.543 |
| BB05 | 457277.38 | 3600852.993 |
| BB06 | 431730.219 | 3621461.54 |
| BB07 | 470106.774 | 3573983.26 |
| BB08 | 469123.409 | 3584645.148 |
| BB09 | 463508.833 | 3623595.085 |
| BB10 | 492143.906 | 3576685.799 |
| BB11 | 440229.46 | 3564725.332 |

| Point | Easting | Northing |
|---|---|---|
| KU01 | 575965.369 | 3596426.681 |
| KU02 | 579114.103 | 3597706.13 |
| KU03 | 575758.753 | 3598612.258 |
| KU04 | 596435.189 | 3560818.244 |
| KU05 | 539415.041 | 3601413.274 |
| KU06 | 516117.845 | 3624827.44 |
| KU07 | 505983.009 | 3640935.061 |
| KU08 | 619690.431 | 3604051.576 |
| KU09 | 479453.451 | 3643259.028 |
| KU10 | 586435.826 | 3663746.448 |

| Point | Easting | Northing |
|---|---|---|
| KA01 | 409340.254 | 3607834.579 |
| KA02 | 408815.477 | 3606646.723 |
| KA03 | 404579.759 | 3612578.621 |
| KA04 | 418611.68 | 3617593.267 |
| KA05 | 421573.083 | 3615541.836 |
| KA06 | 426359.612 | 3601107.469 |
| KA07 | 411972.079 | 3603203.652 |
| KA08 | 357514.788 | 3604916.925 |
| KA09 | 366649.802 | 3592854.586 |
| KA10 | 394084.669 | 3604209.325 |

| Point | Easting | Northing |
|---|---|---|
| DIO1 | 492413.403 | 3539888.076 |
| DIO2 | 491950.505 | 3542593.693 |
| DIO3 | 495408.174 | 3536347.717 |
| DIO4 | 522317.075 | 3546753.54 |
| DIO5 | 482308.449 | 3566668.702 |
| DIO6 | 498397.994 | 3556601.659 |
| DIO7 | 496703.386 | 3509871.494 |
| DIO8 | 466060.44 | 3494694.449 |
| DIO9 | 538292.978 | 3538142.191 |
| DIO10 | 463082.936 | 3512112.57 |

| Point | Easting | Northing |
|---|---|---|
| NJ01 | 438934.689 | 3541957.799 |
| NJ02 | 437908.366 | 3539101.114 |
| NJ03 | 444290.981 | 3544422.876 |
| NJ04 | 443725.27 | 3535673.783 |
| NJ05 | 441521.232 | 3556020.418 |
| NJ06 | 442447.529 | 3540267.75 |
| NJ07 | 431669.35 | 3576034.106 |
| NJ08 | 452933.579 | 3517705.85 |
| NJ09 | 435985.496 | 3543556.703 |
| NJ10 | 452022.621 | 3529910.13 |

MATEC Web of Conferences 162, 03029 (2018) https://doi.org/10.1051/matecconf/201816203029
BCEE3-2017

# TransCad analysis and GIS techniques to evaluate transportation network in Nasiriyah city

*Zaynab.* Qasim [1*], *Abdul-Razzak* Ziboon [1] *and Khaldoon* Falih [2]

[1] Building and Construction Engineering Department, University of Technology, Baghdad, Iraq
[2] Civil Engineering Department, Mazaya University College, Nasiriyah, Iraq

**Abstract.** Transportation is an essential element for the development of countries economically, socially, and culturally. AL-Nasiriyah is one of the highly congested urban centers within Dhi-Qar province. Network in the city center suffers congestion, particularly during peak hours simply because of the existence of vitality centers (educational, commercial and government). The network is used to assess the flow of the current traffic network patterns through several programs such as the (TransCad, GPS, GIS) so collecting different types of data, such as (Traffic volumes and free flow speed) using a device (MSSS), field surveys is done for the work of large-scale map road network.The results of the evaluation showed that most of the roads in the city have a level of service type (B), as the network showed service-level type (F) at which the vehicle relative to the road capacity (v/c) is greater than one in the city center, such as (Habboubi Street, Nasir bridge, Sumer Street, etc.) as illustrated in analysis map. On the basis of that proposed to add new roads to change path of external - external trips and also add new bridges to get rid of the congestion that appeared in the city center.

## 1 Introduction

The purpose of transportation is to provide a mechanism for the exchange of goods, information, and to support economic improvements for society, [1]. Transportation provides the means to travel for purposes of employment, exploration, personal fulfillment and is a necessary condition for human activities, such as commerce, recreation, and defense. Travel patterns within the central part of Dhi-Qar are focused on Nasiriyah which provides strong links and good services between Nasiriyah and the other District centers, Al Shatra, Al Rifaai, Suk Shoyukh and Al Jabayish, which are lack good services locally, [2].

Generally road networks comprise of three classes of roads:

- Major arterial: These connect main towns and governorates, they carry mostly long- distance traffic that is either generated in the towns or collected from the rural areas by the secondary and local roads
- Minor arterial: these connect small towns and groups of villages and link these areas into the primary road network.
- Collector roads: these connect the more lightly-populated rural areas into the secondary and primary road network. They also provide access to individual plots such as farms and homes, usually these roads are unpaved or poorly paved.

## 2 Case study

AL-Nasiriyah is the capital city of Dhi-Qar Province. It is located between latitude (30°40' 00" to 32° 00' 00") north and longitude (45°40' 00" to 47°10' 00") east and is located at zone (38 N) according to UTM (Universal Transverse Mercator) geographic coordinate system. The area of Dhi-Qar is (12900 sq. km) representing 3% of the total area of Iraq. Most of it is very fertile producing various agricultural crops. The governorate is famous for its Marsh area (Haur Al-Hammar). That means the governorate has economic importance in terms of agricultural and tourism potential, the latter including both natural and historical/ archaeological tourism as illustrated in Figure 1, [3, 4].

There are four main roads which link Nasiriyah with the other four Qada centers: (1) Nasiriyah – Ur (recently paved, single carriageway, with no proper shoulders), (2) Nasiriyah Bathaa (moderate condition) parts of these two roads run parallel to the Nasiriyah-Bathaa railway, (3) Nasiriyah Suk Shoyukh (moderate single carriageway with no proper shoulders), (4) Nasiriyah Sayid- Dakhil (dual carriageway, well paved with no traffic signs, part of the road currently damaged due to wars and conflicts).

* E-mail: zaynab.2010@yahoo.com

© The Authors, published by EDP Sciences. This is an open access article distributed under the terms of the Creative Commons Attribution License 4.0 (http://creativecommons.org/licenses/by/4.0/).

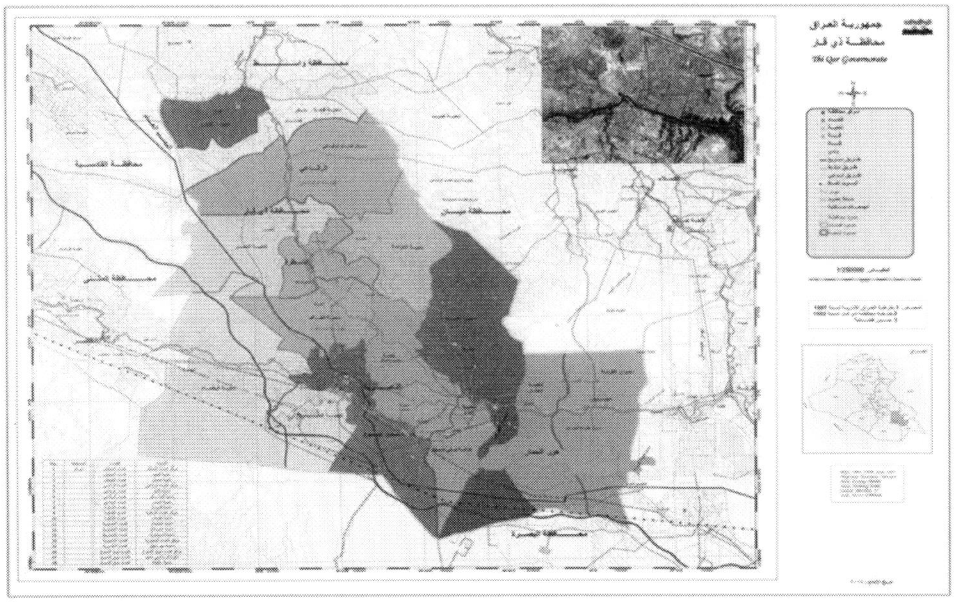

**Fig.1.** The Map of Thi-Qar Governorate

## 3 Problem statement

Al-Nasiriyah city as many other Iraqi cities has no comprehensive published studies in transportation planning, or traffic management plans, [5], taking into account the annual growth in population, employment and car ownership that effects on the daily activities, to become represent a burden increases day after day. Transportation engineers usually face the problem of how to reproduce information from field survey. This process is to create predictive statistical models that describe the behavior and relationship of the phenomenon und reconsideration. Practically, it is not easy to develop a model from field data, because the reality that many phenomena are non-linear and/or collinear or it's a difficult to derive a model.

## 4 Aim of study

Given this context, the aim of this research is to evaluate Al-Nasiriyah city transportation system by applying a transportation planning process and mitigate the challenges mentioned above. The main objectives are:

- Gathering information pertaining to the transportation system in the province to facilitate the tasks of the decision-maker for the development of this network
- Analyzing the existing traffic situation of Al-Nasiriyah City based on a reliable traffic count.
- Modelling and building the road network of Al-Nasiriyah City using a suitable transportation planning software.
- Preparing traffic data base including as follows:
- Vehicle movements survey on both external and internal ambit
- Count and classify vehicles on the internal road network
- Survey peak time for the entrance to the city

## 5 Traffic survey method

Mobile Speed Safety System (MSSS) was used to determine the size of each traffic congestion survey areas of the stations registered within the city boundaries Figure 2 illustrated this device, [6].

Mobile Speed Safety System (MSSS) is based on Sense's' multi-racking radar, capable of tracking multiple vehicles simultaneously, uses a lobe that oversees several lanes up to 150 meters deep. Vehicles moving within that cone are tracked and their movements analyzed.

Speed is determined through Doppler and checked by distance over time, the advantages and benefits of this device are [7]:

- High-resolution digital images
- Continuous speed verification by applying two independent methods
- Provides information about time, date, location, vehicle speed, speed limit
- Uses Senses' unique multi-tracking radar
- Tracks and measures each vehicle more than 20 times per second
- Triggers at an adjustable speed limit and report line

**Fig. 2.** Mobile Speed Safety System (MSSS)

## 6 AL-Nasiriyah city traffic volumes

The most important points in a position were chosen to control the surveys to the center of the city where these points distributed as shown in the Figure 3. These five stations were observed surveys using MSSS device

The major traffic Volume getting into the Al-Nasiriyah city is through the internal zone station EN2, Figure 3 explains the internal zone stations for the traffic survey around the city center especially concerning the passing traffic leaving the center to the other areas, and Table 1 and Figure 4 show the traffic flow. From all of that, it is found that the city center generating more attraction than with ratio 21%, also the traffic flow leaving the city center is about 61% of the whole traffic flow passing the city center of AL Nasiriyah city.

**Fig. 3**. Satellite image (Quick bird resolution 50 cm) illustrating Locations of observing points in the city center

**Table 1.** Traffic Volume through city centre

| Observing points | Out of center (veh.) | To center (veh.) | Total (veh.) |
|---|---|---|---|
| En1 | 6308 | 6267 | 12575 |
| En2 | 24640 | 1106 | 35705 |
| En3 | 5433 | 3406 | 8839 |
| En4 | 7272 | 5221 | 12493 |
| En5 | 4090 | 4857 | 8957 |

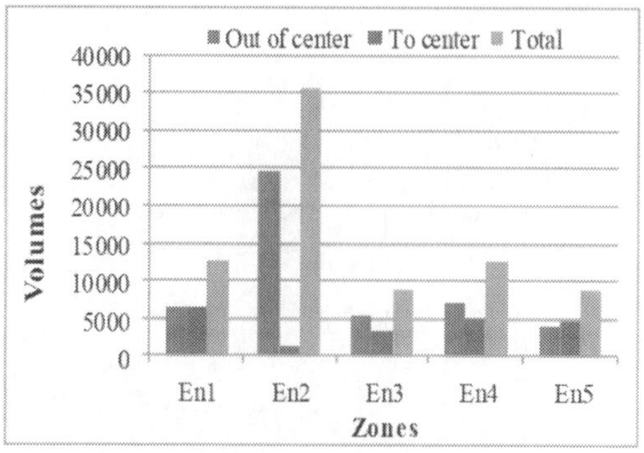


**Fig. 4.** Distribution of traffic flow towards city center

## 7 Type of traffic composition internal traffic volume

Most ratios of vehicle type were nearly kindred and steady through the traffic flow passing every station from the internal zone stations around AL Nasiriyah city center , and the major part is for the saloon private cars with 51% of the passing vehicles in the city center, and about the same ratio for the vehicles leaving the center 50%, followed by the taxis 14% of all the vehicles traffic flow, but the mini buses recorded 13% of the vehicles traffic flow getting into the center , and 19% for the vehicles leaving the center , totally its forms around 16% of the whole volume vehicles traffic flow for the city center , this is shown Table 2 and Figure 5.

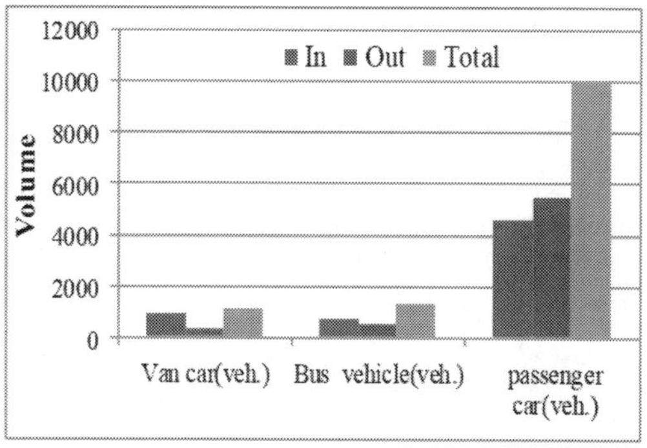


**Fig. 5.** Distribution of vehicles in traffic volume throw internal cordon station (En1) for Example

## 8 Data input

### 8.1 Data Base of Links Shape File

The database of Al-Nasiriyah road network was taken from the Directorate of AL-Nasiriyah Municipality (GIS unit) while the others are obtained by observation or measuring them. Mainly three types of attribute data were collected.

**Table 2.** Distribue infernal Traffic volume by vehicle type

| Entrances | | Passenger car (veh.) | Bus vehicle (veh.) | Van car (veh.) |
|---|---|---|---|---|
| En1 | To center | 4640 | 737 | 890 |
| | Out of | 5474 | 552 | 282 |
| | Total | 10114 | 1289 | 1172 |
| En2 | To center | 8565 | 1691 | 810 |
| | Out of | 16181 | 6913 | 1546 |
| | Total | 24746 | 8604 | 2356 |
| En3 | To center | 4075 | 858 | 500 |
| | Out of | 6711 | 1153 | 975 |
| | Total | 3306 | 937 | 978 |
| En4 | To center | 4494 | 1549 | 1229 |
| | Out of | 7800 | 2486 | 2207 |
| | Total | 2125 | 1680 | 1062 |
| En5 | To center | 2289 | 1219 | 582 |
| | Out of | 4414 | 2899 | 1644 |
| | Total | 21272 | 5340 | 4215 |

The First data were descriptions attributes, which give the descriptions like roads (length, width, number of lane, speed limit, and capacity etc.), the Second data was cost attributes, which play the essential role in the analysis of road to found the optimal route location, the cost of length of road in meters and cost of time (drive time) in minutes.

**Table 3.** Part of Data Set of Links

| ID | Link Id | One Way | Length (m) | Time (min) | Speed (km/hr.) | Link width (m) | Capacity (pcph) | Dir. |
|---|---|---|---|---|---|---|---|---|
| 1 | 78 | FT | 65.788 | 0.131576 | 30 | 12 | 5484 | 1 |
| 2 | 77 | | 68.085 | 0.13617 | 30 | 12 | 5484 | 1 |
| 3 | 671 | FT | 121.415 | 0.24283 | 30 | 12 | 5484 | 1 |
| 4 | 55 | FT | 65.788 | 0.131576 | 30 | 12 | 4113 | 0 |
| 5 | 621 | FT | 509.413 | 0.611296 | 50 | 10 | 5484 | 1 |
| 6 | 841 | | 3.909.19 | 2.606126 | 90 | 12 | 5484 | 0 |
| 7 | 128 | | 3.909.19 | 2.606126 | 90 | 12 | 5484 | 0 |
| 8 | 137 | FT | 82.031 | 0.164061 | 30 | 10 | 5484 | 1 |
| 9 | 139 | | 4925.325 | 3.28355 | 90 | 12 | 4113 | 1 |
| 10 | 4432 | FT | 60.461 | 0.120922 | 30 | 9 | 4113 | 0 |

Attributes

Finally, third was the restriction attributes, which also have the main role accompanying cost attributes. This attribute data include restriction values like one-way and no entry road (closed road). Table 3 shows part of these attributes, [8, and 9]. Account access through each time has been through the equation (time = distance / speed) and speed were taken from the municipality to the specific speed of each road has been the application of the equation, in reality, There was no significant difference, [9].

## 8.2 Data Base of Nodes Shape File

The database of nodes layer contained the number and the geographic coordinate of node (Easting and Northing) this is shown in Table 4. Examples of these were inserted in the attributed table of nodes layer. The necessary attribute of nodes is necessary in network analysis and traffic assignment of road network for the study area

**Table 4.** Part of Nodes and Intersections Attributes

| ID | (E)Coordinate (m) | (N) Coordinate (m) |
|---|---|---|
| 1 | 614230.088148 | 3433700.98101 |
| 2 | 614242.397959 | 3433791.77074 |
| 3 | 614244.155956 | 3433695.99353 |
| 4 | 614244.380534 | 3433697.63914 |
| 5 | 614256.785087 | 3433788.53176 |
| 6 | 614256.892316 | 3433789.32004 |
| 7 | 614261.040399 | 3434889.08932 |

## 9 Performing a traffic assignment

Executing traffic assignments is simple in TransCAD software ver. 4.5. It is preparing the O-D matrix and network files with all the right data included as discussed in previous sections. TransCAD Stochastic User Equilibrium and System Optimum models are utilized for each assignment with a default number of iterations equal 20 per assignment. After each assignment, the output volumes for each link are exported to a new geographic file, [10].

In this study, city has been divided into four zones based on the administrative division of the Department of Municipal and through these divisions was the traffic count field O-D matrix.

### 9.1 Performing System Optimum Method

The System Optimum assignment assigns trips to the network so that the total travel time for all trips in a network is minimized. In this assignment method, individual travel times are not minimized (as in the UE and SUE assignments), only the total travel time. Individual users may be able to choose alternate routes to reduce their own travel time but only at the cost of increasing the travel times of other travelers by a greater amount and thus the total system travel time. Ultimately, the System Optimum yields a solution whereby congestion is minimized.

## 10 Analyses

Trips for each O-D pair are then assigned to the links in the minimum path and the trips are added up for each link. The assigned trip volume is then compared to the capacity of the link to see if it is congested. If a link is congested the travel time is adjusted to result in a longer travel time on that link. Changes in travel time means that the shortest path may change. The whole process is repeated several times (iterated) until there is an equilibrium between travel demand and travel supply. Trips on congested links will be shifted to uncongested links until this equilibrium, condition occurs. Trans CAD automatically joins the results table to the attribute table of the street network file and shows this as a new data view on screen.

### 10.1 Implementation of System Optimum Method in TransCAD

The results of the stochastic user equilibrium assignment model is presented in the Figure (6) and Table (5-A-B) which present part of the of result table that join to the attribute table of the road network.

The v/c ratio is very importance indicators to the level of service of roads. From Figure (6) it can be noticed that: the v/c ratio is divided into ten classes. The links with green color present most of Al-Nasiriyah road network have (v/c) ratio range between (0.00-0.50), level of service A. The yellow and orange colors represent the v/c range between (0.50-0.70) LOS B that includes links approximately 2% from all links, such as (188-521-616-617-628-655-660-724-730-789-791-812-815. etc.).

The links with orange color represent the range between (0.70-0.80) LOS C are (341-651-544-251-285-321-204-709-714-749-693-868-847-849-841). Then there are 1.5% links in Al-Nasiriyah network which has v/c ratio between the range (0.80-0.90) which represent level of service D and the ID for some links (88-91-93-867-980-982-1114-1123) and there are 14 links which have v/c ratio between the ranges (0.90-0.1) which represent LOS E. The Black color represent the range (v/c > 1) which is shown clearly in the city, especially in the center where it appeared in 93 link such as ( 57-171-179-488-508-575-853-966-1001-1121).

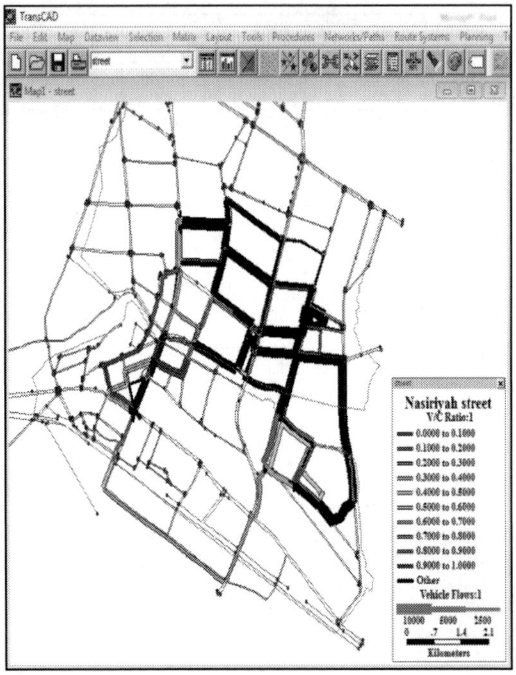


**Fig.6.** System Optimum Method Assignment Model Result

## 11 Conclusions

Through the application of System Optimum method for the analysis of the road network and also through traffic surveys that have been applied to the entrances of the city and some of the points in the city center was observed that the concentration of traffic congestion occurs in the city center which is a crowded market and most of the government state departments and this has resulted through the analysis in the program Trans CAD where black color appeared clearly that represents the ratio (v/c> 1.0) in many links almost 100 links specially in city center and which have the type of services of the type F between emerged after the other of the ratios (v/c) then gradually generally in the road network of the of

Nasiriyah city is the type of service level B which represents the ratio (v/c) of (0.40-0.70).

**Table (5-a).** Part of Data Set of Links Attributes

| ID | DIR | One Way | Length m | Time min | Speed Km/h | Link Width | Capacity | Alpha | Beta |
|---|---|---|---|---|---|---|---|---|---|
| 33 | 1 | -- | 939.61 | 0.96 | 60 | 12.00 | 3984.000 | 0.15 | 4 |
| 43 | 1 | -- | 93.54 | 0.18 | 30 | 12.00 | 3984.000 | 0.15 | 4 |
| 432 | 0 | FT | 281.84 | 0.33 | 50 | 9.00 | 2813.000 | 0.15 | 4 |
| 441 | 1 | -- | 727.69 | 0.87 | 50 | 13.00 | 4241.000 | 0.15 | 4 |
| 542 | 1 | -- | 89.82 | 0.09 | 20 | 9.00 | 2813.000 | 0.15 | 4 |
| 698 | 1 | -- | 898.01 | 0.89 | 60 | 11.50 | 3255.500 | 0.15 | 4 |
| 770 | 0 | FT | 77.53 | 0.19 | 20 | 9.50 | 3341.500 | 0.15 | 4 |
| 814 | 1 | -- | 564.22 | 0.67 | 50 | 11.5 | 3341.500 | 0.15 | 4 |

**Table (5-b).** Part of System Optimum Assignment Model

| ID | AB-Flow | BA-Flow | TOT-Flow | AB-Time | BA-Time | Max-Time | AB-VOC | BA-VOC | MAX-VOC | AB-Speed | BA-Speed |
|---|---|---|---|---|---|---|---|---|---|---|---|
| 33 | 1722.36 | 0 | 1722.36 | 0.944 | 0.939 | 0.944 | 0.432 | 0 | 0.432 | 35.40 | 35.58 |
| 43 | 246.05 | 0 | 246.05 | 0.187 | 0.187 | 0.187 | 0.061 | 0 | 0.061 | 30.00 | 30.00 |
| 432 | 0 | | 0 | 0.338 | | 0.338 | 0.000 | | 0.000 | 21.03 | |
| 441 | 1881.63 | 1957.89 | 3839.52 | 0.142 | 0.142 | 0.142 | 0.443 | 0.461 | 0.461 | 29.85 | 29.82 |
| 542 | 926.526 | 1790.42 | 2716.94 | 0.059 | 0.060 | 0.060 | 0.329 | 0.636 | 0.636 | 20.02 | 19.58 |
| 698 | 3756.89 | 3945.47 | 7702.36 | 1.136 | 1.188 | 1.188 | 1.154 | 1.211 | 1.211 | 47.40 | 45.33 |
| 770 | 0 | | 0 | 0.142 | | 0.142 | 0.000 | | 0.000 | 20.03 | |
| 814 | 182.63 | 112.473 | 295.105 | 0.128 | 0.128 | 0.128 | 0.054 | 0.033 | 0.054 | 20.00 | 20.00 |

## 12 Recommendations

The external- external trips that pass through the network should be changed their path to:

- Rehabilitation of the highway, which passes on the Alholandy Bridge for trips destined for Basra.

- Use the university road for trips destined for the provinces of Al-Muthanna, Karbala and other arterial road as planned as first step and a ring road should be constructed to carry the external trips without pass through the center of the city.

In order to increase the capacity of congested links, on street parking and encroachment should be removed and the carriage way of these links should widened in addition to Traffic Management Measure is required.

New roads should be added to current network and new bridges on the rivers that cross and divide the road network should be constructed as the master plan suggested.

**Fig.7.** The process of taking photos and survey data during the monitoring process

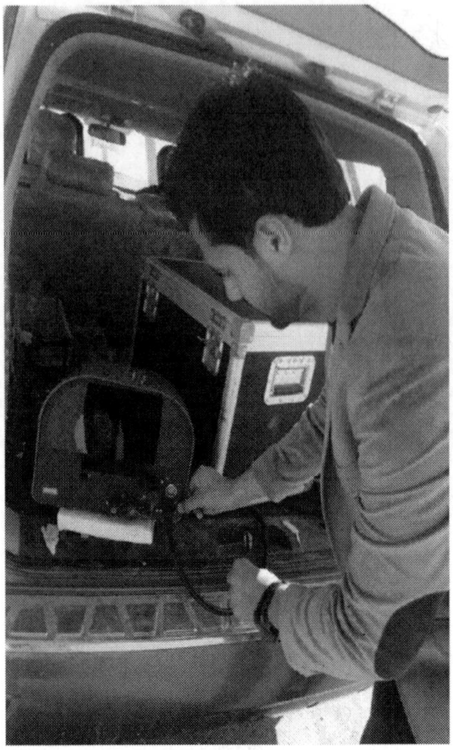

**Fig.8.** The MSSS device during the traffic survey

## References


1. N. Garber, L.Hoel, *Traffic and Highway Engineering*, University of Virginia (2010).
2. COSIT. *Population Estimates of Dhi Qar For the Year 2010*, DPLFS **20**, (2010)
3. K. Falih *The Utilization of Geomatics Techniques for Analysis of Urban Areas Transportation Network*, A Thesis Submitted to the Building and Construction Engineering Department, University of Technology, (2016).
4. *Dhi-Qar Governorate Assessment Report,* UNHCR, (2006).
5. A. Ziboon, Z. I. Qasim, K. Falih, JEASD **21** (2017).
6. H. Weiping, Wu Chi *Urban Road Network Accessibility Evaluation Method Based On GIS Spatial Analysis Techniques* (SG, South China, 2008)
7. SENSYS® Traffic AB (publ) *Advanced sensors and systems for traffic informatics and traffic safety*", Jonkoping Sweden, (2015). www.sensys.se
8. M. Haider *Trip Assignment models in Trans CAD*". Ted Rogers School of Management Ryerson University. (2013)
9. G.I. Qasim *Travel Demand Modeling: AL-Amarah City As a Case Study* thesis submitted to the Engineering College, University of Baghdad. (2015)
10. N. M. Ismael "*A GIS-Assisted Optimal Route Selection Based on Transportation Network Design (Baghdad Metro Case Study)*". College of Engineering, University of Baghdad. (2015)

MATEC Web of Conferences **162**, 03030 (2018) https://doi.org/10.1051/matecconf/201816203030
*BCEE3-2017*

# Preparing a comprehensive geodatabase for Iraq geodetic networks systems

*Oday* Jasim [1,*] *Khalid* Hassoon [2], *and Mazin Hussein*[1]

[1]Building and Construction Engineering Department, University of Technology, Baghdad, Iraq
[2]Ministry of Science and Technology, Baghdad, Iraq

**Abstract.** This paper concerns the subject geodetic networks data in Iraq and the need to put them in specialized spatial databases that can be updated and developed, which makes it much easier for those concerned such as scientific researchers ad surveying engineers to access these data and use them in scientific applications and Engineering projects. The practical aspect of this study was divided into three stages, the first stage was limited to the collection of official data related to the main projects of geodetic networks in Iraq, beginning with the first English network 1934, the second English network 1967, the Polish network 1979, the gravitational networks 1963-1984, and ending with the CORS-HARN networks that have been based in their observations and calculations upon the GPS. The second stage consists of sorting the data obtained from the first stage, and then entering the selected data into spatial databases. The third stage includes the reconnaissance, ground survey and verification of data obtained from the second stage. Furthermore, it also showed that all the English networks, the first 1934 and the second 1967, as well as the points of gravity were almost completely extinct and it became difficult to identify any traces of any of its points within this region.

## 1 Introduction

It is well known that the geodetic networks have fundamental importance in surveying. Generally there are four types of these networks: horizontal, vertical combined 3D System (horizontal and vertical) and gravity, each type has specific trends and applications, in Iraq. From historical point of view, the development of these networks can be categorized as shown in figure 1

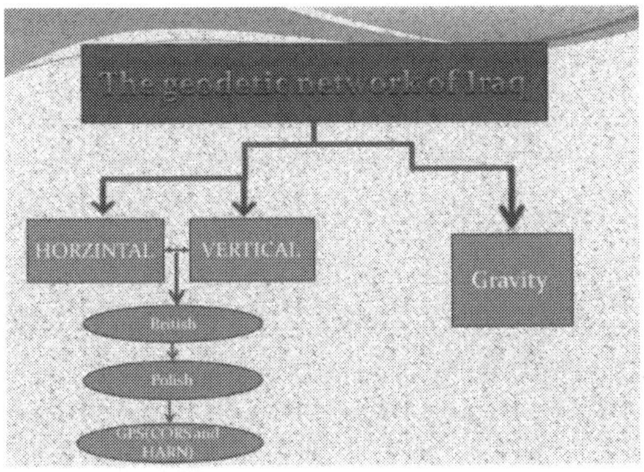


**Fig.1**. Geodetic network in Iraq

## 2 Review historical development for Iraqi geodetic networks

### 2.1 First British network of Iraq1934

The first geodetic network of Iraq was created by the English oil companies in 1934 and called the name (Nahrwan Datum 1934) on this network as a label Nahrawan is the ratio to the city in southeastern Baghdad, where it was relying on the reference in this region point to be a reference to the system of coordinates the adoption of this network in drawing all first property maps of Iraq and used up to the present time which what is known maps Alcadstro in addition to other different maps have been benefit from it in the planning of infrastructure projects. The flatfish surfaces (CLARKE 1880) have been used Theodolite instrument, settlement and painting and planar chain in the creation of this network has been relying on the sea level (M.S.L) [1]The Primary Triangulation of Iraq.-The primary triangulation of Iraq is only of secondary precision and was executed between 1930-1934. It is divided into two series: the North Series arid the South Series. Both the series start with the measured base at Nahrwan (E. of Baghdad). The North Series closes on a measured base at Kirkuk and the South Series at Batha. The measured

* Corresponding author 40004@uotechnology.edu.iq

© The Authors, published by EDP Sciences. This is an open access article distributed under the terms of the Creative Commons Attribution License 4.0 (http://creativecommons.org/licenses/by/4.0/).

values of the bases have been accepted and discrepancies between the triangulated and measured values have been adjusted. The details are given below:

**Table 1.** The measured values of the bases in British first Network of Iraq1934

| no | Name of base | Data of measurement | Length (m) | Probable Error (mm) |
|---|---|---|---|---|
| 1 | Nahrwan | 1930 | 4441.01448 | ±3.00 |
| 2 | Kirkuk | 1930 | 2000.30950 | ±3.34 |
| 3 | Batha | 1930 | 2012.17679 | ±2.40 |

The initial latitude and longitude are the astronomical values of Nahrwin S. End Base:

Latitude 33" 19' 10" .87 ±0" .20 (for N & S. stars)

Longitude 44" 43' 25" .64 ± 0" .69 (from wireless time signals)

## 2.2 Second British network of Iraq 1967

The establishment of a new system in 1967, flat depends surface flatfish (CLARKE 1880) and projected UTM PROJECTION) The reference point in Nahrawan region called locally "binary system .whereas elevation level depends sea level at FAO as basis for elevation system currently called GTS .This system used until 1980 .but the information about this system was little and it is called nahrwan datum 1976 whereas projection that was used (PROJECTION) has Been projected to use the UTM the following specifications [1]

1-Central Meridian = 46°30´00´´E

2-Origion Latitude = 29°01´34.5560´´N

3-Scale Factor = 0.9994

4-False Easting = 800000m

5-False northing =0.00m

## 2.3 Polish geodetic control networks of Iraq 1974-1979

### 2.3.1 Horizontal networks

It is the period from 1974-1979, where the one of the competent international companies in this field (Pole Service Company) establishment of a network adjust the horizontal and vertical first-order at the country level extra to do a lot of work surveying Other, These can be one of the best geodetic networks in the region network considered at that time where it was using the latest instruments and techniques in the development of this network has been chosen flounder CLARCK1880andprojectedUTM.The following specifications

1-The horizontal network consisting of 2778 stations distributed in all over the country

2-50 Laplace azimuth/average mean square error±0.42″

3-8606 distances measured by means of AGA8 /mean square error of distance 5mm+$1.10^{-6}D$

4-316 distances measured by means of tellurometers /mean square error of distance 15mm+$3.10^{-6}D$

5-1267 angle /average mean square error ±4^cc the mean square error of the distance after adjustment is equal to 22mm

6- The distances are ranging from 8 to 35 kilometers, while mean distance is about 15 kilometers.

7- The primary control is countrywide-trilateration net, in which all the distances between consecutive points and also some selected angles are measured.

8- 24 Base line which connects Laplace points[2]

### 2.3.2 Vertical networks

It has been relying on the sea MSL level rate where he was erected two station-Gauge Tide for about two weeks "in the Fow region of the Arabian Gulf and recording the highest elevations readings and took her average per day and record readings lower elevations and took her average per day and then taking those rates and took rate of two per day and then taking the rates of these days and taking the average of the sum of these days to station one and then two rate This is attributed to a zero meters, which depends on the double minute settlement process for the distribution of the vertical grid points, the whole area of the country51 leveling lines from 21 closed perimeters the average length of a line is 180 kilometers, the average length of a perimeter600 kilometers. The bench-marks on precise leveling lines are situated at a distance of about 5 kilometers one from another in rural and desert areas and 2 kilometers in developed and urban areas, which amounts to up to 1984 station bench – marks. The trigonometrically leveling net adjustment comprises 5357 vertical angular observation single or double and was based on leveled point .the mean square error of adjusted height does not exceed 0.5m ,while the average value is 0.3m. [3]

## 2.4 Gravity basic station network of Iraq1963-1984

Gravity basic station are found mental for regional and detailed gravity surveys previously foreign oil companies conducted regional gravity survey using basic station apparently connected to the Syrian net unfortunately on information could be obtained concerning location values and accuracies of the basic station . First document information available were these of six international basic stations established in Iraq by woolard in 1963 and connected with world gravity net of wiscons in university. Arose to create basic station to carryout regional gravity survey in an covered area of Iraq. So in the period between 1969-1977 some 180 basic station were established and connected with woolard . On the initiative of geophysical determent of Som a five year plane to established gravity basic station net. Covering the whole territory of Iraq was projecting .462 basic stations distributed all over country, mostly on triangulation point and benchmarks of polservis were established, measured and al aborted during the years 1979-1984. These basic stations were connected to the previously established ones making a total of 648 gravity basic stations covering the territory of Iraq [4]as in figure 2

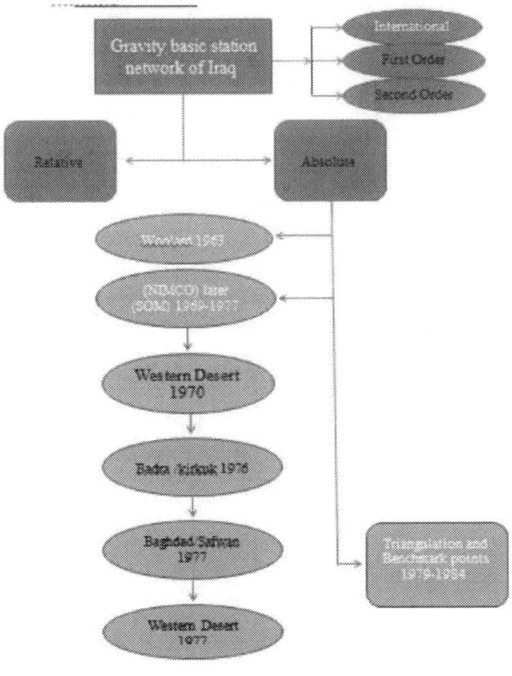


**Fig.2.** Historical development of gravity basic stations network in Iraq .

## 2.5 Iraqi geospatial reference system 2004 - 2016

During the period from 2004 to 2005, the American institution for the National Geodetic Survey, which is considered one of the most prestigious Institutions in the competent Geodesy world, prepared the plan for the project geodesy network development in Iraq. The project has been started and the completion of part of it, but without the participation of the Iraqi side and the project was renamed IRAQI GEOSPATIAL REFRENCE SYSTEM (IGRS).

the work done by this institution has been Distinguished by ber formic two major tasks

1-The first is the monument 6 of reference CORS stations

2-The second is erected 64 adjusted Stations of the three-dimensional first-class or as-called points of HARN in four provinces.(southern )

At the beginning of 2007 the Authority General of surveying started to think about completing the stoppage work and the addition of some other important things and already has been moving and start racing to activate the project, but at a slow pace due to lack of expertise and cadres . In fact such that acts have always been and still carried out by companies and institutions world with experience in this field it has been identified that, the goals of this new project follows as in figure 3

- Adoption of the Universal ITRF system as the basis for a new system of coordinates, according to the recommendation of the US geodesy Authority NGS
- Adoption of flatfish WGS84 Ellipsoid
- The adoption of the Universal projected UTM
- Installed new network is made up of seven reference CORS and world-class associated with the World Wide Web will be the basis for the new system of coordinates of stations and provide a free service for users to fully and be this network is an alternative to the old network because of a stoppage and not to the possibility of re-entry to work in addition to the new network will be controlled by the Directorate General of surveying fully
- Continue to install the ground control stations, three-dimensional HARN (High Accuracy Reference Network) in the provinces and in the same specifications stations which installed previously with some modifications in the site selection process and the establishment of stations. [1]

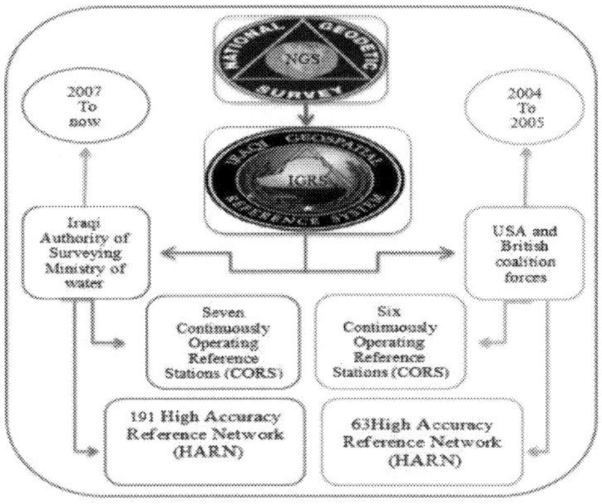


**Fig. 3.** Iraqi geospatial reference system

### 2.5.1 Continuously Operating Reference Stations of USA and British (CORS)

The monument 6 of reference CORS stations and connect them with the institution of the NGS which provide data and processing of data for these stations free of charge to users where these stations regarded as the basis of Iraq's new coordinate system, which depends on the center of the globe as a point of reference have been chosen coordinates of ITRF2000 system as a basis for work and according to international standards accredited [1].The distribution points in the military headquarters of the multinational forces, of the table as follows (table-2)

**Table-2** distribution the USA CORS stations in Iraq

| Code station | Location | Code station | Location |
|---|---|---|---|
| IZ BD | Baghdad | IZTL | Thi Qar |
| IZ BA | Basrah | IZ AD | Anbar |
| IZ BL | Salahuddin | IZQW | Nainawa |

### 2.5.2 High accuracy reference network of USA and British (HARN)

Erected 63 Stations Ground adjust the three-dimensional first-class or as-called points of HARN in four provinces, as shown as follows .It utilized to international standards in the establishment and monitoring points and in accordance with the six reference stations and were prepared. Data bases of these points which have been distributed in the province of Basrah, Muthanna,. Maysan and Thi Qar.. [1]

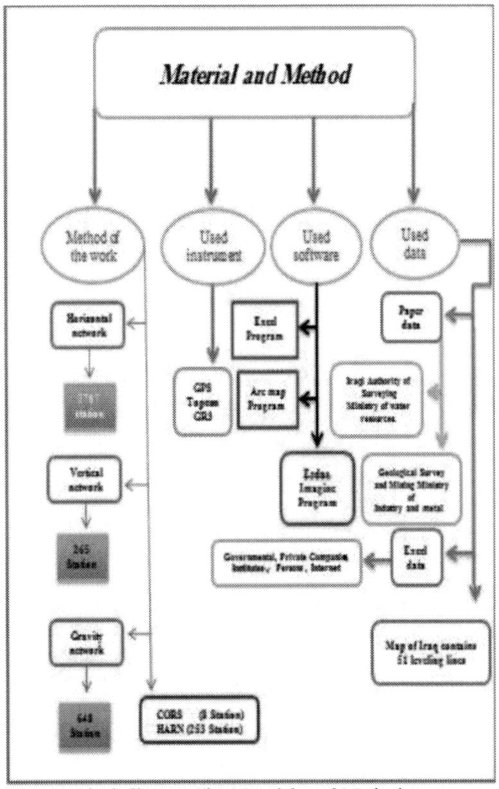


**Fig.4.** Block diagram The Materials and Methods

### 2.5.3 Continuously Operating Reference Stations of Iraqi general authority for surveying (CORS)

Seven stations have been installed during the period from 2008 to the end of 2009 in each of the governorates (Erbil, Salahuddin, Baghdad, Najaf, Kut, Basra and Zakho).

### 2.5.4 High accuracy reference network of Iraqi general authority for surveying (HARN)

The High Accuracy Reference System (HARN )is three-dimensional control points be monitored by system of GPS particular way and depending on the stations of the CORS and be given a central geodetic coordinates and quadratic depending on the global system ITRF surface flatfish WGS84 and projected UTM in addition to providing a national level, depending on the Polish network of vertical and most of the points is done through method double accurate leveling thus can rely on these points in the topographic surveys using GPS RTK Technology System .where During 2005 the institution of NGS installed 64 distributed stations in the provinces (Basra, Maysan, Dhi Qar, Muthanna) but without giving the levels depends on the sea level and is currently completing the work in the installed stations in other provinces are currently working shut down due to lack of financial allocations were So far, 191 Have been installed distributed station in the country's governorates. [1]

## 3 Problem statements

In the past eight decades, different geodetic projects have been planned and implemented within Iraq country, the hypothesis of this thesis based on representing the main problem in these projects was not in their accuracy or in their technical approaches rather than in the official documentation of them. However, two indicators for weak documentation process have been noticed, as follows:

A- Throughout searching in the Internet about the specialized publications of geodetic control networks in Iraq, the acquired documents (technical reports, maps, sketches, theses, papers, datasheets) are too few and do not reflect the reality of these huge geodetic projects that had been implemented in Iraq.

B- If anyone could visit the scientific libraries of the specialized directorates in geodetic surveying, he will not find the standardized archiving for the documents related with geodetic control networks in Iraq.

## 4 Materials and methods

To study in depth the thesis problem, different materials and methods in figure 4 are to be prepared and arranged as follows:

A- The authorized documents including maps, sketches, technical reports, datasheets in both hardcopy and digital formats

B- Utilizing ArcGIS software package to design and implement the required spatial data base for geodetic control networks ( with their different types, i.e.

traditional horizontal, traditional vertical, gravity and CORS-HARN ).

C- Other software packages have been used in minor operations that are required in preparing the geodetic geodatabases, as follows:

1- ERDAS Imagine: has been used in coordinates transformations of geodetic control points within different networks.

2- Microsoft Excel: was required in initial editing of the tabular data that are related with the geodetic geodatabases.

D- Geodetic GPS device (Topcon GR3 model) has been used as ground truth by field re-surveying for selected geodetic control points within two Governorates ( Babil and Kerbaka) .

## 5 Result

According to the availability of the Iraqi geodetic data, a comprehensive geodatabse has been prepared (using the ArcGIS software package), this geodatabse consists of different datasets as shown in figures 5-16 and tables 3-6

### 5.1 Polish geodetic control networks of Iraq 1974-1979

#### *5.1.1 Polish geodetic horizontal networks*

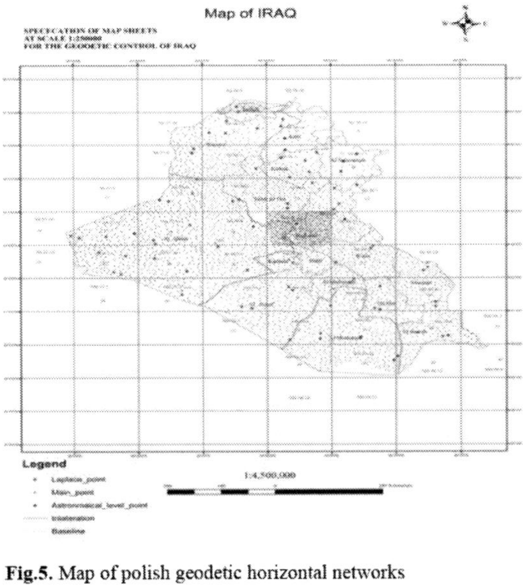


**Fig.5.** Map of polish geodetic horizontal networks

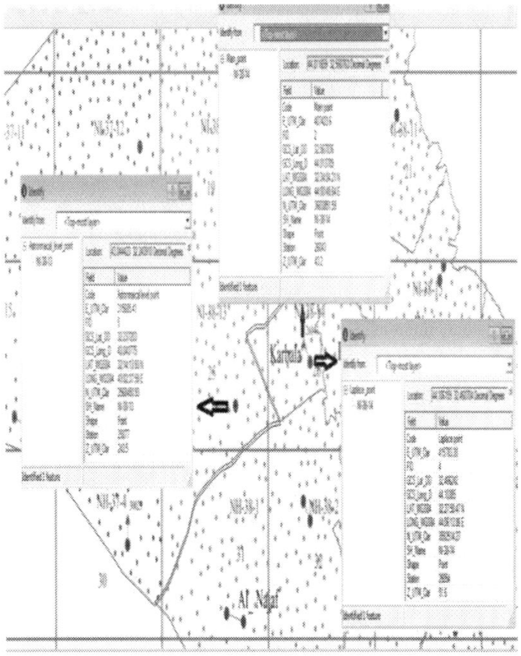


**Fig.6.** Geodatabase for polish geodetic horizontal networks

**Table.3** Geodatabace for polish geodetic horizontal networks

| StNr | E_utm | N_utm | Height | Dist_obsv | Angle_obs |
|---|---|---|---|---|---|
| 1001 | 339329.9 | 4128320 | 2161.6 | AGA_8 | Wild_T3 |
| 1002 | 323013.5 | 4124051 | 1766.2 | AGA_8 | Wild_T3 |
| 1003 | 310354.8 | 4126191 | 1982.8 | AGA_8 | Wild_T3 |
| 1004 | 293071.5 | 4122335 | 870.6 | AGA_8 | Wild_T3 |
| 1005 | 355135.2 | 4118299 | 1548.6 | AGA_8 | Wild_T3 |
| 1006 | 304670.1 | 4115525 | 753.1 | AGA_8 | Wild_T3 |
| 1007 | 319781 | 4114325 | 969.4 | AGA_8 | Wild_T3 |
| 1008 | 336084.7 | 4113060 | 1412.2 | AGA_8 | Wild_T3 |
| 1009 | 345411.7 | 4111424 | 2086.7 | AGA_8 | Wild_T3 |
| 1011 | 275681.8 | 4107531 | 1071.3 | AGA_8 | Wild_T3 |
| 1012 | 292036.2 | 4104023 | 1117.5 | AGA_8 | Wild_T3 |
| 1013 | 321854.3 | 4102324 | 966.9 | AGA_8 | Wild_T3 |
| 1014 | 308403.1 | 4097901 | 1385.8 | AGA_8 | Wild_T3 |
| 1015 | 341059.4 | 4096594 | 1508.4 | AGA_8 | Wild_T3 |

### *5-1-2 Polish geodetic vertical networks*

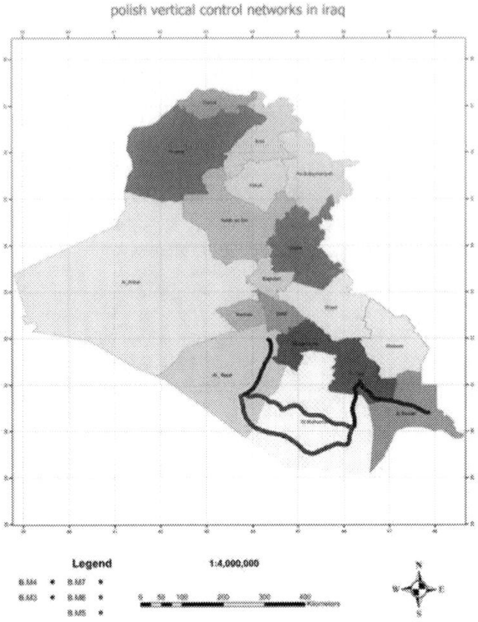


**Fig.7**. Map of polish geodetic vertical networks

**Table.4** Geodatabase for lines bench mark in polish geodetic vertical networks

| B.M_no | Dis_KM | Final_H_M | GCS_Long_Deg_Min_Sec | GCS_Lat_Deg_Min_Sec |
|---|---|---|---|---|
| 1-4 | 4.99 | 3.7432 | 46.359541 | 31.034796 |
| 2-4 | 10 | 4.7711 | 46.376435 | 30.991966 |
| 3-4 | 15.56 | 4.6174 | 46.401419 | 30.947708 |
| 4-4 | 18.45 | 3.8859 | 46.421645 | 30.930338 |
| 5-4 | 21.92 | 3.9189 | 46.446867 | 30.909874 |
| 6-4 | 27.22 | 4.8185 | 46.483749 | 30.878941 |
| 7-4 | 29.91 | 4.5325 | 46.504926 | 30.867044 |
| 8-4 | 32.99 | 2.6631 | 46.528483 | 30.852291 |
| 9-4 | 36.03 | 3.5782 | 46.549422 | 30.833493 |
| 10-4 | 39.07 | 2.8296 | 46.571789 | 30.816599 |
| 11-4 | 43.93 | 4.2632 | 46.59249 | 30.774959 |
| 12-4 | 48.95 | 8.5594 | 46.613191 | 30.731653 |
| 13-4 | 53.03 | 18.0333 | 46.645076 | 30.711665 |
| 14-4 | 57.81 | 18.5569 | 46.685527 | 30.695247 |

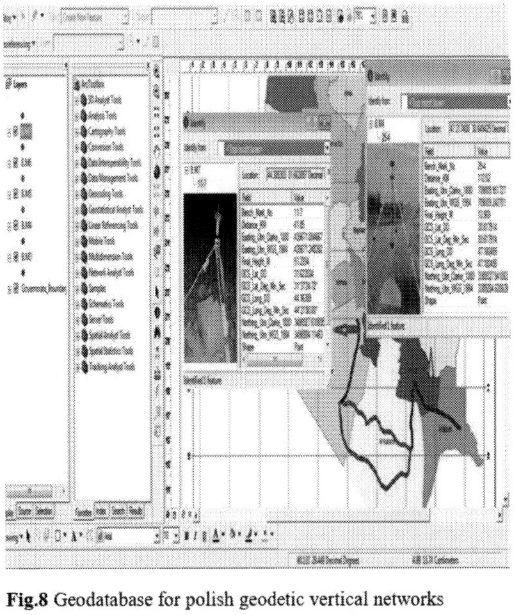

**Fig.8** Geodatabase for polish geodetic vertical networks

## 5.2 Gravity basic station network of Iraq1963-1984

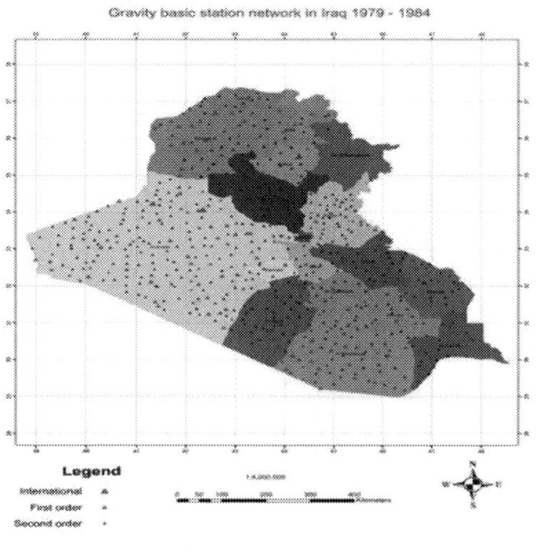


**Fig.9**. Map for gravity basic stations network

**Table.5** Geodatabase for gravity basic stations network.

| Code of Basic Station | Order | Y UTM | X UTM | Elev. UTM | Absolute gravity value (mgal) | Old of Basic Station |
|---|---|---|---|---|---|---|
| NCDa2 | 2nd | 4106113.000 | 267007.000 | | 979 794.27 | Fish Khabour |
| NCDab1 | 2nd | 4113279.000 | 293873.000 | | 979 794.82 | Zacho |
| NCDb2 | 2nd | 4120868.000 | 361428.000 | | 979 572.14 | Kani Masi |
| NCDb1 | 2nd | 4115383.000 | 323417.000 | | 979 660.83 | Batufa |
| NDa3 | 2nd | 4110307.000 | 346518.000 | | 979 590.70 | Bameni |
| 1 | 1st | 4105909.000 | 365726.000 | | 979 579.16 | Amadiya |
| NDa4 | 2nd | 4100033.000 | 352139.000 | | 979 624.78 | Sersank |
| NDb1 | 2nd | 4103002.000 | 380723.000 | | 979 697.75 | Balenda |
| NB1 | 2nd | 4062054.000 | 763469.000 | 377.700 | 979 775.99 | TP 03001 |
| NCDa1 | 2nd | 4097362.000 | 262357.000 | 484.300 | 979 775.73 | TP 04001 |
| NC1 | 2nd | 4084396.000 | 260461.000 | | 979 805.23 | Wadi Swedi |
| 2 | 1st | 4076827.000 | 239781.000 | | 979 780.38 | Rabiaa |
| NBC | 2nd | 4062092.000 | 267189.000 | | 979 792.90 | Tel Awainat |

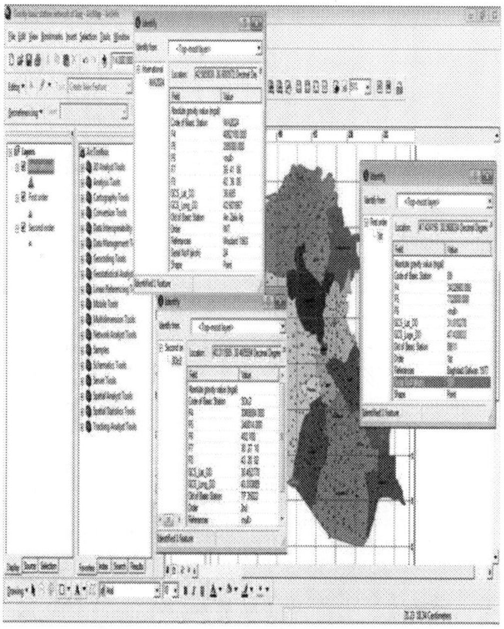

**Fig.10.** Geodatabase for gravity basic stations network

## 5.3 Continuously operating reference stations of USA and British (CORS)

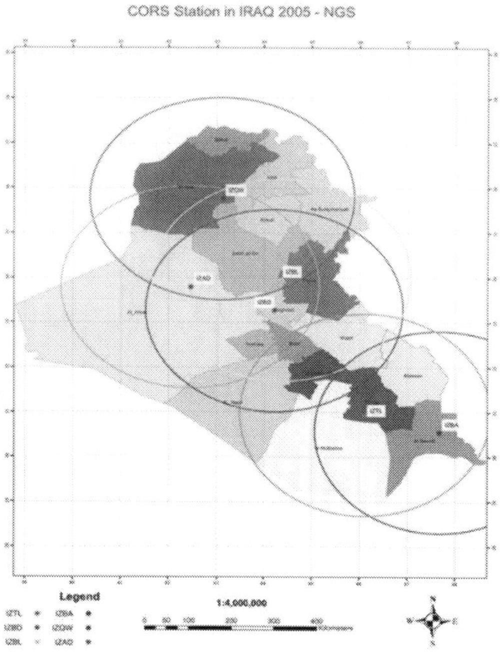


**Fig.11.** Map the (CORS) network 2005 in Iraq – NGS

## 5.4 High accuracy reference network of USA and British (HARN)

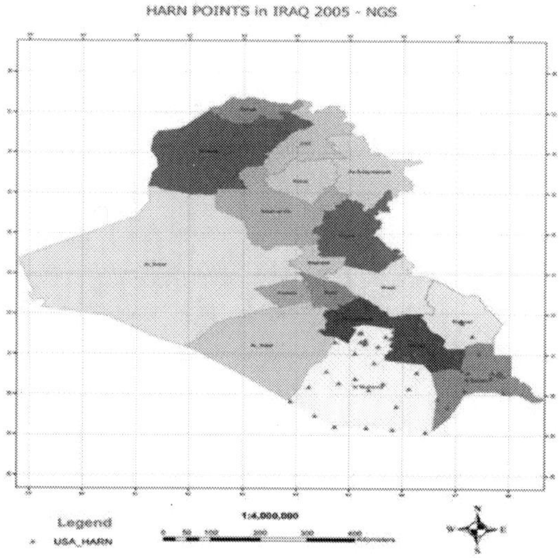


**Fig.12** Map for high accuracy reference network (HARN) in Iraq 2005

### 5.5 Continuously operating reference stations of Iraqi general authority for surveying (CORS)

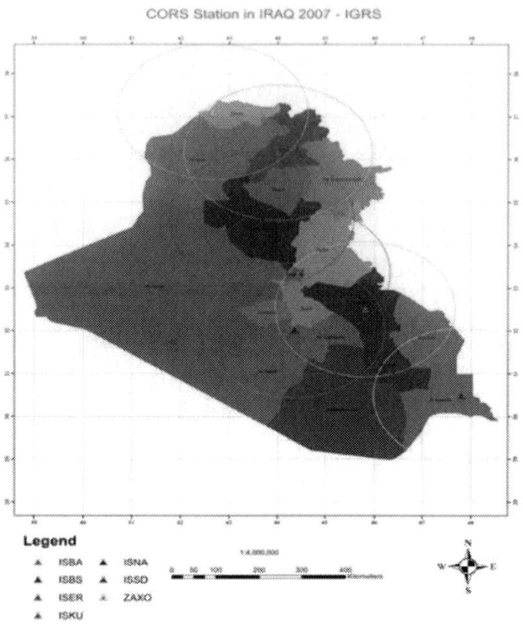


**Fig.13.**Map for (CORS) network in Iraq – IGRS

### 5.6 Continuously operating reference stations of Iraqi general authority for surveying (CORS)

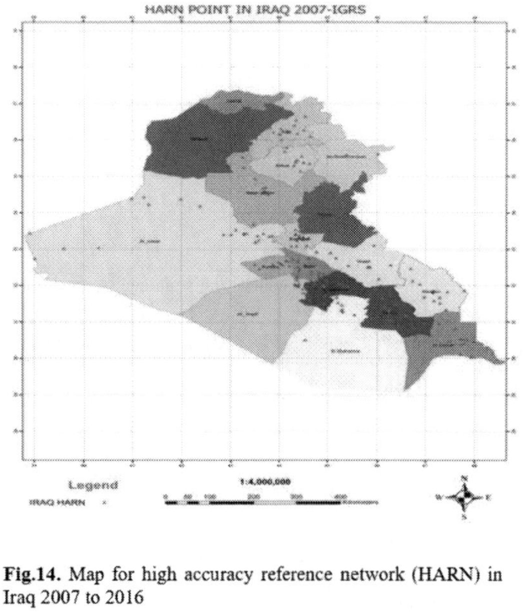


**Fig.14.** Map for high accuracy reference network (HARN) in Iraq 2007 to 2016

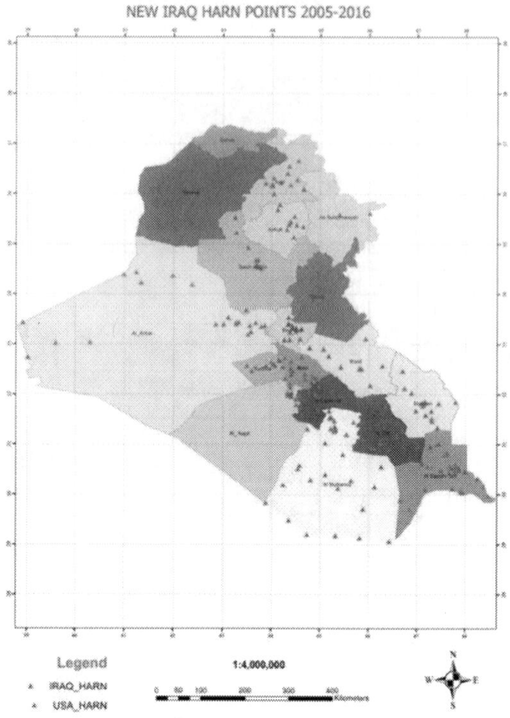


**Fig.15.** map for high accuracy reference network (HARN) in Iraq 2005 to 2016

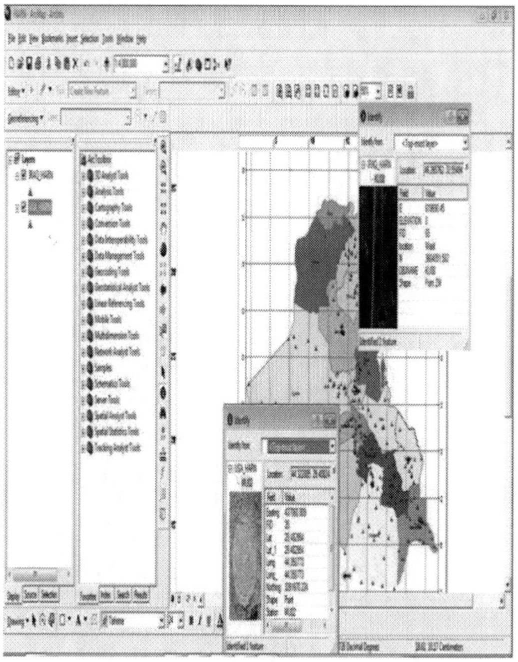

**Fig.16.**Geodatabase for high accuracy reference network (HARN) in Iraq 2005 to 2016

**Table.6** Geodatabase for high accuracy reference network (HARN)

| point | Code | Location | East. | North. | ELEV. |
|---|---|---|---|---|---|
| 01 | DQ01 | Thi Qar | 621452.078 | 3438594.909 | 5.299 |
| 02 | DQ02 | Thi Qar | 618568.200 | 3432954.322 | 6.508 |
| 03 | DQ03 | Thi Qar | 618053.464 | 3435422.901 | 4.180 |
| 04 | DQ04 | Thi Qar | 611159.108 | 3475960.042 | 1.296 |
| 05 | DQ05 | Thi Qar | 618011.963 | 3463616.072 | 1.725 |
| 06 | DQ06 | Thi Qar | 604723.820 | 3503133.510 | 6.023 |
| 07 | DQ07 | Thi Qar | 690793.343 | 3427588.589 | 9.367 |
| 08 | DQ08 | Thi Qar | 664627.161 | 3427978.998 | 9.404 |
| 09 | DQ09 | Thi Qar | 652368.232 | 3448873.044 | 9.091 |
| 10 | DQ10 | Thi Qar | 591378.678 | 3531492.574 | 4.127 |
| 11 | DQ11 | Thi Qar | 638578.686 | 3420464.993 | 6.535 |
| 12 | DQ12 | Thi Qar | 624745.393 | 3435122.974 | 6.025 |
| 13 | DQ13 | Thi Qar | 631242.612 | 3485316.793 | 4.190 |
| 14 | DQ14 | Thi Qar | 584681.384 | 3442576.912 | 5.438 |
| 142 | BS12 | Al Basrah | 761711.744 | 3334676.016 | 10.404 |
| 143 | BS13 | Al Basrah | 781884.881 | 3326610.608 | 9.007 |
| 144 | BS14 | Al Basrah | 255952.921 | 3318774.347 | 0.140 |
| 145 | BS15 | Al Basrah | 247020.348 | 3338235.728 | 3.385 |
| 146 | BS16 | Al Basrah | 760550.287 | 3364410.957 | 7.986 |
| 147 | BS01 | Al Basrah | 772378.358 | 3380094.345 | 11.280 |

## 6 Recommandations

To continue the research work in the future, and to go strongly toward solving the technical and administrative problems that are related with geodetic control networks, especially in the field of preparing their required corrected updated spatial data bases, the following recommendations are to be taken in considerations:

1- Enhancing the courses of Geodetic Control Networks for both undergraduate and postgraduate programs within all departments of colleges that are related with geomatics subject).

2- Preparing and supporting the advanced and practical training courses for scientific researchers and geomatics engineers in the field of Geodetic Control Networks, throughout increasing the scientific coordination between the academic sector ( geomatics wised colleges ) and the applied sector ( specialized directorates in geodetic surveying).

3- Establishing the Iraqi higher council for geodetic and cartographic activities, it could be helpful in the improvement of the technical and financial managements for the geodetic control networks projects.

4- Establishing the Digital Iraqi Gate for Geodetic Control Networks, this gate is specialized web. site, it will manage , control the accessibility and downloading for all geodetic official documents ( datasheets, sketches, photos, technical reports ), this gate could be run by the Iraqi higher council mentioned in the recommendation number 3 above.

## References

1. W. Hussein, National geodetic networks of Iraq Between the past and the present (2011).
2. K. AlKadimi, Gravity control measurement to on basic station along Baghdad Basrah – Safwan road, Baghdad, Som (1977).
3. J. Gazdzicki, H. Kwiatkowshi , New Geodetic Control Network In Iraq Design, Surveys, And Data Processing , (1977).
4. K. AlKadimi , Gravity Basic Station Network of Iraq (1985).

# Using geospatial analysis to determine livability in Najaf city by GWR

*Mustafa* Ebraheem[1,*]

[1] Center of Urban and Regional Planning for Postgraduate, University of Baghdad, Baghdad, Iraq

**Abstract**. Iraqi cities suffer from degradation in many livable elements, sometimes related to an economic situation, social, and environmental, therefore in this research we have applied some indicators relating to livable city by using GIS technique, to determine the trend of Iraqi city in livability term.Also we are cannot depend on any foreign standards, because we did not have enough archive to our cities to explaining it is trends, so in other side the relationships between activities in the city did not have enough exploring depending on complexity model to clarify the connections among these activities and how it depends on livability of cities.We try applying GWR, on socio-economic factors, to study which effective factors on city livability. In this research we select Najaf city as a case study, for its importance as international religious tourism city, and for its stability through last decades, which has helped growth in economic activities, and made it attractive to immigrations.From discussing in literature review we determine some indicators, and finally we found the difficult point that determining Iraqi city trends in terms of livability was incomplete, because it is have a physical structure in many parts of it.

## 1 Introduction

The main problems in the Iraqi cities include decline in environmental, social, and economical circumstances, causing unlivable city, as compared to international standards. So we can make problems statements such as: (1) what are the capabilities required for a livable city experiment in Iraq, and what are obstacles around it, (2) what are the characteristics of a livable city, and (3) How we can plan for a modern city with an ability to face different problems.
Research goal knowing how far Iraqi city is from livability, also how it is near from human activates, that consternate about development elements related to man welfares life, therefore; research hypothesis are being all data for Iraqi city are nearing it to a livable city.

## 2 Livable city

The concept of the livable city is a part of a sustainable city, which constrain on many common principles, (accessibility, equality, participation, infrastructure, and others), so if we distinguish any society have participated in its governorate to take decision for building the city, we can say this is a livable city. When people have many choices of movement and are free to take trips with a variety of ways, we can say there is some indicator for livable city. However; cooperation between participating and accessibility to movement for urban economic activities, resulting in higher income, there is an increase livability level in the city [1]. In other way low rate polluted sources, attractive stability community, are more attractive for investments, visitors, and tourists, making a high level of livability in the city [2].
The livable city is a city for all ages, children have safety in playing, and elders have attractive employment opportunities.
To understand the concepts of livable city and sustainable city we can say that livability is the first stage towards sustainability. In other words, livability corresponds to the present, and sustainability to the future. Sustainability is a goal, logically to reach the goal we must have strong base in the present [3].

### 2.1 Aspects of livable city

There are many aspects, but they are different in their priority, the most important; (1) city must contain many public spaces, to be attractive and leisure, (2) priority in the centers for movement of the pedestrians, (3) having transit stations for multi purposes, and accessibility to any place, (4) diversity in housing, with reasonable costs, (5) economical competitive activities, and (6) decreasing pollution level with time.

## 3 International experiences

There are many experiences related to livability in the world, this study selects two of leader experiences in the world in terms of livable city, one of them in a wealthy country, UAE, Masdar city.

* Corresponding author: dr.mustafa.a.jalil@iurp.uobaghdad.edu.iq

© The Authors, published by EDP Sciences. This is an open access article distributed under the terms of the Creative Commons Attribution License 4.0 (http://creativecommons.org/licenses/by/4.0/).

https://doi.org/10.1051/matecconf/201816203031

The second one in a developing country with fair economic activities, Philippines

### 3.1 City of Masdar

It is one of the UAE cities; we can consider it the first city in the whole world with zero carbon and zero waste, it is the world with Solar Energy, the city build about (17) Km south east Abu Dhabi, near Abu Dhabi international airport, with connectivity to most other airports, and to all cities in the UAE.
Masdar city has been built based on some principles of livable city, and also sustainable city. These principles are: (1) energy management, using solar plate, descending beam, clean electricity station, (2) water management, (3) waste management and recycling, (4) recycling building martial, and (5) transportation systems, through environmental train, personal vehicle, buses and mass transit [4].

### 3.2 Philippines experiences

Philippines is one of the most crowded cities in the world, it is ranked seventh in Asia, and twelfth in the world in terms of population. The principles suitable to make Philippines a livable city are, (1) free living way to all population, (2) diversity and variety with housing, according to types and costs, to suit so everyone's need , (3) accessibility, (4) infrastructure for people with special needs, (5) job opportunities for various skill sets, (5) decreasing pollution rate, and (6) decreasing average crime rate. [5].
From these experiences, we can find out what indicators suitable to our city, and be able to manage it spatially.

## 4 Geospatial analysis

Spatial analysis now is an important tool to find out spatial indicators, so we can recognize and define areas without needing to attribute data, just depending on data related to spatial activities and their locations. The most powerful tool for that were GIS technologies, supported by remote sensing data, so by depending on basic analysis tool in GIS we can express the situation of our city through spatial indictors [6].
We must be rethinking about spatial data; we need to recognize features to find out channels among these features through a certain network. The strength of this network is related to its flow, may be passengers, goods, or any other flow that causes interaction creating interrelationship among features.
With this aim we can manipulate spatial dimensions as different shapes, by ranking the features; therefore, we can create different sizes of catchment areas. By using GIS simple analysis tool, not complicated, to represent actual situation and by explaining spatial dimensions in any location in clear language, we can ensure that GIS analysis functions are used by more people [7].

## 5 Case study

In this research (18) districts in Najaf city, have been selected as sample for finding indictors of livable city. The main reasons to select these districts were: (1) variety at district area, (2) variety of density and population, (3) variety in physical structure, (4) variety of social and economic characteristics, and (5) variety in plot shapes and area.

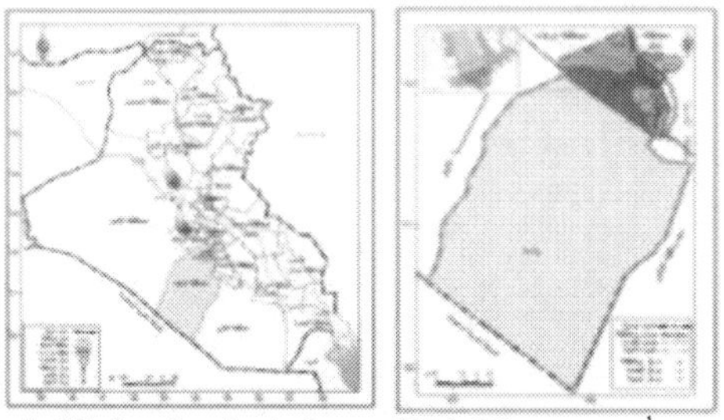

**Fig. 1.** City location due to Iraq and Najaf Governorate[1]

We can see distribution of these districts from image (2) that showing all Najaf city, and districts coding by letters, instead of names, to easy dealing with it. Note we area excluding city center from our analysis because it is represent special case due to holy shrine of Imam Ali, and the functional system is different on all other districts in the city.

### 5.1 Selected indictors

Research tries to build compromised methodology to help it getting results from real data, this compromisation depending on explanation in the previous literature review, because we cannot depending on any other city from the surrounding regions or another place in the world, because the huge different according to economic, social, and natural characteristic. Even in the same region, Iraq, because we did not have similar study or specialized center to depending it is criteria or scale.
Research depend on indicators can be measured, also measuring it in the same unit, to prevent any cumulative or bias errors in the model interpretation later. All seven indicators represent the important indicators in the city, but we must sure there are other unknown indicators effects on the livability of any city; we can see indicators as follow:

[1] All maps produced based on GIS system build from researchers with information collected from field survey, city municipality, directorate of physical planning.

 https://doi.org/10.1051/matecconf/201816203031

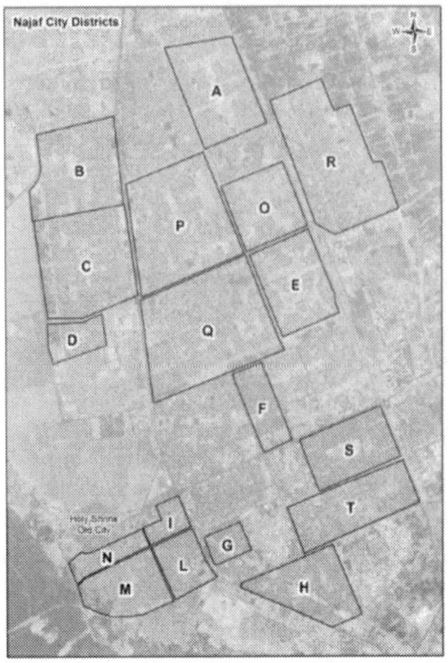


**Fig. 2.** Slected district to field study

Source: researcher depending on image and felid investigations.

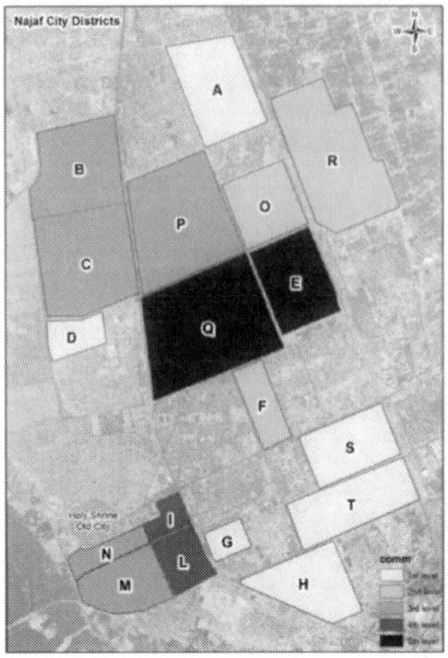


**Fig. 3.** Spatial indicatores trend of Commercial Landuse

Source: researcher depending on image and felid investigations.

### 5.2 Commercial indicators

The commercial land use reflect the economic situations of any district, so it is make this district desire for living, because more commercial land use meaning more chances of work and getting high income. The commercial land use must be attractive indicators to the district, but when the reaching to the critical point, we occurring theme negative indictors because the side effects be decay all districts services.

Fig. (3), showing distribution of commercial land use on districts, we can notice from all five category of distributing commercial land use the districts (Q and E) have the maximum share of this land use, and in the other side the districts (A, D,G, H, S, T,) have the minimum share of this land use.

### 5.3 Green Area indicators

This indicator have positive relationship with livable city, increasing this indicator effecting on all activities in the city, research indicate all active green area in the city districts, we can also depending this indicators as attractive elements to resident in the district. From the Fig. (4), we can recognizing the maximum green area land use in (C) districts, and minimum land use in (A, E, F, G, H, I, L, M, N, O, R, S, and T) districts.

### 5.4 Mass transit indicators

These indicators express on accessibility level to the city through time, also it is positive relationship with livable city, we calculating this indicator by representing all mass transit paths in the city, the mass transit related to paths of public vehicles in the districts, means the long of the paths represent quality of mass transit, usually when mass transit related to private sector only, we'll find it have relation with distracts attractive to resident. We can notice from Fig. (5) that district (C) have maximum value from the five category of mass transport, when in the other side, distracts (B, F, G, M, O, and R) recording minimum value of mass transport.

### 5.5 Population Density indicators

The population density showing to us the reality of district attractive and population willing to get residents, therefore; naturally the districts with high density means there are many good conditions in this district, so be attractive for people, but low density surly that means the district unattractive. Fig. (6), showing us that districts (C, and O) have high proportion of population density, where the districts (A, D, E, F, G, P, Q, R, S, and T) have very low density.

 https://doi.org/10.1051/matecconf/201816203031

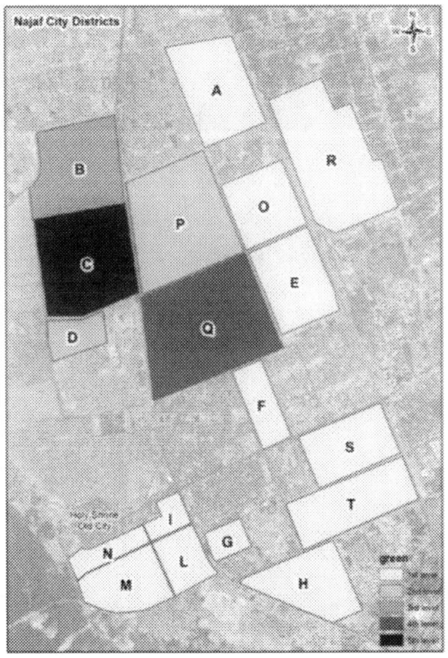


**Fig. 4**. Spatial indicatores trend of Green Area

Source: researcher depending on image and felid investigations.

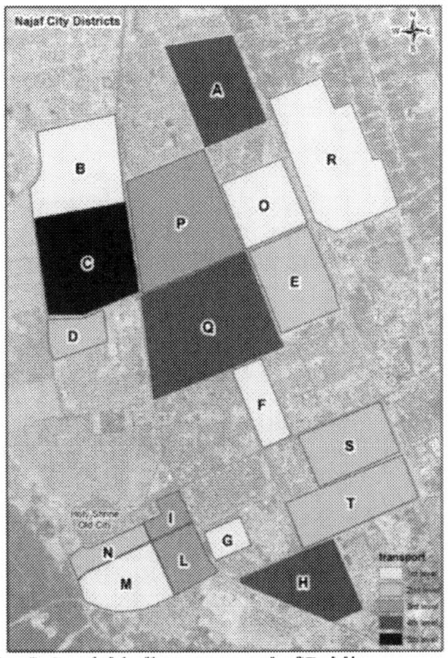


**Fig. 5.** Spatial indicatores trend of Public Transportation

Source: researcher depending on image and felid investigations.

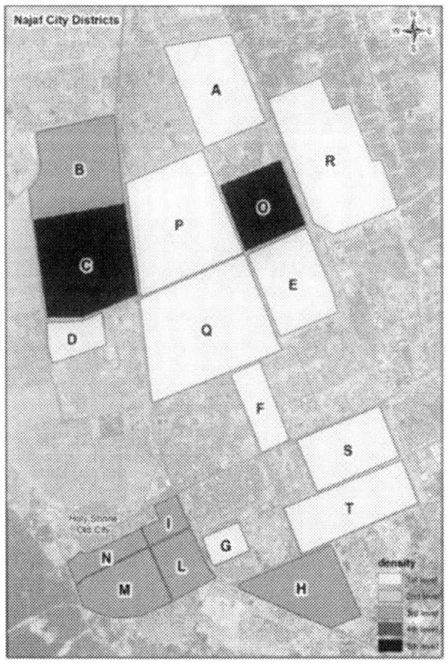


**Fig. 6**. Spatial indicatores trend of Population Density

Source: researcher depending on image and felid investigations.

### 5.6 Road Network indicators

This indicator refer to accessibility in the districts, so the distance of all major roads may showing us how district have accesses, intersections, and links, to be easy the movement among district parts. Fig. (7) clarifies that district (M and R) have maximum value of roads network, and districts (D, F, G, I, and N) have minimum value of it.

### 5.7 Services indicators

This indicator consist of all services provides in district, education, health, cultural, and recreation, mainly if these services are good and available the district will attractive to populations, Fig. (8), showing that high value of availability services in the districts (P and Q), and districts (C, D, G, I, L, M, and N) recorded minimum value of service availability.

### 5.8 District area indicators

We are using this indicator as representative of difficulty in the providing any service, so the area of district almost it causing reversely in attractive population, because when area is high needing more efforts to be suitable to livability, but if be small the services in it, was controlled easier, Fig. (9), showing high districts area in (P, Q, and R), and small area with districts (D, F, I, L, G, and N).

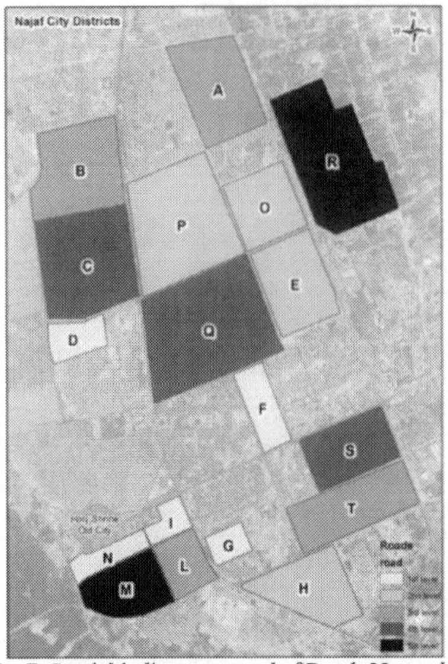


**Fig. 7**. Spatial indicatores trend of Roads Network

Source: researcher depending on image and felid investigations.

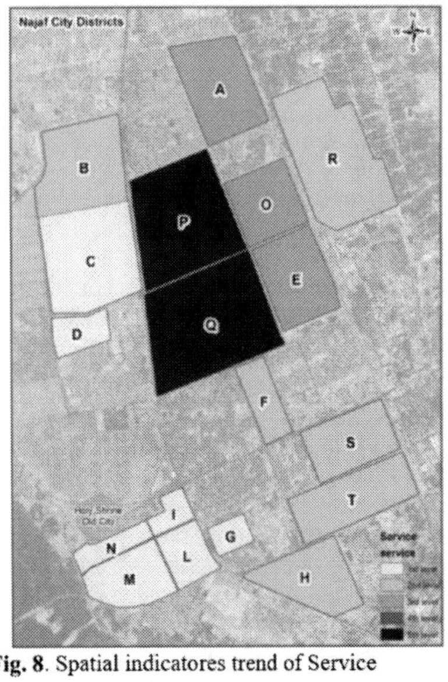


**Fig. 8**. Spatial indicatores trend of Service

Source: researcher depending on image and felid investigations.

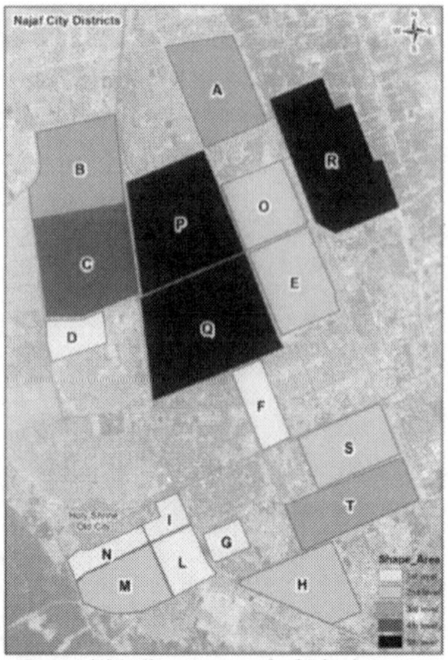


**Fig. 9**. Spatial indicatores trend of Districts Area

Source: researcher depending on image and felid investigations.

### 5.9 Review indicators

From all perviuose review to all indictors, we can shwo in Fig. (10), behaviore of indicitors with districts. We found that district (C and Q) have three maximum value, when districts (P and R) have just two maximum value, and the recent district have one or no maximum value, so logically we found the best districts for livabilty are (C and Q), also table refering to no relation among population density and other indicatores, for that, distric (C and O) have maximum value of population density, but at the same time all other indicatores were in minimum value, espically to (O) district.

## 6 Applying GWR Model

We are must define parameters of GWR model, by determine dependent variable, in our case, we are selecting (Population Density) as occurring it dependent variables. This variable referring to population willing to live in the certain district than other, so the condition system related to live daily circulation in this district, who it is density, is high, sure must be better than other districts, so, the livability of this district is more than others.

For previous discussion we selecting GWR, because it is capable to explain the effect of any variable on livability level separately with remaining in mind these spreading do not mean works alone without interrelationships affecting on and among Model variables.

The independent viarables can may see it in table (1).

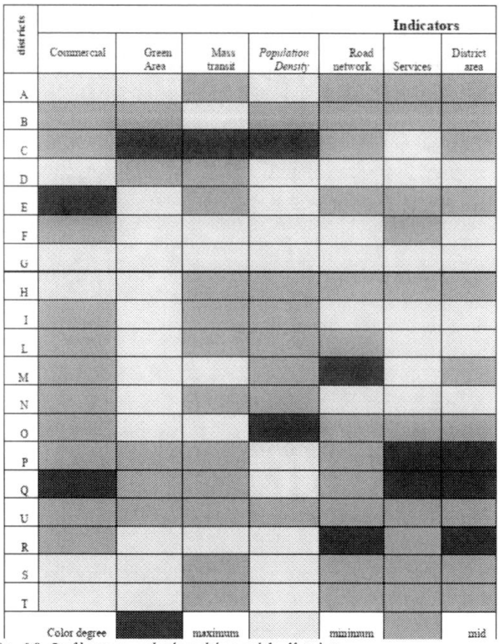


**Fig. 10.** Indicators relationships with districts

Source: researcher.

**Tab 1.** GWR model varibles

| **Variables** | **Type** | **Explain** |
|---|---|---|
| Population density | Dependent | The willing to live in the district |
| Commercial | Independent | Describing availability of works, and income level |
| Green Area | Independent | Referring to good quality of live without pollution and have may chance for recreation |
| Mass transit | Independent | Meaning accessibility in movements |
| Road network | Independent | Infrastructure for movement |
| Services | Independent | The import conditions to stability |
| District area | Independent | The cost of rehabilitations |

Source: researcher.

Firstly, the most important thing after running variables in the model was checking Spatial Validity of model; this will be through applying (Moran I) on the residuals of GWR model.

The (Moran I) will discovering the Spatial Autocorrelation among variables through spatial distribution of them, if apparent ant concentration, meaning the model was no have confidently, which means, there are other variable effective in essential way on model behavior. Fig. (11) Showing the result of (Moran I) to our model, and it is describe behavior of our variables are Spatial Randomly Distributed, meaning our model have true variables and related to logic of our study problem.

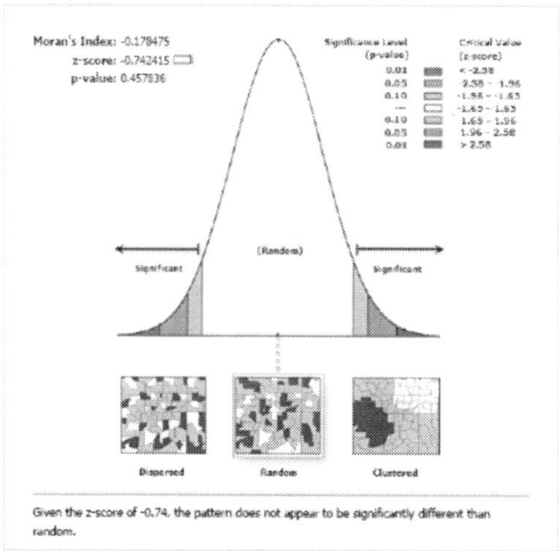


**Global Moran's I Summary**

| | |
|---|---|
| **Moran's Index:** | -0.178475 |
| **Expected Index:** | -0.058824 |
| **Variance:** | 0.025974 |
| **z-score:** | -0.742415 |
| **p-value:** | 0.457836 |

**Fig. 11**. results of applying Moran I on residuals of GWR Model

Source: researcher depending on running Moran I.

Secondly, discussing results of GWR, table (2), are showing the main results of model. The important one, the value of $R^2$, this value define to strength of explaining relationship through model, in our case it was 23%, this value is accepted, when occurring we dealing with man behavior, and surly there are many many systems effecting on the urban life. The result of Sigma is good, and may be bettering it, when it is value minimizes, by taking more variables.

Thirdly, we can from table (3) clarify the behavior of independent variables, when it working together as system to effect on the independent variable, the Population Density.

We distinguish that two variables reversely effective on the dependent variables, District Area and Services, and the other four variables have direct effecting on increasing population density, in the first level we found road network was more effecting than other, then the commercial land use, with mass transit, and lastly the Green Area.

**Tab 2.** results of GWR model.

| subject | value |
|---|---|
| Bandwidth | 90277.505 |
| ResidualSquares | 926758.491 |
| EffectiveNumber | 7.0213 |
| Sigma | 290.542 |
| AICc | 278.446 |
| R2 | 0.23238 |
| R2Adjusted | -0.18862 |

Source: researcher depending on running GWR model

**Tab 3.** Resultes of GWR model equation parameters

| variables | values |
|---|---|
| Intercept | 353.64062481539 |
| Commercial | 0.00118731451 |
| Services | -0.00029769160 |
| Road Network | 0.01217881478 |
| Mass transit | 0.00236091701 |
| Green Area | 0.00084879463 |
| District area | -0.00008052649 |

Source: researcher depending on running GWR model

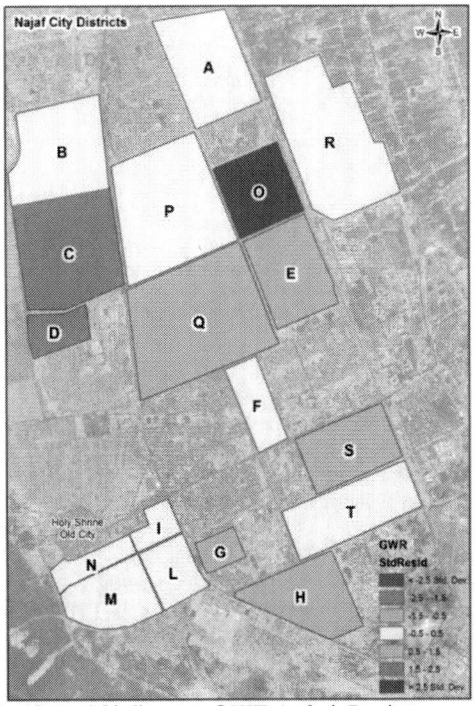


**Fig. 12.** Spatial indicatores of GWR Analysis Results

Source: researcher depending on image and felid investigations.

## 7 Conclusions

1- We can express on livability by some electing indicators, conditioning these indicators suitable with city situation.
2- The difficulties through measuring liaveablity were not collecting data, but comparing the results with base line.
3- Best way to monitor trend of any city livability through comparing it is situation with time.
4- We cannot found main measured variable can express on city livability directly.
5- Measuring levels of livability are clarifying city systems trend work during the time.
6- Livable city characteristics are safety and free with variety choices.
7- Major consideration of the livability related to movement system, with all kinds and types.

## 8 Recommendations

1- We can compromise among group of variables to define indicator can expressing clearly about city livability.
2- Must are represent cities livability yearly by charts and monitoring it.
3- We can maximize confidence by increasing cases and independent variables
4- The logic of building model are most important than measuring variables.

## References

1. E. Momtaz, *in ITAME Conf. Proc.*, (UAE 2015).

2. L. Sabra, Downtown living for families, The Vancouver, BC Urban Liveability Experience and Lessons for Other Cities, M.Sc., (UoW2008).

3. Lowe & others, Place, *HLRP*,1, (2013)

4. A. Yones, *EAJ*, Measuring and analysis welfare in Karkok city and related to distributing income for 2009, **no 83**, (2010).

5. https://dmciphilippines.wordpress.com.

6. K. Kemp, Encyclopaedia GIS, (Sage Publications 2008).

7. B. Burian, Advances in spatial planning, *InTech*, **9**, 51000, (Croatia 2012).

 https://doi.org/10.1051/matecconf/201816203032

# Assessment of land surface temperature variation over Rusafa side of Baghdad city, Iraq

*Ayad* Beg[1,*]

[1] College of Education, Department of Geography, Mustansiriyah University, Baghdad, Iraq

**Abstract.** The current study is concerned with Rusafa side of Baghdad city, which extends on 362 square kilometres and is inhabited by 3738371 people. It suffers from population growth, which led to decline in municipal services and rising of land surface temperature (LST) during the summer season. The current study aims to assess the spatial and temporal variation of LST using different dated satellite images of Landsat 4, 7 and 8 at path 168 and raw 037 during July 16, 1989, April 25, 2000, April 26, 2006, and May 31, 2016. Eight analysis models which were built using ERDAS 2014 software are used to calculate the LST parameters; while the land use/land cover (LULC) types are classified using ENVI v.5.3 software. The maximum surface temperature recorded at barren lands are between 26 and 41 °C, while at vegetation cover ranging from 25 to 34 °C and an inverse correlation are recorded between LST and NDVI ranging from -0.5 to -0.81. The built-up area was increased by 23%, whereas open spaces and vegetation cover are decreased by 16% and 6% respectively all that consequently will make the environment of the study area prone to be warmer.

## 1 Introduction

Evolution of Land Surface Temperature (LST) based on remote sensing data depends on the spatial resolution of the sensors, as well as modelling process requiring a collection of mathematical equations for calculation the solar radiation parameters, vegetation indices and land surface temperature. The spatial analysis, includes three elements: cartographic modelling represent mapping and their related calculations, mathematical modelling, based on the spatial interaction and relationships between the objects in the model according to the mathematical equations, and finally statistical analysis modelling, which use the appropriate spatial statistical analysis [1]. The land surface temperature (LST) derived from low spatial resolution imagery is termed as urban heat island (UHI), and correspond closely with canopy layer heat island [2]. LST is important in direct estimation of long wave fluxes and indirect estimation of latent and sensible heat fluxes, and other applications including geology, hydrology, vegetation monitoring and global circulation model and the use of thermal infrared of satellite imageries are very attractive in global access and uniform estimation of surface temperatures [3]. Land surface temperature (LST) is important in understanding the dynamic of urban thermal environment and it is a function of four parameters i.e., albedo, emissivity, the thermal properties of construction materials, composition and structure of urban canopy, and correspond closely to the spatial variation of land use/land cover [2]. The material properties i.e., heat capacity; thermal conductivity and thermal inertia are the main factors controlling the spatial variation of LST [4]. Nichol [5] referred to the advantage of remote sensing in analysis of urban heat island by providing dense grid of time synchronized surface temperature over the whole city. Uneven distribution of trees and vegetation cover in the city lead to uneven distribution of ecosystem services and community benefits, in fact, the tree cover in the city can be considered as a social justice issue, because a 5% reduction in tree cover can rise air temperature by 1 -2 °C [6].

Rusafa side of Baghdad city is characterized by high population density; population growth and spread of slums after 2003 resulting in illegal expansion on land especially vegetation cover and landscapes, consequently leading to a change in the physical characteristics of the city canopy. In fact, since Iraqi cities are mostly surrounded by barren or semi barren land, so during the summer season, impact of the heat island is imperceptible between the city and its surrounding areas; hence, the land surface temperature is the main factor impacting the city climate. The study area is characterized by high temperature during summer season, induced negative impact on the ecosystem of the city, many citizens are exposed to diseases and stresses caused by rising temperature, as well as the activity of organisms and rodents fertility are increased during summer, accompanied by appearance of viruses that cause diseases. The current study aims to a). Assessment of the temporal and spatial variation of land surface temperature using remote sensing and geographic

* Corresponding author: aafbeg64@uomustansiriyah.edu.iq

© The Authors, published by EDP Sciences. This is an open access article distributed under the terms of the Creative Commons Attribution License 4.0 (http://creativecommons.org/licenses/by/4.0/).

information systems techniques based on Landsat 4, 7 and 8 satellite imageries acquired during four periods between 1989 and 2016, b). Building models using ERDAS 2014 to calculate the spectral radiance, reflectance, albedo, vegetation indices, and land surface temperature, c). Classification of land use/ land covers using ENVI v.5.3 software, d). Production of LULC and LST distribution maps using ArcGIS v 10.3, and e). Point out the statistical relation between LST and NDVI.

## 2 Description of study area

Baghdad is a capital city and located in the central part of Iraq; divided by Tigris river into two sides - Rusafa on East and Karkh on West. The current study focusing on Rusafa side which extends on an area of 362 square kilometers and bounded by latitudes 33° 13′, 33° 27′ N, and longitudes 44° 17′, 44° 33′ E (Fig. 1), and inhabited by about (3738371) million people, with population density of about (10326) people/ sq km. Administratively, the area composed of (8) municipal units and has a road network with a length of 3814 kilometres.

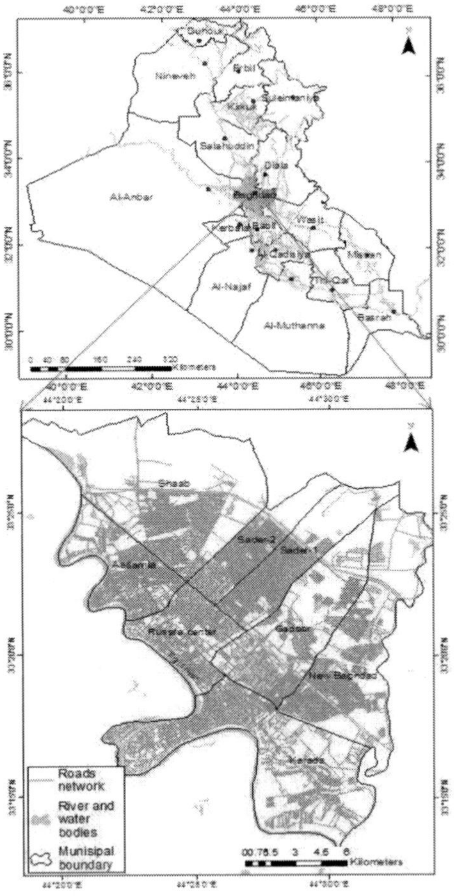


**Fig. 1**. Location of study area.

## 3 Methodology

The analysis method involved the collection of satellite images, selection of necessary software for image processing and presentation of the results, and gathering of the surface temperature above different land use/land cover types by in-situ measurements, collection the mathematical bases required for modelling, building the analysis models to calculate the spectral radiation parameters, vegetation cover indices, and land surface temperature.

### 3.1 Data and processing software

The most important data used in the analysis are remote sensing data of Landsat 4, 7 and 8 satellites (raw 037, path 168) acquired on 16 July 1989, 25 April 2000, 26 April 2006 and 31 May 2016. The satellite imageries are downloaded from the website https://earthexplorer.usgs.gov. To validate the analysis carried out from satellite imageries, field measurements of the contact surface temperature and at a height of half a meter above different types of land use /land cover in the city was carried out. Digital image processing models were prepared using the ERDAS 2014 software, while land use/ land cover types were classified using spectral angle mapper (SAM) method executed on ENVI 5.3 software.

The spatial distribution maps for different variables and calculation of statistical parameters of LSTs according to land cover classes were performed using ArcGIS 10.3. The MICROSOFT EXCEL program was used to create tables. The correlation coefficient and scatterplots of LST against NDVI were carried out using STATISTICA software based on data extracted at random points layer prepared using ArcGIS software.

### 3.2 Mathematical bases of modelling

Mathematical equations are collected and used in building eight models for calculating the values of spectral radiance, surface reflectance, albedo, NDVI, SAVI, LAI, emissivity and LST, as follows:-

#### *3.2.1 Spectral radiance*

The first phase of the land surface temperature calculation requires the conversion of digital number (DN) of satellite data into spectral radiance values. There are two ways of conversion; first is based on the minimum and maximum radiance of each band, taken from the metadata file according to the type of satellite sensor. This method was used in calculation of the spectral radiance of Landsat 4,5 and 7 data based on equation 1 [7]:

$$L_\lambda = \left(\frac{Lmax_\lambda - Lmin_\lambda}{Qcal_{Max}}\right) \times Q_{cal} + Lmin_\lambda \quad (1)$$

Where Lλ spectral radiance (W/(m2. sr. μm)), lminλ, lmaxλ are minimum and maximum spectral radiance for each band from the metadata file, $Q_{cal}$ is the band digital number, Qcalmin= 0.0 before 2004 and 1 after 2004, Qcalmax is 255 (Landsat 7,NASA, 2011).

The second method was used to calculate the spectral radiance of Landsat 8 data according to equation 2 [8]:

$$L_\lambda = ML \times Q_{cal} + AL \qquad (2)$$

Where *ML* and *AL* are radiance Mult_band and Add-band values from the metadata file.

### 3.2.2 Surface reflectance

To remove the errors from spectral radiance; atmospheric correction was carried out by conversion of spectral radiance to surface reflectance [7]:

$$\lambda = \frac{\pi[L_{sat}(\lambda) - L_p(\lambda)] \times d^2}{E_o(\lambda) \times \cos(\theta_z)} \qquad (3)$$

$$L_{p(\lambda)} = L_{min}(\lambda) - L_{1\%}(\lambda) \qquad (4)$$

Where $E_o(\lambda) = E_{SUN}\lambda$ = Exoatmospheric irradianc (W x $m^{-2}$ x $\mu m^{-1}$), *Lsat(λ)* represent spectral radiance at sensor, cos(θz) = Cosine of solar zenith angle calculated as follows:

$$\theta = \text{solar zenith} = 90 - \text{solar elevation} \qquad (5)$$

While $L_{1\%}(\lambda)$ is haze radiance at sensor and calculated as follows[1]:-

$$L_{1\%}(\lambda) = \frac{0.01 \times \cos(\theta_z)^2 \times E_o(\lambda)}{\pi \times d^2} \qquad (6)$$

$$d = 1 + 0.033\cos(DOY \frac{2\pi}{365}) \qquad (7)$$

d is Earth-sun distance [2], DOY is day of year [9, 10].

### 3.2.3 Albedo

The albedo of Landsat 4, 5 and 7 can be calculated according to the following equation [11, 12]:

$$\alpha_{TOA} = (0.356 \times b1 + 0.13 \times b3 + 0.373 \times b4 + 0.085 \times b5 + 0.072 \times b7 - 0.0018) \times 100 \qquad (8)$$

Where b1, b3,...b7 is the reflectance of specific band of the sensor.

For Landsat 8 data, the value of the Albedo is calculated according to the following equation [13]:

$$\alpha_{TOA} = 0.356 \times ref_2 + 0.326 \times ref_3 + 0.138 \times ref_4 + 0.084 \times ref_5 + 0.056 \times ref_6 + 0.041 \times refl_7 \qquad (9)$$

Albedo value at top of atmosphere was converted to surface albedo as follows [10]:-

$$\alpha = \frac{(\alpha_{TOA} - \alpha_{path\text{-}radiance})}{\tau_{sw}^2} \qquad (10)$$

$$\tau_{sw} = 0.75 + 2 \times 10^{-5} \times z \qquad (11)$$

Where *a* is the surface albedo, $\tau_{sw}$ is atmospheric transmissivity.

### 3.2.4 Normalized Difference Vegetation Index (NDVI)

The NDVI is an important index in calculating the land surface temperature ranging between (-1 and 1), where the positive values indicate that there is a vegetation and the negative values mean the barren land [14]. Comparison of LST and NDVI values shows linear and mostly negative correlation [3]. To increase accuracy of NDVI, the values of spectral radiance are converted to atmospherically corrected reflectance values and then used to calculate the values of the NDVI [15].The following equation is used for NDVI [16]:

$$NDVI = (NIR - RED)/(NIR + RED) \qquad (12)$$

Where NIR is the reflectance at near infrared band and RED is the reflectance at red band.

### 3.2.5 Soil Adjusted Vegetation Index (SAVI)

Because of reflectivity properties of soil and their interaction with vegetation cover added to that the chemical characteristics of soil including metallic, chemical composition, percentage of organic matter, surface roughness, tissue and moisture content of soil, therefore, the SAVI need to be calculated [17], and calculated as follows [18]:

$$SAVI = \frac{(1+L) \times (TIR - RED)}{(L + TIR + RED)} \qquad (13)$$

Where L is calibration factor equal to 0.5 [10].

### 3.2.6 Leaf Area Index (LAI)

Leaf area index is calculated as follows [10]:

$$LAI = -(1/a2) * \ln(a0 - SAVI/a1) \qquad (14)$$

Where a0, a1, a2 are constants calculated to be suitable with study area using calibration method [13].

### 3.2.7 Emissivity

The values of vegetation indices are used to calculate surface emissivity, which is defined as the ability of the object or surface to emit radiation compared to the black body at the same temperature [19]. Emissivity can vary greatly depending on the type, intensity and stage of vegetation growth [20]. There is more than one method to calculate the values of surface emissivity according to the variables or boundary conditions used in the selection of the emissivity value. One of the methods; depend on the values of the albedo, NDVI and LAI according to the following equations:

$$\varepsilon = 1.009 + 0.047 \times \ln(NDVI) \qquad (15)$$

If LAI is less than 3 then the following equation is used:

$$\varepsilon = 0.97 + 0.0033 \times LAI \qquad (16)$$

If LAI is greater than or equal to 3, the emissivity is 0.98, whereas if NDVI is less than 0 and albedo is less than 0.47, the emissivity is 0.99 [10, 13, 21].

Another method depends on the NDVI values of vegetation cover and soil and considered boundary conditions, if the value of NDVI is less than 0.2, it means that the surface is free of vegetation cover and the emissivity value is 0.97, but if the value of NDVI is greater than 0.5, the value of emissivity is 0.99, while if the NDVI value is between 0.2 and 0.5, the emissivity value is measured from the following equation [22]:-

$$\varepsilon = \varepsilon_v \times p_v + \varepsilon_s(1 - p_v) + d_\varepsilon \qquad (17)$$

Where ε is emissivity, εv emissivity of vegetation cover, εs is emissivity of soil, dε is the effect of internal reflections and geometric distribution of natural surfaces, *Pv* vegetation cover proportion [22].

$$d_\varepsilon=(1-\varepsilon_s)\times(1-p_v)\times F\times\varepsilon_v \quad (18)$$

$$p_v=\left[\frac{NDVI-NDVI_{min}}{NDVI_{max}-NDVI_{min}}\right]^2 \quad (19)$$

### 3.2.8 Brightness temperature (Tc)

To calculate LST from the thermal bands of Landsat satellite sensor; first, the brightness temperature (Tc) or black body temperature on board of the satellite needs to be calculated as follows [7, 23, 24]:

$$T_c=\frac{K2}{\ln(\frac{K1}{L\lambda}+1)} \quad (20)$$

Where K1 and K2 are constants as given in Table 1, Lλ is spectral radiance of thermal band.

**Table 1**. Constants of Landsat satellite sensors.

| Satellite | K1 | K2 |
|---|---|---|
| LANDSAT_4 (TM) | 671.62 | 1284.30 |
| LANDSAT_7 (ETM+) | 666.09 | 1282.71 |
| LANDSAT_8 (OLI_TIRS) | b10=774.8854<br>b11= 480.8883 | b10= 1321.0789<br>b11= 1201.1442 |

**Source**: USGS, Landsat scene metadata file.

### 3.2.9 Land Surface Temperature (LST)

The land surface temperature is calculated from the brightness temperature or black body temperature corrected according to the emissivity values of the different land use/land cover classes, based on the following equation [2, 21, 23]:

$$T_s=\frac{T_c}{1+(\lambda\times T_c/\rho)\times\ln\varepsilon} \quad (21)$$

Where λ is average wavelength of thermal band, ρ is constant equal to 0.01438 m. kelvin and calculated based on *a* Boltzmann constant, *h* Blank constant, *c* is velocity of light [2]:

$$\rho=h\times {}^{c}/_{a} \quad (22)$$

LST in kelvin is converted to Celsius as follows[3]:

$$T(°C) = T(K) - 273.15 \quad (23)$$

## 4 Results and discussion

The spatial variations of LSTs and land use/land cover (LULC) types are illustrated in distribution maps. During the four periods; LULC are classified into seven types i.e., water bodies (River, Lake, Swamps), cropland, grass and garden, green land (Orchards), bare land, residential area and roads and parks. The statistics of LSTs are summarised according to LULC as shown in Table 2. The results extracted from Landsat 4 TM image acquired on July 16, 1989 (Fig. 2), show that the lowest average LST about 30.6 °C was recorded on water bodies (River, Lake, and Swamps), followed by vegetation cover (Cropland, Grass and Garden, and Green land) with average LST ranging from 32.7 to 37.3 °C, while the highest LST was recorded at barren land (Bare land, residential area, roads and parks) with average values ranging from 38.8 to 40.6 °C and covered about 76.7% of study area, while the vegetation cover was covered about 20.8% of the area and most of them are extended along Tigris River. The LSTs measured from Landsat 7 ETM+ image acquired on April 25, 2000 (Fig. 3), show a decline in temperature of all classes. In fact, the water bodies recorded low temperature of about 21.3 °C, meanwhile, the vegetation cover gives LST ranging from 26.3 to 29.4 °C with approximately the same coverage area of about 21.7%. The barren lands including a residential area, roads, and parks are covered about 76.3% of the area and their average LSTs ranging from 30.7 to 33.2 °C with a relatively high standard deviation of LSTs at roads and parks class, because of mixed pixels with other classes. Also, the study area witnesses an increase in the percentage of built-up area (Residential area, roads and parks) 45% during 1989 up to 50.4% that means increase of impervious surfaces which makes the city canopy to be warmer. On April 26, 2006, Landsat 7 ETM+ image was analyzed for LSTs over the study area (Fig. 4), the results show that water bodies still covering about 2.7% with average LSTs 19.4°C, while vegetation cover remains about 20.6% of the area with average LSTs ranging from 23.6 to 26.7 °C, barren lands cover about 76.7% of the area with average LSTs ranging from 26.1 to 26. 8 °C, meanwhile, built up area (Residential area, roads and parks) are continued to increase among other land use classes and reached up to 54.5%. The results of latest analysis of LSTs from Landsat 8 (OLI_TIRS) image acquired on May 31, 2016 (Fig. 5 and Table 2) was carried out based on thermal bands i.e., b10, b11 and average values of the two bands. The achieved results of LSTs are compared with in-situ measurements carried out on different land use/land cover types during June 24, 2016, at time 10.50-11.20 am listed as 35 °C at building shadow, 42 °C at building roofs, 40 °C at road pavement, 37 °C at vegetation cover and 28 °C at river water. The comparison of in-situ measurements and calculated LSTs; indicates the LSTs measured from b10 and b11 are deviated from in-situ measurement, but they are more comparable with LSTs calculated from average b10 and b11. The LULC classification results indicate the vegetation cover was reduced to 15.3% of study area while barren lands increased up to 82.5% at the same time; residential area witness dramatic increase up to 62.5% compared to 31.8% at 1989, the roads network and parks are reduced to 5.1% because of low maintenance which led to making their spectral signatures mixed with residential area and lost their sharp contrast which was recognized during 1989. In fact, the increase in barren lands especially residential area will lead to rising LSTs and the local climate being prone to be warmer during the summer season. To show the relation between LSTs and vegetation cover; scatterplots of LST against NDVI were plotted and the correlation was measured as shown in scatterplots in Figure 6, the regression lines during the studied periods show inverse correlation between LST and NDVI and the correlation coefficient values ranging from – 0.54 to – 0.81, indicating the importance of vegetation cover as the main factor of land use in reducing the LST in study area. Indeed, the lack of interest in increasing and enlarging tree cover and landscaping in the study area will make the city suffer from warm local climate.

**Table 2.** Statistical parameters of LST of different LULC classes.

| Land use/ land cover type | Area ($km^2$) | % area | Avg. T. (°c) | SD. T. (°c) |
|---|---|---|---|---|
| **16 July 1989** | | | | |
| River, lake & swamps | 8.9 | 2.5 | 30.6 | 3.2 |
| Crop land | 10.3 | 2.8 | 33.9 | 1.8 |
| Grass & garden | 51.5 | 14.2 | 37.3 | 2.0 |
| Green land (Orchards) | 13.7 | 3.8 | 32.7 | 1.5 |
| Bare land | 114.6 | 31.7 | 40.6 | 1.5 |
| Residential area | 114.8 | 31.8 | 38.8 | 1.5 |
| Roads & parks | 47.7 | 13.2 | 38.8 | 1.8 |
| **26 April 2006** | | | | |
| River, lake & swamps | 9.64 | 2.7 | 19.4 | 2.2 |
| Crop land | 2.85 | 0.8 | 23.6 | 1 |
| Grass & garden | 48.27 | 13.3 | 26.7 | 1.6 |
| Green land (Orchards) | 23.35 | 6.5 | 24.9 | 1.5 |
| Bare land | 80.46 | 22.2 | 26.8 | 1.2 |
| Residential area | 140.9 | 39 | 26.1 | 1.1 |
| Roads & parks | 56.12 | 15.5 | 26.1 | 1.5 |
| **25 April 2000** | | | | |
| River, lake & swamps | 7.3 | 2 | 21.3 | 1.05 |
| Crop land | 4.4 | 1.2 | 26.76 | 1.58 |
| Grass & garden | 53.6 | 14.8 | 29.36 | 1.67 |
| Green land (Orchards) | 20.6 | 5.7 | 26.27 | 1.56 |
| Bare land | 93.8 | 25.9 | 33.16 | 1.85 |
| Residential area | 155.4 | 43 | 31.14 | 1.63 |
| Roads & parks | 26.6 | 7.4 | 30.67 | 2.23 |
| **31 May 2016 from band 10** | | | | |
| River, lake & swamps | 7.73 | 2.1 | 30.9 | 2.4 |
| Crop land | 4.15 | 1.1 | 36.2 | 2.2 |
| Grass & garden | 44.61 | 12.3 | 40.2 | 2.2 |
| Green land (Orchards) | 6.81 | 1.9 | 36.1 | 2.6 |
| Bare land | 53.47 | 14.8 | 42.7 | 1.6 |
| Residential area | 225.9 | 62.5 | 41.3 | 1.2 |
| Roads & parks | 18.88 | 5.2 | 40.9 | 2.7 |
| **31 May 2016 from band 11** | | | | |
| River, lake & swamps | 7.73 | 2.1 | 25.6 | 1.9 |
| Crop land | 4.15 | 1.1 | 29.7 | 1.7 |
| Grass & garden | 44.61 | 12.3 | 32.8 | 1.7 |
| Green land (Orchards) | 6.81 | 1.9 | 29.6 | 2.0 |
| Bare land | 53.47 | 14.8 | 34.7 | 1.2 |
| Residential area | 225.9 | 62.5 | 33.7 | 0.95 |
| Roads & parks | 18.88 | 5.2 | 33.4 | 2.1 |
| **31 May 2016 from Avg. b10 &b11** | | | | |
| River, lake & swamps | 7.73 | 2.1 | 28.31 | 2.21 |
| Crop land | 4.15 | 1.1 | 33.04 | 1.95 |
| Grass & garden | 44.61 | 12.3 | 36.55 | 1.95 |
| Green land (Orchards) | 6.81 | 1.9 | 32.88 | 2.32 |
| Bare land | 53.47 | 14.8 | 38.78 | 1.43 |
| Residential area | 225.9 | 62.5 | 37.58 | 1.08 |
| Roads & parks | 18.88 | 5.2 | 37.17 | 2.41 |

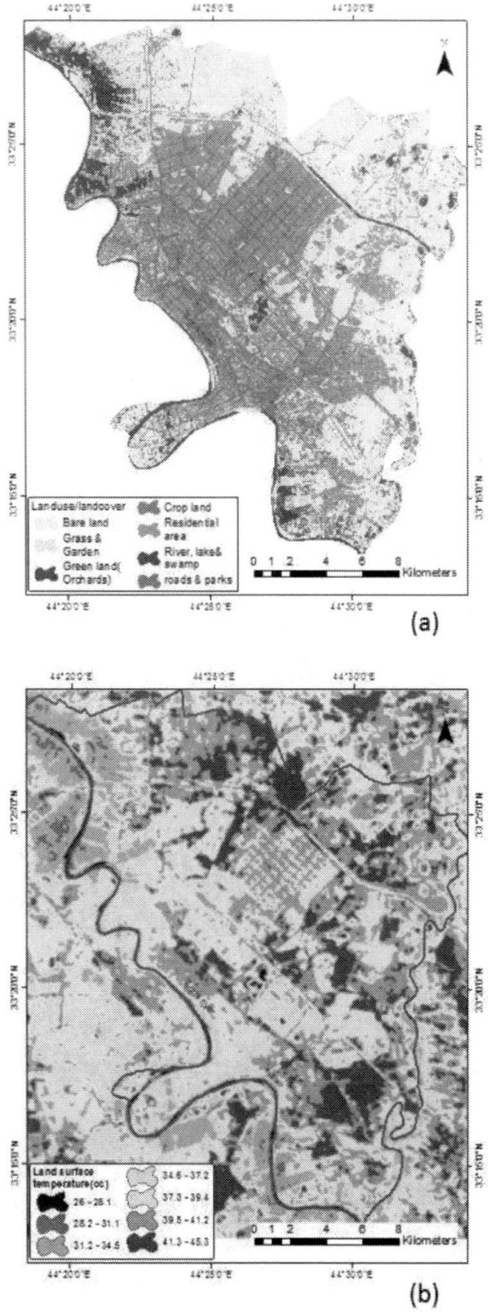


**Fig. 2.** Land use/land cover map during 16 July 1989: a). LULC, b). LSTs

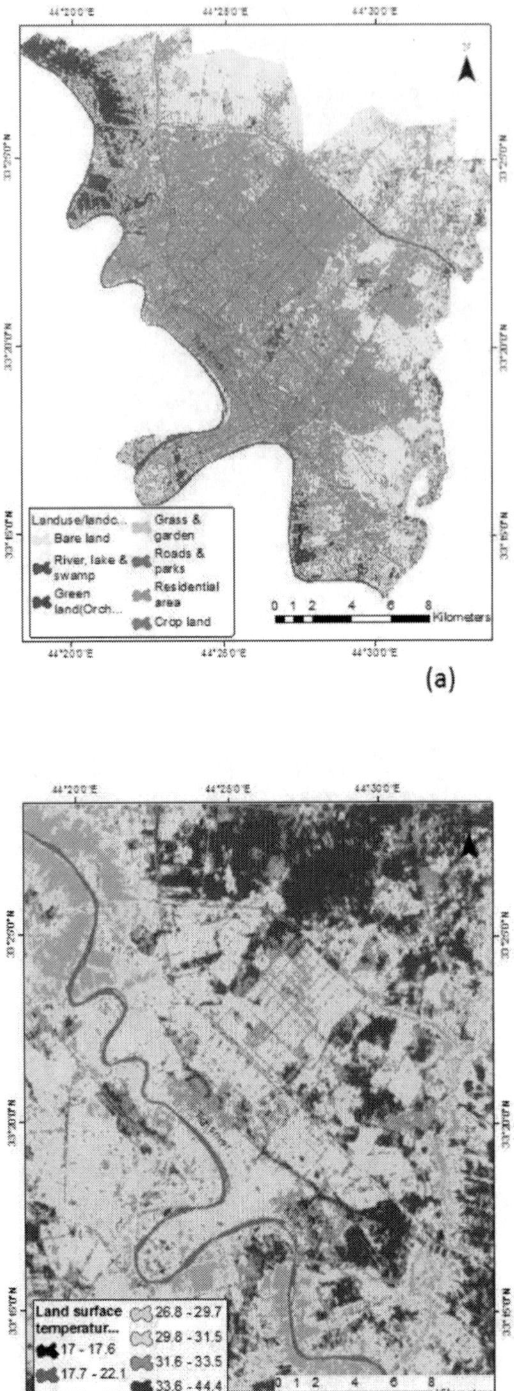


**Fig. 3**. Distribution maps during 25 April 2000: a).LULC, b). LSTs

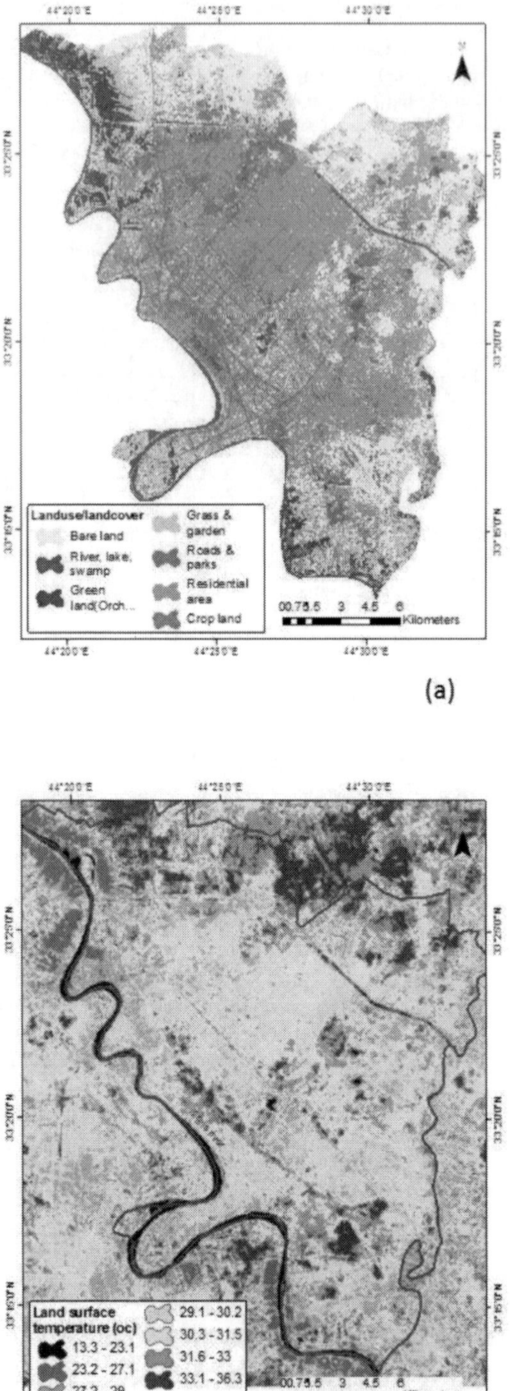


**Fig. 4**. Distribution maps during 26 April 2006:a). LULC, b). LSTs.

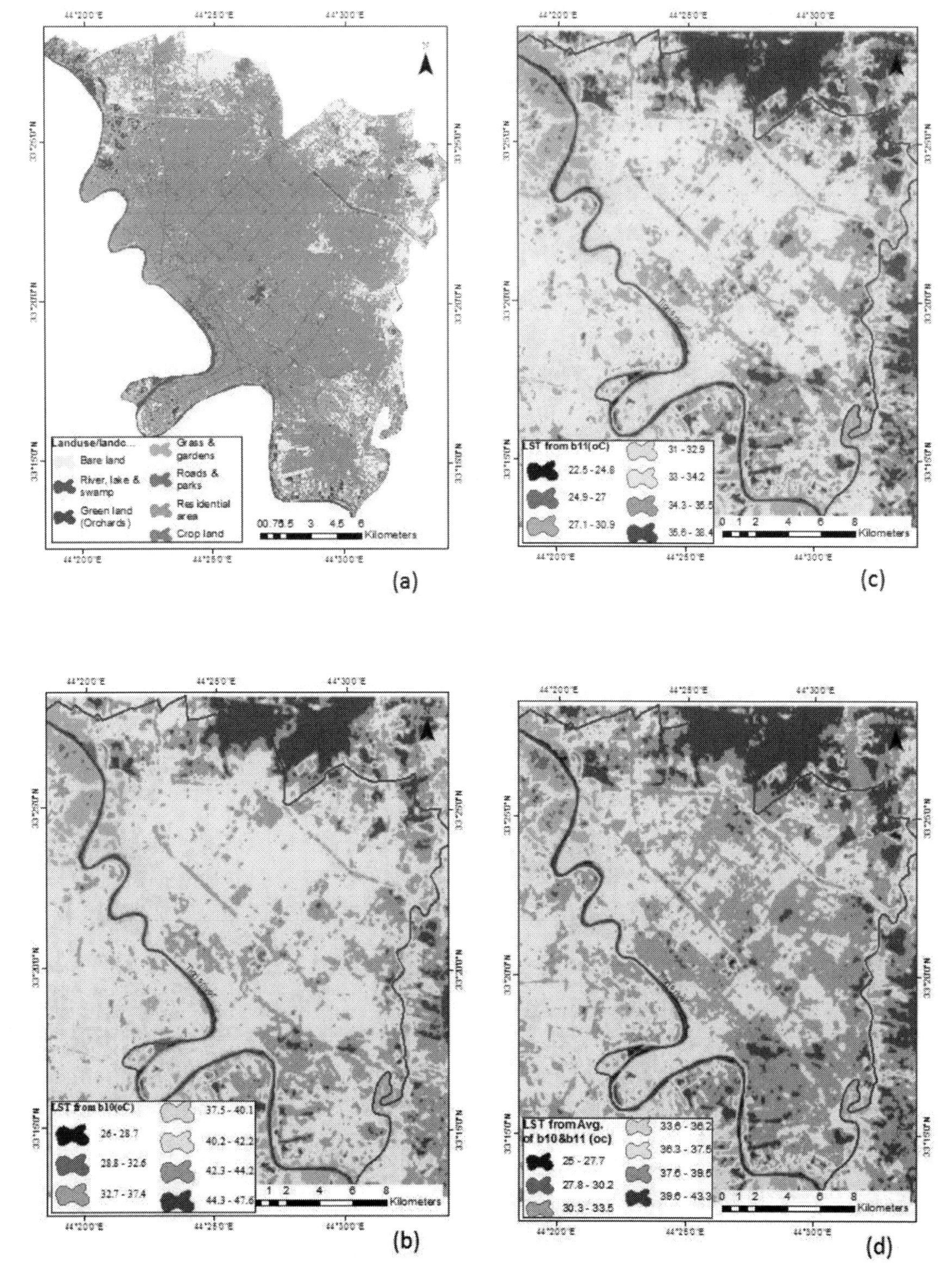


**Fig. 5.** Distribution maps during 31 May 2016: a). LULC, b). LSTs. from B10., c). LSTs. from B11 and d). LSTs. From average of B10&B11.

**Fig. 5.** Continued

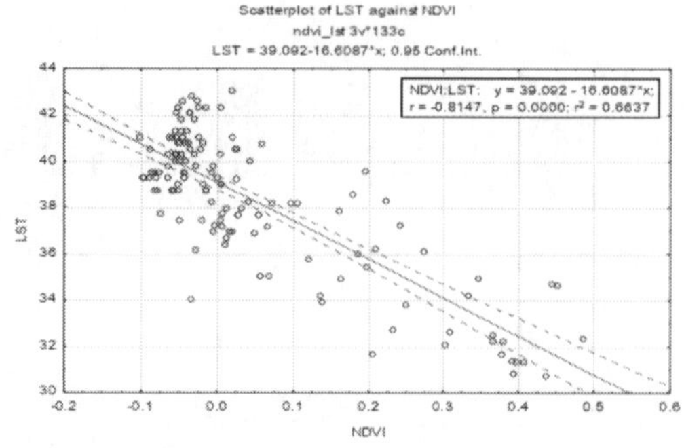


**16 July 1989**

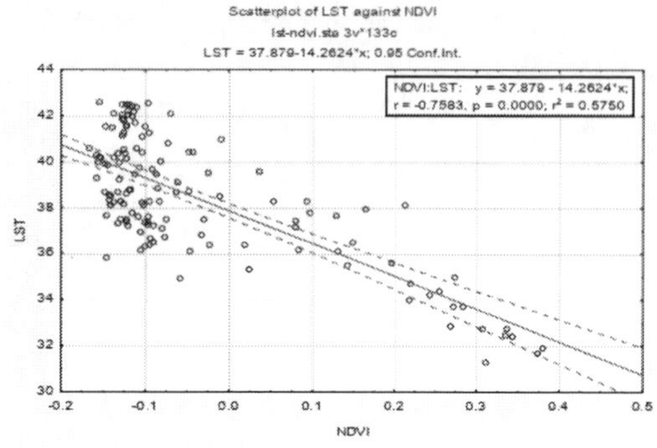


**26 April 2006**

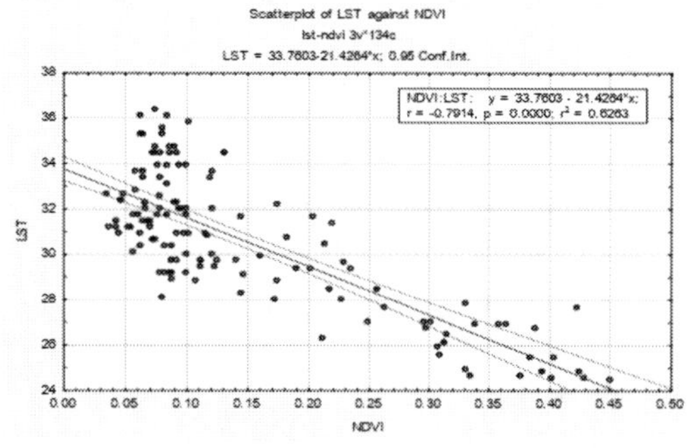


**25 April 2000**

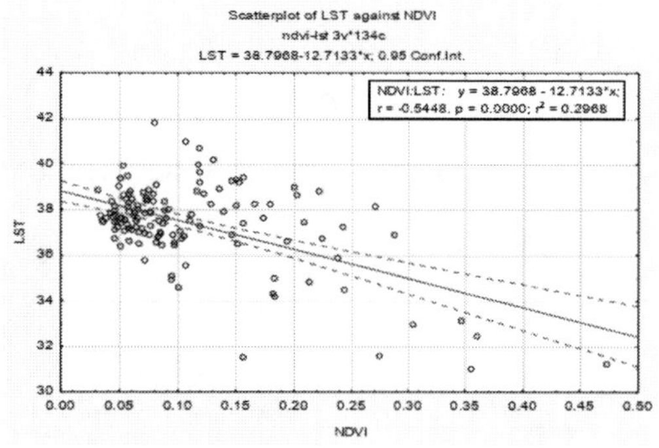


**31 May 2016**

**Fig. 6.** Scatterplots of LST against NDVI.

## 5 Conclusions

The analysis of the LSTs and LULC relationship shows that there are two main land use/ land cover controlling the spatial distribution of LSTs in study area i.e., barren lands (built up area and bare lands) and vegetation cover density. Based on the results, the vegetation cover area shows temperature difference of 3-5ºC less than barren lands during the study period, indicating their positive impact on reducing LST in the study area. The orchards area are the most important vegetation cover in the city for reducing the LSTs, in Rusafa side of Baghdad city the percentage of orchards did not exceed 6% of the study area during 1989, even this percent was reduced to 2% on 2016. While the residential area increased dramatically from 32% during 1989 up to 62% during 2016 resulting in increase in impervious surfaces and pavement areas which consequently lead to rising of LSTs of the city. Roads network faced low maintenance resulting in difficulty in differentiating them from the residential area during digital classification of the images that is clear from comparison of roads extracted from an image acquired during 1989 and that acquired during 2016. The LST and NDVI are correlated inversely with correlation coefficient ranging from -0.54 to -0.81. For Landsat 8; the LSTs measured from average of b10 and b11 is more correlated with the measurement of temperature in-situ.

The author would like to record his thanks to Mustansiriyah University for the encouragement for completion and publishing of this work.

## References

1. Rolf A. D., *Principles of Geographic Information Systems: An Introductory Textbook*. Vol. **1** (International Institute for Geo-Information Science and Earth Observation, 2004)
2. Weng Q., *Remote Sensing and GIS Integration: Theories, Methods, and Applications* (McGraw-Hill New York, 2010)
3. Quattrochi D. A. J. C. Luvall, *Thermal Remote Sensing in Land Surface Processing*. (CRC Press, 2004)
4. G. D. Jenerette, S. L. Harlan, A. Buyantuev, W. L. Stefanov, J. Declet-Barreto, B. L. Ruddell, S. W. Myint, S. Kaplan, and X. Li,. L. Ecol. **31**, 4, 745 (2015)
5. J. E. Nichol, J. Appl. Meteo. **35**, 1, 135 (1996)
6. Parris J. K. M. Parris, *Ecology of Urban Environments* (John Wiley & Sons, 2016)
7. G. Chander and B. Markham. IEEE Trans. Geos. R.S. **41**, 11, 2674 (2003)
8. Usgs, *Landsat 8 (L8) Data Users Handbook*. (2ed. Lsds-1574, EROS Sioux Falls, South Dakota: USGS, 2016)
9. Allen R. G. L. S. Pereira, D. Raes, and M. Smith, *Crop Evapotranspiration-Guidelines for Computing Crop Water Requirements-Fao Irrigation and*

*Drainage Paper 56* (FAO, Rome. **300**, 9, D05109, 1998)
10. Waters R. R. Allen, W. Bastiaanssen, M. Tasumi, and R. Trezza, *Sebal. Surface Energy Balance Algorithms for Land* (Idaho, USA, 2002)
11. Mohamed M., *Summer Land Surface Temperature: Small-Local Variation in Intro-Urban Environment in El Paso* (U. of Texas, El Paso, Tx. 2013)
12. S. Liang, R. S. of Env. **76**, 2, 213 (2001)
13. A. A. F. Beg, A. H. Al-Sulttani, A. Ochtyra, A. Jarocińska, and A. Marcinkowska, J. Geo. Res. E. **6**, 257 (2016)
14. F. Yuan, and M. E. Bauer, R. S. of Env. **106**, 3, 375 (2007)
15. X. Yu, X. Guo, and Z. Wu, R. S., **6**, 9829 (2014)
16. Lillesand T. R. W. Kiefer, and J. Chipman, *Remote Sensing and Image Interpretation*. (John Wiley & Sons, 2014)
17. E. R. Stoner, M. F. Baumgardner, R. A. Weismiller, L. L. Biehl, and B. F. Robinson, S. Sci. Soc. of A. J., **44**, 3, 572 (1980)
18. A. R. Huete, R. S. of Env., **25**, 3, 295 (1988)
19. Watson C., *Analysis of Urban Heat Island Climates Along the I-85/I-40 Corridor in Central North Carolina*. (U. of North Carolina, Greensboro, 2012)
20. Q. Weng, J. of Photo. and R. S., **64**, 4, 335 (2009)
21. A. Faris, and Y. S. Reddy, IJEE., **3**, 3, 332 (2010)
22. A. Asgarian, B. J. Amiri, and Y. Sakieh, U. Ecos., **18**, 1, 209 (2014)
23. Q. Weng, D. Lu, and J. Schubring, R. S. of Env., **89**, 4, 467 (2004)
24. J. Li, C. Song, L. Cao, F. Zhu, X. Meng, and J. Wu, R. S. of Env., **115**, 12, 3249 (2011)

# Evaluating and monitoring of changes in the environmental elements by using remote sensing techniques in Iraq

*Hussein* Al-Tameemi[1,*]

[1]School of Engineering and Informatics, Sussex University, UK

**Abstract.** The primary aim of project highlight was to E&M a universally-applicable suite of elements and techniques that allow remote sensing techniques to the E&M-environmental changes and human interventions and the impact of disaster management systems for the period (2010-2015). The framework allows evaluators to monitor changes in a manner that is rapid, independent and reliable. Such a pattern is required to promote contribution and transparency towards a more systematic understanding of the process of E&M. The factors are based on the use of Very High-Resolution Optical Satellite Imagery and have been developed to fit existing change approaches, such as the In-Disaster Needs Assessment-IDNA, and the Affected Cluster Framework. The elements encompass a range of environmental, climate, economic, health, physical and social factors, which can be analysed to provide a holistic representation of the E&M changes process. The paper begins by addressing the need for a E&M framework and by reviewing current guidelines and systems. Which are continuously developed in this case study. Recommendations are made on: 1- How to use disaster management systems; 2- Which elements should be monitored; and 3- How they can be evaluated, drawing on the experience of environmental scientific researcher, and feedback by the evaluators' academics.

## 1 Introduction

Main issues contributing to changing the outlook to bring the situation to an acceptable level to the evolution in the environment. The impact of the change is often long, and complex, with patterns and factors affected among numerous life sectors and its elements. There is a difference over time and the area of influence due to several factors, including socio-economic factors, and health-environmental, because of the multiplicity of decisions taken at the stage of change and after, which adversely affected it. There are subsequently similarities between change and environmental vulnerability theories, indicating that. There are weak factors, which in themselves are more likely to suffer serious losses, unable to recover with difficulty in the near or distant future as local environmental conditions. Post-disaster change is often conceptualised as a complex, multi-dimensional process. Often we find a reference to the period of the disaster [1].

The primary objective of these elements is often to differentiate emergency impact patterns from long-term change and mitigation. While the effect phase is characterised by changing of natural life tasks, such as high temperature is the first to cause desertification and drought, water scarcity and pollution; impact change is likely to include the climate environment (factors, values and patterns) and contribute to changing natural conditions to abnormal (water, soil thus the plant levels).

Hence, this article highlights the importance of the economic side and the ideological side according to the principles of change, as follows:

A. Environmental changes region: 78% of Iraq is assessed;

B. This region provides more than half of Iraq's production;

C. It is estimated that about 53% of the human population lives in this region;

D. One-third of Iraq's cities have a combined population of more than 32 million. Approximately 90% of the livelihoods of these people has been destroyed, region shown in Figure (1). Evaluating and monitoring the regions suffering change identifies: desertification, disaster-management systems impacts, and environmental pollution levels.

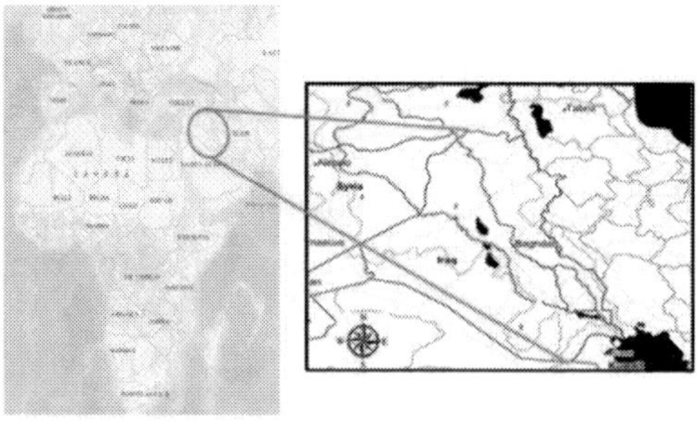

**Fig .1.** Study area of Iraq.

* Corresponding author: *h.al-tameemi@sussex.ac.uk*

© The Authors, published by EDP Sciences. This is an open access article distributed under the terms of the Creative Commons Attribution License 4.0 (http://creativecommons.org/licenses/by/4.0/).

The average impact of radioactive materials is estimated in Iraq's eastern, west, and southern environmental area. The trend of the higher-temperature system (HTS) for MODIS of Iraq [2, 9]. Environmental changes over time and space through sedimentation, and erosion during this dangerous phase of the Tigris-Euphrates Rivers for more than 25 years (1990–2017) is analysed and, shown in Figure (2). The analysis is based on the assessment of environmental-climatic changes in the central and southern cities. Studies conducted on the Shatt al-Arab estuary make comparisons of satellite imagery using MATLAB software. The environmental changes in Basra city, from 2001 to 2016 were estimated [7, 9, 17].

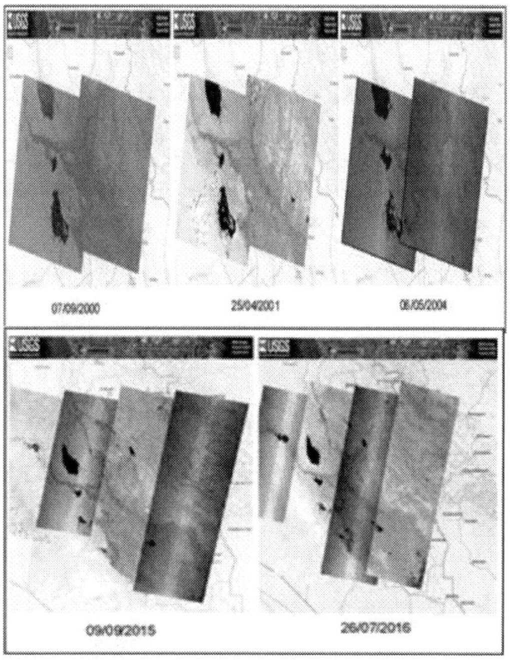


**Fig.2.** Environmental changes shown using satellite images of Iraq.

## 1.1 Methods for E&M of the environmental impact

Iraq's fields are complex and with an ever-changing environment. Remote sensing is therefore important because of its great potential to provide an enormous amount of accurate and rigorous information, for planning and disaster management of Iraq's region. The evaluation of environmental changes with climate, and the analysis of data using the GIS technique establishes priorities [3, 4, 1]. Such, evaluation and monitoring are necessary to improve future aid policy to mitigate environmental impact, to provide time and to assist evaluators' with on-going work on the ground. Although many such research studies have been conducted in the past, with somewhat different ideas action paths: the priority now is to identify issues as they arise to inform decision-making and utilise analytical evaluations to guide environmental disasters management systems to mitigate the effects of climatic changes in Iraq. The evaluators try to assess the damages they are doing with their baseline changes [8]. They require data to provide the analytical basis for their decisions and to ensure that their objectives are being met on time. Much of the work is often pledged to assist people after massive natural disasters. Despite the fact that there is a large effort supporting work to aid E&M, there are currently no standard frameworks or methodologies being applied, it is nevertheless possible to evaluate and monitor the process of change. In 1992, in June, a United Nations Conference on Environment & Development was held in Rio de Janeiro-Brazil, where a discussion was held [17], which identified an urgent need for basic and applied research in the long-term evaluation processes for the years to come. In particular, the absence of a fixed and independent standard was observed to replicate the approach of using remote sensing techniques to monitor and assess the magnitude of impact and changes occurring in Iraqi environment. Figure (3) shows the changes of the environmental elements via temporal coverage from the Landsat 7 and 8 L-Sa-images between 1990-2015 in Iraq [16, 6, 5, 2]. More recently, the Active Environmental Assessment and Monitoring Techniques Association called for "The primary research is in the field of demonstrating the methods of assessment and the development of the most suitable solution for emergency response to disasters and risks".

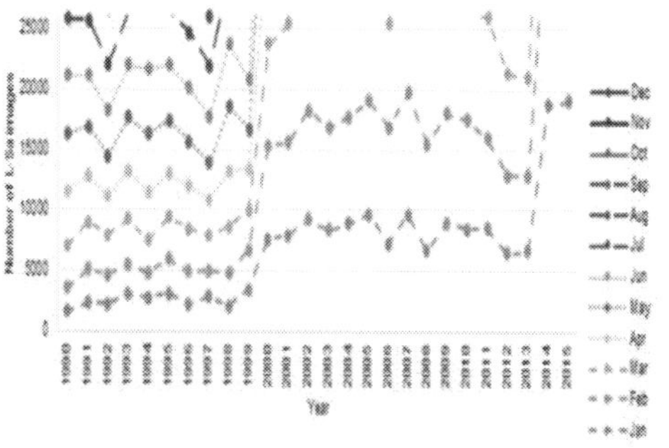


**Fig.3.** Changes to the environmental elements in temporal coverage of the Landsat 7 and 8 L-SA for 1990-2015 of Iraq.

## 1.2 Methodology and analytical concept

The main objective of the E&M changes was to develop a standard approach to monitoring post-disaster environmental factors that are affected by change and instability. This is achieved through the use of high-resolution satellite image datasets and spatial analysis within a Geographical Information System (GIS). The E&M was conducted from 2010 to 2015. Appropriate case study sites were selected, and datasets were acquired and integrated into a single database in the path changes evaluation and monitoring [12, 3, 7]. A standardised monitoring framework was then developed after evaluating affected data levels and relevant values. Appropriate analysis indicators and techniques were developed and applied to the case study sites in the affected areas. This is supported by verification of the results of the analysis in detail, the level of evaluation by using data collected with advanced ground survey techniques and technique-audit methods [14]. A detailed impact-effectiveness analysis was then conducted to identify the weaknesses before the strengths of each element, and to provide evaluations on impact spread and how they may be monitored via remote sensing techniques. A set of Guidelines has been produced to introduce this approach and to demonstrate remote sensing as an appropriate monitoring tool [12]. The following sections describe the methodology in more detail:

#### *1.2.1 Assessment of the regional-local scale of environmental changes*

The evaluators reviewed current regional, local scales frameworks, conducted a usé needs assessment during the evaluation of the affected area. That work resulted in a list of points that encompassed evaluation and analysis for long-term development. The points of joint action will be mentioned later [10, 16]. Results demonstrated that all points received medium-to-high levels of prioritisation suggesting there was a strong view and preference for a comprehensive new approach to monitor and evaluate all changes, but that the points of interest differed according to the nature of the region. Selected techniques helped to highlight these specialised areas of interest. That said "In fact, we lack new and up-to-date information on patterns of damage at this point, which in turn has spread of the impact on the foundations of life and infrastructure (Josef Leitmann-Disaster Management Coordinator-World Bank–Indonesia). So, it would be good to handle overlapping satellite images to counteract these changes and stop the widening of damage in areas" Anne-Cécile Vialle – UNEP, "We need maps of environmental damage, economic loss, health loss, industrial and environmental sites, increased sites of harmful waste, including chemical, toxic, debris of war." Overall, at the local level, increasingly data is using environmental changes evaluation, such as satellite image data, such as figure (4).

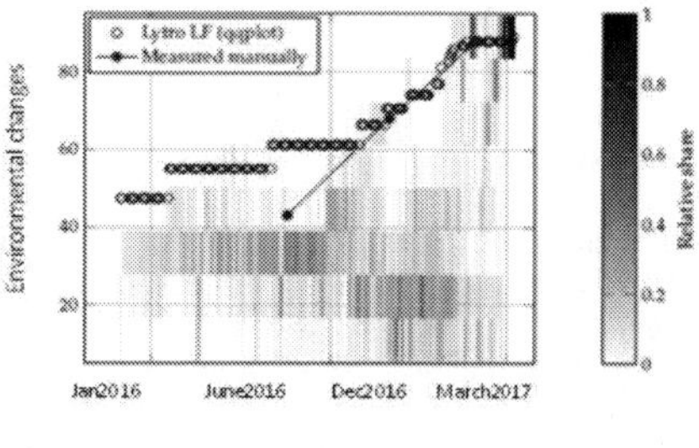


**Fig.4.** Iraq's environnemental changes.

Also, to reduce human intervention to minimise human impact on the lowest possible level and verify the high rate of assessment. It can briefly test by shortening work in two stages: (1) check for changing the numerical and colour values of a pixel at the satellite images during comparison with the previous image and map. (2) Making complete classifications of the pixel change scene, drawing on the principles and patterns of environmental change.

## 2 Describe the analysis of environmental changes for GIS

The VHR satellite imagery was analysed and techniques developed throughout the study period. A range of techniques was developed for each factor that uses both manual and semi-automatic procedures. The methods differ in the time and expertise required to conduct them including changes occurring now in the Iraq area. The remote sensing techniques and methods were first applied to the Al-Tameemi study and then applied to practical research by Al-Tameemi himself to test the transferability of the technical methodology. To summarise, the methodology is comprised of four components: 1. Technical processing 2. Database Creation 3. Data Analysis and 4. Establish full assessment of technology and process processing. Changes were monitored using a time-series of VHR satellite images acquired before the disaster, immediately afterwards and in the months-and-years after that in Iraq. While Maximum Likelihood Supervised Classification was used to create land cover maps marking out the extent of the built and natural environments changes. More advanced extraction techniques, such as Object-Based Image Analysis, were used to extract similar features of the impact and the environmental factors. These features were then integrated into a multi-temporal change Geodatabase as point GIS [13, 10, 11,15]. A new temporal layer of the database was derived from each satellite image, representing a different state in the process of impact changes. Each temporal layer was further divided into 12 thematic sub-layers, each representing one of the main performance factors [15, 17]. Once the features in the imagery were mapped and stored in a GIS database of effect changes were determined by applying change detection analysis to the geodatabase. This technique calculates the rate that various processes were conducted, including temperature rise, salinity level, rivers pollution, desertification expansion compared to green vegetation, pollution areas as black spots, etc.

The method also identified when key features, such as population density, urbanisation, rural-to-urban areas migration [10,3]. The overall progress of E&M changes was then factored by noting the presence or absence of the main features at different points in time. For more detailed analysis, the datasets were disaggregated by geographic boundaries, allowing disparity in the change levels to be assessed [11]. The progress of the effect was further evaluated by normalising the numbers of features (soil, water and vegetation) in each region to baseline statistical data of the satellite images, which were constructed using the pre-disaster image and the objectives of each evaluation- analysis project. This analysis was successful at identifying slow impacts and gaps in the environmental climate damages [15]. Overall, environmental changes and impact of damages were affected by variations in the performance of the economy, the social standard of the population, and unpredictable effects of nature.

This led to the conception of normative analytical systems organised in a detailed assessment on three crucial stages, as follows:

A) Environmental change levels, including evaluation, a high slope at temperature, a wetness index, soil salinity, desertification, lack of water levels and geomorphological types;

B) Water, including low rainfall, lack of river flow, and land migration;

C) Effect of socioeconomic levels, including population density, lack of resources, pollution, health deterioration rate, shown Figure (5). There is a direct and indirect impact on climate change with the environment, through a causal human lifestyle, and unstable development factors [12].

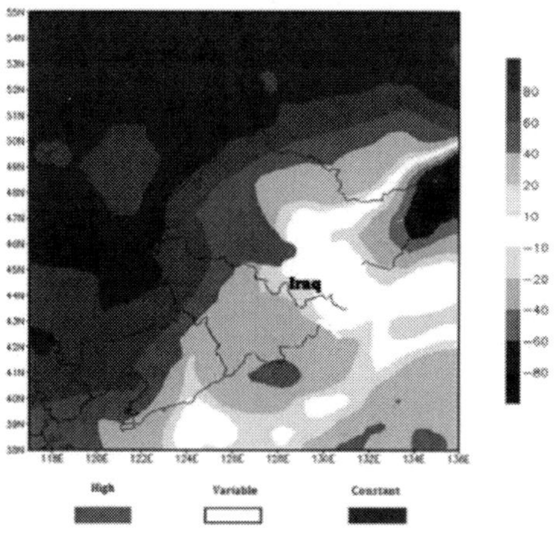


**Fig. 5.** Analytical-environnemental changes with GIS.

## 3 Recycling environmental climate change data based on the remote sensing technique

Balancing the rate of environmental changes causing in permanent differences, according to the selection and determination of the research area. Each region has significant damage that and changes the future life of the community. Additionally, environmental changes, ecological environment evaluations and crop yield assessment in the area. The available data has been used, such as temperature, remote sensing data and a world-class scale of other data sets used [11].

## 4 Results

### 4.1 Use of the statistics-analytical

Each of the statistics provided a rich source of information that could be interpreted independently or integrated to provide a holistic understanding of the progress of evaluation. Analytical data evaluated, according to the principles of remote sensing was found to closely match statistics from key informants, focus group meetings and environmental changes [9, 17]. Also, results were considered to be accurate when they were compared to detailed digital data collected using MATLAB. Moreover, the analysis also relied on many assumptions that weren't always accurate, such as, that all roads were in use and that all effects of damage change was found at the time the satellite images were acquired. These assumptions and errors were minimised in the environmental change knowledge data before the image analysis was begun by identifying likely forms of miss-interpretation and impact changes. In addition to the physical factor analysis, economic, social levels were successfully used to infer information about the evaluation and monitoring strategy [16, 11, 10]. A detailed discussion of these sectors and factors was not possible with remote sensing alone, but the information from satellite imagery was used to complement results acquired with ground surveys and modern techniques of the digital processing images with GIS [12]. Overall, a workflow of research was proposed that shows how the various instruments may be used in a complementary manner. GIS was found to be a suitable system in which to store and analyse the different datasets, including satellite imagery, ground-based observations and survey results data. The GIS also allowed multiple layers of the database to be examined simultaneously, thus providing a greater contextual understanding of the progress on the ground [8]. As an example of the results gained using this approach, a summary of the evaluation and monitoring process in the Al-Tameemi study after 2015 is presented in Table 1 shows a selection of the factors results from four months for 2011-2014.

### 4.2 Case study: Al-Tameemi

In Al-Tameemi, the 2010-2015 Iraqi environmental, climate change had a devastating effect on infrastructure and people's lives. Table 1 shows that over 45% of environmental factors were changed and nearly 35% of climate factors were impacted or in case of the continuous change in the Iraq areas. Also, figures (1-5) show how quickly disaster of change took place in the Al-Tameemi. In the Al-Tameemi study, natural resources were also affected by the changes, with severe damage to large tracts of land and urban green spaces. Livelihoods were disrupted by the impacts of the damage. All these activities were capable of being monitored with remote sensing. For example, image analysis showed an increase in the area effected by damage. This was validated with survey data from the Iraqi government and confirmed by UND and FAO reports between 2010-2015. The Tameemi study has not been completed to finish all changes in Iraq, which dates back to its beginnings in the year of 1990.

Table (1) Shows set of selecte results for the environnemental factor change of the Al-Tameemi study from four months for the period 2011-2014.

**Table 1.** The results for the environnemental factor.

| Change Indicator | Pre-Damage | Post-Damage | Jan2011 | Sep2014 |
|---|---|---|---|---|
| Damaged Population | N/A | N/A | 4.5M | 2.6M |
| Total Number of Areas (City) | 14/18 | 9/18 | 16-13/18 | 11-8/18 |
| Effect Temperature | 10-39C$^0$ | 22-58C$^0$ | 18-24C$^0$ | 46-60C$^0$ |
| Rate Salinity | 1654mg /L | 9875mg/ L | 2623-15416mg /L | 85765-126520mg /L |
| Quantity of Groundwater | 1.4 $m^3$/year | 2.1$m^3$/year | 1.9 $m^3$/year | 3.1 $m^3$/year |
| Rate Desertification | 36-48% | 70-92% | 68% | 93% |
| Rate Pollution | 35-55% | 48-63% | 53% | 76% |
| Disease Prevalence Rate | 3.5-12% | 10.6-20% | 9.76% | 17.99% |
| Effect Soil | 15-25% | 48-60% | 51% | 84% |

## 5. Comparison of analysis

A number of the data patterns are already used by evaluators monitoring the environment. The models are divided into two parts: 1) direct observation (e.g. remote sensing with ground survey) using software techniques/ programs methods (e.g. GIS, Digital processing technology via of the satellite images). The patterns each have their weaknesses and strengths and used to collect different forms of data. The methods are therefore likely to be more appropriate at different phases in the evaluation and monitoring process of the changes. To better understand patterns for each method an effectiveness analysis was conducted based on six or more criteria: a) Temperature; b) Salinity; c) Rivers; d) Desertification; e) Groundwater; and f) Pollution. Additionally, Technical Assessment- Analysis. The weaknesses and strengths of each factor were then examined, and the results used to make recommendations on when each factor may be most appropriately applied.

## 6. Conclusion

The paper presents a list of analytical statistics evaluating the principles of remote sensing. Their shifts at the regional level, add to categorising the study of the stated objective. In solidarity, with the existence of high-resolution data from local data covering the assessment rates of existing environmental change. Also, areas of distinct activity were included for use in monitoring and evaluating changes in the Iraqi environment during the period 2010-2015. It is evident from the results that remote sensing is one of the most appropriate methods to use to evaluate the performance of environmental and climatic disaster reduction in Iraq, such as providing detailed analysis and assessment of the environmental values affected by human activities in changes in the degraded landscape nature also, to help other sectors, which are related to the impact of these developments in the region. Eco-environmental priorities must be developed based on an evolving technical basis, such as for environmental evaluation and analysed the spatiotemporal variation from 2010-2015. Additionally, adopting the quality requirement of the image resolution within the sensor limits of 1.0 m. So, to complete the holistic theory of research in providing a normative and reliable assessment of the quality of analytical work to study the disaster before and after even occurrence. Satellite imagery is available to all and has been shown to be cost-effective. A specific timetable was for analysis was set up at the beginning of practical research. In particular, once the initial mapping was finalised and the database constructed, the time and resources necessary to update and query the signature of the data set was reduced. This work, therefore, provides a scientific explanation on the preservation of data used later, before the disaster that is ideally updated every 1-5 years. Make a copy of the maps with the imagery to assess the impact of the damage, during environmental change monitoring. Overall, to show the superiority of methods and patterns of sensor techniques with GIS to the incidence of the environmental changes. High capabilities and skill focused on a highly efficient assessment of the changes that have been taking place in Iraq for a decade. To activate a comprehensive management system to reduce the damage that causes the deterioration of vital areas of human life.

## References

1. A. WN, Brown K, Fairbass J, Jordan A, Paavola J, Rosendo S. A 35, 1095 (2003).
2. A. D.J. Complex Urban Systems and Extreme Events: towards a theory of disaster recovery (IC Urban Disaster Reduction, 2004).
3. Å. Nilsson, M. Brewin, C. T. Bakwatsa. Monitoring and evaluation system for the environment and natural resources sector in rwanda (Sw 2013).
4. Al-Timimi, Y. K., Monim H. Al-Jiboori. Int J E. E. 4.2, 291 (2013).
5. A. Williams, Options for results monitoring and evaluation for resilience-building operations 2016).
6. B. Indicators for disaster risk and risk management 2011).
7. B. D. Saito, K., Spence, R., Chenvidyakarn, T., Adams, B., Mcmillan, A. and Platt, S. 6th, I. RSDA. 11-12 (2008).
8. C. R., Sîrodoev, I., Koeppel, S., Denisov, N., & Sîrodoev, G. S.W. J. 10.1155 (2013).

9. Cu. Disaster recovery monitoring and evaluation: centre for risk in the Built environment 2010).
10. D. S. I. P. 52.4, 444 (2015).
11. M. Agha, O.M.A., Şarlak, N. IOSR J. A. Ph, 6, 25 (2016).
12. O. M. A. Mahmood Agha, Nermin. Şarlak . Int. J. Adv. 5, 473 (2017).
13. W. D., Svoboda, M., & Hayes, M. Monitoring drought in the United States (Ed 2005).
14. W. R. M., Fazey, I., Stafford Smith, M., Park, S. E., Eakin, H. C., Archer Van Garderen, E. R. M., & Campbell, B. Reconceptualising adaptation to climate change as part of pathways of change and response (GECh 2014).
15. G. H. K., M. Alexander, P. Matczak, M. Pettersson, S. Bruzzone. E. So. 21, 27 (2016).
16. G. M. Borsuk, and R. B. Howarth. E. So. 22, 39 (2016).
17. UNC. (Rio de Janerio-Brazil 1992).

MATEC Web of Conferences **162**, 04001 (2018) https://doi.org/10.1051/matecconf/201816204001
*BCEE3-2017*

# Serviceability limit state of two-way reinforced concrete slab strengthened with different techniques

*Eyad* Sayhood[1,*], *Ammar* Ali[1], and *Zahraa* Sharhan[1]

[1]Building and Construction Engineering Department, University of Technology, Baghdad, Iraq

**Abstract**. The experimental results for service load of sixteen simply supported two way reinforced concrete slabs under the action of concentrated patch load were determined based on the amount of permissible deflections and the crack widths.All the slabs had the same overall dimensions and flexural steel reinforcement. Five types of strengthening were adopted. The first and second methods include applying either near surface mounted (NSM) or near reinforcement mounted (NRM) ferrocement layers. While the third method includes applying a concrete layer reinforced with welded wire fabric mesh of various diameters. The fourth and fifth methods include fixing CFRP rods and laminates, respectively, on the bottom surface of slabs. Strengthening techniques were applied on the bottom surface of fifteen slab specimens. In addition, a control slab specimen without any strengthening was used for the purpose of comparison. The calculated results for ultimate loads based on serviceability requirements (deflection and crack width according to both ACI and BS formulae) were lower than the experimental results.

## 1 Introduction

Serviceability refers to the conditions under which a structure is still considered useful. It refers to conditions other than the structure strength that render the structures unusable. Serviceability limit state design of structures includes factors such as durability, overall stability, fire resistance, deflection, cracking and excessive vibration. past and have been quite popular. Five of these methods are included in this paper: wire mesh reinforced cement mortar (both NSM and NRM ferrocement), external welded wire fabric (WWF) reinforced concrete layer, CFRP bars and CFRP laminates.
The advantages of strengthening techniques are well explained in a previous research [1].

## 2 Experimental program

Sixteen simply supported slab specimens were cast. All slabs have the same dimensions of 700 mm width, 700 mm length and 130 mm thickness. A control slab (without strengthening) was designed to fail in flexure as shown in Fig. 1. The steel bars used in reinforcing slab specimens were of size 6 mm with 150 mm spacing. The steel bars were added at the bottom face of the slabs in both directions. Fifteen of the specimens were strengthened by different types of strengthening techniques using mesh wires, WWFs, CFRP bars and laminates as shown in Fig. 2. The specimens were classified into five groups GA, GB, GC, GD and GE, and slab G0 was the control specimen. The parametric study was designed to classify slabs into five series, as shown in Table1. Many trial mixes were made to gain a suitable strength. The mix was designed to achieve cylinder compressive strength (f'c) of about 30 MPa at 28 days. The details of the normal concrete mix which was used in this study are shown in Table 2. The mortar that was used in groups GA and GB was of cement-sand mix in the ratio of 1:2.

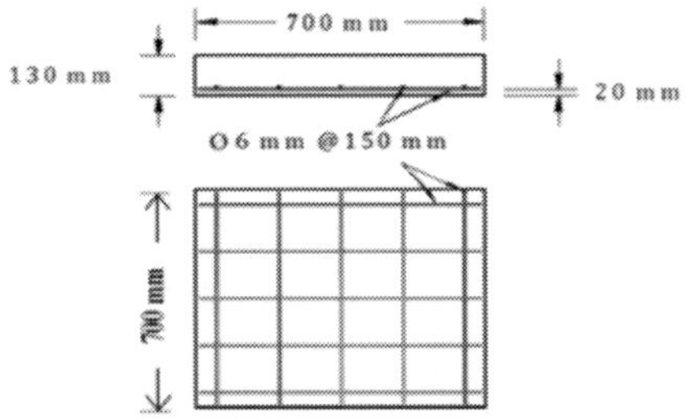


**Fig. 1.** Details and arrangement of reinforcement of a typical slab specimen

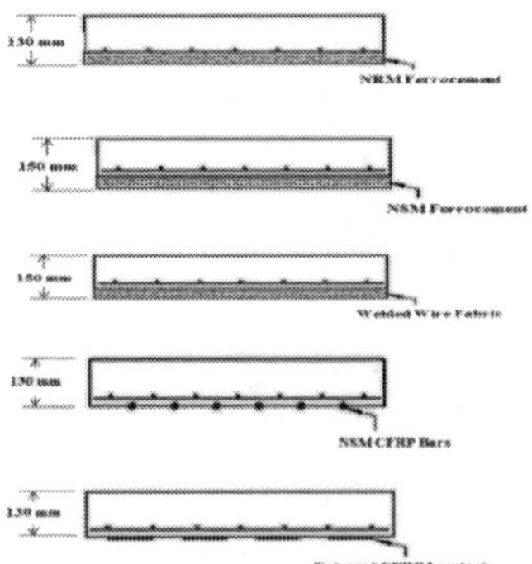


**Fig. 2.** Strengthening techniques

* Corresponding author: dr_eyad_alhachamee@yahoo.com

© The Authors, published by EDP Sciences. This is an open access article distributed under the terms of the Creative Commons Attribution License 4.0 (http://creativecommons.org/licenses/by/4.0/).

## 2.1 Concrete mix and mechanical properties

Each mix batch for all groups were controlled by casting nine standard cylinders of 150x 300 mm, three 150 mm cubes and three prisms of 100x100x400 mm. These specimens were used to determine the average cylinder compressive strength of concrete of 30 MPa, splitting tensile strength ranges between 2.6 to 3.2 MPa, flexural strength ranges between 3.8 to 4.6 MPa and the modulus of elasticity ranges between 26.5 to 28.2 GPa. Details of the tests and results are given in Ref. 1.

All concrete slab specimens were tested using a testing machine of 700 kN capacity. During the test, central deflection was recorded using a dial gauge with accuracy of 0.01 mm. The first crack was observed, also the corresponding cracking load was recorded, and the crack patterns were marked. Steel plate of 200 x 200 x 25 mm was placed at the center of the slab to represent patch load and to prevent local crushing of concrete. The nominal effective areas of strengthening ($A_t$) shown in Table 1 were calculated as:

$$A_t = A_b \times \frac{1000}{S} \qquad \text{.......... (1)}$$

where, $\mathbf{A_b}$ is the area of a single steel wire in mesh reinforcement in groups A to C or CFRP bars or strip in groups D and E in $mm^2$,

And, **S** is the spacing distance between wires, bars or strips, center to center, in mm.

**Table 2.** Normal concrete mix proportions

| Cement ($kg/m^3$) | Water ($kg/m^3$) | Coarse aggregate ($kg/m^3$) | Fine aggregate ($kg/m^3$) | W/C |
|---|---|---|---|---|
| 400 | 193 | 990 | 745 | 0.48 |

## 2.2. Test results and discussion

The cracking loads ($\mathbf{P_{cr}}$), crack widths ($\mathbf{W_{cr}}$), central deflections ($\mathbf{\Delta v}$) and ultimate load ($\mathbf{P_u}$) are shown in Table 3. The control slab was designed to fail in flexure. Deflection is considered an important property of any structure. In this study, a central dial gauge with accuracy 0.01mm was used to determine the deflection. At low load levels all the tested slabs behaved in an elastic manner where the cracks did not appear at any place and the central deflections are small and proportional to the applied load. The first crack appeared around the sides of the steel plate on the tension face of the slab and other cracks formed at the central region of the slab. By increasing the load, these cracks widened and increased in number and spread to the support of the slab through a diagonal direction. Also new cracks appeared starting from the supports. Eventually the cracks were many and one or more of these cracks had the potential to penetrate into the compression zone at the loading positions. Figs. 3 to 7 explain the load-deflection relationship of the tested slabs, clearly shows that the usage of the five strengthening techniques reduces the deflection of the slab. It can be noticed that all specimens have undergone three stages of behavior during the entire load process.

**Table 1.** Details of specimen groups

| Group | Samples | Type of strengthening | Nominal Area of Strengthening* ($mm^2/m$) | Description |
|---|---|---|---|---|
| G0 | G0 | Without strengthening | ----- | Control specimen |
| GA | GA1 | Strengthened by NRM ferrocement | 90 | One layer of wire mesh |
| | GA2 | | 180 | Two layers of wire mesh |
| | GA3 | | 270 | Three layers of wire mesh |
| GB | GB1 | Strengthened by NSM ferrocement | 90 | One layer of wire mesh |
| | GB2 | | 180 | Two layers of wire mesh |
| | GB3 | | 270 | Three layers of wire mesh |
| GC | GC1 | Strengthened by fabric wires | 90 | One layer of fabric wires with Ø4 mm @150mm spacing |
| | GC2 | | 180 | One layer of fabric wires with Ø6 mm @150mm spacing |
| | GC3 | | 270 | Two layers of fabric wires with (Ø4mm and Ø6 mm) @150mm spacing |
| GD | GD1 | Strengthened by CFRP bars | 90 | 3 bars at each direction with Ø6mm |
| | GD2 | | 180 | 5 bars at each direction with Ø6 mm |
| | GD3 | | 270 | 7 bars at each direction with Ø6 mm |
| GE | GE1 | Strengthened by CFRP laminates (1.2mm thickness) | 90 | 5 strips of carbon laminate with width of 15 mm |
| | GE2 | | 180 | 5 strips of carbon laminate with width of 30 mm |
| | GE3 | | 270 | 5 strips of carbon laminate with width of 45 mm |

*From Eq.1

**Table 3.** Experimental results of the tested slabs

| Sample Name | $P_{cr}$ (kN) | $W_{cr}$ (mm) at $P_u$ | Δv (mm) at $P_u$ | $P_u$ (kN) | $P_{cr}/P_u$ | $P_{cr}/P_{cr}$-G0 | $P_u/P_{u,}$ G0 |
|---|---|---|---|---|---|---|---|
| G0 | 72.1 | 1.3 | 11.2 | 170 | 0.42 | 1 | 1 |
| GA1 | 80.5 | 1 | 7.7 | 176 | 0.46 | 1.11 | 1.04 |
| GA2 | 84 | 0.95 | 7 | 183 | 0.46 | 1.17 | 1.08 |
| GA3 | 86 | 0.9 | 7.3 | 205 | 0.42 | 1.19 | 1.21 |
| GB1 | 79.3 | 1 | 8.7 | 190 | 0.42 | 1.1 | 1.12 |
| GB2 | 85 | 0.98 | 8.5 | 207.7 | 0.41 | 1.18 | 1.22 |
| GB3 | 88.1 | 0.97 | 9.4 | 235 | 0.37 | 1.22 | 1.38 |
| GC1 | 98.9 | 1.2 | 10.8 | 242.2 | 0.41 | 1.37 | 1.42 |
| GC2 | 105 | 1.15 | 10 | 264 | 0.4 | 1.46 | 1.55 |
| GC3 | 106 | 1.1 | 9 | 281.2 | 0.38 | 1.47 | 1.65 |
| GD1 | 121.3 | 1 | 11 | 312 | 0.39 | 1.68 | 1.84 |
| GD2 | 128.1 | 0.95 | 10 | 344.1 | 0.37 | 1.78 | 2.02 |
| GD3 | 131.5 | 0.92 | 10 | 375.1 | 0.35 | 1.82 | 2.21 |
| GE1 | 126.9 | 0.83 | 10.2 | 322 | 0.39 | 1.76 | 1.89 |
| GE2 | 139.4 | 0.8 | 9.5 | 350.2 | 0.4 | 1.93 | 2.06 |
| GE3 | 144.6 | 0.78 | 9.2 | 399.4 | 0.36 | 2.01 | 2.35 |

At the first stage, the linear behavior of the load deflection response is considered. This stage covers the region up to the first crack load, below this limit the materials behave elastically and the cracks originating in the tensile regions of the specimens cross section are still stable. After this, the cracks spread and their width increases with the increasing load. Both the reinforcement for all groups and mortar for group GA, GB and GC in compression zone are still elastic.

At the second stage, a nonlinear behavior of the load deflection response is noticed. This stage covers the region beyond the proportional limit. A gradual yielding of steel reinforcement occurs, because several layers of mesh are placed at different depths of the cross section, many carbon rods were also placed near surface and several slices of carbon laminate were applied. At this stage, the increase in the load carrying capacity beyond the proportional limit is due to the increase in the tensile stresses accompanied with a continuous shift in the position of the neutral axis towards the compression zone.

Finally, as the applied load reaches near its ultimate value, the rate of increase in deflection is substantially exceeding the rate of increase in the value of applied load. The crack width versus applied load responses of the slab specimens are shown in Figs. 8 to 12.

**Plate 1.** Slab specimens after testing, crack patteren and mode of failure

https://doi.org/10.1051/matecconf/201816204001

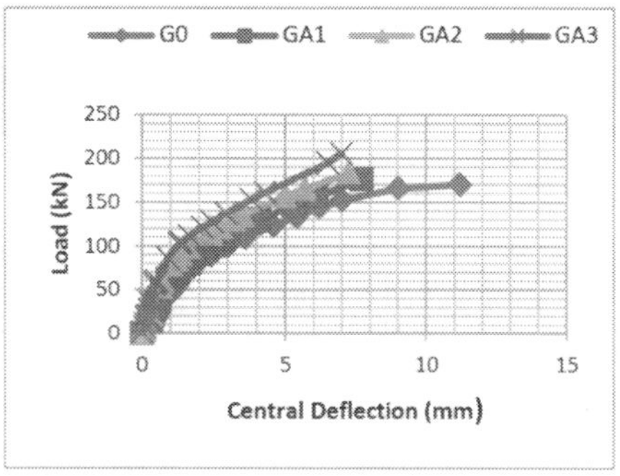


**Fig. 3.** Load – deflection curve for specimens of group GA

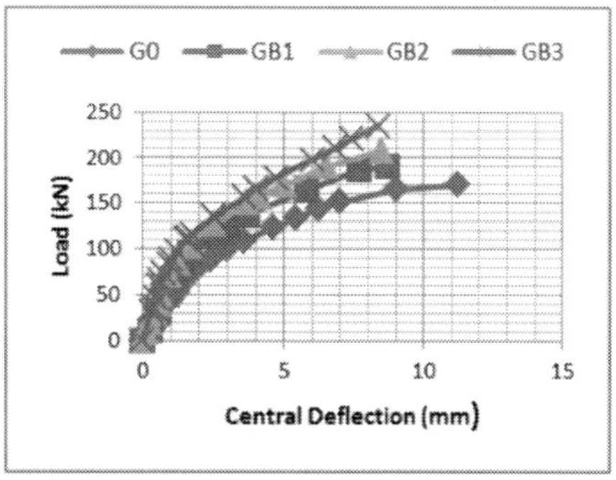


**Fig. 4.** Load – deflection curve for specimens of group GB

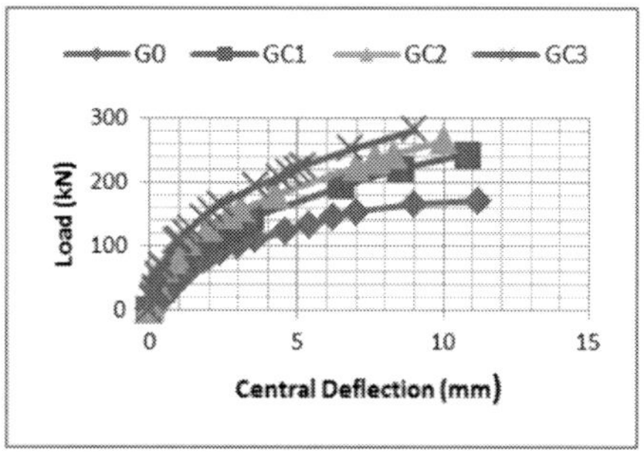


**Fig. 5.** Load – deflection curve for specimens of group GC

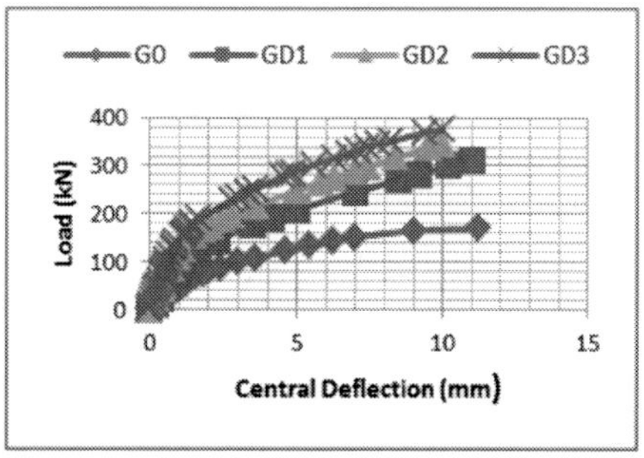


**Fig. 6.** Load – deflection curve for specimens of group GD

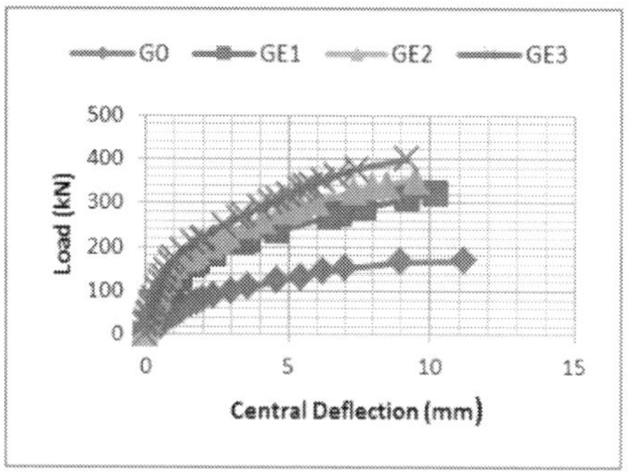


**Fig. 7.** Load – deflection curve for specimens of group GE

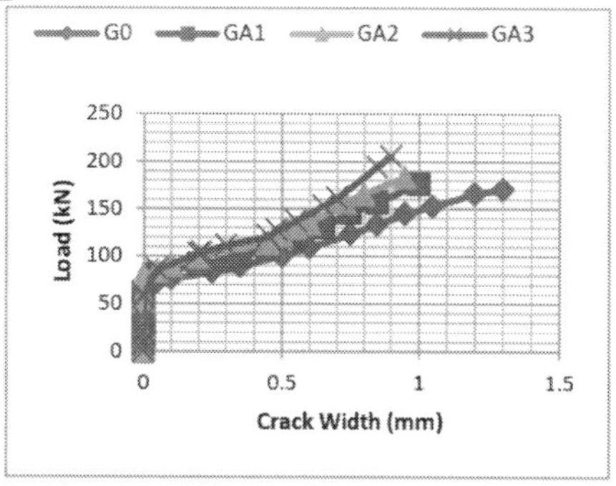


**Fig. 8.** Load – crack width curve for specimens of group GA

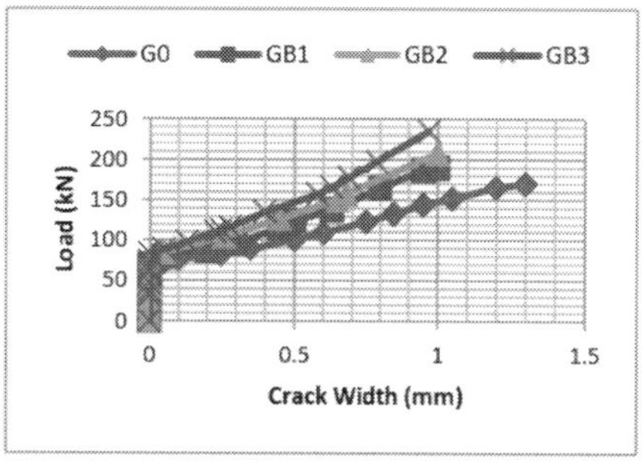


**Fig. 9.** Load – crack width curves for specimens of group GB

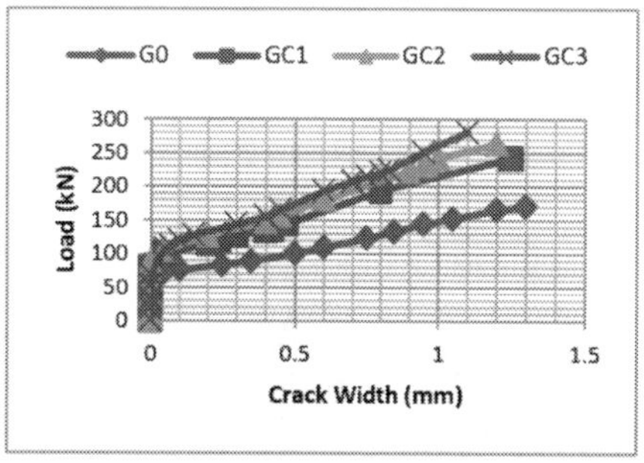


**Fig. 10.** Load – crack width curve for specimens of group GC

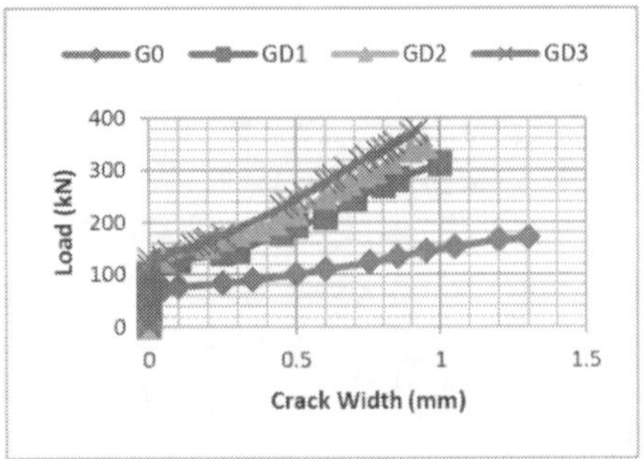


**Fig. 11.** Load – crack width curves for specimens of group GD

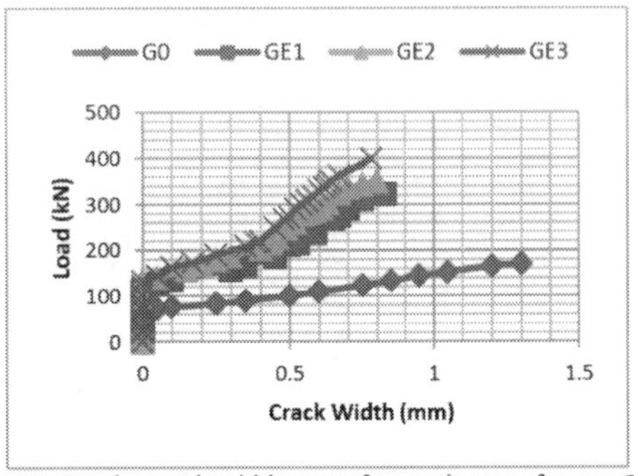


**Fig. 12.** Load – crack width curve for specimens of group GE

## 3 Serviceability requirements

### 3.1 Deflection

The ACI code kindly recheck table to calculate the maximum permissible deflections 318-M-14 for slabs [2]. The deflection limitation obtained from the ACI code for two way reinforced concrete slab is selected to correspond to L/360, where L is the span length. For the slab specimens of 600 mm span length, this limit will give a service deflection of 1.67 mm. By applying this value of deflection on the load-deflection curves given in Fig.3 the service loads of all sixteen specimens have been determined. These values of service load were multipled by a factor of 1.6 to get the factored load as recommended by ACI 318M-14 code [2]. These factored loads were divided by the corresponding experimental ultimate load to get the percentages. Fig. 13 clarifies the experimental ultimate nominal load (Pn) and the calculated factored load (Pu).

While the B.S8110:part 2:1985 standarad[3] applies a limit to calculate the deflection due to vertical load for structural members that are visible, the sag in a member will usually become noticeable if the deflection exceed L\250, where L is the span length which was 600 mm. So the service deflection is 2.4mm, then this value has been applied to the load-deflection curves that are shown in Figs. 3 to 7 to extract the value of the service load of all sixteen specimens. There after these values of service load were multipled by a factor of 1.6 to get the factored load and the last factored values were divided by the corresponding value of experimental ultimate load to get the percentage. Fig. 14 clarifies the experimental ultimate nominal load (Pn) and the calculated service load (Pu).

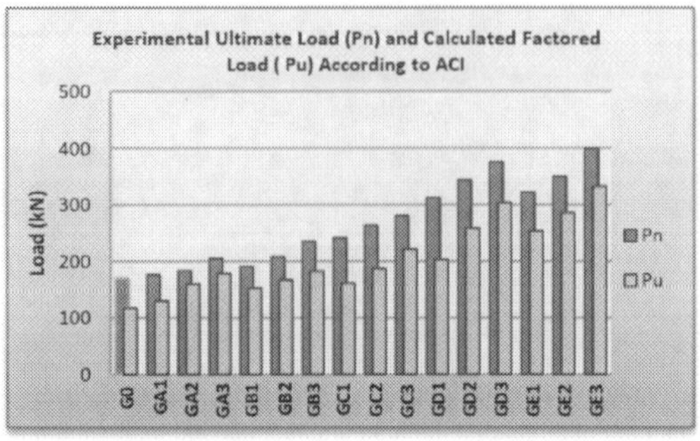


**Fig.13.** Experimental ultimate load (Pn) and calculated factor

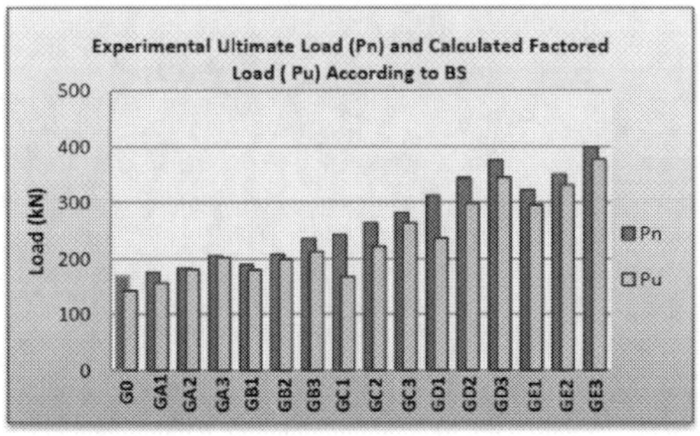


**Fig.14.** Experimental ultimate load (Pn) and calculated factored load (Pu) based on permissible deflection

### 3.2 Crack width

ACI 318-14 code [2] provisions for allowable crack-width limits in steel-reinforced concrete structures correspond to 0.013 in. (0.3 mm) for exterior exposure and 0.016 in. (0.4 mm) for interior exposure. Blair and Nawy [4] have provided the following expression, for predicting the maximum crack width:

$$w = k\beta f_s\sqrt{I} \quad ...(2)$$

where

$$I = \frac{d_{b1}S_2}{\rho_{t1}} = \frac{S_1\, S_2\, dc}{d_{b1}}\frac{8}{\pi}$$

$k$ = fracture coefficient with a value $k = 2.8 \times 10^{-5}$ for uniformly loaded restrained two-way action square slabs and plates. For concentrated loads or reactions or when the ratio of short to long span is less than 0.75 but larger than 0.5, a value of $k = 2.1 \times 10^{-5}$ was applicable. For span aspect ratios less than 0.5, k = 1.6. For simply supported slabs, the value of k should be multiplied by 1.5.

$\beta$ = 1.25 (chosen to simplify calculations, although it varies between 1.20 and 1.35)

$f_s$ = actual average service-load steel (reinforcement) stress level or 40% of the specified yield strength fy, ksi

$d_{b1}$ = diameter of the reinforcement in direction 1 closest to the concrete outer fibers, in.
$S_1$ = the spacing of reinforcement in direction 1, in.
$S_2$ = the spacing of reinforcement in perpendicular Direction 2, in.
$\rho_{t1}$ = the active steel ratio, that is, the area of steel $\frac{A_s}{[12(d_{b1}+2c_1)]}$
where $A_s$ = Area of steel per ft
$c_1$ the clear concrete cover measured from the tensile face of the concrete to the nearest edge of the reinforcing bar in direction 1; and,
w the crack width at face of concrete caused by flexure, in. direction 1 refers to the direction of reinforcement closest to the outer concrete fibers; this is the direction for which crack-control check should be made. Subscripts 1 and 2 pertain to the directions of reinforcement .

The value extracted from the equation above was 0.6 mm and the code mentions a maximum allowable value equal to 0.4 mm which should not exceed. So By dropping this crack value 0.4mm on the Load-crack width curves that are shown in Figs. 8 to 12, the values of service load of all sixteen specimens has been gained, then these values of service load were multiplied by a factor of 1.6 to get the factored loads and the last values were divided by the corresponding value of experimental ultimate load to get the percentage. Fig.15 clarifies the experimental ultimate load (Pn) and the calculated factored load (Pu).

While BS 8110-97[5] standarad recommends that the design surface crack width should not exceed a specific appropriate value. Cracking should not lead to spoil appearance. So for members that are visible, the calculated maximum crack width should not exceed 0.3 mm. Also, cracking should not lead to steel corrosion, so for members in aggressive environment the calculated maximum crack width should not lead to a loss of the performance of the structure. BS 8110-97[5] provisions are based on Beeby [6] empirical equations.

Design surface crack width, $W_d$

$$= \frac{3\ a_{cr}\ \varepsilon_m}{1+2(\frac{a_{cr}-C_{min}}{h-x})} \quad ...(3)$$

where:

$\mathbf{a_{cr}}$ = distance from the point considered to the surface of the nearest longitudinal bar;

$$\mathbf{a_{cr}} = \sqrt{(\frac{S}{2})^2 + (c+\frac{d_b}{2})^2} - \frac{d_b}{2} \quad ....(4)$$

$\varepsilon_m$ = average strain at the level where the cracking is being considered;
**S**: reinforcement spacing (mm)
$\mathbf{d_b}$: diameter of reinforcement (mm)
$\mathbf{c_{min}}$ = minimum cover to the tension steel;
h = overall depth of the member;
x = depth of neutral axis
For cracked section, the value of €m is expressed as:

$$\varepsilon_m = \varepsilon_1 - \frac{\mathbf{b(h-x)(a'-x)}}{\mathbf{3}Es\ As\ (d-x)} \quad .... \ \mathbf{(5)}$$

where: $\varepsilon_1$ = strain at the level considered, calculated ignoring the stiffening effect of the concrete in the tension zone,

$$\varepsilon_1 = \frac{d-x}{h-x} * \varepsilon_s \quad .... \ (6)$$

$\varepsilon_s = \frac{F_s}{E_s}$ *where* $\mathbf{F_s}$ should not exceed 0.8 fy, thus, let.

$$\varepsilon_s = 0.8 * \frac{F_y}{E_s} \quad ....(7)$$

b = width of the section at the centroid of the tension steel.
a' = distance from the compression face to the point at which the crack width is being calculated = h

The value obtained from the equation above was 0.287 mm which was consistent with the specified value in the code. Then this value (0.287mm) has been used in the load-crack width curves that are shown in Figs. 8 to 12 to extract the value of the service load of all sixteen specimens, after that these values of service load were multiplied by a factor of 1.6 to gain the factored load and the last factored values were divided by the corresponding value of ultimate load to get the percentage. Fig.16 clarifies the experimental ultimate load (Pn) and the calculated factored load (Pu).

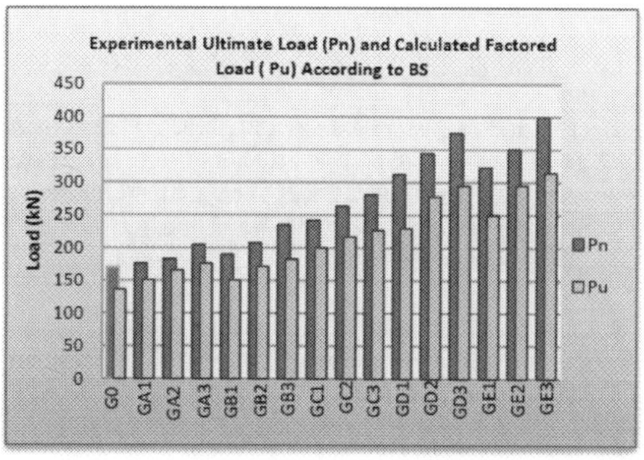


**Fig.15.** Experimental ultimate load (Pn) and calculated factored load (Pu) based on permisible crack width

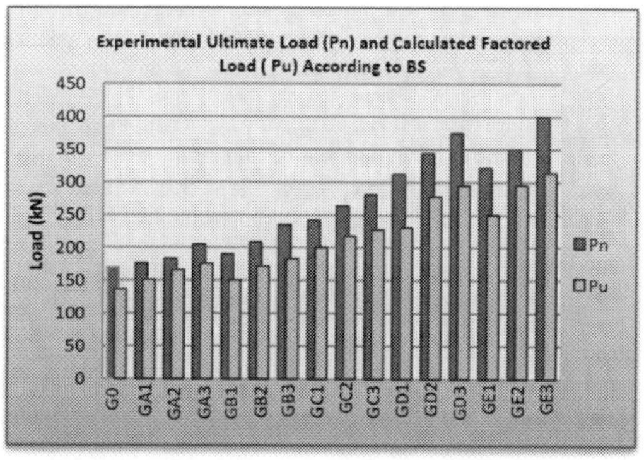


**Fig.16.** Experimental ultimate load (Pn) and calculated factored load (Pu) based on permissible crack width

Table 4 shows the ratio of experimental ultimate load (Pn) and calculated factored load (Pu) to the corresponding values of control (unstrengthend) slab specimen. Also, this table shows that the slab with CFRP laminate strengthening GE3 has the greatest gain percentage of 135% in ultimate load for experimental results. While NRM ferrocement with one layer of steel mesh specimen GA1 gave the lowest gain percentage of 3.5% in ultimate strength. On the other hand, the ultimate loads that have been calculated based on ACI and BS requirements for both deflections and crack widths revealed similar trend. The highest percentages of gains in strength using ACI code formulae for both deflection and crack widths were 183% and 138% while for BS formulae they were164% and 131% respectively. Also, specimen GA1 got the lowest percentages in strengths which were 10% and 12% based on ACI requirement for both deflection and crack widths, respectively. For BS formulae the percentages were 10% and 11%, for both deflection and crack widths formulae, respectively.
Table 5 shows the ratios of the calculated factored loads (Pu) based on permissible deflection and crack width to experimental ultimate load (Pn) according to ACI and BS provisions. The average value of this ratio is 0.8.

## 4. Conclusions

From this experiment, the following conclusions can be drawn:

1. Strengthening ratios within each strengthening technique had a clear effect on the behavior of the strengthened slabs. It was seen that the increase in the ratio of strengthening led to an increase in the ultimate strength in all types of strengthening.

2. The increased in the strengthening ratios of all the five techniques tends to increase the initial cracking and a decrease in the deflection.

3. The calculated results for ultimate loads based on serviceability requirements (deflection and crack width according to both ACI and BS formule) were lower than the experimental results.

4. Generally, the ultimate strength determination using deflection gives more conservative values than that based on crack width.

5. Service load of B.S formula based on permissible deflection gave values higher than ACI formula service load. While for crack width requirements, they gave approximately equal values.

6. NRM ferrocement strengthening technique has the least value of gain in service strength, while the CFRP laminate strengthening gave the highest value.

## References

1. Z.S. Sharhan, Master Thesis, Building and Construction Engineering Department, University of Technology, Baghdad, (2016)

2. ACI 318M-14, 2014. Building Code Requirements for Structural concrete ( ACI 318M-14) and Commentary, USA: American Concrete Institute, (2014)

3. BS 8110: part 2, British Standard use of concrete, code of practice for special circumstances, (1985)

4. E. G. Nawy, K. W. Blair, SP-30, American Concrete Institute, Farmington Hills, Mich., pp. 1-41, (1971)

5. BS 8110-1997. Structural use of concrete, Part 1: Code of practice for design and construction, (1997)

6. A.W. Beeby, SP-20, American Concrete Institute, Detroit, 1971, pp. 55–75 (cited by Makhlouf and Malhas (1996)

**Table 4.** Ratio of experimental load (pn) and calculated factored load (pu) of strengthed slabs to control slab

| Samples Name | Experimental | Calculated | | | |
|---|---|---|---|---|---|
| | | Deflection | | Crack width | |
| | $\frac{P_n}{P_{n.control}}$ | ACI code | B.S standarad | ACI code | B.S standarad |
| | | $\frac{P_u}{P_{u.control}}$ | $\frac{P_u}{P_{u.control}}$ | $\frac{P_u}{P_{u.control}}$ | $\frac{P_u}{P_{u.control}}$ |
| G0 | 1 | 1 | 1 | 1 | 1 |
| GA1 | 1.04 | 1.10 | 1.10 | 1.12 | 1.11 |
| GA2 | 1.08 | 1.36 | 1.27 | 1.19 | 1.22 |
| GA3 | 1.21 | 1.52 | 1.42 | 1.27 | 1.29 |
| GB1 | 1.12 | 1.29 | 1.27 | 1.17 | 1.10 |
| GB2 | 1.22 | 1.42 | 1.40 | 1.31 | 1.26 |
| GB3 | 1.38 | 1.55 | 1.49 | 1.39 | 1.35 |
| GC1 | 1.42 | 1.37 | 1.18 | 1.45 | 1.47 |
| GC2 | 1.55 | 1.59 | 1.55 | 1.58 | 1.60 |
| GC3 | 1.65 | 1.88 | 1.85 | 1.67 | 1.67 |
| GD1 | 1.84 | 1.73 | 1.66 | 1.80 | 1.69 |
| GD2 | 2.02 | 2.20 | 2.10 | 2.08 | 2.04 |
| GD3 | 2.21 | 2.59 | 2.42 | 2.32 | 2.17 |
| GE1 | 1.89 | 2.16 | 2.08 | 1.89 | 1.84 |
| GE2 | 2.06 | 2.43 | 2.32 | 2.19 | 2.16 |
| GE3 | 2.35 | 2.83 | 2.64 | 2.38 | 2.31 |

**Table 5.** Ratio of experimental load (pn) and calculated factored load (pu)

| Samples | Experimental | Calculated factored load (Pu) | | | | | | | |
|---|---|---|---|---|---|---|---|---|---|
| | | Deflection | | | | Crack width | | | |
| | Ultimate Load (Pn) (kN) | ACI Code | | B.S standards | | ACI Code | | B.S standards | |
| | | Pu | Pu/Pn | Pu | Pu/ Pn | Pu | Pu/Pn | Pu | u/ Pn |
| G0 | 170 | 117.54 | 0.69 | 142.4 | 0.84 | 148.27 | 0.87 | 136.15 | 0.80 |
| GA1 | 176 | 129.76 | 0.74 | 156.8 | 0.89 | 165.6 | 0.94 | 151.17 | 0.86 |
| GA2 | 183 | 159.79 | 0.87 | 181.01 | 0.98 | 176.83 | 0.97 | 166.06 | 0.91 |
| GA3 | 205 | 178.27 | 0.87 | 201.71 | 0.98 | 187.73 | 0.92 | 175.91 | 0.86 |
| GB1 | 190 | 152.16 | 0.8 | 180.44 | 0.94 | 173.6 | 0.91 | 150.41 | 0.79 |
| GB2 | 207.7 | 167.26 | 0.81 | 198.87 | 0.96 | 193.52 | 0.93 | 171.63 | 0.83 |
| GB3 | 235 | 182.02 | 0.78 | 212.25 | 0.90 | 206.13 | 0.88 | 183.13 | 0.78 |
| GC1 | 242.2 | 161.13 | 0.67 | 168 | 0.69 | 214.56 | 0.89 | 200.20 | 0.83 |
| GC2 | 264 | 187.11 | 0.71 | 221.34 | 0.84 | 234.24 | 0.89 | 217.25 | 0.82 |
| GC3 | 281.2 | 221.17 | 0.79 | 264 | 0.94 | 247.36 | 0.88 | 227.24 | 0.81 |
| GD1 | 312 | 203.04 | 0.65 | 236.57 | 0.76 | 267.46 | 0.86 | 230.08 | 0.74 |
| GD2 | 344.1 | 258.5 | 0.75 | 298.56 | 0.87 | 307.68 | 0.89 | 278.29 | 0.81 |
| GD3 | 375.1 | 304 | 0.81 | 344.41 | 0.92 | 344.6 | 0.92 | 294.97 | 0.79 |
| GE1 | 322 | 253.74 | 0.79 | 296 | 0.92 | 280.96 | 0.87 | 250.41 | 0.78 |
| GE2 | 350.2 | 286.16 | 0.82 | 330.97 | 0.95 | 324 | 0.93 | 295.50 | 0.84 |
| GE3 | 399.4 | 332.64 | 0.83 | 376.39 | 0.94 | 352.16 | 0.88 | 314.19 | 0.79 |

# Comparison of ultimate strength results from ACI and Eurocode 4 for steel tubular columns filled with SCC

*Asraa* Al-Goody[1, 2,*], *Esra* Mete Güneyisi[2], and *Süleyman* İpek[2]

[1] Building and Construction Department, University of Technology, Baghdad, Iraq
[2] Department of Civil Engineering, Gaziantep University, 27310, Gaziantep, Turkey

**Abstract.** The Concrete Filled Steel Tubular (CFST) columns have several benefits in comparison to the ordinary steel or reinforced concrete ones. Therefore, they have become more commonly acknowledged in the structural applications. In this study, two design codes such as American Concrete Institute (ACI) and Eurocode 4 (EC4) were used for predicting the ultimate axial strength of CFST columns filled with self-compacting concrete (SCC). To evaluate the results, circular steel tube with different diameter to thickness (D/t) ratio of 30, 60, and 90 and steel yielding strength of 185, 275, and 450 MPa were considered as prediction parameters. The wall thickness and length to diameter (L/D) ratio of the steel tubes were kept constant. As an infill material for the steel tubes, 16 different SCC mixtures reported in the literature were studied and their compressive strength results were used to get the code predicted ultimate axial strength of the composite columns. The analysis of the results based on ACI and EC4 were performed and discussed comparatively.

## 1 Introduction

Composite columns that have been widely used in the engineering applications have superior strength and ductility properties. When the traditional reinforced concrete columns are compared with the Concrete Filled Steel Tubular (CFST) columns, significant differences with respect to ductility and energy absorption capacity can be observed. Due to their outstanding static and dynamic characteristics, CFST columns have been commonly utilized in the various types of structures such as bridges, high-rise buildings, subway platforms, etc. [1-7]. The load carrying capacity of the column is significantly influenced by the properties of steel and concrete. Due to the progress in the strength of steel and concrete, cross section of the structural elements gets smaller. Steel in CFST columns is used to confine the concrete that is the structural advantage of CFST and provides contribution in load carrying capacity. In addition, the construction time is also reduced by eliminating the permanent formwork. Also, the utilization of CFST in structures procures an inhibited or delayed local buckling of steel tube due to the concrete core. Briefly, increase in the load carrying capacity, stiffness, and ductility and decrease in the cost and time of construction by use of CFST columns make utilization of such columns attractive [7].

The CFST columns have been experimentally investigated for decades [6-11]. Researches revealed that the properties and geometry of the materials are leading parameters which remarkably affect the behavior and capacity of CFST columns. For the last few years, there have been a plenty number of theoretical and experimental studies on circular CFST columns with several diameter to thickness (D/t) ratios with combinations of various material properties [2,12-17]. Since the experimental investigation of axial load carrying capacity of CFST columns extensively attracts the interests of researchers, the theoretical analysis, therefore, plays a significant role in engineering research and practice.

In this study, the formulas proposed by ACI 318 and Eurocode 4 [18,19] were used for the calculation of the ultimate strength of steel tubular columns infilled with self-compacting concrete (SCC). For this, several parameters were considered such as different D/t ratios and steel yield strength of steel tube and compressive strength of SCC core. A comparison of the analysis results based on both design codes was performed for the composite columns under the action of compressive loading.

## 2 Details of the study

To evaluate the behavior of SCC filled steel tubular columns subjected to axial compression, as previously mentioned, two international design codes were employed. To this aim, three ratios of D/t (30, 60, 90) and three yield strength of steel, $f_y$ (185, 275, 450 MPa) were specified for circular CFST columns. However, a constant length to diameter (L/D) ratio and steel tube thickness of 2.5 and 4 mm was considered. A special procedure was adopted to take the benefit of

* Corresponding author: israayounis@yahoo.com

© The Authors, published by EDP Sciences. This is an open access article distributed under the terms of the Creative Commons Attribution License 4.0 (http://creativecommons.org/licenses/by/4.0/).

compressive strength results of SCC as an infill material on the behavior of CFST columns. For this, the concrete strength was used from the experimental studies of Güneyisi et al. [20] and Al-Goody [21]. According to these works, a total of 16 SCC mixtures were designed with respect to various nanosilica, NS (0-6%) and fly ash, FA (0-75%) contents. Their cubic compressive strength results at 28 days varied from 30 to 78 MPa. More detailed explanations on the properties of the mixtures can be found in the aforementioned study. It should be noted that the cubic compressive strength results were converted to cylindrical compressive strength by using Eurocode 2 [22] and then used in the calculation of ultimate strength of the steel tubular columns regarding to ACI 318 and EC 4 [18,19].

There are a number of expressions given by the standardization institutes of several conditions [18,19]. The formulae used in this paper are proposed by ACI 318 and Eurocode 4. The American Concrete Institute [18] uses the following expression for calculating the ultimate axial capacity of the circular CFST columns.

$$N_u = 0.85 f_c' A_c + f_y A_s \qquad (1)$$

This relation does not contain effectiveness of the concrete confinement and the interaction between the concrete core and steel tube. However, in the relation suggested by Eurocode 4 [19], the contribution of the steel tube and the concrete core as well as confinement effect is taken into account. The coefficients of confinement for the concrete ($\eta_c$) and steel tube ($\eta_a$) are employed. Thus, the concrete strength is increased by $\eta_c$ due to the occurrence of a triaxial state of stress condition while steel tube strength is reduced by $\eta_a$ since the hoop stresses cause a reduction in the effective yield stress of the steel. The following expression is proposed by Eurocode 4 [19]:

$$N_u = \left(1 + \eta_c \frac{t}{D}\frac{f_y}{f_y'}\right) f_c' A_c + \eta_a f_y A_s \qquad (2)$$

$$\eta_c = 4.9 - 18.5\bar{\lambda} + 17\bar{\lambda}^2 \; (\eta_c \geq 0) \qquad (3)$$

$$\eta_a = 0.25(3 + 2\bar{\lambda})(\eta_a \leq 1.0) \qquad (4)$$

$$\bar{\lambda} = \sqrt{\frac{N_{pIR}}{N_{cr}}} \qquad (5)$$

$$N_{pIR} = f_y A_s + f_c' A_c \qquad (6)$$

$$N_{cr} = \frac{\pi^2 (EI)_{eff2}}{l^2} \qquad (7)$$

$$(EI)_{eff2} = E_s I_s + K_e E_{c2} I_c \qquad (8)$$

$$P_{o,EC4} = \left(1 + 4.9 \frac{t}{D}\frac{t_y}{f_c'}\right) f_c' A_c + 0.75 f_y A_s \left(if\ \bar{\lambda} = 0\right) \qquad (9)$$

## 3 Results and discussion

The compressive strength of SCC mixtures at 28 days were used in the calculation of the ultimate strength of the steel tubular columns by using ACI and Eurocode 4 formulae. The relation between ultimate strength (calculated by using ACI and Eurocode 4 formulae) and NS/FA contents are presented in Figure 1 with respect to D/t ratio of 30 and steel yielding strength of 185 MPa. In the other figures, the relations are given with respect to D/t ratio of 60 and 90 and steel yielding strength of 275 and 450 MPa (Figures 2-9). The results indicated that the ultimate strength of steel tubular columns filled with SCC is strongly affected by concrete compressive strength. Utilization of NS till replacement level of 4% increased the SCC compressive strength and an increase in the compressive strength directly increased the ultimate strength of the steel tubular columns. However, at the replacement level of 6%, a decrease in the compressive strength of SCC was observed at all FA contents. This is also caused decreasing in the ultimate strengths of CFST calculated by both formulae. On the contrary to NS, the utilization of FA decreased the compressive strength of SCC. This situation resulted in lower ultimate strength of steel tubular columns determined by both formulae. Moreover, the ultimate strengths determined by ACI formula were lower than that calculated by using Eurocode 4. Calculating the ultimate strength of the steel tubular columns filled with compressive strength at different D/t ratio showed that increasing the concrete surface area increases the ultimate strength. Also, utilization steel tubular with higher yielding strength remarkably increases the ultimate strength of the columns.

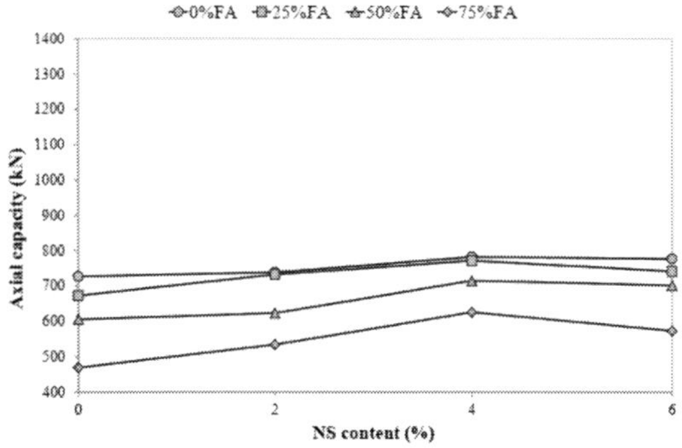


(a)

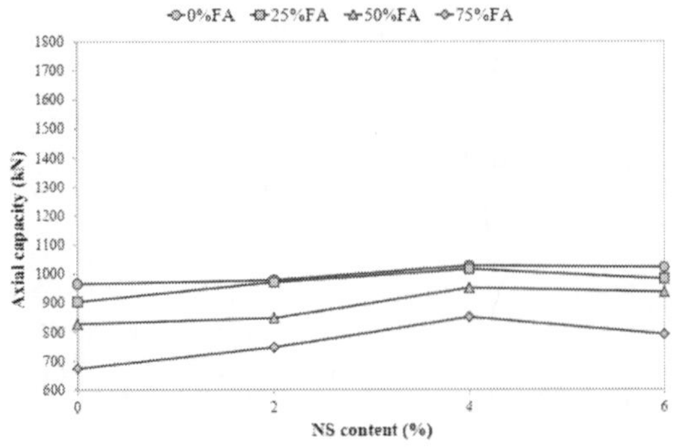


(b)

**Fig. 1.** Axial capacity of CFST columns with D/t value of 30 and *fy* of 185 MPa having different SCC mixtures according to: a) ACI and b) Eurocode 4

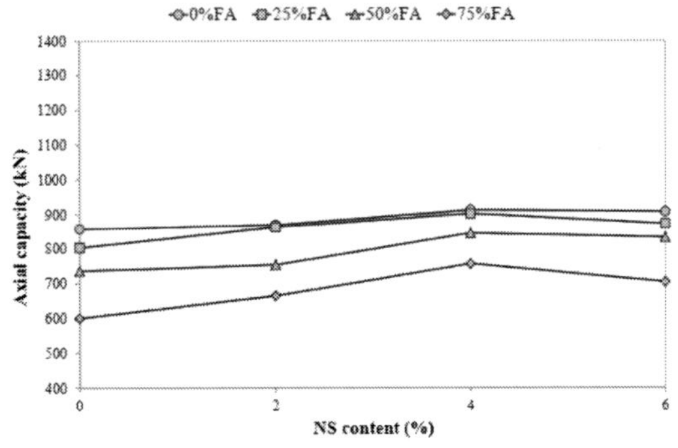


(a)

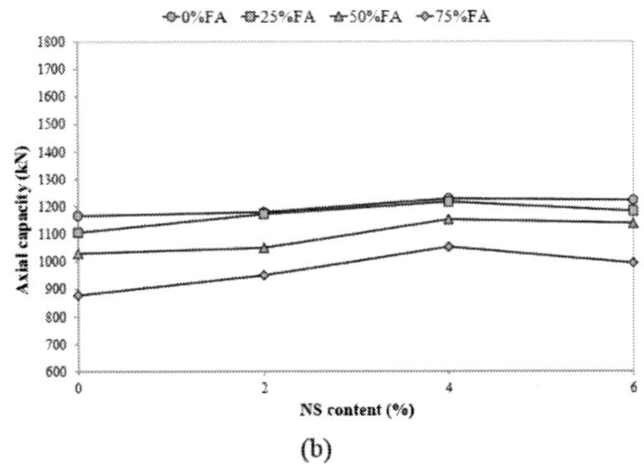


(b)

**Fig. 2.** Axial capacity of CFST columns with D/t value of 60 and *fy* of 185 MPa having different SCC mixtures according to: a) ACI and b) Eurocode 4

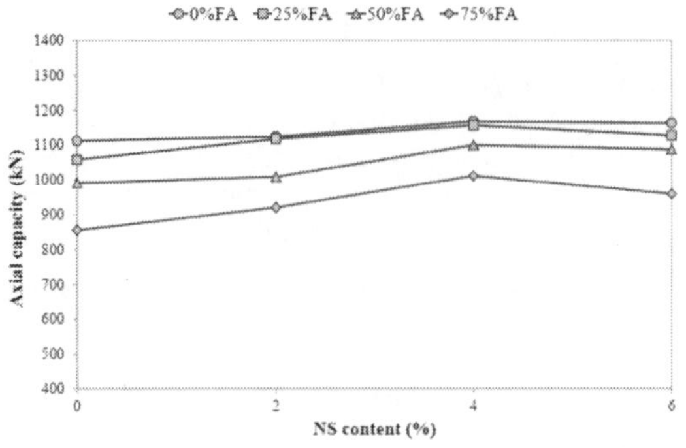


(a)

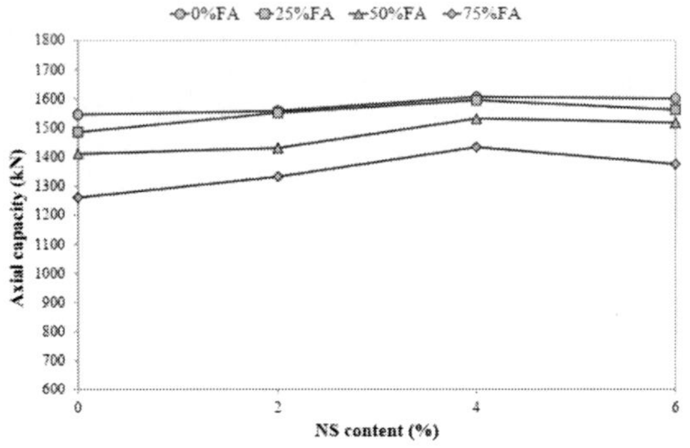


(b)

**Fig. 3.** Axial capacity of CFST columns with D/t value of 90 and *fy* of 185 MPa having different SCC mixtures according to: a) ACI and b) Eurocode 4

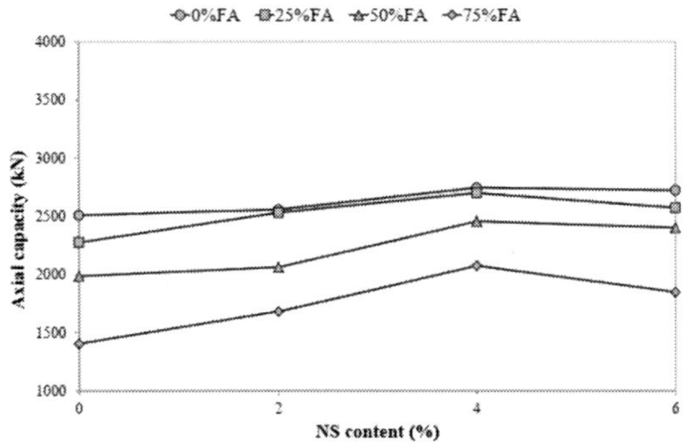


(a)

https://doi.org/10.1051/matecconf/201816204002

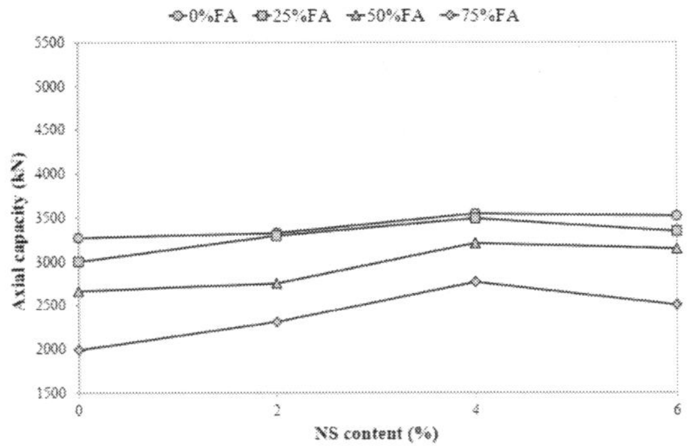


(b)

**Fig. 4.** Axial capacity of CFST columns with D/t value of 30 and *fy* of 275 MPa having different SCC mixtures according to: a) ACI and b) Eurocode 4

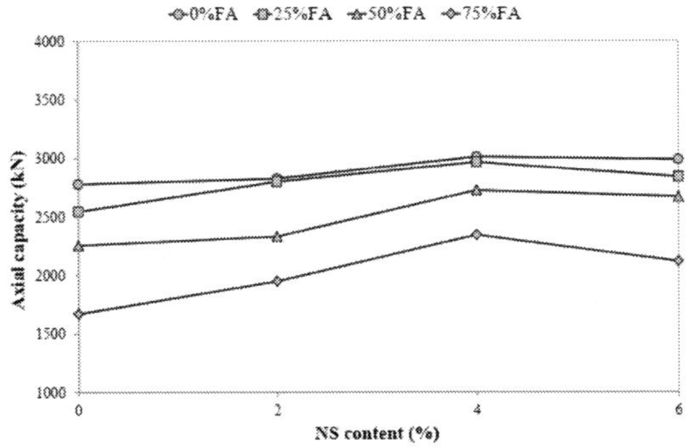


(a)

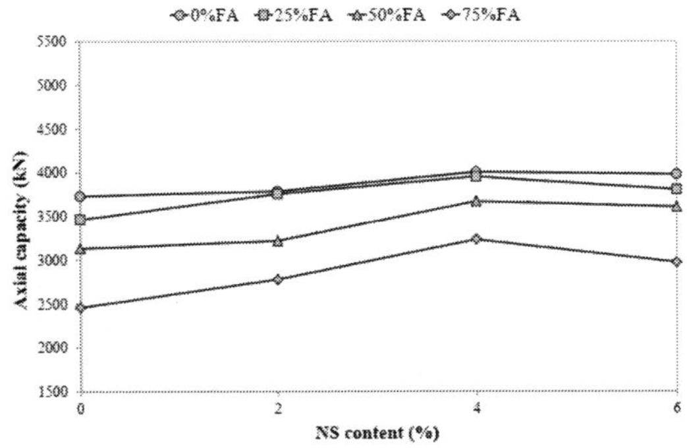


(b)

**Fig. 5.** Axial capacity of CFST columns with D/t value of 60 and *fy* of 275 MPa having different SCC mixtures according to: a) ACI and b) Eurocode 4

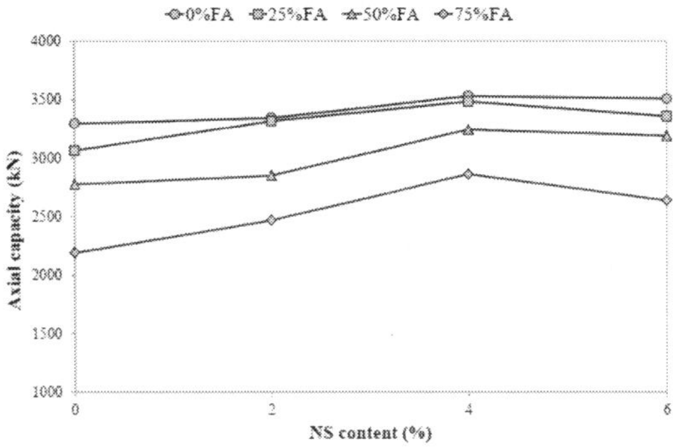


(a)

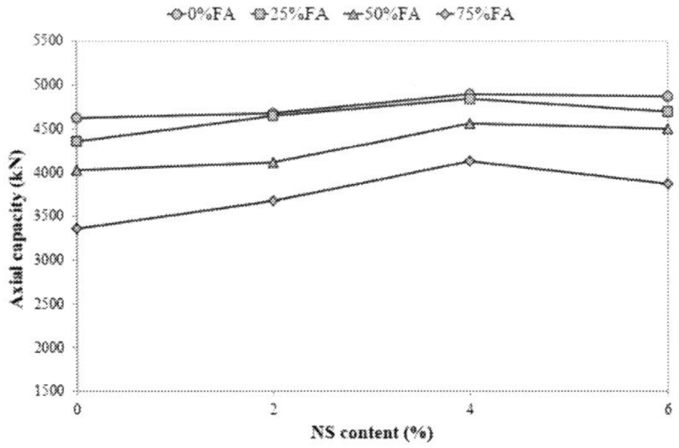


(b)

**Fig. 6.** Axial capacity of CFST columns with D/t value of 90 and *fy* of 275 MPa having different SCC mixtures according to: a) ACI and b) Eurocode 4

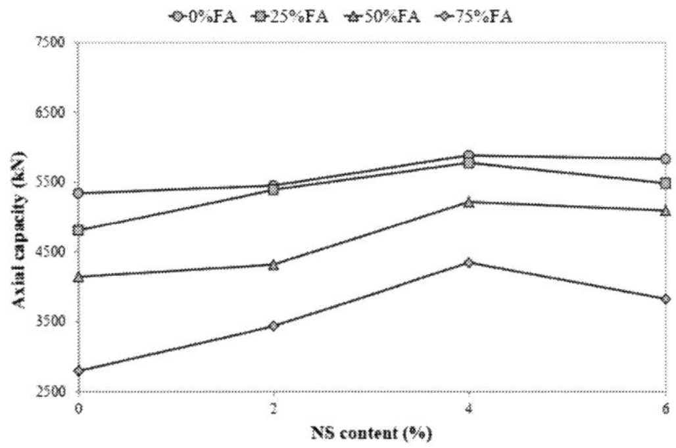


(a)

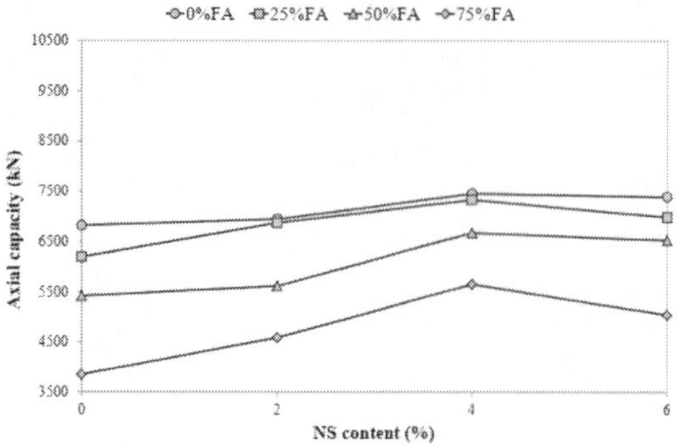


(b)

**Fig. 7.** Axial capacity of CFST columns with D/t value of 30 and *fy* of 450 MPa having different SCC mixtures according to: a) ACI and b) Eurocode 4

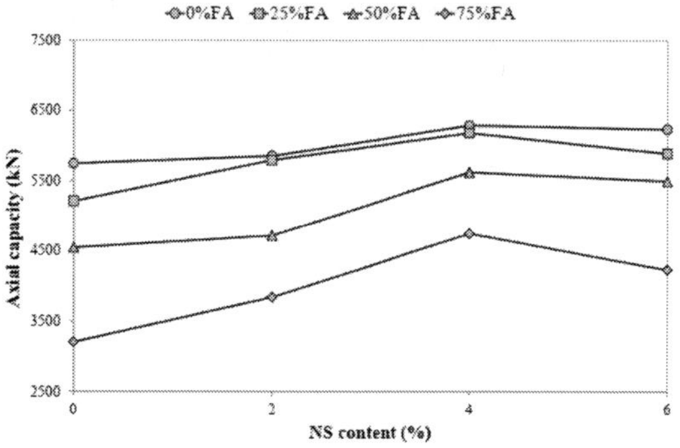


(a)

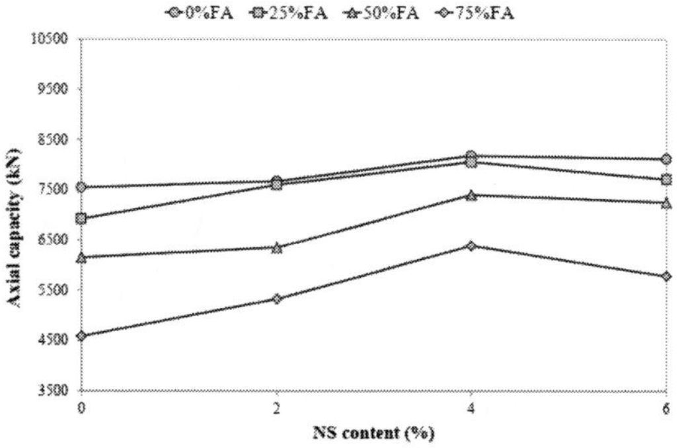


(b)

**Fig. 8.** Axial capacity of CFST columns with D/t value of 60 and *fy* of 450 MPa having different SCC mixtures according to: a) ACI and b) Eurocode 4

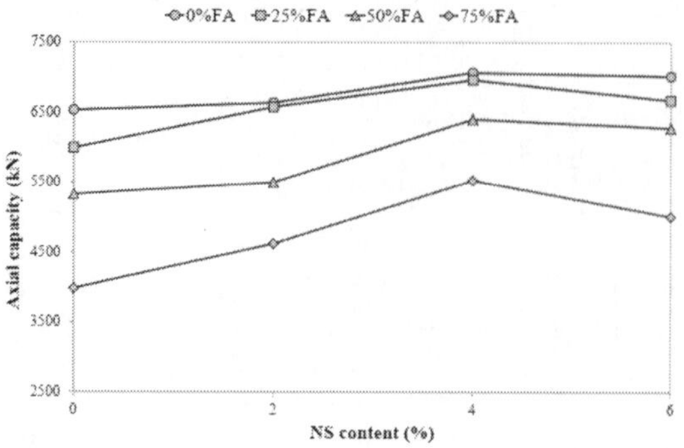


(a)

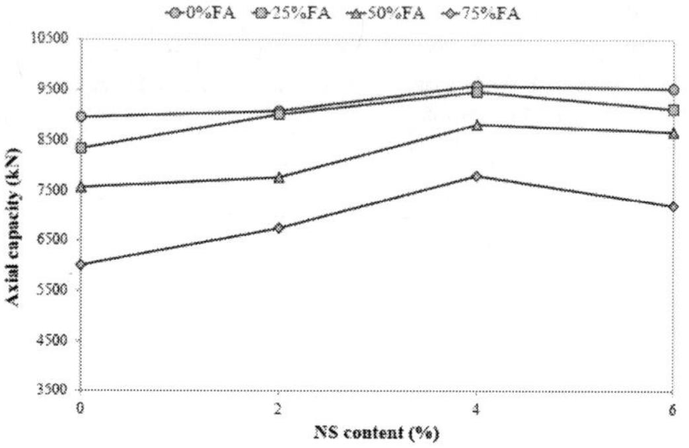


(b)

**Fig. 9.** Axial capacity of CFST columns with D/t value of 90 and *fy* of 450 MPa having different SCC mixtures according to: a) ACI and b) Eurocode 4

In addition, in Figure 10, the correlation between the design codes for the composite sections having different SCC mixtures was given. Results indicated that there is a strong relation between the ultimate strength calculated by ACI and Eurocode 4 formulae.

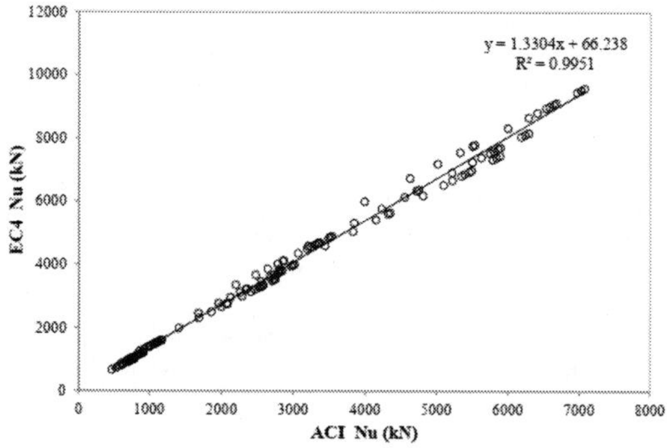


**Fig. 10.** Correlation between the design codes

## 4 Conclusions

Based on the findings of the study, the following conclusions can be drawn.

- The compressive strength of the concrete and the yield strength of steel have an important influence on the ultimate strength of the circular steel tubular columns filled with concrete.
- FA and NS contents significantly affect the strength of SCC and CFST columns.
- Variation in diameter to thickness ratio is observed to be remarkable effects on the ultimate strength of the circular steel tubular columns filled with SCC.
- Ultimate strength calculated by using ACI formula resulted in lower values than that calculated by Eurocode 4.
- There is a strong relation between ultimate strength values obtained by using ACI and Eurocode 4 formulae.

## References

1. K. Tsuda, C. Matsui, Y. Ishibashi, *Fifth Asia-Pacific Conf. on Struc. Eng. and Const.* (EASEC-5) (1995)
2. Z.H. Lu, Y.G. Zhao, J. of Constr. Steel Res. **66**, 850-862 (2010)
3. W.F. Chen, C.H. Chen, Inter. Assoc. for Bridge and Struc. Eng. P. **33(11)**, 37-52 (1973)
4. N.J. Gardner, E.R. Jacobson, American Conc. Ins. J. **64**,7, 404-412 (1967)
5. C.Y. Lin, *Ninth Inter. Spec. Conf. on Cold-Formed Steel Struc.* (St. Louis, Missouri, USA, 1988)
6. J. Zeghiche, MSc Thesis, University of Manchester, (1988)
7. J. Zeghiche, K. Chaoui, J. of Const. Steel Res. **61**, 53-66 (2005)
8. H. Shakir-Khalil, J. Zeghiche, The Struc. Eng. **67**, 346-353 (1989)
9. H. Shakir-Khalil, M. Mouli, The Struc.Eng. **68**,20, 405-413 (1990)
10. N.E. Shanmugam, B. Lakshmi, J. of Const. Steel Res. **57**, 1041-1080 (2001)
11. P.K. Neogi, H.K. Sent, J.C. Chapman, The Struc. Eng. **47**,5, 187-195 (1969)
12. M.D. O'Shea, R.Q. Bridge, J. of Struc. Eng. 1295-1303 (2000)
13. C.D. Goode, R. Narayanan, Seminar of Assoc. for Inter. Coop. and Res. in Steel-Conc. Comp. Struc., (Austria, 1997)
14. M. Saisho, T. Abe, K. Nakaya, J. of Struc. and Cons. Eng. **523**, 133-140 (1999)
15. G. Giakoumelis, D. Lam, J. of Cons. Steel Res. **60**,7, 1049-1068, (2004)
16. K. Sakino, H. Nakahara, S. Morino, I. Nishiyama, J of the Struc. Eng. (ASCE) **130**, 2, 180-188 (2004)
17. L.H. Han,G.F. Yao, X.L. Zhao, J. of Cons. Steel Res. **61**, 1241-1269 (2005)
18. ACI-318R, Building code requirements for structural concrete and commentary. American Concrete Institute, (MI, USA, 2005)
19. Eurocode 4, Design of composite steel and concrete structures. European Committee for Standardization: British Standards Institution, (2004)
20. E. Güneyisi, M.Gesoğlu, A. Al-Goody, S. İpek, Con. Build. Mat. **95**, 29-44 (2015)
21. A. Al-Goody, PhD Thesis, Gaziantep University (2017)
22. Eurocode 2, Design of composite steel and concrete structures. European Committee for Standardization: British Standards Institution, (2004)

# Influence of flange on the shear capacity of reinforced concrete beams

*Kaiss* Sarsam[1,*], *Raid* Khalel[1] and *Nisreen* Mohammed[1]

[1]Building and Construction Engineering Department, University of Technology, Baghdad-Iraq

**Abstract.** In structural engineering (RC, steel, etc.) it is usual to base the shear strength of members on the web only- e.g. in RC the stirrups used are usually called "web reinforcement". Presently all codes, and several researches, base the strength of members on the capacity of the web alone. 93 tests of T-beams failing in shear available from the literature are studied in this work to estimate the influence of flanges on the shear capacity of RC beams. These include 32 ones without web reinforcement and 61 with stirrups. Comparison between test results and theoretical shear capacity show that all available equations conservatively estimate the occurrence of shear failure. In this work an equation for predicting the contribution of the flange to shear capacity in T-beams is presented.

The best available design method obtained from the literature leads to a coefficient of variation (COV) of 17.58% compared to 13.46% for the proposed design method in this work.

## 1 Introduction

There are significantly different methods to estimate shear capacity of reinforced concrete (RC) beams. The main parameters affecting the occurrence of shear failure are: concrete compressive strength, longitudinal reinforcement ratio, shear span to effective depth ratio, and size effect of the member. Bresler and MacGregor [1] indicate that shear failure is commonly initiated by the occurrence of diagonal cracks developing in the shear span, and that the flexural cracks always come before the occurrence of diagonal cracks in rectangular, I, and T- sections. They noted that the shape of the beam (I and T-sections) influences the shear capacity and the behavior including propagation of diagonal cracking due to the different magnitudes of shearing stress developed in the web. However, not much attention has been given to the behavior of RC beams with T-sections.

Swamy and Qureshi [2] proposed a procedure to estimate the ultimate shear strength of the compression zone of T-beams. This procedure was derived by using the concept of biaxial stress criterion and was based on Mohr's theory of failure. However, this method needs several steps of calculation and is difficult to apply in practice.

Ahmed [3] proposed an empirical equation which is simpler than the procedure of reference 2. This equation includes the flange contribution in predicting the ultimate shear capacity of T-sections.

Existing equations for shear capacity of concrete available in the codes do not take into account the influence of the flange in T-sections [4, 5].

In this work, empirical equations of researchers and design codes for shear capacity of beams are compared with test results obtained from the literature to find an equation for predicting the contribution of flange to the shear capacity of concrete in RC beams with T-sections.

## 2 Research significance

Several code and research estimates of shear capacity of RC beams are studied in this work. A data base of 93 tests is used in this work: 32 without stirrups and 61 with stirrups. It is found that including of flange contribution leads to a significantly improved COV for all available 93 tests from the literature. In fact, the proposed equation leads to a COV of 13.46% compared to the range of (17.58% - 42.55%) for the available methods.

## 3 Experimental investigations

The 93 shear tests have been taken from the literature: Taub and Neville [6]; Swamy and Qureshi [7]; Haddadin et al [8]; Placas and Regan [9]; Palaskas et al [10]; Kotsovos et al [11]; and Thamrin et al [12].

All of these RC T- beams involved slender ones (shear span /effective depth ratio$\geq 2.5$) that failed in shear under concentrated loading. Table 1 indicates the range of variables in all 93 tests.

**Table 1.** Range of variables

| Variable | Unit | Range |
|---|---|---|
| $f_c'$ | MPa | 11.99 – 56.95 |
| $a/d$ | ---- | 2.5– 10.4 |
| $\rho_w$ | ---- | 0.49% – 5.2% |
| $\rho_v f_{yt}$ | MPa | 0 – 2.71 |
| $b_w d$ | mm² | 2903 – 76000 |
| $b/b_w$ | ---- | 2– 7 |

* Corresponding author: kaisssarsam@yahoo.com

© The Authors, published by EDP Sciences. This is an open access article distributed under the terms of the Creative Commons Attribution License 4.0 (http://creativecommons.org/licenses/by/4.0/).

where:

$f_c'$ = cylinder compressive strength of concrete, MPa

$a$ = shear span, equal to the distance from center of concentrated load to center of support for simply supported members, mm

$d$ = distance from extreme compression fiber to centroid of longitudinal tension reinforcement, mm

$\rho_w$ = ratio of $A_s$ to $b_w d$

$A_S$ = area of non prestressed longitudinal tension reinforcement, mm$^2$.

$b_w$ = web width, mm

$\rho_v$ = ratio of $A_v$ to $b_w S$

$A_v$ = area of shear reinforcement within spacing S, mm$^2$

$S$ = center- to- center spacing of transverse reinforcement, mm

$f_{yt}$ = specified yield strength of transverse reinforcement, MPa.

$b$ = flange width, mm

## 4 Code estimates of RC beam shear capacity

### 4.1 BS 8110:1997 [13]

$$V_{rBS} = 0.79\left(\frac{100A_s}{b_w d}\right)^{1/3}\left(\frac{400}{d}\right)^{1/4}\left(\frac{f_{cu}}{25}\right)^{1/3} b_w.d/\gamma_m + \frac{0.95\,A_v\,f_{yt}\,d}{S} \quad (1)$$

$$0.15 \le \frac{100A_s}{b_w d} \le 3.0 \quad (2)$$

$$\frac{400}{d} \ge 1.0 \quad (3)$$

where:

$V_{rBS}$ = estimated shear strength per reference 13, N

$\left(\frac{400}{d}\right)$ = size effect, used for enhancement only

$f_{cu}$ = cube compressive strength of concrete, not to exceed 40 MPa. For comparison in this work, $f_{cu} = f_c' / 0.82$

$\gamma_m$ = 1.25

### 4.2 Australian Code; AS 3600:2001 [14]

$$V_{rAS} = 0.7\left[\beta_1\, b_w d\left(\frac{A_s f_c'}{b_w d}\right)^{1/3} + \frac{A_v\, f_{yt}\, d}{S}\right] \quad (4)$$

$$\beta_1 = 1.1\left(1.6 - \frac{d}{1000}\right) \ge 1.1 \quad \text{[size effect]} \quad (5)$$

where:

$V_{rAS}$ = estimated shear strength per reference 14, N

### 4.3 Canadian code; A23.3:2004 [15]

$$V_{rCAN} = 0.65\,\beta\sqrt{f_c'}\,b_w d_v + \frac{0.85\,A_v\,f_{yt}\,d_v \cot 35^o}{S} \quad (6)$$

$\beta = 0.18$, with stirrups (7)

$$\beta = \frac{230}{1000 + d_v}, \text{without stirrups} \quad (8)$$

$$f_c' \le 64\,MPa, \text{upper limit} \quad (9)$$

$$d_v = \begin{matrix} 0.9d \\ 0.72h \end{matrix}\Big| \text{ whichever is greater} \quad (10)$$

where:

$V_{rCAN}$ = estimated shear strength per reference 15, N

$h$ = overall thickness or height of member, mm

### 4.4 Euro code: 2014 [16]

$$V_{rEURO} = [0.12\,k\,(100\,\rho_w f_c')^{1/3}]b_w d + \frac{0.75\,A_v\,f_{yt}\,d}{S} \quad (11)$$

where:

$V_{rEURO}$ = estimated shear strength per reference 16, N

$$k = 1.0 + \sqrt{200/d} \le 2.0 \quad (12)$$

### 4.5 ACI 318M-14 [17]- Simplified method

$$V_{rACI-S} = 0.75\left[0.17\sqrt{f_c'}\,b_w d + \frac{A_v\,f_{yt}\,d}{S}\right] \quad (13)$$

where:

$V_{rACI-S}$ = estimated shear strength per reference 17; based on simplified ACI method, N

### 4.6 ACI 318M-14 [17]- Detailed method

$$V_{rACI-D} = 0.75\left[(0.16\sqrt{f_c'} + 17\rho_w \frac{V_u d}{M_u})\,b_w d + \frac{A_v\,f_{yt}\,d}{S}\right] \quad (14)$$

$$\frac{V_u d}{M_u} \le 1 \quad (15)$$

where:

$V_{rACI-D}$ = estimated shear strength per reference 17; based on detailed ACI method, N

$V_u$ = factored shear force at section, N

$M_u$ = factored moment at section, N.mm

$V_{rS\&A}$ = estimated shear strength per reference 20, N

## 5 Existing research estimates of RC beam shear capacity

### 5.1 Zsutty approach [18]

$$V_{rZsutty} = 2.17 \left(\rho_w f_c' \frac{d}{a}\right)^{1/3} b_w d + \frac{0.75\, A_v\, f_{yt}\, d}{S} \tag{16}$$

where:

$V_{rZsutty}$ = estimated shear strength per reference 18, N

### 5.2 Niwa et al approach [19]

$$V_{rNIWA} = 0.2\ (\rho_w f_c')^{1/3}\ (d)^{1/4} \left(0.75 + 1.4\frac{d}{a}\right) b_w d + \frac{0.75\, A_v\, f_{yt}\, d}{S} \tag{17}$$

where:

$V_{rNIWA}$ = estimated shear strength per reference 19, N

### 5.3 Sarsam and Al-Musawi approach [20]

$$V_{rS\&A} = 1.8 \left(f_c' \rho_w \frac{V_u d}{M_u}\right)^{0.38} b_w d + \frac{A_v\, f_{yt}\, d}{S} \tag{18}$$

$$\frac{V_u d}{M_u} \le 0.5 \tag{19}$$

where:

### 5.4 Kuo et al approach [21]

$$V_{rKUO} = 1.17 \left(\frac{a}{d}\right)^{-0.7} \sqrt{f_c'}\ b_w C + A_v\, f_{yt} \left(\frac{d}{S} - 1\right) \tag{20}$$

$$\frac{C}{d} = \sqrt{2\, \rho_w\, n + (\rho_w\, n)^2} - \rho_w\, n \tag{21}$$

$$n = \frac{E_S}{E_C} \tag{22}$$

$$E_C = 4700 \sqrt{f_c'} \tag{23}$$

where:

$V_{rKUO}$ = estimated shear strength per reference 21, N

$E_S$ = modulus of elasticity of reinforcement = 200000 MPa

$E_c$ = modulus of elasticity of concrete, MPa

## 6 Results and discussion

From the methods used in codes & research proposals, a comparison was made for the ratio of ($V_{exp}/V_r$), where:

$V_{exp}$ = Shear resistance of tested beam, N

$V_r$ = calculated shear resistance based on different methods of prediction, N

Table 2 gives a comparison of the results of the different methods, based on the ratio of ($V_{exp}/V_r$).

**Table 2.** Comparison of the ratio of ($V_{exp}/V_r$) for all 93 beam tests

| Detail | BS [13] | Australian code [14] | Canadian code [15] | Euro code [16] | ACI [17]-simplified | ACI [17]-detailed | Zsutty approach [18] | Niwa approach [19] | Sarsam and Al-Musawi [20] | Kou et al [21] | Proposed method |
|---|---|---|---|---|---|---|---|---|---|---|---|
| Equation used | (1) | (4) | (6) | (11) | (13) | (14) | (16) | (17) | (18) | (20) | (24) |
| Mean | 1.808 | 2.079 | 2.634 | 2.433 | 2.619 | 2.418 | 1.903 | 2.894 | 1.808 | 1.832 | 1.718 |
| Standard deviation | 0.34739 | 0.36546 | 1.12085 | 0.94413 | 0.88921 | 0.69026 | 0.66996 | 1.21593 | 0.32701 | 0.32303 | 0.23112 |
| COV % | 19.215 | 17.580 | 42.547 | 38.797 | 33.957 | 28.545 | 35.203 | 42.021 | 18.088 | 17.636 | 13.464 |
| Max. ratio | 2.669 | 3.066 | 6.063 | 5.354 | 5.341 | 4.608 | 4.002 | 6.194 | 2.970 | 2.724 | 2.348 |
| Min. ratio | 1.131 | 1.184 | 1.246 | 1.164 | 1.148 | 1.166 | 0.889 | 1.241 | 1.213 | 1.033 | 1.130 |
| Range (max/min) | 2.360 | 2.588 | 4.864 | 4.190 | 4.652 | 3.953 | 3.113 | 4.953 | 2.449 | 2.637 | 2.078 |
| *Number <1 | 0 | 0 | 0 | 0 | 0 | 0 | 1 | 0 | 0 | 0 | 0 |

*Number <1 indicates the number of specimens (out of 93) for which $V_{exp}<V_r$.

Regression analysis was performed on all 93 tests. This led to the following equation:

$$V_{r,PROP.} = 1.7 \left(f_c' \rho_w \frac{V_u d}{M_u}\right)^{0.5} \alpha + \frac{\phi_S\, A_v\, f_{yt}\, d}{S} \tag{24}$$

$$\frac{V_u d}{M_u} \le 0.5 \tag{25}$$

$$\alpha = b_w d + 0.7\,(b - b_w)\, h_f \tag{26}$$

where:

$V_{r,PROP.}$ = estimated shear strength by the proposed equation, N

$\alpha$ = parameter taking into account the effect of flange in T-section.

$\phi_S = 0.8$

As can be seen in Eq. (26), the contribution of flange to shear capacity will be eliminated in the case of beams with rectangular cross section ($h_f = 0$).

The last column in Table 2 indicates the various ratios of ($V_{exp}/V_r$) for the proposed method_- Eq. (24). In this table, the COV values range between (17.58% - 42.55%). By using the proposed equation, the COV value decreases to 13.46%.

Figs. 1-5 indicate the trend of the influence of major parameters on the prediction of shear resistance of 4 methods: BS [13], ACI [17] - detailed method, reference 20, and the proposed method. Fig. 1 is plotted in order to examine the effect of wide variation of $f_c'$ on the shear capacity of T-sections. This Fig. shows that the proposed method reveals little change, despite the change in $f_c'$ between (11.99 – 56.95) MPa. This contrasts with other methods- e.g. the ACI [17] method leads to significantly uneconomic prediction of strength with rising $f_c'$. Fig. 2 also shows little change in prediction for the proposed method with $a/d$ between (2.5-10.4). Again, the other methods lead to dropping values of the ratios of ($V_{exp}/V_r$), with the highest in the ACI [17] method.

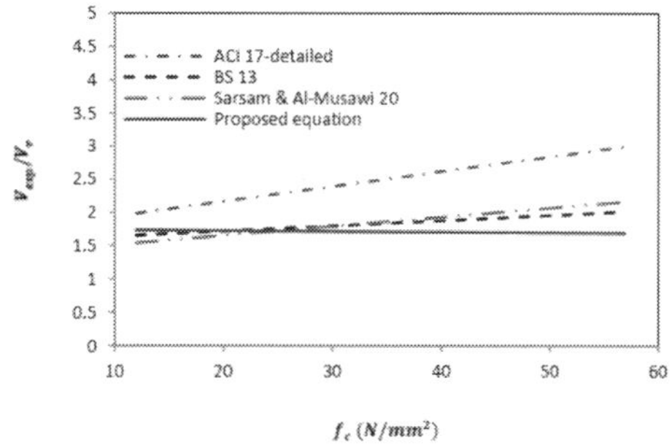


**Fig. 1.** Influence of $f_c'$ on the ratio $V_{exp}/V_r$

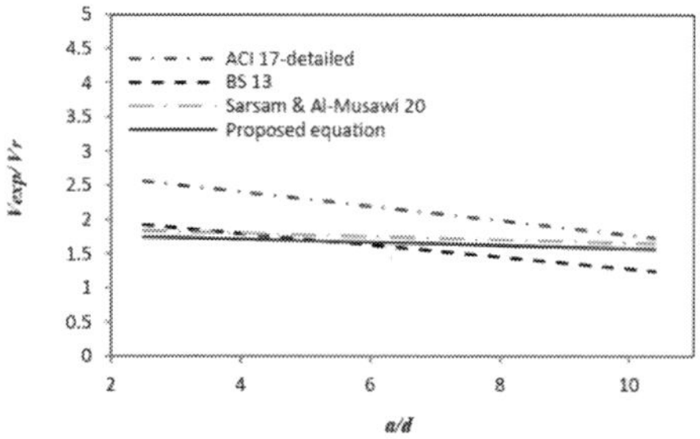


**Fig. 2.** Influence of $a/d$ on the ratio $V_{exp}/V_r$

While Fig. 3 illustrates little change in prediction for the proposed method with $\rho_w$ (0.49% - 5.2%), this is not the case with other methods– especially the ACI [17] method. Similar results are shown in Fig. 4 with significant influence of $\rho_v f_{yt}$ ranging between 0 (without stirrups) – 2.71 MPa. The proposed method shows no trend of $\rho_v f_{yt}$ influence, contrasting with the other methods when the safety factor decreases with rising $\rho_v f_{yt}$. In Fig. 5, the influence of $b/b_w$ ranging between (2-7) on the shear capacity prediction was examined. All the methods show a decrease in the factor of safety with rising $b/b_w$ – especially the ACI [17] method.

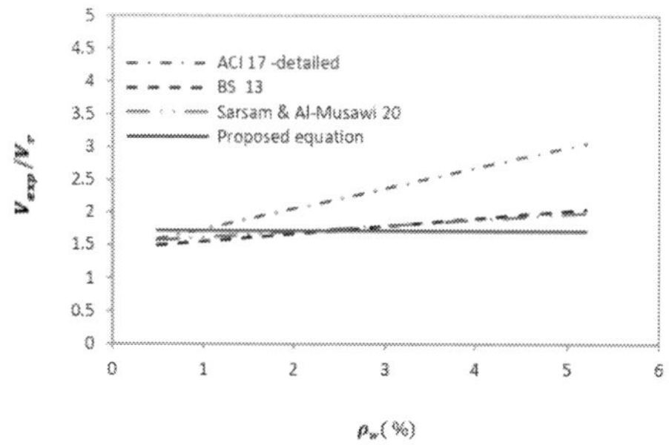


**Fig. 3.** Influence of $\rho_w$ on the ratio $V_{exp}/V_r$

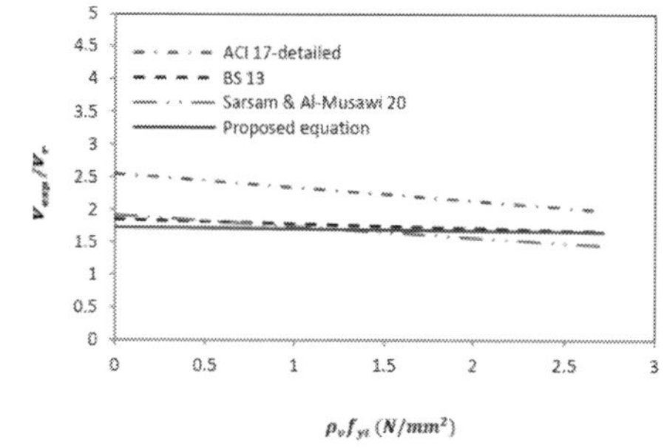


**Fig. 4.** Influence of $\rho_v f_{yt}$ on the ratio $V_{exp}/V_r$

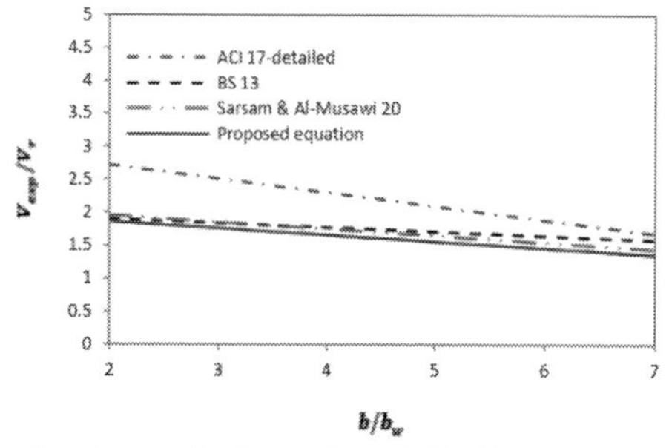


**Fig. 5.** Influence of $b/b_w$ on the ratio $V_{exp}/V_r$

## 7 Conclusions

Based on 93 tests of RC T- beams failing in shear obtained from the literature, the following conclusions are made:

1- Regression analysis of all tests indicates that the ratio of ($V_{exp}$ / $V_r$) drops significantly to 13.46% for the proposed method (which is taking into account the contribution of flange on shear capacity), compared with the COV values of the other methods (17.58% -42.55%).

2- The Australian Code [14] Method gives the lowest COV of all existing methods- 17.58%. It is clear from the proposed method that using $\alpha$- Eq. (26) for flange effect has significantly improved the COV for shear capacity prediction.

3- The proposed method by Kuo et al [21] also leads to a significant drop in COV compared to several code methods. In contrast with this method, the proposed method leads to a much simpler prediction of shear strength.

4- Of the code methods, the Canadian Code [15] Method gives the highest COV of 42.55%. This is because strength is based only on $(f_c^{'})^{1/2}$. However, with $(f_c^{'})^{1/2}$ in the ACI [17], the COV is lower with the simplified method at 33.96%.

5- All the empirical equations used to calculate shear strength conservatively predict the shear capacity of the beams. The proposed method also leads to a conservative prediction for T-beams.

6- In contrast with the proposed method, Figs. 1 and 3 show significant safety factor changes, respectively with increasing $f_c^{'}$ and $\rho_w$ by BS [13], ACI [17], and reference 20.

7- With the proposed design method- Eq. (24), Figs. 1-4 show little change in safety factor with increasing $f_c^{'}$, $a/d$, $\rho_w$, and $\rho_v\, f_{yt}$.

8- In all design methods, there is a decrease in safety factor with increasing $b/b_w$ – Fig. 5.

## Acknowledgment

This work was carried out at the Building and Construction Engineering Department, University of Technology, Baghdad, Iraq. The authors express their gratitude to the department.

## References

1. B. Bresler, J. MacGregor, ASCE Journal, **93,** ST1, 343-372 (1967)
2. R. Swamy, S. Qureshi, Materiaux et Construction **7**, 39, 181-189 (1974)
3. A. Ahmed, M.Sc. thesis, University of Technology, Baghdad, Iraq, 87 pp. (2001)
4. P. Zararis, ACI Structural Journal **103**, 5, 693-700 (2006)
5. A. Tureyen, T. Wolf, R. Frosch, ACI Structural Journal **103**, 5, 656-663 (2006)
6. J. Taub, A. Neville, ACI Journal **57**, 8, 193-220 (1960)
7. R. Swamy, S. Qureshi, *Strength, Cracking, and Deformation Similitude in Reinforced T-Beams Under Bending and Shear*, ACI Journal Proceedings **68**, 3, 187-195 (Mar. 1971)
8. M. Haddadin, S. Hong, A. Mattock, *Stirrup Effectiveness in Reinforced Concrete Beams with Axial Force*, ASCE Proceedings **97**, ST9, 2277-2297 (Sep. 1971)
9. A. Placas, P. Regan, *Shear Failure of Reinforced Concrete Beams*, ACI Journal Proceedings **68**, 10, 763-773 (Oct. 1971)
10. M. Palaskas, E. Attiogbe, D. Darwin, ACI Journal **78**, 6, 447-455 (1981)
11. M. Kotsovos, J. Bobrowski, J. Eibi, The Structural Engineer **65**, B (1), 1-10 (1987)
12. R. Thamrin, J. Tanjung, R. Arianti, O.F. Nur, A. Devinus, Journal of Engineering Science and Technology, Andalas Universiy, Indonesia, 1-15 (2014)
13. BS 8110: (1997), *Structural Use of Concrete, Part 1- Code of Practice for Design and Construction*, British Standards Institution, London, UK, 168 pp. (1997)
14. AS 3600- (2001), *Concrete Structures*, Standards Australia International, Sydney, Australia, 175 pp. (2001)
15. A23.3-04, *Design of Concrete Structures*, Canadian Standards Association, CSA Standard, Ontario, Canada, 214 pp. (2004)
16. British Standards Institution, *Eurocode 2: Design of Concrete Structures, Part 1-1: General Rules and Rules for Buildings*, BSI, London, UK, (2014)
17. ACI Committee 318, *Building Code Requirements for Structural Concrete (ACI 318M-14) and Commentary (ACI 318RM-14)*, American Concrete Institute, Farmington Hills, Michigan, 519 pp. (2014)
18. T. Zsutty, ACI Journal, **65**, 11, 942-951 (1968)
19. J. Niwa, K. Yamada, K. Yokozawa, H. Okamura, Concrete Library International of JSCE, **9**, 65-84 (1987)
20. K. Sarsam, J. Al-Musawi, ACI Structural Journal, **89**, 6, 658-664 (1992)
21. W. Kuo, T. Hsu, S. Hwang, ACI Structural Journal, **111**, 4, 809-818 (2014)

# Experimental comparative study of reactive powder concrete: mechanical properties and the effective factors

*Eyad* Kadhem[1,*], *Ammar* Ali[1], and *Sameh* Tobeia[1]

[1]Building and Construction Engineering Department, University of Technology, Baghdad, Iraq

**Abstract.** Reactive Powder Concrete (RPC) is a type of ultra-high performance concrete, this dense composite material generally characterized by high cement content, high durability, low porosity, low water/cement ratio and in most cases contains steel fibers as new types of concrete appears, further investigation for their mechanical properties are needed. This work aims to give a better understanding of RPC behavior by deriving formulas to calculate the modulus of elasticity and the splitting tensile strength in relation with compressive strength and steel fibers content. This study is based on data obtained from the experimental investigation done in this work and from others pervious works. The parametric study is based mainly on the silica fume content which is used in four different ratios (12 %, 15 %, 20 % and 25 %), the use of micro steel fibers 15 mm in length, 0.2 mm in diameter and aspect ratio of 75 added in ratios of (0 %, 1 %, 1.5 % and 2 %), and water/cement in ratios of (16 %, 18 %, 20 % and 22 %), respectively. The proposed equations show a better behavior in comparison to some available equations that were used in the estimation of modulus of elasticity and splitting tensile strength of reactive powder concrete, the coefficient of variation for the proposed equations (COV) decrease to 10.677% and 10.455% respectively.

## 1 Introduction

In order to make a better understanding of concrete (plain concrete, reinforced concrete and prestressed concrete) behavior and to ensure a more accurate concrete structural design, the modulus of elasticity which represents a fundamental factor for the elastic deformation prediction of concrete members should be studied. The splitting tensile strength is a simple method used to estimate the tensile strength of concrete that considered very difficult to estimate by using the direct tension test due to the brittle nature of concrete. Both modulus of elasticity and splitting tensile strength are usually estimated as a function of concrete compressive strength. [1, 2]

As new types of concrete appear, further investigation for their mechanical properties are needed, one of these types is the reactive powder concrete as a type of ultra high performance concrete having a dense and durable composite mix generally characterized by low water/cement ratio, high cement content and in most cases containing steel fibers [3]. The strength of reactive powder concrete is highly affected by its contents especially the amount of the steel fibers that is usually added as a volumetric ratio, the increase in this ratio will reflect positively on the modulus of elasticity, compressive strength, splitting tensile strength and modulus of rupture [3, 4].

## 2 Static modulus of elasticity and splitting tensile strength of concrete

Concrete is a composite material consisting of materials of different sizes and properties, each material has a different modulus of elasticity, thus the modulus of elasticity of concrete is highly dependent on its components properties.

A conventional basic form Eq. (1) was adopted by many provisions to establish concrete modulus of elasticity as a function of concrete compressive strength.[5]

$$Ec = a\,.(f_c')^b\ \gamma^c \tag{1}$$

Where:

Ec= modulus of elasticity
($f_c'$)= specified concrete compressive strength
$\gamma$ = unit weight of concrete
a, b and c = non-dimensional coefficient

To reach at a final form of this formula which can be applied for concrete design purposes, a regression analysis must be done to establish the values of a, b and c parameters. This equation reflects the effects and the mechanical properties of materials that are used in concrete production as well as air content in the fresh mix, unit weight, curing procedure and heat temperature and the age of test on the modulus of elasticity[5].

* Corresponding author: dr_eyad_alhachamee@yahoo.com

© The Authors, published by EDP Sciences. This is an open access article distributed under the terms of the Creative Commons Attribution License 4.0 (http://creativecommons.org/licenses/by/4.0/).

In the same way, the splitting tensile strength of concrete is mainly dependent on the compressive strength as many provisions and previous researches adopted formulas which can be represented in Eq. (2). This basically reflects the effects of original materials used in concrete production represented in terms of concrete compressive strength [6, 7].

$$f_{ct} = k\ \sqrt{f_c'} \qquad (2)$$

Where:
$f_{ct}$= splitting tensile strength of concrete
$f_c'$= specified concrete compressive strength
k= non-dimensional coefficient may found by regression analysis

These equations used in concrete design calculation in order to estimate the modulus of elasticity and the splitting tensile strength are essentially derived for conventional concrete. In case of ultra-high performance concrete like reactive powder concrete most of these equations need re-examination to make them more suitable and representative for the new concrete types. The use of different types of additives, like cementitious materials, pozolanic materials and steel fibers significantly improve the strength of concrete. The addition of steel fibers mainly enhance the flexural strength and the tensile strength of concrete and in less manner the compressive strength. The presence of steel fibers in concrete has a positive impact on crack resistance, where cracks are prevented from forming and the formation of additional cracks gets delayed after the first crack appears [8].

## 3 Previous provisions

Over the years several specifications and researchers try to estimate the mechanical properties of concrete such as modulus of elasticity, splitting tensile strength and flexural strength. As the concrete manufacturing evolved producing new types of concrete, new formulas were proposed to simulate the new types of concrete.

- The ACI-318M-14[9] estimates the modulus of elasticity and the splitting tensile strength of normal concrete as a function of concrete compressive strength ;Eq.(3) and Eq.(4):

$$E_c = 4700\sqrt{f_c'} \qquad (3)$$
$$f_{ct} = 0.56\sqrt{f_c'} \qquad (4)$$

- As the reactive powder concrete become more recognizable, H. M. Al-Hassani et. al.[10] try to give a better estimation for the modulus of elasticity based on regression analysis as in Eq.(5)

$$E_c = 113.43(f_c') + 31126.74 \qquad (5)$$

- N. Arioglu et. al [7] proposed an equation to calculate the splitting tensile strength for high strength concrete containing different types of additives (silica fume and fly ash), this equation (Eq.(6)) is applicable for concrete compressive strength up to 120 MPa with maximum curing temperature of 30 C°.

$$f_{ct} = 0.387\ (f_c')^{0.63} \qquad (6)$$

Where:

$f_{ct}$ = splitting tensile strength of concrete

## 4 Experimental program

### 4.1 Material properties:

As the properties of the materials have major effects on the reactive powder concrete strength; the types, properties and specification matching these materials can be summarized as follows:

1- Cement: ordinary portland cement type I produced by Mass Iraq Company are used in this work to produce RPC. The physical and chemical properties of cement agreed with the Iraqi specification NO.5/1984[11].
2- Fine sand: Fine quartz sand with particle size of 300 µm - 600 µm from DCP (Don Construction Products) is used in this work. The specific gravity and the sulfate content of this sand are matching the requirements of the Iraqi specification NO.45/1984[12].
3- Tap water is used in RPC production (mixing and curing).
4- Silica fume: Silica fume is an extremely fine powder. Its particles are several times smaller than cement and are usually used to enhance the strength of concrete. Silica fume (MEYCO MS610) from BASF Company is used as a percentage of cement by weight with several different percentages in this work. The physical and chemical properties agreed with ASTM C1240[13] requirements.
5- Superplasticizer: represented by ViscoCrete –5930 provided from Sika is used to enhance the workability of fresh reactive powder concrete paste, the properties of this admixture satisfy the requirements of ASTM C494[14].
6- Steel fibers: Micro steel fiber 15 mm in length, 0.2 mm in diameter, aspect ratio of 75 and tensile strength of 2850 MPa, is used in four different volumetric ratios varying from 0% to 2%.

### 4.2 Concrete: mix design parameters, casting and curing

In this work six mixes of reactive powder concrete has been made, the content of cement and fine quartz sand is constant for all mixes with 900 kg/m3 and 990 kg/m$^3$ respectively. Silica fume is used as a weight percentage of cement in four different ratios (12%, 15%, 20% and 25%). Also, four volumetric ratios (Vf) of the steel fibers (0%, 1%, 1.5% and 2%) are used. The content of water represented by w/c ratio varies from 16% to 22%. A steady content of 6% for ViscoCrete-5930 superplasticizer is used for all mixes.

Cylindrical specimens with diameter of 150 mm and height of 300 mm are made cured and tested to find the compressive strength, the modulus of elasticity and the splitting tensile strength of reactive powder concrete mixes, and all specimens are cured in tap water for 28 days.

### 4.3 Experimental evaluation of compressive strength, modulus of elasticity and splitting tensile strength

The compressive strength and splitting tensile strength of reactive powder concrete were evaluated by tests carried out on cylinders (150mm×300mm), these tests are performed according to ASTM C39[15] and ASTM C496[16] respectively. All cylindrical specimens were loaded up to failure by using hydraulic digital test machine with maximum capacity of 4000 kN.

To estimate the static modulus of elasticity ASTM C469[17] specification was adopted, a percentage of 40% of the predicted ultimate concrete compressive strength was applied during the test of the cylindrical specimens of (150mm × 300mm).

## 5 Tests results and discussion

Several tests were carried out on the cylindrical specimens to find the compressive strength, splitting tensile strength and the modulus of elasticity where the following results are found:

- Compression tests were carried out on cylinders at age of 28 days to estimate the compressive strength of different mixes made with reactive powder concrete, the test results show that the content of silica fume and the w/c ratio have major effects on the compressive strength, and in less manner steel fibers content. Six mixes with three compressive strength targets 70MPa, 80MPa and 90MPa were examined to evaluate the influence of changes in silica fume and steel fibers content as well as w/c ratio on the reactive powder compressive strength. As the silica fume content increased from 20% to 25% (mixes M1-70 and M2-80) the concrete strength increased from 81.292 MPa to 84.3 MPa for constant w/c ratio of 22%. While, in order to get a target compressive strength of 90MPa (mix M3-90) w/c was decreased to 20% which produced the compressive strength of 99.5 MPa. These results indicate that the silica fume and water content share the greatest effects on reaching compressive strength targets. A lesser effect of steel fibers can be noticed, as the steel fibers content increased from 1% to 1.5% and 2% silica fume content can be decreased from 20% to 15% and 12% to keep the strength in the designed target range. The mix M4-0 with no steel fiber has high silica fume content of 25% and a less w/c ratio is used to get a compressive strength of 76.682 MPa. Table (1) shows the results of mechanical properties for all mixes experimentally examined in this work and Fig. (1) shows the cylindrical specimens during and after the test. The effect of silica fume and steel fibers on reactive powder properties were examined, and the gain of silica fume can be noticed not only by its pozzolanic effects but by increasing mix density and enhancing the bond strength between the fine materials and the steel fibers[18].

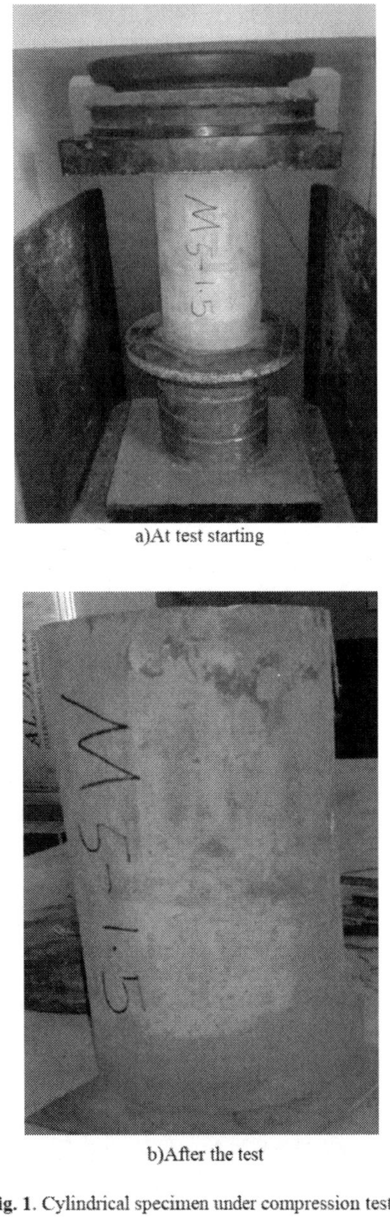


a)At test starting

b)After the test

**Fig. 1.** Cylindrical specimen under compression test

- The splitting tensile strength is highly affected by the changes in the volumetric ratio of steel fibers, as the steel fibers content increases from 0% to 1%, 1.5% and 2% the splitting tensile strength increased from 4.033 MPa to 9.609 MPa, 11.944 MPa and 12.896 MPa respectively. This enhancement in tensile strength with the increase in steel fibers content occurred because steel fibers resist and delay the formation of the cracks, and permit the transmission of stresses through cracks[19]. On the other hand, the changes in silica fume content has a slight effect on the splitting tensile strength, see Table (1). Fig.(2) shows the failure mode for specimen with and without steel fibers respectively.
- The increase in silica fume from 12% to 15%, 20% and 25% provide some noticeable enhancement in modulus of elasticity production, as shown in Table(1).
- The relationship between modulus of elastic and steel fibers content is not clear. Where the increase in steel fibers show a little improvement in the modulus of elasticity (for example mix M4-0 through M6-2) this can be explained from the procedure used in modulus of elasticity predication itself as the percentage of 40% from the ultimate concrete compressive strength is adopted, in most cases the estimation of modulus of elasticity at this ratio ends before the crack stage, while steel fibers act when cracks start to form and beyond[10].

a) With steel fibers

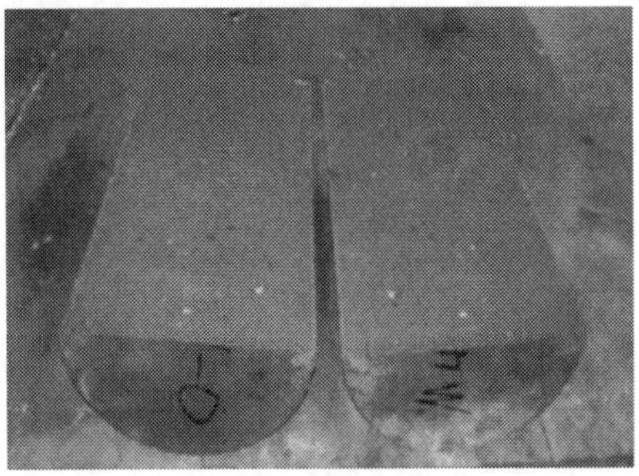

b) Without steel fibers

**Fig. 2** .Splitting tensile strength failure mode

## 6 Data analysis and proposed equations

In this work two equations are proposed to predict the modulus of elasticity and the splitting tensile strength based on data obtained experimentally from this work and data provided from previous works by applying regression analysis:

a) **Static modulus of elasticity:** As the modulus of elasticity mainly depends on the compressive strength, a regression analysis was applied on the basic form Eq.(1) based on the available data from this work and other previous works Table(1) and Table(2) respectively which is made of reactive powder concrete, the proposed equation (Eq.(7)) was found to be more suitable and representative for reactive powder concrete behavior and try to be more reflective of its components properties and adopted a new power for the compressive strength of 0.25.

$$E_c = 14000\ (f_c')^{0.25} \tag{7}$$

The coefficient of variation (COV) used to make a comparison of results, where the proposed equation gives a COV of 10.677 % and a mean value of 1.001, which can be considered better than other approaches discussed in this work, Table(3) show the modulus of elasticity results for all methods studied in this work. Also, Fig.(3) illustrates the relationship between the modulus of elasticity experimentally estimated to the calculated modulus of elasticity (Ec test/Ec calculated) and the concrete compressive strength (MPa). Fig.(4) shows a comparison between the three approaches adopted in this work, where the proposed equation shows the best convergence of the test to predicted modulus of elasticity ratio.

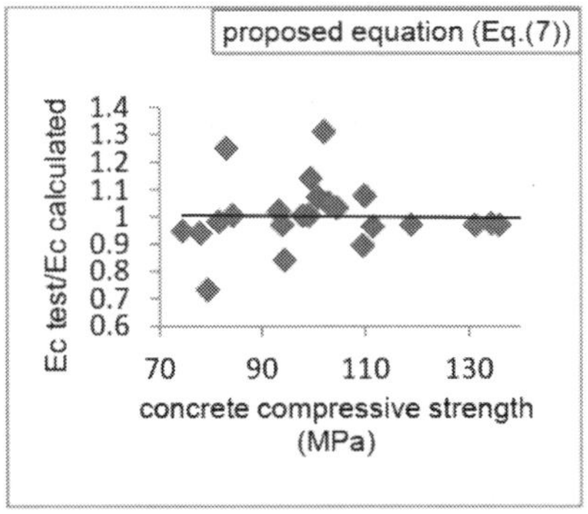


a) Proposed equation (Eq.7)

 https://doi.org/10.1051/matecconf/201816204004

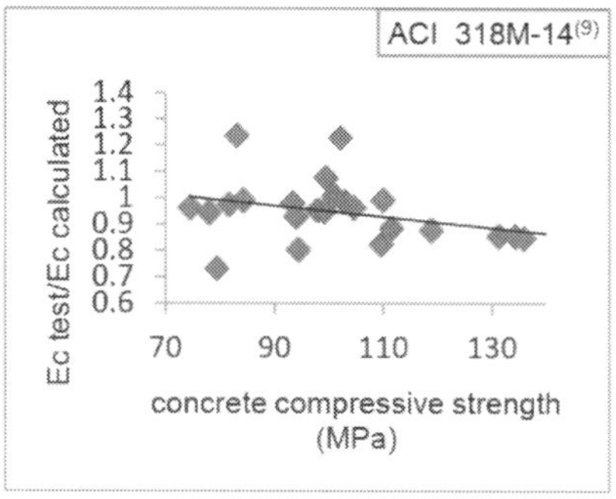


b) ACI 318M – 14

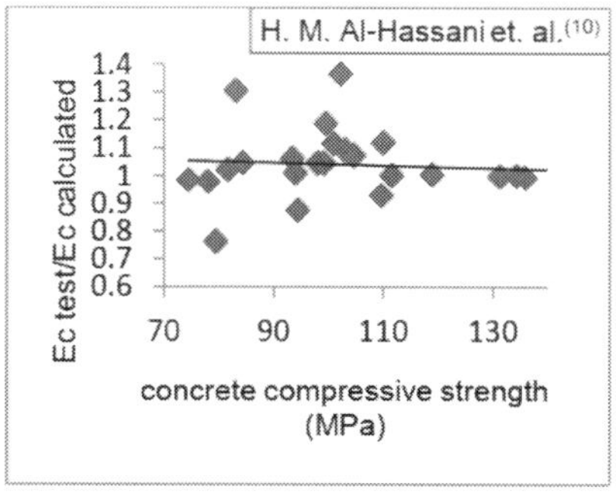


**c)** H. M AL-Hassani et. al. equation

**Fig. 3**. Relationship between (Ec test/Ec calculated) and concrete compressive strength (MPa)

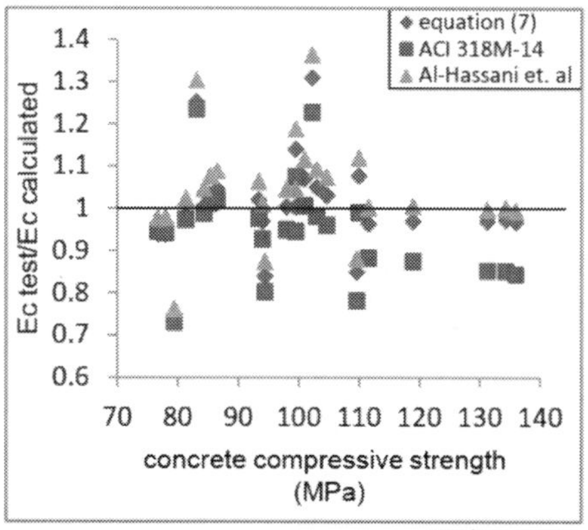


**Fig. 4**. Comparison of (Ec test/Ec calculated) for all approaches

## b) Splitting tensile strength:

In the same way as in the modulus of elasticity estimation, a new formula for splitting tensile strength of reactive powder concrete is derived Eq. (8):

$$f_{ct} = 0.1\,(f_c')^{0.85} \tag{8}$$

Besides the new power of compressive strength, and due to the vital role of steel fibers presence in strengtheingn the reactive powder concrete against the tensile stresses a new term based on the content of the steel fibers is added as in Eq.(9).

$$f_{ct} = 0.1\,(f_c')^{0.85} + 3.6\,V_f \tag{9}$$

The proposed equation gives a COV of 10.455 % and a mean value of 1.204, which is considered much better than the ACI-318M -14 with COV of 26.206 and Arigula equation with COV of 25.828%. Table(4) shows the splitting tensile strength for all approaches and Fig.(5) illustrates the relationship between the ratio of splitting tensile strength taken from test results to the calculated ones and the steel fibers content (Vf %). From this figure, a straight line is provided by the proposed method (Eq.(9)), while the other two methods tend to be highly unconservative as the steel fibers content increased, this behavior occurred due to the negligence of the effects of steel fibers by the other methods. Fig.(6) shows the comparison of the ratio of ($f_{ct}$ test/$f_{ct}$ calculated) for the proposed equation (eq.(9)) and the other methods that discussed in this work, equation (9) shows a good agreement between the experimental and theoretical splitting tensile strength value.

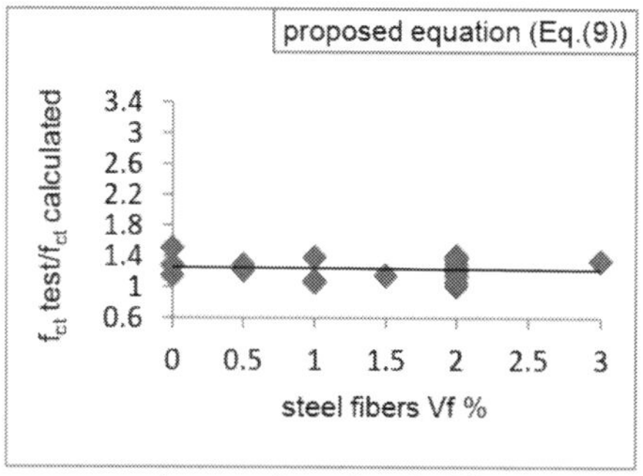


a) proposed equation (Eq.(9))

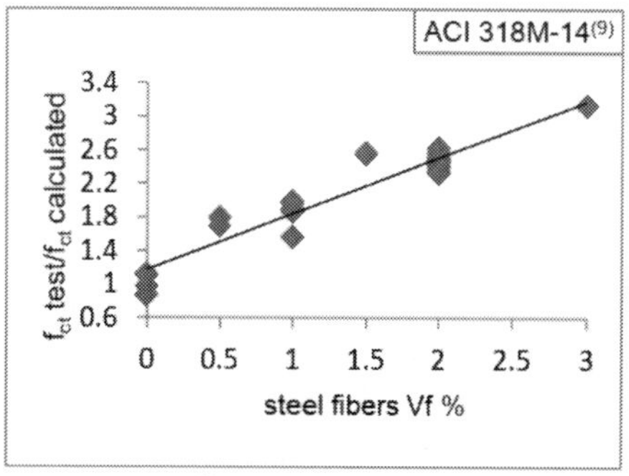


b) ACI 318M – 14

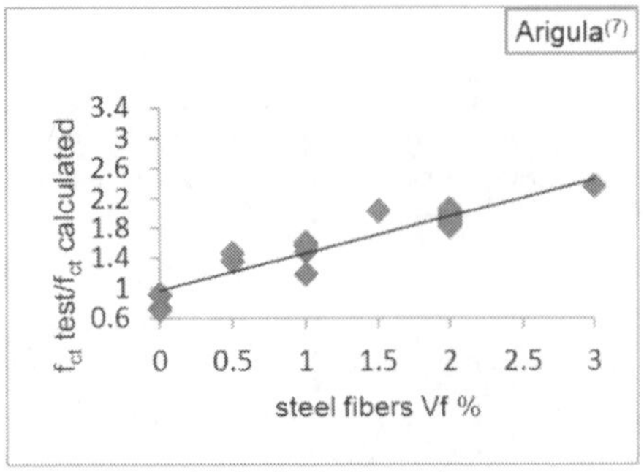


c) Arigula et. al. equation

**Fig. 5 .** Relationship between (fct test/fct calculated) and steel fibers content (MPa)

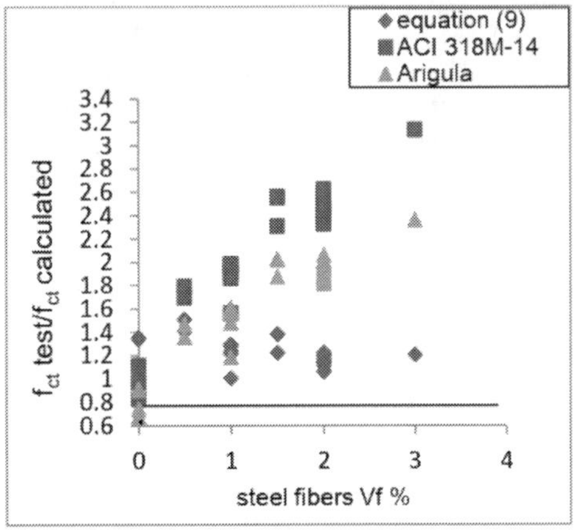


**Fig. 6.** Comparison of ($f_{ct}$ test/$f_{ct}$ calculated) for all approaches

## 7 Conclusions:

From the above results the following facts can be concluded:

1. The splitting tensile strength is highly affected by the presences of steel fibers, the experimental results show that the increase in steel fibers from 0% to 1%, 1.5% and 2% lead to increase in splitting tensile strength from 4.033MPa to 9.609 MPa, 11.944 MPa and 12.896 MPa respectively.
2. Fewer effects on the splitting tensile strength can be noticed as the compressive strength changes with a constant ratio of steel fibers, as the compressive strength increases from 81.292 MPa to 84.3 MPa and 99.5 MPa the splitting tensile strength slightly increases from 9.609 MPa, 10.148 MPa and 10.362 MPa respectively.
3. The proposed equation (Eq. (9)) that was used to predict the splitting tensile strength of reactive powder concrete focused on the effects of steel fibers. This equation shows a good agreement with the experimental test results. The COV provided by this equation is 10.398% and less than the other two approaches examined in this work (ACI-14 and Arigula) with COV of 26.944% and 25.828% respectively.
4. The modulus of elasticity slightly affected by the presence of steel fibers, this behavior probably occurred due to the adopted test procedure where 40% from the ultimate concrete compressive strength is adopted to estimate the modulus of elasticity.
5. Generally, the strength of reactive powder is mainly dependent on the content of steel fibers, silica fume, w/c ration and the finesses of its contents.

## References:

1. B. Haranki, MS.c. thesis, University of florida, (2009)
2. T. Noguchi, F. Tomosawa, K. M. Nemati, B. M. Chiaia, A. P. Fantilli, ACI structural Journal **106**, 5, 690-696 September-October (2009)
3. B. Bae, H. Choi, B. Lee, C. Bang, Advances in Materials Science and Engineering Journal, (2016)
4. H. Yazici, M. Yardimci, S. Aydin A. S. Karabulut, Construction and Building Materials **23**, 1223–1231 (2009)
5. F. Tomosawa and T. Noguchi, Journal of Structural and Construction Engineering **60**, 474, 1-10 (1995)
6. K. A. Gajendran, R. Anuradha G. S. Venkatasubramani, ARPN Journal of Engineering and Applied Sciences **10**, 14, (August 2015)
7. N. Arioglu, Z. C. Girgin, E. Arioglu, ACI Materials Journal **103**, 1, (January-February 2006)
8. M. K. Maroliya, International Journal of Engineering Research and Development **4**, Issue 4, 58-61 (October 2012)
9. ACI Committee 318," Building Code Requirements for Reinforced Concrete and Commentary -ACI 318RM-14", American Concrete Institute, Detroit, pp. 432, 2014.
10. H. M. Al-Hassani, W. I. Khalil, L. S. Danha, Eng. &Tech. Journal **31** Part (A), 16 (2013)
11. Iraqi Specification, No. 5/1984, "Portland Cement", Ministry of Planning, Central Organization for Standardization and Quality Control.
12. Iraqi Specification, No.45/1984, "Aggregate from Natural Sources for Concrete and Construction", Ministry of Planning, Central Organization for Standardization and Quality Control.
13. ASTM C 1240-05, "Standard Specification for the Use of Silica Fume as a Mineral Admixture in Hydraulic Cement Concrete, Mortar, and Grout", Vol. 04.02, 1-7 (2005)
14. ASTM C 494 /C 494M-05a, "Standard Specification for Chemical Admixtures for Concrete", Vol. 04.02, 1-11 (2005)
15. ASTM C39/C39M- 05, "Standard Test Method for Compressive Strength of Cylindrical Test Specimens", Vol. 04.02, 1-9 (2005)
16. ASTM C496/C496M-04, "Standard Test Method for Splitting Tensile Strength of Cylindrical Concrete

Specimens", Vol. 04.02, 1-5 (2005)
17. ASTM C469-02, "Standard Test Method for Static Modulus of Elasticity and Poisson's Ratio of Concrete in Compression", Vol. 04.02, 1-7 (2005)
18. A.M. Neville, Pearson Education Limited, Fourth and Final Edition, p.844 (2005)
19. S.P. Yaakoub, Ph.D. thesis, University of Technology, Baghdad, 2010.
20. M. M. Saeed., Ph.D. thesis, University of Technology, Baghdad, 2010.

**Table 1.** Test results of mechanical properties experimentally examined in this work

| NO. | Mix | Steel fibers $V_f$ | Silica fume | w/c ratio | Compressive strength ($f_c'$) MPa | splitting tensile strength($f_{ct}$) MPa | modulus of elasticity (Ec) MPa |
|---|---|---|---|---|---|---|---|
| 1 | M1-70 | 1% | 20% | 0.22 | 73.214 | 9.609 | 41286.951 |
| 2 | M2-80 | 1% | 25% | 0.22 | 84.301 | 10.148 | 42617.255 |
| 3 | M3-90 | 1% | 20% | 0.20 | 96.500 | 10.362 | 44327.260 |
| 4 | M4-0 | 0% | 25% | 0.16 | 72.462 | 4.033 | 38924.274 |
| 5 | M5-1.5 | 1.5% | 15% | 0.22 | 73.936 | 11.944 | 43924.827 |
| 6 | M6-2 | 2% | 12% | 0.22 | 74.804 | 12.896 | 44568.264 |

**Table 2.** Mechanical properties available from previous works

| No. | Mix | Silica fume % | Steel fibers ($V_f$) % | Compressive strength ($f_c'$) MPa | Splitting tensile strength MPa | modulus of elasticity MPa | Ref. No. |
|---|---|---|---|---|---|---|---|
| 1 | MSP-N | 15 | 2 | 118.91 | 14.78 | 44841 | 10 |
| 2 | MSF0 | 0 | 2 | 111.6 | 15.19 | 43836 | 10 |
| 3 | MSF10 | 10 | 2 | 131.27 | 15.4 | 45900 | 10 |
| 4 | MSF15 | 15 | 2 | 134.33 | 15.9 | 46398 | 10 |
| 5 | MSF20 | 20 | 2 | 146.19 | 16.23 | 47422 | 10 |
| 6 | MSF25 | 25 | 2 | 149.39 | 17.25 | 48295 | 10 |
| 7 | MSF30 | 30 | 2 | 157.48 | 17.55 | 49103 | 10 |
| 8 | MFR0 | 25 | 0 | 135.93 | 6.32 | 46262 | 10 |
| 9 | MFR1 | 25 | 1 | 144.57 | 10.5 | 47363 | 10 |
| 10 | MFR3 | 25 | 3 | 151.62 | 21.59 | 48538 | 10 |
| 11 | M0.0-15 | 15 | 0 | 78 | 5.5 | 39100 | 20 |
| 12 | M0.5-15 | 15 | 0.5 | 94 | 9.2 | 42250 | 20 |
| 13 | M1.0-15 | 15 | 1 | 98 | 11 | 44200 | 20 |
| 14 | M1.5-15 | 15 | 1.5 | 103 | 14.5 | 46800 | 20 |
| 15 | M2.0-15 | 15 | 2 | 110 | 15.4 | 48800 | 20 |
| 16 | M2.0-10 | 10 | 2 | 101 | 14 | 47500 | 20 |
| 17 | M2.0-17 | 5 | 2 | 93.4 | 12.7 | 44370 | 20 |
| 18 | M0 | 15 | 0 | 79.42 | 5.5 | 30578 | 19 |
| 19 | M0.5 | 15 | 0.5 | 83.11 | 9.16 | 52885 | 19 |
| 20 | M1 | 15 | 1 | 99.53 | 11.01 | 50379 | 19 |
| 21 | M1.5 | 15 | 1.5 | 102.2 | 14.49 | 58265 | 19 |
| 22 | M2 | 15 | 2 | 109.65 | 14.44 | 38432 | 19 |
| 23 | Msf10 | 10 | 2 | 104.66 | 14.14 | 46171 | 19 |
| 24 | Msf5 | 5 | 2 | 94.41 | 12.66 | 36618 | 19 |

**Table 3.** Modulus of elasticity calculated by several methods

| Method | COV. % | Ec test/Ec calculated mean (average) | Min. | Max. | Max./Min. |
|---|---|---|---|---|---|
| Proposed equation Eq.(7) | 10.677 | 1.001 | 0.732 | 1.309 | 1.788 |
| ACI 318M-14[(9)] | 12.419 | 0.932 | 0.73 | 1.234 | 1.690 |
| H. M. Al-Hassani[(10)] | 10.904 | 1.038 | 0.762 | 1.364 | 1.790 |

**Table 4.** Splitting tensile strength calculated by several methods

| Method | COV. % | $f_{tsp}$ test/$f_{tsp}$ calculated mean (average) | Min. | Max. | Max./Min. |
|---|---|---|---|---|---|
| Proposed equation Eq.(9) | 10.455 | 1.204 | 0.971 | 1.506 | 1.551 |
| ACI 318M-14[(9)] | 26.206 | 2.125 | 0.867 | 3.131 | 3.611 |
| Arigula[(7)] | 25.266 | 1.673 | 0.716 | 2.359 | 3.295 |

# Influence of ties on the behavior of short reinforced concrete columns strengthened by external CFRP

*Kaiss* Sarsam [1,*], *Raid* Khalel [1], and *Mohammed* Hadi [1]

[1]Building and Construction Engineering Department, University of Technology, Baghdad, Iraq

**Abstract.** An experimental study was carried out to investigate the behavior of normal strength reinforce concrete (RC) circular short column strengthned with "carbon fiber reinforced polymer (CFRP) sheets". Three series comprising totally of (15) specimens loaded until failure under concentric compresion load. Strengthening was varied by changing the number of CFRP strips, spacing and wrapping methods. The findings of this research can be summarized as follows: for the columns without CFRP, the influence of the tie spacing was significant: compared with 130 mm tie spacing, dropping the spacing to 100 mm and 70 mm increased the load carrying capacity by 18% and 26%, respectively. The columns with less internal confinement (lesser amount of ties) were strengthened more significantly by the CFRP than the ones with greater amount of internal ties. As an example of the varying effectiveness of the fully wrapped CFRP, the column with ties at 130 mm was strengthened by 90% with the CFRP. In contrast, the ones with 70 mm spaced ties only increased in strength with CFRP by 66%. Compared with the control specimen (no CFRP), the same amount of CFRP when used as hoop strips led to more strengthening than using CFRP as a spiral strip- the former led to nearly 9% more strengthening than the latter in the case of 130 mm spaced internal steel ties. In the case of 100 mm internal steel ties, the difference (between the hoops & spiral CFRP strengthening) is close to 4%. In contrast, there is no difference between the two methods of strengthening in the heavily tied columns (70 mm tied spacing).

## 1 Introduction

Strengthening of RC structures using CFRP has become a popular technology over the last few years. One of the substantial applications of CFRP-strengthening technology is to enhance the load carrying capacity of RC columns throughout the provision of confining CFRP wraps. The column wrapping technique is especially effective for circular columns because the strength and ductility of concrete in a circular section can be basically increased through lateral confinement [1]. The objectives considered in this study are: (1) the effect of the ties on the load carrying capacity of short RC columns represented as a volumetric ratio of transverse reinforcement; (2) the effect of the external CFRP on the load carrying capacity of short RC columns.

## 2 Experimental work

Through the design of the experimental program, the variables included in this study are focused mainly on the
internal tie spacing, CFRP strips number, spacing and wrapping methods. Detailed description of each variable is presented below.

### 2.1 Specimen's description

A total of 15 RC specimens were designed with a circular section of 150 mm diameter and overall height of 600 mm. 15 mm concrete cover was used in all confined specemens and between the ends of longitudinal steel bar and the bottom and top surface of the column specimens to preclude direct loading on the steel bars.

The column specimens were divided into three groups depending on the spacing of the ties reinforcement, each five specimens were reinforced with (6) longitudinal bars with diameter of (6)mm and of (4) mm diameter for the lateral ties reinforcement with 130, 100 and 70 mm tie spacing respectively. Fig (1) shows the geometry of the Tied column specimens.

* Corresponding author: kaisssarsam@yahoo.com

© The Authors, published by EDP Sciences. This is an open access article distributed under the terms of the Creative Commons Attribution License 4.0 (http://creativecommons.org/licenses/by/4.0/).

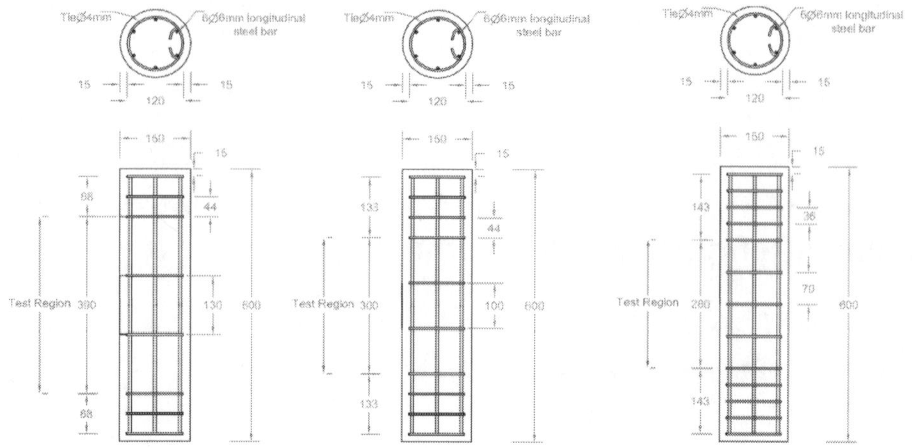


(All dimensions in mm)

(a) 130mm tie spacing column | (b) 100mm tie spacing column | (c) 70mm tie spacing column

**Fig .1.** The geometry of the tied column specimens.

### 2.2 Specimen's identification and strengthening schemes

Strengthening schemes were chosen accurately based on the practiced needs and the field condition. In each specimen group (five columns), the first column specimen (type A) is control column. The second column specimen (type B) is strengthened by 25 mm width CFRP strips
with hoop spacing of 115 mm c/c (CFRP ratio= 25%), the third column specimen (type C) is strengthened by 25 mm width CFRP strip as spiral with pitch of 160 mm (CFRP ratio= 25%), the fourth column specimen (type D) is strengthened by 25 mm width CFRP strips with hoop spacing of 58 mm c/c (CFRP ratio =50%) and the fifth column specimen (type E) is strengthened with full CFRP wrap (CFRP ratio=100%), as shown in Fig (2).

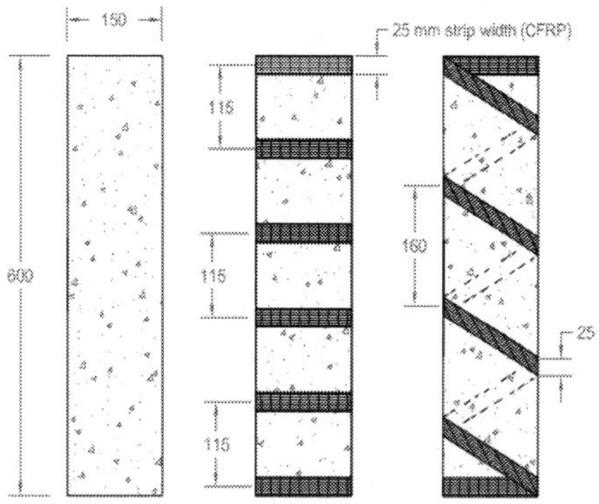


(a) Type A | (b) Type B | (c) Type C

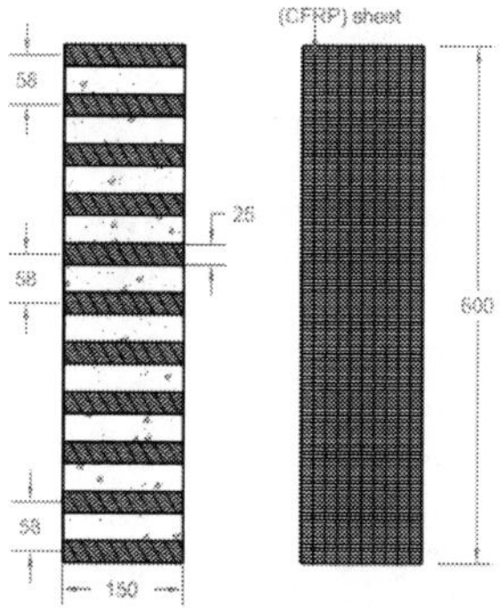


(All dimensions in mm)

(d) Type D | (e) Type E

**Fig .2.** Types of CFRP confinement.

Column specimens are identified with a series of letters and numbers, which refer to the tied columns by letter T. The numbers (70, 100 and 130) represent the lateral ties spacing. While the letters (A, B, C, D and E) refer to the type of column's classification according to the external confinement by CFRP as in Fig.2, Fig. 3. shows the key for the column specimens.

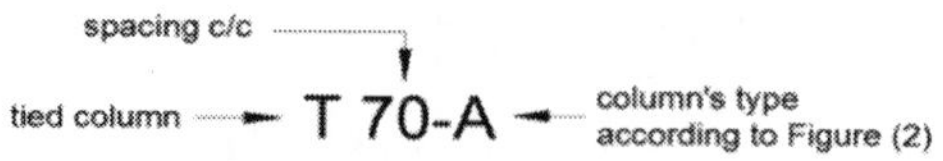


**Fig .3.** Column specimen's key.

### 2.3 Construction Materials

*2.3.1 Cement*

The used cement in this research is (Type I) ordinary Portland cement; Iraqi made with trade mark (Al-Mass). The physical and the chemical analysis properties are in agreement with the Iraqi specification No.5/1984 [2].

*2.3.2 Fine aggrgate*

Al Ukhader natural sand, maximum size 4.75mm was used as fine aggregate through-out this research work. The specific gravity, fineness modulus, sulfate content and absorption of fine aggregate are within the requirements of Iraqi specification No.45/1984 [3].

*2.3.3 Coarse aggrgate*

Natural crushed gravel with maximum size of (10mm) is used throughout this work. The aggregate was washed, stored in air to dry the surface, and then used as saturated surface dry condition. The grading of the coarse aggregate is as per the Iraqi specification No.45/1984[3].

*2.3.4 Water*

Water supplied by the network system (Tap Water) was used for mixing and curing the concrete.

*2.3.5 Reinforcing steel bars*

A 6mm diameter deformed bars were used for longitudinal reinforcement with 513 MPa yield strength and 4 mm diameter deformed bars with 717 MPa yield strength were used for the transverse reinforcement.

*2.3.6 Carbon fiber reinforced polymer (CFRP)*

Sika-Wrap®-300C woven carbon fiber fabrics (mid strength type fiber) was applied as strengthening agent in this work. The roll of carbon fiber is 50cm width and 50 m length, as reported by the manufacturer. This system was supplied by (Sika Switzerland). Table 1 shows the product description of the (CFRP) from Sika-Wrap®-300C.

**Table1.** The product description for the Sika-Wrap®-300C woven carbon fibre fabric.

| | |
|---|---|
| Areal weight | 300g/m$^2$ ± 15 g/m$^2$ |
| Fabric design thickness | 0.166 mm |
| Tensile E-modulus | 230 000 N/mm$^2$ |
| Tensile strength | 3900 N/mm$^2$ |
| Elongation at break | 1.5% (nominal) |

*2.3.7 Bonding materials*

Sikadur®-330 was used as recommended by SikaWrap®-300C Woven carbon fiber fabric manufacturer to bond CFRP to the concrete. Table 2 presents the product description for the Sikadur®-330

**Table 2.** The product description for the Sikadur®-330.

| | |
|---|---|
| Appearance /colors | Part A:white<br>Part B:grey<br>Parts A+B mixed=light grey |
| Mixing ratio | Part A : part B = 4 : 1 by weight |
| Density | 1.30 kg/l + 0.1 kg/l (parts A+B mixed) |
| Tensile strength | 30 N/mm$^2$ (7 days at +23°C) |
| Tensile E-modulus | Flexural: 3800 N/mm$^2$ (7 days at +23°C)<br>Tensile: 4500 N/mm$^2$ (7 days at +23°C) |
| Elongation at break | 0.9% (7 days at +23°C) |

## 2.4 Trial Mixes of Concrete

Several trial mixes were made aiming at cylinder strength of (30MPa). Trial mix No (4) was adopted for this study. Table 3 represents details for trial concrete mixes.

**Table3.** Details for the trial concrete mixes.

| Trial No. | Cement (kg/m$^3$) | Sand (kg/m$^3$) | Gravel (kg/m$^3$) | Water (kg/m$^3$) | W/C Ratio | Mix ratio by weight | Compressive strength (MPa) 28 Day |
|---|---|---|---|---|---|---|---|
| 1 | 330 | 780 | 922 | 205 | 0.62 | (1:2.36:2.79) | 17 |
| 2 | 350 | 710 | 1025 | 190 | 0.54 | (1:2:2.93) | 25 |
| 3 | 370 | 685 | 1020 | 190 | 0.51 | (1:1.85:2.76) | 27.3 |
| 4 | 400 | 740 | 1000 | 192 | 0.48 | (1:1.85:2.5) | 31 |

## 2.5 Concrete Mixes

The concrete is mixed by applying a horizontal rotary mixer with a capacity of (0.1m$^3$). The dry components of cement, sand and gravel are mixed for 4 to 5 minutes, then the water is added to the mixture in quarters and mixed for 2 to 3 minutes.

## 2.6 Specimen's molds

PVC tubes are cut to match the column height. The base of the mold is constructed from plywood, to prevent water from leaking through.

## 2.7 Casting and curing

In total, three batches of concrete were used to cast the columns. Each batch was used to cast five columns with six cylinders (100 ×200) mm [4] to determine the compressive strength of concrete at age of 7 days and age of 28 days, respectively. And six cubes (100 ×100 ×100) mm [5], were used to calculate the compressive strength of concrete at age of 7 days and 28 days for comparison with the cylinder compressive strength. Three cylinders (100×200) mm [6], were used for splitting tensile strength test. Three prisms (100 ×100 ×400) mm[7], were used for flexural strength test(modulus of rupture). In this study the column specimens were cast vertically and vibrated with two stages, by using the electric table vibrator to consolidate the concrete and to remove the air bubbles.

After 24 hours, control specimens are stripped from the molds and cured in water tanks for 28 days as long with the columns.

## 2.8 Preparation for column specimens

The column specimens were taken out of the curing tanks and left to dry for one day. All the specimen's surfaces were grinded by applying the grinder machine in order to eliminate all weak and losses particles. Later the specimens were wash again and left to dry for one or two days.

## 2.9 Installation of CFRP

After drying the column specimen's surface, the installation procedure initially started, by planning the locations of the CFRP strips to ensure that the spacing is distributed along the column according to the design with the CFRP being cut as required. An epoxy resin is prepared by mixing one part type B (hardener) and four parts type A (resin) by weight. A mixing machine is used to obtain a homogeneous light grey color. 1mm thickness of epoxy resin is applied to the concrete surface, while at the same time the carbon fiber sheets were fully saturated with that resin. Then the CFRP sheet is installed to that concrete surface and pressed with a roll in the direction of the CFRP fibers to eliminate the entrapped air, another layer of epoxy was used to coat the CFRP. According to the recommendation of the manufacturer, the overlapping of around 100mm of CFRP is sufficient to provide full strength of CFRP and to prevent de-bonding failure during the test. Finally after 7-days at laboratory temperature, the column specimens will be ready for testing.

## 2.10 RC Column specimens test

A 2500 kN capacity compression testing machine (AVERY) located at structural laboratory of the department is used to apply monotonically compression load to the column specimens. A total of two dial gauges and 8 demec points, 4 in each opposite face were used for each specimen. By using a dial gauge located at the bottom surface of the testing machine to measure the column gross axial shortening. Another gauge is located at the mid height of the column specimens to measure the lateral displacement. Two demec points were stuck in the mid-height of the column to measure the longitudinal compresive strains from two opposite direction of the columns. To measure the lateral strains at two opposite directions of the column, two other demec points are mounted at the mid-height through the column cross section (horizontal axis). Both ends of column specimens were confined by steel collars 50 mm height and 10 mm thick as shown in Fig.4.

All the column specimens were tested under constant loading at an average rate of 20 kN/sec. from the beginning of test up to failure. For each increment of loading dial gauge readings were recorded and longitudinal strain and lateral strain were recorded until the failure of column.

**Fig. 4.** Locations of dial gauges and demec points for the column specimen.

## 2.11 Control specimens

The control specimens were cast from the same concrete batch used for casting the columns. The control specimens were tested immediately after the columns test. Table 4 shows the test results for the compressive strength of concrete cylinders and cubes and for three concrete batches.

**Table 4.** Mechanical properties for the column specimens

 https://doi.org/10.1051/matecconf/201816204005

| Batch No. | column specimen | $f'_c$ (MPa) | $f_{cu}$ (MPa) | $f'_c/f_{cu}$ | Splitting tensile strength $f'_{sp}$(MPa) | Calculated splitting tensile strength according to ACI-Code (MPa) * | Modulus of rupture $f_r$ (MPa) | Calculated modulus of rupture according to ACI-Code (MPa)** |
|---|---|---|---|---|---|---|---|---|
| 1 | T70 | 30.05 | 36.42 | 0.825 | 2.8 | 3.07 | 3.9 | 3.4 |
| 2 | T100 | 30.12 | 36.25 | 0.831 | 2.9 | 3.07 | 4 | 3.4 |
| 3 | T130 | 29.89 | 36.32 | 0.823 | 2.7 | 3.06 | 3.8 | 3.39 |

* $0.56\sqrt{f'_c}$
** $0.62\sqrt{f'_c}$

## 3 Test results and discussion

### 3.1 Experimental Axial Load Capacity for the Tested RC Column Specimens.

Table 5 represents the experimental ultimate load capacity for the tied columns.

**Table 5.** The experimental ultimate load capacity for tied column specimens.

| Column designation | Capacity of tested columns (kN) | Column designation | Capacity of tested columns (kN) | Column designation | Capacity of tested columns (kN) |
|---|---|---|---|---|---|
| T130-A | 495 | T100-A | 585 | T70-A | 625 |
| T130-B | 620 | T100-B | 730 | T70-B | 745 |
| T130-C | 610 | T100-C | 720 | T70-C | 745 |
| T130-D | 735 | T100-D | 855 | T70-D | 869 |
| T130-E | 940 | T100-E | 1025 | T70-E | 1040 |

From the experimental results, it can be seen that the strengthening of RC columns by CFRP strips is significantly effective in increasing the ultimate load capacity. The increases were (25, 48 and 90) % for 130 mm ties spacing strengthened with CFRP ratios of (25, 50 and 100) %, respectively. The corresponding values of increase for 100 mm and 70 mm ties spacing were (25, 46 and 75)% and (19, 39 and 66) % respectively, for the same CFRP ratios.

### 3.2 Failure mode

All specimens were tested until failure. The reference columns showed linear behavior initially, the first crack initiated through the mid height (Test Region) at the respective 55% of its ultimate load. After further loading, the cracks were increased in numbers and width. Finally, the columns failed at mid height by crushing and spalling of the concrete accompanied by local buckling for longitudinal steel bars as shown in Fig 5.

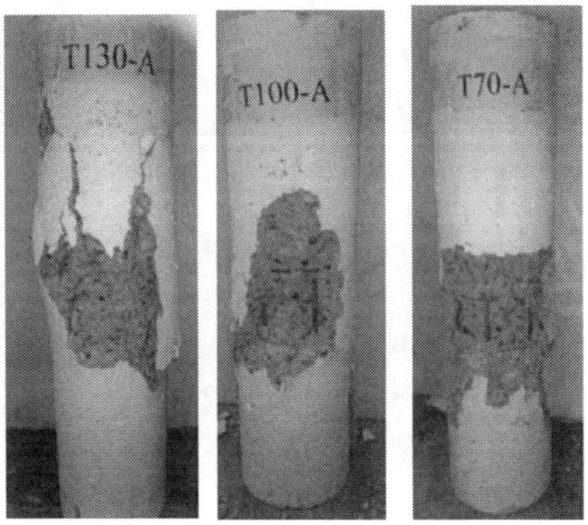


(a) (b) (c)

**Fig 5.** Failure modes for control columns

(a) (b) (c)

**Fig 6.** Failure modes for columns type (B)

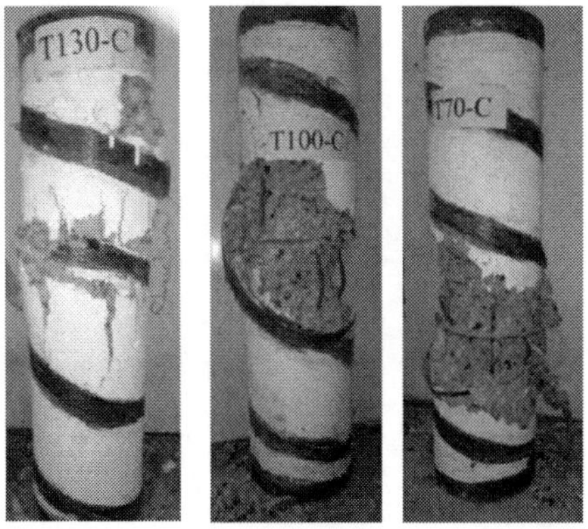


(a) (b) (c)

**Fig 7.** Failure modes for columns type (C)

(a) (b) (c)

**Fig 8.** Failure modes for columns type (D)

For the columns confined with CFRP strips, the first crack was at the mid height for unbounded concrete surface between the CFRP strips. At higher loading the cracks developed further. Finally, the columns failed by rupture of fiber followed by crushing of concrete and buckling for one or more longitudinal steel bar. It was an explosive failure. Figs. (6 to 8) show the failure patterns due to the rupture of CFRP strips accompanied by local buckling between the steel ties.

Columns with fully wrapped CFRP exhibited no visible cracking. Fig. 9 shows the failure mode of columns type (E). Because of Poisson effects of confinement, failure occurred with cracking sound of CFRP-rupture happening at column mid height.

(a) (b) (c)

**Fig 9.** Failure modes for columns type (E)

### 3.3 Load-strain behavior of column specimens

Fig.10. shows that for unconfined CFRP columns the influence of the tie spacing was significant: compared with 130 mm tie spacing, dropping the spacing to 100 mm and 70 mm, the effect of the internal confinement was clear-with the load carrying capacity rising with confinement.

Figs. (11 to 13). show the effect of CFRP confinement for tied columns; compared with the unconfined column the increase in strengthening by the CFRP was more significant in raising load carrying capacity of the column specimens.

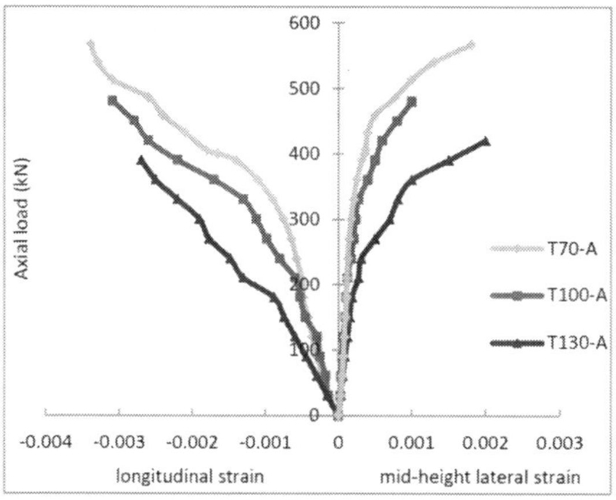


**Fig 10.** Load- strain curves for unconfined CFRP tied columns.

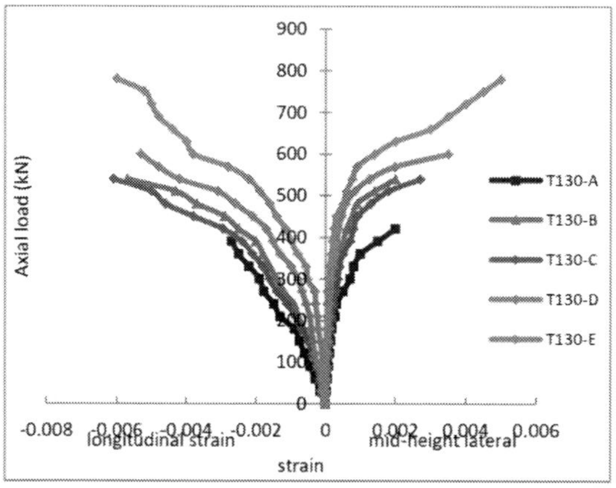


**Fig.11.** Load- strain curves for confined CFRP tied columns-T130.

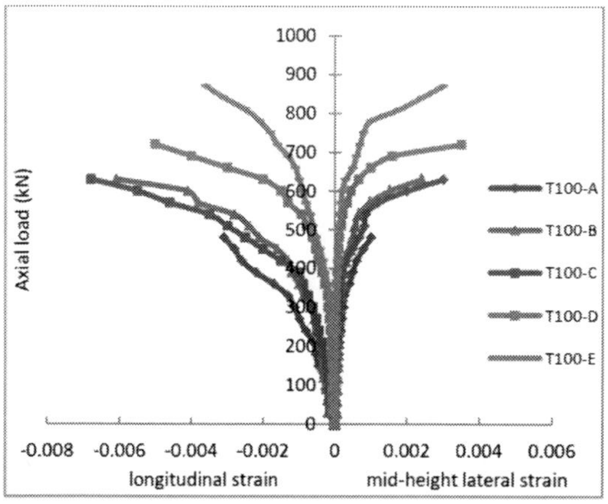


**Fig.12.** Load- strain curves for confined CFRP tied columns-T100

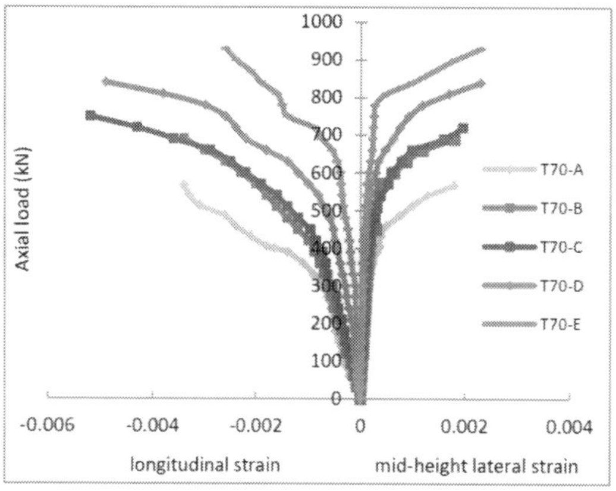


**Fig.13.** Load- strain curves for confined CFRP tied columns-T70.

### 3.4 Load-displacement behavior of column specimens

Fig.14. illustrates the load-longitudinal displacement behavior of unconfined columns (no CFRP) on load carrying capacity. There is a significant differience between these un-confined columns. The column with less internal confinement (lesser amount of ties) shows larger longitudinal displacement than the other columns. This in turn affects the decrease of load carrying capacity of that column. On the other hand there is a reduction of lateral displacement with increase of the internal confinement (increased number of ties) as shown in Fig.15.

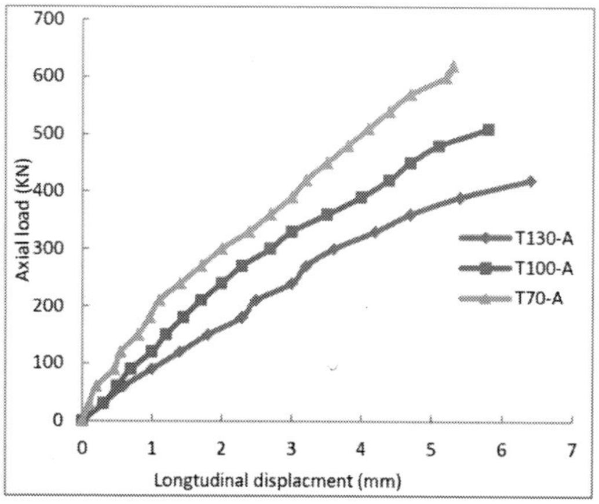


**Fig.14.** Load- longitudinal displacement curves for un-confined tied column.

https://doi.org/10.1051/matecconf/201816204005

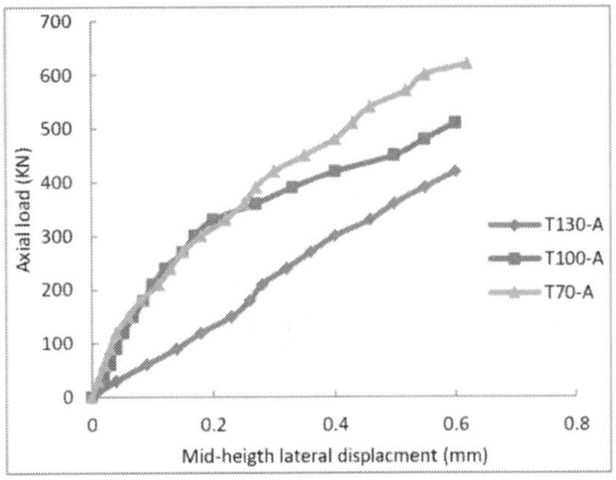


**Fig.15.** Load-lateral displacement curves for un-confined tied column.

Figs. (16 to 18) show the load- longitudinal displacement of columns externally confined with CFRP. Compared with the control specimens (no CFRP). There is an improvement in reducing the longitudinal displacement with the increase of the CFRP confinement. Thus, the lateral displacement will decrease too. This in turn increases the load carrying capacity as shown in Figs. (19 to 21).

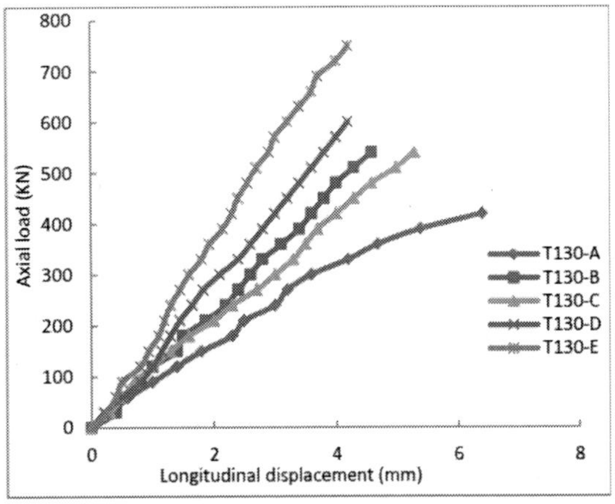


**Fig.16.** Load-longitudinal displacement curves for CFRP confined tied columns with internal tie spacing 130 mm.

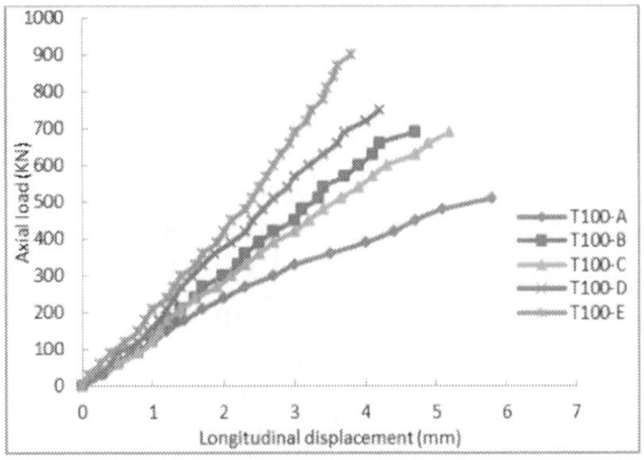


**Fig.17.** Load- longitudinal displacement curves for CFRP confined tied columns with internal tie spacing 100 mm.

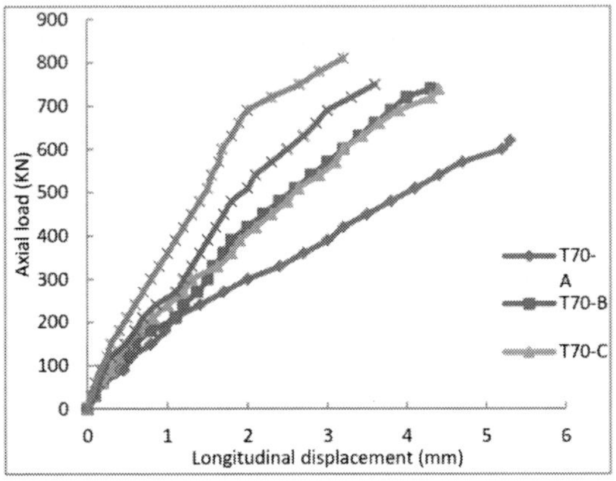


**Fig.18.** Load-longitudinal displacement curves for CFRP confined tied columns with internal tie spacing 70 mm.

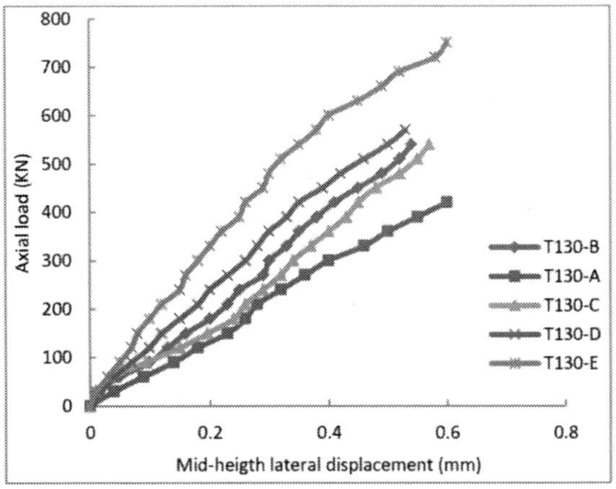


**Fig.19.** Load-lateral displacement curves for CFRP confined tied columns with internal tie spacing 130 mm.

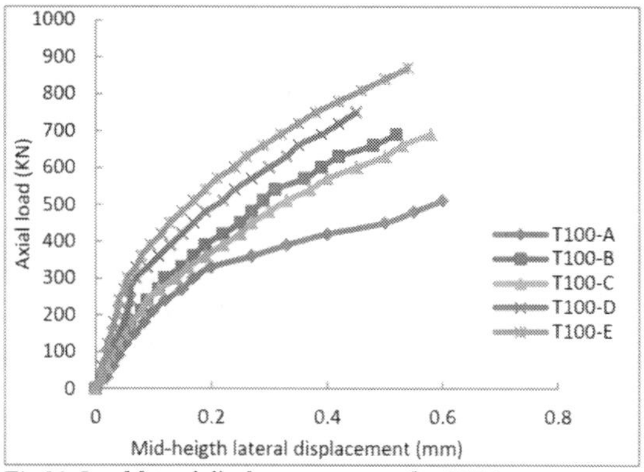


**Fig.20.** Load-lateral displacement curves for CFRP confined tied columns with internal tie spacing 100 mm.

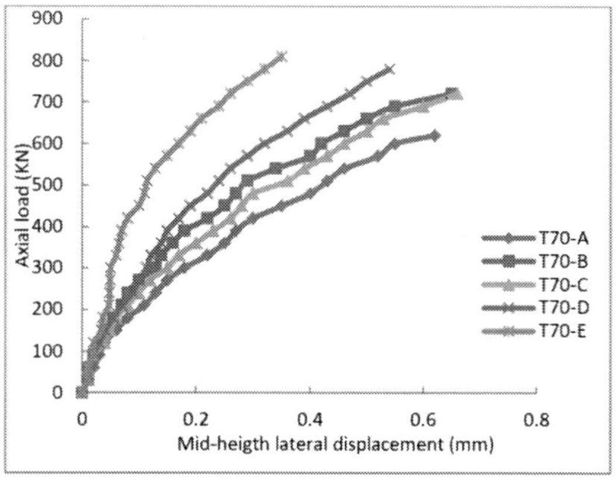


**Fig.21.** Load-lateral displacement curves for CFRP confined tied columns with internal tie spacing 70 mm.

## 3.5 Load capacity

### *3.5.1 Ultimate load capacity for unconfined RC tied columns*

To predict the ultimate axial compressive load for column the American Concrete Institute (ACI 318M-14) [8] has suggested the following equation for the tied column (neglecting the reduction factors):

$$P_o = 0.85 f'_c (A_g - A_{st}) + f_y A_{st} \qquad \text{...... (1)}$$

Where $P_o$ is the maximum axial load, $f'_c$ is the cylinder compressive strength of concrete, $A_g$ is the gross cross sectional area of RC column, $A_{st}$ is the total area of the longitudinal reinforcement, $f_y$ is the yield stress of longitudinal reinforcement.

### *3.5.2 Ultimate load capacity for strengthened RC tied columns*

For a column confined with CFRP under concentric loading, the ultimate load capacity can be calculated according to American Concrete Institute (ACI Committee 440.2R-08) [9] for nonpre-stressed members with tie reinforcement by applying the following equation neglecting reduction factor (0.8Ø).

$$P_o = 0.85 f'_{cc} (A_g - A_{st}) + f_y A_{st} \qquad \text{. ... (2)}$$

The maximum confined concrete compressive strength $f'_{cc}$ is calculated by applying equation (3) (Lam and Teng 2003a,b) [10&11] using the inclusion of additional reductions factor $\Psi f$=0.95.

$$f'_{cc} = f'_c + \Psi_f\, 3.3 K_a f_l \qquad \text{....... (3)}$$

$f'_c$ is the unconfined cylinder compressive strength of concrete, Ka is the efficiency of the geometry of the section($K_a$ =1 for circular section), $f_l$ is the maximum lateral confinement pressure is calculated by Equation (4)(International Federation for Structural Concrete) [1]

$$f_l = \frac{1}{2} K_e \, p_j \, E_j \, \varepsilon_{ju} \qquad \text{........ (4)}$$

$$p_j = 4\, t/D \qquad \text{........ (5)}$$

$p_j$ is the volumetric ratio of the CFRP jacket, $E_j$ is the modulus of CFRP jacket, $\varepsilon_{ju}$ is the effective failure strain of the CFRP wrapping, t is CFRP jacket thickness and D is the diameter of CFRP jacket. For fully wrapped columns $K_e$ =1, use equation (6) [1] to find $K_e$ for partial wrapping, for partial wrapping with fiber orientation use $K_e$ by applying equation (7) [1].

$$K_e = \left[1 - \frac{s'}{2D}\right]^2 \qquad \text{..................(6)}$$

$$K_e = \left[1 + \left(\frac{P}{\pi D}\right)^2\right]^{-1} \qquad \text{..................(7)}$$

Where s' is the clear spacing between the CFRP strips, P is the pitch for CFRP spiral. Table (6) shows the experimental values of the axial compressive strength compared with the calculated values according to references 8 and 9.

**Table 6.** The experimental and calculated values of axial compressive strength for RC tied columns

| Column designation | Tested unconfined concrete compressive strength $f'_c$ (MPa) | calculated confined concrete compressive strength $f'_{cc}$ (MPa) | Experimental values of the axial compressive strength (kN) | Calculated values of the axial compressive strength (kN) | $P_{exp}/P_{cal}$ |
|---|---|---|---|---|---|
| T130-A | 29.89 | ----- | 495 | 527.40 | 0.94 |
| T130-B | 29.89 | 43.15 | 620 | 729.10 | 0.85 |
| T130-C | 29.89 | 41.69 | 610 | 707.45 | 0.86 |
| T130-D | 29.89 | 51.33 | 735 | 850.78 | 0.86 |
| T130-E | 29.89 | 56.95 | 940 | 934.38 | 1.01 |
| T100-A | 30.12 | ---- | 585 | 534.97 | 1.09 |
| T100-B | 30.12 | 43.36 | 730 | 732.22 | 1.00 |
| T100-C | 30.12 | 41.91 | 720 | 710.88 | 1.01 |
| T100-D | 30.12 | 51.54 | 855 | 853.91 | 1.00 |
| T100-E | 30.12 | 57.18 | 1025 | 937.81 | 1.09 |
| T70-A | 30.05 | ----- | 625 | 534.23 | 1.17 |
| T70-B | 30.05 | 43.31 | 745 | 731.48 | 1.02 |
| T70-C | 30.05 | 41.84 | 745 | 709.83 | 1.05 |
| T70-D | 30.05 | 51.49 | 869 | 853.16 | 1.02 |
| T70-E | 30.05 | 57.11 | 1040 | 936.76 | 1.11 |

According to the ACI 318M-14 code [8] limits, the maximum ties spacing is 96 mm. It can be seen that the experimental values of the ultimate load capacity were greater than the calculated ones with 70 mm ties spacing (within the ACI code limits). When the ties spacing increased to 100 mm (close to maximum spacing value of the code), the difference between the experimental and calculated values was decreased. While the calculated values were greater than the tested results (unsafe results) with increasing the ties spacing to 130 mm (greater than $S_{max}$ of the code). Therefore, the ACI 318M-14 code [8] provisions for the ties spacing of the unconfined ties columns can be applied for RC tied columns confined with CFRP strips.

## 4 Conclusions

1-For the columns without CFRP, the influence of the tie spacing was significant: compared with 130 mm tie spacing, dropping the spacing to 100 mm and 70 mm increased the load carrying capacity by 18% and 26%, respectively.

2-Using the CFRP strips to strengthen RC columns are significantly effective in increasing the ultimate load capacity. The increases were (25, 48 and 90) % for 130 mm spaced internal steel ties strengthened with CFRP ratios of (25, 50 and 100) %, respectively. The corresponding values of increase for 70 mm spaced internal steel ties were (19, 39 and 66) % for the same CFRP ratios.

3-The columns with less internal confinement (lesser amount of ties) were strengthened more significantly by the CFRP than the ones with greater amount of internal ties. As an example of the varying effectiveness of the fully wrapped CFRP, the column with ties at 130 mm was strengthened by 90% with the CFRP. In contrast, the ones with 70 mm spaced ties was only increased in strength with CFRP by 66%.

4-Compared with the control specimen (no CFRP), the same amount of CFRP when used as hoop strips led to more strengthening than using CFRP as a spiral strip-the former led to nearly 9% more strengthening than the latter in the case of 130 mm spaced internal steel ties. In the case of 100 mm internal steel ties, the difference (between the hoops & spiral CFRP strengthening) is close to 4%. In contrast, there is no difference between the two methods of strengthening in the heavily tied columns (70 mm tied spacing).

5- The experimental values of the ultimate load capacity were greater than the calculated ones, when the internal steel tie spacing was 70mm (within the limits of the ACI code [8]). When the ties spacing increased to 100 mm (close to maximum spacing according to the code limits), the difference between the experimental and calculated values decreased. However, unsafe results were obtained (experimental/calculated values $< 1.0$) with increasing the ties spacing to 130 mm (greater than $S_{max}$ of the code).

6-For the same load, the increase in strengthening of tied columns externally confined with CFRP led to a decrease of longitudinal strain, mid-height lateral strain.

7-Failure of the RC column specimens under axial compressive load starts by rupture of fiber followed by crushing of concrete at mid height of the column and spalling of concrete cover. For column specimens with low transverse reinforcement ratio, the failure is accompanied with buckling of the longitudinal reinforcement between the transverse ties. While, for the high transverse reinforcement ratio, the failure is accompanied with buckling of the longitudinal reinforcement and rupture for internal ties.

8-The strengthening schemes used in this study were very successful with no CFRP debonding with any tested specimen.

## References

1. Externally bonded FRP reinforcement for RC structures, International Federation for Structural Concrete (fib), Zürich, Switzerland, 130 pp, (2001)
2. Iraqi Specification, No. 5/1984, Portland Cement, وزارة التخطيط, الجهاز المركزي للتقييس و السيطرة النوعية

3. Iraqi Specification, No. 45/1984, Aggregate from Natural Sources for Concrete and Construction, وزارة التخطيط , الجهاز المركزي للتقييس و السيطرة النوعية
4. ASTM C 39/C 39M-05, Standard Test Method for Compressive Strength of Cylindrical Test Specimens, Vol. **04.02**, 9 pp, (2005)
5. B.S. 1881, part 116, Method of Determination of Compressive Strength of Concrete Cubes, British Standards Institution, (1989)
6. ASTM C 496/C 496M-04, Standard Test Method for Splitting Tensile Strength of Cylindrical Concrete Specimens, Vol. **04.02**, 5 pp, (2004)
7. ASTM C 78-02, Standard Test Method for Flexural Strength of Hydraulic - Cement Mortars, Vol. **04.02**, 6 pp, (2002)
8. ACI committee 318, Building Code Requirements for Structure Concrete, (ACI 318M-14) and Commentary (ACI 318RM-14), American Concrete Institute, Farmington Hills, Michigan, 519 pp, (2014)
9. ACI Committee 440, Guide for the Design and Construction of Externally Bonded FRP Systems for Strengthening Concrete Structures (ACI 440.2R-08), American Concrete Institute, Farmington Hills, Michigan, 76 pp, (2008)
10. L. Lam, J. Teng, Construction and Building Materials **17**, 6, 471-489 (2003)
11. L. Lam, J. Teng, Journal of reinforced Plastics and Composites **22**, 13, 1149-1186 (2003)

# Analytical predictions of moment curvature relationship of steel beam columns under fire attack

*Haitham* Al-Thairy*

Civil Engineering Department, College of Engineering, University of Al-Qadisiyah, Diwaniyah, Iraq

**Abstract.** Fire attack is one of the worst scenarios that may cause catastrophic consequences of steel buildings such as progressive collapse and failure. Current design codes and standards have addressed fire as one of the extreme loading conditions to be accounted for in the design of buildings. However, most of the approaches and procedures suggested by these codes and standards still lack accuracy and rationality. The purpose of this paper is to develop an analytical approach to predict the elastic-plastic moment-curvature relationship of steel beam – columns section under elevated temperature. The analytical method was derived based on dividing the steel section to layers and integrating the resistance moment equation of each layer in terms of the section curvature taking into account the effect of elevated temperature on the material properties of the steel by using EC3 reduction factors of the yield stress and modulus of elasticity. The suggested method has been validated against numerical simulation results. Validation results have shown the reliability of the suggested method to predict the resistance moment – curvature relationship of steel beam-column members at different elevated temperatures and under different values of the axial compressive force. The suggested methods may be used to develop more accurate design approaches for steel beam columns under fire condition.

## 1 Introduction

Fire attack is one of the most dangerous incidents that may cause catastrophic consequences to steel buildings leading to progressive collapse and failure. Whilst current design codes and standards have addressed the fire load as one of the extreme loading conditions to account for in the design of steel buildings, most of the approaches and procedures suggested by these codes and standards still lack accuracy and rationality.

Section 3 of Part 1-2 in European Standards, EC1 [1] provides general rules to determine the thermal loads induced by fire to be used for assessment of steel members under fire attack such as determining of temperature – time curves for different fire exposures using heat transfer analysis. In addition, EC1 suggests procedures to determine the mechanical action of the fire based on prescriptive requirements. On the other hand, Part 1-2 of European standards, EC3 [2] have set out design procedures to predict structural resistance against fire. The code suggests that simplified analytical methods can be used for the design of isolated steel members exposed to fire assuming a uniform temperature in the cross-section taking into account the mechanical properties of steel at elevated temperatures. Temperature-dependent material properties for the structural steel have been suggested by the code including modulus of elasticity, effective yield stress and proportional limit along with thermal expansion, thermal conductivity and specific heat [2]. These values

can be used in analytical and numerical analyses of steel structures under elevated temperature. However, if non uniform temperature distribution is encountered, different temperature-dependent material properties must be proposed.

For example, Part 1-1 of EC3 [2] has suggested the following simple alternative method to calculate the reduced plastic bending moment capacity of steel sections at elevated temperature:

$$M_{pfi} = \frac{k_{y,T} M_p}{\kappa_1 \times \kappa_2} \quad (1)$$

Where $M_p$ the plastic moment capacity of the steel is section at normal temperature; $k_{y,T}$ is the reduction factor for the yield strength of steel due to elevated temperature [2]. $\kappa_1$ is a factor used to account for non-uniform temperature distribution in the cross-section; and $\kappa_2$ is a factor used to calculate the support negative bending

Corresponding author: haitham.althairy@qu.edu.iq

© The Authors, published by EDP Sciences. This is an open access article distributed under the terms of the Creative Commons Attribution License 4.0 (http://creativecommons.org/licenses/by/4.0/).

moment capacity when there is non-uniform temperature distribution along the beam [2]. From the equation above, it can be noticed that the effect of the axial compressive load is neglected which may result in overestimated values of the column plastic moment capacity.

The British Standards (BS5950) part 8 [3] has also suggested a bending moment capacity method as an alternative and accurate method for the design of steel structures exposed to non-uniform and uniform fire induced temperature. In this method, the plastic moment capacity of the steel section is calculated according to the reduced strengths of steel at each layer over the steel cross section. However, no guidance was given on how to consider the effect of the compressive axial force on the moment capacity of the section. Moreover, it has been shown by Wang [4] that BS5950 part 8 method would provide over estimated prediction of the section strength as compared to cases where the steel members fail during the elastic or the elastic-plastic phase of their response. However, for steel members subjected to axial load and bending moment BS5950 part 8 has suggested to use the limiting temperature method at which beam column load ratio can be obtained from the interaction equation of the column under combined axial load and bending moment at ambient temperature [3].

A considerable amount of experimental, numerical and analytical research studies have been published during the last decade on how to quantify the effect of elevated temperature on the behaviour and failure of steel members under fire. From these research studies, Dharma and Tan [5, 6] have conducted experimental and numerical studies to investigate the effect of elevated temperature on the rotational resistance of a stocky steel beam under a steady state fire condition. It has been found that fire has considerably reduced the rotational moment resistance of the steel beam section. It has also been shown that the moment–rotation and load-deflection responses of the elevated steel beams have followed same trends of the stress-strain behaviour of the steel under fire suggested by EC3 part 1-2 [2], which is characterized by a reduced yield strength and disappearance of the strain hardening behaviour. It has been concluded [5] that thermal reduction factors of yield stress and modulus of elasticity suggested by EC3 part 1-2 predominate the effect of temperature on rotational capacity whereas, increasing the flange, web and LTB slenderness ratios reduces the rotational capacity. Dharma and Tan [6] have also suggested a bilinear analytical model of the moment–rotation relationship of a compacted steel beams at elevated temperatures. The suggested model was based on numerical integration of moment–curvature linear relationship across the steel section. However, the effect of the axial compression of the moment- rotation relationship has not been considered in the suggested mode

Yin and Wang [7, 8] presented analytical model to investigate the effect of catenary action on behaviour of elastic perfectly plastic steel beams under uniform [7] and no uniform [8] elevated temperatures. The suggested methods use the equilibrium equation of the isolated beams and assume a polynomial deflection shape of the beams under lateral load corresponding to the beam loading and boundary conditions. For beams under non- uniform elevated temperatures, thermal bowing deflection has been accounted for to calculate the total deflection of the beam. The elastic moment resistance was determined using the elastic bending theory taking into consideration the reduction in modulus of elasticity and yield stress due to elevated temperature by using EC3 part 1-2 reduction factors [2]. The plastic moment resistance was calculated using the equilibrium of the forces exerted on the cross section when the steel material reached its yield limit taking into consideration the effect of axial tension force and the reduction in modulus of elasticity and yield stress due to elevated temperature. However, the effect of the axial compression force has also not been accounted for.

Choe et al. [9] have conducted experimental tests to investigate the moment-curvature response of hot rolled steel beam columns under a localized uniform temperature induced by fire. Experimental results have shown that the moment-curvature relationship was affected by the value of the elevated temperature. It has also been shown that the nonlinear behaviour of the moment-curvature relationship gets initiated earlier when the applied temperature is increased. These experimental test results can also be used to set out assumptions and to validate suggested analytical or numerical models to predict the moment-curvature response of steel beam columns under fire attack.

Walz et al. [10] has suggested a closed-form analytical model to determine the moment-curvature- relationship of steel beam columns subjected to uniform elevated temperatures under axial compression load. The suggested model assumes an elastic-perfectly plastic response of the steel material and uses steel material properties as described in Sec. 1-2 of EC3 [2]. MATLAB software was implemented to evaluate the section response using Newton-Raphson iterative procedure. However, inelastic behaviour was not included in the suggested model along with the unloading effects.

The ultimate aim of the current study is to derive an analytical method to predict the moment-curvature relationship of the steel beam column under uniform elevated temperature and different axial load levels. The suggested method was derived by dividing the steel section into numbers of fibres or layers and calculating the moment resistance of each layer taking into account the effect of elevated temperature on the material properties of the steel using EC3 reduction factors. The suggested method is validated against numerical simulation results.

## 2 Methodology

The steel beam -column section shown in Fig. 1 is under combined bending moment and axial compressive load and subjected to elevated temperature. For the elastic-plastic phase of the beam-column deformations, strains at top and bottom layers of the steel section are assumed to exceed the proportional limit of the steel whereas strains at other layers are assumed within elastic range. The depth of yielding over the cross section in compression and tension zones measured from the top and bottom surfaces of the cross section are referred to hereafter, as yc(w) and yt(w) respectively. The strains at the top layer ( $\varepsilon_{c,Top}$ ) and at the bottom layer ( $\varepsilon_{t,Bottom}$ ) of the elastic part of the cross section, corresponding to each value of transverse displacement (w+Δw), can be calculated as a function of the elastic curvature ( $\varphi$ ), total section depth (h), and depths of yielding at compression and tension zones (yc(w) and yt(w)), using the elastic bending theory as follows, see Fig. 1.

$$\varepsilon_{c,top(w+\Delta w)} = \varepsilon_{axial} + \varphi \times (h - y_{c(w)} - y_{t(w)})/2 \leq \varepsilon_f \quad (2)$$

$$\varepsilon_{t,bot.(w+\Delta w)} = \varepsilon_{axial} - \phi \times (h - y_{c(w)} - y_{t(w)})/2 \geq -\varepsilon_f \quad (3)$$

Where (w) is the accumulative value of the transverse displacement at the previous analysis step; Δw is the incremental value of the transverse displacement at the current analysis step; $\varepsilon_f$ is the fracture strain of the steel material; $\varepsilon_{axial}$ is the axial strain due to the axial compressive load calculated by:

$$\varepsilon_{axial} = \frac{P}{E_T A} \quad (4)$$

Where P is the applied axial compression force, $E_T$ is the elastic modulus of elasticity of the steel at elevated temperature calculated using EC3 part 1-2 method (see Fig. 2) [2]; and A is the beam – column cross sectional area.

The elastic curvature of the beam column $\phi$ is calculated for the elastic part of the elastic- plastic section using the elastic bending theory by the following equation:

$$\phi = \frac{\partial^2 w(x)}{\partial x^2} \quad (5)$$

Eq.5 must be evaluated at the location where the elastic plastic moment resistance is to be determined

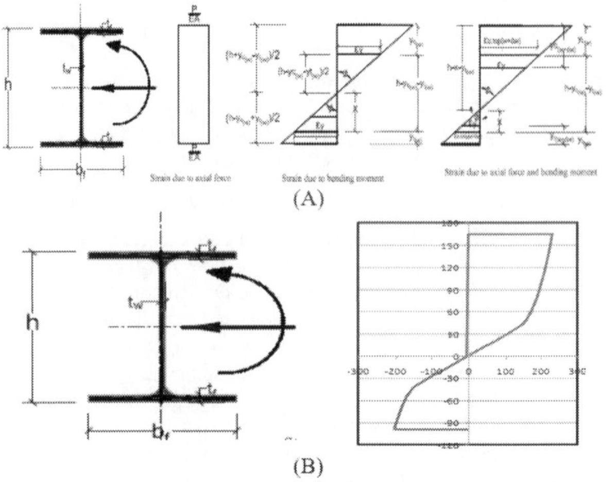

Fig. 1. Strain (A) and stress (B) distribution over a cross section of the steel column when bending is about the major axis.

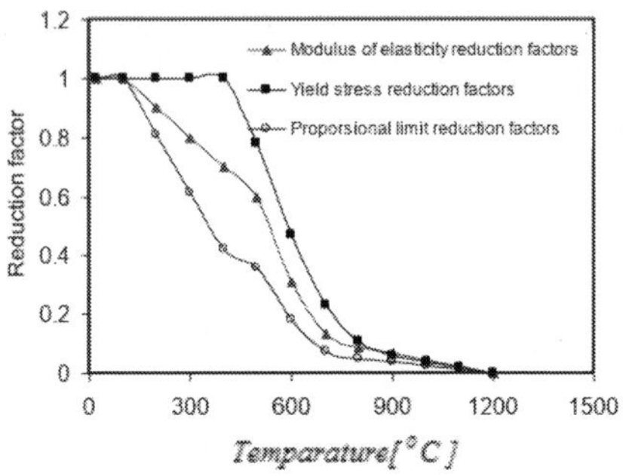


**Fig. 2.** EC3 reduction factors used in calculating the material properties of steel at elevated temperatures [2]

Yielding of the section is initiated when values of $\varepsilon_{c,top(w+\Delta w)}$ and $\varepsilon_{t,bot.(w+\Delta w)}$ calculated from Eqs.2 and 3, exceed the proportional limit strain of steel ( $\varepsilon_p$ ) which can be calculated as:

$$\varepsilon_p = \frac{f_{pT}}{E_T} \quad (6)$$

Where $f_{pT}$ is the proportional limit stress of the steel at elevated temperature determined using the method suggested by EC3 part 1:2 [2].

Values of $yc_{(w+\Delta w)}$, $yt_{(w+\Delta w)}$ are calculated accumulatively according to following incremental equations derived from trigonometric symmetry assuming a linear strain distribution over the elastic part of the section, see Fig. 1(A):

$$y_{c(w+\Delta w)} = y_{c(w)} + (h - y_{t(w)} - x_{(w)}) \times \left( 1 - \frac{\varepsilon_p}{\varepsilon_{c,top\,(w+\Delta w)}} \right) \quad (7)$$

$$y_{t\,(w+\Delta w)} = y_{t(w)} + x_{(w)} \times \left( 1 - \frac{\varepsilon_p}{\varepsilon_{t,bott.(w+\Delta w)}} \right) \quad (8)$$

Where $x_{(w)}$ is the distance from the first layer of the elastic part of tension zone to the neutral axis (N.A.) of the section as shown in Fig. 1 The value of $x_{(w)}$ is calculated from the previous analysis step using trigonometric symmetry as follows:

$$x_{(w)} = \frac{h - y_{c(w)} - y_{t(w)}}{\left( 1 + \frac{\varepsilon_{c,top(w)}}{\left|\varepsilon_{t,bot.w)}\right|} \right)} \quad (9)$$

Eqs. 6 and 7 can be used to capture the gradual spread of yielding over cross section of the steel beam-column when the transverse displacement increases from $(w)$ to $(w+\Delta w)$.

Hence, the elastic-plastic bending resistance of a beam-column section under elevated temperature corresponding to each value transverse displacement $(w+\Delta w)$ can be calculated by integrating the moment resistance of all layers over the section depth as follows, see Fig. 1 (B):

$$M_{El-Pl} = \int_{-\kappa}^{(h-x-y_t-\kappa)} fc_{(T)} \times b_1 \times dh \times [x + y_t + (\kappa + k_1) - \Delta h] + \int_{\kappa}^{(x+t+\kappa)} ft_{(T)} \times b_1 \times dh \times [x + y_t - (\kappa - k_2) + \Delta h] - P \times \frac{h}{2} \quad (10)$$

Where $fc_{(T)}$ and $ft_{(T)}$ are stress at elevated temperature in compression and tension zones respectively. Values of $fc_{(T)}$ and $ft_{(T)}$ can be determined from the stress- strain curve of the steel at elevated temperature suggested by part 1:2 of EC3 [2[ as follows (see Fig. 3)

$fc_{(T)}$ or $ft_{(T)} = (\varepsilon_c$ or $\varepsilon_t) \times E_T$ when $(\varepsilon_c$ or $\varepsilon_t) < \varepsilon_p$ (11)

$fc_{(T)}$ or $ft_{(T)} = fp_{(T)}$ when $(\varepsilon_c$ or $\varepsilon_t) > \varepsilon_p < \varepsilon_y$ (12)

$fc_{(T)}$ or $ft_{(T)} = fy_{(T)}$ when $(\varepsilon_c$ or $\varepsilon_t) > \varepsilon_y$ (13)

Where $\varepsilon_y$ is the yield strain of the steel at elevated temperature which assumed to be equal to 0,2 [2]; $fy_{(T)}$ is the effective yield stress. Values of $fp_{(T)}$ and $fy_{(T)}$ can be determined from the corresponding equations suggested by part 1:2 of EC3 [2] and they are not presented in this paper due to the lack of space. $\varepsilon_c, \varepsilon_t$ are values of strain at the corresponding layer of the compression and tension zones of the section respectively,. Values of $\varepsilon_c, \varepsilon_t$ can be calculated from the following equations assuming a linear strain distribution over the cross section depth as shown in Fig. 1:

$$\varepsilon_c = \varepsilon_{axial} + \phi \times k_1 \leq \varepsilon_f \quad (14)$$

$$\varepsilon_t = \varepsilon_{axial} - \phi \times k_2 \geq -\varepsilon_f \quad (15)$$

Back to Eq.10, Δh is the incremental value of section depth which must be small enough to give accurate results $k_1$ and $k_2$ are distances from N.A. of the section assuming pure bending to the corresponding layer at which strain is to be determined at compression and tension zones respectively; $\kappa$ is the distance from the N.A. of the section at which the strain at the section under combined axial force and pure bending is equal to zero. However, $\kappa$ is measured from the N.A. of the section under pure bending which can be determined by setting the strain equations 14 or 15 equal to zero:

$$0 = \varepsilon_{axial} + \phi \times \kappa \Rightarrow \kappa = \left| \frac{-\varepsilon_{axial}}{\phi} \right| \quad (16)$$

b1 and b2 are widths of the cross section layer in the compression and tension zones respectively and they are determined according to the location of the layer over the cross section as follows, see Fig. 1(A):

$b_1 = t_w$ when $(h - x - yt - \kappa) - t_f \geq k_1 \geq -\kappa$ (17)

$b_1 = b_f$ when $(h - x - yt - \kappa) \geq k_1 \geq (h - x - yt - \kappa) - t_f$ (18)

$b_2 = t_w$ when $(x + y_t + \kappa) - t_f \geq k_2 \geq \kappa$ (19)

$b_2 = b_f$ when $(x + y_t + \kappa) \geq k_2 \geq (x + y_t + \kappa) - t_f)$ (20)

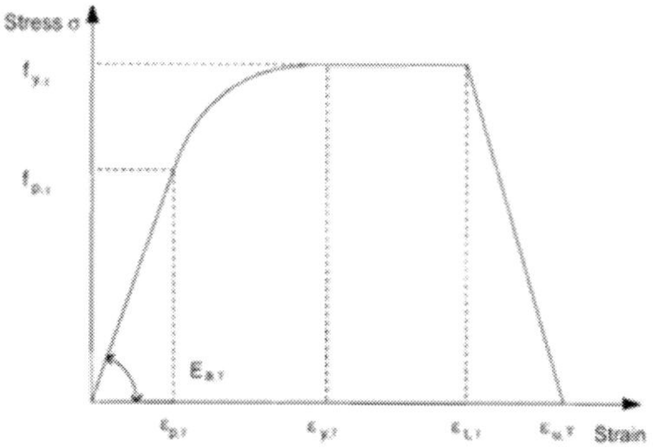


**Fig. 3**. Stress-strain relationship for structural steel at elevated temperatures [2].

## 3 Validation

The accuracy of the suggested analytical procedure is ascertained throughout comparison with numerical

 https://doi.org/10.1051/matecconf/201816204006

simulation results using the financial finite element software ABAQUS. A numerical simulation was carried out on a 1.6 meter length fixed base and free tip A992 W10×68 section steel beam column subjected to combined axial compression force and bending moment under elevated temperature [9]. The boundary and loading conditions of this case of the beam column was arranged to allow direct prediction of the moment-curvature response of steel beam columns under uniform elevated temperature, see Fig. 4. The beam columns were axially compressed to the intended axial load levels which are taken as a percentage of the column squash load (0.3Py and 0.5Py) then subjected to uniform temperature over a 330mm length at the fixed end where maximum moment and plastic hinge were expected to occur. After that, a lateral load was applied at the top end of the beam - columns and monotonically increased until the lateral displacement at this point was approximately equal to 150 mm. Values of the yield and ultimate tensile stress were taken equal to 390 MPa, 500 MPa respectively [9]. Uniform elevated temperatures of 300°C and 500 °C were applied to the beam-column over the outer surfaces of the flanges and both surfaces of the web at the fixed end using a temperature rate (temperature –time curves) as shown in Fig. 5. Table 1 shows the geometry and loading condition of the simulated beam column members.

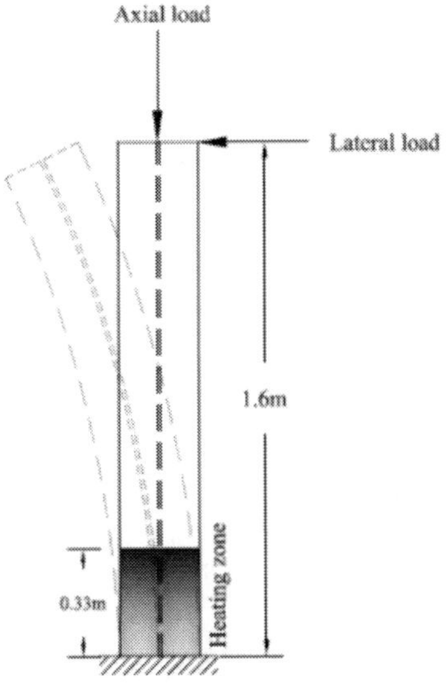


**Fig. 4.** Geometry and loading conditions of the steel beam-column example [9].

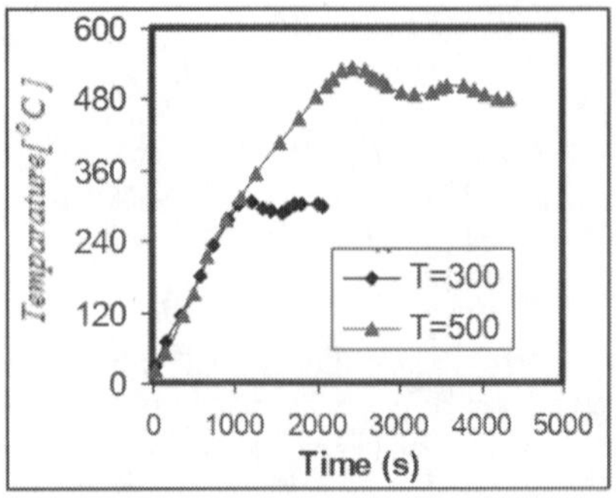


**Fig. 5.** Temperature-time curves used to apply the elevated temperatures [9]

**Table 1.** Geometry and loading properties of the beam-column specimens

| Model Case | Beam-column section | Applied temperature °C | Axial load level (P/Py) | Mu(max) numerical | Mu(max) Analytical | Mu analytical /Mu experimental |
|---|---|---|---|---|---|---|
| 1 | W10×68 | 20 | 0.3 | 499 | 435 | 0.87 |
| 1 | W10×68 | 20 | 0.5 | 382 | 304 | 0.8 |
| 2 | W10×68 | 300 | 0.3 | 522 | 432 | 0.83 |
| 3 | W10×68 | 300 | 0.5 | 477 | 410 | 0.86 |
| 4 | W10×68 | 500 | 0.3 | 383 | 326 | 0.85 |
| 5 | W10×68 | 500 | 0.5 | 272 | 264 | 0.97 |

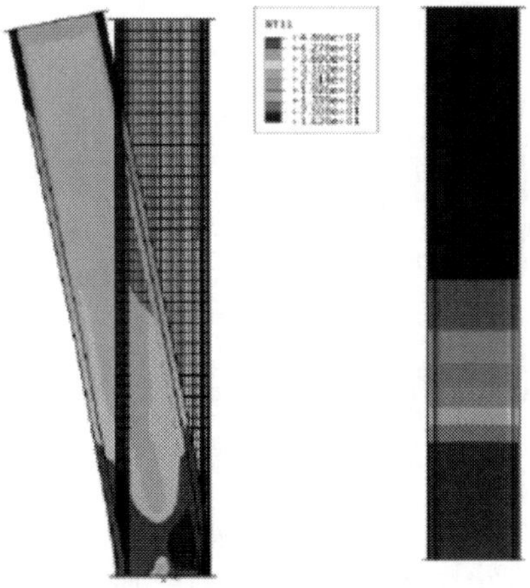

(A) (B)

**Fig. 6.** A) Geometry of the steel beam-column model used in numerical simulations; B) temperature distribution over the steel column obtained from heat transfer analysis

The sequential thermal mechanical analysis procedure available in ABAQUS/Standard was utilized in this study to simulate the behaviour of the steel beam-column members under elevated temperature. In this sequential procedure, a heat transfer analysis was firstly carried out to determine the temperature distribution over the length and over the cross-section of the beam-column member. In the heat transfer analysis, the 8-node linear heat transfer brick element (DC3D8) was used to model the beam-column member using a mesh size of 10 mm at the heated zone and mesh size of 30 mm at the rest of the beam-column as shown in Fig. . The elevated temperature was applied as a boundary condition in the heat transfer analysis step using a tabular temperature time amplitude available in ABAQUS/Standard as shown in Fig.4.

The heat transfer analysis was followed by a mechanical or structural analysis to determine the behaviour of the axially loaded beam-column under fire. In this analysis the 8-node linear brick with reduced integration and hourglass control (C3D8R) was employed to model the geometry of steel members using same mesh size used in the heat transfer analysis. The EC3 stress – strain curves shown in Fig. was used to model the material behaviour of steel under elevated temperatures. Furthermore, EC3 reduction factors of the yield stress, proportional limit, and modules of elasticity were used to account for the effect elevated temperate on the stress-strain response of the steel material as shown Fig. . The fixed end of the steel beam-column members were simulated by restraining all degrees of freedom at the column base.

During the mechanical or structural analysis, the axial compression load was firstly applied as a pressure load in a static analysis step followed by another static analysis step at which the temperature distribution over the steel beam column was restored from the heat transfer analysis procedure using the PREDEFINED FEILED option available in ABAQUS. Finally, a lateral concentrated static force was applied at the column free end during static analysis step until the column experienced large displacement. The moment-curvature relationship was obtained from the numerical load-displacement results at each simulation. The moment at the column base was obtained by multiplying the concentered lateral loads by the distance between the load and the plastic hinge location which is approximately equal to 1.5m. The elastic curvature at the plastic hinge location is obtained from the equation obtained from the elastic bending theory whereas; the elastic- plastic curvature is obtained from the numerical differentiation.

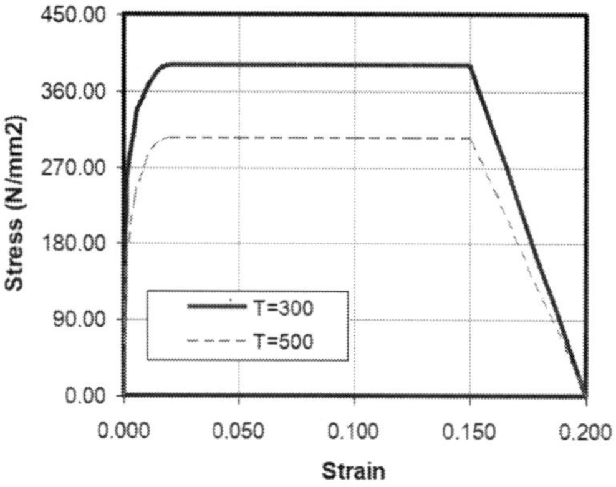


**Fig. 7.** Stress-strain relationship for structural steel at elevated temperatures used in the numerical simulations.

On the other hand, the suggested analytical method has also been used to predict the moment-curvature relationship of the same problem. However, implementation of the procedure described in section II requires determining the elastic curvature equation of the beam - column according to the assumed elastic deformation shape, which must satisfy the loading and the ends support conditions. For a fixed base and free tip column, Timoshenko and Gere [11] have suggested the following equation of the elastic deformation shape:

$$w_{(x)} = W(1 - cos(\frac{\pi x}{2L})) \tag{21}$$

Where W is the amplitude of the elastic deformation shape; L is the length of the beam-column member and x is the distance along the beam -column length at which elastic displacement is to be calculated. Hence, elastic curvature equation can be determined from Eqs. 5 and 21 as:

$$\phi = \frac{\partial^2 w(x)}{\partial x^2} = W(\frac{\pi}{2L})^2 \cos(\frac{\pi x}{2L}) \tag{22}$$

And the curvature at the column's base, where the maximum moment would occur, can be determined by substituting x in Eq. 22 by null as follows:

$$\phi = W(\frac{\pi}{2L})^2 \tag{23}$$

**Fig.** 8, **Fig.** 9 and **Fig.** 10 show the comparison between the numerical and analytical predictions of the moment curvature relationship of the steel beam-column under different axial load levels and subjected to two values of the elevated temperature. As can be seen from the figures, reasonable agreement was obtained between the two sets of results indicating reliability of the suggested method. It can also be noticed from Figs. 8 to 9 that with an increase in the axial load values, the agreement between the proposed analytical and numerical results increases accordingly. Moreover, Figs. 8 and 9 indicate that with an increase in the degree of the elevated temperature, more agreement can be obtained between analytically and numerically predicted results.

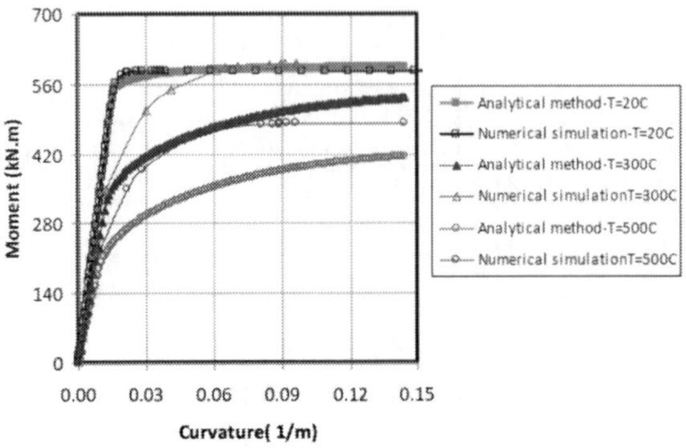


**Fig. 8.** Moment curvature relationships of the steel beam-column section W10×68 subjected to different elevated temperatures with no axial compressive load

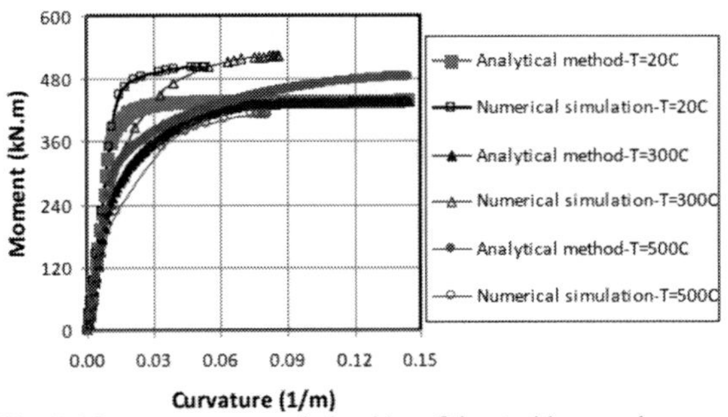


**Fig. 9.** Moment curvature relationships of the steel beam-column section W10×68 subjected to different elevated temperatures under axial compressive load of 0.3Py.

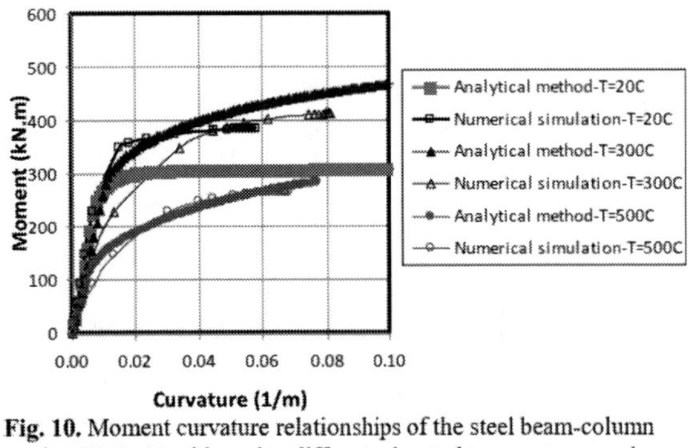


**Fig. 10.** Moment curvature relationships of the steel beam-column section W10×68 subjected to different elevated temperatures under axial compressive load of 0.5Py.
Units

## 4 Parametric study

A parametric study is carried out in this section using the suggested and validated analytical method to investigate the effect of elevated temperature on the moment-curvature response of steel beam-column section under a compression force equal to 50% of the section squash load (Py) and the results are shown in Fig. . As can be seen from this figure, a considerable reduction in the predicted moment of the section is obtained when the elevated temperature is increased. It can also be noticed from Fig. 11 that the linear elastic part of the moment-curvature relationship gets vanished with increasing elevated temperature. This behaviour is conforms to the disappearance of the strain hardening behaviour indicated experimentally in Dharma and Tan study [5] and Choe et al. [9] and the suggested analytical method could predict this behaviour reasonably.

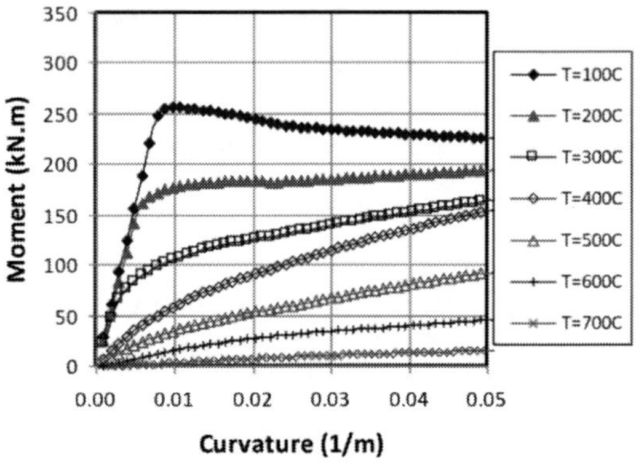


**Fig. 11.** Effect of elevated temperatures on the moment-curvature response of steel beam-column section.

Fig. illustrates the effect of elevated temperature on the stress distribution over the cross section of the steel beam-column subjected to same value of the axial compression force. It is clearly shown from Fig. 12 that with an increase in the elevated temperature, the depths of compression zone increased owing to the reduction in the stresses caused by the elevated temperatures [2]. When the compressive stress decreases, a more compression area is needed to achieve the equilibrium of forces acting on the section.

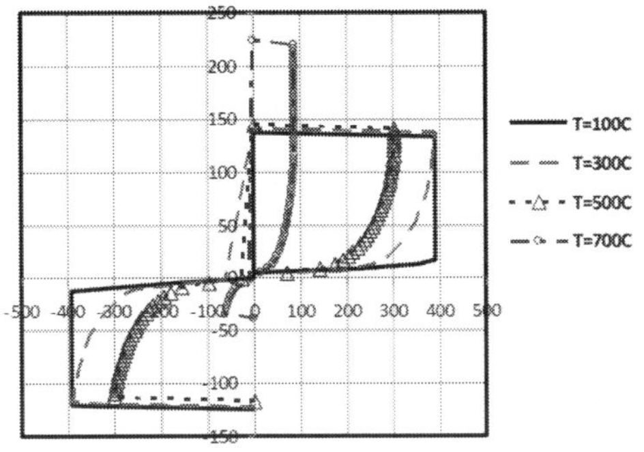


**Fig. 12.** Effect of elevated temperature on the stress distribution over the cross section of the steel beam column

## 5 Conclusions

This paper has presented a development of an analytical method for prediction of the elastic-plastic moment-curvature relationship of steel beam – columns section under elevated temperatures. The suggested analytical approach divides the steel section to numbers of layers and integrates the moment equation of the single layer over the entire section depth taking into account the degradation of the steel material properties due to the effect of elevated temperature using EC3 reduction factors. The suggested analytical method has been validated against numerical simulation results which have shown the reliability of the suggested method. The suggested methods may be used to develop more accurate design approaches for steel beam columns under fire condition.

A parametric study is carried out to investigate the effect of elevated temperature on the moment-curvature response of steel beam-column section showing that a considerable reduction in the predicted moment of the section occurred with reducing the linear elastic part of the moment-curvature relationship when the elevated temperature is increased. It has also been concluded that with an increase in the elevated temperature, the depths of compression zone also increases.

## References

1. EC 1: *Actions on structures - Part 1-2: general actions - Actions on structures exposed to fire*. EN 1991-1-2:2002 (European Committee for Standardization, 2002).
2. EC 3: *Design of steel structures - Part 1-2: General rules - Structural fire design*. EN 1993-1-2:2005 (European Committee for Standardization; 2004).
3. BS5950: *Part 8:1990: Code of Practice for Fire Resistant Design* (The steel construction institute, 1993)
4. 2nd ed (Taylor and Francis e-Library, 2005)
5. R.B. Dharma, K.H. Tan, Eng. Str., **29**, 2391–2402 (2007)
6. R.B. Dharma, K.H. Tan, Eng. Str., **29** , 2403–2418, (2007)
7. Y.Z. Yin, Y.C. Wang, Jour. of Const. Ste. Res., **61**, 183–211 (2005)
8. Y.Z. Yin, Y.C. Wang. Jour. of Con. Ste. Res., **61**, 213-234 (2005)
9. L. Choe; A. H. Varma, A. Agarwal; and A, Surovek, Jour. of Str. Eng., **127**, 954-966 (2001)
10. J. Walzl, A. Surovek, A. Agarwal, L. Choe, A.Varma, (Council Pittsburgh, Pennsylvania, 2011)
11. S. P Timoshenko, J. M. Gere "*Theory of Elastic Stability*" 2nd ed. (McGraw-Hill, NY, 1985)

# FEM performance of concrete beams reinforced by carbon fiber bars

*Hashim* Hasan
Building and Construction Engineering Department, University of Technology, Baghdad, Iraq

**Abstract.** Concrete structures may be vulnerable to harsh environment, reinforcement with Fiber Reinforced Polymer (FRP) bars have an increasing acceptance than normal steel. The nature of (FRP) bar is (non-corrosive) which is very beneficial for increased durability as well as the reinforcement of FRP bar has higher strength than steel bar. FRP usage are being specified more and more by public structural engineers and individual companies as main reinforcement and as strengthening of structures. Steel reinforcement as compared to (FRP) reinforcement are decreasingly acceptable for structural concrete reinforcement including precast concrete, cast in place concrete, columns, beams and other components. Carbon Fiber Reinforcement Polymer (CFRP) have a very high modulus of elasticity "high modulus" and very high tensile strength. In aerospace industry, CFRP with high modulus are popular among all FRPs because it has a high strength to weight ratio. In this research, a finite element models will be used to represent beams with Carbon Fiber Reinforcement and beams with steel reinforcement. The primary objective of the research is the evaluation of the effect of (CFR) on beam reinforcement.

## 1 Introduction

Recently, carbon FRP bars were developed. These CFRP bars provide a comparable performance to that of conventional steel bars. In between of the various types of fiber reinforcement polymer, (Carbon Fiber Reinforcement Polymer) has the highest tensile modulus of elasticity. There are serious problems faced by concrete elements reinforced with normal steel, so CFRP internal reinforcement bars have been developed with the advantages below:

- CFRP bars do not corrode
- CFRP bars are light weight
- CFRP bars are high strength

This material is suitable for structural applications because of high strength to self-weight ratio, low own weight, electromagnetic transparency, high resistance against corrosive agents, along with the other recommend structural and technologic parts. Disadvantages in the global use of these materials are the high cost of initial application, high cost of manufacturing these materials and the main disadvantage of this material is the lack of standard national design codes [1].

## 2 Research significance

Evolution of reinforcement technology is going to be very progressive because engineers design will not employ concrete reinforced with steel. In the last years, one of the very important materials in reinforced concrete which has been suggested was Fiber Reinforced Polymer (FRP). The reinforcement of fiber polymer is very modern and limited to some works on structures, it increases the full capacity of reinforced concrete beams, enhances the tensile strength and reduces deflection but the utilization of carbon fiber reinforcement is very low and it is very expensive. Carbon fiber reinforcement polymer can be produced with higher strength and modulus of elasticity than steel, then the structural members will improve in flexure, shear strength, and deflection. The study aims to investigate the effect of carbon fiber bars on the reinforced concrete beams. Finite element approach was used to simulate the beam reinforced with CFRP bars then compare the results with analytical solution [6].

## 3 Historical review

One of the popular materials for building construction in the world is reinforced concrete. Reinforced concrete has been used as main material construction in many structures such as bridges and buildings. Concrete lacks tensile strength but the common reinforcement in tensile strength use in reinforced concrete is steel. Though performance of steel is excellent in reinforcement of concrete but it has some problems. Steel tends to crumble fast, especially when expose to sea water and extreme coastal weather. In 2013, Norazman Mohamad Nor et al [3], investigated an alternative reinforcement for concrete structures which is non-metal called "Carbon fiber reinforced polymer CFRP". Actually,

© The Authors, published by EDP Sciences. This is an open access article distributed under the terms of the Creative Commons Attribution License 4.0 (http://creativecommons.org/licenses/by/4.0/).

(CFRP) is being used for damaged structures due to long age or due to exposure to bad weather conditions and for structural repair. The aim of their study, is to increase the tensile strength of reinforced concrete, replacing steel reinforcement, completely and using (CFRP) instead. The most important advantage of using (CFRP) in reinforcement is to avoid reinforcement corrosion and reinforcement rusting. In 2002, Padmarajaiah and Ramaswamy [4], illustrated the results of increasing concrete compressive strength by adding fibers. They increased the concrete compressive strength up to (15%) from normal strength by adding fibers up to 1.5% in volume. So, the concrete compressive strength and concrete strain affected by fibers volume. In 2008, Ţăranu Nicolae et al [1], have presented two main issues, using internal reinforcements of FRP composite materials for new members in the structure and strengthening of existing members (externally bonded reinforcements). The advantages, disadvantages, constraints as well as the problems associated are mainly related to concrete members.

## 4 Properties of Materials

In this paper five finite element models were built up using ANSYS software [5] to simulate reinforced concrete beams with traditional reinforcement, CFRP bars, and both were investigated. Model's materials and properties are listed in Table (1). All beams have the same geometry of concrete. The parameters considered are shown in Table (2).

**Table 1.** Materials properties

| Concrete | | Steel bars | | CFRP bars | |
|---|---|---|---|---|---|
| fc̍ | 31 MPa | fy | 414 MPa | $fy_{(CFRP)}$ | 3690 MPa |
| $E_C$ | 26168 MPa | $E_S$ | 200000 MPa | $E_{(CFRP)}$ | 580000 MPa |
| $v_c$ | 0.15 | $v_s$ | 0.3 | $v_{(CFRP)}$ | 0.3 |

Where,
fc̍: Cylinder compressive strength of concrete.
$E_C$: Concrete modulus of elasticity.
$v_c$: Poisson's ratio of concrete.
fy: Steel yield strength.
$E_S$: Steel modulus of elasticity.
$v_s$: Poisson's ratio of steel.
$fy_{(CFRP)}$: Yield stress carbon for fiber reinforcement polymer.
$E_{(CFRP)}$: Modulus of elasticity for carbon fiber reinforcement polymer.
$v_{(CFRP)}$ : Poisson's ratio for carbon fiber reinforcement polymer.

**Table 2.** Types of beams considered in the analysis

| Beam no. | Reinforcement Bars | | | Method of Analysis |
|---|---|---|---|---|
| | Main | Top | Stirrups | |
| B1 | steel | steel | steel | Linear |
| B2 | steel | steel | steel | Nonlinear |
| B3 | CFRP | CFRP | CFRP | Linear |
| B4 | CFRP | steel | steel | Linear |
| B5 | CFRP | CFRP | CFRP | Nonlinear |

## 5 Geometric modelling

(SOLID65) Element is used for the (3-D) modeling of solids (with or without) reinforcement bars. (SOLID65) element is qualified for crushing in compression and cracking in tension. The solid ability of the element may be used for modeling the concrete while the rebar ability is available for modeling the behavior of reinforcement in concrete applications. (SOLID65) shown in (Figure 1) which is defined by eight nodes and each node has three degrees of freedom: translations in the nodal (x, y, and z directions). May be defined up to three different rebar specifications. The treatment of properties for nonlinear material is the most important part in this element. Concrete is capable of cracking in three orthogonal directions, crushing, plastic deformation, and creep. The rebar are not capable of shearing but they are capable of compression and tension. The rebar are also capable of creeping and plastic deformation. Below are the elements which were used to model the concrete.

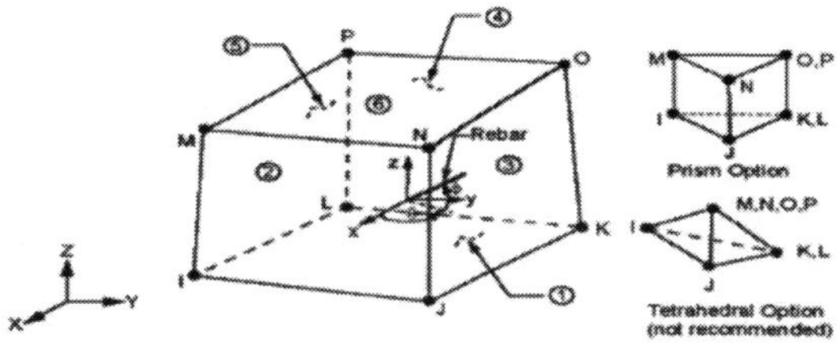


**Fig.1.** SOLID65 element geometry

**Table 3.** Materials Representations

| Element Number | Element Type | Representation |
|---|---|---|
| 1 | SOLID65 | Concrete cross section |
| 2 | LINK8 | Top and bottom longitudinal reinforcement |
| 3 | LINK8 | Stirrups and CFRP bars |

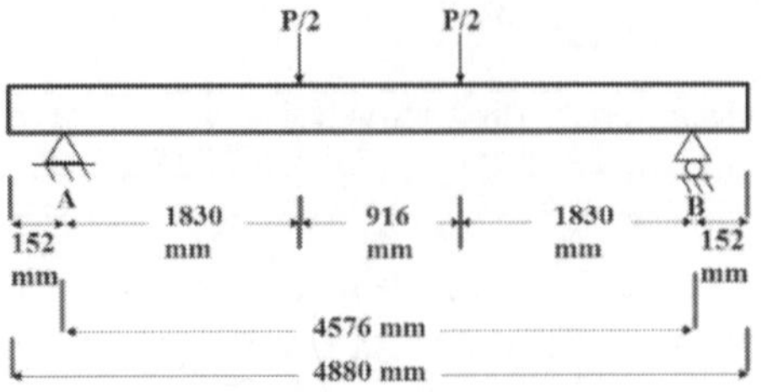


**Fig.2.** Beam dimensions

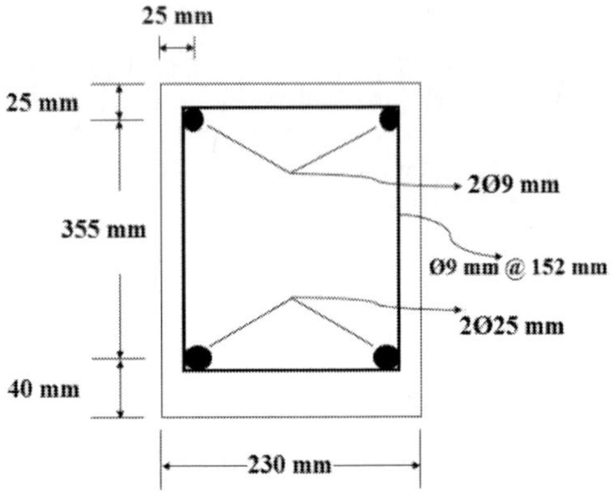


**Fig. 3.** Beam cross section

**Table 4.** Reinforcement bars details

| Reinforcement | Details |
|---|---|
| Bottom (main reinforcement) | 2 bars Ø 25 mm. |
| Top (secondary reinforcement) | 2 bars Ø 9 mm. |
| Stirrups | Ø 9 mm @ 152 mm. center to center |

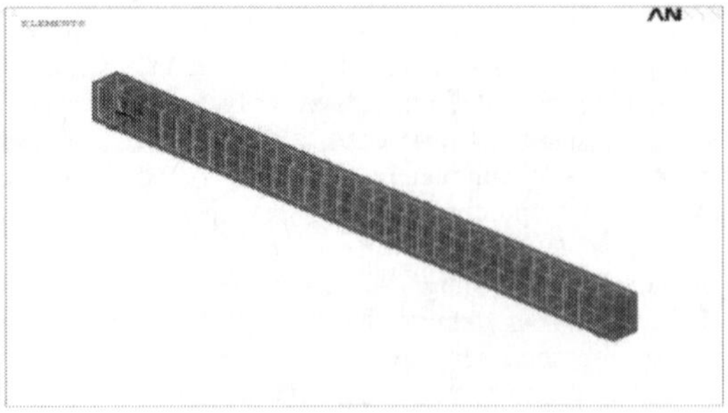

**Fig. 4.** Beam reinforcement ANSYS model

## 6 Beams deflection

The deflection of beams was calculated by taking deflection results of beams tested in the laboratory [2], and then compared these results using the same loading with hand out calculation by applying the theoretical approach described in (ACI 318M -14) [6]. Finite element models were representing beams and calculate the deflection at midspan beam. Table (5) show models deflection results at load level 196 kN.

**Table 5.** Models deflection results

| Beam | Applied Load (kN) | Ultimate Deflection at Midspan (mm) | % Reduction of Deflection |
|---|---|---|---|
| B1 (linear) | 196 | 23.51 | 100 % (reference) |
| B3 (linear) | 196 | 11.21 | 52% |
| B4 (linear) | 196 | 11.67 | 50% |
| B2 (nonlinear) | 196 | 45 | 100% (reference) |
| B5 (nonlinear) | 196 | 11.28 | 74% |

Figures (5, 6 and 7) represent relationship between beams loading and beams deflection, while Figures (8 and 9) show beams deflection and beams cracks captures by ANSYS program.

## 7 Results discussion

- Because of the yielding & modulus of elasticity of CFRP material the deflection are performed % reduction to (52 % linearly and 74 % nonlinearly) instead of the traditional reinforcement steel.

- The (stress-strain) curve for traditional reinforcement steel assumed to be (elastic-perfectly plastic) but CFRP behavior is assumed to be linear according to the assumption recommended by (ACI 440-2R-06) [7].

- In simply supported beam, the main factor affecting deflection is the main reinforcement, modulus of elasticity and the geometry of beam because of it is function of compressive strength. The modulus of elasticity and the geometry of beam assumed to be constant so that in present study the main reinforcement of CFRP is more effective than stirrups and secondary top reinforcement.

- In beam B2, the nonlinear of steel reinforcement assumed elastic – full plastic behavior. Applied load (196 kN) has not reached as a full loading. Because of the cracked zone at tension faced due to increased loading due to the nonlinearity of reinforcement and reached to the plastic zone, the deflection became more so that at 80% of applied loading the deflection 45 mm as shown in Figure (5), but at full load the deflection became 125 mm.

- Analytical solution for steel reinforcement with maximum deflection equal to (23.373 mm) and maximum deflection equal to (23.511 mm) as numerical solution as shown in Figure (10), so that the ratio between them equal to (1.006) and the difference ratio equal to (0.59 %) which is acceptable because it is less than (10%) for the design of the reinforced concrete beam.

- Analytical solution for CFRP reinforcement with maximum deflection equal to (10.261 mm) and maximum deflection equal to (11.215 mm) as numerical solution as shown in Figure (11), so that the ratio between them equal to (1.09) and the difference ratio equal to (8.5 %) which is acceptable because it is less than (10%) for the design of the reinforced concrete beam.

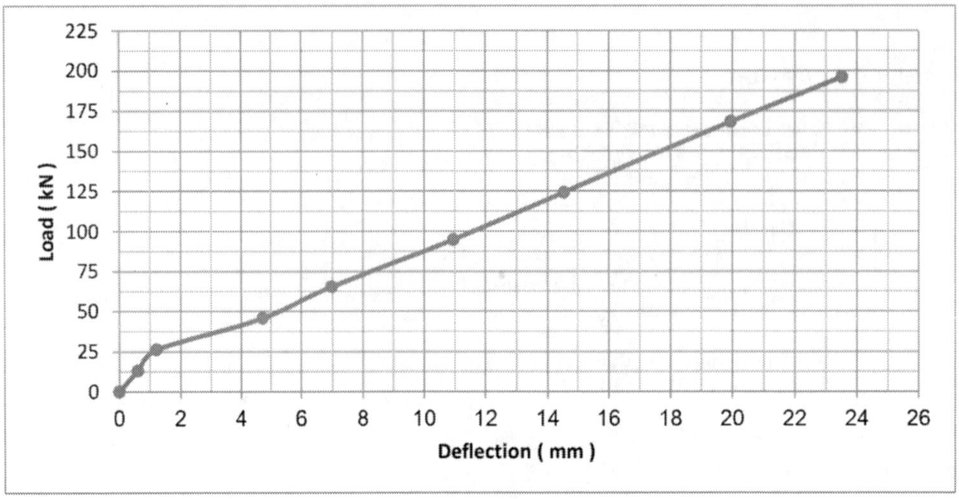


(a) Linear performance of B1 (Steel Reinforcement)

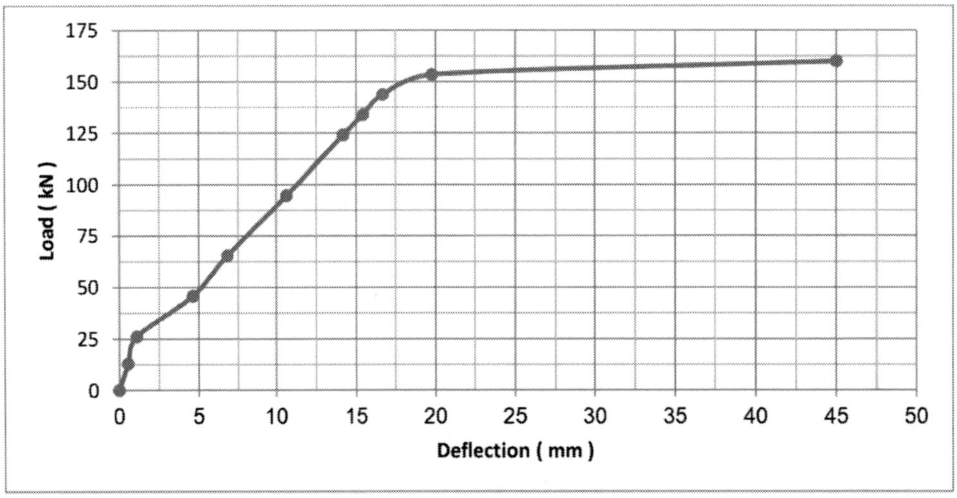


(b) Nonlinear performance of B2 (Steel Reinforcement)

**Fig. 5.** Load-midspan deflection relation of beams with steel reinforcement

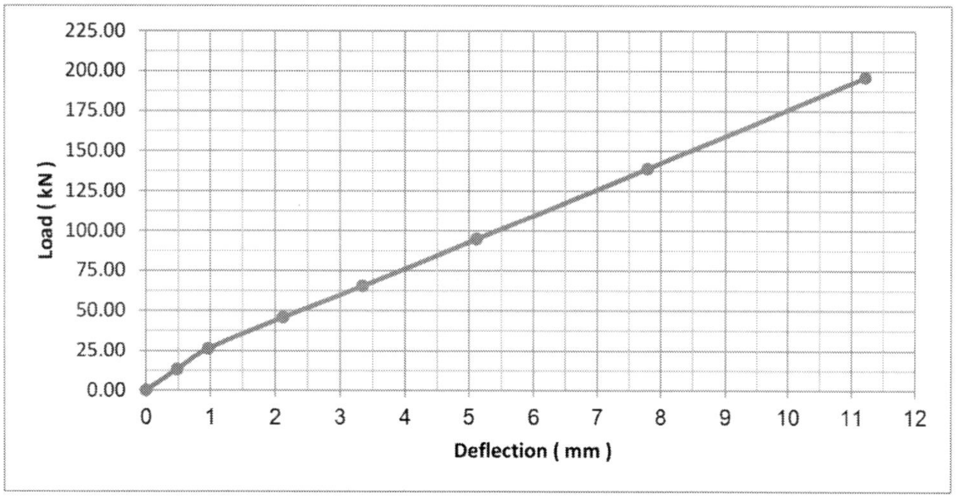


(a) Linear performance of B3 ( CFRP reinforcement)

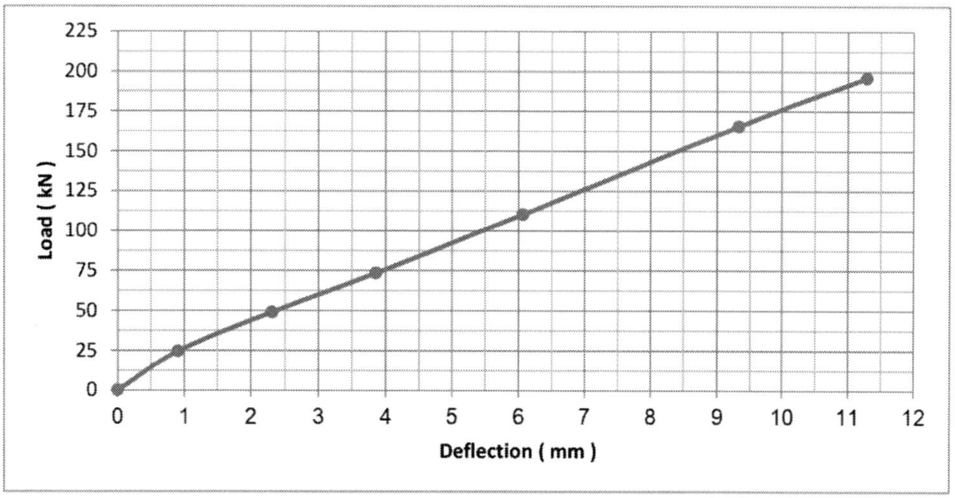


(b) Nonlinear performance of B5 (CFRP reinforcement)

**Fig. 6.** Load-midspan deflection relation of beams with CFRP reinforcement

 https://doi.org/10.1051/matecconf/201816204007

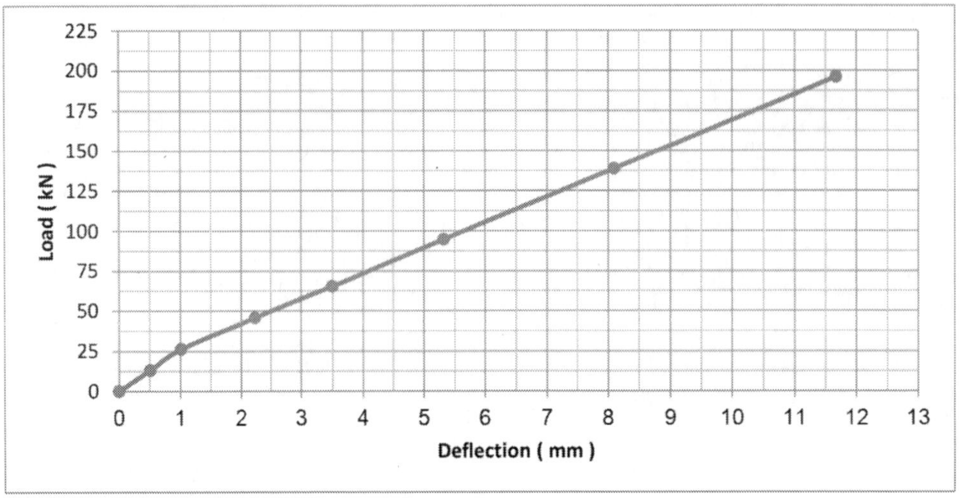


Linear performance of B4 (CFRP main reinforcement, steel top reinforcement and steel stirrups)

**Fig. 7.** Load-midspan deflection relation of beam with Steel and CFRP reinforcement

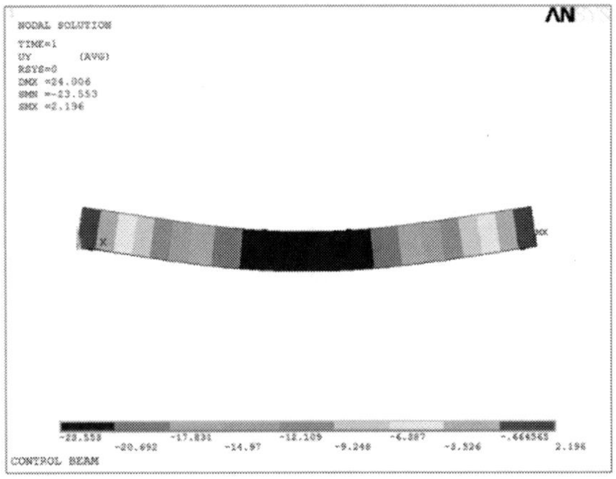


**B1 deflection**

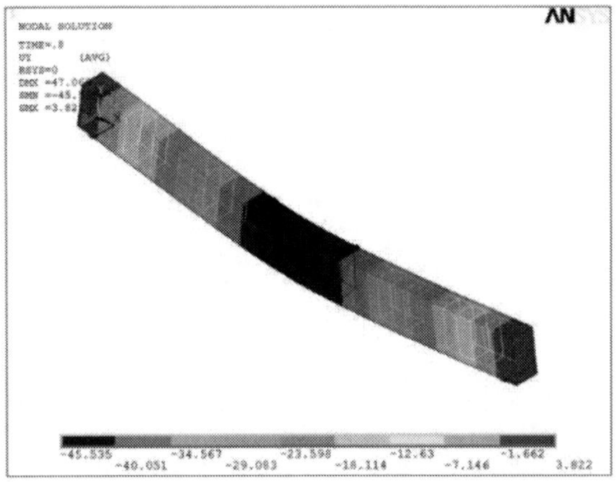


**B2 deflection**

**Fig. 8.** Deflection of beams 1 and 2 captured by ANSYS

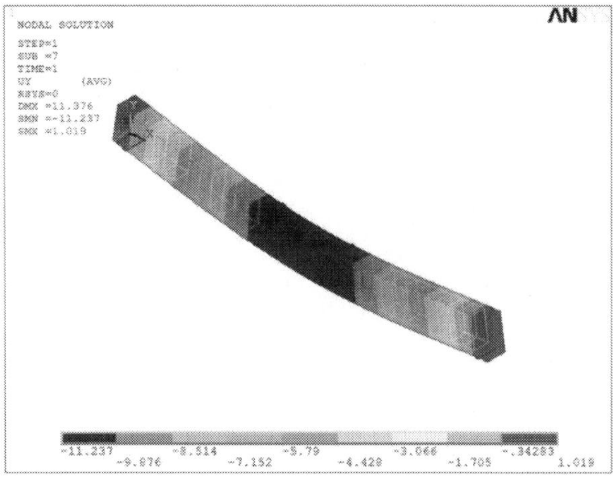


**B3 deflection**

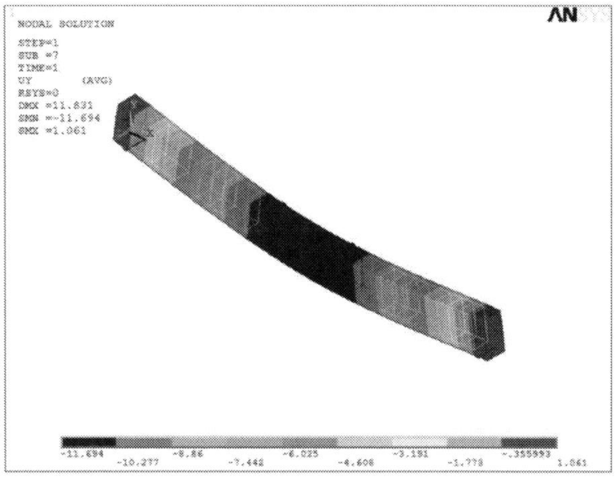


**B4 deflection**

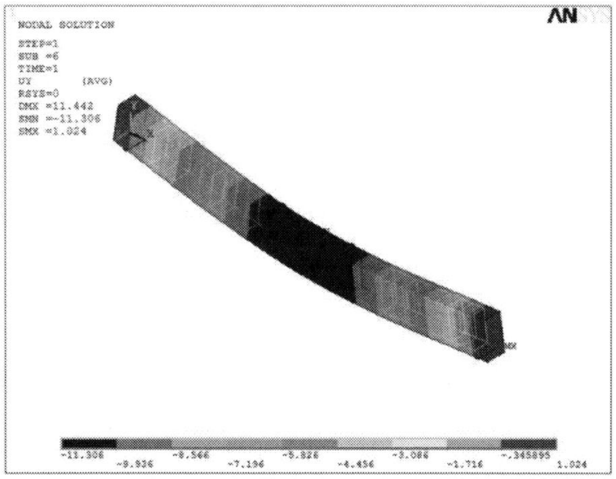


**B5 deflection**

**Fig. 9.** Deflection of beams 3, 4 and 5 captured by ANSYS

## References

1. T. Nicolae, O. Gabriel , I. Dorina, E. Ioana, M. Vlad, B. Cătălin, Universitatea Tehnică, **1**, 7-20 (2008)
2. P. Alagusundaramoorthy, I. E. Harik C. C. Choo, Research report, College of Engineering, University of Kentucky, Frankfort, KTC-02-13/SPR 200-99-1F, (Aug.2002)
3. N. M. Nor, M. H. A. Boestamam and M. A. Yusof, International Journal of Emerging Technology and Advanced Engineering, Malaysia, **3**, 2, 6-10, (Feb.2013)
4. S. K. Padmarajaiah, A. Ramaswamy, Journal of Material in Civil Engineering **14**, 2, 130-136 (Mar.2002)
5. ANSYS: Inc. is a UL registered ISO 9001:2000 Company. Release 11.0SP1 UP20070830. ANSYS Multiphysics. Manual help, USA
6. Building Code Requirements for Structural Concrete and Commentary. ACI 318M-14. Farmington Hills, MI: American Concrete Institute, 503pp. (2014)
7. Guide for the Design and Construction of Structural Concrete Reinforced with FRP Bars. ACI 440.1R-06: American Concrete Institute, 44pp. (2006)

# Tensile strength of 3D textiles reinforced cementitious composites plates

*Ikbal* Gorgis[1], *Waleed* Abbas[1], and *Nadia* Moneen[2,*]

[1] Building and Construction Engineering Department, University of Technology, Baghdad, Iraq
[2] Al- Najaf Technical Institute, Najaf, Iraq

**Abstract.** Tensile plate specimens with dimension of 450×100×40mm were cast with 3D glass fabric having three different thicknesses 6, 10 and 15mm to measure their tensile strength. Plates with one and two layers of chicken wires, as well as micro steel fiber of 0.75% volume fraction were tested under tensile for comparison with references plates. Cement mortar with 61.2MPa cube compressive strength at 28 days was designed for casing the plates. The results indicated that after cracking of the mortar the textile reinforcement adds a strain hardening trajectory, that cause failure to occurs at slightly higher load and a higher strain. The improvement in tensile strength at 28 days ranged between 5 to 30%, and for 90 days between 5 to 60% for the three types of fibres used. Based on the results a significant increase was indicated with micro steel fiber.

## 1 Introduction

The mechanical possibilities and advantages of Textile Reinforced Mortar have been proven, both as strengthening material and as precast material (in this field this composite is also called Textile Reinforced Concrete TRC) [1&2]. As Textile Reinforced Mortar (TRM) usually bears tensile loads, it is very relevant to investigate the behaviour of this material under pure tensile loads.

The tensile strength can be determined approximately by direct tensile specimens. Usually tensile strength of normal brittle concrete is less imperative, but not in the case of steel fibre reinforced concrete, since steel fibres influence the ductility [3]. Tensile testing is performed to understand the material behaviour of fibre reinforce cementations composites in tension and to determine the values necessary for analysis and design of structural elements. To provide accurate material characteristics for design, the geometric relationships between the microstructure and the stresses in the test specimens as well as the boundary conditions must correlate to those in the product to be designed [4]. The load bearing behaviour of textiles as reinforcement can be seen similar to steel reinforced concrete (SRC) since the reinforcement is aligned in the direction of the tensile strength. The main difference in the load bearing behaviour between textile reinforced concrete (TRC) and steel reinforced concrete (SRC) is the bonding between the reinforcement and the concrete matrix. The reinforcement bars in steel reinforced concrete (SRC) are profiled to transfer forces between the reinforcement and concrete. In (TRC) the reinforcement yarns of the textile work on adhesion and friction. When a yarn in the textile structure consist of multiple non-twisted filaments the full tensile strength cannot be utilized, since only a part of the yarns has direct bonding with the concrete matrix (because a part of the yarn gets impregnated with concrete). Tensile test on textile reinforced concrete (TRC) specimen was done. It was found that the behaviour was the same as steel reinforce concrete (SRC)[5]. This behaviour can be described in three stages.

From previous experiments, it was observed that the (TRC) specimen has lower (10 to 30%) stiffness than the textile itself. This has been explained by the failure of a small part of the filaments during the cracking stage. Unlike steel reinforcement, textile reinforcement does not show plastic behaviour. Although the reinforcement does not deform in a plastic way, (TRC) does not fail in a brittle way due to pull-out of the fibres. Figure (1b) shows the stress-strain diagram for multiple types of TRC specimens loaded with a centric tensile force [6].

The influences of geometric patterns of 3D spacer fabric on the tensile behaviour of concrete canvas were investigated [7]. Tensile stress–strain curves and crack propagating patterns of concrete canvas were obtained in warp and weft directions through experiments. The average values of three parallel samples were acquired for different concrete canvas in warp and weft direction, respectively. Experimental results revealed that the concrete canvas samples reinforced by the 3D spacer fabrics exhibited improved tensile behaviours in terms of tensile strength, reinforcing efficiency factor and crack pattern. Moreover, for concrete canvas, 3D spacer fabric reinforcement was a better option than spacer yarns alone.

* Corresponding author: nadia.material@gmail.com

© The Authors, published by EDP Sciences. This is an open access article distributed under the terms of the Creative Commons Attribution License 4.0 (http://creativecommons.org/licenses/by/4.0/).

Experimental program to evaluate the mechanical properties of cement-based composites were carried out [8]. The variables were water to cement (w/cm) ratios, silica fume (SF) percent and volume fraction of steel fibre. Compressive strength, direct tensile strength, and drop weight tests were performed. It was found that cement-based composites containing 10% SF demonstrates better compressive strength, direct tensile strength and abrasion resistance, and worse impact resistance than composites containing 5% SF. The 10% SF specimens with w/cm of 0.35 and 0.65 have 18% and 15% higher compressive strength, 3% and 8% higher direct tensile strength, 83% and 62% lower impact toughness than the control specimen. The addition of steel fibre to SF composites achieve 10% and 9% increase in compressive strength, 31% and 15% increase in direct tensile strength, and 418% and 296% increase in impact toughness. Lightweight concrete with ferrocement subjected to axial tension was studied [9], in which blast furnace slag with sand was used. Steel galvanized woven square mesh of 4 × 20 gauge with 6.35 mm diameter of wire is used. Sand was replaced by blast furnace slag in the range of 0 %, 20 %, 35 %, and 70 %. Five volume fractions had been used for total 6 layers of mix proportion of in ratio 1: 2 with w/c of 0.5. The result showed that the replacement of blast furnace till 20 % cause maximum increase in axial tension strength while any increase in blast furnace slag to 35 % and 70 % shows decrease in strength. Tensile strength of specimen was found to be directly proportional to the number of mesh layers.

Enhancement of the load carrying capacity of mortar of cement sand ratio 1:2 with w/c of 0.45 was investigated [10]. Using wire mesh with diameter of 0.94 mm and grid size of (12×12 mm) welded. Plain concrete with ferrocement jacket provides remarkable lateral confinement pressure which increases axial load carrying capacity.

An experimental study [11] on the direct tensile strength of plates by using different numbers of wire mesh and inducing steel fibres with, keeping constant thickness of plates. Wire mesh layers ranged from (1-6) layers, test specimens used were 300*50mm with a uniform thickness. Steel fibres with volume fraction of 0.5 % of total volume of composite were added. Test result indicates that; direct tensile strength is directly proportional to the number of mesh layers used. Steel fibres usage increases tensile strength by about (10 to 17) % as compared with specimens without steel fibres.

In this work more than 60 plate specimens were cast with 3D glass fabric having three different thicknesses 6, 10 and 15mm and different number of layers to measure their tensile strength. Plates with one and two layers of chicken wires, as well as to plates with micro steel fibre of 0.75% volume fraction were tested under tensile for comparison with references plates. Cement mortar with 61.2MPa cube compressive strength at 28 days was designed for casing the plates.

## 2 Experimental work

An experimental program was devoted in this work to investigate the structural behaviour of self-compact reinforced concrete plates subjected to tensile load. Test Specimens with dimension (450 ×100×40) mm were divided into six croups according to the fibre reinforcement thickness, layers, direction of reinforcement, and type of reinforcement. First group consisted of plates cast with self-compact mortar (SCM) without any reinforcement; the second group was reinforced with 6mm 3D glass textile fibre, reinforced with one layer, two layers. Third group cast with10mm 3D glass textile fibre with one and two layers, while the fourth group contains 15mm 3D glass textile fibre casted with one layer only. The fifth plate group consisted of self-compact mortar reinforced with chicken wire casted with one and two layers, the sixth group reinforced with micro steel fibre of 0.75% volume fraction, three specimens were casted for each test age (28 and 90) days. All details of the specimens are shown in Table (1).

**Table 1.** Details of test specimens.

| **Group No.** | **Mix Symbol** | | **Direct Tensile Test 450×100×40 mm** | |
|---|---|---|---|---|
| | | | **28 days** | **90 days** |
| 1 | Ref. | Without any fiber | 3 | 3 |
| 2 | F6-1 | Glass fiber 6 mm thickness (one layer). | 3 | 3 |
| | F6-2 | Glass fiber 6 mm thickness (two layers). | 3 | 3 |
| 3 | F10-1 | Glass fiber 10 mm thickness (one layer). | 3 | 3 |
| | F10-2 | Glass fiber 10 mm thickness (two layers). | 3 | 3 |
| 4 | F15 | Glass fiber 15 mm thickness (one layer). | 3 | 3 |
| 5 | FS-1 | Mesh chicken wire (one layer). | 3 | 3 |
| | FS-2 | Mesh chicken wire (two layers). | 3 | 3 |
| 6 | M S F | With $V_f$=0.75 micro steel fiber | 3 | 3 |

### 2.1 Materials properties

The properties of materials used in the preparation of the tested reinforced and unreinforced self-compact mortar are described below:

1- Cement: Ordinary Portland cement (type I) from Krasta Factory was used to cast all the specimens, the chemical composition and physical properties are listed in Table (2). It conforms to the requirements of the Iraqi Standard Specification IQS. No. 5 /1984 [12].

2- Fine aggregate: Natural sand from Najaf sea region was used with 0.6 mm maximum size. The results of physical and chemical properties of the sand are listed in Table (3). Test results comply with the requirements of the Iraqi Standard Specification IQS. No.45-1984 [13].

3- Water: Ordinary water tape was used for both casting and curing all specimens.

4- High Range Water Reducing Admixture (HRWRA): A high performance concrete super-plasticizer (also named High Range Water Reduction Agent HRWRA) based on polycarboxylic technology, known commercially as Glenium 54 was used. It is produced by BASF Company and conforms to ASTM C 494 Type F [14]. Table (4) shows its properties.

**Table 2.** Chemical and physical properties of the cement

| Oxide | % | I.O.S. 5: 1984 Limits |
|---|---|---|
| CaO | 66.11 | – |
| $SiO_2$ | 21.93 | – |
| $Al_2O_3$ | 4.98 | – |
| $Fe_2O_3$ | 3.10 | – |
| MgO | 2.0 | < 5.0 |
| $K_2O$ | 0.75 | |
| $Na_2O$ | 0.35 | |
| $SO_3$ | 2.25 | < 2.8 |
| Loss on Ignition (L.O.I) | 2.39 | < 4.0 |
| Lime Saturation Factor | 0.93 | 0.66 - 1.02 |
| (L.S.F) Insoluble residu | 1.29 | < 1.5 % |
| Free lime (F.L) | 0.67 | - |
| **Compound Composition** | **%** | **I.O.S. 5: 1984 Limits** |
| $C_3S$ | 58.16 | – |
| $C_2S$ | 19 | – |
| $C_3A$ | 7.95 | – |
| $C_4AF$ | 9.43 | – |
| **Physical Properties** | **Test Result** | **I.Q.S.5:1984 [11] Limits** |
| Fineness, Blaine, $cm^2/gm$ | 3300 | >2300 |
| Setting Time: Initial hrs.; min Final hrs.; min | 1;08 4;00 | ≥45 min ≤10hrs |
| Compressive Strength ( MPa) 3-days 7-days | 20.0 25.0 | ≥15 ≥23 |

**Table 3.** Properties of fine aggregate.

| Physical properties* | Test results | Iraqi specification. 45/1984[13] |
|---|---|---|
| Specific gravity | 2.65 | - |
| Sulfate content | 0.3 % | Not more than 0.5% |
| Absorption | 1 % | - |
| Bulk density($kg/m^3$) | 1560 | - |

**Table 4.** Typical properties of sp (glenium54)*

| Form | Viscous Liquid |
|---|---|
| Commercial name | Glenium 54 |
| Chemical composition | Sulphonated melamine and naphthaline formaldehyde condensates |
| Appearance | Whitish to straw colored liquid |
| Relative density | 1.07 $gm/cm^3$at 20 °C |
| Chloride content | Nil. |
| pH | 5-8 |
| Storage | Should be stored in original containers and at above 5°C |
| Transport | Not classified as dangerous |
| Labeling | Not hazard label required |
| Alkali content (as $NaO_2$) equivalent) | 0.26% |

*According to manufacturer

## 2.2 Additive or mineral admixture

1- Fly Ash: Class F Fly Ash (FA) produced from Thermal Power plant in Turkey was used as an additive according to ASTM C618 [15], fly ash was used as 20% by weight replacement of cementitious material. The physical and chemical properties are presented in Table (5).

2- Silica Fume: Silica Fume (SF) produced by BASF Company was used as pozzolanic admixture. Cement was replaced by 5% of silica fume by weight of cementitious material. The silica fume used in this work conforms to the requirements of ASTM C-1240-05 [16], ASTM C-311-05[17]. The technical specifications of silica fume are presented in Table (6).

3- 3D textile glass fibre: 3D textile glass fibre woven fabric consists of two bidirectional woven fabric surfaces mechanically connected with vertical woven piles. Two S-shaped piles combine to form pillar, 8-shaped in the warp direction and I- shaped in the weft direction. For the use of a textile as reinforcement in concrete, textile should have an open mesh allowing the mortar or concrete to penetrate the textile for good bond between the materials, for this reason, 19 mm holes at distance 50mm center to center - for both direction were drilled (by tool cutters) for the 3D textile glass fibre glass woven; this process must ensure good penetration of the cement matrix Plats (1) shows the surface, thickness and fibres holes). 3D glass fibre textile woven was export from china. Tables (7) and (8) show typical properties, of different-Fabric Specifications.

4- Mesh chicken wire: The Rhombic shape meshes of reinforcement, fabricated from 0.54mm – nominal diameter steel bars, the opening in the long and short direction (10.6, 7.92mm) respectively.

5- Micro steel fibre: Steel fibres are used in self-compact mortars (SCMs) to enhance some properties and improve the ductility; it was manufactured by Ganzhou Daye Metallic Fibre Co., Ltd, China. (http// www.gzdymf.com), it is subjected to (ISO 9001 /2008), the properties are summarized in Table (9). After pouring the first layer of self-compact mortars; micro steel fibre (0.75%) volume fraction was randomly distributed by hand. See Plate (2) micro steer fibre casted.

**Table 5.** Properties of class (f) fly ash*

| Particular | Fly ash (Class F) | ASTM C 618 Class F fly ash [14] |
|---|---|---|
| Chemical composition | | |
| Silica ($SiO_2$%) | 65.65 | ($SiO_2$ + $Al_2O_3$+ $Fe_2O_3$) ≥ 70 |
| % Alumina ($Al_2O_3$) | 17.69 | |
| Iron Oxide ($Fe_2O_3$) % | 5.98 | |
| Lime (CaO)% | 0.98 | |
| Magnesia (MgO)% | 0.72 | |
| % Sulphur Trioxide ($SO_3$) | 0.19 | Max. 5.0 |
| Loss on Ignition | 3.1 | Max. 6.0 |
| $Na_2O$ | 1.35 | |
| $K_2O$ | 2.98 | |
| Physical properties | | |
| Specific gravity | 2.12 | |
| | | |
| Fineness ($cm^2$/gm) | 3600 | Min. 2250$cm^2$/gm |

**Table 6.** The technical specifications of silica fume

| Structure of material | Silica fume | Limits of ASTM C 1240-05 [15] |
|---|---|---|
| Color | Dark gray | |
| Density | 0.55-0.7 kg/$m^3$ | |
| Chlorine amount | < 0.1 % | |
| Specific surface area ($cm^2$/gm) | > 150000 $cm^2$/g | ≥ 150000 $cm^2$/g |
| $SiO_2$ | > 85 % | ≥ 85 % |
| CaO | < 1 % | |
| Activity index* | 156 % | ≥ 105 % |
| Specific gravity | 2.2 | |

**Table 7.** Glass fiber woven fabrics specifications*.

| Area Weight g/$m^2$ | Core (mm) | Density of Warp (ends /cm) | Density of Weft (ends /cm) | Tensile Strength (MPa) (Warp) | Tensile Strength (MPa) (Weft) |
|---|---|---|---|---|---|
| 900 | 6 | 15 | 10 | 5500 | 9400 |
| 1480 | 10 | 15 | 8 | 6800 | 12000 |
| 1650 | 15 | 12 | 6 | 7200 | 13000 |

*According to manufacturer.

**Table 8.** Properties of 3D textile glass fibre woven fabrics specifications.

| Weight/ area | From 820 to 2580g/$m^2$ |
|---|---|
| Surface Treatment | Silicon Coated |
| Width | 1.3m or Made to order |
| Weave Type | Plain Woven |
| Yarn Type | E-Glass |
| Alkali Content | Alkali Free |
| Standing Temperature | 260 °C |
| Color | White |
| Specific stiffness | Extremely High |
| Woven | parabeam |
| Acoustic insulation | Excellent |
| Wave transmittable | Well |
| Construction | Two layers and one hollow spacer |

**Table 9.** Specification of micro steel fiber*

| Property | Specification |
|---|---|
| Type | WSF 0213 |
| Surface | Brass coated |
| Tensile Strength | 2850 MPa |
| Length | 15mm |
| Diameter | 0.2 mm |
| Aspect ratio | 65 |

* Adopted by manufacturer

## 3 Mixing of self-compacting mortar (SCM)

The proportion of the constituents for the prepared mortar mix is 1:1 (cement sand) by weight of ordinary Portland cement and cementitious materials: fine aggregate with 0.6mm maximum size. The super-plasticizer (SP) had a dosage of 1liters per 100 kg of material (cement and binder), w/b of 0.36 was selected for this investigation. The mortar was prepared using a small mechanical mixer conforming to ASTM C305-04 [18]. The mixing procedure is descried below:

**First**, add cement, fly ash and silica fume in the mixer, operating at low speed for 30 sec. until a uniform distribution is reached,

**Second,** add sand and mix for 1 min. at medium speed,

**Third**, add the first part (2/3) of water and mix thoroughly for 30 sec. at low speed, stop 2 min to clean blades, now, add SP and remaining water and mix for 2 min. at normal velocity, Stop the mixer and wait for 1 min., and then finalize the process by mixing at normal velocity for 3 min.

**Finally**, do the discharge.

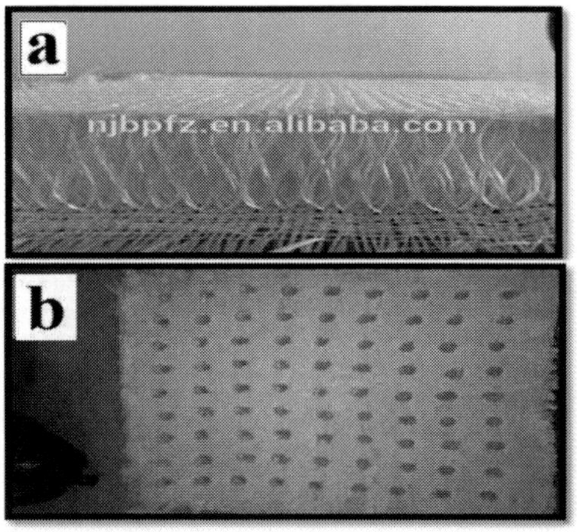


**Plates 1.** 3D textile glass fiber made in china; a) Zoom view. b) Holes made in 3D glass fiber.

**Plate 2.** Specimen casted molds for tensile test.

# 4 Fresh Mortar properties

## 4.1 Determination of slump-flow

The test apparatus for measuring the flow and viscosity of mortar comprises a mini slump cone with a graduated glass plate. Slump cone has 70mm and 100mm top and bottom diameters respectively with height of (59mm). The subsequent diameter of the mortar is measured in two perpendicular directions and the average of the diameters is reported. In this test, the truncated cone mold is placed exactly on the (100 mm) diameter graduated circle marked on the glass plate, filled with mortar and lifted upwards. Fresh properties of mortars were evaluated by the mean value of two perpendicular flow diameters in the spread test. Mortars were prepared manually in a container to observe its behaviour. The procedure for test was followed as described in [19] see Plate 3.

## 4.2 Determination of Flow Time

Flow time was determined in the v-funnel test; the dimension of v- funnel is show in Fig. 1.

# 5 Mechanical properties of hardened self- compact mortar

## 5.1 Compressive strength

This test was done on cubes according to the Standard Specification I.Q.S. No.5/1984[15]. A3000kN capacity testing machine was used for compressive test. The average compressive strength of three cubes (77×77×77mm) was recorded for each testing age (7, 28, 90 days).

## 5.2 Flexural strength

The flexural strength testing is carried out on prisms with (40 × 40 ×160 mm) self-compact mortar prism. The prism is then loaded at its center point until failure. Using three mortar prisms for each age (28, 90) days and the average of three results is adopted. A 20kN capacity Beijing United testing machine was used for this test. The flexural strength is calculated, as follows: -

$$Fr = 2PL/(3bd^2) \quad (1)$$

where: fr is the flexural strength [which is measured in (MPa), P is the maximum load, measured in (N), L is the clear span length, measured in (mm), b is the width of specimen, measured in (mm), Finally, d is the average depth of specimen, measured in (mm). The mortar prisms were prepared according to ASTM C 348-04[20]. Plate (4) shows the flexural test.

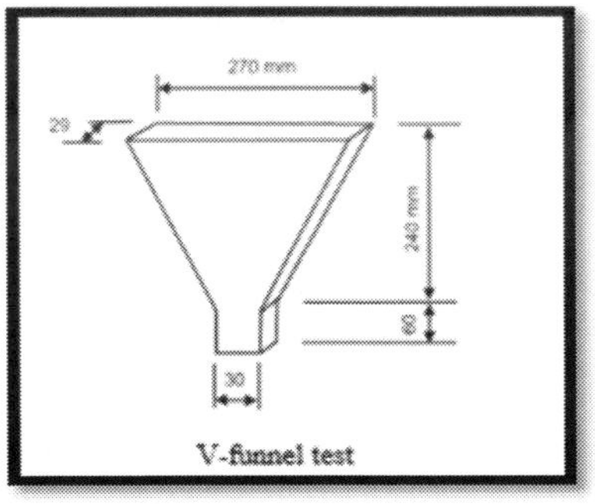


**Fig. 1.** V-funnel apparatus.

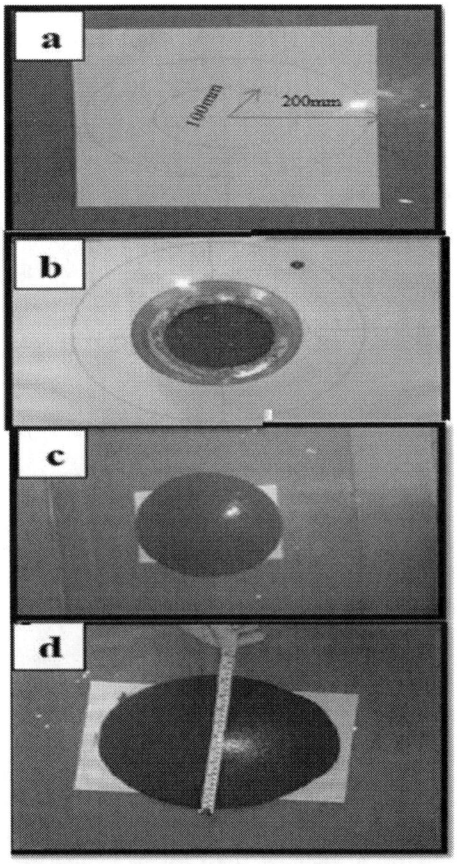


**Plate 3.** Flow test steps : a) Draw two circle (10&20) cm. b) Slump cone filled with mortar. c) Flow of mortar. d) Measured appendicular diameter.

https://doi.org/10.1051/matecconf/201816204008

### 5.3 Tensile strength of mortar

Tensile Strength of mortar test was tested according to B.S 6319-7:1985[21]. Briquette mold was used for this test, Plate (5) shows mortar tensile test. The average of three samples was used.

### 5.4 Tensile strength for plate specimens

The specimens are designed with a dimension (450×100 ×40 mm), hydraulic machine of 1000 kN capacity used for this test. To connect the specimens to the test device a re-usable steel connection was designed. In this connection, the specimen is clamped between two steel plates with a length of 100 mm, leaving a measurement area of 250 mm. The tensile stress in the specimen is uniformly introduced by friction between the clamping plates and the specimen element. The 'friction' load in orthogonal direction is provided by pre-tensioning 4 M 20 bolts between two steel plates that run over the clamping plates. To increase the friction coefficient and to prevent stress concentration on the uneven prism surface of the specimen a rubber layer was placed between the steel plates and the prism specimen, LVDT was clamped at the face of the prism specimen and connected to output screen data. Fig. 2 and Plate (6) shows setup of tensile test that was designed, tensile specimen connects to test device by holes and bolts.

**Plate 4**. Flexural test for SCM.

**Plate 5**. Tensile test for SCM

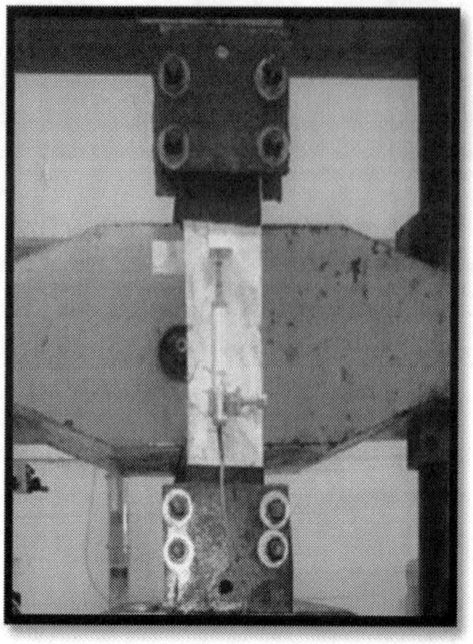

**Plate 6**. Tensile plate specimen test.

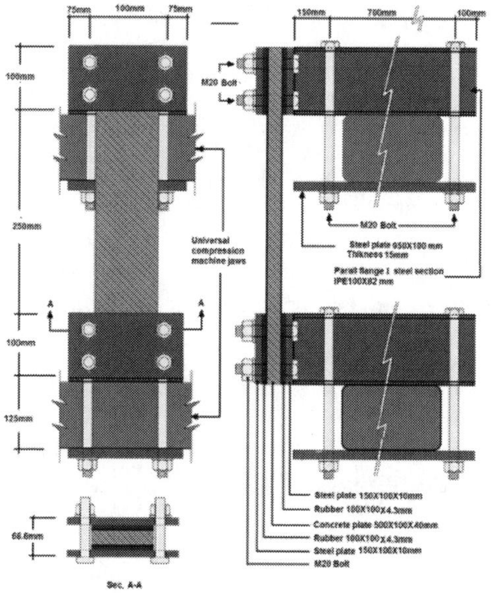


**Fig. 2**. Tensile test setup.

## 6 Results and discussion

### 6.1 Properties of fresh mortar

The test results relevant to the slump flow diameter, V-funnel flow time are presented in Table 10. The result shows the mixture had slump flow diameter, V-funnel flow time conforming to [19].

Where Dm is the mean value of the two perpendicular diameters, measured in (mm); Do is the initial diameter of the base of the cone, measured in (mm), and finally, the (t) represents the time of flow in the v-funnel, which is measured in seconds.

### 6.2 Properties of hardened SCMs

The hardened properties of the mortars were summarized in Table 11. From results shown in Table 11, the strength increased with ages. This development in compressive strength, tensile strength and flexural strength can be attributed to the continuous hydration process (C-S-H), also presence of silica fume tends basically to consume the calcium hydroxide crystals released from the hydration process leading to the formation of further calcium-silicate-hydrate (secondary C-S-H).

### 6.3 Direct tensile strength of plate specimens

The tensile test was conducted at 28 and 90 days after casting of specimens. During the execution of the tensile tests changes were made to the connection device based on the observations during the experiment, the specimen started to slip out of the connection device at a load of 11kN. This indicates that frictional coefficient between these layers was not sufficient. It was decided to glue the rubber layer on the steel clamping plate, making the connection suitable for testing, and by increasing the pretension in the bolts increased the frictional force between the steel, rubber and concrete. After this measurement, the specimens failed before slipping occurred. Table 12 and Figs (3 - 5) show an overview of the results of the tensile tests.

**Table 10.** Fresh properties of SCMs

| SCMs | Slump Flow Diameter (cm) | V-funnel time (s) | $G_m = \left(\frac{Dm}{D_0}\right)^2 - 1$ | $R_m = \frac{10}{t}$ |
|---|---|---|---|---|
| | 25.4 | 9.5 | 5.45 | 1.05 |
| Acceptance criteria of SCMs suggested by [19] | 24-26 | 7-11 | - | - |

**Table 11.** Results of hardened properties of SCMs

| Type of Test | Compressive Strength (MPa) | | | Tensile Strength (MPa) | | Flexural Strength (MPa) | |
|---|---|---|---|---|---|---|---|
| Test Age (days) | 7 | 28 | 90 | 28 | 90 | 28 | 90 |
| | 45.7 | 61.22 | 77.9 | 3.3 | 5.1 | 5.3 | 9.12 |

**Table 12.** Direct tensile strength of plate specimens

| Specimen | $F_t$ ( MPa) 28 days | $F_t$ ( MPa) 90 days |
|---|---|---|
| R | 3.9 | 4.1 |
| F6-1 | 4.1 | 4.63 |
| F6-2 | 4.31 | 5.2 |
| F10-1 | 4.43 | 4.88 |
| F10-2 | 4.45 | 5.28 |
| F15 | 4.3 | 4.85 |
| S-1 | 4.2 | 4.4 |
| S-2 | 4.3 | 4.6 |
| MSF | 5.21 | 6.7 |

The direct tensile strength increases with age for all mixes as shown in Fig. 3, this development in strength can be attributed to the continuous hydration process (C-S-H), also presence of silica fume tends basically to consume the calcium hydroxide crystals released from the hydration process leading to the formation of further calcium -silicate- hydrate (secondary C-S-H) and contributing to interfacial bond strength between diverse types of fiber used in present work and the matrix. All fibrous mixes demonstrated higher tensile strength relative to reference (R) mix at all ages. Direct tensile strength indicated significant increase in strength due to the inclusion of fibers. From the results shown in Table 12, Figs (4 & 5) the percent of increase in tensile strength was found to be increase with the increase in thickness and number of layers of 3D textile, about (5.13 to 13.6%), (12.9 to 19%) for ages (28, 90 days) respectively. Also, the increase in tensile strength when reinforced with two layers, about (10.51 to 14.1%), (26.8 to 28.8%), for mesh chicken wire increased tensile strength (7.7, 7.3%) and for two layers about (10.2, 12.2%), finally in case of micro steel fiber the increment (33.6, 63%) at ages (28, 90 days), this result agree with [12]. This increase in strength is likely due to the bond between the matrix and fibers which increase the possibility of counteracting crack growth and preventing micro-cracks coalescing to form large cracks. The strength increase due to inclusion of steel fibers is attributed to the mechanism of steel fibers in arresting crack progression. Where, the presence of fibers in concrete restrains the development of internal micro-cracks and thus contributes to an increased tensile strength. Thus, the increase in fiber thickness or fiber layers, i.e. increase in fiber content, leads to an increase in the tensile strength of SCMs as shown in Figs (3- 5). On the other hand, with an increase in fiber content, the fibers become more densely spaced, and they may hinder growth of micro-cracks within the brittle matrix and increase the direct tensile strength of the plates.

All the specimen's testes showed the same failure mode; the failure cracks on each specimen were located near the support and clearly progressed from one to the other side of the specimen. This indicated the specimens were not loaded on pure centric tensile, but on a small bending moment as well, when the failure point of the fiber in the crack stabilizing stage was reached it was observed that the width of the cracks would increase rapidly.

The following reasons are identified that could introduce a bending load on the specimens:

1. The specimens were cast in an open mold, this leads to one smooth surface and one rough surface on the specimen. The friction coefficient of concrete and rubber is likely to be dependent on roughness of the surface. Different friction coefficients on each side of the specimen could lead to an uneven introduction of the tensile stress.
2. Using 4 bolts to apply the load on the clamping plates to introduce the tensile stress by friction, causes

unequal pre-tension in these bolts, this could lead to uneven introduction of the friction load in the specimen.

After failing of the reinforcement and with decreasing tensile load, the specimens showed a long strain softening trajectory due to the pull-out (friction) of the inner fibers of the yarn, as indicated in Fig. 3, this behavior conformed to [8].

## 7 Conclusions

Based on the results from experimental tests the following conclusions can be drawn for the self-compact mortars reinforced with diverse types of fibers:

1. The direct tensile strength increases with age for all mixes. All fibrous mixes present higher tensile strength relative to reference mix at all ages.
2. The increase in fiber thickness or fiber layers, (i.e. increase in fiber content) leads to an increase in the tensile strength of SCMs.
3. Increase in strength is likely due to the good bond between the matrix and fibers which increase the possibility of counteracting crack growth and prevent micro-cracks coalescing to form large crack.
4. The tested for reference (R) (unreinforced), also SCM reinforced with mesh chicken wire and micro steel fiber, show the same behavior (failed suddenly and split into two separate parts).
5. 3D textile fibers plates are cracked at failure without separation.
6. The tested specimens showed the same failure mode; the failure cracks on each specimen were located near the support and clearly progressed from one to the other side of the specimen.

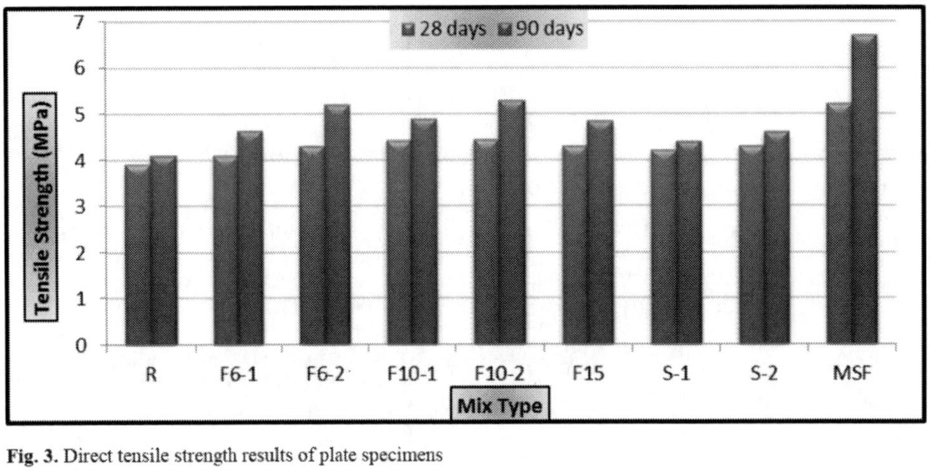


**Fig. 3.** Direct tensile strength results of plate specimens

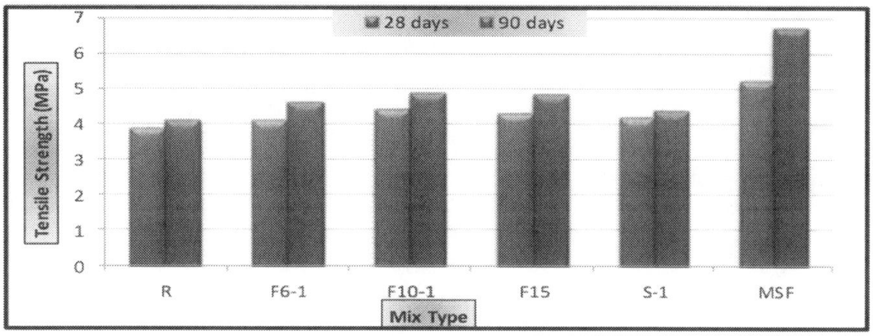


**Fig. 4.** Direct tensile strength results for one layer reinforcement.

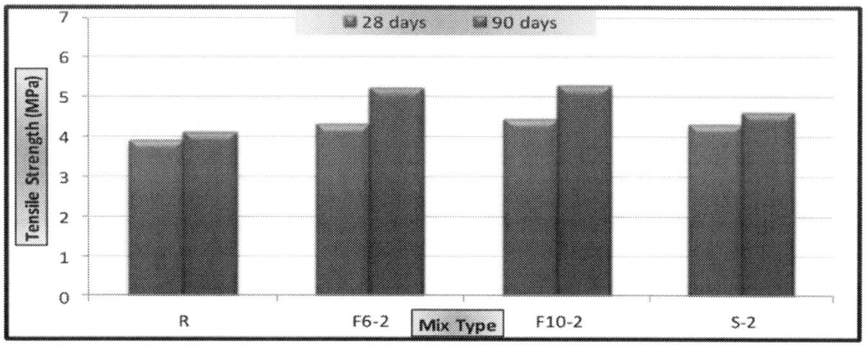


**Fig. 5.** Direct tensile strength results for two-layer reinforcement.

## References

1. T. C. Triantafillou, C. Papanicolau, Mater. & Struct. J. E **39**, 85 (2007)
2. D. Bournas, P. Lontou, C. G. Papanicolau, T. C. Triantafillou, ACI Struct J **104**, 740 (2007)
3. A. L. Dossland, *"Fiber Reinforcement in Load Carrying Concrete Structures"*. Ph.D. Thesis, Norwegian University of Science and Technology, Trondheim, Norway, pp. 254, (March 2008)
4. J. R. Davis, *Tensile testing*, 2nd edition, ASM international (2004)
5. W. Brameshuber, *Textile Reinforced Concrete* - State-of-the-Art, TC 201-TRC. Bagneux, France. RILEM Publications S.A.R.L. (2006)
6. J. Hegger, J. & S. Voss, Eng. Struct. E **30**, 2050 (2008)
7. H. Fangyu, C. Huisu, J. Kefeng, Z. Wulong, L. Tao, Y. Yujie, Const. & Build. Mat. J., E **65**, 620 (2014)
8. W. T. Lin, R. Huang, C. Lee, & H. M. Hsu, Journal of Marine Science and Technology, E **16**,214, (2008)
9. B. M. Gangadharappa, K. E. Prakash, G. S. Suresh, M. N. Shesha, (IJETCAS), E **5**, 239, (2013)
10. S. Eldin, M. Mourad, R. R. No. 47/426 King Saud University, college of engineering research center, (2014)
11. P. S. Parande, D. G. Gaidhankar, (IJRAT), E **2**, 2321 (2014)
12. ISQ. No. 5/1984, *"Portland cement"*, COSQC, Iraq
13. ISQ No. 45/1984, *"Ruins the natural resources used in concrete construction"*, COSQC, Iraq
14. ASTM C-494/C 494M-05, *"Standard Specification for Chemical Admixtures for Concrete"*
15. ASTM C-618-03, *"Standard specification for coal fly ash and raw or calcined natural pozzolan for use in concrete"*
16. ASTM C-1240-07, *"Standard Specification for the Use of Silica Fume as a Mineral Admixture in Hydraulic Cement Concrete, Mortar, and Grout"*.
17. ASTM C-311-05. *"Standard Test Methods for Sampling and Testing Fly Ash or Natural Pozzolans for Use in Portland-Cement Concrete"*
18. ASTM C305-04, *"Standard Practice for Mechanical Mixing of Hydraulic Cement Pastes and Mortars of Plastic consistency"*.
19. EFNARC, *"Specification and Guidelines for Self-Compacting Concrete"*. London", UK: Association House, February (2002)
20. ASTM C348-14, *"Standard Test Method for Flexural Strength of Hydraulic-Cement Mortars"*
21. B.S 6319-7 Part 116,*"Method for Determination of Tensile Strength of Mortar"*, British Standard Institution (1985)

# Torsional strength evaluation of reinforced SCC box beams strengthened internally by opened and closed transverse concrete diaphragms

*Ali* Aziz[1,*], and *Oday* Hashim[1]

[1]Faculty of Engineering, Al-Mustansiriayah University, Baghdad, Iraq.

**Abstract.** There are a few researches about box (hollow core) RC beams under the effect of torque moment; therefore, a torsional failure mechanism is not understood well. In the present paper, the concept (technique) of adding internal concrete diaphragms, in transverse direction, is adopted. Six self-compacting concrete beam specimens having a dimension of (2200x220x350mm) for length, width and height, respectively were poured and tested. Three variables are adopted in the present study, section type, number and type of the internal diaphragms. Experimental results show that the ultimate torque moment increases about (43%-61%) and (89%-94%) for beam specimens strengthened internally by two and four diaphragms respectively. Also, increasing internal diaphragms from two to four, led to increase the ultimate torque moment about (33%-46%). Furthermore, the change of section from box section with four internal diaphragms to solid section, led to increase in ultimate torque moment about (28%-33%). The paper concludes that the increase in torsional capacity is attainable for box (hollow core) beams by using the technique of strengthening by internal diaphragms; Also, the box sections can be made with internal opened diaphragms and possess same capacity or contribution of the closed ones.

## 1 Introduction

A box concrete section (or hollow core beam or box beam or hollow beam) is often used for beam design, particularly for long spans bridges for reducing dead load and saving cost construction. Many structural members such as building and bridge elements are subjected to equilibrium torsional moments. In order to avert failure of these elements at torsional load, adequate reinforcement (longitudinal and transverse), repairing and strengthening are required. Strengthening of concrete members to resist torsional stresses may be achievable by one of the following techniques: (i) increasing the member cross-sectional area, (ii) adding transverse reinforcement, (iii) using externally bonded steel plates, (iv) applying an axial load to the member by external prestressing[1, 2].

Reinforced concrete sections under torsional stresses and externally strengthened by CFRP are interested in several research [3, 4]. Beams reinforced internally with GFRP reinforcements under pure torsion are also interested [5].

Due to rapid development in concrete technology and construction techniques, new concrete types such as SCC has emerged. Many researches showed that SCC is commonly used within precast and cast-in-place construction. Furthermore, it is also used in the thin concrete members and where the steel reinforcement tightness is required. However, other implementations of SCC include caissons, drilled piers, walls and bridge abutments [6].

There are few researches about box reinforced concrete beams under the effect of torque moment; therefore, a torsional failure mechanism is not understood well. The concept of adding internal concrete diaphragms (opened or closed), in transverse direction is a new technique for strengthening reinforced concrete box (hollow core) sections under the effects of torsional stresses and will be adopted in the present study. It may be noted that, additional variables can be added later to the present study as a suggestions for future researches.

## 2 Research Significance

Despite the many investigations on the torsional behavior of reinforced concrete beams with solid or hollow cross section, little has been done on the applications of SCC with a complete absent for strengthening internally by any techniques. So the need to study the torsional behavior of such sections with strengthening by special technique gives more attention. The main objectives of the current study is to investigate, experimentally, torsional behavior of reinforced SCC rectangular box (hollow core) beams strengthening transversely by internal diaphragms with different variables, including, type of section (solid or box), number and type of the internal diaphragms (closed or opened).

* Corresponding author: alihaziz@uomustansiriyah.edu.iq

© The Authors, published by EDP Sciences. This is an open access article distributed under the terms of the Creative Commons Attribution License 4.0 (http://creativecommons.org/licenses/by/4.0/).

## 3 Experimental program

The experimental program consists of cast and test of six reinforced SCC beam specimens. One specimen was control beam; four beams were strengthened using internal diaphragms while the last beam was a solid. Three variables were adopted in the present study, section type (solid or box), number and type of the internal diaphragms (closed or opened). The specimen's dimensions, clear span, concrete grade, load location and reinforcement (longitudinal and transverse) were kept constant thought out this study. The experimental work includes, also, a series of tests carried out on control specimens (cubes and cylinders). Cracking load, ultimate load, mode of failure, torque-angle of twist behavior and torque-strain curve are recorded, discussed and presented in this study.

### 3.1. Beam specimens description

The specimens have a dimension of (2200x220x350mm) for length, width and height respectively. Since, the hollow and solid reinforced concrete beams can be designed directly according to ACI-318 code [7], the longitudinal and transverse reinforcement were calculated based on code requirement for torsion. Each beam was reinforced by (2ϕ12mm) bars at the top and bottom, while, the transverse reinforcement consists of (ϕ10@60mm) stirrups at edges and (ϕ10@120mm) stirrups at the mid, see Figure (1). As mentioned before, the concept of strengthening of hollow sections by transverse internal diaphragms (to resist torsional stresses) are adopted in the present paper. Therefore, lateral (transverse) closed or opened diaphragms with thickness of (50mm) were created (during concrete poured) inside the hollow beam specimens. It may be noted that, all diaphragms were reinforced longitudinally (vertically) and transversely by (6ϕ10mm) and (4ϕ10mm) bars respectively. Each beam specimen was designated in a way to refer to section type (HB= Hollow Beam or SB=Solid Beam), number of internal diaphragms (0, 2 and 4) and type of diaphragm (CD=Closed Diaphragm, OD=Opened Diaphragm). Therefore, for example, the beam specimen (HB-4CD) is a hollow beam made with four transverse closed diaphragms. Description and details of tested beam specimens are shown in Table (1) and Figures (2) to (8).

**Table 1.** Beam Specimens Description

| Beam | Dimensions (mm) | | | Description |
|---|---|---|---|---|
| | L | W | D | |
| HB-0D* | 2200 | 220 | 350 | Hollow |
| HB-2CD | | | | Hollow With Two Closed Diaphragms |
| HB-2OD | | | | Hollow With Two Opened Diaphragms |
| HB-4CD | | | | Hollow With Four Closed Diaphragms |
| HB-4OD | | | | Hollow With Four Opened Diaphragms |
| SB | | | | Solid |

*Reference Beam
Note: HB= Hollow Beam, CD=Closed Diaphragm, OD=Opened Diaphragm, SB=Solid Beam

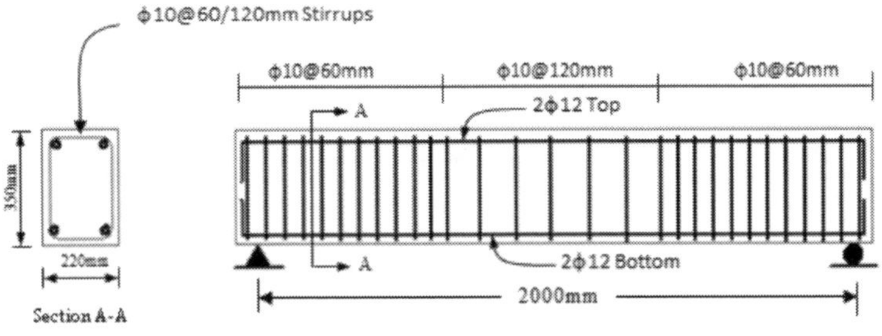


**Fig.1.** Reinforcement Details and Dimensions of the Tested Beams

**Fig. 2.** Reinforcement Details and Diaphragms Location

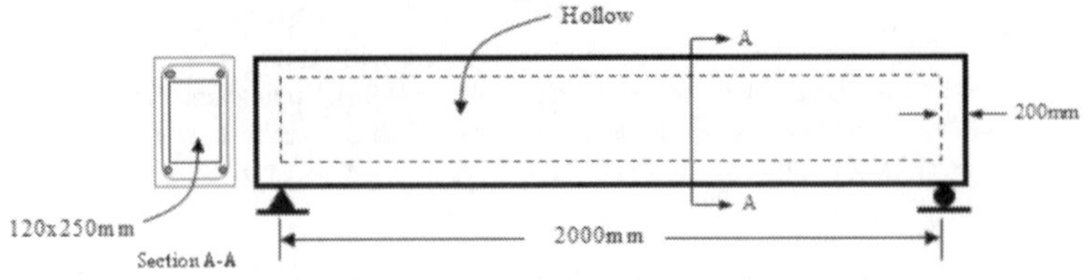


**Fig. 3.** Description of Beam Specimen (HB-0D)

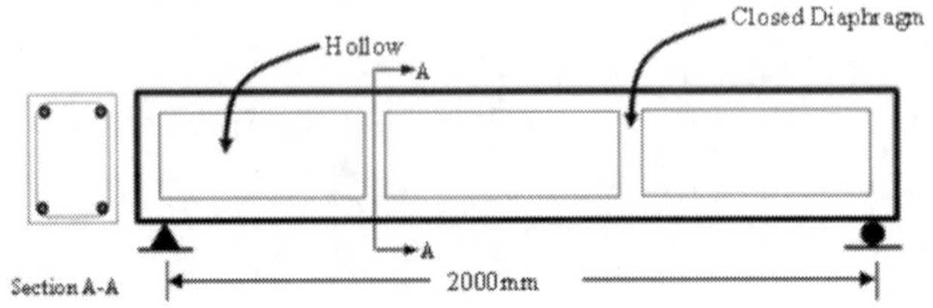


**Fig. 4.** Description of Beam Specimen (HB-2CD)

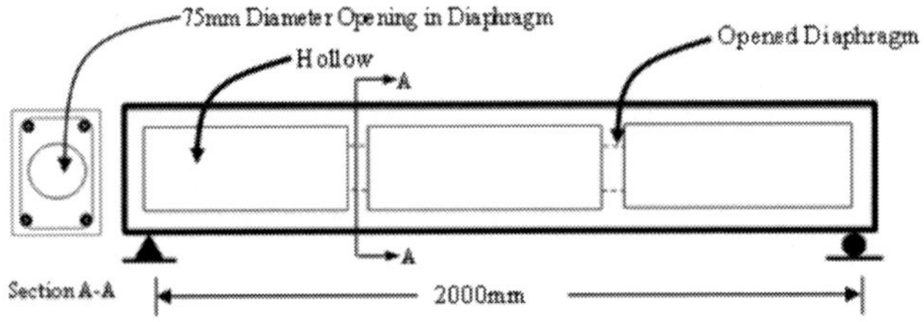


**Fig. 5.** Description of Beam Specimen (HB-2OD)

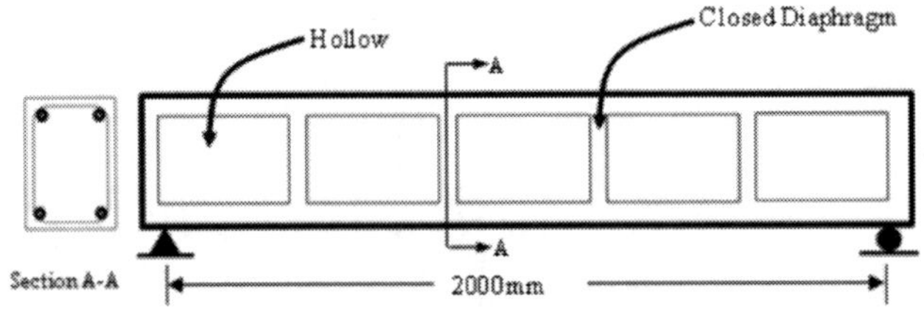


**Fig. 6.** Description of Beam Specimen (HB-4CD)

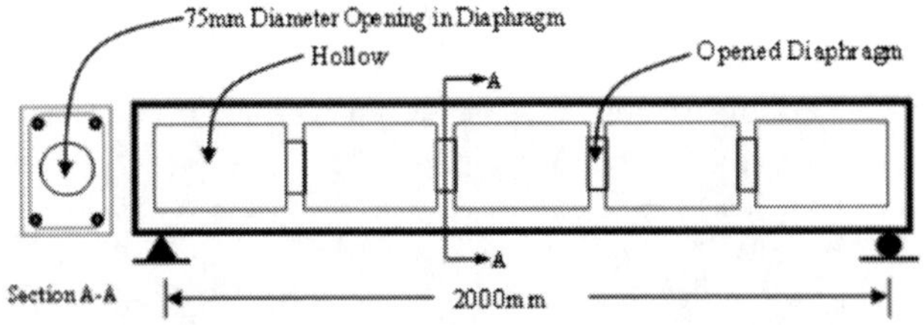


**Fig. 7.** Description of Beam Specimen (HB-4OD)

 https://doi.org/10.1051/matecconf/201816204009

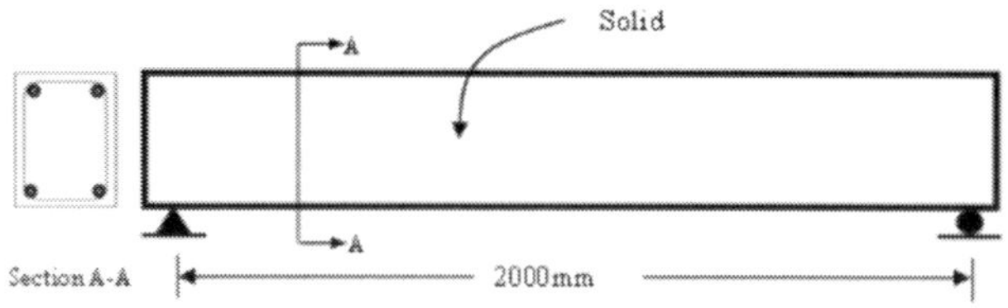


**Fig. 8.** Description of Beam Specimen (SB)

### 3.2. Molds

One wooden mold with (18mm) thickness has been used for casting the beam specimens. The mold consists of a bed and four movable sides, these sides have been fixed to the bed by screws. The clear dimensions of the mold were (2200x220x350mm) for length, width and height respectively. For casting the mold was oiled and the required reinforcement cage was placed inside the mold before casting the concrete batch.

To form the box beam, the cross section contains a polystyrene void that blocks out the center of the beam over specified lengths. The void is included to block out concrete where it does not add much resistance to the section and reduce the self-weight of the beam. For practical reasons, the void (box) in the beam cross-section is obtained by a stay-in-place polystyrene blocks. It may be noted that, the closed diaphragms are created directly when the spacing between Polystyrene blocks are filled by SCC during the pouring process. While, to create opened diaphragms, UPVC pipe of diameter of (75mm) is restricted between polystyrene pieces. For all tested beams, beyond the cells (at the ends), whole beam section (220x350mm) was solid concrete.

### 3.3. Materials

To manufacture the test beams, properties and description of the used materials are reported and presented in Table (2); and mix proportions of concrete (by weight) are reported and presented in Table (3).

## 4 Measurements and Instrumentation

All beams as well as control specimens were tested by using Hydraulic Testing Machine (MFL type) with a maximum range capacity of (3000kN), Figure (9). To ensure free rotation at supports, spherical bearing seats are used as shown in Figure (10). To determine the angle of twist, simple method was used by fixing a dial gauge to the beam ends (at the bottom at a point 45mm from longitudinal axis of the beam) as shown in Figure (11). The gauge (0.01mm/div. accuracy) recorded the vertical displacement to estimate the angle of twist (in radians) at every load stage.

**Table 2.** Properties of Construction Materials

| Material | Descriptions |
|---|---|
| Cement | Ordinary Portland Cement (Type I) |
| Sand | Natural sand with maximum size of (4.75mm) |
| Gravel | Crushed gravel of maximum size (14 mm) |
| Limestone powder | Fine limestone powder of Jordanian origin |
| Superplasticizer | Sika viscocrete-5930 manufactured by BASF Construction Chemicals |
| Reinforcing Bars | ($\phi$12mm) deformed steel bar, having (495 MPa) yield strength ($f_y$)<br>($\phi$10mm) deformed steel bar, having (510 MPa) yield strength ($f_y$) |
| Water | Clean tap water |

**Table 3.** Concrete Mix Proportions

| Material | Quantity | Ratio | Limits |
|---|---|---|---|
| Cement (Kg/m$^3$) | 450 | | 350-600 |
| Sand (Kg/m$^3$) | 750 | 33%[a] | <40% |
| Gravel (Kg/m$^3$) | 900 | 40%[b] | <50% |
| Silica fume (Kg/m$^3$) | 30 | | |
| Limestone (Kg/m$^3$) | 130 | | |
| Super plasticizer (L/m$^3$) | 10 | 1.64%[c] | <2% |
| Water (L/m$^3$) | 190 | | |
| W/P (%) | | 31.1[d] | 28% - 38% |

a-Sand ratio=Sand/(cement+ sand+ gravel+ silica fume+ limestone)
b-Gravel ratio=Gravel/(cement+ sand+ gravel+ silica fume+ limestone)
c-Super plasticizer ratio= Super plasticizer/(cement content +silica fume+ limestone)
d-W/P=Water/(cement content+ silica fume+ limestone)

**Fig. 9.** Test Machine and Beam Setup

**Fig. 10.** Spherical Bearing Seats at Supports

**Fig. 11**. Locations of Dial Gages

## 5 Testing procedure

The testing machine, shown in Figures (9) and (11), was used to test all beam specimens as well as control samples. The vertical load can be applied directly on the beam specimens at any point and the bearings should remain fixed without any rotation around the beam's longitudinal axis. In the present study, the applied loads outside the bed of the test machine were needed in order to get the torsional movement. The test requirements need to:-

a- Create ball bearing (move the supports in a circular manner).

b- Transmitting the load from the center of the test machine to the two external points (moment arm). In this paper, the used clamping frame (loading frame) is shown in Figures (9) and (11). The steel frame consist of two clamps work as an arms for applied torque moment with separated faces to connect them over the beam by means of bolts (for each arm, four bolts were used). The frame is made with two steel shafts of thick steel plate of (20 mm) attached by welding. The final shape of the steel frame is look like to a bracket. The steel arms are capable to provide eccentricity of a maximum of (600 mm) with respect to the beam longitudinal axis. To transmit the applied loads from the center of the test machine to the arms, steel girder of (2 m) length and (250 mm) depth was used. The steel girder was fixed to the machine of test, as shown in Figures (9) and (11).

## 6 Characteristics of SCC

### 6.1. Characteristics of fresh SCC

Four tests to evaluate three characteristics (filling ability, segregation resistance and passing ability) are made to check self-compacting concrete. Table (4) shows the test methods, test results and SCC requirements according to EFNARC [8]. Tests results show that the used SCC confirmed to standard requirements.

### 6.2. Characteristics of hardened SCC

All control specimens (cylinders and cubes) were removed from curing at the age of (28 days). Compressive strength of SCC for cylinders and cubes are measured based on ASTM C39-96 [9] and BS1881-116 [10] respectively. Tests results are reported and presented in Table (5).

**Table 4.** Test Results of Fresh SCC

| Test | Property | Test Result | EFNARC |
|---|---|---|---|
| Slump (mm) | Filling ability | 800 | 650-800 |
| $T_{50}$ (sec) | | 2.81 | 2.0-5.0 |
| V-funnel (sec) | Segregation Resistance | 8.28 | 6.0-12 |
| L-Box | Passing ability | 1.0 | 0.8-1.0 |

**Table 5.** Test Results of Hardened SCC

| Property | Test Result |
|---|---|
| Cube Compressive Strength ($f_{cu}$), MPa | 41 |
| Cylinder Compressive Strength ($f_c'$), MPa | 38 |

## 7 Test results

### 7.1 Cracking and ultimate load

The results of cracking torque (Tc) and ultimate torque (Tu) are shown in Table (6) for all tested beam specimens. The ultimate torque increases about (61%) and (94%) for beam specimens (HB-2CD and HB-4CD) which strengthened internally by two and four closed diaphragms respectively. While, for beam specimens (HB-2OD and HB-4OD) who strengthened internally by two and four opened diaphragms, the ultimate torque increases about (43%) and (89%) respectively. In comparison with the reference beam, (HB-0D), the cracking torque increases about (57%) and (100%) for beam specimens (HB-2CD and HB-4CD) which are strengthened internally by two and four closed diaphragms, respectively. While, for beam specimens (HB-2OD and HB-4OD) who strengthened internally by two and four opened diaphragms, the cracking torque increases about (29%) and (86%), respectively. This means the presences of internal diaphragms improves the torsional resistance and allow higher forces to be carried through internal diaphragms. When the beam section changes from box section (HB-0D) (without internal diaphragms) to solid section (SB), the cracking torque and ultimate torque increase by (43%) and (122%), respectively. For box beam specimen, only the thinned wall (ribs) is contributed to resist both, cracking and ultimate loads. From the other hand, full section of solid beam (SB) is contributed to resist both, cracking and ultimate loads. The results indicated that the increase in torsional capacity is attainable for box (hollow core beams) by using the concept of strengthening by internal opened or closed diaphragms.

**Table 6.** Test Results of Beam Specimens

| Beam | Tc (kN.m) | Increase in Tc* | Tu (kN.m) | Increase in Tu* |
|---|---|---|---|---|
| HB-0D | 5.25 | - | 17.25 | - |
| HB-2CD | 8.25 | 57% | 27.75 | 61% |
| HB-2OD | 6.75 | 29% | 24.75 | 43% |
| HB-4CD | 10.5 | 100% | 33.45 | 94% |
| HB-4OD | 9.75 | 86% | 32.55 | 89% |
| SB | 7.5 | 43% | 38.25 | 122% |

### 7.2. Modes of failure

The progress of cracks provided useful information regarding the failure mechanism of tested specimens. First crack of all specimens occurred at mid span and increased gradually. When the torque moment was increased, cracks appeared on each side and finally took the spiral shape. Figures (12) to (14) show the mode of failure of the tested specimens. The beam specimens were failed by torsional spiral cracks (diagonal concrete cracks). For box beams, the extensive concrete cracking is bounded between diaphragms which ultimately resulted in beam failure.

**Fig. 12.** Beam Specimens Failure Mode

**Fig. 13.** Close-up Mode of Failure of Tested Beam

**Fig. 14.** Failure Mechanism of Tested Beam

### 7.3 Effect of number of internal diaphragms

As shown in Table (6), the ultimate torque increases about (43%-61%) for beam specimens (HB-2OD and HB-2CD) which are strengthened internally by two diaphragms. While, for beam specimens (HB-4OD and HB-4CD) which are strengthened internally by four diaphragms, the ultimate torque increases about (89%-94%). Also, increasing of internal diaphragms from two to four, led to increase the ultimate torque moment about (33%-46%). From the other hand, the change of section from box section with four internal diaphragms to solid section, led to increase in ultimate torque moment by about (28%-33%). This is due to increase in torsional stiffness due to presence of internal diaphragms or full solid section.

### 7.4 Effect of type of internal diaphragms

When the internal diaphragms changed from opened to closed type, the ultimate capacity increased about (3.6%) and (2.76%) for beam specimens containing two and four internal diaphragms respectively. Since the increase is marginal (can be negligible), it can be concluded that the box sections can be strengthened by internal opened diaphragms and possess same capacity or contribution of the closed ones.

### 7.5 Effect of hollowness

When the beam section changes from box section (HB-0D) (without internal diaphragms) to solid section (SB), the cracking torque and ultimate torque increase by (43%) and (122%) respectively, see Table (6). For box beam specimen, only the thinned wall (ribs) is contributed to resist both, cracking and ultimate loads. From the other hand, full section of solid beam (SB) is contributed to resist both, cracking and ultimate loads.

### 7.6 Torque-angle of twist behavior

Figure (15) shows the relationship between torque and angle of twist (T-θ Diagram) for all beams. Ultimate angle of twist (θ) for all tested beam specimens are shown in Table (7). The strengthening using internal diaphragms can obviously change the deformation capacity of box beams. Test results indicated that the angle of twist was approximately constant at about ($15x10^{-2}$ radian) until occurrence of the first crack then increased gradually until failure. Before the first crack, at the same level of torque, the angle of twist of strengthened beams was smaller than the angle of twist of control beams. Compared to the control beam (HB-0D), the ultimate twist angle of beam specimens (HB-2-CD), (HB-2-OD), (HB-4-CD) and (HB-4-OD) are increased by (37%), (43%), (100%) and (144%), respectively. This is may be due the transverse diaphragms that were placed inside the beam and that allowed to higher torque to be carried and as a result, the corresponding angle of twist was increased. Moreover for beam specimen (SB), the ultimate twist angle is increased slightly, about (10%), as the test results indicated. This may be due to high rigidity of this beam (entire beam section will resist torsional stresses and as a result, the corresponding angle of twist decreased). For the control beam (HB-0D), the presence of hollow seems to be more effective in decreasing the crack torque and torsional capacity because the hollow decreased the section rigidity (the weakness extend throughout an entire the whole beam), as a result, the corresponding ultimate twist angle is decreased.

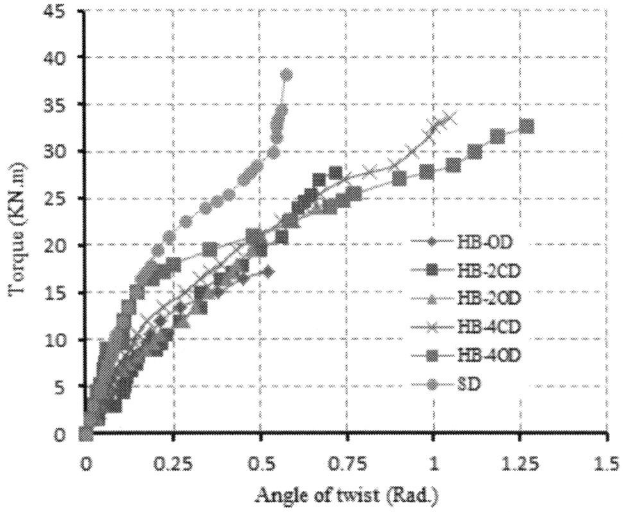


**Fig 15.** Torque-Angle of Twist Behavior

### 7.7 Ductility

The toughness may be considered as a ductility indicator. The toughness was measured as the area under the torque-angle of twist curve for each beam specimen using Microsoft Excel. In Table (7), it is noticed that toughness of the strengthened beam specimens (HB-2-CD), (HB-2-OD), (HB-4-CD) and (HB-4OD) is higher than reference beam (HB-0D). This might be because the strengthening by using internal transverse diaphragms led to increase both, torque and rotational (twisting) capacity and as a result, the toughness of these beams increased. Clearly in Table (7), when the number of internal diaphragms increased from two to four diaphragms, the corresponding toughness increased too. The unstrengthened reference beam (HB-OD) did not provide any restraint against rotation, and recodes lowest toughness. While, the solid beam (SB) provides high restraint against rotation (angle of twist) because entire section resists the torsional stresses and as a result, less angle of twist was recorded, even though the beam torsional capacity is high.

**Table 7.** Toughness of Tested Beam Specimens

| Beam | θ (Rad.) | θi/ θr (%) | Toughness (kN.m. rad) x10$^{-4}$ |
|---|---|---|---|
| HB-0D | 0.5214 | - | 391 |
| HB-2-CD | 0.7162 | 137 | 537 |
| HB-2-OD | 0.7448 | 143 | 559 |
| HB-4-CD | 1.0455 | 200 | 784 |
| HB-4-OD | 1.2718 | 244 | 838 |
| SB | 0.5729 | 110 | 430 |

## 8 Conclusions

Based on experimental results, the following conclusions can be drawn:-

1-The ultimate torque moment increases about (61%) and (94%) for beam specimens which are strengthened internally by two and four closed diaphragms respectively. While, for beam specimens which are strengthened internally by two and four opened diaphragms, the ultimate torque increases about (43%) and (89%), respectively.

2- The cracking torque moment increases about (57%) and (100%) for beam specimens which are strengthened internally by two and four closed diaphragms, respectively. While, for beam specimens that are strengthened internally by two and four opened diaphragms, the cracking torque increases about (29%) and (86%), respectively.

3- When the section changes from the box section, (without internal diaphragms), to solid section, the cracking and ultimate torque increase about (43%) and (122%), respectively.

4- When the internal diaphragms changed from opened to closed type, the ultimate capacity increased about (3.6%) and (2.76%) for beam specimens containing two and four internal diaphragms, respectively. Since the increase is marginal (can be negligible), it can be concluded that the box sections can be made with internal opened diaphragms and possess same capacity or contribution of the closed ones.

5- The toughness of the strengthened beam specimens (by internal closed or opened diaphragms) is higher than the box beam without any strengthening.

## References

1. J. MacGregor, and M. Ghoneim, ACI St. J. **92**, 211-218 (1995)
2. A. Khalil, E. Etman, A. Atta and S. Fayed, IOSR-JMCE, **12**, 30-41 (2015)
3. B. Sahib, A. Hameed, and I. Kh.uder , IJMME **12**, 754-769 (2012)
4. A. Abeer, A. Allawi, and K. Chai, WASET-IJCSE **7**,115-122( 2013)
5. A. Prabaghar, and G. Kumaran, IJCS, **2**, 570-594 (2011)
6. H. Okamura and M. Ouchi, JACT **1**, 5-15 (2003)
7. ACI Committee 318M, ACI (2014)
8. EFNARC, 21 (2002), Available at (www.efnarc.org)
9. ASTM C39-96, ASTM (1996)
10. BS1881-116, BSI (1983)

MATEC Web of Conferences **162**, 04010 (2018) https://doi.org/10.1051/matecconf/201816204010
*BCEE3-2017*

# Effect of replacing the main reinforcement by steel fibers on flexural behavior of one-way concrete slabs

*Muyasser* Jomaa'h [1, *], *Ammar* Khazaal [1], and *Sinan* Ahmed [1]

[1] College of Engineering, Tikrit University, Tikrit, Iraq

**Abstract.** The main objective of the research is to study the preparation of one way slabs of ordinary concrete, and then to prepare concrete slabs by replacing the main reinforcing steel with two kinds of steel fibers (ordinary steel fibers and recycled steel fibers) by fraction volumes of 0.125, 0.250, and 0.375%. Also, study the mechanical properties of the mixtures as a ompressive strength, indirect tensile strength, and flexural strength. Concrete slabs of these mixtures have been prepared with specific geometrical dimensions700 * 300 * 70 mm, exposed to line load, to study the bending moment and maximum failure load of these slabs. A concrete mixture was produced after proportionment based on the ACI and casting six cubes tested at the ages of 7 and 28 days where the strength requirements for design were achieved. The main mixtures of research were produced, three cubes for compressive strength testing, twenty one cylinders for indirect tensile strength testing, and twenty one prisms for modulus of rupture testing. Also sixteen concrete slabs were prepared, two of them were reference slabs without reinforcing steel, the other two with main reinforcing steel only, six of them replaced by the main reinforcing steel with ordinary steel fibers, and the last six replacing the main reinforcing steel with recycled steel fibers. The results showed that both the failure load and the resultant deflection in concrete slabs decreased by (20.09%, 51.91%) for maximum load and by (35.18%, 81.48%) compared to the reference slabs when replacing the main reinforcing steel with steel fiber, also by (25.72%, 38.29%) to failure load and (38.88%, 79.63%) for the deflection compared to the reference slabs when replacing the main reinforcing steel with recycled steel fibers. The best value for maximum load and deflection could be obtained from this study was at 0.125% replacement ratio of the main reinforcing steel with recycled steel fibers, the highest value of ductility was 3.77 at the replacement ratio of 0.250% of the main reinforcing steel with the ordinary steel fibers, also the highest hardness value was 11.67 kN / mm with the replacement ratio of 0.125% of the main reinforcing steel with recycled steel fibers and increased by 61.85% than the hardness of the reference slabs.

## 1 Introduction

Due to the increase in waste that adversely affects the environment, the researchers found many ways to get rid of these wastes and reduce their impact on the environment. It is known that reinforced concrete is used to build structures on a large scale all over the world. The reinforced concrete slabs are the main parts of these facilities. There are many analytical and experimental studies on reinforced concrete slabs. It has taken time and effort for many researchers and for many years. Concrete slabs are used as floors not only in industrial and residential buildings but also as floors in bridges. Portland cement concrete is also known as brittle when it is subjected to tensile stresses, unreinforced concrete cracks under these stresses. Since the middle of the 18th century steel reinforcement has been used to solve this problem. Using it in the concrete as a composite material, reinforcing steel supports all tensile stresses [1]. Concrete containing cement, water, fine and coarse aggregates, non-continuous fibers called as Fibrous Reinforced Concrete (FRC), has been widely used in the construction of roofs, paving, machinery foundations, and Seismic buildings [2]. Fibers are separate intermittent parts made of many materials such as steel, plastic, glass, and other natural materials of different shapes and sizes and the best numerical variable to define is the aspect ratio which is equal to the length of the fiber divided by the length of the equivalent diameter. The value of this ratio ranges (30-150) [3]. Different labels for different types of fiber, as fiber can be divided into two groups according to the elasticity coefficient. The first fiber has a lower elasticity coefficient than the concrete elasticity coefficient, such as cellulose fibers, nylon and polypropylene fibers, and the second has a higher elasticity coefficient than the elasticity of concrete, such as asbestos fibers, fiberglass, iron fiber and carbon fiber [4]. It works to bridge cracks and increases the strength and durability of concrete. The effectiveness of this fiber depends on the pull-out resistance of the concrete mixture. When reinforced concrete is used in reinforced concrete beams then due to tensile strength the high tension of this type of concrete causes high tensile deformities in the longitudinal reinforcement bars. As a result, the reinforced concrete beams are acquired by the longitudinal reinforcing steel

* Corresponding author: muyasserjomaah@tu.edu.iq

© The Authors, published by EDP Sciences. This is an open access article distributed under the terms of the Creative Commons Attribution License 4.0 (http://creativecommons.org/licenses/by/4.0/).

and the steel fibers are highly ductile [5]. As for recycled steel fibers, they are a residue of vehicle tires or steel residues and their management remains a concern for those interested in the environment around the world. Therefore, researchers in Europe have sought to use tire waste as additives to concrete and to study their effect on concrete mixtures [6]. Materials are removed from damaged tires either mechanically or by burning. Mechanical methods produce rubber and steel. They start with the recycling of tire waste, turning tire residues into small pieces of rubber and then separating the raw rubber by freezing the parts to separate the tires. Rubber fracture which will be frosted and thus separating fiber after breaking rubber surrounded by and saved, and the method of burning produces steel including, coal, liquid and gases. It can be used as rubber and iron in concrete reinforcement as it works to increase the strength of concrete. Portability roads have hardness. It also can use the steel in it (with any type of steel fiber) as possible to reduce the amount of processed fiber used to increase construction speed, this can be used as waste tires to improve the energy and materials and reduce the buried material under the ground with the other uses, as the waste management FRAMEWORKS impact on the economy and national factors yield a practical example of this case. The frame of recycled recycling, where the demand is less than a new tire but it contributes to the reduction of prices depending on the level of use.

## 2 Aim of study

The purpose of this research is to study the structural and mechanical properties of reinforced concrete panels with steel fibers when used in wide beams (like one-way slab), and the possibility to take advantage of reinforcing fibers as a substitute for ordinary reinforcement, according to the study parameters and determinants process laboratory, and the possibility to take advantage of the fibers available or re-use of recycled steel fibers for concrete reinforcement with low cost, use of steel fibers from worn tire manufacturer as a substitute for fiber reinforced concrete as an alternative of synthetic steel fibers, it has proposed structural units with high strength and low cost used as concrete roofing.

## 3 Experimental Program

### 3.1 Material preparation

#### *3.1.1 Cement*

An ordinary Portland Cement Type I was used in all concrete mixtures. It conforms to the Iraqi Standard specification (IQS No.5 / 1984) [7]. Tables 1 and 2 show the chemical and physical properties of the cement used.

#### *3.1.2 Fine Aggregate*

The fine aggregate (sand) used was river sand from the short of river Dijla in Samarra area - Salah al-Din Governorate. It confirms to the Iraqi Standard Specification/ zone III IQS NO.45 / 1984. Table 3 shows the grading of fine aggregate, and Table 4 shows the chemical and physical properties.

**Table 1.** Chemical properties of cement.

| Oxides Composition [8] | Content % | Limit of Iraqi Specification No. 5/1984 [7] |
|---|---|---|
| CaO | 60.45 | - |
| $Al_2O_3$ | 4.65 | 8% Max |
| $SiO_2$ | 20.11 | 21% Max |
| $Fe_2O_3$ | 3.62 | 5% Max |
| MgO | 4.1 | 5 % Max |
| SO3 | 2.33 | 2.5 %Max |
| Loss on Ignition, (L.O.I) | 2.72 | 4 %Max |
| Insoluble material | 1.33 | 1.5 %Max |
| Lime Saturation Factor (L.S.F) | 0.89 | (0.66-1.02) |

**Table 2.** Physical properties of cement.

| Physical properties | Test Results | Limit of Iraqi Specification No. 5/1984 |
|---|---|---|
| Specific surface area (Blaine method), ($m^2$/kg) [9] | 308 | (230 $m^2$/kg) lower limit |
| Setting time (Vacate Apparatus) [10]<br>Initial setting, (hrs : min)<br>Final setting, (hrs : min) | 2hrs:15min<br>4hrs:10mn | Not less than 45min<br>Not more than 10 hrs |
| Compressive strength (MPa) [11] For 3-day | 28.7 | Not less than 15 MPa |

**Table 3.** Grading of fine aggregate.

| Sieve Size | Cumulative Retained % | Cumulative Passing % | Limit of IQS No. 45/1984 for Zone No. (3) [12] |
|---|---|---|---|
| 4.75-mm (No.4) | 5.61 | 94.40 | 90-100 |
| 2.36-mm (No.8) | 8.62 | 91.35 | 85-100 |
| 1.18-mm (No.16) | 15.62 | 84.30 | 75-100 |
| 600-μm (No.30) | 23.68 | 76.32 | 60-79 |
| 300-μm (No.50) | 65.69 | 34.25 | 12-40 |
| 150-μm (No.100) | 90.55 | 9.45 | 0-10 |

#### *3.1.3 Coarse Aggregate*

Well rounded river coarse aggregate (gravel), available at natural quarries on the Dijla River, from Samarra area was used with maximum size of aggregate of 14 mm, Table 5 shows the grading of coarse aggregate and confirms to (I.Q.S no.45/1984) [12], and table 6 shows the chemical and physical properties.

**Table 4.** Physical and Chemical Properties of fine aggregate.

| Properties | Specification | Test Results | Limits of Specification |
|---|---|---|---|
| Specific gravity | ASTM C128-01 [13] | 2.63 | |
| Absorption (%) | ASTM C128-01 [13] | 2.00 | |
| Dry loose unit weight ($kg/m^3$) | ASTM C29/C29M/97 [14] | 1600 | |
| Sulfate content (as SO3) (%) | (I.Q.S.) No. 45/1984 [12] | 0.36 | 0.5 (max. value) |
| Material finer than 0.075 mm (%) | (I.Q.S.) No. 45/1984 [14] | 3 | 5 (max. value) |

**Table 5.** Grading of coarse aggregate.

| Sieve size | Cumulative Passing % | Limit of IQS No. [45/1984] [10] |
|---|---|---|
| 12.5-mm | 93.975 | 90-100 |
| 9.5-mm | 85.462 | 85-100 |
| 4.75-mm | 16.699 | 10-30 |
| 2.36mm | 0.164 | 0-10 |

**Table 6.** Physical and Chemical Properties of coarse aggregate.

| Properties | Specification | Test Results | Limits of Specification |
|---|---|---|---|
| Specific Gravity | ASTM C127-01 [15] | 2.62 | - |
| Absorption (%) | ASTM C127-01 [15] | 0.7 | - |
| Dry Loose Unit Weight $kg/m^3$ | ASTM C29/C29M/97 [14] | 1596 | - |
| Sulfate Content (as $SO_3$) (%) | (I.Q.S.) No. 45-1984 [12] | 0.08 | 0.1 (max.value) |

## 4 Mix Water

Ordinary drinking water was used for the mixing, curing and all concrete works.

## 5 Admixtures

### 5.1 Superplasticizer

A plasticizer of type Sika Visco Crete super E4 has been used as a high range water reducer additive that lends positive effects to fresh concrete such as increased slump without need for additional water, reducing the ratio of water to cement (w/c), in addition to improve and facilitating the workability and pumping process of concrete, Table 7 shows the chemical and physical properties of the plasticizer.

**Table 7.** Technical properties of plasticizer.

| Properties of Plasticizer Sika Visco Crete Super E4 | Activity |
|---|---|
| Brown Liquid | Appearance |
| 1.06± 0.02 kg/ltr | Density at 20 C° |
| -6° C | Freezing |
| 6-8 | pH |
| None | Chloride |

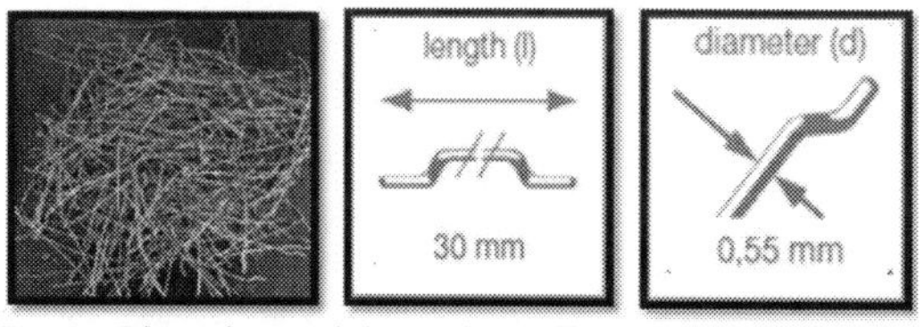


**Fig. 1.** Dimensions and shape of steel fibers Sika Fiber SH 60/30.

### 5.2 Silica fume

SIKA-FUME-HR was used, it confirms to ASTM C1240-03 [16], and Table 8 shows the properties of SIKA-FUME-HR.

**Table 8.** Chemical analysis and activity index of SIKA-FUME-HR

| Pozzolanic activity | Limit of ASTM C1240-03 | Chemical decomposition | | Limit of ASTM C1240-03 [16] |
|---|---|---|---|---|
| | | Oxides | Result% | |
| 121.5% | 105% | L.O.I | 3.89 | 6% Max |
| | | $SiO_2$ | 91.03 | 85% Min |
| | | $Al_2O_3$ | 4.02 | - |
| | | $Fe_2O_3$ | 0.32 | - |
| | | $SO_3$ | 0.73 | - |

## 6 Steel Fibers

Steel fibers of type Sika Fiber SH 60/30 were used as shown in Fig. (1), which is the most commonly used species and works to bridge cracks and increases ductility and toughness of concrete. Fibers were of 30 mm length and diameter of 0.55 mm, fibers were added with replacement ratios of the main reinforcement steel in fraction volumes of (0.125, 0.125, and 0.375) % of concrete volume, Table 9 shows the specifications and characteristics of typical steel fibers used in concrete mixtures.

### 6.1 Recycled Steel Fiber

Recycled steel fibers were used from vehicles tires in the local market in Iraq. The fibers were cut manually at uniform length of 30 mm and 1 mm diameter and aspect ratio of 30, Fig. (2) Shows the recycled steel fibers used

**Table 9.** Pproperties of steel fibers.

| Commerc-ial name | Configuration | Property | Specifications |
|---|---|---|---|
| Sika Fiber SH 60/30 |  Hooked ends | Density | 7860 kg/m$^3$ |
| | | Ultimate strength | 1345 MPa |
| | | Modulus of Elasticity | 210x10$^3$MPa |
| | | Strain at proportion limit | 6.4 x10$^{-3}$ |
| | | Poisson's ratio | 0.28 |
| | | Average length (Lf) | 30 mm |
| | | Nominal diameter(Df) | 0.55 mm |
| | | Aspect ratio ($L_f/D_f$) | 60 |

**Fig. 2.** Recycled steel fiber.

### *7.1 Steel Reinforcement*

Deformed steel reinforcement of 6mm diameter, were used at the bottom of the slabs to resist tension stresses resulting from bending moment. Table 10 shows the Lab. results conducted on this steel bars and confirms to ASTM A615 [17].

### *7.2 Mold Dimensions and Slab Reinforcement Details*

Four steel molds were manufactured with 300 x 700 x 70 mm in dimensions and thick steel sheets were used to ensure fare faces. The slab samples were designed with rectangular section to ensure bending failure before shear failure is reached, Rebars of 6mm Ø were used in both directions as shown in Fig. 3.

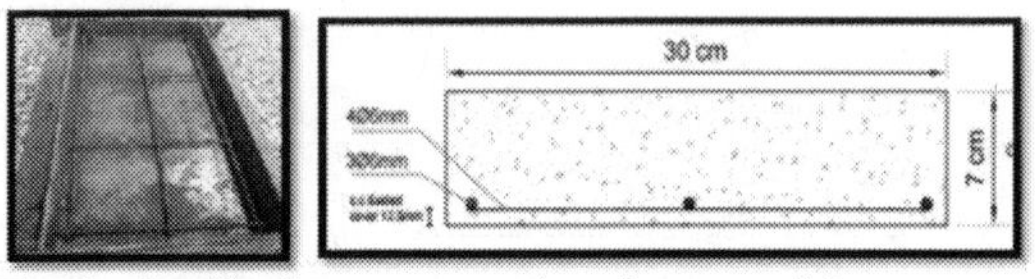


**Fig. 3.** Details of transverse section of one-way slab.

### *7.3 Reference Mix*

The concrete mixture used was designed according to the American Concrete Institute code (ACI 211.4R- 08) [18] to obtain a concrete compressive strength of 35 MPa at the age of twenty-eight days, Superplasticizer was used by 1%, and the silica fume by 10% by weight of cement. Tables 11,12 shows the weight of the material used in preparing the concrete mixture.

**Table 11.** Ingredients of reference concrete mixture.

| Ingredient | Coarse aggregate | Fine aggregate | Cement | Water | Superplasticizer | Silica Fume |
|---|---|---|---|---|---|---|
| Kg/m$^3$ | 898 | 890 | 447 | 137 | 5 | 50 |

**Table 10.** Results of steel bars test.

| Bars diameters (mm) | Yield stress ($f_y$) (MPa) | Ultimate stress ($f_u$) (MPa) | Elongation % |
|---|---|---|---|
| 6mm | 621.81 | 720.34 | 6.4 |

## 7 Mixes, casting, and concrete curing

Steel molds (slabs, cubes, cylinders, and prisms) are prepared, cleaned and oiled with a thin film of oil, then reinforcement is placed in the steel mold, and the mixing process begins with half quantity of coarse aggregate and water, then fibers (steel fibers or recycled steel fiber) were added gradually to avoid congestion. Finally all other mixing ingredients are added to complete the mix. Slump test is conducted on the mix afterward within 75 -125 mm limits. Molds are filled with fresh mix and vibrated at layers of 50mm thickness each [19]. Samples are left for 24 hours, then placed in water for 27 days to have ready for testing of strength. Eight concrete mixes were prepared with two slabs samples from each mix. The eight mixes consist of one mix as a reference mix with no additives and reinforcement. Another mix reinforced but no additives in addition to six mixes with different volumetric ratios of reinforcement as shown in Table XII .

## 8 Laboratory tests

### *8.1 Mechanical tests*

#### *compressive strength*

Compressive strength test was conducted according to British specification (BS 1881: Part 116:1989) for standard cubes samples used of (150 x 150 x 150) mm. An electrical testing machine of capacity 2025 kN was loaded axially. The average compressive strength value for three cubes was adopted at the age of 28 days, as shown in Table 13.

**Table 12.** Details of mixtures used.

| Symbol | Fibers % From steel | W/C | SP % |
|---|---|---|---|
| R1 | 0 | 0.392 | 0.6 |
| R2 | 0 | 0.392 | 0.6 |
| SF1 | 0.125 | 0.392 | 0.8 |
| SF2 | 0.250 | 0.392 | 0.9 |
| SF3 | 0.375 | 0.392 | 1.0 |
| RSF1 | 0.125 | 0.392 | 0.8 |
| RSF2 | 0.250 | 0.392 | 0.9 |
| RSF3 | 0.375 | 0.392 | 1.0 |

https://doi.org/10.1051/matecconf/201816204010

### *Splitting Tensile Strength*

A splitting tensile strength test was conducted on the samples according to (ASTM C 496-96) [20]. A U-Test machine was used to test samples of (150 × 300) mm cylinders at 28 days age. Three samples were used for each mix, for each percentage ratio of steel and recycled steel proportion of the percentages used for steel and recycled steel fiber. In the test, a diagonal compressive load was applied which will cause tensile failure instead of compressive since the location of applied load will be in the triaxial compression and enable sample to resist compression higher than when under compressive testing. Table 14 shows detailed results of increase or decrease in tensile strength. Fig. (5) Shows the effect of hooked reinforced fiber on bridging cracks compared to samples with unhooked fiber reinforcement.

### - Modulus of rupture

Modulus of rupture ($f_r$) test was conducted on concrete samples of (100×100×500) mm with prisms specimens according to ASTM C78-01[21]. Third -point loading method was used at loading rate (1MPa / min). Three prisms were used for each type of mix at 28 age. Fig. (6) Shows the mode of failure of samples. While table 15 shows detailed results of using different ratios of fibers. Some results show that reinforcement fibers didn't work well at certain level because level was exposed to tension and compression stresses at the same time. The different mixtures that contained the two types of fiber were adopted for the number of standard prisms and testing. It noted the increase in the ratio of modulus of rupture to increase the ratio of the hooked fibers compared to the reference sample and the increase rates of (35.41%, 58.20%) for fibers proportion of (0.125% (0.2%), while the decrease in fiber fraction (0.375%) to be (26.41%). This difference may be due to the steel fiber ratio exceeding the normal fraction of the $V_f$. when comparing the same ratios of both types of fiber a difference was noticed in the amount of increase or decrease of resistance, which requires the study of replacement ratios at additional rates to reach the exact limits of the change in rupture resistance.

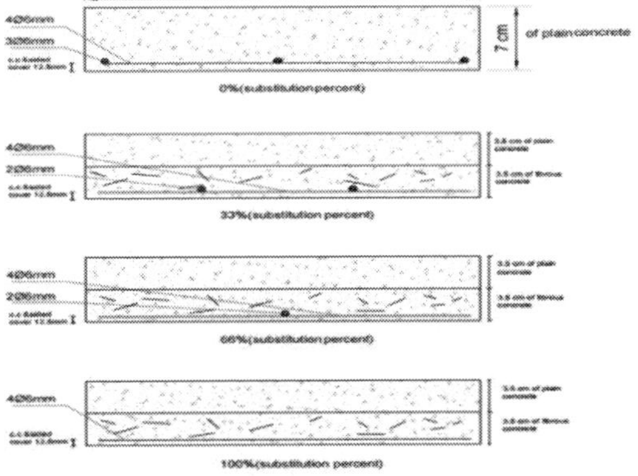


**Fig. 4.** Details of steel ratios.

**Table 13.** Compressive strength of reference concrete mixtures.

| Group No. | Mix No. | Steel Fibers Percentage % ($V_f$) SF (Normal) % | Steel Fibers Percentage % ($V_f$) SF (Recycling) % | W/C | SP % | Silica Fume from Cement % | Comp. - Strength (28) day ($f_{cu}$) MPa | Av. Comp. - Strength (28) day ($f_{cu}$) MPa |
|---|---|---|---|---|---|---|---|---|
| Ref | 1 | 0 | 0 | 0.392 | 0.6 | 10 | 48.7 | 44.89 |
| | 2 | 0 | 0 | 0.392 | 0.6 | 10 | 42.24 | |
| | 3 | 0 | 0 | 0.392 | 0.6 | 10 | 43.73 | |

a-without fiber b- with fiber

**Fig. 5.** Mode of failure of concrete cylindrical samples.

**Fig. 6.** Mode of failure of prisms.

### - Flexural strength test of one way slabs

Slabs of (700 × 300 × 70) mm were tested for flexure at 28 days age. The slabs reinforced identically were simply supported and tested by a flexural testing machine of 200 kN capacity at loading rate (1kN/sec). A uniform load was applied with reading taken from a digital gage machine of one micron (1 μmm) accuracy. Results were based on load-deflection curves at mid-span. Fig. (8 & 9). Table 16 shows many different cases of the effects of changing reinforcement with typical steel fibers and recycled steel fibers.

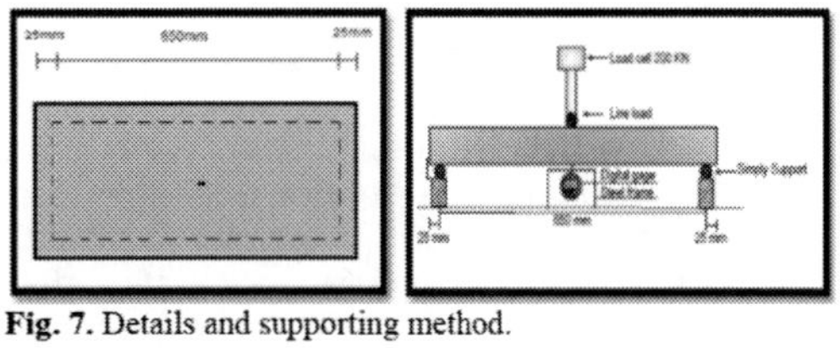

**Fig. 7.** Details and supporting method.

The replacement of the main reinforcing steel with the steel fibers and the recycled fiber has an effect on the

https://doi.org/10.1051/matecconf/201816204010

maximum load of the concrete slabs. Table 16 shows the results obtained from the bending test, for different ratios of steel fibers (0.125%, 0.250%, and 0.375%). The maximum load decreases by 20.09%, 51.91% and 48.26% respectively, compared to the maximum load of the reference slabs. As well as for slump values (35.18%, 37.03%, and 81.48%) compared to the slump values of the reference slabs. This is due to the type of fiber used, as well as the nature of its behavior and the place of addition where it was added to the tensile area only. The maximum load is gradually reduced by 25.72%, 36.41%, and the maximum weight is reduced by 25.72% 38.29%), respectively, compared to the maximum load of slabs (38.88%, 70.37%, 79.63%) compared with the slump values of the reference slabs.

Fig. (10) Shows how slabs with main reinforcement replaced with fibers at (0.125%) will have maximum load higher than those slabs with recycled reinforcement of fiber. The same result goes to slump tests and cracks. This is due to the hooks of the reinforcement of normal steel fiber at their ends which spread widely to tension zone and thus give more resistance. Fig. (11 & 12) show the comparison of slabs with recycled steel fibers and normal steel fibers at (0.25%) and (0.375%) respectively. The latter gives less strength for the same steel fiber and recycled steel fiber slabs as shown in Fig. 10.

### Ductility

The Ductility Index (μd) is defined as the ratio between the deflection at the maximum load of failure to the deflection at the yield of the steel reinforcement [22] which can be determined from the curve (load - deflection) of the slab when the curve moves from the elasticity stage to the plasticity phase, Table 17 shows the values of ductility that were determined from the rate of two slabs for each replacement ratio between the main reinforcement and the ordinary and recycled steel fibers. Based on these results, the highest value of ductility index was 3.77 for the replacement ratio of 0.250% of main steel reinforcement by steel fibers that increases by (18.92%) from the ductility Index for the reference slabs. This is due to the decrease in the deflection for the first cracking or when the reinforcing steel is replaced by the steel fibers instead of the main reinforcing steel which is widely distributed during the mixture, causing a decrease in the value of the deflection. Also, it has been noted that the ductility index has decreased to a large value for a replacement ratios of (0.125%, 0.375%) compared to the reference tile. When replacing the main steel reinforcement with recycled steel fibers, there is an increase in the value of ductility for a replacement ratio of (0.125%) compared to the reference slab, the increase was (4.1%). Gradual decrease in ductility index has been noted compared to reference slabs for replacement ratios (0.375%, 0.25%) by (16.08% and 30.60%) respectively. This is due to the increase in the amount of fibers inside the concrete mixture with a decrease in the amount of reinforcing steel which reduces the strength of the stresses generated in the tensile areas at the beginning of the loading. It results a decrease in the rate of deflection at yield and deflection at maximum load of failure, which greatly affects the value of ductility Index.

### Stiffness

The stiffness of the concrete slabs was calculated based on the maximum load of failure for the slabs by taking 45% of the maximum load and projecting it on the load-deflection curve and determines the value of the deflection corresponding to the load and dividing the calculated load on the resulting deflection [23]. Table 18 shows the stiffness values of concrete slabs. The results show that the highest stiffness value of slabs at a replacement ratio of (0.125%) of the main reinforcement by steel fibers in the second group (G2) which has increased by (61.85%) than the stiffness value of the reference slabs (R2), and the highest stiffness in the first group (G1) at a replacement ratio of (0.125%) of the main reinforcing steel by the steel fibers, which is (18.58%) higher than the stiffness of the reference slabs (R2). The stiffness values of the second group (G2) slabs with reinforcing steel replacement ratios by recycled steel fibers are higher than the stiffness values of the first group (G1) slabs, and their stiffness is also higher than the stiffness of the reference slabs. The stiffness of this group gradually decreases with an increase of fibers content, due to reasons such as the type of fiber used, their size and their distribution. The addition of these fibers in the tensile area only, which greatly affects the maximum deflection values and the deflection at the first crack, as well as the maximum load, giving high stiffness values compared to reference slabs and that contain steel fibers.

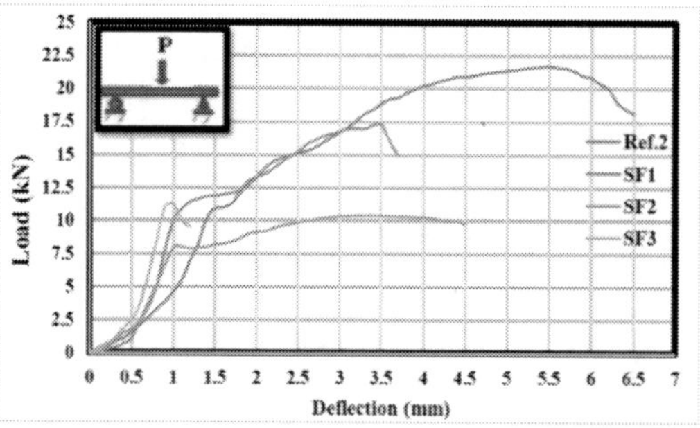


**Fig. 8.** Load-deflection relationship at mid-span of concrete slabs.

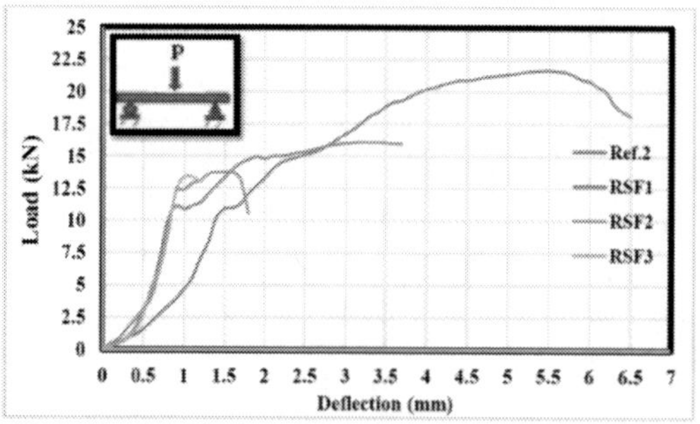


**Fig. 9.** Load- deflection relationship at mid-span of concrete slabs.

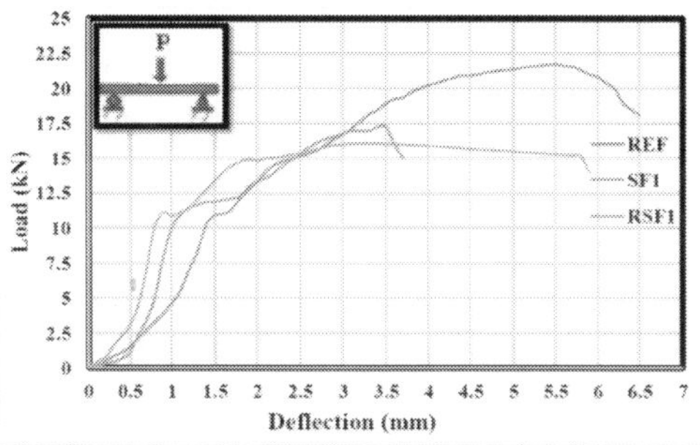


**Fig.10.** load-deflection relationship at mid-span of concrete slabs

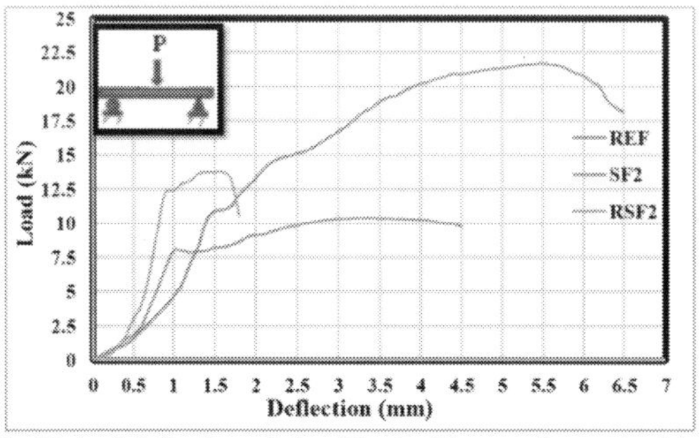


**Fig.11.** load-deflection relationship at mid-span of concrete slabs

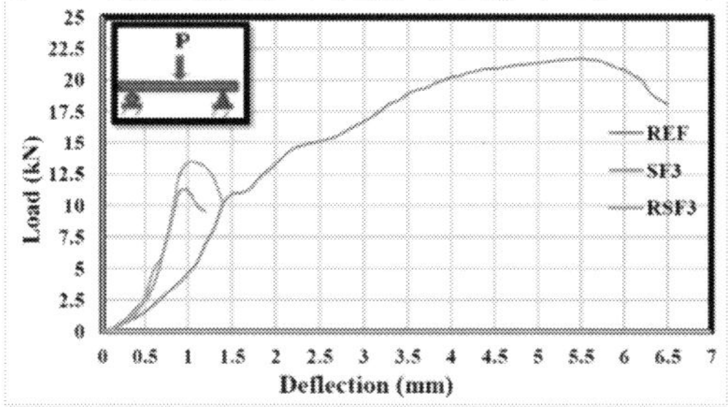


**Fig.12.** load-deflection relationship at mid-span of concrete slabs

## 9 Conclusions

It has been concluded that;

1. The max. Failure load in flexure and deflection has decreased by 20.09% to 51.91% of max. Load and by 35.18% to 81.48% in one-way slabs compared with refrenece slab, upon replacing main steel reinforcement by steel fibers.
2. The max. Failure load in flexure and deflection has decreased by 25.72% to 38.29% of max. Load and by 38.88% to 79.63% in one-way slabs compared with refrenece slab, upon replacing main steel reinforcement by recycled steel fibers.
3. The best values of max. Load and deflection were at replacement ratio of 0.125% of main steel reinforcement by recycled ones.
4. The heighest value of ductility obtained was 3.77 at replacement ratio 0.25% of main steel reinforcement by steel fibers.
5. The highest value of hardness obtained was 11.76 kN/mm at replacement ratio 0.125% of main steel reinforcement by recycled steel fibers, and increased by 61.85% than that of reference slabs.

## References

[1] Brown R., Shukla A., Natarajan K.R., "Fiber Reinforcement of Concrete Structure", University of Rhode, Island, Transportation Center, Sep. 2002

[2] Dhonde, Hemant B. ,Mo, Y.L. ,Thomas,T.C. ,"Fiber Reinforcement in Prestressed Concrete Beams" , University of Hoston , Department of Civil Engineering & Environmental Engineering, 2005.

[3] ACI Committee 544, "State-of-the Art Report on Fibre Reinforced Concrete", Sp-81 ACI Publication, International Symposium on Fiber Reinforced Concrete, pp. 411-431, 1984.

[4] Hannant, D."Fiber Cement and Fiber Concrete", Wiley, Chichester, 1987.

[5] Shah, R. H., & Mishra, S. V. "Crack and Deformation Characteristics of SFRC Deep Beams" Journal of the Institution of Engineers. India. Civil Engineering Division, 85(mai), pp. 44-48, 2004.

[6] Hannant, D.J., "Fibre Cement and Fibre Concrete", Properties in the Hardend State, Chapter 6, John Wiley and Sons, Chichester, 1978, pp. 61-80.

[7] COSQC, Standard Specification for Portland Cement, Iraqi Specif. 5 (1984).

[8] ASTM C114–01 "Standard Test Methods for Chemical Analysis of Hydraulic Cement ", ASTM Standard, vol. 04.02, pp. 1 - 4, 2004.

[9] ASTM C204–00 "Standard Test Method for Fineness of Hydraulic Cement by Air-Permeability Apparatus", ASTM Standards, vol. 04.02, pp. 1 - 9, 2004.

[10] ASTM C191–04 "Standard Test Method for Time of Setting of Hydraulic Cement by Vicat Needle", ASTM Standards, Vol. 04.02, 2004.

[11] ASTM C 109 / C 109M – 07 "Standard Test Method for Compressive Strength of Hydraulic Cement Mortars (Using 2-in. or [50-mm] Cube Specimens)".

[12] Iraqi Standard no. 45, "Natural Aggregate used in Construction Works", Central Directorate of Standardization and Quality Control.

[13] ASTM C128 – 01 "Standard Test Method for Density، Relative Density (Specific Gravity)، and Absorption of Fine Aggregate", ASTM Standards, Vol. 04.02, 2004.

[14] ASTM C29 / C29M – 97 (Re - approved 2003), "Standard Test Method for Bulk Density 15(Unit Weight) and Voids in Aggregate", ASTM Standards, Vol. 04.02, pp. 1-4, 2004.

[15] ASTM C127-01, "Standard Test Method for Density, Relative Density (Specific Gravity), and Absorption of Coarse Aggregate1", ASTM Standards, Vol. 04.02, pp. 1-6, 2004

[16] ASTM C1240 - 03, "Standard Specification for Use of Silica Fume as a Mineral Admixture in Hydraulic-Cement Concrete, Mortar, and Grout" ASTM Standards, 2004

[17] ASTM A615 / A615M - 09, "Standard Specification for Deformed and Plain Carbon-Steel Bars for Concrete Reinforcement".

[18] ACI Committee 211" Guide for Selecting Proportions for High-Strength Concrete Using Portland Cement and Other Cementitious Materials ", ACI 211.4R-08 , 2008.

[19] Alyaa AlAtaar, "Micro structure and Mechanical Properties of Lightweight Aggregate Concrete Containing Fibers".

[20] ASTM C 496 – 96 "Standard Test Method for Splitting Tensile Strength of Cylindrical Concrete Specimens

[21] ASTM C78-02, "Stander test method for flexural strength of concrete (using samples beam with third point loading" ASTM standards, vol. 04.02, 2004.

[22] Elavenil S. , Samuel K., G. M ,"Behavior of Steel Fiber Reinforced Concrete Beams and Plates Under Static Load" , Journal of Research in Science , Computing and Engineering , vol. 4, no. 3, pp. 11-28, 2007.

[23] Al-Jwari, Ruaa S., " Behavior of reinforced concrete beams strengthening by Polymer carbon fibers fixed near top surface, M.Sc Thesis, University of Mosul, 2010.

**Table 14.** Results of splitting tensile strength.

| Group No. | Mix No. | W/C | Steel Fibers Percentage ($V_f$) | | SP% | Silica Fume % | Splitting Tensile Strength (28) day ($f_t$) MPa | Increase or decrease % |
|---|---|---|---|---|---|---|---|---|
| | | | SF (Normal) % | SF (Recycling) % | | | | |
| Ref | R | 0.392 | 0 | 0 | 0.6 | 10 | 3.91 | - |
| G1 | SF1 | 0.392 | 0.125 | 0 | 0.8 | 10 | 4.29 | +9.71 |
| | SF2 | 0.392 | 0.250 | 0 | 0.9 | 10 | 5.35 | +38.83 |
| | SF3 | 0.392 | 0.375 | 0 | 1.0 | 10 | 3.57 | -8.70 |
| G2 | RSF1 | 0.392 | 0 | 0.125 | 0.8 | 10 | 3.13 | 19.95- |
| | RSF2 | 0.392 | 0 | 0.250 | 0.9 | 10 | 3.40 | -13.04 |
| | RSF3 | 0.392 | 0 | 0.375 | 1.0 | 10 | 3.32 | -15.09 |

**Table 15.** Modulus of rupture for different concrete mixes with steel fiber ratios.

| Group No. | Mix No. | W/C | Steel Fibers Percentage % ($V_f$) | | SP% | Silica Fume % | Modulus of Rapture (28) day ($f_r$) MPa | Increase or decrease % |
|---|---|---|---|---|---|---|---|---|
| | | | SF (Normal) % | SF (Recycling) % | | | | |
| Ref | R | 0.392 | 0 | 0 | 0.6 | 10 | 6.89 | - |
| G1 | SF1 | 0.392 | 0.125 | 0 | 0.8 | 10 | 9.33 | +35.41 |
| | SF2 | 0.392 | 0.250 | 0 | 0.9 | 10 | 10.9 | +58.20 |
| | SF3 | 0.392 | 0.375 | 0 | 1 | 10 | 8.71 | +26.41 |
| G2 | RSF1 | 0.392 | 0 | 0.125 | 0.8 | 10 | 10.16 | +47.46 |
| | RSF2 | 0.392 | 0 | 0.250 | 0.9 | 10 | 9.46 | +37.30 |
| | RSF3 | 0.392 | 0 | 0.375 | 1 | 10 | 9.96 | +44.55 |

**Table 16.** Results of maximum load and first crack load of one – way concrete slabs.

| Gr. No. | Slabs No. | W/C | Steel Fibers Percentage % ($V_f$) | | SP% | Silica Fume % | Load at First Crack $P_{cr}$ (kN) | ($\delta_{cr}$) (mm) | Average Failure Load($P_u$) (kN) | decrease in ($P_u$) | ($\delta_u$) (mm) | decrease in ($\delta_u$) % |
|---|---|---|---|---|---|---|---|---|---|---|---|---|
| | | | SF (Normal) % | SF (Recycling) % | | | | | | | | |
| Ref | R1 | 0.392 | 0 | 0 | 0.6 | 10 | - | - | 8.14 | - | | - |
| | R2 | 0.392 | 0 | 0 | 0.6 | 10 | 11.30 | 1.7 | 21.65 | - | 5.4 | - |
| G1 | SF1 | 0.392 | 0.125 | 0 | 0.8 | 10 | 11.87 | 1.4 | 17.3 | 20.09% | 3.5 | 35.18% |
| | SF2 | 0.392 | 0.250 | 0 | 0.9 | 10 | 6.66 | 0.9 | 10.41 | 51.91% | 3.4 | 37.03% |
| | SF3 | 0.392 | 0.375 | 0 | 1 | 10 | 1.90 | 0.4 | 11.2 | 48.26% | 1 | 81.48% |
| G2 | RSF1 | 0.392 | 0 | 0.125 | 0.8 | 10 | 10.88 | 1 | 16.08 | 25.72% | 3.3 | 38.88% |
| | RSF2 | 0.392 | 0 | 0.250 | 0.9 | 10 | 4.01 | 0.6 | 13.77 | 36.41% | 1.6 | 70.37% |
| | RSF3 | 0.392 | 0 | 0.375 | 1 | 10 | 2.63 | 0.5 | 13.36 | 38.29% | 1.1 | 79.63% |

**Table 17.** Ductility Index of Reinforcement concrete slabs.

| Group No. | Mix No. | W/C | Steel Fibers Percentage ($V_f$) | | SP% | Silica Fume % | ($\delta_y$) (mm) | ($\delta_y$) (mm) | Ductility index ($\mu_{d=\delta_u/\delta_y}$) |
|---|---|---|---|---|---|---|---|---|---|
| | | | SF (Normal) % | SF (Recycling) % | | | | | |
| Ref | R2 | 0.392 | 0 | 0 | 0.6 | 10 | 1.7 | 5.4 | 3.17 |
| G1 | SF1 | 0.392 | 0.125 | 0 | 0.8 | 10 | 1.4 | 3.5 | 2.5 |
| | SF2 | 0.392 | 0.250 | 0 | 0.9 | 10 | 0.9 | 3.4 | 3.77 |
| | SF3 | 0.392 | 0.375 | 0 | 1.0 | 10 | 0.4 | 1.0 | 2.5 |
| G2 | RSF1 | 0.392 | 0 | 0.125 | 0.8 | 10 | 1.0 | 3.3 | 3.3 |
| | RSF2 | 0.392 | 0 | 0.250 | 0.9 | 10 | 0.6 | 1.6 | 2.66 |
| | RSF3 | 0.392 | 0 | 0.375 | 1.0 | 10 | 0.5 | 1.1 | 2.2 |

**Table 18.** Results of stiffness of one – way concrete slabs.

| Gr. No. | Slabs No. | W/C | Steel Fibers Percentage % ($V_f$) | | SP% | Silica Fume % | $P_u$ (kN) | (45% $P_u$) (kN) | Deflection @45%(Pu) (mm) | Stiffness (kN/mm) $St = \frac{45\%\ pu}{\delta u\ @\ 45\%\ pu}$ | Increasing %) |
|---|---|---|---|---|---|---|---|---|---|---|---|
| | | | SF (Normal) % | SF (Recycling) % | | | | | | | |
| Ref | R1 | 0.392 | 0 | 0 | 0.6 | 10 | 8.14 | 3.78 | - | - | - |
| | R2 | 0.392 | 0 | 0 | 0.6 | 10 | 21.65 | 9.74 | 1.35 | 7.21 | - |
| G1 | SF1 | 0.392 | 0.125 | 0 | 0.8 | 10 | 17.3 | 7.78 | 0.91 | 8.55 | 18.58 |
| | SF2 | 0.392 | 0.250 | 0 | 0.9 | 10 | 10.41 | 4.68 | 0.75 | 6.24 | -13.45 |
| | SF3 | 0.392 | 0.375 | 0 | 1 | 10 | 11.2 | 5.04 | 0.70 | 7.20 | -0.13 |
| G2 | RSF1 | 0.392 | 0 | 0.125 | 0.8 | 10 | 16.08 | 7.24 | 0.62 | 11.67 | 61.85 |
| | RSF2 | 0.392 | 0 | 0.250 | 0.9 | 10 | 13.77 | 6.19 | 0.60 | 10.31 | 42.99 |
| | RSF3 | 0.392 | 0 | 0.375 | 1 | 10 | 13.36 | 6.01 | 0.60 | 10.01 | 38.83 |

MATEC Web of Conferences **162**, 04011 (2018)
*BCEE3-2017*
https://doi.org/10.1051/matecconf/201816204011

# Shear strength enhancement of lightweight aggregate reinforced concrete deep beams by using CFRP strips

*Ahlam* Mohammad[1,*], *Kaiss* Sarsam[1] and *Nabeel* Al-Bayati [1]

[1]Building and Construction Engineering Department, University of Technology, Baghdad, Iraq

**Abstract.** In this research, results of an experimental investigation on the shear strengthening of lightweight aggregate reinforced concrete deep beams are presented. A total of eight lightweight aggregate deep beams were cast and tested in the experimental work to study the effect of externally bonded CFRP strips in improving their structural behavior, one of them was unstrengthened to serve as a control beam while the remaining seven beams were strengthened in different orientation, spacing and number of layers of CFRP. The locally available natural porcelanite rocks are used to seek the possibility of producing structural lightweight aggregate concrete. The beams were designed to satisfy the requirements of ACI 318M- 14 building code. Results show that the CFRP strips have increased the load carrying capacity for the strengthened deep beams up to 50 % when comparedto the unstrenghtened control one. The diagonal compression strut crack of unstrenghtened control beam is changed to several diagonal cracks in the mid-depth within the shear span of the strengthened beams and exhibited more ductile failure modes. The results also indicate that bonded CFRP system in the shear span was seen to delay the formation of diagonal shear cracks and provided positive restraint to the subsequent growth of cracks. Increasing the amount of CFRP (by increasing the number of layers from one to two layers) results in increase in the ultimate load by about 15%. However, the increase in the spacing between the strips (from 100 to 150mm) led to a decrease in the ultimate load by about 13%.

## 1 Introduction

Reinforced concrete deep beams are used as load distributing structural units like transfer Girderswhich are used in multistory buildings to provide column offsets or as horizontal diaphragms, pile caps, foundation walls, and offshore structures. They are also used in construction of bunkers, water tanks and bins where the walls act as vertical beams spanning between the columns supporting and carrying a part of the floor load [1].

According to ACI 318M-14[2] deep beams could be defined as members loaded on one face and supported on the opposite face, in which compression struts can develop between the supportsand the applied loads. deep beams have either: $l_n$/h ≤ 4.0; or a/h ≤ 2.0, where ($l_n$) is the clear span for distributed loads measured from the face of the support, (h) is the overall depth, and (a) is the shear span length (figure 1).

After inclined cracks formulation in deep beam, it takes on a tied arch behavior, allowing the forces to transfer directly to the supports [3]. Such behavior provides some reserve shear capacity in deep beams which is not found in shallower members. The deep beams shear strength is significantly greater than that of slender beams because of this special capacity of deep beams in redistributing internal forces before failure [4].

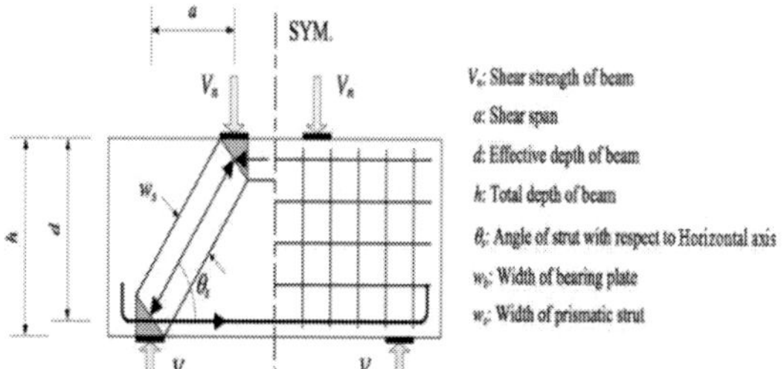


**Fig. 1.** Typical deep beam

The strain is distributed nonlinearly across the depth of the cross section of deep beams and great amount of load is carried to the supports by a compression strut passing through the line joining the applied load and reaction [2, 5].

Shear behavior of reinforced concrete deep beam is very complex, and difficult to predict accurately. This behavior depends on the size of the deep beam,

*Corresponding author: ahlamsader@yahoo.com

© The Authors, published by EDP Sciences. This is an open access article distributed under the terms of the Creative Commons Attribution License 4.0 (http://creativecommons.org/licenses/by/4.0/).

reinforcement type and detail in addition to the type of the applied load and its position. The total shear resistance of beams depends usually on the shear resistance capacity of concrete and the additional contributions of external and/or internal reinforcement of the beams. In particular, the shear resistance of reinforced concrete beams can be enhanced by externally bonding FRP composite materials. This system may be regarded as additional web reinforcement but fixed externally.

Using FRP materials as external shear strengthening gained popularity over steel plates because of their lightweight feature and the ease of application, moreover, its being available in the form of thin sheets which make very little change to the dimension of the existing members. FRP could play a very important role in retrofitting and strengthening of degraded structures or strength deficient structures that are already in existence.

The four main types of FRP used in the construction industry are: aramid (AFRP), carbon (CFRP), basalt, and glass (GFRP) [6]. Within these fiber groups, there are numerous different performance characteristics available. The fibers have a linear elastic response up to ultimate load with no significant yielding as shown in figure (2). A comparison among some of FRP composites (based on fiber area only) and reinforcing steel in terms of stress-strain relationship is illustrated in figure (2).

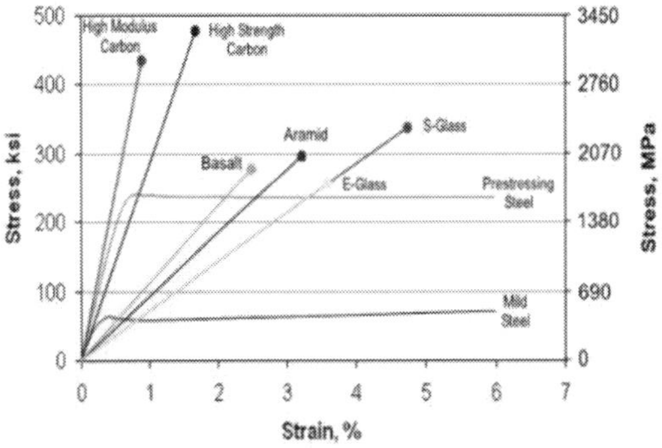


**Fig. 2.** stress-strain relationship comparison between some of FRP composites and reinforcing steel [7].

## 2 Strengthening of beams in shear with FRP sheets

The three types of wrapping schemes used to increase the shear strength beams or columns (As stated in ACI Committee 440.2R-08[8])are illustrated in figure (3). The most efficient wrapping scheme is FRP Completely wrapped around the section on all four sides(which is most commonly used in column). However, total wrap is not practical from constructability standpoint in the case of beam applications (where an integral slab makes it impractical to completely wrap the member). the shear strength could be improved by wrapping the FRP system around three sides of the member (U-wrap) or bonding to the two opposite sides of the beams [8,9].

In all wrapping systems, the installation of the FRPs can be used continuously along the span of the member or as spaced strips (figure 4). The effective alternative in optimizing the amount of material used is the second one.

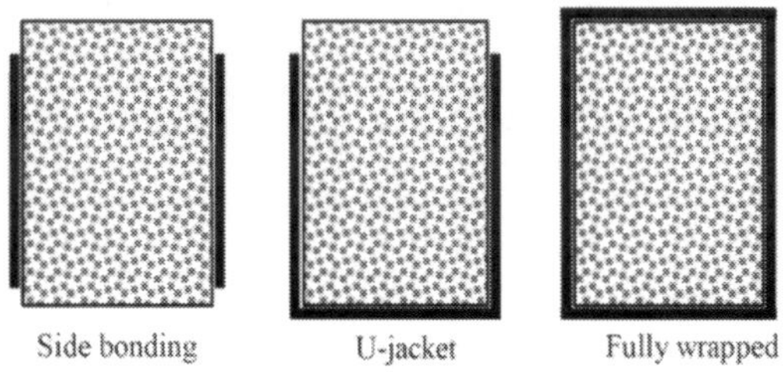


**Fig. 3.** Shear strengthening schemes with FRP composites [8].

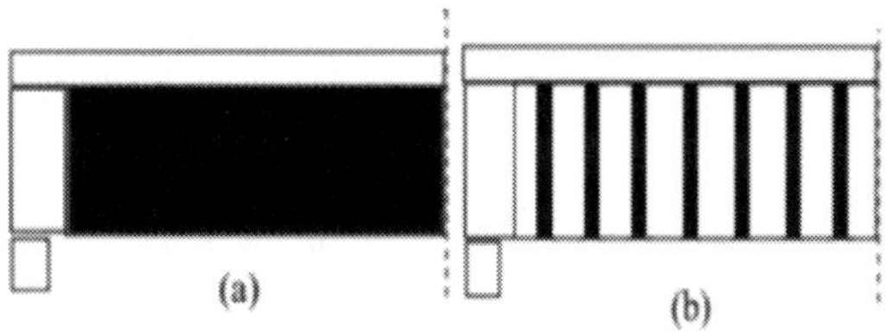


**Fig. 4.** FRP distributions [8, 9].

Figure (5) shows the orientation of FRP. AligningFRP fibers transverse to the axis of the member or perpendicular to the shear cracks paths is more effective in providing additional shear strength [10]. This is achieved by the use of inclined strips.

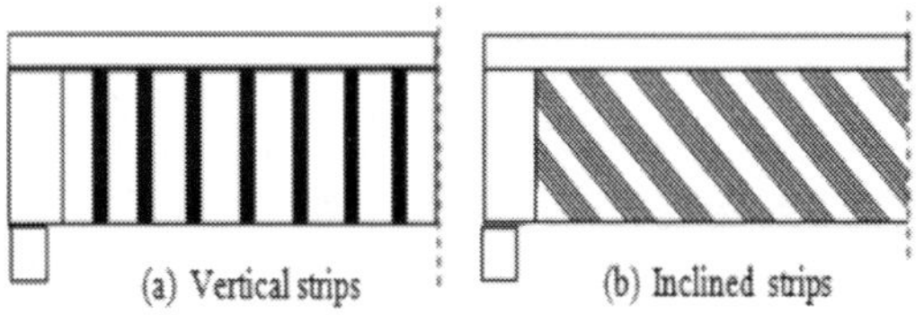


**Fig. 5.** FRP orientations [9].

## 3 Experimental program

### 3.1 Beam specimen description

A total of eightidentical simply supported lightweight aggregate reinforced concrete deep beams were molded and tested to evaluate the effectiveness of the CFRP strengthening schemes on the behavior and shear load carrying capacity of these beams, as well as, studying some parameters related to this field. All beams were strengthened by various configurations of CFRP strips except one which serves as a control beam. The shear span to effective depth ratio (a/d) was chosen to be 1.0.

### 3.2 Beam Specimen Details

The deep beams were with the same dimensions of (1400 mm span length, 150mm width, 400 mm depth. The internal reinforcement is kept constant as (3Φ16mm for main bars and Φ5 at 100 mm for shear reinforcement in bothvertical and horizontal directions).The chosen shear reinforcement ratio ($\rho_v$=$\rho_{vh}$=0.00262) was almost the same to the minimum shear reinforcement ratio recommended by the ACI 318M-2014 Code[2] for both vertical and horizontal directions. Shear reinforcement's yield strength was 476MPa while for flexural reinforcement is 450MPa. Dimensions and details of the internal steel reinforcement for the tested deep beams are shown in Figures6 and 7 respectively. Additional reinforcement cages were provided at the points of applied load and supports which connected tightly to the beams reinforcement to avoid premature local failure at those zones.

### 3.3 Concrete mix design

As recommended by ACI 213.2R-03[11] Committee, a concrete shall be deemed to be a structural lightweight concrete if its compressive strength (*f'c*) is greater than (17 MPa) at (28 days) and the oven dry density less than (2000kg/m$^3$).Hence, several trial mixes were made by the researchers in order to satisfy these two conditions and reach to an acceptable *f'c*of (26.34 MPa) at (28 days) with an oven dry density of about (1950 kg/m$^3$).

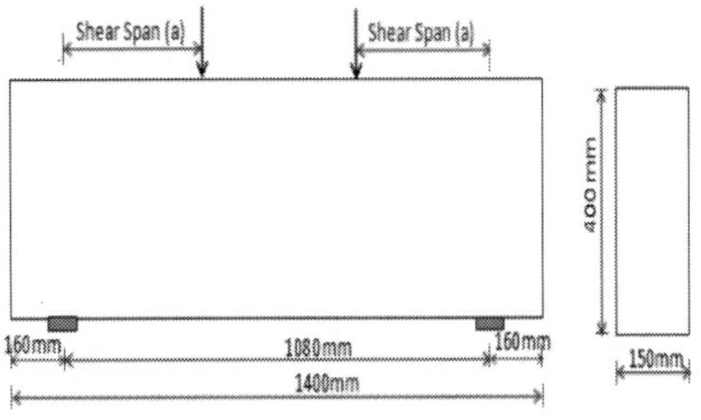


**Fig. 6.** Deep Beam specimen demensions

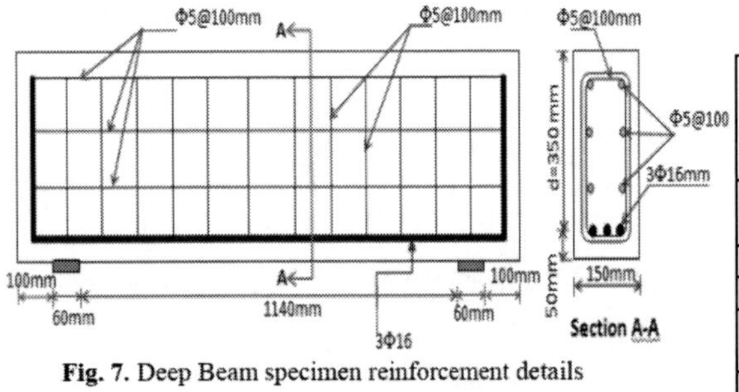


**Fig. 7.** Deep Beam specimen reinforcement details

## 4 Strengthening system

### 4.1 CFRP sheets and adhesive

Only seven of the eight deep beams have been strengthened with CFRP strips. strengthened deep beams were bonded with one or two layersof different orientation and direction of CFRP strips spaced at 100mm center to center. The width of CFRP strips was kept constant at 50 mm with thickness of 0.166 mm. Sika Wrap-300 C/60 unidirectional woven carbon fiber fabric and Sikadore-330 two parts epoxy impregnation resin were used in the present study to strengthen the tested beams. The strengthening process was done after the 28 days of curing of the test beams by about two weeks, when the concrete surface was completely air dried.

Beam specimen strengthening details are illustratedin table1. The names of the deep beams include the short (DB) which refers to deep beam. DB1 was unstrenghtened to serve as a control beam. Deep beams (DB2, DB6, DB7, and DB8) were strengthened with vertical fiber wrap as strips. Each strip was continuous at the bottom face of the beam and exhaustible close to the top surface of beams on both sides. Two layers of CFRP strips were applied, one over the other, for the beams DB6 and DB8. While three strips were used to strengthen DB3 in a horizontal configuration covering all sides of the beam. DB4 was strengthened with both vertical and horizontal strips spaced at (100mm) for both directions. In DB5, four isolated carbon fiber strips of (300 mm length) and (200 mm length) were bonded orthogonallyto the diagonal line connecting the loading and the supports points in a symmetrical manner with the (400 mm) strip at the middle as shown in figure 8. The length of the strips No (2) in figure (8 of DB5) was chosen to be 75% of the length of the middle strip. While the length of strip number (3) was equal to 50% of the middle strip length. Decreasing in the strips length was intended to be analogous to a decrease in the bottle shape representation of the strut and tie region.

**Table 1.** Beams specimen strengthening details

| Beam No. | a/d ratio | a (mm) | Shear Reinf. | CFRP Details | | |
|---|---|---|---|---|---|---|
| | | | | No. of Layers | Spacing (mm) | Orientatin |
| DB1 | 1.0 | 350 | Φ5@ 100 | Ref. beam | | |
| DB2 | | | | 1 | 100 | Vertical |
| DB3 | | | | 1 | 100 | Horizontal |
| DB4 | | | | 1 | 100 | Vertical & Horizontal |
| DB5 | | | | 1 | 100 | Inclined |
| DB6 | | | | 2 | 100 | Vertical |
| DB7 | | | | 1 | 150 | Vertical |
| DB8 | | | | 2 | 150 | Vertical |

 https://doi.org/10.1051/matecconf/201816204011

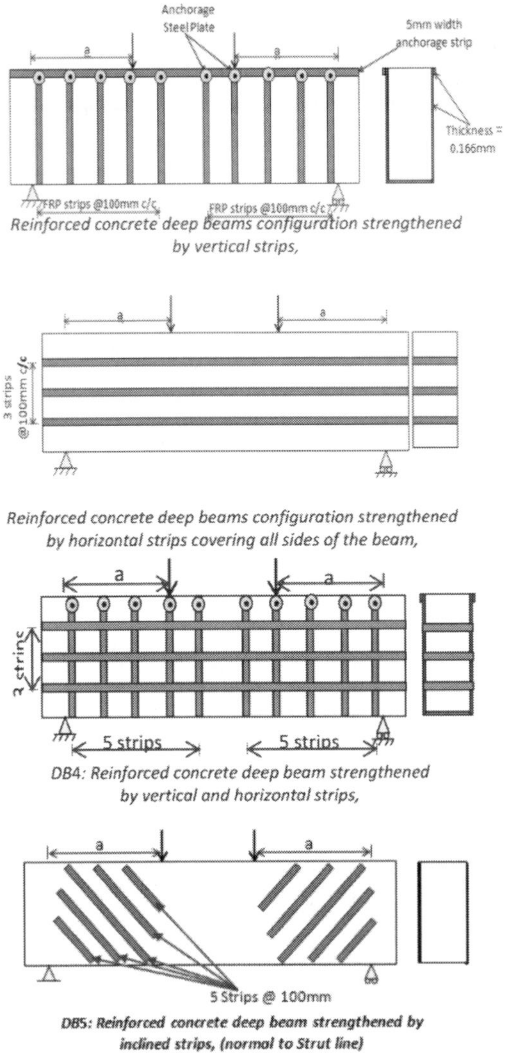


**Fig. 8**. Design drawings and all details of the strengthening pattern of the tested beams.

### 4.2 CFRP strips installations

The strips of carbon fiber were applied by a manual lay-up process to the external surface of concrete. Resin Chemical reaction hardens the material to a strong lightweight product. The resin serves as the matrix for carbon fiber as concrete action for the steel reinforcing bars. The insulation process was done as described in the following steps:

- The concrete surface was abraded at the locations of gluing carbon fiber strips on concrete.
- The side or/and lower angle of the strengthened specimens have been chamfered to provide a radius of approximately (13 mm) (Figure 9) according to ACI Committee 440.2R-08[8] recommendations.
- Just before composites application on concrete beam, the epoxy resin was mixed according to the manufacturer's instructions. Then a thin layer of epoxy resin was applied along the concrete substrate in specified locations where the fiber will installed. The strips are placed on top of the epoxy resin coating and bonded to tensile soffit. Then, the resin is squeezed through the portable of the fabric by a special roller.

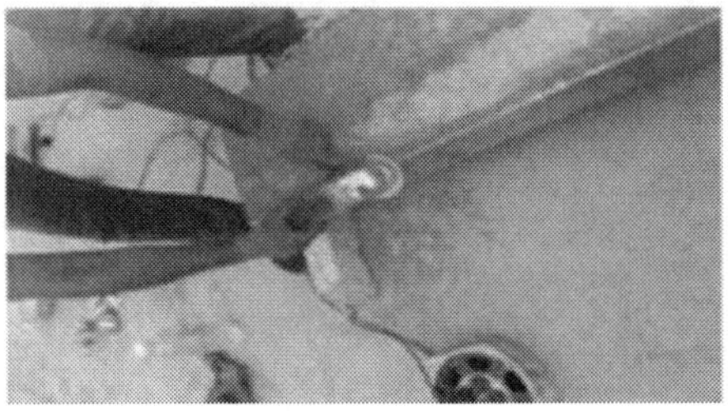

**Fig. 9.** Smoothing the concrete corners prior to fix CFRP strips

## 5 Test setup and loading procedure

The laboratory tests for the all the deep beams were done by a hydraulic testing machine (AVERY Denison testing machine) of (2500 kN) maximum capacity which was available in the structural laboratory of the Building and Construction Engineering Department/ University of Technology (Figure 10).For all tested beams, one dial gauge of (0.01 mm) accuracy capacity was positioned at midspan of each beam to measure the midspan deflection.

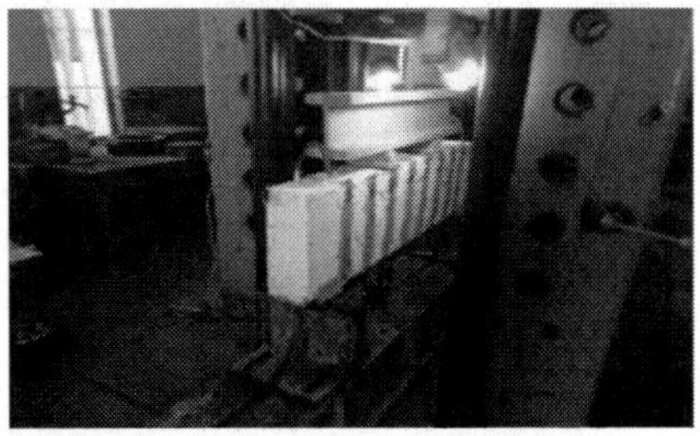

**Fig. 10.** One deep beam specimen under testing

## 6 Results and discussion

### 6.1 General behavior of deep beam specimens

At the early stages of loading, all the testeddeep beams behaved in elastic manner and they were free of cracks. Gradually, with the increase of the loads, small flexural and shear cracks appeared at center of the beam and the shear span regions respectively. The first shear cracks appeared at approximately about (16 to 35%) of ultimate loads followed by long and slightly wide cracks that developed in the shear span zone with further increasing of the loads.

The carbon fiber strips (CFRP) in the strengthened deep beams play their role in loads resisting capacity at later stages of loading. The behavior of the strengthened beams under increasing loads was varied according to the type of strengthening, number of layers, spacing between the strips.

The failure planes in all beams were formed along the diagonal cracks joining the loading and the reaction points. The crack patterns in the strengthened specimens were nearly identical to the control deep beam DB2 except that DB2 shows widely spaced and more cracks as compared to strengthened deep beams.

### 6.2 First cracking and ultimate loads capacity

Test results including first shear cracking and ultimate (maximum) loads gain by carbon fiber strengthening technique are illustrated in table 2. Theelementary or initial cracks developed in the strengthened deep beams at a higher load than their regulate specimen because the CFRP strips provided a good restraint to the cracks growth. Increasing in the first cracking load of the strengthened samples could be attributed to the stiffness increasing due to the CFRP strips suppressing effect. The ratios of ultimate load carrying capacity of the strengthened beams to the control onewere computed and given in Table 2. The highest value was recorded for beam specimen DB4 among all the tested beams, which was equal to 145%.

The experimental investigation shows that the percentage increase in ultimate load of one and twolayers of CFRP is 18% (for DB2) and 27.3% (for DB6) respectively compared to the control sample. The strengthening of DB6 (two layers of CFRP strips, one over the other) results in increase in load carrying capacityabout 9% over DB2 (one layer), both have the same spacing between strips (s=100mm). The same conclusion could be drawn by comparing DB8 (two layers of CFRP) with DB7 (one layer) each with spacing of 150mm. In the latter case, the increase in load carrying capacity is about 16%. According to the experimental observation, the tendency of having brittle failure perceptible was observed to be weaker in CFRP strengthened deep beams than in unstrenghtened beams.

**Table 2.** Experimental results of the tested beams

| Beam No. | CFRP Details | | | $P_{cr(crack)}$ kN | Increasing in crack loads % | $P_u$ kN | Increasing in ultimate load % |
|---|---|---|---|---|---|---|---|
| | No. of Layers | Spacing (mm) | Orientation | | | | |
| DB1 | - | - | - | 100 | - | 440 | - |
| DB2 | 1 | 100 | Vertical | 160 | 60 | 520 | 18 |
| DB3 | 1 | 100 | Horizontal | 120 | 20 | 500 | 14 |
| DB4 | 1 | 100 | Vertical & Horizontal | 160 | 60 | 640 | 45 |
| DB5 | 1 | 100 | Inclined | 140 | 40 | 580 | 32 |
| DB6 | 2 | 100 | Vertical | 120 | 20 | 560 | 27 |
| DB7 | 1 | 150 | Vertical | 140 | 40 | 460 | 5 |
| DB8 | 2 | 150 | Vertical | 140 | 40 | 530 | 14 |

The central deflection at first shear cracking and Ultimate loads of the tested beams is presented in figure (11). The central deflection at ultimate load for the tested deep beams is smaller than that of the control beam DB2 (reinforced by internal shear reinforcement only); this probably could be a result of the externally attached carbon fiber strips.

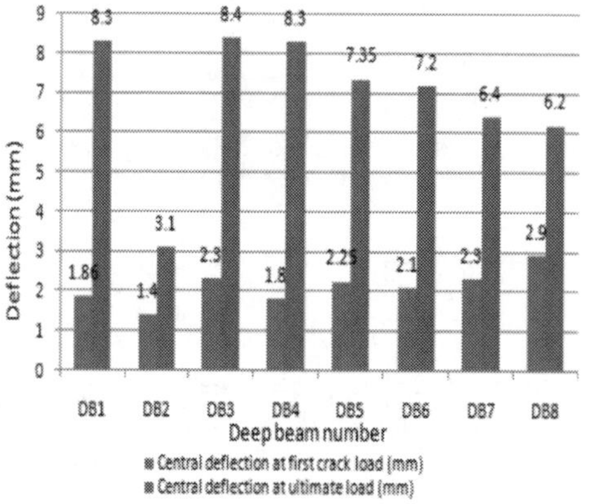


**Fig. 11.** Central deflection at shear crack and ultimate loads of the tested deep beams

### 6.3 Crack pattern and failure modes.

Figure 12 shows the failure mode of the tested beams, the vertical orientation of strips in DB2 showed better performance than horizontal placement in DB3. Also, the presence of carbon fiber in addition to internal stirrups provides a better distribution of diagonal cracks throughout the shear span in DB2.

The mode of failureof DB5 was started due to the lower end deboning of the middle strip from the concrete surface. Shortly after that, the strip nearest to the middle one on the upper end of the beam gave way, consequently causing the complete failure in a same manner to DB1. The using of transverse strips gives an increase of 60% and 32% respectively, in crack load and ultimate load compared to that of the control sample. When compared to DB2 which had one layer vertical strips, this increase in the ultimate load was more by about 12% while there is a decrease in crack load by about 14%.

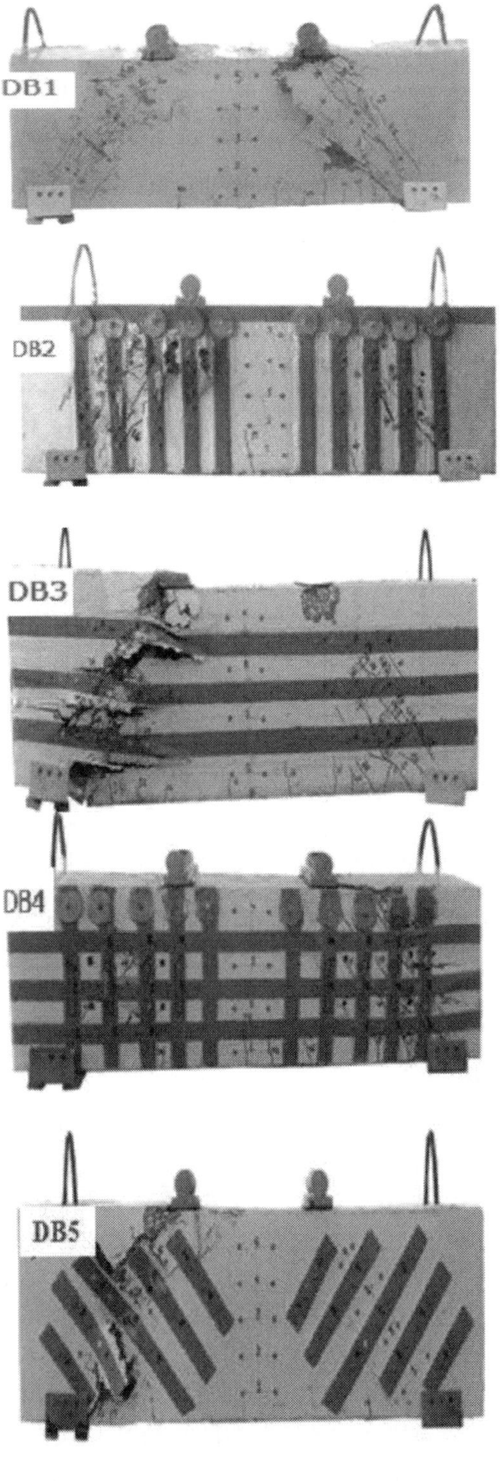


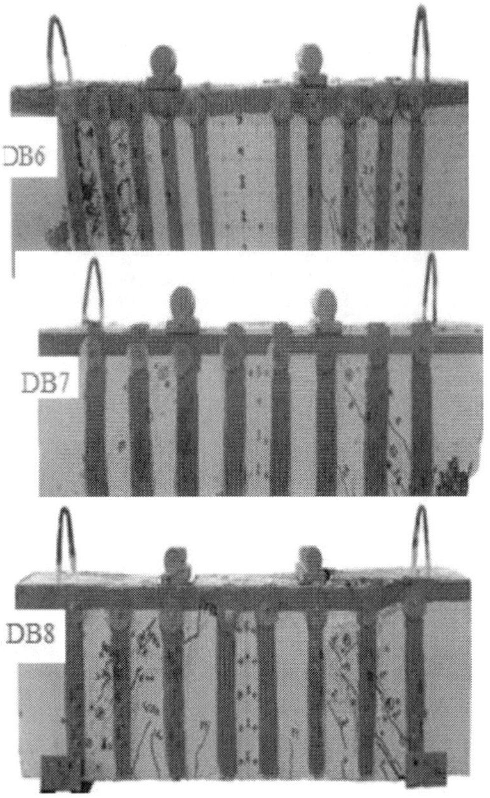


**Fig. 12.** Crack pattern at failure for the tested deep beams

A longitudinal splitting crack was established on the top face of the specimen DB8at about 450 kN as shown in figure 13. The crack started to the right side of the beams (roller support) and extends towards the support. It could be said that the splitting failure of DB8 happened due to the high longitudinal compressive stress developed at the top of the beam that produce aepisodic tension led to splitting failure. DB8 failed at a total ultimate load of 530 kN with an increase of 20% in ultimate capacity compared with DB1 and an increase about 15% compared with DB7

**Fig. 13.** Splitting crack on the top of DB8

### 6.4 Load Mid-span deflection responses

Load-deflection relationship is essential to describe behavior of the beams under various stages of loading. Table (3) demonstrates the midspan deflection for the tested deep beams to show the additional capacity gained by using CFRP strips. The largest deflection was experienced by DB3. For comparison, the deflections of the beams at first crack load of the control beam (DB1) which was of (100kN) are also shown in table (3). The table shows that the use of carbon fiber composite material results in reduced deflections by about (9 to 80%) when compared at the first crack load. This decrease in deflection is dependent on the type of strengthening pattern. While at the ultimate load, the deflection of the strengthened beams was more than the deflection of unstrenghtened beam in companied with increase in shear capacity loading.

**Table 3.** Comparison between failure load and deflection for the deep beams

| Beam No. | Deflection at 100kN | | Maximum deflection at ultimate load | | |
|---|---|---|---|---|---|
| | mm | Ratio to DB1 % | mm | Ratio to DB1 % | % decrease of max. deflection |
| DB1 | 2.18 | 100 | 8.30 | 100 | - |
| DB2 | 0.8 | 37 | 3.60 | 43 | - |
| DB3 | 2.12 | 97 | 8.30 | 100 | - |
| DB4 | 1.1 | 50 | 7.60 | 92 | 9 |
| DB5 | 1.8 | 83 | 7.35 | 89 | 11 |
| DB6 | 1.59 | 73 | 7.20 | 87 | 13 |
| DB7 | 1.72 | 79 | 6.50 | 78 | 28 |
| DB8 | 2.38 | 109 | 7.80 | 94 | 64 |

Figure (14) shows the load-midspan deflection behavior of the deep beams strengthened by vertical carbon fiber strips. The deflection curve of the strengthened deep beams appeared to be identical in shape to their control one (DB1) but with different extremes. Initially, the internal steel reinforcement bars in the strengthened beams carried the majority of the tensile force in the beam cross section. Then, the additional tensile force are carried by the CFRP strips when the internal steel abandon and as a result, an increase of the load carrying capacity of the beams is gained.

Figure (15) shows the curves of load midspan deflection for DB1, DB3, DB4, and DB5which were the control beam and the beams strengthened by horizontal, vertical and horizontal, and inclined CFRP strips, respectively. All curves have somewhat similar character and demonstrated nearly linear response till to the first shear crack load. Response of all strengthened beams of this group shows a typical load-deflection behavior with the sharp drop as soon as their reaching the ultimate loads.

It can be observed from figure (16) that the horizontal strips in DB5 have less effect on the load carrying capacity and the maximum deflection value compared with the unstrengthened control deep beam DB1, while the presence of inclined strips in DB5 have a significant effect in improving the load capacity by about 32% as compared with DB1 and 16% as compared with DB3 respectively. Also, these inclined strips play a noticeable role in decreasing the maximum deflection by about 13% compared with the control beam DB1 and the beam with horizontal strips only (DB3). DB4 was the best strengthened deep beam among the all beams in increasing the load carrying capacity by 45% as compared with the control one.

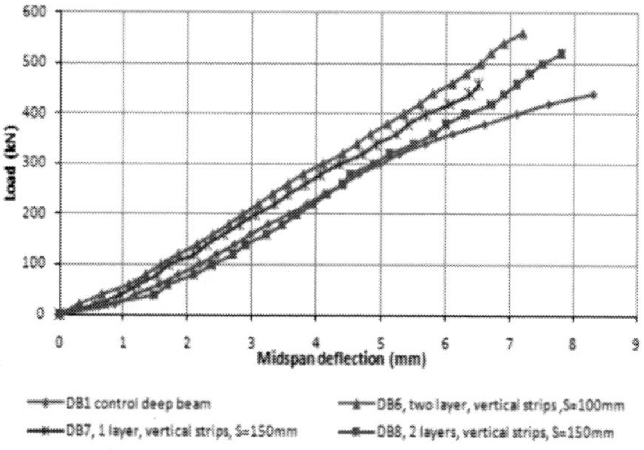


**Fig. 14.** Load-midspan deflection curves for beams with vertical strengthening system

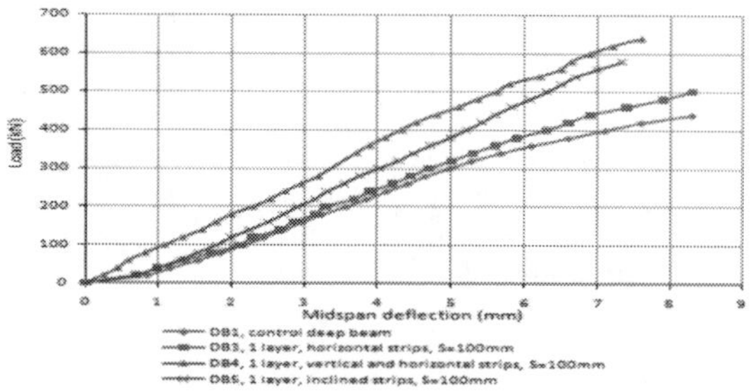


**Fig. 15.** Load midspan deflection curves for beams with horizontal and inclined strengtheningsystem

## 7 Conclusions

The following conclusion could be made based on the experimental results:

1. The use of CFRP strips in the strengthening lightweight aggregate deep beams increases the load carrying capacity by about 45% for the present work. Cracks were smaller and more distributed in the strengthened beams compared with their controls ones.
2. Bonded FRP system in the shear span considerably delayed the formation of diagonal shear cracks and provided positive restraint to the subsequent growth of cracks. The shear crack load of the beams varies from 16 to 35% of their ultimate loads.
3. Compared to control beam, the use of inclined CFRP strips in DB5 gives an increase of 60% and 32% in crack load and ultimate load respectively.
4. Diagonal compression strut crack of unstrenghtened control beams was changed to many small diagonal cracks at mid depth within the shear span of the strengthened beams and exhibited more ductile failure mode with the increased stiffness in tension face.
5. The strengthening accomplished by externally bonded carbon fiber strips, in general, provided restraint to the widening of the diagonal cracks. Thus, it can be concluded that the strip thickness and normal orientation significantly influences the structural performance of the strengthening lightweight deep beams.

## References

1. S.Tain, PhD Thesis, University of Manchester/ Engineering and Physical Sciences, 239pp. (2013)
2. ACI Committee 318, American Concrete Institute, Farmington Hill, MI,524pp. (2014)
3. G.Russo, R.Venir,M.Pauletta, ACI Structural Journal**102,** 3, 429-437 (May-June 2005)
4. E.Skibbe, MSc. Thesis, Kansas State University, Department of Architectural Engineering, College of Engineering, Manhattan, Kansas,98pp. (2010)
5. P.Nagarajan,T.M. Pillai, N. Ganesan, IE (I) Journal-CV **88**, 38-43(July 2007)
6. C.Dolan,H.R. Hamilton,A.Nanni, FHWA Contract, Final Report **1**, 24,133pp (August 2001)
7. A.F. Izzet, PhD Thesis, University of Technology/ Building and Construction Department, Baghdad, Iraq,176pp.(2008)
8. ACI Committee 440.2R-08, American Concrete Institute, Michigan, USA, 80pp.(July 2008)
9. A.M. Khalifa, PhD. Thesis, University of Missouri-Rolla, Center for Infrastructure Engineering Studies (CIES), 155pp. (1999)
10. Y.Sato, T.Ueda, Y.Kakuta, Tanaka, T., *Proceedings of The 2nd International Conference on Advanced Composite Materials in Bridges and Structures, ACMBS-II*, Canada,621-627(1996)
11. ACI Committee 213.2R-03, American Concrete Institute, Farmington Hill, USA, 38pp. (2003)

# The effectiveness of CFRP strengthening of steel plate girders with web opening subjected to shear

*Mohammed* Hamood[1,*], *Wael* AbdulSahib[1], and *Ali* Abdullah[1]

[1]Building and Construction Engineering Department, University of Technology, Baghdad, Iraq

**Abstract.** The present study is experimental by nature; it involves applying shear buckling test on seven steel-plated girders that are subjected to shear loading. Four of these girders represent the strengthened girders whereas the other three are reference girders. One of the latter type of girders has a square web opening; the second has a diamond web opening while the last one has no opening at all. The webs of the strengthened girders were adhesively attached to CFRP sheets of different patterns. This step was carried out to evaluate the most effective strengthening scheme by CFRP composite. The study aims at examining the effect of several parameters on the behavior of web girders. It further examines the percentage of increase in the ultimate shear capacity of the perforated girders that have a fixed location and size where the latter is equal to 40 percent of the web depth. Results have shown that, first, the ultimate shear load of the CFRP-strengthened girders with square web opening is higher than that of the reference girder with a square web opening; ranging from 8.7% to 15.7%. The obtained ranges depend on the orientation of the CFRP strips. Second, the ultimate shear load of the CFRP-strengthened girders with a diamond web opening is higher than the reference girder with a range of 9.8% to 21.5%. Again, the obtained ranges depend on the orientation of the CFRP strips. Analytically speaking, Von Mises stresses have been used to predict the ultimate shear load of girders with square and diamond web openings and without web opening.

## 1 Introduction

Many research studies have been so far conducted on the use of carbon fiber reinforced plastic composite in both strengthening and repairing the webs of girders. These previously conducted studies mainly focused on the application of CFRP to girders with axial or bending stresses. As far as the investigations on strengthening and repairing web girders using CFRP are concerned, they have been reported few.

The standard strengthening techniques used with steel-plated girders include bolting or welding. However, these techniques are not void of disadvantages; cases in point are the effect of corrosion, the fatigue-sensitivity of the repaired system to the stress concentration resulted from either the bolting or welding techniques. Another shortcoming is the long period of service interruption. Recently, alternative girder strengthening techniques have been used, such as that of the external bonded (CFRP) composites. Such techniques help prevent web buckling in steel-plated girders by adding stiffness to the web of girder. In other words, the stiffness of CFRP provides the required high stiffness, tensile strength and corrosion /fatigue resistance to the girders [1, 2, 3, 4].

In modern buildings and bridges, large ducts for air-condition system, sanitary pipes, sewage pipes, electric cables and other services need to be provided. However, to provide such services, an unacceptably large construction depth between storeys is needed. To solve this problem, a web opening needs to be made on each of the web girders. This helps in extending the pipes of services and reducing the floor height. However, such web openings change the distribution of stress at the web of girders and ultimately reduce the shear and bending strength capacity of the girder [2, 3, 4].

To restore the lost strength, the locations of these openings and the lost strength need first to be studied and measured, respectively. Second, the lost strength in these locations and the locations surrounding the openings needs to be strengthened with different patterns of CFRP. In addition to that, one should avoid making such openings in locations where the shear stresses are high [5].

## 2 Material properties

The properties of the materials used in the tested plate girders include the following:

### 2.1 Steel

Table (1), below, shows the details and properties of the steel plates used to form the seven steel-plated girders obtained from the Tensile strength test, as shown in Figure (1). In this study, the modules of elasticity and poison's ratio are assumed to be 200000 MPa and 0.3 respectively:

* Corresponding author: mjh_smm@yahoo.com

© The Authors, published by EDP Sciences. This is an open access article distributed under the terms of the Creative Commons Attribution License 4.0 (http://creativecommons.org/licenses/by/4.0/).

**Table 1.** Properties of Steel Plates Used for Girders

| Component | Thickness (mm) | σy (MPa) | | σu (MPa) | |
|---|---|---|---|---|---|
| Flange/ stiffeners | 10 | 255 | 255 | 400 | 403 |
| | | 255 | | 405 | |
| | | 255 | | 404 | |
| web | 2 | 250 | 260 | 365 | 370 |
| | | 270 | | 375 | |
| | | 260 | | 370 | |

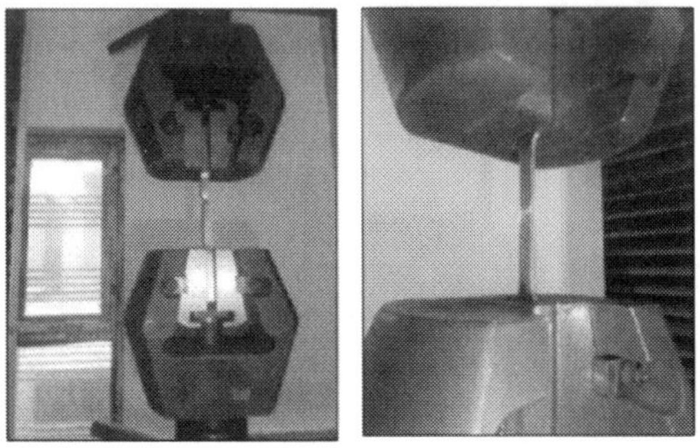

**Fig. 1.** Tensile Strength Test for Steel Plates

## 2.2 CFRP Laminate

The type of CFRP used to strengthen the girders is **sika-warp-230 c** produced by Sika Corporation, as shown in Figure (2) and Table (2), where the latter is about the properties of CFRP provided according to manufacturer data sheet attached to the CFRP laminate.

**Table 2.** CFRP Properties

| Properties | Sika Warp 230-c |
|---|---|
| Tensile Strength (MPa) | 4300 |
| E-Modulus (GPa) | 238 |
| Elongation At Break (Strain)% | 1.8 |
| Thickness (mm) | 1.131 |
| Density (g/cm3) | 1.76 |

**Fig. 2.** CFRP Laminate

## 2.3 Adhesive epoxy

The suitable adhesive material used with the carbon fiber strips is Sikadur®-330, as shown in Figure (3). This type of adhesive consists of two parts: part A, resin, which is white and part B, hardener, which is grey. The main properties of the adhesive are provided according to manufacturer data sheet attached to the adhesive package, as shown in Table (3).

**Fig. 3.** Sikadur®- 330 Container

**Table 3.** Sikadur®- 330 Properties

| Properties | Sikadur®- 330 |
|---|---|
| Density (kg/l) at +23°C | Parts A+B mixed : 1.31 |
| Tensile Strength (MPa) | 30 |
| | Flexural : 3.8 |
| E-Modulus (GPa) | Tensile : 4.5 |
| Elongation At Break (Strain)% | 0.9 |
| Setting Time (Minute) at 35°C | 30 |

# 3 Type and details of the plate girders

Two types of plate girders have been used in this study. The first type is without web opening whereas the second is with a web opening. The second type is further sub-divided into two types: square and diamond web opening. The length of web opening is 240 mm and the shear stresses in the webs are relatively low in comparison to the bending stresses in the flanges. As a result, the 2mm web plate is generally chosen to be much thinner than the flanges which are (6mm) [7]. For more details about the girders, consider Figure (4) and Table (4) where three of the girders are reference

 https://doi.org/10.1051/matecconf/201816204012

whereas the other four are strengthened girders with different patterns of CFRP laminate.

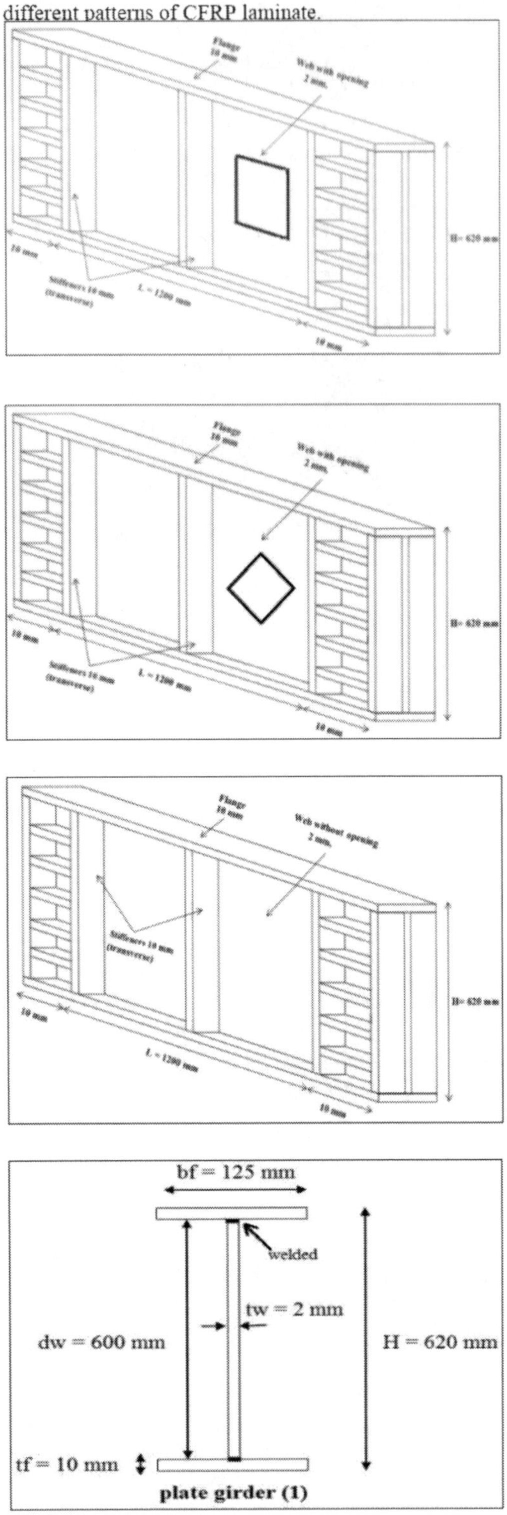


**Fig. 4.** Details of Plate Girders

**Table 4.** Details of Plate Girders in (mm)

| Girder | $t_w$ | $t_s$ | $t_f$ | $b_f$ | L | a | d | $d/t_w$ | a/d | Reinf. |
|---|---|---|---|---|---|---|---|---|---|---|
| T2-0 | 2 | 10 | 10 | 125 | 1200 | 600 | 600 | 300 | 1 | reference |
| T2-S-0 | 2 | 10 | 10 | 125 | 1200 | 600 | 600 | 300 | 1 | reference |
| T2-S-1 | 2 | 10 | 10 | 125 | 1200 | 600 | 600 | 300 | 1 | Type1 |
| T2-S-2 | 2 | 10 | 10 | 125 | 1200 | 600 | 600 | 300 | 1 | Type2 |
| T2-D-0 | 2 | 10 | 10 | 125 | 1200 | 600 | 600 | 300 | 1 | reference |
| T2-D-1 | 2 | 10 | 10 | 125 | 1200 | 600 | 600 | 300 | 1 | Type1 |
| T2-D-2 | 2 | 10 | 10 | 125 | 1200 | 600 | 600 | 300 | 1 | Type2 |

## 4 Types of strengthening

**Type1** includes strengthening the web of girders using CFRP strips of 10mm width from two sides. The first side is parallel to the flanges while the second is perpendicular to flanges, as it is used in (T2-D-1), (T2-S-1) girders shown in figure (5).

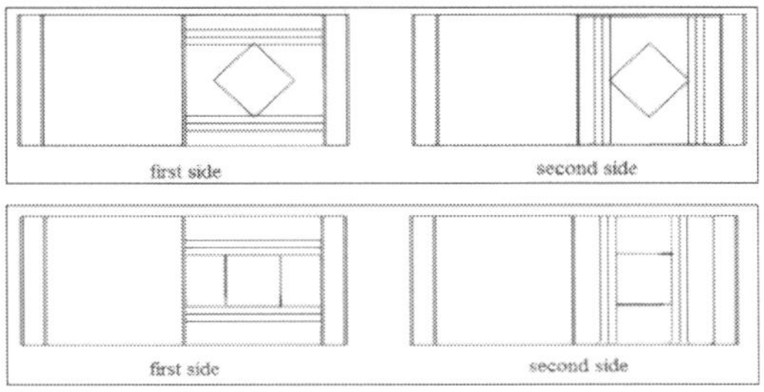


**Fig. 5.** Type (1) of strengthening (T2-D-1) and (T2-S-1) Girders

**Type2** involves strengthening the web of the girders via CFRP strips of 10mm width from two sides. In the first side, the strengthening is diagonally put along the tension at ϴ45 whereas in the second side, it is put diagonally along the compression at ϴ135. Cases in point are the ones used in (T2-D-2) and (T2-S-2) girders; consider Figure (6).

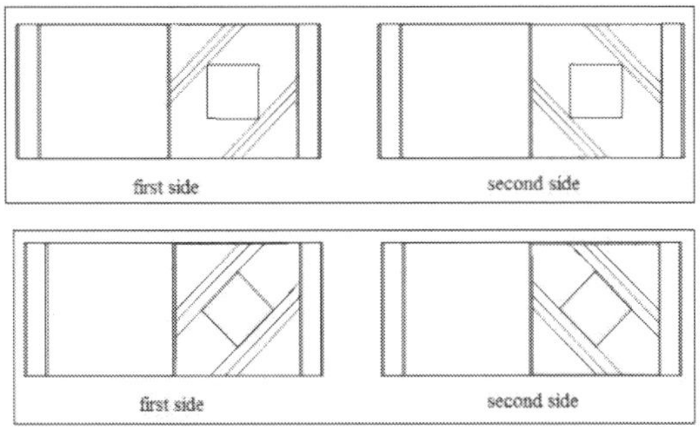


**Fig. 6.** Type (2) of Strengthening (T2-D-2) and (T2-S-2) Girders

## 5 Testing procedure and instrumentation

All plate girders were simply supported by a pin and roller over 1200 mm span. The girders were tested up to the failure mode under the action of applied load at the mid span of the girder. Such an application helps ensure that each of two panels equally have half of the applied load; see Figure (7). The maximum load capacity of the tested machine (Avery) is 120 tons and the deflections of all girders were measured by a dial gauge of 0.01 accuracy installed at the mid span under the girders.

**Fig. 7.** Test Setup for a sample of Plate Girder

## 6 Evaluation of ultimate shear strength

### 6.1 Girders without web opening:

The ultimate shear strength can be obtained as follows:

The critical shear stress ($\tau_{cr}$) in this stage is given through adopting Gardiff Method:

$$\tau_{cr} = K_s * \frac{\pi^2 * E}{12(1-\nu^2)} * \left(\frac{t_w}{d}\right)^2 \qquad (1)$$

Where:

($K_s$): the shear buckling coefficient given by:

$$K_s = 5.35 + 4\left(\frac{d}{a}\right)^2 \quad \textit{where } \left(\frac{d}{a}\right) \geq 1$$

$$K_s = 5.35 * \left(\frac{d}{a}\right)^2 + 4 \quad \textit{where } \left(\frac{d}{a}\right) \leq 1 \qquad (2)$$

Where:

(d): Depth of the girder in mm;

($t_w$): web thickness in mm;

(a): Clear distance of the web plate between the vertical stiffeners in mm;

(E): Modulus of elasticity in MPa; and

(v): poisons ratio.

$$\sigma_{yt} = \sqrt{\sigma_{yw}^2 + \tau_{cr}^2\left(\frac{9}{4}sin^2 2\theta - 3\right)} - \frac{3}{2}\tau_{cr} * sin2\theta \qquad (3)$$

Where:

$$\theta = \frac{2}{3}tan^{-1}\left(\frac{d}{a}\right) \qquad (4)$$

($\sigma_{yw}$): the tensile stress of web in MPa.

$$M_{pf} = \frac{1}{4} * \sigma_{yf} * b_f * t_f^2 \qquad (5)$$

$$M_p^* = \frac{M_{pf}}{d^2 * \sigma_{yw} * t_w} \qquad (6)$$

$$V_{cr} = \tau_{cr} * d * t_w \qquad (7)$$

$$V_{post} = \sigma_{yt} * t_w * sin^2\theta(d * cot\theta - a) + 4d * t_w * sin\theta\sqrt{\sigma_{yw} * \sigma_{yt} * M_{p*}} \qquad (8)$$

$$V_{ult} = V_{cr} + V_{post} \qquad (9)$$

Where:

($\sigma_{yf}$): the tensile stress of flange in MPa;

($b_f$): flange width; and

($t_f$): flange thickness.

### 6.2 Girders with web opening

$$k_o = ks\left(1 - \frac{dh}{d}\right) \qquad (10)$$

Where:

($K_s$): the shear buckling coefficient from equation (2); and

(dh): the opening diameter or length.

$$(\tau_{cr})_o = K_o \frac{\pi^2 * E}{12(1-v^2)} * \left(\frac{tw}{d}\right)^2 \quad (11)$$

$$\sigma_{yt} = \sqrt{\sigma_{yw}^2 + (\tau_{cr})_o^2 \left(\frac{9}{4}\sin^2 2\theta - 3\right)} - \frac{3}{2}(\tau_{cr})_o * \sin 2\theta \quad (12)$$

Where:

$$(V_{cr})_o = (\tau_{cr})_o * d * t_w; \text{ and}$$

$$V_{post} = \sigma_{yt} * t_w * \sin^2\theta(d * \cot\theta - a) + 4d * t_w * \sin\theta * \sqrt{\sigma_{yw} * \sigma_{yt} * M_{p*}} \quad (13)$$

$$(V_{post})_o = 0.5 * V_{post} \quad (14)$$

$$(Vult)_o = (Vcr)_o + (Vpost)_o \quad (15)$$

## 7 Test results

Figure (8) shows the load deflection curves for the tested girders (T2-0) without web opening and (T2-D-0) with diamond web opening as reference girders. Moreover, the (T2-D-1) and (T2-D-2) girders were strengthened by CFRP strips. Results have revealed that the diamond web opening decreases the ultimate shear load resistance to 44%, whereas the girders (T2-D-1) and (T2-D-2) which were strengthened by CFRP strips, as in Type 1 and Type 2, show a higher stiffener and a high load than that of (T2-D-0).

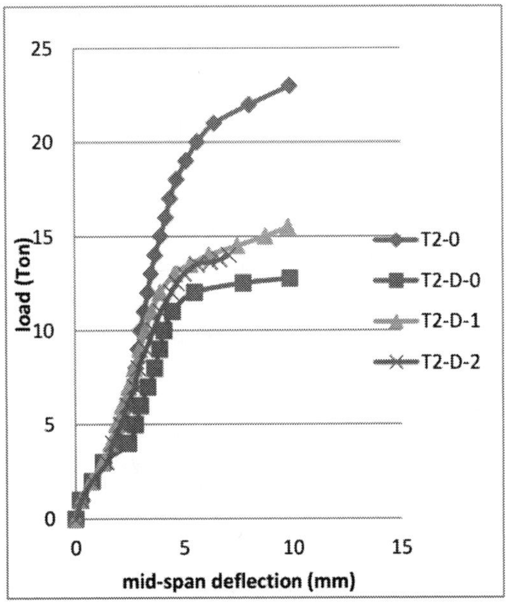


**Fig. 8.** Load –Deflection Curves for (T2-0), (T2-D-0), (T2-D-1), and (T2-D-2)

Figure (9) shows the load deflection curves for the tested girders (T2-0) without web opening and (T2-S-0) with square web opening where both represent the reference girders. In addition, the girders (T2-S-1) and (T2-S-2) were strengthened by CFRP strips. Results have shown that the square web opening decreases the ultimate shear load resistance to 38%, while the girders (T2-S-1) and (T2-S-2), which were strengthened by CFRP strips, as in Type 1 and Type 2, show higher stiffener and a high load than that of (T2-S-0).

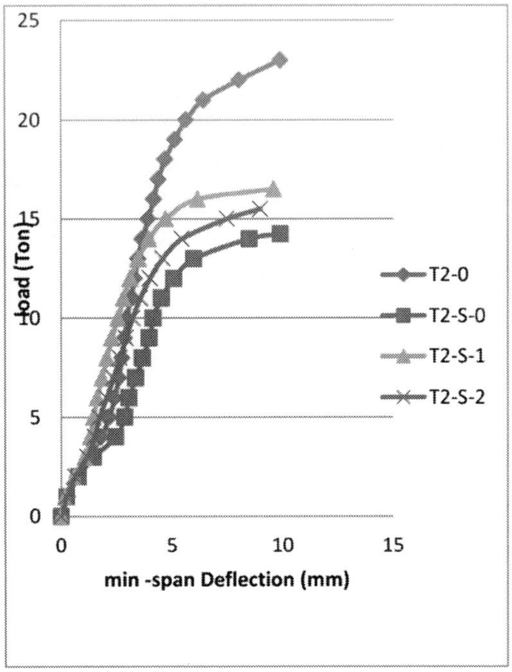


**Fig. 9.** Load –Deflection Curves for (T2-0), (T2-S-0), (T2-S-1), and (T2-S-2)

In the vein, the CFRP strips-strengthened girders were all painted in light grey. Besides, the test was conducted on only one panel of girders whereas the second panel was diagonally strengthened by a wood-made stiffener. This is because the girder is symmetrical in shape and has two panels. As it was previously mentioned, each panel should equally carry half of the applied load. Accordingly, and economically speaking, there is no need for the two panels to be tested. This is because both the first and second panels of the girders share common results and behavior. Figure (10) portrays the failed mode of all girders after the tests. In this respect, the researchers compared between the analytical and experimental results. In addition, the ultimate shear load was compared with its reference girder, as illustrated Table (5).

 https://doi.org/10.1051/matecconf/201816204012

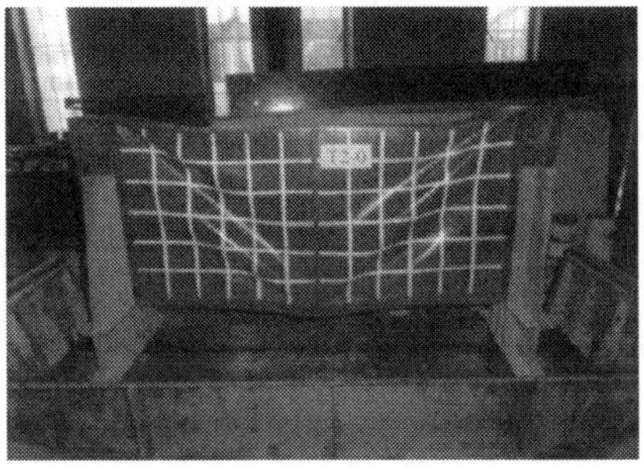

A. A Reference Plate Girder without a Web Opening

**B.** A Reference Plate Girder with a Web Opening (T2-D-0)

**C**. A Plate Girder with a Web Opening (T2-D-1)

D. A Plate Girder with a Web Opening (T2-D-2)

E. A Reference Plate Girder with a Web Opening (T2-S-0)

F. A Plate Girder with a Web Opening (T2-S-1)

G. A Plate Girder with a Web Opening (T2-S-2)

**Fig. 10.** The Tested Plate Girder after the Failure Mode

**Table 5.** The Analytical and Experimental Ultimate Shear Load for the Girders in comparison to their Reference Girders

| Plate Girder | Max Ultimate Shear Load (Ton) (analytical) | Max Ultimate Shear Load (Ton) (experimental) | The Effect of Reinforcing Perforated Girders Compared to their Reference (%) (Experimental) | Ultimate Shear Load Compared to the Reference Girder (T2-0) (%) (Experimental) |
|---|---|---|---|---|
| T2-0 | 25.25 | 23 | - | - |
| T2-D-0 | 12.7 | 12.75 | - | -10.25 |
| T2-D-1 | 14.2 | 15.5 | 21.5 | -7.5 |
| T2-D-2 | 12 | 14 | 9.8 | -9 |
| T2-S-0 | 13.48 | 14.25 | - | -8.75 |
| T2-S-1 | 14.21 | 16.5 | 15.7 | -6.5 |
| T2-S-2 | 9.63 | 15.5 | 8.7 | -7.5 |

## 8. Conclusions

1- The ultimate shear resistance has decreased by about (38%) for the girder with a square opening of 40% side length of the web depth compared to the girder without a web opening;

2- The ultimate shear load has decreased by about (44.5%) for the girder with a diamond opening of 40% side length of the web depth compared to the girder without a web opening;

3- The ultimate shear load for the girders with a square web opening strengthened by CFRP strips was higher than the reference girder with a square web opening. The ranges of the ultimate shear obtained are 8.7% and 15.7%, depending on the orientation of the CFRP strips;

4- The ultimate shear load for the girders with a diamond web opening strengthened by CFRP strips are higher than that of the reference girder with a diamond web opening. The ranges of the ultimate shear obtained are 9.8% and 21.5%, depending on the orientation of the CFRP strips;

5- The experimental results of the plate girders have shown the applicability and effectiveness of the strengthening technique used in girders with a web opening. Furthermore, the technique is shown to increase the shear strength of these girders; and

6- The most effective form used to strengthen the web opening of plate girders was type (1). This type was strengthened by CFRP strips that were parallel and perpendicular to their flanges.

## References

1. K. Abed, MSc Thesis, Structural Engineering Department, University of Technology, Baghdad, pp.3, (2015)
2. M. S. Abdul Gabar, MSc Thesis, Structural Engineering Department, University of Technology, Baghdad, pp.3 (2012)
3. N. C. Hagen, P. K. Larsen, A. Aalberg, Journal of constructional Steel Research **65**, 1, 142-50 (2009)
4. M. j. Hamoodi, I. N. Korkess, K. F. Sarsam, The 6th Engineering Conference, College of Engineering, Baghdad University **1,** 1-17 (April 2009)
5. Web site www.steel-insdag.org/teaching material /chapter28.pdf
6. K. Rockey, H. Evans, D. porter, *A design method for predicting the collapse behavior of plate girders*, proceeding institution of civil engineers, part2, 85-112 (1978)
7. Z. Al-azzawi, T. Stratford, M. Rotter, L. Bisby, Conference: Response of Structures under Extreme Loading- Protect, The University of Edinburgh UK., 1-11 (January 2015)

https://doi.org/10.1051/matecconf/201816204012

## Notations

**a:** The clear of the web plate between vertical stiffeners;
**τ:** Shear stress;
**tw:** Thickness of the web;
**d:** Depth of girder;
**MPf:** Plastic moment capacity of the flange plate;
**τcr:** Critical shear stress;
**dh:** Diameter of the opening;
**σyw:** Tensile yield stress of the web; and
**σyt:** Tensile membrane stress at yield.
**ko:** Buckling coefficient for a perforated web;
**ks:** Shear buckling coefficient;
**θ:** The angle of inclination of the membrane stress;
**σyf:** Tensile yield stress of the flange;
**Vpost:** Post buckling shear force;
**Vult.:** Ultimate shear strength;
**Vcr.:** Critical buckling shear force;
**Mp *:** Non-dimensional flange strength parameter;
**τcrv:** Shear buckling stress of plate girder reinforced by CFRP;
**Dv:** Flexural rigidity of plate girder bonded CFRP laminate;
**σy:** Yield stress of steel material; and
**σu:** Ultimate stress of steel material.

# Behavior of self-compact reinforced concrete deep beams with small shear span to depth ratio

*Hassan* Hassan[1,*], *Mu'taz* Medhlom [1], *Mohammed* Hatem[1]

[1]Faculty of Engineering, Al-Mustansiriyah University, Baghdad, Iraq

**Abstract.** This research is devoted to investigate the experimental and theoretical behavior of deep beams under monotonic two points loading. An experimental program examining six RC deep beams is carried out. The investigated parameters include shear span to depth ratio varying from 1.0 to 0.276. A comparative study is conducted in this paper by using finite element software ANSYS. The experimental and numerical results show that concrete strength and shear span to depth ratio are the two most important parameters in controlling the behavior of RC deep beams. Comparison of experimental results was made with corresponding predicted values using the Strut and Tie procedure presented ACI 318M-11Code and with other procedures mentioned in the literature. It was found that the Strut and Tie procedure presented in ACI 318M-11Code give conservative results as compared with the experimental tested results. The results showed reliability of analysis in predicting deep beams behavior in terms of failure load, failure mode as well as crack propagation.

## 1 Introduction

Reinforced concrete (RC) deep beams are structural members for which the load is applied at a distance from the support so that a substantial proportion of the load is transferred directly to the support by arching action. Common structural applications of deep beams include transfer girders in buildings, bridges, and offshore structures. Acording to ACI 318-14 [1]. "Deep beams are members that are loaded on one face and supported on the opposite face such that strut-like compression elements can develop between the loads and supports and that satisfy (a) or (b): (a) Clear span does not exceed four times the overall member depth $h$; (b) Concentrated loads exist within a distance $2h$ from the face of the support." On the other hand, Eurocode 2 (EC2) [2] defines deep beams as all beams with span to depth ratio smaller than three.

The strength of deep beams is usually controlled by shear rather than flexure [3]. As the shear behavior of RC members is still not well understood and is influenced by many parameters, existing design models rely on empirical equations [3, 4]. Even though such approaches are generally extremely conservative, [5-8] they can also lead to unsafe design solutions [8-10]. Therefore, the provisions of current codes of practice need to be reviewed and improved to account for parameters affecting shear behavior and capacity of RC deep beams.

Self-Compacting Concrete (SCC), which flows under its own weight and does not require any external vibration for compaction, has revolutionized concrete placement. SCC first introduced in the late 1980's by Japanese researchers [11], is highly workable concrete that can flow under its own weight through restricted sections without segregation and bleeding. Such concrete should have a relatively low yield value to ensure high flow ability, a moderate viscosity to resist segregation and bleeding, and must maintain its homogeneity during transportation, placing and curing to ensure adequate structural performance and long term durability [12].

## 2 Research significance

This work aims to provide experimental evidence on the behavior of RC deep beams to enable a better understanding of the effects of shear span-depth ratio, and lead to improved design procedures. The results will also allow an evaluation of the current code provisions and help identify their limitations.

## 3 Experimental program

The experimental program consists of testing 6 simply supported deep beams under two point loads to investigate the behavior of reinforced concrete deep beams. All beams have the same dimensions and flexural reinforcement. They had an overall length of 1000 mm, a width of 100 mm and a height of 400 mm. The amount of flexural reinforcement for all the tested beams was 3Φ16 mm ($\rho$=0.0185), as shown in Figure1. Table (1) show details of the six tested reinforced concrete deep beams.

* Corresponding author: hassanfalah@uomustansiryah.edu.iq

© The Authors, published by EDP Sciences. This is an open access article distributed under the terms of the Creative Commons Attribution License 4.0 (http://creativecommons.org/licenses/by/4.0/).

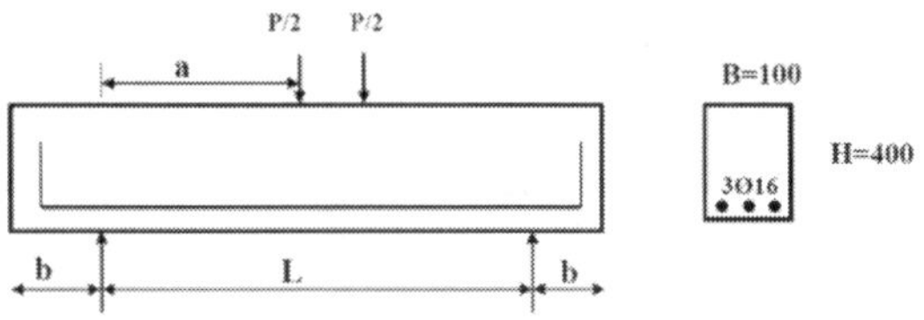


**Fig. 1**. Details of beam.

**Table 1**.Specimen details.

| Beam No. | L mm | a mm | b mm | d mm | ρ % | a/d |
|---|---|---|---|---|---|---|
| S1 | 824 | 362 | 38 | 362 | 1.85 | 1.0 |
| S2 | 700 | 300 | 100 | 362 | 1.85 | 0.828 |
| S3 | 580 | 240 | 160 | 362 | 1.85 | 0.663 |
| S4 | 500 | 200 | 200 | 362 | 1.85 | 0.552 |
| S5 | 400 | 150 | 250 | 362 | 1.85 | 0.414 |
| S6 | 300 | 100 | 300 | 362 | 1.85 | 0.276 |

## 3.1 Materials

Properties and description of used materials are reported and presented in Table (2) and the concrete mix proportions are reported and presented in Table (3).

**Table 2**. Properties of materials.

| Material | Descriptions |
|---|---|
| Cement | Ordinary Portland Cement (Type I) |
| Sand | Natural sand from Al-Ukhaider region with maximum size of (4.75mm) |
| Gravel | Crushed gravel of maximum size (10 mm) |
| Superplasticizer (S.P.) | Sika Visco Crete PC-20 was used as an admixture to produce SCC in this study |
| Limestone Powder (L.P.) | Limestone powder from Al-Mousel district is less than 0.125 mm, which satisfies EFNARC 2002[13] recommendations |
| Reinforcing Bars | (ϕ16mm) deformed steel bar, having (630MPa) yield strength (fy) |
| Water | Clean tap water |

**Table 3**. Mix prportions

| Cement ($Kg/m^3$) | Sand ($Kg/m^3$) | Gravel ($Kg/m^3$) | S.P. ($lit/m^3$) | L.P. ($Kg/m^3$) | Water |
|---|---|---|---|---|---|
| 520 | 790 | 750 | 10 | 150 | 0.35 |

## 3.2 Fresh SCC properties results

Table (4) illustrates the results of these three tests that carried out on SCC mix and the comparisons with the standard limitations are also presented. From this table, one can notice that the results of all test mixes satisfy the requirements of EFNARC [13] specifications.

## 3.3 Hardened SCC mechanical properties results

Table (5) shows test results of mechanical properties obtained for SCC mix. These properties are concrete compressive strength (f'c), splitting tensile strength (ft) and modulus of rupture (fr). Each value presented in this table represents the average value of three specimens.

**Table 4**. Results of fresh SCC

| Mix | Slump flow (mm) | $T_{50}$ (sec) | L – box ($H_2/H_1$) |
|---|---|---|---|
| SCC | 700 | 3 | 0.85 |
| Limits of EFNARC[13] | 650-800 | 2-5 | 0.8-1 |

**Table 5.** Mechanical properties of hardened SCC

| Mix Name | $f'_c$ (MPa) | $f_t$ (MPa) | $f_r$ (MPa) |
|---|---|---|---|
| SCC | 50 | 4.1 | 5.6 |

## 3.4 Test results of deep beams

### *3.4.1 Ultimate and First Cracking Load*

Table (6) summarizes the results of first cracking load (Pcr) and ultimate load (Pu) for all tested beams together with their modes of failure.

**Table 6**. Results of deep beam

| Beam No. | ρ % | a/d | $P_{cr}$ kN | $P_u$ kN | Mode of shear failure |
|---|---|---|---|---|---|
| S1 | 1.85 | 1.0 | 25 | 123 | Diagonal tension failure |
| S2 | 1.85 | 0.828 | 26 | 152 | (Shear +flexural) failure |
| S3 | 1.85 | 0.663 | 40 | 190 | (Shear +flexural) failure |
| S4 | 1.85 | 0.552 | 74 | 245 | Diagonal tension failure |
| S5 | 1.85 | 0.414 | 85 | 320 | Diagonal tension failure |
| S6 | 1.85 | 0.276 | 90 | 330 | Diagonal tension failure |

### *3.4.2 Load-Mid Deflection Relationships*

From the load mid-span deflection relationship shown in Figure 2 for all deep beams, the following three distinct stages are observed:

1. The first stage shows linear behavior with constant slope.
2. In the second stage, vertical flexural cracks were initiated at the tensile face within the maximum bending moment region of the beam, and extend upward, then inclined cracks originated in the shear spans. These cracks developed with increased load, causing a corresponding shift of the neutral axis towards the compression face, and consequently, a continuous reduction in the moment of inertia of the

cracked section. The curve changed from linear to non-linear behavior in this stage.

3. In the third stage, the shape of the load-deflection curve tends to be asymptotic to the horizontal as the beam approached its ultimate load.

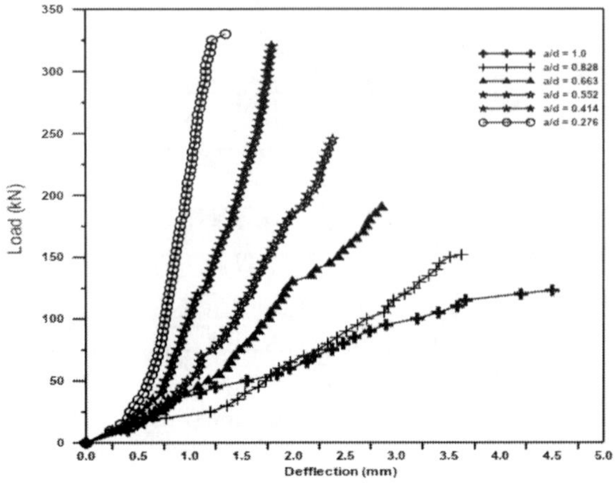


**Fig. 2**. Load mid-span defelection of beams

### 3.5 Failure mode

Figure 3 shows the crack patterns after testing all the beams to failure. This plate shows that the failure mode for most of the deep beams tested was through a diagonal shear crack with different widths extending from the bottom of beam near the support to the loading points at the top with different widths. The cracks were accompanied, in some specimens, by the formation of new inclined cracks parallel to the initial cracks in the shear span. However, three specimens failed by flexural vertical cracks extended to the compression zone. The diagonal cracks extended towards the beam's bottom at or near the supports and the loading points at the top but did not reach both.

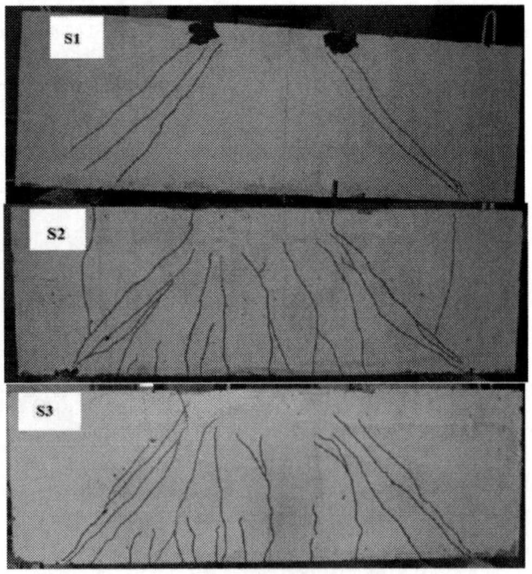


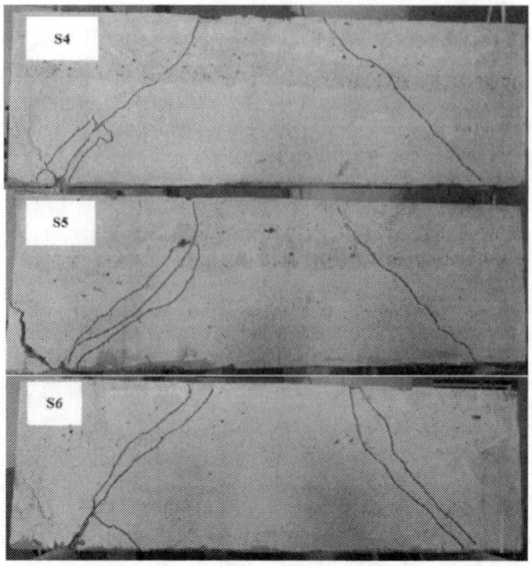


**Fig. 3**. CRACK PATTERNS AND MODES OF FAILURE

## 4 Analytical program

### 4.1 Strut and tie model in design

The strut-and-tie model is a simple equilibrium model based on the lower-bound solution of the plasticity theory and can be used to design D-regions such as deep beams. Design based on the strut-and-tie model (STM) is allowed in all four major international codes: ACI 318-14[1], AASHTO LRFD[14] EC2[2] and Model Code 2010[15] The first step in designing with the STM is to select an appropriate strut-and-tie layout and define the size of each element (Figure 4). Subsequently, the stresses in each element, which should not exceed the maximum allowable stresses, are calculated. Codes of practice give provisions on allowable stresses but do not provide guidance on how to determine the size of the resisting strut and- tie elements, such as struts and nodal regions. As this information is needed to determine the resisting capacity of each element, designers are required to make arbitrary decisions on the size of the elements that transfer the imposed loads to the supports. The STM shown in Fig. 1 can generally be used to analyze all beams with and without shear reinforcement. Sagaseta and Vollum[8] proposed a different STM to account for the effect of shear reinforcement on the strut configuration. This work, however, is not discussed in the current paper, as codes of practice do not suggest the use of different configurations of struts to account for the effect of shear reinforcement. The height of the bottom node (hB) (refer to Figure 4) can be assumed as twice the distance between the center of force of the main longitudinal reinforcement and the tension face. According to Yang and Ashour[16] the height of the top node (hT) can be determined from the equilibrium between the limit of resultant compressive force at the top node (C-C-C), and the limit of resultant tensile force of the bottom node (C-C-T). According to this hypothesis, the height of the top node (hT) is equal to

80%, 88%, 85%, and 75% of the height of the tie according to ACI 318-14[1], AASHTO LRFD[14], EC2[2], and Model Code 2010[15], respectively, while the strut angle (θ) can be calculated from geometry using Eq. (1).

The top and bottom width of the inclined strut (WST and WSB) can be calculated using Eq. (2) and (3)[1] respectively.

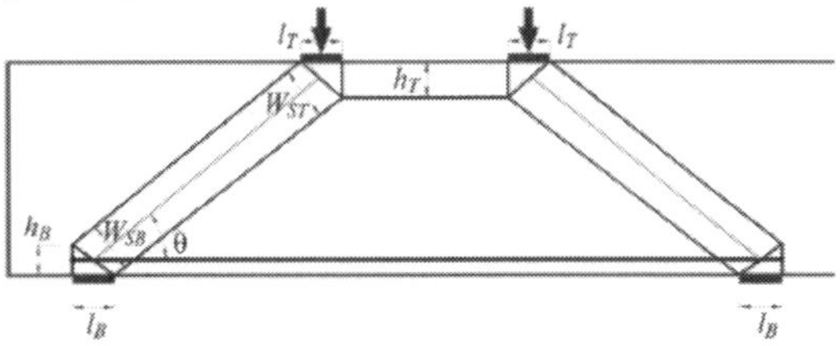


FIG. 4. TYPICAL STRUT-AND-TIE MODEL USED TO PREDICT LOAD CAPACITY OF TESTED BEAMS.

$$\theta = tan^{-1}\left[\frac{d-{}^{h_t}/_2}{a}\right] \quad (1)$$

$$W_{ST} = l_T sin\theta + h_T\, cos\theta \quad (2)$$

$$W_{SB} = l_B sin\theta + h_B cos\theta \quad (3)$$

Cracked RC is an orthotropic material, for which its principal stresses can be assumed to have the same directions of the principal tensile and compressive strains. The presence of lateral tensile strain in the concrete strut, however, can reduce its compressive strength. This is accounted for by using an effectiveness factor to calculate an effective concrete compressive strength. According to ACI 318-14[1], the effective concrete strength (fce) can be calculated using Eq. (4a) while AASHTO LRFD[14], EC2[2], and Model Code 2010[15] use Eq. (5a), Eq. (6a), and Eq. (7a), respectively.

$$f_{ce} = 0.85\beta_s f'_c \quad (4a)$$

Where βs is 0.75 for strut with shear reinforcement satisfying
Eq. (4b); otherwise, βs is taken as 0.6.

$$\sum \frac{A_{si}}{b_s S_i} sin\alpha_i \geq 0.003 \quad (4b)$$

where $A_{si}$ is the area of the reinforcement at spacing si in the
i-th layer of reinforcement crossing a strut at an angle αi to
the axis of the strut.

$$f_{ce} = \frac{f'_c}{0.8+170\varepsilon_i} \leq 0.85 f'_c \quad (5a)$$

Where

$$\varepsilon_i = \varepsilon_s + (\varepsilon_s + 0.002)cot^2\alpha_s \quad (5b)$$

Where αs is the smallest angle between the strut and adjoining
tie; and εs is the tensile strain in the direction of the tie.

$$f_{ce} = 0.6v' f_{cd} \quad (6a)$$

Where v′ can be calculated according to Eq. (6b); and fcd is the design concrete compressive strength.

$$v' = 1 - \frac{f_{ck}}{250} \quad (6b)$$

$$f_{ce} = k_c f_{cd} \quad (7a)$$

$$k_c = 0.55\left(\frac{30}{f_{ck}}\right)^{1/3} \leq 0.55 \quad (7b)$$

The strut-and-tie model shown in Figure 4 with the aforementioned element size definitions will be used for the analysis of the experimental results obtained from the program presented in this paper.

### 4.2 Modified STM theory

Zhang and Tan [17], suggested a modified STM for calculation of shear strength of reinforced concrete deep beams based on a previous fulfillment reported by Tan and Cheng [18]. For simply supported reinforced concrete beams subjected to symmetric two point loads, from the structural analysis it is well known that the ultimate load (P) is equal to twice the shear force at the support.

$$P = 2V_n \quad (8)$$

The expression for calculated the shear strength Vn according to Zhang and Tan [17], is as follows:

$$V_n = \frac{1}{\frac{4\, \sin\theta_s \cos\theta_s}{A_c f_t} + \frac{\sin\theta_s}{A_{str} f'_c}} \quad (9)$$

where;
Vn: shear strength of deep beams (N).
Ac: is the beam effective cross- sectional area in mm2, equals
to bw dc.
dc: effective beam depth (mm).
Astr: cross-sectional area of the concrete diagonal strut in mm2,
equal to ws bw
ws: effective width of the inclined strut (mm).
bw: width of deep beam (mm).
ft: combined tensile strength of reinforcement and concrete
(MPa).

$\theta_s$: angle between the axis of the strut and the horizontal axis
of the member.

It can be noted that the expression ft is the composite tensile strength including contributions from concrete and reinforcement (web and main bars), where;

$$f_t = f_{ct} + f_{st} \quad (10)$$

$f_{ct}$: represents the contribution of concrete tensile strength.
$f_{st}$: represents the contribution of steel reinforcement which
consists of two parts, $f_{sw}$ from the web reinforcement and from the longitudinal reinforcement $f_{ss}$.

The expression fss refers to the contribution of bottom longitudinal steel, it can be obtained according to the following equation:

$$f_{ss} = \frac{4A_s f_y \sin\theta_s}{A_c / \sin\theta_s} \quad (11)$$

Where:
$A_s$: total areas of bottom longitudinal main reinforcement (mm2).
$f_y$: tensile yield strength of main reinforcement (MPa).

The performance of the four codes of practice and Zhang in predicting the shear capacity of specimens with a different shear span-depth ratio, and the prediction values are given in Table 7.

## 5 Theoretical program

To study the structural behavior of the tested beams thoroughly, three dimensional finite element analyses by using ANSYS (version-15) software. The theoretical study includes, in additional to verifications of all experimental beams, modeling and analyzing of twelve additional beam specimens. A nonlinear, eight nodes brick element, (SOLID-65), with three translations DOF at each node is used to model the SCC. For FEM modeling of the steel reinforcement, two nodes, discrete axial element, (LINK-180), with three translations DOF at each node is used. In ANSYS software, the real constants such as cross-sectional area and thickness are needed to represent the geometrical properties of the used elements. While, the material properties are needed to represent behavior and characteristics of the constitutive materials which depends on mechanical properties such as yield stress, modulus of elasticity, Poisson's ratio and stress-strain relationship. However, each element has a number of fundamental parameters that are identified in the element library of ANSYS. The use of a rectangular mesh is recommended to secure good results from the concrete element (Solid-65), therefore, a rectangular meshing was applied to model all beam specimens. In spite of volumes meshing for concrete and while volumetric and real meshing are used for the concrete media, no meshing was needed for LINK-180 elements because individual elements were introduced in the model through the nodes created by volumetric meshing of the concrete.

The meshing of concrete elements for and alignments of the steel reinforcing bars beams is shown in Figure 5.

FIG. 5. CONCRETE MESHING AND MODELING OF STEEL BARS

### 5.1 Ultimate loads

Table 8 shows the comparison between the ultimate loads of the experimental (tested) beams, $V_{exp}$ , and the final loads from the finite element models, $V_{num}$. The final loads for the finite element models are the last applied load steps before the solution starts to diverge due to numerous cracks and large deflections. It can be observed that there is a simulation between the finite element analysis and the experimental about (93%) for ultimate load capacity, and about (86%) for ultimate deflection and these ratios are considered reasonable and accepted.

### 5.2 Crack patterns

Crack patterns obtained from the finite element analysis and the failure modes of the experimental beams agree well, as shown in Figure 6. The appearance of the cracks reflects the failure mode for the beams. The finite element model accurately predicts that the beams fail in shear. The cracks were concentrated in the shear span region and vanish diagonally towards the beam supports.

### 5.3 Parametric study

The parametric study presented here consists of modeling and analyzing of twelve additional beam specimens using ANSYS software. The considered parameters in this study are concrete compressive strength (f'c), the vertical and skin shear reinforcement. The parametric study parameters are reported and presented in Table 9.

#### *5.3.1 Effect of Concrete Compressive Strength*

shown in Table 10. It can be noted that the increase in the value of the compressive strength (*fc'*) from (50MPa) to (100MPa) and due to arching action, concrete compressive strength is a dominant parameter influencing the shear capacity of RC deep beams. Figure 7shows the effect of concrete compressive strength on the ultimate capacity of deep beams. The results show that shear capacity of deep beams increases with increasing concrete compressive strength, and this enhancement is more pronounced for beams with smaller shear span-depth ratios.

**Table 7**: ANALYTICAL SHEAR STRENGTH PREDICTIONS AGAINST EXPERIMENTAL DATA

| Beam No. | $V_{exp}$, kN | ACI 318-14 | AASHTO LRFD | EC2 | Model Code 2010 | Zhang and Tan |
|---|---|---|---|---|---|---|
| | | $V_{exp}/V_{cal}$ | $V_{exp}/V_{cal}$ | $V_{exp}/V_{cal}$ | $V_{exp}/V_{cal}$ | $V_{exp}/V_{cal}$ |
| S1 | 61.5 | 0.65 | 1.31 | 0. 84 | 0.87 | 1.25 |
| S2 | 76 | 0.71 | 1.1 | 0.89 | 0.94 | 1.17 |
| S3 | 95 | 0.8 | 1.06 | 0.92 | 0.95 | 1.09 |
| S4 | 122.5 | 0.84 | 0.84 | 0.98 | 1.05 | 1.00 |
| S5 | 160 | 0.87 | 0.75 | 0.99 | 1.08 | 0.91 |
| S6 | 165 | 0.89 | 0.73 | 1.03 | 1.13 | 0.85 |
| Average | | 0.793 | 0.965 | 0.941 | 1.003 | 1.045 |
| Standard deviation | | 0.08 | 0.2 | 0.06 | 0.09 | 0.14 |
| C.O.V | | 0.12 | 0.24 | 0.075 | 0.098 | 0.15 |

**Table 8**: NUMERICAL AND EXPERIMENTAL RESULTS

| Beam No. | ρ % | a/d | $P_{exp}$ kN | $\Delta_{exp}$ mm | ANSYS | |
|---|---|---|---|---|---|---|
| | | | | | $P_{exp}/P_{num}$ | $\Delta_{exp}/\Delta_{num}$ |
| S1 | 1.85 | 1.0 | 123 | 4.5 | 0.93 | 0.8 |
| S2 | 1.85 | 0.828 | 152 | 3.62 | 0.92 | 0.87 |
| S3 | 1.85 | 0.663 | 190 | 2.86 | 0.91 | 0.84 |
| S4 | 1.85 | 0.552 | 245 | 2.38 | 0.95 | 0.9 |
| S5 | 1.85 | 0.414 | 320 | 1.79 | 0.95 | 0.88 |
| S6 | 1.85 | 0.276 | 330 | 1.4 | 0.93 | 0.87 |
| Average | | | | | 0.93 | 0.86 |
| Standard deviation | | | | | 0.016 | 0.032 |
| C.O.V | | | | | 1.09 | 2.17 |

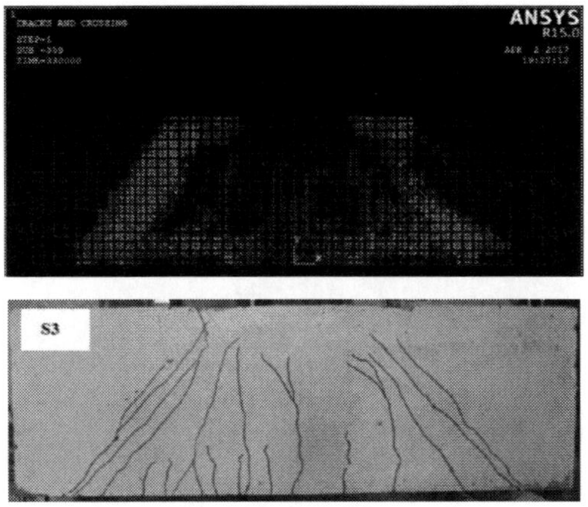


**Fig. 6**. Crack Patterns Compression

Three grades of concrete compressive strength have been used. The numerical results of ultimate load capacity ($P_{num}$) and the corresponding deflection ($\Delta_{num}$), are

### *5.3.2 Effect of Vertical and Skin Shear Reinforcement*

Two ratios of shear reinforcement have been used. The numerical results of ultimate load capacity ($P_{num}$) and the corresponding deflection ($\Delta_{num}$), are shown in Table 11. The increase in shear reinforcement ratio from 0 to 0.4% led to an increase in load capacity. This can be attributed to the fact that the vertical shear reinforcement can carry directly a portion of the applied shear force and provide confinement to the concrete, thus allowing the development of higher compressive stresses. In addition, both vertical and skin reinforcements can reduce crack opening, thus increasing the contribution to shear resistance offered by the concrete through aggregate interlock. Figure 8 shows the effect of shear reinforcement on the ultimate capacity of deep beams.

**Table 9:** EFFECT OF SHEAR REINFORCEMENT ON ULTIMATE LOAD AND DEFLETION

| Beam Designation | a/d | Vertical & Horizontal shear reinforcement % | $p_{num}$ kN | $\Delta_{num}$ mm |
|---|---|---|---|---|
| S1 | 1 | 0 | 114 | 3.6 |
| S1-0.2 | 1 | 0.2 | 155 | 3.5 |
| S1-0.4 | 1 | 0.4 | 190 | 3.46 |
| S2 | 0.828 | 0 | 140 | 3.1 |
| S2-0.2 | 0.828 | 0.2 | 188 | 3.07 |
| S2-0.4 | 0.828 | 0.4 | 245 | 2.99 |
| S3 | 0.663 | 0 | 173 | 2.4 |
| S3-0.2 | 0.663 | 0.2 | 209 | 2.31 |
| S3-0.4 | 0.663 | 0.4 | 289 | 2.26 |
| S4 | 0.552 | 0 | 233 | 2.14 |
| S4-0.2 | 0.552 | 0.2 | 267 | 2.08 |
| S4-0.4 | 0.552 | 0.4 | 310 | 2.02 |
| S5 | 0.414 | 0 | 304 | 1.57 |
| S5-0.2 | 0.414 | 0.2 | 348 | 1.5 |
| S5-0.4 | 0.414 | 0.4 | 399 | 1.42 |
| S6 | 0.276 | 0 | 307 | 1.2 |
| S6-0.2 | 0.276 | 0.2 | 345 | 1.11 |
| S6-0.4 | 0.276 | 0.4 | 410 | 1.03 |

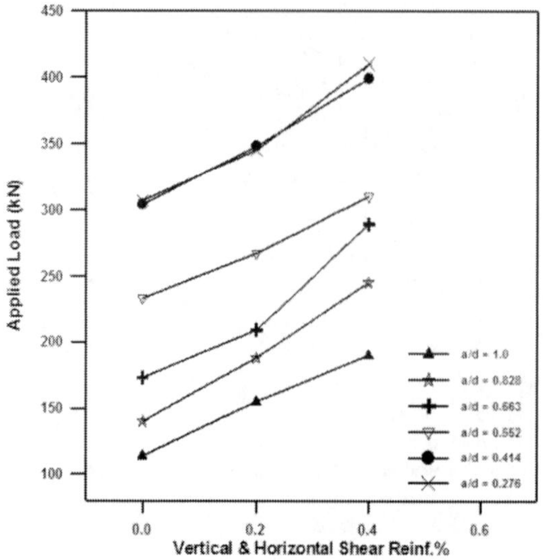


**FIG. 7** EFFECT OF SHEAR REINFORCEMENT ON ULTIMATE FAILURE LOAD.

## 6 Conclusions

On the basis of the experimental and theoretical results presented in this study and the assessment of different design approaches, the following conclusions can be drawn:

1. As a/d ratio, reduces the arch action becomes more dominant whereby loads are transferred directly by arch in compression. Thus the conventional plane-sectional shear approach is inappropriate for deep beams.
2. Concrete compressive strength has more influence on the shear strength of deep beams than shear reinforcement. However, the presence of shear reinforcement is crucial in controlling crack propagation and providing ductility to deep beams.
3. The effectiveness factor is dominated by concrete compressive strength but it is also influenced by the shear span-depth ratio. ACI 318-14 provisions neglect the effect of these two parameters in estimating the effective concrete strength of the inclined strut and, as a result, lead to unconservative predictions.
4. Shear strength predictions by the STM with the provision of AASHTO LRFD generally yield more conservative results among the examined codes.
5. Shear strength predictions by the STM with the provisions of both EC2 and Model Code 2010 are generally conservative; however, their conservatism reduces with increasing shear span-depth ratio because these two codes neglect the effect of shear span-depth ratio on the concrete effectiveness factor.
6. The predicted load by ANSYS in deep beams at various stages was found to be in good agreement with the test data. Where the estimated deflection of beam at ultimate load was much closer to the experimental data.

## References

1. ACI Committee 318, "Building Code Requirements for Structural Concrete (ACI 318M-14) and Commentary (ACI 318RM-14)," American Concrete Institute, Farmington Hills, MI, 520 pp (2014)
2. British Standards Institution, "Eurocode 2: Design of Concrete Structures: Part 1-1: General Rules and Rules for Buildings," British StandardsInstitution, London, UK, 97 pp (2004)
3. K. Smith, A. Vantsiotis, "Shear Strength of Deep Beams," ACI J. Proceedings, **79**, 3, Mar., pp. 201-213 (1982)
4. M. P. Collins, D. Mitchell, E. C. Bentz, The Struc. Eng., **86**, 10, pp. 32-39 (2008)
5. R. Tuchscherer, D. Birrcher, C. Williams, D. Deschenes, O. Bayrak, ACI Struc. J., **111**, 6, Nov.-Dec. (2014), pp. 1451-1460. doi: 10.14359/516869926
6. A. Arabzadeh, A. Rahaie, R. Aghayari, Inter. J. of Civil Eng., pp. 141-153, **7**, 3 (2009)
7. D. Kuchma, S. Yindeesuk, T. Nagle, J. Hart, H. Lee, ACI Struc. J., **105**, 5, pp. 578-589, Sept.-Oct. (2008)
8. J. Sagaseta, R. Vollum, Magazine of Conc. Research, **62**, 4, (2010), pp. 267-282. doi: 10.1680/macr.2010.62.4.267
9. M. P. Collins, E. C. Bentz, E. G. Sherwood, ACI Struc. J., **105**, 5, pp. 590-600, Sept.-Oct. (2008)
10. M. D. Brown, O. Bayrak, ACI Struc. J., **105**, 4, pp. 395-404, July-Aug. (2008)
11. N. Nagamoto, K. Ozawa, "Mixture properties of Self-Compacting, High-Performance Concrete", Proceedings, Third CANMET/ACI Inter. Conf. on Design and Materials and Recent Advances in Concrete Technology, SP-172, V. Malhotra, American Concrete Institute, Farmington Hills, p. 623-637, Mich. (1997)
12. K. H. Khayat, A. Ghezal, "Utility of Statistical models in Proportioning Self-Compacting Concrete", Proceedings, RILEM Inter. Symp. on Self-Compacting Concrete, Stockholm, p. 345-359, (1999)
13. EFNARC: European Federation Dedicated to Specialist Construction Chemicals and Concrete Systems, "Specifications and Guidelines for Self-Compacting Concrete" Association House, 99 West Street, Farnham, Surrey, U.K., 32 pp., Feb. (2002)

**14.** AASHTO, "AASHTO LRFD Bridge Design Specifications," American Association of State Highway and Transportation Officials, Washington, DC, (2012)

**15.** International Federation for Structural Concrete, "Model Code 2010 – first complete draft," fib Bulletin No. 55, 288 pp., (2010)

**16.** K. H. Yang, A. F. Ashour, J. of Struc. Eng., **137**, 10, (2011), pp. 1030-1038. doi: 10.1061/(ASCE)ST.1943-541X.0000351.

**17.** N. Zhang, K.H. Tan, Science Direct, Engineering Structures J., **29**, pp. 2987-3001, March, (2007)

# Investigation on the performance of bridge approach slab

*Amr* Abdelrahman[1], *Mohamed* Tawfik[1] and *A.* El-Saify[1]

[1]Ain Shams University, Cairo, Egypt

**Abstract.** In Egypt, where highway bridges are to be constructed on soft cohesive soils, the bridge abutments are usually founded on rigid piles, whereas the earth embankments for the bridge approaches are directly founded on the natural soft ground. Consequently, excessive differential settlement frequently occurs between the bridge deck and the bridge approaches resulting in a "bump" at both ends of the bridge deck. Such a bump not only creates a rough and uncomfortable ride but also represents a hazardous condition to traffic. One effective technique to cope with the bump problem is to use a reinforced concrete approach slab to provide a smooth grade transition between the bridge deck and the approach pavement. Investigating the geotechnical and structural performance of approach slabs and revealing the fundamental affecting factors have become mandatory. In this paper, a 2-D finite element model is employed to investigate the performance of approach slabs. Moreover, an extensive parametric study is carried out to appraise the relatively optimum geometries of approach slab, i.e. slab length, thickness, embedded depth and slope, that can yield permissible bumps. Different geo-mechanical conditions of the cohesive foundation soil and the fill material of the bridge embankment are examined.

## 1 Introduction

As a result of the current increase in urban development, construction activities have become of essential need in areas of crucial geotechnical conditions, such as construction of highway bridges on soft ground. Whenever a highway bridge has to be constructed on thick layers of soft cohesive subsoil (e.g. soft or very soft clay), the bridge abutments are usually founded on rigid piled foundations. In such a case, the settlement of bridge abutments and, consequently, bridge deck is minimal and frequently considered negligible. The bridge approaches, on the other hand, are commonly constructed of earth embankments that are directly founded on the top of the natural soft ground, and, thus, they are free to settle. The approach embankment settlement is often relatively large 65[1]. Settlement of the approach embankment is attributed mainly to the time-dependent deformations of the soft subsoil.

Consolidation and creep of soft clay, and secondly to the short-term compression of the embankment fill material. In most cases, short-term settlements of the approach embankments are accomplished after construction works have been completed and pavements have been fixed along the bridge. Nevertheless, during the bridge lifetime, consolidation and creep of soft ground below the approach embankments can result in excessive differential settlement between the bridge deck and the adjoining approach pavement. Such a differential settlement emerges as a "bump" at both ends of the bridge deck. The magnitude of the bump is dependent on the bridge specific in-situ conditions. A bridge deck bump not only creates a rough and uncomfortable ride but also represents a hazardous condition to traffic. These defects of the bridge pavements usually require costly maintenance and that often tends to impede the normal flow. Mostly cohesive soil represents a threat instead of cohesion-less soil. Cohesive soil like soft or high plastic clay represents more critical situation.

Cohesive soils are more susceptible to high lateral and vertical movement. Settlements of soil are divided in to three different phases: Initial, primary and secondary consolidation [2]. The initial settlement is a short term settlement when load is applied on the surface of the soil [2].The primary settlement mainly contributes to the total settlement of the soils [3].

The "Bump" can affect drivers varying from feeling uncomfortable to being hazardous to their lives [2].

© The Authors, published by EDP Sciences. This is an open access article distributed under the terms of the Creative Commons Attribution License 4.0 (http://creativecommons.org/licenses/by/4.0/).

Several solutions have been introduced in practices in order to overcome the bump problem, applying a ground improvement technique such as surcharge (or) preloading, dynamic compaction, compaction piles, grouting, and gravel columns are preferred [4, 5] and/or installing approach slabs. The use of approach slabs, however, has been well-recognized as one of the most efficient techniques. To eliminate the effects of the bump, the approach slab must be built to provide a smooth grade transition between these two structures (bridge and roadway). Another function of the approach slab is to keep the magnitude of differential settlement within a control limit [6, 7]. However, in practice it was found that the approach slabs also exceed differential settlements [6, 7]. Approach slabs are reinforced concrete structural slabs that are often constructed at both ends of the bridge. The slab is supported from one end on the bridge abutment and from the other end on the embankment soil or a sleeper slab [4]. Approach slabs do not prevent the approach embankments from settlement. The fundamental function of approach slabs is to provide a smooth grade transition between the bridge deck and the approach pavement, so that the potential differential settlements can be quite diminished. Nevertheless, their use may result merely in moving a controlled bump from the end of the bridge to the end of the approach slab. Furthermore, feasible settlement of the approach embankment can cause the approach slab to lose its contact and support from the underlying embankment soil. Hence, the slab will bend in a concave manner that may cause a sudden change in the slope grade near its ends and may result in a redistribution of loads to the slab both ends; influencing the slab structural performance. Accordingly, profound understanding of the approach slabs performance, as being transition zones, is essentially required.

A number of comprehensive studies have been presented in the literature over the last few years to discuss and identify the sources of bridges bumps and to introduce specific practical solutions including the approach slabs. However, there is still a need for an extensive investigation on the performance of approach slabs and revealing the fundamental affecting factors. Moreover, clear guidelines for approach slabs design considering the approach embankment settlements have become essentially required. In this regard, the numerical analysis technique will be employed in this research study, through a 2-D finite element model, to investigate the performance of bridge approach slabs.

## 2 Numerical model

Finite element analysis has been used as one of the numerical techniques for solving problems in the mechanics of continuous media [8]. Nowadays, the finite element plays an important role in all branches of engineering for the analysis and design of the structures [9]. The analysis is typically performed by transforming the physical problem, an actual structure and structural components, into a mathematical model [9].

The purpose of the numerical analysis is to evaluate the behavior of the approach slabs and give recommendations to design the approach slab according to the international standards. This paper used "PLAXIS" to simulate the behavior of the transition zone, including the bridge abutment, the piles, the approach slab, the embankment and the subsoil material.

### 2.1 Model configuration

One of the most important steps in numerical simulations is to determine where the boundaries should be placed. In this paper, it is assumed that the boundary is located 35m below the bottom of the fill embankment and the horizontal distance from the approach slab to the end of the embankment is 30m. A schematic of the model is shown in Figure (1). Figure (2) shows a finite element model to simulate the bump at the free end of the approach slab.

Using the finite element model described above, this paper simulated several cases. The thickness, the length, the embedded depth and varying thickness of the approach slab were changed to study their influence on the bump at the free end of the approach slab. The different properties of materials which used in this study are shown in Tables (1) and (2).

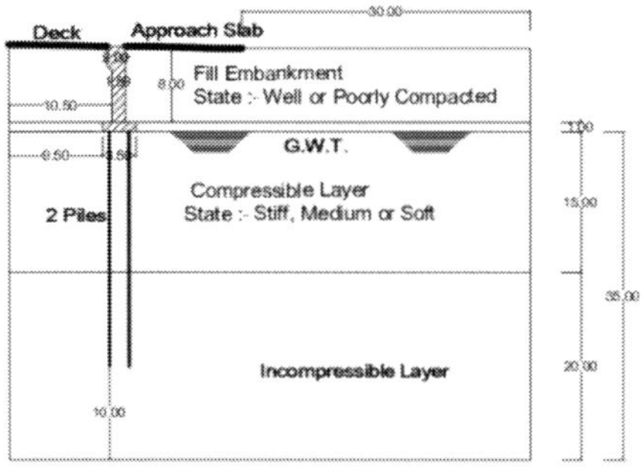


**Fig. 1.** General configuration and parameters the used in analytical model

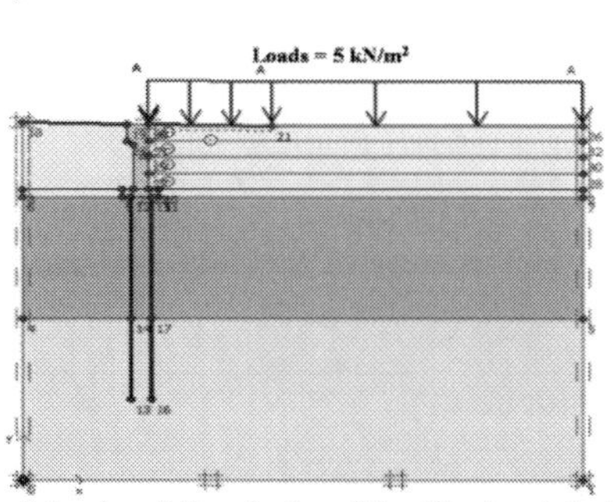


**Fig. 2.** Loads and discretization of the utilized analytical model

**Table 1.** Properties of plates used in the model

| | Approach Slab | Deck | Piles |
|---|---|---|---|
| EA (kN/m) | Variable | 3 X $10^7$ | 1.508 X $10^7$ |
| EI ($kN.m^2/m$) | Variable | 2.5 X $10^6$ | 6.032 X $10^5$ |
| d (m) | 1 | 1 | Pile D=0.8 |
| w (kN/m/m) | 7.5 | 25 | 15 |
| ν | 0.2 | 0.2 | 0.2 |

**Notations:** Where $\gamma_{Unsat}$: Bulk unit weight, $\gamma_{Sat}$: Saturation unit weight, $E_{ref}$: Young's modulus, C: Cohesion, Φ: Friction angle, Ψ: Dilatancy angle, ν: Poission's ratio, $R_{inter}$: Strength reduction factor inter, EA: Normal stiffness, EI: Flexural rigidity, d: Equivalent thickness, w: weight.

The description of the utilized numerical model is presented and properties of soils and plates are carefully defined. A Figure (3) shows the configuration of the basic analytical model used to investigate the performance of the approach slab.

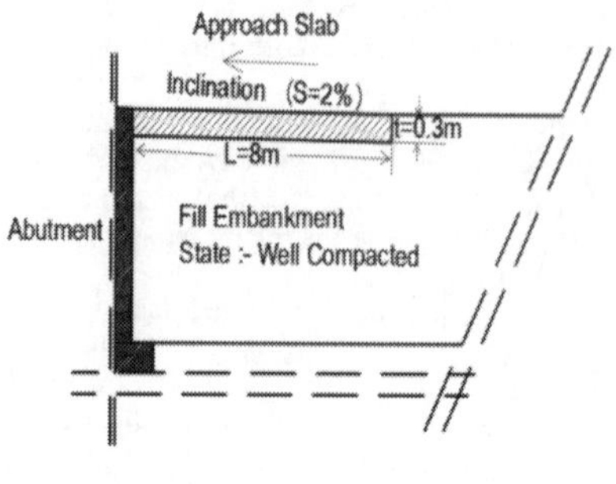


**Fig. 3.** Basic Model

**Table 2.** Properties of concrete, incompressible layer, compressible layer and fill embankment used in the model

| | | | Compressible Layer | | | Fill Embankment | |
|---|---|---|---|---|---|---|---|
| | Concrete | Incompressible Layer | Soft Clay | Medium Clay | Stiff Clay | Well Compacted Fill | Poorly Compacted Fill |
| Material Model | Linear Elastic | Mohr-Coulomb | Mohr-Coulomb | Mohr-Coulomb | Mohr-Coulomb | Hardening Soil | Hardening Soil |
| Material Type | Non-porous | Drained | Drained | Drained | Drained | Drained | Drained |
| $\gamma_{Unsat}$ ($kN/m^3$) | 25 | 22 | 16 | 17 | 17.5 | 20 | 18 |
| $\gamma_{Sat}$ ($kN/m^3$) | | 22 | 16 | 17 | 17.5 | 20 | 18 |
| $E_{ref}$ ($kN/m^2$) | 3 X $10^7$ | 1 X $10^7$ | 1500 | 3000 | 6000 | 1 X $10^4$ | 7 X $10^3$ |
| C ($kN/m^2$) | | 3000 | 15 | 30 | 45 | 0.1 | 0.1 |
| Φ (Degree) | | 30 | 10 | 20 | 25 | 35 | 32 |
| Ψ (Degree) | | 0 | 0 | 0 | 0 | 0 | 0 |
| ν | 0.25 | 0.25 | 0.35 | 0.3 | 0.28 | | |
| Power | | | | | | 0.5 | 0.5 |
| $R_{inter}$ | | | | | | 0.67 | 0.67 |

## 2.2 Behavior of the model

A typical two-dimensional model is shown in Figure (4) using "PLAXIS" software, which contains facilities for automatic mesh generation. The figure illustrates a typical mesh, which is used in the numerical model representing the soil and approach slab.

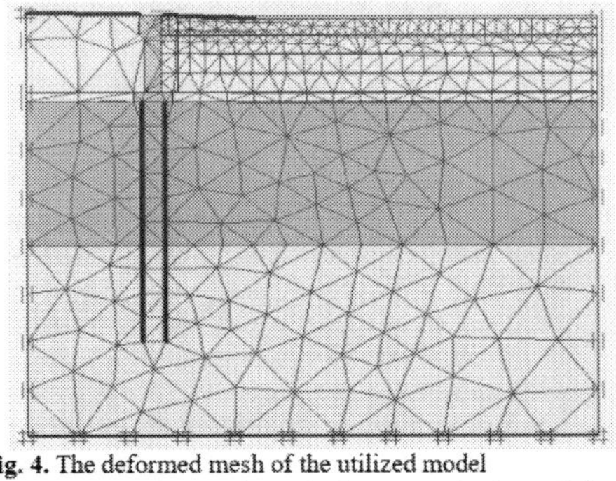

**Fig. 4.** The deformed mesh of the utilized model

The analysis results of the numerical model are represented in Figures (5) through (7), due to actions of the applied loads. Figure (5) shows the total displacements occurring in the embankment and the approach slab. Figures (6) and (7) show the vertical soil displacements and the horizontal soil movements, respectively. The pattern of color shades from the results of the vertical movements is quite similar to the pattern of the results of total movements shown in Figure (5), which indicates that the vertical movement is the predominant factor in this study.

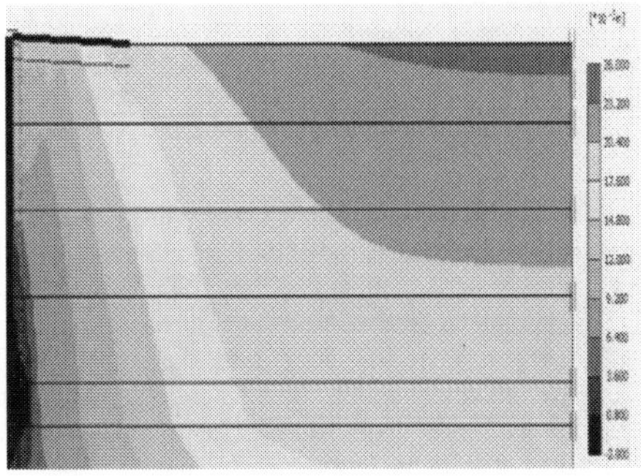

**Fig. 5.** Total displacement of the basic model

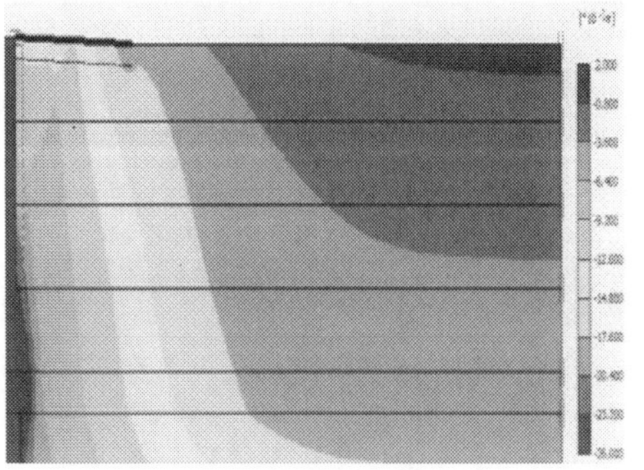

**Fig. 6.** Vertical displacement of the basic model

**Fig. 7.** Horizontal displacement of the basic model

## 2.3 Behavior of the approach slab

When the approach slab is loaded with the applied loads, straining actions for the slab represented in bending moment are shown in Figure (8). The effective normal stress on soil below the approach slab is shown in Figure (9).

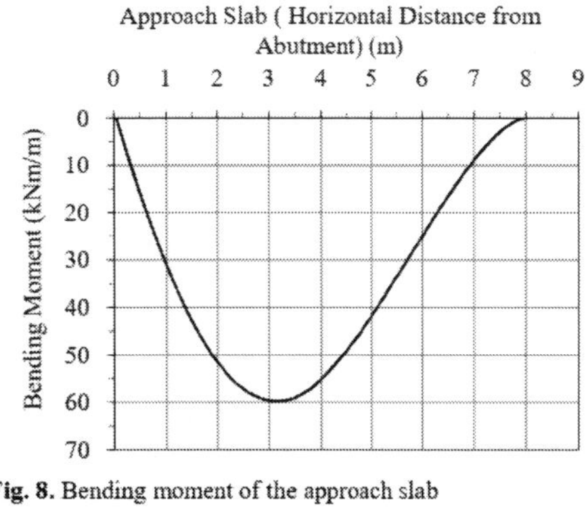


**Fig. 8.** Bending moment of the approach slab

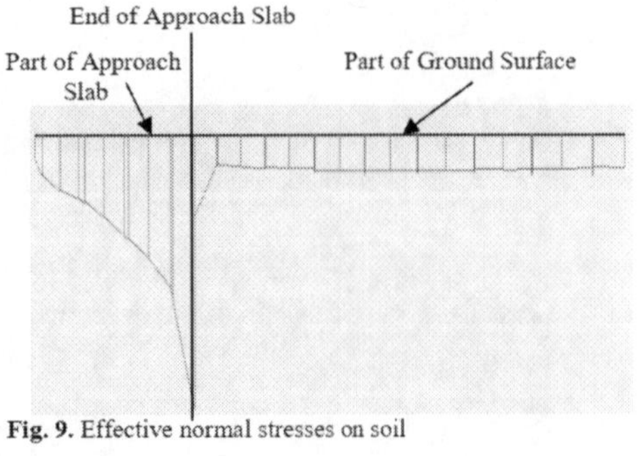


**Fig. 9.** Effective normal stresses on soil

### 2.4 Formation of the bump

Figures (10) and (11) show the settlement of the approach slab and soil. It is evident that a bump is formed at the free end of the approach slab because of the different stiffness between the approach slab and the soil as shown in Figure (9). Total settlement at the free end of the approach slab ($\Delta_0$) and relative bump value defined as the differential settlement in 3m width at the free end of the approach slab ($\Delta_b$), are shown in these Figures.

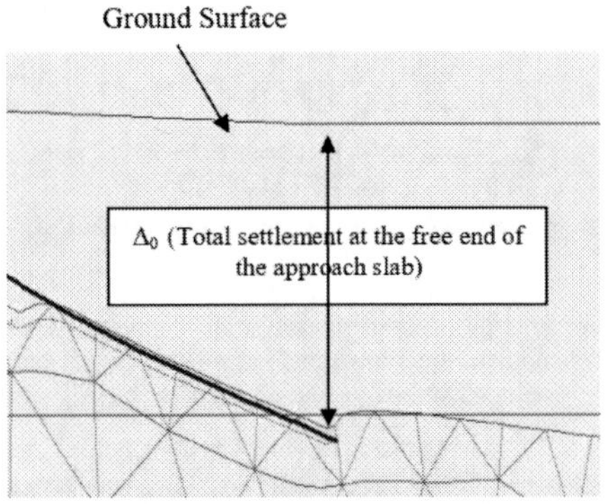


**Fig. 10.** The deformed mesh of the utilized model

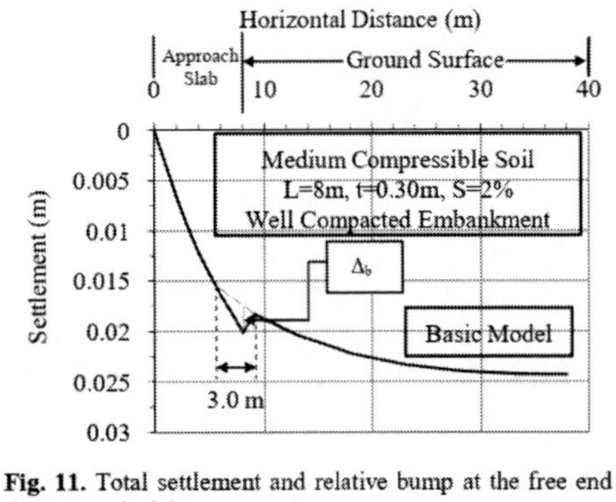


**Fig. 11.** Total settlement and relative bump at the free end of the approach slab

## 3 Effect of the approach slab length

A parametric study was performed to investigate the effect of different factors on the behavior of the approach slab and to examine the mechanism of bump at the end of the approach slab. The examined parameters included different fill embankment and different types of compressible soil. Different lengths (L) of approach slabs starting from 8 up to 20m were studied. The thickness of the slab was 0.30m and the slope was 2%. These analyses were studied with different types of compressible soil, soft, medium or stiff, and different types of fill embankment, well or poorly compacted.

### 3.1 Effect of compressible soil

By analyzing the results from the finite element models using different compressible soil and well compacted embankment, Figure (12) illustrates that an increase in the approach slab length leads to an increase in the total settlement at the free end of the approach slab. In case of soft compressible soil, the total settlement at the free end is higher than that for the medium or stiff soil. The total settlement at the free end of the approach slab is constant regardless of the slab length in case of having medium or stiff soil; however, the settlement is increasing in case of having soft soil with the increase of the slab length.

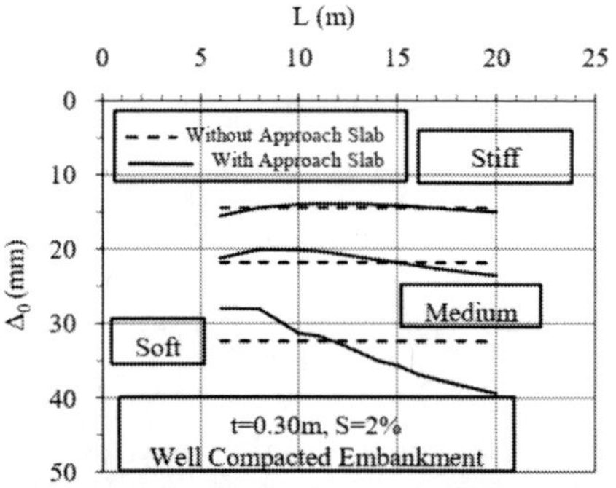


**Fig. 12.** Effect of length of the approach slab (L) on total settlement at the free end of the approach slab ($\Delta_0$)

Figure (13) illustrates that the relative value of the bump at the free end of the approach slab is inversely proportional to the length of the approach slab. For different compressible soils, increasing the length of the approach slab leads to a decrease in the relative bump value within 3m width at its the free end. When the approach slab length increases more than 13m, there is almost no change in the relative bump value within a 3m width at the free end of the approach slab.

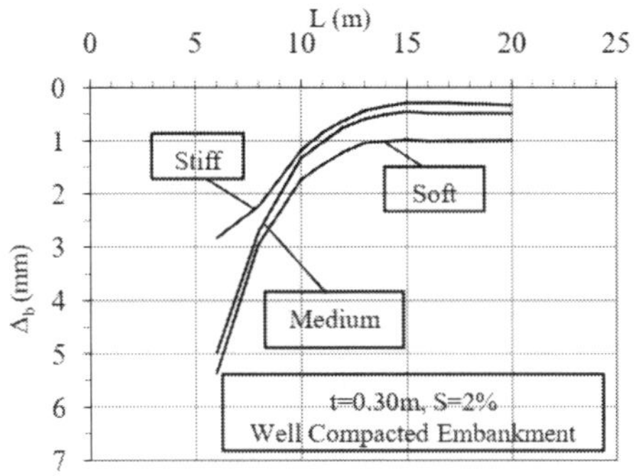


**Fig. 13.** Effect of different compressible soils on the relative bump value at the free end of the approach slab

Figure (14) shows the change in the maximum bending moment in the approach slab for compressible soils. The maximum bending moment on the slab ($M_{max}$) increases with the increase of the slab length. The maximum moment also increases for soft soil compared to medium or stiff soils for the same length. The figure also shows that there is no change in the maximum slab bending moment when the slab length is more than 12, 14 and 16m for the soft, medium and stiff soil, representatively.

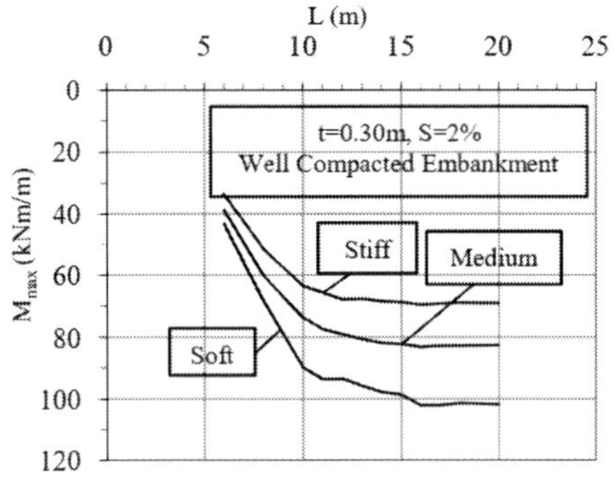


**Fig. 14.** Maximum bending moment of the approach slab ($M_{max}$)

Figure (15) shows the normal stress at the free end of the approach slab. The figure shows that the normal stresses decrease with increase of the slab length. When the approach slab length (L) increases more than 15m for different cases of compressible soils (stiff, medium, soft), there is almost no change in the normal stresses at the free end of the slab ($\sigma_N$).

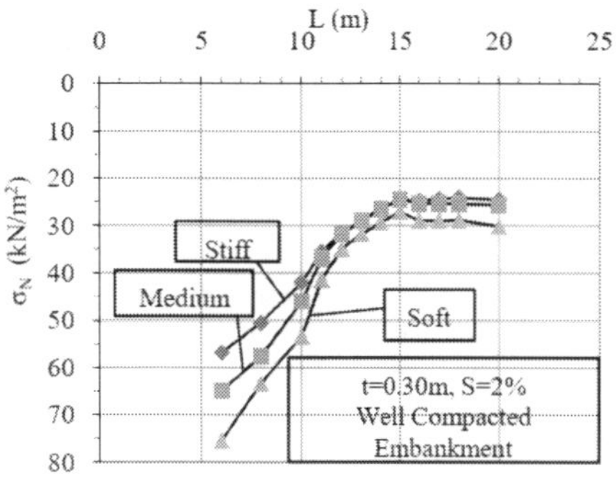


**Fig. 15.** Normal stresses at the free end of the approach slab ($\sigma_N$)

### 3.2 Effect of embankment compaction

Analysis was carried out for different compacted fill embankment and medium compressible soil. Figure (16) shows that the relative bump value within 3m width at the free end of the approach slab ($\Delta_b$) increases for poorly compacted fill compared to the well compacted fill. When the length of the approach slab (L) increases more than (13 and 16) m for the well and poorly compacted embankment, there is almost no change in the relative bump value.

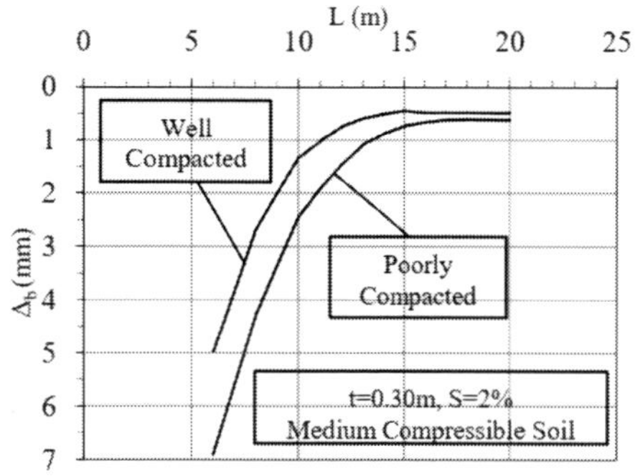


**Fig. 16.** Effect of fill embankment on the relative bump

## 4 Thickness of the approach slab

The parametric study was extended to investigate the effect of slab thickness on the general behavior of the approach slab. In this 2-D analysis, 12m length approach slab was studied. The thickness was 0.15, 0.20, 0.25, 0.28, 0.30, 0.32, 0.35, 0.40 or 0.50m with a slope of 2%. These slab analyses were studied with medium compressible soil and different cases of fill embankment, well or poorly compacted.

By analyzing the results from the finite element, the thickness of the approach slab and the value of the bump at the free end of the approach slab are directly proportional. This means that increasing the thickness of the approach slab leads to an increase in the total settlement at the free end of the approach slab ($\Delta_0$).

Figure (17) illustrates that an increase in the thickness of the approach slab leads to an increase in the relative bump value within a 3m width at the free end of the approach slab ($\Delta_b$) because the increase in the weight of the approach slab and the relative stiffness between the slab and soil.

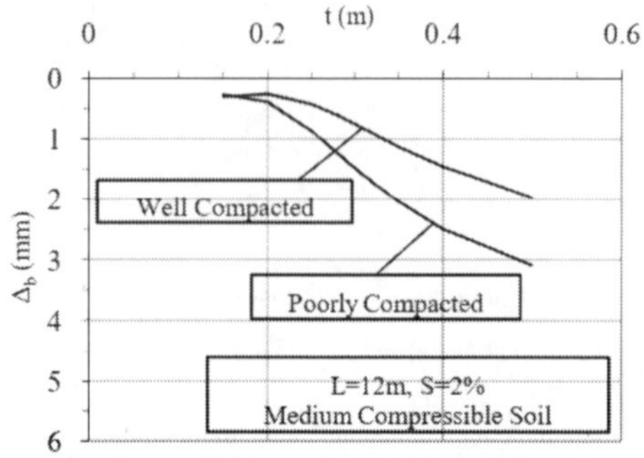


**Fig. 17.** Effect of thickness of the approach slab (t) on the relative bump

## 5 Effect of embedded depth at the free end of the approach slab

A parametric study was performed to investigate the effect of embedding the slab below the ground level. In this 2-D finite element analysis, 1, 1.5, 2, or 2.5m embedded depth at the free end of the approach slab (Z) was studied, as shown in Figure (18). The slab length was 12m with thickness of 0.30m. These analyses were studied with different cases of compressible soil, soft, medium or stiff, and well compacted fill embankment.

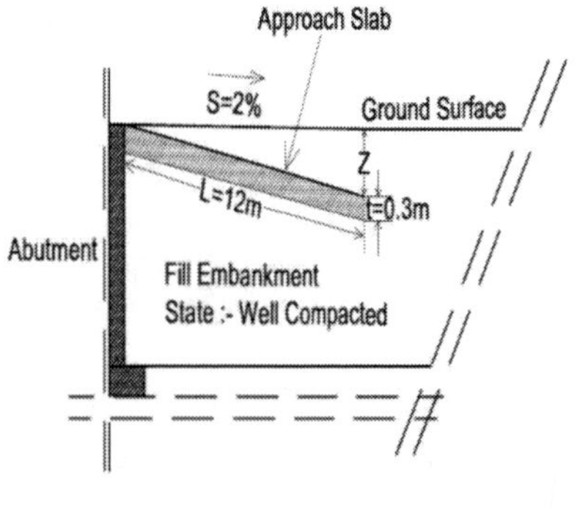


**Fig. 18.** General configuration

By analyzing the results from the finite element analyses, it was found out that the embedded depth at the free end of the approach slab (Z) and the value of the bump at the free end of the approach slab are inversely proportional. This means that an increase the embedded depth leads to a decrease in the relative bump value within 3m width at the free end of the approach slab ($\Delta_b$), as shown in Figure (19).

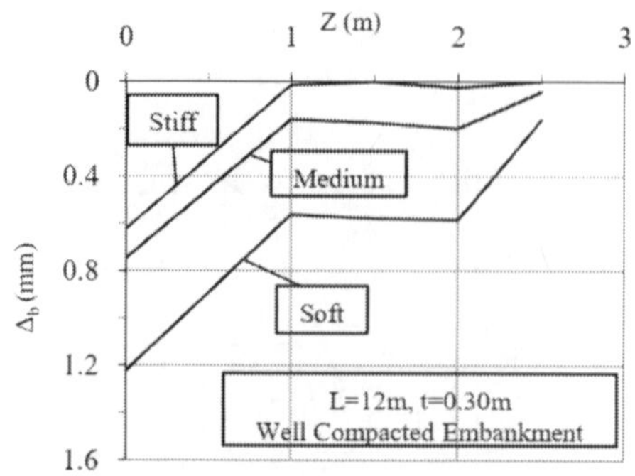


**Fig. 19.** Effect of embedded depth at the free end of approach slab (Z) on relative bump value within 3m width at the free end of the approach slab ($\Delta_b$).

Figure (20) illustrates that an increase in the embedded depth at the free end of the approach slab (Z) leads to an increase in the maximum bending moment on the approach slab ($M_{max}$) because of increasing the applied loads on the approach slab.

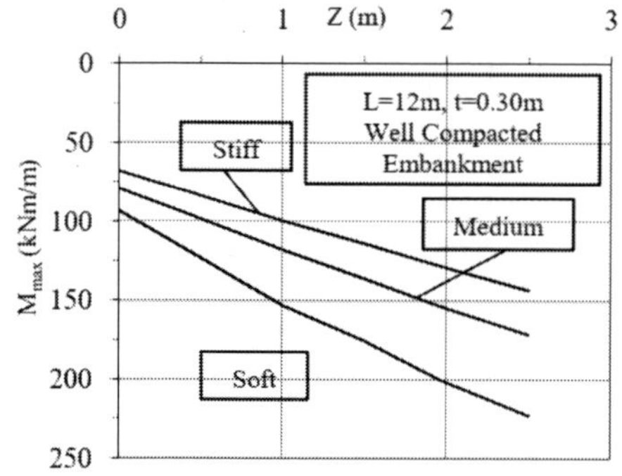


**Fig. 20.** Maximum bending moment on the approach slab ($M_{max}$)

## 6 Effect of variable thickness of approach slab

The parametric study was performed to investigate the relative effect of different factors on the behavior of performance of the approach slab to examine the mechanism of bump between the approach slab and fill embankment that represent different cases of compressible soil. In this 2-D finite element analysis, the length was 12m and the thickness of it was a variance ($t_1$: Thickness of approach slab in the direction of abutment, $t_2$: Thickness at the free end of approach slab), as shown in Figure (21). This analysis was studied with different cases of compressible soil, soft, medium or stiff, and fill embankment was well compacted.

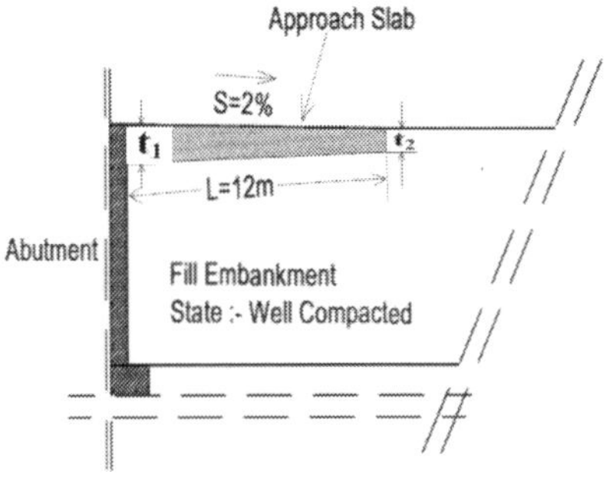


**Fig. 21.** General configuration.

By analyzing the results from the finite element analysis, which the varying thickness of the approach slab has almost no effect on the total settlement at the free end of the approach slab ($\Delta_0$). However, decreasing the varied thickness of the approach slab leads to an increase in the effect normal stresses at the free end of the approach slab ($\sigma_N$). It is clear from Figure (22), that the varying thickness of approach slab has almost no effect on the relative bump value within a 3m width at the free end of the approach slab ($\Delta_b$).

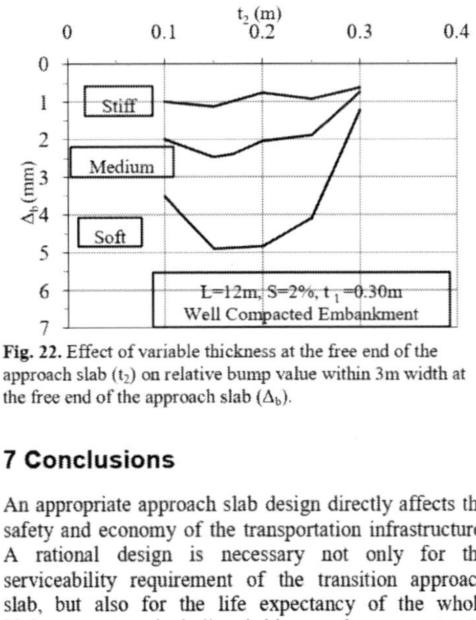


**Fig. 22.** Effect of variable thickness at the free end of the approach slab ($t_2$) on relative bump value within 3m width at the free end of the approach slab ($\Delta_b$).

## 7 Conclusions

An appropriate approach slab design directly affects the safety and economy of the transportation infrastructure. A rational design is necessary not only for the serviceability requirement of the transition approach slab, but also for the life expectancy of the whole highway system including bridges and pavements. As the bump problem has existed for years, the design of the approach slab is still more an art than a science. Engineering calculations of the approach slab are typically not conducted or the approach slab is simply designed as a simply-supported beam since the information about the interaction of the approach slab and the embankment settlement is unknown for a routine office design. There are no AASHTO guidelines for designing approach slabs considering a given embankment settlement.

The present study investigated the effect of different factors on the behavior of approach slab that represent different cases of compressible soil and embankment soil. The investigated factors were as follows: the length of approach slab; the thickness of approach slab; the embedded depth at the free end of approach slab and the varying thickness of approach slab. Total settlement at the free end of the approach slab ($\Delta_0$), the relative bump value within a 3m width at the free end of the approach slab ($\Delta_b$), the normal stresses at the free end of the approach slab ($\sigma_N$) and the maximum bending moment on approach slab ($M_{max}$) were predicted with finite element modeling. The best approach slab recommended in this paper is from 12m to 14m length, 0.3m thickness and the embedded depth with more than 1m.

It is concluded from this study that the most effective design of the approach slab in order to reduce the pumps at the end of the slab can be achieved by the following

1. Total length of the approach slab is more than 12m.
2. Thickness of the approach slab is more than or equal 0.30m.
3. Inclining the approach slab away from the bridge deck to achieve a depth not less than1.0m below ground surface at the end of the slab.

## References

1. J.L.Briaud, R.W. James, S.B.Hoffman, NCHRP synthesis 234: Settlement of Bridge Approaches (the bump at the end of the bridge), Transportation Research Board, National Research Council, D.C.Washington, pp.75(1997).
2. T.C.Hopkins, "Settlement of Highway Bridge Approaches and Embankment Foundations." Rep. No. KYHPR-64-17; HPR-1(4), Kentucky Transportation Center, Lexington, Kentucky(1969).
3. T.C.Hopkins, "Settlement of Highway Bridge Approaches and Embankment Foundations" Rep. No. KYHPR-64-17; HPR-1(8), Kentucky Transportation Center, Lexington, Kentucky, **40**(1973).
4. H.E.Wahls, "Design and Construction of Bridge Approaches"NCHRP Synthesis of Highway Practice No. 159, Transportation Research Board, National Research Council, D.C. Washington (1990).
5. N.Abu-Hejleh, D. Hanneman, D.J. White, I.Ksouri, "Flowfill and MSE Bridge Approaches: Performance, Cost and Recommendations for Improvements" Report No. CDOT-DTD-R-2006-2, ColoradoDepartmentof Transportation, Denver(2006).
6. J.G. Laguros,M.M.Zaman, I.U.Mahmood, "Evaluation of Causes of Excessive Settlements of Pavements Behind Bridge Abutments and their Remedies; Phase II. (Executive Summary)" Rep. No. FHWA/OK 89 (07), Oklahoma Department of Transportation(1990).

7. E.J.Hoppe, "Guidelines for the Use, Design, and Construction of Bridge Approach Slabs" Rep. No. VTRC 00-R4, Virginia Transportation Research Council, VA.Charlottesville, (1999).
8. A.K.Bugrov, "The Finite Element Method in Consolidation Analyses of Water- Saturated Soils" Power Technology and Engineering (Formerly Hydrotechnical Construction), **9**, 7, 661–668(1975).
9. K.J. Bathe,Finite Element Procedures, Prentice-Hall of India, New Delhi(2003).
10. J.Horvath,"Integral Abutment Bridges: Problem and Innovative Solutions Using EPS Geofoam and other Geosynthetics" Research Report No. CE/GE-00-2, Manhattan College, Bronx, New York(2000).
11. J. Kunin, S.Alampalli, "Integral Abutment Bridges: Current Practice in United States and Canada" ASCE Journal of Performance of Constructed Facilities, **14**(3), 104-111(2000).
12. J.Seo, "The Bump at the End of the Bridge: An Investigation" Dissertation submitted in partial fulfillment of the requirements for the degree of the Doctor of Philosophy, Texas, A&M University, College Station, Texas(2003).

# Strengthening of self-compacting reinforced concrete deep beams containing circular openings with CFRP

*Nabeel* Al-Bayati[1,*], *Bassman* Muhammad[1], and *Murooj* Faek[1]

[1]Building and Construction Engineering Department, University of Technology, Baghdad, Iraq

**Abstract.** This paper shows the behavior of reinforced self-compacting concrete deep beams with circular openings strengthened in shear with various arrangements of externally bonded Carbon Fibre Reinforced Polymer (CFRP). Six simply supported deep beams were constructed and tested under two points load up to the failure for this purpose. All tested beams had same geometry, compressive strength, shear span to depth ratio, main flexural and web reinforcement. The variables considered in this study include the influence of fiber orientation, utilizing longitudinal CFRP strips with vertical strips and area of CFRP. The test results indicated that the presence of the circular openings in center of load path reduce stiffness and ultimate strength by about 50% when compared with solid one, also it was found that the externally bonded CFRP can significantly increase the ultimate load and enhance the stiffness of deep beam with openings.

## 1 Introduction

Deep beams are significant structural elements in which most of the applied load transferred directly to supports by tied arch[1] they are used as transfer girder in bridges and buildings, panel beams, foundation walls of rectangular tanks, walls of bunker, load bearing walls in buildings, and pile cap. Because of the deep beam commonly has a small width, big depth and contain heavy web reinforcement, therefore; the use of self-compacting concrete (SCC) for casting of those members more convenient to avoid the usual problems of utilization conventional concrete such as segregation, voids and bleeding. This type of concrete is different from conventional concrete by its ability to spread readily under self-weight only and fill the formwork without need any mechanical compaction compaction [2]. In many cases, transverse openings in such member are often provided for the passage of utility ducts and drainage pipes. These openings cause many problems in behavior of the deep beams particularly when these openings were located to interrupt the load path which define as (the line joining the load bearing edge and support bearing edge)[3-6] as a result of reduction in beam stiffness such as increased cracking and deflection and may be a reduction in beams carrying capacity. Deep beams like other structural members needed strengthening or rehabilitation programs due to extra load or mistake in their design or construction without rise their self-weight. The use of externally bonded Carbon Fiber-Reinforced Polymer (CFRP) for the strengthening of Reinforced Concrete structure (RC) is becoming very common in recent years. Because of their higher strength to corrosion, higher strength to weight ratio, comfort of handling and application, has higher strength and stiffness compared to steel and availability in various lengths and shapes.

## 2 Research significance

No enough research was found deals with the strengthening self-compacting deep beams containing circular openings positioned to interrupt the load path with CFRP fabric. So this paper is prepared to provide experimental information about the shear strength and behavior of this type of structural members strengthening with different configuration of CFRP.

## 3 Experimental program

A total of six self-compacting concrete (SCC) deep beams were constructed and tested under two points loading applied on the top face. All specimens had total long 1400 mm and cross section of 150×400 mm, shear span to depth ratio (a/d) equal to (1), the clear span between the simply support 1070 mm which give L/D ratio equal 2.67 less than 4 as recommended in ACI 318M-14[7] for deep beam requirements, Details of the flexure and shear reinforcement in both vertical and horizontal directions shown in Figure. (1). Two circular openings with diameter 110mm were placed in the center of load path in two sides of the beam in five specimens and one solid beam as the reference beam. Four specimens from

* Corresponding author: 40063@uotechnology.edu.iq

© The Authors, published by EDP Sciences. This is an open access article distributed under the terms of the Creative Commons Attribution License 4.0 (http://creativecommons.org/licenses/by/4.0/).

 https://doi.org/10.1051/matecconf/201816204015

deep beams with openings strengthened in the shear zone with uni-directional CFRP strips of 40 mm wide with different configuration and one specimen unstrengthened were served as the reference beam. Details of the beams are mentioned in Table (1) and Figure (2). The beam notation contained three parts; in the first part the letter (D) refers to deep beam, (O) refers to presence the circular openings, the second part indicates to the shear span to effective depth ratio; and the third part indicates to the type of strengthening. As an example, DO-1-V refers to a deep beam with openings, a/d equals to 1, type of strengthening was vertical strips.

## 4 Material properties

### 4.1 Cement

An Ordinary Portland cement (type I) made in a Mass-Bazian Factory, Sulymania- Iraq, was used in this work. The physical and chemical properties of this cement are corresponding to the requirements of the Iraqi standard specifications IQS NO.5/1984[8] as shown in Table 2 and 3.

### 4.2 Fine sand

Natural sand from Al-akhaidur region was used for this study. It has a fineness modulus of 2.55. The grading and sulfate content of this sand were conformed with the requirement of specification ASTM C33-03[9] and IQS No.45/1984[10] as shown in Table 4 and 5

### 4.3 Coarse aggregate

Crushed river gravel brought from Al-Niba'ee region has (14 mm) maximum size was utilized in this work. The physical and chemical properties and grading of it complies with the Iraqi specification IQS No.45/1984[10] as shown in Table 6 and 7.

### 4.4 Limestone powder (LSP)

This material is commercially known as "Al-Ghubra". Fine limestone was utilized as a filler to increase the amount of fine materials in this mixture to improve segregation resistance and enhance its cohesiveness. Chemical composition of limestone powder shown in Table 8.

### 4.5 Superplasticize

Aqueous solution of modified polycarboxylate commercially named as Sika viscocrete was used to obtain fresh properties of SCC. All information about main properties of this material is summarized in Table 9.

### 4.6 Steel reinforcement

Three deformed steel bars were utilized with a diameter of 16mm as the flexural reinforcement and shear reinforcement made with 5 mm diameter deformed steel wires spaced at 75 mm C/C. Test results indicated that the steel reinforcement of Ø16 conforms to the requirements of the ASTM A615/615M-14[11] for grade 60 bars of minimum yield stress = 420 MPa and elongation = 9%. Furthermore, the deformed wire reinforcement Ø5 complies with the requirements of the ASTM A1064/A1064M–14[12] which recommended minimum values for yield stress and ultimate stress of 515 and 585 MPa, as shown in Table10.

### 4.7 Carbon fiber reinforced polymer (CFRP)

Unidirectional woven carbon fiber fabric known as Sika Wrap-300 C/60 was used for externally strengthening in this study. All information about main properties of this material is summarized in Table 11.

### 4.8 Bonding materials

Sikadur®-330 is the suitable adhesive material for carbon fiber fabric strips comprises of two parts, white part (resin) called part A, and grey part (hardener) called part B. The properties of this type of adhesive are mentioned in Table 12.

## 5 Mixing, casting and curing of concrete specimens

The mix design of Self-Compacting Concrete must achieve the criteria of filling ability and segregation resistance. In the present work the SCC mix was designed according to requirements of European Guidelines for Self-Compacting Concrete [13] to a ccomplish both fresh and hardened properties of SCC. Many trail mixes were made to achieve a minimum compressive strength of 32MPa at 28 days and to ensure the SCC requirements slump flow, T50, L-box V-Funnel tests carried out. Tables 13 and 14 indicate the details of

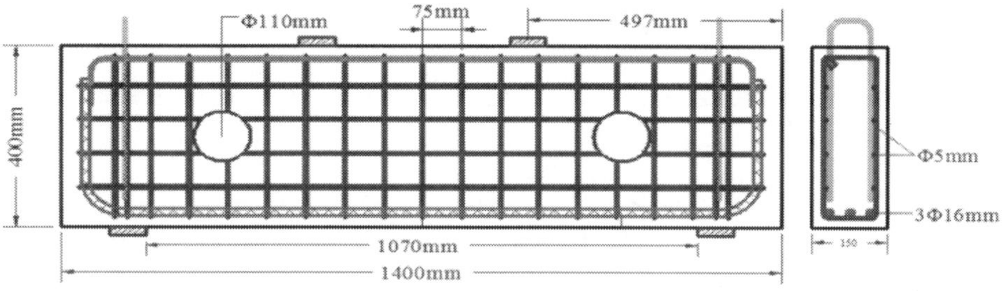


**Fig. 1.** Reinforced concrete deep beam

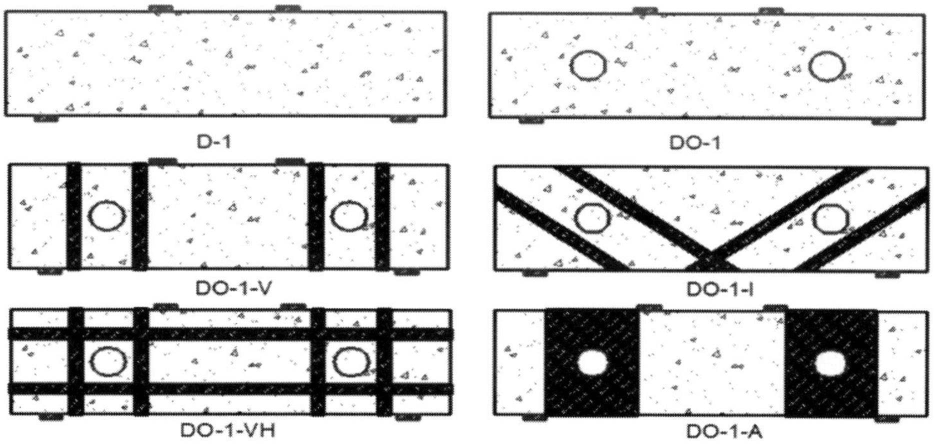


Fig. (2) Strengthening Schemes of the Tested Deep Beams

**Table 1.** Summary of specimens Details

| No | Specimen Symbols | Openings diameter(mm) | a/d | Strengthening details |
|---|---|---|---|---|
| 1 Solid beam | D-1 | - | 1 | No strengthening |
| 2 | DO-1 | 110 | 1 | No strengthening |
| 3 | DO-1-V | 110 | 1 | Two vertical U-shape strips |
| 4 | DO-1-VH | 110 | 1 | Two vertical U-shape strips at two sides of beam and two horizontal strips at two faces of beam |
| 5 | DO-1-I | 110 | 1 | Two inclined U-shape strips |
| 6 | DO-1-A | 110 | 1 | Covering all shear span with vertical U-shape fiber |

**Table 2.** Physical properties of the cement

| Physical Properties | Unit | Test results | IQS 5/1984 |
|---|---|---|---|
| Fineness uses Blaine air permeability apparatus | $m^2/Kg$ | 380 | 230 min |
| Soundness using autoclave method | % | 0.24 | 0.8 max |
| Time of setting (Vicat)<br>Initial time<br>Final time | minutes<br>minutes | 158<br>273 | 45 min<br>600 max |
| Compressive strength of cement paste cube mold (50mm) at:<br>3days<br>7days | MPa<br>MPa | 29.52<br>34 | 15 min<br>23 min |

**Table 3**. Chemical properties of cement

| No. | Compound name | Chemical composition | % by Weight | IQS 5/1984 |
|---|---|---|---|---|
| 1 | Lime | CaO | 61 | - |
| 2 | Silica | $SiO_2$ | 21.1 | - |
| 3 | Alumina | $AI_2O_3$ | 4.64 | - |
| 4 | Iron oxide | $Fe_2O_3$ | 4.2 | - |
| 5 | Magnesium oxide | MgO | 1.51 | 5 max |
| 6 | Sulfur trioxide | $SO_3$ | 2.35 | 2.8max |
| 7 | Loss on ignition | L.O.I | 2.79 | 4max |
| 8 | Insoluble residue | I.R | 0.81 | 1.5max |
| 9 | Lime saturation factor | L.S.F | 0.88 | 0.66-1.02 |
| 10 | Tricalcium aluminates | $C_3A$ | 5.2 | - |
| 11 | Tricalcium silicate | $C_3S$ | 44 | - |
| 12 | Dicalcium silicate | $C_2S$ | 27.39 | - |
| 13 | Tetracalcium | $C_4AF$ | 12.8 | - |

**Table 4.** Physical properties of fine aggregate

| Physical Properties | Test Results | Limits of Iraqi Standard IQS.No.45, 1984 |
|---|---|---|
| Specific Gravity | 2.59 | - |
| Sulfate Content% | 0.12 | $\leq 0.5$ |
| Absorption% | 0.75 | - |

**Table 5.** Grading of fine aggregate

| Sieve size (mm) | % Passing by Weight | IQS45/1984 for zone 2 |
|---|---|---|
| 10 | 100 | 100 |
| 4.75 | 100 | 90-100 |
| 2.36 | 88 | 75-100 |
| 1.18 | 76.9 | 55-90 |
| 0.6 | 57.9 | 35-59 |
| 0.3 | 20 | 8-30 |
| 0.15 | 2 | 0-10 |

**Table 6.** Physical properties and sulfate content of coarse aggregate

| Properties | %Passing by weight | IQ45/1984 for zone 2 |
|---|---|---|
| Specific gravity | 2.63 | - |
| SO3 % | 0.07 | 0.1 max |
| Absorption % | 0.8 | - |

**Table 7.** Grading of coarse aggregate

| Sive size (mm) | % Passing by weight | IQS45/1984 |
|---|---|---|
| 14 | 100 | 90-100 |
| 10 | 80 | 50-85 |
| 5 | 5 | 0-10 |

**Table 8.** Chemical Composition of Limestone Powder

| Oxide Composition | Content by weight % |
|---|---|
| CaO | 54.1 |
| MgO | 0.12 |
| $SiO_2$ | 1.4 |
| $Fe_2O_3$ | 0.13 |
| $Al_2O_3$ | 0.7 |
| $SO_3$ | 0.21 |
| Loss on ignition | 42.57 |

**Table 9.** Properties of Superplasticizer

| | |
|---|---|
| From | liquid |
| Color | Turbid liquid |
| Specific Gravity | 1.08Kg/l |
| pH | 7.0-9.0 |
| Normal dosage | 0.8-2% |
| Chorides | Free from choride |
| Transport | Non-hazardous |
| Labeling | No hazard label required |

**Table 10.** Properties of steel bars

| Bar Diameter (mm) | Yield Stress (MPa) | Ultimate stress (MPa) | Elongation % |
|---|---|---|---|
| 16 | 590 | 692 | 15.16 |
| 5 | 554 | 617 | 5 |

**Table 11.** Properties of CFRP

| Properties | Sika Warp 300c/60 |
|---|---|
| Tensile strength (MPa) | 3900 |
| Density (g/cm2) | 1.79 |
| Elongation at break% | 1.5 |
| Tensile modulus (MPa) | 230000 |
| Weight (g/m2) | 300 |
| Thickness (mm) | 0.166 |

**Table 12.** Properties of Epoxy Resin

| Properties | Sikadur-330 |
|---|---|
| Density (kg/l) | Mixed Res-in:1.31 |
| Mixing ratio by weight A:B | 4:1 |
| Tensile strength (MPa) | 30 |
| Flexural modulus (MPa) | 3800 |
| Elongation at break(%) | 0.9 |
| Setting time (minute) | 30 |
| Full cure (day) | 7 |

**Table 13.** SCC mix proportion by weight

| | |
|---|---|
| Super plasticizer (L/m$^3$) | 4.9 |
| Fine aggregate (Kg/m$^3$) | 797 |
| Coars aggregate (Kg/m$^3$) | 767 |
| Water/ Powder (by Weight) | 0.34 |
| Water (Kg/m$^3$) | 180 |
| Limestone powder (Kg/m$^3$) | 175 |
| Cement (Kg/m$^3$) | 350 |

**Table 14.** Fresh SCC test results

| Test method | Results | EFNARC (2002) Limits[14] | ACI237R-07 Limits[2] |
|---|---|---|---|
| Slump flow (mm) | 710 | 650-800 | 450-760 |
| T500(Sec) | 3 | 2-5 | 2-5 |
| L-box (H2/H1) | 0.9 | 0.8-1 | 0.8-1 |
| V-Funnel (Sec) | 10 | 6-12 | - |

the adopted SCC mixture and test results respectively. Furthermore concrete compressive strength (*f'c*) and splitting tensile strength (*fct*) was determined by testing three 150×300 mm cylinders, also modulus of rupture (*fr*) was determined utilizing three 100 × 100 × 400 mm prisms. The test results are shown in Table 15.

**Table 15.** Test results of hardened SCC

| Beams | D-1 DO-1 | DO-1-V DO-1-VH | DO-1-I DO-1-A |
|---|---|---|---|
| $f'_c$ (MPa) | 34.14 | 33.5 | 34.3 |
| *fct* (MPa) | 3 | 3.38 | 3.35 |
| *fr* (MPa) | 4.5 | 4.42 | 6.2 |

Each beam was cast utilizing two batches of SCC mix in vertically positioned steel formworks then cured for 28 days utilizing wet burlaps after removing the formwork. Four deep beams with openings were strengthened by one layer, U-shaped of CFRP strips that have constant width (40mm). The locations of CFRP strips were chosen carefully based on the failure mode of the reference beams. The strengthening schemes of specimens are described in detail in Table 1 and Figure 2.

## 6 Devices and testing procedure

The specimens were tested utilizing the universal testing machine (AVERY) of 3000 kN capacity at the Structural Laboratory of Building and Construction Department of the University of Technology/ Baghdad /Iraq. Loading was applied in small increase (20 kN). At each load level, midspan deflection was recorded utilizing dial gage of accuracy of 0.01 mm. The first cracking load was noted, and the crack patterns were marked. To avoid the failure at applied load regions and to make sure that bearing stress was uniformly distributed at these locations, a steel plate with dimensions of (70,150, and 20 mm) soldered to a cylindrical shaft of (Ø 40 mm and along of 150 mm) were utilized at load points. In addition to imitate the boundary conditions of the simply supported beam, the pin and the roller end supports have been used as per Al-Bayati [15] were used.

## 7 Experimental results and discussion

### 7.1 Effect of web openings

For solid deep beams most of the applied load is transmitted directly from loading point to the reaction point along the "load path". The cracking process in (D-1) started with the formation of flexural cracks at about (160 kN) as shown in Table 16, while the shear cracks appeared at about 200 kN nearly in the mid-depth of the specimen in the direction of diagonal compression strut and thereafter expanded towards the load and the support points after the width of the diagonal cracks became more widening, this beam failed by shear mode with crushing of concrete near the loading point at a load of 760 kN. When load path was intercepted by the opening as in beam (DO-1) the cracking and ultimate load decrease by 50% as-compared with solid beam (D-1). The presence of openings raised the midspan deflection of the hollow deep beam when compared with solid beam (D-1) at the same load levels, such rise in deflection values was caused by decrease in moment of inertia as shown in fig.3.

### 7.2 Effect of strengthening with CFRP

#### *7.2.1 Cracking process and failure load*

The crack patterns of RC deep beams are shown in Plates (1). Generally, in strengthened beams after elastic stage the first shear crack, appeared in shear zone from the diagonal edges of the openings then expanded towards the load and the supports plates. With a further increment in the load, flexure cracks formed in the beams soffit as in control beam (DO-1). The first visible shear cracking and ultimate loads for all beams are summarized in the Table 16.The test results illustrated that the appearance of shear cracks were belated for the strengthened specimen somewhat more than reference beam and this may be refer to the increase of stiffness because the strips restraining effect. At the front face of beam, shear cracks appeared at load of 120 kN for beams (DO-1-VH, DO-1-I) with increasing 20% in cracking load compared with reference beam (DO-1). While in beam (DO-1-V) shear crack formed at the same loading level for beam (DO-1). Table 16 also shows that beam (DO-1-A) has the highest ultimate load (560 kN) among other strengthened beams in this group. The percentage of increase in ultimate load for this beam was about 47.4% as compared to the reference beam. This deep beam specimen was strengthened by covering the shear zone with vertical fibers. That reflects the advantage of CFRP strips in enhancing the ultimate strength of SCC deep beams, because they lead to delaying in the appearance of the shear cracks and gave good restraint to the next expansion of cracks. The failure type in this deep beam specimen was crushing of concrete along the line joining the edge of load points and opening tangents opposite to the load points under the CFRP which led to debonding the fiber and rupture it near the load point as shown in plate 1. The strengthening system in beam (DO-1-V) was two vertical (U) shape CFRP strips on each side of this beam to investigate the influence of strengthening by using these strips on behavior of deep beam with openings. The Failure of this beam occurred by debonding of CFRP strip near the load point and rupture of CFRP strip near the support poin combined with splitting the specimen along the path connecting the bearing points and openings tangents opposite to load point at (420 kN) as shown in plate 1.The strengthening of (DO-1-V) results in about 10.5% increase in the ultimate load over control beam (DO-1). Beam (DO-1-I) was strengthened by two inclined CFRP strips (U) wrapped (orthogonal to the load path) mirrored on each side of the beam to discuss the influence of the fiber orientation on behavior of deep beams. The failure of this specimen was due to separation of the two strips from the concrete surface after diagonal crack became wider at ultimate load (530 kN) as shown in Plate 1. Strengthening of this beam gave a (39.5%) and (26%) increase in the ultimate loads as compared to (DO-1) and (DO-1-V) respectively. As a concluding remark, the inclined orientation of strips in (DO-1-I) showed better performance than vertical placement in (DO-1-V) both with respect to first crack, ultimate strength of SCC deep beams as can be observed from Table 16. Beam (DO-1-VH) was constructed and tested to discuss the effect of adding horizontal CFRP strips to vertical CFRP strips when strengthening the deep beams. This beam was strengthened by two vertical CFRP strips (U) wrapped on each side and two horizontal CFRP strips on two faces of the beam. With increasing load, the shear crack extended towards the load and the support points and became wider under CFRP strips causing de-bonding of the vertical CFRP strip near load point and rupture of vertical CFRP strip near the support point and horizontal CFRP strips finally, the failure occurred at ultimate load (500kN) with (31.5%) and (19%) increase in the ultimate load as compared to the beam (DO-1) and (DO-1-V) respectively.

**Table 16.** Test results of all beams

| Deep beam No. | a/d | CFRP details | Shear crack force | | Flexural crack Force | | Ultimate Force | | Increase In Ultimate load % | $Pcr_s$/ Pu | Mode of Failure |
|---|---|---|---|---|---|---|---|---|---|---|---|
| | | | $Pcr_s$ | $\Delta cr_s$ | $Pcr_f$ | $\Delta cr_f$ | Pu | Δu | | | |
| D-1 | 1 | - | 200 | 1.55 | 160 | 1.26 | 760 | 7.5 | - | 26.3 | DS+CC |
| DO-1 | 1 | - | 100 | 1.15 | 140 | 1.75 | 380 | 4.55 | - | 26.3 | DS |
| DO-1-V | 1 | V | 100 | 1 | 220 | 2.32 | 420 | 4.65 | 10.5 | 23.8 | FRPD +FRPT |
| DO-1-VH | 1 | V&H | 120 | 1.07 | 180 | 1.66 | 500 | 4.8 | 31.6 | 24 | FRPD +FRPT |
| DO-1-I | 1 | I | 120 | 0.98 | 240 | 2.06 | 530 | 4.7 | 39.5 | 22.6 | FRPD |
| DO-1-A | 1 | A | - | - | 200 | 1.55 | 560 | 4.85 | 47.4 | - | FRPD +FRPD+CC |

Where: V: vertical orientation, H: horizontal orientation, I: inclined orientation, Pcrs: the load at initiated first shear crack (kN), Pcrf : the load at initiated first flexural crack(kN), Pu: failure load (kN), Δcrs: vertical deflection at initiated first shear crack, Δcrf: vertical deflection at initiated first flexural crack, Δu: vertical deflection at failure stage (mm), DS: diagonal splitting failure, FRPD: shear induced FRP Debonding, FRPT: FRP tearing.

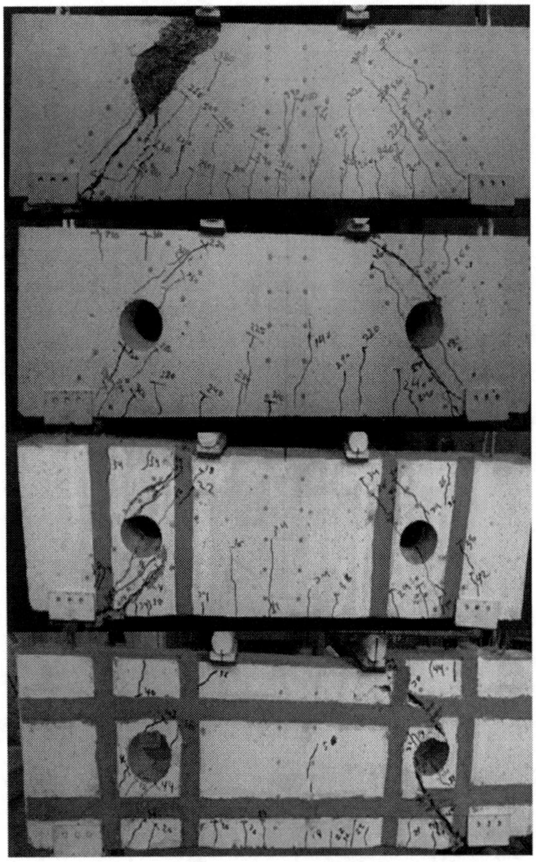

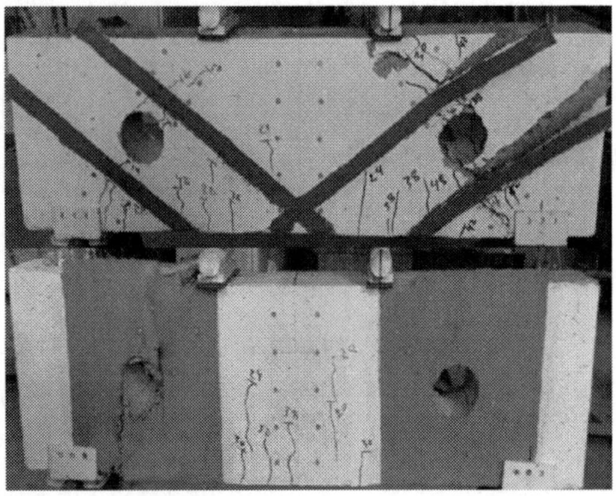

**Plate 1.** All deep beams after failure

### *7.2.2 Load midspan deflection response*

The load-deflection relationship for all specimens is mentioned in Fig.4. The mid span deflection curve of the strengthened specimens is similar in shape with the reference beam (DO-1) but with different extremes. From fig.4 it can be seen that the largest deflection was recorded in (DO-1-A) at ultimate load and show the highest stiffness in post cracking stage. In the other hand, the vertical strips in (DO-1-V) have less effect on carrying capacity and the stiffness of this beam as compared with other specimens, while the using vertical and horizontal strips when strengthening (DO-1-VH) led to significant increase in the beams stiffness in post cracking stage as compared with (DO-1) and (DO-1-V). The deflection at ultimate load Of (DO-1-I) and (DO-1-V) are close to each other despite the difference in their ultimate loads

(530 and 420 kN) respectively, this mean that the using inclined strips improve load carrying capacity and the beams stiffness more than vertical strips.

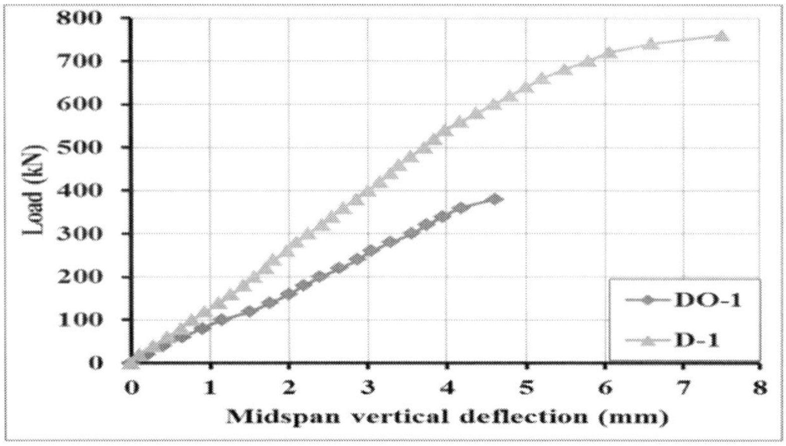


**Fig 3.** Relationship between load -deflection for deep beams with and without openings.

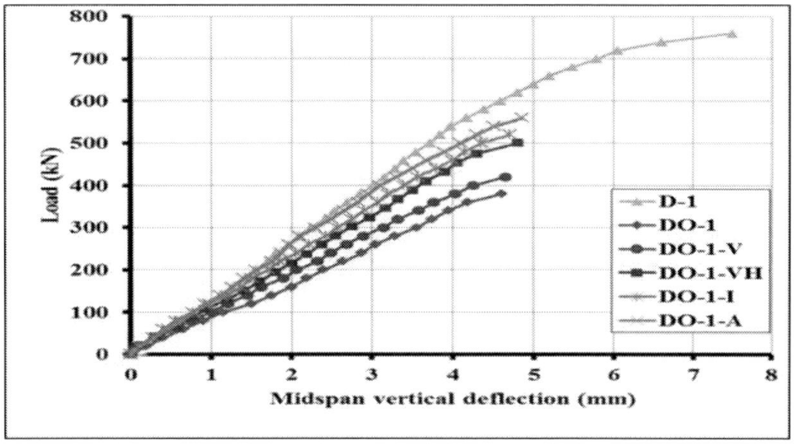


**Fig4**. Relationship between load-deflection for strengthening deep beams.

## 8 Conclusions

1. In this study, All deep beams failed in shear. This failure in the solid deep (D-1) was distinguished by a diagonal splitting type, which happened along path load between the load and support points. While, in deep beams with openings the failure occurred by splitting along the diagonal crack joining the load and support bearing edges with opening tangents opposite to them.
2. The presence of the circular openings in center of load path had a considerable influence on the load carring capacity of the concrete deep beams where a decreased in beam (DO-1) by about 50% as compared with corresponding solid one (D-1). While, the deflection increased in beam (DO-1) when compared with solid beam (D-1) at the same load levels.
3. Bonded CFRP system in the shear zone delayed the appearance of diagonal shear cracks slightly and gave good restraint to the next expansion of them. The shear crack load of the strengthened beams varies from (22.6 to 24%) of their ultimate strength.
4. Generally, the strengthening self-compacting RC deep beam by CFRP strips reduces the deflection if compared at same loads levels and increases the load carrying capacity.
5. Using inclined strips of CFRP was more effective in upgrading the stiffness and shear strength of self-compacting RC deep beams than the vertical one as could be seen in Table 16.
6. When combining CFRP strips in longitudinal direction of deep beam followed by vertical CFRP strips, the results showed that there was an additional increase ultimate load by (19%) as compared with (DO-1-V).
7. Covering the shear span by vertical U-shape CFRP woven increased the ultimate load up to 47 % and increases the stiffness.

## References

1. D.M. Rogowsky, and J.G. MacGregor, Structural Engineering Report, 110, Department of Civil Engineering, University of Alberta, Edmonton, 178 pp. (1983)
2. ACI Committee 237R, Reported by Emerg-ing Technology Series, 30 pp. (2007)
3. F. K. Kong, and G. R. Sharp, The Structural Engineer **51**, 8, 267-275 (1973)
4. M. A. Mansur, and W. A. M. Alwist, The International journal of Cement Composites and Lightweight Concrete **6**, 4, 263- 271 (1984)
5. A. F. Ashour, and Rishi, G., ACI Structural Journal **97**, 3, 418-426 ( 2000)
6. K. H. Yang, H. C. Eun, and H. S. Chung, Engineering Structures Journal **28**, 13, 1825–1834 ( 2006)
7. ACI Committee 318, "Building Code Requirements for Structural Concrete (ACI 318M-14) and Commentary," American Concrete Institute, Farmington Hills, MI 48331, USA, 520 pp. (2014)
8. IQS No. 5/1984 "*Portland Cement*" Central Agency for stand-ardization and Quality Control, Planning Council, Baghdad, IRAQ, (in Arabic)
9. ASTM C 33-03, "Standard Specification for Concrete Aggregate Annual Book of ASTM Standards, Vol.04.02 Concrete and Aggregates, West Conshohocken, PA, United States, , pp. 11, (2003)
10. IQS No. 45/1984, Aggregate from Natural Sources for concrete" Central Agency for standardization and Quality Control, Planning Council, Baghdad, IRAQ,(in Arabic)
11. ASTM A615/A615M-14, "*Standard Specification for Deformed and Plain Carbon-Steel Bars for Concrete Reinforcement*," ASTM International, West Conshohocken", PA, 7 pp (2014)
12. ASTM A1064/A1064M-14, "*Standard Specification for Carbon-Steel Wire and Welded Wire Reinforcement, Plain and Deformed for Concrete*," ASTM International, West Conshohocken", PA, 11 pp. (2014)
13. Bibm, CEMBUREAU, EFCA, EFNARC, ERMCO, "*The European Guidelines for Self-Compacting Concrete*," http://www.efnarc.org. (2005)
14. EFNARC: European Federation Dedicated to Specialist Construction Chemicals and Concrete Systems, Association House, 99 West Street, Farnham, Surrey, U.K, 32 pp. (2002)
15. N. A. J. Al-Bayati, Ph. D. Thesis, Building and Construction Engineering Department, University of Technology, Baghdad, Iraq, 298 pp. (2012)

# Retrofitting of shear strength self-compacting reinforced concrete deep beams with FRP

*Nabeel* Al-Bayati[1,*], *Bassman* Muhammad[1] and *Sarah* Sadkhan[1]

[1]Building and Construction Engineering Department, University of Technology, Baghdad, Iraq

**Abstract.** Experimental program were carried out to investigate the behavior of self-compacting reinforced concrete deep beams retrofitting with carbon fiber reinforced polymer (CFRP). Six simply supported deep beams were tested under symmetrically two point loads, three beams were tested up to failure without strengthening as a control beams with different shear span to effective depth ratio (a/d) while the other two beams were loaded up to 60% from the ultimate load of control beams for each a/d ratio and then retrofitted by the same configuration of CFRP to study the effect of a/d ratio on the properties of deep beams retrofitted. a/d for tested beams were (0.8, 1, 1.2). Study was focused on determining failure loads, cracking loads, failure modes, load midspan deflection. All the beams had the same compressive strength, overall dimensions and flexural and shear reinforcement. It was concluded that using this retrofitted method is very efficient and a gain in the ultimate load capacity of the deep beams was obtained also the results showed that when a/d ratio increase from 0.8 to 1.2, the ultimate load was decrease by 25% and midspan deflection was increased approximately at all load stages for control and retrofitted beams.

## 1 Introduction

A reinforced concrete deep beam is a structural element which has useful applications in many structures such as transfer girders and pile caps. Deep beams usually have a large depth, narrow width and contain congested shear reinforcement so that, the use of Traditional Vibrated Concrete in such beams may results in many problems like voids, segregation, and holes on the surface of the member. Therefore, it is expected that the self-compacting concrete (SCC) is suitable for casting of deep beams. This particular type of concrete is different from vibrated concrete by its ability to fill completely the formwork and achieve full compaction due to self-weight only without segregation, even in the existence of congested reinforcement. During the service life, reinforced concrete structures often experience improvement and modification of their performance. The main contributing factors are change in their use, new design standards etc. In such circumstances there are two possible solutions: replacement, or strengthening or retrofit. The replacement of the full structure may have many disadvantages such as higher labor and material costs. When possible, it is better to strengthen or repair the structures (Mohammed, 2016).There are many ways to rehabilitation of reinforced concrete deep beams. Adding new concrete jacket with additional reinforcement, using external steel sections, and wrapping the section with Fiber-reinforced polymers, FRP are the most popular methods of strengthening and retrofitting concrete beams. A Fiber Reinforced Polymer (FRP) composite is defined as a polymer matrix that is reinforced with a fiber with a sufficient aspect ratio to provide a distinguishable reinforcing function. In the last decade, the investigation on the behavior of reinforced concrete structures retrofitted by FRP has become a very important research field. The shear behavior depends on the detailing and type of shear reinforcement, size of the element, the position and type of applied loading and the shear resistance capacity of concrete. External shear strengthening using FRP composites has become very popular due to several reasons including ease of application, lightweight feature and aesthetic appearance (Rasheed 2016).

## 2 Experimental program

* Corresponding author: 40063@uotechnology.edu.iq

© The Authors, published by EDP Sciences. This is an open access article distributed under the terms of the Creative Commons Attribution License 4.0 (http://creativecommons.org/licenses/by/4.0/).

In this study, six deep beams were tested to investigate the influence of a/d ratio on the behavior of SCC deep beams retrofitted with CFRP with special attention to cracking and ultimate load, vertical mid-span deflection and modes of failure. All the specimens had a length of 1400 mm, height h of 400 mm and width b of 150 mm were constructed and tested under symmetrically two points top loading as shown in Figure 1. To investigate the effect of a/d ratio, all the beams retrofitting with the same configuration of CFRP, consisting of one layer of five carbon fiber strips and orthogonal to the load path that joining load and support points as presented in Figure 2

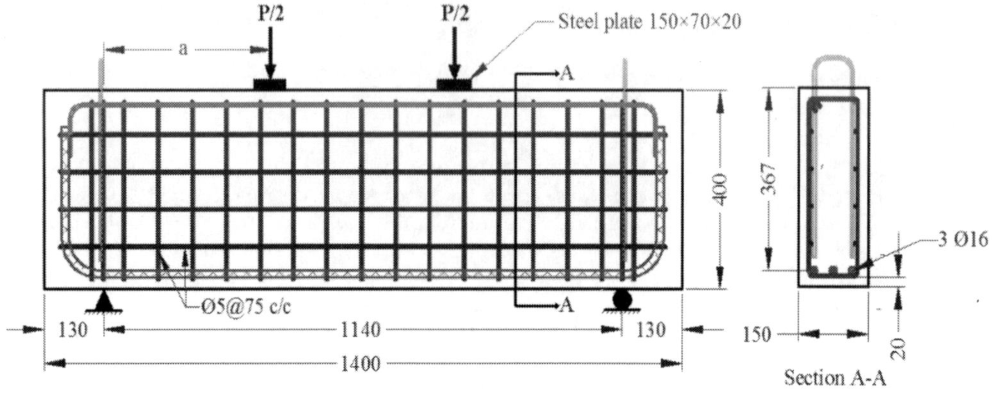


**Fig. 1** Specimens details and arrangement of reinforcement, note: All dimensions in mm

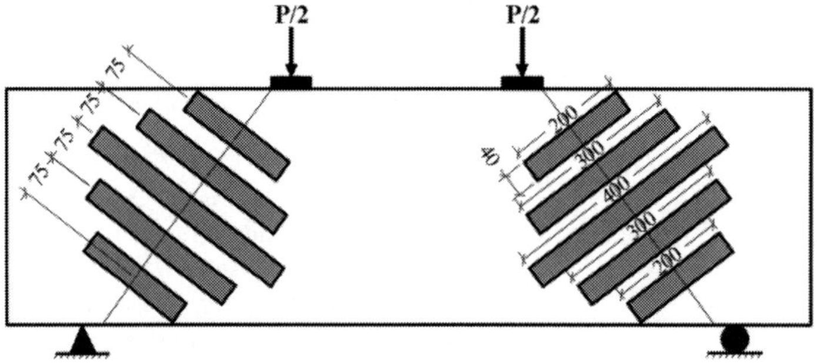


**Fig. 2** CFRP details

## 3 Concrete mixes

Several trial mixes have been made in accordance with the recommendations of (EFNARC, 2002). Mix design of SCC must meet the specification of segregation resistance and filling ability. Details of the adopted SCC mixture are shown in Table 1. To ensure the used concrete is self-compacting, Slump Flow, T50, L-box and V-funnel tests were performed. Furthermore, the results were compared with the limitation of (EFNARC, 2002) and (ACI 237R-07) as indicated in Table 2.

**Table 1** SCC mix proportions

| **Cement ($kg/m^3$)** | **Limestone Powder ($kg/m^3$)** | **Water ($kg/m^3$)** | **W/p By weight** | **Coarse aggregate ($kg/m^3$)** | **Fine aggregate ($kg/m^3$)** | **Super plasticizer ($l/m^3$)** |
|---|---|---|---|---|---|---|
| 350 | 175 | 180 | 0.34 | 767 | 797 | 4.9* |

*1.4 liter/100 kg cement

https://doi.org/10.1051/matecconf/201816204016

**Table 2** Fresh SCC test results

| Test method | Result | EFNARC (2002) Limits | ACI 237R-07 Limits |
|---|---|---|---|
| Slump flow(mm) | 720 | 650-800 | 450-760 |
| $T_{500}$ (sec) | 3 | 2-5 | 2-5 |
| L-box ($H_2/H_1$) | 0.9 | 0.8-1 | 0.8-1 |
| V-funnel (sec) | 10 | 6-12 | - |

## 4 Materials

**1. Cement:** an ordinary Portland cement (type I) was used throughout this research. Test results complying with ASTM Iraqi standards specifications (IQS No.5, 1984). Table 3 and 4 show chemical composition and Physical properties of cement respectively. These tests were conducted by the National Center for Constructional Laboratories and Research.

**2. Fine Sand:** Natural sand was used. The results obtained showed that the sand used was within the requirement of specifications and (IQS No.45, 1984). Table 5 and 6 show Grading and Physical properties of fine aggregate. These tests were performed by the researcher at the laboratories of the University of Technology and National Center for Constructional Laboratories and Researches respectively.

**3. Coarse aggregate:** Crushed gravel with a maximum size of 14 mm used throughout this work. Test results showed that the sulfate content and grading of coarse aggregate conforms to the Iraqi specification (IQSNo.45, 1984). Table 7 and 8 show Grading and Physical properties of fine aggregate. These tests were performed by the researcher at the laboratories of the University of Technology and National Center for Constructional Laboratories and Researchers respectively.

**4. Limestone Powder:** Limestone powder (LSP) was used as a filler material for self- compacting concrete. The increased of fine materials will improve its cohesiveness and enhance the resistance to segregation. Table 9 show chemical composition of limestone powder. These tests were conducted by the National Center for Constructional Laboratories and Research.

**5. Superplasticizer:** to produce, self- compacting concrete, a superplasticizer which named commercially as SikaVicoCrete-5930 was used. SikaVicoCrete-5930 is appropriate with all Portland cement types and adding it to the mix leads to a significant increase in slump without segregation. Table (10) shows Properties of superplasticizer according to manufacturer data sheet.

**6. Carbon, Fiber (CFRP):** beams were externally retrofitted using SikaWrap®-300C/60 woven carbon fiber fabric for strengthening of structures. The orientation angle of the fabric is equal to 0°, unidirectional. Table (11) indicates the technical description of CFRP. According to manufacturer data sheet.

**7. Bonding Materials:** Sikadur®-330 is the most adequate adhesive material for carbon fiber fabric is which consists from two parts. Table (12) shows the Properties of the used adhesive material according to manufacturer data sheet.

**8. Steel reinforcement:** Deformed steel bars of a diameter 16 mm and 5mm were used for tension reinforcement and shear reinforcement (vertical and horizontal) respectively. Three samples were prepared for each type for tensile tests to obtain their yield strength, ultimate strength and elongation percentage. The test was conducted at the Laboratories of University of Technology. Test results obtained are given in Table 13. The results showed that the steel reinforcement meet to the requirements of the ASTM A1064-14 and ASTM A615- 16. Figure1 show the reinforcement details for the beams.

**Table 3** Chemical Composition of Cement

| Compound composition | Chemical composition | % By Weight | IQS 5/1984 |
|---|---|---|---|
| lime | CaO | 61 | - |
| Silica | $SiO_2$ | 21.1 | - |
| Alumina | $Al_2O_3$ | 4.64 | - |
| Iron Oxide | $Fe_2O_3$ | 4.2 | - |
| Magnesia | MgO | 1.51 | $\leq 5$ |
| sulfate | $SO_3$ | 2.35 | $\leq 2.8$ |
| Loss on ignition | L.O.I | 2.79 | $\leq 4$ |
| Insoluble residue | I.R | 0.5 | $\leq 1.5$ |
| Lime saturation factor | L.S.F | 0.88 | 0.66-1.02 |
| Tricalcium Silicate | $C_3S$ | 44 | - |
| Dicalcium Silicate | $C_2S$ | 27.4 | - |
| `Tricalcium Aluminate | $C_3A$ | 5.2 | - |
| Tetracalcium Alumminoferrite | $C_4AF$ | 12.77 | - |

**Table 4** Physical Properties of the Cement

| Physical Properties | Units | Test Result | IQS 5/1984 |
|---|---|---|---|
| Fineness using Blaine air permeability apparatus | $m^2/kg$ | 384 | ≥230 |
| Setting Time (Vicat's Method) | | | |
| Initial | Minutes | 120 | ≥ 45 min |
| Final | Minutes | 250 | ≤ 10 hrs |
| Soundness using autoclave method | % | 0.25 | ≤ 0.8 |
| Compressive strength for cement paste cube mold (50 mm) at: | | | |
| 3 days | MPa | 22 | ≥ 15 |
| 7 days | MPa | 26.5 | ≥ 23 |

**Table 5** Grading of Fine Aggregate

| Sieve Size (mm) | % passing by weight | IQS 45/1984 Zone (2) |
|---|---|---|
| 10 | 100 | 100 |
| 4.75 | 97.5 | 90-100 |
| 2.36 | 92.7 | 75-100 |
| 1.18 | 77.8 | 55-90 |
| 0.60 | 48.7 | 35-59 |
| 0.30 | 19 | 8-30 |
| 0.15 | 4.2 | 0-10 |

**Table 6** Physical Properties of Fine Aggregate

| Physical Properties | Test Results | IQS 45/1984 |
|---|---|---|
| Specific Gravity | 2.58 | - |
| Sulfate Content % | 0.14 | ≤ 0.5 |
| Absorption % | 0.75 | - |

**Table 7** Grading of Coarse Aggregate

| Sieve Size (mm) | % passing by weight | IQS 45/1984 |
|---|---|---|
| 20 | 100 | 100 |
| 14 | 100 | 90-100 |
| 10 | 78 | 50-85 |
| 5 | 3.8 | 0-10 |

**Table 8** Physical Properties of Coarse Aggregate

| Physical Properties | Test Results | IQS 45/1984 |
|---|---|---|
| Specific Gravity | 2.64 | - |
| Sulfate Content % | 0.05 | ≤ 0.1 |
| Absorption % | 0.72 | - |

**Table 9** Chemical Composition of Limestone Powder

| Oxide Composition | % By Weight |
|---|---|
| CaO | 54.3 |
| $SiO_2$ | 1.40 |
| $Al_2O_3$ | 0.71 |
| $Fe_2O_3$ | 0.14 |
| MgO | 0.12 |
| $SO_3$ | 0.23 |
| L.O.I | 42.64 |

**Table 10** Properties of Superplasticizer

| Form | Liquid |
|---|---|
| Color | Turbid liquid |
| Relative Density | 1.08 Kg/l ±0.005 |
| Boiling | 100 °C |
| Specific Gravity | 1.08± 0.005 Kg/l |
| pH | 7-9 |
| Normal dosage | 0.8-2% |
| Chlorides | Free from chloride |
| Transport | Non-hazardous |

**Table 11** The Technical Properties of CFRP

| Properties | Sika Warp®-300 C/60 |
|---|---|
| Tensile Strength (MPa) | 3900 |
| E-Modulus (MPa) | 230000 |
| Elongation At Break (strain)% | 1.5 |
| Thickness (mm) | 0.166 |
| Areal Weight ($g/m^2$) | 300±15 |
| Density ($g/cm^3$) | 1.79 |

**Table 12** Properties of the Used Adhesive Material

| Properties | Sikadur®-330 |
|---|---|
| Density (kg/l) at+23°C | Parts A+B mixed : 1.31 |
| Tensile strength (MPa) | 30 |
| E-Modulus (GPa) | Flexural : 3.8<br>Tensile : 4.5 |
| Elongation At Break (strain)% | 0.9 |
| Setting Time (Minute) at 35°C | 30 |
| Mixing ratio | Part A:part B = 4 : 1 by weight |

**Table 13** Properties of the Reinforcing Steel Bars

| Bar Diameter (mm) | Yield Stress (MPa) | Ultimate Stress (MPa) | Elongations (%) |
|---|---|---|---|
| 5 | 558 | 643 | 5.6 |
| 16 | 575 | 680 | 17 |

## 5 Results and discussion

### 5.1 Mechanical Properties of Hardened SCC

Compressive, Strength (f 'c): in order to perform a compressive strength test, ASTM C39/C39M-05 procedure was followed by using three cylinders of 150×300 mm at 28 days from each mix batch. Average compressive strength tests results were shown in Table 14.

Splitting, Tensile Strength, (fct): this test was carried out using three cylinders of 150×300 mm for each batch mix according to ASTM C496/C496M -04 at 28 day. The concrete cylinder fails when a longitudinal crack breaks the cylinder into two halves test. Test results of average splitting tensile strength were shown in Table 14.

Flexural Strength (fr): The flexural strength of concrete had been determined according to ASTM C78-10. Three prisms of 100 × 100 × 400 mm was tested. Table 14 shows test results of average flexural strength.

**Table 14** Mechanical Properties of SCC

| Deep beam Symbols | $f'_c$ (MPa) | $f_{ct}$ (MPa) | $f_r$ (MPa) |
|---|---|---|---|
| D-Control-0.8 and DB-0.8 | 34.2 | 3.61 | 4.43 |
| D-Control-1 | 33.38 | 3.59 | 4.57 |
| DB-1 | 32.75 | 3.72 | 4.6 |
| D-Control-1.2 and DB-1.2 | 33.8 | 3.82 | 4.62 |
| average | 33.5 | 3.68 | 4.55 |

Procedure and Instruments of Testing: The beams were tested using the universal testing machine (AVERY) of 2500 KN capacity. midspan vertical deflection was recorded during the test. Also, First crack load and crack patterns were observed during the test. To prevent local crushing of concrete, bearing plates, placed at the loading points.

Behavior of Control Beams: The crack pattern and failure mode of the control beams are depicted in plate 1. They were failed along the splitting line. The first shear cracks appeared in D-Control-0.8, D-Control-1, D-Control-1.2 at a load 200 kN, 160 kN and 160 kN respectively. After that many flexural and shear cracks were observed and the cracks become wider and longer with increasing of load. The collapse happened at a load of 800 kN for D-Control-0.8, 680kN for D-Control-1 and 640kN for D-Control-1.2. The results showed that the ultimate load decrease by 25% when a/d ratio increased from 0.8 to 1.2. The control beam D-Control-0.8 shows maximum load and higher central deflection value at failure compare with other control beams. Figure 3 show the ultimate load and central deflection for control beams with different a/d ratios (0.8, 1 and 1.2). This figure demonstrates that the deflection increase with the increase of a/d ratio.

Behavior of Pre-cracking Beams: The beams D1-0.8, D3-1 and D1-1.2 were loaded to 60% of the ultimate load of control beams D-Control-0.8 D-Control-1and D-Control-1.2 respectively. Many shear and flexural cracks were noticed in both faces (front and rear) of beams as shown in plate 2. The central deflection of these beams shows in Figure 4. They shows that their behavior similar to control beams behavior and the beam with a/d ratio equal to 1.2 have higher deflection than other beams. Table 15 shows first crack load in shear and flexural with the corresponding deflection. Also, it presents deflection at a load of 60% from control beam for pre-cracking deep beams.

**Table 15** Experimental results for Pre-cracking beams

| Deep beam symbols | $P_{cr)s}$ kN | $\Delta_{cr)s}$ mm | $P_{cr)f}$ kN | $\Delta_{cr)f}$ mm | $\Delta_{60\%}$ mm |
|---|---|---|---|---|---|
| D-control-0.8 | 200 | 1.57 | 180 | 1.43 | 3.72 |
| D1-0.8 | 200 | 1.61 | 160 | 1.25 | 3.73 |
| D-control-1 | 160 | 1.37 | 160 | 1.37 | 3.56 |
| D3-1 | 140 | 1.16 | 160 | 1.35 | 3.34 |
| D-control-1.2 | 160 | 1.67 | 200 | 2.35 | 3.75 |
| D1-1.2 | 140 | 1.4 | 120 | 1.18 | 3.72 |

### 5.2 Behavior of the retrofitted beam (D1R-I-0.8, D3R-I-1, D1R-I-1.2)

**Crack pattern and failure modes:** The crack pattern and failure mode of the retrofitted beams are depicted in plates 3. The retrofitting beams show a rise in ultimate load compare with reference beams for all a/d ratios. The beams D1R-I-0.8, D3R-I-1, D1R-I-1.2 failed at loads of 980kN, 820kN and 720kN respectively which gives increase in ultimate load by 22.5%, 20.5 and 12.5 respectively compared with control beams. For the retrofitting beams, the reduction in ultimate load capacity with increase a/d ratio from 0.8 to 1.2 was about 36%. The failure loads of the beams in third group were higher than those of the corresponding beams of other groups because of the shorter shear span length. The mode failure of beam D1R-I-0.8 was by de-bonding of CFRP strips at the roller side and crashing the concrete near the load point. While D3R-I-1 failed by tearing the middle strip at the right side of the beam followed by de-bonding the other strips in the same side.

It can be noticed that the failure mode D1R-I-1.2 was shear compression failure happened by de-bonded of CFRP as well as crushing of concrete at compression zone so a large piece of concrete was pulled up at this zone and take off the upper carbon fiber strip at the left side of the beam. Test results of beam specimens are shown in Table 16.

**Table 16** Experimental results of beam specimens

| Beam symbols | a/d | Shear force | | Flexural Force | | Ultimate force | | % Increasing in Ultimate Load | % $P_{cr)s}$ / $P_u$ | Modes of failure |
|---|---|---|---|---|---|---|---|---|---|---|
| | | $P_{cr)s}$ kN | $\Delta_{cr)s}$ mm | $P_{cr)f}$ kN | $\Delta_{cr)f}$ mm | $P_u$ kN | $\Delta_u$ mm | | | |
| D-Control-0.8 | 0.8 | 200 | 1.57 | 180 | 1.43 | 800 | 7.17 | - | 25 | DS |
| D1R-I-0.8 | 0.8 | 180 | 1.28 | 200 | 1.44 | 980 | 7.62 | 22.5 | 18 | FRPD+CC |
| D-Control-1 | 1 | 160 | 1.37 | 160 | 1.37 | 680 | 6.53 | - | 23 | DS |
| D3R-I-1 | 1 | 200 | 1.7 | 160 | 1.35 | 820 | 7.19 | 20.5 | 24 | FRPD+FRPT |
| D-Control-1.2 | 1.2 | 160 | 1.67 | 200 | 2.35 | 640 | 6.8 | - | 25 | DS |
| D1R-I-1.2 | 1.2 | 160 | 1.32 | 180 | 1.5 | 720 | 7.15 | 12.5 | 22 | FRPD+CC |

**Load mid-span deflection responses:** The ultimate load and mid-span vertical deflection for all the beams are illustrated at Table 16. The load-deflection relationship for the control and retrofitting beam for the three groups are shown in Figure 5 to 7. it is noticeable from this figures the effect of CFRP on decreasing the deflection value for retrofitting beams in compare with control ones. Figure 8 shows the deflection of the retrofitting beams with different a/d ratio. It present that the deflection increase with the increase of a/d ratio. Similar to control beams, the retrofitted beam D1R-I-0.8 shows maximum load and higher central deflection value at failure compare with other retrofitted beams. The adopted retrofitting arrangement is identical in the three beams to allow for studying the effect of composite action on the retrofitting of reinforced concrete deep beams in case of changing a/d ratios.

## 6 Conclusions

According to the experimental results of retrofit deep beams by using CFRP with different a/d ratio, the following conclusion can be recorded:

1. All the tested beams failed in shear, the failure mode of control beams was diagonal splitting failure along the line joining the load and support while retrofitted beams failed by combination many modes of failure such as deponding or tearing of CFRP as well as crushing of concrete.
2. All retrofitted deep beams given an increase in ultimate strength with represents to reference model.
3. For All a/d ratios, retrofitted deep beams showed reduction in deflection values in compare with control beams.
4. In general, the behavior of beams tested influenced by a/d ratio. It was found that the increase of a/d ratio from 0.8 to 1.2 decreases the failure load for control and retrofitted beams by about 25% and 36% respectively.
5. midspan deflection decrease with increase a/d ratio for control and retrofitted beams.

 https://doi.org/10.1051/matecconf/201816204016

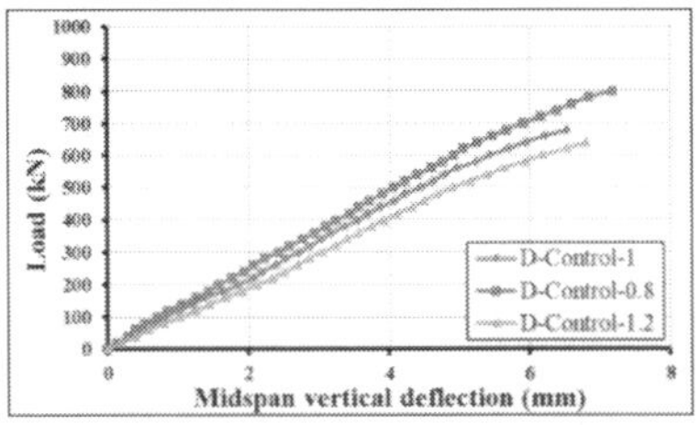


**Figure 3** Load-deflection relationship for control deep beams

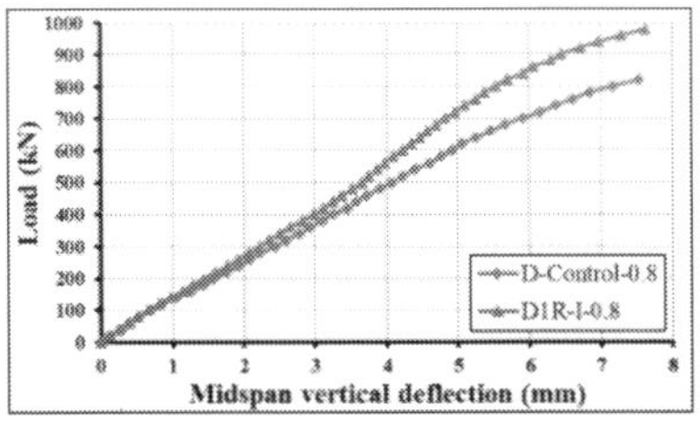


**Figure 4** Load-deflection relationship for pre-cracking deep beams

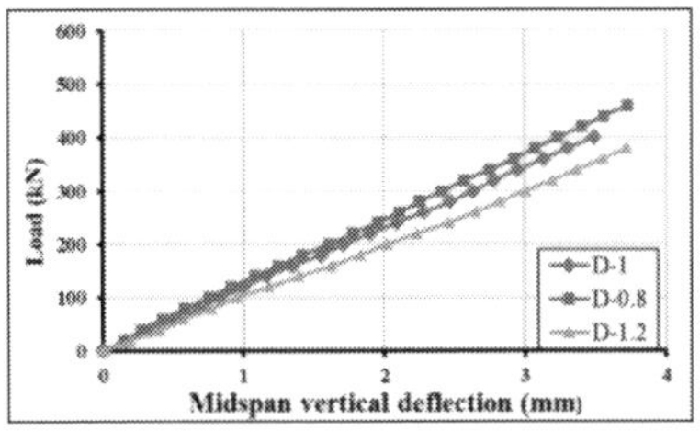


**Figure 5** Load-deflection relationship for D-Control-0.8 and D1R-R-0.8

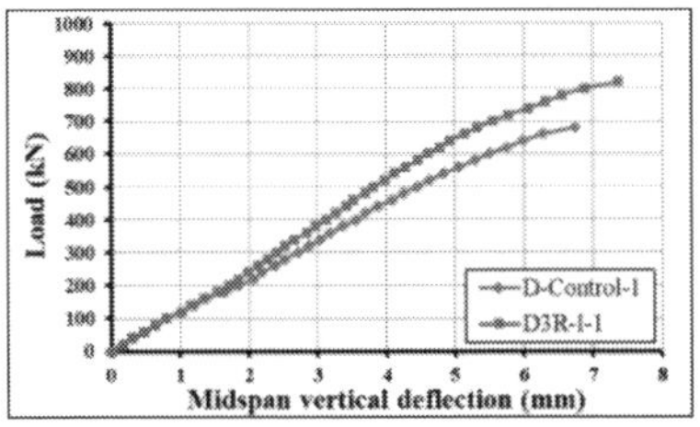


**Figure 6** Load-deflection relationship for D-Control-1 and D3RI-I-1

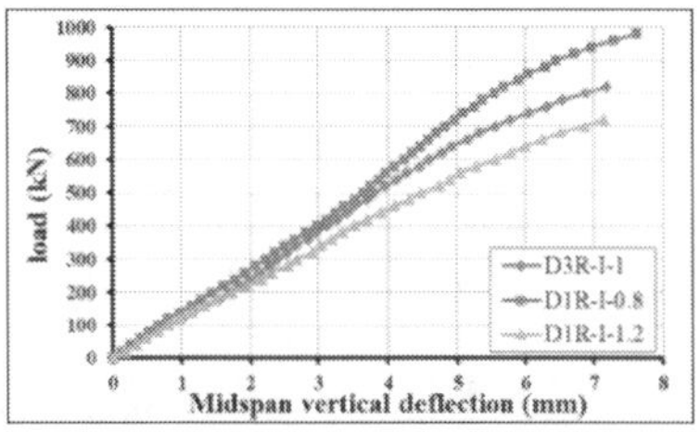


**Figure 7** Load-deflection relationship for D-Control-1.2 and D3RI-I-1.2

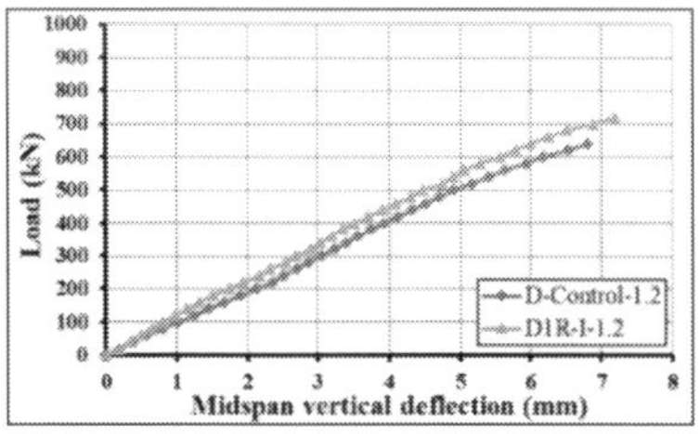


**Figure 8** Load-deflection relationship for D-Control-1 and D3RI-I-1

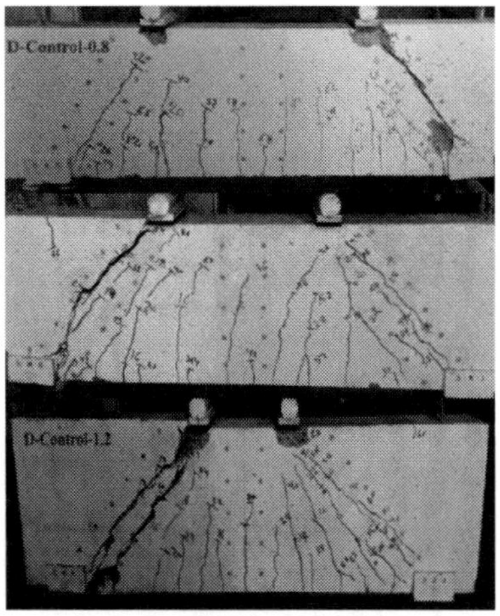


**Plate 1** Crack pattern of control beams

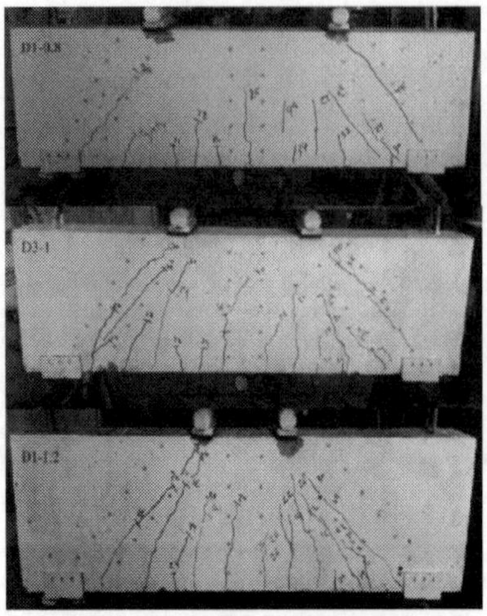

**Plate 2** Crack pattern of pre-cracking beams

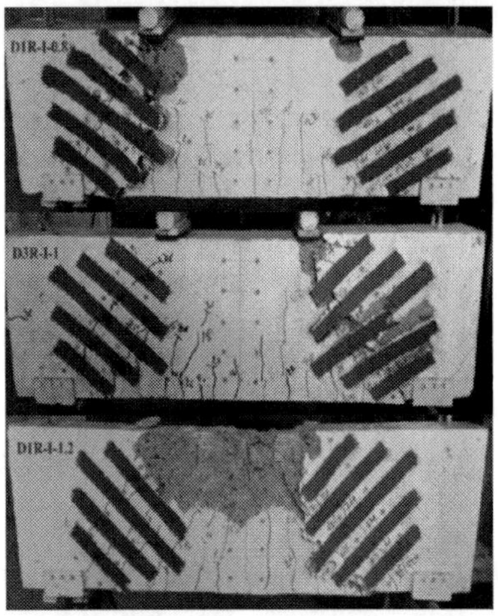

**Plate 3** Crack pattern of retrofitted beams

## References

1. ASTM C496 / C496M-04, West Conshohocken, PA: ASTM Int.
2. ASTM C 39 / C 39M-05. Con. and Agg., West Conshohocken, PA, US.
3. ACI 237R-07. ACI Committee 237: Farmington Hills, MI, USA
4. ASTM C78-10, 2010. West Conshohocken, PA: ASTM Int.
5. ASTM A1064 / A1064M−14. West Conshohocken, PA: ASTM Int.
6. ASTM A615 / 615M-14. West Conshohocken, PA: ASTM Int.
7. EFNARC, 2002. **99** West Street, Farnham, Surrey GU9 7EN, UK.
8. Iraqi Standard **5**, (1984). Agg. from Natural Sources for Concrete, Baghdad, Iraq.
9. IS (45), (1984). Por. Cement, Central Agency.
10. Stand. and Quality Control, Planning Council, Baghdad, Iraq.
11. Mohammed, A. S., (2016). Build. and Const. Eng. Depart., Uni. of Tech., Baghdad, Iraq, 247 pp.
12. Rasheed, M.M., (2016). Civil and Env. Research, **8**, 5, 2, pp. 6-14

# The role of cover and rebar characteristics on load-slip behavior of reinforced concrete members in compression

*Tareq* al-Attar[1,*], *Qais* Hassan[1], *Sura* Mejbel[1], and *Hussein* Dawood[2]

[1] University of Technology, Baghdad, Iraq
[2] Private Sector, Baghdad, Iraq

**Abstract.** This paper describes a part of an extensive research work devoted to evaluate the bond strength between rebars and concrete through different testing procedures. Main parameters in this part are the concrete cover and rebar diameter. The tested specimen consisted of a single bar embedded in a concrete block with square cross-sectional area and is being tested under compressive load. Three concrete block sizes were cast to offer three different cover for the embedded rebars. The dimensions of these blocks were; 150×150×135, 100×100×135 and 200×200×185 mm. Three bar diameters, 12, 16 and 20 mm, were investigated. The specimens were water-cured and tested at the ages of 7 and 28 days. A new proposed test set-up was used to monitor the load-slip behavior of the specimens. The test results showed that there is no significant difference in bond energy between the two curing ages, 7 and 28 days. The concrete cover has a significant effect on the bond strength between rebar and concrete. By increasing the cover, the confinement offered by concrete increases, bond strength increases, and slip increases. Based on the present results, a concept of effective cover was developed. This concept showed a high correlation with the mode of failure for the tested specimens.

## 1 Introduction

The bond between reinforcing bars and concrete has been known as a key parameter to a proper performance of reinforced concrete structures. For many years, bond strength was represented in the term of shear stress at the interface between bar and surrounding concrete. The transfer of axial force from a reinforcing bar to the surrounding concrete results in the development of tangential stress components along the contact surface. The stress acting parallel to the bar along the interface is called the bond stress. The term bond is represented as a structural property [1]. In reinforced concrete structures the real value of force transferred between bars and concrete is required for optimal design [2]. The transfer of forces is depending on the chemical adhesion between bar and concrete, frictional forces and bearing of the front key (concrete between two ribs) against shear forces [3]. The bond behavior is decisively determined by the behavior of concrete close to the rib [4]. The simplest model representing the stress transfer between steel and concrete is so called " friction concept " whereby the shear stress that develops along the lateral surface (bond stress) is a function of the normal confining pressure exerted by the surrounding concrete on the bar surface and concrete cover. Higher the confining pressure higher the frictional force required for bond failure and higher the strength reserves of splitting failure [5]. When the first concrete key fails, there is a sudden drop in bond strength due to the formation of longitudinal splitting cracks which are visible from the surface of the specimen. Therefore, if there is an adequate confinement (cover) available, the tension stiffening effect reduces the sudden drop in bond strength. As the concrete ages, the compressive strength increases, so the bond stress differs from early ages to later ages. Increasing the concrete compressive strength can improve bond performance. In other words, higher adhesion and friction force between concrete and rebar can be expected for concrete with higher compressive strength [6]. For a given bonded length, larger bars require larger forces to cause either a splitting or pullout failure [1]. For bars not confined by transverse reinforcement, the total force developed at bond failure is not only an increasing function of concrete cover, bar spacing, and bonded length, but also of bar area [7, 8]. The bond strength (stress) can be expressed as:

$$\tau = P_{max} / (\pi \, d_b \, l_d) \quad (1)$$

Where:
$\tau$: bond strength (stress), MPa.
$P_{max}$: the maximum force at failure, N.
$d_b$: diameter of bar, mm.
$l_d$: bonded length, mm.

* Corresponding author: 40076@uotechnology.edu.iq

© The Authors, published by EDP Sciences. This is an open access article distributed under the terms of the Creative Commons Attribution License 4.0 (http://creativecommons.org/licenses/by/4.0/).

## 2 Experimental program

### 2.1 Materials

Ordinary Portland cement conforming to the ASTM C150 [9] was used throughout this study. The used fine aggregate was natural sand. The fineness modulus, specific gravity (SSD) and sulfate content of this sand were 2.36, 2.6 and 0.09 percent respectively. Crushed gravel with a maximum size of 12.5 mm is the coarse aggregate. The unit weight, specific gravity (SSD) and sulfate content of this gravel were 1640 kg/m$^3$, 2.7, and 0.01 percent respectively. Hot rolled deformed bars with nominal diameters 12, 16 and 20 mm having yield strength of582, 552 and 559 MPa respectively, were used for bond test specimens.

### 2.2 Concrete mix

According to the recommendations of the ACI Committee 211 [10], a concrete mix, with proportions of 1: 1.88: 2.63: 0.45 (cement: fine aggregate: coarse aggregate: W/C ratio) by weight, was used to cast the studied specimens. The cement content of this mix was 400 kg/m$^3$. This mix was designed to attain an average compressive strength of 30 MPa at 28 days. Two standard cylinders of 150mm in diameter and 300mm in height were used to determine the compressive strength of concrete and the measured compressive strength values were 23 and 32 MPa at 7 and 28 days respectively.

### 2.3 Test specimen

The test specimen consisted of a single bar embedded in a concrete block with square cross-sectional area and is being tested under compressive load. Three concrete block sizes, A, B and C were cast to offer three different cover lengths for the embedded rebars. The dimensions of these blocks were; 150×150×135, 100×100×135and 200×200×185 mm. Size A blocks were reinforced with a single vertically embedded bar with a diameter of 12, 16 and 20 mm ($A_{12}$, $A_{16}$ and $A_{20}$). Meanwhile, size B and C blocks were reinforced with only 16 mm diameter bars ($B_{16}$ and $C_{16}$). Figure 1 shows a schematic representation of used test specimen. The bottom end of the rebar was free to move for 15mm to allow the movement of the bar during the test, while the loaded end (top end) was jutted out for $2.5d_b$. PVC sleeves with smooth outer and inner surfaces were used as bond breakers to achieve the required embedded (bonded) length for the steel bar. PVC tubes do not restrain the slip of the bars, and do not permit the transfer of force between bar and concrete. The concrete blocks were tested by a universal testing machine in the compression mode applied on the steel bar (push-off mode), and the load-deformation (slip) curves were recorded by the data acquisition system of the UTM.

### 2.4 Fabrication of test specimen

Wooden molds were fabricated to cast the specimens. Lubricating oil was applied on all the inner sides of the mold. The bar was fixed vertically by using a steel plate at the top and a squared wooden piece at the bottom as shown in figure 2. After the specimens were prepared with bars, the concrete was poured and compacted by a vibrating table machine according to ASTM C192 [11]. After 24 hours, de-molding was carried out and the specimens were placed in a water tank for curing process until the age of 7, or 28 days.

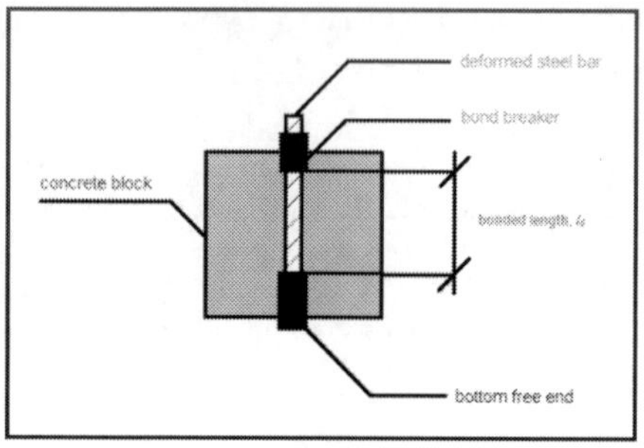


**Fig.1.** Schematic representation of test specimen

**Fig.2.** Wooden molds used for specimen fabrication

### 2.5 Set-up and testing

Displacement control system was used to test the specimens by pushing-off the bar inside the block which was set in the machine on a steel plate with a hole in the middle to permit the free end of the bar to slip down. This bottom bearing plate had a Rockwell hardness of 60 HRC and horizontal dimensions conforming to ASTM C 39 [12]. The effective embedded (bonded) length was kept constant at 100mm. The load was applied vertically on the cross section of the bar (load direction was parallel to the longitudinal axis of the bar) as shown in figure 3. Monotonic load was applied by the actuator and the stroke was maintained at a constant speed of 1.5 mm/min. Each test result represents the average of three tested specimens.

## 3 Results and discussion

### 3.1 Basic considerations

Push-off test was carried out on thirty concrete blocks. Strain rate was kept at 1.5 mm/minute. The specimens were tested under push-off test at two ages, 7 and 28 days.

**Fig.3.** Test set-up of the reinforced concrete block

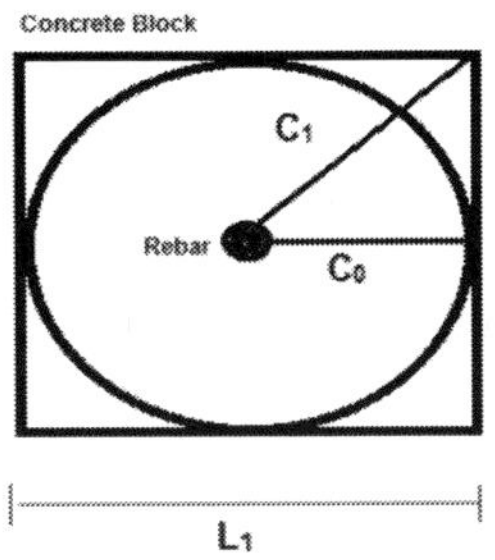


**Fig.4.** Schematic representation of the rebar covers $C_0$ and $C_1$.

According to figure 4, two distances could be considered as rebar cover, $C_0$ and $C_1$. Stress-wise, $C_0$ is considered a more critical stress path than $C_1$. This consideration was based on the work of Tepfers [13], to focus on prediction of bond strength for deformed reinforcement. Tepfers was the first to propose an analytical model in which the concrete surrounding a single reinforcing bar is characterized as a thick-walled cylinder subjected to internal shear and pressure. In this analogy the internal shear and pressure correspond respectively to the bond and radial stresses developed at the concrete-steel interface. Thus, it follows that the radial force transfer at the concrete-steel interface determines the tensile hoop stress developed in the concrete surrounding the bar and thus the critical load. Tepfers proposes that bond strength is determined by the capacity of the concrete surrounding the reinforcing bars to carry the hoop stresses, as shown in figure 5. Moreover, the failure modes of most of the tested specimens have shown that the splitting of concrete block was occurring through $C_0$ stress path.

The $C_0$ length was calculated as follows:

$$C_0 = (L_1 - d_b)/2 \quad (2)$$

Where:

$C_0$: rebar cover, mm.

$L_1$: lateral dimension of the concrete block, mm.

$d_b$: nominal diameter of the rebar, mm.

In the present work, the ratio of $C_0/d_b$ is nominated as the *effective cover*. Table 1 lists the values of this ratio for the tested groups.

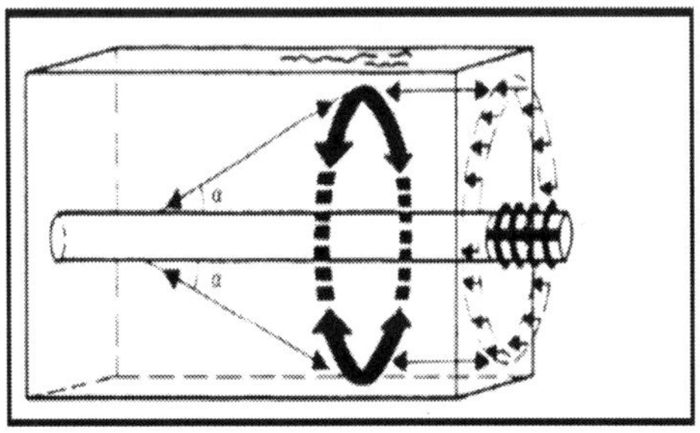

**Fig.5.** Schematic representation of internal forces within a reinforced concrete specimen [13]

**Table 1.** The effective cover for the tested groups

| Group | $L_1$, mm | $d_b$ (mm) | $C_0$, mm | $C_0/d_b$ |
|---|---|---|---|---|
| A | 150 | 12 | 69 | 5.75 |
| | | 16 | 67 | 4.19 |
| | | 20 | 65 | 3.25 |
| B | 100 | 16 | 42 | 2.63 |
| C | 200 | 16 | 92 | 5.75 |

### 3.2 Load-slip relationship

Table 2 lists the maximum load, $P_{max}$, maximum slip at $P_{max}$, bond stress (according to Eq. 1) and the bond energy (area under the load-slip curve) which was calculated using Microsoft excel program and trapezoidal rule at ages of 7 and 28 days for specimens $A_{12}$, $A_{16}$, $A_{20}$, $B_{16}$ and $C_{16}$ respectively. Figures 6 -10 show the load-slip curves for the abovementioned specimens.

The bond energy could be considered as an indication to bond development with curing age. It was observed that there is no significant difference in energy between the two curing ages, 7 and 28 days. The only exception to that is the case of specimens $A_{16}$.

Increasing the bar diameter has a positive effect on the maximum load and associated slip and negative effect on the bond energy for the same age.

The confinement that was offered by the concrete block size (cover) also had a positive effect on load-slip relationship and on the bond-energy values.

**Table 2.** Push-off test results for the studied specimens at 7 and 28 days.

| Group | Specimen | Age, days | $P_{max}$, kN | Bond stress, MPa | Max. Slip*, mm | Mode of failure** | Bond energy, N.mm | % increase in bond energy |
|---|---|---|---|---|---|---|---|---|
| A | $A_{12}$ | 7 | 28.20 | 5.16 | 1.01 | P | 108.2 | 8.3 |
| | | 28 | 33.27 | 6.62 | 1.30 | P | 117.2 | |
| | $A_{16}$ | 7 | 53.64 | 10.67 | 1.41 | PS | 79.8 | 19.3 |
| | | 28 | 64.72 | 12.88 | 1.53 | PS | 95.2 | |
| | $A_{20}$ | 7 | 59.55 | 11.84 | 1.63 | S | 84.6 | 6.3 |
| | | 28 | 81.34 | 16.18 | 1.70 | S | 89.9 | |
| B | $B_{16}$ | 7 | 47.96 | 12.72 | 1.84 | S | 34.5 | 6.7 |
| | | 28 | 56.72 | 15.04 | 2.09 | S | 36.8 | |
| C | $C_{16}$ | 7 | 64.16 | 10.21 | 2.58 | P | 164.9 | 4.1 |
| | | 28 | 70.47 | 11.22 | 3.21 | P | 171.7 | |

*: the max. slip was recorded by a computer program implemented in the testing machine
**: P: push-off, PS: combined, S: splitting

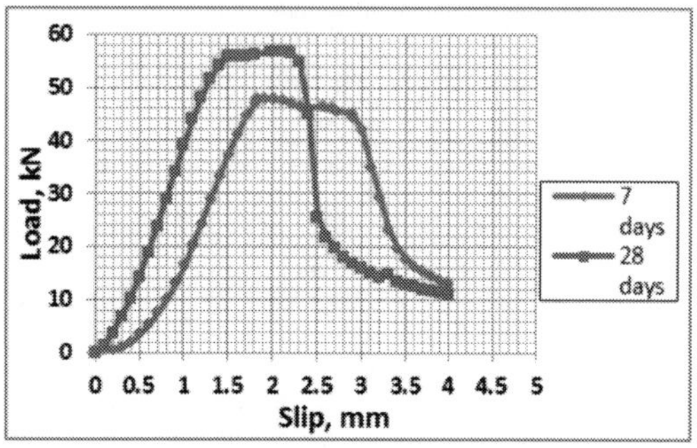


**Fig.6.** Load-slip curve for specimen $A_{12}$

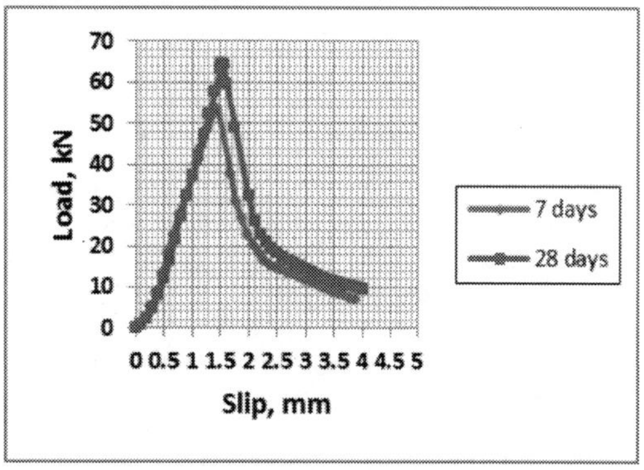


**Fig.7.** Load-slip curve for specimen $A_{16}$

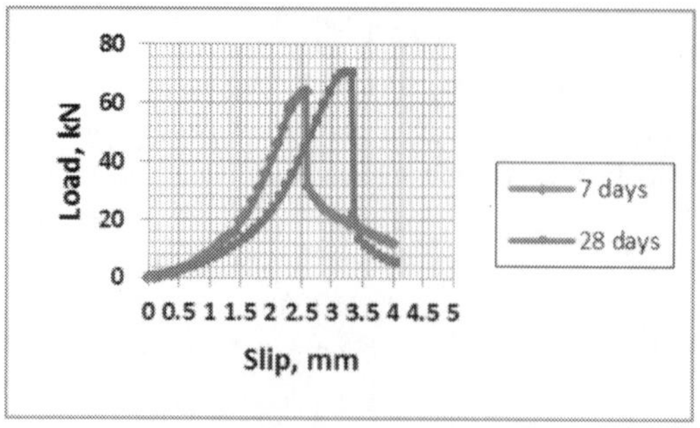


**Fig.8.** Load-slip curve for specimen $A_{20}$

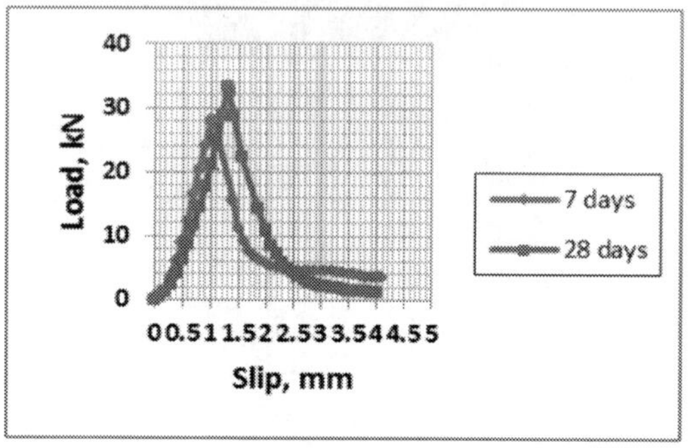


**Fig.9.** Load-slip curve for specimen $B_{16}$

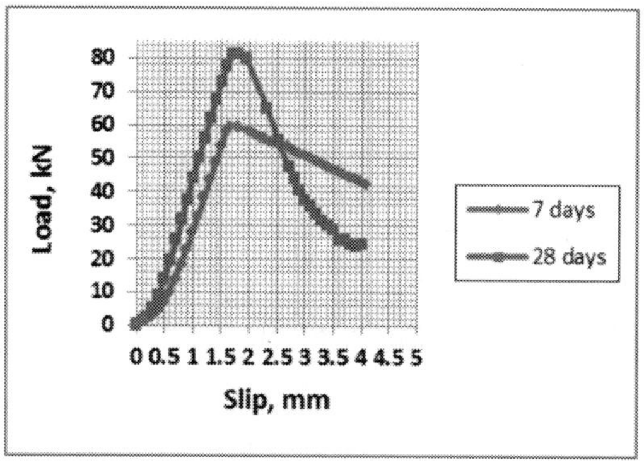


**Fig.10.** Load-slip curve for specimen $C_{16}$

### 3.3 Failure modes

Figure 11 displays the observed modes of failure for the push-off test specimens. They failed by the following three modes of failure; push-off failure, P, with pure slippage of the rebar, push-off with splitting of concrete block, PS and only splitting of concrete, S. The push-off failure mode occurred when the concrete block size (cover) provided an adequate confinement, thus preventing a splitting failure of the test specimen. The push-off failure was observed for $A_{12}$ and $C_{16}$ specimens. The combined failure was observed only for $A_{16}$ specimens. The splitting failure was diagnosed in $A_{20}$ and $B_{16}$specimens. This behavior may be explained by that bond strength is inversely proportional to the surface area of the bonded length of the rebar, as shown in Eq.1.

For the smaller one, 12mm, most of the shearing force in the interface is applied on the rebar and that leads to pure slippage of the rebars. Increasing the bar diameter to 16 or 20 mm for the same concrete block size would raise the share of concrete in load carrying thus the failure will change to PS or S. Therefore, it could be concluded that the mode of failure is highly dependent to the bar diameter and concrete confinement (cover).
When relating the mode of failure to the effective cover, $C_0/d_b$, it could be concluded that higher effective cover causes pure slippage of rebars and lower $C_0/d_b$ values are companions to the concrete splitting mode.

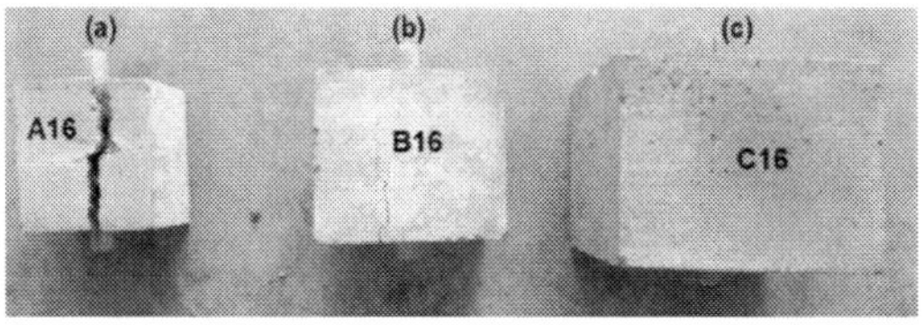


**Fig.11.** Failure modes a: splitting failure in group A, b: combined failure in group B, c: push-off failure in group C

### 3.4 The relationship between effective cover and $P_{max}$ and slip

Figure 12 illustrates the variation of $P_{max}$ with the effective cover at the age of 28 days for specimens of different sizes, A, B and C, but with the same bar diameter, 16mm.The increase in the block size makes the concrete stronger and sharing higher percentage of load. In addition to that the confinement becomes higher when the block size increases [14].

Figure 13 shows the variation of $P_{max}$ with effective cover at 28 days for specimens of A size reinforced with different bar diameters, 12, 16 and 20mm. Increasing the bar diameter would cause the effective cover to be reduced but this will improve the bond and makes the rebar to withstand higher load. The change in rebar diameter has the same effect on bond slip as shown in figure 14.

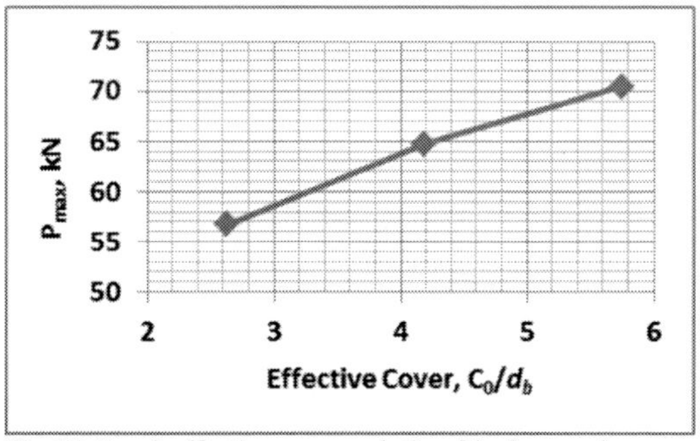


**Fig.12.** Load-effective cover relationship for specimens $A_{16}$, $B_{16}$ and $C_{16}$at the age of 28 days

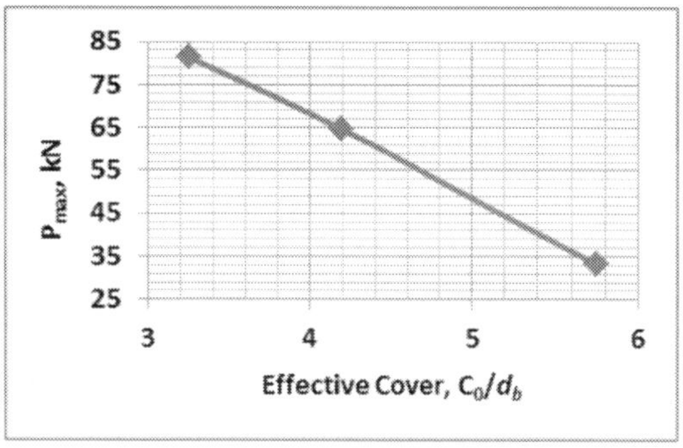


**Fig.13.** Load-effective cover relationship for specimens $A_{12}$, $A_{16}$ and $A_{20}$ at the age of 28 days

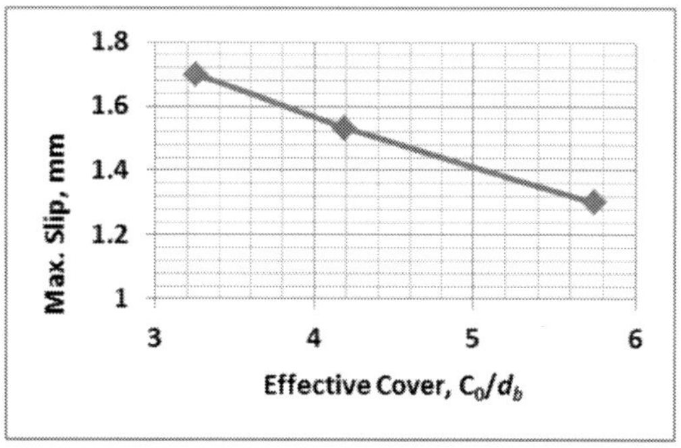


**Fig.14.** Slip-effective cover relationship for specimens $A_{12}$, $A_{16}$ and $A_{20}$ at the age of 28 days

## 4 Conclusions

The following are the conclusions reached throughout this research work:

1. For all studied cases, the aging of concrete caused the load at failure and the related slip to increase. This is attributed to the fact that concrete compressive strength increases with time leading to an increase in tensile strength, which means more confining for the steel bar by the surrounding concrete.
2. The concrete cover has a significant effect on the bond strength between rebar and concrete. By increasing the $C_0/d_b$ ratio, the confinement increases, bond strength increases, and slip increases.
3. When relating the mode of failure to the effective cover, $C_0/d_b$, it could be concluded that higher effective cover causes pure slippage of rebars and lower $C_0/d_b$ values are companions to the concrete splitting mode.
4. The increase in the surrounding concrete size makes the concrete stronger and share higher percentage of load. Meanwhile, increasing the bar diameter would cause the effective cover to be reduced but this will improve the bond and make the rebar to withstand higher load.

## References

1. ACI 408R-03, ACI Manual of Concrete Practice, (2015)
2. K. Ahmed, A. Al Ragi, U. Kausar, A. Mahmood, International Journal of Advancements in Research & Technology, **3**, 1, (2014)
3. K. Ahmed, Z. Siddiqi, M. Ashraf; A. Ghaffar, Pakistan Journal of Engineering and Applied Science, **2**, (2008)
4. J. S. Martin, MSc Thesis, University of Washington, (2006)
5. S. Tastani, and S. Pantazopoulou, Bond in concrete-from research to standards, (2002)
6. G. Appa Rao, International Journal of Applied Engineering and Technology, **4**, (2014)
7. D. Darwin, M. L. Tholen, E. K. Idun, J. Zuo, , ACI Structural Journal, **93**, 1, (1996)
8. D. Darwin, S. L. McCabe, E. K. Idun, S. P. Schoenekase, ACI Structural Journal, **89**, 6, (1992)
9. ASTM C150, Annual Book of ASTM Standards, (2015).
10. ACI 211.1-91, ACI Manual of Concrete Practice, (2015)
11. ASTM C192, Annual Book of ASTM Standards, (2015)
12. ASTM C39, Annual Book of ASTM Standards, (2015)
13. R. Tepfers, Magazine of Concrete Research, **31**, (1979)
14. M. N. S. Hadi, Open Civil Engineering Journal, **2**, (2008)

MATEC Web of Conferences **162**, 04018 (2018) https://doi.org/10.1051/matecconf/201816204018
*BCEE3-2017*

# Behaviour of normal reinforced concrete columns exposed to different soils

*Laith* Rasheed[1, *], *Shakir* Salih[2], *Zahraa* Hanash[1]

[1]College of Engineering, University of Kerbala, Kerbala, Iraq
[2]Building and Construction Engineering Department, University of Technology, Bagdad, Iraq

**Abstract.** Concrete resistance to sulfate attack is one of the most important characteristics for maintaining the durability of concrete. In this study, the effect of the attack of sulfate salts on normal reinforced concrete column was investigated by burying these columns in two types of soils (sandy and clayey) in two pits at a depth of 3 m in one of the agricultural areas in the holy city of Karbala, one containing sandy soil (SO3 = 10.609%) and the other containing clayey soil with (SO3 = 2.61%). The tests were used (pure axial compression test of reinforced concrete columns, compressive strength test, and splitting tensile strength test, absorption, voids ratio and finally density). It's found that the strength of RC columns decreasing by (12.51%) for age (240 days), for columns buried in clayey soil, where the strength increased by (11.71%) for the same period, for columns buried in sandy soils, with respect to the reference column.

## 1 Introduction

Sulfate ions are found in groundwater and soil in southern parts of Iraq. Concrete structures in the regions of this part subjected to attack from sulfates may suffer from two types of damage; loss of strength of the matrix due to degradation of calcium-silicate-hydrate (C-S-H), and volumetric expansion due to formation of gypsum or ettringite that leads to cracking [1]. Protecting against sulfate attack requires reducing the permeability of concrete, whereby low permeability characteristics of concrete can prolong the service life of a structure that is subjected to severe exposure conditions [2]. The present study includes an attempt to produce reinforced concrete mix that can stand up against sulfate attack investigating, Normal reinforced concrete was used to study the strength and durability performance of this type of concrete. The main aim of the present work is to evaluate the effect of sulfate salts in groundwater and soil on strength and durability characteristics of normal reinforced concrete columns under pure axial compression force. Through this investigation (7) reinforced concrete columns are tested, where a (42) cubes are performed to cover a compressive strength, absorption, density, and voids tests. Splitting tensile strength are undertaken on (21) cylinders. The specimens are buried in sandy and clayey soils, for up to (240) days, in aggressive solution containing ($Cl^- + SO_4^=$) at concentration identical to those present in soil and groundwater in the southern parts of Iraq.

## 2 Experimental work

### 2.1 Experimental program

All groups of columns, were identical in size, (150x150) mm square section and overall height (1080 mm). They were casted in a vertical direction and hardened in fresh water conditions until 28-days. After that, they were buried and hardened in soils with different concentrations of sulfate and chloride solutions for a period of 60, 150, and 240 days. Then, all groups of columns were tested under axial static load by mechanical machine with electrical seal load in period 60, 150, and 240 days as shown in *Figure 1* and mechanical and electronic dial gauges. The capacity of machine was (1000 kN), it had a rigid structure frame, and the capacity of mechanical gauges (2 cm) and (5 cm) for electronic dial gaugeTest started with the application of 7 $kN/m^2$ load to set and check dial gauge. Then initial reading of dial gauges is obtained. The load was applied in stages. At each load increment, observations of crack development on the concrete columns traced. Also, at each test, the first visible cracking load and reading of dial gauges were recorded. The process of reading the gauges and crack observations took about five to ten minutes. Once this process got completed, loading

* Corresponding author: laith.alqarawee@gmail.com

© The Authors, published by EDP Sciences. This is an open access article distributed under the terms of the Creative Commons Attribution License 4.0 (http://creativecommons.org/licenses/by/4.0/).

resumed to the next load step. The same procedure was followed. The load was continued until ultimate load that recorded. All columns were divided into two groups, one buried in clayey soils and other buried in sandy soil, the main details of the mix used throughout this investigation in Table 1. The columns have the same amount of longitudinal reinforcement of 4 Ø 8 mm and stirrups of 7 Ø 6mm/m ($A_s$=0.00893 $A_g$), as shown in *Figure 1,* which is less than the smallest percentage allowed in the ACI code by (0.0011) to observe the effect of sulfates.

**Table 1.** The main details of the mixes used throughout this investigation.

| *Mix Designation* | *Cement (kg/m³)* | *Sand (kg/m³)* | *Gravel (kg/m³)* | *W/C* |
|---|---|---|---|---|
| Normal | 336 | 760 | 1044 | 0.55 |

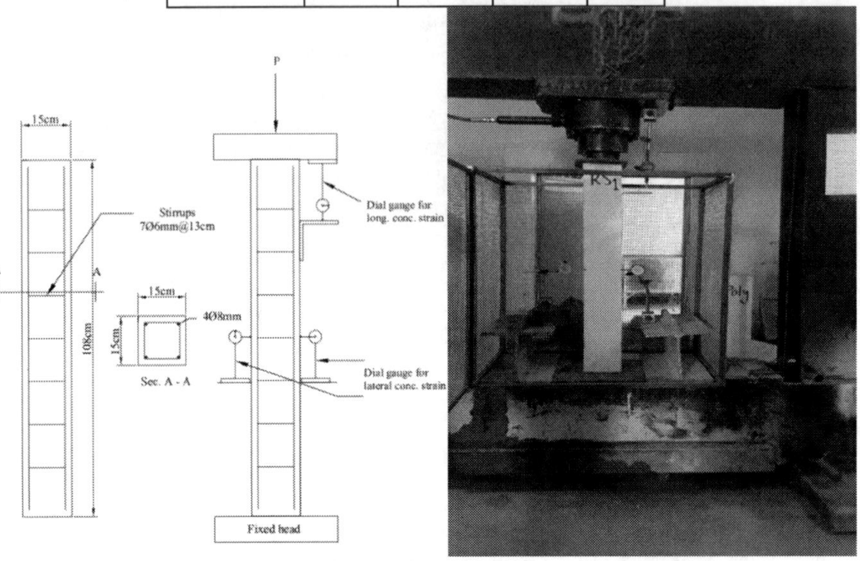


**Fig. 1.** Details of RC. tested columns and method of measuring their deformations.

## 2.2 Materials

In this research Portland Sulfate Resist Cement (ASTM-Type V) [3] manufactured by United Cement Company commercially known (AL-GESR) was used. The adopted cement confirms to the Iraqi Specifications No.5/1984 [4]. The fine aggregate from Al-Ekadir region was used throughout this work, it was within the requirements of (IQS No.45/ 1984) [5]. Black crushed stones of 20 mm maximum size from Al-Nebai quarry were used as a coarse aggregate in all mixes. which confirms to the Iraqi specification (IQS No.45/ 1984) [5]. The water used in the mix design was potable water from the water-supply network system; so, it was free from suspended solids and organic materials, which might have affected the properties of the fresh and hardened concrete. Deformed steel bars (8 and 6) mm diameter were used as reinforcement to test column specimens obtained from Turkish production. The steel reinforcement was tested according to ASTM-A615/A-615M-05a.

## 2.3 Soils

Two types of soil were used to bury the specimens and RC columns models in clayey soil and sandy soil. Two pits were excavated in an agricultural land located to a location (1.5 km) west of Karbala centre. The pit`s dimension was (2×2×3) m. The reinforced concrete columns models and their attached specimens were buried in the pits, then one pit was filled with the sandy soil, while the other pit was filled with the clayey soil, see *Table 2*. The ground water at August appeared in the pits in level 1.75 m, which had the properties shown in the *Table 3*.

**Table 2.** Physical and chemical properties for clayey and sandy soils.

| Soil Type | $SiO_2$ % | $Fe_2O_3$ % | $Al_2O_3$ % | CaO % | MgO % | $SO_3^{=}$ % | $Na_2O$ % | $K_2O$ % | $Cl^-$% | O.M % | PH | GS |
|---|---|---|---|---|---|---|---|---|---|---|---|---|
| Sandy | 18.31 | 2.95 | 4.55 | 26.86 | 5.87 | 10.60 | 4.24 | 0.64 | 1.95 | 0.62 | 8.15 | 2.6 |
| Clayey | 33.79 | 6.40 | 12.43 | 12.49 | 7.79 | 2.61 | 2.20 | 1.35 | 0.78 | 0.83 | 8.3 | 2.5 |

**Table 3.** Chemical properties for ground water

| | |
|---|---|
| $SO_4^{=}$ ppm | 4675.5 |
| $Cl^-$ ppm | 11182.5 |
| PH | 8.45 |

## 3 Results and discussion

### 3.1 Results of hardened concrete properties

The compressive strength of specimens decreased in clayey soil with time and increased in sandy soil. The tensile strength of concrete cylinders showed that it has increased over time, this may be due to the salt affected to improve bonds between the concrete particles. The absorption has decreased with time, the maximum increment in the absorption was recorded in cubes buried in the sandy soil for (240 days) by 5.09%, and in cubes buried in the clayey soil by 9.83 % for (240 days) compared with the reference cubes. It was also noted that the percentage of voids in the cubes had also decreased with time in all concrete mixes, specifically in the polymer mix, where the increase was 4.57% in cubes buried in sandy soil and 7.79% in cubes buried in the clayey soil for (240 days). A decrease in the percentage of voids and the absorption ratio in the normal mix was observed. This is due to the crystallization in the normal mix because of the presence of salts, which led to the closure of pores and reduced the percentage of voids and absorption in this mix and it was expected over time that the crystallization would increase the stresses in concrete, thus increase absorption rate and voids again *Table 6* provides a summary of cubes and cylinders tests such as compression strength, tensile strength, absorption, voids ratio and density.

**Table 4.** Percentage of sulfates and chlorides in clayey and sandy soil over time.

| *Soil Type* | *Salt (%)* | *Time of buried (days)* | | | |
|---|---|---|---|---|---|
| | | *0* | *60* | *150* | *240* |
| *Clayey* | *Sulfate ($SO4^{=}$)* | 2.61 | 4.41 | 3.50 | 3.04 |
| | *Chloride ($Cl^-$)* | 0.78 | 0.78 | 1.99 | 0.78 |
| *Sandy* | *Sulfate ($SO4^{=}$)* | 10.6 | 9.5 | 9.0 | 9.6 |
| | *Chloride ($Cl^-$)* | 1.95 | 1.95 | 1.24 | 1.07 |

**Table 5.** Results of the Experimentally Tested Columns

| *Group Name* | *Soil Type* | *Designation Name* | *Time of Test (days)* | *Cracking Load (Pc), (kN)* | *Failure Load (Pu), (kN)* | *Long. Def. (mm)* | *Lateral. Def. (mm)* |
|---|---|---|---|---|---|---|---|
| Nc | Ref. | RS1 | 28 | 142.92 | 495 | 4.8 | 0.61 |
| | Clayey | N2 | 60 | 327.24 | 490.32 | 3.45 | 1.63 |
| | | N3 | 150 | 440.23 | 509.76 | 4.79 | 0.525 |
| | | N4 | 240 | 254.16 | 433.08 | 3.75 | 0.98 |
| Ns | Sandy | NS2 | 60 | 158.4 | 439.2 | 7.43 | 1.56 |
| | | NS3 | 150 | 111.24 | 529.56 | 5.23 | 0.74 |
| | | NS4 | 240 | 552.96 | 552.96 | 3.25 | 1.28 |

**Table 6.** Results of the Experimentally Tested Cubes and Cylinders.

| *Group Name* | *Soil Type* | *Time of Test (days)* | *Comp. Strength (Cube), (MPa)* | *Tensile Strength (Cylinder), (MPa)* | *Absorption %* | *Voids %* | *Density Mg/m³* |
|---|---|---|---|---|---|---|---|
| Nc | Ref. | 28 | 27.75 | 2.712 | 5.2061 | 11.566 | 2.3457 |
| | Clayey | 60 | 27.365 | 2.034 | 3.1967 | 7.3703 | 2.383 |
| | | 150 | 26.723 | 3.3502 | 4.6073 | 9.7224 | 2.2076 |
| | | 240 | 24.48 | 3.3815 | 4.6942 | 10.665 | 2.3755 |
| Ns | Sandy | 60 | 27.767 | 2.0865 | 3.3125 | 7.5891 | 2.3665 |
| | | 150 | 27.013 | 2.7194 | 4.3789 | 9.3165 | 2.2212 |
| | | 240 | 29.75 | 3.9142 | 4.9412 | 11.037 | 2.3435 |

### 3.2 Effect of aggressive solution of clayey soil

To understand the effect of sulfate and chloride on normal reinforced concrete columns, specimens buried for 60, 150, and 240 days in clayey soil to investigate its behavior. First visible crack was observed in the Normal Reinforced concrete columns group (Nc) at top and bottom of these columns at applied load of (142.92, 327.32, 440.32, and 254.16) kN for columns (RS1, N2, N3, and N4) respectively, as shown in *Figure 2*. The cracks propagated and grew, through the load progressing. At the final stages of loading the columns reached ultimate loads at (495, 490.32, 509.76, and 433.08) kN for columns (RS1, N2, N3, andN4) respectively. The failure occurred gradually with increased loads in the developed ages while the longitudinal deflection decreased, thus decrease in the area under the curve of load - deformation, which led to the columns became less tough and ductile with time, *Figure 3* shows that the deflection was decreased with increasing buried time for the columns tested under the same loading, which means decreasing the ductility and toughness. As seen in the *Figure 4* the lateral deflection increased over time at the same load for example (400 kN), these increments were the reflection of the decrements in the longitudinal direction. From *Table 5,* it can be seen that, the ultimate load lightly enhanced until 150 days of buried time, while the harmful effected of sulfate and chloride was seen after this time, where the cracking load, and failure load of N2, N3 and N4 with respect to RS1 were (229, 308, and 171 %), and (99, 103 and 87%) respectively.

**Fig. 2.** First cracks for group (Nc) (* the load factored by 3.6)

A possible reason for this increment in the lateral and longitudinal deflection is the effect of the sulfate and chloride in clayey soil on the strength of columns during the buried time.

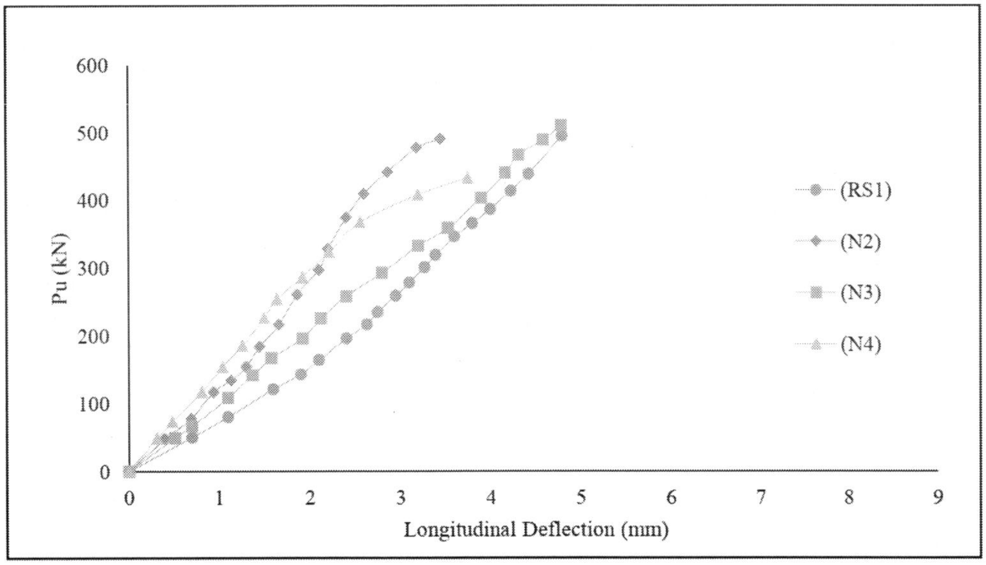


**Fig. 3.** Load – Longitudinal Deflection behaviour of **Normal** RC. column buried in clayey soil

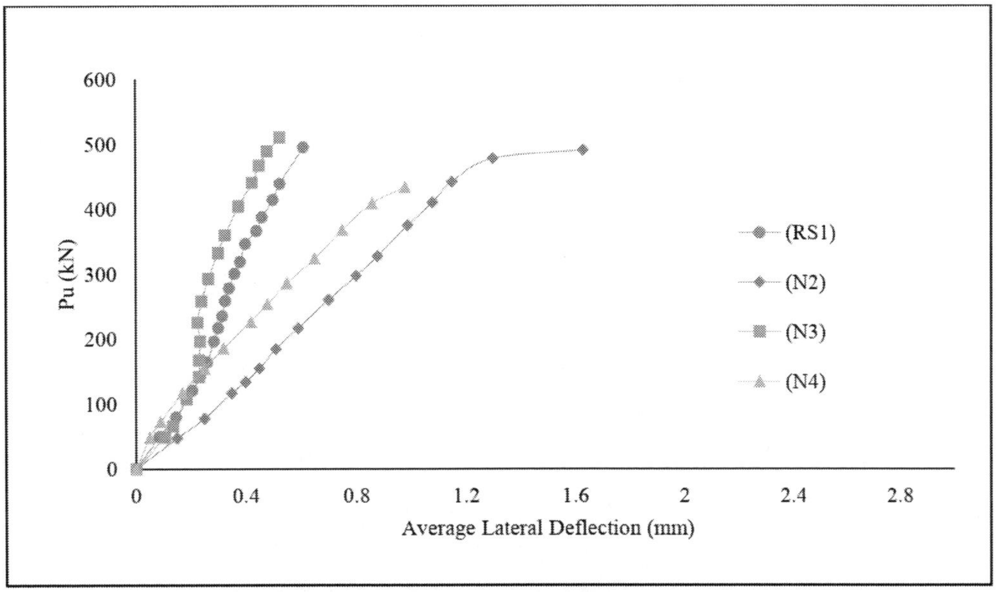


**Fig. 4.** Load – Lateral Deflection behaviour of **Normal** RC. column buried in clayey soil

After 60 days columns in the clayey soil, the normal mix remained conservative on its behavior and was not affected by sulfate salts during duration of 60 days. Visually there is no penetration of sulfate inside columns in this age. At the age of 150 days buried in the soil it notes an improvement in the strength of the concrete columns of the normal mix with a decrease in ductility, this may be due to the low proportion of sulfates in the soil at this age. The percentage of increase of normal mix strength was 2.98% with respect to reference column (RS1). By visual vision there is a penetration of sulfates in the normal column for depth (3 cm). At the age of 240 days of burial in the clayey soil, the concrete specimen's strength decreasing due to the high percentage of sulfates in the soil as in *Table 4.* The decreasing percentage of the strength in normal columns was 12.51% with respect to reference columns. Visually there is a penetration of sulfates for depth (3.5 cm).

### 3.3 Effect of Aggressive Solution of Sandy Soil

The columns were buried for 60,150, and 240 days in sandy soil to understand the effect of sulfate on reinforced concrete columns. The first visible crack in the normal Reinforced concrete columns group (Ns) was observed at top and bottom of these columns at applied load of (142.92, 158.4, 111.24, and 552.96) kN for columns (RS1, NS2, NS3, and NS4) respectively, NS4 failed suddenly without appearing cracks, with load sequence cracks grow and became wider. With continuous loading, the columns reached ultimate loads at (495, 439.2, 529.56, and 552.96) kN for columns (RS1, NS2, NS3, andNS4). See *Figure 5*.

Gradually failure occurred except column (NS4) which failed suddenly, it was observed for 60 days age of the column the compressive strength decreased and then it increased in the developed ages to reach its large strength at 240 days age. This means there was improvement in strength with 6.98% and 11.71% in 150 and 240 days ages respectively compared with reference strength, these happened because the concentration of sulfate decreased in sandy soil at 150 and 240 days as shown in *Table 4*. In the developed ages deflection decreased in longitudinal directions with time, due to which the columns became less tough and ductile because of the decrease in the area under the curve of load - deformation, so at last age it was found that for the same loading values, there is a decrease in the longitudinal deflection due to decrease in the toughness and ductility, these conclusions were covered by *Figure 6*. As seen in the *Figure 7* the lateral deflection increases over time at the same load for example (380 kN), these increments were the reflection the decrements in the longitudinal direction. From *Table 5*, it can be seen that, cracking load, and failure load of NS2, NS3 and NS4 with respect to RS1 are (118, 77, and 386 %), and (88, 106 and 111%) respectively. The effect of the sulfate in sandy soil on the strength of columns during the time of buried is a possible reason for this increment in the lateral and longitudinal deflection.

**Fig. 5.** Failure pattern for group (Ns) (* the load factored by 3.6)

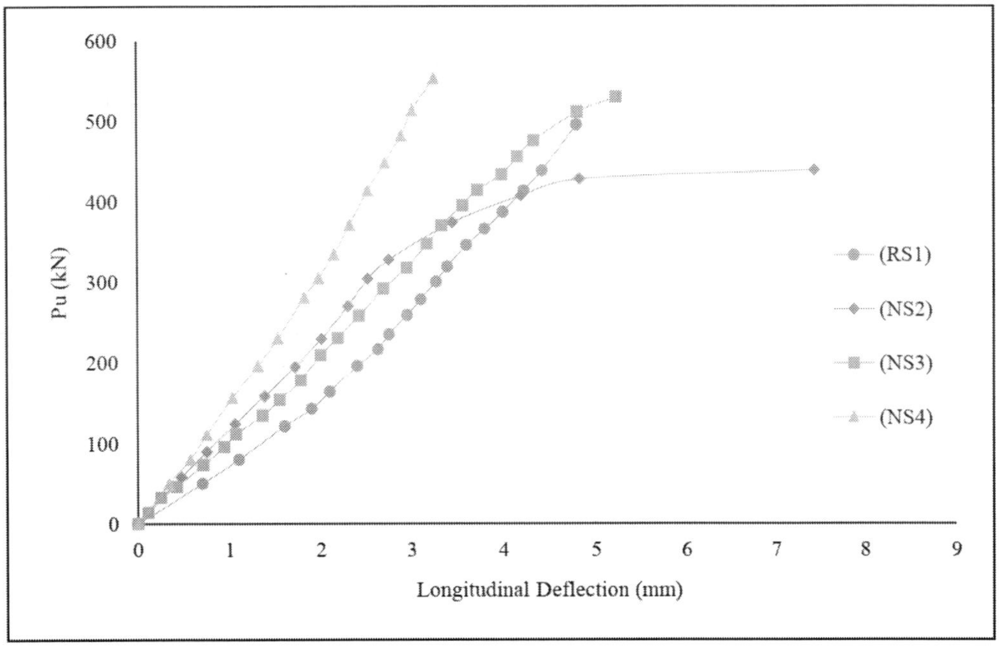


**Fig. 6.** Load – Longitudinal Deflection behaviour of **Normal** RC. column buried in sandy soil

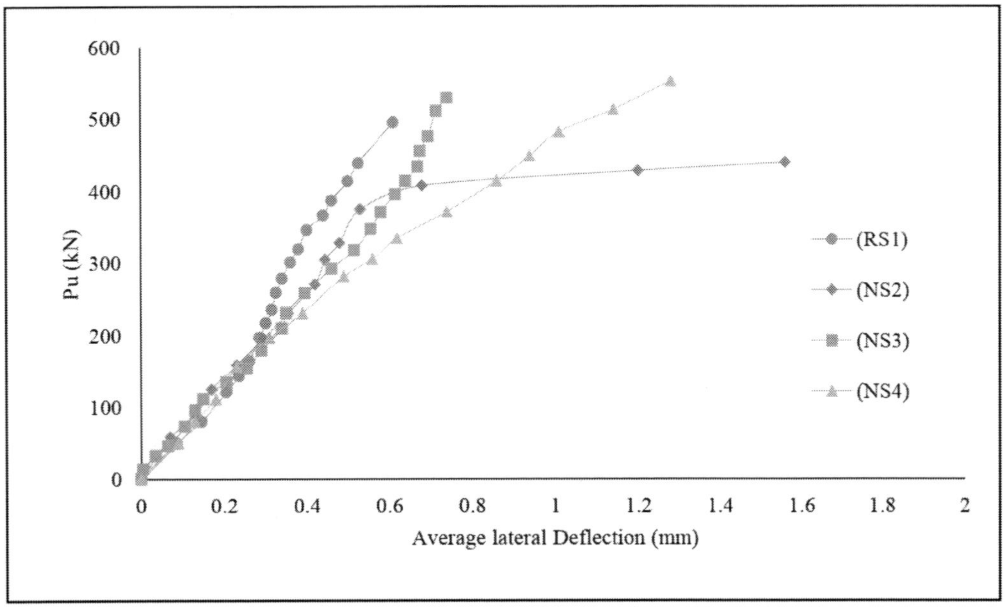


**Fig. 7.** Load – Lateral Deflection behaviour of **Normal** RC. column buried in sandy soil

At test age of 60 days high strength decreasing was recorded for normal reinforced concrete columns which reached to 11.27% compared to the reference strength, the failure was gradual in concrete columns. There is no penetration of sulfate inside columns in this test age. At 150 days the strength was enhanced, the increase became 6.98 %. From figures of load-deformation curves it can be seen the failure happened gradually. There was penetration for sulfate to the columns and its value was (4) cm. Finally, at 240 days the increasing in strength was 11.71%, the failure was suddenly, and the penetration was (5 cm).

### 3.4 The Effect of Different Soils

The columns in sandy soil were more ductile than the columns in clayey soil but, generally the behavior of columns was relatively similar with a difference in the value of deformation and compressive strength recorded in each test, these are shown in *Figure 8 to 10*.

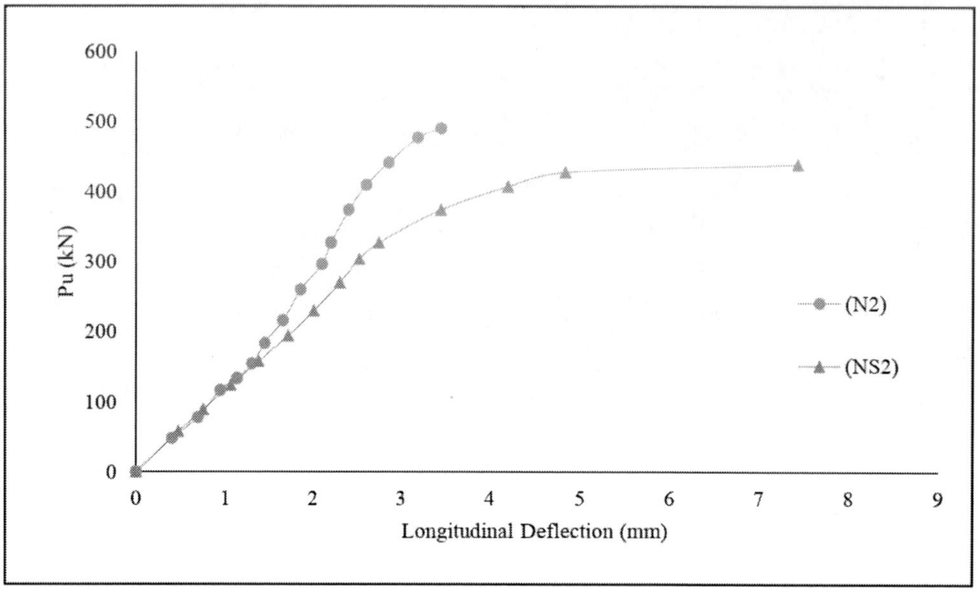


**Fig. 8.** Load – Longitudinal Deflection behavior of **Normal** RC. column buried in clayey and sandy soil for **60 days**

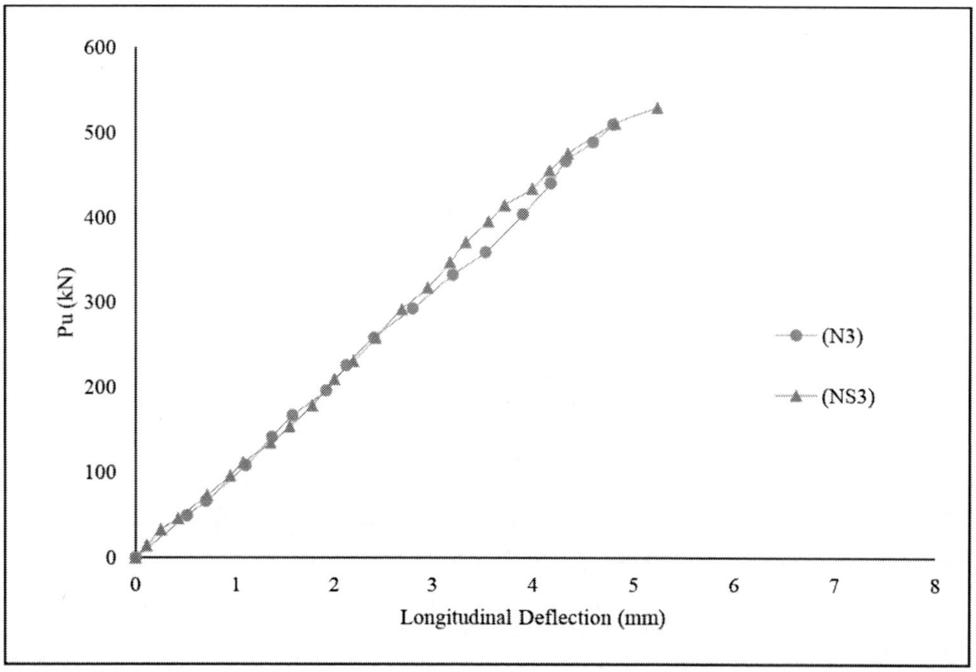


**Fig. 9.** Load – Longitudinal Deflection behavior of **Normal** RC. column buried in clayey and sandy soil for **150 days**

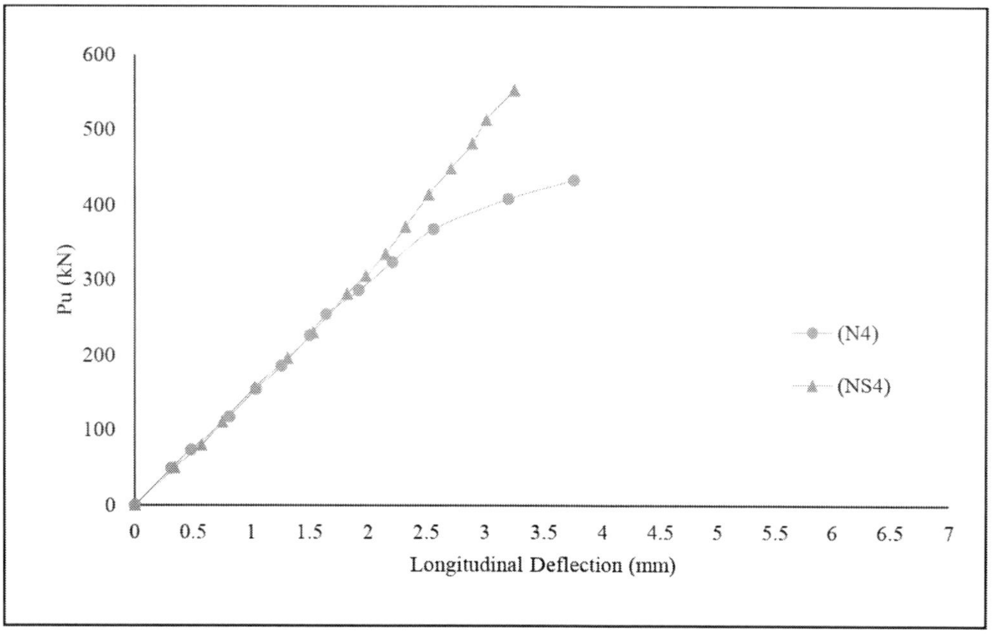


**Fig. 10.** Load – Longitudinal Deflection behaviour of **Normal** RC. column buried in clayey and sandy soil for **240 days**

## 4 Conclusions

Based on the test results obtained from experimental program and the dissection of the results, the following conclusions can be drawn:

1. Ductility decreased with time, for models buried in clayey and sandy soils.
2. Strength of normal RC. columns decreased by (12.51%) for age (240 days) and increased it by (2.98%) for age (150 days), while it`s not effected at age (60 days), all these changes were for columns buried in clayey soil, where the strength decreased by (11.27) % for age (60) days while it enhanced in age (150, 240) days (6.98, 11.71) % respectively, compared with reference columns, for columns buried in sandy soils.
3. The failure was gradual failure in normal RC. columns.
4. Decreasing the percentage of absorption ratio with time in normal concrete cubes, the highest ratio recorded at (240 days) age were (5.09, 9.83) % for sandy and clayey soils respectively, compared with reference cubes.
5. Decreasing the voids ratio with time in normal concrete cubes, the highest ratio recorded at (240 days) age were (4.57, 7.79) % for sandy and clayey soils respectively, compared with reference cubes.

## References

1. M.D. Cohen, B. Mather, Materials Journal 88, 1, 62-69 (1991)
2. K. Obla, C. Lobo, L. Lemay, Specifying concrete for durability. National Ready Mix Concrete Association, USA, (2006)
3. ASTM/C150, Standard Specification for Portland Cement, in American Society for Testing and Materials, ASTM International West Conshohocken, Pa (1999)
4. Iraqi/Specification, No.5 Portland Cement (1984)
5. Iraqi/Specification, No. 45, Aggregates from Natural Sources for Concrete and Construction, Central Organization for Standardization andQuality Control COSQC (1984)

# Evaluation of p-delta effect in structural seismic response

*Sardasht* Sardar[1, 2, *], and *Ako* Hama[1]

[1] Komar University of Science and Technology, Sulaymaniyah, Iraq
[2] Housing Research Center, Faculty of Engineering, UPM, Malaysia

**Abstract.** Numerous recent studies have assessed the effect of P-Delta on the structures. This paper investigates the effect of P-Delta in seismic response of structures with different heights. For indicating the effect of P-Delta, nonlinear static analysis (pushover analysis) and nonlinear dynamic analysis (Time history analysis) were conducted by using finite element software. The results showing that the P-Delta has a significant impact on the structural behavior mainly on the peak amplitude of building when the height of the structures increased. In addition, comparison has been made between concrete and steel structure.

## 1 Introduction

One of the major parameters in tall building analysis is the structural height. As the height of building increases, the structure requires more attention to be given. Several studies indicated that the structural height has a significant effect on P-Delta value. The studies showed that P-Delta effect on low and medium column height is lesser than slender columns [1].

Gaiotti and Smith, (1989) proposed a method to analyze the effect of P-Delta in tall buildings. It has been shown that computing P-delta is crucial. In particular, for low structural stiffness and tall buildings [2]. It was found that ignoring the effect of P-Delta leads to a loss of 16% of load carrying capacity when the structure get subjected to repeated loading. Furthermore, the difference between static and repeated loading was 13% of load carrying capacity due to P-Delta effect [3]. Therefore, Non-linear Dynamic Analysis can be used to determine the contribution of P-Delta in the structural analysis. Kwak et al. (2007) investigated the effect of P-Delta in slender reinforced concrete columns based on the principle of the layered section method. The correlation between the influence of axial force, P-Delta effect, and the effect of vertical earthquake have been assessed. The method was successfully verified [4]. Davidson et al. (1992) indicated that the strength required to support influence of P-Delta is up to 50% of the basic design strength of the beams at lower third of the building. To calibrate the effect of P-Delta two factors must be investigated before incorporating P-Delta effect into design, which are: structural ductility demand and ground motion period [5]. For this purpose, Malaysian Earthquake ground motion has been used in this study. Malaysian Earthquake ground motion is considered as low to medium seismicity region [6]. However, there are many limitations which might be implemented in order to consider the effect of P-Delta such as: the total story drift of three one-fourth scale reinforced concrete frame structure models exceeds 1/85 rad [7]. Yen (1993) introduced a method for direct analysis of nonlinear behavior of slender beam-columns based on the strain control techniques. A strain control technique was developed to obtain the updated flexural stiffness directly for lateral deflection estimation [8]. Fenwick et al. (1992) reviewed thousands of inelastic time history analyses on single degree of freedom to examine the effect of P-Delta. It is shown that the influence factors are ductility, ground motion duration, level of damping, and period of structure [9].

Few experimental research works have been conducted on the effect of P-Delta. Consequently, it is very difficult to show the effect of P-Delta accurately and efficiently. In particular, lack of literature regarding the influence of P-Delta in relation to structural height was another barrier in order to identify its impact on P-Delta effect. The main aim of this paper is to determine to what extent height variation of building is adverse to building design.

## 2 Finite element modeling

Ten different concrete and steel models (different height) were created by using finite element software. Two methods of seismic analysis were used, which were nonlinear dynamic procedure (NDP) and nonlinear static procedure (time history analysis and pushover analysis respectively). All models were subjected to Malaysian earthquake excitation. Furthermore, the performance level of plastic hinges was examined to show the impact of P-Delta. The models were analyzed and simulated according to ASCE 7-05 [10], and FEMA [11].

* Corresponding author: sardasht.sardar@komar.edu.iq

© The Authors, published by EDP Sciences. This is an open access article distributed under the terms of the Creative Commons Attribution License 4.0 (http://creativecommons.org/licenses/by/4.0/).

## 3 Nonlinear dynamic procedure (time history analysis)

At least three ground motions should be used to compute the acceptance spectra because the calculated response is very sensitive. In addition, mathematical model for time history analysis are identical to those of spectra response analysis. Consequently more than one ground motion has been used. The time history analysis allows seeing the response of the structure in time dependent history (see Fig.1).

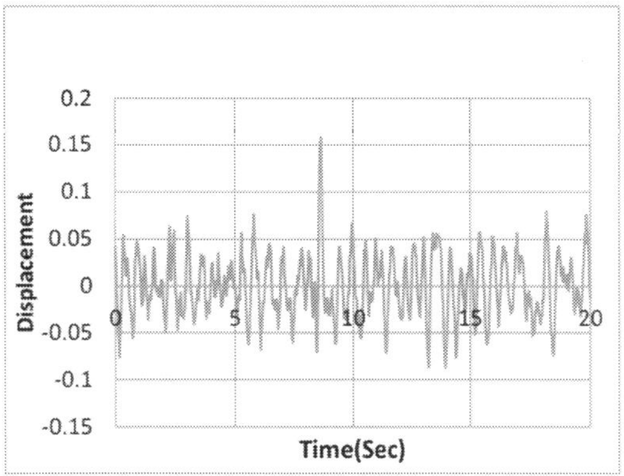


**Fig. 1.** Malaysian time history

## 4 Nonlinear static procedure (pushover analysis)

Impulse force was applied to the highest joint A1 (see Fig. 2) of each model. The steps of the increased loading are due to the characteristics of structural element. At the beginning, plastic hinges happens. As the applying force continues, the number of plastic hinge occurrence increases until the last stage of plastic hinge performance occurs. P-Delta effect was included since pushover analysis is an indicator of structural health and safety.

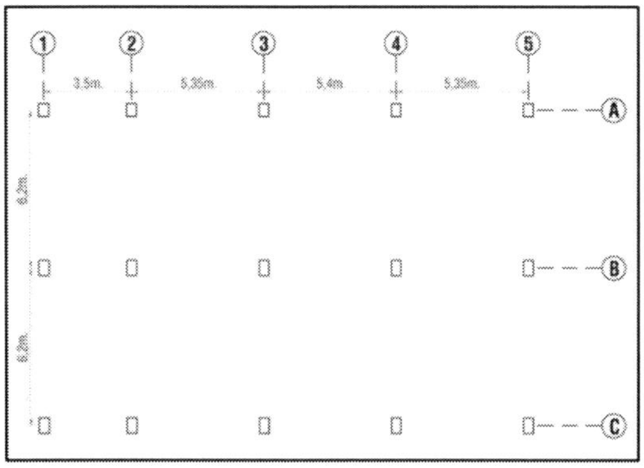

**Fig. 2.** Site Plan of the mode

## 5 Concrete models

Five concrete models have been developed with different heights. The main variable parameter was the number of story levels. The models were subjected to First, Malaysian ground acceleration, and Second, Pushover force. The physical properties of all concrete models were same. Columns and beam properties have been shown in Table 1.

**Table 1.** Concrete section properties

| Sections | Size(mm) | Steel bars #9 |
|---|---|---|
| All beams are 600 x 400 reinforced with 6#9 | | |
| Column B & G | 600x500 | 6bars |
| Column 1st ,2nd and the rest | 600x500 | 6bars |

### 5.1 Time history analysis

The result of Non-linear Dynamic Procedure (NDP) can be divided into five subdivisions to obtain the result of each model and interpret the effect of P-Delta on amplitude. For all models, structural response was examined by applying Malaysian ground motion once with P-Delta and another without P-Delta to measure its effect on structural response. Fig.3 shows time history displacement of FE model of 5, 10, 15, 20 story concrete models. Time history displacement of 25 story model was not following the same pattern; therefore it has been excluded from almost all the analyses.

### 5.2 Peak amplitude

FE model analysis showed that including P-Delta effect was very severing for 5 story model. The peak amplitude escalated to 68.28mm from 39.27mm (see Fig. 4a). It can be seen the amplitude variation was doubled due to the contribution of P-Delta effect. As the height of the building increased, the effect of P-Delta decreased. For 10 story concrete model peak amplitude changed to 100mm. The effect of P-Delta was 33%. It was observed that due to weight height ratio (see Fig.4 a, and b), the influence of P-Delta has been reduced. As the number of story levels increased, the peak amplitude increased as well. The peak amplitude of 15 and 20 story concrete models became 128mm, and 251mm, respectively.

The results were shown that P-Delta effect contribution leads to increase in the amplitude by 30% and 19.92%, respectively (see Fig.4 c, and d). The amplitude value of 25 story concrete models was dropped to less than 185mm. This is mainly due to catastrophic failure which could be observed from the mode shape (see Fig.5).

https://doi.org/10.1051/matecconf/201816204019

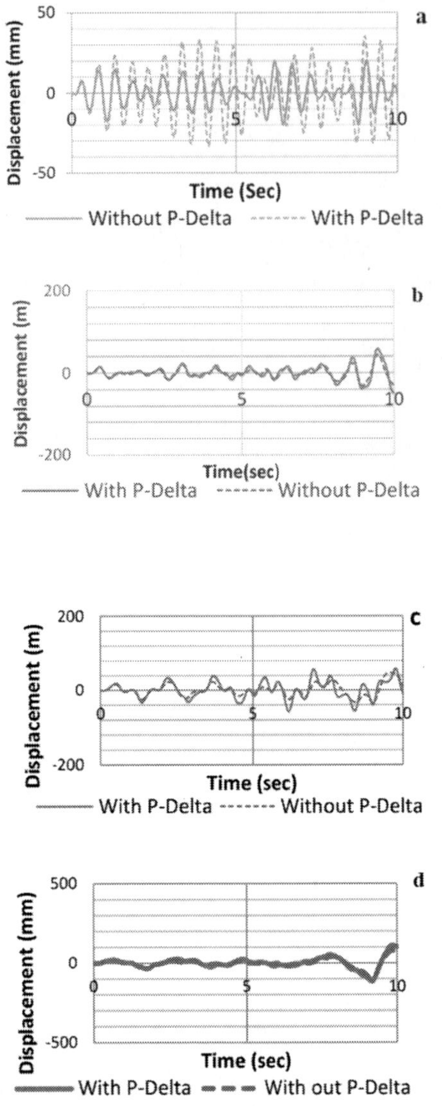


**Fig.3.** Time History Analysis Displacement; a. 5 story level, b. 10 story level, c. 15 story level, d. 20 story level

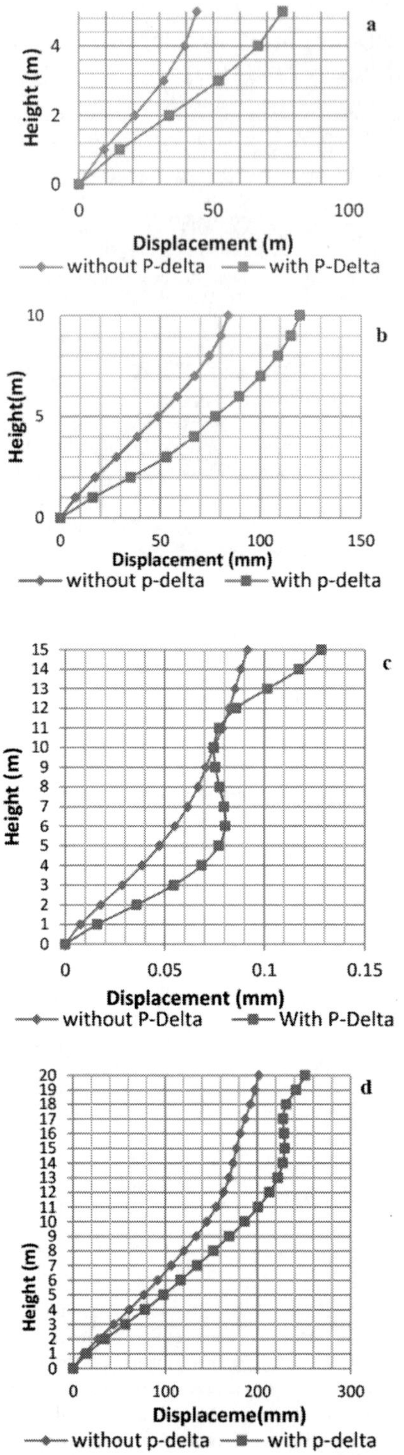


**Fig.4.** Peak amplitude; a. 5 story level, b. 10 story level, c. 15 story level, d. 20 story level

https://doi.org/10.1051/matecconf/201816204019

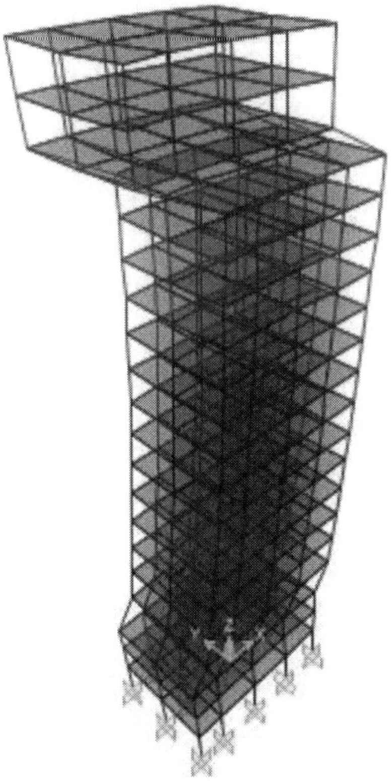

**Fig.5.** 25 story levels 3D mode shape

### 5.3 Pushover analysis

In this section three different concrete models have been investigated to determine the effect of structural height on P-Delta effect. P-Delta effect contribution resulted in increasing number of plastic hinges occurrence from 0 to 56 for 5 story concrete model (see Fig. 6). The result shows that the effect is sever and should not be overlooked without proper design consideration. Apparently, the amplitude was affected by the influence of P-Delta. The effect of P-Delta leads to increase the amplitude by three times (see Fig.7). Fig. 8 shows the effect of P-Delta on the number of plastic hinge occurrence for 15 story concrete model. The study observed that the number of plastic hinges escalated from 16 to 121. The amplitude was also increased by 39.8% due to the effect of P-Delta (see Fig.9).

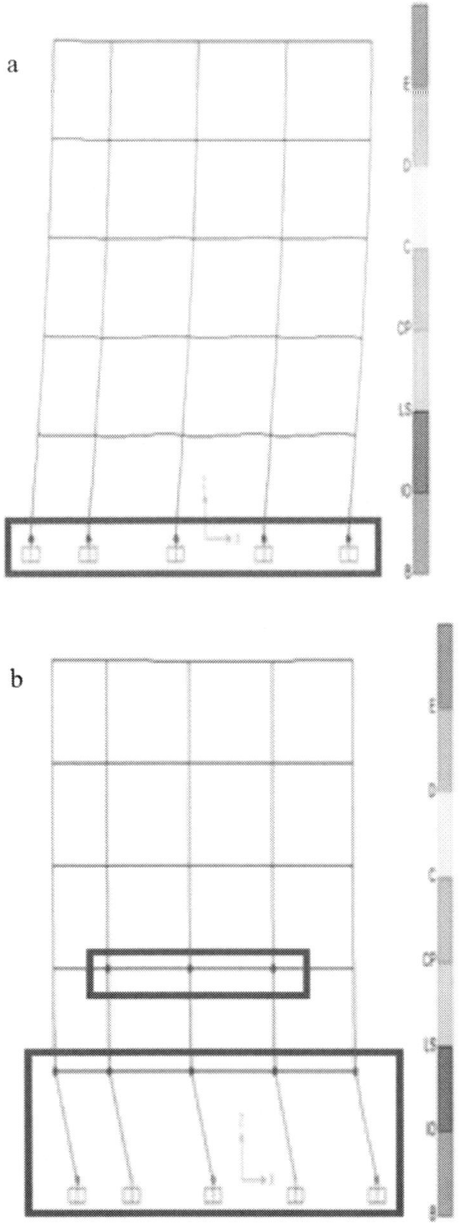


**Fig. 6.** Deformed shape and Plastic Hinge Occurrence. a. Without P-Delta effect, b. With P-Delta effect

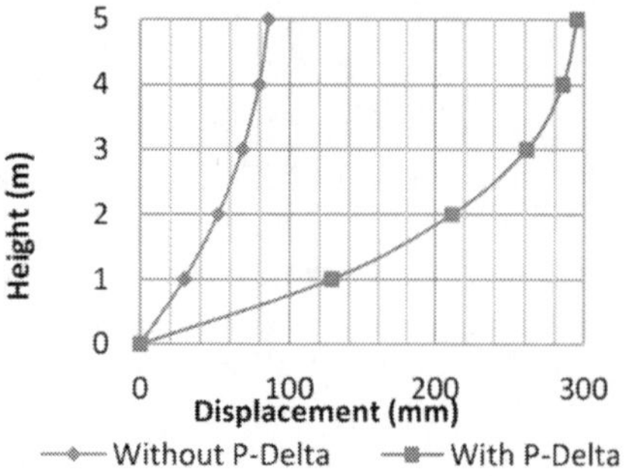


**Fig. 7.** P-Delta effect on peak amplitude

a

b

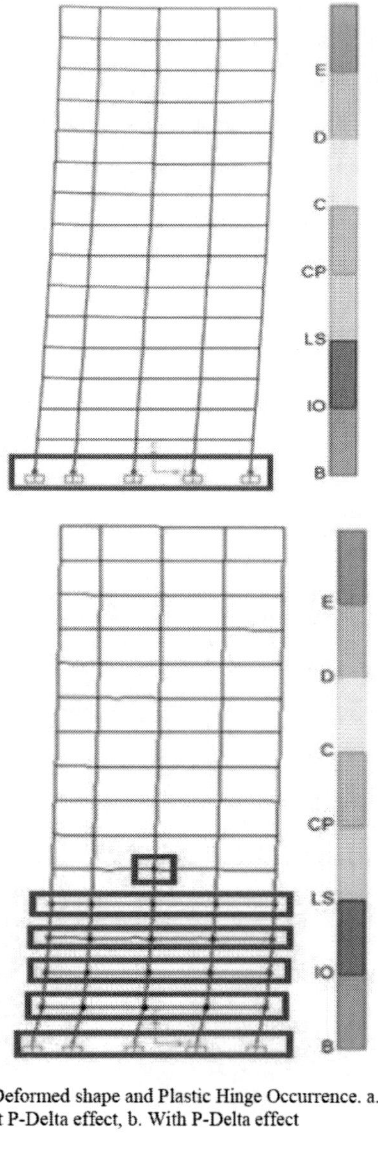


**Fig.8.** Deformed shape and Plastic Hinge Occurrence. a. Without P-Delta effect, b. With P-Delta effect

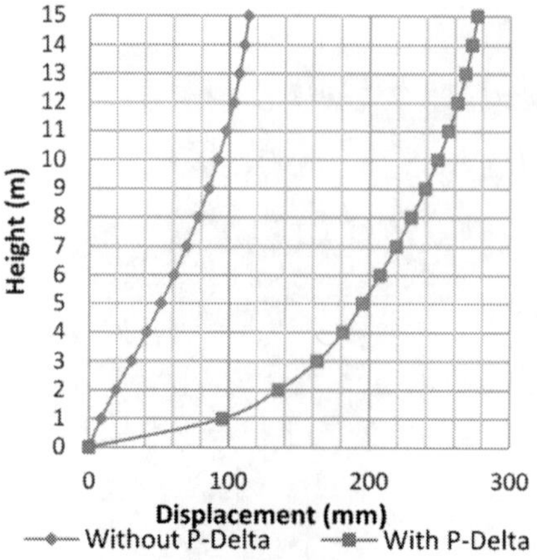


**Fig. 9.** P-Delta effect on Amplitude

Numbers of plastic hinge happening of 25 story concrete models were increased to 206 from 23 due to the effect of P-Delta (see Fig.10). The amplitude is also affected and it increased to 490mm from 345mm (see Fig.11).

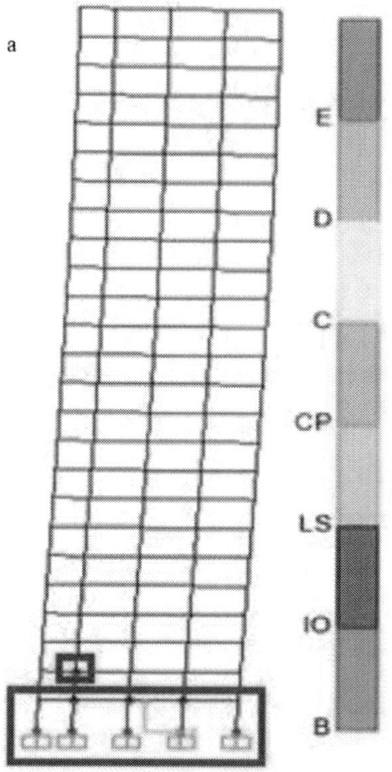

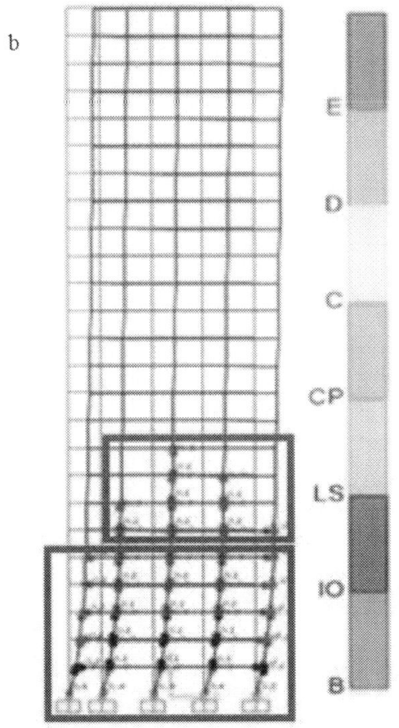


**Fig.10.** Deformed shape and Plastic Hinge Occurrence. a. Without P-Delta effect, b. With P-Delta effect

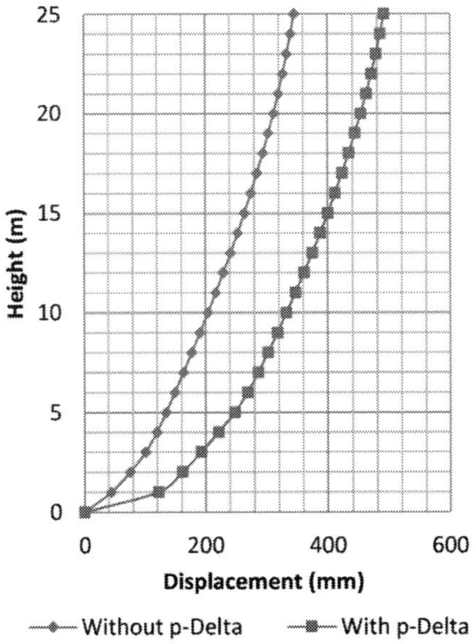


**Fig. 11.** P-Delta effect on Amplitude

## 6 Steel models

Five steel models have been developed with different heights. The main variable parameter was the height. The steel models were subjected to First, Malaysian ground acceleration, and Second, Pushover force. The physical properties of all steel models were same. Both beams and columns sections have different section properties as shown in Table 2.

**Table 2.** Steel section properties

| Sections | Type | Size |
|---|---|---|
| Beams | W | 533X210 |
| Columns | W | 533X310 |

### 6.1 Time history analysis

Non-linear Dynamic Procedure (NDP) has been applied on all the steel models. It was indicated that the time history displacement for all different height models was following the same pattern. The result of each model has been determined and shown briefly to present the structural response. Fig.12 shows time history displacement of all five steel models.

### 6.2 Peak amplitude

It's indicated that peak amplitude for 5 story steel model was higher than 5 story concrete models. For 5 story steel structures the amplitude is 40.5mm, while contributing P-Delta effect to the analysis leads to increase the amplitude to 69.8mm (see Fig.13 a). The effect of P-Delta on the amplitude for 10 story steel model was significantly low (4%) (see Fig.13 b). Eventually, the effect was increased to 21% for 15 story steel model. Furthermore, the peak amplitude was raised to 51.7% which was increased to 317mm from 164.1mm (see Fig.13 c and d). On the other hand, 25 story steel model has failed to provide logical result due to paramount collapse (see Fig.14). The same failure pattern for 25 story concrete model can be observed.

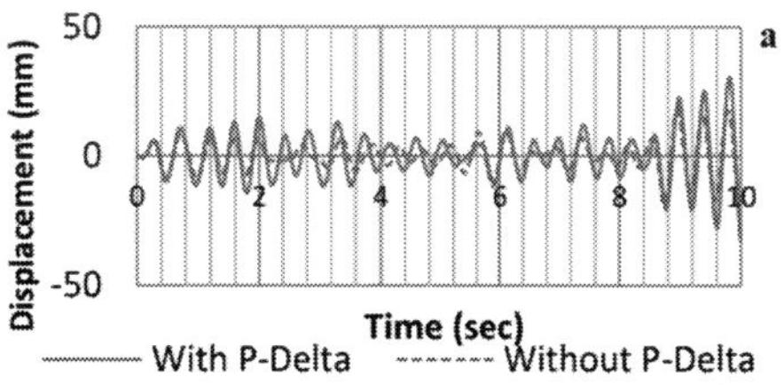

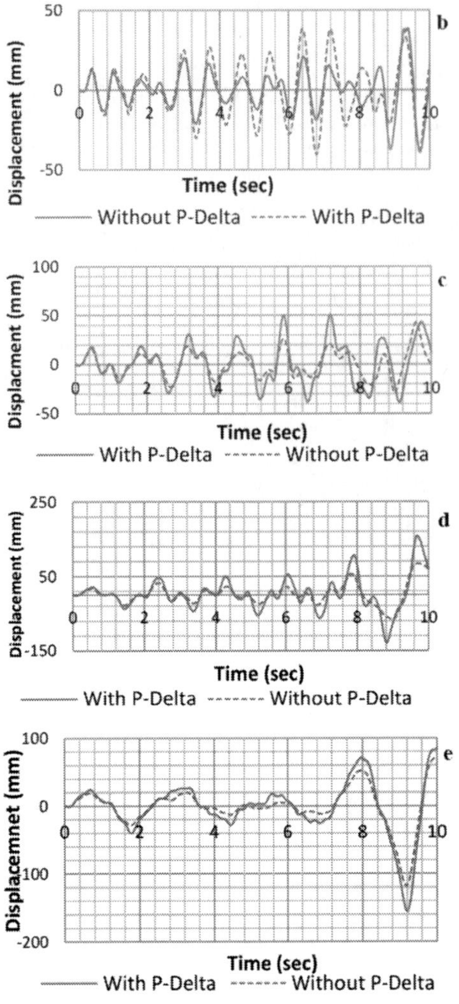


**Fig. 12.** Time History Analysis Displacement; a. 5 story level, b. 10 story level, c. 15 story level, d. 20 story level, e. 25 story level

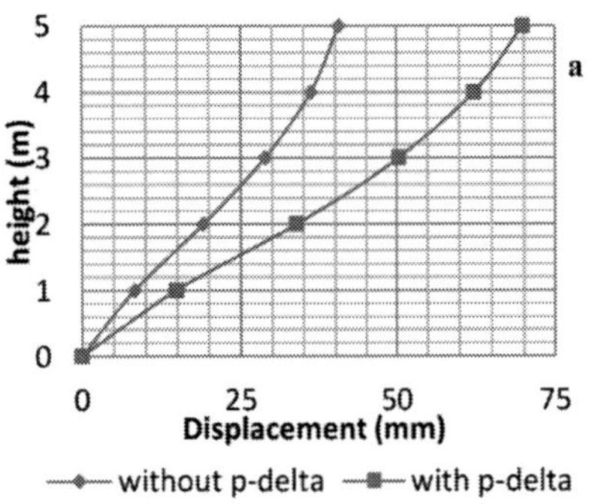


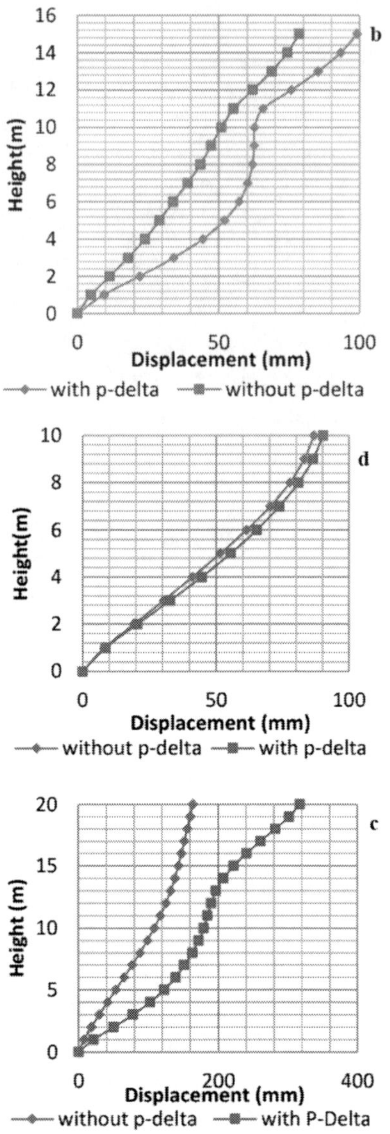


**Fig.13.** Peak amplitude; a. 5 story level, b. 10 story level, c. 15 story level, d. 20 story level

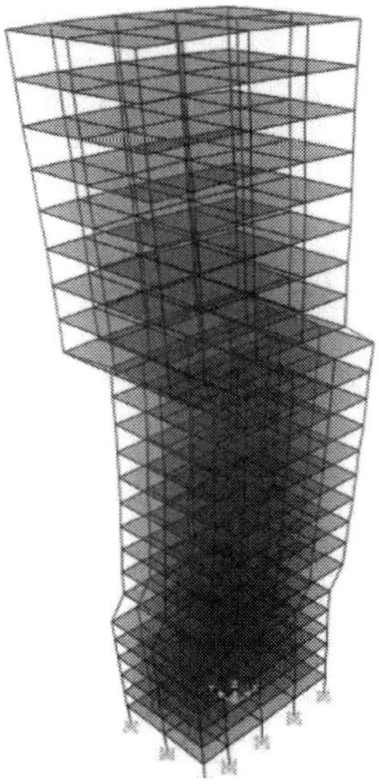

**Fig. 14.** 25 Story steel model which is failed due to excessive height

## 6.3 Pushover analysis

Ignoring P-Delta effect resulted in reducing 6 plastic hinges for 5 story steel model. While for 10 story steel model the difference was increased to 66 plastic hinge occurrence. Further height increment leads to change the performance level of plastic hinge formation from very slight (elastic range) to slight (collapse prevention) according to FEMA-365 description (see Fig.15). Very little changes in the number of plastic hinges occurrence has been observed for 15, 20, and 25 story steel models. It was investigated that the effect of P-Delta was insignificant to the quantity of plastic hinges formation, but plastic hinges performance level moved to beyond strong level (IO-Immediate Occupancy).Fig.16 shows plastic hinge formation distribution throughout the height of 15, 20, and 25 story steel models.

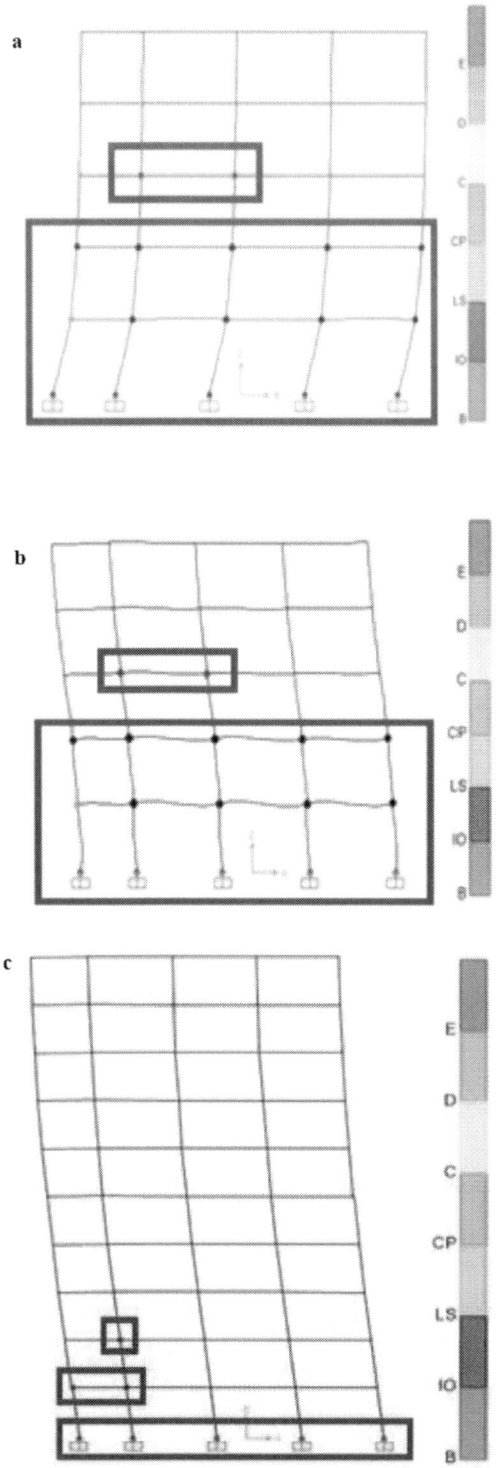

 https://doi.org/10.1051/matecconf/201816204019

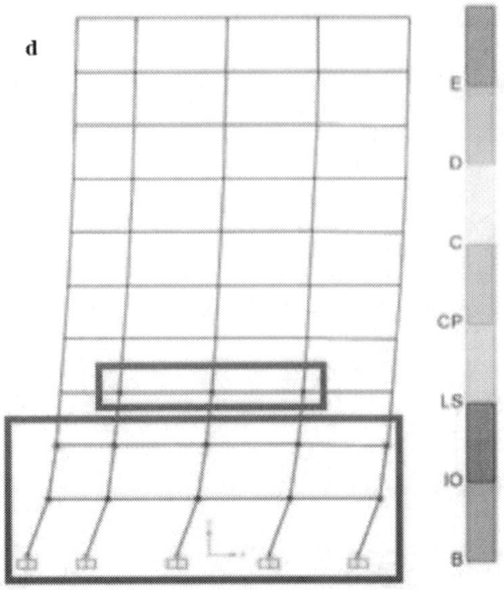


**Fig.15.** Plastic Hinge Occurrence, a. 5 story steel model without P-Delta, b. 5 story steel model with P-Delta, c. 10 story steel model without P-Delta, d. 10 story steel model with P-Delta

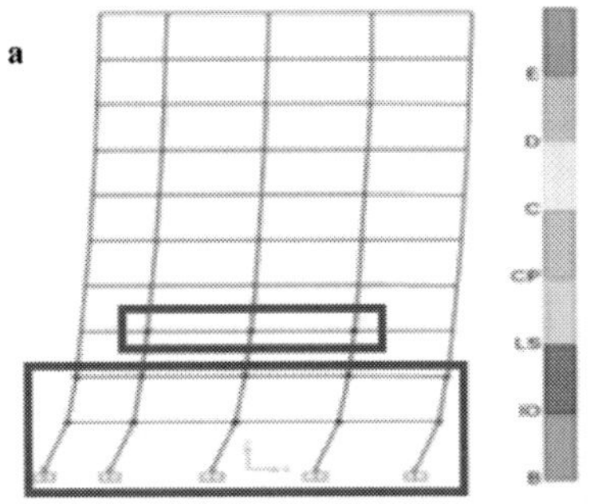


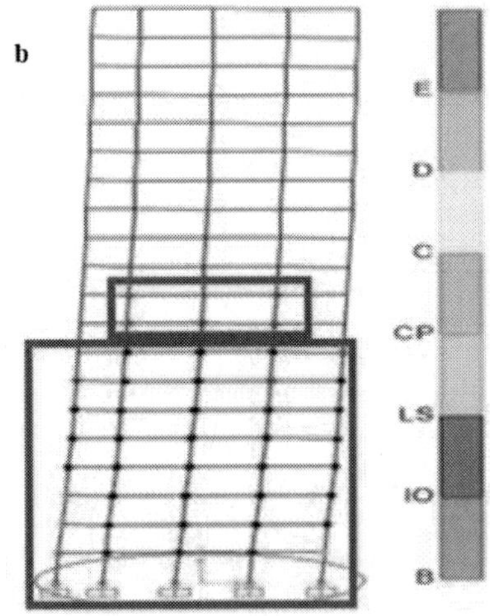


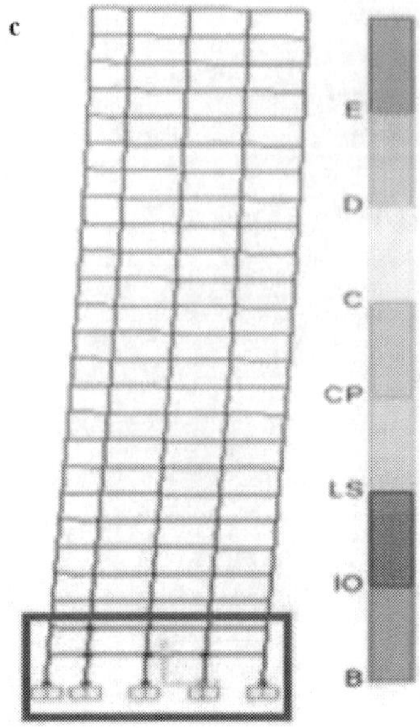


**Fig.16.** Plastic hinges formation, a. 10 story steel model, b. 15 story steel models, c. 20 story steel mode

Amplitude is one of the parameters which might be affected by P-Delta in pushover analysis. The impact of P-Delta was sever and cannot be overlooked without proper design consideration. For instance, for 5 story steel model the effect of P-Delta leads to increase the amplitude by 60mm (see Fig.17 a). It can be seen that the effect of P-Delta was reduced to 22mm, when the height of the model increased to 10 story level (see Fig.17 b). The results showed that further increasing the number of story levels did not affect the amplitude. This might be due to the performance level of plastic hinges occurrences which was beyond immediate occupancy according to FEMA description.

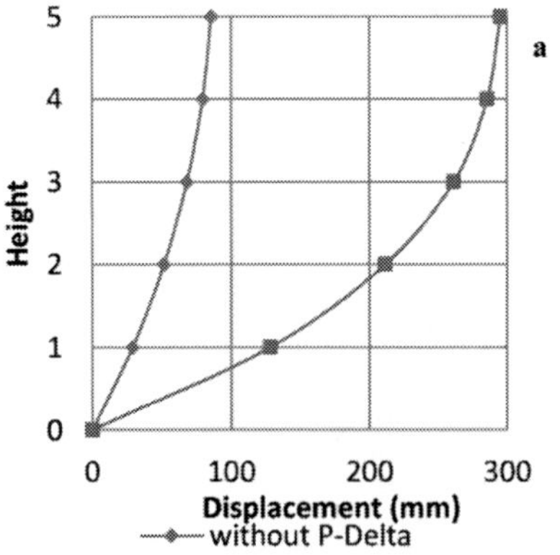

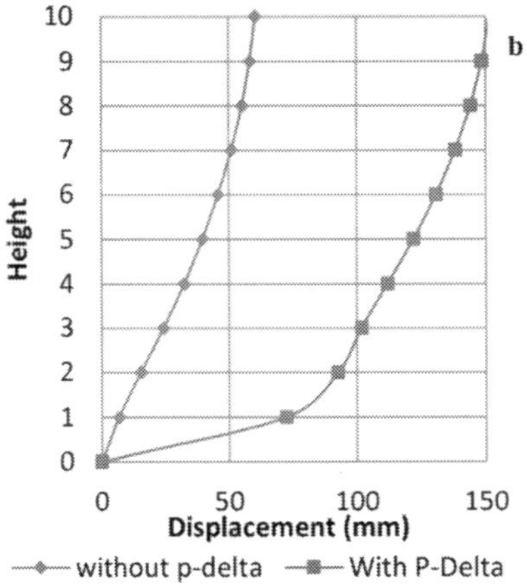


**Fig.17.** Peak amplitude, a. 5 story steel model, b. 10 story steel model,

## 7 Results and comparison

The effect of P-Delta on the amplitude of 5 story steel model and 5 story concrete model was 42%, while the effect of P-Delta on the amplitude of 10 story steel and concrete models were 4.4%, and 33%, respectively. The effect of P-Delta for 15 and 20 story concrete models were decreased to 30%, and 20%, respectively. On the other hand, time history results shown that the effect of P-Delta in 15 and 20 story steel models were 21%, and 52%, respectively. It was observed that the effect of P-Delta due to the height variation of steel structures is more effective than the concrete structures. Structural weight has a major contribution to the effectiveness of P-Delta. In fact, concrete models weight were greater than the weight of steel models, as a result, P-Delta influence became lesser for concrete models. Ultimately, 25 steel and concrete models failed to give rational results regards the effect of P-Delta. This was mainly due to calamities failure of the structure. P-Delta effect had a very clear impact on the quantity and performance level of plastic hinges formation. For 5, 10, 15, 20, and 25 story concrete models the effect of P-Delta were 70%, 60%, 59%, 46%, and 30%, respectively. It was indicated that the P-Delta effect became lesser for higher story levels. Again the reason for this is the height weight ratio. Steel models recorded different outcomes. The height increment is too sever on the quantity of plastic hinges occurrence of steel models. The effect of P-Delta on plastic hinges performance was jumped from 6.67% to 76.7% when the height increased from 5 to 10 story levels. Further height increment leads to structural elements failure. Therefore, the structural elements of 15, 20, and 25 story steel models failed to carry the excessive loads due to P-Delta effect.

## 8 Conclusions

The second order effect (P-Delta) has been investigated to determine the correlation between structural height and P-Delta effect. Two variables have been included in the analysis, which are; Frist- Number of story levels. Second- Different construction materials have been utilized in the analysis such as steel and concrete. The study utilized two types of analysis, time history analysis and pushover analysis. In both analyses, peak amplitude was a reliable indicator to obtain the effect of P-Delta. Furthermore, plastic hinge formation was found to be a good measurement to observe structural performance. To sum up, the height of the building should be considered as a critical factor in P-Delta effect analysis. The study indicates that steel structures are more sensitive towards the height than concrete structures. In addition, performance level of plastic hinges occurrence in concrete models influenced by the height inversely, while failures have been recorded when the height is increased up to fifteen story levels for steel models.

## References

1. H. Scholz, *J. Struct. Eng.* **113**, 3, 534–545 (Mar. 1987)
2. B. R. Gaiotti, S. Smith, *ASCE J. Struct. Eng.* **115**, 4, 755–770 (1989)
3. B. H. Scholz, **116**, 8, 2070–2082 (1991)
4. H.-G. Kwak, J.-K. Kim, *Eng. Struct.* **29**, 11, 3121–3133 (Nov. 2007)
5. B. J. Davidson, R. C. Fenwick, B. T. Chung, *Earthquake Engineering ,Tenth World Conference Balkema,Rotterdam*, 3847–3852 (1992)
6. S. Engineering, T. Division, Gathering of Views and Opinions on Seismic Investigations in Peninsular Malaysia – Report on the IEM Workshop on Earthquake ( Part 1 ), (October 2011)
7. M. A. A. Mollick, *J. Civ. Eng.* **CE 25**, 2, 107–119 (1997)
8. B. J. Y. R. Y. I, *ASCE J. Struct. Eng.* **119**, 3, 871–890 (1993)
9. R. C. Fenwick, B. J. Davidson, B. T. Chung, *Bull. New Zeal. Natl. Soc. Earthq. Eng.* **25**, 1, 56–69 (1992)
10. S. a. James R. Harris, *ASCE J. Struct. Eng* (2010)
11. E. Zeller,*FEMA273* (1997)

 https://doi.org/10.1051/matecconf/201816204020

# The use of modal parameters in structural health monitoring

*Ali* Al-Ghalib[1,*], and *Fouad* Mohammad[2]

[1]College of Engineering, Al-Mustansiriyah University, Baghdad, Iraq
[2]School of Architecture, Design and the Built Environment, Nottingham Trent University, Nottingham, UK

**Abstract.** The concrete is liable to damage due to various stresses which compensate its adequacy and safety. The estimation of remaining strength in reinforced concrete beams when subjected to increased loading action utilizing vibration parameters is investigated. For this reason, three beams are loaded statically close to failure in various increasing load steps and then repaired. The beams are all of same dimensions, but are different in strength and range of defects introduced to each sample. Following each loading step, the experimental modal testing is utilized to collect the vibration parameters (natural frequency, damping ratio and mode shapes) of each beam when tested under free support boundary conditions. The use of vibration parameters for the purpose of damage identification are known to be an elaborate and lengthy process. On the other hand, they are successful for the structural health monitoring given that they are able to provide global on-site automated continuous monitoring. The paper features post analysis procedures for experimental modal measurements of three concrete samples to obtain and correlate the basic modal parameters (natural frequency, modal damping and mode shapes). The results of the extracted modal parameters and their combination are exploited in this research as quantified identification parameters. This paper concludes that modal parameters are successful in determining the location and quantity of structural degradation, when holistic approach considered through a system.

## 1 Introduction

Thenumber of aged and damaged civil engineering infrastructure has increased significantly in the past few decades, and therefore, the subject of structural health monitoring (SHM) becomes the focus of many researchers. Much research concentrates on applying an automated monitoring method that replaces or complements the current onerous visual inspection and other local non-destructive testing (NDT) methods. Important part of research has particularly given to study the vibration-based paradigms for damage identification purposes [1 and 2].

The vibration-based monitoring systems are developed and constantly adapted to the monitoring of civil engineering structures. These systems are advantageous of global and automated nature. Quantitative analysis on any type of structure can be performed by using the assembled measurements from these systems. It is widely accepted that these are the only methods can present a mathematical model for the actual structure [3]. The possible combination of the theoretical model together with the experimental measurements makes these vibration-based methods more enticing. The current wide interest in the use of monitoring systems in the SHM has been attributed to[1] the advancement and low expenses of measurement facilities [4].

The concept of arranging the damage identification assessment in four levels, namely: detection, localization, quantification and prediction was first proposed by Rytter in his doctoral study back in 1993 [4 and 5].

The identification of damage in RC beams inflicted with progressive cracking behavior by utilizing vibration data is considered difficult and is still unpredictable in the published literature. Therefore, this research endeavors to explore the extent of some vibration parameters for damage identification when applied to reinforced concrete (RC) beams. For this purpose, three RC beams of laboratory size were cast and set to various damage situations. For each particular state condition the Experimental Modal Analysis (EMA) was implemented for the beam under free-free support condition. An appropriate post analysis procedure for EMA measurement of the three RC beams was conducted in order to obtain the natural frequencies, damping ratios and mode shapes. The study exploits the modal parameters and their combinations in order to interpret changes in vibration properties with the corresponding drop in bending stiffness. The facts established in this research can be widened to involve large scale structures. Having experienced the advantage of the vibration-based techniques in SHM, the author wishes to share the concepts of these prominent systems with the engineering community in Arab and Middle East regions.

* Corresponding author: ali.alghalib@uomustansiriyah.edu.iq

© The Authors, published by EDP Sciences. This is an open access article distributed under the terms of the Creative Commons Attribution License 4.0 (http://creativecommons.org/licenses/by/4.0/).

## 2 Laboratory tests

### 2.1 Characteristics of concrete beams

The degree of damage was captured and evaluated through testing three RC beams modeled with the following levels of faults.

All the three test beam samples were of the same dimensions, which are 2.0m long, 140mm deepand 90mm wide. The first beam sample (Beam A) has been investigated under seven loading cases: intact, 3kN load, 6kN, 10kN, 13kN, 16kN and externally repaired. Second sample (Beam B) has been investigated under same loading conditions as for Beam A. However, the third sample (Beam C) has been investigated under three different statuses, that is: intact, 10kN and 14.5kN load. The load status and hence, the related cracking conditions have beencreated by using the procedure of four-point static load. Table 1 shows the dimensions and physical properties of the three beams. The experimental investigations of the beams as well as the dynamic test results represent some outcomes of the author's PhD thesis [6].

### 2.2 Vibration test

One of the main issues could be faced in vibration tests is the interaction of support conditions and its influence on the measurements. Therefore, in small scale samples, a freely supported testing model is usually chosen so as to remove the effect of support rigidity [7]. According to this point, the RC beams have been tested in a free-free support condition, as shown in Fig. 1. In this procedure, each beam has been hung by the means of elastic cords attached at 0.46m from each end. Three accelerometers have been fixed beginning from the one end at stations 0.2m apart. These three accelerometers then moved to fresh three points to cover the whole length with a total of 11 measurement points. The beam has been hit and translation accelerations considered normal to the direction of the elastic cords. The Brüel&Kjaerimpact hammer type 8208 generated a dynamic force with a sensitivity of 0.0002V/N containing frequency components up to 1.0kHz. The first four modes been covered with the force excitation. The response measurements collected using Brüel&Kjaer 4514 dynamic acceleration sensor with sensitivity of 0.00985V/g and a frequency range up to 10000Hz. Both output (response) and input (force) signals have been captured using a data acquisition system of 4100 samples, taken at a rate of 8180 samples per second.

### 2.3 Experimental modal analysis

The EMA is a dynamic test whereby it is generally utilised to obtain the specifications of the structural elements with regard to the modal parameters. In this particular study and for each measurement point, the amplification, filtration and transformation into frequency domain have been conducted to independent four force and response time signals.

The accuracy of Frequency Response Function (FRF) from four averages was checked through constant checking of features, including repeatability and reciprocity.

**Table 1**. Specifications of The Three Test RC Beams

| Test Beam | Cube strength $N/mm^2$ | Mass $kg/m^3$ | $E_d$* $10^3$ MPa* | $G_d$* $10^3$MPa† |
|---|---|---|---|---|
| Beam A | 20.4 | 2240 | 19.1 | 8.13 |
| Beam B | 29.0 | 2270 | 27.5 | 11.5 |
| Beam C | 27.3 | 2320 | 25.0 | 10.4 |

*$E_d$: Dynamic (tangent) modulus of elasticity.

†$G_d$: Dynamic (tangent) shear modulus of elasticity.

**Fig. 1**. Freely supported RC beam under dynamic test.

The coherence functions, which measure the correlation between the output and the input signals [8], and are used as a dataverification procedure, were exploited in this study for the FRFs of the system.

In the post-test analysis stage, FRF measurements are basically analysed so as to find the modal parameters of the structures. The simplest modal post-test analysis method is known as peak-picking method, and it is used along with the single degree of freedom systems to obtain the modal parameters. In this method, the response measurements are apparently controlled by resonant modes at their corresponding natural frequency [9], which is the case in this study.

## 3 Damage identification using modal parameters

### 3.1 Effect of damage on natural frequency

The frequency corresponding to the peak amplitude of the FRF is taken as the resonance frequency (ω) of a particular vibration mode. For each one of the three beam samples and for all damage conditions, the results of the natural frequencies for the first bending modes were found. The progressive of cracks in concrete beams will cause a drop in the stiffness, and hence, the drop stiffness is manifested as a decrease in the natural frequencies.

For Beams A, B, and C, the outcomes of frequency variation of every loading status from the reference frequency of the intact status are shown in Fig. 2. The variation in frequencies for the first mode is noteworthy compared to the variation in higher modes. This observation suggests that the first mode of vibration is influenced more by the crack pattern compared to the higher modes.

For instance in Beam A, close to the complete failure stage (P=16kN), the first four bending frequencies dropped respectively by 20.6%, 11.8%, 9.6% and 11.6% from the intact frequency. The first bending frequency corresponding to 6kN load case drops about 6% from the intact first frequency, even though the beam shows no signs of visual cracks. The first bending frequency resumed 7.2% of frequency variation when retrofitting work was done to the beam.

The modal frequencies of sample B were considerably dropped with the development of cracks of the increased loading compared with beforehand beam sample (beam A), as shown in Fig. 2. The number and width of cracks driven through the loading steps were ample. This behaviour was manifested as bigger drops in frequencies.

When the beam was loaded with 6kN, the first four frequencies decreased by 8.8%, 3.8%, 4.6%, and 3.3%, respectively compared to the initial intact state. The big differences in natural frequencies conclude that vibration-based methods and natural frequencies (in particular) are able to detect faults at its early stage, without any apparent visible change. In additions, the first four natural frequencies of the four bending modes decreased respectively by 30.5%, 18.9%, 14.6% and 16.5, when the beam was loaded by (P=15.5kN).
For Beam C, the discrepancy of frequencies with growth of cracks is shown in Fig. 2. The measured frequencies of the first four modes decreased continuously responding to the increasing damage.

The abovementioned results can be generalized to draw threshold by which the structure is considered risky when the variation in frequencies of first modes drop to a certain level. It is fortunate that results similar to those of present study were reported in other research employed to real size RC structures [10, 11, and 12].

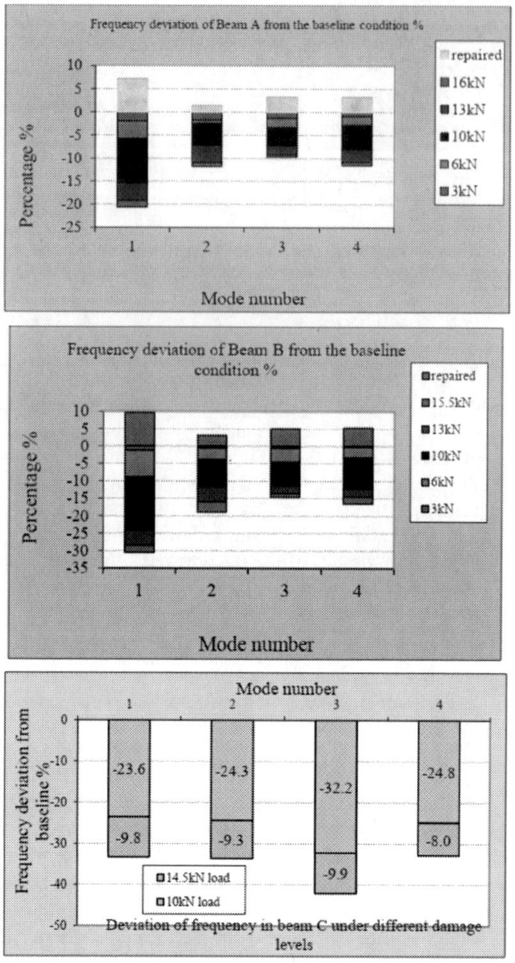


**Fig. 2.** Variation of natural frequencies for beams A, B and C for different damage conditions.

### 3.2 Effect of damage on modal damping ratio

Of the various modal parameters, damping ratio is supposedly the less reliable quantity to estimate because it is quite sensitive to the background noise. The modal damping ratios of the current study were estimated from the width of the resonance peak, because the beams are lightly damped elements. The half-power bandwidth technique is utilized herein because it is common, quick and widely practiced [13].

The usual presence of cracks in structural elements often increases the amount of the energy dissipation. This phenomenon will affect and control the height and width of peaks of the FRF measurements.

After a round of experiments and calculations, the normalised modal damping ratios for Beam A for the four modes are given in Fig. 3. Although the evaluation of damping ratios is error-prone, consistent results for modal damping were found. For the consecutive loading steps, changes in modal damping ratios were regular as they increased with the magnitude of load. The modal damping ratios gone higher by 39%, 6.3%, 33%, and 78.1% for the four modes of vibration, respectively compared to the intact beam when the beam was loaded to 16kN load. The non-regular rounds of application of impact forces in magnitude, direction and position could as well cause unexpected behavior in damping results. It might seem illogical for the rate of change of damping with the gradual development of cracks appeared non-uniform because the damping ratios were estimated assuming linear structure fitted to non-linear vibration frequency functions.

The Beam B showed clearer changes than the case of Beam A. This beam exhibited obvious increase in modal damping ratios with the developed damage. In Fig. 3, the normalised damping ratios for the first four bending modes with evolution of damage are given in reference to the intact beam. At the final loading stage whereby P=15.5kN, the modal damping increased in the order of 143%, 96%, 114%, and 141% for the four modes compared to the virgin beam stage. Similar behaviour to that described in section A, trends of modal damping for the first mode increased significantly responding to the increasing damage.

Obviously, for the retrofitted beam, the first four damping ratios decreased by 40%, 33%, 23% and 31% from the severely damaged (15.5kN) load stage for the four modes, respectively. The effect of modification added to the repaired beam was sufficiently exposed.

When a four-point static load of 10kN was carried out to beam C, intrinsic distributed cracks was experienced, and this increased the damping ratios significantly, as shown in Fig. 3. For this load step, the modal damping ratios were increased by 60%, 8%, 213% and 555% from the virgin beam reference line for the four modes, respectively.

### 3.3 Effect of damage on mode shape

The third outcome in the modal analysis campaign that is the mode shapes of vibration was further assessed as damage identification factor. Prior to the identification process, the main mode shapes have to be taken out from FRF measurements. One of the approaches used for extracting the mode shapes, where a little effort is needed, is the approach known as the (Quadrature Picking). In this approach a light pairing between the modes is assumed [14]. The observed mode shapes of the three beams and for all loading steps were generated all through in this work.

In view of state deterioration of a structure, the change in the amount and/or shape of a mode shape reflects change in strength. The evaluation of various mode shapes represent different state conditions with respect to one of virgin model, qualitative information is likely to achieve regarding the site and scale of structural changes [13]. Meanwhile, the effect of the environmental conditions has little effect on the mode shapes compared to their effect on frequencies [13].

As anexample to exercise the ability of mode shapes to localize the damage, Fig. 4 shows the overlaid first four flexural mode shapes for beam A.The damage was very noticeable in the first and second mode shapesin comparison to the third and fourth mode shapes.

As well, different damage statuses were sufficiently recognized by those mode shapes. The mode shape for the retrofitted beam managed to place the associated status within its reasonable place. Even though damage (change in condition) was light, the change of mode shape was light as well but subtle.

The changes are liable to become larger with damage level. However, the mode shapes of the first bending mode shape for the beam A produced the most substantial changes in the magnitude of mid span where visual inspection gave no indication of cracking at early loading stages. In addition, the changes of mode shape were not restricted at the damaged area of the beams, but were distributed over the entire length. Results of mode shapes from investigations on pre-stressed concrete bridge girder performed by Huth et al. [15] and results of this study are alike.

## 4 Damage quantification via stiffness calculation

### 4.1 Theory

The work of Maeck and De Roeck[16] was concluded that the location and size of damage in one-dimensional element could be estimated by using the curvature of mode shape instead of the deflection curve. The method, which is named: Direct Stiffness Calculation (DSC), exploits two modal parameters to identify the distributed rigidity in an element, which are natural frequency and mode shape. The basic bending-curvature relationship of the beam element is used for this objective. Certainly, this technique is applicable for small amount of deflection in each section, and can be defined mathematically as [15]:

$$\frac{d^2y}{dx^2} = y'' = \frac{M}{EI} \quad (1)$$

In which the bending moment at any position is denoted by M, E is modulus of elasticity, I is the second moment of area and $y''$ is the second derivative (curvature) of the deflection mode shape function.

In order to supplement the analysis for free vibration loading under damping-free condition, a static simulated inertia loading system may be expressed in the following form:

$$K.y = \omega_i^2\, m\, y \quad (2)$$

The right-hand side of the preceding equation represents the modal inertia force for specific mode shape (i) affecting abeam of rigidity (K) with the purpose of generating a displacement vector (y).

The flexural moment at section position x (i+1) of a beam using the particular curvature mode shape and the inertia loading condition is derived and given in the following Equation 3, [16, 17]:

$$M_{i+1} = M_i + V_i(\Delta x) - \frac{1}{2}\int_{x_i}^{x_{i+1}} \omega_i^2.\gamma.A.y(x).(\Delta x).dx \qquad (3)$$

Where γ is the mass density, the cross-sectional area is (A), V and M represent the shear and bending moment, respectively.

### 4.2 Damage quantification

By using the calculated modal bending moment distribution utilizing Equation 3, the dynamic bending stiffness (EI) can be evaluated using Equation 1. From these computations, thriving insight on the location of spread damage followed by computation of its severity can be achieved.

Fig.5 shows the consecutivedrop in flexural dynamic stiffness with the applied load in beam A, as an example of applying the direct stiffness approach. The trend was approximatelysteady. A comprehensive insight on this case study has shown that the dropinflexural stiffness gonefurther from the load points to reach the supports. As a good sign of right approach, the theoretical dynamic stiffness (EI) of the starting point (intact) beam was in good agreement with the measured one.Whereas the theoretical stiffness was 840kN.m2, the measured stiffness is 870kN.m2. Contrary to what is visually seen,in the load step corresponding to a vertical load of 6kN, the mid span stiffnessdecreased by almost 18%,where no bending cracks were developed.

At the final load, when the beam was practically falling apart, the mid span strength reduced by 47% compared withstrength of the reference case. The beam in this stage was gone throughsevere cracking conditions.Furthermore, when the cracking area was retrofitted, the beam recovered35% of its strength compared with the final damaged step.

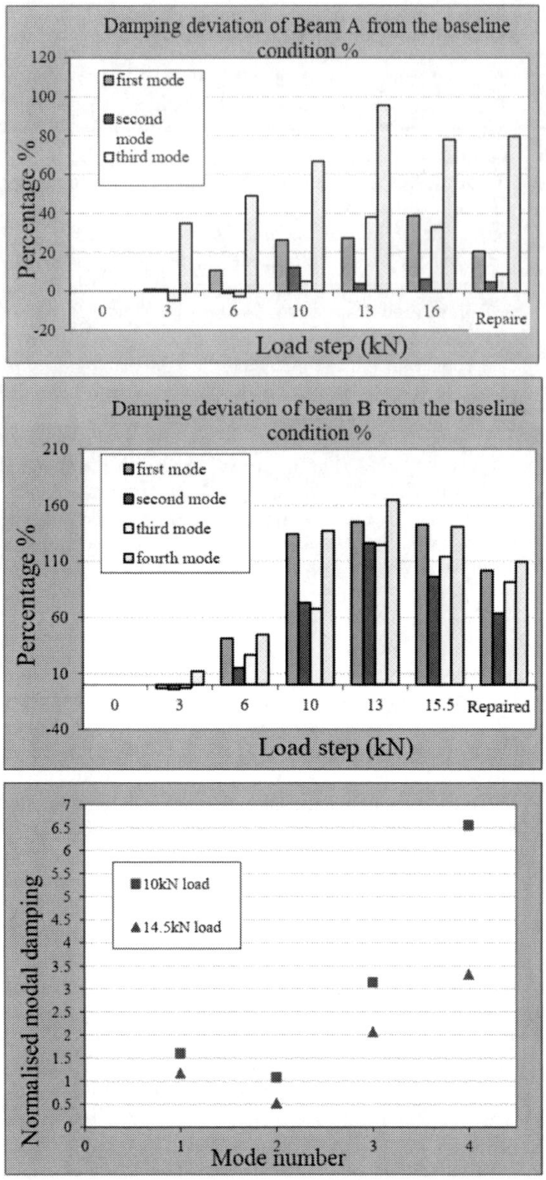


**Fig. 3**. Variationof modal damping for beams A, B and C for different damage conditions.

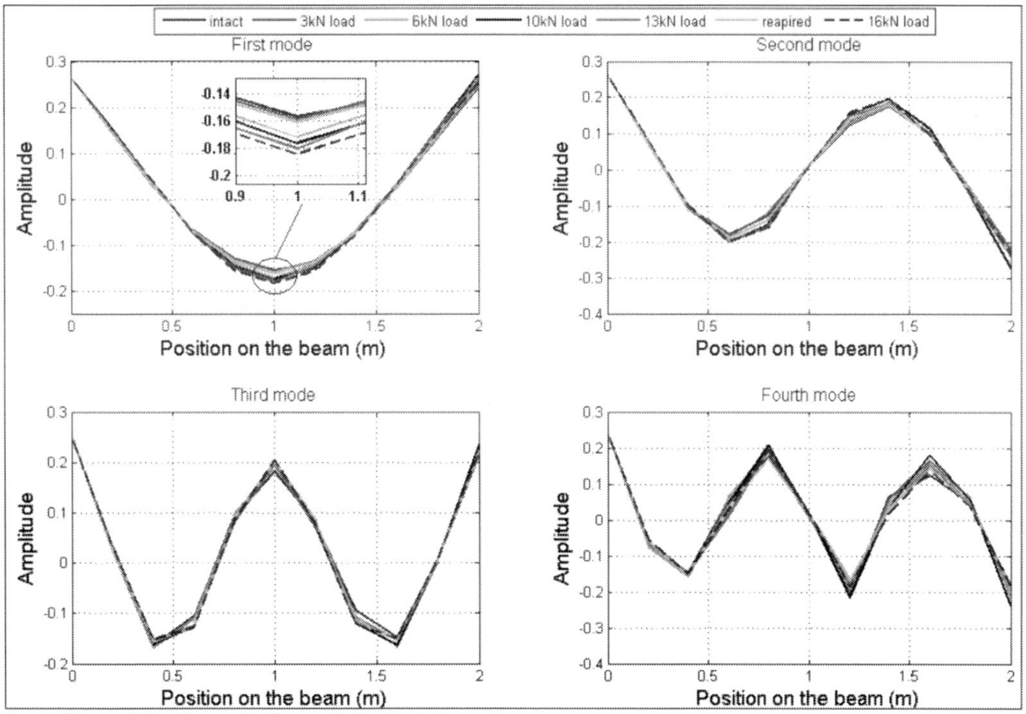


**Fig. 4**. Observed first four mode shapes of vibration in beam A.

## 5 Summary

SHM is now a deep-rooted area of both investigation and application along with a decent solid knowledge and research literature. There are many examples on successful applications on using vibration-based monitoring and identification systems in various disciplines. This technique can simply be applied to real-world civil engineering structure where clear and deep understanding for the structure can be gained. Of specific significance for this research is to determine the location and extent for structural degradation through combining frequencies and mode shapes of a system in a proper analytical model. A comprehensive modal parameter survey has been drawn for three RC beams deliberately damaged by active static loading. The damping parameters are usually estimated through fitting the non-linear dynamic behaviour to a linear model of FRF. Therefore, the consistencyof outcomes may be defeated, as it has been seen in the damping ratios from this study.

When the complete deterioration was looming, the biggest resonance change in a RC beam was about 30% compared with the intact state condition. This drop was accompanied with maximum reduction in the flexural stiffness of about 52%. Because the model of produced damage was mostly contributed to the first mode, the biggest decrease in strength and frequency basically occurred in this mode. Although such results give no concrete insight on degradation, they underpin the understanding on abnormal performance and point to faulty behaviour to correct and rectify error.

In the subject of estimating the quantity of damage, the direct stiffness method is one of the useful techniques can be advantageous used in beam structures. Mode shapes outperforms their rival modal parameters for damage sites localization, especially for the first two modes, given that the amplitude increased with the load in relation to the initial virgin state.

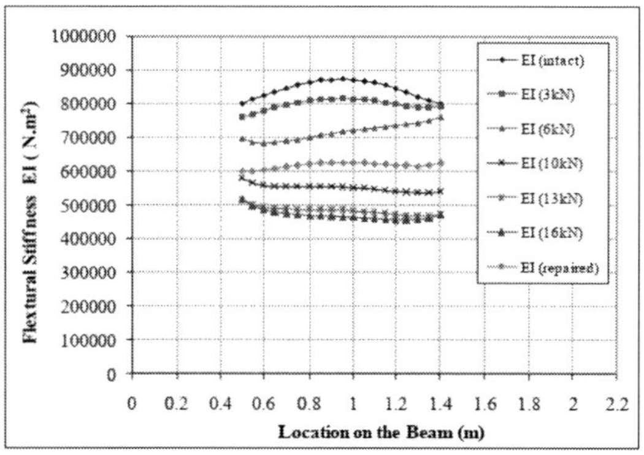


**Fig. 5.** Flexural stiffness distribution for beam A featured from first mode shape.

## References

1. C. Farrar, D.A. Jauregui, Journal of Smart Material Structures **7**, 704-719 (1998)

2. C. Farrar, S. Doebling, D. Nix, Philosophy Transactions Royal Society **359**, 131-149 (2001)
3. L. Zhengsheng, J. Swanson , A. Helmicki, V. Hunt, ASCE, Journal of Bridge Engineering **10**, 2 (2005)
4. C. Farrar, K. Worden, Philosophy Transactions of Royal Society **365**, 303-315 (2007)
5. E.P. Carden, P. Fanning, Journal of Structural Health Monitoring **3**, 4, 355–377 (2004)
6. A.A. Al-Ghalib, PhD. Thesis, Nottingham Trent University, Nottingham, UK (2013)
7. D. Ewins, Research Studies Press Ltd., Baldock, Hertfordshire, England (2000)
8. E. Figueiredo, G. Park, J. Figueriras, C. Farrar, K. Worden, Report LA-14393, Los Alamos National Laboratory, US (2009)
9. N. Haritos, J. Owen, Journal of Structural Health Monitoring **3** 2, 141-163 (2004)
10. A. Mordini , K. Savov, H. Wenzel, Structural Engineering International **17**, 4, 352–358 (2007)
11. W. X. Ren, G. De Roeck, ASCE, Journal of Structural Engineering **128**, 1, 96-104 (2002)
12. J. Unger, A. Teughels, G. De Roeck, The Journal of Structural Engineering **132**, 11, 1691-1698 (2006)
13. H. Wenzel, *Health Monitoring of Bridges*, John Wiley & Sons Ltd, West Sussex, United Kingdom (2009)
14. S. Gade, H. Herlufsen, H. Konstantin-Hansen, Application Note 3560, Brüel & Kjær, Denmark (2005)
15. O. Huth, G. Feltrin, J. Maeck, N. Kilic, M. Motavalli, Journal of Structural Engineering, ASCE **131**, 12, 1898-1910 (2005)
16. J. Maeck, G. De Roeck, Journal of Sound and Vibration **225**, 1, 153-170 (1999)
17. A. Al-Ghalib, F. Mohammad, M. Rahman, J. Chilton, *Damage and Repair Quantification in Reinforced Concrete Beams Using Vibration Data*, Proceedings of the 4th International Conference on Concrete Solutions, September 2011, Dresden, Germany ( 2011)

# Behaviour of hybrid concrete columns under axial compression loads

*Ali* Resheq *

Building and Construction Engineering Department, University of Technology, Baghdad, Iraq

**Abstract.** Hybrid Concrete Column (HCC) is composite column consists of two types of concrete, Normal Concrete (NC) and Self Compacted Concrete (SCC). The main objective of this work is to study the effect of outer to inner diameter ratio of hybrid concrete on the ultimate strength of column under the axial loading. Ten scaled columns (150 mm diameter and 600 mm length) were fabricated and cast with different ratio of NC to SCC and tested under axial loading. The test results were presented in term of loading and the axial and lateral strains at mid span of column length. It was found that the ultimate strength of column increased from 281 kN to 605 kN (215%). Also it was found that the strength of column increased when the SCC is in inner thickness rather than in outer thickness of the column and the strength increased with increasing the inner thickness of SCC.

## 1 Introduction

Hybrid concrete columns can be defined as a compression member consisting of composite SCC and NC. Hybrid columns have been used to carry high loading with small cross section area as an alternative to ordinary reinforced concrete during the past decades. Concrete columns can be reinforced by wound composites, which are providing circumferential confinements forces as well as higher compressive strength [1, 2]. In last years, using of self compacting concrete SCC gave the structural engineers very favourably. Due to SCC properties many advantages mentioned and developed structural applications can be realizing. Elchalakani, et. Al. was reported that CFDSST column has almost the same advantage as normal CFST member. It has lighter weight, higher bending stiffness's and has best cyclic performances [3]. Tomii found that the axial loads capacity of concrete filling steel tube was better than the nominal capacity, which defined as the total of compressive strength of the unconfined steel and concrete [4]. Wei et. al., was verify the nominal strength of CFDSST column got from experimental work. He had find that the confinement effect of inner and outer tube, the strength from experimental tests are greater than the numerical as shown in figure (1). [5].

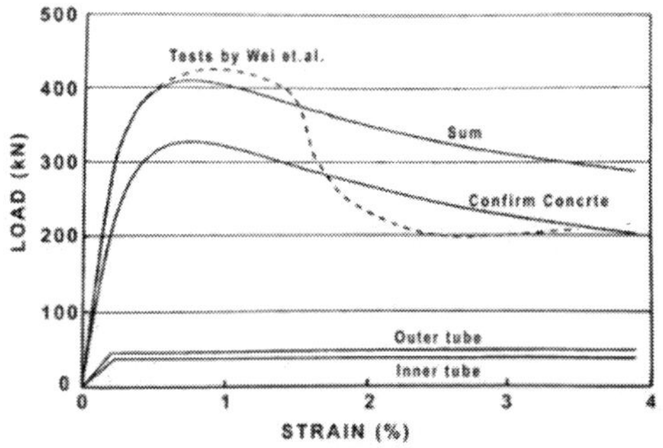


**Fig. 1.** The Comparison between analysis and test results of strength for CFSSDT column, [5].

## 2 Experimental works

The main objective of this experimental work is preparing and testing 10 HCC columns divided to two groups having different ratio of SCC to NC as inner and outer cross section area. Both groups were tested under axial loading. All columns have same length 600 mm and same outer diameter 150 mm and the inner diameter was changed between NC and SCC. Table (1) and Figure (5) show the details of the specimens.

* Corresponding author: dr.ali_sadiq@yahoo.com

© The Authors, published by EDP Sciences. This is an open access article distributed under the terms of the Creative Commons Attribution License 4.0 (http://creativecommons.org/licenses/by/4.0/).

**Table 1.** HC specimen Details.

| Group | no. | Column designation | Inner Material (mm) | Inner Diameter (mm) | Di/Do Ratio |
|---|---|---|---|---|---|
| 1 | 1 | C-NC150-SCC0 | NC | 150 | 1 |
| | 2 | C-NC125-SCC25 | NC | 125 | 0.83 |
| | 3 | C-NC100-SCC50 | NC | 100 | 0.67 |
| | 4 | C-NC75-SCC75 | NC | 75 | 0.5 |
| | 5 | C-NC50-SCC100 | NC | 50 | 0.33 |
| 2 | 6 | C-SCC50-NC100 | SCC | 50 | 0.33 |
| | 7 | C-SCC75-NC75 | SCC | 75 | 0.5 |
| | 8 | C-SCC100-NC50 | SCC | 100 | 0.67 |
| | 9 | C-SCC125-NC25 | SCC | 125 | 0.83 |
| | 10 | C-SCC150-NC0 | SCC | 150 | 1 |

Where,
C: Column
NC: Normal Concrete
SCC: Self Compacting Concrete.
Di: Inner Diameter.
Do: Outer Diameter (Do=150mm).

## 3 Materials

### 3.1 Cement

Ordinary Portland cement was used in the experimental works, the chemical composition and physical properties are tested and found comply with the Iraqi Specifications Standards Specifications IQS No.5/1984[6].

### 3.2 Fine aggregate

Natural Sands brought from Al-Akhaider Zone was used in concrete mixes. The fine aggregate had fineness modulus of (2.78) and shaped rounded and smooth texture. The lab results showed that fine aggregate is within Iraqi Specifications Standards Specifications IQS No.5/1984[6] and ASTM C33 specification [7].

### 3.3 Course aggregate

Crushed gravel of Al-Nibay Zone was used in this experimental works. The lab results showed that Course aggregate is within Iraqi Specifications Standards Specifications IQS No.5/1984[6] and ASTM C33 specification [7].

### 3.4 Water

Ordinary tap water was used for both mixes and curing of all specimens used in this experimental work. It was free from oil and organic material.

### 3.5 Limestone powder (LSP)

This material is a white grinding material coming from lime stone in the north of Iraq, usually used for the construction process and as filler of concrete production. Fine limestone powder was used in this experimental work, grinded by bowling technical. Size of particles of the limestone powder was less than 125x10-3 mm, and it was conformed to EFNARC [8].

### 3.6 Superplasticizer

Sulphated Melamine Formaldehyde (SMF) was used in this experimental works. (SMF) was free from chlorides and comply with ASTM C494 [9].

## 4 Mixing procedure for NC and SCC

### 4.1 Normal concrete NC

The laboratory mixing used in this work was according to Neville [10] for Normal Concrete. Table (2) presents the quantities by weight for material used per cubic meter of this mix.

**Table 2.** Mix Properties for NC/$m^3$.

| no. | Material | Weight (kg) |
|---|---|---|
| 1 | Cement | 400 |
| 2 | Fine aggregate | 740 |
| 3 | Course aggregate | 990 |
| 4 | Water | 180 |
| 5 | Superplasticizer | 0.45 |

### 4.2 Self-compacting concrete SCC

The laboratory mixing used in this work was according to Emborg [11] and modified by Al-Jabiri [12] for self-compacting concrete. Table (3) presents the quantities by weight for material used per cubic meter of this mix.

**Table 3.** Mix Properties for SCC/m3.

| no. | Material | Weight (kg) |
|---|---|---|
| 1 | Cement | 400 |
| 2 | Fine aggregate | 797 |
| 3 | Course aggregate | 767 |
| 4 | Limestone powder | 170 |
| 5 | Water | 180 |
| 6 | Superplasticizer | 4.9 |

## 5 Fresh SCC tests

SCC is approached on the ability to flow under own weight without any vibration and the ability to gain the homogeneity without any segregation of aggregate. Slump flow, V-Funnel and L-box tests are done to show the fresh properties of SCC. All results were within Specifications.

## 6 Casting and curing of HC specimens

All columns were cast and cured in the laboratory. The Steel mould was containing 2 parts of half circular section and fastens to the base plate with bolts. Then a steel tube with dimeter equal to the inner diameter of the experimental works was fastening inside the steel mould. 6 bars with diameter 8mm was used as longitudinal reinforcement and bar with diameter 6mm was used as stirrup with spacing 100mm. At first the outer material was casting and the inner material cast later.
The curing of the specimens was done by submerge the specimen in the water poll for 28 days. Six cubes and six cylinders were also cast and cured with same procedure. Figures (2, 3, 4 and 5) show the preparation of the specimens and Table (4) present the properties of NC and SCC mixes.

**Fig. 2.** Cylinder and Cubic specimens under compressive strength tests.

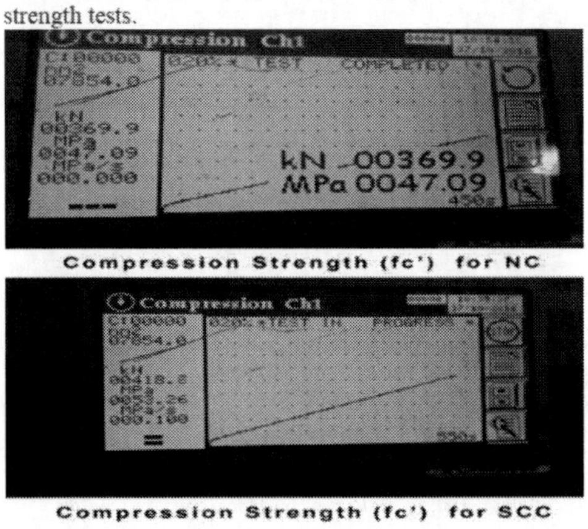


**Fig. 3.** Results of compressive strength tests for NC and SCC.

**Table 4:** NC and SCC Properties

| Type | fc'(Mpa) | ft (Mpa) | fr (Mpa) |
|---|---|---|---|
| NC | 47 | 3.45 | 4.25 |
| SCC | 53 | 3.65 | 4.5 |

**Fig. 4.** Reinforcement Details of the columns.

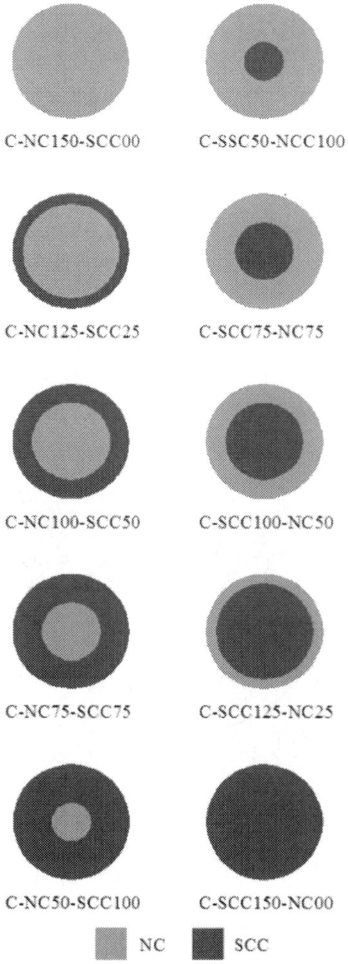


**Fig. 5.** HC Columns Details.

https://doi.org/10.1051/matecconf/201816204021

## 7 Results and discussion

Respecting to the experimental works the ten specimens were tested under uniaxial compressive loadings. Table (5) shows the summary of the experimental results.

**Table 5:** Summary of experimental test results

| no. | Column designation | Ultimate Load (kN) | Axial Strains at Ultimate Load |
|---|---|---|---|
| 1 | C-NC150-SCC0 | 281 | 0.00063 |
| 2 | C-NC125-SCC25 | 320 | 0.00064 |
| 3 | C-NC100-SCC50 | 354 | 0.00066 |
| 4 | C-NC75-SCC75 | 405 | 0.00072 |
| 5 | C-NC50-SCC100 | 480 | 0.00089 |
| 6 | C-SCC50-NC100 | 503 | 0.00120 |
| 7 | C-SCC75-NC75 | 532 | 0.00085 |
| 8 | C-SCC100-NC50 | 550 | 0.00075 |
| 9 | C-SCC125-NC25 | 573 | 0.00069 |
| 10 | C-SCC150-NC0 | 605 | 0.00062 |

A study of the structural behaviour of HCC is conducted, the effect of SCC to NC ratio on the axial strength of the column is considered. All columns were failed in compression mode of failure, figures (6, 7, and 8).

**Fig. 6.** Failure of Column C-SCC50-NC100

**Fig. 7.** Failure of Column C-SCC75-NC75

**Fig. 8.** Failure of Column C-SCC150-NC00

Figures from (9 to 18) show the relationship between the loading and the axial deformation in the columns.

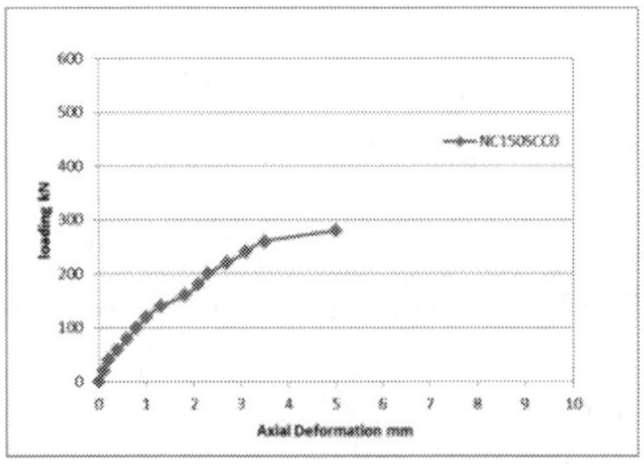


**Fig. 9.** Load-deformation curve for C-NC150-SCC0.

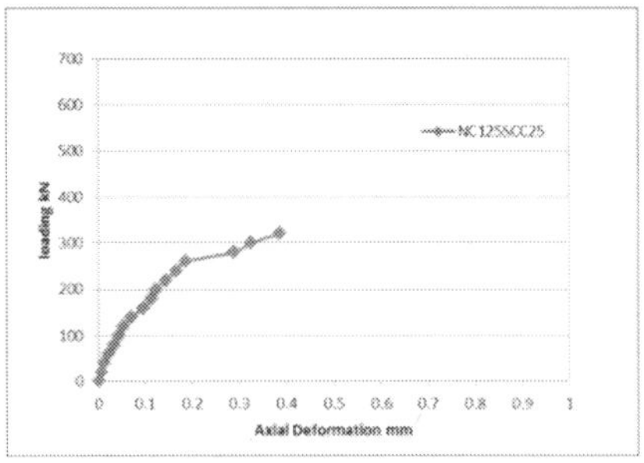


**Fig. 10.** Load-Deformation curve for C-NC125-SCC25.

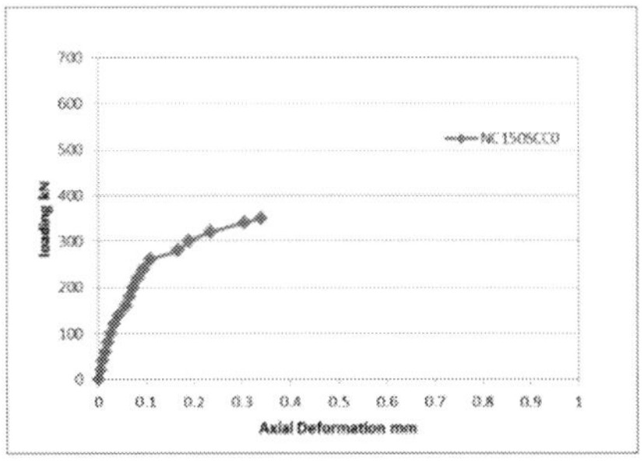


**Fig. 11.** Load-Deformation curve for C-NC100-SCC50.

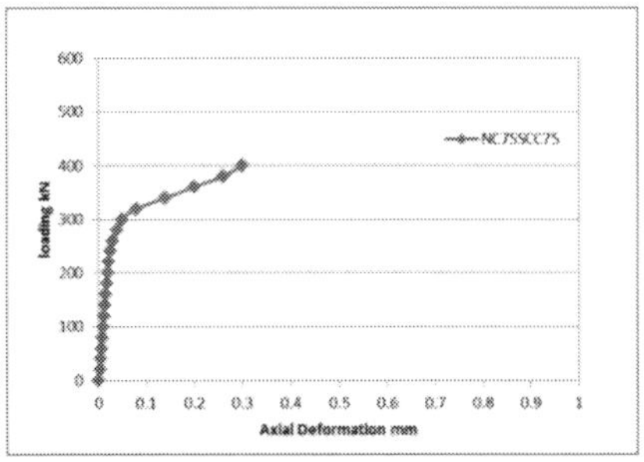


**Fig. 12.** Load-Deformation curve for C-NC75-SCC75.

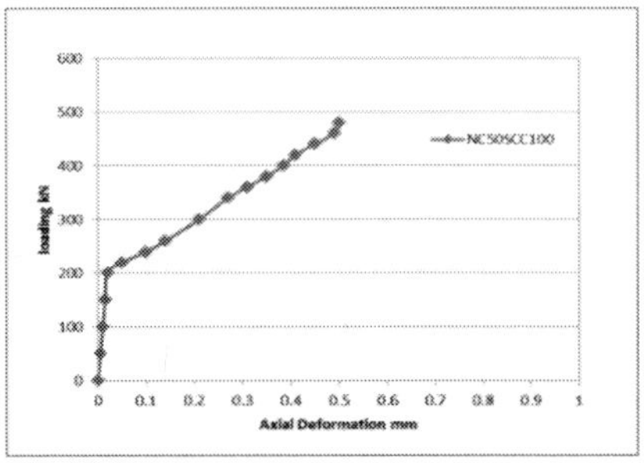


**Fig. 13.** Load-Deformation curve for C-NC50-SCC100.

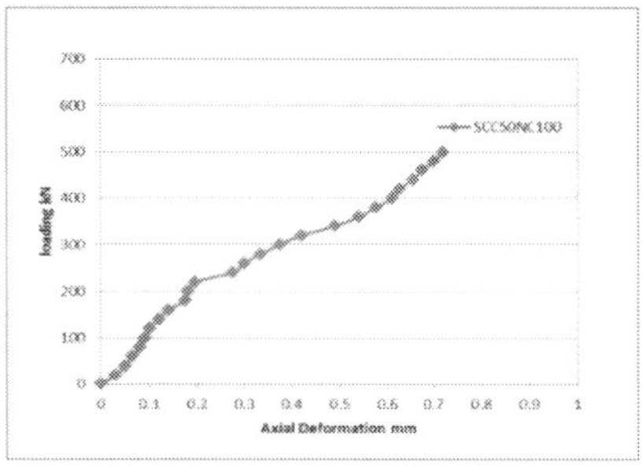


**Fig. 14.** Load-Deformation curve for C-SCC50-NC100.

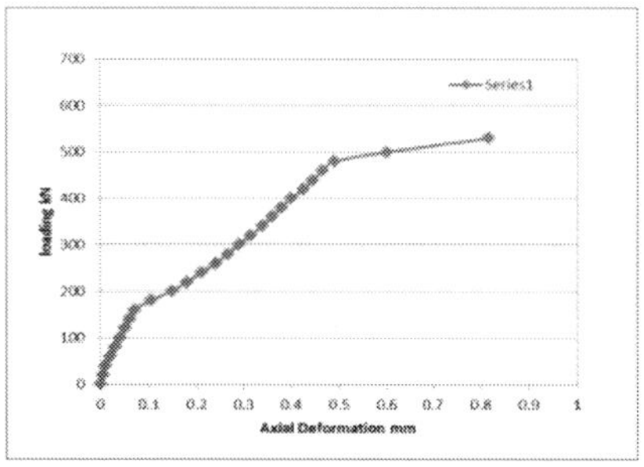


**Fig. 15.** Load-Deformation curve for C-SCC75-NC75.

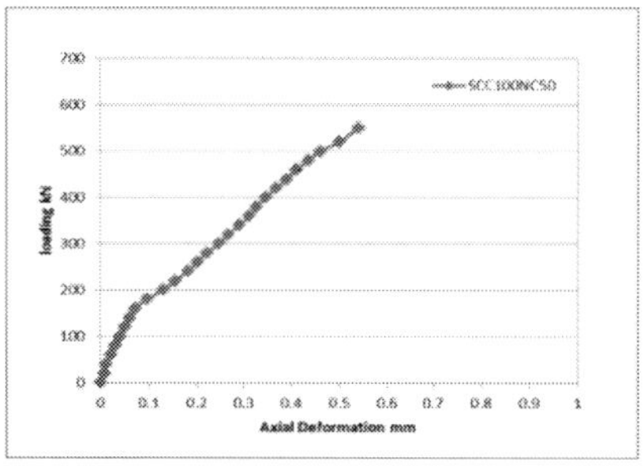


**Fig. 16.** Load-Deformation curve for C-SCC100-NC50.

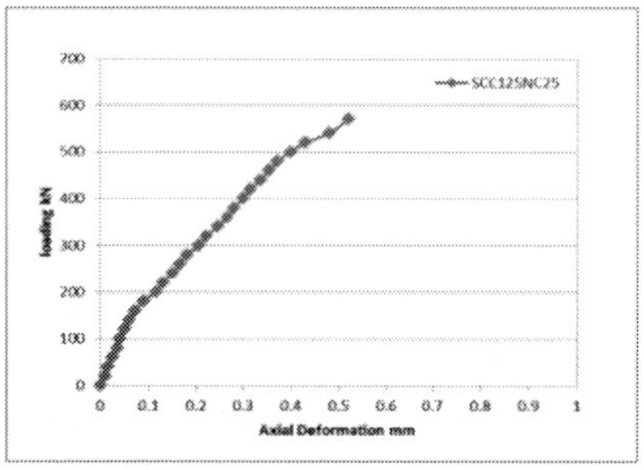


**Fig. 17.** Load-Deformation curve for C-SCC125-NC25.

https://doi.org/10.1051/matecconf/201816204021

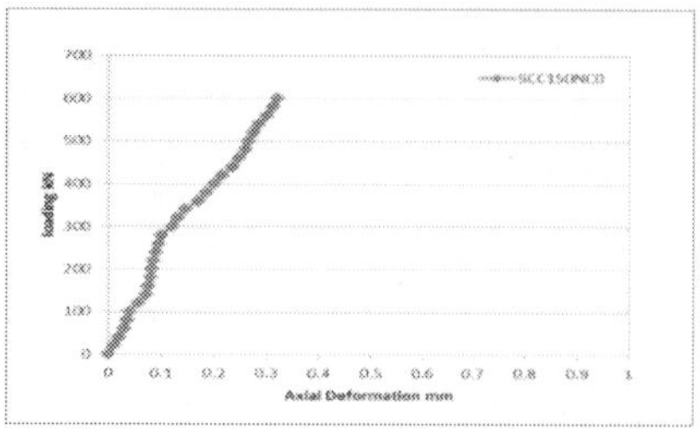


**Fig. 18.** Load-Deformation curve for C-SCC150-NC0.

## 7.1. Effect of SCC/NC ratio on HCC under axial compressing loads

### *7.1.1 Group 1 (NC inner and SCC outer)*

As the ratio of inner diameter to outer diameter (Di/Do) decreased, the load capacity of column increased from 281 kN to 480 kN and the axial deformation was increase also from 0.00063 to 0.00089 as shown in figure (19).

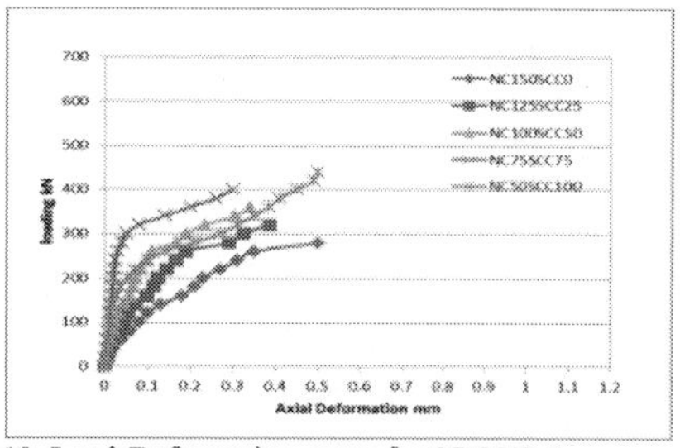


**Fig. 19.** Load-Deformation curve for GROUP 1.

### *7.1.2 Group 2 (SCC inner and NC outer)*

As the ratio of inner diameter to outer diameter (Di/Do) increase, the load capacity of column was increased from 503 kN to 605 kN and the axial deformation was decreased from 0.0012 to 0.00062 , as shown in figure (20).

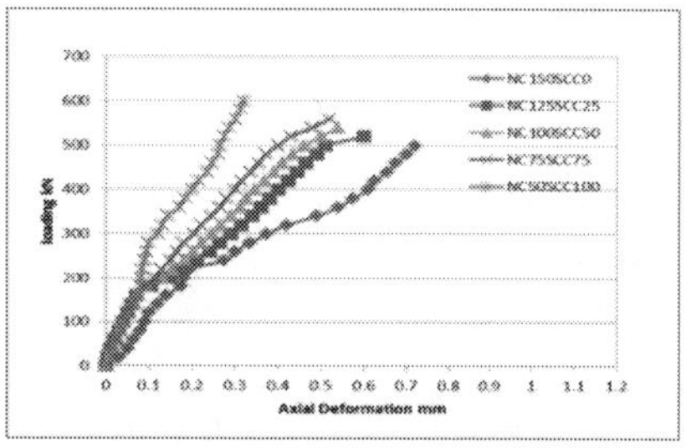


**Fig. 20.** Load-Deformation curve for GROUP 2.

### *7.1.3 Group 3 (Same ratio of Di/Do)*

For the same ratio of inner diameter to outer diameter (Di/Do) for NCC and SCC , the load capacity of column was increased from 354 kN to 550 kN and the axial deformation also increased from 0.00066 to 0.00075.

The load capacity of column was increased from 281 kN to 605 kN when you change it from fully NC to fully SCC and the axial deformation almost equal

## 8 Conclusions

In this study, the experiment investigation of behaviour of Hybrid Concrete Column under axial compressing loads. The testing of specimens considers the effect of outer to inner diameter ratio on the capacity of the columns.

The experimental program yields following conclusions:

1. All columns were failed in compression mode of failure.
2. The load capacity of column was increased 70% when the NCC inner diameter decrease and SCC outer dimeter increased.
3. The load capacity of column was increased 20% when the SCC inner diameter decrease and NC outer dimeter increased.
4. it's better to use SCC as outer material because the its work as confinement layer and increasing of load capacity compared with control column.

## References

1. Hsien-Kuang Liu, Nyan-Hwa Tai, Zhi-Cheng Chen. Composite Part B, 1999.M. Ben Rabha, M.F. Boujmil, M. Saadoun, B. Bessaïs, Eur. Phys. J. Appl. Phys. (to be published)
2. Nyan-Hwa Tai, Hsien-Kuang Liu, Zhi-Cheng Chen. Polymer composite, 1999.
3. Elchalakani M ;Zhao XL, and Grzebieta RH. Thin-WallStruct.;40(5) :415-41,2002.
4. Tomii, M.proceedings of the Third International Conference on Steel-Concrte Composite Structures, 26-29 Septem ber, Fukuoka, Japan, 39-67, 1991.
5. Wei, S., Mau, S. T., Vipulanadan, C. and Mantrala, and S.K. J. Struct. Eng. ASCE, Vol. 121(12), 1806-1814, 1995.
6. Iraqi Specifications No. (5), for Portland Cement, 1984.
7. ASTM-C33-03, ASTM International, 2003.
8. EFNARC, European Federation Dedicated To Specialist Construction Chemicals and Concrete Systems, 2005.
9. ASTM-C494-05ASTM International, 2005.
10. Neville, A.M., Fifth and Final Edtion, Wiley, New York and Longmont, London, 844, 2010.
11. Emborg, M., Final Report of Task 8.1, Betongindustri AB, Birte EuRam, Sweeden, 2000.
12. Al-Jabri, L. A., M.Sc. Thesis, Al-Mustansirya University, Baghdad, Iraq, 2005.

# Behavior of reinforced concrete slender rectangular columns under uniaxial loading

*Bassman* Muhammad*

Building and Construction Engineering Department, University of Technology, Baghdad, Iraq

**Abstract.** Two proposed formulae are presented to obtain the reinforced concrete slender column capacity. Reinforced concrete slender columns are used more nowadays than in the past 30 years because of development of engineering materials. Structural engineers that adopted the ACI 318M Code know that the design of this type of columns is more tedious than short ones. This research proposes two formulae that reflect the ultimate strength of cross section of slender columns, one for the axial and the second for the bending moment. Case study of 656250 columns is generated to cover most variable properties of columns, analyzed using MATLAB (R2017a) Program to obtain the ultimate strength of cross section then compared to the strength of slender columns.

Notations:

| | | |
|---|---|---|
| $h_x$ (h) | = | The cross sectional column side in the direction parallel with the eccentricity ($e_x$), mm. |
| $h_y$ (b) | = | The cross sectional column side in the direction perpendicular to the eccentricity ($e_x$), mm. |
| $f'_c$ | = | Specified compressive strength of standard cylinder concrete at 28 days, MPa. |
| $\alpha_1$ | = | Correction intensity cylinder compressive strength, 0.85. |
| $f_y$ | = | Specified yield strength for longitudinal steel reinforcement, MPa. |
| $A_s$ | = | Total area of the longitudinal column reinforcement, $mm^2$. |
| P | = | Column total gross steel ratio, ($A_s$ / $A_g$). |
| Γ | = | End steel lever arm column cross sectional factor. |
| $\beta_1$ | = | Equivalent relating rectangular compressive stress block to depth of neutral axis. |
| C | = | The neutral compression axis depth, mm. |
| $e_x$ | = | The applied eccentricity of the load about x – axis, mm. |
| $\beta_d$ | = | Creep reduction factor of the column. |
| $k\ell$ | = | Effective length of the column, mm. |
| R | = | Column cross section radius of gyration = 0.3 h, mm. |
| $\varepsilon_{cu}$ | = | The strain limit of extreme concrete fiber in compression = 0.003 mm / mm. |
| $P_o$ | = | The ultimate concentric axial load capacity, kN. |
| $Ø_s$ | = | ACI 318M-14 Euler Load factor of safety, 0.75. |
| $P_c$, $P_s$ | = | Cross Sectional capacity for cross section and column ultimate axial load capacity (for slender column), respectively, (kN). |
| $M_c$, $M_s$ | = | Cross Sectional bending moment capacity for cross section and column ultimate bending moment capacity (for slender column), respectively, (kN.m). |

* Corresponding author: bassman.riyadh@gmail.com

© The Authors, published by EDP Sciences. This is an open access article distributed under the terms of the Creative Commons Attribution License 4.0 (http://creativecommons.org/licenses/by/4.0/).

## 1 Introduction

In last decades, the high rise buildings, stronger materials tend to construct taller buildings with less cross sectional element.[1]

The slender column is an important element in any structural building and its calculations tend to be tedious because of its properties like the flexural rigidity (EI), Euler load buckling ($P_{cr}$) to obtain the magnification bending moment factor ($\delta$). In 1993, MacGregor[4] proposed a new method for analysis of slender columns, he developed the magnification factor ($\delta$) and proposed magnifications of moments applied on the columns for sway then magnify the overall sway and nonsway.

This provision was adopted in the ACI 318M-95[4] and is still accepted up to now. Many researchers studied this method, compared with experimental columns results, such as Lloyd and Rangan[5] and found good correlation between results.

## 2 Research significance

Two proposed formulae were obtained using statistical nonlinear best fit regression analysis for 656250 slender columns. These formulae are derived statistically to obtain the strength of the slender column from the strength of its cross section.

Also, study of the behavior of slender columns for all its mechanical and geometrical properties are presented.

## 3 Strength of slender columns

The ACI 318M-14[2] provides magnification method for analysis and design of the slender columns. To study this method, 656250 columns were generalized with the following parameter properties:

- $f'_c$ = 20 to 80 MPa; (10 MPa increments).
- $f_y$ = 200 to 600 MPa; (100 MPa increments).
- $(e / h)_x$ = 0.1 to 5; (0.1 increments).
- $\gamma$ = 0.5 to 0.9; (0.1 increments).
- $\rho$ = 0.01 to 0.05; (0.01 increments).
- $k \ell / r$ = 20 to 60; (10 increments).
- $\beta_d$ = 0 to 0.4; (0.2 increments).

In this research, constant values of some parameters according to the ACI 318M-14 depended. The strain limit of extreme concrete fiber in compression ($\varepsilon_{cu}$) = 0.003 mm/mm; Correction intensity cylinder compressive strength ($\alpha_1$) = 0.85; the modulus of elasticity of reinforcement ($E_s$) = $2 \times 10^5$ MPa; the strength reduction stability factor ($\emptyset_s$) = 0.75.

MATLAB (R2017a) was used to generate and analyze and the statistical results were computed in an incremental process program. The ratios between the ultimate strength of axial load for the slender column to the ultimate strength of its cross section is shown in Fig. (1).

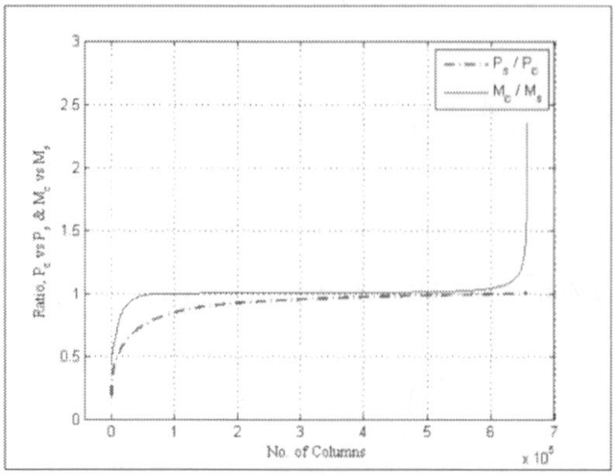


**Fig. 1.** The relation between ultimate strength of the column vs its cross section for generated Columns.

Noting that; the minimum and maximum ratios between ($P_s$ / $P_c$) are 0.1922; 1.0000; and the minimum and maximum ratios between ($M_c$ / $M_s$) are 0.4300; 2.3442.

The minimum ultimate load reduction for the axial force (column number = 562515) has the following properties:

| | | |
|---|---|---|
| $f'_c$ | = | 80 (MPa) |
| $f_y$ | = | 200 (MPa) |
| $(e / h)_x$ | = | 0.1 |
| $\gamma_x$ | = | 0.5 |
| $\rho_x$ | = | 0.01 |
| $(k \ell / r)_x$ | = | 60 |
| $\beta_d$ | = | 0.4 |
| $P_s / P_c$ | = | 0.192 |
| $M_c / M_s$ | = | 1.098 |
| $P_n / f'_c\, b\, h$ | = | 0.691 |
| $\rho f_y / f'_c$ | = | 0.025 |
| $P / P_o$ | = | 0.798 |

In addition, the ratios between cross sectional bending moment and the column moment stagger above and below unity.

## 4 MATLAB (R2017a) program

Incremental of the compression zone technique was used to solve each column twice, first for the cross section capacity then second for the slender column with full geometrical and mechanical properties. The tolerance dependence in solution of each column was 0.001.

A huge results data for input and output (array's size that treatment is 1312500 by 26). Therefore, statistical and graphical methods were used to study these results.

The ACI 318M-14 cross sections and magnification columns procedure was used in the programme.

## 5 Nonlinear regression for the axial load ($P_n$)

A nonlinear curve fitting programed in MATLAB (R2017a) was used to obtain a compatible representation with the true column behavior. The variables studied with many equations to obtain a good relation with the curve shown in Fig. (2), these independent variables are:

1. $P_n / f'_c\, b\, h$;
2. $\rho f_y / f'_c$;
3. $(e / h)$;
4. $(k\, \ell / r)$;
5. $P / P_o$;
6. $\beta_d$; and
7. $\gamma$.

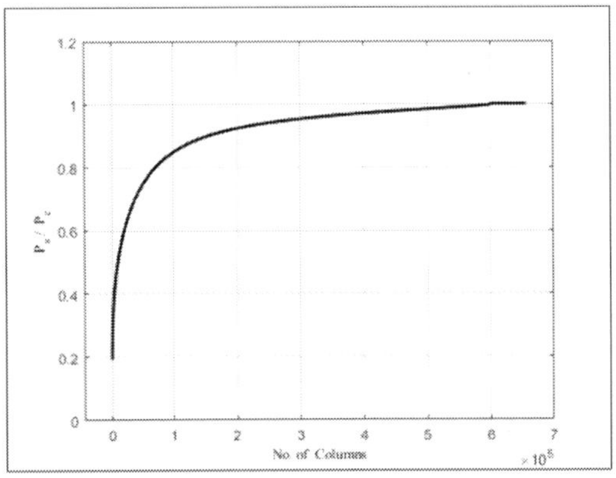


**Fig. 2.** The relation between ultimate strength of the axial load columns vs its cross section capacity for generated Columns.

Statistically, Eq. (1) is obtained to represent $R_p$.

$$R_p = 0.872 + \frac{0.8\, P_n}{f'_c\, b\, h} - \frac{2\, \rho\, f_y}{30\, f'_c} + \frac{e}{42\, h} - \frac{P_n}{5\, P_o}\sqrt{\frac{k\, \ell}{r}} - \frac{3\, \beta_d}{55} + \frac{\gamma}{14.8} \leq 1.0 \qquad (1)$$

In which; $R_p = P_s / P_c$; $P_n$ is the column capacity for cross section and $P_o$ is the nominal axial strength at zero eccentricity for the cross section, $P_o = 0.85 f'_c (A_g - A_{st}) + f_y A_{st}$.

Graphically, this equation has a good accuracy with the axial load behavior. The statistical results are:
Multiple R = 0.875.
R Square = 0.76.
Adjusted R Square = 16.86.
Mean values = 0.92.
Coefficient of variation = 8.16 %.
Fig. (3) shows the relation between the results of ($P_s / P_c$) vs the proposed Eq. (1).

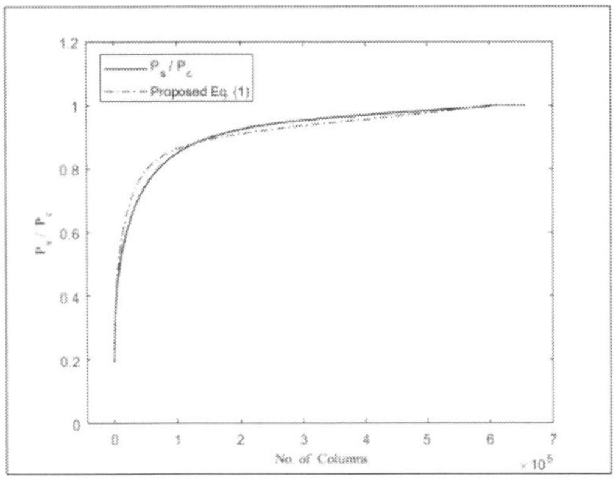


**Fig. 3.** Comparison between the proposed load formulae with the exact results.

## 6 Nonlinear regression for the bending moment ($M_n$)

Again, nonlinear curve fitting was programmed using MATLAB (R2017a) to best fit the behavior of column in bending moment. This behavior is divided into three parts to obtain statistical results, this division was decided because the relation between ultimate strength of the bending moment for the columns vs its cross section moment capacity for generated columns behaves in three parts - semi-vertical, semi-horizontal and semi-vertical, Fig. (4). Each part of the curve was analyzed separately. The statistical results (squared r) show no valuable differences between these three parts.

For this reason, one formula is proposed to formulate the relation between the cross sectional bending capacity and the bending capacity of slender ones. To make the formula practical for this research, the same independent variables were intentionally used in Eq. (1).

From Fig. (4), $R_m$ (which equals $M_c$ / $M_s$) is safe when the inspected value is less than the real ratio, so that, to make the formula of bending moment prediction more accurate, statistical results were made for 600000 columns only, i.e. the third part of Fig. (4) was excluded from the statistical analysis. The rest of the columns (56250 columns makes about 8.5% from the total columns studied) have no major effect on the results.

From some trials of regression analyses, Eq. (2) is selected to predict the slender bending moment capacity of the slender columns.

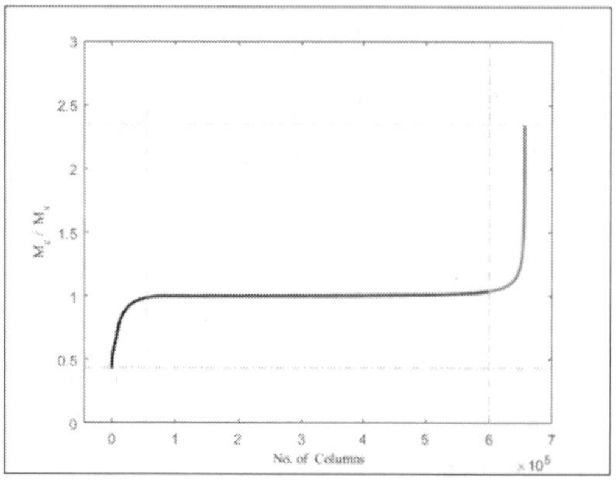


**Fig. 4.** The relation between ultimate strength of the bending moment for the columns vs its cross section moment capacity for generated columns.

$$R_m = 1.05 + \frac{P_n}{74\, f_c'\, b\, h} + \frac{\rho\, f_y}{52\, f_c'} - \frac{e}{70\, h} - \frac{2\, P_n}{31\, P_o}\sqrt{\frac{k\,\ell}{r}} + \frac{\gamma}{50} \qquad (2)$$

$R_m = M_c / M_s$ and $P_n$ is the column capacity for cross section.

The statistical results of Eq. (2) are:
Multiple R = 0.885.
R Square = 0.784.
Adjusted R Square = 8.503.
Mean values = 0.993.
Coefficient of variation = 3.659 %.

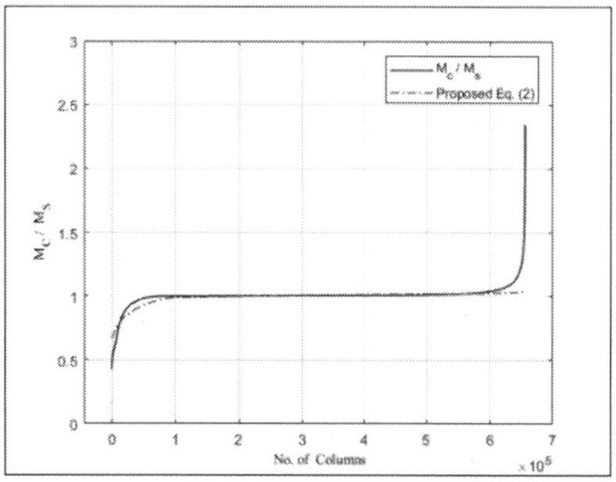


**Fig. 5.** Comparison between the proposed bending moment formulae with the exact results.

From Fig. (5), it can be seen that the proposed formula for bending moment capacity of the column has a good representation of the real behavior. Note: the last part (less than 8 % of the total number of studied columns) predict a conservative value of a ratio equals 2.101, since the prediction is less than these analyzed columns. Columns under large eccentricity, results a mall axial strength and large moment capacity, especially when the column steel force is small in value ($A_s$ $f_y$ is small) with stronger concrete compressive strength (large $f'_c$ value).

## 7 Conclusions

A theoretical investigation for 656250 RC slender columns with various geometrical and mechanical properties analyzed according to the ACI 318M-14, two formulae are proposed in this research; Eq. (1) and Eq. (2). For all columns, the ultimate axial strength was less than the cross sectional ultimate strength, the results show that these ratios ranged between (0.1922 – 1.0). While for bending moments ultimate strength the ratios were scattered above and below 1.0, the ratios between the cross sectional bending moment to the column bending moment strength were between (0.43 - 2.3442).

Design columns using these proposed equations are simpler using the interaction diagram by calculating the mechanical properties of columns, then obtaining the factors $R_p$ and $R_m$ and finally obtaining the required results according to the new load values ($R_p \times P_n$ and $M_n / R_m$).

## References

1. J. Wight, J. MacGregor, Rein. Con. Mech. and Design, Pearson Edu., Inc., Upper Saddle River, New Jersey (2012)
2. ACI 318M-14 and ACI 318RM-14, Detroit (2014)
3. J. MacGregor, ACI St. J. **90**, 3, 302-309, May-June (1993)
4. ACI 318-95 and ACI 318R-95, Detroit (1995)
5. N. A. Lloyd, B. V. Rangan, ACI St. J, **93**, 6, 631-638, Nov. - Dec. (1996)

## Appendix (Numerical Example):

For a RC column (600×800) mm; $f'_c$ = 35 MPa; $f_y$ = 400 MPa; total longitudinal steel area ($A_{st}$) = 6000 mm$^2$ ($\rho_s$ = 0.0125); $\gamma$ = 0.8; $e_x$ = 240 mm (i.e. $e_x$ / h = 240 / 800 = 0.3); effective length ($\ell_u$) = 7.2 m (k $\ell_u$ / r = 30) and the creep factor ($\beta_d$) = 0.4.

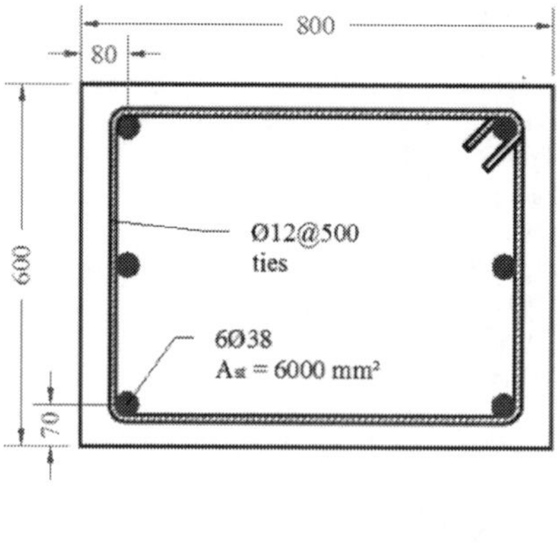

**Fig. A-1.** The cross section of the column

According to the ACI 318M-14, the ultimate strength of the column is as follows: (note that the nominal cross section capacity can be obtained using any interaction diagram instead of manual iteration method).

**Part I - Cross Sectional Analysis (short column):**

After some trials; the compression depth (c) = 538.91 mm; therefore, $P_n$ = 8201.53 kN and $M_n$ = 1968.3 kN.m.

**Part II – Slender Column Analysis:**

After some trials; the compression depth (c) = 457.88 mm; therefore, $P_n$ = 6618.81 kN and $M_n$ = 2103.05 kN.m. The total eccentricity ($e_x$) = 317.74 mm (i.e. $\delta$ = 1.324).

Therefore:
$R_{p\ ACI\ 318M\text{-}14} = P_{n\ slender} / P_{n\ short}$ = 6618.81 / 8201.53 = 0.807.
$R_{m\ ACI\ 318M\text{-}14} = M_{n\ short} / M_{n\ slender}$ = 1968.3 / 2103.05 = 0.936.

**Part III – Proposed Equations of this Research:**

$P_{no}$ = [0.85×35 (800×600 – 6000) + 400×6000] × $10^{-3}$ = 16501.5 kN.
From Eq. (1),

$$R_p = 0.872 + \frac{0.8\,P_n}{f_c'\,b\,h} - \frac{2\,\rho\,f_y}{30\,f_c'} + \frac{e}{42\,h} - \frac{P_n}{5\,P_o}\sqrt{\frac{k\,\ell}{r}} - \frac{3\,\beta_d}{55} + \frac{\gamma}{14.8} \le 1.0$$

$R_p = P_s / P_c$ = 0.872 + 0.8 × {8201.53 × $10^3$ / (35 × 600 × 800)} – 2 × 0.0125 × 400 / (30 × 35) + 0.3 / 42 - 8201.53 / (5 × 16501.5) × $(30)^{0.5}$ – 3 × 0.4 / 55 + 0.8 / 14.8
$R_p$ = 0.751. ($R_p \le 1$).
From Eq. (2),

$$R_m = 1.05 + \frac{P_n}{74\,f_c'\,b\,h} + \frac{\rho\,f_y}{52\,f_c'} - \frac{e}{70\,h} - \frac{2\,P_n}{31\,P_o}\sqrt{\frac{k\,\ell}{r}} + \frac{\gamma}{50}$$

$R_m = M_c / M_s$ = 1.05 + {8201.53 × $10^3$ / (74 × 35 × 600 × 800)} + 0.0125 × 400 / (52 × 35) – 0.3 / 70 – 2 × 8201.53 / (31 × 16501.5) × $(30)^{0.5}$ + 0.8 / 50
$R_m$ = 0.895.

Tabulated the results (Table (A.1) and comparing with the ACI 318M-14 results, it is shown that the results was a good prediction for the ultimate strength capacity.

**Table A.1.** Numerical Example Results

| | ACI 318M-14 | Proposed | Diff % |
|---|---|---|---|
| $R_p$ | 0.807 | 0.751 | 93 % |
| $R_m$ | 0.936 | 0.895 | 96 % |

From Table A.1, $P_n$ predicted for the slender column = 0.751 × 8201.53 = 6159.35 kN ($P_{n\text{-}ACI\ 318M\text{-}14\ slender}$ = 6618.81 kN - accuracy = 93.1 %)). $M_n$ predicted for the slender column = 1968.3 / 0.895 = 2200 kN.m ($M_{n\text{-}ACI\ 318M\text{-}14\ slender}$ = 2103.05 kN.m – accuracy = 95.6 %).

# Reinforcement bars tying types in reinforced concrete

*Marawan* Hamid [1,*], *Ali* Joudah[1], *Alyaa* Mohammed[1]

[1]University of Technology, Baghdad, Iraq

**Abstract.** In huge reinforced concrete structures, it is often difficult to distribute the reinforcing bars, especially at the connection regions, to keep the spacing between bars within the minimum acceptable limits as recommended in the global specifications standard, by using the common methods such as, overlapping or welding, therefore, a new method for bonding was introduced, using mechanical joints and electro fusion bonding method, and comparing these methods with the maximum strength of reinforcing bars without joints and the failure location within maximum strength limits outside the joints.

## 1 Introduction

While constructing structures with huge concrete frames, especially within the construction stage, the structural engineer faces some difficulties in applying the structural plans prepared by the designers based on the global specifications standard, because of the minimum limits for spacing between reinforcing bars. Therefore, the constructing engineer and his technical team must follow other procedures than overlapping or welding such as mechanical joints or electro fusion bonding methods. The solution is connecting and overlapping between the steel bars in different ways and different factors will affect this like the reinforced steel bar diameter, quantity of the reinforcement steel, grade of steel and high compressive strength of concrete.

## 2 Objectives

The main objective of this research is to investigate the optimum connection between steel bars reinforcement connection such as mechanical connection or butt welding in heavy structure sections.

## 3 Research significance

The knowing of deferent type steel bars reinforcement connection and when we can use it on deferent member in structure to give more surety according to the tests from this study.

## 4 Types of tying and overlapping between steel bars: Tying by overlapping between steel bars

Splicing by wire: is a simple and common way, give the user flexibility in the work so that he can separate the steel (before pouring) whenever he wants, as shown in figure 1. Formation depends on reinforced steel diameter, steel grade, compressive strength of concrete, the lap length between pieces is according to the following table 1 and plate 1.

**Table 1.** Overlap in steel bar by splicing by wire

| Bar Type | Concrete Strength grade | | | | |
|---|---|---|---|---|---|
| | C20 | C25 | C30 | C35 | >C40 |
| HPB 235 | 31 d | 27 d | 24 d | 22 d | 20 d |
| HPB 335 | 39 d | 34 d | 30 d | 27 d | 25 d |
| HPB 400 | 46 d | 40 d | 36 d | 33 d | 30 d |

- (d) the diameter for steel bar

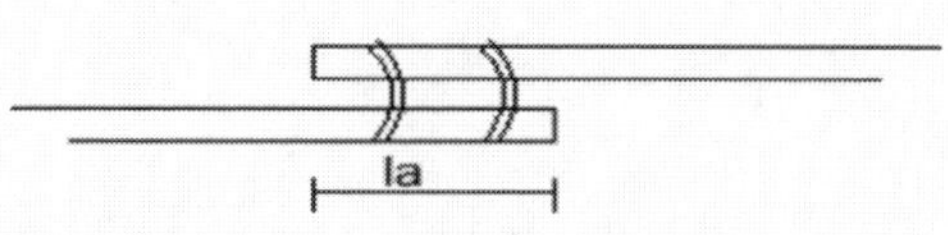


**Plate 1.** Shows the distance of (la)

* marawancivil@gmail.com

© The Authors, published by EDP Sciences. This is an open access article distributed under the terms of the Creative Commons Attribution License 4.0 (http://creativecommons.org/licenses/by/4.0/).

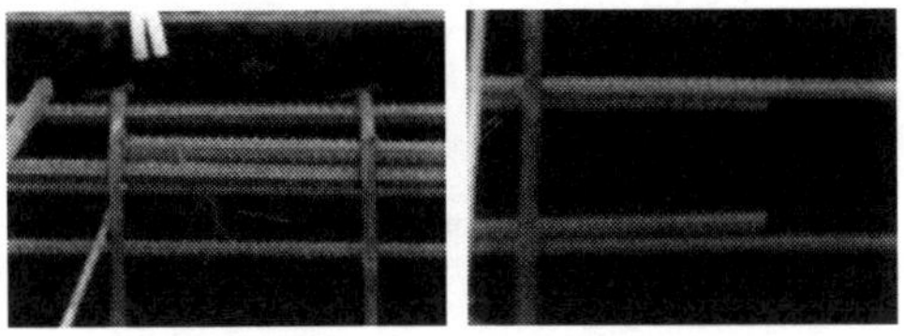

**Fig. 1.** Splicing by wires

Tying by welding: this type is used for strong steadiness in the structural parts wherein the steel reinforcement network is subjected to movement during the work or if the design requirements assume that. Like reinforcement of columns, retaining walls and Pile reinforcement. For example: in the situation of pouring long piles, after the completion of excavation, the steel box is installed first before pouring concrete, the steel box is installed by pieces and each piece is welded with another one. Using the previous method, the wire between steel pieces will not bear the heavy weight of the steel box and will get separated, thus, welding is used in this scenario. See Figure 2.

**Fig. 2.** Shows a steel cage for piling

In case of welding, the type of welding wire should be taken into account so that the cooling process happens gradually for the welded pieces:

### 4.1. Welding

#### *4.1.1Welding one side of the steel bar*

Shown in plate 2, 3 and figure 3.

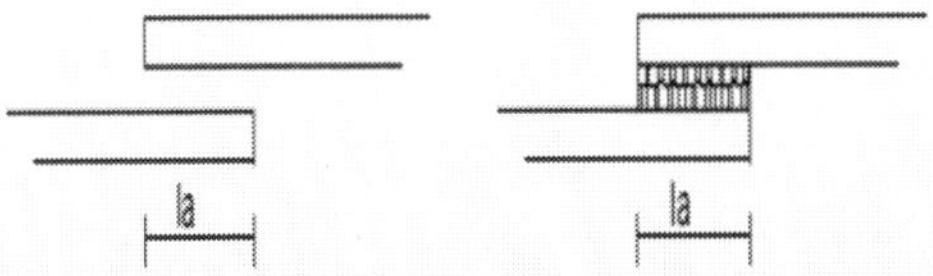


**Plate 2.** Shows both the faces of steel bar welding

Where: la= 10 * d [1]
(d) diameter for steel bar

* sometimes welding one side of the steel bar equals to (la=15*d).

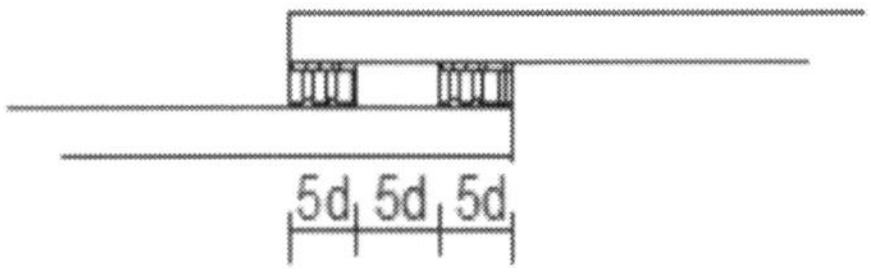


**Plate 3.** Shows the welding one side of the steel bar equal to 15* d

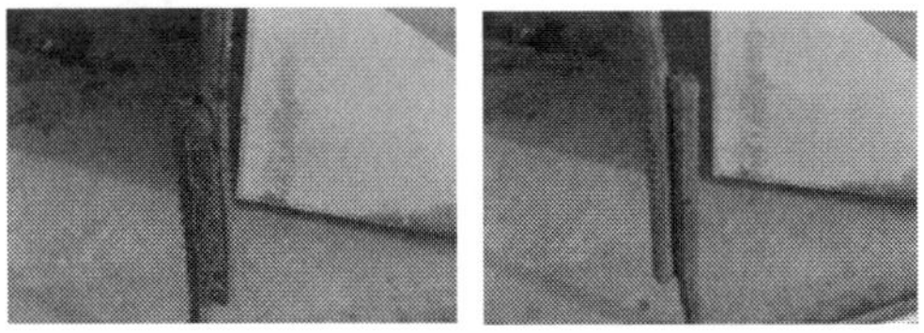

**Fig. 3.** Shows welding one side of the steel bar and the other side without welding

#### *4.1.2 Welding for two sides*

Welding distance for two sides is less than the distance for welding one side but in this case a consideration should be given for modulus of elasticity of steel as welding two sides will cause melting locations behind the welding part and also cooling these locations when they get in contact with the surrounding air thus, reducing modulus of elasticity in it.as shown in plate 4 and plate 5.

**Plate 4.** The front and back face for welding two sides of steel bar

Where: la= 5 * d [1]
(d) Diameter for steel bar

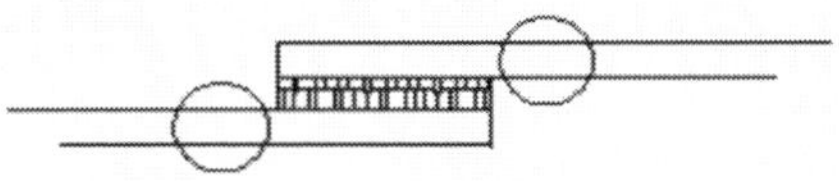

**Plate 5**. The area that are put in a circle show the weaken points

Tensile tests were performed for these two types of welding (one side and two sides) and the results were successful as no failure occurred in the connection area or it is surrounding, and as shown in figure 4 and according to the attached table 2, which represents the average of six samples that were taken for testing with different diameters.

**Table 2.** Results for testing steel bars were welding together by two ways for connect

| Nominal steel bar diameter (mm) | Single side lap welding | | Double side lap welding | |
|---|---|---|---|---|
| | Yield strength* ($f_y$) MPa | Tensile strength ($f_t$) MPa | Yield strength* ($f_y$) (MPa) | Tensile strength ($f_t$) (MPa) |
| 25 | 430 | 585 | 440 | 590 |
| 22 | 430 | 570 | 430 | 570 |
| 20 | 445 | 585 | 440 | 585 |
| 18 | 425 | 560 | 430 | 570 |
| 12 | 435 | 590 | 425 | 580 |

* tested steel bars reinforcement conformed to requirements of grade 60 (60 ksi), min fy=420 MPa, min. min. elongation = 9% according to ASTM A 615 -05[2] for bars dai. 12 mm, 18 mm, 20mm, 22mm and 25 mm.

## 5 Materials and experimental work

The steel bars ore samples of Turkey made with various diameters of bars reinforcement (20,22,25 and 32 mm) were taken as the most commonly used in large concrete and heavy reinforcement sections. These sections were subjected to tensile and compressive strength testing in the laboratories of MAS International Investment Company / Iraq Sulaymaniyah and its comparison with the strain of submission and elongation allowed within the standard specifications of the manufacturer of reinforcing steel, the types of linkage conducted within this work are detailed mechanical bonding and butt welding.

### 5.1Butt welding

It is an economic method and is used in the areas with heavy reinforcements (bridge with high loads): so that the distance between reinforced steel is the minimum distance while when using other type of steel welding, we can't get the adequate distance between the reinforced steel.as shown in plate (6) and Figure 4.

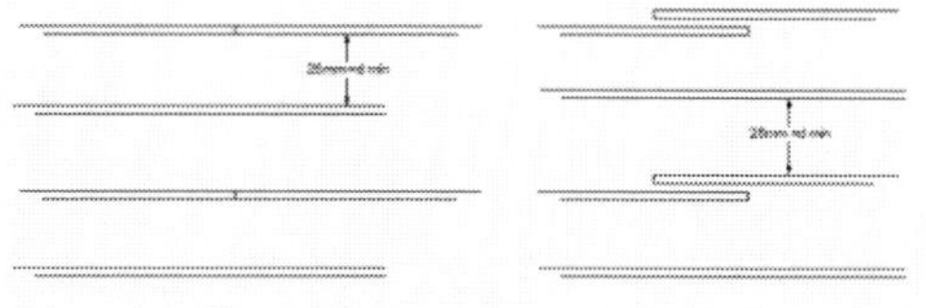

**Plate 6.** Shows the difference between type one and the butt welding

**Fig. 4.** Shows the joint in the butt welding

Welding between two pieces of steel is done by special equipment that forms a cathode for the first one and electrode for the second one and when connecting between the two, an electrical discharge occurs between the two poles and steel melts and then welding takes place. Tensile test was performed for this type of welding and it was found successful as no failure in the welded area or it's nearby had occurred and also Yield strength test was performed as shown in Figure 4 and according to Table 3 that shows results for six testing samples with different diameters.

**Table 3.** The result for steel bar connected with butt welding

| Nominal steel bar diameter (mm) | Yield strength* ($f_y$) (MPa) | Tensile strength ($f_t$) (MPa) | Elongation |
|---|---|---|---|
| 25 | 450 | 620 | 0.00244 |
| 20 | 455 | 590 | 0.00229 |
| 18 | 420 | 580 | 0.00247 |
| 12 | 525 | 610 | 0.00263 |

* tested steel bars reinforcement conformed to requirements of grade 60 (60 ksi), min fy=420 MPa, min. min. elongation = 9% according to ASTM A 615 -05[2] for bars dai. 12 mm, 18 mm, 20mm and 25 mm.

This method is used outside the forms and not like the previous ones which means the steel should be welded before installing it into the structure as shown in plate 7 and figure 5 and 6.

**Fig. 5.** Shows how to connect in the butt welding spark

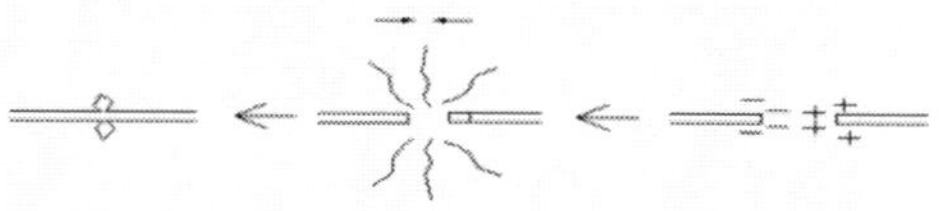

**Plat 7.** Shows the interaction with the emission of heat and

**Fig. 6.** Shows steel bars stacks

## 5.2 Mechanical joint

It a new method that has been used for almost 15 years now. It is used for reinforcement diameter of more than (a 20 mm). This method depends on welding on friction force between the steel (the same as in galvanised steel pipes), Plate 8.

**Plat 8.** Tying two pieces of steel bars by mechanical joint

This method is used in different places like foundations, columns, beams and walls, where it is on a large distance and is difficult to use the previous welding and connecting methods, especially in embedded reinforcement steel bar .as shown in Figure 7 and 8 and Plat 9.

**Fig. 7.** Shows mechanical joint

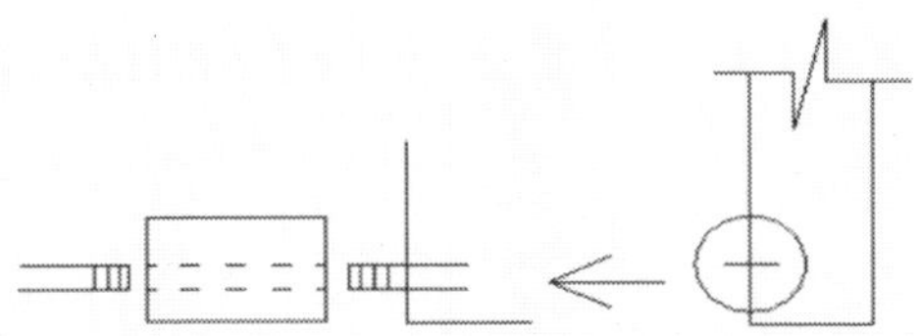

**Plate 9.** Shows embedded reinforcement steel bar

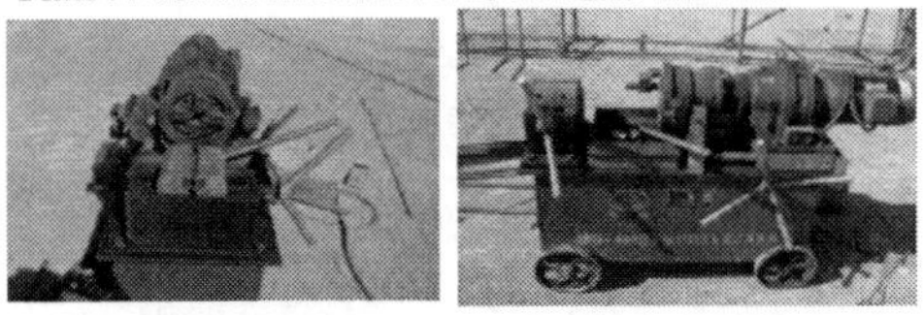

**Fig.8.** the machine that make the mechanism of mechanical joint

**Fig. 9.** The production by the machine

Procedure to produce the mechanical joint from the steel bars:

Outside the structure (form) the machine starts cleaning the bump in the reinforced steel in the area where the serrated part is wanted in an automatic way (depending on the bar diameter), Figure 10.

**Fig. 10.** The machine at the work

And then making serrated in a cone shape (the angle is very small) Figures 11and 12.

**Fig. 11.** Shows the machine at work

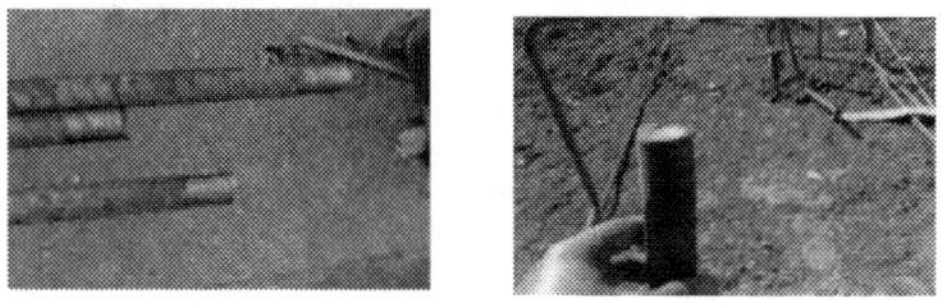

**Fig. 12.** Shows the steel bar after making it serrated

Tensile tests for several mechanical joint samples have proved the possibility of using the technique. It was found that this welding has the same normal yield stress for steel and the same compression even in the structures that have dynamic loads and vibrations and it was widely used in the United States, Germany, and China. Also, this type of welding was used in Kurdistan /Iraq in the cement factories and in all the fields of foundations, columns, bridges, walls and ceilings, and has proved its creditability in both lab and field tests as several samples were taken directly and without modification for testing for this type of welding for the purpose of this testing as it's shown in Figure 13. The result was successful as no failure occurred in the welded area or its surroundings (Table 4) shows average results for six samples that was taken for testing for different diameters.

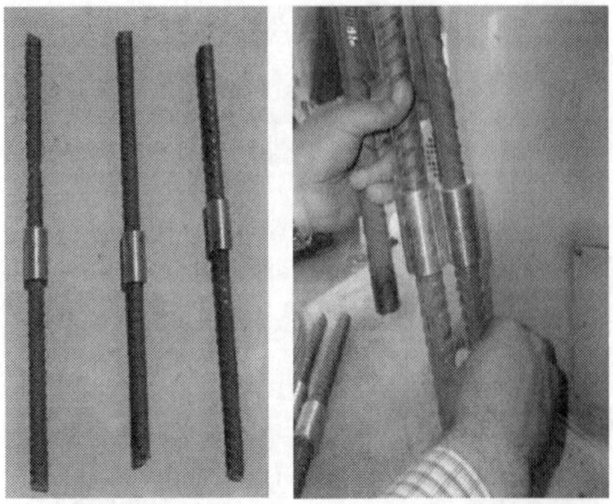

**Fig. 13.** Shows the steel bar that connected with mechanical joint after tasted in the laboratory

In Germany steel is made in different way as the nut is fixed in the bar from the factory (poured with it) directly and according to the DIN standard) each type is according to the location of the steel bar if it was tension or compression and so on. As shown in Figure 14 and in the same way is used in the United States. We can see this type of welding in ACI Journal and according to the shapes, types and the method of welding as shown in Figures 14, 15 and 16.

**Fig. 14.** Shows a German catalogue that deals with mechanical joint

**Fig. 15.** Shows concrete structures constructed by using mechanical joints in the steel reinforcement [7]

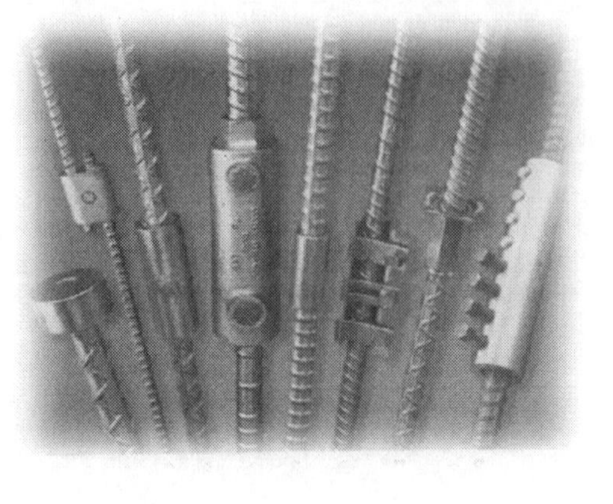

**Fig. 16.** Shows some shapes and types for mechanical joints [7]

ICC Evaluation Service, Inc. has performed tensile and compression (Static and Dynamic) on the mechanical joint its different types when the steel nut is poured with it in the field, and the results from these tests referred to the success and readiness for these samples and the possibility of many uses in buildings and different manufactural structures. The nuts are made in different locations in the world like United Arab Emirates as there are factories for steel manufacturing by using the welding method.

## 6 Test results

As mentioned before, the mail aim of this research is to study failure behaviour for tested steel bars reinforcement and comparison it with the same main bars without connection. This tables and figure shows results for each tested deferent steel bar connections as shows in tables 4, 5, 6 and 7and the figure 17 shows the summary.

**Table 4.** For Steele bar 20mm diamètres

| Semple no. | Nominal dia. mm | Mechanical joint tensile strength test MPa | Complete bar tensile strength test MPa | *Elongation % for mechanical joints | Compare between Mechanical joint and Complete bar tensile strength |
|---|---|---|---|---|---|
| 1 | 20 | 560 | 550 | 0.00280 | close |
| 2 | 20 | 540 | 570 | 0.00295 | close |
| 3 | 20 | 555 | 570 | 0.00270 | close |
| 4 | 20 | 550 | 540 | 0.00285 | close |
| 5 | 20 | 565 | 535 | 0.00300 | close |
| 6 | 20 | 550 | 540 | 0.00270 | close |

* Min. elongation = 9% according to ASTM A 615 -05[2] for bars dai. 20 mm

**Table 5.** For Steele bar 22mm diamètres

| Sample no. | Nominal dia. mm | Mechanical joint tensile strength test MPa | Complete bar tensile strength test MPa | *Elongation % for mechanical joints | Compare between Mechanical joint and Complete bar tensile strength |
|---|---|---|---|---|---|
| 1 | 22 | 580 | 590 | 0.00267 | close |
| 2 | 22 | 570 | 570 | 0.00266 | close |
| 3 | 22 | 595 | 570 | 0.00260 | close |
| 4 | 22 | 600 | 560 | 0.00265 | close |
| 5 | 22 | 560 | 545 | 0.00255 | close |
| 6 | 22 | 550 | 565 | 0.00266 | close |

* Min. elongation = 9% according to ASTM A 615 -05[2] for bars dai. 22 mm

**Table 6.** For steel bar 25mm diameter

| Sample no. | Nominal dia. mm | Mechanical joint tensile strength test MPa | Complete bar tensile strength test MPa | *Elongation % for mechanical joints | Compare between Mechanical joint and Complete bar tensile strength |
|---|---|---|---|---|---|
| 1 | 25 | 610 | 620 | 0.00226 | close |
| 2 | 25 | 600 | 595 | 0.00222 | close |
| 3 | 25 | 595 | 580 | 0.00240 | close |
| 4 | 25 | 610 | 585 | 0.00227 | close |
| 5 | 25 | 590 | 640 | 0.00213 | close |
| 6 | 25 | 615 | 600 | 0.00216 | close |

* Min. elongation = 9% according to ASTM A 615 -05[2] for bars dai. 25 mm

**Table 7.** For steel bar 32mm diameter

| Sample no. | Nominal dia. mm | Mechanical joint tensile strength test MPa | Complete bar tensile strength test MPa | *Elongation % for mechanical joints | Compare between Mechanical joint and Complete bar tensile strength |
|---|---|---|---|---|---|
| 1 | 32 | 580 | 600 | 0.00234 | close |
| 2 | 32 | 570 | 590 | 0.00258 | close |
| 3 | 32 | 615 | 600 | 0.00249 | close |
| 4 | 32 | 590 | 610 | 0.00248 | close |
| 5 | 32 | 580 | 610 | 0.00283 | close |
| 6 | 32 | 630 | 580 | 0.00199 | close |

* Min. elongation = 9% according to ASTM A 615 -05[2] for bars dai. 32 mm

## 7 Conclusions and recommendations

Based on the results of the experimental tests the following are obtained:

1. The results were very close for the reinforcing bar specimens with mechanical joints and those without joints, for the same type of reinforcing bars. Therefore, from the researcher's point of view, it is safer to be used within the required global specifications, based on the ratio of specimen length to the bar diameter according to the American standards.

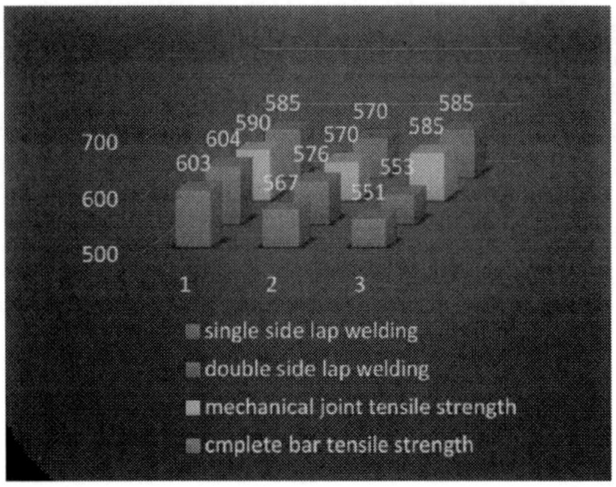


**Fig.17.** shows deferent results for each tested steel bar connections

2. The mechanical bonding method is relatively easy and quick to implement and configure.
3. The cost of this method is relatively low because it saves some quantities of reinforcing bars.
4. It saves areas and distances between reinforcing bars in concrete walls and lintels with congested reinforcement.

## References

1. National Standard of the People's Republic of china, Code of Design of concrete Structures (GB 50010-2002).
2. ASTM A615/A615M-05a, "Standard Specification for Deformed and Plain Billet-Steel Bars for Concrete Reinforcement," ASTM Committee A01 on Steel, Stainless Steel, and Related Alloys, West Conshohocken, United States, 2005, 5 pp.
3. Steel designers manual 5th edition chapter 24 p.677 bolt welds
4. Building Code Requirement for Structural Concrete and Commen Tory ACI-318m-05chapter12 p.211
5. Building Code Requirement for Structural Concrete and Commen Tory ACI-318m-05 chapter21 p.311
6. Design of concrete structures edition 13th chapter 5 p.195.
7. The Magazine of the American Concrete Institute – International technical society, December 2013/vol. 35 No. 12.

# Harmonic dynamic loading response of reinforced concrete column

*Saba* Sabih*

University of Technology, Baghdad, Iraq

**Abstract.** A reinforced concrete column is classified as compression structural element mostly analyzed and designed due to the applied combinations of dead and live loading with other considered loadings. Industries of considerable or relatively great size, production and electrical utilities are very concerned about the presence of dynamic loads in their electrical power systems. This behavior provides current with different components that are multiples of the fundamental frequency of the system which are called harmonics. Reinforced concrete elements such as column must be checked for the strength capacity and the response due to applied harmonic loading after completed the static analysis and design. In present article evaluations of reinforced concrete columns under the effects of dynamic harmonic loadings are studied. The main parameters are the reinforcement ratio and harmonic ranged loadings. Finite elements approach was adopted to analyze the columns by ANSYS software and all models are simulated in three dimensions. The analysis results indicated that the square cross sections with that rectangular of the same cross sectional area are closed in performance against static and dynamic loadings.

## 1 Introduction

The response of reinforced concrete columns to dynamic loadings are very important to check out the static design of columns and to make the structural building safe in case of presence of dynamic loading. Many parameters that the reinforced concrete columns effecting in case of dynamic loading such as type of dynamic loading, amount of main reinforcements, tie reinforcements, type of concrete and the geometry of column. In 2004, Hideaki Nishida and Shigeki Unjoh [1], studied the performance of reinforced concrete columns (RCC) under the effects of dynamic loading such as earthquake. Different cross sections of columns were tested and the results showed the excitation of the column significant as nonlinear behavior of the column. In 2004, Siddik Sener et al [2], investigated the effects of RCC sizes due to axially loading. The square RCC with pin ended boundary condition was explored taking into account different slenderness ratios. The test results showed that stress at failure load decreased as the column size increased. In 2010, T. Yu et al [3], studied the non-uniform confinement of circular, square and other columns types using FEA by ABAQUS software. The analysis results indicated that close as compared with the test results by others researchers. In 2011, Zhishen Wu et al [4], explored the damaged that occurs in RCC under the effects of progressive seismic at the final stage. The test results be visible that the increased in progression of cracks due to reverse dynamic loads. In 2011, Hugo Rodrigues et al [5], looked out on the performance of RCC subjected to different histories as uniaxial and biaxial loading. The test results showed that a higher energy misspending in case of uniaxial than biaxial loading. In 2013, Libo Yan et al [6], investigated the performance RCC in presence of fiber reinforced polymer (FRP) under the effects of seismic loading. The test results showed that the presence of FRP enhancing the performance of RCC subjected to different loadings such as uniaxial and biaxial loading. In 2013, Yuanfeng Wang and Xiaoran Li [7], studied the hysteretic of RCC in case of presence of FRP subjected to cyclic loading. The nonlinear damping analysis using numerical method was adopted and the results indicated that the dynamic response of RCC more effectively reduced in case of taking into account the nonlinear damping analysis. In 2014, Ahmed Samir [8], studied the effects of blast loading on the RCC using finite elements approach FEA by ABAQUS software. Many parameters such as reinforcement ratios and charge weights were considered in study. The analysis results indicated that the residual displacement became more significant.

In present paper, the response of reinforced concrete

* Corresponding author: Saba.sabeeh@yahoo.com

© The Authors, published by EDP Sciences. This is an open access article distributed under the terms of the Creative Commons Attribution License 4.0 (http://creativecommons.org/licenses/by/4.0/).

columns under the effects of dynamic harmonic loading with different parameters such as frequency ranges, longitudinal and transverse reinforcement will be discussed and evaluate the strength capacity of RCC using FEA by ANSYS software

## 2 Column strength capacity

The reinforced concrete column is the compression member that reinforced by steel reinforcements to prevent buckling and to reducing the long term effects of creep in concrete. The tied column strength capacity based on the ACI – 318 – 2014 [9] as follow:

$$\emptyset P_{n,max} = 0.8\emptyset\{0.85f_c'(A_g - A_s) + f_y A_s\} \quad \text{.........(1)}$$

In which, Pn,max is the nominal axial strength, ϕ safety factor equal to (0.8), f 'is the compressive strength of concrete, Ag and As are the gross sectional area of concrete and area of main reinforcement steel respectively, fy is the yielding tensile strength of steel bar. The same material properties for the analyses were used as actual height and geometry of the columns. The main column is square with dimensions (600x600x3000 mm) that transfer to circular column with diameter around to ( 650 mm) and rectangular with dimensions as (450x800 mm). The main reinforcements adopt in present study based on the minimum requirements of ACI – 318 – 2014 [9] that (1.11%) of gross sectional area of concrete. The main reinforcement for all cross sectional area is (20ϕ16) and the tied reinforced of column is (ϕ10@200 mm) center to center along the columns. The design compressive strength of concrete is (28 MPa) and the yield strength of reinforcement is (400 MPa). The ultimate strength capacity of columns lists in Table (1). Figure (1) show the cross sectional area for all columns.

**Table 1:** Columns strength capacity

| Column mark | Geometry (mm) | Column strength capacity (kN) Equation (1) |
|---|---|---|
| Square Column (CS) | Square 600x600 | 6450 |
| Rectangular Column (CR) | Rectangular 450x800 | 6450 |
| Circular Column (CC) | Circular 650 | 6450 |

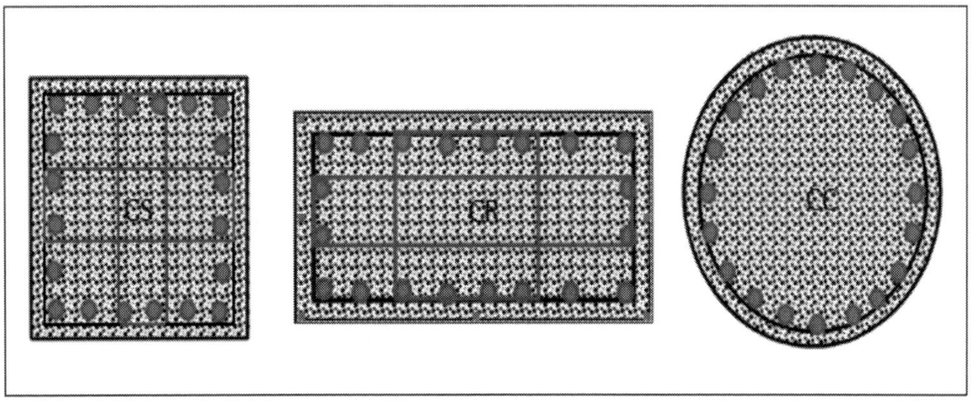


**Fig. 1.** The cross sectional area for all columns

https://doi.org/10.1051/matecconf/201816204024

## 3 Harmonic response analysis

The applied continuing for an extended period or without interruption cyclic loading will produce a sustained cyclic response (harmonic response) in any type of structural system. Harmonic response analysis gives the ability to predict the sustained dynamic behavior of any structural elements such as the reinforced concrete columns, so that in this case enabling to verify whether or not the reinforced concrete columns design will successfully overcome resonance or other damaging effects of forced vibrations.The steady-state response of the reinforced concrete columns (RCC) subjected to

$$[M\}\{\ddot{u}\}+[c]\{\dot{u}\}+[k]\{u\}=F_o\sin(\omega t)............(2)$$

sinusoidal with time and find out the response of the RCC at frequency range. The general equation of motion as follow: In which, M is mass matrix, c, is the damping matrix, k, is the stiffness matrix, ü, u. and u are the acceleration, velocity and displacement respectively. The solutions of the right hand side with amplitude force equal to the ultimate static loading and with frequency range as low (0-40) and medium (40-400).

proper material models, and boundary condition simulation for columns at the ends, convergence criteria and solution method are selected. The solver that used for static analysis was Newton – Raphson method and for dynamic analysis was Frontal solver with displacement convergence (5%). The loading applied with sub-steps in case of static loading up to the ultimate loading. The dynamic analysis, first the modal analysis run to determine the frequencies that make the model resonance range for modal analysis. ANSYS [10] program is adopted to build all reinforced concrete columns that subjected to static with harmonic dynamic analysis. The solver for dynamic analysis that adopted in present study is sparse solver that takes into accounts the real and imaginary roots during solution of equations to find the unknown variables.

## 4 Results

The analysis results for reinforced concrete columns using finite elements approach by ANSYS are presented in Figures below. Figures (2) to (7) show that the longitudinal and horizontal displacements for all columns. Figures (8) to (10) represents the static analysis of models as longitudinal, horizontal displacements and longitudinal stress. Figures (11) to (16) show the performance of models under the

## 5 Numerical analysis (finite elements approach)

The numerical analysis is approximation approach as distinguished from discrete mathematics that used in present work as finite elements approach (FEA) by ANSYS [10] to simulate the reinforced concrete column models. The structural composed elements connected together at a finite number of nodes. Compatibility of displacement at any connections nodes between steel section and concrete will be satisfied due to the full bond between them. The full connections between the interior nodes of steel with the exterior nodes of concrete make the displacement same for the two materials at common nodes The numerical analysis adopted by FEA by ANSYS to simulate the composite columns relay on the suitable selection of elements type, so, SOLID65 [10], for concrete, SOLID185 for steel plate above the column that used to distributes the applied loading. In additions to the element types, number of elements,

applied harmonic loading as low and medium and presents the results of longitudinal, horizontal displacements and longitudinal stress.

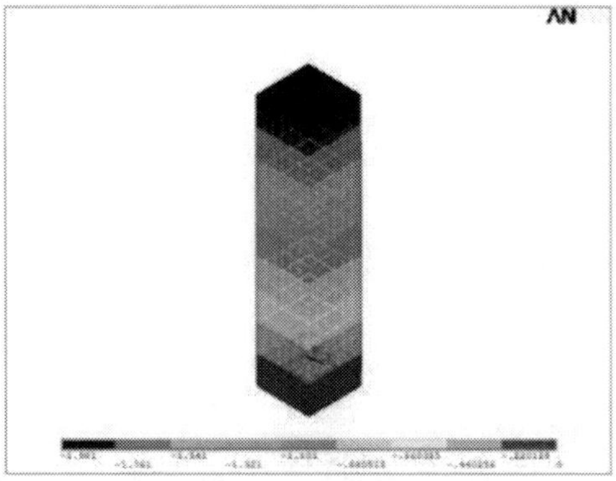


**Fig. 2.** Longitudinal displacement of CS (mm)

 https://doi.org/10.1051/matecconf/201816204024

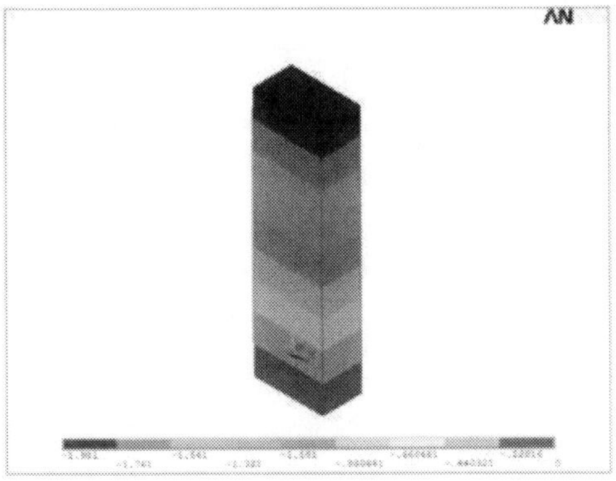

**Figure 3:** Longitudinal displacement of CR (mm)

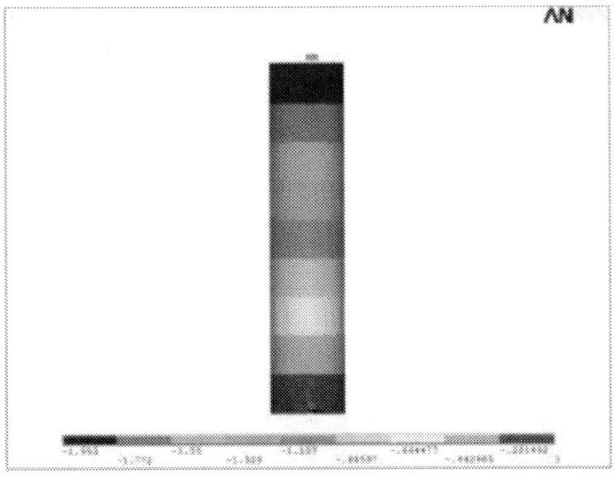

**Figure 4:** Longitudinal displacement of CC (mm)

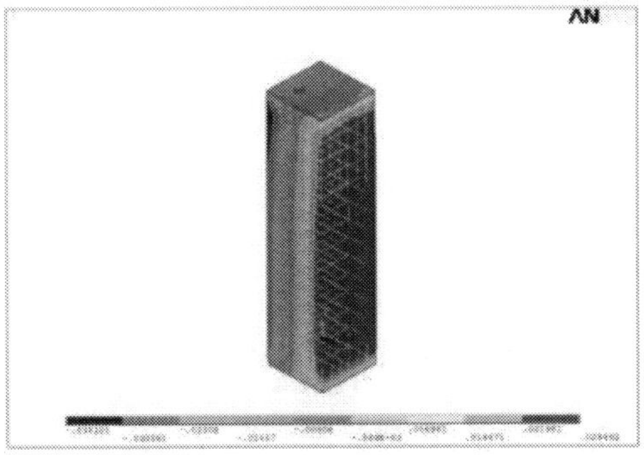

**Figure 5:** Horizontal displacement (ux) of CS (mm)

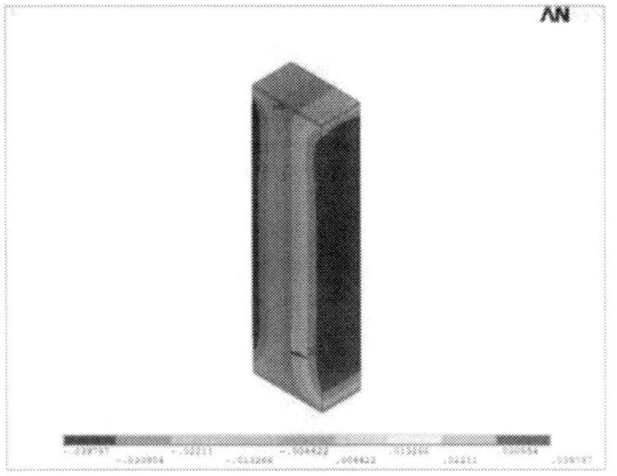

**Figure (6):** Horizontal displacement (ux) of CR (mm)

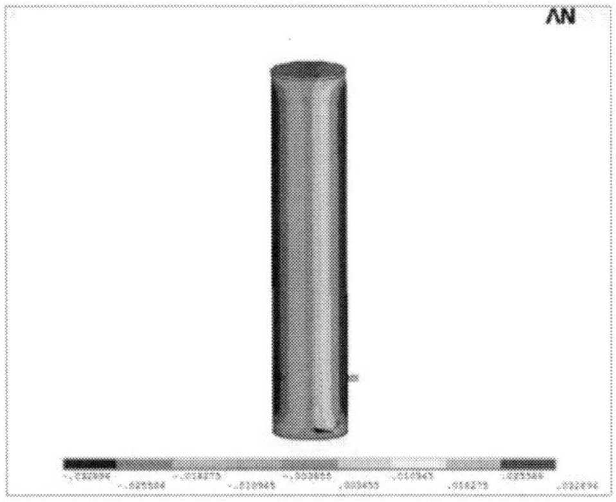

**Figure 7:** Horizontal displacement (ux) of CR (mm)

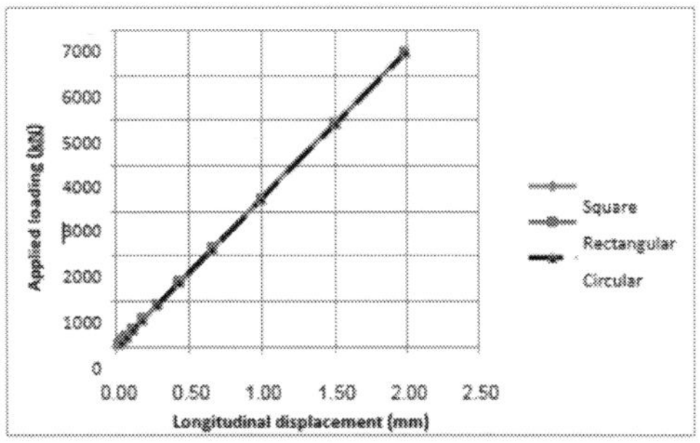


**Figure 8:** Applied loading vs. longitudinal displacement for all columns

 https://doi.org/10.1051/matecconf/201816204024

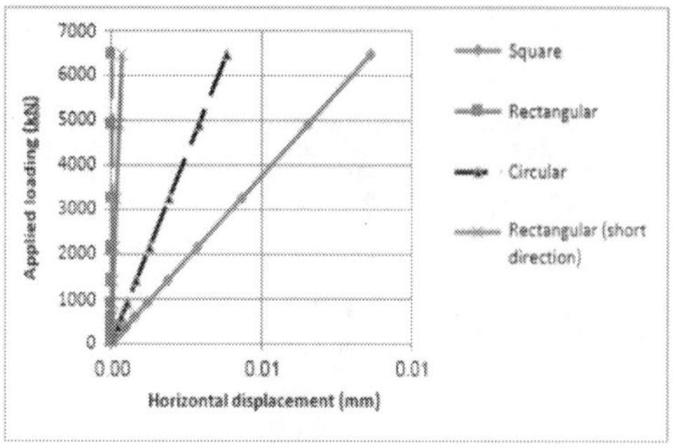


**Figure 9**: Applied loading vs. horizontal displacement for all columns

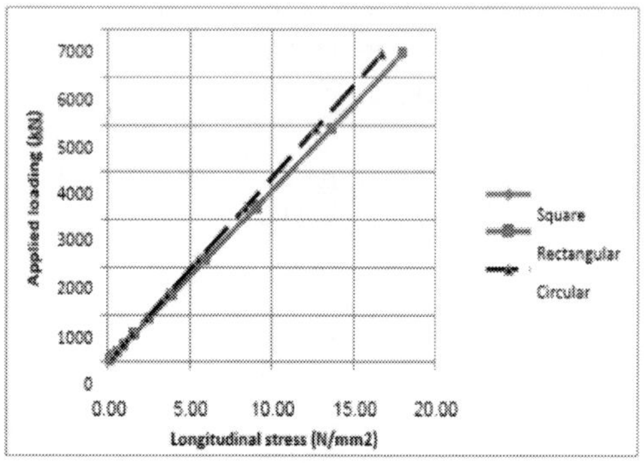


**Figure 10**: Applied loading vs longitudinal stress for all columns

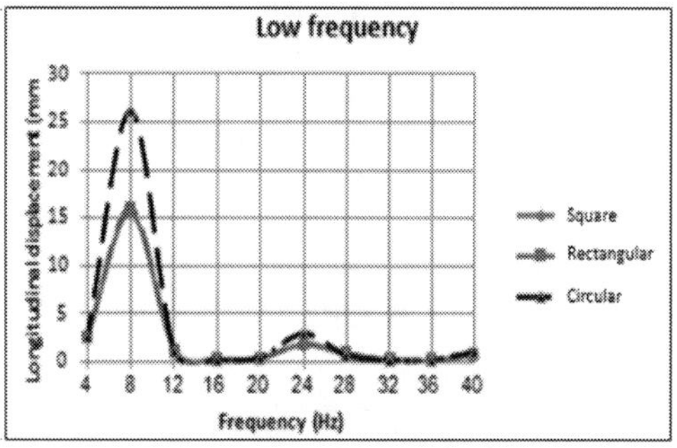


**Figure 11**: Longitudinal displacement for all columns vs. frequency (low)

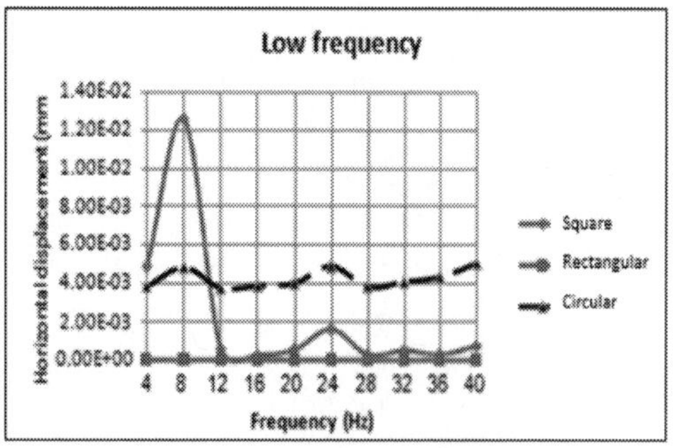


**Figure 12**: Horizontal displacement for all columns vs. frequency (low)

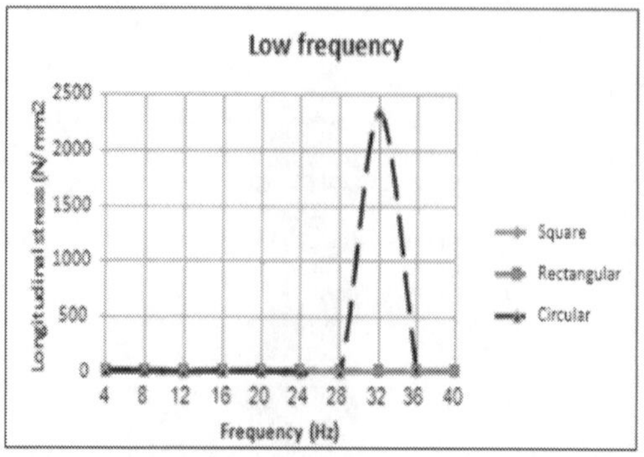


**Figure (13)**: Longitudinal stress for all columns vs. frequency (low)

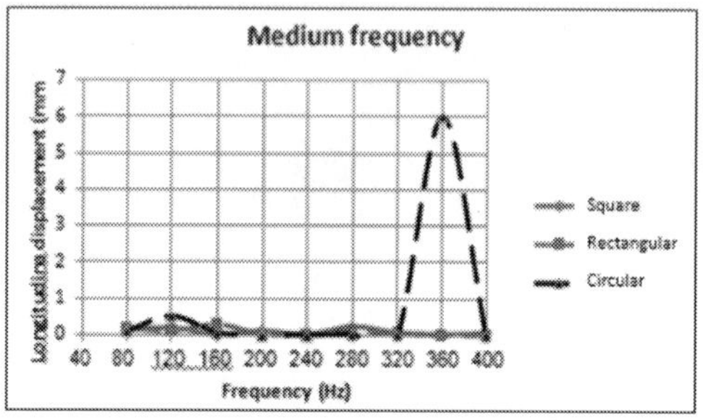


**Figure 14**: Longitudinal displacement for all columns vs. frequency (medium)

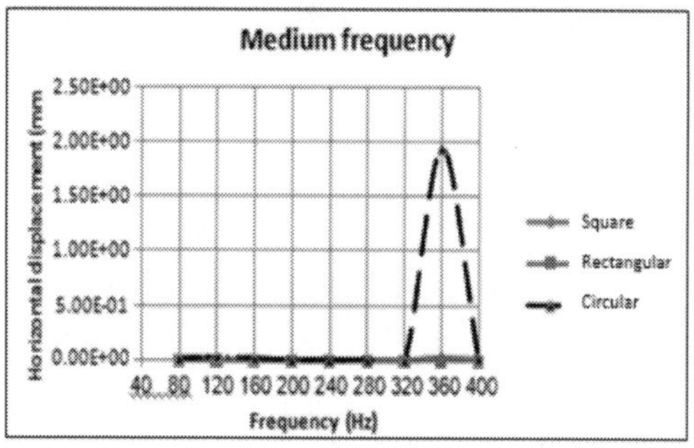


**Figure 15**: Horizontal displacement for all columns vs. frequency (medium)

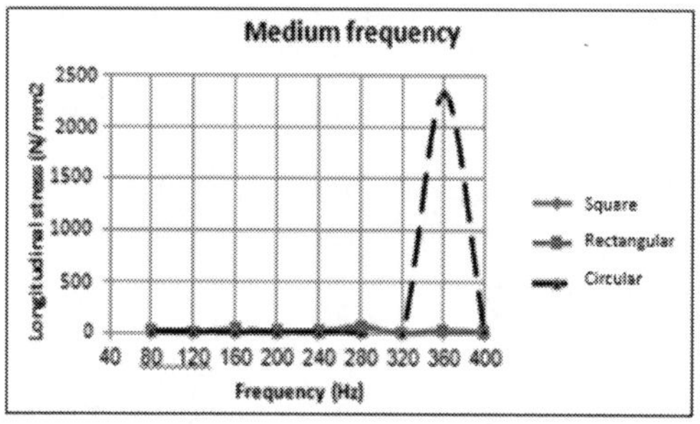


**Figure 16**: Longitudinal stress for all columns vs. frequency (medium)

## 6 Discussions

Reinforced concrete columns were designed based on the ACI-318-2014 code under static loading as a square column first then make equivalent area to rectangular and circular columns with minimum steel reinforcements and then analyzed using finite elements approach and after that a harmonic analysis for all models are adopt to check out the full performance of the columns under dynamic loading Following points was concluded from study:

1.The static analysis results as longitudinal displacement approximately same for all columns cross section and closed with the axial displacement based on the strength of material principle that equal to (2.10 mm), for square (1.98 mm), for rectangular (1.98 mm) and for circular (1.99 mm) so that finite elements simulation deflection results by ANSYS software gives closed results.

2. The maximum horizontal displacement for square is (0.0428 mm), for rectangular is (0.039 and 0.0251 mm) and for circular is (0.0328 mm). the rectangular column gave the lowest value of displacement as compared with other cross sections this is because of having smaller dimension in that direction

3. The longitudinal stress performance developed due to applied loading same for all sections because of the applied same load on the same area.

4. In case of low and medium frequency, the longitudinal displacement for circular section gave value more than other sections but at same frequency (8 and 360 Hz) respectively, because of the stiffness of the circular column is less due to an approximate equivalent area rather than rectangular that gave same results for square section.

5. Stress developed due to the effects of applied harmonic loading are more than that produced due to static loading because the frequency range that applied on the models caused a stress-concentration at the point of maximum longitudinal displacement (points in which the axial displacement occur) that is mean at the point of applied loading. The frequency of circular section gave more stress concentration as compared with others cross section in case of low and medium frequency range.

6. The harmonic analysis must be taken into account to check out the reinforced concrete columns in case of presence of this type of loading because of the performance of the column is different under this type of loading.

7. Square and the equivalent rectangular cross section gave approximately same results and performance under static and dynamic loading.

8. No resonance will occur throughout the low and medium frequency ranges because of there are no infinite result.

## References

1. H. NISHIDA,13th World Conference on Earthquake Engineering , August 1-6, Paper No. 576, (2004).
2. S.sener et al, J.Stru.Eng, **volume 130** issue 4, april (2004).
3. T. Yu et al,*ELSEVIER Engineering Structures* , 680_691 ,PP.681-691, **32** (2010)
4. Z. wu et al, J. Earth. Tsun, **Volume 05**, Issue 02, June (2011).
5. H. Rodrigues et al, ELSEVIER Engineering Structures , 149–164 ,pp.150, 164, **35** (2012)
6. L. Yan et al, Australian Earthquake Engineering Society , Nov 15-17, Hobart, Tasmania ,pp.1-7.
7. Y. Wang Xiaoran Li, *ELSEVIER Engineering Structures* ,289–295 ,pp.289-294, **57** (2013)
8. A. Samir Eisa, International Journal of Civil and Structural Engineering , **Volume 5 No 2**, pp. 155-164..
9. ACI Committee 318M-14, *"Building Code Requirements for Structural Concrete and Commentary "*, American Concrete Institute
10. ANSYS, Inc. Help Manual: is a UL registered ISO 9001:2000 Company. Version (16.2), USA, (2017).

# Resisting punching shear stress in reinforced concrete slabs

*Ebada* Ahmed[1], *Boshra* Eltaly[1], *Fatma* El-Zhraa[1] and *Magdy* Tayel[1]

[1]Faculty of Engineering, Minoufia University, Minoufia, Egypt

**Abstract.** Punching shear failure is a major problem encountered in the design of reinforced concrete flat slabs. The utilization of shear reinforcement via shear studs or other means has become a choice for improving the punching shear capacity. In this study, a new alternative of reinforcement modalities were tested and demonstrated the effect of self-compact concrete on the punching shear capacity, beside that compared between the difference codes to identify the suitable one for determining the position of critical section of punching shear. Nevertheless, in this investigation, the proposed reinforcement system is examined for interior columns only. An experimental work consisting of six specimens: five of them were cast with normal reinforced concrete and one was cast with self-compact strength concrete. The obtained results indicate that the proposed shear reinforcement system has a positive effect in the enhancement of the punching shear capacity of interior slab–column connection of self-compact strength concrete.

## 1 Introduction

Punching shear is a critical design factor of reinforced concrete flat slabs since it is associated with brittle failure. The punching shear strength and deformation capacity are strongly influenced by the type and characteristics of the shear reinforcing system. Ruiz and Muttoni [1] carried out a series of a six full scale slab tests (3.0 x 3.0 x 0.25 m) with the same flexural and shear reinforcing ratio. The slabs were different punching shear reinforcing systems; e.g., separated stirrups, continuous stirrups, bonded reinforcement with anchorage plates, vertical studs and inclined studs. The improvement in punching shear strength and ductility as a result of these systems increased with the same order as they have been mentioned, with the vertical and inclined studs giving the best results; 77% and 119% improvement, respectively. Bent-up bars also improve the punching shear strength and deformation capacity as reported by Tassinari [2]. Lips [3] and Lips and Muttoni [4] demonstrated the positive effect of both shear studs and continuous stirrups on the punching shear strength and deformation capacity of slab–column connection. The same conclusion was achieved by Pilakoutas [5] for inclined shear band reinforcement. The shear strength is proportional to the flexural reinforcement ratio; in contrast, the rotation capacity is inversely proportional to the flexural reinforcement [6]. This study explored the possibility of enhancing the punching shear strength by introducing horizontal mesh reinforcement at the middle of the depth of slab–column connection zone. This reinforcement arrangement is easy to apply and economic in comparison with the other reinforcement types. Kueres et al[7], evaluated the punching shear design provisions according to Eurocode 2 by means of comparisons with test results. To verify the changes in the current design provisions, the new design method was evaluated using large databanks for flat slabs and column bases without and with shear reinforcement as well as systematic test series. The evaluation of the large databanks indicates a well-balanced level of safety of the proposed design method for all investigated types of members. Mabrouk et al., [8] asses the contribution of horizontal flexural reinforcement and vertical shear reinforcement on the punching behavior of reinforced concrete flat slabs. The result from this study showed that the failure of all the tested specimens was a brittle punching failure, adding vertical shear reinforcement in the form of stirrups improved the punching capacity of the slabs, increasing the width of stirrups caused a slight increase in the punching strength, also as the flexural reinforcement ratio was increased, the punching capacity was increased. Finally the current Egyptian code is more conservative than the ACI code in estimating the punching shear capacity.

Self-compacting concrete (SCC) is one of the new types of concrete that possesses High flow-ability, passing ability and stability. It is a highly workable concrete that can flow through congested reinforcing under its own weight and sufficiently fill voids without tending to segregation or excessive bleeding and without the need for Vibration to consolidate it [9-11]. In addition, self-compacting concrete should maintain its flow-ability for a reasonably long time [12].

The provisions for calculating the ultimate punching shear capacity recommended by different building codes are reviewed in the following:

© The Authors, published by EDP Sciences. This is an open access article distributed under the terms of the Creative Commons Attribution License 4.0 (http://creativecommons.org/licenses/by/4.0/).

- In the Egyptian Code of Practice, ECP-203 [13], the critical shear perimeter is located at a distance 0.5d from the column face, it does not account for the flexural reinforcement effect and the concrete strength is limited to 40 MPa. Besides, it does not take into consideration the contribution of punching shear reinforcement.
- According to the ACI 318-11 [14], the critical perimeter is assumed at 0.5d from the perimeter of the loaded area.
- The critical section adopted by the British Standard, BS-8110 [15], lies at 1.5d from the column face. If shear reinforcement is considered, then it should be provided on at least two perimeters within the punching zone. The first perimeter of reinforcement should be located at approximately 0.5d from the face of the loaded area and should contain not less than 40% of the calculated area of reinforcement. The spacing of perimeters of reinforcement should not exceed 0.75d and the spacing of the shear reinforcement around any perimeter should not exceed 1.5d. The shear reinforcement should be anchored round at least one layer of tension reinforcement. The shear stress should then be checked on successive perimeters at 0.75d intervals until a perimeter is reached which does not require shear reinforcement.
- The critical section adopted by the Eurocode 2, EC2 [16], lies at 2d from the column face, In case of shear reinforcement, a control perimeter set at a distance of 1.5d from the outer most perimeter of shear reinforcement

The main objective of the current work is to compare the punching shear strength of six flat slabs; one made from self-compact concrete, and the remaining made with normal strength concrete, four of these normal slabs were strengthened by bent bar at different location from the column face and one without strengthened to be used as a control slab.

# 2 Experimental work

## 2.1 Test specimens

An experimental program were consisting of six square flat slab specimens. One specimen was cast with self-compact concrete and the others specimens were cast with normal strength concrete (the control specimen and four specimens strengthened against the shear punching). The specimens ABBS1, ABBS2 and ABBS3 were strengthened by bent bar started at 0.5 d, 1.0 d and 1.5 d, respectively from the column face in two directions, the remaining specimen ABBS4 was strengthened by bent bar in which the middle of it was at 0.5 d from the column face. Where d is the distance between the column face and bent bar. The details of test specimens are indicated in Table 1 and in Figs. 1&2. All the test specimen consisted of square flat slabs 1200 mm length and 140 mm thick with 200 mm square reinforced concrete column stubs extending 200 mm above the slabs. The column stub was cast monolithically with the slab. All the slabs were identical in dimensions. The reinforcement was distributed uniformly throughout the width of the slab as shown in Figs. 1 and 2.

## 2.2 Properties of materials

Natural siliceous sand was used as the fine aggregate throughout the current research. The sieve analysis was done on the sand and its result is presented in Table 2. Natural dolomite from Suez zone was used as a coarse aggregate. The dolomite has a nominal maximum size of (20 mm). The particles were smooth in texture with 80 percent of them angular shape. The sieve analysis for the dolomite used is shown in Table 3. Ordinary Portland cement was used and its chemical and physical properties were analyzed according to E.S.S. (2011) for concrete works. Fresh drinking water and free from impurities was used for mixing and curing of the test specimens.

For the concrete strength; standard cubes have been tested for strength control. Table 4 shows the average strength of all specimens designated as 25 MPa for normal slab concrete and 40 MPa for self-compact concrete slab. Based on the results of different trail mixes, the concrete mix proportions illustrated in Table 5 were selected to cast the test specimens with normal strength concrete. The concrete mix proportions illustrated in Table 6 were selected to cast the test specimen with self-compact concrete.

High tensile steel bars with 10 mm diameters were used as main and secondary reinforcement in the two directions of tested slabs. The concrete columns were reinforced with normal mild steel as stirrups with 8 mm diameters and high tensile steel bars with 10 mm diameters as main reinforcement. Table 7 illustrates mechanical properties of steel reinforcement.

## 2.3 Preparation of test specimens

Six square wooden forms for slab specimens 120×120×14 cm with stub columns 20×20x20 cm at the center of each slab specimen were prepared. The reinforcement was then placed in their right position in the forms (Fig. 3). Wooden forms for the stub columns were installed at the center of the specimens. A 50 liter capacity mixer was used with a speed of about 50 revolutions per minute. Materials were added in the following sequence: coarse aggregate, sand and cement. The dry constituents were mixed without water for two minutes, then the water was added and the operation continued for an additional three minutes to ensure adequate mixing as shown in Fig. 4. The forms were coated with oil before casting. The concrete mix was charged out from the mixer bowl. A mechanical vibrator was used in placing the concrete around the reinforcing bars together with the hand tamping and rodding to ensure full compaction. The slabs specimens were left in the forms for 24 hours after which the sides of the forms were stripped. The specimens were cured by water sprinkling twice a day for 28 days.

### 2.4 Test set-up

The specimens were tested in Al-Menoufia University Reinforced Concrete Lab. The models were tested under concentrated loads at the center of the column stub up to failure. Loads were applied in increments using a hydraulic jack of 100 ton maximum capacity. Dial gauges of 0.01 mm accuracy and total capacity of 25 mm were used for measuring deflections, first cracking loads and ultimate failure loads were recorded. Propagation of cracks was marked after each load increment up to failure. The description of loading system is shown in Fig. 5, while Fig .6 shows the arrangements of the dial gauges for different specimens.

## 3 Results and discussion

The behavior of the test specimens in terms of load-deflection relationship and mode of failure are illustrated and discussed in the current sections. Load-deflection curves for all specimens are presented in Fig. 7. Table 8 shows the experimental results of all the tested slabs in the terms of ultimate and first crack load, the deflection at the failure and at the first crack load, ductility ratio and energy absorption. The ductility ratio was calculated as ratio of the mid span deflection at the ultimate load to that at the first cracking load. The energy absorption was obtained by calculating the area under the load-deflection curve for each slab. Fig. 8 shows the ultimate and the first crack loads for all tested slabs. The average ductility ratio and energy absorption for all tested slabs are shown in Fig. 9.

### 3.1 Effect of concrete strengthening

Fig. 10 shows the comparison between load-deflection curves at the center (D1) of the control slab S025 and specimen S040, and at the midpoint (D3) between the center of slab and the support. The ultimate load and the first crack load for self-compact concrete (S040) was higher than the control slab (S025) by 30% and 25%, respectively.

### 3.2 Effect of bent bars position

Fig. 11 illustrates comparison between load-deflection curves at D1 and D3 for control slab (S025) and Group C (ABBS1, ABBS2 and ABBS3). The results shows that the ABBS1 has the highest ultimate load and it was double load of the control slab, the ABBS2 was higher than the control slab by about 70% for ultimate load and by about 50% for first crack load, the lowest load for experimental groups was associated with ABBS3, it was 60% and 20% higher than the control slab for ultimate load and the first crack load, respectively.

### 3.3 Effect of start and middle of bent bar at 0.5 d

Comparison between load-deflection curves at D1 and D3 for the specimen ABBS1 and ABBS4 in which the specimens was strengthened at 0.5 d by bent bar at start and at the middle from the column face in two directions; respectively was shown in Fig. 12. The results show that ABBS4 has higher ultimate and first crack loads (375 KN and 175 KN) than control group (250 kN and 100 kN), but it is lower than that for ABBS1 (500 kN and 200 kN).

### 3.4 Crack observations

For all the specimens, surface cracks on the tension side of reinforcement, had been observed and marked during the test. The initial crack development in all specimens followed a similar pattern. Diagonal cracks were getting wider and tangential cracks had been developed. Fig. 13 shows the crack pattern in the tension side for all specimens. For the compression side, cracks had been developed around the column in all specimens.

## 4 Conclusions

The main goal of the current research is examining the punching shear resistance of flat slab by using self-compact concrete and using additional modalities of bent bar reinforcement, to compare between different codes in order of identifying the suitable code for determining the position of critical section of punching shear. From the experimental results, the following conclusions could be drawn as below:

1-The increase of strength concrete lead to increase of ultimate load of slab, and this explain the high resistance of punching shear strength of self-compact concrete.

2-The critical parameter of punching shear lay at 0.5 d from the column face, so the Egyptian code is the best international code for determining the critical section of punching shear.

3-Strengthening of slab at position more or less than 0.5 d leads to reduce of punishing shear resistance of slab.

## References

1. Ruiz F. M. and Muttoni A. "Performance and design of punching shear reinforcing systems", 3rd Fib International Congress, 2010.
2. Tassinari L., Lips S., Muttoni A., and Ruiz MF, "Applications of bent-up bars as shear and integrity reinforcement in R/C slabs", Session 2B-5: construction Technology, 2011.
3. Lips S., "Punching of flat slabs with large amounts of shear reinforcement", Ecole Polytechnique Federale de Lausanne, 2012.
4. Lips S., and Muttoni A., "Experimental investigation of reinforced concrete slabs with punching shear reinforcement", 8th Fib PhD Symposium, Lyngby, Denmark, 2010.
5. Pilakoutas K., and Li X. "Alternative shear reinforcement for reinforced concrete flat slabs", Journal of Structural Engineering, Vol. 129, No. 9pp.1164–72, 2003.
6. Kinnunen S., and Nylander H., "Punching of concrete slabs without shear reinforcement", Transactions of the

Royal Institute of Technology, No. 158, Stockholm, Sweden, pp. 112, 1960.
7. Kueres D., Siburg C., Herbrand M., Classen M., and Hegger J. "Uniform Design Method for punching shear in flat slabs and column bases", Engineering Structural, Vol. 136,149–64, 2017.
8. Mabrouk R. T. S., Bakr A., and Abdalla H. "Effect of flexural and shear reinforcement on the punching behavior of reinforced concrete flat slabs", Alexandria Engineering Journal, 2017.
9. Okamura H., and Ouchi M. "Self-compacting concrete, development, present use and future", Proceeding of the 1st international RILEM symposium on self- compacting concrete, stochkolm, RILEM Publications, 1999.
10. Yahia A., Tanimura M., Shimabukuro A., and Shimoyama Y. "Effect of rheological parameters on self compactibility of concrete containing various mineral admixtures" Proceedings of the 1st international RILEM symposium on self-compacting concrete, Stockholm, RILEM Publications, 1999.
11. Ozawa K., Sakata N., and Okamura H., "Evaluation of self-compact ability of fresh concrete using the funnel test", Concrete Library JSCE, 1995.
12. Okamura H., and Ozawa K. "Mix design for self-compacting concrete", Concrete Library JSCE, 1995.
13. Egyptian Code Committee 203, Egyptian Code of Practice for Design and Construction of Reinforced Concrete Structures, Housing and Building Research Centre, Cairo, 2007 (in Arabic).
14. American Concrete Institute, ACI-318, Building Code Requirements for Structural Concrete, ACI 318-11, ACI Committee 318, pp. 503, Detroit, USA, 2011.
15. British Standards Institution, Structural Use of Concrete, BS- 8110: Part 1-Code of Practice for Design and Construction, London, 1997.
16. Eurocode 2, Design of Concrete Structures, Part 1–6, General Rules and Rules for Buildings, European Prestandard, European Committee for Standardization, 1992.

**Table 1.** Details of the test specimens

| Group | Specimen Code | Fcu (MPa) | Conditions | Bottom RFT | Top RFT | Bent Bar |
|---|---|---|---|---|---|---|
| (A) | S025 | 25 | Control | 6 Ø 10 | 8 Ø 10 | - |
| (B) | S040 | 40 | Self-compact concrete | 6 Ø 10 | 8 Ø 10 | - |
| (C) | ABBS1 | 25 | Strengthen by bent bar started at 0.5 d from the column face | 6 Ø 10 | 8 Ø 10 | 4 Ø 10 |
| | ABBS2 | 25 | Strengthen by bent bar started at 1.0 d from the column face | 6 Ø 10 | 8 Ø 10 | 4 Ø 10 |
| | ABBS3 | 25 | Strengthen by bent bar started at 1.5 d from the column face | 6 Ø 10 | 8 Ø 10 | 4 Ø 10 |
| (D) | ABBS4 | 25 | strengthen by bent bar in which the middle of it was at 0.5 d from the column face | 6 Ø 10 | 8 Ø 10 | 4 Ø 10 |

**Table 2.** Grading of fine aggregate according to (ASTM C33).

| Sieve size (mm) | 9.5 | 4.75 | 2.36 | 1.18 | 0.61 | 0.31 | 0.16 |
|---|---|---|---|---|---|---|---|
| % passing (ASTM C33) | 100 | 95-100 | 80-100 | 50-85 | 25-60 | 5-30 | 0-10 |
| % passing used dolomite | 100 | 100 | 96 | 75 | 45 | 20 | 1 |

**Table 3.** Grading of Coarse Aggregate According to (ASTM C33) and Grading of Natural Dolomite Used.

| Sieve size (mm) | 20 | 12.5 | 9.5 | 4.75 | 2.36 |
|---|---|---|---|---|---|
| % passing (ASTMC33) | 100 | 90-100 | 40-70 | 0-15 | 0-5 |
| % passing used dolomite | 100 | 96 | 85 | 53.2 | 7.3 |

**Table 4.** Used material properties

| Cubes number | F cu (MPa) | Average (MPa) |
|---|---|---|
| 1 | 27.32 | Normal strength concrete 25 |
| 2 | 26.89 | |
| 3 | 26.49 | |
| 4 | 24.65 | |
| 5 | 39.81 | Self-compact concrete 40 |
| 6 | 43.65 | |
| 7 | 42.21 | |

**Table 5.** Concrete mix design (normal strength concrete 25 MPa)

| Cement | Water | Coarse aggregate | Fine aggregate |
|---|---|---|---|
| 300 kg | 150 kg | 1290 kg | 645 kg |

**Table 6.** Concrete mix design (self-compact concrete 40 MPa)

| Cement | Water | Coarse aggregate | Fine aggregate | Flyash | ViscoCrete |
|---|---|---|---|---|---|
| 425 kg | 148 kg | 838 kg | 686 kg | 85 kg | 17 kg |

**Table 7.** Mechanical properties of steel reinforcement

| Steel Type | Yield Stress (kg/cm2) | Tensile Strength (kg/cm2) | Elongation % | Modulus of Elasticity (t/cm2) |
|---|---|---|---|---|
| High Tensile | 3675 | 5350 | 16.0 | 2100 |
| Normal Mild | 2550 | 3750 | 24.30 | 2050 |

**Table 8.** Experimental results

| slab code | Load (KN) | | Deflection at Ultimate Load (mm) | Deflection at First Crack (mm) | Ductility Ratio | Energy Absorption (KN.mm) |
|---|---|---|---|---|---|---|
| | Ultimate | First Crack | | | | |
| S025 | 250 | 100 | 6.50 | 2.6 | 2.50 | 78.12 |
| S040 | 325 | 125 | 8.20 | 2.15 | 3.81 | 1528.00 |
| ABBS1 | 500 | 200 | 9.0 | 2.04 | 4.41 | 3027.50 |
| ABBS2 | 425 | 150 | 8.10 | 1.49 | 5.44 | 2209.50 |
| ABBS3 | 400 | 120 | 7.60 | 1.12 | 6.79 | 1917.40 |
| ABBS4 | 375 | 175 | 8.40 | 3.42 | 2.46 | 1721.50 |

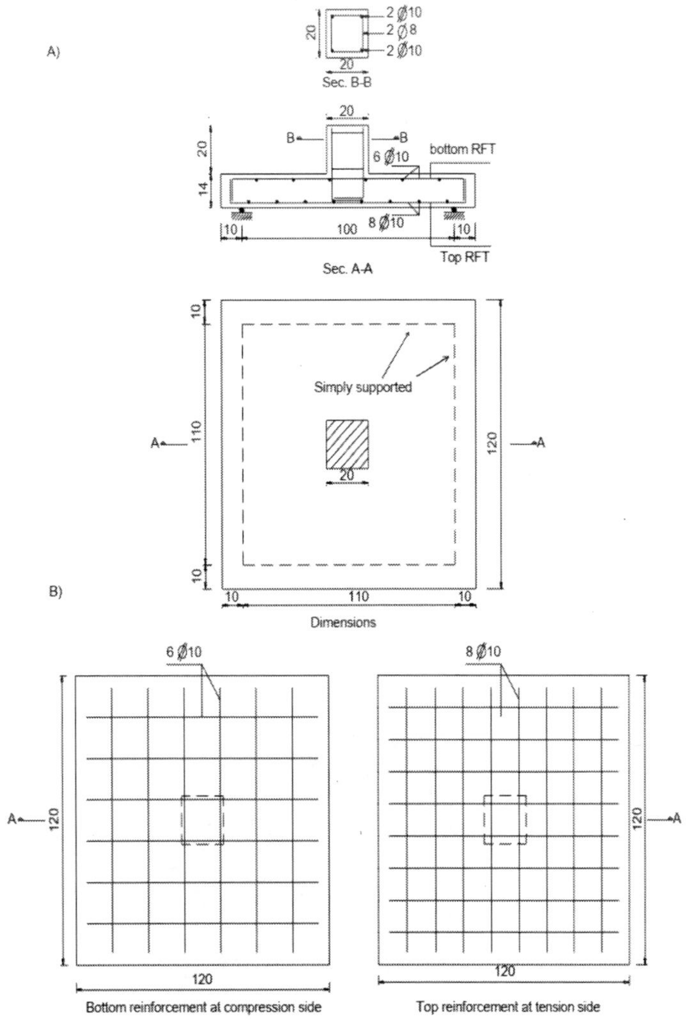


**Fig. 1.** A) Dimensions of slabs, B) Details of reinforcement of reference S025, S040, Dimensions in cm.

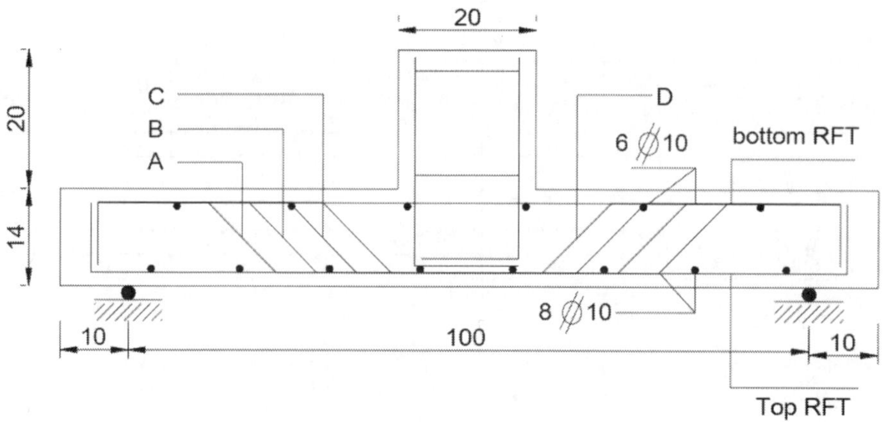


**Fig. 2.** Typical steel arrangement details of slabs: A) Start of bent bar at 1.5 d from column face, B) Start of bent bar at 1.0 d from column face, C) Start of bent bar at 0.5 d from column face, D) Middle of bent bar at 0.5 d from column face, Dimensions in cm.

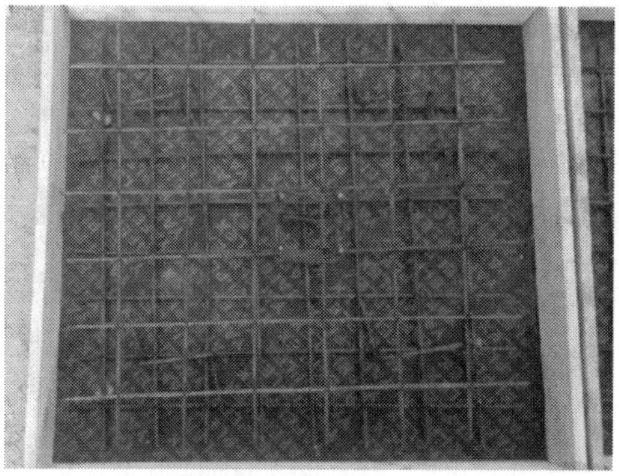

**Fig. 3**. Wooden forms and reinforcement.

https://doi.org/10.1051/matecconf/201816204025

**Fig. 4.** Mixing of materials

**Fig. 5.** Description of the loading system.

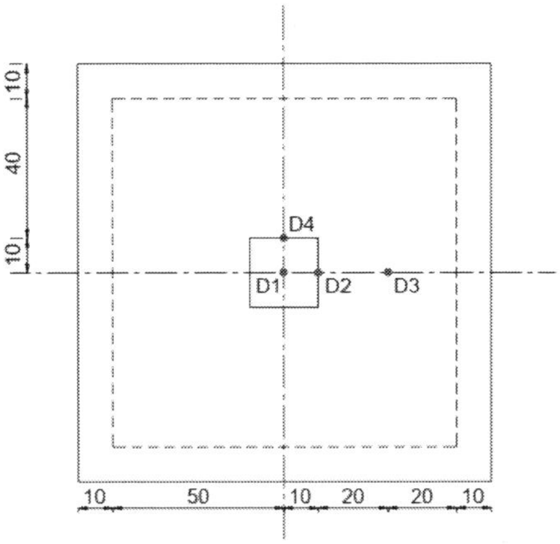


**Fig. 6.** Arrangements of dial gauges for different specimens, Dimensions in cm

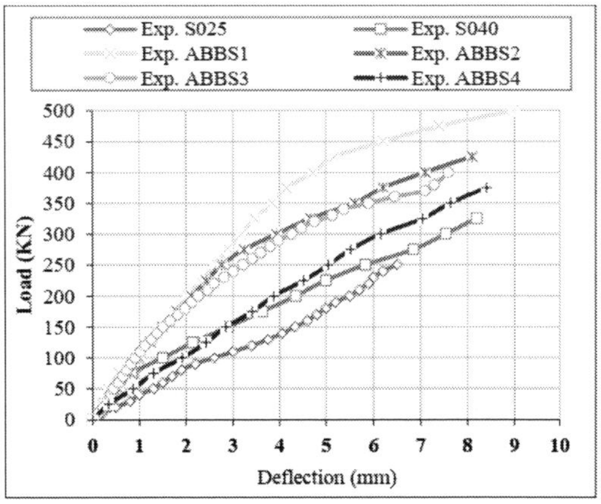

https://doi.org/10.1051/matecconf/201816204025

**Fig.7.** Load-deflection curves of all test specimens at D1

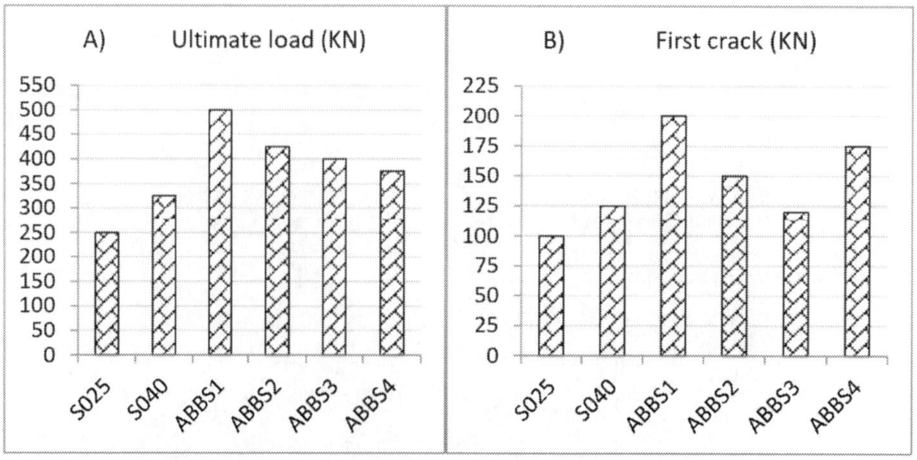


**Fig. 8.** A) Ultimate load for all tested slabs, B) First crack load for all tested slabs

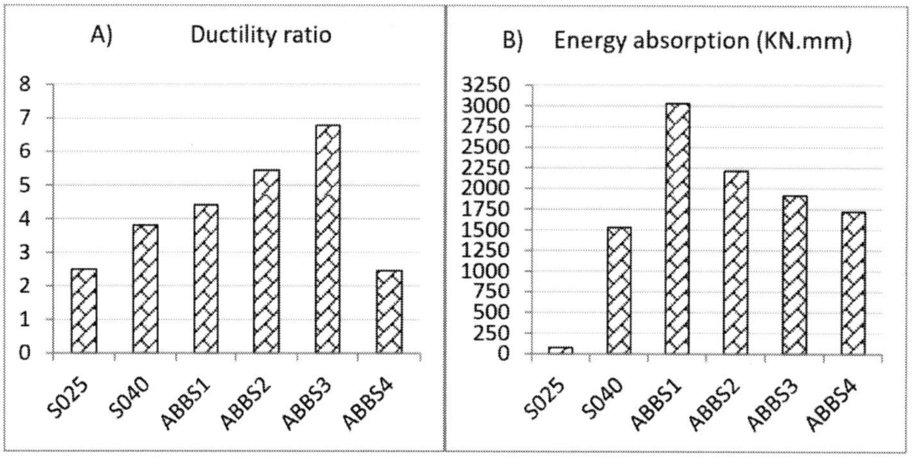


**Fig. 9.** A) The average ductility ratio for all tested slabs, B) Energy absorption for all tested slabs

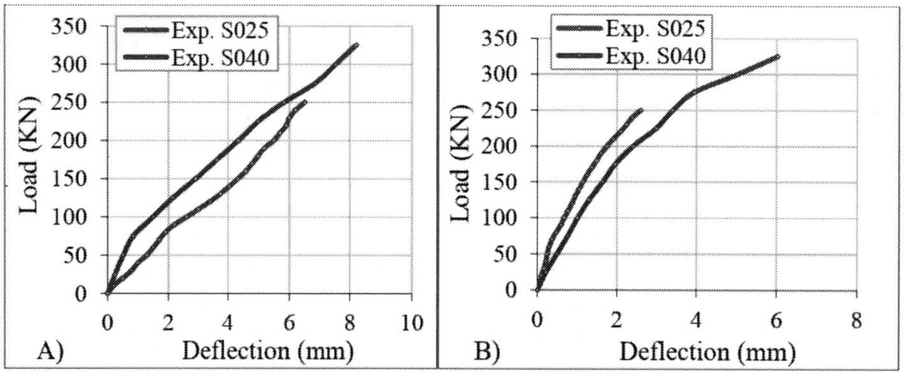

**Fig. 10.** Comparison between load-deflection curves of the specimens S025 and S040, A) load-deflection at D1, B) load-deflection at D3.

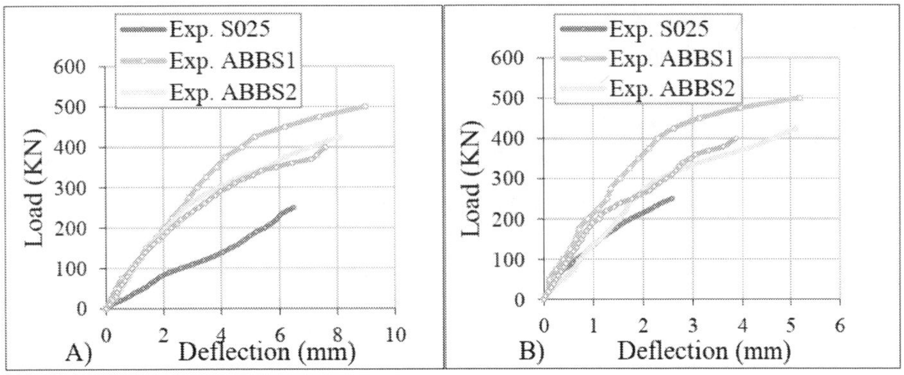


**Fig. 11.** Comparison between load-deflection curves of the specimens S025 and all specimens of Group D: A) load-deflection at D1 and B) load-deflection at D3.

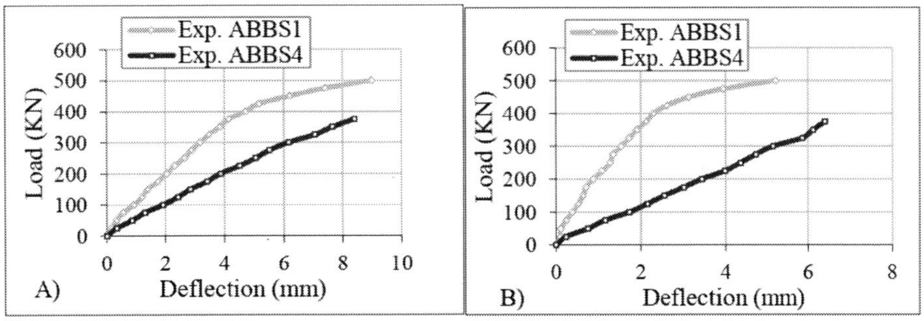


**Fig. 12.** Comparison between load-deflection curves of the specimens ABBS1 and ABBS4; A) load-deflection at D1, B) load-deflection at D3.

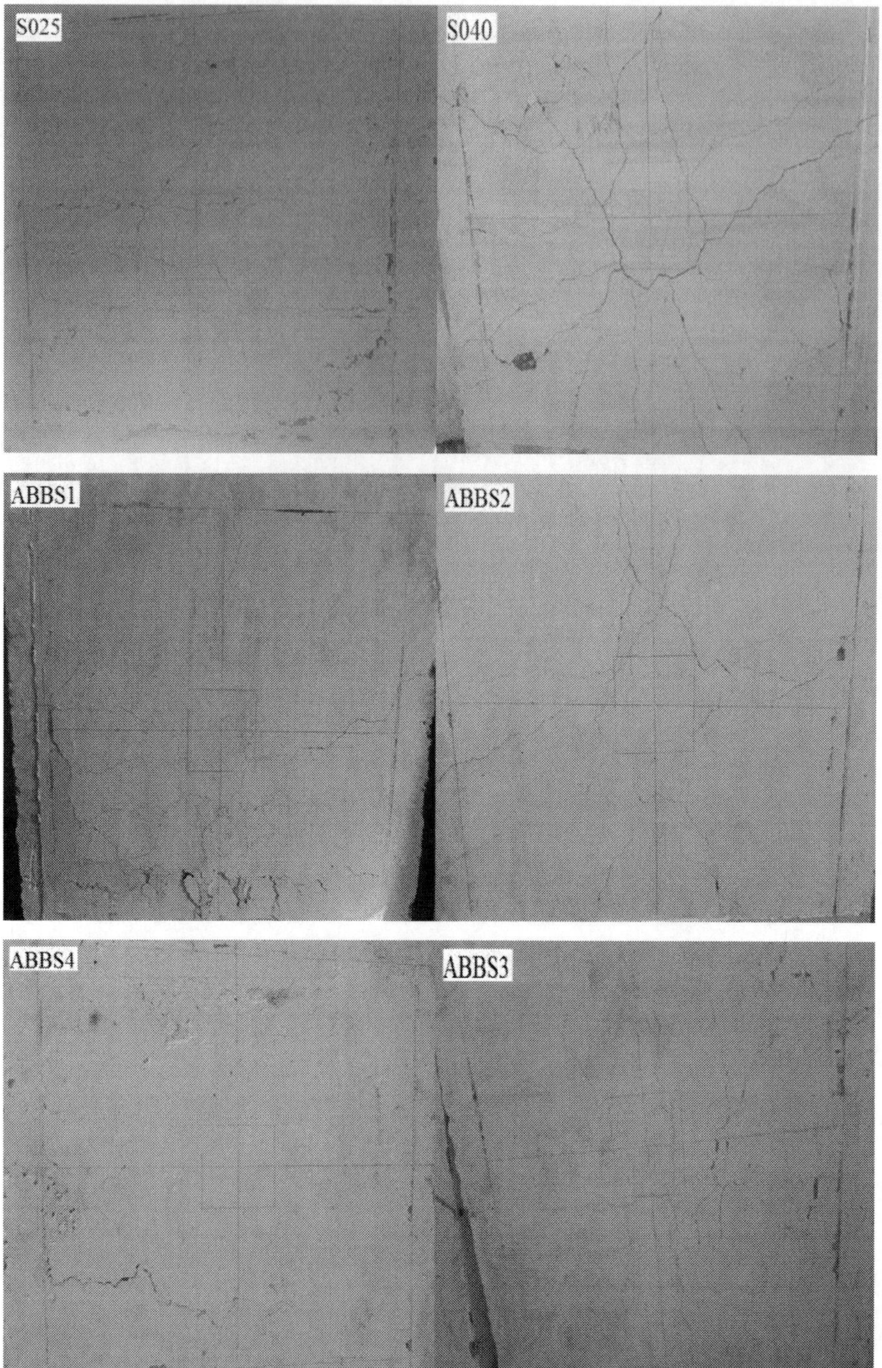


**Fig. 13.** The crack pattern (propagation) in tension side for all specimens

# Evaluation of soil-structure interaction for structures subjected to earthquake loading with different types of foundation

*Mohammed* Elwi[1, *], *Bassman* Muhammed[1] and *Nada* Alhussiny[1]

[1]Building and Construction Engineering Department, University of Technology, Baghdad.

**Abstract.** However though the structures are supported on soil, most of the designers do not consider the soil structure interaction and its subsequent effect on structure during an earthquake. Different soil properties can affect seismic waves as they pass through a soil layer. When a structure is subjected to an earthquake excitation, it interacts the foundation and soil, and thus changes the motion of the ground. It means that the movement of the whole ground structure system is influenced by type of soil as well as by the type of structure. Tall buildings are supposed to be of engineered construction in sense that they might have been analyzed and designed to meet the provision of relevant codes of practice and building bye-laws. IS 1893: 2002 "Criteria for Earthquake Resistant Design of Structures" gives response spectrum for different types of soil such as hard, medium and soft. An attempt has been made in this paper to study the effect of Soil-structure interaction on multi storeyed buildings with various foundation systems. Also to study the response of buildings subjected to seismic forces with Rigid and Flexible foundations. Multi storeyed buildings with fixed and flexible support subjected to seismic forces were analyzed under different soil conditions like hard, medium and soft. The buildings were analyzed by Response spectrum method using software SAP2000. The response of building frames such as Lateral deflection, Story drift, Base shear, Axial force and Column moment values for all building frames were presented in this paper.

## 1 Introduction

As waves from an earthquake reach a structure, they produce motions in the structure. These motions depend on the structure's vibrational characteristics and the layout of structure. For the structure to react to the motion, it needs to overcome its own inertia force, which results in an interaction between the structure and the soil. The extent to which the structural response changes the characteristics of earthquake motions observed at the foundation level depends on the relative mass and stiffness properties of the soil and the structure. Thus the physical property of the foundation medium is an important factor in the earthquake response of structures supported on it. The process in which the response of the soil influences the motion of the structure and the motion of the structure influences the response of the soil is termed as soil-structure interaction (SSI) [1].

## 2 Literature review

Anand and Mightraj [2] studied the seismic behavior of RCC buildings with and without shear wall under different soil conditions. One to fifteen storied space frames with and without shear wall were analyzed using ETABS software for different soil conditions (hard, medium, soft). The values of Base shear, axial force and lateral displacement were compared between two frames. Lateral displacement, base shear, axial force and moment in the column value increases when the type of soil changes from hard to medium and medium to soft for all the building frames. It was concluded that the soil structure interaction must be suitably considered while designing frames for seismic forces.

Pandey et al. [3] studied the seismic soil-structure interaction of buildings on hill slopes. Static pushover analysis and response spectrum analysis (RSA) was conducted on five buildings i.e. three step back buildings and two step back-set back buildings with varying support conditions. These buildings had been analyzed for different soil conditions (hard, medium and soft soils) using SAP2000 software idealized by equivalent springs. The response parameters, i.e. total base shear (V), displacement from pushover analysis ($\delta$ performance point), displacement from RSA ($\delta$ elastic) and response correction factor (R) had been studied with respect to fixed base analysis to compare the effect of soil springs. In general it was found that response reduction factor decreases with increasing time period, but it was expected to be constant beyond a certain value of time period.

Amin and Hasan [4] studied the effect of soft soils on multi-storey reinforced concrete building frame. Four

* Corresponding author: civillaw80@gmail.com

© The Authors, published by EDP Sciences. This is an open access article distributed under the terms of the Creative Commons Attribution License 4.0 (http://creativecommons.org/licenses/by/4.0/).

building models (3, 6, 9 and 12 storey) with identical building plan were considered to investigate the effect of soft soil for multistoried reinforced concrete building frame.

## 3 Procedure

Multi storeyed building with fixed and flexible base subjected to seismic forces were analysed under different soil conditions like hard, medium and soft. The buildings were analysed using response spectrum method by the software SAP2000 Expert. Seismic dissection might have been conveyed crazy eventually Tom's perusing accompanying IS1893:2002-PartI [6]. Diverse reaction outcomes were found.

### 3.1 Input- information

Measure of the building 30m X 30m, sort from claiming structure : RCC Multi story frames, seismic zone v , reaction decrease variable - 3, essentialness component – 1 , tallness of the building- 40m, no about storey–10, tallness from claiming floor – 4m, forced load – 4 kN/m$^2$, Materials – M25(beam), M30(column), Fe415, profundity of the piece – 150mm, unit weight for RCC – 25kN/m$^3$, kind of soil – difficult , medium and soft, reaction spectra – may be 1893(Part I) 2002 , damping 5% , profundity of establishment - 1. 5m, divider thickness-230mm. Table 1 indicates the quality from claiming soil firmness in parallel (x, z and y) headings.

**Table 1.** Soil stiffness values for buildings with flexible base

| Type of soil | Soil Stiffness (kN/m) | | |
|---|---|---|---|
| | $K_x$ | $K_y$ | $K_z$ |
| hard | 8000 | 100000 | 8000 |
| medium | 4000 | 50000 | 4000 |
| soft | 1500 | 25000 | 1500 |

## 4 Analysis of results

Ten storeyed building frames with fixed and flexible base analysed and designed to understand the behaviour were under seismic forces. Various seismic responses were compared for both the type of building frames. Fig. 1 indicates the parallel redirection of fabricating frames subjected will seismic powers. Fig 1 parallel redirection of a building outline with settled Fig. 2 indicates the conveyance from claiming seismic powers all around those stature of the building frame.

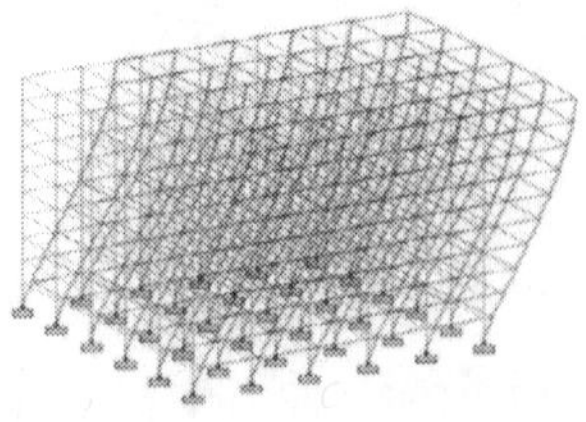

**Fig. 1.** Parallel redirection of a building outline with altered base.

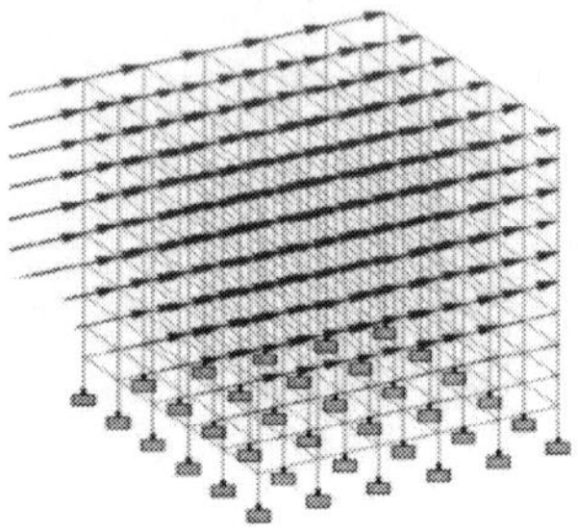

**Fig. 2.** Seismic stacking of a building outline with altered base.

Table 2 shows the values of Lateral deflection of building frames with fixed base for different soil condition.

**Table 2**. Lateral deflection of building frames with fixed base

| Storey | Lateral Deflection (mm) | | |
|---|---|---|---|
| | Hard soil | Medium soil | Soft soil |
| Tenth | 58.93 | 79.19 | 96.72 |
| Ninth | 56.60 | 76.12 | 93.05 |
| Eighth | 52.92 | 71.17 | 87.01 |
| Seventh | 48.01 | 64.55 | 78.89 |
| Sixth | 42.15 | 56.61 | 69.15 |
| Fifth | 35.56 | 47.69 | 58.22 |
| Fourth | 28.45 | 38.09 | 46.47 |
| Third | 21.03 | 28.09 | 34.23 |
| Second | 13.46 | 17.90 | 21.78 |
| First | 5.98 | 7.88 | 9.54 |

Table 3 indicates the values for story float about fabricating frames with settled build to diverse dirt condition.

**Table 3**. Storey float qualities for fabricating with settled base.

| Story | Story Drift (mm) | | |
|---|---|---|---|
| | Hard soil | Medium soil | Soft soil |
| Base | 0 | 0 | 0 |
| First | 5.10 | 6.94 | 8.52 |
| Second | 7.38 | 10 | 12.33 |
| Third | 7.60 | 10.34 | 12.69 |
| Fourth | 7.48 | 10.18 | 12.49 |
| Fifth | 7.19 | 9.77 | 12 |
| Sixth | 6.69 | 9.11 | 11.18 |
| Seventh | 5.98 | 8.13 | 9.99 |
| Eighth | 5 | 6.80 | 8.35 |
| Ninth | 3.72 | 5.06 | 6.22 |

Table 4 shows the values of Lateral deflection of building frames with flexible base for different soil condition

**Table 4.** Lateral deflection values of building with flexible base

| Storey | Lateral Deflection (mm) | | |
|---|---|---|---|
| | Hard soil | Medium soil | Soft soil |
| Tenth | 92.27 | 131.38 | 218.76 |
| Ninth | 89.97 | 128.62 | 214.63 |
| Eighth | 86.48 | 124.31 | 208.56 |
| Seventh | 81.88 | 118.88 | 201.14 |
| Sixth | 76.45 | 112.64 | 192.79 |
| Fifth | 70.44 | 105.91 | 183.87 |
| Fourth | 64.07 | 98.95 | 174.76 |
| Third | 57.50 | 92.01 | 165.70 |
| Second | 50.85 | 85.19 | 156.82 |
| First | 43.88 | 78.46 | 147.92 |

Table 5 indicates the values for story float about fabricating frames with settled build to diverse dirt condition.

**Table 5.** Story float values about building with adaptable base.

| Story | Story Drift (mm) | | |
|---|---|---|---|
| | Hard soil | Medium soil | Soft soil |
| Base | 0 | 0 | 0 |
| First | 16.25 | 18.77 | 26.47 |
| Second | 8.54 | 11.13 | 15.45 |
| Third | 7.47 | 9.75 | 13.62 |
| Fourth | 7.19 | 9.33 | 13.12 |
| Fifth | 6.89 | 8.94 | 12.62 |
| Sixth | 6.43 | 8.36 | 11.89 |
| Seventh | 5.77 | 7.55 | 10.83 |
| Eighth | 4.87 | 6.44 | 9.40 |
| Ninth | 3.71 | 5.02 | 7.56 |

Table 6 indicates diverse reactions from claiming edifices with altered Furthermore adaptable base to Different soil state.

**Table 6.** Examination of reactions about fabricating span for altered Furthermore adaptable base under diverse soil Condition for adaptable build.

| Parameter | Fixed base | | | Flexible base | | |
|---|---|---|---|---|---|---|
| | Hard soil | Medium soil | Soft soil | Hard soil | Medium soil | Soft soil |
| Lateral deflection (mm) | 58.93 | 79.19 | 96.72 | 92.27 | 131.38 | 218.76 |
| Storey drift (mm) | 7.38 | 10 | 12.33 | 16.25 | 18.77 | 26.47 |
| Base shear (kN) | 1744.16 | 2391.97 | 3093.24 | 1763.33 | 2452.24 | 3140.66 |
| Axial force (kN) | 5640 | 5770 | 5810 | 5520 | 5330 | 5120 |
| Column moment (kNm) | 103.54 | 133.59 | 151.95 | 218.23 | 316.86 | 393.99 |
| Seismic weight (kN) | 71190.54 | 71831.44 | 75629.53 | 71972.79 | 73640.94 | 76788.80 |

## 5 Conclusions

Lateral deflection, Storey drift, Base shear and Moment values increases when the type of soil changes from hard to medium and medium to soft for fixed and flexible base buildings. Lateral deflection, Storey drift, Base shear and Moment values of fixed base building was found to be lower as compared to flexible base building. Hence suitable foundation system considering the effect of Soil stiffness has to be adopted while designing building frames for seismic forces.

## References

1. J. P., Wolf, Prentice Hall, Englewood Cliffs, N.J., 1985.
2. N. Anand, and C. Mightraj, Proceedings of the Indian Geotechnical Conference on GEOtrendz, (2010)
3. A.D. Pandey, P. Kumar and S. Sharma, Journal of Civil and Structural Engineering, (2011)
4. M. R., Amin, and P. Hasan, Proceedings of 4th Annual Paper Meet and 1st Civil Engineering Congress, (2011)
5. R. M., Priyanka, and N. Anand, Proceedings of the INCACMA Conference on Advances in Construction, Manufacturing and Automation Research, (2012)
6. IS 1893-Part 1: Indian Standard Criteria for Earthquake Resistant Design of Structures, Part 1–General Provisions and Buildings (Fifth Revision), Bureau of Indian Standards, (2002)
7. IS 456: Indian standard Criteria for Plain and Reinforce Structure, Bureau of Indian Standards, (2000)

# Experimental and numerical study for the post buckling behaviour of plate girders subjected to bending and shear

*Wael* Shawky[1*], and *Ghaidaa* Nabil[2]

[1]University of Technology, Baghdad, Iraq
[2]Ashoor General Constructive Contract Company, Baghdad, Iraq

**Abstract.** This research offers experimental and numerical study for the ultimate strength analysis and post buckling behaviour of plate girders subjected to shear and bending stresses. Two Plate girders of length 2.5m are designed consisting from three separate structural steel plates welded together to form I – section. The dimensions of the first girder section is 512 mm × 120 mm while the second girder section is 412mm × 120 mm. Mechanical properties for plate girders components were found by testing three samples of each plate. The ultimate shear strength was found by examining two plate girders in the laboratory under concentrated load applied at the middle span. A numerical study of the tested plate girders was carried out by using the software program (ANSYS) to study the behaviour of girders steel under the influence of loads application. The results showed that the ratio of the experimental ultimate load strength to the numerical ultimate load strength is 97% 94% for plate girders 1 and 2 , respectively. For the part of using theoretical study. The results obtained from the equations , it was observed that the ratio of the experimental ultimate load strength to the theoretical ultimate load strength is 83% for plate girders1 and 90% for the plate girder 2. Through comparison between experimental, numerical and theoretical results it was observed good and acceptable agreement. Thus confirming the accuracy of the modelling between this methods and obtain convergent results in all cases. Also it was found when an increase in the depth of the web has effect in increasing the ultimate shear strength for plate girder by about 19%.

## 1 Introduction

A girder bridge is likely the most common and most basic bridge. Plate Girders Bridge became popular in the late 19$^{th}$ century, when the plate girders were used for the construction of railroad bridges. In order to obtain plate girders, plates were connected together by rivets and angles. Bolted and riveted plate girders were replaced by using welded plate girders due to their better economy, quality and aesthetics. In general, Modern plate girders are fabricated by welding two steel flanges, a steel web and a series of transverse steel stiffeners. The main function for using the bottom and top steel flange plates is to resist the bending moments by developing axial tensile and compressive stresses from the bending action. Shear stresses will be resisted by the steel webs. The deep and thin steel web is subjected to a large shear force and thus subjected to shear buckling, the shear buckling strength is always depending on plate thickness, aspect ratio, material properties and imperfections. In order to delay shear buckling, it is possible to use a rigid transverse steel stiffeners, they are constructed to limit the extent and separate the buckling regions. Theoretically [1], buckling is caused by a bifurcation in the solution to the equations of static equilibrium. Under an increasing load and at a certain stage, the further loads are able to be resisted in one of two equilibrium states: first is the purely compressed state (with no lateral deviation) while the second is the laterally-deformed state. The buckling is characterized by the sudden lateral failure of a structural member exposed to high compressive stresses, where the ultimate compressive stress that the material can withstand is larger than the compressive stress at the point of failure. As an applied load is increased on a member, such as a column, it will ultimately become large enough to cause the member to become unstable and is said to have buckled. In 1969 Chern and Ostapenko [2] presented a paper for the ultimate strength analysis of the plate girder panels subjected to shear. The method is applicable to homogeneous, hybrid, symmetrical, and unsymmetrical plate girders. The ultimate shear strength was assumed to be equal to the sum of the strengths of the tension field action, the frame action and the beam action. While computing the web plate buckling for the beam action, the web plate was assumed to be simply supported at the stiffeners and fixed at the flanges. The tension field developed by the pattern web after buckling different than those proposed by others. At the same time, the frame action (a panel mechanism with plastic hinges) will be developed by flanges at each corner of the girder panel. A comparison was done with the available 33 test results indicates a more consistent and better accuracy of the proposed method than of the

* Corresponding author: waelalmuhsin@yahoo.com

© The Authors, published by EDP Sciences. This is an open access article distributed under the terms of the Creative Commons Attribution License 4.0 (http://creativecommons.org/licenses/by/4.0/).

methods suggested by other researchers. In 1971 Rockey, Skaloudb[3] presented a plastic method of design for the plate girder Webs. The method allows for the influence of the rigidity of flange upon the webs post buckled behaviour. The design procedure for the study was based on a the behaviour of 40 plate girders tested by the authors, Results were compared to other available experimental data in the Technical Press. Because the design procedure allows for the influence of flange stiffness upon the shear webs post buckled behaviour and according to authors, the method was more accurate than existing design methods. In 1978 Evans, H.R., Porter, D.M. and Rockey, K.C.[4] presented a method to calculate the collapse load for a steel plate girder loaded in both shear and bending. Authors presented an accurate and quick design method which is capable to predict the failure loads of steel plate girders when loaded in bending, shear or by bending and shear. The design process was applied to a large number of test steel girders. Eighty-eight girders, tested by various researchrs were considered. The result of the study showed that a mean value for the ratio of the predicted failure load to the experimental load is 0.997 while the Standard deviation is 0.064. The method can be accurately applied for girders reinforced by both transverse and longitudinal stiffeners. In 2005 Carlos Graciano, and Euro Casanova[5] presented a numerical study to predict the ultimate strength of I-steel girder webs under the effect of combined action of patch loading and bending moment. The nonlinear finite element analysis was used to conduct the study. Plastic material behaviour, Initial geometrical imperfections, and large deformation effects were taken into account in the model. Experimental results taken from the literature were used to validate the finite element model . A parametric study was carried out to investigate the influence of the relative location of the stiffener and the magnitude of the bending moment on the ultimate strength to patch loading. In 2007 Hassan H. Abbas, Richard Sause, and Robert G. Driver [6] presented experimental, theoretical, and finite-element analysis results for the linear elastic behavior of corrugated web I-girders under in plane loading. A typical corrugated web steel girder formed by two steel flanges welded to one corrugated steel web. A theoretical method, the fictitious load method, was used as an analytical tool for quantifying flange transverse bending in corrugated web I-girders. For validating the method, four experimental large-scale corrugated web I-girder were used. To get additional insight, the finite element analysis results for the tested girders were presented, it is then compared to both the theoretical and experimental results. In 2008 F. Shahabian T.M. Robert [7] presented a theoretical predictions for the ultimate resistance of slender plate girders subjected to shear loading, based on existing formulas and theories, an interaction formula for the ultimate resistance of slender plate girders subjected to combined shear and bending loading was proposed. Results show satisfactory correlation with the available theories. The proposed interaction equation covers web panel aspect ratios $(b_w/d_w)$ from 1 to 2 and slender ratios $(d_w/t_w)$ from 150 to 300. In 2008 Donald W. White, and Michael G. Barker[8] a paper to evaluate the accuracy and ease for the use of twelve of the most promising models for the shear resistance of transversely stiffened I-girders. Well established models that are used in civil engineering practice in addition to a number of other recently proposed models also were considered. the paper focuses on the merits and limitations of the alternative models relative to Basler's model. Statistical analyses were conducted on the predictions by using various models and an updated data set from 129 experimental shear tests, samples including 11 horizontally curved and 30 hybrid I- steel girders. The results of the study support the conclusion that the form of Basler's model implemented in 2004 AASHTO LRFD Bridge Design Specifications and 2005 AISC Specification for Structural Steel Buildings gives the best combination of simplicity and accuracy to calculate the shear resistance of transversely stiffened steel I-girders. In 2013 by Dong-Ku Shin Eun-Young Cho and Kyungsik Kim [9] conducted a study for the ultimate flexural strengths of bisymmetric and monosymmetric I-girders subjected to local bend-buckling in webs by nonlinear finite element analysis to include post-buckling behaviours. High performance steel (HPS) was used to fabricate the plate girders. Conventional steel were modelled three-dimensionally with thin shell elements and ultimate strength analyses were performed utilizing nonlinear techniques provided by a commercial finite element analysis package program, ABAQUS. The ultimate flexural strengths of girders with various slenderness of web were evaluated from the nonlinear FEA and compared with values predicted by the specifications in AASHTO LRFD and Eurocode 3. In 2016, Young Bong Kwon, and Seung Wan Ryu [10] presented a paper for a series of shear test results of plate girders fabricated by using steel plates with a yield stress of 315 MPa for flanges and 235 MPa for the web. The performance of end web panels and the ultimate shear strength with a single double-sided transverse stiffener at support of plate girders undergoing a significant action of tension field were investigated theoretically and experimentally. According to the results of shear test, post-buckling strength in the shear buckling mode is of a significant effect on the ultimate shear strength of end web panels of plat girders. Based on shear test results, a design shear strength formulae for the direct strength method (DSM) for end web panels of plate girders were presented. According to the results, the DSM shear curves can accurately predict the ultimate shear strength of web panels of steel plate girders showing a significant post-buckling behaviour in the shear buckling mode.

## 2. Ultimate plate girder resistance

When a web panel of a plate girder subjected to shear loading, compressive and tensile stresses are developed as shown in Fig.1. Any additional shear loading is carried by an inclined tensile membrane stress field. When increase in load on the girder, the membrane stresses coupled with buckling stress reach yield.

Extensive yielding in the web leads to failure of the girder with plastic hinges formed in the flanges. The failure mechanism under bending and shear is similar to that for girders loaded in pure shear only. Thus, failure will occur when the plastic hinges have formed in the flanges, which together with the formation of a yield zone in the web plate, form a plastic mechanism.

Four important additional factors must be taken into account when the effects of bending moment are considered [4].

-The presence of the bending stresses will reduce the buckling stress of the web plate.

-The influence of the bending stresses upon the magnitude of the membrane stresses required to produce yield in the web.

-The influence of load shedding from the web to the flanges as a consequence of buckling.

-The reduction in the plastic moment capacity of the flanges by the axial stresses arising from the bending moment.

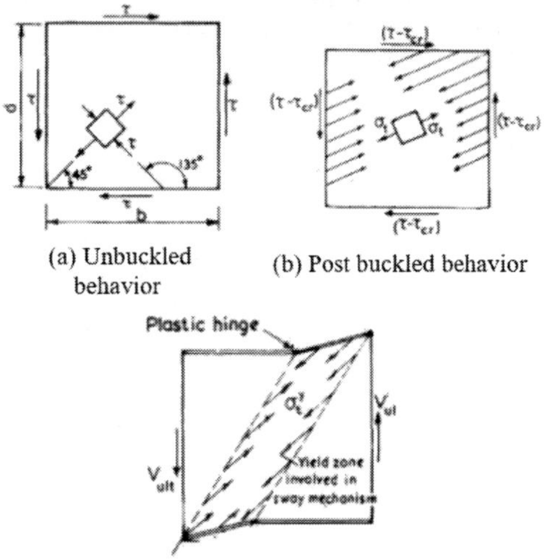


(a) Unbuckled behavior (b) Post buckled behavior

(c) Collapse behaviour

**Fig.1**. Shear mechanism of web panel

**Table 1.** Dimensions of plate girders

| Sample | Flange(mm) | | Web (mm) | | |
|---|---|---|---|---|---|
| | bf (mm) | tf (mm) | dw (mm) | tw (mm) | Stiffeners spacing (mm) |
| Girder 1 | 120 | 6 | 500 | 2 | 575 |
| Girder 2 | 120 | 6 | 400 | 2 | 575 |

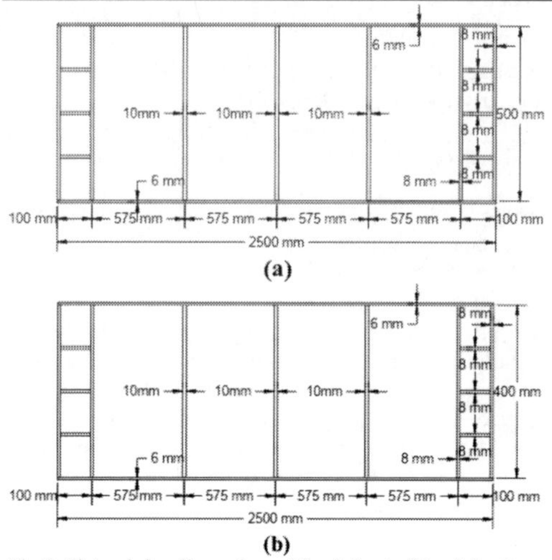


**Fig.2.** Plate girder dimensions, (a)- girder 1, (b)- girder 2

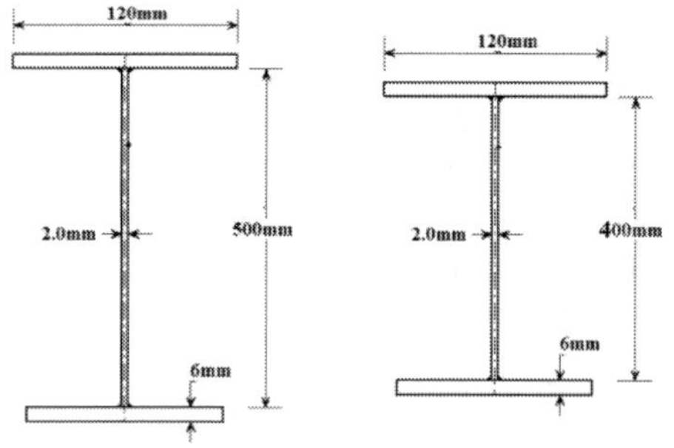


**Girder(1)** **Girder (2)**

**Fig. 3.** The geometry of the tested girders

## 3. Experimental work

Two specimens of steel plate girders were designed. The only difference between the two girders is the web depth. Each girder was built up from three plates will act as one unit by welding vertical web plate to the horizontal flange plates(one top and one bottom flange) to form I-shaped section. To increase the strength of the girder it was used transverse and longitudinal stiffeners are welded between the web panel and the flanges from the both sides of the plate girders. Figs.2 (a, b) and Fig. 3. Show details for the girders.

### 3.1 Dimensions of plate girders

The plate girder dimensions for both samples are listed in Table 1 and Figs.2 (a, b). The span of each plate girder is (2500 mm). The external concentrated load was applied to the specimens at its mid span.

### 3.2 Steel properties

Six steel tensile specimens were used to test the mechanical properties for the (6mm) thickness flange steel plate and for (2mm) thickness of web steel. Tensile tests of the steel were carried out according to (ISO6892).The detailed mechanical properties are summarized in Table 2

**Table 2.** Mechanical properties of the girders

| Component | Thickness (mm) | Yield stress $\sigma_y$ (MPa) | | Ultimate tensile strength $\sigma_u$ (MPa) | |
|---|---|---|---|---|---|
| Flange | 6 | 171 | 181 | 286 | 296 |
| | | 197 | | 311 | |
| | | 175 | | 290 | |
| Web | 2 | 426 | 423 | 558 | 553 |
| | | 458 | | 581 | |
| | | 385 | | 520 | |

### 3.3 Test setup and equipment

Tests for to two girders was carried out using a testing machine "AVERY" at the structural laboratory in the University of Technology. The maximum load capacity of the machine is (1500 KN). The testing machine is shown in Plate 1, load can be applied by hydraulic pressure. The I-section girders were placed on a horizontal base of the simply supported and the load was applied to the mid-span of the plate girders, were tested under concentrated load. The applied load is automatically shown on a large dial contained in the indicator cabinet of the testing machine. The deflections for two girders were measured using dial gauge of 0.01mm increments. The dial gauge was installed under the bottom flange at the mid-span of girder. The deflection was read at every increment of load of 10 KN for girder 1 and 5 KN for girder 2. The test was terminated when the girders had shown collapse and large deformation had occurred at the web panel and flanges without any ability for the load to increase. After the girders had failure, the load was removed.

**Plate 1.** Testing machine

**Plate 2.** Test Setup

### 3.4 Experimental results

The two plate girders were examined by applying load gradually until reaching failure. The ultimate load value for the first girder is (175 KN) while the second girder is(146 KN). The post buckling and failure of each plate girders are shown in Plates 3 and plate 4. The load-deflection curves for the girders are presented in Fig.4. The relationship was approximately linear in the major part of the loading, and then the curve starts to bend slightly due to the occurrence of deformation in the web panel of the girder. All stages of the steel girders deformations were noted, first, the web buckled , then plastic hinges form on the flanges.

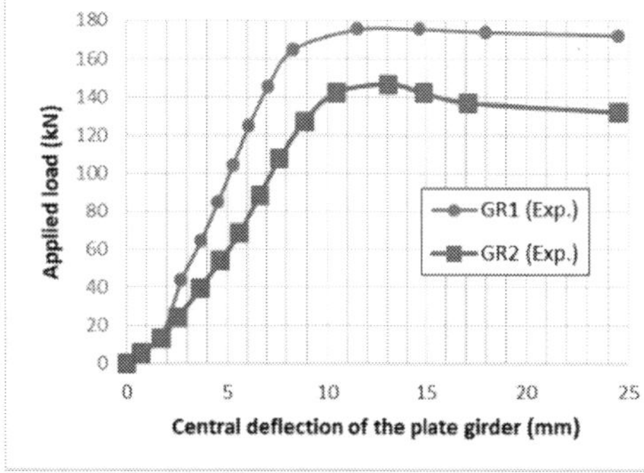


**Fig.4** Comparison load-deflection curve between girders 1 and 2

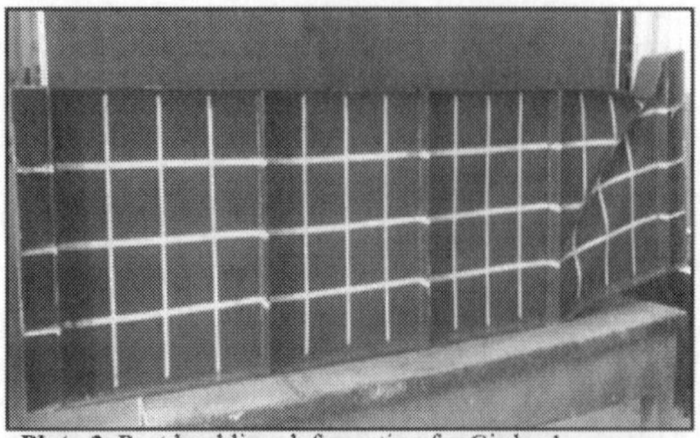

**Plate 3**. Post buckling deformation for Girder 1

**Plate 4**. Post buckling deformation for Girder 2

### 3.5 Numerical results

The finite element analysis software ANSYS is used to perform the numerical part of the study. The two steel plate girders loaded in shear and bending were analysed. In order to build the model of the girders, the shell elements 43 from the code element library was adopted. Perfectly elastoplastic behaviour for the material was considered. Poisson's ratio was set to $v = 0.3$. The method of solving an Eigen value problem was used to obtain the value of t cr whereas V ult was obtained by using the inelastic buckling analysis. The inelastic buckling analysis is essentially nonlinear static analysis that incorporates both material and geometrical nonlinearities with the assumption of an elastic-perfectly plastic constitutive relation. The Von-Mises yield criterion was employed to account for material nonlinearity. To initiate out-of-plane deformations, and since the out-of-plane deformations cannot occur for a perfect flat plate, an initial imperfection was applied to the plate girders having the same shape of the first buckling mode. The results showed that the ultimate load value for the first girder is (180kN) while the second girder is (155.26kN). The deflection and failure results of each plate girder are shown in Figs.5, 6 and 7.

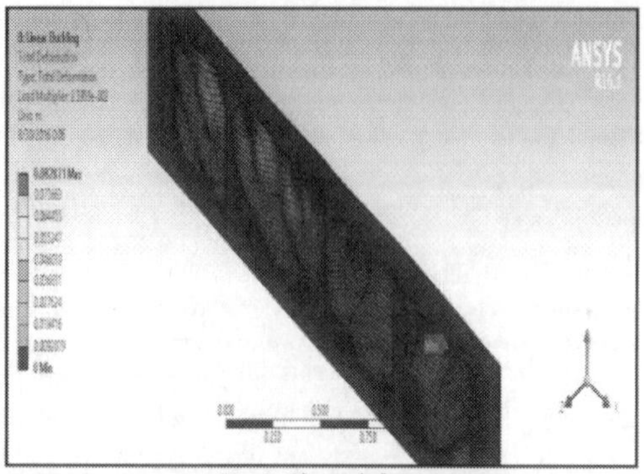


**Fig.5**. Plate girder 1

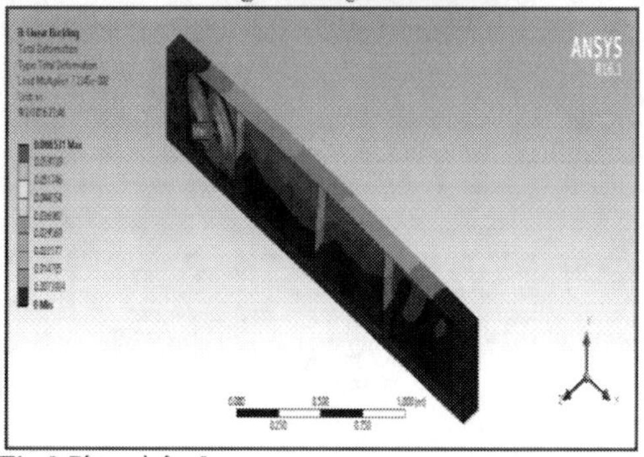


**Fig.6**. Plate girder 2

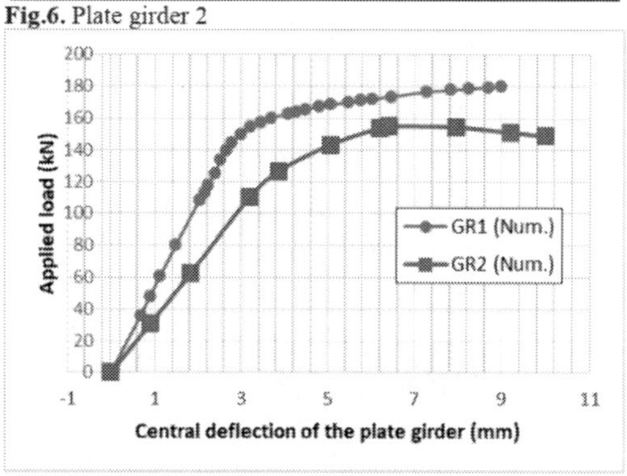


**Fig.7**. Load-deflection curve for girder 1

### 3.6 Theoretical equation results

The theoretical equations derived by Porter et al.[11] were used to calculate the ultimate shear strength for the two plate girders. Mechanical properties of the steel plates shown in table 2 were used. An average yield stress ($\sigma_{yw}$) for web is 423 MPa and an average yield stress ($\sigma_{yf}$) for flange is 181 MPa were used. Results are listed in Table 4.

**Table 3.** Mechanical properties of the girders

| Girder | Depth of the Girder (mm) | $2V_{ult}$ kN |
|---|---|---|
| GR 1 | 500 | 210.9 |
| GR 2 | 400 | 162.13 |

## 4. Discussion

Two different comparisons were conducted , the first was between the results obtained from experimental work ($V_{EXP}$) and numerical study ($V_{ANS}$). The second was with the ultimate shear load results which obtained from theoretical equations ($2V_{ult}$). Results are listed in Table 4.

**Table 4.** Comparison between experimental , numerical and theoretical equation results.

| Models | $V_{EXP}$ (KN) | $V_{ANS}$ (KN) | $V_{EXP}$ / $V_{ANS}$ | $2V_{ult}$ (KN) | $V_{EXP}$ / $2V_{ult}$ |
|---|---|---|---|---|---|
| GR 1 | 175 | 180 | 0.972 | 210.9 | 0.83 |
| GR 2 | 146 | 155.26 | 0.94 | 162.13 | 0.90 |

From this comparison (comparable in both cases) it can be observed the ratio $V_{EXP}$ / $V_{ANS}$ show reasonable agreement within (6%). While the ratio $V_{EXP}$ /$2V_{ult}$ refers to the approximation in the experimental and theoretical results within (17%). Equations used for the theoretical part of the study was derived for shear only while the study was based on shear and bending , therefore some differences have appeared.

## 5. Conclusions

Two plate girders were tested until failure considering the combined effect of shear and bending for studying the behaviour of the ultimate shear strength and buckling. Conclusions based on the experimental, numerical and theoretical equations results were obtained. The results showed that the ultimate shear strength of web panel can be affected significantly by the size of the web and the span/depth ratio. The samples were capable of simulating the behaviour of the web panel and provided very good results. Theoretical study depend on the equations proposed by previous researchers shows an acceptable accuracy. In general, there are acceptable and good agreement between the results that have been obtained from the experimental , numerical and theoretical study.

The ANSYS model was used to perform a study for investigating the effect of geometrical and mechanical properties on the ultimate resistance of the plate girder. Experimental works confirm the general equations that are used for calculating the failure load in the plate girders and confirms the ANSYS model results. As a conclusion ,the increase in the web depth form 400 mm to 500 mm will give an increase in the ultimate load failure by approximately 19 %.

## References

1. J. Newman, *Physics Of The Life Sciences* (Springer, New York, 2008).

2. C. Chern, A. Ostapenko, Fritz Laboratory Reports. Paper 264 (1969).

3. K.C. Rockey, M Skaloud, The Structural Engineer, **50**, 313 (1972).

4. H. R Evans, D.M. Porter, K.C. Rockey, *IABSE Proceedings* (1978).

5. C. Graciano, E. Casanova E., Journal of Constructional Steel Research,**61(1)**, 93 (2005).

6. H. Abbas, R. Sause, R. Driver, Engineering Structures, **29(11)**, 2816 (2007).

7. F. Shahabian, T.M. Roberts, Scientia Iranica, Sharif University of Technology, **15(1)**, 16 (2008) .

8. D.W. White, M.G. Barker, J. Struct. Eng., **134(9)**, 1425 (2008).

9. K.S. Dong, Y.C. Eun, K. Kyungsik, International Journal of Steel Structures, **13(2)**, 291 (2013)

10. Y.B. Kwona, and S.W. Ryu, Thin-Walled Structures, **98(B)** , 578 (2016).

11. D.M. Porter, K.C. Rockey, H.R. Evans, The Structural Engineer, **53(8)**, 313 (1975).

# Evaluation water quality of Diyala River in Iraq using Bhargava method

*Nawar* Al-Musawi*

College of Engineering, University of Baghdad, Baghdad, Iraq

**Abstract**. Diyala River is a tributary of Tigris River, it is one of the important rivers in Iraq. It covers a total distance of 445 km (275 miles). 32600 km$^2$ is the area that drains by Diyala River between Iraqi-Iranian borders. This research aims to evaluate the water quality index WQI of Diyala River, where three stations were chosen along the river. These stations are D12 at Jalawlaa City at the beginning of Diyala River, the second station is D15 at Baaquba City at the mid distance of the river, and the third station is D17 which is the last station before the confluence of Diyala River with Tigris River at Baghdad city. Bhargava method was used in order to evaluate the water quality index for both irrigation and drinking uses. The results indicated that Diyala river water quality at its beginning was excellent for irrigation and good for drinking, while at the mid distance of the river, it was good for irrigation but heavily polluted and unsafe for drinking. Water quality of the river at the third site was acceptable for irrigation but again severely polluted and unsafe for drinking.

## 1 Introduction

Diyala River is one of the important water resources and one of the main tributaries of Tigris River in Iraq. It covers a total distance of 445 km Diyala weir controls the river floods and irrigates the area northeast of Baghdad. Diyala city depends mainly on Diyala River for water supply for domestic, municipal, agriculture and other purposes. **[1]**. Many towns are located on the river banks. Wastes from agricultural and industrial activities in these towns are drained straightway to the river. The river covers an area of 32600 km$^2$ . The river catchment is widely varied through the entire catchments area from semi-arid plain north of Baghdad to mountainous area of western Iran [2]. The catchments of this river are divided into four parts; the first part is above Derbendikhan, the second part is Upper Diyala, the third part is Middle Diyala and the fourth part is down Diyala, each of these parts has different characteristics and different contribution to the main river flow. From the geological point of view, river catchments have different geological units. Down Diyala is covered mainly by recent alluvium and lies within the unfolded zone **[3]**. Derbendikhan and Hemrin are two dams constructed on the river, (360 and 188 km upstream the confluence with the Tigris River south Baghdad respectively. In addition to the two dams, the Diyala wier was constructed on the river (11 km downstream from Hemrin Dam) and it distributes the outflow of Hemrin reservoir to the Lower Diyala River and the irrigation canals (Al-Khalis canal and Sadr Al-Mushtarak). **[4]**.
A research on Evaluation of Water Quality of Diyala River For Irrigation Purposes , comprised suitability evaluation of water quality of Diyala River for irrigation. The researcher depends on four stations along Diyala River which are Jalawlah station (DI2) before Jalawlah Bridge, Saadiyah station (DI3) before inter to Hamren Dam, Muqdadia station (DI4) after Diyala Dam, and Baqubah station (DI5) at Iron Bridge, besides wand river station in Khanaqin (DIW) Before Dam. The parameters studied were pH, electric conductivity (EC), TDS, $Na^+$, $K^+$, $Ca^{+2}$, $Mg^{+2}$, $HCO^{-3}$, $CO_3^{-2}$, $SO_4^{-2}$, $Cl^-$, and $NO_3$ The study were done during January to December 2010, the study showed that water samples of DIW and DI5 fall in the class of high salinity with low sodium water, which can be used for irrigation on almost all types of soil, while water samples of other stations fall in the class of medium salinity with low sodium water, which can be suitable to salt tolerant plants with probability developed permeability. The results of this study revealed that the quality of Diyala River water can be classified as suitable for irrigation with few exceptions[1].
Suspended and Solute Loads on the Lower Diyala River is a research done on Suspended sediment and solute loads associated with the Lower Diyala River were evaluated for the period from June 1984 to May 1985. During the studied period the mean discharge reached 128 cumecs and the maximum and minimum were 505 and 1.72 cumecs respectively. The annual suspended load reached 64589 tonnes. The rate of transport was not uniform throughout the studied period, where most of the load (51.02 percent of the suspended load and 54.68 percent of the solute load) was transported during the Winter [4].

* Corresponding author: nawaromran@yahoo.com

© The Authors, published by EDP Sciences. This is an open access article distributed under the terms of the Creative Commons Attribution License 4.0 (http://creativecommons.org/licenses/by/4.0/).

A research dealing with the Heavy Metals in Diyala River, aimed to investigate the status of Diyala River water quality in Iraq with respect to the concentrations of its heavy metals by preparing the heavy metal pollution index. The locations selected were along of Diyala River, from Kalar district to the meeting with Tigris River. The study indicates that Diyala River is polluted with heavy metals. The study showed the impact of anthropogenic sources on the pollution load of the river water. [5]

A research on Spectral Analysis of Some Selected Hydrochemical Parameters of Diyala River In Iraq , has Some hydrochemical variables of Diyala River to analyze their periodicities quantify and qualify surface water and investigate the behavior of these variables.

The results appeared the variation of the factors affecting the water quality of the river in both hydrological and meteorological conditions also the impact of human activities through the area of the river basin **[3]**.

Bhargava in 1983 studied WQI to evaluate the water quality for several activities in Ganga River in India and Saigon River in Vietnam using the sensitivity function method. The method transfers the values of variables according to a standard specification (0 – 1) scale. The WQI was used to classify rivers into five groups of different water activities depending on the variables which affected the WQI.

Bhargava 1983 studied the potable Water Quality Index(PWQI ) using the sensitivity function which were represented as a curve showing that the ratio of pollutant concentration to the maximum permissible concentration level had the range of (0 – 1). He divided the variables into four groups as: biological, heavy metals, physical and organic and non- organic variables. He suggested that, for potable water supply, the WQI should be greater than 90.

Bhargava Method is one of the modern methods which is used in many countries, and it is easy for dealing with parameters for different uses by sensitivity functions curves which take the value between zero to one. The results are accumulated using the geometric mean.

## 2 Area of study

Diyala River is one of the main water resources of Iraq and one of the most important tributaries of Tigris River in Iraq. The river passes through Diyala province northeast of Baghdad as shown in Figure 1. It covers a total distance of 445 km (275 miles). The catchment of Diyala River is located between Latitude (33° 13′00" N- 35° 50′ 00" N) and Longitude (44° 30′ 00" E- 44° 50′ 00" E),

Many towns are situated on its banks, agricultural and industrial wastes are discharged in to Diyala river. The river drains 32600 km$^2$ of the area between Iraqi-Iranian borders . North of Baghdad the catchment is varied from semi-arid plain while mountainous area at western Iran **[2]**.

## 3 Methods and calculations

In this research, Diyala River was studied for the period 2011-2016.Three stations were chosen in order to evaluate the Water Quality Index of Diyala River. The first station was D12 which is at Jalawlaa City at the beginning of the river after the Iraqi-Iranian borders. The second station was D15 at Baquba City in the middle of the river, while the third station was D17 south of Baghdad City before the confluence of Diyala River with Tigris River. The data were collected from the Operation Department [6].

For calculating the Water Quality Index (WQI), Bhargava method was used to evaluate Diyala River for irrigation and drinking uses.

The sensitivity functions can be calculated by the impact of the parameter value on water quality as in Figs 2, 3.

The sensitivity functions curves are used to evaluate the quality of river water. For instance, when the concentration of sulfate ($SO_4^{-2}$) is 400 ppm, the s nsitivity function will be very low which makes water worse for drinking use, while the same concentration value can give sensitivity function equal to 0.6 for irrigation use which means it is acceptable 60%. **[7]**

The relative parameters for irrigation use are: TDS, pH, $SO_4^{-2}$, SAR, EC, and $Cl^{-1}$, while the relative parameters of drinking use are TDS, T.H, $Cl^{-1}$, $SO_4^{-2}$, BOD, pH, and $Ca^{+2}$. Finally; the relative parameters for industrial use are EC, T.H, and Tu.

This index is suitable for classifying rivers into five classes as shown inTable 1 and to determine the WQI for each use of different water uses depending on the variables which affect that use by geometric mean formula.

The geometric mean formula is shown below:

$$WQI = \left[\pi_{i=1}^{n} fi\left(P_i\right)\right]^{1/n} \quad *100 \qquad \ldots\ldots\ldots\ldots 1$$

Where:

$f_i$ ($P_i$) is the sensitivity function for each variable including the effect of variable weight concentration which is related to a certain activity and varies from (0 – 1]. **[7]**

## 4 Results and discussion

After applying the Bhargava Method for the period from 2011 to 2016, the results indicated that the first station D12 which is at Jalawlaa city after the Iraqi-Iranian borders has an excellent water quality for irrigation and good water quality for drinking as shown in Table 2 and Table3. The second station D15 which is at mid distance of Diyala River in Baquba city, the results indicated that the river has good water quality for irrigation but severely polluted water for drinking as shown in Table 4 and Table 5. D17 the last station , located at Baghdad city before the meeting of Diyala River with Tigris River, the Tables 6 and 7 indicated that the river has acceptable water quality for irrigation but severely polluted water for drinking.

Figures 4,5,6 show that the river is suitable for irrigation but unsuitable for drinking except for station D12 in Jalawlaa City. Figure 7 shows that the variation of WQI with respect to time for the three stations for irrigation uses, it is clear that the best station is D12 then D15 and D17 Figure 8 showes the variation of WQI with respect to time for the three stations for drinking uses, it is clear that only station D12 is suitable for drinking uses, and that is because of the high values of BOD concentration that was discharged to Diyala River from the mid distance of river at Baquba city until its end at Baghdad city. The zero values for the WQI shown in Tables 5 and 7 refer to the parameters concentrations were out of the range suggested by Bhargava. In other words the parameters concentration exceeded the permissible limits that were used in this method.

## 5 Conclusions

From this research we can conclude that:

1- The first station at Jalawlaa city after the Iraqi-Iranian borders has excellent water quality for irrigation and good water quality for drinking, because it is the nearest point to the river origin and few population beside it.

2-The second station at the mid distance of Diyala River in Baquba city, has good water quality for irrigation but severely polluted water for drinking, this city considered as Diyala province center and many agricultural and industrial wastes discharged directly to the river.

3-The last station located at Baghdad city before the meeting of Diyala River with Tigris River, the river has acceptable water quality for irrigation but severely polluted water for drinking, we can conclude that the river is heavy polluted with various wastes and high concentration of BOD values affected on river water quality.

4-The heavy population and uncontrolled discharging wastes to the river affected the river water quality.

## 6 References

**1.** Hamza, Nahida H, Evaluation Of Water Quality Of Diyala River For Irrigation Purposes, Diyala Journal of Engineering Sciences, Vol. 05, No. 02, pp. 82-98 , December 2012, Iraq.(2012)

**2.** Al-Ansari, N. and Al-Jabbari, M, Hemrin reservoir, geological and hydrological investigation, J. of Water Res., Special Publ. No. 2, Sector 2, (1987).

**3.** Aqeel et al., S, Qusay A. Al-Suhail, Spectral Analysis of Some Selected Hydrochemical Parameters of Diyala River In Iraq , Fourteenth International Water Technology Conference, IWTC14 2010 Cairo, Egypt, (2010)

**4.** Al-Ansari, et al., Suspended and Solute Loads on the Lower Diyala River , Lulea University of Technology, (2014)https://www.researchgate.net/publication/253518411.

**5.** Abdullah, Enaam, Evaluation of Surface Water Quality Indices for Heavy Metals of Diyala River-Iraq, Journal of Natural Sciences Research,ISSN 2224-3186 (Paper) ISSN 2225-0921 (Online), Vol.3, No.8, Iraq.(2013)

**6.** Documented data, Operation department, Water Resources Ministry, Baghdad, Iraq.(2016)

**7.** Bhargava, D. S., Use of a water quality index for river classification and zoning of the Ganga River. *Environmental Pollution*, B6, 51–67.(1983)

**Table 1.** Bhargava classification for Water quality

| A. *Class* | B. *WQI Value* | C. *Water Quality* |
|---|---|---|
| I | 100 – 90 | Excellent |
| II | 89 – 65 | Good |
| III | 64 – 35 | Acceptable |
| V | 34 – 11 | Polluted |
| IV | Less than 10 | Severe Polluted |

Fig. 1. Diyala River map

https://doi.org/10.1051/matecconf/201816205001

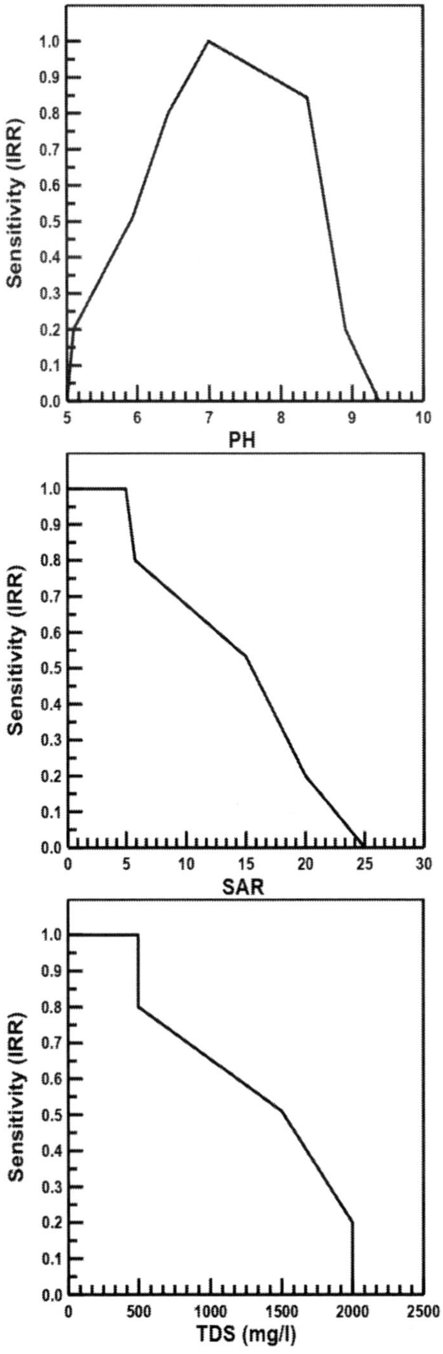


**Figure 2** Sensitivity functions curves according to Bhargava for irrigation use

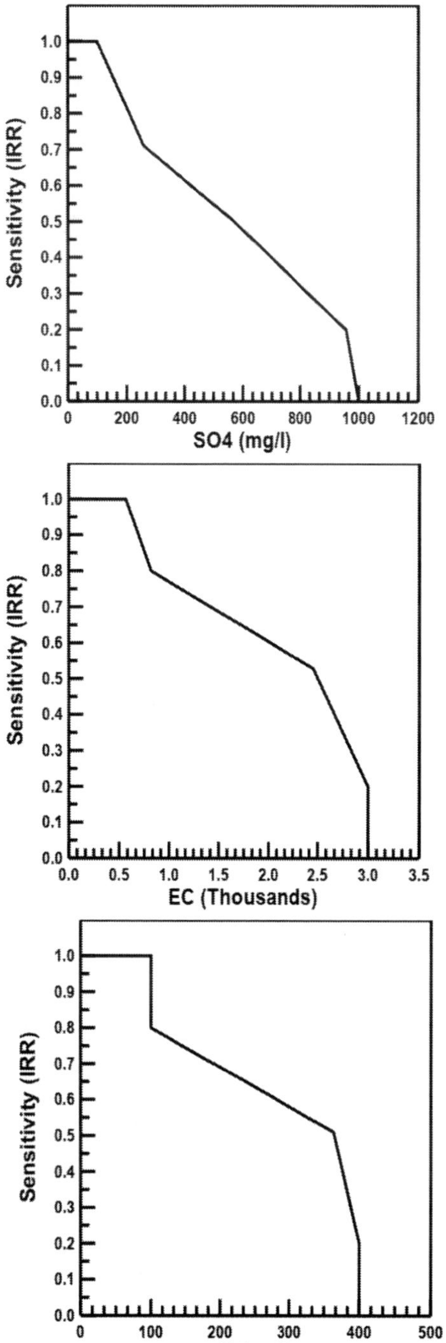


**Figure 2** Sensitivity functions curves according to Bhargava for irrigation use continued

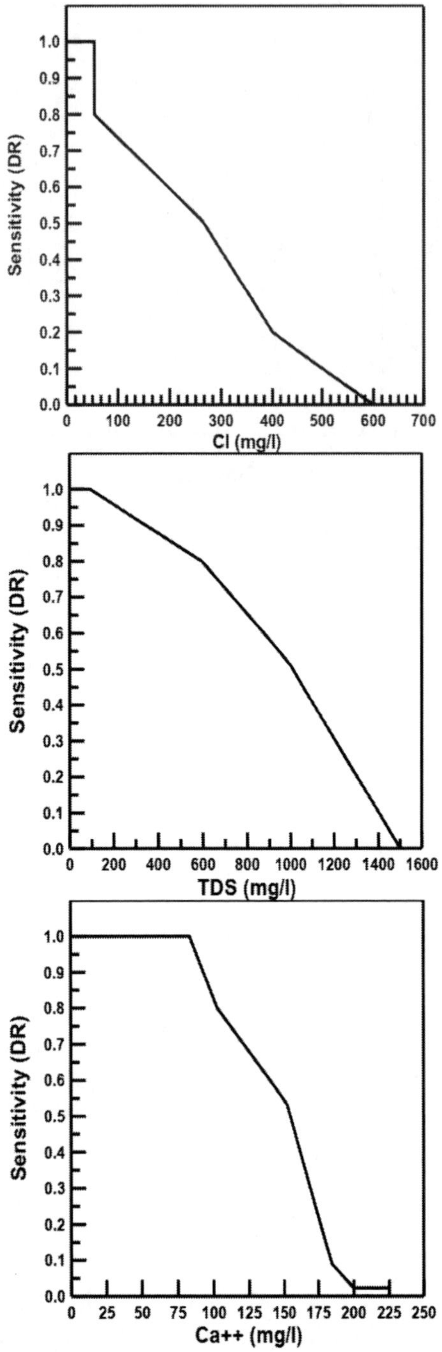


Figure 3 Sensitivity functions curves according to Bhargava for drinking use

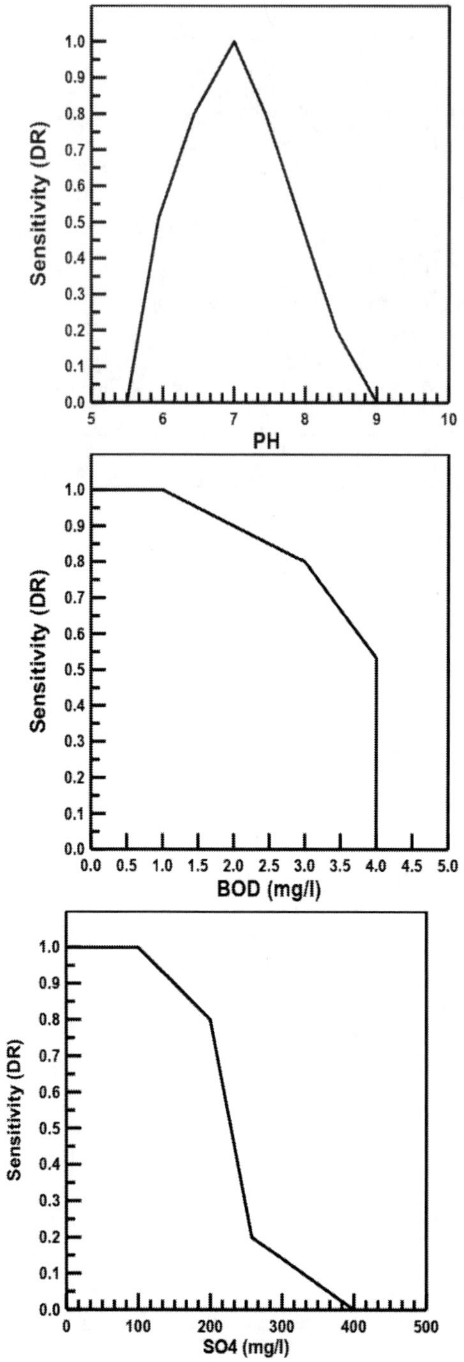


Figure 3 Sensitivity functions curves according to Bhargava for drinking use continued

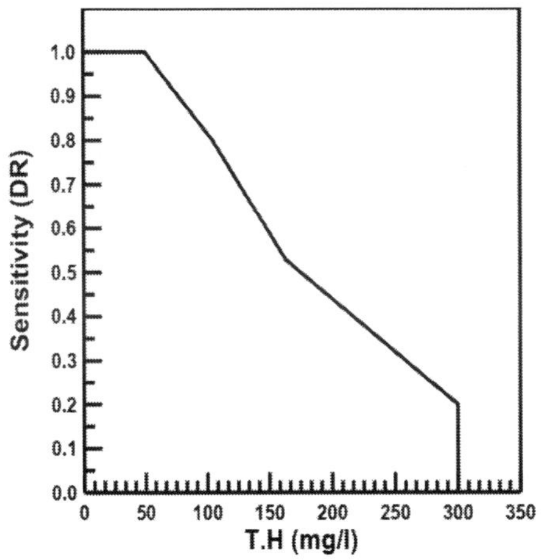


Figure 3 Sensitivity functions curves according to Bhargava for drinking use continued

**Table 2** The relative sensitivity functions for the parameters of irrigation uses for station D12

| Year | SAR | EC | TDS | Cl | pH | SO4 | WQI | Class |
|---|---|---|---|---|---|---|---|---|
| 2011 | 1 | 1 | 1 | 1 | 0.95 | 1 | 99.15 | 1 |
| 2012 | 1 | 1 | 1 | 1 | 0.94 | 1 | 98.98 | 1 |
| 2013 | 1 | 1 | 1 | 1 | 0.9 | 1 | 98.27 | 1 |
| 2014 | 1 | 0.95 | 1 | 1 | 0.9 | 0.9 | 95.74 | 1 |
| 2015 | 1 | 1 | 1 | 0.761 | 1 | 1 | 95.55 | 1 |
| 2016 | 1 | 1 | 1 | 1 | 1 | 0.733 | 94.95 | 1 |

**Table 3** The relative sensitivity functions for the parameters of drinking uses for station D12

| Year | Ph | TH | TDS | Ca | Cl | SO4 | BOD | WQI | Class |
|---|---|---|---|---|---|---|---|---|---|
| 2011 | 0.89 | 0.24 | 0.9 | 1 | 1 | 0.85 | 0.53 | 70.48 | 2 |
| 2012 | 0.75 | 0.4 | 0.9 | 1 | 1 | 1 | 0.55 | 76.13 | 2 |
| 2013 | 0.75 | 0.4 | 0.9 | 1 | 1 | 0.95 | 0.54 | 75.38 | 2 |
| 2014 | 0.7 | 0.3 | 0.94 | 1 | 1 | 0.9 | 0.56 | 71.89 | 2 |
| 2015 | 0.79 | 0.3 | 0.9 | 1 | 1 | 0.9 | 0.55 | 72.66 | 2 |
| 2016 | 0.88 | 0.23 | 0.9 | 1 | 1 | 0.9 | 0.54 | 70.87 | 2 |

**Table 4** The relative sensitivity functions for the parameters of irrigation uses for station D15

| Year | SAR | EC | TDS | Cl | pH | SO4 | WQI | Clas s |
|---|---|---|---|---|---|---|---|---|
| 2011 | 1 | 0.7 | 0.75 | 0.8 | 0.94 | 0.9 | 84.17 | 2 |
| 2012 | 1 | 0.68 | 0.69 | 0.7 | 0.94 | 0.69 | 77.29 | 2 |
| 2013 | 0.79 | 0.73 | 0.69 | 0.69 | 0.95 | 0.68 | 74.79 | 2 |
| 2014 | 0.78 | 0.7 | 0.68 | 0.72 | 0.94 | 0.67 | 74.32 | 2 |
| 2015 | 0.65 | 0.65 | 0.65 | 0.69 | 0.95 | 0.6 | 69.02 | 2 |
| 2016 | 0.65 | 0.59 | 0 | 0.61 | 0.94 | 0.59 | 68.97 | 2 |

**Table 5** The relative sensitivity functions for the parameters of drinking uses for station D15

| Year | pH | TH | TDS | Ca | Cl | SO4 | BOD | WQI | Class |
|---|---|---|---|---|---|---|---|---|---|
| 2011 | 0.7 | 0 | 0.71 | 0.58 | 0.79 | 0.88 | 0 | 0 | 5 |
| 2012 | 0.7 | 0 | 0.55 | 0.49 | 0.78 | 0.2 | 0 | 0 | 5 |
| 2013 | 0.75 | 0 | 0.54 | 0.49 | 0.6 | 0.09 | 0 | 0 | 5 |
| 2014 | 0.7 | 0 | 0.56 | 0.49 | 0.78 | 0.09 | 0 | 0 | 5 |
| 2015 | 0.75 | 0 | 0.35 | 0.45 | 0.78 | 0 | 0 | 0 | 5 |
| 2016 | 0.75 | 0 | 0 | 0.1 | 0.5 | 0 | 0 | 0 | 5 |

**Table 6** The relative sensitivity functions for the parameters of irrigation uses for station D17

| Year | SAR | EC | TDS | Cl | pH | SO4 | WQI | Class |
|---|---|---|---|---|---|---|---|---|
| 2011 | 0.25 | 0.28 | 0.24 | 0.55 | 0.95 | 0.45 | 39.68 | 3 |
| 2012 | 0.25 | 0.29 | 0.25 | 0.54 | 0.95 | 0.39 | 39.21 | 3 |
| 2013 | 0.46 | 0.29 | 0.3 | 0.54 | 0.94 | 0.44 | 45.57 | 3 |
| 2014 | 0.45 | 0.5 | 0.6 | 0.56 | 0.95 | 0.43 | 56.03 | 3 |
| 2015 | 0.2 | 0.56 | 0.62 | 0.63 | 0.94 | 0.43 | 51.05 | 3 |
| 2016 | 0.55 | 0.6 | 0.64 | 0.71 | 0.9 | 0.45 | 62.71 | 3 |

**Table 7** The relative sensitivity functions for the parameters of drinking uses for station D17

| Year | pH | TH | TDS | Ca | Cl | SO4 | BOD | WQI | Class |
|---|---|---|---|---|---|---|---|---|---|
| 2011 | 0.8 | 0 | 0 | 0.58 | 0.33 | 0 | 0 | 0 | 5 |
| 2012 | 0.8 | 0 | 0 | 0.34 | 0.31 | 0 | 0 | 0 | 5 |
| 2013 | 0.75 | 0 | 0 | 0.36 | 0.31 | 0 | 0 | 0 | 5 |
| 2014 | 0.8 | 0 | 0 | 0.44 | 0.41 | 0 | 0 | 0 | 5 |
| 2015 | 0.75 | 0 | 0 | 0.05 | 0.15 | 0 | 0 | 0 | 5 |
| 2016 | 0.65 | 0 | 0.24 | 0.58 | 0.63 | 0 | 0 | 0 | 5 |

 https://doi.org/10.1051/matecconf/201816205001

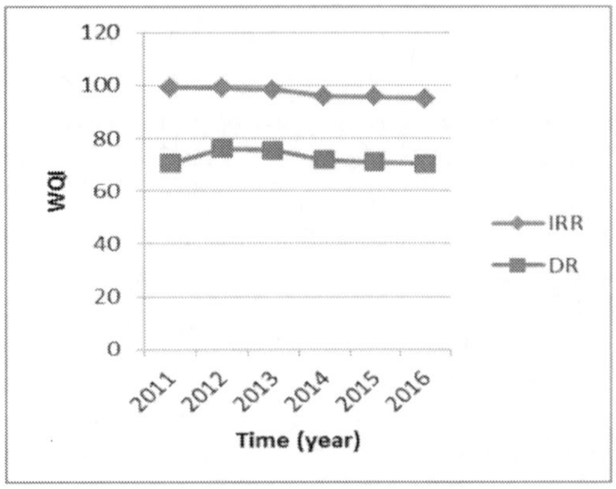


**Figure 4** Variation of WQI with respect to time for both irrigation and drinking uses for station D12

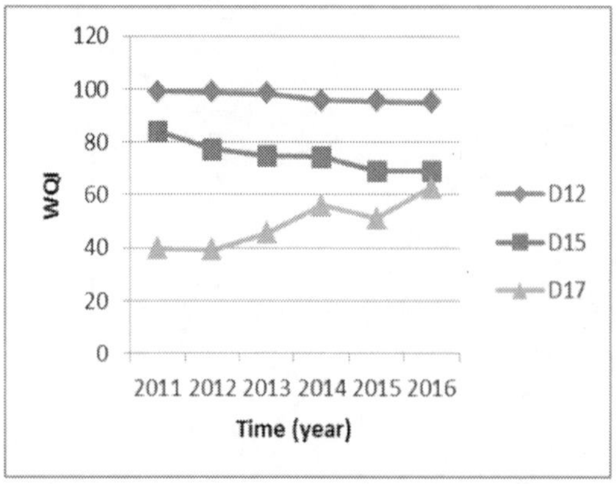


**Figure 7** Variation of WQI with respect to time for irrigation uses for the three stations

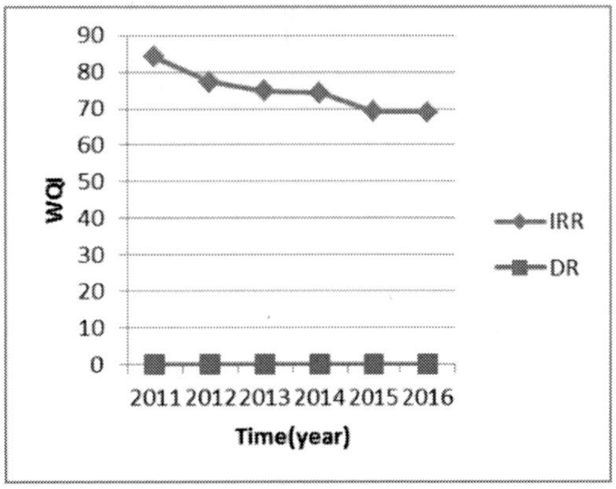


**Figure 5** Variation of WQI with respect to time for both irrigation and drinking uses for station D15

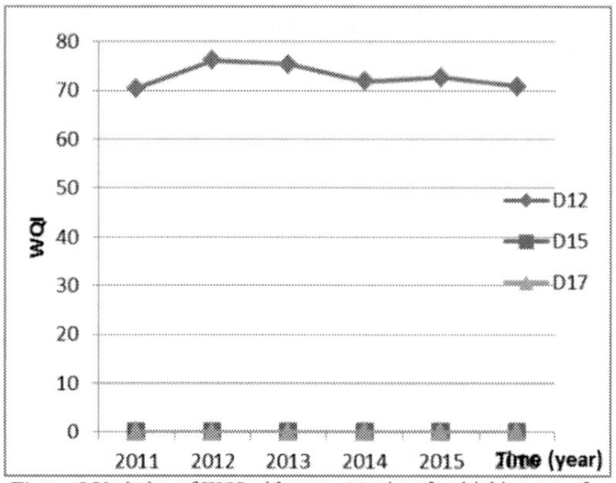


**Figure 8** Variation of WQI with respect to time for drinking uses for the three stations

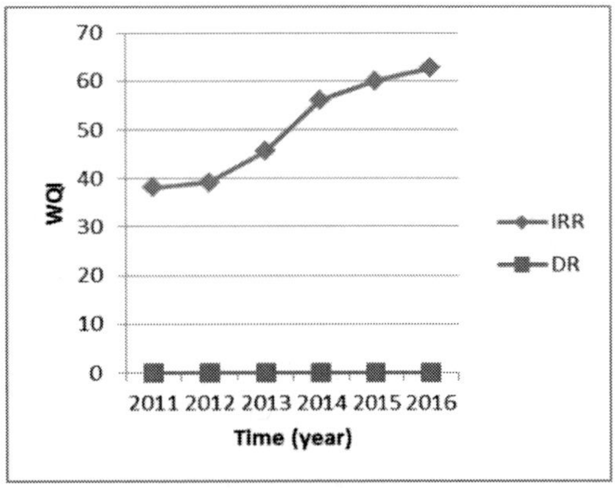


**Figure 6** Variation of WQI with respect to time for both irrigation and drinking uses for station D17

# Hydrochemical properties of ground water of Rahaliya-Ekhedhur region, west Razzaza lake, iraq

*Moutaz* Al-Dabbas[1], *Qusai* Al-Kubaisi[1], *Tariq* Hussein[2] and *Shahad* Al- Qaraghuli[3]

[1]College of Science, University of Baghdad, Baghdad, Iraq
[2]Building and Construction Engineering Department, University of Technology, Baghdad, Iraq
[3]Al Karkh University of Science, Baghdad, Iraq.

**Abstract.** Rahaliya – Ekhedur region is located to the west of Razzaza Lake (M.IRAQ). It is within the Salman – Rutba tectonic zones characterized by the existence of Abu Jir and Imam Ahmad bin Hashim Fault Systems. The major geological formations in the studied area consists of Tertiary and Quaternary deposits described from the oldest to the youngest as : Dammam, Euphrates, Nfayil and Injana, formations, and the slope of beds is less than one degree in the direction of east and northeast towards Al-Razzaza lake. To evaluate the hydrochemical properties and the quality groundwater samples were collected from 16 wells at different depths 80-120 m, during April and October 2013. These samples have been chemically analyzed and the results indicated that the ground water is of mixed origin, very hard (TH ranging from 227.5 to 1032.8 ppm), having high pH values, while the Total Dissolved Solids (TDS) ranging from 1700 to 2750 mg/l so that the water is of brackish type. On the basis of the major cation and anion concentration the groundwater during April 2013 was of mixed Mg and Na -SO4 group having two families which are Na2SO4 family with Na+>Ca2+>Mg2+-SO42- >Cl->HCO3- and MgSO4 water family with Mg2+>Ca2+>Na+-SO42- >Cl->HCO3-. While, the groundwater during October 2013 is of two groups (Na -SO4) having two families which are (Na2SO4) family with Na+>Ca2+>Mg2+-SO42- >Cl->HCO3-and (MgSO4) water family with Mg2+> Ca2+>Na+-SO42- >Cl->HCO3-. The second group was Na - Cl with Na+> Mg2+>Ca2+ - Cl->SO42- > HCO3- and Na+>Ca2+>Mg2+ - Cl->SO42- > HCO3- . Therefore, the ground water in the region is not good for drinking due to high salinity and it is of Doubtful to unsuitable-Unsuitable for irrigation but it can be used to irrigate sensitive crops in areas.

## 1 Introduction

Water resources systems need to be operated to cope with variability of climate changes, mainly the expected changes in temperature and precipitation. There is great need for an emergency or water resources risk management practices. These water systems, traditionally, are designed on the assumption that the statistical characteristics of the hydro-meteorological processes are almost expected annually and on the long run. Several studies were conducted concerning the hydraulic characteristics and the hydrochemical analysis properties of the groundwater of Rahhalia, Shithatha or Ekhedhur [1], [2], [3]. But no one investigated all these locations together. Therefore, the area from Rahhalia to Ekhedhur is chosen for further deals of the hydrochemical analyses and future utilization of the groundwater (Fig. 1).

The chosen region is in central Iraq between Al-Anbar and Karbala governorates, to the west of the Razzaza Lake. The region is extending from Rahhalia to the Ekhedhur between latitudes 32° 25\`-32° 50\` North and longitudes 43°15\`- 43° 40\` East (Fig. 1). The region is characterized by arid climate of hot dry summer, cold dry winter with annual rainfallof 90 mm mainly during January to April and annual evaporation of 2954 mm (Table 1) [4]. The geology of study area represented by the tertiary sediments of different ages and rock units such as Dammam, Euphrates, Nfayil, Injana and Dibdiba formations. Quaternary sediments cover vast area near Al-Razzaza Lake which represented by gypcrete, inland sabkha, depression fill, flood plain and Aeolian sediments (Figure 1). The studied area has two fault systems, Abu Jir and Imam Ahmad bin Hashim which are deepening more than 150 m for the period before

© The Authors, published by EDP Sciences. This is an open access article distributed under the terms of the Creative Commons Attribution License 4.0 (http://creativecommons.org/licenses/by/4.0/).

https://doi.org/10.1051/matecconf/201816205002

Miocene and its main trend is to the northwest - southeast [5]. It is believed that the groundwater related to the same line, which extends to more than 10 km in width, [6]. Topography of the area has gentle slope terrains to the east towards Al-Razzaza Lake and the dip of the beds is less than one degree to the east and northeast with elevations of 65-35 m. above mean sea level [7], (Figures 1 and 2). Because of geological nature of the study area, it is characterized by distinguished hydrogeology through the presence of the Abu Jir and Imam Ahmad bin Hashim fault systems. The hydrogeology situation depends on the nature of the structural, geological, type of formations, nature of the water bearing rocks and cavities [8]. The ground water aquifers are represented by three aquifers, (Tayarat – UmmRadhuma), (Dammam – Euphrates) and Nifayil formations [2], [9].

The studied area is considered as a discharge area to Al-Razzaza Lake, and the study area is considered to be one of the important areas due to the existence of several economic projects [10]. Two hydrogeological units were defined in the study area, unconfined and confined aquifers. The unconfined aquifer in the studied area is represented by the Euphrates - Nifayil formations [9], [11]. The Euphrates aquifer is characterized by the total dissolved solids ranging between 4000–6000 mg/l and the electrical conductivity ranges between 5000-7000 µs/cm, with sulphatic water type at Shithatha and Rahaliya localities [11].

The confined aquifers represented by the Dammam Formation which is one of the most vital aquifers in western Iraq. It is composed of variable carbonate rocks mainly limestone, dolomatic limestone and dolomite, with secondary It is characterized by the presence of cavities and canals in addition to fractures, fissures and joints, which cause the formation to have highest transmissivity and permeability, in most areas [11]. The Dammam aquifer is considered to be the main source of water in the area [12]. The aims of this study are to investigate the hydrochemical analyses and to evaluate the quality of ground water suitability for different purposes at Rahhalia - Ekhedhur area during April and October 2013.

## 2 Materials and methods

Sixteen wells have been chosen for this study distributed over all the area (Figure 1). Groundwater samples were selected from sixteen boreholes during April and October 2013 from Rahaliya to Ekhedhur region (Figure-1). The analysis of the concentration of cations ($K^+$, $Na^+$, $Mg^{2+}$, $Ca^{2+}$) and anions ($Cl^-$,$HCO_3^-$, $SO_4^{2-}$) in addition to (Electric conductivity EC, Total Dissolved Salts, TDS and pH) have been done in the chemical laboratory of the General Commission for Groundwater. The results of major ions, TDS, EC and pH were shown in Tables 1 and 2.

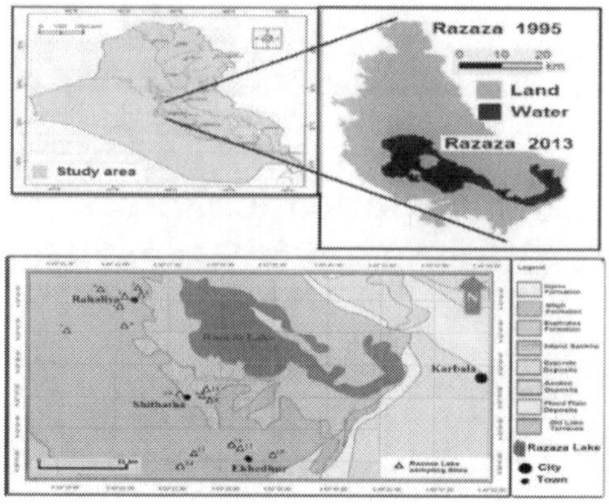

Fig. 1: Location and geological map of the study area.

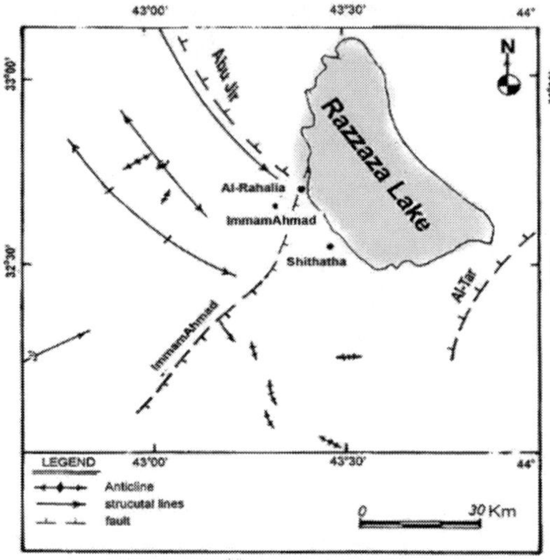

Fig. 2: Tectonic map of the study region (modified after [13]).

## 3 Results and discussion

The TDS of the groundwater is ranging from 1700 to 2420 ppm during April 2013 while October 2013 ranging from 1740 to 2750 ppm (Tables 1 and 2).The EC value is ranging from 2520 to 3710 µs/cm during April 2013 while during October 2013 ranging from 2560 to 3850 µs/cm .

Applying three classifications of water for salinity values for both periods reflect slightly-brackish water (Tables 1 and 2) [4], [8], [14]. The pH values ranging from 7.08 to 7.61 during April 2013 while during October 2013 ranging from 7.11 to 7.80.

The total hardness (TH) was ranging from 227.5 to 772.9 ppm during April 2013 while in October 2013 it was ranging from 338.4 to 1032.8 ppm) and indicated very hard water.

Previous study [14] examined the relationship between electrical conductivity (EC μS/cm) and water mineralization. It can be concluded that the type of spring and Al-Razzaza Lake waters indicate to excessively mineralized water.

Comparison of TDS values for both periods with three classifications of water was classified as slightly-brackish water not useful for drinking [4], [14]. Applying Sulin's [15] classification, the ground water samples fall in the zone of A< 1 that represents marine water in semi confined aquifer and in the zone of B<1 that represent meteoric origin of Na+K - Sulphate .

According to Schoeller [4], classification the groundwater samples shows that the groups are $SO_4$ and Cl and the families are Na-$SO_4$, Mg-$SO_4$ , Ca-$SO_4$, Na - Cl for wet and dry periods. While according to Piper's [4], classification the groundwater samples falls in e – class and some of these samples fall in g – class [4]. The application of Piper's diagram [4] shows that all water samples have non carbonate hardness (secondary salinity >50%). Moreover, all the water samples have total hardness value > 300 and according to Todd's classification [4]. So, all the water samples were not useful for drinking. The result of using Hassan, et al, [16], method to classify the ground water of Rahaliya – Ekhedhur area during April and October 2013, and identifying their water types are shown in Tables 1 and 2. The Rahaliya and Ekhedhur groundwater have clearly indicated that the sulphate and chloride groups are dominant. During April 2013, the sulphate group with two families (sulphate -sodium and sulphate-magnesium) where the first family with one major water type rNa>.rCa>rMg; SO4> rCl, while the second family with two water types rMg> rCa> rNa; rSO4> rCl and rMg> rNa >rCa; rSO4> rCl water type. The chloride group contains one major family (chloride -sodium family) and one water type which is rNa >rMg> rCa; rCl>rSO4 water type. While during October 2013, the sulphate group with two families (sulphate -sodium and sulphate- calcium) the first family with two water types rNa>.rCa>rMg; SO4> rCl and rNa>rMg>rCa; rSO4>Cl, while the second family with one major water type rCa>rMg>rNa; rSO4>Cl water type. The chloride group contains one major family (chloride -sodium family) and with two water types rNa >rMg> rCa; rCl>rSO4 and rNa>.rCa>rMg; rCl>SO4.

This variation can be attributed to the lithological and mineralogical contents of the geological formation [16]. Al-Basrawi, [2], concluded that the main groundwater recharge sources at the west and south west of Al-Razaza lake are from the subsurface water flow from the west and south west catchment of the studied area which flow through fractures and subsurface channels that affected by the fault systems of the area represented by Abu Jeir and Emam Ahmad faults (Fig.2). Moreover, the ground water is similar to the deep groundwater of Tayarat aquifer of Chloride group with Chloride-Sodium family that affect the Dammam and Euphrates aquifers of Sulphate group with Sulphate- Sodium and Sulphate- Magnesium families. The rainfall represents water availability element which is vital for the water surplus occurred in the study area and it is the most important climate elements in the water balance and aquifer recharge. The increase in precipitation has an impact on the groundwater and transmission of material from the upper parts to the lower parts. The average mean annual rainfall in mm for ten years intervals indicate that there were a remarkable decrease in the amount of rainfall for years 1992-2013 that reflected the regional climatic change as shown for Karbala meteorological station, the average mean annual rainfall in mm for ten years intervals decrease from 90 mm for the period 1992-2001 to about 71 mm for the period 2002- 2013 [4].

### 3.1 Groundwater suitability for different purposes

The groundwater chemical analyses of two periods reflected unsuitability for drinking according to Iraqi standards and World Health Organization standard [4]. The ground waters are good or acceptable for animal drinking according to Altoviski , Crist and Lowery and Ayers and Westcot [4], but they were not good for industrial purposes according to Hem [4]. Also, they are suitable for building purposes according to Altoviski [4] and doubtful according to Ayers and Westcot [4] and Don [8] classifications for irrigation purpose.

## 4 Conclusions

1- The examined groundwater is classified as slightly-brackish, very hard water and can be concluded that the type water indicate excessively mineralized water.

2- The groundwater is of mixed Mg and Na - $SO_4$ group having two families which are ($Na_2SO_4$) and ($MgSO_4$).

3- The results indicate two types, the first is marine water and the second is meteoric water. Marine origin indicates deep origin because of existence of a different pressure caused to make a vertical flow of groundwater.

4- The ground water in the region were not good for drinking and it is Doubtful to unsuitable-Unsuitable for irrigation whereas it can be used to irrigate sensitive crops. Also, it is not suitable for industrial and suitable for building purposes.

## References:

1. Consrtium, 1977. Water development projects, Western Desert Block 7; Hydrogeological explorations and Hydrotechnical works, Climatology and Hydrology, Yugoslavia Final Report Vol.1, Republic of Iraq, Directorate of Western Desert Development Projects, (Unpub.).
2. Al-Basrawi, N H., 1996, Hydrogeology of Razzaza Lake, Unpub. Ph.D. Thesis, College of Science ,University of Baghdad. Iraq. (in Arabic). P.128.
3. Jassim R. and Al-Zubaidi A. 2013,Hydrochemical assessment of water resources southern part of Al-Razzazah Lake and surroundings, Karbala area, central Iraq, Iraqi bulletin of Geology and Mining, Vol.9, No.2, pp 81-95.
4. Al- Qaraghuli S. A. ,2014, Hydrogeological and Hydrochemical Study of Al-Rahaliya –Al-Ekhedhur Area, Unpub. M.Sc. Thesis, College of Science, University of Baghdad.
5. Barazanji, M.A., and Yasi, A.I., (1987): Geophysical Study of Habbania- Razzaza area, J. Water Reso., Vol. 6, No.2.
6. Abbass, M,J:, and AL-Khatib, H.H;(1982) : Complex Geophysical Interpertation of Hit-Abu-Jir, Shithatha area, SOM Lib, Rep. No. 1247, 34P.
7. Al-Meheidi, H., Vejlupek, M., and Yacoub, S. Y., 1975: The regional geological mapping of Shithatha - Habbaniyah area. GEOSURV, int. rep. No. 697.
8. Don, C.M., 1995: A grows guide to water quality. University college station, Texas. P. 601-609.
9. Al- Fatlawi , A.N. (2000): Hydrochemical study of Shithatha area under white valley. M.Sc. Thesis. Collage of science. University of Baghdad. (In Arabic). P.113.
10. Hassan, A. H. and al Kubaisi, Q. Y., 2002: The fourth axis and the third subaxis, natural recharge for the ground water in the western desert. Second part. PP. 308-310.
11. Al-Jiburi, H. K., and Al-Basrawi, N.H., 2007: Geology of the Iraqi western desert, Iraqi Bulletin of Geology and Mining, (GEOSURV, Iraq), P. 169.
12. Al-Jiburi, H.K., and Al-Basrawi, N. H., 2014: Hydrogeological map of Iraq, scale 1:1000000, 2nd edition, Iraqi Bulletin of Geology and Mining, (GEOSURV, Iraq), vol.10, No.2, P. 29.
13. Sissakian, V., Ibrahim, E., Ibrahim, F., & Al-Ani, N., 2000.Geologic Map of Iraq, 3rd.edition, GEOSURV, Baghdad, Iraq.
14. Detay, M., 1997: Water Wells- implementation, maintenance and restoration, John Wiley and Sons, London, P. 379.
15. Sullin, V.A., 1946, Oil water in the system of natural groundwater, Gostopichezdat, Moscow USSR, 215p.vol. 30, No. l, P. 37-45.
16. Hassan, A.H., Al-Din, T.S. and Al-Sayyab, A., (1988) The(SS) hydrochemical model, Iraqi J. Sci.,V. 29 ,No. 1&2 pp. 109-121.

**Table 1.** Groundwater properties in the wet period April, 2013.

| Well No | Na+K epm (%) | Mg epm (%) | Ca epm (%) | Cl epm (%) | SO4 epm (%) | HCO3 epm (%) | TDS ppm | EC µs/cm | pH | T.H (ppm) |
|---|---|---|---|---|---|---|---|---|---|---|
| 1 | 61.57 | 19.96 | 18.45 | 54.67 | 32.24 | 13.08 | 2137 | 3710 | 7.31 | 648.9 |
| 2 | 69.27 | 14.13 | 16.58 | 32.31 | 61.09 | 6.59 | 1740 | 2570 | 7.12 | 276.9 |
| 3 | 72.09 | 12.50 | 15.38 | 36.31 | 47.86 | 15.82 | 1848 | 2760 | 7.08 | 320.8 |
| 4 | 78.90 | 8.38 | 12.7 | 36.85 | 42.35 | 20.78 | 1860 | 2560 | 7.15 | 227.5 |
| 5 | 45.97 | 18.37 | 35.64 | 38.04 | 48.23 | 13.71 | 1755 | 2520 | 7.30 | 604.3 |
| 6 | 32.03 | 35.99 | 32.58 | 35.42 | 58.96 | 5.61 | 1932 | 2650 | 7.40 | 655.1 |
| 7 | 32.7 | 43.97 | 32.32 | 35.27 | 59.09 | 5.61 | 1820 | 2600 | 7.50 | 648.5 |
| 8 | 31.41 | 36.5 | 32.08 | 35.33 | 59.49 | 5.17 | 1932 | 2550 | 7.61 | 687.2 |
| 9 | 31.26 | 36.59 | 32.16 | 35.36 | 59.4 | 5.23 | 1700 | 2560 | 7.16 | 687.2 |
| 10 | 34.22 | 34.37 | 31.39 | 35.59 | 59.14 | 5.55 | 1919 | 2650 | 7.31 | 652.6 |
| 11 | 42.2 | 27.93 | 29.79 | 41.67 | 45.25 | 13.07 | 1872 | 2540 | 7.15 | 772.9 |
| 12 | 47.51 | 17.84 | 34.63 | 37.43 | 47.58 | 14.98 | 1723 | 2560 | 7.34 | 604.3 |
| 13 | 74.40 | 11.57 | 14.01 | 40.25 | 42.47 | 17.27 | 2000 | 2670 | 7.34 | 320.9 |
| 14 | 55.47 | 24.29 | 20.24 | 49.38 | 39.13 | 11.47 | 2420 | 3000 | 7.31 | 685.8 |
| 15 | 42.00 | 18.61 | 35.38 | 37.5 | 47.52 | 14.96 | 1812 | 3020 | 7.21 | 620 |
| 16 | 74.59 | 11.39 | 14.02 | 40.44 | 42.31 | 17.24 | 2172 | 2920 | 7.19 | 320.9 |

**Table 2.** Groundwater properties in the dry period October, 2013

| welll No | Na+K epm (%) | Mg epm (%) | Ca epm (%) | Cl epm (%) | SO4 epm (%) | HCO3 epm (%) | TDS ppm | EC µs/cm | pH | T.H (ppm) |
|---|---|---|---|---|---|---|---|---|---|---|
| 1 | 56.37 | 21.59 | 22.03 | 55.61 | 31.28 | 13.09 | 2350 | 3850 | 7.20 | 839.4 |
| 2 | 69.42 | 13.3 | 17.26 | 36.56 | 51.29 | 12.13 | 1740 | 2560 | 7.21 | 339.8 |
| 3 | 70.00 | 14.23 | 15.37 | 40.46 | 42.39 | 17.13 | 2204 | 2900 | 7.11 | 389.5 |
| 4 | 69.5 | 12.47 | 17.53 | 36.55 | 51.19 | 12.24 | 2226 | 2950 | 7.22 | 338.4 |
| 5 | 37.07 | 31.90 | 31.01 | 37.37 | 56.75 | 5.67 | 2100 | 2800 | 7.21 | 632.5 |
| 6 | 45.1 | 23.58 | 31.40 | 41.13 | 44.84 | 14.02 | 2305 | 2800 | 7.80 | 978.1 |
| 7 | 45.46 | 23.30 | 31.22 | 40.46 | 44.53 | 14.49 | 2220 | 2800 | 7.3 | 962.4 |
| 8 | 44.34 | 23.26 | 32.37 | 33.39 | 51.62 | 10.97 | 2238 | 3670 | 7.36 | 1032.8 |
| 9 | 71.81 | 12.65 | 15.53 | 40.25 | 42.47 | 17.27 | 2000 | 3670 | 7.40 | 366.4 |
| 10 | 27.33 | 33.31 | 39.34 | 43.50 | 51.52 | 4.96 | 2260 | 2750 | 7.20 | 852.1 |
| 11 | 38.99 | 31.5 | 29.49 | 53.96 | 37.35 | 8.68 | 2228 | 2700 | 7.20 | 979.6 |
| 12 | 36.91 | 31.99 | 31.09 | 35.09 | 59.14 | 5.55 | 2000 | 2750 | 7.30 | 632.4 |
| 13 | 41.95 | 24.58 | 33.53 | 35.74 | 53.18 | 11.07 | 2450 | 3480 | 7.22 | 972.5 |
| 14 | 48.24 | 22.24 | 29.51 | 29.75 | 46.32 | 23.05 | 2750 | 3260 | 7.25 | 813.7 |
| 15 | 44.21 | 23.11 | 33.47 | 40.42 | 45.17 | 14.4 | 2113 | 2750 | 7.25 | 802 |
| 16 | 70.94 | 13.79 | 15.24 | 40.46 | 42.39 | 17.13 | 2476 | 3260 | 7.11 | 389.6 |

**Table 3.** Rahaliya– Shithatha groundwater classification for the wet period April, 2013 according to according to [16]

| | Family | Group | Index | Water type | Spring No. | % |
|---|---|---|---|---|---|---|
| 1 | Sulphate-Sodium | Sulphate | 23; 32 | rNa>.rCa>rMg;rSO4>rCl | 2,3,4,5,11,12, 13,15,16 | 56.25% |
| 2<br>3 | Sulphate-Magnesium | | 53; 32 | rMg>rCa>rNa; rSO4>Cl | 6,8,9 | 18.75% |
| | | | 33; 32 | rMg>rNa>rCa;rSO4>Cl | 7,10 | 12.5 % |
| 4 | Chloride-Sodium | Chloride | 13;12 | rNa>rMg>rCa;rCl> SO4 | 1,14 | 12.5 % |

**Table 4.** Rahaliya– Shithatha groundwater classification for the dry period October, 2013 according to [16]

| | Family | Group | Index | Water type | Spring No. | % |
|---|---|---|---|---|---|---|
| 1 | Sulphate-Sodium | Sulphate | 13; 32 | rNa>rMg>rCa;rSO4> Cl | 5 ,12 | 12.5% |
| 2 | | | 23; 32 | rNa>.rCa>rMg;rSO4> rCl | 2,3,4,6,7,8,9,13,14,15,16 | 68.75% |
| 3 | Sulphate-Magnesium | | 63;32 | rCa>rMg>rNa; rSO4>Cl | 10 | 6.25% |
| 4 | Chloride-Sodium | Chloride | 13;12 | rNa>rMg> rCa; rCl>SO4 | 11 | 6.25% |
| 5 | | | 23;12 | rNa>.rCa>rMg; rCl>SO4 | 1 | 6.25% |

# Toluene concentrations reduction by using photocatalytic coating methods for cementitious materials

*Ayat* Hussein [1,*], *Riyad* Al Anbari [1] and *Maan* Hassan [1]

[1]Building and Construction Engineering Department., University of Technology (10066), Baghdad, Iraq.

**Abstract.** Volatile organic compounds (VOC) generated from various sources like car combustion is one of the most surrounding pollutants, which can be transferred from one form to another in the presence of sunlight. In the present work the ability of the reduction of VOC have been conducted on cementitious samples surfaces by using toluene as a pollutant gas and TiO2 as an environmentally friendly photocatalyst . Two coating methods have been used (dip and spray) with two types of aqueous solution, prepared by spreading 3g/L micro and nano TiO2 powder with deionized water. A laboratory test procedure was adopted to assess the performance of the coated specimens. The specimens were subjected to toluene gas and there efficiency in gas removal was monitored with time. Furthermore, contact angle for the coated samples was also examined to investigate the hydrophilicity of the coated substrate. The results showed that TiO2 give high activity and work as a good photocatalyst for mitigation of air pollutants, and that it could be used in different ways and concentrations to obtain better air quality and aesthetic building appearance. The observed coating efficiency in reducing the toluene was more pronounced in higher than lower gas concentrations for the micro coated samples with removal efficiency equals to 13% and 10% for dipping and spraying method, respectively. While the higher removal was about 20% for both nano dipping and nanospray methods for the inlet gas concentration 3ppm. This behavior reflects the beneficial effects of TiO2 coating procedure in highly pollutant environments, like Iraq.

## 1. Introduction

Air pollution is one of the most harmful environmental problems that effects human health (Boyd 2006). Recent studies showed that photocatalytic materials can be used with concrete building through surface coatings or mixing with building materials. In the presence of UVlight and photocatalyst material, oxidation of the pollutants started and precipitated on the surface of the material. The products of the reaction are washed from the surface by rain. (Beeldens 2006).

Semiconductors used for the degradation of pollutants are usually metal oxides or metal sulfides. In searching for the ideal photocatalyst in remediation of organic pollutants by using sunlight, several considerations must be taken;chief among them are energy of the band gap and oxidation potential (Hathway 2009).
$TiO_2$ considered one of the most active photocatalyst which has three main crystal forms: anatase, rutile, and brookite. Anatase and rutile are the main two crystals which show photocatalytic activity. For photocatalytic applications, anatase has been considered to be more active than rutile because the conduction band location for anatase is more favorable for driving conjugate reactions involving electrons.(Zhao & Yang 2003)

The widely use of $TiO_2$ as a photocatalytic with building materials is due to the following characteristics(Chen & Poon 2009): a) Effective, inexpensive, safe, and have chemical stability, b) It has high photocatalytic activity compared with other photocatalysts metal oxide, c) Compatible with building materials.
The utilization of the photocatalytic building materials began from 1990s(Fujishima & Zhang 2006). Recently, one of the most effective ways to apply $TiO_2$ is by adding a thin coating to the weathering layer. If this layer is partly used by bad weather conditions a new coating of titanium dioxide particles could be added to the surface (Beeldens 2006). There are various types of substrates as supporting materials that could be used for applying $TiO_2$ to reduce air pollutants. More interests are growing for using cementitious

*Corresponding author : 40128@uotechnology.edu.iq

© The Authors, published by EDP Sciences. This is an open access article distributed under the terms of the Creative Commons Attribution License 4.0 (http://creativecommons.org/licenses/by/4.0/).

materials (e.g. mortar and concrete or cement paste) because of their porous structures and high binding property(Chen et al. 2011).

Studies reported lower photocatalytic activity of $TiO_2$ when using in immobilized form while the suspended form was more efficient (Rachel et al. 2002).

Finishing sprayed coating for wide horizontal and vertical surfaces, like ceiling, walls, tunnel archs and multi-story car parks were began to be used significantly(Marco et al. 2013).

Coatings must be protective barrier against chemical attack, abrasion, in addition to aesthetics aspects. Interface performance between film and substrate material takes an important part in the durability of the whole system. (Balaguru & Chong 2008).

Applying coating solutions are interesting because of the surface photocatalysis reaction, furthermore lead to lower consumption of $TiO_2$ (Martinez et al. 2011) and (Granqvist 2013). (Martinez et al. 2011) stated that there is a relation with the optimal $TiO_2$ content corresponded to the maximal photocatalytic surface area accessible to polluting substances.

Most solids are naturally rough; however, superhydrophilic surface coating could be obtained by chemical treatment or deposited in sub-microscopic particles. Films of nanoparticles are often deposited on substrates from solutions/suspension, ink-jet printing, by a sol-gel dip technique, spin coating or through sputtering and spray coatings (Chen et al. 2011) and (Drelich et al. 2011). Other researchers, (Guo & Poon 2012) compared 5% nano-$TiO_2$ intermixed mortars with dip-coated mortars in nano-$TiO_2$ solution; dip-coated mortars displayed a highly improved photocatalytic activity.

(Faraldos et al. 2016) stated that all of the sols (aqueous suspensions of nanoparticles) coatings showed very high affectivity, even when the coatings were extremely thin or diluted, independently on the deposition method (spraying or dip coating).

Time scale for dip coating method is slow, also this method has a mechanical and environmental stability, while spray coating method is rapid, have moderate stability, fast and easy reparability (Liu et al. 2016).

(Ramirez et al. 2010) concluded high toluene removal efficiencies (up to 86%) obtained with the dip-coated samples, solution of $TiO_2$ was by using a suspension of ethanol and $TiO_2$ (0.05 g/ml).

(Hassan et al. 2012) compared removal efficiency for three types of coatings: thin film of concrete surface with 3% and 5% $TiO_2$ mixtures, nano $TiO_2$ powder sprinkled to fresh concrete surface before hardening, and spraying a water based $TiO_2$ on a hard concrete surface. Results obtained appears that 5% $TiO_2$ mixture and spraying water based $TiO_2$ were the most efficient in removing pollutants from air stream.

Studies showed that the photocatalytic degradation of VOCs is dependent on both the air humidity and the initial concentration of pollutant (Martinez et al. 2011) . Toluene removal has been investigated by (Demeestere et al. 2008), using $TiO_2$ coatings with particle sizes of 30–60 nm and a specific surface area larger than 250 $m^2/g$ , demonstrating the effect of increasing RH (1–77%) with increasing [TOL]in concentrations. Results revealed lower toluene removal efficiencies at higher RH and [TOL]in, whereas a positive effect was observed with increased t.

The aim of this research work is to evaluate the ability of nano scale $TiO_2$ for the removal of toluene from surrounding air by applying two economic surface coating methods (spraying and dipping) that could be used even for previously constructed buildings. The performance of the coated cementitious specimens were assessed and then compared with the control, non-coated specimens.

## 2. Materials and experimental procedure

Concrete substrates were prepared according to ASTM C305 standards (ASTM C305 2011) standards. Cement used was type I ordinary portland cement, sand was graded according to ASTM C778 (ASTM C778 2002). Two grain size of commercial titanium dioxide powder was used; the first type is micro size range (150-200 nm) from Tianjin Zhi Yuan ReagentCo.,Ltd.-Chinese market with specific surface area of 6.9 $m^2/g$ , the second type is nano size PC105 (20$\pm$5 nm) with specific surface area of 85.6 $m^2/g$ from Cristal Active Millennium SAS-France.

Pollutant gas used was from mesa specialty gases and equipment -USA, California. Concentration of toulene was 400ppm which diluted by another air cylinder (80% $N_2$ and 20% $O_2$).

Gas detector device was from RAE system by Honeywell-USA, San Jose, detection limits for VOCs were from 1 to 1000 ppm. Parts for reactor details were collected from local market and designed according to International Organization for Standardization (ISO 22197-1:2007) with some modification according to test requirements. Two inlet toluene gas concentrations were applied on the coated samples 3ppm and 5 ppm in order to study the effect of low and high gas concentrations on the removal efficiency. Tests parameters (humidity, temperature and light intensity) were the same for all samples. Minimum residence time of 60 min was used for all samples to investigate toluene concentration reduction. If reduction happened for a sample another 60 min was added to note if there is any more reaction happened.

### 2.1 Mortar (subatrate) preparation

Sand was sieved by using 4 sets of sieves and standard sand gained by using the retained on sieve 600 μm according to ASTM C778.

Mixing procedure was conducted according to ASTM C305 standards with cement to sand ratio of 1:2.75 and w/c of 0.484. Flow test was made to ensure workability for the mixture (ASTM:C1437-13 2013). All mortar specimens were casted in 9 cm diameter petri dish to form 6 mm thickness samples and let harden for 24 hr at room temperature then cured for 28 days at 100% relative humidity, then dried and stored in sealed bags.

### 2.2 Aqueous solution preparation

Samples have been coated by using two methods (spray and dip coating).Two types of titanium powder were used nano and micro, aqueous solutions for both types prepared on the same concentration. A suspension was prepared by spreading 3g $TiO_2$ powder in 250 ml ethanol and 750ml deionized water stirred for (30 min.) then ultrasonicated for 1 hour to obtain solution homogeneity.

In spray method, substrates were sprayed for 5 minutes by using spray pyrolysis equipment, (Fig. 1).

**Fig. 1**. Spray pyrolysis equipment

In dip coating method substrates were dipped in the same concentration of the aqueous solution for 10 minutes, and then dripped for 3 minutes (Fig. 2). Finally, for both methods samples were oven-dried at 105°C for 1 hour, and kept sealed till test (Fig. 3).

### 2.3 Reactor setup

Rectangular chamber used with dimensions of about (30 cm x10 cm), as shown in Fig. 4. All Tests were carried out at room temperature with applying a continuous air flow, gas flow used was about 1.6$\pm$0.2 l/min. Two 6 W UV lamps have been used with intensity of 19 W/m$^2$ for each lamp and wave

**Fig. 2**. Dipping method for spraying micro and nano $TiO_2$ aqueous solutions

length range (300-400 nm). To control humidity inside the reactor a humidifier has been used with control valves. Test procedure was as follows:

1- Calibration for the gas detector must be checked.

2- Putting the sample inside chamber and close tightly.

3- A continuous air flow ejected inside the reactor for about 1 hr to persuade cleaning of the reactor especially VOC with UV lamps turned off and dark conditions.

4- Inlet gas pollutant (toluene) was opened and calibrated till reached the desired concentration 3 ppm and 5 ppm.

5- Humidity was calibrated through control valves to obtain RH of about 50%.

6- Waiting for about 25 min to ensure chamber saturation and readings stabilization after this UV source is turned on.

7- The reaction continued till reaching a steady state.

8- After this UV is turned off, close the pollutant gas valve and again only air flow is ejected inside chamber.

**Fig. 3**. Samples after spraying and dried (right) and after dipping and drying (left)

**Fig. 4.** Lab. Photocatalytic Reactor

## 3. Results and discussion

Tests were made to evaluate the suitability of the samples for VOC mitigation and the hydrophilicity of the coated samples was also investigated.

### 3.1 Low VOC concentration reduction results by using dipping method

Air flow meter was adjusted in order to obtain gas concentration equals to 3ppm. There was no removal observed for toluene concentration from micro-dipped samples (Fig. 5). The removal for toluene for other samples was one value from the inert concentration 3ppm to 2 ppm with differences in the time required for the reduction. Nano-dipped samples showed overall removal efficiency of 20.62% and the reduction happened after 37 min. from starting the reaction (Fig. 6).

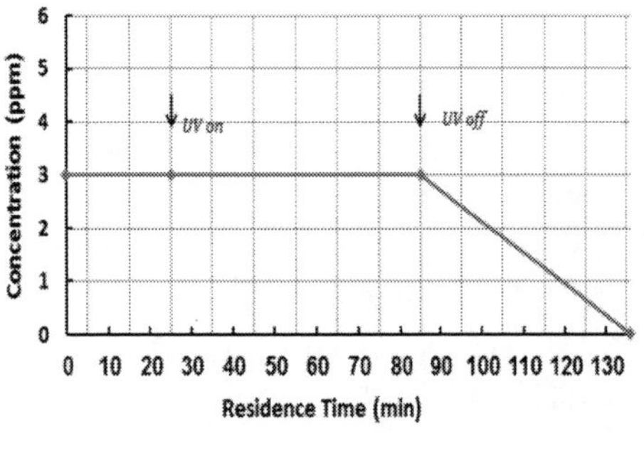


**Figure 5**- Micro-dipped sample reduction for low gas concentration

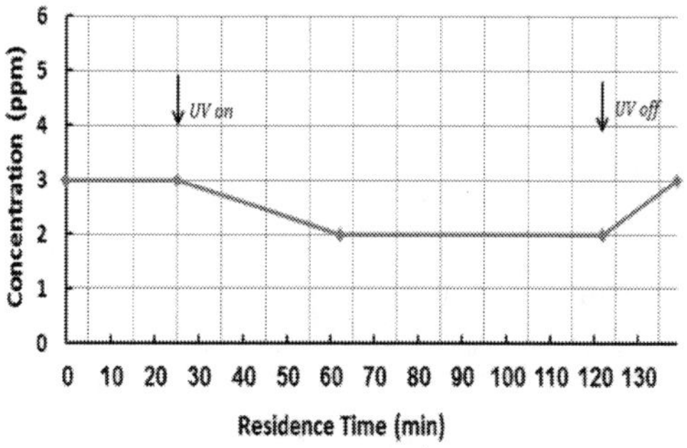


**Figure 6**- Nano-dipped sample reduction for low gas concentration

### 3.2 High VOC concentration reduction results by using dipping method

Adjusting flow meter for an inert gas concentration of 5ppm, the removal for toluene was one value from the inert concentration 5ppm to 4 ppm with differences in the time required for the reduction. Removal efficiency was 13.48 % for micro-dipped samples after 29 min from reaction and 15% within 20 min for nano-dipped samples which indicates increasing susceptibility for the high gas concentrations rather than low gas concentrations as stated by (Martinez et al. 2011) especially for the micro-dipped samples; Figs. 7 and 8 show the obtained results for micro and nano-dipped samples, respectively.

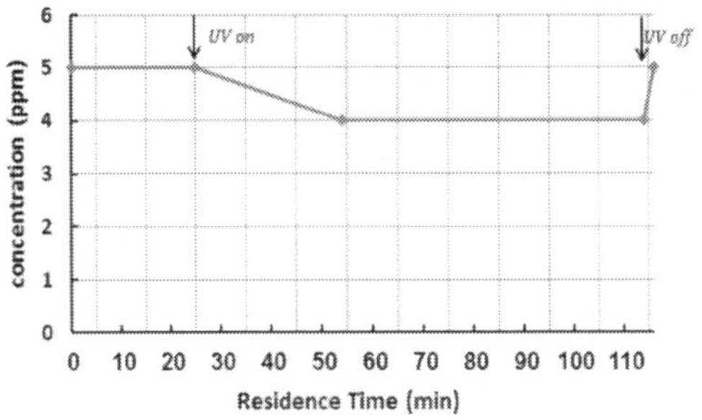


**Figure 7**- Micro-dipped samples reduction for high gas concentration

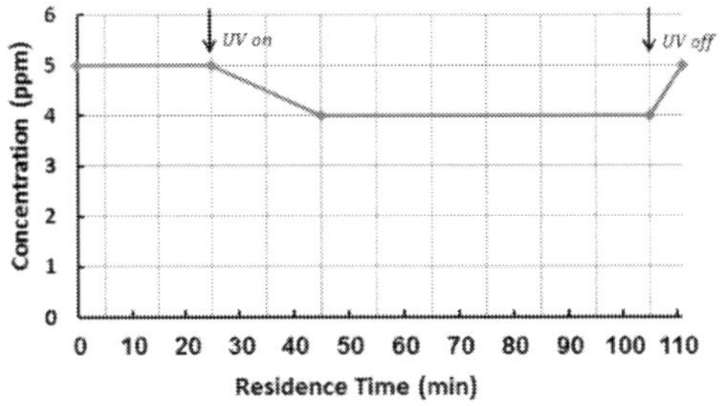


**Figure 8**- Nano-dipped samples reduction for high gas concentration

From the above results it could be concluded that the same range of removal happened but within different times, that is depending on the type of the material used. Nano coated samples required shorter time than micro coated sample. This result is expected due to the high surface area for the nano coated samples, which results to reduce time required for the reduction especially for high gas concentrations as stated by (Demeestere et al. 2008) and (Martinez et al. 2011) .

### 3.3 Low VOC concentration reduction results by using spraying method

The same test procedure in dipping method was applied on the previously micro and nano sprayed samples for 5 min spraying time.

Ejecting low gas concentration equals to 3ppm on micro-sprayed samples showed no gas removal. Nano-sprayed samples showed removal of 20.62%, the reduction happened after 37 min from starting the reaction as shown in Fig. 9

### 3.4 High VOC concentration reduction results by using spraying method

5 min micro-sprayed samples showed removal of 10.16 %, the reduction happened after 58min from

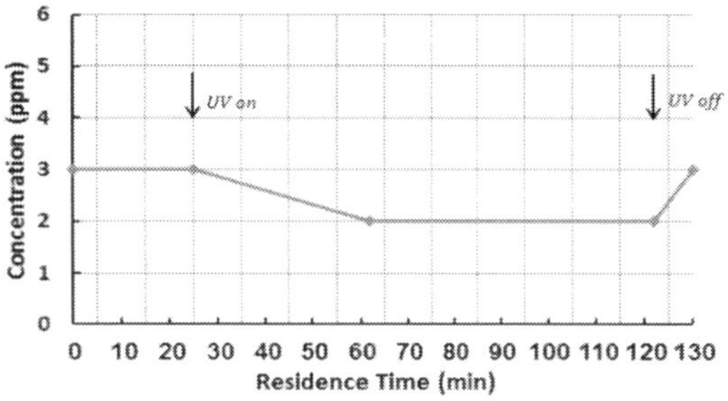


**Figure 9**- (5 min) Nano-sprayed samples reduction for low gas concentration

starting the reaction as shown in Fig.10. The removal requires less time for the 5 min nano-sprayed sample that is equal to 15.79 % and the reduction happened within 16 min, Fig. 11. These results are expected due to high reactivity of nano coated samples.

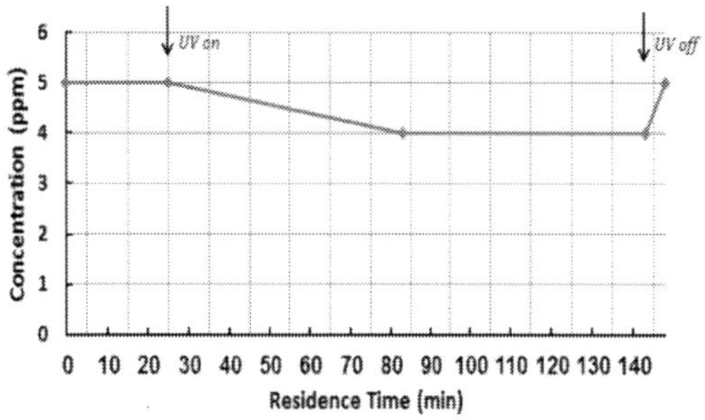


**Figure 10**- (5 min) Micro-sprayed samples reduction for high gas concentration

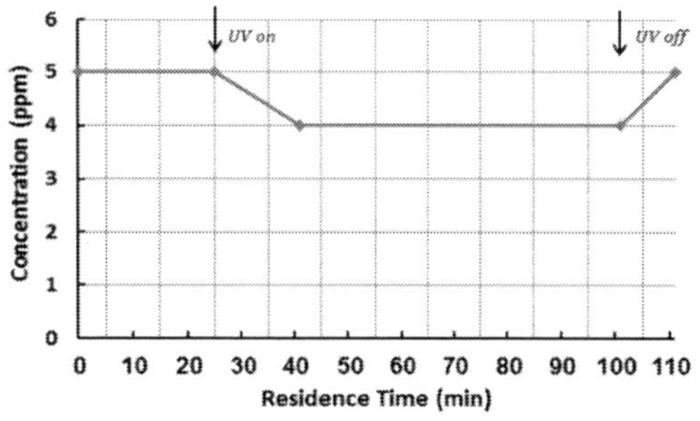


**Figure 11**- (5 min) Nano-sprayed samples reduction for high gas concentration

### 3.5 Comparison between coating methods

The removal of low toluene gas concentrations were not reduced by both micro-dipped and micro-sprayed samples, however, high gas concentrations were reduced.

Using the micro$TiO_2$ solution for removing toluene concentration by dipping method need less time with better removal than spraying method this could be explained due to the high amounts with large agglomerations of the particles that could be absorbed and bind to the substrate surface area, while in spray method some amounts of the solution adsorbed to the inner part of mortar substrate due to the small liquid particle volume generated by the spryer that could be inbounded by cement based material. This results in reducing the effective part for the coating and increasing time required for the removal.

For both nano-dip and 5 min nano-spray methods, the same removal for low toluene concentration was achieved within the same time (37 min).

For high toluene concentration removal, using nano$TiO_2$ solution also seems to be almost within the same time for each spray and dip method 16 min and 20 min, respectively. Fig. 12 illustrates the stated results.

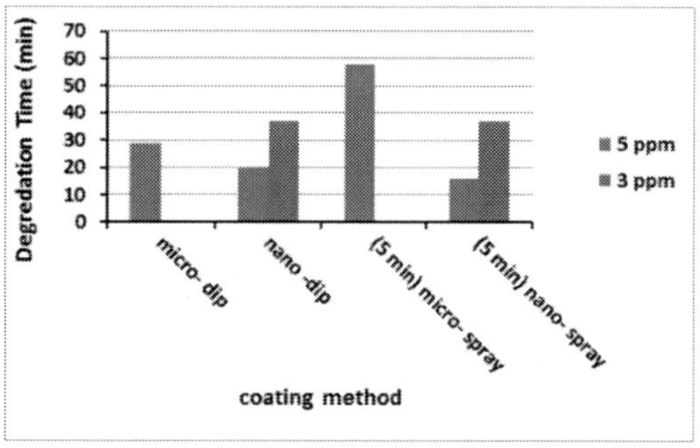


**Figure 12**- Comparison between coating method and time elapsed for toluene degradation

### 3.6 Contact angle results

Titanium dioxide have a hydrophilic specification, micro and nano dip coatings showed some agglomerations on the surface as expected, therefore substrates have a contact angle of about 34.7° and 16.20° for the micro and nano dip, respectively, also 5 min nano-sprayed samples have a contact angle equals to 14° as shown in Fig. 13, other coating types showed high hydrophilicity (superhydrophilic) surfaces with 0° contact angle, Table 1 summarize the obtained results. Decreasing the hydrphilicity of the cementious material may helps in removing solid pollutants that adhere on surfaces by washing in addition to the photochemical reaction.

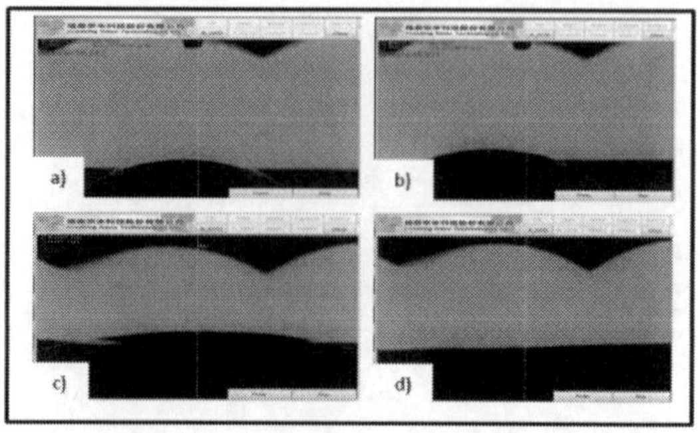


**Figure13**- Contacts angle image for various substrates; a) Micro –dipped substrate, b)Nano–dipped substrate) 5 min nano-sprayed substrate, d) Substrates with 0° contact angle.

**Table 1**- Summary of contact angle test for the coated samples

| Sample Type | Contact Angle | Description |
|---|---|---|
| Reference mortar sample | 0 | hydrophilic |
| Micro dip | 34.7° | hydrophilic |
| Nano dip | 16.20° | hydrophilic |
| (5 min) micro-spray | 0 | hydrophilic |
| (5 min) nano-spray | 14.04° | hydrophilic |

## 4. Conclusions

In this study, simple methods of spray or dip coating were adopted to assess the self-cleaning performance of the produced cementitious specimens. Based on the experimental work results in this investigation, the following conclusions can be drawn:

1. Coating cementitious materials with an aqueous solution contains nano titanium dioxide was effective in reducing toluene concentration.
2. Coatings with micro titanium dioxide were less efficient in the removal especially to the low gas concentration.
3. Both dip and spray coatings have the same activity in the removal when nano aqueous solution was used.
4. Coatings give a hydrophilic surface for the cementious material this helps in removing solid pollutants in addition to the photochemical reaction.

## References


1. ASTM:C1437-13, 2013. Standard Test Method for Flow of Hydraulic Cement Mortar 1. *ASTM International*, pp.1–2.
2. ASTM C305, 2011. Standard Practice for Mechanical Mixing of Hydraulic Cement Pastes and Mortars of Plastic Consistency. *ASTM International*, pp.1–3.
3. ASTM C778, 2002. Standard Specification for Standard Sand. *ASTM International*, Vol 14.04.(Reapproved), pp.1–4.
4. Balaguru, P. & Chong, K., 2008. Nanotechnology and concrete: research opportunities. *ACI Special Publication*, pp.15–28. Available at: http://www.concrete.org/Publications/GetArticle.aspx?m=icap&pubID=20208.
5. Beeldens, A., 2006. Environmentally Friendly Concrete Pavement Blocks : Air Purification in the Centre of Antwerp. *8th International Conference on Concrete Block Paving*, pp.277–284.
6. Boyd, D.R., 2006. The Air We Breathe: An International Comparison of Air Quality Standards and Guidelines. *A REPOR T PREPARED FOR THE DAVID SUZUKI FOUNDATION HEALTH AND ENVIRONMENT SERIES BY.*
7. Chen, J., Kou, S. cong & Poon, C. sun, 2011. Photocatalytic cement-based materials: Comparison of nitrogen oxides and toluene removal potentials and evaluation of self-cleaning performance. *Building and Environment*, 46(9), pp.1827–1833. Available at: http://dx.doi.org/10.1016/j.buildenv.2011.03.004.
8. Chen, J. & Poon, C. sun, 2009. Photocatalytic construction and building materials: From fundamentals to applications. *Building and Environment*, 44(9), pp.1899–1906. Available at: http://dx.doi.org/10.1016/j.buildenv.2009.01.002.
9. Demeestere, K. et al., 2008. Heterogeneous photocatalytic removal of toluene from air on building materials enriched with TiO2. *Building and Environment*, 43(4), pp.406–414.
10. Drelich, J. et al., 2011. Hydrophilic and Superhydrophilic Surfaces and Materials*. *Electronic Publishing*, 7(21), pp.9804–9828.
11. Faraldos, M. et al., 2016. Photocatalytic hydrophobic concrete coatings to combat air pollution. *Catalysis Today*, 259(x), pp.228–236. Available at: http://dx.doi.org/10.1016/j.cattod.2015.07.025.
12. Fujishima, A. & Zhang, X., 2006. Titanium dioxide photocatalysis: present situation and future approaches. *Comptes Rendus Chimie*, 9(5–6), pp.750–760.
13. Granqvist, C.G., 2013. 8 - Thin films and nanostructured coatings for eco-efficient buildings. In A. N. F. Pacheco-Torgal, M. V. Diamanti & and C.-G. Granqvist, eds. *Nanotechnology in Eco-Efficient Construction*. p. 161–187a. Available at: http://www.sciencedirect.com/science/article/pii/B9780857095442500087.
14. Guo, M.Z. & Poon, C.S., 2012. An Effective Way to Incorporate Nano-TiO 2 in Photocatalytic Cementitious Materials. *Third International Conference on Sustainable Construction Materials and Technologies*. Available at: http://www.claisse.info/Proceedings.htm.
15. Hassan, M.M. et al., 2012. Methods for the application of titanium dioxide coatings to concrete pavement. *International Journal of Pavement Research and Technology*, 5(1), pp.12–20.
16. Hathway, T.L., 2009. *Titanium dioxide photocatalysis : studies of the degradation of organic molecules and characterization of photocatalysts using mechanistic organic chemistry*. Iowa State University.
17. Liu, H. et al., 2016. Recent progress in fabrication and applications of superhydrophobic coating on cellulose-based substrates. *Materials*, 9(3), pp.1–37.
18. Marco, T. De et al., 2013. Use of photocatalytic products for sustainable construction development. *Third International Conferences on Sustainable Construction Materials and Technologies*.
19. Martinez, T. et al., 2011. Degradation of NO using photocatalytic coatings applied to different substrates : Dégradation de NO en utilisant des revêtements photocatalytiques appliqués sur différents substrats. *Building and Environment*, 46(9), pp.1808–1816. Available at: http://www.sciencedirect.com/science/article/pii/S0360132311000722%5Cnhttp://www.sciencedirect.com/science?_ob=MImg&_imagekey=B6V23-52CYKGN-1-11&_cdi=5691&_user=636532&_pii=S0360132311000722&_origin=search&_coverDate=09/30/2011&_sk=999539990&view=c&wchp=d.
20. Rachel, A., Subrahmanyam, M. & Boule, P., 2002. Comparison of photocatalytic efficiencies of TiO2 in suspended and immobilised form for the photocatalytic degradation of nitrobenzenesulfonic acids. *Applied Catalysis B: Environmental*, 37(4), pp.301–308.
21. Ramirez, A.M. et al., 2010. Titanium dioxide coated cementitious materials for air purifying purposes: Preparation, characterization and toluene removal potential. *Building and Environment*, 45(4), pp.832–838.
22. Zhao, J. & Yang, X., 2003. Photocatalytic oxidation for indoor air purification: A literature review. *Building and Environment*, 38(5), pp.645–654.

# Removing chromium and lead metals using phytoremediation technique

*Riyad* Al-Anbari [1], *Abdul-Hameed* Al-Obaidy [2], *Mahmoud* Al-Khafaji[1]* and *Tiba* Al-Imari [1]

[1]Building and Construction Engineering Dept, University of Technology, Baghdad, Iraq
[2]Environment Research Center, University of Technology, Baghdad, Iraq

**Abstract.** Phytoremediation technique uses plants parts to remove, extract, and absorb heavy or toxic matter from soil and water. In the present study, Catharanthusroseus (Periwinkle) and Nerium Oleander (Oleander) were used for removing Chromium (Cr) and Lead (Pb) metals. These plant species were seeded in polyethylene pots containing 8kg of soil. Each pot was irrigated with wastewater for four months (May, June, July and August) and accumulation of the considered metals was analyzed after every month for leaf, stem and root by using Atomic Absorption Spectrophotometer (AAS). This experimental work was carried out in the laboratories of Water Desalination Researches Unit - Building and Construction Engineering Department and Environmental Research Centre at the University of Technology in Baghdad City, Iraq. The concentration of Cr was found to be increased with time. High Cr concentration, 20.34 mg/kg, was recorded at August in leaf of Periwinkle and 19.61 mg/kg in root of Oleander in case of using 100% wastewater (WW). While, for Pb, the maximum concentration, 22 mg/kg, was recorded in June in leaf of Periwinkle and 19.5 mg/kg in steam of Oleander. Accordingly, Oleander has the maximum removal efficiency.

## 1 Introduction

Water, air and soil pollution is the main environmental issue faced by the modern society. While, soils are considered as the major sink for toxic metals released to the environment by the mention anthropogenic activities [1]. Thus, contaminated soils and waters pose a major environmental and human health problem [2]. The heavy metals may come from natural sources, leached from rocks and soils according to their geochemical mobility or come from anthropogenic sources, as a result of human land occupation and industrial pollution [3]. Heavy metals are found soluble in water, so that, it can enter in ecosystems by water pathways [4].
So that, many researchers and environmentalists have developed cheap and good technology to remediate the pollutant through "Phytoremediation" [5]. Phytoremediation is an emerging technology that uses various plants to degrade, extract, contain, or immobilize contaminants from soil and water. This technology has received much attention lately as an innovative, cost-effective alternative to the more established treatment methods used at hazardous waste sites [6]. There are several factors which are considered to select the plant for phytoremediation, one of them is that the plant should have acceptable degree of tolerance to the site conditions to grow well and should be able to accumulate numerous metal contaminants and the most important factor is that the plant species should grow quickly and should be easy to harvest [7]. Several types of phytoremediation are being used to-day Phytoextraction, Phytostabilazation, Phytovolatilization, rhizofiltration and phytodegradation.
The objective of this research is to determine the ability of Catharanthusroseus (Periwinkle) and Nerium oleander (Oleander) plants to remove Chromium (Cr) and Lead (Pb) ions from soil by irrigating with wastewater (WW) and to determine the effectiveness of these plants in removing these metals.

## 2 Experimental works

In order to evaluate the effect of ornamental plants, such as Periwinkle and Oleander, in the removal of Cr and Pb from soil irrigated with WW, Experimental pots were conducted during the period from May 2015 to August 2015. This experimental work was carried out in the laboratories of Water Desalination Researches Unit (WDRU) - Building and Construction Engineering Department (BCED) and Environmental Research Centre (ERC) at the University of Technology (UOT), Baghdad, Iraq. The experiment was arranged for four months to evaluate the effect of ornamental plants; Periwinkle and Oleander as shown in Plate 1, in the removal of Pb and Cr from soil irrigated with (WW). Four levels of water were used for irrigation including tap water (control) (T1), 50% of WW + 50% of tap water (50% WW) (T2), 75% WW + 25% tap water (75% WW) (T3) and 100% WW (100% WW) (T4). The selected

* Corresponding author: 41100@uotechnology.edu.iq

© The Authors, published by EDP Sciences. This is an open access article distributed under the terms of the Creative Commons Attribution License 4.0 (http://creativecommons.org/licenses/by/4.0/).

plants were planted in plastic pots (6 plant per pot; 25 cm length × 25 cm width × 25 cm height) filled with 8 kg of sandy loam soil. The pots were kept in the greenhouse of WDRU - BCED at UOT.

**Plate 1.** Tested plants.

The WW samples were collected from sewage nearby the experimental site in acid washed polyethylene containers. Potable water also collected from tap water nearby the experimental site. WW, tap water, soil and plant samples were monthly collected to determine the concentration of Cr and Pb.

The soil was air-dried, powdered, sieved through a 2 mm sieve and Cr and Pb concentration was measured. The total concentration of soil Pb and Cr was measured using wet digestion method. The WW samples were collected in acid washed polyethylene containers. 100 ml of WW sample has been filtered using Whatman filter paper (0.45 μm pore size) for estimation of dissolved heavy metals (Cr and Pb). Another 100 ml of the sample have been digested with (3: 1) of Hydrochloric acid (HCl) and Nitric acid (HNO3) to determine total heavy metals. Procedures followed for analysis have been in accordance with the standard methods for examination of water and WW [8]. Analysis of the total and dissolved of heavy metal was performed with a flame atomic absorption spectrophotometer (AA6300, Shimadza, Japan).

## 3 Results and discussion

The concentration of Cr and Pb heavy metals that were found in the parts of plants which were irrigated with WW during the experimental period (four months) is shown in Tables I and II. The measured concentrations of Cr in Periwinkle and Oleander plant parts are listed in Table I. The minimum concentration, 1.35 mg/kg, was found in the stem of Oleander plant in May in the control treatment and the maximum concentration, 19.855 mg/kg, was observed in the leaf of the Oleander plant at 100%WW in May. While the minimum concentration of Cr, 1.37 mg/kg, was found in the roots of Periwinkle plants at a control treatment in May and maximum concentration was measured in Periwinkle, 20.34 mg/kg, in August T 100%WW. However, the experimental results indicated that the Periwinkle plant can be considered as an effective plant to remove the Cr from soil irrigated with WW.

Table II demonstrates the results of concentration of Pb in leaf, stem and root of Periwinkle and Oleander. The results show that the minimum concentration, 4.7 mg/kg, was measured in May at the control treatment in the leaf of Periwinkle plant and maximum concentration, 22.00 mg/kg, was observed in the leaf of Periwinkle plant in June at 100% WW. For Oleander plant, the maximum and minimum concentration of Pb was observed in leaf at 25%WW in May and stem of plant at 100%W) in June. The highest and lowest value was 4.02 and 19.5 mg/kg, respectively.

The percent of removal efficiency of each heavy metal was calculated as follows:

$$\text{Removal Efficiency (\%)} = \left(\frac{C_i - C_f}{C_i}\right) \times 100 \qquad (1)$$

Where $C_i$ is the initial concentration of heavy metal (mg/kg) and $C_f$ is the final concentration of heavy metal (mg/kg).

Figures 1 to 4 show the removal efficiency in the level of heavy metals in plants for all treatment. The percentage of removal efficiency of Cr and Pb in the Periwinkle and Oleander for all the treatment was calculated for every month.

Figures 1 and 2 show that the maximum removal efficiency of Periwinkle to remove Pb from soil that was irrigated with WW was 91.08% which occurs in August. While for Oleander it was 95.96%. The minimum removal efficiency of Periwinkle and Oleander for removing Pb was 6.89 % and 6.08 %, respectively which occurs in May.

Figures 3 and 4 illustrated that the maximum removal efficiency to remove Cr from soil that was irrigated by WW of Periwinkle and Oleander was 94.36% and 94.92% which occurs in August. While the minimum was 1.82 % and 2.64 %, respectively, this occurs in May. Generally, the removal efficiency increases with time and both Periwinkle and Oleander are good plant to remove Pb and Cr from soil irrigated with WW.

## 4 Conclusions

In this research and from the obtained results, it can be concluded that both Periwinkle and Oleander observed in this study were very effective in removing Cr and Pb because of their acceptable percentage of removal efficiency. The efficiency of removing Pb for both plants, (94.36 % for Periwinkle and 94.92 % for Oleander), is better than that of Cr which is 91.08% and 95.96% for Periwinkle and Oleander, respectively.

Accordingly, it can be concluded that Periwinkle and Oleander have hyper accumulation properties and are good to remove heavy metals.

**Table 1.** Concentration of Cr in plant parts during the experimental period.

| Type of plants | 0% Leaf | 0% Stem | 0% Root | 25% Leaf | 25% Stem | 25% Root | 75% Leaf | 75% Stem | 75% Root | 100% Leaf | 100% Stem | 100% Root |
|---|---|---|---|---|---|---|---|---|---|---|---|---|
| **May** (Percent of WW) | | | | | | | | | | | | |
| **Periwinkle** | 1.78 | 2.80 | 1.37 | 5.7 | 5.01 | 6.34 | 6.37 | 6.12 | 7.83 | 7.35 | 6.145 | 9.24 |
| **Oleander** | 3.56 | 1.35 | 3.73 | 10.32 | 9.04 | 11.32 | 12.21 | 11.74 | 13.46 | 19.86 | 14.50 | 15.16 |
| **June** | | | | | | | | | | | | |
| **Periwinkle** | 5.16 | 2.72 | 8.91 | 7.31 | 8.21 | 6.11 | 9.95 | 10.98 | 9.87 | 10.13 | 9.23 | 10.98 |
| **Oleander** | 7.59 | 5.72 | 3.1 | 7.59 | 5.87 | 7.94 | 8.78 | 8.03 | 9.28 | 8.98 | 9.34 | 10.35 |
| **July** | | | | | | | | | | | | |
| **Periwinkle** | 7.02 | 6.97 | 7.05 | 9.21 | 8.98 | 9.01 | 13.89 | 12.12 | 9.87 | 15.9 | 13.79 | 13.99 |
| **Oleander** | 7.39 | 6.39 | 8.66 | 8.01 | 8.78 | 9.11 | 13.11 | 12.21 | 11.87 | 14.87 | 13.98 | 15.88 |
| **August** | | | | | | | | | | | | |
| **Periwinkle** | 9.48 | 11.25 | 9.46 | 11.43 | 11.12 | 10.28 | 11.8 | 11.8 | 11.47 | 20.34 | 18.45 | 19.35 |
| **Oleander** | 10.51 | 9.62 | 11.40 | 11.73 | 12.17 | 11.05 | 13.28 | 13.63 | 12.68 | 16.93 | 15.99 | 19.61 |

**Table 2.** Concentration of Pb in plant parts during the experimental period.

| Type of plants | 0% Leaf | 0% Stem | 0% Root | 25% Leaf | 25% Stem | 25% Root | 75% Leaf | 75% Stem | 75% Root | 100% Leaf | 100% Stem | 100% Root |
|---|---|---|---|---|---|---|---|---|---|---|---|---|
| **May** (Percent of WW) | | | | | | | | | | | | |
| **Periwinkle** | 6.78 | 5.8 | 5.37 | 4.7 | 5.01 | 4.34 | 5.37 | 6.12 | 5.83 | 11.5 | 9 | 10.5 |
| **Oleander** | 6.56 | 6.35 | 5.73 | 4.02 | 4.04 | 4.32 | 4.21 | 4.74 | 5.46 | 10.5 | 9.5 | 11.5 |
| **June** | | | | | | | | | | | | |
| **Periwinkle** | 10.16 | 9.72 | 10.10 | 13.7 | 14.01 | 13.34 | 14.37 | 16.12 | 17.83 | 22 | 19.5 | 15.5 |
| **Oleander** | 7.59 | 8.72 | 9.095 | 13.02 | 13.04 | 15.32 | 15.21 | 14.74 | 16.46 | 15.7 | 19.5 | 16.5 |
| **July** | | | | | | | | | | | | |
| **Periwinkle** | 6.155 | 6.72 | 5.905 | 12.32 | 11.01 | 12.34 | 13.37 | 14.12 | 13.83 | 16 | 15.5 | 17 |
| **Oleander** | 5.59 | 6.72 | 6.095 | 11.89 | 13.21 | 10.32 | 16.21 | 15.74 | 12.46 | 19.4 | 18.5 | 15.5 |
| **August** | | | | | | | | | | | | |
| **Periwinkle** | 7.155 | 8.12 | 9.105 | 5.32 | 7.01 | 8.34 | 13.32 | 11.01 | 11.34 | 17.5 | 15.5 | 17 |
| **Oleander** | 7.01 | 6.72 | 8.095 | 7.23 | 7.21 | 8.32 | 12.89 | 12.21 | 10.32 | 17 | 18.5 | 16 |

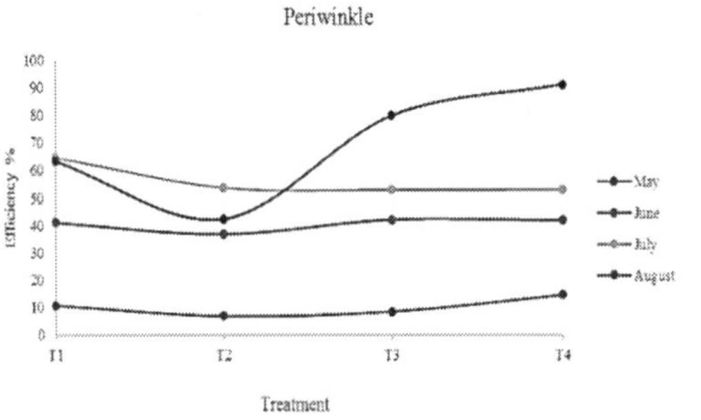


**Fig. 1.** Removal efficiency (%) of Pb for Periwinkle.

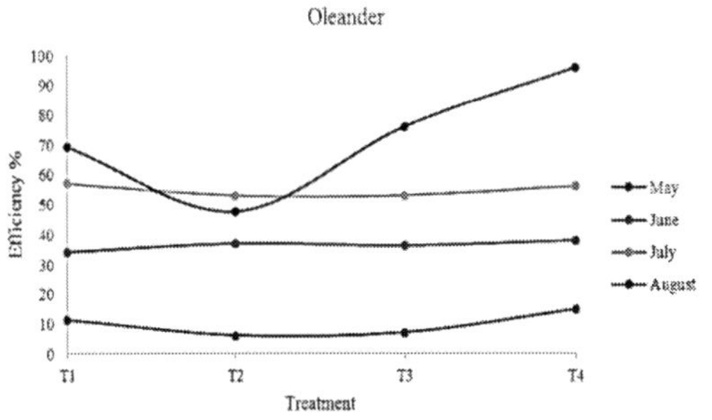


**Fig. 2.** Removal efficiency (%) of Pb for Oleander.

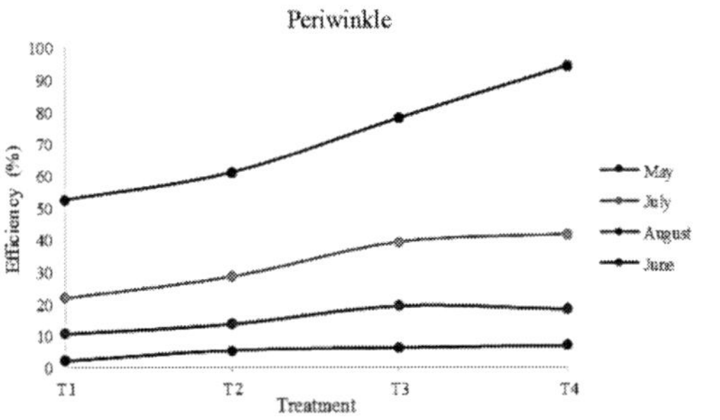


**Fig. 3.** Removal efficiency (%) of Cr for Periwinkle.

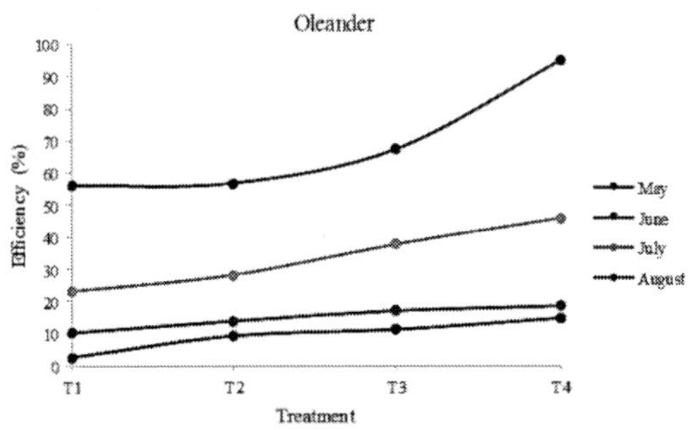


**Fig. 4.** Removal efficiency (%) of Cr for Oleander.

## Acknowledgment

The Authors thank and acknowledge the Ministry of Higher Education and Scientific Research, Research and Development Directorate for their support to construct the greenhouse in the University of Technology campus. We also cordially thank the Environmental Research Centre, University of Technology, Baghdad, Iraq for their scientific support. Special thanks to Mrs. Athmar Al-Mashhady, Environmental Research Centre, for her efforts in carrying out the laboratory analysis.

## References

1. M. M. Rahman, S. M. Azirun, A.N. Boyce, "Enhanced Accumulation of Copper and Lead in Amaranth (Amaranthuspaniculatus), Indian Mustard (Brassica Juncea) and Sunflower (Helianthus annuus)" PLoS ONE, 8(5), e6294, (2013)..
2. A. Mojiri, "Phytoremediation of Heavy Metals from Municipal Wastewater by Typhadomingensis" African Journal of Microbiology Research, 6(3), 643-647, (2011).
3. O. Nathan, K. P. Njeri, O.E. Rang'ondi, C. J. Sarima, "The potential of Zea mays, Commelinabengelensis, Helianthus annuus and Amaranthushybridus for phytoremediation of waste water". An Interdisciplinary Journal of Applied Science: vol. 7, no.3, 2012.
4. A. H. M. J. Al-Obaidy, N. H. Hamza, E. Shakir, A. A. M. Al-Mashhady, "Assessment of some atmospheric heavy metals in selected sites within Baghdad city". Mesopotamia Environmental Journal, vol.2, no.3: pp. 42-46, (2016).
5. S. Sahay, A. Inam, S. Iqbal, "Effect of Wastewater Irrigation on Soil, Metal Tolerance and its Remediation by Four Oil Yielding Brassica Cultivars" Environmental Journal of Environmental Sciences, 4(2), 158-170, (2013).
6. EPA Environmental Protection Agency, "Introduction to Phytoremediation". Cincinnati, Ohio. (2000).
7. A.M. Mathew, "Phytoremediation of Heavy Metal Contaminated Soil." M.Sc., Thesis, Bachelor of Technology Cochin University of Science and Technology Cochin, Kerala, India, (2001).
8. S. Marin, S. Lacrimioara, R. Cecilia, "Evaluation of Performance Parameters for Trace Elements Analysis in Perennial Plants using ICP-OES Technique", J. Plant Develop, 18, 87-93, 20, (2011).

# Assessment study of water quality index (WQI) for Shatt Al-arab River and its branches, Iraq

*Ahmed* Hamdan[1,*], *Ammar* Dawood[1], and *Douaa* Naeem[1]

[1]College of Engineering, University of Basrah, Basrah, Iraq

**Abstract.** Shatt Al-Arab River (in Basrah province South of Iraq) is approximately 192 km long. It plays a key role in providing water for domestic purposes, irrigation, manufacturing, in addition to shipment. Recently the river suffers from increasing pollution, due to wastes from industries, domestic sewage and agricultural activities that find their way into water sources and result in large scale deterioration of water quality. Investigating the river size and significance, becomes necessary to perform a study to understand the water quality of this river that is considered by some experts as one of the most contaminated in Iraq. This work uses the Water Quality Index (WQI) to describe the pollution level of the river and by using Geographic Information System (GIS) to create WQI map. This study also determines the critical pollutants affecting the river water quality throughout its course. WQI has been formulated making use of several water quality parameters such as pH, temperature, Dissolved Oxygen (DO), Biological Oxygen Demand (BOD5), Chemical Oxygen Demand (COD) , Nitrate (NO3-2), Phosphate (PO4-3), Total Dissolved Solids (TDS), Total Suspended Solids (TSS), Turbidity (Tur), and Electrical Conductivity (E.C) which were measured at 37 sites along the river. Bad water quality was observed at the sites of the river branches, near the center of Basrah governorate. Furthermore, it was discovered that the main reason for river pollution was due to the high sewage water discharged into the river, especially river branches and illegal discharges of industrial effluent and sewage.

## 1 Introduction

Rivers are regarded as one of the main resources for supplying water for different purposes, such as for drinking, agricultural and industrial uses. Additionally, these resources are employed at some place for discharge of industrial wastewater, sewage, and additionally agricultural drainage[1].

Conventional reports on water quality are usually too detailed and technical, confronting monitoring data on single water quality variables, without supplying a comprehensive assessment of the water quality. To meet this gap, different water quality indices (WQIs) have been formulated to integrate water quality variables [2]. The WQI is regarded to be considered as one criterion for the classifications of surface water, on the basis of the usage of standard parameters for water evaluation and characterization[3]. The results associated with the WQI enable the classification of water for the objective of numerous uses and supply a benchmark for assessing administration strategies [4]. The WQI is a number that is certainly unitless ranging between 0 and 100 with the higher value, suggesting higher quality of water [5].
WQI regarded to be considered as one criterion for the classifications of surface water, on the basis of the usage of standard parameters for water evaluation and characterization.

In Iraq, social and industrial advancements replaced the qualitative attributes of the river water and contributed to increased pollution that is excessive.

The main objectives of this study are to assess the physicochemical properties of the river water, to determine the water quality of the Shatt Al -Arab River and the main branches depending on WQI, to create WQI map based on Geographic Information System (GIS) and to gatherknowledge of the resources of pollutants discharged into the river so that it can suggest the concern of actions that should be supposed by the local authority.

## 2 Study area

The Shatt Al-Arab river is formed by the confluence of the Tigris and the Euphrates Rivers at Qurna town located in the southern part of Iraq (Fig.1), then it discharges its water to the Arabian Gulf, with a total length of about 192 km. There are many branches affected by Shatt Al-Arab River during tide and ebb cases due to wedge intrusion from Arabian gulf, the

* Corresponding author: ahmed_n_ahmed2005@yahoo.com

© The Authors, published by EDP Sciences. This is an open access article distributed under the terms of the Creative Commons Attribution License 4.0 (http://creativecommons.org/licenses/by/4.0/).

main branches of which are Al Saraji, Al Khora, Al Ashar, Al khandaq, Al Ribat, Al Mufteah and Al Jubailah (see Fig. 2). The study area in this research includes the region between the confluence of Tigris and Euphrates Rivers and to a distance of about 5 km downstream of Basrah center.

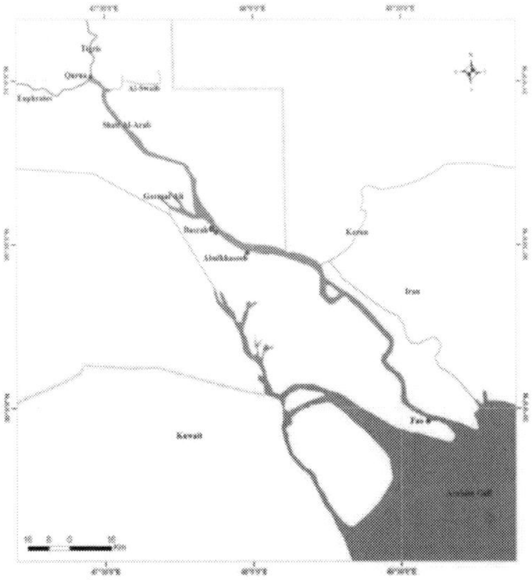

**Fig. 1.** Study area

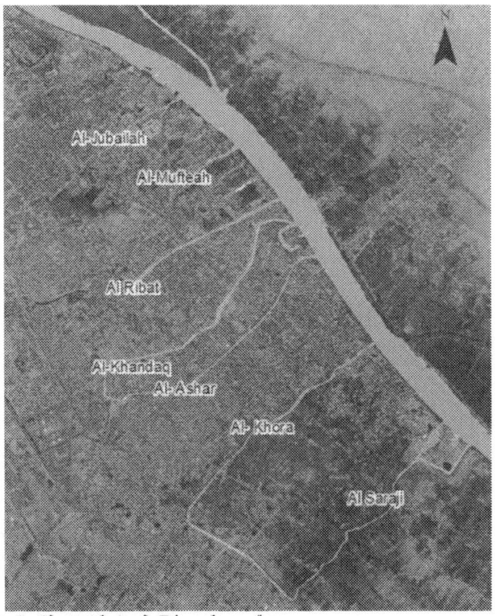


**Fig.2.** Shatt Al-Arab River branches

## 3 Collection of samples and analysis procedures

The sampling of physical, chemical and biological parameters was performed for thirty-seven sites along Shatt Al-Arab River and its branches in December, 2016 (see Fig.3 and Fig.4).

The environment parameters surveyed included pH, water temperature (T), Total Dissolved Solids (TDS), turbidity (Tur), Electrical Conductivity (EC), Dissolved Oxygen (DO), Biochemical Oxygen Demand ($BOD_5$), Chemical Oxygen Demand (COD), Total Suspended Solids (TSS), phosphate ($PO_4^{-3}$), and nitrate ($NO_3^{-2}$).

Some of water parameters were measured directly at site such as temperature, pH, TDS, EC, DO and turbidity, for the rest, samples were collected in polyethylene bottles of 1.5 liter volume, and kept in the cool box until transportation to the laboratory for the tests by using the method illustrated by (APHA, 2005; Ell-Amin et al., 2012), [6].

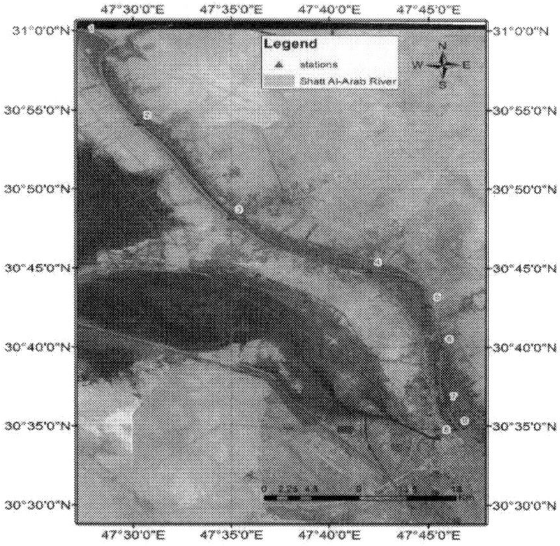


**Fig. 3**: Water sampling locations in the Shatt Al-Arab River in the upper direction

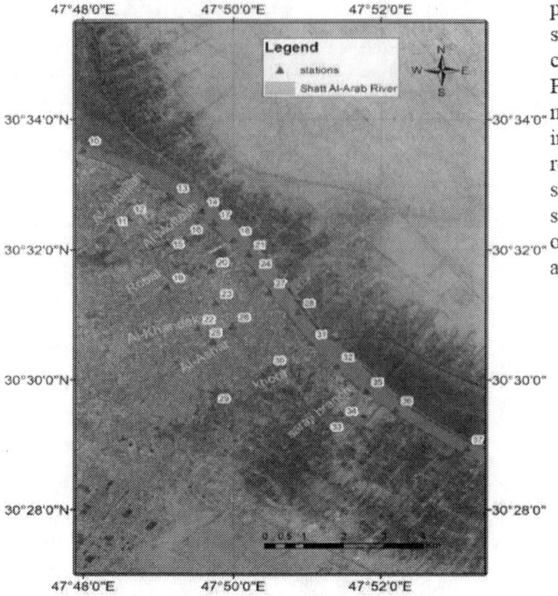


**Fig. 4**: Water sampling locations in the Shatt Al-Arab River and its branches in the lower direction

Physicochemical analyses were performed at the University of Basrah and Basrah Environment Directorate following the standard and recommended methods (APHA 2005). pH, water temperature were measured on site using a pH meter type (SD 300), DO test was measured at site by using a membrane electrode (DO meter type Lovibond (Oxi 200)), turbidity test was carried out at the site by using device of type Lovibond (TB 300 IR), TDS and EC, were measured at site using UBANTE device. The $BOD_5$ test was measured in the laboratory, based on five days incubation at 20°C, COD test was carried out in the laboratory using device type Lovibond (RD 125). In addition, $NO_3^{-2}$ and $PO_4^{-3}$ were tested in laboratory according to the standards methods [6].

## 4 Water quality index (WQI)

There are different indices which tend to be appropriate for water quality indices. Among these indices is National Sanitation Foundation Water Quality Index (NSFWQI). NSFWQI had been chosen due to its high accuracy, simplicity and accessibility to the parameters that are required [7]. The index of water quality calculated by NSFWQI is one of the most extensively used indicators which includes primary parameters such as pH, water temperature, DO, $NO_3^{-2}$, $PO_4^{-3}$, $BOD_5$, TDS, and turbidity. NSFWQI is obtained by the sum of multiplication of $W_i$ by $I_i$, [8].

$$NSFWQI = \sum W_i I_i \quad .................(1)$$

In this equation, Ii is the quality of the ith parameter (a number between 0 and 100 read from the appropriate subindex graph) and Wi is the weight factor of the ith parameter. After measuring the above characteristics, the sub-index of each of them is obtained from conversion curves (Wilcox Standard Graphs)[9], as shown in Figs.(5-12), these curves convert the parameters into measures that range from 0 to 100. To calculate the final index in this method, each sub-index obtained from the related curves is multiplied by the weight factor (as shown in Table 1), and the final index is obtained by summing them, according to Equation 1. Classification of river pollution intensity and water description according to NSFWQI are presented in Table 2[10].

**Table (1)**: weighing factors of water quality parameters

| Parameters | Weight factor |
|---|---|
| DO | 0.17 |
| $BOD_5$ | 0.11 |
| pH | 0.11 |
| Nitrate | 0.10 |
| Phosphate | 0.10 |
| Temperature | 0.10 |
| Turbidity | 0.08 |
| Total Dissolved Solid | 0.07 |

**Table 2.** Water quality index (WQI) legend

| Range | Class | Quality |
|---|---|---|
| 91-100 | A | Excellent |
| 71-90 | B | Good |
| 51-70 | C | Medium |
| 26-50 | D | Bad |
| 0-25 | E | Very Bad |

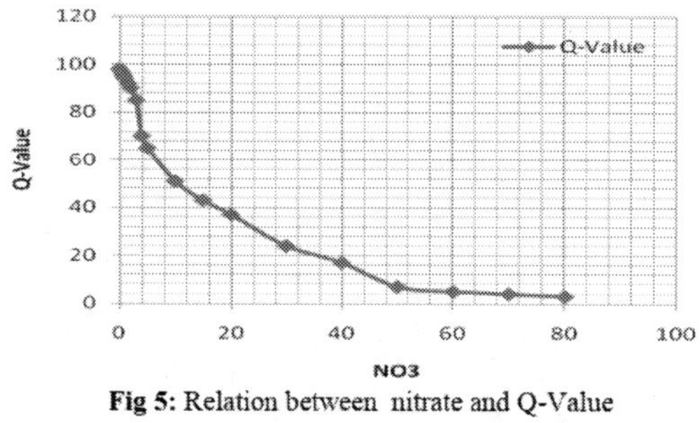


**Fig 5:** Relation between nitrate and Q-Value

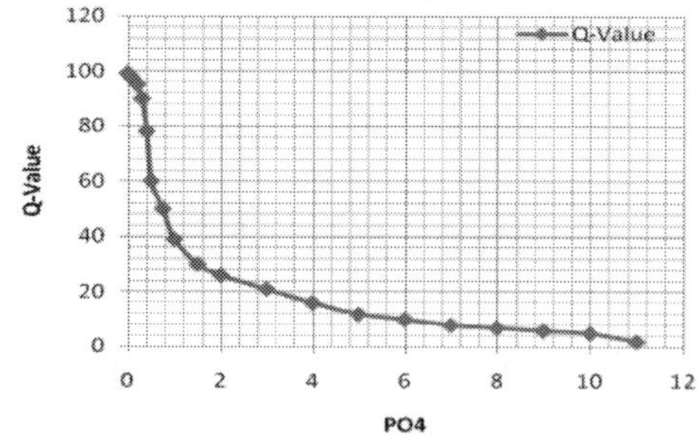


**Fig 6:** Relation between Phosphate and Q-Value

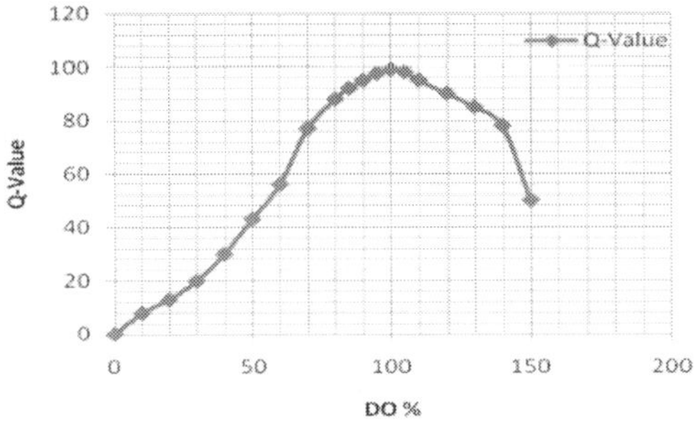


**Fig 7:** Relation between DO and Q-Value

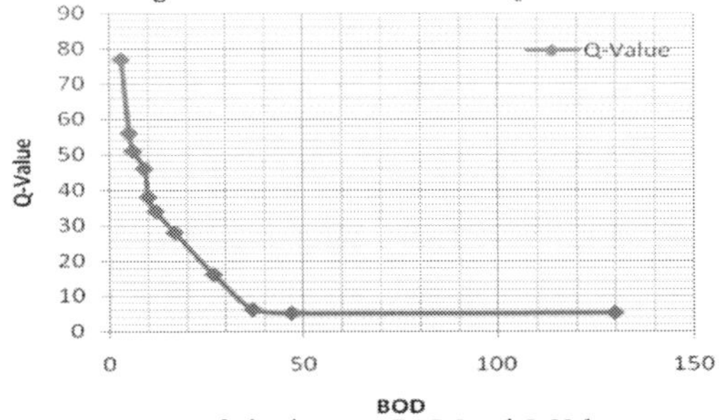


**Fig 8:** Relation between BOD5 and Q-Value

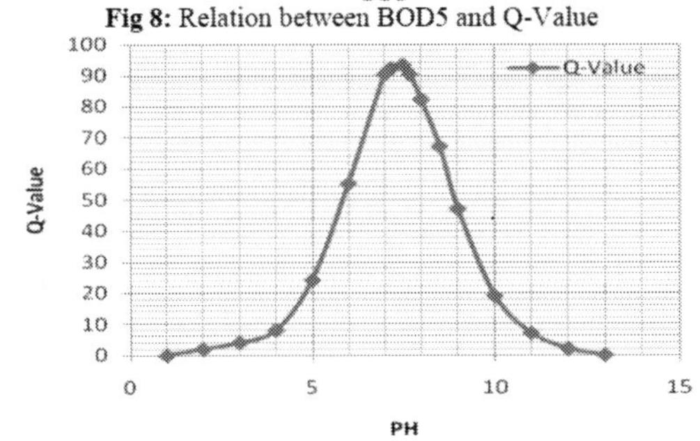


**Fig 9:** Relation between PH and Q-Value

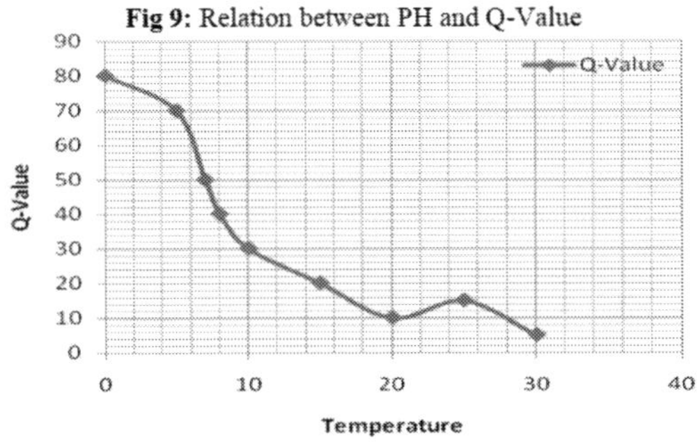


**Fig 10:** Relation between Temprature and Q-Value

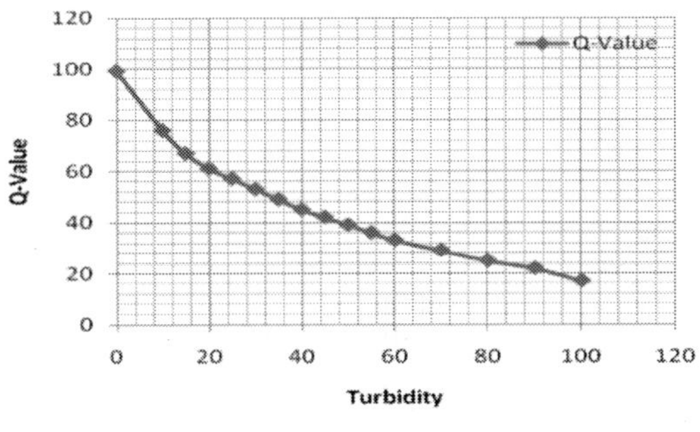


**Fig 11:** Relation between Turbidity and Q-Value

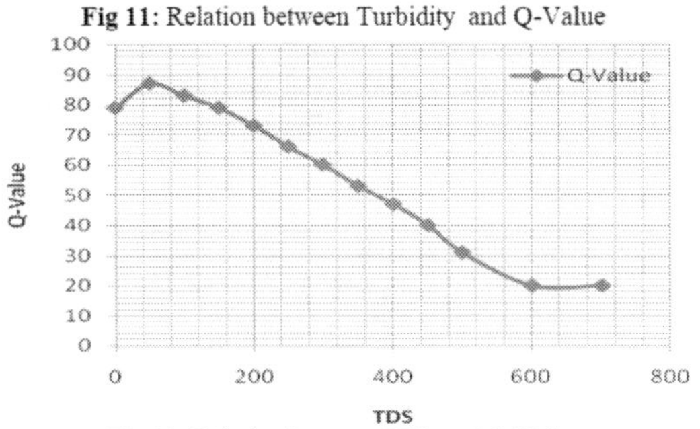


**Fig 12:** Relation between TDS and Q-Value

## 5 Results and discussion

The descriptive statistics of the parameters of the studied sites in the Shatt Al-Arab River and its branches are shown in Table 3.

**Table 3. Descriptive statistics for study area's parameters**

| Parameter | Min. | Max. | Mean | Std. Deviation |
|---|---|---|---|---|
| T (C°) | 13.40 | 22.70 | 16.78 | 2.12 |
| DO (mg/l) | 4.71 | 9.85 | 7.64 | 1.43 |
| EC (μs/cm) | 1375.00 | 7718.00 | 3721.22 | 1325.63 |
| pH | 6.90 | 8.15 | 7.60 | 0.28 |
| TDS (mg/l) | 805.00 | 3859.00 | 1956.27 | 630.50 |
| TSS (mg/l) | 48.00 | 974.00 | 365.95 | 266.10 |
| $NO_3$ (mg/l) | 1.56 | 10.24 | 3.97 | 1.98 |
| $PO_4$ (mg/l) | 0.25 | 2.47 | 0.712 | 0.55 |
| BOD (mg/l) | 3.00 | 123.00 | 30.06 | 32.84 |
| COD (mg/l) | 12.90 | 546.00 | 112.71 | 126.57 |
| Tur (NTU) | 8.83 | 206.00 | 34.75 | 44.15 |
| WQI | 41.00 | 71.00 | 58.7838 | 9.88 |

The water temperature values in this study varied from the lowest value equal to 13.4 °C at site 1 and the highest value which was 22.7 °C at site 19. It also showed a high temperature beside the industry locations such as Paper mill factory (beside site 4 where the temperature was 20.2°C) and Al Hartha Power Station (located in Hartha region 10 km upstream of Basrah governorate, in the middle distance between sites 5 and 6, where the temperature exceeds 21 °C), where this station withdraws water from Shatt Al-Arab River and uses it for refrigerating the power station plant, after some RO treatments, the wastewater, at high temperature, is discharged back into Shatt Al Arab River that leads to an increase in the river water temperature near this site.

Generally, optimal pH range for sustainable aquatic life is pH 6.5-8.5. pH of an aquatic system is an important indicator of the water quality and the extent of pollution in the watershed areas. The maximum pH value for the water samples was 8.15 which was recorded at site 5 while the minimum was 6.9 which was recorded at site

11, this range indicates that water is slightly alkaline. All water samples were within the permissible limits prescribed by (WHO standards for drinking water in 2004) [12], EEC 464/76 standards for surface water quality [13] and Iraqi standards for water quality in 1998 (as shown in Table. 4).

**Table 4:** Comparison between water quality parameters for Iraqi and international standards

| Parameter | WHO standards for drinking water in 2004 | EEC 464/76 standard for surface water Quality Tebbutt,1997 | Iraqi standards for water quality in 1998 |
|---|---|---|---|
| TDS, mg/l | 500-1500 | 500-1500 | 1000 |
| Turbidity,NTU | 0-50 | 0-25 | 25 |
| pH | 6.5- 8.5 | 6.5- 8.5 | 6.5- 8.5 |
| DO, mg/l | > 5 | > 5 | > 5 |
| $BOD_5$, mg/L | <3 | < 5 | <3 |
| COD, mg/l | 10 | | |
| $NO_3^{-2}$, mg/l | 0 – 45 | 0 – 40 | 25-50 |
| $PO_4^{-3}$, mg/l | 0.1 | 0.4 | 0.4 |

DO, $BOD_5$, and COD had been tested in this research which tend to be an indicator of organic pollutants.

In this study it was shown the lowest value of DO recorded was at site 31 which was 4.71 mg/l that may be assigned to raw sewage discharge and highest value was 9.85 mg/l at site 2. The concentration of dissolved oxygen increases when the temperature decreases because the fact that the levels of oxygen saturation are temperature dependent, the average DO concentration in the river is 7.64 mg/l. Furthermore, the key factor that is main in contributing to changes in the level of DO may be the accumulation of organic wastes. Apart from the other parameters, the lower the DO concentration the poorer the water quality. When the DO concentration is less than 2 mg/L, it indicates that the river water is not safe.

The obtained $BOD_5$ data showed that the maximum value was recorded at site 29 which was 123 mg/L and minimum value which was 3 mg/l at site 8, from whole area of study it can be shown that all the sites the water quality parameter is beyond the range of WHO standards, and Iraqi standards, as well as most of the water samples were beyond the range of EEC 464/76 standards, the areas of high dangerwere in all branches, the high $BOD_5$ value was probably linked to the level of organic matter load from sewage, industrial or urban discharges.

The obtained COD data showed that the maximum value was recorded at site 29 which was 546 mg/L and minimum value which was 12.9 mg/l at site 24, From the whole area of study it can be shown that all the sites were beyond the range of WHO, EEC 464/76 standards and Iraqi standards, and the areas of high danger were in all branches.

The lowest value of TDS which was 805 mg/l was shown at site 1, and the highest value was 3859 mg/l which was shown at site 29. Generally the highest values of TDS were in branches.The high value of TDS recorded could be related to the increase in the load of soluble salts, increase in the urban and fertilizer runoff, wastewater, septic effluent, decaying plants, and erosion of river banks. Hence, it was shown that a decrease of TDS in upstream is due to a decrease in the effect of saline wedge intrusion from Arabian gulf and the sources of pollution.

Being an important factor that limits growth, phosphorus has been focused on studies that deal with water quality assessement. The obtained $PO4^{-3}$ data showed that the maximum value was recorded at site 29 which was 2.47 mg/L and minimum value which was 0.25 mg/l in site 6. From the whole area of study it can be shown that all sites exceed the maximum limits for $PO4^{-3}$ as per the WHO standards, but as per the EEC 464/76 standards and Iraqi standards most of the samples which were downstream of site 10 exceeded the limit. Increasing levels of $PO4^{-3}$ in Shatt Al-Arab River may be due to high concentration of detergents in sewage, in addition to the high density of agricultural areas, where phosphate fertilizers were used. That might be the cause of high concentrations of phosphate. High levels of phosphorus in nature can create algal blooms causing eutrophication or the premature "aging" of a water body. This process decreases sunlight and oxygen levels (hypoxia) thus, affecting fish and other aquatic life.

The obtained $NO_3^{-2}$ data showed that the maximum value was recorded at site 29 which was 10.24 mg/l and minimum value which was 1.56 mg/l at site 7. From whole area of study, it can be shown that all sites had acceptable values for $NO_3^{-2}$ as per WHO standards for drinking water in 2004, EEC 464/76 standards for surface water quality (Tebbutt, 1998) and Iraqi standards for water quality in 1998.

The highest value of turbidity recorded was 206 NTU, which was recorded in site 23. While the lowest value of turbidity was 8.83 NTU which was recorded at site 16. In the main channel of the river, turbidity was shown to be within the acceptable limits of standards and the highest values of turbidity was shown in branches of the river, especially in Al Khora, Al Ashar, and Al Khandek. This is due to the effect of banks erosions, domestic wastewater discharge , and appearance of phytoplankton. These branches showed the excess limits of standards.

Final WQI map of the river was prepared using Geographic Information System (GIS) and by using Inverse Distance Weighted (IDW) interpolation technique and presented in Fig.13 and 14, the observed range of WQI was 41–71 for Shatt Al-Arab River and

https://doi.org/10.1051/matecconf/201816205005

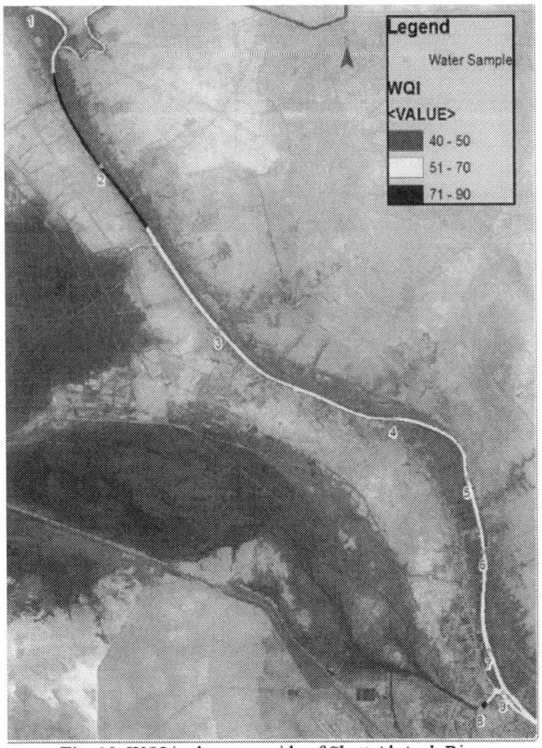


**Fig. 13.** WQI in the upper side of Shatt Al-Arab River

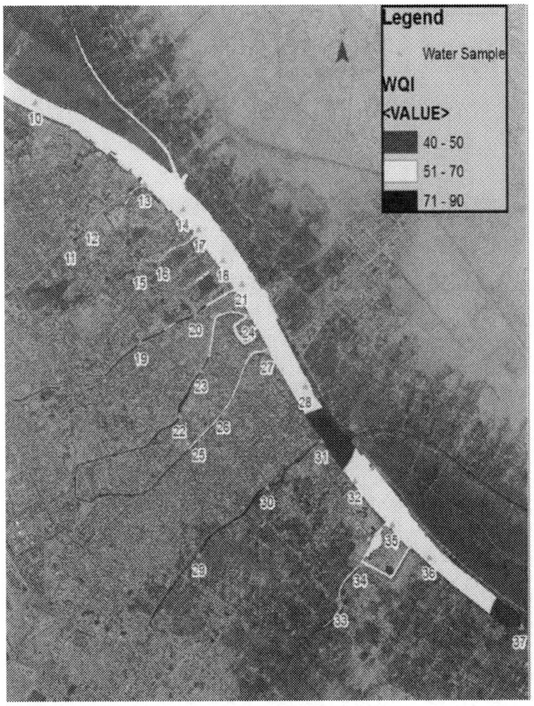


**Fig. 14.** WQI in the lower side of Shatt Al-Arab River and its branches

its branches and ranged between Bad and Good (as shown in Table 5). The WQI distribution map shows that the water of Shatt Al-Arab River was medium in most sites of the main channel, and bad in the Al-Ribat, Al-Khandaq, Al-Ashar and Al-Khora branches. The deterioration of water quality was mainly due to the input of municipal and industrial wastes and/or agricultural activities discharge at the bank of the river (especially in branches). It is also noticed that the quantity of impurities get diluted when branches poured into the main river and it becomes medium water quality. According to calculations, the highest mean effective weights value belongs to DO and pH parameters with 27.63% and 20.76%, respectively and these parameters are the most effective parameters in the WQI calculations.

## 6 Conclusions

The water quality of Shatt Al-Arab River was evaluated based on the water quality analysis data for December in 2016. The main conclusions are as follows:

- Results from this work presented that water quality of the Shatt Al-Arab River was generally influenced by the activities that occur along its watershed. The worst water quality of the river was found at the branches, whereas most of the Shatt Al-Arab River water quality was medium.
- The deterioration of water quality is mainly due to the input of municipal and industrial wastes and/or agricultural activities discharge at the bank of the river (especially in branches).
- The highest mean effective weights value belongs to DO and pH parameters with 27.63% and 20.76%, respectively and these parameters are the most effective parameters in the WQI calculations.
- It can also be concluded that WQI is an effective tool for classifying the water of the river for its different beneficial uses and gives an accurate idea about the pollution load in the Shatt Al-Arab River.
- The results revealed that GIS was a good and efficient technique to create WQI map and to give an expected picture to help the decision maker to take the appropriate actions to improve the river water quality.

## References

1. C. G. Cude, JAWRA Journal of the American Water Resources Association, Oregon water

quality index a tool for evaluating water quality management effectiveness. 37(1): 125-137 (2001)

2. S. M. Liou, S. L. Lo, et al., Environmental Monitoring and Assessment. A generalized water quality index for Taiwan, 96(1): 35-52 (2004)
3. A. Said, D. K. Stevens, et al., Environmental management. An innovative index for evaluating water quality in streams, 34(3): 406-414 (2004)
4. A. A.Bordalo, R. Teixeira, and W. J. Wiebe, Environmental Management. A water quality index applied to an international shared river basin: The case of the Douro River, 38, 910–920 (2006)
5. P. Debels, R. Figueroa, R. Urrutla, R. Barra, and X. Niell, Environmental Monitoring and Assessment. Evaluation of water quality in the Chillan river (Central Chile) using physicochemical parameters and modified water quality index, 110, 301–322 (2005)
6. APHA, American Water Works Association/ Water Environment Federation, Standard Methods for the Examination of Water and Wastewater, 21st ed, American Public Health Association, Washington, DC, USA, (2005)
7. P. Zandbergen, and K. Hall, Journal of Water Quality. Analysis of the British Columbia Water Quality Index for watershed manager: a case study of two small watersheds, 33, pp. 519–525 (1988)
8. A. Shamsai, S. Urei, and A. Sarang, Journal of Water and Wastewater. Comparative of qualitative indexes and qualitative zoning of Karoon river and Dez river, 16, pp. 88–97 (2006)
9. Pur, S. Ebrahim, Mohammad Zadeh, H. and A. Mohammadi, paper presented at the 4th Iranian Water Resource Management Conference , Qualitative investigation of Zarivar lake water and it's zoning according to OWQI and NSFWQI by GIS, University of Amirkabir, Tehran, Iran (2012)
10. D. Fabiano, S. Santos, B. Altair, M. Sonia, G. Nobre, and J. Maria, Ecological Indicators. Water quality index as a simple indicator of aquaculture effects on aquatic bodies, 8, pp. 476–484 (2008)
11. E. Sanchez, Ecological Indicators. Use of the water Quality index and dissolved oxygen deficit as simple indicators of watershed pollution, 7, pp. 315–328 (2007)
12. World Health Organization. Guidelines for drinking-water quality. Vol. 1, (2004)
13. Tebbutt, Thomas Hugh Yelland, Butterworth-Heinemann. Principles of water quality control, (1997)

**Table 5.** WQI values and water types of the samples

| Sam ple Num | Longitude | Latitud e | W Q I | Water type |
|---|---|---|---|---|
| 1 | 47.841274 | 30.5224 | 7 | Mediu |
| 2 | 47.837922 | 30.5275 | 7 | Good |
| 3 | 47.836753 | 30.5323 | 7 | Mediu |
| 4 | 47.833374 | 30.5359 | 6 | Mediu |
| 5 | 47.828996 | 30.5403 | 6 | Mediu |
| 6 | 47.826236 | 30.5433 | 6 | Mediu |
| 7 | 47.819451 | 30.5468 | 6 | Mediu |
| 8 | 47.799987 | 30.5589 | 7 | Good |
| 9 | 47.772195 | 30.5804 | 6 | Mediu |
| 10 | 47.757204 | 30.5715 | 6 | Mediu |
| 11 | 47.763017 | 30.6066 | 4 | Bad |
| 12 | 47.759519 | 30.6672 | 5 | Mediu |
| 13 | 47.751258 | 30.7118 | 6 | Mediu |
| 14 | 47.699864 | 30.7482 | 6 | Mediu |
| 15 | 47.582071 | 30.8036 | 5 | Mediu |
| 16 | 47.504144 | 30.9033 | 6 | Mediu |
| 17 | 47.457534 | 31.0002 | 6 | Mediu |
| 18 | 47.857669 | 30.4914 | 6 | Mediu |
| 19 | 47.841116 | 30.5028 | 4 | Bad |
| 20 | 47.833216 | 30.5137 | 4 | Bad |
| 21 | 47.82934 | 30.5198 | 6 | Mediu |
| 22 | 47.818594 | 30.5238 | 4 | Bad |
| 23 | 47.822561 | 30.5362 | 4 | Bad |
| 24 | 47.809999 | 30.5414 | 7 | Mediu |
| 25 | 47.88627 | 30.4824 | 5 | Mediu |
| 26 | 47.870069 | 30.4924 | 5 | Mediu |
| 27 | 47.863405 | 30.4972 | 6 | Mediu |
| 28 | 47.856822 | 30.5035 | 5 | Mediu |
| 29 | 47.850880 | 30.5094 | 4 | Bad |
| 30 | 47.848036 | 30.5174 | 4 | Bad |
| 31 | 47.854305 | 30.4856 | 4 | Bad |
| 32 | 47.828713 | 30.4930 | 5 | Mediu |
| 33 | 47.828853 | 30.5098 | 7 | Mediu |
| 34 | 47.825413 | 30.5132 | 6 | Mediu |
| 35 | 47.828421 | 30.5279 | 6 | Mediu |
| 36 | 47.818544 | 30.5349 | 5 | Mediu |
| 37 | 47.806254 | 30.5385 | 4 | Bad |

# Synthesis and photocatalytic activity for $TiO_2$ nanoparticles as air purification

*Adawiya* Haider [1], *Riyad* Al- Anbari[1], *Ghadah* Kadhim[1], *Zainab* Jameel[1]

[1]University of Technology, Baghdad, Iraq

**Abstract.** In the present work, titanium dioxide (TiO2) nanoparticles (NP's) were prepared using sol-gel process from Titanium Tetrachloride (TiCl4) as a precursor with calcinations at two temperatures (500 and 900) °C. The effect of calcinations temperatures on the structural, optical, morphological and Root Mean Square (roughness) properties were investigated by means of Scanning Electron Microscopy, X-ray Diffraction (XRD), and Atomic Force Microscopy (AFM). Bacterial inactivation was evaluated using TiO2-coated Petri dishes. A thin layer of photocatalytic TiO2 powder was deposited on glass substrate in order to investigate the self-cleaning effect of TiO2 nanoparticles in indoor and outdoor applications. Ultra-hydrophilicity was assessed by measuring the contact angle and it evaluated photolysis properties through the degradation of potassium permanganate (KMnO4) under direct sunlight. XRD analysis indicated that the structure of TiO2 was anatase at 500 °C and rutile at 900 °C calcination temperatures. As the calcination temperature increases, the crystallinity is improved and the crystallite size becomes larger. Coated films of TiO2 made the has permeability, low water contact angle and good optical activity. These are properties essential for the application of the surface of the self-cleaning. The final results illustrate that titanium dioxide can be used in the build materials to produce coated surfaces in order to minimize air pollutants that are placed in microbiologically sensitive circumference like hospitals and the food factory.

## 1 Introduction

Ecological pollution, such as dirty water or contaminated air, has become a universal problem that affects human health. Typical polluting sources are toxic organic molecules or exhausted gas compounds that are released from household waste, livestock waste and local industries [1, 2, 3, 4, 5]. Growth and deposition of microbes on surfaces such as walls, windows and tables also pose a substantial threat to human health. Continual exposure to these pollutants can lead to varied health troubles like irritations and toxic effects, and other various respiratory or skin diseases. Most studies on photocatalysts show that it is a better technique for biodegradable organic pollutants using clean solar energy without generating any harmful products [6]. Photocatalytic technique is an effective purification process capable of disrupting a wide range of harmful microorganisms in various media. It is safe, non-toxic, and relatively inexpensive method of disinfection which allows the ability to adapt to use in many applications such as air purification, water purification, and treatment of hazardous waste, water purification [7, 8, 9]. This technique can be envisaged as one of the most promising Advanced Oxidation Process (AOPs) due to its specific advantages, such as bland reaction conditions, the possibility of using molecular oxygen as oxidant species, the total mineralization of pollutants into substances innocuous to the environment. Photocatalysis is based on the interaction between semiconductor materials and light by considering that 'free' light from the sun, electron will be excited from the valance band to the conduction band thus creating an electron-hole pair, the holes (h+) and hydroxyl radicals (OH•) generated in the valence band, also electrons and superoxide anions (O2 •-) generated in the conduction band, irradiated TiO2 photocatalysts can decompose and mineralize organic compounds by a series of oxidation reactions leading to carbon dioxide [10, 11]. The idea of using solar light energy as resource to clean up the environment is an ideal and promising approach.

*Corresponding Author: 100081@uotechnology.edu.iq

© The Authors, published by EDP Sciences. This is an open access article distributed under the terms of the Creative Commons Attribution License 4.0 (http://creativecommons.org/licenses/by/4.0/).

Among various oxide photocatalysts, the preparation, characterization and applications of semiconductor titanium dioxide (TiO2) have become a research hot spot. The photocatalytic oxidation (PCO) using titanium dioxide is an efficient technology for the degradation of many pollutants [12, 13, 14, 15]. It has also been known that the TiO2 shows the hydrophilicity due to the photocatalytic nature, the water tends to sheet across this surface rather than forming droplets [16]. This makes the surface anti-fogging and easily washable. TiO2 photocatalyst has good chemical stability, relative low price, wide availability, non-toxicity and it has strong oxidizing power [17, 18, 19, 20, 21, 22].

Super-hydrophilicity is a phenomenon that occurs when TiO2 coated surface is illuminated by UV radiation. The hydrophilic surface has a very small contact angle ($\theta \leq 5o$) that allows water to sheet across this surface rather than forming droplets. This makes the surface anti-fogging and easily washable [23].

In this work, the synthetic method of nanostructure, a Titanium dioxide (TiO2) based on TiCl4, has been synthesized by sol-gel method and thermally treated at two temperatures. The effect of thermal treatment on the structural and morphological properties of the material was investigated by means of SEM, XRD and AFM and measuring the contact angle of water drop on the films by spin coating technique to evaluate the hydrophilicity property.

## 2 Experimental section

### *2.1 Synthesis of $TiO_2$ nanoparticles*

$TiO_2$ Nanaparticles was prepared by Sol-Gel method using Titanium Tetrachloride ($TiCl_4$) as precursor of $TiO_2$ as described in (Adawyia et al. 2017).

### *2.2 Characterization*

The prepared nanoparticles were characterized for the crystalline phase by X-ray diffraction using (XRD-6000 2kw type), at room temperature, operating at 40 kV and 30 mA, Cu (1.54060 A). The morphological changes of prepared $TiO_2$ nanoparticles at different calcination temperatures were analyzed using Scanning Electron Microscopy (SEM, Tescan VEGA3 SB). Zeta potential (ZetaPlus) was used to measure Zeta potential of the particles in liquid suspension using Electrophoretic Light Scattering (ELS) technique with the range between -250 mV to 250 mV (sample dependent). The size of the prepared $TiO_2$ nanoparticles was measured by using (Particle size analyzer, 90Plus). Fourier Transform Infrared spectroscopy (FTIR) has been extensively used for identifying the various functional groups.

## 3 Results and discussion

### 3.1 X-ray diffraction (XRD)

Precise metal composition and grain size of $TiO_2$NP's were determined by XRD.

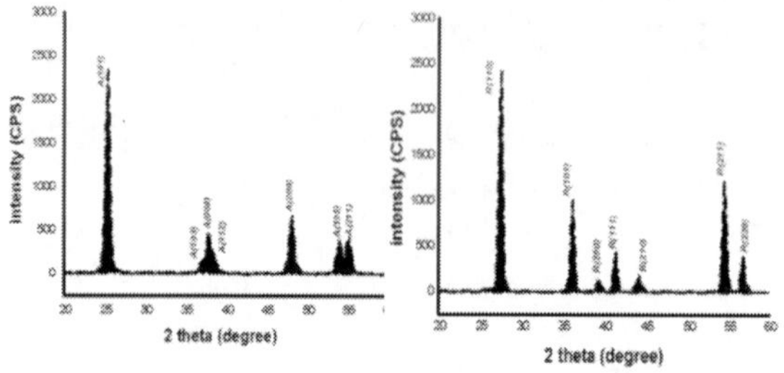


Fig- 1 presents the XRD pattern of TiO2 nanoparticles obtained by sol gel method and calcined at different temperatures.

Crystallite size was obtained by Scherrer's formula given by equation:

$$D = K\lambda / (\beta\cos\theta) \qquad (1)$$

Where K is a constant that depends on the crystallite shape (0.9, with the assumption of spherical particles) [24], $\lambda$ is the X-ray wavelength, $\beta$ is the full width at half maximum of the selected peak and $\theta$ is the Bragg's angle of diffraction for the peak.

The XRD patterns results indicate that the structure of $TiO_2$ was anatase at calcination temperatures at 500 °C, rutile at 900 °C. The anatase phase was identified at $2\theta$ values of 25.4°, 38.1°, 48.2°, 53.9° and 55.1°, corresponds to the crystal planes of (101), (004), (200), and (105) respectively, which are in agreement with the standard XRD pattern (JCPDS files No. 21-1272). Rutile phase was also identified at $2\theta$ values of 27.5°, 36.1°, 41.3°, 44.2° and 55.5° corresponding to the crystal planes of (110), (101), (200), (111), (210), (211), and (220) respectively,in agreement with the standard XRD pattern (JCPDS files No. 21-1276).

Table 1.Effect of calcination temperatures on the crystallite sizes and phase content of TiO2 nanoparticles

| Calcination temperature (°C) | Phase [a] | $2\theta$ (Deg.) | $\beta$(FWHM) (Deg.) | Grain sized [b] (nm) | WR % |
|---|---|---|---|---|---|
| 500 | A | 25.4005 | 0.6209 | 12.5 | 0 |
| | | 48.1266 | 0.6902 | 12.45 | |
| | | 37.9087 | 0.8796 | 11 | |
| 900 | R | 27.5443 | 0.5243 | 16.9 | 100 |
| | | 54.5425 | 0.5611 | 16.7 | |
| | | 36.2899 | 0.4991 | 18 | |

[a] A stands for anatase and R stands for rutile. [b] The crystallite size was calculated by using Sherrer's equation.

## 3.2 Scanning Electron Microscopy (SEM)

The micrographs SEM images of TiO2 nanoparticles were prepared at different calcination temperatures as shown in Figure 2. (Figure 2a) shows TiO2 nanoparticles calcined at 500ºC have roughly spherical spongy shape with small size nanoparticles around the range of less than 10 nm. When the temperature increases, the sizes become bigger and the agglomeration becomes significant. Further increase in temperature up to 900ºC, $TiO_2$ nanoparticles exhibited non-uniform particles shape due to the agglomeration of initial particles with increase in crystalline size. These results are in agreement with XRD results that showed that the particles size of the anatase phase is smaller than rutile phase due to aggregation of nanoparticles with increased calcination temperature.

Fig-2 SEM images of TiO2 nanoparticles calcined at (a): 500ºC, (b) 900ºC

## 3.3 Zeta potential surface charge for TiO2 nanoparticles

Zeta potential surface charge for TiO2 nanoparticles is illustrated in Figure (3). According to ASTM Standard (D 4187–82), illustrated in Table (3), TiO2 collides prepared at 500ºC have good stability. An excellent stability has been found with TiO2 collides prepared at 900ºC.

Table 2. List of Zeta potential, Frequency, Frequency shift, and Mobility data for TiO2 nanoparticles calcined at different calcination temperatures

| Tio2 Samples Calcined At | Zeta Potential (Mv) | Frequency (Hz) | Frequency Shift (Hz) | Mobility (μ/S)/(V/Cm) |
|---|---|---|---|---|
| 500 ºC | -52.95 | 151.29 | -98.71 | -12.80 |
| 900 ºC | -76.12 | 147.96 | -102.04 | -13.47 |

Table 3. ASTM Standard (D 4187–82)

| Zeta potential mV | Stability behavior of the colloid |
|---|---|
| from 0 to ±5, | Rapid coagulation or flocculation |
| from ±10 to ±30 | Incipient instability |
| from ±30 to ±40 | Moderate stability |
| from ±40 to ±60 | Good stability |
| more than ±61 | Excellent stability |

## 3.4 Particle Size Analyzer

Particle size analysis illustrated in Table 4 shows that when the calcination temperature was increased the crystallites tend to agglomerate leading to the formation of bigger particles. This result is in agreement with XRD and SEM results.

## 3.5 Atomic force microscopy (AFM)

The surface morphology was studied by means of AFM (Atomic force microscopy). Figure 3 shows two- and three-dimensional AFM images of the $TiO_2$ coatings calcined at 500 and 900 °C, respectively. It was observed that, at different calcination treatment temperatures, the surface morphologies and roughness of the $TiO_2$ surface are obviously different. As it can be seen from Figure 3a (coating treated at 500 °C), a granular microstructure and a flat texture, with the lowest surface roughness, and is composed of large particles diameter of spherical shapes. According to previous XRD results, these range on the surface of $TiO_2$ coating are aggregates of many small $TiO_2$ crystallites [25].

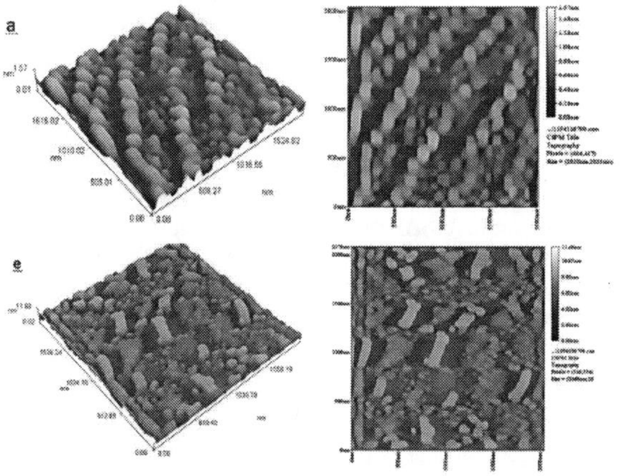

Fig-3 Two- and three-dimensional AFM image of Titanium dioxide (TiO2 nanoparticles calcined at: (a) 500ºC, (b) 900ºC.

Give images AFM analysis also surface roughness values, as shown in Table 2. means the root of the square roughness (RQ) of the surface of less than (1 nm) and increases with increasing temperature calcinations , which shows the grain of normal forms on the surface. When the temperature increases, the changes in topography and roughness of the paint $TIO_2$ are largely due to the phase shift of titanium Anatase to rutile and the growth of the crystal $TiO_2$ [27, 28]. The table shows (1) that the roughness values increase with increasing temperature calcination. After calcined up to 900 ° C, it observed a significant increase in grain size and RQ to 1.22 nm. It is because the shift from titanium Anatase phase peak toward Rutile phase peak went through a total of smaller particles and the formation of larger particles (J. Yu Yu, is, Jiang, 2002), which then increased its grain size. This result is in agreement with Jimmy C Yu, *et al.* [29] and Jiaguo Yu, *et al.* [30] references.

Table 4. The relationship between roughness values and calcination temperature.

| Calcination temperature. (℃) | *Sq. (Root mean square) (nm)* |
|---|---|
| 500 | 0.206 |
| 900 | 1.22 |

### 3.6 Fourier Transform Infrared (FTIR) Analysis

FTIR spectrum shows a wide band range (3400-4000) $cm^{-1}$ which is contribution to the H-O-H stretching mode of the surface adsorbed $H_2O$ molecule and also due to the reaction between the O-H groups of $TiO_2$ [31]. This $OH^-$ adsorbate plays an important role in photoreceptor stimulation by trapping charge carriers to produce OH reaction radicals, which are the driving force behind the process. Furthermore, they act as adsorbents and activating sites for reacting molecules or contaminants [32]. Another band of around (1620-1640) $cm^{-1}$ is attributed to the H-O-H bending mode. These results agreed with Huang *et al.* [33] and Thangavelu *et. al.* [34]. FTIR results of $TiO_2$ nanoparticles calcined at (400, 500 and 600) °C showed a broad band at 3419 $cm^{-1}$ and 1634 $cm^{-1}$ due to the stretching and deformation vibrations of surface adsorbed water and hydroxyl groups in the surface of $TiO_2$. Whereas, $TiO_2$ calcined at 900 °C did not show any aforesaid two broad peaks and it completely disappeared at high calcination temperature. That led to reduced surface area, pore volume and adsorbed water molecule at high calcinations temperature.

All the samples showed the foremost vibration bands around (500-900) $cm^{-1}$ due to the Ti-O-Ti bond-stretching mode, which indicates the formation of a Titanium oxide [35].

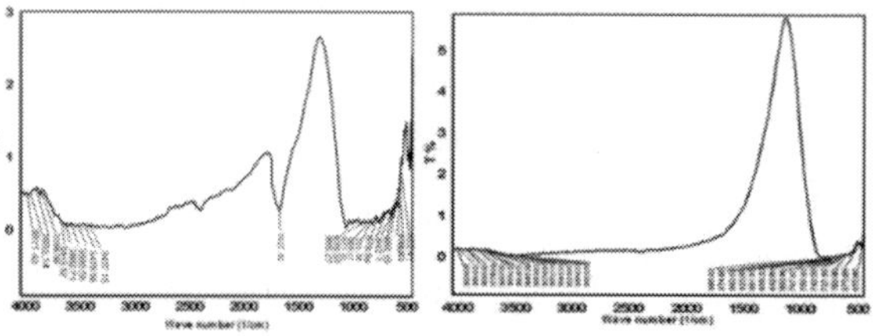

Fig-4 FTIR spectra of the TiO2 nanoparticles calcined at, a) 500°C, b) 900°C

### 3.7 Photocatalysis activity

The photocatalytic activity was evaluated by photocatalytic oxidation of Potassium permanganate ($KMnO_4$) under direct sunlight irradiation, by measuring the degradation of Potassium permanganate ($KMnO_4$) under direct sunlight irradiation. UV-Visible Spectrophotometry demonstrated the absorbance value for different film samples with Potassium permanganate after exposure to UV- irradiation from sunlight. Spectrum shown in Figure 5 demonstrates the relationship between the absorbance values and exposure time of irradiation in (hr.). The color of Potassium permanganate dye gradually reduces and finally it completely disappears which means the dye is completely degraded .

The results illustrate calcination temperate effect on the degradation of the dye [36]. It demonstrates that $TiO_2$ with pure anatase phase exhibit higher photocatalytic activity compared with $TiO_2$ with pure rutile that needs more time to degrade the dye. This perhaps is due to the higher adsorptive affinity of anatase surface and lower recombination rate of electron–hole pairs as compared to rutile [37]. At high calcination temperatures (900°C), the rutile crystallite size increased due to aggregation under direct sunlight irradiation. The electron generated was permanently trapped in the larger rutile crystallites aggregates, which results in increasing electron–hole recombination rate and decreasing photocatalytic activity. In addition, the growth of $TiO_2$ crystallites at higher calcination temperature, results in a significant decrease in the surface area. $TiO_2$ calcined at 800°C shows the highest photocatalytic activity. This is because its' composed of both anatase and rutile phases. The photo-generated electrons transfered from the anatase to the rutile phase may have reduced the rate of electron–hole combination and thus enhanced the photoactivity.

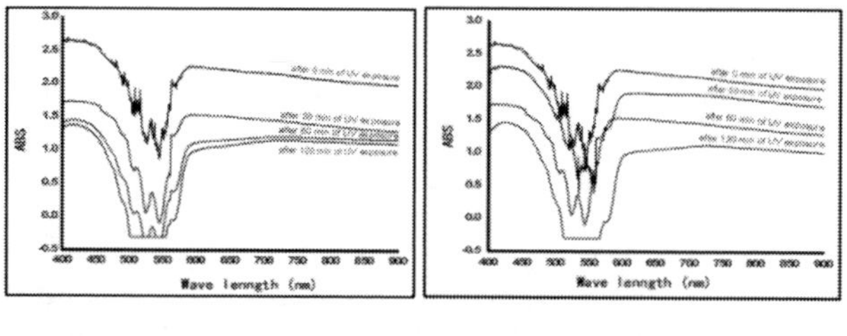

Fig-5: Illustrates the relationship between absorbance and wavelength for TiO2 nanoparticle calcined at a) 500 °C, b) 900 °C, measured befor irradation by sunlight (0 min) and after 30, 60, and 120 min exposure to direct sunlight.

## 4 Conclusions

$TiO_2$ NP's were synthesis by sol- gel technique at various calcinations temperatures, and then deposited on a glass substrate by spin coating method. The XRD pattern of calcined $TiO_2$ nanoparticles shows the sharp picks that reveals the anatase crystalline phase at 500 °C, and rutile phase at 900 °C.

Self-cleaning properties of the prepared nanoparticles were characterized by Water Contact Angle (WCA) and photocatalysis activity. WCA show that the anatase and mixed phase are super-hydrophobic surfaces with WCA of zero value, while rutile phase shows good hydrophilicity with WCA of 7.775.

## Acknowledgement

The authors would like to thank the University of Technology, Department of Building and Construction Engineering, and Nanotechnology and Advanced Materials Research Center (Iraq) for help to work on this search.

## References

1. Edgerton, S. A., Holdren, M. W., Smith, D. L., & Shah, J. J. (1989). Inter-urban comparison of ambient volatile organic compound concentration in U.S. cities. Journal of the Air Pollution Control Association (JAPCA), 39(5), 729–32.
2. Sweet, C. W., & Vermette, S. J. (1992). Toxic volatile organic compounds in urban air in Illinois. Environmental Science & Technology, 26(1), 165–173. http://doi.org/10.1021/es00025a020
3. Kostiainen, R. (1995). Volatile organic compounds in the indoor air of normal and sick houses. Atmospheric Environment, 29(6), 693–702.
4. Mukund, R., Kelly, T. J., & Spicer, C. W. (1996). Source attribution of ambient air toxic and other VOCs in Columbus, Ohio. Atmospheric Environment, 30(20), 3457–3470. Retrieved from 10.1016/1352-2310(95)00487-4
5. Alberici, R. M., & Jardim, W. F. (1997). Photocatalytic destruction of VOCs in the gas-phase using titanium dioxide. Applied Catalysis B: Environmental, 14(1–2), 55–68. http://doi.org/http://dx.doi.org/10.1016/S0926-3373(97)00012-X.
6. Chun, H., Park, S., You, S., Kang, G., Bae, W., Kim, K., Park, J. E., Öztürk, A., Shin, D. (2009). Preparation of a transparent hydrophilic TiO 2 thin film photocatalyst. Journal of Ceramic Processing Research, 10(2), 219–223.
7. Adawiya J. Haider, Zainab N. Jameel, Samar Y. Taha (2015). Synthesis and Characterization of TiO2 Nanoparticles via Sol-Gel Method by Pulse Laser Ablation. Eng. & Tech. Journal Vol.33 part (B ).No.5, p. 761-771.
8. Yu, J. (2003). Preparation and characterization of highly photoactive nanocrystalline TiO2 powders by solvent evaporation-induced crystallization method. Science in China Series B, 46(6), 549. http://doi.org/10.1360/03yb0012
9. Fujishima, A., Rao, T. N., & Tryk, D. a. (2000). Titanium dioxide photocatalysis. Journal of Photochemistry and Photobiology C: Photochemistry Reviews, 1(1), 1–21. http://doi.org/10.1016/S1389-5567(00)00002-2
10. Zainab N. Jameel, Adawiya J. Haider,Samar Y. Taha (2014).Synthesis of TiO2 Nanoparticles by Using Sol-Gel Method and its Applications as Antibacterial Agents. Eng. & Tech. Journal Vol. 32. Part (B).No.3, p.418-426.
11. Verdier, T., Coutand, M., Bertron, A., & Roques, C. (2014). Antibacterial Activity of TiO2 Photocatalyst Alone or in Coatings on E. coli: The Influence of Methodological Aspects. Coatings, 4(3), 670–686. http://doi.org/10.3390/coatings4030670
12. Guo, H., Lee, S. C., Li, W. M., & Cao, J. J. (2003). Source characterization of BTEX in indoor microenvironments in Hong Kong. Atmospheric Environment, 37(1), 73–82. http://doi.org/10.1016/S1352-2310(02)00724-0
13. Finlayson-Pitts, B. J., & Jr., J. N. P. (1997). Tropospheric air pollution: ozone, airborne toxics, polycyclic aromatic hydrocarbons, and particles. Science, 276(5315), 1045–1052.
14. Monod, A., Sive, B. C., Avino, P., Chen, T., Blake, D. R., & Rowland, F. S. (2001). Monoaromatic compounds in ambient air of various cities: a focus on correlations between the xylenes and ethylbenzene. Atmospheric Environment, 35(1), 135–149.
15. Convertino, A., Leo, G., Striccoli, M., Di Marco, G., & Curri, M. L. (2008). Effect of shape and surface chemistry of TiO2 colloidal nanocrystals on the organic vapor absorption capacity of TiO2/PMMA composite.Ba-abbad, M. M., Kadhum, A. A. H., Mohamad, A. B., & Takriff, M. S. (2012). Synthesis and Catalytic Activity of TiO2 Nanoparticles for Photochemical Oxidation of Concentrated Chlorophenols under Direct Solar Radiation. International Journal of ELECTROCHEMICAL SCIENCE, 7, 4871–4888. Retrieved from www.electrochemsci.org
16. Zainab N. Jameel, Adawiya J. Haider, Samar Y. Taha , Shubhra Gangopadhyay and Sangho Bok. (2016) .Evaluation of Hybrid Sol-gel Incorporated with Nanoparticles as Nano Paint., AIP Conf. Proc. 1758, 020001-1–020001-14; http://doi: 10.1063/1.4959377.
17. McCullagh, C., Robertson, J. C., Bahnemann, D., & Robertson, P. J. (2007). The application of TiO2 photocatalysis for disinfection of water contaminated with pathogenic micro-organisms: a review. Research on Chemical Intermediates, 33(3-5), 359–375. http://doi.org/10.1163/156856707779238775
18. Sirimahachai, U., Phongpaichit, S., & Wongnawa, S. (2009). Evaluation of bactericidal activity of TiO 2 photocatalysts: a comparative study of laboratory-made and commercial TiO 2 samples. Songklanakarin J Sci Technol, 31(5), 517–525.
19. Chen, L., Graham, M. E., Li, G., Gentner, D. R., Dimitrijevic, N. M., & Gray, K. A. (2009). Photoreduction of CO2 by TiO2 nanocomposites synthesized through reactive direct current magnetron sputter deposition. Thin Solid Films, 517(19), 5641–5645. http://doi.org/http://dx.doi.org/10.1016/j.tsf.2009.02.075
20. Fujishima, A., Zhang, X., & Tryk, D. A. (2008). TiO 2 photocatalysis and related surface phenomena. Surface Science Reports, 63(12), 515–582.
21. Gao, K., Zhou, S., & Zhao, X. (2011). Preparation and Characterization of Zn-Containing Hydroxyapatite/TiO2 Composite Coatings on Ti Alloys. Materials Science Forum, 685, 367–370. http://doi.org/10.4028/www.scientific.net/MSF.685.367
22. Sung-Suh, H. M., Choi, J. R., Hah, H. J., Koo, S. M., & Bae, Y. C. (2004). Comparison of Ag deposition effects on the photocatalytic activity of nanoparticulate TiO2 under visible and UV light irradiation. Journal of Photochemistry and Photobiology A: Chemistry, 163(1-2), 37–44.
23. Taga, Y. (2009). Titanium oxide based visible light photocatalysts: Materials design and applications. Thin Solid Films, 517(10), 3167–3172. http://doi.org/http://dx.doi.org/10.1016/j.tsf.2008.11.087
24. Zieli, B., Grzechulska, J., Grzmil, B., & Morawski, A. W. (2001). Photocatalytic degradation of Reactive Black 5 A comparison between TiO 2 -Tytanpol A11 and TiO 2 -Degussa P25 photocatalysts, 35, 3–9.
25. Jimmy, Y., Yu, J., & Zhao, J. (2002). Enhanced photocatalytic activity of mesoporous and ordinary TiO2 thin films by sulfuric acid treatment. *Applied Catalysis B: Environmental*, *36*(1), 31–43.
26. Malnieks, K., Mezinskis, G., & Pavlovska, I. (2015). Optical , Photocatalytical and Structural Properties of TiO2 -SiO2 Sol-Gel Coatings on High Content SiO2 Enamel Surface. Materials Science, 21(1), 100–104. Retrieved from http://dx.doi.org/10.5755/j01.ms.21.1.5188
27. Reli, M., Kočí, K., Matějka, V., Kovář, P., & Obalová, L. (2012). Effect of Calcination Temperature and Calcination Time on the Kaolinite/TiO2 Composite for Photocatalytic Reduction of Co2. GeoScience Engineering, 58(4), 10–22. http://doi.org/10.2478/v10205-011-0022-2
28. Zhang, H., & Banfield, J. F. (2000). Understanding Polymorphic Phase Transformation Behavior during Growth of Nanocrystalline Aggregates: Insights from TiO2. The Journal of Physical Chemistry B, 104(15), 3481–3487. http://doi.org/10.1021/jp000499j
29. Yu, J., Yu, J. C., Ho, W., & Jiang, Z. (2002). Effects of calcination temperature on the photocatalytic activity and photo-induced super-hydrophilicity of mesoporous TiO2 thin films. New Journal of Chemistry, 26(5), 607–613. http://doi.org/10.1039/B200964A
30. Yu, J. C., Yu, J., Ho, W., & Zhao, J. (2002). Light-induced super-hydrophilicity and photocatalytic activity of mesoporous TiO2 thin films. Journal of Photochemistry and Photobiology A: Chemistry, 148(1-3), 331–339. http://doi.org/10.1016/S1010-6030(02)00060-6

31. Bessekhouad, Y., Robert, D., & Weber, J. V. (2003). Synthesis of photocatalytic TiO2 nanoparticles: optimization of the preparation conditions. Journal of Photochemistry and Photobiology A: Chemistry, 157(1), 47–53. http://doi.org/http://dx.doi.org/10.1016/S1010-6030(03)00077-7
32. Maira, A. ., Coronado, J. ., Augugliaro, V., Yeung, K. ., Conesa, J. ., & Soria, J. (2001). Fourier Transform Infrared Study of the Performance of Nanostructured TiO2 Particles for the Photocatalytic Oxidation of Gaseous Toluene. Journal of Catalysis, 202(2), 413–420. http://doi.org/10.1006/jcat.2001.3301
33. Huang, C., Bai, H., Huang, Y., Liu, S., Yen, S., & Tseng, Y. (2012). Synthesis of neutral SiO 2/TiO 2 hydrosol and its application as antireflective self-cleaning thin film. International Journal of Photoenergy, 2012. http://doi.org/10.1155/2012/620764
34. Thangavelu, K., Annamalai, R., & Arulnandhi, D. (2013). Preparation and Characterization of Nanosized TiO2 Powder by Sol-Gel Precipitation Route. International Journal of Emerging Technology and Advanced Engineering, 3(1), 636–639.
35. Kusmahetiningsih, N., & Sawitri, D. (2012). Application of TiO2 for Self Cleaning in Water Based Paint with Polyethylene Glycol (PEG) as Dispersant. International Conference on Chemical and Material Engineering, 5–10.
36. Liu, G., Zhang, X., Xu, Y., Niu, X., Zheng, L., & Ding, X. (2005). The preparation of Zn2+-doped TiO2 nanoparticles by sol–gel and solid phase reaction methods respectively and their photocatalytic activities. Chemosphere, 59(9), 1367–1371.
37. Riegel, G., & Bolton, J. R. (1995). Photocatalytic Efficiency Variability in TiO2 Particles. The Journal of Physical Chemistry, 99(12), 4215–4224. http://doi.org/10.1021/j100012a050

# Experimental investigation and thermodynamic study of heavy metal removal from industrial wastewater using pomegranate peel

*Thamer* Mohammed[1], *Raheek* Ibrahim[2,*], and *Alia* Naji[1]

[1]Chemical Engineering Department, University of Technology, Baghdad, Iraq
[2]Electromechanical Engineering Department, University of Technology, Baghdad, Iraq

**Abstract.** The objective of the study is to treat industrial wastewater and perform a thermodynamic experimental study for more economical and efficient adsorption process. This has been done by biosorption and chemisorption of chromium ions from synthetic wastewater into already prepared pomegranate peel as a natural adsorbent. Pomegranate peel proved to be an economic solution since it is available locally and it has an elevated adsorption capacity. Adsorption has been done in a fixed bed column (10 cm inside diameter, and 30 cm height) with continuous flow rate of 1 ml/min. The characteristic properties of adsorption for the pomegranate peel, analyzed by FTIR and SEM, proved that pomegranate peel is a good adsorbent because of its effective combinations and holes on its exterior. Also, comparing between adsorption ability of pomegranate peel together with Langmuir and Freundlich models indicates that Langmuir is preferable in fitting the experimental data. However, the elimination efficiency of the pomegranate peel was 90%, and the greatest adsorption capability was 9.45 mg/g. While, thermodynamic study results showed that endothermic adsorption was approved from positive values of $\Delta H^{\circ}$, also the negative values of $\Delta G^{\circ}$ proved that the adsorption is unprompted and convenient.

## 1 Introduction

Water pollution with poisonous heavy metals by industrial wastewater discharge is an environmental problem through the world [1]. Fast manufacturing and growth sent out heavy metals into the water. Industries like mining, metal plating and battery manufacturing result into a release of heavy metals into aquatic ecosystem [2,3]. Biosorption is an environment friendly process for metal removal since it can be considered as cost effective and efficient adsorbents for heavy metal removal. Many biomasses can be used as biosorbent like waste agriculture materials, and different processes have been utilized in the treatment, this includes biological method through biosorption. Biosorption advantages are: low cost, high efficiency and availability during most of the year [4]. Local fruit wastes are successfully applied for the treatment of wastewater polluted with heavy metals, but it needs a further understanding of biosorption mechanism [5]. Pomegranate is a common fruit; it includes an edible part, seed, and peel. The pomegranate peel (PP) forms 5 % to 15% of its total weight [6]. Laterally, several researches were achieved for employing a pomegranate peel to eliminate Cr(VI), Ni(II), Pb(II), and Cu(II) from water samples [7, and 8]. Thermodynamic study was performed using adsorption measurements [9]. The aim of the present work is to remove chromium ions from synthetic wastewater using pomegranate peels as a natural biomass with an efficient, low cost, and environmental friendly adsorption process. In addition to performing a thermodynamic study to explain the impact of temperature on feasibility of the adsorption process, the characterization of PP with FTIR and SEM are done to present the impact of chromium ion binding with the functional combinations in the exterior surface of PP as natural adsorbent. Also, the adsorption isotherms study is performed.

## 2 Materials and method

### 2.1 Adsorbent preparation

*Corresponding author: doctorraheek@yahoo.com

© The Authors, published by EDP Sciences. This is an open access article distributed under the terms of the Creative Commons Attribution License 4.0 (http://creativecommons.org/licenses/by/4.0/).

https://doi.org/10.1051/matecconf/201816205007

The pomegranate peels bought locally from a marketing center, washing it using tap water and distilled water. After that peels were cut off to small portions then dried by sun light for three days. The dried portions were dried again using an oven at 80°C for 36 hrs. The dried materials were grinded and sieved to give 1 mm particle size.

## 2.2 Cr (VI) solution preparation

Cr (VI) solution was prepared by dissolving the nitrates salts in deionized water and diluted to a desired metal concentration.

## 2.3 Analysis method

Atomic absorption instrument type (AA250, China) used to find Cr(VI) concentration.

## 2.4 Thermodynamic experiments

Thermodynamic experiments were done using 25 ml of chromium solution of 10 mg/L concentration, PP dose of 0.01 g/25 ml at different temperatures (30, 40, 50 °C) for contact time of (1hr). The amount of Cr(VI) adsorbed per unit of pomegranate peel (PP) and removal percentages were calculated according to mass balance of chromium concentration as:

$$q_e = \frac{(C_i - C_f)V}{m} \quad \text{................(1)}$$

$$\%\text{Removal} = \frac{C_i - C_f}{C_i} \times 100 \quad \text{..........(2)}$$

## 2.5 Column tests

In this experiment 25 gm of PP powder were placed in 30 cm high, 10 cm i.d. glass column as shown in Fig. 1. After that100 ml of the heavy metal solution were inserted into the column. The out flowing samples of 5.0 ml were assembled, labeled and analyzed. The samples were filtered using filter paper No. 40 and the quantity of chromium ions was found by atomic absorption.

**Fig. 1**. Experimental apparatus used in the experiments.

# 3 Results and discussion

## 3.1SEM test of pomegranate peel

The porous composition of (50 μm) particle size of pomegranate peel was noticed with 500X resolution (Figs. 2, and 3). The test display several holes in the surface that are able to uptake chromium ions. The test result showed that there is a noticeable variation in surface structure of PP before and after Cr(VI) adsorption.

**Fig. 2**, SEM image of pomegranate peel before adsorption.

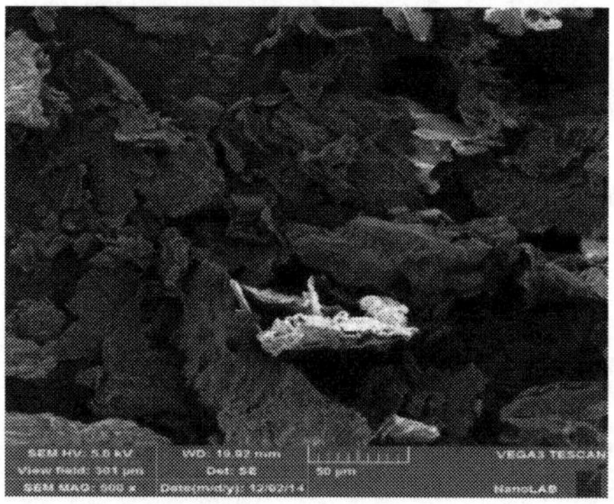

**Fig. 3**, SEM image of pomegranate peel after adsorption.

## 3.2 FTIR characterization of pomegranate peel

The FTIR analysis of PP before and after Cr (VI) binding is offered by Figure 4. FTIR analysis was utilized to show the variation in the functional groups before and after adsorption. The test demonstrates several peaks, the peak at 2360.67 and 2341.7 $cm^{-1}$ has been seen which could be as a result of chromium ion binding and adsorbing in the surface of the pomegranate peel.

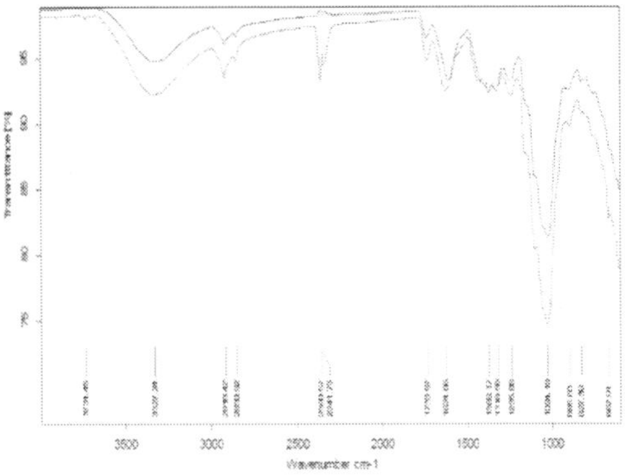

**Fig.4,** FTIR analysis of pomegranate peel before and after adsorption (blue color is PP before adsorption, and red color is PP after adsorption).

## 3.3 Thermodynamic analysis

The thermodynamic properties like change in free energy ΔG°, enthalpy ΔH°, and entropy ΔS° have been obtained using:

$$K_C = \frac{C_{AC}}{C_e} \quad \text{.................(3)}$$

$$\Delta G^O = -RT \ln K_C \quad \text{...........(4)}$$

$$\log K_C = \frac{\Delta S^O}{2.303R} - \frac{\Delta H^O}{2.303RT} \quad \text{...............(5)}$$

Vant Hoff equation (No.5) was utilized to evaluate ΔH° and ΔS°. From the intercept and slope for log $K_c$ vs. 1/T as shown in Fig.5, the results are available in Table 1.

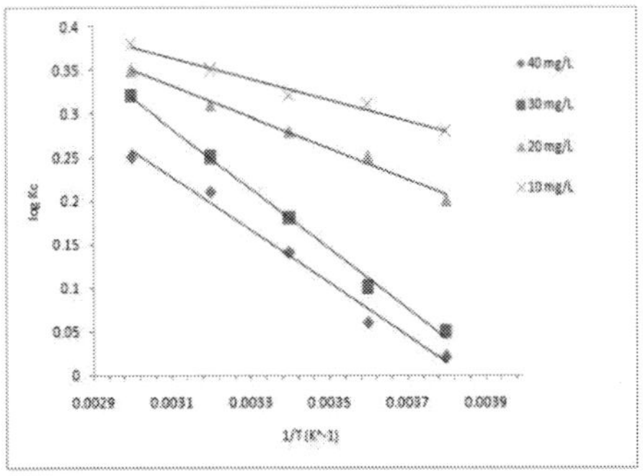


**Fig. 5,** Aplot of Vant Hoff equation for Cr(VI) adsorption onto PP at different Cr(VI) concentrations.

The values of ΔH° lie between 4.65 to 15.26 $kJmol^{-1}$ which proved the biosorption, the results indicate that biosorption is most effective. The positive values of ΔH° indicate the endothermic type of adsorption and it decides the potential of physical adsorption. Since in the adsorption process the rising of system temperature tends to enhance adsorption of Cr (VI), these principles are different from chemisorption. ΔG° values are gained from Eq. 4, the negative values of ΔG° shown in Table 1, suggest it is highly appropriate and unprompted. Furthermore, it was noticed that negative values of ΔG° increased as the system temperature was increased, which indicates that the adsorption is more superior at high temperatures. The positive values of ΔS° are shown in Table 1, proved the rising in disturbance at the interface of chromium ion with PP.

**Table 1.** Thermodynamic properties of Cr(II) adsorption on Pomegranate Peels.

| Cr(VI) conc. | ΔH° | ΔS° | ΔG° kJ/mol | | | | |
|---|---|---|---|---|---|---|---|
| mg/l | kJ/mol | J/mol | 20°C | 30°C | 40°C | 50°C | $R^2$ |
| 10 | 4.65 | 18.86 | -1.56 | -1.73 | -1.85 | -2.11 | 0.998 |
| 20 | 7.32 | 26.38 | -1.34 | -1.43 | -1.75 | -1.93 | 0.978 |
| 30 | 14.42 | 48.53 | -0.83 | -0.99 | -1.55 | -2.21 | 0.954 |
| 40 | 15.26 | 49.49 | -0.06 | -0.68 | -1.31 | -1.43 | 0.943 |

## 3.4 Adsorption isotherms for the adsorption of Cr(VI) onto PP

The adsorption results were tested using Freundlich and Langmuir models at different temperatures. The parameters of each adsorption isotherm were obtained. The linear form of Langmuir isotherm is given by Eq. 6 [11]:

$$1/q_e = (1/q_m) \times (1/b) \times (1/C_e) + (1/q_m) \quad ..(6)$$

It was demonstrated that the plot of $1/q_e$ versus $1/C_e$ displays a straight line (Fig.6), and the values of $b$ and $q_m$ are documented in Table 2. Also, the values of $q_e$ estimated from the model as $q_e(cal.)$ is quite around the experimental values as $q_e(\exp.)$. Thus, the Langmuir isotherm fits well the experimental data. The linear form of Freundlich isotherm is given by Eq. 7 [11].

$$\log q_e = \log K + (1/n) \log C_e \quad \text{.......(7)}$$

It is observed that the log $q_e$ versus log $C_e$ gives a straight line relation as shown in Fig.7. The values of Freundlich constants are given in Table 2, the Freundlich isotherm is also complied because $q_e(\exp.)$ values are near $q_e(cal.)$. In addition to that, the high correlation coefficient $(R^2)$ for Cr (VI) adsorption by PP refers that the system undergoes both Langmuir and Freundlich models, with better fitted to Langmuir model.

Table 2 shows that $n$ value of Cr (VI) is greater than 1; it is proved that adsorption is convenient for removal of Cr (VI) by pomegranate peel.

**Table 2**, Isotherms for Cr(II) removal by pomegranate peels.

| Langmuir coefficients | | | Freundlich coefficients | | |
|---|---|---|---|---|---|
| qmax. (mg/g) | b (l/mg) | $R^2$ | K (mg/g) | n | $R^2$ |
| 9.45 | 0.2 | 0.970 | 3.15 | 1.2 | 0.953 |

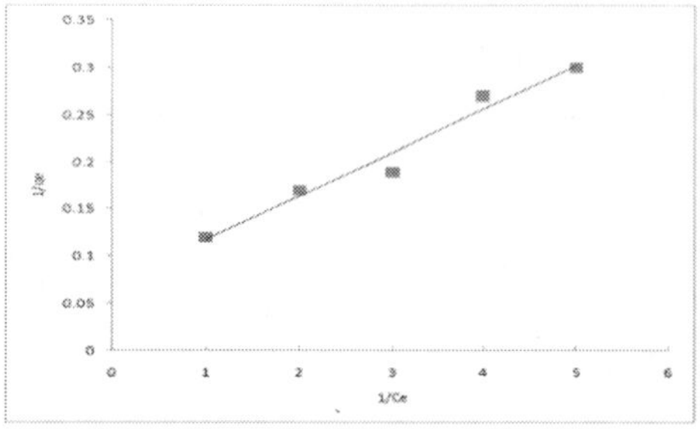

**Fig. 6,** Langmuir Isotherm for adsorption of Cr(VI) on PP.

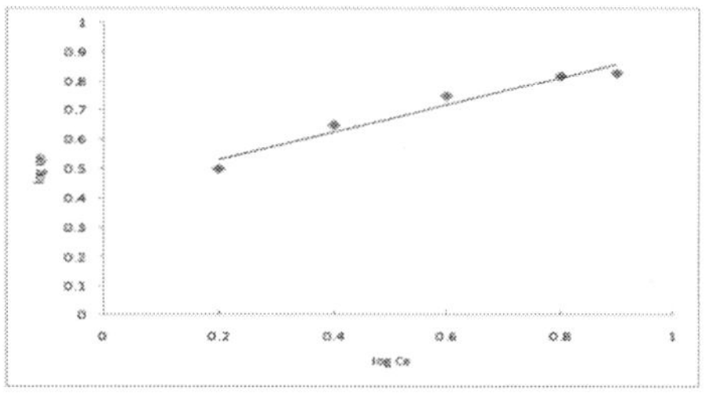

**Fig. 7,** Freundlich Isotherm for adsorption of Cr(VI) on PP.

## 4 Conclusions

From the experimental results of this work, it can be concluded that pomegranate peel proved to be an efficient and economical natural adsorbent for remediation of Cr (VI) ions from industrial wastewater. The elimination efficiency of the pomegranate peel was 90%, and the highest possible adsorption capability was 9.45 mg/g. The FTIR and SEM analyses proved that pomegranate peel is a strong adsorbent because of complex structure and holes on its surface. Although, the adsorption capacity of pomegranate peel in comparison with Langmuir and Freundlich models refers that Langmuir is well fitted with the experimental data. Also, thermodynamic investigation demonstrates an endothermic adsorption as well as the process is unprompted and convenient.

## List of symbols

| Symbol | Definition (unit) |
|---|---|
| b | Constant (-) |
| $C_{AC}$ | Concentration of metal ion in the adsorbent (mg/L) |
| $C_e$ | Equilibrium concentration of metal ion (mg/l) |
| $C_i$ | Initial concentration (mg/L) |
| $C_f$ | Final concentration (mg/L) |
| $\Delta G°$ | Change in Gibbs free energy (kJ/mol) |
| $\Delta H°$ | Change in enthalpy (kJ/mol) |
| K | Freundlich constant (-) |
| $K_C$ | Equilibrium constant (-) |
| m | Mass of adsorbent (gm) |
| n | Freundlich constant (-) |
| $q_e$ | Amount of adsorbed metal per unit weight (mg/g) |
| $q_m$ | Maximum adsorption capacity (mg/g) |
| R | Gas constant (Kj/mol K) |
| $\Delta S°$ | Change in entropy (J/mol) |
| T | Temperature (K) |
| V | Volume of solution (mL) |

## References

1. N. Othman, S. Mohd-Asharuddin and Azizul-Rahman M-F-H. (2013).An Overview of fruit waste as sustainable adsorbent for heavy metal removal.
2. Masoudrohanimoghadam,NavidNasirizadeh, ZienabDashti and EsmeilBabanezhad. Removal of Fe(II)from aqueous solution using pomegranate peel carbon : equilibrium and kinetic studies.International journal of industrial chemistry (IJIC)a springeropen journal 2013 .4:19.
3.E.S.Z.ELAshtoukhy.,N.K.Amin,andO.Abdelwahab.(2008).Removal of lead (II)and copper(II)from aqeuous solution using pomegranate peel as new adsorbant .Desalination.223,162-173.
4. T.V ,Ramachandra, N Ahalya., and R.D Kanamadi.(2008).Biosorption:Techniques and mechanisms .CES Technical Report 110.
5. K.Vijayaraghavan. and Y-S Yun.(2008).Bacterial biosorbent and biosorption.Biotechnology Advances.26,266-291.
6. H.K.Alluri.,S.R. Ronda.,S.SVigaya.,S.B Bondili.,V. Suryanarayana. And P Venkateshwar.(2007) Biosorption: An eco-friendly altervative for heavy metal removal .African journal of Biotechnology Vol.6(25) ,2924-2931.
7. EikaniMH,GolmohammadF,Homami SS (2012) Extraction of pomegranate (punicagranatuml) seed oil using superheated hexane .Food Biprod process 91:32-36.
8. EL Nemr A (2009) Potential of pomegranate peel carbon for Cr(VI) removal from wastewater ,kinetic and isotherm studies ,J Hazard Mater 161:132-141.
9. Chandra Sekhra K, Kamala CT,Chary NSN ,Anjaneyulu Y (2003) Removal of heavy metals using a plant biomass with reference environmental control .Int J Miner Process 68:37-45.
10. Babel S ,kurniawan CT ,Chary NSN ,Anjaneyulu Y (2003) Removal of heavy
Metals uptake from contaminatedwater: a review .J Hazard Mater 97:219-243.
11. Thamer J. Mohammed, and Raheek I. Ibrahim (2016), Remediation of Cu(II) from well water of Iraq using cortex of fruit and agriculture waste. Arabian J. for Sci. and Eng., 41:345–355.

# Study of the properties of drinking water for Kirkuk city and its pollution in some areas

*Iman* Hussein Zainulabdeen

Kirkuk Technical Institute, Kirkuk, Iraq

**Abstract.** This study was conducted to demonstrate the difference between drinking water quality for seven different areas of Kirkuk city with the treatment water in the main and three sub storages tanks, affected with drinking water pipe lines, supplying system Alternate and continuous and also pipes across the construction work's areas. The physical and chemical properties (turbidity, pH, conductivity, Alkalinity, Total hardness, calcium, magnesium, chloride, sulfate, TDS (Total dissolve solids), TSS (Total suspended solids), sodium and potassium) for raw water and treated water in the Storages in addition to the supplying water from seven different areas in Kirkuk were analyzed, samples taken at a rate nine times during the month of March. Results showed that Drinking water of Dumez, Wasti, Shorja, Imam Qasim and Arubaa polluted with turbidity with an average levels of (8.68, 6.04, 7.41, 6.95, 7.64) NTU respectively, over the limit allowed by IBWA (International Bottled Water Association), 2012 and WHO (World Health Organization), 2006. The turbidity in the main and sub storages have reached (4.04, 4.09, 4.47, 4.16) NTU respectively, while other physical and chemical properties were within the allowed limits for human use. The pollution in those areas caused by the alternative supplying system of treatment water and it has negative impact on the pipelines, it includes constriction work in areas also where pipelines exist.

## 1 Introduction

In large drinking water distribution networks , water needs to move in a large distances from the treatment units to the storage units and then to the consumer through a major and sub distribution networks , the problem is that this process will affect the stomach drinking water quality , which is an important issue that has attracted the attention of many governments and centers research in the world [1] .This problem may occur due to rarity of water pressure inside the pipelines, [2] it indicates that the increase of transferred water's pressure in the pipeline and non – stability causes Water Turbidity and erosion in the pipes . The corrosion or decomposition in the pipe results , an increase in the proportion of metals and organic substances in the water, which in turn causes an increase in the number of microorganisms exceeding the allowable limit (1-10 times) and poses a threat to consumer health [3].

Though the amount of corrosion also depends on the quality of the materials used in water pipeline and storage tanks. [4] They have studied this kind of effects using pipes made of PVC (Polyvinyl Chloride), HDPE (High Density Polyethylene) and GRP (Glass Fiber Reinforced Polyester) in addition to the effect of small bottle waters ( plastic ) on the quality of drinking water, they noted the possibility of leaks of some organic ingredients causing a change in taste, color and smell as well as the increase in bacterial growth in it, [5] e Comparing the effect of the two types of metal and plastic water pipes quality in Honduras, they found an increase in the proportion of iron at the very first flow of water and an increase in the microorganisms in the metal pipe when compared to plastic pipes. it showed an increase in concentrations of manganese, lead and dark colours.

In another research, [6] study water samples for 40 sites in Dhahran in Saudi Arabia, he noted an increase of lead's component in the water plastic pipelines and increase of the copper component in the copper water pipes, [7] He pointed a change in water taste and smell to an unacceptable degree to the consumer because of the interaction of water with piping components . In another study, [8] he observed the discoloration of water that has been collected from copper pipes (with ages serving more than six years) It was blue-green colored due to the corrosion.

It has been observed that the breakage and cracking in buried pipelines under various depths from the surface

© The Authors, published by EDP Sciences. This is an open access article distributed under the terms of the Creative Commons Attribution License 4.0 (http://creativecommons.org/licenses/by/4.0/).

of the soil and defect in the joints of the pipes cause water pollution as will as leakage of large amounts of water 17% in 1997 in large and medium-sized cities [9]. The contaminated soil material surrounding the pipe increase the pollution in the rainy days at construction work place, the passage of networks near the sewage and industrial area networks [10]

the time period of using pipeline contributes the cracks occurrence in the walls [11]. It indicates that the age of plastic pipes affects the extent of shock resistance to the walls of the pipes caused by the movement of water within the pipes , this effect has been shown since the first time usage , Vulnerability is up to the amount of 80% compared with normal .

In general, [12] he studied failure rate in the pipes used to deliver water in the United Kingdom Building through 40 years ( which compares the types of pipes metal , aspect , plastic P.V.C. and ductile plastic P.E.) , he noted that the highest percentage occurred in the metal pipe compared with PVC pipe .

Various pipe-related incidents that could occur in urban areas are mostly related to the lack of information on pipelines and underground management. With regard to water supply, the torn pipes resulting from these accidents lead to different leaks. These fatal accidents lead to increase in direct and indirect injuries[13].Kirkuk province is situated to the north-east of the city of Baghdad, with a population of 940,000 thousand people. Almost 65% of them live in the city of Kirkuk, according to unofficial statistics of 2009, the city of Kirkuk is equipped with clean water from the Kirkuk general water treatment project based on Bottom Zab River as well as water wells adopted in some modern areas that have not been linked to a network pipeline after water treatment units when there are high levels of pollution in the water of those wells .

## 2 Materials and methods

Water samples collected from 12 different location including main water source for project of liquidation ( Zab Al-asfal river ), main water storage tank (SGT1) after treatment, water storage tanks (SGT2, SGT3, SGT4) located to the north east of the city of Kirkuk and seven residential areas distributed within the city of Kirkuk ( Al-muhafatha, Imam Qasim, Shorja ) which are fed from SGT2, ( Arubaa , Tesien ) which are fed from SGT3, ( Dumez, Al-wasti ) that are fed from SGT4 (Figure1,2). Water samples collected nine times repeatedly and all tested at different dates, areas starting from (01/03/2014 - 02/04/2014).

Since the supply of water to most residential areas in the city of Kirkuk is periodic from day to day, it was considered that samples will be taken 15 minutes after the beginning of the flow of water in the pipeline in anticipation of not collecting samples of stagnant water inside pipes after the interruption of flow in the previous processing period , 2 liters of each sample were taken in pre-sterilized glass containers in the laboratory, samples transferred to the laboratory and stored at a temperature of 4 C° away from light.

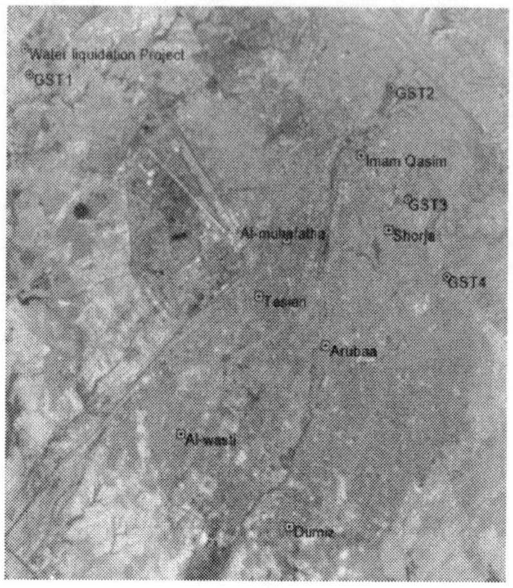


Figure 1 : Distribution of sampling areas within the city of Kirkuk

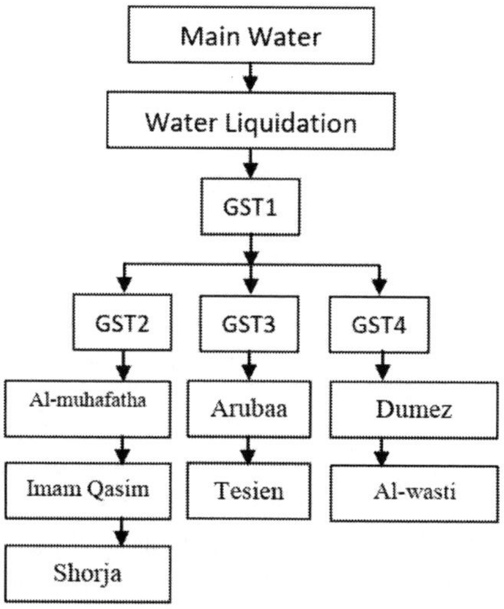


Figure 2 : Distribution plan of the water pipeline inside the city of Kirkuk

Some physical and chemical analysis were conducted in the laboratories of the Directorate of Water in Kirkuk, where measurements were taken as indicators of the amount of pollution in the water, such as (Turbidity, pH, ECu, Total Dissolved Solid (T.D.S.), Total

Soluble Substances (T.S.S.), Hardness, Alkalinity as $CaCo_3$, Calcium, Magnesium, Chloride, Sulfate, Sodium and potassium).

PH and E.C.u were measured using a pH meter and a conductivity meter depending on the method used by [14].

1. Positive ions: The method used was adopted before [15] to estimate calcium, sodium, potassium and magnesium.
2. Negative ions: These include sulphates, carbonates and chlorides, It was estimated by following the method used before [15].
3. Total severity: It was estimated using the method described by [15].
4. Total dissolved salts: It was calculated by the sum of negative and positive ions.
5. Inverter: It was measured using a Hach2100.
6. Suspended Materials: It was calculated from the results obtained for total solids and total soluble substances by the following equation:

TSS = Total solids - Total dissolved

## 3 Results and discussion

Table 1,2 shows differences in the values of physical and chemical analysis in the tested water samples for all the study area inside the city of Kirkuk, in addition to the main water tanks GTS1 , GTS2 , GTS3 and GTS4 Which distributes water networks to different areas of study.

Table 1: Average values of physical analysis of water samples tested

| Sampling area | Turbidity in NTU | PH | E.C.u in Mmohs/cm | T.D.S mg/l | T.S.S. mg/l | T.Hardness as CaCo3 mg/l |
|---|---|---|---|---|---|---|
| Zab river | 29.9 | 7.59 | 393.5 | 245.5 | 65.22 | 172 |
| GST1 | 4.04 | 7.53 | 373.3 | 239.1 | 47.25 | 164 |
| GST2 | 4.09 | 6.55 | 381.1 | 237.3 | 43.25 | 172 |
| Almuhafatha | 4.69 | 7.07 | 380.7 | 240.2 | 45.87 | 170 |
| Imam Q. | 6.95 | 7.07 | 372.3 | 238.2 | 47.33 | 171 |
| Shorja | 7.41 | 6.78 | 388.7 | 253.8 | 47.22 | 170 |
| GST3 | 4.47 | 7.30 | 383.6 | 234.8 | 43.25 | 171 |
| Arubaa | 7.64 | 6.91 | 381.6 | 226.7 | 41.62 | 170 |
| Tesien | 4.09 | 6.95 | 380.6 | 231.8 | 44.12 | 164 |
| GST4 | 4.16 | 7.30 | 375.7 | 237.0 | 36.62 | 171 |
| Dumez | 8.68 | 6.85 | 392.7 | 236.3 | 39.50 | 172 |
| Al-wasti | 6.04 | 6.81 | 400.3 | 233.5 | 46.12 | 168 |
| IBWA-WHO | 5.00 | 6.5-8.5 | 70-300** | 500 | 1000 | 500 |

[16](WHO,1996)=World Health Organization
*[17](IBWA)=International Bottled Water Stander Quality
**[18](FAO)=Food & Agriculture Organization

Table 2: Average value of chemical & Percentage of component ratios of water samples tested

| Sampling area | Alkalinity as $CaCo_3$ mg/l | Calcium as $Ca^+$ mg/l | Magnesium Mg mg/l | Chloride $Cl^-$ mg/l | Sulphate $So_4^=$ mg/l | Sodium $Na^+$ mg/l | Potassium $K^+$ mg/l |
|---|---|---|---|---|---|---|---|
| Zab river | 133 | 42.3 | 16.4 | 15.8 | 44.8 | 7.3 | 1.6 |
| GST1 | 130 | 42.7 | 15.0 | 16.5 | 48.5 | 7.3 | 1.5 |
| GST2 | 129 | 41.0 | 16.7 | 15.6 | 46.1 | 7.3 | 1.5 |
| Al-muh. | 133 | 42.8 | 15.6 | 17.3 | 41.3 | 7.4 | 1.5 |
| Imam Q. | 134 | 41.5 | 16.4 | 16.3 | 39.5 | 7.2 | 1.5 |
| Shorja | 139 | 43.5 | 15.6 | 18.9 | 42.9 | 7.1 | 1.6 |
| GST3 | 130 | 41.5 | 15.8 | 15.7 | 41.5 | 7.5 | 1.5 |
| Arubaa | 132 | 42.0 | 15.2 | 16.7 | 42.0 | 7.0 | 1.5 |
| Tesien | 131 | 40.5 | 16.6 | 17.8 | 40.0 | 6.6 | 2.0 |
| GST4 | 130 | 42.5 | 14.7 | 16.6 | 47.5 | 7.3 | 1.5 |
| Dumez | 143 | 40.6 | 17.6 | 16.4 | 39.6 | 7.0 | 1.7 |
| Al-wasti | 133 | 40.9 | 15.8 | 16.9 | 41.4 | 7.2 | 1.5 |
| IBWA-WHO | 125-200* | 50 | 50 | 250 | 250 | 200 | 10* |

The highest value of turbidity was found in the areas of (Dumez, Al-Wasati, Shorja, Imam Qassem and Arubaa), with a ratio of ( 8.68, 6.04, 7.41, 6.95, 7.64 ) mg / L respectively, exceeding the limit of [16]&[17] . Note that the amount of turbidity in the collection tank of the liquidation project and three major feeder reservoirs of the city were between ( 4.04 - 4.47 ) NTU , on the other hand when compared turbidity between sampling dates, ( Figure 3 ).

This discrepancy may be due to the periodic water distribution system within the residential areas in the city of Kirkuk, which depends on supplying the residents from one day to another, where the interruption of the flow of water inside the pipes leads to the deposition of the contents of the insoluble and cumulatively on the one hand, and the variation of water pressure moment of flow within the tubes and the sediments contained in them increase the rate of turbidity in the supplied water, which is clearly visible in the first supply of the water flow to the consumer and may continue for a few hours, and the turbidity ratios in the areas of Tesien and the province have recorded low rates and within the limit of the consumer (4.09 and 4.69) NTU, it may be happen, that both regions are provided water per day without the adoption of deliberation system.

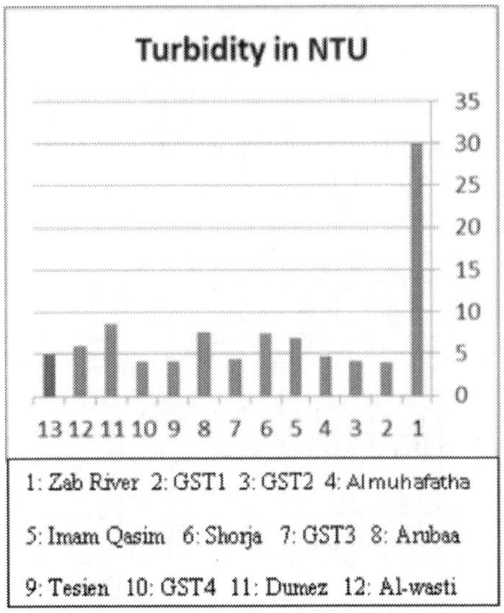


Figure 3 : Turbidity in NTU for water samples tested.

The highest rate of turbidity in the Dumez area, which is supplied daily with water, exceeded the limit of 8.68 NTU. When asked about the problem from the specialists and the residents of the area, the water pipes were found damaged in the area due to the construction work in the area. two years ago.

PH values in Table (1) showed no significant differences between the sampling sites and ranged between ( 6.3 - 8.1) . They were within the permissible range of human use and were not affected during the transfer of water from the project to the consumer during the pipeline.

Also, it was noted in (Tables 1) that the electrical conductivity values E.C.u rates have exceeded the upper limit allowed by the Food and Agriculture Organization FAO in its report at 1989 and all study areas, including the main water reservoirs as values ranging from ( 372.3 Imam Qasim area and 400 Wasti area) Mmohsm/cm ( Figure 4 ) in with a decline slight declination from what water source of values shown for the delivery of electrical proving a defect in the treatment of electrical conductivity of water intended for human consumption. It indicates the researcher [19] that high electrical conductivity values may cause diseases to its users Electrical conductivity values are closely correlated with dissolved solids, suspended materials and chlorides .

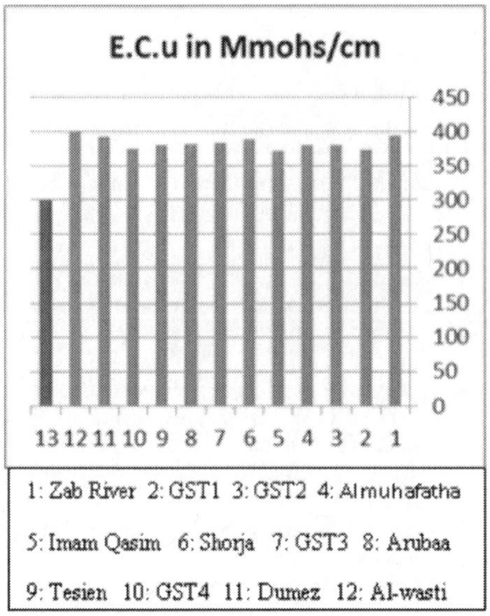


Figure 4 : E.C.u in Mmohs/cm for water samples tested.

The Values of T.D.S. for Source water were 245.5 mg / l, in the main collection tank 239.1 mg / L, the values of the residential sampling areas for T.D.S. were not significantly different from the values of the main collection tank water after treatment and source water this may be due to the lack of salinity treatment units, or lack of focus on the fact that the water is not salty and well below the permissible limit allowed by ( IBWA and WHO ) .

The percentage of total suspended solids in raw water was 47.25 mg / l with a significant difference of 17.97 mg / l from its share in the source water of 65.22 mg / l . This is due to the disappearance of these particles. It was observed that this percentage was reduced in the collection tanks at most of sampling the times as a result of deposition.

Table 2 shows a difference in the ratio of suspended materials between regions. This may be due to the rapid flow of water inside the pipes, on the one hand, and the rapid supply of water to the residential areas from the collection tanks without leaving a time period on the other hand. The presence of dilutions in the joints of the conveyor tubes may be another cause of the variation.

Total Hardness known as a group of total salts of carbonates, bicarbonates, sulfates, chlorides, calcium nitrate and magnesium. Tables ( 2 ) showed that the total hardness rates did not exceed the permissible limit for the water consumed and for all the sampling areas. The value ranged from (163.7 to 172.1) mg / l . It was noticed that no significant change in the value of the source water after entering the treatment unit and then

to the reservoirs of the main city of Kirkuk and to the consumer except for some fluctuations that may be resulted from the interaction of water with the contents of the pipes from the material of sediment or contact with the external soil in the sites of discord within the conveyor tubes and this is consistent with [9] and [13] There is an additional cost for it.
The results of Alkalinity as $CaCo_3$ Table (2) showed alow rates within the specified percentage of water content allowed for human consumption. No significant differences were observed between the sampling sites of the residential areas and the water reservoirs and the main source of feed for the project ( Figure 5 ).

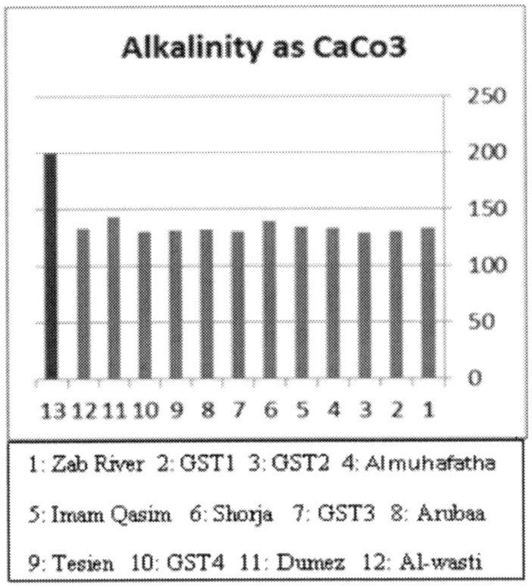

Figure 5 : Alkalinity as $CaCo_3$ for water samples tested

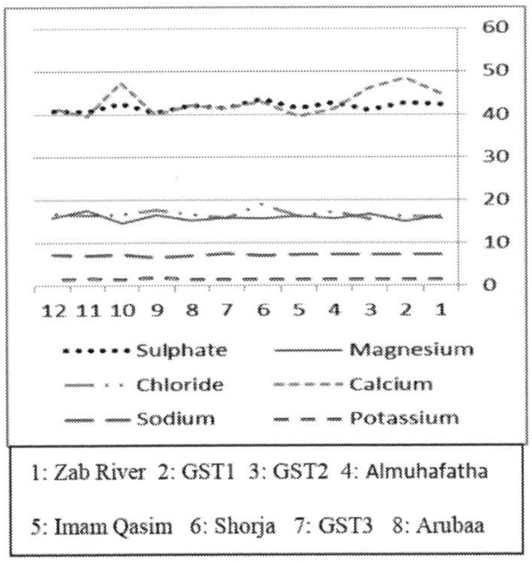

Figure 6 : $Ca^{+}$,$Mg^{+}$, $Cl^{-}$ , $So_4^{=}$, $Na^{+}$ and $K^{+}$ rates for water samples tested .

Positive ions ( Magnesium, Chloride, Sulphate, Sodium and potassium ) all showed low rates within the specified percentage of the permitted water content for human consumption ( IBWA & WHO ), also there are no significant differences observed between the sampling sites of the residential neighborhoods and the water reservoirs and the main source of feed for the project.

## 4 Conclusions

The water samples taken from the collection tanks for the treated water are pure and confirm the approved standard specifications. It should be noted here that the raw water is of acceptable quality in most physical properties and does not require the material cost and time for treatment.
The periodic distribution of water in most areas of Kirkuk led to a high rate of turbidity and caused damage to the distribution networks, which in turn reflected on the quality of water transferred.
Pollution in the water supplied to some areas are due to the existence of construction and excavations and lack of coordination with the competent authorities on this subject.
Non-maintenance and replacement of old pipes has a reflection on the quality of water supplied to those areas.

## 5 Recommendations

Based on the results of this study, the following are recommended:
- Maintenance and cleaning of liquefaction networks and reservoirs on a regular basis.
- Replacement of some old and old-fashioned pipes with high-quality annealing pipes.
- Pre-coordination between companies and departments responsible for excavation works and Kirkuk water department to prevent the occurrence of excesses on the pipelines.
- Communicate the areas equipped with clean drinking water within the approved standards of non-consumption of cash to buy bottled water and reduce the use of excess water.

## References

1. Weifeng Li, Wencui Ling, Suoxiang Liu, Jing Zhao, Ruiping Liu, Qiuwen Chen, Zhimin Qiang, and Jiuhui Qu ( 2011) . Development Of Systems For Detection, Early Warning, And Control Of Pipeline Leakage In Drinking Water Distribution. A case study . Journal of Environmental Sciences 2011, 23(11) 1816–1822.

2. Hossein Shamsaei , Othman Jaafar, Noor Ezlin Ahmad Basri ( 2013 ) . Disadvantage Pressure Changes on the Decline of Water Quality in Water Distribution Systems . Engineering, 2013, 5, 97-105.

3. Grabińska A. Łoniewska , T. Konilłowicz-Kowalska , G. Wardzyńska and K. Boryn (2007) . Occurrence Of Fungi In Water Distribution System . Polish J. of Environ. Stud. Vol. 16, No. 4 (2007), 539-547.

4. Hem Lars J. and Ingun Skjevrak ( 2002). Potential Water Quality Deterioration Of Drinking Water Caused By Leakage Of Organic Compounds From Materials In Contact With The Water . Proceedings, 20 th NoDig conference, Copenhagen May 28-31 2002.

5. Cerrato José Manuel ( 2005 ) . Impact Of Piping Materials On Water Quality In Tegucigalpa, Honduras . Master of Science in Environmental Engineering . September 5, 2005 , Blacksburg, Virginia . USA.

6. M. Sadiqa, T.H. Zaidia, H. Al Muhannaa & A.A. Miana ( 1997 ) . Effect Of Distribution Network Pipe Material On Drinking Water Quality . Journal of Environmental Science and Health . Part A: Environmental Science and Engineering and Toxicology .Volume 32, Issue 2, 1997.

7. Timothy Heim , ( 2006 ). Impact Of Polymeric Plumbing Materials On Drinking Water Quality And Aesthetics . (master thesis) , April 26th, 2006 . Blacksburg, Virginia , USA.

8. Ahn Jeahwan , Hyondong Lee , Seoku Kim , Pilijae Kwak , and Jooyeon Lee ( 2010 ). Corrosion Control Of Drinking Water Copper Pipe By Corrosion Inhibitor . materials science forum . vol. 658 . 2010 pp. 320 – 323.

9. Chen Q W, Qu J H, Liu R P, Li W F.(2008) . Rule-Based Model For Aging-Induced Leakage From Water Supply Pipe Network In Beijing City. China Water & Wastewater, 24(11): 52–56.

10. Nkanga Enefiok J. ( 1980 ). Investigations Of Pipe-Borne Pollution In Benin City, Nigeria . Water, Air, and Soil Pollution . December 1980, Volume 13, Issue 4, pp 425-437.

11. Scholten Frans and Mannes Wolters ( 2011 ) . Physical Ageing Of U-P.V.C. Gas And Water Pipes . PVC 2011, 12-14 April 2011, Brighton, UK, British Plastics Federation.

12. Rubeiz C. ( 2010 ) , Impact Off Potable Water Disinfectant Of PE Pipe . Jana Laboratories Inc. 280B industrial Pkwy S Arora Ontario , L4G , 3T9 . www.janalab.com.

13. Avhad, Priyanka T. ( 2016 ) . To Study Management and Monitoring of Water Pipeline. International Journal of Innovative Research in Computer and Communication Engineering . Vol. 4, Issue 4, April 2016.

14. Page, A.L. ( 1982 ). Methods Of Soil Analysis , Agron.9, part 2:Chemical and Mineralogical Properties , 2nd ed.,Am. Soc. Agron , Madison, WI, USA(1982).

15. Wahab. Yousri Shehab, Abtasam Hoppe, Amal Francis, Ali Zubair Khalaf (2000). Analytical Guide To The Water Testing Unit. North Oil Company Publications - Fields Authority.

16. IBWA ( 2012 ) . Bottled Water Code of Practice , International Bottled Water Association 1700 Diagonal Road, Suite 650 Alexandria, VA 22314 , (703) 683-5213 , http://www.bottledwater.org.

17. WHO ( 2006 ) . A Compendium Of Drinking-Water Quality Standards In The Eastern Mediterranean Region. World Health Organization Regional Office for the Eastern Mediterranean Regional Centre for Environmental Health Activities -CEHA.

18. FAO ( 1989 ) . Water Quality For Agriculture . Food and Agriculture Organization of the United Nations Rome, 1985 © FAO.

19. Al-Hayani AG (2009) . Evaluation Some Groundwater Wells Of Khaffjiyia Village In Al-Anbar Governorate. J. Al-Anbar Univ. Pure Sci 3, (no2):30–38, Arabic.

# Biodegradation of phenolic components in wastewater by micro algae: a review

*Muthanna Al-Dahhan*[1], *Faris* Al-Ani[2] and *Afaf* Al-Saned[2, *]

[1]Missouri University of Science and Technology, Rolla, MO 65409, USA
[2]Building and Construction Engineering Department, University of Technology, Baghdad, Iraq

**Abstract.** The biodegradation of phenolic contaminant present in wastewater by micro algae has been investigated in the current research. The study focused on algae growth requirements and factors affecting algae growth in contaminated water leading to the best treatment. Different techniques used to functionalize microalgae in the treatment such as batch and continuous operation modes have been further investigated. Mechanisms for microalgae pollutants degradation and microalgae growth kinetics were also highlighted to present the required knowledge in phenol contaminant biodegradation field by microorganisms

## 1 Introduction

Among the most essential environmental issues, pollution of surface water has become a standout problem among the most imperative environmental issues. Two sorts of substantial and long-lasting pollution dangers can be perceived at the worldwide level. Organic contamination prompting by high organic content in aquatic ecosystems and eutrophication. Polluted water can affect water quality in a manner, of confining utilization of water, bodies for some reasons.

Organic pollution happens when huge amounts of organic compounds from numerous sources are discharged into running waters, lakes and seas. Organic pollutants are derived from farm water, industrial effluents, urban runoff and domestic sewage. Organic pollution could adversely influence the water quality in numerous ways.

one of the 129 specific significance chemicals is Phenol, which is considered toxic under the 1977 modifications to the Clean Water Act and for which the US Environmental Protection Agency (EPA) has issued water quality standards [23]. Phenol is an aromatic hydrocarbon containing a hydroxyl group (OH) tied to the benzene ring. It is a basic structural unit for a diversity of artificial organic compounds. As an organic substance, phenol is soluble in most organic solvents and it is slightly soluble in water at room temperature [1]. Phenols are divided either as natural or synthetic nonaromatic compounds in different environmental sites as major pollutants. Due to the aromatic structure of phenol, it is renitent to biodegradation naturally and phenolic compounds have been reported to have high stability because of the difficulty of cleaving the benzene ring.

Effective treatment methods are available for the removal of hydrocarbons and phenols such as activated carbon adsorption, ion exchange, liquid–liquid extraction, and chemical oxidation; however, they often suffer from dangerous drawbacks such as high cost. In addition, most of these techniques do not degrade organic compounds, but rather transfer it to another phase, which lead to the realization of dangerous byproducts (secondary pollution). On the other hand, biodegradation is considered a more environmental friendly and cost-effective alternative. So, biological treatment has been an increasingly important process in pollution control). However, a lot of microorganisms can tolerate phenol and using it as a carbon and energy source. The biological degradation is done through benzene ring cleavage using the enzyme presents in the microorganism [14].

## 2 Biological treatment method

The biological treatments are globally preferred when they are compared to physiochemical methods due to the possibility of complete mineralization of the toxic compounds from aqueous solutions while producing innocuous end products, maintaining concentrations of pollutants below the toxic limit, cost competitive, there are no chemicals involved and no secondary problems (toxic by-products) in the effluent. Relying on the microbial ability of growing in specific conditions, organic material can be degraded anaerobically or aerobically. Though aerobic and anaerobic, microorganisms can degrade phenol and other organic compounds; commonly aerobic processes are preferred. For degrading toxic compounds, aerobic microorganisms are more effective because they grow faster and commonly transform organic compounds to inorganic

* Corresponding author: jadaan_64@yahoo.com

© The Authors, published by EDP Sciences. This is an open access article distributed under the terms of the Creative Commons Attribution License 4.0 (http://creativecommons.org/licenses/by/4.0/).

compounds ($H_2O$, $CO_2$). Aerobic processes are also preferred because of the low cost related with this option large number of microorganisms including bacteria, fungi, and algae are capable of degrading phenol [25].

## 3 Microalgae

Microalgae are microscopic photosynthetic microorganisms. They are a type of eukaryotic cells and they contain similar organelles such as chloroplasts, nucleus, etc. Microalgae are generally, more efficient than land plants in utilizing sun light, CO2, water and other nutrients which are translated to higher biomass yields and higher growth rates. They can crown and grow in a different aquatic environment. Microalgae can be crowned without the use of fertilizers and pesticides which results in less waste and pollution.

Algae are included in the plant kingdom and are distinguished from other Chlorophyllous plants on the basis of sexual reproduction. There are seven divisions of algae, divisions that include the larger visible algae are: Cyanophyta (blue-green algae), Chlorophyta (green algae), Rhodphyta (red algae) and Phaeophyta (brown algae). These divisions are subdivided into orders, which are subsequently divided into families and then into genus and species. The main differences between these types of algae are in cell wall, where sorption takes place [33].

## 4. Algae growth requirements

Microalgae have a particular set of requirements for growth identical to that of other photosynthetic plants. Basic inputs for microalgae growth include water, sunlight, nutrients and an acceptable range of environmental conditions particular to the algal species. Environmental conditions like temperature, pH, salinity, and dissolved gases may influence the growth characteristics of microalgae. Growth inputs have an optimum range of supply, and providing the optimum growth conditions produce the largest algal population. Generally speaking, larger algal populations lead to greater amounts of algal biomass [9].

Successful treatments of wastewater with microalgae require a good growth, and understanding of the factors that influence growth is therefore essential. The growth rate of algae and cyanobacteria is affected by biological, chemical and physical factors.

### 4.1 Light "photo" requirements

Microalgae require a dark period as well as a light period for growing which is known as a photoperiod. The dark period allows the algae to respire while the light period permits photosynthesis to occur. The photosynthesis reaction is shown below as Equation 1 [10].

$$6CO_2 + 6H_2O + light \leftrightarrow C_6H_{12}O_6 + 6O_2 \quad (1)$$

Chloroplasts absorb light energy, and in the presence of $CO_2$ and water transform the captured energy into potential chemical energy In this way photosynthesis converts light energy along with $CO_2$ and water into chemical energy in the form of carbohydrates and releasing oxygen in the process. Through the dark phase, respiration follows the same equation proceeding in the opposite direction.

The mathematical model showed that degradation of phenolic compounds is improved by increasing the light intensity in a broad range of light intensities; which is mainly attributable to increases in photosynthetic oxygen production. Both biodegradation rate and microalgae growth can be improved under increased light intensities, by addition of inorganic carbon sources such as CO2 and sodium bicarbonate (NaHCO3) [26].

### 4.2 Nutrient requirements

Carbon is an essential nutrient for the growth of microalgae. CO2 is a main ingredient driving photosynthesis and is the primary GHG to be sequestered.

The carbonic acid (H2CO3) results from injection CO2 into water, therefore lowering the water pH. The following set of equilibrium equations clarify how $H_2CO_3$ behaves in an aqueous solution and recognize the carbonic species that are consumed by microalgae according to the pH [42].

$$CO_2(gas) + H_2O(aq) \rightarrow H_2CO_3(aq) \quad (2)$$

$$H_2CO_3 \leftrightarrow H^+ + HCO_3^- \quad (3)$$

$$HCO_3^- \leftrightarrow H^+ + CO_3^{2-} \quad (4)$$

However, since the ambient atmospheric concentration (0.033 %) is far below optimum for algal growth, supplying of extra carbon dioxide may be necessary. This can be achieved by providing the cultures with air enriched with 1 to 5% $CO_2$ [20].

Besides carbon, the second basic nutrients are nitrogen, phosphorous, and potassium. The essential macro nutrients required for all plants are potassium (K), phosphorous (P) and Nitrogen (N). Usually plant fertilizers are classified by their N-P-K ratios.

Different micro nutrients such as calcium (Ca), iron (I), magnesium (Mg), sulfur (S), zinc (Zn), and different trace elements are also basic for a full satisfaction to the nutrient requirements of microalgae. Because microalgae are suspended in water, there is a great availability of nutrient, and thus maintaining enough nutrient levels is fundamental for optimum algal growth. Atmospheric carbon dioxide may be provided to algal cultures by means of aeration [28].

### 4.3 Environmental requirements

Environmental conditions play an important role in governing the microalgae growth. Environmental parameters such as salinity, pH, temperature and dissolved oxygen ultimately have an impact on the success of algal cultivation. The microalgae population is around of 100,000 recognized species. Therefore, the ideal environmental conditions are particular to the different algal species [11].

## 5. Microalgae for wastewater treatment

The biological treatment of wastewaters is based on the capacity of naturally occurring microorganisms to degrade pollutants presented in the aquatic environment. Most studies on biodegradation to date have focused on bacteria or fungal mediated processes. The idea of using microalgae and/or cyanobacteria in decontamination processes, such as in the treatment of domestic wastewater was initially proposed in the late 1950's. Since that period, there have been number of reports on microalgae use in biological treatment strategies.

Their uses in such applications have the main advantage of being able to generate microbial biomass from a cheap and most abundant energy source (sunlight). Another benefit is that additional pollution is not generated when the biomass is harvested and efficient recycling of nutrients is possible [29].

Photosynthetic oxygen-evolving cyanobacteria are ideal for the treatment of some effluents containing xenobiotic since unlike heterotrophic organisms; they can undertake biodegradation simultaneously with oxygenation Some marine cyanobacteria have been successfully used in the treatment of solid wastes and wastewaters containing pesticide , phenols , aromatic hydrocarbons , textile dyes , removing uranium from mining wastewater and detergents .This is primarily due to their capacity to metabolize these compounds as nitrogen, phosphorus, carbon and sulfur sources. They have also been used to reduce the content of nitrogen, ammonia and phosphorous in sewage and various agricultural wastes. Microalgae Scenedesmus Chlorella, cyanobacteria Phormidium and Oscillatoria are the most frequently used genera in wastewater treatment systems mediated by microalgae and/or cyanobacteria. The use of these microorganisms led to a progressive reduction of chemical oxygen demand (COD) and biological oxygen demand (BOD), to values below the disposal limits [36].

Microalgae have also been successfully employed in the reduction of heavy metal content in industrial wastewaters, both in batch or in continuous systems. Additionally, the acidic nature of some industrial wastewater can be buffered by algal growth. The growth of algae in wastewater can balance the acidity which can help reducing harmful degradation of waste infrastructure [22].

## 6. Degradation of phenolic compounds by microalgae

Till now, researches on phenol biodegradation have almost exclusively concentrated on the use of bacteria and fungi. However, these bacteria and fungi may further aggravate the pollution problem because of anaerobic degradation of the aromatic amino acids exist in the wastewater that may further result in release of phenol and cresol Even in low concentration, phenolic compounds are so toxic to organisms. However, some microorganisms are capable of utilizing them as carbon and energy source [34]

Many phenols show acute toxicity to algae, both cyanobacteria and eukaryotic microalgae are able to bio-transforming aromatic combinations, including phenols [36].

Semple et al. [36] examined the catabolic versatility of the microalgae Ochromonasdanica (CCAP 933/2B) to degrade mixtures of phenols. It was observed that after growth of algae, all phenol had disappeared from a phenolic mixture in the incubation media in 2 hours.

Shashirekha et al. [38] reported inhibitory effects of phenol on the growth of phormidium photosynthetic conditions up to 100 mg/ L.

Ishihara and Nakajima [19] reported the removal of bis-phenol from a medium spiked with 40 um of bis-phenol by two marine microalga strains of nannochloropsis sp. And chlorella gracilis up to a level of 34 % and 53% respectively over a period of 6 days under light.

. Pinto et al. [30] reported the removal of more than 70% of phenol by two green microalgae, ankistrodesmusbraunii and scenedesmusquadricauda from olive-oil mill wastewaters within 5 days of treatment. This was similar to those obtained with fungi and bacteria.

D and N [12] tested the capability of the microalgae Chlamydomonasreinhardtii, Chlorella vulgaris Beyerinck, Scenedesmusquadricauda and blue green microalgae Anabaena ambigua, Nostocmuscorum, Oscillatoria animals and Oscillatoria sancta to use phenol as carbon source on agar salt medium. The study demonstrated better growth of algae and they were capable of degrading all phenol levels in culture medium. However,

## 7. Culturing techniques

Generally, there are two systems for mass photosynthetic microorganism production, either open or enclosed system. The open systems, i.e., raceway ponds and lakes, are usually illuminated naturally (outdoors) and open to the environment. They are cost-effective but intrinsically hard to control. Hence, they are usually used to grow resistant strains, such as Spirulina platens and Dunaliela Salina. Those strains can resist hard environmental conditions (such as high pH for Spirulina and high salinity) that prevent contaminations by other strains.

The enclosed systems, usually called photo bioreactors (PBR) have numerous possible photo bioreactor (PBR) configurations (e.g., tubular, panel, internally illuminated, airlift bubble column reactors) as shown in Figure (1).

https://doi.org/10.1051/matecconf/201816205009

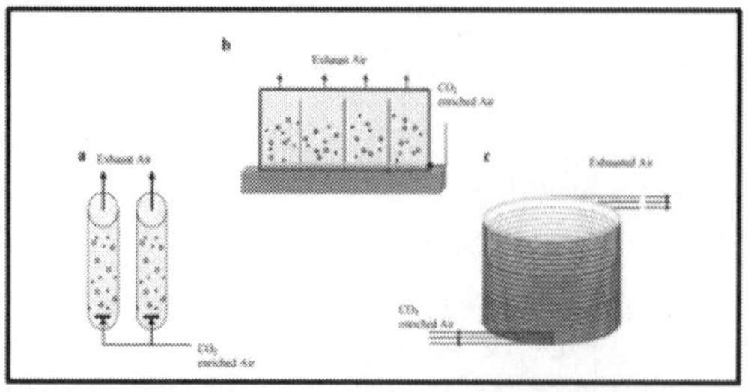

**Fig. 1.** Schematic presentation of a vertical column reactor (a), a flat plate reactor (b), and tubular photo bioreactor arranged in spiral [27].

Table 1 shows a summary of researches performed in different PBR configurations using various microorganisms.

Table 1. Summary, of Continuous and, Semi-Continous Cultivation of Microalgae

| Author | [15] | [39] | [41] | [16] | [18] | [43] |
|---|---|---|---|---|---|---|
| Species | Chorella vulgaris | Thermosynechococcussp | Chlorella minutissima, D.tertiolecta | Scenede-smus obliquus | Cyano-bium sp. | Chlorella vulgaris |
| Type | | Cyanobacte-rium | freshwater | Fresh water | | Fresh Water |
| Media | Artificial Waste Water | Fitzgerald media | Modified Bold 3N | Jaworski | BG11 | Diluted Manure |
| Media Supplement | Glucose NH4CL | 15%CO2 absored by NaOH solution | 3xReg.NaNO3 | | 0.4-1.6g/l NaHCO3 | |
| Working Vol.(L) | 2 | 1 | 3-6 | 500 | 1.8 | 1 |
| PBR Design | Tubular | Tubular | CSTR | Fence+ Tank | Tubular | 4L flasks |
| Light source | ------ | ------- | Circular LED | Sunlight | ---- | Fluorescent lamp |
| Tem.(°C) | 30 | | 25 | | 30 | 25±2 |
| %CO2 | | | 4 | 3 (control pH) | | 2 (for pH control) |
| Growth rate (d-1) | | | | 0.03-0.23 | | |
| Dilution rate | 0.5-1.5 | 1.4d-1 | 0.076-0.64d-1 | Not specified | Not constant | 0.05-0.3d-1 |
| pH | 8-10 | 10-11.5 | 7.8 (initial) | 7.0±0.2 | | |

Table (2) shows various tubular photo bioreactor characteristics, focusing on the algal species employed, the main parameters useful to the biomass growth (temperature and pH), the adopted mechanism of circulation, the methods of light supply and the maximum illumination flux provided, the system of CO2 supply and the materials employed in the construction of the re actors. [42].

Photobioreactor can be either illuminated naturally or synthetically (indoors) and permit axenic cultures under full control. These enclosed systems, therefore, are suitable for growing strains for high value products, as these strains usually need protection from the exterior environment and should be maintained at suitable temperature, pH value, salinity, etc [17].

Among all sorts of photo bioreactors studied, airlift column reactors have emerged as one of the most promising photo bioreactors for their excellent mixing intensity. These reactors have been widely used in the chemical, biochemical fermentation and biological wastewater treatment industries where high mass transfer and good mixing are required. These reactors possess numerous advantages that are especially appropriate for mass microalgae/cyanobacteria culture, such as simple construction, long liquid phase residence time, and low shear stress while maintaining high turbulence intensity. Therefore, exploring the possibility of applying airlift column reactors in mass autotrophic microorganism production and advancing the understanding. In air lift column, photo bioreactors, air (or any other gas) is spargered into the column from the riser leading to a continuous cyclical recirculation. of liquid and solid (if any, e.g., cells) phases in the reactor. The advantages of the airlift reactor are [31]:

1- Stable.
2- Little power consumption.
3- High mass transfer.
4- High heat transfer characteristics.
5- Good solid suspending.
6- Homogeneous shear.
7- Easy to control.
8- Operational flexibility.
9- Rapid mixing.

The homogeneity inside the reactor can further provide a consistent metabolic state to the cells and the light usage efficiency is high. Furthermore, the scalability of the airlift reactors is better than the other kinds of reactor mentioned above [21].

Kanhaiya.et al. [24] have examined the suitable design of photo bioreactors for $CO_2$ sequestration addressing global warming using green algae and cyanobacteria. The Biological fixation of $CO_2$ culture density can produce renewable energy carriers such as carbon free energy hydrogen, bioethanol, biodiesel and other valuable biomolecules. The results showed that Airlift reactor seems the most promising for the $CO_2$ sequestration.

A pilot plant designed by Sudarsan et al [40] to study the removal of ammonia Cal nitrogen by a biological system. The nitrification process was carried out in a sequential batch reactor and removal of nitrate from the system generated during the oxidation of ammonia Cal nitrogen was carried out in a photo bioreactor using algae. Results showed that 95% ammonia Cal nitrogen removal was done and 98% nitrate removal was done also by using algal photo bioreactor.

The cultivation of two algal strains; Chlorella vulgaris and Tetraselmis sp. in lab scale air lift photo bio reactors to estimate the feasibility of using flue gas as a carbon source for microalgae growth was conducted by Bryan

[6]. The study demonstrated that the both microalgae species were capable of growing under flue gas dosing. And they are able to fixate $CO_2$ from coal-fired flue gas.

Deviram, et al. [13], developed a novel alternate method for wastewater treatment by using a special constructed column treatment plant. Algal species like Ulva sp, Cladophorasp and Chlorella sp used for the treatment of waste water. the treatment method were done in three stages, free cell manner, batch and continuous flow, continuous flow process with using chlorella sp showed better results. the percentage of COD and BOD removal were 52.1% and 50.8% respectively.

Rothemal [35] investigated the probability of coupling a conventional wastewater treatment system with an algal photo bio reactor to remove of phosphate and nitrate from wastewater and production of renewable resource was tested. The study showed whole nutrients were removed.

Amer [3] investigated the growth of Scenedesmus algae, cultivated in a split-column airlift bioreactor under three various light intensities and two values of superficial gas velocities. The obtained results indicate that the growth rate and chlorophyll content of the Scenedesmus algae directly proportional to light intensity and superficial gas velocity.

## 8 Microalgae growth kinetics

In order to estimate growth rates, a series of measurements at various times that allow the calculation of the rate of change in biomass concentration should be available. Cell number must be counted either through a direct method, as through light microscopy with a hemacytometer, or indirectly through biomass concentration as dry weight or optical density, as long as this measurement correlates linearly with the number of cells [2].

In a usual homogenous, batch regime (in a closed system), microalgae will pass through the following. growth phases as illustrated in Figure 2.

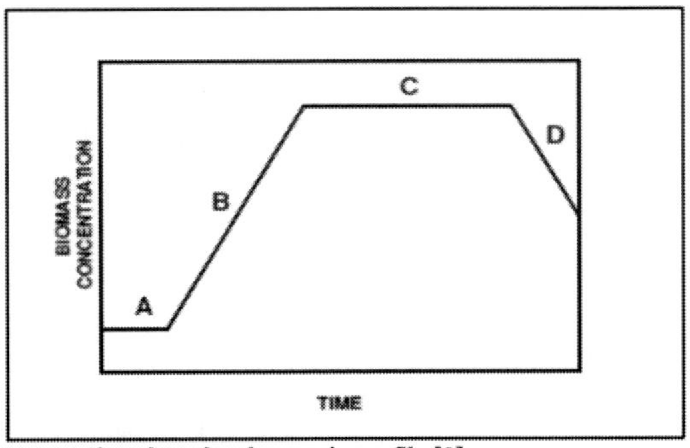


**Fig. 2.** Microalgae batch growth profile [2].

A. Adaptation (lag phase).
B. Exponential growth phase (log phase).
C. Stationary phase.
D. Logarithmic death phase.

The specific stages illustrated in Figure 2 are not always defined in a good manner. Their incline or length may differ according to the situations of culture. For the period of (phase A), the adaptation of microalgae cells to the new environments will accrue. When lag phase end, the cells are well adapted and then begin to increase rapidly, now this is the exponential phase (phase B), where the number of the living cells doubles in a regular tone with time in this phase [5].

Equations (5) and (6) describe cell growth:

$$\frac{dn}{dt} = \mu n \quad or \quad \frac{1}{n}\frac{dn}{dt} = \mu \tag{5}$$

$$n = n_o \quad or \quad t = t_o \tag{6}$$

Where n and $n_0$ are the concentrations of cells at various times $t$ and $t_0$, where:

$\mu$ is the specific growth rates and $t_0$ is referred as the initial time.

The following relation cab be generated from the integrated form of equation (5):

$$ln\frac{n}{n_o} = \mu(t - t_o) \quad or$$

$$\mu = \frac{ln^n/_{n_o}}{t - t_o} \tag{7}$$

From equation (7) it can deduce that the time required for cell population to double ($t_d$) is given by:

$$t_d = \frac{ln^2}{\mu} \tag{8}$$

The state of the cell population can be characterized during the exponential phase using μ or $t_d$.

Media of fresh nutrients In open systems was supplementary in at the same ratio that medium is thoughtful from the culture; this saves the constant level of both growth parameters and cell concentration. Mass balance is applicable to any component of the System:

*rate of addition to system − rate of removal from system + rate of production within system = 0* (9)

The above can, be reformulated as:

$$fx_o - fx + V r_x = 0 \tag{10}$$

Where;

$f$ = volumetric. flow rate.

$x_o$ = the concentration of biomass or cell in the feed stream.

$x$ = the cell concentrations in the system.

V = the cultures volume.

$r_x$ = the rate of cells formation.

If it is supposed that the growth rates is much higher than cell death rate and cell death is, neglected, if μ denotes the specific growth rate ($r_x/x$), then it, can be substituted in equation (10) to obtain the following [7]:

$$Dx_o = (D - \mu)x \tag{11}$$

 https://doi.org/10.1051/matecconf/201816205009

The ratio of the volumetric, flow rate, at which fresh medium is added to the culture, is referred to as the dilution rate $D$, so that:

$$D = f/V \tag{12}$$

$D$ = dilution rate (time $^{-1}$)
$f$ = flow rate
$V$ = reactor volume

It could be seen that, the dilution rates were inverse of residence time. Thus, increase the dilution rate increases the flow rate, and inversely decreases the residence time.

equation (13) and (14) are satisfied represented the the steady state of the system when the concentration of algae cell remain constant in the system [6].

$$\frac{dx}{dt} = 0 \tag{13}$$

$$\mu = D \tag{14}$$

When $D > \mu$ the nutrients are flowing through the reactor faster than the algae can divided in other, words. The organisms are being washed out of the reactors and only those cells attached, to a surface will stay and the cells cannot, grow fast enough to reach steady-state, washout will occur. as shown in figure (3).

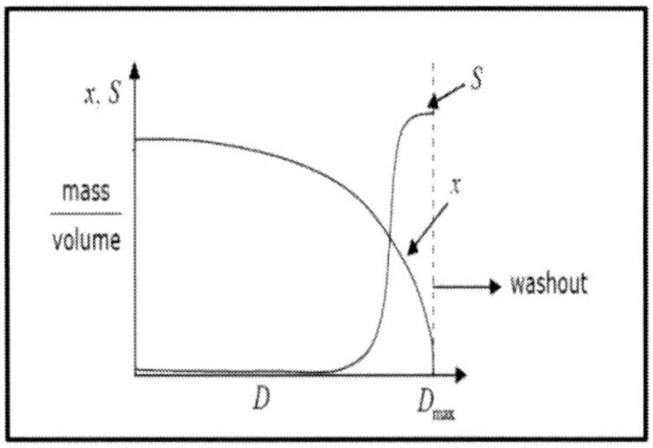


**Fig. 3.** Biomass / volume versus dilution rate beyond the maximum dilution rate, washout occurs. satisfied [6].

## 9 Mechanisms for microalgae degradation of pollutants

Microalgae and cyanobacteria are able to rapid uptake of nutrients and $CO_2$ and possess quicker cell growth. They have been studied for their capability of degrading wide ranges of organic chemicals including pesticides. In the case, of some microalgae, the presence of molecular oxygen is indispensable. to initiates the enzymatic attacked on the aromatic ring and biodegradation of phenols only occurred under aerobic conditions. A typical pathway for metabolizing phenol is to hydroxylate the rings to form catechol that can be degraded through ortho or meta oxidationtion [36]. The reaction will be additional explained in "Enzymatic aspects. of the degradation processes by microalgae and, cyanobacteria and illustrated in the Figure (4).

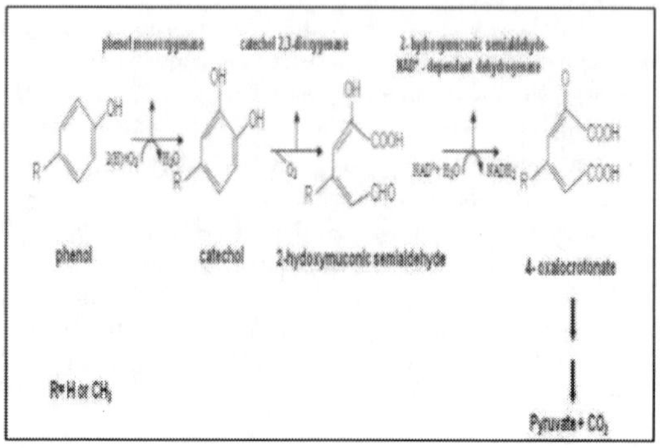

**Fig. 4.** Phenol biodegradation by meta cleavage pathway, described in the microalgae Ochromonasdanica [36].

## 10 Conclusions

In the current study, biodegradation of phenolic contaminate presented in wastewater by microalgae had been examined. The significant impact of microalgae in treating contaminated wastewater had been shown exploring the dominant factors that affected the biomass growth. Culturing techniques were further investigated to show the present development in such field. Different reactors set up like air left reactor were investigated to highlight the progress and benefits of culturing process under specific conditions. It had been noticed that studying microorganism kinetics and mechanisms possesses a great benefit in understanding the whole process leading to more development in the current field.

## References

1. Agarry, S. E., Durojaiye, A. O., Yusuf, R. O., and Aremu, M. O. International Journal of Environment and Pollution, **Vol.32**, Pages:3–11. (2008)
2. Andersen R. AAlgal culturing techniques. 1st Edition Elsevier/Academic Press, Burlington, Mass. pp. 578. (2005).
3. Amer D. ZmatAlbdiri, Al-Qadisiyah" Journal For Engineering Sciences; **Vol. 8**, 410-421 (2015).
4. Bailey, J.E. and D.F. Ollis. *Biochemical engineering fundamentals*. McGraw-Hill, Montreal (1977).
5. Becker, E.W. *Microalgae Biotechnology and Microbiology*. Cambridge University Press, New York. (1994).
6. Bryan Daniel Brooker, *Investigation of microalgae growth kinetics using coal-fired flue gas as a carbon*. A Thesis Presented to The Faculty of California Polytechnic State University, San Luis

Obispo. In Partial Fulfillment of the Requirements for the Degree Master of Science in Engineering, Specialization in Bioengineering. (2011).

7. Bailey, J.E. and D.F. Ollis. *Biochemical engineering fundamentals*. McGraw-Hill, Montreal. ,(1977).
8. Becker, E.W., *Microalgae Biotechnology and Microbiology*, Cambridge University Press, New York. (1994).
9. Chisit . Y and Moo.Young. M. Jchm Tech Biotech, **Vol.42**, 211-219 (1988).
10. Carvalho, P. A., Silva, O. S., Baptista, M. Jo., Malcata, F. X; **Vol.89**, 1275 –1288 (2011).
11. D. Samanthakamani and N. Thangaraju. Ind. J. Sci. Res. and Tech. **Vol. 3**, 9-12 (2015).
12. Deviram GVNS, Pradeep K. V and R GyanaPrasuna.European Journal of Experimental Biology; **Vol. 1**, 216-222 (2011).
13. El-Naas, M., Al-Zuhair, S., and Makhlouf, S., Journal of Industrial and Engineering Chemistry; **Vol.16**, 267–272 (2010).
14. Feng, Y., C. Li and D. Zhang, Bioresour. Technol. **Vol.102,** 101-105 (2011).
15. Hulatt, C.J. and D.N. Thomas, "Energy efficiency of an outdoor microalga lphotobioreactor sited at mid-temperate latitude," Bioresour. Technol; **Vol.102**, 6687-6695 (2011).
16. Huang, G.H., Chen, F., Wei, D., Zhang, X.W., Chen, G., Appl. Energy; **Vol.87**,38–46(2010).
17. Henrard, A.A., M.G. de Morais and J.A.V. Costa,. Technol;**Vol. 102**, 4897-4900 (2011).
18. Ishihara. K.and Nakajima. N. J. Mo Catal B-Enzym; **Vol. 23**, 411–417(2003).
19. Juang, R. S., and Tsai, S. Y. Biochemist Eng. J.; **Vol.31**, 133–140(2006).
20. Janssen, M., Tramper, J., Mur, L.R., Wijffels, R.H., Biotechnology. Boeing; **Vol.81**,193-210 (2003).
21. Kalin M., Wheeler W.N., Meinrath G., J. Environ. Radioactivity; **Vol.78**, 151– 177 (2005).
22. Kumar, A., Kumar, S., and Kumar, S. Biodegradation kinetics of phenol and catechol using Pseudomonas putrid. MTCC 1194 Biochemist Eng. J; **Vol.22**, 151–159(2005).
23. Kanhaiya Kumar a, Chitralekha Nag Dasgupta a, Bikram Nayak a, Peter Lindblad b, and Debabrata Das Bio resource Technology; **Vol.102** ,4945–4953. Elsevier Ltd (2011).
24. Lika, K., Papadakis, I. A. J. Sea Res. **Vol.62**, 135–146 (2009).
25. Lam, M.K. and K.T. Lee, "Potential of using organic fertilizer to cultivate Chlorella vulgaris for biodiesel production", Appl. Energy; **Vol.94**, 303-308(2012).
26. Lee, J.S., Lee, J.P., Biotechnology Bioprocess E. **Vol.8**, 354-259 (2003).
27. Luo Hu-Ping, Muthanna H. Al-Dahhan, Biotechnology and Bioengineering; **Vol. 85**, (2004).
28. Pérez-Garcia, O.; Escalante, F.M.E.; de-Bashan, L.E.; and Bashan, Y. Water Res; **Vol.45**, 11-36. (2011).
29. Pinto, G.; Pollio, A.; Previtera, L.; Stanzione, M. and Temussi, F.Journal of Biotechnology; **Vol.25**,1657-1659 (2003).
30. Petersen E. and Margaritis A. Crit. Rev. Biotechnol., **Vol.21**(4), 233-294 (2001).
31. Ruiz-Ordaz, N., Ruiz-Lagunez, J. C., Casta˜n´on-Gonz´alez, J. H., Hern´andez- Manzano, E., Cristiani-Urbina, E., and Gal´ındez-Mayer, J. Rev. Latino am. Microbial; **Vol.43**, 19–25 (2001).
32. Romera, E., Gonzalez, F., Ballester, A., Blazquez, M.J., Crit. Rev. Biotechnology; **Vol.26**, 223-235(2006).
33. Radwan, R.H. Al-Hasan, H.M. Mahmoud and M. EliyasJournal of applied miocrobiology; **Vol.103** (2007).
34. Rothermel, Monica Christine, *Coupling the Wastewater Treatment Process With An Algal Photo bioreactor For Nutrient Removal And Renewable Resource Production*, Master's Thesis, University of Pittsburgh (2011).
35. Semple, K. T., & Cain, R. B. Applied and Environmental Microbiology; **Vol.62**, 1265–1273 (1996).
36. Semple, K.T.; Cpain, RB.; FEMS Microbial Lett. **Vol.170**, Pages: 291-300 (1999).
37. ShashirekhaS,Uma L. Subramanian G. J. Ind Microbiol Biotechnol ;**Vol.19** pages:130-133 (1997).
38. Su, C.M., H.T. Hsueh, H.H. Chen and H. Chu, Chemosphere. **Vol.88**, 706-711 (2012).
39. Sudarsan.J.S1 , Renganathan.K2 , and Ann Christy. international journal of environmental sciences; **Vol.2**, No. 1 (2011).
40. Tang, H., M. Chen, K.Y. Simon Ng and S.O. Salley. Biotechnol. Bioeng j. (2012).
41. Van Langenhove H, Kumar A, Ergas S, Yuan X, Sahu A, Zhang Q. [Internet]. 1st ed (2014).
42. Ugwu, C.U., Aoyagi, H., Uchiyama, H. Bio resource Technology; **Vol.99**(10) p. 4021–8. (2008).
43. Wang L, Min M, Li Y., and Chen P., Chen Y. Cultivation of green algae chlorella sp. in different wastewaters from municipal wastewater treatment plant. Appl Biochemist; **Vol.**162, 1174-1186 (2010).

 https://doi.org/10.1051/matecconf/201816205009

**Table 2.** Different tubular photobioreactor systems

| Tubular Photo bioreactor | Algae | Approximate capacity volume [l] | T[°c] | pH | Circulating mechanism | Light supply method and illumination flux | $CO_2$ Supply |
|---|---|---|---|---|---|---|---|
| Indoor | Spirulina sp. Scenedesmus obliquus | - | 30 | Determined by a digital pH-meter | Agitation and aeration using air from a compressor and a sintered sparger | A 12h dark/light photo period with 3200 lux of illumination provided by 40 W daylight type fluorescent lamp | $CO_2$ added to the air at a rate of 0.3 wm for 15 min every 2h during the 12h light period |
| | Chicrell asorckiniana | 58 | Controlled by sprinking the reactor surface with tap water | | | Natural illumination by solar light energy | $CO_2$ added to the air a rate of 0.25 wm |
| Outdoor: consisting of 4m tall airlift section with a degassed zone | P. tricornutum | 200 | | | Airlift device | Solar illumination | |
| Semi-continuous | Chlorlla | | 26±1 | Directly determined by an ISFET pH – meter KS723 | | Continuous cool white fluorescent illumination at the intensity of 300 μmol $m^{-2}s^{-1}$ measured at the photobioreactor using a Basic Quantum Meter | Air of different $CO_2$ concentration produce mixing air and pure $CO_2$ at rate of 0.25wm |
| Channel chambers later connected by means of horizontal baffles attached alternately to the front and the back of the larger flat faces of the reactor | [illegible]rella [illegible]aris | 3 | Controlled by circulating cooling water through a transparent jacket located at the front of the reactor | 6.8±1 measured by a sensor placed in the upper part of the reactor and controlled by injecting $co_2$ by demand | By a peristaltic pump | Illumination from the both side of the tubes with 4 OsramFluora lamps at the intensity of 120 $\mu Em^{-2}s^{-1}$ | $Co_2$ added to the air a rate of 0.45 wm |

# Turbidity and oil removal from oilfield produced water, middle oil company by electrocoagulation technique

*Thamer* Mohammed[1], *Esraa* Abbas[1] and *Thabit* Ahmed[1]

[1]Chemical Engineering Department, University of Technology, Baghdad, Iraq

**Abstract.** Huge quantity of produced water is salty water trapped in the oil wells rock and brought up along with oil or gas during production. It usually contains hydrocarbons as oil and suspended solids or turbidity. Therefore the aim of this study is to treat produced water before being discharge to surface water or re_injected in oil wells. In this paper experimental results were investigated on treating produced water (which is obtained from Middle Oil Company-Iraq), through electrocoagulation (EC). The performance of EC was investigated for reduction of turbidity and oil content up to allowable limit. Effect of different parameters were studied; (pH, current density, distance between two electrodes, and electrolysis time). The experimental runs carried out by an electrocoagulation unit was assembled and installed in the lab and the reactor _was made of a material Perspex, with a capacity of approximately 2.5 liters and dimensions were 20 cm in length, 14 cm in width and 16 cm height. The electrodes employed were made of commercial materials. The anode was a perforated aluminum rectangular plate with a thickness of 1.72 mm, a height of 60 mm and length of 140 mm and the cathode was a mesh iron. The current was used in the unit with different densities to test the turbidity removing efficiency (0.0025, 0.00633, 0.01266 and 0.0253 A/cm2).The experiment showed that the best turbidity removing was (10, 9.7, 9.2, 18 NTU) respectively. The distance between the electrodes of the unit was 3cm. The present turbidity removing was 92.33%. A slight improvement of turbidity removing was shown when the distance between the electrodes was_changed from 0.5 to 3 cm with fixation of current density. The best turbidity removing was 93.5% , (7.79 NTU) when the distance between the electrodes were 1 cm. The experimental results found that concentration of oil had decreased to (10.7, 11.2, 11.7, 12.3) mg/l when different current densities (0.00253, 0.00633, 0.01266, 0.0253) A/cm2 were used, respectively with the distance between the electrodes was 3 cm. The best result of oil content decreasing was 10.7 mg/l with current density 0.0253 A/cm2. These results are within allowable limit to provide the possibility of reuse the water and can be injected in oil wells

© The Authors, published by EDP Sciences. This is an open access article distributed under the terms of the Creative Commons Attribution License 4.0 (http://creativecommons.org/licenses/by/4.0/).

## 1 Introduction

During the process of producing oil or gas from underground formations, trapped water is brought to surface along with oil or gas. This water is known as produced water in upstream oil and gas operations. On a global spectrum, it is found that 3 barrels of water produced for every barrel of crude oil. The ratio of water to crude oil is about 10:1 in North America, [1]. Lee et al., [2] found that for every 1 billion of oil, 7 bbl of water is produced. Therefore treatment and disposal of produced water is becoming a leading economic factor in viability assessment of the asset and it discharges within allowable limit of environmental regulations. This major problem in the 21st century effects on quality of water because of increasing pollution from point and non-point sources and become a crucial problem, particularly for the Third-World Countries, [3]. The promising technical methods based on electrochemicals are electrocoagulation, electro flotation [4], electro decantation, and others [4-6].

But in recent years the treatment by electrocoagulation (EC) has been widely used for the treatment of wastewater having oil wastes, turbidity or suspended solids [7-8], organic matter from landfill leachate [9], heavy metals [10], phenolic waste. This technique has multi benefits [5]: such as compatibility, amenability to automation, cost effectiveness energy efficiency, without use coagulants, safety, and versatility. This process has capability to overcome the disadvantages of other treatment methods [1]. The literature surveys indicated that EC has the capability of removing turbidity and most of the water contaminant in oily wastewater and produced water [11-13]. Electrocoagulation (EC) is a complicated process involving many chemical and physical phenomena that use sacrificial electrodes, such as Al, Fe and others, to supply ions into the water [14]. In the EC process, the coagulants are generated in situ by dissolving electrically the consumable electrodes (Fe/Al). The metal ions generation takes place at the anode;

hydrogen gas is released from the cathode [15]. In the solution, the metal cations resulted from the anode oxidation combine with hydroxyl ion ($OH^-$) resulting from water to form highly charged coagulant. In the case of Aluminum anode, the $Al^{3+}$reacts with $H_2O$ to form $Al(OH)_3$, and in the case of magnesium anode, the $Mg^{2+}$reacts with $H_2O$ to form $Mg(OH)_2$. The basic process can be summarized in Fig. 1. The aim of the present work is to remove suspended solid and oil from produced water which is obtained from Middle Oil Company by EC and study different parameters effect on removal efficiency.

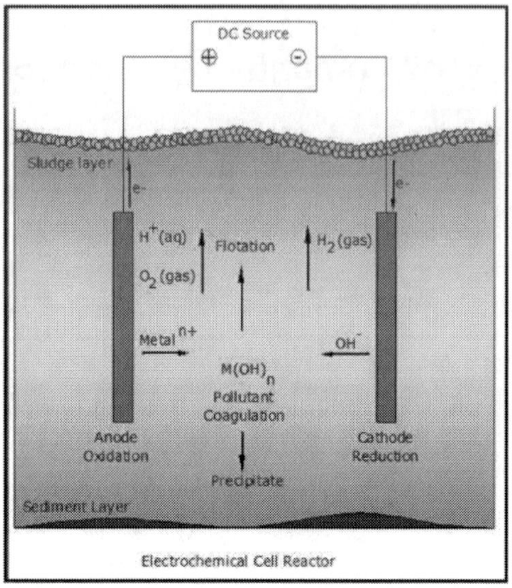


Fig. 1 Interactions occurring within an electrochemical reactor

## 2 Experiment

### 2.1 Characteristics of oilfield produced water

Oilfield produced water used in the present study was kindly provided by Petroleum Research and Development Center staff from Middle Oil Company.

Its chemical and physical characteristics include pH around (7), EC (144300μs/cm), Turbidity (120 NTU), oil content (46.6 mg/l), TDS (133477mg/l) and TSS (90mg/l).

### 2.2 Materials and methods

Monopolar aluminum and iron electrodes were used in the electro coagulation cell to form coagulants for the purpose of oil and turbidity removing from produced water. Experimental work was conducted at the laboratories of the Chemical Engineering Department of Technology University in Baghdad._The electrodes used were made with commercial materials. The anode was a perforated aluminum rectangular plate with a thickness of 1.72 mm, a height of 60 mm and length of 140 mm; these were perforated in order to increase the contact surface area, which increases the electron transfer. The cathode was constructed with iron mesh, which provides a large area for the reaction. For the preliminary study of the process variables (applied current, distance between the electrodes and electrolysis time) were taken.. The reactor is made of a material Perspex, with a capacity of approximately 2.5 liters. The dimensions were 20 cm in length, 14 cm in width and 16 cm height shown in Fig. 2. The electrodes used in this cell were variable distance.

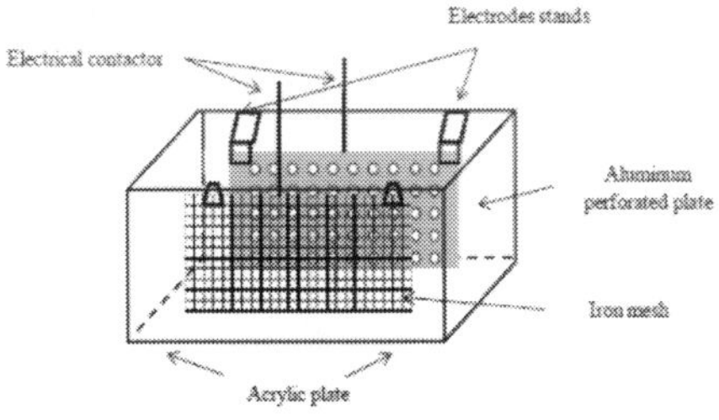


Fig.2 : Electrocoagulation cell

Determination of the effect of current density on the reaction of electro- coagulation was carried out in a parallel-plate cell. In these experiments, 2 L of the sample produced water were transferred to the electrochemical cell with parallel plates as shown in Fig. 3. Once the water to be treated was transferred, the magnetic stirrer was at slow speeds), the electrodes were connected to the power supply by which the work circuit current (ranged **0.2 A - 0.5 A – 1 A -2 A**) was established. The initial turbidity was determined, and the system was then allowed to react, measuring turbidity at different time intervals. For experiments with different currents, the distance between electrodes was kept fixed. Once the experiment was completed, the procedure described above was repeated but with a change in the distance between electrodes ranged (0.5 cm, 1cm, 2cm, 3cm).

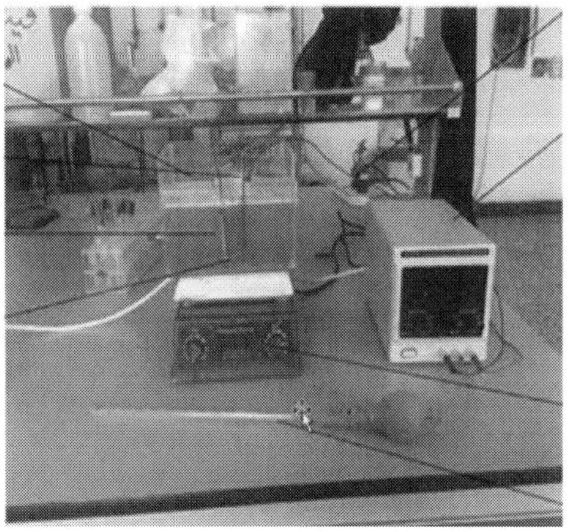

Fig. 3 : photographic picture of EC cell at Chemical. Eng. Labs, Technology Unv , Baghdad.

## 3 Results and discussion

### 3.1 Effect of current density on the turbidity

Four current densities (**I=0.00253, 0.00633, 0.01266, and 0.0253 A/cm2**) were applied to examine the effect of current on the electrocoagulation treatment of produced water. The effect of current density on the turbidity removal efficiencies is depicted in Fig. 4 to 7.The removal efficiency increases with the applied current density. The removal efficiencies of turbidity were 91.6, 91.9, 92.3 and 85% with residual turbidity (10, 9.7, 9.2, and 18) of current densities, respectively. It has been found that the best value is where the current density (**0.01266** A/cm$^2$) which gives the best results for removal efficiency (92.3%). The effectiveness of current density on increasing removal efficiency slows down at higher current densities. The rate of anodic dissolution of aluminum is increased at higher current densities, resulting in a greater amount of coagulant and precipitant production. Consequently, this results in a higher removal efficiency of turbidity. Bubble size decreases, while bubble generation rate increases with current density, which results in higher removal efficiency of turbidity via H2 flotation, in addition to this the effect of coagulation were agreed with this result of [16].

https://doi.org/10.1051/matecconf/201816205010

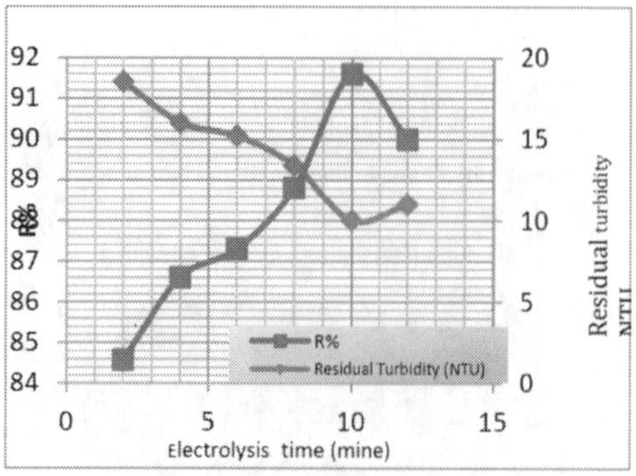


Fig. 4: Effect of the current density (0.00253 A/cm$^2$) on the turbidity

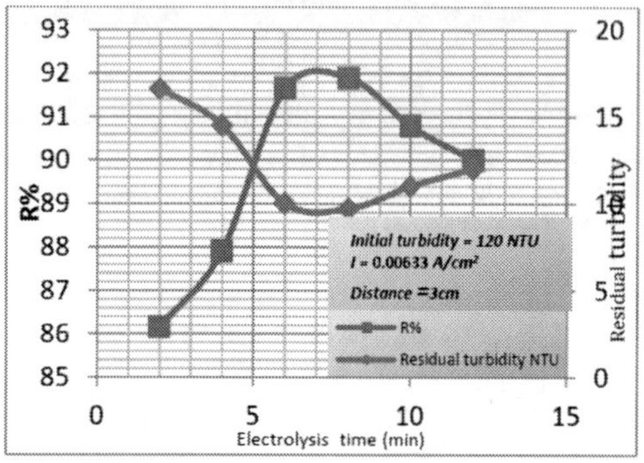


Fig. 5: Effect of the current density (0.00633 A/cm$^2$) on the turbidity

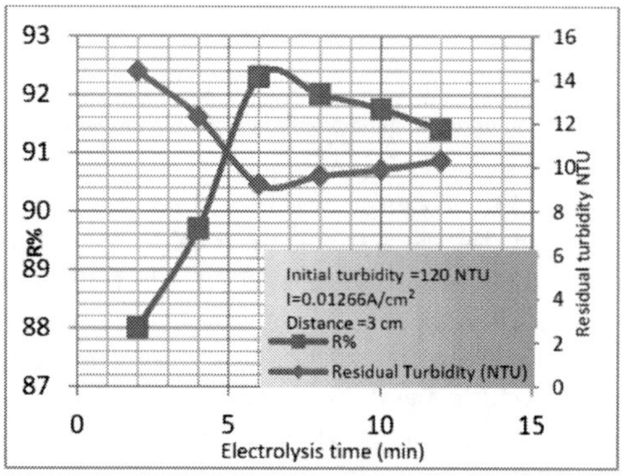


Fig. 6: Effect of the current density (0.01266 A/cm$^{2}$) on the turbidity

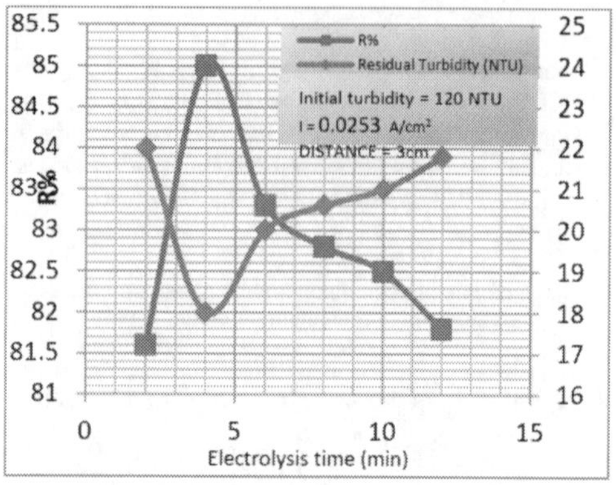


Fig. 7: Effect of the current density (0.0253 A/cm$^{2}$) on the turbidity

## 3.2 Effect of distance between the electrodes on the turbidity

Four experiments were performed under fixed operational condition (pH = 6.98, current density **I** = **(0.01266 A/cm2**, and electrolysis time (2-12 minutes) and altering distances (0.5, 1, 2 and 3 cm) between electrodes. The results show in Fig. 8 that the turbidity decreased slightly when the distance was increased from **0.5** to **1** cm (from 7.9 to 7.79) NTU, but the turbidity increased to 9 NTU when the distance between electrodes was kept 2 and 3 cm. Thus, the optimal distance was shown to be approximately 1 cm. This can be explained that decreasing the space between electrodes results in low resistance through the solution which in turn results in increasing the rate of Aluminum dissolution and $Al^{+3}$ releases and consequently leads to more turbidity removal from the solution. On the other hand, decreasing the space could enhance the flotation process by limiting the generated bubbles in a narrow space which results in higher removal efficiencies.

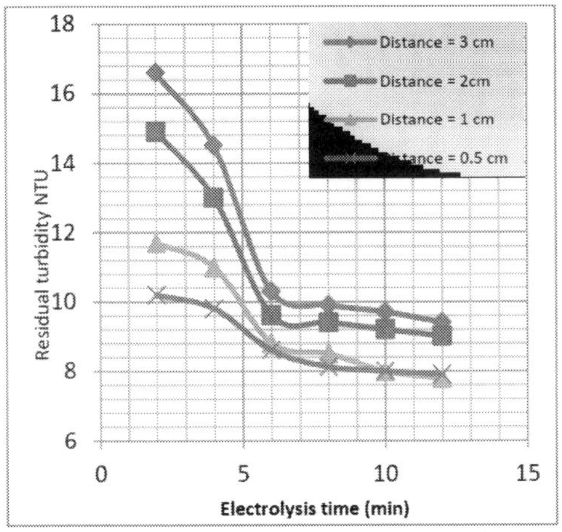


Fig. 8 Effect of distance between electrodes on the turbidity ( t 2-12) minutes, pH = 6.98, $I$ = (0.01266 A/cm$^2$)

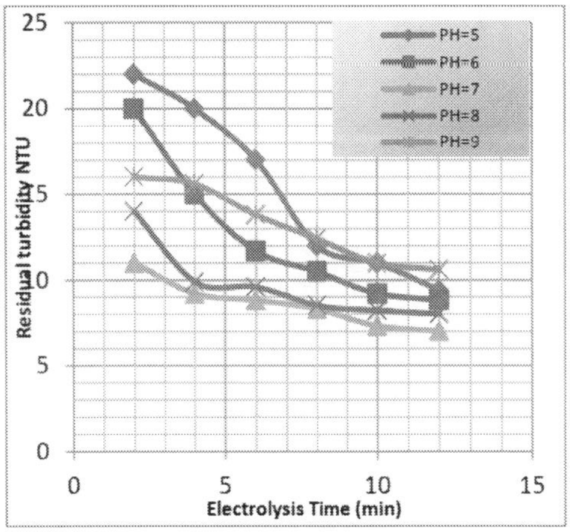


Fig. 9 Effect of initial pH on the turbidity (t 2-12) minutes distance = 1 cm, $I$ = (0.01266A/cm$^2$). Content Distance = 3 cm and initial oil content 46.6 mg/l

### 3.3 Effect of Initial pH

Aluminum ions may be found in different forms and phases, depending on the pH and chemical characteristics of the solution. At pH values less than 4, aluminum ions that are released from electrodes can be found in the form of **$Al(H_2O)_6^{+3}$**, while at a pH range of 5 to 6, aluminum m the pH rises to higher values (5 to 8), aluminum changes to the form Al $(OH)_3$. pH values greater than 8.8 may cause dissolution of aluminum as ions again. Based on the effect of pH on the chemical form of aluminum, it is expected that electro coagulation efficiency would be dependent on pH, to some extent. Five experiments at initial pH values of 5, 6, 7, 8 and 9 were performed to investigate this effect. The results presented in Fig.9 shows that the optimal residual turbidity is achieved at pH value of 7.

### 3.4 Effect of the current density on the Oil Content Distance = 3 cm and initial oil content 46.6 mg/l

The results shown in Fig 10 depict the effect of current density on the oil content removal .It was found that oil content removal increases with increasing current density. It was found that the best removal achieved when the current was (**0.0253** A/cm2) where the oil content reached (10.7 mg/l) after 10 min. The higher (**0.0253** A/cm2) current density has no significant effect on the removal efficiency of oil. This is ascribed to the fact that at higher current densities, the dissolution of anode electrode increases according to Faraday's law; the resulting metal hydroxides produce more sludge.

### 3.5 Effect of distance between the electrodes on the oil content

Four experiments performed under fixed operational condition (pH = 6.98, current density I = 0.0253 A/cm2, electrolysis time (2-12 minutes) and altering distances (0.5, 1, 2 and 3 cm) between electrodes. The results shown in Fig 11 display that the oil content decreased slightly with the distance from 0.5 to 10.2 mg/l. Thus, the best distance was shown to be approximately 0.5 cm.

 https://doi.org/10.1051/mateconf/201816205010

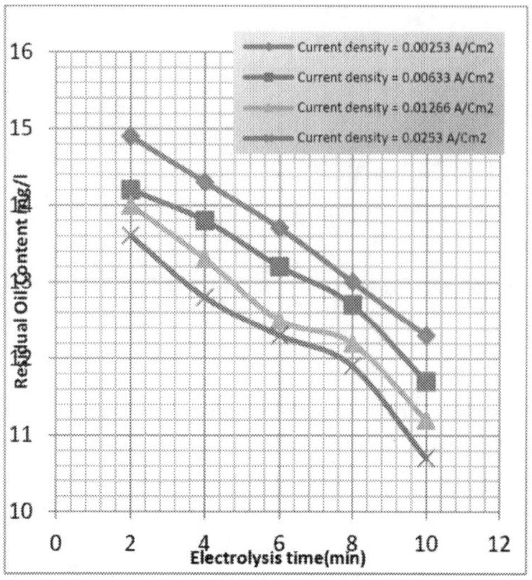


Fig. 10 Effect of current density on the Oil Content Distance = 3 cm and Initial oil content = 46.6 mg/l (pH = 6.98)

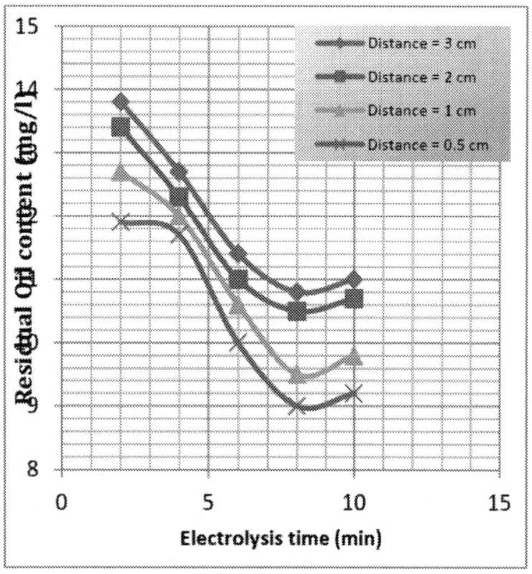


Fig. 11 Effect of distance between the electrodes on the oil content at (I=0.0253 A/cm2) (t=2-10 min)

## 4 Comparisons between experimental and theoretical sacrificial Aluminum anode consumption

A comparison between experimental and theoretical sacrificial aluminum anode consumption has been achieved by calculating the theoretical Al loss from the anode using Faraday's equation and compared with those obtained experimentally [17].

$$w = \frac{itM}{Nf}, \qquad \text{Faraday's law}$$

$N$ = 3 is the number of electrons corresponding to aluminum oxidation, $M$ *is* the molecular weight (mg/mol) and $f$ is Faraday's constant (96 500 C/mole), i=current (A), t= time (sec), and the second value is determined from the aluminum anode consumption by weighing the electrode before and after experiment. The results shown in Fig 12 and Table (1.1) describe the amount of Al generated from anode always exceeded by those calculated theoretically from Faraday's equation. The reason for this may be due to the fact that Faraday's law does not take into account the impact of salt in PW on Metal dissolution while the salt has a strong effect on the rate of dissolution by increasing the conductivity of the solution.

## 5 Conclusions

**a.** In Electrocoagulation method, by using different current densities (**0.00253, 0.00633, 0.01266,** and **0.0253 A/cm2**) it is found that the best current density is (0.01266 **A/cm2**) to reduce turbidity removal efficiency and residual turbidity up to (92.3%), (9.2 NTU), respectively at constant distance between electrodes 3cm.

**b.** Experiments carried out using different distances between electrodes (0.5, 1, 2, 3 cm) showed that the best distance to reduce removal turbidity up to (7.79NTU) is 1cm.

**c.** The experimental results observed the effect of initial pH on turbidity removal efficiency with electrocoagulation treatment. The best value is PH =7 to obtain turbidity removal efficiency (94.1%).

**d.** In the electrocoagulation methods, the oil content could be decrease to (10.8, 10.5, 9.5, 9) mg/L with different distances between electrodes at constant current (**0.0253A/cm2).**

Table (1.1):- Experimental and Theoretical Sacrificial Aluminum Anode Consumption

| Current density A | Al consumption based on Faraday's law (gm Al) | Al consumption (gm) experimental |
|---|---|---|
| 0.2 | 0.0111 | 0.01566 |
| 0.5 | 0.0279 | 0.039 |
| 1 | 0.0559 | 0.071 |
| 2 | 0.1119 | 0.17948 |

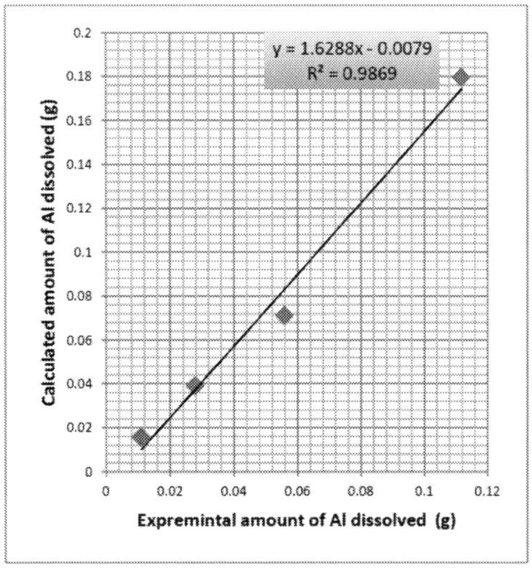


Fig. 12 : Measured Vs. calculated amount of Al released from the anode

## References


1. BakerCorp "Treatment Of Produced Water By Electrocoagulation" water treatment solution.(2014).
2. Lee, R., Seright, R., Hightower, M., Sattler, A., Cather, M., Mcpherson, B., Wrotenbery, L., Martin, D., and Whiteworth, M., "Strateries for Produced Water Handling in New Mexico" Conference, Colorado Springs, Co. 16-17, Oct. (2000).
3. Satish, I., Chaturredi, "Electrocoagulation; A Novel Wastewater Treatment Method" I. JMER, Vol.3, Issue, 1, Jan-Feb. PP. 93-100, (2013).
4. Rajeshwar, K., Ibanez, J.G. and Swain, G.M., " Electrochemistry and the environment",J. APL. Electrochem., 24, 1077, (1994).
5. Mollah,M.Y.A., Schennach, R., Parga, J.R. and Cocke, D.L, " Electrocoagulation (EC)--science and applications.," J. Hazzard Matter, 84, 29-41, (2001).
6. M. Y. A. Mollah, P. Morkovsky, J. A. G. Gomes, M. Kesmez, J. Parga, and D. L. Cocke, "Fundamentals, present and future perspectives of electrocoagulation," Journal of Hazardous Materials, vol. 114, no. 1–3, pp. 199–210, (2004).
7. Donini, J.C.; Kan, J.; Szynkarczuk, J.; Hassan, T.A.; Kar, K.L. Operating cost of electrocoagulation. *Can. J. Chem. Eng.*, 72 (6): 1007.. (1994).
8. Abuzaid, N.S., Bukhari, A.A., Al-Hamouz, Z.M., ". Removal of bentonite causing turbidity by electrocoagula-tion". J. Environ. Sci. Health, Part A A33, 1341–1358, (1998).
9. CT Tsai, ST Lin, YC Shue, PL Su "Electrolysis of soluble organic matter in leachate from landfills" Water Res., 31, 3073, (1997).
10. V.L. Pogrebnaya, A.A. Klimenko, T.N. Bokovikova, E.P. Tsymbal, N.P. Pronina, "Purification of waste water of heavy metals by electrocoagulation", Chem. Petrol. Eng. 31 , 280, (1995).
11. Maria Elektorowicz, Shiva Habibi, Rozalia Chifrina , " Effect of electrical potential on the electro-demulsification of oily sludge" Journal of Colloid and Interface Science 295 535–541, (20006).
12. McBeth, I., Reddy, K.J., Skinner, Q.D. "Chemistry of Trace Elements in Coalbed Methane Produced Water" Water Research, 27,884-890, (2003).
13. Tir, M., Moulal-Mostefa, N., "Optimization of Oil Removal from Oily Wastewater by EC Using RSM" J. of Hazards Material, 158, 107-115, (2008).
14. *Liu*, H., Zhao, X., and Qu, J., "Electrocoagulation in Water Treatment." From: *Electrochemistry for the Environment*, Chapter 10. Comninellis, C and Chen, G. (eds.), Springer.(2010).
15. *Lawrence K.*, Yung Tse, Nazih K.,,"Advanced Physicochemical Treatment Technologies", Humana Press, Totowa, New Jersey, p 64-65. (2007).
16. Song, S.; Yao, J.; He, Z.; Qiu, J.; Chen, J. Effect of Operational Parameters on the Decolorization of C.I. Reactive Blue 19 in Aqueous Solution by Ozone-Enhanced Electrocoagulation. *J. Hazard. Mater.* 152, 204-210. (2008).
17. 8- Christos C. and G. Chen, , "Electrochemistry for the Environment, Springer", New York, pp. 245-262., (2010).

 https://doi.org/10.1051/matecconf/201816205011

# Synthesis of antibacterial additive for metal working fluids application

*Ezzat* Hamed[1], *Nagy* Saker[2], *Shawky* ElShazly[1], *Tarek* Fahmy[3], and *Yasser* Aboulazm[2]

[1]Faculty of science, Alexandria University, Alexandria, Egypt
[2]Research Center, Misr Petroleum Company, Egypt
[3]Alexandria Specific Petroleum Products Company, Alexandria, Egypt

**Abstract.** Metalworking fluids, the class of lubricants most susceptible to microbial attack, metalworking fluid must also be safe for human use and exposure, The failure of the fluid to perform any of its functions has the potential to result in operational problems, process shutdowns, decreased tool life, and product-quality issues, all of which will result in increased costs. Perhaps one of the most common and controllable complications is microbial degradation specially standing from the fact that metalworking fluids contain the nutrients that can permit unchecked microbial growth. In this work we prepared Antimicrobial metal working fluid additive which fulfills the criteria of successful local production in Egypt, as the commercial availability of starting components, effectiveness of the prepared additive, easiness of addition and homogenization with other MWF additives, good environmental profile and biodegradability. The effectiveness of the prepared additive was evaluated by standard method ASTM E645-97 test. It was revealed that; the formic acid which condensate with the polyethylene glycol 400 and ethylene glycol had excellent antibacterial action, it could also greatly reduce growth of bacteria.

## 1 Introduction

Microbial contamination may cause some health problems for workers such as skin irritation and dermatitis, and respiratory problems. Hypersensitivity pneumonitis (HP), the most severe condition, has been attributed to the presence of Mycobacterium immunogenum in contaminated fluid mists [1]. Endotoxins, which are cell wall fragments of gram-negative bacteria that may be found in metalworking fluids, may cause respiratory impairment, and mycotoxins, produced by fungi in contaminated fluids. These situations continue to be explored; Passman [2] has very recently provided a detailed discussion of worker health issues. The American Society for Testing and Materials (ASTM) Subcommittee E34.50 on Health and Safety Standards for Metal Working Fluids have produced standard test methods for enumeration of mycobacteria (E2564-07) [3] and determination of endotoxin concentration (E2250-02) [4] in metalworking fluids.

Formaldehyde-condensate biocides exhibit a broad spectrum of antimicrobial activity; they are by far the most popular and proven biocide chemistries for metalworking fluid applications. They control microbiological growth through their ability to generate formaldehyde in situ. Its antimicrobial activity is derived from an electrophilic active group that reacts with nucleophilic cell sites. These targets include amino acids or proteins, which are often important components of enzymes or other functional proteins critical to cell function [5, 6].

Our aim of the work was to prepare effective, local market available and affordable components, with good environmental profile.

In order to achieve that esterification reaction between the glycols(s) and formic acid using para-toluene sulfonic acid as catalyst, Xylene as an azeotropic solvent (as shown in equation 1). The using of Azeotropes to remove water and removing it, drives the equilibrium in favor of the ester product. [7]

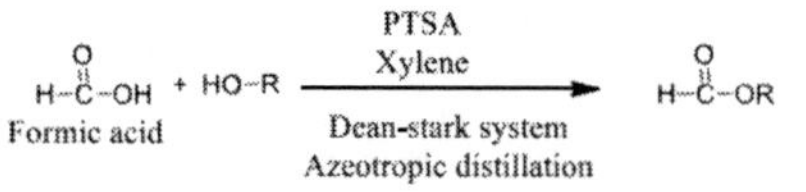

(1)

Then the obtained esters were tested using standard method ASTM E645-97 test, showed excellent antibacterial results for both condensates.

## 2 Materials and methods

### 2.1 Materials

© The Authors, published by EDP Sciences. This is an open access article distributed under the terms of the Creative Commons Attribution License 4.0 (http://creativecommons.org/licenses/by/4.0/).

The IR spectra were recorded using Perkin-Elemer FT-IR spectrophotometer. Reactions were routinely followed by thin layer chromatography (TLC) using Merck Kiesel gel using eluent n-Hexane (8): Ethyl acetate (2); 60F254 pre-coated plastic plates and the spots were detected by UV light absorption, the antimicrobial studies were performed at the microbiological laboratory, Misr Petroleum Company, Egypt. Chemicals were obtained from the local market.

### 2.2 Reaction method

In a two necked round bottom flask equipped with Stark-Dean separation adaptor equivalent amount of Ethylene glycol, or Poly ethylene glycol were placed, a thermometer was placed in a neck while the Stark-Dean adaptor was placed into the other neck using Xylene as azeotropic solvent, the system was placed on a hot plate equipped with magnetic stirrer. Catalytic amount of Para Toluene Sulfonic Acid (PTSA) was added (one gram for each hundred mille of reactants). The reaction mixture left to reflux with stirring at temperatures of 90°C, reaction progress was traced by TLC. After completion of reaction noticed from the acidity test and the separated water, the reaction mixture was washed with 20ml of 10% aqueous sodium carbonate and 100 ml water, extracted using Benzene, dehydrated using anhydrous sodium sulphate and left for solvent evaporation. Spectral analysis was used to identify and proof the structure.

2.2.1. *Reaction of Ethylene glycol with Formic acid (compound 1)*: 56.4 ml of Ethylene glycol (1 mol.) with 115.075 ml of Formic acid (2.2 mol.) were refluxed for 7 hours at 90°C to produce ethane-1,2-diyl diformate (1). IR υ $cm^{-1}$: 2954.72 (CH Aliphatic), 1721.96 (C=O).

2.2.2. *Reaction of poly ethylene glycol with Formic acid (compound 2)*:: 44.32 ml of poly ethylene glycol (0.125 mol.) with 10.153 ml of Formic acid (0.1875 mol.) were refluxed for 5 hours at 90°C to produce polyethylene glycol formate (2). IR υ $cm^{-1}$: 2875.21 (CH Aliphatic), 1719.55 (C=O).

2.2.3. *Reaction of Ethylene glycol with acetic anhydride (compound 3)*: 14.05 ml of Ethylene glycol (0.25 mol.) with 35 ml. (excess amount) of acetic anhydride were refluxed for 5 hours at 110-120 °C to produce ethane-1,2-diyl diacetate (3). IR υ $cm^{-1}$: 2961.81 (CH Aliphatic), 1746.42 (C=O).

2.2.4. *Reaction of poly ethylene glycol with acetic anhydride (compound 4)*: 44.326 ml of poly ethylene glycol (0.125 mol.) with 14.318 ml. (1.1 equivalents) of acetic anhydride were refluxed for 8 hours at 160 °C to produce polyethylene glycol acetate (4). IR υ $cm^{-1}$: 2873.85 (CH Aliphatic), 1740.21 (C=O).

## 3 Evaluation tests

We used the following tests to evaluate the action of the prepared additives and also the performance and effect on other final product properties.

2.3.1. *IP 311*: This method covers the determination of the thermal stability of water mix metal working fluids over the range of temperatures at which the fluids would normally be stored. We perform this test to insure that the prepared additive don't affect the thermal stability of the final product after mixing with other components.

2.3.2. *IP 312*: This method assesses the frothing characteristics of water mix metal working fluids used in the form of aqueous dilutions.

2.3.3. *IP 263*: Water mix metal working fluids used in the form of aqueous dilutions should remain stable in use. This method is accordingly designed to assess the stability of water mix metal working fluids used in the form of aqueous dilutions.

2.3.4. *IP 125*: This standard specifies a procedure for the assessment of the corrosion characteristics of water-mix metal working fluids, in the form of aqueous dilutions, when in contact with cast iron. It is applicable to pre-mixed aqueous dilutions or to supply concentrates which are diluted in the manner described in IP 263. Metal working fluids should not permit the corrosion of metals with which they come into contact, and cast iron is chosen as a typical sensitive example of such metals.

2.3.5. *ASTM E645-97*: this test method outlines a procedure for evaluating the efficacy of microbicides (algaecides, bactericides, and fungicides) that will be used for controlling microbial growth in cooling water systems. The microbicides will be evaluated using simulated or real cooling tower water against microbes from cooling water, microbiological deposits (biofilms) from operating cooling systems, or microorganisms known to contaminate cooling water systems, or combination thereof.

## 4 Results

The antibacterial investigation using the ASTM E645-97 method results shown in the following table:

Table 1. The antibacterial analysis ASTM E645-97 results.

| Percentage added | compound 1 | compound 2 | compound 3 | compound 4 |
|---|---|---|---|---|
| 0.1 | Not visible | Growth | Over growth | Over growth |
| 0.2 | Not visible | Growth | Over growth | Over growth |
| 0.3 | Not visible | Growth | Over growth | Over growth |
| 0.4 | Not visible | Growth | Over growth | Over growth |
| 0.5 | Not visible | Slight growth | Over growth | Over growth |
| 1 | - | - | Over growth | Over growth |

https://doi.org/10.1051/matecconf/201816205011

Table 2. The evaluation tests result for effective biocides

| Test | compound 1 | compound 2 |
|---|---|---|
| IP 311 | No separation | No separation |
| IP 312 | Cream, 0.1 | Cream, 0.1 |
| IP 263 | Oil, Nil | Oil, Nil |
| IP 125 | 0/1-1 | 0/1-1 |

## 5 Discussion and conclusions

The esterification reaction has taken place between organic acids (formic) or acid anhydride (acetic anhydride) and glycols (ethylene glycol and/or poly ethylene glycol) using the azeotropic conditions with the Lewis acid (PTSA) gave successfully the corresponding esters as shown in Equation1. The reactions of formic acid shown in equations 2,3 giving the effective biocide compounds 1 and 2. The 2 equivalent molecules of formic acid were used for the condensation with the two hydroxyl groups of the glycol molecule, Para-toluene sulfonic acid was used as Lewis acid for the successful catalysis of the esterification reaction. While the resulted condensate water was separated using xylene as azeotrope in the system equipped with Dean-Stark adaptor.

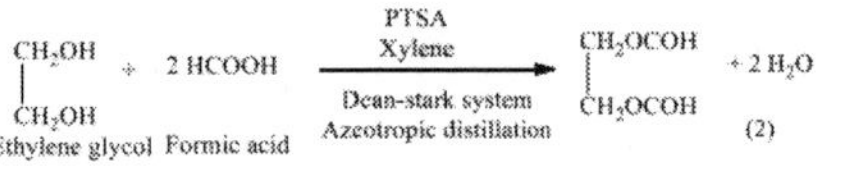

The results stated in Table 1 shows that the reaction between ethylene glycol and formic acid gave the best product (ethane-1,2-diyl diformate) with antibacterial effect, while the same reaction with polyethylene glycol 400 gave less potent antibacterial compound (polyethylene glycol formate).

While preparation of antibacterial compounds using acetic anhydride gave almost negative results with both ethylene glycol (ethane-1,2-diyl diacetate); and polyethylene glycol (polyethylene glycol acetate).

That advance of the formyl condensates may be attributed to the high electrophilic active group that reacts with nucleophilic cell sites of the bacteria affecting the cell activity and wall.

The relative higher antibacterial effect of the mono ethylene glycol condensate over that of the poly ethylene glycol condensate may be attributed to the easy dissociation of the formyl group from the mono ethylene glycol over that of the bulky polyethylene glycol where steric hindrance and molecular size may affect the dissociation ability of the formyl group.

The evaluation tests were performed to the already successful compounds 1 and 2. They showed no effect on the thermal stability (IP 311) or the emulsion stability as shown with the IP 312 and IP 263, respectively. Also IP 125 showed that the prepared compounds don't permit the corrosion of metals.

## References

1. Rossmoore, H.W., L. Rossmoore, D. Bassett. Life and death of mycobacteria in the metalworking environment. *Lubes'N'Greases* 21–27, April (2004).
2. Passman, F.J. Metalworking fluid microbes: What we need to know to successfully understand causeand- effect relationships. *Tribol Trans* **51**(1): 107–117, (2008).
3. ASTM E2564-07, Standard test method for enumeration of mycobacteria in metalworking fluids by direct microscopic counting (DMC) method, *Annual Book of Standards,* **vol. 11.03**, Philadelphia, (2007).
4. ASTM E2250-02, Standard method for determination of endotoxin concentration in water miscible metal working fluids, *Annual Book of Standards*, **vol. 11.03**, Philadelphia, (2002).
5. Sondossi, M., H.W. Rossmoore, J.W. Wireman. Observations of resistance and cross-resistance to formaldehyde and a formaldehyde condensate biocide in *Pseudomonas aeruginosa*. *Int Biodetn* **21**: 105–106, (1985).
6. Sondossi, M., H.W. Rossmoore, R. Williams. Relative formaldehyde resistance among bacterial survivors of biocide-treated metalworking fluids. *Int Biodetn* **25**: 423–437, (1989).
7. L. H. Horsley in R. F. Gould, ed., Azeotropic Data-III, American Chemical Society, Washington, D.C., (1973).

# Determination of greenhouse gasses in al-Nahrawan suburban by geomatics technique

*Abdul-Razzak* Ziboon[1], *Ali* Resheq*[1], *Abbas* Hamza[2], and *Mays* Abass[1]

[1]Building and Construction Engineering Department, University of Technology, Baghdad, Iraq
[2]Architectural Engineering Department, University of Technology, Baghdad, Iraq

**Abstract.** Climate change is now more important than before, based on many evidence, humans were affecting on changing Earth's climate. The atmosphere and oceans have warmed, by sea-level rise, a strong decline in Arctic ice sea, and other climate-related changes. The aim of this study was to calculate some important greenhousse gasses concentration such as $CO_2$, $CH_4$, $N_2O$ in AL-Nahrawan suburban – Baghdad city -Iraq. Geographic Information System (GIS) was utilized to map greenhouse gasses dispersion in AL-Nahrawan bricks factory. From GIS distribution maps for $CO_2$, $CH_4$, $N_2O$, it was found that the value of these gasses were changed from one location to another according to the quantity of fuel used in bricks factory, Where the value of emitted $CO_2$ ranged from (695 -854 tones), value of $N_2O$ ranged from (1.905 - 2.318 tones), and finally value of $CH_4$ ranged from (0.286 - 0.347 tones).

## 1 Introduction

For a deeper understanding of climate change, it is essential to distinguish between weather and climate as they are mutually exclusive events that affect in a complex way human presence and activities on the earth. Weather is the state of the atmosphere above a given place at a specific time. It is the day-to-day state of the atmosphere in terms of air temperature and moisture, cloud covering, relative humidity. Weather is derived from the chaotic nature of the atmosphere and is unstable as it is affected by perturbations.

Climate on the other hand is described as average weather over a defined time period. Weather is a scientific concept. It deals with statistics such as the average of all-weather events, or over a long period of time (normally 30 years). Weather has a very limited predictability effect and could be directly perceived by people while climate cannot [1]. Climate is what you expect and weather is what you get. Climate varies from place to place, depending on latitude, distance to the sea vegetation, presence or absence of mountains or other geographic factors. Climate also varies in time, from season to season, year to year, decade to decade or on much longer scales such as the ice ages. The statistical significant variations of the mean state of the climate or of its variability, typically persisting for decades or longer have been referred to as "climate change" [2].

### 1.1 Climate change

Climate Change Refers to any change in climate over time, whether due to natural variability or as a result of human activity. [3] Climate change is nowadays recognized as one of the most challenging and complex problem facing the globe. [4] Climate change may be due to natural internal processes or external forcing such as modulations of the solar cycles, volcanic eruptions, and persistent anthropogenic changes in the composition of the atmosphere or in land use. [5] Without the natural greenhouse effect, the average temperature at Earth's surface would be below the freezing point of water. Thus, Earth's natural greenhouse effect makes life as we know it possible. However, human activities, primarily the burning of fossil fuels and clearing

*Corresponding author: dr.ali_sadiq@yahoo.com

© The Authors, published by EDP Sciences. This is an open access article distributed under the terms of the Creative Commons Attribution License 4.0 (http://creativecommons.org/licenses/by/4.0/).

of forests, have greatly intensified the natural greenhouse effect, causing global warming. Global climate change has resulted primarily from the accumulation of Carbon Dioxide ($CO_2$) and other greenhouse gases (GHGs) in the atmosphere over the last 150 years [6].

## 2 Greenhouse gasses

The most important greenhouse gasses are:

### 2.1 Carbon dioxide ($CO_2$)

Carbon Dioxide ($CO_2$) is the most common G HG emitted by human activities in terms of the quantity released and total impact on global warming. The main effects of Carbon Dioxide in the atmosphere are its contribution in the greenhouse effect, global warming, acid rain, and maintaining the PH of rain water. It is one of the end products of burning of fossil fuels like coal, natural gas and petroleum [7].

### 2.2 Methane ($CH_4$)

Methane is emitted by natural sources such as wetlands, as well as human activities such as leakage from natural gas systems and the raising of livestock. Natural processes in soil and chemical reactions in the atmosphere help remove $CH_4$ from the atmosphere. Methane's lifetime in the atmosphere is much shorter than Carbon Dioxide ($CO_2$), but $CH_4$ is more efficient at trapping radiation than $CO_2$ [8]. The IPCC Fifth Assessment Report determined that Methane in the Earth's atmosphere is an important greenhouse_gas with a global warming potential of 34 compared to $CO_2$ over a 100-year period.

### 2.4 Nitrous oxide ($N_2O$)

Nitrous Oxide ($N_2O$) is nonflammable, colorless gas with pleasant, sweet odor and taste. Also called Dinitrogen Monoxide or more commonly laughing gas. When inhaled, it produces relaxation, and a reduced sensibility to pain [9] Nitrous Oxide is generated as a byproduct during the production of Nitric Acid, which is used to make synthetic commercial fertilizer, and in the production of adipic acid, which is used to make fibers, like Nylon, and other synthetic products [10] .

### 2.3 Geographic information system

Geographical Information System (GIS) is a modern powerful tool that facilitates linking spatial data to non-spatial information, obviously, the use of GIS has become essential in providing boundary conditions to the air quality models [11]. GIS technology allows us to locate where pollutants are coming from and monitor those areas for change to conserve the quality of our air, the major goal of GIS ito represent the complicated geographic phenomena in digital model [12]

## 3 Materials and methods

### 3.1 Study area

Baghdad city is the political capital of Iraq, situated on the banks of Tigris River at the northwest end of the alluvial plain. It covers 4555km$^2$ which represented 1% from total area of Iraq (435052 km$^2$) [13] Fig (1) AL-Nahrawan suburban is located in the south – east part of Baghdad city.it lies about 65 km from it .It was created in 1988 and its population is about 150000, its originally an agricultural area that transformed lately into city. [14] The study area contain (172) factory of brick which considered the main reason for air pollution in that area.

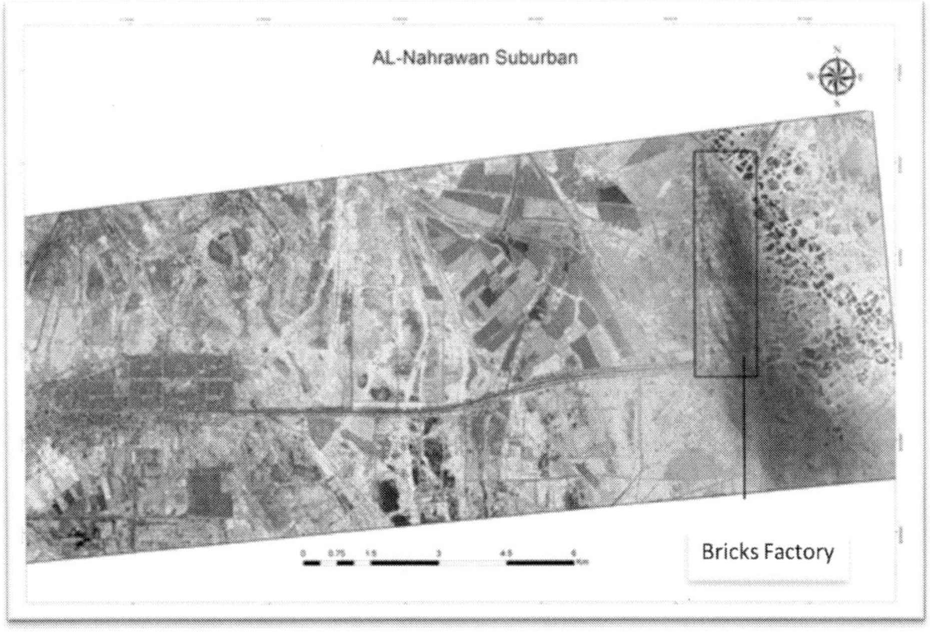


**Figure (1):** Study area

**Mathematical model of climate change**

$$Eij = \frac{Q_i \times EC_i \times EF_{ijoxec}}{1000}$$

Where: -

$E_{ij}$ is the emissions of gas type (j), (Carbon Dioxide, Methane or Nitrous Oxide, from fuel type (i) ($CO_2$-e tons).

$Q_i$ is the quantity of fuel type (i) (kiloliters) combusted for stationary energy purposes.

$EC_i$ is the energy content factor of fuel type (i) (gigajoules per kiloliter) for stationary energy purposes, according to Table (1). If $Q_i$ is measured in gigajoules, then $EC_i$ is 1.

$EF_{ijoxec}$ is the emission factor for each gas type (j) (which includes the effect of an oxidation factor) for fuel type (i) (kilograms $CO_2$-e per gigajoule) according to **Table 1**

**Table 1:** Standard for Fuel combustion emission factors - liquid fuels and certain petroleum based products for stationary energy purposes [15].

| Fuel Combusted | Energy content factor (GJ/kL unless otherwise indicated | Emission factor kg CO2-e/GJ (relevant oxidation factors incorporated) | | |
|---|---|---|---|---|
| | | $CO_2$ | $CH_4$ | $N_2O$ |
| Petroleum based oils (other than petroleum based oil used as fuel, eg lubricants | 38.8 | 27.9 | 0.0 | 0.0 |
| Petroleum based greases | 38.8 | 27.9 | 0.0 | 0.0 |
| Crude oil including crude oil condensates | 45.3 GJ/t | 68.9 | 0.06 | 0.2 |
| Other natural gas liquids | 45.3 GJ/t | 60.4 | 0.06 | 0.2 |
| Gasoline (other than for use as fuel in an aircraft) | 34.2 | 66.7 | 0.2 | 0.2 |
| Gasoline for use as fuel in an aircraft (avgas) | 33.1 | 66.3 | 0.2 | 0.2 |
| Kerosene (other than for use as fuel in an aircraft) | 37.5 | 68.2 | 0.01 | 0.2 |
| Kerosene for use as fuel in an aircraft (avtur) | 36.8 | 68.9 | 0.01 | 0.2 |
| Heating oil | 37.3 | 68.8 | 0.02 | 0.2 |
| Diesel oil | 38.6 | 69.2 | 0.1 | 0.2 |
| Fuel oil | 39.7 | 72.9 | 0.03 | 0.2 |
| Liquefied aromatic hydrocarbons | 34.4 | 69.0 | 0.02 | 0.2 |
| Solvents if mineral turpentine or white Spirits | 34.4 | 69.0 | 0.02 | 0.2 |
| Liquefied petroleum gas | 25.7 | 59.6 | 0.1 | 0.2 |
| Naphtha | 31.4 | 69.0 | 0.00 | 0.02 |
| Petroleum coke | 34.2 GJ/t | 90.8 | 0.06 | 0.2 |
| Refinery gas and liquids | 42.9 GJ/t | 54.2 | 0.02 | 0.03 |
| Refinery coke | 34.2 GJ/t | 90.8 | 0.06 | 0.2 |
| Petroleum based products other than mentioned in the items above | 34.4 | 69.0 | 0.02 | 0.2 |
| Biodiesel | 34.6 | 0.0 | 0.06 | 0.2 |
| Ethanol for use as a fuel in an internal combustion engine | 23.4 | 0.0 | 0.06 | 0.2 |
| Biofuels other than those mentioned in the items above | 23.4 | 0.0 | 0.06 | 0.2 |

## 4 Results

For the purposes of computing emissions of greenhouse gases (Carbon Dioxide, Methane and Nitrous Oxide) in tones of $CO_2$-e is estimated as follows; (the fuel used in computations is fuel oil). And the quantity of fuel used different from brick factory to another and we have four groups of quantity used, so by applying the equation that mention in the above section the emission of ($CO_2$, $CH_4$, $N_2O$) shown in the tables (2, 3, 4).

Table (2, 3, 4) show the result of greenhouse gasses emission for each quantity of oil

**Table 2**: Emission of $CO_2$ for each quantity of oil

| Quantity of oil ($Q_i$) (kiloliters) | Emission of $CO_2$ (t $CO_2$-e) |
|---|---|
| 292 | 854.086 |
| 270 | 781.415 |
| 247 | 714.850 |
| 240 | 694.591 |

**Table 3**: Emission of $N_2O$ for each quantity of oil

| Quantity of oil ($Q_i$) (kiloliters) | Emission of $N_2O$ (t $CO_2$-e) |
|---|---|
| 292 | 2.3185 |
| 270 | 2.1438 |
| 247 | 1.96118 |
| 240 | 1.9056 |

**Table 4**: Emission of $CH_4$ for each quantity of oil

| Quantity of oil ($Q_i$) (kiloliters) | Emission of $CH_4$ (t $CO_2$-e) |
|---|---|
| 292 | 0.3477 |
| 270 | 0.3215 |
| 247 | 0.29417 |
| 240 | 0.28584 |

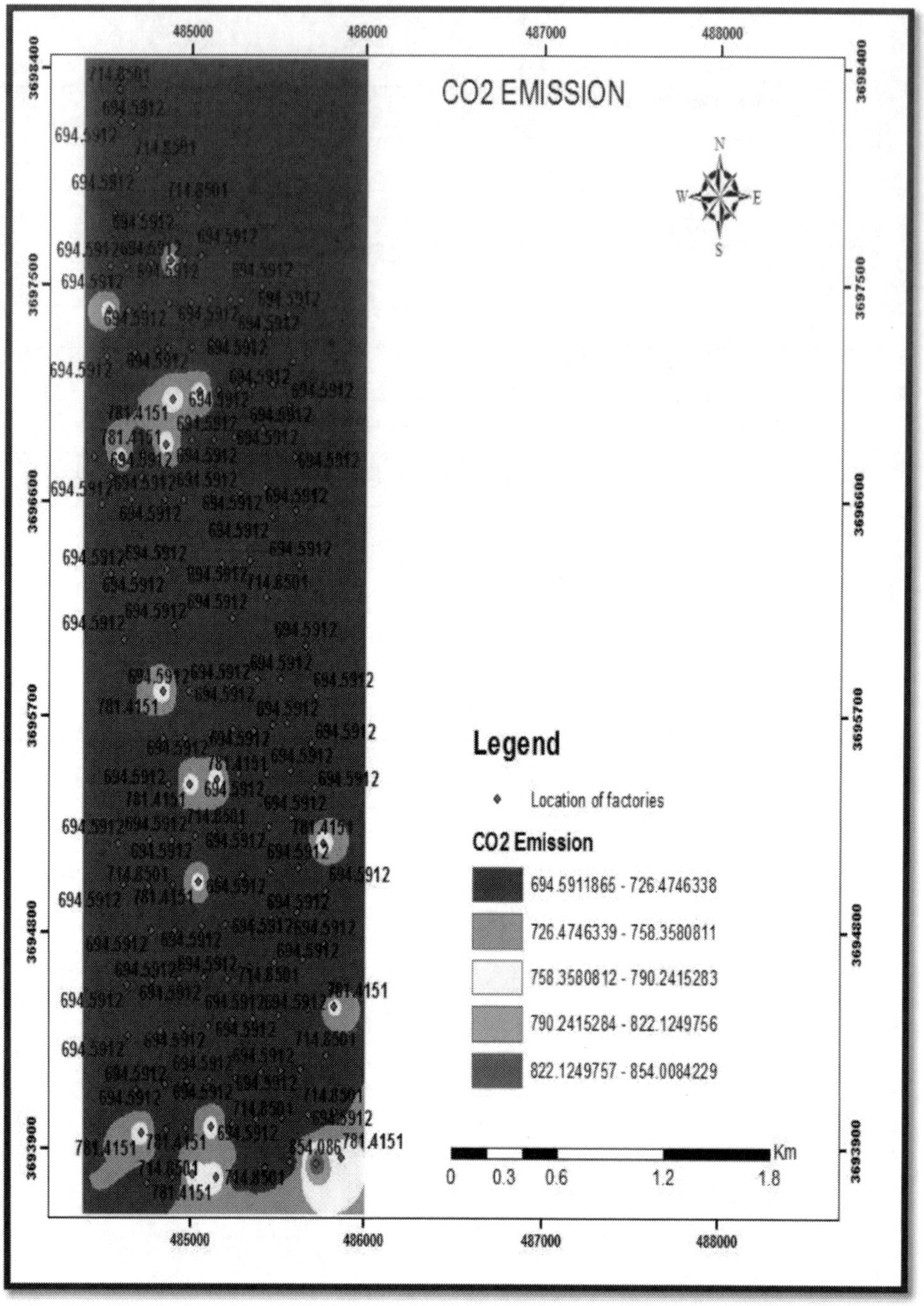


**Figure (2):** distribution of brick factory with emission factor for $CO_2$.

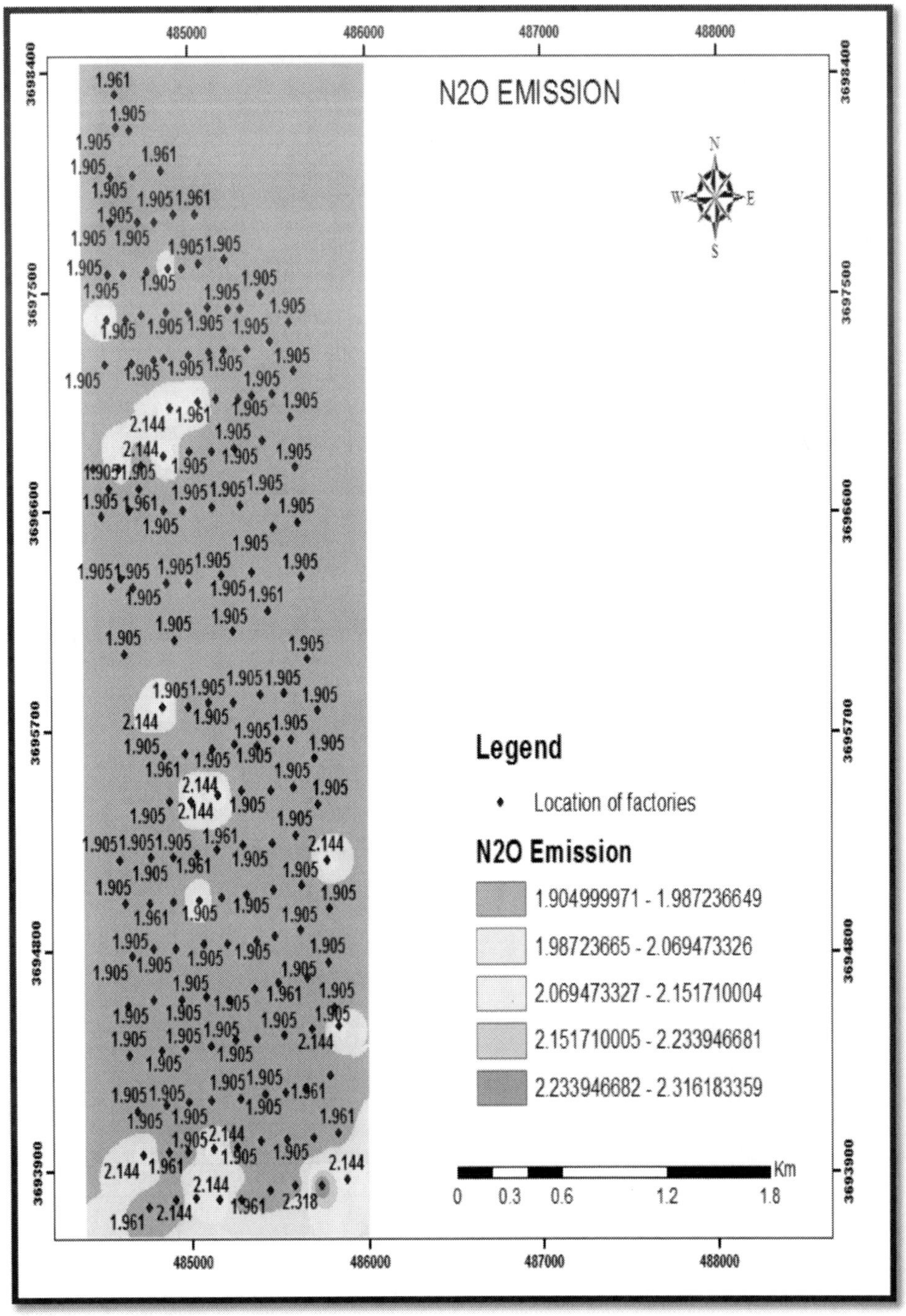


**Figure (3):** distribution of brick factory with emission factor for $N_2O$

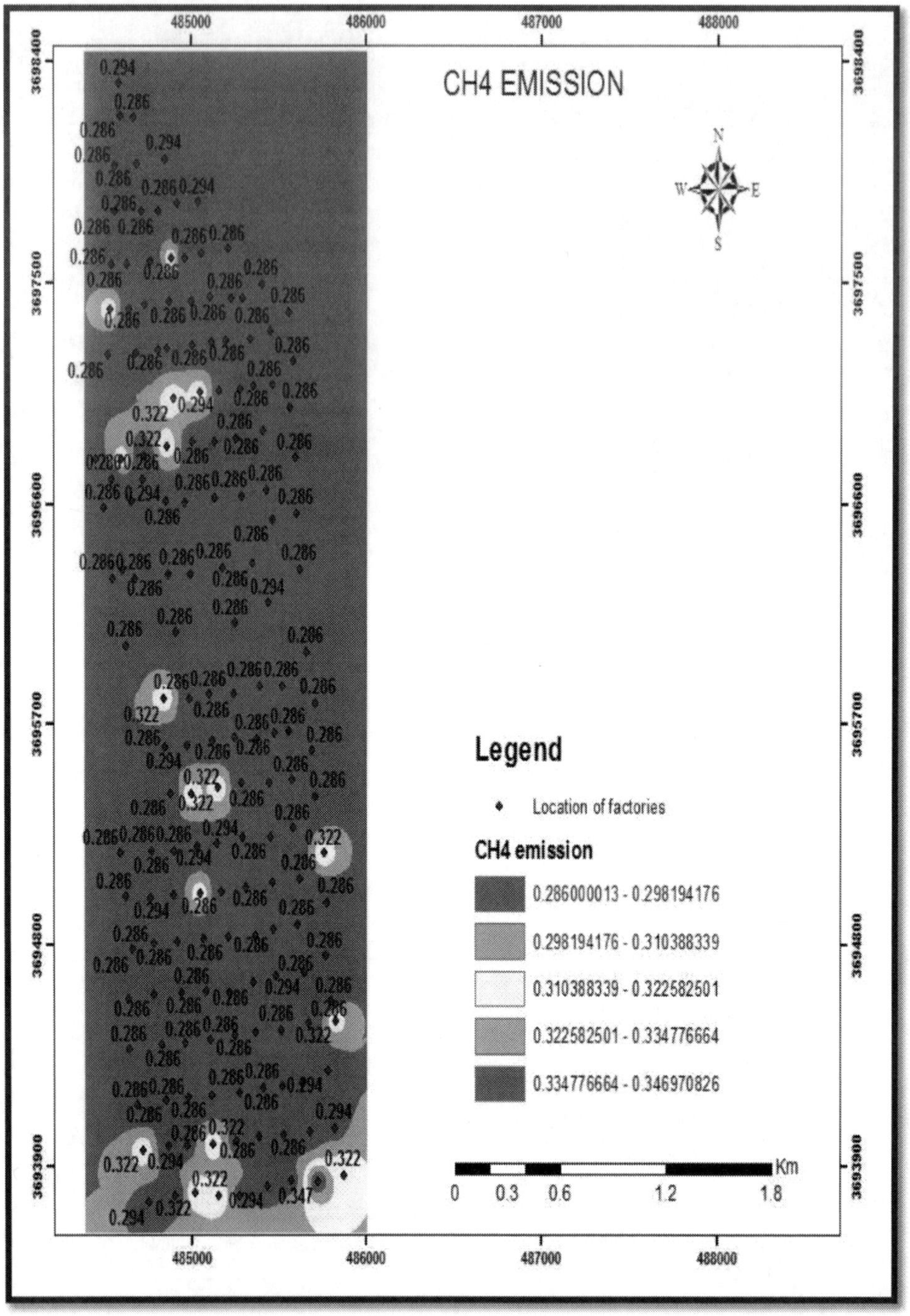


**Figure (4):** distribution of brick factory with emission factor for $CH_4$

## 5 Conclusions

By using geomatics technique (remote sensing then identifying areas of influence by using GIS is proved to be a good efficiency in the quick identification of contaminants and prepared it for environmental decision maker. There are a lot of gasses that influence in the climate change but here we discussed the most important gasses influence on the change in climate. The values of climate change gasses ($CO_2$, $N_2O$, and $CH_4$) that have been calculated from the formula of emission have different value according to the quantity of fuel used in each brick factory. Where the value of $CO_2$ ranged from (695 - 854 tones), value of $N_2O$ ranged from (1.905 - 2.318 tones), and finally value of $CH_4$ ranged from (0.286 -0.347 tones).

## References

1. Kropp J. and Scholze M. (2009) Climate change information for effective adaptation: A practitioner's manual. Eschborn: Deutsche Gesellschaft für Technische Zusammenarbeit (GTZ) GmbH.

2. Baede APM, Ahlonsou E, Ding Y Schimel D (2001) The Climate System: An Overview. In Houghton, J. T. et al. (eds.), Climate Change 2001: The Scientific Basis. Cambridge University Press, Cambridge, UK, pp. 85-98.

3. IPCC TAR (2001) IPCC Third Assessment Report.Climate Change 2001. Working Group II: Impacts, Adaptation and Vulnerability. Chapter1.

4. Watson RT (2001) Climate Change 2001.Synthesis Report. Third Assessment Report of the Intergovernmental Panel on Climate Change (IPCC). Cambridge University Press, UK.

5. IPCC 2014: Climate Change 2014: Impacts, Adaptation, and Vulnerability, IPCC WGII AR5 Summary for Policymakers.
6. Kuylenstierna Johan, Kevin Hicks., 2008 : Benefits of Integrating Air Pollution and Climate Change Policy Published by: Stockholm Environment Institute University of York
7. Ahluwalia, V.K., Malhotra,S., 2006: Environmental Science, an a book, Pvt. Ltd. Gopaljee Enterprises, India
8. EPA 2010: Methane and nitrous oxide emission from natural source U.S environmental protection agency, Washington, DC, USA.
9. Kaznosky James., 2015: Nitrous Oxide: Hazards and Proper Use Manager, Env. & Occ. Safety Environmental Health and Safety Version: 2.1.
10. EPA 2010: Methane and nitrous oxide emission from natural source U.S environmental protection agency, Washington, DC, USA.
11. Matejicek L., 2005: Spatial modelling of air pollution in urban areas with GIS: a case study on integrated database development. Advances in Geosciences, 4, 63–68
12. Zhiewei Cao, Mordechai (Muki) Haklay and Tao Cheng :IDW implementation in map calculus and commercial GIS products ,accuracy 2010 sympsium ,july 20-23 leicester ,UK
13. Saleh, S.A. H., "Impact of Urban Expansion on Surface Temperature in Baghdad, IRAQ using Remote Sensing and GIS Techniques", University of al-Nahrain, Baghdad, Iraq, Canadian Journal on Environmental, Construction and Civil Engineering,2 (8) 2011.
14. INVESTOR GUIDE OF BAGHDAD NOVEMBER 2011: This report was produced for review by the U.S. Agency for International Development (USAID). It was prepared by The Louis Berger Group, Inc
15. Australian National Greenhouse Account, 2014: National Greenhouse Accounts Factors, Published by the Department of the Environment.

# Production of fuel gases from reed by using downdraft gasifier

*Abdul-Hameed* Al-Obaidy[1], *Riyad* Al-Anbari[2], *Hussein* Kazim[2]*

[1]Environmental Research Center, University of Technology, Baghdad, Iraq,
[2]Department of Building and Construction Engineering, University of Technology, Baghdad, Iraq

**Abstract.** This study was conducted to investigate the ability of reed as a Production of gases via gasification technique by using special gasifier made for this purpose and to evaluate the produced gases (CO, $H_2$, $CO_2$ and $CH_4$). The results showed that the percentage of production gases was 15.6%, 10.2%, 7.81% and 1.69% for carbon monoxide, carbon dioxide, hydrogen and methane respectively. In addition, it showed that the highest temperature up to the gasifier was 700°C and the time required to complete gasification process was 23 min. The results indicated that 5 kg of reed was capable of producing 10 m3 of gases which produces about 1 HP-hr. that equivalent to 3.75 kWh.

## 1 Introduction

Reeds are found all over the world except Antarctica, but there are many area in Europe, the Middle East and America [1], and the reed is called either 'first generation or second generation.. The first generation is when be source of biomass from food crop (e.g maize) and the second generation is when the source of biomass from non-food crops (e.g reed).[2] There are three ways to use the reed as a source of energy which are combustion or gasification, biogas and biofuel.[3]

There are two ways to convert solid waste into fuel; where the first method is biochemical, which is achieved by the decomposition of organic matter by organisms. The disadvantages of this method are the requiring expensive enzymes, a large amount of feed preconditioning, and producing a single product (e.g. ethanol) [4,5]. The second way is to convert solid waste into fuel by thermochemical; this method is carried happening by heating the biomass in the presence of steam, oxygen or air to a temperature between 500 and 850°C. This method is called gasification in case oxygen presence or pyrolysis in the absence of oxygen .High temperature allows breaking of the cellulosic material into a mixture of gases mainly hydrogen and carbon monoxide known as synthesis gas ("syngas").
Gasification is usually defined as the conversion of carbon into gases produced by partial oxidation. The resulting gas is known as syngas, which contains hydrogen and carbon monoxide mainly, as well as methane and carbon dioxide. This gas is later used in more than one field after passing through many processes [6]. Coal is the first material used in the process of gasification but due to demand of sustaining the earth resources, other types of feedstock are being used; which include reeds, corn residues and sawdust…etc. [7].Chemical reaction in the gasification process where some of which are exothermic and some are endothermic [8]. The main advantages of this technique are the low cost of waste used in addition to recycle this waste. Gasification is similar to combustion, but it is considered as partial combustion process, and it has smaller carbon dioxide (greenhouse gas) emission [9].

## 2 Materials and methods

### 2.1 The gasifier model: Overview and Specifications

The gasifier GEK system (version 4) was constructed from ALL Power Labs Company, and given to the University of Technology by the US Embassy in Baghdad then building in Ministry of Science and Technology, Baghdad, Iraq.. It is essential to work on all the required specifications for gasification systems as shown in Table 1. The GEK Systems are designed for raw biomass and organic-based feedstock's. Experimental feedstock's lying outside of the suggested ranges may require modification of the equipment and testing of the gas quality produced.

*Corresponding Author: Hussein.a.kazim@gmail.com

© The Authors, published by EDP Sciences. This is an open access article distributed under the terms of the Creative Commons Attribution License 4.0 (http://creativecommons.org/licenses/by/4.0/).

The capacity of gases produced by this model is about 2 m2 per 1 kg biomass which produces about 0.75 kWh with biomass consumption rate (kg/hr.) about 12 kg. Dimension of this model is 61 cm height and 43 cm diameter.

Table 1: Solid waste requirement

| | |
|---|---|
| Size of particle (in) | 5 – 1.5 |
| Content of moisture (% by dry weight) | Less than 25 |
| Fixed to volatile ratio | More than 0.25 |
| Content of ash (%) | More than 5 |

## 2.2 Preparation of feedstock

Feedstock selection and preparation are the most important variable toward successful operation of gasifier. Biomass fuel does not come in standard forms. Shape, size, moisture levels and mineral contents vary greatly. Understanding the impacts of these variables on gasifier operation, and preparing fuel accordingly, are most important assignment for operating gasifier model. Fresh reed was harvested from the Tigris river in Baghdad at Al-Kadhemia section, the reed samples were dried by sun light for 5 days to obtain moisture content required for gasification process (15-20 % is usually ideal) then used as feedstock to gasifier operation. High moisture content leads to reduce the temperature in the combustion zone and thus not to get a process of gasification completed [10]. After drying process, the reed was milling into small particles using blender 0.5 – 1.5 in, where larger particles can form bridges which avoid the feed moving down, while smaller particles lead to block the available air void age, leading to shutdown of the gasifier [11].The physical and chemical properties were measured at the Ministry of Science and Technology according to the standard method ASTM. Figure 1 shows the steps followed in the process of gasification. After the process of harvested reeds, the preparation of raw materials through drying and reduce the size to the specifications required in addition to the procedure of preliminary tests, such as moisture content, density and size. The second step is to transfer the raw materials to the gasifier and then to measure the internal and external temperature as well as the gases produced by sensors prepared for this purpose.

## 2.3 Operation of gasifier

As shown in Figure 2, after the process of sizing and drying, the raw material was transferred to the reactor through a hole located at the top of the gasifier, and after filling reactor with raw material, the compressor was connected to the reactor through a controlled pipe. Then we burn the reeds by torch with insert a small amount of crude oil to accelerate the ignition, after few minutes.. The hot gas exits the reactor and passes through both a cyclone and filter to separate char particulates. The temperature and concentration of gas measured by sensors proposed to this purpose.

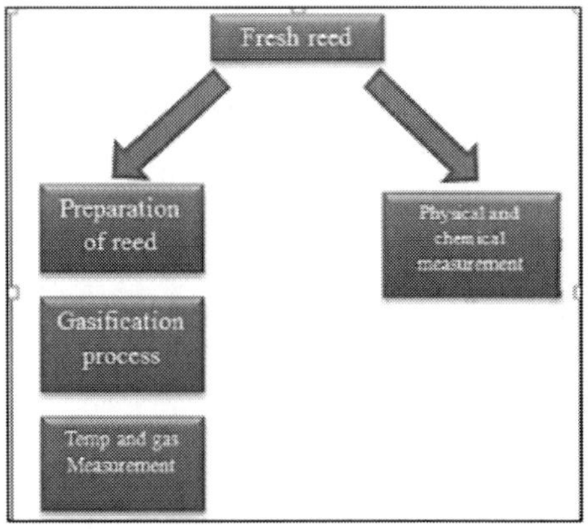


Figure 1: The route of gasification process

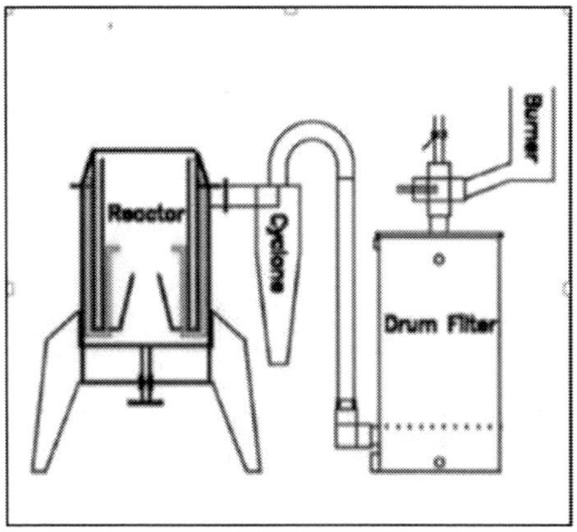


Figure 2: Diagram of downdraft gasifier

# 3 Results

The physical and chemical properties were measurement before gasification process, table below shows properties of reed.

Table 2: Analysis of physical and chemical the reed used [Ü. Kask]

| | |
|---|---|
| Moisture content (dry basis %) | 17 |
| Ash content (%) | 6.5 |
| CV(MJ/kg) | 18.5 |
| C | 46.5 |
| H | 6.2 |
| O | 40.7 |

The following figure shows that the chemical composite of the gas produced from reeds versus the reactor temperature, where it shows percentage of gases (CO, $CO_2$, $H_2$ and $CH_4$) at a temperature up to 700°C.

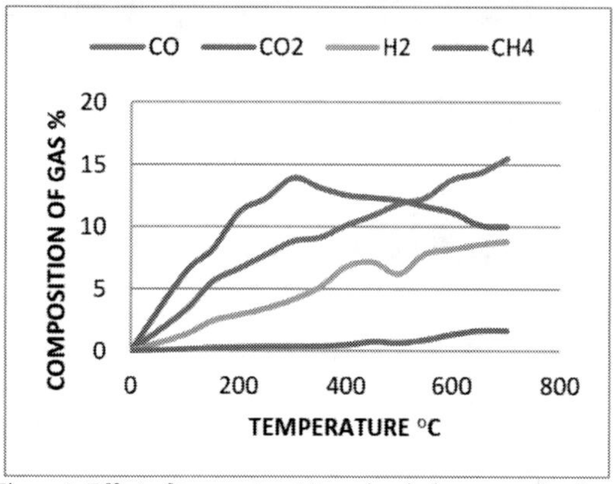


Figure 3: Effect of temperature °C on chemical compounds of gas produced from reeds (%)

Lower heating value (the amount of heat released during the combustion of a specified amount of it) was calculated based on the resulting of gas compounds and according to the following equation [Reza A. Moghadom]: LHV (MJ/M3) = (107.98*H2+126.36*CO+358.18*CH4+56*C2H2)/1000.

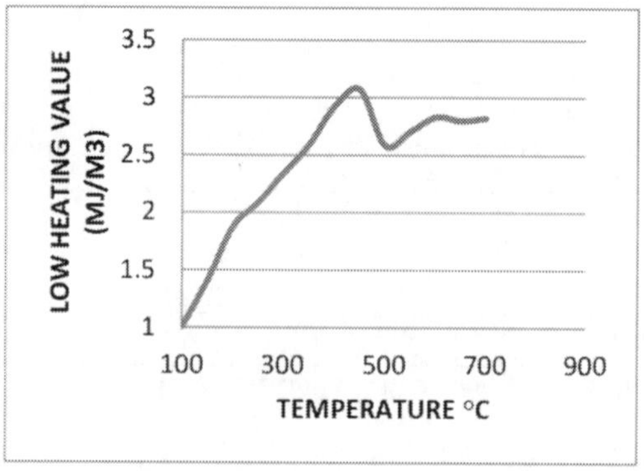


Figure 4: The relationship between heat values (MJ/M3) with temperature °C

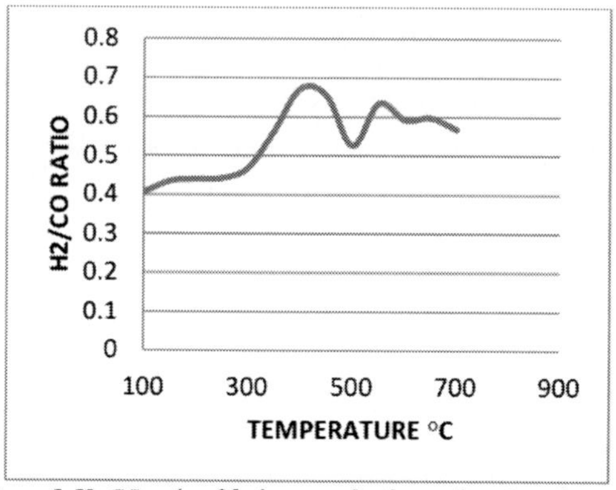


Figure 5: $H_2$:CO ratio of fuel gas production as function of reactor temperature °**C**

Figure 6 illustrates the relationship between the chemical compounds of the yield gas and the rate of consumption of solid waste. Where it was noted that the time of consumption of solid waste is 23 minutes.

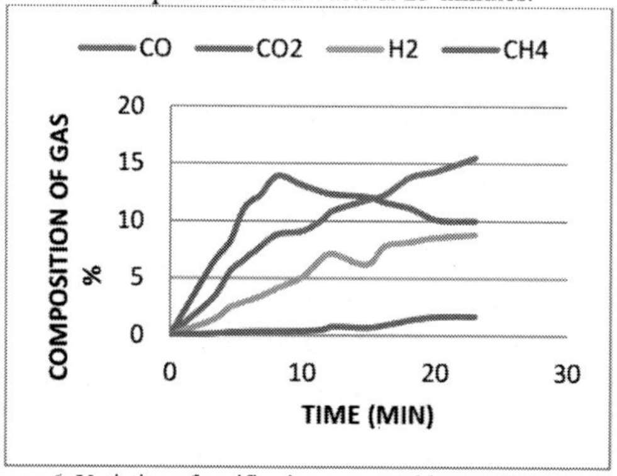


Figure 6: Variation of gasification output with operation time (min)

The Figure (5) shows the relationship between the reactor temperature and the time required to consumption the reed, this figure also illustrates the thermal behavior of reed used as raw material to gasification process.

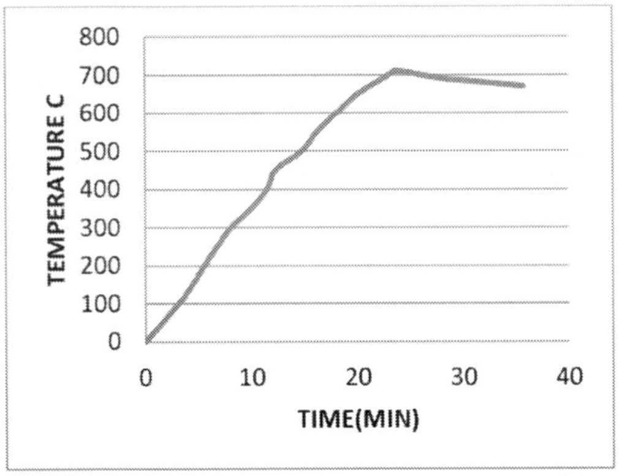


Figure 7: Thermal behavior of reed with operation time (min)

Table 3: The results of the experimental gasification

| Characterizes | Min | Max | Average | STD |
|---|---|---|---|---|
| Temp °C | 100 | 710.6 | 493.26 | ±2120.29 |
| CO% | 3.23 | 15.6 | 9.99 | ±3.63 |
| $CO_2$% | 6.27 | 10.5 | 9.10 | ±1.04 |
| $H_2$% | 1.31 | 7.81 | 5.39 | ±2.3 |
| $CH_4$% | 0.12 | 1.67 | 0.71 | ±0.52 |
| LHV (MJ/M3) | 1 | 2.53 | 2 | ±0.54 |

## 4 Discussion

From the results shown in Figure 3, the highest produced gas was CO with percentage of 15.6 % at 700.4°C, followed by carbon dioxide with content of 10.5% at 300.5°C, then hydrogen and methane with 7.81% and 1.67% at 550.2°C, 700.4°C respectively. The high percentage of CO gas was due to the high carbon content found in the reed as shown in Table 2.

It was also observed that carbon monoxide and hydrogen were rapidly increased with increasing temperature through reaction (6 and 7)[lv et 2004] while carbon dioxide was decreased at 300.5 °C. Methane was found to have increased slowly as the temperature increased.

Figure 4 shows an increase in the value of LHV with increasing temperature due to the increase of concentration of CO in syngas [Moni et 2012], where LHV depends on the flammable gases and then decrease it at a temperature of 450°C because of the apparent increase of both hydrogen gas and methane compared to slightly increase of gas.

As for Figure 5, it was found that the highest value of $H_2$:CO at 450 °C temperature due to the increase of hydrogen gas compared to CO. the increase concentration of CO compared to decreasing amount of $H_2$ with increased temperature caused decreasing in $H_2$:CO ratio[Moni et 2012]

The study of the effect of time on the concentration of gas produced is to know the maximum operating time until the gasifier needs to stop the supply syngas. Figure 6 shows that the highest value was for gas CO at a time of 23 minutes after steady climb and then experiencing a drop towards the end of operation. Also for H2 showed almost the same pattern.

Increasing the temperature of the reed regularly indicates the efficiency of the gasifier system as illustrated in Figure 7 Also the highest value heart temperature (T1) is approximately 700.3 °C at 23 minutes. The heart temperature stabilized at 600 °C due to depletion of the batch of fuel fed to reactor.[17]

The thermodynamic changes during the gasification process are due to the chemical reaction obtained within the gasifier. Increased temperature for reed is the result of energy released from exothermic the reactions as show below [14]

| Reactions |
|---|
| $2C + O2 = 2CO$ partial oxidation (1) |
| $C + O2 = CO_2$ complete oxidation, (2) |
| $C + 2H_2 = CH_4$ (hydro-gasification/methanation reaction), (3) |
| $CO + H_2O = CO2 + H_2$ water-gas shift reaction,(4) |
| $CH_4 + H_2O = CO + 3H_2$ steam reforming reaction, (5) |
| $C + H_2O = CO + H_2$ water gas reaction, (6) |
| $C + CO_2 = 2CO$ bounded reaction (7) |

## 5 Conclusions

The reed is found at high quantities in Iraq especially in the south area, this paper has dealt with the reed plant and tested by using gasification technology to assess how much energy can be produced from them. From the these results, it is clear that reeds have the ability to produce gas by gasification technique and the higher gas obtained was CO then $CO_2$, H2 and $CH_4$ respectively, This is due to chemical composition and thermal behavior of reed. As well 5 kg of reed used as raw material in this paper is capable of producing 10 m3 of combustible gases which equivalent about 3.75 kWh.

## References


1. S.M. Haslam "A Book of Reed: *(Phragmitesaustralis (Cav.) Trin. ex Steudel, Formerly Phragmites communis Trin.),"* Cardigan, GB: Forrest Text, pp 254, 1993.
2. J.F Kobbing, N. Thevs and S. Zerbe *"The utilisation of reed (Phragmitesaustralis),"* Mires and peat. **Vol.13**, pp.1-14, 2013.
3. Ecology Land and People (ELP) and N. Ash, "New Opportunities for the Sustainable Management of Fens: Reed Pelleting, Composting and the Productive Use of Fen Harvests," in Ecology, Land and People (ELP) for the Broads Authority, Norwich, UK, pp.66, 2010.
4. L.R. Lynd, J.H. Cushman, R.J. Nichols and C.E. Wyman *"Fuel Ethanol from Cellulosic Biomass,"* Science, 251, 4999, 1318-1323, 1991.
5. Lynd, L.R., W.H. van Zyl, J.E. McBride and M. Laser "Consolidated Bioprocessing of Cellulosic Biomass: An Update. Current Opinion in Biotechnology," Curr Opin Biotechnol., 16, 5, 577-83, 2005.
6. Gasification Technologies Council *"Gasification: Redefining Clean Energy,"* available online: http://www.gasification.org/ [accessed Oct 2010]
7. M.R. Bricka, D.C. Swalm *"Energy Crop Gasification and Gasification Issues,"* School of Chemical Engineering Mississippi State University.
8. C. Higman, M. van der Burgt, *"Gasification,"* Gulf Professional Publishing, Second edition, 12-15, 33-36, 2008.
9. L. Zhang, C. (Charles) Xu, and P. *Champagne "Overview of Recent Advances in Thermo-Chemical Conversion of Biomass,"* Energy Conversion and Management, **51**, 969–982, 2010.
10. W. Simpson and A. Tenwolde, *"Physical Properties and Moisture Relations of Wood,"* in Wood Handbook: Wood as an Engineering Material. Madison, WI: USDA Forest Service, Forest Products Laboratory, 1999. General technical report FPL; GTR-**113**, pp 3.1-3.24, 1999.
11. P. McKendry, *"Energy Production from Biomass (Part3): Gasification Technologies,"* Bioresource Technology, **83, 1**, 55–63, 2002.
12. Ü. Kask, L. Kask and S. Link, *"Combustion characteristics of reed and its suitability as a boiler fuel"* Mires and Peat, **Volume 13** (2013/14)
13. G. Xian Sheng, *" Biomass Gasifiers: From Waste to Energy Production"* Biomass 20 (1989) 3-12 14. J. F. V´elez, F. Chejne, C. F. Vald´es, E. J. Emery, and C. A. Londo˜no, *"Cogasification of Colombian coal and biomass in fluidized bed: an experimental study,"* Fuel, **vol. 88**, no. 3, pp. 424–430, 2009
15. R.Moghdam, S.Yusup, W.Azlin, S.Nehzati, A.Tavasoli,*''Investigation on syngas production via biomass conversion through the integration of pyrolysis and air–steam gasification processes''*Energy Conversion and Management, **Volume 87**, November 2014, Pages 670–675
16. M.N.Z.Moni,S.A.Sulaiman*"Downdraft Gasification of Oil Palm Frond,"*journal of applied science **12(24)**:2547-2579.2012
17. Samson mikbeb antaw *' study on tar generated from downdraft gasification of oil palm fronds '* the scientist world journal ,**volume 2014**

# Evaluation of Combined Sewage Trunk System (East Bank Trunk) in Al-Rusafa Area, Baghdad City, Iraq

*Faris* Hammodi Al-Ani, *Sataa* A.F. Al- Bayati and, *Ali* N. Jasim*

Building and Construction Engineering Department, University of Technology, Baghdad, Iraq

**Abstract.** This study came because of problems that recently occurred in the sewerage trunk in Baghdad city. In Rusafa, there are two main trunks (East Bank Trunk and Zublin Trunk) for the carry of sewage to the wastewater treatment plant in Rustamiyaha, some parts of Baghdad city were flooded due to high rainfall intensity in year 2013 and 2014 during winter for several days, the results obtained from the analysis showed that the design capacity of the transmission trunk of Baghdad does not meet the maximum operating conditions. It was also found that the hydraulic gradient is very low, which made the flow velocity less than the minimum recommended Self-cleansing velocity. This caused the deposition of particles that cause the blockage of the transmission main trunks.

***Keywords:*** East Bank Trunk, Discharge of wastewater, Self-cleansing velocity

## 1. Introduction

Sanitary sewer systems are composed of piping that is used to transport wastewater consisting of residential, commercial, and industrial waste. Some amount of storm water, surface water, and groundwater may also be present in sanitary sewer systems.

Storm sewer systems are composed of those piping systems that carry only storm water, surface water, and other waters that are drained into the storm sewer system. They do not carry residential, commercial, or industrial wastes [7].

A combined sewer system consists of a combination of a sanitary sewer system and a storm sewer system. Thus, a combined sewer system carries both storm water as well as wastewater. Many existing sewer systems collect both wastewater and stormwater. In such systems, stormwater, up to a design limit, is channeled with sanitary wastewater to the treatment plant. When combined flow exceeds the sewer's capacity, combined sewer overflows (CSOs) occur and wastewater, along with stormwater, is discharged to receiving waters. When stormwater is transported with wastewater to the treatment plant, pumping and treatment costs are increased and treatment problems may occur. Combined sewer systems are no longer designed except as limited extensions or replacements for existing combined systems [1].

Studies by the U.S. Public Health Service (USPHS 1964) and the U.S. Environmental Protection Agency [5] show that combined stormwater and wastewater overflows introduce large quantities of polluting materials into the nation's waters. As a result of these issues and other concerns, the EPA in 1994 adopted a CSO control policy [12]. This policy includes minimum controls that sharply limit the use of combined sewer systems and essentially preclude their use in new construction.

Sewer systems are generally designed as gravity flow systems with a free water surface [9]. This means that the sewer pipe may run full or partially full so that there is an air space above the water level. This is known as open-channel flow. The advantage of open-channel flow includes ventilation of the sewer and maintenance of good velocities at low flow rates for cleaning the sewers. Pumps are also used to provide the lift necessary from deep sewer locations to force the sewage to a higher elevation from which point gravity flow can continue [9].

In general terms, older networks in urban area which have been based upon water courses culverted during urban development end to be on the combined system, while new networks are more commonly designed on the separate systems [8]. The design of sewers tends to be based upon minimum gradients and velocities to minimize excavation and number of pumping Stations. With a conventional design of combined system therefore self-cleansing velocities may only be reached once per day or at worst during the winter months only storm conditions. From this point of view therefore, a separate system is to be preferred [8].

Manholes are structures installed in gravity sewers to provide access for repairing, maintenance and cleaning of

*Corresponding author: alinaeem46@gmail.com

© The Authors, published by EDP Sciences. This is an open access article distributed under the terms of the Creative Commons Attribution License 4.0 (http://creativecommons.org/licenses/by/4.0/).

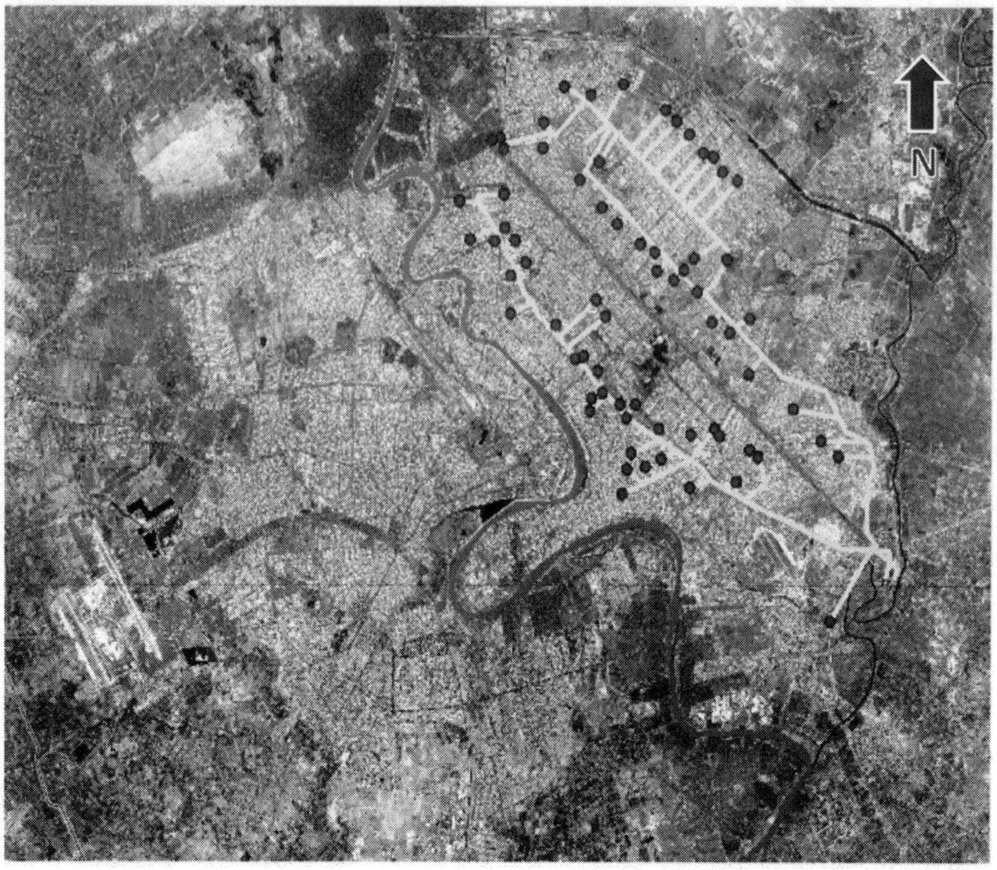


**Fig. 1.** Baghdad Map

the sewer line. Typical cleaning equipment are 90-150 meters long, which naturally causes limiting factors for manhole spacing. Typically, no manholes should be placed approximately over 120 meters apart in straight runs of pipe [10].

Lift stations are needed because there are limitations how deep one can install the pipe in. When this point is reached, the water must be pumped to a higher altitude to proceed with the gravitational flow [4]. These facilities that do the pumping are called the pumping/lift stations. Common components found in pumping stations include: pumps (centrifugal pumps, pneumatic ejectors or screw pumps), hardware, bar racks, valves, electrical system, alarms, motor control center, hour recorders and pump controls [13]. Lift stations are subject to mechanical and electrical failures due to many components involved [6].

## 2. Study Area

Flows to the existing sewer works at Rustamiyaha rise primarily from the existing (East Bank Trunk) Sewer network which at present serve a total area of 5300 hectares 3600 ha.net . Its catchment area includes the central area between the Army canal and Tigris River, and the Karada. There are two main sewer trunks in al-Rusafa

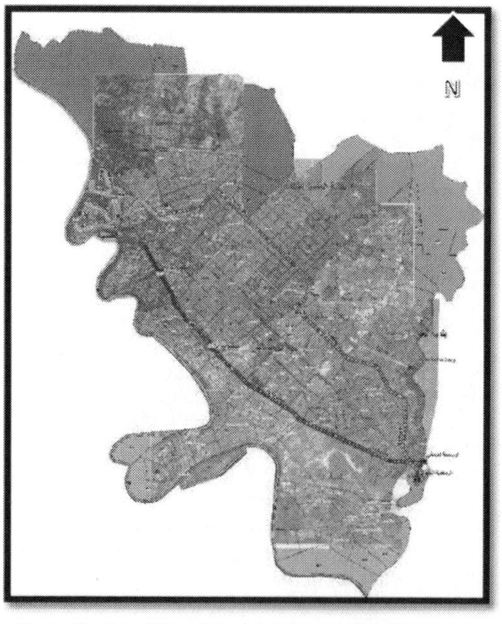


**Fig. 2.** Municipalities of Baghdad Rusafa Side [2]

side/ Baghdad (East Trunk Sewer System (ET) and Zublin Trunk). ET trunk served (Al-Adamiah, Al-Rusafa center, Part of Al- Gadeer, Al-Karada, Part of Baghdad Al Jadidah) Municipalities.

The East Trunk sewer commences at manhole ET 139 in 760 mm dia. pipe increasing to 1100 mm dia. At Wazariyah pumping Station the diameter increases from 1100 mm to 1600 mm. from Ghazali Pumping Station to the junction with Karrada submain trunk at ET 44 the diameter is 1850 mm. from ET 44 to the sewage treatment works the diameter increases from 2150 to 2300 mm in terminal section [2]. The east trunk sewer is likely to be overloaded at future design flows in its lower lengths allowing for the sewerage of any mashing smaller areas within its existing catchment; It is therefore recommended that no further major areas should be connected [8].

## 3. Methodology

Data Collection: The data collected from municipalities - sewage department, these Data include profiles of line (East trunk bank), maps that show municipalities zones and pump stations within areas served and discharge of these Pump Stations.

Data Analysis: The Manning equation is widely used and is one of the best open-channel hydraulics equations. Over time, it has largely replaced the Chézy and Kutter equations in engineering practice because of its relative simplicity. [9]. The value of roughness coefficient use, which depends on the type and internal condition of the pipe [1]. For older sewer pipes, a value of 0.015 may be used [9]. The general form of the Manning equation for open-channel flow is as follows:

$$v = \frac{1}{n} R^{2/3} S^{1/2} \qquad (1)$$

where

- V = average velocity of flow, m/s
- n = roughness coefficient, dimensionless
- R = hydraulic radius = (wetted cross-sectional area/ wetted perimeter), m [for a circular pipe flowing full) $R = {}^{D}/_{4}$
- S = slope of hydraulic energy gradient, m/m

The general form of the Manning equation for circular pipe flow is as follows:

$$Q = \frac{0.312}{n} D^{8/3} S^{1/2} \qquad (2)$$

where

- Q = flow rate, m3/s
- n = roughness coefficient, dimensionless
- D = inside diameter of pipe, m
- S = slope of hydraulic energy gradient, m/m

Self-Cleansing Velocity: Since sanitary sewers contain suspended solids that may deposit on the pipe wall, some minimum velocity is desirable to keep the solid particles suspended and in motion. This velocity that is necessary to prevent deposition of solids is known as the self-cleansing velocity. When a sanitary sewer system is flowing full, minimum velocities range from (0.6 to 0.75 m/s). Storm sewers generally have a minimum velocity range of (1.0 to 1.2 m/s) [11]. The velocity of flow ensures the solids will remain in suspension and move with the water. There is also a maximum velocity that must not be exceeded to prevent erosion of the sewer pipe. The maximum velocity is in the range of (3 to 3.5 m/s) for both sanitary sewers and storm sewers [9].
The flow from pump stations calculated as flowrate from final pump station that connected with ET line and 60% as efficiency [3].

Capacity requirements determine the sewer slope, often sanitary sewers through 15 inches (375 mm) in diameter are designed to flow half-full. Larger sanitary sewers have been typically designed to flow three-quarters full [1]. Hence, Calculate the Y/D ratio by using figure That Should be equal to 75% for Diameter Larger than 700 mm [11].

Represent the Data calculation in figure 3 to Show the variation of Sanitary Load through the Sections and figure 4 to show variation of flow velocity.

**Table 1.** Data Calculations; MH: manhole, D(m): diameter, Qa(l/s): actual flow rate (from P.S)[2].

| sections | MH | D (m) | Qa (l/s) | P.S |
|---|---|---|---|---|
| 1 | ET139-ET135 | 0.76 | 147.6 | Al jawazat (322) |
| 2 | ET134-ET118 | 0.84 | 114 | Al-Shamasia (318) |
| | | 0.84 | 127.2 | Aswaq Zohair (316) |
| 3 | ET118-ET96 | 1.10 | 33 | Al Moror Al Sarie (307) |
| | | 1.10 | 612 | Al Mojama Al Ilmi (304) |
| | | 1.10 | 66 | Al Waziriah Buildings (305) |
| 4 | ET96-ET88 | 1.30 | 114 | Al Sinaa (303) |
| | | 1.30 | 81.6 | Al Muror (301) |
| | | 1.30 | 99.96 | 58 (147) |
| | | 1.30 | 16.8 | Bueeot Al-Ama (137) |
| 5 | ET88-ET71 | 1.60 | 99.96 | Birute (503) |
| | | 1.60 | 15 | Al-Sekak (503) |
| | | 1.60 | 19.8 | 135 |
| | | 1.60 | 1500 | Al Gazali (135) |
| 6 | ET71-ET44 | 1.85 | 16.8 | AL-Maardh/1 (147) |
| | | 1.85 | 16.8 | Al-Jumlla Al-Asabia (105) |
| | | 1.85 | 30 | 8B (103) |
| | | 1.85 | 30 | 9A (101) |
| | | 1.85 | 58.2 | 9ب (702) |
| | | 1.85 | 50.4 | Al-Qaser Al-Abeeadh (103) |
| | | 1.85 | 27.6 | 702 (702) |
| | | 1.85 | 37.8 | 10 B (103) |
| | | 1.85 | 50.4 | 27 (708) |
| | | 1.85 | 30 | 11 (904) |
| | | 1.85 | 37.8 | 12 (906) |
| | | 2.15 | 414 | Al-Hindiah (903)+14 B (409)+13 (904) |
| 7 | ET44-ET1 | 2.15 | 177.6 | Zaiona 3 + Zaiona 2+Sommer (704) |
| | | 2.30 | 216 | 16 B (910) |
| | | 2.30 | 10.8 | 701 |
| | | 2.30 | 99.6 | Arafa (701) |
| | | 2.30 | 241.8 | Pepsi +No. 3 (703) |
| | | 2.30 | 6.6 | 713 |
| | | 2.30 | 399.6 | SE 1 (954) |

## 4. Results and Discussions

The ET line (East Bank Trunk) was analyzed by using the Manning equation for the design discharge (full flow rate) calculation according to the pipe diameters (D) and the hydraulic gradient (S) using n= 0.015, depending on the operational conditions from the pumping stations including the number of pumps operated and the discharge of each pump with the efficiency of the pump 60%. The result of analysis shows the ability of (East Bank Trunk) to quantity of discharge incoming from pumping stations (assuming there are no sediments) except the end of section 7 overflow appearance at ET7 from SE1 pump station that served Al- Zafarania city from Karada Municipal as shown in Table (1).

The Y/D ratio should not exceed 75% in large diameter, The Figure (3) shows this ratio during the sections and shows that some stations have exceeded this 75% as in Aswaq Zohair (316) (section 2), Al Waziriah Buildings (305) (section 3) in Al-Adhamiha Maniciple,5B (147) (section 4) in Rusafa Municipal , Arafa (701),Pepsi +No. 3 (703) (section 7) in Baghdad al-Jadidah municipal and finally SE 1 (954) (section 7) in Al-Karada municipal.

**Table 2.** Results : MH : manhole Qt(l/s) :total flow rate ,Qd (l/s): design flow rate , V(m/s): flow velocity, 75%: partial flow rate ratio

| sections | MH | Qt (l/s) | Qd (l/s) | V(m/s) | QT/QD % | 75% | Y/D |
|---|---|---|---|---|---|---|---|
| 1 | ET139-ET135 | 147.6 | 340.41 | 0.32 | 43 | 75 | 46 |
| 2 | ET134-ET118 | 261.6 | 412.92 | 0.47 | 63 | 75 | 58 |
| | | 388.8 | 412.92 | 0.70 | 94 | 75 | 77 |
| 3 | ET118-ET96 | 421.8 | 775.93 | 0.44 | 54 | 75 | 53 |
| | | 612 | 725.81 | 0.64 | 84 | 75 | 70 |
| | | 678 | 725.81 | 0.71 | 93 | 75 | 77 |
| 4 | ET96-ET88 | 792 | 981.35 | 0.60 | 81 | 75 | 68 |
| | | 873.6 | 981.35 | 0.66 | 89 | 75 | 73 |
| | | 973.56 | 981.35 | 0.73 | 99 | 75 | 81 |
| | | 990.36 | 1365.79 | 0.49 | 73 | 75 | 63 |
| 5 | ET88-ET71 | 1090.32 | 1527.00 | 0.54 | 71 | 75 | 62 |
| | | 1105.32 | 1763.23 | 0.55 | 63 | 75 | 57 |
| | | 1125.12 | 1763.23 | 0.56 | 64 | 75 | 58 |
| | | 1500 | 2097.15 | 0.56 | 72 | 75 | 63 |
| 6 | ET71-ET44 | 1516.8 | 2097.15 | 0.56 | 72 | 75 | 63 |
| | | 1533.6 | 2169.07 | 0.57 | 71 | 75 | 62 |
| | | 1563.6 | 2169.07 | 0.58 | 72 | 75 | 63 |
| | | 1593.6 | 2169.07 | 0.59 | 73 | 75 | 64 |
| | | 1651.8 | 2169.07 | 0.61 | 76 | 75 | 65 |
| | | 1702.2 | 2169.07 | 0.63 | 78 | 75 | 67 |
| | | 1729.8 | 2169.07 | 0.64 | 80 | 75 | 67 |
| | | 1767.6 | 2338.33 | 0.66 | 76 | 75 | 65 |
| | | 1818 | 2338.33 | 0.68 | 78 | 75 | 66 |
| | | 1848 | 2418.07 | 0.69 | 76 | 75 | 65 |
| | | 1885.8 | 2514.39 | 0.70 | 75 | 75 | 65 |
| | | 2299.8 | 2837.65 | 0.63 | 81 | 75 | 68 |
| 7 | ET44-ET1 | 2477.4 | 2837.65 | 0.68 | 87 | 75 | 72 |
| | | 2693.4 | 3177.37 | 0.65 | 85 | 75 | 71 |
| | | 2704.2 | 3177.37 | 0.65 | 85 | 75 | 71 |
| | | 2803.8 | 3177.37 | 0.68 | 88 | 75 | 73 |
| | | 3045.6 | 3177.37 | 0.73 | 96 | 75 | 78 |
| | | 3052.2 | 3177.37 | 0.74 | 96 | 75 | 79 |
| | | 3451.8 | 3177.37 | 0.83 | 109 | 75 | 100 |

Figure (4) shows the flow velocity through (East Bank Trunk) compared to the minimum self-cleaning velocity that shows the velocities are less than minimum self-cleaning velocity leading to cause sedimentation of particles.

## 5. Conclusion

The ET trunk has low slope that caused low self-cleaning velocity in trunk leading to sedimentation of particles and discharge of more than the design capacity of trunk during extreme conditions.

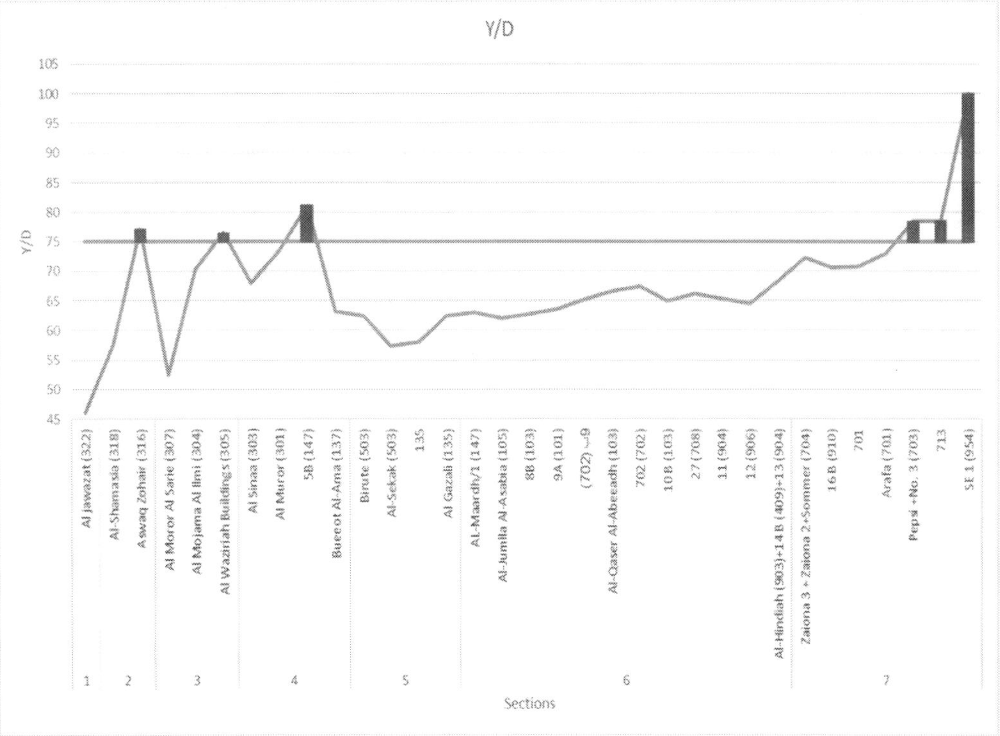


**Fig. 3.** Total flow rate compared with 75% Y/D ratio

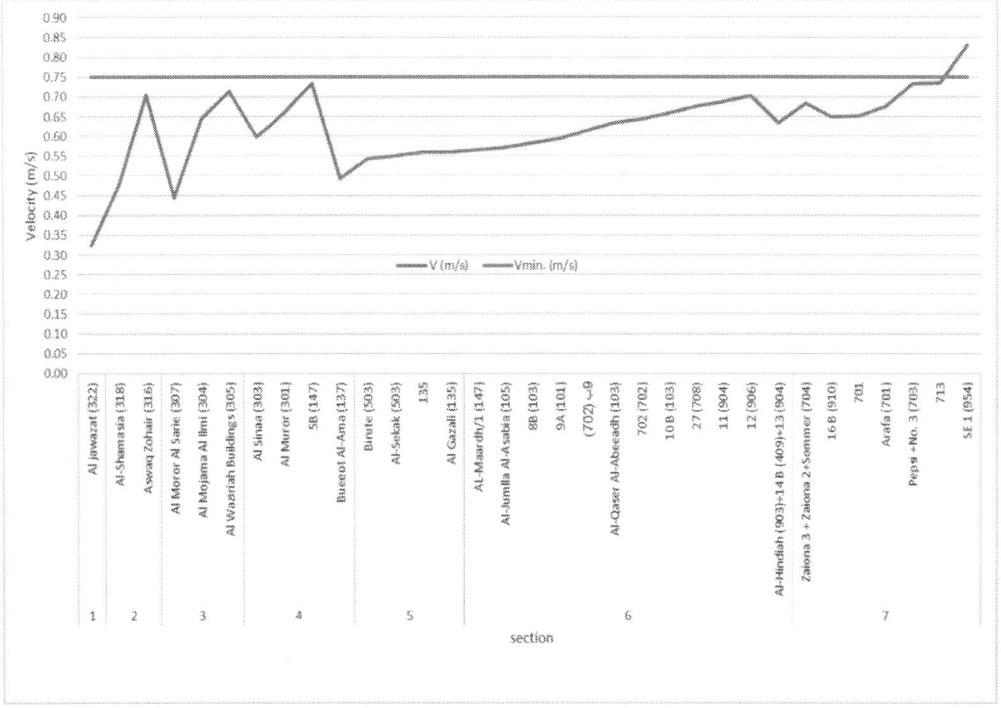


**Fig. 4.** Flow velocities with minimum self-cleaning velocity

## References

1. American Society of Civil Engineers (ASCE) "*Gravity sewer design and construction.*" **2nd** Edition (1982).
2. Baghdad Mayoralty, Design department (2016)
3. Baghdad Mayoralty, Pump Station department (P.S), workers . (2016)
4. J.R.. Buchanan, *Cost of Decentralized Unit Processes*. pp. 45-55 (2010). Available at: http://ndwrcdp.werf.org/documents/DEC2R08/DEC2R08web.pdf.
5. Corporation Central Engineering Laboratories. A Flushing System for Combined Sewer Cleansing, 11020DNO03/72, NTIS PB 219 858, U.S. Environmental Protection Agency, Office of Research and Monitoring, Washington,D.C. (1970)
6. M.J.. Delzingaro *There's No Use Crying Over Spilled Sewage: Using Standby Lift Station Pumps to Prevent SSOs before They Happen*. pp. 1-3 (2006). Available at: http://www.environmentalexpert.com/Files%5C5306%5Carticles%5C8984%5C168.pdf.
7. R. H. French, "*Hydraulics of open channel flow*." Stormwater collection systems design handbook, L. W. Mays, ed., McGraw-Hill, New York, with permission. (2001)
8. Haiste and Partiners. *Greater Baghdad Sewerage Study*. Volume **1** (1981)
9. Menon, E. Shashi. "*Piping Calculations Manual*" New York, McGraw-Hill (2005)
10. F. Ragsdale (n.d.). "*Wastewater System Operator's Manual*". Chapter **13**. Available at: http://www.ragsdaleandassociates.com/js2524/index.php/tmat/wastewater-systemoperator-s-manual.
11. E.W. Steel, & T.J. McGhee, " *water supply & sewerage*", **5th** Ed., McGraw-Hill (1979)
12. U.S. EPA (n.d.). *Combined Sewer Overflow (CSO)*. pp. 2-6. Available at: http://www.epa.gov/nrmrl/wswrd/wq/stormwater/cso.pdf.
13. WUTAP. *Wastewater Systems Operator Certification Study Manual* (2007). Available at: http://www.nmrwa.org/sites/nmrwa.org/files/WastewaterOperatorStudyManual.pdf.

# Assessment of air quality for Kerbala city using field survey

*Riyad* Al-Anbari [1], *Zainab* Mohammed *[1], *Muayad* Ismail[1], *Ali* Resheq [1]

[1] Building and Construction Engineering Department, University of Technology, Baghdad, Iraq

**Abstract**. The main objectives of this work are monitoring and analyzing air pollutant gases in Karbala city like (O3, SO2, H2S, VOCS, TSP, CO, CO2, NO2, PM1, PM2.5, PM7 and PM10). In this study, 11 deferent locations with 22 monitoring site has been selected to measuring the concentrations of air pollutants. The measuring period was six months starting from December (2015) to May (2016). The concentration results of TSP were more than Iraqi and world standard during April and May. The concentrations of CO recorded high values at the north of the city. The maximum concentration of NO2 was exceeding (1.5 ppm) at April 2016, while the minimum value was reached to (0.1 ppm) in many locations. The values of O3 were within international and Iraqi standard limits while the SO2 concentrations were above the International standard limits. High concentration of CO2 were founded in the north site of Karbala. There were high H2S concentrations in all sites, H2S values were between (0.006 ppm – 1.81 ppm) during the period study, which were exceeding acceptable limits value of the international. There is variation in the volatile organic compounds (VOCs), the highest concentration for VOCs was about (20 ppm) at December 2015. The minimum concentrations of VOCs were recorded during March 2016. In all sites the concentrations were within the acceptable limits (3 ppm). The results of field tests of air sites show that the highest pollutants concentrations were in December, April and May months, whereas lower concentrations of pollutants are observed during the remainder of the study, especially in March.

## 1. INTRODUCTION

Throughout the world, air pollution is a matter of concern at all levels. The worldwide epidemiological study on the effect of air pollution had revealed that gaseous pollutants and particulate matter had enough potential to cause severe health effect like respiratory, cardiovascular diseases and cardio pulmonary mortality **[1, 2]**. Being a serious matter of concern now-a-day, a systematic monitoring program all over the world especially in urban cities are urgently needed as the level of air pollution is increasing rapidly in many areas of mega cities of the developing world **[3]**. It was found that the modernization and industrialization of developing countries had led to the increase use of fossil fuels and their derivatives. As such, developing countries were confronted with the great challenge of controlling the atmospheric pollution especially in the rapidly growing mega cities. Concern about air pollution in urban regions is receiving increasing importance worldwide, especially pollution by gaseous and particulate trace metals **[4, 5]**. The urban centers might be viewed as dense sources of enormous anthropogenic emissions of pollutants, which could alter the atmospheric composition, chemistry and life cycles in its downwind regimes, extending over several hundred kilometers **[6]**. It had been found that world motor vehicle population growth had reached 700 million in the year 2000 **[7]**. Petrol and diesels engines of motor vehicles were found to emit a wide variety pollutants, principally, oxide of nitrogen (NOx) which had an increasing impact on urban air quality **[8]**.

## 2. OBJECTIVE

The main objective of this work are measuring and analyzing air pollutant gases in Karbala city like ($O_3$, $SO_2$, $H_2S$, $VOC_S$, $TSP$ $CO$, $CO_2$, $NO_2$, $O_3$, $SO_2$, $H_2S$, $VOC_S$, $TSP$, $PM_1$, $PM_{2.5}$, $PM_7$ and $PM_{10}$)

## 3. STUDY AREA

Karbala is an Iraqi city, located about 100 km southwest of Baghdad. Karbala is the capital of Karbala Governorate, the geographic location of the city is **32.61°N, 44.08°E**, and has an estimated population of (1,151,200) people (Statistical Center Organization in Karbala, 2014). Karbala was one of Iraq's wealthiest cities, profiting from both religious visitors and its agricultural produce, especially dates. It split into two districts, "Old Karbala", the religious center, and "New Karbala", the residential district containing Islamic

© The Authors, published by EDP Sciences. This is an open access article distributed under the terms of the Creative Commons Attribution License 4.0 (http://creativecommons.org/licenses/by/4.0/).

schools and government buildings. Figures (1,2) below show the location of Karbala city to Iraq and the map of Karbala city.

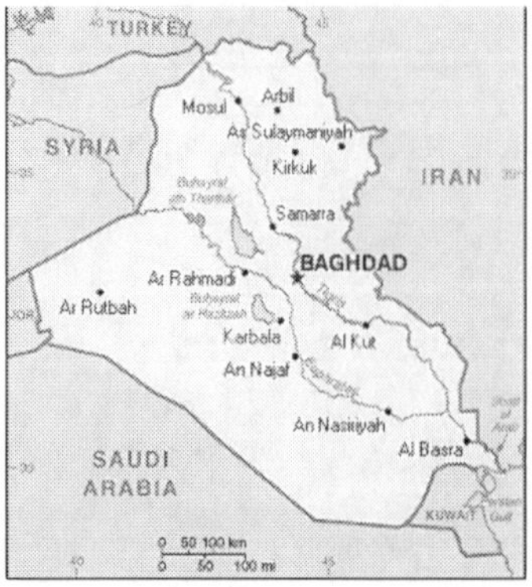


Figure (1): Map of Iraq show the location of Karbala governorate

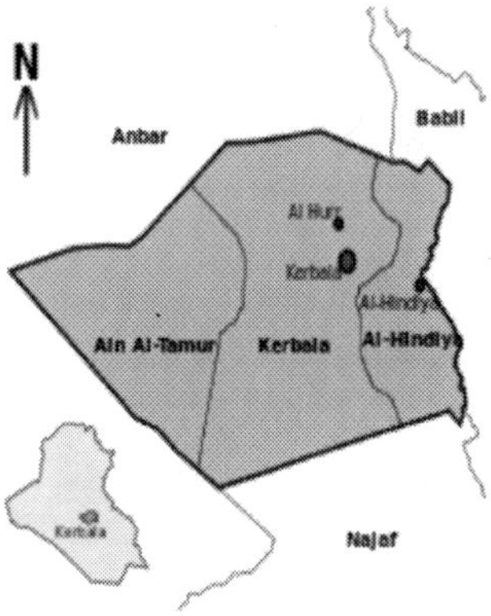


**Figure (2) the map of Karbala city**

## 4. METHODOLOGY AND SAMPLING

In this study, 11 deferent locations with 22 monitoring site has been selected to measuring the concentrations of air pollutants. The distribution of selected sites was to cover nearly all the area of Karbala city, including the mainly crowded intersections and different land classifications (residential, commercial, vegetation, barren and industrial areas). The measurement of concentrations has been done by using different devices to monitoring the concentrations of air pollutants such as (AEROCET 531, GIG450 devices, NOVA600 devices, MiniRAE 3000, Eco Sensors UV-100) as shown in Figure (3), which measure (TSP, $PM_1$, $PM_{2.5}$, $PM_7$, $PM_{10}$, $H_2S$, $NO_2$, $SO_2$, $CO_2$, CO, VOCs, $O_3$). The site locations were located by using Garmin GPS72. The measuring period was six months starting from December (2015) to May (2016).

## 5. RESULTS AND ANALYSIS

The results of this study deals with air pollutants (TSP, $PM_1$, $PM_{2.5}$, $PM_7$, $PM_{10}$, $H_2S$, $NO_2$, $SO_2$, $CO_2$, CO, VOCs, $O_3$) for six months starting from December 2015 to May 2016 and analysis their results. The average gases, TSP and PM concentrations measurements of air in the sites of the study area for six months are shown in Table (1).

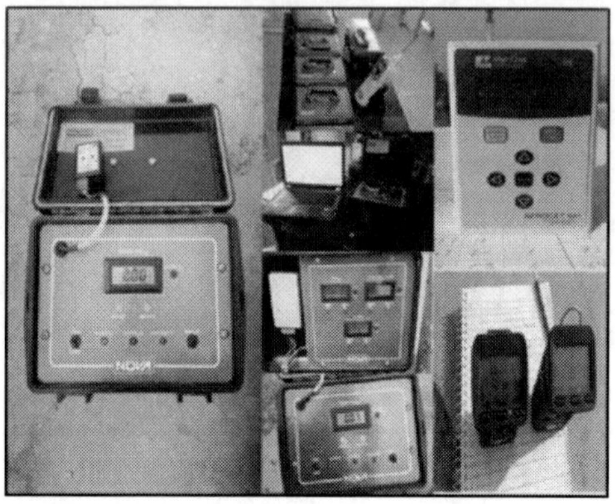

Figure (3): The equipment used in the fieldwork to measure the concentrations of air pollutants in Karbala city

Table (1): Average gases, TSP and PM concentrations measurements of air in the sites of the study area for six months

| Site Name | Site No. | RH % | Temp. °C | VOCs ppm | $H_2S$ ppm | $CO_2$ ppm | $O_3$ ppm | $NO_2$ ppm | $SO_2$ ppm | CO ppm | $PM_1$ µg/m³ | $PM_{2.5}$ µg/m³ | $PM_7$ µg/m³ | $PM_{10}$ µg/m³ | TSP µg/m³ |
|---|---|---|---|---|---|---|---|---|---|---|---|---|---|---|---|
| [illegible] | Site 1 | [illegible] | [illegible] | [illegible] | [illegible] | [illegible] | [illegible] | [illegible] | [illegible] | [illegible] | [illegible] | [illegible] | [illegible] | [illegible] | [illegible] |
| | Site 2 | [illegible] | [illegible] | [illegible] | [illegible] | [illegible] | [illegible] | [illegible] | [illegible] | [illegible] | [illegible] | [illegible] | [illegible] | [illegible] | [illegible] |
| [illegible] | Site 3 | [illegible] | [illegible] | [illegible] | [illegible] | [illegible] | [illegible] | [illegible] | [illegible] | [illegible] | [illegible] | [illegible] | [illegible] | [illegible] | [illegible] |
| | Site 4 | [illegible] | [illegible] | [illegible] | [illegible] | [illegible] | [illegible] | [illegible] | [illegible] | [illegible] | [illegible] | [illegible] | [illegible] | [illegible] | [illegible] |
| [illegible] | Site 5 | [illegible] | [illegible] | [illegible] | [illegible] | [illegible] | [illegible] | [illegible] | [illegible] | [illegible] | [illegible] | [illegible] | [illegible] | [illegible] | [illegible] |
| | Site 6 | [illegible] | [illegible] | [illegible] | [illegible] | [illegible] | [illegible] | [illegible] | [illegible] | [illegible] | [illegible] | [illegible] | [illegible] | [illegible] | [illegible] |
| [illegible] | Site 7 | [illegible] | [illegible] | [illegible] | [illegible] | [illegible] | [illegible] | [illegible] | [illegible] | [illegible] | [illegible] | [illegible] | [illegible] | [illegible] | [illegible] |
| | Site 8 | [illegible] | [illegible] | [illegible] | [illegible] | [illegible] | [illegible] | [illegible] | [illegible] | [illegible] | [illegible] | [illegible] | [illegible] | [illegible] | [illegible] |
| [illegible] | Site 9 | [illegible] | [illegible] | [illegible] | [illegible] | [illegible] | [illegible] | [illegible] | [illegible] | [illegible] | [illegible] | [illegible] | [illegible] | [illegible] | [illegible] |
| | Site 10 | [illegible] | [illegible] | [illegible] | [illegible] | [illegible] | [illegible] | [illegible] | [illegible] | [illegible] | [illegible] | [illegible] | [illegible] | [illegible] | [illegible] |
| [illegible] | Site 11 | [illegible] | [illegible] | [illegible] | [illegible] | [illegible] | [illegible] | [illegible] | [illegible] | [illegible] | [illegible] | [illegible] | [illegible] | [illegible] | [illegible] |
| | Site 12 | [illegible] | [illegible] | [illegible] | [illegible] | [illegible] | [illegible] | [illegible] | [illegible] | [illegible] | [illegible] | [illegible] | [illegible] | [illegible] | [illegible] |
| [illegible] | Site 13 | [illegible] | [illegible] | [illegible] | [illegible] | [illegible] | [illegible] | [illegible] | [illegible] | [illegible] | [illegible] | [illegible] | [illegible] | [illegible] | [illegible] |
| | Site 14 | [illegible] | [illegible] | [illegible] | [illegible] | [illegible] | [illegible] | [illegible] | [illegible] | [illegible] | [illegible] | [illegible] | [illegible] | [illegible] | [illegible] |
| [illegible] | Site 15 | [illegible] | [illegible] | [illegible] | [illegible] | [illegible] | [illegible] | [illegible] | [illegible] | [illegible] | [illegible] | [illegible] | [illegible] | [illegible] | [illegible] |
| | Site 16 | [illegible] | [illegible] | [illegible] | [illegible] | [illegible] | [illegible] | [illegible] | [illegible] | [illegible] | [illegible] | [illegible] | [illegible] | [illegible] | [illegible] |
| [illegible] | Site 17 | [illegible] | [illegible] | [illegible] | [illegible] | [illegible] | [illegible] | [illegible] | [illegible] | [illegible] | [illegible] | [illegible] | [illegible] | [illegible] | [illegible] |
| | Site 18 | [illegible] | [illegible] | [illegible] | [illegible] | [illegible] | [illegible] | [illegible] | [illegible] | [illegible] | [illegible] | [illegible] | [illegible] | [illegible] | [illegible] |
| [illegible] | Site 19 | [illegible] | [illegible] | [illegible] | [illegible] | [illegible] | [illegible] | [illegible] | [illegible] | [illegible] | [illegible] | [illegible] | [illegible] | [illegible] | [illegible] |
| | Site 20 | [illegible] | [illegible] | [illegible] | [illegible] | [illegible] | [illegible] | [illegible] | [illegible] | [illegible] | [illegible] | [illegible] | [illegible] | [illegible] | [illegible] |
| [illegible] | Site 21 | [illegible] | [illegible] | [illegible] | [illegible] | [illegible] | [illegible] | [illegible] | [illegible] | [illegible] | [illegible] | [illegible] | [illegible] | [illegible] | [illegible] |
| | Site 22 | [illegible] | [illegible] | [illegible] | [illegible] | [illegible] | [illegible] | [illegible] | [illegible] | [illegible] | [illegible] | [illegible] | [illegible] | [illegible] | [illegible] |

### 5.1 Total Suspended Particulate (TSP)

The concentration of total suspended particulate (TSP) from (22) locations were obtained as shown in **Figure (4)** below. The figure demonstrate high values of TSP above WHO standards in the April and May due to the increase the numbers of the cars.

### 5.2 Particulate Matter (PMs)

There is appositive relationship between (TSP) and particular matter **Figures (5a and 5b)** shows the concentrations of PM10 and PM2.5 were withen the WHO standards.

### 5.3 Carbon Monoxide (CO)

It was notice that highest concentrations of CO was in the north of city at locations No. (10, 11, 12, 13, 14, and 15) during the study period as shown in **Figure (6)** below. Although these value are high but within acceptable limits of Iraqi standards that is (35 ppm) and exceeding international standards, which is (9 ppm). The remaining sites ranging between (0.15 ppm – 14 ppm) and were within acceptable standards.

The increase in the car number and using bad-quality of benzene led to increase the CO emission which generates carbon monoxide gas instead of carbon dioxide gas, and the high temperature increase the value of CO.

High values of CO were founded in locations (No.12, 14) with values (9 ppm and 14 ppm respectively) during February 2016 to March 2015 due to the traffic volume and the presence of garage and petrol station. minimum concentration was (0.11 ppm) at (No.22) In December and January, as shown in **Figures (6),** this location is in the west north of city and conceder as agriculture area with low traffic volume.

### 5.4 Nitrogen Dioxide ($NO_2$)

Figure (7) shown that the concentration of $NO_2$ was higher than International limits which is about (0.11 ppm).

site (No.16) shows maximum concentration of $NO_2$ with value exceeding (1.5 ppm) at April 2016, While many locations through several months shows the minimum value reaching to (0.1 ppm).

Graphically, It can noticed that in December 2015, the maximum value of $NO_2$ was (1.35 ppm) which detected in site (No.20), while the minimum value was (0.1 ppm) in site (No.10). In the next two month, the values were gradually decrease in the most of the sites. While in April and May gradually increase in concentrations for all locations of the study area.

Rising $NO_2$ concentrations shown in the north and middle of city. Which represented by locations No. (1, 2, 3, 15, 16, 17, 18, 19, and 20).

### 5.5 Ozone ($O_3$)

**Figure (8)**. Showed that the concentrations of $O_3$ ranging between (0.01-0.08 ppm) not exceeded the Iraqi and international standard, which are (0.08 ppm), and (0.12 ppm) high values of $O_3$ were in site (No.14) measured in December While the minimum value of ozone gas was in many locations through different periods.

### 5.6 Sulfur Dioxide ($SO_2$)

The concentrations of $SO_2$ were exceeding the International standard limits in almost locations for the city as shown in **Figure (9)** .

In December, the minimum value (0.18 ppm) was founded at location (No.20) south of city. On the other hand, the maximum value of (0.9 ppm) founded in the location of (No.16). While the remaining sites, their concentrations ranging between (0.18 -0.8 ppm).

Finally, it can be notice that the concentration of $SO_2$ increased due to industrial activity in this area.

### 5.7 Carbon Dioxide ($CO_2$)

the values of $CO_2$ were highly recorded in the north and the middle of city due to the main road that connect with Baghdad, also the main intersection in Karbala city. High concentrations were resulting from passing a significant number of big trucks that produce a large quantities of carbon dioxide and contaminated the surrounding air.

The highest concentration was in December 2015 (380 ppm) and this value was above the WHO limits (250 ppm) as shown in **Figure (10)** at site No.6 and the lowest concentration was (300 ppm) in site No.3 during study period.

The minimum concentrations during the study period were during March that witnessed decreasing in concentrations in all sites during this month. The next two months witnessed gradually increase in $CO_2$ concentrations.

### 5.8 Hydrogen Sulfide ($H_2S$)

The concentrations of $H_2S$ ranging between 0.006-1.81ppm and these values above the acceptable limits 0.047ppm, as shown in **Figures (11)**.

highest concentration recorded at May 2016 in all sites with concentrations ranging between (0.3 ppm) to (1.81 ppm). While the lowest concentration was (0.004 ppm) at many locations during several months as shown in **Figures (11)**.

The lowest concentration of $H_2S$ was founded at almost months while The high concentrations of $H_2S$ recorded in the north of the city at may 2016 due to industrial activity in the area ,destruction of sewer networks.

### 5.9 Volatile Organic Compounds (VOCs)

**Figure (12)** show, the highest concentration for VOCs was in site No.14 with value about (20 ppm) at December 2015, due to high traffic volume in this area and the proximity from petrol station and textile factory. While the lowest values of VOCs measured to be zero ppm in many sites during several periods.

https://doi.org/10.1051/matecconf/201816205015

The concentrations of VOCs at locations (No.14, 18, 19, 20, 21, and 22) exceeding the acceptable limit standards. The presence of petrol stations in middle and north of the city is the main source to increase the values of VOCs.

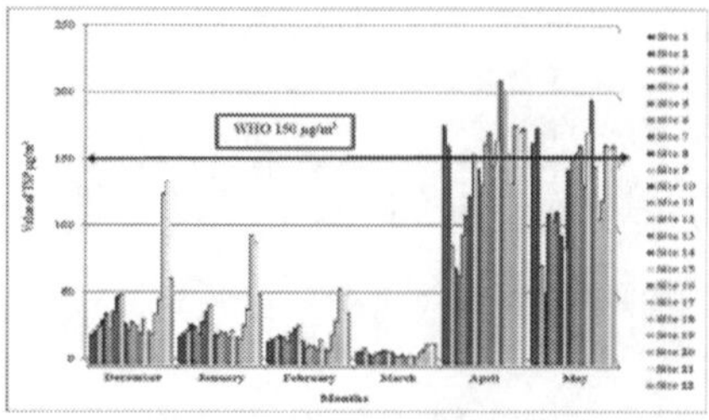


Figure (4): Concentration of TSP in the air of Karbala city for Six months.

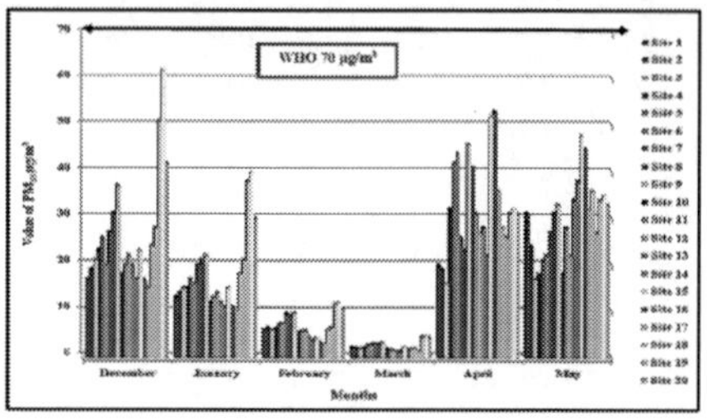


Figure (5, A): Concentration of $PM_{10}$ in the air of Karbala city for Six months.

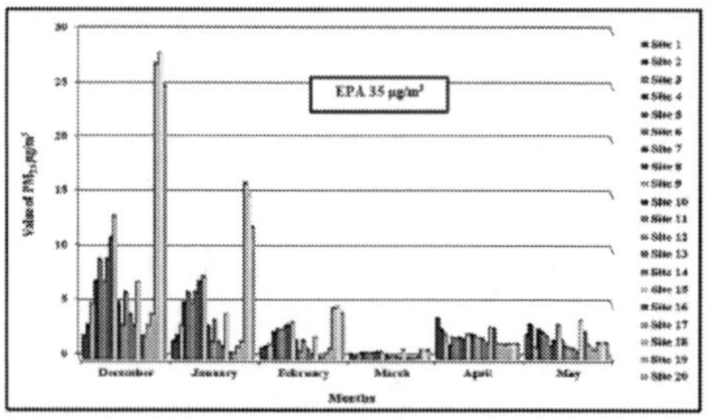


Figure (5, B): Concentration of $PM_{2.5}$ in the air of Karbala city for Six months.

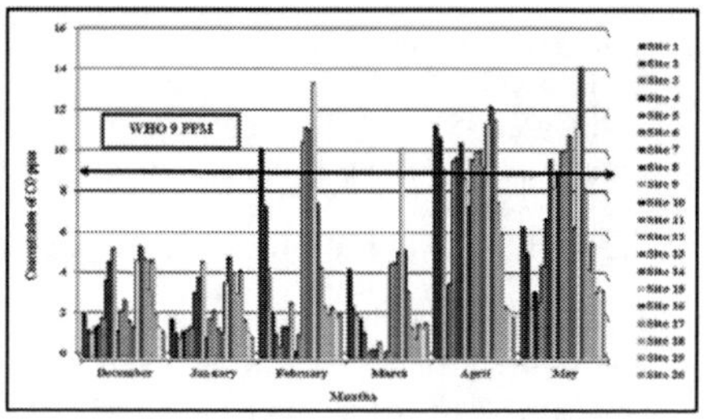


Figure (6): Concentration of CO in the air of Karbala city for Six months.

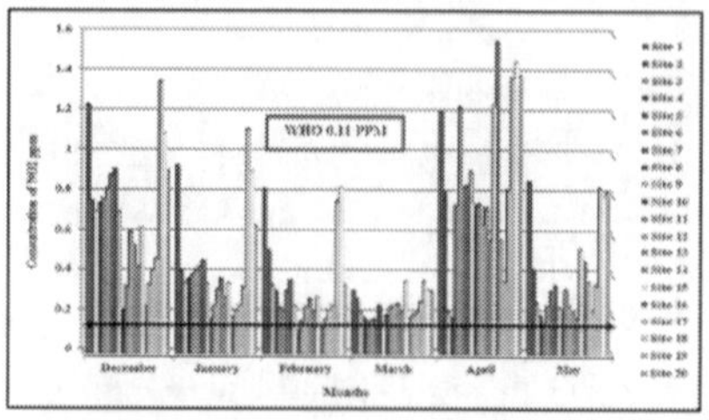


Figure (7): Concentration of $NO_2$ in the air of Karbala city for Six months.

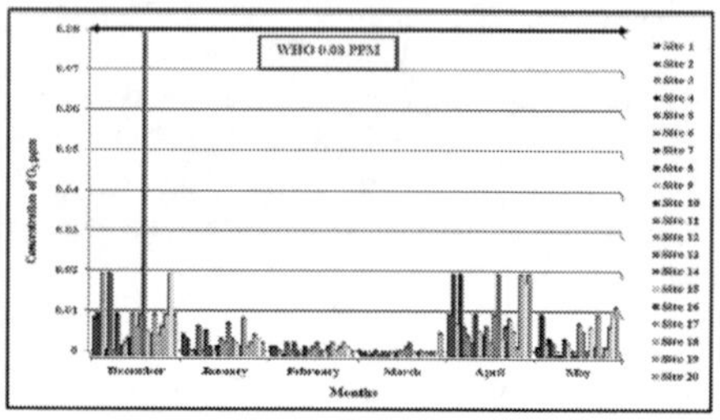


Figure (8): Concentration of $O_3$ in the air of Karbala city for Six months.

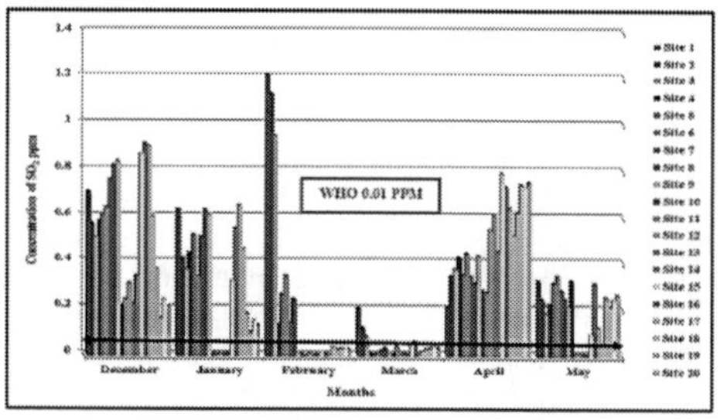


Figure (9): Concentration of $SO_2$ in the air of Karbala city for Six months.

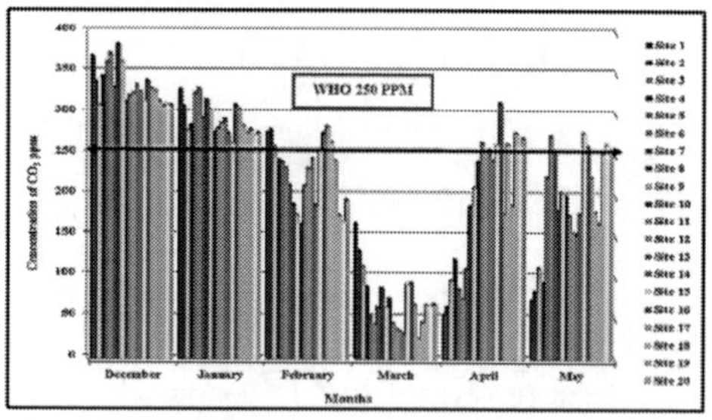


Figure (10): Concentration of $CO_2$ in the air of Karbala city for Six months.

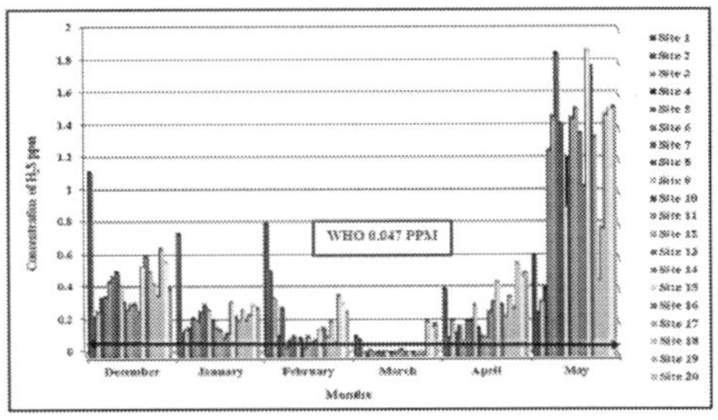

Figure (11): Concentration of $H_2S$ in the air of Karbala city for Six months.

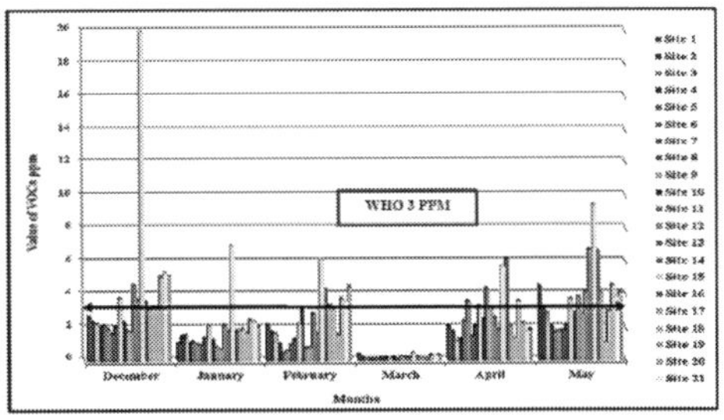

Figure (12): Concentration of $VOC_S$ in the air of Karbala city for Six months.

## 6. CONCLUSIONS

- High pollution by gases has increased in summer seasons for all gases while decreased in wintertime.
- The results of field tests of air sites show that the highest pollutants concentrations were in December, April and May months, whereas lower concentrations of pollutants are observed during the remainder of the study and especially in March. The study took into consideration the rainfall and its impact on the air clearness of pollutants, as well as the effect of wind direction and speed in the transfer of pollutants.
- values of (TSP) were above the permissible limits during April and May and high concentrations of PMs was founded in the same sites contain high values of TSP.
- highest concentrations of CO was in the north of city at locations No. (10, 11, 12, 13, 14, and 15). the concentration of $NO_2$ was higher than International limits
- The concentrations of $SO_2$ were exceeding the International standard limits in almost locations for the city. the concentrations of $O_3$ ranging between (0.01-0.08 ppm) not exceeded the Iraqi and international standard.
- the values of $CO_2$ were highly recorded in the north and the middle of city. The concentrations of $H_2S$ ranging between 0.006-1.81ppm and these values above the acceptable limits and the highest concentration for VOCs was in site No.14 with value about (20 ppm) at December 2015.

## REFERENCES

**[1]** D. W. Dockery, C. A. Pope, X. Xu, J. D. Splender, J. H. Ware, M. E. Fay, B. G. Ferris and F. E. Speizer, "An Association between Air Pollution and Mortality in Six US cities, New England," *Journal of Medicine*, Vol. 329, No. 24, 1993, pp. 1753 1759.

**[2]** P. J. Koken, W. T. Piver, F. Ye, A. Elixhauser, L. M. Olsen and C. J. Portier, "Temperature, Air Pollution and Hospitalization for Cardiovascular Diseases among Elderly People in Denver," *Environmental Health Perspectives*, Vol. 111, No. 10, 2003, pp. 1312-1317.

**[3]** United Nations Environment Programme, "Global Environment Outlook," Earthscan, London, 1999.

**[4]** A. K. Azad and T. Kitada, "Characteristics of the Air Pollution in the City of Dhaka, Bangladesh in winter," *Atmospheric Environment*, Vol. 32, No. 11, 1998, pp. 1991-2005.

**[5]** H. Cahier, A. F. ulagnier, R. Sarda, F. Gautier, P. Masclet, J. L. Besombes, *et al.*, "Aerosol Studies during the ESCOMPTE Experiment: An Overview," *Atmospheric Research*, Vol. 74, No. 1-4, 2005, pp. 547-563.

**[6]** A. K. Gupta, K. Karar, S. Ayoob and K. John, "Spatio- Temporal Characteristics of Gaseous and Particulate Pollutants in an Urban Region of Kolkata, India," *Atmospheric Research*, Vol. 87, No. 2, 2008, pp. 103-115.

**[7]** M. K. Ghose, R. Paul and S. K. Banerjee, "Assessment of the Impacts of Vehicular Emissions on Urban Air Qualit and its Management in Indian Context: The Case of Kolkata (Calcutta)," *Environmental Science and Policy*, Vol. 7, No. 4, 2004, pp. 345-351.

**[8]** D. Mage, G. Ozolins, P. Peterson, A. Webster, R. Orthofer, V. Vandeweerd and M. Gwynne, "Urban Air Pollution in Mega Cities of the World," *Atmospheric Environment*, Vol. 30, No. 5, 1996, pp. 681-686.

**[9]** Leksmono, N. S., Longhurst, J. W. S., Ling, K. A., Chatterton, T. J., Fisher, B. E. A., Irwin, J. G. "Assessment of the relationship between industrial and traffic sources contributing to air quality objective exceedances": a theoretical modelling exercise, Environmental Modelling & Software, 21, (4), pp. 494-500. (2006).

# Assessment of pollution in water quality of oil and grease for Tigris River-Iraq by GIS mapping (2010-2016)

*Ibtihal* Mawlood[1], *Abdul-Razzak* Ziboon[1] and *Abdul-Hammed* Al-Obaidy[2]

[1]Building and Construction Engineering Department, University of Technology, Baghdad, Iraq
[2]Environmental Research Center, University of Technology, Baghdad, Iraq

**Abstract.** In this study have followed the oils and grease in Tigris River within Baghdad city during the period 2010-2015 by using ArcGIS 10.2 program. Two satellite image of each year were used which were taken in as same time as collected samples and painted the mapping which show the locations of pollution the river by oil and grease. This study showed the Tigris river is pollution by the oil and grease, and the values were up limit allowable depend on the world health organization (who) and Iraqi criteria.

## 1 Introduction

Remote sensing techniques play increasingly important role over recent decades in both problems of global climate change and frequent deterioration of the status of aquatic ecology, driven by the ever-increasing needs of growing populations for drinking water, polluted by overland runoff from point and non-point sources, as well as fish and other seafood [1]. With the advent of new sensor technologies, it is possible to monitor land cover / land change in large area simultaneously and quickly [2, 3]. Remote sensing techniques have been widely used in water quality assessment [4, 5 and 6].

Many documents describe water quality monitoring using different satellite sensor [4, 7]. The spectrum characteristics of water and pollutants are essential to water quality monitoring and assessment. The spectral characteristics of the signal received from water are a function of hydrological, biological and chemical characteristics of water, and other interference factor [8]. Suspended sediments increase the radiance emergent from surface waters in the visible and near infrared proportion of the electromagnetic spectrum [9] so it is promising and feasible to detect water

Pollutants using spectral signatures in the visible and near infrared band. Tigris River is course water alone in Bagdad city in addition The Tigris has long been an important transport route in a largely desert country. Shallow-draft vessels can go as far as Baghdad.

The Tigris is heavily dammed in Iraq and Turkey to provide water for irrigating the arid and semi-desert regions bordering the river valley. Damming has also been important for averting floods in Iraq, to which the Tigris has historically been notoriously prone following April melting of snow in the Turkish mountains.

Recent Turkish damming of the river has been the subject of some controversy, for both its environmental effects within Turkey and its potential to reduce the flow of water downstream.

Water from both countries is used as a means of pressure during conflicts. [10]

In 2014 a major breakthrough in developing consensus between multiple stakeholder representatives of Iraq and Turkey on a Plan of Action for promoting exchange and calibration of data and standards pertaining to Tigris river flows was achieved. The consensus which is referred to as the "Geneva Consensus on Tigris River" was reached at a meeting organized in Geneva by the think tank Strategic Foresight Group. [11]

All that, Lead to save the river from any problem expose on it, and monitor the river periodically.

## 2 Study Area

The Tigris River is one of the largest rivers in Iraq, and is considered the main source of drinking water for Baghdad, which is the largest city in the country and the second largest city in the Arab world with a population estimated by 7.5 million [4,11].

© The Authors, published by EDP Sciences. This is an open access article distributed under the terms of the Creative Commons Attribution License 4.0 (http://creativecommons.org/licenses/by/4.0/).

https://doi.org/10.1051/matecconf/201816205016

The River enters Baghdad city at coordinates 44°24' E, 33°36' N and divides the city into two parts (Karkh & Risafa). The length of the river along Baghdad city is about 110 km. The samples collected from seven sites in the river as the show in table (1) and the figure (1) in the same time took the satellite image.

Table 1. Location of samples in study area

| site | Location | Coordinate | |
|---|---|---|---|
| Site 1 | Al-Muthanna Bridge | Lon: 44$^{0}$20'46" | Lat: 33$^{0}$25'32" |
| Site2 | Al-A'imma bridge | Lon: 44$^{0}$21'19" | Lat: 33$^{0}$22'28" |
| Site 3 | Al-Shuhada Bridge | Lon: 44$^{0}$23'16" | Lat: 33$^{0}$20'17" |
| Site 4 | Al- Jumhuiya Bridge | Lon: 44$^{0}$26'31" | Lat: 33$^{0}$19'28" |
| Site 5 | Al- Jadriah Bridge | Lon: 44$^{0}$2$^{3}$'09" | Lat: 33$^{0}$17'26" |
| Site 6 | Al- Rasheed w.w | Lon: 44$^{0}$22'55" | Lat: 33$^{0}$15'43" |
| Site 7 | Al- Doura w.w | Lon: 44$^{0}$27'15" | Lat: 33$^{0}$16'17" |

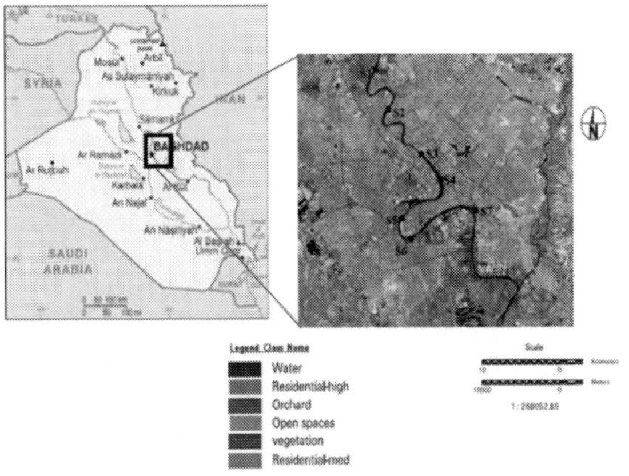


**Fig 1:** shows the site of sample in Tigris River

## 3 Materials and methods

The satellite images from the Landsat archived on United States Geological Survey (USGS) server (http://earthexplorer.usgs.gov/) were used in this study. The Landsat Enhanced Thematic Mapper plus (ETM+) image was acquired in (2010-2016).All images employed in this study are level-1 products of the USGS. They are referenced in the World Geodetic System (WGS84) datum, in GeoTiff format, and are projected using the Universal Transverse Mercator system (zone UTM 38 North).

Geographical Information system (GIS) is a computer-based technology for handling geographical data in digital form. It is designed to capture, store, manipulate, analyze, and display diverse sets of spatial or geo-referenced data [12, 13]. Coordinates of sampling points were recorded by mobile GPS (Etrax). The results of the chemicals analysis (mean value per sampling point) were then used as input data in ArcGIS 10.2. The sampling locations were integrated with the water data for the generation of spatial distribution maps. The present study used the Inverse Distance Weighted (IDW) method for spatial interpolation of water parameters. IDW determines cell values using a liner-weighted combination set of sample points. The weight assigned is a function of the distance of an input point from the output cell location. The greater the distance, the less influence the cell has on the output value [14]

The Analyzing based on contaminant concentration change in the river from the low concentration to high concentration, where this indicates that the presence of contamination points in that region and Search for these points sources of pollution of the river.

### 3.1 Oils and grease (O&G)

The concentration of single oils and grease (OG) is an important limitation for water quality and safety. OG in water can reason surface films and shoreline deposits leading to environmental degradation, and can induce human health risks when discharged in surface or ground waters. Also, OG might interfere with aerobic and anaerobic biological processes and lead to declined wastewater treatment efficiency. Regulatory bodies worldwide set limits in order to control the amount of OG incoming natural bodies of water or reservoirs through industrial discharges, and to boundary the amount present in drinking water.

OG in water is normally determined by extraction into a non-polar, hydrocarbon free solvent followed by quantity of the infrared absorption spectrum of the extract. The absorption between 3000 and 2900 cm-1 by C-H sets in the OG is connected to the concentration of OG. There are several standard check protocols based around this methodology, 1–4 most usually using 1, 1, 2-trichloro-1, 2, 2-trifluoroethane (CFC-113) or tetra-chloromethane.

Though, these solvents are identified ozone-depleting compounds, and under the Montreal Protocol, the use of CFC-113 WAS PHASED OUT BY 1996 AND THE USE OF TETRA-chloromethane will develop illegal in 2010[15].

## 4 Results and Discussion

Figures from (2) to (9) have shown the classification of Tigris river according to O&G during the period 2010-2016.The Iraqi Ministry of health Stipulated on maximum concentration of O&G is up stream of (0.1mg/l).all maps show increased about concentration up of the limit in all

 https://doi.org/10.1051/matecconf/201816205016

sites, as seen in figure (11)

Source of pollution in Tigris River: in through compare between the satellite images and determined the change in points from low concentration to high concentration, it can be determined the most important of source pollution in Tigris River depend on the drawing maps in this study and information from Field survey of the study area and the Department of Environment Baghdad and the Iraqi Ministry of Water Resources. As seen in Figure (10) and table (2)

Table2. Source of pollution in Tigris River

| STATION | COORDINATES | RESPONSIBLE | DISCHARGE L/SEC |
|---|---|---|---|
| RAINS STATION AL-MASBAH | N 330 17′ 54.6″ E 0440 25′55.1 | Al-karrada municipality | 360 |
| RAINS STATION AL-JADRIAH | N 330 17′ 32.5″ E 0440 23′ 28.2″ | Al-karrada municipality | 1500 |
| RAINS STATION AL-HURIYA | N 330 24′ 37.2″ E 044026′ 48.4″ | Al-karrada municipality | 420 |
| RAINS STATION AL-RASHEED | N 330 17′41.4″ E 0440 27′11.6″ | Al-karrada municipality | 4500 |
| RAINS STATION SHIP | N 33022′02.2″ E 0440 21′ 20.6″ | Al-adhamiya municipality | 285 |
| RAINS STATION A1 | N 330 23′ 28.2″ E 0440 21′ 35.7″ | Al-adhamiya municipality | 3200 |
| RAINS STATION AL-KUWAITY | N 330 24′10.5″ E 0440 20′ 24.2″ | Al-adhamiya municipality | 390 |
| RAIN STATION T1 | N 330 21′ 05.7″ E 0440 22′ 29.0″ | Municipality rusafa center | 15000 |
| RAIN STATION PN-AL-SAIDIYA | N 330 16′ 05.5″ E 0440 22′20.5″ | Rasheed municipality | 18000 |
| RAIN STATION ABDUL MOHSEN AL-KAZEMI PUMPING | N 330 21′ 59.2″ E 0440 20′ 43.8″ | Kadhimiya municipality | 9000 |

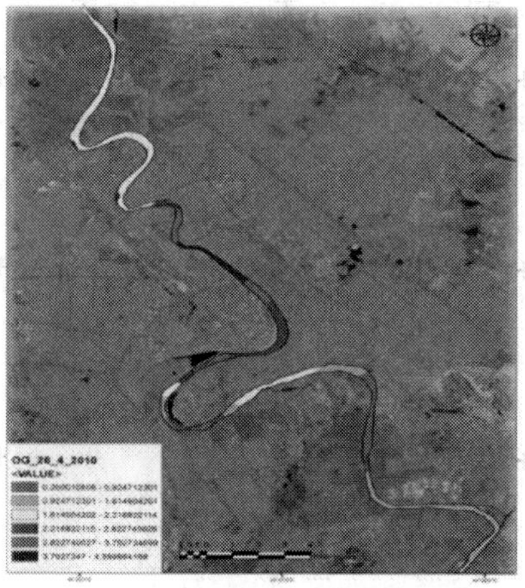


**Fig 2:** shows the classification of Tigris River by oils and grease in April 2010

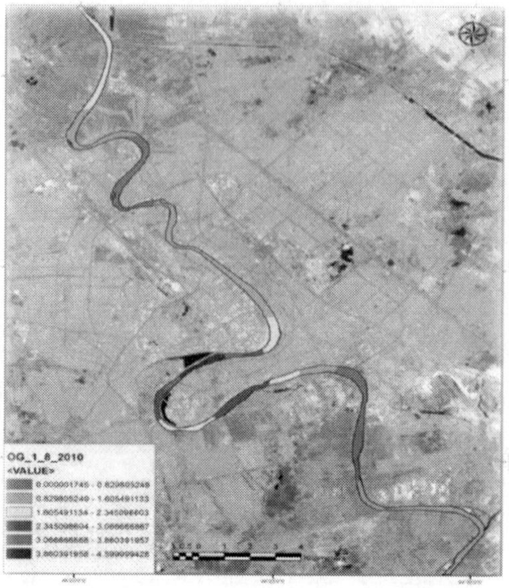


**Fig 3:** shows the classification of Tigris River by oils and grease in Aug. 2010

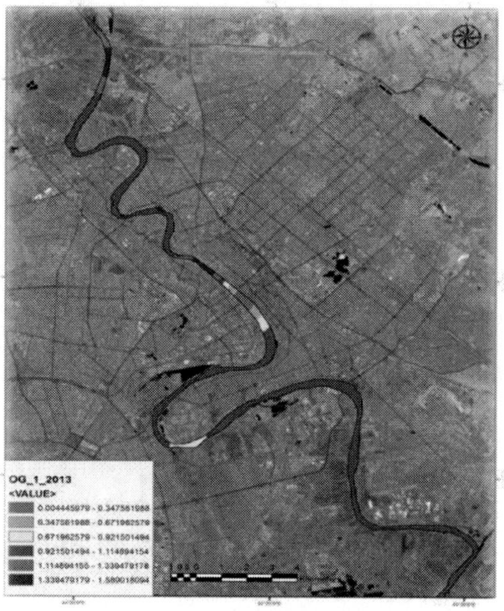


**Fig 4:** shows the classification of Tigris River by oils and grease in Janu 2013

https://doi.org/10.1051/matecconf/201816205016

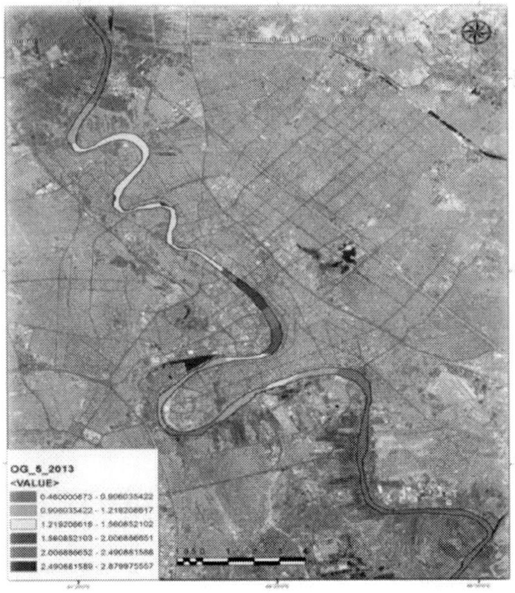


**Fig 5:** shows the classification of Tigris River by oils and grease in May 2013

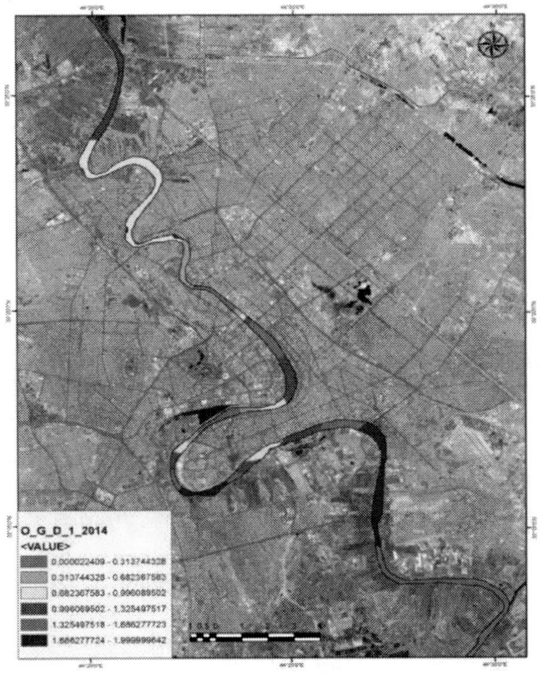


**Fig 6**: shows the classification of Tigris River by oils and grease in Janu 2013

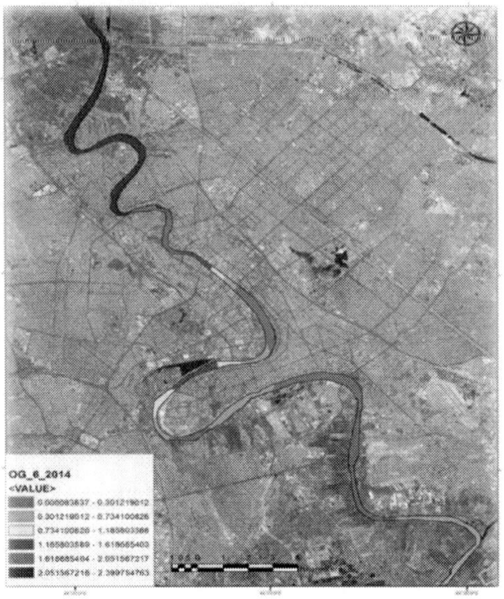


**Fig 7:** shows the classification of Tigris River by oils and grease in Jun 2014

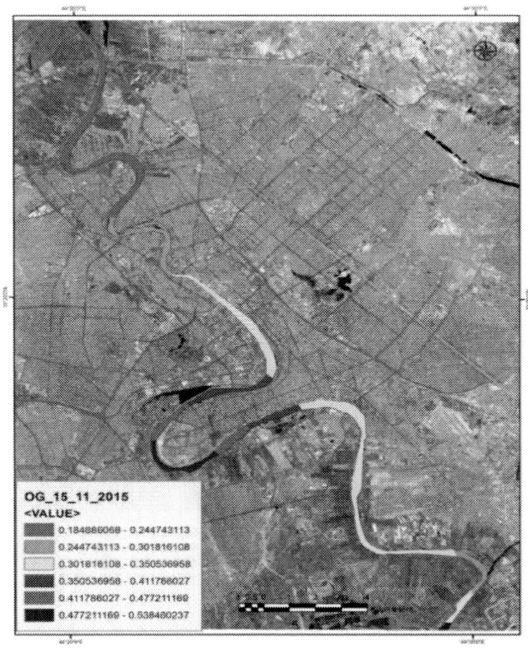


**Fig 8:** shows the classification of Tigris River by oils and grease in Nov. 2015

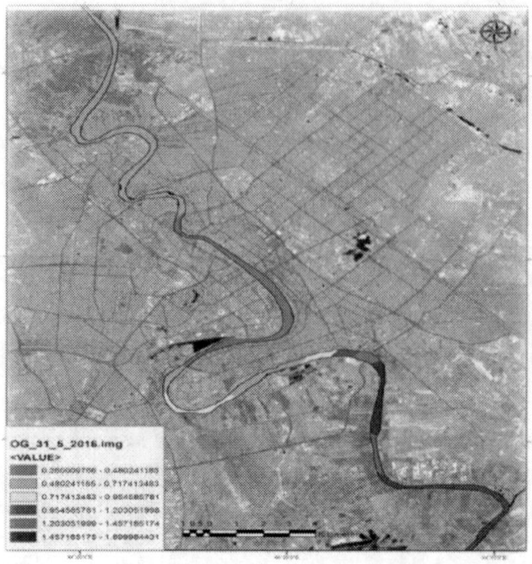


**Fig 9:** shows the classification of Tigris River by oils and grease in May 2016

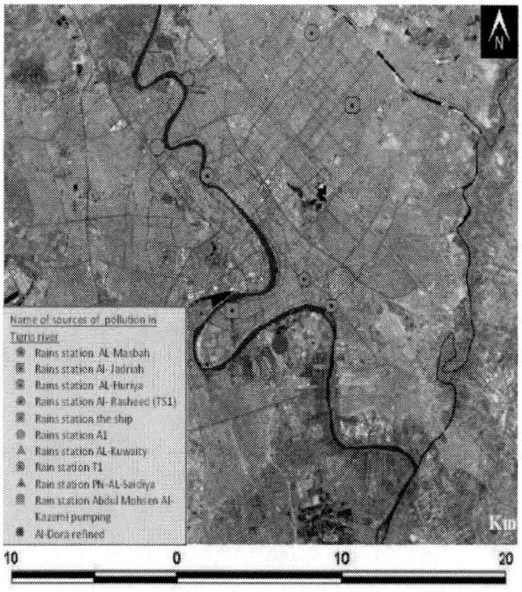


**Fig 10:** shows the Source of pollution by oils and grease in Tigris River

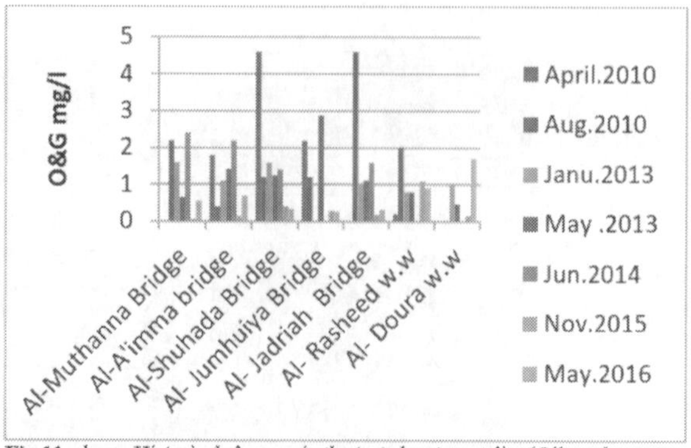


**Fig 11:** shows Historical changes in the tested water quality (Oils and Grease) in Tigris River

## References

[1]Pozdnyakov, D., Shuchman, R., Korosov, A. & Hatt, C. (2005). Operational algorithm for the retrieval of water quality in the Great Lakes. Remote Sensing of Environment, Vol. 97, No. 3, 352-370.

[2] Wen, X. & Yang, X. (2009a). Change detection from remote sensing imageries using spectral change vector analysis. Asia-Pacific Conference on Information Processing (APCIP2009), 189-192, IEEE Computer Society, Shenzhen, China

[3] Wen, X. & Yang, X. (2009b). A new change detection method for two remote sensing images based on spectral matching. International Conference on Industrial Mechatronics and Automation (ICIMA 2009), 89-92, IEEE Computer Society, Chengdu, China

[4]Alparslan, E., Ayd○ner, C., Tufekci, V. & Tüfekci, H. (2007). Water quality assessment at Omerli Dam using remote sensing techniques. Environmental Monitoring and Assessment, Vol. 135, No. 1, 391-398

[5]Brando, V. E. & Dekker, A. G. (2003). Satellite hyperspectral remote sensing for estimating estuarine and coastal water quality. Geoscience and Remote Sensing, IEEE Transactions on, Vol. 41, No. 6, 1378-1387.

[6]Chen, Q.; Zhang, Y. & Hallikainen, M. (2007). Water quality monitoring using remote sensing in support of the EU water framework directive (WFD): A case study in the Gulf of Finland. Environmental Monitoring and Assessment, Vol. 124, No. 1, 157-166.

[7]Boken, V. K. (2007). Linking landuse and groundwater quality in the Mississippi delta Using MODIS satellite data. IEEE International Geoscience and Remote Sensing Symposium, 5025-5027,

[8]Seyhan, E. & Dekker, A. (1986). Application of remote sensing techniques for water quality monitoring. Aquatic Ecology, Vol. 20, No. 1, 41-50.

[9]Ritchie, J.; Zimba, P. & Everitt, J. (2003). Remote sensing techniques to assess water quality. Photogrammetric Engineering and Remote Sensing, Vol. 69, No. 6, 695-704.

[10]Vidal, John. "Water supply key to outcome of conflicts in Iraq and Syria, experts warn" The Guardian, 2 July 2014.
[11]"Analysis & Water Agenda". ORSAM. Retrieved 2015-11-28
[12]V. P. Singh and M. Fiorentino, "Geographic Information System in Hydrology," Kluwer Academic Publishers, Dordrecht, 1996.
[13]Y. Saeedrashed and A. Guven, "Estimation of Geomor- phological Parameters of Lower Zab River-Basin by Us-ing GIS-Based Remotely Sensed Image," *Water Resour- ces Management*, Vol. 27, No. 1, 2013, pp. 209-219. http://dx.doi.org/10.1007/s11269-012-0179-x
[14]C. Childs, "Interpolation Surfaces in ArcGIS Spatial Analyst," ESRI Education Services, 2004. www.seri.com
[15] Chapman. D., 1996. Water Quality Assessments – A Guide to Use of Biota, Sediments and water in Environmental Monitoring. Second ed. Great Britain at the University Press, Cambridge: UNESCO/WHO/UNEP.

MATEC Web of Conferences **162**, 05017 (2018) https://doi.org/10.1051/matecconf/201816205017
*BCEE3-2017*

# Suitability of the Iraqi natural clay for the preparation of Al-Fe pillared- clays

*Faris* Al-Ani[1], *Ghayda* Al-Kindi[1,*]

[1]Building and construction Engineering Department, University of Technology, Iraq

**Abstract.** Some organic pollutant in aqueous wastewater inhabitant biological treatment, catalyst wet air oxidation is a method to treat this waste. Al-Fe pillared Iraqi clays with different Fe content were prepared. Aluminum was mixed with iron in different molar ratio Fe/Al (1:9- 2:8- 3:7- 4-6- 5:5) for each regions. These clays collected from different location in Iraq (Anbar, Erbil, Mosel, Baghdad, and Basra), the results of some properties studied for natural and pillared clay, from X-Ray-Fluorescence (XRF) show that iron cation replaces some Ca in the clay, and from X-Ray Diffraction (XRD) show the Calcite types of Iraqi clays. constructs clusters of active ingredient which dramatically clear on the catalytic surface for all catalyst types show through photos of catalysts by Scanning Electron Microscopy (SEM), the best molar ratio of Fe/Al in which give strong interaction between Fe and Al in the pillared were (0.5-0.4-0.3-0.3 and 0.2) for (Mosul-Basra –Anbar-Baghdad and Erbil) respectively. This ratio gives high surface area, and pore size in mesopore. It is concluded that clay and silty clay (Mosul clay) consider the best catalyst due to higher activity and stability through experiment of CWAO in batch reactor, with operation condition 150°C, 2.5 Mpa, Ccat equal to 7 mg/l.

## 1 Introduction

Pillared is one of the methods of clay modification, which have used the ion material in application of catalysts potential such as adsorption and separation processes [1]. That widespread delayed because in pilot scale difficulty obtaining a uniform and reproducible product because of several variables contribute to the preparation process and influence on the PILCs' properties [2]. The minerals clay properties tendency to improve the pillared such as, environmental compatibility, reusability and operational simplicity, low cost, high selectivity, porosity and stability of these materials [3], so that clay support catalyst exemplify of the largest group of catalysts (heterogeneous) which are important particularly in petroleum and petrochemical industries [4], as well as for its wastewater treatment.

The clay minerals structure, arrangement in two groups tetrahedrally and octahedrally in each group the atoms arranged in sheets and layers with the anion mainly O, OH, about the cations mainly Mg, Al, and Fe in Octahedral position and Si, and Al, in tetrahedral positions together in same minerals with interlayer cation S such as K, Na, Ca, Mg, The clays mostly used as catalysts are smectite, halloysite, sepiolite palygorskite and kaolin, [5]. The largest use of kaolin is in catalyst substrates in the catalytic cracking of petroleum [6], and in the synthesis of zeolites [7]. In addition, kaolin has been identified as a potential raw material for the production of alumina. The three layer stack clay structure know Montmorillonite. The cation connection in the intervening by natural attraction and Vander Waal forces, this soil have exhibited high shrinkage and swelling characteristics when have a large amount of the mineral, [8] montmorillonite are complex layers of $SiO_2$ and $ALO_4$ tetrahedral, also containing small amount of MgO and $Fe_2O_3$ [9]. The most smectite minerals commonly used are Na-montmorillonite, Ca-montmorillonite, saponite (Mg), nontronite (Fe), hectorite (Li) and beidellite (Al) [10]. Illite clay have Basic structure of this clay is the same as the one of montmorillonite. However, potassium ion ($K^+$) are filled in between facing $O^{2-}$ and $O^{2-}$surfaces of silica sheets. The clay characteristics are been classified between those of montmorillonite and kaolinite [11], these clays in granulated and powder form are the most important clay catalysts in the world. The small amount of active materials are added to surface of porosity solid support in the preparation of the solid catalyst, Most important of these oxides solids silica gel, $TiO_2$, zeolite, alumina oxides, aluminum silicates, activated carbon, $ZrO_2$, MgO, and ceramics [12]. The classical pillaring method includes two steps, physical and chemical. In physical stage swelling in clay and exchange of the interlayer cations by partially hydrated polymeric, but in second step thermal treatment of the intercalated clay mineral.

Some attempts were successful in preparing large pore in pillared clays using rare element /aluminum. The ratio of OH/Al in the pillaring solution was a

* Corresponding author: alkindi.ghydaa@gmail.com

© The Authors, published by EDP Sciences. This is an open access article distributed under the terms of the Creative Commons Attribution License 4.0 (http://creativecommons.org/licenses/by/4.0/).

controlling factor of increasing basal spacing [13]. While calcing was a controlling factor on surface area, various types of solids have been employed to catalyst for oxidation of organic compounds in wastewaters: transition-metal and noble metals, the noble metals have been extensively used as active components [14]. Fe used in prepared pillared clay due to high oxidation state and high stability [15], the advantage of used combined two or more metal oxide catalysts, which can improved selectivity, catalytic activity and increasing the rate of the processes [16]. Such as increase thermal stability and catalytic properties when used Fe as second cation with Al-pillared clays [17], To increase the efficiency of process of wet oxidation, which has been modified or has been used a suitable combination of processes [16].

The goal of this study is to prepare AL-Fe pillared Iraqi clay to be used in removal of some pollutants of wastewater. The clays were collected from five regions. Which have different type of clays. Anbar (from west 40Km from H3), Mosul (from north Smehlla Quarries 36°31´25N, 43°53´52E), Baghdad (from central Quarries brick factory Mohamed Adnan in AL Nahrawan 33°24´6N, 44° 51´ 0E), and Basra (from south brick factory for Abu AL Khaseeb E:782004, N:3373832 from Shatt Al-Arab) Collected to represent the topography of all Iraq, and only those available at the General Authority for Geological Survey, in addition to the samples collected from Erbil in the Kurdistan Iraq from Khlefan mountains, to choose the best catalyst depending on surface area, pore size, activity and stability for removal phenol and TOC from synthetic wastewater by CWO.

## 2 Experimental work

### 2.1 Selecting the best condition

In the beginning, according to XRD test Anbar soil was selected to find the best conditions as calcing temperature, and molar ratio OH/(Al +Fe), the scheme diagram which is used to find the stable operation condition show in Figure (1).

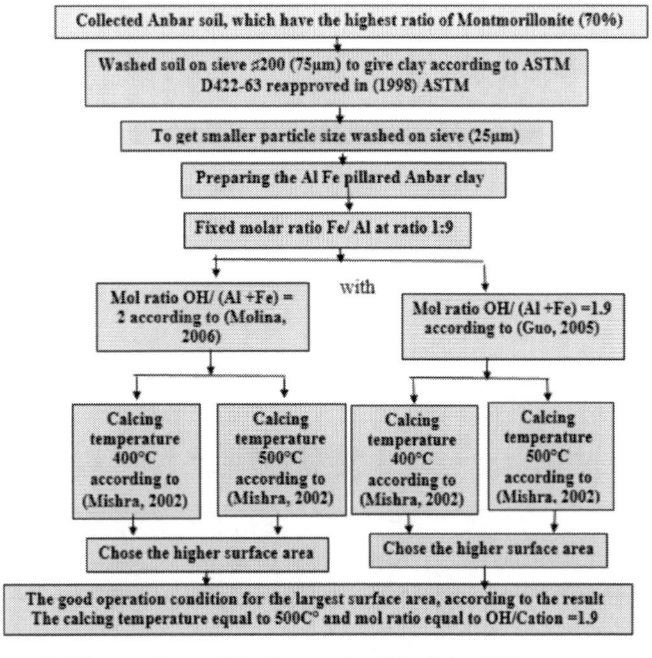


**Fig. 1.** Scheme diagram to find the operation condition for preparing pillared clays [18].

### 2.2 Preparation of pillared clay

From collected stack soil the preparation has done according to Professor N. Papayannakos process (Analysis and Plant Design Group at National Technical University of Athens, NTUA) as prepared by Guo, 2005 [17]. Figure (2) shows schematic representation of the conventional process of preparation pillared clay.

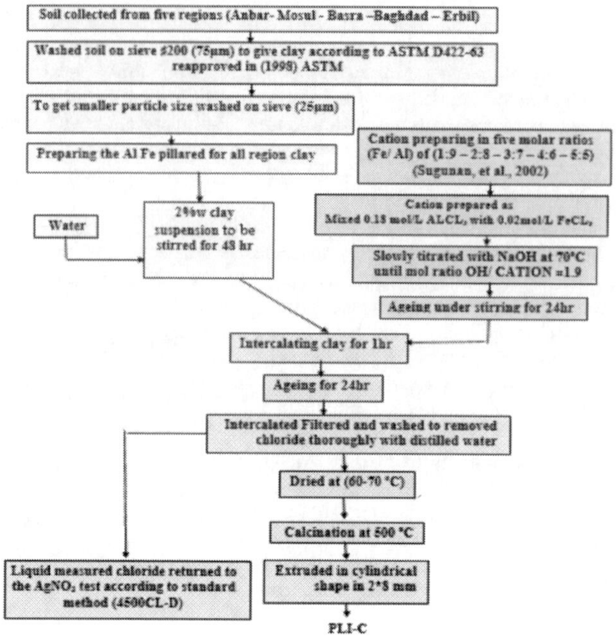


**Fig. 2.** Schematic representation of the conventional process of pillared clay [19]

Many tests required to study the characterization of soil and pillared clay many tests were done such as hydrometer test, Blumenauer-Emmer-Teller (BET), X-ray Diffraction XRD, X-ray florescent (XRF), Scanning Electron Microscope (SEM).

## 2.3 Activity measurement

From the result of previous experimental test, the maximum operating condition are temperature and pressure of 500 C° - 5 MPa respectively, in Basket Stirred Batch Reactor (BSBR), made from stainless steel designed for experimental work. The experimental scheme is depicted in Figure (3).

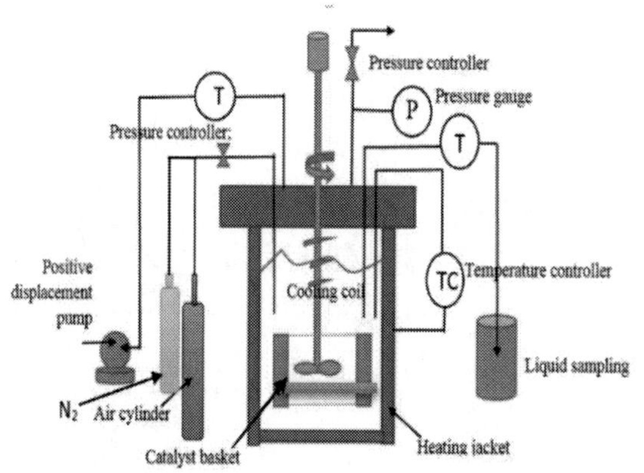


**Fig. 3.** Diagram of the experimental setup.

The 500mg of phenol has been diluted with distilled water, add sulfuric acid (1N) to reach pH 3.9-4, putting in 600ml Vessel. For each catalyst region, and putting in a basket inside the reactor, the rotation speed was kept at 800 rpm. Nitrogen pressured to 1Mpa to inert the reactor, purified air (21% pure oxygen mixed with 79% pure nitrogen) was continuously fed into the reactor. Many experiments have been done for each catalytic region, the operating conditions of experimental works are temperature range (90-150 Co), total pressure of (2.5 MPa), and catalyst loading (7mg/l). Once the set temperature reached (defined as time zero), liquid sample was withdrawn all 30min, each sample collected rapidly cooled, and filtered (nylon filter) 0.2 μm pore size. The samples were stored in the freezer to the time of test in order to stop the reaction [17].

The air was manually set to the required partial pressure for the duration of the experiment. The airflow rate was maintained constant [9]. Every batch was washed to remove all fines, then dried overnight prior to other use.

The filtered samples were analyzed by Germany High Performance Liquid Chromatography (HPLC) YL 9100 HPLC System, YL 9101 Vacuum Degasser. And YL 9110 Quaternary pump, YL 9120 UV/Vis Detector with C18 column. A mobile phase $H_2SO_4$ of variable composition was programmed at a 1.2 ml/min flow rate. The wave length between 450-550λ detected phenol and low molecular weight Intermediates compound. The (UV –VV-1100, UEBO9025) Spectrophotometer was used to detect phenol concentration. The total organic carbon (TOC) has been measured by Shimadzu TOC-LCSH/CSN Standalone Model TOC analyzer. Leaching Fe concentration in the wastewater from reaction media was also analyzed by Atomic Absorption.

## 3 Result and Discussion

Hydrometer tests of fine-grained soils, (silts and clays), which gives the particle size distribution for the Iraqi's soil collected from five areas. The hydrometer test result, for Basra refer to sand, silt, and clay in varying percentages of 50 % silt, 30 %clay, and 20 % sand while, Baghdad, Mosul proved it is a clay (60 % ) and silt (40 % ) approximately. Anbar and Erbil did not give any result because swelling occurred as shown in Figure (4).

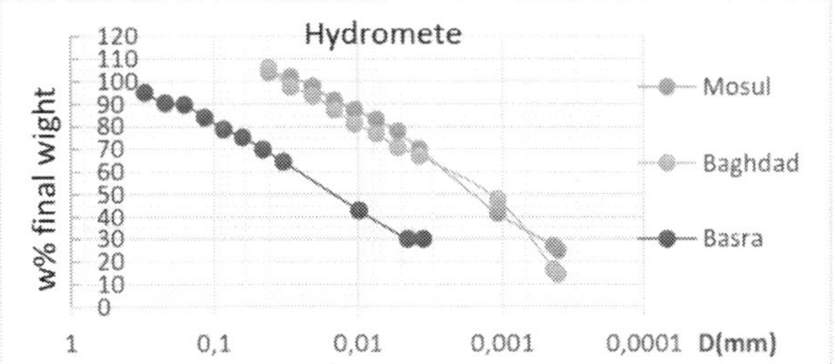


**Fig. (4):** Hydrometer of collected Iraqi soil

(XRD) X-ray diffraction is a technique used to research about characteristics of crystalline structure and to study the mineralogy of finer grained clays [20]. According to X-ray diffraction patterns, the following phases were identified in the investigated catalyst samples according to (standard powder diffraction data 1990) [21], typical XRD patterns of Al-Fe clay before and after pillaring with mixed Fe and Al polyoxocation (Fe-Al-CTE). The d (001) peak for the ordinary clay corresponds to a 3.0202, 3.02516, 3.0400, 3.03019, 3.02524 Å for Anbar- Mosul- Baghdad- Basra- Erbil clays respectively, spacing, and is increased to d (001) spacing of 3.03165, 3.3286, 3.30064, 3.33455, 3.33879 Å, for Anbar- Mosul- Baghdad- Basra- Erbil clays respectively, which indicates the extent of the pillaring process. Also show the Iraqi clays are calsumy have two layers of silica and one layer of alumina this conforming to result of Aibert, 1980 [22].The result of this test is shown in Figure (5)

**XRD for ordinary Anbar clay**

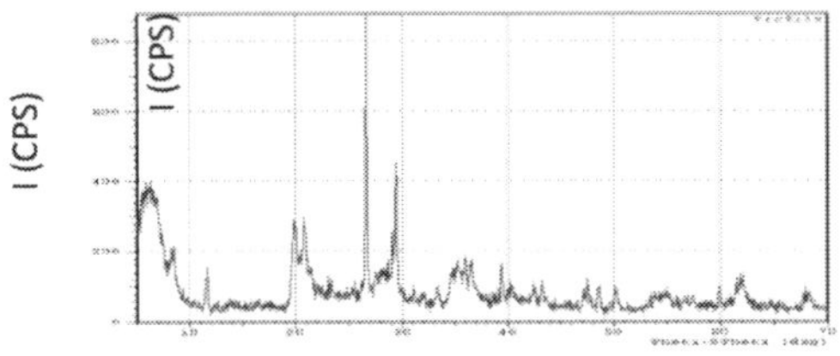


**XRD for catalyst 0.3 Anbar**

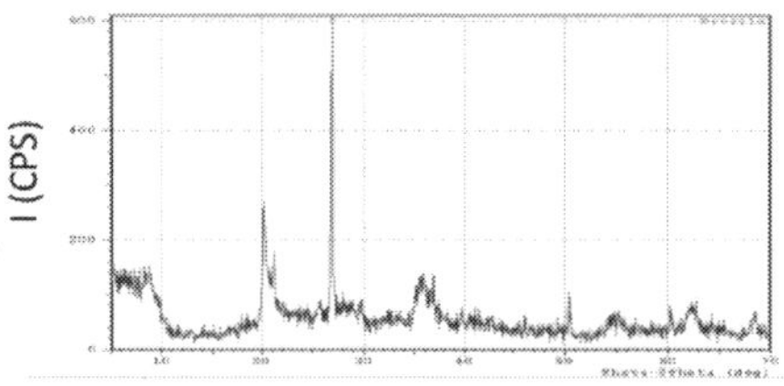


**XRD for ordinary Mosul clay**

**XRD for catalyst 0.5 Mosul**

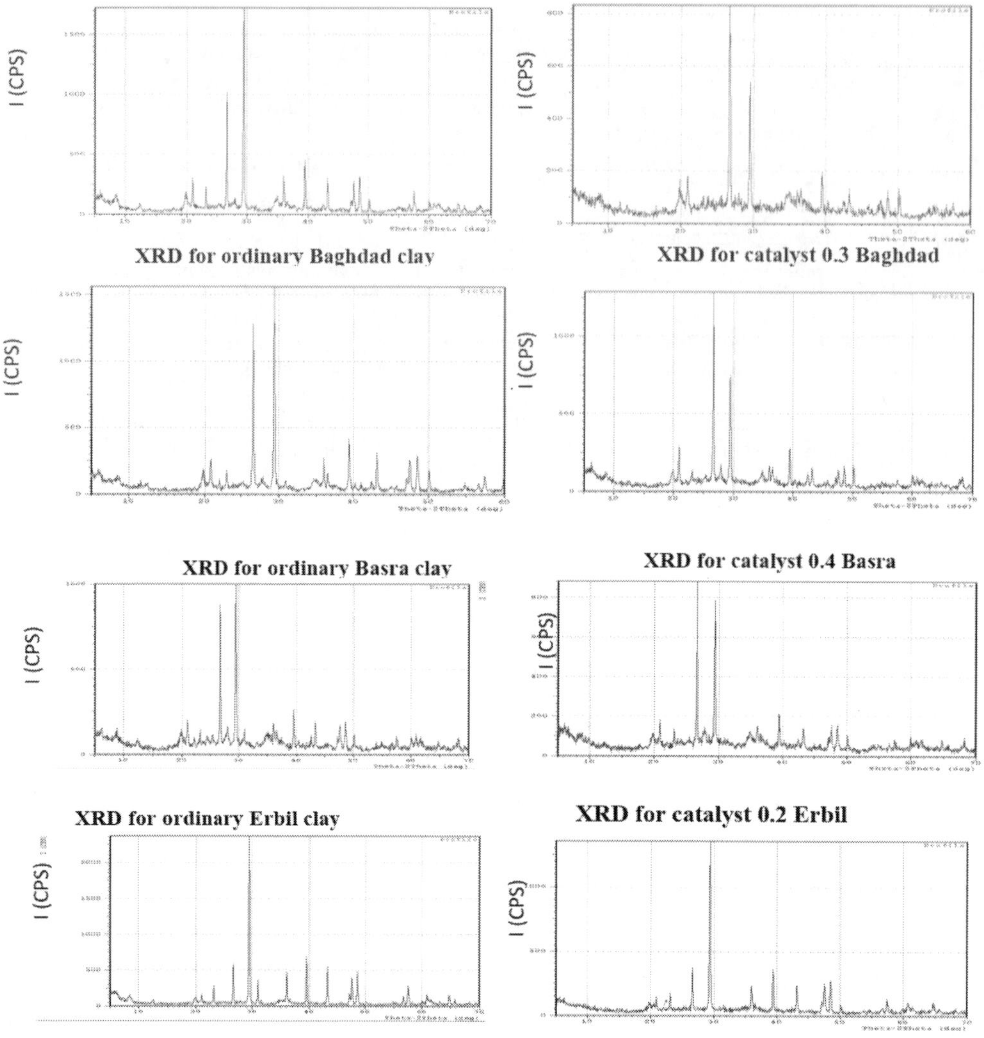


**Fig. (5):** XRD result for ordinary and Al -Fe pillared clay for Erbil, Mosul, Anbar, Baghdad, Basra

The results of XRF analyses for ordinary clay and Al - Fe pillared clays (catalyst) indicate that ratio of the $Fe_2O_3$ % and $Al_2O_3$ % in pillared clay is higher than ordinary clay, Baghdad clay shows increase in $Al_2O_3$ % than other pillared clay, followed by Basra, Erbil, Anbar, and Mosel, but the ratio of calcium oxide in all pillared clay degrease in ratio than ordinary clay due to exchange between cation and Ca. Also the ratio of $Si_2O_3$ has degrease, these result shown in Table (1).

**Table 1:** main composition of the ordinary and catalys

| Ratio | Anbar Ordinary clay | Anbar catalyst | Mosul Ordinary clay | Mosul catalyst | Baghdad Ordinary clay | Baghdad catalyst | Basra Ordinary clay | Basra catalyst | Erbil Ordinary clay | Erbil catalyst |
|---|---|---|---|---|---|---|---|---|---|---|
| $SiO_2$% | 23.75 | 20.26 | 36.38 | 32.35 | 41.38 | 39.04 | 34.18 | 28.7 | 57 | 55.36 |
| $Fe_2O_3$% | 3.93 | 7.98 | 5.42 | 9.82 | 5.2 | 7.8 | 6.2 | 14.81 | 5.5 | 9.6 |
| $AL_2O_3$% | 5.82 | 15.37 | 9.64 | 14.35 | 9.76 | 29.49 | 9.77 | 21.73 | 13.5 | 23.4 |
| CaO% | 33.43 | 21.44 | 17.53 | 11.3 | 17.56 | 5.21 | 18.14 | 9.28 | 8 | 6.51 |
| Type of clay | calsumy | | calsumy | | calsumy | | calsumy | | calsumy | |

The Blumenauer-Emmer-Teller (BET) Method give description and more explanation of surface area for each region according to above conditions shown in Tables (2 to 6). The final result of surface area of catalyst for all types which was significantly higher than ordinary clay, showns in Table (7). The results of experimental works show that the best ratio used in preparing the pillared clay were (0.3, 0.5, 0.3, 0.4, and 0.2) for (Anbar, Mosul, Baghdad, Basra, and Erbil), respectively. These ratios were selected depending on result of surface area. And these results of surface area are in agreement with results obtained by Bankovic, 2009 [23]. The constructing an adsorption isotherm in Figure (6) contributed an understanding of the pore size and surface area of a porous media. That figure represents the typical nitrogen adsorption isotherm obtained for the Al-Fe pillared Iraqi clay catalyst. The linear portion of the adsorption curve in the $P/P_0$ (relative pressure, correlated to the amount of adsorbed nitrogen) show the range between 0.04-0.3

The pore size being at upper mesoporous, but from 0.3-0.99 the pore size being at the low macropore range as reference by Gregg and Sing, 1983 [24], The Al -Fe pillared Iraqi clay for five regions range between (0.04-0.3) that can be considered mesopore, after this value (0.3-0.99) it is small macropore as reference by Gregg and Sing, 1983 (24). Also the pores size is characteristic of mesopore according to the International Union of Pure and Applied Chemistry (IUPAC) classification of pore size (Å ) as macropore (≥ 500), mesopore (20 to 500) supermicropore (7-20) and ultramicropore (≤ 7) [25]. From the results Baghdad have the highest value of pore size width 150.4 Å, but Basra have lowest value 75.1 Å, all regions are mesopore size so that all pillared suitable to be catalyst as reference by Lastoskie, 1993 [26].

Regarding pore volume which depends on the type of clay and calcing , Mosul has the largest pores volumes, followed Erbil, Basra, Anbar, and Baghdad, where the major part of the surface area contributed by pores, where internal chemical reaction gates.

**Table (2)** show result of BET surface area of Anbar Traphawy at 500C° for 1h

| samples | Surface area $m^2/gm$ | Pore volume $cm^3/gm$(Å) |
|---|---|---|
| Stack clay | 75.2158 | 0.08635 |
| pillared 0.1 | 92.1518 | 0.08198 |
| pillared 0.2 | 88.1734 | 0.0508 |
| pillared 0.3 | 97.8157 | 0.0962 |
| pillared 0.4 | 73.8363 | 0.0611 |
| pillared 0.5 | 63.6507 | 0.0607 |

**Table (3)** show result of BET surface area of at AL Mosul Smahalla 500C° for 1hr

| samples | Surface area $m^2/gm$ | Pore volume $cm^3/gm$(Å) |
|---|---|---|
| Stack clay | 78.8355 | 0.0911 |
| pillared 0.1 | 55.2218 | 0.0999 |
| pillared 0.2 | 80.9344 | 0.1320 |
| pillared 0.3 | 63.3475 | 0.0989 |
| pillared 0.4 | 74.9119 | 0.1214 |
| pillared 0.5 | 93.6052 | 0.1430 |

**Table (4)** show result of BET surface area Basra Abu AL-Khaseeb at 500C° for 1hr

| samples | Surface area $m^2/gm$ | Pore volume $cm^3/gm$(Å) |
|---|---|---|
| Stack clay | 41.475 | 0.05325 |
| pillared 0.1 | 42.235 | 0.0746 |
| pillared 0.2 | 53.317 | 0.0894 |
| pillared 0.3 | 56.973 | 0.0907 |
| pillared 0.4 | 49.1577 | 0.8385 |
| pillared 0.5 | 54.5577 | 0.7921 |

**Table (5)** show result of BET surface area of of Baghdad AL Nahrawan at 500C° for 1hr

| samples | Surface area $m^2/gm$ | Pore volume $cm^3/gm$(Å) |
|---|---|---|
| Stack clay | 51.117 | 0.084 |
| pillared 0.1 | 47.00 | 0.108 |
| pillared 0.2 | 43.28 | 0.096 |
| pillared 0.3 | 46.45 | 0.098 |
| pillared 0.4 | 61.3711 | 0.1109 |
| pillared 0.5 | 60.28 | 0.101 |

**Table (6)** show result of BET surface area Erbil Khleaphan at 500C° for 1hr

| samples | Surface area $m^2/gm$ | Pore volume $cm^3/gm$(Å) |
|---|---|---|
| Stack clay | 47.57 | 0.06236 |
| pillared 0.1 | 85.527 | 0. 1201 |
| pillared 0.2 | 100.742 | 0.1221 |

**Table 7:** physical properties of various catalyst sample

| Samples Region | Ratio Fe/(AL+Fe) Used to prepare AL-Fe pillars clay | (BET) Surface area $m^2/g$ | Average Pore volume $m^3/g$ | Pore size width Å |
|---|---|---|---|---|
| Anbar CLAY | | 75.215 | 0.08635 | |
| Anbar Catalyst | 0.3 | 97.815 | 0.0962 | 78.4 |
| Erbil CLAY | | 47.57 | 0.06236 | |

| | | |
|---|---|---|
| pillared 0.3 | 99.412 | 0.1332 |
| pillared 0.4 | 96.6770 | 0.1242 |
| pillared 0.5 | 99.3033 | 0.0131 |

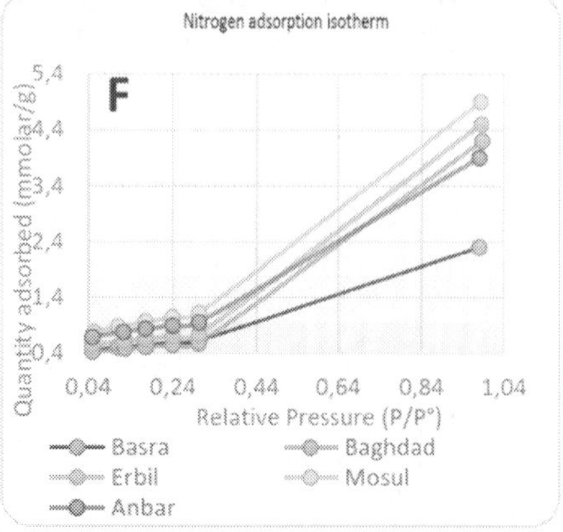


**Fig.6:** Nitrogen adsorption isotherms liner plot for Fresh Al-Fe pillared Iraqi clay for (a) Anbar (b) Erbil (c) Mosul (d) Baghdad (e) Basra

| | | | | |
|---|---|---|---|---|
| Erbil Catalyst | 0.2 | 100.74 | 0.1221 | 124.2 |
| Baghdad CLAY | | 41.475 | 0.05325 | |
| Baghdad Catalyst | 0.3 | 56.973 | 0.0907 | 150.4 |
| Mosul CLAY | | 78.835 | 0.0911 | |
| Mosul Catalyst | 0.5 | 93.605 | 0.1430 | 89.7 |
| Basra CLAY | | 51.117 | 0.084 | |
| Basra Catalyst | 0.4 | 61.371 | 0.1109 | 75.1 |

To understand the morphology of ordinary clay and catalyst visualization of surface reaction scanning electron microscopy was used as shown in Figure (7) which give the image of ordinary clay and catalyst, these figures show the pores and surface roughness in different magnification and show boundary of crystals regions on the catalyst image described in a circle around it.

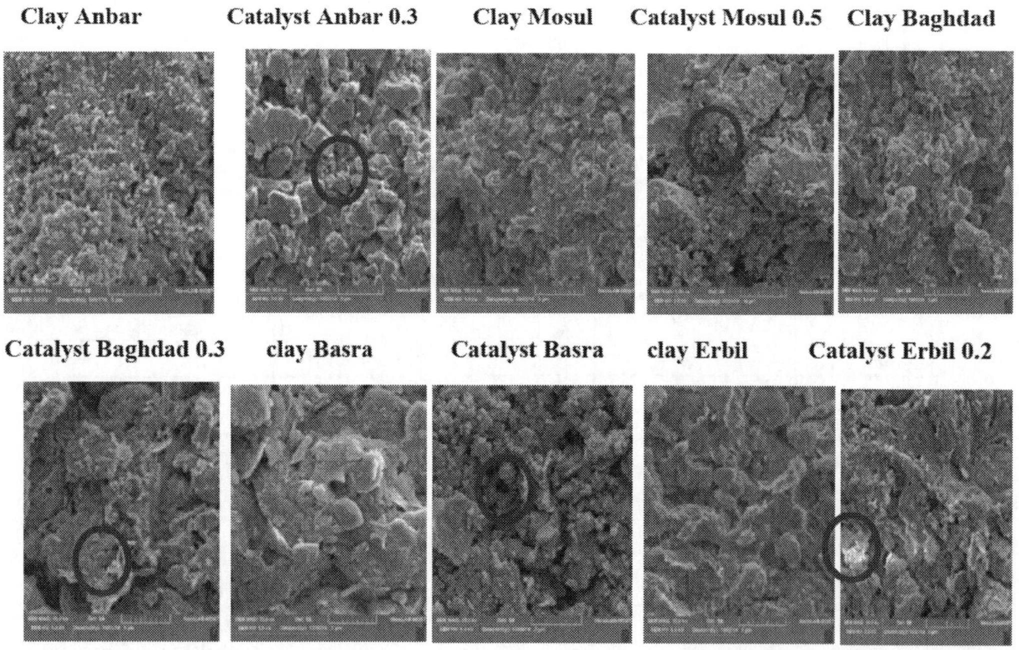


**Fig. 7:** SEM images of Fresh Al-Fe pillared clay catalyst with magnification of (5 μm scale) crystals marked by circle

## 3.1 Chemical properties of clay catalyst

From the results of experimental works, the pH increases in catalyst than clay due to $AlOH_3$ and

$FeOH_3$ obtained from the reaction of hydroxide with $FeCl_3$ and $AlCl_3$ and remained after calcing. The value of chloride in catalyst is less than that in clay because of washing during preparing. The needle shape sulfate in catalyst is considered impurity. If the sulfate ratio was between 1-2% that important to solid catalyst due to increase in acidity, and causes deactivation when the sulfate contain in catalyst >3% as stated by McNaught et al., (2006) [27]. The value of sulfate concentration in catalyst was reduced but still >3% so that it reasoned of deactivation. Carbonation causes the problems due to its low solubility in water and increases the cohesion of clay, which reduces the surface area, in this study the carbonation concentration was reduced slightly after preparing the catalyst, but considered high due to the natural Iraqi clay which have a high amount of sulfate and carbonite, these results shown in Table (8) (for stack clay) and in Table (9) (for catalyst).

**Table (8):** Measuring the percentage of ions in clay

| Measuring of Catalyst | | | | | |
|---|---|---|---|---|---|
| **test** | **Anbar** | **Erbil** | **Mosul** | **Baghdad** | **Basra** |
| **pH** | **9.81** | **9.53** | **9.82** | **9.39** | **9.22** |
| **EC mv/cm** | **166.3** | **149.3** | **165.4** | **141.6** | **130.9** |
| **$CL^-$ %** | **0.007** | **0.056** | **0.015** | **0.039** | **0.097** |
| **$SO_3^-$ %** | **2.058** | **2.06** | **6.86** | **3.43** | **3.43** |
| **$SO_4^-$ %** | **4.424** | **4.425** | **14.75** | **7.37** | **7.37** |
| **$CO_3^-$ %** | **0.63** | **0.73** | **0.68** | **0.61** | **0.69** |
| **TSS %** | **0.006** | **0.0067** | **0.006** | **0.007** | **0.0076** |
| **T C°** | **25** | **25** | **25** | **25** | **25** |

**Table (9):** Measuring the concentration of ion and salt in catalyst

| Measuring of CLAY | | | | | |
|---|---|---|---|---|---|
| **test** | **Anbar** | **Erbil** | **Mosul** | **Baghdad** | **Basra** |
| **pH** | **7.98** | **8.82** | **8.55** | **8.35** | **8.54** |
| **EC mv/cm** | **57.7** | **107.7** | **91.7** | **79.5** | **90.7** |
| **$CL^-$ %** | **0.74** | **0.055** | **0.143** | **0.264** | **0.172** |
| **$SO_3^-$ %** | **6.86** | **13.7** | **13.6** | **13** | **20.58** |
| **$SO_4^-$ %** | **14.75** | **29.46** | **29.24** | **27.95** | **44.25** |
| **$CO_3^-$ %** | **0.65** | **0.6** | **0.65** | **0.72** | **0.65** |
| **TSS %** | **0.0173** | **0.093** | **0.012** | **0.013** | **0.011** |
| **T C°** | **25** | **25** | **25** | **25** | **25** |

The activity of the heterogeneous catalytic system was also studied for all catalyst regions in stirred basket batch reactor, at different temperatures between (90-150 °C), while the constant for other parameter pressure 2.5MPA, catalyst loading $C_{Cat}$ equal to 7mg/l. Figures (8) shows phenol concentration at different temperatures for different catalyst regions. Figures (9) shows phenol conversion profiles of reactions performed at temperatures 90, 110, 130 and 150C, the result phenol conversion 96.2, 95.44, 93.7, 80, 76.3% for Erbil, Mosul, Anbar, Baghdad, Basra, respectively. The temperature at 150°C is necessary to obtain phenol conversion along the process because mass transfer of oxygen from the gas phase to the liquid phase is required. Also the temperature between (100-150 C) is considered ideal for most of catalyst so that the activated energy increase with increase temperature reach to 50kj/molar than increase 1°C in temperature increase activated energy 10 kj/molar therefore causing double conversion rate lead to disequilibrium state [28]. The reaction rate is increasing gradually with time especially at a temperature of 150°C. Erbil has the highest value of phenol degradation rate and followed by Mosul, Anbar, Baghdad and Basra respectively. This is expected in some catalyst types which reached to complete minimization ($H_2O$ and $CO_2$), because in the first step of phenol oxidation corresponds to the formation organic acid and in further oxidation reach to $H_2O$ and $CO_2$ [29]. Figure (10) shows TOC concentration at different temperatures for different catalyst regions. Figures (11) shows TOC conversion, the higher TOC conversion for Anbar catalyst reach to 88% than Erbil 79.28%, Mosul 78.2%, Baghdad 66.5% and Basra 66.46%. The deactivation appears on the surface of pillared Anbar and Erbil clay. The pathway reaction for all sample catalysts were shown in first step phenol oxidation the benzoquinone formation in some samples while other have Malic acid, were in subsequent period all samples formation acetic acid.

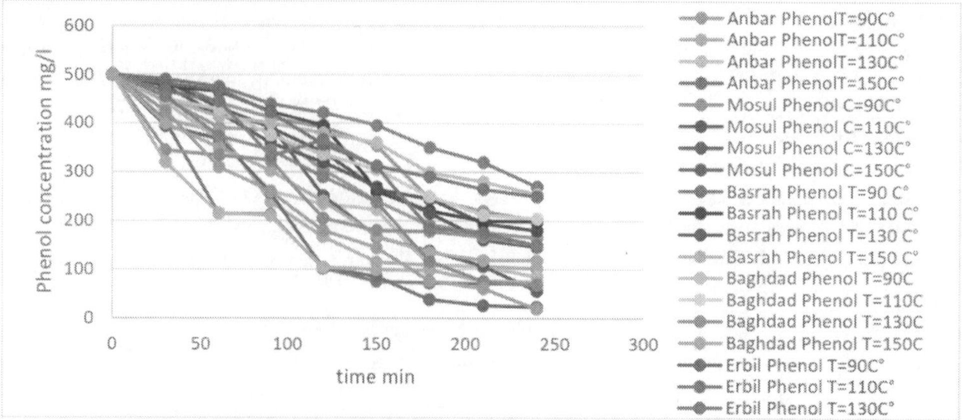


**Fig (8):** Phenol concentration during 4hr in catalyst loading 7mg/l, pressure 2.5Mpa, 500mg/l phenol concentration.

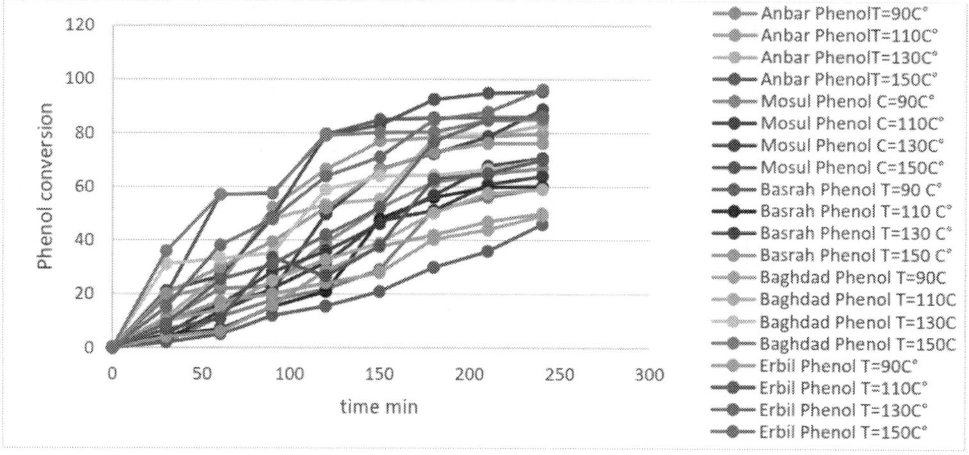


**Fig (9):** Phenol conversiontion during 4hr in catalyst loading 7mg/l, pressure 2.5Mpa, 500mg/l phenol concentration.

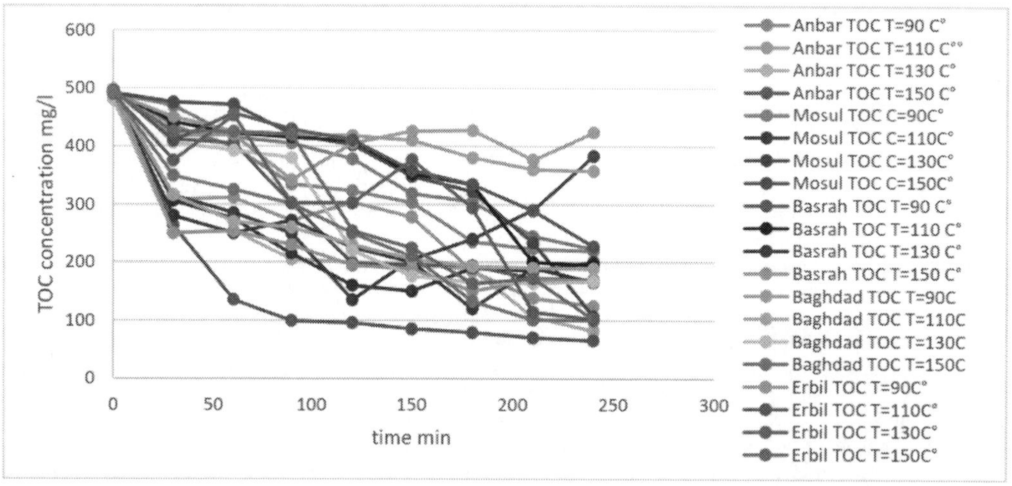


**Fig (10):** TOC concentration during 4hr in catalyst loading 7mg/l, pressure 2.5Mpa, 500mg/l phenol concentration.

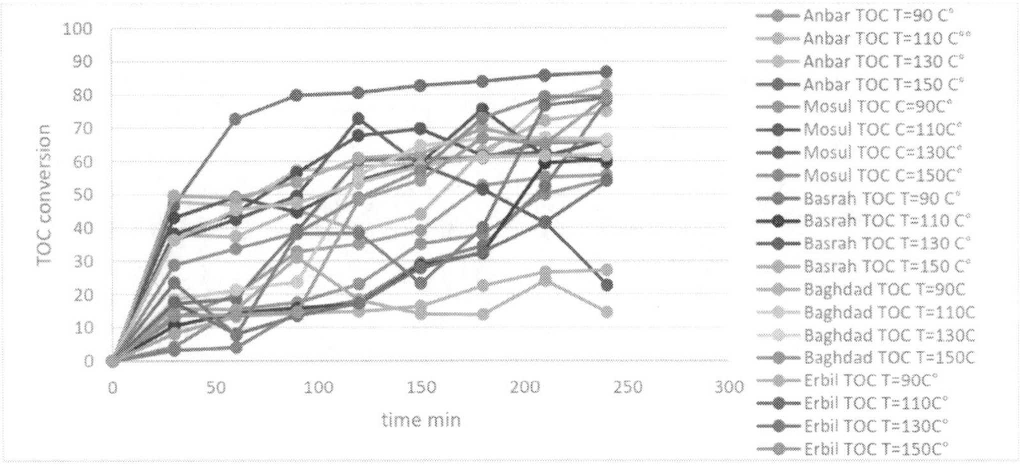


**Fig (11):** TOC concversion during 4hr in catalyst loading 7mg/l, pressure 2.5Mpa, 500mg/l phenol concentration

During the experiment of catalyst activity, the stability is studied too, during the leaching, and blocking, in batch reactor, the atomic adsorption equipment was used for leaching test, observed the trace Fe in the liquid effluent samples that collected every 30min, low values result leaching so that these catalyst can be continuously operated. The result of leaching for Mosul an example is shown in Figure (12). While Figure (13) show the catalysts after used. The blocking is also measured by changing of surface area (BET). The surface area after blocking equal to (15.34, 23.65, 37.26, 79.04, 80.17 $m^2/g$) for (Basra- Baghdad- Erbil- Anbar- and Mosul) respectively. The higher blocking was shown in Basra and Baghdad equal to 75, 70% respectively, Erbil catalyst equal to (63%), then Anbar catalyst (19.2%), while AL-Mosul catalyst (14.35%).

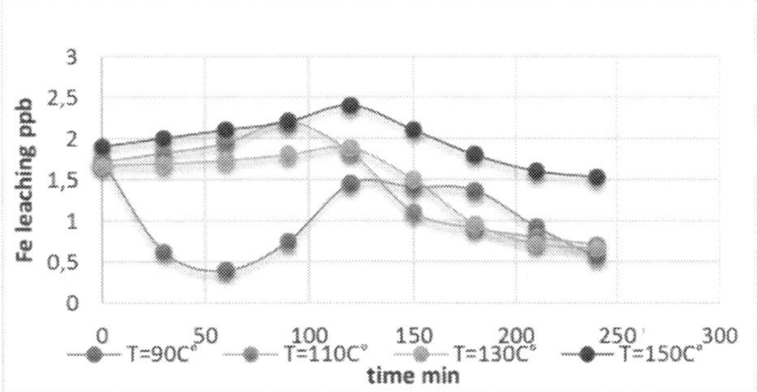


**Fig. (12):** Fe concentration with different temperature in batch reactor Mosul change temperature Fe leaching

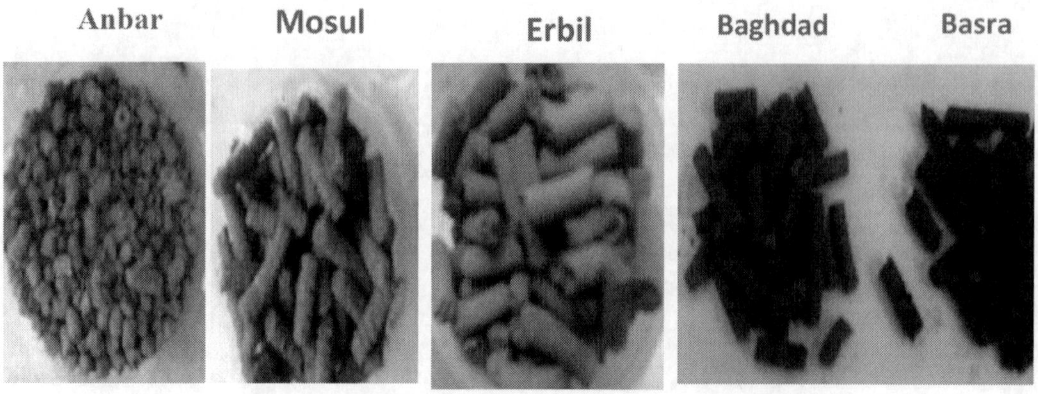


**Fig. (13):** Deactivation in all catalytic regions during phenol oxidation in batch reactor.

## 4 Conclusions

All the tests were conducted over extrudates that prepared Fe-Al pillared Iraqi clay catalyst from five regions (Anbar- Erbil- Mosul- Basra- and Baghdad) with dimension (2 × 8 mm). All catalyst regions are stable and maintain activity during the experimental process, various characterization methods were employed for fresh and aged pillared clay catalysts previously used in batch reaction operation.

Surface analysis showed that the combined Al and Fe species were homogeneously dispersed between the silicate layers. During batch reactor deactivation appeared on Anbar clay, some time on Erbil and Mosul pillared cay, and the trace amount in (ppb) of Fe were detected in the liquid effluent, which means there is no leaching. The lowest blocking were shown in Mosul, and followed by Anbar, Erbil, Baghdad and Basra clay. From above it can be concluded that clay and silty clay (Mosul clay) are considered the best catalysts due to higher activity and stability through experiment of CWAO in batch reactor, with operation condition 150°C, 2.5 Mpa, $C_{cat}$ equal to 7mg/l.

## References

1- Nam, H.J, I.-S. S. W. Ham, and S. B. Hong, Catal. Today **68**, 31 .Cross REF. (2001)
2- Figueras F. *Catal. Rev.-Sci. Eng.* **30**, 457-499. (1988),
3- Molina, C.B. J.A. Casas, A.H. Pizarro, J.J. Rodriguez, Facultad de Ciencias, pp. 435-474Top » Catalog » Books » Materials Science » Clay: Types, Properties and Uses Chapters Universidad Autónoma de Madrid, Madrid, Spain (2004).
4- Buker, R. A. and S.A.M. AL- Mallah, (2006) J. Educ. And Sci., **17**(4), 42.Cited by Buker, R. A. J.Edu. & Sci., **25** (1), 26-33. (2012)
5- Rodrigues M G F. Cerâmica, 49: 146(2003)
6- Murray H H. Applied clay mineralogy. 1st edition, Copyright , Elsevier B.V., ISBN-**13**: 978-0-444-51701-2 (2007)
7- Oghenejoboh K M, Ohimor E O. European J Sci Res, **56**(4): 445, (2011)
8- Wilson M. J. clay minerals 37, 7-25. (1999)
9- James, T .Richardson, book Library of congress cataloging in publication data .NEW YORK AND LONDON. (1989)
10- Naman S A, Maher K A, Ibrahim R P. Journal of Petroleum and Gas Exploration Research, 2(4): 80. (2012)
11- Wilson M. J. clay minerals 37, 7-25. (1999)
12- Stones, F.S. (1990) J. Mol. Cat., 59, 147. Cited by Buker, R. A., J.Edu. & Sci., **25**, (1). (2012)
13- Buker, R. A., J.Edu. & Sci., **25** (1), 26-33. (2012)
14- Fajerwerg, K. H. Debellefontaine, Appl. Catal. B: Environ. 10 L229–L235. (1996)
15- Biomorgi, J., Pérez Zurita, M. J., Scott, C., de Navarro, C. U., Microsc Microanal 9, 434-435 (2003).
16- Suranjana Chaliha & Krishna G Bhattacharyya. Journal of Chemical Technology vol 13, pp 499-504. (2006)
17- Guo, J., PhD .thesis –school of chemical Engineering, university of Washington. (2005).
18- Mishra, B. G.; Rao, G. R. Bull. *Mater. Sci25*, 155. (2002),
19- Sugunan S and N. K. Renuka. *Ind. J. Chem.*, **41A**, 304. (2002)
20- Poppe, L.J. Paskevich, V.F. Hathaway, J.C. and Blackwood, D.S., USGS U.S. Geological Survey Open-File Report 01-041. (2002)
21- International Center for Diffraction Data Joint Committee on 1990 "Powder Diffraction

Standards, Powder Diffraction Data", Swarthmore (PA).
22- Aibert, F. Cotton, John Wiley and Sons New York Inc. (1980)
23- Bankovic, P., A. Milutionovic., Nikolic, A. Rosic., N. Jovic – Jovicic, and Jovanovic, Journal of physical chemistry, vol 83 No. 9 PP 1485-1489. (2009)
24- Gregg S. J. and Sing K. S. W., 2nd ed., Academic: London. (1983),
25- Nicoleta Platon, ANA-Maria ROSU, Vasilica Alisa ARUS, Denisa Ileana Nistor, Ilie Siminiceanu, Journal of Engineering Studies and Research – Volume 19 No. 4, PP. 52-58. (2013)
26- Lastoskie, Christian, Keith E. Gubbins, and Nicholas Quirke. journal physical and chemical, 97, 4786-4796. (1993)
27- McNaught, A. D. and A. Wilkinson. Blackwell Scientific Publications, Oxford http://goldbook.iupac.org (2006) created by M. Nic, J. Jirat, B. Kosata; updates compiled by A. Jenkins. ISBN 0-9678550-9-8. doi:10.1351/goldbook. Last update: 2014-02-24; version: 2.3.3. http://goldbook.iupac.org/C00876.html (1997).
28- Rob. J. Berger, E. Hugh Stitt*, Guy B. Marin, Freek Kapteijn, Jacob A. Moulijn, , EUOKEN Volume 5, no. 1, USA (2001).
29- Guo, J. and Al-Dahhan, M.H., Chem. Eng. Sci, 60, 735-746. (2005b)

# Removal of iron and manganese from groundwater: a study of using potassium permanganate and sedimentation

*Mohamed* Elsheikh[1,] *, *H.* Guirguis[1], *A.* Fathy[1]

[1]Civil Engineering Department, Faculty of Engineering, Menoufia University, Egypt

**Abstract.** Experiments are done for different combinations of Fe+2 and Mn+2 concentrations. The obtained results show that Potassium Permanganate (PP) gives good results in iron and manganese removal. By using PP dose near to half of the theoretically required one, it can remove up to 100 % and 90 % of iron and manganese, respectively over different tested concentrations at pH=7.0. Increasing rate of filtration influences the Mn+2 removal process obviously. Sedimentation is required when combined concentrations of iron and manganese are greater than 5.0 ppm to reduce filter rapid clogging. Using conventional treatment with adding alum, flocculation, sedimentation and filtration can remove up to 97% and 18% of iron and manganese, respectively. Using PP in addition to alum enhances manganese removal but decreases iron removal. However, using alum with increasing pH to 10 leads to 100 % and 95 % of $Fe^{+2}$ and $Mn^{+2}$ removal and increases filter working period.

## 1. INTRODUCTION

Groundwater sources in Egypt can be divided into renewable aquifers like Nile valley and Delta and non-renewable aquifers like the aquifer of the western desert in the Nubian sandstone **[1]**. Iron and manganese are usually present in groundwater as dissolved minerals or associated with other components **[2]**. Existence of iron and manganese in water causes many problems like water coloring and taste, clothes staining and encouraging bacterial growth in water distribution networks which affect the pipes transfer efficiency **[3]** but in general they do not cause health problems **[4]**. The secondary maximum contaminant levels for iron and manganese are 0.3 mg/l and 0.05 mg/l, respectively **[5]**. The most common methods used to remove iron and manganese include oxidation by aeration, chlorine, chlorine dioxide, potassium permanganate and/or ozone followed by filtration alone or sedimentation and filtration **[6,7]**. There are other methods that can also be used like using filters with special media like green sand, using ion exchange, biological methods or membrane filtration **[8]**.

Potassium Permanganate (PP) is considered a stronger oxidant and has many advantages over other oxidants. Oxidation chemistry of iron and manganese by PP can be described as follows **[9]**:

$$3Mn^{+2} + 2KMnO_4 + 2H_2O \longrightarrow 5MnO_2 + 2K^+ + 4H^+ \quad \text{....(1)}$$

$$3Fe^{+2} + KMnO_4 + 7H_2O \longrightarrow 3Fe(OH)_3 + K^+ + MnO_2 + 5H^+ \quad \text{....(2)}$$

To oxidize one mg of iron and one mg of manganese, about 0.94 mg and 1.92 mg of PP are required, respectively **[10]**.

When iron and manganese exist in high concentrations, the filter runs less than 24 hours so a clarification step is needed before filtration to increase the filtration period **[11]**. So, treatment method would there include flocculation, sedimentation and filtration stages. **[12]**

This paper studies using PP followed by filtration only or using PP and/or alum (Al) followed by flocculation, sedimentation and

*Corresponding author: mshafy2@yahoo.com

© The Authors, published by EDP Sciences. This is an open access article distributed under the terms of the Creative Commons Attribution License 4.0 (http://creativecommons.org/licenses/by/4.0/).

filtration for $Fe^{+2}$ and $Mn^{+2}$ oxidation and removal.

## 2. MATERIALS AND METHODS:

### 2.1. Study Method:

Simulated groundwater was prepared by adding salts of iron and manganese to tap water. The study consists of three groups of experiments: the first one discusses the different factors (e.g. dosages, detention time, pH) that affect the oxidation of $Fe^{+2}$ and $Mn^{+2}$ by using PP followed by direct filtration. Experiments are done for $Fe^{+2}$ and $Mn^{+2}$ = 1.50 and 1.0 mg/l, respectively which represent iron and manganese concentrations in the Delta region, Egypt. The second group investigates the results of the first one on other combined concentrations of $Fe^{+2}$ and $Mn^{+2}$ by using direct filtration alone. The third group discusses the efficiency of using sedimentation if high concentrations of $Fe^{+2}$ and $Mn^{+2}$ exist. This phase includes flocculation, sedimentation and Filtration.

### 2.2. Pilot plant:

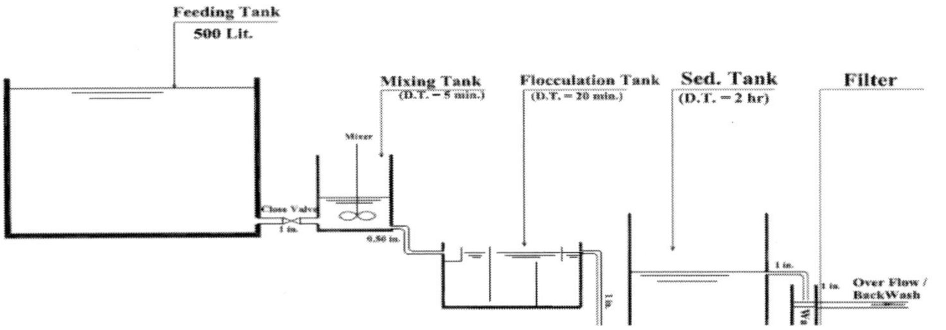


**Fig. (1):** Pilot Plant with mixing, flocculation, sedimentation and filtration tanks. When direct filtration mode is used, the flocculation and sedimentation tanks are removed.

Figure (1) shows the pilot plant, which is constructed for the study. It consists of feeding tank, process tank for adding and mixing chemicals, flocculation and sedimentation tanks and a Rapid Sand Filter (RSF). For 1st and 2nd phase experiments, the flocculation and sedimentation tanks are not included. The RSF is made of PVC pipe of internal diameter = 100 mm and includes 35 cm gravel layer with diameter 6:25 mm and 75 cm of coarse sand with diameter 1.18:1.60 mm. The Rate of Filtration is obtained by controlling the outlet filter valve.

### 2.3. Chemicals:

The simulated groundwater containing iron and manganese is made by adding ferrous sulfate heptahydrate ($FeSO_4.7H_2O$) and manganese sulfate mono-hydrate ($MnSO_4.H_2O$) to tap water. Potassium permanganate ($KMnO_4$) with 99.90% purity and aluminum sulphate hexadecahydrate $Al_2(SO_4)_3 16H_2O$) were used as sources of PP and alum. Sodium hydroxide (NaOH) was used to adjust pH.

### 2.4. Devices and Analyses:

Iron and manganese measure devices were used to measure iron and manganese concentrations (Hanna, USA). Portable pH device was used to measure water pH. The devices were calibrated before the study.

## 3. RESULTS AND DISCUSSIONS:

### 3.1. Group 1: study using Potassium Permanganate:

In this group, the experiments were done for $Fe^{+2}$ and $Mn^{+2}$ concentrations = 1.50 and 1.0 mg/l, respectively at constant ROF= 150 $m^3/m^2/d$, discuss the removal efficiency of iron and manganese using potassium permanganate under different conditions like different dosages of PP, contact time, effect of pH and ROF effect. The experiments lasted for 6 hours before taking samples.

***Effect of PP dosages:*** **Fig. (2)** shows the results of using different PP dosages for $Fe^{+2}$ and $Mn^{+2}$ removal. Using a dosage of 1 ppm of PP can remove up to 97% of iron just after 10 minutes. For manganese, using PP enhances the removal process greatly at pH=7.0. Using PP dose = 2.0 ppm -which is near to half of the theoretically calculated dose- can remove 66 % of manganese after 10 minutes. Using doses near to the theoretical dose remove up to 80% of manganese in just 10 minutes. Increasing PP dosage more than the theoretical one has a bad effect on water. When the dosage of 4 ppm is used, the water is colored pink from the effect of increased PP dosage. The increased PP quantity contains manganese according to the following equation **[13]**:

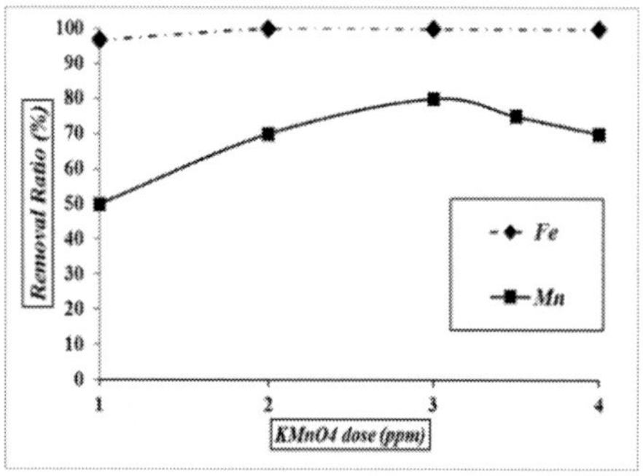


***Fig. (2): Effect of using different PP doses on $Fe^{+2}$ and $Mn^{+2}$ removal (Initial $Fe^{+2}$ and $Mn^{+2}$ are 1.50 and 1.0 mg/l, pH=7.0, ROF=150 m/d and RT=10 min.)***

$$KMnO_4 + 8H^+ + 5e^- \longleftrightarrow Mn^{+2} + K^+ + 4H_2O \quad (3)$$

Therefore, a special care should be taken into consideration for choosing the PP dosage. These results agree with other studies that recommend using PP dosages near the theoretical ones **[13]**.

***Effect of Retention Time:*** The results for using PP dose =2.0 ppm for RT=5 to 20 minutes are shown in **Fig. (3)**. The oxidation process of iron using PP happens very fast. Complete iron oxidation needs less than 5 minutes. The oxidation process of manganese using PP also happens fast and is enhanced by increasing RT. At RT less than 5 minutes, about 66% of manganese is oxidized, however, at RT equals to 20 minutes increases the RR to 90%.

***Effect of pH:*** **Fig. (4)** shows results of pH effect on $Fe^{+2}$ and $Mn^{+2}$ oxidation by PP dose = 2.0 ppm and RT = 10 min. It is shown that iron is oxidized at pH greater than 7.0 and much of manganese oxidation by using PP happens also at pH near to 7.0. Increasing pH to 8 and 9, increases the RR to 75% and 85%, respectively. Therefore, it is concluded that pH affects the process of $Fe^{+2}$ and $Mn^{+2}$ oxidation by PP slightly or in other words, they need pH greater than 7.0.

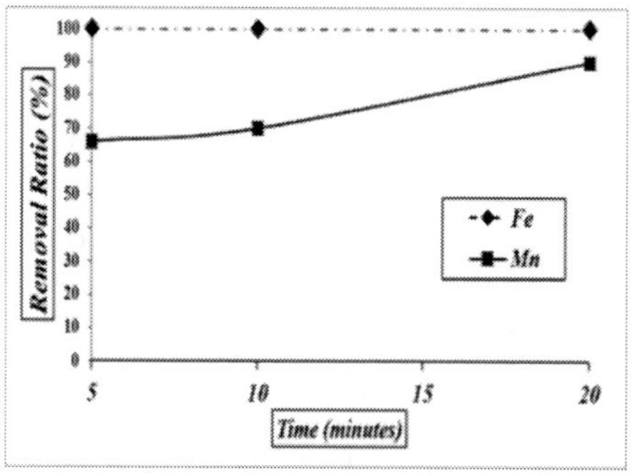


***Fig. (3): Effect of retention time on $Fe^{+2}$ and $Mn^{+2}$ removal by PP (Initial $Fe^{+2}$ and $Mn^{+2}$ are 1.50 and 1.0 mg/l, pH=7.0, PP=2.0 ppm and ROF=150 m/d)***

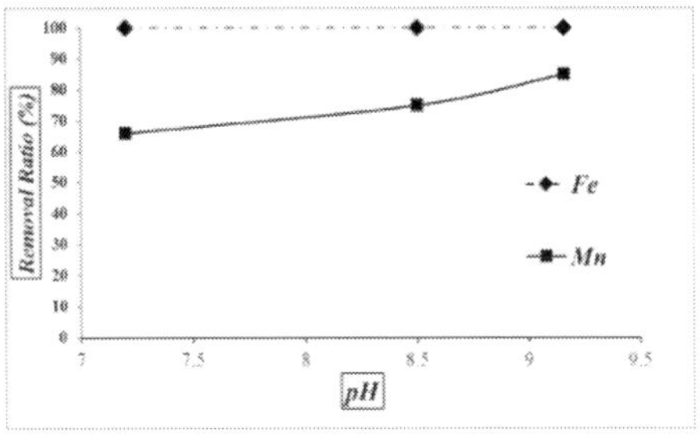


*Fig. (4): Effect of pH on $Fe^{+2}$ and $Mn^{+2}$ removal by PP (Initial $Fe^{+2}$ and $Mn^{+2}$ are 1.50 and 1.0 mg/l, PP=2.0 ppm ROF=150 m/d and RT=10 min.)*

The obtained results agree with other studies which found that the required dose of PP to oxidize $Mn^{+2}$ was less than that indicated by stoichiometry. It is thought that when $Mn^{+2}$ is separated on the filter, it starts to coat the filter sands and convert it to work like green sand filter so the required dose becomes smaller **[10]**. The oxidation time ranges from 5 to 10 minutes, provided that the pH is over 7.0 **[12]**. On the other hand, some studies found that the required dosage is slightly more than the required theoretical dose at pH less than 8.0 **[13]** and so, the required dose should be determined accurately.

***Rate Of Filtration (ROF) effect:*** **Fig. (5)** shows the results of experiments that discuss the effect of increasing ROF on $Fe^{+2}$ and $Mn^{+2}$ removal. PP dose = 2.0 ppm was used for RT = 20 min. For iron: increasing ROF from 150 to 180 m/d does not affect the removal efficiency. At ROF=220 m/d, RR becomes

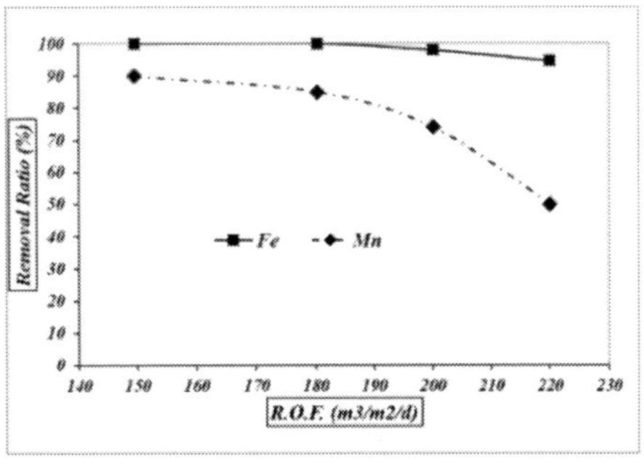


***Fig. (5): Effect of R.O.F. on $Fe^{+2}/Mn^{+2}$ removal (Initial $Fe^{+2}/Mn^{+2}$ conc. is 1.50/1.0 mg/l, pH=7.0, P.P.=2.0 ppm and D.T.=20 min)***

***Table (1): $Fe^{+2}/Mn^{+2}$ concentrations of phase 2 experiments***

| $Fe^{+2}$ (mg/l) | $Mn^{+2}$ (mg/l) | Combined Conc. (mg/l) | PP dose (ppm) |
|---|---|---|---|
| 1.5 | 1 | 2.5 | 2 |
| 3 | 2 | 5 | 3.5 |
| 5 | 3 | 8 | 5.5 |
| 6 | 4 | 10 | 7 |

95%. But in general, it is seen that iron removal is slightly affected by increasing ROF. For manganese: the biggest RR is obtained at ROF equals 150 m/d. At ROF equals 220 m/d, the removal ratio becomes 50% only, therefore, it is obvious that manganese removal depends on ROF in contrast to iron. Therefore, ROF of 150-180 m/d -(6.25-7.5 m/h) is recommended for $Mn^{+2}$ and 150-220 m/d (6.25-9.17) for iron. However, it is reported that the recommended ROF for $Mn^{+2}$ removal is about 15-18 m/hr where for iron is 6-7.5 which is considered totally different from our study **[12]**.

### 3.2. Group 2: Study results of Phase 1 on other $Fe^{+2}$ and $Mn^{+2}$ concentrations:

In this group, experiments were done for different concentrations of iron and manganese with using half of the required stoichiometric dose of PP. Table (1) shows these concentrations and the used PP dose. The experiments started with ROF equals to 180 $m^3/m^2/d$, pH=7.0 and RT equals to 20 minutes. The experiments lasted for 24 hours.

***Effect of using PP:*** **Fig. (6)** shows results of using half of the stoichiometric PP dose on the removal of $Fe^{+2}$ and $Mn^{+2}$ different concentrations. The figure shows that using PP is very efficient to oxidize and remove $Fe^{+2}$ and $Mn^{+2}$ at different concentrations, where using half of stoichiometric dose of PP can remove about 98% of $Fe^{+2}$ and more than 90% of $Mn^{+2}$ despite increasing combined concentrations to 10 mg/l.

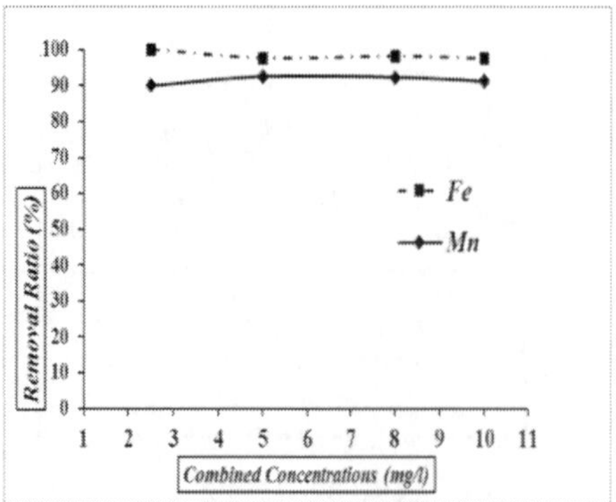


***Fig. (6): Effect of using half dose of PP on removal of different concentrations of $Fe^{+2}$ and $Mn^{+2}$ (Initial pH=7.0, RT=20 min. Samples are taken and analyzed after 8 hours from filter run start).***

***Filter clogging rate:*** **Fig. (7)** shows the change in ROF with time (in hours) for experiments of group 2. The change in ROF expresses the velocity of filter clogging at different concentrations. The figure shows that the ROF decreases with time as expected. This occurs as removed iron and manganese accumulate on filter media and start to clog the filter. It is shown that increasing $Fe^{+2}$ and $Mn^{+2}$ concentrations lead to rapid filter clogging. When concentrations of iron and manganese exceed 5.0 mg/l, the removed particles cause rapid clogging of filters and decrease the period of filter run to be less than 12 hours. This period is not practical nor economical to be applied to water treatment plants as the ratio of backwash water will increase. And so, it is recommended that when these concentrations exceed 5.0 mg/l to use sedimentation before filtration. Other studies suggest using sedimentation when combined concentrations of $Fe^{+2}$ and $Mn^{+2}$ exceed 8.0 mg/l **[8]** or when iron concentration is greater than 5.0 mg/l **[12]**.

### 3.3. Group 3: Study using conventional treatment to remove High concentrations of $Fe^{+2}$ and $Mn^{+2}$:

This group of experiments deals with using conventional treatment to remove $Fe^{+2}$ and $Mn^{+2}$

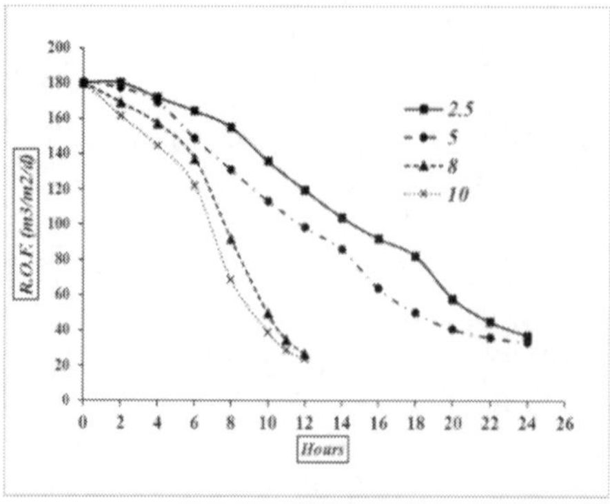


***Fig. (7): Change in ROF of different concentrations of $Fe^{+2}$ and $Mn^{+2}$ (Initial pH=7.0, RT=20 min. All experiments Started at ROF=180 m/d)***

when found in high concentrations. The treatment consists of adding alum (with/without other chemicals), mixing with water for 5 minutes, flocculation for 25 min., sedimentation for 2.0 hrs and then filtration. This set discusses using alum alone, alum with PP and alum with increasing pH.

**Fig. (8)** shows the results of these experiments. It is shown that using alum alone can remove high ratio of $Fe^{+2}$. By using alum dose=60 ppm, about 97% of $Fe^{+2}$ can be removed. But, RR of $Mn^{+2}$ does not exceed 18%. Using PP with alum enhances $Mn^{+2}$ RR to reach about 63% but decreases $Fe^{+2}$ RR to 79% when PP dose=5.0 ppm used with alum dose=40 ppm. using alum alone with increasing water pH to over 10 leads to complete removal of iron and about 95% of $Mn^{+2}$.

Another study found that $Fe^{+2}/Mn^{+2}$ removal by coagulation and clarification were about 18:75% and 8:24% for $Fe^{+2}/Mn^{+2}$, respectively for different concentrations of both. However, when PP was used, $Fe^{+2}$ RR was about 99% which is greater than our study results and $Mn^{+2}$ RR was 72% which are like the above results **[14]**.

## 4. CONCLUSIONS:

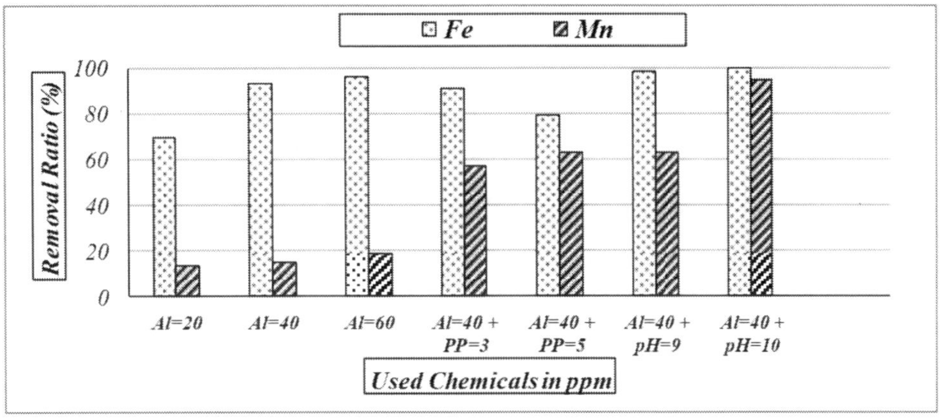


**Fig. (8):** Results of using conventional treatment for removal of $Fe^{+2}$=5.0 mg/l and $Mn^{+2}$=3.0 mg/l (Initial pH=7.0, All experiments Started at ROF=150 m/d and samples are taken after 10 hours from filtration start).

Iron and manganese can be removed from groundwater by using oxidation by potassium permanganate followed by filtration which gives good results. By using doses near to half of the theoretically calculated dose, it can remove up to 100% and 90% of iron and manganese at pH=7.0 over tested concentrations. Sedimentation is required when combined concentrations of iron and manganese are greater than 5.0 ppm to increase the filtration period. Using conventional treatment of adding alum, flocculation, sedimentation and filtration can remove up to 97% and only 18% of iron and manganese, respectively. Using PP in addition to alum enhances manganese removal but decreases iron removal. However, using alum with increasing pH to 10 leads to 100% and 95% of $Fe^{+2}$ and $Mn^{+2}$ removal and increases filter working period.

## RECOMMENDATIONS:

***Using PP:*** PP is considered a very good selection for the process of iron and manganese removal. The experiments show that using dosages near to half of the required theoretical dose at normal pH and retention time of 20 minutes can remove iron completely and 90% of manganese. However, the dose should be determined very accurately to prevent water coloring.

***Using sedimentation:*** sedimentation is required when iron and manganese combined concentrations exceed 5.0 mg/l to allow longer filtration periods.

***Using Alum:*** using alum with sedimentation alone cannot remove high ratio of manganese. However, using PP with alum affect $Fe^{+2}$ removal. Using alum while raising pH to 10 leads to remove high ratio of both $Fe^{+2}/Mn^{+2}$ concentrations.

## REFERENCES

1. Ahmed R., E. Saafb, Mohamed A. Dawoud: 'Desalination of brackish groundwater in Egypt', Desalination, 152, pp. (19–26), 2002.
2. Qingfeng C., Lichao N, Linlin B., et al.: 'Interactions between ammonia, iron and manganese removal using pilot-scale biofilters' Journal of Water Supply: Research and Technology, AQUA, in press, 2017.
3. Demir, Neslihan Manav. "Experimental Study of Factors that Affect iron and manganese Removal in Slow Sand Filters and Identification of Responsible Microbial Species." Polish Journal of Environmental Studies 25.4: 1453-1465, 2016.
4. Azrin M., Mohd N., Mohd R., et al: 'Removal of iron and manganese Using Cascade Aerator and Limestone Roughing Filter' Materials Science Forum, 1662-9752, Vol. 857, pp 509-513, 2016.
5. R. El Araby, S. Hawash, G. El Diwani.: 'Treatment of iron and manganese in simulated groundwater via ozone technology', Desalination, 249, pp. (1345–1349), 2009.
6. Wen-Hsiang C., Yung-Hsu H., Chih-Chao W., et al: 'The on-site feasibility study of iron and manganese removal from groundwater by hollow-fiber microfiltration', Journal of Water Supply: Research and Technology -AQUA, pp. (391-401), 2011.
7. Jutta K. Piispanen & Jarmo T. Sallanko: 'Mn(II) removal from groundwater with manganese oxide coated filter media', Journal of Environmental

Science and Health, Part A: Toxic/Hazardous Substances and Environmental Engineering, 45:13, 1732-1740,2010.
8. John C., Mark T.: 'iron and manganese removal handbook', 2nd edition, American Water Works Association, 2015.
9. P. Roccaro, C. Barone, G. Mancini, et al: 'Removal of manganese from water supplies intended for human consumption: a case study', Desalination, 210, pp. (205–214), 2007.
10. Piaw P., Jatuporn W., Wen-Hsiang C., et al: 'Removal of manganese(II) and iron(II) from synthetic groundwater using potassium permanganate', Desalination and Water Treatment, 52, p. 5942–5951, 2014.
11. Lee H. Odell: 'Treatment technologies for Groundwater', 1st edition, American Water Works Association, 2010.
12. K. M. Johnson, D. D. Ratnayaka, M. J. Brandt: 'Twort's Water Supply', 7th edition, Elsevier Ltd CO., 2016.
13. G. Khadse, P. Patnil, P. Labhasetwar: 'Removal of iron and manganese from drinking water supply', Sustain. Water Resour. Manag. 1:157–165, 2015.
14. Zogo D., Bawa L., Soclo H. and Atchekpe D.: 'Influence of pre-oxidation with potassium permanganate on the efficiency of iron and manganese removal from surface water by coagulation-flocculation using aluminium sulphate: Case of the Okpara dam in the Republic of Benin', Journal of Environmental Chemistry and Ecotoxicology Vol. 3(1), pp. 1-8, January 2011.

# Phytoremediation of soil polluted with Iraqi crude oil using grass plant

*Abdul Hameed* Al-Obaidy[1], *Riyad* Al-Anbari[2], *Sarah* Hassan[2]

[1]Environmental research Center, University of Technology, Baghdad, Iraq
[2]Department of Building and Construction Engineering, University of Technology, Baghdad, Iraq

**Abstract.** Remediation technology is a promising technique that decreases pollutants like hydrocarbons from the environment. An experimental work was done at green house of University of Technology in order to study the effect of crude oil on the plant growth and to measure the decrement which happened on shoot height, germination rate and the reduction of total petroleum hydrocarbon (TPH), which resulted by this phytoremediation technique. The samples of soil were measured for TPH reduction and removal by Horiba model OCMA – 350. Five doses were used in this experiment (0 control, 10x103, 30 x103, 50 x103, 75 x103) (mg crude oil / kg soil) or (ppm). The greater efficiency was obtained in the treatment 50 x103 ppm seeded with cotton, in which cotton removed 50.66% of the primary TPHs from soil. Results showed that the employed vegetate species were promising and effective in reducing and removing TPHs from freshly polluted soil.

## 1. Introduction

Pollution that had been found in the used agricultural lands is serious trouble contributed by the processing and allocation of crude oil and refined petroleum output in many oil producing countries [1]. As these problems are becoming bigger day by day, the solution must be found to clean up the pollution and this has led to exploration of many remedial approaches including physical and chemical methods to reduce the pollution. However it was found that such techniques aggravated the problem rather than eliminating it. Biodegradation method is used now a days because its being a good option for the remediation of polluted sites mainly because it uses inexpensive equipment, friendly to the environment and simple.

One of the most effective techniques in the biodegradation is phytoremediation where it employed vegetates and microbes for the remediation of contaminated sites.

Many research studies evaluated the phytoremediation technique to remediate petroleum polluted lands [2],[3],[4]. Many vegetates were also studied to know if they have phytoremediation potential to remove and decrease petroleum contaminants from soil [5],[6],[7].

In this study grass was used for the phytoremediation of the crude oil. This research shows the effect of different concentrations of crude oil on the plant development and germination.

## 2. Material and methods

The experimental soil was chosen and collected from location without oil contamination. Then, Iraq crude oil, which was supplied and analyzed, was used to contaminate the selected soil. The following treatments were used in the experimental study:

T1: Unpolluted soil (control) sown with selected plant seeds and without seeded plant.

T2: Polluted soil with crude oil (10000 ppm, 30000 ppm, 50000 ppm and 75000 ppm) sown with seeded plants and without seeded plant.

© The Authors, published by EDP Sciences. This is an open access article distributed under the terms of the Creative Commons Attribution License 4.0 (http://creativecommons.org/licenses/by/4.0/).

Control soil was used to show the degradation of crude oil by microorganisms without the plant effect and compared it with planted pots to see the effect of plant on degradation. Petroleum hydrocarbon pollution have negative impact on the plant growth, which was discover by this comparison. In this research sieve of 2mm was used to sieve soil. After mixing the soil homogeneously , the weighed soils were transmitted to PVC pots (7 kg of soil per pot).

Plants were seeded in contaminated soil as well as unpolluted soil. Then the TPH concentration was measured in control and contaminated soil with one vegetation species was employed in this research. However, like we know, natural processes can be used to reduce concentration of TPHs such as microbial action in soil. It appears that, even if the contaminated soil is ignored and neglected without any treatment or remediation effort, TPH level will reduce after some time. The effect of phytoremediation itself should be separated from the natural processes, which is called "Natural Attenuation". Therefore, in this study, natural attenuation (NA) treatment was also considered for TPHs reduction. Grass was seeded in soils T1 and T2 for 5 month duration in a greenhouse. Plants were vegetated in a height of 1 - 1.5 cm in the upper layer of soil in the each pot. The pots were put into the greenhouse beneath the sunlight. The plants were irrigated two times a week to keep sufficient and constant humidity level and to decrease the outgoing of leachate [11].

The temperature was between 25°C and 35°C. Leachate was collected by PVC pans kept beneath each pot. The collected water from the pans was put in the next irrigation to avoid the loss of the petroleum crude oil. However, only 0.02% of the TPH in soil was leaked out from the container with the excessive water that got down to the pans [12]. The number of developed vegetates or the upper layer of soil density was monitored to see the germination rate in the first weeks. The shoot height was monitored and measured.

Core sampler (inner diameter = 10 mm) was used to take soil samples from the 5cm depth of the container in each month of study. For TPH measurement, soil samples were air dried by putting it at normal room temperature and sieved with a 2 mm sieve. There are various procedures, which will accurately measure TPH in soil. This method is provided as a simple procedure for Horiba model OCMA – 350 oil concentration monitoring and analysis. Samples of 5g of soil were taken and put in 40 ml clean vial. 1g of anhydrous sodium sulfate, $Na_2SO_4$ was added to dry the soil samples.

30 ml of solvent (infrared spectrometer grade of Horiba S-316) were added to the vials. An extraction of oil in soil to oil in solvent was performed by shaking the vials vigorously for one or more minutes. After shaking the vials were placed in its upright position and waited for at least one min to allow setting of soil particles. 11cm No 40 Whatman filter paper was put in glass funnel. 2 gm of conditioned silica gel was added above the filter paper. The solvent / extract was filtered through the filter into a clean beaker. Pipette about 6 ml of extract into the OCMA-350 cell. Placed the cell into the OCMA -350 and measured.

## 3. Results

A promising behavior is shown by the vegetated species that was employed in this phytoremediation research in removing petroleum hydrocarbon-contaminated soil. Because of the oil pollution plant development was depressed to some extent. Shoot heights of the studied plants at the beginning, middle and end of the study are showed in Figures 1.

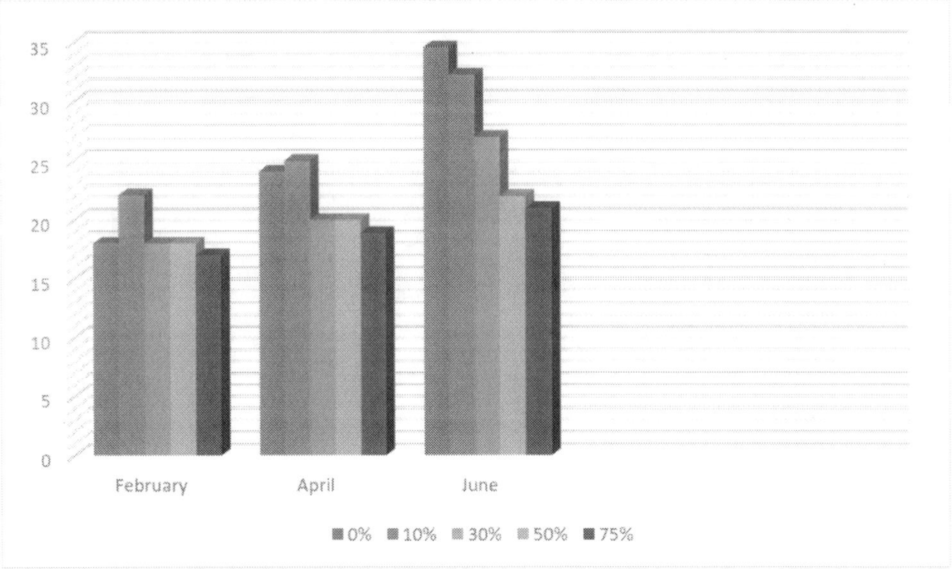


**Fig. 1. Shoot height monitoring during phytoremediation.**

Seedling germination of grass was reduced by the petroleum hydrocarbons present in the soil (Figure 2). The plant shoot heights vegetated in oil-polluted soil have shorter height in compare with the shoot heights of the plants grown in clean soil.

The decrease of TPH concentration was shown in table (1). The impact of the observed plant on petroleum hydrocarbon elimination at specific period of times was monitored. The seeded plant caused a significantly higher petroleum hydrocarbon dispersion in comparison with un-vegetated soil. Natural attenuation capable of minimize the polluted degree by TPH in soil by (76.19%, 86.5%, 86.21%, and 77.8%) for ($75x10^3$, $50 x10^3$, $30 x10^3$, $10 x10^3$) ppm respectively crude oil concentrations when the study ended. The highest effectiveness of phytoremediation was achieved for grass, in which plant existed in comparison with natural attenuation reduced the TPH level by (90.47%, 90%, 89.73%, 84.42%) for ($75 x10^3$, $50 x10^3$, $30 x10^3$, $10 x10^3$) ppm respectively. Table (3) presented the TPH removal rate at one-time intervals. TPH dispersion in S2 with the existing plant was greater in the whole five months. This drift was also studied for natural attenuation. The least TPH depression rate was observed in the final month.

Table 1. TPH removal rate in different treatments (mg kg-1).

| Different concentration of crude oil with plant | TPH concentration reduction by soil seeded with plant For five months (mg/kg) | Concentration of TPH Removed by plant (mg/kg ) | Removal rate by plant (%) |
|---|---|---|---|
| %0 | 0 | - | - |
| %10 | 610-105 | 292 | %47.8 |
| %30 | 994-39 | 496 | %49.9 |
| %50 | 3876-786 | 1964 | %50.6 |
| %75 | 4368-1610 | 1130 | %25.8 |
| Natural Attenuation (soil without plant ) | TPH concentration reduction by soil For five months (mg/kg) | TPH Removed by soil (mg/kg) | Removal rate by soil (%) |
| %0 | 0 | - | 0 |
| %10 | 610-397 | 213 | %34.9 |
| %30 | 994-535 | 459 | %46.1 |
| %50 | 3876-2750 | 1126 | %29 |
| %75 | 4368-2740 | 1628 | %37.2 |

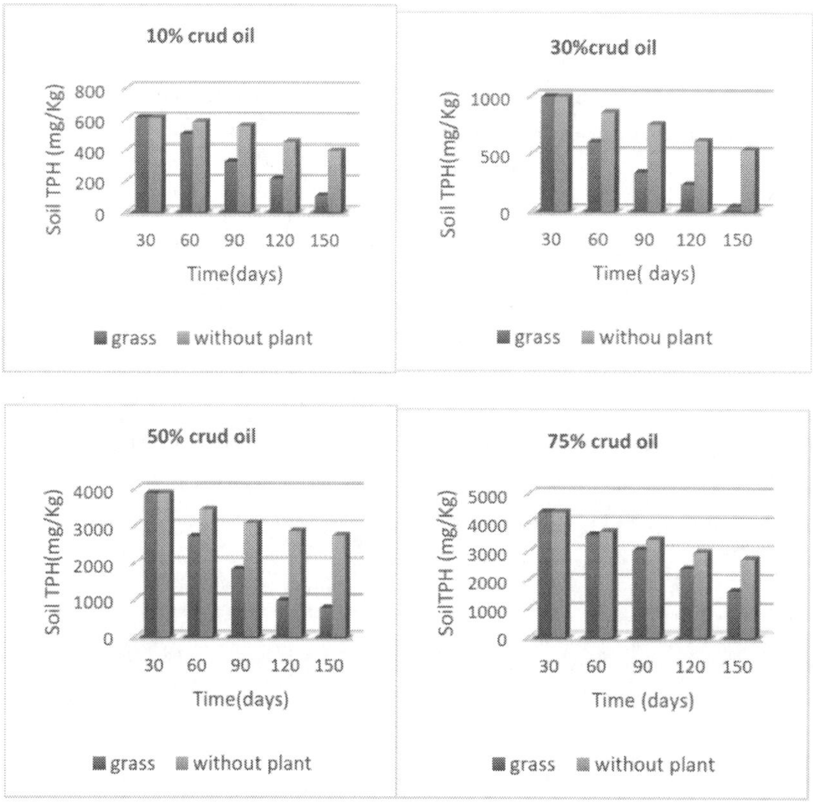


**Fig. 2. TPH concentration for five months.**

## 4. Discussion

A promising behavior was showed by the plant species, which were used in the current phytoremediation study in petroleum hydrocarbon-polluted area. However, the plant development was affected in the oil polluted soil to some extent. The most substantial point in plant establishment is the development rate. Germination Sensitivity is also the first developmental step of seeding plant can influence the efficiency of phytoremediation. A relationship between poor emergence and subsequent poor development in hydrocarbon-polluted area was mentioned by some researchers [10].

An adverse effect and delay caused by the increase of petroleum hydrocarbon concentration on the development of the observed plant species; in addition, the subsequent evolution was frustrated greatly in most cases by petroleum hydrocarbon contamination.

The soil employed in this research was fresh soil, refers to the fresh oil pollution. The water repellent characteristic of hydrocarbons may be the cause of development delay of the studied plants. Delaying or preventing the access of water and oxygen to seeds may be caused by the Hydrocarbons, act as physical barrier [8].

The presence of petroleum hydrocarbon made a remarkable reduction of shoot height, which was observed for cotton plants. The effects of toxicity of petroleum crude oil caused an inhibition of plant growth. Small molecules of hydrocarbons can enter and pass cell membranes leading to reduced membrane integrity or even to the death of the cell [13].

Plant health can be indicated by plant height and shoot biomass; however, more shoot biomass observed in the study is not necessarily an indicative of enhanced remediation efficiency [9]. Extensive root elongation is likely to be associated with more root biomass within the soil.

It was observed that the studied plants increase the removal rate of petroleum hydrocarbon at varied sampling times. The seeded plant made significantly higher petroleum hydrocarbon dispersion in contrast with unplanted soil.

Based on the obtained results, cotton is a promising species for phytoremediation of fresh petroleum hydrocarbon-contaminated soils. However, plants growth was depressed by petroleum hydrocarbon contamination. Cotton as we know, it is easy to vegetate in many countries. Remediation is well known as inexpensive and requires minimum maintenance technology which is why it is used in Iraq and most other developing countries.

## References

**1.** J. Ayotamuno, R. Kogbara, M. Taleat, *Bioremediation of a petroleum-hydrocarbon polluted agricultural soil at different levels of water application in Port Harcourt*, Nigeria. J Food, Agric. Environ. **4** (3&4): 214-217, (2006)

**2.** N. Merkl., R. Schutze-Kraft, C. Infante, *Phytoremediation in the tropics – influence of heavy crude oil on root morphology characteristics of graminoids*. Environ. Poll., **138**(1): 86-91(2005).

**3.** I. Issoufi,, R.L. Rhykerd, K.D. Smiciklas, *Seedling growth of agronomic crops in crude oil contaminated soil*. J. Agron. Crop Sci., **192**: 310 (2006).

**4.** E. Diab., *Phytoremediation of Oil Contaminated Desert Soil Using the Rhizosphere Effects*. Global J. Environ. Res., **2**(2): 66-73 (2008).

**5.** C.M. Frick, R.E. Farrell, J.J. Germida, *Assessment of Phytoremediation as an in situ technique for cleaning oil-contaminated sites*. Petroleum Technology Alliance Canada, Calgary. (1999). http://www.rtdf.org/pub/phyto/phylinks.htm.

**6.** K. Njoku, M. Akinola, B. Oboh, *Growth and Performance of Glycine max L. (Merrill) Grown in Crude Oil Contaminated Soil* Augmented With Cow Dung. Nat. Sci., **6**(1): 46-56 (2008).

**7.** K. Njoku, M. Akinola, B. Oboh, *Phytoremediation of crude oil contaminated soil: the effect of growth of Glycine max on the physico-chemistry and crude oil contents of soil.* Nat. Sci., **7**(10): 79-87. (2009).

**8.** G. Adam, H. Duncan, *Influence of diesel fuel on seed germination*. Environ Pollut **120**: 363-370. (2002).

**9.** M. Banks, P. Schwab, B. Liu, , Kulakow, P., Smith, J.S., Kim, R, *The effect of plants on the degradation and toxicity of petroleum contaminants in soil*: A field assessment. Adv Biochem Eng Biotech **78**: 75-96. (2003).

**10.** C. Chaineau, J. Morel, J. Oudot, *Phytotoxicity and plant uptake of fuel oil hydrocarbons*. J Environ Qual **26**: 1478- 483. (1997)

**11.** J.Tiba .,J. Abdul Hameed .,A. Riyad, *The Use of Phytoremediation Technique to Remove Cr and Pb Ions from Soil Irrigated with Wastewater* . (2013)

**12.** S. Hutchinson, M. Banks, A. Schwab, *Phytoremediation of aged petroleum sludge: Effect of inorganic fertilizer*. J Environ Qual **30**: 395-403. (2001)

**13.** N. Merkl , R. Schultze-Kraft.,C. Infante, *Phytoremediation in the tropics-the effect of crude oil on the growth of tropical plants*. Biorem J **8**: 177-184. (2004).

# Evaluation of raw water quality in Wassit governorate by Canadian water quality index

*Abdulkhaleq* Mahmood

Water Resource Department, Kut Technical Institute, Middle Technical University, Kut, Iraq

**Abstract.** In this work an attempt has been done to evaluate the raw water quality in Wassit Governorate by using Canadian Council of Ministers of Environment Water Quality Index (CCME WAI). Six stations along Tigris River were located and the field work was conducted during one year from October 2015 to September 2016 in collecting data. Twelve water parameters were used to evaluate the water quality index (pH, turbidity, total dissolved solid, total alkalinity, total hardens, nitrate, calcium, sodium, potassium, magnesium, chloride, and sulfate). The raw water quality in Wassite Governorate has been ranged between (65-79) based on CCME WQI results, which means that the river has not been in its good condition, and need to be manage to control the sources of pollutants by monitoring to keep them in their natural condition.

## 1 Introduction

One of the essential part of any environmental monitoring program is the reporting of results to both general public and the managers. This task is so complex in water quality monitoring because of the large number of measured variables required analyzing. One of the suggested solution of this problem is employing an (water quality index), that will arithmetically combine all water variable parameters measures and provide a general understood description of water. This index is a good tool for ranking the suitability of water use by humans, wild life, aquatic life, etc. [1]. Water quality index was creating by the Canadian Council of Ministries of Environment (CCME) in 1997. The index value can range between 0 and 100. Increasing values are indicative of increasing water quality in which CCME WQI value between 95-100 means that the water quality is excellent while CCME WQI values between 0-44 means that the water quality is poor. In Iraq, many attempts has been done to evaluate the water quality index for both Tigris and Euphrates rivers in different locations. Shaymma and Ayad, 2012 studied the water quality index for Euphrates river from Al-Qaim to Al-Qurna for both drinking and irrigations water. The results shows that the water quality for overall drinking and irrigation uses in the period between April 1998 to April 2001 is mostly rated between good to very poor [2]. Alhassan, 2013 make an attempt to calculate the overall water quality index for Tigris river as a case study using six water parameters in Baghdad city. The results indicated that the Tigris water quality are ranged between poor and slightly polluted [3]. Layla, 2015 used CCME QWI to evaluate the water quality index of three adjacent water treatment stations in Shat Al-Hila River. The results showed that all evaluated parameters of treated water were within Iraqi standard except turbidity and the water quality is good [4]. Salam, 2016 studied the water quality of Al-Gharraf river in south of Iraq using CCME WQI. The result shows that the water of Al-Gharraf classified to poor for aquatic life and fair for irrigation with seasonal overall WQI values of 30-39 and among stations was 38-29.

The objective of the present research is to evaluate the raw water quality of Tigris river in Wassit Governorate using CCME WQI.

## 2 Data and samples collection

To assess water quality conditions relative to water quality objectives, twelve water parameters were tested during one year from October 2015 to September 2016. Six locations were used for tacking samples over 160 km long on Tigris River within Wassite governorate. The sampling locations distributed between Al-Swiara and Al-Kut City. Water samples were tested twice in month and the average values for each parameters were recorded for WQI evaluation.

© The Authors, published by EDP Sciences. This is an open access article distributed under the terms of the Creative Commons Attribution License 4.0 (http://creativecommons.org/licenses/by/4.0/).

Figure 1 shows the aerial photo for sampling location within Wassite governorate, while table 2.1 shows the UTM coordinate for sampling locations.

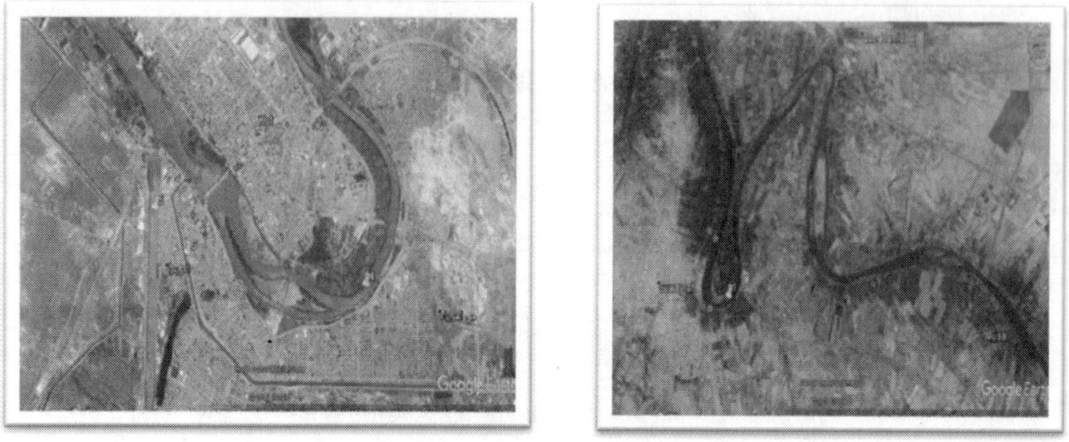

**Fig. 1.** Aerial photo for sampling locations within Wassite governorate.

**Table 1.** The UTM coordinates for sampling locations

| Sample no. | N | E |
|---|---|---|
| 1 | 3642820 | 481467 |
| 2 | 3602139 | 539368 |
| 3 | 3597279 | 575568 |
| 4 | 3594828 | 579097 |
| 5 | 3595648 | 575885 |
| 6 | 3595150 | 576391 |

### 2.1 Water quality index measurement

The selection of appropriate water quality variables is necessary to yield meaningful water quality index results. This related to a professional judgment of the observer. After collecting the data for the twelve parameters for raw water in Wassite governorate during one years, the water quality index were calculated using CCEM WQI. The values of water quality index factors (F1, F2, and F3) are calculated as flow, F1, and F2 are calculated directly, while F3 are need some additional steps.

$$F1 = \frac{Number\ of\ variables\ that\ failed}{Total\ number\ of\ variables} * 100 \quad (1)$$

Where F1, is the percentage of variables that not meet the objectives at least one during the period of observation failed variables).

$$F2 = \frac{Number\ of\ failed\ test}{Total\ number\ of\ tests} * 100 \quad (2)$$

Where F2, is the percentage of tests that not meet the objectives (failed test). For calculation the factor F3, which represented by the equation

$$F3 = \frac{nse}{0.01\ nse + 0.01} \quad (3)$$

In which, F3 represent the amount by which failed tests values don't meet their objectives, and nse is the normalized sum of excursion, which are calculated by

$$nse = \frac{\sum_{i=1}^{n} excursion}{\#of\ tests} \quad (4)$$

equation are calculated by dividing the sum of excursion that represented the amount by which each failed test are out of complains with relative of all tests( sum of tests meet and not meet objectives).

The excursion are calculated either by

$$\text{Excursion} = \left(\frac{\text{test values failed}}{\text{objectives}}\right) - 1 \qquad (5)$$

This equation is used when the test values must not exceed the objectives.

$$\text{Excursion} = \left(\frac{\text{objectives}}{\text{test values failed}}\right) - 1 \qquad (6)$$

This equation is used when the test values is fall below the objectives.

After calculating the three factors ( F1,F2, and F3) the

$$\text{CCME Water Quality Index} = \frac{\sqrt{F1^2+F2^2+F3^2}}{1.732} \qquad (7)$$

The CCME.WQI results ranged between 0 and 100 as shown in table 2.

**Table 2**. Shows the ranking of water categories due to CCME results

| Degree | Water categories |
|---|---|
| 95 -100 | Excellent ( natural water, no threat detected) |
| 80 - 94 | Good ( rarely depart from natural condition) |
| 65 - 79 | Fair ( sometimes depart from natural levels) |
| 45 - 64 | Marginal ( often depart from natural and desirable level) |
| 0 - 44 | Poor ( usually depart from natural or desirable level) |

## 3 Results and discussion

The average of some of the physical and chemical properties (turbidity, alkalinity, and calcium) in all six sample locations during study period are greater than the permissible level recommended by the Iraqi standard for drinking water, respectively greater than 5 NTU, 125 mg/ L , and 100 mg/L. Figure 2 shows the variation in turbidity values during study period in two locations samples in Kut city, while the sulfate values recorded are greater than recommended levels ( 250mg/L) in all locations during some specific months. Figure 3 shows the variation of sulfate also in these two locations in kut city. The results obtained by this investigation showed that the values of other parameters are within Iraqi standard. See tables in appendix A, while the permissible levels for all water parameters in the study are tabulated in table 3 for Iraqi, WHO, and CCME.

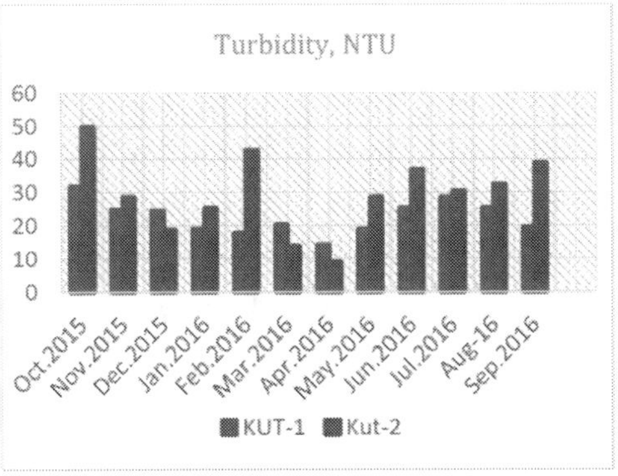


**Fig.2.** Variation of turbidity values in two locations samples in Kut City.

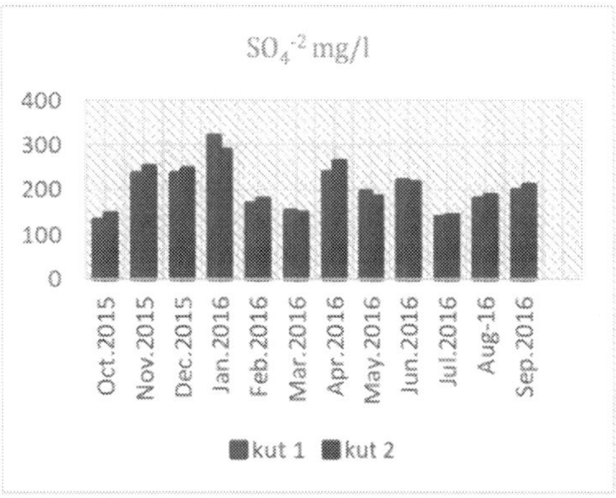


**Fig.3.** Shows the values of sulfate recorded during one year in two locations in Kut city

**Table 3.** Shows the limits of some raw water parameters for Iraq, WHO, and CCME

| parameter | pH | TDS mg/l | Turbidity (NTU) | Alkalinity mg/l | Nitrate mg/l | T.H as $caco_3$ | $Ca^{+2}$ mg/l | Mg mg/l | K mg/l | Na mg/l | Cl mg/ L | $SO_4$ mg/L |
|---|---|---|---|---|---|---|---|---|---|---|---|---|
| Iraqi standard | 6.5-8.5 | 1000 | 5 | 125 | 50 | 500 | 50 | 50 | 10 | 200 | 250 | 250 |
| WHO | 6.5-9.5 | 500 | 5 | 100 | 50 | 250 | 100 | 50 | 12 | 200 | 250 | 250 |
| CCME | 6.5-9.0 | 500 | 5 | - | 48.2 | 250 | - | 50 | 12 | 100 | 110 | 125 |

The estimated values of water quality index by CCME method for all locations are tabulated in table 4, the values are ranged between 70.4 and 72.0 .

The water quality of Tigris river and its branches( Al- Djila and Ghraff )in Wassite governorate are ranged in fair class ( 65-79), that means the river required to be managed, and take care to be under consideration to control the source of pollutant that thrown to the river to keep them in its natural conditions.

**Table 4.** Shows the values of CCME WQI for raw water in Wassite governorate.

| location | F1 | F2 | F3 | CCMEWQI |
|---|---|---|---|---|
| Al-Swaira | 33.3 | 25.7 | 23.3 | 72.0 |
| Al- Noamania | 33.3 | 25.7 | 27.5 | 70.8 |
| Before Kut Barrage | 33.3 | 25.7 | 27.0 | 71.1 |
| After Kut Barrage | 33.3 | 27.0 | 27.0 | 70.4 |
| Al-Djiala river | 33.3 | 25.7 | 28.0 | 70.8 |
| Al-Gharaff river | 33.3 | 25.7 | 27.0 | 70.5 |

## 4 Conclusions

Water quality index is a useful tools for the assessment of raw water states and can be used in formulating the pollution control strategies in terms of required treatment.

The calculated water quality index results showed that the Tigris raw water quality in Wassit governorate in all six sampling location were in fair condition with accordance to CCME WQI categories.

The raw water of Tigris river in Wassite governorate required to be observed and managed well by the local authorities to keep it in its natural condition and prevent degradation of its properties.

## References

1- CCME, Canadian Council of Ministers of the Environment (2001). *Water Quality Index 1.0 Technical Report".* In Canadian Environmental Quality Guidelines.
2- S. Alhashimi, A. Mustafa. *International Journal of Environment and Water (2012)*
3- H. Ismail. J. Al-Taqani, *vol.***26**, *no.* **2**. (2013)
4- L. Abdulkareem. J. Appl. Science, Engineering and Technology10(**11**)- 1343-1346 (2015)
5- S. Ewaid. J. of Science, Vol. **57**, No. 24, pp: 878-885(2016)
6- K. Husain, J. *Environmental Monitoring and Assessment,* **88**: 221–242 (2003).
7- UNEP/GEMS,. http://www.gemswater.org/.
8- L. Boemo, A.Musso. j. Water SA., 33(**4**): 479-485 (2007)
9- S. Gholamreza , A. Masoud, j. *Environmental Science*, 6, (**4**):19-28(2009).
10- A. Al-Obaidy. J. Water Resource and Protection, **2**: 629-635(2010)
11- Z. Al-janabi, j. Al-Nahrain Univ., 15(**1**) 119-126(2012)
12- Word Health Organization (WHO)," *Guidelines for Drinking Water Quality* " 4th Edition Word Health Organization ( WHO), Geneva(2011)

## APPENDIX A

**Table A1**. Chemical test results for raw water at Swiara city station.

| parameter | pH | TDS mg/l | Turbidity (NTU) | Alkalinity mg/l | Nitrate mg/l | T.H as $caco_3$ | $Ca^{+2}$ mg/l | Mg mg/l | K mg/l | Na mg/l | Cl mg/ L | $SO_4$ mg/L |
|---|---|---|---|---|---|---|---|---|---|---|---|---|
| Oct.2015 | 7.31 | 684 | 16 | 152 | 4.1 | 286 | 74 | 25 | 4.21 | 86.8 | 107.8 | 129 |
| Nov.2015 | 7.26 | 824 | 29.2 | 160 | 5.4 | 324 | 88 | 25 | 5.2 | 105 | 137.2 | 240 |
| Dec.2015 | 7.06 | 786 | 18.1 | 150 | 4.6 | 316 | 80.6 | 26 | 3.8 | 105.5 | 132.3 | 216 |
| Jan.2016 | 7.21 | 864 | 14.9 | 180 | 4.6 | 356 | 88.2 | 34 | 4.2 | 120.5 | 135.2 | 336 |
| Feb.2016 | 7.76 | 636 | 9.44 | 150 | 5.4 | 290 | 72 | 24.2 | 4.3 | 98.5 | 117.6 | 162 |
| Mar.2016 | 7.15 | 602 | 14.4 | 146 | 4.2 | 308 | 70.4 | 28.8 | 3.2 | 92.5 | 109.7 | 173 |
| Apr.2016 | 7.62 | 696 | 19.7 | 156 | 3.2 | 286 | 67.4 | 30.2 | 3.2 | 90 | 123.5 | 208 |
| May.2016 | 7.6 | 522 | 23.6 | 140 | 3.6 | 286 | 72 | 24 | 3.4 | 86.4 | 107.8 | 176 |
| Jun.2016 | 7.36 | 630 | 17.7 | 150 | 4.2 | 286 | 80.4 | 21.2 | 3.1 | 86.5 | 111.7 | 200 |
| Jul.2016 | 7.6 | 590 | 31.5 | 152 | 4.4 | 280 | 73.2 | 25 | 3.4 | 95 | 113.7 | 132 |
| Aug 2016 | 7.73 | 596 | 32.4 | 156 | 3.4 | 274 | 75 | 20.8 | 4.2 | 80.5 | 107.8 | 162 |
| Sep.2016 | 7.2 | 658 | 10.2 | 166 | 3.4 | 310 | 82 | 26.6 | 2.8 | 95.5 | 129.3 | 200 |
| objective | 6.5-8.5 | 1000 | 5 | 125 | 50 | 500 | 50 | 50 | 10 | 200 | 250 | 250 |

F1 = 33.3 , F2 = 25.7, F3= 23.3, **WQI = 72**

**Table A2**. Chemical test results for raw water at Noamania city station.

| parameter | pH | TDS mg/l | Turbidity (NTU) | Alkalinity mg/l | Nitraite mg/l | T.H as $caco_3$ | $Ca^{+2}$ mg/l | Mg mg/l | K mg/l | Na mg/l | Cl mg/ L | $SO_4$ mg/L |
|---|---|---|---|---|---|---|---|---|---|---|---|---|
| Oct.2015 | 7.29 | 710 | 27.8 | 160 | 2.84 | 300 | 78 | 23.5 | 5.16 | 92 | 113 | 140 |
| Nov.2015 | 7.22 | 820 | 23.6 | 166 | 5.8 | 320 | 86 | 28 | 5.6 | 115 | 139 | 248 |
| Dec.2015 | 7.08 | 792 | 16.8 | 165 | 4.2 | 328 | 83.4 | 27.8 | 3.4 | 110 | 135 | 224 |
| Jan.2016 | 7.32 | 860 | 20.6 | 170 | 5.2 | 340 | 86.2 | 36 | 4.8 | 115 | 137 | 314 |
| Feb.2016 | 7.81 | 654 | 40.2 | 156 | 5.8 | 298 | 74.2 | 26.3 | 4.8 | 105 | 124 | 148 |
| Mar.2016 | 7.32 | 674 | 13.5 | 160 | 4.4 | 316 | 72 | 28.2 | 3.8 | 98 | 115 | 162 |
| Apr.2016 | 7.7 | 796 | 23.8 | 160 | 4.8 | 300 | 71 | 28.6 | 3.4 | 96 | 127 | 227 |
| May.2016 | 7.4 | 564 | 21.8 | 156 | 3.4 | 300 | 80 | 23.4 | 4.1 | 91.5 | 115 | 166 |
| Jun.2016 | 7.57 | 710 | 20.3 | 170 | 4.6 | 300 | 80 | 23.5 | 3.4 | 95 | 121 | 183 |
| Jul.2016 | 7.4 | 610 | 36.8 | 150 | 4.8 | 294 | 78.6 | 23.6 | 3.6 | 98.5 | 119 | 128 |
| Aug 2016 | 7.81 | 630 | 26.7 | 160 | 3.6 | 286 | 78 | 22.5 | 3.4 | 85.2 | 117 | 172 |
| Sep.2016 | 7.6 | 666 | 32.6 | 172 | 3.2 | 320 | 86 | 25 | 3.2 | 90.5 | 132 | 186 |
| objective | 6.5-8.5 | 1000 | 5 | 125 | 50 | 500 | 50 | 50 | 50 | 200 | 250 | 250 |

F1 = 33.3, F2 = 25.69 F3 = 27.53, **WQI =70.97**

Table A3. Chemical test results for raw water at Kut city station. Before Kut Barrage section 1.

| parameter | pH | TDS mg/l | Turbidity (NTU) | Alkalinity mg/l | Nitrite mg/l | T.H as $caco_3$ | $Ca^{+2}$ mg/l | Mg mg/l | K mg/l | Na mg/l | Cl mg/ L | $SO_4$ mg/L |
|---|---|---|---|---|---|---|---|---|---|---|---|---|
| Oct.2015 | 7.3 | 692 | 31.8 | 146 | 2.84 | 294 | 80 | 21.4 | 3.82 | 90.5 | 107.8 | 132 |
| Nov.2015 | 7.2 | 826 | 24.8 | 178 | 5.2 | 336 | 91 | 30 | 4.8 | 98.8 | 133.3 | 236 |
| Dec.2015 | 7.04 | 826 | 24.5 | 160 | 4.8 | 310 | 84 | 22.6 | 4.2 | 100.8 | 139 | 236 |
| Jan.2016 | 7.64 | 986 | 19 | 200 | 6.8 | 380 | 98 | 38.4 | 5.6 | 138.6 | 152 | 320 |
| Feb.2016 | 7.58 | 720 | 17.8 | 148 | 4.6 | 300 | 75.6 | 30 | 3.6 | 96.4 | 113.7 | 168 |
| Mar.2016 | 7.23 | 620 | 20.3 | 155 | 5.2 | 294 | 68.7 | 30 | 4.6 | 105 | 120.5 | 152 |
| Apr.2016 | 7.62 | 824 | 14.2 | 150 | 3.6 | 290 | 68.6 | 30 | 4.4 | 100.5 | 122.5 | 238 |
| May.2016 | 7.8 | 618 | 18.9 | 144 | 2.8 | 282 | 77.4 | 26.4 | 3.2 | 85.5 | 109.8 | 196 |
| Jun.2016 | 7.42 | 626 | 25.2 | 160 | 3.8 | 284 | 76 | 22.4 | 2.8 | 100 | 115.6 | 220 |
| Jul.2016 | 7.5 | 616 | 28.6 | 166 | 3.8 | 266 | 70.6 | 21.8 | 4.2 | 105 | 127.4 | 138 |
| Aug 2016 | 7.9 | 664 | 25.2 | 156 | 4.2 | 264 | 74.2 | 21.2 | 3.6 | 90.5 | 125.4 | 180 |
| Sep.2016 | 7.4 | 654 | 19.4 | 160 | 4.4 | 300 | 80.2 | 27.5 | 2.8 | 100 | 127.4 | 198 |
| objective | 6.5-8.5 | 1000 | 5 | 125 | 50 | 500 | 50 | 50 | 50 | 200 | 250 | 250 |

F1= 33.33, F2 = 25.7, F3 = 26.84, **WQI = 71.19**

**Table A4**. Chemical test results for raw water at Kut City station. Section 2. After Kut Barrage

| parameter | pH | TDS mg/l | Turbidity (NTU) | Alkalinity mg/l | Nitrate mg/l NO3 | T.H as $caco_3$ | $Ca^{+2}$ mg/l | $Mg^{+2}$ mg/l | $K^{+1}$ mg/l | $Na^{+1}$ mg/l | $Cl^{-1}$ mg/ L | $SO_4^{-2}$ mg/L |
|---|---|---|---|---|---|---|---|---|---|---|---|---|
| Oct.2015 | 7.28 | 718 | 49.6 | 166 | 3.66 | 310 | 84 | 22 | 4.81 | 98.8 | 122 | 146 |
| Nov.2015 | 7.19 | 832 | 28.6 | 184 | 6.6 | 342 | 90 | 32 | 6.4 | 126 | 147 | 252 |
| Dec.2015 | 7.11 | 818 | 18.7 | 156 | 5.6 | 330 | 90 | 28 | 4.8 | 120 | 143 | 246 |
| Jan.2016 | 7.52 | 876 | 25.2 | 176 | 4.8 | 336 | 96 | 36 | 5.2 | 125 | 139 | 288 |
| Feb.2016 | 7.78 | 766 | 42.6 | 166 | 5.8 | 312 | 80 | 28 | 4.8 | 115 | 131 | 178 |
| Mar.2016 | 7.08 | 604 | 13.8 | 166 | 4.8 | 310 | 71.5 | 29 | 3.8 | 100 | 117 | 148 |
| Apr.2016 | 7.84 | 838 | 8.91 | 166 | 5.2 | 310 | 72 | 31 | 4.2 | 110 | 135 | 263 |
| May.2016 | 7.5 | 632 | 28.6 | 160 | 4.6 | 318 | 82 | 24.6 | 4.6 | 105 | 121 | 184 |
| Jun.2016 | 7.64 | 642 | 36.8 | 170 | 4.2 | 296 | 81 | 21.6 | 3.2 | 105 | 125 | 216 |
| Jul.2016 | 7.7 | 622 | 30.4 | 170 | 4.2 | 290 | 75 | 25 | 4.8 | 110 | 125 | 142 |
| Aug 2016 | 7.93 | 668 | 32.4 | 166 | 3.8 | 276 | 71.5 | 23 | 3.2 | 95 | 129 | 187 |
| Sep.2016 | 7.7 | 686 | 38.8 | 180 | 3.8 | 328 | 82 | 28.8 | 4.2 | 105 | 137 | 210 |
| objective | 6.5-8.5 | 1000 | 5 | 125 | 50 | 500 | 50 | 50 | 50 | 200 | 250 | 250 |

F1 = 33.33, F2 = 27, F3 = 27, **WQI = 70.42**

**Table A5**. Chemical test results for raw water at Kut City station. Al- Dijialy section

| parameter | pH | TDS mg/l | Turbidity (NTU) | Alkalinity mg/l | Nitrate mg/l NO3 | T.H as caco3 | $Ca^{+2}$ mg/l | $Mg^{+2}$ mg/l | $K^{+1}$ mg/l | $Na^{+1}$ mg/l | $Cl^{-1}$ mg/ L | $SO_4^{-2}$ mg/L |
|---|---|---|---|---|---|---|---|---|---|---|---|---|
| Oct.2015 | 7.25 | 734 | 30.4 | 180 | 2.85 | 330 | 82 | 30 | 3.6 | 100 | 128 | 156 |
| Nov.2015 | 7.17 | 816 | 34.4 | 156 | 5.2 | 310 | 82 | 24 | 5.8 | 105 | 137 | 246 |
| Dec.2015 | 7.14 | 828 | 21.4 | 160 | 5.3 | 328 | 88 | 25 | 3.6 | 115 | 140 | 228 |
| Jan.2016 | 7.26 | 870 | 8.47 | 180 | 4.2 | 350 | 87 | 35 | 4.3 | 132 | 127 | 276 |
| Feb.2016 | 7.43 | 732 | 8.4 | 160 | 4.2 | 308 | 75 | 28 | 3 | 109 | 123 | 172 |
| Mar.2016 | 7.17 | 616 | 5.71 | 140 | 4.6 | 286 | 69 | 26 | 4.4 | 116 | 127 | 137 |
| Apr.2016 | 7.72 | 882 | 6.51 | 150 | 4.2 | 326 | 80 | 32 | 3.2 | 92 | 124 | 221 |
| May.2016 | 7.4 | 642 | 34.5 | 156 | 3.8 | 305 | 80 | 25 | 4.2 | 100 | 137 | 168 |
| Jun.2016 | 7.64 | 746 | 15.4 | 176 | 5.1 | 310 | 82 | 26 | 4.1 | 116 | 129 | 250 |
| Jul.2016 | 7.8 | 630 | 39.6 | 176 | 4.6 | 272 | 72 | 22 | 3.7 | 100 | 124 | 152 |
| Aug 2016 | 7.84 | 660 | 16.3 | 150 | 3.2 | 292 | 80 | 22 | 3.4 | 87 | 131 | 176 |
| Sep.2016 | 7.8 | 668 | 52.4 | 170 | 4.8 | 308 | 82 | 28 | 3.4 | 95 | 132 | 188 |
| objective | 6.5-8.5 | 1000 | 5 | 125 | 50 | 500 | 50 | 50 | 50 | 200 | 250 | 250 |

F1 = 33.3, F2 = 25.7, F3 = 28, **WQI= 70.82**

**Table A6**. Chemical test results for raw water at Kut City station. Al- ghraf section

| parameter | pH | TDS mg/l | Turbidity (NTU) | Alkalinity mg/l | Nitrate mg/l NO3 | T.H as caco3 | $Ca^{+2}$ mg/l | $Mg^{+2}$ mg/l | $K^{+1}$ mg/l | $Na^{+1}$ mg/l | $Cl^{-1}$ mg/ L | $SO_4^{-2}$ mg/L |
|---|---|---|---|---|---|---|---|---|---|---|---|---|
| Oct.2015 | 7.21 | 726 | 60 | 176 | 2.46 | 324 | 80 | 31.5 | 3.44 | 105 | 122.5 | 156 |
| Nov.2015 | 7.16 | 808 | 32.6 | 148 | 5.4 | 300 | 84 | 21.4 | 5.4 | 100 | 135 | 240 |
| Dec.2015 | 7.08 | 842 | 26.2 | 170 | 5.8 | 340 | 92 | 26.5 | 4.4 | 110 | 137 | 248 |
| Jan.2016 | 7.66 | 852 | 30.9 | 166 | 6.4 | 330 | 84 | 30.8 | 4.2 | 105 | 140 | 310 |
| Feb.2016 | 7.52 | 728 | 19.6 | 155 | 4.8 | 296 | 70.6 | 30 | 3.8 | 110 | 127 | 184 |
| Mar.2016 | 7.21 | 586 | 6.25 | 146 | 4.8 | 300 | 65 | 28.2 | 4.2 | 96.5 | 111 | 128 |
| Apr.2016 | 7.82 | 834 | 21.8 | 146 | 3.6 | 296 | 70 | 31.5 | 3.5 | 95 | 119 | 250 |
| May.2016 | 7.7 | 626 | 22.6 | 140 | 4.1 | 300 | 79 | 27.2 | 3.8 | 96 | 118 | 179 |
| Jun.2016 | 7.72 | 660 | 43.6 | 180 | 5.4 | 268 | 80.5 | 26 | 4.2 | 110 | 132 | 210 |
| Jul.2016 | 7.3 | 624 | 29.6 | 160 | 3.8 | 264 | 70.8 | 23.2 | 3.8 | 116 | 132 | 148 |
| Aug 2016 | 7.82 | 666 | 28.7 | 152 | 4.2 | 320 | 72.8 | 22.4 | 3.2 | 88.5 | 122 | 178 |
| Sep.2016 | 7.3 | 672 | 22.6 | 166 | 3.6 | 325 | 88 | 25.6 | 3.2 | 105 | 129 | 192 |
| objective | 6.5-8.5 | 1000 | 5 | 125 | 50 | 500 | 50 | 50 | 50 | 200 | 250 | 250 |

F1 = 33.3, F2 = 25.7, F3 = 27, WQI= 70.5

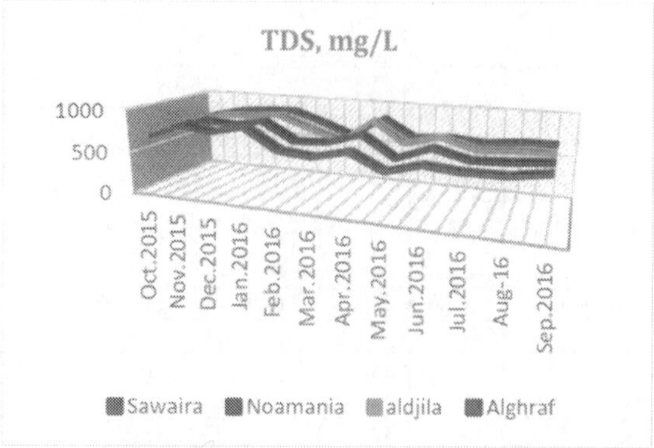


**Fig.** A1 Shows the variation of TDS in four location during study period.

# Removal of cadmium Cd(ii) and silver Ag(i) from aqueous solutions by nano activated alumina. part i: batch adsorption experiments

*Alaa* Wadie[1,*], and *Enas* Al-Khawaja[2]

[1] College of Engineering, Babylon University, Hilla, Iraq
[2] Najaf Directorate for Health, Najaf, Iraq

**Abstract.** The extent of removal of heavy metal ions cadmium Cd(II) and silver Ag(I) in single and binary system by adsorption on alumina has been investigated. Adsorption experiments were performed in batch technique from synthetic solutions using high purity nano activated alumina prepared in laboratory as adsorbent for heavy metals remove. Several experimental parameters that affect the extent of adsorption of the metal ions of interest have been investigated such as adsorbent dosage, concentration of the adsorbate, contact time, agitation speed and pH of the system under study. The equilibrium nature of the adsorption of the metal ions at different concentrations has been followed by Freundlich and Langmuir adsorption isotherms. Correlation coefficients ($R^2$) derived from these plots are (0.965) for cadmium Cd(II) and (0.966) for silver Ag(I). This work proposes a cost-effective method for the efficient removal of Cd(II) and Ag(I) from aqueous solutions. The conditions required for effective removal have been optimized.

## 1 Introduction

Water pollution due to the disposal of industrial effluents enriched with toxic metallic species is a matter of concern [1]. The metals, namely chromium, cadmium, arsenic, nickel, zinc, copper, lead and mercury, are regularly used in various industries. Furthermore, these heavy metals are non-biodegradable and exist for a long time in the environment. They have a tendency to enter into living tissues. These metals are reported to be harmful even at low concentrations. Thus, industrial effluents must be treated prior to their discharge into water resources [2].

Alumina ($Al_2O_3$) is a fine white powder which resembles common salt [3]. Activated alumina is manufactured from aluminum hydroxide in a way that produces a highly porous material; this material can have a surface area significantly over 200 $m^2$ /gm. The compound is used as a desiccant (to keep things dry by absorbing water from the air) and as a filter of fluoride, arsenic and selenium in drinking water. It is made of aluminum oxide (alumina, $Al_2O_3$), the same chemical substance as sapphire and ruby (but without the impurities that give those gems their color). It has a very high surface-area-to-weight ratio, due to the many "tunnel like" pores that it has [4].

Various technologies such as ion exchange, reverse osmosis, electrolytic removal, reduction, adsorption, precipitation, membrane filtration and flocculation have been reported for the removal of heavy metals from industrial effluents. But these technologies have some disadvantages such as high operational and maintenance cost, expensive equipment, incomplete metal removal, high energy requirement and generation of toxic residual metal sludge. Disposal of toxic sludge is another major problem in most methodologies [2].

The presence of heavy metals in wastewater will contaminate the water environment.

*Corresponding Author: Eng.alaa.husaeen@uobabylon.edu.iq

© The Authors, published by EDP Sciences. This is an open access article distributed under the terms of the Creative Commons Attribution License 4.0 (http://creativecommons.org/licenses/by/4.0/).

Furthermore, the wastewater which has not been handled properly will pose an admitted hazard to human health and the environment. Hence, heavy metals must be removed from wastewater to bring their concentrations down to below the prescribed legal limit [5].

Nowadays the importance of nano-particles and their uses in different industries have attracted many researches. The materials in nano-scale show different characteristics in comparison with their bulk state. Nano-materials have potential applications in optoelectronics, catalysis, and membranes. In this paper Nano-size γ-alumina was successfully synthesized by sol-gel method [6].

## 2 Materials and methods

### 2.1 Material

Cadmium nitrate ($Cd(NO_3)_2$) and silver nitrate $AgNO_3$ were used for preparing synthetic solutions of heavy metal ions in the required concentration. Alumina ( $Al_2O_3$ ) of 99.9% prepared in laboratory was used as the adsorbent for the study. $Al(NO_3)_3.9H_2O$ (46.9 g) was dissolved in 125mL deionized water and 125 mL ethanol. then add glucose surfactant. Under continues magnetic stirring added a drop of ammonium hydroxide untie the solution mixture was turn to gel and then filtered to remove any insoluble impurity. The gel was draying at 110°C for around 12 h. The pH of the solution mixture was initially 2. As the reaction proceeds the pH gradually increased and rose sharply from 2 to 5 producing alumina gel. The draying gel was heated for 2 h at 700°C.

## 2.2 Batch adsorption experiments

A definite volume of heavy metal ion stock solution with a known initial concentration was stirred with a definite amount of activated alumina for the stipulated time at a fixed temperature in a mechanical shaker. The un-adsorbed heavy metal ions which remained in solution were filtered and the amount of heavy metal ions was estimated after attainment of equilibrium. The extent of retention of the different metal ions by alumina was studied by varying several parameters such as the initial concentration of adsorbate (50, 100, 150 and 200) mg/L, adsorbent dosages (0.1, 0.5, 1, 1.5 and 2) g, solution temperature (25°C), pH (3.5, 5, 6.5, 8 and 9.5) and contact time (20, 30, 60, 120 and 180) min.. Freundlich and Langmuir isotherms were also studied for varying concentrations of the different metal ions at constant temperature.

## 3 Results and discussion

### 3.1 Equilibrium Time

For batch experiments, the contact time should be fixed at value ensure reaching equilibrium concentrations. Figures (1, 2 and 3) show the effect of contact time on $Cd^{+2}$ and $Ag^{+1}$ removal using 1.5 gm of alumina added to 100 mL of metal solution for batch tests at 25°C. These figures demonstrate that $Cd^{+2}$ and $Ag^{+1}$ removal percentage significantly increased with an increase in contact time. It is clear that the sorption rate was rapid at the initial stage and gradually slowed down afterwards. The slower sorption was likely due to the decrease in sorption sites on the surface of the alumina. About 98.6% $Cd^{+2}$ were removed in 120 min; while 99% $Ag^{+1}$ were removed in 60 min for single system. While 98% of $Cd^{+2}$ and 97% of $Ag^{+1}$ was removed in 60 min for binary system. The kinetic data show that $Cd^{+2}$ and $Ag^{+1}$ removal mainly occurred within 120 and 60 min, respectively.

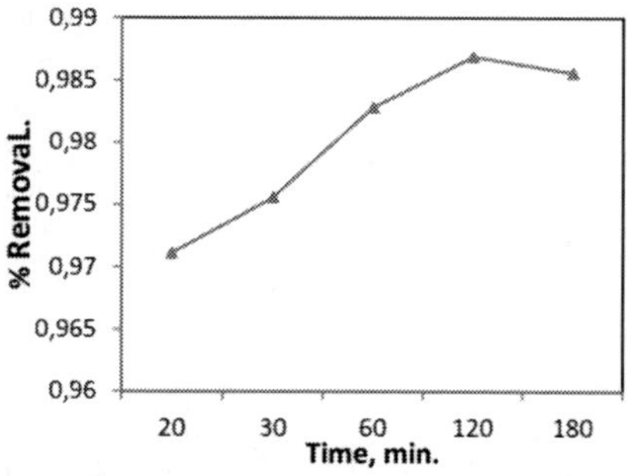


**Fig. 1**. Removal efficiency of cadmium $Cd^{+2}$ on nano activated alumina as a function of contact time, ( $C_0$ = 50 mg/L; pH= 6.5; dose=1.5 gm; agitation speed= 250 rpm).

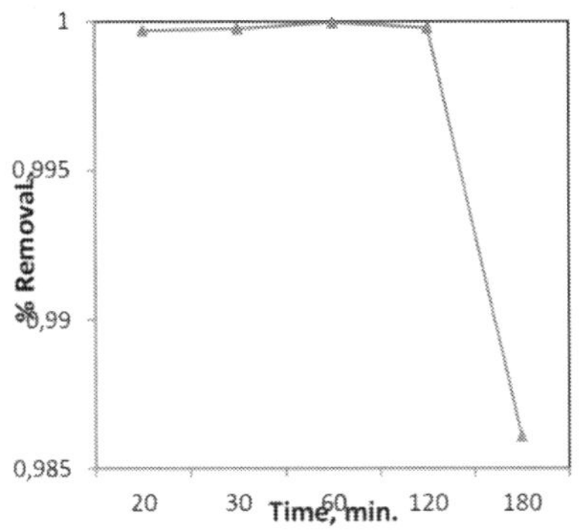


**Fig. 2.** Removal efficiency of silver $Ag^{+1}$ on nano activated alumina as a function of contact time ($C_0$= 50 mg/L; pH= 6.5; dose=1.5 gm; agitation speed= 250 rpm).

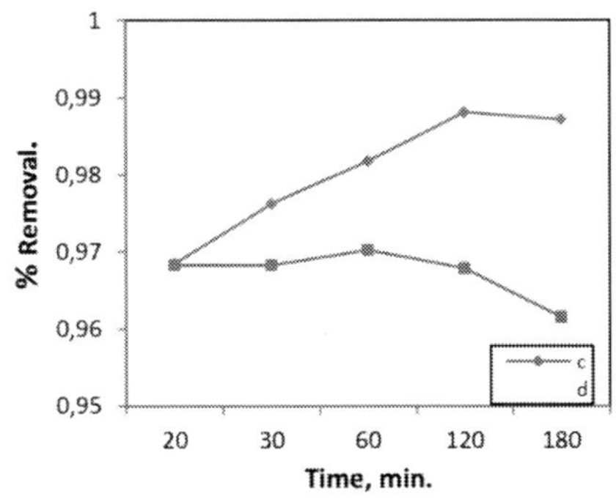


**Fig. 3.** Removal efficiency of binary system cadmium and silver on alumina as a function of contact time ($C_0$= 50 mg/L; pH= 6.5; dose=1.5 gm; agitation speed= 250 rpm).

### 3.2 Initial pH of the solution

The solution pH is an important parameter which influences most of the solid/liquid sorption processes. The sorption of $Cd^{+2}$ and $Ag^{+1}$ on the alumina was examined at different pH ranging from 3.5 to 9.5 with an initial metal concentration of 50 mg/L as shown in Figures (4, 5 and 6).

The affinity of sorption is generally determined by pH at point of zero charge $pH_{PZC}$. The $pH_{PZC}$ is the pH at which a positive charge by protonation exists in approximately equal numbers with negative charges caused by the development of deprotonation on the alumina surface. Under an acidic condition or pH below $pH_{PZC}$, protons are sorbed on the functional group that causes the alumina surface to have a net positive charge and, thus, inhibit cation sorption. At above pHpzc, the oxygen atom stays deprotonized and the surface prevails to have a net negative charge, thus, enhancing cation sorption [7].

At pH < pHpzc = 6.5 (Figures 4 to 6), the surfaces of $AL_2O_3$ having a net positive charge hinder the $Cd^{+2}$ ion adsorption due to the electrostatic repulsion. Also, stiff competitions between $H^+$ and $Cd^{+2}$ion for the active sites will also decrease the $Cd^{+2}$ ion adsorption. At pH> pHpzc, hydroxyl groups were progressively deprotonated and a net negative charge was presented on the surfaces of $AL_2O_3$, which contributed to the $Cd^{+2}$ ion adsorption through the formation of metal - ligand composite complexes with cationic $Cd^{+2}$ ion .

Over the pH range from 8-9.5, the adsorption capacity of $Cd^{+2}$ ion increased from 0.965 to 0.99 mg/gm, which was attributed to the critical pH value for $Cd^{+2}$ ion hydrolysis formation of $Cd(OH)^+$ and $Cd_2(OH)_3^+$ and precipitation ( $Cd(OH)_2$ ≥8) through the electrostatic interaction and deposition, [Jie, et al., 2015]. Meanwhile, the electrostatic attraction between the pairs of electrons on the oxygen atoms of alumina and the positive cationic $Cd^{+2}$ ion also facilitated the adsorption of $Cd^{+2}$ ion. Moreover, surface precipitation and complexation between the carboxylic group, hydroxyl and $Cd^{+2}$ ion also contributed to the adsorption of $Cd^{+2}$ ion. Finally, the physical property of the $AL_2O_3$ and the Vander vales interactions occurring between the hexagonally arrayed carbon atoms in the $Cd^{+2}$ ion could also conduce to the adsorption of $Cd^{+2}$ ion [8].

The experimental results showed that $Cd^{+2}$ and $Ag^{+1}$ can be removed at different pH values. It is clear from figures mentioned above that the time required to reach equilibrium in batch systems is equal to 120 and 60 min for $Cd^{+2}$ and $Ag^{+1}$, respectively, although the initial pH of the solution was varied from 3.5 to 9.5. Also, the maximum removal efficiency of these metals was achieved at initial pH of 6.5.

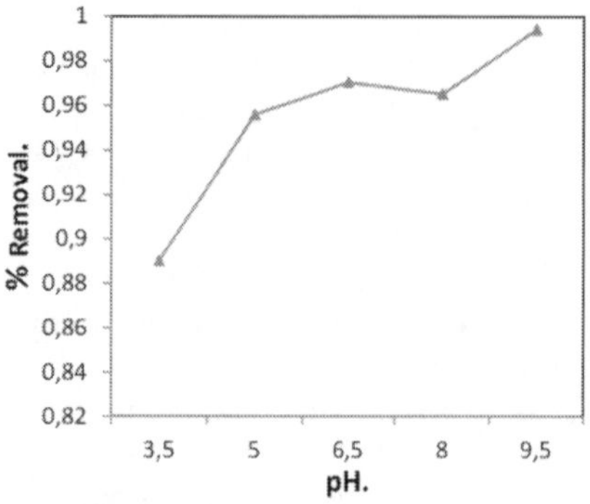


**Fig. 4.** Removal efficiency of $Cd^{+2}$ ions on alumina at different pH solution; Reaction conditions: (Co= 50 mg/L, dose=1.5 gm, agitation speed= 250 rpm, time= 180 min).

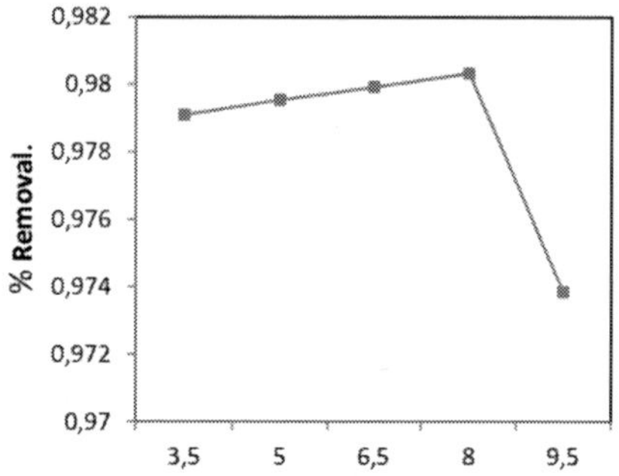


**Fig. 5.** Removal efficiency of $Ag^{+1}$ ions on alumina at different pH solution; Reaction conditions: (Co= 50 mg/L, dose=1.5 gm, agitation speed= 250 rpm, time= 180 min).

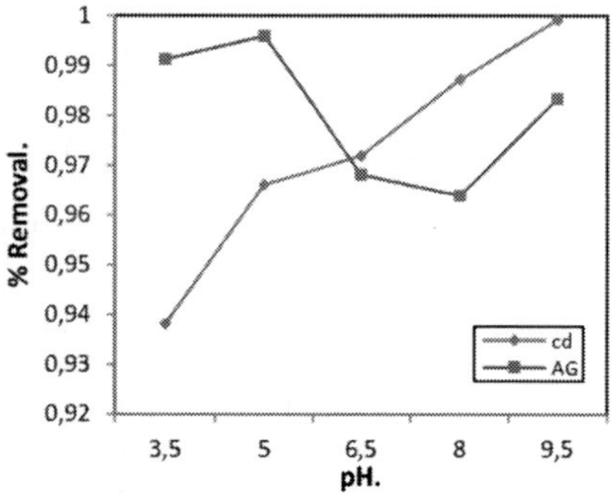


**Fig. 6.** Removal efficiency of binary system at different pH solution; Reaction conditions: (Co= 50 mg/L, dose= 1.5 gm , 250 rpm, time 180 min).

### 3.3 Optimizing alumina weight

Figures (7, 8, and 9) present the $Cd^{+2}$ and $Ag^{+1}$ removal efficiencies as a function of different amounts of alumina. It can be observed that removal efficiency improved with increasing adsorbent dosage from 0.1 gm to 2 gm for $Cd^{+2}$ and $Ag^{+1}$ at a fixed initial metal concentration. This was expected due to the fact that the higher dose of sorbents in the solution, the greater availability of sorption sites. This means that the more alumina particles are able to provide more alumina surface-active sites for collision with metal molecules to accelerate the metal removal efficiencies.

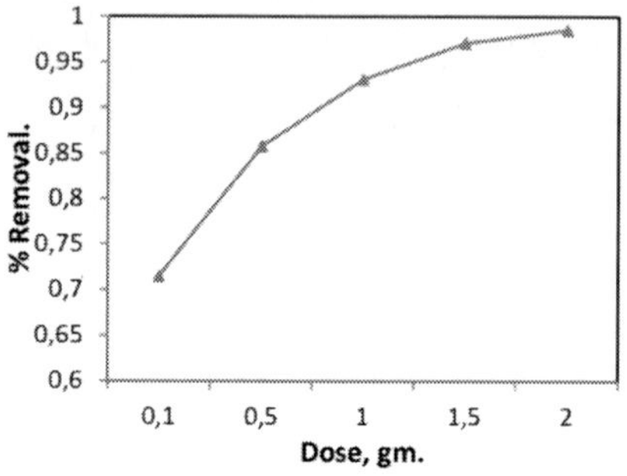


**Fig. 7.** Effect of alumina dosage on removal efficiencies of $Cd^{+2}$(Co=50mg/L; pH=6.5, time=120min, agitation speed= 250 rpm).

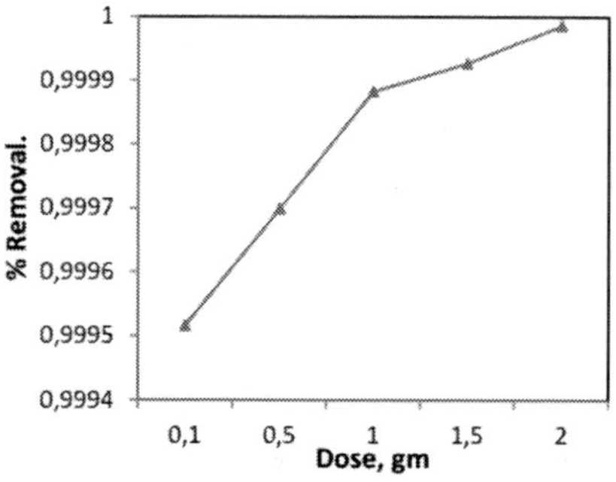


**Fig. 8.** Effect of alumina dosage on removal efficiencies of $Ag^{+1}$ (Co=25mg/L; pH=6.5, time=60min, agitation speed= 250 rpm).

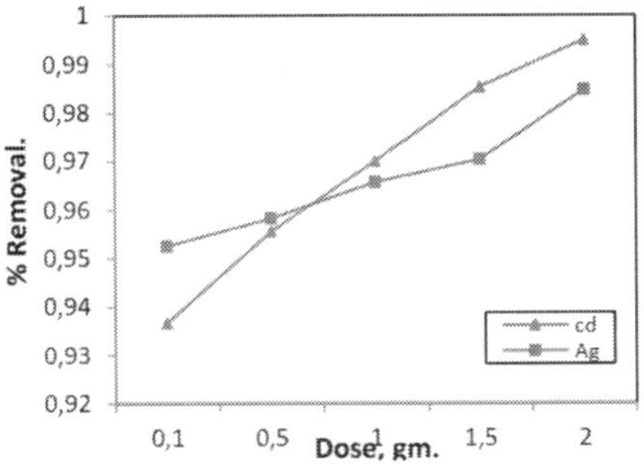


**Fig. 9.** Effect of alumina dosage in binary system of $Cd^{+2}$ and $Ag^{+1}$ (Co=50 mg/L; pH=6.5; agitation speed= 250 rpm, time=60min).

To investigate the effect of initial $Cd^{+2}$ and $Ag^{+1}$ concentration, the batch experiments were conducted at various concentration of metal from (25 to 200 mg/L). Figures (10 to 12) explains that the removal efficiency decreased inversely with increased the initial metal concentration. This plateau represents saturation of the active sites available on the alumina samples for interaction with metal ions. These results indicate that energetically less favorable sites become involved with increasing metal concentrations in the aqueous solution.

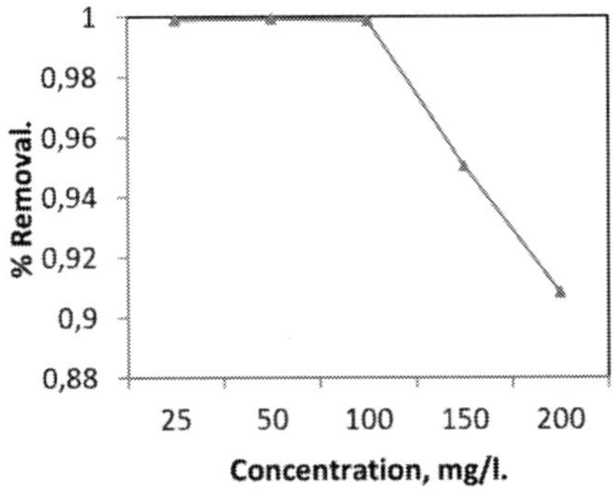


**Fig. 10.** Effect of initial concentration on removal efficiency of $Cd^{+2}$ on alumina (pH=6.5; dose=2gm; agitation speed= 250 rpm ,time=120min).

### 3.4 Agitation speed

The effect of agitation speed on removal efficiency of metal ions adopted here was studied by varying the speed of agitation from (50 to 250 rpm) and keeping other parameters equal to best ones obtained in the previous steps.

Figures (13 to 15) show that removal of $Cd^{+2}$ and $Ag^{+1}$ ions were increases with the increase in shaking rate. The results also indicate that a shaking with 250 rpm is sufficient to ensure that all the surface binding sites are made readily available for metal uptake. These results can be associated to the fact that the increase in the agitation speed improves the diffusion of metal ions towards the surface of the reactive media. Thus, proper contact is developed between metal ions in solution and the binding sites, which promotes effective transfer of sorbet ions to the sorbent sites.

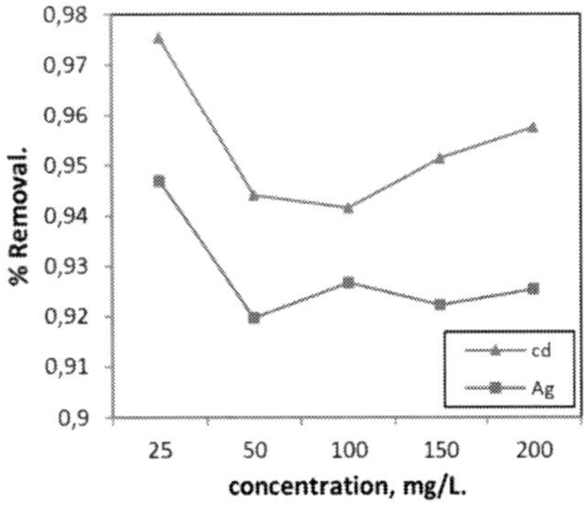


**Fig. 11.** Effect of initial concentration on removal efficiency of $Ag^{+1}$ on alumina (pH=6.5; dose=2gm; agitation speed= 250 rpm ,time=60min).

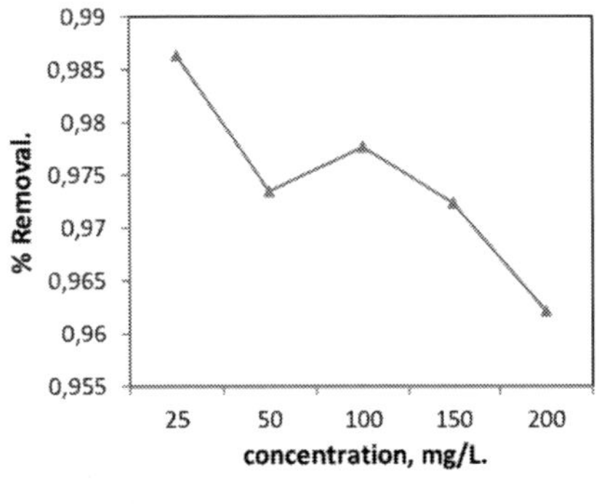


**Fig. 12.** Effect of initial concentration on removal efficiency in binary system of $Cd^{+2}$ and $Ag^{+1}$ on alumina (pH=6.5; dose=2gm ;agitation speed= 250 rpm ;time=60min).

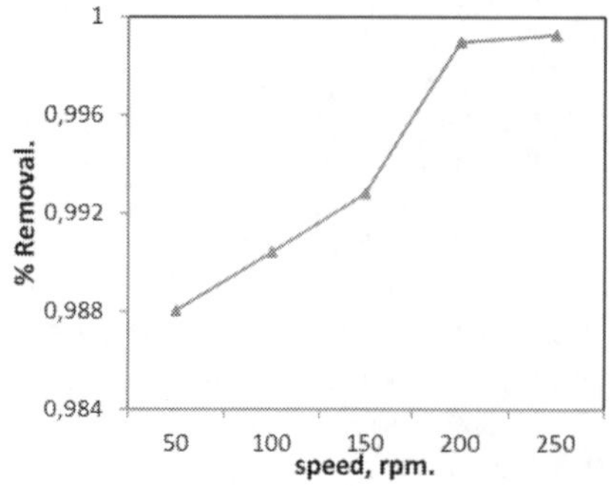


**Fig. 13.** Effect of agitation speed on percentage removal of $Cd^{+2}$ (Co=50 mg/L; pH=6.5, dose=2gm , time=120min).

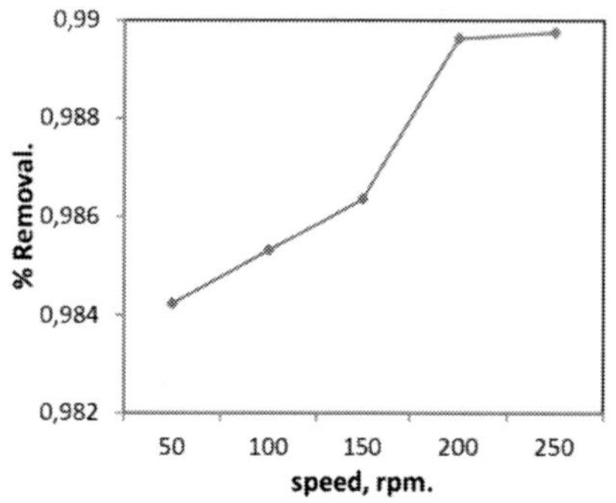


**Fig. 14.** Effect of agitation speed on percentage removal of $Ag^{+1}$ (Co=25 mg/L; pH=6.5, dose=2gm , time=60min).

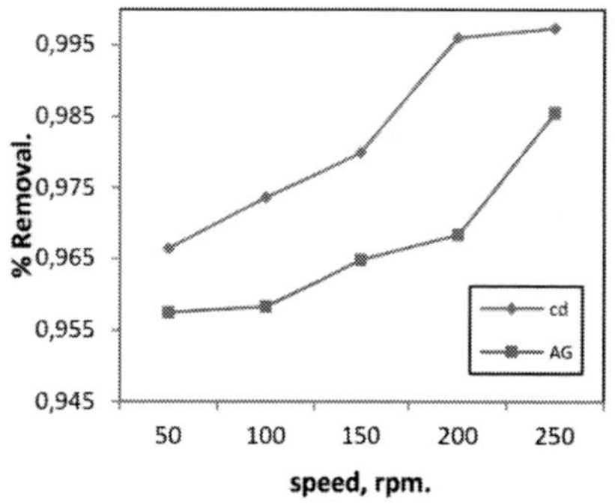


**Fig. 15.** Effect of agitation speed on percentage removal of metals in binary system(Co=25mg/L, pH=6.5, time=60min, dose=2gm).

## EXPERIMENTS

The isotherm is represented by the linear plot of specific adsorption ($C_e/q_e$) against the equilibrium concentration ($C_e$). Figures (16 to 18) show that the adsorption obeys the Langmuir model. The Langmuir equation is represented by Jenan, et al.[9]:

$$q_e = \frac{abC_e}{1+bC_e} \quad ...... (1)$$

And the equation may be linearized as:

$$\frac{C_e}{q_e} = \frac{1}{a}C_e + \frac{1}{ab} \quad ....... (2)$$

The constants b and $q_e$ relate to the energy of adsorption and maximum adsorption capacity, and their values are obtained from the slope and interception of the plot and are presented in Table1. The Freundlich isotherm is an empirical equation that is very useful because it accurately describes much adsorption data. The Freundlich equation is represented by Jenan, et al. [9]:

$$q_e = K_f C_e^{1/n} \quad ...... (3)$$

The linear form of the equation obtained by taking the logarithm as follows:

$$\log q_e = \log K_f + \frac{1}{n}\log C_e \quad .... (4)$$

The Freundlich isotherm Figures (17 and 19) are introduced as an empirical model, where $q_e$ represents the amount adsorbed per amount of adsorbent at the equilibrium (mg/gm), $C_e$ represents the equilibrium concentration (mg/L), and $K_f$ and $n$ are parameters that depend on the adsorbate and adsorbent[Reed, and Matsumoto, 1993]. Freundlich equilibrium constants were determined from the plot of $\ln q_e$ versus $\ln C_e$. From Figures (17 and 19) on the basis of the linear form of Freundlich equation, the $n$ value indicates the degree of nonlinearity between solution concentration and adsorption as follows: if $n$ =1, then adsorption is linear; if <1, then adsorption is a chemical process; if $n$ >1, then adsorption is a physical process. The situation $n$>1 is most common and may be due to a distribution of surface sites or any factor that causes a decrease in adsorbent-adsorbate interaction with increasing surface density [10]

https://doi.org/10.1051/matecconf/201816205021

and the values of $n$ within the range of 1–10 represent good adsorption [11] [12]. In the present study, since $n$=(1.92, 1.91) for $Cd^{+2}$ and $Ag^{+1}$, respectively lies between 1 and 10 it indicates the physical adsorption of $Cd^{+2}$ and $Ag^{+1}$ ions onto alumina and represent favorable adsorption condition. In both cases, linear plots were obtained, which reveal the applicability of these isotherms on the ongoing adsorption process.

Figures (16 to 19) exhibit Langmuir and Freundlich plots, respectively, for the adsorption of $Cd^{+2}$ and $Ag^{+1}$ ions onto alumina, and different Langmuir and Freundlich constants and coefficient of determination ($R^2$) derived from these plots are presented in Table 1. To find the most appropriate model for the $Cd^{+2}$ and $Ag^{+1}$ ions adsorption; data were fitted to Langmuir and Freundlich isotherm models. Results revealed that Freundlich adsorption isotherm was the best model for the $Cd^{+2}$ and $Ag^{+1}$ ions adsorption onto alumina with $R^2$ of 0.9657 and 0.9664, respectively.

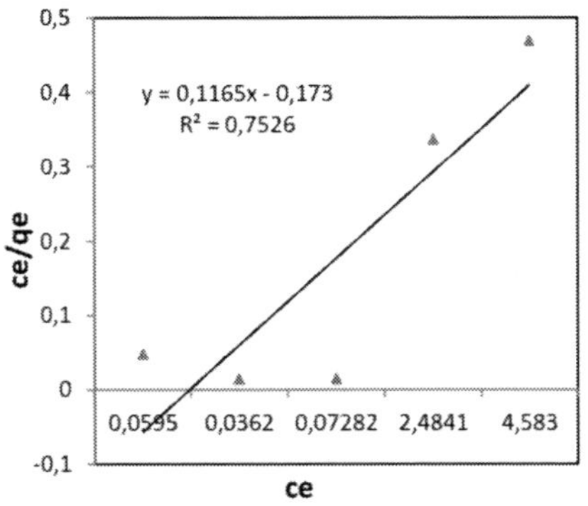


**Fig. 16.** Langmuir isotherm for sorption of Cd $^{+2}$ (pH=6.5; dose=2gm; agitation speed=250rpm; contact time = 120min.).

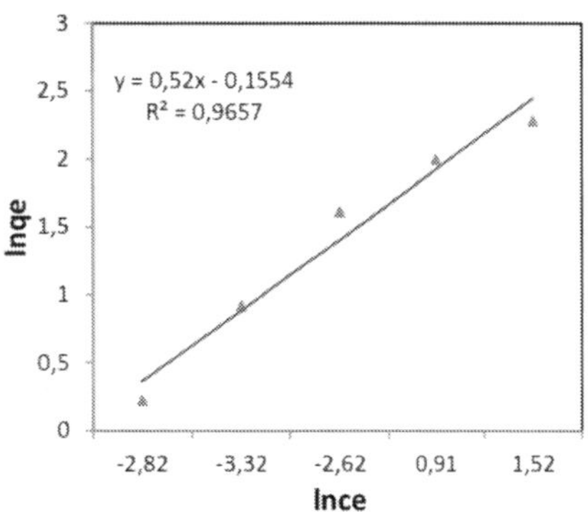


**Fig. 17.** Freundlich isotherm for sorption of $Cd^{+2}$ (pH=6.5; dose=2gm; agitation speed=250 rpm; contact time=120min.).

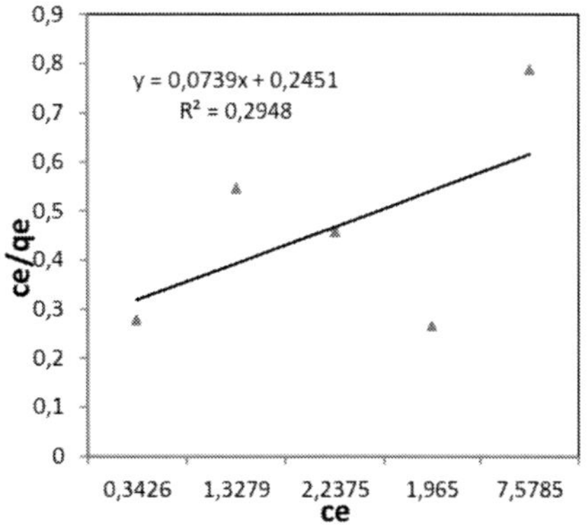


**Fig. 18.** Langmuir isotherm for sorption of $Ag^{+1}$ (pH=6.5; dose=2gm; agitation speed=250 rpm; contact time=60min.).

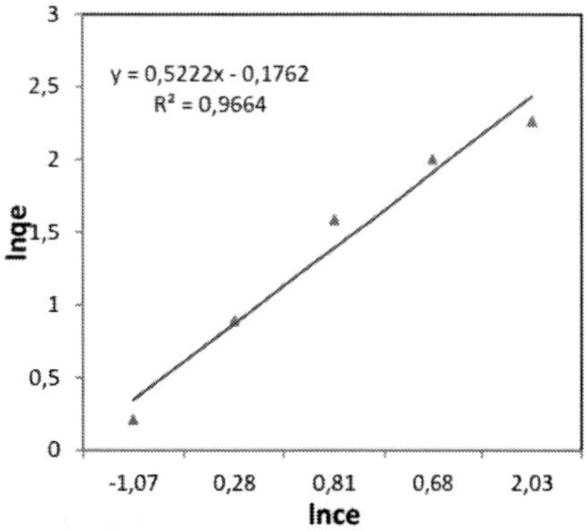


**Fig. 19.** Freundlich isotherm for sorption of $Ag^{+1}$ (pH=6.5, dose=2gm, agitation speed=250 rpm, contact time=60min.).

**Table 1:** Sorption isotherm constants for cadmium and silver on alumina reactive material. It is clear that the Freundlich isotherm model provided the best correlation in compared with Langmuir isotherm model for $Ag^{+1}$ and $Cd^{+2}$ sorption on the alumina.

| Pollutant | Langmuir constants | | | Freundlich constants | | |
|---|---|---|---|---|---|---|
| | a, mg/gm | b, L/mg | $R^2$ | $K_F$, mg/gm | 1/n | $R^2$ |
| $Cd^{+2}$ | 8.583 | -0.673 | 0.752 | 0.856 | 0.52 | 0.965 |
| $Ag^{+1}$ | 13.5318 | 0.301 | 0.294 | 0.839 | 0.522 | 0.966 |

## 4 Conclusions

The retention of Cd(II), and Ag(I) onto nano activated alumina has been carried out. Batch adsorption experiments were used for analyzing the effects of the following variables: such as concentration, adsorbent dosage, pH, agitation speed and contact time. The % of adsorption increases on changing the pH of the medium from acidic towards neutral. The time taken for attainment of equilibrium for all the metal ions has been established. The removal yield increases with increasing the adsorbent dosage, agitation speed and decreases with increase initial metal ion concentration. The effective removal of the heavy metal ions is mainly dependent on the ratio of adsorbent dosage to the concentration of metal ions, which is of course a compromise between cost effectiveness of the process and the desired retention percentage. The mathematical description for the adsorption behavior fit well to Freundlich isotherm model with good coefficient of determination.

## References

1. Sharma Y. C., V. Srivastava, A. K. Mukherjee., 2010. Synthesis and Application of Nano-Al2O3 Powder for the Reclamation of Hexavalent Chromium from Aqueous Solutions. Department of Applied Chemistry, Institute of Technology, Banaras Hindu University, Varanasi-221005, India
2. Sharma Y.C., V. Srivastava, V.K. Singh, S.N. Kaul and C.H. Weng, 2008. Nano-adsorbents for the removal of metallic pollutants from water and wastewater. MIT College of Engineering, Kothrud, Pune, India.
3. Sheeba S., R. Rajkumar, 2013. Removal of Cr(VI), Cu(II), Pb(II) and Ni(II) from Aqueous Solutions by Adsorption on Alumina. India, Research Journal of Chemical Sciences, ISSN 2231-606X, Vol. 3(8), 44-48, Res. J. Chem. Sci.
4. Metcalf and Eddy, 2003. Ch.1Wastewater Engineering 4thed.
5. Tingyi L., Zhong-Liang W., Yanqiu S., 2015. Manipulating the morphology of nanoscale zero-valent iron on pumice for removal of heavy metals from wastewater. Laboratory of Water Resources and Environment, Tianjin Normal University, Tianjin 300387, China.
6. Omid R., Ahmad S., Mohammad R., 2012. New Method for Synthesis Nano Size γ-Al2O3 Catalyst for Dehydration of Methanol to Dim-ethyl Ether. International Journal of Chemical Engineering and Applications, Vol. 3, No. 2.
7. Rangsivek, 2010. Removal of dissolved metals from storm water runoff by zero-valent iron. Ph.D. thesis, University of Berlin.
8. Gao Z., T.J. Bandosz, Z. Zhao, M. Han, J. Qiu, 2009. Investigation of factors affecting adsorption of transition metals on oxidized carbon nanotubes. Journal of Hazard Material.
9. Jenan A., Asawer A., Ramzi H., Abbas H., 2014. Study the Feasibility of Alumina for the Adsorption of Metal Ions from Water. Chemical Engineering Department, College of Engineering, University of Baghdad, Baghdad, Iraqi Journal of Chemical and Petroleum Engineering, Vol.15 No.3, 37-50, ISSN: 1997-4884.
10. Reed, B.E., Matsumoto, M.R., 1993. Modeling cadmium adsorption by activated carbon using the Langmuir and Freundlich isotherm expressions". Separation Science and Technology,vol.28, no. 13-14, pp. 2179–2195.
11. McKay, G., Blair, H.S., Gardner, J.R., 1982. Adsorption of dyeson chitin. Journal of Applied Polymer Science,vol.27,no.8, pp.3043–3057.
12. Ozer, A., Pirinc, H.B., 2006. The adsorption of Cd(II) ions on sulphuric acid-treated wheat bran. Journal of Hazardous Materials, vol.137, no.2, pp.849–855.

# Pollution assessment of Euphrates River in three governances of Iraq

*Alaa* Al-Fatlawi[1*], *Mohammed* Al-Jassani[2], and *Jawad* Al-Rifaie[3]

[1] Engineering College, Babylon University, Hilla, Iraq
[2] Al-Alamain Institute for Higher Education, Hilla, Iraq
[3] Al-Furat Al-Awsat University, Karbala, Iraq

**Abstract.** This study was concerned with assessment the water quality of Euphrates River in the middle of Iraq. Water quality assessment is the overall process of evaluation of the physical, chemical and biological nature of the water. Water pollution for Euphrates River in Middle of Iraq (Babylon, Al-Najaf, and Al-Diwaniya governorates) occurs in both rural and urban areas. In rural areas, drinking water from natural sources such as rivers and streams is usually polluted by organic substances from upstream users who use water for agricultural activities. To protect the water resources from pollution and deterioration which caused by natural pollutants or human activities, an environmental database was constructed and applied. A Visual Basic (version 6.0) Software program was constructed to evaluate the pollutant concentrations using regression models obtained by Data Fit Software program (version 8.0) to make the monitoring system. The result of the program is verified with data of year 2008 which is not included in regression model. This verification shows a good agreement with coefficient of determination ranged between "0.914 to 0.994"To assess water body for compliance with standards that have been set to define requirements for various functions and uses, the water body concerned, were compared with the Iraqi, and WHO standards for domestic and irrigation purposes, for determining whether water quality at sampling sites exceeds water quality standards. A comparison shows that these parameters in some stations are within the Iraqi and WHO standards (i.e. total dissolved solids (TDS): in Al-Musaib, Al-Kufa, Al-Shamya station), and (calcium (Ca): in Al-Hindya Barrage station).

## 1 Introduction

Water resource form about 15% from area of Iraq which found in different forms like rivers, lakes and streams etc. These resources are necessary to lasting and assurance belonging of human, domestic requirements and another uses like irrigation and industrial uses.

The monitoring and assessment of water quality is based, ultimately, upon the fundament physical, chemical and biological properties of water. However, water quality monitoring and assessment is a process of analysis, interpretation and communication of those properties within the wider context of human activity and use, and the conservation of the natural environment. It is not a fixed process, therefore, but is adapted in the light of local, national or international needs. The final aim is to provide information useful for management. Styles and strategies of management vary greatly, depending on institutions, resources and priorities.

Many researchers have been interested in studying the water quality parameters. These parameters are analyzed either by applying different statistical model or by qualities comparison with available standards.

Al-Masri and Ali [1], Wasfi [2] evaluated some pollution concentrations such as total hardness, sulphates, alkalinity, chlorides, calcium and magnesium in Tigris River through Baghdad city. The result of the study indicates that concentrations of total hardness, sulphates and calcium exceed the Iraqi permissible limits of drinking water.

*Corresponding Author: Eng.alaa.husaeen@uobabylon.edu.iq

© The Authors, published by EDP Sciences. This is an open access article distributed under the terms of the Creative Commons Attribution License 4.0 (http://creativecommons.org/licenses/by/4.0/).

They noticed that the time and site of sampling were of high importance in effecting the measured concentrations.

Al-Delaimy [3] performed a study on the effect of some environmental factors and their correlation with the inlet of river in Baghdad city (7-Nissan water treatment plant), and at the outlet (Al-Rasheed water treatment plant). He predicated models to simulate the effect of these environmental factors (temperature, turbidity, DO, pH and electrical conductivity) on the bacterial contents in river water by the adoption of single linear regression analysis test.

Al-Malikey [4] studied the determination of Tigris river water quality. He chose three stations for this purpose located at north, middle and south of Baghdad city. The result of the study shows that Tigris river water of good quality to be treated at north of Baghdad city.

Al-Husseini [5] evaluated some of water quality parameters of Shatt Al-Hilla river in four main water treatment plants abstracting their water from the sites situated along the river were selected as a case study. He found there is a noticeable increase in concentrations of chloride and sulphate especially at Al-Hsien water treatment plant during the different periods of the year and there were clear effects of municipal wastes of Al-Hilla city on levels of turbidity and bacteria.

Al-Fatlawi [6] studied some physical and chemical parameters in addition the qualitative and quantitative study for phytoplankton. Four trace elements were determined (Fe, Zn, Cu, Pb) in five stations which distributed along the Euphrates river from Al-Hindya reservoir to Al- Kifil town in the middle of Iraq. The results showed that the concentration of $NO_3$ (1.56 – 360.62), $NO_2$ (0.38 -0.84), $SiO_2$ (85.15 – 597.71) , $PO_4$ (N.D. – 0.73 )μg/l, the dissolved (Fe, Zn, Cu, Pb) (0.1– 404) , (N.D. – 1.2) , (N.D. – 0.7),(N.D. – 1.2), (N.D. – 1) ppm respectively.

Kalel [7] made a study for alteration of the following concentration parameters, turbidity, electrical conductivity, total hardness, calcium, magnesium, chloride, alkalinity, total dissolved solids, suspension solids and pH-value. These concentrations were compared to those of World Health Organization (WHO). For both branches the quality of water was checked, conducting hypothesis test at 95% level of confidence for where T-distribution had been used. It has been found that water quality for both branches did not exhibit any difference.

Fahd [8] studied some of physical and chemical characteristics of Al-Masab Al-Aam river. Results show monthly variation in air temperature ranged 14 – 43 ċC and the light penetration was 38 – 64 cm. Dissolved oxygen concentrations and free $CO_2$ inversely correlated with water temperature. The total alkalinity of water was due to the bicarbonate and carbonate. Salinity between 4.20 – 8.40 g/L. Significant positive relation were found between water temperature, alkalinity and salinity, while negative relation between dissolved oxygen and temperature.

## 2 Data processing

To gauge the success of the program, data of water quality of the Euphrates river are being analyzed monthly, and the pollution levels are being determined. The program depend five sampling stations along the river including (Al-Musaib, Al-Hindya Barrage, Al-Kufa, Al-Shamya and Al-Shenafya) as shown in Fig. (1).

### 2.1 The independent variables

They are also named as explanatory variables or predictors. Each one of the five sampling stations have specified independent variables which are assumed to have a reliable impact on water quality assessment.

### 2.2 The dependent variable

In the present work, the total dissolved solids in Al-Musaib, Al-Kufa, Al-Shamya and Al-Shenafya stations, and calcium in Al-Hindya Barrage stations were assumed to be the dependent variable and it varied greatly.

Table (1) shows the independent and dependent variables for Euphrates river used in these stations. In this study, water samples were analyzed at Al-Musaib station for the period between 2005 to 2007, Al-Hindya Barrage station for the period between 2000 to

2007, and period between 2005 to 2007 for Al-Kufa, Al-Shamya and Al-Shenafya stations, to construct the monitoring system along Euphrates river.

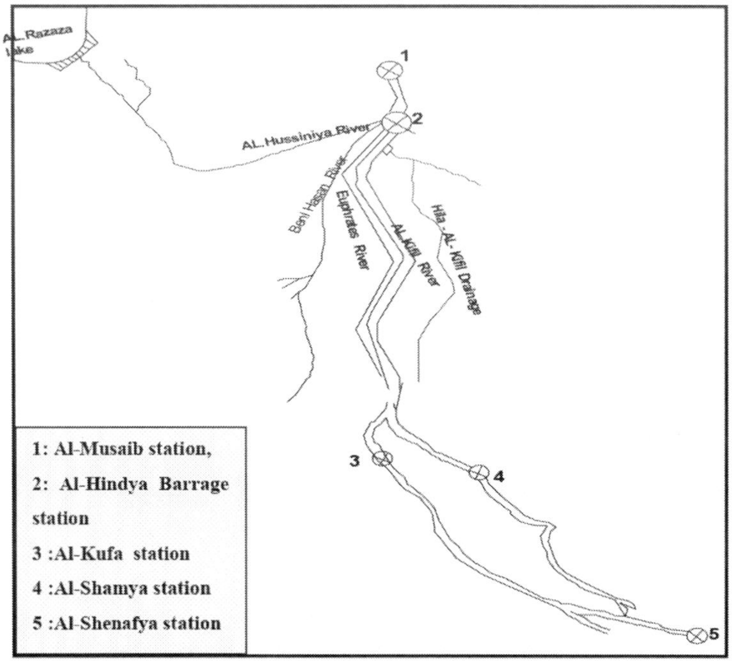


Fig. 1: Map of the studying area

Table 1: Description of independent and dependent variables of Euphrates river at different stations.

| Stations | Variables | | | | | | | | | |
|---|---|---|---|---|---|---|---|---|---|---|
| | Independent | | | | | | | | | Dependent |
| | $x_1$ | $x_2$ | $x_3$ | $x_4$ | $x_5$ | $x_6$ | $x_7$ | $x_8$ | $x_9$ | y |
| Al-Musaib | pH | Ec | Ra | T | - | - | - | - | - | TDS |
| Al-Hindya Barrage | Ec | TDS | $So_4$ | TH | Mg | Cl | Ra | Q | T | Ca |
| Al-Kufa | pH | Ec | Ra | T | RD | SSD | SD | - | - | TDS |
| Al-Shamya | pH | Ec | Ra | T | RD | SSD | SD | - | - | TDS |
| Al-Shenafya | pH | Ec | Ra | T | RD | SSD | SD | - | - | TDS |

Where: Ec : electrical conductivity (μs/cm),
TDS : total dissolved solids (mg/L),
$So_4$ : sulphate (mg/L),
TH : total hardness (mg/L),
Mg : magnesium (mg/L),
Cl : chloride (mg/L),
Q : discharge ($m^3$/sec),
Ca : calcium (mg/L),
pH : hydrogen ion concentration,
Ra : monthly rainfall totals (mm),

SSD : no. of days of suspended dust,
RD : no. of days of rising dust, SD : No. of days of storm dust, and
T : ambient temperature ($^0$C).

## 3 Statistical analyses and regression models

In standard regression analysis it is generally assumed that the determining (independent) variable is measured without error. All errors of measurement are then assigned to the determined (dependent) variable. The relationship between two normally distributed variables x and y can be expressed mathematically in a variety of ways.

In the present study, multiple non-linear regression models in three forms were used for each design requirements to choose which form gives the best fitting of data. The regression models that were proposed and investigated can be seen in table (2).

**Table 2: The proposed models.**

| Rank | Equation Description |
|---|---|
| A | $y = \exp(b_1 x_1 + b_2 x_2 + ... + b_k x_k + G)$ |
| B | $y = b_1 x_1 + b_2 x_2 + ... + b_k x_k + G$ |
| C | $y = b_1 x_1 + b_2 x_2 + ... + b_k x_k$ |

Where; y = dependent variables; $x_1, x_2, \ldots, x_k$ = the independent variables.
$b_1, b_2, b_3, \ldots, b_k$ = are model coefficients; and G = model constant term.

Table (3), (4), (5), (6) and (7) show the correlation matrix in Al-Musaib, Al-Hindya Barrage, Al-Kufa, Al-Shamya and Al-Shenafya station respectively. while plot models of these stations were shown in Figs. 2 to 6.

**Table 3: Correlation Matrix of Euphrates river in Al-Musaib station**

| | $x_1$ | $x_2$ | $x_3$ | $x_4$ | y |
|---|---|---|---|---|---|
| $x_1$ | 1 | -0.109 | 0.312 | -0.466 | -0.124 |
| $x_2$ | -0.109 | 1 | 0.323 | -0.280 | -0.389 |
| $x_3$ | 0.312 | 0.323 | 1 | -0.563 | -0.572 |
| $x_4$ | -0.466 | -0.280 | -0.563 | 1 | 0.931 |
| y | -0.124 | -0.389 | -0.572 | 0.931 | 1 |

**Table 4: Correlation matrix of Euphrates river in Al-Hindya Barrage station**

| | $x_1$ | $x_2$ | $x_3$ | $x_4$ | $x_5$ | $x_6$ | $x_7$ | $x_8$ | $x_9$ | y |
|---|---|---|---|---|---|---|---|---|---|---|
| $x_1$ | 1 | | | | | | | | | |
| $x_2$ | 0.825 | 1 | | | | | | | | |
| $x_3$ | 0.564 | 0.574 | 1 | | | | | | | |
| $x_4$ | 0.489 | 0.501 | 0.810 | 1 | | | | | | |
| $x_5$ | 0.256 | 0.306 | 0.571 | 0.544 | 1 | | | | | |
| $x_6$ | 0.603 | 0.617 | 0.810 | 0.748 | 0.432 | 1 | | | | |
| $x_7$ | 0.233 | 0.167 | 0.365 | 0.380 | 0.120 | 0.299 | 1 | | | |
| $x_8$ | -0.010 | -0.076 | -0.003 | -0.152 | -0.092 | -0.084 | 0.045 | 1 | | |
| $x_9$ | -0.135 | -0.082 | -0.257 | -0.188 | 0.013 | -0.228 | -0.541 | -0.085 | 1 | |
| y | 0.565 | 0.640 | 0.841 | 0.897 | 0.423 | 0.907 | 0.390 | -0.170 | -0.241 | 1 |

**Table 5: Correlation Matrix of Euphrates river in Al-Kufa station.**

| | $x_1$ | $x_2$ | $x_3$ | $x_4$ | $x_5$ | $x_6$ | $x_7$ | $x_8$ | y |
|---|---|---|---|---|---|---|---|---|---|
| $x_1$ | 1 | | | | | | | | |
| $x_2$ | -0.282 | 1 | | | | | | | |
| $x_3$ | -0.085 | -0.100 | 1 | | | | | | |
| $x_4$ | -0.151 | 0.188 | -0.395 | 1 | | | | | |
| $x_5$ | 0.023 | -0.186 | 0.660 | -0.594 | 1 | | | | |
| $x_6$ | -0.033 | -0.369 | 0.689 | -0.422 | 0.726 | 1 | | | |
| $x_7$ | 0.136 | -0.331 | 0.552 | -0.363 | 0.629 | 0.796 | 1 | | |
| $x_8$ | 0.120 | -0.194 | 0.101 | 0.141 | 0.044 | 0.195 | 0.456 | 1 | |
| y | -0.182 | 0.637 | -0.143 | 0.569 | -0.252 | -0.579 | -0.454 | -0.333 | 1 |

**Table 6: Correlation Matrix of Euphrates river in Al-Shamya station.**

| | $x_1$ | $x_2$ | $x_3$ | $x_4$ | $x_5$ | $x_6$ | $x_7$ | y |
|---|---|---|---|---|---|---|---|---|
| $x_1$ | 1 | | | | | | | |
| $x_2$ | 0.168 | 1 | | | | | | |
| $x_3$ | -0.267 | -0.011 | 1 | | | | | |
| $x_4$ | 0.355 | -0.097 | -0.562 | 1 | | | | |
| $x_5$ | 0.089 | -0.263 | -0.324 | 0.625 | 1 | | | |
| $x_6$ | -0.109 | -0.268 | -0.224 | 0.482 | 0.809 | 1 | | |
| $x_7$ | -0.378 | -0.147 | 0.168 | 0.034 | 0.270 | 0.505 | 1 | |
| y | 0.295 | 0.906 | -0.176 | 0.035 | -0.404 | -0.450 | -0.306 | 1 |

**Table 7: Correlation Matrix of Euphrates river in Al-Shenafya station.**

| | $x_1$ | $x_2$ | $x_3$ | $x_4$ | $x_5$ | $x_6$ | $x_7$ | y |
|---|---|---|---|---|---|---|---|---|
| $x_1$ | 1 | | | | | | | |
| $x_2$ | 0.244 | 1 | | | | | | |
| $x_3$ | -0.224 | 0.085 | 1 | | | | | |
| $x_4$ | 0.465 | -0.123 | -0.562 | 1 | | | | |
| $x_5$ | 0.352 | -0.056 | -0.324 | 0.625 | 1 | | | |
| $x_6$ | 0.180 | -0.157 | -0.224 | 0.482 | 0.809 | 1 | | |
| $x_7$ | -0.009 | -0.052 | 0.168 | 0.034 | 0.270 | 0.505 | 1 | |
| y | 0.309 | 0.959 | -0.034 | -0.031 | 0.027 | -0.165 | -0.199 | 1 |

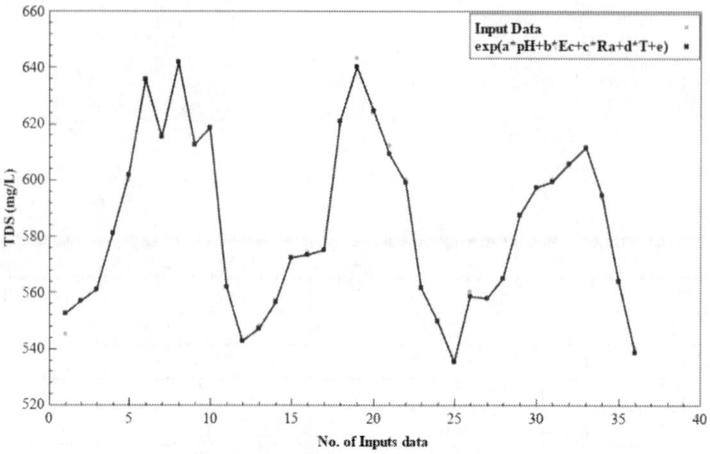


Fig. 2: Plot model of Euphrates river in Al-Musaib station

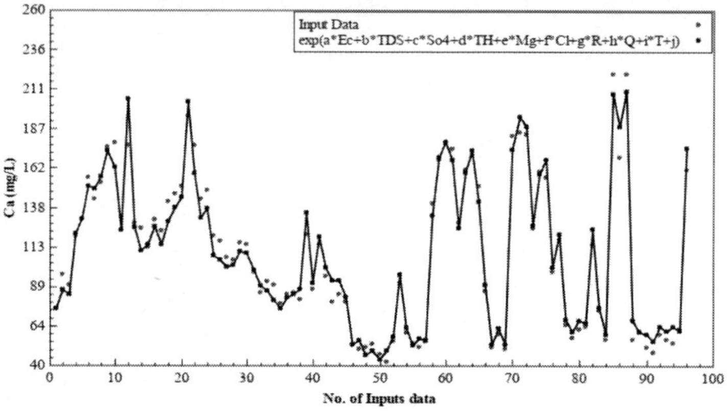


Fig. 3: Plot model of Euphrates river in Al-Hindya Barrage station.

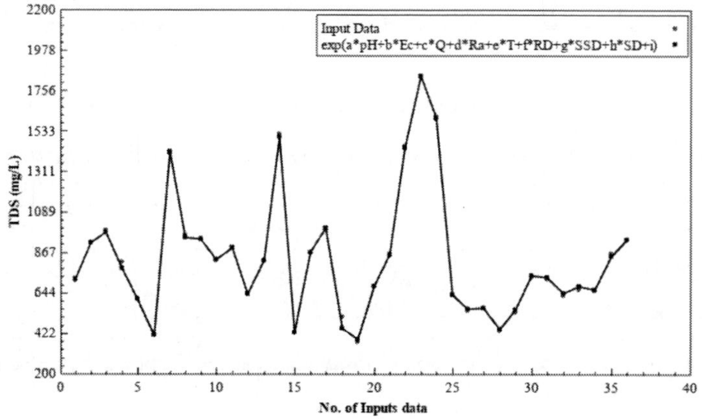


Fig. 4: Plot model of Euphrates river in Al-Kufa station

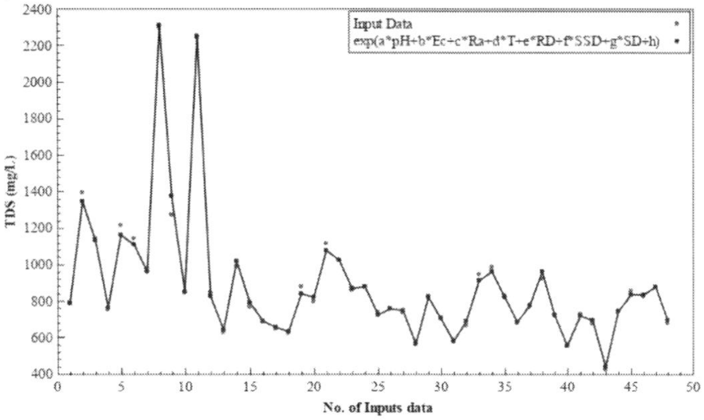


Fig. 5: Plot model of Euphrates river in Al-Shamya station

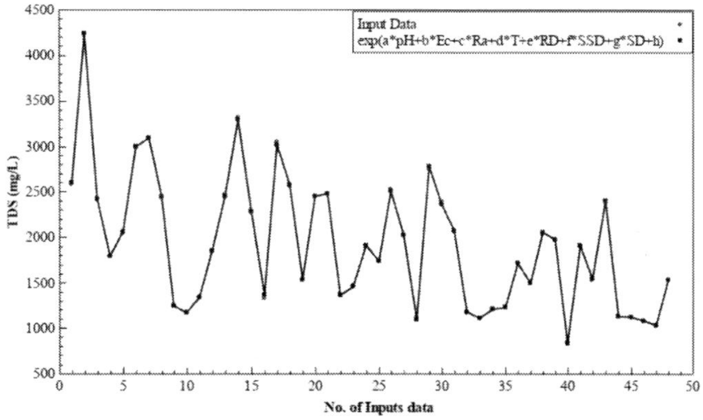


Fig. 6: Plot model of Euphrates river in Al-Shenafya station

## 4 Verification of regression model

Verification of the obtained model can be made by plotting the data of the year 2008 which were not included in the building of the regression model versus modeling data. The obtained model was also calibrated by the following equation:

$$y = ax^{b}$$

Verification of regression model is shown in Figs. (7 to 11). These Figs. showed that the experimental data are correlated well with modeling data. This verification shows a good agreement with coefficient of determination ranged between "0.914 to 0.994".

https://doi.org/10.1051/matecconf/201816205022

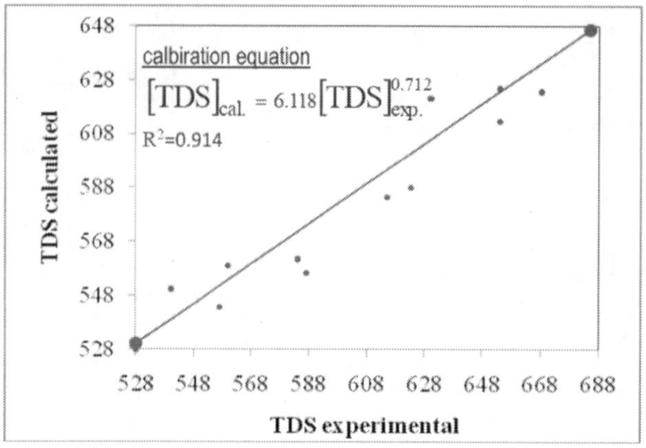


Fig.7: Verfication of regression model for Euphrates river in Al-Musaib station.

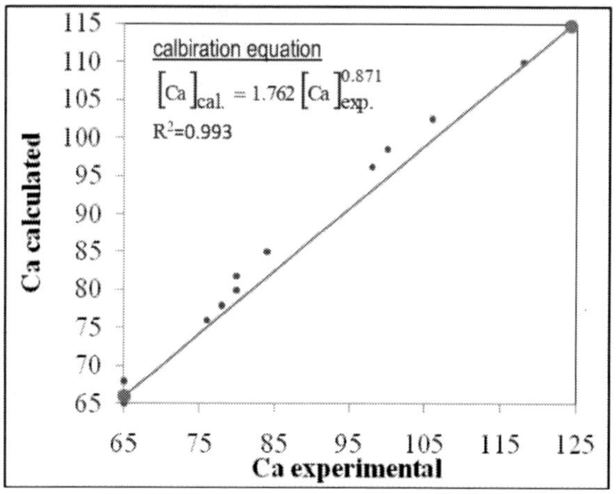


Fig. 8: Verfication of regression model for Euphrates river in Al-Hindya Barrage station.

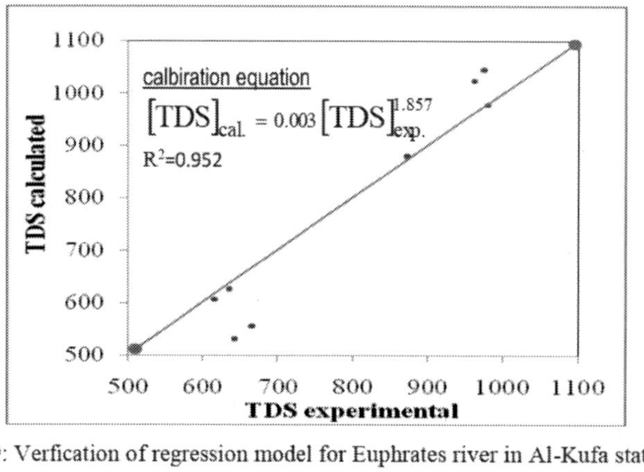


Fig. 9: Verfication of regression model for Euphrates river in Al-Kufa station.

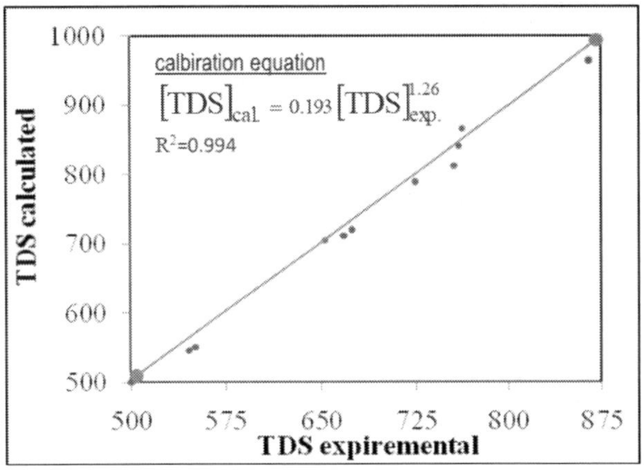


Fig.10: Verfication of regression model for Euphrates river in Al-Shamya station.

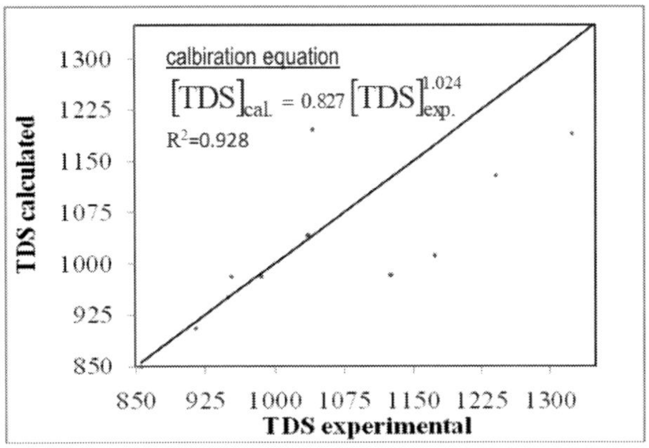


Fig. 11: Verfication of regression model for Euphrates river in Al-Shenafya station.

## 5 Comparison of the program output data

A Comparison of the program output data (TDS, Ca) are compared with the Iraqi and WHO standards for domestic and irrigation purposes (table 8). A Comparison shows that these parameters in some stations are within the Iraqi and WHO standards for domestic and irrigation purposes (table 9) (i.e. TDS: in Al-Musaib, Al-Kufa, Al-Shamya station), and (Ca: in Al-Hindya Barrage station).

Table 8: Allowable limits of water quality parameters in surface water body used as domestic and irrigation water source according to Iraqi and WHO standards [9],[10]

| Parameter | Unit | Domestic water standards | | Irrigation water standards |
|---|---|---|---|---|
| | | Iraqi | WHO | WHO |
| pH | | 6.5-8.5 | 6.5-8 | 6-8.5 |
| Ec | μs/cm | 2000 | - | <250 Excellent, 250-750 Good and 750-2000 Permissible |
| Ca | mg/L | 200 | 75-200 | 0-200 |
| Cl | mg/L | 200 | 250-600 | 0-300 |
| Mg | mg/L | 50 | 30-150 | 0-50 |
| TA | mg/L | 170 | 5-200 | - |
| TDS | mg/L | 1500 | 500-1000 | 0-700 Excellent, 700-2000 Good and >2000 Unsuitable |
| TH | mg/L | 500 | 100-500 | |
| Tur | NTU | <10 | 5-25 | |
| $So_4$ | mg/L | 200 | 200-400 | |

**Table 9: Qualitative properties of Euphrates river for drinking and irrigation purposes.**

| Station | Raw water suitability for following purposes | | |
|---|---|---|---|
| | Demotic | | Irrigation |
| | Iraqi | WHO | WHO |
| Al-Musaib | Suitable | Suitable | Excelent |
| Al-Hindya Barrage | Suitable | Suitable | Suitable |
| Al-Kufa | Suitable | Suitable | Good |
| Al-Shamya | Suitable | Suitable | Good |
| Al-Shenafya | Suitable | Suitable | Good |

## 6 Description of computer program

A computer program in a Visual Basic (software 6.0) was designed to construct a monitoring system that works with the data gathered to calculate the pollutant concentrations for river water and compare with the Iraqi and WHO standards for domestic and irrigation purposes.

The program was divided into three main parts: first named gathering of general information, second: calculation of the pollutant concentrations and third: data updating. The outputs data from statistical program (Data Fit version 8.0) were entered to the monitoring system designed as follow and shown in Figs. 12 to 18:

### 6.1 Part 1: Gathering of the general information

In this part, general information about the selected surface water is shown in the monitor forms. The user can select the water resource according to:

1. The type of surface water.
2. The name of surface water.
3. The station according to location of surface water.

### 6.2 Part 2: Calculation of the pollutant concentrations

This part of program calculates the concentrations of pollutant which polluted the surface water, then compared these concentrations with Iraqi and WHO for domestic and Irrigation purposes. The calculation depends on the database information about the surface water. The following steps were applying on Euphrates river in Al-Hindya Barrage station:

1. From the front page of the program in Fig. (12) click (Assessment Surface Water in the Middle Euphrates Region) button.
2. From the Combo Box in Fig. (13) select surface water type.
3. From Comb Box in Fig. (14) select the station of the surface water which selected in step (2).
4. From Fig. (15) select the regression model rank A, B or C.
5. Click the (Calculation) button in Fig. (16).
6. Click the (Comparison) button in Fig. (17).
7. Fig. (18) gives the final result which can be used by decision makers.

### 6.3 Part 3: Data updating

Data updating is permitted for all information in the database. Updating includes:

1. Adding.
   A. New water resource and corresponding information.
   B. New station and corresponding information.
2. Updating of the existing data.

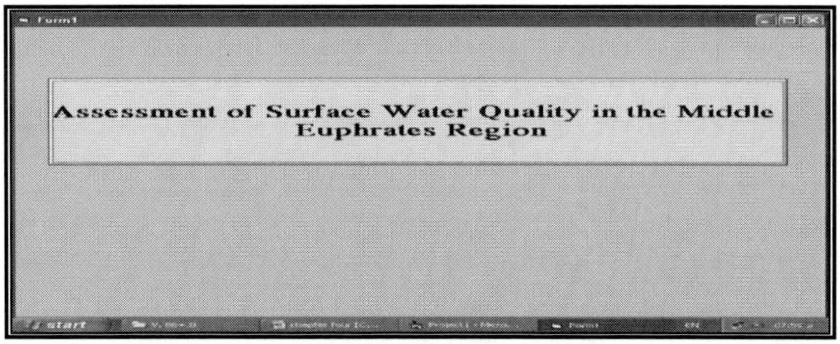


**Fig. 12: Front page of program.**

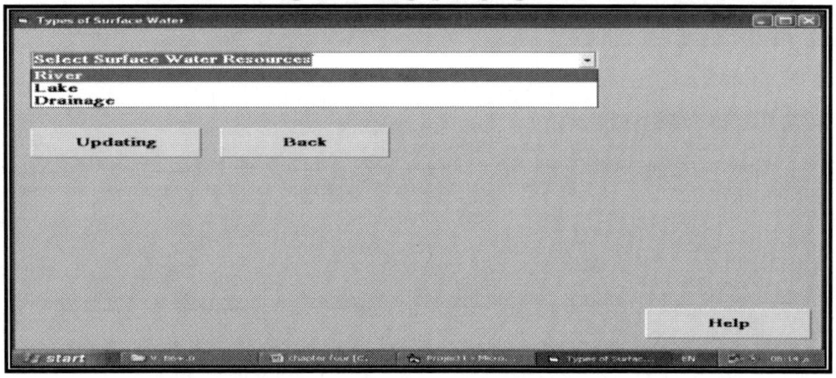


Fig. 13: Page to select the type of surface water.

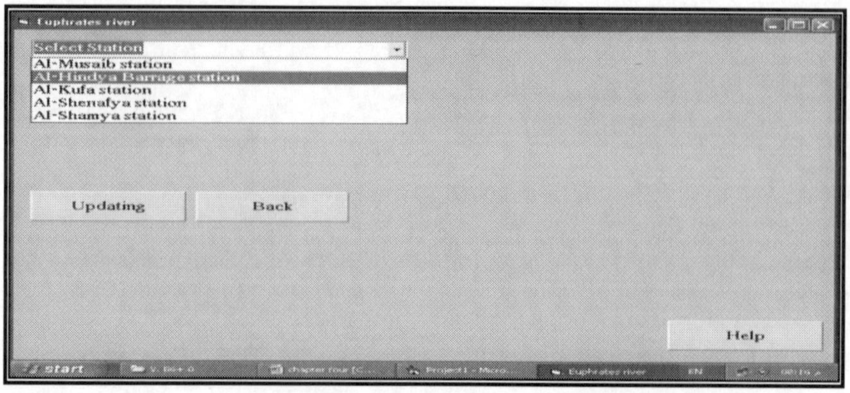


Fig. 14: Page of inputs data of Euphrates river in Al-Hindya Barrage station.

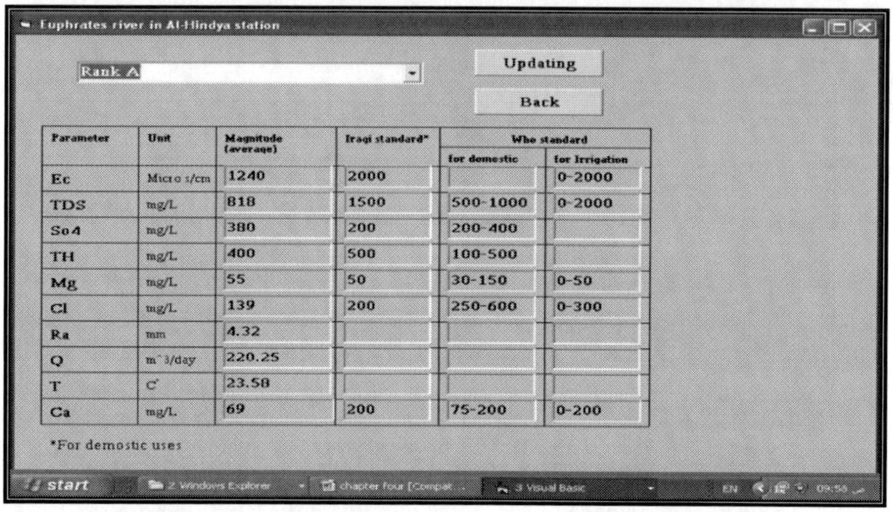


| Parameter | Unit | Magnitude (average) | Iraqi standard* | Who standard | |
|---|---|---|---|---|---|
| | | | | for domestic | for Irrigation |
| Ec | Micro s/cm | 1240 | 2000 | | 0-2000 |
| TDS | mg/L | 818 | 1500 | 500-1000 | 0-2000 |
| So4 | mg/L | 380 | 200 | 200-400 | |
| TH | mg/L | 400 | 500 | 100-500 | |
| Mg | mg/L | 55 | 50 | 30-150 | 0-50 |
| Cl | mg/L | 139 | 200 | 250-600 | 0-300 |
| Ra | mm | 4.32 | | | |
| Q | m^3/day | 220.25 | | | |
| T | C° | 23.58 | | | |
| Ca | mg/L | 69 | 200 | 75-200 | 0-200 |

*For demostic uses

Fig. 15: Page of inputs data of Euphrates river in Al-Hindya Barrage station.

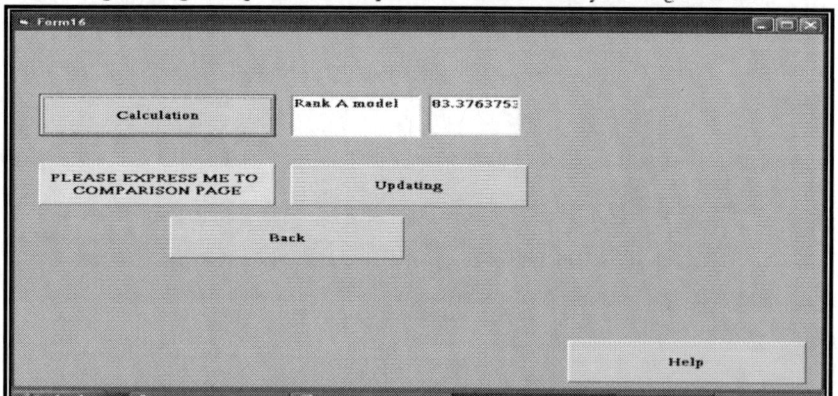


Fig. 16: Page of output data of Euphrates river in Al-Hindya Barrage station.

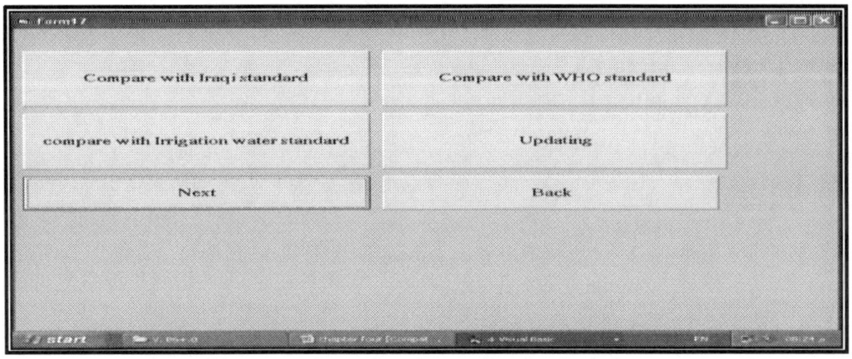


Fig. 17: Page of the comparison of Euphrates river in Al-Hindya Barrage station.

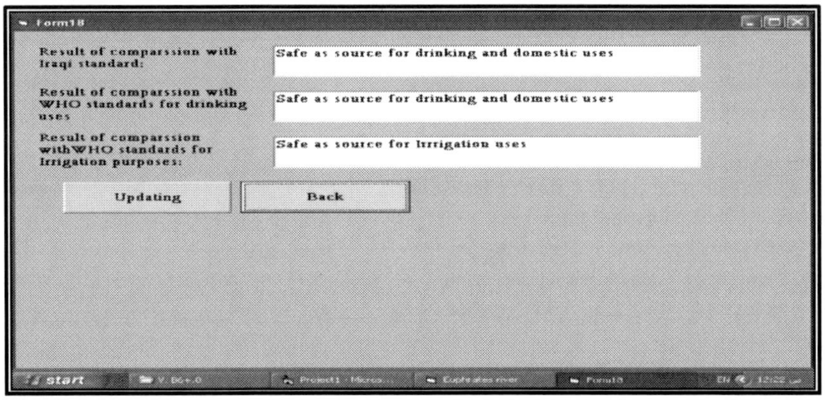


Fig. 18: Page of comparison of Euphrates river in Al-Hindya Barrage station.

## 7 Conclusions

From this study the following conclusions are obtained:

1. Environmental Assessment procedure is necessary to further characterize water quality condition of Euphrates river in the middle of Iraq.
2. Environmental Assessment approaches and techniques have much to offer in improving performance in water resources management and planning given the significant, and widely divergent, demands put on water resources.
3. The multiple regressions are found more suitable to evaluate the pollutants concentration.
4. Improved surface water quality in the middle of Iraq leading to improve the drinking water quality, reduced chemical treatment, reduce the rate of disease and improved hygienic conditions in the area.
5. For Euphrates river positive relation was found between total dissolved solids and temperature, while negative relation between total dissolved solids, hydrogen ion concentration, electrical conductivity and monthly rainfall totals in Al- Musaib station, also positive relation was found between total dissolved solids, electrical conductivity and monthly rainfall totals, while negative relation between total dissolved solids, hydrogen ion concentration, discharge, temperature, no. of days of rising, storm and suspended dust in Al-Kufa station.
6. The verification models show a good agreement with coefficient of determination ranged between "0.914 to 0.994".

7. The computer program can be considered as a guide to the application of rapid technique in the inventory stage in each tool of the environmental management.
8. A Comparison of the program output data (TDS, Ca) shows that these parameters in some stations are within the Iraqi and WHO standards for domestic and irrigation purposes (i.e. TDS: in Al-Musaib, Al-Kufa, Al-Shamya station), and (Ca: in Al-Hindya Barrage station).

## References

1. Al-Masri, N., and Ali, A. A., 1985, "Effect of Domestic and Industrial Waste Water Discharge on Pollutant Variation in River Tigris", Proc. Of Iraqi conference on Engineering, ICE, 467-474, sited in Al-Husseini.
2. Wasfi, A. A., 1986, "Evaluation of Tigris River Water Quality of Different Uses ", M. Sc. Thesis, University of Mosul
3. Al-Delaimy, H. H., 1989, "Effect of Tigris River Pollution on the Operation of Al-Rasheed Water Treatment Plant", M. Sc. Thesis, College of Engineering, University of Baghdad, sited in Al-Husseini.
4. Al-Malikey, S. J. B., 1993, "Effect of Tigris River Pollution on the Performance of Water Treatment Plants in Baghdad City", M. Sc. Thesis, College of Engineering, University of Baghdad, sited in Al-Husseini.
5. Al-Husseini, H. H., 1999, "Evaluation of Shatt Al-Hilla Water quality for Domestic Use and the Suitability of Main Water treatment Plants on it", M. Sc. Thesis, College of Engineering, University of Babylon, sited in Al-Husseini
6. Al-Fatlawi, H. J., 2005, "Ecological Study for Euphrates River between Al-Hindya Reservoir and Al-Kifil Town/ Iraq ", M. SC. Thesis, College of Science, University of Babylon, In Arabic
7. Kalel, B., 2005, "Comparative Study in Water Quality of Euphrates between Shatt - Al-Kufa and Shatt -Al-Abasiya ", Journal of Babylon University, Engineering Science, 10 (5): 1011-1021.
8. Fahd, K. K., 2006, "Some of Physico-Chemical Characteristics of Al-Masab Al-Aam River at Al-Nasiriyia City", Journal of Engineering Science, 10 (5): 1011-1021
9. Abbawi, S., A. and Mohsen M., S., 1990, "Environmental Practical Engineering-Water Testing", Muosel University, In Arabic.
10. WHO, 2006," Guidelines for Drinking _ Water Quality", First Addendum to Third Edition, WHO Library Cataloguing, Publication Data, ISBN 924154696 4, Vol. 1, Recommendations, Third Edition, Geneva.

# Urban air quality evaluation over Kut city using field survey and Geomatic techniques

*Zainab* Mohammed[1], *Abdulrazzak.* Ziboon[1,*], *Ali* Kamal[1] and *Mahdi* Alfaraj [2]

[1]Building and Construction Engineering Department, University of Technology, Baghdad, Iraq
[2]Architectural Engineering Department, University of Technology, Baghdad, Iraq

**Abstract.** Air pollution is caused by various sources such as cars exhaust, energy sources, petrol stations, industrial activities, and other sources. The aim of this study was to measure some air pollutants gases, representing the results by Arc GIS maps over AL-Kut city and finding the ways for reducing them. Twenty samples have been taken using Global Positioning System (GPS) for measuring the main air pollutants (sulfur dioxide ($SO_2$), nitrogen dioxide ($NO_2$), carbon monoxide (CO), ozone ($O_3$), total suspended particle (TSP), and particulate matters (PMs). The measurements were carried out by several specialized portable equipment at afive months starting from November 2015 to March 2016. The distribution maps resulted from ArcGIS 10.2, display that the highest concentrations of air pollutants were in the center and north of the study area. The highest concentrations of pollutants were in November and March, whereas lower concentrations observed during January period. The average monthly concentrations of (TSP) measured during the study period was (504.4 μg/m3, 359.5 μg/m3, 32.2 μg/m3, 324.8 μg/m3, and 392.45 μg/m3) from November 2015 to March 2016, these results were significantly higher than the permissible allowable limits of the Iraqi national (350 μg/m3) and international allowable limits (150 μg/m3), except for January is founded (32.2 μg/m3) within the acceptable limits due to heavy rainfall. the averages monthly concentrations of (CO), (6.567, 5.165, 4.08, 5.454 and 7.25 ppm) were lower than standards limits for five months, comparing with the Iraqi National limits (35 ppm) and WHO Limits (9 ppm). Pollutant concentrations such as ($O_3$), were within acceptable limits of Iraqi and WHO limits at all period of study, while other air pollution gasses such as ($SO_2$ and $NO_2$), their concentrations over the study area were within the Iraqi national limits but slightly over the WHO limits.

## 1 Introduction

Air is essential to all life forms. All depends on its existence; without it, nothing breathes or lives. For these reasons, air quality monitoring should be one of the most important priorities. While recognizing that air quality has such a significant impact on the quality of life, the European Union placed air quality at the top of its thirteen quality of life indicators list [1]. Air is always unclean and always contaminated with gases such as carbon dioxide $CO_2$, nitrogen dioxide $NO_2$, and sulfur dioxide $SO_2$, which can be toxic in nature and divided into solid, liquid and smog particles. Air becomes polluted due to the presence of pollutants mentioned above. The presence of these contaminants in the air called air pollution and the materials, which pollute the air, called air pollutants. In fact, the pollution of air (or atmosphere) is an undesirable change in air quality caused by the emission of gases from the burning of fossil fuels, volcanic eruptions and the discharge of particulate matter into the atmosphere by industrial activities and soil erosion [2].

The impact of air pollution on public health depends on many factors, such as the chemical composition of pollutants, the level of concentration, the presence of other pollutants, the existing health of individuals, and periods of exposure. Other than air pollution, the most likely risk to public health in the present car dominated transport system is road accident deaths and injuries worldwide.

The transport sector is a major contributor to climate change caused by greenhouse emissions. Transportation is one of the basic components in modern human daily life. However, it has both productive effects on human development and detrimental effects on public health. The number of motor vehicles is estimated to be over 800 million worldwide and is increasing almost everywhere at higher rates than human population, and road traffic may be growing even more rapidly. The number of private cars worldwide rose to 500 million in 1990 from 50 million in 1950 [3]. Transport emissions comprise about 20% of the total greenhouse gas emissions [4]. The transport sector, as an important source of emissions, adds a wide range of gaseous air

* Corresponding author: zbm_asr@yahoo.com

© The Authors, published by EDP Sciences. This is an open access article distributed under the terms of the Creative Commons Attribution License 4.0 (http://creativecommons.org/licenses/by/4.0/).

pollutants and particulate matter of different sizes and compositions to the environment.

In the new global economy, human population has become a central issue. As the population on the earth grows, vehicle and other emissions increase, pollutants, which contain toxic substances such as heavy metals, polycyclic aromatic hydrocarbons (PAH), volatile organic compounds (VOC) and particulates ($PM_{10}$), have an increasing impact on urban air quality. In addition, photochemical reactions resulting from the action of sunlight on nitrogen dioxide ($NO_2$) and VOCs from vehicles lead to the formation of ozone. This secondary long-range pollutant sometimes affects population far away from the original emission site. In addition, acid rain is another long-range pollutant caused by vehicle nitrogen oxides and sulfurous emissions, and it could therefore concluded that traffic pollution problems are worsening worldwide [5].

## 2 Objectives

The main objectives of this thesis work are:

1. Monitor and analyze of urban air pollution dispersion in Al-Kut city using ground measurements of some pollutant such as (CO, $NO_2$, $O_3$, $SO_2$, TSP, $PM_{2.5}$, and $PM_{10}$).

2. Investigate the relationship between in site urban air pollutants with land use classes and some metrological factors (temperature, humidity, wind speed).

## 3. Study area

Al-Kut is an Iraqi city located on the banks of the Tigris River, which is the administrative center of the Wasit province of Iraq. Which it built by Al-Hajjaj ibn Yusuf Althaqafy at 78 AH and completed in 86 AH. Kut, is a city with distinct features are on a peninsula shape surrounded by the Tigris River from the east, west and south sides, and containing Kut Barrage built by the city's sons under the supervision of the English in the first half of the twentieth century (1938), which divided them rivers of Dujail and Garraf, as well as containing Al-Karama bridge which was built in the era leader Abdul Karim Kassem, founder of the Republic of Iraq in 1958. The well-known sites in the city: the textile factory and Alhorh Street, Almasaif hotel, and the University of Wasit, which is one of the most important landmarks in the city as shown in Figure. 1.

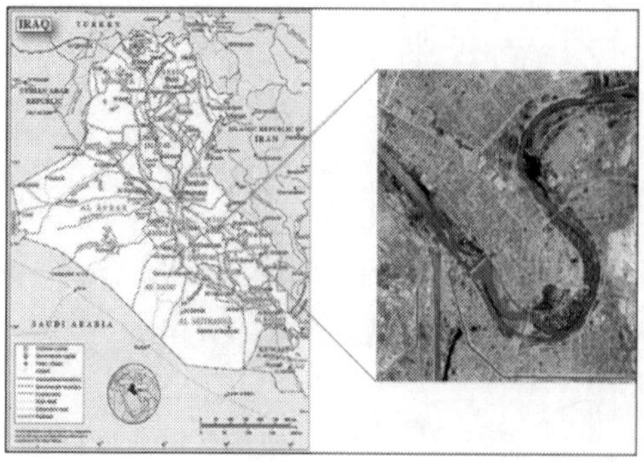


**Fig.1.** Map of Iraq showing the location of Al-Kut city.

## 4. Methodology and sampling

In this study, 20 deferent site locations has been selected to monitoring the concentrations of air pollutants. The distribution of selected sites was to cover nearly all the area of AL_Kut city center, including the mainly crowded intersections and different land classifications (residential, commercial, vegetation, barren and industrial areas). Figure (2) shows AL_Kut map with samples location. The measurement of concentrations has been done by using different devices to monitoring the concentrations of air pollutants such as (AEROCET 531, GIG450 devices, NOVA600 devices, MiniRAE 3000, Eco Sensors UV-100) as shown in Figure (3) below, Which measure (TSP, $PM_1$, $PM_{2.5}$, $PM_7$, $PM_{10}$, CO, $NO_2$, $SO_2$, $O_3$). The site locations were located by using Garmin GPS72. The measuring period was five months starting from November (2015) to March (2016).

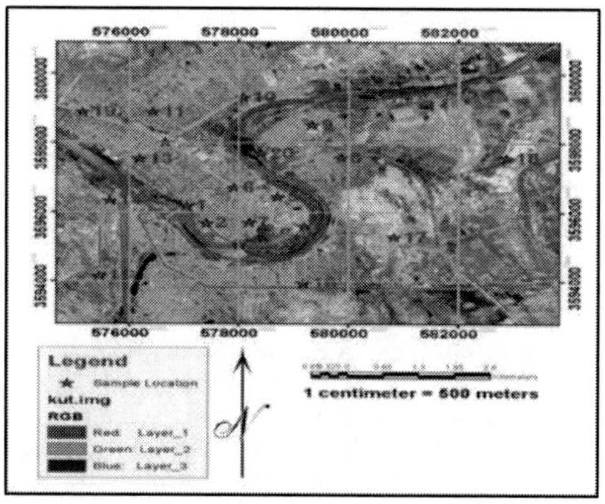


**Fig. 2.** Map of AL_Kut city with sample locations.

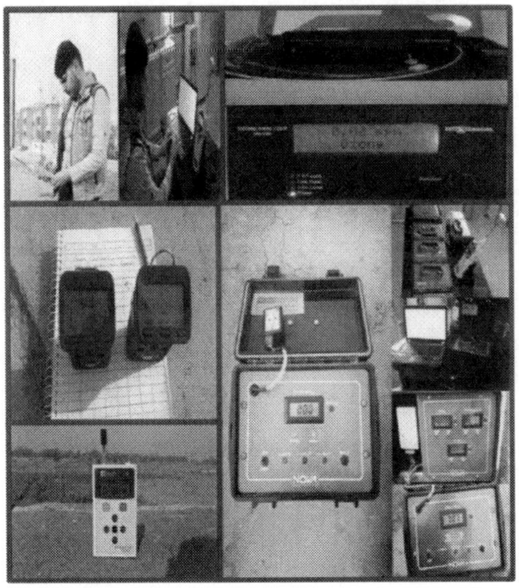

**Fig. 3.** Some devices that use to measuring air pollutants during the study period.

## 5. Results

The results of this study were including two parts: the first one deals with air pollutants (TSP, PMs, CO, $NO_2$, $SO_2$, $O_3$) for five months starting from November 2015 to March 2016 and the analysis of their results. Secondly, express the relationship between land use and air pollution.

### 5.1 Total suspended particulate (TSP)

Total suspended particulate data from (20) locations were imported to geographical information system to drive TSP dispersion maps for the center of AL-Kut city as shown in Figure (4) below. This figure show that concentration of total suspended particulate pollutants is higher than the permissible limits of the Iraqi National limits (350 µg/m3) and the world limits (150 µg/m3), which are mostly in traffics intersections.

Very high concentration of TSP in most of the sites and along the study period was due to several causes, such as very high traffic congestion in these sites, the uncleaned streets specially streets shoulders which filled with dust, in addition to wind that led to blowing dust from streets to the air. Excluding the period during January 2016, where the concentrations of TSP was at minimum values and within the permissible limits of Iraqi and international standards because of the rains almost continuously during this month. Which led to reduce concentration of TSP from air and stabling on the ground. The great difference between the periods reflects the great effect of the seasonal changes specially rains and temperature. Figure (5) below shows some of the intersections and the uncleaned streets that filled with dust and causes high concentration of TSP in most of the study area.

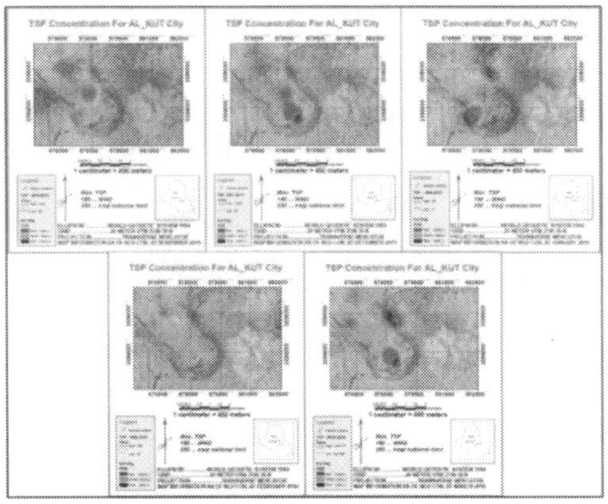

**Fig. 4.** Distribution of TSP concentration over AL-Kut city from November 2015 to March 2016.

**Fig. 5.** Some intersections and streets of the study area.

### 5.2 Particulate matter (PMs)

Figures (6) demonstrate the sites which has high concentrations of PMs was the same sites of high concentrations of TSP, that means both of them have the same causes and sources. Which means that PM2.5 and PM10 were results due to traffic intensity, uncleaned streets that filled with dust, furthermore winds that led to blowing dust from streets to the air.

By using geomatics technique (remote sensing then identifying areas of influence by using GIS is proved to be a good efficiency in the quick identification of contaminants and prepared it for environmental decision maker. There are a lot of gasses that influence in the climate change but here we discussed the most important gasses influence on the change in climate. The values of climate change gasses (CO2, N2O, and CH4) that have been calculated from the formula of emission have different value according to the quantity of fuel used in each brick factory. Where the value of CO2 ranged from (695 -854 tones), value of N2O ranged from (1.905 - 2.318 tones), and finally value of CH4 ranged from (0.286 -0.347 tones).

## 5.3 Carbon monoxide (CO)

From a filed measurement in the study area, it has been found that the highest concentrations of CO was in the north of city at locations No. (10, 11, and 12) during the study period (five months) as shown in Figure (7) below. The concentrations of these locations were high but within acceptable limits of Iraqi standards that is (35 ppm) and exceeding international standards, which is (9 ppm). Other sites were within acceptable standards ranging between (0.15 ppm – 12 ppm).

The high values for CO in the north of city were due to the intersections with huge numbers of vehicles furthermore the using of bad-quality benzene by the vehicles that led to incomplete combustion, which generates carbon monoxide gas instead of carbon dioxide gas.

The concentration of CO gradually increased with increasing air temperature as shown in all sites. For example CO concentration was (22 ppm) in November 2015 which decrease to become (16 ppm) in January 2016 due to the decreasing in air temperature and gradually increased to be (25 ppm) in March 2016 due to the increasing temperature in this month.

Highest values of CO were recorded in location (No.11) with values (25 ppm and 22 ppm) during March 2016 and November 2015 sequentially and that was due to the heavy traffic and locate near a garage and petrol station. In December and January, as shown in Figures (7), the minimum concentration was (0.11 ppm) in (No.20) which located in west-north of the city and conceder as agriculture area with low traffic volume.

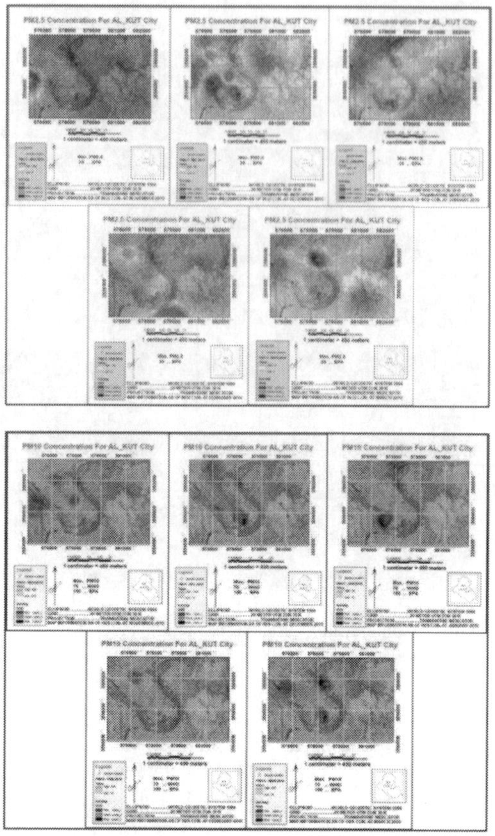

**Fig. 6.** Distribution of PMs concentration over AL-Kut city from November 2015 to March 2016.

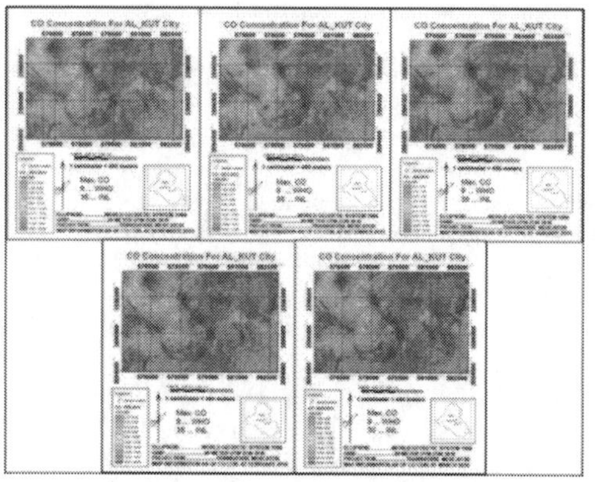

**Fig. 7.** Distribution of Co concentration over AL-Kut city from November 2015 to March 2016.

## 5.4 Nitrogen dioxide ($NO_2$)

Figure (8) shown below describes the distribution of NO2 in the air of AL-Kut city. It obvious that, for mostly locations and during all period that the concentration of NO2 was within the Iraqi national limits which is about

(0.25 ppm). Expect one location was slightly exceeding the limits by (0.01 ppm – 0.02 ppm), that location was (No.11) with concentration of (0.26 ppm) at November 2015 and (0.27 ppm) at March 2016.

The maximum concentration of NO2 was in site (No.11) with value of (0.27 ppm) at March 2016, While the minimum value was reached to (0.01 ppm) in many locations through several months.

Graphically, It can noticed that in November 2015, the maximum value of NO2 was (0.26 ppm) which detected in site (No.11), while the minimum value was (0.01 ppm) in site (No.16). In the next two month, the values were gradually decrease in the most of the sites. While February and march gradually increase in concentrations for all locations of the study area.

The rise in NO2 concentrations in major intersections of the city was due to heavy traffic jam that led to emitted high quantities of NO2 into the air as shown in the north and middle of city. Which represented by locations No. (6, 7, 10, 11, and 12).

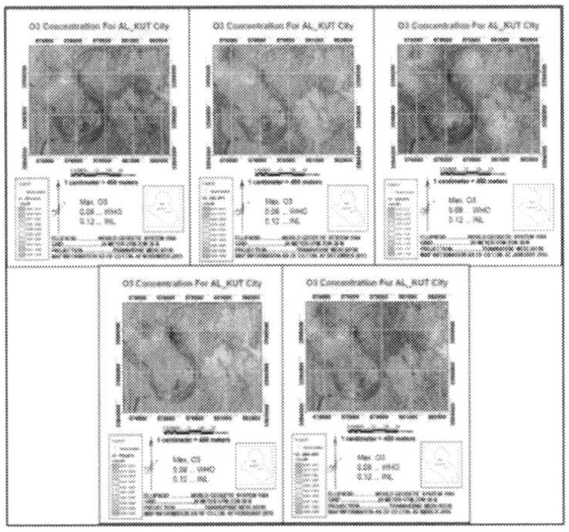

**Fig. 8.** Distribution of No2 concentration over AL-Kut city from November 2015 to March 2016.

## 5.5 Ozone ($O_3$)

Ozone (O3) measured several times in (20) different sites distributed in the study area aiming to understand the sources and causes of this gas as shown in Figure (9). In general, the measurement showed that the concentrations of O3 not exceeded the international and Iraqi standard limits, which are (0.08 ppm), and (0.12 ppm) sequentially. During the period November 2015 to March 2016, the maximum values of O3 were in site (No.12) measured in November and March with values reached to (0.05 ppm). In addition, site (No.11) measured in March where the concentrations reached to (0.05 ppm). While the minimum value of ozone gas was (0.01 ppm) in many locations through different periods.

In November, the highest reading of O3 was recorded to be (0.05 ppm) that demonstrated the city is not polluted by Ozone because all readings were below the acceptable Standard limit, which is (0.08 ppm). In the next month, the concentration of O3 declined slightly in all sites of the study area. However, the lowest values of ozone recorded in January due to the decrease in air temperature during this month. While the next two months witnessed gradually rise in concentrations.

Finally, it can be notice that the major reasons of O3 formation is that sunlight causes photochemical reactions involving NOx and VOCs. Therefore, high O3 concentrations founded in a crowded intersections in the north of the study area due to high concentrations of NO2 and VOCs.

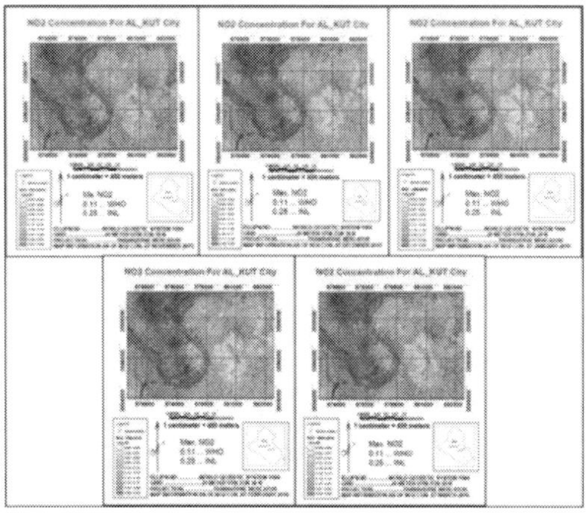

**Fig. 9.** Distribution of O3 concentration over AL-Kut city from November 2015 to March 2016

## 5.6 Sulfur dioxide ($SO_2$)

Broadly, the SO2 concentrations were within the Iraqi standard limits in almost locations for the city. Figure (10) shows one location was exceeded the standard limit (0.14 ppm), during the study period; this location is (No.12) were concentrations ranged between (0.2 ppm) to (0.25 ppm) due to high traffic volume that led to significant quantities of SO2 through vehicles exhausts according the process of fuel combustion. Additionally, it is located adjacent to a textile factory, garage, and a petrol station.

In November, the minimum concentrations were (0.11 ppm) that recorded in (No.4, No.14 and No.19) which locates in the south of city. On the other hand, the maximum value found in the location of (No.12) with a value of (0.22 ppm). While the remaining sites, there concentrations ranging between (0.02 ppm) to (0.14 ppm). The concentrations of SO2 were slightly inclined in December and January in all locations. The next two months which represented by February and March witnessed gradually increased in SO2 concentrations.

Finally, it can be notice that the concentration of SO2 spreads widely in areas that crowded by cars and neighboring industrial areas specifically in the north of the study area.

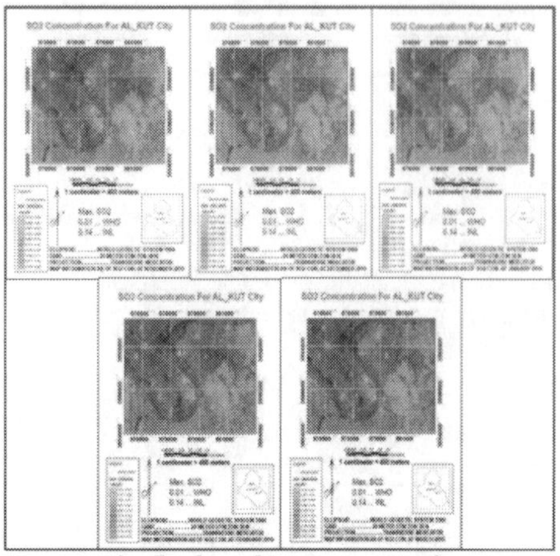

**Fig. 10.** Distribution of SO2 concentration over AL-Kut city from November 2015 to March 2016

### 5.7 Land use relationship with air pollution

Urban areas are associated with sources of a variety of air pollutants, and regional pollution problems such as acid rain and photochemical smog. Cities are also major contributors to global air pollution related to ozone depletion and carbon dioxide (CO2) warming. Within an urban area, the level of pollution varies with the distance to pollution sources, including both stationary and mobile sources (e.g., vehicles).

Land use and land cover changes due to urbanization impact and effect on air quality in the city are important. The impacts of land use predicted to have substantial impacts on gases emissions and directly effects on the urban environment. Therefore, land use of the studied area was investigate and studied with the gases concentration patterns. Figure (11) shows the land use of AL-Kut city with samples locations. By comparing the land use map with the gases concentrations, it can noticed that approximately all gases concentrations were widely distributed in urban areas and nearly to industrial areas especially in the north and middle of the city. Locations (10, 11, 12, 6 and 7) have the maximum values during the study period. On the other hand, the minimum concentrations were widely distributed near the vegetation and barren areas. Finally, the land use impacts on the TSP and PMs founded in the barren and urban areas especially in the intersections and roads of the city. The TSP and PMs Distribution were widely founded in stations of (6, 7, 10, 11, 12 and 18).

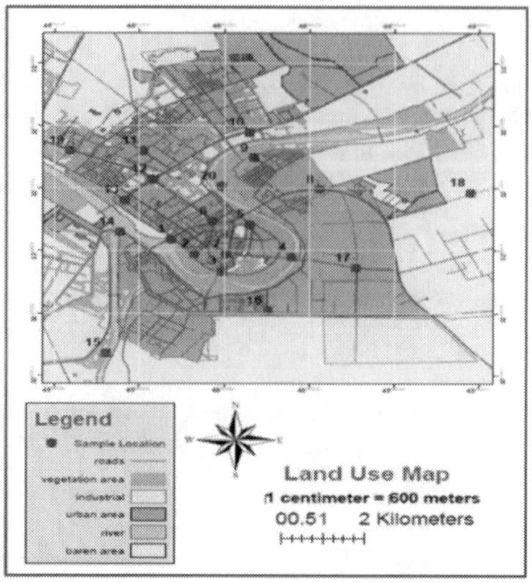


**Fig. 11.** Land Use map classification with sample locations

## 6 Conclusions

1. Arc GIS analysis distribution of air pollution maps showed that the highest concentration of pollutants were in the center and north of the city, where important intersections with heavy traffic volume and petrol stations are locate, and near to the industrial areas.
2. TSP and PM concentrations of AL-Kut city during (November, December, February, and March) were higher than Iraqi standards, while during January were within acceptable standards. The TSP concentrations were high in those locations due to high traffic volume and due to Non-regular; streets filled with dust that volatile on the air because of vehicles movement and wind gusts.
3. In terms of gases, the research showed that CO, and O3 concentrations were not exceeded the Iraqi standards. However, in the north of city, a high concentration had detected, the reason was due to the heavy traffic of the area.
4. NO2 and SO2 concentrations were higher than the allowable limits during March 2016 in north of AL-Kut city. The increment of concentrations found to be due to the fuel combustion by vehicles.
5. Anthropogenic sources affecting on the city's air pollution through a variety of plants and factories that surround the city from all destinations, along with the presence of more than an industrial area in the center and north of the city and near the residential neighborhoods that content of laboratories and industrial workshops.
6. The exhausts in all forms of transportation affecting on the city's air pollution. which has increased their numbers in recent years as a result of improved living conditions and the economic level of the city's people, with the survival of a large collection of old vehicle with damaged motors that operating on the

emission of black smoke into the air without censorship and sense of responsibility.

## References

1. Tulloch, M. and Li Jonathan, 2004, Environmental Information Archives, Volume 2 (2004), 846-854.
2. Ahluwalia, V.K., Malhotra, S., Environmental Science, a book, Pvt. Ltd. Goalie Enterprises, India, (2006).
3. Kamerman, S. B. and OECD, Social policies, family types and child outcomes in selected OECD countries. Organization for Economic Cooperation and Development. Employment, Labor and Social Affairs Committee (2003).
4. Beer, T., Grant, T., Williams, D. and Watson, H, Atmospheric Environment, 36, (4), pp. 753-763, (2002).
5. Leksmono, N. S., Longhurst, J. W. S., Ling, K. A., Chatterton, T. J., Fisher, B. E. A., Irwin, J. G.:a theoretical modelling exercise, Environmental Modelling & Software, 21, (4), pp. 494-500. (2006).

# Energy saving in air conditioning of buildings

*Mousa* Mohamed[1,*] and *Mohammed* Almarshadi[2]

[1]Faculty of Engineering, Minoufia University, Shebin El kom, Egypt
[2]Faculty of Meteorology, King Abdel Aziz University, Jeddah, Saudi Arabia

**Abstract.** The external heat load of residential buildings in summer is the dominant parameter of the required cooling load and refrigeration capacity of air conditioning systems. The consumed energy of air conditioning system is proportional to the outside conditions and intensity of solar radiation. The maximum heat load of building may occur at 3 O'clock PM, although the peak of solar radiation occurs at noon. The construction materials of building is playing an important rolls of heat transmission through buildings outside walls and glazing windows. The walls thermal insulation can be effective in energy conservation by reducing the cooling load and required electrical energy. The building is constructed from common materials with 0~12 cm thermal insulation in outside walls, ceilings, and double layers glazing windows. The building heat loads are calculated for two models of walls. The optimum thickness of thermal insulation is also determined and is found between 6~8 cm for insulation of thermal conductivity of 0.039 *W/m.K* the energy saving is 50.45% at 6 cm insulation thickness.

## 1 Introduction

The buildings heat load is proportional to the heat transmission through outside walls, ceiling, floor, and glasses windows. The heat transmission is positive in summer and negative in winter to the internal heat load. The thermal insulation is an effective tool to achieve the minimum heat transmission in buildings. The heat transmission is expressed in U-values or R-values which related to the thickness of thermal insulation. On the other hand, increasing the thickness of thermal insulation will increase the investment cost. Thus, it is inevitable to determine an optimum insulation thickness by considering economic analysis.
Recently, many authors have studied the minimum thermal insulation thickness [1-10]. Most of studies estimated the transmission load through the exterior walls into room by using the degree-time concept (degree-day/degree hour) which is one of the simplest methods that applied in static conditions [1-4]. A few studies estimated the transmission load through the exterior wall into room by using numerical method [5-6], which neglects the effect of moisture transfer on heat transmission. Furthermore, the optimum insulation thickness of the external wall for different energy sources and different insulation materials was found [11-13]. Moreover, the energy sources and energy savings were used to obtain the optimum insulation thickness [14-16]. The results showed that the increase of insulation thickness has a significant effect on the building heat load. The analysis indicates that the building energy savings are different for a given wall insulation and building conditions. A numerical solution of transient heat transfer through a multilayer walls submitted to the average outdoor temperature and solar radiation specific has been presented [17]. The inside surface heat flux of two common uninsulated walls was predicted. The results presented for the representative day of the hottest month of the considered climate showed that the significant effect of walls orientation and solar shading on the thermal performance of the two walls.
The objective of the present study is to determine the optimum thermal insulation thickness of building walls. The analysis is conducted using one type of insulation material with various thickness and two type walls of building with and without insulation. The hourly buildings heat load with cooling load temperature deference, CLTD method are estimated at various thermal insulation thickness. The optimum insulation thickness and percentage decrease of the building heat loads are also investigated.

## 2 Building models

The building heat load is calculated according to the buildings specification and construction materials. Fig. 1 shows the layout of a residential buildings we used as shown in [18-19]. The building is constructed from a common construction materials and single layer glass

* Corresponding author: mousamohamed@yahoo.com

© The Authors, published by EDP Sciences. This is an open access article distributed under the terms of the Creative Commons Attribution License 4.0 (http://creativecommons.org/licenses/by/4.0/).

windows with wooden doors. Two walls Models, A and B are used as Fig. 2.

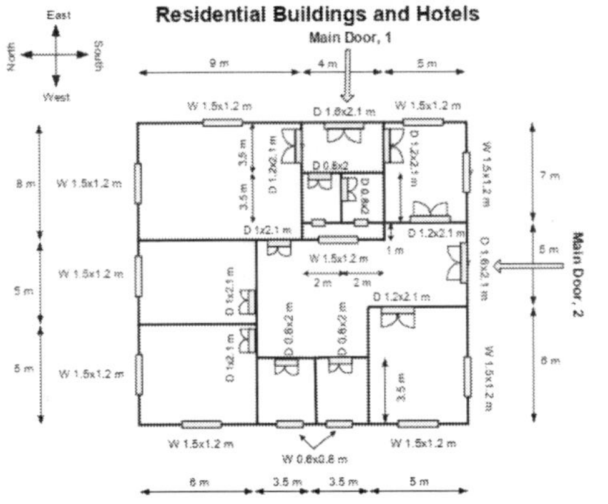


Fig. 1 Layout of residential buildings

The building is constructed from common materials; hollow bricks, heavy concrete, single layer clear glazing windows, and wooden doors as illustrated in Fig. 2. The walls Model B is typical as Model A of construction materials in addition of 0~12cm thermal insulation of 0.039 W/m.K in outside walls, floor, and ceiling. The glazed windows are double layers clear glass with air gab of 1~3 cm [19].

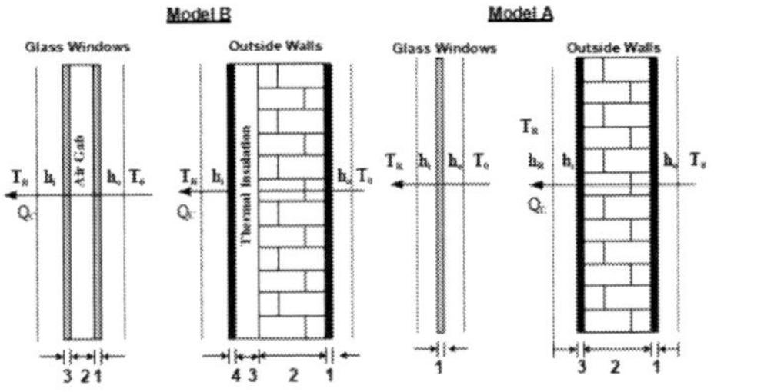


Fig. 2 Walls Models A and B with common construction materials and thermal insulation

The building heat load is the amount of heat removed or added to the conditioned space to maintain the thermal comfort. The building heat load is divided in two parts, the first is the sensible heat load which is a function of temperature difference between outside and inside condition, and radiation heat through glazing windows. The second is the latent heat load which is a function of humidity ratio difference between outside and inside condition, and moisture content of conditioned space. The building heat load in summer is the summation of external and internal heat loads. But, the building heat load in winter is the difference between the external and internal heat loads. The building heat load in summer is calculated using appropriate cooling load temperature difference, CLTD method [20]. A computer program is prepared and examined under the constraints of the building location of four directions, and latitude and longitude. The data needed according to CLTD method for outside environment and building specification are inserted as subroutine to get the final results of hourly building heat load.

The building heat load is the amount of heat removed or added to the conditioned space to maintain the thermal comfort. The building heat load is divided in two parts, the first is the sensible heat load which is a function of temperature difference between outside and inside condition, and radiation heat through glazing windows. The second is the latent heat load which is a function of humidity ratio difference between outside and inside condition, and moisture content of conditioned space. The building heat load in summer is the summation of external and internal heat loads. But, the building heat load in winter is the difference between the external and internal heat loads. The building heat load in summer is calculated using appropriate cooling load temperature difference, CLTD method [20]. A computer program is prepared and examined under the constraints of the building location of four directions, and latitude and longitude. The data needed according to CLTD method for outside environment and building specification are inserted as subroutine to get the final results of hourly building heat load.

## 3 Walls heat transmission model

The outside walls are subjected to variations of temperature $T_o(t)$ and solar radiation $I(t)$. The inside faces of walls comes in contact with the indoor air which maintained at a fixed temperature of $T_i$ to have a better thermal comfort. The one dimensional transient heat conduction equation for this problem is as follows:

$$\frac{\partial T(x,t)}{\partial t} = \frac{k}{\rho c} \frac{\partial^2 T(x,t)}{\partial x^2} \tag{1}$$

The outdoor boundary condition is,

$$-k\frac{\partial T(x,t)}{\partial x}\bigg|_{x=0} = h_o\left(T_o(t) - T_{x=0}(t)\right) \tag{2}$$

The indoor boundary condition is,

$$-k\frac{\partial T(x,t)}{\partial x}\bigg|_{x=L} = h_i\left(T_{x=L}(t) - T_i\right) \tag{3}$$

Where, $h_o$, $h_i$ are the film heat transfer coefficient of outside and inside walls of building, and their values are $h_o = 22 W/m^2.K$ and $h_i = 9 W/m^2.K$. The instantaneous transmission heat transfer rate through walls from outside to inside is defined as,

$$Q_{trans}(t) = AU\left(T_o(t) - T_i\right) \tag{4}$$

Where, $T_o(t)$ is the instantaneous outside temperature, and $T_i$ is the indoor temperature. In steady state condition, the temperature deference $\left(T_o(t) - T_i\right)$ is replace by $\Delta T$ as mention in [20] according to cooling load temperature deference, CLTD method and is defined as,

$$\Delta T = (CLTD + LM).K + (25.5 - T_i) + (T_o - 29.4) \quad (5)$$

Where, $T_o$ is hourly outside temperature, $A$ walls area, and $U$ is the overall heat transfer coefficient as,

$$\frac{1}{U} = \frac{1}{h_i} + \sum \frac{\Delta x}{k}\bigg|_{\text{constructon material}} + \frac{\Delta x}{k}\bigg|_{\text{thermalinsulation}} + \frac{1}{h_o} \quad (6)$$

To investigate the effect of thermal insulation thickness on the building heat load, we assumed all parameters are constant except insulation thickness, and the inside conditions are maintained at $24\,^\circ C$ and $50\,\%$ relative humidity.

## 4. Building cooling load profile

The hourly outside conditions of dry bulb temperature and relative humidity are used in this analysis as illustrated in [19]. The average data of dry bulb temperature, and relative humidity for each season are illustrated in Figs. 3 and 4. The dry bulb temperature is increased hourly from sunrise to maximum value at 15:00 O'clock and decreased to initial value at night as shown in Fig. 3. Opposite trend of dry bulb temperature, the relative humidity decreased from sunrise to minimum value at 15:00 O'clock and increased to maximum value at night as shown in Fig. 4.

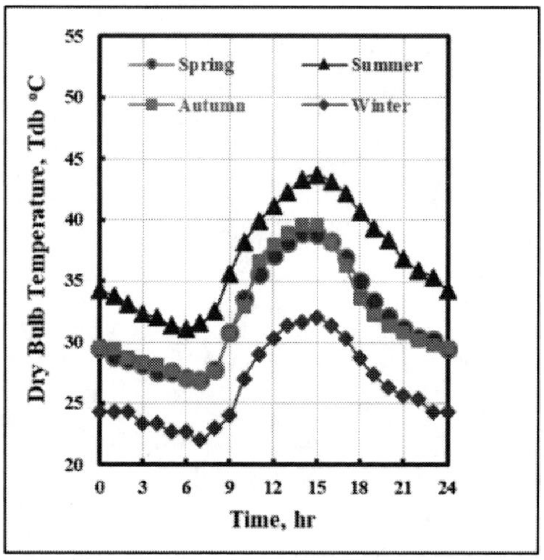


Fig. 3 Average of dry bulb temperature [19]

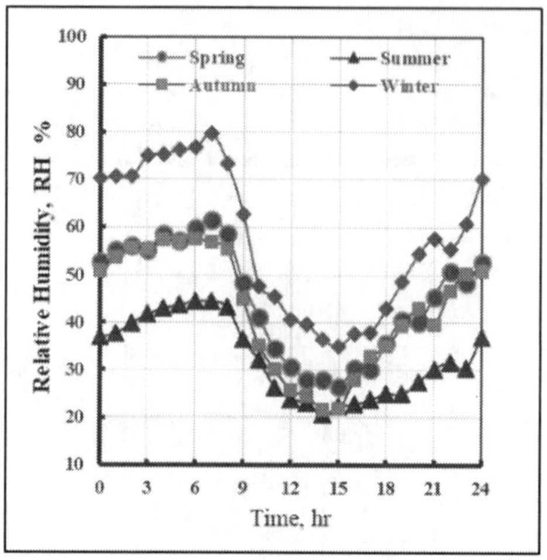


Fig. 4 Average of relative humidity [19]

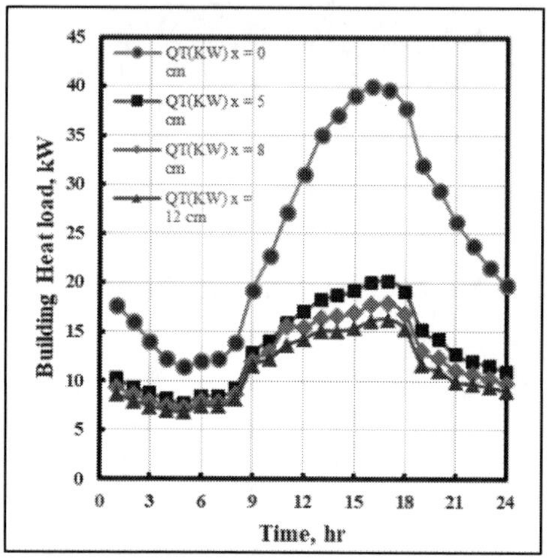


Fig. 5 Building heat load at various thermal insulation thickness

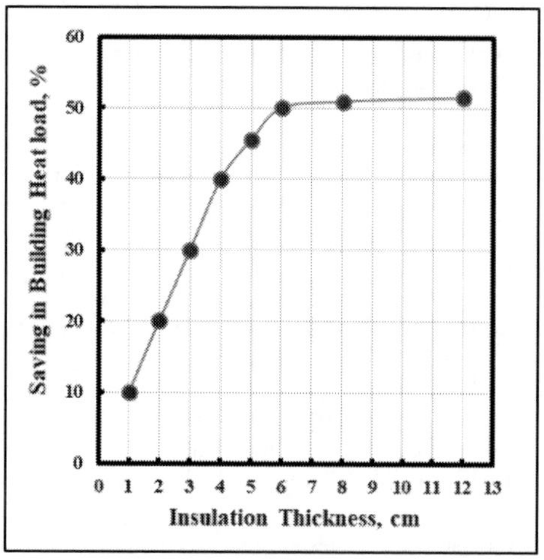


Fig. 6 saving energy in building heat load

## 5 Optimum thermal insulation thickness

The total building heat load consists of external heat load and internal heat load. The external heat load contains the heat transmission and direct solar radiation across the building envelope from outdoor to indoor conditions. The total building heat load with constant internal heat load are calculated at various thermal insulation thickness of 0 ~ 12 cm and the results are shown in Fig. 5. We can see a clear effect of thermal insulation thickness on the building heat load similar to the trend of outside dry bulb temperature in Fig. 3. The heat transmission through building walls decreased with increasing of insulation thickness.

The percentage saving in building heat load is the difference between building heat load without insulation and with insulation to the value without insulation as,

$$\left[\text{Saving energy} = \frac{Q_t\,(\text{without insulation}) - Q_t\,(\text{with insulation})}{Q_t\,(\text{without insulation})}\right] \quad (7)$$

Figure 6 shows the trend of energy saving in the residential buildings as a function of thermal insulation thickness. The energy saving increases with increasing of insulation thickness until 6 cm. Small increase in energy saving until 8 cm is found. So, the optimum thickness of thermal insulation is found between 6 ~ 8 cm and the energy saving my 50.46 %. Also, similar results are illustrated in [15-16] and the optimum thickness of extruded polystyrene (XPS) is between 0.053 and 0.069m and the optimum thickness of expanded polystyrene (EPS) is between 0.081 and 0.105m.

## 6 Conclusions

The external heat load of residential buildings in summer is the dominant parameter of the required cooling load and refrigeration capacity of air conditioning systems. The maximum heat load of buildings may occur at 3 O'clock PM, although the peak of solar radiation occurs at noon. The building is constructed from common materials; hollow bricks, heavy concrete, single layer clear windows, and wooden doors. The walls Model B is typical as Model A of construction materials in addition of 0~12 cm thermal insulation of 0.039 *W/m.K* in outside walls, floor, and ceiling. The glazing windows are double layers of clear glass with air gab of 1~3 cm. The optimum thickness of thermal insulation is also determined and is found between 6~8 cm for insulation of thermal conductivity of 0.039 *W/m.K*, and the energy saving is 50.46 % compared to the building heat load without thermal insulation.

## References

1. Satman A. Yalcinkaya N. Heating and cooling degree-hours for Turkey. Energy, Vol. 24, pp. 833-40, 1999.
2. Buyukalaca O, Bulut H, Yilmaz T. Analysis of variable-base heating and cooling degree-days for Turkey. Applied Energy, Vol. 69, pp.269-283, 2001.
3. Zedan MF, Mujahid AM. An efficient solution for heat transfers in composite walls with periodic ambient temperature and solar radiation. Int J Ambient Energy, Vol. 14, pp.83-98, 1993.
4. Zedan MF, Mujahid AM. An efficient solution for heat transfers in composite walls with periodic ambient temperature and solar radiation. Int J Ambient Energy, Vol. 14, pp.83-98, 1993.
5. Daouas N, Hassen Z, Ben Aissia H. Analytical periodic solution for the study of thermal performance and optimum insulation thickness of building walls in Tunisia. Applied Thermal Engineering, Vol.30, pp.319-326, 2010.
6. A. Hasan. Optimizing insulation thickness for buildings using life cycle cost. *Applied Energy*, Vol. 63, pp.115-124, 1999.
7. Durmayaz A. Kadioglu M, Sen Z. An application of the degree-hours method to estimate the -residential heating energy requirement and fuel consumption in Istanbul. *Energy*, Vol. 25, pp. 1245–1256, 2000.
8. Sarak H, Satman A. The degree-day method to estimate the residential heating natural gas consumption in Turkey: a case study. *Energy*, Vol. 28, pp.929-939, 2003.
9. Granja AD, Labaki LC. Influence of external surface color on the periodic heat flow through a flat solid roof with variable thermal resistance. *Int J Energy Res*, Vol. 27, pp.771-779, 2003.
10. Yumrutas R, Unsal M, Kanog-lu M. Periodic solution of transient heat flow through multilayer walls and flat roof by complex finite Fourier transform technique. *Built Environment*, Vol. 40, pp.1117-1125, 2005.
11. Dombayci, O. A., Golcu, M., Pancar, Y., Optimization of insulation thickness for external walls using different energy-sources. *Applied Energy*, Vol. 83 No. 9, pp 921–928, 2006.
12. O. A. Dombaycı, M. Golcu, Y. Pancar. Optimization of insulation thickness for external walls using different energy-source. *Applied Energy*, Vol. 83, pp.921-928, 2005.
13. K.C, Omaklı, B. Yuksel. Optimum insulation thickness of external walls for energy saving. *Applied Thermal Engineering, Vol.* 23, pp.473–479, 2003.
14. A. Bolatturk. Determination of optimum insulation thickness for building walls with respect to various fuels and climate zones in Turkey. *Applied Thermal Engineering*, Vol. 26, pp.1301–1309, 2006.
15. Zhang Lianying, Numerical Study of Effects of Wall's Insulation Thickness on, Energy Performance for Different Climatic Regions of China, *Energy Procedia*, Vol. 75, pp. 1290 – 1298, 2015.
16. Xiangwei Liu, Determination of Optimum Insulation Thickness of Exterior Wall with Moisture Transfer in

Hot Summer and Cold Winter Zone of China, Procedia Engineering, Vol. 121, pp. 1008 – 1015, 2015.
17. Modeste K. N., A comparative study on optimum insulation thickness of walls and energy savings in equatorial and tropical climate, International Journal of Sustainable *Built Environment*, Vol. 6, pp. 170–182, 2017.
18. Mousa M. Mohamed, Mohammed H. Almarshadi, Simulation of District Cooling Plant and Efficient Energy, 1st International Conference on New Trends for Sustainable Energy, Pharos University, Alexandria, Egypt. 1-3 October, 2016.
19. Mousa M. Mohamed, Mohammed H. Almarshadi, Simulation of District Cooling Plant and Efficient Energy Air Cooled Condensers (Part I), Journal of Electronics Cooling and Thermal Control, JECTC, Vol. 7, pp. 45-62, 2017.
20. ASHRAE Fundamentals Handbook, SI, Chapter 27, 1997.

https://doi.org/10.1051/matecconf/201816205025

# Effect of urban geometry and green area on the formation of the urban heat island in Baghdad city

*Younis* Mohammed [1] *and Aws* Salman [1,*]

[1]Department of Architecture Engineering, University of Technology, Baghdad, Iraq

**Abstract.** With the growth of cities, the ambient air temperatures (Ta) inside the urban areas are expected to be higher compared to the surrounding rural areas, creating urban heat island (UHI) phenomenon. The city of Baghdad is an example of a hot dry climate cities and during summer, the UHI intensity is significantly affected by the extreme direct solar radiation and leads to outdoor thermal discomfort. Also it causes an increase in energy consumption and air pollution. This research work focuses on the effect of urban geometry and green area in the formation of heat island through a study of two different fabrics of residential neighbourhoods. The height to width ratio (H/W) and vegetation are adopted while the materials of buildings were unified in all study cases. Three-dimensional numerical software Envi-met 4.1 was utilized to analyze and assess the studied parameters including: ambient air temperature (Ta), street surface temperature (Ts) and mean radiant temperature (Tmrt). This study has given a better understanding of the role of urban geometry and green area on forming the UHI that influence on the microclimatic conditions in hot dry climate of the city of Baghdad. So that helped to generate guidelines of urban design and planning practices for a better thermal performance in hot and dry cities.

Day after day the cities growth and became bigger and bigger and the ambient air temperatures ($T_a$) inside the urban cores are expected to be higher than that in their surrounding countryside, creating urban heat island (UHI) phenomenon[1-4]. The bad effect of the urban heat island (UHI) phenomenon, particularly on energy use, human health and air quality as well as its significant effect on human comfort, many researchers are focusing on the possible strategies for mitigating UHI and many of them are interested to investigate at the scale of neighbours [4-7].

There are a number of mitigation strategies. One of them is related with the use of trees or vegetation in urban areas. Vegetation can contribute significantly to park cool island (PCI) influence and consequently can reduce air temperature up to 3 to 4°C in hot summer periods [8-10]. Vegetation can make $T_a$ reduction possible through several mechanisms. First, higher albedo value compared with construction materials such as asphalt. Second, through evaporation and transpiration. Third, with lower specific heat capacity greeneries absorb less heat. Finally, with shading, leaves and branches reduce the amount of solar radiation that reaches the urban surfaces below the canopy layer [9-15].

The effect of vegetation on the conditions of microclimate and thermal comfort can be evaluated by computer simulation software programs for the environment. This helped to understand many of the current environmental problems where it let to quantify the influences of zone changes (land coverage) on microclimate conditions and human thermal comfort, through microclimate models such as Envi-met [5], [16].

Also, this study is highlight on the role of urban geometry as a potential mitigation strategy [8]. Investigation that studies the impact of urban configurations variables in hot dry climate (density, urban canyons, and orientation) in forming the UHI effect in cities is still required [2], [17]–[20]. Huge quantities of solar radiations are mainly stored and reradiated in urban areas due to massive construction material and canyon effect, which is more relevant in case of taller buildings [20]. Density of buildings can significantly impact the UHI by reducing or increasing the areas of exposed surfaces. The geometry of urban canyons can influence the speed of airflow for natural ventilation system (NVS) and also can shape shaded paths for pedestrians. This geometry is expressed by the ratio of building height (H) and street width (H/W), or the aspect ratio [16], [20].

The current study focuses on the city of Baghdad, the largest urban area in Iraq [21]. This city is characterized by a variety of urban forms: These urban patterns include attached and detached. The patterns include the areas that have developed primarily in the modern suburbs in the middle of the 20th century; and the areas that have been shaped primarily in the last quarter of the 20 century and are distributed in the city edges. These patterns have been developed on the basis of modern planning practices. These practices were imported from western contexts, where cold and

* Corresponding author: aws_m_s2007@yahoo.com

© The Authors, published by EDP Sciences. This is an open access article distributed under the terms of the Creative Commons Attribution License 4.0 (http://creativecommons.org/licenses/by/4.0/).

moderate climates are prevalent without consideration to extreme local climates in hot and arid regions [20].

It was decided to investigate how urban form and green area influences the quality of outdoor spaces of two common residential neighbourhoods in the city of Baghdad. A comparison have been made to find out the impact of these UHI mitigation strategies on thermal comfort in the selected area.

The aim of the study presented is to investigate the influence of several variables, which co-cause the Urban Heat Island effect, on temperature distribution ($T_a$, $T_s$, and $T_{mrt}$) and on outdoor thermal comfort. In particular, it tries to provide a better understanding of how the geometry of the urban canyon and vegetation contributes to the formation of the UHI. A simulation tool, ENVI-met Version 4.1, is used to forecast the microclimatic changes within urban environment.

# 1 Methodology

## 1.1 Selected urban areas

Two locations of the city were selected to represent the attached, and detached urban configurations. The case studies represent a variety of urban forms with regards to the density, urban canyons, and orientation Fig. (1). The key determinant factor of choosing such configurations is that they all represent urbanized areas and vary significantly in their design, layout, and characteristics. First, the modern attached case includes rectangular long urban blocks, gridded street systems, and attached buildings from three sides and is laid out perpendicularly on the streets. Second, even though it is derived from the previous pattern, the modern detached case has wider streets aligned with detached buildings that occupy small footprints and typically are placed in the middle of plots. To establish comparable cases and to assure that the urban configuration and vegetation are the dominant factor in the analysis, streets in all cases are oriented northeast-southwest and materials for all buildings surfaces are identical.

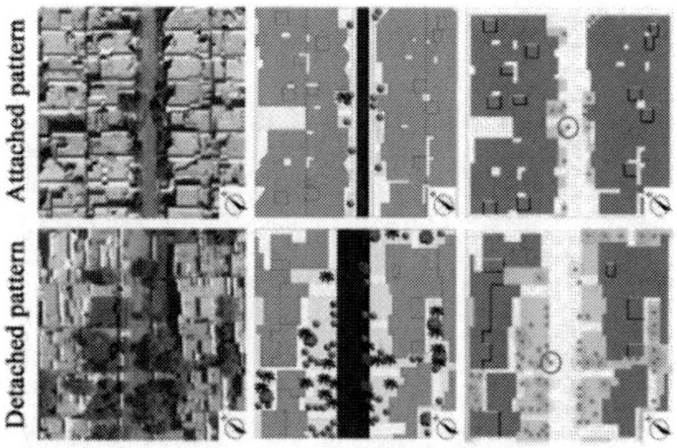


**Fig. 1.** Representation of the selected two study cases of urban form in the city of Baghdad. Red point in (Input File) column represented the measurement point in each area.

Building height and the land use in each area are shown in Table 1. Also, this table shows the building surface cover in terms of building density and average height. The two selected location covering an area of 100×100 m were used for the investigation.

**Table 1.** Building height and land use properties in each area.

| | | Detached | Attached |
|---|---|---|---|
| Land surface cover (%) | Building | 39.75 | 75.6 |
| | Road asphalt | 16 | 6 |
| | Pavement | 12.85 | 9.5 |
| | Soil | 0 | 2.9 |
| | Grass | 31.4 | 6 |
| Building height (m) | Elevations height | 3 or 6 | 6 |
| | Highest point | 8 | 8 |

## 1.2 Urban environmental simulation

With the aim to forecast the microclimatic changes within urban environment, a simulation tool, ENVI-met Version 4.1, was used. ENVI-met, a three-dimensional computer software that simulates micro-scale conditions in urban environments. ENVI-met is designed to simulate the surface-plant-air interactions. Typically, Envi-met model has spatial resolution from 0.5m to 10m, and a temporal resolution of 10 seconds. The best time to start a simulation is at night, so that the simulation can follow the solar radiation increase during the day. In hot and dry climate simulation was carried out for more than 24 hours and results are taken after the first 24 hours. This gives more accurate results. A detailed description of microclimate model Envi-met and further information can be accessed through its official website [22]–[25]. Envi-met can be used to evaluate several aspects of urban canyons and the effects of vegetation on outdoor comfort and urban heat island mitigation [26].

In this research, typical summer day are selected for simulation. (Table 2) illustrates the input data for July 15th, 2015, which has the highest recorded air temperature in the city of Baghdad. ENVI-met 4.1 is using to evaluate the $T_a$, $T_s$, and $T_{mrt}$ of the two selected cases. In order to compare the effects of different scenarios in the two urban locations, three models for each location were considered (Table 3):

- Base model: it represents the existing current location;
- No green model: it represents the existing current location without green area and vegetation;
- Green model: it represents the existing current location with a vegetation coverage higher than in the base model.

**Table 2.** Details of initialization input parameters for the simulations in ENVI-met.

| Data Type | Parameter | Value | |
|---|---|---|---|
| General | Date of start the simulation | 15 July 2015 | |
| | Simulation starting time | 5:00 A.M. | |
| | Simulation duration | 44 hours | |
| Initial data | Wind Speed in 10 m ab. Ground [m/s] Wind | 2.0 | |
| | Direction (0:N..90:E..180:S..270:W..) | 315 | |
| | Roughness Length z0 at Reference Point [m] Specific | 0.1 | |
| | Humidity in 2500 m [g Water/kg air] Factor of | 7.0 | |
| | shortwave adjustment (0.5 to 1.5) | 0.9 | |
| | Initial Temperature Atmosphere [K] | 315.250 | |
| Soil data | Initial temperature/RH Upper Layer (0-20 cm) | 310.0 K / 10% | |
| | Initial temperature/RH Middle Layer (20-50 cm) | 300.0 K / 50% | |
| | Initial temperature/RH Deep Layer (below 50cm) | 293.0 K / 60% | |
| Building Material Properties | | Walls: | Roof: |
| | Albedo (%) | 0.3 | 0.3 |
| | Thermal conductivity (W/m.K) | 1.3 | 1.9 |
| Surface Material | Albedo (%) | Asphalt Street: 0.2<br>concrete pavement gray: 0.5 | |
| [LBC-TYPES] | LBC for T and q (1:open, 2:forced, 3:cyclic) | 2 | |
| | LBC for TKE (1:open, 2:forced, 3:cyclic) | 2 | |

In this study, each case was modelled with a domain dimension of 50 x 50 x 20 and a resolution of 2m x 2m x 1m, making a total model area of 100m x 100m x 20m. Six different scenarios were created viz. DN, DB, DG, AN, AB and AG. The first letter represents detached (D) or attached (A) pattern while the second letter represents the selected scenario which is no green (N), base (B) or green (G). Table 3 shows the number of trees and the type of trees in each scenario. Big tree have a height of 7 meters and a width of 9 meters, small tree have a height of 5 meters and a width of 3 meters and Plame have a height of 8 meters and a width of 6 meters.

**Table 3.** Details of number and type of trees in each scenario.

| Scenarios | Big tree | Small tree | Plame |
|---|---|---|---|
| DN | 0 | 0 | 0 |
| DB | 10 | 39 | 21 |
| DG | 21 | 50 | 31 |
| AN | 0 | 0 | 0 |
| AB | 0 | 13 | 1 |
| AG | 0 | 60 | 1 |

## 2 Results and discussion

### 2.1 Air temperature ($T_a$)

In the detached pattern (Fig2), the hottest values of $T_a$ were found at 3:00 PM which were 50.0°C, 49.9°C and 49.65°C in no green, base and green scenarios, respectively. The lowest values of $T_a$ were found at 5:00 AM which were 35.3°C, 34.7°C and 34.67°C in no green, base and green scenarios, respectively. The average values of $T_a$ were of 42.9°C 42.6°C and 42.4°C in no green, base and green scenarios, respectively. During daytime periods, the difference in average was 0.18°C between no green and base scenarios while it was 0.46°C between no green and green scenarios. During night time periods, the maximum difference in $T_a$ was found in 5:00 AM which was 0.6°C between no green and base scenarios while its 0.63°C between no green and green scenarios. The average difference at night time periods was 0.45°C between no green and base scenarios

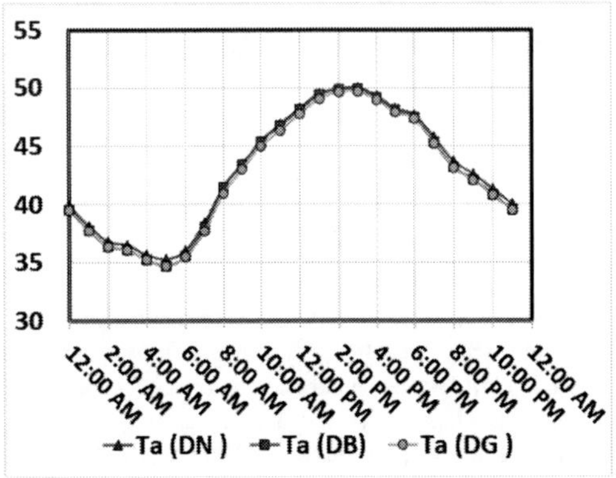

**Fig. 2.** Variations of $T_a$ during the day at pedestrian walkway of detached pattern at a height of 1.5m from the ground.

while it was 0.46°C between no green and green scenarios.

The results have shown that the green area is more effective during night time periods. The reason for this is that the larger area of asphalt and cement surfaces in no green scenario stores more heat during daytime and radiate at night time which causes higher values of $T_a$ during night time periods. However, green area effects on $T_a$ values was small.

In the attached pattern (Fig 3), the hottest values of $T_a$ were found at 3:00 PM which were 49.6°C, 49.5°C and 49.5°C in no green, base and green scenarios, respectively. The lowest values of $T_a$ were found at 5:00 AM which were 35.3°C, 35.2°C and 35.1°C in no green, base and green scenarios, respectively. The average values of $T_a$ were 42.8°C, 42.7°C and 42.6°C in no green, base and green scenarios, respectively. During daytime periods, the difference in average was 0.07°C between no green and base scenarios while its 0.18°C between no green and green scenarios. During nighttime periods, the maximum difference in $T_a$ was found in 5:00 AM which was 0.1°C between no green and base scenarios while it was 0.2°C between no green and green scenarios. The difference on an average during nighttime periods was 0.08°C between no green and base scenarios while its 0.15°C between no green and green scenarios.

The results have shown that the green area has low effects during daytime and nighttime periods. The reason for this is the small area of vegetation and green surfaces in both base and green scenarios.

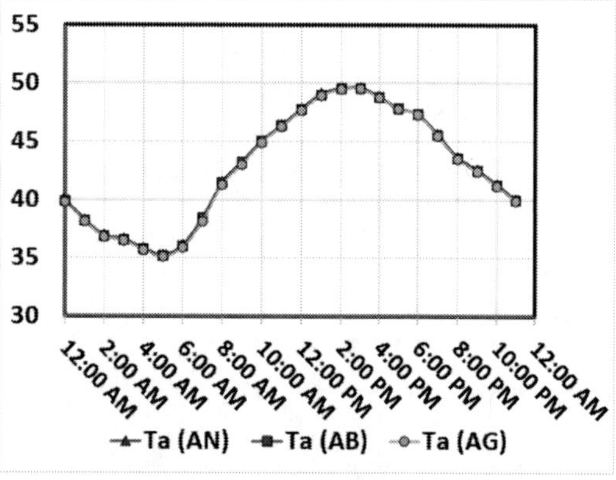


**Fig. 3.** Variations of $T_a$ during the day at pedestrian walkway of attached pattern at a height of 1.5m from the ground.

Figure 4 compares between average values of $T_a$ in different scenarios. In no green scenarios, the results have shown that the detached pattern has higher value of $T_a$ than attached by 0.18°C. The maximum difference in $T_a$ during daytime periods was found in 3:00 PM which was 0.4°C. The difference in average during daytime periods was 0.3°C. During nighttime periods the detached pattern have lower value of $T_a$ than attached by 0.07° C in average. It is obvious that the H/W ratio in the shallow detached street enhances the performance of dissipating and cooling by winds and breezes during night, while the H/W ratio in the attached street helps to minimize the accessibility of solar radiation into the canyons during the daytime and maximize the shaded areas, which help the surfaces to remain cool.

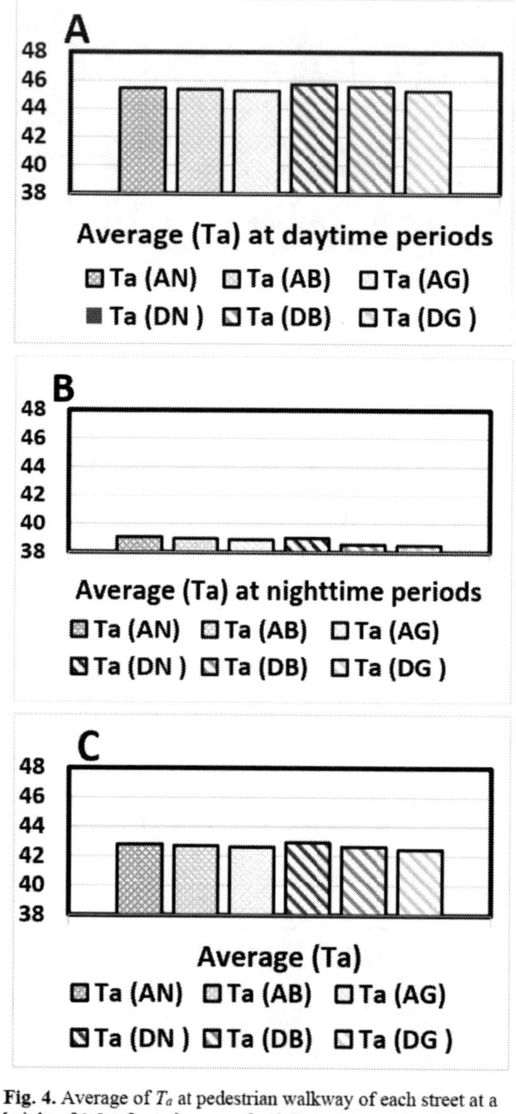


**Fig. 4.** Average of $T_a$ at pedestrian walkway of each street at a height of 1.5m from the ground. A) Represent average of total $T_a$ during daytime periods; B) represent average of $T_a$ during night time periods; C) represent average of total $T_a$ during the daytime and night time periods.

In base scenarios, the results have shown that the

detached pattern have higher value of $T_a$ during daytime periods. The maximum difference in $T_a$ during daytime periods was found at 3:00 PM which was 0.4°C. The difference in average during daytime periods was 0.16°C. During nighttime periods the detached pattern have lower value of $T_a$ than attached by 0.4°C in average. The maximum difference in $T_a$ during nighttime periods was found at 5:00 AM which was 0.47°C. The H/W ratio in the attached street with trees helps to minimize the accessibility of solar radiation into the pedestrian walkway during the daytime and maximize the shaded areas, which helped the surfaces to remain cool. In the detached pattern the buildings play a lesser role in the shading because it is far in the distance from the pedestrian walkway. During nighttime, the large green area in the detached pattern compared with the attached pattern as well as the wind plays important role in reduce $T_a$.

In green scenarios, the results have shown that the two areas have almost equal value of $T_a$ during daytime periods. The maximum difference in $T_a$ during daytime periods was found in 4:00 PM which was 0.01°C. During nighttime periods the detached pattern have lower value of $T_a$ than attached by 0.4°C in average. Wider and denser green spaces in the detached pattern contribute more to reducing $T_a$, however, the attached pattern needs fewer trees to achieve almost equal value of $T_a$ during daytime periods compare with detached pattern.

The best scenario of $T_a$ in all cases was found in detached pattern with green scenario while the worst scenario of $T_a$ in all cases was found in detached pattern with no green scenario. That show us how vegetation densities can affect the microclimate conditions and contribute to its improvement. The results also have shown that attached pattern have more balanced value of $T_a$ and needs fewer trees to achieve almost equal value of $T_a$ during daytime periods compare with detached pattern.

### 2.2 Ground surface temperature ($T_s$)

In the detached pattern (Fig 5), the hottest values of $T_s$ was found at 3:00 PM which were 62.1°C, 58.5°C and 51.1°C in no green, base and green scenarios, respectively. The lowest values of $T_s$ were found at 5:00 AM which were 35.7°C, 35.8°C and 35.9°C in no green, base and green scenarios, respectively. The average values of $T_s$ were 48.3°C, 47.5°C and 43.9°C in no green, base and green scenarios, respectively. During daytime periods, the difference in average was 1.4°C between no green and base scenarios while it was 7.3°C between no green and green scenarios. During nighttime periods, the maximum difference in $T_s$ was found in 5:00 AM which was 0.1°C between no green and base scenarios while it was 0.2°C between no green and green scenarios. The difference in average during nighttime periods was 0.05°C between no green and base scenarios while it was 0.4°C between no green and green scenarios. The results have shown that the green area is more effective during daytime periods. The reason for this is the shade from trees which increases with increasing tree density and causes a reduction in values of $T_s$.

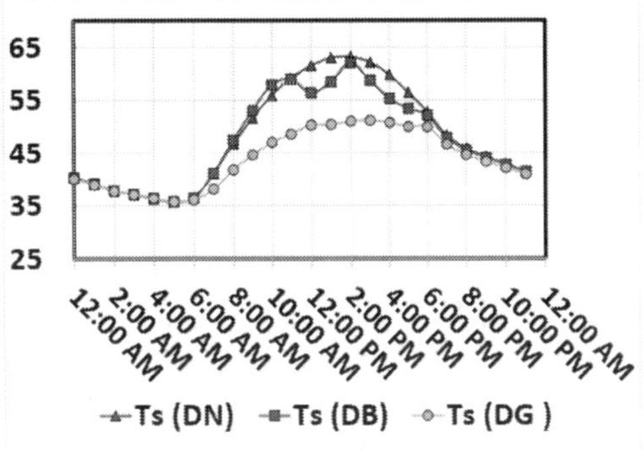


**Fig. 5.** Variations of $T_s$ during the day at pedestrian walkway of detached pattern.

In the attached pattern (Fig. 6), the hottest values of $T_s$ were found at 3:00 PM which were 56.0°C, 53.0°C and 53.6°C in no green, base and green scenarios, respectively. The lowest values of $T_s$ were found at 5:00 AM which were 30.6°C, 30.6°C and 30.9°C in no green, base and green scenarios, respectively. The average values of $T_s$ were 45.2°C, 43.5°C and 43.0°C in no green, base and green scenarios, respectively. During daytime periods, the difference in average was 2.7°C between no green and base scenarios while it was 3.6°C between no green and green scenarios. During nighttime periods, the difference in $T_s$ values were small and it's ineffective.

The results have shown that the green area is more effective during daytime periods. The reason for this is the shade from trees. This effectiveness is increases with increasing tree density and causes a reduction in values of $T_s$.

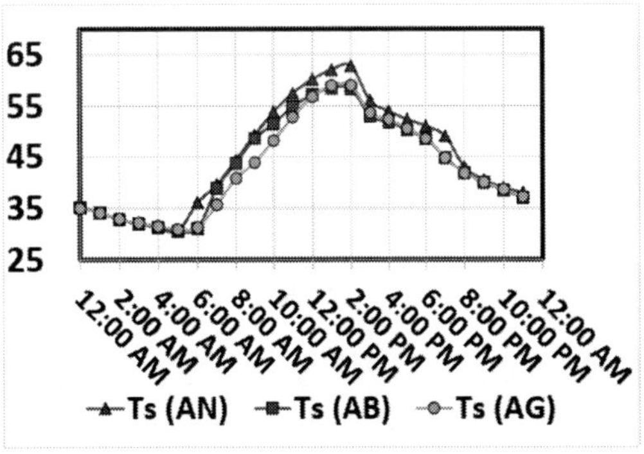


**Fig. 6.** Variations of $T_s$ during the day at pedestrian walkway of attached pattern.

 https://doi.org/10.1051/matecconf/201816205025

Figure 7 compares the average values of $T_s$ in different scenarios. In no green scenarios, the results have shown that the detached pattern have higher value of $T_s$ than attached by an average of 3.0°C. The maximum difference in $T_s$ during daytime periods was found at 3:00 PM which was 6.1°C. The difference in average during daytime periods was 2.1°C. During nighttime periods, the Attached pattern has lower value of $T_s$ than Detached by an average of 4.4°C. The reason behind it may be the nearly building and trees in Attached pattern were provided more shade on walkways and lead to reduce $T_s$ in it.

In base scenarios, the results have shown that the detached pattern have higher value of $T_s$ than attached by 3.7°C on an average. This is more pronounced during the nighttime where the difference in $T_s$ is 4.7°C while it was 3.0°C during daytime. The reason behind it may be the nearly building and trees in Attached pattern were provided more shade on walkways and lead to reduce $T_s$ in it.

In green scenarios, the results have shown that the detached pattern has higher value of $T_s$ than attached by 0.9°C on an average. During daytime the detached pattern has lower value of $T_s$. The maximum difference in $T_s$ during daytime periods was found at 2:00 PM which was 7.0°C. The difference in average during daytime periods was 1.5°C. The reason of that was the shade from trees which has higher density in the detached pedestrian walkway. During nighttime periods the detached pattern has higher value of $T_s$ than attached. The maximum difference in $T_s$ during nighttime periods was found at 5:00 AM which was 4.9°C. The difference in average during nighttime periods was 4.2°C. This may be because of two reasons: first, the quality and albedo of material in pedestrian walkway and second, the wind speed. The material in detached pedestrian walkway is asphalt which has lower albedo than cement surfaces in attached pedestrian walkway. Asphalt stores more heat during daytime and radiate during nighttime. Wind speed can also be less due to high density of trees in detached pedestrian walkway. Less wind and breeze leading to slower $T_s$ loss.

The best scenario of $T_s$ in all cases was found in attached area with green scenario while the worst scenario of $T_s$ in all cases was found in detached pattern with no green scenario. The trees cover and the material played an important role in influencing the value of $T_s$.

### 2.3 Mean radiant temperature ($T_{mrt}$)

In the detached pattern (Fig. 8), the hottest values of $T_{mrt}$ was found at 3:00 PM which was 88.4°C, 69.2°C and 61.2°C in no green, base and green scenarios, respectively. The lowest values of $T_{mrt}$ was found at 5:00 AM which was 28.6°C, 29.8°C and 30.7°C in no green, base and green scenarios, respectively. The average values of $T_{mrt}$ was 58.0°C, 52.5°C and 46.0°C in no green, base and green scenarios, respectively. During daytime periods, the difference in average was 9.9°C between no green and base scenarios while it was 21.3°C between no green and green scenarios. During nighttime periods, the maximum difference in $T_{mrt}$ was found at 5:00 AM which was 1.2°C between no green and base scenarios while it was 2.1°C between no green and green

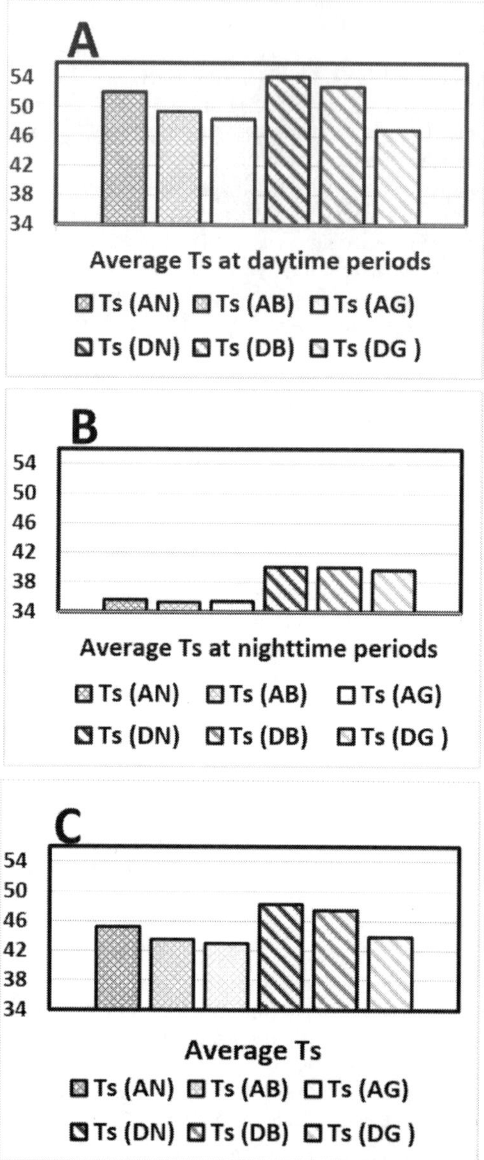


**Fig. 7.** Average of $T_s$ at pedestrian walkway of each street. A) Represent average of total $T_s$ during the daytime periods; B) represent average of $T_s$ during the night time periods; C) represent average of total $T_s$ during the daytime and night time periods.

scenarios. The difference in average during nighttime periods was 0. 5°C between no green and base scenarios while it was 1.0°C between no green and green scenarios. The results have shown that the green area is more effective during daytime periods. The reason for this is the shade from trees which increases with increasing tree density and causes a reduction in values of $T_{mrt}$.

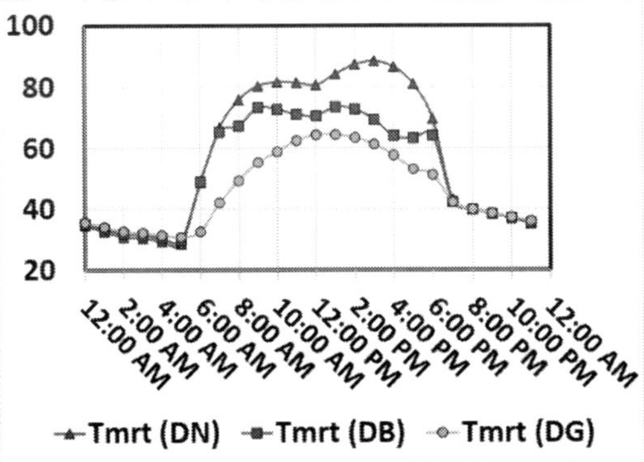


**Fig. 8.** Variations of $T_{mrt}$ during the day at pedestrian walkway of detached pattern at a height of 1.5m from the ground.

In the attached pattern (Fig 9), the hottest values of $T_{mrt}$ was found at 3:00 PM which were 85.4°C, 72.5°C and 69.5°C in no green, base and green scenarios, respectively. The lowest values of $T_{mrt}$ was found at 5:00 AM which were 27.9°C, 29.2°C and 29.6°C in no green, base and green scenarios, respectively. The average values of $T_{mrt}$ were 53.4°C, 49.8°C and 48.0°C in no green, base and green scenarios, respectively. During daytime periods, the difference in average was 6.8°C between no green and base scenarios while it was 10.1°C between no green and green scenarios. During nighttime periods, the maximum difference in $T_{mrt}$ was found in 5:00 AM which was 1.3°C between no green and base scenarios while it was 1.7°C between no green and green scenarios. The difference in average at nighttime periods was 0.9°C between no green and base scenarios while it was 1.0°C between no green and green scenarios. The results have shown that the green area is more effective during daytime periods. The reason for this is the shade from trees which increases with increasing tree density and causes a reduction in values of $T_{mrt}$.

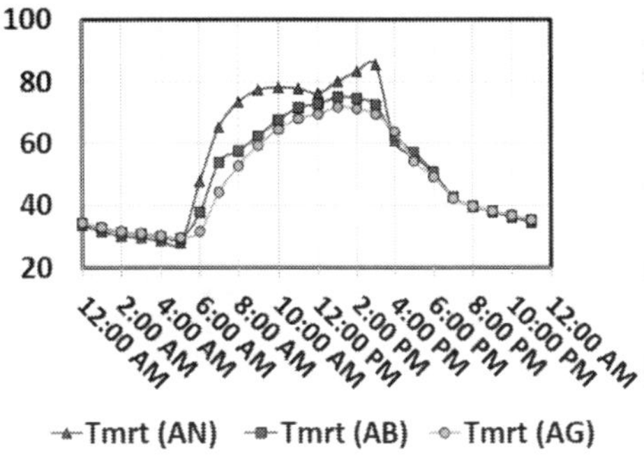


**Fig. 9.** Variations of $T_{mrt}$ during the day at pedestrian walkway of attached pattern at a height of 1.5m from the ground.

Figure 10 compares between average values of $T_{mrt}$ in different scenarios. In no green scenarios, the results have shown that the detached pattern has higher value of $T_{mrt}$ than attached by 4.5°C on an average. The maximum difference in $T_{mrt}$ during daytime periods was found in 4:00 PM which was 25.8°C. The difference on an average during daytime periods was 7.2°C. During nighttime periods the two area have almost equal value of $T_{mrt}$. The maximum difference in $T_{mrt}$ during nighttime periods was found in 0:00 AM o'clock which was 1.1°C. The average difference during nighttime periods was 0.8°C.

In base scenarios, the results have shown that the detached pattern has higher value of $T_{mrt}$ than attached by 2.7°C on an average. The maximum difference in $T_{mrt}$ during daytime periods was found in 6:00 PM which was 13.4°C. The difference on an average during daytime periods was 4.2°C. During nighttime periods the two area have almost equal value of $T_{mrt}$. The maximum difference in $T_{mrt}$ during nighttime periods was found in 0:00 AM which was 0.9°C. The difference on an average during nighttime periods was 0.5°C. It is noticed that green area works to create a greater thermal balance between the two patterns.

In green scenarios, the results have shown that the detached pattern have lower value of $T_{mrt}$ than attached by 2.0°C on an average. The maximum difference in $T_{mrt}$ during daytime periods was found in 1:00 PM which was 7.4°C. The difference on an average during daytime periods was 3.9°C. During nighttime periods, the two area have almost equal value of $T_{mrt}$. The maximum difference in $T_{mrt}$ during nighttime periods was found in 0:00 AM which was 1.1°C. The difference on an average during nighttime periods was 0.8°C. It was noticed that the higher density of trees in the detached pedestrian walkway greatly reduce $T_{mrt}$ than the lower density of trees in the attached pedestrian walkway.

The best scenario of $T_{mrt}$ in all cases was found in detached pattern with green scenario while the worst scenario of $T_{mrt}$ in all cases was found in detach area

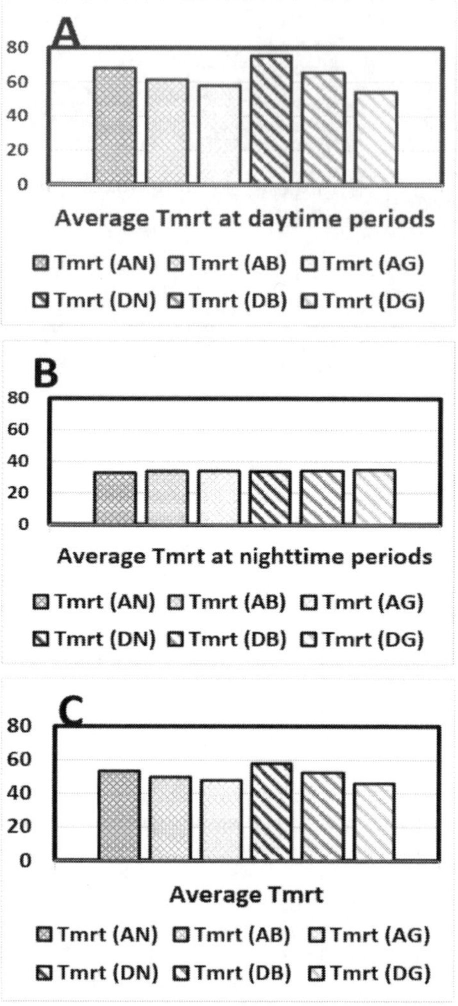


**Fig. 10.** Average of $T_{mrt}$ at pedestrian walkway of each street at 1.5m height. A represent average of total $T_{mrt}$ during the daytime periods. B represent average of $T_{mrt}$ during the nighttime periods. C represent average of total $T_{mrt}$ during the daytime and nighttime periods

with no green scenario. The trees cover and its density played an important role in influencing the value of $T_{mrt}$.

## 3 Conclusions

Simulation studies have been carried out for evaluating the effect of geometric properties of different patterns of urban canyons and green area on temperature distribution (ambient air temperature ($T_a$), street surface temperature ($T_s$) and mean radiant temperature ($T_{mrt}$) in two locations in the city of Baghdad.

The study reached the following conclusions:

-First, at the pedestrian level scale, the urban geometry and green area contribute significantly to reduce of mean radiant temperature ($T_{mrt}$) and street surface temperature ($T_s$) values while there was no significant effect on the ambient air temperature ($T_a$) values.

-Second, changes in the values of $T_s$ and $T_{mrt}$ were during daylight hours, while there was no significant effect during night time.

Accordingly, this study recommends reviewing the practices and standards of planning and urban design in the city of Baghdad, such as the height of the building or the distance between them, in accordance with the hot and dry extreme external microclimate conditions of the city. This can be achieved by providing the greatest amount of shading in the pedestrian walkway using the urban geometry or vegetation or both.

## References

[1] T. R. Oke, "Urban climates and global environmental change," Applied Climatology: Principles & Practices. New York, NY: Routledge, 1997, pp. 273–287.

[2] E. Johansson, "Influence of urban geometry on outdoor thermal comfort in a hot dry climate: A study in Fez, Morocco," *Build. Environ.*, vol. 41, no. 10, pp. 1326–1338, 2006.

[3] S. Shooshtarian, "Urban heat island," pp. 4–6, 2014.

[4] Q. S. R. Alshammary, "The Role of Green Areas in the Environmental System Efficiency for the City," University of Technology, 2013.

[5] C. O'Malley, P. Piroozfar, E. R. P. Farr, and F. Pomponi, "Urban Heat Island (UHI) mitigating strategies: A case-based comparative analysis," *Sustain. Cities Soc.*, vol. 19, no. August 2016, pp. 222–235, 2015.

[6] Y. Wang and H. Akbari, "Development and application of 'thermal radiative power' for urban environmental evaluation," *Sustain. Cities Soc.*, vol. 14, pp. 316–322, 2015.

[7] Y. Wang, U. Berardi, and H. Akbari, "The Urban Heat Island Effect in the City of Toronto," *Procedia Eng.*, vol. 118, pp. 137–144, 2015.

[8] M. Z. Targhi and S. Van Dessel, "Potential Contribution of Urban Developments to Outdoor Thermal Comfort Conditions: The Influence of Urban Geometry and Form in Worcester, Massachusetts, USA," *Procedia Eng.*, vol. 118, pp. 1153–1161, 2015.

[9] A. Dimoudi and M. Nikolopoulou, "Vegetation in the Urban Environment," *Energy Build.*, vol.

35, pp. 69–76, 2003.

[10] Aws M. Salman and Younis M. Mohammed, "Impact of Microclimate Improvement Strategies on the Quality of External Spaces in Baghdad City," in *Scientific Conference of the Ministry of Construction, Housing, Municipalities and Puplic works*, 2017.

[11] T. R. Oke, *Boundary layer climates*. Routledge, London.

[12] H. Taha, S. Douglas, and J. Haney, "Mesoscale meteorological and air quality impacts of increased urban albedo and vegetation," *Energy Build.*, vol. 25, no. 2, pp. 169–177, 1997.

[13] D. J. Sailor, "A review of methods for estimating anthropogenic heat and moisture emissions in the urban environment," *Int. J. Climatol.*, vol. 31, no. 2, pp. 189–199, 2011.

[14] L. S. Pereira, A. Perrier, R. G. Allen, and A. and I. A. Member, "Evapotranspiration:concepts and future trends," no. APRIL, pp. 45–51, 1999.

[15] U. S. Epa, "Reducing Urban Heat Islands: Compendium of Strategies," *Heat Isl. Reduct. Act.*, pp. 1–23, 2008.

[16] K. Perini and A. Magliocco, "Effects of vegetation , urban density , building height , and atmospheric conditions on local temperatures and thermal comfort," *Urban For. Urban Green.*, vol. 13, no. 3, pp. 495–506, 2014.

[17] D. Alobaydi, H. Mohamed, and H. Attya, "The Impact of Urban Structure Changes on the Airflow Speed Circulation in Historic Karbala, Iraq," *Procedia Eng.*, vol. 118, pp. 670–674, 2015.

[18] M. M. Elnahas, "Energy Simulation of Urban Dwellings i n Temperate Climates," no. June 2013, pp. 37–41, 2011.

[19] D. Robinson, "Urban morphology and indicators of radiation availability," *Sol. Energy*, vol. 80, no. 12, pp. 1643–1648, 2006.

[20] D. Alobaydi, M. A. Bakarman, and B. Obeidat, "The Impact of Urban Form Configuration on the Urban Heat Island: The Case Study of Baghdad, Iraq," *Procedia Eng.*, vol. 145, pp. 820–827, 2016.

[21] The National Committee for Population Policies, "The Situation of Iraq 's Population in 2010, The first national report on the situation of the population as part of the recommendations of the Cairo Conference on Population and Development Goals," *Cent. Stat. Organ. Minist. Plan.*, 2011.

[22] M. Bruse, "ENVI-met website," 2014. [Online]. Available: http://www.envi-met.com/.

[23] M. Bruse, "The influences of local environmental design on microclimate – development of a prognostic numerical model ENVI-met for the simulation of Wind, temperature and humidity distribution in urban structures," University of Bochumو, Germany, 1999.

[24] A. J. B. I. Ozkeresteci, K. Crewe, "Use and evaluation of the envi-met model for environmental design and planning : an experiment on linear parks," *Proc. 21st Int. Cartogr. Conf.*, no. August, pp. 10–16, 2003.

[25] B. M. S. Alznafer, "The Impact of Neighbourhood Geometries on Outdoor Thermal Comfort and Energy Consumption From Urban Dwellings," 2014.

[26] H. Bruse, M., Fleer, Simulating surface–plant–air interactions inside urban environments with a three dimensional numerical model. 1998.

# Investigation of thermal behaviour for various exterior walls materials and their impact on rationalization of energy consumption of Kirkuk city, Iraq

*Naseer* Alharbawee*

Engineering Technical College, Kirkuk, Iraq

**Abstract.** The present study aims to make a comparison of the thermal behaviour for various exterior Packaging materials (Local or Imported) or a new proposed (a phase change material), which are used in covering the facades of buildings in Kirkuk city/Iraq. And thus know how much those materials effect on energy conservation (rationalization in energy consumption). In this study the focus was on making comparison of heat dissipation and power consumption (Save energy) when using the conventional and proposed materials and which of them has the greatest ability to withstand various environmental conditions. The study was carried out (from the beginning of January 2017 until the end of February, 2017) which this period represents the cold months in the year. (Where temperatures ranged from $-2\ to\ 15$ °C) which within those months all facilities, buildings need to be heated. The thermal conductivity of various classical covering materials samples was measured and compared with the values of the other certified project sources. It was found through analysis of the results that the use of proposed material which is a phase change material (Paraffin Wax) exceeds in terms of energy reduction than other classical covering materials. It was also shows that the use of classical and proposed covering materials reduces heat loss through the exterior walls (reducing demanded heating load in the comfort zone), thus minimizing the consumption of (save) expended energy within the comfort.

## 1 Introduction

In light what the world is witnessing of the enormous revolution in the architecture field and especially what our beloved country and particularly what our beloved country witnesses of large urban renaissance because of population growth, and there for the demand have increased for building and facilities, and this increase was accompanied by consumption of fuel for needed generation power electricity for air-conditioning and heating those buildings and facilities which use high-quality materials in construction and therefore it requires a relatively high cost. It has been therefore the subject of keeping the thermal energy in the buildings of the topic most important at the present time due to the increased consumption of energy in buildings as a result of the use of means of heating and air conditioning to ensure a comfortable and healthy atmosphere to humans where is the using of heat insulation materials in buildings is most important economic means used to reduce the growing energy consumption, as it provides a saving in energy consumption for heating and cooling, which in turn ease the growing burden on the citizen, and especially those who have limited income and in return the construction materials have far away developed which is the packaging materials which is use to cover the exterior walls is one of them and which characterized of enduring the different environmental conditions. But there is some of the packaging materials have drawbacks which are related to its thermal properties, for example there are some of them which own a very high thermal conductivity , and its dissipating heat rapidly, therefore using them in building construction without adding and heating insulation or air-conditioning will be uncomfortable for the human comfort, on contrary of that in case of using a primitive construction material like clay or stone which characterize by their high ability to store the heating energy for the surrounding environment moreover

*Corresponding Author: nassir.towfeek79@gmail.com

© The Authors, published by EDP Sciences. This is an open access article distributed under the terms of the Creative Commons Attribution License 4.0 (http://creativecommons.org/licenses/by/4.0/).

their huge slow in heat conductivity compared to the modern Tech. which are not thermally isolated. At the present time, there is a lot of packaging imported materials like from (China, Turkey) or local which are used in covering the buildings. Therefore, researchers noticed it necessary to develop the method of building and use modern materials to get the utmost usage from energy through improving the thermal properties of the construction materials in order to reduce the possible dissipated heat through it. But improving the thermal properties of the construction materials require new style of available local materials which are cheap and can accomplish the same purpose [1]

Thermal energy is transferred through the solid medium represented by the wall by thermal conductivity, and thermal energy conductivity is the very huge and profound forked study. So thermal energy conduction is very important in engineering application which depends on it. So this science occupied a large part in engineering mechanics and physics science which study it and need some application, for example in increasing heat transfer in a possible big way, like in evaporative cooling or in residential heating or even reject heat gained in electric and computers devices, whereas in other applications we need to maintain heat energy inside it, like for example in residential water heating or inside the houses in winter season [2].

There are a lot of researches and studies which discussed to analyze and study different types of packaging materials for exterior classical walls (imported and local materials). All of those studies aims is economic feasibility through saving more energy by using packaging materials with the capability of saving high thermal energy for the surrounding environment in addition to the great slow in its thermal conductivity with the comparison to the unisolated modern Tech.

Reference [3] Had carried out a study and made a comparison in thermal performance between modern and classical Iraqi building walls. Where results showed that the amount rate in load reduction when using interior packaging walls is ranging (1%- 12%) which is the highest comparison in a cording to in reducing the cost of packaging materials which didn't exceed (2%). and the study had declared the necessity of using the thermal isolation during wall covering for making Iraqi thermal properties wall within the international measurements.

Reference [4] also carried out a study about the wall thermal behavior (implemented currently and suggested for implementation) within the climate conditions for Baghdad city (North 33.2 degree latitude) with the presence and without the presence of the thermal isolation which is thickness (40 mm), and the study was carried out through the day work for one day (21) of July and for a wall faced to the west. The researcher was able scheduling the results, which included total thermal energy coefficient, the weight of each wall per area, and the difference in temperature with finishing interior wall materials and an air temperature of the designed room according to hours changing of day and their daily average.

Reference [5] Carried out a study about energy conservation by using Polyurethane material as an isolation material which characterized with a low thermal conductivity in comparing with other isolating material used, and through the laboratories tests the thermal conductivity was (0.024 watts per meter. Percentage ) whereas these results were assured through experimental tests, the difference was ranged between two values (experimental and measured ) ranged ( 5-7%) also the researchers study thermal loads of faculty of dentist in Tishreen/Syria university, without the thermal isolation and with thermal isolation by using Polyurethane , the results showed that using this isolating material(Polyurethane ) caused in decreasing in thermal loads ranged about 46% and therefore ensuring energy saving consumed with rate about of 46%.

Reference [6] Carried out a study to evaluate the artificial stone product, which one of the Iraqi local factories to produce it and compare it with natural Granit stone, to explore the involved material to produce this material and knowing their costs and their suitability for packaging. The researcher carried out the experimental test, which included (total absorption test, density broken criteria ), the following results were obtained: absorption (5.182 %, compression 2.1 N/mm$^2$, density 2159.92 Kg/m$^3$, broken criteria's 6.32 N/m$^2$). Those results was so far good and within the limits of American description for Granit stone which has the number C99-87 and C217-85 and C170-87 , and the researchers found out it was more close to half (B) according to Iraqi description which has the number 1387-1989 for the Granit stone in all tests except the compression test, whereas the thickness of the stone is 2.5 cm, therefore, the researcher depend on the cement tile gradual compression and it was successful in that test, and from the side of the cost it is half cost of the natural Granit stone.

The present study is an extension of previous studies. In the present research a thermal behavior comparison were held in the winter for several different packaging exterior classical walls ( local and ready- made imported ) which are used in packaging facades of Kirkuk / Iraq city or a suggested unused material as PCM (a paraffin wax), and therefore knowing the amount of impact using those materials for energy saving ( conservation in energy consumption ) in this study it was focused on making comparison between heat dissipation and energy consumption when using those materials ( classical and suggested ) and which of them have the large ability to withstand the different environmental condition. The study was carried out from the beginning of (January 2017 until the end of February 2017) which those months represent the cold months in the year (temperature is ranged from ($-2$ *to* 15 °C ) which are needed to heat the building in the winter season.

## 2 Practical part

In order to study the thermal behavior of different exterior wall packaging materials (classical and suggested), a practical test station was designed which simulate the process of heat dissipation for rooms facing the south side. A packaging materials were installed on the southern wall (building façade), whereas most buildings are surrounded by three walls of other buildings (to the both sides of the building and the rear side of the building), time of the study was carried out in winter and according to Kirkuk / Iraq city climate. this station is consisting of five rooms (50x50x50cm) five different packaging materials were installed on the exterior walls for each room a packaging material, in addition to the measurement and control unit. As it shown in " Fig."(1).

Testing rooms (five sample rooms) were designed and constructed and were placed in second floor in the building of air-conditioning and refrigeration department in Kirkuk technical college, on 4m high of ground level, in order to expose those rooms to be exposed to the environment all the day and faced to the southern side Kirkuk city with dimension (5ox50x50 cam) and in order to reduce heat transfer ( limited transferring ) through room spaces in five sides ( eastern wall, western wall, northern wall, ground and ceiling), the five sides had been packed with three layers of ( a, b, c ) as it shown in the " Fig." (2) as following:

a. Wooden layer type of (MDF) thickness of (18mm).
b. Layer of trapped air (inert gas) with thickness of (2cm).
c. Wooden layer of (MDF) with thickness of (4mm).

as for the southern wall (sixth) its consists of pure Aluminum with thickness (0.9 mm) of high thermal conductivity value in order to allow a large amount of heat to transfer through this wall (wall in process of study) which represent the source or the influential side in a big manner on the comfort level inside each room with comparison with other walls for the same room. Different Packaging materials were placed (classical and suggested) on the southern wall for each sample, as shown in " Fig."(3).

Types of packaging materials used in the current study, as shown in " Fig."(4):

a. Classical packaging materials, it was divided as follow, shown in " Fig."(4) The first pattern ($R_1$): installing (Limestone) with a dimension of (5x50x50 cm). The second pattern ($R_2$): installing Granit stone with a dimension of (2x50x50x cm). The third pattern ($R_3$): installing Ceramic packaging materials front side with a dimension of (1x 50x50 cm). The fourth pattern ($R_4$): installing kind of cork (Styrofoam) with dimension of (4x 50x50 cm).

The fifth / suggested material (R5): Installation packaging material type phase change material (PCM) called Paraffin wax. The wax installed adjacent to the southern Aluminum wall and was trapped with transparent glass of a thickness (4mm) to inform a rectangular basin made of transparent glass with a thickness of (2cm) only the side which is adjacent room to be from Aluminum as for the dimension of this a rectangular box its (2x50x50 cm) as it shown in " Fig."(5).

## 2.1 Control and measurement unit

An electronic control panel was constructed for the control and measurement unit and for each room has an independent control panel which consists of:

a) Temperature sensor placed by a socket on a programmable electronic panel in order to adjust temperature degree according to inside rooms requirement and sensor bulb was placed in the middle of the room from above on a high of (45cm), electronic panel is fed with electrical power of (12v), through an electrical depressor transformer to transform electrical power from (220 v) to (12v). The main function of the electronic panel is to control the electrical load's work. (electrical bulb of thermal wire which represent the heat source simulate the heating system inside testing room placed beneath the rear wall for the room in the middle, supplied with an electrical source of (220 v) and electrical power of (100 watts). Which when the temperature room reaches the demanded degree( for example 32 °C) the programmable electronic panel will command to stop (open) the electrical unit which supply the electrical load (turn off), whereas if the temperature is decreased below the demanding temperature then it will give a command to connect the electrical unit which supplies the electrical load (turn on ).

b) A four intake temperature measurement was used supplied with thermocouples type of [k] with a diameter of (0.3mm) to measure temperature degree in different spots for the testing room ( the exterior and interior surface testing southern wall (sixth wall)) those couples were calibrated according to the criteria and of a range of temperature (0 – 100 °C).

c) Power meter device multi-function with (watt/h) units was imported from outside of Iraq (USA) its purpose is for knowing the amount of energy per hour which will be consumed (spend) inside each room through a certain time (heating load) where measures each of (current in amperage units, voltage in volt unit, electrical power in watt units, energy in watt per h) the device is fed with (220v) electrical power. This device is consisting of inductive loop its function to measure the value of the electrical current which passes through it. The electrical unit was placed inside a box made of plastic kind of (PVC) with dimension of (7x1.5x1.5) as it's shown in " Fig."(6)

## 2.2 Control and measurement units work mechanism

The electronic circle is connected as it's mentioned above, for example, a temperature sensor is adjusted on (32 °C) inside testing room in the winter season (likely with residential air conditioning units). So when the temperature of the room reaches the demanded temperature the programmable board of the temperature degree sensor will issue a command to cut the electrical source which supplies the electrical bulb with the thermal wire (turn off), but in case the temperature is lowered from the demanding one the mechanism will connect the circle which supplies the bulb with electricity (turn on). The difference in degree between activation and shutting down the circuit was adjusted (2) degree. The feeding wire is passed to the electrical bulb to the negative side (-) from inside of inductive looping for the power meter, which will take its role and measure the passing electric current, and the power which the electrical load consumed (electrical bulb with thermal wire) and measure the energy which will the electrical bulb consumed in each hour in (watt/h), which is the amount of heating load in each hour the room consumed it.

# 3 Research plan

Plans to carry out experiments on the system for the interval from (the beginning of January 2017 until the end of February20117) which those months represent the cold months of the year temperature degree were ranged from (-2 to 15 °C ) which is needed to heat the buildings in the winter season.

Measurements in this research were carried out on stages:

**First stage**

In the first stage primer tested was conducted on the suggested design without adding packaging materials on the southern wall (layer of pure Aluminum with the thickness of 0.9 mm) to evaluate the performance and the possibility of the amendment in the design in order to reach the best design.

**Second stage**

This stage included studying the thermal performance and the amount of consumed energy in each hour for the five patterns in a similar working condition and then comparing the results obtains from those patterns (suggested and classical) with each other.

Tested were conducted for the stages in different days and in a uneven environmental condition ( sunny day, cloudy day, partial cloudy, rainy day, snowy day ) and number of tests became (30) test, all the reading were registered in every hour for all variables ( electrical current, voltage, electrical power, energy for each hour, temperature of the room and temperature of the surrounding area, temperature of the exterior testing wall surface with the interior wall of the room) all the tests begin at 9 am and continue to several hours in a day and several days to acknowledge the capability of each packaging material to store and lose the thermal energy moreover and the extension of ability to endure all the different environmental conditions.

## 4 Theoretical part

The wall has been analyzed using the accurate analytical method (Exact solution) in order to calculate the overall heat transfer coefficient (U) and the amount of heat dissipation (rate of heat transfer through the constructed placed area), to know the amount of power needed to supply inside the comfort zone [7], on the following way:

### 4.1 The heat transfer through a composite wall

The thermal properties of the structure of architectural building are accounted the most important basics which any air-conditioning system (heating and cooling) rely on it. The heat is transferred from inside the heated building to outside the same building in winter and in opposite way in summer by two ways through exterior structural covering material for the building. First of them is heat transferred through the walls of the air from one of its sides to the air of the other side, secondly is through leaking (infiltration) the cold air to outside the building. Therefore, if its demand to shrink the heat transfers we must improve the walls thermal properties and reduce air infiltration through the cracks and gaps through the doors and windows. And the heat transfer occurs between two bodies their temperature by conduction or convection or radiation or more than one method in one time.

And the heat is transfer in conduction by direct contact between the molecular without displacing them from their places. And the first scientist who put the heat transfer conduction is J.B. Fourier.

$$\frac{dQ}{d\theta} = -KA\frac{dt}{dx} \quad \ldots (1)$$

Where ( $dQ/d\theta$) is the rate of heat transfer per time and (A) is the cross-sectional area which heat transfers occur through it and (dt) is the deferent in temperature which causes heat to transfer and (dx) the path which through the heat passes and the thermal conductivity. And it's seen that the rate of heat transfer inversely proportional to the thickness of the material, which it means heat transfer is reduced with the increasing the thickness of the material and vice versa. And if the temperature changed with time and location once, therefore, it's called unsteady state. But if the temperature changed only with the location it will call steady state and in the steady state, the amount ( $dQ/d\theta$ ) will remain constant and which its symbol is (q) with watt units. And the negative sign in equation (1) means that temperature degree is moving in a bounded direction from high-temperature side to the lowest temperature side, therefore the equation will be put on this way:

$$q = K\frac{A}{-x}(t_1 - t_2) \quad \ldots (2)$$

Where (A) is in square meters, $(t_1 - t_2)$ is the temperature difference in Celsius degrees and $(K)$ is the thermal conductivity of the material with $(W/m.°C)$ units.

The idea of thermal resistance is used to find heat transfer across the composite walls, where equation (2) can be rearranged as follows:

$$q = \frac{(t_1 - t_2)}{x/KA} = \frac{1}{R}(t_1 - t_2) \quad \ldots (3)$$

Where $(x/KA)$ is called thermal resistance with $(°C/W)$ units.

An experimental test was conducted for the classical packaging materials used in this research in the Kirkuk/ Technical Institute laboratory (Laboratory for measuring solid materials thermal conductivity) Japanese source, in order to reach to the real value of the thermal conductivity (K) for each material and then making comparison for those values and dependent values in books and previous sources.

Table (1) indicated values of thermal conductivity of the classical packaging materials for exterior which is trusty in the sources and previous research [3,7], also the values thermal conductivity of packaging materials during the laboratory test. The value of the thermal conductivity of the suggested packaging material for the sixth pattern (a wall of Aluminum layer with thickness of 0.9mm + Paraffin wax packaging material with a thickness of 2cm + glass thickness of 4mm) as following: the value of pure Aluminum thermal conductivity layer is $(202 W/m.°C)$ and for the regular glass is (0.78 $W/m.°C$ ) [2], and for Paraffin wax is $(0.21 W/m.°C)$ [8].

And compound wall case (structural wall consists of several materials) whereas the heat is flowing in constant rate in the steady state through the neighboring wall. Also there is an additional resistance resulted from a thin fluid layer (Air adjacent to the neighboring wall from inside and outside) this semi-static layer try to stick to the wall and hinder heat transferring and thickness of this layer depends on the condition of the heat transfer with load naturally of forced and also depends on the roughness and the inclination of the surface. And it was the Coefficient of heat transfer through the film thinks layer indicated with the letter $(f)$ and its unit in $W/m.°C$ and its thermal resistance as follows:

$$R_f = \frac{1}{f} \quad (4)$$

As customary when the heating load need to be calculated a general value is given for the coefficient $(fi)$considering the inside air is static, and giving two values each of them for the cretin wind velocity which represent winters condition and the second for summer season and those values are

Static air
$fi = 9.37\ W/m^2.°C$

In the winter – wind with velocity $24\ km\ /\ h$
$fo = 34.1\ W/m^2.°C$

In the summer – wind with velocity 24 $km\ /\ h$
$fo = 22.7\ W/m^2.°C$

And now it is possible to write a general equation for thermal resistant for any compound wall contains thermal resistant for the air layers (inside and outside the space) and the homogenous and not –homogenous for walls material as following:

$$R_t = \frac{1}{f_i} + \frac{x_1}{k_1} + \frac{x_2}{k_2} + \cdots + \frac{x_n}{k_n} + \frac{1}{f_o} \quad (5)$$

*4.1.1 The overall heat transfer coefficient*

The total inverted resistant known as the overall heat transfer coefficient in units $W/m^2.°C$ as following:

$$U = \frac{1}{R_t} \quad (6)$$

And therefore for to determine heating load (Heat dissipation) for the structural area ( walls, ceiling, and floor) within the American Society of heating, refrigeration and air-conditioning Engineers criteria (Ashrae), the amount of heating load for the structural area was calculated for the structural are ( thermal loss through compound structure ) as following:

$$q = U.A.(t_i - t_o) \quad (7)$$

Table (2) indicated values of the overall heat transfer coefficient and Heat dissipation through wall in process of study (sixth wall).

But thermal loss through the other structural area for the designed room (remain walls in addition to the ceiling and floor) is very small compared with the sixth wall which is assigned for study, the total thermal loss was found through other structural area (30 watts) , and the amount of the thermal loss because of the air infiltration through the cracks and holes around the doors and windows wasn't calculated because there isn't any doors or windows in the design room. It was found through results analyzing the followings as it shown in " Fig."(7), using suggested material (Paraffin wax) from the side view of reducing energy consumption the material exceed some of the packaging materials and also found the packaging materials (classical and suggested) reduces the amount of thermal loss through the exterior walls (reducing the heating load inside the comfort zone) and therefore reducing energy consumption inside the comfort zone.

In the practical side the design room was supplied the heat energy (heating load) through electric bulb with thermal wire (100 watts for each hour) this value was measured with a power meter multi-function. And the thermal loss (the consuming energy for each hour) for the design room with and without packaging material for the exterior wall and for all the five included patterns ($R_1, R_2, R_3, R_4, R_5$) for the study the results were analyzed.

# 5 Results and discussion

## 5.1 Results

In this study, the results were obtained through conducting (30) test and a long two months and according to Kirkuk /Iraq environment. The experiments was conducted on the system for the time interval (from the beginning of January 2017 until the end of February, 2017), those months represents the cold months in the year ( the temperature were ranged from -2 to 15Celsius degree) where heating is needed to warm the building, in the winter, tests were conducted on different days and uneven circumstances (sunny day, cloudy day, partially cloudy, rainy day, snowy day, tests were conducted through 24hrs day &night), tests were registered each hour for all variables which are (electrical current, voltage, electrical power, consuming energy for each hour, temperature inside room and temperature of the climates surrounding design room, temperature of the exterior and interior surface of the room) all the experiments start at nine am and continue several hours of a day and sometimes several days in order to know the capability of each packaging material on storing and losing the thermal energy and their extensibility on enduring all the different variable condition , the study included five patterns of design room without packaging materials for with them for the five patterns ($R_1, R_2, R_3, R_4, R_5$) and with same operating conditions, in the analytical solution it was calculated the amount of the thermal resistances and the overall thermal conductivity coefficient through the sixth designed southern wall for the fifth study patterns, and also the amount of the thermal loss through those walls . and through the practical results we found out, that using the suggested material (Paraffin wax) exceeds some of the classical packaging materials from the side of reducing the consumption of energy, and it was also noticed that using classical and suggested packaging materials reducing the rate of thermal loss through the exterior walls (reducing the demanded heating load inside the comfort zone), and therefore reducing the consumption of electrical energy disbursed inside the comfort zone.

### 5.2 Discussion

Fig. (8) illustrates the relation between the time for each hour (from nine am until five pm) and energy consumed (watt per hr.) inside the five design rooms (R1, R2, R3, R4, R5) and with same operating condition, and the time of study was a sunny day. And through the practical results it was noticed that the pattern (R2) is the least energy consumed and this is because of the falling solar radiation impact on the polished Granit stone surface which in turn leads raise its temperature (heated it) and therefore a reverse conduction occurred in the direction of the thermal energy from outside to inside (adding a thermal load inside the room), and because of the chemical composition of the stone which makes him thermal energy reservoir in a very good way, therefore the designed temperature degree inside the room will be semi –stable (at 32 °C) and this, in turn, leads to reduce the consuming of energy inside the room. Following it in less consuming energy the pattern (R5) which includes the packaging material (Paraffin wax), and that is because of the falling solar radiating impact on Paraffin through the glass (absorbing thermal energy from reaching solar radiation on the surface of the Paraffin) and in turn, leads to change its phase gradually (phase change) from the solid state to liquid state (storing a potential energy) and therefore a reverse transferring occurred in the direction of the energy from outside to inside (adding heating load in the room), therefore the design temperature (at 32 °C) inside the room will be semi- stable which it means reducing energy consumption inside the room. Following it in less energy consumption the pattern (R4) which has a high thermal isolation of Styrofoam, and after it the pattern (R1), but the highest energy consumption is the pattern (R3) because of the highly thermal loss through packaging material (Ceramic facade) which have a very high thermal conductivity.

Fig. (9) illustrates the relation between the time for each hour (from nine am until five pm) and energy consumed (watt per hr.) inside the five design rooms (R1, R2, R3, R4, R5) and with same operating condition, and the time of study was a partially cloudy day. And through the practical results it was noticed that the pattern (R2) is the least energy consumed and this is because the intermittent effect of the falling solar radiation impact on the polished Granit stone surface which in turn leads raise its temperature (heated it) and therefore a reverse conduction occurred in the direction of the thermal energy from outside to inside ( adding a thermal load inside the room), and because of the chemical composition of the stone which makes him thermal energy reservoir in a very good way, therefore the designed temperature degree inside the room will be semi –stable (at 32 °C) and this, in turn, leads to reduce the consuming of energy inside the room. Following it in less consuming energy the pattern (R5) which includes the packaging material (Paraffin wax), and that is because the intermittent effect of the falling solar radiating impact on Paraffin through the glass (absorbing thermal energy from reaching solar radiation on the surface of the Paraffin) and in turn, leads to change its phase gradually (phase change) from the solid state to liquid state (storing a potential energy) and therefore a reverse transferring occurred in the direction of the energy from outside to inside (adding heating load in the room ),therefore the design temperature (at 32 °C) inside the room will be semi- stable which it means reducing energy consumption inside the room. Following it in less energy consumption the pattern (R4) which has a high thermal isolation of Styrofoam, and after it the pattern (R1), but the highest energy consumption is the pattern (R3) because of the highly thermal loss through packaging material (Ceramic facade) which have a very high thermal conductivity.

Figures (10, 11, and 12) illustrate the relation between time in each hour (from nine am until five afternoons) and the consuming energy (watt per hour) inside the five design rooms (R1, R2, R3,R4,R5) and in the same operating condition, time of the study was during (cloudy day, rainy day, snowy day). Through the practical experiments we noticed that the less energy consuming is the pattern (R5) because of the effect of thermal energy transferred in conduction manner from the heat source inside the deign rooms through the Aluminum on the packaging suggested material (Paraffin wax) and therefore, the Paraffin wax gradually turning in from the solid phase to the liquid phase, and after changing his phase to a liquid it started to inform an adjacent thermal layer adjoined to the Paraffin wax for inside ( Aluminum layer side) and which acts as an isolating layer prevent heat transferring from inside toward paraffin wax, and then the temperature inside the design rooms then stabilize at (32 °C), following it the less energy consumption the pattern (R4) because of the high thermal isolation of the Styrofoam following the Styrofoam the pattern (R2) and for the same mention reason in fig (2) , following it the pattern (R1), but the more energy consumption is the pattern (R3) because of the high thermal loss through packaging material (Ceramic facade ).

## 6 Conclusions and recommendations

Through the results which were obtained, we concluded the followings:

1. It was noticed through analyzing the results that using the suggested materials (Paraffin wax) from the view of reducing energy consumption it exceeds some of the classical packaging materials, it is noticed that the consuming energy for each room of the five rooms along five days in a different environmental circumstances and uneven( sunny day, cloudy day, partial cloudy day, rainy day, snowy day )and the results as following: for the room(R1) which was included by Limestone as a packaging material , the consuming energy was (57453 watt per hr) for the room (R2) which included (Granit stone ) as a packaging material, the consumed energy was rated to (566397 watt per hr), for the room (R3) which was the Ceramic the facade was as a packaging material the consuming energy was (55853 watt per hr) for the room (R4) which was packed with (Styrofoam facade core) the consuming energy reached (52351watt per hour), but for the room (R5) which was the (paraffin wax ) as a packaging material the consumed energy reached ( 50140 watts per hr)

2. It was noticed also the using the suggested and classical packaging materials are help to reducing the thermal loose the exterior walls (reducing the missing thermal energy) and therefore reducing the consumption (saving) energy the consuming energy inside the comfort zone.

3. The possibility of using two phase material (Paraffin wax) as an isolated material inside the walls for areas and structure.

4. We can observe, that through the practical results there are some of the packaging materials for the exterior wall like Granit stone is very suitable for use in winter and cold areas because it added a thermal load for the comfort zone attributed to his ability for storing energy. Therefor we advise not to utilize it in the hot regions which are exposed to sunlight because it will add in the summer a heat load to the comfort zone.

5. It is advised to use Paraffin wax within the walls (which it means not exposed to the sunlight and the outside climate environmental condition ), for using the paraffin was in energy storing at the time of loads and rejected them to the surrounding area in shadow and night times, and also it can be used for thermal isolation.

## References

[1] A. A. Hassan. (2012) On the Environmental Effect Reduction On Internal Space Temperature By Internal Wall Covering, [Online]. 10(2), pp. 35-48. Available: http://www.iasj.net/iasj?func=fulltext&aId=63390

[2] J.P. Holman. (2010) Heat Transfer (10th ed.) [Online]. Available: http://www.academia.edu/6340170/Heat_Transfer_J.P_Holman_Ed.10

[3] A. K. Kalef, A.A. Wahab, and N.F. Hussein(2013) On the Comparison of the thermal performance between modern packaging materials and traditional for walls of buildings in Baghdad, Journal of Babylon University, [Online]. 21(4), pp. 1450-14630. Available: http://iasj.net/iasj?func=fulltext&aId=78323

[4] A. A. Hasan. (2012) On the Thermal Behavior of Present and Future Iraqi Constructed Walls(An Experimental Study), Anbar Journal for Engineering Sciences, [Online]. 5(1), pp. 140-164. Available: http://iasj.net/iasj?func=fulltext&aId=41148

[5] Y. A. Kabili " Energy conservation by improving the specifications of some thermal insulation materials" M.s. Thesis, Department of Mechanical and Electrical Engineering. Tishreen University. Latakia, Syria. 2009.

[6] I. F. Nassir.(2011) On the Evaluation of Artificial Stone (Local Tile) and Compared with Natural Stone, Iraq Academic Scientific Journals, [Online]. 24(8), pp. 17-35. Available: http://www.iasj.net/iasj?func=issueTOC&isId=1721&uiLanguage=ar

[7] K. A. Al-joudy "Principles of Air Conditioning Engineering and Refrigeration" NA Edition, Published by Philadelphia University, 1991, pp. 192-233.

[8] M. T. Chaichan, S. H. Kamel, A. N. Ajeely, (2015) On the Thermal Conductivity Enhancement by Using Nano-Material in Phase Change Material for Latent Heat Thermal Energy Storage System, Journal Saussure, [Online]. 5(6), pp. 48-55, Available: http://saussurea.org/wp-content/uploads/2015/08/Saussurea-Aug-04-2015.pdf.

https://doi.org/10.1051/matecconf/201816205026

Fig.1. Test station

Fig. 2. Testing rooms

Fig. 3 Southern wall (sixth)

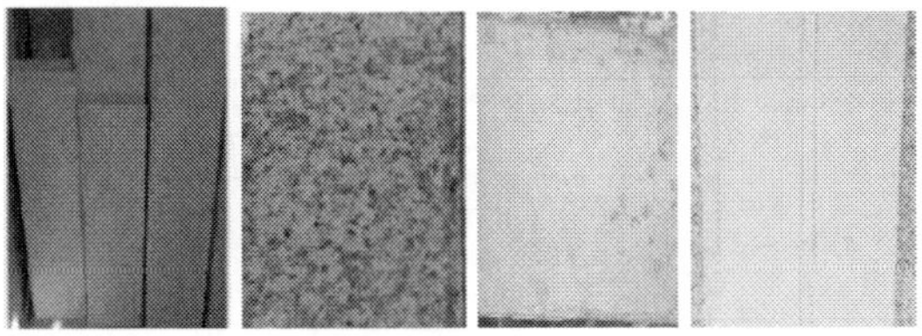

R1 R2 R3 R4

Fig.4 Classical packaging materials

Fig. 5. Suggested material / R5

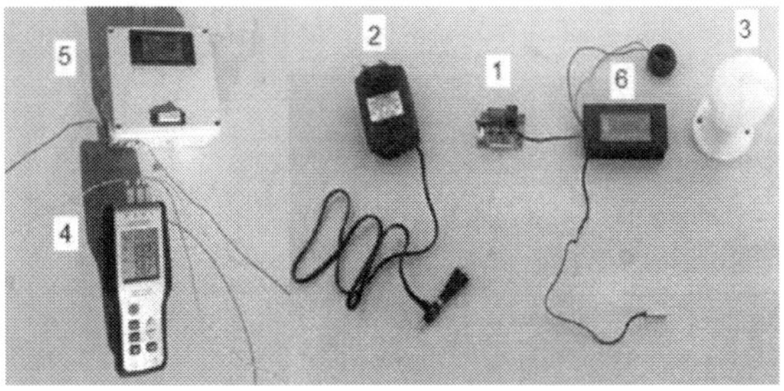


Fig. 6. Control and measurement unit

Table (1) the values of thermal conductivity (sources and previous research & during the laboratory test.)

| packaging materials for exterior walls | Thermal conductivity which is trusty in the sources and previous research ($W/m.$℃) | Thermal conductivity during the laboratory test |
|---|---|---|
| Limestone | ١,٣ | ١ |
| Granit stone | ٣ | ٢ |
| Ceramic | ١ | 1.2 |
| Styrofoam | ٠,٠٣٣ | 0.025 |

Table (2) indicated values of the overall heat transfer coefficient and Heat dissipation through wall in process of study (sixth wall)

| *No* | Type wall | Overall heat transfer coefficient ($W/m^2.°C$) | Heat dissipation through wall in process of study ($W$) |
|---|---|---|---|
| ١ | Model $R_0$ without packaging materials for exterior walls (a wall of Aluminum with thickness 0.9 mm) | 7.3 | 49.6 |
| ٢ | Model $R_1$ with packaging materials for exterior walls (+ wall of Aluminum with thickness 0.9 mm+ wall of Limestone 5 cm) | 5.37 | 36.29 |
| ٣ | Model $R_2$ with packaging materials for exterior walls( wall of Aluminum with thickness 0.9 mm + wall of Granit stone 2 cm ) | 6.84 | 46.23 |
| ٤ | Model $R_3$ with packaging materials for exterior walls( wall of Aluminum with thickness 0.9 mm+ wall of Ceramic 1 cm ) | 6.92 | 46.76 |
| ٥ | Model $R_4$ with packaging materials for exterior walls( wall of Aluminum with thickness 0.9 mm+ wall of Styrofoam 4 cm ) | 0.71 | 4.74 |
| ٦ | Model $R_5$ with packaging materials for exterior walls (wall of Aluminum with thickness 0.9 mm+ wall of Paraffin wax 2 cm + wall of transparent glass 4 mm) | 3.64 | 24.59 |

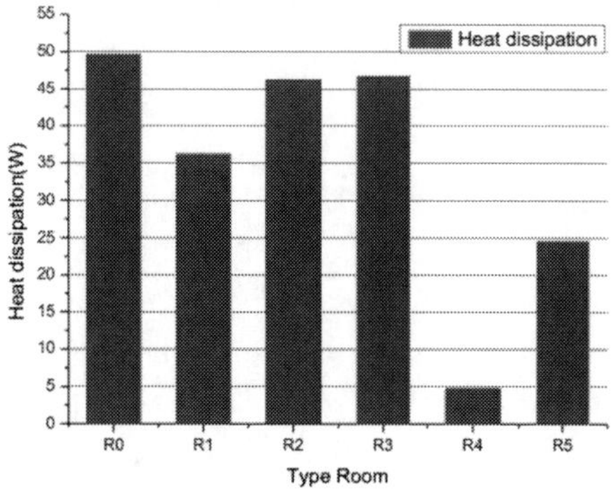


Fig.7. Energy consumption inside the comfort zone for the six patterns

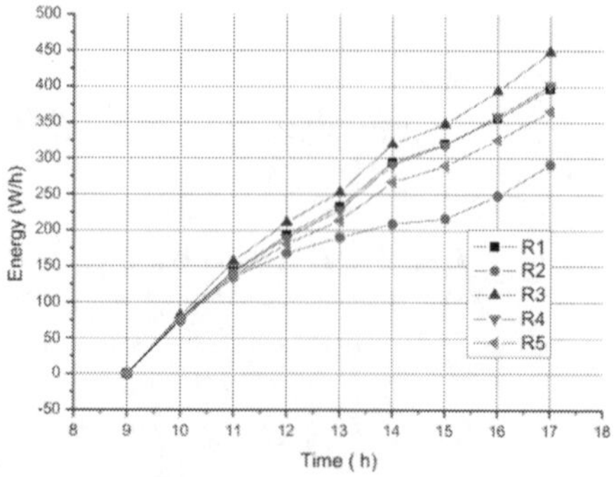


Fig.8. Relation between the time for each hour and energy consumed for a fifth patterns in a sunny day

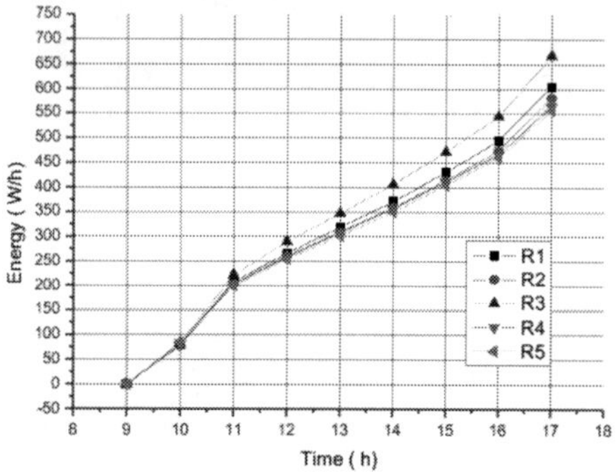


Fig.9. Relation between the time for each hour and energy consumed for a fifth patterns in a partially cloudy day

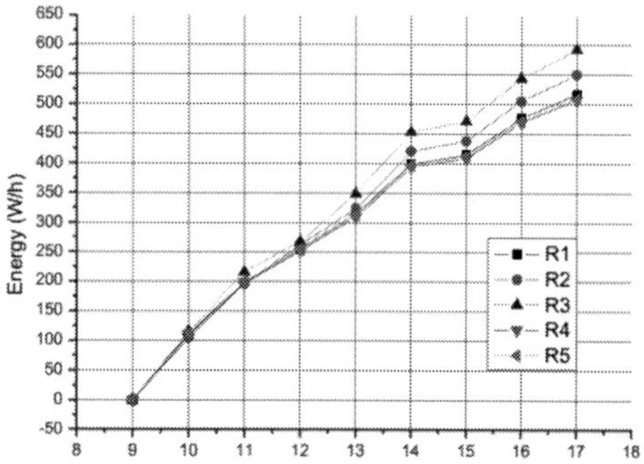


Fig.10. Relation between the time for each hour and energy consumed for a fifth patterns in a cloudy day

 https://doi.org/10.1051/matecconf/201816205026

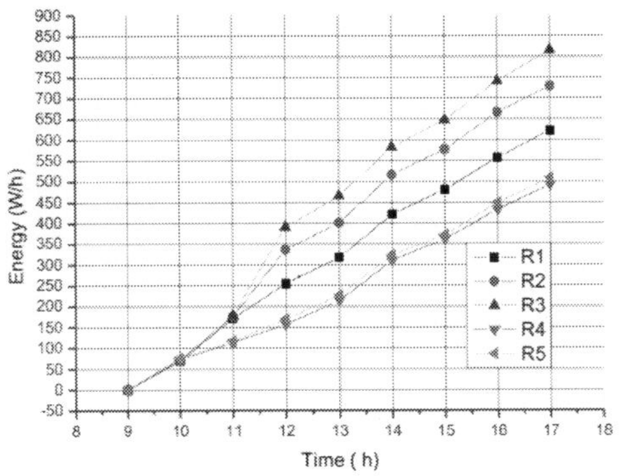


Fig.11. Relation between the time for each hour and energy consumed for a fifth patterns in a rainy day

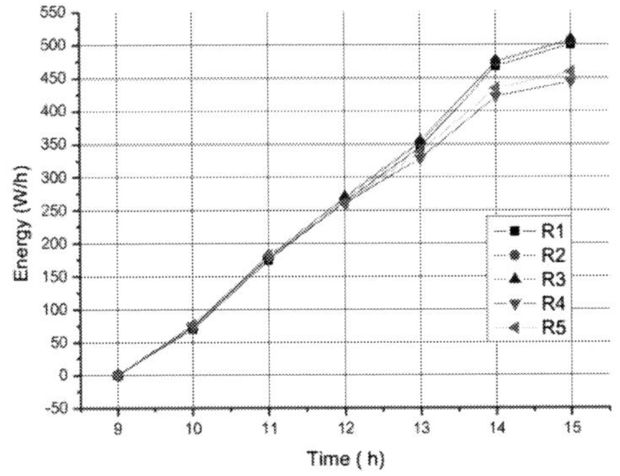


Fig. 12. Relation between the time for each hour and energy consumed for a fifth patterns in a snowy day

# Colored asphalt and street print are decorating paving in public spaces

*Khawla* Kawther*

Department of Building and Construction Engineering, University of Technology, Baghdad, Iraq

**Abstract.** Asphalt is the main material used to finish the internal and external streets of cities. Newly emerged in some countries, the use of colors which is added to the asphalt, it would be possible to use in the paving of this material widely in walking pavements, bicycle paths, parks and open spaces. By analyzing the international experiences as well as practical study, by a survey that was done by Iraqi professors in the specialized field of civil and road engineering, as well as architects, urban designer, and landscapers. Non-proliferation of Colored Asphalt and Street Print applications locally, this application has not been used in Iraq till now. There is a lack of knowledge on this subject, hence, the studies and research need to be carried out on this aspect, and find out their suitability for the local environment. The reasons that it is not used in Iraq are: cost of devices, staff, lack of experience, unavailability of materials, laws - culture- traffic engineering and lack of desire for development. It was found that, it is necessary to legislate engineering laws related to traffic engineering and educate vehicle owners about this new culture.

## 1 Introduction

Asphalt applications are very effective in residential and commercial evolution. Through colored asphalts can be obtained texture and style, as well as add safety on the roads, adding architectural and aesthetic values, less expensive than Al-Muqarnas, natural stone or other types of industrial flooring, changing the color of road surfaces reduces the temperature in urban areas.
There are two types to get the colored asphalt paving in Public Spaces; Colored Asphalt and Street Print. The colors can be added to the mixture of asphalt during the process of making, or decoration can be done during the tiling process, or the work of decoration can be done after the tiling process. There are many methods to get colored asphalt, using colorful aggregates, recycled glass, iron, slag, through these red, brown yellow, orange and green colors can be obtained in colored asphalt, and it can also add various colors to asphalt through the cladding of surface material, such as printing on the streets. These materials break through the gaps and act as adhesives to the asphalt.

In the 1990's, Highway Safety Research Center - North Carolina University was tested with color options for bike lane identification in Portland city, their study provided a broad analysis of material durability. These aspects are summarized in Table1.

## 2 Colored asphalt

Colored asphalt is used in many urban areas around the world to define bus and bicycle lanes. Such as London, New York, Auckland, Ottawa, and, Sydney, all of these cities were marked with bus corridors. See Figure 1. [1]

**Fig.** 1. Red bus lane Colored asphalt Manhattan -United States / W. Carry, (2012).

Their researchers expected that the most durable solution would be a dyed asphalt-wearing course; however, this was not tested, the implementation cost in Portland installed test sections of painted and thermoplastic colors and found that while the painted material wore off after the first winter, the thermoplastic proved to still be in good condition after one year". [2]

The asphalt uses have many benefits، but the black color can sometimes give a bad impression, Color asphalt was produced as an alternative to black asphalt in 1992. The new product is aesthetically pleasing also it maintains the asphalt qualities that are unparalleled. The new

* Corresponding author:Khawla_kareem@yahoo.com

© The Authors, published by EDP Sciences. This is an open access article distributed under the terms of the Creative Commons Attribution License 4.0 (http://creativecommons.org/licenses/by/4.0/).

product consists of two types, integral color for hot mix and integral color dry sealant mix. The color is added to the hot mix and mixed well, and then mix with the hot asphalt in the factory, Colored asphalt is delivered to the project site after it is fully colored. [3]

**Table 1**. Material Considerations for the City of Portland's Bike lane coloring research (Birk et al., 1999) [4]

| Material | Known Vendors | Approximate Materials Cost | Durability | Colors, Availability |
|---|---|---|---|---|
| Paint | A Local paint supply stores | \$.04-\$.10/ lineal Foot | Poor | Wide variety |
| Thermo plastic | Flint Trading (Premark®) 336-475-6600 | \$2.66/ sq. ft. | Good. Withstands Significant volume &Turning movements | Blue, red, Yellow, white |
| Methyl Methacrylat based marking | Morton Traffic Markings (Dura Stripe®) 800-835-3357 | \$.50-\$.60/ lineal Foot | Potentially good | Yellow, White, Red |
| Cold plastic | 3M (Stamark®) 800-362-3455 | \$4.50 sq. ft. | Durable with inlay, not as good with existing asphalt; Unlikely to hold up to heavy turning volumes | Blue, red, yellow. White |
| Dyed asphalt | Asphacolor® 800-258-7679 | Very costly. Must apply fresh, treated asphalt. | Excellent | Earth-tones |
| Imprinted and sealed asphalt | Integrated Paving Concepts (Street Print) 800-688-5652 | Costly. Must Apply fresh Asphalt. | Unknown | Earth-tones |
| Colored acrylic Coating | Traffic Safety Systems (Zebraflex®) 407-348-2624 | Unknown | Potentially good | Blue, Green, red, yellow |

## 3 Street print

"Street Print is an innovative property application approach, means the ground coverage of public space with asphalt decorates, similar to stone, bricks or other materials. Street coloring is done by using special phosphoric pigments, or by adding the dye to the asphalt mix, the goal is to give the function of corridors or spaces and categorize them with different colors, this method may be used for traffic and road engineering to determine traffic density and speed. This style is characterized by flexibility, strength, low cost and high durability.
"It does not need maintenance". See Figure 2. [5]

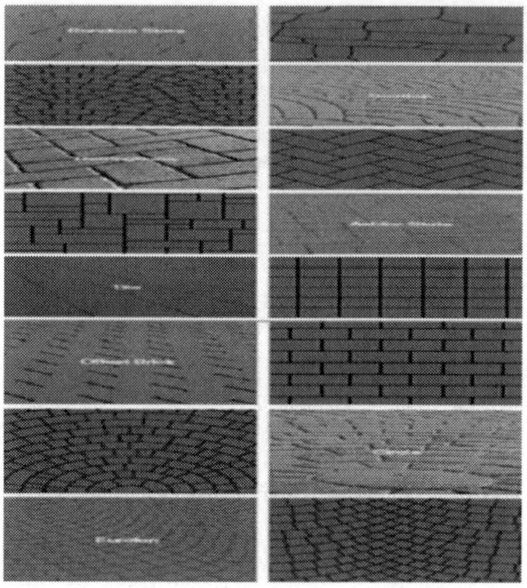

**Fig.** 2. Different colors & patterns.
**http://cjasphaltpaving.com/StreetPrint.pdf.**

Street Print Pavement Texturing is " a decorative paving system, which use special templates designed to work on a variety of decorative forms of asphalt، the installation of colored pieces on the surface called the connecting of the streets is intended to preserve the surface.In the end decorative architecture pattern will show through repetition form".[6]
"Printing of Street gives multiple possibilities and creativities and enables architects and designers to achieve more with less cost in comparison to brick or stone". See Figure 3. [7]
Printing of Street is fast, easy to be done, needs fewer workers, but requires special equipments for printing and installing decorative paving surfaces. [8]

Fig. 3. Application of street print,
**http://cjasphaltpaving.com/StreetPrint.pdf.**

### 3.1. Working steps

1. Heating the asphalt to SR60 temperature.
2. Printing by Woven Metal Molds on the Streets, the warm asphalt above the ground is pressed using the compressor plate and then lifted.
3. Apply Paint bonds, high performance street using printing by the highway sprayer. See Figure 4.

**Fig.4.** Working method
http://cjasphaltpaving.com/StreetPrint.pdf.

### 3.2. Benefits of street printing

**Re-decoration:** It can combine a variety of styles and colors, to produce a unique output and interesting designs. See Figure 5.

**Fig. 5**. Benefits of street printing
http://asphaltcementprinting.com/streetprint/WhyStreetPr.

**Endurance:** the base of asphalt paving is without joints or layers, so it is preventing the changes that are induced by time, weeds can not grow, its ability to resist various forces and environmental conditions. [9]
**Low cost:** Methods of effective synthesis, the availability of working costs.
**Maintenance**: Surface street printing has less maintenance requirements, thus costs less than others.
**Seamless Repairs:** coating systems enable easy repairs to "good as new" condition.
**Increase life**: Printing street coating systems prevents oxidation and asphalt base immunization which leads to a better appearance, durability. The possibility of creating diversity in design and patterns, the pictures below are some of the examples implemented and their locations: See Figure 6. 7. 8. 9. 10. 11.

## 4 Main characteristics

There are basic properties of architectural paving, that allow to evaluate sidewalk when using of stamped asphalt based on cost, aesthetics, application constraints, and environmental impact.[10]
**1.** Environmental Factors: Surface materials, unique characteristics and chemical compositions, have different installation; Geography plays a key role in selection for the best user installation [11]
**2.** Application Factors: downtime, multiple textures, hardness
**3.** Aesthetic Factors: initial appearance & design, aesthetic effects of wear & repair. [12]
**4.** Cost Factors: Large continuous coverage, cost of printing by concrete costs twice as much as asphalt, according to a tile company, it charges 5.25 \$- 6.25 \$ per sq. ft. While 11\$ per sq. foot for printed concrete [13]

**Fig. 6**. Pedestrian crossing in Thuringowa- Australia
**Fig**. **7**. Kern Bros. Drive, Thuringowa- Australia
http://asphaltcementprinting.com/photos/images/acp_p1.j

**Fig. 8.** Footpath outside Cairns, High School
http://asphaltcementprinting.com.
**Fig. 9.** Location of Cairns High School- cairns city
**http://asphaltcementprinting.com/photos/map10.html.**

**Fig. 10.** Pedestrian crossing - Car Parks in Townsville.
**Fig. 11.** location, Townsville City
**http://asphaltcementprinting.com/photos/map5.html**.

## 5 Projects with colored asphalt applications

### 5.1. Dammam, Saudi Arabia

Dammam city began an experiment in 2016, using colored asphalt, on the main streets of Dammam، a professor and supervisor of traffic safety، at the University of Dammam, stated that asphalt coloring has nothing to do with the power of the vehicle، and that the purpose of using it in Dammam is only aesthetic، he pointed that the cost of colored exceeded the normal، which reduced the chances of using it on a large area، and that the coloring should be using pigments that are highly resistant to vehicle movement, and weather conditions. See Figure 12.

**Fig 12.** the red asphalt plant in Dammam (Makkah)
**http://makkahnewspaper.com/article/138757.**

He explained that way occupy between 35 - 50% of the cities، when the temperature increases, it absorbs heat, and contributes to hot weather. The cold roads work in contrast to 30-50% of the sun's heat. While drawing that states applicable the red asphalt experience was designed to achieve three objectives; absorb the sun urban heat mitigation، where it can reduce the reflection of the sun's heat to 5% Beautification of city entrances. Identify specific paths, and distinguish them from the path of vehicles، such as bicycle, bus or pedestrian routes. And select two methods for asphalt coloring:
1. Mixing a chromosome with asphalt material processor to convert it to any color required.
2. Covering black asphalt layer coated with colored material, it is the most widespread method. [14]

### 5.2. Streetscapes in Scottsdale, Arizona

In 2003, municipal officials made the decision to use Street Print, construct, paving 12077.4 m2 to create, traffic movement domains and urban landscape.[15] The quality of the previous paving influenced in their decisions to use the street printing in a very large project. See Figure 13.

**Fig. 13**. Streetscapes in Scottsdale, AZ
**http://www.asphalt-impressions.com.**

### 5.3. Waterfront revitalization in Toledo, Ohio

The plot of land in the center of Toledo has been redeveloped; the pavements were paved with unique and beautiful walkways like the cobblestone paths.
Budget constraints made it impossible to choose a hand-laid pavers for the 3716.12 m2 area along the Maumee River. By using Printing of Street the project team was able to replicate the look of brick and cobblestone and completed the project within budget. See Figure 14.

**Fig. 14.** Waterfront Revitalization, Toledo, OH.
**http://www.asphalt-impressions.com.**

### 5.4. Campus at (North Carolina) Greensboro

Colored asphalt was used to enhance the appearance and image of the campus at the same time capitalizing the low cost of the application of the decorative pavement.
The University of North Carolina wanted to find a beautiful image of the campus streets in Greensboro. Chose to print the street to achieve the architectural integrated value therefore of the advantage of the low cost of printing from the street as decorative pavement. The project was 2043.87 m2 consisting of 9 intersections and 3 corridors. It used Brick Patterns as a statistical style. See Figure 15.
With Printing of Street the University was able to create the unique visual image as well as a fraction of the cost of pavers. [16]

**Fig. 18**. Bus ways finding -Ontario - S. Varamini, (2016).

Asphalt paving is achieving safe, resistant to environmental conditions, permanent, recyclable, and more economic, as far as maintenance, asphalt is the quickest and easiest to repair, more viable, marketable material. [18] See Figure 19. 20. 21. 22. 23. 24.

**Fig. 15.** Campus of the U of NC, Greensboro.
**http://www.asphalt-impressions.com.**

### 5.5. New Jersey walkways

Within the process of urban renewal, which was once the healthy landfill for city waste, 2229.67 m2 of hiking routes was paved. See Figure 16.

**Fig. 16**. Walkways for the City of Carteret, NJ

This project is located in the Arthur Hill waterway which leads to the Atlantic Ocean, The workers and technicians needed more than four weeks to complete this application. This project included a park, fishing docks and launch boats.

## 6 Advantages of colored asphalt

Safety: can be achieved easily by using colored asphalt in transportation line, because of the use of colors, it is possible to differentiate between bus corridors, pedestrian paths and bike paths. [17] See Figure 17. 18.

**Fig. 17.** bus lane in Seoul. Colors help ensure safe connections.

**Fig. 19.** shapes of light colors. Playground in Seoul. Colors designate safe environments-**www.lanxess.com**

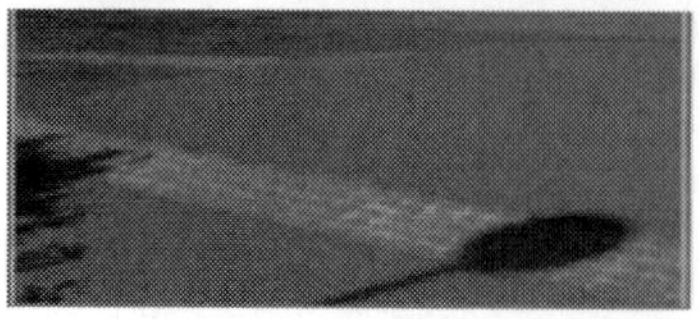

**Fig. 20**. Pedestrian zone in Marseilles. Gently curving **www.lanxess.com**

**Fig. 21** Marketplace in Ales. Open space and meeting point with South-of-France charm. **www.lanxess.com**

**Fig. 22.** Determination of directivity

**Fig. 23**. State Garden shows ٢٠٠٥ in Leverkusen: on colored trails through a world of fascination. **www.lanxess.com**

**Fig. 24.** "Forum" exhibition Adding color and cultural center in Barcelona. Color lends character to open spaces. **www.lanxess.com**

While it may be the beginning to get colored streets instead of blacks and grays, researchers at the US Department of Energy's "Berkeley Lab" laboratory has been trying to change the color of road surfaces to reduce urban temperatures. See Figure 25. 26.

**Fig. 25**. Lawrence Berkeley National Laboratory, Cold pavement, Case study, Researcher, J. Woods, Works on measuring Levels of solar reflection

**Fig. 26.** Device, White counter. Credit Lawrence Berkeley National Laboratory War Over Cool Pavements Heats Up, 21/3/2014 - LAM Staff, by Arthur Allen https://landscapearchitecturemagazine.org/2014/03/21/war-over-cool-pavements-heats-up/.

The researchers have devised grounds work the same refrigeration ceiling theory keep the air indoor cold by reverse sun rays، cold working ways unlike 30-50% of the heat of the sun compared to 5% reflected by new asphalt and 10-20% is reflected by old asphalt.
Currently the research group is working (Hit Island Group) on the conversion of a sector from the parking space of new temporary vehicles in the laboratory "Birklly Lab" to refrigerated floors to review its work, assess their effectiveness over time, The arena will provide an opportunity to test a range of coatings, which is placed directly on the cobbled surfaces [*19*]
There are many companies specialized in colored asphalt produce and develop of colored powder and strengthening pluralism copolymer synthetic resin emulsion liquid preparation, it has lasting and rich colors, excellent performance, longer life and the construction more convenient, inexpensive, both can improve the physical characteristics of asphalt pavement (with sufficient mechanical strength, a certain flexibility and ability to wake up deformation), dust, easy to clean and wash, easy maintenance and other characteristics, but also improve the ordinary black asphalt pavement monotony, with the surrounding public spaces with each other, coordination, more suitable for construction applications in the city. [20]
Colored Asphalt product used for the "new green building materials" than the bleaching of color asphalt healthier and environmentally friendly.
Specific applications in urban streets, squares, scenic areas, parks and tourist attractions such as roads may be by Colored Asphalt.
Some product of colored asphalt is a color enhancement powder and special multi-copolymer resin emulsion in the field with an electric mixer to fully stir into the slurry, the slurry will be poured into a good mixture of ordinary black asphalt pavement to be paved with a special scraper Evenly spread, shop scraping the thickness of control in 2-3 mm, flat surface, 24 hours after the surface layer of brushing protective agent, beautiful color asphalt pavement can be presented in front.

## 7 Practical study

To measure Available possibilities, The survey was done by the Iraqi professors in the specialization field of

Civil Engineering and road engineers, because they are dealing with the uses of asphalt in paving, as well as architects, urban designer, and landscaper, where 40 engineers were interviewed، the questionnaire survey was prepared and its objective were based on the hypotheses of this study, in order to prove those assumptions، analysis، and try to propose appropriate solutions، and to indicate the relationship between available resources، and appropriate alternative.

• At first data like gender, certificate, specialization, experience were collected: The results were as follows. See Figure 27. 28. 29. 30.

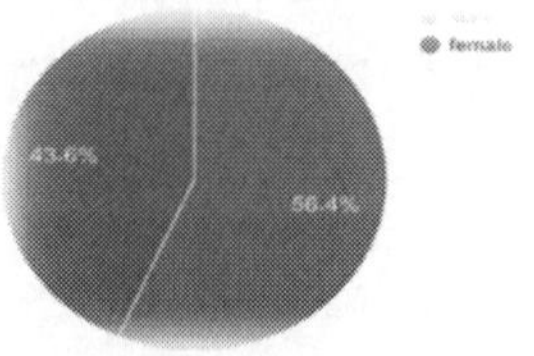


**Fig. 27.** The Gender

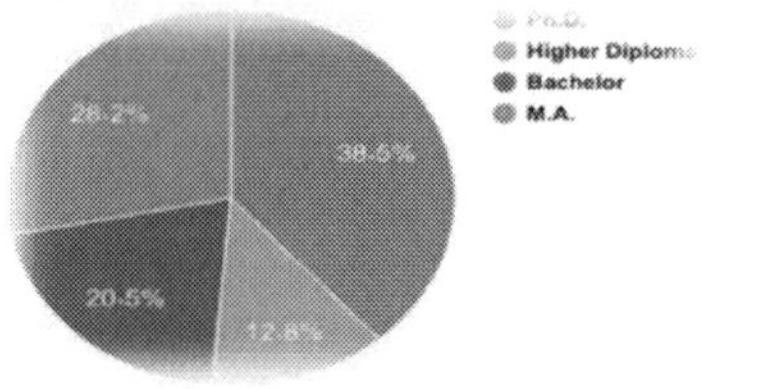


**Fig. 28.** Certificate

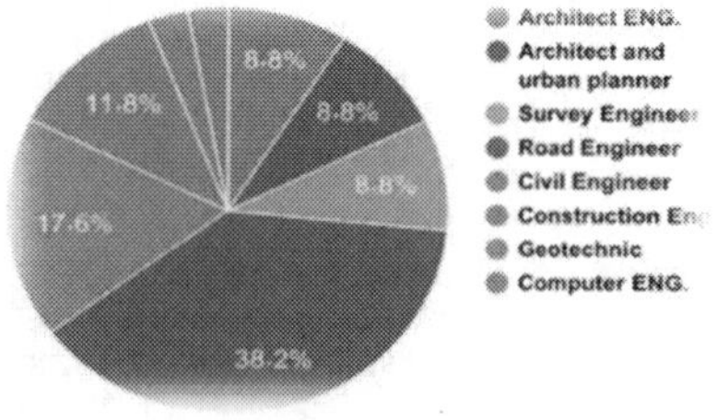


**Fig. 29.** Specialization

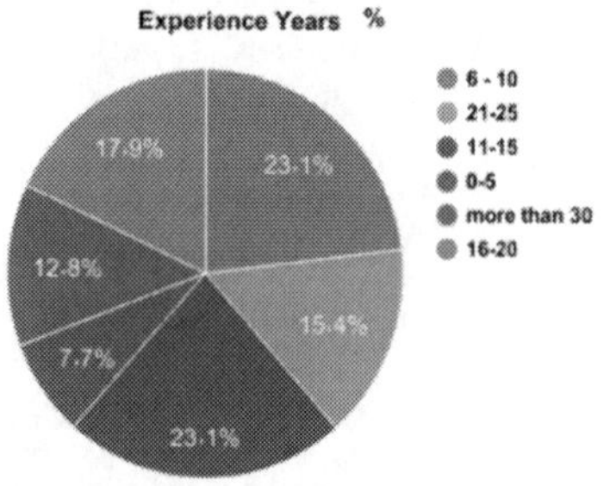


**Fig. 30.** Experience Years

The best building material used for (open space pavement), Paving by colored asphalt with decorations 22.5%, Paving by colored asphalt without decorations 20%, see Figure 31.

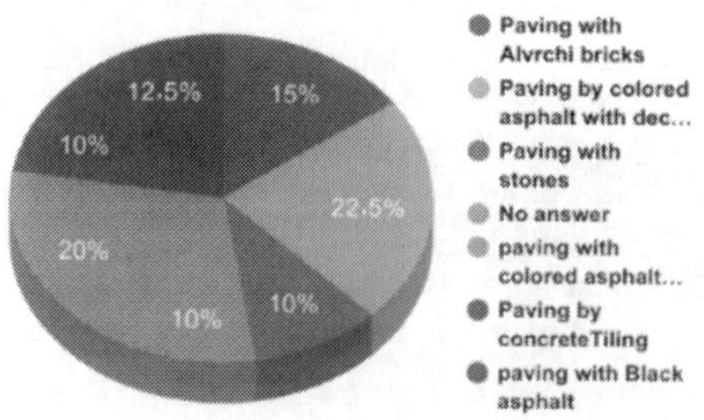


**Fig. 31.** best building material used for (open space paving)

❖ The extent of knowledge about colored asphalt applications only 17.9% they know, see Figure 32.

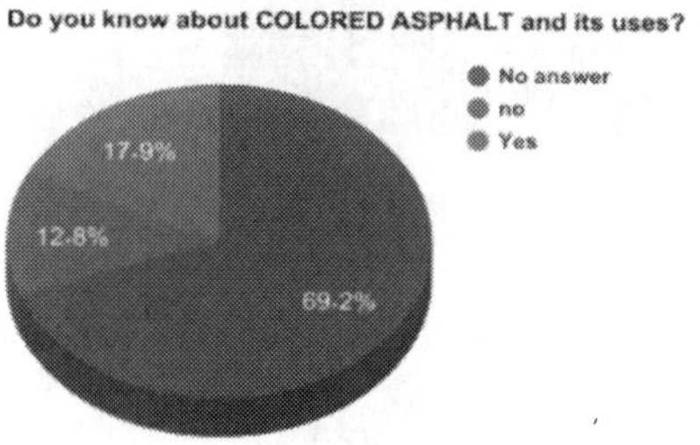


**Fig. 32.** Knowledge about colored asphalt.

❖ Only 25% they know Street Print applications, see Figure 33. this application not used in Iraq till now. See Figure 34. 35.

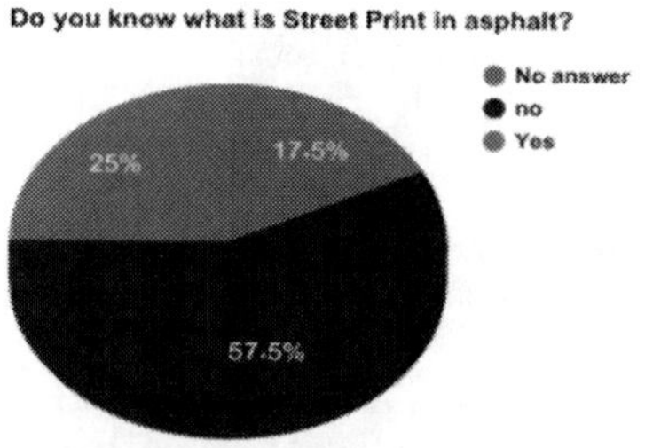


**Fig. 33.** Knowledge about Street Print.

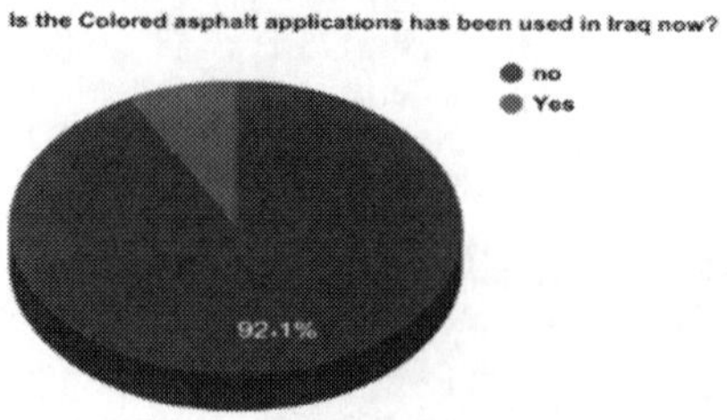


**Fig. 34.** Colored asphalt not available in Iraq

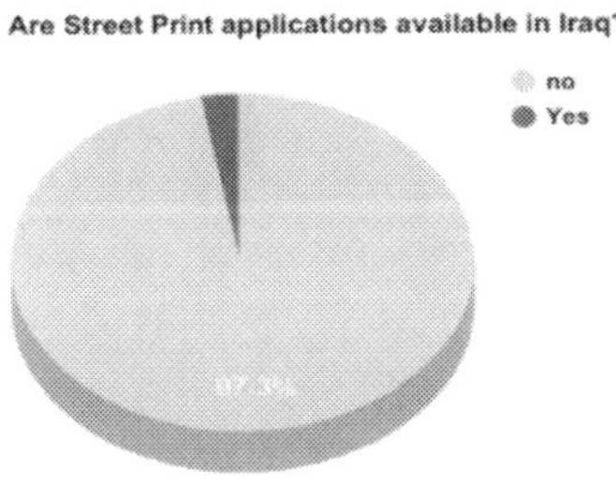


**Fig. 35.** street print not available in Iraq

❖ The meaning of street print by asphalt:

When asked about what ***street print*** means to the Engineers in this survey, the percentage of responses was 25.6 %, see Figure 36. their answers were as follows:

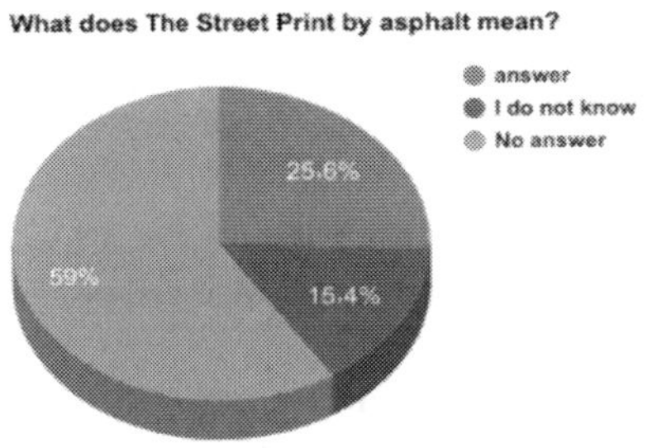


**Fig. 36.** The meaning of the street print

- Make decorations on asphalt no scratches similar to stone or bricks or other materials.
- It means a street covered with asphalt.
- Street coloring using special phosphoric pigments, or add the dye to the asphalt mix, The goal is to give the function of corridors or spaces and categorized them with different colors. This method may be used for traffic and road engineering to determine traffic density and speed.
- Adding different paints and colors, not an aesthetic addition to the paving, or may mean a name for a particular region or for a particular effectiveness, or may be added by different colors and this adds aesthetic and vitality and joy.
- Use engravings and colors to decorate the forms of paving asphalt.
- Using colored asphalt mixtures.
- The meaning of street printing is the work of two or three-dimensional graphics. The correct term is street coloring, which means the work of a specific and distinctive road surface layer (including concrete paving and asphalt) by adding dyes, or materials colored or additions that cause the paving surface to appear in a certain color after execution.
- Changing the street color with other colors for safety purposes and traffic regulation.
- Making decorations on the street, work in different forms and patterns.

❖ The best color which is preferred less absorption of heat is an important consideration:

Depending on the location of use of the road , white can be used in some methods, flexible pavement, and low-speed, and high-load, while the yellow color is preferred to distinguish the corridors of some roads, and determine their use, it was for buses, bicycles or pedestrians, it is not preferable to use white. For foreign expressways, because it causes a high reflection of the light that coming cars from the opposite direction and it can result in serious conditions, if the process of blocking those lights is not observed from the other direction. See Figure37.

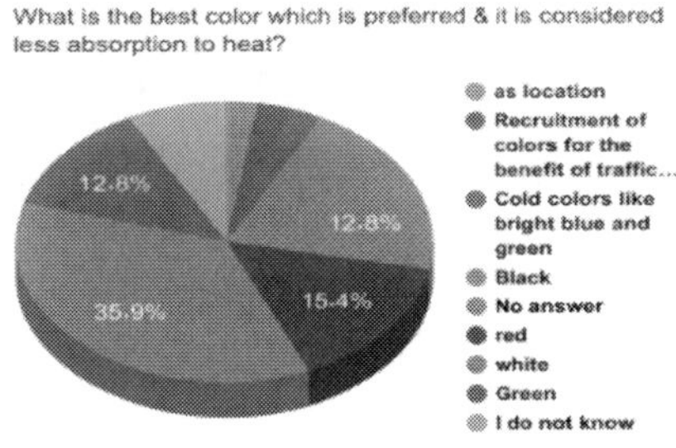


**Fig. 37**. best color- less absorption of heat.

The methods to add colors to asphalt emulsions:
Various answers: see Figure 38.

- There are specialized companies in this field
- During the manufacture of asphalt mix by adding it to the prime coat, it is a collection of compositions that are not mixed with each other.
- Add colors to the asphalt emulsions as desired and available.
- After the paving process, it can add color to the roads or you can also, Add color during the mixing process.
- During the mixing process asphalt, add certain materials.
- There are many aspects on which the desired color depends First, on the type of paving, (Solid or flexible), Second, in addition to the time period required, to resist atmospheric factors, and climate change and weather (rain, snow, temperature, etc.)

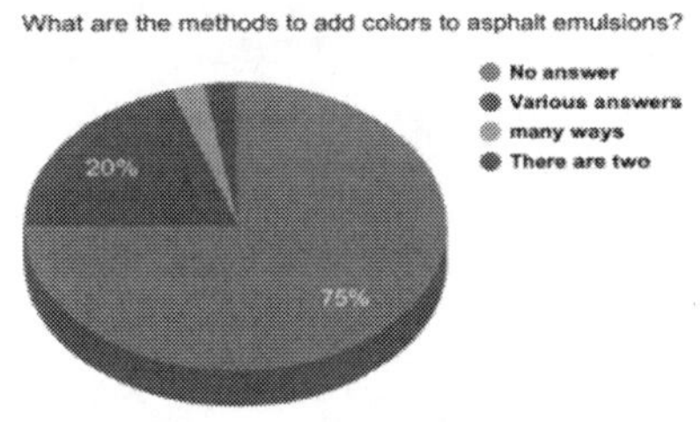


**Fig. 38.** Various methods to add colors

❖ Adding colors increases safety. See Figure 39.

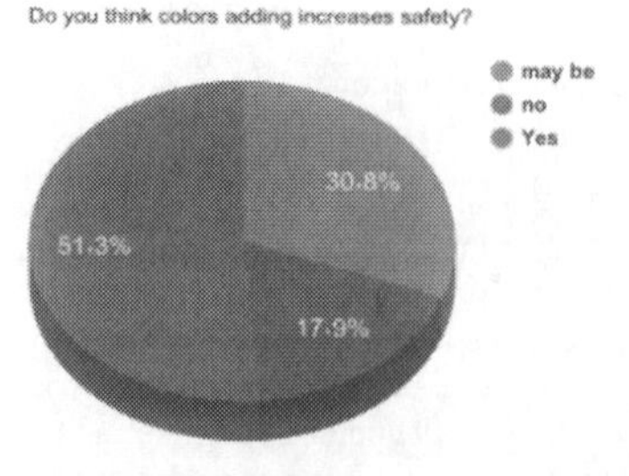


**Fig. 39.** Colors, adding increases safety

❖ There is a relationship between the colors and city temperatures that the colors increase temperature. See Figure 40.

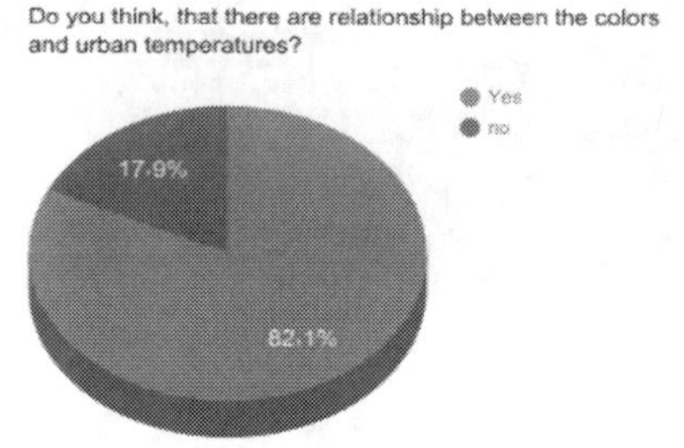


**Fig. 40.** Colors decreased temperature

❖ There are refrigerant colored and asphalt surface treatment & floors cooling, see Figure 41.

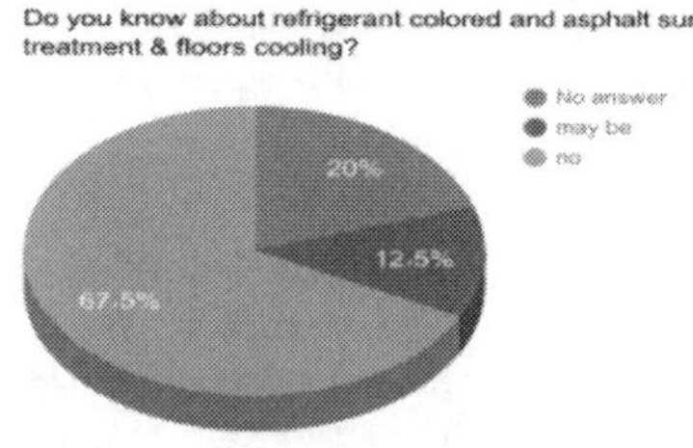


**Fig. 41.** Knowledge about refrigerant colored

❖ Using colors in order to determine the type of passing vehicles, such as a designated bus street, another passage for small vehicles and so on. See Figure 42.

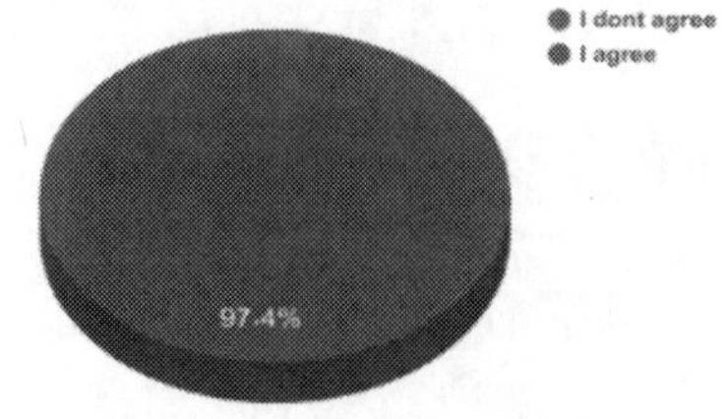


**Fig. 42.** Colors & type of passing

❖ The relationship between the road lighting and the colors used for the ground, see Figure 43.
❖ Solar heat effect on the color of asphalt, negatively, see figure 44.
❖ Pavement materials which are less expensive locally in Iraq), see Figure 45.

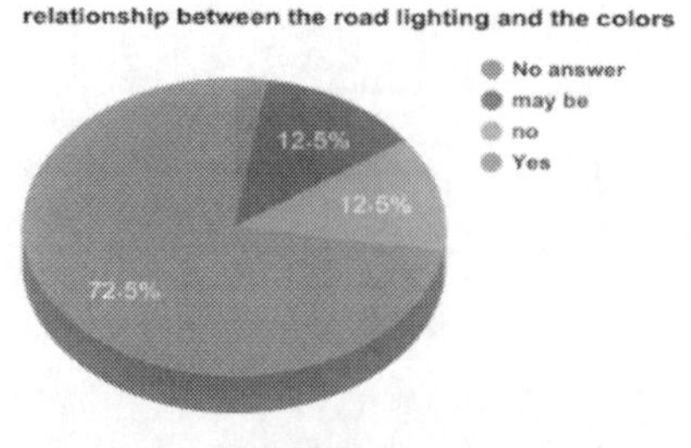


**Fig. 43.** Colors & road lighting

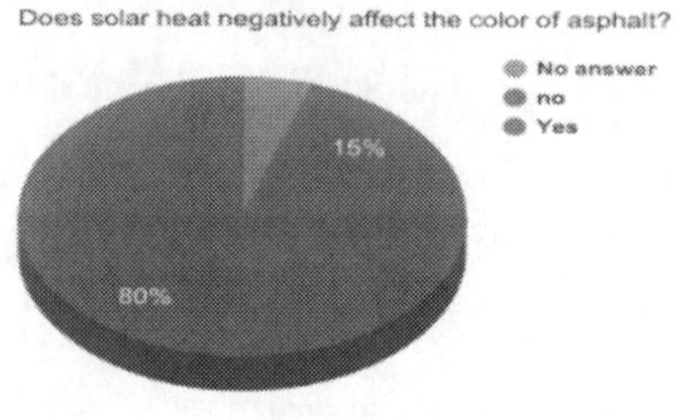


**Fig. 44.** Lighting & Solar heat effect

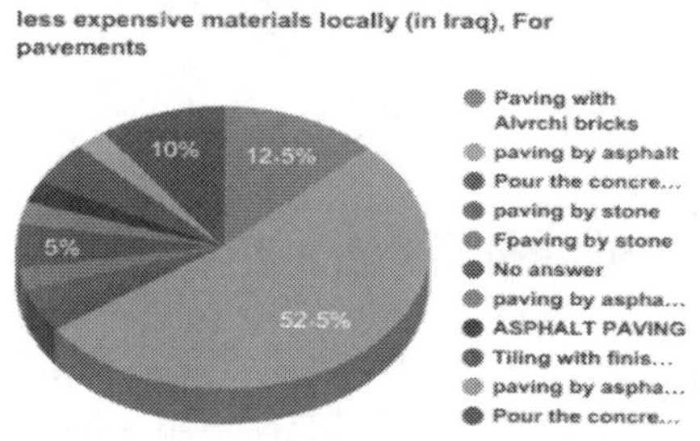


**Fig. 45.** Less expensive Pavements Materials

❖ The best material properties of water absorption in paving, flooring in public spaces:
❖ To resist the absorption of water and should not reduce the permeability of black asphalt or stained (can be used as a type of no pigment or materials and no colored spaces that increase the resistance of water permeability), if it is intended, no more preferred in terms of speed of discharge and permeability within the layer and the stone is the best, see Figure 46.

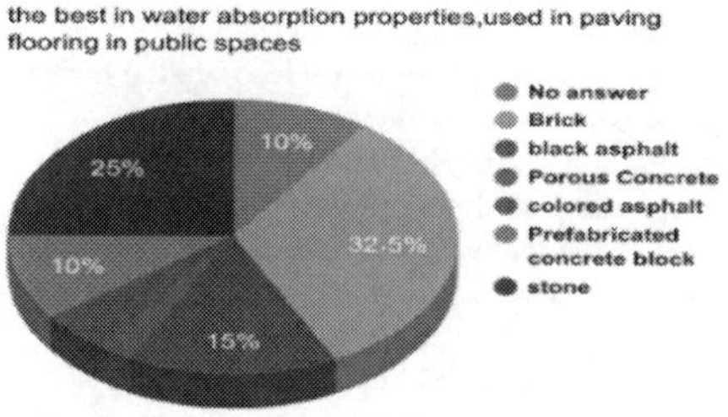


**Fig. 46.** Water absorption Properties

❖ The Rain changes the asphalt color negatively, see Figure 47.

 https://doi.org/10.1051/matecconf/201816205027

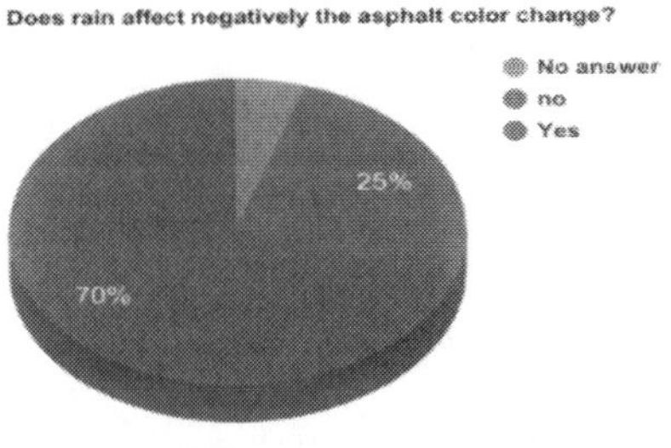


**Fig. 47.** Rain Effects

❖ The rough texture is more preferred in the floors of public spaces on this basis, the asphalt 32.5% and the stone 32.5%, the best floor surfaces, see Figure 48.

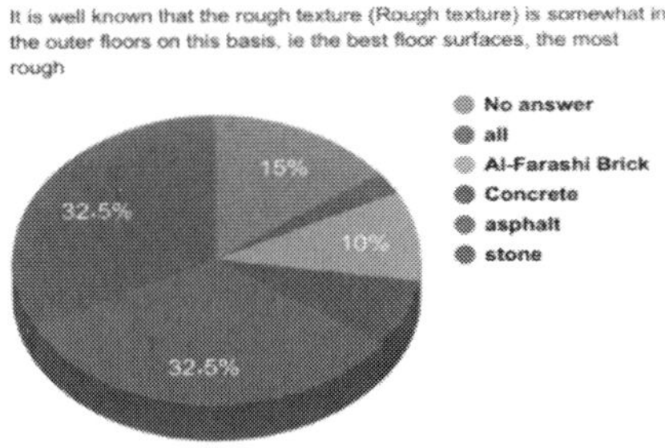


**Fig. 48.** Best material as a texture

❖ The reasons that color asphalt is isn't in Iraq, see Figure 49.

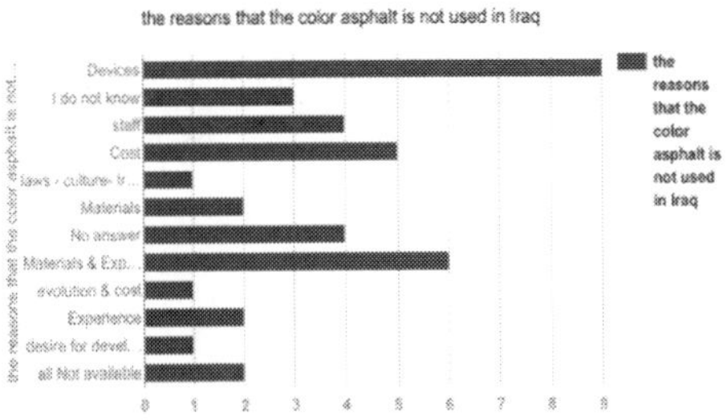


**Fig. 49**.The reasons that color asphalt is not used in Iraq

❖ Reasons that street print isn't used in Iraq see Figure 50.

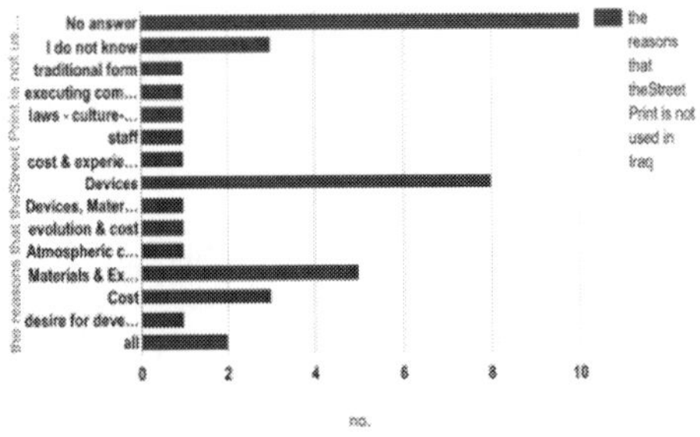


**Fig. 50.** The reasons that street print is not used in Iraq.

❖ Before using colored asphalt, it is necessary to legislate laws and engineering equations related to traffic engineering and educate vehicle owners about this new culture.

❖ There is not enough awareness among the majority of citizens about this new development, because it is essential to provide road services and infrastructure, to educate the citizen before thinking about applying such methods.

## 8 Conclusions

- It is necessary to legislate laws and engineering equations related to traffic engineering and educate vehicle owners about this new culture.
- The majority of citizens is not aware about this new development, it is essential to provide road services and infrastructure, to serve the citizens.
- Colored Asphalt and Street Print has not been used locally in Iraq until now, there is a lack of knowledge on this subject, the need to provide studies and research on this aspect, and find out their suitability for the local environment
- The best building material used for pavement of open spaces, is Paved with colored asphalt with decorations.

The definition of street print by asphalt:

- Means: street covered with asphalt, using colored asphalt mixtures.
- Making decorations on asphalt that are similar to stone or bricks or other materials.
- Street coloring using special phosphoric pigments, or adding the dye to the asphalt mix, the goal is to give the function of corridors or spaces and categorize by color. This method may be used for traffic and road engineering to determine traffic density and speed.
- Adding different paints and colors, is not an aesthetic addition to the paving, or may mean a name for a particular region or for a particular effectiveness, or may be added by different colors and this adds aesthetic, vitality and joy.
- Use engravings and colors to decorate the form of paving asphalt.
- The meaning of street printing is the work of two or three-dimensional graphics. The correct term is street coloring, which means the work of a specific and distinctive road surface layer (including concrete paving and asphalt) by adding dyes, or colored materials or additions that cause the paving surface to appear in a certain color after execution.
- Change the street color with other colors for safety purposes and traffic regulation.
- Making decorations on the street, work in different forms and patterns.
- The best color which is preferred & considered is the one which absorbs less heat:
- Depending on the location of the road, white can be used in some methods, flexible pavement, and low-speed, and high-load. While the yellow color is preferred, in distinguishing the corridors of some roads, and determine their use, it was used for buses, bicycles or pedestrians. It is not preferable to use

white, for foreign expressways, because it causes a high reflection of lights of cars coming from the opposite direction, therefore it is very dangerous.

❖ The methods to add colors to asphalt emulsions:

- There are specialized companies in this field, during the manufacturing of asphalt mix the colors by adding it to the prime coat.
- During the mixing process asphalt, add certain materials.
- Add colors to the asphalt emulsions as desired and available.
- After the paving process, we can add color to the roads or you can also, add color during the mixing process.
- After the paving process, it can add color to the roads or you can also, add color during the mixing process.
- There are many methods, of which depends on the desired color, First, based on the type of paving, (Solid or flexible), Second, in addition to the time period required, to resist atmospheric factors, and climate change, weather (rain, snow, temperature, etc.)
- Adding colors increases safety.
- Colors & types of passings: using colors in order to determine the type of passing vehicles such as a designated bus street, another passage of small motor vehicles and so on. There is the relationship between the colors and city temperature, some colors can increase the temperature.
- There are refrigerant colored and asphalt surface treatment & floors cooling.
- There is a relationship between the road lighting and the colors used for the ground.
- Solar heat affects the color of asphalt, negatively.
- It may be beginning to get colored streets instead of blacks and grays, change the color of road surfaces to reduce urban temperatures.
- Pavement materials of asphalt are less expensive locally (in Iraq).
- The best properties of the material considering its water absorption in paving, flooring in public spaces:
- Resist the absorption of water and do not reduce the permeability of black and colored asphalt, if it is intended, no more preferred in terms of speed of discharge and permeability within the layer and the stone is the best.
- Rain change the asphalt color negatively.
- The rough texture is more preferred in public spaces' floors based on this both the asphalt and the stones are preferred

❖ The reasons that color asphalt & street print is not used in Iraq:

- Devices not available
- Staff not available
- The cost
- Laws - culture- traffic engineering
- Materials not available
- Lack of experience
- Not to keep pace with development, lack of the desire for development.

## Acknowledgment

Acknowledgments to all Iraqi professors and architects and civilians engineers, who are shared their experience in this survey and practical study of this search.

## References

1 W. Carry *"Red Bus Lane Treatment Evaluation"*, Institute of Transportation Engineers Annual Meeting and Exhibit, California (2012).
2 M. Birk, R. Burchfield, J. Flecker *"Improved Safety through Enhanced Visibility"*. City of Portland, Office of Transportation, Oregon, (1999).
3 http://www.asphacolor.com/index_main.html.
4 S. Varamini, Technical, Economic and Environmental Evaluation, of Warm Mix Asphalt and Coloured Asphalt for Usage in Canada, Doctor thesis, Civil Engineering, Waterloo, Ontario, Canada, 32 (2016).
5Alternative Paving Concepts Inc. "Street Print Making Asphalt Beautiful". http://www.alternativepaving.com/downloads/Product_Overview.pdf
6 http://asphaltcementprinting.com/questions/index.html
7What is Street Print. http://www.asphalt-impressions.co.za/what%20is%20streetprint.htm.
8 Ibid. 5
9 Asphalt Applications. *Construction and Surfacing of Footways and Cycle Ways Using Asphalt*, Information Sheet 11, Mineral Products Association, London, 1 (2009).
10 Blaine, *"Integrated Paving Concepts, The Paving Surface Selector,* a Guide to The Applications and Economics of Commercial and Municipal Grade Paving Solutions" Peace Portal. http://www.streetprint.com/ http://asphaltcementprinting.com/publications/384.pdf
11 Ibid.
12 Mclaughlin, Mary-Beth, *oncrete or Asphalt, Choice is on Surface*, The Blade. All rights reserved, 2003. http://www.toledoblade.com/Real-Estate/2003/07/20/Concrete-or-asphalt-Choice-is-on-surface.print.
13 Asphalt Applications. *Decorative and Coloured Finished for Asphalt Surfacing*, Information Sheet 4, Mineral Products Association, London (2009).
14 http://makkahnewspaper.com/article/138757.
15 W. Frank, Colored Asphalt Is Decorating Paving Public Spaces", Asphalt Institute, Triangle logo and Asphalt magazine are all trademarks of the Asphalt Institute, 2017.
16 K.K. kawther, The Visual Image Quality for the Future City, the Engineering and Technology Journal, University of Technology, **33**, 1-36, 2015.
17 Lee, H., & Kim, Y. *Laboratory Evaluation of Color Polymer concrete Pavement with Synthetic Resin Binder for Exclusive Bus Lanes*. Transportation Research Record (1991), (2007).
18. S. Varamini, M. Farashah S. L. Tighe, *Coloured*

*Asphalt Bus Rapid Transit Lanes in The Regional Municipality of York*: Integrating Laboratory Performance Testing Into Sustainable Pavement Asset Management, Resilient Infrastructure, (2016).

19http://www.mbc.net/ar/programs/mbc-news/articles/ 2012/9/12.

20 Asphalt Applications, *Construction and surfacing of car parking areas including private drives*, Information Sheet 1, Mineral Products Association, London, (2009).

21. K. K. kawther, M.Sc. thesis, University of Technology, (2014).

22. NACTO, *Urban Street Design Guide*, National Association of City Transportation Officials, 2013.

# Iraqi architecture in mogul period

*Shatha* Hasan[1,*] and *Yasameen* Rasheed[1]

[1]Department of Architecture - College of Engineering, University of Baghdad, Baghdad, Iraq

**Abstract**. Iraqi architecture have many periods passed through it until now, each on from these periods have it is architectural style, also through time these styles interacted among us, to creating kind of space forming, space relationships, and architectural elements (detailed treatments), the research problem being from the multi interacted architectural styles causing some of confused of general characteristic to every style, that we could distinguish by it. Research tries to study architecture style through Mogul Conquest to Baghdad. Aim of research follow main characteristic for this architectural style in the Mogul periods on the level of form, elements, and treatments. Research depending on descriptive and analytical all buildings belong to this period, so from analyzing there style by, general form for building, architectural elements, and it architectural treatment, therefore; repeating this procedures to every building we get some similarities, from these similarities we can making conclusion about pure characteristic of the style of these period. Other side, we also discover some Dissimilar in the building periods, these will lead research to make what interacting among styles in this period, after all that we can drew clearly main characteristic of Architectural Style for Mogul Conquest in Baghdad

## 1 Introduction

There many periods through any civilization, every period must be have comment characteristics that we can recognize from it, the most important and significant characteristic that be permanent and durable with time, like physical structure, when we wont to research about any records to any periods, firstly we look on architecture of this period, so we can imagination period facts when we intersect documented information with building architecture of this period.

In this research we look for very important period in history of Baghdad, Mogul Period, many sources referring to this period as dark period, may be the hard circumstance clear some facts in this way, but there are other way referring to ability of society to live and resistance ant conditions, from that we have many evidence in many Iraqi Cities showing create building architecture in this period.

Research build it is methodology on clarify the building belongs to Mogul Period, and then try to comparing among them, to reach define common architectural characteristics that be occurring the main character of this period.

* Corresponding author: shatha_arch2010@yahoo.com

## 2 Selecting building

Research put criteria for selecting building depends on:

1. History of building must be clear.
2. Structure of building was obvious and without modification.

So, the buildings will be analyzed in levels as:

1. The general form
2. Architectural elements
3. Treatment

In this way, we can identify the characteristics of the era style, Table (1), shows selected buildings to apply comparative analysis among them.

Table 1: Buildings in Case Study

| Building name | Dates | Using |
|---|---|---|
| Omer Al Shrawrdy Shrine | 1334 | Shrine |
| Minaret Alaqoly Mosque | 1327 | Mosque |
| Minaret Alkhukafa'a Mosque | 1271 | Mosque |
| Merjan School | 1357 | School |
| Merjan Khan | 1358 | Motel |

© The Authors, published by EDP Sciences. This is an open access article distributed under the terms of the Creative Commons Attribution License 4.0 (http://creativecommons.org/licenses/by/4.0/).

## 3 Buildings descriptive

Research try in this paragraph to get general descriptive to all building we'll discuss it:

### 3.1 Omer Al Shrawrdy shrine

Shrine location is in historical city center of Baghdad city, Rusafa side, building nearly 1334; the shape is simple, rectangular, with conic tomb. The tomb based on square base, length of it about (4.9m), creating mosque shape beside the rectangular.

Figure 1: Omer Al Shrawrdy Shrine

#### *3.1.1 Architectural elements*

The building of entrance is having the same shape without modification, with writing ornament by brick. The tomb dome consist of (10) layers of Muqarnas, direction of it to inside. There are four opens in the dome, shape of them as vault, using for ventilation and lighting. Mosuqe minaret is having three lyres of Muqarnas, supporting Muezzin balcony, this minaret was built in the Ottomans era.

### 3.2 Manarate Alaqoly mosque

The Aqoly mosque located in the middle of Rasheed Street, nearing Hayderkhana mosque, built at 1327, then renovated at 1802, also after 50years renovated too. (Editors, 2013, p.250)

Figure 2: Minaret Alaqoly Mosque

Source: field survey

#### *3.2.1 Architectural elements*

Mosque minaret is having wide body, based on rectangular shape, with six layers of Muqarnas, minaret height was nine meter, and proportion of it is height to width is (1:3.6). (Sid Mahmoud, 1346, p.47). The entrance of minaret is simple without any ornament, and the basic building unit is brick, with white color due to coated by gypsum. (Editors, 2013, p.250).

### 3.3 Minaret Al Khukafa'a mosque

The minaret is having cylindrical shape, with huge base, and having two balconies. It was built at 1179, and then after it is damaged, renovated at 1271, in the time of Alaa Al Dein Aljouany in the Mogul era. (Sid Mahmoud, 1346, p.38)

Figure 3: Minaret Alkhukafa'a Mosque

Source: field survey

#### *3.3.1 Architectural elements*

The original height of minaret was 35 meter, with (8) meter diameter, the proportion of height to width is

(1:4). All faces of minaret body decorated with patterns of lettering as ornament by bricks, and we can recognize Koufi style in these letters. (Refa'at, 2014, p.82), the Muezzin balcony based on Muqarnas without decoration; just consist of composed four lyres of Muqarnas. (Tariq, 1982, p.95) main basic material was brick, (Yousif, 1982, p.397).

Figure 4: Muezzin balcony in the Minaret Alkhukafa'a

Source: field survey

### 3.4 Merjan School

It is called by Merjan according to persons who built it, Ammen Al Deen Merjan, at 1367; it is located in the last part of Rasheed Street, nearing Tigress River. (Sid Mahmoud, 1346, p.65)

#### *3.4.1 Architectural elements*

The building is consisting of two floors, it is height about ten meter, with (40) meter in width, and (18) meter long, it is proportion of width to height it was (1:4). Building was divided into two parts, school and mosque, having central court, and three big domes. (Yousif, 1982, p.520).

Figure 5: Merjan School and Mosque

Source: field survey

The school have huge gate in it is entrance, decorated by cylindrical ornament column, it is shape tapered vault. The main dome above shrine is based on long cylinder of bricks, decorated by geometric pattern from brick also. (Khalid, 2013, p.142)

Figure 6: Gate of Merjan Mosque

Source: field survey

The relationship between mosque and school are through three big gates, with tapered vaults, (Khalid, 2013, p.180), the middle one is the bigger from the others, and decorated by geometric and lettering ornament patterns. (Yousif, 1982, p.522).

Figure 7: Decoration inside Merjan Mosque

Source: field survey

### 3.5 Merjan Khan

The function of Khan is commercial, like motel, it used by traveler especially merchants among cities. So Merjan Khan related to Merjan School, occurring one of funded sources to the school, it is also named Ortema Khan. (Sid Mahmoud, 1346, p.66)

Figure 8: Merjan Khan

#### *3.5.1 Architectural elements*

Khan shape is rectangular, (11.4) meter width, with (30.25) meter long, the proportion of width to long was (1:2.7). (Ali, 2007).

Figure 9: Roofing of Merjan Khan

Source: field survey

The Khan consist of two floors, the first one, having continues indoor balcony, viewed on ground floor, on it is court. This balcony based on supported wall, length of each one about (1) meter.

Figure 10: Balcony in Merjan Khan

Source: field survey

The court is roofing by (8) intersected vaults, width of each one is (2.5) meter, height point about (14) meter.

Figure 11: Vaults in Merjan Khan

Source: field survey

Windows opens in the ground floor have small horizontal rectangular, and thus in the first floor it was bigger and with vaults. (Khalid, 2013, p.121)
They are no more decoration in the khan, except the main gate and it is entrance, was decorated by geometric pattern, consist of octagonal stars, main building material was brick only with wood. (Ahmed, 2013, p.20)

Figure 12: Gate of Merjan Khan

Source: field survey

## 4 Analysis of the general form

1. The rectangular shape is prevailing on the building's general form with the ratio of length to width in average of (1: 3)
2. The presence of the central courtyard or – yard (Sahan) in the roofed buildings - which the space or different rooms open toward it by a structural ratio of (1: 2.5).
3. The overall height of buildings is two stories with average of (10 m).
4. The entrance is prominent with a height of two stories and may be higher than the building itself, ornamented, with a structural ratio of (1: 1.5).
5. The external facades are solid and if there are openings, they are either at the level of the ground floor, small or elevated horizontal.
6. Buildings are free of adornments and ornaments, except in the entrances and domes.
7. The horizontal lengthiness of the building ration is of (1: 5).

### 4.1 Architectural elements

#### *4.1.1 The entrance*

1. The entrances with a height of two stories with a ration of (1: 1.5).
2. Higher than the main building.
3. The entrance mainly consists of a rectangular external frame.
4. The presence of decorative rectangle above the threshold of the door.
5. The decorations differ:

• Some were limited to radial decoration, similar to floral compositions, while others were limited to geometric motifs.

• The presence of twisted columns in Merjan School only after it was first found in decorations of the Abbasid palace columns and has been absent until it reappeared here.

#### *4.1.2 The dome*

1. Seljuk style domes continue to present as a continuation of the Abbasid era.
2. The Dome (onion shaped) with a neck.
3. The emergence of triangles corner for the first time in the domes of this period.
4. The domes were not devoid of some simple inscriptions and decorations at times.
5. The presence of small openings for functional purposes in the domes

#### *4.1.3 Minaret*

1. The minarets height differed but the structural ratio remained close to 1: 3.5.
2. The minarets unified by a candle onion shaped striated dome despite the difference in final treatment.
3. The decorations and ornaments appeared only on the minaret of the Khulafaa Mosque in crossing stripes.
4. The minarets are based on the adjacent building or on an independent base.

5. The minarets balcony is based on six layers of muqarnas rows in the minarets of Merjani School and Aqouliya mosque.
6. The great difference Khulafaa Mosque minaret from the rest of the minarets, probably due to the Abbasid origin.

See Minaret of al - Sahrawardi tomb, minaret of the Aqouliya mosque, minaret of the mosque of the Khulafaa, minaret of Merjan mosque.

Figure 13: Cylindrical Decoration in Gate of Merjan Mosque

Source: field survey

#### *4.1.4 Arches*

1. The dominant type of arces is Tudor arches, although continuous appearance of pointed arch.
1. The different ratios were used in arches construction and ranged from (1: 1.5-1: 2).
2. The arches are decorated with geometric motifs, if any. See Suhrawardi, Khan Merjan , Merjan School

**Figure** 14: Comparative Dome with Omer Al Shrawrdy Shrine Dome

#### *4.1.5 The vault*

1. The vault used for Khan roof, a phenomenon appeared for first time.
2. Cross-sectional vaults with half-circular section were used.

#### *4.1.6 The iwan*

1. The obvious formalistic similarity with the Abbasid era's Iwan, and its length to the width ratio (1: 2).
2. Detailing treatments differed from the previous era by the introduction of inscriptions and facing with marble.
3. Iwan remains an important element in the building formation and di the height of the building has not been increased or become salient.

#### *4.1.7 The gallery*

1. The continuation of gallery element appearance in the case of buildings and mosques.
2. The used arches are Tudor arches.

#### *4.1.8 The courtyard*

1- The courtyard did not have a specific shape. The Merjan school's courtyard is similar to the Abbasid palace one with its semi-square shape before the adding of the mosque's gallery, while the Khan has a rectangular courtyard surrounded by four sides and a roof.

2- The structural ratio of courtyards is proximate at (1: 2.5).

3- The floor is paved with the square bricks.

#### *4.1.9 The openings*

1- The rectangular openings are purely for the functional purposes in the minarets and are ornaments free, and they are a continuation of openings of Abbasids era.

1- Ornaments free Tudor arches, a type that appeared in more than one example: an opening with a Tudor arch enclosed by a rectangular frame containing the inscriptions and it is similar to the buildings of the Abbasid era in terms of the decorated rectangular shape.

2- The structural ratio for openings of this era is (1: 1.70)

### 4-2 Analysis of architectural treatments

#### *4.2.1 Ornaments and inscription*

1. Plant decoration is the dominant decorative element, especially in floral form.
2. The geometric ornaments appearance, but only a few and at Merjan School only.
3. The Twisted frame appears once again at Merjan School only.

Figure 15: Demolished part from Merjan School

*Muqarnas*

1- The Muqarnas are different forms in this era in terms of form and number, they are serrated.

2- The decorative filling of the Muqarnas have geometric shapes, which distinguished it from the Abbasid era buildings.

*Buildings inscription treatments*

1. The Thulth calligraphy used in the inscription of Mughal era.

*Building materials*

1. The basic material in construction during this era was the baked brick
2. Inlaying of Kashani (glazed brick) in some places for aesthetic purposes.
3. Using the timber in the windows.

*4.2.2 Buildings color*

1. The predominant color for the buildings of this era yellow of the baked brick.
2. The buildings are not free from some color treatments such as in Muqarnas of Khan Merjan or color inlying of Merjan School.

# 5 Discussion and conclusions of the Mogul era buildings

## 5.1 The building form

### *5.1.1 The shape*

1- The rectangular shape prevailed on the general form of the building with a ratio of (1: 3).

2- The presence of open or roofed courtyard. The overall height of the buildings is (10 m).

3- External facades are solid and, if there are openings, they are either small located at the level of the ground floor or elevated horizontal.

4- The buildings have no decoration except in the entrances and domes.

## 5.2 Architectural elements

### *5.2.1 The entrance*

1. The entrances to the height of two stories with a structural ratio of (1: 1.5).
2. They exceed the height of the original building.
3. The entrance consists mainly of a decorated rectangular external frame with geometric or floral motifs.

### *5.2.2 The dome*

1- The onion shaped domes emerged and belong to the Abbasid era.

1. The emergence of triangles corners for the first time in domes of this era.
2. The domes are free of inscriptions and decorations that adorn them.

## 5.3 Detailed treatments

### *5.3.1 Ornaments*

1- Plant ornaments are the dominant decorative element, particularly in floral form.

2- The little appearance of geometric decorations.

3- The twisted appears again at Merjan School only, after it was used only in the Abbasid palace, but in Merjan School the reefs is older than the decorations of the Abbasid palace, because the detailed treatment of the Abbasid palace did not reach to the present.

## 5.4 Architectural elements

### *5.4.1 The minaret*

1. The structural ratio (1: 3.5).
2. The minarets unified with an onion shaped dome.
3. The decorations and inscriptions only appeared on the minaret of the Khulafaa Mosque in the form of cross ribbons affected by the Abbasid head.
4. The minarets are based on the ceiling of the prayer hall.
5. The minaret balconies are based on muqarnas layers of six rows in the minarets of Merjan School and Aqouliya Mosque.

https://doi.org/10.1051/matecconf/201816205028

*5.4.2 Arches*

a. The predominant type of arches is the Tudor arch, despite the continuity of the pointed arch.

1. Different ratios were used in arches construction and ranged between (1: 1.5-1: 2).

2. The Arches were decorated with geometrical ornaments, if any decorations.

#### 5.5 Detailed treatments

***5-5-1 Muqarnas***

1- The forms of Muqarnas differ in this era in terms of form and number.

2- The muqarnas have no decorative fillings.

*5.5.2 Inscription treatments*

1- The inscription in Thulth calligraphy is the characteristic of the era.

2- Used in the Koran and historical inscriptions

Figure ١٦: Alaqoly Mosque Minaret

Source: field survey

#### 5.6 Architectural elements

*5.6.1 The courtyard*

1- The structural ratio of both courtyards is similar at (1: 2.25).

2- The floor paved with baked brick square tiles.

*5-6-2 The hallway*

1- The hallway is with flat roofing.

2- The hallway appeared on the upper floor.

*5-6-3 The iwan*

1- The element of Iwan was emphasized and its length to width ratio is (1: 2).

2- Rises for two stories.

3- The Iwan facade was decorated.

***5-6-4 The openings***

1- The rectangular openings were purely for functional purposes in minarets and free of ornaments

2- The predominant openings were the Tudor arches ones.

3- Slotted arch openings have also emerged.

#### 5-7 Detailing treatments

*5-7-1 Building materials*

The baked brick was used in buildings of this era.

*5-7-2 Color treatments*

The yellow color of baked brick is the predominant color on buildings of this era.

## 6 Final conclusions

After the analysis of the buildings which were built up under the Mughal occupation and the submitting buildings still remains to be inferred, the research reached to several important points that may not have been aware of because the historical overlaps abundance, and the various factors rotation on city of Baghdad and its buildings...

Research, over selected period of time, investigated the possibility of diagnosing its style.

The research comes out with several important points that can be summarized as follows:

1- The buildings of the Mughal era (Mughal style) its architecture witnessed proximities to Abbasid era architecture:

2- The horizontal orientation is the prevailing trend in the buildings, despite the characterized immensity.

3- The attention in entrances, and they were very similar to the entrances of Abbasid era, with emergence of new types of decorations.

4- The central courtyard continued to be the mainstay of the buildings and the roofed central courtyard appeared and was a new phenomenon.

5- The dome of this era was characterized by being flat with the use of Abbasid domes with long neck with it.

6- The arches used during that period were the pointed arches.

7- No clear style was discovered for this era minarets, which were different, but generally have a thick body compared to the height.
8- The use of muqarnas in the minaret reaches to six layers.
9- The Khulafaa Mosque minaret (Minaret of Suq -Ilgazil) is a double-style minaret. It was built in a transition era between the Abbasid and Mughalian eras. It has the base of Abbasid era, so it is a hybrid style minaret.

The emergence of corner curves for the first time in the buildings of this era to roof the domes.

## References

1. Al Alousy, Sid Mahmoud Shukry, Mousuqe History in baghdad and it is effect, Dar Alsalam Press, (1346 H).
2. Al Janabi, Tariq, (Studies in mediaeval Iraqi architecture, Dar AlHuriaa press, (1982).
3. Al Saffar, Refa'at Marhoon, (Older Baghdadi neighborhood in the memory), Wisdom Dar, London, (2014).
4. Hamoudy, Khalid Khalil, Mierjan school in Baghdad, historical, architecture, and art study, Publisher Scientific Books Dar, Ministry of Culture, 1st edition, (2013).
5. Researchers and Historian, 2013, Mustensiryah School History in Baghdad, Publisher Scientific Books Dar, Ministry of Culture.
6. Shiref, Yousif, Art history of Iraqi architecture, in many ages, Dar Al Rasheed for Publshing, Ministry of Culture, (1982).
7. Susa, Ahmed, Jouad, Mustafa, Detail guide for Baghdad map in Baghdad plans, new and old, Mesopotamia press, 2nd edition, (2013).
8. Thouiny, Ali, Civilized conversation, Researches and Studies in History, Language, and Heritage, no.1940, (2007).

https://doi.org/10.1051/matecconf/201816205029

# Green technologies in sustainable urban planning

*Suaad* Laffta[1,] * and *Areaj* Al-rawi[1]

[1]Center of urban and regional planning for post graduate studies, Baghdad University, Baghdad, Iraq

**Abstract.** The world is facing many environmental and economic problems resulting from rapid urbanization and over-exploitation of available resources as well as due to the dramatic increase of the population and corresponding increases in the number of cities. Technological and industrial development plays a big role in these problems. As a result, the term green technology became as a tool to solve these problems with an orientation towards sustainable development at all levels. , It seems to be very effective tool in modern urban planning which incorporates of all aspects of planning such as infrastructure and industry, energy, telecommunications, transportation and other vital areas in cities. These technologies are environment friendly inventions that often involve - energy efficiency, recycling, safety and health concerns, renewable resources, and more. Green technologies include several forms of technology that help to minimize negative effects on the environment and create new ways to achieve sustainable development. This paper is about green technology and its role in sustainable urban planning, it aims to clarify the most important areas for the application of this technology and the benefits provided by these technologies in improving the lives of the population in cities and a development orientation towards sustainability.

## 1 Technologies applied to urban sustainable development

Cities face a lot of problems and challenges resulting from the significant expansion due to population growth, which increases pressure on the infrastructure and results in depletion of natural resources, as a result, the concept of sustainability came up to address these challenges and guide urban development, to ensure the sustainability of resources and reduce environmental problems. Using modern technology can overcome a lot of problems in cities and guide the planning of these cities towards sustainability. These techniques help cities to achieve an appropriate standard of living and at the same time maintain natural ecosystems and resources. Technology used to make production processes more efficient, finding solutions to various "threats" that may affect the ability of cities to compete, such as the use of soil and urban transport, waste management in the city, quality of air, cultural heritage of cities, urban information systems, sustainable energy, agriculture, and new building materials applied to urban development and sustainable water management, [1].

These techniques are known as green technologies which are just tools to helpe make cities sustainable and provide inhabitants with a good quality of life. For example in large and important cities, the integration of technologies into urban planning policies has improved many of the major problems, [2]. In the past, the technological factor has not been as important as it is now to achieve sustainability through urban planning. Currently, the challenges are more difficult, and technology is needed not only to solve technical issues related to energy, water, materials and construction, but also to develop important sectors of the city such as transportation, infrastructure, communications and housing, to make these sectors sustainable through the application of green technology, [3].

Green technologies are environment friendly inventions that often involve - energy efficiency, recycling, safety and health concerns, renewable resources, and more, [4].

Green technologies reduce waste and contaminants, resource consumption and increase the efficiency of services in the city. The various aspects of green technology that can be incorporated into the spatial planning process help find new ways to achieve sustainable development by reducing the negative impacts of various economic and human activities on the environment and ecosystems and guiding development towards adoption of green and eco-friendly ways of life in cities and urban areas. For example, promoting green transport can enhance access to services and help reduce pollution levels and health inequities of the city's population; the use of treated wastewater in public water and green gardens help to reduce overall water consumption. Cities should look at circular development models that recycle water and waste and produce energy in them, so that sewage can be used, for example, to grow fish, plants or recycled solid waste to create methane. These initiatives help to reduce harmful

*Corresponding author: dr.suaad.j@iurp.uobaghdad.edu.iq

© The Authors, published by EDP Sciences. This is an open access article distributed under the terms of the Creative Commons Attribution License 4.0 (http://creativecommons.org/licenses/by/4.0/).

environmental impacts improve the city's landscape and improve quality of life, [5].

"The greening of cities will require some, or preferably all, of the following, [6]:

- "Reduction of chemical and physical hazards".
- "Control over environmental impacts on health".
- "Creation of quality environments for all"
- "Minimized ecological footprints outside the urban area".
- "Ensured sustainable consumption".
- "Adaptation to climate change impacts".

Buildings, energy, and transport are the largest sources of CO2 emission and through design strategies and technologies to improve building construction, urban transport efficiency, and technology to generate energy, water management and waste management, the amount of emissions can be reduced. Depending on the green technologies, all this will reduce the rate of energy and natural resources consumption.

"There are four eco-technologies that may be utilized by city planners and other professions the urban planning and design: “Environmental Technologies (ET)” encompass a broad range of technologies related to energy, water, and waste, while “Information Technologies (IT)” include a broad range of computer-based hardware and software and environmental sensing technologies for gathering of environmental data. “Geographic Information systems (GIS)” provide for the storage, transformation, manipulation, management, visualization, updating, querying, and reporting in related databases of geo-referenced data in a tabular format, and incorporates an Urban Information System that facilitates land use analyses, the preparation of development and environmental plans, etc. Finally, “Communication Technologies (CT)” are used for the transfer of environmental data, information, knowledge, and decisions in wired or wireless environments", [7] .

These green technologies are one of the most important and effective tools for achieving sustainable development in cities by contributing to the greening of economic, service and urban sectors and preserving natural resources because they enable planners to choose the best pathways to minimize the negative effects of development and maximize the benefits.

## 2 Green technology and efficiency of energy

Energy is one of the most important elements upon which life depends in modern cities, and is the main engine for all core activities in these cities. It also represents one of the main causes of environmental problems in cities because it is produced from non-sustainable sources, therefore, the production of energy in cities by using green technology is very important in order to achieve sustainability of these cities. Investments in renewable energy today are likely to gain benefits in the coming years. Renewable technologies are usually more labour-intensive than fuel-based power generation projects, they have positive effects of being clean and non-polluting energy and rely on non-exhaustive sources, [8].

Unplanned urban expansion increases energy demand and carbon emissions, thereby distorting the ecosystems and making cities socially divisive places to live. This process is called the “negative ecological footprint”. The key to meeting growing future energy demand is obtaining energy from clean, non-traditional sources. Diversification of renewable energy supply must extend into solar, wind, geothermal, ocean wave, and tidal energy sources, [9].

The investment in low-carbon technology is expected to extends significantly in the future, more than 460 billion $ will be invested in renewable energy projects globally by 2030, [10]. Achieving benefits from using clean energy sources without losing social and economic benefits of urban areas is the key goal for the future, [11].

In several urban cities, the traditional grid model of electricity generation is being re-examined. The smart grid, which changes the urban energy landscape, is being used. These intelligent networks operate efficiently in cities and will play an important role in changing the way power is generated and distributed in the future, [8].

## 3 Sustainable and green infrastructures.

Infrastructure is the basis on which modern cities, activities and population depend, It represents the basic structure of these cities which urban growth and economic development and the life quality of its population depend.

Due to global climate change, which leads to increased frequency of storm, rising temperatures and rising sea levels, some countries have resorted to green infrastructure as a hedge against the risks of climate change, [12].

Recently green infrastructure and technologies have been identified as ‘best practices’ at the local level when combined with traditional grey and it focuses on prevention of unnecessary consumption of natural resources (especially non-renewable ones) and mitigation of harmful emissions, [13].

The US Environmental Protection Agency (EPA) defines sustainable infrastructure as a range of practices that "encourage utilities and their customers to address existing needs so that future generations will not be left to address the approaching wave of infrastructure needs that will result from aging infrastructure", [14].

The term green infrastructure is a new term, but it is not a new idea that began a hundred and fifty years ago through planning and preservation efforts through two concepts:

- Link parks and open areas.

Conservation and linkage of natural areas to biodiversity, [15]

The importance of green infrastructure can be summarized, [16]

- Sustainable management of land and water resources, including production (eg energy and food crops),

pollution control, climate improvement, increased land cover porosity,
-Biodiversity.
- Recreation - through the corridors or green roads and the use of methods to contribute to improving the quality of life and public health.
- Aesthetic - because of the green spaces and corridors of beautiful landscapes.

The sustainability of human habitats is directly related to the sustainability of natural habitats, so when planning sustainable infrastructure to balance them, it is essential to adopt green infrastructure, [17].

Green infrastructure approaches help adapt to climate change. Their ability to mitigate the effects of heavy precipitation and temperature reduction, better management of storm water runoff, and water retention and conservation, [12].

The use of these approaches in the planning process for sustainable cities will support the resilience of cities to the effects of climate change and adaptation to climate change and help reducing the impact of climate change on people's lives and natural resources.

Green infrastructure helps to preserve arable land and reduce the consumption of this valuable resource, in addition to conserving natural green areas. These green structures also help reduce hazardous waste and thus improve the quality of life and public health.

## 4 Sustainable planning of housing

Due to population growth, urbanization, poverty, inequality, pollution and overcrowding, which necessitated UN-Habitat to develop new approaches and standards and urban planning systems to address them.
To achieve these goals they put five principles, [18 ]:
- The allocation of sufficient and efficient space for the street network, and 30% of the master plan, and not less than 18 km per km 2.
- Achieve a high population density of 15,000 people / km2, ie 150 people / ha.
- Adopt the principle of mixed use of 40% of the land area and be allocated for commercial use.
-Social integration, through the provision of housing with different holdings of private ownership, rent, 50% each, and provide (20-50) % of low-income.

## 5 Green technology for water treatment

Water is one of the most valuable resources, and it must be free of pollutants to avoid risks to humans and our environment, Water treatment methods exist in traditional ways but are not sufficient to eliminate the pollutants completely which requires the development of alternatives to water purification, solar energy, which is a green technologies to get rid of water pollution.

There are four methods for solar powered water purification, [19 ,20]:
-Solar water treatment systems:-The water purification system is based on the use of battery operated on solar energy, and the battery can work for three days even without sunlight.

-Solar water disinfection: Water is sterilized with plastic bottles and exposed to sunlight for six hours
-Solar water pasteurization : Bacteria are killed in solar-heated stoves.
- Solar distillation: It is used to desalinate sea water and make it drinkable.

## 6 The information technology as a tool for sustainable urban planning

"Geographic Information Systems (GIS) are systems of spatial location making use of geographic and statistical databases, where the characteristics and trends of a model of geographic development can be visualized", [21].

GIS is used to apply geo-knowledge to solve problems. In the case of green government decisions, geographic information systems are essential in identifying land uses and zoning.

For decades, urban and regional planners have used GIS to plan better communities, developing smart growth of cities and controlling urbanization.

Using GIS on the Internet, we can publish maps and help citizens understand the proposed plans, public awareness of plans and discussions is essential for development initiatives and for improving the ability of planners to serve the community.

The success of planners in combating chronic urban problems is largely determined by their ability to utilize effective tools and planning support systems.

The planners use GIs as a tools to achieve their goals of creating viable communities in a pollution-free environment and achieving economic development, as well as its use for analysis and modeling, [23].

There are four steps of GIs system, [23]:
-Firstly: Data collection, input and correction.
-Secondly: Data storage and retrieval.
-Third: Data Manipulation and Analysis Phase.
-Fourth: Data Display and Reporting Phase.

## 7 Green roads

Road network design has traditionally been concerned with vehicle traffic and road safety. Now, road network design is concerned with many things such as aesthetics, environment and community interests. The landscaping of roads and green spaces gives advantages to transport corridors, [24].

### 7.1 Principles of green roads concept

The green road concept is an approach that refers to a sound and appropriate methodology either cost (low cost), environmental, technical, work-based or road-based.

The Green Road Concept focuses on conserving the delicate mountain ecology, in particular the protection of vegetation cover as means of soil conservation. Actually, Green Road is a low-cost, low-volume, fair-weather earthen road. They are usually village roads or district roads under the rural road network and not a new class of roads in itself.

Greater emphasis is given to the selection of a smooth longitudinal alignment that allows for progressive upgrading as traffic increases. This approach is most appropriate for rural hill road construction having low traffic volume. But some aspect of Green Roads could be used in Feeder and Roads and Highways too,[24,25].

# 8 Countries that have the greatest potential to produce and commercialize green technology

## 8.1 Ireland

Ireland is a strong competitor in clean technology, focusing on wind and tidal energy. There is considerable innovation progress, supported by prominent science and engineering institutions.

Clean jobs in 2020 are expected to reach 80,000, and wind farm projects for power generation of 138 megawatts, and these figures are likely to be incremental ,[ 26 ].

## 8.2 Germany

Germany continues to be a leader in the renewable energy market and occupies one-third of the world's installed solar energy, despite its often cloudy climate, and has also put plans to reduce carbon emissions by 80-95%, [26].

## 8.3 Switzerland

Switzerland has high innovative inputs, supportive government policies, and medium-sized energy and transport infrastructure, all of which support the clean technology industry, [26].

## 8.4 Canada

The clean technology industry in Canada is still average today, although many innovative factors are in their early stages, which means that the country can make huge profits in several years, [26].

## 8.5 UK

The UK is a strong competitor in green technology innovation, currently outperforming the wind sector, and has the world's largest wind market.

And that there is a new innovator Cyber Hawk in the wind sector acts as an observer of the air infrastructure, [26].

## 8.6 Denmark

Denmark ranks first in the wind energy investment list and one of the two highest clean technology budgets with the other country Finland.

And looking for partnerships to invest in places like China and this attracts investors both domestically and abroad, [26].

## 8.7 Sweden

Sweden leads the list of clean technology innovation programs and has a number of clean technology companies as well as a major consumer of renewable energy, [26].

## 8.8 USA

There are some green initiatives, such as a 30% reduction in carbon emissions by 2030, and an increase of 30% in the fiscal year 2014 for development research and clean energy to $ 7.9 billion, [26].

## 8.9 Finland

There are several factors that encourage clean technology in Finland, such as the lack of fossil fuels, severe climate and the possibility of innovation. So it operates the second largest clean technology industry, and the industry provides jobs for up to 50,000 people. This figure will double by 2020, [26].

# 9 The Eco-town concept

Eco-Town is a town where most of its activities depend on green technologies It is based on several concepts. which include the 3R (Reduce, Reuse and Recycling) concepts as well as building an economy based on the life-cycle approach and accumulation of recycling facilities. Eco-Town concepts also include extended producer responsibility, socially responsible investment, green procurement, green consumerism, industrial ecology , and integrated waste management, global reporting initiative, corporate social responsibility, green labelling, EMS and ISO 14001, [27,p1-2],see Fig.1 below .

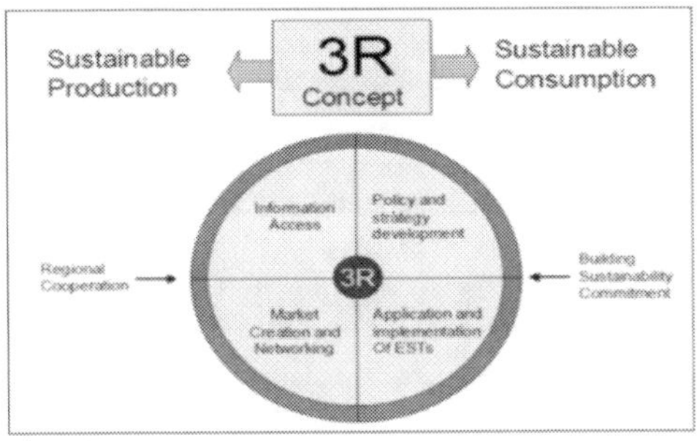


Fig. 1. Eco-town concept.

The concept of Eco-Town has been expanded, from focuses mainly on individual systems related to 3R to becoming part of the eco-city concept and to include industrial environmental parks, industrial solidarity and concentration on collective areas, focus generally on urban planning, urban ecosystems and other spatial planning branches. See fig.2

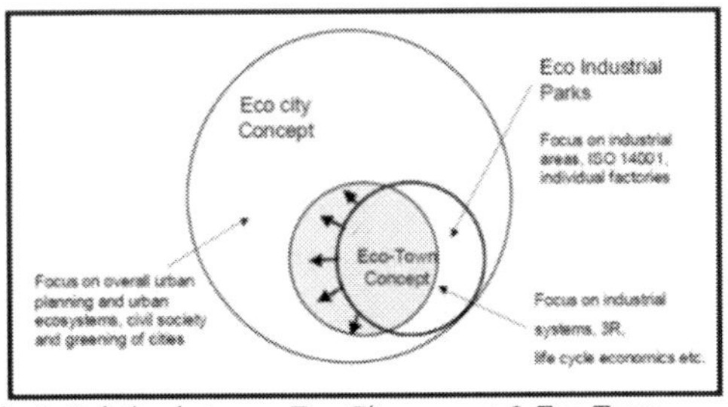


Fig. 2 Relation between Eco-City concept & Eco-Town concept

### 9.1 Kawasaki Eco-town in Japan.

Kawasaki is the first environmental city in Japan. It is located in the coastal area of Kawasaki. It was founded in 1997 and includes several economic activities based on green technologies including plastic waste recycling system for use as blast furnace raw materials, paper recycling facility, waste recycling, Ammonia recycling system [27,p14].

The Kawasaki Eko-Toon plan aims to define a wide area to extend the entire coast of Kawasaki, Aspiring to make the community rely on recycling resources, and seek to develop the coastal zone, and to achieve these goals it plans to develop zero-emission industrial zone. Individual companies in the industrial zone will not only reduce their emissions but also effectively use or recycle emissions from other facilities there to new resources that could be used as raw materials for industry. Other recycling facilities around the industrial zone will be built To promote cooperation, where resources are recycled with existing companies and thus the production process will be circular with zero waste, [27].

## 10 The Indian experience in green technology

The emphasis will be on wind power and the road network as a green technique.

### 10.1 Wind energy.

Wind energy plays a prominent role as renewable energy in India and supports a number of means and legislation in the development of wind energy technology, Because of this support, India ranks fifth in the wind energy sectors after China, the United States, Germany and Spain. Four phases were proposed during the period (1981-2020) and the energy started from (2.2MW) during the first phase to (23444Mw) in 2020. Wind energy (70% of the rest of the sources), as shown in the Figure, F.3 below.

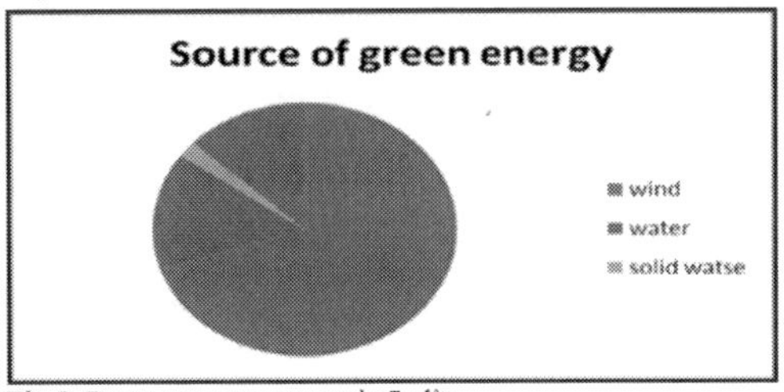


Fig.3 Green energy sources in India.

India has the world's second largest road network, the fourth rail network and the ninth civil aviation. The transport sector is expected to grow more than any other sector over the next 30 years,[28]. So a plan has been developed for the transition to green technology agencies:

### 10.2 UK electric smart vehicles.

Due to the increase in population growth and increased demand for energy, leading to increased demand for energy, and to reduce dependence on conventional fuels, emission limits and Standards have been proposed to compare with European standards by a plan of 5-10 years depending on the quality of the vehicle.

There are three companies in the Indian market:

-Riva Electric Vehicle Company

-TATA Motors.

-Bharat Heavy Electric Company.

"A plan has been put in place to produce 600,000 electric vehicles by 2020 and (40 -50 million) two-wheeled vehicles and bicycles".

## 11 Suggestions and recommendations for application of green technologies in Iraq

The research will attempt to present some suggestions and recommendations for application of green technologies in Iraq through the use of these techniques in sustainable spatial planning to achieve sustainable development.

Sustainable Green Technology is one of the most important tools for sustainability in spatial planning. We will review here the most important areas in which green technology can be used to achieve sustainability in Iraqi cities:

-Energy Sector: This sector is an important sector for the lives of the population in Iraqi cities, as well as its essential role in the economic and service sector. The production of electricity in Iraq depends on the thermal stations based on oil and gas as fuel, Iraq has good sources of alternative and clean energy that can be used to generate electricity and thus eliminate environmental problems associated with the use of fossil fuels for power generation.

-Transport Sector: It is important to apply green technology in the field of transport to minimize the possible negative impacts of this sector on the environment and to achieve sustainable development. These technologies are applied through the use of mass transport based on electric trains and transport modes using electric power

-Green infrastructure: The adoption of green infrastructure in the planning of cities in Iraq is very important to address the environmental problems resulting from climate change and the acceleration of desertification of agricultural land and the high temperatures in these cities.

-Housing sector: The use of green building techniques and green architecture in the housing sector will greatly help achieve the sustainability of the housing sector in Iraq by reducing the consumption of natural resources such as land, in addition to water and energy consumption.

-Agriculture is one of the main economic activities in Iraq. The use of green technologies in agricultural sector is a necessary step to activate this sector and to overcome the problems that result from using unsustainable methods in agriculture.

- Industrial sector : industrial sector in Iraq suffers from stumbling as a result of circumstances experienced by Iraq in recent years in addition to the old methods used in the industrial sector, so the adoption of green and sustainable technologies in this sector are essential to develop this sector, and to reduce environmental, social and health problems resulting from this sector.

## References

1. S. Moreno, *Curent technologies appli to urban sustinable development*, Faculty of Architecture and Design, Autonomous University of the State of Mexico Cerro de Coatepec s/no., Ciudad Universitaria,C.P.50100 Toluca, Estado de México, México
2. X. Chen, *Urban planning management system in Los Angeles: an overview*, Theoretical and Empirical Researches in Urban Management Journal, 2, (2009), pp. 50-63.
3. Commissie Lange Termijn Milieu-Beleid [CLTM]. (1990). Het Milieu denkbeelden voor, 21 ste eeuw, Kerkebosch, Zeist.
4. GH. Soni, *Advanteges of green technology*, Social Issues and Environmental Problems, Vol.3 (Iss.9:SE): Sep, (2015).
5. Investing in sustainable Cities, *Challenges and Opportunities*, International Development Finance Club Special Interest Group on Sustainable Urban Development, October (2014).
6. D. Satterthwaite, *Sustainable citie s or cities that contribute to sustainable development?Urban Studies*, 34 (10), (1997), 1667–1691.
7. O. Ercoskun, *Green and ecological technologies for urban planning: creating smart cities* ,(2012)
8. .S. Hammer, *Cities and green growth: A conceptual framework*, OECD Regional Development Working Papers(2011), OECD Publishing. http://dx.doi.org/10.1787/5kg0tflmzx34-en
9. Rees, W. Revisiting Carrying Capacity, *Area-Based Indicators of Sustainability*. Population and Environment. (3), ( 1996).
10. Bloomberg New Energy Finance (BNEF). , *Global renewable energy market outlook*, Executive Summary. BNEF ,http://bnef.com/ free publications / white-papers/ (accessed 5 April 2012).
11. UNEP, *Decoupling natural resource use and environmental impacts from economic growth*, Paris,(2011).
12. J. Foster, A. Lowe, S. Winkelman, , *The value of green infrastructure for urban climate adaptation*, The Center for Clean Air Policy February (2011).
13. R. Howes, H. Robinson, *Infrastructure for the built environment—global procurement strategies*; Elsevier: Oxford, UK,( 2005).
14. US Environemental Protection Agency, *Sustainable infrastructure for water and wastewater*, http://www.epa.gov/waterinfrastructure. Accessed: March 12, (2009).
15. A. Mark, T. Edward, *Green infrastructure: smart conservation for the 21st century*, The Conservation Fund, S P R A W L W A T C H C L E A R I N G H O U S E M O N O G R A P H S E R I E S.
16. R. Davies, MacFarlan, C. McGloin,More,, *Green infrastructuer planning guide* https://www.researchgate.net/profile/Maggie_R

oe/publication/265012095_GREEN_INFRASTRUCTURE_PLANNING_GUIDE_Authors/links/54f087a70cf24eb87940c8c1.pdf.

17. Planning Infrastructure to Sustain America , *Next generation concepts to guide the community, design and infrastructure professions*, ), p.10 , (2010).
18. unhabitat for better planning future, *A new strategy of sustainable neighborhood planning: Five Principles*.
19. P. Kalt, C. Birzer , H. Evans, A. Liew, M. Padovan, and Michael Watchman, *A Solar Disinfection Water Treatment system for remote communities*, The School of Mechanical Engineering, The University of Adelaide, Adelaide, 5005, Australia.
20. Dip, L. Bauer, *Solar-driven water treatment – experiences in India*, Project Engineer, Solar Spring, Intersolar Europe, (2004).
21. C. Tomlin, *Geographic information systems and cartographic modeling*, Prentice Hall, N J,USA. (1990).
22. ESRI, *GIS for Green Government Providing Sustainable Solutions*.
23. ESRI, *GIS for Urban and Regional Planning*, January,(2011).
24. Harmonizing rural road development with mountain environment, *Green roads in nepal harmonizing rural road development with mountain environment*, Green Roads in Nepal Hare Ram Shrestha Director, SIDeF, Nepal Vice-President, SCAEF Nepal,( 5/30/2009).
25. Green roads: *Building environmentally friendly, low maintenance rural roads through local participation* Frances Klatzel GTZ Food for Work, Nepal 2000 FfW@gtz.org.np.
26. http://www.energydigital.com/top10/3708/Top-10-Countries-with-Cleantech-Innovators
27. Global Environment Centre Foundation, *Eco-Towns in Japan -Implications and Lessons for Developing Countries and Cities*, June(2005).

# AUTHOR INDEX

# AUTHOR INDEX

# AUTHOR INDEX

# AUTHOR INDEX